Mechanical Engineer's Reference Book

Mechanical Engineer's Reference Book

Twelfth edition

Edited by

Edward H. Smith BSc, MSc, PhD, CEng, FIMechE

Head of Computing Services,
University of Central Lancashire

With specialist contributors

Published by:
Society of Automotive Engineers, Inc.
400 Commonwealth Drive
Warrendale, PA 15096-0001

Butterworth-Heinemann Ltd
Linacre House, Jordan Hill, Oxford OX2 8DP

$\mathcal{R}$ A member of the Reed Elsevier group

OXFORD   LONDON   BOSTON
MUNICH   NEW DELHI   SINGAPORE   SYDNEY
TOKYO   TORONTO   WELLINGTON

First published as *Newnes Engineer's Reference Book* 1946
Twelfth edition 1994

© Butterworth-Heinemann Ltd 1994

British Library Cataloguing in Publication Data
A catalogue record for this book is available from the
British Library

ISBN 1 56091 450 5
SAE order number R - 136

Library of Congress Cataloguing in Publication Data
A catalogue record for this book is available from the
Library of Congress

Typeset by TecSet Ltd, Wallington, Surrey
Printed in Great Britain by The Bath Press, Avon

Contents

Preface

List of contributors

1 Mechanical engineering principles

Status of rigid bodies · Strength of materials · Dynamics of rigid bodies · Vibrations · Mechanics of fluids · Principles of thermodynamics · Heat transfer · References

2 Electrical and electronics principles

Basic electrical technology · Electrical machines · Analogue and digital electronics theory · Electrical safety · References · Further reading

3 Microprocessors, instrumentation and control

Summary of number systems · Microprocessors · Communication standards · Interfacing of computers to systems · Instrumentation · Classical control theory and practice · Microprocessor-based control · Programmable logic controllers · The z-transform · State variable techniques · References · Further reading

4 Computers and their application

Introduction · Types of computer · Generations of digital computers · Digital computer systems · Categories of computer systems · Central processor unit · Memory · Peripherals · Output devices · Terminals · Direct input · Disk storage · Digital and analogue input/output · Data communications · Computer networks · Data terminal equipment · Software · Database management · Language translators · Languages

5 Computer-integrated engineering systems

CAD/CAM: Computer-aided design and computer-aided manufacturing ·Industrial robotics and automation · Computer graphics systems · References · Further reading

6 Design standards

Standardization in design · Drawing and graphic communications · Fits, tolerances and limits · Fasteners · Ergonomic and anthropometric data · Total quality – a company culture · References

7 Materials, properties and selection

Engineering properties of materials · The principles underlying materials selection · Ferrous metals · Non-ferrous metals · Composites · Polymers · Elastomers · Engineering ceramics and glasses · Corrosion · Non-destructive testing · References · Further reading

8 Mechanics of solids

Stress and strain · Experimental techniques · Fracture mechanics · Creep of materials · Fatigue · References · Further reading

9 Tribology

Basic principles · Lubricants (oils and greases) · Bearing selection · Principles and design of hydrodynamic bearings · Lubrication of industrial gears · Rolling element bearings · Materials for unlubricated sliding · Wear and surface treatment · Fretting · Surface topography · References · Further reading

10 Power units and transmission

Power units · Power transmission · Further reading

11 Fuels and combustion

Introduction · General fuel types · Major property overview · Major fuel groupings · Combustion · Conclusions · References

12 Alternative energy sources

Introduction · Solar radiation · Passive solar design in the UK · Thermal power and other thermal applications · Photovoltaic energy conversion · Solar chemistry · Hydropower · Wind power · Geothermal energy · Tidal power · Wave power · Biomass and energy from wastes · Energy crops · References

13 Nuclear engineering

Introduction · Nuclear radiation and energy · Mechanical engineering aspects of nuclear power stations and associated plant · Other applications of nuclear radiation · Elements of health physics and shielding · Further reading

14 Offshore engineering

Historical review · Types of fixed and floating structures · Future development · Hydrodynamic loading · Structural strength and fatigue · Dynamics of floating systems · Design considerations and certification · References

15 Plant engineering

Compressors, fans and pumps · Seals and sealing · Boilers and waste-heat recovery · Heating, ventilation and air conditioning · Refrigeration · Energy management · Condition monitoring · Vibration isolation and limits · Acoustic noise · References

16 Manufacturing methods

Large-chip metal removal · Metal forming · Welding, soldering and brazing · Adhesives · Casting and foundry practice · References · Further reading

17 Engineering mathematics

Trigonometric functions and general formulae · Calculus · Series and transforms · Matrices and determinants · Differential equations · Statistics · Further reading

18 Health and safety

Health and safety in the European Community · Health and safety at work – law and administration in the USA · UK legislation and guidance · The Health and Safety at Work etc. Act 1974 · The Health and Safety Executive · Local Authorities · Enforcement Notices · Control of Substances Hazardous to Health Regulations 1988 · Asbestos · Control of lead at work · The Electricity at Work Regulations 1989 · The Noise at Work Regulations 1989 · Safety of machines · Personal protective equipment · Manual handling · Further reading

19 Units, symbols and constants

SI units · Conversion to existing imperial terms · Abbreviations · Physical and chemical constants · Further reading

Index

Preface

I was delighted when Butterworth-Heinemann asked me to edit a new edition of *Mechanical Engineer's Reference Book*. Upon looking at its predecessor, it was clear that it had served the community well, but a major update was required. The book clearly needed to take account of modern methods and systems.

The philosophy behind the book is that it will provide a qualified engineer with sufficient information so that he or she can identify the basic principles of a subject and be directed to further reading if required. There is a blurred line between this set of information and a more detailed set from which design decisions are made. One of my most important tasks has been to define this distinction, so that the aims of the book are met and its weight is minimized! I hope I have been able to do this so that the information is neither cursory nor complex.

Any book of this size will inevitably contain errors, but I hope these will be minimal. I will be pleased to receive any information from readers so that the book can be improved.

To see this book in print is a considerable personal achievement, but I could not have done this without the help of others. First, I would like to thank all the authors for their tremendous hard work. It is a major task to prepare information for a book of this type, and they have all done a magnificent job. At Butterworth-Heinemann, Duncan Enright and Deena Burgess have been a great help, and Dal Koshal of the University of Brighton provided considerable support. At the University of Central Lancashire, Gill Cooke and Sue Wright ensured that the administration ran smoothly.

I hope you find the book useful.

Ted Smith
University of Central Lancashire, Preston.
Christmas Eve, 1993

Contributors

Dennis H. Bacon BSc(Eng), MSc, CEng, MIMechE
Consultant and technical author

Neal Barnes BSc, PhD
Formerly Manager, Pumping Technology, BHR Group Ltd

John Barron BA, MA(Cantab)
Lecturer, Department of Engineering, University of Cambridge

Christopher Beards BSc(Eng), PhD, CEng, MRAeS, MIOA
Consultant and technical author

Jonh S. Bevan IEng, MIPlantE, ACIBSE
Formerly with British Telecom

Ronald J. Blaen
Independent consultant

Tadeusz Z. Blazynski PhD, BSc(Eng), MIMechE, CEng
Formerly Reader in Applied Plasticity, Department of Mechanicakl Engineering, University of Leeds

James Carvill BSc(MechE), BSc(ElecEng)
Formerly Senior Lecturer in Mechanical Engineering, University of Northumbria at Newcastle

Trevor G. Clarkson BSc(Eng), PhD, CEng, MIEE, Senior Member IEEE
Department of Electronic and Electrical Engineering, King's College, University of London

Paul Compton BSc CEng, MCIBSE
Colt International Ltd, Havant, Hants

Vince Coveney PhD
Senior Lecturer, Faculty of Engineering, University of the West of England

Roy D. Cullum FIED
Editor, *Materials and Manufacture*

A. Davies
National Centre of Tribology, Risley Nuclear Development Laboratory

Raymond J. H. Easton CEng, MIMechE
Chief Applications Engineer, James Walker & Co Ltd

Philip Eliades BSc, AMIMechE
National Centre for Tribology, UKAEA, Risley, Warrington

Duncan S. T. Enright BA, MA(Oxon), CertEd, GradInstP
Commissioning Editor, Butterworth-Heinemann, Oxford

Charles J. Fraser BSc, PhD, CEng, FIMechE, MInstPet
Reader in Mechanical Engineering

Eric M. Goodger BSc(Eng), MSc, PhD, CEng, MIMechE, FInstE, FInstPet, MRAeS, MIEAust
Consultant in Fuels Technology Training

Edward N. Gregory CEng, FIM, FWeldI
Consultant

Dennis R. Hatton IEng, MIPlantE
Consultant

Tony G. Herraty BTech, MIMechE, CEng
SKF (UK) Service Ltd, Luton, Bedfordshire

Martin Hodskinson BSc, PhD, CEng, FIMechE, MIED, REngDes
Senior Lecturer, Department of Engineering and Product Design, University of Central Lancashire

Allan R. Hutchinson BSc, PhD, CEng, MICE
Deputy Head, Joining Technology Research Centre, School of Engineering, Oxford Brookes University

Jeffery D. Lewins DSc(Eng), FINucE, CEng
Lecturer in Nuclear Engineering, University of Cambridge and Director of Studies in Engineering and Management, Magdalene College

Michael W. J. Lewis BSc, MSc
Senior Engineering Consultant, National Centre of Tribology, AE Technology, Risley, Warrington

R. Ken Livesley MA, PhD, MBCS
Lecturer Department of Engineering, University of Cambridge

J. Cleland McVeigh MA, MSc, PhD, CEng, FIMechE, FInstE, MIEE, MCIBSE
Visiting Professor, School of Engineering, Glasgow Caledonian University

Gordon M. Mair BSc, DMS, CEng, MIEE, MIMgt
Lecturer, Department of Design, Manufacture and Engineering Management, University of Strathclyde

Fraidoon Mazda MPhil, DFH, DMS, MIMgt, CEng, FIEE
Northern Telecom

Bert Middlebrook
Consultant

John S. Milne BSc, CEng, FIMechE
Professor, Department of Mechanical Engineering, Dundee
Institute of Technology

Peter Myler BSc, MSc, PhD, CEng, MIMech
Principal Lecturer, School of Engineering, Bolton Institute

Ben Noltingk BSc, PhD, CPhys, FInstP, CEng, FIEE
Consultant

Robert Paine BSc, MSc
Department of Engineering and Product Design, University
of Central Lancashire

John R. Painter BSc(Eng), CEng, MRAes, CDipAF
Independent consultant (CAD/CAM)

Minoo H. Patel BSc(Eng), PhD, CEng, FIMechE, FRINA
Kennedy Professor of Mechanical Engineering and Head of
Department, University College, London

George E. Pritchard CEng, FCIBSE, FInst, FIPlantE
Consulting engineer

Donald B. Richardson MPhil, DIC, CEng, FIMechE, FIEE
Lecturer, Department of Mechanical and Manufacturing
Engineering, University of Brighton

Carl Riddiford MSc
Senior Technologist, MRPRA, Hertford

Ian Robertson MBCS
Change Management Consulatnt, Digital Equipment
Corporation

Roy Sharpe BSc, CEng, FIM, FInstP, FIQA, HonFInstNDT
Formerly Head of National Nondestructive Testing Centre,
Harwell

Ian Sherrington BSc, PhD, CPhys, CEng, MInstP
Reader in Tribology, department of Engineering, and
Product Design, University of Central Lancashire

Edward H. Smith BSc, MSc, PhD, CEng, FIMechE
Head of Computing Services, University of Central
Lancashire

Keith T. Stevens BSc(Phy)
Principle scientist

Peter Tucker BSc(Tech), MSc, CEng, MIMechE
Formerly Principal Lecturer, Department of Mechanical and
Production Engineering,Preston Polytechnic

Robert K. Turton BSc(Eng), CEng, MIMechE
Senior Lecturer in Mechanical Engineering, Loughborough
University of Technology and Visiting Fellow, Cranfield
University

Ernie Walker BSc CEng, MIMechE
Formerly Chief Thermal Engineer, Thermal Engineering
Ltd

Roger C. Webster BSc, MIEH
Roger Webster & Associates, West Bridgford, Nottingham

John Weston-Hays
Managing Director, Noble Weston Hays Technical Services
Ltd, Dorking, Surrey

Leslie M. Wyatt FIM, CEng
Independent consultant and technical author

1

Mechanical engineering principles

Robert Paine (Sections 1.1–1.4.2)
Christopher Beards (Section 1.4.3)
Peter Tucker (Section 1.5)
Dennis H. Bacon (Sections 1.6 and 1.7)

Contents

1.1 Statics of rigid bodies 1/3
1.2 Strength of materials 1/3
1.3 Dynamics of rigid bodies 1/4
   1.3.1 Basic definitions 1/4
   1.3.2 Linear and angular motion in two dimensions 1/6
   1.3.3 Circular motion 1/7
   1.3.4 Linear and angular motion in three dimensions 1/7
   1.3.5 Balancing 1/8
   1.3.6 Balancing of rotating masses 1/8

1.4 Vibrations 1/9
   1.4.1 Single-degree-of-freedom systems 1/9
   1.4.2 Multi-degree-of-freedom systems 1/14
   Further reading 1/15
   British Standards 1/15
   1.4.3 Random vibrations 1/15
   Further reading 1/18

1.5 Mechanics of fluids 1/18
   1.5.1 Introduction 1/18
   1.5.2 Fluid statics 1/18
   1.5.3 Fluid flow 1/22

1.5.4 Flow measurement 1/25
1.5.5 Open-channel flow 1/27
1.5.6 Boundary layer flow 1/28
1.5.7 Pressure transients (water hammer) 1/30
1.5.8 Gas flow 1/31
1.5.9 Ideal fluid flow 1/33
1.5.10 Conclusion 1/35
Further reading 1/35

1.6 Principles of thermodynamics 1/35
   1.6.1 Introduction 1/35
   1.6.2 The laws of thermodynamics 1/36
   1.6.3 Thermoeconomics 1/37
   1.6.4 Work, heat, property values, process laws and combustion 1/37
   1.6.5 Cycle analysis 1/37

1.7 Heat transfer 1/40
   1.7.1 Introduction 1/40
   1.7.2 Basic principles of heat transfer 1/40
   1.7.3 Analysis of heat transfer 1/43
   1.7.4 Use of computers 1/47
   1.7.5 Heat transfer: nomenclature 1/48

References 1/48

In general, the study of mechanics may be divided into two distinct areas. These are *statics*, which involves the study of bodies at rest, and *dynamics*, which is the study of bodies in motion. In each case it is important to select an appropriate mathematical model from which a 'free body diagram' may be drawn, representing the system in space, with all the relevant forces acting on that system.

1.1 Statics of rigid bodies

When a set of forces act on a body they give rise to a resultant force or moment or a combination of both. The situation may be determined by considering three mutually perpendicular directions on the 'free body diagram' and resolving the forces and moment in these directions. If the three directions are denoted by x, y and z then the sum of forces may be represented by ΣF_x, ΣF_y and ΣF_z and the sum of the moments about respective axes by ΣM_x, ΣM_y and ΣM_z. Then for equilibrium the following conditions must hold:

$$\Sigma F_x = \Sigma F_y = \Sigma F_z = 0 \tag{1.1}$$

$$\Sigma M_x = \Sigma M_y = \Sigma M_z = 0 \tag{1.2}$$

If the conditions in equations (1.1) and (1.2) are not satisfied then there is a resultant force or moment, which is given by

$$F = [(\Sigma F_x)^2 + (\Sigma F_y)^2 + (\Sigma F_z)^2]^{1/2}$$

$$M = [(\Sigma M_x)^2 + (\Sigma M_y)^2 + (\Sigma M_z)^2]^{1/2}$$

The six conditions given in equations (1.1) and (1.2) satisfy problems in three dimensions. If one of these dimensions is not present (say, the z direction) the system reduces to a set of coplanar forces, and then

$$\Sigma F_x = \Sigma M_x = \Sigma M_y = 0$$

are automatically satisfied, and the necessary conditions of equilibrium in a two-dimensional system are

$$\Sigma F_x = \Sigma F_y = \Sigma M_z = 0 \tag{1.3}$$

If the conditions in equation (1.3) are not satisfied then the resultant force or moment is given by

$$F = [(\Sigma F_x)^2 + (\Sigma F_y)^2]^{1/2}$$

$$M = \Sigma M_z$$

The above equations give solutions to what are said to be 'statically determinate' systems. These are systems where there are the minimum number of constraints to maintain equilibrium.[1]

1.2 Strength of materials

Weight: The weight (W) of a body is that force exerted due to gravitational attraction on the mass (m) of the body: $W = mg$, where g is the acceleration due to gravity.

Centre of gravity: This is a point, which may or may not be within the body, at which the total weight of the body may be considered to act as a single force. The position of the centre of gravity may be found experimentally or by analysis. When using analysis the moment of each element of weight, within the body, about a fixed axis is equated to the moment of the complete weight about that axis:

$$\bar{x} = \Sigma \delta mg \cdot x/\Sigma \delta mg, \quad \bar{y} = \Sigma \delta mg \cdot y/\Sigma \delta mg,$$

$$\bar{z} = \Sigma \delta mg \cdot z/\Sigma \delta mg$$

where δm is an element of mass at a distance of x, y or z from the respective axis, and $\bar{x}$, $\bar{y}$ and $\bar{z}$ are the positions of the centres of gravity from these axes. Table 1.1 shows the position of the centre of gravity for some standard shapes. (See reference 2 for a more comprehensive list.)

Shear force and bending moment: If a beam subject to loading, as shown in Figure 1.1, is cut, then in order to maintain equilibrium a shear force (Q) and a bending moment (M) must be applied to each portion of the beam. The magnitudes of Q and M vary with the type of loading and the position along the beam and are directly related to the stresses and deflections in the beam.

Relationship between shear force and bending moment: If an element of a beam is subjected to a load w then the following relationship holds:

$$\frac{d^2M}{dx^2} = \frac{dF}{dx} = -w$$

Table 1.2 shows examples of bending moments, shear force and maximum deflection for standard beams.

Bending equation: If a beam has two axes of symmetry in the xy plane then the following equation holds:

$$M_Z/I_Z = E/R_Z = \sigma/y$$

where M_Z is the bending moment, R_Z is the radius of curvature, I_Z the moment of inertia, E the modulus of elasticity, y the distance from the principal axis and σ is the stress.

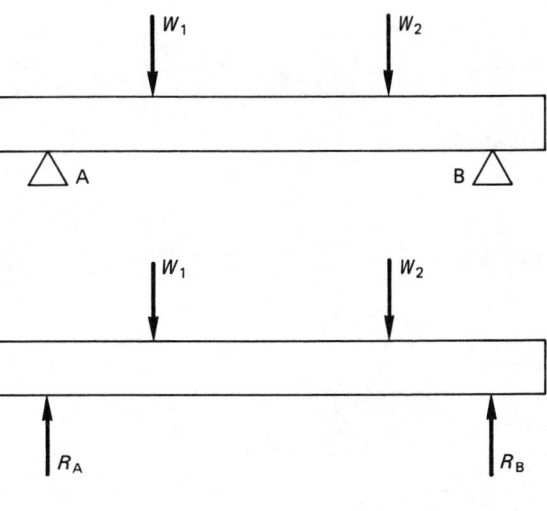

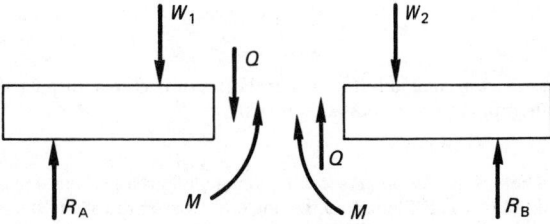

Figure 1.1

Table 1.1 Centres of gravity and moments of inertia or second moments of area for two-dimensional figures

Shape	G	I
Triangular area	$\bar{y} = h/3$	$I_{GG} = bh^3/36$
		$I_{XX} = bh^3/12$
Rectangular area		$I_{GG} = bh^3/12$
		$I_{XX} = bh^3/3$
Circular sector	$\bar{x} = \dfrac{2r}{3}\dfrac{\sin\alpha}{\alpha}$	$I_{XX} = \dfrac{r^4}{4}\left(\alpha - \dfrac{1}{2}\sin^2\alpha\right)$
		$I_{YY} = \dfrac{r^4}{4}\left(\alpha + \dfrac{1}{2}\sin^2\alpha\right)$
Slender rod		$I_{XX} = I_{ZZ} = ml^2/12$
Rectangular prism		$I_{XX} = m(b^2 + c^2)/12$
		$I_{YY} = m(c^2 + a^2)/12$
		$I_{ZZ} = m(a^2 + b^2)/12$
Thin disc		$I_{XX} = mr^2/2$
		$I_{YY} = I_{ZZ} = mr^2/4$
Circular cylinder		$I_{XX} = mr^2/2$
		$I_{YY} = I_{ZZ} = m(3r^2 + L^2)/12$
Sphere		$I_{XX} = I_{YY} = I_{ZZ} = 2mr^2/5$
Hemisphere	$\bar{y} = 3r/8$	$I_{YY} = 2mr^2/5$
Circular cone	$\bar{x} = h/4$	$I_{XX} = 3mr^2/10$
		$I_{YY} = \dfrac{3mr^2}{20} + \dfrac{mh^2}{10}$

Torsion equation: If a circular shaft is subject to a torque (T) then the following equation holds:

$$T/J = \tau/r = G\theta/L$$

where J is the polar second moment of area, G the shear modulus, L the length, θ the angle of twist, τ the shear stress and r the radius of the shaft.

1.3 Dynamics of rigid bodies

1.3.1 Basic definitions

1.3.1.1 Newton's Laws of Motion

First Law   A particle remains at rest or continues to move in a straight line with a constant velocity unless acted on by an external force.

Table 1.2

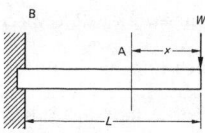

One concentrated load W
M at A $= Wx$, Q at A $= W$
M greatest at B, and $= WL$
Q uniform throughout
Maximum deflection $= WL^3/3EI$
at the free end.

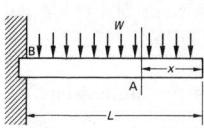

Uniform load of W
M at A $= Wx^2/2L$
Q at A $= Wx/L$
M greatest at B $= WL/2$
Q greatest at B $= W$
Maximum deflection $= WL^3/8EI$
at the free end.

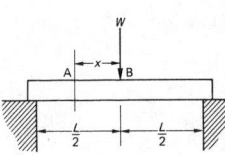

One concentrated load at the
centre of a beam

$$M \text{ at A} = \frac{W}{2}\left(\frac{L}{2} - x\right),$$

Q at A $= W/2$

M greatest at B $= WL/4$
Q uniform throughout
Maximum deflection $= WL^3/48EI$
at the centre

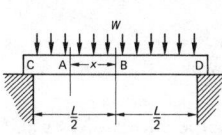

Uniform load W

$$M \text{ at A} = \frac{W}{2L}\left(\frac{L^2}{4} - x^2\right)$$

Q at A $= Wx/L$
M greatest at B $= WL/8$
Q greatest at C and D $= W/2$
maximum deflection at
B $= 5WL^3/384EI$

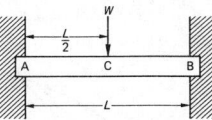

Beam fixed at ends and loaded at
centre.
M is maximum at A, B and C
and $= WL/8$.
Maximum deflection at
C $= WL^3/192EI$

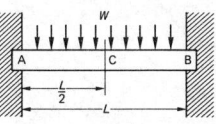

Beam fixed at ends with uniform
load.
M maximum at A and B
and $= WL/12$
Maximum deflection at
C $= WL^3/384EI$

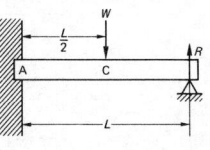

One concentrated load W
Reaction $R = 5W/16$
M maximum at A, and $= 3WL/16$
M at C $= 5WL/32$
Maximum deflection is $L/\sqrt{5}$ from
the free end, and $= WL^3/107EI$

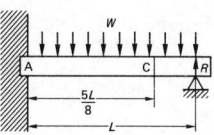

Uniform load W
Reaction $R = 3W/8$
M maximum at A, and $= WL/8$
M at C $= 9WL/128$
Maximum deflection is $3L/8$ from
the free end, and $= WL^3/187EI$

Second Law The sum of all the external forces acting on a particle is proportional to the rate of change of momentum.

Third Law The forces of action and reaction between interacting bodies are equal in magnitude and opposite in direction.

Newton's law of gravitation, which governs the mutual interaction between bodies, states

$$F = Gm_1m_2/x^2$$

where F is the mutual force of attraction, G is a universal constant called the constant of gravitation which has a value 6.673×10^{-11} m³ kg^{-1} s^{-2}, m_1 and m_2 are the masses of the two bodies and x is the distance between the centres of the bodies.

Mass (m) is a measure of the amount of matter present in a body.

Velocity is the rate of change of distance (x) with time (t):

$$v = dx/dt \text{ or } \dot{x}$$

Acceleration is the rate of change of velocity (v) with time (t):

$$a = dv/dt \text{ or } d^2x/dt^2 \text{ or } \ddot{x}$$

Momentum is the product of the mass and the velocity. If no external forces are present then the momentum of any system remains constant. This is known as the Conservation of Momentum.

Force is equal to the rate of change of momentum (mv) with time (t):

$$F = d(mv)/dt$$

$$F = m \cdot dv/dt + v \cdot dm/dt$$

If the mass remains constant then this simplifies to $F = m \cdot dv/dt$, i.e. Force = mass × acceleration, and it is measured in Newtons.

Impulse (I) is the product of the force and the time that force acts. Since $I = Ft = mat = m(v_2 - v_1)$, impulse is also said to be the change in momentum.

Energy: There are several different forms of energy which may exist in a system. These may be converted from one type to another but they can never be destroyed. Energy is measured in Joules.

Potential energy (PE) is the energy which a body possesses by virtue of its position in relation to other bodies: $PE = mgh$, where h is the distance above some fixed datum and g is the acceleration due to gravity.

Kinetic energy (KE) is the energy a body possesses by virtue of its motion: $KE = \frac{1}{2}mv^2$.

Work (W) is a measure of the amount of energy produced when a force moves a body a given distance: $W = F \cdot x$.

Power (P) is the rate of doing work with respect to time and is measured in watts.

Moment of inertia (I): The moment of inertia is that property in a rotational system which may be considered equivalent to the mass in a translational system. It is defined about an axis xx as $I_{XX} = \Sigma \delta mx^2 = mk^2_{XX}$, where x is the perpendicular distance of an element of mass δm from the axis xx and k_{XX} is the radius of gyration about the axis xx. Table 1.1 gives some data on moments of inertia for standard shapes.

Angular velocity (ω) is the rate of change of angular distance (θ) with time:

$$= d\theta/dt = \dot{\theta}$$

Angular acceleration (α) is the rate of change of angular velocity (ω) with time:

$$= d\omega/dt \text{ or } d^2\theta/dt^2 \text{ or } \ddot{\theta}$$

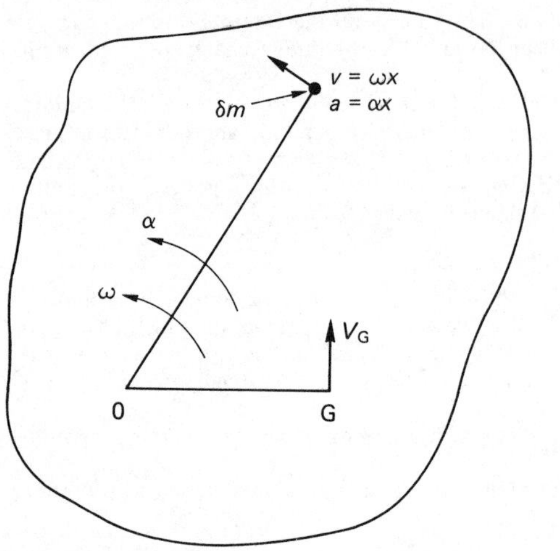

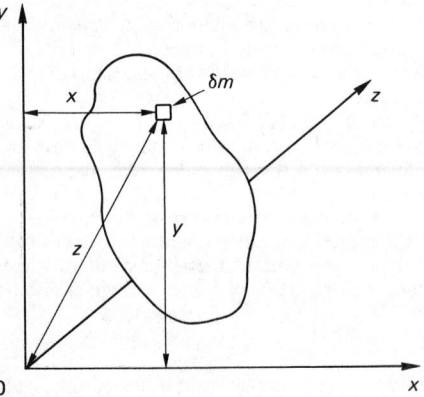

Figure 1.2

Both angular velocity and acceleration are related to linear motion by the equations $v = \omega x$ and $a = \alpha x$ (see Figure 1.2).

Torque (T) is the moment of force about the axis of rotation:

$$T = I_0\alpha$$

A torque may also be equal to a *couple*, which is two forces equal in magnitude acting some distance apart in opposite directions.

Parallel axis theorem: if I_{GG} is the moment of inertia of a body of mass m about its centre of gravity, then the moment of inertia (I) about some other axis parallel to the original axis is given by $I = I_{GG} + mr^2$, where r is the perpendicular distance between the parallel axes.

Perpendicular axis theorem. If I_{XX}, I_{YY} and I_{ZZ} represent the moments of inertia about three mutually perpendicular axes x, y and z for a plane figure in the xy plane (see Figure 1.3) then $I_{ZZ} = I_{XX} + I_{YY}$.

Angular momentum (H_O) of a body about a point O is the moment of the linear momentum about that point and is ωI_{OO}. The angular momentum of a system remains constant unless acted on by an external torque.

Angular impulse is the produce of torque by time, i.e. angular impulse $= Tt = I\alpha \cdot t = I(\omega_2 - \omega_1)$, the change in angular momentum.

Angular kinetic energy about an axis O is given by $\frac{1}{2}I_O\omega^2$.

Work done due to a torque is the product of torque by angular distance and is given by $T\theta$.

Power due to torque is the rate of angular work with respect to time and is given by $Td\theta/dt = T\omega$.

Friction: Whenever two surfaces, which remain in contact, move one relative to the other there is a force which acts tangentially to the surfaces so as to oppose motion. This is known as the force of friction. The magnitude of this force is μR, where R is the normal reaction and μ is a constant known as the coefficient of friction. The coefficient of friction depends on the nature of the surfaces in contact.

1.3.2 Linear and angular motion in two dimensions

Constant acceleration: If the accleration is integrated twice and the relevant initial conditions are used, then the following equations hold:

Linear motion	Angular motion
$x = v_1 t + \frac{1}{2}at^2$	$\theta = \omega_1 t + \frac{1}{2}\alpha t^2$
$v_2 = v_1 + at$	$\omega_2 = \omega_1 + \alpha t$
$v_2^2 = v_1^2 + 2ax$	$\omega_2^2 = \omega_1^2 + 2\alpha\theta$

Variable acceleration: If the acceleration is a function of time then the area under the acceleration time curve represents the change in velocity. If the acceleration is a function of displacement then the area under the acceleration distance curve represents half the difference of the square of the velocities (see Figure 1.4).

Curvilinear motion is when both linear and angular motions are present.

If a particle has a velocity v and an acceleration a then its motion may be described in the following ways:

1. *Cartesian components* which represent the velocity and acceleration along two mutually perpendicular axes x and y (see Figure 1.5(a)):

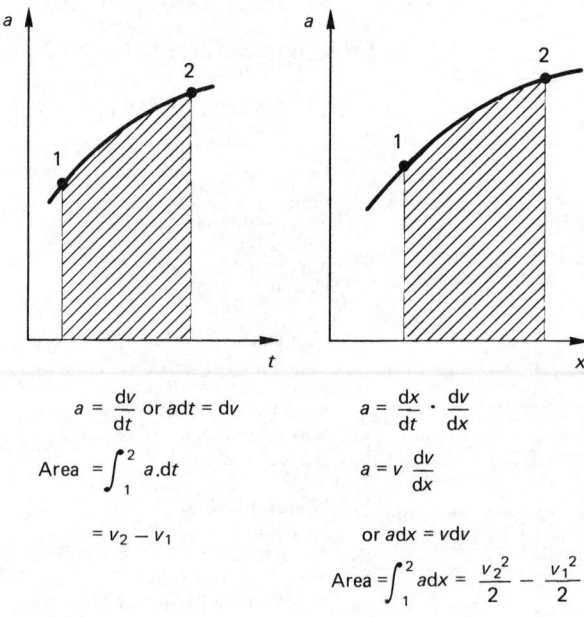

$$a = \frac{dv}{dt} \text{ or } adt = dv$$

$$\text{Area} = \int_1^2 a.dt$$

$$= v_2 - v_1$$

$$a = \frac{dx}{dt} \cdot \frac{dv}{dx}$$

$$a = v\frac{dv}{dx}$$

$$\text{or } adx = vdv$$

$$\text{Area} = \int_1^2 adx = \frac{v_2^2}{2} - \frac{v_1^2}{2}$$

Figure 1.3

Figure 1.4

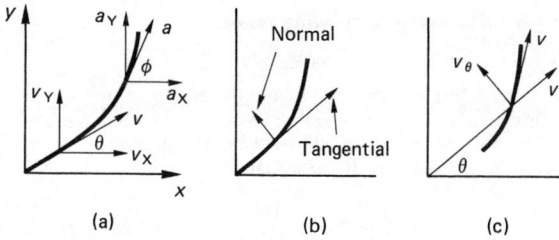

(a)   (b)   (c)

Figure 1.5

$$v_X = v \cos \theta, \, v_Y = v \sin \theta, \, a_X = a \cos \phi,$$
$$a_Y = a \sin \phi$$

2. *Normal and tangential components*: see Figure 1.5(b):

$$v_t = v = r\dot\theta = r\omega, \, v_n = 0$$
$$a_t = r\ddot\theta + \dot r\dot\theta = r\alpha + \dot r\omega,$$
$$a_n = v\dot\theta = r\omega^2$$

3. *Polar coordinates*: see Figure 1.5(c):

$$v_r = \dot r, \, v_\theta = r\dot\theta$$
$$a_r = \ddot r - r\dot\theta^2, \, a_\theta = r\ddot\theta + 2\dot r\dot\theta$$

1.3.3 Circular motion

Circular motion is a special case of curvilinear motion in which the radius of rotation remains constant. In this case there is an acceleration towards the cente of $\omega^2 r$. This gives rise to a force towards the centre known as the *centripetal force*. This force is reacted to by what is called the *centrifugal reaction*.

Velocity and acceleration in mechanisms: A simple approach to determine the velocity and acceleration of a mechanism at a point in time is to draw velocity and acceleration vector diagrams.

Velocities: If in a rigid link AB of length l the end A is moving with a different velocity to the end B, then the velocity of A relative to B is in a direction perpendicular to AB (see Figure 1.6).

When a block slides on a rotating link the velocity is made up of two components, one being the velocity of the block relative to the link and the other the velocity of the link.

Accelerations: If the link has an angular acceleration α then there will be two components of acceleration in the diagram, a tangential component αl and a centripetal component of magnitude $\omega^2 l$ acting towards A.

When a block slides on a rotating link the total acceleration is composed of four parts: first, the centripetal acceleration towards O of magnitude $\omega^2 l$; second, the tangential acceleration αl; third, the acceleration of the block relative to the link; fourth, a tangential acceleration of magnitude $2v\omega$ known as Coriolis acceleration. The direction of Coriolis acceleration is determined by rotating the sliding velocity vector through 90° in the direction of the link angular velocity ω.

1.3.4 Linear and angular motion in three dimensions

1.3.4.1 Motion of a particle in a moving coordinate system

xyz is a moving coordinate system, with its origin at O which has a position vector **R**, a translational velocity vector $\dot{\mathbf{R}}$ and an angular velocity vector $\boldsymbol\omega$ relative to a fixed coordinate system XYZ, origin at O′. Then the motion of a point P whose position vector relative to O is $\boldsymbol\rho$ and relative to O′ is **r** is given by the following equations (see Figure 1.7):

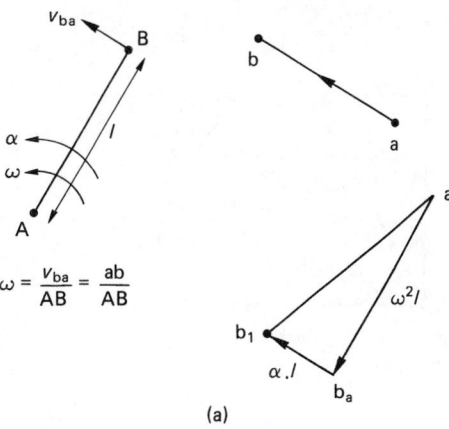

$$\omega = \frac{v_{ba}}{AB} = \frac{ab}{AB}$$

(a)

E is on the link
F is on the slider

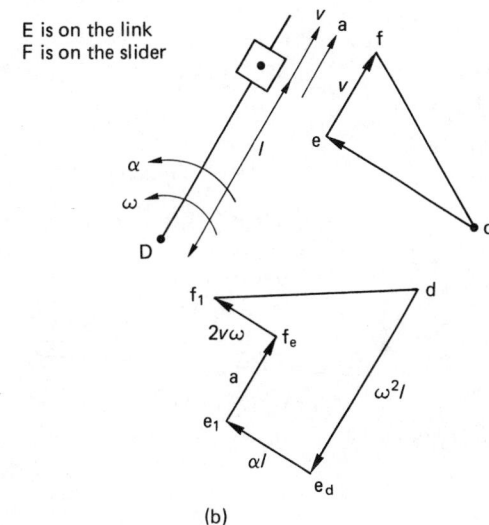

(b)

Figure 1.6

$$\dot{\mathbf{r}} = \dot{\mathbf{R}} + \dot{\boldsymbol\rho}_r + \boldsymbol\omega \times \boldsymbol\rho$$

where $\dot{\boldsymbol\rho}_r$ is the velocity of the point P relative to the moving system xyz and $\boldsymbol\omega \times \boldsymbol\rho$ is the vector product of $\boldsymbol\omega$ and $\boldsymbol\rho$:

$$\ddot{\mathbf{r}} = \ddot{\mathbf{R}} + \dot{\boldsymbol\omega} \times \boldsymbol\rho + \boldsymbol\omega \times (\boldsymbol\omega \times \boldsymbol\rho) + 2\boldsymbol\omega \times \dot{\boldsymbol\rho}_r + \ddot{\boldsymbol\rho}_r$$

where $\ddot{\boldsymbol\rho}_r$ is the acceleration of the point P relative to the moving system. Thus $\dot{\mathbf{r}}$ is the sum of:

1. The relative velocity $\dot{\boldsymbol\rho}_r$;
2. The absolute velocity $\dot{\mathbf{R}}$ of the moving origin O;
3. The velocity $\boldsymbol\omega \times \boldsymbol\rho$ due to the angular velocity of the moving axes xyz.

and $\ddot{\mathbf{r}}$ is the sum of:

1. The relative acceleration $\ddot{\boldsymbol\rho}_r$;
2. The absolute acceleration $\ddot{\mathbf{R}}$ of the moving origin O;
3. The tangential acceleration $\dot{\boldsymbol\omega} \times \boldsymbol\rho$ due to the angular acceleration of the moving axes xyz;
4. The centripetal acceleration $\boldsymbol\omega \times (\boldsymbol\omega \times \boldsymbol\rho)$ due to the angular velocity of the moving axes xyz;
5. Coriolis component acceleration $2\boldsymbol\omega \times \dot{\boldsymbol\rho}_r$ due to the interaction of coordinate angular velocity and relative velocity.

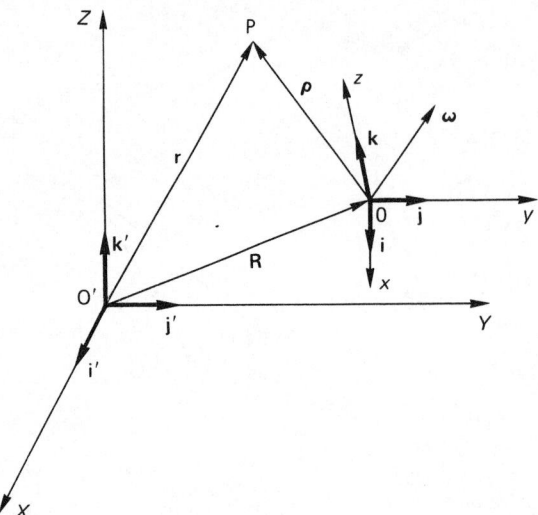

Figure 1.7

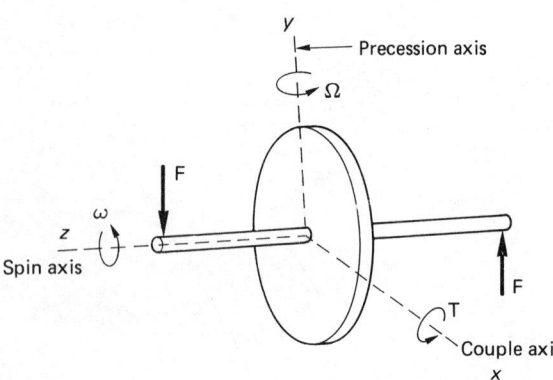

Figure 1.8

In all the vector notation a right-handed set of coordinate axes and the right-hand screw rule is used.

1.3.4.2 Gyroscopic effects

Consider a rotor which spins about its geometric axis (see Figure 1.8) with an angular velocity ω. Then two forces F acting on the axle to form a torque T, whose vector is along the x axis, will produce a rotation about the y axis. This is known as precession, and it has an angular velocity Ω. It is also the case that if the rotor is precessed then a torque T will be produced, where T is given by $T = I_{XX}\omega\Omega$. When this is observed it is the effect of gyroscopic reaction torque that is seen, which is in the opposite direction to the gyroscopic torque.[4]

1.3.5 Balancing

In any rotational or reciprocating machine where accelerations are present, unbalanced forces can lead to high stresses and vibrations. The principle of balancing is such that by the addition of extra masses to the system the out-of-balance forces may be reduced or eliminated.

1.3.6 Balancing of rotating masses

1.3.6.1 Single out-of-balance mass

One mass (m) at a distance r from the centre of rotation and rotating at a constant angular velocity ω produces a force $m\omega^2 r$. This can be balanced by a mass M placed diametrically opposite at a distance R, such that $MR = mr$.

1.3.6.2 Several out-of-balance masses in one transverse plane

If a number of masses ($m_1, m_2, \ldots$) are at radii ($r_1, r_2, \ldots$) and angles ($\theta_1, \theta_2, \ldots$) (see Figure 1.9) then the balancing mass M must be placed at a radius R such that MR is the vector sum of all the mr terms.

1.3.6.3 Masses in different transverse planes

If the balancing mass in the case of a single out-of-balance mass were placed in a different plane then the centrifugal force would be balanced. This is known as *static balancing*. However, the moment of the balancing mass about the

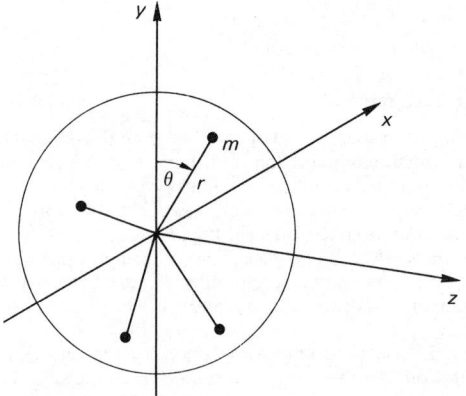

$$\Sigma F_X = \Sigma m\omega^2 r \sin\theta = 0$$
$$\Sigma F_Y = \Sigma m\omega^2 r \cos\theta = 0$$

Figure 1.9

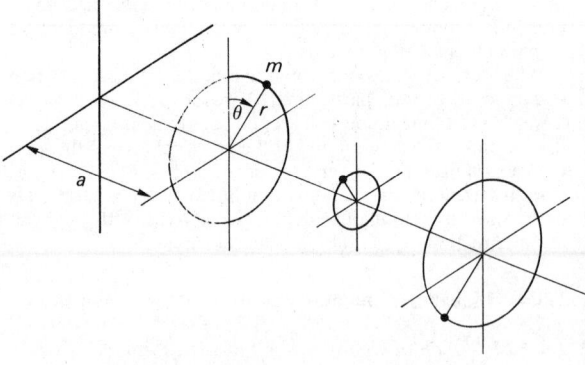

$\Sigma F_X = \Sigma m\omega^2 r \sin\theta = 0$ and $\Sigma F_Y = \Sigma m\omega^2 r \cos\theta = 0$
as in the previous case, also
$\Sigma M_X = \Sigma m\omega^2 r \sin\theta . a = 0$
$\Sigma M_Y = \Sigma m\omega^2 r \cos\theta . a = 0$

Figure 1.10

original plane would lead to what is known as *dynamic unbalance*.

To overcome this, the vector sum of all the moments about the reference plane must also be zero. In general, this requires two masses placed in convenient planes (see Figure 1.10).

1.3.6.4 Balancing of reciprocating masses in single-cylinder machines

The acceleration of a piston as shown in Figure 1.11 may be represented by the equation[5]

$$\ddot{x} = -\omega^2 r[\cos\theta + (1/n)\cos 2\theta + (1/4n)$$

$$(\cos 2\theta - \cos 4\theta) + \ldots]^*$$

where $n = 1/r$. If n is large then the equation may be simplified and the force given by

$$F = m\ddot{x} \simeq -m\omega^2 r[\cos\theta + (1/n)\cos 2\theta]$$

The term $m\omega^2 r\cos\theta$ is known as the *primary force* and $(1/n)m\omega^2 r\cos 2\theta$ as the *secondary force*. Partial primary balance is achieved in a single-cylinder machine by an extra mass M at a radius R rotating at the crankshaft speed. Partial secondary balance could be achieved by a mass rotating at 2ω. As this is not practical this is not attempted. When partial primary balance is attempted a transverse component $M\omega^2 R\sin\theta$ is introduced. The values of M and R are chosen to produce a compromise between the reciprocating and the transverse components.

1.3.6.5 Balancing of reciprocating masses in multi-cylinder machines

When considering multi-cylinder machines account must be taken of the force produced by each cylinder and the moment of that force about some datum. The conditions for primary balance are

$$F = \Sigma m\omega^2 r \cos\theta = 0, \quad M = \Sigma m\omega^2 r\cos\theta \cdot a = 0$$

where a is the distance of the reciprocating mass m from the datum plane.

In general, the cranks in multi-cylinder engines are arranged to assist primary balance. If primary balance is not complete then extra masses may be added to the crankshaft but these will introduce an unbalanced transverse component. The conditions for secondary balance are

$$F = \Sigma m\omega^2 (r/n) \cos 2\theta = \Sigma m(2\omega)^2 (r/4n) \cos 2\theta = 0$$

and

$$M = \Sigma m(2\omega)^2 (r/4n) \cos 2\theta \cdot a = 0$$

The addition of extra masses to give secondary balance is not attempted in practical situations.

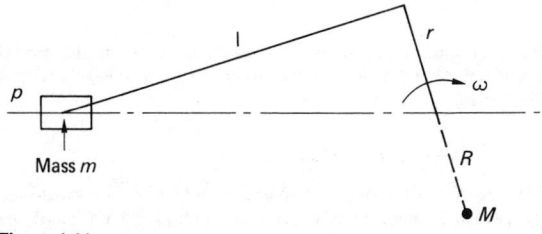

Figure 1.11

* This equation forms an infinite series in which higher terms are small and they may be ignored for practical situations.

1.4   Vibrations

1.4.1   Single-degree-of-freedom systems

The term *degrees of freedom* in an elastic vibrating system is the number of parameters required to define the configuration of the system. To analyse a vibrating system a mathematical model is constructed, which consists of springs and masses for linear vibrations. The type of analysis then used depends on the complexity of the model.

Rayleigh's method: Rayleigh showed that if a reasonable deflection curve is assumed for a vibrating system, then by considering the kinetic and potential energies* an estimate to the first natural frequency could be found. If an inaccurate curve is used then the system is subject to constraints to vibrate it in this unreal form, and this implies extra stiffness such that the natural frequency found will always be high. If the exact deflection curve is used then the natural frequency will be exact.

1.4.1.1   Transverse vibration of beams

Consider a beam of length (l), weight per unit length (w), modulus (E) and moment of inertia (I). Then its equation of motion is given by

$$EI \frac{d^4 y}{dx^4} - w\omega^2 y/g = 0$$

where ω is the natural frequency. The general solution of this equation is given by

$$y = A \cos\beta x + B \sin\beta x + C \cosh\beta x + D \sinh\beta x$$

where $\beta^4 = w\omega^2/gEI$.

The four constants of integration A, B, C and D are determined by four independent end conditions. In the solution trigonometrical identities are formed in β which may be solved graphically, and each solution corresponds to a natural frequency of vibration. Table 1.3 shows the solutions and frequencies for standard beams.[6]

Dunkerley's empirical method is used for beams with multiple loads. In this method the natural frequency (f_1) is found due to just one of the loads, the rest being ignored. This is repeated for each load in turn and then the natural frequency of vibration of the beam due to its weight alone is found (f_0).

* Consider the equation of motion for an undamped system (Figure 1.13):

$$m \cdot \frac{d^2 x}{dt^2} + kx = 0 \qquad (1.4)$$

but

$$\frac{d^2 x}{dt^2} = \frac{d}{dt}\left(\frac{dx}{dt}\right) = \frac{dx}{dt} \cdot \frac{d}{dx}\left(\frac{dx}{dt}\right) = \frac{1}{2} \cdot \frac{d}{dx}\left(\frac{dx}{dt}\right)^2$$

Therefore equation (1.4) becomes

$$\frac{d}{dx}\left[\frac{1}{2}m\left(\frac{dx}{dt}\right)^2\right] + kx = 0$$

Integrating gives

$$\frac{1}{2}m\left(\frac{dx}{dt}\right)^2 + \frac{1}{2}kx^2 = \text{Constant}$$

the term $\frac{1}{2}m(dx/dt)^2$ represents the kinetic energy and $\frac{1}{2}kx^2$ the potential energy.

Table 1.3

	End conditions	Trig. equation	Solutions $\beta_1 1$	$\beta_2 1$	$\beta_3 1$
	$x = 0, y = 0, y' = 0$ $x = 1, y = 0, y' = 0$	$\cos \beta 1 \cdot \cosh \beta 1 = 1$	4.730	7.853	10.966
	$x = 0, y = 0, y' = 0$ $x = 1, y'' = 0, y''' = 0$	$\cos \beta 1 \cdot \cosh \beta 1 = -1$	1.875	4.694	7.855
	$x = 0, y = 0, y'' = 0$ $x = 1, y = 0, y'' = 0$	$\sin \beta 1 = 0$	3.142	6.283	9.425
	$x = 0, y = 0, y' = 0$ $x = 1, y = 0, y'' = 0$	$\tan \beta 1 = \tanh \beta 1$	3.927	7.069	10.210

Then the natural frequency of vibration of the complete system (f) is given by

$$\frac{1}{f^2} = \frac{1}{f_0^2} + \frac{1}{f_1^2} + \frac{1}{f_2^2} + \frac{1}{f_3^2} + \ldots \ldots \frac{1}{f_n^2}$$

(see reference 7 for a more detailed explanation).

Whirling of shafts: If the speed of a shaft or rotor is slowly increased from rest there will be a speed where the deflection increases suddenly. This phenomenon is known as whirling. Consider a shaft with a rotor of mass m such that the centre of gravity is eccentric by an amount e. If the shaft now rotates at an angular velocity ω then the shaft will deflect by an amount y due to the centrifugal reaction (see Figure 1.12). Then

$$m\omega^2(y + e) = ky$$

where k is the stiffness of the shaft. Therefore

$$y = \frac{e}{(k/m\omega^2 - 1)}$$

When $(k/m\omega^2) = 1$, y is then infinite and the shaft is said to be at its critical whirling speed ω_c. At any other angular velocity ω the deflection y is given by

$$y = \left(\frac{\omega^2}{\omega_c^2 - \omega^2}\right) . e$$

When $\omega < \omega_c$, y is the same sign as e and as ω increases towards ω_c the deflection theoretically approaches infinity. When $\omega > \omega_c$, y is opposite in sign to e and will eventually tend to $-e$. This is a desirable running condition with the centre of gravity of the rotor mass on the static deflection curve. Care must be taken not to increase ω too high as ω might start to approach one of the higher modes of vibration.[8]

Torsional vibrations: The following section deals with transverse vibrating systems with displacements x and masses m. The same equations may be used for torsional vibrating systems by replacing x by θ the angular displacement and m by I, the moment of inertia.

1.4.1.2 Undamped free vibrations

The equation of motion is given by $m\ddot{x} + kx = 0$ or $\ddot{x} + \omega_n^2 x = 0$, where m is the mass, k the stiffness and $\omega_n^2 = k/m$, which is the natural frequency of vibration of the system (see Figure 1.13). The solution to this equation is given by

$$x = A \sin(\omega_n t + \alpha)$$

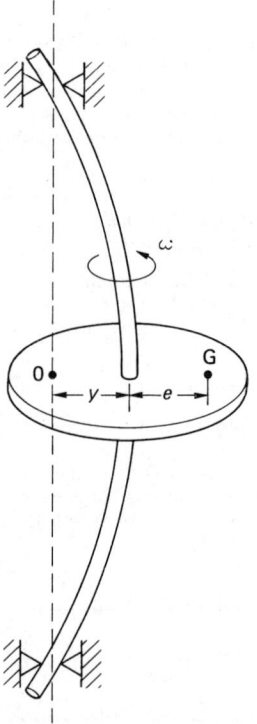

Figure 1.12

where A and α are constants which depend on the initial conditions. This motion is said to be *simple harmonic* with a time period $T = 2\pi/\omega_n$.

1.4.1.3 Damped free vibrations

The equation of motion is given by $m\ddot{x} + c\dot{x} + kx = 0$ (see Figure 1.14), where c is the viscous damping coefficient, or $\ddot{x} + (c/m)\dot{x} + \omega_n^2 x = 0$. The solution to this equation and the resulting motion depends on the amount of damping. If $c > 2m\omega_n$ the system is said to be overdamped. It will respond to a disturbance by slowly returning to its equilibrium posi-

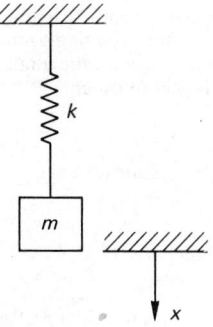

Figure 1.13

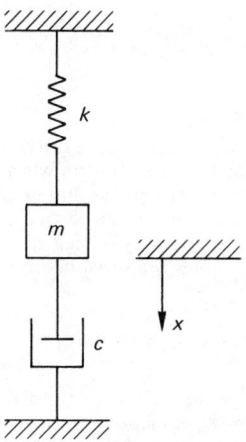

Figure 1.14

tion. The time taken to return to this position depends on the degree of damping (see Figure 1.15(c)). If $c = 2m\omega_n$ the system is said to be critically damped. In this case it will respond to a disturbance by returning to its equilibrium position in the shortest possible time. In this case (see Figure 1.15(b))

$$x = e^{-(c/2m)t(A+Bt)}$$

where A and B are constants. If $c < 2m\omega_n$ the system has a transient oscillatory motion given by

$$x = e^{-(c/2m)t}[C \sin(\omega_n^2 - c^2/4m^2)^{1/2}t + D(\cos \omega_n^2 - c^2/4m^2)^{1/2}t]$$

where C and D are constants. The period

$$T = \frac{2\pi}{(\omega_n^2 - c^2/4m^2)^{1/2}}$$

(see Figure 1.15(a)).

1.4.1.4   Logarithmic decrement

A way to determine the amount of damping in a system is to measure the rate of decay of successive oscillations. This is expressed by a term called the *logarithmic decrement* (δ), which is defined as the natural logarithm of the ratio of any two successive amplitudes (see Figure 1.16):

$$\delta = \log_e(x_1/x_2)$$

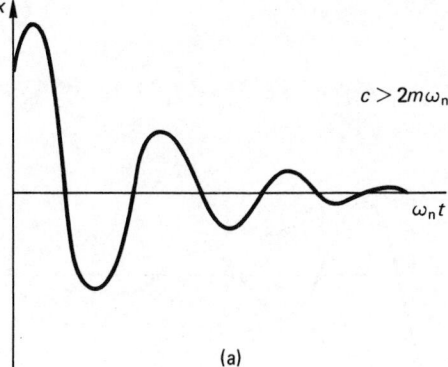

(a)

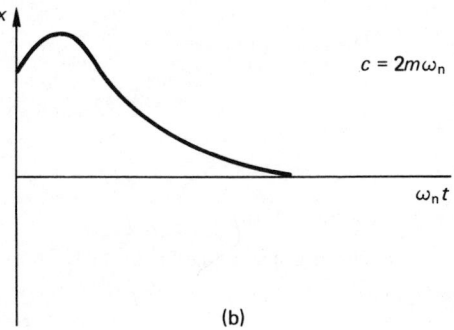

(b)

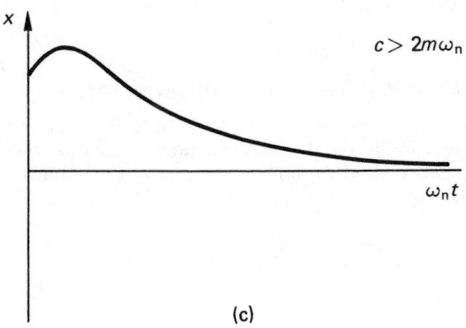

(c)

Figure 1.15

where x is given by

$$x = e^{cT/m} \sin\left[\left(\omega^2 - \frac{c^2}{4m^2}\right)^{1/2} + \phi\right]$$

Therefore

$$\delta = \log_e(e^{-cT/2m}/e^{-c(T+\tau)/2m})$$

$$= c\tau/2m$$

where τ is the period of damped oscillation.

If the amount of damping present is small compared to the critical damping, τ approximates to $2\pi/\omega$, and then

$$\delta = c\pi/m\omega_n$$

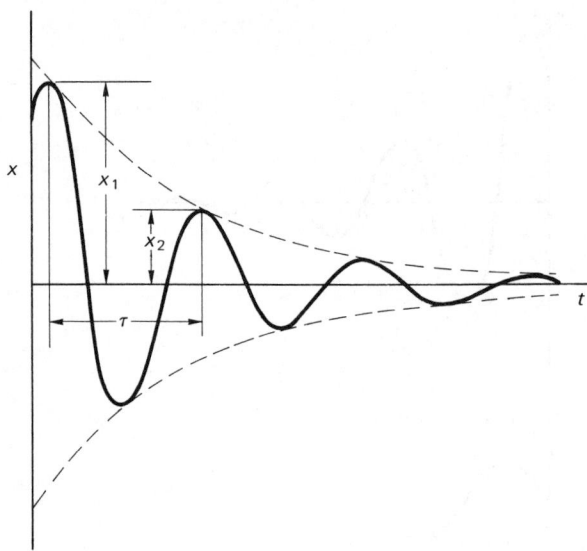

Figure 1.16

1.4.1.5 Forced undamped vibrations

The equation of motion is given by (see Figure 1.17)

$$m\ddot{x} + kx = F_0 \sin \omega t$$

or

$$\ddot{x} + \omega_n^2 = (F_0/m) \sin \omega t$$

The solution to this equation is

$$x = C \sin \omega_n t + D \cos \omega_n t + F_0 \cos \omega t/[m(\omega_n^2 - \omega^2)]$$

where ω is the frequency of the forced vibration. The first two terms of the solution are the transient terms which die out, leaving an oscillation at the forcing frequency of amplitude

$$F_0/[m(\omega_n^2 - \omega^2)]$$

or

$$\frac{F_0}{k}\left(\frac{\omega_n^2}{\omega_n^2 - \omega^2}\right)$$

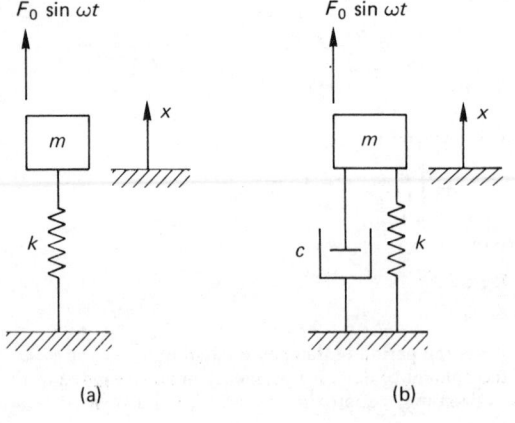

$F_0 \sin \omega t$ $F_0 \sin \omega t$

(a)   (b)

Figure 1.17

The term $\omega_n^2/(\omega_n^2 - \omega^2)$ is known as the dynamic magnifier and it gives the ratio of the amplitude of the vibration to the static deflection under the load F_0. When $\omega = \omega_n$ the amplitude becomes infinite and resonance is said to occur.

1.4.1.6 Forced damped vibrations

The equation of motion is given by (see Figure 1.17(b))

$$m\ddot{x} + c\dot{x} + kx = F_0 \sin \omega t$$

or

$$\ddot{x} + (c/m)\dot{x} + \omega_n^2 = (F_0/m) \sin \omega t$$

The solution to this equation is in two parts: a transient part as in the undamped case which dies away, leaving a sustained vibration at the forcing frequency given by

$$x = \frac{F_0}{m} \frac{1}{[(\omega_n^2 - \omega^2)^2 + (c(\omega/m)^2]^{1/2}} \sin(\omega t - \alpha)$$

The term

$$\frac{\omega_n^2}{[(\omega_n^2 - \omega^2)^2 + (c\omega/m)]^2]^{1/2}}$$

is called the dynamic magnifier. Resonance occurs when $\omega \simeq \omega_n$. As the damping is increased the value of ω for which resonance occurs is reduced. There is also a phase shift as ω increases tending to a maximum of π radians. It can be seen in Figure 1.18(a) that when the forcing frequency is high compared to the natural frequency the amplitude of vibration is minimized.

1.4.1.7 Forced damped vibrations due to reciprocating or rotating unbalance

Figure 1.19 shows two elastically mounted systems, (a) with the excitation supplied by the reciprocating motion of a piston, and (b) by the rotation of an unbalanced rotor. In each case the equation of motion is given by

$$(M - m)\ddot{x} + c\dot{x} + kx = (me\omega^2) \sin \omega t$$

The solution of this equation is a sinusoid whose amplitude, X, is given by

$$X = \frac{me\omega^2}{\sqrt{[(K - M\omega^2)^2 + (c\omega)^2]}}$$

In representing this information graphically it is convenient to plot MX/me against ω/ω_n for various levels of damping (see Figure 1.20(a)). From this figure it can be seen that for small values of ω the displacement is small, and as ω is increased the displacement reaches a maximum when ω is slightly greater than ω_n. As ω is further increased the displacement tends to a constant value such that the centre of gravity of the total mass M remains stationary. Figure 1.20(b) shows how the phase angle varies with frequency.

1.4.1.8 Forced damped vibration due to seismic excitation

If a system as shown in Figure 1.21 has a sinusoidal displacement applied to its base of amplitude, y, then the equation of motion becomes

$$m\ddot{x} + c\dot{x} + kx = ky + c\dot{y}$$

The solution of this equation yields

$$\frac{x}{y} = \sqrt{\left[\frac{k^2 + (c\omega)^2}{(k - m\omega^2)^2 + (c\omega)^2}\right]}$$

where x is the amplitude of motion of the system.

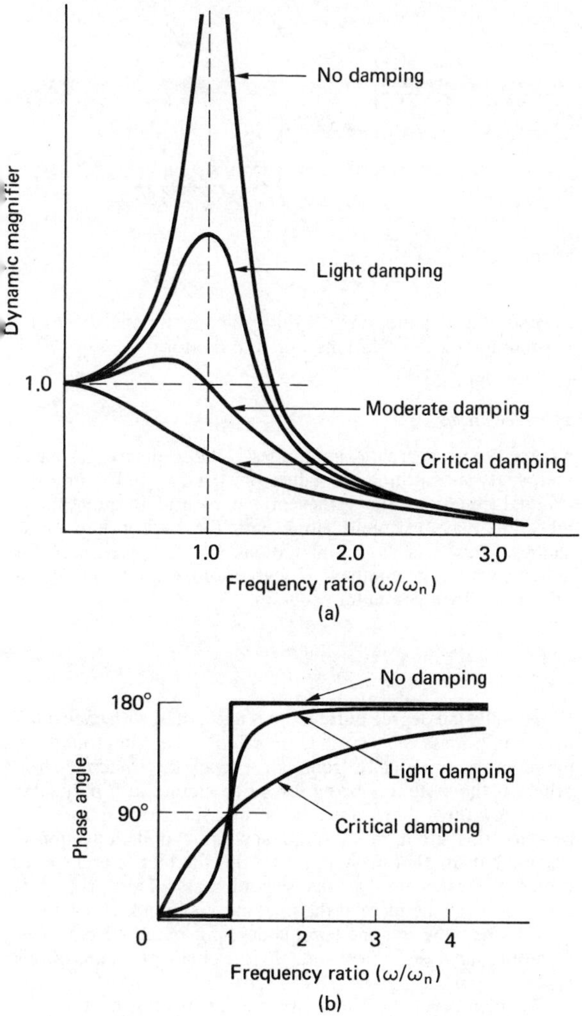

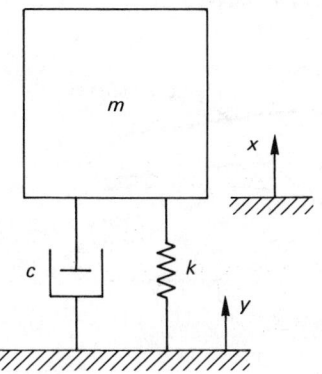

Figure 1.18

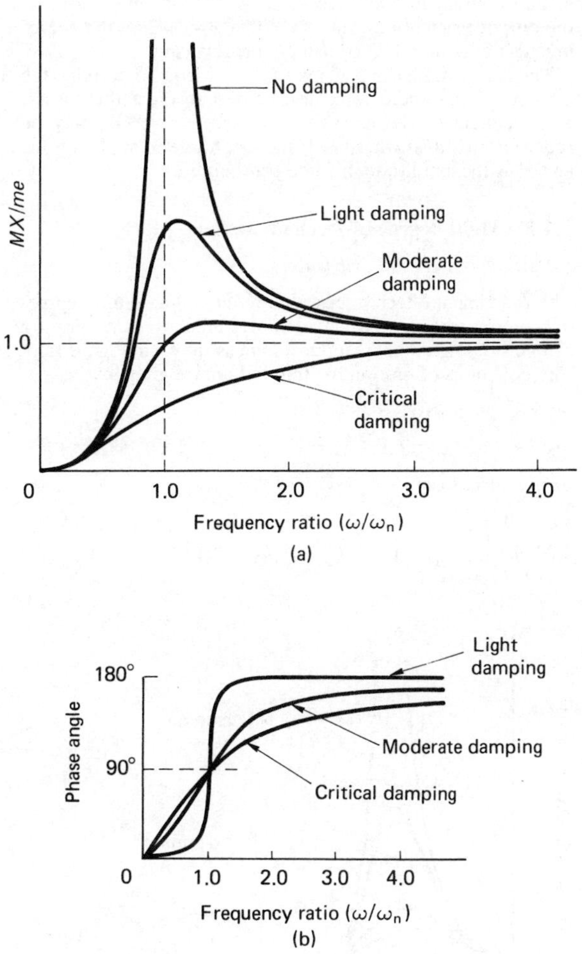

Figure 1.20

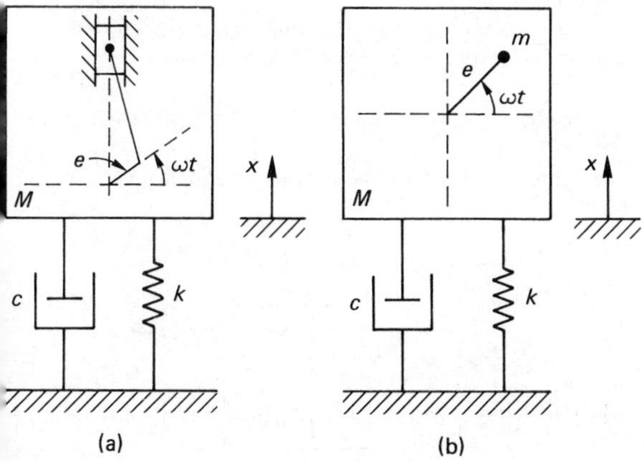

Figure 1.19

Figure 1.21

When this information is plotted as in Figure 1.22 it can be seen that for very small values of ω the output amplitude X is equal to the input amplitude Y. As ω is increased towards ω_n the output reaches a maximum. When $\omega = \sqrt{2}\,\omega_n$ the curves intersect and the effect of damping is reversed.

The curves in Figure 1.22 may also be used to determine the amount of sinusoidal force transmitted through the springs and dampers to the supports, i.e. the axis (X/Y) may be replaced by (F_t/F_0) where F_0 is the amplitude of applied force and F_t is the amplitude of force transmitted.

1.4.2   Multi-degree-of-freedom systems

1.4.2.1   Normal mode vibration

The fundamental techniques used in modelling multi-degree-of-freedom systems may be demonstrated by considering a simple two-degree-of-freedom system as shown in Figure 1.23. The equations of motion for this system are given by

$$m_1 \ddot{x}_1 + (k_1 + k_2)x_1 - k_2 x_2 = 0$$

$$m_2 \ddot{x}_2 + (k_3 + k_2)x_2 - k_2 x_1 = 0$$

or in matrix form:

$$\begin{bmatrix} m_1 & 0 \\ 0 & m_2 \end{bmatrix} \begin{Bmatrix} \ddot{x}_1 \\ \ddot{x}_2 \end{Bmatrix} + \begin{bmatrix} (k_1 + k_2) & -k_2 \\ -k_2 & (k_3 + k_2) \end{bmatrix} \begin{Bmatrix} x_1 \\ x_2 \end{Bmatrix} = \begin{Bmatrix} 0 \\ 0 \end{Bmatrix}$$

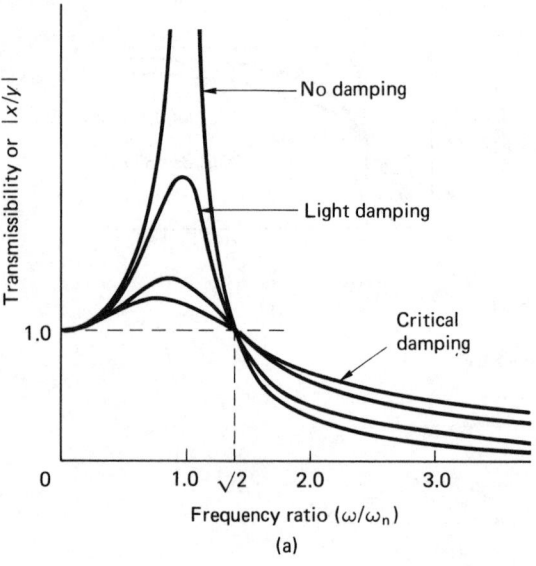

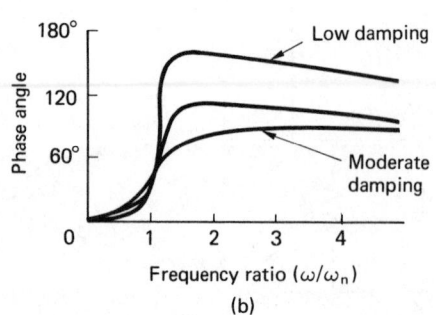

Figure 1.22

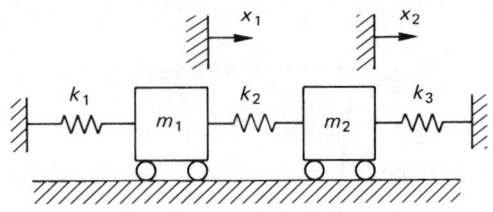

Figure 1.23

Assuming the motion of every point in the system to be harmonic then the solutions will take the form

$$x_1 = A_1 \sin \omega t$$

$$x_2 = A_2 \sin \omega t$$

where A_1 and A_2 are the amplitudes of the respective displacements. By substituting the values of x_1, x_2, $\ddot{x}_1$ and $\ddot{x}_2$ into the original equations the values of the natural frequencies of vibration may be found along with the appropriate mode shapes. This is a slow and tedious process, especially for systems with large numbers of degrees of freedom, and is best performed by a computer program.

1.4.2.2   The Holtzer method

When only one degree of freedom is associated with each mass in a multi-mass system then a solution can be found by proceeding numerically from one end of the system to the other. If the system is being forced to vibrate at a particular frequency then there must be a specific external force to produce this situation. A frequency and a unit deflection is assumed at the first mass and from this the inertia and spring forces are calculated at the second mass. This process is repeated until the force at the final mass is found. If this force is zero then the assumed frequency is a natural frequency. Computer analysis is most suitable for solving problems of this type.

Consider several springs and masses as shown in Figure 1.24. Then with a unit deflection at the mass m_1 and an assumed frequency ω there will be an inertia force of $m_1\omega^2$ acting on the spring with stiffness k_1. This causes a deflection of $m_1\omega^2/k_1$, but if m_2 has moved a distance x_2 then $m_1\omega^2/k_1 = 1 - x_2$ or $x_2 = 1 - m_1\omega^2/k_1$. The inertia force acting due to m_2 is $m_2\omega^2 x_2$, thus giving the total force acting on the spring of stiffness k_2 as $[m_1\omega^2 + m_2\omega^2 x_2]/k_2$. Hence the displacement at x_3 can be found and the procedure repeated. The external force acting on the final mass is then given by

$$\sum_{i=1}^{n} m_i \omega^2 x_i$$

If this force is zero then the assumed frequency is a natural one.

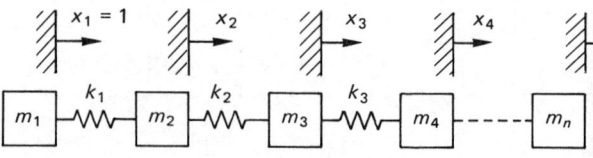

Figure 1.24

Further reading

Johnston, E. R. and Beer, F. P., *Mechanics for Engineers*, Volume 1, *Statics*; Volume 2, *Dynamics*, McGraw-Hill, New York (1987)

Meriam, J. L. and Kraige, L. G., *Engineering Mechanics*, Volume 1, *Statics*, second edition, Wiley, Chichester (1987)

Gorman, D. J., *Free Vibration Analysis of Beams and Shafts*, Wiley, Chichester (1975)

Nestorides, E. J., *A Handbook of Torsional Vibration*, Cambridge University Press, Cambridge (1958)

Harker, R., *Generalised Methods of Vibration Analysis*, Wiley, Chichester (1983)

Tse, F. S., Morse, I. E. and Hinkle, R. T., *Mechanical Vibrations: Theory and Applications*, second edition, Allyn and Bacon, New York (1979)

Hatter, D., *Matrix Computer Methods of Vibration Analysis*, Butterworths, London (1973)

Nikravesh, P. E., *Computer Aided Analysis of Mechanical Systems*, Prentice-Hall, Englewood Cliffs, NJ (1988)

British Standards

BS 3318: Locating the centre of gravity of earth moving equipment and heavy objects

BS 3851: 1982: Glossary of terms used in mechanical balancing of rotary machines

BS 3852: 1986: Dynamic balancing machines

BS 4675: 1986: Mechanical vibrations in rotating and reciprocating machinery

BS 6414: 1983: Methods for specifying characteristics of vibration and shock absorbers

1.4.3 Random vibrations

1.4.3.1 Introduction

If the vibration response parameters of a dynamic system are accurately known as functions of time, the vibration is said to be *deterministic*. However, in many systems and processes responses cannot be accurately predicted; these are called *random processes*. Examples of a random process are turbulence, fatigue, the meshing of imperfect gears, surface irregularities, the motion of a car running along a rough road and building vibration excited by an earthquake (Figure 1.25).

A collection of sample functions $x_1(t), x_2(t), x_3(t), \ldots, x_n(t)$ which make up the random process $x(t)$ is called an *ensemble* (Figure 1.26). These functions may comprise, for example, records of pressure fluctuations or vibration levels, taken under the same conditions but at different times.

Any quantity which cannot be precisely predicted is non-deterministic and is known as a *random variable* or a *probabilistic quantity*. That is, if a series of tests are conducted to find

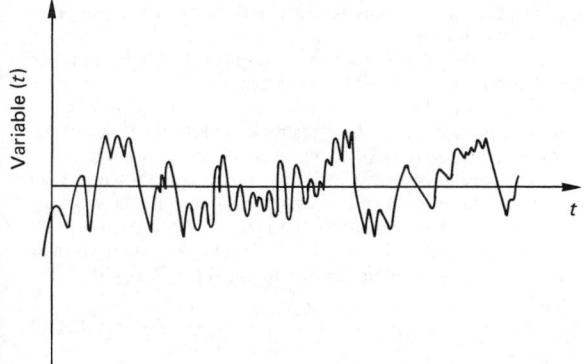

Figure 1.25 Example random process variable as $f(t)$

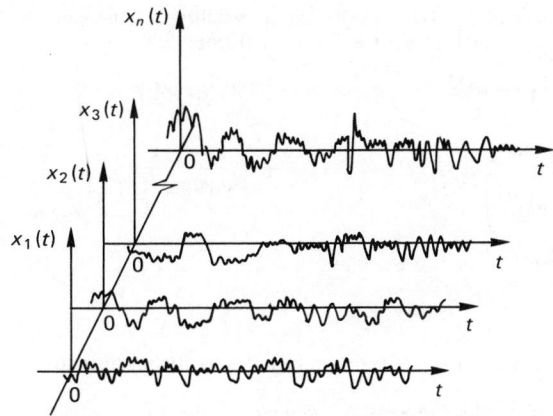

Figure 1.26 Ensemble of a random process

the value of a particular parameter, x, and that value is found to vary in an unpredictable way that is not a function of any other parameter, then x is a random variable.

1.4.3.2 Probability distribution

If n experimental values of a variable x are $x_1, x_2, x_3, \ldots, x_n$, the probability that the value of x will be less than x' is n'/n, where n' is the number of x values which are less than or equal to x'. That is,

$$\text{Prob}(x \leqslant x') = n'/n$$

When n approaches ∞ this expression is the probability distribution function of x, denoted by $P(x)$, so that

$$P(x) = \underset{n \to \infty}{\text{Lt}} (n'/n)$$

The typical variation of $P(x)$ with x is shown in Figure 1.27. Since $x(t)$ denotes a physical quantity,

$$\text{Prob}(x < -\infty) = 0, \text{ and } \text{Prob}(x < +\infty) = 1$$

The *probability density function* is the derivative of $P(x)$ with respect to x and this is denoted by $p(x)$. That is,

$$p(x) = \frac{dP(x)}{dx}$$

$$= \underset{\Delta x \to 0}{\text{Lt}} \left[\frac{P(x + \Delta x) - P(x)}{\Delta x} \right]$$

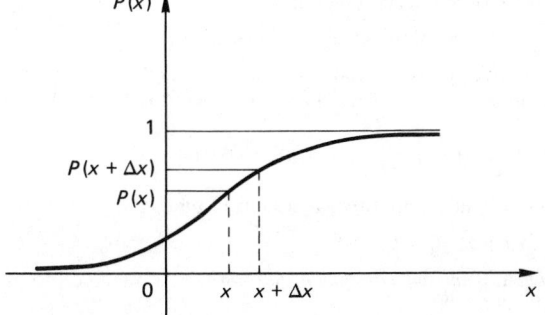

Figure 1.27 Probability distribution function as $f(x)$

where $P(x + \Delta x) - P(x)$ is the probability that the value of $x(t)$ will lie between x and $x + \Delta x$ (Figure 1.27). Now

$$p(x) = \frac{dP(x)}{dx}$$

so that

$$P(x) = \int_{-\infty}^{x} p(x)dx$$

Hence

$$P(\infty) = \int_{-\infty}^{\infty} p(x)dx = 1$$

so that the area under the probability density function curve is unity.

A random process is *stationary* if the joint probability density

$$p(x(t_1), x(t_2), x(t_3), \ldots)$$

depends only on the time differences $t_2 - t_1$, $t_3 - t_2$ and so on, and not on the actual time instants. That is, the ensemble will look just the same if the time origin is changed. A random process is *ergodic* if every sample function is typical of the entire group.

The expected value of $f(x)$, which is written $E[f(x)]$ or $\overline{f(x)}$ is

$$E[f(x)] = \overline{f(x)} = \int_{-\infty}^{\infty} f(x)p(x)dx$$

so that the expected value of a stationary random process $x(t)$ is

$$E[x(t_1)] = E[x(t_1 + t)]$$

for any value of t.

If $f(x) = x$, the expected value or *mean value* of x, $E[x]$ or $\bar{x}$, is

$$E[x] = \bar{x} = \int_{-\infty}^{\infty} xp(x)dx$$

In addition, if $f(x) = x^2$, the *mean square value* of x, $\bar{x}^2$ is

$$E[x^2] = \bar{x}^2 = \int_{-\infty}^{\infty} x^2 \cdot p(x)dx$$

The *variance* of x, σ^2 is the mean square value of x about the mean, that is,

$$\sigma^2 = E[(x - \bar{x})^2] = \int_{-\infty}^{\infty} (x - \bar{x})^2 p(x)dx = \overline{(x^2)} - (\bar{x})^2$$

σ is the *standard deviation* of x, hence

Variance = (Standard deviation)^2

= {Mean square − (Mean)^2}

If two or more random variables x_1 and x_2 represent a random process at two different instants of time, then

$$E[f(x_1,x_2)] = \int_{-\infty}^{\infty}\int_{-\infty}^{\infty} f(x_1,x_2)p(x_1,x_2)dx_1dx_2$$

and if t_1 and t_2 are the two instants of time,

$$E[x(t_1), x(t_2)] = R(t_1,t_2)$$

which is the *autocorrelation function* for the random process (Figure 1.28).

For random processes which are stationary,

$$E[x(t_1), x(t_2)] = R(t_1,t_2) = R(t_2 - t_1) = R(\tau), \text{ say,}$$

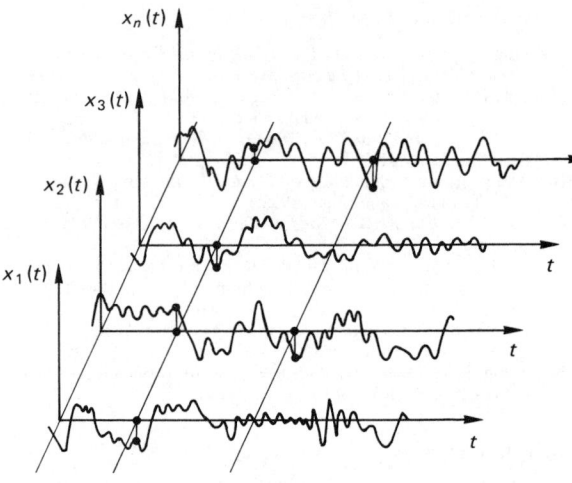

Figure 1.28 Random processes

since the average depends only on time differences. If the process is also ergodic, then

$$R(\tau) = \underset{T \to \infty}{\text{Lt}} \frac{1}{2\pi} \int_{-T}^{T} x(t)x(t + \tau)dt$$

It is worth noting that

$$R(0) = E[x(t)^2] = \underset{T \to \infty}{\text{Lt}} \frac{1}{2T} \int_{-T}^{T} x^2(t)dt$$

which is the average power in a sample function.

1.4.3.3 Random processes

The most important random process is the *Gaussian* or *normal random process*. This is because a wide range of physically observed random waveforms can be represented as Gaussian processes, and the process has mathematical features which make analysis relatively straightforward.

The probability density function of a Gaussian process $x(t)$ is

$$p(x) = \frac{1}{\sqrt{2\pi}\,\sigma} e^{-1/2[(x-\bar{x})/\sigma]^2}$$

where σ is the standard deviation of x and $\bar{x}$ is the mean value of x. The values of σ and $\bar{x}$ may vary with time for a non-stationary process but are independent of time if the process is stationary.

One of the most important features of the Gaussian process is that the response of a linear system to this form of excitation is usually another (but still Gaussian) random process. The only changes are that the magnitude and standard deviation of the response may differ from those of the excitation.

A Gaussian probability density function is shown in Figure 1.29. It can be seen to be symmetric about the mean value $\bar{x}$, and the standard deviation σ controls the spread.

The probability that $x(t)$ lies between $-\lambda\sigma$ and $+\lambda\sigma$, where λ is a positive number, can be found since, if $\bar{x} = 0$,

$$\text{Prob}\{-\lambda\sigma \leq x(t) \leq +\lambda\sigma\} = \int_{-\lambda\sigma}^{+\lambda\sigma} \frac{1}{\sqrt{2\pi}\,\sigma} e^{-1/2(x^2/\sigma^2)}dx$$

Figure 1.30 shows the Gaussian probability density function with zero mean. This integral has been calculated for a range

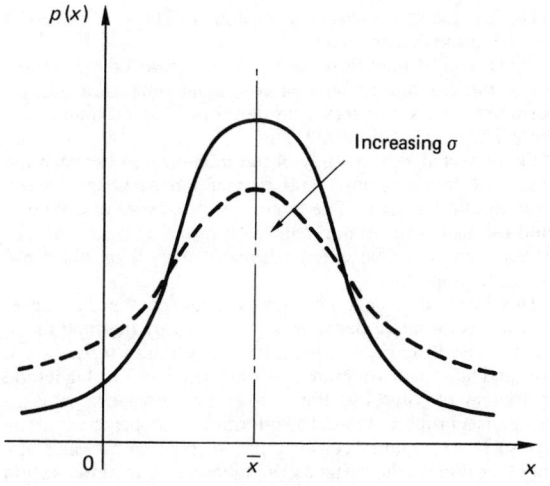

Figure 1.29 Gaussian probability density function

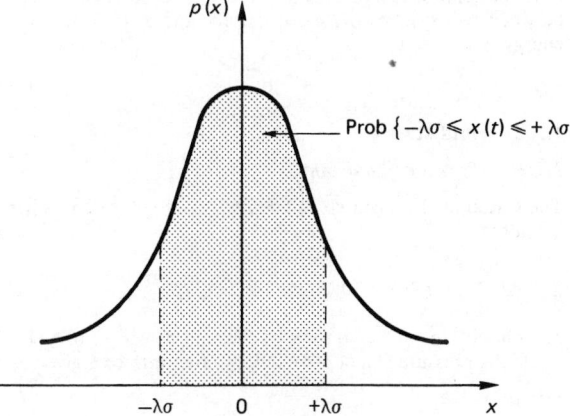

Figure 1.30 Gaussian probability density function with zero mean

Table 1.4

Value of λ	Prob$[-\lambda\sigma \leqslant x(t) \leqslant \lambda\sigma]$	Prob$[1x(t)/ > \lambda\sigma]$
0	0	1.0000
0.2	0.1585	0.8415
0.4	0.3108	0.6892
0.6	0.4515	0.5485
0.8	0.5763	0.4237
1.0	0.6827	0.3173
1.2	0.7699	0.2301
1.4	0.8586	0.1414
1.6	0.8904	0.1096
1.8	0.9281	0.0719
2.0	0.9545	0.0455
2.2	0.9722	0.0278
2.4	0.9836	0.0164
2.6	0.9907	0.0093
2.8	0.9949	0.0051
3.0	0.9973	0.0027
3.2	0.9986	0.00137
3.4	0.9993	0.00067
3.6	0.9997	0.00032
3.8	0.9998	0.00014
4.0	0.9999	0.00006

Prob$[-\lambda\sigma \leqslant x(t) \leqslant +\lambda\sigma]$

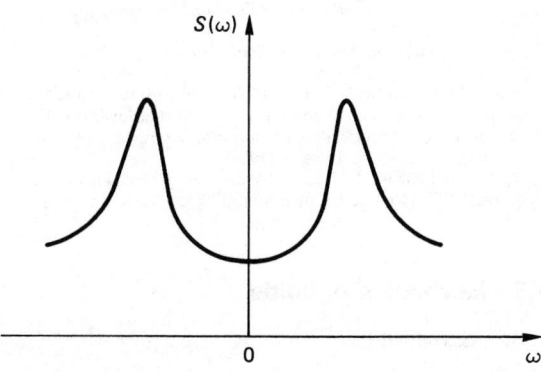

That is, the mean square value of a stationary random process x is the area under the $S(\omega)$ against frequency curve. A typical spectral density function is shown in Figure 1.31.

A random process whose spectral density is constant over a very wide frequency range is called *white noise*. If the spectral density of a process has a significant value over a narrower range of frequencies, but one which is nevertheless still wide compared with the centre frequency of the band, it is termed a *wide-band process* (Figure 1.32). If the frequency range is narrow compared with the centre frequency it is termed a *narrow-band process* (Figure 1.33). Narrow-band processes frequently occur in engineering practice because real systems often respond strongly to specific exciting frequencies and thereby effectively act as a filter.

of values of λ and the results are given in Table 1.4. The probability that $x(t)$ lies outside the range $-\lambda\sigma$ to $+\lambda\sigma$ is 1 minus the value of the above integral. This probability is also given in Table 1.4.

1.4.3.4 Spectral density

The spectral density $S(\omega)$ of a stationary random process is the Fourier transform of the autocorrelation function $R(\tau)$, and is given by

$$S(\omega) = \frac{1}{2\pi} \int_{-\infty}^{\infty} R(\tau)e^{-i\omega\tau}d\tau$$

The inverse, which also holds true, is

$$R(\tau) = \int_{-\infty}^{\infty} S(\omega)e^{-i\omega\tau}d\omega$$

If $\tau = 0$

$$R(0) = \int_{-\infty}^{\infty} S(\omega)d\omega$$

$$= E[x^2]$$

$S(\omega)$

Figure 1.31 Typical spectral density function

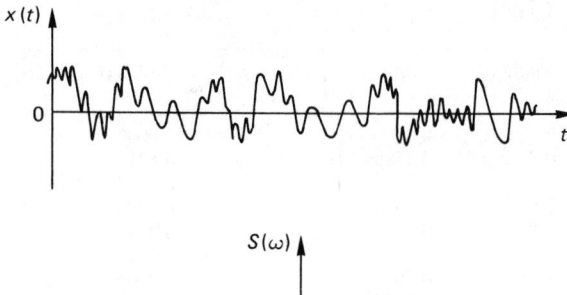

Figure 1.32 Wide-band process

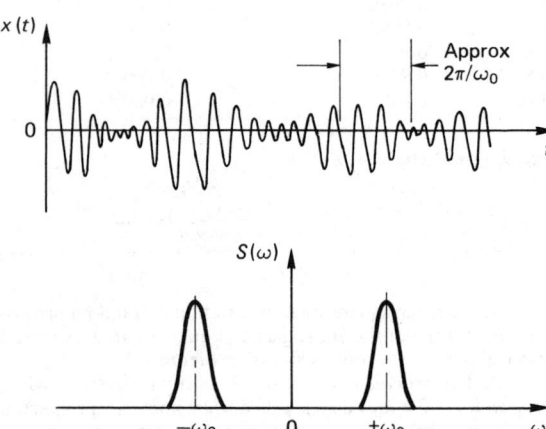

Figure 1.33 Narrow-band process

Further reading

Crandall, S. H., *Random Vibration*, Technology Press and John Wiley, Chichester (1958)

Crandall, S. H. and Mark, W. D., *Random Vibration in Mechanical Systems*, Academic Press, London (1963)

Robson, J. D., *An Introduction to Random Vibration*, Edinburgh University Press (1963)

Davenport, W. B., *Probability and Random Processes*, McGraw-Hill, New York (1970)

Nigam, N. C., *Introduction to Random Vibrations*, MIT Press (1983)

Newland, D. E., *An Introduction to Random Vibrations and Spectral Analysis*, second edition, Longman, Harlow (1984)

Helstrom, C. W., *Probability and Stochastic Processes for Engineers*, Macmillan, London (1984)

Piszek, K. and Niziol, J., *Random Vibration of Mechanical Systems*, Ellis Horwood, Chichester (1986).

1.5 Mechanics of fluids

1.5.1 Introduction

Fluid is one of the two states in which matter can exist, the other being solid. In the fluid state the matter can flow; it will,

in general, take the shape of its container. At rest a fluid is not able to sustain shear forces.

Some 'solids' may flow over a long period (glass window panes thicken at the base after a long time in a vertical position). The substances considered in fluid mechanics are those which are continously fluid.

Fluid mechanics is a study of the relationships between the effects of forces, energy and momentum occurring in and around a fluid system. The important properties of a fluid in fluid mechanics terms are *density, pressure, viscosity, surface tension* and, to some extent, *temperature*, all of which are intensive properties.

Density is the mass per unit volume of the substance. *Pressure* is the force per unit area exerted by the fluid on its boundaries. *Viscosity* is a measure of the fluid's resistance to flow and may be considered as internal friction. The higher the coefficient of viscosity, the greater the resistance. *Surface tension* is a property related to intermolecular attraction in the free surface of a liquid resulting in the apparent presence of a very thin film on the surface. The meniscus at the intersection of a liquid and its container wall and capillarity are further examples of intermolecular attraction.

Temperature is more relevant to thermodynamics than to fluid mechanics. It indicates the state of thermal equilibrium between two systems or, more loosely, the level of thermal energy in a system.

1.5.2 Fluid statics

1.5.2.1 Pressure at a depth

The variation of pressure p and depth h in a fluid of density ρ is given by

$$\int_{p_1}^{p_2} dp = \int_{h_1}^{h_2} \rho g dh \tag{1.5}$$

Most liquids are assumed to be of constant density ρ. In such a liquid the pressure at a depth h below a free surface is given by

$$p = p_0 + \rho g h \tag{1.6}$$

where p_0 is the pressure above the free surface.

For gases equation (1.5) may be solved only if the relationship between ρ and h is known. A typical case is the atmosphere, where the relationship may be taken as polytropic or isothermal, depending on the altitude. Tables relating the properties of the atmosphere to altitude are readily available as the International Atmosphere (Rogers and Mayhew, 1980).

1.5.2.2 Pressure measurement

Pressure may be expressed as a pressure p in Pa, or as a pressure head h in m of the fluid concerned. For a fluid of density ρ, $p = \rho g h$. There are various instruments used to measure pressure.

(a) Manometers Manometers are differential pressure-measuring devices, based on pressure due to columns of fluid. A typical U-tube manometer is shown in Figure 1.34(a). The difference in pressure between vessel A containing a fluid of density ρ_A and vessel B containing fluid of density ρ_B is given by

$$p_A - p_B = \rho_B g Z_B + (\rho_m - \rho_B)gh - \rho_A g Z_A \tag{1.7}$$

where h is the difference in the levels of the manometer fluid of density ρ_m and $\rho_m > \rho_A$ and $\rho_m > \rho_B$. If $\rho_A = \rho_B = \rho$, then the difference in pressure head is

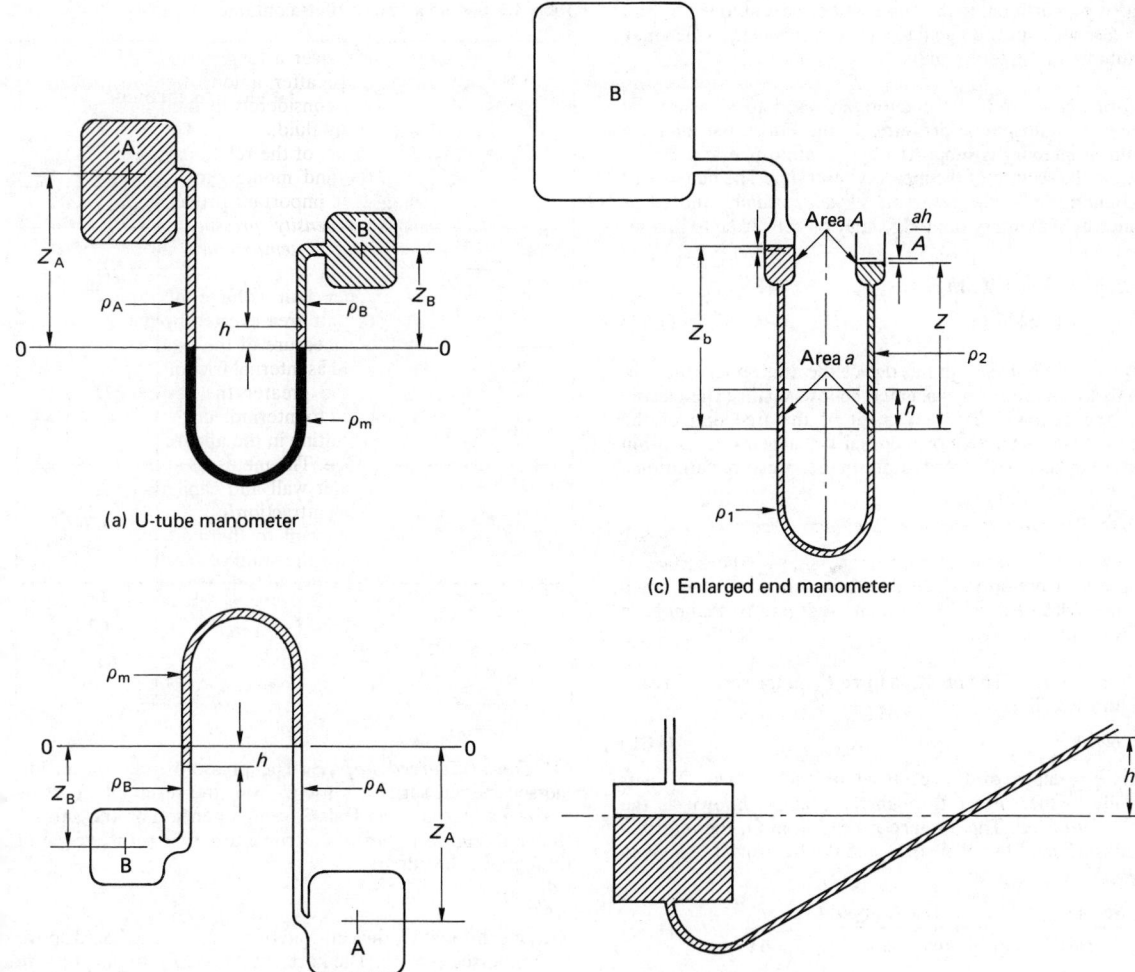

(a) U-tube manometer

(b) Inverted U-tube manometer

(c) Enlarged end manometer

(d) Inclined limb manometer

Figure 1.34

$$\frac{p_A - p_B}{\rho g} = Z_B - Z_A + \left(\frac{\rho_m}{\rho} - 1\right)h \qquad (1.8)$$

If $\rho_m < \rho_A$ and $\rho_m < \rho_B$ then an inverted U-tube manometer is used as shown in Figure 1.34(b). In this case the pressure difference is

$$p_A - p_B = \rho_A g Z_A + (\rho_B - \rho_m)gh - \rho_B g Z_B \qquad (1.9)$$

and if $\rho_A = \rho_B = \rho$ then the difference in pressure head is

$$\frac{p_A - p_B}{\rho g} = Z_A - Z_B + \left(1 - \frac{\rho_m}{\rho}\right)h \qquad (1.10a)$$

of if $\rho_m \ll \rho$ (the manometer fluid a gas and A and B containing liquid),

$$\frac{p_A - p_B}{\rho g} = Z_A - Z_B + h \qquad (1.10b)$$

The accuracy of a U-tube manometer may be increased by sloping one of the legs to increase the movement of the fluid interface along the leg for a given difference in vertical height. This may be further enhanced by replacing the vertical leg by a

reservoir and the inclined leg by a small-bore tube (Figure 1.34(d)).

Another method is to increase the cross-sectional area of the ends of the legs (or one of the legs), as shown in Figure 1.34(c), so that a small movement of the free surfaces in the enlarged ends results in a large movement of the surface of separation.

(b) Dial gauges Most pressure dial guages make use of a *Bourdon tube*. This is a curved tube with an oval cross section. Increase in pressure causes the tube to straighten, decrease makes it bend. The movement of the free end turns a pointer over a scale, usually via a rack and pinion mechanism. The scale may be calibrated in the required pressure units.

(c) Diaphragm gauges In these gauges the pressure changes produce a movement in a diaphragm which may be detected by a displacement transducer, or by the output from strain gauges attached to the diaphragm surface.

(d) Piezoelectric transducers A piezoelectric crystal produces a voltage when deformed by an external force. This induced

charge is proportional to the impressed force and so the output can be used to supply a signal to a measuring device which may be calibrated in pressure units.

(e) Fortin barometer   Barometers are used to measure the ambient or atmospheric pressure. In the Fortin barometer a column of mercury is supported by the atmospheric pressure acting on the surface of the mercury reservoir. The height h of the column above the reservoir surface, usually quoted as millimetres of mercury (mm Hg), may be converted to pressure units p_0 by

$$p_0 = \rho g h = 13.6 \times 9.81h$$

$$= 133.42h \text{ Pa} \qquad (1.11)$$

(f) Aneroid barometer   In this device the atmospheric pressure tends to compress an evacuated bellows against the elasticity of the bellows. The movement of the free end of the bellows drives a pointer over a dial (or a pen over a drum graph) to indicate (or record) atmospheric pressure variations.

1.5.2.3   Force due to pressure on an immersed surface

These forces are only relevant if one side of the surface is exposed to a pressure which does not depend on the depth (e.g. the sides of a vessel, an immersed gate or manhole, a dam wall, etc.).

(a) Plane surface   The pressure force F_p on the surface area A in Figure 1.35 is

$$F_p = \rho g \bar{h} A \qquad (1.12)$$

where $\bar{h}$ = depth of the centroid of the surface. F_p acts normally to the surface through the point C known as the *centre of pressure*. The distance x_c of C from O, the intersection of the line of the plane of A and the free surface, is given by

$$x_c = \frac{\text{Second moment of area A about O}}{\text{First moment of area A about O}} = \frac{I_0}{A\bar{x}} \qquad (1.13)$$

The depth of the centre of pressure $h_c = x_c \sin \theta$.

The force F_p does not include the pressure above the free surface p_0, since this is often atmospheric and may also act on the opposite side of the immersed surface to F_p. If this is not the case $F_p = (\rho g \bar{h} + p_0)A$.

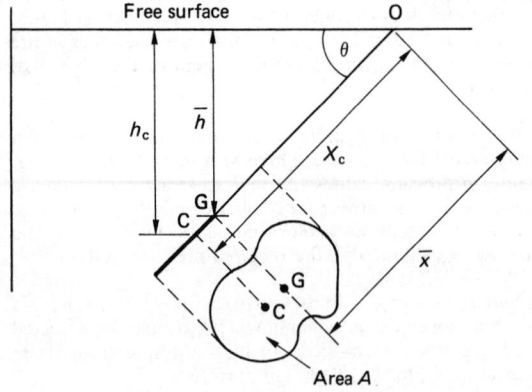

Figure 1.35   Immersed surface (G is centroid, C is centre of pressure)

Table 1.5   Second moments of area

	Area A	I_G
Parallel axis theorem $I_X = I_G + A\bar{x}^2$		
	BD	$\dfrac{BD^3}{12}$
	$\dfrac{BH}{2}$	$\dfrac{BD^3}{36}$
	$\dfrac{\pi R^2}{2}$	$0.1102R^4$
	$\dfrac{\pi D^2}{4}$	$\dfrac{\pi D^4}{64}$

(b) Concave curved surface   The pressure force on an immersed curved surface is found from the resultant F_R of its horizontal F_H and vertical F_V components. For the surface shown in Figure 1.36(a) the vertical force F_V = the weight of fluid above the curve

$$= \rho g AB \qquad (1.14)$$

and acts through G_v the centroid of the volume of liquid above the immersed surface. The horizontal force F_H = the pressure force on the projected area of the immersed surface in the vertical plane

$$= \rho g \bar{h}_{C_P} A_P \qquad (1.15)$$

and acts through the centre of pressure C_P of the projected area.

The resultant pressure force on the curved surface F_R is given by

$$F_R = (F_H^2 + F_V^2)^{0.5} \qquad (1.16)$$

The angle of inclination α of F_R to the horizontal is given by

$$\alpha = \tan^{-1}\left(\frac{F_V}{F_H}\right) \qquad (1.17)$$

or

$$F_R = \bar{F}_H + \bar{F}_V \qquad (1.18)$$

(c) Convex curved surface   This is as a concave surface, except that F_V is the *buoyancy* force of the displaced volume of liquid above the immersed surface and acts vertically upwards through the centre of buoyancy (see Figure 1.36(b)).

1.5.2.4   Buoyancy

When a body is immersed in a fluid the difference in pressure over the depth of the body produces a displacement force on

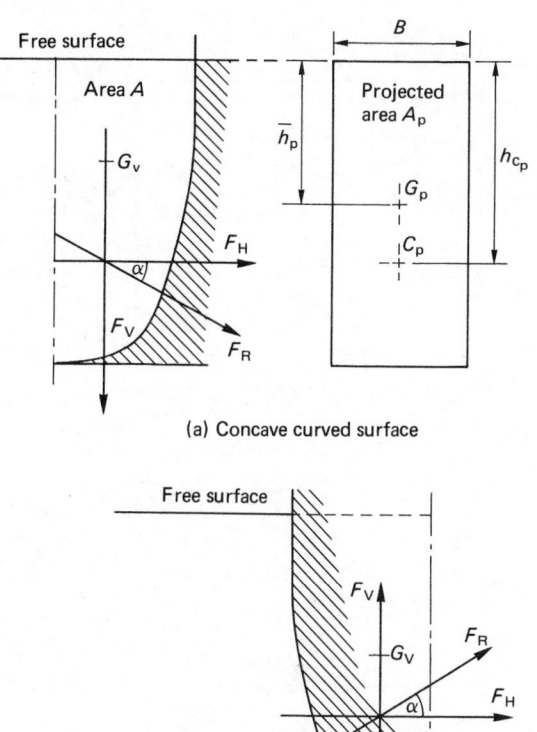

(a) Concave curved surface

(b) Convex surface

Figure 1.36

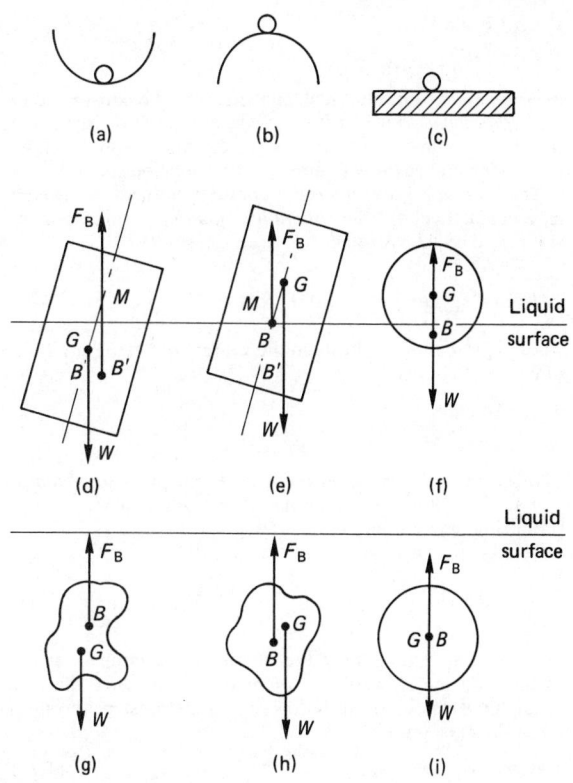

Figure 1.37 Stability

the body. The first recognition of this is attributed to Archimedes.

(a) Displacement force The buoyancy or displacement force F_B on a body fully or partially immersed in a fluid is equal to the weight of the volume of the fluid equivalent to the immersed volume of the body (the weight of the displaced volume V_D of the fluid):

$$F_B = \rho g V_D \qquad (1.19)$$

This buoyancy force acts vertically upwards through the centroid of the displaced volume, which is known as the *centre of buoyancy (B)*. If the buoyancy force is equal to the weight of the body then the body will float in the fluid. If the weight of the body is greater than the buoyancy force then the body will sink. If the buoyancy force is greater than the weight of the body then the body will rise.

In a liquid, for example, a body will sink until the volume of liquid displaced has a weight which is equal to that of the body. If the body is more dense than the liquid then the body will not float at any depth in the liquid. A balloon will rise in air until the density of the air is such that the weight of the displaced volume of air is equal to the weight of the balloon.

(b) Stability of a floating body Figure 1.37 shows bodies in various stages of equilibrium. A body is in *stable* equilibrium if a small displacement produces a restoring force or moment as for the ball in the saucer in Figure 1.37(a) or the floating

bodies in (d) and (g). A body is in *unstable* equilibrium if a small displacement produces a disturbing force or moment as for the ball in Figure 1.37(b) or the floating bodies (e) and (h). A body is in *neutral* equilibrium if a small displacement produces no force or moment as for the ball in Figure 1.37(c) or the floating bodies in (f) and (i).

For a partially immersed body, the point at which the line of action of the buoyancy force F_B cuts the vertical centre line of the floating body in the displaced position is known as the *metacentre (M)*. For a floating body to be stable M must lie above the body's centre of gravity, G. If M lies below G the body is unstable; if M lies on G the body is in neutral equilibrium. The distance GM is known as the *metacentric height*. The distance of the metacentre above the centre of buoyancy

$$BM = \frac{I}{V_D} \qquad (1.20)$$

where I = second moment of area of the body at the water line (liquid surface) about its central axis normal to the direction of displacement.

(c) Period of oscillation of a stable floating body A floating body oscillates with the periodic time T of a simple pendulum of length k^2/GM, where k is the radius of gyration of the body about its axis of rotation. The periodic time is given by

$$T = 2\pi \left(\frac{k^2}{GMg} \right)^{0.5} \qquad (1.21)$$

1.5.3 Fluid flow

1.5.3.1 Definitions

(a) Continuity For almost all analysis, a fluid is considered to be a *continuum*, that is, with non-discontinuities or cavities in the flow stream. Cavitation, two-phase flow, 'bubbly' flow, etc. are special cases with non-standard relationships.

Therefore for a continuum, by considering the flow through an elemental cuboid the *continuity equation* in three dimensions may be shown to be

$$\frac{\partial}{\partial x}(\rho v_x) + \frac{\partial}{\partial y}(\rho v_y) + \frac{\partial}{\partial z}(\rho v_z) = 0 \tag{1.22}$$

where v_x is the fluid velocity in the x direction, etc. For a fluid of constant density

$$\frac{\partial v_x}{\partial x} + \frac{\partial v_y}{\partial y} + \frac{\partial v_z}{\partial z} = 0 \tag{1.23}$$

That is, the velocity of an incompressible fluid flow cannot increase in all three directions at the same time without producing discontinuity or cavitation.

For two-dimensional flow:

$$\frac{\partial v_x}{\partial x} + \frac{\partial v_y}{\partial y} = 0 \tag{1.24}$$

For one-dimensional flow the continuity equation may be linked with the *conservation of mass*, which states that for steady flow conditions mass flow rate, $\dot{m}$, is constant throughout a flow system:

$$\dot{m} = \rho A v \tag{1.25}$$

where A is the cross-sectional area normal to the direction of flow.

(b) Circulation Γ Circulation is defined as the line integral of the tangential velocity around a closed contour:

$$\Gamma = \oint v_s ds \tag{1.26}$$

Γ is positive if the closed contour is on the left.

(c) Vorticity ζ Vorticity is defined as the circulation per unit area, and by considering the circulation around the element in Figure 1.38(a) it can be shown that

$$\zeta = \frac{\Gamma}{\partial x \partial y} = \frac{\partial v_y}{\partial x} - \frac{\partial v_x}{\partial y} \tag{1.27}$$

(d) Rotation ω Rotation is defined as the instantaneous mean angular velocity of two mutually perpendicular lines in a plane of the flow field. By considering the angular velocities of the two lines OA and OB in Figure 1.38(b) it can be shown that

$$\omega = \frac{1}{2}\left(\frac{\partial v_y}{\partial x} - \frac{\partial v_x}{\partial y}\right) \tag{1.28}$$

or the rotation is equal to half the vorticity.

(e) Stream lines The stream line is a line drawn in a flow stream which is everywhere tangential to the direction of flow. A family of stream lines may be described mathematically by a stream function ψ, where $\psi = \text{fn}(x,y)$. Each stream line has the same function with a value of ψ peculiar to that line.

(f) Stream tubes Since a line has no thickness, there can be no flow along a stream line. The stream tube is a concept introduced to enable flow along a stream line to be studied. It

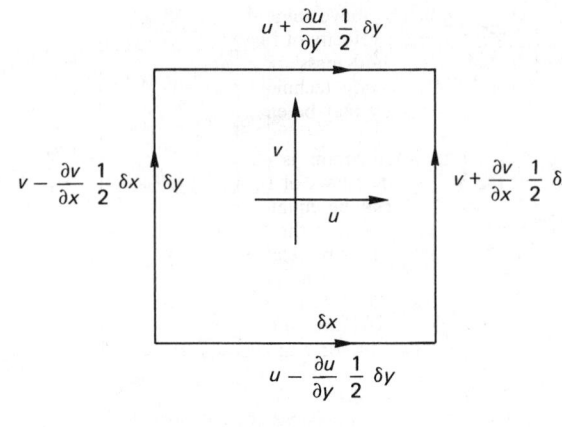

(a) Vorticity

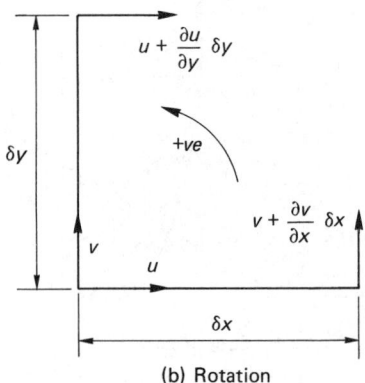

(b) Rotation

Figure 1.38

is a tube of infinitely small cross section with a stream line as its axis.

(g) Energy Energy is the stored form of *heat* and *work*. The basic concepts applied in fluid mechanics are:

- The conservation of energy
- That energy is transferred only as heat or work
- That energy in a fluid flow system is stored only as *internal energy*, *kinetic energy* or *potential energy*.

Other forms of energy (electrical, magnetic, chemical, etc.) may have to be taken into account in some circumstances, but are not usually included in general fluid mechanics relationships.

Enthalpy and *entropy* need to be considered for gas flow analysis (see Section 1.5.8). The basic energy-flow equation is the *steady-flow energy equation*:

$$\dot{Q} + \dot{W} = \dot{m}\Delta\left(h + \frac{v^2}{2} + gZ\right) \tag{1.29}$$

where $\dot{Q}$ is the rate of heat transfer,
$\dot{W}$ is the rate of work transfer (power),
h is the specific enthalpy (if e is the specific internal energy, p the pressure and ρ the fluid density, then $h = e + (p/\rho)$),

Z is the height above some datum,
v is the mean velocity of flow.
Specific means 'per unit mass'. For non-steady flow conditions, either *quasi-steady* techniques or the integration of infinitely small changes may be employed.

(h) Momentum Momentum is the product of mass and velocity (mv). Newton's laws of motion state that the force applied to a system may be equated to the rate of change of momentum of the system, in the direction of the force. The change in momentum may be related to time and/or displacement. In a steady flow situation the change related to time is zero, so the change of momentum is usually taken to be the product of the mass flow rate and the change in velocity with displacement. Hence the force applied across a system is

$$F = \dot{m}\Delta v \tag{1.30}$$

where Δv is the change in velocity in the direction of the force F.

For flow in two or three dimensions the resultant force may be obtained by resolving the forces in the usual way. The flow round an expanding bend shown in Figure 1.39 is a typical example. The force in the x direction, F_x, and the force in the y direction, F_y, are given by

$$F_x = p_1A_1 + \dot{m}v_1 - (p_2A_2 + \dot{m}v_2)\cos\theta \tag{1.31a}$$
$$F_y = -(p_2A_2 + \dot{m}v_2)\sin\theta \tag{1.31b}$$

from which the resultant force F_R and its angle of inclination α can be found:

$$F_R = \{F_x^2 + F_y^2\}^{0.5} \tag{1.32}$$

and

$$\alpha = \tan^{-1}\left(\frac{F_y}{F_x}\right) \tag{1.33}$$

In pipe flow the pressure forces pA must not be forgotten.

1.5.3.2   Equations of motion

Application of the momentum equation in three dimensions to an irrotational, inviscid fluid flow leads to the *Euler equation*:

$$-\frac{Dv}{Dt} = \frac{1}{\rho}\nabla p + g\nabla h \tag{1.34}$$

which for steady flow along a stream tube becomes:

$$\frac{1}{\rho}\,dp + v\,dv + g\,dZ = 0 \tag{1.35}$$

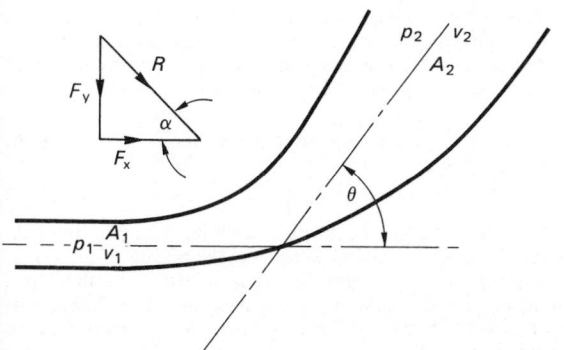

Figure 1.39   Expanding bend

Integration for a constant-density fluid gives:

$$\frac{p}{\rho} + \frac{v^2}{2} + gZ = \text{constant} \tag{1.36}$$

These energy per unit mass terms may be converted to energy per unit weight terms, or heads, by dividing by g to give:

$$\frac{p}{\rho g} + \frac{v^2}{2g} + Z = \text{constant} \tag{1.37}$$

which is the *Bernoulli* (or constant head) equation.

These equations are the generally more useful simplifications of the *Navier–Stokes* equation:

$$\frac{Dv}{Dt} = \rho B - \nabla p + \nabla\{u(\nabla v + \nabla \epsilon)\} \tag{1.38}$$

where B is the body force and ϵ the rate of expansion.

1.5.3.3   Incompressible pipe flow

(a) Flow regimes The two major flow regimes are *laminar* and *turbulent*. Laminar flow may be fairly accurately modelled mathematically. The fluid moves in smooth layers and the velocity is everywhere tangential to the direction of motion. Any perturbations are quickly dampened out by the fluid viscosity.

In turbulent flow the mathematical models usually need to be empirically modified. Viscous damping may not be sufficient to control the perturbations, so that the fluid does not move in smooth layers and the instantaneous velocity may have components at an angle to the direction of motion.

The ratio of inertia forces to viscous forces in a fluid flow is known as *Reynolds' Number (Re)*. In a pipe diameter D, with a fluid of density ρ and dynamic viscosity η flowing with velocity v, Reynolds' number $Re = \rho Dv/\eta$.

A high value of $Re > 2300$ indicates relatively low damping, predicting turbulent flow. A low value of $Re < 2000$ indicates relatively high damping, predicting laminar flow. These values were suggested in an historical experiment by Osborne Reynolds.

(b) Pipe losses (friction) Liquids (and gases under small pressure changes) flowing through pipes usually behave as incompressible fluids. Within the flow there is a relationship between the shear stress in the fluid and the gradient of the change of velocity across the flow. In most light liquids and gases, the relationship approximates to the *Newtonian* one:

$$\tau = \eta\frac{dv}{dy} \tag{1.39}$$

where τ is the shear stress in the fluid, dv/dy the gradient of the velocity distribution across the pipe and η the dynamic viscosity.

The viscosity of the fluid produces not only the velocity variation across the flow but also a loss of energy along the pipe usually regarded as a friction loss. The force associated with this loss of energy appears as a shear force in the fluid at the pipe wall. A relationship between the shear stress at the pipe wall τ_0 and the *friction coefficient*, f is:

$$\tau_0 = \frac{1}{2}\rho v^2 f \tag{1.40}$$

where v is the average flow velocity.

For use in pipe flow problems with viscous fluids the Bernoulli equation (1.37) may be adapted to incude a head loss term, h_L. Applied between two positions (1) and (2) in a pipe, the head equation gives:

$$\frac{p_1}{\rho g} + \frac{v_1^2}{2g} + Z_1 = \frac{p_2}{\rho g} + \frac{v_2^2}{2g} + Z_2 + h_L \tag{1.41}$$

where the head loss term h_L is the loss of energy per unit *weight* of fluid flowing.

Note that if a pump, say, is introduced between (1) and (2) an energy gain per unit weight term h_w, equivalent to the output of the pump written as a head, should be added to the left-hand side of the equation to give

$$\frac{p_1}{\rho g} + \frac{v_1^2}{2g} + Z_1 + h_w = \frac{p_2}{\rho g} + \frac{v_2^2}{2g} + Z_2 + h_L \tag{1.42}$$

The relationship used to determine the head loss in a pipe depends on the flow regime in operation as well as the type and surface finish of the pipe wall.

A mathematical analysis of laminar flow may be used to obtain an expression for the head loss along a pipe in terms of the fluid properties, pipe dimensions and flow velocity. Relating the pressure change along a length, L, of pipe of diameter, D, to the change in shear force across the flow produces *Poiseuille's equation*:

$$h_L = 32 \frac{\eta v L}{\rho g D^2} \tag{1.43}$$

If the flow regime is turbulent, then the relationships in the flow cannot be easily described mathematically, but the head loss may be derived by equating the shear force at the pipe wall to the change in pressure force along the pipe. This gives the *D'Arcy* equation:

$$h_L = \frac{4fL}{D} \frac{v^2}{2g} \tag{1.44}$$

This relationship may also be established using dimensional analysis.

Unfortunately, the friction coefficient, f is *not* a constant but depends on the type of flow and the roughness of the pipe walls. There are general relationships between f and Re which may be expressed as equations of varying complexity or as charts. For smooth pipes:

$$\frac{1}{\sqrt{f}} = 4 \log_{10}(2R_e\sqrt{f}) - 1.6 \tag{1.45}$$

For rough pipes with a roughness size k this becomes:

$$\frac{1}{\sqrt{f}} = 4 \log_{10}\left(\frac{D}{2k}\right) + 3.48 \tag{1.46}$$

The *Colebrook and White* equation is a general or universal friction equation:

$$\frac{1}{\sqrt{f}} = 3.47 - 4 \log_{10}\left(\frac{2r}{D} + \frac{9.35}{Re\sqrt{f}}\right) \tag{1.47}$$

It is, however, usually more useful to obtain values of f from a chart such as Figure 1.40. (*Note*: the value of f used in American equations for head losses is *four times* that used in the United Kingdom, so if values of f are obtained from American texts they should be moderated accordingly or the corresponding American equation used.)

An empirical relationship widely used in water pipe work is the *Hazen–Williams* equation, usually written as:

$$v = 1.38 \, C \, m^{0.63}\left(\frac{h_L}{L}\right)^{0.54} \tag{1.48}$$

where m is the ratio of the cross-sectional area of flow to the wetted perimeter known as the *hydraulic mean diameter* and C is a coefficient which depends on the condition of the pipe wall.

(c) Pipe losses (changes in section) When a fluid flows through a sharp (sudden) change in the cross section of a pipe, energy is dissipated in the resulting turbulent eddies at the edge of the flow stream, producing a loss of head (or energy per unit weight). If the flow is from a smaller area to a larger one (sudden enlargement) the head loss is

$$h_L = \frac{(v_1 - v_2)^2}{2g} \tag{1.50}$$

When the flow is from a larger area to a smaller area (sudden contraction) the narrowed flow stream entering the smaller pipe is known as a *vena contracta*. The loss of head is assumed to be that due to a sudden enlargement from the vena contracta to the full area of the smaller pipe:

$$h_L = \frac{v_2^2}{2g}\left(\frac{1}{C_c} - 1\right)^2 \tag{1.51}$$

The *contraction coefficient*, C_c, is the ratio of the vena contracta area to that of the smaller pipe area. A typical value of C_c is 0.6, which gives

$$h_L \simeq = 0.5 \frac{v_2^2}{2g} \tag{1.52}$$

which is also the head loss at the sharp entry to a pipe from a reservoir. Energy dissipation at changes in section, and pipe entry and exit, may be reduced by making the changes smooth and gradual, though this may be relatively costly.

Other pipe fittings, such as valves, orifice plates and bends, produce varying values of head loss, usually quoted as a fraction of the velocity head ($v^2/2g$).

(d) Pipe networks A system of pipes may be joined together either in series (one after the other) or parallel (all between the same point). The friction head loss across a system of pipes in series is the sum of the losses along each pipe individually. The flow rate through each pipe will be the same. Using D'Arcy's head loss equation:

$$h_L = 4f_1\frac{L_1v_1^2}{D_12g} + 4f_2\frac{L_2v_2^2}{D_22g} + \ldots 4f_n\frac{L_nv_n^2}{D_n2g} \tag{1.53}$$

and

$$\dot{V} = v_1A_1 = v_2A_2 = \ldots = v_nA_n \tag{1.54}$$

If the system of pipes is connected in parallel the head loss across the system is equal to the head loss along any one of the pipes, when the flow has settled down to steady. The flow rate through the system is the sum of the flow rates along each pipe. Again using the D'Arcy equation:

$$h_L = 4f_1\frac{L_1v_1^2}{D_12g} = 4f_2\frac{L_2v_2^2}{D_22g} = \ldots = 4f_n\frac{L_nv_n^2}{D_n2g} \tag{1.55}$$

$$\dot{V} = v_1A_1 + v_2A_2 + \ldots v_nA_n \tag{1.56}$$

In addition, the rate of flow into each junction of a network, either in series or parallel, is equal to the rate of flow out of it.

Pipe network problems are thus solved by setting up a number of such equations and solving them simultaneously. For a large number of pipes a computer program may be needed to handle the number of variables and equations. An example of a pipe network computer solution is given in Douglas *et al.* (1986).

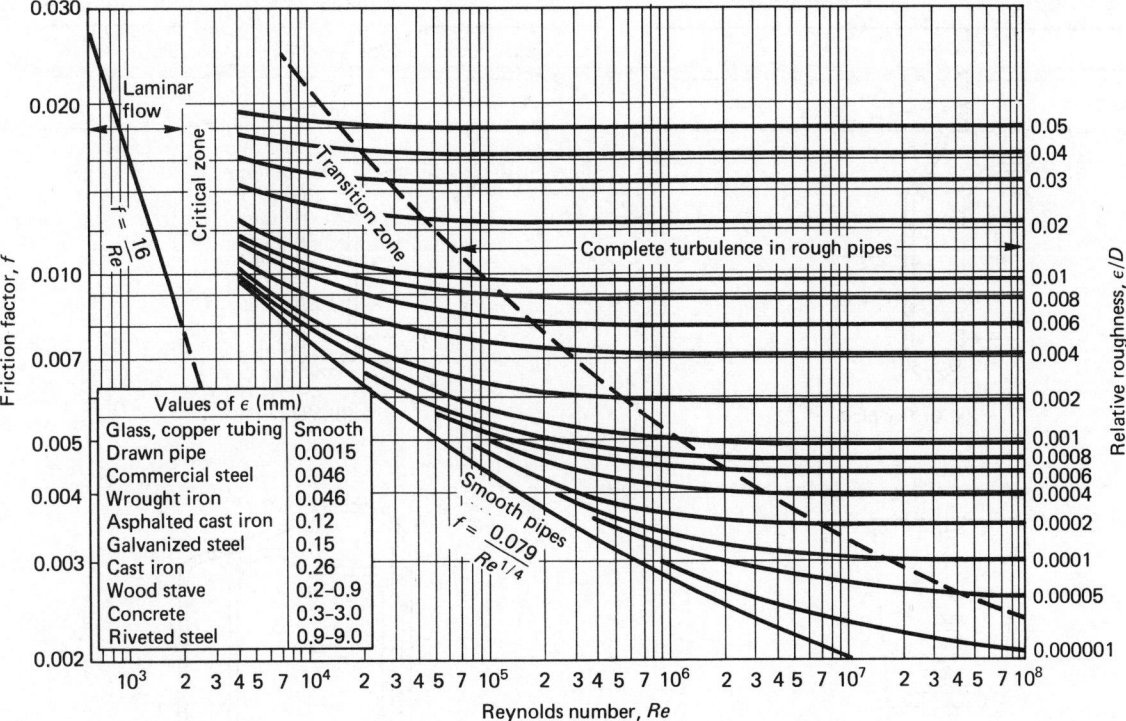

Figure 1.40

1.5.4 Flow measurement

1.5.4.1 Pipe flow

One very accurate measure of flow rate is to catch the discharge in a bucket over a known time and then weigh it. This method, made more sophisticated by the electronic timing of the balancing of a tank on a weighbridge, is often used to calibrate other devices, but may not always be acceptable.

(a) Orifices and nozzles (see Figure 1.41(a)) Another basic flow measurement technique is to introduce some restriction into the flow passage and calibrate the resulting pressure changes against known flow rates.

Often the restriction in a pipe is in the form of an *orifice plate* (a plate with a hole) or a *nozzle*. A simple application of the Bernoulli equation may be used for the design calculations, but it is always advisable to calibrate any measurement device in conditions as close to the required operating conditions as possible.

Bernoulli and the continuity equations give the flow rate:

$$\dot{V} = C_d A_o \left\{ \frac{2(p_p - p_o)}{\rho[1 - (A_o/A_p)^2]} \right\}^{0.5} \quad (1.57)$$

where A_o is the orifice (or nozzle throat) area,

A_p is the upstream pipe area,

p_p is the upstream pressure,

p_o is the pressure at the orifice or the nozzle throat, and

C_d is a discharge coefficient which takes account of losses and contraction of the flow stream through the device.

Recommended orifice and nozzle dimensions, values of C_d and methods of operation are contained in BS 1042. It is most important to place the orifice or nozzle so that its operation is not affected by perturbations in the upstream flow caused by valves, bends or other pipe fittings.

(b) Venturi meters (see Figure 1.41(b)) The introduction of any restriction, particularly a sharp-edged orifice or nozzle, in a pipe will result in a loss of head (energy). If it is required to keep this loss to a minimum, a *venturi meter* may be used. The flow passage in a venturi is gradually and smoothly reduced to a throat followed by a controlled expansion to full pipe section. In this way the head loss across the meter is greatly reduced, but the cost of producing a venturi meter is much higher than that of an orifice. Equation (1.57) may be used to calculate the flow rate $\dot{V}$, but the value of C_d will now be approximately 0.98 for a well-designed venturi meter. Again, BS 1042 should be consulted for recommended dimensions, values of C_d and methods of operation.

(c) Rotameter or gap meter (see Figure 1.41(c)) If, somewhere within the system, it is acceptable to tolerate flow up a vertical section of piping, then a *rotameter* or *gap meter* may be used. This instrument depends on the balancing of the weight of a rotating float in a tapered glass tube with the drag forces in the annular passage surrounding the float. The drag forces depend on the flow rate and the corresponding area of the annulus. As the flow rate increases, the annulus area which will produce a drag force equal to the weight of the float also increases. Therefore the float moves up the tapered tube until the annulus area is such that the forces again balance. As the flow rate decreases the float descends to a reduced annulus area to again achieve a balance of forces.

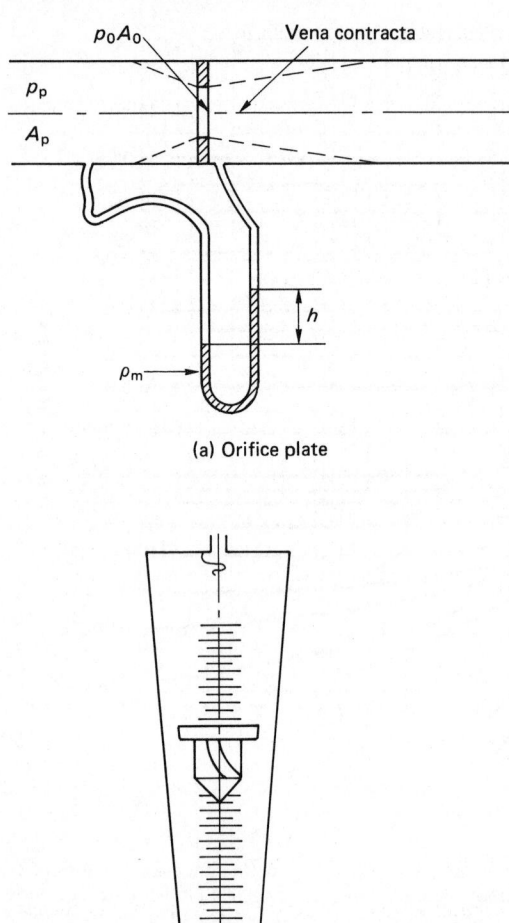

(a) Orifice plate

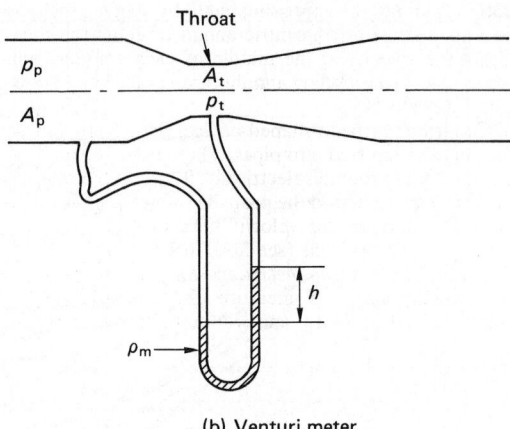

(b) Venturi meter

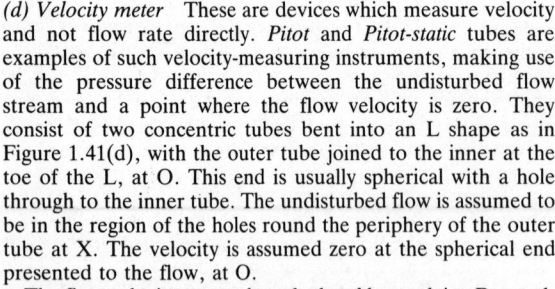

(c) Rotameter or gap meter

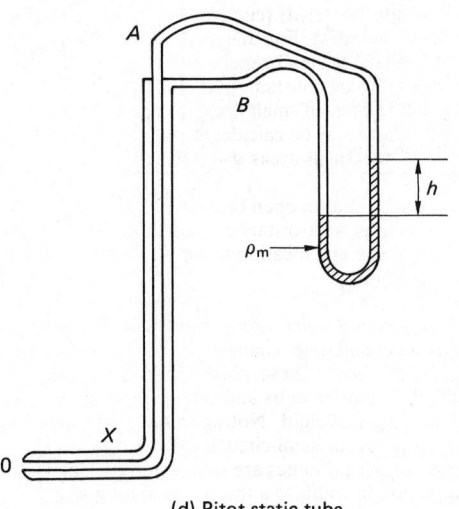

(d) Pitot-static tube

Figure 1.41 Flow meters

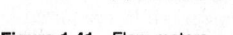

(d) Velocity meter These are devices which measure velocity and not flow rate directly. *Pitot* and *Pitot-static* tubes are examples of such velocity-measuring instruments, making use of the pressure difference between the undisturbed flow stream and a point where the flow velocity is zero. They consist of two concentric tubes bent into an L shape as in Figure 1.41(d), with the outer tube joined to the inner at the toe of the L, at O. This end is usually spherical with a hole through to the inner tube. The undisturbed flow is assumed to be in the region of the holes round the periphery of the outer tube at X. The velocity is assumed zero at the spherical end presented to the flow, at O.

The flow velocity, v may be calculated by applying Bernoulli's equation between the two points O and X to give

$$v = C_v \left[2 \frac{(p_o - p_x)}{\rho} \right]^{0.5} \quad (1.58)$$

where p_o is connected to O via the inner tube to the tapping at A, p_x is connected to X via the outer tube to the tapping at B and C_v is a coefficient to cater for losses and disturbances not

accounted for in Bernoulli's equation. C_v is often taken to be unity.

The pressure difference may be measured using a manometer and then written into equation (1.58) as a head, h, to give

$$v = \left[2gh\left(\frac{\rho_m}{\rho} - 1\right) \right]^{0.5} \quad (1.59)$$

As usual, it is advisable to calibrate the tube and obtain a calibration curve or an accurate value for C_v. BS 1042 should be consulted for operational instructions and placement advice.

Care should be taken when a pitot-static tube is used to measure pipe flow, since the velocity will vary across the pipe. As a rough guide to the flow rate the maximum velocity, which is at the centre of the pipe, may be taken to be twice the average velocity. Alternatively, the velocity at half the radius may be taken to be equal to the average velocity in the pipe. For an accurate evaluation the velocity distribution curve may be plotted and the flow rate through the pipe found by integration. This may be approximated to by dividing the cross section into a series of concentric annuli of equal thickness, measuring the velocity at the middle of each annulus, multiplying by the corresponding annulus area and adding to give the total flow rate.

Current meters, torpedo-shaped devices with a propeller at the rear, may be inserted into pipes. The number of rotations of the propeller are counted electrically. This number together with coefficients peculiar to the propeller are used in empirical equations to determine the velocity. These meters are more often used in open channels (see Section 1.5.4.2).

Velometers, *vaned anemometers* and *hot wire anemometers* are not usually used to measure the velocities of incompressible fluids in pipes, and will be discussed in Section 1.5.8.

1.5.4.2 Open-channel flow

(a) Velocity meters In channels of regular or irregular cross section the flow may be measured using the velocity meters described in Section 1.5.4.1(d) (current meters are often used in rivers or large channels). For this method the cross section is divided into relatively small regular areas, over which the velocity is assumed to be constant. The velocity meter is then placed at the centre of each small area, and from the velocity and area the flow rate may be calculated. Adding together the flow rates for all the small areas gives the flow rate for the channel.

It should be noted that in open channels the velocity varies with depth as well as with distance from the channel walls. Selection of the shape and location of the small areas need to take this into account.

(b) Notches, flumes and weirs As in pipe flow, flow rates in channels may be related to changes in head produced by obstructions to the flow. These obstructions may be in the form of notches, flumes or weirs and change in head observed as a change in depth of fluid. Notches may be rectangular, V-shaped, trapezoidal or semi-circular. Weirs may be sharp-edged or broad-crested. Flumes are similar to venturis, with a controlled decrease in width to a throat followed by a gradual increase to full channel wdith. They are often known as venturi flumes. For most of these devices there is a simplified relationship between the flow rate $\dot{V}$ and the upstream specific energy e:

$$\dot{V} = K e^n \tag{1.60}$$

where K is a coefficient which may be constant for a particular type of device (and for a specific device). The index n is approximately 1.5 for rectangular notches, weirs and flumes, and 2.5 for V-notches. The specific energy e is the sum of the depth and the velocity head:

$$e = \frac{v^2}{2g} + D \tag{1.61}$$

In many applications, particularly at the exit of large tanks or reservoirs, the upstream (or approach) velocity may be negligible and e becomes equal to either the depth D or the head above the base of the notch or weir H.

For a V-notch of included angle 2θ:

$$\dot{V} = 2.36C_d(\tan \theta)H^{2.5} \tag{1.62}$$

For a 90° notch:

$$\dot{V} = 2.36C_d H^{2.5} \tag{1.63}$$

For a rectangular notch of width B:

$$\dot{V} = 2.953C_d BH^{1.5} \tag{1.64}$$

The empirical *Francis* formula may be applied to sharp-edged weirs and rectangular notches:

$$\dot{V} = 2.953C_d(B - 0.2H)\left[e^{1.5} - \left(\frac{v^2}{2g}\right)^{1.5} \right] \tag{1.65}$$

or neglecting the approach velocity:

$$\dot{V} = 2.953C_d(B - 0.2H)H^{1.5} \tag{1.66}$$

For a venturi flume of throat width B, or a broad-crested weir of width B, the same equation applies:

$$\dot{V} = 1.705C_d B\, e^{1.5} \tag{1.67}$$

Since the value of e depends on the approach velocity v, which in turn depends on the flow rate $\dot{V}$, equations (1.65) and (1.67) are usually solved by an iterative method in which the first estimation of the approach velocity v is zero. Successive values of v are found from the upstream flow cross-sectional area and the preceding value of $\dot{V}$, the resulting value of e is then used in equation (1.67) for $\dot{V}$. This is repeated until there is little change in the required values. The discharge coefficient C_d in each of the flow equations (1.62) to (1.67) has a value of about 0.62.

As before, it is much more accurate to calibrate the device. For convenience, the calibration curves often plot the flow rate against the upstream depth.

(c) Floats In large rivers, where it is inconvenient to install flumes or weirs, or to use velocity meters, floats may be used. The timing of the passage of the floats over a measured distance will give an indication of the velocity. From the velocity, and as accurate a value of cross-sectional area as possible, the flow can be estimated.

(d) Chemical dilution In large, fast-flowing rivers chemical dilution may be the only acceptable method of flow measurement. The water is chemically analysed just upstream of the injection point and the natural concentration C_1 of the selected chemical in the water established. The concentration of the chemical injected is C_2 and the injection rate is R_i. Analysis of the water again at some distance downstream of the injection point determines the new concentration C_3 of the chemical. The flow rate $\dot{V}$ along the river may be estimated from

$$\dot{V} = R_i\left(\frac{C_2 - C_3}{C_3 - C_1}\right) \tag{1.68}$$

1.5.5 Open-channel flow

An open channel in this context is one containing a liquid with a free surface, even though the channel (or other duct) may or may not be closed. A pipe which is not flowing full is treated as an open channel.

1.5.5.1 Normal flow

Normal flow is steady flow at constant depth along the channel. It is not often found in practice, but is widely used in the design of channel invert (cross section) proportions.

(a) Flow velocity The average velocity, v, of flow in a channel may be found by using a modified form of the D'Arcy head loss equation for pipes, known as the *Chezy* equation:

$$v = C(mi)^{0.5} \tag{1.69}$$

C is the Chezy coefficient, a function of Reynolds' number Re and the friction coefficient f for the channel wall and i is the gradient of the channel bed. C may be obtained from tables or from the *Ganguillet and Kutter* equation or (more easily) the *Bazin* formula:

$$C = \frac{86.9}{1 + km^{-0.5}} \qquad (1.70)$$

where k is a measure of the channel wall roughness, typical values are shown in Table 1.6. m is the ratio of the cross-sectional area of flow to the wetted perimeter (the length around the perimeter of the cross section in contact with liquid), known as the *hydraulic mean depth*.

A widely used alternative modification of the D'Arcy equation is the *Manning* equation:

$$v = Mm^{0.67}i^{0.5} \qquad (1.71)$$

where M is the Manning number which depends, like the Chezy coefficient, on the condition of the channel walls. Values of M are tabulated for various channel wall materials (see Table 1.6). Some texts use Manning number $n = 1/M$.

The Chezy coefficient, C, the Manning number, M, and the roughness factor k used in equations (1.69)–(1.71) are *not* dimensionless. The equations and the tables are written in SI units and they must be modified for any other system of units.

(b) Optimum dimensions In order to produce the maximum flow rate in normal flow with a given cross-sectional area, the optimum channel shape is semi-circular. However, particularly for excavated channels, a semi-circular shape may be expensive to produce. It is easier and much cheaper to dig a rectangular or trapezoidal cross section. The optimum dimensions are: for the rectangular channel, when the width is twice the depth; for the trapezium, when the sides are tangential to a semi-circle. In both cases the hydraulic mean depth m will be equal to half the liquid depth, as for the semi-circular section.

The *maximum flow rate* through a circular pipe not flowing full will occur when the depth of liquid at the centre is 95% of the pipe diameter. The *maximum average velocity* will be achieved when the depth of liquid at the centre is 81% of the pipe diameter.

1.5.5.2 Non-uniform flow

In most instances of real liquids flowing in real channels the depth D of the liquid will vary along the length L of the channel with the relationship

Table 1.6

$$\frac{dD}{dL} = \frac{i - j}{1 - (v^2/gD)} \qquad (1.72)$$

where j is the slope of the *total energy line* (the plot of the total energy per unit weight against length). The total energy per unit weight is $(e + Z)$ at any point in the channel where the bed is at a height Z above the datum. Therefore if $i = j$ the depth is constant with L, normal flow conditions obtain and the slope of the channel i is the same as the slope of the energy line. v^2/gD is a dimensionless quantity known as *Froude's number* (Fr).

If $v = (gD)^{1/2}$, then $Fr = 1$, and from equation (1.72) the rate of change of depth with length (dD/dL) becomes infinite, which is the required condition for a *standing wave* or *hydraulic leap* to be formed in the channel (see Figure 1.42). The standing wave is a sudden increase in depth as the flow velocity is reduced from fast to slow (*supercritical* to *subcritical*), usually by channel friction or some obstruction such as a weir. The *critical velocity* v_c and the *critical depth* D_c are those which correspond to a Froude number of unity.

This phenomenon may also be explained by considering a graph of specific energy e against depth D (Figure 1.43). At the minimum value of e on the graph there is only one value of D, namely D_c, the critical depth. For a particular flow rate in a given channel it can be seen that any value of e above the minimum corresponds to two values of D. The higher value of D represents slow flow, the lower value represents fast flow. As the flow changes from fast to slow it passes through the critical value and a standing wave is formed (Figure 1.42).

The ratio of the downstream depth D_2 to the upstream depth D_1 across the standing wave is given by

$$\frac{D_2}{D_1} = \left(0.25 + \frac{2v_1^2}{gD_1}\right)^{0.5} - 0.5 \qquad (1.73)$$

and the loss of energy per unit weight or head loss by

$$h_L = \frac{(D_2 - D_1)^3}{4D_1D_2} \qquad (1.74)$$

1.5.6 Boundary layer flow

When a fluid flows over a solid boundary there is a region close to the boundary in which the fluid viscosity may be assumed to have an effect. Outside this region the fluid may be assumed inviscid. The viscous effect within the region is evidenced by a reduction in velocity as the boundary is approached. Outside the region the velocity is constant. The region is known as a *boundary layer*.

It is usual to assume that at the solid surface the fluid velocity is zero and at the boundary layer outer edge it is equal to the undisturbed flow velocity v_s. This defines the boundary layer thickness δ. (In practice, δ may be taken to be the distance from the boundary surface at which the velocity is 99% of the undisturbed velocity, or 0.99 v_s.)

Type of channel	Manning number, M	Bazin roughness factor, k
Smooth cement	91	0.06
Smooth wood	83	0.08
Concrete	77	0.10
Brick	67	0.15
Cast iron	67	0.16
Riveted steel	53	
Corrugated iron	44	
Rubble	40	0.50
Earth	37	1.00
Gravel	33	1.30
Rough, weedy earth	10	1.50

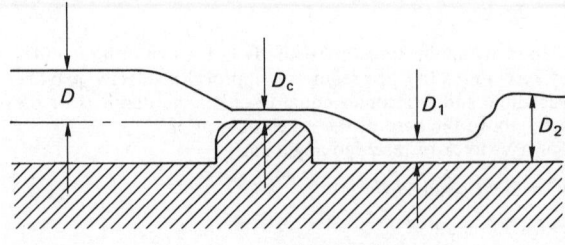

Figure 1.42 Broad-crested weir and standing wave

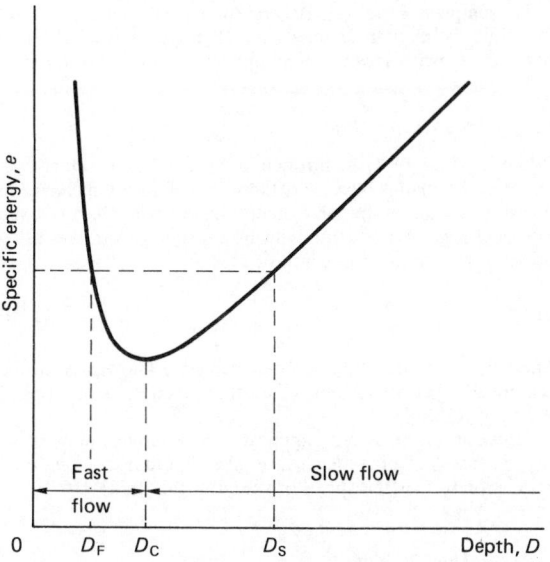

Figure 1.43   Graph of specific energy versus channel depth

When a flow stream at a velocity v_s passes over a flat plate the boundary layer thickness δ is found to increase with the distance x along the plate from the leading edge. Near the leading edge the flow inside the boundary layer may be assumed to be laminar, but as x increases the flow becomes turbulent and the rate of increase of δ with x also increases, as shown in Figure 1.44.

Within even a turbulent boundary layer there is a narrow region close to the plate surface where the flow is laminar. This is known as the *laminar sublayer* and has thickness δ_b. The reduction in velocity across the boundary layer is associated with a shear force at the plate surface, usually known as the *drag force*.

Application of the momentum equation produces *Von Karman's momentum integral*, in which the drag force per unit width, F_D, becomes

$$F_D = \rho v_s^2 \int_0^\delta \frac{v}{v_s}\left(1 - \frac{v}{v_s}\right) dy \qquad (1.75)$$

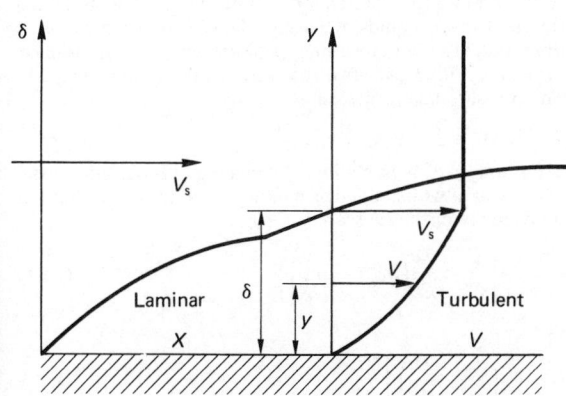

Figure 1.44   Boundary layer

where v is the velocity within the boundary layer at a distance y above the plate surface. (The integral

$$\int_0^\delta \frac{v}{v_s}\left(1 - \frac{v}{v_s}\right) dy$$

may be defined as the *momentum thickness* (θ) and the integral

$$\int_0^\delta \left(1 - \frac{v}{v_s}\right) dy$$

as the *displacement thickness* (δ^*) so that

$$F_D = \rho v_s^2 \theta.) \qquad (1.76)$$

In order to solve the Von Karman integral equation (1.75) or equation (1.76) it is necessary to know the value of δ and the relationship between v and y, the velocity distribution. Both of these are dependent on each other and the flow regime, laminar or turbulent, within the boundary layer.

1.5.6.1   Laminar boundary layers

A laminar boundary is normally assumed if $Re_x < 500\ 000$. (Re_x is Reynolds' number based on x or $\rho v_s x/\eta$.) For laminar boundary layers various simplified velocity distribution relationships may be used, such as linear, sinusoidal or cosinusoidal. The generally accepted most accurate relationship is, however, that obtained by the reduction of a four-term polynomial, which gives

$$\frac{v}{v_s} = \frac{3y}{2\delta} - \frac{1}{2}\left(\frac{y}{\delta}\right)^3 \qquad (1.77)$$

From this the shear stress at the plate surface, τ_0, may be found for Newtonian fluids:

$$\tau_0 = \eta \left(\frac{dv}{dy}\right)_{y=0} \qquad (1.78)$$

The shear force

$$F_D = \int_0^x \tau_0 dx \qquad (1.79)$$

Substitution from equation (1.77) in equations (1.75) and (1.78) and equating F_D from equations (1.75) and (1.79) leads via a separation of variables technique to

$$\frac{\delta}{x} = 4.64\ Re_x^{-0.5} \qquad (1.80)$$

The drag force is usually quoted in terms of a *drag coefficient*, C_D:

$$C_D = \frac{2F_D}{\rho v_s^2 x} \qquad (1.81a)$$

or

$$F_D = \frac{1}{2} C_D \rho v_s^2 x \qquad (1.81b)$$

By manipulation of the above equations an equation for C_D for a laminar boundary over the whole length L of the plate:

$$C_D = 1.29 Re_L^{-0.5} \qquad (1.82a)$$

is obtained. The drag force on the whole plate surface of area A is found from:

$$F = \frac{1}{2} C_D v_s^2 A \qquad (1.82b)$$

1.5.6.2 Turbulent boundary layers

For $Re_x > 500\,000$, the boundary is assumed to be turbulent. In a turbulent boundary layer the velocity distribution is often written in a power form:

$$\frac{v}{v_s} = \left(\frac{y}{\delta}\right)^{1/n} \tag{1.83}$$

The index n varies between 6 and 9, depending on Re_x.

Because of the presence of the laminar sublayer, the turbulent regime is not continuous down to the plate surface, and $(dv/dy)_{y=0}$ does not give a useful result.

The equation used for τ_0 is

$$\tau_0 = 0.0225\rho v_s^2 R_\delta^{-0.25} \tag{1.84}$$

based on work on smooth pipes by *Blasius*.

Taking $n = 7$ and using the same techniques as for laminar boundary layers gives:

$$\frac{\delta}{x} = 0.37\, Re_x^{-0.2} \tag{1.85}$$

and

$$C_D = 0.072\, Re_L^{-0.2} \tag{1.86}$$

This result assumes that the turbulent boundary layer obtains over the whole length of the plate to L.

Prandtl suggested a more realistic expression which takes into account the presence of a laminar boundary layer near the leading edge:

$$C_D = 0.074\, Re_L^{-0.2} - 1700\, Re_L^{-1} \tag{1.87}$$

This may be used for $5 \times 10^5 < Re_L < 10^7$. For $10^7 < Re_L < 10^9$ Schlichting (1960) suggests a logarithmic velocity distribution and

$$C_D = 0.44(\log_{10} Re_L)^{-2.58} = 3.91(\ln Re_L)^{-2.58} \tag{1.88}$$

Again, equation (1.82b) may be applied to find the drag force on the whole plate.

1.5.6.3 Laminar sublayers

The analyses in Sections 1.5.6.1 and 1.5.6.2 above assume that the plate surface is smooth or at least *hydraulically smooth*. A surface is regarded as hydraulically smooth if the average roughness height k is less than the laminar sublayer thickness δ_b. For a turbulent layer with a velocity distribution power index of $n = 1/7$, the laminar sublayer thickness at a point at a distance x along the plate from the leading edge is given by

$$\frac{\delta_b}{\delta} = 199\, Re_x^{-0.7} \tag{1.89}$$

Thus δ_b may be compared with the roughness height, k, if the boundary layer thickness, δ, is known.

1.5.7 Pressure transients (water hammer)

Water hammer is the common name for the rattling which occurs in water pipes as result of pressure transients. This phenomenon is due either to the collapse of cavitation bubbles or to rapid valve closure. The former is not uncommon in domestic water pipe work as an irritating vibration or noise on valve closure or opening (usually of the hot taps). It can be much more serious on a larger scale, where high-pressure rises over short periods may cause severe damage. Similar effects due to valve closure can be analysed on different levels of sophistication.

The simplest is the *rigid column* theory, which assumes that the fluid is incompressible, and that the valve is closed relatively slowly. This is often applied to water flow in pipes.

1.5.7.1 Slow valve closure

When a fluid flowing through a pipe with a velocity v_0 undergoes a change in velocity there is an associated change in pressure. Equating the force due to the pressure change to the rate of change of momentum during closure gives the resulting pressure rise Δp over a length of pipe L:

$$\Delta p = -\rho L \frac{dv}{dt} \tag{1.90}$$

The solution to this equation depends on a knowledge of the relationship between v and t (the valve closure rate in terms of the flow velocity).

Equation (1.90) is only applicable to relatively slow valve closure rates in which the closure time should not be less than $2L/C$ (where C is the speed of sound in the fluid).

1.5.7.2 Time to establish flow

The rigid column theory is also often used to calculate the time required to establish flow in a pipe on opening a valve. The theory implies that the time required to fully establish the flow is infinite and so the time t to achieve 99% of the final velocity v_0 is usually accepted:

$$t = 2.646 \frac{Lv_0}{gH} \tag{1.91}$$

where H is the supply head to the pipe entrance. The time t_x required to reach $x\%$ of the final velocity is given by

$$t_x = \frac{Lv_0}{2gH} \ln\left(\frac{1 + 0.01x}{1 - 0.01x}\right) \tag{1.92}$$

1.5.7.3 Rapid valve closure

When a fluid is brought to rest *instantaneously* from a velocity of v_0 by the closure of a valve at the exit of a pipe of diameter D there will be a relatively high pressure rise at the valve. If the valve closure time is *less than* $2(L/c)$ then the resultant pressure rise is as if it were instantaneous; c is the speed at which the pressure wave travels through the fluid, which is the sonic velocity.

On such a rapid valve closure the kinetic energy of the flow is converted into strain energy in both the pipe material and the fluid (even liquids are acknowledged as compressible in this context). The resulting pressure wave is transmitted through the fluid away from the valve as shown in Figure 1.45. The pressure rise produced is

$$\Delta p = \rho c v_0 \tag{1.93}$$

For a fluid of bulk modulus G, in a pipe of wall thickness x, of a material with a Young's modulus E and Poisson's ratio σ, the velocity of the pressure wave is

$$c = \left[\rho\left\{\frac{1}{G} + \frac{D}{Ex}\left(1.25 - \sigma\right)\right\}\right]^{-0.5} \tag{1.94a}$$

or

$$c = \left[\rho\left(\frac{1}{G} + \frac{D}{Ex}\right)\right]^{-0.5} \tag{1.94b}$$

if longitudinal stress is small compared to hoop stress.

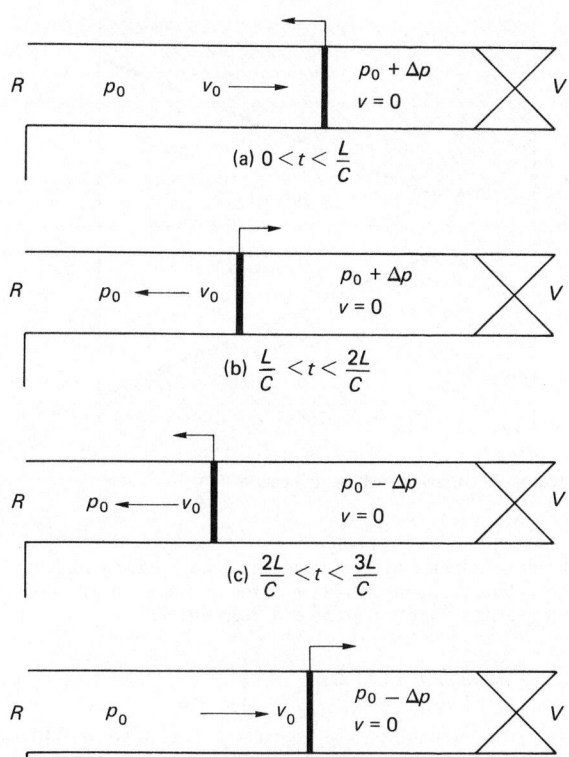

Figure 1.45 Progress of a pressure wave

Equation (1.94a) is often written as $c = [G_e/\rho]^{0.5}$, where G_e is the effective bulk modulus of the fluid and pipe combinations:

$$\frac{1}{G_e} = \frac{1}{G} + \frac{D}{Ex}(1.25 - \sigma) \qquad (1.95)$$

If the valve is at the entrance to the pipe, then rapid valve closure results in a rarefaction (pressure drop) at the valve. In other words, the pressure change is

$$\Delta p = -\rho c v_0 \qquad (1.93')$$

1.5.7.4 The progress of a pressure wave

Assuming no friction and no cavitation in the fluid, the progress of a pressure wave along a pipe between a valve and a reservoir following valve closure is as shown in Figure 1.45. The fluid in the pipe is successively brought to rest by the passage of the pressure wave.

At a time $t = L/c$ after valve closure, the pressure wave reaches the reservoir. The whole of the fluid in the pipe is at rest at a pressure $p = p_0 + \Delta p$, which at the reservoir end instantaneously drops to reservoir pressure p_0. The resulting pressure wave travels along the pipe towards the valve and the fluid at the higher pressure in the pipe flows towards the reservoir at its initial velocity v_0.

At $t = 2L/c$, the situation is the same as for a rapid closure of a valve downstream of the flow, producing an instantaneous pressure drop to $p_0 - \Delta p$ and a rarefaction which travels towards the reservoir. The passage of the rarefaction successively brings the fluid to rest along the pipe.

At $t = 3L/c$, the rarefaction reaches the reservoir and the pressure instantaneously rises to reservoir pressure p_0. The resulting pressure wave travels towards the valve and fluid flows away from the reservoir at velocity v_0.

At $t = 4L/c$, the situation is the same as when the valve first closed at $t = 0$, and the cycle is repeated.

In practice, friction quickly dampens out the pressure waves and cavitation reduces the pressure decrease during the rarefactions.

A typical plot of pressure against time at a valve following rapid valve closure is superimposed on the theoretical plot in Figure 1.46.

1.5.8 Gas flow

1.5.8.1 General relationships

The behaviour of gases during processes involving thermal energy interactions and exchanges fits more properly into a study of thermodynamics. However, if only the flow mechanics are considered, the thermal and temperature effects may be restricted to those mainly relating to pressure and density.

The most straightforward approach is to consider zero thermal energy transfer (heat transfer) to or from the fluid, or adiabatic flow. If, in addition, the changes in the fluid's properties are assumed to be reversible, then the flow becomes isentropic and the relationship between pressure and density the simple and well-known one for an ideal gas:

$$\frac{p}{\rho^\gamma} = \text{constant} \qquad (1.96)$$

Also, the *ideal gas law*:

$$\frac{p}{\rho} = RT \qquad (1.97)$$

applies. Other useful relationships are:

1. The ratio of specific heats $\gamma = \dfrac{c_p}{c_v}$ $\qquad (1.98)$

2. The gas constant $R = c_p - c_v$ $\qquad (1.99)$

3. The *universal gas constant* $R_o = R \times$ molecular mass

$$= 8.3143 \text{ kJ kg}_{mol}^{-1} \text{ K}^{-1}$$

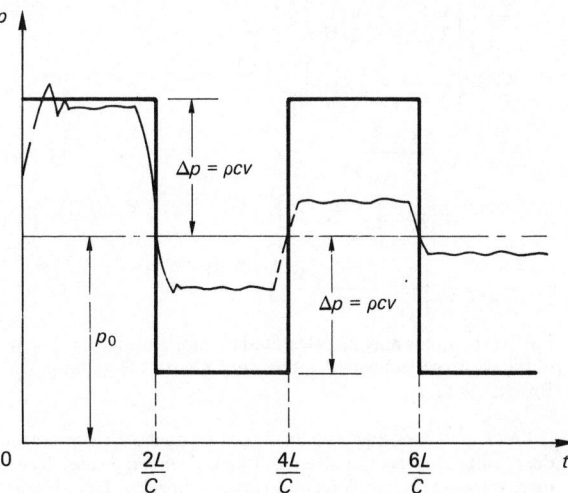

Figure 1.46 Pressure versus time at a valve

The terms *stagnation* or *total* temperature T_0 and pressure p_0 are often applied as the datum temperature and pressure of a fluid flow, even when stagnation conditions (zero velocity) do not exist in the particular situation under consideration. In gas flow the relationships between T_0 and the temperature T and p_0 and the pressure p at some point in the flow is often given in terms of the *Mach number* (M), the ratio of the flow velocity v to that of sound c, i.e.

$$M = \frac{v}{c} \tag{1.100}$$

and

$$c = \left[\gamma \frac{p}{\rho}\right]^{0.5} = [\gamma RT]^{0.5} \tag{1.101}$$

In these terms T_0 and p_0 may be found from Euler's equation to be

$$T_0 = T\left[1 + \frac{(\gamma - 1)}{2}M^2\right] \tag{1.102}$$

$$p_0 = p\left[1 + \frac{(\gamma - 1)M^2}{2}\right]^{\gamma/(\gamma-1)} \tag{1.103}$$

1.5.8.2 Flow in ducts and nozzles

(a) Ducts The analysis of gas flow in ducts is based on the Euler equation (1.34) and the one-dimensional continuity equation (1.25). Consideration of the differential forms of these equations will demonstrate that for subsonic flow ($M < 1$) the velocity will increase as the cross-sectional area of the duct decreases (in the converging entrance to a convergent/divergent nozzle, for example). For supersonic flow ($M > 1$) the velocity will increase as the cross-sectional area increases (in the diffuser of the convergent/divergent nozzle).

The properties of the fluid at a position in the flow stream where the local Mach number is unity are often denoted by a superscript * (p^*, ρ^*, T^*) and used as a datum, so that

$$M^* = 1; \text{ and } v^* = c^* = [\gamma RT^*]^{0.5} \tag{1.104}$$

The ratios of the properties at any position in the flow stream to those at the * position are:

1. $\dfrac{p}{p^*} = \left[\dfrac{\gamma + 1}{2 + (\gamma - 1)M^2}\right]^{\gamma/(\gamma-1)}$ (1.105)

2. $\dfrac{\rho}{\rho^*} = \left[\dfrac{\gamma + 1}{2 + (\gamma - 1)M^2}\right]^{1/(\gamma-1)}$ (1.106)

3. $\dfrac{T}{T^*} = \dfrac{\gamma + 1}{2 + (\gamma - 1)M^2}$ (1.107)

4. $\dfrac{v}{v^*} = M\left[\dfrac{\gamma + 1}{2 + (\gamma - 1)M^2}\right]^{1/2}$ (1.108)

5. $\dfrac{A}{A^*} = \dfrac{1}{M}\left[\dfrac{2 + (\gamma - 1)M^2}{\gamma + 1}\right]^{(\gamma+1)/2(\gamma-1)}$ (1.109)

For air the ratios may be calculated by substituting $\gamma = 1.4$, or obtained from published tables and charts (Houghton and Brock, 1961).

(b) Nozzles A nozzle is an example of a duct with a smoothly decreasing cross-sectional area, followed in some cases by an increasing area (convergent/divergent nozzle) (see Figure 1.47). Since the velocity in the throat (minimum cross section) is often sonic, the approach velocity may be negligible. The

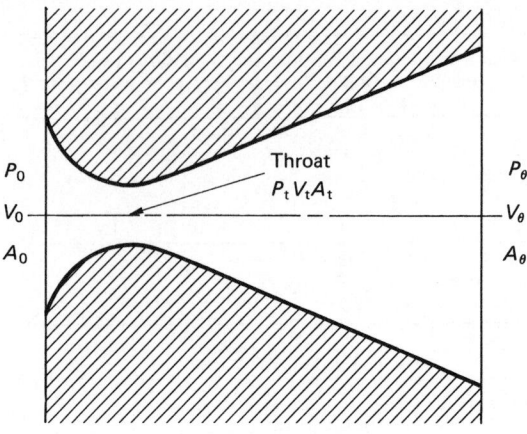

Figure 1.47 Convergent/divergent nozzle

throat velocity v_t and the exit velocity v_e are found by applying the Euler equation (1.34) between the upstream (entry conditions suffix $_0$) and throat and exit, respectively:

$$v_t = \left\{\frac{2}{\gamma - 1}\frac{p_0}{\rho_0}\left[1 - \left(\frac{p_t}{p_0}\right)^{(\gamma - 1)/\gamma}\right]\right\}^{1/2} \tag{1.110}$$

For v_e the pressure term p_e replaces p_t in equation (1.110). The mass flow rate through the nozzle is usually found at the throat by substituting v_t in the mass flow equation (1.25) to give

$$\dot{m} = C_d A_t \left\{\frac{2}{\gamma - 1}\frac{p_0}{\rho_0}\left[\left(\frac{p_t}{p_0}\right)^{2/\gamma} - \left(\frac{p_t}{p_0}\right)^{(\gamma + 1)/\gamma}\right]\right\}^{1/2} \tag{1.111}$$

where C_d is a discharge coefficient which depends on the nozzle design. For a well-designed nozzle C_d will be close to unity.

The mass flow rate will be the same at the exit as at the throat. It may be calculated from the exit conditions by substituting A_e and p_e for A_t and p_t, respectively, in equation (1.111).

Nozzles are usually designed for maximum mass flow rate. This will occur when the throat velocity is sonic ($v_t = c$). The pressure ratio which produces this situation is known as the *critical pressure ratio*, given by

$$\left(\frac{p_t}{p_0}\right)_{\text{crit}} = \left(\frac{2}{\gamma + 1}\right)^{\gamma/(\gamma - 1)} \tag{1.112}$$

For many light diatomic gases such as air, where γ is approximately 1.4, $(p_t)/(p_0)_{\text{crit}} = 0.528$.

The throat area will be that which gives the required mass flow rate through the throat at sonic velocity for critical pressure ratio. The exit area will be that which gives the calculated exit velocity for the given mass flow rate at the exit conditions. For convergent nozzles the throat also becomes the exit.

If nozzles, orifices or venturi meters are used to measure gas flow rates through a pipe then the approach velocity may be signficant and the mass flow rate given by

$$\dot{m} = C_d A_t \rho_0\left\{\frac{2(p_0/\rho_0)[p_t/p_0]^{2/r} - (p_t/p_0)^{(\gamma+1)/r}}{(\gamma - 1)[1 - (p_t/p_0)^{2/r}(p_t/p_0)^2]}\right\}^{1/2}$$
$$\tag{1.113}$$

1.5.8.3 Shock waves

Under normal design conditions the flow in the nozzle downstream of the throat will be supersonic. The velocity of the gas at exit will depend on the external pressure p_b (back pressure). If the back pressure is greater than the theoretical exit pressure $p_e (p_b > p_e)$ then *shock waves* will be set up in the nozzle. These are discontinuities similar to standing waves in open-channel flow. The shock waves set up in such a way are *normal shock waves*, normal to the direction of flow.

If $p_b < p_e$ then the expansion will continue outside the nozzle (over-expansion).

If conditions upstream of a normal shock wave are denoted by suffix 1 and downstream by suffix 2, then it can be shown that the product of the up- and downstream velocities is equal to the square of the sonic velocity at $M = 1$:

$$v_1 v_2 = C^{*2} \tag{1.114}$$

and since $M_1 > 1$, then $M_2 < 1$. Also,

1. $\quad M_2 = \left[\dfrac{2 + (\gamma - 1)M_1^2}{2\gamma M_1^2 - (\gamma - 1)} \right]^{0.5}$ $\qquad$ (1.115)

2. $\quad \dfrac{p_2}{p_1} = \dfrac{2\gamma M_1^2}{(\gamma + 1)} - \dfrac{\gamma - 1}{(\gamma + 1)}$ $\qquad$ (1.116)

3. $\quad \dfrac{p_2}{p_1} = \dfrac{v_1}{v_2} = \dfrac{(\gamma + 1)M_1^2}{2 + (\gamma - 1)M_1^2}$ $\qquad$ (1.117)

These are known as the *Rankine–Hugoneot* relationships. Values for air may be obtained by putting $\gamma = 1.4$ or by the use of published tables (Houghton and Brock, 1961).

The strength of a shock wave may be defined as the ratio of the pressure change across the wave to the upstream pressure, or in terms of the upstream Mach number:

$$\frac{\text{Shock wave}}{\text{strength}} = \frac{p_2 - p_1}{p_1} = \left(\frac{2}{\gamma + 1} \right)(M_1^2 - 1) \tag{1.118}$$

$$= 1.167(M_1^2 - 1), \text{ for air.}$$

Oblique shock waves, at an angle β to the upstream flow direction, are produced when a supersonic gas flow is turned through an angle θ by an obstruction such as an aircraft's nose, wing or tail, the inside walls of a duct, etc. The relationships between the up- and downstream Mach numbers and the angles β and θ are published in tables and charts (Houghton and Brock, 1961).

In some cases both the up- and downstreams will be supersonic and subsequent shock waves produced, for example, at the leading and trailing edges of a wing. The effects of such shock waves produced by aircraft in flight, say, may be noted at ground level as *sonic booms*.

1.5.8.4 Gas flow measurement

Gas flow rates through ducts will normally be measured using devices and techniques similar to those used for incompressible fluids, namely orifice plates, venturi meters and nozzles. However, for gases the flow rate is usually quoted as a mass flow rate. Equations (1.110) and (1.112) may be used with orifices and venturis as well as nozzles. Relevant values of C_d for each device will be found in BS 1042, in addition to operational advice.

When pitot-static tubes are used to measure gas flow velocities equation (1.58) may be acceptable for low flows with low pressure differences. At high velocities the compressibility must be taken into account and equation (1.110) used with stream conditions at X replacing those at the throat t.

For accurate velocity measurement with little disturbance to the flow *hot wire anemometers* may be used. The resistance of an electrically heated wire is related to the temperature of the wire, which in turn is related to the velocity of the fluid flow past the wire. The wire resistance measured on a bridge may be calibrated against a known velocity, to give either direct readout or (more usually) a calibration curve. The fine wire of the anemometer is suceptible to fluid contamination.

Other velometers and anemometers (depending on the relationship between the speed of rotation of a set of blades and the velocity (or speed) of the gas flow) may be used in very large cross-sectional ducts or to measure wind speed in the open air. They may depend on the rotation to generate a small electrical current, which can be calibrated as a speed, or the number of revolutions may be inserted into an empirical formula. A typical example is the three-vaned meteorological anemometer.

1.5.9 Ideal fluid flow

The concept of using idealized conditions to establish the shape of the mathematical models of real situations is common in engineering science studies. These models may then be modified to accommodate observed relationships, for application to real situations.

Ideal fluid (or *potential*) flow is such a concept. It may be used to set up flow patterns in the region of a flow stream outside the boundary layers described in Section 1.5.6. The combination of ideal flow and the boundary layer effects may be used to predict the performance of a real situation, so long as the limitations of both are recognized. The fluid is assumed to be inviscid and the flow steady, continuous and irrotational, as defined in Section 1.5.3.1. This means that there are no cavities or discontinuities in the flow stream, and that the fluid particles do not rotate about their own axes, even though flow may be circular.

The continuity equation (1.23) applies, and may be modified to

$$\frac{\partial v_x}{\partial x} + \frac{\partial v_y}{\partial y} = 0 \tag{1.119}$$

if required for two-dimensional flow. For irrotation in two dimensions equation (1.28) becomes

$$\frac{\partial v_x}{\partial y} - \frac{\partial v_y}{\partial x} = 0 \tag{1.120}$$

1.5.9.1 The stream function

From the definitions of the stream line and stream function in Section 1.5.3.1(e) the equation to a stream line may be shown to be

$$v_x \mathrm{d}y - v_y \mathrm{d}x = 0 \tag{1.121}$$

and since the stream function ψ is an equation which describes a family of stream lines, then, for example,

$$\psi = 2x - y$$

represents a family of parallel straight lines with a variable intercept ψ.

For unit thickness in the z direction the volumetric flow rate $\dot{V}$ between two stream lines 1 and 2 is

$$\dot{V} = \psi_1 - \psi_2 \tag{1.122}$$

or

$$\mathrm{d}\dot{V} = \mathrm{d}\psi \tag{1.123}$$

Stream functions may be *superposed* so that if

$\psi_1 = \text{fn}(x,y)$ describes flow pattern A,

and

$\psi_2 = \text{fn}(x,y)$ describes flow pattern B,

then $\psi = \psi_1 + \psi_2$ describes the flow pattern produced by the combination of A and B.

The x and y components of the flow velocity v are given by

$$v_x = -\frac{\partial \psi}{\partial y}; \text{ and } v_y = \frac{\partial \psi}{\partial x} \qquad (1.124)$$

or

$$v_x = \frac{\partial \psi}{\partial y}, \text{ and } v_y = -\frac{\partial \psi}{\partial x}$$

depending on sign convention. In *polar coordinates* the components are

$$\text{Radial velocity } v_r = \frac{-\partial \psi}{r \partial \theta} \qquad (1.125)$$

$$\text{Tangential velocity } v_\theta = \frac{\partial \psi}{\partial r} \qquad (1.126)$$

1.5.9.2 The velocity potential

In a gravitational field there is a property the change in which is independent of the path of the change: *potential energy*. In a continuous, irrotational flow field there is also a property the change in which is independent of the path of the change. This property is the *velocity potential* (ϕ). It can be shown that

$$v_x = -\frac{\partial \phi}{\partial x}; \text{ and } v_y = -\frac{\partial \phi}{\partial y} \qquad (1.127)$$

or

$$v_\theta = -\frac{\partial \phi}{r \partial \theta}; \text{ and } v_r = -\frac{\partial \phi}{\partial r} \qquad (1.128)$$

Lines of constant ϕ are known as *velocity potential* lines with an equation

$$v_x dx + v_y dy = 0 \qquad (1.129)$$

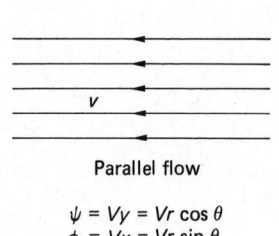

Parallel flow

$$\psi = Vy = Vr \cos \theta$$
$$\phi = Vx = Vr \sin \theta$$

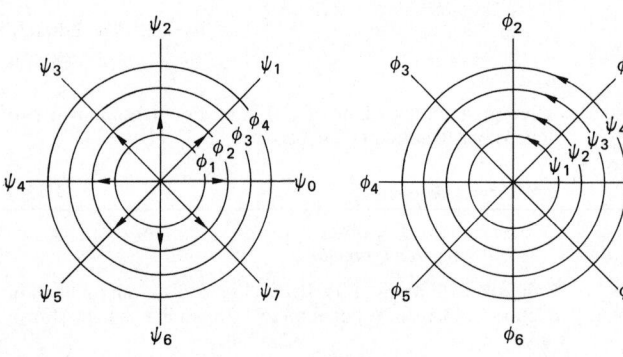

Source $\psi = -\dfrac{m\theta}{2\pi}$

$$\phi = -\frac{m}{2\pi} \ln r$$

Vortex $\psi = \dfrac{\Gamma \ln r}{2\pi}$

$$\phi = -\frac{\Gamma 0}{2\pi}$$

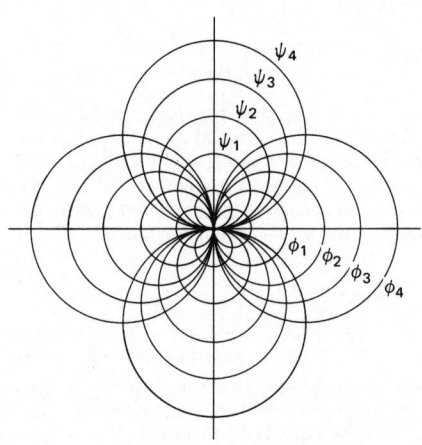

Doublet $\psi = -\dfrac{\mu \sin \theta}{2\pi r}$

$$\phi = \frac{\mu \cos \theta}{2\pi r}$$

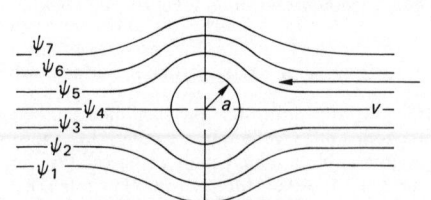

Doublet in parallel flow

$$\psi = V\left[r - \frac{a^2}{r}\right] \sin \theta$$

$$\phi = V\left[r + \frac{a^2}{r}\right] \cos \theta$$

Figure 1.48 Simple flow patterns

which intercept the stream lines at right angles to form an orthogonal network of characteristic pattern for each flow field.

From equations (1.124) and (1.127) it can be seen that

$$\frac{\partial \psi}{\partial y} = \frac{\partial \phi}{\partial x}; \text{ and } \frac{\partial \psi}{\partial x} = -\frac{\partial \phi}{\partial y} \qquad (1.130)$$

which are the *Cauchy–Rieman* equations. In such flow fields the *Laplace* equations for ϕ and ψ must both be satisfied:

$$\nabla^2 \phi = 0; \text{ and } \nabla^2 \psi = 0 \qquad (1.131)$$

Because the fluid is ideal, the Bernoulli equation (1.37) may be readily applied between points in the field, both along and across the stream lines.

1.5.9.3 Flow patterns

Examples of simple flow patterns are shown in Figure 1.48 with the equations to their stream functions and velocity potentials. A *source* is mathematically a point at which fluid appears and flows radially outwards. A *sink* is a negative source at which fluid flows radially inwards to disappear at a point (similar to the plug hole in a domestic sink, where, however, there is a vortex superposed to produce a spiral vortex or whirlpool). A *vortex* is flow in concentric circles with no radial flow. A *doublet* is the superposition of a source and a sink of equal strength m, initially a distance $2a$ apart brought infinitely close together so that the product of their strength and the distance between them remains a constant k. k is the strength of the doublet and is equal to $2am$.

This is a mathematical concept which is apparently impractical but yields a useful flow pattern. It is often used in combination with other simple patterns.

1.5.9.4 Modelling

Since there can be no flow across a stream line and the fluid is assumed inviscid, any stream line in a flow pattern may be replaced by a solid surface with no effect on the rest of the pattern. A stream line forming a closed contour may be replaced by a solid body to model the flow pattern around a body of the same shape. This provides a method of writing mathematical models to describe the flow streams around various shapes.

The limitations of the model must be understood, and the effects of rotation and viscosity particularly considered, when applying the analysis to *real* situations. For example, the flow round a cylinder may be modelled by a combination of doublet and parallel flow. The drag and lift forces calculated by integrating the resulting pressure forces at the cylinder surface appear to be zero. This is obviously not correct.

In the real situation the viscosity of the fluid produces a boundary layer at the cylinder surface, which, because of curvature, separates from the surface to form a wake. The presence of the wake disturbs the downstream flow pattern and the form drag force is a consequence. Viscosity also introduces a surface friction drag.

The shedding of vortices from the cylinder surface into the wake also produces alternate positive and negative lift forces, which are not predicted by the ideal flow analysis, although telegraph wires may often be observed vibrating in the wind.

1.5.10 Conclusion

Fluid mechanics is often regarded as an empirical subject which makes use of formulae based only on observed experimental results. This misconception is further compounded by

the extensive use of 'coefficients' (discharge especially) to account for effects which are difficult to model mathematically. However, almost all fluid mechanics equations in common use are based on the conservation of energy, the conservation of momentum or the fact that the rate of change of momentum may be equated to an applied force, usually a pressure force. The experimental checking and empirical amendment to derived formulae is just *good engineering practice*.

This section does not provide rigorous derivations of the various equations quoted. Some appreciation of such derivations may be required in order to establish the limitations and modifications necessary for the application of the equations, especially to non-standard situations. For this, standard textbooks on fluid mechanics as listed below should be consulted.

Further reading

General fluid mechanics
Douglas, J. F., Gasiorek, J. M. and Swaffield, J. A., *Mechanics of Fluids*, third edition, Pitman, London (1986)
Walshaw, A. C. and Jobson, D. A., *Mechanics of Fluids*, third edition, Longman, London (1979)
Ireland, J. W., *Mechanics of Fluids*, Butterworths, London (1971)
Reference tables
Houghton, E. L. and Brock, A. E., *Tables for the Compressible Flow of Dry Air*, Edward Arnold, London (1961)
Rogers, G. F. C. and Mayhew, Y. R., *Thermodynamic and Transport Properties of Fluids*, third edition, Blackwell, Oxford (1980)
Additional computer solutions
Smith, P. D., *BASIC Hydraulics*, Butterworths, London (1982)
Boundary layer flow
Schlicting, H., *Boundary Layer Theory*, fourth edition, McGraw-Hill, New York (1960)
Ideal fluid flow
O'Neill, M. E. and Chorlton, F., *Ideal and Incompressible Fluid Dynamics*, Ellis Horwood, Chichester (1986)
For students preparing for examinations
Douglas, J. F., *Solutions of Problems in Fluid Mechanics*, Parts 1 and 2, third edition, Pitman, London (1987)
Brasch, D. J. and Whyman, D., *Problems in Fluid Flow*, Edward Arnold, London (1986)

1.6. Principles of thermodynamics

1.6.1 Introduction

Thermodynamics is concerned with energy transfers in processes. Two modes of transfer are recognized: *work* (transfer) and *heat* (transfer). From the mechanical engineer's viewpoint the main interest is the production of shaft power (rate of work transfer) by a machine to which the energy is supplied by heat transfer obtained directly or indirectly from the combustion of a hydrocarbon fuel in air. Before proceeding, the terminology used in thermodynamics must be defined.

The particular part of the *working substance* under consideration is called the *system*, and this is separated from the *surroundings* by a *boundary*. In the *closed* or *non-flow* system the mass of working substance is constant, but in the *open* or *flow* system there is a mass flow rate across the boundary. Some processes in reciprocating plant may be considered by non-flow analysis but in steam plant, for example, most are considered by flow analysis. The *state* of a system is defined by the *properties* (pressure, temperature, etc.). Properties are normally expressed *specifically* (i.e. per unit mass) to enable charts or tables to be used. The state of simple substances can be described by two independent properties, but complex ones such as mixtures need more definition. A *change of state* is achieved by a *process* which is idealized as *reversible* with no losses. Reversible processes can be described by mathematical equations and enable analysis to be made to give answers for ideal situations. Real processes have losses and are described as *irreversible*, and the ideal results are multiplied by a coefficient or efficiency (based on measurement or experience) to predict real performance.

1.6.2 The laws of thermodynamics[9,10]

1.6.2.1 The first law of thermodynamics

This is a law of energy conservation. When applied to a process we write

$$Q - W = \Delta E \text{ or } q - w = \Delta e$$

where Q is the heat transfer (kJ) or q is the specific heat transfer (kJ/kg), W is the work transfer (kJ) or w is the specific work transfer (kJ/kg), and ΔE is the energy change (kJ) or Δe is the specific energy change (kJ/kg).

The change symbol Δ means final value minus initial value. ΔE embraces all forms of energy but in the non-flow process it is usual to find that the only significant change is in the internal energy (U, u) and we write the non-flow energy equation

$$Q - W = \Delta U \quad \text{or} \quad q - w = \Delta u$$

For the *steady* flow system we write

$$\dot{Q} - \dot{W}_x = \dot{m}\Delta\left(h + \frac{V^2}{2} + gz\right)$$

$$\text{or} \quad q - w_x = \Delta\left(h + \frac{V^2}{2} + gz\right)$$

where $\dot{Q}$ and $\dot{W}_x$ are the energy transfer rates and $\dot{m}$ is the steady mass flow rate across the boundary (in and out), Δh is the change in specific enthalpy ($h = u + pv$), $\Delta V^2/2$ is the change in specific kinetic energy and Δgz is the change in specific potential energy. The suffix x is used on the work transfer to denote that this is the useful work from the system as the flow work is included in the enthalpy term. In flow problems it will also be necessary to use the continuity equation

$$\dot{m} = \rho A V$$

where ρ is the density and A is the area normal to the velocity V. Analysis of non-steady flow may also be made, in which case energy terms to allow for the storage of energy in the system will be added.

Warning: A sign convention for work and heat is built into the equations above. Positive work means work obtained from the system and positive heat means heat put into the system. Care should be taken to be clear about the symbol V, which may appear as velocity or volume in many equations.

In order to allow continuous energy transfers a *cycle* is defined in which a series of processes brings the working substance back to the initial state so that the cycle can be repeated continuously. If we apply the first law to a cycle it follows that ΔE is zero and

$$\sum_{\text{cycle}} Q = \sum_{\text{cycle}} W$$

1.6.2.2 The second law of thermodynamics

It might be thought that the first law of thermodynamics permits all the heat transfer to a cycle to be returned as work transfer, but unfortunately the second law places restraints on the achievement of this desirable situation. The restraint takes the practical form of demanding that some of the heat transfer to the cycle *must* be rejected as a heat transfer to a lower temperature. Thus when we build a *heat engine* it has to exchange heat with (at least) two reservoirs in order to produce work (Figure 1.49). Since work is the objective, the amount produced per unit heat input is vital information and we define the *thermal efficiency* of a heat engine as

$$\eta_{\text{thermal}} = \frac{\text{Net work transfer from the cycle}}{\text{Heat transfer to the cycle}} = \frac{W}{Q_1}$$

Since the first law states $Q_1 - Q_2 = W$ we see that efficiency is less than unity.

The second law makes further investigations and determines the maximum possible efficiency of a heat engine using reversible isothermal processes to transfer heat from two reservoirs as

$$\eta_{\text{thermal maximum}} = 1 - (T_{\text{min}}/T_{\text{max}})$$

This efficiency is known as the Carnot efficiency and is not attainable due to losses. It is also found that constant temperature processes, except during phase change, are not practical and real processes of heat transfer take place at approximately constant volume or constant pressure. One positive product of the second law is that it tells the engineer that thermal efficiency will be increased by making the maximum cycle temperature as high as possible (a materials constraint) and by making the minimum cycle temperature as low as possible (ambient conditions)

The losses in a real cycle are due to internal fluid friction and the necessity of having a temperature difference to cause a heat transfer. The larger the temperature difference, the greater the losses. The fluid friction losses in a work-producing process are defined by the process efficiency

$$\eta_{\text{process}} = \frac{\text{Actual work produced}}{\text{Ideal work produced}}$$

which is inverted for work-absorbing processes (compression). The most common application of this efficiency is in steady flow *adiabatic* processes. Ideally, these are processes with no heat transfer which are often used as models for real processes in which the heat transfers are negligible compared with the work transfers (turbine expansion). In such processes the

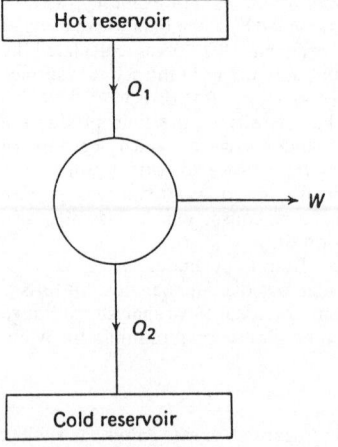

Figure 1.49 A heat engine

efficiency is known as the isentropic efficiency because the ideal adiabatic process has constant entropy (qv).

Losses due to fluid friction and losses due to heat transfer across finite temperature differences are found to result in an increase in the value of the entropy that would be expected in a reversible process. Thus an expected increase would be larger and an expected decrease would be smaller. It is not easy to define entropy except mathematically. In practical use as the abscissa of charts it enables work transfers in ideal adiabatic processes to be represented as vertical lines if enthalpy is used as an ordinate, and in this guise is a valuable visual method of presentation.

1.6.3 Thermoeconomics[11,12]

When a more detailed study of a flow process is made by the second law of thermodynamics is is found that specific entropy ($kJ.kg^{-1}K^{-1}$) appears as part of a property known as availability. In a flow process we write $b = h - T_0 s$, in which b is the specific *availability* function, h is the specific enthalpy, T_0 is the temperature (absolute) of the surroundings and s is the specific entropy. The second law shows that the maximum work potential or *exergy* of any state in surroundings at state 0 is given by $b - b_0$. Thus for a change of state in a flow process from 1 to 2 the maximum specific work obtainable is given by the exergy change, $w_{x\max} = (b_1 - b_0) - (b_2 - b_0) = (b_1 - b_2) = -\Delta b$. If we measure or predict by analysis the actual work achieved it is possible to determine numerically the lost work or irreversibility in the process. If engineering plant is to be designed to the best advantage it is clear that processes should be chosen to minimize this loss. The lost work may be associated with costs and we move into the developing field of thermoeconomics. Clearly, this is a complex subject but it is important in that it unites thermodynamics with costs and can help in the design of long-life expensive plant.

1.6.4 Work, heat, property values, process laws and combustion[9,10]

To deploy the laws of thermodynamics outlined above we need more information. To perform simple cycle analysis the data below is vital.

1.6.4.1 Work

In a non-flow process work transfer can be determined from $w = \int p dv$. The mathematical relation for the process is known as the process law (qv). In most flow processes used in engineering cycles the adiabatic approximation is used so that the steady flow energy equation, neglecting changes in kinetic and potential energy, gives

$$w_x = \Delta h$$

1.6.4.2 Heat

This is usually an unknown quantity and is found by the application of the energy equation. As stated earlier, many processes are approximately adiabatic so that heat transfer is zero and in others heat transfer is obtained from combustion data or, if a heat exchange process, by heat exchanger efficiency. In heating plant such as boilers which do no useful work the steady flow energy equation shows $q = \Delta h$.

1.6.4.3 Property values[13]

These are found in tables or from charts for common substances. Computer formulations are also available.

1.6.4.4 Process laws

This is a particularly important step in thermodynamic analysis because an idealized reversible process has to be chosen to represent as closely as possible the real process in order to calculate energy changes. When the working substance is a gas it is convenient in an elementary analysis to use perfect gas laws with the process calculation. These are

$$pv = RT \qquad \text{or} \qquad pV = mRT$$
$$u = c_v \Delta T \qquad \text{and} \qquad h = c_p \Delta T$$

where c_v and c_p are the specific heat capacities at constant volume and constant pressure, respectively, which are related as follows:

$$c_p - c_v = R \qquad \text{and} \qquad c_p/c_v = \gamma$$

where R is the specific gas constant and γ is the isentropic index.

Ideal processes commonly used are *constant pressure*, *constant volume*, *constant temperature* (which for a perfect gas becomes $pv = $ constant) together with two other more general relations: the *adiabatic process*, $pv^k = $ constant (which for a perfect gas becomes $pv^\gamma = $ constant) and the *polytropic process*, $pv^n = $ constant. The last process is a general relation between pressure and volume which is used if none of the other clearly special cases are considered valid. Usually $1 < n < 1.4$.

It is possible (by using the gas laws) in adiabatic and polytropic gas processes to rearrange the relations to involve pressure and temperature or temperature and volume to yield very useful relations:

$$\frac{T_1}{T_2} = \left(\frac{p_1}{p_2}\right)^{(\gamma-1)/\gamma}; \frac{T_1}{T_2} = \left(\frac{V_2}{V_1}\right)^{\gamma-1}; \frac{T_1}{T_2} = \left(\frac{p_1}{p_2}\right)^{(n-1)/n};$$
$$\frac{T_1}{T_2} = \left(\frac{V_2}{V_1}\right)^{n-1}$$

Processes may be represented on property diagrams to enable cycle visualization (Figures 1.50–1.54).

1.6.4.5 Combustion

To avoid involving complex chemical equations, engineers often use the calorific value of a fuel coupled with a combustion efficiency to estimate the energy transfers in combustion processes. Thus the rate of energy input by combustion is

$$\dot{E} = \dot{m}_f \cdot CV \cdot \eta_{comb}$$

where $\dot{m}_f$ is the fuel mass flow rate, CV the calorific value of the fuel and η_{comb} the combustion efficiency.

1.6.5 Cycle analysis

One example will be given of the simple analysis of the ideal Joule cycle for a gas turbine plant (Figure 1.55). The cycle consists of four flow processes described in Table 1.7 and analysed by the steady flow energy equation.

From the data in the table it can be seen that the specific work $w = c_p(T_3 - T_4) - c_p(T_2 - T_1)$ and the thermal efficiency

$$\eta_{thermal} = \frac{c_p(T_3 - T_4) - c_p(T_2 - T_1)}{c_p(T_3 - T_2)}$$

If allowance is made for the isentropic efficiency of the compression and expansion processes the cycle diagram is changed to show the associated entropy increases but the expressions for work and efficiency above are still valid with

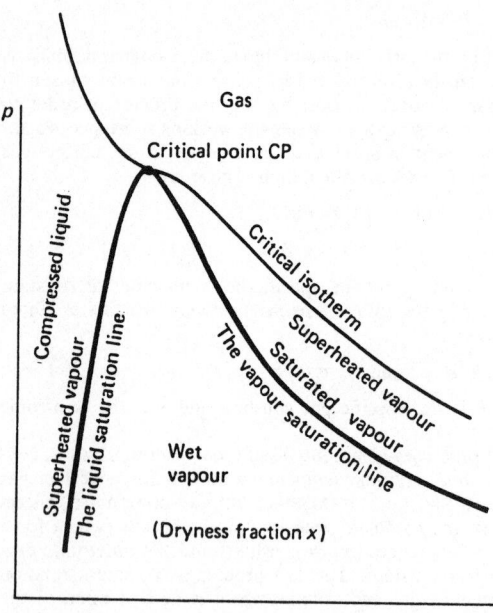

Figure 1.50 Substance phases and definitions

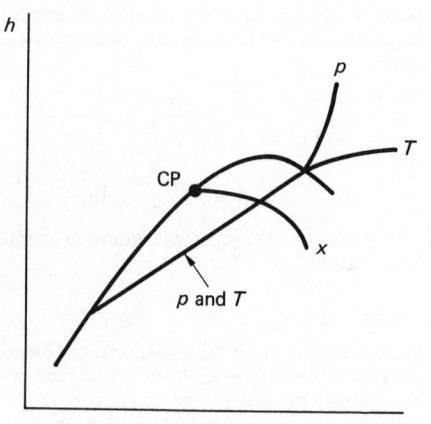

Figure 1.51 Steam processes on an h–s diagram

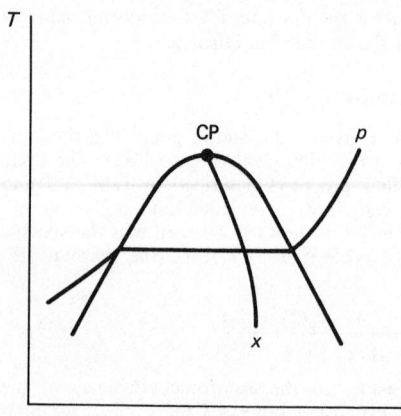

Figure 1.52 Steam processes on a T–s diagram

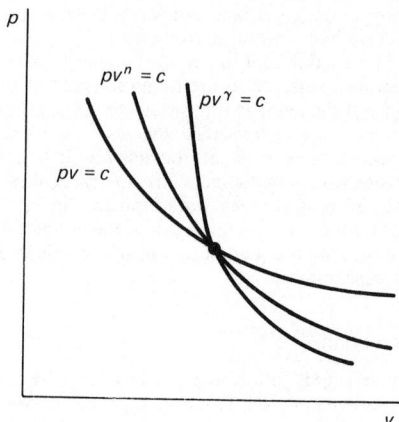

Figure 1.53 Gas processes on a p–v diagram

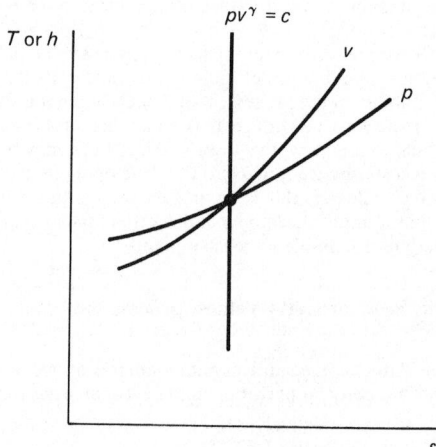

Figure 1.54 Gas processes on a T–s diagram

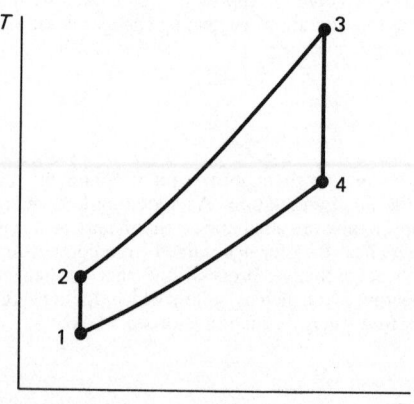

Figure 1.55 The Joule cycle

Table 1.7

Process	Description	$q - w_x = h$
1 to 2	Reversible adiabatic compression pv^γ = constant	$-w_x = h_z - h_1$ $\quad = c_p(T_2 - T_1)$
2 to 3	Reversible constant pressure heat transfer *to* the cycle	$q = h_3 - h_2$ $\quad = c_{[p}(T_3 - T_2)$
3 to 4	Reversible adiabatic expansion pv^γ = constant	$-w_x = h_4 - h_3$ $w_x = c_p(T_3 - T_4)$
4 to 1	Reversible constant pressure heat transfer *from* the cycle	$q = h_1 - h_4$ $\quad = c_p(T_1 - T_4)$

the changed values of T_2 and T_4 (Figure 1.56). These values are determined from the use of the reversible adiabatic process relation and the isentropic efficiency as

$$T_2 - T_1 = T_1(r_p^{(\gamma-1)/\gamma} - 1)/\eta_c$$

and

$$T_3 - T_4 = \eta_t T_3(1 - 1/r_p^{(\gamma-1/\gamma)})$$

where r_p is the cycle pressure ratio,

$$\eta_c = \frac{T_2' - T_1}{T_2 - T_1}$$

and

$$\eta_t = \frac{T_3 - T_4}{T_3 - T_4'},$$

η_c and η_t being the isentropic efficiencies of compression and expansion. If these values are substituted into the work and thermal efficiency expressions they become more useful in that they involve the thermodynamically significant maximum and minimum cycle temperatures which are fixed by material and ambient conditions respectively, so that the only variable is the cycle pressure ratio (Figure 1.57). If the expressions are differentiated with respect to this pressure ratio it is possible to

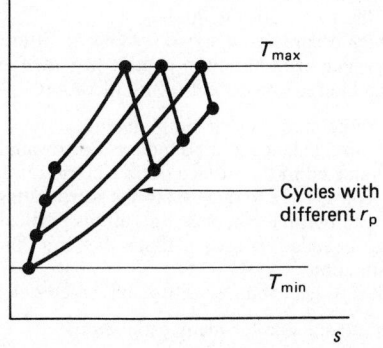

Figure 1.57 The effect of pressure ratio in a cycle with fixed T_{max} and T_{min}

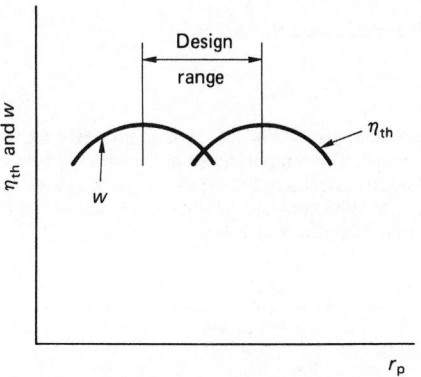

Figure 1.58 Thermal efficiency and specific work transfer variation in a Joule cycle with allowance for isentropic process efficiency

find the pressure ratio for maximum work and that for maximum efficiency. The cycle designer then has a choice, depending on the proposed application and Figure 1.58 shows that it would be expected that the chosen ratio would fall between these two maxima. Obviously, this simple approach is not the complete answer to gas turbine cycle analysis but it illustrates the use of the laws of thermodynamics, and similar work may be done for other plant cycles.

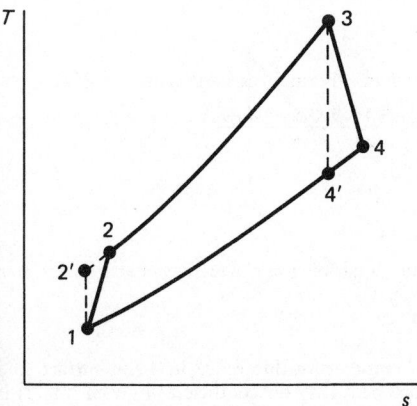

Figure 1.56 The effect of isentropic process efficiency on the Joule cycle

1.7   Heat transfer

1.7.1   Introduction

Whenever a temperature difference occurs there is an energy flow from the higher temperature to the lower. A study of heat transfer is concerned with the determination of the instantaneous rates of energy flow in all situations. We determine heat transfer rates in watts. These rates will be constant in situations where the temperature difference remains constant but variable (transient) when the temperature difference varies either due to the heat transfers or to other energy changes such as internal chemical reaction.

There are three modes of heat transfer:

1. *Conduction*, which is of greatest interest in solid bodies but also occurs in fluids where it is often overshadowed by convection;
2. *Convection*, which occurs in fluids when energy is transferred due to the motion of the fluid:
3. *Radiation*, which occurs between two systems at different temperatures which need not be in contact provided any intervening medium is transparent to the radiation.

In practice, all three modes occur simultaneously and it is necessary to draw up a balance at a boundary. For example, energy may be conducted to the surface of an electric storage heater and is then convected and radiated to the surroundings. Thus calculations can become complex, and in this particular case where energy is added at certain times this is a continuously varying situation.

Three approaches to heat transfer will be discussed below:

1. A simple method suitable for many estimations;
2. A more detailed appraisal of the field;
3. Comments on the use of computers.

1.7.2   Basic principles of heat transfer[14-16]

1.7.2.1   Conduction

Fourier's law for conduction states

$$\dot{Q}'' = -k\frac{dT}{dx}$$

The thermal conductivity $k(\mathrm{Wm^{-1}K^{-1}})$ is a property of the material which varies with temperature but for small temperature ranges is usually considered constant. Typical values are shown in Table 1.8. With constant k Fourier's equation can be integrated for four common situations.

Table 1.8

Substance	Thermal conductivity, $\mathrm{Wm^{-1}\ K^{-1}}$, at 20° C
Aluminium	204
Iron	52
Facing brick	1.3
Water	0.597
Air	0.026 (100 kPa)
Glass wool	0.04

Plane surfaces   Integration gives

$$\dot{Q}'' = k_{12}\frac{(T_1 - T_2)}{x_2 - x_1} = \Delta T \Big/ \left(\frac{\Delta x}{k}\right)_{12}$$

(see Figure 1.59). The quantity $(\Delta x/k)$ is known as the thermal resistance in $\mathrm{m^2\ W^{-1}}$. Thermal resistances can be added in a

Single layer Plane surface

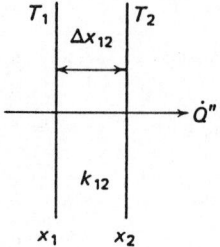

Figure 1.59   One-dimensional conduction through a single-layer plane wall

Multi layer plane surface

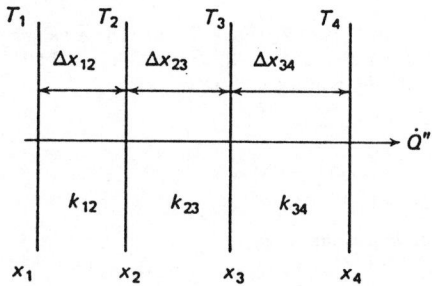

Figure 1.60   One-dimensional conduction through a multi-layer plane wall

similar way to electrical resistances so that for a multilayer plane surface there are a number of resistances in series (Figure 1.60). Thus we can write

$$\dot{Q}' = (T_1 - T_4)\Big/\sum\frac{\Delta x}{k}$$

Cylindrical surfaces   For tubes it is more convenient to evaluate heat transfer rates per unit length, and integration gives

$$\dot{Q}' = (T_1 - T_2)\Big/\frac{\ln r_2/r_1}{2\pi k_{12}} = \Delta T\Big/\frac{\ln r_2/r_1}{2\pi k_{12}}$$

(see Figure 1.61), and in this case the thermal resistance is $\left(\dfrac{\ln r_2/r_1}{2\pi k_{12}}\right)$ in m K $\mathrm{W^{-1}}$.

For a multilayer tube, thermal resistances are added to give

$$\dot{Q}' = (T_1 - T_4)\Big/\sum\left(\frac{\ln (r_{\text{outer}}/r_{\text{inner}})}{2\pi k}\right)$$

(see Figure 1.62).

1.7.2.2   Convection

The fundamental equation for convective heat transfer at a solid–fluid interface is

$$\dot{Q}'' = h\theta$$

where θ is the temperature difference between surface and fluid. The surface heat transfer coefficient h (W $\mathrm{m^{-2}\ K^{-1}}$) is not a property of the fluid or the surface but depends on the flow pattern, the fluid properties and the surface shape. The coefficient has to be determined for each situation and can

Single layer cylindrical surface

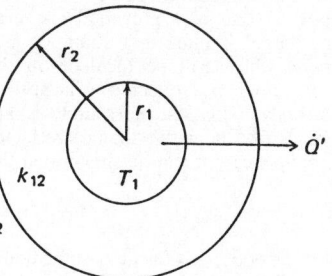

Figure 1.61 Conduction through a single-layer cylindrical surface

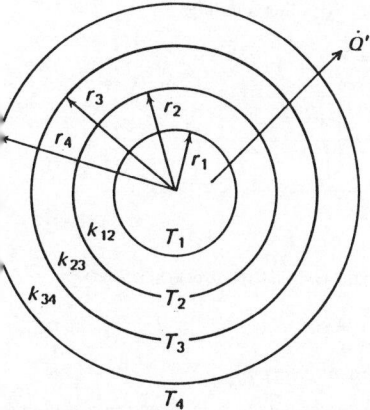

Figure 1.62 Conduction through a multi-layer cylindrical surface

...ary considerably (Table 1.9). Although the determination of ... is crucial to convection calculations it is an extremely difficult process, and accurate prediction of convective heat ...ransfer is not always possible.

If we express the convection equation in a thermal res-...tance form suitable for plane surfaces

$$\dot{Q}'' = \theta \Big/ \left(\frac{1}{h}\right)$$

... can be seen that the thermal resistance is $(1/h)$ m^2 K W^{-1}. ...or tubular surfaces it is again more convenient to work per ...nit length, so that

$$\dot{Q}' = \theta \Big/ \left(\frac{1}{2\pi rh}\right)$$

...nd the thermal resistance is $(1/2\pi rh)$ m K W^{-1}.

Table 1.9 Range of values of surface heat transfer ...oefficient (W m^{-2} K^{-1})

Free convection	Gases	0.5 to 500
	Liquids	50 to 2000
Forced convection	Gases	10 to 700
	Liquids	100 to 10 000

7.2.3 Overall heat transfer coefficients

. common heat transfer situation is a solid wall separating two ...uids, and for this problem the thermal resistances for conduc-...tion and convection can be added to enable the heat transfer rate to be determined in terms of the two fluid temperatures.

For a plane surface (Figure 1.63)

$$\dot{Q}'' = (T_{f_1} - T_{f_2}) \Big/ \left(\frac{1}{h_1} + \sum \frac{\Delta x}{k} + \frac{1}{h_2}\right)$$

For a tubular surface (Figure 1.64)

$$\dot{Q}' = (T_{f_1} - T_{f_2}) \Big/ \left(\frac{1}{2\pi r_1 h_1}\right) + \sum \left(\frac{\ln(r_{outer}/r_{inner})}{2\pi k}\right) + \frac{1}{2\pi r_2 h_2}$$

It can be seen that the added resistances may be inverted to give an overall *conductance* which is known as a '*U*-value' or overall heat transfer coefficient.

For a plane surface

$$\frac{1}{U} = \frac{1}{h_1} + \sum \frac{\Delta x}{k} + \frac{1}{h_2}$$

For a tubular surface

$$\frac{1}{U'} = \frac{1}{2\pi r_1 h_1} + \sum \left(\frac{\ln(r_{outer}/r_{inner})}{2\pi k}\right) + \frac{1}{2\pi r_2 h_2}$$

The heat transfer rate is then simply written

For a plane surface

$$\dot{Q} = UA(T_{f_1} - T_{f_2}) \text{ where } A \text{ is the area}$$

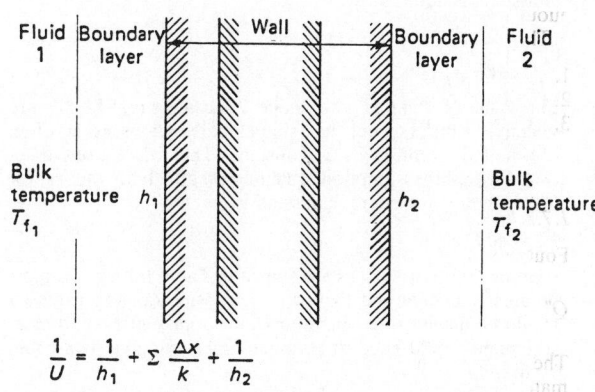

Figure 1.63 *U*-value for a plane surface

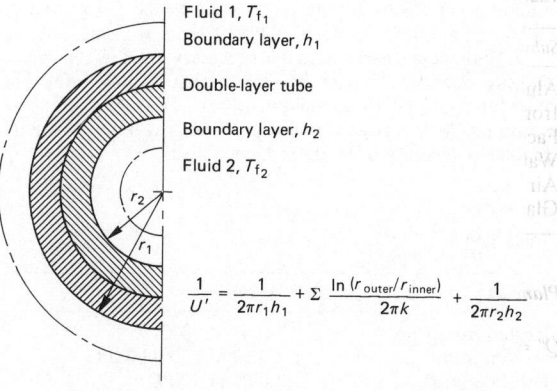

Figure 1.64 *U'*-value for a cylindrical surface

For a tubular surface

$$\dot{Q} = U'l(T_{f_1} - T_{f_2})$$ where l is the length

The situations in which this technique is commonly used include walls of buildings, double glazing and hot water pipes, and the method is also used in heat exchanger design. It should suffice for simple calculations provided suitable values of the surface heat transfer coefficients for convection can be obtained. Values of U can be found for buildings, but care should be taken to ensure that the quoted figures include convection effects, as some only account for conduction and will need the effect of convection to be added.[17] If the temperature difference is not constant then a mean value should be used. A suitable equation for a mean can be found in the heat exchanger section which follows.

1.7.2.4 Radiation

Heat transfer due to radiation effects is of increasing importance as temperature increases because the rate of energy emitted by an ideal black body is given by the Stefan–Boltzmann law, in which the *absolute* temperature (Celsius + 273) is raised to the fourth power:

$$\dot{E}''_b = \sigma T^4$$

where σ is the Stefan–Boltzmann constant 5.67×10^{-8} W m^{-2} K^{-4} and the subscript b refers to the ideal black body. Real bodies emit less radiation and the monochromatic emissivity is defined by

$$\varepsilon_\lambda = \left[\frac{\dot{E}''_\lambda}{\dot{E}''_{b\lambda}} \right]_T$$

The value of ε varies with λ and T because real bodies are selective emitters, but for simple calculations it is often assumed that emissivity is constant. The calculations associated with this assumption are based on grey body theory, for which the rate of energy emission is given by

$$\dot{E}''_g = \varepsilon \sigma T^4$$

Some emissivity values are shown in Table 1.10, but it must be clearly understood that with such wide-ranging values it would be unwise to estimate unknown emissivities, and measurements would need to be made unless suitable data could be found.

Radiation incident on a body may be absorbed, reflected or transmitted. Thus we write $\alpha + \rho + \tau = 1$ where α, ρ and τ are the absorbtivity, reflectivity and transmissivity, respectively. Ideal black bodies absorb all incident radiation but real bodies do not. Gases are often assumed to transmit all radiation, but this is not always true, particularly with hydrocarbon combustion products. Solids have a transmissivity of zero. With these simple ideas it is necessary to know the values of only α and ρ. It can be shown that a grey body has absorbtivity equal to emissivity, $\alpha = \varepsilon$. Thus, provided the transmissivity is zero, a knowledge of the grey body emissivity enables reflectivity to be determined, since

$$\rho = 1 - \varepsilon$$

The only simple radiation problem that can be solved with the simple approach above is that of a grey body in *large* surroundings. The word 'large' implies that radiation not incident on the grey body which will be incident on the surroundings and will therefore be reflected will not be re-incident on the grey body. Thus the surroundings are effectively black. (This might be true with a linear size factor greater than 10.) In this simple case it can be shown that the heat transfer rate is

$$\dot{Q} = \varepsilon \sigma A(T^4 - T_s^4)$$

where ε is the emissivity of the body, A is the area of the body, T is the temperature of the body in K and T_s is the temperature of the large (black) surroundings in K.

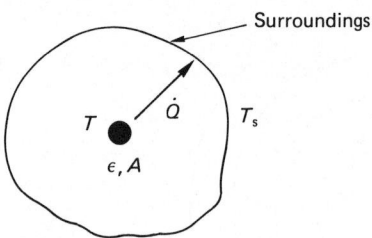

Figure 1.65 A small grey body in large (black) surroundings

1.7.2.5 Simple transient problems

If a body is being cooled or heated by convection or radiation and the thermal conductivity is large so that the rapid heat transfer rates within the body enable it to be assumed that the body temperature distribution is uniform, then the situation is known as a lumped capacity system. For such a system the complex methods of transient heat transfer are not required and a simple energy balance equation may be drawn up and integrated. The most common case is quenching, a convective boundary problem for which in time dt a small heat transfer δQ occurs when the body temperature changes from T by an amount dT (Figure 1.66). Thus

$$\delta Q = \rho c_p V dT = -hA(T - T_f)$$

where T_f is the fluid temperature, A the body surface area, V the body volume, ρ the body density, c_p the body specific heat and h is the surface heat transfer coefficient. Integration gives

$$\frac{\theta}{\theta_0} = e^{-(hAt/\rho c_p V)}$$

where θ_0 is the initial temperature difference between fluid and body and θ is the temperature difference at any future time t. The quantity $(\rho c_p V/hA)$ may be regarded as the time constant of the system.

Table 1.10

Material	Emissivity
Rusted iron plate	0.69 at 19°C
Molten iron	0.29 at 1300–1400°C
Polished brass	0.06 at 100°C
Asbestos board	0.96 at 21°C

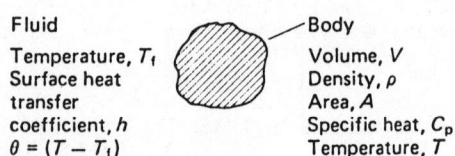

Fluid
Temperature, T_f
Surface heat transfer coefficient, h
$\theta = (T - T_f)$

Body
Volume, V
Density, ρ
Area, A
Specific heat, C_p
Temperature, T

Figure 1.66 A lumped-capacity system with convection

1.7.3 Analysis of heat transfer[14-16]

1.7.3.1 Conduction

By considering the thermal equilibrium of a small, three-dimensional element of solid, isotropic material it can be shown that for a rectangular coordinate system

$$\frac{\partial T}{\partial t} = \alpha \left[\frac{\partial^2 T}{\partial x^2} + \frac{\partial^2 T}{\partial y^2} + \frac{\partial^2 T}{\partial z^2} \right] + \frac{\dot{Q}'''}{\rho c_p}$$

where $\partial T/\partial t$ is the rate of change of temperature with time, α is the thermal diffusivity of the material $\alpha = k/\rho c_p$ and $\dot{Q}'''$ is the internal heat generation rate per unit volume, which may be due, for example, to electric current flow for which $\dot{Q}''' = i^2 r$, where i is the current density and r the resistivity. The solution to this equation is not easy, and numerical approximation methods are often used. One such method is the finite difference technique in which continuously varying temperatures are assumed to change in finite steps. Consider the three planes shown a distance Δx apart (Figure 1.67).

At A $\dfrac{\partial T}{\partial x} = \dfrac{T_1 - T_2}{\Delta x}$ and at B $\dfrac{\partial T}{\partial x} = \dfrac{T_2 - T_3}{\Delta x}$

so that at C $\dfrac{\partial^2 T}{\partial x^2} = \dfrac{T_1 + T_3 - 2T_2}{\Delta x^2}$. Similarly, $\partial T/\partial t$ may be

written $\dfrac{T_{n,1} - T_{n,0}}{\Delta t}$ where $T_{n,1}$ is the temperature at layer n

at time 1 and $T_{n,0}$ is the temperature at layer n at time 0.

For steady state situations $\partial T/\partial t = 0$ and a two-dimensional plane surface will be used for illustration requiring a solution of

$$\alpha \left[\frac{\partial^2 T}{\partial x^2} + \frac{\partial^2 T}{\partial y^2} \right] + \frac{\dot{Q}'''}{\rho c_p} = 0$$

Consider the surface to be divided by a grid (Figure 1.68). It is then found that the solution for any point in the plane for steady state conduction without heat generation is

$$T_1 + T_2 + T_3 + T_4 - 4T_0 = 0$$

or with heat generation is

$$T_1 + T_2 + T_3 + T_4 - 4T_0 = -\frac{a^2 \dot{Q}'''}{k}$$

At the boundary of the plane conditions are usually isothermal, in which case $T = $ constant, or convective when an energy balance yields for a straight boundary (Figure 1.69)

$$\frac{T_1}{2} + T_2 + \frac{T_3}{2} + \frac{ha}{k} \cdot T_f - T_0 \left(2 + \frac{ha}{k} \right) = 0$$

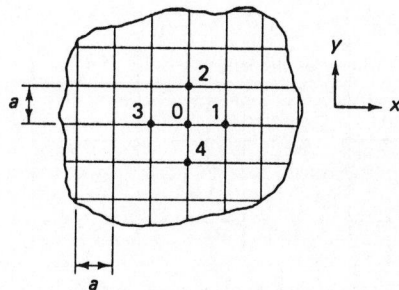

Figure 1.68 The two-dimensional grid concept

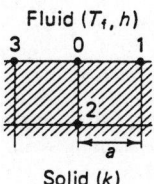

Figure 1.69 Convective surface nomenclature

where h is the surface heat transfer coefficient, T_f is the fluid temperature and a is the grid size. Similar expressions can be derived for corners, curves, etc. at the boundary.

For a large number of grid points, a large number of simultaneous equations are obtained which can be solved by iteration or Gaussian elimination. Computer programs may be used to advantage. The solution obtained will be the temperature distribution in the plane, and the heat transfer rates may be found at the boundary (Figure 1.70):

With an isothermal boundary $\dot{Q} = \Sigma k(T_m - T_{wall})$

With a convective boundary $\dot{Q} = \Sigma ha(T_f - T_m)$

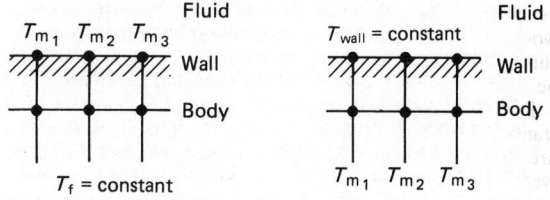

(a) Convective boundary   (b) Isothermal boundary

Figure 1.70 Calculation of heat transfer rate

For transient heating or cooling for which $\partial T/\partial t \neq 0$, a one-dimensional illustration is used. The equation to be solved is

$$\frac{\partial T}{\partial t} = \alpha \frac{\partial^2 T}{\partial x^2}$$

when there is no heat generation (Figure 1.71). When expressed in finite difference form this becomes

$$T_{n,1} = F\left(T_{n-1,0} + T_{n+1,0} + T_{n,0}\left(\frac{1}{F} - 2 \right) \right)$$

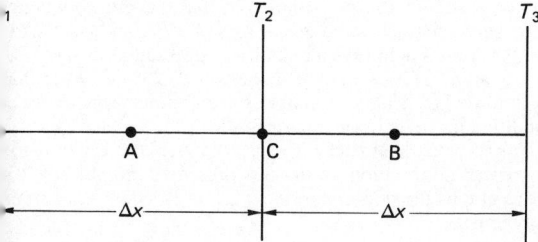

igure 1.67 One-dimensional finite difference formulation

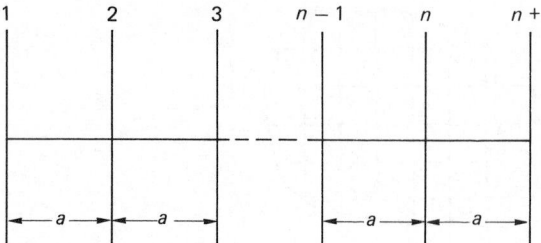

Figure 1.71 One-dimensional transient conduction formulation

where F is the non-dimensional grid size Fourier number $F = \alpha\Delta t/a^2$. The only unknown in this equation is $T_{n,1}$, the temperature at layer n after one time internval Δt. Thus from a knowledge of the initial conditions successive temperatures in each layer can be found directly for each time interval. This is the *explicit* method and is used for tabular or graphical (Schmidt method) solutions. If $F > 0.5$ the solution is unstable and in three dimensions the criterion becomes severe. The boundary conditions may be isothermal or convective and in the latter case the solution is

$$T_{n,1} = 2F \cdot T_{n-1,0} + T_{n,0}[1 - 2F - 2F \cdot B] + 2F \cdot BT_{\mathrm{f}}$$

where B is the non-dimensional grid Biot number, $B = ha/k$. For this case the solution is unstable if $(F + FB) > 0.5$. The solutions obtained give the temperature distribution in the one-dimensional plane and the heat transfer is found at the boundary

$$Q = \sum_t kA \left(\frac{\partial T}{\partial x}\right)_{1,t} \Delta t$$

or from the temperature profile

$$Q = \sum_{\mathrm{layer}} mc_{\mathrm{p}}(T_{\mathrm{final}} - T_{\mathrm{initial}})$$

The stability problems of the explicit method can be overcome by the use of *implicit* methods for which there is no direct solution, but a set of simultaneous equations are obtained which may be solved by Gaussian elimination. A computer program may be used to advantage. A satisfactory implicit method is that due to Crank and Nicolson. The importance of a stable solution is that if the choice of F is limited then the grid size and time interval cannot be freely selected, leading to excessive calculations for solution. The implicit method releases this constraint but care is still needed to ensure accuracy.

Although the finite difference method has been chosen for demonstration because the method is easy to understand most modern computer programs are based on the finite element technique. However, the mathematical principles are involved, and would not lend themselves to simple programming. Before the availability of computer software analytical solutions were obtained and presented as graphs of transient solutions for slabs, cylinders and spheres. These graphs enable solutions for other shapes to be obtained by superposition methods. Such methods should be used to avoid or validate computer solutions.

Warning: If fibre-reinforced materials are used in which the lay-up is arranged to give directional structural strength it will be found that the thermal conductivity has directional variation and the methods above will need considerable amendment.

1.7.3.2 Convection

A knowledge of the surface heat transfer coefficient h is essential in determining heat transfer rates. Fluid flow over a solid surface is a boundary layer problem, and the heat transfer depends on boundary layer analysis. This analysis may be by differential or integral approach, but solution is difficult and the modelling of turbulence is complex. Computer solutions based on numerical approximations may be used to advantage, but simple approaches have been used for many years and are still extremely useful. These methods are based on Reynolds' analogy (modified by later workers) and dimensional analysis backed by experimentation.

Convection may be free or forced. In *forced convection* it is found that the heat transfer coefficient can be included in a non-dimensional relation of the form

$$Nu = \phi(Re, Pr) = \text{constant} \cdot Re^a \cdot Pr^b$$

where Nu is the Nusselt number ($Nu = hl/k$), Re is the Reynolds number ($Re = \rho Vl/\mu$) and Pr is the Prandtl number ($Pr = \mu c_{\mathrm{p}}/k$). In these relations l is a representative length dimension (diameter for a pipe and some chosen length for a plate), V is the bulk or free stream velocity outside the boundary layer. The values of the constants a and b depend on whether the flow is laminar or turbulent and on the geometry of the situation, and are usually found by experiment.

The determination of whether flow is laminar or turbulent is by the value of the Reynolds number;

For plates, $Re < 500\,000$, flow is laminar: $Re > 500\,000$, flow is turbulent

For tubes, $Re < 2000$, flow is laminar: $Re > 4000$, flow is turbulent

(between these two values there is a transition zone). There are many relations to be found in texts which allow for entry length problems, boundary conditions, etc. and it is not feasible to list them all here. Two relations are given below which give average values of Nusselt number over a finite length of plate or tube in forced, turbulent flow with Mach number less than 0.3 using total plate length and diameter for representative length dimension. Care must be taken in any empirical relation to use it as the author intended.

Plate; $Nu = 0.036Re^{0.8}Pr^{0.33}$

In this relation fluid properties should be evaluated at the film temperature, $T_{\mathrm{film}} = (T_{\mathrm{wall}} + T_{\mathrm{bulk}})/2$.

Tube; $Nu = 0.023Re^{0.8}Pr^{0.4}$

In this relation fluid properties should be evaluated at the bulk temperature, $0.6 < Pr < 160$ and $(l/d) > 60$.

It should be noted that the index of Reynolds number of 0.8 is characteristic of turbulent flow; in laminar flow 0.5 is found.

It must be emphasized that reference to other texts in all but these simple cases is essential to estimate heat transfer coefficients. It should also be pointed out that the values obtained from such relations could give errors of 25%, and a search of the literature might reveal equations more suited to a particular situation. However, an estimate within 25% is better than no knowledge, and is a suitable starting point which may be modified in the light of experience.

For complex heat exchange surfaces such as a car radiator empirical information is usually presented graphically (on these graphs the non-dimensional group St (Stanton number) may appear:

$$St = \frac{Nu}{RePr} = \frac{h}{\rho V c_{\mathrm{p}}}$$

In *free convection* the relationship used is $Nu = \phi(Pr \cdot Gr)$ where Gr is the Grashof number, $\rho^2\beta g\theta l^3/\mu^2$ in which β is the coefficient of cubical expansion of the fluid and θ is a temperature difference (usually surface to free stream temperature). The transition from laminar to turbulent flow is determined by the product $(Pr \cdot Gr)$ known as the Rayleigh number, Ra. As a simple example, for plane or cylindrical, vertical surfaces it is found that;

For $Ra < 10^9$, flow is laminar and $Nu = 0.59(Pr \cdot Gr)^{0.25}$

For $Ra > 10^9$, flow is turbulent and $Nu = 0.13(Pr \cdot Gr)^{1/3}$

The representative length dimension is height and the resulting heat transfer coefficients are average values for the whole height. Film temperature is used for fluid properties. Warnings similar to those given for forced convection apply to the use of these equations.

Phase change convection heat transfer (condensing and evaporation) shows coefficients that are, in general, higher than those found in single-phase flow. They are not discussed here but information may be found in standard texts.

1.7.3.3 Radiation

When an emitting body is not surrounded by the receiver the spatial distribution of energy from the radiating point needs to be known. To determine this distribution the intensity of radiation i_ϕ is defined in any direction ϕ as

$$i_\phi = \left(\frac{d\dot{E}''}{d\omega}\right)_\phi$$

where $d\omega$ is a small solid angle subtended at the radiating point by the area intercepting the radiation, $d\omega = dA/r^2$ (Figure 1.72) (the solid angle represented by a sphere is 4π steradians). Lambert's law of diffuse radiation states that $i_\phi = i_n \cos\phi$ where i_n is the normal intensity of radiation which can be determined for black and grey bodies;

Black $i_n \, \sigma T^4/\pi$ Grey $i_n \, \varepsilon\sigma T^4/\pi$

With this knowledge of the radiation intensity in any direction it is only necessary to determine the amount that any body can see of any other body to calculate the heat transfer rate. For this purely geometric problem mathematical analysis (Figure 1.73) suggests a quantity variously known as the geometric, configuration or shape factor, which is defined as

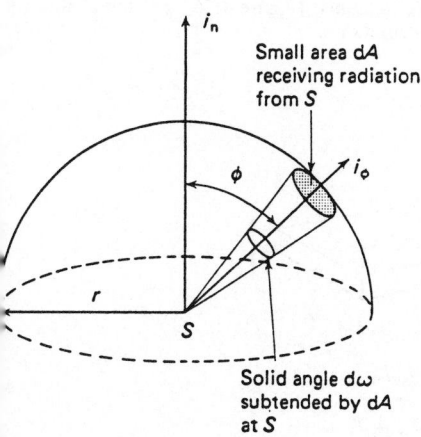

Figure 1.72 Spatial distribution of radiation

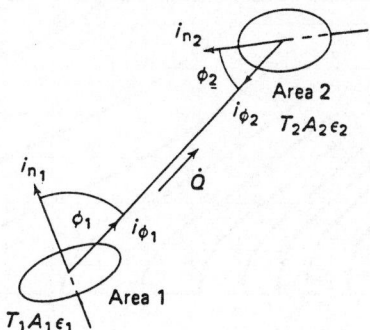

Figure 1.73 Heat transfer by radiation between two arbitrarily disposed grey surfaces

the fraction of the energy emitted per unit time by one surface that is intercepted by another surface. The geometric factor is given by

$$F_{12} = \frac{1}{A_1} \int_{A_1}\int_{A_2} \frac{\cos\phi_1 \, \cos\phi_2 \, dA_1 dA_2}{\pi x^2}$$

It can be seen that $A_1F_{12} = A_2F_{21}$, a useful reciprocal relation. It is also clear that the equation will require skill to solve in some situations, and to overcome this problem geometric factors are available for many situations in tables or on graphs (Hottel charts). The charts can give more information than anticipated by the use of shape factor algebra, which enables factors to be found by addition, subtraction, etc. (Figure 1.74).

Having established the intensity of radiation and the geometric factor, problems may be solved by an electrical analogy using the radiosity of a surface. Radiosity is defined as the total emitted energy from a grey surface:

$$\dot{J}'' = \dot{E}_g'' + \rho\dot{G}''$$

where $\dot{J}''$ is the radiosity and $\rho\dot{G}''$ the reflected portion of the incident radiation $\dot{G}''$. Since $\dot{E}_g'' = \varepsilon\dot{E}_b''$ the net rate of radiation leaving a grey surface of area A becomes

$$\frac{\dot{E}_b'' - \dot{J}''}{\rho/A\varepsilon}$$

which may be envisaged as a potential difference, $\dot{E}_b'' - \dot{J}''$, divided by a resistance, ρ/AE. A similar geometric resistance of $1/AF$ can be established to enable complete circuits to be drawn up. Thus for a three-body problem we may sketch the analogous electrical circuit (Figure 1.75) and apply Kirchhoff's electric current law to each $\dot{J}''$ node to obtain three simultaneous equations of the form

$$\frac{\dot{E}_{b_1}'' - \dot{J}_1''}{\rho_1/A_1\varepsilon_1} + \frac{\dot{J}_2'' - \dot{J}_1''}{1/A_1F_{12}} + \frac{\dot{J}_3'' - \dot{J}_1''}{1/A_1F_{13}} = 0$$

If there are more than three bodies sketching becomes complex and the equation above can be rearranged and generalized. For N surfaces ($j = 1$ to N) there will be N equations, the ith of which ($i = 1$ to N) will be

$$\dot{J}''_i - (1 - \varepsilon_i)\sum_{j=1}^{N} F_{ij}\dot{J}_j'' = \varepsilon_i\dot{E}''_{b_i}$$

When $j = i$, F_{ii} will be zero unless the the surface is concave and can see itself. This set of N simultaneous equations may be solved by Gaussian elimination for which a computer program may be used. The output of the solution will be N values of $\dot{J}''$

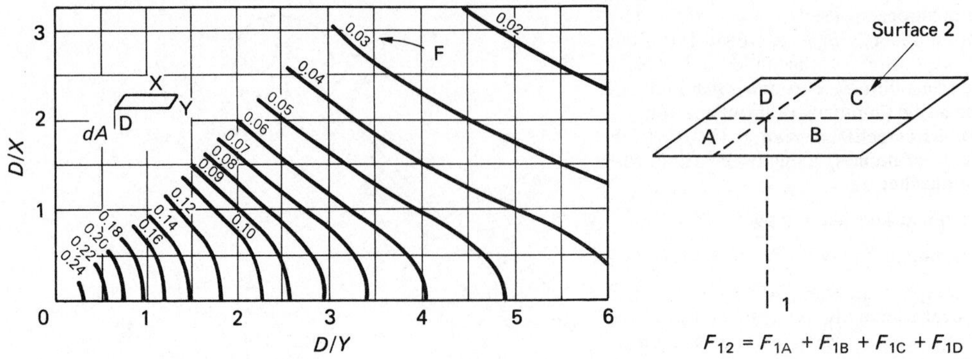

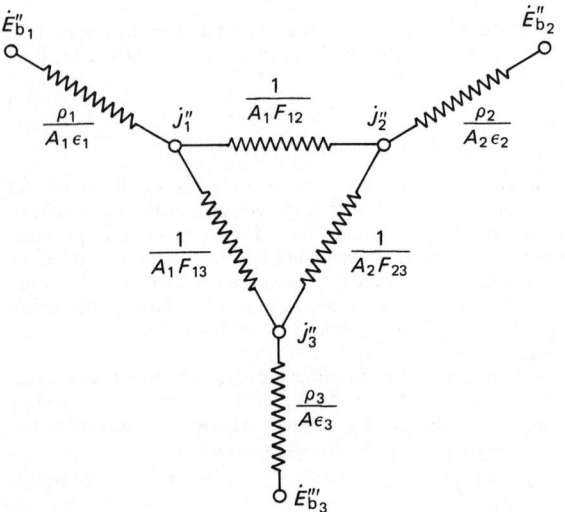

$$F_{12} = F_{1A} + F_{1B} + F_{1C} + F_{1D}$$

Figure 1.74 The geometric factor chart

Figure 1.75 The electrical analogy for grey-body radiation problems

and any required heat flux can be found. In the three-body problem, for example, the heat transfer from 1 to 2

$$\dot{Q}_{12} = \frac{\dot{J}_1'' - \dot{J}_2''}{1/A_1 F_{12}}$$

or the total heat transfer from body 1 is

$$\frac{\dot{E}_1'' - \dot{J}_1''}{\rho_1/A_1\epsilon_1}$$

Special cases are:

1. If $N - 1$ bodies are in large surroundings then for this Nth body $\dot{J}'' = \dot{E}_b''$;
2. An insulated or refractory surface has no black body potential but contributes to the heat transfer by taking up an equilibrium temperature T given by $\sigma T^4 = \dot{J}''$;
3. Radiation shield problems will show six resistances in series rather than the series-parallel circuits used previously.

In all the discussion on radiation above no account has been taken of the selective emitter for which emissivity is not constant. Additional techniques are required to solve these real problems and care should be exercised if widely varying

emissivity is encountered. In new situations it will be necessary to determine emissivity by experiment before proceeding. No account has been taken of intervening media for which transmissivity is not unity. Gas absorption and radiation needs further information. For gases such as oxygen, nitrogen and hydrogen with symmetric, diatomic molecules the above work is adequate, but asymmetric molecular structures cause problems. In particular, hydrocarbon fuel combustion products (H_2O), CO_2, CO, SO_2) are important in engineering calculations and account must be taken of their radiation properties. Solar radiation problems also need special consideration.

1.7.3.4 Finned surfaces

In many heat exchange problems involving the determination of a U-value it is found that the surface heat transfer coefficient on one side of the solid interface is much smaller than that on the other. The smaller coefficient will dictate the heat transfer rate achieved, and in order to overcome this problem fins may be added to this poor convection surface to increase the area for heat transfer. This problem will occur in liquid to gas exchangers on the gas side. The addition of fins will alter the flow pattern so that a new coefficient should if possible be determined. It is also possible that there may be variation of coefficient over the fin surface.

Simple fin theory in which conduction along the fin is balanced with convection from the surface can be used to determine the temperature distribution and heat transfer rate of the fin. For example, when a long fin of constant cross-sectional area is examined (Figure 1.76) it is found that the temperature distribution is

$$\frac{\theta}{\theta_0} = \frac{\cosh(m(l - x))}{\cosh(ml)}$$

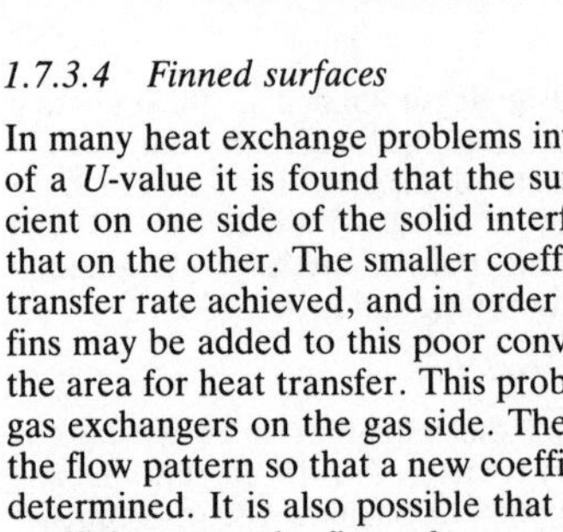

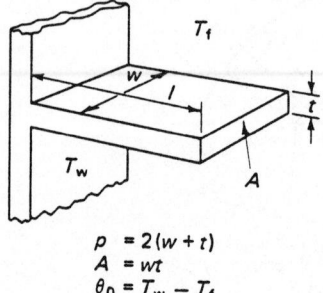

$$p = 2(w + t)$$
$$A = wt$$
$$\theta_0 = T_w - T_f$$

Figure 1.76 Simple rectangular fin nomenclature

and the heat transfer rate is

$$\dot{Q} = mkA\theta_0 \tanh(ml)$$

where θ is the temperature difference between fin and fluid and θ_0 is the difference at the fin root, $m = hp/kA$, p is the fin perimeter and A is the cross-sectional area.

A fin efficiency is then defined to compensate for the varying temperature difference along the fin as

$$\eta_{fin} = \frac{\text{Actual heat transfer rate}}{\text{Heat transfer rate if whole fin were at the wall temperature}}$$

which for the simple case above is $\eta_{fin} = \tanh(ml)/ml$. Fins are usually fitted in arrays and the efficiency of a fin system can be established in the form of an area weighted fin efficiency, η'

$$\eta' = \eta_{fin}\beta + 1 - \beta$$

where

$$\beta = \frac{\text{total fin area}}{\text{total area including fins}} = \frac{A_{fin}}{A}$$

The U value based on the enhanced area A is then

$$\frac{1}{U_A} = \frac{1}{(A_1/A)h_1} = \sum\frac{\Delta x}{R} + \frac{1}{\eta' h_2}$$

For complete surfaces β is supplied by the manufacturer.

When fins of more complex shape are used (tapered fins or annular fins) the cross-sectional area is not constant and fin efficiency data are obtained from graphs. Care should be taken in the interpretation of such graphs since the equations above may not agree with the definitions used for the graphs.

1.7.3.5 Heat exchangers[18]

It is possible to 'design' a heat exchanger with the information above and obtain a basic idea of size and configuration of simple tubular structures. It is first necessary to realize that temperature differences change along a heat exchanger, and that flow may be parallel or counter in direction. The latter is to be preferred, as it leads to smaller sizes. To allow for the changing temperature difference a log mean temperature difference is used (Figure 1.77):

$$\theta_{LMTD} = \frac{\theta_1 - \theta_2}{\ln(\theta_1/\theta_2)}$$

and to allow for varying flow patterns (which are neither counter nor parallel flow) graphs are available to give a factor F to modify the (usually) counterflow value of mean temperature difference. Thus for any heat exchanger

$$\theta_{mean} = F,\theta_{LMTD}$$

If an estimated mean U-value for the surface is then determined the heat exchange equation is

$$\dot{Q} = UAF\theta_{LMTD}$$

Thus the area is determined. There are many solutions to this equation to satisfy all the constraints which will include:

Energy for each stream $\dot{Q} = (\dot{m}\Delta h)_{hot} = (\dot{m}\Delta h)_{cold}$

$$(\Delta h = \text{enthalpy change})$$

Continuity of each stream $\dot{m} = \rho A V$

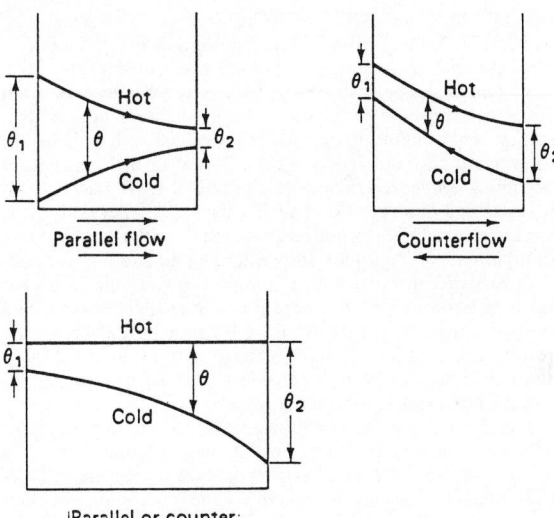

Figure 1.77 The mean temperature difference in heat exchangers

Heat transfer area $A = \pi dl$

The above equations will involve options with varying tube numbers and diameters which will affect the determination of heat transfer coefficients (and hence the U-value), and a number of solutions will be obtained. The optimum choice will involve allowable pressure drops, velocities and exit temperatures. It is not a difficult calculation, but, because of the choice, a computer program may be used. There are short cuts to this approach based on the interrelation between pressure drop and heat transfer (modified Reynolds' analogy) to determine heat transfer coefficients and a method based on graphs of effectiveness (E), capacity ratio (c) and number of transfer units (NTU) is sometimes used.

By whatever method a 'design' is achieved, it will, unless the application is very simple, be necessary to consult a professional heat exchanger designer with experience and full computer programs.

1.7.4 Use of computers

Computers may be used as an aid to heat transfer calculations at various levels of skill. At the simplest level, computer programs may be written for a home computer to determine, for example, heat losses in buildings, heat inputs from pipes and radiators, etc. This will save repetitive calculations and build a small library of useful programs. To aid those whose heat transfer and computing skills are slight Bacon[19] gives a simple approach to problems. At a higher level of skill Adams and Rogers[20] is a considerable advance in both programming and heat transfer technique. Both books use BASIC and the latter emphasizes the finite difference technique. Both books include programs. Supplementary reading from other heat transfer books would be advantageous. For more detail on the mathematics of finite difference techniques in heat transfer Myers[21] is useful.

An alternative approach to numerical approximations in computer work is to use finite element methods. For heat transfer applications Myers[21] is again useful. Very few people will find the need for finite element programming skills as there is an ever-increasing range of software which is aimed at the solution of many engineering problems, including heat transfer. Many of these are mainframe or PC packages in which there is integration between design, drawing, manufacture and analysis of stress, vibration, heat transfer, etc.

Whenever numerical approximation techniques are used large-scale analysis leads to considerable demand on data storage and computing time, which is expensive. It is therefore essential to do as much as possible with simple methods before becoming committed to large-scale finite element packages. If it is decided that the use of such a package is necessary it is vital to be sure of the requirements of the problem, for volume of input data and output results will be large and unpalatable. To assist with this problem, preprocessing packages are used for data input and mesh generation with graphic display, and post-processing packages are used for graphic display of the results. For example, the temperature distribution in a combustion chamber or piston displayed by colour graphics enables easier identification of problem areas.

Finally, it must be emphasized that computing is not a substitute for understanding heat transfer problems. It is an aid to enable a more detailed investigation to be achieved and presented in a fashion to enable engineers to improve their designs.

1.7.5 Heat transfer: nomenclature

A	area
B	Biot number
c_p	Specific heat capacity at constant pressure
$\dot{E}_b''$	Black-body emissive power
$\dot{E}_g''$	Grey-body emissive power
F	Fourier number, geometric factor, mean temperature factor
g	gravitational constant
Gr	Grashof number
h	surface heat transfer coefficient
i_ϕ	intensity of radiation in direction ϕ
i_h	normal intensity of radiation
$\dot{j}''$	radiosity
k	thermal conductivity
l	length
m	mass, fin parameter
$\dot{m}$	mass flow rate
Nu	Nusselt number
p	perimeter
Pr	Prandtl number
Q	heat transfer
$\dot{Q}$	heat transfer rate
$\dot{Q}'$	heat transfer rate per unit length
$\dot{Q}''$	heat transfer rate per unit area
$\dot{Q}'''$	heat transfer rate per unit volume
r	radius
Ra	Rayleigh number
Re	Reynolds number
St	Stanton number
T	temperature
t	time
U	overall heat transfer coefficient per unit area
U'	overall heat transfer coefficient per unit length
V	velocity
x,y,z	rectangular coordinates

Greek letters

α	thermal diffusivity, absorbtivity
β	coefficient of cubical expansion
Δ	change in
η	efficiency
ε	emissivity
λ	wavelength
μ	viscosity
ϕ	angle
ρ	density, reflectivity
σ	Stefan–Boltzmann constant
Σ	summation
θ	temperature difference
τ	transmissivity
ω	solid angle

References

1. Mathesun, J. A. L., *Hyperstatic Structures: an introduction to the theory of statically indeterminate structures*, Butterworths, London (1971)
2. Meriam, J. L. and Kraige, L. G., *Engineering Mechanics*, Volume 2, *Dynamics*, second edition, John Wiley, Chichester (1987)
3. Anvoner, S. *Solution of Problems in Mechanics of Machines*, Volume 2, Pitman, London (1972)
4. Scarborough, J., *The Gyroscope: theory and applications*, Interscience, New York (1958)
5. Wilson, W. K., *Vibration Engineering: A practical treatise on the balancing of engines, mechanical vibration and vibration isolation*, Griffin, London (1959)
6. Young, D. and Feglar Jr, R. P., *Tables of Characteristic Functions Representing Normal Modes of Vibration of a Beam*, The University of Texas Publication Number 4913 (July 1949)
7. Cole, E. B., *Theory of Vibrations*, Crosby Lockwood, London (1950)
8. Thompson, W. T., *Theory of Vibrations with Applications*, second edition, George Allen & Unwin, London (1983)
9. Rogers, G. F. C. and Mayhew, Y. R., *Engineering Thermodynamics, Work and Heat Transfer*, Longman, Harlow (1986)
10. Bacon, D. H., *BASIC Thermodynamics and Heat Transfer*, Butterworths, London (1983)
11. Kotas, T. J., *The Exergy Method of Thermal Plant Analysis*, Butterworths, London (1985)
12. Bejan, A., *Entropy Generation Through Heat and Fluid Flow*, Wiley, Chichester (1982)
13. Haywood, R. W., *Thermodynamic Tables in SI (Metric) Units*, University Press, Cambridge (1976)
14. White, F. M., *Heat Transfer*, Addison-Wesley, Reading, MA (1984)
15. Özişik, M. N., *Heat Transfer*, McGraw-Hill, New York (1985)
16. Kreith, F. and Bohn, M. S., *Principles of Heat Transfer*, Harper & Row, New York (1986)
17. *CIBS Guide*, Section A3 Thermal properties of building structures; Section A5, Thermal response of buildings, London (1980)
18. Kern, D. Q., *Process Heat Transfer*, McGraw-Hill, New York (1984)
19. Bacon, D. H., *Basic Heat Transfer*, Butterworths, London (1989)
20. Adams, J. A. and Rogers, D. F., *Computer-aided Heat Transfer Analysis*, McGraw-Hill, New York (1973)
21. Myers, G. E., *Analytical Methods in Conduction Heat Transfer*, McGraw-Hill, New York (1971)

2

Electrical and electronics principles

Charles J. Fraser

Contents

2.1 Basic electrical technology 2/3
   2.1.1 Flux and potential difference 2/3
   2.1.2 Simple resistive circuits 2/3
   2.1.3 Electromotive force and potential difference 2/3
   2.1.4 Power and energy 2/3
   2.1.5 Network theorems 2/3
   2.1.6 Double-subscript notation 2/4
   2.1.7 Electrostatic systems 2/4
   2.1.8 Simple capacitive circuits 2/4
   2.1.9 Charging a capacitor 2/5
   2.1.10 Dielectric strength 2/5
   2.1.11 Electromagnetic systems 2/5
   2.1.12 Magnetic field of a toroid 2/5
   2.1.13 Permeability 2/6
   2.1.14 Faraday's law 2/6
   2.1.15 Self-induced e.m.f. 2/7
   2.1.16 Energy stored in an inductor 2/7
   2.1.17 Mutual inductance 2/7
   2.1.18 Hysteresis in magnetic circuits 2/8
   2.1.19 Eddy current loss 2/8
   2.1.20 Kirchhoff's laws and the magnetic circuit 2/8
   2.1.21 Alternating quantities 2/9
   2.1.22 Relationship between voltage and current in R, L and C elements 2/9
   2.1.23 RL and RC circuits under transient switching conditions 2/9
   2.1.24 Steady-state alternating currents 2/10
   2.1.25 Phasor diagrams 2/11
   2.1.26 Complex notation 2/11
   2.1.27 The parallel RLC circuit 2/11
   2.1.28 Power and power factor in a.c. circuits 2/12
   2.1.29 Frequency response of circuits 2/12
   2.1.30 Semiconductors 2/13
   2.1.31 Doped semiconductors 2/13
   2.1.32 pn junction diode 2/13
   2.1.33 A.C. rectification 2/14
   2.1.34 The Zener diode 2/15

2.2 Electrical machines 2/15
   2.2.1 The d.c. generator 2/15
   2.2.2 Methods of connection 2/17
   2.2.3 The separately excited generator 2/17
   2.2.4 The shunt-wound generator 2/19
   2.2.5 The series-wound generator 2/19
   2.2.6 The compound-wound generator 2/19

2.2.7 The d.c. motor 2/19
2.2.8 The shunt-wound motor 2/20
2.2.9 The series-wound motor 2/20
2.2.10 The compound-wound motor 2/20
2.2.11 Starting d.c. motors 2/20
2.2.12 Speed control of d.c. motors 2/21
2.2.13 Efficiency of d.c. machines 2/23
2.2.14 Three-phase circuits 2/23
2.2.15 Generation of three-phase e.m.f.'s 2/23
2.2.16 Star and delta connections 2/23
2.2.17 Three-phase voltage and current relations 2/24
2.2.18 Power in three-phase circuits 2/24
2.2.19 Three-phase alternators 2/24
2.2.20 Synchronous motors 2/25
2.2.21 Induction motors 2/26
2.2.22 Starting induction motors 2/26
2.2.23 Braking induction motors 2/27
2.2.24 Speed control of induction motors 2/27
2.2.25 Single-phase induction motors 2/28
2.2.26 The d.c. permanent magnet (PM) motor 2/28
2.2.27 The stepper motor 2/29
2.2.28 Brushless d.c. motors 2/30
2.2.29 Transformers 2/30
2.2.30 Basic transformer action 2/31
2.2.31 Transformer voltage equation 2/31
2.2.32 Transformer losses 2/31
2.2.33 Determination of transformer losses 2/31
2.2.34 Referred values 2/31
2.2.35 Transformer efficiency 2/32
2.2.36 Voltage regulation 2/32
2.2.37 Three-phase transformers 2/32
2.2.38 Auto-transformers 2/33

2.3 Analogue and digital electronics theory 2/33
   2.3.1 The bipolar (or junction) transistor 2/33
   2.3.2 Common-base characteristics 2/33
   2.3.3 Common-emitter characteristics 2/34
   2.3.4 The transistor in a circuit 2/35
   2.3.5 The field effect transistor (FET) 2/36
   2.3.6 Integrated circuits 2/36
   2.3.7 The thyristor 2/37
   2.3.8 The triac 2/37
   2.3.9 Amplifiers 2/37
   2.3.10 Effect of feedback on amplifiers 2/38
   2.3.11 Noise and distortion in amplifiers 2/38
   2.3.12 Amplifier frequency response 2/38
   2.3.13 Positive feedback and stability 2/39

2.3.14   The operational amplifier   2/39
2.3.15   The differential amplifier   2/41
2.3.16   Instrumentation amplifier   2/41
2.3.17   Power supplies   2/41
2.3.18   Analogue and digital systems   2/42
2.3.19   Boolean algebra   2/43
2.3.20   Digital electronic gates   2/44
2.3.21   Gate symbols   2/44
2.3.22   Logic systems using simple gates   2/44
2.3.23   Logic systems using NAND and NOR gates
         only   2/45
2.3.24   Unused inputs   2/48
2.3.25   Latches   2/48
2.3.26   The Karnaugh map   2/49
2.3.27   Minimization of Boolean expressions   2/50
2.3.28   Positive and negative logic   2/51
2.3.29   Tri-state logic   2/51
2.3.30   Sequential logic circuits   2/51

2.3.31   Registers and counters   2/53
2.3.32   Timers and pulse circuits   2/54
2.3.33   Digital computers and microprocessors   2/55
2.3.34   Application-specific integrated circuits
         (ASICs)   2/55

2.4   Electrical safety   2/55
2.4.1   Electric shock   2/56
2.4.2   Electric burn   2/56
2.4.3   Rescue   2/56
2.4.4   Protection   2/56
2.4.5   Earthing   2/57
2.4.6   Doube insulation   2/57
2.4.7   Low-voltage supplies   2/57

References   2/57
Further reading   2/57

2.1 Basic electrical technology

2.1.1 Flux and potential difference

The concept of flux and potential difference enables a unified approach to be adopted for virtually all the 'field' type of problems. Generally, the flowing quantity is termed the flux and the quantity which drives the flow is called the potential difference. This consistency of method is equally applicable to problems in fluid flow, heat transfer, electrical conduction, electrostatics and electromagnetism, to name but a few.

In general terms, the flux may be written as

$$\text{Flux} = \frac{(\text{Field characteristic}) \times (\text{Cross-sectional area}) \times (\text{Potential difference})}{(\text{Length})} \tag{2.1}$$

In specific terms, for the flow of an electric current through a conducting medium, equation (2.1) takes the form:

$$I = \frac{\sigma a V}{l} \tag{2.2}$$

where I is the current in amperes (A),
σ is the conductivity of the medium (siemens/m), i.e. the field characteristic,
a is the cross-sectional area of the medium (m^2),
l is the length of the medium (m), and
V is the applied potential difference, or voltage (V).

The group $(\sigma a / l)$ is termed the conductance, denoted by G and measured in siemens, thus:

$$I = GV \tag{2.3}$$

The reciprocal of conductance is referred to as the resistance, R and is measured in ohms (Ω). Hence

$$I = V/R \tag{2.4}$$

Equation (2.4) is the familiar 'Ohm's law', which defines a linear relationship between voltage and current in a conducting medium. If the resistance, R, varies with the magnitude of the voltage, or the current, then the resistance is non-linear. Rectifiers constitute one particular class of non-linear resistors.

Comparing equations (2.4) and (2.2) gives:

$$R = l/(\sigma a) \tag{2.5}$$

It is more usual, however, to quote the 'resistivity' as opposed to the conductivity, and resistance is generally written as:

$$R = \rho l/a \tag{2.6}$$

where ρ is the resistivity of the conductor in ohm-metres.

The resistance of all pure metals is temperature dependent, increasing linearly for moderate increases in temperature. Other materials, including carbon and many insulators, exhibit a decreasing resistance for an increase in temperature.

2.1.2 Simple resistive circuits

The effective total resistance of a series arrangement is the algebraic sum of all the resistances in series, i.e.

$$R_t = R_1 + R_2 + R_3 \tag{2.7}$$

where R_t is the total resistance of the circuit.

For resistors in parallel the effective total resistance obeys an inverse summation law, i.e.

$$\frac{1}{R_t} = \frac{1}{R_1} + \frac{1}{R_2} + \frac{1}{R_3} \tag{2.8}$$

2.1.3 Electromotive force and potential difference

In a metallic conductor which has a potential difference applied across opposite ends, free electrons are attracted to the more positive end of the conductor. It is this drift of electrons which constitutes the electric current, and the effect is simply Nature's attempt to redress an energy imbalance. Although the negatively charged electrons actually drift towards the positive end of the conductor, traditional convention gives the direction of the current flow from positive to negative. There is nothing really at issue here, since it is only a simple sign convention which was adopted long before the true nature of the atom and its associated electrons were postulated.

A current of 1 A is associated with the passage of 6.24×10^{18} electrons across any cross-section of the conductor per second. The quantity of charge is the coulomb, Q, and

$$Q = I \cdot t \tag{2.9}$$

where 1 coulomb of charge is passed when a current of 1 A flows for a period of 1 s.

The electromotive force (e.m.f.) is that which tends to produce an electric current in a circuit, and is associated with the energy source. Potential difference is simply the product of current and resistance across any resistive element in a circuit, irrespective of the energy source. For circuit elements other than purely resistive, the potential difference across the element becomes a time-dependent function.

2.1.4 Power and energy

Power is the rate at which energy is expended, or supplied. The potential difference across any two points in a circuit is defined as the work done in moving unit charge from a lower to a higher potential. Thus the work done in moving Q coulombs of charge across a constant potential difference of V volts is:

$$W = Q \cdot V \tag{2.10}$$

Therefore

$$\text{Power} = \frac{dW}{dt} = \frac{dQ}{dt} V$$

From equation (2.9) $(dQ/dt) = I$. thus

$$\text{Power} = IV \tag{2.11}$$

Using Ohm's law, the power dissipated across a simple resistive circuit element is

$$\text{Power} = IV = I(I \cdot R) = I^2 R \tag{2.12}$$

2.1.5 Network theorems

A network consists of a number of electrical elements connected up in a circuit. If there is no source of electromotive force in the circuit it is said to be passive. When the network contains one or more sources of electromotive force it is said to be active.

A number of well-established theorems have been developed for the analysis of complex resistive networks [1-3] and are listed below:

Kirchhoff's first law: The algebraic sum of the currents entering (+ve) and leaving (−ve) a junction is zero.
Kirchhoff's second law: The algebraic sum of potential differences and e.m.f.'s around any closed circuit is zero.
Superposition theorem: In a linear resistive network containing more than one source of e.m.f. the resultant current in any branch is the algebraic sum of the currents that would be

produced by each e.m.f. acting on its own while the other e.m.f.'s are replaced with their respective internal resistances. *Thevenin's theorem*: The current through a resistor R connected across any two points in an active network is obtained by dividing the potential difference between the two points, with R disconnected, by $(R + r)$, where r is the resistance of the network between the two connection points with R disconnected and each e.m.f. replaced with its equivalent internal resistance. An alternative statement of Thevenin's theorem is: 'Any active network can be replaced at any pair of terminals by an equivalent e.m.f. in series with an equivalent resistance.' The more concise version of Thevenin's theorem is perhaps a little more indicative of its power in application. *Norton's theorem*: Any active network can be replaced at any pair of terminals by an equivalent current source in parallel with an equivalent resistance. It may appear that Norton's theorem is complementary to Thevenin's theorem and both can be equally well used in the analysis of resistive networks.

Other useful network analysis techniques include 'mesh analysis', which incorporates Kirchhoff's first law, and 'nodal analysis', which is based on Kirchhoff's second law. Mesh and nodal analysis are also essentially complementary techniques.

2.1.6 Double-subscript notation

To avoid ambiguity in the direction of current, e.m.f. or potential difference, a double-subscript notation has been adopted. Figure 2.1 shows a source of e.m.f. which is acting from D to A. The e.m.f. is therefore E_{da}. The current flows from A to B, by traditional convention, and is designated I_{ab}. From this simple circuit it is apparent that $I_{ab} = I_{bc} = I_{cd} = I_{da}$.

The potential difference across the load R is denoted V_{bc} to indicate that the potential at B is more positive than that at C. If arrow heads are used to indicate the potential difference, then they should point towards the more positive potential.

2.1.7 Electrostatic systems

Electrostatic systems are quantified by the physical behaviour of the 'charge'. Fortunately, the unified field approach lends itself well to the quantification of electrostatic systems.

Figure 2.2 shows two parallel, conducting metal plates separated by an evacuated space. A potential difference is applied across the plates such that they become charged at equal magnitude but opposite sign. For the electrostatic system, equation (2.1) is written

$$Q = \frac{\varepsilon_0 a V}{l} \tag{2.13}$$

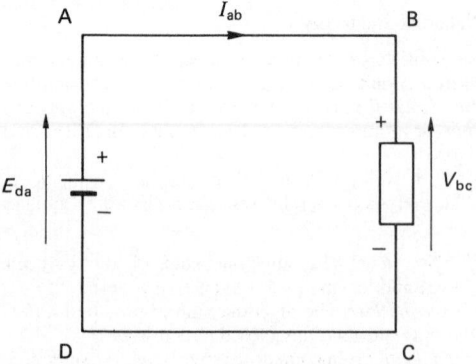

Figure 2.1 Double-subscript notation

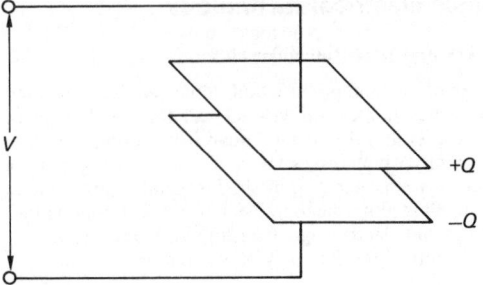

Figure 2.2 Electrostatic system

where Q is the total charge in coulombs,
  ε_0 is the permittivity of free space in Farads/m, i.e. the field characteristic,
  a is the cross-sectional area of the plates,
  l is the distance separating the plates, and
  V is the applied potential difference.
The group $(\varepsilon_0 a/l)$ is termed the capacitance of the system. It is usually denoted by C, and is measured in farads (F). Thus

$$Q = C \cdot V \tag{2.14}$$

It is more common to use the microfarad (μF) or the picofarad (pF) as the unit of measurement.

NB: $1 \ \mu F = 10^{-6}$ F: 1 pf $= 10^{-12}$ F

If the plates are separated by an insulating medium other than free space, then these so-called dielectric media have a different value of permittivity. The actual permittivity is related to the permittivity of free space by the relative permittivity of the dielectric, i.e.

$$\varepsilon = \varepsilon_0 \cdot \varepsilon_\tau \tag{2.15}$$

where ε_τ is the relative permittivity of the dielectric. The permittivity of free space, ε_0, is numerically equal to $(1/36\pi) \times 10^{-9}$. The relative permittivity of some of the more common dielectric materials are listed in Table 2.1.

2.1.8 Simple capacitive circuits

For three capacitors connected in a simple parallel arrangement, the equivalent total capacitance is given as the algebraic sum of all the capacitances in the circuit, i.e.

$$C = C_1 + C_2 + C_3 \tag{2.16}$$

where C is the total capacitance. For a series capacitance arrangement of three capacitors, the total equivalent capacitance is related through the inverse summation given as

$$\frac{1}{C} = \frac{1}{C_1} + \frac{1}{C_2} + \frac{1}{C_3} \tag{2.17}$$

Table 2.1 Relative permittivities of some typical dielectric materials

Material	Relative permittivity
Air	1
Paper	2–2.5
Porcelain	6–7
Mica	3–7

Equations (2.16) and (2.17) can be used to reduce series and parallel capacitor circuits to a single equivalent capacitor.

Composite capacitors, involving different dielectric media, may also be treated in the same manner as a series capacitor arrangement.

2.1.9 Charging a capacitor

Figure 2.3 shows a parallel plate capacitor which is connected in series with a resistor to a source of e.m.f. (say, a battery) through a switch. Initially, the capacitor is uncharged before the switch is closed. When the switch is closed a charging current will flow until such time that the potential difference across the capacitor is equal to the e.m.f. available from the source. The charging process consists of taking electrons from plate A and transferring them through the external wiring to plate B. The energy required to do this is derived from the battery. The build-up of electrons from the negative terminal of the battery to plate B of the capacitor induces a dielectric flux between the plates and a balancing positive charge is developed on plate A. As long as the dielectric flux is changing, a current will flow externally. Eventually a state of equilibrium will be reached. Note that no electrons can pass through the dielectric since it is an insulator.

The instantaneous current during charging is

$$i = dQ/dt$$

From equation (2.14), this may be written for a capacitor as

$$i = dQ/dt = C(dv/dt) \qquad (2.18)$$

where v is the instantaneous voltage. The instantaneous power is therefore

$$p = iv = Cv(dv/dt)$$

The energy supplied over the time period, dt, is

$$Cv(dv/dt)dt = Cvdv$$

Hence the total energy supplied is

$$\int_0^V Cvdv = \frac{1}{2}CV^2 \qquad (2.19)$$

2.1.10 Dielectric strength

If the potential difference across opposite faces of a dielectric material is increased above a particular value, the material breaks down. The failure of the material takes the form of a small puncture, which renders the material useless as an insulator. The potential gradient necessary to cause break-down is normally expressed in kilovolts/millimetre and is termed the 'dielectric strength'. The dielectric strength of a given material decreases with increases in the thickness. Table 2.2 gives approximate values for some of the more common dielectric materials.

Table 2.2 Dielectric strength of some common insulators

Material	Thickness (mm)	Dielectric strength (kV/mm)
Air	0.2	5.75
	0.6	4.92
	1.0	4.36
	10.0	2.98
Mica	0.01	200
	0.10	115
	1.00	61
Waxed paper	0.10	40–60

2.1.11 Electromagnetic systems

The magnetic field can be defined as the space in which a magnetic effect can be detected, or observed. An obvious magnetic field is observable around a straight length of conductor carrying a current. In particular, exactly the same magnetic field as that produced by a bar magnet is observed when the current-carrying conductor is formed into a helical type coil. The equipotential loops describe the path of the magnetic flux, ϕ, and although the flux lines have no physical meaning, they provide a convenient vehicle to quantify various magnetic effects.

The direction of the magnetic flux is governed by the so-called 'right-hand screw rule'. This states that the direction of the magnetic field produced by a current corresponds with the direction given by turning a right-hand screw thread. The direction of the current corresponds with the translational movement of the screw.

2.1.12 Magnetic field of a toroid

Figure 2.4 shows a toroidal coil, of N turns, which is wound round an annular former. A resultant magnetic flux, shown as broken lines in the figure, is generated when the coil carries a current. For the magnetic field, equation (2.1) takes the general form:

$$\phi = \frac{\mu a F}{l} \qquad (2.20)$$

where ϕ is the magnetic flux(in webers),
μ is the permeability of the medium (in henrys/m),
a is the cross-sectional area of the flux path in the toroid,
l is the length of the flux path, and
F is the magnetic potential difference, or magneto-motive force (in amperes).

The magnetomotive force (m.m.f.) is equal to the product of the number of turns on the coil and the current carried, i.e.

$$F = IN \qquad (2.21)$$

Note that the m.m.f. is descritively expressed in 'ampere-turns'. Since the number of turns is already a dimensionless quantity, the accepted unit of magnetomotive force is the ampere (A).

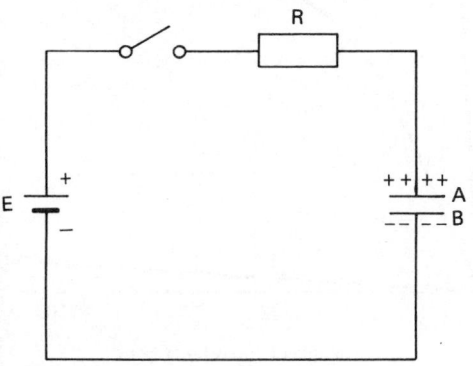

Figure 2.3 Charging a capacitor

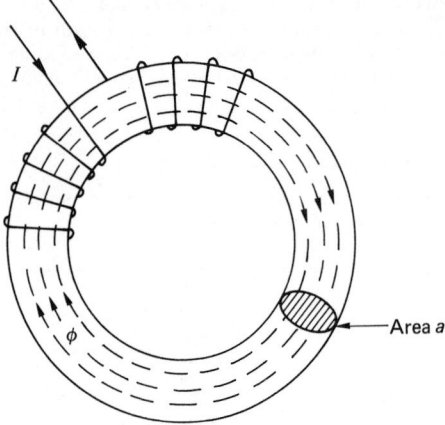

Figure 2.4   Toroid

The group ($\mu a/l$) is termed the permeance and the inverse of permeance is the reluctance, S. Thus equation (2.20) may be rewritten as

$$\phi = F/S \qquad (2.22)$$

Equation (2.22) represents an electromagnetic version of Ohm's law.

Alternatively, equation (2.20) can be expressed as

$$\frac{\phi}{a} = \mu \frac{F}{1}$$

or

$$B = \mu H \qquad (2.23)$$

where $B = \phi/a$ is the magnetic flux density (in webers/m², or Tesla (T)) and $H = F/l$ is the magnetic intensity (in A/m).

2.1.13   Permeability

The permeability of free space, μ_0, is numerically equal to $4\pi \times 10^{-7}$. The absolute permeability of other materials is related to the permeability of free space by the relative permeability, i.e.

$$\mu = \mu_0 \cdot \mu_r \qquad (2.24)$$

For air and other non-magnetic materials, the absolute permeability is the same constant. For magnetic materials, absolute permeability is not a fixed constant but varies non-linearly with the flux density. The non-linear variation of permeability is conveniently displayed as a functional plot of magnetic flux density, B, against magnetic intensity, H. Figure 2.5 illustrates a number of B–H curves for some common materials.

Also shown in Figure 2.5 is the B–H curve for air, the only straight-line relationship in the diagram. It is apparent that for an applied magnetic intensity, the magnetic flux developed in a coil with a ferrous core is many times greater than that through a similar coil with an air core. In most practical systems, therefore, a ferrous core is normally used, since it greatly facilitates the establishment of a magnetic flux.

2.1.14   Faraday's law

Faraday's law states that the e.m.f. induced in a magnetic circuit is equal to the rate of change of flux linkages in the circuit, and is given as

$$e = N(d\phi/dt) \qquad (2.25)$$

where e is the instantaneous induced e.m.f.

Equation (2.25) forms the basis of all electrical power generation machines and is a statement of the fact that an electric current can be produced by the movement of magnetic flux relative to a coil. In all rotating electrical generators it is actually the coil which is moved relative to the magnetic field. The net result, however, is exactly the same.

The direction of the induced e.m.f. is always such that it tends to set up a current to oppose the motion (or the change of magnetic flux) which was responsible for inducing the e.m.f. This is essentially a statement of Lenz's law. In many texts, therefore, the right-hand side of equation (2.25) is often shown as a negative quantity.

The motion, or change of flux, is associated with the application of a mechanical force which ultimately provides the torque required to drive the electric generator. Figure 2.6 shows a single conductor of length l metres, carrying an induced current I and lying perpendicular to a magnetic field of flux density, B T.

The force applied causes the conductor to move through a distance dx metres. The mechanical work done is therefore $F \cdot dx$. The electrical energy produced is given as the product of the power developed and the time duration, i.e. $e \cdot I \cdot dt$.

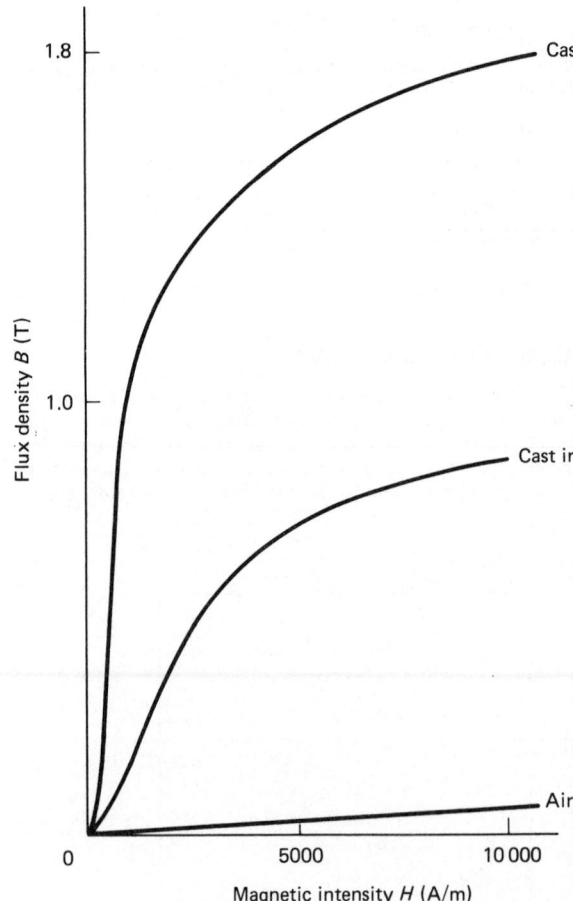

Figure 2.5 B–H curves for some common materials

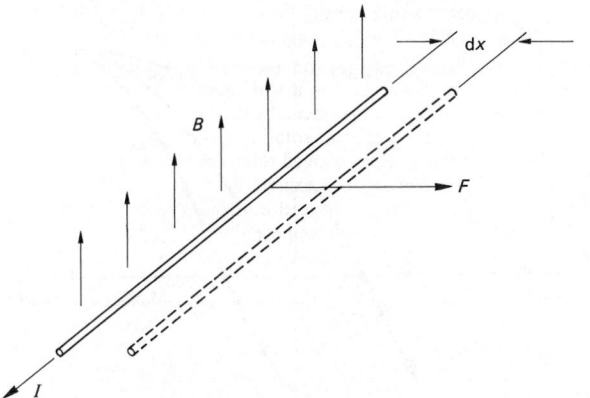

Figure 2.6 Generation of e.m.f.

For no external losses, the mechanical work done is converted into electrical energy. Thus

$$e \cdot I \cdot dt = F \cdot dx \qquad (2.26)$$

Using equation (2.25) the induced e.m.f. is equal to the rate of change of flux linkage. For a single conductor, $N = 1$, and in consequence

$$e = (B \cdot l \cdot dx)/dt$$

Therefore

$$(B \cdot l \cdot dx/dt) \cdot I \cdot dt = F \cdot dx$$

i.e.

$$F = B \cdot l \cdot I \qquad (2.27)$$

Equation (2.27) relates the applied force to the corresponding current generated in a conductor moving through a magnetic field. The equation applies equally to an electric generator or, conversely, to a motor, in which case the electrical power supplied is converted into a mechanical torque via the electromagnetic effect.

2.1.15 Self-induced e.m.f.

If a current flows through a coil a magnetic flux links that coil. If, in addition, the current is a time-varying quantity, then there will be a rate of change of flux linkages associated with the circuit. The e.m.f. generated will oppose the change in flux linkages.

When dealing with electric circuits it is convenient if the voltage across individual elements can be related to the current flowing through them. Figure 2.7 shows a simple circuit comprising a coil having N turns and resistance R, connected in series with a time-varying voltage. The voltage drop across the terminals A and B can be split into two components. First, there is the voltage drop due solely to the resistance of the coiled element. Second, there is a voltage drop which is a consequence of the self-induced e.m.f. generated through the electromagnetic effect of the coil. Thus

$$v = vr + vl$$

$$= iR + N\frac{d\phi}{dt} \qquad (2.28)$$

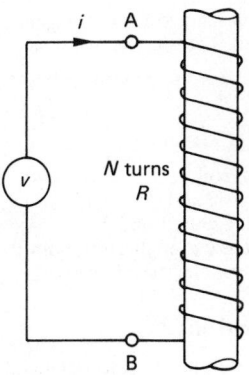
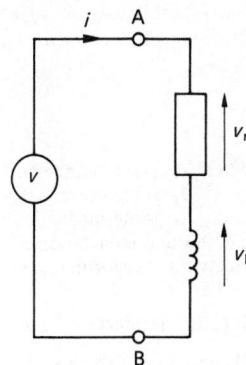

Figure 2.7 Self-induced e.m.f.

From equations (2.20) and (2.21)

$$\phi = \frac{\mu aF}{l} = \frac{\mu aiN}{l}$$

Therefore

$$v = iR + N\frac{d}{dt}\left[\frac{\mu aiN}{l}\right]$$

$$= iR + N^2\left[\frac{\mu a}{l}\right]\frac{di}{dt} \qquad (2.29)$$

The group $N^2(\mu a/l)$ is called the 'self-inductance' of the coil and is denoted by L. The unit of self-inductance is the henry (H). Therefore

$$v = iR + L\frac{di}{dt} \qquad (2.30)$$

By comparing equations (2.28) and (2.30) it is apparent that

$$L\frac{di}{dt} = N\frac{d\phi}{dt}$$

Integration then gives

$$L = N\phi/i \qquad (2.31)$$

The nature of the self-induced e.m.f. (i.e. Ldi/dt) is such that it will oppose the flow of current when the current is increasing. When the current is decreasing the self-induced e.m.f. will reverse direction and attempt to prevent the current from decreasing.

2.1.16 Energy stored in an inductor

Instantaneous power $= vi$

$$\text{Energy stored} = W = \int_0^t vi\,dt$$

$$= \int_0^t L\frac{di}{dt}i\,dt$$

$$= L\int_0^I i\,di = \frac{1}{2}LI^2 \qquad (2.32)$$

2.1.17 Mutual inductance

Two coils possess mutual inductance if a current in one of the coils produces a magnetic flux which links the other coil.

Figure 2.8 shows two such coils sharing a common magnetic flux path in the form of a toroid. The mutual inductance between the two coils is

$$M = \frac{N_2\phi}{I_1} \text{ (H)} \tag{2.33}$$

where N_2 is the number of turns on coil 2,
$\quad\quad I_1$ is the current through coil 1, and
$\quad\quad \phi$ is the magnetic flux linking coils 1 and 2.
The mutual inductance effect finds great application both to electrical transformers and to rotating electrical machines.

2.1.18 Hysteresis in magnetic circuits

Hysteresis can be described with reference to a toroidal coil wound on an iron core (see Figure 2.4). The current supplied to the coil can be imagined to be taken through a cyclic process where it is increased from 0 to $+I$ A, back through 0 to $-I$ A and again back through 0 to $+I$ A. Measurement of the flux density in the core, as the current varies, results in a B–H curve as depicted in Figure 2.9.

The behaviour of the B–H relationship is termed a 'hysteresis loop'. This behaviour is typical for ferrous cores and is an illustration of the fact that all the electrical energy supplied to magnetize an iron core is not returned when the coil current is reduced to zero. The loss of energy is called 'hysteresis loss', and it is manifested as heat in the iron core.

Hysteresis is characterized by two parameters, which are the 'remanent flux density' (or 'remanence') and the 'coercive force'. The remanent flux density is the flux density which remains in the core when the magnetic intensity (i.e. the coil current) has been reduced to zero. The remanent flux density is represented by line OA in Figure 2.9. The coercive force is the magnetic intensity required to reduce the remanent flux density to zero, and is represented by line OC in Figure 2.9.

2.1.19 Eddy current loss

Faraday's law (equation 2.25)) shows that a time-varying magnetic flux will induce an e.m.f. in a coil. If the ends of the coil are connected and form a closed circuit, then the induced voltage will circulate a current around the closed loop. Consider now an iron core, in which a time-varying magnetic flux exists. Since iron is a conductor then there will be a multitude of arbitrary closed paths within the iron matrix. These closed paths constitute effective conduction routes and the varying magnetic flux will generate a flow of current round them. The currents are called 'eddy currents' and, because of the ohmic resistance of the core, the end result is an energy loss as the eddy currents are dissipated as heat.

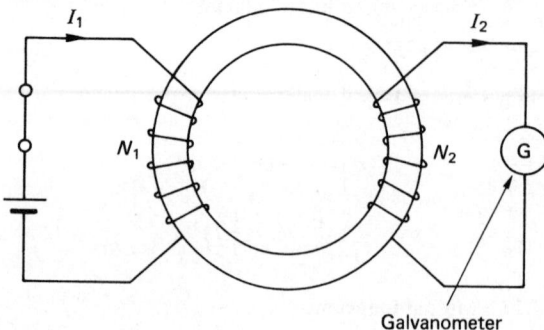

I_1 $\quad\quad\quad\quad\quad\quad\quad\quad\quad\quad$ I_2

N_1 $\quad\quad\quad\quad\quad\quad\quad\quad$ N_2 $\quad$ (G)

Galvanometer

Figure 2.8 Mutual inductance

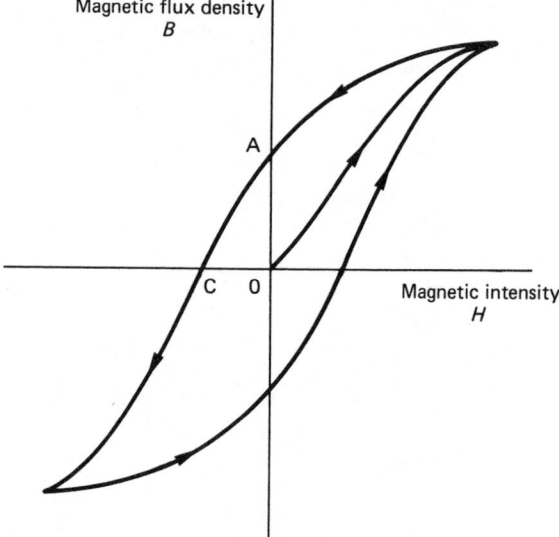

Figure 2.9 Hysteresis loop for an iron-cored toroid

Eddy current losses can be greatly reduced by building the iron core in the form of laminations which are insulated from one another. The laminated assembly confines the path lengths for the eddy currents to each respective lamination. The cross-sectional area of the eddy current path is also reduced and the eddy current loss is approximately proportional to the square of the thickness of the laminations. A practical minimum thickness for any lamination is about 0.4 mm. Increasing manufacturing costs could not justify the use of much thinner laminations.

2.1.20 Kirchhoff's laws and the magnetic circuit

Figure 2.10 shows a magnetic circuit in which a magnetizing coil is wound on one of the limbs and another limb incorporates the usual feature of an 'air gap'. Using the analogy between the magnetic and the conduction circuits, the magnetic circuit can be represented in terms of an energy source (or m.m.f.) and each limb of the magnetic circuit is written in terms of the appropriate reluctance, S. This is illustrated in Figure 2.11.

Given all the relevant dimensions and material properties, the problem is resolved to one of calculating the current required to establish a prescribed magnetic flux density in the air gap. The solution invokes the use of Kirchhoff's laws as they apply to magnetic circuits:

First law: At any instant in time, the sum of the fluxes flowing into a node is equal to the sum of the fluxes flowing out.
Second law: Around any closed magnetic circuit the total magnetomotive force is equal to the sum of all the m.m.f.'s round the circuit.

Manipulation of equations (2.20) and (2.24) then yields the required solution. The self-inductance of the coil (if required) may be calculated from equation (2.31), or from the definition

$$L = N^2(\mu a/l) = N^2/S \tag{2.34}$$

It has already been shown that the lowest permeability is that of air, and that the m.m.f. required to produce a flux density in air is many times greater than that required to produce the same flux density in a ferrous material. It may

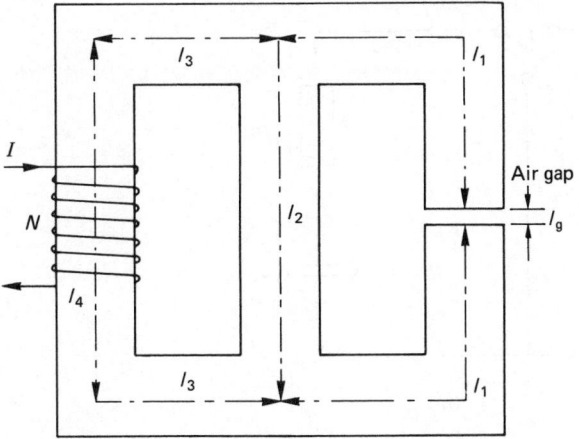

Figure 2.10 Magnetic circuit

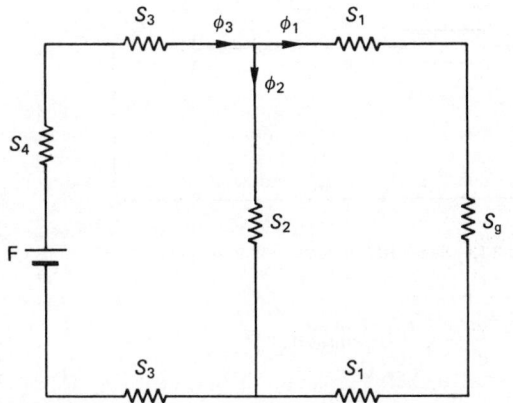

Figure 2.11 Representation of the magnetic circuit of Figure 2.10

reasonably be questioned therefore why air gaps are used at all in iron-cored magnetic circuits. The only function of the air gap is to provide a measure of linearity to the magnetic system such that the inductance remains reasonably constant over a range of operating currents.

2.1.21 Alternating quantities

If an electrical quantity varies with time, but does not change its polarity, it is said to be a direct current (d.c.) quantity. If the quantity alternates between a positive and a negative polarity, then it is classified as an alternating current (a.c.) quantity.

The period, T, is the time interval over which one complete cycle of the alternating quantity varies. The inverse of the period is the frequency, f, in Hertz (Hz). Circular frequency, ω, in radians per second is also commonly used.

Instantaneous values of the quantities encountered in electrical systems are usually denoted by lower-case letters. Since the instantaneous values are difficult to measure and quantify, a.c. quantities are usually expressed as 'root mean square' (r.m.s.) values. For a periodically varying a.c. quantity, the r.m.s. value is given by

$$\text{r.m.s.} = \left[1/t \int_0^t (\text{quantity})^2 dt \right]^{1/2} \tag{2.35}$$

Many electrical quantities vary in a sinusoidal manner and it can easily be shown that the r.m.s. value is simply related to the maximum value by

$$\text{r.m.s.} = \text{max}/(\sqrt{2}) = 0.707 \text{ max} \tag{2.36}$$

2.1.22 Relationship between voltage and current in R, L and C elements

For a simple resistive element, current is directly proportional to voltage. The current waveform will therefore be essentially the same shape as the voltage waveform.

For an inductive coil with negligible resistance, the relation between voltage and current is given by equation (2.30), i.e.

$$v = L \frac{di}{dt}$$

Thus

$$i = \frac{1}{L} \int v \, dt \tag{2.37}$$

The relation between voltage and current for a capacitive element is given by equation (2.18), i.e.

$$i = C \frac{dv}{dt}$$

For the capacitive element it can be seen that a current will flow only when the voltage is changing. No current can flow if the voltage is constant since dv/dt will then be equal to zero. The capacitor then, will block any steady d.c. input and indeed is sometimes used for this express purpose.

2.1.23 RL and RC circuits under transient switching conditions

Circuits involving a single resistor, capacitor or inductance are rare. It is more usual to find circuits involving some or other combination of these elements in both d.c. and a.c. applications. Figure 2.12 illustrates two simple RL and RC circuits.

2.1.23.1 RL circuit

With the switch open there is no flow of current in the circuit. At the instant of switching, the current will rise and eventually reach a steady-state value of V_s/R. The transient period is governed by equation (2.30), which represents a first-order, ordinary differential equation in i. The solution of equation (2.30) involves separating the variables to allow integration. The general solution is

$$i = I[1 - \exp(-Rt/L)] \tag{2.38}$$

Equation (2.38) shows that the current growth in the circuit will rise exponentially to reach a steady state value as time, t, increases. It may also be shown that

$$v = L \frac{di}{dt} = V_s \cdot \exp(-Rt/L) \tag{2.39}$$

The 'time constant', T, for the RL circuit is L/R.

2.1.23.2 RC circuit

In Figure 2.12(b), with the switch open there is zero potential difference across the capacitor. On closing the switch the

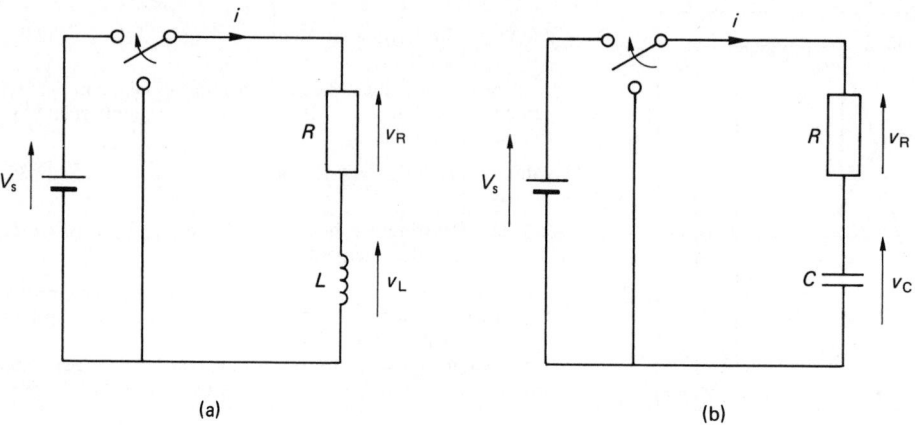

Figure 2.12 Simple *RL* and *RC* circuits under transient switching conditions

voltage across the capacitor will rise in an asympototic manner, reaching a steady-state value of V_s. From Kirchhoff's second law:

$$V_s = iR + v_c \qquad (2.40)$$

where v_c is the instantaneous voltage across the capacitor.

From equation (2.18) we can write:

$$V_s = RC \frac{dv_c}{dt} + v_c \qquad (2.41)$$

Equation (2.41) shows that the instantaneous voltage across the capacitor also conforms to a first-order system. The solution gives

$$v_c = V_s[1 - \exp(-t/RC)] \qquad (2.42)$$

The time constant for the simple *RC* circuit is

$$T = RC \qquad (2.43)$$

Both the simple *RL* and *RC* circuits are first-order systems with a generalized form of transient behaviour. In circuits containing both inductive and capacitive elements the transient behaviour is governed by a second-order ordinary differential equation. The transient behaviour of these circuits is, however, less important than their response to sinusoidally varying inputs.

2.1.24 Steady-state alternating currents

In most practical applications in electrical engineering the voltages and currents are sinusoidal. A simple series *RLC* circuit is depicted in Figure 2.13. Since the current is common to each of the circuit elements, it is used for reference purposes. The instantaneous current is defined as

$$i = I_m \sin(\omega t) \qquad (2.44)$$

where I_m is the maximum (or peak) value of the current and ω is the angular, or circular frequency in radians/s.

The voltage drop across the resistor is

$$v_R = iR = I_m R \sin(\omega t) \qquad (2.45)$$

Equation (2.45) indicates that the voltage drop across the resistor is in phase with the current. In other words, v_R reaches a positive maximum at the same instant as the current, i. The voltage drop across the inductor is

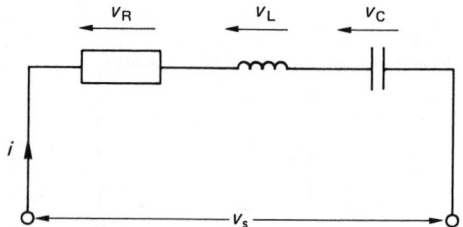

Figure 2.13 Series *RLC* circuit

$$v_L = L \frac{di}{dt} = L \frac{d}{dt} [I_m \sin(\omega t)]$$

$$= LI_m \omega \cos(\omega t)$$

$$= \omega L I_m \sin(\omega t + 90) \qquad (2.46)$$

The relationship between current and voltage drop across the inductor is shown in Figure 2.14.

It can be seen that there is a phase difference between the voltage drop and the current through the inductor. In fact, v_L reaches a positive maximum 'before' i and v_L is said to 'lead' the current by 90°. For the capacitor, the voltage drop is given by

$$v_c = 1/C \int i\, dt = 1/C \int I_m \sin(\omega t)\, dt$$

$$= -\frac{I_m}{\omega C} \cos(\omega t)$$

$$= +\frac{I_m}{\omega C} \sin(\omega t - 90) \qquad (2.47)$$

The voltage drop across the capacitor therefore reaches its positive maximum after that of i. In general terminology, v_c 'lags' i by 90°.

Equations (2.45)–(2.47) are all of similar form in that they can be expressed as

Voltage drop = constant × current

In equation (2.46) the constant ωL is termed the 'inductive reactance' and is denoted by X_L. In equation (2.47) the

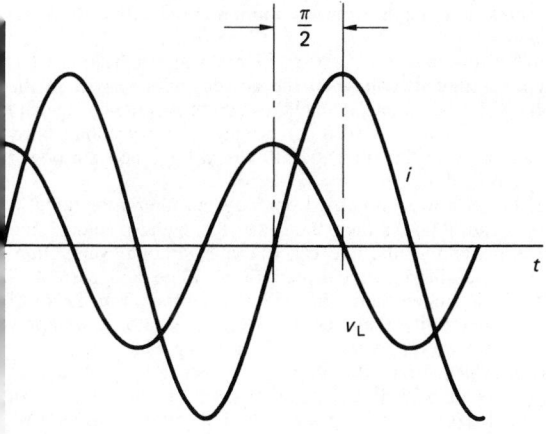

Figure 2.14 Current and voltage drop across an inductor

constant $(1/\omega C)$ is the 'capacitive reactance', which is denoted as X_c. Both of these reactances have units of ohms.

The total voltage drop across the three circuit elements is

$$v = v_R + v_L + v_C$$

$$= iR + L\frac{di}{dt} + \frac{1}{C}\int i\,dt$$

Therefore

$$v = I_mR\sin(\omega t) + \omega LI_m\sin(\omega t + 90°) + \frac{I_m}{\omega C}\sin(\omega t - 90°) \tag{2.48}$$

While equation (2.48) defines the total instantaneous voltage drop in mathematical terms, it is rather cumbersome to deal with. To simplify the analysis, the addition of a.c. voltages is conveniently performed using a graphical technique involving 'phasors'.

2.1.25 Phasor diagrams

Any sinusoidally varying quantity can be represented as a phasor, which is a vector quantity. The length of the phasor is proportional to the magnitude of the product of the reactance and the maximum current. The direction of the phasor is determined by the phase angle and its relation to some common reference.

For the RLC circuit of Figure 2.13 the voltage drop across the inductance may be arbitrarily assumed greater than that across the capacitor. The total voltage drop in the circuit is then given as the phasor addition of the three individual potential difference components. This is illustrated in Figure 2.15.

The vector addition of the three phasors shows that the source voltage leads the current by an angle of ϕ degrees, i.e.

$$\overline{V} = V_m\sin(\omega t + \phi) \tag{2.49}$$

The circuit is therefore essentially inductive and, using standard notation, the total phasor voltage is designated by a capital letter with an overbar.

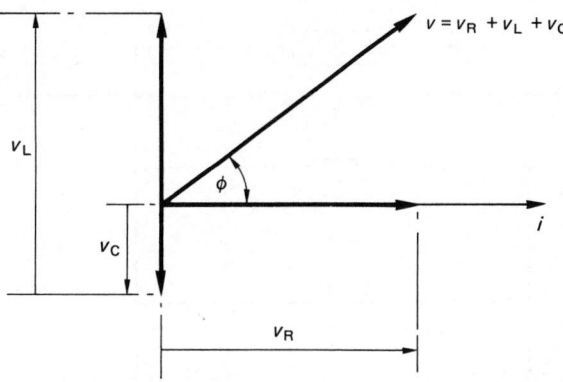

Figure 2.15 Phasor diagram for series *RLC* circuit

2.1.26 Complex notation

Since inductive and capacitive elements in a.c. circuits involve a phase shift of $+90°$ and $-90°$, respectively, the complex number notation is used extensively to manipulate phasor quantities. The complex operator j (defined as $\sqrt{-1}$) is a unit operator which, when multiplying a phasor, shifts it by 90° in an anti-clockwise direction. Thus for the series RLC circuit

$$\overline{V}_R = \overline{I}R, \; \overline{V}_L = j\overline{I}X_L \text{ and } \overline{V}_c = -j\overline{I}X_c$$

where $\overline{I}$ can be taken as the r.m.s. value of the current. The voltage drop across the complete circuit can then be written as

$$\overline{V} = \overline{I}R + j\overline{I}X_L - j\overline{I}X_C$$
$$= \overline{I}[R + j(W_L - X_C)] \tag{2.50}$$

The term in the square brackets is called the 'impedance' of the circuit and is denoted by $\overline{Z}$. Thus

$$\overline{V} = \overline{I} \cdot \overline{Z} \tag{2.51}$$

Equation (2.51) represents Ohm's law for a.c. circuits. The phase angle between the source voltage and the current is

$$\phi = \tan^{-1}[(X_L - X_C)/R] \tag{2.52}$$

2.1.27 The parallel *RLC* circuit

A parallel RLC circuit is shown in Figure 2.16. The applied voltage is common to all the circuit elements and it is therefore chosen as the reference.

Using Ohm's law, the currents through each of the circuit elements are

$$\overline{I}_R = \overline{V}/R; \; \overline{I}_L = \overline{V}/X_L; \; \overline{I}_C = \overline{V}/X_C$$

Applying Kirchhoff's first law, the total current is the vector sum of the three currents $\overline{I}_R$, $\overline{I}_L$ and $\overline{I}_C$. The magnitude and phase of the total current may subsequently be determined from a phasor diagram, or calculated using the complex number notation. Using the latter and noting that the current through an inductor lags the voltage while the current through a capacitor leads the voltage, it may be shown that

$$\overline{I} = \overline{I}_R + \overline{I}_L + \overline{I}_C$$
$$= \overline{V}\left[\frac{1}{R} - j\left(\frac{1}{X_C} - \frac{1}{X_L}\right)\right] \tag{2.53}$$

and the phase angle,

$$\phi = \tan^{-1}\left[\frac{R(X_L - X_C)}{X_L \cdot X_C}\right] \tag{2.54}$$

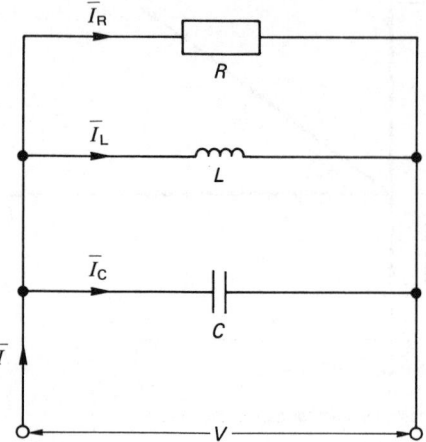

Figure 2.16   Parallel *RLC* circuit

2.1.28   Power and power factor in a.c. circuits

Denoting the phase angle between the voltage and the current as ϕ, it may be shown[2] that the average power is

$$P_{av} = \frac{V_m}{\sqrt{2}} \frac{I_m}{\sqrt{2}} \cos(\phi)$$

In terms of r.m.s. values:

$$P_{av} = VI \cos(\phi) \qquad (2.55)$$

where $\cos(\phi)$ is called the 'power factor'.

Power factor is an important parameter when dealing with electrical transformers and generators. All such machines are rated in terms of kilo-volt amperes (kVA), which is a measure of the current-carrying capacity for a given applied voltage. The power that can be drawn depends both on the kVA rating and the power factor of the load. Figure 2.17 shows the relationship between kVA, kilowatts (kW) and power factor. sometimes referred to as the power triangle. It can readily be seen that

$$kW = kVA \cos(\phi) \qquad (2.56)$$

and

$$kVA_R = kVA \sin(\phi) \qquad (2.57)$$

where kVA_R is the reactive power. Thus knowing the kVA rating and the power factor of a number of various loads, the

power requirements from a common supply may be determined.

When quoting power factor in practical applications it is usual to state the phase of the current with respect to the voltage. For an inductive load the current lags the voltage and the power factor is said to be lagging. For a predominantly capacitive load the current leads the voltage and the power factor is leading.

If the power is supplied from, say, an alternator rated at 400 V and 1000 A, then these are the highest voltage and current that the machine can tolerate without overheating. The phase difference between the voltage and current is entirely dependent upon the load. Thus if the power factor of the load is unity then the 400 kVA alternator can supply 400 kW of power to the load. Neglecting losses, the prime mover which drives the alternator must also be capable of supplying 400 kW. If, on the other hand, the power factor of the load is 0.5, then the power supplied will only be 200 kW. This means that although the generator will be operating at its rated kVA, the prime mover which drives the genrator will be operating at only half of its capacity.

An alternative way of looking at this phenomenon is to consider a load of, say, 100 kW, with a lagging power factor of 0.75. If the supply voltage is 50 V, then the required current, from equation (2.55), is 2.67 A. If, however, the power factor of the load were to be increased to unity, then the required current would be reduced to 2 A. This means that the conducting cables, in supplying a reduced current, may have a correspondingly reduced cross-sectional area.

In general, the size of an electrical system including transmission lines, switchgear and transformers is dependent upon the size of the current. It is economically viable therefore to ensure that the current is minimized. As a further incentive to industrial consumers, the electricity supply authorities normally operate a two-part tariff system. This consists of a fixed rate depending on the kVA rating of the maximum demand and a running charge per unit kilowatts consumed per hour.

For these reasons it is advantageous to try to increase the power factor such that it is close to (but not quite) unity. A unity power factor is in fact avoided, because it gives rise to a condition of resonance (see Section 2.1.29). In practice, capacitors connected in parallel are often used to improve the power factor of predominantly inductive loads such as electric motors. For large-scale power systems, a separate phase advance plant is used.

2.1.29   Frequency response of circuits

The 'frequency response' of a circuit is usually presented as a plot of the ratio of output over input against the frequency as base. The ratio plotted could be one of voltages, currents or powers. Since the range of frequencies involved may be quite large, a logarithmic scale is normally employed. A logarithmic scale is also usually adopted for the vertical axis and the output/input ratio quoted in decibels (dB), i.e.

$$\text{Voltage ratio in dB} = 20 \log_{10}\left[\frac{V_{out}}{V_{in}}\right] \qquad (2.58)$$

Considering the series *RLC* circuit shown in Figure 2.13 and taking the voltage across the resistor as an output,

$$V_{out} = IR$$

$$V_{in} = I[R + j(\omega L - 1/\omega C)]$$

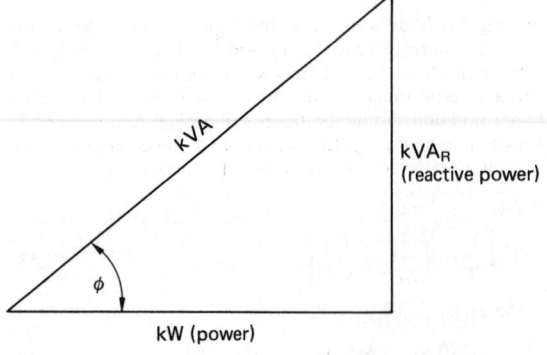

Figure 2.17   Power triangle

Therefore

$$\frac{V_{\text{out}}}{V_{\text{in}}} = \frac{R}{R + j(\omega L - 1/\omega C)}$$

Using the complex conjugate and calculating the modulus of the voltage ratio gives

$$\left| \frac{V_{\text{out}}}{V_{\text{in}}} \right| = \frac{R}{[R^2 + (\omega L - 1/\omega C)^2]^{1/2}} \qquad (2.59)$$

The phase angle

$$\phi = -\tan^{-1}\left[\frac{(\omega L - 1/\omega C)}{R} \right] \qquad (2.60)$$

The voltage ratio will have a maximum value of unity when the frequency

$$\omega = \frac{1}{\sqrt{LC}} \qquad (2.61)$$

Equation (2.61) defines the 'resonance' condition at which the inductive and capacitive reactances are equal and self-cancelling. The resonant frequency is usually denoted ω_0 and it is the frequency at which the power transferred through the circuit is maximum. At any other frequency above or below ω_0 the power transferred is reduced.

The impedance of the circuit is given by

$$\overline{Z} = R + j(X_L - X_C) \qquad (2.62)$$

At the resonant frequency the total reactance is zero and the circuit behaves as if only the resistive element were present.

The general variation of the voltage ratio (or amplitude ratio) and phase angle with frequency is illustrated in Figure 2.18. Also shown in the figure are the two frequencies, ω_1 and ω_2, at which the amplitude ratio is -3 dB. The -3 dB amplitude ratio is chosen because it corresponds to a halving in the power transmitted.

The 'bandwidth' is the frequency range between ω_1 and ω_2. A quality parameter, used with respect to resonant circuits, is the so-called 'Q factor', which is defined as the ratio of the resonant frequency to the bandwidth.

2.1.30 Semiconductors

The materials commonly used for semiconductors are germanium and silicon. In recent times silicon has all but replaced germanium as a semiconductor material. These materials have a crystalline structure such that each atom is surrounded by equally spaced neighbours. The basic structure can be visualized as a two-dimensional grid where the node points represent the central nucleus and the inner shell electrons, while the connecting lines of the grid represent the four valence electrons associated with each nucleus. This grid concept is adequate to describe an intrinsic (or 'pure') semiconductor.

At absolute zero temperature the crystalline structure is perfect and the electrons are all held in valence bonds. Since there are no current carriers available, the crystal behaves as a perfect insulator. As the temperature rises above absolute zero an increasing number of valence bonds are broken, releasing pairs of free electrons and their associated 'holes'. In the absence of an applied field the free electrons move randomly in all directions. When an electric field is applied the electrons drift in a preferential direction to oppose the field and a net flow of current is established.

The covalent bond, with a missing electron, has a large affinity for electrons such that an electron from a neighbouring bond may easily be captured. This will leave the neighbouring

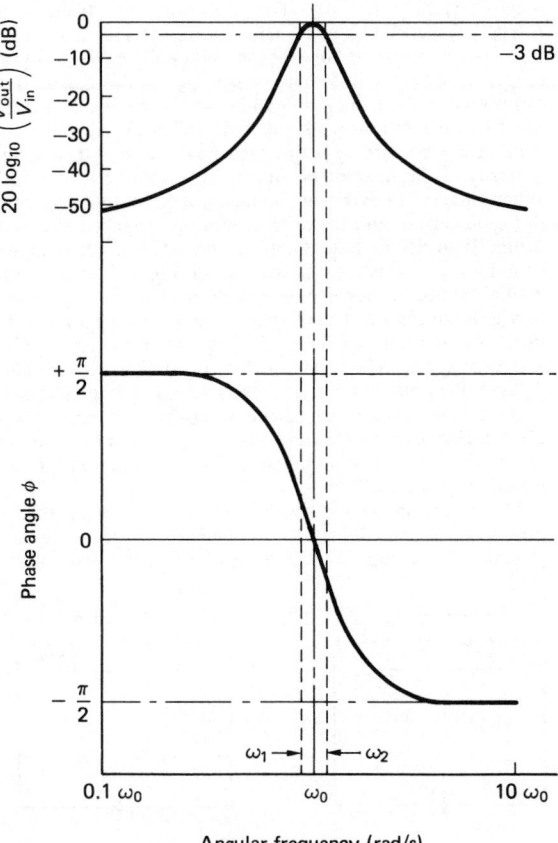

Figure 2.18 Voltage ratio and phase angle versus frequency (series RLC)

atom depleted of electrons and the flow of electrons is generally associated with a counterflow of so-called holes. The mobile hole, to all intents and purposes, is essentially a simple positive charge.

2.1.31 Doped semiconductors

Doped semiconductors are those in which an impurity has been introduced into a very pure intrinsic silicon. The nature of the impurity depends on the type of semiconductor required:

1. *n-type*: Impurities with five valence electrons can be added to produce a negative type of semiconductor. These impurities are referred to as 'donors', since the additional electron is very easily freed within the matrix. In the *n*-type semiconductor the free electrons are the dominant current carriers.
2. *p-type*: the *p*-type semiconductor is one in which the added impurities have only three valence electrons. Such impurities are called 'acceptors' and they produce a positive type of semiconductor within which hole conduction is the dominant current carrier.

2.1.32 *pn* junction diode

A *pn* junction is formed by doping a crystal in such a way that the semiconductor changes from *p*- to *n*-type over a very short length (typically 10^{-6} m). The transition zone from p- to n-type is called the 'carrier depletion layer' and, due to the high concentration of holes on one side and electrons on the other, a potential difference exists across this layer. The diffusion of holes from p to n and electrons from n to p is the majority carrier movement, called the 'diffusion current'. The drift of electrons from p to n and holes from n to p is the minority carrier movement, referred to as the 'drift current'. When there is no externally applied potential difference, the diffusion current and the drift current are balanced in equilibrium. If an electric field is applied across the device then two situations can exist, as illustrated in Figure 2.19. Figure 2.19(a) shows the reverse-bias mode in which the potential barrier is increased. The diffusion current is reduced while the drift current is barely altered. Overall, the current is negative and very small. When forward bias is applied, as in Figure 2.19(b), the potential barrier is reduced and a large diffusion current flows. Overall, the current is positive and large. These general characteristics are the basis of a semiconductor diode which displays the typical current/voltage relationship depicted in Figure 2.20.

This figure shows clearly that a very high impedance is presented by the diode to an applied voltage of reverse polarity. A low impedance is presented to a forward polarity

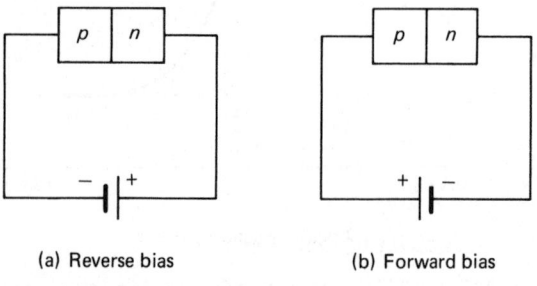

(a) Reverse bias   (b) Forward bias

Figure 2.19 *pn* junction with applied potential difference

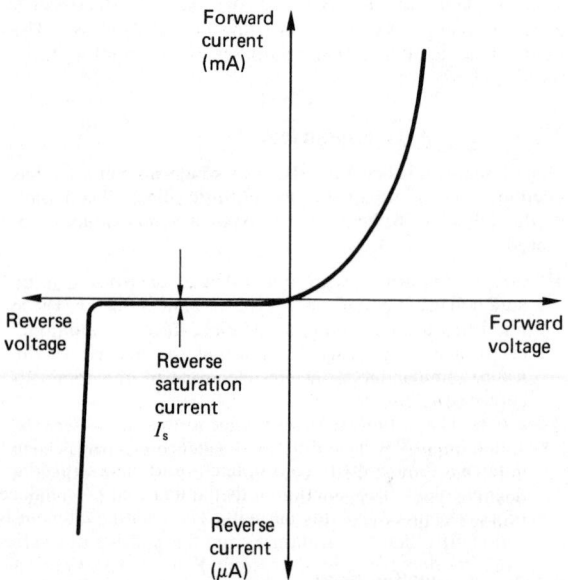

Figure 2.20 Current/voltage relationship for a *pn* semiconductor diode

voltage. In simple terms, the diode accommodates a forward flow of current but greatly inhibits a reverse flow. The diode may be likened therefore to a switch which is activated 'on' for forward voltages and 'off' for reverse voltages. The reverse saturation current, I_s, is typically of the order of a few nano-amperes and can sensibly be regarded as zero.

The general characteristic also shows that the reverse voltage has a critical limiting value at which a 'breakdown' occurs. Depending upon the diode construction, the breakdown (or 'Zener' voltage) may range from as low as one volt to as much as several thousand volts. Up to the breakdown voltage, the reverse saturation current is independent of the reverse voltage.

Since the current/voltage relationship for a diode is a non-linear exponential function, the analysis of circuits involving diodes can become complicated. A simple awareness of the diode's practical function as a rectifier is perhaps more important than a proficiency in analysing circuits involving diode elements.

2.1.33 A.C. rectification

Figure 2.21 shows an a.c. circuit with a diode in series with a load resistor. When the diode is forward biased a current will flow in the direction indicated by the arrowhead. No current can flow when the diode is reverse biased, provided that the applied voltage does not exceed the breakdown value. The resultant current waveform through the resistor, for a sinusoidal voltage input, will therefore consist of positive only half sine waves. Since the output waveform is positive only, then it is, by definition, a d.c. voltage. It can be shown that the r.m.s. voltage across the resistor is

$$V = \frac{V_m}{2} \cdot \frac{R_L}{R_L + R_F} \qquad (2.63)$$

where R_L is the load resistance, R_F is the diode forward resistance and V_m is the peak input voltage. Determination of R_F is problematic, however, and models of varying complexity are used to simulate the diode in the circuit.

The single-diode circuit results in half-wave rectification. To obtain full-wave rectification a diode bridge circuit can be used. The diode bridge is shown in Figure 2.22. When A is positive with respect to B then diodes D_1 and D_3 are conducting. When B is positive with respect to A then diodes D_2 and D_4 are conducting. The circuit arrangement ensures that the current, which consists of a continuous series of positive half sine waves, is always in the same direction through the load R_L.

With full-wave rectification there are twice as many half sine pulses through the load than there are with half-wave rectification. In addition, there are always two diodes effectively in series with the load. The resultant r.m.s. voltage across the load resistor for the full-wave diode bridge rectification circuit is

$$V = \frac{V_m}{\sqrt{2}} \cdot \frac{R_L}{(R_L + 2R_F)} \qquad (2.64)$$

The 'peak inverse voltage' (PIV) is defined as the maximum reverse-biased voltage appearing across a diode. When used as a rectifier the diodes must have a sufficiently high reverse voltage rating in excess to the peak inverse voltage that the circuit can generate. For both the half- and the full-wave rectification circuits considered, the peak inverse voltage is equivalent to the maximum supply voltage, V_m. Additional manufacturers' diode specifications would normally include the maximum power rating and the maximum allowable forward current.

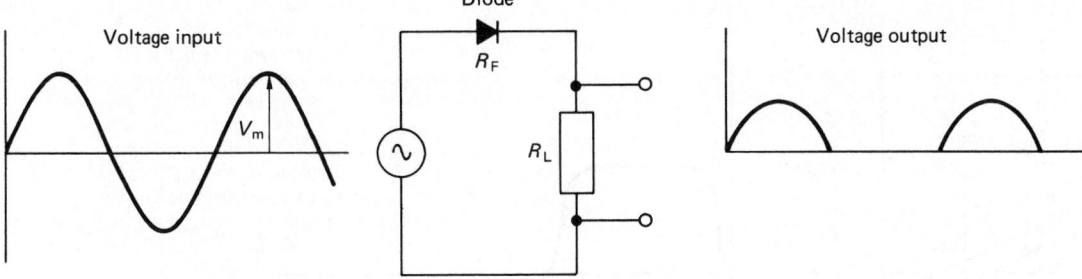

Figure 2.21 Half-wave rectification circuit

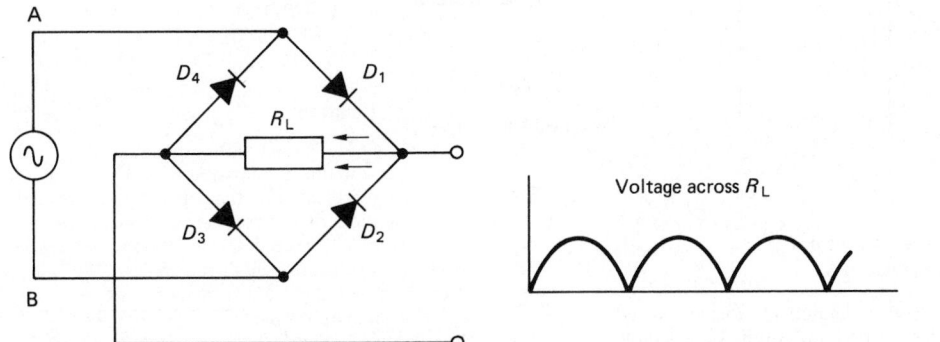

Figure 2.22 Full-wave rectification with a diode bridge

2.1.34 The Zener diode

The diode breakdown effect is also used in a variety of circuits to provide a stabilized reference voltage. Special diodes which are designed to operate continuously in the reverse bias mode are called 'Zener diodes'. These diodes are manufactured with a range of breakdown voltages from between 3 to 20 V. Figure 2.23 shows a Zener diode being used in a circuit to give a stable voltage which is essentially independent of the current flowing through the device. The series resistor in the circuit is included to limit the reverse current through the diode to a safe value.

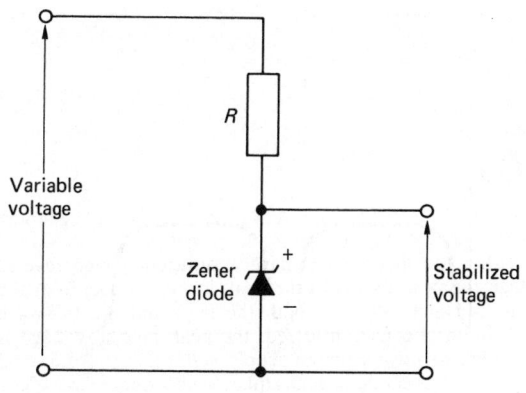

Figure 2.23 Zener diode as a reference voltage source

2.2 Electrical machines

The function of a rotating electrical machine is to convert mechanical power into electrical power, or vice versa. The conversion from mechanical to electrical power is made with a 'a generator' and the conversion of electrical to mechanical power with a 'motor'. Electrical machines may be further sub-divided into a.c. or d.c. machines. The major part of all electrical energy generated in the world today is produced by a particular type of a.c. machine called an 'alternator'. The applications of electric motors are no less substantial and they are used in a great variety of industrial drives. It is usually the mechanical features of a particular application which determines the type of electric motor to be employed, and the torque–speed characteristics of the machine are therefore very important.

2.2.1 The d.c. generator

All conventional electrical machines consist of a stationary element and a rotating element which are separated by a air gap. In d.c. machines – generator or motor – the stationary element consists of salient 'poles' which are constructed as laminated assemblies with coils wound round them to produce a magnetic field. The function of the laminations is to reduce the losses incurred by eddy currents. The rotating element is traditionally called the 'armature', and this consists of a series of coils located between slots around the periphery of the armature. The armature is also fabricated in laminations which are usually keyed onto a locating shaft. A very simple form of d.c. generator is illustrated in Figure 2.24.

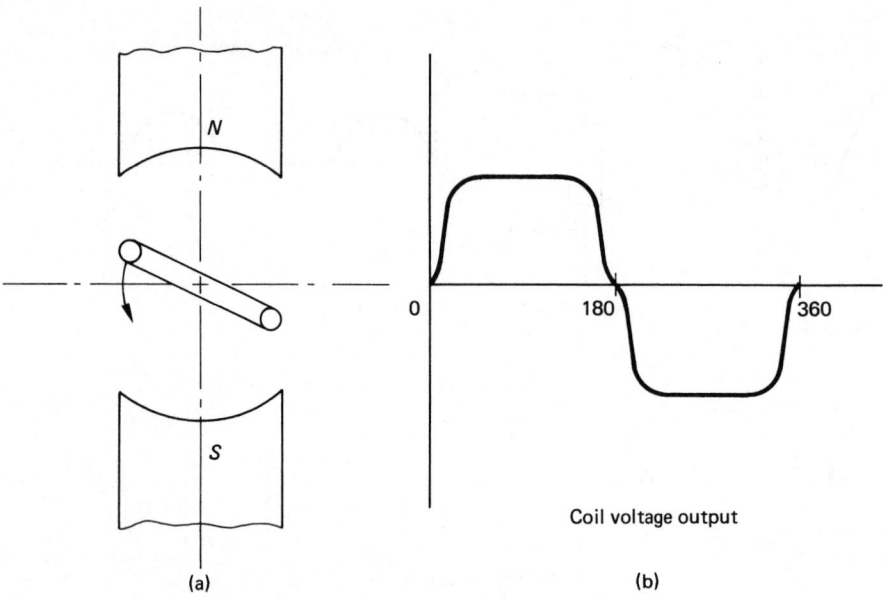

Figure 2.24 Single-coil, two-pole d.c. generator

In the figure the single coil is rotated at constant speed between the opposite poles, north and south, of a simple magnet. From Faraday's law (equation (2.25)) the voltage generated in the coil is equal to the rate of change of flux linkages. When the coil lies in the horizontal plane there is maximum flux linking the coil but a minimum rate of change of flux linkages. On the other hand, when the coil lies in the vertical plane there is zero flux linking the coil but the rate of change of flux linkages is a maximum. The resultant variation in generated voltage in the coil, as it moves through one revolution, is shown in Figure 2.24(b). It is apparent that the generated voltage is alternating with positive and negative half-cycles. To change the a.c. output voltage into a d.c. voltage, a simple yet effective mechanical device called a 'commutator' is used. The commutator (Figure 2.25) incorporates brass segments separated by insulating mica strips. External connection to the armature coil is made by stationary carbon 'brushes' which make sliding contact with the commutator. Referring to Figures 2.24(a) and 2.25(a), as the coil rotates from the horizontal plane through 180° the right-hand side of the coil is under the north pole and is connected via the commutator to the upper brush. Meanwhile, the left-hand side of the coil is under the south pole and is connected to the lower brush. A further 180° of rotation effectively switches the coil sides to the opposite brushes. In this manner the coil side passing the north pole is always connected to the positive upper brush, while the coil side passing the south pole is always connected to the negative lower brush. The resultant output voltage waveform is shown in Figure 2.25(b).

If two coils, physically displaced by 90°, are now used, the output brush voltage becomes virtually constant, as shown in Figure 2.26. With the introduction of a second coil, the commutator must have four separate segments. In a typical d.c. machine there may be as many as 36 coils, which would require a 72-segment commutator.

The simple d.c. generator of Figure 2.24 can be improved in perhaps three obvious ways. First, the number of coils can be increased, second, the number of turns on each coil can be increased and third, there is no reason why another pair of

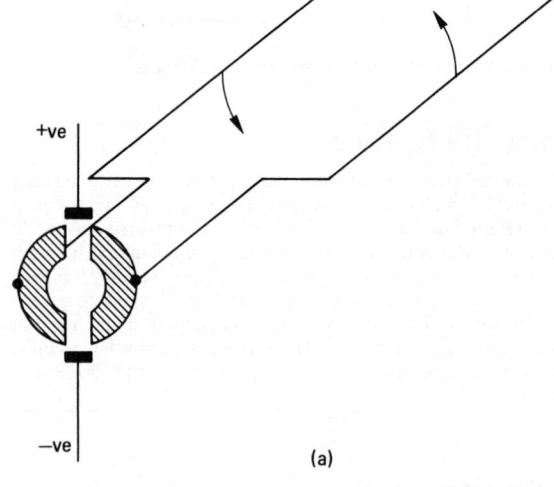

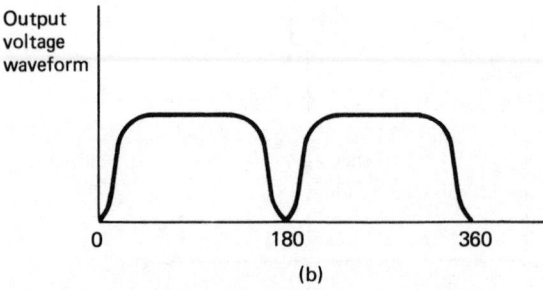

Figure 2.25 Commutator connections to armature

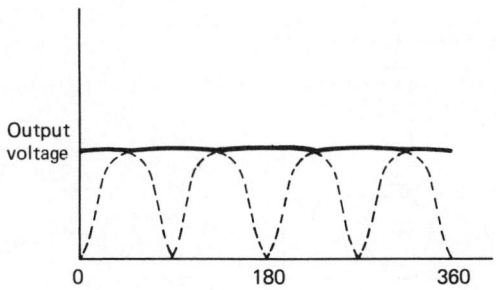

Figure 2.26 Two-coil, two-pole d.c. generator output voltage

poles cannot be introduced. A typical d.c. machine would therefore normally incorporate four poles, wired in such a way that each consecutive pole has the opposite magnetic polarity to each of its neighbouring poles. If the e.m.f.'s generated in the armature coils are to assist each other then while one side of the coil is moving under a north pole, the other side must be moving under a south pole. With a two-pole machine the armature coils must be wound such that one side of the coil is diametrically opposite the other. With a four-pole machine the armature coils can be wound with one side of the coil physically displaced 90° from the other. The size of the machine will generally dictate how many coils and the number of turns on each coil that can be used.

2.2.1.1 Armature e.m.f.

If a conductor cuts flux then a voltage of 1 V will be induced in the conductor if the flux is cut at the rate of 1 Wb/s. Denoting the flux per pole as Φ and the speed in revolutions per second as N, for the single-turn coil and two-pole generator of Figure 2.24(a) the e.m.f. induced in the coil is

$$E_{coil} = \frac{\text{Flux per pole}}{\text{Time for half revolution}} = \frac{\Phi}{1/(2N)} = 2N\Phi$$

For a machine having Z_s armature conductors connected in series, i.e. $Z_s/2$ turns, and $2p$ magnetic poles, the total induced e.m.f. is

$$E = 2N\Phi \frac{Z_s}{2} 2p = 2N\Phi Z_s\, p \text{ volts} \tag{2.65}$$

Z_s depends on the type of armature winding, and the two main types are 'lap-wound' and 'wave-wound'.

The lap winding is characterized by the fact that the number of parallel paths through the winding is equal to the number of poles. In the alternative wave winding the number of parallel paths through the winding is always equal to two. If Z denotes the total number of armature conductors then for the lap winding

$$Z_s = \frac{Z}{\text{Number of parallel paths}} = \frac{Z}{\text{Number of poles}} = \frac{Z}{2p} \tag{2.66}$$

and for the wave winding

$$Z_s = \frac{Z}{\text{Number of parallel paths}} = \frac{Z}{2} \tag{2.67}$$

Lap windings are generally used in low-voltage, heavy-current machines and wave winding in all other cases.

2.2.1.2 Armature torque

The force on a current-carrying conductor is given by equation (2.27), i.e.

$$F = BlI$$

The torque on one armature conductor is therefore

$$T = Fr = B_{av}lI_a r \tag{2.68}$$

where r is the radius of the armature conductor about the centre of rotation,
  I_a is the current flowing in the armature conductor
  l is the axial length of the conductor, and
  B_{av} is the average flux density under a pole. Note that

$$B_{av} = \frac{\Phi}{(2\pi rl)/2p}$$

The resultant torque per conductor is

$$T = \frac{\Phi 2plI_a r}{2\pi rl} = \frac{\Phi pI_a}{\pi}$$

For Z_s armature conductors connected in series the total torque on the armature is

$$T = \frac{\Phi pI_a Z_s}{\pi} \text{ Newton-metres} \tag{2.69}$$

2.2.1.3 Terminal voltage

Denoting the terminal voltage by V, the induced e.m.f. by E and the armature resistance by R_a,

$$V = E - I_a R_a \quad \text{(for a generator)} \tag{2.70}$$

$$V = E + I_a R_a \quad \text{(for a motor)} \tag{2.71}$$

For the motor, the induced e.m.f. is often called the 'back e.m.f.'.

2.2.2 Methods of connection

The methods of connecting the field and armature windings may be grouped as follows:

1. *Separately excited* – where the field winding is connected to a source of supply independently of the armature supply;
2. *Self-excited* – which may be further sub-divided into:
    (a) *Shunt-wound* – where the field winding is connected across the armature terminals;
    (b) *Series-wound* – where the field winding is connected in series with the armature winding;
    (c) *Compound-wound* – which is a combination of shut and series windings.

The four alternative methods of connection are illustrated in Figure 2.27.

2.2.3 The separately excited generator

Consider the separately excited generator, shown in Figure 2.27(a), running at a constant rated speed with no load across the output. It is assumed that initially the poles were completely de-magnetized. If the field current, and hence the magnetic field, is gradually increased then a plot of terminal voltage against field current takes the form shown in Figure 2.28.

As the field current increases, the iron poles begin to saturate and the proportionality between the flux and the field current no longer exists. If the field current is then reduced,

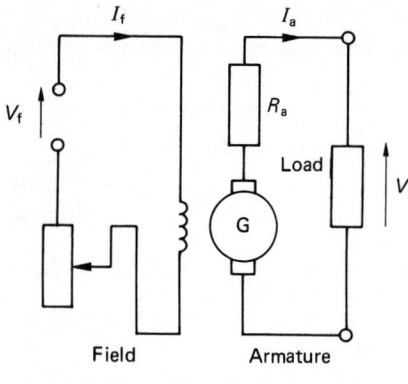

(a) Separately excited

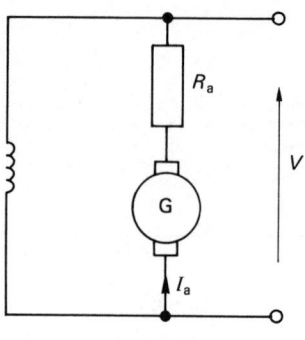

(b) Shunt-wound

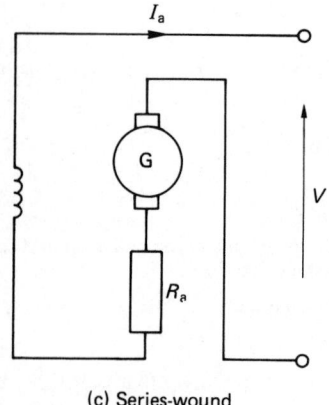

(c) Series-wound

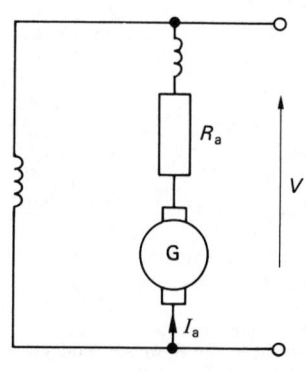

(d) Compound-wound

Figure 2.27 Methods of field connection

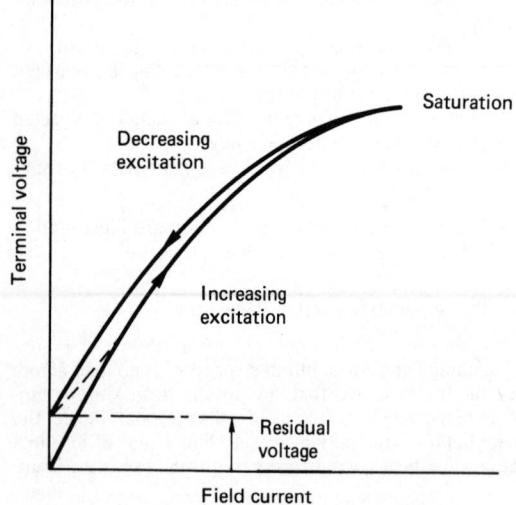

Figure 2.28 Open-circuit characteristics of a separately excited generator

the magnetic hysteresis causes the terminal voltage to have a slightly greater value than that obtained when the field current was being increased. When the field current is reduced to zero, a 'residual voltage' remains. On increasing the field current once more, the curve follows the broken line to merge with the original lower curve. These curves are termed the 'open-circuit characteristics' of the machine.

If the generator is now connected to a variable external load and driven at constant speed with a constant field current, I_f, the terminal voltage variation with armature current is as shown in Figure 2.29. The decrease in terminal voltage with increase in load is due mainly to the voltage drop across the armature resistance, R_a. Additionally, the decrease in terminal voltage is attributed to a decrease in flux caused both by the de-magnetizing ampere-turns of the armature and the magnetic saturation in the armature teeth. These effects are collectively known as 'armature reaction'. Figure 2.29 is referred to as the 'load characteristic' of the generator.

The separately excited generator has the disadvantage inherent with a separate source of direct current required for the field coils. They are, however, used in cases where a wide range in terminal voltage is required.

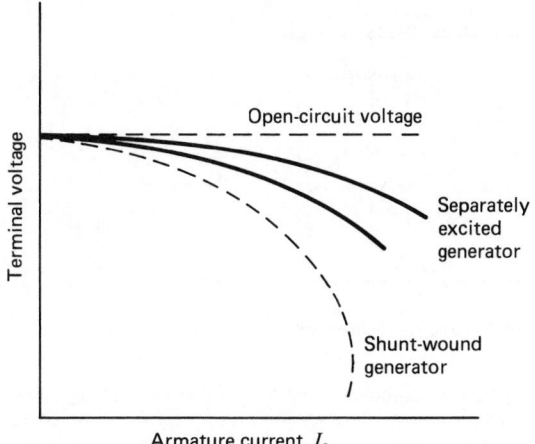

Figure 2.29 Load characteristic of a separately excited generator

2.2.4 The shunt-wound generator

The field winding in the shunt-wound generator is connected across the armature terminals as shown in Figure 2.27(b) and is therefore in parallel (or 'shunt') with the load. A shunt generator will excite only if the poles have some residual magnetism and the resistance of the shunt circuit is less than some critical value.

If, when running at constant speed, the field is disconnected from the armature, the voltage generated across the armature brushes is very small and entirely due to the residual magnetism in the iron. When the field is connected, the small residual voltage generates a flow of current in the field winding. The total flux in the field winding will gradually build up and the final terminal voltage will depend on the resistance of the field winding and the magnetization curve of the machine. The general characteristic is shown in Figure 2.30.

When connected to an external load the shunt-wound generator exhibits a drop in terminal voltage as the armature current is increased (see Figure 2.29). The drop in voltage in the shunt-wound generator is much greater than that in the separately excited generator. This stems from the fact that, as the terminal voltage drops, the field current also reduces, which causes a further drop in terminal voltage.

The shunt-wound machine is the most common type of d.c. generator employed. The load current, however, must be limited to a value well below the maximum value to avoid excessive variation in terminal voltage.

2.2.5 The series-wound generator

For the series-wound generator the field winding is connected in series with the armature terminals as shown in Figure 2.27(c). The armature current therefore determines the flux. The constant speed load characteristic (Figure 2.31) exhibits an increase in terminal voltage as the armature (or load) current increases.

At large values of load current the armature resistance and reactance effects cause the terminal voltage to decrease. It is apparent from Figure 2.31 that the series-wound generator is totally unsuitable if the terminal voltage is required to be reasonably constant over a wide range of load current.

2.2.6 The compound-wound generator

The compound-wound generator (Figure 2.27(d)) is a hybrid between the shunt- and the series-wound generators. Normally, a small series field is arranged to assist the main shunt field. This is termed 'cumulative compounding'. The shape of the load characteristic (Figure 2.32) depends upon the number of turns on the series winding. If the series field is arranged to oppose the main shunt field ('differentially compounded') a rapidly falling load characteristic is obtained. The number of turns on the series coil can be varied to give an over-compounded, level-compounded or an under-compounded characteristic as shown in Figure 2.32.

2.2.7 The d.c. motor

There is no difference in basic construction between a d.c. generator and a d.c. motor. The only significant distinction between the two machines is quantified by equations (2.70) and (2.71). These illustrate the fact that, for a d.c. generator, the generated e.m.f. is greater than the terminal voltage. For the d.c. motor, the generated e.m.f. is less than the terminal voltage.

Equation (2.65), which gives the relationship between the induced e.m.f. and the speed of a d.c. generator, applies

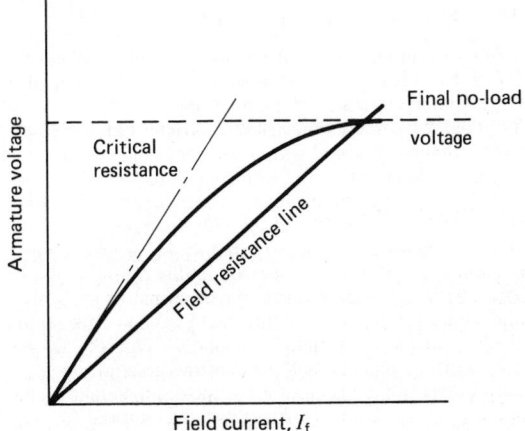

Figure 2.30 No-load characteristic of a shunt-wound generator

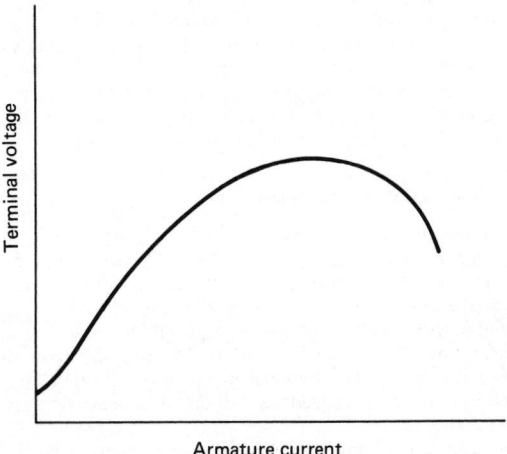

Figure 2.31 Constant speed load characteristic for the series-wound generator

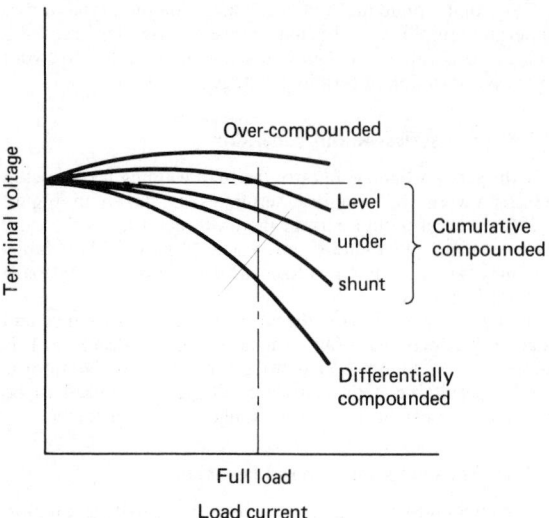

Figure 2.32   Load characteristic for the compound-wound generator

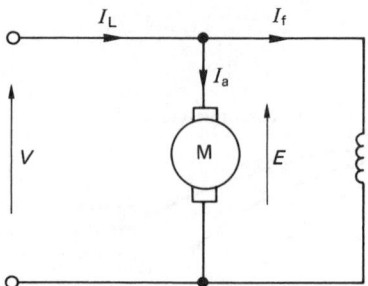

Figure 2.33   The shunt-wound motor

equally well to the d.c. motor. Since the number of poles and number of armature conductors are fixed, a proportionality relationship can be derived to relate speed as a function of induced e.m.f. and flux, i.e.

$$N = E/\phi \qquad (2.72)$$

or, using equation (2.71),

$$N = (V - I_a R_a)/\Phi \qquad (2.73)$$

The value of $I_a R_a$ is usually less than about 5% of the terminal voltage such that, to a reasonable approximation,

$$N \approx V/\Phi \qquad (2.74)$$

Similarly, equation (2.69), which gives the armature torque on a d.c. generator, also applies to the d.c. motor. A proportionality relationship for the d.c. motor torque is therefore

$$T = I_a \Phi \qquad (2.75)$$

Equation (2.74) shows that the speed of a d.c. motor is approximately proportional to the voltage applied to the armature and inversely proportional to the flux. All methods of controlling the speed of d.c. motors are based on these proportionality relationships. Equation (2.75) indicates that the torque of a given d.c. motor is directly proportional to the product of the armature current and the flux per pole.

2.2.8   The shunt-wound motor

The shunt-wound motor is shown schematically in Figure 2.33. Under normal operating conditions the field current will be constant. As the armature current increases, however, the armature reaction effect will weaken the field and the speed will tend to increase. The induced voltage will decrease due to the increasing armature voltage drop, and this will tend to decrease the speed. The two effects are not self-cancelling, and, overall, the motor speed will fall slightly as the armature current increases.

The motor torque increases approximately linearly with the armature current until the armature reaction starts to weaken the field. These general characteristics are shown in Figure 2.34, along with the derived torque–speed characteristic.

Figure 2.34(a) shows that no torque is developed until the armature current is large enough to supply the constant losses in the machine. Since the torque increases significantly for a slight decrease in speed, the shunt-wound motor is particularly suitable for driving equipment such as pumps, compressors and machine tool elements, where the speed must remain 'constant' over a wide range of load.

2.2.9   The series-wound motor

The series-wound motor is shown in Figure 2.35. As the load current increases, the induced voltage, E, will decrease due to reductions in the armature and field resistance voltages. Because the field winding is connected in series with the armature the flux is directly proportional to the armature current. Equation (2.74) therefore suggests that the speed/armature current characteristic will take the form of a rectangular hyperbola. Similarly, equation (2.75) indicates that the torque/armature current characteristic will be approximately parabolic. These general characteristics are illustrated in Figure 2.36 along with the derived torque–speed characteristic.

The general characteristics indicate that if the load falls to a particularly low value then the speed may become dangerously high. A series-wound motor should therefore never be used in situations where the load is likely to be suddenly relaxed.

The main advantage of the series-wound motor is that it provides a large torque at low speeds. These motors are eminently suitable, therefore, for applications where a large starting torque is required. This includes, for example, lifts, hoists, cranes and electric trains.

2.2.10   The compound-wound motor

Compound-wound motors, like compound generators, are produced by including both series and shunt fields. The resulting characteristics of the compound-wound motor fall somewhere in between those of the series- and the shunt-wound machines.

2.2.11   Starting d.c. motors

With the armature stationary, the induced e.m.f. is zero. If, while at rest, the full voltage is applied across the armature winding, the current drawn would be massive. This current would undoubtedly blow the fuses and thereby cut off the supply to the machine. To limit the starting current, a variable external resistance is connected in series with the armature. On start-up the full resistance, is connected in series. As the machine builds up speed and increases the back e.m.f., the external resistance can be reduced until the series resistance is disconnected at rated speed.

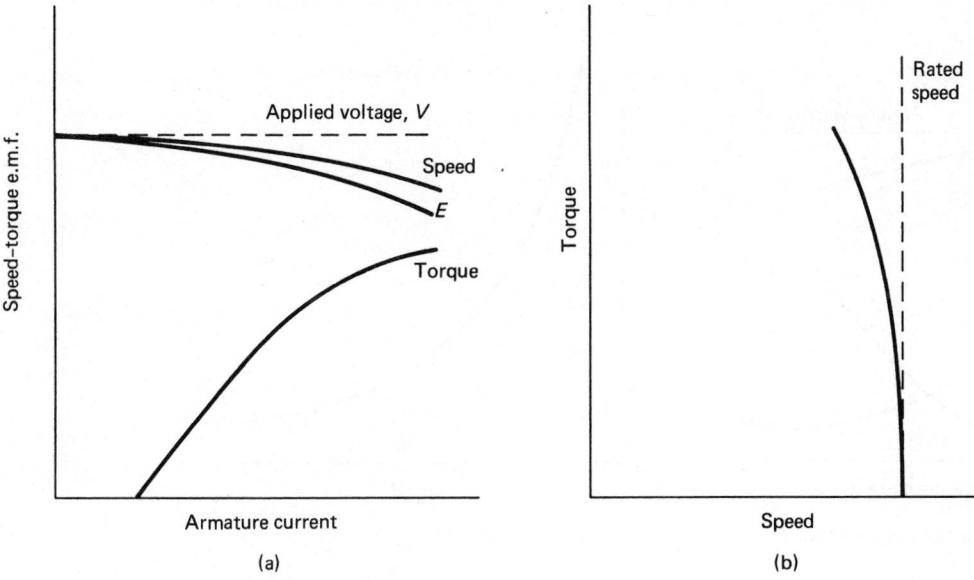

Figure 2.34 The shunt-wound motor load characteristics

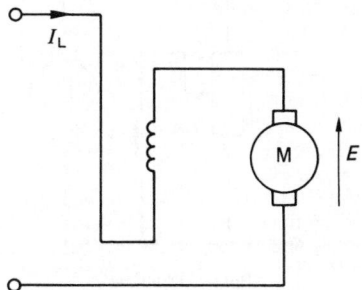

Figure 2.35 The series-wound motor

Variable-resistance 'starters' are also usually equipped with a return spring and an electromagentic 'catch plate'. The latter keeps the starter in the zero resistance position while the machine is running at its rated speed. The electromagnet is powered by the field current and, in the event of a supply failure, the electromagnet is de-energized and the return spring pulls the starter back to the full-resistance 'off' position. This ensures that the full starting resistance will always be in series with the armature winding when the machine is re-started.

An overload cut-out switch is another normal feature incorporated into the starter mechanism. The overload cut-out is another electromagnetic switch which this time is powered by the supply current. The overload switch is normally 'off', but if the supply current becomes excessive, the switch is activated and it short-circuits the supply to the electromagnetic catch plate. This, in turn, de-energizes the catch plate and the return spring takes the starter back to the 'off' position. Figure 2.37 illustrates the essential features of a starter device for a shunt-wound motor.

2.2.12   Speed control of d.c. motors

Equation (2.74) shows that the speed of a d.c. motor is influenced both by the applied voltage and the flux. A variation in either of these parameters will therefore effect a variation in the motor speed.

2.2.12.1   Field regulator

For shunt- and compound-wound motors a variable resistor, called a 'field regulator', can be incorporated in series with the field winding to reduce the flux. For the series-wound motor the variable resistor is connected in parallel with the field winding and is called a 'diverter'. Figure 2.38 shows the various methods of weakening the field flux for shunt-, compound- and series-wound motors.

In all the above methods of speed control the flux can only be reduced, and from equation (2.74) this implies that the speed can only be increased above the rated speed, and may, in fact, be increased to about three or four times the rated speed. The increased speed, however, is at the expense of reduced torque, since the torque is directly proportional to the flux which is reduced.

2.2.12.2   Variable armature voltage

Alternatively, the speed can be increased from standstill to rated speed by varying the armature voltage from zero to rated value. Figure 2.39 illustrates one method of achieving this.

The potential divider, however, carries the same current as the motor, and this limits this method of speed control to small machines. Additionally, much of the input energy is dissipated in the controller, which consequently renders the system inefficient.

2.2.12.3   Ward Leonard drive

In this case the variable d.c. voltage for the speed-controlled motor is obtained from a separate d.c. generator which is itself driven by an induction motor (see Figure 2.40). The field coil for the d.c. generator is supplied from a centre-tapped potential divider. When the wiper arm is moved from O to A the armature voltage of the d.c. motor is increased from zero and the motor speed will rise. In moving the wiper from A to O

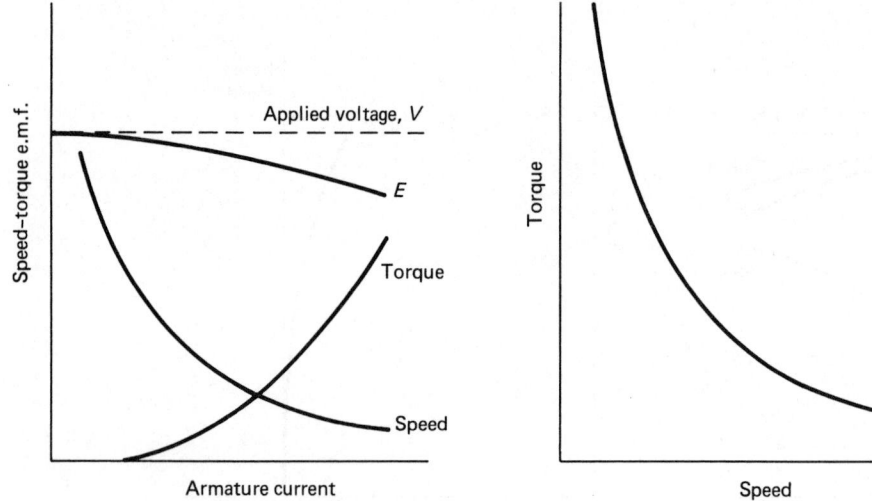

Figure 2.36 The series-wound motor load characteristics

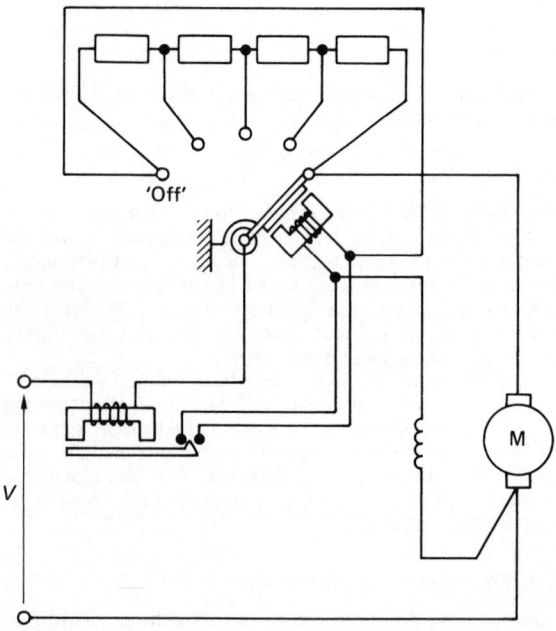

Figure 2.37 Starter-device for d.c. motors

and on through to B the motor will decelerate to a standstill and then increase in speed again, but in the opposite direction of rotation. The Ward Leonard drive is smooth and accurate in either direction and also provides for very responsive braking. Its complexity, however, makes it a very expensive system, and it is only used in high-quality applications.

2.2.12.4 Chopper control

Figure 2.41 shows a thyristor circuit connected in series with the armature of a d.c. motor. The thyristor circuit is triggered such that it operates essentially as a high-speed on/off switch. The output waveform across the armature terminals is depicted in Figure 2.42. The ratio of time on to time off (i.e. the

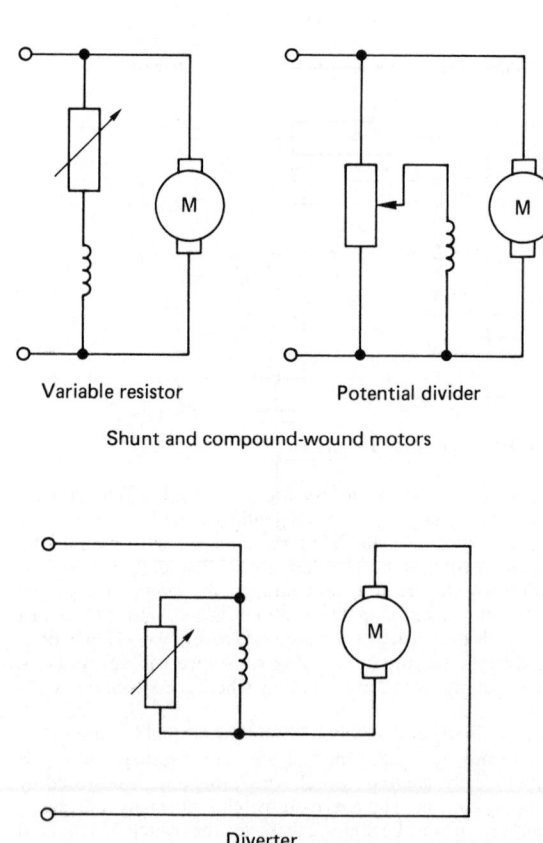

Shunt and compound-wound motors

Series-wound motor

Figure 2.38 Speed control for flux reduction

'mark/space ratio') can be varied, with the result that the average voltage supplied to the armature is effectively between zero and fully on. The frequency of the signal may be up to about 3 kHz and the timing circuit is necessarily complex.

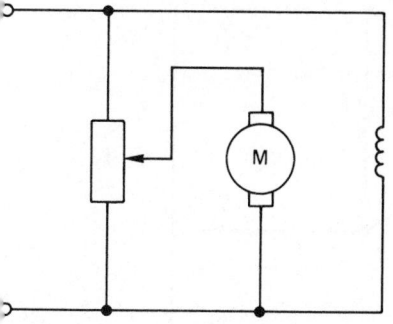

Figure 2.39 Speed control by varying armature voltage

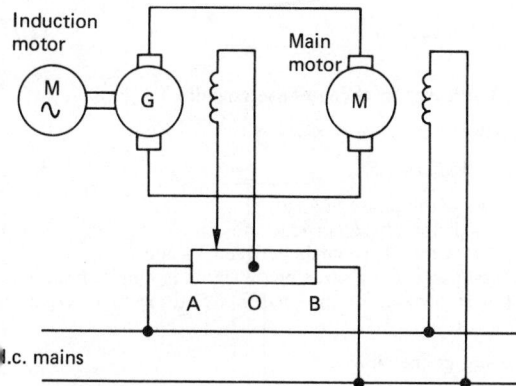

Figure 2.40 Ward Leonard drive

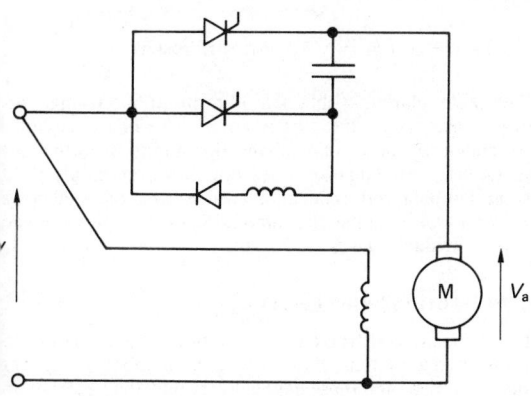

igure 2.41 Speed control using thyristors

peed control of d.c. motors using thyristors, is, however, ffective and relatively inexpensive.

.2.13 Efficiency of d.c. machines

he losses in d.c. machines can be generally classified as:

Armature losses,
ron loss,
'ommutator losses,
:xcitation loss, and
'earing friction and windage

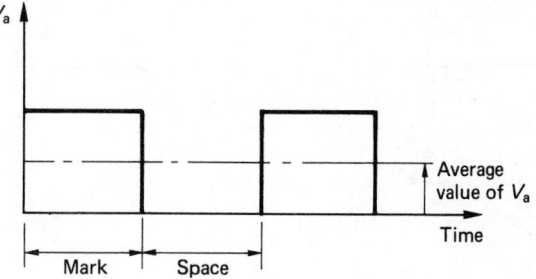

Figure 2.42 Voltage across armature terminals

Despite the variety and nature of the losses associated with d.c. machines, they have, nonetheless, a very good performance with overall efficiencies often in excess of 90%.

2.2.14 Three-phase circuits

Since a.c. machines are generally associated with three-phase systems it is necessary to consider some aspects of three-phase circuits before a meaningful discussion of a.c. machines can be undertaken. The limiting factor of a d.c. machine is related to the commutator which restricts the maximum voltage that can be generated. Because of their efficiency and performance, three-phase machines have emerged as the dominant type of electrical generator and motor and, on a worldwide basis, three-phase electrical distribution networks are the norm.

2.2.15 Generation of three-phase e.m.f.'s

Figure 2.43 shows three similar coils displaced at 120° relative to each other. Each loop terminates in a pair of 'slip-rings' and if the coils are to be isolated from one another, then six slip-rings are required in total. If the three coils are rotated in the anti-clockwise direction at constant speed, then each coil will generate a sinusoidally varying e.m.f. with a phase shift of 120° between them.

2.2.16 Star and delta connections

The three coils shown in Figure 2.43 can be connected together in either of two symmetrical patterns. These are the 'star' (or 'wye') connection and the 'delta' (or 'mesh') connection. The two types of connection are shown in Figure 2.44.

The star pattern is made by joining R_0, Y_0 and B_0 together. This connection point is referred to as the 'neutral point'. The delta pattern is formed by connecting R_0 to Y_1, Y_0 to B_1 and B_0 to R_1.

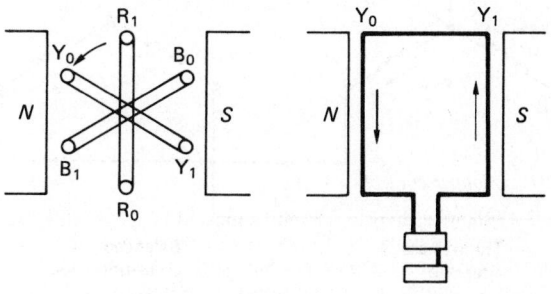

Figure 2.43 Generation of three-phase e.m.f.'s

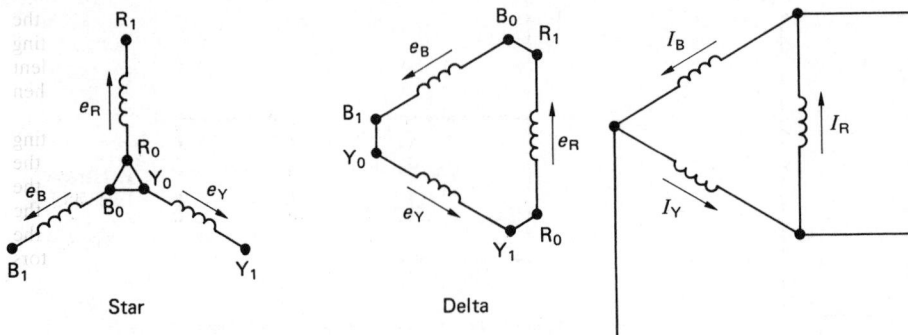

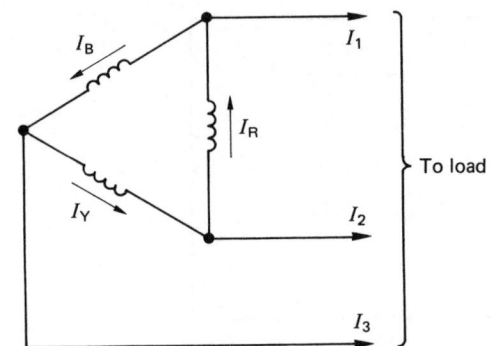

Figure 2.44   Star and delta connections for three-phase systems

Figure 2.46   Alternator windings in delta connection

2.2.17   Three-phase voltage and current relations

Figure 2.45 shows a three-phase star connected alternator supplying currents I_R, I_Y and I_B to a balanced (or equal) resistive–inductive load. This gives the usual 'four-wire' star-connected system. Since there are only four transmission cables involved, the alternator connected in a star pattern will only require four slip-rings.

For a balanced system the phase voltages V_{RN}, V_{YN} and V_{BN} are all equal in magnitude and equally displaced by a phase angle of 120°. The currents I_R, I_Y and I_B are also equal in magnitude and equally displaced in phase angle but they all lag their respective phase voltages by some angle ϕ. Phasor addition of the currents shows that the neutral current, I_N, is zero.

The voltages between the transmission cables are called the 'line' voltages. If the phase voltages are all equal then phasor addition shows that the line voltages are given by

$$V_{line} = 2V_{phase} \cos(30)$$

or

$$V_L = \sqrt{3} \times V_P \qquad (2.76)$$

For the star connection, the line currents, I_L, are equal to the phase currents, I_P. Figure 2.46 shows the alternator windings connected in the delta pattern. In this pattern the line voltages are equal to the phase voltages. Phasor addition of the currents shows that if the phase currents are equal then the line currents are given by

$$I_L = \sqrt{3} \times I_p \qquad (2.77)$$

2.2.18   Power in three-phase circuits

The power per phase is given by

$$P_{phase} = V_P I_P \cos(\phi) \qquad (2.78)$$

where V_P is the phase voltage,
   I_P is the phase current, and
   ϕ is the phase angle between V_P and I_P.
The total power for a three-phase circuit is simply three times the power for one of the phases, i.e. three times equation (2.78).

For a star connection:

$$P = 3 \frac{V_L}{\sqrt{3}} I_L \cos(\phi) = \sqrt{3} \times V_L \times I_L \cos(\phi) \qquad (2.79)$$

For a delta connection:

$$P = 3V_L \frac{I_L}{\sqrt{3}} \cos(\phi) = \sqrt{3} \times V_L \times I_L \cos(\phi)$$

The same relation is obtained. In terms of line voltages and currents therefore, the power in a three-phase circuit is independent of the winding connection and is given by equation (2.79). This equation does not, however, apply if the system is unbalanced. In an unbalanced system the total power can only be obtained as the summation of the powers in each of the individual phases.

2.2.19   Three-phase alternators

Alternators are constructed with a stationary a.c. winding and a rotating field system. This reduces the number of slip-rings required to two, and these have to carry only the field-exciting current as opposed to the generated current. The construction is thereby simplified and the slip-ring losses are minimized. In addition, the simpler arrangement enables heavier insulation to be used and, in consequence, much higher voltages can be generated. The robust mechanical construction of the rotor also means that higher speeds are possible and substantially higher power outputs can be generated with an alternator. A simple form of three-phase generator is depicted in Figure 2.47.

The three coils on the stator are displaced 120° and the rotor, which is a salient pole type, is supplied via the two slip-rings with a d.c. current. As the rotor is driven by some form of prime mover, a rotating magnetic field is established and the e.m.f.'s generated in the coils will be displaced with

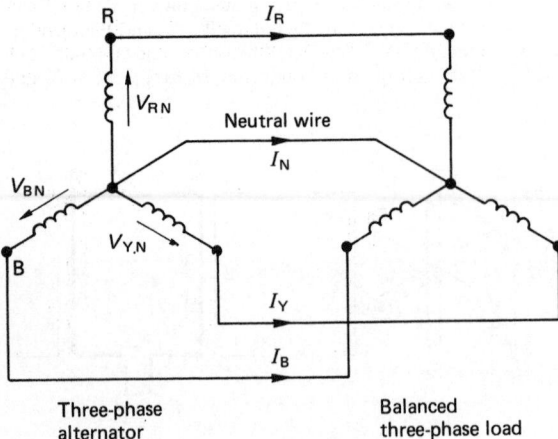

Three-phase
alternator

Balanced
three-phase load

Figure 2.45   Three-phase supply connections

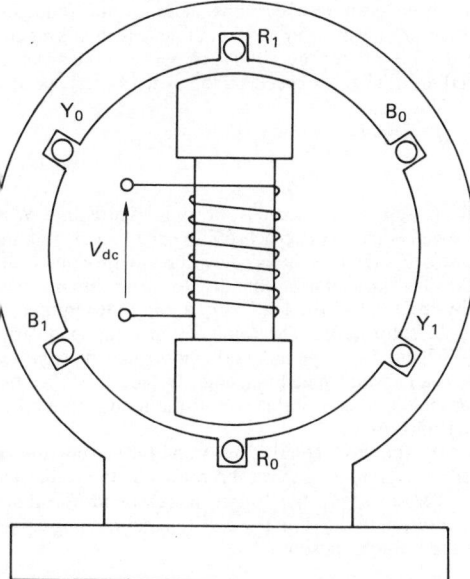

igure 2.47   Simple three-phase generator

where N_s is the speed of the field (rev/min) and f is the frequency of the supply currents. The speed of the rotating field is termed the 'synchronous speed' and for an equivalent single pair of poles (i.e. three coils) this is 3000 rev/min when the frequency of the supply currents is at 50 Hz.

The use of a.c. excited rotor coils to produce the rotating magnetic field simplifies the mechanical construction of the rotor and greatly facilitates the dynamic balancing of the machine. An added advantage is that the waveform of the generated voltage is improved. The a.c. method of exciting the field is used extensively in large alternators. Salient pole rotors are normally restricted to the smaller machines.

2.2.20   Synchronous motors

Synchronous motors are so called because they operate at only one speed, i.e. the speed of the rotating field. The mechanical construction is exactly the same as the alternator shown in Figure 2.47. The field is supplied from a d.c. source and the stator coils with a three-phase current. The rotating magnetic field is induced by the stator coils and the rotor, which may be likened to a permanent bar magnet, aligns itself to the rotating flux produced in the stator. When a mechanical load is driven by the shaft the field produced by the rotor is pulled out of alignment with that produced by the stator. The angle of misalignment is called the 'load angle'. The characteristics of synchronous motors are normally presented in terms of torque against load angle, as shown in Figure 2.48. The torque characteristic is basically sinusoidal, with

$$T = T_{max} \sin(\delta) \tag{2.81}$$

where T_{max} is the maximum rated torque and δ is the load angle.

It is evident from equation (2.81) that synchronous motors have no starting torque and the rotor must be run up to synchronous speed by some alternative means. One method utilizes a series of short-circuited copper bars inserted through the outer extremities of the salient poles. The rotating magnetic flux induces currents in these 'grids' and the machine accelerates as if it were a cage-type induction motor (see

hase shift of 120°. The magnitude of the generated voltages re dependent on the flux produced by the rotor, the number f turns on the stator coils and the speed of rotation of the otor. The rotor speed will also dictate the frequency of the enerated voltage.

The no-load and load characteristics of an alternator are ery similar to those of the d.c. separately excited generator Figures 2.28 and 2.29, respectively). In constant speed opera- ion the terminal voltage exhibits a drooping characteristic, vhere the decrease in terminal voltage is due to 'armature' esistance and reactance effects. For an alternator, the term armature' is taken to imply the stator windings.

As the load on an alternator is increased, the speed of the rime mover will drop. This is an unacceptable situation, ecause the speed controls the frequency of the generated oltage. To maintain a constant frequency, the prime mover nust be governed to run at constant speed over the entire ange of expected loads. This is particularly important where nany alternators are to be run in parallel to supply a distribu- ion system such as the National Grid. In such cases the prime novers are always speed controlled and the output voltage is egulated to comply with the rated values. In the UK, lternators are usually two-pole machines driven at 3000 rev/ nin to produce the rated frequency of 50 Hz. In the USA a reat deal of the electrical power consumed is generated from ydroelectric power stations. The water turbines used in these nstallations are fairly low-speed machines and the alternators, vhich are directly driven, are equipped with multiple poles to roduce the rated frequency of 60 Hz. An alternator running t 240 rev/min, for example, must have 30 poles to give the ated output frequency.

The production of the rotating magnetic field may also be ctivated using three, 120° displaced, rotor coils supplied with hree-phase current. The rotational speed of the field is elated to the frequency of the currents, i.e.

$$_s = \frac{f \times 60}{\text{Number of pole pairs}} \tag{2.80}$$

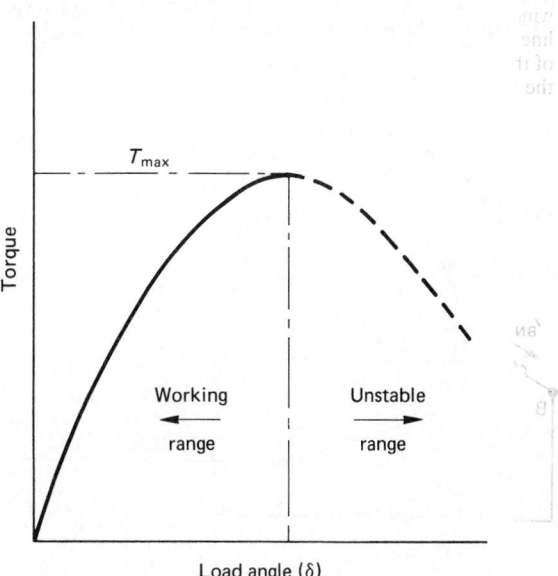

Figure 2.48   Torque characteristic for a synchronous motor

Section 2.2.21). A second method uses a wound rotor similar to a slip-ring induction motor. The machine is run up to speed as an induction motor and is then pulled into synchronism to operate as a synchronous motor.

The advantages of the synchronous motor are the ease with which the power factor can be controlled and the constant rotational speed of the machine, irrespective of the applied load. Synchronous motors, however, are generally more expensive and a d.c. supply is a necessary feature of the rotor excitation. These disadvantages, coupled with the requirement for an independent starting mode, make synchronous motors much less common than induction ones.

2.2.21 Induction motors

The stator of an induction motor is much like that of an alternator and, in the case of a machine supplied with three-phase currents, a rotating magnetic flux is produced. The rotor may be either of two basic configurations: the 'squirrel-cage' or the slip-ring type. In the squirrel-cage motor the rotor core is laminated and the conductors consist of uninsulated copper (or aluminium) bars driven through the rotor slots. The bars are brazed or welded at each end to rings or plates to produce a completely short-circuited set of conductors. The slip-ring machine has a laminated core and a conventional three-phase winding, similar to the stator and connected to three slip-rings on the locating shaft.

Figure 2.49 shows a schematic representation of an induction motor having three stator coils displaced by 120°. If the stator coils are supplied with three-phase currents a rotating magnetic field is produced in the stator. Consider the single-rotor coil shown in the figure. At standstill the rotating field will induce a voltage in the rotor coil since there is a rate of change of flux linking the coil. If the coil forms a closed circuit then the induced e.m.f. will circulate a current in the coil. The resultant force on the current-carrying conductor is a consequence of equation (2.27) and this will produce a torque which will accelerate the rotor. The rotor speed will increase until the electromagnetic torque is balanced by the mechanical load torque. The induction motor will never attain synchronous speed because, if it did, there would be no relative motion between the rotor coils and the rotating field. Under these circumstances there would be no e.m.f. induced in the rotor

coils and subsequently no electromagnetic torque. Induction motors therefore always run at something less than synchronous speed. The ratio of the difference between the synchronous speed and the rotor speed to the synchronous speed is called the 'slip', s, i.e.

$$s = \frac{N_s - N}{N_s} \qquad (2.82)$$

The torque–slip characteristic is shown in Figure 2.50. With the rotor speed equal to the synchronous speed, i.e. $s = 0$, the torque is zero. As the rotor falls below the synchronous speed the torque increases almost linearly to a maximum value dictated by the total of the load torque and that required to overcome the rotor losses. The value of slip at full load varies between 0.02 and 0.06. The induction motor may be regarded therefore as a constant-speed machine. In fact the difficulties of varying the speed constitutes one of the induction motor main disadvantages.

On start-up, the slip is equal to unity and the starting torque is sufficiently large to accelerate the rotor. As the rotor runs up to its full-load speed the torque increases in essentially inverse proportion to the slip. The start-up and running curves merge at the full-load position.

2.2.22 Starting induction motors

As with d.c. motors, the current drawn during starting of a.c. motors is very large, up to about five times full-load current. A number of devices are therefore employed to limit the starting current but they all involve the use of auxiliary equipment which is usually quite expensive.

2.2.22.1 Star–delta starter

The star–delta switch (Figure 2.51) is the cheapest and most common method employed. With the machine at standstill and the starter in the 'start' position, the stator coils are connected in the star pattern. As the machine accelerates up to running speed the switch is quickly moved over to the 'run' position, which reconnects the stator windings in the delta pattern. By this simple expedient the starting supply current is reduced to one third of what it would have been had the stator windings been connected in the delta pattern on start-up.

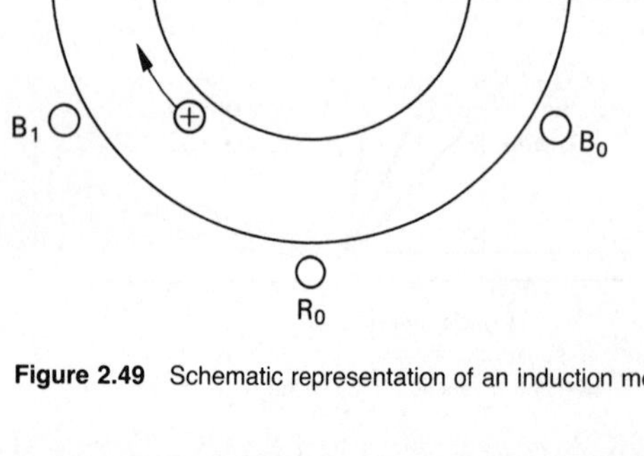

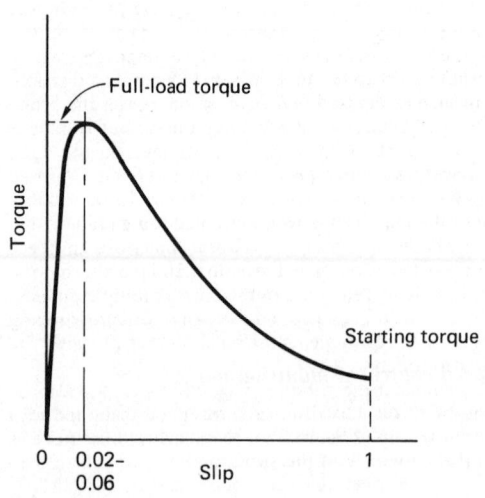

Figure 2.49 Schematic representation of an induction motor

Figure 2.50 Torque–slip characteristic for an induction motor

hree-phase supply

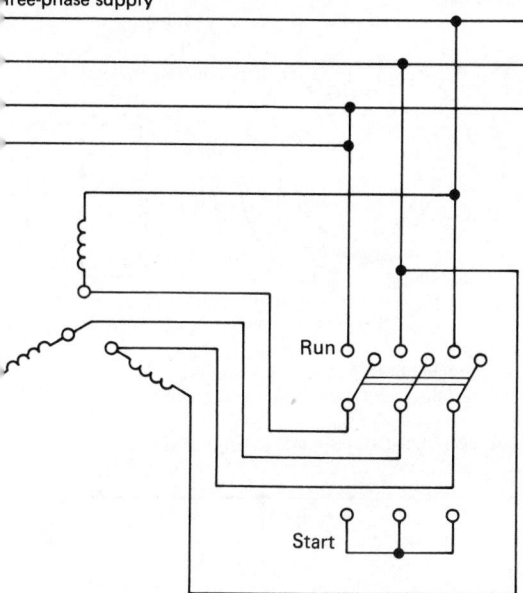

Run

Start

ure 2.51 Star–delta starter

2.22.2 *Auto-transformer starter*

e auto-transformer represents an alternative method of ducing the starting current drawn by an induction motor.

2.22.3 *Rotor resistance*

ith slip-ring induction motors it is possible to include ditional resistance in series with the rotor circuit. The clusion of extra resistance in the rotor provides for reduced rting current and improved starting torque.

2.23 Braking induction motors

duction motors may be brought to a standstill by either ugging' or 'dynamic braking':

Plugging: This refers to the technique where the direction of the rotating magnetic field is reversed, and is brought about by reversing any two of the supply leads to the stator. The current drawn during plugging is, however, very large and machines which are regularly plugged must be specially rated.

Dynamic braking: In this technique the stator is disconnected from the a.c. supply and reconnected to a d.c. source. The direct current in the stator produces a stationary uni-directional field and, as the rotor will always tend to align itself with the field, it will come to a standstill.

2.24 Speed control of induction motors

der normal circumstances the running speed of an induc-n motor will be about 94–98% of the synchronous speed, pending on the load. With the synchronous speed given by ation (2.80) it is clear that the speed may be varied by nging either the frequency of the supply current or the nber of poles.

2.2.24.1 *Change of supply current frequency*

Solid state variable-frequency drives first began to appear in 1968. They were originally applied to the control of synchronous a.c. motors in the synthetic fibre industry and rapidly gained acceptance in that particular market. In more recent times they have been used in applications such as pumping, synchronized press lines, conveyor lines and, to a lesser extent, in the machine-tool industry as spindle drives. Modern a.c. variable-frequency motors are available in power ratings ranging from 1 kW to 750 kW and with speed ranges from 10/1 to 100/1.

2.2.24.2 *Change of number of poles*

By bringing out the ends of the stator coils to a specially designed switch it becomes possible to change an induction motor from one pole configuration to another. To obtain three different pole numbers, and hence three different speeds, a fairly complex switching device would be required.

Changing the number of poles gives a discrete change in motor speed, with little variation in speed over the switched range. For many applications, however, two discrete speeds are all that is required and changing the number of poles is a simple and effective method of achieving this.

2.2.24.3 *Changing the rotor resistance*

For slip-ring induction motors additional resistance can be coupled in series with the rotor circuit. It has already been stated that this is a common method used to limit the starting current of such machines. It can also be employed for marginal speed control. Figure 2.52 shows the torque characteristics of a slip-ring induction motor for a range of different resistances connected in series with the rotor windings. As the external resistance is increased from R_1 to R_3 a corresponding reduction in speed is achieved at any particular torque. The range of speeds is increased at the higher torques.

The method is simple and therefore inexpensive, but the decrease in speed is accompanied with a reduction in overall efficiency. Additionally, with a large resistance in the rotor circuit (i.e. R_3) the speed changes considerably with variations in torque.

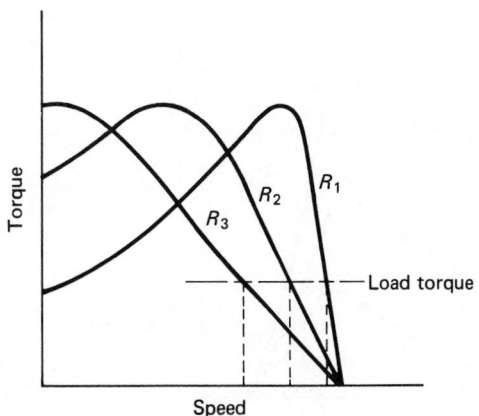

Figure 2.52 Torque–speed characteristics for various rotor resistances

2.2.24.4 Reduced stator voltage

By reducing the applied stator voltage a family of torque–speed characteristics are obtained, as shown in Figure 2.53. It is evident that as the stator voltage is reduced from V_1 to V_3, a change in speed is effected at any particular value of torque. This is provided, of course, that the torque does not exceed the maximum load torque available at the reduced stator voltage. This latter point is obviously a limiting factor which places a constraint on this method of speed control. Generally, only very small speed ranges can be obtained using a variable stator supply voltage.

2.2.25 Single-phase induction motors

The operation of an induction motor depends upon the creation of a rotating magnetic field. A single stator coil cannot achieve this, and all the so-called single-phase induction motors use some or other external means of generating an approximation to a two-phase stator supply. Two stator coils are therefore used and these are displaced by 90°. Ideally, the currents which supply each coil should have a phase difference of 90°. This then gives the two-phase equivalent of the three-phase induction motor.

2.2.25.1 The shaded-pole motor

The stator of the shaded-pole motor consists of a salient pole single-phase winding and the rotor is of the squirrel-cage type (see Figure 2.54). When the exciting coil is supplied with alternating current the flux produced induces a current in the 'shading ring'. The phase difference between the currents in the exciting coil and the shading ring is relatively small and the rotating field produced is far from ideal. In consequence, the shaded-pole motor has a poor performance and an equally poor efficiency due to the continuous losses in the shading rings.

Shaded-pole motors have a low starting torque and are used only in light-duty applications such as small fans and blowers or other easily started equipment. Their advantage lies in their simplicity and low cost of manufacture.

2.2.25.2 The capacitor motor

A schematic layout of a capacitor motor is given in Figure 2.55. The stator has two windings physically displaced by 90°.

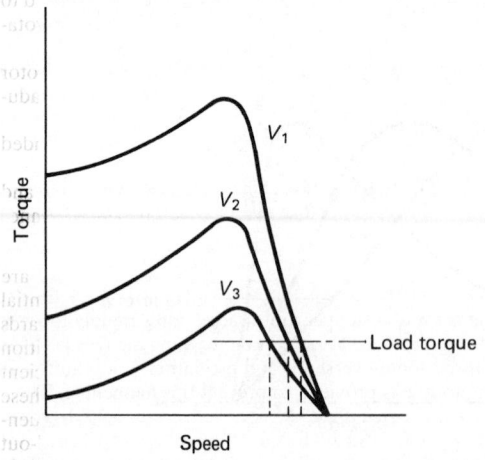

Figure 2.53 Torque–speed characteristics for various stator voltages

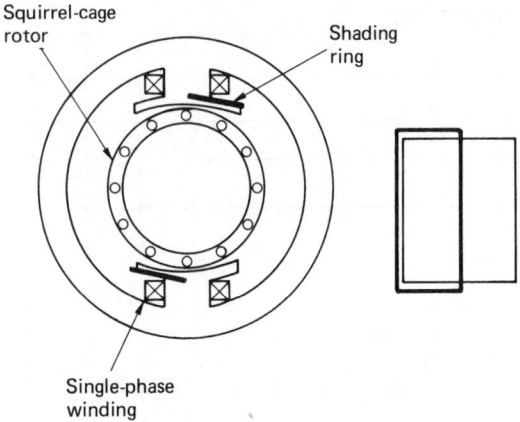

Figure 2.54 Shaded pole motor

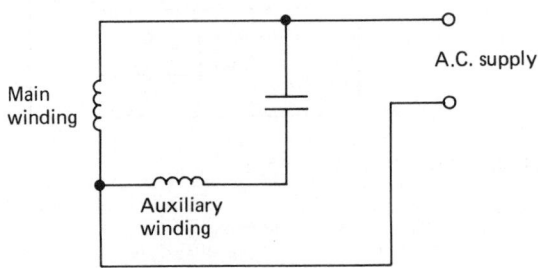

Figure 2.55 Capacitor motor

A capacitor is connected in series with the auxiliary winding such that the currents in the two windings have a large phase displacement. The current phase displacement can be made to approach the ideal 90°, and the performance of the capacitor motor closely resembles that of the three-phase induction motor.

2.2.25.3 The universal motor

These are small d.c. series-wound motors which operate at about the same speed and power on direct current, or on single-phase current with approximately the same root mean square voltage. The universal (or plain-series) motor is used mainly in small domestic appliances such as hair dryers, electric drills, vacuum cleaners, hedge trimmers, etc.

2.2.26 The d.c. permanent magnet (PM) motor

The d.c. permanent magnet (PM) motor is a continuous rotation electromagnetic actuator which can be direct coupled to its load. Figure 2.56 shows the schematic representation of a d.c. PM motor. The PM motor consists of an annular brush ring assembly, a permanent magnet stator ring and a laminated wound rotor. It is particularly suitable for servo systems where size, weight, power and response time must be minimized and where high position and rate accuracies are required.

The response times for PM motors are very fast and the torque increases directly with the input current, independent of the speed or the angular position. Multiple-pole machines maximize the output torque per watt of rotor power. Commercial PM motors are available in many sizes from 35 mm

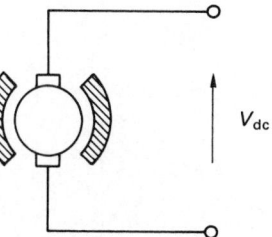

Figure 2.56 D.C. permanent magnet motor

Newton-metres at about 25 mm diameter to 13.5 Newton-metres at about 3 m diameter.

Direct-drive rate and position systems using PM motors utilize d.c. tachogenerators and position sensors in various forms of closed-loop feedback paths for control purposes.

2.2.27 The stepper motor

A stepper motor is a device which converts a d.c. voltage pulse train into a proportional mechanical rotation of its shaft. The stepper motor thus functions both as an actuator and as a position transducer. The discrete motion of the stepper motor makes it ideally suited for use with a digitally based control system such as a microcomputer.

The speed of a stepper motor may be varied by altering the rate of the pulse train input. Thus if a stepper motor requires 48 pulses to rotate through one complete revolution then an input signal of 96 pulses per second will cause the motor to rotate at 120 rev/min. The rotation is actually carried out in finite increments of time, but this is visually indiscernible at all but the lowest speeds.

Stepper motors are capable of driving a 2.2 kW load with stepping rates from 1000 to 20 000 per second in angular increments from 45° down to 0.75°. There are three basic types of stepper motor:

Variable reluctance: This type of stepper motor has a soft iron multi-toothed rotor with a wound stator. The number of teeth on the rotor and stator, together with the winding configuration and excitation, determines the step angle. This type of stepper motor provides small to medium-sized step angles and is capable of operation at high stepping rates.

Permanent magnet: The rotor used in the PM-type stepper motor consists of a circular permanent magnet mounted onto the shaft. PM stepper motors give a large step angle, ranging from 45° to 120°.

Hybrid: The hybrid stepper motor is a combination of the previous two types. Typically, the stator has eight salient poles which are energized by a two-phase winding. The rotor consists of a cylindrical magnet which is axially magnetized. The step angle depends on the method of construction and is generally in the range 0.9–5°. The most popular step angle is 1.8°.

The principle of operation of a stepper motor can be illustrated with reference to a variable-reluctance, four-phase machine. This motor usually has eight stator teeth and six rotor teeth (see Figure 2.57).

If phase 1 of the stator is activated alone then two diametrically opposite rotor teeth align themselves with the phase 1 teeth of the stator. The next adjacent set of rotor teeth in the clockwise direction are then 15° out of step with those of the stator. Activation of the phase 2 winding on its own would cause the rotor to rotate a further 15° in the anti-clockwise

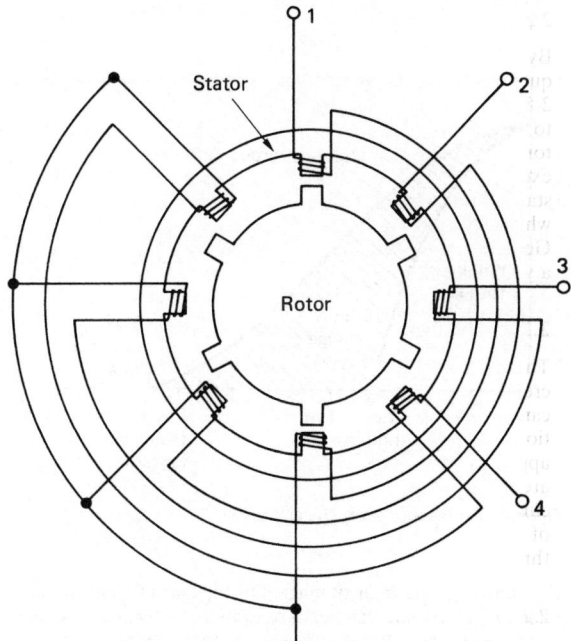

Figure 2.57 Variable-reluctance stepper motor

direction to align the adjacent pair of diametrically opposite rotor teeth. If the stator windings are excited in the sequence 1, 2, 3, 4 the rotor will move in consecutive 15° steps in the anti-clockwise direction. Reversing the excitation sequence will cause a clockwise rotation of the rotor.

2.2.27.1 Stepper motor terminology

Pull-out torque: The maximum torque which can be applied to a motor, running at a given stepping rate, without losing synchronism.

Pull-in torque: The maximum torque against which a motor will start, at a given pulse rate, and reach synchronism without losing a step.

Dynamic torque: The torque developed by the motor at very slow stepping speeds.

Holding torque: The maximum torque which can be applied to an energized stationary motor without causing spindle rotation.

Pull-out rate: The maximum switching rate at which a motor will remain in synchronism while the switching rate is gradually increased.

Pull-in rate: The maximum switching rate at which a loaded motor can start without losing steps.

Slew range: The range of switching rates between pull-in and pull-out in which a motor will run in synchronism but cannot start or reverse.

The general characteristics of a typical stepper motor are given in Figure 2.58. During the application of each sequential pulse the rotor of a stepper motor accelerates rapidly towards the new step position. However, on reaching the new position there will be some overshoot and oscillation unless sufficient retarding torque is provided to prevent this happening. These oscillations can cause rotor resonance at certain pulse frequencies, resulting in loss of torque, or perhaps even pull-out conditions. As variable-reluctance motors have very little inherent damping they are more susceptible to resonances

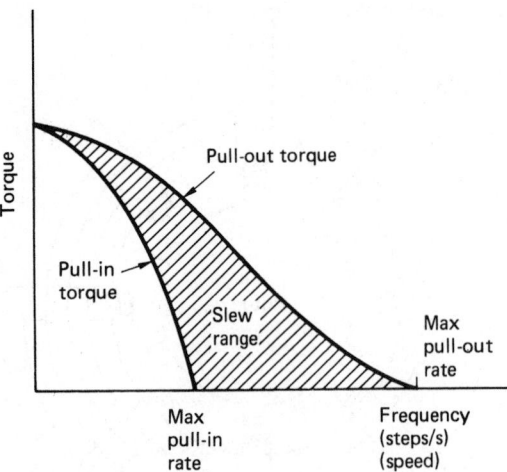

Figure 2.58   Stepper motor characteristics

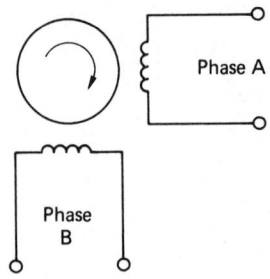

Figure 2.59   Two-phase brushless motor

than either the permanent magnet or the hybrid types. Mechanical and electronic dampers are available which can be used to minimize the adverse effects of rotor resonance. If at all possible, however, the motor should be selected such that its resonant frequencies are not critical to the application under consideration.

Because of their unique characteristics, stepper motors are widely used in applications involving positioning, speed control, timing and synchronized actuation. They are prevalent in X-Y plotters, punched-taped readers, floppy disc head drives, printer carriage drives, numerically controlled machine tool slide drives and camera iris control mechanisms.

By far the most severe limitation on the purely electric stepper motor is its power-handling capability. Currently, this is restricted to about 2.25 kW.

2.2.28   Brushless d.c. motors

These motors have position feedback of some kind so that the input waveforms can be kept in the proper timing with respect to the rotor position. Solid-state switching devices are used to control the input signals, and the brushless d.c. motor can be operated at much higher speeds, with full torque available at those speeds. The brushless motor can normally be rapidly accelerated from zero to operating speed as a permanent magnet d.c. motor. On reaching operating speed the motor can then be switched over to synchronous operation.

The brushless motor system consists of a wound stator, a permanent magnet rotor, a rotor position sensor and a solid-state switching assembly. The wound stator can be made with two or more input phases. Figure 2.59 gives the schematic representation of a two-phase brushless motor. The torque output of phase A is

$$T_A = I_A(Z\Phi/2\pi) \sin(p\theta/2) = I_A K_T \sin(p\theta/2) \tag{2.83}$$

where I_A is the current in phase A,
$K_T = (Z\Phi/2\pi)$, is the torque constant of the motor,
p is the number of poles, and
θ is the angular position of the rotor.
In the expression for the torque constant, Z is the total number of conductors and Φ is the magnetic flux.
Similarly, the torque output of phase B is

$$T_B = I_B K_T \cos(p\theta/2) \tag{2.84}$$

If the motor currents are arranged to be supplied in the following relationships:

$$I_A = I \sin(p\theta/2) \text{ and } I_B = I \cos(p\theta/2)$$

then the total torque for a two-pole motor becomes

$$T = T_A + T_B = IK_T[\sin^2(\theta) + \cos^2(\theta)]$$
$$= IK_T \tag{2.85}$$

Equation (2.85) shows that if all the above conditions are satisfied then the brushless d.c. motor operates in a manner similar to the conventional d.c. motor, i.e. the torque is directly proportional to the armature current. Note that the armature current in this context refers to the stator windings.

Excitation of the phases may be implemented with sinusoidal or square-wave inputs. The sine-wave drive is the most efficient, but the output transistors in the drive electronics must be capable of dissipating more power than that dissipated in square-wave operation. Square-wave drive offers the added advantage that the drive electronics can be digitally based.

The brushless d.c. motor will duplicate the performance characteristics of a conventional d.c. motor only if it is properly commutated. Proper commutation involves exciting the stator windings in a sequence that keeps the magnetic field produced by the stator approximately 90 electrical degrees ahead of the rotor field. The brushless d.c. motor therefore relies heavily on the position feedback system for effective commutation. It might also be apparent that the brushless motor as described is not strictly a d.c. machine but a form of a.c. machine with position feedback.

The further development of the brushless d.c. motor will depend to a large extent upon future advances in semiconductor power transistor technology. It is likely, however, that within the next decade the true brushless d.c. motor, using solid-state swiching, will become commercially viable and will progressively dominate the d.c. servosystem market.

This brief discussion of rotating electrical machines is in no way comprehensive. A fuller discourse on a.c. and d.c. machines is given by both Gray[4] and Sen.[5] Orthwein[6] presents an interesting practical discussion on the mechanical applications of a.c. and d.c. motors and Kenjo and Nagamori provide a detailed in-depth study of permanent-magnet d.c. motors.

2.2.29   Transformers

One of the major advantages of a.c. transmission and distribution is the ease with which an alternating voltage can be increased or decreased. Common practice in the UK is to generate voltages at 11–22 kV and then transform up to 33 kV (or 132 kV) for transmission on the National Grid to the consumer centres. At these centres, the voltages are transformed back down to 415 V (or 240 V) and then distributed for industrial and domestic use.

2.2.30 Basic transformer action

Figure 2.60 illustrates a simple single-phase transformer in which two separate coils are wound onto a ferrous core. The coil connected to the supply is called the 'primary winding' and that connected to the load is the 'secondary winding'. The ferrous core is made in laminations, which are insulated from one another, to reduce eddy current losses.

If a sinusoidal voltage, V_1, is applied across the primary winding a current, I_1, in the coil will induce a magnetic flux, ϕ, in the core. From Faraday's law (equation (2.25)) the induced e.m.f. in the primary coil is

$$E_1 = N_1(d\phi/dt) \tag{2.86}$$

Since the magnetic flux is common to both coils the e.m.f. induced in the secondary winding is

$$E_2 = N_2(d\phi/dt) \tag{2.87}$$

Hence

$$\frac{E_1}{E_2} = \frac{N_1}{N_2} \tag{2.88}$$

The ratio of primary coil turns to secondary turns, N_1/N_2, is called the 'transformation ratio'. The primary and secondary winding impedances, Z_1 and Z_2, respectively, are both very small such that when the secondary winding is on open circuit, then $V_1 \doteqdot E_1$ and $V_2 \doteqdot E_2$. Therefore

$$\frac{V_1}{V_2} = \frac{N_1}{N_2} \tag{2.89}$$

When a load is connected across the secondary winding a current, I_2, will flow in the secondary winding. From Lenz's law this will set up a flux which will tend to oppose the main flux, ϕ. If the main flux is reduced then E_1 would be correspondingly decreased and the primary current, I_1, would then increase. This increased primary current would tend to produce a flux to oppose that induced by the secondary current. In this manner the main flux is generally maintained. In steady state the ampere-turns in the primary and secondary windings are balanced, i.e.

$$I_1N_1 = I_2N_2$$

or

$$\frac{I_1}{I_2} = \frac{N_2}{N_1} \tag{2.90}$$

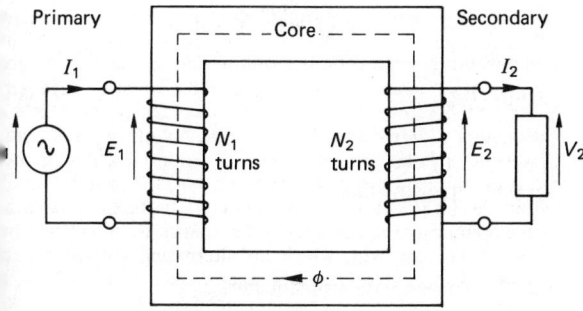

Figure 2.60 Single-phase transformer

2.2.31 Transformer voltage equation

In normal operation the flux may be considered to be a sinusoidally varying quantity, i.e.

$$\phi = \Phi \sin(\omega t) \tag{2.91}$$

The induced e.m.f., from Faraday's law, is

Primary side, $e_1 = N_1(d\phi/dt) = N_1\Phi\omega\cos(\omega t)$

The r.m.s. value of the induced e.m.f. is

$$E_1 = \frac{2\pi f N_1 \Phi}{\sqrt{2}} = 4.44\, fN\Phi \tag{2.92}$$

Similarly, for the secondary side,

$$E_2 = 4.44\, fN_2\Phi$$

2.2.32 Transformer losses

Equations (2.89) and (2.90) define the ideal transformer in which there are no resistive or inductive losses. An actual transformer, of course, does involve some losses, which are:

1. *Copper losses*: These are associated with the I^2R loss in both of the coils. They may be represented therefore as a resistance in series with each coil.
2. *Iron loss*: These are associated with magnetic hysteresis effects and eddy current losses in the iron core. The iron losses are essentially constant for a particular value of supply voltage. Iron losses can be represented as a resistor in parallel with the primary coil.
3. *Flux leakage*: The useful (or main) flux is that which effectively links both coils. In practice, some of the flux will escape, or otherwise fail to link both coils. The e.m.f.'s produced by the leakage fluxes are proportional to (and lead the fluxes by) 90°. The effect of flux leakage may be likened therefore to having an additional inductive coil in series with the primary and secondary coils. In practice, the flux leakage loss is usually lumped together with the iron loss.

2.2.33 Determination of transformer losses

2.2.33.1 Open-circuit test

The secondary coil is on open-circuit and the full-rated voltage is applied to the primary winding. The transformer takes a small no-load current to supply the iron loss in the core and the copper losses are essentially zero. Since the normal voltage and frequency are applied, a wattmeter connected to the primary side will give a measure of the iron loss. The iron loss can then be taken as a constant, irrespective of the load.

2.2.33.2 Closed-circuit test

With the secondary winding short-circuited the transformer requires only a small input voltage to circulate the full-load current. The wattmeter on the primary side then gives an indication of the full-load copper losses. If the load is expressed as a fraction of the full load, the copper losses at reduced loads are proportional to the load squared. At half load, for example, the copper losses are one quarter of the full-load value.

2.2.34 Referred values

In dealing with transformers it is usual to base all calculations on one side of the transformer. Parameters on the neglected side are accounted for by 'referring' them over to the side on which the calculation is to be based. The transformation ratio is used to scale the equivalent values. For example, the copper loss on the secondary side, $I_2^2 R_2$, can be referred to the primary side through the relation

$$I_2'^2 R_2' = I_2^2 R_2 \tag{2.93}$$

where the prime denotes the referred values. Using equation (2.90) the referred resistance becomes

$$R_2' = \{N_1/N_2\}^2 R_2 \tag{2.94}$$

Thus equation (2.94) gives an equivalent resistance, R_2', in the primary side which accounts for the actual resistance, R_2, of the secondary winding. Reactances may be similarly referred to one or other side of the transformer for calculation purposes.

2.2.35   Transformer efficiency

The transformer efficiency, as with any machine, is the ratio of the output power to the input power. The difference between the output and the input power is the sum of the losses, which, for the case of a transformer, is the copper and the iron losses, i.e.

$$\eta = \frac{\text{Output}}{\text{Input}} = \frac{\text{Output}}{\text{Output} + \text{copper loss} + \text{iron loss}}$$

Therefore

$$\eta = \frac{V_2 I_2 \cos(\theta_2)}{V_2 I_2 \cos(\theta_2) + I_2^2 \cdot R_e + F_e} \tag{2.95}$$

Note that R_e represents an equivalent resistance, which consists of the resistance of the secondary winding and that of the primary winding referred over to the secondary side, i.e.

$$R_e = R_2 + (N_2/N_1)^2 R_1 \tag{2.96}$$

The iron loss, F_e, is assumed to be constant and $\cos(\theta_2)$ is the load power factor, also assumed constant.

By dividing the numerator and the denominator of equation (2.95) by I_2, then differentiating the denominator with respect to I_2, and equating the result to zero, it can be shown that for maximum efficiency, $I_2^2 \cdot R_e = F_e$. Maximum transformer efficiency then occurs when the copper loss is equal to the iron loss. The general efficiency characteristics for a transformer are shown in Figure 2.61.

Equation (2.95) also shows that the output will be influenced by the load power factor. At unity power factor the output (and hence also the efficiency) is maximized. As the power factor decreases, the transformer efficiency also reduces proportionally.

2.2.36   Voltage regulation

As the load current drawn from a transformer is increased, the terminal voltage decreases. The difference between the no-load output voltage and the output voltage on load is called the 'regulation'. The percentage regulation is defined as

$$\frac{\text{No-load voltage} - \text{load voltage}}{\text{No-load voltage}} \times 100 \tag{2.97}$$

Figure 2.62 shows the two voltages in terms of phasors referred to the primary side. In the figure V_1 is the no-load primary voltage and V_2' is the secondary-side voltage referred to the primary. R_e and X_e denote the equivalent resistance and reactance, respectively, including the referred secondary values. Since δ is very small, then, to a reasonable approximation,

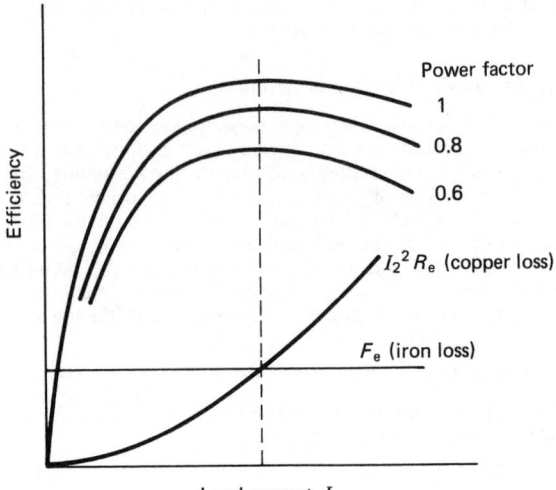

Figure 2.61   Transformer efficiency characteristics

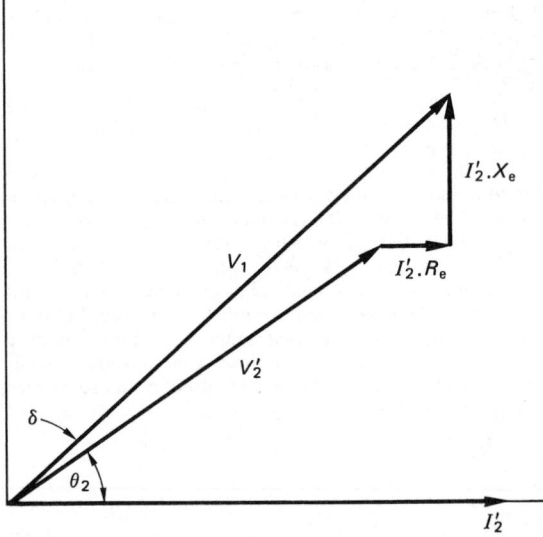

Figure 2.62   Phasor diagram for a transformer with a lagging power factor load current

$$V_1 = V_2' + I_2' \cdot R_e \cdot \cos(\theta_2) + I_2' \cdot X_e \cdot \sin(\theta_2) \tag{2.98}$$

The percentage regulation is therefore

$$(100/V_1)[I_2'R_e \cos(\theta_2) + I_2'X_e \sin(\theta_2)] \tag{2.99}$$

Equation (2.99) is based on the assumption that the load power factor is lagging, and this is the normal situation. If, however, the load power factor is leading, the plus operator within the term in square brackets must be replaced with a minus operator.

2.2.37   Three-phase transformers

Modern large three-phase transformers are usually constructed with three limbs as shown in Figure 2.63. In the figure

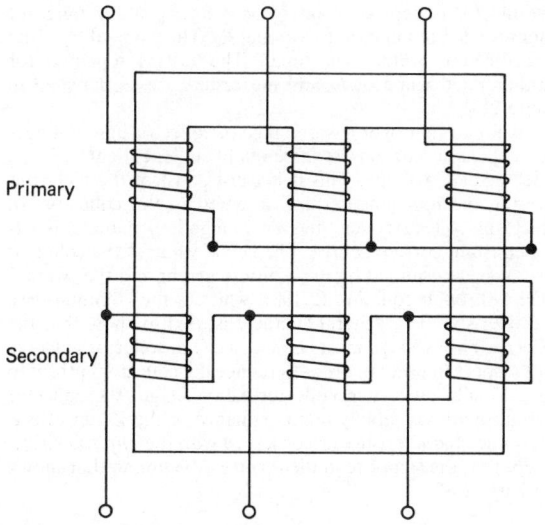

Figure 2.63 Three-phase transformer

the primary windings are star-connected and the secondary windings are delta-connected. In fact, the primary and secondary windings can be connected in any pattern, depending upon the conditions under which the transformer is to operate. It is important, however, to know how the three-phase transformer is connected, particularly when two or more transformers are to be operated in parallel. It is essential, for instance, that parallel operation transformers belong to the same main group and that their voltage ratios are perfectly compatible.

2.2.38 Auto-transformers

The auto-transformer is characterized by having part of its winding common to both the primary and secondary circuits (see Figure 2.64). The main application of auto-transformers is to provide a variable voltage, and it is used, for example, to limit the starting current drawn by an induction motor (see Section 2.2.22).

A major disadvantage of the auto-transformer is that the primary and secondary windings are not electrically isolated from one another. This presents a serious risk of shock, and therefore auto-transformers cannot be used for interconnecting high- and low-voltage systems.

2.3 Analogue and digital electronics theory

2.3.1 The bipolar (or junction) transistor

The term 'transistor', derived from 'transfer resistor', describes a device which can transfer a current from a low-resistance circuit to a high-resistance one with little change in current during the process. The junction transistor consists of two *pn* diodes formed together with one common section, making it a three-layer device (see Figure 2.65).

Current flow in the transistor is due to both electron and hole conduction. The common central section is referred to as the 'base' and is typically of the order of 25 μm in length. Since the base can be made either an *n*-type or a *p*-type semiconductor, two basic configurations are possible. These are the *npn* and the *pnp* types, as illustrated in Figure 2.65. The two other terminals are called the 'emitter' and the 'collector'. An arrowhead is traditionally shown between the emitter and the base to indicate the conventional direction of the current flow in that part of the circuit.

A brief description of the physical operation of the junction transistor can be made with respect to the *npn* type. The mode of operation of the *pnp* type is the same as that of the *npn* type, except that the polarities of all applied voltages, currents and charge carriers are reversed.

In normal use, as a linear amplifier, the transistor is operated with the emitter to base junction forward biased and the collector to base junction reversed biased. For the *npn* transistor, the emitter is therefore negative with respect to the base while the collector is positive with respect to the base (see Figure 2.66). The junction n_1p is forward biased such that the free electrons drift from n_1 to p. On the other hand, junction n_2p is reverse biased and it will collect most of the electrons from n_1. The electrons which fail to reach n_2 are responsible for the current at the base terminal, I_B. By ensuring that the thickness of the base is very small and that the concentration of impurities in the base is much lower than either that of the emitter or the collector, the resultant base current will be limited to some 2% of the emitter current. The basic transistor characteristic is therefore

$$I_C = h_{FB} \cdot I_E \tag{2.100}$$

where I_C is the collector current, I_E is the emitter current and h_{FB} is the current gain between the collector and the emitter. Normally, h_{FB} would range between 0.95 and 0.995 for a good-quality transistor.

2.3.2 Common-base characteristics

Figure 2.67 shows an *npn* transistor connected in a circuit to determine its static common-base characteristics. The emitter

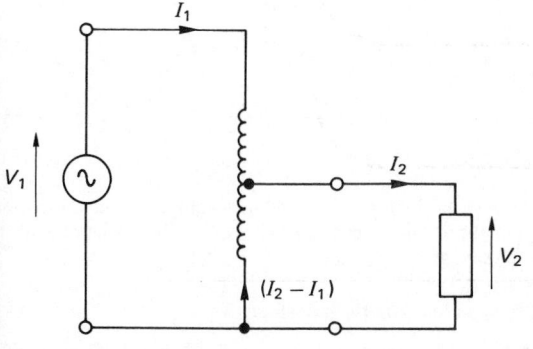

Figure 2.64 Auto-transformer

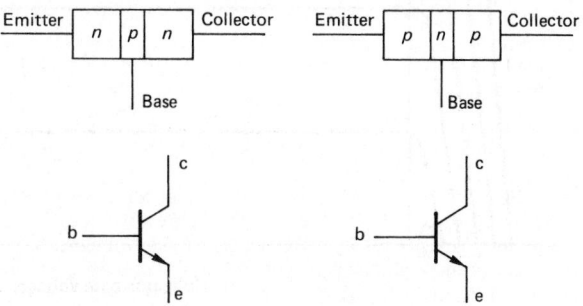

Figure 2.65 *npn* and *pnp* junction transistors

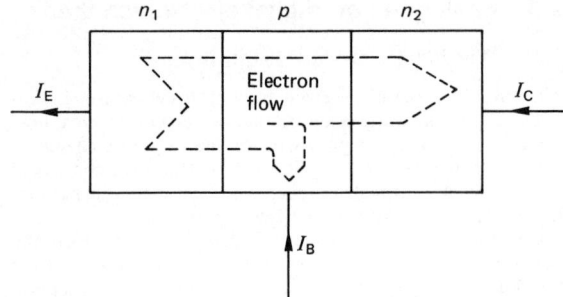

Figure 2.66 *npn* transistor in normal operation

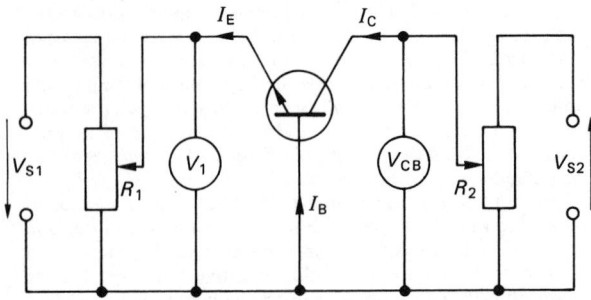

Figure 2.67 *npn* transistor in common base circuit

current, I_E, is kept constant by varying R_1 and a range of values for I_C are imposed by varying R_2. The value of V_{CB}, the collector-base voltage, is noted. The test is repeated for another fixed value of I_E and the results are as depicted in Figure 2.68.

It is found that over a wide range of collector-base voltages the collector current is essentially independent of the collector-base voltage. This is because most of the electrons entering the *npn* junction are attracted to the collector. In effect, the collector circuit has a very high impedance and acts as a constant current source. The actual value of the collector current is determined by the emitter current and the two are related through equation (2.100), which is the common-base characteristic. The general characteristics also show that the collector-base voltage must be reversed (i.e. collector negative with respect to base) in order to reduce the collector current to zero. Finally, at a high collector-base voltage, the collector current increases rapidly in consequence of the Zener effect. The same characteristics are observed with the *pnp* transistor, except that the signs are in the reverse direction to that shown in Figure 2.67.

2.3.3 Common-emitter characteristics

Figure 2.69 shows the *npn* transistor with its emitter terminal connected to both the base current and the collector current circuits. Using the same test procedure as before, the resulting characteristics are as shown in Figure 2.70. The first significant observation is that the collector-emitter voltage, V_{CE}, must be positive to produce a positive collector current. At low values of V_{CE} the collector current I_C is also low, but when V_{CE}

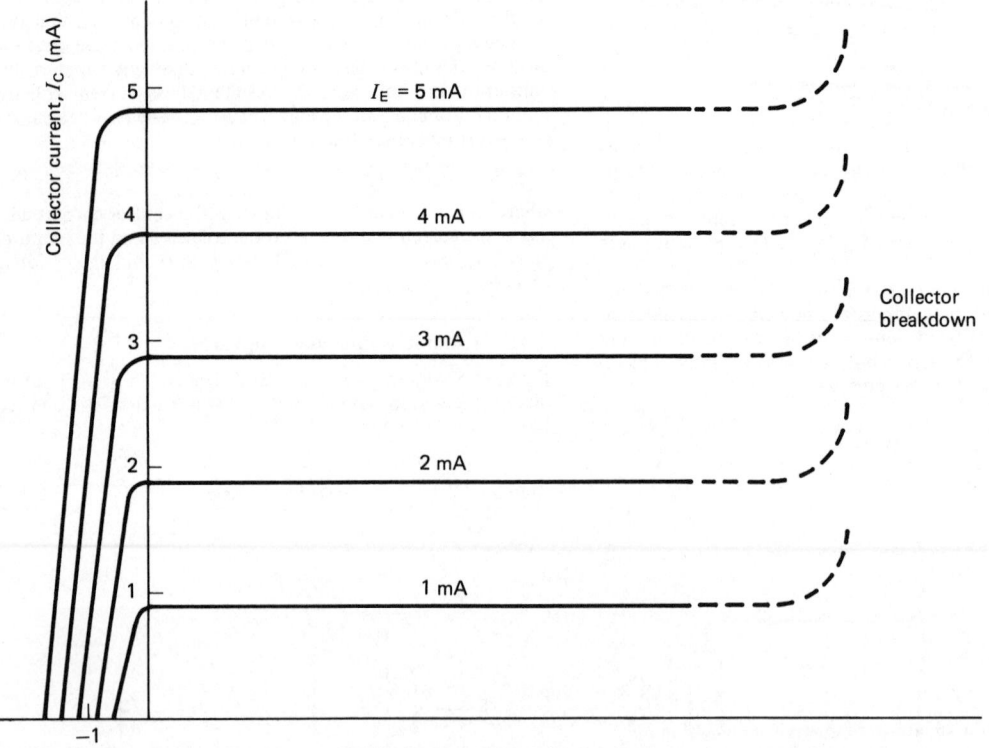

Figure 2.68 Common-base characteristics

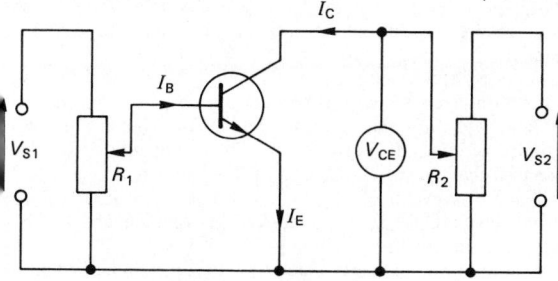

Figure 2.69 *npn* transistor in common emitter circuit

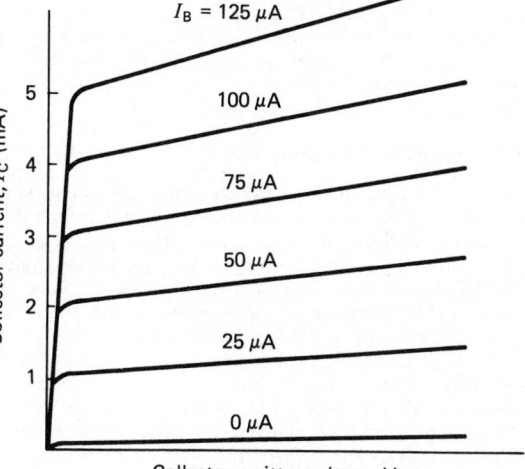

Figure 2.70 Common-emitter characteristics

exceeds the so-called 'knee' voltage the characteristic assumes a linear relationship. The gradient of the linear region is generally much higher than that for the common-base configuration and the collector impedance is therefore lower than that for the common-base circuit. When the base current is zero the collector current still has a positive finite value.

The common-emitter characteristic is generally written as

$$I_C = h_{FE} \cdot I_B \qquad (2.101)$$

where h_{FE} is the current gain between the collector and base. Application of Kirchhoff's first law to the common-emitter circuit gives

$$I_E = I_C + I_B$$

Using equation (2.100) and eliminating I_E, it can be shown that

$$\frac{I_C}{I_B} = \frac{h_{FB}}{1 - h_{FB}} = h_{FE} \qquad (2.102)$$

For a transistor with a steady-state current gain in common base of 0.95 the common-emitter gain is

$$h_{FE} = \frac{0.95}{1 - 0.95} = 19$$

If, due to some temperature effect, h_{FB} undergoes a minor change to, say, 0.96, the new value of h_{FE} becomes 24. It is clear therefore that the common-emitter gain, h_{FE}, is much more sensitive to small-order effects than the common-base gain, h_{FB}.

For a *pnp* transistor the characteristics of the common-emitter circuit are the same, except that the polarity of all voltages and currents are again in reverse order to that shown in Figure 2.69.

2.3.4 The transistor in a circuit

In most practical applications transistors are operated in the common-emitter mode where the emitter terminal forms the common connection between the input and output sections of the circuit (see Figure 2.71).

The transistor collector characteristics are shown again in Figure 2.72. The load line for the resistor, R_C, is superimposed and the operating point is given by the intersection of the load line with the collector characteristic. The operating point will therefore be dependent on the base current, since this controls the collector characteristic. Also shown in Figure 2.72 is the maximum power dissipation curve (broken line), which represents the locus of the product of collector current and collector-emitter voltage. The maximum power dissipation curve represents a physical limitation and the operating point must be constrained to lie below the curve at all times.

As the base current is reduced the operating point moves down the load line. When I_B reaches zero the collector current will be minimized and the transistor is said to be 'cut-off'. Alternatively, as the base current is increased the operating point moves up the load line and eventually reaches a maximum value at which the transistor is said to be 'bottomed', or 'saturated'. When saturated, the collector-emitter voltage is at a minimum of about 0.1–0.2 V and the collector current is a maximum. The two extremes between cut-off and saturation represent a very high and a very low impedance state of the transistor, respectively. These extremes have great practical application to rapid, low-power switching, and transistors operating between cut-off and saturation are frequently used in digital electronics circuitry. The low-impedance state represents a switch closed (or on) and the high-impedance state represents the switch open (or off). When operating as a linear

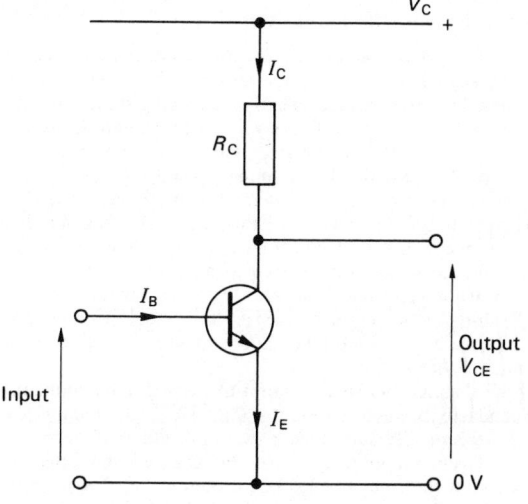

Figure 2.71 *npn* transistor in a practical common-emitter circuit

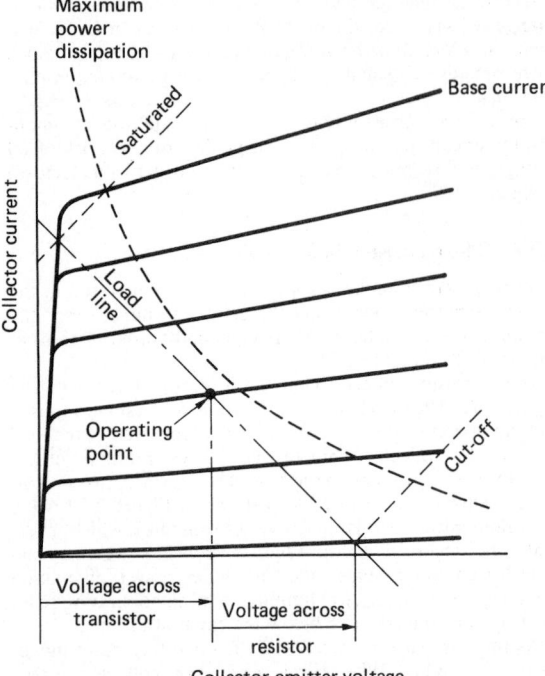

Figure 2.72 Common-emitter characteristics with superimposed load line

current amplifier the operating point is ideally located in the centre of the active region of the characteristic.

The analysis of circuits involving transistors is conveniently dealt with by representing the transistor in terms of an equivalent circuit and using the conventional current flow direction from positive to negative. Consideration of the charge carriers (i.e. holes or electrons) is only necessary to describe the internal physical operation of the transistor. Fully detailed worked examples are particularly informative, and these are usually provided in all standard textbooks on electrical and electronics technology.

2.3.5 The field effect transistor (FET)

Field effect transistors (or FETs) are a much more recent development than bipolar transistors and they operate on a substantially different mechanism in achieving signal amplification. Operationally, FETs are voltage-controlled devices as opposed to the bipolar transistor, which is current-operated. FETs are often described as unipolar, since conduction in the FET is the result of only one predominant charge carrier.

The junction field effect transistor (JFET) consists of a thin bar of semiconductor which forms a channel between its two end-connections that are referred to as the 'source' and the 'drain'. If the semiconductor used in the construction of the FET is *n*-type, the device is called an '*n*-channel'. Conversely, a FET made from a *p*-type semiconductor is called a '*p*-channel' device.

If the channel consists of a uniformly doped semiconductor, the conductivity will be constant and the FET will function as a linear resistor. By introducing two opposite type semiconductor layers on either side of the channel the effective thickness of the channel (and hence the current flow) can be controlled. The opposite type layers are denoted as 'gates' and in normal operation they are reverse biased by a d.c. poten-

tial, V_{GS}, referred to as the 'gate source voltage'. The reverse bias ensures that no current can flow between the two gates and the gate inputs have an extremely high impedance. By using a lightly doped semiconductor for the channel the gate depletion layer, which is determined by V_{GS}, can be made to extend well into the channel width. This controls the resistance of the path between the source and the drain. The general characteristics of such a FET are shown in Figure 2.73.

For a given value of V_{GS} an increase in drain-source voltage from zero initially gives a linear rise in drain current. Further increases in drain-source voltage result in a so-called 'pinch-off' in the drain current, which then becomes independent of the drain-source voltage. Finally, at a particular limiting value of drain-source voltage a breakdown is initiated. The similarities between Figures 2.73 and 2.70 or 2.72 are clear, and it is evident therefore that the bipolar junction transistor and the unipolar FET can perform essentially a similar function in any given application. Many other types of transistor (for example, the metal oxide semiconductor FET, or MOSFET) use alternative means to control the resistance of the source to drain channel. The general characteristics of these devices, however, are all very similar to that shown in Figure 2.73.

2.3.6 Integrated circuits

While transistor-based amplifiers are still found as individual elements in many working circuits, the modern trend is towards the development of integrated circuits, where all the circuit elements are housed within a single silicon wafer. MOSFET technology is predominant in this area, since the number of components on a single silicon chip can be packed up to twenty times more densely than with bipolar technology.

The integrated circuit components include diodes and transistors which may be either bipolar junction type or FETs. Resistors can be deposited on top of the wafer in the form of tantalum, which is a poor conductor, or built into the wafer as 'pinch' resistors, which are partially turned-off FETs. Capacitors can also be produced within the silicon wafer. Capacitive elements may be formed when a *pn* junction diode is reverse biased. The *p*- and *n*-type layers form the plates of the capacitor and the carrier-depletion layer acts as a dielectric. The capacitance is, however, limited to a few picofarads.

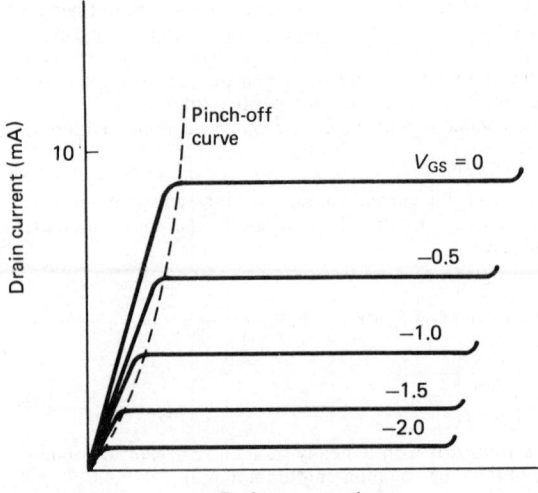

Figure 2.73 Characteristics of a FET

is no microelectronic equivalent for an inductor, but most circuit designs can generally avoid the requirement for coiled inductive elements.

When the integrated circuit is complete it is usually encapsulated as a 'dual-in-line' (DIL) package. This is the normal form in which the integrated circuit is sold. An eight-pin DIL package may contain a relatively simple circuit, but a 40-pin DIL could easily contain all the electronics associated with a central processing unit (CPU) for a computer system. These latter devices contain a very large number of transistors and diodes (approaching 10 000 on a chip of less than 10 mm square). The technology to produce this density of integration is commonly called 'very large-scale integration', or VLSI.

2.3.7 The thyristor

Both the bipolar transistor and the FET can be utilized for switching operations. These devices, however, are usually associated with low-power switching. For switching very large currents and voltages a special device called a 'thyristor' (formerly known as a silicon-controlled rectifier, SCR) is normally used. The thyristor is a four-layer, unidirectional semiconductor device with three connections referred to as the anode, cathode and the control gate (see Figure 2.74).

The current flow is from the anode to the cathode only and, with the cathode positive with respect to the anode, the device has a very high impedance. Under normal circumstances the thyristor will fail to conduct current in any direction. If a voltage is applied such that if the thyristor were a diode it would conduct in the forward-biased direction, then application of a very small current between the gate and the cathode will cause the thyristor to abruptly change from non-conducting to conducting mode. The turn-on is rapid (within a few microseconds) and, once turned on, the thyristor will remain on, even if the gate current is removed.

Once triggered into conduction the thyristor will turn off again only when the current flowing through it is reduced below a critical value. This minimum conducting current is called the 'holding current' and may range between a few microamps to a few tens of milliamps. Thyristors are additionally connected in series with a resistor, which serves to limit the current to a safe value. The basic thyristor function is that of a power-control device, and thyristors are used extensively for switching mains electricity and as speed controllers for d.c. motors.

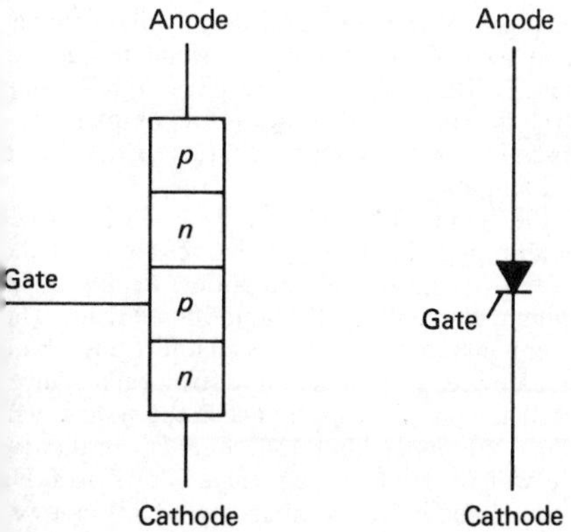

2.3.8 The triac

The triac (or bidirectional thyristor) is similar in operation to the thyristor but differs in that it can be switched into conduction in either direction. In essence, the triac is equivalent to two thyristors mounted back to back. Triacs find application to switching in full-wave alternating power supplies.

2.3.9 Amplifiers

In general, electronic amplifiers are supplied with energy from a d.c. source. An input signal to the circuit controls the transfer of energy to the output, and the output signal should be a higher-power version of that supplied to the input. The amplifier does not, however, function as some magical source of free energy. The increased power across the amplifier is invariably drawn from the supply.

The term 'amplifier' is actually a shortened form for the complete specification 'voltage amplifier'. This has transpired because most amplifiers are intended to magnify voltage levels. Any other type of amplifier is normally prefixed with the name of the quantity which is amplified (e.g. current amplifier, charge amplifier or power amplifier).

Amplifiers may be broadly classified with reference to the frequency range over which they are designed to operate. In this respect there are two general categories: 'wide-band' and 'narrow-band' amplifiers. The names are self-explanatory in that the wide-band amplifier exhibits a constant power gain over a large range of input signal frequencies. The narrow-band (or 'tuned') amplifier, on the other hand, provides a power gain over a very small frequency range. This gain is usually expressed in decibels and is defined by equation (2.58).

The bandwidth of an amplifier is used in the same context as in Section 2.1.29, i.e. to define the operating frequency range. In this respect the -3 dB amplitude ratio is used consistently to define the upper and lower input signal frequencies at which the power transferred across the amplifier is halved.

Using the system model, the amplifier can be represented as shown in Figure 2.75. In the figure the amplifier is shown enclosed within the broken lines. There is a single input, a single output and one common connection. The amplifier also features an internal input impedance, shown as resistance R_i, and an internal output impedance, shown as resistance R_o. In fact, the input and output impedances could have both inductive and capacitive components as well as the simple resistances, as shown in the figure.

Connected to the input stage of the amplifier is a voltage source, V_s, and its associated internal resistance, R_s. This could be taken to represent some form of transducer having a

Figure 2.74 Thyristor device and circuit symbol

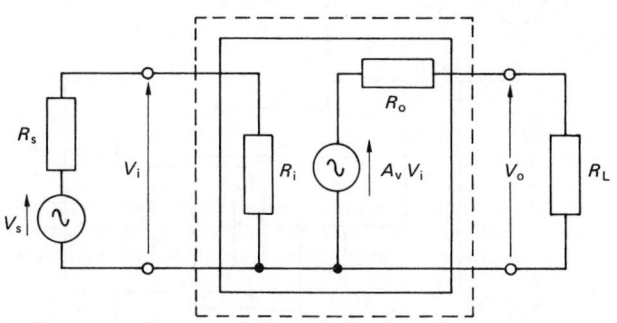

Figure 2.75 System representation of an amplifier

low-voltage output in the millivolt range. At the output stage the amplifier acts as a voltage source where A_v is the voltage gain. The output is shown connected to an external load, R_L, which might be considered to be a recording instrument such as a digital voltmeter.

Considering the input stage, it may be shown, from Ohm's law, that

$$V_i = \frac{V_s}{(1 + R_s/R_i)} \qquad (2.103)$$

This equation indicates that the voltage applied to the amplifier input stage, V_i, will approach the source voltage, V_s, only when R_i tends to infinity. The amplifier should therefore ideally have a very large input impedance to prevent serious voltage attenuation at the input stage. By a similar argument, the output impedance, R_o, should be very small in comparison to the load resistance, R_L, for maximum voltage gain.

2.3.10 Effect of feedback on amplifiers

The amplifier illustrated in Figure 2.75 is specified by its input and output impedances and its open-circuit gain, A_v, this gain being obtained when the load resistance is infinite. These parameters are not fixed, but will vary with ambient temperature, power supply voltage and variation with age. The adverse effects of these variabilities can be minimized through the application of 'negative feedback'.

One particular method of obtaining negative feedback is the so-called 'series voltage' method (see Figure 2.76). The feedback system in Figure 2.76 is applied by connecting a potentiometer across the output terminals and tapping off a fraction, β, of the output signal. This fraction is connected in series with the input and with a polarity which will always oppose the input signal. Assuming both that the input impedance of the amplifier is very large in comparison to the internal resistance of the voltage source and that the resistance of the potentiometer is very large in comparison with the output impedance of the amplifier,

$$V_i = V_s - \beta \cdot V_o \qquad (2.104)$$

Since, $V_o = A_v \cdot V_i$, then $V_o = A_v \cdot V_s - \beta \cdot A_v \cdot V_o$. The overall gain of the system with feedback, A_f, is

$$A_f = \frac{V_o}{V_s} = \frac{A_v}{1 + \beta A_v} \qquad (2.105)$$

Equation (2.105) shows that the feedback loop has reduced the original gain by the factor $(1 + \beta \cdot A_v)$. If, in addition, the original gain A_v was in itself very large such that $\beta \cdot A_v \gg 1$, then

$$A_f = A_v/(\beta \cdot A_v) = 1/\beta \qquad (2.106)$$

Under the above circumstances the overall gain of the system with feedback is essentially dependent only on the feedback fraction, β. Any changes therefore which alter the original gain, A_v, of the amplifier will not affect the gain of the overall system with feedback.

Consideration of the system with and without the feedback loop shows that the effect of series voltage negative feedback is to increase the input resistance by the factor $1 + \beta \cdot A_v$ and to reduce the output resistance by the same factor. Both these effects are of benefit to the operation of the system. These comments refer only to a negative feedback system using the series voltage method. Other methods of obtaining negative feedback can be used, including series current feedback, shunt current and shunt voltage feedback. These alternative methods have different effects on the overall gain and on the input and output impedances of the amplifier.

2.3.11 Noise and distortion in amplifiers

Noise is inherently present in all electronic amplifier systems. The source of the noise is due to a number of effects, which include the random charge movements within solid-state devices, thermoelectric potentials, electrostatic and electromagnetic pick-up and interference from the standard 50 Hz or 60 Hz mains power supply. The noise is fairly evenly distributed across the whole frequency spectrum and appears superimposed upon the amplified input signal. If the noise is generated at the input stage of the amplifier then the 'signal-to-noise' ratio is not improved by feedback. This ratio can, however, be improved if an intermediate amplifying stage, free from noise effects, can be included in the system.

Distortion is another undesirable feature which arises when the amplifier input/output (or transfer) characteristic deviates from an ideal linear relationship. If the transfer characteristic is linear then the output signal will be a faithful amplified replica of the input. A non-linear characteristic will give a distorted output, and a non-sinusoidal output will be generated from a sinusoidal input. Distortion is usually associated with a high level of input signal, which overextends the linear operating range of the amplifier.

2.3.12 Amplifier frequency response

The frequency response of an amplifier is usually illustrated as a plot of the gain in decibels against the input signal frequency. The graph is called a 'Bode plot' and the phase relationship between the output and input is also shown for completeness. Figure 2.77 illustrates the frequency characteristics for a typical wide-band amplifier.

In the figure the bandwidth between the −3 dB cut-off frequencies is determined either by the characteristics of the active devices used to make the amplifier or by other frequency-dependent elements in the amplifier circuit. The upper limiting frequency is fixed by the charge transit time through the active device. In practice, any stray capacitance which is manifested as a parallel capacitance in the system, will considerably reduce the upper limiting frequency. In theory the active device will respond to frequencies down to 0 Hz but, because of the variabilities due to ageing effects, a lower cut-off frequency is often imposed by including series capacitors on one or both of the input connections.

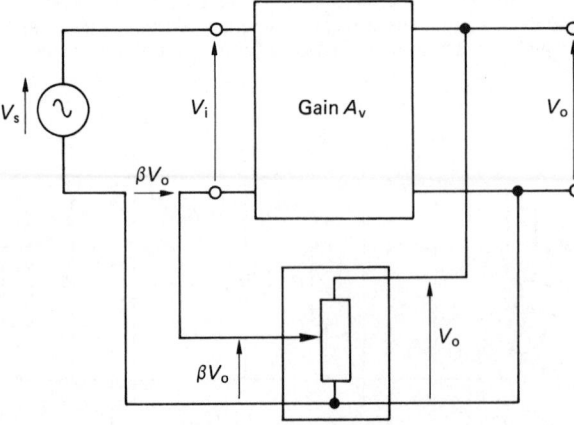

Figure 2.76 Series voltage method of negative feedback

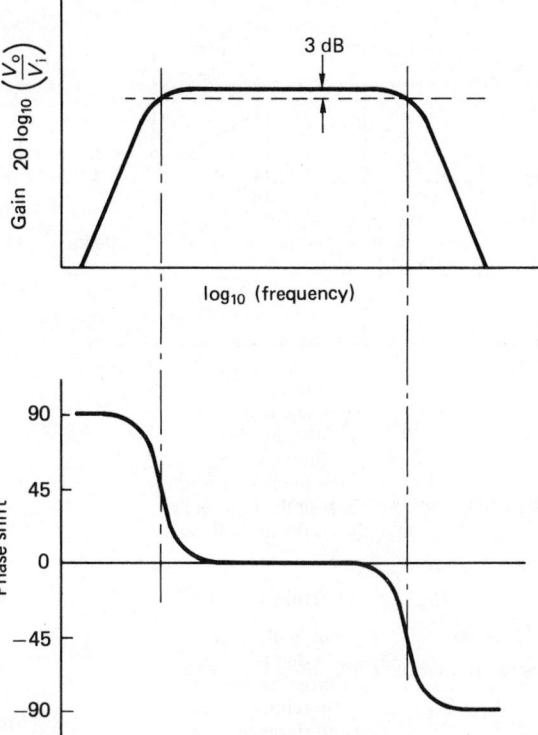

Figure 2.77 Frequency response for a wide-band amplifier

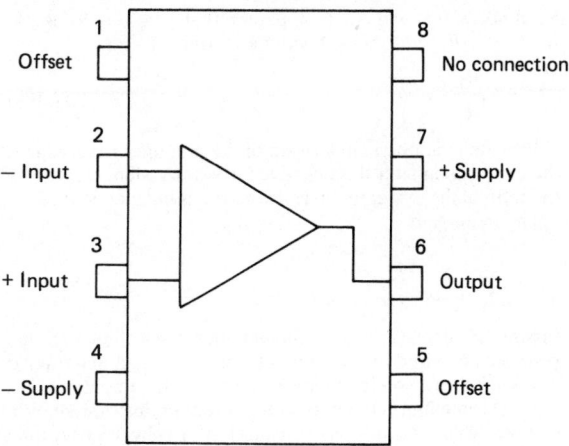

Figure 2.78 SN72741 operational amplifier

2.3.13 Positive feedback and stability

In Figure 2.76 a negative feedback signal is produced by using a series voltage. If the phase of the series voltage was changed such that the feedback signal augmented the input, then the nature of the feedback loop would become positive. With this positive feedback system the overall gain would then become

$$A_f = A_v/(1 - \beta \cdot A_v) \qquad (2.107)$$

Positive feedback therefore increases the overall system gain. If indeed the product $\beta \cdot A_v$ is made equal to unity then the overall gain becomes infinite. Positive feedback, however, is inherently unstable, since the output signal tends to increase indefinitely in an uncontrolled manner. Systems with positive feedback are found, nonetheless, in oscillator circuits where the amplifier produces its own input signal via a positive feedback loop.

2.3.14 The operational amplifier

Modern amplifier systems rely less on discrete active devices such as transistors and much more on the large range of integrated circuits which are readily available. One of the most prevalent operational amplifiers based on integrated circuit technology is the generic type SN72741, or, as it is often abbreviated, the 741. The 741 is available as an eight-pin DIL package and internally consists of 20 bipolar transistors, 11 resistors and one capacitor. The DIL package takes up less area than a small postage stamp and costs less than a cup of coffee. Figure 2.78 shows the usual representation of the 741 operational amplifier (or 'op-amp') in its DIL form.

The internal circuitry is quite complex but is conveniently reduced to the basic schematic form shown in the figure. The operational amplifier consists of an output, an inverting input and a non-inverting input. The circuitry in addition, requires a bipolar power supply which may range between ±3 to ±18 V. There is also provision for an offset null on connection pins 1 and 5. For the most part, the offset pins can be ignored.

The operational amplifier has a high input impedance, a low output impedance and a very high open-circuit gain, A. Ideally, the gain should be infinite. The bandwidth should also be infinite but the 741, for example, has an effective band-width limited between 0 Hz and about 1 MHz.

For operational amplifiers such as the 741 there are a number of standard circuits which are used routinely to perform specific functions.

2.3.14.1 Inverting amplifier

Figure 2.79 shows an op-amp wired up for an inverted output. The input current i_1 is given as V_1/R_1 and, because the amplifier input impedance is very high, the current flowing into the input terminal is approximately zero. This is equivalent to having the potential available at point E equal to zero. For this reason, E is referred to as a 'virtual earth'. From

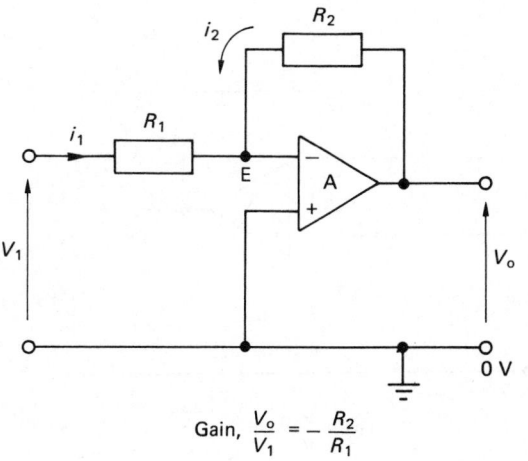

$$\text{Gain, } \frac{V_o}{V_1} = -\frac{R_2}{R_1}$$

Figure 2.79 Inverting amplifier

Kirchhoff's first law, it is apparent that $i_1 = -i_2$. Thus $V_1/R_1 = -V_o/R_2$, and the gain can be written as

$$\frac{V_o}{V_1} = -\frac{R_2}{R_1} \qquad (2.108)$$

Provided the open-circuit gain of the amplifier is very high, the overall gain with this negative feedback system is given by the ratio of the two external resistors and is independent of the open-circuit gain.

2.3.14.2 Unity gain amplifier

Figure 2.80 depicts a unity gain amplifier in which no external resistors are wired into the circuit. The unity gain amplifier is also known as a voltage follower or as a buffer amplifier. This type of amplifier circuit is often used in instrumentation systems where the internal resistance of a voltage-generating transducer and that of the voltage-recording instrument are so poorly matched that the transducer voltage is seriously attenuated. This situation arises when the transducer internal resistance is large in comparison to that of the recording instrument. Since the buffer amplifier has a large input impedance and a low output impedance it can be interfaced between the transducer and the recording instrument to provide optimum impedance matching. This gives a low source impedance and high destination impedance between both the transducer and amplifier and also between the amplifier and the instrument.

Summing the voltages round the amplifier in Figure 2.80 gives

$$V_1 + V_a = V_o$$

Since the internal impedance of the amplifier is very large then V_a is effectively zero and the gain is

$$V_o/V_1 = 1 \qquad (2.109)$$

2.3.14.3 Non-inverting amplifier

Figure 2.81 shows the operational amplifier connected up for a non-inverting output. Assuming that the currents through resistors R_1 and R_2 are equal and that point E is a virtual earth,

$$\frac{V_i}{R_1} = \frac{V_o - V_i}{R_2}$$

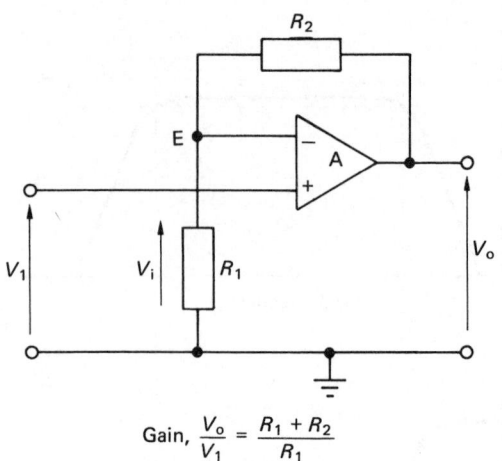

$$\text{Gain,} \quad \frac{V_o}{V_1} = \frac{R_1 + R_2}{R_1}$$

Figure 2.81 Non-inverting amplifier

Hence

$$\frac{V_o}{V_i} = \frac{R_2 + R_1}{R_1}$$

Since E is a virtual earth, then $V_i = V_1$ and

$$\frac{V_o}{V_1} = \frac{R_2 + R_1}{R_1} \qquad (2.110)$$

If, in addition, $R_2 \gg R_1$,

$$\frac{V_o}{V_1} \doteqdot \frac{R_2}{R_1} \qquad (2.111)$$

2.3.14.4 Summing amplifier

The summing amplifier is shown in Figure 2.82. As point E is a virtual earth, then

$$-i_4 = i_1 + i_2 + i_3$$

Therefore

$$\frac{V_o}{R_4} = -\left[\frac{V_1}{R_1} + \frac{V_2}{R_2} + \frac{V_3}{R_3}\right]$$

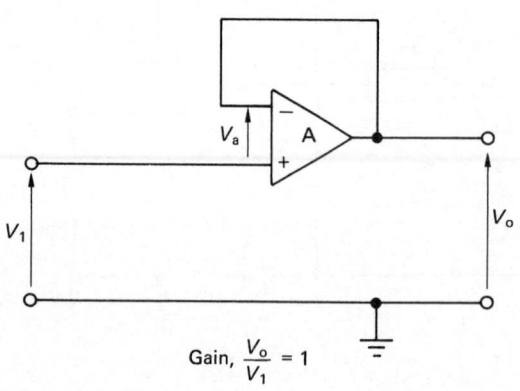

$$\text{Gain,} \quad \frac{V_o}{V_1} = 1$$

Figure 2.80 Unity gain amplifier

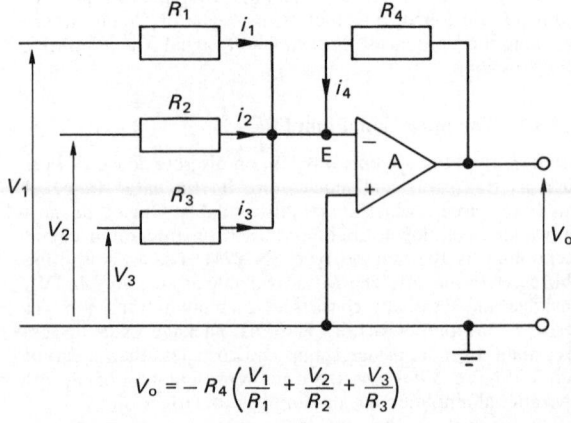

$$V_o = -R_4\left(\frac{V_1}{R_1} + \frac{V_2}{R_2} + \frac{V_3}{R_3}\right)$$

Figure 2.82 Summing amplifier

or

$$V_o = -R_4\left[\frac{V_1}{R_1} + \frac{V_2}{R_2} + \frac{V_3}{R_3}\right]$$ (2.112)

If the resistances used in the circuit are all of equal value, the output voltage will be equivalent to the summation of all the input voltages and with a reversed sign. Subtraction of any of the voltages can be performed by reversing its polarity, i.e. by first passing the voltage through a unity gain inverting amplifier before it is passed on to the summing amplifier.

2.3.14.5 Integrating amplifier

The integrating amplifier uses a capacitor, as opposed to a resistor, in the feedback loop (see Figure 2.83). The voltage across the capacitor is

$$1/C \int_0^t i_2 dt$$

Since E is a virtual earth then $i_1 = -i_2$. therefore $i_2 = -(V_1/R_1)$. The voltage across the capacitor, which is, in effect, V_o, is

$$V_o = -(1/C) \int_0^t (V_1/R_1)dt = -(1/CR_1) \int_0^t V_1 dt$$ (2.113)

Thus the output voltage is related to the integral of the input voltage.

Apart from various mathematical processes, operational amplifiers are also used in active filtering circuits, waveform generation and shaping, as a voltage comparator and in analogue-to-digital (A/D) and digital-to-analogue (D/A) conversion ICs.

2.3.15 The differential amplifier

The differential amplifier (or subtractor) has two inputs and one output, as shown in Figure 2.84. The differential amplifier yields an output voltage which is proportional to the difference between the inverting and the non-inverting input signals. By applying the superposition principle, the individual effects of each input on the output can be determined. The cumulative effect on the output voltage is then the sum of the two separate inputs. It can be shown therefore that

$$V_o = (R_2/R_1)[V_2 - V_1]$$ (2.114)

The input signals to a differential amplifier, in general, contain two components; the 'common-mode' and 'difference-

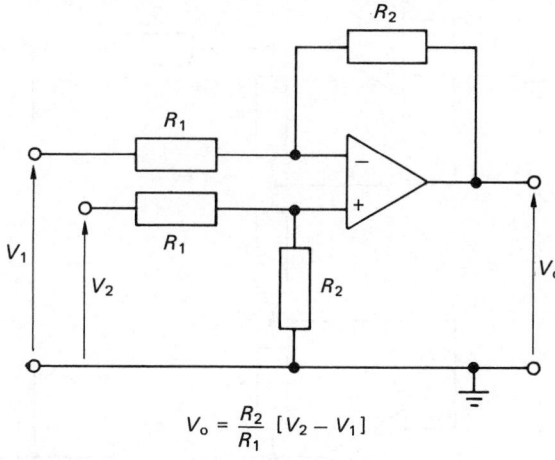

$$V_o = \frac{R_2}{R_1}[V_2 - V_1]$$

Figure 2.84 The differential amplifier

mode' signals. The common-mode signal is the average of the two input signals and the difference mode is the difference between the two input signals. Ideally, the differential amplifier should affect the difference-mode signal only. However, the common-mode signal is also amplified to some extent. The common-mode rejection ratio (CMRR) is defined as the ratio of the difference signal voltage gain to the common-mode signal voltage gain. For a good-quality differential amplifier the CMRR should be very large.

Although particularly important to the differential amplifier, the common-mode rejection ratio is a fairly general quality parameter used in most amplifier specifications. The 741 op-amp has a CMRR of 90 dB and the same signal applied to both inputs will give an output approximately 32 000 times smaller than that produced when the signal is applied to only one input line.

2.3.16 Instrumentation amplifier

Instrumentation amplifiers are precision devices having a high input impedance, a low output impedance, a high common-mode rejection ratio, a low level of self-generated noise and a low offset drift. The offset drift is attributable to temperature-dependent voltage outputs. Figure 2.85 shows the schematic representation of a precision instrumentation amplifier.

The relationship between output and input is

$$V_o = (R_4/R_3)[1 + 2(R_2/R_1)]V_1$$ (2.115)

The first two amplifiers appearing in the input stage operate essentially as buffers, either with unity gain or with some finite value of gain.

A number of instrumentation amplifiers are packaged in IC form and these are suitable for the amplification of signals from strain gauges, thermocouples and other low-level differential signals from various bridge circuits. Kaufman and Seidman[8] give a good practical coverage on the general use of amplifiers.

2.3.17 Power supplies

In Section 2.1.33 the use of *pn* junction diodes were illustrated as a means of a.c. voltage rectification. Both the half-wave and full-wave rectification circuits give outputs, which, although varying with respect to time, are essentially d.c. in that there is

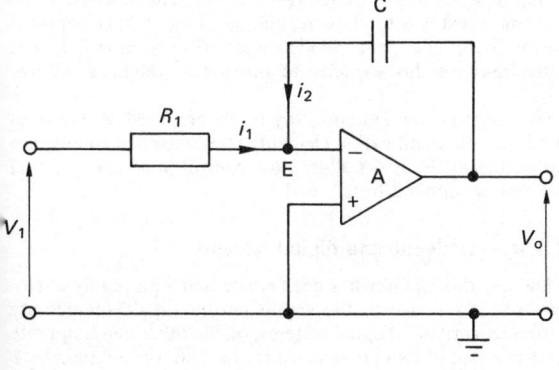

Figure 2.83 Integrating amplifier

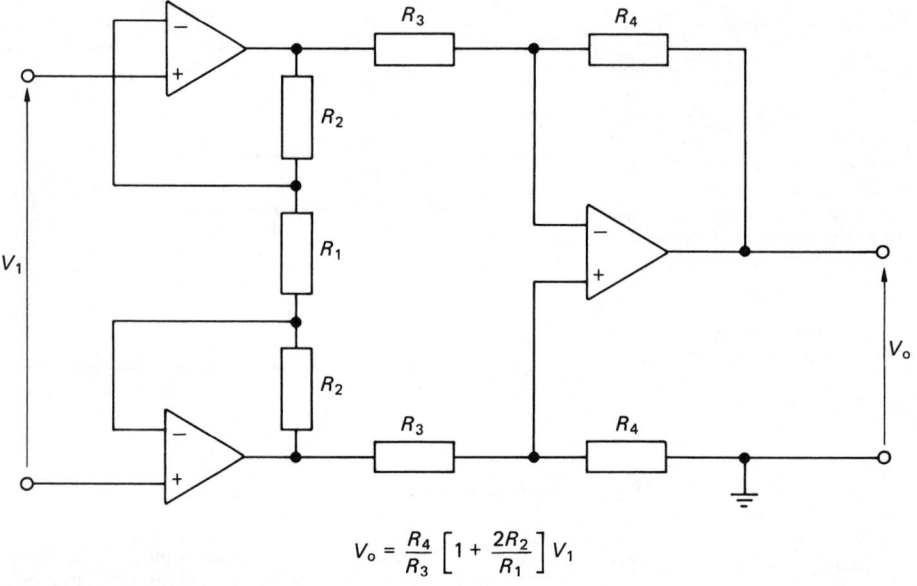

$$V_o = \frac{R_4}{R_3}\left[1 + \frac{2R_2}{R_1}\right]V_1$$

Figure 2.85 Precision instrumentation amplifier

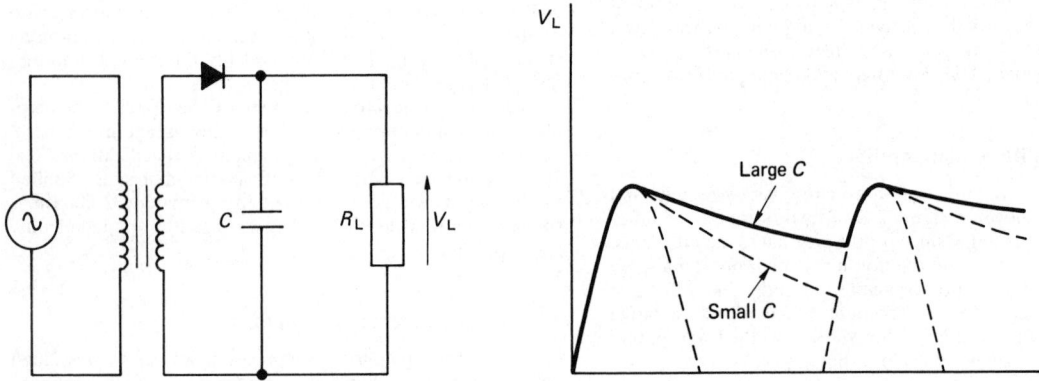

Figure 2.86 Half-wave rectification circuit with reservoir capacitor

no change in polarity. These rectification circuits provide a first stage in the production of a steady d.c. voltage from an a.c. power supply. Some further refinements are, however, added to the circuits to reduce the variation (or 'ripple') in the d.c. output voltage. The ripple factor can be greatly reduced by adding a 'reservoir capacitor', as shown in Figure 2.86, which is connected in parallel with the load.

A further reduction in ripple can be achieved by using a full-wave rectification circuit, since there are then twice as many voltage pulses and the capacitor discharge time is halved. The reservoir capacitor is, of necessity, quite large, and electrolytic capacitors are often used in this application. A leakage resistor is also frequently connected in parallel with the reservoir capacitor as a safety feature. In the event that the load is disconnected leaving the reservoir capacitor fully charged, the leakage resistor will dissipate the charge safely. For applications where the reservoir capacitor still cannot reduce the ripple to an acceptable level an additional ripple filtering circuit may be added.

Further enhancement might include a variable resistor either in series or in parallel with the load. The function of the variable resistor is to allow regulation of the voltage supplied to the load. The Zener diode discussed in Section 2.1.34 is often used in this capacity to provide a stabilized voltage supply.

For high-power systems, thyristors are used in place of diodes as the rectification element. The controlled conduction properties of thyristors allow close control to be exercised on the power supplied to the load.

2.3.18 Analogue and digital systems

Thus far, this chapter has been concerned with purely analogue systems in which the circuit currents and voltages are infinitely variable. Digital systems, on the other hand, operate between one of two possible states, i.e. 'off' or 'on' (conducting or not conducting) and, as such, digital systems are essentially discrete in their operation.

2.3.19 Boolean algebra

The basic rules of Boolean algebra are conveniently described with reference to simple manually switched circuits. In the binary notation a '0' denotes that the switch is off and a '1' that the switch is on. The '0' and '1' can also be taken to represent the absence or presence, respectively, of a voltage or a current.

2.3.19.1 Logical AND

Figure 2.87 shows a simple AND circuit. Obviously, the lamp will light only when both switches A AND B are closed. Writing this as a Boolean expression

$$F = A \text{ AND } B \tag{2.116}$$

where A, B and F are Boolean variables denoting switches A, B and the lamp, respectively. The logical operator AND is denoted by a dot, thus:

$$F = A.B$$

or

$$F = AB \tag{2.117}$$

2.3.19.2 Logical OR

Figure 2.88 shows the simple OR circuit. It is clear that the lamp will light in the OR circuit when either switch A OR switch B is closed. As a Boolean expression, the OR function is written

$$F = A \text{ OR } B$$

i.e. $F = A + B$ (2.118)

The + sign is used to denote the logical OR and must not be confused with the arithmetical meaning.

The AND and the OR are the basic logical functions, and quite complex switching circuits can be represented by them in Boolean form.

2.3.19.3 Logical NOT

The NOT function is the inverse complement, or negation of a variable. The negation of the variable A is $\overline{A}$. Thus if $A = 1$, then $\overline{A} = 0$ and vice versa.

2.3.19.4 Logical NAND

The NAND function is the inverse of AND.

2.3.19.5 Logical NOR

Similarly, the NOR is the inverse of OR.

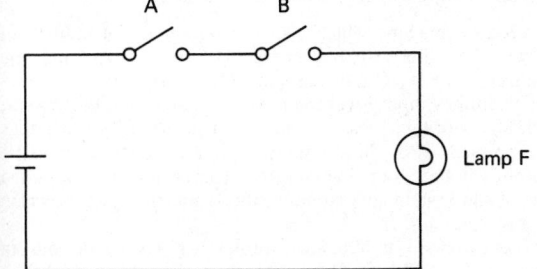

Figure 2.87 Simple AND circuit

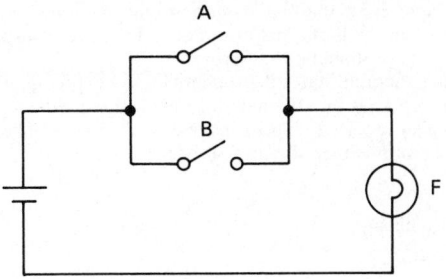

Figure 2.88 Simple OR circuit

2.3.19.6 Exclusive OR

In Figure 2.88 it can be seen that the lamp will also light when both switches A and B are closed. The exclusive OR is a special function which does not enable an output when both switches are closed. Otherwise the exclusive OR functions as the normal OR operator.

The logical functions may also be represented in a tabular form known as a 'truth table'. This table indicates the output generated for all possible combinations of inputs, and is illustrated in Figure 2.89 for the AND and NAND operators with three inputs A, B and C.

Using the basic logical functions, the Boolean identities are specified in Table 2.3.

Table 2.3 Boolean identities

$A + 0 = A$	$A \cdot A = A$
$A + 1 = 1$	$A + \overline{A} = 1$
$A \cdot 0 = 0$	$A \cdot \overline{A} = 0$
$A \cdot 1 = A$	$\overline{\overline{A}} = A$
$A + A = A$	$\overline{A + B} = \overline{A} \cdot \overline{B}$
	$\overline{A \cdot B} = \overline{A} + \overline{B}$

In this table a '0' can be taken to represent an open circuit while a '1' represents a short circuit.

Using a truth table, it is easy to prove the validity of various logical expressions by evaluating both sides, e.g.

$$A(B + C) = AB + AC$$

$$(A + B).(A + C) = A + BC$$

$$A + \overline{A}B = A + B \qquad \text{etc.}$$

A	B	C	A.B.C	$\overline{A.B.C}$
0	0	0	0	1
0	0	1	0	0
0	1	0	0	0
0	1	1	0	0
1	0	0	0	0
1	0	1	0	0
1	1	0	0	0
1	1	1	1	0

Figure 2.89 Truth table for AND and NAND operators with three inputs

The first example shows that the parentheses may be removed by multiplying out, as in normal arithmetic. The second two examples have no arithmetic counterpart.

De Morgan's theorem states that, in any logical expression, AND can be replaced by OR and vice versa, provided that each term is also replaced with its inverse complement. The resulting expression is then the inverse of the original.

Example 1
From ABC we negate to

$$\overline{ABC} = \overline{A} + \overline{B} + \overline{C}$$

Hence

$$ABC = \overline{\overline{A} + \overline{B} + \overline{C}}$$

Example 2
From $F = AB + CD$ we negate to

$$\overline{F} = (\overline{A} + \overline{B}) + (\overline{C} + \overline{D})$$

Applying De Morgan again,

$$F = \overline{(\overline{A} + \overline{B}).(\overline{C} + \overline{D})}$$

The equivalence of the original and the final expressions in the above two examples may be checked by using a truth table.

2.3.20 Digital electronic gates

The principles of Boolean algebra have been considered with respect to manually switched circuits. In modern digital systems the switches are formed with transistors for speed of operation, and they are generally referred to as 'gates'. Over the years, various technologies have been developed in the manufacture of logic gates. The earliest forms of electronic gate were based on the unidirectional conduction properties of diodes. Diode logic gates have now been superseded by transistor–transistor logic gates (TTL) or the more recent CMOS family of logic gates.

The internal construction and operation of modern logic gates may be quite complex, but this is of little interest to the digital systems designer. Generally, all that the designer need to know is the power supply voltages, the transient switching times, the 'fan out' and the 'fan in'. Fan out refers to the number of similar gates which can be driven from the output of one gate. Fan in, on the other hand, denotes the number of similar gate outputs which can be safely connected to the input of one gate.

2.3.20.1 TTL

The TTL family is based on the bipolar junction transistor, and was the first commonly available series of logic elements. TTL logic gates are rapid-switching devices (the SN7400, for example, takes just 15 ns to change state). The standard power supply is 5 V with a low tolerance band of ±0.25 V. This, in turn, necessitates a reliable power supply regulation which is reasonably facilitated through the great variety of supply regulators which are now available in IC form. For the SN74 series TTL ICs, the fan out is about 10.

A TTL-based system can draw quite large instantaneous loads on a power supply, and this can result in substantial interference 'spikes' in the power lines. Since the spikes can upset the normal operation of the system it is common practice to connect small capacitors directly across the power lines, as close to the TTL ICs as possible. One capacitor, 0.1–10 μF, per five ICs is sufficient in most instances.

TTL circuits are continually being improved and a major recent advance has been the introduction of the low-power 'Schottky' TTL circuits. These use the same generic code numbers as the standard series, but have 'LS' inserted before the type code (e.g. SN74LS00). The operating speed is about twice as high and the power consumption is about 20% of the standard series. Schottky devices are, however, slightly more expensive.

2.3.20.2 CMOS

The problematic features of the power supply associated with the TTL family of logic devices has been largely responsible for the growth of its major competitor, CMOS. CMOS ICs are based on the field effect transistor and can operate off a range of power supply voltages between ±3 V to ±18 V. CMOS devices dissipate very little power, are very cheap and are simple in operation. The fan out is about 50 and they have a far greater immunity to power supply noise. The noise immunity of CMOS devices means that there is no requirement for smoothing capacitors to the extent that they are generally found in TTL circuitry.

There are also some disadvantages associated with CMOS devices, the main one being that CMOS is slower than TTL, roughly about one tenth of the equivalent TTL circuit. CMOS ICs are also very sensitive to electrostatic voltages. Manufacturers do build in some safety features to reduce the electrostatic sensitivity, but CMOS devices must still be handled with due care. Table 2.4 gives a brief comparison between TTL and CMOS devices.

Table 2.4 Comparison between TTL and CMOS devices

Property	TTL	CMOS
Power supply	5 V ± 0.25 V	3 V to 18 V d.c.
Current required	Milliamps	Microamps
Input impedance	Low	Very high
Switching speed	Fast – 10 ns	Slow – 300 ns
Fan out	10	50

2.3.21 Gate symbols

Having defined a system output in terms of a Boolean expression, the actual circuit can be constructed using the required gates selected from the logic family chosen. Generally, the design will be centred round the more readily available NAND and NOR logic gates. In laying out a gate interconnection diagram, standard symbols are used to represent the individual gates. Unfortunately, no universal set of symbols has emerged, and several systems are in current use. Figure 2.90 summarizes the most common gate symbol systems.

2.3.22 Logic systems using simple gates

A vending machine which dispenses either tea or coffee can serve as an illustrative example. The logic circuit may be realized using AND gates as shown in Figure 2.91.

The money input is common to both gates, and the system, although workable, has a minor fault in that if both buttons are pressed, after the money criterion is satisfied, then the output will be both tea and coffee. This fault can be designed out of the system by extending the logic circuit as shown in Figure 2.92.

The extended system incorporates a NAND gate and an additional AND gate. If both buttons are now pressed then the output from G_3 will be 0. With the output 1 from G_1, the output from G_4 will be 0 and the machine will dispense tea. On

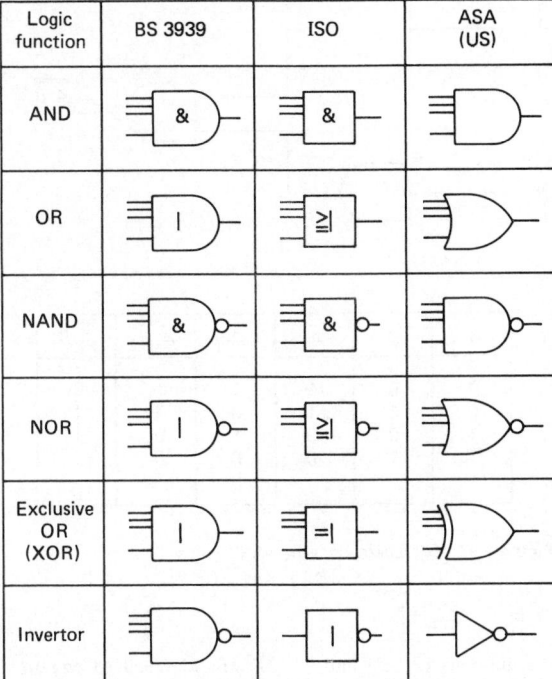

Figure 2.90 Gate symbol systems in current use

Logic function	BS 3939	ISO	ASA (US)
AND			
OR			
NAND			
NOR			
Exclusive OR (XOR)			
Invertor			

Inputs			Outputs	
C	M	T	Coffee	Tea
0	0	0	0	0
0	0	1	0	0
0	1	0	0	0
0	1	1	0	1
1	0	0	0	0
1	0	1	0	0
1	1	0	1	0
1	1	1	0	1

Figure 2.93 Truth table for drinks-vending machine

where C, T and M represent the coffee button, tea button and money input, respectively, and the overbar represents the inverse complement as usual.

Using De Morgan's theorem the system may alternatively be written as

$$\text{Coffee} = \overline{(\overline{C + M})} + \overline{(\overline{C} + \overline{T})} \tag{2.121}$$

$$\text{Tea} = \overline{\overline{T} + \overline{M}} \tag{2.122}$$

Thus the same logic system can be implemented using one OR and three NOR gates, as shown in Figure 2.94.

The validity and equivalence of equations (2.119)–(2.122) may easily be checked using a truth table. Four logic gates are again required but the circuit operates with inverted input signals. This means that three inverters are also required in the circuit as shown.

It is apparent that the logical function can be realized in several different ways, e.g.

$$\text{Coffee} = \overline{(\overline{C + M})} . (\overline{C} + \overline{T}) \text{ and Tea} = TM$$

Using the above realization, the circuit takes the form shown in Figure 2.95.

2.3.23 Logic systems using NAND and NOR gates only

Logic gates are packaged as arrays of the same type in IC form. A typical example is SN7408, which is a 14-pin DIL package containing four separate two-input AND gates. Because the logic gates are marketed in this particular form it is advantageous to design the logic circuit using only one type of gate. This normally minimizes the number of IC packages required. Figure 2.96 shows a two-input NAND gate driving into a single-input NAND gate.

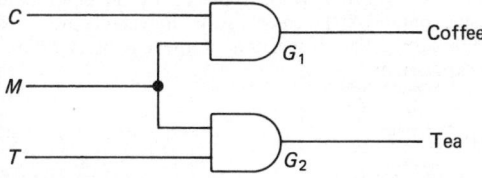

Figure 2.91 Logic circuit for drinks-vending machine

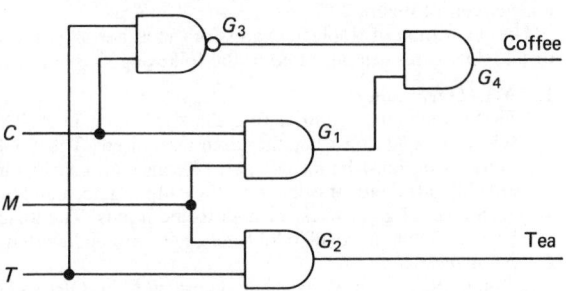

Figure 2.92 Extended logic circuit for drinks-vending machine

pressing either button on its own and satisfying the money input criterion the correct drink will be output. The operation of the extended system is verified in the truth table shown in Figure 2.93.

By inspection of Figure 2.92, the system can be represented in Boolean expressions as

$$\text{Coffee} = (CM) . (\overline{CT}) \tag{2.119}$$

$$\text{Tea} = TM \tag{2.120}$$

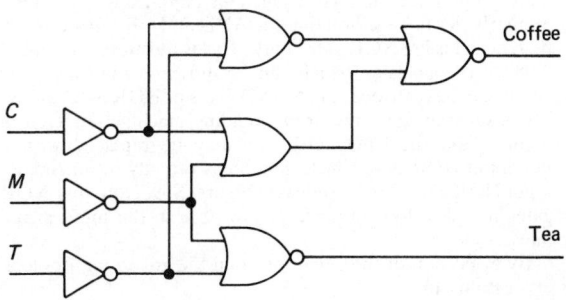

Figure 2.94 Logic circuit for drinks-vending machine using OR and NOR gates

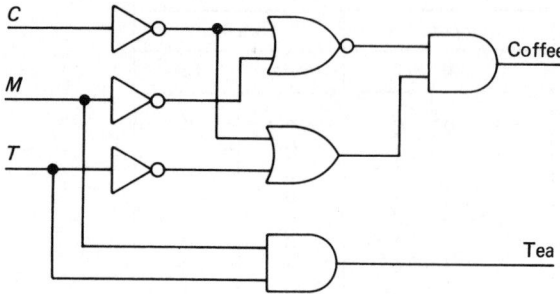

Figure 2.95 Alternative logic circuit for drinks-vending machine

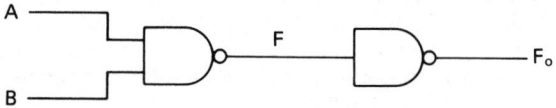

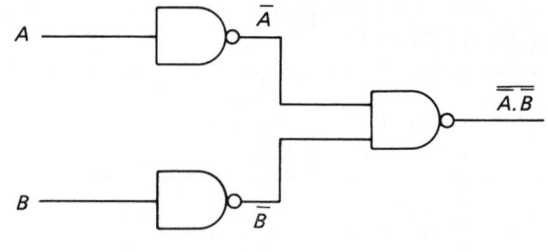

A	B	A.B	$\overline{A.B}$	$\overline{A}.\overline{B}$
0	0	0	1	0
0	1	0	1	0
1	0	0	1	0
1	1	1	0	1

Figure 2.96 AND realization using NAND gates

For the two-input NAND gate, the Boolean expression is

$$F = \overline{A \cdot B}$$

Since F is then fed into a single-input NAND gate, which operates as an inverter, then the final output is

$$F_o = \overline{\overline{F}} = \overline{\overline{A \cdot B}} = A \cdot B$$

It is apparent therefore that the circuit given in Figure 2.96, using NAND gates, performs the same function as the logical AND operator.

Figure 2.97 shows two single-input NAND gates with their outputs driving into a two-input NAND gate. Following through the truth table it can be seen that the circuit performs the logical OR function. If the output F is then fed to another single-input NAND gate (not shown in the figure) then the function performed will be a logical NOR. It can be seen, therefore, that suitable combinations of NAND gates can be made to perform the logical functions AND, OR and NOR. Similarly, it can be shown that the AND and OR functions can be realized using NOR gates only. This is illustrated in Figure 2.98. The conclusion which can be drawn is that any logic circuit can be realized using NAND gates or NOR gates alone.

Considering again the drinks-vending machine depicted in Figure 2.94, the single OR gate may be replaced with a two-input NOR gate which then feeds directly into a single-input NOR gate. This is shown in Figure 2.99. Note that NOR gates are also used in place of invertors in the input signal lines.

By inspection of the circuit diagram the governing Boolean expressions are

$$\text{Coffee} = \overline{(\overline{C} + \overline{M}) + \overline{(\overline{C} + \overline{T})}}$$
$$= (\overline{C} + \overline{M}) + (\overline{C} + \overline{T}) \tag{2.123}$$

Figure 2.97 OR realization using NAND gates

A	B	$\overline{A}$	$\overline{B}$	$\overline{A.\overline{B}}$	$\overline{\overline{A}.\overline{B}}$
0	0	1	1	1	0
0	1	1	0	0	1
1	0	0	1	0	1
1	1	0	0	0	1

$$\text{Tea} = \overline{\overline{T} + \overline{M}} \tag{2.124}$$

Equations (2.123) and (2.124) are identical to equations (2.121) and (2.122), respectively. This, of course, must be true, since the circuits from which the expressions were deduced perform identical logical functions.

Similarly, the circuit in Figure 2.92, involving one NAND and three AND gates, may be replaced by an equivalent circuit using only NAND gates. This equivalent circuit is shown in Figure 2.100. Inspection of the circuit gives the Boolean expressions

$$\text{Coffee} = \overline{\overline{(C \cdot M)} \cdot \overline{(C \cdot T)}}$$
$$= (C \cdot M) \cdot \overline{(\overline{C \cdot T})} \tag{2.125}$$
$$\text{Tea Q} = \overline{\overline{(T \cdot M)}} = TM \tag{2.126}$$

Perhaps as expected, the Boolean expressions are identical to equations (2.119) and (2.120), which were deduced from the logic circuit of Figure 2.92.

The realization of Boolean expressions in either all NAND or all NOR gates can be stated in the following simple rules:

1. *NAND realization*
   First, obtain the required Boolean expression in AND/OR form and construct the circuit required. The final output gate must be an OR gate. Replace all gates with NAND gates and, starting with the output gate, number each level of gates back through to the inputs. The logic level at the inputs to all 'odd' level gates must be inverted.

2. *NOR realization*
   Obtain the required Boolean expression in OR/AND form. The final output gate must be an AND gate. Replace all gates with NOR gates and number each level of gates from the output back through to the input. The logic level at all inputs to 'odd' level gates must be inverted.

Application of these rules is best illustrated by e.g.

NAND realization of $F = AB + C(D + E)$

Figure 2.101 shows the realization of the function in AND/OR form. As inputs D and E appear at an odd level of gate input they must be inverted. In terms of the actual circuit this will

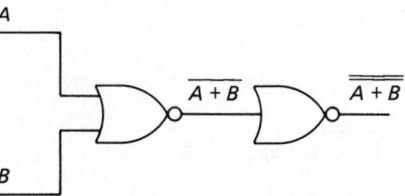
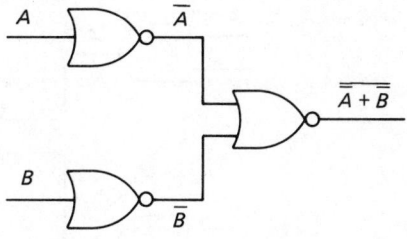

A	B	$\overline{A+B}$	$\overline{\overline{A+B}} = A+B$
0	0	1	0
0	1	0	1
1	0	0	1
1	1	0	1

A	B	$\overline{A}$	$\overline{B}$	$\overline{\overline{A}+\overline{B}} = A.B$
0	0	1	1	0
0	1	1	0	0
1	0	0	1	0
1	1	0	0	1

Figure 2.98   OR and AND realizations using NOR gates only

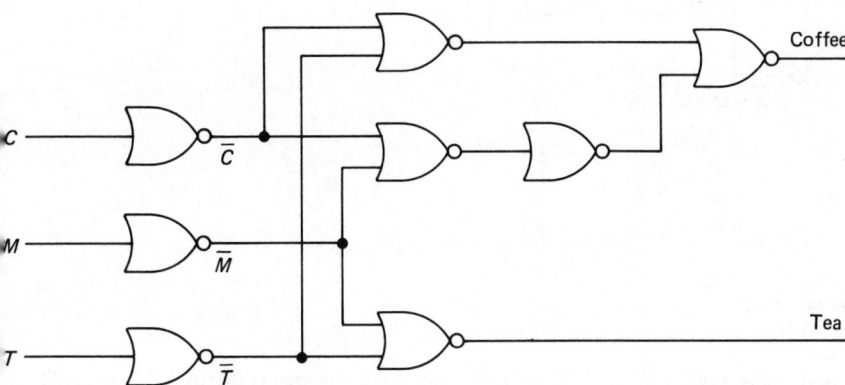

Figure 2.99   Logic circuit using NOR gates only for drinks-vending machine

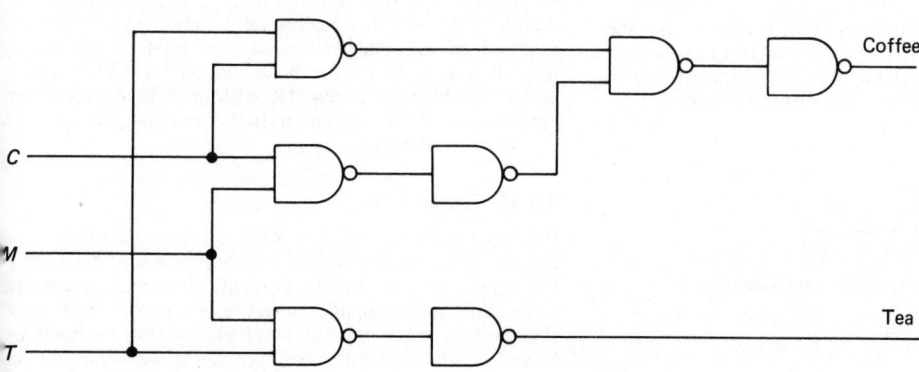

Figure 2.100   Logic circuit using NAND gates only for drinks-vending machine

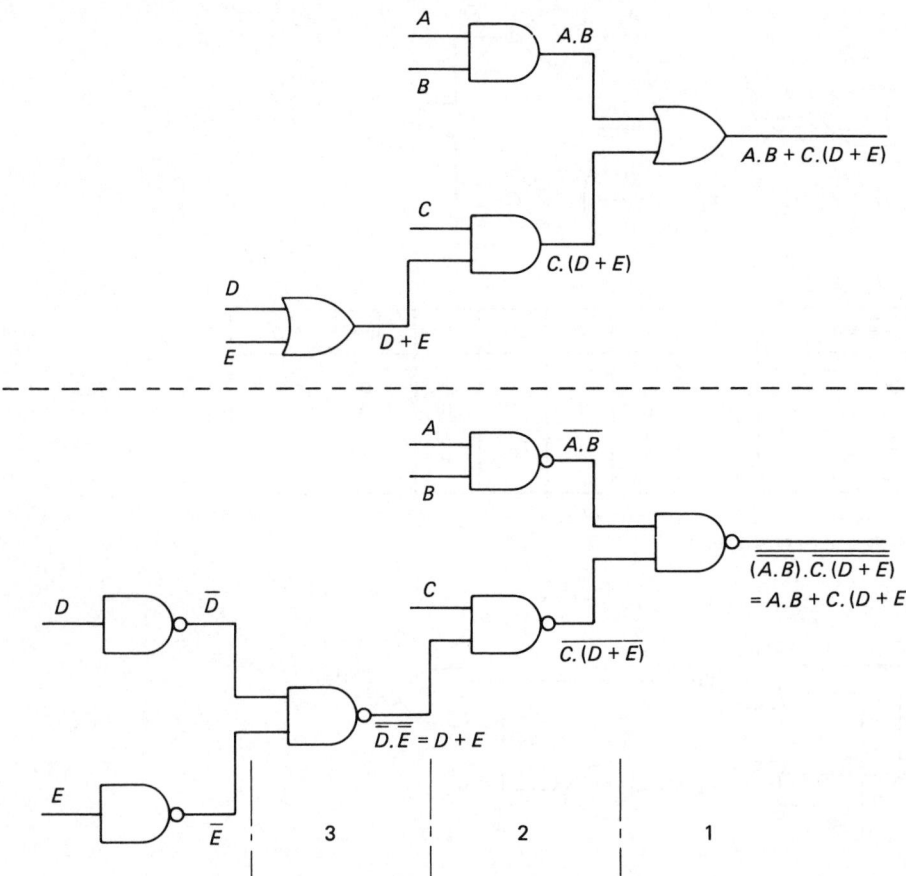

Figure 2.101 NAND realization of a Boolean function

mean that inputs D and E are inverted, using NAND gates, prior to entering the NAND gate at level 4.

A similar procedure is adopted for a NOR realization of a Boolean expression. The exclusive-OR function serves as an interesting example. Written as a Boolean expression, the exclusive-OR is

$$F = A \cdot \overline{B} + \overline{A} \cdot B \qquad (2.127)$$

For the NOR realization, however, it is necessary that the final output gate is an AND. The exclusive-OR function must therefore be manipulated such that the final logical function in the expression is an AND. Using De Morgan's theorem,

$$F = \overline{(A \cdot \overline{B}) \cdot (\overline{A} \cdot B)}$$

$$= \overline{(\overline{A} + B) \cdot (A + \overline{B})}$$

Multiplying out this expression gives

$$= \overline{\overline{A} \cdot A + \overline{A} \cdot \overline{B} + A \cdot B + B \cdot \overline{B}}$$

Since $A \cdot \overline{A} = B \cdot \overline{B} = 0$, the expression simplifies to

$$= \overline{\overline{A} \cdot \overline{B} + A \cdot B}$$

$$= \overline{(\overline{A} \cdot \overline{B}) \cdot (\overline{A \cdot B})}$$

Using De Morgan again gives

$$F = (A + B) \cdot (\overline{A} + \overline{B}) \qquad (2.128)$$

The realization of this equation is shown in Figure 2.102.

2.3.24 Unused inputs

Multi-input gates are also commonly available. In practical circuits, however, it is important that any unused inputs are tied, i.e. they are connected either to the positive voltage supply or to the zero voltage supply. The unused inputs are therefore set at either logic level 1 or at logic level 0, as required. In connecting an unused input to the positive supply the connection should be made through a 1 kΩ resistor. Failure to connect any unused inputs can result in intermittent malfunction of the circuit or in harmful oscillations with attendant overheating.

2.3.25 Latches

It is often useful to 'freeze' a particular binary sequence, and devices called 'latches' are used for this purpose. A latch has four inputs and four outputs. Normally the outputs assume the same state as the inputs. However, when a control signal (known as a 'strobe' input) is taken to logic '1' the outputs are locked in whatever state they were in at the instant of the strobe input going high. This enables the binary sequence to be 'captured' without affecting the ongoing processes, whatever they may be. The latch therefore serves as a temporary

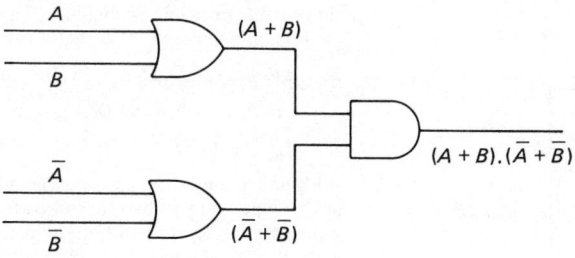

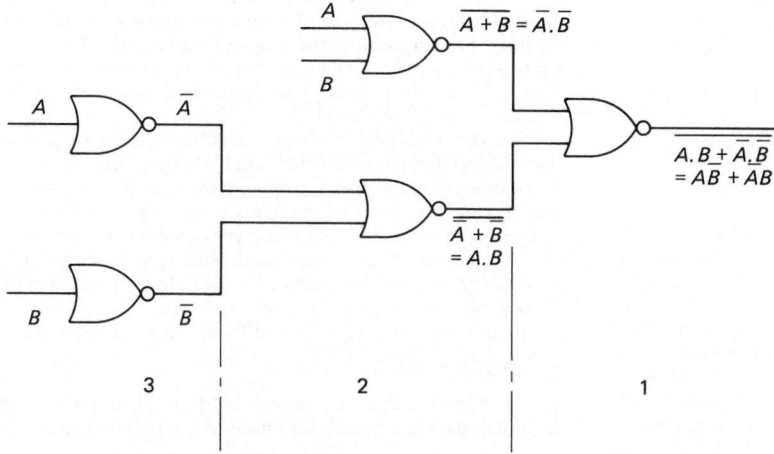

Figure 2.102 NOR realization of the exclusive-OR function

state-recording device which may subsequently be referred to during various interrupt operations.

2.3.26 The Karnaugh map

The Karnaugh map provides an alternative representation of a Boolean expression for all possible Boolean input combinations. In some respects the Karnaugh map is like a truth table in that identical logical expressions display an identical pattern on a Karnaugh map. The Karnaugh map, however, also has a great utility in simplifying Boolean expressions in a systematic manner.

The Karnaugh map consists of a set of boxes in which each box represents one possible combination of the Boolean input variables. The boxes are assigned either a '1' or a '0' to indicate the value of the Boolean expression for the particular combination of input variables that the box represents. The number of boxes required is 2^n, where n is the total number of input variables. Although any number of input variables can be represented, a practical limitation is about seven. Figure 2.103 shows the Karnaugh map for a four-input system. Within each box the unique Boolean input combination is represented by assigning each variable the logic values indicated along the horizontal and vertical axes. These values conform to the binary Gray code in which adjacent consecutive characters differ only in one variable. This imparts a property to the Karnaugh map in that the adjacent squares (vertically or horizontally) differ only in one variable.

As an example, The Boolean expression, $F = \overline{A}BCD + A\overline{B}C\overline{D} + \overline{A}B\overline{C}D$ is represented by the Karnaugh map given

CD \ AB	00	01	11	10
00	$\overline{A}\,\overline{B}\,\overline{C}\,\overline{D}$	$\overline{A}\,B\,\overline{C}\,\overline{D}$	$A\,B\,\overline{C}\,\overline{D}$	$A\,\overline{B}\,\overline{C}\,\overline{D}$
01	$\overline{A}\,\overline{B}\,\overline{C}\,D$	$\overline{A}\,B\,\overline{C}\,D$	$A\,B\,\overline{C}\,D$	$A\,\overline{B}\,\overline{C}\,D$
11	$\overline{A}\,\overline{B}\,C\,D$	$\overline{A}\,B\,C\,D$	$A\,B\,C\,D$	$A\,\overline{B}\,C\,D$
10	$\overline{A}\,\overline{B}\,C\,\overline{D}$	$\overline{A}\,B\,C\,\overline{D}$	$A\,B\,C\,\overline{D}$	$A\,\overline{B}\,C\,\overline{D}$

Figure 2.103 Karnaugh map for a four-input system

in Figure 2.104. The maps are drawn up by placing a '1' in each box for which the combination of input variables makes the logical expression have a value of 1. All the other boxes represent the combination of input variables which make the expression have a logical value of 0. Usually the '0' is not entered in the box.

A second example for consideration is

$$F = \overline{A}\overline{B}CD + AC + \overline{C}D$$

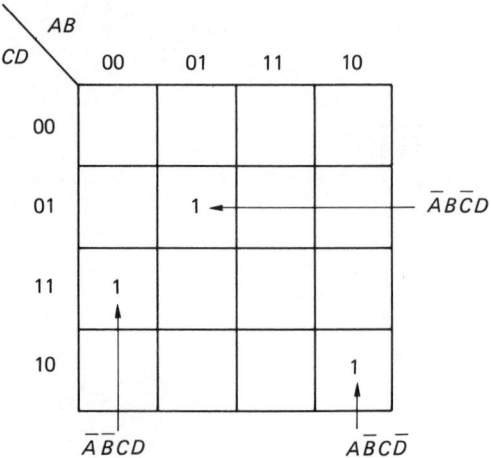

Figure 2.104 Karnaugh map for a Boolean expression (1)

The Karnaugh map for this expression is shown in Figure 2.105. It can be seen that the term AC includes all four squares in which both A and C are included. Similarly, the term $\overline{C}\overline{D}$ also encompasses four squares on the map. It may be concluded that in a four-variable expression any term which contains the four variables will occupy one square on the Karnaugh map. Any term which contains only three of the variables will occupy two squares and any which contains only two of the variables will occupy four squares. A term containing only one of the variables will occupy eight squares in the Karnaugh map. The Karnaugh map may be used in a reverse mode to deduce the Boolean expression.

2.3.27 Minimization of Boolean expressions

The principle of minimization is based on the Boolean identity $A + \overline{A} = 1$. Thus

$$F = ABCD + ABC\overline{D} = ABC(D + \overline{D}) = ABC \qquad (2.129)$$

The grouping of squares along any axis therefore enables the minimization which is typified by equation (2.129). An exten-

sion of this principle is shown in Figure 2.106. The Boolean expression depicted in this figure can be written as

$$F = AB\overline{C}\overline{D} + ABC\overline{D} + \overline{A}\overline{B}CD + A\overline{B}CD$$
$$= AB\overline{D}(C + \overline{C}) + \overline{B}CD(\overline{A} + A)$$
$$= AB\overline{D} + \overline{B}CD$$

Minimization in the above example reduces the four terms in the expression to two, each involving three variables. The groupings in the example are akin to the idea of rolling the map into a cylinder about either axis to complete the two groupings as shown.

In extending the minimization principle to five variables the number of squares required is $2^5 = 32$. This is best handled as two sets of 16 squares in a top and bottom arrangement. The 16-square layers represent the first four-input variables and each layer accommodates the two possible input combinations for the fifth variable. Higher numbers of input variables can be dealt with, but the map becomes increasingly more difficult to handle.

In certain situations involving a number of input variables, particular combinations of the variables never actually occur in practice. Under these circumstances the output which would occur with these combination of variables is irrelevant. The output can therefore have any value, since it is a situation that never occurs. Such input combinations are called 'don't care' conditions, and they can be incorporated into a system to allow a simpler circuit realization.

The principle can be illustrated by means of an example:

$$F = \overline{A}B\overline{C}D + AB\overline{C}D + \overline{A}BCD$$

It is stated that the combination $ABCD$ will never occur. Including the don't care condition in the expression gives

$$F = \overline{A}B\overline{C}D + AB\overline{C}D + \overline{A}BCD + \{ABCD\}_x \qquad (2.130)$$

The don't care combination is usually enclosed within parentheses and subscripted with either x or 0.

The Karnaugh representation for the expression is shown in Figure 2.107 and the don't care condition is clearly indicated in the figure. By ignoring the don't care condition, minimization of the expression results in

$$F = B\overline{C}D + \overline{A}BD$$

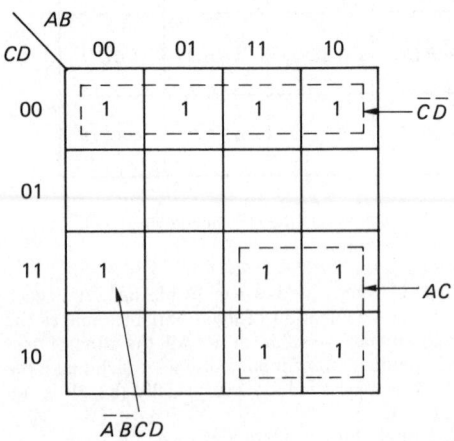

Figure 2.105 Karnaugh map for a Boolean expression (2)

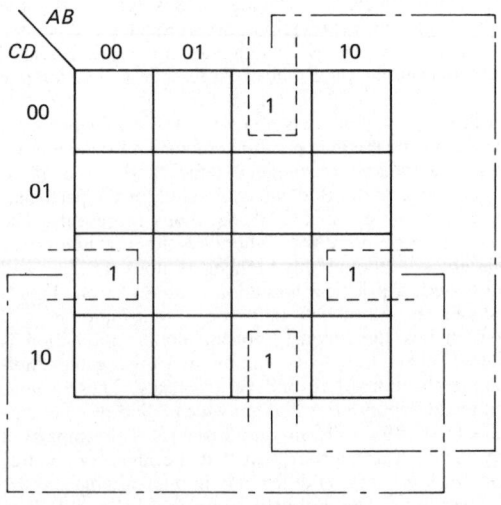

Figure 2.106 Extended minimization principle

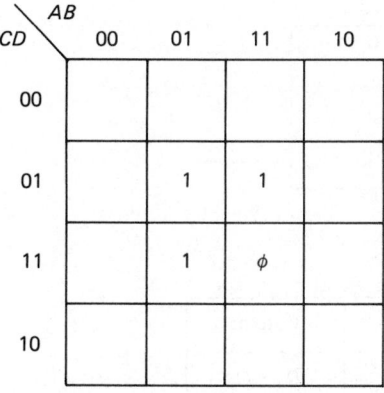

Figure 2.107 Karnaugh representation for equation (2.130)

If the network output is allowed to be 1 for the don't care condition, the minimization yields

$F = BD$

The example evidently shows that considerable savings in the realization of an expression can be made by including a relevant don't care condition.

It is also worth bearing in mind that although the Karnaugh map can yield a minimum gate solution to a given problem, it might not be an optimum solution. In the real world other considerations may well dictate in terms of parts, design, assembly costs and the number of IC packages required.

2.3.28 Positive and negative logic

In considering the digital logic systems so far, no mention has been made of the significance of the logic levels in terms of the actual voltages applied. Two possibilities exist to differentiate between logic 1 and 0. In a positive logic system, logic level 1 is represented by a more positive voltage level than logic level 0. Both logic voltage levels could actually be negative, but many digital systems operate with a voltage between 0 V and 0.8 V, denoting logic level 0 and a voltage between 2.4 V and 5 V, denoting logical level 1. This standard is used in the TTL and CMOS series of logic devices.

In a negative logic system, logic level 1 is represented by a less positive voltage than logic level 0. This standard applies to data-transmission interfaces where a voltage in the range −3 V to −15 V denotes logic 1 and a voltage in the range +3 V to +15 V logic 0. The large differentiation between 0 and 1 ensures good immunity to electrical noise. These voltages, however, are not compatible with TTL and CMOS devices, and interconversion ICs are required within the data-transmission interface.

As an alternative to using the terms logic 1 and logic 0, 'high' and 'low' are often substituted. In a positive logic system a transition from logic 0 to logic 1 can be termed a transition from low to high.

The logic level definitions also influence the function of the logic device. Figure 2.108 shows two types of two input NOR gates. In Figure 2.108(a) the inputs are negative logic and the output is positive logic. The NOR gate therefore performs the logical AND function. In Figure 2.108(b) the inputs are positive logic while the output is negative logic. This NOR gate therefore performs the logical OR function.

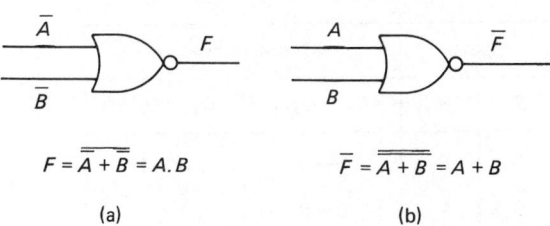

$$F = \overline{A} + \overline{B} = A . B$$

(a)

$$\overline{F} = \overline{\overline{A} + \overline{B}} = A + B$$

(b)

Figure 2.108 NOR gates using positive and negative logic input/output systems

2.3.29 Tri-state logic

Tri-state logic does not represent three logic levels but denotes three states which may be logic 1, logic 0, or 'unconnected'. A separate 'enable' input determines whether the output behaves as a normal output or goes into the third (open-circuit) state. Tri-state devices are used in applications where different logic devices are required to be connected into output lines which are common to other logic devices (for example, computer data buses). While one set of logic devices is transmitting signals the other set is temporarily disconnected or disabled.

2.3.30 Sequential logic circuits

The logic circuits considered so far are all examples of combinational logic systems where the output is determined by the combination of input variables present at that time. Sequential logic circuits are those in which the outputs depend upon the sequence of prior inputs. The main difference between sequential and combinational logic systems is that the former circuits must possess some semblance of 'memory'. The basic memory element in sequential logic systems is provided by one of several 'bistable' gates, so called because of the two different but stable outputs which the gates produce.

2.3.30.1 The SR bistable (flip-flop)

The term 'flip-flop' is traditionally used with respect to basic memory elements, and in the SR flip-flop the 'S' denotes Set and 'R' denotes Reset. The SR flip-flop was an early development, commonly constructed using discrete transistors. The internal operation, in which two transistors alternate between the cut-off and saturated states, is of less importance than the external function which the device performs.

Using the systems approach, the SR flip-flop can be represented as shown in Figure 2.109. The system shows the two inputs S and R and the two output lines traditionally denoted as Q and $\overline{Q}$. For sequential circuits the truth table is more usually called a state table. The state table for the SR flip-flop is given in Figure 2.110. Each set of input variable values is considered for both possible states of the output. This is necessary because the output values depend not only on the

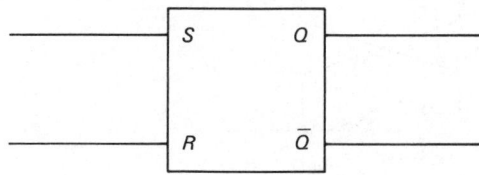

Figure 2.109 The SR flip-flop

Inputs		Output changes	
S	R	$Q_n \rightarrow Q_{n+1}$	$\overline{Q}_n \rightarrow \overline{Q}_{n+1}$
0	0	$0 \rightarrow 0$	$1 \rightarrow 1$
0	0	$1 \rightarrow 1$	$0 \rightarrow 0$
0	1	$0 \rightarrow 0$	$1 \rightarrow 1$
0	1	$1 \rightarrow 0$	$0 \rightarrow 1$
1	0	$0 \rightarrow 1$	$1 \rightarrow 0$
1	0	$1 \rightarrow 1$	$0 \rightarrow 0$
1	1	Not available	
1	1		

Figure 2.110 State table for the SR flip-flop

input variable values but also on the current values of the outputs themselves.

The operation of the SR flip-flop may be summarized as follows:

1. With $S = 0$ and $R = 0$, the output is not affected and remains as it was.
2. With $S = 1$ and $R = 0$, the output will change to $Q = 1$ if previously Q was 0. Q will remain at 1 if previously Q was 1.
3. With $S = 0$ and $R = 1$, the output will change to $Q = 0$ if previously Q was 1. Q will remain at 0 if previously Q was 0.
4. In all cases considered, the output $\overline{Q}$ will be the inverse complement of Q.

The SR flip-flop may be constructed using cross-coupled NOR or NAND gates, as shown in Figure 2.111.

2.3.30.2 The T (Trigger) flip-flop

The T flip-flop is another bistable circuit having two outputs, Q and $\overline{Q}$, but only one input, T. The T flip-flop changes state on every T input signal and then remains in that state while the T input remains low.

2.3.30.3 The JK flip-flop

The JK flip-flop uses integrated-circuit technology and, since it can perform both the SR and T flip-flop functions, it has become the most common flip-flop in current use. Figure 2.112 gives the state table and logic symbol for the JK flip-flop. The state table is identical to the SR flip-flop with the exception that the input condition $J = 1$, $K = 1$ is allowed. For these latter inputs the JK flip-flop functions as a T flip-flop

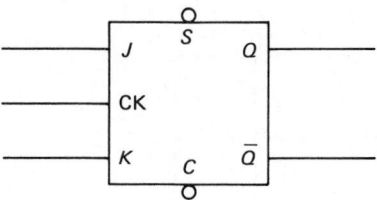

Inputs		Outputs
J	K	$Q_n - Q_{n+1}$
0	0	$0 \rightarrow 0$
0	0	$1 \rightarrow 1$
0	1	$0 \rightarrow 0$
0	1	$1 \rightarrow 0$
1	0	$0 \rightarrow 1$
1	0	$1 \rightarrow 1$
1	1	$0 \rightarrow 1$
1	1	$1 \rightarrow 0$

Figure 2.112 JK flip-flop and corresponding state table

using an input clock signal, in the form of a pulse train, as the trigger.

The JK flip-flop operates in a clocked or synchronous mode. In synchronous mode, the J and K inputs do not in themselves initiate a change in the logic outputs but are used to control inputs to determine the change of state which is to occur. A pulsed input to the clock terminal (CK) then determines the timing of the state changes. The clocked mode allows for precise timing of the state changes in a sequential circuit.

JK flip-flops may also be provided with additional Set, S, and Clear, C, inputs which can be used to set output Q to 1, or clear output Q to 0 at any time. Muliple J and K inputs are also commonly available to enable logical ANDing of multiple-input signals.

A slightly more complicated flip-flop arrangement is the JK master–slave flip-flop. This consists of a pair of SR flip-flops connected together by various logic gates as shown in Figure 2.113. The JK master–slave flip-flop differs from the simpler arrangement in that if the clock pulse is at logic 1, a logic 1 applied to either J or K will not set the outputs. The new data, however, are accepted by the 'master'. When the clock pulse returns to 0, the master is isolated from the inputs but its data are transferred to the slave, with the result that Q and $\overline{Q}$ can then change state. In a circuit involving many such flip-flops,

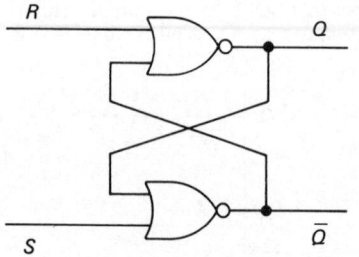

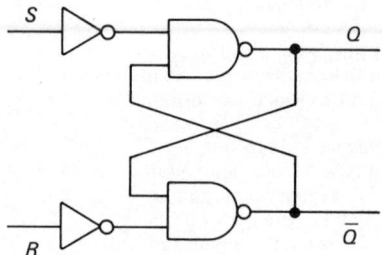

Figure 2.111 SR flip-flops using cross-coupled gates

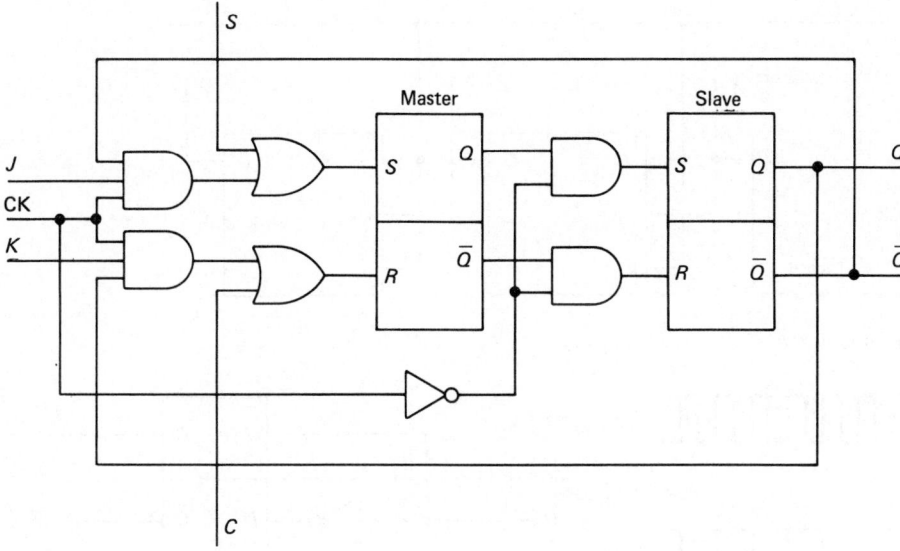

Figure 2.113 The JK master–slave flip-flop

the advantage of the master–slave arrangement is that it can allow for synchronization of all the output state changes.

2.3.31 Registers and counters

In the previous section it was shown that a logic level of 1 on a particular input line to a flip-flop can set an output line to 1. In this way the flip-flop can perform an elementary memory function. For a binary signal of length n bits, n flip-flops are required to construct a memory device for the n-bit input signal. A group of flip-flops used together in this manner constitutes a 'register'.

Data may be entered into the register in a serial or a parallel manner. In the parallel method the n-bit binary 'word' is available on n input lines. Each line is connected to its own flip-flop and the n data bits are entered simultaneously into the register. In the serial entry method the data are available on only one input line in a time sequence. They are entered consecutively and are timed into the register by a system clock. Serial entry registers are also called 'shift' registers, since the data bits are entered into the first flip-flop and moved consecutively along into the next flip-flop as the next data bit arrives at the first flip-flop, and so on. The serial method of data entry requires as many shift and store operations as the number of bits in the binary word. This means that the serial entry method is much slower than the parallel method. Serial entry, however, is also much less expensive than parallel entry.

Yet another type of register is the counting register. This consists of a number of flip-flops arranged to store a binary word which is representative of the number of input pulses applied at the input terminal. Using n flip-flops, a total count of 2^n can be made.

Counting registers (or 'counters') may be synchronous in which the state changes in all the flip-flops occur simultaneously, or asynchronous in which the state changes in various flip-flops do not occur at the same time. Figure 2.114 illustrates an asynchronous, 3-bit binary counter composed of JK flip-flops.

All J and K inputs are held at logic level 1 and the input signal consists of a pulse train fed to the clock input of the first flip-flop. In this mode the JK flip-flop is operating as a T flip-flop. The output Q from the first flip-flop provides an input for the clock of the second flip-flop, and so on through the network. The outputs Q also form the binary representation of the counter, where A is the least significant bit, increasing through to C, which represents the most significant bit. The state table and timing diagram for the counter are shown in Figure 2.115.

The state table shows that each flip-flop changes state when the next less significant flip-flop output changes from 1 to 0. The output signal from each flip-flop, moving through the network, is at half the frequency of that of the previous flip-flop. These output signals thus provide the correct binary count of the number of input pulses applied to the input. The 3-bit binary counter can count up to a maximum of 8 decimal. If a ninth pulse is applied at the input, the count reverts back to the initial zero setting and the count continues again as normal for further input pulses.

Asynchronous counters are also referred to as 'ripple' counters because of the way that the changes of state ripple through the network of flip-flops.

A synchronous version of the counter can also be realized using a network of JK flip-flops. The synchronous counter additionally uses the outputs Q and $\overline{Q}$ of each flip-flop, in logic gate networks, to produce the necessary control signals for the J and K inputs. This ensures that all flip-flops change state correctly to the desired state table for each clock pulse. The synchronous counter alleviates the problems associated with transient operation inherent in the asynchronous counter.

There are, of course, many other types of flip-flop available, but the only one of significant practical importance is the D flip-flop (see Figure 2.116), where the 'D' refers to Data.

In the D-type flip-flop the D input is fed directly into the J input line and the inverse complement of D is fed to the K input line. This ensures that J and K are always the inverse complement of one another. A logic 0 or 1 on the data input will then flip (or flop) the outputs when the clock pulse is at logic 0.

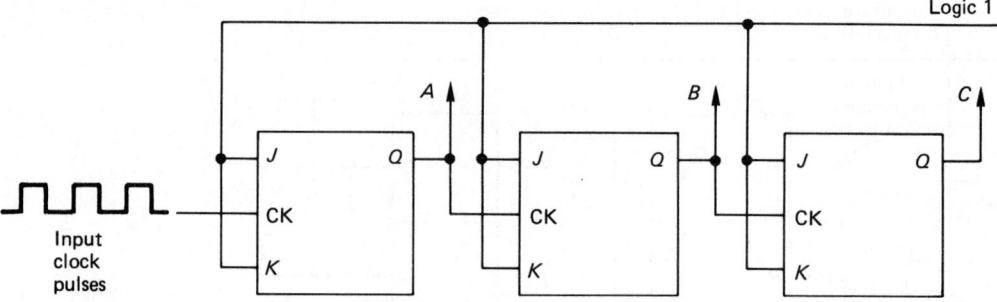

Figure 2.114 Asynchronous, 3-bit binary counter

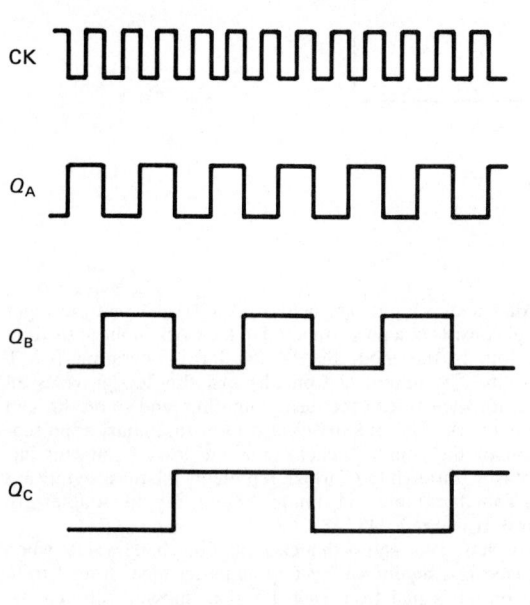

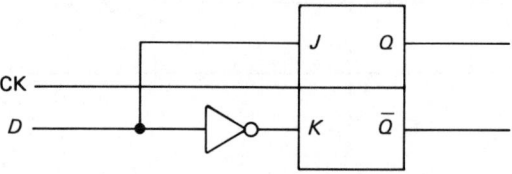

Figure 2.116 D-type flip-flop

manufactured as an 8-pin DIL package. The CMOS equivalent to the TTL-based NE555 is the ICM7555. These packages are essentially identical, and ICM7555 can be used to replace NE555, although the converse is not always applicable. The so-called '555' timer is very versatile and can be used in either 'monostable' or 'astable' mode.

2.3.32.1 Monostable

Figure 2.117 shows the 555 wired up for monostable operation. When the 'trigger' is taken from +5 V to 0 V (i.e. high to low) the output will go high for a period determined by the values selected for R and C. The length of the output pulse is given by $1.1RC$.

Clock pulses	Flip-flop		
	C	B	A
0	0	0	0
1	0	0	1
2	0	1	0
3	0	1	1
4	1	0	0
5	1	0	1
6	1	1	0
7	1	1	1
0	0	0	0

Figure 2.115 State table and timing diagram for a 3-bit binary counter

2.3.32 Timers and pulse circuits

An essential feature of the flip-flop circuits described in the previous two sections was the provision of a pulsed clock signal. Although timers can be designed using discrete components, it is normal to design round the commonly available timers which are already available in IC form. The most prevalent timer currently in use is the NE555, which is

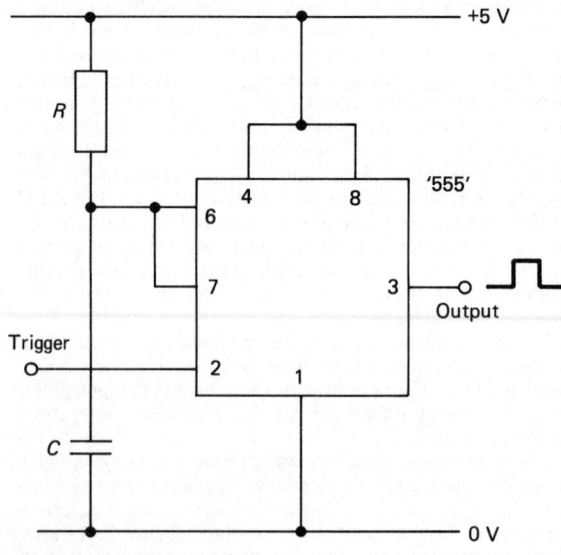

Figure 2.117 555 timer in monostable operation

The timer can deliver currents of more than 100 mA, and it can therefore be used to drive a DIL reed relay directly. When such a relay is switched off, however, the back e.m.f. generated by the relay coil could damage the timer. As a precaution, a diode is normally connected in parallel with the relay coil, in the opposite direction to the current flow, to absorb the high induced voltage.

2.3.32.2 Astable

Figure 2.118 depicts the 555 wired up for astable operation. The 100 nF capacitor is only required for TTL-based timers. In astable operation the output is a continuous pulse train. The ON and OFF times can be controlled independently within certain limitations with

$$\text{ON time} = 0.693(R_1 + R_2)C \qquad (2.131)$$

$$\text{OFF time} = 0.693(R_2)C \qquad (2.132)$$

Obviously, the ON time can only be equal to or greater than the OFF time. The output signal, however, can always be inverted if, in a particular application, short-duration positive pulses are required.

The maximum operating frequency for the 555 timer is about 500 kHz and the minimum frequency, limited by the leakage of the capacitor, is about one cycle per several hours.

Alternative pulsed output circuits can be constructed using TTL or CMOS gates (see Kaufman and Seidman[8] and Watson[9]).

2.3.33 Digital computers and microprocessors

No coverage of digital electronics, however brief, can fail to give some cognizance to the impact of the digital computer and its associated microprocessor. The modern digital computer, although a complex digital system, consists of no more than the basic logical sub-systems previously discussed. This includes AND, OR, NAND and NOR gates, registers, counters and communication interfaces. Space limitations do not allow a detailed description of computer systems and microprocessors in this chapter. However, Chapter 3 of this book contains specific details related to microprocessor tech-

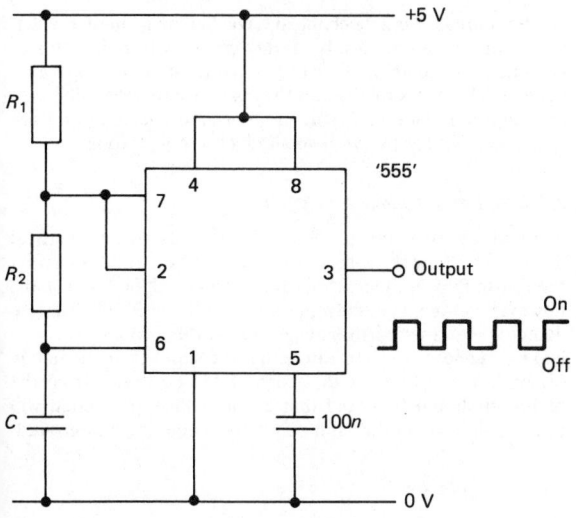

Figure 2.118 555 timer in astable operation

nology, number systems and interfacing techniques for digital computers.

The main advantages of the microprocessor-based system are that the logical functions for a particular application can be developed and implemented in software, as opposed to electronic hardware. In many instances the microprocessor-based system may actually be the cheaper alternative to a hardwired logic gate circuit. The software is easy to alter in the event of incorrect system operation and the complete system can be tested as a simulation before being committed.

For relatively small logical switching applications (up to, say, 32 inputs) the single-card (or single-chip) microcomputer represents an ideal low-cost solution (see Milne and Fraser[10]). These micro-systems can be used as dedicated devices where all the system components reside on a single card or a single chip, respectively. The major applications for these devices are in the high-volume production markets such as automotive electronics, washing machines, bus ticket dispensers and time-attendance recorders.

2.3.34 Application-specific integrated circuits (ASICs)

Application-specific integrated circuits are programmable logic devices (PLDs) which have their internal logic configuration determined by the user, as opposed to the manufacturer. The systems design engineer therefore customizes the actual silicon building blocks to meet the requirements of the system. Such customization provides for performance, reliability, compactness, low cost and design security. PLDs are available in both TTL and CMOS technology. The latter are erasable and can be reprogrammed almost indefinitely. PLDs represent the fastest-growing segment of the semiconductor industry in recent times, and it can be expected that they will play an increasingly important role in the design of digital logic systems in the future.

Internally, PLDs consist of an array of AND gates connected to an array of OR gates, with input and output blocks containing registers, latches and feedback options. Figure 2.119 shows the general architecture of a programmable logic device.

In customizing the PLD the user essentially determines which of the interconnections between the gate arrays will remain open and which will be closed. The customization procedure, however, requires additional development 'tools' which consists of:

1. A word processor to generate the source code;
2. Development software to transform the high-level language source code into a fuse pattern for the PLD. The code which is generated is referred to as a 'JEDEC' file;
3. A PLD programmer to implement the program within the device.

The PLD programmer, connected to the parallel printer port of an IBM-PC or a true compatible programs the PLD by 'burning' the fuse pattern in the memory array of the device. When returned to its normal operating mode, the PLD then performs the customized logic function. Horowitz and Hill[11] provide a reasonably detailed coverage on applications of programmable logic devices.

2.4 Electrical safety

The definitive document on electrical safety in the UK is The Electricity at Work Regulations 1989, which came into force on 1 April 1990. The full text of the regulations is set out in

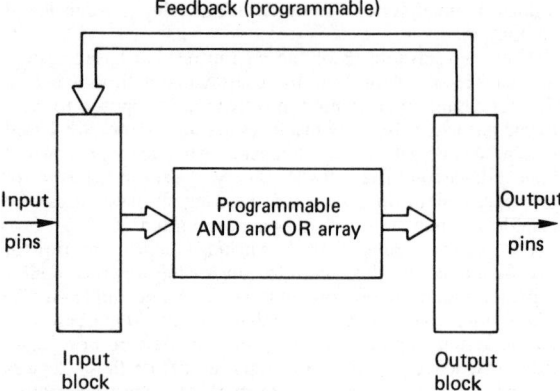

Feedback (programmable)

Input pins

Programmable
AND and OR array

Output pins

Input block

Output block

Figure 2.119 General architecture of a PLD

Statutory Instrument 1989, No. 635, available from HMSO. The notes included here are based on the Health & Safety series booklet HS(R)25, which forms part of the Electricity at Work Regulations.[12] The regulations are made under the Health and Safety at Work Act 1974, and the Act imposes the responsibility for people at work on both the employer and the employee. It is partly the individual's responsibility therefore to ensure that safe working practices are observed in the workplace.

2.4.1 Electric shock

Most serious injuries and deaths from electric shock occur from contact with the mains electricity supply. In the UK, the mains supply is about 240 V a.c. and in the USA it is about 110 V a.c. The live wire (colour coded brown) is at a higher potential with respect to the earth. If a person is in contact with the ground, an electric shock can be sustained by touching the live wire only. When a shock is received the passage of the electric current through the body may cause muscular contractions, respiratory failure, fibrillation of the heart, cardiac arrest or injury from internal burns. Any one of these can be fatal. The greatest danger occurs when the current flows across the chest. This can happen either when the current flows from one arm to the other or when it flows from one arm to the opposite leg.

The magnitude of an electric shock depends on the strength of the current, which in turn depends on the voltage and the ohmic resistance of the body. The resistance of the human body varies in different persons and is primarily dependent on the resistance of the skin. This variability in skin resistance means that a 'safe' voltage cannot be readily specified, and all voltages in excess of about 50 V must therefore be regarded as being ppotentially lethal. If the skin is damp, either from water or from perspiration, then the skin resistance is considerably reduced. Under such circumstances the chances of an electric shock proving to be fatal are greatly increased.

Injury can also be caused by a minor shock, not serious in itself but which has the effect of contracting the muscles sufficiently to result in a fall.

2.4.2 Electric burn

Burns can be caused either by the passage of heavy current through the body or by direct contact with an electrically heated surface. They may also be caused by the arcing across a short circuit or as a result of an electrically originated fire or explosion.

Any circuit brought near an induction heater will receive energy and heat will be generated extremely rapidly. No rings or other metal objects should therefore be worn when in the vicinity of an induction heater, neither should any metal be held in the hands. In general, no part of the body should come within close proximity (about a metre) to an induction heater.

2.4.3 Rescue

To render assistance to a person undergoing an electric shock the rescuer should first attempt to isolate the circuit by switching off the supply. If the rescuer cannot isolate the supply, he or she should try to break the victim's contact with the live apparatus by using insulating material. This material is essential to prevent the rescuer from becoming a second victim.

If the victim is unconscious the rescuer should send for medical assistance and start artificial resuscitation. This assumes, of course, that the rescuer is trained in modern methods of such resuscitation.

2.4.4 Protection

2.4.4.1 Insulation

Electrical cables consist of one or more metal cores which may be single wires but are more usually stranded wire and surrounded by insulation. The insulation serves to contain the flow of current and prevent a person from touching the live metal and thereby receiving a shock.

2.4.4.2 Fuses

A fuse is a device which will melt when the current exceeds a predetermined value. In operation, fuses serve as current-limiting devices, and they are used for overload protection of electrical equipment. Two types of fuses are generally available: rewirable and cartridge.

Fuses are specified by the maximum current that they can transmit and the correctly rated fuse must be used at all times. A high-rating fuse must never be substituted for a low-rating one.

2.4.4.3 Circuit breakers

A circuit breaker is a mechanical device in the form of a switch which opens automatically if the circuit which it controls becomes overloaded. Circuit breakers may be operated magnetically or thermally and they can also be manually reset and adjusted. Plug-in circuit breakers are available and are recommended for use with small electric power tools.

2.4.4.4 Earth-leakage protection

Normally there is no net flow of electricity to an electrical device. The flow in from the live wire is exactly balanced by the return flow in the neutral cable. If an earth fault develops, however, a leakage current will result, and this can be detected by the earth-leakage apparatus. Modern earth-leakage devices are so sensitive that the supply is immediately disconnected before a lethal current can be drawn from the mains. It should be noted that earth leakage protection will operate only when the fault occurs between line and ground.

2.4.4.5 Isolation

Effective means of disconnecting cables or apparatus from the source of supply must be provided so that maintenance, repair or alteration may be carried out safely. This is achieved by isolating switches which have no automatic features. Various circuits and motors should not 'share' an isolation switch unless it is clear that under no circumstances will it be necessary, or convenient, to use one circuit while the other is being serviced.

Isolation switches should be capable of being locked in the 'off' position but not in the 'on' one. If the isolation switch cannot be locked it should be possible to remove the fuse on the power line so that the line cannot be energized by inadvertent closing of the isolation switch.

2.4.5 Earthing

The external metal casing of electrical apparatus and cables must be earthed for three reasons:

1. To prevent the casing rising to a dangerous voltage if there is a fault such as a short circuit between the conductor and the casing;
2. To conduct any current away by a safe path;
3. To ensure that the faulty circuit is automatically disconnected from the supply by drawing sufficient current to blow the protective fuse or operate the circuit breaker.

Earthing consists of connecting the metal casing by means of a conductor to an earth electrode. The earth electrode may be a buried pipe or other such conductor which is known to be making an effective connection to the 'general mass of the earth'. Where the earth connection to a casing is made with a nut and bolt a spring washer or other similar locking device must be used. Earthing is a legal requirement and must be effective at all times.

2.4.6 Double insulation

Although the electricity regulations require all portable apparatus used at normal mains voltage to have an earthing conductor, these can introduce their own hazards. As a result 'double-insulated' or 'all-insulated' apparatus is made which does not require earthing. Double insulation means what its name says, and all live conductors are separated from the outside world by two separate and distinctive layers of insulation. Each layer of insulation would adequately insulate the conductor on its own, but together they virtually negate the probability of danger arising from insulation failure. Double insulation avoids the requirement for any external metalwork of the equipment to be protected by an earth conductor.

2.4.7 Low-voltage supplies

Portable tools (particularly hand inspection lamps) can be a source of danger because they are subject to severe wear and tear and are likely to be used in confined spaces where the skin

resistance could easily be reduced by damp conditions. In cases where work is carried out within confined metal enclosures, mains voltage equipment must not be used. A double-wound transformer with a secondary centre tap to earth is allowable in these cases. This transformer gives 50 V for lighting and 100 V for portable tools.

These few notes on general electricity safety are by no means extensive or authoritative. Reference should always be made to the full guide to the regulations.[12] Further recommended reading on electrical safety guidelines can be obtained from references 13 and 14.

References

1. Hughes, W., *Electrical Technology*, sixth edition, revised by I. McKenzie Smith, Longman, Harlow (1987)
2. Bell, E. C. and Whitehead, R. W., *Basic Electrical Engineering & Instrumentation for Engineers*, third edition, Granada, St Albans (1987)
3. Bell, D. A., *Fundamentals of Electric Circuits*, third edition, Reston Publishing, New York (1984)
4. Gray, C. B., *Electrical Machines and Drive Systems*, Longman Scientific & Technical, Harlow (1989)
5. Sen, P. C., *Principles of Electric Machines and Power Electronics*, John Wiley, Chichester (1989)
6. Orthwein, W., *Machine Component Design*, West Publishing, New York (1990)
7. Kenjo, T. and Nagamori, S., *Permanent-Magnet and Brushless dc Motors*, Monographs in Electrical and Electronic Engineering, Clarendon Press, Oxford (1985)
8. Kaufman, M. and Seidman, A. H., *Handbook of Electronics Calculations for Engineers and Technicians*, second edition, McGraw-Hill, New York (1988)
9. Watson, J., *Mastering Electronics*, Macmillan, London (1983)
10. Milne, J. S. and Fraser, C. J., 'Development of a mechatronics learning facility', in *Mechatronic Systems Engineering*, Vol. 1, pp. 31–40, Kluwer Academic, New York (1990)
11. Horowitz, P. and Hill, W., *The Art of Electronics*, second edition, Cambridge University Press, Cambridge (1989)
12. Health and Safety Executive, Memorandum of Guidance on the Electricity at Work Regulations, 1989, Health & Safety series booklet HS(R)25, HMSO (1989)
13. Imperial College of Science and Technology, *Safety Precautions in the Use of Electrical Equipment*, third edition, Imperial College Safety Booklets (1976)
14. Reeves, E. A., *Handbook of Electrical Installation Practice*, Vol. 1, *Systems, Standards and Safety*, Ed. E. A. Reeves, Granada, St Albans (1984)

Further reading

Fitzgerald, A. E., Higginbottom, D. G. and Grabel, A., *Basic Electrical Engineering*, fifth edition, McGraw-Hill, New York (1981)
Nasar, S. A., *Handbook of Electric Machines*, McGraw-Hill, New York (1987)
Say, M. G., *Alternating Current Machines*, Pitman, London (1983)

3

Microprocessors, instrumentation and control

Charles J. Fraser
John S. Milne (Sections 3.1–3.4, 3.5.9 and 3.6–3.9)
Ben Noltingk (Sections 3.5.1–3.5.8)

Contents

3.1 Summary of number systems 3/3
   3.1.1 ASCII code 3/3
   3.1.2 Gray code 3/3

3.2 Microprocessors 3/4
   3.2.1 System architecture 3/4
   3.2.2 Bus structure 3/4
   3.2.3 Memory devices 3/5
   3.2.4 Input/output (I/O) structure 3/5
   3.2.5 Memory map 3/6

3.3 Communication standards 3/7
   3.3.1 Serial communication 3/7
   3.3.2 Parallel communication 3/7

3.4 Interfacing of computers to systems 3/8
   3.4.1 Digital interfacing 3/8
   3.4.2 Controller output interface hardware 3/9
   3.4.3 Analogue interfacing 3/12
   3.4.4 Multiplexing 3/15
   3.4.5 Machine tool interfaces 3/15
   3.4.6 Robot control interfaces 3/15
   3.4.7 Signal conditioning 3/18
   3.4.8 Analogue and digital filtering 3/19

3.5 Instrumentation 3/21
   3.5.1 Introduction 3/21
   3.5.2 Dimensional/geometrical measurements 3/23
   3.5.3 Volume and level 3/27
   3.5.4 Measurement of vibration 3/28
   3.5.5 Force/weight measurement 3/28
   3.5.6 Pressure 3/29
   3.5.7 Flow 3/31
   3.5.8 Temperature measurement 3/34
   3.5.9 Bar code readers 3/39

3.6 Classical control theory and practice 3/40
   3.6.1 Introduction 3/40
   3.6.2 Mathematical models of systems – time-domain analysis 3/43
   3.6.3 Laplace notation for differential equations – frequency-domain analysis 3/44
   3.6.4 Stability criteria 3/47
   3.6.5 Control strategies 3/49

3.7 Microprocessor-based control 3/58
   3.7.1 Direct digital control 3/59
   3.7.2 Hardware requirements 3/59
   3.7.3 Software considerations 3/60
   3.7.4 Sampling frequency in digital control loops 3/62
   3.7.5 PID digital control algorithm 3/62
   3.7.6 Speed control 3/63
   3.7.7 The PC as a controller 3/65

3.8 Programmable logic controllers 3/65
   3.8.1 The PLC in automation systems 3/65
   3.8.2 The PLC versus the microcomputer 3/66
   3.8.3 Ladder logic programming 3/67
   3.8.4 Controlling pneumatic and hydraulic systems 3/68
   3.8.5 Safety 3/69
   3.8.6 Networking of PLCs 3/69

3.9 The z-transform 3/70
   3.9.1 Representation of discretely sampled data 3/71
   3.9.2 The z-transform of a closed-loop system 3/72
   3.9.3 Proportional control using digital techniques 3/73
   3.9.4 The z-transform for a PID controller 3/74
   3.9.5 A $P + I$ strategy using digital techniques 3/74
   3.9.6 Stability in discrete time systems 3/75

3.10 State variable techniques 3/76
   3.10.1 State variable representation of systems 3/76
   3.10.2 Application to a first-order system with a $P + I$ controller 3/77
   3.10.3 Application to a second-order system with a $P + I$ controller 3/78
   3.10.4 Non-linear system elements (method of isoclines) 3/78
   3.10.5 Sampled-data systems 3/83
   3.10.6 State variable transformations 3/84
   3.10.7 The state transition matrix 3/86

References 3/87

Further reading 3/87

3.1 Summary of number systems

In the manipulation of data within a computer a two-state numbering system is used. This is termed the binary system, and it is based on a simple ON/OFF principle. For the semiconductor integrated circuits which make up the computer system, 5 V denotes ON (or logic level '1') while 0 V denotes OFF (or logic level '0'). In practice, a tolerance band is adopted, with 2.4–5 V representing logic '1' and 0–0.8 V logic '0'.

The microelectronic devices in the system handle the transfer of information in 1's and 0's which are referred to as 'BITs', being a short form for BInary digiT. A group of eight bits is termed a 'byte' and a number of computer systems are based on 8-bit technology with the handling of data codes as 8-bit 'words'. Sixteen- and 32-bit machines are also available.

The computer operates with three numbering systems, which are decimal, binary and hexadecimal (often simply called 'hex'). Numerical data would normally be entered by a human operator in decimal form since this is the most familiar number system. The computer, however, must ultimately convert the decimal number into a binary code, since this is the eventual form in which the number will be processed and stored. The hexadecimal system is an in-between state and represents a particularly compact method of handling binary numbers as groups of four bits.

In binary representation the only possible logic levels are 0 and 1. The base is chosen as 2 and integer numbers can be represented using 8-bit codes as shown below:

bit number	7	6	5	4	3	2	1	0
	2^7	2^6	2^5	2^4	2^3	2^2	2^1	2^0
	128	64	32	16	8	4	2	1

Most significant bit (MSB) Least significant bit (LSB)

The conversion from binary to decimal is illustrated as follows:

binary number	1	0	1	1	1	0	0	1

giving $128 + 0 + 32 + 16 + 8 + 0 + 0 + 1$
$= 185$ decimal

Conversion from decimal to binary is the reverse process to the above.

It is apparent that the highest number which can be accommodated in 8-bit binary notation is 1111 1111, which is equivalent to 255 decimal. Generally, therefore, computer systems handle integer numbers in four consecutive bytes, i.e. as 32-bits. The most significant bit is used to denote the sign of the number and the resulting range of integer numbers is

$-2^{31} - 1$ to 2^{31}, or $-2, 147, 483, 648$ to $2, 147, 483, 647$

Real numbers are handled in five bytes with the most significant byte representing an exponent and a sign bit and the other four bytes the mantissa and a sign bit. The resultant range of real numbers is 2^{-128} to 2^{127}, with either a positive or a negative sign for the mantissa.

The handling of numbers in binary notation is extremely cumbersome for a human and a shorthand notation is adopted for convenience. This is the hex system in which the binary number is arranged into groups of four bits. Four bits, which is half of a byte, is called a 'nibble'. A byte therefore consists of an upper and a lower nibble.

Since there are only ten unique symbols in the decimal numbering system, the first six letters of the alphabet are used to denote the additional six symbols in the hexadecimal system, i.e.

Decimal   0 1 2 3 4 5 6 7 8 9 10 11 12 13 14 15
Hexadecimal   0 1 2 3 4 5 6 7 8 9 A B C D E F

Using hex notation then, 8-bit binary numbers may be replaced by two hex symbols, e.g.

167 decimal = 1010 0111 binary = A7 hex

Higher numbers are similarly handled:

6836 decimal = 0001 1010 1011 0100 binary = 1AB4 hex

3.1.1 ASCII code

In the interchange of information between the constituent parts of a computer or a peripheral device a binary code is used to represent the alphanumeric characters. The most commonly used code for digital communication links is the American Standard Code for Information Interchange (ASCII, pronounced Askey). ASCII is a 7-bit code which can accommodate 128 definable characters.

When communication takes place in a serial fashion the ASCII code is extended to 8-bits, usually by inserting a zero in the most significant bit. Additionally, one or two start bits, a parity bit and a stop bit are also included. The start bit(s) inform the receiving device that a character code follows. The parity bit provides a check that no bits have been corrupted during transmission, by ensuring that the sum of all the 1's in the ASCII group give either an even number for 'even parity' or an odd number for 'odd parity'. The stop bit, set to logic '1', terminates the transmission of the character.

The transmission rate in bits/second is termed 'baud'. Since there are 11 bits associated with the transmission of one character, a speed of 2400 baud corresponds to 2400/11 = 218 characters per second.

3.1.2 Gray code

The Gray code is but one of many binary codes in which only one of the digits change between successive consecutive numbers. The main application is in the sensing of rotational and translational position in mechanical systems (see Section 3.5.2.7).

In converting from Gray to binary code the most significant bit of the binary number, B, is equal to the most significant bit of the Gray code, G. For all other bits, the relationship between binary and Gray is given by

$$B(n) = G(n) \oplus B(n + 1) \qquad (3.1)$$

where n denotes the bit reference number and $\oplus$ is an exclusive-OR logic comparison.

The conversion of 1101 Gray to binary is shown below:

bit number	3	2	1	0
Gray code	1	1	0	1

binary number   1 = 0 = 0 = 1
i.e. 1101 Gray = 1001 binary

In practical position-sensing applications the conversion process can be programmed in the software or implemented in a hardwired logic circuit using logic gates.

3.2 Microprocessors

3.2.1 System architecture

Since the mid-1970s there are currently a number of different microprocessor designs available in several versions. The 'popular' designs are produced by a few manufacturers, and these include such companies as Intel, Motorola, Rockwell, Texas Instruments and Zilog.

Microprocessor-based systems require additional family support chips and, in true digital form, all microelectronic components which constitute a microcomputer are designated numerically rather than by name. Some of the more popular microprocessors available are as follows:

1. *8-bit*: Binary data handled in a word, 8-bits wide, defining the accuracy of the number handling representation, e.g.
   Intel 8080 and 8085
   Motorola 6800 series
   Rockwell 6502 series
   Zilog Z80
2. *16-bit*: Binary data handled in a word of 16-bits width, e.g.
   Intel 8086
   Motorola 68000
   Zilog Z8000

The 8086 is one of the most powerful and versatile 16-bit microprocessors available, and it has been widely adopted by industry. Further enhancements include the 32-bit versions – 80286, 80386 and 80486 – which provide increased processing power.

Although it is unnecessary for the user of the technology to understand in detail how each individual chip actually functions, it becomes essential to have at least a working knowledge of the logical organization of the system hardware and how each component relates to each other. The composition of this hardware structure is known as the system architecture.

A digital computer system comprises three main constituent parts: the *microprocessor*, the *memory* and the *input/output*. Digital signals which have a common function are transmitted between the main components by a group of wires or conduction tracks, termed a bus. In a microcomputer there are three buses, i.e. the data bus, the address bus and the control bus. The interconnection between the basic hardware components in a microcomputer is illustrated in Figure 3.1.

The microprocessor is a very large-scale integrated circuit (VLSI), which is the brain of the microcomputer system and acts as the central processing unit (CPU). Integrated circuits are generally classified according to the number of components on the silicon chip, and VLSI has tens of thousands. The

Intel 8086 microprocessor, introduced in 1979, has 29 000 transistors packed on the 225 mm^2 chip.

The main feature of the microprocessor is the arithmetic and logic unit (ALU). The ALU allows the arithmetical manipulation of data, addition and subtraction with 8-bit systems and multiplication with 16-bit systems. Logical operations (AND, OR, etc.) can also be performed. In addition to the ALU, the CPU contains a number of temporary data storage registers to fetch, decode and execute instructions, and to time and control circuits to manage the internal and external operation of the complete microprocessor system.

The processing power of the CPU is influenced by such factors as word length, instruction set, addressing modes, number of available registers and information transfer rates. For word processors or the manipulation of large quantities of data as in CAD packages, 16- or 32-bit microprocessors are essential. In the field of measurement and control, 8-bit systems are usually adequate.

The system clock, accurately controlled by a quartz crystal maintaining a constant frequency, acts as the heartbeat for the system and provides all the timing reference points.

All the basic components (CPU, memory and I/O) and their interconnections may reside in a complete microcomputer system encompassing the traditional keyboard, monitor, etc. Alternatively, they may reside on a single card, or even on a single chip to give a single-chip microcomputer.

3.2.2 Bus structure

As outlined in Section 3.2.1, the connection between the system components is made by an arrangement of three buses:

1. *The data bus* transmits the words in binary form representing the program instructions for the CPU to execute. It can also carry the information transmitted between the CPU and memory or I/O devices. Although the popular PCs use a 16-bit data bus, 8-bit data transfer operations remain the norm in data acquisition and control applications.
2. *The address bus* transmits the memory address related to the data bus information. In 8-bit systems this bus commonly has 16 lines to give 64K of addresses. PCs usually have an effective 20 lines to give 1M byte of available addresses, although software limitations often restrict this to 640K.
3. *The control bus* transmits the collection of timing and control signals which supervise the overall control of the system operation.

The physical format of a bussing system is basically a circuit board with a number of connectors. Different types of microprocessors require different hardware interfaces, and to alleviate the problems, standard bus structures have been developed in order to facilitate the connection of hardware components. In industrial-type systems, cards for various microcomputer functions such as processor, memory, digital and analogue I/O, power switching, etc. slot into a standard backplane or motherboard rack. This offers the advantage of being able to plug any specific card, designed to the bus standard, into a free slot in the rack to build up the system as required.

The physical form of the bus is represented by its mechanical and electrical characteristics. Such information as card dimensions, input and output pin-out connections, signal levels, loading capability and type of output gates must be known.

Standard buses are compatible with cards from different manufacturers, and the most popular bus structures include Multibus, S-100 Bus, STD Bus and the STE Bus.

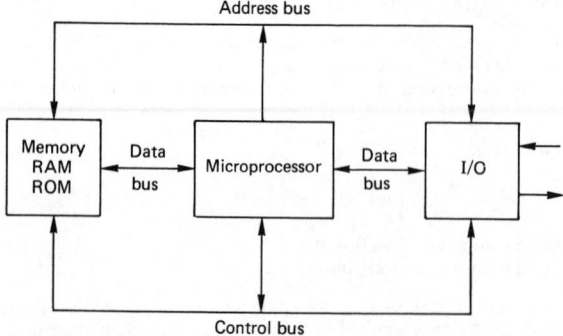

Figure 3.1 Basic components in a microcomputer

3.2.3 Memory devices

Memory devices consist of those used to store binary data, which represents the user program instructions, and those which are necessary for the user to operate the system. Memory takes the form of one or more integrated circuits. These basically hold locations capable of storing a binary word. Each location is assigned a unique address within the system and data can be selected through the address bus. As a binary code is deposited by the CPU on the address bus, defining a specific location in memory, the contents of that location are selected and placed on the data bus. The appropriate piece of memory hardware and specific location is selected by means of an address-decoding circuit built up from logic gates within the microcomputer system. The end result is a highly flexible data manipulation arrangement.

In an 8-bit microcomputer (i.e. 8-bit data bus) the address bus is 16-bits wide. This enables $2^{16} = 65\,536$ locations to be addressed, and the total memory capacity of the machine is said to be 64K. The memory is further sub-divided into pages with the high-order byte of the address denoting the page number and the low-order byte indicating one of the 256 locations available on each page.

In machines with 8086/8088 microprocessors an additional four bits are effectively made available on the address bus. This theoretically constitutes one megabyte of addressable memory. Microcomputers such as the IBM-PS2, employing the 80286 and 80386 microprocessors, have an address bus which is 24-bits wide and can address up to 16M bytes of physical memory. This releases new levels of processing power to accelerate the processing speed in measurement and control applications.

The types of memory chips built into the system basically divide into two categories:

1. *Random Access Memory (RAM)*, where data can be read from or written to any specified location. RAM is more correctly defined as read/write memory and data retention is dependent upon power being applied to the device. This type of memory is normally employed for the temporary storage of the computer programs, at the editing or execution stage, or the storage of data from measuring transducers prior to permanent storage as a disc file. In a number of systems available, the RAM is made non-volatile by providing battery back-up.

2. *Read Only Memory (ROM)*, where data are held in a secure manner and can be read in any specified sequence. Once the chip is configured it cannot be overwritten, and the programs which specify the system operation (termed the monitor program) are 'burnt' into ROM when they are known to operate in a satisfactory manner. Basic ROM is inflexible since the software contained therein is developed by the system manufacturer. It is often useful, however, to have all programs which are to be permanently stored in the microcomputer in a non-volatile form, held in an Erasable and Programmable Read Only Memory (EPROM). This is undoubtedly the most popular type of ROM used because the write process is reversible. These chips are available in popular memory capacities of 2K, 4K, 8K, 16K and 32K, and they are respectively designated as 2716, 2732, 2764, 27128 and 27256. The numbers following the '27' indicate the number of kilobits of memory available within the device.

EPROMs are supplied in an uncommitted form with each location holding FF hex. They are configured using an EPROM programmer which 'burns' or 'blows' the required data, in machine code form, onto the chip. If an error in the data exists, or an alteration is to be made, then the complete EPROM can be returned to its uncommitted state by exposing the small 'window' in the device to intense ultraviolet light for about 20–30 min. EPROM erasers are available for this purpose. Once programmed as required, it is usual to cover the window with opaque material. If uncovered, it would normally take some months before program corruption was experienced through the effects of natural sunlight.

A similar type of memory device is an Electrically Erasable Read Only Memory (EEPROM or E^2PROM). This is essentially similar to the EPROM but enables the user to alter any particular byte of data rather than wiping the entire chip. E^2PROM is not so popular as the EPROM for economic reasons.

3.2.4 Input/output (I/O) structure

With the microprocessor acting as the brain of the microcomputer system and the memory chips storing the system-operating software and application programs, the other essential hardware required is that associated with the input and output of data in essentially binary form. Interface support chips associated with the various microprocessor families are available to enable communication with such hardware essentials as keyboards, display monitors, disc drives and printers.

The same I/O interface circuits are used in measurement and control applications and the main functions required of the devices are:

1. Digital I/O logic lines which can be read or set by the microprocessor;
2. Data direction register to configure lines as either input or output;
3. Handshake lines to supervise data transfer via the I/O lines;
4. Timing and counting facilities.

The software used for controlling the communication between the microcomputer and other external devices is dependent upon the I/O interfacing technique employed. The two most common methods are either 'memory mapped' or 'dedicated port addressed'.

3.2.4.1 Memory mapped I/O

In this method the I/O chip is connected into the system in the same way as the memory illustrated in Figure 3.1. The I/O lines are contained in groups of 8-bits termed a 'port' and this byte is addressed in the same manner as any other location in memory. The port is accessed using memory transfer instructions like PEEK and POKE in high-level BASIC, or LDA and STA in low-level 6502 assembly language.

Since the interface is connected into the bus structure in exactly the same way as the RAM and ROM, no additional decoding hardware is required. Memory addresses are, however, used up for I/O, and as a result, communication is slower than the port addressed alternative.

3.2.4.2 Dedicated port addressed I/O

This method involves a second dedicated I/O data bus as shown in Figure 3.2. When data are to be input or output the necessary control signals are sent from the CPU to the I/O interface chip and the port data are transmitted via the dedicated I/O data bus. This does not effect the addressing of memory within the system and results in faster data transfer than with the memory mapped technique. The ports (or channels) are assigned unique addresses (numbers) on the dedicated bus and are accessed using the additional software

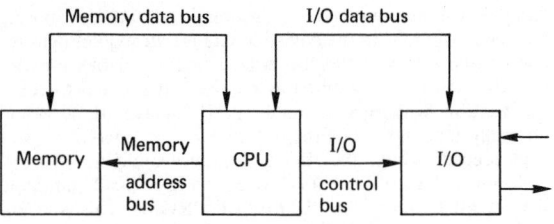

Figure 3.2 Port addressed I/O

instructions of IN (or INP) and OUT in both low- and high-level programming languages.

Although a number of I/O support chips are available, there are essentially two which figure prominently with the memory mapped and port addressed techniques. These are the 6522 versatile interface adapter, (VIA), usually associated with the memory mapped 6502 microprocessor systems and the 8255 programmable peripheral interface, (PPI), associated with such processors as the 8080, Z80 and 8086 in port addressed systems.

3.2.4.3 The 6522 VIA

This is a general interface chip which provides such interface functions as two 8-bit parallel bi-directional ports, each with a pair of handshake lines and two 16-bit counter timers. The ports, often designated as data registers A and B ('DRA' and 'DRB') each have an associated data direction register ('DDRA' and 'DDRB'), which is used for setting a bit on a port as either an input or an output. The addresses follow the sequence PORTB, PORTA, DDRB and DDRA.

If a.c. or d.c. loads such as solenoids, motors or lamps are to be driven from the port logic signal levels, then a power scaling interface, such as a Darlington Driver, compatible with the microprocessor VIA must be used (see Section 3.4.1).

The VIA control lines CA1, CA2, CB1 and CB2 can be set to operate in various read/write modes. This is achieved through the peripheral control register (PCR) in the VIA. CA1 has no output capability, but CA2, CB1 and CB2 can all be used as either input or output. These lines are incapable, however, of switching a power-scaling device.

The two programmable timers within the 6522 are generally referred to as T1 and T2. These are 16-bits wide and implemented as two 8-bit registers with a low-byte/high-byte arrangement. The modes of operation, selected by writing the appropriate code to the auxiliary control register (ACR), are:

1. Generate a single time interval.
2. Generate continuous time intervals (T1 only).
3. Produce a single or continuous pulses of bit 7 of DRB (T1 only).
4. Count high to low transitions on bit 6 of DRB (T2 only).

3.2.4.4 The 8255 PPI

All microprocessor families have parallel I/O interfaces and these are designed for use with the particular type of CPU. The 8255PPI is used basically with Intel 8 and 16-bit devices such as the 8080 and 8086/8088. The 8255PPI provides three 8-bit bi-directional ports which may be operated in three modes. No other functions such as timing or additional handshaking are available. The ports are designated as A, B and C, and data direction is specified by writing to a write-only Control Register.

If hardware timing is required then a separate counter/timer device must be used. One commonly adopted with an 8255PPI

8255PPI is the Intel 8253 chip, which provides three independent 16-bit counters, each with a count rate of up to 2.6 MHz. The 8253 has various modes of operation but works basically on the same principle as the timers in the 6522VIA.

3.2.4.5 Direct Memory Access (DMA)

In data-acquisition systems involving analogue and digital signals suitably conditioned for inputting to a microcomputer there is a limitation of about 100 kHz on the sampling rate when using direct program control to transfer data to memory. If it was necessary to acquire the maximum amount of data at the highest speed, using the maximum amount of the computer's resources, then the DMA technique might be employed.

This is a hardware technique which causes the microprocessor to momentarily abandon control of the system buses so that the DMA device can directly access the memory. The DMA controller, connected to the I/O interface, needs to know how many bytes are to be transferred and where in memory the input data are to be stored. The data transfer rate is much faster than in an interrupt servicing method and data-sampling rates of the order of 1 MHz are possible for most microcomputers.

3.2.5 Memory map

The memory locations in RAM and ROM, which the processor can address, must accommodate space for such requirements as system monitor and utilities, user software and input/output. The manufacturer of the microcomputer assigns an area of memory for each functional requirement and provides the necessary information in a system memory map.

In 8-bit systems, with 64K of addressable memory, the memory map is usually composed of 32K of RAM and 32K of ROM or EPROM. The ROM holds the operating system software and normally some space is available in EPROM form for user firmware. In addition to providing space for user programs, the RAM area contains the system stack and the visual monitor data storage. The I/O facilities are also assigned an area of memory in a memory mapped system.

In a 6502 or 6800 based system the RAM is usually low down in memory and the ROM is high up. A typical memory map is shown in Figure 3.3. The I/O is accommodated

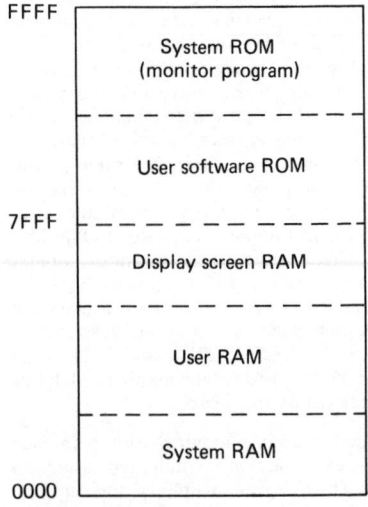

Figure 3.3 System memory map

anywhere within the above structure and varies from one manufacturer to another.

The display screen RAM is often fitted with a moveable boundary in order to set variable resolution modes for graphics. Typically, the screen memory need only be 1K for a text mode to give a 40 × 25 character display. This would be insufficient for computer graphics and a high resolution would require 20K of memory to MAP the screen. It should be noted that this greatly reduces the amount of RAM available to the user.

8080- and Z80-based systems have a distribution of memory similar to that as shown in Figure 3.3. The ROM, however, is usually low down and the RAM high up in the memory map. The 16-bit PCs, with a 20-bit address bus, have 1M byte of addressable memory with the RAM comprising the first three quarters and ROM occupying the last quarter. A general memory map showing the distribution of RAM/ROM for a PC is given in Figure 3.4.

A familiarity with the memory map of the system to be used in any data-acquisition application is essential, since it indicates the areas reserved for the operating system. The programmer can then knowledgeably determine the locations available for data storage and machine-code programs.

3.3 Communication standards

Various standards have been drawn up to define the protocol for the transmission of binary data from within the microcomputer bus structure to external devices such as display monitors, printers and other peripheral equipment. Most microcomputers are equipped with this facility and manufacturers of data measurement and control instrumentation usually offer an external communication port as an extra.

The most commonly accepted standards are those defined by the American Electronic Industries Association (EIA) and the Institute of Electrical and Electronics Engineers (IEEE). The standards fall into the two categories of serial and parallel data communication. The difference between the two relates to the number of bits of information transmitted simultaneously between the devices. The serial method is the slower

of the two, with the bits denoting the characters of information travelling sequentially along a single path. In the parallel method the data word is sent as a parallel code, invariably 8-bits wide, resulting in a 'bit parallel, byte serial' transmission of information.

3.3.1 Serial communication

Serial communication is the most common method used for the interconnection of a microcomputer to the relatively slow peripheral hardware, or between two computers, when transferring a low volume of information. The (EIA) RS232C, or its successors the RS422 and RS423, is the most widely adopted standard employed and connection between devices is made via a standard 25-pin connector. This allows communication with one peripheral device only. Twenty-one of the signal lines are defined in the standard although only five (or even three) are all that are usually required.

The three main connections are 'transmitted data' (pin 2), 'received data' (pin 3) and 'signal ground or common return' (pin 7). These would normally be connected as shown in Figure 3.5. For communication in both directions, i.e. full duplex, the two handshaking control lines – 'request to send' (pin 4) and 'clear to send' (pin 5) – are also required.

The standard applies to data transmission interchange usually at rates between 110 and 9600 baud. A logic '1' is represented by a voltage in the range of −3 to −15 V and a logic '0' by a range of +3 to +15 V. This large differential between '1' and '0' ensures good immunity against electrical noise. However, the voltages used are not compatible with the TTL logic semiconductor family and interconversion chips are required within the interface.

The RS232C is limited to short communication links of about 30 m, although the RS422 and RS423 standards, succeeding the RS232, have extended communication distances and increased transmission speeds. The RS423, which is compatible with the RS232, has superior driving and receiving interfaces, allowing communication over distances of up to 1500 m at 9600 baud, or 15 m at 100K baud.

It should be noted that while the voltages and signal connections for the plug are defined in the standard, the data protocol is not identified. This must be known for the devices which are to be connected and can be set accordingly by software. The requirements are:

1. Baud rate;
2. Number of bits in the ASCII group defining the character being transmitted;
3. Odd, even or no parity;
4. Number of stop bits.

3.3.2 Parallel communication

The RS232 serial standard for communication was developed essentially for the connection of microcomputers via a telephone link. The parallel standard emerged from the need to

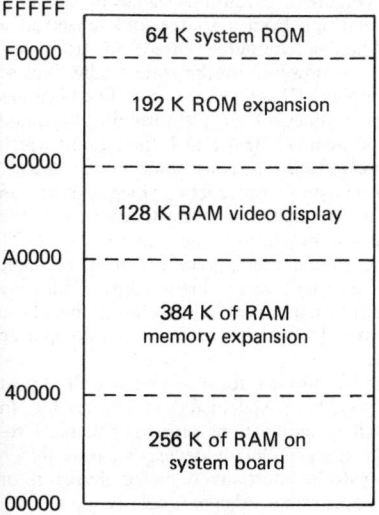

Figure 3.4 General memory map for a PC

FFFFF

| 64 K system ROM |
F0000

| 192 K ROM expansion |

C0000

| 128 K RAM video display |

A0000

| 384 K of RAM memory expansion |

40000

| 256 K of RAM on system board |

00000

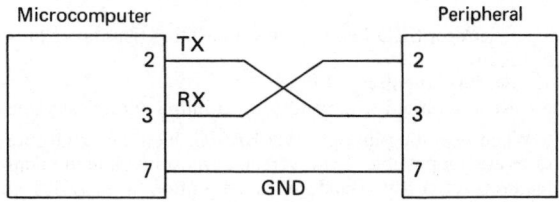

Figure 3.5

establish a means of interfacing a variety of instruments for data-logging applications. The most common standard for the integration of automated test systems, developed by Hewlett-Packard, is referred to as the IEEE-488 interface bus, and has achieved wide recognition among instrument manufacturers since the start of the 1980s.

The bus consists of 24 lines, accommodated within standard stacked type connectors. The eight bi-directional data lines carry information as 7-bit ASCII codes between the microcomputer (controller), and an instrument (listener) on the bus. The roles may be reversed when data are being logged. To process the information on the data bus, up to eight control and status signals are available.

The bus is designed to interface with up to 15 instruments, within a localized area, involving a total cable length of not more than 20 m. Each instrument is uniquely numbered within the range of 0–30 and the overall activity is controlled by one of the devices, termed the Controller. This is usually the microcomputer with an appropriate interface. Each device number is switch selectable within the instrument. Other functional aspects of the devices on the bus are that they must be capable of acting as a 'Listener' or a 'Talker'. A 'Listener' is a device that can receive data over the bus and a 'Talker' is capable of transmitting data. There may be several Listeners active on the bus at any one time, but there can only be one Talker. Most devices, including the microcomputer Controller, can act as either Listeners or Talkers.

When setting up an instrument to measure some physical variable, codes devised by the instrument manufacturer are sent on the bus, in ASCII format, as a data string to the numbered device. In the case of a multichannel DVM, this could take the form of the channel number to be monitored, voltage range to be selected and a terminating character. An example of the corresponding string to be put on the bus is

'C9R2T'

which denotes channel 9, range number 2 (say, 0–10 V) and 'T' is the string terminating character recognized by the instrument.

Manufacturers of add-on cards, to give IEEE-488 facilities with microcomputers, usually supply software for initializing the bus, setting it up for transmitting data from controller to instrument and returning data from instrument to controller. The measured quantity is also sent to the computer in the form of an ASCII string from which the actual numerical value can be extracted.

One of the most important management control lines is the service request (SRQ). This is a type of interrupt line that is activated low by a device residing on the bus and needing service from the controller. It is used as a means of indicating that the instrument is ready to transmit the current reading onto the bus.

Thus, a typical software sequence for implementing the control of an instrument on the IEEE-488 bus for data acquisition is:

1. Initialize bus and set instrument as a Listener.
2. Put control string on the bus to set up the instrument as required.
3. Check for SRQ line to go low indicating that data can be read.
4. Set instrument as a Talker.
5. Read returned string and convert into a numerical value.

When operating in high-level BASIC, high data-collection rates are not possible. However, since most instrument manufacturers offer the standard as an option it provides an intelligently controlled flexible arrangement for test and measuring instruments.

3.4 Interfacing of computers to systems

The serial and parallel communication standards are the basic interfacing links between computers and their associated peripheral devices, or between computers and a comprehensive range of measurement instrumentation. In general computer control applications, however, two other common interfaces are fundamental. These are the digital interface which implements the controller output and the analogue interface associated with the measured variable input.

3.4.1 Digital interfacing

The computer output port may be used to transmit control signals on any one of the available lines by writing the appropriate number to the port address. When a line (or bit) is set 'high' (i.e. a logic level of 1) the voltage on the line is approximately 5 V. The current available, however, is fairly minimal (of the order of 1 mA) and no load can be connected directly to the port. There is in fact a danger of causing extensive damage to the computer by connecting a load directly to the port. An interface must therefore be provided to enable the computer to switch in power loads using the logic level control signals from the output port. The most common interface device used for this purpose is the power transistor. In typical applications the power transistor, operating on logic level control signals, switches in a mechanical relay which in turn switches in the load.

The 'Darlington Driver' is a popular power transistor available as an integrated circuit and which normally includes a number of separate stages. Figure 3.6 shows the wiring diagram for a single stage in a Darlington Driver. The Darlington Driver can switch up to 500 mA at 50 V and each stage is diode protected for the switching of inductive loads.

To provide a total isolation from high voltages, port output signals can be coupled through an 'opto-isolator', which, interfaced between the computer output port and the power control device, is not an essential element in the digital interface. Opto-isolation, however, ensures that no hardwired connections are made between the computer and the power device. An added advantage is that the opto-isolator acts as a buffer to spurious noise signals which can corrupt the digital logic values being transmitted on the buses.

The opto-isolator (Figure 3.7) transmits signals by means of infrared radiation, emitting from a source and sensed at a photo-transistor. When a computer based on transistor/transistor logic (TTL) is powered up the state of the lines of the output port 'float high'. That is, each output line becomes set to a logic value of 1. Since a logic 1 is normally associated with the function of switching a device ON, then a port which floats high could inadvertently activate some power device. Obviously, this is a dangerous precedent, which requires an additional element in the digital interface to counteract the effect. The device commonly used is the 'inverter' (or NOT gate) which has the simple function of inverting all logic signals from 1 to 0 and vice versa (Figure 3.8). Following power up, a logic 0 must then be sent to the relevant line of the output port, to become a logic 1 after inversion and to operate the control function.

The composite digital interface for a computer output port suitable for power switching is depicted in Figure 3.9. In high-frequency switching applications electromechanical relays are not suitable. Semiconductor devices such as silicon controller rectifiers (SRCs, alternatively called thyristors or triacs) may be more appropriate. Also particularly suitable are the various solid-state relays which can operate directly from logic level signals.

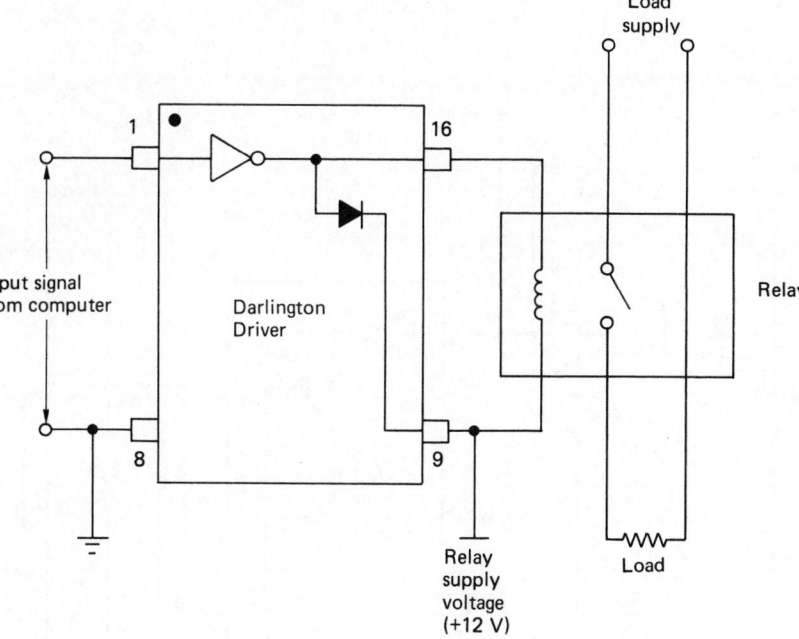

Figure 3.6 Power switching with a Darlington Driver

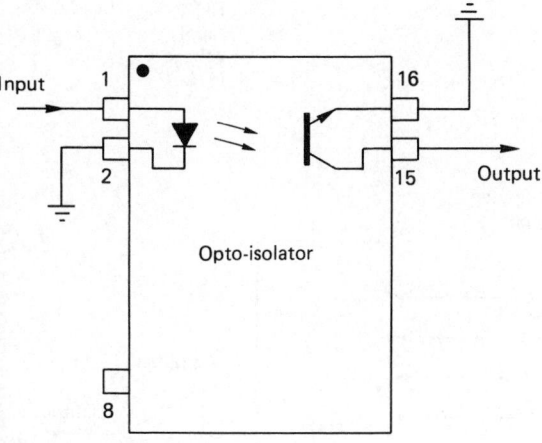

Figure 3.7 Opto-isolator

3.4.2 Controller output interface hardware

The digital interfaces discussed above are suitable for switching in power loads in an ON/OFF control system. For a digital control algorithm based on a PID strategy, some means is required of discretely varying the output power supplied to the controlled device (Figure 3.10). A number of different methods are used to supply variable power to the system and these include the following.

3.4.2.1 The digital-to-analogue converter (DAC)

The required control effort value, U, is calculated in the program according to the control strategy employed. This value is converted to an equivalent binary number and output

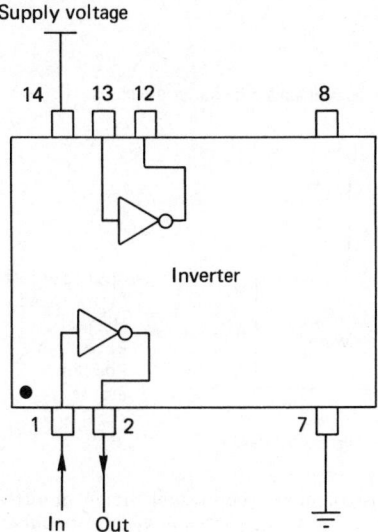

Figure 3.8 Inverter integrated circuit

to the computer port where it is then transmitted to a DAC. The DAC converts the binary input into a proportional output voltage which may then be suitably amplified to drive the controlled device. The controlled device could, for example, be a d.c. motor whose speed is directly related to the supply voltage. The interface is illustrated in Figure 3.11.

Two basic types of DAC are available: the adder converter and the ladder converter. The adder converter can be illustrated as a simple example of Ohm's law. A 4-bit adder type DAC is shown in figure 3.12. The resistance value of the line resistors are halved for each consecutive increasing 'bit' and

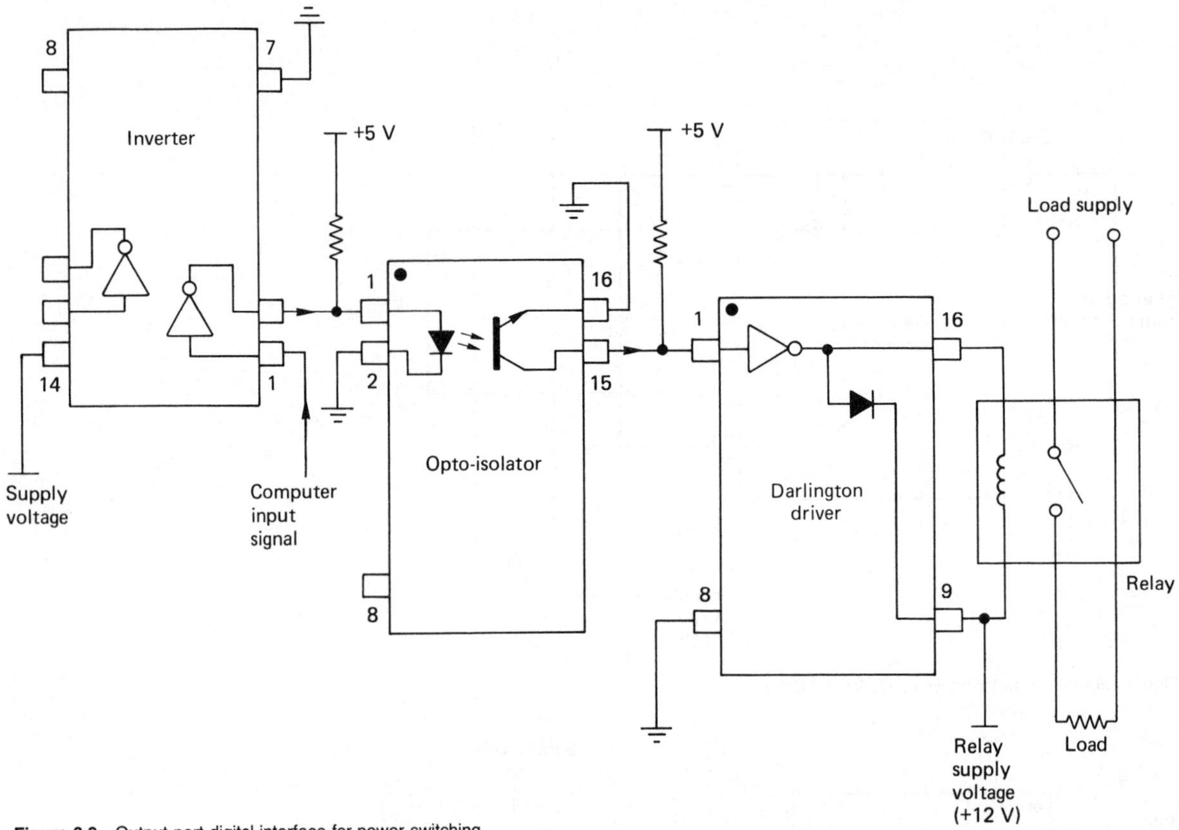

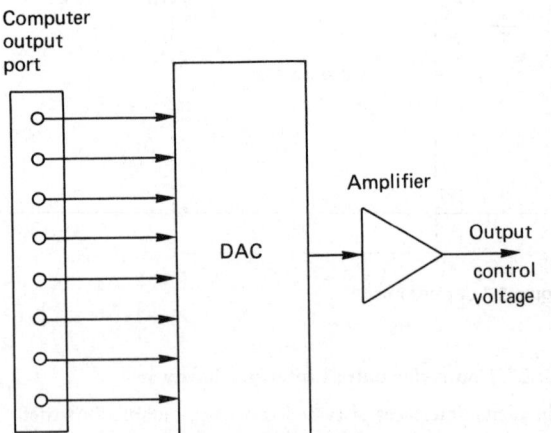

Figure 3.9 Output port digital interface for power switching

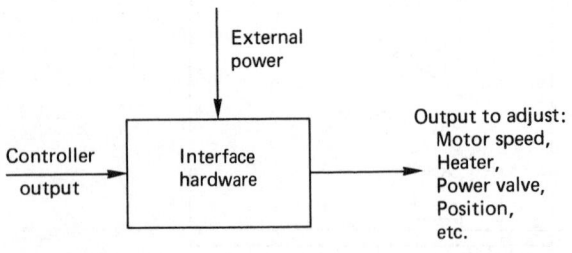

Figure 3.10 Variable power output device

the supply voltage is common. The current drawn through each line, if it is connected, is 1, 2, 4 and 8 mA, corresponding with bits 0, 1, 2 and 3. The summation of the currents at output therefore equates on a decimal to binary basis with the input. Figure 3.12 shows a digital or switched input of 1011 binary. This gives a current summation of 11 mA, equivalent to the corresponding decimal number. The output, a proportional current, can be converted to a proportional voltage through an operational amplifier. Because of the range of resistor values required, the adder converter is less popular than the ladder converter, which uses only two resistor values, i.e. R and 2R.

The ladder converter DAC (Figure 3.13) must be analysed using the network theorems of Thevenin and Norton. The end result is similar, however, with a proportional voltage output corresponding to the digital switched input on a decimal-to-binary basis.

Figure 3.11 Variable power output using a DAC

3.4.2.2 Pulse Width Modulation (PWM)

Output power from the interface hardware can be varied by sending ON/OFF pulses to the power device. The frequency range is normally between 2 and 10 kHz. If the time ON and the time OFF periods are equal, then rapid switching of the power supply will transmit 50% of the total power available. Due to the relatively high switching frequencies and the levels

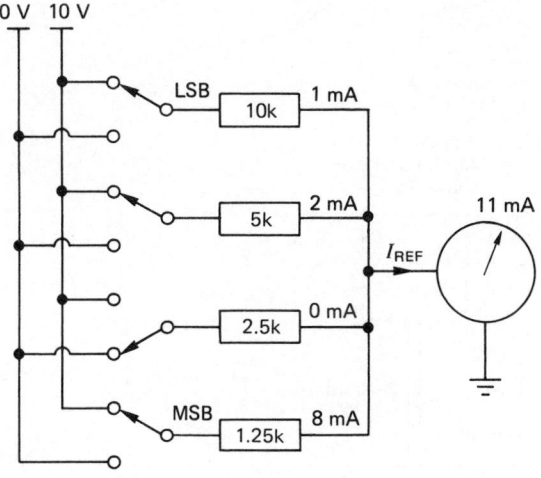

Figure 3.12 4-bit, adder-type DAC

implemented over the remainder of the sync pulse. Two separate timing loops are used to control the MARK and SPACE times, respectively. Due to the relatively high frequency of the output signal, the control software must be written in assembly language for speed. An alternative to this is to use hardware support chips to output the PWM signal under the control of the computer's CPU.

3.4.2.3 Controlling a.c. power by control of thyristor phase angle

Various applications, such as temperature control, require a.c. power adjustment and solid-state relays can be used effectively to vary output power between 0% and 100%. The power is controlled by varying the phase angle between the supply voltage and that which appears across the load when current conduction begins. A phase control device operating off a 0–5 V signal from a DAC can be used to alter the phase angle between voltage and current in the range 0–180°. The control of the power output to the load is non-linear, but linearization between output and input can be accomplished in the software.

3.4.2.4 Controlling flow control valves

In level-control systems, fluid flow rates are controlled by varying the degree of opening of a gate-type valve. Because of the forces involved, a pneumatic 'actuator' is normally employed, working off a controlled pressure in the range between 3–15 pounds per square inch (i.e. psi). (*Note*: The industry still favours the Imperial unitary system but the approximately equivalent pressures in the metric system are 20–100 kN/m^2.) An applied pressure of 3 psi is equivalent to the valve being fully open, 15 psi corresponds to the valve being fully closed.

Most flow control valves are fitted with a 'positioner', which operates off a current signal in the range 4–20 mA. The current range generally corresponds on a linearly proportional basis to the pressure range. Since the computer interface usually involves a DAC, then an additional element (a voltage-to-current converter) is required to interface between

of power transmitted, the solid-state relays are more suitable for this application. The time ON is called the 'MARK' and the time OFF is the 'SPACE'. The MARK/SPACE ratio can be evaluated from the controller output value calculated in the software. PWM can be achieved either by keeping the MARK fixed and varying the SPACE (i.e. varying the signal frequency) or by varying both the MARK and the SPACE within a constant period.

If the PWM output is supplied by the controller, then it is not possible to allow the MARK+SPACE period to extend over the complete control loop cycle. There must be some time allowed to sample the process transducer and to calculate the required controller output value. The problem is overcome by using a synchronization ('sync') pulse signal with a longer period than the MARK+SPACE time. The excess time during the sync pulse is used to read the transducer output and to process the information. The MARK/SPACE ratio is then

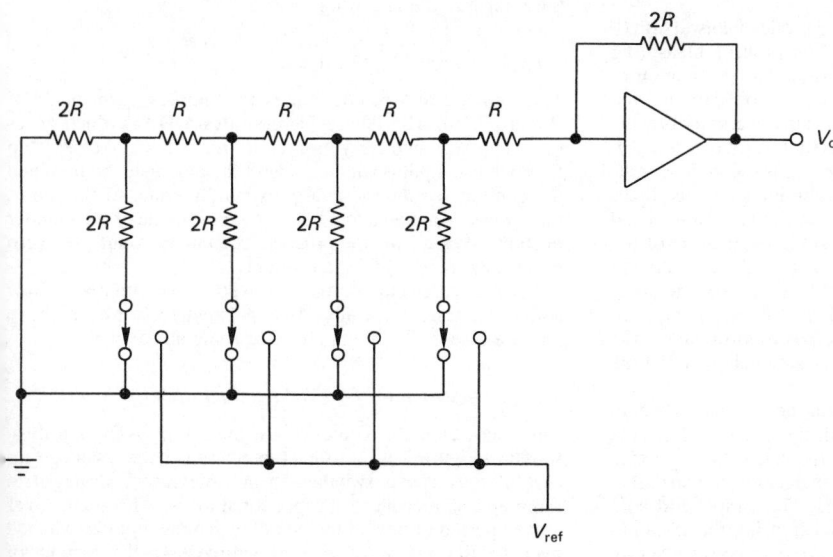

Figure 3.13 Ladder-type DAC

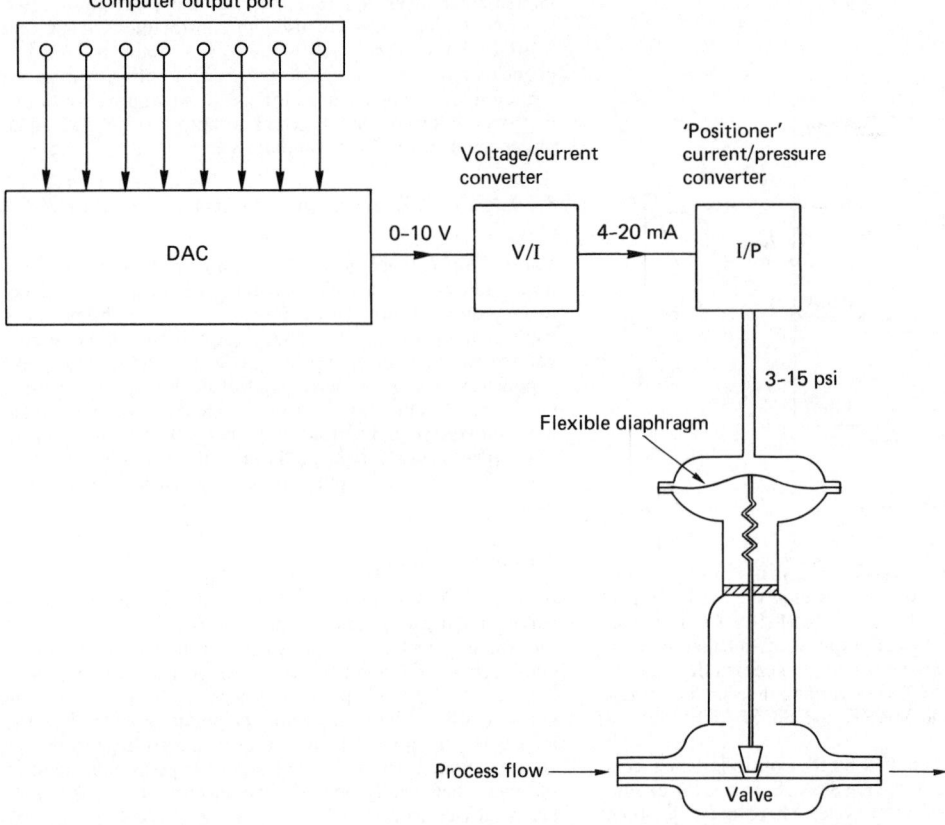

Figure 3.14 Digital interface for a flow control valve

the DAC and the valve positioner. The complete interface is shown in figure 3.14.

3.4.3 Analogue interfacing

The basic role of the analogue interface is one of conversion of the continuous analogue signals, from process measuring transducers to the digital representation that the computer requires to operate on. In all practical applications, the monitoring and acquisition of the data is the necessary precursor to the subsequent control functions that might be actioned.

The process variables are ultimately represented as voltages. Using the appropriate signal-conditioning circuits, these voltages would ideally be processed to range between zero and some reference value. The final task is the digitization of the analogue signal, which is accomplished through an analogue-to-digital converter (ADC). The ADC samples the analogue signal, performs the conversion and outputs a digitally encoded binary number which is directly proportional to the magnitude of the input voltage. The essential elements in the signal train are shown in Figure 3.15.

This figure indicates a sample and hold (S/H) element between the signal conditioner and the ADC. Since the analogue input may be varying while the conversion is taking place, there is a degree of uncertainty in deciding the instant in time which the output code represents. The sample and hold element removes this uncertainty by capturing the instantaneous 'snapshot' of the input for the ADC to convert before moving on to the next sample. The S/H element is only

essential if the input signal is varying very rapidly. The ADC and S/H functions are often packaged within a composite integrated circuit.

ADCs are available in a number of different forms and these include the following.

3.4.3.1 Staircase and comparator

The staircase and comparator is the simplest form of ADC (Figure 3.16). The device incorporates a DAC which generates a voltage increasing in small steps as shown. At each step the staircase input is compared with the analogue input. When the generated staircase is approximately equal to the input, the process is halted and a binary count is made of the number of steps taken during the process. The binary count from zero represents the coded digital output.

The staircase and comparator ADCs have relatively slow conversion times (typically, 20 ms). They are, however, cheap and are essentially immune to electronic noise.

3.4.3.2 Integrating type ADC (or dual slope)

The major elements comprising a dual-slope ADC are illustrated in Figure 3.17. At the start of conversion a voltage-to-current converter is switched to the integrator, causing it to ramp up a slope which is proportional to V_{in}. This occurs over a fixed period of time at the end of which the input is switched over to the reference current source. At the instant of switching the integrator output voltage is proportional to V_{in},

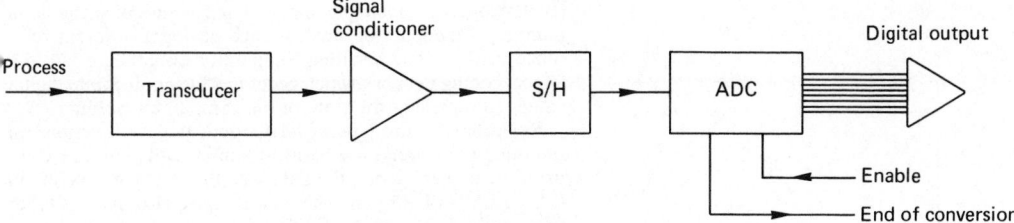

Figure 3.15 Analogue-to-digital conversion

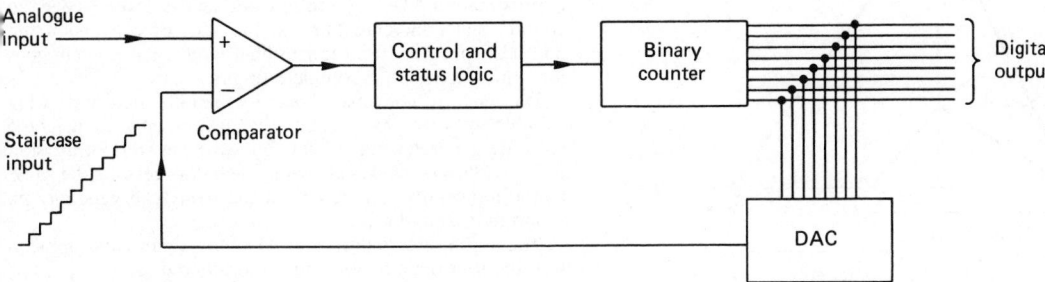

Figure 3.16 Staircase and comparator-type ADC

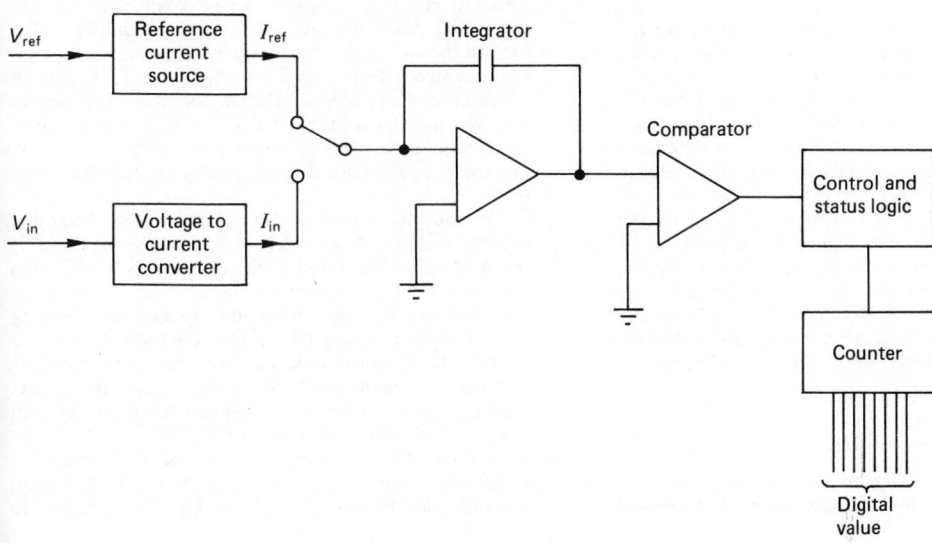

Figure 3.17 Dual-slope ADC

a counter is enabled and counting begins at a rate set by the internal clock. Meanwhile, the reference current causes the integrator to ramp down at a slope which is proportional to V_{ref} (i.e. a constant slope). When the integrator output again reaches ground the comparator switches the counter off and the counter then contains a digitally encoded value proportional to V_{in}. Figure 3.18 shows the voltage variation at the integrator output.

From this figure it can be seen that there are two similar triangles such that

$$V_{in} = V_{ref}(T_v - T_f)/(T_{max} - T_f) \qquad (3.2)$$

T_v is directly proportional to the counter output and with T_{max}, T_v and V_{ref} all known, the input voltage, V_{in}, is determined by proportion.

The integrating types of ADC have similar operating characteristics and conversion times to that of the staircase and comparator types. For faster analogue-to-digital conversion, the 'successive' or 'counter' type are generally employed.

3.4.3.3 Successive approximation type ADC

In this ADC the input signal is compared with a number of standard reference voltages, generated from a DAC, until the

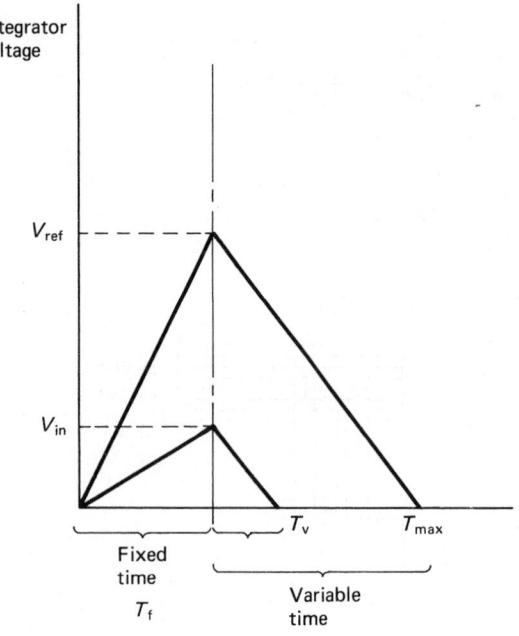

Figure 3.18 Integrator voltage variation

combination of standard voltages required to make up the input value has been determined. The main components of the converter are a clock, a counter, a comparator and a DAC.

When an analogue signal is input to the converter the counter starts a count and passes a digital value to the DAC. The DAC generates a voltage to represent the most significant bit and the comparator assesses this against the analogue input. If the analogue signal is greater than the voltage from the DAC then the logic 1 in the MSB is retained. If the analogue signal is smaller then a logic 0 is assigned to the MSB. This process is then repeated on the next most significant bit and so on for all the other bits down to the LSB. The conversion time for these types of converters may be of the order of 10–25 μs, but this will depend upon the hardware design. Figure 3.19 outlines the essential features of a successive approximation ADC.

3.4.3.4 Parallel conversion type ADC

The parallel type ADC has by far the fastest conversion time (at about 1 μs) but it is also the most expensive. With parallel conversion, the analogue input is fed simultaneously to a number of comparator circuits, each having a different reference voltage. The resulting comparator outputs are fed to a logical coding network which generates the appropriate digital values to represent the state of the comparator outputs.

Regardless of the type of ADC used, the pin functions on the integrated circuit are basically similar and generally comprise the power supply, the data bits, the start conversion pin ($\overline{SC}$ or $\overline{CONVERT}$) and the end of conversion pin ($\overline{EOC}$ or $\overline{STATUS}$). The overbar signifies that the pin is active low.

The conversion is software initiated by sending a 'pulse' (logic 0, followed by logic 1) to the $\overline{CONVERT}$) pin. On the negative edge of this pulse the counter in the successive approximation ADC is set to zero and on the positive edge the counter starts incrementing. At the start of conversion the $\overline{STATUS}$ pin goes from low to high, and when it again goes low, the conversion is complete (Figure 3.20).

The end of conversion may be readily detected using suitable software. As an alternative, it is possible to include a software-generated time delay following the start conversion pulse to allow conversion to complete before reading the value at the input port. The length of the delay can generally be found by trial and error.

In choosing the appropriate ADC for a particular application the four main features to be considered are:

1. *Conversion time*: The conversion time is a measure of the operating speed of the converter and is the time taken for the complete translation of an analogue signal to digital form. In many of the staircase and comparator and the integrating types of ADC this may be dependent on the level of the analogue input signal. Faster conversion times are obtained with low-level inputs due to the manner in which the conversion is completed. Successive approximation and parallel conversion types of ADC have a fixed conversion time. This is because the same conversion process is performed, regardless of the analogue input level.

   The conversion time of the ADC does not, however, indicate the fastest rate at which data can be captured. If the data are to be stored in the computer's RAM, then this must be done in a sequential and ordered manner. This involves setting a base address and incrementing various registers to step the storage addresses of each byte of data placed in memory. Further time delays could be accrued in a sample and hold device. The minimum data capture period is often therefore many times greater than the specified conversion time of the ADC.

2. *Resolution*: The resolution of an ADC is the number employed to represent the digital output in binary form. For example, the resolution of an 8-bit ADC is limited to

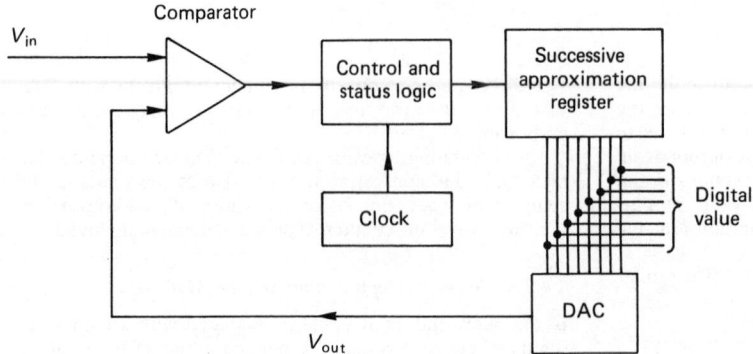

Figure 3.19 Successive-approximation ADC

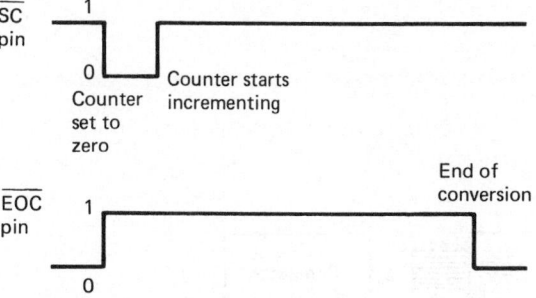

Figure 3.20 Start conversion and end of conversion pin signals

one part in 256 of the maximum voltage corresponding to the full-scale setting. An improvement in resolution can be obtained with a 12-bit converter, with one part in 4096. Table 3.1 summarizes the relation between the number of bits and the resolution.

Table 3.1

n-bits	2^n	Resolution (%)
8	256	0.4
10	1 024	0.1
12	4 096	0.025
16	65 536	0.0015

3. *Accuracy*: The accuracy is related to linearity defects, zero error and calibration defficiencies in the electronics of the converter and should not be confused with the resolution.
4. *Cost*: Cost will depend on the quality required in the three areas previously described and on the means of conversion employed. It is closely associated with the speed of the conversion and with the resolution and accuracy. Cost generally rises with increases in all or either of the three other variables.

3.4.4 Multiplexing

In applications where a number of transducers are to be sampled, a multiplexer (MUX) can be used to switch in various channels as and when required to a single ADC. The switching is software controlled from the computer and Figure 3.21 illustrates the basic principle.

The multiplexer and ADC often form an integral part of a complete system. In some cases, even the signal conditioning can be software controlled, with all the necessary hardware mounted on a single 'card' and plugged directly into the computer's bus system. Multiplexers (or analogue switches) are available with various numbers of input channels.

Minimum cost conditions usually dictate whether multiplexing will be implemented or not, but the reduced cost must be balanced against an inevitable reduction in sampling rate. Figure 3.22 shows three possible arrangements of signal conditioning, multiplexing and conversion for analogue interfaces.

System A is the most common, while B and C can provide for virtually simultaneous sampling. System C gives the most representative snapshot at a particular period in time, but it is also the most costly.

3.4.5 Machine tool interfaces

The control system for a machine tool slide is shown in Figure 3.23. Typically, there are two negative feedback loops, one for position and one for velocity in a cascade arrangement as shown in the figure. The position sensor is usually an optical grating device (or an inductosyn) and the speed sensor a tachometer.

The CNC interface initially has to decode the manual or control tape input data. This consists of a sequence of commands, including feed and speed data, essential dimensional reference points and other constraints to be observed by the machine during its operation. In operation, the interface is required to monitor the slide position and speed and check various limit switch settings for compliance with the sequential program instructions. The transducer input signals would normally be switched in through a multiplexer prior to digitization with a fast conversion type ADC. Limit switches would also be checked or set through additional digital I/O lines. If any errors are detected, the interface must be able to indicate these and take the appropriate action. The interface includes a real-time clock which generates an interrupt every few milliseconds. The clock acts as a monitor of operator actions, enables the output of error signals to the machine servos and checks all current signals from each of the feedback sensors.

For a typical CNC milling machine there are three independent axes, and each would have the same monitoring and control functions applied to them. In addition, the spindle speed would be monitored and controlled and the machine might also incorporate a tool-changing facility based on a simplified robot arm.

Further refinements could include a load transducer in an additional feedback loop to measure the cutting forces during machining. Force sensing may be used as the basis for an adaptive control loop. In the context of machine tools, adaptive control is usually associated with the alteration of feed rates and cutting speeds to maximize the cutting power. Figure 3.24 shows an adaptive control option on an NC turning machine.

The adaptive loop can optimize the cutting operations, prevent spindle overload, maximize tool life, reduce time loss in 'air cuts' and simplify the programming. The additional sensors and their protection in the harsh machining environment means, however, that the adaptive loop is much more costly to implement. The adaptive control interface, which has no manual input data facility, is also necessarily complex and requires considerable memory capacity.

3.4.6 Robot control interfaces

The machine tool interface described in the previous section can be programmed to perform a series of operations which might be described as 'sequenced automation'. Many of the simpler robots (e.g. pick-and-place machines) use the same technology and perform essentially similar tasks. These machines are not, however, robots in the strictest sense. The essential feature of a true robot is its capability of exercising independent control in each of its axes, or rotating joints, such that its 'hand' can reach any position and any orientation within the working volume.

Each joint on the robot has an actuator, an associated position sensor and a velocity sensor. Six actuators are required for full flexibility in position and orientation, although in most cases only five or less are used. The computer must at all times be able to ascertain the current and desired locations of the hand. The position sensor data processing therefore involves the manipulation of various coordinate transforma-

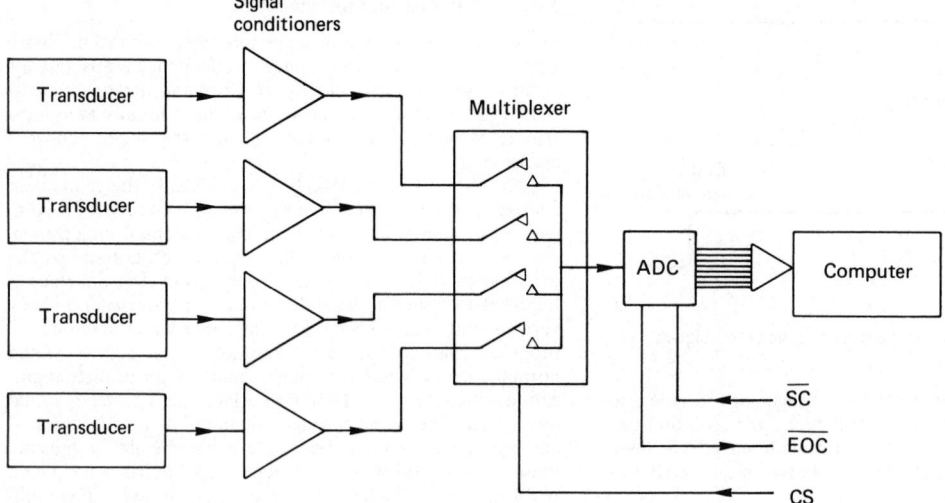

Figure 3.21 Multiplexer for multiple inputs

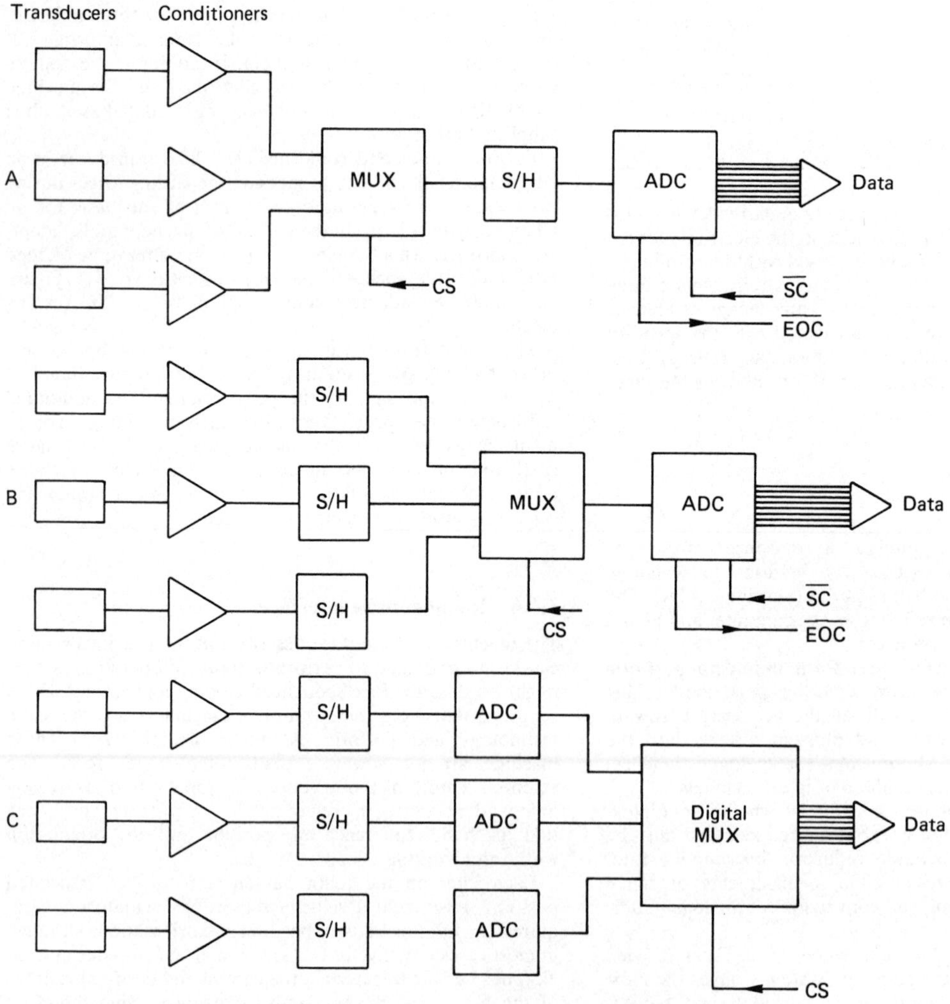

Figure 3.22 Multi-input systems

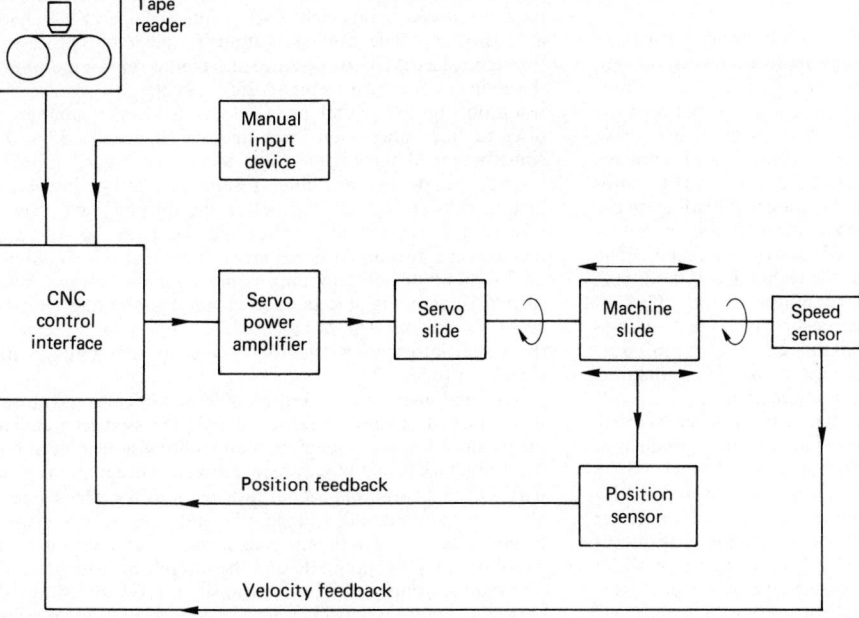

Figure 3.23   Machine tool slide control interface

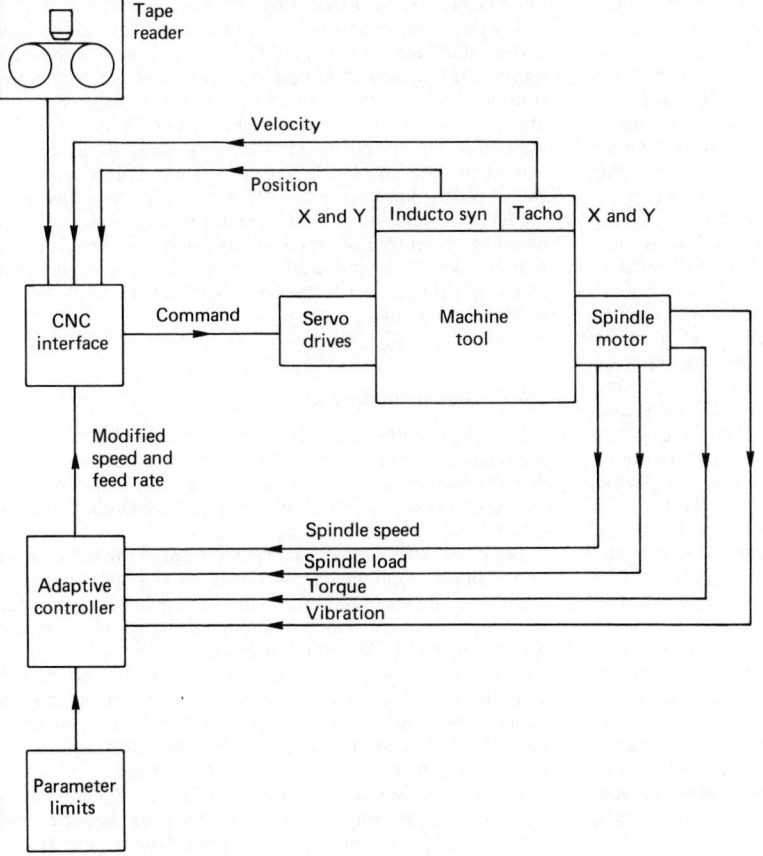

Figure 3.24   Adaptive control on a NC turning machine

tion matrices in the definition and control of the hand location.

Force sensing within the 'gripper' is commonly featured in many industrial robots. The obvious application is to prevent the proverbial 'vice-like' grip on some fragile object. Force sensing is also incorporated by inserting a sensor between the gripper and the wrist. Many wrist force sensors are strain gauge based and can respond to applied forces and moments in each of the three Cartesian directions. The wrist sensor provides feedback signals for static and inertial loading on the robot arm. Robots which operate a control strategy based on wrist force sensor data are termed 'active compliant'. The active compliant robot system has the capability of modifying the joint positions and motion to minimize the effect of external forces. Many wrist force sensors have been successfully developed for robot manipulators engaged in pin insertion type operations. Whitney[1] reports on some detailed investigations on wrist force sensor applications.

The sensors producing signals related to position, velocity and force form the main feedback loops in many industrial robot assemblies. These signals are converted through ADCs and the digital control algorithm, often based on a PID strategy, is then solved to determine the required controller output values at that given time. Power is subsequently output to the actuators either through DACs or using pulse width modulation techniques. In most respects the above process is exactly the same as that which would be used to control a CNC machine tool with an adaptive loop. The robot arm motion, however, is much more complex and the computation involved is far more extensive than that required for the CNC machine tool. The computer used in robotic manipulators therefore requires a large memory capacity and a fast processing speed to accommodate the multiplicity of tasks associated with the control functions. The control algorithm is itself complicated by the multiple feedback loops for position, velocity and force, and a 'hybrid' control strategy is normally adopted. This strategy allows force to be controlled along certain Cartesian axes, while position is controlled along the remaining degrees of freedom. Velocity is controlled along the same degrees of freedom as position. The computational effort is significant due to the many coordinate transformations required, and hybrid control can only be implemented when the robot is moving slowly.

Recent advances in robot technology are centred on the development of tactile sensors, range finders and machine vision. Tactile sensors ideally simulate the human sense of touch. They should therefore be able to detect presence, shape, location and orientation. In addition, the ideal tactile sensor should be able to respond to applied forces and moments in terms of magnitude, direction and location. Varying degrees of success have been achieved with tactile sensors (see Bejczy,[2] Harmon[3] and Rebman and Trull[4]).

Many range-finding devices are based on optical proximity sensors or on ultrasonic echo-sounding transducers in applications to short and long range finding, respectively. The function of the rangefinder is twofold; either to locate the gripper in the correct position relative to the object to be picked up or to avoid potential collisions with other objects. Like tactile sensors, however, many range-finding devices have not yet been developed to the stage where they are consistently reliable for routine use.

Perhaps the most useful additional sensory attribute that a robot can have is that of vision. Industrial machine vision is currently the subject of considerable worldwide research and development and a number of effective systems are already available.

The first problem with a machine vision system is the formation of a digital image. There are a number of different digital imaging devices in use but they all have the common function of converting light energy into voltage (see Ballard and Brown[5]). The voltage output is inevitably processed through a fast ADC to produce the digital representation of the light intensity, in terms of 256 possible grey levels. The resolution of the digital image depends on the number of discrete lines and points per line which are used in the construction of the picture. The use of 512 lines, with 512 discrete points in each line, results in 262 144 individual picture cells (or 'pixels') to define the image. Each pixel is represented as an 8-bit number and these can be stored in memory as a two-dimensional array. Note that this would use up 262 144 bytes of computer memory for the storage. With such large amounts of data to be handled, a separate 'preprocessor' is often used in the system. The preprocessor interprets the visual information and then transmits the result to the robot controller.

Since the image can be represented numerically and stored in memory, it can also be retrieved and processed according to the particular requirements. In many robot vision applications the vision task is that of part recognition and orientation of the part. Often a two-dimensional projection of the part shape is all that is required, and a single solid-state camera can be used to generate the two-dimensional image. Part recognition is based on the part silhouette and the image-processing technique requires only a brightness threshold level to distinguish between 'object' and 'background'. The level of vision is crude but can be sufficient for many automated assembly processes. Control of the lighting arrangements can greatly enhance the vision system but, in general, machine vision still leaves much to be desired at its current stage of development.

The three major development areas of tactile sensing, range finding and machine vision are all being actively researched at present. It is apparent that these sensory functions all involve considerable amounts of additional input data and this imposes further limitations on the system in general. The limitations are related to the data-processing speed and the communication interfaces. Recent new microprocessors (e.g. Intel's 80486) operate at much higher speeds than any of its predecessors and future developments in communications are likely to be centred on applications using light signals transmitted along fibre-optic cables. These advances should open up the possibilities for further enhancements in robot sensory perception systems.

3.4.7 Signal conditioning

Although the output signal from many sensors and transducers is already in the necessary voltage form, it is only very rarely that the voltage is high enough to be directly interfaced to a measurement system. In most instances the transducer output will require some form of amplification.

The term 'amplifier' is a shortened form of the full description 'voltage amplifier'. Most amplifiers amplify voltage but other types are encountered and these are normally given their full description (e.g. current amplifier, charge amplifier and power amplifier). Amplifiers are further classified as a.c. or d.c. The d.c. amplifier will accept a.c. inputs but the a.c. amplifier will block any d.c. input. The input signal may be either single ended with one, or differential with two active signal lines. Finally, the amplifier may be 'inverting', with a reversal of sign at the output, or 'non-inverting'.

The most common amplifier configuration embodies a differential input, single-ended output voltage amplifier with either inverted or non-inverted output. These type are referred to as 'operational amplifiers'. Figure 3.25 illustrates some typical idealized operational amplifier circuits.

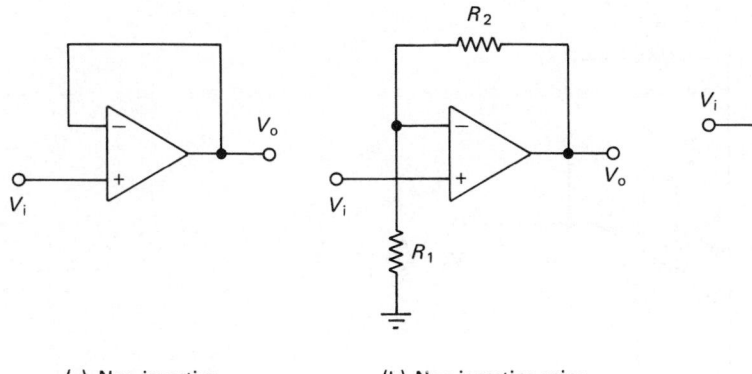
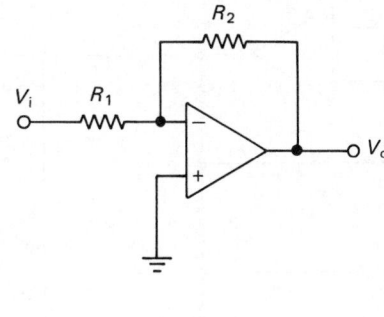

(a) Non-inverting
   unity gain

(b) Non-inverting gain:
$$V_o = \left(\frac{R_1 + R_2}{R_1} \right) V_i$$

(c) Inverting gain:
$$V_o = - \left(\frac{R_2}{R_1} \right) V_i$$

Figure 3.25 Operational amplifier circuits

The important characteristics of an operational amplifier are its gain and its bandwidth. The gain (which can be as high as 10^6) is related to the combination of passive resistors which make up the external circuit. The gain is constant only over a restricted range of input signal frequencies. Outside this range the gain is attenuated as shown in Figure 3.26.

The 3 dB 'cut-off' frequencies are used to define the bandwidth of the amplifier. The product of gain and bandwidth (GBW) is quoted by some manufacturers as a quality para-

meter. Other descriptive parameters associated with amplifiers are:

1. *Common-mode rejection ratio*: Ability of an amplifier to reject differential input gain variation.
2. *Offset voltage*: Voltage output attributed to input and output voltages generated by component variations.
3. *Offset drift*: Temperature-dependent voltage output.
4. *Non-linearity*: Departures from linear input/output characteristics.
5. *Distortion*: Frequency-dependent non-linearities.

Popular and inexpensive general-purpose operational amplifiers are those in the so-called '741' family. For high-accuracy measurements, where low drift and low noise are also essential, an instrumentation amplifier with a high input impedance and high common-mode rejection ratio would normally be required (Figure 3.27).

The problem of impedance mismatch is frequently encountered in instrumentation systems. This occurs when a transducer having an internal resistance and functioning as a small voltage source is coupled to a recording instrument which also has an internal resistance. If the internal resistance of the transducer is many times smaller than that of the recording instrument then the problem is insignificant. If, on the other hand, the internal resistances are of similar orders of magnitude then the error in the measurement and the loss of the signal can be quite large. The standard solution to the impedance mismatch problem is to interface an amplifier with a high-input impedance and a low-output impedance between the transducer and the recording equipment. Such amplifiers are usually configured to have unity gain since their primary function is the prevention of serious signal loss.

For many measurement applications, custom-built special-purpose amplifiers are available in integrated circuit form. These greatly ease and facilitate the signal-conditioning circuitry. Fraser and Milne[6] give a practical guide to many of the signal-conditioning interfaces required for the amplification of signals from strain bridges, thermocouples and other low-level voltage output sensors.

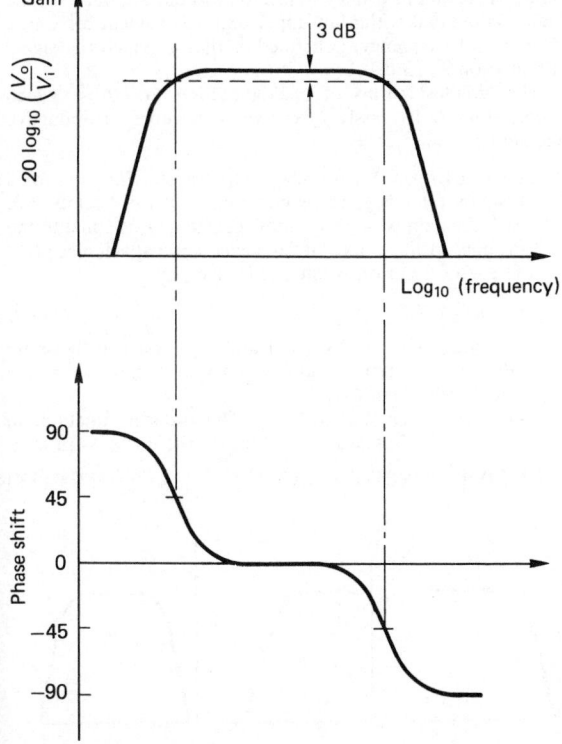

Figure 3.26 Gain and phase shift characteristics of an operational amplifier

3.4.8 Analogue and digital filtering

Noise is inherently present in all physical systems where measurements are made. In sampled data systems (i.e. when an ADC is employed) the effect of noise, illustrated in Figure

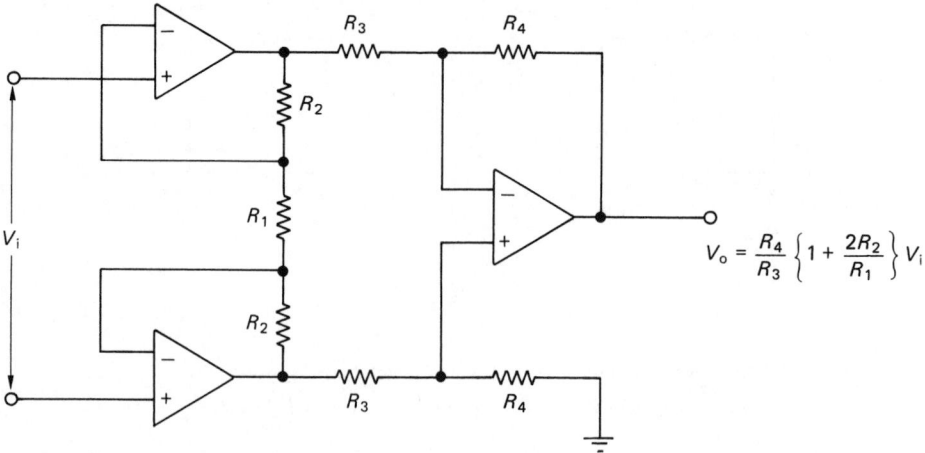

Figure 3.27 Instrumentation amplifier

$$V_o = \frac{R_4}{R_3}\left\{1 + \frac{2R_2}{R_1}\right\}V_i$$

3.28, can give rise to further misinterpretation in the form of 'aliases' (see also Section 3.7.4).

Discrete sampling (shown in Figure 3.28) results in an output signal which suggests that the measured variable is increasing linearly but with a superimposed sinusoidal fluctuation. The apparent sinusoidal variation is entirely the effect of background noise, and it is obviously good practice to try to eliminate noise in the measurement system. It is perhaps fortuitous in mechanical systems that background noise is generally manifested at much higher frequencies than that associated with the primary variable of interest.

The sources of noise are varied and may originate from thermoelectric effects, electrochemical action, electrostatic and electromagnetic pick-up, self-generated component noise, offset voltages and common earth loops. If the frequency

content of the signal to be measured is known beforehand, then positive steps can be taken to eliminate most of the unwanted effects of noise by the inclusion of suitable filters. Filters exist in three broad categories: (1) lowpass, (2) highpass and (3) bandpass. The gain characteristics for each type are shown in Figure 3.29.

A lowpass filter is one which allows the transmission of signals below a particular cut-off frequency. Signals whose frequencies are above the selected cut-off are progressively attenuated. The highpass filter, in contrast, transmits only that part of the signal whose frequencies are above the cut-off value. The bandpass filter transmits, without attenuation, the signal contained within an upper and a lower cut-off value. The cut-off frequency, is defined as that at which the signal attenuation is -3 dB.

The simplest forms of analogue filter are those which incorporate only passive resistive, capacitive or inductive elements:

1. *Lowpass filter* (Figure 3.30): The transfer function for a lowpass filter is given as equation (3.18) in Section 3.6, with the gain, $k = 1$. The time constant, τ, is equal to the product of the values of resistance and capacitance, RC. The -3 dB cut-off frequency is given by

$$f = 1/(2\pi RC) \tag{3.3}$$

A suitable choice of resistor and capacitor can therefore allow any desired cut-off frequency to be imposed in the signal-conditioning train.

2. *Highpass filter* (Figure 3.31): The transfer function, in terms of the Laplacian variable, for the highpass filter is

$$(V_o/V_i) = (sRC)/(1 + sRC) \tag{3.4}$$

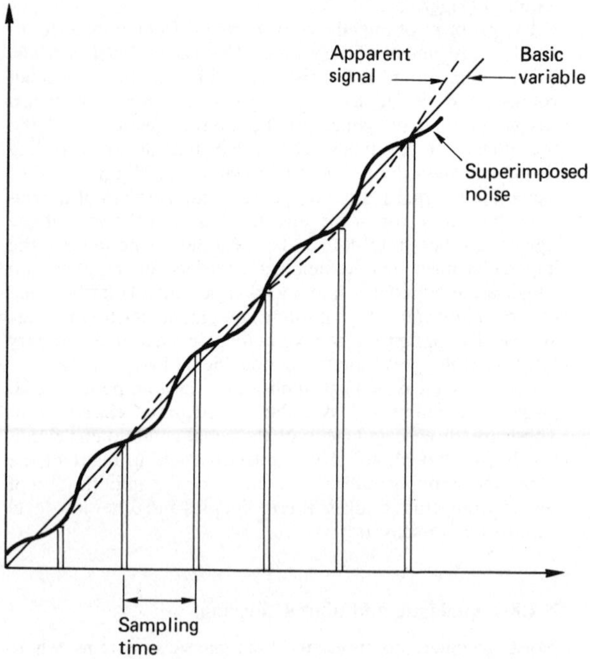

Figure 3.28   Noise generated 'aliases' in a sampled-data signal

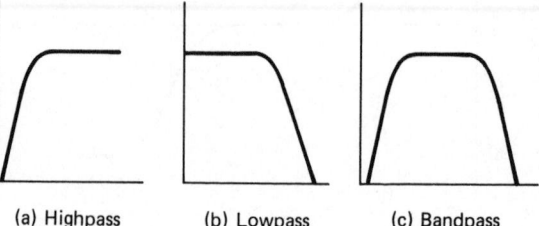

(a) Highpass   (b) Lowpass   (c) Bandpass

Figure 3.29   Filter performance curves

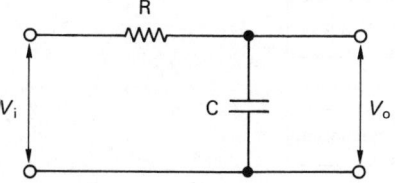

Figure 3.30   Lowpass filter

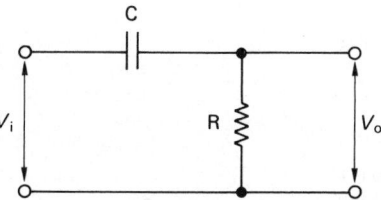

Figure 3.31   Highpass filter

The cut-off frequency is similarly selected through a judicious choice of resistor and capacitor.

Bandpass filters might be thought of as a series arrangement of a lowpass and a highpass filter. With the appropriate combinations of resistors and capacitors, the low- and high-frequency noise components in the signal can be suitably attenuated as required.

Filters which include an amplifier in the circuit are referred to as active and the relationship between input and output is a much more complex function of time. Williams[7] gives a comprehensive discourse on filter design.

Figures 3.30 and 3.31 illustrate the simplest forms of passive analogue filter which are used to suppress background noise. The governing equations (3.4) and (3.18) may equally well be expressed in terms of finite differences. In finite difference form, the equations can be used to action the filtering process on a discretized version of the input signal. This is the basis of a digital filter which can be implemented in software and requires no external hardwired components.

The setting of the cut-off frequencies in the digital filter are achieved through adjustment of the constants appearing in the finite difference approximating function. These numerical constants are simply related to the physical time constant in the equivalent analogue filter and also the digital sampling rate.

The advantages that the digital filter has over its analogue counterpart are the ease with which the cut-off frequencies can be adjusted. The -3 dB cut-off frequency can also be set exactly, since no hardwired components, with physical tolerance bands, are used. The digital signal may also be filtered any number of times simply by processing the data repetitively through the filtering algorithm. The disadvantage incurred is that digital filtering takes longer in real time to perform. The results of digital filtering on the signal are exactly the same as would be obtained using an analogue filter. That is, any time-varying signal whose frequency is outside the cut-off value is subject to attenuation with a corresponding phase shift.

3.5   Instrumentation

3.5.1   Introduction

Many aspects of mechanical engineering depend essentially on the ability to make measurements of relevant quantities. In some areas, such as non-destructive testing or experimental stress analysis, the techniques for making the measurements are linked so closely to the rest of the subject that they are better dealt with in their particular context. Here, we consider the more general measurements encountered throughout the whole field and the instrumentation that makes them possible.

In later sections, techniques are described under the headings of the quantities to be measured – sometimes referred to as measurands. First, it is useful to discuss topics that are of importance in many instrumentation systems.

Sometimes the process of measurement is a simple one, with the human operator playing a direct part: putting a rule alongside a component to determine a length, or measuring time using a stopwatch. More often a more complex system is used, which can be understood by thinking of the elements that make it up. Such a system is shown in Figure 3.32. Many variations are possible.

The measurand acts on a transducer and produces a signal corresponding to its value. This output signal is commonly electrical, but may be pneumatic or optical. The signal can then be transmitted as needed. We show the transmission or telemetry as a distinct part of the system in Figure 3.32 to indicate its importance. Some processing of the signal is often needed before it is displayed or recorded. Often a computer is used to control the whole operation, probably covering many transducers, partly in the light of the information given by the measurements.

For instance, temperatures might need to be determined at many points on a plant. Thermocouples could be used as transducers, other equipment being some distance away, more accessible and in a less severe environment. Temperatures should be recorded at regular intervals; maxima, minima and averages being extracted and an alarm given if any temperature exceeds a pre-chosen limit. Such a system might be part of the control of operational plant, signals being available as inputs for actuators. Alternatively, it might be logging the results of an experiment.

Most data processing and computing use digital signals, as does complex telemetry; simple, short-range telemetry can be analogue. Most transducers have analogue outputs, so an analogue-to-digital converter (ADC) is needed at the input to data processors. If the transducer gives an output in the form of frequency, the ADC can operate simply by counting the number of cycles in a given time.

Communicating to operators the results of measurement is a distinct phase of the whole. It may amount to recording permanently for study later, possibly simply on paper, more probably in some form suitable for computer input. Alternatively, an indication (called Display) may be needed to show immediately the output of the instrumentation system. This is all part of the human/machine interface that is dealt with in Chapter 4.

3.5.1.1   Accuracy

The accuracy of any instrumentation system is, of course, of great importance. When making a measurement it is possible to distinguish systematic and random errors. When a measurement of what should be the same thing is repeated it will be found that the readings are not identical; the spread of readings corresponds to random errors between individuals. The final error can be reduced by repetition and averaging

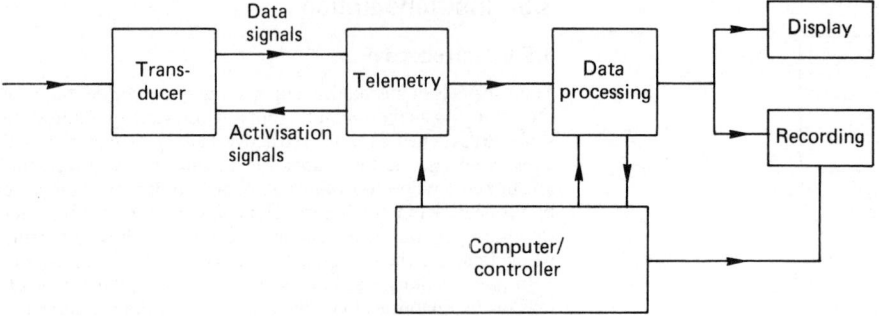

Figure 3.32   Typical instrumentation system

(provided time allows). However, even the average is not necesssarily correct, because it contains the systematic error. Systematic errors can be greatly reduced by careful calibration under conditions resembling the operating ones as closely as possible. Calibration is in fact an important feature of all systems where it is required to relate the readings taken back to absolute values of the quantities concerned. The word 'traceability' is used to describe building up links in this chain. Absolute values of measurands are not always important. Sometimes it is 'repeatability' – the capability of giving the same reading (even if it is the wrong one!) under the same conditions – that matters. The 'precision' of a reading relates to the smallest difference in value that can be detected.

When considering how faithfully an instrument represents what it is supposed to be measuring, the idea of 'influence quantities' should be taken into account. The resistance of a strain gauge, for instance, varies with strain, and so is used to measure it; but it also varies with temperature, so a one-to-one correspondence with strain will only hold provided temperature is constant. Temperature is an influence quantity. Reducing the errors introduced by influence quantities is facilitated by making 'bridge measurements'. This idea is most familiar in electrical circuits, when two components are connected so that changes in them affect the output in opposite directions. It can then be arranged that influence quantities affect them equally and so cancel, while the measurand either only affects one component or affects them oppositely.

Another concept that is helpful in understanding transducer performance is that of 'gauge factor'. This is the ratio of the fractional change in output to the fractional change in measurand. Again it is easily understood in the frequently quoted example of the resistance strain gauge, where it is given by the formula

$$\delta R/R = \delta l/l$$

However, the concept is more generally applicable, with the expectation that a transducer with a higher gauge factor will be more immune to influence factors.

'Hysteresis' is observed in many instruments. This is the name given to the defect of indicating a different output according to whether the quantity measured is increasing or decreasing (see Figure 3.33). Non-linearity is also shown in Figure 3.33. This is the situation when the relation between input and output cannot be exactly represented by a straight line.

3.5.1.2   Reliability

The reliability of any equipment is often of the utmost importance, and instrumentation systems are no exception. Attempts have been made to develop the concept of reliabil-

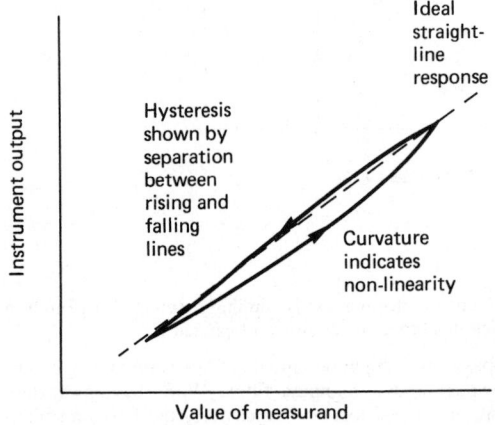

Figure 3.33   Hysteresis and non-linearity errors

ity, putting it on a semi-quantitative – though necesarily probabilistic – basis, so allowing the chances of failure to be calculated. While carrying out such an exercise, lessons can be learned of features that have been identified as significant. With large systems, the configuration in which different items are connected can make an unexpected difference. The schedule of servicing plays a large part in the down time that is to be expected. The consequences of any particular failure, which may be very diverse, should be analysed. The idea of 'independence' may be developed; this implies that it is safer for two measurements, if they are intended to corroborate one another, to be made with essentially different components – or even techniques – because they are then less likely to suffer from a common fault. In all considerations of instruments and their installations due weight must be placed on reliability. High accuracy has little value if it cannot be counted on, or perhaps it may be put that the possibility of a very large error can be more damaging than the probability of a small one.

3.5.1.3   Environmental conditions

In a broad way, it is always recognized that there are limits to the conditions under which instruments can be used. The working temperature range may well be specified. Users should be aware of other significant conditions. The atmosphere can be harmful, from moisture or other corrosive effects; dust in the atmosphere is an enemy to many instruments. Subjection to mechanical abuse (notably excessive vibration) must also be guarded against. More sophisticated hazards include electrical interference and nuclear radiation. The capabilities of particular items to withstand all these are often not spelt out. Common sense is often called for to judge what is reasonable to expect. Sometimes precise limits are quoted by suppliers, but it should be noted that approaching these limits closely may reduce accuracy and especially reliability.

3.5.1.4 Frequency coverage

A simple description of what an instrument does may imply that there is an indefinite amount of time in which to make a measurement – the steady-state or d.c. behaviour. Sometimes this is not the whole relevant story. There are certain devices that cannot be used for steady-state measurements but only for varying ones. There is always an upper limit to the frequency at which an instrument will operate faithfully. This may restrict the number of readings that can be taken in a limited time and is very often of concern when the measurement process is part of a closed control loop (see Section 3.7). It is therefore of great importance to take account of this feature of an instrument's performance.

3.5.2 Dimensional/geometrical measurements

The lengths that people have been interested in and hence wanted to measure range from perhaps 10^{-15} to 10^{18} metres. At the extremes, the interest is mainly from nuclear physicists and astronomers, and we shall here concentrate more on the middle distances that feature more in mechanical engineering.

Of course, there are simple, manual instruments that will continue to be used: rules, micrometers, calipers. A 'dial gauge' incorporates gears to provide magnification so that a movement of even a fraction of a millimetre gives an observable rotation of a pointer. Gauge blocks should be mentioned – elements with accurately parallel faces a precise distance apart; they can be thought of as coming into a calibration exercise, or as forming references allowing the dimensions of other parts to be compared with them.

However, a large part of modern dimensional instrumentation involves converting lengths into electrical (or, occasionally, optical or pneumatic) signals. The three electrical quantities – resistance, inductance and capacitance – are all used in transducers for this purpose.

3.5.2.1 Resistance transducers

In one form of transducer a slider moves over an extended resistance element, which may be either wire-wound or made of some non-metal. The resistance between the contact of the slider and either end of the element varies with the position of the former. Such devices are simple and only need simple electrical equipment. They can be made to obey some non-linear law if that should be wanted. However, they are not of the highest accuracy, and, because rubbing and friction occur, cannot be expected to have indefinitely long lives.

A different form of resistance transducer is the strain gauge. These are described in detail in Chapter 8. In essence, a small element is bonded to a structural unit and changes its resistance with the dimensional changes brought about by the latter being strained. Two gauges being installed are shown in Figure 3.34. The length changes are small (typically a fraction of a per cent) and the resistance changes are generally only a little larger. However, because it is possible to measure electrical resistance very precisely, strain gauges can still be used in accurate work. When directly measuring strain, they are, of course, measuring very small movements – and needing a large force to act on them. The range can be increased

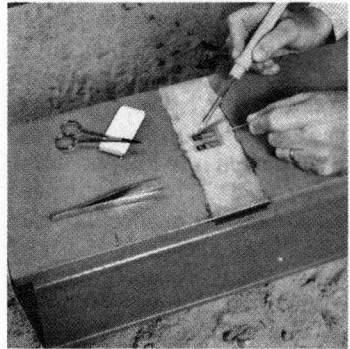

Figure 3.34 Strain gauges being installed (courtesy HBM)

and the force reduced by having the movement to be measured applied to a flexible strip on which the gauge is mounted, as shown in Figure 3.35.

3.5.2.2 Inductance transducers

The inductance of a coil is given approximately by

$$L = N^2 \frac{A}{l} \mu$$

where N is the number of turns, A the cross-sectional area, μ the effective permeability and l the length of the magnetic path. For an iron-cored coil with an air gap as shown in Figure 3.36 two very different values of μ arise, and the formula becomes

$$L = \frac{N^2 A}{(l_{iron}/\mu_{iron}) + (l_{air}/\mu_{air})}$$

Even for small values of the air gap, $(l_{iron}/\mu_{iron}) \ll (l_{air}/\mu_{air})$ and L is very sensitive to changes in l_{air}. Many inductance transducers make use of this principle.

If extreme sensitivity is not required, the 'iron circuit' need not be so nearly closed. An example of an instrument of this

Figure 3.35 Strain gauge on flexible strip

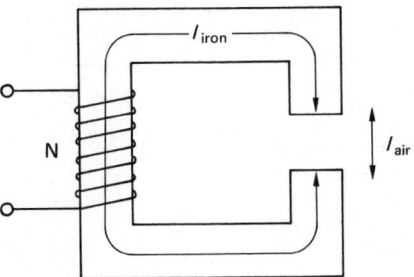

Figure 3.36 Coil with air gap

type is the linear variable differential transformer (LVDT), where there are two further refinements shown in Figure 3.37:

1. A differential system is used so that the inductance of one winding increases at the same time as that of another decreases; and
2. By having a further winding, a mutual inductance or transformer replaces the self-inductance.

These additions increase the magnitude and linearity of the output and give an inbuilt bridge system, so increasing the effective gauge factor. Figure 3.38 shows how small these devices can be made.

While most practical devices employ ferromagnetic (iron) cores, it is, in principle, possible to have an air-cored inductor serving as a transducer.

3.5.2.3 Capacitance transducers

The electrical capacitance in farads between a pair of parallel plates (as shown in Figure 3.39) is

$$C = \varepsilon_0 \varepsilon \frac{A}{d}$$

where ε_0 is the permittivity of free space ($= 8.9 \times 10^{-12}$ F/m),
  ε is the relative permittivity of the material between the plates,
  A is the area of either plate, or of their overlap if they are not exactly equal and opposite, and
  d is the separation between the plates.

C can therefore be changed by changing either A or d. Since d can be a millimetre or less, while the lengths involved in A are likely to be a centimetre or more, and the percentage change in C equals the percentage change in A or d, it can be seen that variable-d transducers are more sensitive than variable-A ones, i.e. those in which the overlap is changed. On the other hand, the proportionality between C and A makes the second type inherently linear, unlike the inverse C/d relationship when the gap is varied. A differential arrangement, in which the same movement increases one capacitance and decreases another, is often used and can improve linearity.

A variety of constructions have been used. One example is shown in Figure 3.40, where it can be seen that movement of an earthed screen alters the capacitance between the central, cylindrical electrode and one of the outer, co-axial electrodes,

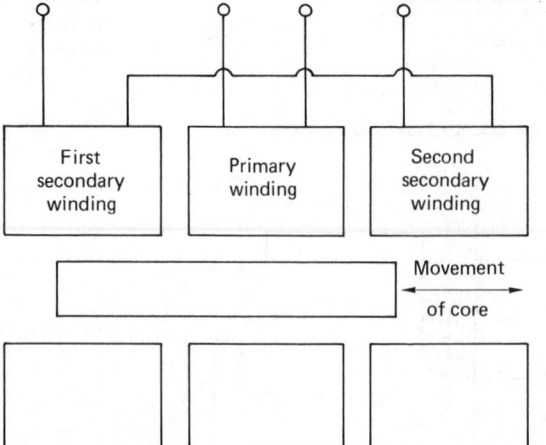

Figure 3.37 Linear variable differential transformer (LVDT)

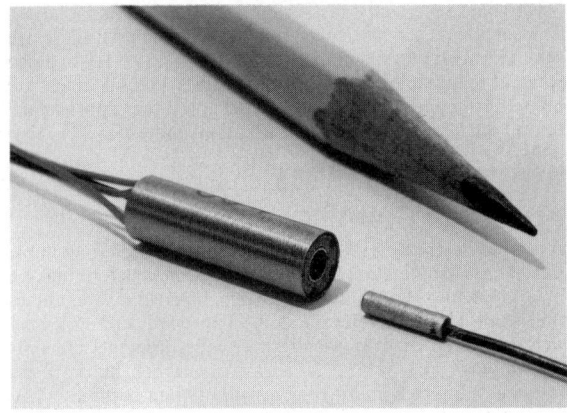

Figure 3.38 A LVDT (courtesy Schaevitz)

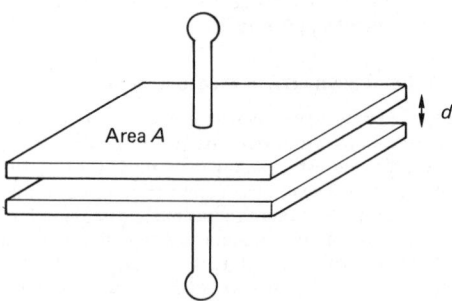

Figure 3.39 Parallel plate capacitor

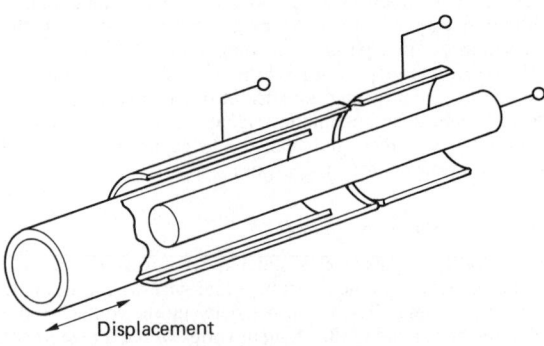

Figure 3.40 Cylindrical form of variable-area capacitance transducer (courtesy Automatic Systems Laboratories)

while the reference capacitance to the other electrode remains constant. Figure 3.41 shows an embodiment of this principle, the super-linear variable capacitor (SLVC).

3.5.2.4 Electrical circuits and comparison of techniques

Resistance measurements are simple and straightforward for the large changes involved in sliding contact devices; the basic accuracy of the transducer may not be high enough to justify elaborate circuitry. The much smaller changes in devices using strain gauges call for the use of bridge circuits, and some form of bridge configuration is usual for inductance and capacitance

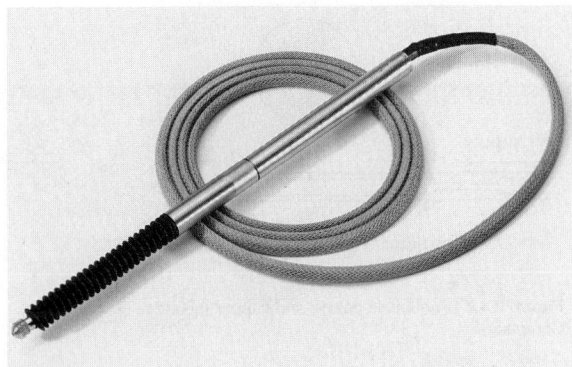

Figure 3.41 Super-linear variable capacitor (SLVC) (courtesy ASL)

transducers. Audiofrequency (AF) power supplies are used for resistance and inductance measurements, though d.c. is, of course, also effective for resistance.

Circuits used for capacitance measurement must take account of the stray capacitance that occurs between nearby conductors unless they are specifically screened from each other. Thus in Figure 3.42 capacitance variations between its lead and earth (either within or outside the screened cable) are indistinguishable (to the measuring circuit) from transducer capacitance changes. Various arrangements can be adopted to overcome this problem. The relatively small capacitance in most transducers corresponds to a very high impedance at lower frequencies and this is an argument for working at higher frequencies, but in fact AF bridge systems with very high sensitivities are available and give good performance when spurious effects are eliminated.

Many considerations come into the choice of transducer technique. As a very simple summary, it may be suggested that resistance devices are simple and inexpensive, inductance devices, while tending to be larger and more complicated, have a long history of development and mass production. Capacitance devices, simple and sensitive in principle, need more elaborate circuitry, but may well give the best approach for particulalrly onerous requirements. Sometimes the force needed to move a transducer element is important. In general, the force is less for capacitors than for inductors, while with variable resistors it may be less repeatable.

3.5.2.5 Optical methods of position measurement

Some classical experiments in physics depend on optical interference. If two coherent light beams are superposed they reinforce or cancel each other, according to whether they are in or out of phase, and this phase difference depends on the different lengths of the paths they have travelled. If they travelled the same distance or their paths differ in length by an integral number of wavelengths then they reinforce, while if

their paths differ by an odd number of half wavelengths they cancel.

Figure 3.43 shows how this can be used for an accurate measurement of movement. As the mirror M moves, the light intensity changes from maximum to minimum and back for successive distances of half a wavelength – a fraction of a micrometre, making the system highly sensitive. Refinements are needed to determine the direction of motion, and to give general stability; a corner cube reflector instead of a simple mirror eliminates the otherwise high sensitivity to the angle of the mirror. A laser is a convenient source of coherent radiation. The output signal, going through a succession of peaks, is essentially digital.

Moiré fringes are sometimes used to measure movement. Figure 3.44 shows two adjacent gratings as seen from above. If they are positioned as in (a) light can pass through, but if one is moved by half a 'wavelength' as in (b) the path is blocked and the combination appears dark. The 'wavelength' or pitch can be very short, as small as a few micrometres (the name Moiré comes from the silk weave in which the effect can be observed), giving a high potential accuracy. Again there is a basically digital output and the need to determine the direction of movement.

As shown in Figure 3.44, the interrogating light is transmitted through the gratings; it is, of course, possible to have a mirror system when the light source and detector are both on the same side of the gratings. The gratings may be at an angle to each other (Figure 3.45), when the alternate bright and dark areas form fringes perpendicular to the gratings; the fringes move bodily with linear displacement of either grating, while the separation between them depends on the angle between the two.

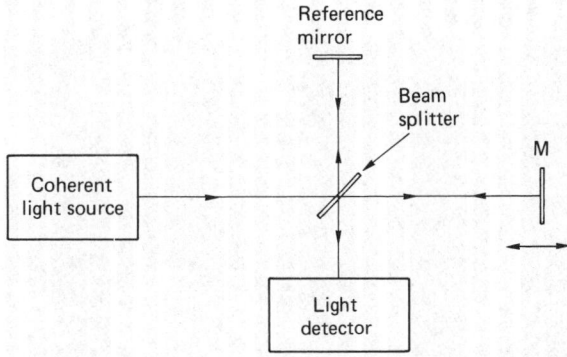

Figure 3.43 Movement measured by optical interference

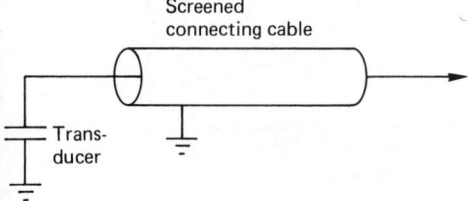

Figure 3.42 Effects of stray capacitance

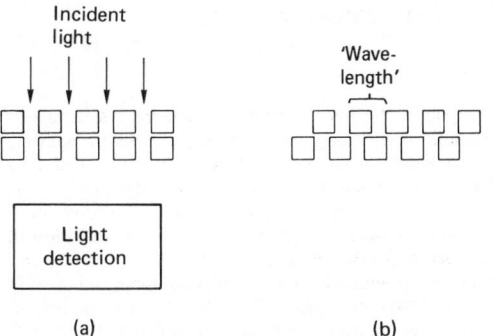

Figure 3.44 Principle of Moiré fringes

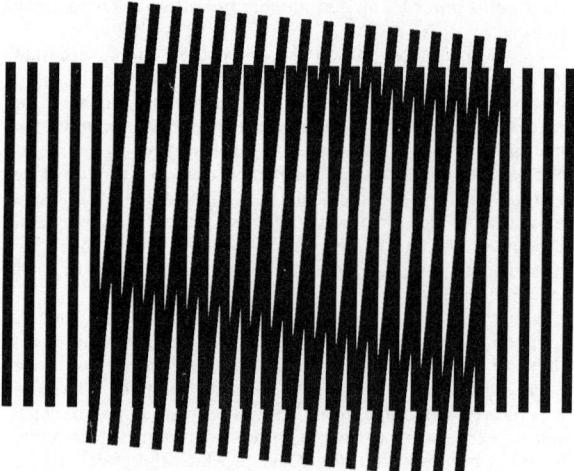

Figure 3.45 Moiré gratings at an angle

If the gratings do not have quite the same pitch, there are fringes parallel to the grating elements (Figure 3.46). This principle is sometimes used in strain measurement, when the strain to be measured is arranged to alter the pitch. In all these arrangements there is an effective magnification, so that small movements, on the scale of the small pitch of the gratings, give rise to much larger movements of the fringes.

Figure 3.46 Unequally spaced Moiré gratings

3.5.2.6 Pneumatics

Currently, pneumatic instrumentation systems are used less than electronic ones. They have the drawbacks of needing somewhat delicate mechanical devices and of introducing significant delays when signals are transmitted over long distances. However, they are by no means extinct and have the great safety advantage that there need be no question of their introducing electric sparks.

The heart of a pneumatic instrument is a flapper adjacent to a nozzle as shown in Figure 3.47. As the separation, d, between these is changed, the air flow through the nozzle changes markedly and hence also the pressure drop across the 'series' restrictor. A typical relation between d and the pressu-

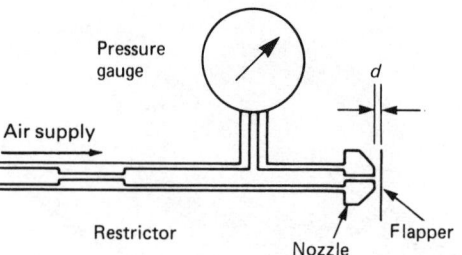

Figure 3.47 Pneumatic nozzle and flapper (courtesy Foxboro Company)

re P just upstream of the nozzle is shown in Figure 3.48. The effect can be amplified by the introduction of further elements in the shape of valves and a pressure-sensitive diaphragm. The primary behaviour is inherently non-linear, but the use of a pressure-feedback device with levers and a spring-controlled bellows allows a movement of the order of a millimetre to give a proportional pressure change of some tens of kilopascals.

3.5.2.7 Angular displacement

The synchro – sometimes called a Magslip or Selsyn – is widely used in the measurement of angles. If a.c. is applied to the central element (rotor) of such a device (left-hand side of Figure 3.49) then the voltages induced in the three circumferential windings depend on the angular position of the rotor

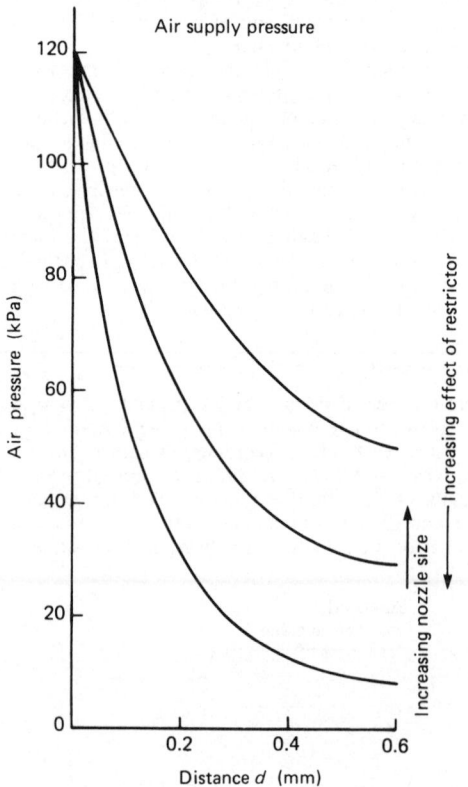

Figure 3.48 Relation between pressure and gap for pneumatic device (courtesy of Foxboro Company)

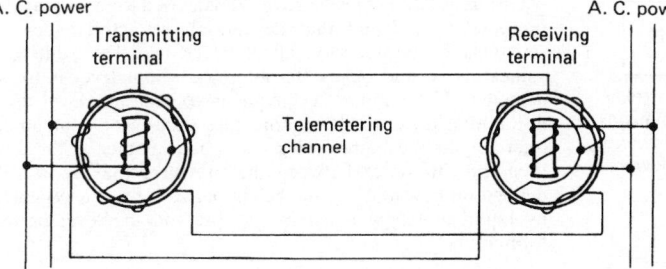

Figure 3.49 Principle of synchro

relative to them. This system has the particular advantage that if a second, identical unit is connected appropriately (right-hand side of Figure 3.49) forces will act within it until the two rotors take up identical angles. This is a robust and widely used technique for telemetering an angular position.

Capacitive transducers with variable overlap readily give a measurement of angle. The arrangement is in fact just that of the orthodox variable capacitor.

Encoders are used to give a digital signal corresponding to angular position. Moving clockwise round the disc shown in Figure 3.50, it can be seen that successive positions 1,2,3 . . correspond to successive binary numbers if black and white areas give digits 1 and 0, respectively, for powers of 2 starting at the largest radius. Black can be distinguished from white using six optical beams in the example shown, or a single beam can be traversed radially across the encoder. Alternatively, the distinction can be between conducting and insulating material, detected electrically.

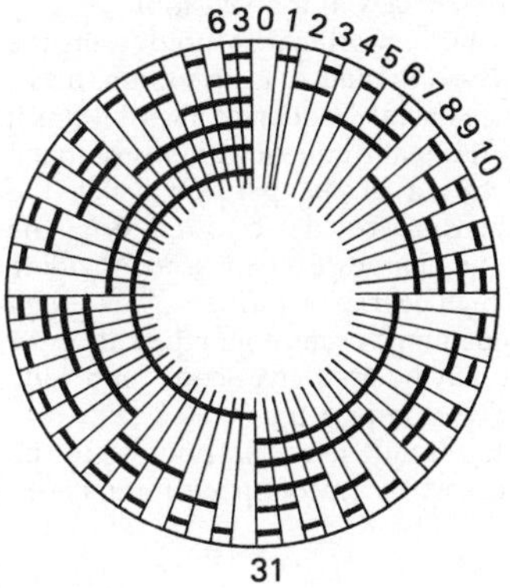

Figure 3.50 Encoding for angular position

A difficulty with this form of coding follows from imperfections of manufacture. Considering, for instance, the move from position 7 to position 8, if the outermost black should turn white slightly before the others, the configuration will momentarily correspond to position 6, while premature changes of the other blacks would indicate 5 or 3, respectively. The problem arises from the need for simultaneous changes at more than one radius, and to overcome this, codes have been devised in which only one change occurs at a time. As indicated previously, small changes of angle can also be detected with Moiré fringes.

3.5.2.8 Velocity measurement

Angular velocity is commonly measured employing electrical induction. Using the fundamental law that induced voltage is proportional to rate of change of flux, generators, either d.c. or a.c., can be made for which output voltage gives a direct measure of the speed of rotation. Under a completely different principle, a technique is to mount markers on the circumference of a rotor and count the number passing a stationary point in a given time, or alternatively, the time lapse between successive passages, which can be detected optically, magnetically or electrostatically. This system, of course, provides a digital output; it requires a finite time to give an indication.

Linear velocity is sometimes deduced from angular velocity as in a car's speedometer. It can also be calculated as the rate of change of position or as the integral of acceleration, and this is particularly relevant to vibration studies (Section 3.5.4). Measurement of fluid velocity is discussed under Flow in Section 3.5.7.

3.5.3 Volume and level

Volume, as such, is a quantity that is seldom measured. Instrumentation for rate of change of volume (or flow) is widely applied and can be integrated to give total volume; this is dealt with in a later section. Volume and mass are simply related through density and mass can be measured as weight. Again, the volume of material in a container can be inferred from the level it reaches, and this is a common measurement.

Measuring level, we can distinguish between continuous, normally analogue methods and digital techniques, in which the action is really detection rather than measurement. The presence or absence of the material in question at a particular level is indicated. The second category can be used, as shown in Figure 3.51, to move a 'follower' outside the container under study so that it remains opposite the internal interface, allowing the height to be measured in a more accessible place.

The level of a liquid conductor can be found from resistance measurement. Figure 3.52 shows two resistive wires that are effectively short circuited where they enter the liquid, so that the resistance seen at their terminals decreases as the level rises.

For an insulating liquid, capacitance measurement is appropriate. With the arrangement of Figure 3.53, capacitance increases as the level rises and a larger area of the overlapping plates is separated by a dielectric of higher permittivity. A sonar-ranging system can also be used in which the time taken for an echo to return from the surface being studied gives an indication of its position (Figure 3.54).

A sophisticated single-point technique involves passing gamma rays through the container. These will be more attenuated if there is a denser material in their path, so the intensity of radiation received at the detector shows whether liquid (or solid) rather than just gas is present. In Figure 3.55 it can be recognized that the detector output will be larger if the level of liquid in the container falls below the line from source to detector.

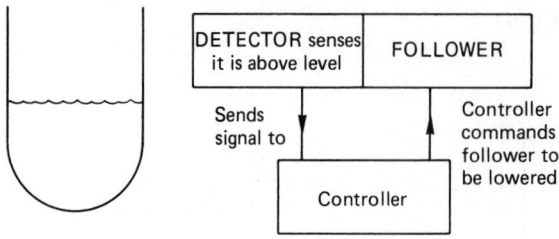

Figure 3.51 Level measurement with a follower

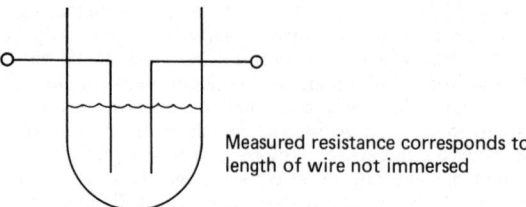

Measured resistance corresponds to
length of wire not immersed

Figure 3.52 Level measurement using resistance

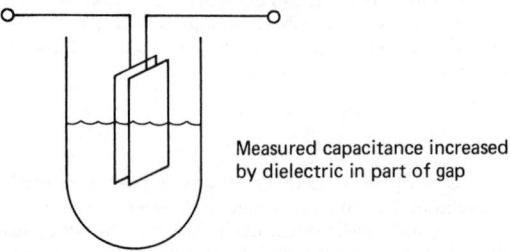

Measured capacitance increased
by dielectric in part of gap

Figure 3.53 Level measurement using capacitance

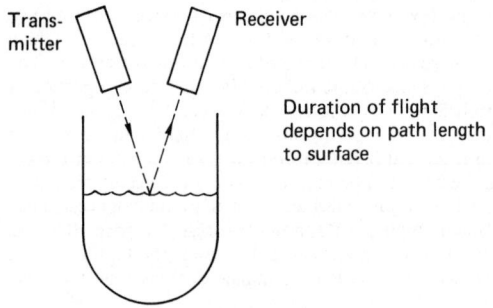

Duration of flight
depends on path length
to surface

Figure 3.54 Level measurement by sound ranging

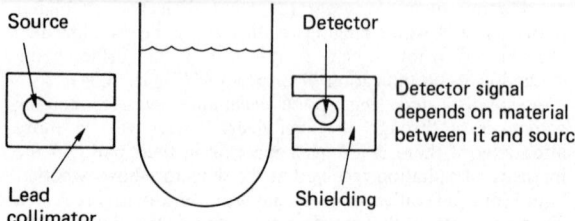

Detector signal
depends on material
between it and source

Figure 3.55 Level measurement by gamma rays

Different types of probe have been devised for detecting the presence of a liquid that depend on refractive index or resistivity or permittivity, all of which may have different values above and below the interface whose level is to be detected. Yet another technique is to measure level as a differential pressure. If one pressure transducer is mounted internally at the bottom of a vessel and another at the top, then the difference between the pressures they show will depend on how much of the height between them is occupied by liquid and how much by gas (i.e. on the level of the former).

3.5.4 Measurement of vibration

Wider aspects of vibration, which may be thought of as movement over small distances at comparatively high frequencies, are discussed in Chapter 1. Here we touch briefly on techniques of measurement.

In sinusoidal motion at frequency ω, linear amplitude s, velocity v and acceleration a are simply related as

$$a = \omega v = \omega^2 s$$

$$v = \omega s$$

Displacement can be measured using techniques described in Section 3.5.2, or velocity with a generator, commonly a coil moving in the field of an electromagnet. If a nearby point is known to be stationary, either measurement can be relative to this. Alternatively, part of the transducer can be an element with sufficient inertia not to move, when the measurement can be made relative to that. The criterion whether the inertia is large enough is that the resonant frequency of the element – decided by its mass and its flexible mounting – should be much lower than any frequencies in the vibration.

Vibrational accelerations are very often measured, using the associated force F, where $F = ma$, and m is an inertial mass. In this case, the element's resonant frequency must be much greater than the highest vibrational frequency to ensure that a is the same as the acceleration of the part on which the transducer is mounted. A piezoelectric device often forms the link to the intertial mass, the charges excited in it providing the output signal, while its high degree of stiffness gives a high resonant frequency. The wide range of mass and the variety of piezoelectric elements that may be used give scope for a wide range of applicability for vibration pick-ups.

We may want to know the frequencies contributing to the vibration studied. For this, some form of spectrum analyser will be desirable.

3.5.5 Force/weight measurement

This field of instrumentation gives good examples of some general principles mentioned in Section 3.5.1. Concerning speed of response, in many instances the measurement called for is a static one, but sometimes a quickly varying force is to be studied, and this calls for a different approach. The potential accuracy varies more than a thousandfold – with corresponding price ranges for equipment. Sensitivity to extraneous influences is also a factor in accurate force measurement, where errors from temperature, wrong location of the force and other things must be guarded against. Weight, of course, is a force, and, in general, it is measured in similar ways to other forces, though the measurement is always a static one.

Lever-type instruments, such as the classical analytical balance, are basically devices for comparing forces – often the weights of different masses. Unequal lever arms allow widely different forces to be compared; an arm of variable length allows a precise ratio to be established without the need for an adjustable force. The spring balance is the most familiar member of a large family of instruments in which the force to be measured is balanced by the reaction from an elastically strained member whose distortion can be measured.

The proving ring illustrated in Figure 3.56 is a refined form of spring balance. The applied force, which may be compressive or tensile, distorts the ring from its intially circular shape. The change in diameter (measured mechanically with a dial gauge or electronically) indicates the force. Much smaller movements are measured than with a coil-spring spring balance, making the total system more compact.

In a strain-gauge load cell the process is taken rather further, the elastic strains in a member being directly measured with strain gauges, allowing very compact devices to be constructed. The principle of a simple, columnar load cell is shown in Figure 3.57; the four strain gauges are connected into the four arms of a bridge. Structures in which shear strains are measured are also widely used; their readings are less dependent on the position where the load is applied. Two instruments of this type are shown in Figure 3.58. In hydraulic load cells the unknown force alters the pressure in a liquid system, allowing it to be measured as a pressure. Load cells can be built into the supports of hoppers to weigh their contents, or included in weighbridges.

Systems of particular value for dynamic measurement of quickly changing forces include piezoelectric elements, mentioned in connection with vibration instrumentation. Force balance systems are also used. Figure 3.59 shows how the displacement produced by a force to be measured can control the restoring force in a coil, the current in which gives a direct indication of the first force provided the gain is large and the displacement small – with due attention paid to stability.

Table 3.2 summarizes the features of different ways of measuring force.

3.5.6 Pressure

Pressure is easily measured from the difference in level of the liquid in two arms of a U-tube (Figure 3.60):

$$P_1 - P_2 = h\rho$$

where ρ is the density of the liquid. Mercury is commonly used as the working liquid. Its high density means that a large pressure difference can be measured without the equipment becoming too big in order to accommodate a large h. There is

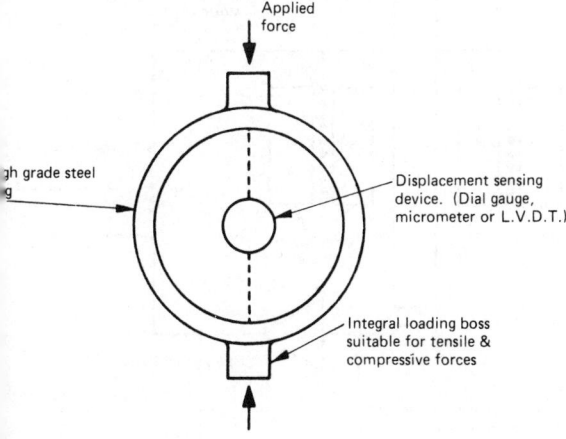

Figure 3.56 Proving ring

greater sensitivity if mercury is replaced by water. Further movement for a given pressure difference can be achieved if such a manometer is at a small angle to the horizontal instead of being held vertical. Care must be taken that the liquid can move freely over the inside surface of the tube (otherwise there will be hysteresis) and that any distortion of the surface from surface tension is the same in both limbs.

Note that it is pressure *differential*, P_1-P_2, that is of concern. Three situations should be distinguished:

1. Absolute pressure, where P_2 is zero, corresponding to a vacuum;
2. Gauge pressure, where P_2 is the atmospheric pressure in the neighbourhood of the equipment;
3. Differential measurements where both P_1 and P_2 may vary but it is their difference that is significant.

Pressure can be measured in terms of its fundamental definition of force per unit area. In Figure 3.61 if the cross-sectional area of the cylinder with its piston is known, then the pressure, P, is directly given by the force, F (usually a weight), needed to balance it. This method can give high accuracy, but there are practical complications, notably to ensure that the piston can move freely without the liquid leaking, so it is used mainly to calibrate other pressure gauges.

Neither of these approaches uses compact equipment or leads directly to an output signal, so they are often replaced by the use of transducer elements. Various configurations change shape with pressure and the consequent displacement can be used for measurement.

A Bourdon tube has an elliptical (or otherwise unsymmetrical) cross-section and is bent into a circular arc (Figure 3.62). If the pressure inside the tube increases, it tends to make the cross-section more nearly circular, and this in turn straightens the arc. With some further mechanical amplification, the movement is large enough to be read against a scale.

A metal diaphragm distorts according to the difference in pressure of the fluid on either side of it. The sensitivity varies widely with the dimensions. A wide, thin diaphragm moves appreciably under small pressures, and the danger of its rupturing under overload can be greatly reduced by the provision of 'stops'. The movement can be detected pneumatically or by capacitive or inductive devices.

An alternative approach is to measure strain in the diaphragm. Since different parts are strained in different senses, strain gauges connected in different arms of an electrical bridge can be mounted on a diaphragm so that their pressure-induced outputs sum while the spurious changes from, for example, temperature variation cancel each other out. A development from this is to have the strain gauges integral with the diaphragm. Using appropriate fabrication techniques, a silicon member can serve as diaphragm and can have certain parts modified and electrically insulated so that their strain-sensitive properties can be used to give an electrical output.

Stiff diaphragms – with a high natural frequency – allow rapidly changing pressures to be measured. Other devices have much slower responses, but often the measurement required is only of quasi-static pressure. Pressure transducers including piezoelectric force measurement have a quick response but cannot be used statically.

3.5.6.1 Vacuum

Some widely different methods are available for measuring vacuum (i.e. a pressure less than atmospheric). As conditions approach the zero of absolute vacuum, measurements become increasingly difficult.

Table 3.2

Method	Type of loading	Force range, N (approx.)	Accuracy % (approx.)	Size
Lever balance	Static	0.001 to 150 k	Very high	Bulky and heavy
Force balance	Static/dynamic	0.1 to 1 k	Very high	Bulky and heavy
Hydraulic load cell	Static/dynamic	5 k to 5 M	0.25 to 1.0	Compact and stiff
Spring balance	Static	0.1 to 10 k	Low	Large and heavy
Proving ring	Static	2 k to 2 M	0.2 to 0.5	Compact
Piezoelectric transducer	Dynamic	5 k to 1 M	0.5 to 1.5	Small
Strain-gauge load cell	Static/dynamic	5 to 40 M	0.01 to 1.0	Compact and stiff

Figure 3.57 Principle of load cell

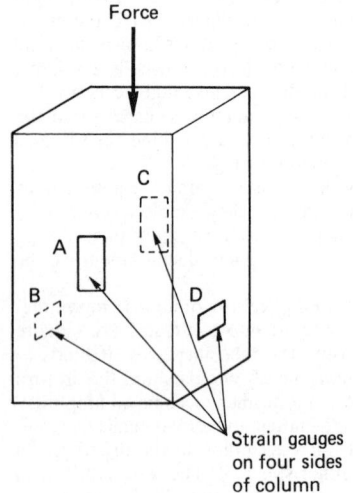

Figure 3.58 Photograph of load cells (courtesy Defiant Weighing)

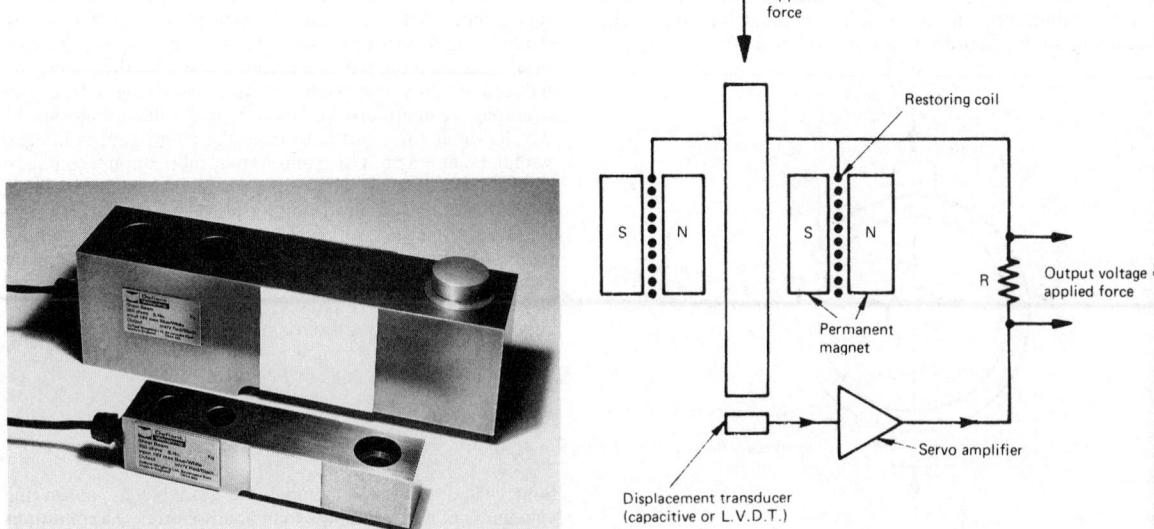

Figure 3.59 Force balance system

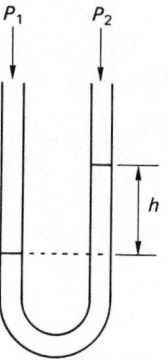

Figure 3.60 U-tube manometer

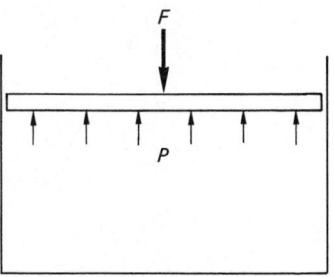

Figure 3.61 Absolute pressure measurement

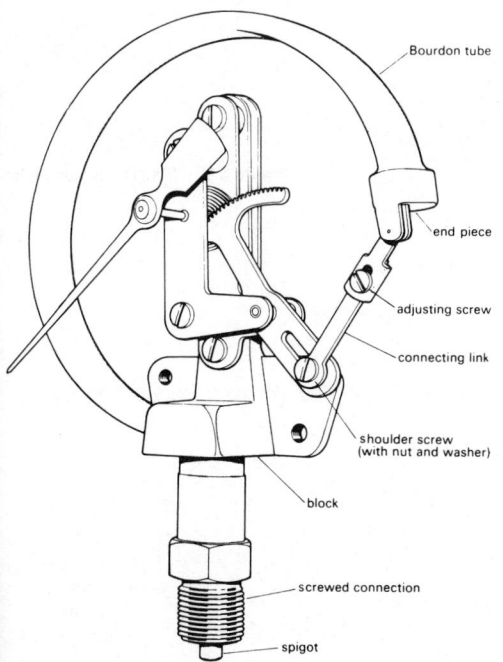

Figure 3.62 Bourdon tube pressure gauge (courtesy Budenberg Gauge Company)

In a low vacuum (i.e. an appreciable fraction of an atmosphere) instruments described in the previous section can be used. The McLeod gauge is a development from the U-tube manometer, in which a sample of gas is compressed by a known amount before its pressure is measured; this allows much lower initial pressures to be measured. Two other broad techniques are used for high vacua.

With thermal conductivity instruments (notably the Pirani) use is made of the fact that the larger number of molecules in a gas at higher pressure increase its heat transfer, so that measurement of the temperature of a heated member can indicate the degree of vacuum surrounding it. In ionization instruments (Buckley, Penning, Bayard-Alpert) the current resulting from ions in the vacuum is measured. This gives the population density of ions and hence the pressure.

The ranges over which different techniques can be used are shown in Table 3.3.

3.5.7 Flow

An important and widely applied field of instrumentation is the measurement of fluid flow. Sometimes the concern is to measure velocity at a 'point'. More often the requirement is for a single measurement representing the total volume of fluid passing along a pipe or other container – though this can be achieved by integrating from 'point' values. Instantaneous readings for flow are of primary interest; often their time integral (i.e. the total volume that has passed) needs to be known. Gases, as well as liquids, come under study. Occasionally, the main interest is with the mass rather than the volume that is passing.

Conceptually, the most direct form of instrumentation for flow measurement is the positive displacement meter, in which it is arranged that a known volume is repeatedly filled and emptied and the number of times that this takes place is counted. Of this type is the common, domestic gas meter, whose operation is illustrated in Figure 3.63. The capacities of chambers A and B are altered by a known volume as the diaphragm between them moves between the limits of its travel. A suitably phased slide valve ensures that the two chambers are connected alternately to the inlet manifold and to the outlet. For smoother operation, A and B are duplicated by C and D, running out of phase with them.

In other types, the volumes that are alternately filled and emptied are defined by rotating parts, either on a single axis or by two meshing rotors as shown in Figure 3.64. Care must be taken that sealing – usually by a liquid – is effective while still allowing free movement under the small forces associated with gas at low pressure.

3.5.7.1 Positive displacement for liquid flow

Similar devices are available as flow meters for liquids. A rotating piston, mounted eccentrically in a larger cylinder, is a common arrangement. Reciprocating pistons are also used as well as the sort of rotary systems described for gases. A turbine meter (Figure 3.65) may be thought of as a positive displacement instrument, having been designed so that the angle its bladed rotor turns through is proportional to the volume of liquid that has passed (axially) through the meter.

With all these meters, the number of rotations or excursions must be counted, the flow rate, of course, being given by the number occurring in a particular time. Information about internal movements must be conveyed through a container wall to the outside, and this is often done by the passage of permanent magnetic poles past external pick-ups.

Table 3.3 Comparison of vacuum gauge techniques

Technique	Pressure range (Pa)	Accuracy (±%)	Cost[a]	Advantages	Limitations
Bourdon tube	10^5–10^2	10	A	Simple, robust	Poor accuracy below 100 Pa
Diaphragm	10^5–10	5	B	Good general-purpose gauge	Zero setting varies
Liquid manometer	10^5–10^2	5–10	A	Simple, direct reading	Vapour may contaminate vacuum
McLeod	10^5–10^{-3}	5–10	C	Wide range. Used for calibration	Intermittent. Measures *gas* pressures only
Thermal conductivity	10^3–10^{-2}	10–20	C	Can be robust, with fast response	Risk of zero variation
Ionization	10^2–10^{-8}	20	D	Sensitive, fast response	Care needed in use

[a] Scale of costs: A less than £100, B £100–200, C £200–400, D £400–600.

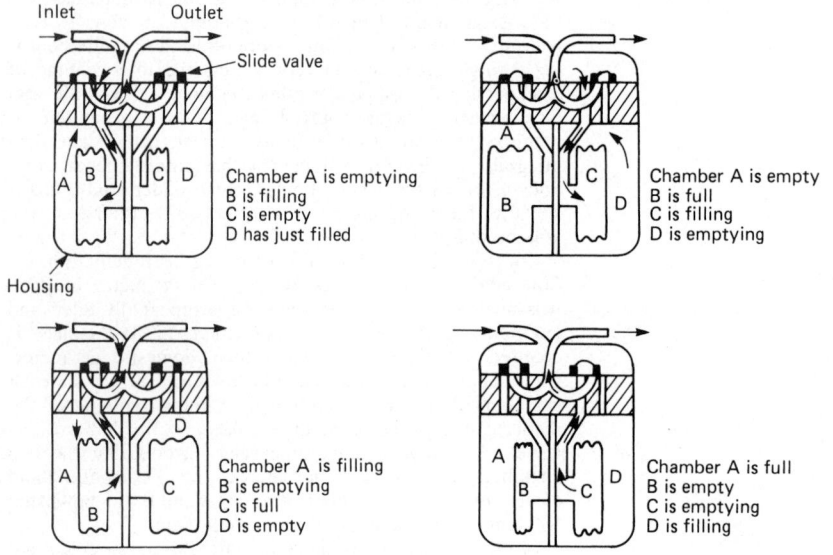

Chamber A is emptying
B is filling
C is empty
D has just filled

Chamber A is empty
B is full
C is filling
D is emptying

Chamber A is filling
B is emptying
C is full
D is empty

Chamber A is full
B is empty
C is emptying
D is filling

Figure 3.63 Principle of gas meter

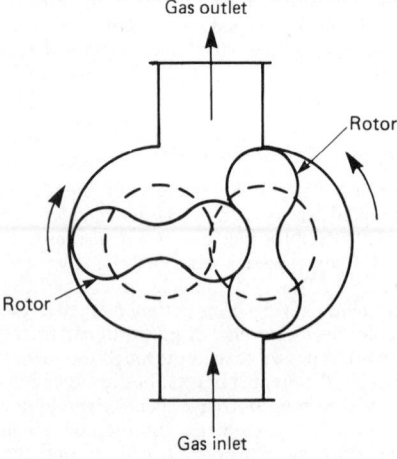

Figure 3.64 Positive displacement flow meter

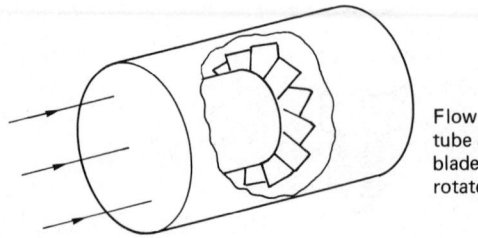

Flow along tube acts on blades to rotate turbine

Figure 3.65 Turbine meter

3.5.7.2 *Differential pressure*

Where the cross-sectional area of a pipe changes, so does the pressure of a liquid flowing in the pipe, and the magnitude of the pressure change depends on the flow rate. This is often used as the principle of a flow meter. Two configurations of changing cross section may be distinguished: the Venturi throat and the orifice plate.

In the former, a smooth profile serves to reduce the area (Figure 3.66); in the latter, changes are more abrupt (Figure 3.67). The Venturi has the advantage that less energy is absorbed, but at the cost of greater size and expense. Profiles different from either of these are sometimes used. In the Venturi, the difference in pressure between the throat and a point upstream is measured. With an orifice plate, the two relevant pressures are simply those upstream and immediately downstream of the plate, since for some distance downstream the effective area is still that of the orifice.

There is a square-law relation between the flow-rate Q (m^3/s) and the differential pressure, Δp:

$$Q^2 \propto \Delta p \quad \text{or} \quad Q = k\sqrt{\Delta p}$$

In some circumstances this is an inconvenience.

One of the factors in the proportionality is the area A at the point of restriction:

$$Q = cA\sqrt{\Delta p}$$

and in one type of flow meter, the Rotameter, Δp is kept constant and A made variable. This is achieved as shown in Figure 3.68 by having the liquid flow up through a tapered tube in which is placed a plummet whose weight causes the differential pressure. Increasing flow carries the plummet to a point where the annular area around it is such as to satisfy the equation.

Superficially similar to differential pressure types is the Target Flowmeter, in which the force exerted on a body obstructing the flow is used as an index of that flow. This again follows a square-root law.

3.5.7.3 *Open channels*

The measurement of flow in an open channel (or a closed duct that is not completely filled with liquid) calls for a different

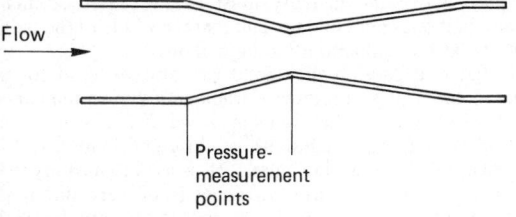

Flow →

Pressure-measurement points

Figure 3.66 Venturi throat

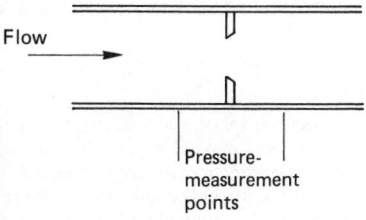

Flow →

Pressure-measurement points

Figure 3.67 Orifice plate

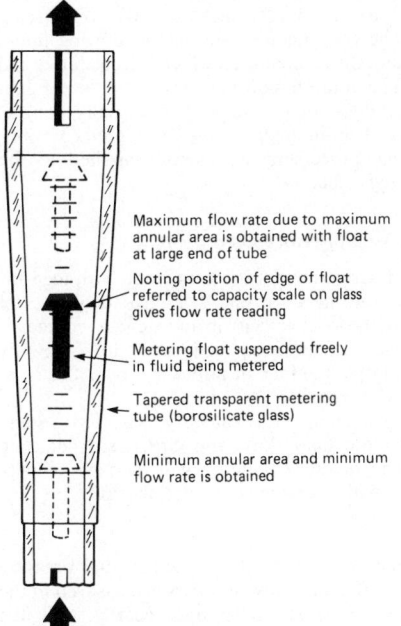

Maximum flow rate due to maximum annular area is obtained with float at large end of tube

Noting position of edge of float referred to capacity scale on glass gives flow rate reading

Metering float suspended freely in fluid being metered

Tapered transparent metering tube (borosilicate glass)

Minimum annular area and minimum flow rate is obtained

Fluid passes through this annular opening between periphery of float head and I.D. of tapered tube. Of course, flow rate varies directly as area of annular opening varies

Figure 3.68 Rotameter (courtesy Fischer and Porter)

approach. Flow over a weir may be measured by noting the level to which the liquid rises in a 'notch' in the weir. If there is not an adequate head of liquid to allow construction of a weir, the channel may have a 'flume' built into it; this is a construction similar to the throat of a Venturi tube, and again the pressure readings at appropriate points allow the flow rate to be calculated. In either case, the water level corresponding to a particular point in the flow is measured in another chamber – a 'stilling well' – connected by a small pipe.

3.5.7.4 *Newer flow-meter principles*

Three techniques may be mentioned that have more recently been developed to measure flow. When an obstruction is mounted in a pipe, the flow can be disturbed so that vortices are shed alternately from its opposite sides, and the frequency of this shedding is accurately proportional to flow rate. Sensitive detectors are needed to detect the vortices, commonly using their pressure or cooling effects, or their modulation of an ultrasonic beam. The method has the advantage of not being dependent on the exact sensitivity of the detector; in fact it uses a digital signal, namely frequency.

Electromagnetic flow meters use the principle of Faraday's law of electromagnetic induction. This states that a conductor moving in a magnetic field will give rise to an electromotive force (i.e. potential or voltage). The field, the movement and the potential are all mutually perpendicular. In a conventional electrical generator the conductor is a wire, but it can equally be a conducting liquid such as water. All that is needed is to provide a magnetic field – commonly non-sinusoidal at a low frequency – and suitably insulated electrodes in contact with the liquid.

A time-of-flight ultrasonic flow meter depends on the fact that sound pulses are transmitted more quickly downstream than upstream. The transmission time across flowing liquid which is the medium thus depends on which is transmitter and which receiver. The transmission path must not be straight across a pipe, but it does not have to be strictly axially along it, and an arrangement as in Figure 3.69 is adopted. Various forms of electronic processing are usefully applied to the primary time-of-flight data.

3.5.7.5 Measurement of velocity at a point

The flow meters described so far are concerned with measuring the total flow rate in a stream. It is also sometimes of interest to measure the local velocity in an extended volume of fluid. This approach can be used to measure total flow by having a representative number of points to cover the complete cross section of a stream.

The Pitot tube comes into this category. This has a small orifice facing into the fluid flow, and closed so that fluid cannot escape. The pressure build-up will then be velocity dependent, in fact with a square-root relationship

$$v \propto \sqrt{p}$$

The Pitot can thus be thought of as a differential pressure device, with p the difference between what is measured in the tube and the static pressure in the fluid nearby, only the primary measurement is of point velocity rather than total flow rate, as is the case with an orifice plate.

The hot-wire anemometer uses the fact that the cooling effect of a flowing fluid increases with the fluid's velocity. A wire is heated and the temperature excess above its surroundings is measured. With a fine wire, the instrument can respond very quickly, particularly if a feedback system is introduced so that the heating power supplied is altered to maintain a nearly constant excess temperature.

The turbine meter is used to measure 'point' velocities, particularly in surveys of large stretches of water.

Doppler techniques, where the frequency of some radiation is changed at reflection from a moving object, should also be mentioned under this heading. The ultrasonic Doppler flow meter is shown in Figure 3.70. A frequency, f_t, is sent out from the transmitter and the receiver registers a frequency f_r. The flow velocity V is then given by

$$V = \frac{c(f_t - f_r)}{2f_t \cos\theta}$$

where c is the velocity of sound in the flowing fluid – which must contain some discontinuities to give reflections. As described, this refers to reflection from a single point; in

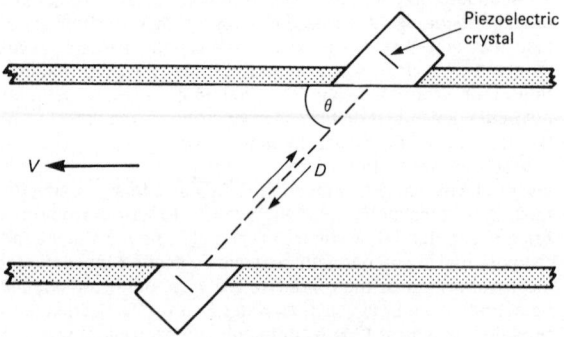

Figure 3.69 Time-of-flight ultrasonic flow meter

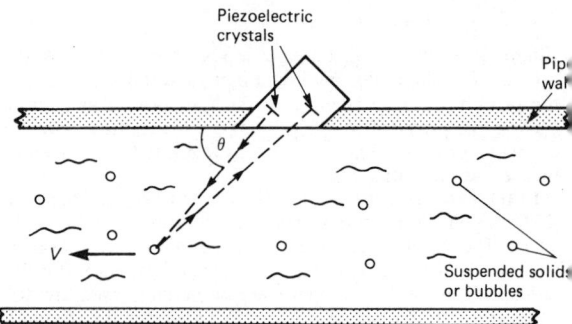

Figure 3.70 Ultrasonic Doppler flow meter

practice the method is used for total flow measurement taking the mean f_r as corresponding to the average of the different values of V across a pipe.

Laser Doppler techniques are a development from this, using electromagnetic instead of acoustic radiation. The closer positional control that is possible with laser beams allows more precise location of the point of reflection – at the cost of more complicated equipment.

3.5.8 Temperature measurement

Although we daily experience its effects, the scientific concept of temperature is an involved one, linked with energy on an atomic scale and not directly accessible. To measure it we therefore need some property that varies consistently as temperature varies.

3.5.8.1 Thermal expansion

A simple effect is thermal expansion. Mercury-in-glass thermometers depend on the expansion coefficient (defined as the fractional change in volume for one degree change in temperature) of mercury being larger than that of glass. The well-known shape of a bulb opening into a fine capillary tube allows the change in relative volumes to show as a change in position of the top of the thread of mercury in the capillary. Note that the space above the mercury must be evacuated so that pressure in it does not build up, and that the whole of the bulb should be at the temperature to be measured.

The linear expansion of a solid can also be used for a thermometer. Unless a reference frame at a known temperature should happen to be available, it will again really be a question of relative expansions of different materials. A 'bimetal strip', as shown in Figure 3.71, is used to magnify the movement. In such a strip, two metals of very different expansion coefficients, e.g. brass and Invar, are bonded together, when their combined curvature will alter as they warm up in order to equalize tensile and compressive forces in the strip. A direct pointer thermometer can be made by coiling the bimetal into a suitable spiral or helix as shown in Figure 3.72.

3.5.8.2 Thermo-couples

Other temperature-dependent properties are electrical. Thermocouples are widely used. They depend on an e.m.f. being set up in the circuit if two different metals are connected in series and their two junctions are not at the same temperature. Knowing one temperature and the controlling law, the other temperature may be deduced.

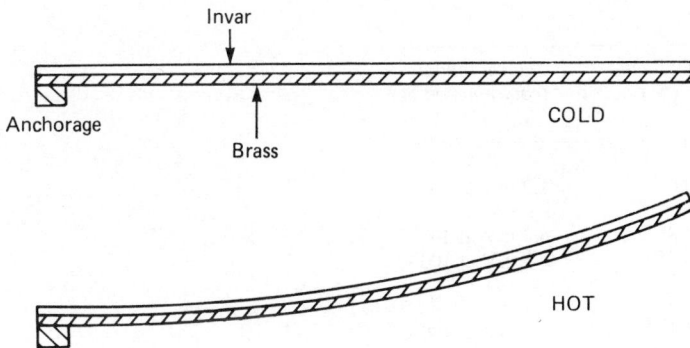

Figure 3.71 Bimetal strip

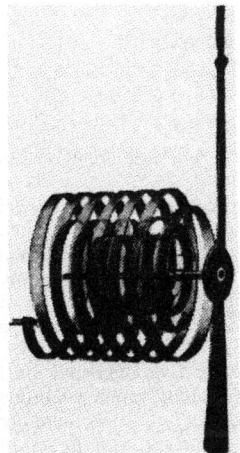

Figure 3.72 Bimetal thermometer

The provision of measuring circuitry is greatly simplifed because additional series junctions at intermediate temperatures (but with the same pair of metals) do not affect the aggregate e.m.f., and also further metals can be included provided all their junctions are at the same temperature. As shown in Figure 3.73, a parabolic relation generally holds, for a given cold junction, between e.m.f. and hot junction temperature. If the peak of the parabola is far enough away from the temperatures to be measured (as is the case for some metal pairs) the characteristic will be approximately linear. However, a firm limit for the working range is set by the reduced sensitivity in the neighbourhood of the peak and, worse, by the ambiguity that arises for temperatures beyond the peak, because two temperatures then correspond to the same voltage output.

Questions of corrosion also limit the type of thermocouple that can be used at high temperatures. Base metals are satisfactory below perhaps 1000°C; the more expensive noble metals and alloys are necessary to cover higher ranges. Some characteristics of common thermocouples are given in Table 3.4.

Potentials generated by thermocouples are never more than a small fraction of a volt. To measure this so as to give an acceptable accuracy calls for a precision of a few microvolts,

but great sensitivity is easily achieved in electrical measurements so that problems arise more from the errors introduced by spurious effects than from the absolute low signal level.

Calculations are often based on the 'cold' junction being held at 0°C. Rather than do this physically with a thermostat, it may be more convenient to introduce compensation from a component whose resistance varies with temperature in a known way (see Figure 3.74). Since it is only necessary to cover a small temperature range, changes can be thought of as linear.

If the e.m.f. of a thermocouple were translated into a current, the resistance of the leads in the circuit would come into the equation and unknown variations along their length, caused by the temperature changes there, would cause errors. This is eliminated if the working current is reduced to zero, either by using the 'null' technique of a potentiometer (when an equal and opposite potential opposes the output of the couple) or by having a detector with a high-input impedance.

With base-metal couples or short cable runs it is not impossible to use the same metals from hot junction to cold, and the additional circuitry can be at nearly constant, room temperature. The expense of a long run of noble metal can be avoided by replacing it, for most of its length, by a cheaper alloy, chosen so that its thermoelectric behaviour matches that of the noble metal over the limited temperature range to which most of the cable is subjected.

The two metals making up the thermocouple are commonly in the form of wires, forming leads as well as a junction. Bare wires welded or even twisted together can make an effective device, having, in fact, the advantage of a quick response. Note that if there are several points of contact it will be the coldest that is measured – the higher potential of the hotter junction being short-circuited through the colder. It is often convenient, however, for the couple to be supplied in a sheath having appropriate internal insulation and the whole forming a robust, replaceable unit. Metallic sheaths are the most common, but at the highest temperatures a ceramic construction gives greater protection against corrosion.

A mineral-insulated form of construction (MI) is widely used. In this, wires of the two different materials are located within a metal sheath and insulated from it and from each other by ceramic powder. It has been established that good insulation and stability are maintained even when the whole combination is drawn down to a very small cross section, perhaps as little as 1 mm overall. Figure 3.75 shows how the junction may either be insulated or welded to the tip of the sheath; the latter gives a quicker thermal response but consequences for the electrical circuit may be undesirable. Some of

Table 3.4 Characteristics of some common thermocouples

Type	Composition	Approx. temp range (°C)	Typical output (cold junction at 0°C)
B	Platinum – 30% rhodium/ platinum – 6% rhodium	0–1500	1.2 mV at 500°C
E	Nickel – chromium/ constantan	−200–850	6.3 mV at 100°C
J	Iron/constantan	−200–850	5.3 mV at 100°C
K	Nickel–chromium/nickel –aluminum (called chromel/alumel)	−200–1100	4.1 mV at 100°C
R	Platinum – 13% rhodium/ platinum	0–1500	4.5 mV at 500°C
S	Platinum – 10% rhodium/ rhodium	0–1500	4.2 mV at 500°C
T	Copper/constantan	−250–400	4.3 mV at 100°C
	Rhodium – iridium/rhodium	0–2000	6 mV at 1200°C
	tungsten – 5% rhenium/ tungsten – 26% rhenium	0–2300	9 mV at 500°C
	tungsten/molybdenum	1250–2600	5 mV at 2000°C

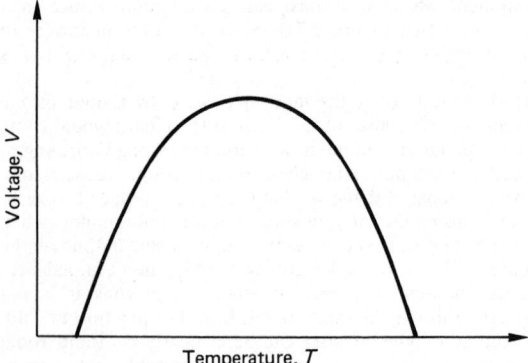

Figure 3.73 Thermoelectric potential

the wide variety of housings that are available are shown in Figure 3.76.

Spot checks on metal temperatures can be made with special devices. In one of these a two-pronged fork has its sharpened prongs made from the two different couple materials. When the prongs are pressed against the metal surface whose temperature is to be measured, it follows from the law of intermediate materials that the thermal e.m.f. will be that of a couple at the required temperature.

To measure temperatures as high as those of molten steel (which is often called for) a permanently protected probe would be both expensive and slow-acting. It is therefore economic to have an expendable hot-junction, with the rest of the probe arranged for easy replacement after a reading has been taken.

3.5.8.3 Resistance thermometers

One of the important properties that varies with temperature and so is used in thermometers is electrical resistance. This

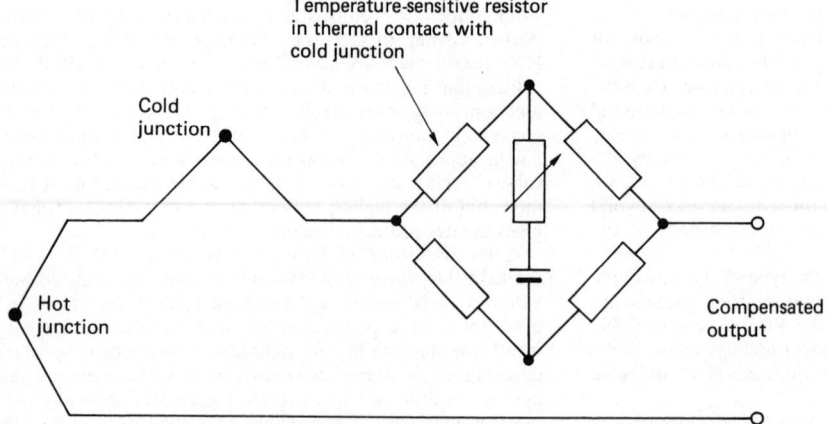

Figure 3.74 Cold-junction compensation

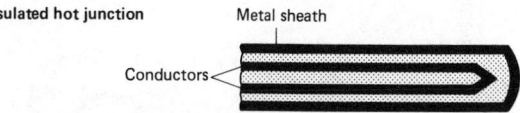

Insulated hot junction

Metal sheath

Conductors

The conductors are welded together and insulated from the sheath

Advantages: insulation resistance can be checked before and after installation

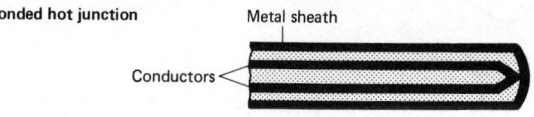

Bonded hot junction

Metal sheath

Conductors

The conductors and sheath are welded together

Advantages: very fast response and very accurate

Figure 3.75 Thermocouple construction

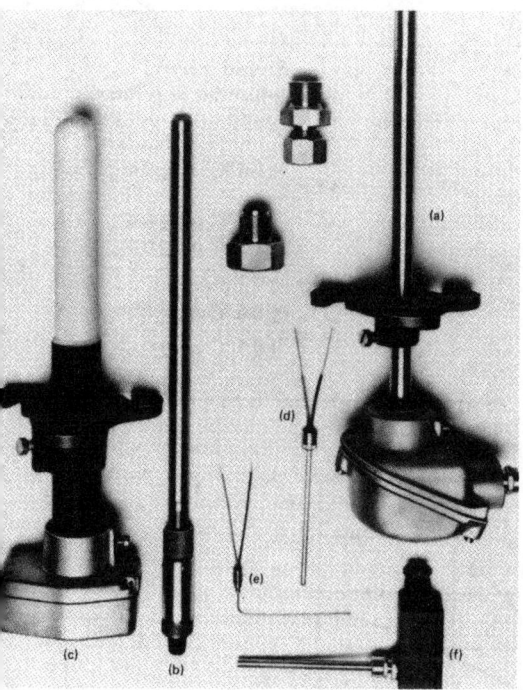

Figure 3.76 Housings for thermocouples (courtesy Kent Industrial Measurements)

pares with thermocouples the advantage of leading into the convenience and precision of electrical measurements.

The electrical measurement of a resistance is easier when it not too small – 100 Ω is a convenient value for a resistance thermometer. That means that comparatively long, fine wire ill be needed. A consequent danger is the susceptibility to corrosion, because even a very shallow, surface attack would make a proportionately large change in cross section, making the resistance rise:

$$= \frac{\rho l}{A}$$

i.e. resistance is given by resistivity times length, divided by area, and we are looking for changes in ρ, not spurious changes in A.

Because of these considerations, platinum is the favourite material for accurate resistance thermometry. A further advantage is that platinum can be made very pure, allowing reproducible characteristics, even though resistivities are highly susceptible to small impurities. For less accurate, lower-temperature operation, nickel may be used.

It is important to avoid unpredictable stressing of the element – which would alter the resistance. Several ways of doing this are shown in Figure 3.77. Metallized film tracks on glass or ceramic are alternatives. The protective outer housing that is added means that, to outward appearance, there is not much difference between thermocouple and resistance transducers.

Conventional bridge and other circuits are used with resistance thermometers. Whereas with thermocouples it can be arranged that lead resistance does not play a large part, when the active element is itself a resistance, the lead resistance cannot be ignored, and unless the associated circuitry is very close, compensation must be introduced for variations in lead resistance due to unknown temperature changes along the length. It can be achieved by introducing one or two extra wires (according to the bridge configuration) in the cable that connects the transducer to the measuring circuit. A.C. circuits have the advantage that they avoid introducing errors from thermal e.m.f.s in leads or temperature elements but with them some care must be taken over stray inductance or capacitance in the transducer.

In the act of measurement a finite current flows through a resistance thermometer element. This generates heat and raises its temperature above that of its surroundings – which, of course, is what is really to be measured. This error, which depends on the thermal insulation of the element, can generally be neglected if the power dissipation is below 10 mW. Film-type resistors have the advantage of introducing particularly small errors of this type.

3.5.8.4 Thermistors

Semiconductors, with a different physical basis for their electrical conduction, can show a much greater change of resistance with temperature than metals do. This is exploited in thermistors, whose behaviour is illustrated in Figure 3.78.

Thermistors can be supplied as beads or in rod, disc, washer or film form; they have the advantage that they can be very small. Their characteristic is given approximately by the law

$$\alpha = -B/T^2$$

where α is the temperature coefficient (ohms per Kelvin), B is a constant and T is absolute temperature. Their high sensitivity makes thermistors attractive for many applications, though an individual's characteristics cannot be predicted to a tight tolerance and may show a drift equivalent to the order of 0.1 K over a year.

The characteristic of a different device, the 'switching thermistor', is also shown in Figure 3.78. It is used for protection purposes rather than continuous control. The switching device, unlike a conventional thermistor, has a positive temperature coefficient. Over a small span of temperature its resistance increases a thousandfold, so drastically reducing the current flowing through it. The critical temperature at which this happens can be chosen, for instance, to prevent electrical insulation being burnt out.

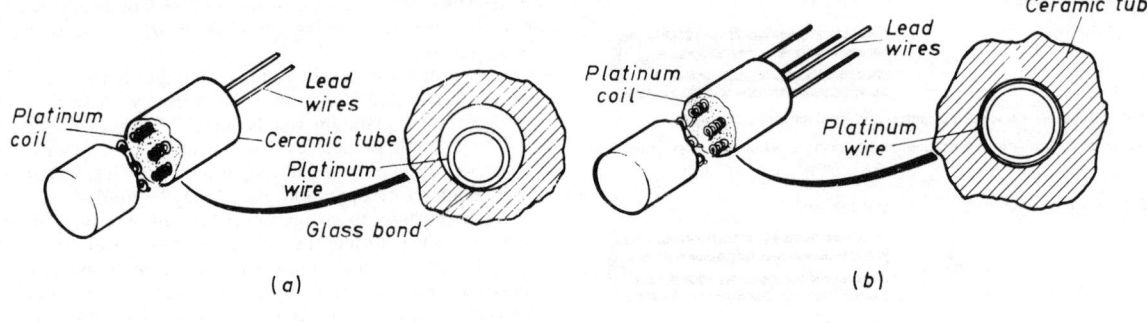

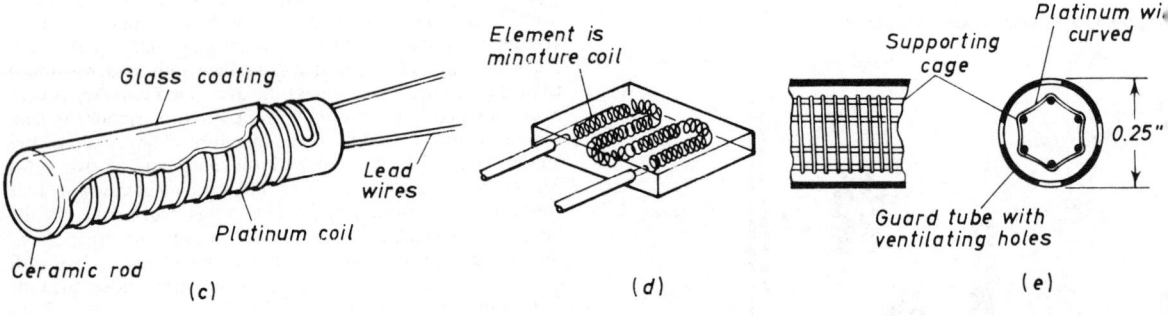

Figure 3.77 Construction of resistance thermometers (courtesy Rosemount Engineering)

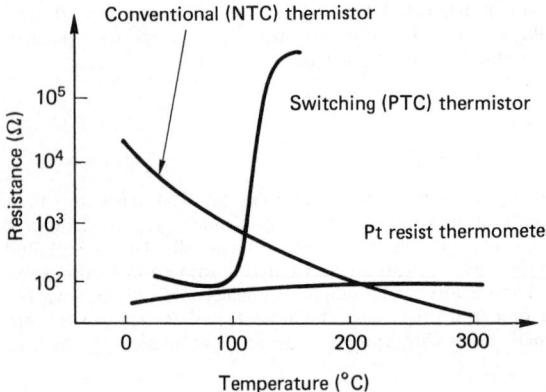

Figure 3.78 Thermistor characteristics

3.5.8.5 Radiation thermometers

Radiation thermometers (which were formerly known as pyrometers) are not based on any change of property with temperature but use the electromagnetic radiation from a body to be measured. As the body warms up, the total radiation it emits increases rapidly (with the fourth power of the absolute temperature) and the spectral distribution shifts to shorter wavelengths. The temperature can thus be deter-

mined by measuring the radiation, and there is the clear advantage that all the detecting equipment is remote from the hot body. Limitations to the technique are that it is more difficult to measure lower temperatures, where the energy emitted is much less, and that the emissivity of the radiating surface comes into the equation as well as its temperature.

In this type of thermometer the radiation is focused on a detector. A lens may be used for this purpose (it must be made of a material that transmits the appropriate radiation) or sometimes a mirror to give complete spectral coverage. A thermopile, consisting of a number of thermoelectric junctions connected in series to increase their output, may be used as detector. Alternatively a pyroelectric device may be employed; in this, charges are liberated as the temperature changes. These latter devices do not respond to steady-state signals, so the radiation must be 'chopped', which is commonly effected by having a segmented disc rotating in its path. The semiconductor photodiode is another detector that is sometimes used at shorter wavelengths.

Surface emissivity is much less important when the radiation to be measured has emerged from a 'window' in a hollow body. This makes the technique particularly applicable to furnaces. Dependence on emissivity is also reduced in the arrangement shown in Figure 3.79. If the reflectivity of the hemi-spherical mirror there approaches unity, the effective emissivity of the surface also tends to unity, though with this set-up the advantage of having all equipment remote from the hot surface is, of course, sacrificed.

Sometimes the measurement is of total radiation, sometimes of that within a particular band of wavelengths, which are chosen from considerations of detector sensitivity and material transmission. Shorter wavelengths are appropriate for hotter bodies, longer for colder. By working in the far

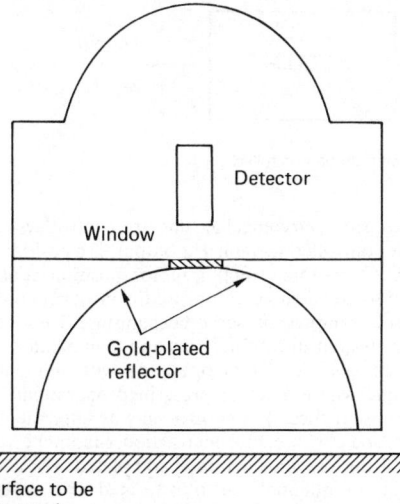

Surface to be
measured

Figure 3.79 Arrangement to reduce emissivity dependence

material transmission. Shorter wavelengths are appropriate for hotter bodies, longer for colder. By working in the far infrared (at 30 μm wavelength) temperatures as low as $-50°C$ have been measured, but applications are much more common upwards from 50 to 100 K higher.

When, as is often the case, radiation thermometers are used in dusty atmospheres, an air purge will be desirable to keep the front optical surface clean. In some designs, the detector is kept further outside a hostile environment by using optical fibres as links.

3.5.8.6 Gas and vapour thermometers

In pneumatic instrumentation systems there is an advantage in having the temperature signal, not in an electrical form, but as something more immediately compatible with pneumatics. This is an attraction of gas and vapour thermometers.

With both of these a sealed bulb is situated where temperature is to be measured and connected by capillary tubing to a pressure-sensitive device. The bulb is either completely filled with a permanent gas or partially filled with a suitable liquid, which means that the pressure in the bulb changes with bulb temperature, according either to the gas laws or to the liquid's vapour pressure. Pressure measurements can thus indicate bulb temperature.

3.5.8.7 Practical considerations

With such a multiplicity of methods for measuring temperature the choice between them depends on many factors. The following line of thought may be helpful as a first, crude guide:

1. Use gas or vapour techniques if and only if they have to feed into a pneumatic system.
2. Failing them, use thermocouples (base metals for lower temperatures, noble metals for higher) unless
3. The highest accuracies are needed, when resistance elements have attractions or
4. Contact with the object studied is difficult or impossible, when radiation thermometers are the solution.
5. Remember that the large signals from thermistors may be an advantage if their limited range is acceptable.

6. For the highest temperatures, radiation techniques may prove much cheaper than thermocouples.

Speed of response is sometimes an important matter, but it depends more on details of construction than on the basic type of thermometer.

When deciding the details of an installation, several points should be borne in mind. The requirement is often to measure the temperature of a fluid, but when a transducer is immersed in a fluid its equilibrium temperature is decided not only by the conductive and convective heat exchange with the fluid but also (if the fluid is transparent, as are most gases) by radiative heat exchange with the walls. Therefore the transducer takes up a temperature that is intermediate between fluid temperature and wall temperature. The error so caused can be greatly reduced by introducing 'radiation shields'. A further error arises when, as is normal, the transducer is mounted from the walls of the container, so providing a heat-conducting path through the mounting. The tendency to bring the transducer's temperature closer towards that of the walls is especially marked if it is housed inside a more permanent pocket (the American term is 'Thermowell') in order to facilitate replacement.

There can be 'sampling errors' when measuring temperature. For instance, if the fluid of concern is flowing inside a pipe and has a temperature different from that of its surroundings, there will be a 'temperature profile' which relates the local fluid temperature to its (radial) position in the pipe. As shown in Figure 3.80, at only one radius will the temperature correspond strictly to that of the mean. However, such errors can be positive or negative, and it is sometimes possible to make them offset the radiation and conduction errors referred to above.

3.5.9 Bar code readers

A form of optical data input which is finding increasing application is the bar code reader. The code consists of a series of black and white vertical lines which are printed onto the object (Figure 3.81).

The code is read by an optical sensor which incorporates a lamp, a phototransistor and a number of optical focusing lenses. The decoding software is, however, necessarily complex, since the speed at which the code is read can vary.

A typical bar code might consists of a start pattern, 101, five 7-bit characters, a check sequence, a second group of five characters and an end pattern. Two consecutive black bands

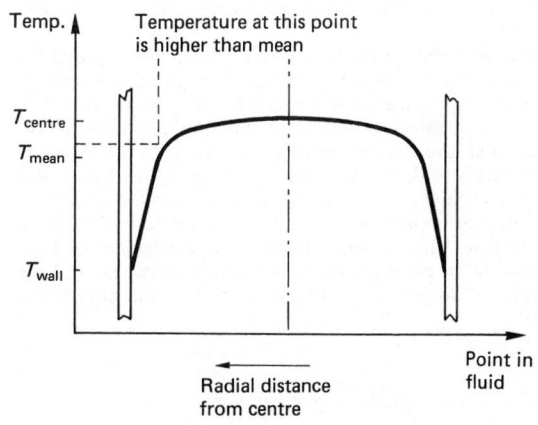

Figure 3.80 Errors caused by temperature profile

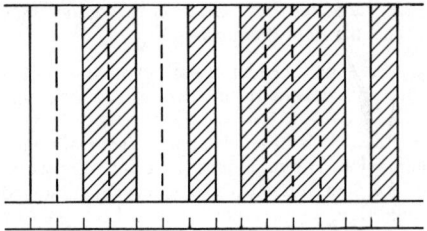

Figure 3.81 Section of a bar code

represent a bit value of 1, while two consecutive white bands represent a bit value of 0. The code is designed such that every character starts with a white band and ends with a black band. This ensures that every character starts and ends with a 1-0 transition. Additionally, every character code includes at least one 1-0 transition within the code.

The decoding program must include a timing loop to determine the speed of reading. The timing is usually based on a count of the 1-0 transitions. With the reading speed established, the code can then easily be translated. Bar code readers are very much evident in large supermarkets and libraries, but they have applications to manufacturing stock control and automatic assembly lines for component counting and identification purposes.

3.6 Classical control theory and practice

3.6.1 Introduction

Control engineering is based on the linear systems analysis associated with the development of feedback theory. A control system is constituted as an interconnection between the components which make up the system. These individual components may be electrical, mechanical, hydraulic, pneumatic, thermal or chemical in nature, and the well-designed control system will provide the 'best' response of the complete system to external, time-dependent disturbances operating on the system. In the widest sense, the fundamentals of control engineering are also applicable to the dynamics of commercial enterprise, social and political systems and other non-rigorously defined concepts. In the engineering context, however, control principles are more generally applied to much more tangible and recognizable systems and sub-systems.

Invariably, the system to be controlled can be represented as a block diagram, as in Figure 3.82. The system is a group of physical components combined to perform a specific function. The variable controlled may be temperature, pressure, flow rate, liquid level, speed, voltage, position, or perhaps some combination of these. Analogue (continuous) or digital (discrete) techniques may individually (or simultaneously) be employed to implement the desired control action. In more recent times the advances made in microelectronics have resulted in an emphasis towards digital techniques, and the majority of modern control systems are now microprocessor based.

3.6.1.1 Classification of control systems

Engineering control systems are classified according to their application, and these include the following:

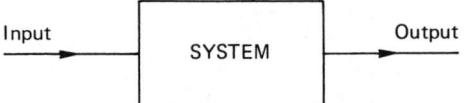

Figure 3.82 System to be controlled

1. *Servomechanisms*: Servomechanisms are control systems in which the controlled variable (or output) is a position or a speed. D.C. motors, stepper motor position control systems and some linear actuators are the most commonly encountered examples of servomechanisms. These are especially prevalent in robotic arms and manipulators.

2. *Sequential control*: A system operating with sequential control is one where a set of prescribed operations are performed in sequence. The control may be implemented as 'event based', where the next action cannot be performed until the previous action is completed. An alternative mode of sequential control is termed 'time based', where the series of operations are sequenced with respect to time. Event-based sequential control is intrinsically a more reliable 'fail-safe' mode than time based. Consider, for example, an industrial process in which a tank is to be filled with a liquid and the liquid subsequently heated. The two control systems are depicted in Figure 3.83.

   The time-based sequential control system is the simplest. The pump is switched on for an interval which would discharge enough liquid into the tank to fill it to approximately the correct level. Following this, the pump is switched off and the heater is switched on. Heating is similarly allowed to continue for a preset time, after which the liquid temperature would approximately have reached the desired value. Note that the control function is inexact and there are no fail-safe features. If the drive shaft between the motor and the pump becomes disengaged or broken, the heater will still come on at the prescribed time, irrespective of whether there is liquid in the tank or not. The event-based sequential control system has fail-safe features built in and is much more exact. In operation the pump is switched on until the liquid-level sensor indicates that the tank is filled. Then (and only then) is the pump switched off and the heater switched on. The temperature of the liquid is also monitored with a sensor and heating is applied until such time that the temperature reaches the desired value.

   Obviously, with two additional sensors, the event-based system is the more expensive. The advantages it offers over the time-based system, however, far outweighs its disadvantages and event-based sequentially controlled systems are by far the most common. Time-based systems do exist, nonetheless, and they are found in applications where the results of malfunction would be far less potentially catastrophic than those occurring in the example described. The essential difference between the two systems is that event-based sequential control incorporates a check that any operation has been completed before the next is allowed to proceed. The modern automatic washing machine and automatic dishwasher are good examples of sequentially controlled systems.

3. *Numerical control*: In a system using numerical control the numerical information, in the form of digital codes, is stored on a control medium which may be a paper tape, a magnetic sensitive tape or a magnetic sensitive disc. This information is used to operate the system in order to control such variables as position, direction, velocity and speed. There are a large variety of manufacturing opera-

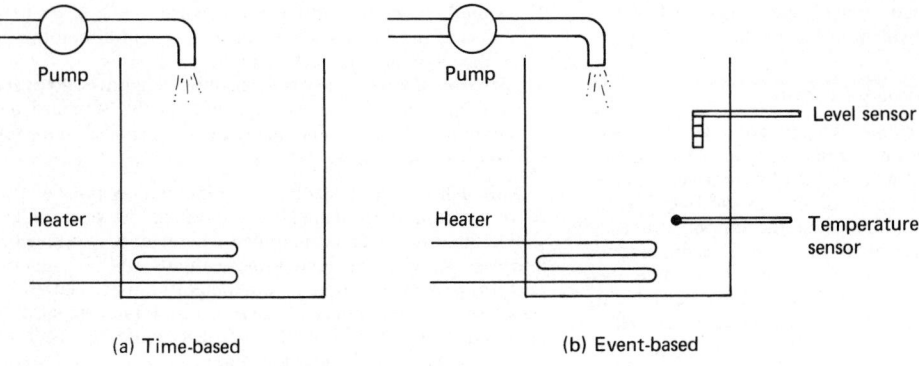

(a) Time-based   (b) Event-based

Figure 3.83 Simple sequential control systems

tions involving machine tools which utilize this versatile method of control.

4. *Process control*: In this type of control the variables associated with any process are monitored and subsequent control actions are implemented to maintain the variables within the predetermined process constraints. The word 'process' is all-encompassing and might include, for example, electrical power generation. The generation of 'electricity' can be considered as a manufacturing process where the 'product' is kilowatt hours. In the control of power generation, the variables which are measured include temperature, pressure, liquid-level, speed, flow-rate, voltage, current and a range of various gas concentrations. This is further complicated by the need to satisfy the power demand, and it is apparent that the control of such a system is necessarily complex. Similarly complex examples exist in the oil and paper-making industries, in automative assembly plants and in any entity which aspires to the designation of a 'flexible manufacturing system'.

3.6.1.2 Open- and closed-loop control

The basic open-loop system is shown in Figure 3.82 and is extended in Figure 3.84 to illustrate a more complete picture. The input element supplies information regarding the desired value, X, of the controlled variable. This information is then acted on by the controller to alter the output, Y.

External disturbances are fed in as shown and will cause the output to vary from the desired value. The open-loop system may be likened to the driving of a vehicle where the driver constitutes the input element. Essentially, two variables are controlled by the driver – the speed and the direction of motion of the vehicle. The controller, in the case of speed, is the engine throttle valve and in the case of direction, is the steering system.

In order that the system becomes closed-loop, two further elements must be added:

1. A monitoring element, to measure the output, Y;
2. A comparing element, to measure the difference between the actual output and the desired value, X.

The monitoring and comparing elements are connected through the 'feedback' link as shown in Figure 3.85.

It can be argued that the driver in the previous example also performs the functions of monitoring and comparing. The vehicle driver, therefore, if considered to be part of the complete system, constitutes a closed-loop feedback control. For the purpose of definition, however, any system which incorporates some form of feedback is termed closed-loop. With no feedback mechanism, the system is categorized as open-loop. For the most practical engineering purposes, control systems are of the closed-loop variety to take advantage of the benefits of feedback, which may be either 'positive' or 'negative'. A positive feedback signal aids the input signal. It is possible therefore to have output with no input when using a positive feedback signal and, since this is detrimental to the control function, positive feedback systems are very rare.

3.6.1.3 Linear and non-linear control systems

For a control system to be linear it must satisfy both the amplitude proportionality criteria and the principle of super-position. If a system output at a given time is $Y(t)$ for a given input $X(t)$, then an input of $kX(t)$ must produce an output of $kY(t)$ if amplitude proportionality is satisfied. Similarly, if an input of $X_1(t)$ produces an output of $Y_1(t)$, while an input of $X_2(t)$ produces an output of $Y_2(t)$, then if an input of $(X_1(t) + X_2(t))$ produces an output of $(Y_1(t) + Y_2(t))$ the superposition principle is satisfied. Non-linear systems do not necessarily satisfy both these criteria, and generally these

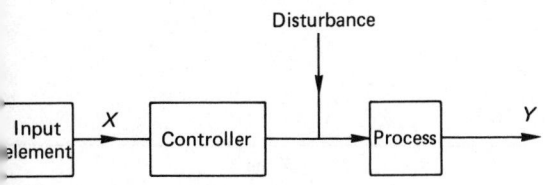

Figure 3.84 Open-loop control system

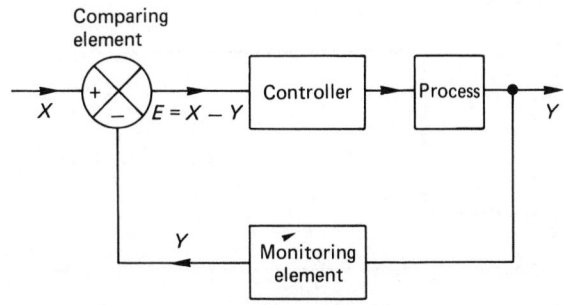

Figure 3.85 Closed-loop feedback control system

systems are 'compensated' such that their behaviour approaches that of an equivalent linear system.

3.6.1.4 Characteristics of control systems

The characteristics of a control system are related to the output behaviour of the system in response to any given input. The parameters used to define the control system's characteristics are stability, accuracy, speed of response and sensitivity.

The system is said to be 'stable' if the output attains a certain value in a finite interval after the input has undergone a change. When the output reaches a constant value the system is said to be in steady state. The system is unstable if the output increases with time. In any practical control system, stability is absolutely essential. Systems involving a 'time delay' or a 'dead time' may tend to be unstable and extra care must be taken in their design to ensure stability. The stability of control systems can be analysed using various analytical and graphical techniques. These include the Routh-Hurwitz criteria and the Bode, Nichols and Nyquist graphical methods.

The accuracy of a system is a measure of the deviation of the actual controlled value in relation to its desired value. Accuracy and stability are interactive and one can in fact be counter-productive to the other. The accuracy of a system might be improved, but in refining the limits of the desired output, the stability of the system might be adversely affected. The converse also applies.

The speed of response is a measure of how quickly the output attains a steady state value after the input has been altered.

Sensitivity is an important factor and is a measure of how the system output responds to external environmental conditions. Ideally, the output should be a function only of the input and should not be influenced by undesirable extraneous signals.

3.6.1.5 Dynamic performance of systems

The dynamic performance of a control system is assessed by mathematically modelling (or experimentally measuring) the output of the system in response to a particular set of test input conditions:

1. *Step input*: This is perhaps the most important test input, since a system which is stable to a step input will also be stable under any of the other forms of input. The step input (Figure 3.86) is applied to gauge the transient response of the system and gives a measure of how the system can cope with a sudden change in the input.

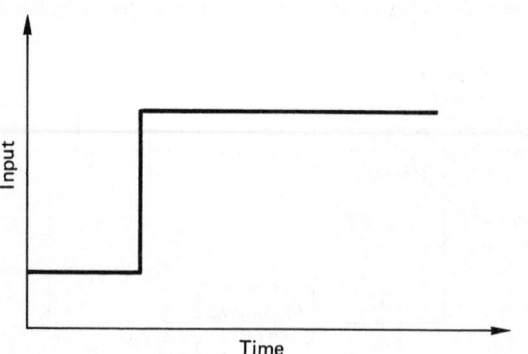

Figure 3.86 Step input

2. *Ramp input*: A ramp input (Figure 3.87) is used to indicate the steady-state error in a system attempting to follow a linearly increasing input.
3. *Sinusoidal input*: The sinusoidal input (Figure 3.88) over a varying range of input frequencies is the standard test input used to determine the frequency response characteristics of the system.

Although the three standard test inputs may not be strict representations of the actual inputs to which the system will be subject, they do cover a comprehensive range. A system which performs satisfactorily under these inputs will, in general, perform well under a more natural range of inputs. The system response to a parabolically varying test input can also be analysed or measured, but this is a less commonly used test signal compared to the previous three.

3.6.1.6 Time domain and frequency domain

The time domain model of a system results in an output $Y(t)$ with respect to time, for an input $X(t)$. The system model is

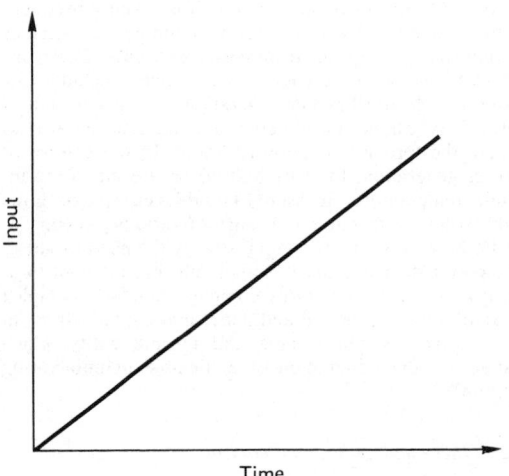

Figure 3.87 Ramp input

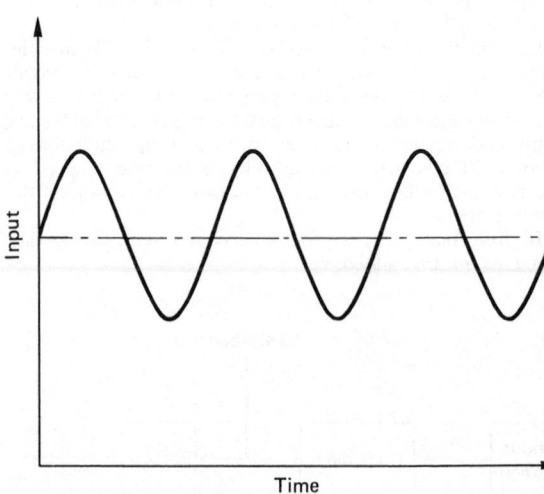

Figure 3.88 Sinusoidal input

expressed as a differential equation, the solution of which is displayed as a graph of output against time.

In contrast, a frequency domain model describes the system in terms of the effect that the system has on the amplitude and phase of sinusoidal inputs. Typically, the system performance is displayed in plots of amplitude ratio, $(Y(t)/X(t))$ or $20 \log_{10} Y(t)/X(t))$, and phase angle, against input signal frequency.

Neither system model has an overriding advantage over the other and both are used to good effect in describing system performance and behaviour.

3.6.2 Mathematical models of systems – time domain analysis

Differential equations are used to model the relationship between the input and output of a system. The most widely used models in control engineering are based on first- or second-order linear differential equations.

3.6.2.1 First-order systems

Some simple control systems (which includes the control of temperature, level and speed) can be modelled as a first-order linear differential equation:

$$\tau \frac{dY}{dt} + Y = kX \tag{3.5}$$

where X and Y are the input and output, respectively. τ denotes the system time constant and k is the system gain. When the input X is a step of amplitude A then the solution to equation (3.5) gives the result shown in Figure 3.89. The solution curve shown in this figure has the analytical form

$$Y(t) = kA[1 - e^{-t/\tau}] \tag{3.6}$$

Equation (3.6), which is the time-domain solution, is an exponential function which approaches the value (kA) as t approaches infinity. Theoretically, the output never reaches kA) and the response is termed an exponential lag. The time constant τ represents the time which the output would take to reach the value (kA) if the initial rate of response were maintained. This is indicated by the broken line which is tangent to the solution curve at time, $t = 0$. For practical purposes the final steady-state output is taken to have been reached in a time of about (5τ).

If the input is a ramp function then the response of a first-order system is as shown in Figure 3.90. The ramp input is simulated by making the right-hand side of equation (3.5) a

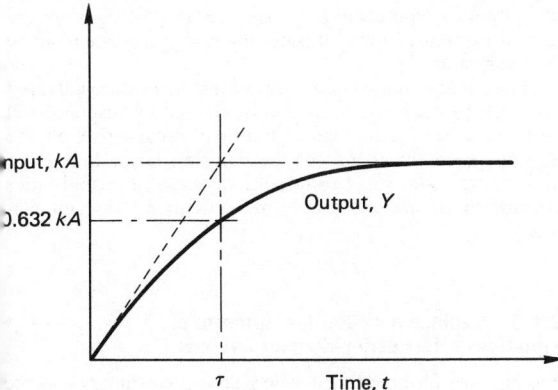

Figure 3.89 Response of a first-order system to a step input

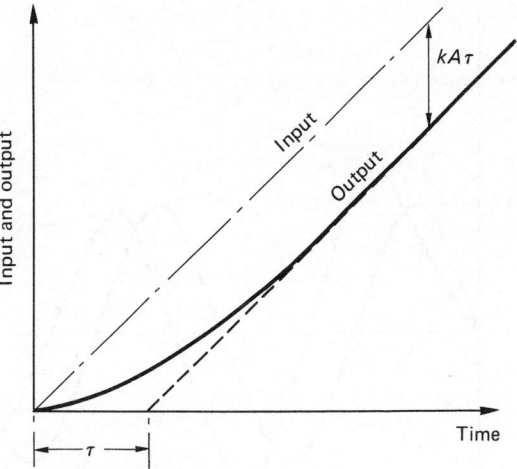

Figure 3.90 First-order system response to a ramp input

linear function of time, i.e. kAt. With this input, the time domain solution becomes

$$Y(t) = kA[t - \tau(1 - e^{-t/\tau})] \tag{3.7}$$

The solution equation shows that as t becomes large the output tends to $kA(t - \tau)$. The output response is asymptotic therefore to a steady-state lag $(kA\tau)$.

The response of a first-order system to a sinuosidal input can be obtained by setting the right-hand side of equation (3.5) equal to $kA\sin(\omega t)$, where ω is a constant circular frequency in radians/second. The time-domain solution yields

$$Y(t) = \frac{kA}{\sqrt{(1 + \tau^2\omega^2)}} [\sin\alpha \cdot e^{-t/\tau} + \sin(\omega t - \alpha)] \tag{3.8}$$

where $\alpha = \tan^{-1}(\tau\omega)$

The response is shown in Figure 3.91. The output response exhibits a decaying transient amplitude in combination with a steady-state sinuosidal behaviour of amplitude, $kA/[\sqrt{1 + \tau^2\omega^2}]$ and lagging the input by the angle α.

3.6.2.2 Second-order systems

While some control systems may be adequately modelled as a first-order linear differential equation, many more practical systems, including position control, are more conformably represented by a differential equation of the second order. The second-order differential equation has the general form:

$$\frac{d^2Y}{dt^2} + 2\zeta\omega_n\frac{dY}{dt} + \omega_n^2Y = kX \tag{3.9}$$

where ζ is termed the damping ratio and is defined as the ratio of the actual damping in the system to that which would produce critical damping. ω_n is the undamped natural frequency of the system and k, again, is the system gain.

The time-domain solution depends on the magnitude of ζ and three solutions for a step input are possible:

1. *Light damping, $\zeta < 1$.*

$$Y(t) = \exp(-\zeta\omega_n t)[A\cos(\omega_n\sqrt{(1 - \zeta^2)}t)$$
$$+ B\sin(\omega_n\sqrt{(1 - \zeta^2)} \cdot t)] \tag{3.10}$$

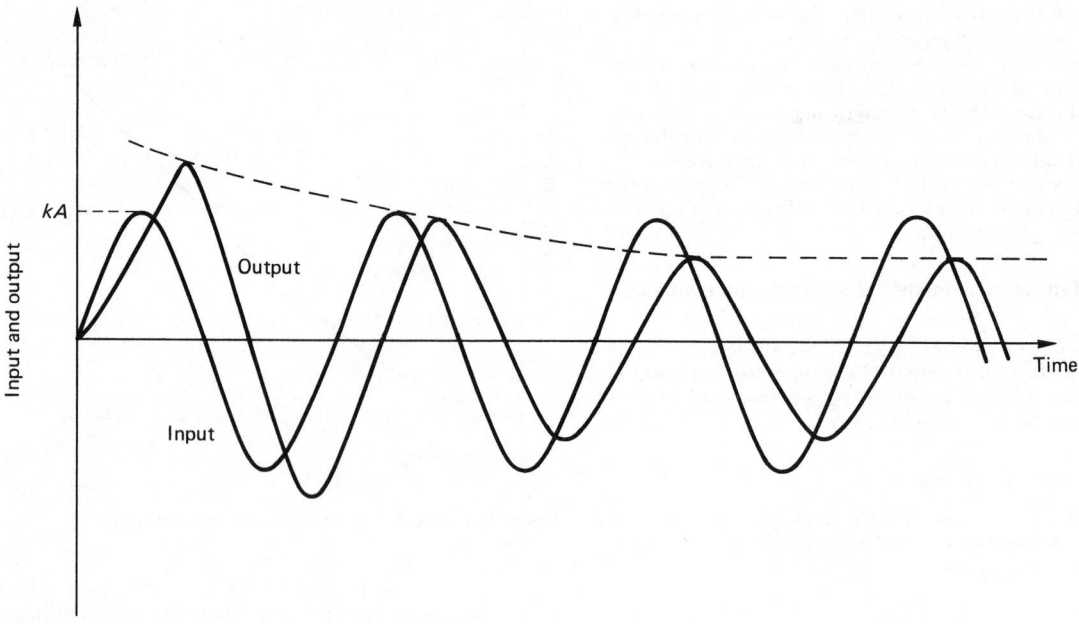

Figure 3.91 First-order system response to a sinusoidal input

2. *Critical damping, $\zeta = 1$.*

$$Y(t) = \exp(-\zeta\omega_n t[At + B] \tag{3.11}$$

3. *Heavy damping, $\zeta > 1$.*

$$Y(t) = A\exp\{-\zeta\omega_n t - [\omega_n\sqrt{(\zeta^2 - 1)}]t\} \\ + B\exp\{-\zeta\omega_n t + [\omega_n\sqrt{(\zeta^2 - 1)}]t\} \tag{3.12}$$

The three possible solutions for a step input are shown in Figure 3.92. The output, $Y(t)$, is plotted as a percentage of the step input, X, against the parameter, $(\omega_n t)$.

For ζ equal to unity, the system is critically damped and the steady-state value is attained in the shortest possible time without any oscillatory response. With ζ greater than unity, the system is overdamped and the response curve is again exponential in form. Overdamped systems may have an undesirably sluggish response. Indeed, since the effect of $\zeta > 1$ simply delays the response to the steady-state value there is no real advantage to be gained in using high ζ values.

For cases where ζ is less than unity, the system is said to be underdamped and the response curve is oscillatory with an exponential decay. A number of performance measures are used to describe the response of an underdamped system to a step input, and these are illustrated in Figure 3.93.

The speed of the response is reflected in the rise time, τ_R, and the peak time, τ_P. For underdamped systems, the rise time is the time taken for the output to reach 100% of the step input. The peak time is that taken to the first maximum in the output response. For critically damped and overdamped systems, the time taken for the output to change between 10% and 90% of the input is used alternatively as a measure of the speed of the response.

The degree in which the actual output response matches the input is measured by the percentage overshoot, *PO*, and the settling time. The percentage overshoot is defined as

$$PO = \frac{M_{PT} - 100}{100} \tag{3.13}$$

where M_{PT} is the peak value of the output.

It may further be shown that the percentage overshoot is given analytically as

$$PO = 100\ e^{-\zeta\pi/\sqrt{(1-\zeta^2)}} \tag{3.14}$$

Another useful relation is derived from the ratio of successive peaks, i.e.

$$\ln\left(\frac{A_n}{A_{n+1}}\right) = \frac{2\pi\zeta}{\sqrt{(1 - \zeta^2)}} \tag{3.15}$$

where n is an integer to denote the peak number (i.e. first, second, etc.). Equation (3.15) is referred to as the 'logarithmic decrement'. The settling time, τ_S, is the time taken for the oscillatory response to decay below a percentage of the input amplitude, δ, often taken as $\pm 2\%$. Finally we have the steady-state error, ε_{ss}, which is self-explanatory.

The response of the second-order system to a ramp input is shown in Figure 3.94. The form of the response curves again depends on the value of the damping ratio, but in each case the output asymptotes to a steady-state lag. The lag is not the same in each case, however, since this is also dependent on the damping ratio.

The response of a second-order system to a sinusoidal input may also be considered. Generally, the output response will lag behind the input with a transient decaying amplitude depending on the nature of the damping ratio. It is more informative, however, to study the response of second-order systems to sinuosoidal inputs using frequency-domain methods.

3.6.3 Laplace notation for differential equations – frequency-domain analysis

For analyses in the frequency domain it is customary to write the differential equation in terms of the Laplace operator, s. This gives rise to the system 'transfer function' which is formed

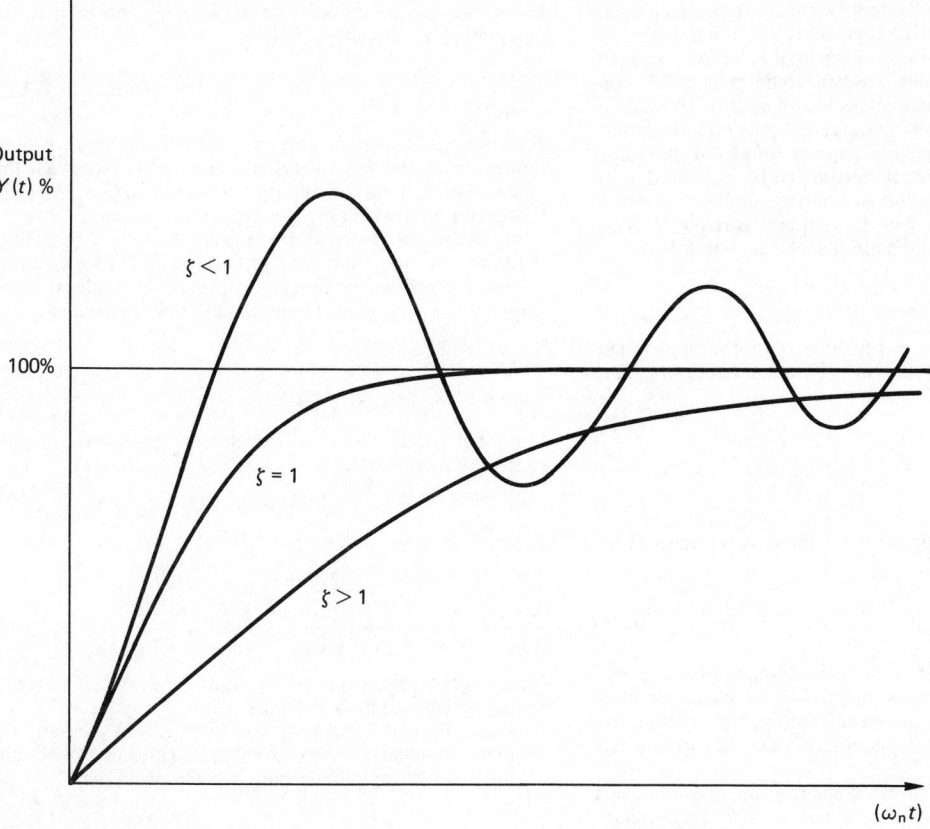

Figure 3.92   Response of a second-order system to a step input

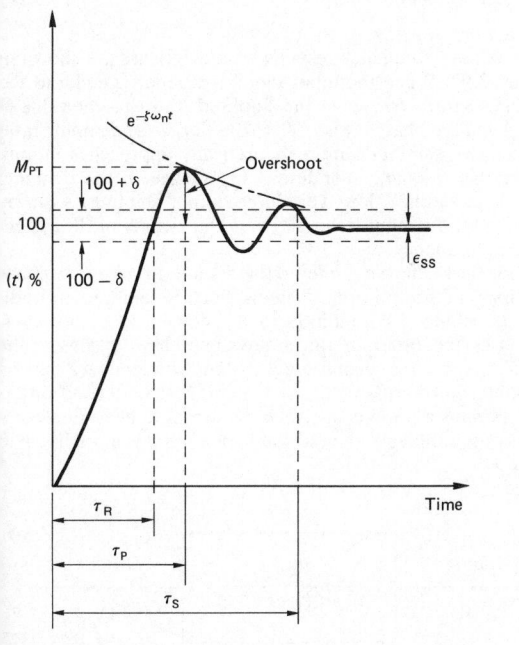

Figure 3.93   Resspponse curve for an underdamped system to a step input

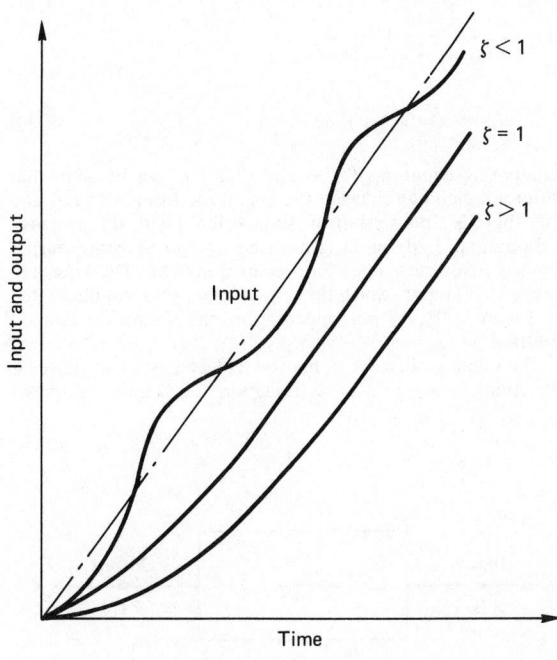

Figure 3.94   Response of a second-order system to a ramp input

by replacing the input and output (X and Y, respectively) with their corresponding Laplace transforms, $X(s)$ and $Y(s)$. The method applies only to linear differential equations. In practice, many systems would contain some degree of non-linearity, and various assumptions would have to be made to simplify and approximately linearize the governing equation.

The advantage in using the Laplace transform method is that it allows the differential equation to be expressed as an equivalent algebraic relation in s. Differentiation is represented by multiplication with the Laplace variable, s. Thus dY/dt becomes $sY(s)$ and d^2Y/dt^2 is replaced with $s^2Y(s)$.

3.6.3.1 First-order systems

The governing equation is rewritten with the appropriate Laplace transforms replacing the differential operators. Thus equation (3.5) becomes

$$\tau \cdot s \cdot Y(s) + Y(s) = kX(s) \tag{3.16}$$

Hence

$$Y(s)[1 + \tau s] = kX(s) \tag{3.17}$$

The system transfer function is defined as the ratio of the output to the input and is

$$\frac{Y(s)}{X(s)} = \frac{k}{1 + \tau s} \tag{3.18}$$

Equation (3.18) enables the convenient facility of incorporating the transfer function within the usual block structure representation of a control system. Thus a first-order, open-loop control system can be systematically depicted as shown in Figure 3.95.

For analyses in the frequency domain we are predominantly concerned with the system response to sinusoidal inputs. Differentiation (or integration) of a sinusoidal function does not alter the shape or frequency. There is simply a change in amplitude and phase, e.g.

$$\text{Input} = A \sin(\omega t) \tag{3.19}$$

$$\frac{d}{dt} (A \sin(\omega t)) = \omega A \cos(\omega t)$$

$$= \omega A \sin\left(\omega t + \frac{\pi}{2}\right) \tag{3.20}$$

Comparing equations (3.19) and (3.20) it can be seen that differentiation has changed the amplitude from A to ωA and that there is a phase shift of 90° associated with the process.

Equation (3.20), in fact, describes the steady-state output from a first-order, open-loop control system. The transient part of the output, which the time-domain solution illustrated in Figure 3.91, is not apparent in the frequency-domain solution.

The Laplace operator, s, may be replaced with $j\omega$, where j is the complex operator $\sqrt{-1}$. Equation (3.18) then becomes

$$\frac{Y}{X} = \frac{k}{1 + j\omega\tau}$$

Using the complex conjugate it may be shown that the modulus of the amplitude ratio is

$$\left|\frac{Y}{X}\right| = \frac{k}{\sqrt{(1 + \omega^2\tau^2)}} \tag{3.21}$$

Equation (3.21) shows how the output amplitude will be influenced by the input sinusoidal frequency. Note that this agrees with the time domain solution (equation (3.8)) for large values of t after which the steady state is achieved.

The technique shown above is general and may be used to determine the amplitude ratio for any second- or higher-order system. *Note*: Common practice, especially in graphical representations, is to express the amplitude ratio in decibels:

i.e. $dB = 20 \log_{10}(Y/X)$ (3.22)

3.6.3.2 Second-order systems

Using the Laplace transfer operator the governing equation (3.9) may be rewritten as

$$s^2 \cdot Y(s) + 2\zeta\omega_n \cdot s \cdot Y(s) + \omega_n^2 \cdot Y(s) = kX(s) \tag{3.23}$$

$$\therefore Y(s)[s^2 + 2\zeta\omega_n \cdot s + \omega_n^2] = kX(s) \tag{3.24}$$

The system transfer function is

$$\frac{Y(s)}{X(s)} = \frac{k}{s^2 + 2\zeta\omega_n \cdot s + \omega_n^2} \tag{3.25}$$

Thus a second-order, open-loop control system can be represented schematically as in Figure 3.96.

For a sinusoidal input of the form $X = X_0 \sin(\omega t)$, the frequency-domain analysis gives the following steady-state solutions for the amplitude ratio and the phase lag:

$$\left|\frac{Y}{X}\right| = \frac{1}{\sqrt{[(1 - r^2)^2 + (2\zeta r)^2]}} \tag{3.26}$$

and

$$\phi = \tan^{-1}[(2\zeta r)/(1 - r^2)] \tag{3.27}$$

where $r = (\omega/\omega_n)$.

The above frequency response characteristics are shown in Figure 3.97. When the input signal frequency is equal to the system's natural frequency the amplitude ratio has the value of $(1/2\zeta)$ and the phase lag is $-90°$. Note that if the damping ratio is zero the amplitude ratio theoretically approaches infinity under this resonance condition. In practice, if the damping ratio is moderately low, very large output amplitudes can be expected if the input frequency is in the vicinity of the system natural frequency.

Thus far we have considered the open-loop system response for first- and second-order systems. Such systems are unconditionally stable. The addition of a feedback loop, however, increases the order of the system and there is always the possibility that the second-order system with feedback may be unstable. Furthermore, if any system, first or second order, incorporates a 'time delay' (also known as a 'dead time' or a 'transportation lag'), then unstable operation is more likely to occur.

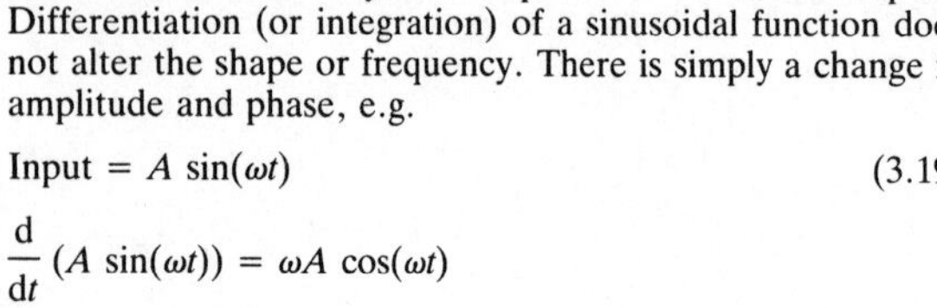

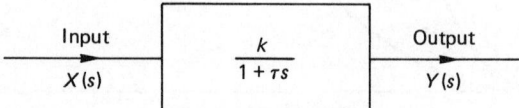

Figure 3.95 First-order open-loop control system

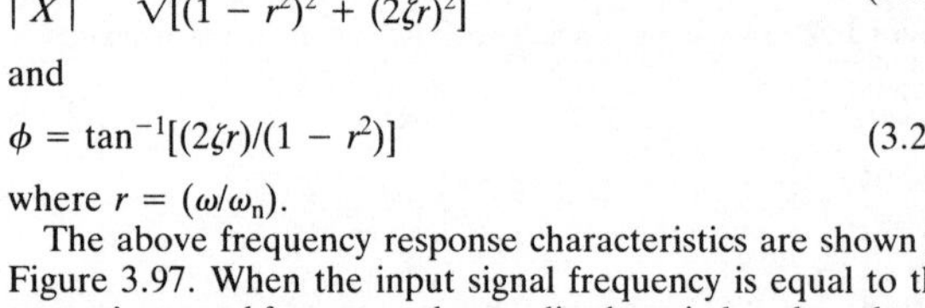

Figure 3.96 Second-order open-loop control system

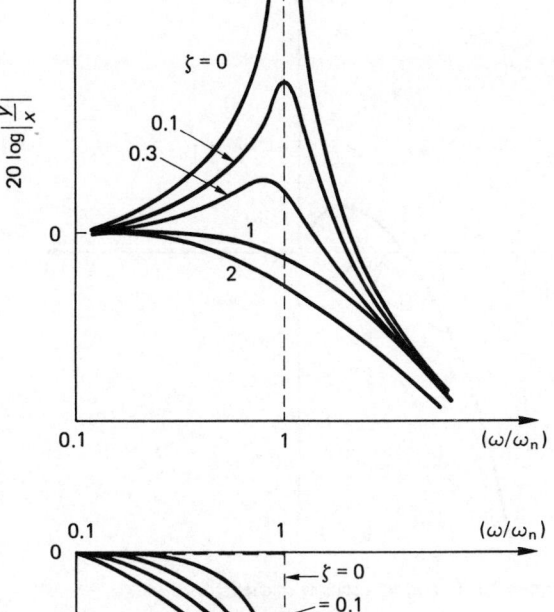

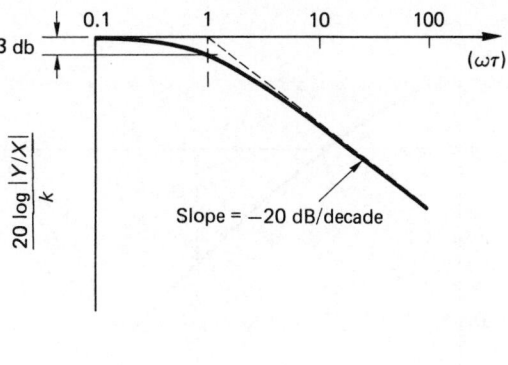

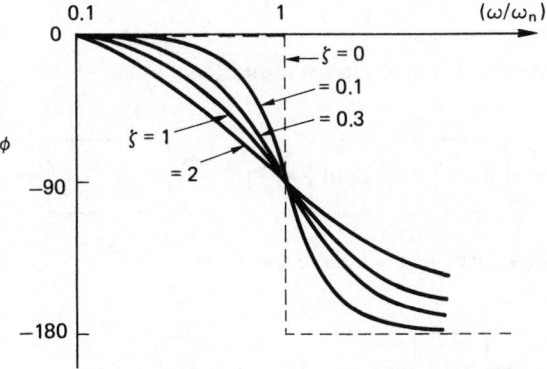

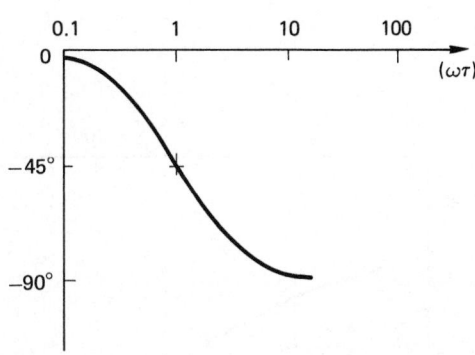

Figure 3.98 Bode plot for an open-loop first-order system

Figure 3.97 Second-order open-loop control system frequency response

3.6.4 Stability criteria

Time delays are very difficult to handle mathematically when they occur in differential equations and the inclusion of multiple feedback loops can greatly increase the order of the governing equation. For these two reasons solutions in the time domain become extremely difficult, and frequency domain methods are almost exclusively used to assess the behaviour of the more complex control systems. The main consideration in frequency-domain analyses is the stability of the system and how it can be adjusted if it happens to be unstable. Various graphical methods are used and these include the Bode and Nyquist plots.

The Bode plot is a graph of amplitude ratio and phase angle variation with input signal frequency. The resulting normalized plot for an open-loop first-order system is shown in Figure 3.98. Note that when the input frequency is equal to the inverse of the system time constant, the output amplitude has been decreased (or attenuated) by 3 dB. The phase lag at this point is $-45°$. This is characteristic of first-order systems.

The Nyquist plot represents the same information in an alternative form. The plot is in polar coordinates and combines the amplitude ratio and phase lag in a single diagram. Figure 3.99 shows the Nyquist plot for the open-loop, first-order system.

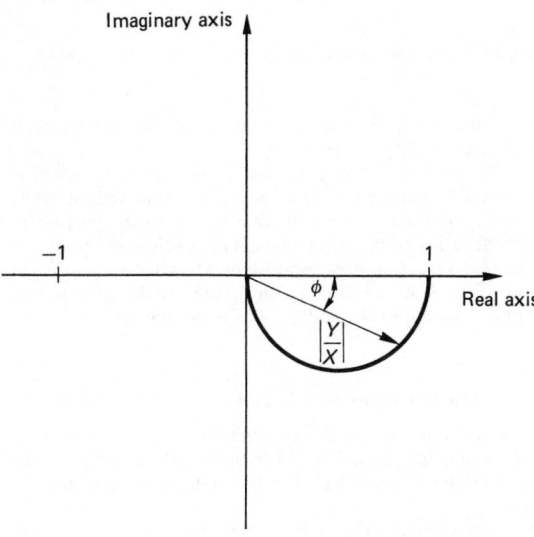

Figure 3.99 Nyquist plot for an open-loop first-order system

3.6.4.1 Bode and Nyquist stability criteria

The Bode[8] criterion for stability is that the system is stable if the amplitude ratio is less than 0 dB when the phase angle is $-180°$. This is illustrated graphically in Figure 3.100. This figure represents a stable system since Bode's criterion is satisfied. The 'gain margin' (GM) and 'phase margin' (PM) are used as measures of how close the frequency response

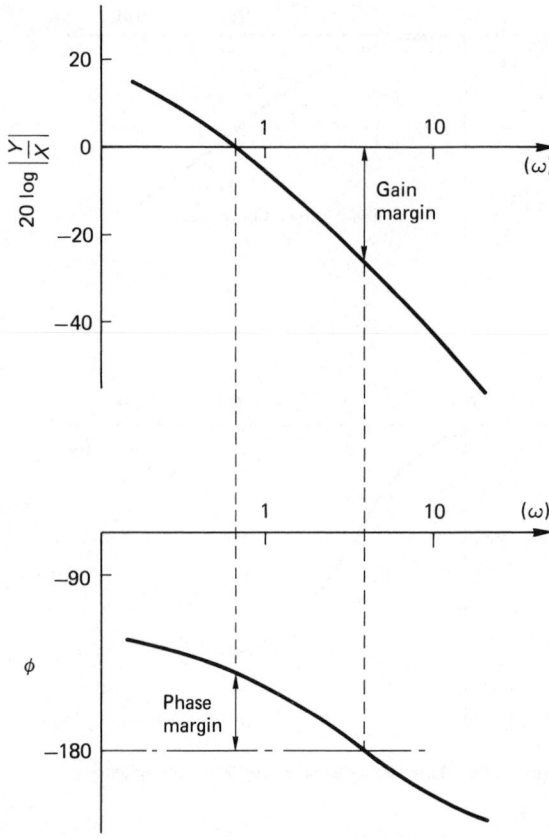

Figure 3.100 Bode's stability criterion

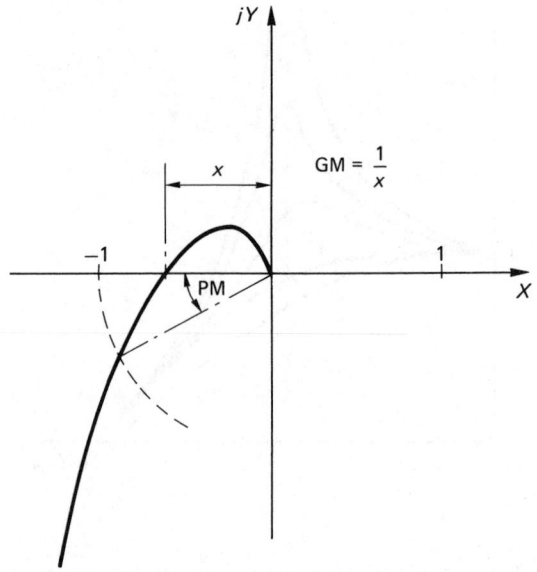

Figure 3.101 Nyquist's stability criterion

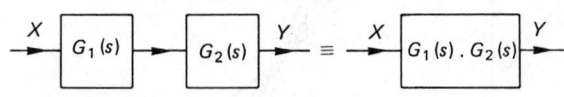

Figure 3.102 Transfer functions in series

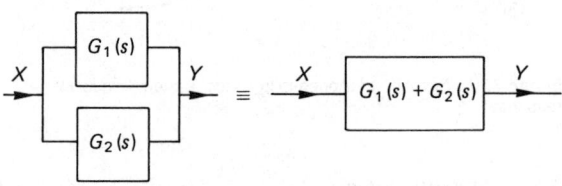

Figure 3.103 Transfer functions in parallel

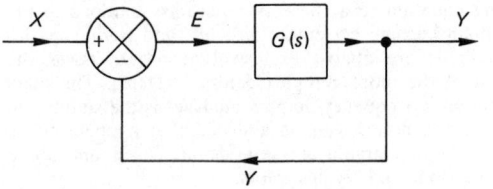

Figure 3.104 Control system with unity feedback

curves are to the 0 dB and $-180°$ points and are indicative of the relative stability of the system.

The Nyquist[9] criterion for stability is that the system is stable if the amplitude ratio is greater than -1 at a phase angle of $-180°$. In effect, this means that the locus of the plot of amplitude ratio and phase angle must not enclose the point -1 on the real axis. A stable response curve is shown plotted in Figure 3.101. Also indicated in this figure are the gain margin and phase margin in the context of the Nyquist plot.

3.6.4.2 System stability with feedback

In a closed-loop system the transfer function becomes modified by the feedback loop. The first task therefore is to determine the overall transfer function for the complete system.

For simple open-loop systems the transfer functions are combined according to the following rules:

1. For elements in series, the overall transfer function is given by the product of the individual transfer functions (see Figure 3.102).
2. For elements in parallel, the overall transfer function is given by the sum of the individual transfer functions (see Figure 3.103).

For a system with feedback, the overall transfer function can be evaluated using a consistent step-by-step procedure. Series and parallel control elements are combined in the manner as shown above to reduce the system to a single block, which then represents the overall transfer function.

Consider the simple control system depicted in Figure 3.104. Since the feedback line does not include any transfer function it is termed a 'unity feedback' system, i.e. the output is compared directly with the input to produce the error signal. The closed-loop transfer function is obtained as follows:

$$Y = G(s) \cdot E = G(s)[X - Y]$$

Thus

$$\frac{Y}{X} = \frac{G(s)}{1 + G(s)} \qquad (3.28)$$

If the element whose open-loop transfer function, $G(s)$, is a first-order sub-system, then $G(s)$ may be replaced with the expression given in equation (3.18). The closed-loop transfer function may then be written as

$$\frac{Y}{X} = \left(\frac{k}{1 + \tau s}\right) \Big/ \left(1 + \frac{k}{1 + \tau s}\right)$$

$$= \frac{k}{1 + k + \tau s} \qquad (3.29)$$

Dividing top and bottom by $(1 + k)$ results in

$$\frac{Y}{X} = \left(\frac{k}{1 + k}\right) \Big/ \left(1 + \frac{\tau s}{1 + k}\right) \qquad (3.30)$$

We define the following terms as

$$k_c = \frac{k}{1 + k} \qquad (3.31)$$

and

$$\tau_c s = \frac{\tau s}{1 + k} \qquad (3.32)$$

where k_c is the closed-loop gain and τ_c is the closed-loop system time constant. The final closed-loop transfer function may be expressed as

$$\frac{Y}{X} = \frac{k_c}{1 + \tau_c \cdot s} \qquad (3.33)$$

Equations (3.31) and (3.32) show, respectively, that both the closed-loop system gain and the time constant are less than those associated with the open-loop system. This means that the closed-loop response is faster than the open-loop one. At the same time, however, the closed-loop gain is reduced.

Using the procedures outlined in the example, any other complex control system may be similarly analysed to determine the closed-loop transfer function. Thus knowing the gain constants and other characteristics of the elements which make up the system, the frequency response may be obtained. The stability of the system may then be assessed and any corrective measures taken as necessary. In practice, it is often found that the gain of some of the system elements must be altered in order to ensure stable operation. Another commonly applied corrective measure is to add a phase advance circuit into the system. The procedure might also be operated in reverse, where, starting with a desired response, a suitable control system can be configured and adjusted to meet the response.

The practising control engineer will use many techniques to assess system stability. These might include the numerical Routh–Hurwitz criterion, which determines only whether a system is stable or not. Alternative graphical methods include the use of Hall charts, Nichols charts, Inverse Nyquist plots and Root Locus plots. The graphical methods additionally indicate the relative stability of a system.

Numerous commercial computer packages are available[10–12] to assist the designer of control systems. These include the usual graphical representations and can be obtained from the suppliers whose addresses are given in the references. The reader is also referred to the Further Reading at the end of this chapter for a more comprehensive coverage of these methods and techniques.

3.6.4.3 Effect of transport delay

The influence of a transport (or time) delay on the response of an underdamped second-order system to a step input is shown in Figure 3.105. Although it is virtually impossible to account for a time delay in a differential equation it is simply accommodated in the frequency-domain model as an additional element in the system block diagram. In the frequency-domain model, the time delay effects a phase shift of $-\omega T$ and can be expressed as

$$\text{Time delay} = e^{-sT} = e^{-j\omega T} \qquad (3.34)$$

Consider the open-loop response of a first-order system incorporating a time delay as illustrated in Figure 3.106. The open loop transfer function becomes

$$\frac{Y}{X} = \frac{Y(s)}{X(s)} = G(s) = \frac{ke^{-sT}}{1 + \tau s} \qquad (3.35)$$

$$\text{Amplitude ratio} = \left|\frac{Y}{X}\right| = \frac{k}{\sqrt{(1 + \omega^2 \tau^2)}} \qquad (3.36)$$

$$\text{Phase lag} = \phi = -[\omega T + \tan^{-1}(\omega \tau)] \qquad (3.37)$$

The first-order, open-loop system response with a time delay is shown in Figure 3.107.

Similarly, a time delay in a system with feedback does not alter the amplitude ratio but adds a phase shift to the frequency response. The system therefore becomes less stable, and in some cases it may be necessary to reduce the closed-loop gain in order to obtain a stable response. The penalty to be paid for increasing the stability in this manner is an increase in the steady-state error.

3.6.5 Control strategies

The basic closed-loop system with common symbol representation is given in Figure 3.108. The nomenclature used in this figure is defined as follows:

$SP(s)$ is the set point (required value, $r(t)$, is sometimes used);

$PV(s)$ is the process value (corrected value, $c(t)$, is sometimes used);

$E(s)$ is the error signal, which is the difference between SP and PV;

$U(s)$ is the control effort output from the controller to the process;

$C(s)$ is the controller transfer function;

$G(s)$ is the process transfer function.

The transfer function for the closed loop system is obtained as before:

$$PV(s) = C(s) \cdot G(s) \cdot E(s) \qquad (3.38)$$

$$= C(s) \cdot G(s) \cdot [SP(s) - PV(s)]$$

Hence

$$\frac{PV(s)}{SP(s)} = \frac{C(s) \cdot G(s)}{1 + C(s) \cdot G(s)} \qquad (3.39)$$

3.6.5.1 ON/OFF control

In many applications a simple ON/OFF strategy is perfectly adequate to control the output variable within preset limits. The ON/OFF control action results in either full or zero power being applied to the process under control. A mechanical type of thermostat provides a good example of an ON/OFF-based controller. The ON/OFF control strategy results in an output

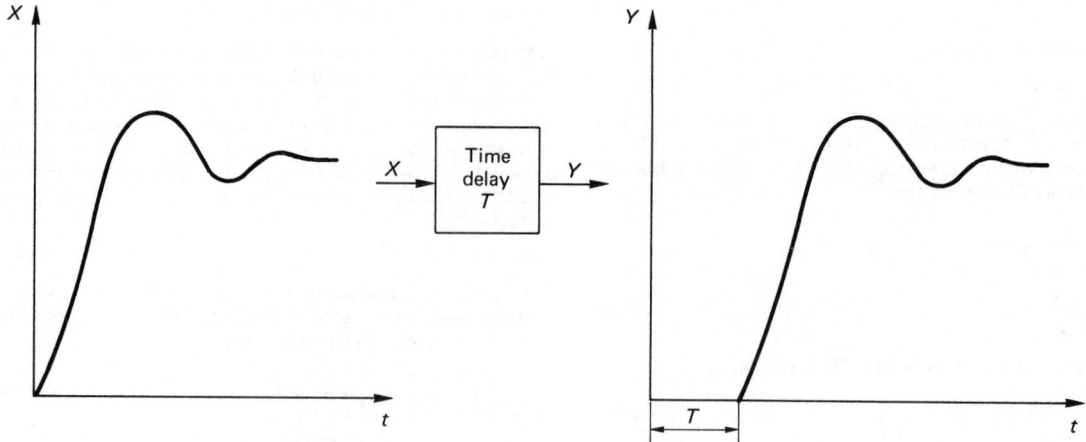

Figure 3.105 Effect of transport delay

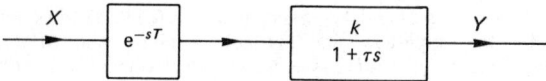

Figure 3.106 First-order system with a time delay

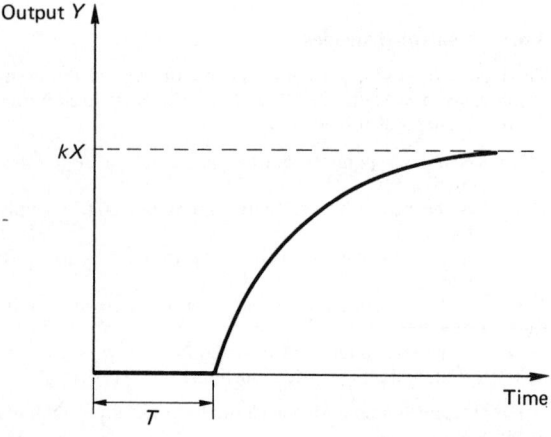

Figure 3.107 First-order open-loop system response with a time delay

which fluctuates about the set point as illustrated in Figure 3.109.

ON/OFF controllers usually incorporate a 'dead band' over which no control action is applied, which is necessary to limit the frequency of switching between the ON and OFF states. For example, in a temperature-control system, the ON/OFF control strategy would be:

If temperature $< T_{min}$, the heater is to be switched ON;
If temperature $> T_{max}$, then heater is to be switched OFF.

The dead band in the above case is $(T_{max} - T_{min})$ and while the temperature remains within the dead band no switching will occur. A large dead band will result in a correspondingly large fluctuation of the process value about the set point. Reducing the dead band will decrease the level of fluctuation but will increase the frequency of switching. The simple ON/OFF control strategy is mostly applicable to processes and systems which have long time constants and in consequence have relatively slow response times (e.g. temperature and level control).

While being simple in concept, ON/OFF control systems are, in fact, highly non-linear and they require some complex non-linear techniques to investigate their stability characteristics.

3.6.5.2 Three-term or PID control

Since complicated transfer functions can be very difficult to model, the most common strategy used to define the con-

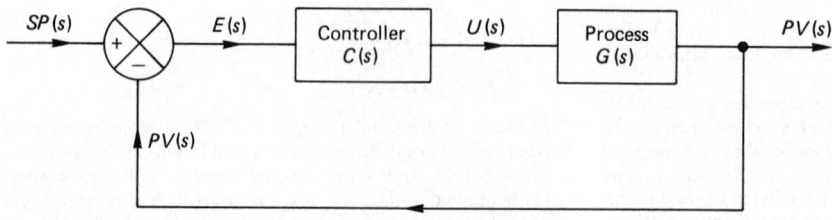

Figure 3.108 Basic closed-loop control system

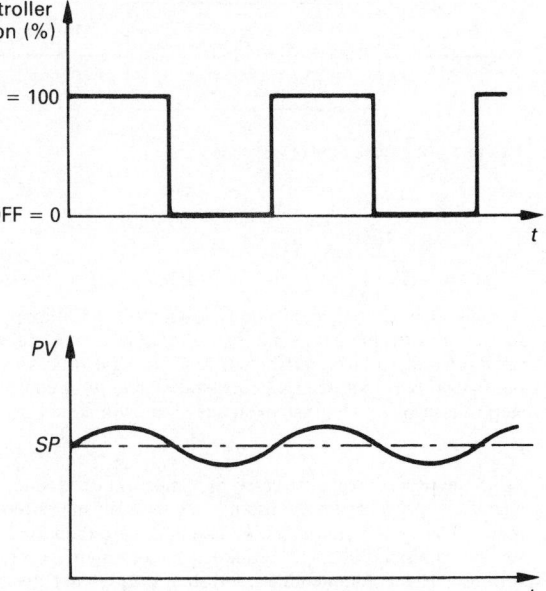

Figure 3.109 Output variation with ON/OFF control

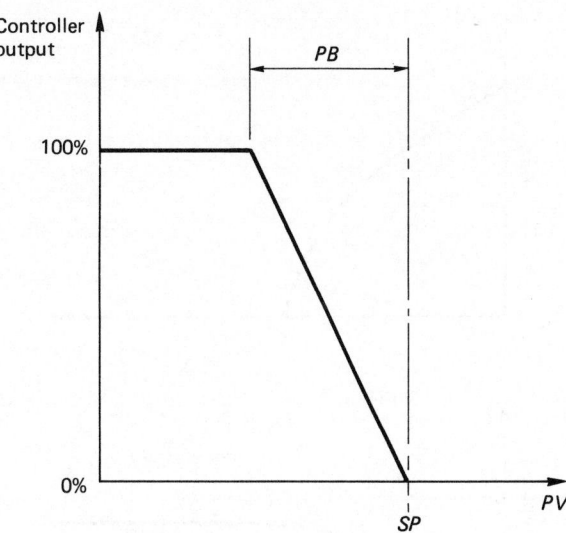

Figure 3.110 Illustration of the proportional band

troller transfer function is the so-called 'three term' or PID controller. PID is the popular short form for Proportional, Integral and Derivative. The three elements of the controller action, U, based on the evaluated error, E, are as follows.

(a) Proportional action

Controller output $= K \cdot E$ (3.40)

where K is the controller gain. Manufacturers of three-term controllers tend to favour the parameter 'proportional band' (PB) in preference to gain, K. The proportional band represents the range of the input over which the output is proportional to the input. The PB is usually expressed as a percentage of the input normalized between 0 and 100% (see Figure 3.110).

To illustrate the concept of proportional band a temperature-control application can be considered where the set point is, say, 80°C and the proportional band is set to, say, 5% over a measured temperature span of 0–100°C. The actual proportional band is therefore 5°C and proportional action will apply over the temperature range between 75°C and 80°C. If the temperature is below 75°C then 100% of the available power will be supplied to the heating device. Between 75°C and 80°C, a proportion of the available power will be applied to the heating device as shown in Figure 3.110. For temperatures in excess of 80°C, 0% of the available power is supplied.

It should be apparent that proportional band is a more meaningful term than gain. The two parameters are, however, very simply related, i.e.

PB% $= 100/K$ (3.41)

It is also apparent from Figure 3.110 that as the proportional band is decreased, the control action is tending towards an ON/OFF strategy. A very large proportional band will result in a somewhat sluggish response.

It must also be noted that for proportional control only, there must always be an error in order to produce a control action. From equation (3.39) proportional control only gives a transfer function of the form

$$\frac{PV(s)}{SP(s)} = \frac{KG(s)}{1 + KG(s)} = \frac{1}{(1/KG(s)) + 1} \quad (3.42)$$

For steady-state conditions, s tends to 0 and $G(s)$ tends to a constant value. Equation (3.42) shows therefore that the gain must theoretically tend to infinity if $PV = SP$ and the steady-state error is to approach zero.

This is simply another manifestation of the classical control problem, i.e. stability at the expense of accuracy and vice versa. With a very high gain (i.e. low proportional band) the steady-state error can be very much reduced. A low proportional band, however, tends to ON/OFF control action and a violent oscillation may result in sensitive systems.

(b) Integral action The limitations of proportional control can be partly alleviated by adding a controller action which gives an output contribution that is related to the integral of the error value with respect to time, i.e.

Controller output $= K_i \int E \cdot dt$ (3.43)

where K_i is the controller integral gain ($= K/T_i$) and T_i is the controller 'integral time' or 'reset'.

The nature of integral action (equation (3.43)) suggests that the controller output will increase monotonically as long as an error exists. As the error tends to zero the controller output tends towards a steady value. The general behaviour of the controller output with integral action is shown in Figure 3.111.

If T_i is very large, the integral action contribution will be low and the error may persist for a considerable time. If, on the other hand, T_i is too small the magnitude of the integral term may cause excessive overshoot in the output response. Unstable operation is also possible when T_i is too small and the controller output value then increases continuously with time.

(c) Derivative action The stability of a system can be improved and any tendency to overshoot reduced by adding derivative action. Derivative action is based on the rate of change of the error, i.e.

Controller output $= K_d \cdot \dfrac{dE}{dt}$ (3.44)

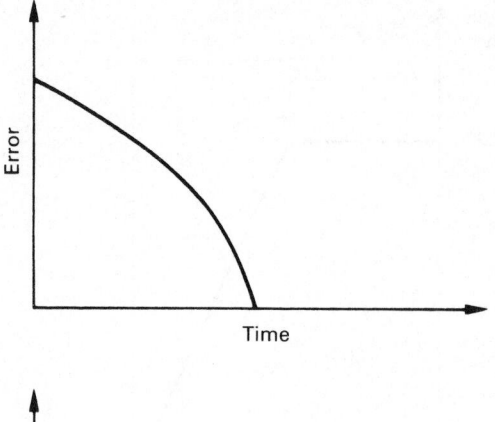

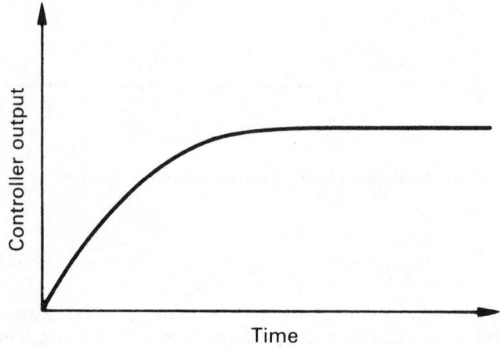

Figure 3.111   Controller output with integral action

where K_d is the controller derivative gain $(= K \cdot T_d)$ and T_d is the controller 'derivative time' or 'rate'.

Equation (3.44) indicates that the derivative action is dependent on how quickly or otherwise the error is changing. Derivative action tends therefore only to come into operation during the early transient part of a system's response.

The full three-term control strategy may be written as

$$K\left[E + \frac{1}{T_i} \int E \cdot dt + T_d \cdot \frac{dE}{dt} \right] \qquad (3.45)$$

To summarize, the proportional action governs the speed of the response, the integral action improves the accuracy of the final steady state and the derivative action improves the stability. Note that derivative action may result in poor performance of the system if the error signal is particularly noisy. In Laplace notation, the three term controller transfer function is as shown in Figure 3.112.

3.6.5.3   Empirical rules for PID controller settings

A simple and still popular technique for obtaining the controller settings to produce a stable control condition is due to Ziegler and Nichols.[13] The method is purely empirical and is based on existing or measurable operating records of the system to be controlled.

(a) Open-loop 'Reaction Curve' method   The process to be controlled is subjected to a step-input excitation and the system open-loop response is measured. A typical open-loop response curve is shown in Figure 3.113. Any system which has a response similar to that given in the figure has a transfer function which approximates to a first-order system with a time delay, i.e.

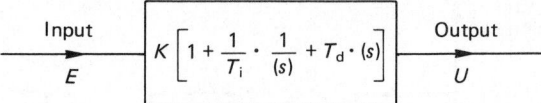

Figure 3.112   Three term or PID control

$$G(s) = \frac{k e^{-sT_1}}{1 + T_2 \cdot s} \qquad (3.46)$$

In general industrial applications, oscillatory open-loop responses are extremely rare and Figure 3.113 is in fact representative of quite a large number of real practical processes. In the figure, N is the process steady-state value for a controller step output of P. The system steady-state gain is

$$k = N/P \qquad (3.47)$$

From the process response curve the 'apparent dead time', T_1, and the 'apparent time constant', T_2, can be measured directly. The three parameters, k, T_1 and T_2, are then used in a set of empirical rules to estimate the optimum controller settings. The recommended controller settings are given in Table 3.5.

In fast-acting servomechanisms, where T_1 may be very small, the method is none too successful. For moderate response systems, however, the method will yield very reasonable first-approximation controller settings.

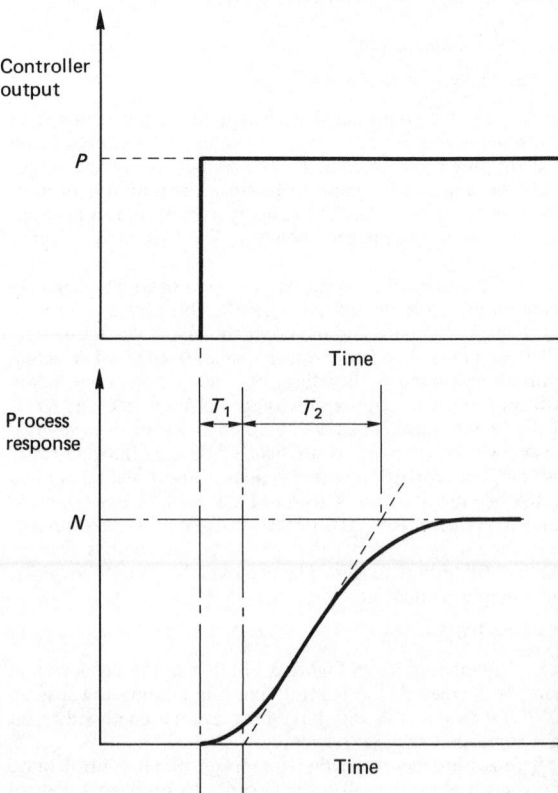

Figure 3.113   Open-loop system response to a step input

Table 3.5 Optimum controller settings according to Ziegler and Nichols

Control action	K	T_i	T_d
P	$T_2/(T_1 k)$	–	–
P + I	$(0.9T_2)/(T_1 k)$	$T_1/0.3$	–
P + I + D	$(1.2T_2)/(T_1 k)$	$2T_1$	$0.5T_1$

(b) Closed-loop 'Continuous Cycling' method The process to be controlled is connected to the PID controller and the integral and derivative terms are eliminated by setting $T_d = 0$ and $T_i = \infty$. In some industrial controllers the integral term is eliminated with $T_i = 0$. A step change is introduced and the system run with a small controller gain value, K. The gain is gradually increased for a step input until constant-amplitude oscillations are obtained as illustrated in Figure 3.114.

The gain, K_u, which produces the constant-amplitude condition is noted and the period of the oscillation, T_u, is measured. These two values are then used to estimate the optimum controller settings according to the empirical rules listed in Table 3.6.

For a temperature-control system, typical values of T_u are about 10 s for a tungsten filament lamp, 2 min for a 25 W soldering iron and from 10 to 30 min for a 3 kW heat treatment furnace.

The PID settings obtained according to the methods of Ziegler and Nichols are approximate only, and some 'fine tuning' would almost certainly be required in practice.

Table 3.6 Optimum controller settings according to Ziegler and Nichols

Control action	K	T_i	T_d
P	$0.5K_u$	–	–
P + I	$0.45K_u$	$T_u/1.2$	–
P + I + D	$0.6K_u$	$T_u/2$	$T_u/8$

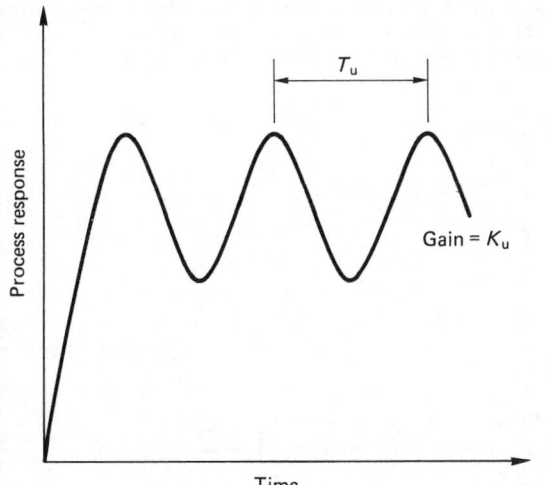

Figure 3.114 Continuous cycling method

3.6.5.4 Three-term controller with a first-order system

The block diagram of the system is depicted in Figure 3.108 and equation (3.39) defines the closed-loop transfer function. If a $P + I$ controller is to be used (i.e. no derivative action) the controller transfer function is

$$C(s) = K\left[1 + \frac{1}{T_i s}\right] \tag{3.48}$$

The process is modelled as a first-order system and its open-loop transfer function is given by equation (3.18).

Substituting equations (3.18) and (3.48) into equation (3.39) results, after some manipulation, in

$$\frac{PV(s)}{SP(s)} = \frac{(kK/\tau T_i)(1 + sT_i)}{\{s^2 + [(1/\tau) + (kK/\tau)]s + (kK/\tau T_i)\}} \tag{3.49}$$

Comparing the denominator with that for the generalized second order system (i.e. equation (3.25)) it can be shown that

$$2\zeta\omega_n = \left(\frac{1}{\tau} + \frac{kK}{\tau}\right) \tag{3.50}$$

and

$$\omega_n^2 = (kK/T_i\tau) \tag{3.51}$$

For the system being controlled, both k and τ are known either via a mathematical model or an open-loop test. The controller settings, K and T_i, can then be calculated for a chosen damping ratio, ζ, and natural frequency, ω_n. Alternatively, a controller gain can be imposed and the corresponding natural frequency evaluated.

For full PID control, an initial value of $T_d = T_i/4$ can be used. Other systems can be similarly handled to obtain the approximate PID controller settings. In all cases some fine adjustment would probably be necessary to obtain the optimum output response.

3.6.5.5 Disturbance sensitivity

The main problem with the classical single-loop control system is that it is not truly representative of the natural environment in which the system operates. In an ideal single-loop control system the controlled output is a function only of the input. In most practical systems, however, the control loop is but a part of a larger system, and it is therefore subject to the constraints and vagaries of that system. This larger system, which includes the local ambient, can be a major source of disturbing influences on the controlled variable. The disturbance may be regarded as an additional input signal to the control system. Any technique therefore which is designed to counter the effect of the disturbance must be based on a knowledge of the time-dependent nature of the disturbance and also its point of entry into the control system. Two methods commonly used to reduce the effect of external disturbances are 'feedforward' and 'cascade' control.

(a) Feedforward control The principle of a feedback loop is that the output is compared with the desired input and a resultant error signal acted upon by the controller to alter the output as required. This is a control action which is implemented 'after the fact'. In other words, the corrective measures are taken after the external disturbance has influenced the output. An alternative control strategy is to use a feedforward system where the disturbance is measured. If the effect of the disturbance on the output is known, then, theoretically, the corrective action can be taken before the disturbance can significantly influence the output. Feedforward can be a

practical solution if the external disturbances are few and can be quantified and measured. The block diagram illustrating the feedforward concept is shown in Figure 3.115.

Feedforward control can be difficult to implement if there are too many or perhaps unexpected external disturbances. In Figure 3.115 the path which provides the corrective signal appears to go back. The strategy is still feedforward, however, since it is the disturbance which is measured and the corrective action which is taken is based on the disturbance, and not the output signal. Some control systems can be optimized by using a combination of feedforward and feedback control.

(b) Cascade control Cascade control is implemented with the inclusion of a second feedback loop and a second controller embodied within a main feedback loop in a control system (see Figure 3.116). The second feedback loop is only possible in practice if there is an intermediate variable which is capable of being measured within the overall process. Cascade control generally gives an improvement over single-loop control in coping with disturbance inputs. The time constant for the inner loop is less than that for the component it encloses, and the undamped natural frequency of the system is increased. The overall effects of cascade control are an increase in the system bandwidth and a reduction in the sensitivity to disturbances entering the inner loop. Disturbances entering the outer loop are unaffected. Cascade control works best when the inner loop has a smaller time constant than the outer one.

3.6.5.6 Direct digital control

Most of the standard texts on control engineering are centred on the mathematical modelling of systems and processes and, subsequently, the stability considerations of these entities. This approach requires detailed knowledge of the system constituent parts to enable the formulation of a suitable differential equation to describe the dynamic behaviour. It is often only in the idealized world of servomechanisms that adequate models can be derived. For many real processes an adequate system model can be difficult (if not impossible) to obtain. The modern emphasis is therefore on an application of computer-based control strategies which can be made to work with real systems.

The recent developments in microelectronics, particularly microprocessors, has made microcomputer devices the natural choice as controllers for many systems. The microcomputer provides the ability to implement such functions as arithmetic and logic manipulation, timing and counting. With many analogue input/output modules available to interface to the

microcomputer, the overall 'intelligence' of the system is greatly enhanced.

The basic elements of the computer-based control system include the microprocessor, memory, an input interface to measure the process variable and an output interface to supply power to the controlled process. The control effort output to the process is determined by the control strategy, which takes the form of an algorithm incorporated within the computer software. The fundamental digital-based control system is depicted in Figure 3.117.

In the generalized layout given in this figure the microcomputer performs a number of tasks which would require separate elements in an equivalent analogue system. The two inputs to the microcomputer are the desired set point and a signal from the process via a feedback loop. The term 'process' is being used in a quite arbitrary sense in this context.

The microcomputer first performs the function of comparing the process value with the set point to establish the error. The control strategy is then applied to the error value to determine the corrective action necessary. The microcomputer subsequently outputs the appropriate signal to the process via other additional elements in the system. These include the input/output interfaces between the digital-based computer and the otherwise analogue-based control system. Digital-to-analogue converters (DACs) and analogue-to-digital converters (ADCs) are featured in Section 3.4. The transducer, which provides the link between the physical world and the electronically based control system, is covered in its various forms in Section 3.5. The essential fundamentals of microprocessor technology are outlined in Section 3.1 and the applications of microcomputer-based control are described in Section 3.7.

3.6.5.7 Adaptive and self-tuning control

The concept of adaptive control is based on the ability to measure the system behaviour at any time and to alter the controller settings automatically to provide an optimum system response. Adaptive control has been a very active research topic over the last years, but it is only recently that practical applications using adaptive controllers have appeared.

The simplest approach to adaptive control is the so-called 'gain scheduling' method (Figure 3.118). The principle of gain scheduling is that some relevant external parameter is measured and an appropriate value of gain is then selected for the controller. Gain scheduling was first developed for aileron control in high-altitude aircraft. The low air density at high

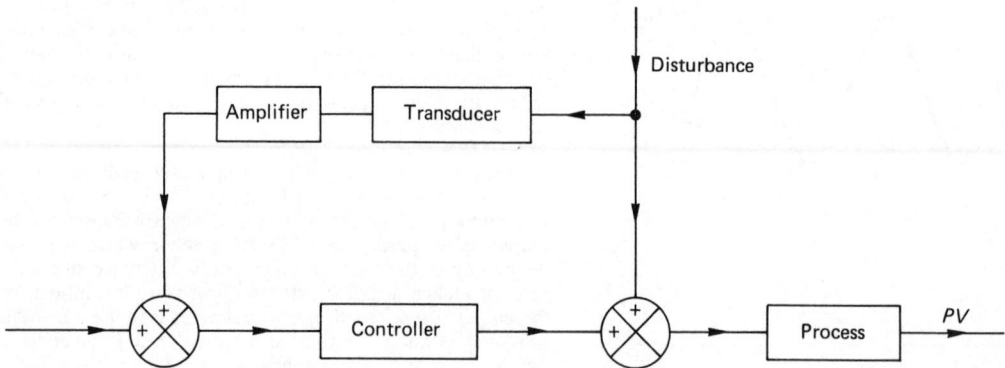

Figure 3.115 Feedforward control system

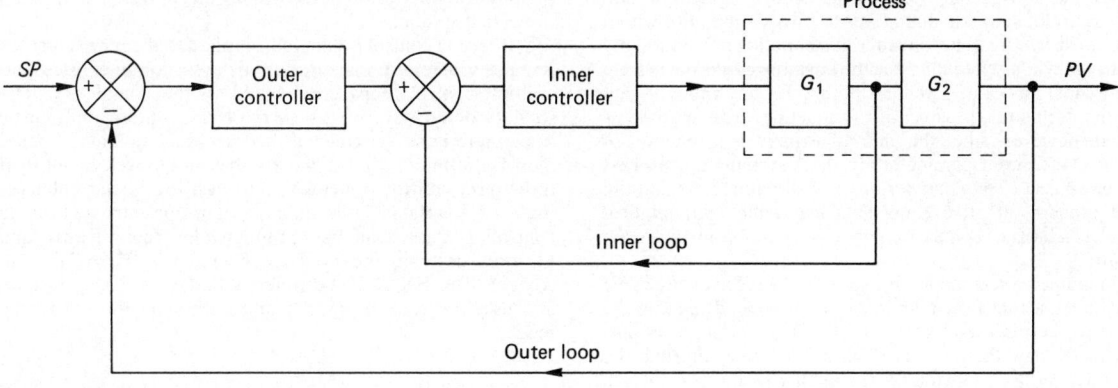

Figure 3.116 Cascade control system

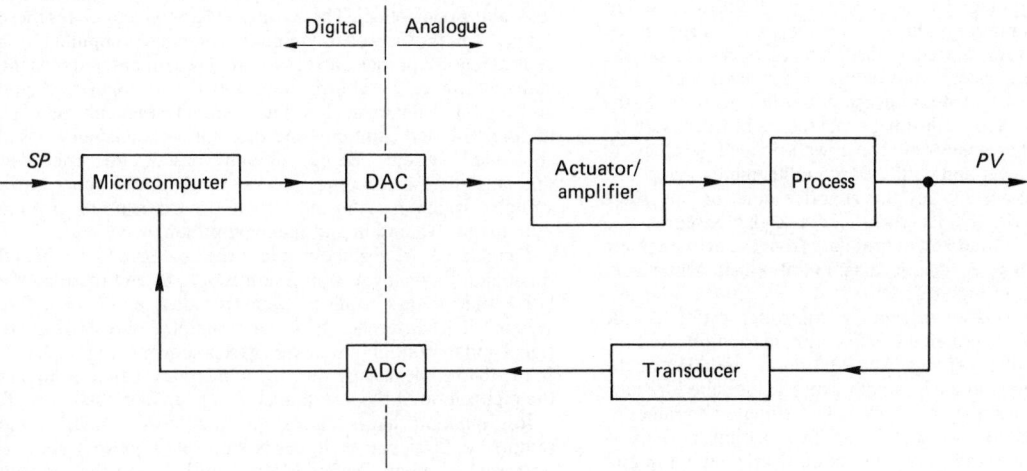

Figure 3.117 Fundamental digital-based control system

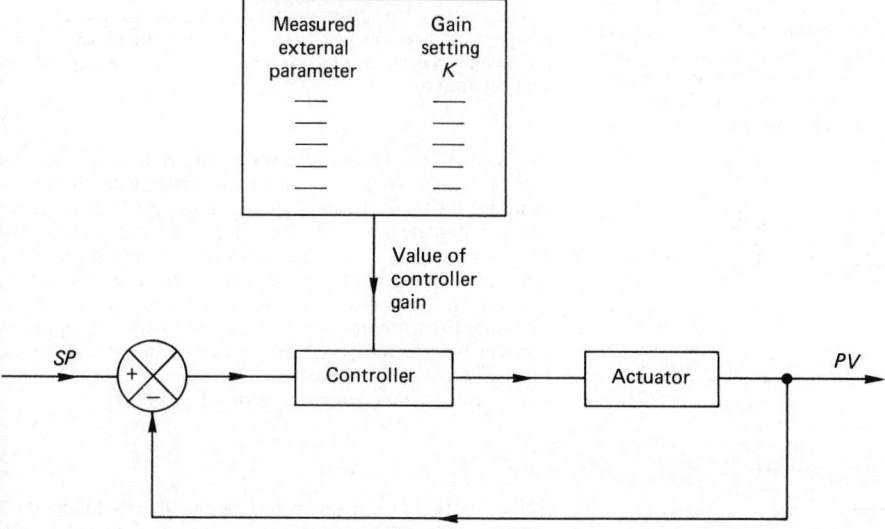

Figure 3.118 Adaptive control by gain scheduling

altitude has a profound effect on the in-flight dynamics, and the purpose of gain scheduling was to provide the pilot with a more consistent 'feel' for the aircraft's handling independently of altitude. Gain scheduling has the advantage that the system stability margins can be well established for any value of gain and the technique is generally fast acting. The method is limited, however, since the gain adjustment is a function of only one measured parameter. In most systems the process may be subject to any number of external parameters and the more modern adaptive controllers use some mathematical model as a basis of comparison with the actual control system (Figure 3.119).

The mathematical model in Figure 3.119 receives the same input as the actual system and an error is created relating the difference between the actual and the model system output. The error may then be used as a basis for altering the controller settings. Obviously, the quality of the control will depend on how well the model reflects the actual system. The usual implementation of model reference adaptive control is illustrated in Figure 3.120.

It is worth noting that the original feedback loop is left intact such that failure of the adaptive loop will not render the system inoperative. External disturbances operating on the actual plant will change the actual/model error signal and provide the basis for re-tuning the controller settings via the adaptive loop. The adjustment of the controller settings implies that there must be some well-defined strategy to determine the level and nature of the adjustments made.

Self-tuning control takes the adaptive concept one stage further in that the mathematical model of the system is also updated as more input and output data from the actual system are acquired. The schematic diagram of a self-tuning controller is shown in Figure 3.121.

The computer-based self-tuning controller estimates the system dynamics and then uses this estimate to implement the optimum controller settings. The continuous updating of the system parameters at each sampling interval is called 'recursive parameter estimation'. Previously estimated parameters are also available, and these can be used in perhaps a 'least-squares' method to provide some overall smoothing of the control function. With the latest system parameters available, the self-tuning controller then goes through a 'design' procedure to optimize the controller settings. This 'design' is usually based on the desired output response of the system. One particular design procedure is based on the root locus method for stability analysis. By adjustment of gains and time constants in the control algorithm, the method seeks to tune the transfer function and thereby govern the output response. Other procedures are often based on the rules of Ziegler and Nichols.[13] The final process in the self-tuning control cycle is

the physical imposition of the optimized controller settings on the actual system.

Self-tuning control is generally applied to the more complex processes where transportation delays, non-linearities and multiple-control loops greatly add to the complexity. The stability of such systems is, in most cases, non-deterministic since there is no generalized theory available. Traditionally, most self-tuning controllers are based on well-established three-term control principles, but with the added enhancement of adaptability. A number of proprietary self-tuning controllers are available commercially (e.g. Kraus and Myron[14] describe the Foxboro Company's 'EXACT' controller). The EXACT controller is based on PID principles and uses the rules of Ziegler and Nichols in the self-tuning mode.

3.6.5.8 Sampled-data systems

The two previous sub-sections gave an overview of direct digital control and the natural progression to adaptive and self-tuning controllers. The common factor which relates these concepts is the use of a computer (or microcomputer) as a central feature of the control system. The computer acts as the compensator in the control loop and the analogue-to-digital and digital-to-analogue interfaces provide the link between the digital-based computer and the otherwise analogue-based controlled system. Being digitally based, the computer operates in discrete time intervals and indeed the control strategy, which exists in the software, must also take a finite time for its evaluation and implementation.

Time delays are also inevitable in the analogue to digital and the digital to analogue conversion processes and these cumulative time delays result in what is called a 'sampled-data system'. The difference between a sampled-date or discrete signal and its continuous counterpart is shown in Figure 3.122. In the figure, the closure time, q, is the time taken to complete the digitization of the instantaneous signal. Generally, $q \ll T$.

It is apparent that much less information is available in the sampled-data signal as it exists only as a pulse train, interspaced with gaps in the information between the sampled points. If the sampling frequency is high enough then this need not be troublesome (see also Section 3.7.4). The inevitable additional time delays in a sampled-data system, however, have implications regarding the overall stability of the system.

The Laplace transform method cannot be used to analyse a sampled-data system but there is a related transform which is applicable to discrete time systems known as the z-transform. The relation is

$$z = e^{sT} \qquad (3.52)$$

The symbol z is associated with a time shift in a difference equation in the same way that s is associated with differentiation in a differential equation. Equation (3.52) then gives a conformal mapping from the s-plane to the z-plane and provides the means for the analysis of discrete time systems. The general method of solution involves the derivation of the closed-loop transfer function in terms of the Laplace variable. The equivalent discrete time system is then represented by introducing a 'zero-order hold' to account for the additional time delays in the discrete system (Figure 3.123).

The transfer function for a zero-order hold is

$$\frac{1 - e^{-sT}}{s} \qquad (3.53)$$

The zero-order hold is simply included in the evaluation of the closed-loop transfer function for the discrete time system. The next step is to replace the Laplace transforms with their

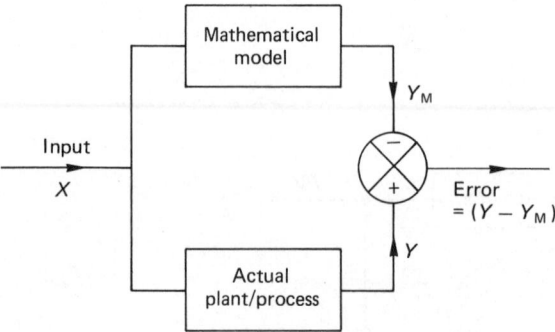

Figure 3.119 Model/actual error generation

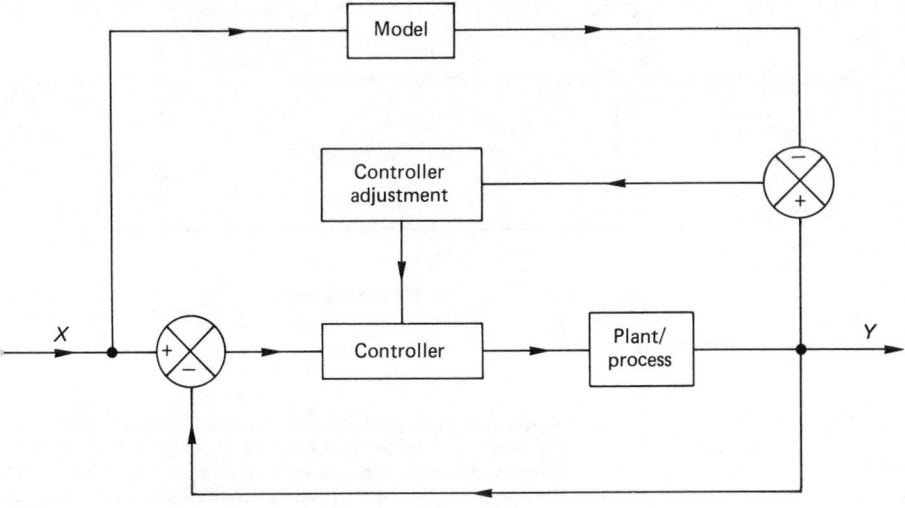

Figure 3.120 Model reference adaptive control

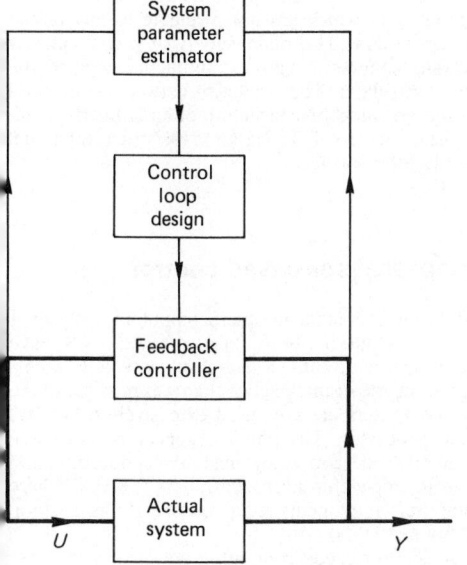

Figure 3.121 Self-tuning controller

equivalent *z*-transforms. The resulting transfer function in terms of *z*-transforms can then be analysed for stability in much the same manner as the root locus method is used for continuous systems. Sampled-data systems and the application of the *z*-transform method is considered in Section 3.9. A comprehensive coverage of *z*-transform techniques and their application to the stability analysis of sampled-data systems is given by Leigh.[15]

3.6.5.9 *Hierarchical control systems*

The ultimate aim in industrial optimization is the efficient control of complex interactive systems. Recent hardware developments and microprocessor-based controllers with extensive data-handling power and enhanced communications have opened up the possibilities for the control of interlinked

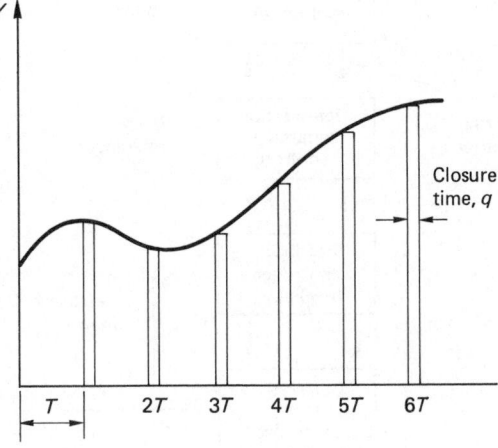

Figure 3.122 Digitization of a continuous signal

systems. What is required (but not yet realized) is a theoretical framework on which to base the analysis of such systems. Nonetheless, and in the absence of theory, hierarchical control systems do exist and are currently being used effectively in the control of various large-scale plant and processes.

The usual approach adopted is to sub-divide the complex system into a number of more manageable parts. This is the concept of hierarchical control which might be thought of as a sub-division in decreasing order of importance. Hierarchical control exists in two basic forms; multi-layer and multi-level. Multi-layer control is that in which the control tasks are sub-divided in order of complexity. Multi-level control, on the other hand, is that where local control tasks are coordinated by an upper echelon of supervisory controllers.

Multi-layer control is illustrated concisely in an elaborate adaptive-type controller, and the hierarchy is depicted in Figure 3.124. The first level is that of regulation, which is characterized by the classical single closed-loop control system. Moving up the hierarchy, we have optimization of the controller parameters. Optimization is representative of the basic adaptive controller, using simple gain scheduling or a

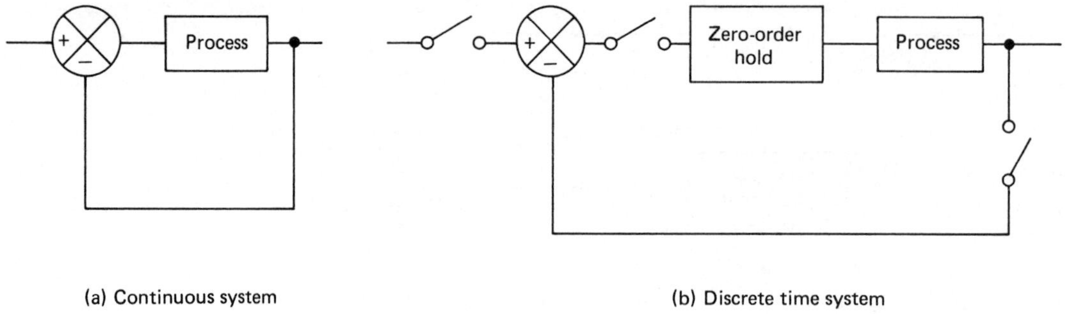

(a) Continuous system (b) Discrete time system

Figure 3.123 Continuous and discrete closed-loop control systems

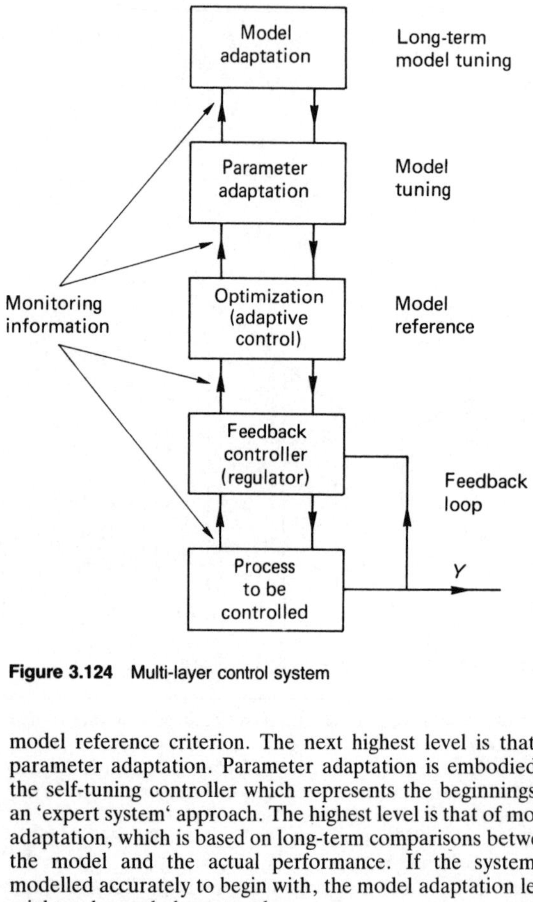

Figure 3.124 Multi-layer control system

model reference criterion. The next highest level is that of parameter adaptation. Parameter adaptation is embodied in the self-tuning controller which represents the beginnings of an 'expert system' approach. The highest level is that of model adaptation, which is based on long-term comparisons between the model and the actual performance. If the system is modelled accurately to begin with, the model adaptation level might only rarely be entered.

Multi-level control is characterized as local controllers whose actions are governed by higher levels of supervisory controllers. The local controllers operate independently to achieve local targets. The function of the supervisory controller is to reconcile the interaction of the local controllers to achieve the 'best' overall performance. The multi-level concept has some similarity with cascade control but is not so amenable to analysis.

Multi-level control gives rise to a pyramid-like structure, typified by that in Figure 3.125. At the base of the pyramid are the local controllers, monitoring and adjusting individual parameters in the overall process. At the next highest level the

supervisory controllers 'oversee' a more complete picture of the process. The intermediate supervisory controllers have more input data to contend with, and they might perhaps relax the control of one of the process variables while tightening up on another. This, of course, would only be done to benefit the process overall.

The highest level of supervisory controller has the responsibility for the entire process. This controller may have access to additional input data which are not available to any of the lower-level controllers. The main supervisory controller is then in overall 'command' and can influence any of the 'subordinate' controllers. The similarity between multi-level control and the organizational structure of an industrial company is not just coincidental. The latter is the structural model upon which the former is based.

3.7 Microprocessor-based control

Technological developments in microcomputers with their associated input/output hardware and software tools have enabled the designers of automatic control systems to incorporate a higher degree of intelligence than was possible in the past. Digital computers are now used extensively to control machines and processes. The physical appearance of these controllers vary considerably according to the application, and may range from single-chip microcontrollers (SCMs), where all microcomputer components reside on one IC, to desktop personal computers (PCs).

SCMs provide very cheap computing power and they are mainly associated with high-volume applications such as washing machines, automotive electronics, taxi meters, ticket machines and time-attendance recorders. They can just as easily, however, be used in the control of manufacturing processes in the same way as PLCs, industrial rack-based controllers and PCs.

Since its first appearance in 1981 the IBM PC and its associated compatibles has been adopted as an industry standard. In addition to an increase in processing power, there are a number of advantages in using a PC-based control system.

This integration of the disciplines of microelectronics, computer science and mechanical engineering is the basis of the developing technology of mechatronics. It has been defined as the synergetic combination of mechanical engineering, microelectronics and systems thinking in the design of products and processes.

Typical examples of mechatronic products include robots, CNC machine tools, automatic guided vehicles, video recorders, automatic cameras and autoteller machines.

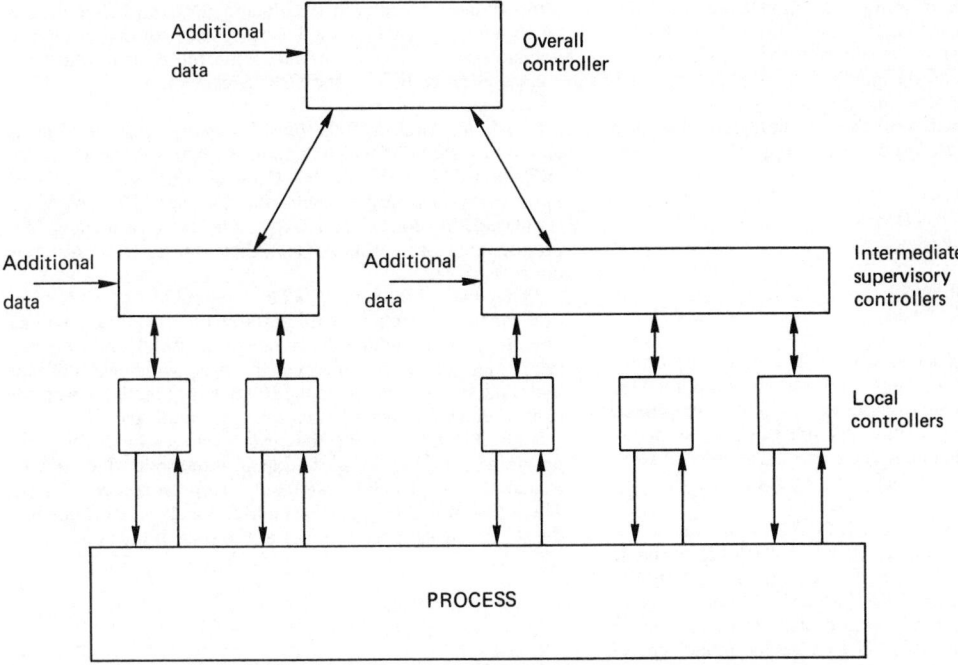

Figure 3.125 Multi-level control system

3.7.1 Direct digital control

Direct digital control (DDC), as outlined in Section 3.6, is employed in systems where such physical quantities as temperature, pressure, flow, position and speed are to be constantly monitored and regulated. The design and operation of a DDC system incorporates formal control theory with computer-related hardware and software to achieve a chosen control strategy. The end product of a real-time control system is a computer-based system which runs the plant efficiently and safely under all operating conditions.

In addition to providing the necessary control functions related to the direct manipulation of valves, drives and other actuators, there are a number of non-control functions available from the system, i.e.:

1. Logging and storage of data;
2. Processing and display of data;
3. Communicating with operators through graphic displays of the plant;
4. Informing of abnormal operating conditions through alarm facilities.

The benefits of DDC are directly identifiable with more precise control, which leads to increased production, improved quality of product and the efficient use of energy and raw materials. There is also a reduction in maintenance requirements and a savings in capital investment through the minimization of recorders, instruments and other control elements. The intangible benefits include an improvement in the handling of engineering and accounting data, increased knowledge of the process from the data collected and displayed, and improved safety.

3.7.2 Hardware requirements

Figure 3.117 in Section 3.6.5 displays the elements of a real-time direct digital control system which constitutes a microprocessor-based controller with the associated input/output interfaces. Digital and analogue interfacing techniques for connecting to external sensors and actuators are described in Section 3.4.

In industrial applications the transducers used for measuring the loop variable normally transmit a current in the range 4–20 mA in proportion to the measured value. This is in preference to using a voltage due to the attenuation over long distances. Also, current signalling can offer a better performance than voltage signalling methods in rejecting electrical noise. At the immediate controller input, the 4–20 mA current signal is converted into a voltage prior to the digital conversion. In control loops the actuators may also operate on a current standard. Since a variable voltage is usually generated through a D/A convertor then voltage to current, (V/I), conversion is also required.

The function of the controller is to monitor the operating conditions of the process and to evaluate, according to a specified strategy, the necessary output control action to ensure that the controlled system operates in a safe and efficient manner. A large variety of digital control devices are available to the system designer, ranging from single-chip microcontrollers, single-card computers and programmable logic controllers (see Section 3.8) to complete personal computer systems (see Section 3.7.7).

3.7.2.1 Low-level control devices

Following the inception of the microprocessor in the mid-1970s the logical progression was to increase component density and incorporate memory and input/output interface facilities onto the one device. These so-called single-chip microcomputers (SCMs) are usually referred to as microcontrollers, and they provide the intelligence required for such applications as measurement and industrial control.

The architecture of a typical SCM is similar to that of a traditional 'single-card computer' and comprises:

1. A CPU;
2. Memory – RAM and ROM;
3. Parallel and Serial Ports;
4. Timers;
5. Hardware Interrupt lines;
6. A/D converters.

There are a large number of SCMs now available and some display all the above features. This greatly simplifies the designer's task, since many of the functions that previously necessitated separate chips in a microprocessor system are now included on a single integrated circuit. Examples of some popular devices currently available include the following.

(a) The Mitsubishi series 740 of 8-bit CMOS microcontrollers These are ultra-low-cost devices which are available in a variety of different forms. For example, a Mitsubishi designer's kit based on the M50734 SCM including a user's manual and software manual, with designer's notes, can be purchased for less than £40. This particular model comes complete with such facilities as four 8-bit I/O ports, a four-channel multiplexed 8-bit A/D converter, timers and counters, a pulse-width modulation output and a single-channel full-duplex Universal Asynchronous Receiver Transmitter (UART) with a built-in baud rate generator.

There is, unfortunately, no internal memory and external RAM and EPROM must be memory mapped through appropriate decoding chips into the system. With a 16-bit address bus, 64K bytes of memory can be accessed, but an area is specified for special function registers associated with the I/O availability, system stack and interrupt vectors. A typical arrangement for the memory map in a control application would be 2K bytes of RAM at the bottom end of memory (i.e. 0000-07FF) and an 8K EPROM at the top end (i.e. C000-FFFF).

The additional hardware required to make the microcontroller functional is a 5 V ±10% power supply, a crystal clock oscillator at 8 MHz and a RESET control. The architecture is based on six main registers similar to those used in the popular 6502 microprocessor and the instruction set is upwards compatible but includes additional mnemonic coding.

An alternative to the Mitsubishi M50734 is the M50747-PGYS, which has an 8K EPROM, mounted piggyback style onto the top of the SCM. Since 192 bytes of internal RAM are available, no external memory with associated decoding is necessary. This model has a number of bi-directional 8-bit I/O ports and timers but has no A/D conversion facilities.

(b) The Motorola MC 68705R3 microcontroller[16] This is a powerful SCM which comprises an 8-bit CPU with 112 bytes of RAM, about 4K bytes of EPROM, four 8-bit I/O ports, a timer and a four-channel A/D converter. The IC which makes this possible requires more on-board transistors than the 16-bit microprocessor used in the Apple Macintosh microcomputer.

The EPROM facility for the user software is not of the piggyback variety, but is embedded into the package which has the disadvantage of requiring a specialized software development system. The inherent A/D converter is extremely useful in control systems which involve analogue measurements. It uses the successive approximation technique with an internal sample-and-hold circuit and the conversion time is 30 μs with a 4 MHz crystal clock. The external reference voltage which is to correspond to the full 8-bit resolution is connected directly onto the SCM package.

(c) The Rockwell R6500/1 one-chip microcomputer This is an easy-to-use SCM which contains 64 bytes of RAM and 2K (R6500/1EB1) or 3K (R6500/1EB3) of addressable EPROM fixed piggyback fashion onto the package. This allows a standard EPROM such as a 2716 or 2732 to be easily removed, programmed or reprogrammed, then reinserted as often as desired.

Thirty-two bi-directional TTL compatible I/O lines from four ports are available with a 16-bit timer/counter that can operate in four modes. This makes a useful and versatile sequential controller which can be used with pneumatic or hydraulic systems. To make the IC package functional a power supply, clock and reset switch must be connected.

The EPROM contains the reset vectors which direct the program counter to the start address of the code which is to be executed in sequence to carry out the specified control task. The RAM available must also be used for stack operation and the stack pointer must be set up at the start of the applications software.

3.7.2.2 Industrial controllers

These are usually modular industrial microcomputers which are built to internationally recognized bus standards such as STD or EURO and selected cards are held in a standard rack. The ability of these systems to acquire data, control equipment and analyse the logged data make them particularly suited to process-control applications where performance monitoring is required. The system also lends itself well to laboratory data acquisition where the function is to collect data from a range of different instruments.

The development of applications software, usually in some form of real-time high-level language, is now usually carried out by communication between the controller and an IBM PC or compatible. This makes life easier for the system builder by adding enhanced editing and data-storage facilities. These rack-mounted industrial controllers are generally more expensive than PLCs although they offer more flexibility.

Since most practical industrial processes tend to be poorly defined and it is often difficult to derive an accurate mathematical model of the system, the control strategy usually adopted is the 'three-term' one. This has led to the emergence of microprocessor-based PID controllers, whose control output is based on the error signal (evaluated from the declared set point and measured process variable) and the chosen settings of gain, integral and derivation time. Self-tuning PID controllers are now available.[14] Some of these employ a process 'pattern-recognition' technique with an 'expert system' approach based upon the tuning rules usually employed by skilled control engineers.

3.7.3 Software considerations

The power in the digital control of a system is in the software, and any controller requires a real-time language. This is one which can synchronize its operations with events occurring in the so-called 'real world' and thereby respond to and control interfaced mechanisms and processes. The facilities required of a real-time software language are:

1. Ease of switching external devices on and off using software;

2. Ease of deciding whether external devices are on or off at any time;
3. Ease of timing the duration of a process;
4. Ease of making a process run for a predetermined time;
5. Ease of making the program respond to things that happen in the outside world.

Although high-level languages such as BASICA, QUICK-BASIC, TURBOBASIC, FORTRAN and PASCAL can, to some extent, meet the above requirements there are a variety of real-time languages which have been specifically designed for control applications.

3.7.3.1 Real-time high-level languages

BASIC This is a programming language commonly used for engineering applications, and real-time interpreted dialects such as CONTROL BASIC and RTBASIC[17] figure prominently. These can be used with both memory mapped and port addressed I/O organization (see Section 3.2.4) and provide keywords which enable the individual bits on a digital port to be read or written to for switching devices on or off. A/D and D/A converters are also easily operated through keywords not usually available in ordinary dialects of BASICs. Timing operations can also be easily implemented.

In addition to the digital and analogue I/O keywords available to assist in the development of control software, floating-point arithmetic is included for data-processing requirements.

FORTH[18] This is an interpreted threaded language developed specifically for control applications. The instructions which constitute an application are stored as a list of previously defined routines. This list is threaded together during the entry of source code from either the computer keyboard or the mass-storage buffers. The process of producing the list is often termed compilation, but this is not strictly correct since the result of the true compiling of source code produces pure machine code.

FORTH is a most unusual language since arithmetic calculations do not follow traditional methodologies. Before calculations can be made it is first necessary to understand how the stack operates in FORTH. Most high-level languages use one or more stacks for their internal operations, but languages such as PASCAL and FORTRAN are designed so that the user does not have to understand how the stack functions. FORTH allows the user full control of the values stored in the stack and their manipulation.

One of the main features of FORTH (and one of the most powerful) is that, once the program has been written, a single word which defines the complete program can be entered into the FORTH dictionary as a command. This dictionary contains words defining, say, routine control operations, and a single word could trigger a series of actions as its threads its way throughout the dictionary.

Programs can be typed in at the keyboard and executed directly, but to save programs for execution at a later date the mass-storage buffers and the FORTH editor must be used. Unlike other high-level languages, FORTH performs only a very limited number of error checks. The errors detected are those which are most likely to cause the system to crash if allowed to pass undetected. The main reason for the lack of error checking is that it would slow FORTH down, and since the main use of FORTH is in time-critical control situations, this would be prohibitive.

C This is a general-purpose programming language which cannot be truly classed as either high-level or low-level. It has all the features expected of a high-level language but allows the programmer to access the computer's hardware and has the high performance usually expected of low-level languages.

Although the language was developed during the 1970s, it is now gaining rapid popularity in a range of industrial applications which include real-time control. There are a number of versions of C available but the accepted standard for the programming language can be found in the book by Kernighan and Ritchie.[19]

Accessing the input and output of data at external ports is easily accomplished and another exceptionally important feature is the use of timing operations which have a resolution of the computer system clock period.

Using C, it is possible to construct concise, well-structured and well-documented programs that can include a variety of useful library functions such as are needed for DOS input/output calls for hardware, screen handling and real-time clock-timing facilities.

3.7.3.2 Low-level languages

At the machine level the program is stored in the memory in a binary format. It is conceivable to write the program immediately in machine code, but this would entail an unjustifiable amount of labour. It is easier to use mnemonic programming, where each instruction is accorded a symbolic name close to colloquial language. The name is easier to remember and the resulting program is much easier to read than pure machine code. This so-called assembly language program can be translated into the necessary machine code for a particular microprocessor by using an assembler program.

The main attraction of using machine code for a particular application is its speed of operation, and if a single-chip microcontroller is to be used the problem of software development must, nonetheless, be faced. This requires the writing of a program into the system memory when no keyboard, display monitor or operating system software exists.

The inexpensive approach is to choose a SCM which supports an external EPROM, in piggyback fashion. The software code for the specific task may then be developed using an appropriate assembler program which operates on an IBM PC or compatible. The resulting code is then used to blow the EPROM. This method can, however, be very time consuming, since the debugging process will require the continual updating of the EPROM data as the program is edited to achieve the specified requirements.

An alternative is to use an EPROM emulator inserted into the target system to temporarily hold the program which is transmitted from the host computer memory. This enables the user to easily alter the program as required and then finally blow the EPROM for permanent use. EPROM emulation should not be confused with the technique of In Circuit Emulation, which involves the removal of the microprocessor from the target system. The In Circuit Emulator then takes complete control by emulating in real time all the functions of the removed microprocessor it replaces.

A software development system should therefore comprise the following:

Hardware
A host microcomputer with keyboard and display monitor;
A printer for a hardcopy of listings and disassemblies;
Disc drives for the permanent storage of data and programs;
An EPROM emulator;
An EPROM programmer.
Software
An editor;
Translation tools such as an assembler or cross-assembler, disassembler, debugger and linker.

Traditional cross-assemblers will only assemble for one microprocessor, and it can be very expensive if a number of upgrades or different types are to be accommodated. These very sophisticated and dedicated development systems are generally too expensive and complicated for producing the operational programs to control a fairly basic and ordered sequence of events, as usually occurs in mechanical systems.

An alternative is to employ an IBM PC or compatible with an assembler which can handle multiple instruction sets. The processor to be used is specified in the source code and the assembler adapts automatically to the correct format.

Tailoring a general-purpose and readily available machine such as the IBM PC to a semi-dedicated role requires a relatively low investment and low software production costs. For one-offs this approach is the only real economic proposition.

3.7.4   Sampling frequency in digital control loops

A DDC loop contains both hardware and software contributions. Transducer measurements of the process variable must be regularly made and the invariably analogue values converted into sequences of numbers that can be handled by the controller. The measured value is compared with some set condition and a control algorithm implemented in software evaluates the necessary control effort. This effort is calculated as a numerical value in the computer, and a conversion of the output is necessary to obtain a form which is suitable to drive the required control elements or actuators. A sampling rule for the measured variable is used to determine the rate at which the sampling is performed.

Digital sampling gives a sequence of 'snapshots' of an analogue variable. The controller only holds representations of the variables at the times when the samples are taken as illustrated in Figure 3.126. The sampling rate must obviously be matched to the rapidity of the variations in the process variable. High-performance control systems require the sampling interval, Δt, to be short, although very rapid sampling will increase the computational load. Δt is usually specified in terms of other system parameters.

The classic reference related to digital sampling is that due to Shannon and Weaver,[20] which states that a signal that is band-width limited can be recovered without any distortion provided samples are taken at a rate of twice that of the highest frequency. 'Twice' may be regarded as an absolute minimum, and a value of five to ten times the highest frequency produces a more realistic digital representation of a sinusoidally varying continuous signal.

Real industrial plant have limited bandwidth, and in applications with long time constants a sampling frequency of twice the highest plant frequency is adequate. In practice, for the great majority of control loops involving the control of variables as temperature, pressure, flow and level, sampling rates of 0.2–1 s usually prove to be fast enough. These are the rates normally fixed in the commercially available industrial PID controllers.

Fast-acting electromechanical servosystems require much shorter sampling intervals (down to a few milliseconds). Shannon's sampling theorem cannot strictly be applied if the system maximum frequency is unknown. However, practical experience and simulation have produced useful empirical rules for the specification of the minimum sampling rate. One such rule is based on the dominant time constant (see Section 3.6.2), which can be obtained from an open-loop test on the process. A suitable sampling interval for use in a closed-loop control digital algorithm is

$$\Delta t < (\tau/10) \tag{3.54}$$

This may, however, prove to be unsatisfactory with systems which have a large time constant when a fast closed-loop response is forced by the controller settings.

3.7.5   PID digital control algorithm

A mathematical model which produces a transfer function (see Section 3.6.3) for the process to be controlled can usually be derived for servomechanisms, and specifications in terms of damping ratio, natural frequency and bandwidth can realistically be defined. This is not the case, however, for the

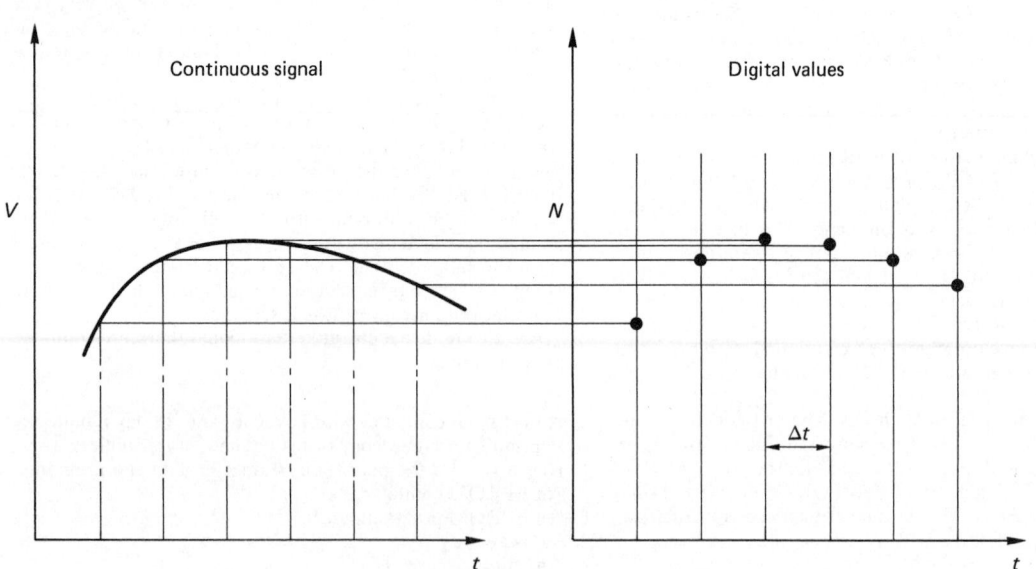

Figure 3.126   Digital sampling of a continuous signal

majority of industrial processes. For this reason, a three-term controller implementation (see Section 3.6.5) is preferred in practice.

The five essential steps required to achieve good process control are:

1. The variable, PV, that best represents the desired condition of the final product must be measured. Measurements such as temperature, pressure, flow, position and speed are commonly used.
2. This measurement must then be compared with the desired set-point, SP, value of the variable to yield an error signal, E, where $E = SP - PV$.
3. This error is then applied as an input to the controller.
4. The controller output, U, must then be applied to a final control device such as a powerstat, a valve or a motor-drive system.
5. A source of energy or material flow is then adjusted in response to the controller output until the deviation between PV and SP will be as near zero as the sensitivity of the system permits.

This is the basic principle of negative feedback as used in most automatic control systems. The sequence of operations at a chosen sample rate is measure, compare and correct.

The function of a digital controller is to apply a control algorithm based on the error value. An algorithm is a computer procedure that performs mathematical functions on a given set of variables. The computational procedure that converts $(SP - PV)$ into a controller output is commonly called the PID algorithm. This algorithm is robust and performs well in practice. Although it may appear inferior as a scientific method, it is difficult to improve on significantly.

The controller output from a PID control strategy is as given in equation (3.45):

$$U = K[E + (1/T_i) \int E dt + T_d(dE/dt)] \qquad (3.45)$$

For implementation on a digital controller this must be transformed into the appropriate software for the system being controlled. This transformation from a continuous to a discrete form can be achieved by either the use of z-transforms or difference equations. The latter are easier to understand and implement, and the steps in deriving any digital algorithm by the difference equation method are:

1. Express the output requirement in a differential equation form.
2. Replace the equation in difference form using the discrete digital approximation:

$$dY/dt = (Y_i - Y_{i-1})/\Delta t \qquad (3.55)$$

where Δt is the sampling interval.
3. Solve for the present value of the variable Y_i from the previous value Y_{i-1}.

Applying these steps to the three-term controller yields:

$$U_i = K[E_i + (\Delta t/T_i) \cdot \Sigma E_i + (T_d/\Delta t) \cdot (E_i - E_{i-1})] \qquad (3.56)$$

The error at any particular time, i, is best evaluated as a percentage of the transducer span, and the above algorithm would give the controller output as a percentage value which must be contained within the range of 0-100%.

This PID digital algorithm can be easily programmed directly in a high-level language and the framework for a program in BASIC is as follows:

1. The values for the chosen set point and the controller setting requirements K, T_i and T_d must first be input to the program.

2. A sampling time Δt is often chosen and specified in the program.
3. Since a summation of the error value is required, a variable termed, say, SUM must be set to zero. This also applies to the 'previous error' value E_{i-1} in order to start the control loop.
4. Numerical constants can be calculated, e.g.

$$IC = \Delta t/T_i \text{ and } DC = T_d/\Delta t$$

5. The program listing continues:

```
100  REM a routine is required to measure the process
     variable value
200  E=100*(SP-PV)/SPAN
300  SUM=SUM+E
400  OUTPUTI=IC*SUM
500  OUTPUTD=DC*(E-EP)
600  U=K*(E+OUTPUTI+OUTPUTD)
700  IF U<0 THEN U=0
800  IF U>100 THEN U=100
900  REM the output U must then be output from the
     controller in a form
1000 REM which is appropriate to the power requirement
     of the process
1010 REM being controlled.
1020 REM e.g. an A/D converter or PWM (see Section
     3.4.2)
1030 EP=E
1040 REM repeat control loop from line 100
```

Note: It is essential that the sampling interval Δt is greater than the time taken to complete the control algorithm. The above routine would generally be called from a main program which contained a time adjustment such that the complete sequence was contained exactly within the specified sampling time Δt.

The program reveals the general method of programming PID control algorithms for use with a computer and forms the basis of the software incorporated into various commercially available controllers.

3.7.6 Speed control

Many industrial processes require variable speed drives using electric motors. Electromechanical methods of motor speed control, such as Ward Leonard motor/generator sets, have largely been superseded by using digital controllers interfaced to power semiconductor devices. The common types of motor are those which are operated by either d.c. or a.c. Alternating machines are also referred to as synchronous motors.

3.7.6.1 D.C. drives

D.C. motors can rotate in either a clockwise or an anti-clockwise direction, depending upon the direction of current flow to the coils via the brushes. The speed is load dependent and it can be varied by altering the supply voltage. Arcing at the brushes generates interference which may cause problems when using computer control through corruption of logic levels representing input/output data. It is essential that the spikes generated at the high interference frequencies are filtered out. This can be achieved by fitting a capacitor across the contacts close to the motor and adding an inductance in series with the power supply. A typical direct digital control loop is illustrated in Figure 3.127.

Details of typical methods of measuring rotational speed are given in Section 3.5. The digital controller invariably implements a three-term strategy as outlined in Section 3.7.5 and

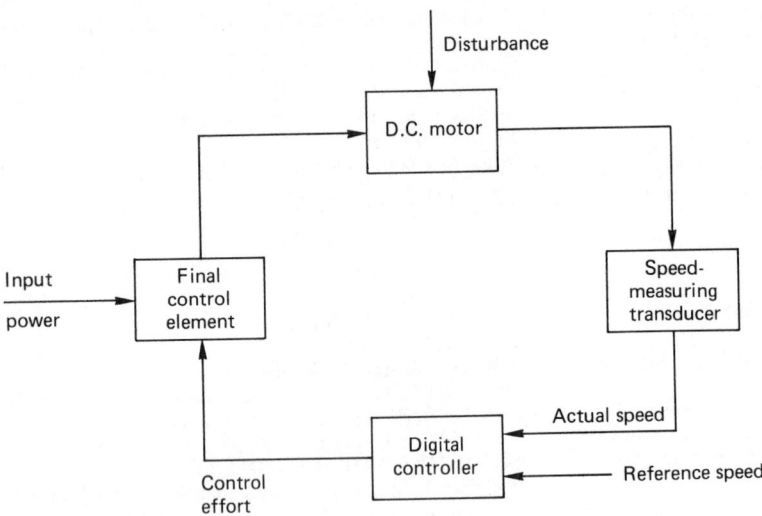

Figure 3.127 Closed-loop control of a d.c. motor

the numerical value calculated must be converted into a motor input through a final control element. This may take the form of a D/A converter and a power amplifier. D.C. servomotor control modules are commercially available for a range of motors, and with a variable input from the D/A converter the motor speed can be varied over its complete operating range. The speed and nature of response to an external disturbance such as a change in load are dependent upon the controller settings.

3.7.6.2 A.C. drives

The a.c. motor does not have brushes and hence the problem with interference when using digital control does not exist. They are also less expensive, more robust and require less maintenance, since they do not have a commutator. However, being a synchronous device, the speed is locked to the frequency of the supply. To control the speed of an a.c. induction motor over a wide range the frequency as well as the amplitude of the applied voltage must be varied. The classic method of acheving this is to use a d.c./a.c. inverter in which a positive and negative d.c. supply is alternatively switched to the motor. This switching arrangement is shown in Figure 3.128.

In order to avoid short-circuiting the d.c. supply any one of the two switches must be off before the other is switched on. The switching may be achieved by using either transistors or

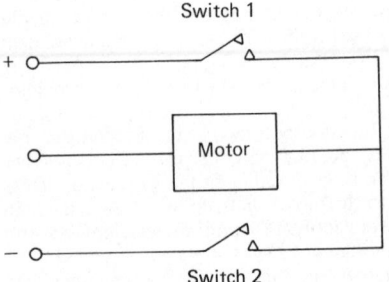

Figure 3.128 Principle of a d.c./a.c. inverter

thyristors which can operate with a turn-on/turn-off time of typically 5 μs. Thyristors have been developed for this purpose and have current- and voltage-handling capabilities of 1000 A and 4000 V, respectively. The corresponding figures for power transistors are 400 A and 800 V.

The control loop for the a.c. drive is identical in principle to that shown in Figure 3.127, where the final control element constitutes the high-frequency switching of an inverter circuit to produce a pulse width modulation (see Section 3.4.2) technique for varying the input power to the machine. The transistor or thyristor based on a.c. drive at powers up to 500 kW is a viable proposition and has the advantages of close speed control in both directions, controlled acceleration or deceleration and high efficiency with a good power factor.

3.7.6.3 Stepper motors

An alternative method of obtaining a variable speed is to use a stepper motor. This is a power device which converts a d.c. voltage pulse train into a proportional mechanical movement of its shaft. There are a range of commercially available motors which are designed for a variety of operating conditions.

The motor is designed with a number of phases which must be supplied with current in a specified preset continuous sequence according to the number of phases. Reversing the sequence causes the motor to rotate in the opposite direction. The motor shaft rotates by a finite amount for each pulse received. Thus if a motor is designed for 48 pulses/revolution then a pulse train received at the rate of 96 pulses/second will cause the motor to rotate at a speed of 120 rev/min. Stepper motors having outputs in the order of kilowatts are available and they are used for applications involving accurate position or speed control such as $X-Y$ plotters, numerically controlled machine tool slide drives and in carburettors for internal combustion engines.

Interfacing a stepper motor to a digital computer is relatively straightforward and a number of integrated circuits and power driver cards are available for a variety of popular digital controllers.

3.7.7 The PC as a controller

The IBM PC and its compatibles are firmly established as the standard for industrial computing throughout the UK and Europe. They are relatively cheap, plentiful and fairly easy to use. As they are integrated into the industrial environment it is evident that they have a high potential as control devices on the factory floor.

In addition to an increase in processing power, there are many other advantages in using a PC-based control system:

1. There is large choice of software which is not available for dedicated controllers.
2. There is a good selection of available tools to produce applications software efficiently.
3. The PC is available in a variety of forms ranging from a single card, a portable, a desktop and an industrial version for use on the factory floor.
4. Expansion plug-in slots to the PC bus structure are available and a large range of cards for digital or analogue I/O have been produced by a number of manufacturers.
5. The PC-based controller is more flexible than the dedicated or minicomputer system and can be easily configured indefinitely to suit different applications.

Data acquisition and control add-ons for PCs are either external rack-mounted systems or plug-in boards. The external box approach usually involves attaching a separate rack-type enclosure with power supply to the host PC. The connection is via either the included serial or the parallel data communication link. Various modules based on a standard card format (such as the Eurocard) can be plugged into the enclosure housing as required.

There are two options for capturing data with PCs. The first is to use an analogue-to-digital converter card that plugs directly into the host computer's backplane. The cards are generally 'port addressed' and may be driven by any language having IN/OUT commands. The base address is usually switch selectable on the card. The second option uses instruments such as digital voltmeters and frequency meters which have an interface board that enables data transfer from or to the controlling PC. The most common standard is the IEEE-488 (GPIB), where data in bit-parallel, byte-serial manner is transmitted from the PC (as an ASCII string) to the instrument, informing it of the settings required for the measurement to be made. Once read, a control signal is sensed and a string representing the measured value is returned to the PC. It is, of course, necessary to have an IEEE-488 card installed into an available slot in the PC. Up to 15 instruments can be accessed in this way from one card.

The quickest, easiest and least expensive way to get measured data into the PC or control signals out is to use I/O cards. These are available for many applications and include:

1. Multi-channel digital I/O with opto-isolation and Darlington Driver facilities;
2. Pulse counting and timing;
3. Multiplexed analogue-to-digital conversion with programmable gain;
4. Digital-to-analogue conversion;
5. Thermocouple input.

Recently there have been rapid advancements made in the software available for data capture and control. Packages are now available which provide the user with a development system having an environment of windows and pull-down menus. The National Instruments[21] 'Lab Windows', for use on PCs, is a library of function modules for programming specific instruments with the IEEE-488 interface. These modules are accessed in the development program via function panels to interactively set up and acquire data from instruments. In addition, a suite of programs are included for data presentation, analysis and formatting in an interactive environment using either QUICKBASIC or C.

A number of manufacturers of rack-mounted microcomputer controllers and programmable logic controllers (PLCs) have united their product with a PC and provide user-friendly software to assist in the development of the required control programs. Such aids are invaluable for displaying, storing and printing PLC ladder relay diagrams as an alternative to the hand-held programmer and EPROM blower. Logged data can also be transferred from the controller to the PC where they can be displayed on a mimic diagram of the plant with animated symbols, updated values of controlled parameters, alarm messages and bar graphs.

Software development for PCs has at last begun to catch up with the hardware, and the data acquisition and control market has been long overdue for reliable development systems.

3.8 Programmable logic controllers

Automation systems generally involve the application of such microprocessor-related equipment as CNC machine tools, pick-and-place machines and industrial robots. As shown in Table 3.7, there has been a greater involvement in the application of programmable logic controllers (PLCs) in manufacturing industry than any other automation devices.[22] The figures in Table 3.7 have been obtained from a survey of all UK manufacturing industries employing 20 or more people.

Table 3.7 Usage of automation equipment[22]

Equipment used	1983 ('000)	1985 ('000)	1987 ('000)	1989 ('000: estimate)
CNC machine tools	17	19	27	42
PLCs	34	59	73	110
Pick-and-place machines	4	11	10	15
Robots	1	2	3	6

PLCs are mostly employed in the relatively straightforward control of a single process or piece of equipment. They are particularly common in the food processing, chemical and automotive industries. PLCs first appeared in the 1970s in place of relay circuits, and they have been continually developed as a result of the rapid progress in microelectronics. Some of the current more powerful PLCs overlap with microcomputers or process computers, and it is often impossible to distinguish between them.

3.8.1 The PLC in automation systems

The PLC is particularly useful for controlling manufacturing processes which constitute a sequence of operations that are either event or time driven. PLCs have had a significant impact on industrial control because of their ruggedness, versatility and reliability. They have virtually replaced hardwired relay, counter and timer logic systems due to their cost, flexibility and relative size. Additionally, the number of applications in which PLCs are being used has increased substantially in recent times.

The principal criteria for determining size is the input/output availability and they can broadly be divided into the following categories:

1. Small PLCs, up to 128 I/O lines, with typically 12 inputs and eight outputs in the basic form. These are designed as stand-alone items and can usually be expanded with respect to digital I/O requirements. A typical system with the associated hand-held controller is shown in Figure 3.129. Such systems basically perform logic, counting and timing operations but have no arithmetical manipulation capabilities.
2. Medium-size systems (128 to 1000 I/O lines), which are generally contained in a rack arrangement. These offer more extensive I/O including analogue-to-digital and digital-to-analogue options with some enhanced programming features.
3. More powerful rack-mounted systems, with over 1000 I/O lines.

The latter PLCs offer data communication for operation within a complete computer-integrated manufacturing arrangement. They embody quite sophisticated man/machine interfaces which employ real-time operating systems and advanced computer graphics. Three-term PID control, digital filtering and vision capabilities can be implemented in many modern PLCs and a variety of operator interfaces are available to allow easy entry of desired settings and process variables.

Recent developments have seen a move from the automation of single machines towards that of the whole manufacturing process. The PLC is a typical representative. The modern concept is the automated factory, where computer-integrated manufacturing (CIM) combines the basic functions of production into a single system. The system incorporates the highest level of automation economically feasible for each function involved. The PLC constitutes one of the building blocks in such a distributed hierarchical control system which is based on a 'pyramid' structure. The peak of such an arrangement contains the mainframe computer, which is responsible for handling the databases for production scheduling, sales and other management needs. The PLC or microcomputer immediately precedes the machine and process applications located at the base of the pyramid.

However, although the small single-purpose basic PLC is relatively secure in its role in the market, the medium and large PLCs face considerable competition from microcomputers. This is particularly the case in applications where much arithmetical manipulation must be performed. It is now possible to get IBM PCs and other microcomputers in a 19-inch rack format which can be mounted within standard

control panels. This represents a serious competitor to many PLC-based systems.

3.8.2   The PLC versus the microcomputer

The PLC and the microcomputer contain the same basic constituent components such as central processing unit, memory devices, input/output interfaces, decoding logic and a connecting bus structure. They are, however, entirely different in physical appearance and operation.

3.8.2.1   The case for the microcomputer

The dominance of the International Business Machines Company (IBM) in the microcomputer market has led to an accepted industry standard machine in the form of the IBM PC. Since its appearance several manufacturers have produced so-called compatibles, which are based on the associated 16- and 32-bit technology. The latest processors now allow the programmer to write programs that use more memory than is actually available in a given system. This is done by exchanging data between the main memory and secondary storage devices. It is this processing power, memory availability and peripheral hardware that has made a significant impact on the engineer's acceptance of PCs for control applications.

In addition to this increase in processing power, there are many other advantages in using a PC-based control system:

1. The large choice of available storage ranging from high-level language interpreters and compilers to sophisticated man/machine software interfaces which include a range of selectable data processing and graphical display routines;
2. The selection of available tools to produce applications software efficiently;
3. The variety of forms available, which range from the small portable type for use in the field to the rugged industrial version for the factory floor;
4. The large range of plug-in cards available for use with the PC bus structure to provide both digital and analogue I/O facilities;
5. The ease with which the system can be reconfigured to enable, for example, the application of word processing, spreadsheets or databases to be carried out. There is no need to purchase a new system for every new application.

However, in order to truly harness the flexibility that accompanies the personal computer some knowledge of microprocessor technology is required, and the ability to understand or develop applications software is essential.

3.8.2.2   The case for the PLC

The alternative to the microcomputer for control applications is the PLC, which may be described as less intelligent but is ideally suited for carrying out logical sequential operations to produce outputs which are conditional on input states. The advantages of using a PLC for controlling machines and processes may be summarized as follows:

1. The PLC is relatively inexpensive compared with a microcomputer.
2. The construction is exceptionally rugged and it requires very little space in the control cabinet.
3. Reliability is high, immunity to electrical noise is good and maintenance is low.
4. Memory is used economically due to the method adopted for the processing of the data. The PLC processor is

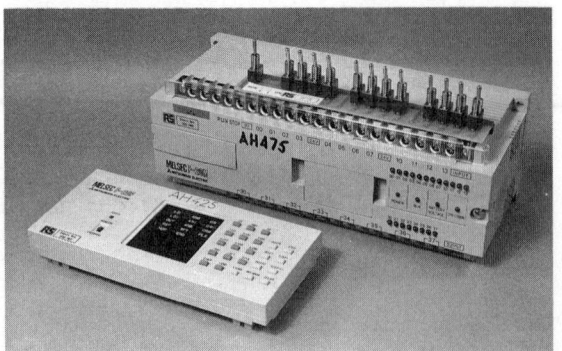

Figure 3.129   A typical small PLC

tailor-made to execute logical operations (AND, OR, etc.) as they occur in the control program.
5. The operating speed is fast.
6. An in-built interface provides for the easy connection to a variety of input/output devices connected directly to the machine or process.

The PLC is, however, less interactive than the microcomputer, partly due to the fact that relay ladder logic is generally used as the programming language because of its alleged wide acceptance by those who design, operate and maintain control systems. During program execution the ladder is continuously scanned and outputs are set in accordance with input conditions. It is reputed that ladder logic is straightforward and easily understood, but software development can be a problem for the non-specialist in systems which involve a number of timer and counter requirements.

In conclusion, there is probably an equal support, at present, for both PC and PLC systems. Both systems, then, should continue to play an important part in manufacturing operations for the foreseeable future.

3.8.3 Ladder logic programming

The most commonly used programming language for use with a PLC is the ladder diagram. For the popular small PLCs it is usually possible to enter the input, output, timer and counter instructions directly into the controller using a hand-held programmer. The mnemonic codes are displayed on a small screen within the programmer prior to conversion into machine language and transfer into the PLC battery back-up memory. Although the program can be edited, the main disadvantage of using this method is that there is no way of saving and printing the developed program.

A preferable alternative is to use a PC with an appropriate software package available from the PLC manufacturer to develop and display the ladder diagram on the PC monitor. Use is made of the PC drives and printer as required. Once completed, the resulting file is transmitted from the PC to the PLC for execution. This is by far the best method of developing ladder logic software.

In a ladder diagram the power source is represented by the two vertical rails of the ladder and the various instructional requirements which represent the control circuits make up the rungs. The symbolic ladder circuit layout is constructed using standard graphic logic symbols to represent input contacts, output loads, timers and counters, as shown in Figure 3.130. When using a PC to develop the program, comments can be added to each instruction rung. This makes the program easier to understand and facilitates fault finding.

Each rung on the ladder starts with an input condition and finishes with an output to a load, timer or counter. When used to automate a machine, the ladder is continuously scanned in a sequential manner and sets the outputs on or off according to the requirements based on the sensed input conditions.

3.8.3.1 Inputs

The most common symbol used in programming a PLC is the input contact, which may be active in either the normally open (NO) or normally closed (NC) mode of operation. The 'NO' contact symbol is a request to examine if the switch is ON in order to proceed along the run. For the 'NO' condition to be satisfied the contact must be OFF.

A popular real-world input device is the switch. This is basically used to open and close an electrical circuit. The PLC supplies the voltage across the switch which is usually either 24 V d.c. or 100 V a.c. A large number of switch designs are

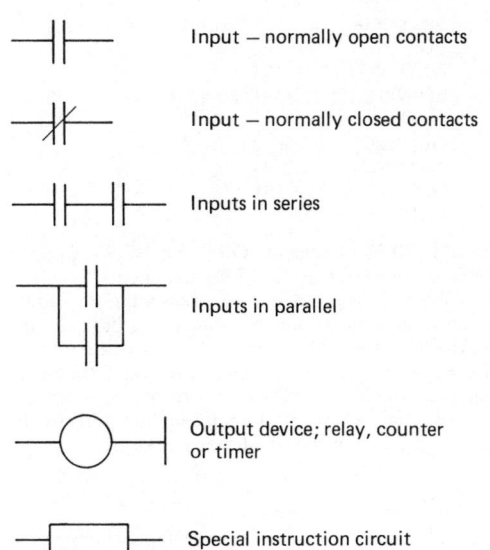

Figure 3.130 Instruction set ladder logic symbols

used to control systems. The limit switch is one such example, and it is designed to open and close when a machine part reaches a specified position or a physical variable obtains a required value. Once the limit switch condition has been sensed, the PLC takes appropriate action as defined in the ladder program.

3.8.3.2 Outputs

The most common output devices are the electromagnetic relay and the solenoid. The relay provides isolation of the load voltage and current from the PLC and is ideal for the switching of high loads via the controller's in-built I/O interface. Alternatively, the solenoid, which consists of wire surrounding a movable plunger, allows electrical control of any mechanical device attached to the plunger. This arrangement is applied in electropneumatic systems to control the position of a valve spool in order to direct the air path through a pneumatic system.

Due to the nature of the sensed input and the switched output load, the PLC I/O module normally provides a safe electrical separation between internal and external circuits. This is effectively carried out by an opto-isolator, which ensures that there is no physical hard-wired connection between the PLC and the external I/O device.

A typical rung in a ladder diagram including I/O with logic operations is shown in Figure 3.131. Thus rung may be written for a particular PLC as:

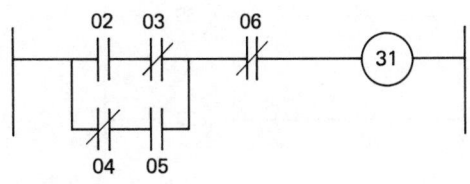

Figure 3.131 A typical ladder rung

Step	Instruction	I/O number
000	LD	02
001	AND NOT	03
002	LD NOT	04
003	AND	05
004	OR BLK	
005	AND NOT	06
006	OUT	31
007	END	

Note: the 'OR BLK' statement 'ORs' the blocks grouped under the preceding two load (i.e. LD) statements.

Different PLCs use various ways of translating the ladder programs into mnemonics, but the principle is basically the same as that illustrated. This translation is necessary if the program is to be entered into the controller via a hand-held programmer. However, if a PC-based software development system is used the ladder diagram is entered directly by pressing the appropriately specified function keys.

3.8.3.3 Timers

Time delays are commonly used in control applications and in a PLC program a timer is used to activate an output following an elapsed time as specified by a value stored in the memory. The timing base units are usually tenths or hundredths of a second.

There is no standard way of specifying timer routines and manufacturers use different methods, but the same basic principles of operation apply. This entails an enabling input condition to start the timer. At the end of the timed period the timer must be reset by disenabling the input.

It must be noted that during the program execution the ladder is *continuously* scanned rung by rung and outputs set according to input conditions. There can be no hold on any particular rung. However, in implementing timing operations the timer output can be used as an input on any rung, thus making outputs dependent upon the specified time period to have expired. This is illustrated in Figure 3.132. In this figure the timer number 50 is set for a timing operation of 1 s and is enabled by switching input number 05 to an ON state. In the next rung the output number 30 will not be energized until the specified time has elapsed. The timer must then be reset by setting the output 05 to OFF.

3.8.3.4 Counters

A counter is used to activate an output after a predetermined number of counts as stored in the PLC memory. It operates in much the same manner as a timer but, rather than counting time, the counter counts the number of times that an event occurs. The event is some specified input condition in the process. The counter is reset from another input switch and it can be used as an input to any other rung in the ladder. This is illustrated in Figure 3.133. Each time that switch 02 is acti-

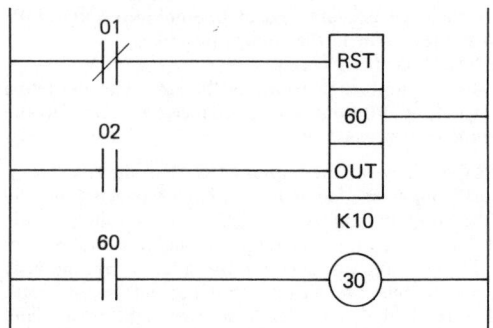

Figure 3.133 Example of a counter operation

vated a count is produced up to the specified value of 10. Once the count is completed the output load number 30 is switched ON. The counter is then reset by switching input 01 OFF. An incrementing or decrementing mode for the counter operation is usually possible with most PLCs.

3.8.3.5 Auxiliary relays

Most PLCs are equipped with various special function input/output facilities that exist in the development of safe and efficient control programs. The auxiliary relay is one such device which operates internally and is assigned an address that is different from that of any other real output device in the system. It is basically a 'dummy relay' which is used as an aid to developing the program. For example, consider an output condition which is dependent upon, say, six inputs and a rung is restricted to four inputs. The auxiliary control relay can be used to connect two rungs into a logical equivalence of the requirement as shown in Figure 3.134.

3.8.3.6 Input/output numbering

Different PLC manufacturers use different numbering systems for the controller's available I/O. The addresses assigned for the small Mitsubishi type F-20M PLC are typical:

12 inputs:	00–07; 10–13
8 outputs:	30–37
8 timers with a range of 0.1–99 s:	50–57
8 counters with a range of 1–99 counts:	60–67
48 auxiliary relays:	100–157

(Note the octal numbering system adopted)

3.8.4 Controlling pneumatic and hydraulic systems

Consider the sequencing of a double-acting pneumatic cylinder with digital sensing devices of mechanical, optical or

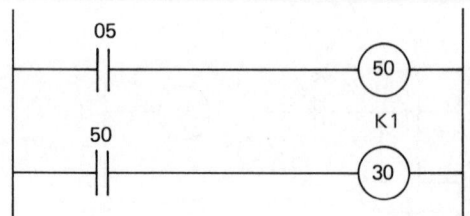

Figure 3.132 Example of a timing operation

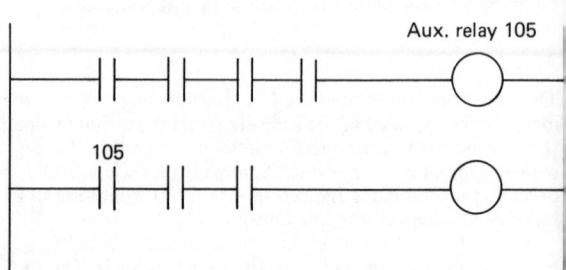

Figure 3.134 Use of an auxiliary relay

magnetic construction to detect the end-of-stroke condition. The direction of motion is controlled by an electropneumatic five-port spool valve which is solenoid/pilot operated. The air supply to the system manifold is enabled by an ON/OFF switch connected to a solenoid-operated three-port valve.

The control program is required to switch on the air, extend the cylinder rod to the end of the stroke, wait 5 s then retract the rod and repeat to complete 10 cycles. The main ON/OFF switch is to be used to reset the counter and it must also stop the cycling at any time. The input/outputs are assigned as follows:

INPUTS
ON/OFF switch   – 00
Sensor to detect retracted position   – 01
Sensor to detect extended position   – 02
OUTPUTS
Air supply valve   – 30
Direction control valve solenoid to extend cylinder rod – 31
Direction control valve solenoid to retract cylinder rod – 32
TIMER – 50 and COUNTER – 60

Figure 3.135 shows the ladder diagram to satisfy the specified sequence. Note that de-activating the ON/OFF switch will stop the system at any time and that the rod will not extend once the counter has completed the required number of cycles.

With solenoid/pilot valves the solenoid need only be pulsed since the pilot line in the valve supplies air to move the spool, and it is usually unnecessary to maintain the solenoid in an energized condition. This is evident in the ladder diagram with inputs 01 and 02 from the end-of-stroke sensors returning to the OFF state once the cylinder rod moves from the end condition.

With solenoid only operated valves, however, it is necessary to latch the output load ON in order to maintain the rod motion. The ladder relay logic for rung 4 to latch on solenoid

31 in order to extend the cylinder rod is illustrated in Figure 3.136. Once the rod has left the sensor 01, power is transmitted to the output via the switched output acting as an input. This latches the control solenoid ON and it is switched OFF once the end of stroke is reached and sensor 02 is switched ON.

3.8.5 Safety

Since the equipment generally associated with PLCs often involve high voltages and currents, electrical safety procedures must be followed at all times. In addition, when testing a developed program to perform a specific task it is is essential that a simulation is first carried out in order to avoid the incorrect movement of large pieces of machinery which would result in dangerous operating conditions. Most PLCs contain a test-mode facility where outputs can be simulated by LED indication for a switched input requirement. The process is simulated by the manual manipulation of switches at specific times. The effects of the control program on the process may then be observed by the status of the PLC LED indicators on the I/O modules. If necessary, the program can be edited in the program mode and then retested prior to connecting to the actual system which is to be controlled.

3.8.6 Networking of PLCs

Computer-integrated manufacturing incorporates the highest level of automation economically feasible from each computer-based activity within a complete system. The overall control task is divided into a number of individual routines assigned to specific control elements such as CNC machine tools, robots and PLCs. The PLC thus constitutes one of the building blocks of the automated factory pyramid concept. This requires a suitable network to enable the various elements to exchange information and utilize a control database.

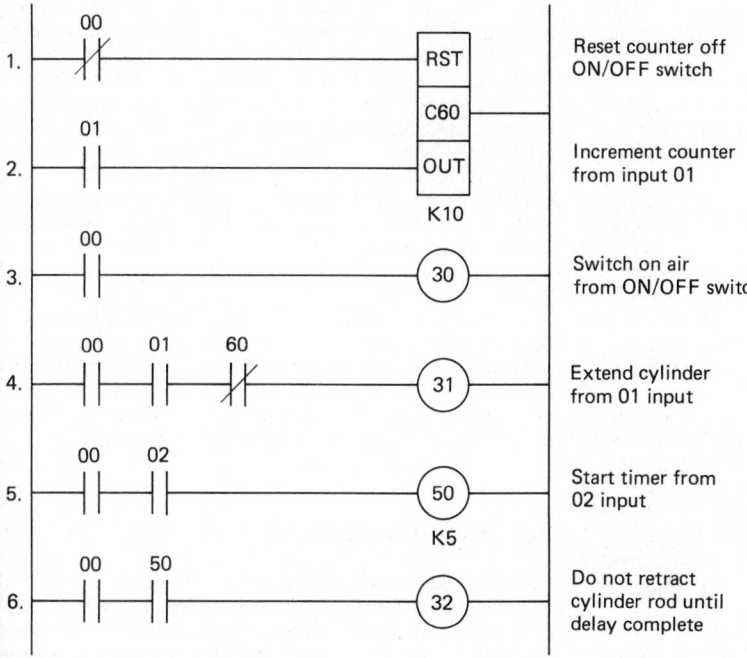

Figure 3.135 Ladder diagram for a specified example

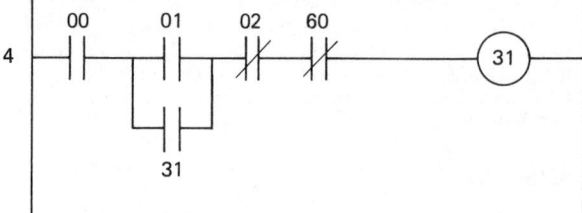

Figure 3.136 Latching of output

The manufacturers of PLCs have addressed the problem of networking, but the major disadvantage is that each manufacturer adopts their own standards. Communicating with a variety of other control devices is not a strength of PLC networks due to the number of data-transmission protocols and data formats adopted. However, most manufacturers provide a dedicated Local Area Network (LAN) for communication between controllers within their own product range and examples are given in Table 3.8.

Table 3.8 Typical PLC networks

Manufacturer	Network
Allen Bradley	Allen Bradley Data Highway
Festo	PC-IFS
General Electric	GE Net Factory LAN
Mitsubishi	Melsec-NET
Texas Instruments	TI-WAY

The basic concepts relating to the standards of communication in digital systems are covered in Section 3.3. In a decentralized network structure the sensing, actuating and processing associated with each controlled element in a system is connected to a supervisory or master computer. This evaluates signals supplied by each component and returns the necessary control requirements. A PLC, PC, single-board or single-chip computer could be used as the controller for the controlled element and a process computer, PC, minicomputer or mainframe can be employed in the supervisory role.

3.8.6.1 Network structures

Although there are various methods adopted for the networking of computer-based products they all possess certain common features. Each element must be equipped with a suitable interface associated with the LAN chosen and each element in the system must be connected into the network by means of wire or fibre-optic cabling to transmit the data from one station to another. Software is also necessary to handle all data transfer within the system correctly.

The physical arrangement of the elements or stations on the network are usually of a 'star' or 'ring' pattern. In the star-shaped network, as shown in Figure 3.137, the stations are connected on a line parallel to one another and connected to a central computer referred to as the 'file server'. With this method, each user station must decide on whether data sent by the file server are for itself or another station. The cable length is limited in this system, but signal amplifiers can be used if required.

In the ring structure as shown in Figure 3.138 the file server master computer transmits data to the first station on the network. These data are checked, evaluated and passed on to the second station if not required. The data are hence tran-

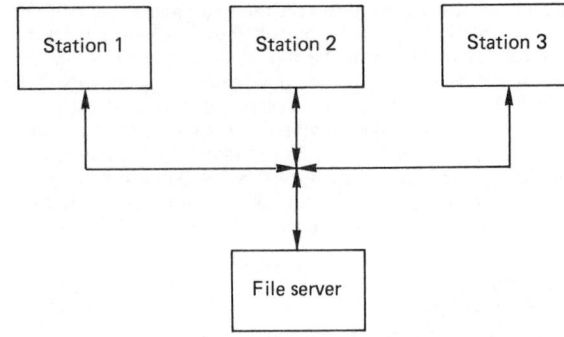

Figure 3.137 A star-shaped network

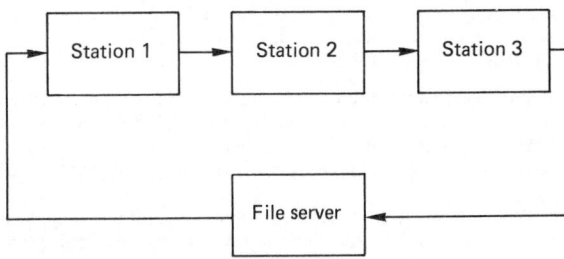

Figure 3.138 A ring network

smitted from one station to another until the user for whom they are intended is found. Data can therefore be passed around great distances but the failure of any one station causes the system to break down.

The integration of all intelligent elements into an automation system requires that all devices communicate with each other. Due to the different suppliers having different communication specifications, international communication standards based on the Open System Interconnection Model (OSI) have evolved. Widespread adoption of such standards will make it easier and cheaper to link devices together.

One such standard is the Manufacturing Automation Protocol (MAP), which was initiated by General Motors to integrate all levels of control systems such as PLCs, robots, welding systems, vision systems, etc, irrespective of the manufacturer. Another very popular LAN standard adopted for industrial and commercial application is ETHERNET, which was created by Digital Equipment Corporation, Intel and Xerox.

Users of automation systems will make their purchasing decisions based on the amount of software packages and support services that the distributed control system vendors can provide.[23] Only those vendors who are capable of supplying such products and services to provide a complete solution to a control problem will be successful in the future.

3.9 The z-transform

As outlined in Section 3.7, the availability of inexpensive and continually improving microprocessors has led to their application to direct digital control techniques in the closed-loop feedback control of processes. With the introduction of these sampled-data systems mathematical techniques have been developed specifically to provide a design method for digital controllers. The technique generally employed is similar to the Laplace transform used in continuous systems.

The Laplace transform, through the s operator, is used to transform linear differential equations into a linear algebraic form which can then be appropriately manipulated to produce open- and closed-loop transfer functions for a system. This is illustrated in Section 3.6.3 and the method forms a basis for the design of continuous controllers.

In a discretely sampled system the z-transform is used to provide an algebraic way of representing a sequence of digital data with respect to time measurements. The symbol z for the digital system is analogous to s in the continuous system. The z-transform is associated with the time shifting in a difference equation as s is associated with differentiation in a differential equation.

The operators may be mathematically defined as follows:

$s = \mathrm{d}/\mathrm{d}t = j\omega$

$z = \mathrm{e}^{sT}$

where T is the discrete time sampling interval and z^{-1} is a delay operator which represents a delay of one sampling period.

3.9.1 Representation of discretely sampled data

The characteristics of a digital control loop are related to the sampling process associated with the conversion of transducer analogue data into a digital format and the converse between the controller output and the system being controlled. The sequence of events are as follows:

1. The process variable (PV) is measured and the corresponding analogue signal is converted into a digital value using an A/D converter.

2. This value is compared with the declared set point (SP) to give on error value (E) on which the controller transfer function operates to produce a control effort (U) as a numerical value preferably expressed as a percentage within the range 0–100%.
3. The control effort derived is then converted by means of a final control element into the actual power supplied to the plant or process. This element invariably involves a D/A converter which holds the value of the impulse received until the next value arrives in a staircase fashion. This is referred to as a 'zero-order-hold' device.

The overall time for the sequencing cycle is the sampling frequency

$f_\mathrm{s} = 1/T$

where T is the sampling interval.

It is most important to appreciate the physical nature of the variables at each stage in the loop and this is illustrated in Figure 3.139. The discretely sampled data associated with the process variable, PV, and its related derived values of error and control effort are displayed at time intervals of T. These are suffixed by an asterisk to indicate that they are snapshots of the plant state at a particular time. The z-transform provides an algebraic way of mathematically presenting such sequences of digital information.

Consider in detail the variation in the derived error signal, E, as shown in Figure 3.140. Numerical values, quoted as percentages of the range of the measuring transducer, are inserted to illustrate the mathematical form of the z-transform equation to represent the variation of E with time. The sequence is shown as:

$E(n) = 80, 60, 40, 25, 12, 6, 0, 0, \ldots$

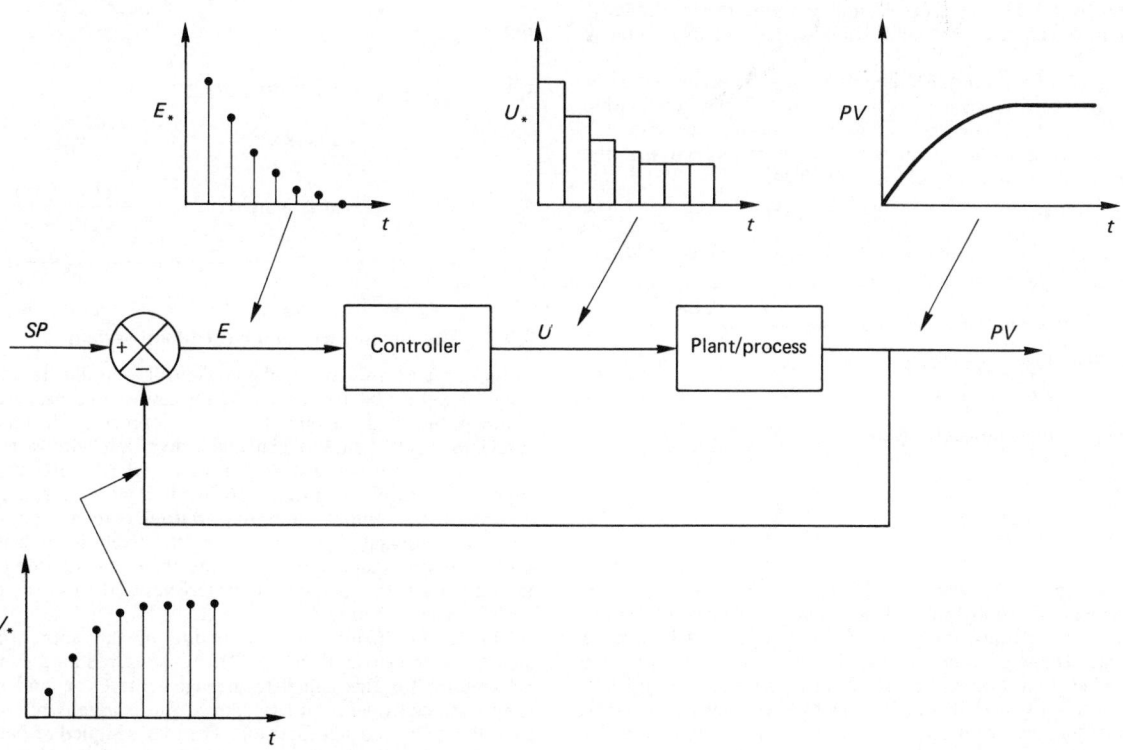

Figure 3.139 Nature of the variables in a digital control loop

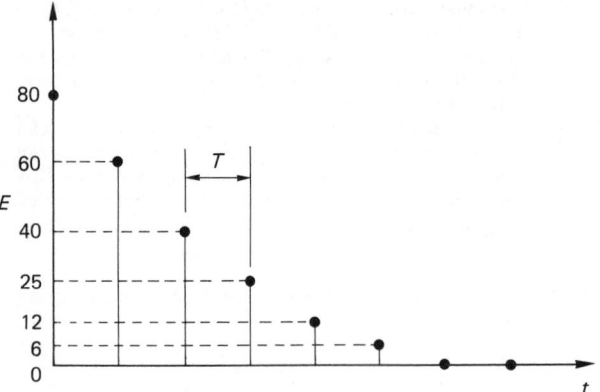

Figure 3.140   Variation of discrete error values with time

and the corresponding z-transform of the sequence is defined by the sum

$$E(z) = 80 + 60z^{-1} + 40z^{-2} + 25z^{-3} + 12z^{-4} + 6z^{-5}$$

This method of representation is particularly useful for handling time delays. For example, if the sequence was delayed by one sampling interval then it becomes

$$0, 80, 60, 40, 25, 12, 6, 0, 0, \ldots$$

and the corresponding z-transform

$$80z^{-1} + 60z^{-2} + 40z^{-3} + \ldots$$

which is the first sequence multiplied by z^{-1}. Thus multiplication by z^{-1} denotes a one sampling period delay. Similarly, multiplication by z^{-2} would denote a delay of two sampling periods.

Generally, z-transform models of digitally sampled data or its derivatives can be obtained provided the relationship between the continuous variation of the data with time is known in a mathematical form. For example, say that the error value can be expressed in the form

$$E(t) = 100e^{-t}$$

which for the data sampled at intervals of T would yield:

$$E(z) = 100(1 + e^{-T}z^{-1} + e^{-2T}z^{-2} + e^{-3T}z^{-3} + \ldots)$$

This is the sum of a geometric series given by

$$SUM = 100/(1 - e^{-T}z^{-1})$$
$$= 100z/(z - e^{-T})$$

which is the z-transform $E(z)$.

If the data were sampled at a frequency of 10 Hz (i.e. $T = 0.1$ seconds) then

$$E(z) = \frac{100z}{(z - 0.905)}$$

In order to aid the design process associated with digital techniques, z-transforms of common sampled functions are tabulated in a way similar to Laplace transforms. Examples of some common time functions with their corresponding Laplace transform and z-transforms are given in Table 3.9.

When applying digital techniques the z-transform, $G(z)$, must usually be derived from the corresponding transfer function for the operation expressed in terms of s, $G(s)$. It should be noted that it is not simply a case of substituting z for

s in the function. The basic approach is generally to obtain the partial fraction expansion of $G(s)$ to yield individual components which translate directly into z-transforms using a standard table such as Table 3.9.

For example, consider the transfer function $G(s)$ given by

$$G(s) = 1/s(1 + 2s)$$

Although this form appears explicitly in Table 3.9 it can be used to illustrate the technique of partial fraction expansion. Using partial fractions it can be assumed that

$$G(s) = A/s + B/(1 + 2s)$$

from which $A = 1$ and $B = -2$. Hence $G(s) = 1/s - 2/(1 + 2s)$. Rearranging the second term to conform to the appropriate Laplace transform in the table gives

$$G(s) = 1/s - 1/(s + 0.5)$$

The constituent parts produce the z-transform from the table as

$$G(z) = z/(z - 1) - z/(z - e^{-0.5T})$$

and after algebraic manipulation gives

$$G(z) = z(1 - e^{-0.5T})/(z - 1)(z - e^{-0.5T})$$

This is as expected from the Laplace transform, $a/s(s + a)$, in Table 3.9.

Table 3.9   Table of z-transforms

Time function	Laplace transform	z-transform
Unit step input	$1/s$	$z/(z - 1)$
t (ramp input)	$1/s^2$	$Tz/(z - 1)^2$
$t^2/2$	$1/s^3$	$\dfrac{T^2 z(z + 1)}{2(z - 1)^3}$
e^{-at}	$1/(s + a)$	$z/(z - e^{-aT})$
te^{-at}	$1/(s + a)^2$	$\dfrac{Tze^{-aT}}{(z - e^{-aT})^2}$
$1 - e^{-at}$	$a/s(s + a)$	$\dfrac{z(1 - e^{-aT})}{(z - 1)(z - e^{-aT})}$

3.9.2   The z-transform of a closed-loop system

Before z-transforms for the digital control loop illustrated in Figure 3.139 can be derived, it is first necessary to consider the nature of the input and output to the plant/process. The input, associated with digital-to-analogue conversion hardware, is basically a 'zero-order-hold' element which effectively supplies a control effort to the process in a pulsed form. The nature of the output from the controller relates to the analogue-to-digital conversion which effectively is indicative of the discrete sampling process. Thus the input is a step and the output sets the sampling characteristics of the loop on the basis of a sample time T.

This arrangement, with the nature of the appropriate signals, is shown in Figure 3.141. What is required is the z-transform for this complete arrangement, $G(z)$, and not simply the equivalence of $G(s)$ alone. This is usually referred to as the 'pulse transfer function'. The total sampled response consists of the difference between the positive stepped input to the process and the same value delayed by one sample period

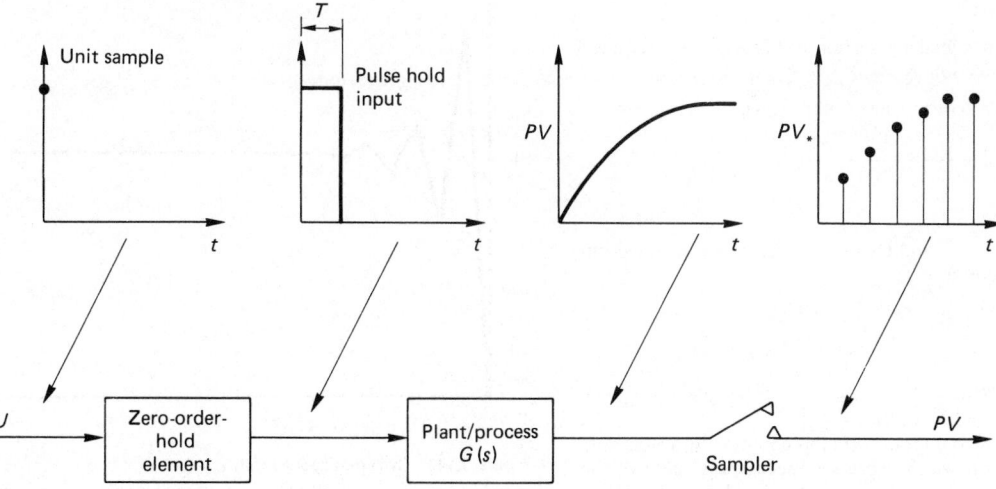

Figure 3.141 Discrete sampling in a control loop

This will give the required pulse transfer function, and it can be written as

$$G(z) = \frac{1}{s} \cdot G(s) - e^{-sT}\frac{1}{s} G(s)$$

$$= (1 - z^{-1}) * \text{ the } z\text{-transform of } \left(\frac{1}{s} \cdot G(s)\right)$$

or

$$G(z) = (z - 1)/z * \text{ the } z\text{-transform of } \left(\frac{1}{s} G(s)\right) \qquad (3.57)$$

which is the z-transform equivalence of all the elements shown in Figure 3.141.

It should be noted that the discrete sampling process effectively introduces a phase lag into the system with its associated inherent stability problems. The closed-loop transfer function in terms of the z-transform is derived in a way similar to that for a continuous system, with $G(z)$ representing the pulse transfer function associated with the hold, plant/process and sampling routine.

Consider the basic loop as shown in Figure 3.142. The forward path transfer function is

$$PV(z) = G(z)U(z)$$

$$= G(z)C(z)E(z)$$

and the closed-loop transfer function is

$$PV(z)/SP(z) = (C(z)G(z))/(1 + C(z)G(z)) \qquad (3.58)$$

which has a form similar to that for the transfer function of a system with a continuous feedback control loop.

The procedure for evaluating the response of a plant/process to a step input with digital control using a sample time of T is as follows:

1. Obtain the pulse transfer function $G(z)$ from the known form for $G(s)$:

   $$G(z) = (z - 1)/z * \text{ the } z\text{-transform of } [G(s)/s]$$

2. Express the controller transfer function $C(s)$ in z-transform format.
3. Evaluate the closed-loop transfer function for $PV(z)/SP(z)$.
4. Algebraically manipulate this expression to yield a discrete time-ordered equation which formulates the current value of PV (say, PV_i) in terms of the past values of PV and SP. This is obtained using the relationships:

   $$PV(z)z^{-1} = PV_{i-1} \text{ where } i - 1 \text{ denotes one period back in time}$$

   $$PV(z)z^{-2} = PV_{i-2}$$

   etc.

3.9.3 Proportional control using digital techniques

For a proportional control strategy as outlined in Section 3.7.5 the controller transfer function $C(s)$ and the corresponding z-transform is simply the gain, K. Taking as an example the plant transfer function to be that for a first-order system of the form

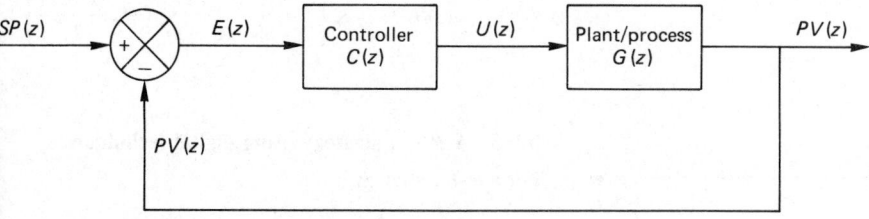

Figure 3.142 Closed loop in terms of a z-transform

$G(s) = 1/(1 + \tau s)$

the procedure outlined in Section 3.9.2 gives the required pulse transfer function as

$G(z) = (z - 1)/z *$ the z-transform of $[1/s(1 + \tau s)]$

$$= \frac{(z - 1)}{z} \left[\frac{z}{(z - 1)} \cdot \frac{(1 - e^{-T/\tau})}{(z - e^{-T/\tau})} \right]$$

$$= C_1/(z - C_2)$$

where $C_1 = 1 - e^{-T/\tau}$ and $C_2 = e^{-T/\tau}$. Hence the z-transform for the closed loop is

$$PV(z)/SP(z) = (C(z)G(z))/(1 + C(z)G(z))$$
$$= ((KC_1)/z - C_2))/1 + KC_1/(z - C_2))$$
$$= a/(z + b)$$

where $a = KC_1$ and $b = (KC_1 - C_2)$.

It is now necessary to transform this equation into a discrete time-stepped form which expresses the current value of the process variable, PV_i, as a function of past values. This can be obtained by dividing the numerator and denominator by z to the power corresponding to the maximum value in the denominator. In this example the dividing factor is simply z, i.e.

$$PV(z)/SP(z) = az^{-1}/(1 + bz^{-1})$$

or

$$PV(z) = aSP(z)z^{-1} - bPV(z)z^{-1}$$

Writing this as a time-difference equation yields

$$PV_i = aSP_{i-1} - bPV_{i-1} \tag{3.59}$$

For a step input, $SP = 1$ for $i > 1$ and starting values are, say,

$i = 1; t = 0; SP = 1$ and $PV = 0$

Equations of this form are easily programmed on a computer to produce a graphical time response of PV to a step input for declared values of system time constant τ, controller gain K and sample time T.

Figures 3.143 and 3.144 illustrate the effect of the sample time, T, on the predicted response of the system using the z-transform technique outlined with $\tau = 1$ and $K = 10$. These are clearly illustrative of the introduction of a delay into the system due to the incorrect sampling of the process, and leads to a transient behaviour or, eventually, an unstable condition. The choice of a satisfactory sampling interval, T, is covered in Section 3.7.4, but it should be noted that it is dependent upon

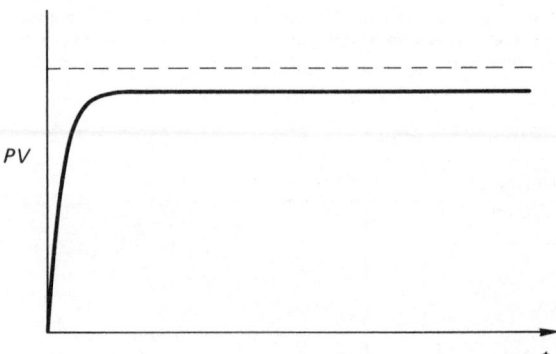

Figure 3.143   System response with $T = 0.01$ s

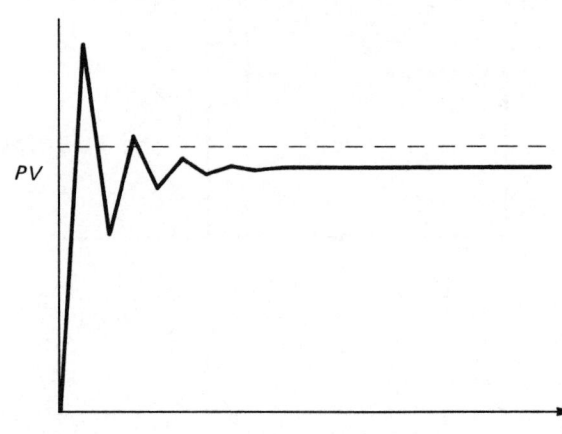

Figure 3.144   System response with $T = 0.15$ s

the system time constant and the controller settings. Hence one advantage of the z-transform technique in obtaining a simulation of the performance of a mathematically modelled system is the ability to investigate the effect of the sample rate for chosen controller settings.

3.9.4   The z-transform for a PID controller

The PID control law for a continuous controller as outlined in Section 3.7.5 is

$$U = K[E + 1/T_i \int E dt + T_d dE/dt]$$

The integral term can be approximated numerically as a summation of rectangular elements which at any time t can be expressed as

$$I_i = I_{i-1} + E_i T$$

where I_{i-1} is the value of the integral term up to the instant $t - T$, or in z-transform notation:

$$I(z) = I(z)z^{-1} + E(z)T$$

i.e. $I(z) = (Tz/(z - 1))E(z)$

Similarly, the derivative term can be approximated as the slope of the line joining the current E value and the previous value, i.e.

$$D_i = (E_i - E_{i-1})/T$$

or $D(z) = (E(z) - E(z)z^{-1})/T$

i.e. $D(z) = [(z - 1)/Tz]E(z)$

The integral and derivative terms in z-transform notation can now be substituted into the full three-term expression to give $C(z)$:

$$C(z) = U(z)/E(z) = K \left[1 + \frac{1}{T_i} \cdot \frac{Tz}{(z - 1)} + T_d \frac{(z - 1)}{Tz} \right]$$

$$\tag{3.60}$$

3.9.5   A $P + I$ strategy using digital techniques

For a $P + I$ strategy

$$C(z) = K \left[1 + \frac{1}{T_i} \cdot \frac{Tz}{(z - 1)} \right]$$

Applied to a first-order system, $G(s) = 1/(1 + \tau s)$ gives a corresponding pulse transfer function of

$$G(z) = C_1/(z - C_2)$$

with $C_1 = (1 - e^{-T/\tau})$ and $C_2 = e^{-T/\tau}$.

Substituting for $C(z)$ and $G(z)$ into the closed-loop transfer function and manipulating the result in a manner similar to that illustrated in Section 3.9.3 yields the time-sequence equation as:

$$PV_i = aSP_{i-1} - bSP_{i-2} + CPV_{i-1} + dPV_{i-2} \qquad (3.61)$$

where:

$$a = K(1 - e^{-T/\tau})(1 + T/T_i)$$

$$b = K(1 - e^{-T/\tau})$$

$$c = (1 + e^{-T/\tau}) - K(1 - e^{-T/\tau})(1 + T/T_i)$$

$$d = K(1 - e^{-T/\tau}) - e^{-T/\tau}$$

For $t < 0$ then $SP = 0$. At $t = 0$: $PV = 0$ and $SP = 1$ for a unit step input.

Figure 3.145 illustrates the response of a first-order system with $\tau = 1$ when a $P + I$ control strategy is applied using a gain K of 2 and an integral time T_i of 0.1. When the discrete sample time T is taken as 0.01 seconds, the result compares favourably with that for continuous control. If, however, the time step is increased to 0.35 seconds then the plant responds in a much more oscillatory manner for the same controller settings. This is illustrated in Figure 3.146. When T is increased beyond the value of 0.35 then the system behaves in an unstable manner. With digital control the sampling time, T, which would cause the system to go unstable depends upon the value of the system time constant and the controller settings.

3.9.6 Stability in discrete time systems

As seen in the previous sections, an incorrectly chosen sample time can result in a system becoming unstable due to the imposed delay. This initial value of T can be obtained by considering the representation of the poles of the pulse transfer function $G(z)$ in the z-plane.

If the closed-loop transfer function in z-transform notation is written in terms of a numerator $N(z)$ and a denominator $D(z)$ then the poles are defined as the roots of the equation $D(z) = 0$. These roots may contain a real and imaginary component and take the form $a + jb$ with the resulting plot termed the 'z-plane'. Systems whose poles are all inside the

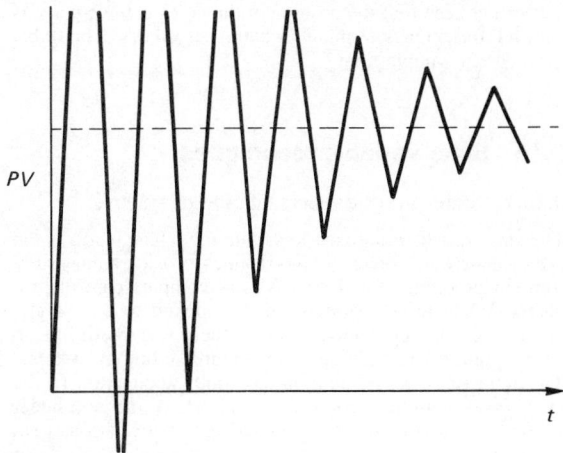

Figure 3.146 System response with $K = 2$, $T_i = 0.1$ s and $T = 0.35$ s

unit circle in the z-plane are stable and those with any poles outside it are unstable. The position of the poles within the unit circle are indicative of the degree of the resulting oscillatory motion which may result following a step input. A z-plane plot is shown in Figure 3.147.

When using the z-transform method of analysis for the digital control of a system the position of the poles in the z-plane are dependent upon the magnitude of the sample time taken. For example, consider a process having a transfer function $G(s) = 1/s$ and controlled using a proportional control strategy. Using the method outlined in Section 3.9.2 the corresponding pulse transfer function and closed-loop z-transfer function are, respectively,

$$G(z) = T/(z - 1)$$

and $PV(z)/SP(z) = KT/(z - 1 + KT)$

i.e. $D(z) = z - 1 + KT$

For the loop to be stable the poles of the real axis should lie in the range of $-1 < z < +1$, which means, by putting $z = -1$, that $KT = 2$ for the limit of stability. If $K = 5$ then the sample time T must be less than 0.4 for the system to remain stable.

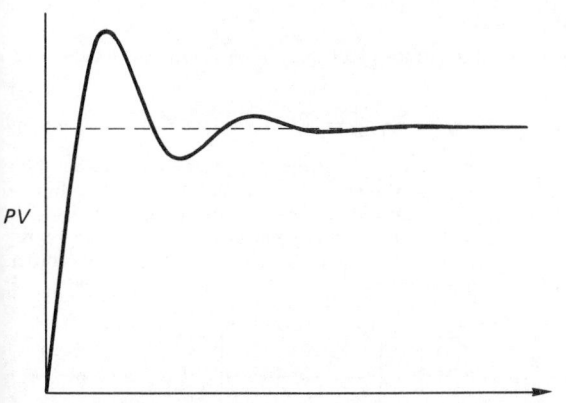

Figure 3.145 System response with $K = 2$, $T_i = 0.1$ s and $T = 0.01$ s

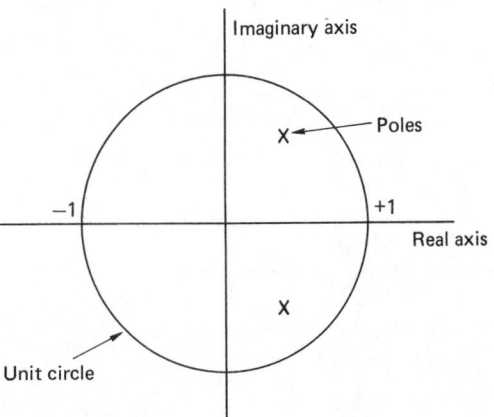

Figure 3.147 The z-plane

It can be seen that the position of the poles in the z-plane is a useful concept in determining whether a system will be stable for a chosen sample time.

3.10 State variable techniques

3.10.1 State variable representation of systems

The state variable approach to system modelling involves the use of matrix and vector methods which provide a consistent solution procedure for the analysis of complex control problems. While these methods can be applied to any single-input–single-output control system, their real power lie in their application to multiple-input–multiple-output systems. The state space representation also finds application to the analysis of non-linear control systems where the non-linear elements of a system are accounted for in a self-consistent manner.

In order to generate a system of state variable equations the system must be adequately defined in terms of the governing differential equations in a continuous system or of the difference equations in a discrete one. A convenient method of representing the mathematical operations and equations is by means of a block diagram made up of the appropriate sequence of operators. In a continuous control system the most common operators are shown in Figure 3.148. The simple damped spring and mass system serves as a useful illustrative example (see Figure 3.149).

The relation which governs the motion of the mass as a function of time is given as a second-order linear differential equation, i.e.

$$\frac{d^2y}{dt^2} + 2\xi\omega_n \frac{dy}{dt} + \omega_n^2 y = u \qquad (3.62)$$

or

$$\ddot{y} + 2\xi\omega_n\dot{y} + \omega_n^2 y = u \qquad (3.63)$$

Solving for the highest derivative gives

$$\ddot{y} = u - 2\xi\omega_n\dot{y} - \omega_n^2 y \qquad (3.64)$$

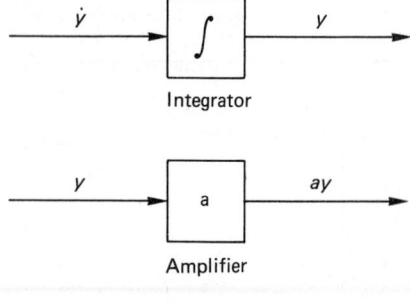

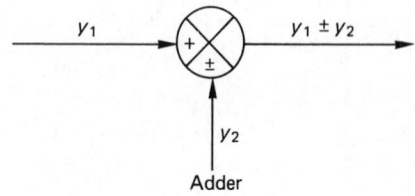

Figure 3.148 Mathematical operators in continuous-control systems

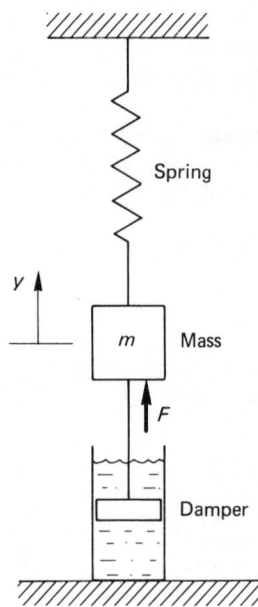

Figure 3.149 Simple damped spring and mass system

where u is the input forcing function.

The block diagram representation of equation (3.64) may be developed by first integrating y twice, as shown in Figure 3.150. The loop may then be closed by satisfying the requirement of equation (3.64) as shown in Figure 3.151.

The quantities y and $\dot{y}$ are called state variables. These are not unique and there is an endless range of possible combinations which may be chosen. It is normal, however, to select as state variables those quantities which are significant to the problem under consideration. In this respect the output, the first derivative of the output and any higher-order derivatives of the output would make the most suitable choice of state variables. In choosing the set of state variables it should be recognized that the state variables must be related through a differential or a difference equation. In a continuous-control system the normal practice is to select the output from any integrator as an appropriate state variable.

For the damped spring mass system the following state variables are selected:

$$X1 = y: X2 = \dot{y}$$

Consideration of the block diagram of Figure 3.151 shows that

$$\dot{X}1 = X2 \qquad (3.65)$$

$$\dot{X}2 = -\omega_n^2 X1 - 2\xi\omega_n X2 + u \qquad (3.66)$$

It transpires therefore that the state variable representation reduces the original second-order differential equation to two simultaneous first-order differential equations. Similarly, a third- or higher-order differential equation could be represented as a set of three or a higher number of first-order differential equations.

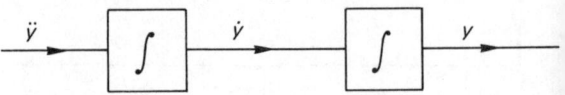

Figure 3.150 Double-integration process

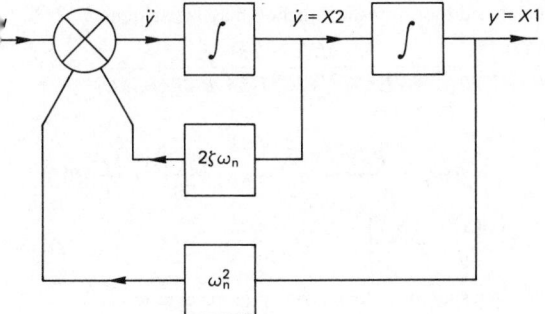

Figure 3.151 Block diagram representation of equation (3.64)

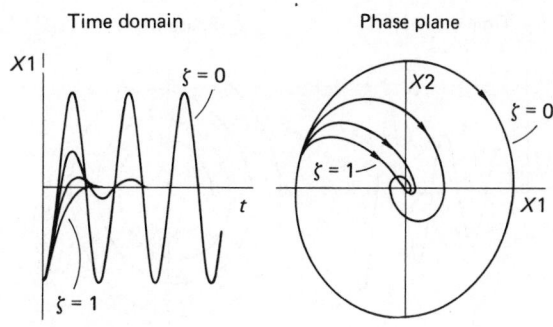

Figure 3.152 Time-domain and state-space representation of the damped spring mass system

Writing equations (3.65) and (3.66) in matrix notation gives

$$\begin{bmatrix} \dot{X}1 \\ \dot{X}2 \end{bmatrix} = \begin{bmatrix} 0 & 1 \\ -\omega_n^2 & -2\xi\omega_n \end{bmatrix} \begin{bmatrix} X1 \\ X2 \end{bmatrix} + u \begin{bmatrix} 0 \\ 1 \end{bmatrix} \tag{3.67}$$

and

$$y = \begin{bmatrix} 1 & 0 \end{bmatrix} \begin{bmatrix} X1 \\ X2 \end{bmatrix} + u[0] \tag{3.68}$$

Any suitable matrix method can be used to solve for $X1$ which is equivalent to the system output (i.e. the position of the mass as a function of time) for any given input u. Alternatively equations (3.65) and (3.66) may be written as simple finite difference approximations and solved simultaneously over an appropriate time increment. For free vibration following a step disturbance to the system, $u = 0$, the finite difference approximation may be written as

$$\Delta X2_i = \Delta t[-2\xi\omega_n X2_{i-1} - \omega_n^2 X1_{i-1}] \tag{3.69}$$

$$X2_i = X2_{i-1} + \Delta X2_i \tag{3.70}$$

$$\Delta X1_i = \Delta t[X2_i] \tag{3.71}$$

$$X1_i = X1_{i-1} + \Delta X1_i \tag{3.72}$$

With suitable values assigned to ξ and ω_n and initial values designated to $X1$ and $X2$ at time = zero, the solution can march forward in time using a suitably small time increment Δt. A plot of $X1$ (which is the system output) against time gives the solution to the equation in the time domain. The more usual state variable representation, however, is shown as a plot of the state variables $X2$ against $X1$. For the damped spring mass system considered, this state space plot is a record of the instantaneous values of the velocity of the mass against its position with time as parameter. Figure 3.152 shows a set of results for selected values of ξ ranging between 0 and 1, with initial conditions of $X1(0) = -0.25$ and $X2(0) = 0$.

The set of curves which result from the solutions of the state variable equations are referred to as trajectories in the state space. For a second-order system, the plot is more generally called the phase plane. The values of the state variables are said to represent the state of the system. With a knowledge of future inputs and the governing state space equations, the phase plane trajectory can be predicted and hence also the behaviour of the system in general. Figure 3.153 shows another set of phase-plane trajectories for the damped spring mass system. In these plots the system is subject to heavy damping and the trajectories start from a range of different initial conditions. Plots such as these are referred to as phase-plane portraits.

One further example is included where a negative damping factor of -1 has been imposed and the computation started

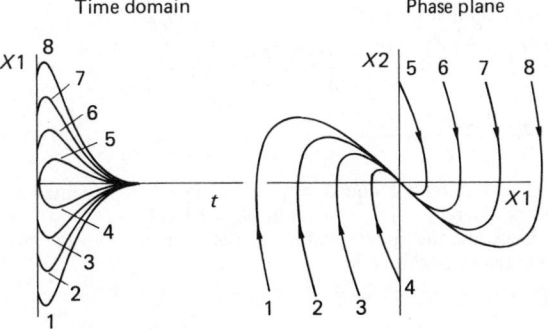

Figure 3.153 Phase-plane characteristic for the spring mass system with very heavy damping

from $X1(0) = -0.10$ and $X2(0) = 0$. The resultant phase-plane plot is shown in Figure 3.154.

Negative damping renders the system unstable and the amplitude of the oscillation increases with time as shown in the time-domain plot. The phase-plane plot indicates a trajectory which is spiralling outwards, and this is the general state-space representation of an unstable system.

Other information which may be easily gleaned from the phase-plane trajectory are an indication of whether the system reponse is over- or underdamped and the extent of any system overshoots.

3.10.2 Application to a first-order system with a P + I controller

Figure 3.155 shows the control block diagram for a process which has an open-loop transfer function which can be modelled as a first-order differential equation. The control effort, U, is determined through a controller which utilizes a combination of proportional and integral action as shown. The closed-loop transfer function may be written as

$$\frac{PV(s)}{SP(s)} = \frac{K[1 + 1/(T_i s)][k/(1 + \tau s)]}{1 + K[1 + (1/T_i s)][k/(1 + \tau s)]} \tag{3.73}$$

Some algebraic manipulation results in

$$\frac{PV(s)}{SP(s)} = \frac{(kK/\tau T_i)(1 + T_i s)}{s^2 + [(1/\tau) + (kK/\tau)]s + (kK/\tau T_i)} \tag{3.74}$$

Time domain   Phase plane

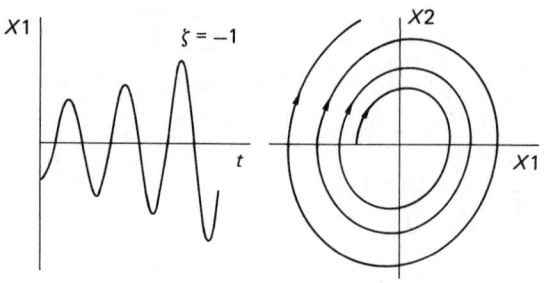

Figure 3.154 Phase-plane trajectory for the spring mass system with negative damping

Cross-multiplication gives

$$\frac{d^2(PV)}{dt^2} + \left[\frac{1}{\tau} + \frac{kK}{\tau}\right]\frac{d(PV)}{dt} + \frac{kK}{\tau T_i}(PV)$$
$$= \frac{kK}{\tau}\left[\frac{(d(SP))}{dt} + \frac{(SP)}{T_i}\right] \tag{3.75}$$

The differential equation (3.75) may be represented in block diagram form as shown in Figure 3.156.

Choosing the state variables as the outputs from the two integrators gives

$$X1 = (PV) : X2 = (P\dot{V})$$

Thus

$$\dot{X}1 = X2 \tag{3.76}$$

$$\dot{X}2 = \frac{-kK}{\tau T_i}X1 - \left[\frac{1 + kK}{\tau}\right]X2 + \frac{kK}{\tau}\left[S\dot{P} + \frac{SP}{T_i}\right] \tag{3.77}$$

The state variable equations can be written again as finite-difference approximations and a solution marched forward in time from specified initial conditions. The step response of the system for various settings of system gain, controller gain, integral time setting and system time constant are illustrated in Figure 3.157.

3.10.3 Application to a second-order system with a P + I controller

If the process to be controlled can be represented as a second-order differential equation then the open-loop transfer function for the process is

$$G(s) = \frac{k}{s^2 + 2\xi\omega_n s + \omega_n^2} \tag{3.78}$$

The closed-loop transfer function may be written as

$$\frac{PV(s)}{SP(s)} = \frac{kK(1 + T_i s)}{(T_i s^3 + 2\xi\omega_n T_i s^2 + \omega_n^2 T_i s + kK T_i s + kK)} \tag{3.79}$$

or

$$\frac{d^3(PV)}{dt^3} + 2\xi\omega_n \frac{d^2(PV)}{dt^2} + (\omega_n^2 + kK)\frac{d(PV)}{dt} + \frac{kK}{T_i}(PV)$$
$$= kK\left[\frac{d(SP)}{dt} + \frac{SP}{T_i}\right] \tag{3.80}$$

The corresponding state variable representation is

$$\dot{X}1 = X2 \tag{3.81}$$

$$\dot{X}2 = X3 \tag{3.82}$$

$$\dot{X}3 = -2\xi\omega_n X3 - (\omega_n^2 + kK)X2 - \frac{kK}{T_i}X1$$
$$+ kK\left[(S\dot{P}) + \frac{(SP)}{T_i}\right] \tag{3.83}$$

where $X1 = PV$, $X2 = (\dot{P}V)$ and $X3 = (\ddot{P}V)$.

The three ordinary differential equations involving the three state variables may be solved simultaneously and a phase-plane trajectory plotted as before. Strictly, the trajectory should be plotted, for this example, in three-dimensional state space. This would involve a plot of the trajectory in terms of the state variables $X1$, $X2$ and $X3$ with time as parameter. This would be difficult to plot and for still higher-order systems might be difficult even to visualize. Note, however, that the simple two-dimensional phase plane still has some significance and can be used as an indication of the system's general stability.

3.10.4 Non-linear system elements (method of isoclines)

In many real control systems there is often some form of non-linear element within the control loop which eliminates the use of linear theory to predict the closed-loop system response. In some cases, a linear approximation can be devised for the non-linear element but the approximation will always imply that there is some degree of error involved in the system model.

The phase plane represents one viable method of accounting for those non-linearities which cannot be either ignored or approximately linearized. It is not always necessary to plot the phase-plane trajectory in its entirety but rather to approximate the curve by estimating its slope at a number of locations in the phase plane. The approximate trajectory can then be fitted to match the known slopes in the phase plane. This is the so-called method of isoclines, which is best illustrated by an example.

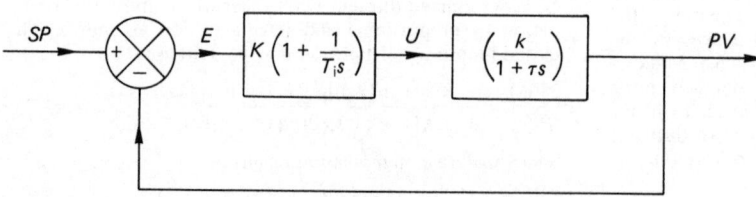

Figure 3.155 First-order system with P + I controller

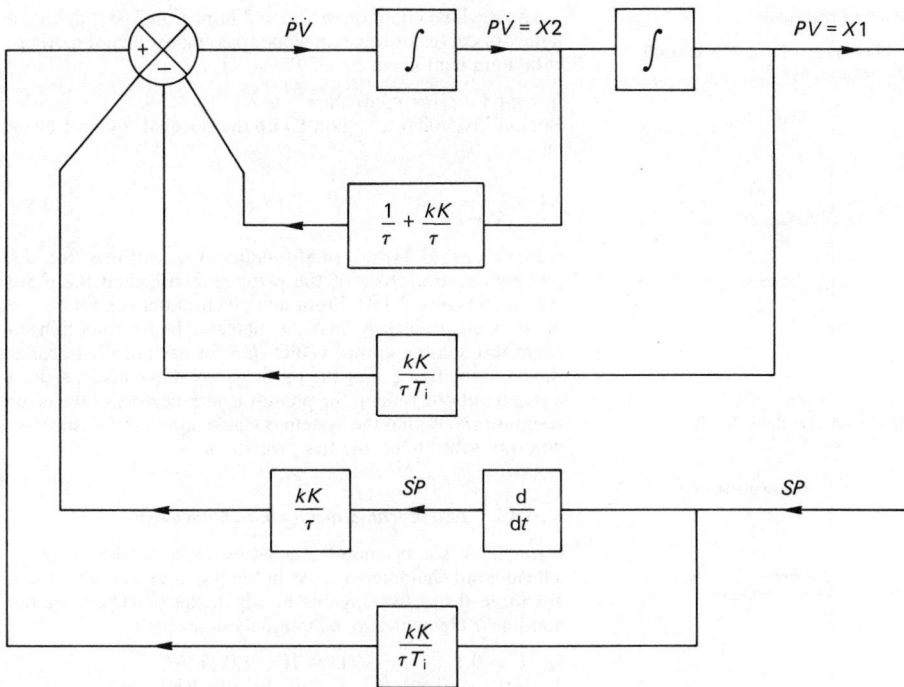

Figure 3.156 Block diagram representation of equation (3.75)

3.10.4.1 Saturation characteristics

Saturation can occur, for example, when an amplifier is operated outside its linear range. This non-linear effect can be depicted as shown in Figure 3.158.

There are three distinct regions over which the amplifier has a significantly different characteristic:

1. $E \geqslant 0.5$; $U = 0.5$
2. $-0.5 < E < 0.5$; $U = E$
3. $E \leqslant -0.5$; $U = -0.5$

In all regions

$$PV = 10U/[s(0.5s + 1)]$$

or $\quad 10U = (0.5s^2 + s)PV = (0.5s^2 + s)(SP - E)$

Thus

$$10U = \frac{1}{2}\frac{d^2(SP)}{dt^2} + \frac{d(SP)}{dt} - \frac{1}{2}\frac{d^2E}{dt^2} - \frac{dE}{dt}$$

For a step input at time $t > 0$,

$$\frac{d^2(SP)}{dt^2} = \frac{d(SP)}{dt} = 0$$

The equation reduces to

$$\frac{d^2E}{dt^2} + 2\frac{dE}{dt} + 20U = 0$$

In region (i), (U = constant = 0.5)
Thus

$$\frac{d^2E}{dt^2} + 2\frac{dE}{dt} + 10 = 0$$

Choosing as a suitable state variable, $X1 = dE/dt$,

$$\dot{X1} = -10 - 2X1$$

Defining the slop S as

$$S = \frac{\dot{X1}}{X1} = \frac{-10 - 2X1}{X1} = -2 - \frac{10}{X1}$$

Alternatively,

$$X1 = \frac{-10}{S + 2} \tag{3.84}$$

Equation (3.84) is the isoclinal equation which applies in the region $U > 0.5$. The isoclinal equation defines a set of curves along which the slope of the trajectory in the phase plane of dE/dt against E will be constant. In selecting a suitable range of values for S, the isoclines may then be plotted in the phase plane. Equation (3.84) shows that these isoclines will be manifested as a set of lines which would be parallel to the horizontal axis.

In region 2 (U = E)
The governing equation becomes

$$\frac{d^2E}{dt^2} + \frac{2dE}{dt} + 20E = 0$$

Again choosing $X1 = dE/dt$ gives

$$S = \frac{\dot{X1}}{X1} = \frac{-20E - 2X1}{X1} = \frac{-2 - 20E}{X1}$$

i.e. $X1 = \dfrac{-20E}{S + 2}$ $\tag{3.85}$

First-order system with a *P* + *I* controller

System time constant, $\tau = 1$ Controller integral time, $T_i = 1E + 09$
System gain, $k = 1$ Controller gain, $K = 2$

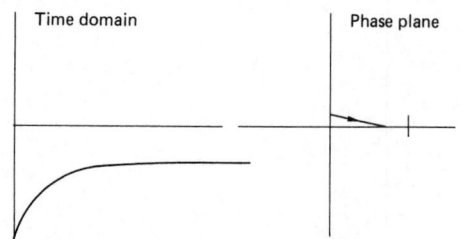

System time constant, $\tau = 1$ Controller integral time, $T_i = 0.1$
System gain, $k = 1$ Controller gain, $K = 2$

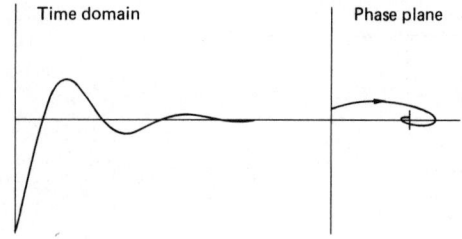

System time constant, $\tau = 1$ Controller integral time, $T_i = 0.01$
System gain, $k = 1$ Controller gain, $K = 2$

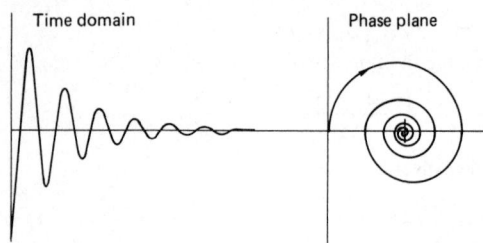

Figure 3.157 System response to a step input

The isoclinal equation in region 2 (equation 3.85)) defines a family of curves of constant slope, passing through the origin, with a gradient given by $-20/(S + 2)$.

In region 3 (U = constant = −0.5)
Region 3 is similar to region 1 with the isoclinal equation given by

$$X1 = \frac{+10}{S + 2} \tag{3.86}$$

Equations (3.84) to (3.86) define the isoclines for the phase-plane trajectory of the error response, and these are shown in Figure 3.159. From any given initial conditions the phase-plane trajectory may be patched in by smoothing a curve through the known values of S for each of the isoclines shown in the figure. For the particular system constants given a step input of 1 radian, the phase-plane trajectory of the error response shows that the system is stable and that the amplifier does not saturate on the first overshoot.

3.10.4.2   System with a dead-band characteristic

Figure 3.160 shows a non-linear system element which exhibits a dead-band characteristic. As in the previous example, there are three distinctive regions in which the behaviour of the non-linear element may be uniquely described:

1. $E > 0.1$, $U = 1(E - 0.1)$
2. $-0.1 < E < +0.1$, $U = 0$, i.e. the dead band
3. $E < -0.1$, $U = 1(E + 0.1)$

Following a similar approach to that adopted earlier in all regions

$$\frac{d^2E}{dt^2} + 2\xi\omega_n \frac{dE}{dt} + U\omega_n^2 = 0$$

Choosing the state variable as $X1 = dE/dt$,

$$X1 = \frac{-U\omega_n^2}{2\xi\omega_n + S} \tag{3.87}$$

In region 1 the isoclinal equation becomes

$$X1 = \frac{-\omega_n^2(E - 0.1)}{S + 2\xi\omega_n} \tag{3.88}$$

Equation (3.88) defines a family of isoclines as a set of constant slope lines with origin (+0.1,0). Region 3 is similar to region 1 and the isoclinal equation defines a family of constant

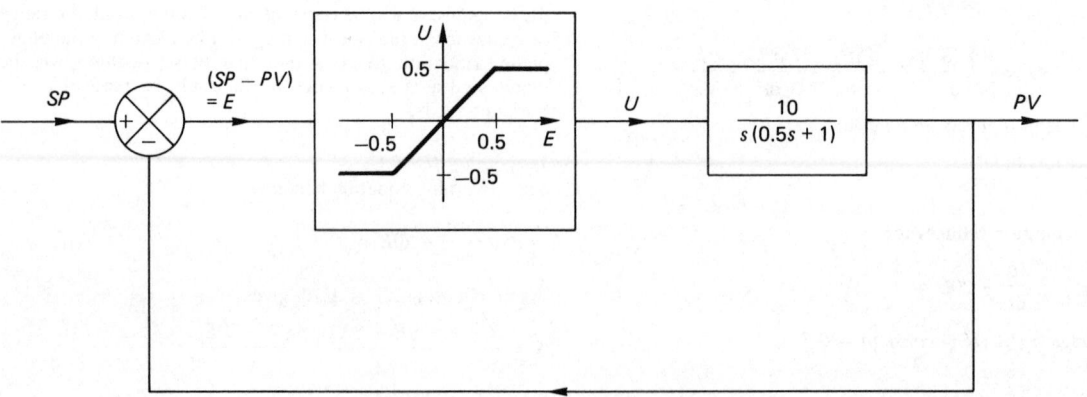

Figure 3.158 Position-control servo with saturation characteristic

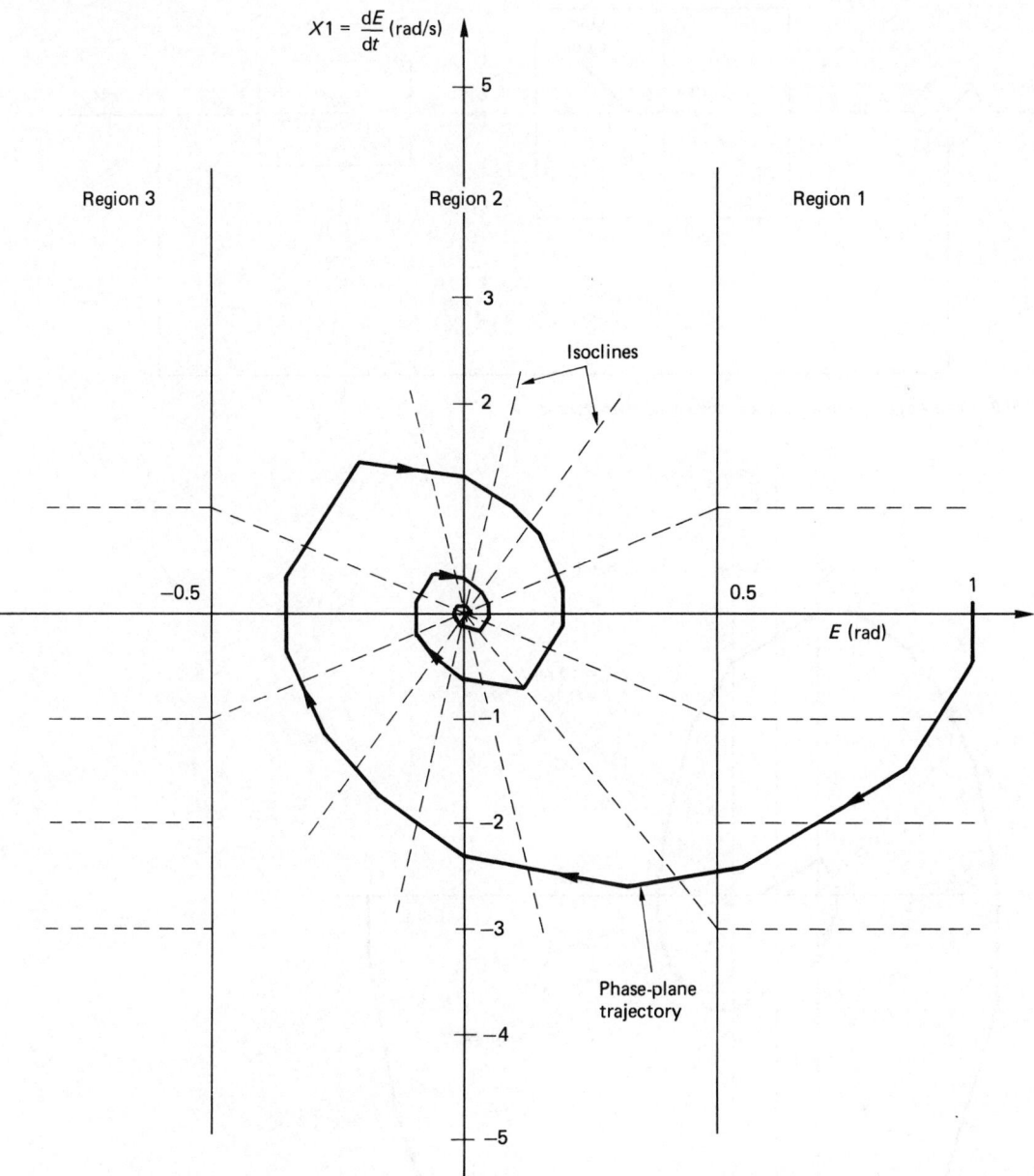

Figure 3.159 Phase-plane trajectory for saturation characteristic

slope with origin $(-0.1, 0)$. In the dead-band region (2) $U = 0$ and the slope becomes

$$\text{Slope} = S = \frac{-2\xi\omega_n X1}{X1} = -2\xi\omega_n$$

The resulting phase-plane trajectory takes the form shown in Figure 3.161.

Other system non-linearities may be handled similarly. These may include, for example, the on/off characteristics of a relay or hysteresis effects as typified by backlash in geared systems. The general representation and typical phase-plane trajectories for some of the common types of system non-linearities are shown in Figure 3.162.

The two examples which include hysteresis effects result in a phase-plane trajectory which traces out a limit cycle. Limit cycles are characterized by isolated closed curves in the phase plane and all stable near trajectories approach the limit cycle as time approaches infinity. The stable limit cycle corresponds to a stable periodic motion in the physical system. Limit cycles may also be unstable, in which case the near trajectories move away from the closed curve.

Unfortunately, the phase-plane approach is generally limited to unforced second-order systems. In theory, the

SP
+
−
E
U
Unity slope
−0.1
0.1
E
Unity slope
U
$\frac{\omega_n^2}{s(s + 2\zeta\omega_n)}$
PV
ω_n = 10 rad/s
ζ = 0.15

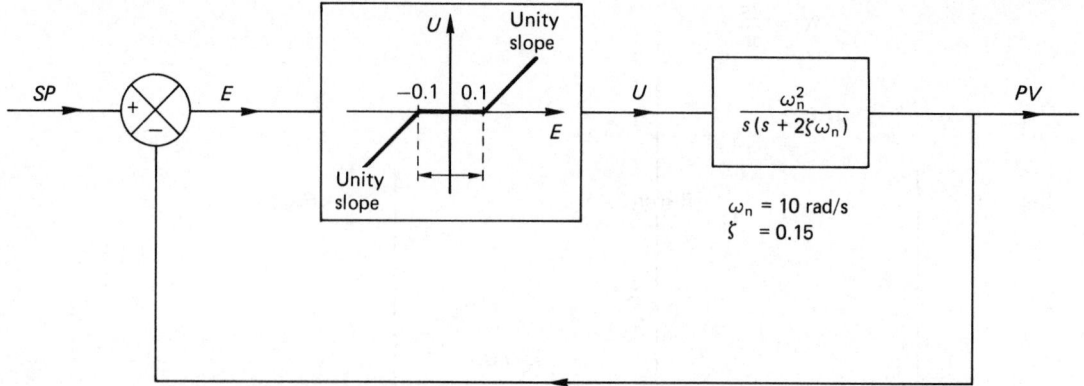

Figure 3.160 Control system with a non-linear dead-band characteristic

$\frac{dE}{dt}$ (rad/s)
4
Dead band
ω_n = 10 rad/s
ζ = 0.15
−0.5
0.5
1
E (rad)
Isoclines
Phase-plane trajectory
−7

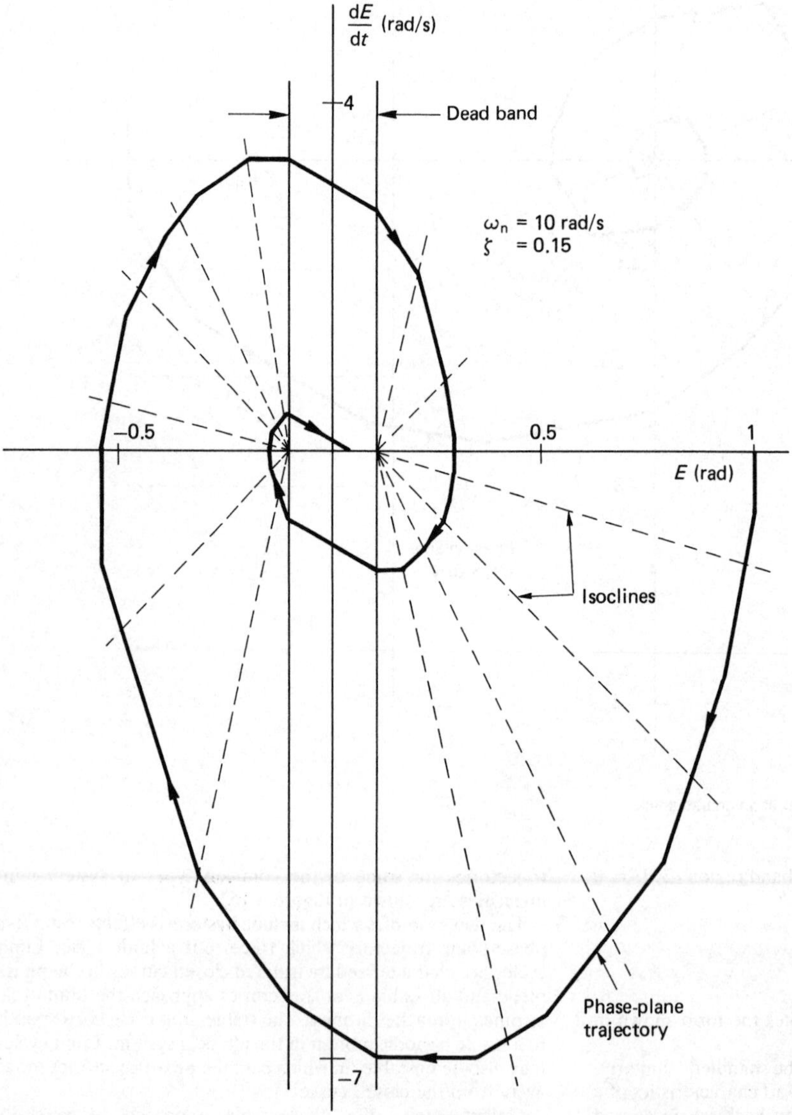

Figure 3.161 Phase-plane trajectory for a system with a dead-band characteristic

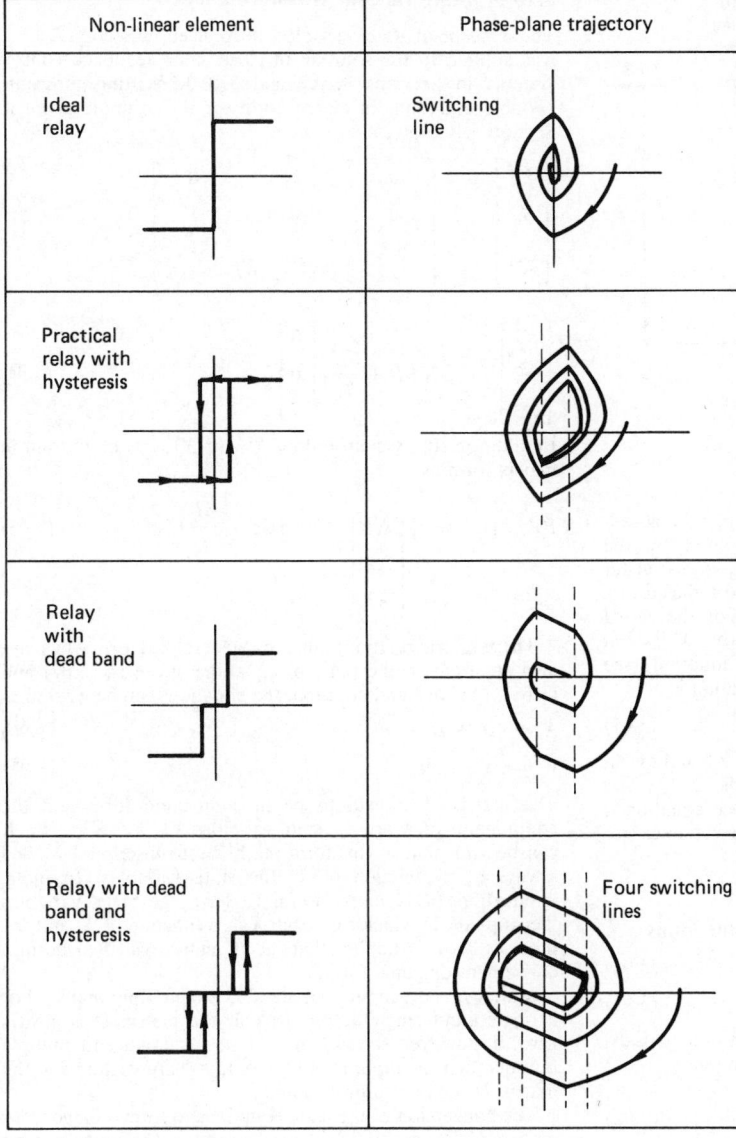

Figure 3.162 Common system non-linearities and corresponding phase-plane trajectories

phase-plane concepts may be extended to higher-order state space. This, however, has not been successful, due to the problems of determining, presenting and visualizing trajectories in n-dimensional state space.

3.10.5 Sampled-data systems

In a discretely sampled system the relationship between output and the input can be represented in terms of z-transforms or, ultimately, as a time-difference equation. In any system described by linear difference equations a block diagram can be similarly constructed using the basic building elements of the adder, the amplifier and the unit time delay. The unit time delay, which is analogous to the integrator in continuous systems, is depicted in Figure 3.163. In the figure the input to the unit delay appears at the output one period later, or delayed by a time T.

If the difference equation has, for example, the form given in equation (3.89) the block diagram for the system is represented as that shown in Figure 3.164, i.e.

$$y_i + ay_{i-1} + by_{i-1} = x_i \tag{3.89}$$

or $y_i = x_i - ay_{i-1} - by_{i-2}$

The system may also be handled in z-transform notation and a block diagram constructed to represent the system in terms of the appropriate operators.

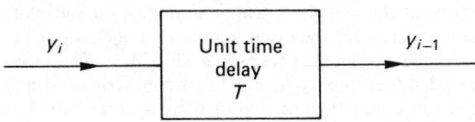

Figure 3.163 The unit time delay

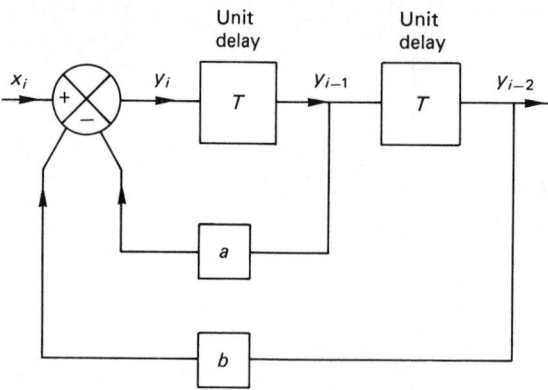

Figure 3.164 Block representation of equation (3.89)

Following the procedures outlined in Section 3.9, a closed-loop digital control system can be represented by the appropriate z-transform operators. The closed-loop transfer operator may subsequently be manipulated and reduced to a finite-difference or time-sequence equation. For the $P + I$ control strategy used in conjunction with a process which can be represented as a first-order differential equation (see Section 3.9.5) the resulting time-sequence equation is

$$PV_i = aSP_{i-1} - bSP_{i-2} + cPV_{i-1} + dPV_{i-2} \qquad (3.61)$$

where a, b, c and d are as defined in Section 3.9.5 and i is an integer variable used to denote the time periods.

The block diagram to represent the difference equation is shown in Figure 3.165. The appropriate state variables are chosen as

$$X1 = PV_{i-2}: X2 = PV_{i-1}$$

The state variable equations can be written in the forms

$$X1_{k+1} = X2_k \qquad (3.90)$$

$$X2_{k+1} = aSP_k - bSP_{k-1} + cX2_k + dX1_k \qquad (3.91)$$

where k is an integer variable used to denote the time periods. In matrix notation the state variable representation is

$$\begin{bmatrix} X1_{k+1} \\ X2_{k+1} \end{bmatrix} = \begin{bmatrix} 0 & 1 \\ d & c \end{bmatrix} \begin{bmatrix} X1_k \\ X2_k \end{bmatrix} + (aSP_k - bSP_{k-1}) \begin{bmatrix} 0 \\ 1 \end{bmatrix} \qquad (3.92)$$

$$PV_i = \begin{bmatrix} 0 & 1 \end{bmatrix} \begin{bmatrix} X1_{k+1} \\ X2_{k+1} \end{bmatrix} + (aSP_k - bSP_{k-1})[0] \qquad (3.93)$$

With any given initial conditions, equations (3.90) and (3.91) may be solved simultaneously for discrete time steps in T. Figure 3.166 shows the response to a step input when $K = 2$, $T_i = 0.1$ and T is set at 0.2 s. The phase-plane trajectory is shown in two forms in the figure. In one form the plot is given of $X2$ against $X1$, these being the chosen state variables. This trajectory shows that the system is stable under the imposed conditions. An alternative representation is shown where the parameter $(X2 - X1)/T$ is plotted against $X2$. In terms of the actual control system, this second plot is equivalent to a graph of the discrete time rate of change of the process variable against the process variable itself. The second plot is included therefore to give a direct comparison with the equivalent continuous system outlined in Section 3.10.2. In either plot, however, the stability of the system is adequately depicted in graphical form.

3.10.6 State variable transformations

The concept of state variables, introduced in Section 3.10.1, was applied to the solution of closed-loop feedback control systems in Sections 3.10.2 and 3.10.3. Writing the state variable equations in matrix form for the example given in Section 3.10.3 gives

$$\begin{bmatrix} \dot{X}1 \\ \dot{X}2 \\ \dot{X}3 \end{bmatrix} = \begin{bmatrix} 0 & 1 & 0 \\ 0 & 0 & 1 \\ \dfrac{-kK}{T_i} & -(\omega_n^2 + kK) & -2\xi\omega_n \end{bmatrix}$$

$$\begin{bmatrix} X1 \\ X2 \\ X3 \end{bmatrix} + kK\left(\dot{S}P + \dfrac{SP}{T_i}\right) \begin{bmatrix} 0 \\ 0 \\ 1 \end{bmatrix} \qquad (3.94)$$

In addition, the system output PV, or $X1$, can be written in matrix form as

$$PV = \begin{bmatrix} 1 & 0 & 0 \end{bmatrix} \begin{bmatrix} X1 \\ X2 \\ X3 \end{bmatrix} + kK\left(\dot{S}P + \dfrac{SP}{T_i}\right)[0] \qquad (3.95)$$

These matrix relations fall into a generalized form which are also apparent in the pairs of equations given as (3.67) and (3.68). In shorthand notation, the equations can be written as

$$\dot{X} = AX + Bu \qquad (3.96)$$

$$y = CX + Du \qquad (3.97)$$

The matrix A is said to be in companion form and the components of X are the state variables $X1$, $X2$, $X3$,...X_n. It can be seen that in this form the highest subscripted $\dot{X}_n$ will always be a function of all the state variables. In many applications it is more useful to have the state variables 'decoupled', in which case each $\dot{X}_i$ is a function of X_i and the input only. This implies that the A matrix will then be in a convenient diagonal form.

Matrix B is the input matrix and C is the output matrix. For a single-input–single-output system, the matrix D is always zero. If, however, the system has multiple inputs and multiple outputs then the input u and the output y are vectors and the matrix D then has a finite value.

The conversion of the state equations to form a diagonal A matrix can be accomplished using a variety of techniques. One method is included here for example.

Consider the transformation where

$$X = Mq \qquad (3.98)$$

Substituting this transformation into equations (3.96) and (3.97) gives

$$M\dot{q} = AMq + Bu$$

and $y = CMq$

or $\dot{q} = M^{-1}AMq + M^{-1}Bu$

i.e. $\dot{q} = A_1q + B_1u \qquad (3.99)$

and $y = C_1q \qquad (3.100)$

where $A_1 = M^{-1}AM$, $B_1 = M^{-1}B$ and $C_1 = CM$.

The procedure involved in transforming the state equations are best illustrated in a simple example. Suppose a particular dynamical system can be represented by the differential equation

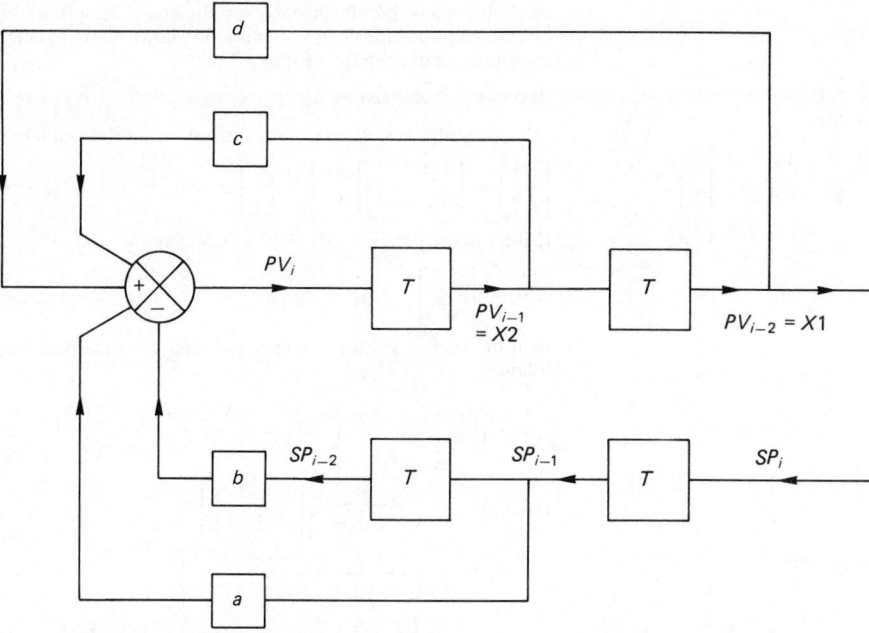

Figure 3.165 Block diagram representation of equation (3.61)

First-order system with a $P + I$ controller

System time constant, $\tau = 1$ Controller integral time, $T_i = 0.1$
System gain, $k_p = 1$ Controller gain, $K = 2$

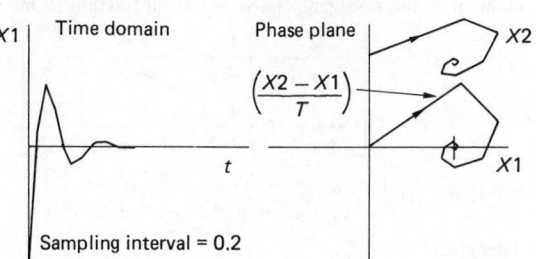

Sampling interval = 0.2

Figure 3.166 System response to a step input

$$\dddot{y} - 2\ddot{y} + \dot{y} - 2y = u \qquad (3.101)$$

The state equations in companion form are

$$\begin{bmatrix} \dot{X1} \\ \dot{X2} \\ \dot{X3} \end{bmatrix} = \begin{bmatrix} 0 & 1 & 0 \\ 0 & 0 & 1 \\ 2 & -1 & 2 \end{bmatrix} \begin{bmatrix} X1 \\ X2 \\ X3 \end{bmatrix} + \begin{bmatrix} 0 \\ 0 \\ 1 \end{bmatrix} u \qquad (3.102)$$

and

$$y = \begin{bmatrix} 1 & 0 & 0 \end{bmatrix} \begin{bmatrix} X1 \\ X2 \\ X3 \end{bmatrix} \qquad (3.103)$$

The characteristic equation of A is

$$[\lambda I - A] = \lambda^n + a_{n-1}\lambda^{n-1} + a_{n-2}\lambda^{n-2} + a_1\lambda^1 + a_0 = 0 \qquad (3.104)$$

where I is the unit matrix, λ_i are the eigenvalues of A and a_i are the coefficients associated with the left-hand side of equation (3.101). Thus

$$\lambda^3 - 2\lambda^2 + \lambda - 2 = 0$$

i.e. $(\lambda - 2)(\lambda^2 + 1) = 0$

Hence $\lambda_1 = 2$, $\lambda_2 = i$ and $\lambda_3 = -i$.

where $i = \sqrt{-1}$

Defining the modal matrix, M, such that

$$AM = A_1 M$$

where

$$A_1 = \begin{bmatrix} \lambda_1 & 0 & 0 \\ 0 & \lambda_2 & 0 \\ 0 & 0 & \lambda_3 \end{bmatrix}$$

Then $A_1 = M^{-1}AM \qquad (3.105)$

It can be shown that

$$M = \begin{bmatrix} 1 & 1 & 1 & - & - & 1 \\ \lambda_1 & \lambda_2 & \lambda_3 & & & \lambda_n \\ \lambda_1^2 & \lambda_2^2 & \lambda_3^2 & - & - & \lambda_n^2 \\ - & - & - & - & - & - \\ \lambda_1^{n-1} & \lambda_2^{n-1} & \lambda_3^{n-1} & - & - & \lambda_n^{n-1} \end{bmatrix} \qquad (3.106)$$

Thus, for this example,

$$M = \begin{bmatrix} 1 & 1 & 1 \\ 2 & i & -i \\ 4 & -1 & -1 \end{bmatrix}$$

and

$$M^{-1} = \frac{1}{10} \begin{bmatrix} 2 & 0 & 2 \\ 4+2i & -5i & -1+2i \\ 4-2i & +5i & -1-2i \end{bmatrix}$$

$$\therefore B_1 = M^{-1}B = \frac{1}{10} \begin{bmatrix} 2 & 0 & 2 \\ 4+2i & -5i & -1+2i \\ 4-2i & +5i & -1-2i \end{bmatrix} \begin{bmatrix} 0 \\ 0 \\ 1 \end{bmatrix}$$

$$\therefore B_1 = \frac{1}{10} \begin{bmatrix} 2 \\ -1+2i \\ -1-2i \end{bmatrix}$$

And, $C_1 = CM$,

$$\therefore C_1 = [1 \ 0 \ 0] \begin{bmatrix} 1 & 1 & 1 \\ 2 & i & -i \\ 4 & -1 & -1 \end{bmatrix}$$

$$= [1 \ 1 \ 1]$$

The transformed state equations become

$$\begin{bmatrix} \dot{q}1 \\ \dot{q}2 \\ \dot{q}3 \end{bmatrix} = \begin{bmatrix} 2 & 0 & 0 \\ 0 & i & 0 \\ 0 & 0 & -i \end{bmatrix} \begin{bmatrix} q1 \\ q2 \\ q3 \end{bmatrix} + \frac{1}{10} \begin{bmatrix} 2 \\ -1+2i \\ -1-2i \end{bmatrix} u \quad (3.107)$$

and

$$y = [1 \ 1 \ 1] \begin{bmatrix} q1 \\ q2 \\ q3 \end{bmatrix} \quad (3.108)$$

Equations (3.107) and (3.108) represent a decoupled system of state equations due to the fact that the A matrix is diagonal. These are referred to as the normal form for the state equations, and they enable full computational use to be made of the diagonality of A.

3.10.7 The state of transition matrix

Given the vector matrix form for a linear system of equations (i.e. equation (3.96)), this can be rearranged and premultiplied by the matrix e^{-At}, i.e.

$$e^{-At}(\dot{X} - AX) = e^{-At}Bu$$

$$\therefore \frac{d}{dt}(e^{-At} \cdot X) = e^{-At}Bu$$

On integration this becomes

$$e^{-At}X(t) - X(0) = \int_0^t e^{-A\tau}Bu(\tau)d\tau \quad (3.109)$$

The matrix e^{At} is termed the state transition matrix and is denoted by $\Phi(t)$, i.e.

$$\Phi(t) = e^{At} \quad (3.110)$$

For the unforced system, when $u(t) = 0$, the solution of equation (3.109) becomes

$$X(t) = \Phi(t)X(0) \quad (3.111)$$

Equation (3.111) shows that $\Phi(t)$ transforms the system from its state $X(0)$ at some initial time, $t = 0$, to the state $X(t)$ at some subsequent time, $t > 0$.

The calculation of the state transition matrix may be performed in several different ways. The most convenient me-

thod, for many problems, involves the use of Laplace and inverse Laplace transforms. It may be shown that the state transition matrix can be written as

$$\Phi(t) = e^{At} = \mathcal{L}^{-1}\{[sI - A]^{-1}\} \quad (3.112)$$

Suppose that a system is characterized by the state equation

$$\begin{bmatrix} \dot{X}1 \\ \dot{X}2 \end{bmatrix} = \begin{bmatrix} 0 & 2 \\ -1 & -3 \end{bmatrix} \begin{bmatrix} X1 \\ X2 \end{bmatrix} + \begin{bmatrix} 0 \\ 1 \end{bmatrix} u \quad (3.113)$$

If the forcing function $u(t) = 1$ for $t > 0$ and

$$X(0) = \begin{bmatrix} 1 \\ -1 \end{bmatrix}$$

then the state X of the system at any time t is determined as follows:

$$[sI - A] = \begin{bmatrix} s & -2 \\ 1 & s+3 \end{bmatrix}$$

$$[sI - A]^{-1} = \frac{1}{s(s+3)+2} \begin{bmatrix} s+3 & 2 \\ -1 & s \end{bmatrix}$$

$$= \frac{1}{(s+1)(s+2)} \begin{bmatrix} s+3 & 2 \\ -1 & s \end{bmatrix}$$

$$= \begin{bmatrix} \dfrac{s+3}{(s+1)(s+2)} & \dfrac{2}{(s+1)(s+2)} \\ \dfrac{-1}{(s+1)(s+2)} & \dfrac{s}{(s+1)(s+2)} \end{bmatrix}$$

The matrix $[sI - A]^{-1}$ is called the resolvent matrix. To evaluate the inverse transforms it is necessary to express each element of the resolvent matrix in partial fraction form:

$$\therefore [sI - A]^{-1} = \begin{bmatrix} \dfrac{2}{s+1} - \dfrac{1}{s+2} & \dfrac{2}{s+1} - \dfrac{2}{s+2} \\ -\dfrac{1}{s+1} + \dfrac{1}{s+2} & -\dfrac{1}{s+1} + \dfrac{2}{s+2} \end{bmatrix}$$

$$\mathcal{L}^{-1}\{[sI - A]^{-1}\} = \begin{bmatrix} 2e^{-t} - e^{-2t} & 2e^{-t} - 2e^{-2t} \\ -e^{-t} + e^{-2t} & -e^{-t} + 2e^{-2t} \end{bmatrix}$$

This gives

$$\Phi(t)X(0) = \begin{bmatrix} 2e^{-t} - 2e^{-2t} & 2e^{-t} - 2e^{-2t} \\ -e^{-t} + e^{-2t} & -e^{-t} + 2e^{-2t} \end{bmatrix} \begin{bmatrix} 1 \\ -1 \end{bmatrix}$$

$$= \begin{bmatrix} e^{-2t} \\ -e^{-2t} \end{bmatrix} \quad (3.114)$$

The above is the complementary function which applies to the unforced system. On consideration of the particular integral it can be noted that

$$\mathcal{L}[u(t)] = \frac{1}{s}$$

Thus

$$[sI - A]^{-1}Bu(s) = \frac{1}{(s+1)(s+2)} \cdot \frac{1}{s} \begin{bmatrix} s+3 & 2 \\ -1 & s \end{bmatrix} \begin{bmatrix} 0 \\ 1 \end{bmatrix}$$

$$= \begin{bmatrix} \dfrac{2}{s(s+1)(s+2)} \\ \dfrac{1}{(s+1)(s+2)} \end{bmatrix}$$

Expanding the above into partial fractions gives

$$\begin{bmatrix} \dfrac{1}{s} - \dfrac{2}{s+1} + \dfrac{1}{s+2} \\[2mm] \dfrac{1}{s+1} - \dfrac{1}{s+2} \end{bmatrix}$$

Taking the inverse Laplace transform gives

$$\begin{bmatrix} 1 - 2e^{-t} + e^{-2t} \\ e^{-t} - e^{-2t} \end{bmatrix}$$

The complete solution is given as the sum of the complementary function and the particular integral, i.e.

$$X(t) = \begin{bmatrix} e^{-2t} \\ -e^{-2t} \end{bmatrix} + \begin{bmatrix} 1 - 2e^{-t} + e^{-2t} \\ e^{-t} - e^{-2t} \end{bmatrix}$$

$$\begin{bmatrix} X1(t) \\ X2(t) \end{bmatrix} = \begin{bmatrix} 1 - 2e^{-t} + 2e^{-2t} \\ e^{-t} - 2e^{-2t} \end{bmatrix} \qquad (3.115)$$

Equation (3.115) defines the state of the system at any time t and provides a direct solution to the state equations.

Results obtained from the solution of equation (3.115) and that obtained from a finite-difference approximation of the state equation (3.113) show no significant difference in the numerical results generated.

An alternative approach to the evaluation of $\Phi(t)$ is to expand e^{At} as a series, and to truncate the series at an appropriate point. This is well suited to a computer.

Further consideration of the state-variable approach is essentially beyond the scope of this chapter. The reader is therefore referred to the texts by DeRusso et al[24] and by Burghes and Graham[25] for comprehensive, in-depth coverage.

References

1. Whitney, D. E., 'Quasi-static assembly of compliantly supported rigid parts', *Jour of Dynamic Systems, Measurement & Control*, March, 65–77 (1982)
2. Bejczy, A. K., 'Effect of hand-based sensors on manipulator control performance', *Mechanism & Machine Theory*, Vol. 12, pp. 547–567, Pergamon Press, New York (1977)
3. Harmon, L. D., 'Automated tactile sensing', *Int. Jour. of Robotics Research*, **1**, No. 2, 3–32 (1982)
4. Rebman, J. and Trull, N., 'A robot tactile sensor for robot applications', ASME Conf. 'Computers in Engineering', Chicago (1983)
5. Ballard, D. H. and Brown, C. M., *Computer Vision*, Prentice-Hall, Englewood Cliffs, NJ (1983)
6. Fraser, C. J. and Milne, J. S., *Microcomputer Applications in Measurement Systems*, Macmillan Education, Basingstoke (1990)
7. Williams, A. B., *Electronic Filter Design Handbook*, McGraw-Hill, New York (1981)
8. Bode, H. W., 'Relations between attenuation and phase in feedback amplifier design', *Bell Systems Tech. Jour.*, July (1940), 421–454; also in *Automatic Control: Classical Linear Theory*, Ed. G. J. Thaler, Dowden, Hutchinson & Ross (1974), pp. 145–178
9. Nyquist, H., 'Regeneration theory', *Bell Systems Tech. Jour.*, January (1932), 126–147; also in *Automatic Control: Classical Linear Theory*, Ed. G. J. Thaler, Dowden, Hutchinson & Ross (1974), pp. 105–126
10. Golten & Verwer Partners, *Control System Design and Simulation for the PC, (CODAS)*, Golten & Verwer Partners, Cheadle Hume, Cheshire (1988)
11. Cambridge Control Ltd, *SIMBOL 2 – Control System Design and Simulation on IBM-PC or PS/2*, Cambridge Control Ltd, Cambridge (1988)
12. Arthur F. Saunders, 'Laplace Systems analysis program', Laplace Systems–Arthur F. Saunders, Woolavington, Bridgewater (1989)
13. Zeigler, J. G. and Nichols, N. B., 'Optimum settings for automatic controllers', *Trans ASME*, **64**, 759 (1942)
14. Kraus, T. W. and Myron, T. J., 'Self-tuning PID controller uses pattern recognition approach', *Control Eng.*, June (1984)
15. Leigh, J. R., *Applied Digital Control*, Prentice-Hall, Englewood Cliffs, NJ (1985)
16. Cahill, S. J., *The Single Chip Microcomputer*, Prentice-Hall, Englewood Cliffs, NJ (1987)
17. Control Universal Ltd, *Real Time BASIC*, Cube Technical Manual, 138 Ditton Walk, Cambridge, CB58QF, UK
18. Brodie, L., *Starting FORTH*, Prentice-Hall, Englewood Cliffs, NJ (1981)
19. Kerninghan, B. W. and Ritchie, D. M., *The C Programming Language*, Prentice-Hall, Englewood Cliffs, NJ (1988)
20. Shannon, C. E. and Weaver, W., *The Mathematical Theory of Communication*, University of Illinois Press, Urbana (1972)
21. National Instruments, 21 Kingfisher Court, Hanbridge Road, Newbury RG14 5SJ, UK
22. Northcott, J. and Walling, A., *The Impact of Microelectronics – Diffusion, Benefits and Problems in British Industry*, PSI Publications (1988)
23. Babb, M., 'Implementing distributed control into the 1990's', *Control Engineering*, August, 2–4 (1989)
24. DeRusso, M. P., Roy, R. J. and Close, C. M., *State Variables for Engineers*, John Wiley, Chichester (1965)
25. Burghes, D. and Graham, A., *Introduction to Control Theory, Including Optimal Control*, Ellis Horwood series – *Mathematics and its Applications*, Ellis Horwood, Chichester (1980)

Further reading

Bannister, B. R. and Whitehead, D. G., *Transducers and Interfacing*, Van Nostrand Reinhold, Wokingham (1986)

Barney, G. C., *Intelligent Instrumentation*, Prentice-Hall, Englewood Cliffs, NJ (1985)

Bollinger, J. G. and Duffie, N. A., *Computer Control of Machines and Processes*, Addison-Wesley, Reading, MA (1988)

Burr Brown, *The Handbook of Personal Computer Instrumentation – for Data Acquisition, Test Measurement and Control*, Burr Brown Corporation, January (1988)

Cassel, D. A., *Microcomputers and Modern Control Engineering*, Reston Publishing, New York (1983)

Cluley, J. C., *Transducers for Microprocessor Systems*, Macmillan, London (1985)

Doebelin, E. O., *Control System Principles and Design*, John Wiley, Chichester (1985)

Dorf, R. C., *Modern Control Systems*, third edition, Addison-Wesley, Reading MA (1980)

Gayakwad, R. and Sokoloff, L., *Analog and Digital Control Systems*, Prentice-Hall, Englewood Cliffs, NJ (1988)

Hunt, V. D., *Mechatronics: Japan's newest threat*, Chapman & Hall, London (1988)

Ismail, A. R. and Rooney, V. M., *Microprocessor Hardware and Software Concepts*, Collier-Macmillan, New York (1987)

Kafrissen, E. and Stephans, M., *Industrial Robots and Robotics*, Reston Publishing, New York (1984)

Kief, H. B, Olling, G. and Waters T. F., *Flexible Automation – The International CNC Reference Book*, Becker Publishing Co. (UK) Ltd (1986)

Meadows, R. and Parson, A. J., *Microprocessors: Essentials, Components and Systems*, Pitman, London (1985)

Scott, P. B., *The Robotics Revolution*, Blackwell, Oxford (1984)

Sharon, D., Harstein, J. and Yantian, G., *Robotics and Automated Manufacturing*, Pitman, London (1987)

Snyder, W. E., *Industrial Robots, Computer Interfacing and Control*, Prentice-Hall, Englewood Cliffs, NJ (1985)

Warnock, I. G., *Programmable Controllers: Operation and Application*, Prentice-Hall, Englewood Cliffs, NJ (1988)

4

Computers and their application

Ian Robertson

Contents

4.1 Introduction 4/3

4.2 Types of computer 4/3
    4.2.1 Analogue computers 4/3
    4.2.2 Hybrid computers 4/3
    4.2.3 Digital computers 4/3

4.3 Generations of digital computers 4/4
    4.3.1 First generation 4/4
    4.3.2 Second generation 4/4
    4.3.3 Third generation 4/4
    4.3.4 Fourth generation 4/4

4.4 Digital computer systems 4/4
    4.4.1 Central processor unit 4/4
    4.4.2 Input/output 4/5
    4.4.3 Peripherals 4/5
    4.4.4 BUS paths 4/5

4.5 Categories of computer systems 4/5
    4.5.1 Pocket computers 4/5
    4.5.2 Laptop computers 4/5
    4.5.3 Microcomputers 4/5
    4.5.4 Minicomputers 4/6
    4.5.5 Superminis 4/6
    4.5.6 Mainframes 4/6
    4.5.7 Combination technology 4/6

4.6 Central processor unit 4/7

4.7 Memory 4/7
    4.7.1 Memory organization 4/7
    4.7.2 Memory technology 4/8
    4.7.3 MOS RAM 4/8
    4.7.4 ROM 4/8
    4.7.5 Bubble memory 4/8
    4.7.6 Core memory 4/8
    4.7.7 Registers 4/8
    4.7.8 Memory addressing 4/8
    4.7.9 Memory management 4/9
    4.7.10 Multi-programming 4/9
    4.7.11 Virtual memory 4/9
    4.7.12 Instruction set 4/9
    4.7.13 CPU implementation 4/11

4.7.14 CPU enhancements 4/11
4.7.15 Fixed and floating-point arithmetic hardware 4/12
4.7.16 Array processors 4/13
4.7.17 Timers and counters 4/13
4.7.18 Input/output 4/13
4.7.19 Input/output bus 4/13
4.7.20 Types of input/output transactions 4/14
4.7.21 Interrupts 4/14

4.8 Peripherals 4/15
    4.8.1 Interactive 4/15
    4.8.2 Storage 4/15
    4.8.3 Communication 4/16

4.9 Ouput devices 4/16

4.10 Terminals 4/16
    4.10.1 Dot matrix printers 4/16
    4.10.2 Letter-quality printers 4/17
    4.10.3 Visual display unit 4/17
    4.10.4 High-speed printers and plotters 4/18

4.11 Direct input 4/19
    4.11.1 Character recognition 4/19
    4.11.2 Writing tablets 4/20

4.12 Disk storage 4/20
    4.12.1 Drive motor 4/20
    4.12.2 Disk medium 4/20
    4.12.3 Head mechanism 4/20
    4.12.4 Electronics 4/20
    4.12.5 Drive control 4/20
    4.12.6 Floppy disk 4/21
    4.12.7 Cartridge disk 4/21
    4.12.8 Disk pack 4/21
    4.12.9 Winchester drive 4/21
    4.12.10 Magnetic tape 4/22
    4.12.11 Industry-standard tape drives 4/22
    4.12.12 Cartridge tape 4/22
    4.12.13 Tape streamer unit 4/22

4.13 Digital and analogue input/output 4/23
    4.13.1 Digital input/output 4/23

4.13.2   Analogue input   4/23
4.13.3   Analogue output   4/23
4.13.4   Input/output subsystems   4/23

4.14   Data communications   4/23
    4.14.1   Introduction   4/23
    4.14.2   Data communications concepts   4/24
    4.14.3   Multi-line interface   4/26
    4.14.4   Modem   4/26
    4.14.5   Fibre-optic cable   4/27
    4.14.6   Laser   4/27
    4.14.7   Microwave   4/27
    4.14.8   Transmission techniques   4/27
    4.14.9   Transmission types   4/27
    4.14.10   Direction of transmission   4/28
    4.14.11   Error detection and correction   4/28
    4.14.12   Communications protocols   4/29

4.15   Computer networks   4/31
    4.15.1   Ethernet   4/32
    4.15.2   Open Systems Interconnect (OSI)   4/32
    4.15.3   Network types   4/32
    4.15.4   Network concepts   4/34
    4.15.5   Network design   4/35

4.15.6   Standard network architecture   4/36

4.16   Data terminal equipment   4/36

4.17   Software   4/36
    4.17.1   Introduction   4/36
    4.17.2   The operating system   4/37
    4.17.3   Batch processing   4/37
    4.17.4   Interactive processing   4/37
    4.17.5   Transaction processing   4/38
    4.17.6   Real time   4/38
    4.17.7   Common concepts   4/38

4.18   Database management   4/40

4.19   Lanaguage translators   4/41
    4.19.1   Assemblers   4/41
    4.19.2   Compilers   4/43

4.20   Languages   4/43
    4.20.1   Introduction   4/43
    4.20.2   Third-generation languages (3GLs)   4/43
    4.20.3   Fourth-generation languages   4/44
    4.20.4   Programmer emphasis   4/45

4.1 Introduction

Although the advent of computers in our everyday lives may seem very recent, the principles of the modern computer were established before the existence of any electronic or electro-mechanical technologies as we know them today, and electronic computers were beginning to take shape in laboratories in 1945.

The work of Charles Babbage, a Cambridge mathematician of the nineteenth century, in attempting to build an 'analytical engine' from mechanical parts, remarkably anticipated several of the common features of today's electronic computers. His proposed design, had he been able to complete it and overcome mechanical engineering limitations of the day, would have had the equivalent of punched-card input and storage registers, the ability to branch according to results of intermediate calculations, and a form of output able to set numeric results in type.

Many purely mechanical forms of analogue computer have existed over the last few centuries. The most common of these is the slide rule, and other examples include mechanical integrators and even devices for solving simultaneous equations.

Much of the development leading to modern electronic computers, both analogue and digital, began during World War II with the intensified need to perform ballistics calculations. The development of radar at this time also provided the stimulus for new forms of electronic circuits which were to be adopted by the designers of computers.

A further development of momentous importance to the technology of computers, as it was for so many branches of electronics, was that of the transistor in 1949. Continued rapid strides in the field of semiconductors have brought us the integrated circuit, which allows a complete digital computer to be implemented in a single chip.

4.2 Types of computer

Although there are two fundamentally different types of computer, analogue and digital, the former remains a somewhat specialized branch of computing, completely eclipsed now, both in numbers of systems in operation and in breadth of applications, by digital computers. While problems are solved in the analogue computer by representing the variables by smoothly changing voltages on which various mathematical operations can be performed, in a digital computer all data are represented by binary numbers held in two-state circuits, as are the discrete steps or instructions for manipulating the data, which make up a program.

4.2.1 Analogue computers

An analogue computer consists of a collection of circuit modules capable of individually performing summation, scaling, integration or multiplication of voltages, and also function-generating modules. On the most up-to-date systems these modules contain integrated-circuit operational amplifiers and function generators. Several hundred amplifiers are likely to be used on a large analogue computer.

To solve a given problem in which the relationship of physical quantities varying with time can be expressed as differential equations, the inputs and outputs of appropriate modules are interconnected, incorporating scaling, feedback and setting of initial conditions as required, with voltages representing the physical quantities. In this way, single or simultaneous equations can be solved in such applications as engineering and scientific calculation, modelling and simulation.

The interconnection and setting of coefficients and initial conditions required is normally done by means of a patch panel and potentiometers. An analogue computer may also have a CRT display or chart recorder for the display or recording of the results of a computation. Where an analogue computer is being used for design work, the designer may choose to observe immediately the effect of changing a certain design parameter by varying the appropriate potentiometer setting, thus altering the voltage representing that parameter. In a simulation application, outputs from the analogue computer may be used as input voltages to another electrical system.

Except in some specialized areas, the work formerly done by analogue computers is now most likely to be carried out on the modern high-speed, cost-effective digital computer, using numerical methods for operations such as integration. Digital computers can work to high precision extremely accurately, whereas using analogue techniques, there are inherent limits to precision, and accuracy suffers through drift in amplifiers.

4.2.2 Hybrid computers

The analogue part of a hybrid computer is no different in its circuitry and function to that of a stand-alone analogue system. However, the task of setting up the network required to solve a particular problem is carried out by a digital computer, linked to the analogue machine by analogue/digital converters and digital input/output. Thus, the digital computer sets potentiometers and can read and print their values, and can monitor and display or log voltages at selected parts of the analogue network.

Linking the two technologies in this way brings the advantage of programmability to the mathematical capabilities of the analogue computer. Set-up sequences for the analogue circuitry can be stored and quickly and accurately reproduced at will. Tasks can be carried out in the part of the system best adapted to performing them. In particular, the ability of the digital computer to store, manipulate and present data in various ways is particularly advantageous.

The use of hybrid systems in, for example, the simulation of aircraft and weapon systems, has helped to perpetuate analogue computing as a technique applicable in certain narrow areas.

4.2.3 Digital computers

Digital computers in various forms are now used universally in almost every walk of life, both in business and in public service. In many cases unseen, computers nonetheless influence people in activities such as travel, banking, education and medicine. There is also a very wide range of computers designed for use by people, without technical skills or training, in their own homes. These are generally used for some form of entertainment or for introduction to the wider subject of computing, but some are large enough and powerful enough to support small businesses. There is also a growing range of truly portable computers now available that will fit into briefcases or even pockets.

As can be imagined from the variety of applications, computers exist in many different forms, spanning a range of price (from the smallest personal system to the largest supercomputer) of more than 1 million:1. Yet there are certain features which are common to all digital computers:

1. Construction from circuits which have two stable states, forming binary logic elements;

2. Some form of binary storage of data;
3. Capability to receive and act on data from the outside world (input) and to transmit data to the outside world (output);
4. Operation by executing a set of discrete steps or instructions, the sequence of which can be created and modified at any time to carry out a particular series of tasks. This ability to be programmed, with a program stored in the system itself, is what gives great flexibility to the digital computer. Recent advances have enabled changes to be made to a program dynamically while it is in operation. In addition, modern techniques permit processing to be carried out on the same set of discrete steps on more than one CPU at the same time within the same 'logical' computer. This technique is referred to as 'parallel processing'.

The computer has also given rise to a whole new series of professions – those of the computer programmer, systems analyst, data specialist, business analyst, computer operator – as well as the industry of designing, building and maintaining computers themselves.

4.3   Generations of digital computers

Beginning with circuits consisting of relays, the history of the digital computer can be seen as having fallen into four generations between the 1940s and today.

4.3.1   First generation

These computers were built with valve circuits and delay-line storage, physically very massive, taking up complete rooms, requiring very large amounts of electricity with corresponding high heat dissipation, and low overall reliability, requiring extensive maintenance often resulting in engineers being on-site 24 hours per day. Input/output was rudimentary (teleprinters, punched cards) and programming very laborious, usually in a binary form that the machine could understand without further interpretation.

4.3.2   Second generation

Developed during the 1950s with transistorized circuits, these computers were faster, smaller and more reliable than the first generation, but still large by today's standards. Magnetic core main stores with magnetic drums were used with tapes as back-up, and line printers were employed for faster printed output. Programming language translators emerged, resulting in the widespread use of Assembler-type languages.

4.3.3   Third generation

Developed during the mid-1960s these computers were heralded by the integrated circuit, allowing more compact construction and steadily improving speed, reliability and capability. The range and capabilities of input/output and mass storage devices increased remarkably. In the software area, high-level languages (e.g. FORTRAN, COBOL, BASIC) became common and manufacturers offered operating system software developed, for example, to manage time-sharing for a large number of computer users or real-time process control.

Most significantly, a trend of downward cost for given levels of performance was established. The minicomputer, aimed at providing a few or even one single user with direct access to and control over their own computing facilities, began to gain in numbers over the large, centrally managed and operated computer system.

4.3.4   Fourth generation

While a great many third-generation computer systems are in use, and will remain so for some time, the semiconductor technology of large-scale integration (LSI) of the 1970s brought complete computers on a chip, known as microprocessors, allowing further refinement and enhancement of third-generation equipment.

Semiconductor memory has almost completely replaced core memory, and the continuing reduction in size and cost has brought the 'personal computer', numbered in hundreds of thousands of units supplied, truly within the reach of individuals in their own homes or offices.

The 1980s saw the arrival of very large scale integration (VLSI) and ultra large scale integration (ULSI) applied to semiconductor memory and processor circuit design. This has resulted in a scalar increase in the density of memory capacity that can be packed onto a single board, as well as the proliferation of multi-processors on a single board. Many computers of power and memory capacity equal to or greater than that of the earliest minis can now be found on a single printed circuit board.

The turn of the decade also saw the introduction of 'surface mount technology', which broke through the previous barrier of physical limitation of the number of connections that could be built between the memory or processor chip and the outside world with which they communicated. This has also permitted closer packing of chips onto each board.

4.4   Digital computer systems

A digital computer system is a collection of binary logic and storage elements combined in such a way as to perform a useful task. Any computer system, whether a microcomputer held in one hand for data collection, or a large data-processing system consisting of many cabinets of equipment housed in a specially built air-conditioned computer room, invariably contains a combination of the parts described in detail in the following sections and shown in Figure 4.1.

4.4.1   Central processor unit

The CPU is where instructions forming the stored program are examined and executed, and is therefore in control of the operation of the system. Instructions and data for immediate processing by the CPU are held in main memory, which is linked directly to the CPU. This general term covers the units that perform logic decisions and arithmetic, collectively known as the Arithmetic and Logic Unit (ALU).

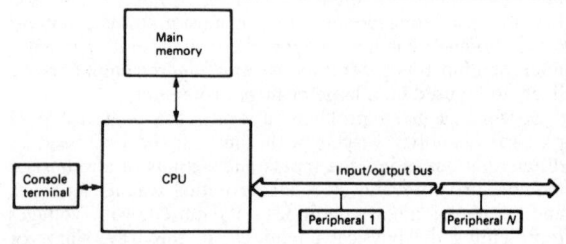

Figure 4.1   Components of a computer system

In recent years considerable progress has been made on computers that contain more than one discrete CPU. These are referred to as 'Multiple processor computers'. The main area of difficulty in the development of these machines has not been in the physical co-location of processors but rather in the design of the operating system. It has to know which instructions can be operated upon simultaneously and which require prior instructions to be completed. It should also be noted that multiple processor computers can operate in two distinct modes. In 'parallel' processing each processor is operating on the same instruction at the same time. This is particularly useful in fail-safe or non-stop critical applications. In 'multi' processing, each processor is operating on a different instruction and each instruction is processed once only. This does not give the 'fail-safe' advantages, but does provide large gains in speed of processing overall.

4.4.2 Input/output

This is the structure which provides optimum communication between the CPU and other parts of the system.

4.4.3 Peripherals

These are the devices external to the CPU and memory which provide bulk storage, human/machine interaction and communication with other electronic systems.

4.4.4 BUS paths

These act as the vehicle for the passing of data and program instructions between all other parts.

4.5 Categories of computer systems

In going from microcomputer to large-data processing systems there are many differences in complexity, technology, system cost or performance. However, there are many similarities between systems from different manufacturers.

Several distinct categories of computer system can be identified. Going from least to most comprehensive and powerful, these are described in subsequent sections.

4.5.1 Pocket computers

These derive their name from their ability to literally fit into their owners' pockets! Though some early prototypes were available to a limited market, they first made a general appearance in 1990. They are typically of two-piece construction with one section folding over the other. Normal facilities include an LCD screen, a reduced-size keyboard (though with a full key set) and removable self-powered memory cards that act as logical disk drives. Many also include an internal modem or the ability to interface with an external one. They provide most of the features of laptop or personal computers without the size or weight penalties. They are ideal for carrying in a briefcase and for use on long journeys on public transport.

4.5.2 Laptop computers

These have the ability to fit onto their owners' laps while they are being used! They are also sometimes referred to as 'portable' computers, though the earliest models required considerable human strength and stamina to enable their portability'.

They were first available on a general basis in the late 1980s and were initially scaled-down versions of existing personal computers. In the last few years they have benefited from a design approach aimed at ease of use and true lightweight portability. Recent models have been reduced to A5 size when opened, leading to smaller laptops referred to as 'notebook' computers.

They are typically of two-piece construction, with one section folding over the other. Normal facilities include an LCD screen, a full-size keyboard, a standard IBM format $3\frac{1}{2}$-inch floppy disk, non-removable $5\frac{1}{4}$-inch Winchester disks of up to 100 megabyte capacity each and a 'windows' type operating system such as Microsoft Windows. Most also include an internal modem and additional facilities for the connection of external peripherals such as lightweight bubble printers or fax equipment. While they cannot fit into a normal briefcase, their shape, size and weight means that they can be carried *as* one.

4.5.3 Microcomputers

The first microcomputers were appearing in the mid-1970s and were mainly intended for use by specialist engineers requiring computing power remote from central resources for large and intricate calculations (an example being the Olivetti Pnnn series). However, they were still large and expensive by modern standards and so did not come into general use.

The second half of the 1970s saw the introduction of mass-produced (and hence cheap) microcomputers, often with an integral CRT, such as the APPLE, NASCOM and the COMMODORE PET. These were aimed initially at the personal market and great numbers were sold on the basis of video games provided for them. As the novelty wore off, the personal computer market did not grow to the size predicted and the manufacturers had to look to another market sector to finance continued growth and, in some cases, just to be able to maintain production at current levels. This led to microcomputers being developed for and marketed in the business sector. These possess memories and facilities far exceeding those of early minicomputers and, in some cases, on a par with smaller mainframes.

Many microcomputers have been specially adapted to cater for a specific market sector such as CAD/CAM and graphics use by a single person while offering any networking and multi-user options as required. A large number are covered by the generic term 'Workstation' and are intended to permit persons to carry out many different functions without having to leave their desks, examples being SUN Microsystems, Hewlett Packard, COMPUTERVISION and the Digital Equipment Corporation VAX stations.

Many of these workstations now run under a 'windows' operating system. The principle behind this is that the screen is split into virtual windows that act as display areas for different applications running at the same time on the same terminal or workstation. Indeed, windows may be related to applications that are running on an entirely separate computer or terminal from the one that is displaying the required information.

Windows can overlap each other, be shrunk or re-sized, can be recalled and need not even be running under the same operating system. The technique used to support this is to separate the management of the display hardware from the management of the hardware that is actually running the application. Users may have as many windows on their display screens as can be physically supported, though they may only be 'active' in one at any one moment. The 'active' window is selected using a mouse to indicate which window is required on the display. Thus users may have windows from systems running UNIX, VMS, MS-DOS or MVS on their screens at the same time.

4.5.4 Minicomputers

Since its introduction as a recognizable category of system in the mid-1960s, with machines such as the Digital Equipment Corporation PDP8, the minicomputer has evolved rapidly. It has been the development which has brought computers out of the realm of specialists and large companies into common and widespread use by non-specialists.

The first such systems were built from early integrated-circuit logic families, with core memory. Characteristics were low cost, ability to be used in offices, laboratories and even factories, and simplicity of operation allowing them to be used, and in many cases programmed, by the people who actually had a job to be done, rather than by specialist staff remote from the user. These were also the first items of computer equipment to be incorporated by original equipment manufacturers (OEMs) into other products and systems, a sector of the market which has contributed strongly to the rapid growth of the minicomputer industry.

Applications of minicomputers are almost unlimited, in areas such as laboratories, education, commerce, industrial control, medicine, engineering, government, banking, networking, CAD/CAM, CAE and CIM. There is also a growing use of minicomputers combined with artificial intelligence for problem solving that benefits from deduction and backward chaining as opposed to predefined procedural stepping.

With advancing technology, systems are now built using large-scale and more often very large scale integrated circuits, and memory is now almost entirely semiconductor. While earlier systems had a very small complement of peripherals (typically, a teleprinter and punched paper-tape input and output) there has been great development in the range and cost-effectiveness of peripherals available. A minicomputer system will now typically have magnetic disk and tape storage holding thousands of millions of characters of data, a printer capable of printing up to 2500 lines of text per minute, 16–512 CRT display terminals (often in colour), low-cost matrix printers for local or operator's hardcopy requirements and a selection of other peripherals for specialist use such as a graphic colour plotter or a laser printer for high-quality output.

4.5.5 Superminis

The word 'supermini' has been coined to describe a type of system that has many similarities in implementation to the minicomputer, but, by virtue of architectural advances, has superior performance. These advantages include:

1. Longer word length. The amount of information processed in one step, or transferred in a single operation between different parts of the system, is usually twice that of a minicomputer.
2. As well as increasing the rate of information handling, the longer word length makes it possible to provide a more comprehensive instruction set. Some common operations, such as handling strings of characters or translating high-level language statements into CPU instructions, have been reduced to single instructions in many superminis.
3. Longer word length provides larger memory addressing. A technique called virtual memory (see Section 4.7.11) gives further flexibility to addressing in some superminis.
4. Higher data transfer speeds on internal data highways, which allow faster and/or larger numbers of peripheral

devices to be handled, and larger volumes of data to be transmitted between the system and the outside world.

Despite providing substantial power, even when compared with the mainframe class of system described below, superminis fall into a price range below the larger mainframes. This is because they have almost all originated from existing minicomputer manufacturers, who have been able to build on their volume markets, including, in most cases, the OEM market.

4.5.6 Mainframes

The mainframe is the class of system typically associated with commercial data processing in large companies where a centralized operation is feasible and desired, and very large volumes of data are required to be processed at high processor speeds, or where a large user base (often in excess of 500 simultaneous users) requires immediate responses during interactive sessions. Today's mainframes, all products of large, established companies in the computer business (except for systems which are software-compatible emulators of the most popular mainframe series) are the successors to the first and second generation as described in Section 4.3. They inherit the central control and location, emphasis on batch processing and line printers, third- and fourth-generation programming and the need for specialized operating staff.

Mainframes are capable of supporting very large amounts of on-line disk and magnetic tape storage as well as large main memory capacity, and data communications capabilities supporting remote terminals of various kinds. Although some of the scientific mainframes have extremely high operating rates (over 100 million instructions per second), most commercial mainframes are distinguished more by their size, mode of operation and support than by particularly high performance.

4.5.7 Combination technology

There have also been some significant developments in methods of combining computing resources to provide more security and faster processing. These fall into the following categories.

4.5.7.1 Tightly coupled systems

In the first instance, certain parts of the system are duplicated and operate in parallel, each mirroring the work performed by the other. This provides security of availability of resource and of data by redundancy, the ultimate being total duplication of every part of the system, with the system designed to continue should one of anything fail. TANDEM and STRATUS machines are early examples of this being applied, though most manufacturers have since offered machines of this type.

The second technique is to have more than one processor within the same CPU, with each one performing different tasks from the others, but with one having overall control. This provides security of availability should a processor fail, since the system will continue automatically with any remaining processor(s) without any human intervention, albeit with a reduced processing capacity overall.

4.5.7.2 Loosely coupled systems

This method employs the technique of sharing all resources across a common group of machines, sometimes known as 'clustering'. Each CPU within the cluster is an independent unit but it knows of the existence of other members of the cluster. It is generally not necesssary to stop processing ('bring down') the cluster in order to add or remove a new member. Each CPU can have its own peripherals attached to it which it decides whether or not to share with other members. The cluster itself will also own peripherals that are available for use by any member. It is also possible to arrange for communications controllers to monitor the load of each member in order to attempt to spread the workload as evenly as possible across the cluster. Should any item, including a CPU, fail within the cluster, then processing can be continued on another part of the cluster. Through careful planning of these clusters it is now common to find a logical system (i.e. a cluster) that rarely has to be stopped for hardware maintenance, operating system upgrades, hardware or software failure, etc. The management of all members of the cluster can be carried out from any member machine.

4.6 Central processor unit

This part of the system controls the sequence of individual steps required for the execution of instructions forming a program. These instructions are held in storage, and when executed in the appropriate order, carry out a task or series of tasks intended by the programmer.

Within any particular computer system, the word length is the fixed number of binary digits of which most instructions are made up. Arithmetic operations within the CPU are also performed on binary numbers of this fixed word length, normally 8, 16, 24, 32, 36, 64 binary digits or bits. The CPU is connected via a memory bus, as in Figure 4.2, to a section of memory organized as a number of randomly accessible words, each of which can be written to or read from individually. The time for reading one word from or writing one word into main memory is typically in the range of 0.05 to 0.6 μs, depending on CPU and memory BUS speed. Each word or location of memory can contain either an instruction or data. Apart from simple systems, some form of magnetic tape or disk memory peripheral is present on a system as file storage and back-up to main memory.

Control and timing circuits in the CPU enable instructions and data to be fetched from memory for processing and the results of any instructions which require to be stored for further processing to be written into memory. The program counter holds the memory address of the next instruction to be fetched after each instruction has been processed. Frequently, the next instruction is held in the next location in memory and the counter need simply be incremented by one. However, some systems work by placing one or more parameters for the instruction immediately after it in memory, thereby causing the next instruction to be displaced further down-memory. At other times, the sequence of the program dictates that a new value be written into the program counter. Instructions which alter the sequence of a program calculate and insert a new value into the program counter for the next instruction.

In order to start the CPU when no programs are already in memory, the program counter is loaded with a predetermined address, usually by the action of switching power onto the system, and a simple loader program held in ROM is loaded into memory. Its function is to load a comprehensive general-purpose loader, which automatically loads user or system programs. This process is known as bootstrapping, or booting the system, and the initial ROM program is known as the bootstrap loader. However, today most systems perform all these functions as a result of switching on the power automatically and are ready for use almost immediately, only requesting date and time if they are not held in any form of battery backed-up memory.

4.7 Memory

In order to provide storage for program instructions and data in a form where they can be directly accessed and operated upon, all CPUs have main memory which can be implemented in a variety of technologies and methods of organization.

4.7.1 Memory organization

Memory is organized into individually addressable words into which binary information can be loaded for storage and from which the storage data pattern can be read. On some systems, memory is arranged in such a way that more than one word at a time is accessed. This is done to improve effective memory access rates, on the basis that by accessing, say, two consecutive words, on most occasions the second word will be the one which the CPU requires next. This is generally referred to as 'interleaved' memory.

A memory controller is required between the memory arrays and the CPU, to decode the CPU requests for memory access and to initiate the appropriate read or write cycle. A controller can only handle up to a certain maximum amount of memory, but multiple controllers can be implemented on a single system. This can be used to speed up effective memory access, by arranging that sequentially addressed locations are physically in different blocks of memory with different controllers. With this interleaved memory organization, in accessing sequential memory locations the operation of the controllers is overlapped, i.e. the second controller begins its cycle before the first has completed. Aggregate memory throughput is thus speeded up.

In some more complex computer systems, all or part of the memory can be shared between different CPUs in the multiprocessor configuration. Shareable memory has a special form of controller with multiple ports, allowing more than one CPU access to the memory.

It is sometimes appropriate to implement two types of memory in one system: random access or read/write memory (RAM) and read-only memory (ROM). Programs have to be segregated into two areas:

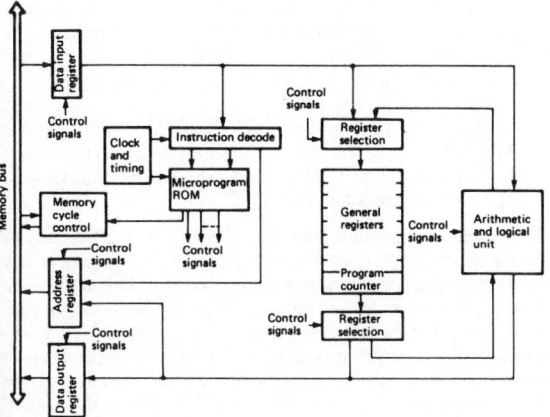

Figure 4.2 CPU block diagram

1.   Pure instructions, which will not change, can be entered into ROM;
2.   Areas with locations which require to be written into (i.e. those containing variable data or modifiable instructions) must occupy RAM.

Read-only memory is used where absolute security from corruption of programs, such as operating system software or a program performing a fixed control task, is important. It is normally found on microprocessor-based systems, and might be used, for example, to control the operation of a bank's cash dispenser.

Use of ROM also provides a low-cost way of manufacturing in quantity a standard system which uses proven programs which never require to be changed. Such systems can be delivered with the programs already loaded and secure, and do not need any form of program-loading device.

4.7.2   Memory technology

The most common technologies for implementing main memory in a CPU are described in the following sections.

4.7.3   MOS RAM

This technology is very widely used, with the abundant availability, from the major semiconductor suppliers, ranging in size from 4 Kbytes to 128 Kbytes. In the latter form, very high density is achieved, with up to 128 megabytes of memory available on a single printed circuit board.

Dynamic MOS RAMs require refresh circuitry which, at intervals, automatically rewrites the data in each memory cell. Static RAM, which does not require refreshing, can also be used. This is generally faster, but also more expensive, than dynamic RAM.

Semiconductor RAMs are volatile, i.e. they lose contents on powering down. This is catered for in systems with back-up storage (e.g. disks) by reloading programs from the back-up device when the system is switched on, or by having battery back-up for all or part of the memory.

In specialized applications requiring memory retention without mains for long periods, battery operations of the complete CPU and CMOS memory can be used. The latter has a very low current drain, but has the disadvantage of being more expensive than normal MOS memory. Where it is essential to use CMOS, circuit boards with on-board battery and trickle charger are now available.

4.7.4   ROM

Read-only memories, used as described in Section 4.7.1, can be either erasable ROMs or a permanently loaded ROM such as fusible-link ROM.

4.7.5   Bubble memory

Bubble memory has not produced the revolution in memory that it seemed to promise at the start of the 1990s. It remains at a comparatively higher price-to-performance ratio than semiconductor memory and is not used on any large scale on a commercial basis.

4.7.6   Core memory

Core memory remains in some applications, but although it has come down substantially in cost under competition from semiconductor memory, more recently MOS RAMs of higher capacity have been much cheaper and have largely taken over.

4.7.7   Registers

The CPU contains a number of registers accessible by instructions, together with more that are not accessible but are a necessary part of its implementation. Other than single-digit status information, the accessible registers are normally of the same number of bits as the word length of the CPU.

Registers are fast-access temporary storage locations within the CPU and implemented in the circuit technology of the CPU. They are used, for example, for temporary storage of intermediate results or as one of the operands in an arithmetic instruction. A simple CPU may have only one register, often known as the accumulator, plus perhaps an auxiliary accumulator or quotient register used to hold part of the double-length result of a binary multiplication.

Large word length, more sophisticated CPUs typically have eight or more general-purpose registers that can be selected as operands by instructions. Some systems such as the VAX use one of its 16 general-purpose registers as the program counter, and can use any register as a stack pointer. A stack in this context is a temporary array of data held in memory on a 'last-in, first-out' basis. It is used in certain types of memory reference instructions and for internal housekeeping in interrupt and subroutine handling. The stack pointer register is used to hold the address of the top element of the stack. This address, and hence the stack pointer contents, is incremented or decremented by one at a time as data are added to or removed from the stack.

4.7.8   Memory addressing

Certain instructions perform an operation in which one or more of the operands is the contents of a memory location (for example, arithmetic, logic and data-movement instructions). In most sophisticated CPUs various addressing modes are available to give, for example, the capacity of adding together the contents of two different memory locations and depositing the result in a third.

In such CPUs instructions are double operand, i.e. the programmer is not restricted to always using one fixed register as an operand. In this case, any two of the general-purpose registers can be designated either as each containing an operand or through a variety of addressing modes, where each of the general-purpose registers selected will contain one of the following:

1.   The memory address of an operand;
2.   The memory address of an operand, and the register contents are then incremented following execution;
3.   The memory address of an operand, and the register contents are then decremented following execution;
4.   A value to which is added the contents of a designated memory location (this is known as 'indexed addressing');
5.   All of the above, but where the resultant operand is itself the address of the final operand (known as 'indirect' or 'deferred' addressing).

This richness of addressing modes is one of the benefits of more advanced CPUs, as, for example, it provides an easy way of processing arrays of data in memory, or of calculating the address portion of an instruction when the program is executed. Further flexibility is provided by the ability on many processors for many instructions to operate on multiples of 8 bits (known as a byte), on single bits within a word and, on some more comprehensive CPUs (such as that of the Digital Equipment VAX series), on double- and quadruple-length words and also arrays of data in memory.

4.7.9 Memory management

Two further attributes may be required of memory addressing. Together, they are often known as memory management.

4.7.9.1 Extended addressing

This is the ability, particularly for a short word length system (16 bits or less), for a program to use addresses greater than those implied by the word length. For example, with the 16-bit word length of most minicomputers the maximum addresses that can be handled in the CPU is 65,536. As applications grow larger this is often a limitation, and extended addressing operates by considering memory as a number of pages. Associated with each page at any given time is a relocation constant which is combined with relative addresses within its page to form a longer address. For example, with extension to 18 bits, memory addresses up to 262,144 can be generated in this way. Each program is still limited at any given time to 65,536 words of address space, but these are physically divided into a number of pages that can be located anywhere within the larger memory. Each page is assigned a relocation constant, and as a particular program is run, dedicated registers in the CPU memory management unit are loaded with the constant for each page (Figure 4.3).

Thus many logically separate programs and data arrays can be resident in memory at the same time, and the process of setting the relocation registers, which is performed by the supervisory program, allows rapid switching between them in accordance with a time-scheduling scheme that is usually based upon resource usage quota, time allocation or a combination of both. This is known as multi-programming. Examples of where this is used are a time-sharing system for a number of users with terminals served by the system, or a real-time control system where programs of differing priority need to be executed rapidly in response to external events.

4.7.9.2 Memory protection

As an adjunct to the hardware for memory paging or segmentation described above, a memory-protection scheme is readily implemented. As well as a relocation constant, each page can be given a protection code to prevent it being illegally accessed. This would be desirable, for example, for a page holding data that are to be used as common data among a number of programs. Protection can also prevent a program from accessing a page outside of its own address space.

4.7.10 Multi-programming

Memory addressing and memory management are desirable for systems performing multi-programming. In such systems the most important area to be protected is that containing the supervisory program or operating system, which controls the running and allocation of resources for users' programs.

4.7.11 Virtual memory

Programmers frequently have a need for a very large address space within a single program for instructions and data. This allows them to handle large arrays, and to write very large programs without the need to break them down to fit a limited memory size.

One solution is known as virtual memory, a technique of memory management by hardware and operating systems software whereby programs can be written using the full addressing range implied by the word length of the CPU, without regard to the amount of main memory installed in the system. From the hardware point of view, memory is divided into fixed-length pages, and the memory management hardware attempts to ensure that pages in most active use at any given time are kept in main memory. All the current programs are stored in a disk backing store, and an attempt to access a page which is not currently in main memory causes paging to occur. This simply means that the page concerned is read into main memory into the area occupied by an inactive page, and that if any changes have been made to the inactive page since it was read into memory then it is written out to disk in its updated form to preserve its integrity.

A table of address translations holds the virtual physical memory translations for all the pages of each program. The operating system generates this information when programs are loaded onto the system, and subsequently keeps it updated. Memory protection on a per-page basis is normally provided, and a page can be locked into memory as required to prevent it being swapped out if it is essential for it to be immediately executed without the time overhead of paging.

When a program is scheduled to be run by the operating system, its address translation table becomes the one in current use. A set of hardware registers to hold a number of the most frequent translations in current use speeds up the translation process when pages are being repeatedly accessed.

4.7.12 Instruction set

The number and complexity of instructions in the instruction set or repertoire of different CPUs varies considerably. The longer the word length, the greater is the variety of instruc-

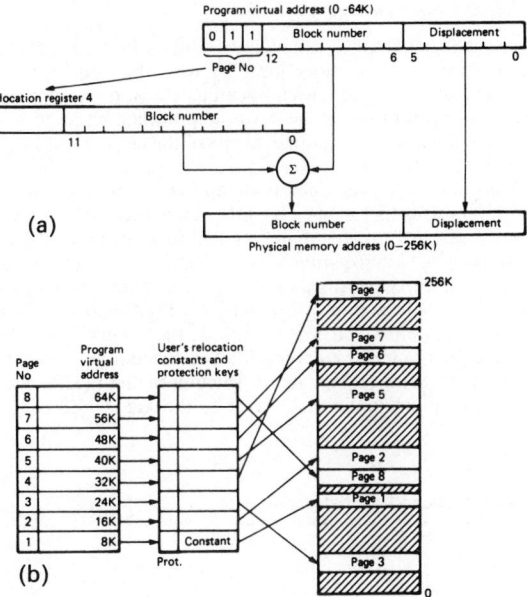

(a)

(b)

Figure 4.3 Memory management for a 16-bit CPU. (a) Generation of a physical address in the range 0 to 256K by combination of user's program virtual address in the range 0 to 64K with a relocation constant for the page concerned. Memory is handled in 64-byte blocks, with eight relocation registers, giving segmentation into eight pages located anywhere in up to 256K of physical memory. (b) The user's program is considered as up to eight pages, up to 8K bytes each. Relocation constants for that program map these pages anywhere in up to 256K bytes of physical memory. Protection per page can also be specified

tions that can be coded within it. This means, generally, that for a shorter word length CPU a larger number of instructions will have to be used to achieve the same result, or that a longer word length machine with its more powerful set of instructions needs fewer of them and hence should be able to perform a given task more quickly.

Instructions are coded, according to a fixed format, allowing the instruction decoder to determine readily the type and detailed function of each instruction presented to it. The general instruction format of the Digital Equipment Corporation VAX is shown as an example in Figure 4.4. Digits forming the operation code in the first (sometimes also the second) are first decoded to determine the category of instruction, and the remaining bytes interpreted in a different way, depending into which category the instruction falls.

There are variations to the theme outlined above for CPUs from differing manufacturers, but generally they all employ the principle of decoding a certain group of digits in the instruction word to determine the class of instruction, and hence how the remaining digits are to be interpreted.

The contents of a memory location containing data rather than an instruction are not applied to the instruction decoder. Correct initial setting of the program counter (and subsequent automatic setting by any branch instruction to follow the sequence intended by the programmer) ensures that only valid instructions are decoded for execution. In the cases where operands follow the instruction in memory, the decoder will know how many bytes or words to skip in order to arrive at the next instruction in sequence.

Logic and arithmetic instructions perform an operation on data (normally one or two words for any particular instruction) held in either the memory or registers in the CPU. The addressing modes available to the programmer (see Section 4.7.8) define the range of possible ways of accessing the data to be operated on. This ranges from the simple single operand type of CPU (where the accumulator is always understood to contain one operand while the other is a location in memory specified by the addressing bits of the instruction) to a multiple-operand CPU with a wide choice of how individual operands are addressed.

In some systems such as the VAX, instructions to input data from (and output data to) peripheral devices are the same as those used for manipulating data in memory. This is achieved by implementing a portion of the memory addresses at the high end as data and control registers in peripheral device controllers.

There are certain basic data transfer, logical, arithmetic and controlling functions which must be provided in the instruction sets of all CPUs. This minimum set allows the CPU to be programmed to carry out any task that can be broken down

and expressed in these basic instructions. However, it may be that a program written in this way will not execute quickly enough to perform a time-critical application such as control of an industrial plant or receiving data on a high-speed communications line. Equally, the number of steps or instructions required may not fit into the available size of memory. In order to cope more efficiently with this situation (i.e. to increase the power of the CPU) all but the very simplest CPUs have considerable enhancements and variations to the basic instruction set. The more comprehensive the instruction set, the fewer are the steps required to program a given task, and the shorter and faster in execution are the resulting programs.

Basic types of instruction, with the examples of the variations to these, are described in the following sections.

4.7.12.1 Data transfer

This loads an accumulator from a specified memory location and writes the contents of the accumulator into a specified memory location. Most CPUs have variations such as adding contents of memory location to the accumulator and exchanging the contents of the accumulator and memory locations.

CPUs with multiple registers also have some instructions which can move data to and from these registers, as well as the accumulator. Those with 16-bit or greater word lengths may have versions of these and other instruction types which operate on bytes as well as words.

With a double operand addressing mode (see Section 4.7.8) a generalized 'Move' instruction allows the contents of any memory location or register to be transferred to any other memory location or register.

4.7.12.2 Boolean logical function

This is a logical 'AND' function on a bit-by-bit basis between the contents of a memory location and a bit pattern in the accumulator. It leaves ones in accumulator bit positions which are also one in the memory word. Appropriate bit patterns in the accumulator allow individual bits of the chosen word to be tested.

Many more logical operations and tests are available on more powerful CPUs, such as 'OR', exclusive 'OR', complement, branch if greater than or equal to zero, branch if less than or equal to zero, branch if lower or the same. The branch instructions are performed on the contents of the accumulator following a subtraction of comparison of two words, or some other operation which leaves data in the accumulator. The address for branching to is specified in the address part of the instruction. With a skip, the instruction in the next location should be an unconditional branch to the code which is to be

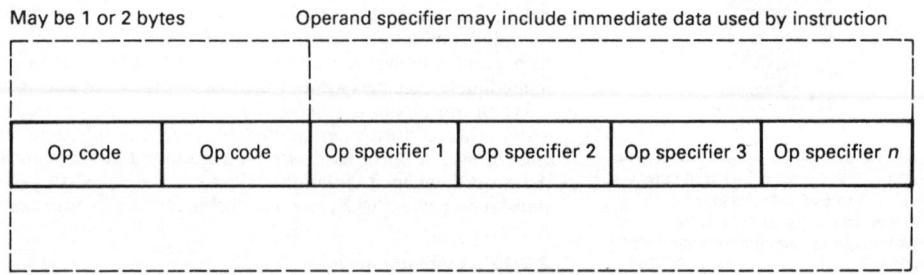

May be 1 or 2 bytes   Operand specifier may include immediate data used by instruction

| Op code | Op code | Op specifier 1 | Op specifier 2 | Op specifier 3 | Op specifier *n* |

Total length of the instruction in bytes – variable according to instruction type

Figure 4.4 General VAX instruction set format

followed if the test failed, while for a positive result, the code to be followed starts in the next but one location.

Branch or skip tests on other status bits in the CPU are often provided (e.g. on arithmetic carry and overflow).

4.7.12.3 Input/output

CPUs like the VAX, with memory mapped inout/output, do not require separate instructions for transferring data and status information between CPU and peripheral controllers. For this function, as well as performing tests on status information and input data, the normal data transfer and logical instructions are used.

Otherwise, separate input/output instructions provide these functions. Their general format is a transfer of data between the accumulator or other registers, and addressable data, control or status registers in peripheral controllers. Some CPUs also implement special input/output instructions such as:

1. *Skip if 'ready' flag set.* For the particular peripheral addressed, this instruction tests whether it has data awaiting input or whether it is free to receive new output data. Using a simple program loop, this instruction will synchronize the program with the transfer rate of the peripheral.
2. *Set interrupt mask.* This instruction outputs the state of each accumulator bit to an interrupt control circuit of a particular peripheral controller, so that, by putting the appropriate bit pattern in the accumulator with a single instruction, interrupts can be selectively inhibited or enabled in each peripheral device.

4.7.12.4 Arithmetic

1. Add contents of memory location to contents of accumulator, leaving result in accumulator. This instruction, together with instructions in category (2) for handling a carry bit from the addition, and for complementing a binary number, can be used to carry out all the four arithmetic functions by software subroutines.
2. Shift. This is also valuable in performing other arithmetic functions, or for sequentially testing bits in the accumulator contents. With simpler instruction sets, only one bit position is shifted for each execution of the instruction. There is usually a choice of left and right shift, and arithmetic shift (preserving the sign of the word and setting the carry bit) or logical rotate.

Extended arithmetic capability, either as standard equipment or a plug-in option, provides multiply and divide instructions and often multiple-bit shift instructions.

4.7.12.5 Control

Halt, no operation, branch, jump to sub-routine, interrupts on, interrupts off, are the typical operations provided as a minimum. A variety of other instructions will be found, specific to individual CPUs.

4.7.13 CPU implementation

The considerable amount of control logic required to execute all the possible CPU instructions and other functions is implemented in one of two ways.

4.7.13.1 Random logic

Random logic uses the available logic elements of gates, flip-flops, etc., combined in a suitable way to implement all the steps for each instruction, using as much commonality between instructions as possible. The various logic combinations are invoked by outputs from the instruction decoder.

4.7.13.2 Microcode

This is a series of internally programmed steps making up each instruction. These steps or micro-instructions are loaded into ROM using patterns determined at design time, and for each instruction decoded, the micro program ROM is entered at the appropriate point for that instruction. Under internal clock control, the micro-instructions cause appropriate control lines to be operated to effect the same steps as would be the case if the CPU were designed using method (1).

The great advantage of microcoded instruction sets is that they can readily be modified or completely changed by using an alternative ROM, which may simply be a single chip in a socket. In this way, a different CPU instruction set may be effected.

In conjunction with microcode, bit-slice microprocessors may be used to implement a CPU. The bit-slice microprocessor contains a slice or section of a complete CPU, i.e. registers, arithmetic and logic, with suitable paths between these elements. The slice may be 1, 2 or 4 bits in length, and, by cascading a number of these together, any desired word length can be achieved. The required instruction set is implemented by suitable programming of the bit-slice microprocessors using their external inputs controlled by microcode.

The combination of microcode held in ROM and bit-slice microprocessors is used in the implementation of many CPU models, each using the same bit-slice device.

4.7.14 CPU enhancements

There are several areas in which the operating speed of the CPU can be improved with added hardware, either designed in as an original feature or available as an upgrade to be added in-field. Some of the more common areas are described below.

4.7.14.1 Cache memory

An analysis of a typical computer program shows that there is a strong tendency to access repetitively instructions and data held in fairly small contiguous areas of memory. This is due to the fact that loops (short sections of program re-used many times in succession) are very frequently used, and data held in arrays of successive memory locations may be repetitively accessed in the course of a particular calculation. This leads to the idea of having a small buffer memory, of higher access speed than the lower-cost technology employed in main memory, between CPU and memory. This is known as cache memory. Various techniques are used to match the addresses of locations in cache with those in main memory, so that for memory addresses generated by the CPU, if the contents of that memory location are in cache, the instruction or data are accessed from the fast cache instead of slower main memory. The contents of a given memory location are initially fetched into cache by being addressed by the CPU. Precautions are taken to ensure that the contents of any location in cache which is altered by a write operation are rewritten back into main memory so that the contents of the location, whether in cache or main memory, are identical at all times.

A constant process of bringing memory contents into cache (thus overwriting previously used information with more currently used words) takes place completely transparently to the user. The only effect to be observed is an increase in execution speed. This speeding up depends on two factors: hit rate (i.e. percentage of times when the contents of a required location are already in cache) and the relative access times of main and cache memory. The hit rate, itself determined by the size of cache memory and algorithms for its filling, is normally better than 90%. This is dependent, of course, on the repetitiveness of the particular program being executed. The increased speed is achieved by using faster, more expensive memory (sometimes core memory). The additional expense for the relatively small amount of memory being used is more than offset by the speed advantage obtained.

4.7.14.2 RISC computers

Most computers require an instructions set of considerable size and complexity in order to provide all the facilities contained in the operating systems that support and manage them. This arrangement has many advantages, especially for commercial organizations, but also suffers from a distinct disadvantage—the more complex the instruction set, the more processor time and effort is required to decode and carry out each instruction. This can (and does) lead to significant reductions in overall processor performance for very large and complex operating systems such as MVS and VMS.

Research into ways of solving this problem began in the late 1970s, principally in the USA, but it was not until 1984 that the first commercially available computer with a 'reduced' instruction set was sold by Pyramid Technology. This design gave rise to the term 'reduced instruction set computer', or RISC as it is more commonly referred to today. In order to distinguish between these processors and the normal complex instruction set computers that preceded them, the term 'complex instruction set computer' (or CISC) was also brought into general use.

Within a RISC processor all superfluous or little-used instructions are removed from the operating system. All instructions will generally be of the same length and take the same amount of time to process. Both of these characteristics enable pipelining and other techniques to be used to effect savings in the time taken to execute each instruction. Typically, all instructions are hardwired in the processor chip (also faster than resorting to microcode). Much use is made of a higher number of registers than normal, thus many more instructions address registers as opposed to main memory. Where memory is addressed, it is often within the very large cache memories that are another feature of RISC processors. All these characteristics contribute to the faster processing speed *per instruction* with a RISC architecture. However, since the instructions are simpler and microcode is not used, then some fucntionality requires many more instructions on RISC than on CISC processors. Overall, there appears to be savings in the region of 25–30% of RISC over CISC for computer-intensive applications. Note that direct comparisons of MIPs (million of instructions per second) between RISC and CISC processors are not a good guide to overall performance that will be obtained.

Most RISC processors run under the UNIX operating system (or one of its clones), since this system is simpler and easier to gain entry to than most proprietary operating systems. Two important players in the RISC arena are Sun Microsystems Inc. and MIPS Computer Systems Inc., both in the USA, the former for its open SPARC (Scalable Processor ARChitecture) RISC architecture, the latter for the fact that all its efforts as a corporation are aimed at the development and sale of RISC-based technology. It is likely that the use of RISC technology will grow over the next decade, though the extremely large existing investments in current operating systems and CISC technology mean that this progress will not be as rapid and as widespread as some of the players in the RISC game would hope for. All the major computer manufacturers have already undertaken research in this area or have announced their intention to do so in the near future.

4.7.15 Fixed and floating-point arithmetic hardware

As far as arithmetic instructions go, simpler CPUs only contain add and subtract instructions, operating on single-word operands. Multiplication, of both fixed and floating-point numbers, is then accomplished by software subroutines, i.e. standard programs which perform multiplication or division by repetitive use of the add or subtract instructions, which can be invoked by a programmer who requires to perform a multiplication or division operation.

By providing extra hardware to perform fixed-point multiply and divide, which also usually implements multiple place-shift operations, a very substantial improvement in the speed of multiply and divide operations is obtained. With the hardware techniques used to implement most modern CPUs, however, these instructions are wired in as part of the standard set.

Floating-point format (Figure 4.5) provides greater range and precision than single-word fixed-point format. In floating-point representation, numbers are stored as a fraction times 2^n, where n can be positive or negative. The fraction (or mantissa) and exponent are what is stored, usually in two words for single-precision floating-point format or four words for double precision.

Hardware to perform add, subtract, multiply and divide operations is sometimes implemented as a floating-point processor, an independent unit with its own registers to which floating-point instructions are passed. The floating-point processor (sometimes called co-processor) can then access the operands, perform the required arithmetic operation and signal the CPU, which has meanwhile been free to continue with its own processing until the result is available.

An independent floating-point processor clearly provides the fastest execution of these instructions, but even without that, implementing them within the normal instruction set of the CPU, using its addressing techniques to access operands in memory, provides a significant improvement over software subroutines. The inclusion of the FPP into 'standard' CPUs is becoming almost standard.

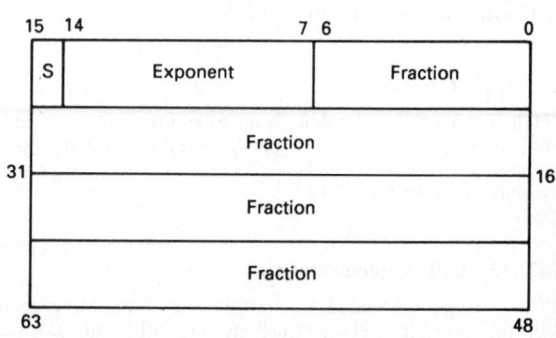

Figure 4.5 32-bit double floating-point format

4.7.16 Array processors

Similar to an independent floating-point processor described above, an optional hardware unit which can perform complete computations on data held in the form of arrays of data in memory, independent from the CPU and at high speed, is known as an array processor. These are used in specialized technical applications such as simulation, modelling and seismic work. An example of the type of mathematical operation which would be carried out by such a unit is matrix inversion. The ability of these units to perform very high-speed searches based upon text keys has also led to a growing use of them for the rapid retrieval of data from large data banks, particularly in areas such as banking, where real-time ATM terminals require fast response to account enquiries from very large data sets.

4.7.17 Timers and counters

For systems which are used in control applications, or where elapsed time needs to be measured for accounting purposes (as, for example, in a time-sharing system where users are to be charged according to the amount of CPU time they use), it is important to be able to measure intervals of time precisely and accurately. This measurement must continue while the system is executing programs, and must be 'real time', i.e. related to events and time intervals in the outside world.

Most CPUs are equipped with a simple real-time clock which derives its reference timing from 50–60 Hz mains. These allow a predetermined interval to be timed by setting a count value in a counter which is decremented at the mains cycle rate until it interrupts the CPU on reaching zero.

More elaborate timers are available as options, or are even standard items on some CPUs. These are driven from high-resolution crystal oscillators, and offer such features as:

1. More than one timer simultaneously;
2. Timing random external events;
3. Program selection of different time bases;
4. External clock input.

The system supervisory software normally keeps the date and time of day up to date by means of a program running in the background all the time the system is switched on and running. Any reports, logs or printouts generated by the systems can then be labelled with the date and time they were initiated. To overcome having to reset the data and time every time the system is stopped or switched off, most CPUs now have a permanent battery-driven date and time clock which keeps running despite stoppages and never needs reloading once loaded initially (with the exception of change to and from Summer Time).

Counters are also useful in control applications to count external events or to generate a set number of pulses (for example, to drive a stepping motor). Counters are frequently implemented as external peripheral devices, forming part of the digital section of a process input/output interface.

4.7.18 Input/output

In order to perform any useful role, a computer system must be able to communicate with the outside world, either with human users via keyboards, CRT screens, printed output, etc. or with some external hardware or process being controlled or monitored. In the latter case, where connection to other electronic systems is involved, the communication is via electrical signals.

All modern computer systems have a unified means of supporting the variable number of such human or process input/output devices required for a particular application, and indeed for adding such equipment to enhance a system at the user's location. As well as all input/output peripherals and external mass storage in the form of magnetic tape, compact disk and disk units, some systems also communicate with main memory in this common, unified structure. In such a system (for example, the VAX) there is no difference between instructions which reference memory and those which read from and write to peripheral devices. The benefits of a standard input/output bus to the manufacturer are:

1. It provides a design standard allowing easy development of new input/output devices and other system enhancements;
2. Devices of widely different data transfer rates can be accommodated without adaptation of the CPU;
3. It permits development of a family concept.

Many manufacturers have maintained a standard input/output highway and CPU instruction set for as long as a decade or more. This has enabled them to provide constantly improving system performance and decreasing cost by taking advantage of developing technology, while protecting very substantial investments in peripheral equipment and software. For the user of a system in such a family, the benefits are:

1. The ability to upgrade to a more powerful CPU while retaining existing peripherals;
2. Retention of programs in moving up- or down-range within the family;
3. In many cases the ability to retain the usefulness of an older system by adding more recently developed peripherals, and in some cases even additional CPU capacity of newer design and technology (see Section 4.5.7).

4.7.19 Input/output bus

The common structure for any given model of computer system is implemented in the form of an electrical bus or highway. This specifies the number, levels and significance of electrical signals and the mechanical mounting of the electrical controller or interface which transforms the standard signals on the highway to ones suitable for the particular input/output or storage device concerned. A data highway or input/output bus needs to provide the following functions.

4.7.19.1 Addressing

A number of address lines is provided, determining the number of devices that can be accommodated on the system. For example, six lines would allow 63 devices. Each interface on the bus decodes the address lines to detect input/output instructions intended for it.

4.7.19.2 Data

The number of data lines on the bus is usually equal to the word length of the CPU, although it may alternatively be a sub-multiple of the word length, in which case input/output data are packed into or unpacked from complete words in the CPU. In some cases data lines are bi-directional, providing a simpler bus at the expense of more complex drivers and receivers.

4.7.19.3 Control

Control signals are required to synchronize transactions between the CPU and interfaces and to gate address and data signals to and from the bus. Although all the bits of an address or data word are transmitted at the same instant, in transmission down the bus, because of slightly different electrical characteristics of each individual line, they will arrive at slightly different times. Control signals are provided to gate these skewed signals at a time when they are guaranteed to have reached their correct state.

4.7.20 Types of input/output transactions

Three types of transaction via the input/output bus between CPU and peripheral device are required, as described below.

4.7.20.1 Control and status

This type of transfer is initiated by a program instruction to command a peripheral device to perform a certain action in readiness for transferring data or to interrogate the status of a peripheral. For example, a magnetic tape unit can be issued with a command to rewind, the read/write head in a disk unit to be positioned above a certain track on the disk, the completion of a conversion by an analogue-to-digital converter verified, or a printer out of paper condition may be sensed.

Normally, a single word of control or status information is output or input as a result of one instruction, with each bit in the word having a particular significance. Thus multiple actions can be initiated by a single control instruction, and several conditions monitored by a single status instruction. For the more complex peripheral devices, more than one word of control or status information may be required.

4.7.20.2 Programmed data transfer

For slow and medium-speed devices (for example, floppy disk units or line printers) data are input or output one word at a time, with a series of program instructions required for every word transferred. The word or data are transferred to or from one of the CPU registers, normally the accumulator. In order to effect a transfer of a series of words forming a related block of data (as is normally required in any practical situation) a number of CPU instructions per word transferred are required. This is because it is necessary to take the data from (or store them into) memory locations. As a minimum, in a simple case, at least six CPU instructions are required per word of data transferred.

In a system such as the VAX, where instructions can reference equally memory locations, peripheral device registers and CPU registers, the operation is simplified since a MOVE instruction can transfer a word of data directly from peripheral to memory without going through a CPU register. This applies equally to control and status instructions on the VAX, with a further advantage that the state of bits in a peripheral device status register can be tested without transferring the register contents into the CPU.

The rate of execution of the necessary instructions must match the data transfer rate of the peripheral concerned. Since it is usually desired that the CPU continue with the execution of other parts of the user's program while data transfer is going on, some form of synchronization is necessary between CPU and peripheral to ensure that no data are lost. In the simplest type of system, the CPU simply suspends any other instructions and constantly monitors the device status word, awaiting an indication that the peripheral has data ready for input to

the CPU or is ready to receive an output from it. This i wasteful of CPU time where the data transfer rate is slow relative to CPU instruction speeds, and in this case the use of 'interrupt' facilities (see Section 4.7.21) provides this synchronization.

4.7.20.3 Direct memory access

For devices which transfer data at a higher rate (in excess o around 20 000 words per second) a different solution i required. At these speeds, efficiency is achieved by giving the peripheral device controller the ability to access memory autonomously without using CPU instructions. With very fas tape or disk units which can transfer data at rates in excess of (million bytes per second, direct memory access (DMA) is the only technique which will allow these rates to be sustained.

The peripheral controller has two registers which are loaded by control instructions before data transfer can begin. These contain:

1. The address in memory of the start of the block of data
2. The number of words which it is desired to transfer in the operation.

When the block transfer is started the peripheral controller using certain control lines in the input/output bus, sequentiall accesses the required memory locations until the specifie number of words has been transferred. The memory addresse are placed on address lines of the input/ouput bus, togethe with the appropriate control and timing signals, for each word transferred. On completion of the number of words specifie in the word-count register the peripheral signals to the CPU that the transfer of the block of data is completed.

Other than the instructions required initially to set the star address and word-count registers and start the transfer, DMA transfer is accomplished without any intervention from the CPU. Normal processing of instructions therefore continues. Direct memory access (more than one peripheral at time can be engaged in such an operation) is, of course competing with the CPU for memory cycles, and the process ing of instructions is slowed down in proportion to th percentage of memory cycles required by peripherals. In th limit, it may be necessary for a very-high-speed peripheral to completely dominate memory usage in a burst mode c operation, to ensure that no data are lost during the transfe through conflicting requests for memory cycles.

4.7.21 Interrupts

The handling of input/output is made much more efficien through the use of a feature found in varying degrees c sophistication on all modern systems. This is known as 'auto matic priority interrupt', and is a way of allowing periphera devices to signal an event of significance to the CPU (e.g. i some systems a terminal keyboard having a character read for transmission, or completion of DMA transfer) in such way that the CPU is made to suspend temporarily its curren work to respond to the condition causing the interrup Interrupts are also used to force the CPU to recognize an take action on alarm or error conditions in a peripheral (e. printer out of paper, error detected on writing to a magneti tape unit).

Information to allow the CPU to resume where it wa interrupted (e.g. the value of the program counter) is store when an interrupt is accepted. It is necessary also for th device causing the interrupt to be identified, and for th program to branch to a section to deal with the conditio which caused the interrupt (Figure 4.6).

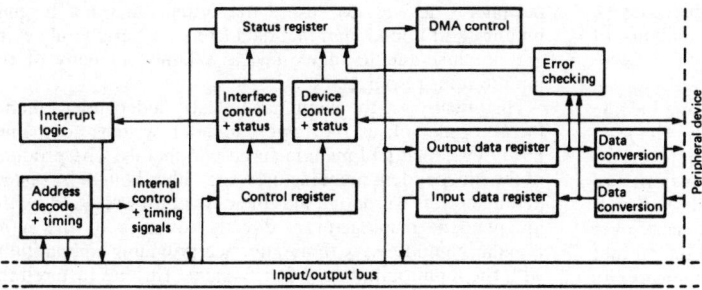

Figure 4.6 Block diagram of peripheral interface

Examples of two types of interrupt structure are given below, one typical of a simpler system such as an 8-bit microprocessor or an older architecture minicomputer, the other representing a more sophisticated architecture such as the VAX. In the simpler system, a single interrupt line is provided in the input/output bus, onto which the interrupt signal for each peripheral is connected. Within each peripheral controller, access to the interrupt line can be enabled or disabled, either by means of a control input/ouput instruction to each device separately or by a 'mask' instruction which, with a single 16-bit word output, sets the interrupt enabled/disabled state for each of up to 16 devices on the input/output bus. When a condition which is defined as able to cause an interrupt occurs in a peripheral, and interrupts are enabled in that device, a signal on the interrupt line will be sent to the CPU. At the end of the instruction currently being executed this signal will be recognized.

In this simple form of interrupt handing the interrupt servicing routine always begins at a fixed memory location. The interrupt forces the contents of the program counter (which is the address of the next instruction that would have been executed had the interrupt not occurred) to be stored in this first location and the program to start executing at the next instruction. Further interrupts are automatically inhibited within the CPU, and the first action of the interrupt routine must be to store the contents of the accumulator and other registers so that on return to the main stream of the program these registers can be restored to their previous state.

Identification of the interrupting device is done via a series of conditional instructions on each in turn until an interrupting device is found. Having established which device is interrupting, the interrupt-handling routine will then branch to a section of program specific to that device. At this point or later within the interrupt routine an instruction to re-enable the CPU interrupt system may be issued, allowing a further interrupt to be received by the CPU before the existing interrupt-handling program has completed. If this 'nesting' of interrupts is to be allowed, each interruptable section of the interrupt routine must store the return value of the program counter elsewhere in the memory, so that as each section of the interrupt routine is completed, control can be returned to the point where the last interrupt occurred.

A more comprehensive interrupt system differs in the following ways from that described above:

1. Multiple interrupt lines are provided, and any number of devices can be on each line or level.
2. The CPU status can be set to different priority levels, corresponding to different interrupt lines. Only interrupt on a level higher than the current priority are immediately serviced by the CPU. This provides a more adaptable way

of dealing with a wide range of devices of different speeds and with different degrees of urgency.
3. When an interrupt is accepted by the CPU the interrupting device sends a vector or pointer to the CPU on the input/output bus address lines. This points to a fixed memory address for each device, which holds the start address of its interrupt routine, and in the following memory word, a new status word for the CPU, defining its priority level and hence its ability to respond to other levels of interrupt during this interrupt routine. By avoiding the need for the CPU to test each device until it finds the interrupting one, response to interrupts is much faster.
4. The current value of the program counter and processor status word are automatically placed on a push-down stack when an interrupt occurs. A further interrupt accepted within the current interrupt routine will cause the program counter and status word to be stored on the top of the stack, and the existing contents to be pushed down into the stack. On return from an interrupt routine, the program counter and status word stored when that interrupt occurred are taken from the top of the stack and used by the CPU, allowing whatever was interrupted to continue as before. This can take place for any number of interrupts, subject only to the capacity of the stack. Thus 'nesting' to any level is handled automatically without the need for the programmer to store the program counter at any stage.

4.8 Peripherals

Peripheral devices fall into the following three categories.

4.8.1 Interactive

These are designed to allow humans to interact with the system by outputting information in the form of voice, readable alphanumeric text or graphics, either on paper or on a display screen, and accepting information from humans through manual devices such as keyboards, voice-recognition devices or by scanning printed text or images. The general function performed by devices in this class is sometimes referred to as human–machine interaction.

4.8.2 Storage

These act as a back-up form of storage to supplement the main memory of the system. The most simple of these (now largely superseded) was punched paper tape or cards, and the most complex now ranges up to very large disks and magnetic tapes, each capable of holding up to thousands of millions of characters of information. Peripherals of this type are generally known as mass-storage devices.

Recent developments in the field of mass storage include both compact disks (CDs) and solid-state disks. The CD used is essentially the same material and technique employed for commercial recording of music, etc., with the majority of equipment able to operate on a 'read only' basis. The advantage is the considerable increase in data-storage density over conventional magnetic tape or disk. Solid-state disks are not really physical disks at all, but rather fast large-capacity CMOS memory, which the operating system treats logically as if it were a disk drive. The advantage is speed over conventional media, but at a cost premium. Solid-state disks are still not yet large enough for long-term data storage or archiving, but this will change as memory technology improves and the unit costs fall in relation to disk technology. Units holding up to 1 gigabyte of information which can then be transferred at a speed of 36 megabytes per second have already been developed.

4.8.3   Communication

These are interfaces between the computer system and other systems or electronic devices. Analogue/digital converters, digital input/output and communication line interfaces are good examples.

An example of progress made with respect to communication principles between the computer and its environment is a device often called the 'laser glove'. This technique combines many of the traditional input/output principles into a single new dimension. In some areas of industry or research (such as nuclear physics) it is undesirable for parts of the human body to come into contact with certain substances such as radioactive material. Yet in many cases a degree of precision of movement that is difficult to predefine is called for, thus eliminating the use of pre-programmed robotics. In such instances the combined skills of the human hand, eye and brain are still required. The laser glove is able to project a hologram of the material being worked on in front of the operator. The operator wears a tight-fitting glove that is able to sense every minute movement of the hand. Any such movement is instantly replicated by a robotic device located in the unsafe environment. Thus the operator is able to carry out tasks on a hologram, and have those same tasks carried out safely by a robot in what, for a human, would be an unsafe environment.

4.9   Output devices

Either interactive or communications peripherals (or both) are required in every system. The existence of storage depends on the need for additional storage over and above main memory. All peripheral devices in a system are connected via the input/output structure (as described in Section 4.13) to the CPU, memory, and in some systems to a separate input/output processor.

The throughput rates and flexibility of the input/output structure determine the number and variety of peripheral devices which can be handled in a system before the input/output requirements begin to saturate the system and prevent any processing of instructions being done by the CPU. In deciding on the configuration of a particular system it is important to analyse the throughput requirement dictated by

peripheral devices, to ensure the system does not become input/output bound, and that data from any peripheral device are not lost due to other devices taking too many of the input/output resources.

Historically, in the computer industry independent manufacturers as well as the large computer systems companies have developed and manufactured peripherals. The products of the independent manufacturers are either bought by system manufacturers for design into their systems or are sold by the independent manufacturers directly to users of the more popular computers, with an interface providing compatibility with the input/output bus of the system. This has fostered the development of many of the widely used, cost-effective peripherals available today, such as floppy disks and printers.

Certain storage devices with removable storage media where the format recording data on the media have been standardized, can be used for exchanging data between systems from different suppliers. This is important where data may be gathered on one system and need to be analysed on a different one more suitable for that purpose. However, due to the very large growth of networking in the 1980s, even between equipment from different manufacturers, the moving of data from one machine to another is most commonly achieved by file transmission, though for very large files magnetic tapes are used for this purpose.

4.10   Terminals

A data terminal is essentially an output device at which a computer user sits, either to receive data in alphanumeric or graphic form or to input data through a keyboard or other form of manual input, or both. Terminals range in cost and complexity from a 100 characters per second serial printer to a graphics terminal with a large colour CRT screen. Most are connected to the CPU by data-communications interfaces and can therefore be situated at some distance from the system within a building (e.g. on a LAN – local area network), or indeed remote from the computer site communicating over the public telephone network or over a company's private network (WAN – wide area network).

4.10.1   Dot matrix printers

Only a small proportion of low-speed printers are now supplied with any means of input such as a keyboard. This is due to the widespread use of interactive terminals for data input and to the fact that many of these printers are themselves attached in some way to a terminal to act as a slave by producing hardcopy output through the terminal.

The most common use for printers with keyboards is as operators' consoles on large computers, where a typical speed is in the range 100–250 characters per second. The print head consists of seven or more needles held in a vertical plane in the head assembly which is positioned with the needles perpendicular to the paper and spaced a short distance from it, with a carbon ribbon interposed between. Each needle can be individually driven by a solenoid to contact the paper through the ribbon, thus printing a dot. A complete character is formed by stepping the head through five or more positions horizontally and at each position energizing the appropriate solenoids. The head is then stepped onto the position for the next character.

When the end of a line is reached the paper is advanced one line and the print head either returns to the left margin position or, in some faster printers, prints the next line from right to left. This is possible where the printer is provided with storage for a line or more of text, and the characters can be xtracted from this store in the reverse order. Throughput peed is improved where this technique is used, by saving edundant head movement. The 7×5 dot matrix within which this type of printer forms each character allows an cceptable representation of alpha and numeric characters. Better legibility, particularly of lower-case characters with descenders, can be achieved by using a larger matrix such as 9×7, i.e. a head with nine needles stepping through seven positions for each character. Manufacturers can now supply 24-pin dot matrix printers, in order to provide higher-quality print and a larger character set.

Character codes are received for printing, sent from the keyboard in asynchronous serial form, on a line driven by a data-communications interface in the computer, whose transmission speed determines the overall printing and keying throughput. Buffer storage for up to 4 million characters is provided in printers which use serial data communications, to make the most efficient use of communication lines, and in printers with built-in intelligence, to allow look-ahead so that the print head can skip blanks and take the shortest route to the next printable character position.

Character sets can be readily changed by replacing the ROM chip which contains the dot patterns corresponding to each character code or, more usually, by sending the character patterns over the network to the printer from the host CPU, often referred to as 'downline loading'. The latter method has the advantage of character selection at any time under program control and without any human intervention. Some printers already contain in-built multiple character sets.

4.10.2 Letter-quality printers

The earliest models of dot matrix printers could not form and print characters of a complex nature or produce a high-quality print image. To overcome this, printers which used a metal wheel, with the characters pre-formed on the end, were used for occasions where the quality of the image was paramount (e.g. business correspondence). These became known as the daisywheel printers. Since dot matrix printers have continued to make large improvements in this respect, some offering different quality of print aacording to requirement and selectable by program control, these letter-quality printers have exceeded in sales and use due to their higher cost of construction and maintenance.

Other types of serial printer which give high print quality are the ink jet printer and the laser printer.

4.10.3 Visual display unit

A VDU is a terminal in which a CRT is used to display alphanumeric text. It is normally also equipped with a keyboard for data input, and occasionally a printer may be slaved to the VDU to produce a permanent paper copy of the information displayed on the screen at a particular time.

The format for the layout of text on the screen is most commonly 24 lines of text (sometimes with a twenty-fifth line reserved for system messages to the terminal user), each with up to 80 character positions. Most VDUs also allow 132 columns to be displayed in a line, using a special compressed character set. The latter feature is useful for compatibility with computer printouts that normally have up to 132 columns of print. Displaying of new text on the screen takes place a character at a time, starting in the top left-hand corner and continuing line by line. When the screen is full, the page of text moves up one line, allowing a new line to be added at the bottom.

Characters are caused to be displayed either by outputting data from the CPU through a serial communications interface or by the operator typing on a typewriter-like keyboard.

VDUs are classified depending on which of the following modes of message composition they use:

1. *Block mode.* A full screen of text is composed and (if need be) edited by the operator, and the corresponding character codes held in a buffer store in the VDU. Transmission of the text from the buffer store to the CPU is then done as a continuous block of characters, when a 'transmit' key is pressed.
2. *Character mode.* Each character code is transmitted to the CPU as the corresponding key is pressed. Characters are 'echoed' back from the CPU for display on the screen. This function, plus editing of entered data, is performed by program in the CPU.

To assist with character positioning on the screen, a cursor is displayed showing the position in which the next character will appear. On most VDUs its position can be altered by control characters sent by program to the VDU or by the operator using special keys, so that inefficient and time-consuming use of 'space' and 'new line' controls is unnecessary.

Other functions, some or all of which are commonly provided depending on cost and sophistication, are:

1. Blinking;
2. Dual intensity;
3. Reverse video (black characters on a white or coloured background);
4. Underline;
5. Alternate character sets;
6. Protected areas.

Each of these can be selected on a character-by-character basis. Additionally, the following are attributes of the whole screen:

1. Bi-directional scrolling;
2. Smooth scrolling (instead of jumping a line at a time);
3. Split-screen scrolling;
4. Enlarged character size;
5. Compressed characters.

The most commonly used colours for displayed characters on monochrome VDUs are white, green and amber, all on a dark background, or the reverse of these where reverse video is available. Almost all VDUs currently produced have keyboards separate from the body of the VDU for ease of use and the comfort of the user, as well as some form of graphics capability ranging from simple graphics to very high-resolution colour monitors that produce an image that is difficult to distinguish from a fine-grain colour photograph. The level of sophistication on 'standard' VDUs has risen considerably since the start of the 1990s and high-resolution multi-choice colour terminals are no longer confined to specialist areas such as CAD/CAM, television and engineering.

Industries and professions such as aircraft and vehicle manufacture, electronics, and structural and civil engineering use VDUs in the design of items such as car bodies, printed circuit boards and building frameworks, by creating and modifying two- or three-dimensional representations of these objects on the CRT screen. Keyboards and devices such as joysticks, rolling balls and light pens are provided for modification of displayed information, as shown in Figure 4.7.

A graphics terminal normally has its own graphics processor to interpret picture information output from the CPU and held in a picture store. Thus, the program in the CPU need only supply parameters such as the start and end points of a vector, or the centre and radius of a circle, and the necessary

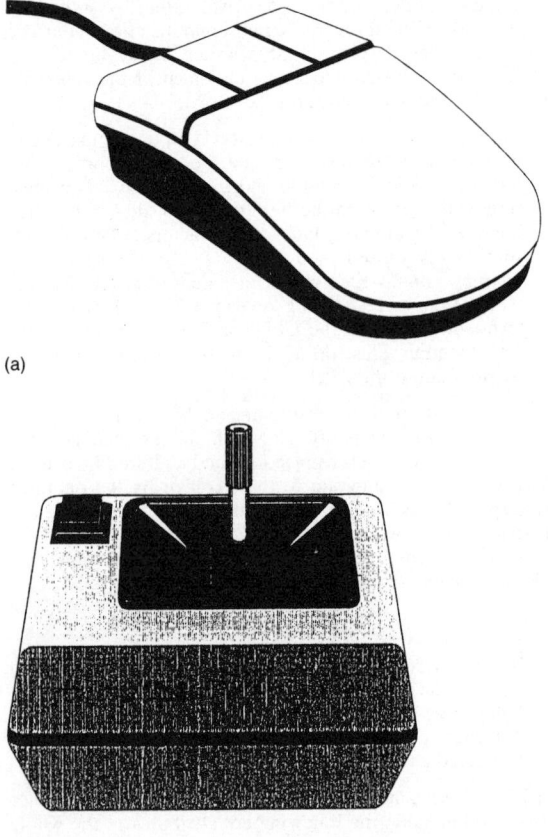

(a)

(b)

(c)

Figure 4.7 Interactive devices. (a) Mouse. (b) Light pen. (c) Joystick

processing is done within the terminal to display them on the screen. As with dot matrix printers, characters are formed as a series of dots within a matrix.

The development of visual display units with very powerful graphics processors, networking capabilities and large-capacity disk-storage facilities has given rise to a new term to describe this revolutionary concept. Such a combination of equipment is commonly referred to as a 'workstation', from the assumption that operatives would have everything that they required to work within one 'station'. Initially, this expression covered the desk and other communications equipment (such as telephones) that was used, but today the term 'workstation' is usually confined to the computer equipment. Though this explanation is given under the heading 'Visual Display Unit', it should be appreciated that workstations are blurring the dividing line between the different types of peripherals attached to computers, with much more emphasis on integrating them into single units.

Specialized terminals are common in areas such as:

1. Shops, as point-of-sale terminals, i.e. sophisticated cash registers linked to an in-store computer which is often adjusting a stock control system at the same time as registering the sale and debiting the customer's bank or credit card account. These are referred to as EFTPOS (Electronic Funds Transfer at Point Of Sale) terminals;

2. Banking for customer cash dispensing, enquiries and other transactions, or for teller use, including the ability to print entries in passbooks or to read a card's magnetic strip containing details of the customer's account and so eliminate the need for completion of a cheque or paying-in slip;

3. Manufacturing for shopfloor data collection and display.

These typically use features found in the terminals described above, and, in addition, may have the capability to read magnetic stripes on credit cards, punched plastic cards or identity badges, or bar codes on supermarket goods or parts in a factory.

4.10.4 High-speed printers and plotters

4.10.4.1 Line printer

For greater volume of printed output than can be achieved with serial printers, line printers that can produce a whole line of characters almost simultaneously are available. Using impact techniques, speeds up to 3500 full lines (usually 132 characters each) per minute are possible. Continuous paper in fan-fold form, which may be multi-part to produce copies, is fed through the printer a line at a time by a transport system consisting of tractors which engage sprocket holes at the edge of the paper to move it upwards and through the printer from front to rear. A paper tray at the rear allows the paper to fold up again on exit from the printer.

As well as advancing a line at a time, commands can be given to advance the paper to the top of the next page, or to advance a whole page or line. This is important, for example, where pre-printed forms are being used.

Two types of line printer are in common use: drum printers and band printers. Both use a horizontal row of hammers, one per character position or, in some cases, shared between two positions. These are actuated by solenoids to strike the paper through a carbon film against an engraved representation to print the desired character. In a drum printer a print drum the length of the desired print line rotates once per print line. In each character position the full character set is engraved around the circumference of the drum. A band printer has a horizontal revolving band or chain of print elements, each with a character embossed on it. The full character set is represented on the band in this way. To implement different character fonts involves specifying different barrels in the case of a drum printer, whereas a change can be made readily on a band printer by an operator changing bands, or individual print elements in the band can be replaced.

The printer has a memory buffer to hold a full line of character codes. When the buffer is full (or terminated if a short print line is required) a print cycle is initiated automatically. During this print cycle the stored characters are scanned and compared in synchronism with the rotating characters on the drum or band. The printer activates the hammer as the desired character on the drum or band approaches in each print position.

4.10.4.2 Laser printers

These are available to meet three different types of printing requirements:

1. Very high volumes of output at speeds exceeding 200 pages per minute. They are normally used by those requiring a constant high-volume printing service, since this equipment is expensive to buy, run and service.
2. Departmental printing requirements, usually consisting of medium to high volumes on an *ad hoc* basis. This equipment would normally be networked to many CPUs and shared by a group of common users. They print at speeds of up to 40 pages per minute.
3. Desktop printing uses laser printers small enough to fit on an individual's desk, designed for intermittent low-volume personal printing requirements.

However, the technology used is common to all three. The principle used is that of the everyday photocopier, the difference being that the image to be copied is set up according to digital signals received from the host CPU instead of from a photoscan of the document to be copied. The main advantage of this form of output is the clarity and quality of the image printed. It is so good that it is possible not only to print data out also to print the 'form' or 'letterhead' of the paper at the same time, thus avoiding the cost of pre-printed stationery. The disadvantage is that it is currently not possible to print multiple copies simultaneously.

4.10.4.3 Pen plotter

Another form of hard copy output is provided by pen plotters. These are devices aimed primarily at high-complexity graphics with a limited amount of text. Their uses range from plotting graphs of scientific data to producing complex engineering drawings in computer-aided design applications such as drawings used in integrated-circuit chip design.

The plotter has one or more pens held vertically above a table on which the paper lies. These can be of different colours, and as well as being raised or lowered on to the paper individually by program commands, they can be moved in small steps, driven by stepping motors. They plot in the X and Y directions, or achieve control in one axis by moving the paper back and forth between supply and take-up rolls under stepping motor control, and in the other axis by pen movement. Diagonal lines are produced by combinations of movements in both axes.

With step sizes as small as 0.01 mm, high accuracy plots can be produced (in multiple colours where more than one pen is used) and annotated with text in a variety of sizes and character sets. Supporting software is usually provided with a plotter. This will, for example, scale drawings and text and generate alphanumeric characters.

4.10.4.4 Electrostatic plotter

The objectives of the electrostatic plotter are the same as those of the pen plotter, the production of high-quality graphics in hardcopy form. However, electrostatic plotters achieve their output by setting an electrostatic charge on the paper in the same pattern as the required output image, and then attracting and retaining ink particles according to that pattern. This ink is then fused onto the paper in order to make a permanent image. This can even be accomplished in colour with almost unlimited ability to recreate the spectrum. The advantage of this approach over conventional pen plotters is speed, with electrostatic plotters achieving speeds up to 50 times faster. There is also considerably less movement of paper and equipment parts. However, electrostatic plots incur heavy production costs when compared to pen plotters, and have not come into large-scale general use for this reason.

4.11 Direct input

Other forms of direct input, eliminating the need for typing on the keyboard of a teleprinter or VDU, are described in this section.

4.11.1 Character recognition

This technique offers a high-speed method for the capture of source data as an alternative to keyboard input and for processing documents such as cheques. Several types of device exist, with varying capabilities and functions:

1. Page and document readers, with the capability to read several special fonts, plus, in some cases, lower-quality print including hand printing and hand marked forms as opposed to written or printed documents. Most character readers have some form of error handling, allowing questionable characters to be displayed to an operator for manual input of the correct character. A wide range of capabilities and hence prices is found, from simple, low-speed (several pages per minute) devices handling pages only, to high-speed readers for pages and comments, the former at up to two pages per second with the latter several times faster.
2. Document readers/sorters which read and optionally sort simple documents such as cheques and payment slips with characters either in magnetic ink or special font. These are

geared to higher throughputs (up to 3000 documents per minute) of standard documents.
3. Transaction devices, which may use both document reading and keyboard data entry, and where single documents at a time are handled.

4.11.2   Writing tablets

Devices using a variety of techniques exist for the conversion of hand-printed characters into codes for direct input to a CPU. The overall function of these is the same – the provision of a surface on which normal forms (typically, up to A4 size) can be filled in with hand-printed alphanumeric characters, using either a normal writing instrument in the case of pressure-sensitive techniques or a special pen. The benefits of this type of device include:

1. Immediate capture of data at source, avoiding time-consuming and error-prone transcription of data;
2. By detecting the movements involved in writing a character, additional information is gained compared with optical recognition, allowing characters which are easily confused by Optical Character Recognition (OCR) to be correctly distinguished.

4.12   Disk storage

Even though computer main memories are very large when compared to those produced during the 1970s (512 million characters is not uncommon), this is still nowhere near approaching the total memory storage capability required to retain and process all data. Backing storage provides this capability and has the added advantage in that it can be copied and stored away from the CPU for security and safekeeping. For backing storage to be most suitable it should be price-effective, reliable and easy to exchange and access. It is for this reason that the use of disk storage has grown considerably over the last ten years.

Disks are connected to CPU by a controller which is normally a DMA device attached to the input/otuput bus or to a high-speed data channel, except in the case of the slowest of disks, which may be treated as a programmed transfer device. The controller is generally capable of handling a number of drives, usually up to 16 or 32. Having multiple-disk drives on a system, as well as providing more on-line storage, allows copying of information from one disk to another and affords a degree of redundancy since, depending on application, the system may continue to function usefully with one less drive in the event of a failure. A disk controller is relatively complex, since it has to deal with high rates of data transfer, usually with error code generation and error detection, a number of different commands and a large amount of status information.

Four types of disk drive will be described in the following sections: floppy disk, cartridge disk, removable pack disk and 'Winchester Technology' disk. The following elements are the major functional parts common to all the above types of drive, with differences in implementation between the different types.

4.12.1   Drive motor

This drives a spindle on which the disk itself is placed, rotating at a nominally fixed speed. The motor is powered up when a disk is placed in the drive, and powered down (normally with a safety interlock to prevent operator access to rotating parts) until it has stopped spinning, when it is required to remove the disk from the system.

4.12.2   Disk medium

The actual recording and storage medium is the item which rotates. It is coated with a magnetic oxide material, and can vary from a flexible diskette of less than one megabyte capacity recording on one surface only (single-sided floppy disk) to an assembly of multiple disks stacked one above the other on a single axle (disk pack) holding thousands of megabytes of data).

4.12.3   Head mechanism

This carries read/write heads, one for each recording surface. The number of recording surfaces ranges from only one on a single-sided floppy disk to ten or more for a multi-surface disk pack. In the latter case, the heads are mounted on a comb-line assembly, where the 'teeth' of the comb move together in a radial direction between the disk surfaces.

During operation, the recording heads fly aerodynamically extremely close to the disk surface, except in the case of floppy disks, where the head is in contact with the surface. When rotation stops, the heads either retract from the surface or come to rest upon it, depending on the technology involved.

The time take for the read/write head to be positioned above a particular area on the disk surface for the desired transfer of data is known as the access time. It is a function partly of the rotational speed of the disk, which gives rise to what is known as the average rotational latency (i.e. one half of the complete revolution time of the disk). Out of a number of accesses, the average length of time it is necessary to wait for the desired point to come below the head approaches this figure. The second component of access time is the head-positioning time. This is dependent upon the number of tracks to be traversed in moving from the current head position to the desired one. Again, an average figure emerges from a large number of accesses. The average access time is the sum of these two components. In planning the throughput possible with a given disk system the worst-case figures may also need to be considered.

4.12.4   Electronics

The drive must accept commands to seek (i.e. position the head assembly above a particular track) and must be able to recover signals from the read heads and convert these to binary digits in parallel form for transmission to the disk controller. Conversely, data transmitted in this way from the controller to the disk drive must be translated into appropriate analogue signals to be applied to the head for writing the desired data onto the disk.

4.12.5   Drive control

Various other functions concerned with control of the drive and sensing of switches on the control panel are performed. On some more advanced drives, much of the operation of the drive and electronics can be tested off-line from the system, allowing fault diagnosis to be performed without affecting the rest of the system.

Information is recorded in a number of concentric, closely spaced tracks on the disk surfaces, and in order to write and thereafter read successfully on the same or a different drive it must be possible to position the head to a high degree of accuracy and precision above any given track. Data are recorded and read serially on one surface at a time, hence transfer of data between the disk controller and disk surface involves conversion in both directions between serial analogue and parallel digital signals. A phase-locked loop clock system s normally used to ensure reliable reading by compensating for variations in the rotational speed of the disk.

Data are formatted in blocks or sections on all disk systems, generally in fixed block lengths pre-formatted on the disk medium at the time of manufacture. Alternatively, 'soft sectoring' allows formatting into blocks of differing length by program. The drive electronics are required to read sector headers, which contain control information to condition the read circuitry of the drive, and sector address information, and to calculate, write and check an error-correcting code – normally a cyclic redundancy check – for each block.

Finding the correct track in a seek operation, where the separation between adjacent tracks may be as little as 0.01 mm, requires servo-controlled positioning of the head to ensure accurate registration with the track. All rigid-disk systems have servo-controlled head positioning, either using a separate surface pre-written with position information and with a read head only or with servo information interspersed with data on the normal read/write tracks being sampled by the normal read/write head. Floppy disk systems, where the tolerances are not so fine, have a simpler stepping motor mechanism for head positioning.

4.12.6 Floppy disk

The floppy disk, while having the four elements described above, was conceived as a simple, low-cost device providing a moderate amount of random access back-up storage to microcomputers, word processors and small business and technical minicomputers. As the name implies, the magnetic medium used is a flexible, magnetic oxide-coated diskette, which is contained in a square envelope with apertures for the drive spindle to engage a hole in the centre of the disk and for the read/write head to make contact with the disk. Diskettes are of three standard diameters, approximately 203 mm (8-inch), 133 mm ($5\frac{1}{4}$-inch) and 89 mm ($3\frac{1}{2}$-inch). The compactness and flexibility of the disk makes it very simple to handle and store, and possible for it to be sent by post.

One major simplification in the design of the floppy disk system is the arrangement of the read/write head. This runs in contact with the disk surface during read/write operations and is retracted otherwise. This feature and the choice of disk coating and the pressure loading of the head are such that, at the rotational speed of 360 rev/min, the wear on the recording surface is minimal. Eventually, however, wear and therefore error rate are such that the diskette may have to be replaced, copying the information onto a new diskette.

Capacities vary from the 256 kilobytes of the earliest drives, which record on one surface of the diskette only, to a figure of over 2 megabytes on more recent units, most of which use both surfaces of the diskette. Access times, imposed by the rather slow head-positioning mechanism using a stepping motor, are in the range of 100–500 ms. Transfer rates are below 300 kilobytes per second.

Another simplification is in the area of operator controls. There are generally no switches or status indicators, the simple action of moving a flap on the front of the drive to load or removing the diskette being the only operator action. The disk motor spins all the time that a disk is present.

4.12.7 Cartridge disk

This type of disk system is so called because the medium – one or two rigid disks on a single spindle, of aluminium coated with magnetic oxide and approximately 350 mm in diameter – is housed permanently in a strong plastic casing or cartridge. When the complete cartridge assembly is loaded into the drive, a slot opens to allow the read/write heads access

to the recording surfaces. As well as providing mechanical mounting, the cartridge provides protection for the disk medium when it is removed from the drive.

Drives are designed either for loading from the top when a lid is raised or from the front when a small door is opened allowing the cartridge to be slotted in. Power to the drive motor is removed during loading and unloading, and the door is locked until the motor has slowed down to a safe speed. On loading and starting up, the controller cannot access the drive until the motor has reached full speed. Operator controls are normally provided for unload, write protection and some form of unit select switch allowing drive numbers to be re-assigned on a multiple-drive system. Indicators typically show drive on-line, error and data transfer in progress.

Access times are normally in the region of 30–75 ms, aided by a fast servocontrolled head-positioning mechanism actuated by a coil or linear motor, the heads being moved in and out over the recording surface by an arm which operates radially. Heads are lightweight, sprint loaded to fly aerodynamically in the region of 0.001 mm from the surface of the disk when it is rotating at its full speed (usually 2400 or 2600 rev/min). Because of the extremely small gap, cleanliness of the oxide surface is vital, as any particle of debris or even smoke will break the thin air gap, causing the head to crash into the disk surface. In this rare event, permanent damage to the heads and disk cartridge occurs. Positive air pressure is maintained in the area around the cartridge, in order to minimize the ingress of dirt particles. Care should be taken to ensure cleanliness in the handling and storage of cartridges when not mounted in the drive.

The capacity of cartridges is in a range up to 100 megabytes, with data transfer rates in the region of 1 megabyte per second. Up to 16 drives can be accommodated per controller, and, because of the data transfer rate, direct memory access is necessary for transfer of data to or from the CPU. Progress made with other disk forms have almost eliminated the use of cartridge disks.

4.12.8 Disk pack

The medium used in this type of drive has multiple platters (five or more) on a single spindle, and is protected when removed from the drive by a plastic casing. When loaded on the drive, however, the casing is withdrawn. The drives are top loading, and, unlike cartridge disks which can generally be rack mounted in the cabinet housing the CPU, are free-standing units.

Other than this difference, most of the design features of disk pack drives follow those of cartridge units. The significant difference is the larger capabilities (up to 1000 megabytes) and generally high performance in terms of access times (25–50 ms) and transfer rates (in the region of 2.5 megabytes per second).

4.12.9 Winchester drive

So-called from a name local to the laboratory in the USA where it was developed, this is a generic name applied to a category of drive where the disk medium itself remains fixed in the drive. The principal feature of the drive – the fixed unit – is known as a head disk assembly (HDA). By being fixed and totally sealed, with the read/write heads and arm assembly within the enclosure, the following benefits are realized:

1. Contaminant-free environment for the medium allows better data integrity and reliability, at the same time

having less stringent environmental requirements. Simpler maintenance requirements follow from this.

2. Lighter-weight heads, flying to tighter tolerances closer to the recording surface, allow higher recording densities. Since the disk itself is never removed, instead of retracting, the heads actually rest on special zones of the disk surface when power is removed.

3. The arrangement of read/write heads is two per surface, providing lower average seek times by requiring less head movement to span the whole recording area.

The head-positioning arrangement differs mechanically from that of the drives previously described by being pivoted about an axis outside the disk circumference.

Three general types of Winchester drive exist, with approximate disk diameters of 133, 203 and 355 mm, providing capacities from 25 megabytes to over 1000 megabytes. Performance, for the reason described above, can exceed that for disk cartridge or pack drives of corresponding capacity.

The smallest versions of Winchester drive are becoming popular as the storage medium for microcomputers and smaller configurations of minicomputer, offering compact size with very competitive prices and fitting above the top end of the floppy disk range. Operationally, the fact that the disks are not removable from the drive means that a separate form of storage medium which is removable must be present on a system using a Winchester drive. Back-up and making portable copies is done using this separate medium, which is usually another type of disk drive, or a magnetic tape system matched to the disk speed and capacity.

4.12.10 Magnetic tape

Reliable devices for outputting digital data to and reading from magnetic tape have been available for a considerable time. The use of this medium, with agreed standards for the format of recorded data, has become an industry standard for the interchange of data between systems from different manufacturers. In addition, low-cost magnetic tape cartridge systems exist providing useful minimal-cost large-scale back-up storage plus a convenient medium for small-volume removable data and the distribution of software releases and updates.

4.12.11 Industry-standard tape drives

These allow reels of 12.7 mm wide oxide-coated magnetic tape, which are normally 731 m in length on a 267 mm diameter reel (or 365 m on a 178 m reel) to be driven past write and read head assemblies for writing, and subsequent reading, at linear densities from 800 to 6250 bits per inch. Tapes are written with variable-length blocks or records with inter-record gaps in the region of 12.7 mm. Each block has lateral and longitudinal parity information inserted and checked, and a cyclic redundancy code is written and checked for each block. The latter provides a high degree of error-correction capability. The tape motion and stop-start characteristics are held within precise limits by a servocontrolled capstan around which the tape wraps more than 180° for sufficient grip.

Correct tape tension and low inertia is maintained by motors driving the hubs of the two tape reels in response to information on the amount of tape in the path between the two reels at any time. One of the following forms of mechanical buffering for the tape between the capstan and reels is used:

1. *Tension arm*. This uses a spring-loaded arm with pulleys over which the tape passes, alternating with fixed pulleys

such that, when loaded, the tape follows a W-shaped path. The position of the arm is sensed and the information used to control the release and take-up of tape by the reel motors. This technique has almost univerally been overtaken by the vacuum method described below.

2. *Vacuum chamber*. This technique, used on modern higher-performance tape drives, has between each reel and the capstan a chamber of the same width as the tape, into which a U-shaped loop of tape of around 1 to 2 m is drawn by vacuum in the chamber. The size of the tape loops is sensed photo-electrically to control the reel motors.

To prevent the tape from being pulled clear of the reel when it has been read or written to the end or rewound to the beginning, reflective tape markers are applied near each end of the reel. These are sensed photo-electrically, and the resulting signal used to stop the tape on rewind, or to indicate that forward motion should stop on reading or writing.

Three different forms of encoding the data on the tape are encountered, dependent upon which of the standard tape speeds is being used. Up to 800 bits per inch, the technique is called 'non-return to zero' (NRZ), while at 1600 bits per inch 'phase encoding' (PE) and at 6250 bits per inch 'group code recording' (GCR) are used. Some drivers can be switched between 800 bits per inch NRZ and 1600 bits per inch PE. Very few systems below 1600 bits per inch are now manufactured.

Block format on the tape is variable under program control between certain defined limits, and as part of the standard, tape marks and labels are recorded on the tape and the inter-block gap is precisely defined. Spacing between write and read heads allows a read-after-write check to be done dynamically to verify written data. Writing and reading can only be carried out sequentially. These tape units do not perform random access to blocks of data, though those units that permit selective reverse under program control do make it possible for the application to access data other than by sequential read of the tape. However, this requires prior knowledge by the application of the layout and contents of the tape and is particularly slow and cumbersome, such applications being far better serviced by a disk-storage device.

Up to eight tape drives can be handled by a single controller. For PE and GCR, a formatter is required between controller and drive to convert between normal data representation and that required for these forms of encoding.

Tape drives can vary in physical form from a rack-mountable unit that is positioned horizontally to a floor-standing unit around 1.75 m in height.

Operator controls for on-line/off-line, manually controlled forward/reverse and rewind motion, unit select and load are normally provided. To prevent accidental erasure of a tape containing vital data by accidental write commands in a program, a write protect ring must be present on a reel when it is to be written to. Its presence or absence is detected by the drive electronics. This is a further part of the standard for interchange of data on magnetic tapes.

4.12.12 Cartridge tape

Low-cost tape units storing many megabytes of data on a tape cartridge are sometimes used for back-up storage.

4.12.13 Tape streamer unit

The emergence of large-capacity, non-removable disk storage in the form of Winchester technology drives has posed the problem of how to make up copies of complete disk contents or security or distribution to another similarly equipped system. An alternative to tape cartridges is a tape drive very similar to the industry-standard units described in Section 4.12.11 but with the simplification of writing in a continuous stream, rather than in blocks. The tape controller and tape motion controls can, therefore, be simpler than those for the industry-standard drive. Many modern tape units are able to operate in both 'block' and 'streamer' mode, according to operater or program selection, but not on the same tape. A streamer unit associated with a small Winchester drive can accept the full disk contents on a single reel of tape.

4.13 Digital and analogue input/output

One of the major application areas for minicomputers and microcomputers is direct control of and collection of data from other systems by means of interfaces which provide electrical connections directly or via transducers to such systems. Both continuously varying voltages (analogue signals) and signals which have discretion or off states (digital signals) can be sensed by suitable interfaces and converted into binary form, for analysis by programs in the CPU. For control purposes, binary values can also be converted to analogue or digital form by interfaces for output from the computer system.

In process and/or machine control and monitoring, data acquisition from laboratory instruments, radar and communications (to take some common examples) employ computer systems equipped with a range of suitable interfaces. They may be measuring other physical quantities such as temperature, pressure and flow converted by transducers into electrical signals.

4.13.1 Digital input/output

Relatively simple interfaces are required to convert the ones and zeros in a word output from the CPU into corresponding on or off states of output drivers. These output signals are brought out from the computer on appropriate connectors and cables. The output levels available range from TTL (± 5 V) for connection to nearby equipment which can receive logic levels, to over 100 V d.c. or a.c. levels for industrial environments. In the former case, signals may come straight from a printed circuit board inside the computer enclosure, while in the latter, they require to go through power drivers and be brought out to terminal strips capable of taking plant wiring. The latter type of equipment may need to be housed in separate cabinets.

Similarly, for input of information to the computer system, interfaces are available to convert a range of signal levels to logic levels within the interface, which are held in a register and can be input by the CPU. In some cases, input and output are performed on the same interface module.

Most mini and micro systems offer a range of logic level input/output interfaces, while the industrial type of input and output equipment is supplied by manufacturers specializing in process control. Optical isolators are sometimes included in each signal line to electrically isolate the computer from other systems. Protection of input interfaces by diode networks or fusible links is sometimes provided to prevent damage by overvoltages. In industrial control, where thousands of digital points need to be scanned or controlled, interfaces with many separately addressable input and output words are used.

Although most digital input and output rates of change are fairly slow (less than 1000 words per second), high-speed interfaces at logic levels using direct memory access are available. These can, in some cases, transfer in burst mode at speeds up to 3 million words per second. High transfer rates are required in areas such as radar data handling and display driving.

4.13.2 Analogue input

Analogoue-to-digital converters, in many cases with programmable multiplexers for high- or low-level signals and programmable gain preamplifiers covering a wide range of signals (from microvolts to 10 V), allow conversion commands to be issued and the digital results to be transferred to the CPU by the interface. Industrial-grade analogue input subsystems typically have a capacity of hundreds of multiplexer channels, low-level capability for sources such as thermocouples and strain gauges, and high common-mode signal rejection and protection. As with digital input/output, this type of equipment is usually housed in separate cabinets with terminal strips, and is supplied by specialized process control equipment or data logger manufacturers

For laboratory use, converters normally have higher throughput speed, lower multiplexer capacity and often direct cable connection of the analogue signals to a converter board housed within the CPU enclosure. Where converters with very high sampling rates (in the region of 100 000 samples per second) are used, input of data to the CPU may be by direct memory access. Resolution of analogue-to-digital converters used with computer systems is usually in the range 10–12 bits, i.e. a resolution of 1 part in 1024 to 1 part in 4096. Resolutions of anything from 8 to 32 bits are, however, available. Where a programmable or auto-ranging preamplifer is used before the analogue-to-digital converter, dynamic signal ranges of 1 million:1 can be handled.

4.13.3 Analogue output

Where variable output voltages are required (for example, to drive display or plotting devices or as set points to analogue controllers in industrial process control applications), one or more addressable output words is provided, each with a digital-to-analogue converter continuously outputting the voltage represented by the contents of its register. Resolution is normally no more than 12 bits, with a usual signal range of ± 1 V or ± 10 V. Current outputs are also available.

4.13.4 Input/output subsystems

Some manufacturers provide a complete subsystem with its own data highway separate from the computer system input/output bus, with a number of module positions into which a range of compatible analogue and digital input/output modules can be plugged. Any module type can be plugged into any position to make up the required number of analogue and digital points.

4.14 Data communications

4.14.1 Introduction

Since 1980 there has been a large growth in the use of data communications between different types and makes of equipment both within a physical location or building and between different buildings situated anywhere in the world. Even when this communication appears to take place between two points on earth it has very often done so by means of a geo-stationary satellite positioned in orbit. The creation and maintenance of such 'networks' is now nearly always the role of network managers and their staff, a function that is separate from (though working closely with) the traditional computer departments.

The requirement for the communication of data is not, of course, new, but what has changed is the basis for that requirement. Previously, the only other means available for the transfer of data between machines was a copy by magnetic media (such as tape or disk) or to key in the data again, with the consequent high risk of error and increased time taken. It was seen that data transmission would be faster and more accurate than both of these methods. Interestingly, data communication was not regarded as a replacement for the data in hardcopy form. Today, more emphasis is being placed on eliminating hardcopy transactions, such as the growing use of EDI (Electronic Data Interchange) to replace paper as the medium for moving order information between companies.

Large-scale integration and consequent lower costs have made very powerful computers much more readily available, which can contain the sophisticated software required to handle complex networks and overcome complex problems such as finding alternative routes for messages when a transmission line is broken. With the computer 'space' thus available, programmers can produce complex programs required to transmit data from a terminal connected to one computer to a program running in another, without the operator being aware of the fact that two or more computers are involved. Interface devices between the computer and the data network are very intelligent and powerful, and are usually minicomputers themselves. Thus they relieve the main computer of much of the previous load that it historically handled for data communications.

Computers have always been able to communicate with their peripheral devices such as card readers, mass-storage devices and printers, but in the 1960s it was not typical for the communications to extend beyond this. Data were transcribed onto 'punching documents' by functional departments within an organization. Now, the widespread use of interactive VDUs by users at their desks has eliminated almost all these departments. Even the very large traditional data-entry organizations such as the utility companies are now introducing data capture at source using hand-held terminals or OCR techniques.

However, in the late 1960s and 1970s the development of both hardware and software technology made it increasingly attractive to replace these terminals with more intelligent remote systems. These systems varied in their sophistication. At one end of the spectrum were interactive screen-based terminals that could interrogate files held on the central computer. Greater sophistication was found in data-validation systems which held sufficient data locally to check that, for example, part numbers on a customer order really existed, before sending the order to the computer for processing. More sophisticated still were complete minicomputers carrying out a considerable amount of local data processing before updating central files to be used in large 'number-crunching' applications such as production scheduling and materials planning.

From these systems have grown a whole range of requirements for data communications. We have communications between mainframe computers, between minis, between computers and terminals, between terminals and terminals and so on.

4.14.2 Data communications concepts

Computers communicate data in binary format, the bits being represented by changes in current or voltage on a wire, or, more recently, by patterns of light through an optic-fibre cable. There are various ways that characters are represented

in binary format. One of the earliest of these was the 5-bit Baudot code, invented towards the end of the nineteenth century by Emile Baudot for use on telegraphic circuits. Five bits can be used to represent 32 different characters, and while this was adequate for its purpose, it cannot represent enough characters for modern data communications. Nonetheless, Baudot gave his name to 'baud', the commonly used unit of speed, which although strictly meaning signal events per second, is frequently used to denote 'bits per second'.

Nowadays, one of the most commonly used codes is the ASCII (American Standard Code for Information Interchange) code (Figure 4.8). This consists of seven information bits plus one parity (error checking) bit. Another is EBCDIC (Extended Binary Coded Decimal Interchange Code), an 8-bit character code used primarily on IBM equipment (Figure 4.9).

Within the computer and between the computer and its peripheral devices such as mass-storage devices and line printer, data are usually transferred in parallel format (Figure 4.10). In parallel transmission a separate wire is used to carry each bit, with an extra wire carrying a clock signal. This clock signal indicates to the receiving device that a character is present on the information wires. The advantage of parallel transmission is, of course, speed, since an entire character can be transmitted in the time it takes to send one bit. However, the cost would prove prohibitive where the transmitter and receiver are at some distance apart. Consequently, for sending data between computers and terminal devices and between computers which are not closely coupled, serial transmission is used.

Here a pair of wires is used, with data being transmitted on one wire while the second acts as a common signal ground. As the term implies, bits are transmitted serially, and so this form of transmission is more practical for long-distance communication because of the lower cost of the wiring required. In addition, it is simpler and less expensive to amplify signals rather than use multiple signals in order to overcome the problem of line noise, which increases as the distance between the transmitter and receiver grows. Data transmission frequently makes use of telephone lines designed for voice communication, and since the public voice networks do not consists of parallel channels, serial transmission is the only practical solution.

Parallel data on multiple wires are converted to serial data by means of a device known as an interface. In its simplest form, an interface contains a register or buffer capable of storing the number of bits which comprise one character. In the case of data going from serial to parallel format, the first bit enters the first position in the register and is 'shifted' along thereby making room for the second bit (Figure 4.11). The process continues until the sampling block which is strobing the state of the line indicates that the correct number of bits has been received and that a character has been assembled. The clock then generates a signal to the computer which transfers the character in parallel format. The reverse process is carried out to convert parallel to serial data.

This 'single-buffered' interface does have limitations, however. The computer effectively has to read the character immediately, since the bit of a second character will be arriving to begin its occupation of the register. This makes no allowance for the fact that the computer may not be available instantly. Nor does it allow any time to check for any errors in the character received.

To overcome this problem, a second register is added creating a 'double-buffered' interface (Figure 4.12). Once the signal is received indicating that the requisite number of bits have been assembled, the character is parallel transferred to the second (or holding) register, and the process can continue

CONTROL CHARACTERS

CHAR	OCTAL	BINARY
NUL	000	0000000
SOH	001	0000001
STX	002	0000010
ETX	003	0000011
EOT	004	0000100
ENQ	005	0000101
ACK	006	0000110
BEL	007	0000111
BS	010	0001000
HT	011	0001001
LF	012	0001010
VT	013	0001011
FF	014	0001100
CR	015	0001101
SO	016	0001110
SI	017	0001111
DLE	020	0010000
DC1	021	0010001
DC2	022	0010010
DC3	023	0010011
DC4	024	0010100
NAK	025	0010101
SYN	026	0010110
ETB	027	0010111
CAN	030	0011000
EM	031	0011001
SUB	032	0011010
ESC	033	0011011
FS	034	0011100
GS	035	0011101
RS	036	0011110
US	037	0011111
DEL	177	1111111

CONTROL CHARACTER KEY

NUL = All zeros
SOH = Start of heading
STX = Start of text
ETX = End of text
EOT = End of transmission
ENQ = Enquiry
ACK = Acknowledgement
BEL = Bell or attention signal
BS = Back space
HT = Horizontal tabulation
LF = Line feed
VT = Vertical tabulation
FF = Form Feed
CR = Carriage return
SO = Shift out
SI = Shift in
DLE = Data link escape
DC1 = Device control 1
DC2 = Device control 2
DC3 = Device control 3
DC4 = Device control 4
NAK = Negative acknowledgement
SYN = Synchronous/idle
ETB = End of transmitted block
CAN = Cancel (error in data)
EM = End of medium
SUB = Start of special sequence
ESC = Escape
FS = Information file separator
GS = Information group separator
RS = Information record separator
US = Information unit separator
DEL = Delete

PRINTABLE CHARACTERS

CHAR	OCTAL	BINARY
A	101	1000001
B	102	1000010
C	103	1000011
D	104	1000100
E	105	1000101
F	106	1000110
G	107	1000111
H	110	1001000
I	111	1001001
J	112	1001010
K	113	1001011
L	114	1001100
M	115	1001101
N	116	1001110
O	117	1001111
P	120	1010000
Q	121	1010001
R	122	1010010
S	123	1010011
T	124	1010100
U	125	1010101
V	126	1010110
W	127	1010111
X	130	1011000
Y	131	1011001
Z	132	1011010

PRINTABLE CHARACTERS

CHAR	OCTAL	BINARY
a	141	1100001
b	142	1100010
c	143	1100011
d	144	1100100
e	145	1100101
f	146	1100110
g	147	1100111
h	150	1101000
i	151	1101001
j	152	1101010
k	153	1101011
l	154	1101100
m	155	1101101
n	156	1101110
o	157	1101111
p	160	1110000
q	161	1110001
r	162	1110010
s	163	1110011
t	164	1110100
u	165	1110101
v	166	1110110
w	167	1110111
x	170	1111000
y	171	1111001
z	172	1111010

PRINTABLE CHARACTERS

CHAR	OCTAL	BINARY
SP	040	0100000
!	041	0100001
"	042	0100010
#	043	0100011
$	044	0100100
%	045	0100101
&	046	0100110
'	047	0100111
(	050	0101000
)	051	0101001
*	052	0101010
+	053	0101011
,	054	0101100
-	055	0101101
.	056	0101110
/	057	0101111
0	060	0110000
1	061	0110001
2	062	0110010
3	063	0110011
4	064	0110100
5	065	0110101
6	066	0110110
7	067	0110111
8	070	0111000
9	071	0111001
:	072	0111010
;	073	0111011
<	074	0111100
=	075	0111101
>	076	0111110
?	077	0111111
@	100	1000000

Figure 4.8 ASCII codes

Bit Positions 4,5,6,7	Hex	0 (0000)	1 (0001)	2 (0010)	3 (0011)	4 (0100)	5 (0101)	6 (0110)	7 (0111)	8 (1000)	9 (1001)	A (1010)	B (1011)	C (1100)	D (1101)	E (1110)	F (1111)
0000	0	NUL	DLE			SP	&	RHY						2	3	½	0
0001	1	SOH	DC1			RSP		/		a	j		°	A	J	NSP	1
0010	2	STX	DC2	SYN						b	k	s		B	K	S	2
0011	3	ETX	DC3	WUS	IRT					c	l	t		C	L	T	3
0100	4									d	m	u		D	M	U	4
0101	5	HT	NL	LF						e	n	v		E	N	V	5
0110	6	RCR	BS	ETB	NBS					f	o	w		F	O	W	6
0111	7	DEL		ESC	EOT					g	p	x		G	P	X	7
1000	8				SBS					h	q	y		H	Q	Y	8
1001	9	SPS			IT			±		i	r	z		I	R	Z	9
1010	A	RPT	UBS	SW	EOP	¼	§	¾	#	:				SHY			
1011	B			CU2		.	£	.	#								
1100	C	FF				<	*	%	@								
1101	D		IGS	ENQ	NAK	(	)	−	'								
1110	E		IRS			+	;	>									
1111	F		ITB	BEL			μ	?	"								

Options

BS — Backspace
CRE — Carrier return
DEL — Delete
CU2 — MCII Format control
HT — Horizontal tab
INX — Index
IRT — Index return
IT — Indent tab
NBS — Numeric backspace
NSP — Numeric space
PE — Page end
RCR — Required carrier return
RHY — Required hyphen
RPT — Repeat
RSP — Required space
SBS — Subscript
SHY — Syllable hyphen
SP — Space
SPS — Superscript
STP — Stop
SW — Switch
UBS — Unit backspace
WUS — Word underscore
PRE — Prefix

Figure 4.9 EBCDIC code table for word processing. EBCDIC is the internal code of IBM mainframes. There is no overall EBCDIC standard, the version of the code depends on the system in which it is implemented. The above is a word processing version for UK keyboards

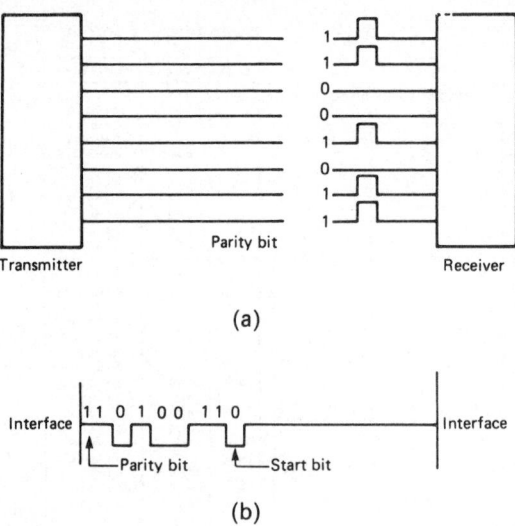

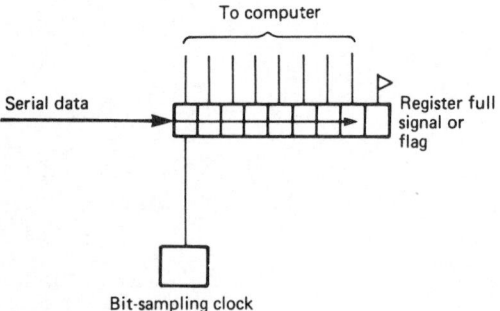

Figure 4.10 Data transmission; (a) parallel, (b) serial

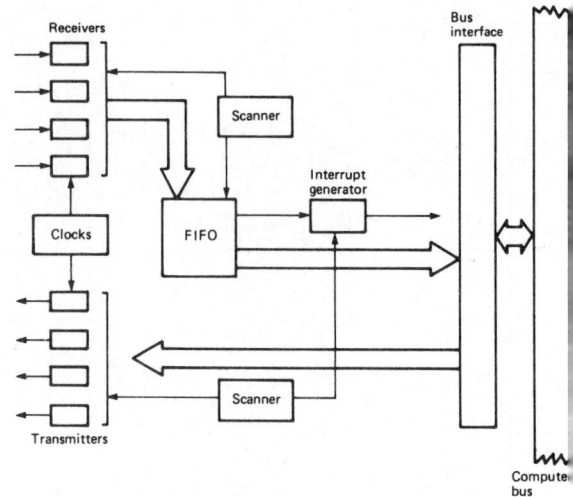

Figure 4.13 Schematic diagram of a multi-line interface

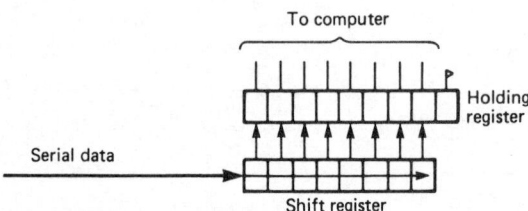

Figure 4.11 Serial-to-parallel interface

Figure 4.12 Double-buffered interface

The computer now has as much time as it takes to fill the shift register in order to check and transfer (again in parallel format) the character.

4.14.3 Multi-line interface

With the development of technology, the transmitter and receiver functions are now carried out by an inexpensive chip. Therefore the major costs in the interface are those of the mechanism used to interrupt the CPU when a character has been assembled and the connection to the computer's bus used

to transmit the received data to the CPU, or, in some cases, direct to memory. The interrupt mechanism and the bus interface are not heavily used. Indeed, they function only when a character is received or transmitted. These facilities are shared in a multi-line interface, sometimes (though not strictly correctly) known as a 'multiplexor' (Figure 4.13). To achieve this the device has several receivers and transmitters and a first-in, first-out (FIFO) buffer for received characters. The receivers are scanned and when a flag is found indicating that a character has been received the character is transmitted into the FIFO buffer, along with its line number. An interrupt tells the CPU that there are characters in the buffer and they are communicated over the bus to the computer. Similarly, a scanner checks the transmitters and when it discovers a flag indicating that a transmitter buffer is empty, it interrupts the CPU. Typically, the number of lines supported by a multi-line interface increases by powers of two for convenient binary representation, 4, 8, 16, 32, 128, 256 being common. The economies of scale in such an interface mean that further sophistications can be included such as program-selectable formats and line speed, and modem control for some or all of the lines.

However, the term 'multiplexing', strictly, actually refers to the function of sharing a single communications channel across many users. There are two commonly used methods of achieving this. One is a technique called time-division multiplexing (TDM), which consists of breaking down the data from each user into separate messages which could be as small as one or two bytes and meaningless when taken individually. The messages, together with identifying characters, are interleaved and transmitted along a single line. They are separated at the other end and the messages reassembled. This is achieved by use of devices known as concentrators or multiplexors. The second technique used to achieve this objective of making maximum use of a communication line is frequency division multiplexing. The concept is similar to that of time-division multiplexing. It is achieved by transmitting complete messages simultaneously but at different frequencies.

4.14.4 Modem

A significant complication of using public voice networks to transmit data is that voice transmission is analogue whereas

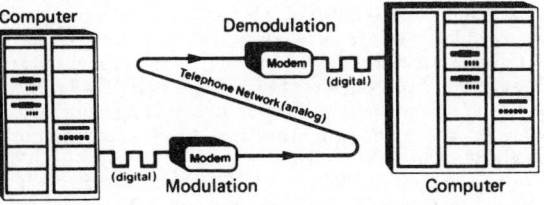

Figure 4.14 The use of modems in a communications link

data generated by the computer or terminal are digital in format. Thus an additional piece of equipment is required between the digital sender/receiver and the analogue circuit. This device modulates and demodulates the signal as it enters and leaves the analogue circuit, and is known by the abbreviated description of its functions, as a modem (Figure 4.14). Modems are provided by the common carrier such as British Telecom or by private manufacturers. In the latter case, however, they must be approved by the carrier and must contain or be attached to a device which provides electrical isolation.

4.14.5 Fibre-optic cable

Cabling for transmissions has traditionally been constructed of a copper-based core, this being a viable compromise between cost and conductivity for anything other than the very shortest communication paths. It is difficult to imagine a gold cable being laid from London to Birmingham and remaining in place for very long! However, copper has its own limitations, such as weight, resistance, noise, etc. The development of fibre-optic cable to the stage where a set light pattern can be sustained over long distances without distortion and then be sensed and interpreted at the other end is signalling the beginning of the end of copper as a standard communication medium. The main advantages of fibre-optic are:

1. Very lightweight;
2. Greater communication capacity (number and speed of channels) for the same size;
3. Digital transmission;
4. Immunity from most causes of interference and noise associated with copper;
5. Cost decreasing as volumes increase.

4.14.6 Laser

This works on exactly the same principle as fibre-optic except that the light signal is passed between two laser/receivers on a point-to-point 'line-of-sight' basis. It is ideal, therefore, in situations where communications are required between two different buildings but neither party owns or controls the land between them. The only other method would be to use a common carrier, resulting in a higher cost and probably lower speed and quality of communication.

4.14.7 Microwave

Where an organization requires extremely large volumes of data to be transmitted, or a very high speed to be achieved, then it is sometimes viable for it to set up its own microwave network. It should be stressed that this is an extremely costly operation and specialist advice should be sought prior to embarking upon it.

4.14.8 Transmission techniques

There are two techniques commonly used to transmit data on serial lines. One varies the current and the other varies the voltage in order to indicate the presence or absence of bits on the line.

4.14.8.1 Current variable

The 'current-based' technique communicates binary data by turning on and off a 20 mA current flowing through both the transmitter and receiver. Current on indicates a 'mark' or '1' bit and current off signifies a 'space' or '0' bit. This technique of turning a current on and off is less susceptible to noise than the technique of varying the voltage. However, it does have some drawbacks. Optical isolators are needed to protect logic circuits from the high voltages which may be required to drive the loop. Since there is one current source, an active interface and a passive interface are required, and finally, since a 20 mA system cannot carry the necessary control information, it cannot be used with modems.

4.14.8.2 Voltage variable

The EIA (Electronic Industries Association) and CCITT (Comité Consultatif Internationale de Telegraphie et Telephone) systems contain specifications and recommendations for the design of equipment to interface data terminal equipment (computers and terminals) to data communication equipment (modems). The specific EIA standard to which most modem equipment is designed is RS232C. The CCITT, being formed by the United Nations to consider all aspects of telecommunications across several national boundaries, was unable to publish firm standards, and instead produced a list of recommendations. Its equivalent of RS232C is known as 'V.24 – List of Definitions of Interchange Circuits Terminating Equipment'. The EIA/CCITT systems communicate data by reversing the polarity of the voltage; a '0' is represented by a positive voltage and a '1' by a negative voltage.

The signals in the EIA/CCITT specifications are not recommended for use over distances greater than 50 feet (15.5 m). Consequently, the modem and interface should not be more than 50 feet apart, though in practice distances in excess of 1000 feet (300 m) have been operated without problems.

4.14.9 Transmission types

Different communications applications use one of two types of transmission: asynchronous or synchronous. Slower devices such as VDUs and low-speed printers typically use asynchronous (or 'start–stop') transmission in which each character is transmitted separately. In order to tell the receiver that a character is about to arrive, the bits representing the character are preceded by a start bit, usually a zero. After the last data bit and error checking bit the line will return to the 1-bit state for at least one bit time – this is known as the stop bit.

Asynchronous transmission has the advantage that it requires relatively simple and therefore low-cost devices. It is, however, inefficient, since at least two extra bits are required to send eight data bits, and so would not be used for high-speed communication.

In synchronous transmission, characters are assembled into blocks by the transmitter and so the stream of data bits travels along the line uninterrupted by start and stop bits. This means that the receiver must know the number of bits which make up a character so that it can re-assemble the original characters from the stream of bits. Preceding the block of data bits, synchronization characters are sent to provide a timing signal for the receiver and enable it to count in the data characters. If the blocks of data are of uniform length, then this is all that is required to send a message. However, most systems would include some header information which may be used to indicate the program or task for which the data are destined and the amount of data in the block. In addition, if the messages are of variable length, some end-of-message characters will be required.

Because it does not contain start and stop bits for every character, synchronous transmission is more efficient than asynchronous. However, it can be inappropriate for some character-oriented applications since there is a minimum 'overhead' in characters which can be high relative to small transmitted block sizes, and the equipment required to implement it is more expensive.

4.14.10 Direction of transmission

There are three types of circuit available for the communication of data and, correspondingly, three direction combinations: simplex, half-duplex and full duplex. However, it is possible to use a channel to less than its full potential.

Simplex communication is the transmission of data in one direction only, with no capability of reversing that direction. This has limitations and is not used in the majority of data-communications applications. It can be employed, however, for applications that involve the broadcasting of data for information purposes in, for example, a factory. In this instance, there is neither a need nor a mechanism for sending data back to the host. The simplex mode of operation could not be used for communication between computers.

Half-duplex, requiring a single, two-wire circuit, permits the user to transmit in both directions, but not simultaneously. Two-wire half-duplex has a built-in delay factor called 'turnaround time'. This is the time taken to reverse the direction of transmission from sender to receiver and vice versa. The time is required by line-propagation effects, modem timing and computer-response time. It can be avoided by the use of a four-wire circuit normally used for full duplex. The reason for using four wires for half-duplex rather than full duplex may be the existence of limitations in the terminating equipment.

Full-duplex operation allows communication in both directions simultaneously. The data may or may not be related, depending on the applications being run in the computer or computers. The decision to use four-wire full-duplex facilities is usually based on the demands of the application compared to the increased cost for the circuit and the more sophisticated equipment required.

4.14.11 Error detection and correction

Noise on most communications lines will inevitably introduce errors into messages being transmitted. The error rates will vary according to the kind of transmission lines being used. In-house lines are potentially the most noise-free since routing and shielding are within user control. Public switched networks, on the other hand, are likely to be the worst as a result of noisy switching system and dialling mechanisms, though this problem is being addressed by many common carriers by the introduction of digital switching exchanges that themselves use computers to perform switching and routing instead of electromechanical switching devices.

Whatever the environment, however, there will be a need for error detection and correction. Three systems are commonly used: VRC, LRC and CRC.

VRC, or vertical redundancy check, consists of adding a parity bit to each character. The system will be designed to use either even or odd parity. If the parity is even, the parity bit is set so that the total number of ones in the character plus parity is even. Obviously, for odd parity the total number will be odd. This system will detect single bit errors in a character. However, if two bits are incorrect the parity will appear correct. VRC is therefore a simple system designed to detect single bit errors within a character. It will detect approximately nine out of ten errors.

A more sophisticated error-detection system is LRC (longitudinal redundancy check), in which an extra byte is carried at the end of a block of characters to form a parity character. Unlike VRC, the bits in this character are not sampling an entire character but individual bits from each character in the block. Thus the first bit in the parity character samples the first bit of each data character in the block. As a result, LRC is better than VRC at detecting burst errors, which affect several neighbouring characters.

It is possible to combine VRC and LRC and increase the combined error detection rate to 99% (Figure 4.15). A bit error can be detected and corrected, because the exact location of the error will be pinpointed in one direction by LRC and the other by VRC.

Even though the combination of LRC and VRC significantly increases the error-detection rate, the burst nature of line noise means that there are still possible error configurations which could go undetected. In addition, the transmission overhead is relatively high. For VRC alone, in the ASCII code, it is 1 bit in 8, or 12.5%. If VRC and LRC are used in conjunction it will be 12.5% plus one character per block.

A third method which has the advantage of a higher detection rate and, in most circumstances, a lower transmission overhead is CRC (cyclic redundancy check). In this technique the bitstream representing a block of characters is divided by a binary number. In the versions most commonly used for 8-bit character format, CRC-16 and CRC-CCITT, a 16-bit remainder is generated. When this calculation has been completed, the transmitter sends these 16 bits – two characters – at the end of the block. The receiver repeats the calculation and compares the two rremainders. With this system, the error detection rises to better than 99.9%. The transmission overhead is less than that required for VRC/LRC when there are more than 8 characters per block, as is usually the case.

The disadvantage with CRC is that the calculation overhead required is clearly greater than for the other two systems. The

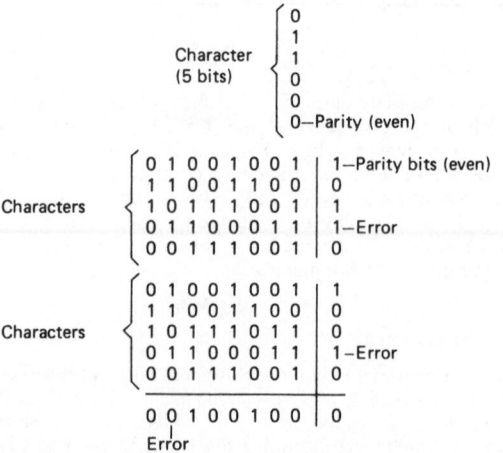

Figure 4.15 VRC, LRC and VRC/LRC combined (with acknowledgements to Digital Equipment Co. Ltd)

heck can be performed by hardware or software but, as is usually the case, the higher performance and lower cost of hardware is making CRC more readily available and commonly used.

Once bad data have been detected, most computer applications require that they be corrected and that this occurs automatically. While it is possible to send sufficient redundant data with a message to enable the receiver to correct errors without reference to the transmitter, the effort of the calculation required to achieve this in the worst possible error conditions means that this technique is rarely used. More commonly, computer systems use error-correction methods which involve retransmission. The two most popular of these are 'stop and wait retransmission' and 'continuous retransmission'.

'Stop and wait' is reasonably self-explanatory. The transmitter sends a block and waits for a satisfactory or positive acknowledgement before sending the next block. If the acknowledgement is negative, the block is retransmitted. This technique is simple and effective. However, as the use of satellite links increases, it suffers from the disadvantage that these links have significantly longer propagation times than land-based circuits and so the long acknowledgement times are reducing the efficiency of the network. In these circumstances, 'continuous retransmission' offers greater throughput efficiency. The difference is that the transmitter does not wait for an acknowledgement before sending the next block, it sends continuously. If it receives a negative acknowledgement it searches back through the blocks transmitted and sends it again. This clearly requires a buffer to store the blocks after they have been sent. On receipt of a positive acknowledgement the transmitter deletes the blocks in the buffer up to that point.

4.14.12 Communications protocols

The communications protocol is the syntax of data communications. Without such a set of rules a stream of bits on a line would be impossible to interpret. Consequently, many organizations – notably computer manufacturers – have created protocols of their own. Unfortunately, however, they are all different, and consequently, yet another layer of communications software is required to connect computer networks using different protocols. Examples of well-known protocols are Bisync and SDLC from IBM, DDCMP from Digital Equipment Corporation, ADCCP from the American National Standards Institute (ANSI) and HDLC from the International Standards Organization (ISO). The differences between them, however, are not in the functions they set out to perform but in the way they achieve them. Broadly, these functions are as follows.

4.14.12.1 Framing and formatting

These define where characters begin and end within a series of bits, which characters constitute a message and what the various parts of a message signify. Basically, a transmission block will need control data, usually contained in a 'header' field, text – the information to be transmitted – held in the 'body', and error-checking characters, to be found in the 'trailer'. The actual format of the characters is defined by the information code used such as ASCII or EBCDIC.

4.14.12.2 Synchronization

This involves preceding a message or block with a unique group of characters which the receiver recognizes as a synchronisation sequence. This enables the receiver to frame subsequent characters and field.

4.14.12.3 Sequencing

This numbers messages so that it is possible to identify lost messages, avoid duplicates and request and identify retransmitted messages.

4.14.12.4 Transparency

Ideally, all the special control sequences should be unique and, therefore, never occur in the text. However, the widely varied nature of the information to be transmitted, from computer programs to data from instruments and industrial processes, means that occasionally a bit pattern will occur in the text which could be read by the receiver as a control sequence. Each protocol has its own mechanism for preventing this, or achieving 'transparency' of the text. Bisync employs a technique known as 'character stuffing'. In Bisync the only control character which could be confusing to the receiver if it appeared in the text is DLE (data link escape). When the bit pattern equivalent to DLE appears within the data a 'second' DLE is inserted. When the two DLE sequences are read, the DLE proper is discarded and the original DLE-like bit pattern is treated as data. This is 'character stuffing'. SDLC, ADCCP and HDLC use a technique known as 'bit stuffing' and DDCMP employs a bit count to tell the receiver where data begin and end.

4.14.12.5 Start-up and time-out

These are the procedures required to start transmission when no data have been flowing and recovering when transmission ceases.

4.14.12.6 Line control

This is the determination, in the case of half-duplex systems, of which terminal device is going to transmit and which to receive.

4.14.12.7 Error checking and correction

As described in Section 4.14.11, each block of data is verified as it is received. In addition, the sequence in which the blocks are received is checked. For data accuracy all the protocols discussed in this section are capable of supporting CRC (cyclic redundancy check). The check characters are carried on the trailer or block check character (BCC) section.

4.14.12.8 RS232C

This is a standard issued by the United States Electronic Industries Association (EIA) to define the interface between Data Circuit-terminating Equipment (DCE) and Data Terminal Equipment (DTE). In plain language these are usually referred to as the 'modem' and 'terminal' respectively. The 'C' at the end of the standard designation indicates the latest revision of this standard that is applicable. This standard is in widespread use in the United States and formed the basis for the European CCITT standard V.24, which defines the interchange circuits and their functionality. Thus V.24 can be considered a subset of the full RS232C standard. In Europe the other components of RS232C are covered by other standards, CCITT V.28 for the electrical characteristics and ISO 2110 for the pin connector allocations. The terms RS232C and V.24 are often interchanged, and for practical purposes an

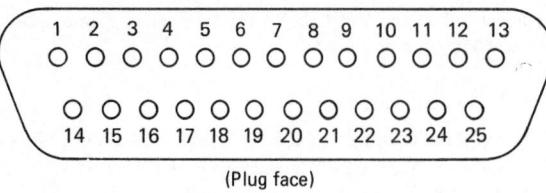

Figure 4.16 D-type connector pin assignments

interface that is said to be 'V.24-compliant' means that it also complies with RS232C.

The full interface specification deals with more than 40 interchange circuits, though, in practice, this number is almost never used. The most common form of connection is the 'D type' connector, so called because of the shape of the male and female plugs used to terminate the cable. A schematic of this connector is shown in Figure 4.16. These interchange circuits are collated into two distinct groups. The '100' series are used for data, timing and control circuits, whereas the '200' circuits are used for automatic telephone calling.

The principle of operation is simple in that both the modem and the terminal are able to indicate their readiness or not to accept/transmit data by adjusting the voltage on a predetermined circuit. A positive voltage represents a binary 0 or logical 'OFF' condition and a negative voltage a binary 1 or logical 'ON' condition. This change in voltage level can then be detected by the other end of the interface. Some circuits are kept constantly in a defined state (usually ± 12 V) at all times during transmission to indicate that a piece of equipment continues to be available. Once readiness to transmit data has been achieved, then other circuits are used to pass data to/from each end of the interface. This is carried out by raising or lowering voltage levels on the send or receive circuits phased according to a clock source, which may be external to the modem or internal to it. Both instances use different circuits for the timing signals, and they may not be used together.

The physical arrangement of the connectors can vary, but the female connector (socket) is usually found on the modem, whereas the male connector (plug) is on the terminal. The connector design itself does *not* form part of the standards but the 'D type' is in such widespread use throughout the world that it has, in practice, become a standard in its own right. The pin connections are defined in ISO 2110 and are shown in Table 4.1. Note that some pin allocations are left to the discretion of national bodies and thus complete compatibility is never certain, though this is not generally a problem in practice.

There are many instances in computing where it is desirable to connect terminals directly to computer or other equipment without physically routing through a modem device. This can be achieved through the use of a special 'switch-over' device or, more simply, by crossing over some of the connections at either end. This cross-over pattern is shown in Figure 4.17. Earlier, such devices were often referred to as 'null modems' and cables wired in this way are still called 'null modem cables'.

4.14.12.9 FDDI

In the mid 1980s it became apparent that the existing high-speed network technology such as Ethernet (see later) would, in the future, become the limiting factor in the transmission of data. Contrast this with the advance that Ethernet gave initially over the then-existing hardware and computer techno-

Table 4.1 RS232C/V. 24 pin/circuit assignments

V.24 (RS232) Circuit	Name	Pin Number
101 (AA)	Cable screen	1
102 (AB)	Signal ground or common return	7
103 (BA)	Transmitted data	2
104 (BB)	Received data	3
105 (CA)	Request to send	4
106 (CB)	Ready for sending (Clear to send)	5
107 (CC)	Data set ready	6
108 (CD)	Connect data set to line/Data terminal ready	20
109 (CF)	Data channel received line signal detector	8
110 (CG)	Data signal quality detector	–
111 (CH)	Data signalling rate selector (DTE source)	23
113 (DA)	Transmitter signal element timing (DTE source)	24
114 (DB)	Transmitter signal element timing (DCE source)	25
115 (DD)	Receiver signal element timing (DCE source)	17
116	Select standby	(24)
118 (SBA)	Transmitted backward channel data	14
119 (SBB)	Received backward channel data	16
120 (SCA)	Transmit backward channel line signal	19
121 (SCB)	Backward channel ready	13
122 (SCF)	Backward channel received line signal detector	12
125 (CE)	Calling indicator	22
126	Select transmit frequency	11
140	Remote loopback for point-to-point circuits	21
141	Local loopback	18
142	Test indicator	25

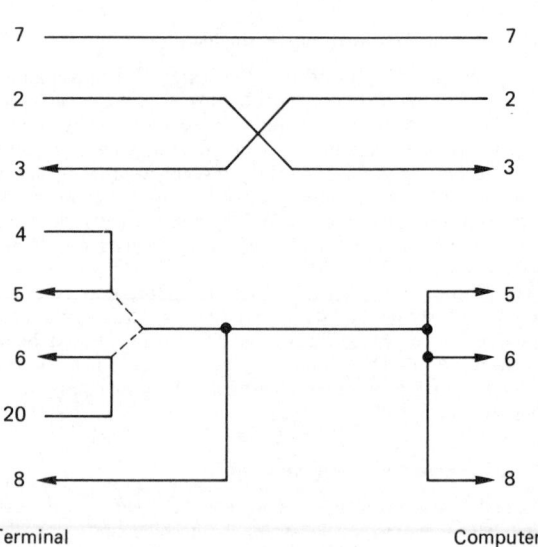

Figure 4.17 Null modem cable connections

logy, which was the limiting factor. It was anticipated that in the 1990s the network would reverse roles and start to limit transmission capabilities. At the time of writing there is already evidence of Ethernet LANs being overloaded from the volume of data now moved between computers and other networks.

In order to overcome these limitations a new network standard was developed, based upon fibre-optic cables, called the Fibre Distributed Data Interface (FDDI). Most of the components of this standard are agreed under ANSI Standard X3T9.5. Much of the basics of FDDI were originally based upon the IEE 802.5 standard for Token Ring networks.

The network design and functionality has proved to be ideal for real-time process control and voice due to the small minimum packet size of 9 bytes, but also efficient for large file/data transfers such as bit-imaged graphics due to the maximum packet size of 4500 bytes.

Gateways can be provided to/from FDDI to PBX, ISDN, ETHERNET, Token Ring and other communication protocols. It is also particularly good for military applications such as ships because of its high fault tolerance rate, lack of electrical radiation for eavesdropping and absence of fire risk on breaking the circuit.

Within the Open Systems Interconnect FDDI equates to the first two layers (Physical and Data Link) of the OSI communications model. In terms of functionality this network standard has been designed for connecting CPUs, high-speed storage devices, workstations, file servers, data terminals, multi-speed circuits and other external services such as the Public Switched Telephone Network (PSTN).

One of the main aims of the network design is to provide rapid and automatic recovery from failure of one or even multiple points in the network. Up to 500 connections (referred to as 'stations') can be on a single LAN, with a maximum cable length of up to 100 km. Stations can be located up to 2 km apart with optical links connecting them.

The network topology is based upon a ring structure with dual circuits, transmitting data at the rate of 100 Mbs. Stars (or spurs) are permitted to hang off the ring, but stations on such spurs do not benefit from the fail-safe facilities that stations on the ring provide. In ring operation one station, designated as the Cycle Master (CM), is responsible for the generation of the cycle structure onto the ring. The CM also acts as a buffer to match data arrival and dispatch rates.

Two types of station exist: Dual Attachment and Single Attachment stations. They are termed Classes 'A' and 'B' stations, respectively. Class A stations have dual-ring connections with traffic moving in opposite directions on each ring. Therefore two physical rings exist: Primary and Secondary. Primary is usually considered the 'live' ring. A problem on the Primary ring causes adjacent stations to switch that leg of the circuit to the Secondary ring, thus making up a complete logical ring from the remaining working components of both rings.

Each Class A station has a passive switching capability that enables a station to be taken out of service without disrupting traffic or lowering service capability on the ring as a whole. It is this combination of a redundant physical ring, alternate routing and station switch-out that gives FDDI a very high degree of fault tolerance. Note that Class B stations must be attached to the network through a Class A station (usually in clusters). Class B stations form part of the logical ring of the network, but not the physical one (all their traffic must pass to/from them via a Class A station).

The right to transmit is gained by 'capturing' a token that is circulating on the ring. Stations detect a token that is 'free', which is then removed from the ring by that station. The station then transmits its packet(s) and finally places a free token on the ring to replace the one that it took in the first place.

Packets are put onto the ring by an 'originating' station, intended for another 'destination' station. Each station regenerates the packet for onward transmission, but the destination station copies it into its own internal buffer as it passed it on.

When the packet arrives back at the originating station, then it is deleted from the ring. Security is assured through a 32-bit Frame Check Sequence cyclic redundancy check at the end of each packet.

Since transmission timing is independent for each and every section of the network, all stations require a phase-lock loop to control receive functions and autonomous transmit clocks for re-clocking the output data stream. The transmit clock runs at 125 Mbs and is used to measure the cycle time of the incoming data and adjust it if required to plus or minus 3 bits.

4.15 Computer networks

In the early days of data communications information travelled along a single, well-defined route from the remote computer to the 'host'. The reason for this was that the remote computer was fairly restricted in its computing and data-storage capabilities and so the 'serious' computing was carried out at the data centre. Most large organizations have retained their large data-processing centres but have changed emphasis on the use to which they are put. They are used principally for batch processing of data where either the volume is too large to be processed by the remote systems or where the processing itself is not time-critical. The advent of very powerful 'mini' computers (some much more powerful than earlier 'mainframes'), coupled with the marked increase in the reliability and speed of networks, has moved much of the data processing out to the world of the user onto the shop floor, into the laboratory, within an office department and even to individuals on the desks in their own homes.

In the motor industry, for example, the European headquarters of a US corporation would have its own designs and engineering department with a computer capable of processing, displaying and printing design calculations. However, it may still require access to the larger US machine for more complex applications requiring greater computer power. In addition, there may be a number of test units, testing engines and transmissions, each controlled by its own mini and supervised by a host machine. If there is a similar engineering department in, for example, Germany, it may be useful to collect and compare statistical data from test results. Also, since people must be paid, it may be useful to have a link with the mainframe computer in the data centre for the processing of payroll records. So it goes on. The demand for the linking of computers and the sharing of information and resources is increasing constantly.

A communications network may exist within a single site. Previously, it was not possible to connect buildings or sites divided by a public thoroughfare without using the services of a common carrier. However, with the advent of laser 'line-of-sight' devices it is possible to do so provided that 'line-of-sight' can be obtained between the points to be connected. Thus an organization may connect its systems together to form its own internal network. Because of the cost and disruption associated with laying cables, many companies are using internal telephone circuits for data communications.

For the factory environment many computer manufacturers offer proprietary networks for connecting terminal equipment to circuits based on 'tree' structures or loops. Connections to the circuit may be from video terminals for collection of, for example, stores data, special-purpose card and badge readers used to track the movement of production batches, or transducers for the control of industrial processes.

Throughout the 1980s the growth of multi-vendor sites, where computers and other equipment from different manufacturers are required to communicate with each other, has highlighted the need for a common means of doing so that is independent of any one manufacturer. Of the various systems initially developed for this purpose, one has become very widespread in its use, with more than 80% of all Local Area Network (LAN) installations using it. The system is 'Ethernet', originally promoted by Xerox, Digital Equipment and Intel.

4.15.1 Ethernet

Ethernet was developed in its experimental form at the Xerox Palo Alto Research Center in 1972. By the early 1980s a revised and more practical version was produced as a result of the cooperative venture between Digital Equipment Corporation, Intel and Xerox. This formed the basis for the standard Ethernet in use today. The prime objective of this system is to enable high-speed communication between computer equipment and other hardware, irrespective of the make or design of that equipment. Until the arrival of Ethernet most inter-machine communication, except that between equipment from the same manufacturer, was limited in practice to around 4800 bps on twisted pairs.

Ethernet is a multi-access communications system for transporting data between distributed computer systems that reside in close proximity to each other. The technique used to transfer data under controlled conditions is packet switching, whereby data are composed into discrete 'packets' for onward transmission without regard to their 'logical' use within an application. There is no central point of management in an Ethernet system. Each station may attempt to transmit when it needs to, and control of packet reception is ensured by the use of unique addresses for every Ethernet device ever manufactured. Only if the packet address matches its own address will a station pick up and use a packet on the network.

Communication occurs on a shared channel that is managed through a concept known as 'carrier sense multiple access with collision detect', or CSMA/CD for short! There is no pre-defined or pre-allocated time slots or bandwidth. Stations wishing to initiate a transmission attempt to 'acquire' control of the communications channel (which is often referred to as the 'Ether') by sensing the presence of a carrier on the network. If so, then the station delays its transmission until the channel is 'free', at which point transmission begins. A station that has detected collision will also jam the channel for a very brief period to ensure that all stations have detected and reacted to the collision it has itself detected.

During transmission the station will listen in to ensure that no other station has started to transmit at the same time. Should this be the case (i.e. a collision has been detected), then both stations will stop transmitting for a randomly generated delay period (called the 'collision interval'). Since all stations will wait a different period of time before attempting to retransmit, the chances of further collision are considerably reduced. It is important that the collision interval is based upon the round-trip propagation time between the two stations on the network that are furthest apart. Software is available that will monitor the collision level on the network and advise on capacity planning and physical network structure to ensure maximum throughput. A CRC check is applied to all packets on transmission and is checked by the receiver before handing the packet over to the station for further processing. Damaged packets are generally retransmitted. Maximum theoretical speed on the network is 10 Mbs but collisions, framing, CRCs, preambles, etc. reduce the level of 'usable' data available to the connected computer systems to 40–60% of this in practice.

4.15.2 Open Systems Interconnect (OSI)

In the past few years much emphasis has been placed on the concept of a standard that would permit equipment from any manufacturer or supplier to communicate with any other equipment, irrespective of the supplier. This would mean the ability to interconnect between systems in a completely 'open' manner, which led to the name Open Systems Interconnect (OSI).

The concept breaks down the whole business of communicating between systems into seven different 'layers'. Thus the problem of physical connection is separated from the method of controlling the movement of data along that connection. Each layer is subject to an individual standard compiled by the ISO. Some of these standards also incorporate earlier standards issued by other bodies such as the IEEE. The seven layers are as follows:

Layer 1 Application. The traditional computer program (application) that determines what need is to be met, what data are to be processed or passed by whom to whom for what purpose.
Layer 2 Presentation. Interfaces between the Application and other layers to initiate data transfer and establish data syntax.
Layer 3 Session. Manages communication session connection/severance, synchronization and reports on exception conditions.
Layer 4 Transport. Manages end-to-end sequencing, data flow control, error recovery, multiplexing and packeting.
Layer 5 Network. Maintains the availability and quality of the overall network, and manages network flow and logical division.
Layer 6 Data Link. Detects and attempts to correct physical errors, and manages data linkages, station identification and parameter passing.
Layer 7 Physical. Provides the actual physical mechanical and electrical services required to establish, maintain and use the physical network.

4.15.3 Network types

There are a number of network types.

4.15.3.1 Point to point

This is the simplest form of network and involves the connection of two devices – two computers or a computer and a terminal. If the communication line goes down for any reason then the link is broken, and so it is usual to back up leased lines with dial-up facilities (Figure 4.18).

4.15.3.2 Multi-point

As the name implies, Multi-point describes the connection of several tributary stations to one host. It is usual for the host to 'poll' the tributary stations in sequence, requesting messages, and for the network to be based on leased lines. In the case of one 'spur' being disconnected, the tributary station will dial into the host using a port reserved for that purpose (Figure 4.19).

4.15.3.3 Centralized

Also known as a 'star' network, in this type of network the host exercises control over the tributary stations, all of which are connected to it. The host may also act as a message-switching device between remote sites (Figure 4.20).

4.15.3.4   Hierarchical

A hierarchical structure implies multiple levels of supervisory control. For example, in an industrial environment special-purpose 'micros' may be linked to the actual process equipment itself. Their function is to monitor and control temperature and pressure. These 'micros' will then be connected to supervisory 'minis' which can store the programs and set points for the process computers and keep statistical and performance records (Figure 4.21). The next link in the chain will be the 'resource management computers', keeping track of the materials used, times taken, comparing these with standards, calculating replenishment orders, adjusting forecasts and so on. Finally, at the top of the network, the financial control system records costs and calculates the financial performance of the process.

4.15.3.5   Fully distributed

Here a station may be connected to several others in the network. The possibility then exists to share resources such as

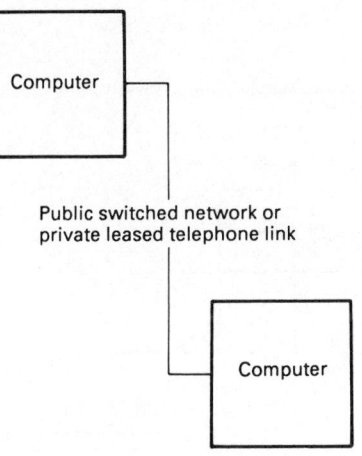

Figure 4.18   A point-to-point link

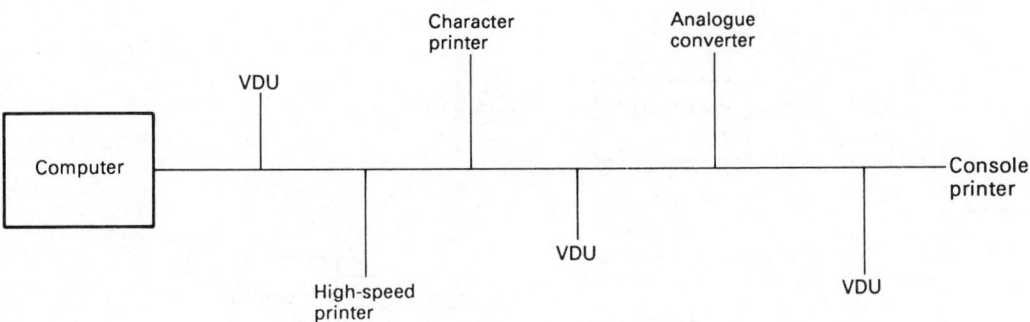

Figure 4.19   A multi-point communications network

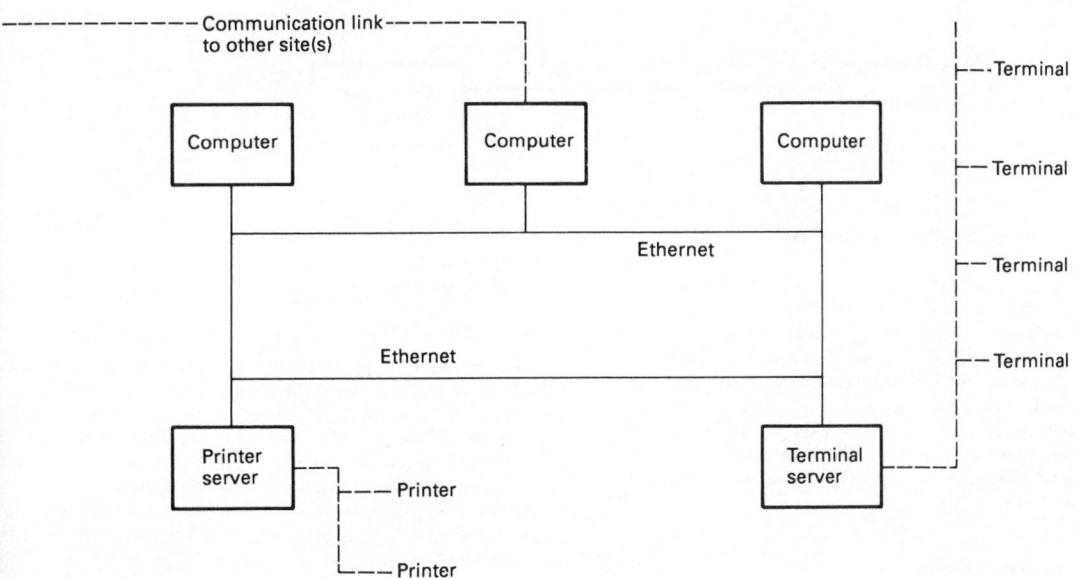

Figure 4.20   A typical centralized single-site network

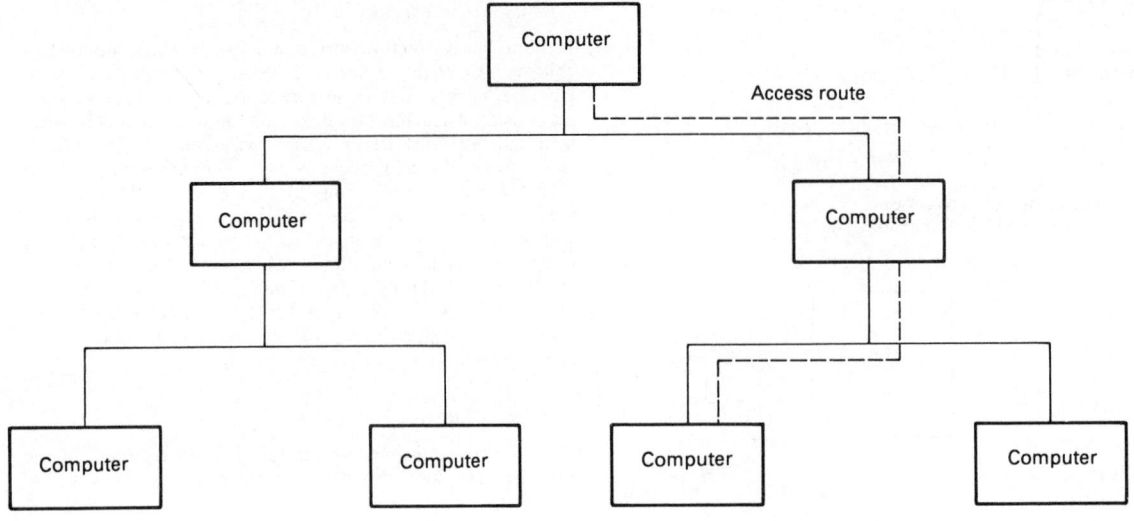

Figure 4.21 A hierarchical network

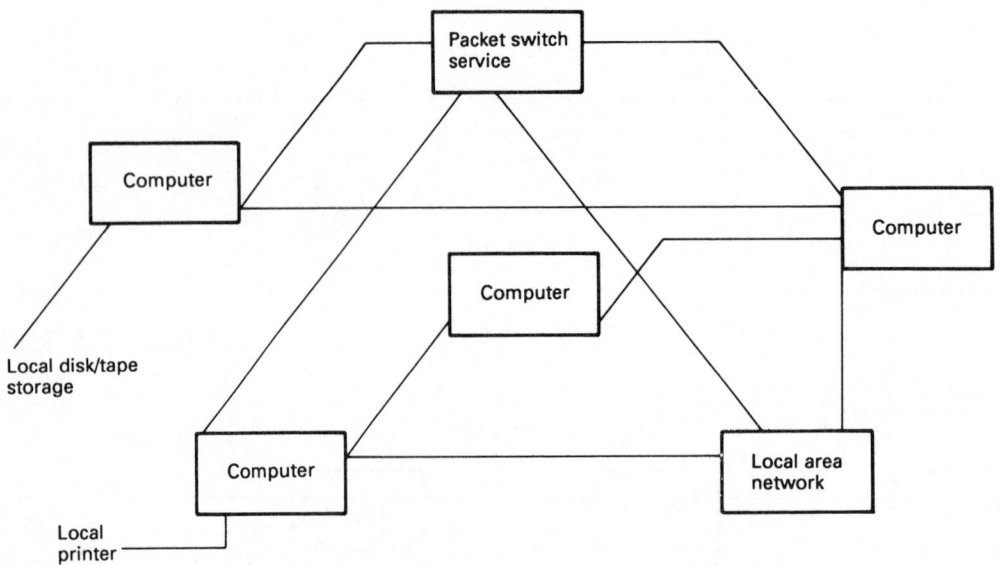

Figure 4.22 A fully distributed wide area network

specialized peripheral devices or large memory capacity and to distribute the database to the systems that access the data most frequently. It also provides alternative routes for messages when communication lines are broken or traffic on one link becomes excessive (Figure 4.22). However, the design of such systems requires sophisticated analysis of traffic and data usage, and even when set up is more difficult to control than less sophisticated networks.

4.15.4 Network concepts

Whatever the type of network, there are a number of concepts which are common.

4.15.4.1 File transfer

A network should have the ability to transfer a file (or a part file) from one node to another without the intervention of programmers each time the transfer takes place. The file may contain programs or data, and since different types (and possibly generations of computers) and different applications are involved, some reformatting may be required. This requires a set of programs to be written to cover all foreseen transfer requests and a knowledge of all local file access methods and formats.

One good example of the need for this is the application known as archiving. This involves the transmission of copies of files held on computer to another system in another location.

In the event of original files being lost as a result of fire, the files can be re-created using the archived information.

4.15.4.2 Resource sharing

It may be more cost-effective to set up communication links to share expensive peripheral devices than to duplicate them on every computer in the network. For example, one computer may have a large sophisticated flatbed printer/plotter for producing large engineering drawings. To use this, the other computers would store the information necessary to load and run the appropriate program remotely. This would be followed by the data describing the drawing to be produced.

4.15.4.3 Remote file access/enquiry

It is not always necessary or desirable to transfer an entire file, especially if only a small amount of data is required. In these circumstances what is needed is the ability to send an enquiry from a program (or task) running in one computer and remotely load, to the other system. This enquiry program will retrieve the requisite data from the file and send them back to the original task for display or processing. This comes under the broad heading of 'task-to-task communications'.

4.15.4.4 Logical channels

Users of a computer network will know where the programs and data, which they want to access, exist. They do not want to concern themselves with the mechanics of how to gain access to them. They expect there to be a set of pre-defined rules in this system which will provide a 'logical channel' to the programs and data they wish to reach. This logical channel will use one or more logical links to route the user's request and carry back the response efficiently and without errors. It may be that there is no direct physical link between the user's computer and the machine he or she is trying to access. In these circumstances the logical channel will consist of a number of logical links. The physical links, in some cases, may be impossible to define in advance, since in the case of 'dial-up' communication using the public switched network the route will be defined at connection time.

4.15.4.5 Virtual terminal

This is a very simple concept, and describes a terminal physically connected to computer A but with access (via A) to computer B. The fact that one is communicating via A should be invisible to the user. Indeed, to reach the ultimate destination, the user may unknowingly have to be routed through several nodes.

The use of common systems such as Ethernet and the promotion of common standards such as Open Systems Interconnect has bred a new concept in connecting terminals to computers, with the emphasis placed more on the 'service' that a user requires. Whereas previously the user had only to know to where the connection was required and not how to get there, with Ethernet-based 'servers' he or she need only know the name of the service that is required and no longer needs to specify where it resides. The terminal will be connected to Ethernet through a computer acting as a router. The server will know on which machine or machines the service required is currently available, and needs to know if the service has been moved, whereas the user does not. Furthermore, if the service is available on more than one machine, then the server will be capable of balancing the terminal workload given to each machine, all without the user even having to know or being aware of from where the service is being provided.

Many terminal servers are even capable of running more than one terminal to computer 'sessions' simultaneously on the same terminal, enabling the user to switch between them as desired without the host computer thinking that the session has been terminated. Workstations are able to carry out this 'sessions' service for themselves.

In all these examples the terminal is considered to be 'virtual' by any of the host machines to which it is connected via the terminal server. This concept and the facilities that it offers is quickly eroding many of the problems associated with previous methods of connecting terminals to computers, and the 'switching' and physical 'patching' that was required to connect a terminal to a new machine.

4.15.4.6 Emulator

As the name implies, this consists of one device performing in such a way that it appears as something different. For example, a network designer wrestling with the 'virtual terminal' concept may define that any terminal or computer to be connected to this network should be capable of looking like a member of the IBM 3270 family of video terminals for interactive work and the IBM 2780/3780 family for batch data transfer. In other words, they must be capable of 3270 and 2780/3780 emulation. Indeed, along with the Digital VT200/300, these two types of emulation have been among the most commonly used in the computer industry.

4.15.4.7 Routing

As soon as we add a third node, C, to a previously point-to-point link from A to B, we have introduced the possibility of taking an alternative route from A to B, namely via C. This has advantages. If the physical link between A and B is broken, we can still transmit the message. If the traffic on the AB link is too high we can ease the load by using the alternate route.

However, this does bring added complications. The designer has to balance such factors as lowest transmission cost versus load sharing. Each computer system has to be capable of recognizing which messages are its own and which it is required merely to transmit to the next node in the logical link. In addition, when a node recognizes that the physical link it was using has, for some reason, been broken, it must know what alternative route is available.

4.15.5 Network design

Network design is a complicated and specialist science. Computer users do not typically want to re-invent the wheel by writing from scratch all the network facilities they require. They expect their supplier to have such software available for rent or purchase, and, indeed, most large computer suppliers have responded with their own offerings.

There are two main network designs in use in the computer industry today.

4.15.5.1 SNA

IBM's Systems Network Architecture (SNA) is a hierarchical network that dominates the many networks hosted by IBM mainframes throughout the world. It is a tried and trusted product developed over a number of years.

4.15.5.2 DNA

Digital Equipment Corporation's Digital Network Architecture (DNA), often referred to by the name of one of its components (DECNET), is a peer-to-peer network, first announced in 1975. Since that time, as with SNA, it has been subject to constant update and development, with particular emphasis recently on compliance with OSI.

4.15.6 Standard network architecture

The fact remains, of course, that there is yet still no standard network architecture in common use permitting any system to talk to any other, hence the need for emulators. However, the PTTs (Post, Telephone and Telegraph operators) of the world have long recognized this need and are uniquely placed as the suppliers of the physical links to bridge the gap created by the computer manufacturers. They have developed the concept of public 'packet-switched networks' to transmit data between private computers or private networks.

4.15.6.1 Public packet-switched networks (PPSNs)

Package switching involves breaking down the message to be transmitted into 'packages' that are 'addressed' and introduced into the network controlled by the PTT. Consequently, the user has no influence over the route the packets take. Indeed, the complete contents of a message may arrive by several different routes. Users are charged according to the volume of data transmitted, giving generally greater flexibility and economy. The exception is the case where a user wants to transmit very high volumes of data regularly between two points. In this instance, a high-speed leased line would probably remain the most viable option.

What goes on inside the network should not concern the subscriber, provided the costs, response times and accuracy meet expectations. What does concern the user is how to connect to the network. There are basically two ways of doing this:

1. If one is using a relatively unintelligent terminal one needs to connect to a device which will divide the message into packets and insert the control information. Such a device is known as a PAD (package assembler/disassembler) and is located in the local packet-switching exchange. Connection between the terminal and the PAD may be effected using dedicated or dial-up lines.
2. More sophisticated terminals and computer equipment may be capable of performing the PAD function themselves, in which case they will be connected to the network via a line to the exchange, but without the need to use the exchange PAD.

4.15.6.2 CCITT X25

The CCITT has put forward recommendation X25 ('Interface between data terminal equipment for terminals operating in the packet mode in public data networks') with the aim of encouraging standardization. X25 currently defines three levels within its recommendations:

1. The physical level defines the electrical connection and the hand-shaking sequence between the data terminals equipment (DTE; see Section 4.16) and the data communications equipment (DCE; e.g. a computer).
2. The link level describes the protocol to be used for error-free transmission of data between two nodes. It is based on the HDLC protocol.
3. The package level defines the protocol used for transmitting packets over the network. It includes such information as user identification and charging data.

Packet-switched networks are now available in most countries in Europe and North America, as well as a smaller but growing number in other parts of the world. Most PTTs offer an international as well as national service under agreements with other PTTs.

4.16 Data terminal equipment

The most basic all-round terminal is the teleprinter which has been almost completely superseded by the video terminal. The most widely used terminal is the video display or VDU. VDUs may be clustered together in order to optimize the use of a single communications line. In this instance a controller is required to connect the screens and printers to the line (Figure 4.23).

Batch terminals are used when a high volume of non-interactive data is to be transmitted. Most commonly, the input medium is punched cards with output on high-speed line printers. As with VDUs, it is quite feasible to build intelligence into batch terminals in order to carry out some local data verification and local processing. However, the middle of the 1980s saw the large-scale introduction of very small but powerful free-standing micros and these are now in commmmon use as local pre-processors in communication with larger processors at remote sites. The advantage of this method is that the raw data, usually in punch card or magnetic tape format, can be read onto the local machine, verified, reformatted if required and then transmitted to the central site and processing initiated; all automatically done by the local micro.

In addition to these commonly found terminals, there are a host of special-purpose devices, including various types of optical and magnetic readers, graphics terminals, hand-held terminals, badge readers, audio response terminals, point of sale terminals and more.

Finally, of course, computers can communicate directly with each other without the involvement of any terminal device.

4.17 Software

4.17.1 Introduction

Software is the collective name for programs. Computer hardware is capable of carrying out a range of functions represented by the instruction set. A program (the American spelling is usually used when referring to a computer program) simply represents the sequence in which these instructions are to be used to carry out a specific application. However, this is achieved in a number of ways. In most cases, the most efficient

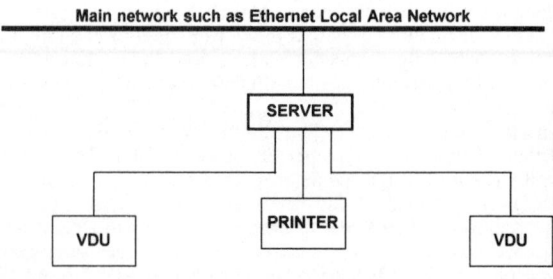

Figure 4.23 Typical network 'server' structure

method of using the hardware is to write in a code that directly represents the hardware instruction set. This is known as machine code and is very machine-dependent. Unfortunately, it requires a high level of knowledge of the particular type of computer in use, and is time consuming. In practice, therefore, programmers write in languages in which each program instruction represent a number of machine instructions. The programs produced in this high-level 'language' clearly require to be translated into code that can operate upon the computer's instruction set.

It would be possible, of course, to buy computer hardware and then set out to write every program one needed. However, this would take a very long time indeed. Most users require their system to perform the same set of basic functions such as reading, printing, storing and displaying data, controlling simultaneous processes, translating programs and many others. Consequently, most computers are supplied with prewritten programs to carry out these functions, and these fall into four basic categories:

1. Operating systems
2. Data-management systems
3. Language translators
4. Windows

4.17.2 The operating system

The operating system sits between the application program designed to solve a particular problem and the general-purpose hardware. It allocates and controls the system's resources such as the CPU, memory, storage and input/output, and allocates them to the application program or programs. Part of the operating system will be permanently resident in main memory and will communicate with the operator and the programs that are running. The functions it will carry out will typically be:

1. The transfer into memory of non-resident operating system routines;
2. The transfer into memory of application programs or parts of them. In some cases, there is insufficient memory to hold an entire program and so little-used portions of the program are held on disk and 'overlaid' into memory as they are required;
3. The scheduling of processor time when several programs are resident in memory at the same time;
4. The communication between tasks. For ease of programming, a large program can be broken down into sections known as tasks. In order to complete the application it may be necessary to transfer data from task to task;
5. Memory protection, ensuring that co-resident programs are kept apart and are not corrupted;
6. The transfer of data to and from input and output devices;
7. The queueing of input/output data until the appropriate device or program is ready to accept them.

There are several ways to use the resources of a computer system and each makes different demands on an operating system. The four main distinctions are as follows.

4.17.3 Batch processing

This was the original processing method and is still heavily used where large amounts of data have to be processed efficiently without a major emphasis on timing. Data are transcribed onto some input medium such as punched cards or magnetic tape and then run through the system to produce, typically, a printed report. Classical batch jobs include such applications as payroll and month-end statement runs.

Batch-operating systems require a command language (often known as JCL – job control language) that can be embedded between the data and that will load the next program in the sequence. Jobs are frequently queued on disk before being executed, and the operating system may offer the facility of changing the sequence in which jobs are run, as a result of either operator intervention or pre-selected priorities. Many operating systems are now capable of running multiple-batch 'streams' at the same time, and even of selecting a batch stream in which a particular job should run. Thus the person submitting the job is instructing the computer to run it under the best possible circumstances without necessarily knowing in advance where it will be run. This technique is particularly effective in a clustered environment, where batch streams may run across an entire cluster, and the operating system will not only choose the best stream but will also select the best processor on which it can run. Many current JCLs are almost programming languages in their own right, with great flexibility offered to the person submitting the job. However, there is a cost to pay for this flexibility, since the language is translated into machine code at the time of running (this is referred to as an 'interpretive language'), which is much slower than executing a pre-compiled language. Generally, though, the ratio of instructions to data to be processed is low, and this disadvantage is not considered significant.

The advantage of batch processing is its efficiency in processing large amounts of data. The major disadvantage is that once a user has committed a job he or she must wait until the cycle is completed before any results are received. If they are not correct one must re-submit the job with the necessary amendments. Some operating systems, however, do permit intermediate 'break points' in a job, so that results so far can be obtained and, if suspended, the job restarted without any loss of data. Others allow a batch job to submit data to another batch job for processing, which is very useful if the other batch stream exists to serve a printer, since intermediate results can then be printed without suspending or affecting the running of the original job submitted.

4.17.4 Interactive processing

This involves continuous communication between the user and the computer – usually in the form of a dialogue. The user frequently supplies data to the program in response to questions printed or displayed on the terminal whereas in batch processing all data must be supplied, in the correct sequence, before the job can be run. Where an operating system does permit a batch job to seek data during the running of the job, human attendance is required, which reduces the benefits of the batch stream principle.

A single person using a keyboard does not use the power of a computer to any more than a fraction of its capacity. Consequently, the resources of the system are usually shared between many users in a process known as 'time sharing'. This should not be apparent to the individual user who should receive a response to a request in one or two seconds under normal loading of the CPU and other resources. Time sharing, as the name suggests, involves the system allotting 'time slices', in rotation, to its users, together with an area of memory. Some users may have a higher priority than others, and so their requests will be serviced first. However, all requests will be serviced eventually.

Requirements of interactive time-sharing operating systems are efficient system management routines to allocate, modify and control the resources allocated to individual users (CPU time and memory space) and a comprehensive command language. This language should be simple for the user to understand and should prompt the inexperienced operator while allowing the experienced operator to enter commands swiftly and in an abbreviated format.

There are many situations today where the use of an interactive system provides an ideal solution to business and administrative problems. An area very close to the heart of computing is the development of programs and systems to run on them. In the earliest days of computing, engineers sat at large consoles and laboriously keyed in binary machine code instructions using toggle swiches. Punched cards and paper tape as a means of input for programmers were quickly adopted due to the time saved. Then it would have taken 20–30 minutes to compile and check a program of average length. Today, most machines perform the same task on much larger programs in a few minutes and sometimes in seconds. The proportion of time spent keying onto punched cards or tape became too high, and the person keying in was rarely the programmer. Therefore delays occurred while the program coding was written out longhand by the programmer, passed to data preparation, keyed in, verified and then sent back to the programmer. This process often took days, leading to very long development time and unproductive programmers. The solution was simple – get the programmers to key in directly themselves. The developments in interactive computing have made this possible, and indeed were for the most part driven by the needs highlighted by this problem.

To overcome the lack of typing and formatting skills of the average programmer, a new tool has been developed called the 'Language Sensitive Editor' (LSE). This checks what the programmer is keying in as part of a program's coding as he or she keys it in for spelling, syntax and format, and highlights any errors at the time of entry. It can even offer a pre-formatted statement framework for the programmer to fill in.

This is just one of the many uses of interactive computing but there are many others such as order input and enquiry, warehouse control, flight planning and booking, Automated Teller Machines, etc. Note that these are not 'real-time' applications in the strict sense of the phrase, since instant response to an event is not guaranteed, and requests for information and resource usage are queued and only seem to be instant.

4.17.5 Transaction processing

This is a form of interactive processing which is used when the operations to be carried out can be pre-defined into a series of structured transactions. The communication will usually take the form of the operators 'filling out' a form displayed on the terminal screen, a typical example being a sales order form. The entered data are then transmitted as a block to the computer which checks them and sends back any incorrect fields for correction. This block method of form transmission back to the computer is very efficient from a communications perspective, but can be inefficient from the point of view of the terminal operator if there are many fields in error, or if the validation of any of the fields is dependent on the contents of other fields on the same form. Some systems, therefore, send back the input character by character and are able to validate any field immediately, and not let the operator proceed past a field until it is correct. The options available to the operator will always be limited and he or she may select the job to be performed from a 'menu' displayed on the screen.

Typical requirements of a transaction-processing operating system are as follows:

1. Simple and efficient forms design utilities;
2. The ability to handle a large volume of simultaneous interactive users;

3. Efficient file-management routines, since many users will be accessing the same files at the same time;
4. Comprehensive journalling and error recovery. Journalling is a recording of transactions as they occur, so that in the event of a system failure the data files can be updated to the point reached at the moment of failure from a previously known state of the system (usually a regular back-up).

4.17.6 Real time

This is an expression sometimes used in the computer industry to refer to interactive and transaction-processing environments. Here it means the recording and control of processes. In such applications, the operating system must respond to external stimuli in the form of signals from sensing devices. The system may simply record that the event has taken place, together with the time at which it occurred, or it may call up a program that will initiate corrective action, or it may pass data to an analysis program.

Such a system can be described as 'event' or 'interrupt' driven. As the event signal is received it will interrupt whatever processing is currently taking place, provided that it has a higher priority. Interrupt and priority handling are key requirements of a real-time operating system. Some operating systems may offer the user up to 32 possible interrupt levels, and the situation can arise in which a number of interrupts of increasing priority occur before the system can return to the program that was originally being executed. The operating system must be capable of recording the point reached by each interrupted process so that it can return to each task according to its priority level.

4.17.7 Common concepts

There are some concepts that are common to most operating systems.

4.17.7.1 Foreground/background

The simplest form of processing is 'single user', either batch or interactive. However, a more effective use of a computer's resources is to partition the memory into two areas. One (background) is used for low-priority, interruptable programs, the other (foreground) is occupied by a program requiring a faster response to its demand. The latter will therefore have higher priority. The recent increases in both the memories and power of many mini and micro systems has relegated the use of this technique to the smaller end of the micro range, particularly home-based PCs.

4.17.7.2 Multiprogramming

This is an extension of foreground/background in which many jobs compete for the system's resources rather than just two. Only one task can have control of the CPU at a time. However, when it requires an input or output operation it relinquishes control to another task. This is possible because CPU and input/output operations can take place simultaneously. For example, a disk controller, having received a request from the operating system, will control the retrieval of data, thus releasing the CPU until it is ready to pass on the data it has retrieved.

4.17.7.3 Boostrapping (booting)

The operating system is normally stored on a systems disk or on a Read Only Memory (ROM) chip. When the computer is started up, the monitor (the memory resident portion of the operating system) must be read from storage into memory. The routine which does this is known as the 'bootstrap'.

4.17.7.4 System generation (sysgen)

When a computer is installed or modified, the general-purpose operating system has to be tailored to the particular hardware configuration on which it will run. A sysgen defines such items as the devices attached to the CPU, the optional utility programs that are to be included and the quantity of memory available, and the amount to be allocated to various processes.

It is unlikely that any single operating system can handle all the various processing methods if any of them is likely to be very demanding. An efficient batch-processing system would not be able to handle the multiple interrupts of a real-time operating system. There are, however, multi-purpose systems that can handle batch interactive and real time.

4.17.7.5 Data-management software

Data to be retained are usually held in auxiliary storage rather than in memory, since if they were held in memory without long-term power back-up they would be lost when the system was turned off. To write and retrieve the data quickly and accurately requires some kind of organization, and this is achieved by data-management software. This is usually provided by the hardware manufacturer, although independent software houses do sell such systems which, they claim, are more efficient or more powerful or both.

The most commonly used organizational arrangement for storing data is the file structure. A file is a collection of related pieces of information. An inventory file, for example, would contain information on each part stored in a warehouse. For each part would be held such data as the part number, description, quantity in stock, quantity on order, and so on. Each of these pieces of data is called a 'field'. All the fields for each part form a record and, of course, all the inventory records together constitute the file.

The file is designed by the computer user, though there will usually be some guidelines as to its size and structure to aid swift processing or efficient usage of the storage medium. With file-management systems the programs using the files must understand the type of file being used and the structure of the records with it. There are six types of file organization:

1. Sequential
2. Chain
3. Direct
4. Physical
5. Relative
6. Indexed

4.17.7.6 Sequential file organization

Before the widespread use of magnetic storage devices, data were stored on punched cards. The program would cause a record (punched card) to be read into memory, the information was updated and a new card punched. The files thus created were sequential, the records being stored in numeric sequence. A payroll file, for example, would contain records in employee-number sequence.

This type of file organization still exists on magnetic tapes and disks. However, the main drawback is that to reach any single record, all the preceding records must be read. Conse-

quently, it is efficient only when the whole file requires to be processed from beginning to end, and random enquiries to individual records are rarely made.

4.17.7.7 Relative file organization

Relative files permit random access to individual records. Each record is numbered according to its position relative to the first record in the file and a request to access a record must specify its relative number. Unfortunately, most user data, such as part number, order number, customer number and so on, does not lend itself to such a simplistic numbering system.

4.17.7.8 Physical file organization

Another version of the Relative technique is used to retrieve a specific 'block' of data relative to the first block in a file from disk. This is done irrespective of where the actual data records reside in the block, and it would be the responsibility of the application program, not the operating system, to separate out individual records (unpacking). Consequently, situations where this method is advantageous are rare, but if the record size equals that of a physcial block on disk then this technique offers considerable advantages in speed of retrieval of the data, particularly if the file is in a physically continuous stream on the disk. This type of file is often referred to as a 'physically direct' file.

4.17.7.9 Chain file organization

This is, in effect, a file that is required to be read sequentially but where not all the data are available at one time. Earlier file systems did not permit the extension of a sequential file once it was written, and adding data to a file meant reading the whole file, writing it out to a new file as it was read and then adding the new data onto the end of the new file.

To overcome this limitation, the chain file technique was introduced. Each record was written to the file using relative file techniques with the application specifying to where each record was to be written. However, each record contained a pointer to the location of the next record in logical (not physical) sequence in the file, or some method of indicating that there were no more records in the chain (usually a zero value pointer). This then enabled the application program to read the file in sequence, irrespective of where the data resided on disk or when the data were put there. The widespread use of sequential files that can be extended coupled with a considerable improvement in database and indexed file techniques has largely made this technique redundant.

4.17.7.10 Direct (hashed) file organization

This is a development of the relative file organization and is aimed at overcoming its record-numbering disadvantage. The actual organization of the file is similar. However, a hashing algorithm is introduced between the user number identifying a particular record and the actual relative record number which would be meaningless to the user. The algorithm is created once and for all when the system is designed and will contain some arithmetic to carry out the conversion.

This file organization permits very fast access but it does suffer from the disadvantage in that most algorithms will occasionally arrive at the same relative record number from different user record-identification numbers, thus creating the problem of 'synonyms'. To overcome this problem, the file management software must look to see if the record position indicated by the algorithm is free. If it is, then a new record can be stored there. If it is not, then a synonym has occurred and the software must look for another available record position. It is, of course, necessary to create a note that this has occurred so that the synonym can subsequently be retrieved. This is usually achieved by means of points left in the original position indicating the relative record number of the synonym.

The user-numbering possibilities permitted with direct files may be more acceptable to the user since they are not directly tied to the relative record number. However, the need for an algorithm means that these possibilities are limited. In addition, the design of the algorithm will affect the efficiency of recording and retrieval since the more synonyms that occur, the slower and more cumbersome will be these operations.

4.17.7.11   Indexed file organization

The indexed method of file organization is used to achieve the same objectives as direct files, namely, the access of individual records by means of an indentifier known to the user, without the need to read all the preceding records. It uses a separate index which lists the unique identifying fields (known as keys) for each record together with a pointer to the location of the record. Within the file the user program makes a request to retrieve part number 97834, for example. The indexed file management software looks in the index until it finds the key 97834 and the pointer it discovers there indicates the location of the record. The disadvantage of the system is fairly apparent; it usually requires a minimum of three accesses to retrieve a single record and is therefore slower than the direct method (assuming a low incidence of synonyms in the latter). However, there are a number of advantages:

1. It is possible to access the data sequentially as well as randomly, since most data-management systems chain the records together in the same sequence as the index by maintaining pointers from each record to the next in sequence. Thus we have indexed sequential or ISAM (indexed sequential access method) files.
2. Depending on the sophistication of the system, multiple keys may be used, thus allowing files to be shared across different applications requiring access from different key data (Figure 4.24).
3. Additional types of keys can be used. Generic keys can be used to identify a group of like records. For example, in a payroll application, employee number 7439 may identify K. Jones. However, the first two digits (74) may be used for all employees in the press shop. It is therefore possible to list all employees who work in this department by asking the software to access the file by generic key.
4. Another possibility is that of asking the system to locate a particular record that contains the key value requested, or

the next highest, if the original cannot be found. This is known as using approximate keys.
5. It is also possible to retrieve records within a given range of keys.
6. Most computer manufacturers provide multi-key ISAM systems and so the users do not need to concern themselves about the mechanics of data retrieval.

4.18   Database management

Files tend to be designed for specific applications. As a result, the same pieces of information may be held several times within the same system, and are often held many times within the same business or organization. This has many disadvantages:

1. It is wasteful of space and effort.
2. It is very difficult to ensure that the information is held in its most recent form in every location.
3. Security maintenance is much more difficult with multiple dispersed copies than it is with a single copy.

It is, of course, possible to share files across applications. However, a program usually contains a definition of the formats of the data files, records and fields it is using. Changes in these formats necessitated by the use of the data within new programs will result in modifications having to be made in the original programs.

The database concept is designed to solve these problems by separating the data from the programs which use them. The characteristics of a database are:

1. A piece of data is held only once.
2. Data are defined so that all parts of the organization can use them.
3. It separates data and their description from application programs.
4. It provides definitions of the logical relationships between records in the data so that they need no longer be embedded in the application programs.
5. It should provide protection of the data from unauthorized changes and from hardware and software.

The data definitions and the logical relationships between pieces of data (the data structures) are held in the schema (Figure 4.25).

The database is divided into realms – the equivalent of files – and the realms into logical records. Each logical record contains data items that may not be physically contiguous.

Records may be grouped into sets which consist of owner and member records. For example, a customer name and address records may be the owner of a number of individual sales order records.

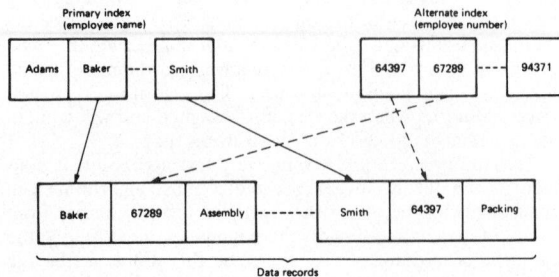

Figure 4.24   Multi-key 'ISAM' file organization

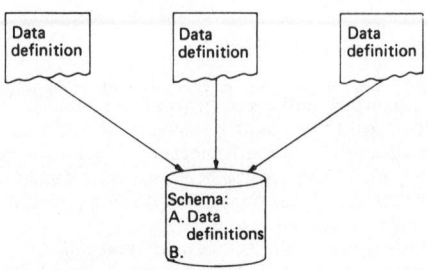

Figure 4.25   The schema

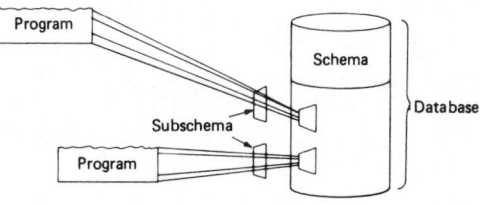

Figure 4.26 The subschema

When an application is developed a subschema is created defining the realms to be used for that application. The same realm can appear in other subschemas for other applications (Figure 4.26).

There are four major definitions of the logical relationships between the data (Figure 4.27):

1. *Sequential* (sometimes known as chain). Here each record is related only to the immediately preceding and following records.
2. *Tree or hierarchical*. In this structure each record can be related to more than one record following it. However, records along separate 'branches' are not directly linked with each other and the relationship can be traced only by travelling along the branches.
3. *Networks*. These are the most complex structures. They are effectively groups of trees where records can be related across branches. Any record can, in fact, be related to any other.
4. *Relational*. A Relational Database stores data in two-dimensional, tabular arrays. Each table (file) of the database is referred to as a relation, and each row of the table (record) is referred to as a tuple. Through the use of normalization (the successive breaking down of data into groups of two-dimensional arrays where each group's data are functionally dependent upon the group's key) the data are defined in a logical format suitable for use with a Relational Database. The result is a totally 'flat' file which, using a Relational Database Management System, has the flexibility to dynamically create new relations from extracts of one or more existing relations.

Because, within a database management system, data are separated from the programs which use them, the data are regarded as a corporate asset. Management of this asset is in the hands of a database administrator. He or she maintains the schema and works with application programmers to define the parts of the database to which they may have access and to help them create subschemas for their particular applications.

Codasyl began to take an interest in developing database standards in 1965 when the database task group was formed. There now exist Codasyl standards for database design.

4.19 Language translators

A programming language is a convention comprising words, letters and symbols which have a special meaning within the context of the language. However, programs have to be translated into the binary language understood by computers. The programmer writes a 'source' program which is converted by the language translator into an 'object program'. Usually, during this process checks are made on the syntax of the source program to ensure that the programmer has obeyed the rules. Any errors discovered will be noted, usually in two categories – terminal and warning. Terminal errors indicate that the translator has found a definite error that is so serious as to either prevent translation from completing or indicate that it is not worth doing so. A warning is generated when an item is encountered and the translator would have expected something different, but that it may not actually be an error. Program errors are known as 'bugs' and the process of removing them, as 'debugging'. Bugs are sometimes humorously referred to as 'planned features'!

Programs are normally stored in both their original code format (known as 'source') and in their final format that is understandable and executable directly by the computer (known variously as 'object', 'binary' or 'executable' formats). On disk, programs are often stored as individual files since they can be accessed directly and quickly. However, to store individual programs on individual magnetic tapes would use up too many tapes and have computer operators constantly loading and unloadding them. To overcome this limitation, the 'library' principle was introduced. Using this method, many different program files are combined into a single file called a library, and facilities exist within the operating system to extract whichever program is next needed from the library without changing the magnetic tape.

The working program is the object program, but when changes have to be carried out, these will be made to the source program which will then be translated to produce a new object program.

There are two kinds of language translators: assemblers and compilers.

4.19.1 Assemblers

An assembler is a language processor designed for use on a particular type of computer. In assembly language there is generally a one-to-one relationship between most of the language mnemonics (abbreviated instruction codes which can be read and interpreted by humans) and the computer binary instructions, although pre-defined sets of instructions can be 'called' from the assembly program.

There are four parts to a typical assembly language instruction:

1. *Label*. This is a name defined by the programmer. When he or she wants to refer to the instruction this can be achieved by means of the label. It is this facility that enables a programmer to alter the sequence in which instructions within a program are obeyed by 'jumping' from the current position to another identified by its label.
2. *Operation code*. This will contain a 'call' or an instruction mnemonic. If a call is used the assembler will insert a pre-defined code during the assembly process. If the programmer used a mnemonic this will define the operation to be carried out.
3. *Operand*. This represents the address of the item to be operated on. An instruction may require one or two operands.
4. *Comments*. This is an optional field used for ease of interpretation and correction by the programmer.

Assembly languages are generally efficient and are consequently used for writing operating systems and routines that require particularly rapid execution. However, they are machine-dependent, slow to write and demanding in terms of programmer skills.

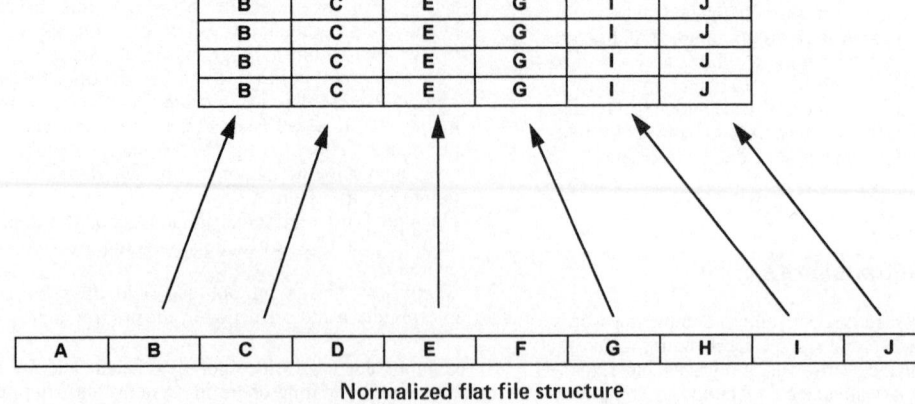

(a)

(b)

(c)

RELATION

TUPLE

(d)

Normalized flat file structure

Figure 4.27   Types of logical data relationships. (a) Sequential. (b) Hierarchical. (c) Network. (d) Relational

4.19.2 Compilers

These are used to translate high-level languages into binary code. These languages are relatively machine-independent, though some modifications are usually required when transferring them from one type of computer to another. The instructions in the high-level language do not have a one-for-one relationship with the machine instructions. Indeed, in some languages a single statement could generate many hundreds of machine instructions.

Most compilers read the entire source program before translating. This permits a high degree of error checking and optimization. An incremental compiler, however, translates each statement immediately into machine format. Each statement can be executed before the next is translated. Although it does not allow code optimization, it does check syntax immediately and the system prompts the programmer to correct errors as they occur. These incremental compilers are often referred to as 'interpreters', since they interpret and then act upon each instruction.

4.20 Languages

4.20.1 Introduction

All computers work with and understand instructions in the same format – binary. The content of an instruction to achieve the same objective may well differ from one machine to another, but the instruction will be coded in a binary format. This first format of computer language is called the first generation. It is almost unheard of today to find anyone other than compiler writers who work at this level.

Programming in this first generation of languages was complex, lengthy, skilful and extremely prone to error. It became obvious that improvements were essential and that assistance was required. What better tool to assist than the very computer that was being programmed? If a code system could be developed that was easier to write and read, and which could then be translated by the computer into machine instructions, then programming would become easier. These languages formed the second generation of computer languages and are referred to as 'assembly' languages (from the action of the translator of assembling everything together to validate it). Many of the original languages have long since gone or have been replaced by easier and more powerful alternatives, but assembly languages are still the best language for writing programs where flexibility and speed of execution are paramount (such as the operating systems themselves).

4.20.2 Third-generation languages (3GLs)

These languages were developed to bring the nature and structure of the instructions much nearer to the programmer's native language, principally to speed up the programming process and reduce the level of skill required. Since many of these languages were developed in the UK or the USA, the 'native' language was English, and this is still the predominant, almost universal, language of computing today. Some examples of third-generation languages are given below.

4.20.2.1 BASIC (Beginners' All-purpose Symbolic Instruction Code)

This is an easy-to-learn, conversational programming language which enables beginners to write reasonably complex programs in a short space of time. The growth in the popular-ity of time-sharing systems has increased its use to the point where it is used for a whole range of applications, from small mathematical problems, through scientific and engineering calculations and even to commercial systems.

A BASIC program consists of numbered statements which contain English words, symbols and numbers such as 'LET', 'IF', 'PRINT', 'INPUT', * (multiple), + and so on.

BASIC was developed at Dartmouth College in the USA, and while there is a standard there are many variations developed by different manufacturers. There are available both BASIC interpreters and compilers.

4.20.2.2 Fortran (FORmula TRANslation)

This originated in the 1950s and was the first commercially available high-level language. It was designed for technical applications and its strengths lie in its mathematical capabilities and its ability to express algebraic expressions. It is not particularly appropriate when the application requires a large amount of data editing and manipulation.

In 1966 an attempt was made by ANSI to standardize the Fortran language. However, manufacturers have continued to develop their own extensions.

A Fortran program consists of four types of statement:

1. Control statements (such as GOTO, IF, PAUSE and STOP) control the sequence in which operations are performed.
2. Input/output statements (such as READ, WRITE, PRINT, FIND) cause data to be read from or written to an input/output device.
3. Arithmetic statements, such as * (multiplication), ** (exponentiation), / (division), perform computation.
4. Specification statements define the format of data input or output.

The language has become more structured with the introduction of FORTRAN 77 and further revisions are being planned.

4.20.2.3 COBOL (COmmon Business-Oriented Language)

This is the most frequently used commercial language. The first Codasyl (Conference of Data Systems Languages) specifications for Cobol were drawn up by 1960, the aims of which were to create a language that was English-like and machine independent.

Cobol is a structured language with well-defined formats for individual statements. A Cobol program consists of four divisions:

1. *Identification*, which names and documents the program;
2. *Environment*, which defines the type of computer to be used;
3. *Data*, which names and describes data items and files used;
4. *Procedures*, which describes the processing to be carried out.

The 'sentences' within Cobol can contain 'verbs' such as ADD, SUBTRACT, MULTIPLY and DIVIDE and are readable in their own right. For example, in an invoicing program you may find the line:

IF INVOICE TOTAL IS GREATER THAN 500 THEN GO TO DISCOUNT ROUTINE

As a result, by intelligent use of the language the programmer can produce a program which is largely self-documenting. This is a significant advantage when modifications have to be made subsequently, possibly by a different programmer.

A general-purpose, structured language like Cobol is not as efficient in terms of machine utilization as assembly language or machine code. However, in a commercial environment, programmer productivity and good documentation are generally the most important factors. The calculations are usually not complex and therefore do not require great flexibility in terms of number manipulation.

4.20.2.4 PL1 (Programming Language 1)

Primarily an IBM language, PL1 was introduced to provide greater computational capabilities than Cobol but better file handling than Fortran.

4.20.2.5 APL (A Programming Language)

Introduced through IBM, APL is generally used interactively. At one end of the scale it allows the operator to use the terminal as a calculator. At the other, it enables one to perform sophisticated operations such as array manipulation with the minimum of coding.

4.20.2.6 RPG (Report Program Generator)

This language was originally introduced by IBM for 360 and System 3 machines. A commercially orientated very structured language, it processes records according to a fixed cycle of operations, although developments have been made in the language to make it more suitable for transaction-processing environments.

4.20.2.7 PASCAL

This is a strongly typed, block structured, procedural third-generation programming language, which is becoming more popular. This may be due to its extensive use in educational establishments in recent years. It is now widely used for a variety of differing applications in commercial environments.

PASCAL lends itself readily to modern design techniques (including top down, stepwise refinement, etc. . . .) and has built into it the necessary building blocks to make the most of structured programming. Additionally, because of its strong enforcement of data type, scope and syntax rules, PASCAL is a language less prone to programmer errors than many others.

All of the above points assist programmers in writing well-designed programs and produce very supportable code.

4.20.2.8 The 'C' language

The C language is a general purpose language. It has a close association with the UNIX operating system which is written in C. However, not being tied to any operating system or machine has made it a popular choice for the development of portable programs that can be run on a variety of hardware. This has been enhanced with the publication of an ANSI standard for the C language, based on the original C reference manual, which includes a definition of the language itself and also of a set of library routines for accessing the operating system such as file operations, memory allocation and string manipulation.

The basic data types available in C are characters, integers and floating point numbers. More complex data types can be created using structures, unions and arrays. A fundamental data type is the pointer. Use of pointers allows programs to be written to take full advantage of dynamic memory allocation and for those programs to be independent of the underlying machine architecture.

The basic language constructs are 'if-else' for decision making, 'for', 'while' and 'do' loops and switch statements for selecting from of a number of possible cases. Statements can be enclosed within brackets '{}' to group them together. There are the usual assignment operators such as '=' and additionally operators such as '++' for auto-increment and '−−' for auto-decrement. There are also a number of bitwise operators such as a logical AND (&) and shift (<<).

Much of the C language is based upon a language called BPCL, developed by Martin Richards. BPCL was the influence behind the language B, written by Ken Thompson in the early 1970s. Following on from B, C itself was designed by Dennis Ritchie, originally for the UNIX operating system running on a DEC PDP-11.

4.20.2.9 ADA

This language was developed primarily to support the development and implementation of complex, real-time software for embedded (i.e. usually contained in ROM) software in military hardware and applications. Its history can be traced back to a US DOD decision in January 1975 to set up the High Order Language Working Group to pursue the development of such a language. In April 1979 a language called 'Green' was selected as the new language for the DOD. It had been developed at a French subsidiary of the Honeywell Corporation and was renamed ADA when adopted by the DOD.

ADA uses PASCAL techniques for control and data typing. It also permits the segregation of regions within a program that enables separate development of each other. Routines can be written in a generic sense (say, for sorting) and then rapidly customized to perform the same operation on different data. It is able to do this since explicit data typing is not required initially. ADA supports multi-tasking in the same program, but the sequence of operation is not guaranteed to the programmer. Exception handling is possible under programmer control; even to the extent of permitting re-entry into routines that under normal circumstances would have crashed the system! Each compiler created for ADA under licence has to be periodically revalidated against a defined ADA standard, in an attempt to ensure that this standard is not compromised. However, the standard ADA 'functionality' is very large and has been a disincentive for many organizations to support development on its machines under ADA.

It has now been adopted as the standard preferred language by many Federal and international organizations including the US Department of Defense, the Federal Aviation Authority, NASA, NATO and the ministries of defence in many western European countries. Despite these advances, the use of ADA is not widespread and is mainly limited to defence contracts, with limited uses in commercial environments.

4.20.3 Fourth-generation languages

These are the closest yet to native language, enabling a person with very little knowledge or skill in a programming language to write instructions performing complex and lengthy tasks on a computer. They do not generally produce code that is as 'efficient' as previous generations, and the amount of machine resource that is required to support their use is sometimes either misunderstood by those using them or misrepresented by those selling them, and very often both.

Unlike previous generations of programming languages, few 4GLs demand that the instructions given to them need to follow the exact procedure and sequence in which they are required to execute. They are capable of breaking complex English-like statements down into many component instruction parts and then deducing the optimum order in which to carry them out. For this reason they are often referred to generically as 'non-procedural' languages.

These languages are most powerful when used in conjunction with some form of database management system, since they can simplify the interface between the programmer/program by undertaking all necessary communication with the database system. Many of these language/data interfaces have developed so far that the user need not know, and will often be unaware of, how the data are organized or formatted within the system. Some examples of these languages are POWERHOUSE, FOCUS, SQL, DATATRIEVE and QUERY.

4.20.4 Programmer emphasis

In broad terms, the increased speed of processing and data handling currently available and the low cost of memory have reduced the pressures on programmers to code for maximum speed and efficiency. It is frequently more economic to spend more money on hardware than to allow programmers to spend time optimizing the performance of their programs.

This, coupled with the shortage of trained programmers has resulted in increased emphasis on simple languages, good program development tools and a general emphasis on programmer productivity.

Computer-integrated engineering systems

5

John R. Painter
(Section 5.1)

Gordon M. Mair
(Section 5.2)

Trevor G. Clarkson
(Section 5.3)

Contents

5.1 CAD/CAM: Computer-Aided Design and
Computer-Aided Manufacturing 5/3
  5.1.1 Introduction 5/3
  5.1.2 Scope 5/3
  5.1.3 Hardware 5/4
  5.1.4 Software 5/7
  5.1.5 Modelling techniques 5/10
  5.1.6 Applications 5/12
  5.1.7 Implementation 5/16
  5.1.8 The future 5/18

5.2 Industrial robotics and automation 5/19
  5.2.1 Industrial robotics 5/19
  5.2.2 Industrial vision systems 5/26

5.3 Computer graphics systems 5/27
  5.3.1 Introduction 5/27
  5.3.2 Creating a computer graphics image 5/28
  5.3.3 Computer graphics systems 5/28
  5.3.4 Applications 5/32
  5.3.5 Workstations 5/40
  5.3.6 Three-dimensional concepts 5/41
  Acknowledgement 5/43
  Appendix: Bresenham's line algorithm 5/43

References 5/43
Further reading 5/43

5.1 CAD/CAM: Computer-Aided Design and Computer-Aided Manufacturing

5.1.1 Introduction

Computer-Aided Design and Computer-Aided Manufacturing (CAD/CAM) is now one of the most important applications of computer science, affecting everybody who makes or buys engineering products. It has a very diverse range of uses and is frequently treated as a computer discipline in its own right. Many of the specialists in the field are not computer scientists but rather experts in the engineering functions who have become involved with the technology. Knowedge of the engineering disciplines is often a prerequisite for understanding CAD/CAM applications.

5.1.1.1 Engineering design

Design and drawing have been an integral part of modern man's development from the cave paintings of 30 000 years ago. The Romans, well known for their engineering achievements, appreciated the need for design and planning before embarking on major engineering projects, and the engineering drawings of Leonardo da Vinci are renowned. Modern engineering design owes much to the nineteenth-century engineers such as Brunel and Stephenson, whose drawings for their engines, ships and bridges are often considered works of art.

Today, product engineering is an essential part of the manufacturing process. Many products are the result of a great deal of engineering effort to develop the optimum design. Over the years, engineers have always been keen to take advantage of any tools available to enable them to put more 'Engineering' into the product. The development of CAD/CAM owes much to their desire to exploit the power of computer systems for this purpose.

5.1.1.2 Early CAD/CAM development

One of the earliest developments in CAD/CAM was Automatically Programmed Tools (APT), a language for programming numerically controlled (NC) machine tools, developed in the mid-1950s. It is regarded as one of the first application languages for digital computers.

The Finite Element Method for structural analysis was conceived in the nineteenth century, but it was only with the advent of digital computers that it would be properly exploited, and the 1960s saw the growth of computer systems for this purpose.

The history of interactive CAD can be traced back to Sutherland and Hanratty, who experimented with systems for graphics and drawing during the early 1960s. Much of the early work was undertaken in large aerospace and automotive companies, with the development of computer systems to mathematically model surfaces for aircraft and car bodies.

By the late 1960s companies such as Calma, Applicon and Computervision began to specialize in the supply of 'turnkey' computer systems for CAD. Today, CAD/CAM is a multi-billion pound industry, actively supported by major computer companies, specialist suppliers and manufacturing firms.

5.1.2 Scope

CAD/CAM embraces a broad range of applications in the fields of engineering design and manufacturing, and systems vary greatly. CAD/CAM software is available for microcomputers at 'give-away' prices, but a large installation may require a multi-million-pound investment.

5.1.2.1 What is CAD/CAM?

The CAD/CAM industry has, like many others, developed a language of its own. Its acronyms and vocabulary are often misunderstood outside the industry as sometimes the meanings are context dependent. CAD/CAM can be defined as the application of computer systems technology to the design and manufacturing process, where CAD refers to the design process and CAM to the preparation of data for the planning and control of manufacturing process.

CAD is often thought of only in terms of drawing or draughting. However, it should encompass the whole of the engineering design process, including preliminary scheming and engineering analysis.

CAM includes manufacturing functions, ranging from production planning and scheduling to process monitoring and control. Some CAM applications involve conventional data processing, and are often treated outside the realm of CAD/CAM.

Computer-Aided Engineering (CAE) is often used to describe the functions of design, analysis and testing, together with those CAM applications closely associated with design and engineering, but excluding conventional DP applications. Thus it encompasses those areas that are usually embraced by integrated CAD/CAM systems.

The integration of CAD and CAM systems (referred to as *CIE* and *CIM* – Computer-Integrated Engineering and Computer-Integrated Manufacturing) is of particular interest. The need is for systems that are able to communicate and exchange information which is essential for realizing many of the benefits of CAD/CAM.

The application areas Within a single company there is usually a requirement for a variety of different CAD/CAM applications. Table 5.1 gives a list of the applications that would be of interest to a company involved with the production of electrical/mechanical components. With such a range of applications, many of which depend on each other for their data, the integration of CAD/CAM systems is of a high priority.

Table 5.1 CAD/CAM for an electromechanical engineering company

Conceptual Design	Design scheming/layout Performance analysis Two- and three-dimensional modelling
Mechanical Design	Three-dimensional geometric modelling Finite element analysis Simulation and kinematics Computer-aided draughting
Electronic Design	Circuit diagram schematics Electronic simulation/analysis PCB design and draughting Harness design/development
Production Engineering	Process planning NC machine programming Jig and tool design
Production Planning	Manufacturing resource planning Shopfloor simulation Production scheduling/control
Manufacturing	Shopfloor scheduling/control Process monitoring and control Quality measurement/control

Other computer systems in a manufacturing company (for example, in finance and administration) may also need an interface to the CAD and CAM systems. Purchasing and stock control systems require information on components and cost control systems need data on processes.

5.1.2.2 The users

The principal users of CAD/CAM systems are engineers, designers, draughtsmen and production engineering staff. They are not usually experienced computer users. Rather, they are trained engineers who understand about the design and/or manufacture of products in their particular field. Until comparatively recently, CAD/CAM applications have generally required an understanding of computer systems in order to make best use of them. While it is evident that future generations of engineers with a greater degree of computer literacy are likely to join the profession, current users need tools that minimize the reliance on the user being able to understand the technology.

Fortunately, with the advent of new computer technologies, particularly in the development of user interfaces, this is happening. Companies are also reorganizing in order to take advantage of CAD/CAM and ease its integration into the design and manufacturing departments.

5.1.3 Hardware

CAD/CAM systems have been made available on many different types of hardware. Some, especially those supplied by specialist vendors, are provided on hardware tailored to suit the needs of particular CAD applications. Others have been developed for standard computer systems ranging from microcomputers to powerful mainframes. A particular feature of the majority of CAD/CAM systems is that they employ 'interactive computer graphics'.

5.1.3.1 The CAD/CAM system

A typical CAD/CAM system provides a 'workstation' which will have either an on-board computing capability or be connected to a time-sharing computer system which runs the application software. The user's workstation will include a screen capable of displaying both text and graphical information, input devices such as a mouse or a digitizing tablet and a keyboard (which is becoming less essential with the latest systems). The cost of a system may range from a few pounds for some application software to run on a 'home computer' to a six-figure sum for a powerful 'engineering workstation' with wide-ranging functionality.

Early systems were often developed on mainframe computers with a user interface provided through graphics terminals employing storage tubes. Later, specialist vendors provided systems with customized CAD/CAM workstations, attached to a minicomputer. Some of the latest systems use powerful engineering workstations together with high-resolution screens to provide the optimum level of performance.

Major manufacturing companies have large CAD/CAM installations, comprising many individual systems connected through a computer network, perhaps serving several hundred users with a large number of different applications.

5.1.3.2 Processors for CAD/CAM

There are still a number of applications in the field of CAD/CAM that run on mainframe computer systems, although many of these tend to be on the border of CAD/CAM and conventional data-processing tasks. Mainframe-based systems are most suited to applications where there is a need for access to a large database, or there is a substantial I/O requirement, and the amount of on-line interactive processing required is small. Typical examples include manufacturing planning and scheduling and the engineering analysis of complex structures.

For a network of CAD/CAM systems, a mainframe computer is often an essential part of the installation, providing for the storage and management of the data. It may also provide the engineering users with access to other systems and other parts of the company through national and international networking facilities. However, for the basic applications in design and manufacturing, the majority of current systems are either stand-alone workstations or minicomputers serving a small number of users.

Minicomputer-based systems A few of the systems in industry are based on the 'small' minicomputer. They are typical of CAD/CAM systems supplied through the late 1970s to middle 1980s which were particularly appropriate for 'turnkey' systems, i.e. complete systems comprising hardware and software supplied as a package by a single vendor. Figure 5.1 shows an example of a typical system. Some installations in large automotive and aerospace plants include 30 or more processors providing for over 200 CAD/CAM workstations.

Figure 5.1 A mini-computer-based CAD/CAM system

A typical CAD/CAM minicomputer-based system included a processor together with a magnetic tape drive (for ½-inch tape), and one or more fixed and/or exchangeable disk drives (typically, 300 Mbyte). Such a system may support four to ten users, the response and performance of such systems being dependent on the type of work and the number of users. To improve performance, some turnkey systems incorporated special graphics processors to handle vector arithmetic and to speed up graphical operations.

The main attraction of such systems was that they were usually simple to implement and operate. They could be installed in a design office and be managed by the design staff. Thus, the processor was dedicated to the CAD system and the user department had control of the facility.

The engineering workstation Stand-alone computer workstations are becoming the current standard for CAD/CAM systems. They provide a great deal of flexibility plus the power to handle some of the very demanding tasks involved in CAD/CAM.

Figure 5.2   CAD/CAM on an engineering workstation

These workstation-based systems are usually dedicated to the needs of a single user with specific software applications. The stand-alone computer system generally comprises a processor with integral fixed disk drives and a cartridge tape drive for software loading and data back-up (Figure 5.2).

Currently, a typical system is based on a RISC-based processor such as the SUN SPARCstation™ offering 40–80 MIPS. However, the hardware technology is advancing rapidly, and more powerful workstations are becoming available. The manipulation of graphically displayed information is an important feature of such systems, and some workstations may incorporate additional hardware to speed up operations such as the scrolling and rotating of three-dimensional models of components.

Improved networking facilities have also enabled systems to be developed which employ 'disk-less' workstations to provide the user interface but enable resources such as disk drives to be shared from a 'file-server', thus providing more economically the benefits of the workstation.

The great advantage of stand-alone workstations is their flexibility. Each workstation can be tailored to the specific needs of particular users, and they do not suffer from the same performance degradation as do time-shared systems. The latest workstations, employing powerful processors using RISC architectures, are able to offer systems which satisfy many of the current requirements for CAD/CAM. The systems can also be run in an office environment, without air conditioning. Thus they can be installed in a location convenient for the user.

The personal computer   Personal computer systems have revolutionized many areas of computing. In CAD/CAM they have provided cheaper applications for a number of design and manufacturing tasks, enabling CAD/CAM techniques to be applied cost-effectively to many engineering problems.

Basic PC-based CAD systems will run on PCs such as the 'Intel 80286™'-based systems, with an EGA screen to provide a sufficient level of resolution and 20–40 Mbytes of hard disk to allow adequate space for the software and CAD data. For the more demanding CAD applications, the PC will need high-resolution graphics, a maths co-processor and additional disk space. The more powerful PCs based on processors such

as the 'Intel 80386™' provide some good CAD/CAM tools very economically.

PC-based systems do not provide the performance and functionality of the more powerful workstation systems but they do make CAD/CAM available to almost anyone who needs it. They are of particular value in education, enabling schools and colleges to introduce CAD/CAM into the curriculum. They are also very useful for first-time users, enabling them to explore the possibilities for cost-effective CAD/CAM in their organization.

5.1.3.3   Input devices – the CAD/CAM seat

A user accesses a CAD/CAM system through a workstation or 'CAD/CAM seat'. The computer hardware is sometimes built into a desk or table which also provides additional space for drawings and papers. Such stations have been the subject of a considerable amount of ergonomic design since they are becoming the designers' principal workplace.

There are almost as many variations on the CAD/CAM seat as there are systems. However, they have one important feature – a screen capable of displaying information and allowing the user to interact directly with the graphic display. Communication of commands to the system may be via a keyboard or a keypad, or by selections from menus on a tablet or a screen. In addition, there may also be devices to digitize information directly from existing drawings. The user's workstation may also include a small printer and/or plotter.

Screens for CAD/CAM   The provision of a suitable screen is most important for many applications. It must generally be capable of supporting high-resolution graphics to enable drawings to be displayed with reasonable detail.

Some early systems used direct view storage tubes (DVSTs) which displayed graphical information by stroking vectors across the screen. They provided high-resolution, high-quality graphics, particularly suitable for curves and surfaces and for shape design in the aerospace and automotive industries. Storage tubes could display very complex drawings, but when the drawing was edited the screen had to be 're-painted', which could take some time.

Initially, raster scan screens generated the image from a vector file and maintained it by continually refreshing the screen. Drawings displayed on the screen could easily be edited but their complexity was limited by the time taken to refresh the image.

Decreasing hardware costs have enabled the development of raster scan screens to generate the image from a pixel map of the drawing. However, if the resolution of the raster display is too low, diagonal lines appear jagged, as was the case with early versions. Raster scan screens now provide the very high resolution required, typically of the order of 1024 × 1024 pixels, and are also able to display the high-quality pictures needed for some CAD/CAM tasks.

A colour screen is useful for many applications. It enables the designer to use colours to separate and analyse component characteristics and features, and greatly simplifies the design task. Indeed, in some cases it is almost essential. Colour screens today can display very high-quality images with subtle changes in shade and colour, enabling computer-generated images to almost mimic a photograph. For many tasks in the design of complicated mechanical components, and in architectural design, this high level of visualization is very necessary.

Some systems have made use of multiple screens, a separate screen being used for the display of text, and/or a second graphical image, so that it does not interfere with the graphics on the first screen. However, with the sophisticated windowing capabilities provided by modern engineering workstations, the trend is towards the use of a single high-resolution screen that can display the graphical image, textual information and commands generated by the system, and icons for interacting with menu-driven user interfaces.

Keyboards, keypads, pens and mice   There are many different ways in which the user communicates commands to the system. The traditional method has been via a standard keyboard on which the user types commands. This is still an essential part of many systems, and is a convenient way for entering text and keying in menu selections from a list. Some systems also incorporate a 'keypad', a rectangular array of keys which enables the user to select a command from a menu provided by an overlay on the keypad.

An essential requirement of most CAD/CAM applications is for the user to be able to interact with the graphical display. This may be to enable a feature on the drawing to be identified or to indicate a position in the displayed image. Communication through the screen may also be used to allow the user to select the next command or command parameter from a set of options displayed on the screen.

Graphical displays provide a cursor in the form of a cross, or horizontal and vertical crossed lines, which can be moved around the screen by the user through devices such as a joystick, rollerball, a pair of thumbwheels or a mouse. Once the cursor has been positioned, a key depression or a button signals to the system that the position on the screen is to be noted and the appropriate action taken, which, depending on context, could be to record the location in the drawing, identify a graphical entity, or action a command.

Other devices for indicating graphical positions include the 'digitizing pen' and tablet, and 'light pen'. A light pen is pointed at the screen to identify a position or to make menu selections. Digitizing pens are used in conjunction with a tablet – a flat rectangular pad which varies in size from about 30 × 30 cm upwards and is sensitive to the position of the pen. The ergonomics of light pens versus digitizing pens has been much debated. A light pen, used to point to a position on the screen, was thought to be more natural for the designer, but most users found that being able to rest the pen on a tablet was more comfortable.

The most usual way of communicating through the screen on modern systems is by a mouse, which is now generally considered to be the most comfortable and user-friendly way to position the cursor. However, to satisfy the exacting needs of CAD/CAM applications it is important that the mouse has a sufficiently high displacement resolution.

Digitizers and scanners   There is frequently a need to trace or capture information from an existing drawing. This can be achieved by using either a digitizing tablet, which is essentially a manual technique, or a scanner if an automatic drawing capture system is required.

In digitizing, the drawing is placed on a digitizing tablet and positions on the drawing translated to cursor positions on the screen, and hence to computer-stored geometry, with a 1:1 (or a suitable scaling factor) correspondence with the original drawing. This method is an extension of the use of the digitizing pen and tablet for input, and many systems equipped with digitizing tablets for command input can also use them to digitize existing drawings. For specialist applications there are large purpose-built digitizers, some capable of digitizing three-dimensional information and generating computer-based geometry from full-size models.

If there is a need to enter existing paper drawings into a CAD system automatically then a scanner is required to 'read' a drawing and generate a computer-based representation of it.

The drawing is created as a raster image, i.e. each line on the drawing is represented by a row of pixels, because the scanner reads the drawing as a pixel image. Although the resultant computerized drawing can be plotted directly, it would have to be converted to a vector format if there was a need to manipulate it through the CAD system. Scanner resolution are typically 100–400 dpi (dots per inch).

Scanners are useful for entering large numbers of drawing automatically but are susceptible to error. Extra marks on the drawing, dirt and fold marks may also be reproduced and some drawings may not be easy to scan because the lines are too faint, although scanners can usually be set to read at a threshold level. Also, the vector conversion of the raster file may not always generate the original design intent. If a digitizer is used a vector is generated directly, and the user can control precisely what is entered, but it is considerably more time consuming.

5.1.3.4   Output devices, plotters and printers

All CAD/CAM systems need some ability to output hard-copy information, typically a plotter and/or printer. In addition some CAM systems may need the ability to generate punched tape to drive numerically controlled machines, and there are occasionally requirements for more exotic output devices.

Plotters for CAD/CAM   Plotting is an important function since, in many organizations, communication of information is still largely paper based. The plotting facility for most current CAD/CAM installations is provided by either a pen plotter or an electrostatic plotter.

The traditional drawing is most closely mimicked by a pen plotter, since the finished drawing can be provided as an inked plot, providing a very high-quality result. Pen plotters vary from small A4 size plotters (approx. 210 × 300 mm) to large draughting machines of 4 m or more, capable of drawing to a tolerance within 0.05 mm. These are essential in aerospace shipbuilding and automotive applications where the need is for large-scale drawings of cars, cross sections of aircraft wings ships' hulls, etc. Pen plotters can often be driven directly from the drawing file on the CAD system. Many pen plotters provide multiple pens, a necessary requirement now because many drawings require colour and/or varying pen thickness.

To output a drawing via an electrostatic plotter the drawing must first be rasterized and then transferred to the paper as individual dots (the ink is transferred by electrostatic charge to the specially treated paper). In recent years the quality of plots from electrostatic plotters has increased substantially, with the development of colour plotters and resolutions of the order of 400 dpi, and they now provide a satisfactory output for most applications, even those requiring a high standard of drawing. However, the CAD data must usually be converted to a raster file, and this can be time consuming for a large and complex drawing. Electrostatic plotters vary in size from about A4 up to A0 (approx. 1200 × 840 mm).

Electrostatic plotters are usually preferred for high-volume drawing because of the simplicity of operation and the fact that they can be left to run unattended (they can be used for plotting overnight if necessary). Pen plotters have to be monitored, and ink flow can frequently be a particular problem (draft plots are often produced using ballpoint). For some of the modern CAD applications, involving the display of shaded images, electrostatic plotters are the only practical solution, and some of the latest plotters provide very high quality shaded colour drawings.

The technology for plotting is developing rapidly, and there have been considerable improvements in the quality of all plotters in the last few years. The latest development is the 'thermo-transfer' plotter, which is similar to an electrostatic plotter, working with a rasterized drawing, but it builds up the image by using heat to transfer the ink. It has the advantage that it is smaller and more portable and it does not require specially treated paper.

Very often some form of textual output is required, and in many cases this can be output via the plotter. However, for larger systems, and applications involving a substantial amount of textual output, access to a conventional printer is preferable.

There are several more exotic devices which satisfy the need for output of data for special tasks. The following are particularly appropriate for some applications in a design engineering environment.

Large-screen displays Plotting is a frequent activity during the course of design and design problems are often discussed around a copy of the current drawing. Large-screen displays enable several engineers to study the design simultaneously and hence eliminate some of the need for plotting.

Computer output to microfilm Many design and engineering offices store their libraries of drawings on microfilm. Direct computer output to microfilm (COM) from a CAD database eliminates the need for plotting the drawings first.

Photoplotters For some electronic applications in printed circuit board design, photoplotters may be linked to the CAD system to generate photographic negatives (for PCB artwork, for example) direct from CAD data.

5.1.3.5 Networking and communications

While there are still many cases where the need is for a single stand-alone CAD/CAM workstation or system, the more usual requirement now is for systems that can communicate with each other and with other computer systems and equipment through a network. The development of computer networking over the last few years has been particularly beneficial to CAD/CAM.

Networked computer systems are needed in CAD/CAM to satisfy a number of specific requirements.

Workstation-to-workstation communication With the development of workstation-based systems there is often a need for users to be able to pass CAD/CAM information to each other and/or to share library data. This is particularly important if the systems run different applications. For example, one system may create component designs to be passed to other systems for engineering analysis and/or NC programming.

Workstation to file-server In larger design offices there is usually a need to provide for the central storage of CAD/CAM databases and libraries. This is often handled by one system which acts as a 'file-server' providing a substantial amount of on-line storage on disk.

Local area systems to mainframes Mainframes are often used to provide for the large-volume storage of engineering databases. The CAD/CAM users may also need to extract information from mainframe-based systems and/or generate data to be passed to other company systems. They may also use the mainframe to run engineering analysis packages and other specialist applications.

CAD/CAM systems to manufacturing NC programmers generate data to drive machine tools. Direct NC (DNC) provides the means to transfer NC data directly from a CAM system to the machine tool. Networking of design and manufacturing systems also enables the integration of CAD/CAM with the manufacturing processes and allows data to be interchanged between the divisions. The ability to be able to put CAD/CAM systems on a company's computing network is a key requirement for many manufacturing organizations.

5.1.3.6 Networking standards for CAD/CAM

To satisfy the networking requirements of CAD/CAM a number of standard networking protocols are used, of which the following are particularly significant.

Local Area Networks (LAN) Many systems based on Unix (a trademark of AT&T), which are becoming the standard for CAD/CAM Engineering Workstations, provide a LAN facility to IEEE 802.3 Standard using Ethernet^{TM} (a trademark of the Xerox Corporation). NFS (Network File System), developed by Sun^{TM}, provides a powerful and versatile network facility permitting the exchange of data and remote logging into other systems, often a necessary requirement to properly exploit a system network for CAD/CAM applications.

Transmission Control Protocol/Internet Protocol (TCP/IP) In many CAD/CAM installations TCP/IP is used as the standard for communicating between different types of processor, enabling them to exchange data and establish host/terminal relationships. It is well established and widely supported by many vendors.

Manufacturing Automation Protocol (MAP) The need for a standard for factory networking systems has led to the development of MAP. This specifies a standard for communication and control between computer systems, and machines and other equipment on the shopfloor, and has been developed to satisfy the needs of large-scale manufacturing organizations.

Technical and Office Protocols (TOP) This provides a set of international OSI (Open Systems Interconnection) standards to enable different computer systems in engineering to exchange data across a network intelligently. It specifies standards for both physical connections and transactions.

5.1.3.7 A CAD/CAM installation

While a number of CAD/CAM installations are single stand-alone systems running on PCs and engineering workstations, many large companies have major installations comprising several individual systems, perhaps from more than one vendor, and networked with the company's mainframe-based installation. Figure 5.3 shows an example of the configuration for a typical installation to provide a variety of applications for different parts of an engineering organization.

5.1.4 Software

Some turnkey systems comprise several million lines of code, the result of several thousand man-years of program development. Software development for CAD/CAM systems covers many different aspects of computing, ranging from databases, screen manipulation and real-time systems to highly specific engineering applications.

5.1.4.1 The multi-level CAD/CAM machine

A CAD/CAM system involves program development at several levels. The concept of the 'multi-level machine', used for current computer design strategies, represents the com-

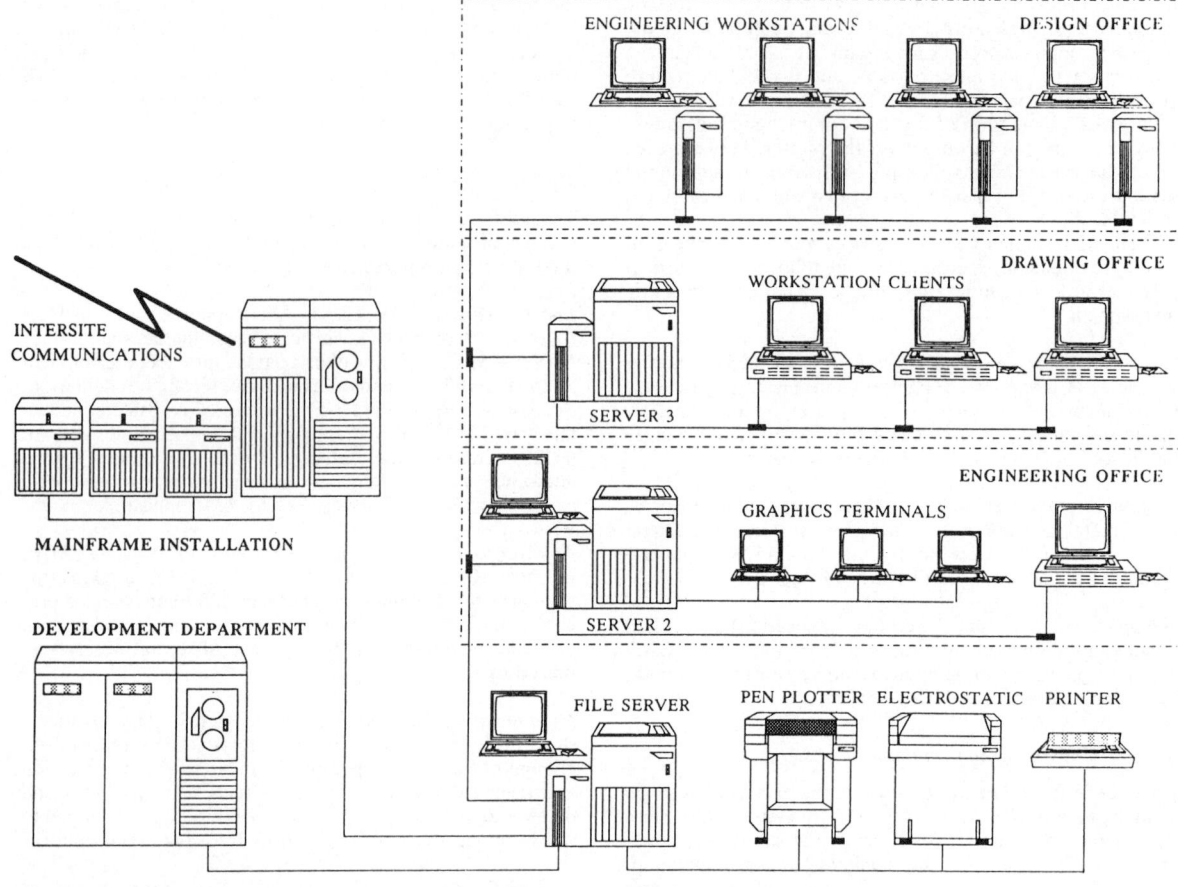

Figure 5.3   An integrated CAD/CAM installation

Table 5.2   The multi-level machine concept extended for CAD/CAM

Level 9	User-tailored software for specific company requirements to automate engineering tasks
Level 8	High-level application software for specific design and manufacturing tasks (e.g. draughting)
Level 7	Problem-oriented tools appropriate to CAD/CAM applications, such as geometric modelling
Level 6	Basic CAD/CAM functionality, such as graphics, vector manipulation and database handling
Level 5	The problem-oriented language level – high-level languages (e.g. Fortran)
. . .	
Level 0	The digital logic level

puter at a series of levels from Level 0 (the Digital Logic) to Level 5 (the Problem-oriented Language).[1] The applications at Level 6 and above can be used to model a CAD/CAM system, as shown in Table 5.2.

The levels provided with the typical CAD/CAM systems are as follows. *Level 5* represents the programming language (typically, Fortran and 'C' for many CAD/CAM applications). *Level 6* includes tools to handle the following:

1. The User Interface, to decode commands, or menus;
2. Graphics Manipulation, to handle screen graphics;
3. Database Handling, to store and retrieve data;
4. Basic routines to provide standard mathematical and geometric functionality.

The basic software functionality required to support CAD/CAM applications is further discussed in Section 5.1.4.2. *Levels 7 and 8* encompass the application-oriented features of a CAD/CAM system, provide engineering users with the software functionality necessary for their work, and are further discussed in Sections 5.1.5 and 5.1.6. *Level 9* represents the level for system tailoring and bespoke software development, which build on CAD/CAM system tools already provided to automate specific processes for particular company needs and is further discussed in Section 5.1.7. The model of CAD/CAM as a multi-level computer application helps in the design of CAD/CAM systems and in the understanding of their operation.

5.1.4.2   *CAD/CAM software fundamentals*

The fundamental components of a CAD/CAM system are the user interface, the graphics manipulation and the database-handling facilities. The design and development of effective software requires a good understanding of the appropriate computer science disciplines, unlike the higher levels, which rely heavily on an in-depth knowledge of the applications.

The user interface The functionality of a CAD/CAM system is usually driven by the user through the input of commands. The two most usual techniques for inputting commands to the system are through a command language or by menu selections.

Command languages Many systems are driven by 'English-like' command statements which are typically of the form:

INSERT LINE X1Y1 X2Y2

This would be interpreted as 'create a line a between the locations given by the coordinates x1y1 & x2y2'. The command statements may include code words, coordinate locations and references to existing geometry. They may be issued to the system in a purely textual form, or by using a combination of text, and screen digitized locations and references to displayed geometry. There are many well-established techniques for developing comman language-driven systems, the main requirement being that the command syntax and complexity of the language are able to reflect the functionality required of the system.

Hierarchical menus Command input by menu selection has been implemented in a number of different ways. In the earliest systems, selection by number from a list of options enabled a simple keyboard interface. However, the development of improved graphics capabilities allowed direct selection by highlighting the command required in a list displayed on the screen, and many of the latest systems provide menu selection through 'icons'. Operating systems for workstations and PCs often provide software tools to assist in the development of menu systems.

Menu-based systems are attractive because they lead the user through the system. However, command language-based systems tend to offer a more flexible approach. They can be adapted to off-line input techniques and support the creation of programs and procedures to automate the tasks.

The provision of sophisticated tools for user interfaces has enabled menu-driven interfaces to be developed for essentially command language-driven systems, thus offering the benefits of both.

CAD/CAM graphics Much of the success of CAD/CAM systems is dependent on the ability of these systems to manipulate the graphics and allow the user to interact with it. They have provided routines to draw lines, circles and other standard shapes and perform operations such as limiting or 'clipping' the displayed graphics within a 'window' on the screen, translating and rotating the screen image, and locating drawn entities from a digitized position.

There is less need for such functionality in some CAD/CAM systems as developments have provided the ability to perform many basic graphics functions through hardware. As a result, CAD/CAM systems are beginning to standardize their graphics output, using standards such as PHIGS (Programmers' Hierarchical Interactive Graphics Standard), to be able to take advantage of these new hardware features as they become available.

CAD/CAM databases The primary purpose of most CAD/CAM systems is to create and/or manipulate engineering information. Thus the organization and structure of the data which represent that information is most important.

Much of CAD/CAM is about the handling of geometric data. For the CAD representation of components, systems need to handle the following types of geometrical entities:

1. Simple linear – points, lines, arcs and circles, conics, free-form curves;
2. Surface – cylindrical, conical and spherical surfaces, complex double-curvature surface geometry;
3. Solid geometry – cuboids, conical solids, complex swept solids.

Simple systems for drawing usually only need to store two-dimensional geometric data while more sophisticated systems which allow complex component design must hold the data as three-dimensional geometry. Since such information is the 'core' data on which many operations are performed, efficient methods for handling it are required.

CAD/CAM databases may also have to store information which is specific to particular applications such as the dimensions on drawings and the names of electronic components. A very flexible database structure is required to enable this. For example, dimensions require information such as the location of leader lines, the shape of arrowheads and additional dimensional text such as tolerances, etc. Additional data to describe the line fonting may also be stored in order to describe the way in which entities are to be displayed.

Intelligent databases are required for some applications to enable the connectivity of data to be stored. In electronic design, for example, there is a need to know how individual components are connected. The use of intelligence in CAD/CAM applications is discussed later.

Information to describe the way in which a three-dimensional object is to be displayed on the screen and the views of it that are required for drawing must also be recorded in the database. Other CAD/CAM systems that output data for post-processing may need to hold instructions to support the re-processing of those data.

5.1.4.3 Languages for CAD/CAM

The development process for CAD/CAM software is similar to that for other computer applications, involving systems analysis, software engineering, development and test. However, CAD/CAM is both a multi-level and a multi-discipline application. Hence the development problems are often compounded by the need to use a variety of different programming languages.

Conventional languages Fortran is probably the most common high-level application language for CAD/CAM, being the one most widely used in scientific and engineering applications. However, there is often a need for some operations to be performed in a very efficient way, and thus low-level assembler code is often 'incorporated'. In recent years, with the use of Unix-based systems, the C language has become popular.

Operating system tools With the growth of workstation-based systems, tools provided by the operating system have assisted in the development of user interfaces and other functions required for CAD/CAM systems. The development of graphics standards also helps the development of output interfaces.

Relational database management systems A number of CAD/CAM systems have specific database-handling requirements that may be best satisfied by a 4GL RDBMS application. If the application is one part of an integrated CAD/CAM system then care is needed in the design of the user interface to ensure that a consistent approach is adopted.

Artificial intelligence   Many engineering tasks rely greatly on the engineers' knowledge and experience. Thus, there is considerable potential for the use of AI and Expert System tools in CAD/CAM applications, although their use has so far been somewhat limited. However, as AI techniques become more proven, their use in CAD/CAM will almost certainly increase.

5.1.5   Modelling techniques

CAD/CAM systems provide the computer tools for a wide range of engineering applications. Central to many of the tasks is the computerized model of a product replacing a traditional engineering representation.

The use of computer-based systems for engineering design has introduced the concept of 'geometric modelling'. Traditionally, a description of an object has been developed and conveyed using two-dimensional drawings on paper, but the use of a computer system enables designers to create and present the necessary information in an entirely different way. Computer technology offers scope for providing not only a better description of a product but also one which can, with the use of other computer-based tools, automate some of the downstream processes.

The modelling techniques in CAD/CAM are not restricted to geometric ones. There are many applications for which some form of 'schematic modelling' is required (e.g. for plant design and electronics).

5.1.5.1   Geometric modelling

Geometric modelling is the CAD tool for generating a computer representation of a 'solid' component. All CAD/CAM systems in mechanical engineering must provide some facility for different levels of geometric modelling from 'wire-frame' to 'surface' and 'solid' modelling, and at each level different techniques are required.

Wire-frame modelling   A 'wire-frame' model is the simplest form of geometric modelling and components may be modelled in two or three dimensions. Component descriptions consist of geometric entities such as lines, arcs, conics and curves.

At the simplest level, two-dimensional geometry is the computerized equivalent of a drawing, and many components can be adequately described in this way. It suits 'flat' objects such as those made from plate or sheet metal and also axi-symmetric (round) components, which can be adequately represented for many purposes by a cross-section through the component.

Additional data may be required to provide enough information about the object. This may be an additional two-dimensional drawing showing another view or further textual information providing details of the material. Although the representation is simple it can still provide sufficient information, and it is a practical solution for many downstream applications (for example, planning and NC programming, or where the required end result is simply a drawing of the component).

Many CAD systems provide a facility for creating wire-frame geometry in three dimensions. The component is still represented by the same type of geometric entities but is defined in three-dimensional space. They are used typically to represent the hard edges of a component, or where the shape of the component changes from a flat surface to a curved one. Free-form curves in three-dimensional space are used in the design of complex surfaces.

Wire-frame models in three dimensions provide a great deal of information about the component but they do not give a complete description of it and the shape between the edges must be inferred. In many cases the shape of the faces is obvious, particularly those that are flat, but complex surface changes cannot be adequately described and the models can be misinterpreted.

Although users can perform quite complex tasks with wire-frame geometry (such as calculations of volume), the operations cannot easily be performed automatically by the system, and it is sometimes very difficult to accurately visualize the component.

Surface modelling   Surface modelling enables complex faces of a component to be accurately described. In a complete surface model each of the component faces will be represented by one or more surfaces which meet at coincident edges.

There have been many different algorithms for mathematically modelling surfaces. Surface modelling was initially developed in the mid-1960s, to satisfy the requirements of the aerospace, automotive and shipbuilding industries for more accurate ways of defining complex surfaces. Many of the early systems were based on the 'Coons patch', developed by Coons in the early 1960s, and some derivatives of these systems are still in use today.

Curve- and surface-fitting algorithms attempt to mimic an 'engineering spline'. The algorithms were initially based on piecewise polynomials. For most applications, fourth-order polynomials were usually sufficient to maintain continuity of slope and curvature across boundaries and provide smooth surfaces, the minimum requirement for most engineering applications. Some applications requiring very smooth flowing curves, such as automotive body design, require higher-order polynomials.

Modern algorithms for curve and surface fitting are based on the B-spline curve fitting algorithm, a derivative of the work by Bezier. This is now preferred because it is mathematically more robust than the polynomial-based curves that were used earlier. The latest systems use NURBs (Non-Uniform Rational B-splines (and B-surfaces)) which can precisely model a wide range of curve functions and give, for example, exact representations of arcs and conics.

Surface modelling provides more information about the component than wire-frame geometry and is of value in describing the complex shapes of aircraft surfaces, ships' hulls and car bodies, as well as for the design of a wide range of general products, such as glass bottles. Surfaces also enable the model to be more easily visualized, and are often provided as an adjunct to the wire-frame representation for additional information.

Modern surface modelling systems, in conjunction with high-technology display devices, provide high-quality surface shading, enabling a very realistic rendering of an object and excellent visualization. Thus many aesthetic design tasks can be performed on the computer screen rather than by building actual models.

Solid modelling   The most complete approach to geometric modelling is through solid modelling. The ability to generate solid models has been available for a number of years, but because of the heavy processing requirement it is only as a result of the recent developments in hardware technology that it has been possible to exploit it in a practical way. Solid models can provide all the geometric information required to describe a component and there have been a number of different techniques for representing a solid, but the majority of current systems make use of one of two methods.

In constructive solid geometry (CSG) the solid model is represented as a set of primitive solids such as cuboids and cylinders, together with operators to combine or subtract them, i.e. a circular hole through a solid is the subtraction of a cylinder from the solid. The user creates and develops the solid by adding and subtracting the appropriate primitives. However, the representation is restricted to solids that can be composed from the given set. It is a compact method of representing a solid but can be time consuming to process for subsequent operations.

In Boundary representation (B-rep) the solid model is represented by a set of 'faces'. These are surface descriptions referencing 'boundary edges' which are made up of lines, arcs and curves meeting at 'vertices' (the faces, edges and vertices relating to those of the object being modelled). The solid may be constructed by Boolean operations on primitives as with the CSG method as well as by sweeping or rotating an 'edge', thus permitting the construction of the highly complex faces required for some solid objects. Processing and downstream applications can exploit the direct availability of the edges but some B-rep models have a large data-storage requirement.

Some solid modellers use a 'faceted' model to represent the solid, rather than an exact mathematical representation of the faces, to speed up processing. The calculations are fast and robust, but may be inaccurate in some instances. For an exact representation of the faces the system must handle all the different curve types, and the routines require a high degree of accuracy in mathematical calculations. It is possible to unify the algorithms by using Non-Uniform Rational B-surfaces and satisfy most modelling requirements.

Solid modelling has distinct benefits over other types of modelling since it models the real solid component and it enables the object to be properly visualized. The model can be shaded and edges hidden, and very realistic pictorical representations can be obtained as the examples in Figures 5.4 and 5.5 show. In addition, mass properties (e.g. volumes and centre of gravity) can be calculated automatically, and models for structural analysis may also be generated.

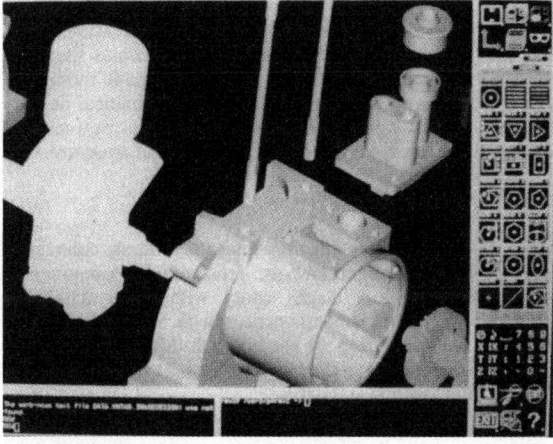

Figure 5.5 The same model as a pictorial image

Feature modelling In the real world, components are actually made up of features such as faces, grooves and holes. Some CAD/CAM applications need to know not just the geometric representation of the component but the features it contains. For example, a cylinder may represent a hole, but it is the characteristics of the hole such as depth, diameter and direction and its surface finish and tolerances that are important when it is manufactured.

Feature-based modelling addresses this problem. It is another representation of the component, which provides a set of related features that reference the geometric entities in the conventional geometric model.

Feature-based modelling offers advantages to the designer. The model can be described in engineering and/or functional terms, and properties can be attached to the features to provide additional information for manufacturing purposes.

Other advantages are that the model conveys the design intent, but more importantly, because the purpose of the geometry is described, it can enable automatic processing for NC programming and process planning. The use of knowledge engineering techniques and artifical intelligence are needed to make full use of feature-based modelling in CAM applications.

Finite element modelling The finite element method of analysis is a technique used for stress analysis and other analytical problems. It involves modelling a component as a matrix of small elements and then determining the effect of the applied loading on each of the individual elements. The method involves the solution of a large number of individual equations, hence its suitability for computer resolution.

The finite element model is a mesh of individual elements, typically brick- and/or wedge-shaped blocks connected at grid points, which model the component geometry. This model is an approximation to the geometry for the purpose of analysis. The finer the mesh, the more accurate the analysis, but the more time consuming is the computation. The mesh is generally designed to reflect the changes in stress in the component, highly stressed areas and areas where there is a marked change in the geometry being given a finer mesh. (The analysis results in Figure 5.8 have been displayed on the original finite element model.)

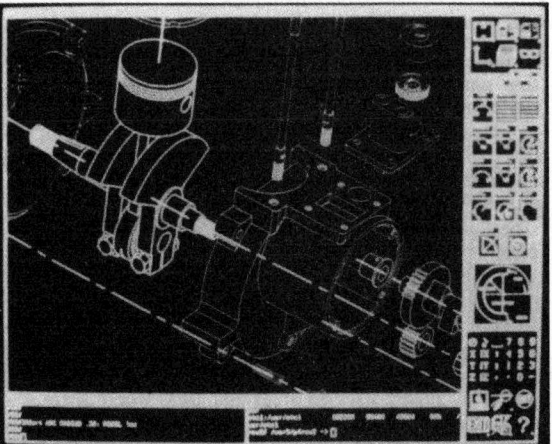

Figure 5.4 A solid model showing hidden line removal

5.1.5.2 Schematic modelling

In some design tasks there is a requirement for creating circuit diagrams and schematic modelling. In electronics they are essential for the design description, but schematic modelling also has applications in plant layout and mechanical design. Schematic modelling may be purely diagrammatic or it may be used to represent a complex three-dimensional structure.

Schematic diagrams A schematic diagram comprises a set of symbols connected together by lines. The symbols, representing components, may be stored as a single entity within the database, referencing a library component which is used for the graphical representation of the symbol. Many applications require a standard set of symbols developed to meet company or even international standards.

Some systems support connectivity within the database, so that the line joining a symbol or a particular component within the symbol (e.g. a pin or a transistor) is represented in the database by entities that are connected. Text displaying component names and other information may also be linked to the component within the database. Thus the database is an intelligent representation of the diagram, enabling the connectivity of the model to be automatically processed.

The use of intelligent databases for schematic diagrams enables information to be extracted from the diagram for use by downstream processes as discussed in Section 5.1.6.2 on CAD electronic design.

Schematic geometry A number of applications, such as a wiring harness or plant design, have a requirement for a schematic representation of the geometry, in which the model is not a true geometric one but rather a diagrammatic representation showing the location of major components and the way they are interconnected.

The components themselves (for example, valves and tanks) are often symbolically represented and the pipes joining them are shown by lines. The schematic representation offers a quick and easy method for design, and permits connectivity and other information to be extracted. Many such systems also provide software to convert the diagrammatic representation into a more realistic geometric model for visualization, as shown in Figure 5.6, a picture generated from a plant schematic.

Figure 5.6 A pictorial image from a plant design schematic

5.1.5.3 CAD/CAM modelling practice

Modelling is an essential part of the design process, and in CAD/CAM computer-based models replace the traditional engineering techniques, which in some cases involved building full-size physical representations of the product.

The CAD/CAM model may often be the ultimate reference for the finished component and it must accurately reflect the actual component, to the level required for its subsequent use, which may be purely geometric, or in some cases also functional.

5.1.6 Applications

Many applications in engineering design and manufacturing are addressed by CAD/CAM systems. They may supply the tools for specific applications and be dedicated to providing 'niche solutions', or answers to particular engineering problems, or a large variety of tools capable of handling many different applications within a single integrated system.

The following sections give a brief overview of the different areas of application together with the CAD/CAM tools to satisfy their requirements. It is not a review of any particular CAD/CAM system but rather of the system features that are required to support the applications.

5.1.6.1 The design to manufacturing process

CAD/CAM systems assist in automating tasks within design and manufacturing. The overall design to production process is broadly similar in most manufacturing establishments, and is illustrated in Figure 5.7.

The different functional areas and the different disciplines in those areas often require specific CAD/CAM solutions to satisfy the needs of the tasks involved. In an integrated CAD/CAM environment information must be passed from step to step in a computer-based form. The process is often an iterative one, and the need to be able to cycle through a series of tasks is an important aspect of the systems.

5.1.6.2 CAD applications

The purpose of engineering design is to define the product and generate information for manufacturing. The tasks involved vary considerably in different engineering disciplines, and hence CAD systems for one discipline may provide very different facilities to those for another.

CAD systems are employed for a number of applications outside the design office. For example, other parts of a company may use CAD for the preparation of drawings for marketing and product support, and illustrations for manuals for input to a desktop publishing system. Many other organizations outside engineering also use CAD as a general tool to create drawings.

The following sections provide an overview to the CAD application requirements for various tasks in a number of different design disciplines.

Mechanical design engineering Mechanical engineering encompasses a wide variety of products ranging from aircraft and cars to zip fasteners. The design of mechanical components involves developing a geometric description of a component to satisfy the needs and functionality of the product, which very often includes a large amount of analytical work. The resultant CAD output may include manufacturing drawings for individual components and assemblies, parts lists, and a variety of other documentation to support the product.

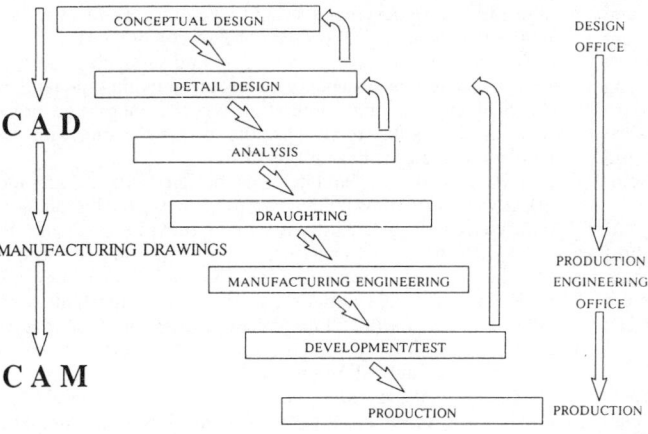

Figure 5.7 The design to manufacturing process

A range of CAD/CAM tools is required to support a mechanical design engineering process from initial scheming, through detailed design and analysis to the preparation of manufacturing drawings.

The initial product design and scheming is likely to make use of schematic design systems, simple CAD tools for two-dimensional draughting and geometric construction, and more powerful systems for three-dimensional modelling of solids and complex surfaces. Ease of manipulation of the computer-based data is a key feature for conceptual design since many different ideas may be tried and modified as the designer optimizes the product.

Conceptual design often involves the use of special programs to generate information to meet pre-specified design criteria (for example, to produce aerofoil sections to meet aerodynamic characteristics). Thus the ability to integrate CAD systems with other computer-based applications is often important.

A powerful geometric modelling tool is a fundamental requirement for the mechanical engineering of new, high-technology products. A high degree of visualization is often required together with the ability to easily perform analytical operations on the component being modelled, such as the calculation of volumes and weights. Many products are the result of the assembly of a number of smaller items. Thus the ability to bring together models of different components to check interfaces and the fit of the product is essential.

Many mechanical design problems can only be resolved by simulating the mechanism. Computer simulation may be used in both conceptual and detail design. The systems enable the component to be modelled as a mechanism, and are able to determine loadings and displacements of linkages, springs and other moving parts, as well as helping the designer to confirm the required mechanical operation of the system.

Many mechanical components are highly stressed, and must be optimized to minimize weight. Hence structural analysis is often an integral part of the design process. There are many finite element analysis packages available. Some of the latest systems make use of high-resolution colour graphics to display the results of the analysis visually, such as that shown in Figure 5.8, as well as providing detailed listings of the results.

Although CAD systems are able to provide computer-based data for the production departments directly, the preparation of manufacturing drawings is still a requirement in many engineering companies. Most CAD systems for mechanical design provide a draughting capability. Computer-aided draughting is discussed below.

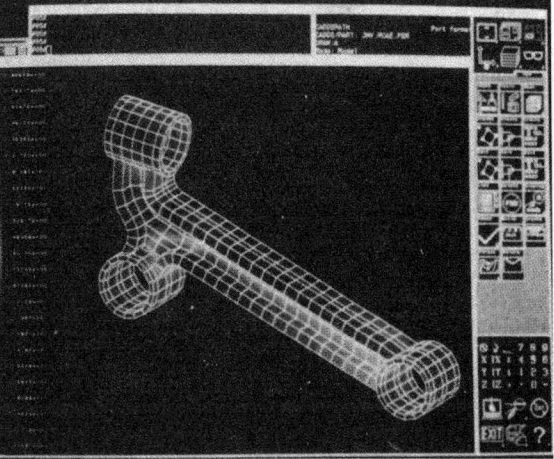

Figure 5.8 A finite element model of a component

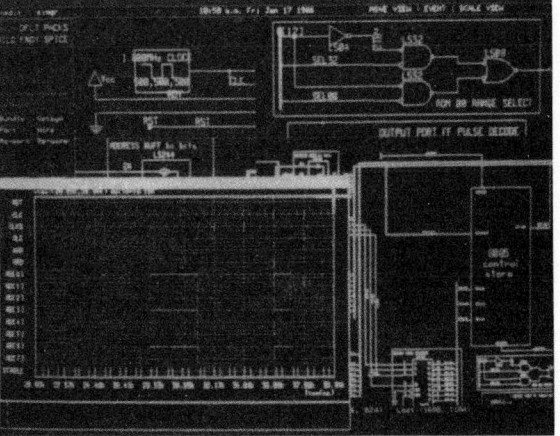

Figure 5.9 Electronic circuit design with analysis results

Mechanical design is frequently an iterative process and therefore the ability to integrate the tools for these different tasks is an important aspect of CAD. This field of engineering is very diverse and, because of its nature and the long traditions in the industry, it is difficult to develop standard techniques. Thus turnkey systems which support a wide range of mechanical design applications must be very flexible and easily able to adapt to the different requirements of engineering companies.

Electronic engineering Electrical/electronic design requires CAD systems with features very different to those for mechanical engineering. The design process is similar but involves circuit design, electrical analysis of the circuit, design of PCBs and/or ICs, and preparation of information for production.

A good schematic modelling package is the basic requirement for the design of electronic circuits. It should provide a component symbol library to suit the company's products. Systems may also provide tools for design rule checking of the circuit as it is created. Additionally, an intelligent schematic database is also necessary to enable automatic generation of data for output.

Systems are available to perform an electronic analysis of the circuit, providing both analogue and digital simulation. Such systems provide the ability to define the electronic inputs, measure and display the outputs, and generate test data. The interfacing of electronic simulation with circuit design is an important part of an integrated system for electronic design.

Systems to assist in the design of PCBs have become increasingly more automated in recent years. The job involves positioning components on the board followed by routing of the tracks. The routing can be a particularly difficult problem, but CAD systems provide automatic routers that are now able to accomplish the task on complex multi-layer boards with nearly 100% success. Other features provided include the facility to perform thermal analysis of the PCB board.

Systems are available to assist in the design of integrated circuits. They enable IC designers to build up masks for IC manufacture by positioning components in the CAD drawing. CAD tools are now essential for IC design, as the number of circuit elements to be interconnected has become so large.

Like mechanical design, the integration of systems for electronic design is also important. The interfacing of circuit schematic systems with systems for simulation enables logic design to be tackled as one integrated process. Figure 5.9 shows an example of such a system. Intelligent databases enable the generation of 'net lists' to assist in the automatic generation of PCBs from circuit diagrams and automatic updating of PCB layouts to reflect changes in the circuit design. The output to manufacturing may include data for silk screening for printing and etching the boards, and NC drilling information.

The implementation of CAD for electronic design is often easier than for mechanical design. Since the design process has developed along with the CAD tools, it is more formalized and thus more easily adapted to automatic processing. As electronic systems have become more sophisticated so have the CAD systems required to design them, and CAD is now an essential tool for many electronic design applications.

Civil and architectural engineering There is some similarity between mechanical design and particular aspects of civil engineering and plant design, and they require similar tools. However, for many areas of civil engineering, specialized CAD features are needed. One significant difference is the way in which the product is modelled. Many civil engineering applications are developed around schematic geometry rather than a true geometric representation of components.

Site engineering involves mapping and modelling the terrain. CAD systems provide a means of modelling the land in three-dimensions, and generating contour maps and cross sections. The ability to visualize the terrain together with the construction is an important feature.

CAD systems for architectural design need to provide effective tools to draw buildings, and to enable the buildings to be accurately visualized. Systems may provide the facility to lay in patterns of bricks and tiles and other decorative features to generate very realistic pictures. As well as providing the tools for design and visualization, it is also important to be able to create traditional architectural drawings, a particular feature of which is the cross sections which use different styles of hatching to show different materials.

The design of structural steelwork is often treated initially as a schematic problem. Structural analysis is important and therefore a link to an appropriate finite element analysis package is necessary. Systems for structural engineering must also enable the generation of materials lists from the schematic models of the structures, manually a difficult and time-consuming task.

Plant design provides the classic example of a 'geometric schematic' system. Systems enable a plant layout to be developed as a three-dimensional schematic, providing symbolic representations of components. These can then be converted to models, providing a high level of visualization such as that shown in Figure 5.6 (Section 5.1.5).

The use of civil engineering CAD applications is widespread. As well as the obvious candidates in the construction and marine engineering industries, many companies in mechanical engineering and manufacturing use the systems for plant design and factory layout.

Mapping itself, although originally regarded as a CAD application is now becoming a more highly specialized application area. The current needs are for the more wide-ranging 'Geographic Information Systems' (GIS), which, as well as providing a mapping capability, also supply a versatile geographical database.

Computer-aided draughting Probably the most widespread use of CAD is in the creation of drawings using computer-aided draughting techniques. Indeed, a very large number of systems are acquired for just this purpose, to provide an 'electronic drawing board'.

Computer-aided draughting systems require a basic capability to create two-dimensional geometry, and provide a set of entities which include points, lines, arcs, circles, conics and free-form curves. A draughting system which is integrated with a three-dimensional modeller must provide the ability to generate two-dimensional views, cross sections and scrap views from the three-dimensional model. The creation of geometry to represent the component is usually the least difficult part of the problem. Having created the geometry, the draughtsman's task is to dimension and annotate the drawing.

One of the most complex tasks is dimensioning. The major problem that most systems face is handling the variety of standards to meet different company requirements. These may include BS (British), ISO (European/International), DIN (German), ANSI (American) and JIS (Japanese), together with variations and restrictions to suit individual companies. Different disciplines also have different dimensioning standards (for example, the dimensioning style of architectural drawings is quite different to that for mechanical components).

Dimensioning often requires 'creativity'. Dimensioning a large drawing of a complex engineering component so that all the dimensions do not excessively overlap is very difficult. Draughtsmen often resort to 'draughting licence'. Thus it can be a difficult task to automate fully.

Drawings often contain as much text as geometry, and many engineering disciplines have symbolic codes which must be incorporated into the text. A drawing is usually presented on a company standard 'form sheet', which also needs to be available to the CAD system. Some computer-aided draughting applications do not require the sophisticated functionality of integrated turnkey systems, and a number of draughting systems are available to run on PC-based computer systems.

5.1.6.3 CAM applications

Computer-aided manufacturing encompasses applications ranging from planning and scheduling to process monitoring and control. In the context of CAD/CAM, however, it is appropriate to restrict this review to those areas of manufacturing which are directly associated with CAD, and in particular those tasks that can be closely integrated with the CAD applications. In this context the most significant CAM applications are those related to 'manufacturing engineering' (see Figure 5.7). The production engineering office is generally responsible for taking information from design, such as drawings and parts lists, and generating the information required for the factory floor. It encompasses the functions of process planning, NC programming and tool design, and is the main interface between design and manufacturing. In an integrated CAD/CAM environment it is important for production engineering to be able to receive information in a computerized form that it is able to use to generate data for the production departments.

Process planning Process planning is the task of determining the production processes required to manufacture a component or a product. Traditionally, it is a task undertaken by engineers with a detailed knowledge of the engineering processes in the factory, and able to determine those that are appropriate for manufacturing the product within the company's capabilities.

Computer-aided process planning (CAPP) offers many advantages over traditional manual methods. It reduces the time and effort that it takes to produce a plan and ensures a greater degree of conformity between planners. CAPP systems have developed along two main directions, variant and generative.

Variant planners are based on the premise that existing plans (or previously prepared partly completed plans for a similar component) can be retrieved and then edited to create new plans. Many variant planners make use of 'group technology' (GT).

GT is the concept of classifying and coding parts according to geometric and other attributes of the component. The value of GT is that, if two parts have the same or similar classification, then they are likely to require a similar manufacturing process. Thus GT can be used to retrieve existing plans for editing. GT also has applications in a design office to avoid repetitious designs.

Generative planners create a new plan from scratch for each component. They make use of a database of processes and rules that a planner would use and, by examining the features of the component and using in-built knowledge, determine the appropriate manufacturing processes for the given component.

Generative process planners require the use of planning knowledge and need to make intelligent decisions in much the same ways as a planner. Thus the development of generative planners is seen as one of the key applications for knowledge engineering and artificial intelligence in manufacturing.

Generative planners enable the process planning task to become more automated, since they can replace most of a planning engineer's task. However, they require the development of computer tools and the use of techniques which are themselves still being researched. The development of feature-based modelling is important to generative planning systems, since generative planners need to understand the component as a set of features rather than as a geometric model.

Computer-aided process planning is a key element in developing computer systems for manufacturing because it captures some of the basic information required for the manufacturing tasks. It can also be a prerequisite to NC programming, one of the key areas for the application of CAD/CAM.

NC programming NC programming is the task of preparing data to drive NC (numerical control) machine tools. The instructions to an NC machine are input in the form of codes which drive the cutting tool and control machine functions. The data are typically provided via eight-track magnetic tape.

Traditionally, NC machines have been programmed directly, the codes being entered into the machine controller. However, the development of NC programming languages such as APT (Automatically Programmed Tools) enabled component geometry to be defined and hence cutting paths in terms of that geometry. NC programming integrated with a CAD/CAM system allows the NC toolpath to be defined with reference to the CAD model geometry.

NC programming systems vary in functionality according to the capability of the machines for which they prepare data. Numerically controlled machines include drills, lathes and machining centres. They may be classed according to the number of axes, or degrees of freedom, in which the cutter may move simultaneously, the most common being 2½-, 3- and 5-axis NC machines.

In 2½-axis NC the cutter may move simultaneously in two axes, typically x and y, but there are positional movements in the z direction. This permits the machining of mainly orthomorphic components (with orthogonal faces, edges and axes) and those with constant angle faces. These make up more than 80% of the components in mechanical engineering.

In 3-axis NC machining in all three coordinate axes allows the machining of complex surfaces as well as orthogonal faces. An example of 3-axis NC programming is shown in Figure 5.10. It can satisfy the machining requirement of many complex components, but may restrict the cutting angle and ease of access for some of the more complex surface shapes.

In 5-axis NC machining in five-axis also permits the machine tool to be rotated, as well as providing all translation movements. It can, for example, maintain a constant cutting angle with the complex surface being machined, a feature which is particularly important in the manufacture of aerofoil shapes and press tools for the automotive industry, where there is a need for a very high-quality finish on complex surface shapes. NC systems for 3- and 5-axis machining usually need to be integrated with a CAD/CAM system which provides a powerful geometric modelling tool to enable the design of complex surfaces.

There are some applications suited to 4-axis machining, especially in the aerospace industry. NC is also used to drive punching machines for sheet metal work and other manufacturing processes.

Most NC systems employ a language that enables the user to create a program of NC machining statements to drive the cutter and perform the operations required, together with an editor to enable the program to be modified or corrected

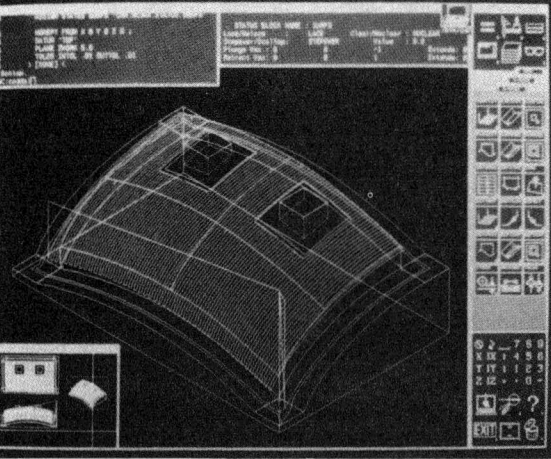

Figure 5.10   An NC 3-axis machining toolpath

easily. Integrated CAD and NC allows references to the CAD data to be included within the program.

Post-processing   Individual machine/controller combinations are driven by specific machine codes. NC programming systems often generate a neutral file output, such as a 'cutter location (CL) path', which can then be post-processed to generate the codes appropriate to a particular machine. DNC, mentioned in Section 5.1.3.5, is an important development, enabling a direct link to an NC machine to download the data and eliminate the need for transfer by tape. NC is one of the most widespread of CAM applications. Many products and processes now rely on NC machining techniques to meet design and manufacturing criteria.

Robotics   Robots are becoming more widespread in the manufacturing industry. Their use, in conjunction with flexible manufacturing systems (FMS), is one of the key ways in which productivity and performance on the factory floor can be improved.

The programming of a robot is similar in some respects to that of other machine tools. However, because of the complexity of the robot's movements, the ability to simulate them, to handle collision detection, and present a visual check are usually prerequisites for robot programming. Thus robotics systems require high-quality graphics and a considerable amount of computing power.

Robotics offers potential applications for artificial intelligence. It may be used to determine the most efficient movements, given the relatively large number of degrees of freedom in a typical robot. Techniques of 'computer vision' may also be used to help the robot to locate components and check their orientation.

Tooling and specialist applications   Tool design tends to be a specialized form of mechanical design. Indeed, for some complex tools there is as much design and analysis effort involved in the tool as in the original component. In addition, there are particular tool design problems which require special application software, of which the following are examples.

The design of moulds requires specialist software to help determine material flow and calculate cooling rates and the allowances for shrinkage of the component. A particularly difficult problem is the determination of the correct shape of the forge tools at the various stages in a forging process, which

must be carefully designed to ensure that the component is manufactured correctly.

The flattening of sheet metal components involves calculation of bending allowances which can be applied to a three-dimensional model as it is unfolded. General solutions can only be applied to components which have straight uniform bends. The flattening of complex doubly curved surfaces requires an understanding of the method of production, since uneven deformation is involved.

Many of the tasks in manufacturing engineering require knowledge and experience of the company's production processes. The rules are often empirical in nature, and thus there is much potential for the use of artificial intelligence.

5.1.7   Implementation

A satisfactory implementation of CAD/CAM may be crucial to the future success of a company. CAD/CAM installations need careful planning and management if they are to achieve the desired results.

5.1.7.1   *People, methods, organization and the technology*

CAD/CAM technology affects three aspects of a company, and each must be addressed during the implementation.

People   In many other applications of computer technology the direct involvement of the users with the system is minimal. However, CAD/CAM systems actively involve the users. This is because the systems, although complete in themselves, can only provide a set of tools to do the work, and hence must rely on the users to apply them to the job in hand in an appropriate way.

Methods   CAD/CAM frequently results in changes to the company's working methods. Although it often happens that an existing process is merely computerized, in order to reap the benefits of the technology it is sometimes necessary to change entirely the way in which a job is tackled.

Organization   The changes as a result of CAD/CAM often embrace several different areas within the organization. The complete integration of these areas may be necessary to enable the company to gain all the advantages of the CAD/CAM installation.

In order for the system to be effective, it frequently relies on the engineering staff who are to use the system becoming fully conversant with it. It also depends upon the company understanding the best way of applying the technology to its product process and exploiting it to advantage.

5.1.7.2   *System implementation*

The initial implementation of CAD/CAM in a company can be treated as a two-stage process. The first step is the installation and system introduction. This is followed by a period of system tailoring which may involve the development of specialist system features, together with appropriate software enhancements.

Introducing the system   When a CAD/CAM system is first introduced into a company it is often important that it is up and running and in productive use as soon as possible. There are three things that must be addressed – the installation itself, the management and administration of the system, and training and support for the users.

Installation Many modern workstation-based CAD/CAM systems can be located in standard design office environments. Although, nowadays, there may not be the need for environmental control (disk drives are usually more sensitive to temperature fluctuation than other devices), it is often beneficial for the system administration if the main units are installed in a purpose-built room.

Careful consideration should be given to the location of workstations. It is usually preferable to install them in the engineering offices close to the users' workplace, so that they can be readily accessed by the users and become well integrated within the office. The move towards desktop systems is of great benefit in this respect.

Systems management In many cases the management and administration of a CAD/CAM system is handled by design office staff who do not have a background in computer systems. Fortunately, much of the day-to-day systems administration can be enacted by well-established procedures, many of which can be automated, thus minimizing the need for specialist personnel.

It is usually beneficial to employ experienced staff, who have some understanding of the operating system, to develop company procedures and provide systems support for the users. Large installations comprising several systems, and a network and communications with a mainframe, usually require staff who are computer literate to operate the systems.

The engineering data which are to be created on the system are a valuable asset to the company and must be properly managed. In particular, careful consideration must be given to the aspects of security and integrity.

Training and support Proper training for the staff is essential if all the benefits of CAD/CAM are to be gained. To exploit its full potential, the system users need to understand its capabilities as well as be able to drive it. A training programme should provide for basic training, following by advanced training and courses in special features appropriate to their work.

The education programme should also include 'awareness training' for managers and other staff who are indirectly affected by the system. CAD/CAM has far-reaching consequences for the company, therefore it is crucial that all staff are kept informed, especially those who are not directly involved with the system.

Tailoring the system CAD/CAM systems provide a standard 'toolkit' for design and engineering offices, and these are sufficient to perform all the basic tasks. However, since all companies have different procedures and standards, the system becomes more efficient if it is further tailored to the specific needs of the organization. There are a number of ways in which this may be achieved.

System parameters Some systems provide versatile commands to allow different standards to be met for tasks such as dimensioning. The setting of system parameters enables the command defaults to be set to meet the company's own standards. In addition, standard company forms should be established for the users as pro formas for their work.

Library components Many applications require a library of components comprising industry standard items supplemented by company-specific parts. The provision of effective component libraries, and frequently used features of components, is essential to save continual re-creation of the same data and ensure that the system is used effectively. The tailoring of the system to automate a particular process or company procedure will usually require the development of special software.

There are two levels at which customized software development may be undertaken.

Use of 'macro' languages Many CAD/CAM systems provide specialist languages, often as an integral part of the system, but incorporating additional programming commands for arithmetic, data input/output, etc. These languages enable users to develop procedures to automate processes and provide complete frequently used command sequences with a single command.

Bespoke software development To satisfy specific processing needs will usually involve the development of software using one of the standard programming languages available for the machine. Macro languages are often tailored for ease of use so that, with a little training, engineering staff can develop their own procedures. Bespoke software development would normally be undertaken by experienced computer personnel.

Tailoring the system to improve the performance and increase the productivity of the users is an essential part of the longer-term development plan for a CAD/CAM system. Such tailoring should be coupled with the development of internal procedures and working practices to make the most effective use of the system.

5.1.7.3 Systems integration

As has already been observed, CAD/CAM encompasses a wide range of activities within the design and manufacturing environment, although many systems only address specific areas. The integration of computer systems throughout the engineering departments is the third stage in the implementation of CAD/CAM. The objective should be for all systems to be able to exchange data, and make use of the data created by other applications where appropriate.

For many applications the flow of information is one way, and in some cases a satisfactory system interface can be implemented quite easily. However, CAD/CAM applications often need a two-way communication of geometric data. It may also be necessary for different applications to be able to access the same geometric database. In these instances a totally integrated system is required.

Interfacing of systems A simple level of interface can generally be provided with a suitable piece of software that extracts the required information from one system and delivers it to the application that needs it, preferably in a form suited to the receiver. Updates to the data may be a particular problem, especially if they need to be incremental (usually the case if some further processing has already taken place on the data). To exchange geometric data between different CAD systems a standard for a neutral file format has been established.

International Graphics Exchange Standard (IGES) IGES provides a mechanism for exchanging graphical information by specifying the format of a neutral file.[2] Any CAD system with the ability to read and write data to the neutral format can pass graphics information to another CAD system. One problem is that only a common subset of the two systems can be successfully exchanged. Often it is the intelligent information in the database that is difficult to transfer. Drawings, too, are sometimes difficult because of the different standards for annotation and dimensioning. Pure geometric information can usually be passed successfully.

To overcome the limitations of IGES, and meet the needs of companies for more powerful data-exchange mechanisms, newer standards such as Product Design Exchange Specification (PDES) are being developed.

5.1.8 The future

The future of CAD/CAM will be determined by changes in current technology and in the way in which the changes are applied.

5.1.8.1 Hardware developments

Many CAD/CAM applications are very CPU-intensive, particularly tasks such as solid modelling and engineering analysis. With increased processing power, solid modelling is a practical solution to many design problems and is affordable. Thus many more design applications are likely to be centred around the use of solid modelling.

Declining hardware prices will make it cost effective for companies to move towards fully computerized engineering offices. As costs decrease and companies can provide a workstation for every engineer, all the tasks can be accomplished through a single computing facility. Networking and communications will enable access to the information required for all the tasks of the user.

The quality of graphic displays will have a considerable impact on CAD/CAM since visual representation is fundamental to many applications. The improved resolution will enable more designers to use these displays for studying results, thus eliminating the need for paper-based output. Applications that may particularly benefit are those that involve aesthetic design of shape, such as automotive body design.

The development in data-storage devices will also benefit CAD/CAM. Many applications have a substantial data-handling requirement. The ability to store data more compactly will enable CAD/CAM applications to benefit from large amounts of on-line data. The permanent storage of data is important in many engineering applications and more reliable methods for the archiving of computer-based data will help to eliminate traditional paper-based storage methods.

5.1.8.2 Software techniques

The development of 4GL languages and expert systems will have a significant impact on future CAD/CAM solutions, as will the use of object-oriented programming techniques. Existing applications are continually being enhanced as better mathematical algorithms and improved programming methods are developed to overcome limitations in the software.

The particular benefits of 4GL languages are likely to be felt in the development of specially tailored software. They will provide better tools for customizing software, thus making it easier and quicker to develop solutions for problems that were previously too costly or time consuming.

However, in the long term, the greater impact on CAD/CAM is likely to be through the use of artificial intelligence and knowledge engineering. Much of the work of engineers is based on experience and knowhow acquired through years of practice. Existing systems often only automate the engineering processes, but AI techniques enable knowledge and experience to be captured and retained for future users. Systems which make use of AI will be able to make the decisions that could previously only be made by experienced designers.

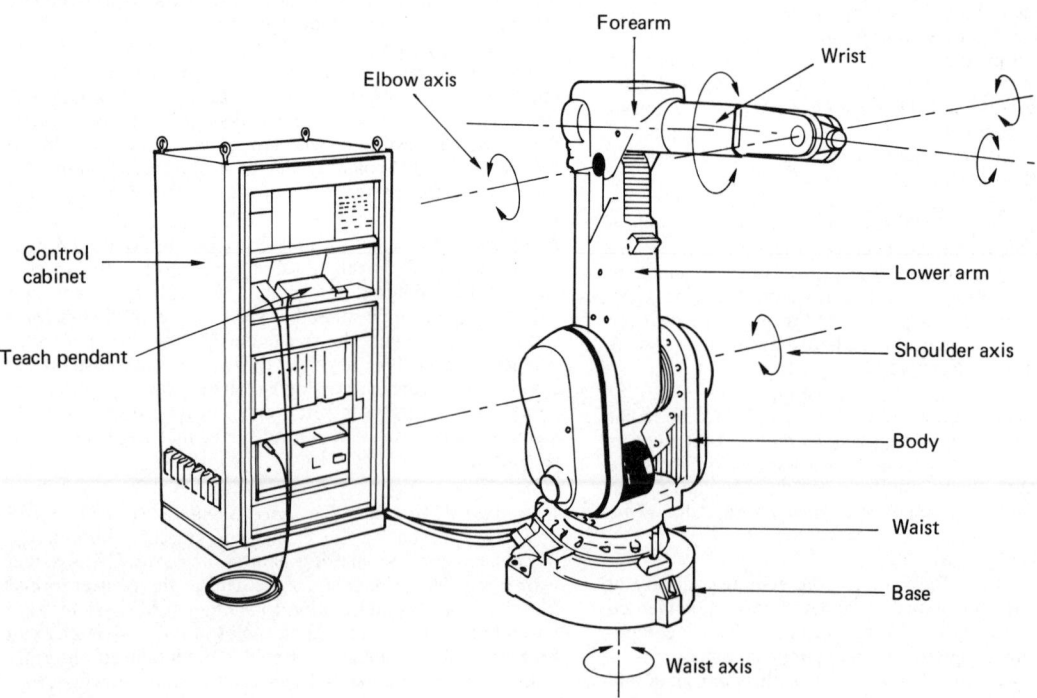

Figure 5.11 ASEA IRB 2000 industrial robot (courtesy ABB)

5.2 Industrial robotics and automation

This section examines the 'robotic' equipment used in manufacturing industry and includes reprogrammable automation equipment. NC machine tools and automated assembly equipment are excluded except where industrial robots are involved. Similarly, 'hard' automation (e.g. cam-controlled mass-production equipment) is omitted. Thus the main subjects of interest will be industrial robotics and artificial vision.

5.2.1 Industrial robots

5.2.1.1 Definition

It is important, in an industrial context, to have an unambiguous definition of a robot. All the major industrialized countries have adopted definitions with some more complex than others. The British Robot Association definition is succinct and is therefore selected for use here, i.e. 'An industrial robot is a reprogrammable device designed to both manipulate and transport parts, tools or specialized manufacturing implements through variable programmed motions for the performance of specific manufacturing tasks'. Industrial robots are therefore *reprogrammable manipulators* of tools, materials or components. A typical robot is shown in Figure 5.11. They should be easily reprogrammed to carry out work on new tasks and they should have a degree of dexterity. Thus an NC machine tool is not a robot since, although it is easily reprogrammable, it is not designed to do anything other than cut material. Neither is the type of arm used to manipulate radioactive material in the nuclear industry; these devices are constantly under the control of a human operator and are therefore remotely controlled, i.e. they are not programmed to operate autonomously.

5.2.1.2 Reasons for using industrial robots

In modern manufacturing there are many advantages in using robots rather than human labour or hard automation. In relation to human labour:

- Robots work to a constant level of quality.
- Waste, scrap and rework is minimized.
- They can work in areas that are hazardous or unpleasant to humans.
- No jobs are boring, tiring or stressful to robots.
- Continuous 24-hour production is possible for many days.
- They are a single investment; salaries do not have to be paid each year at increasing rates, and there are no burden costs such as pension and insurance schemes, holidays, sick pay, etc.
- Investment in a robot involves a once-only capital expenditure, whereas human labour requires an ongoing salary cost that increases annually.
- Robots are advantageous where strength is required, and in many applications they are also faster than humans.

Also, in relation to special-purpose dedicated equipment, robots are more easily reprogrammed to cope with new products or changes in the design of existing ones. Dedicated equipment usually requires expensive strip-down and rebuild in these situations and often has to be discarded as obsolete.

As well as these obvious reasons there are other, indirect, advantages to be gained from robotization:

- When changing from manual to robotic methods, the product components will often have to be redesigned to provide simplicity of presentation, positive gripping points, unambiguous orientation and location, adoption of the stacking principle for assembly, and ease of location for screws, etc. This usually results in a simplified, better and cheaper design for the product.
- Quality will be improved in many areas as automatic inspection techniques are adopted.
- Design changes can be implemented more quickly and new products introduced efficiently.
- Lead times can be reduced.
- Work in progress can be reduced.
- In comparison to dedicated equipment, smaller batch sizes can be handled and downtime between product changeovers is reduced.

It should be remembered that robots are simply one alternative and that human labour and dedicated special-purpose equipment also have their place in the manufacturing environment. Although much simplified, Figure 5.12 shows the cost-to-volume region in which robots are most attractively employed. Generally, unless there are severe environmental or hazardous conditions, human labour is suitable for low-volume high-variety work. Conversely, for very-high-volume/low-variety work dedicated equipment, or 'hard' automation, is probably the most cost-effective.

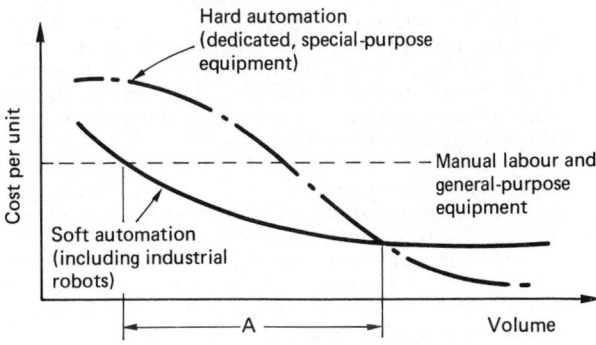

Figure 5.12 Product-volume region 'A' is most suited to robots

5.2.1.3 The construction of industrial robots

Essentially an industrial robot consists of two elements – the manipulator (or 'arm') and the robot controller. The controller contains the microprocessor system and the power control units. Hydraulic and pneumatic robots also have pump and compressor units, respectively. The particular geometry of the arm will provide an associated work envelope; the arm will be powered by electrical, hydraulic or pneumatic means; it will be non-servo or servo controlled; it will be programmed on-line, off-line or both; and will be capable of point-to-point, point-to-point with coordinated path, or continuous path movement. The following is a brief explanation of each of these terms.

The control unit This unit interfaces with the robot's internal and external sensors, drive units, peripheral equipment, and the programmer and operator. It is therefore usually capable of handling serial and parallel data transmission at various rates, and can carry out digital-to-analogue and analogue-to-digital conversion as necessary. Communication with the programmer is via a visual display unit and keyboard, or a teach pendant. There will also be floppy disk drives for loading and saving programs and a printer for hard copy.

Program interpretation will be carried out within the controller. Some robots are capable of using more than one language. In this case the language and operating system is usually loaded from disk at the beginning of the programming session. Within the unit there will be ample memory space to store all necessary data for coordinate transformation, trajectory computation, monitoring and decision making.

The unit will also perform the functions necessary for full servo control. In some robots each axis has its own microprocessor system supervised by a 'master' system. There may also be a system dedicated to handling the sensory data input. Thus in a six-axis robot there may be eight integrated microprocessor systems within the control unit.

Control of the power units is effected by the control unit. In electric-drive robots, low-voltage control signals are sent out to the motor drive amplifiers, one for each axis, to produce power for the motors. Servo control is maintained by monitoring feedback from internal sensors. In fluid power systems, solenoids and servo-valves are also controlled in this way.

Robot geometry The robot arm is composed of links and joints. The joints, also referred to as articulations or kinematic pairs, will normally each have only one degree of freedom. In robots this means that a joint will probably be (1) revolute (i.e. rotation about one axis), (2) prismatic (i.e. linear movement along one axis) or (3) screw (i.e. a combination of linear and rotational movement along one axis with translation defined by the screw pitch and the rotational displacement). Other types of joint with more than one degree of freedom (e.g. a ball joint) are more difficult to control.

The number and relationship of these link and joint arrangements defines the dexterity of the robot arm. There is a maximum number of six degrees of freedom available to any free body in space. For many industrial applications (e.g. arc welding, fettling and spray painting) six degrees of freedom are desirable. With a minimum of six joints it is possible to achieve this. However, the design of the robot must be carefully considered as there will be 'no-go' regions into which the robot arm will not be able to reach due to its physical limitations. Also, even though a robot may have a large number of joints, it may not have the equivalent number of degrees of freedom. For example, an arm with six revolute joints whose axes are all parallel will only have three degrees of freedom.

Usually a robot arm has three major axes providing three degrees of freedom. These axes allow the robot to position its end effector or gripper at any point in space within its work envelope. In addition to the major axes there will also be one, two or three additional axes, normally in the form of a 'wrist' at the extremity of the arm. These will allow the robot to orientate its end effector, which will be fixed to the wrist, about any point in space. As mentioned previously, the number of degrees of freedom required depends on the task to be performed. For example, population of a printed circuit board will only require four degrees of freedom. This operation is essentially a pick-and-place type, with one vertical, two horizontal and one rotational movement for component orientation. However, for arc welding a complex three-dimensional seam (say, at the intersection of two cylinders) a robot with the full six degrees of freedom wil be necessary. This is to ensure that the welding gun is constantly maintained at the correct orientation to the work as welding proceeds. Indeed, this particular task may demand additional axes to be employed by clamping the work on a multi-axis servo-motor controlled worktable whose movements wil be integrated with those of the robot and controlled by the same control unit. Some typical robot configurations and work envelopes are shown in Figure 5.13.

Robot drive and control systems Many of the first industrial robots were hydraulically driven. However, most robots now produced are electrically powered. The previous definition of a robot implies relatively sophisticated control, and this is verified by the fact that nearly all industrial-quality robots are fully servo controlled. Non-servo controlled robots, just on the borderline of the definition between robots and simple pick-and-place units, are usually driven pneumatically. Some very light-duty devices, and often the orientation axes on SCARA robots, use non-servoed stepping motors.

Electric robots are usually driven by d.c. permanent magnet servo motors, brushless motors or, occasionally, stepping motors. Electric-drive systems are relatively clean and quiet when compared to fluid power machines. They are easily maintained and repaired and are well suited to electronic control. Recent developments in the use of rare-earth materials for permanent magnets mean that power-to-size ratios are increasing, and the use of brushless motors reduces maintenance costs. Brushless motor drives can also be used in areas such as clean rooms, since contamination particles are reduced, and in situations where there would previously have been a fire risk due to the possibility of brush arcing. Unless incorporating direct-drive motors, electrically driven robots do have the disadvantage of requiring transmission systems. These add cost and weight, and also reduce precision due to gear backlash and other unwanted movement.

Hydraulically powered robots still have some advantages. For example, they have very good power-to-size ratios and hydraulic force can be applied directly at the desired point without the need of a transmission system. Hydraulic fluid is incompressible and therefore there are no backlash problems. Assuming the power pack, which contains the electrically driven hydraulic pump, is located remotely, then the robot can be used in high fire-risk areas. This is because only very low voltages for control and feedback purposes will be present on the actual robot arm. Because of their necessarily sturdy construction due to the high hydraulic pressures experienced, they can withstand higher shock loads than other robot types. However, it is their disadvantages that have led to their reduction in popularity. A noisy power pack, even when protected by an acoustic muffler, makes them environmentally unattractive. Historically, they have tended to be less reliable than electric robots with leakages occurring which contaminate work areas and cause performance loss. Servo control of hydraulics is not as simple as that for electrics and availability of skilled personnel is more scarce. The viscosity of the hydraulics fluid can be affected by temperature and this can cause variations in performance. Finally, cost is not directly proportional to size. Smaller hydraulic robots tend to be much more expensive than their electric counterparts.

Pneumatic powered robots are the cheapest and least sophisticated type. They are usually not servo controlled but will be able to carry out complex movement sequences if necessary. They are fast, simple, reliable and easily understood by most factory technicians and maintenance personnel. They also have the advantage of being intrinsically safe and can therefore be used in explosive atmospheres. The major disadvantage of pneumatic robots is that precise servo control is not practical. This is due to the compressibility of air, particularly when moving heavier loads. Thus pneumatic robots are usually found in limited-sequence, light-load fixed-speed applications.

5.2.1.4 Path control

The application will influence the choice of the robot path control system. Robots with simple point-to-point control are suitable for assembly, palletizing and other materials-handling

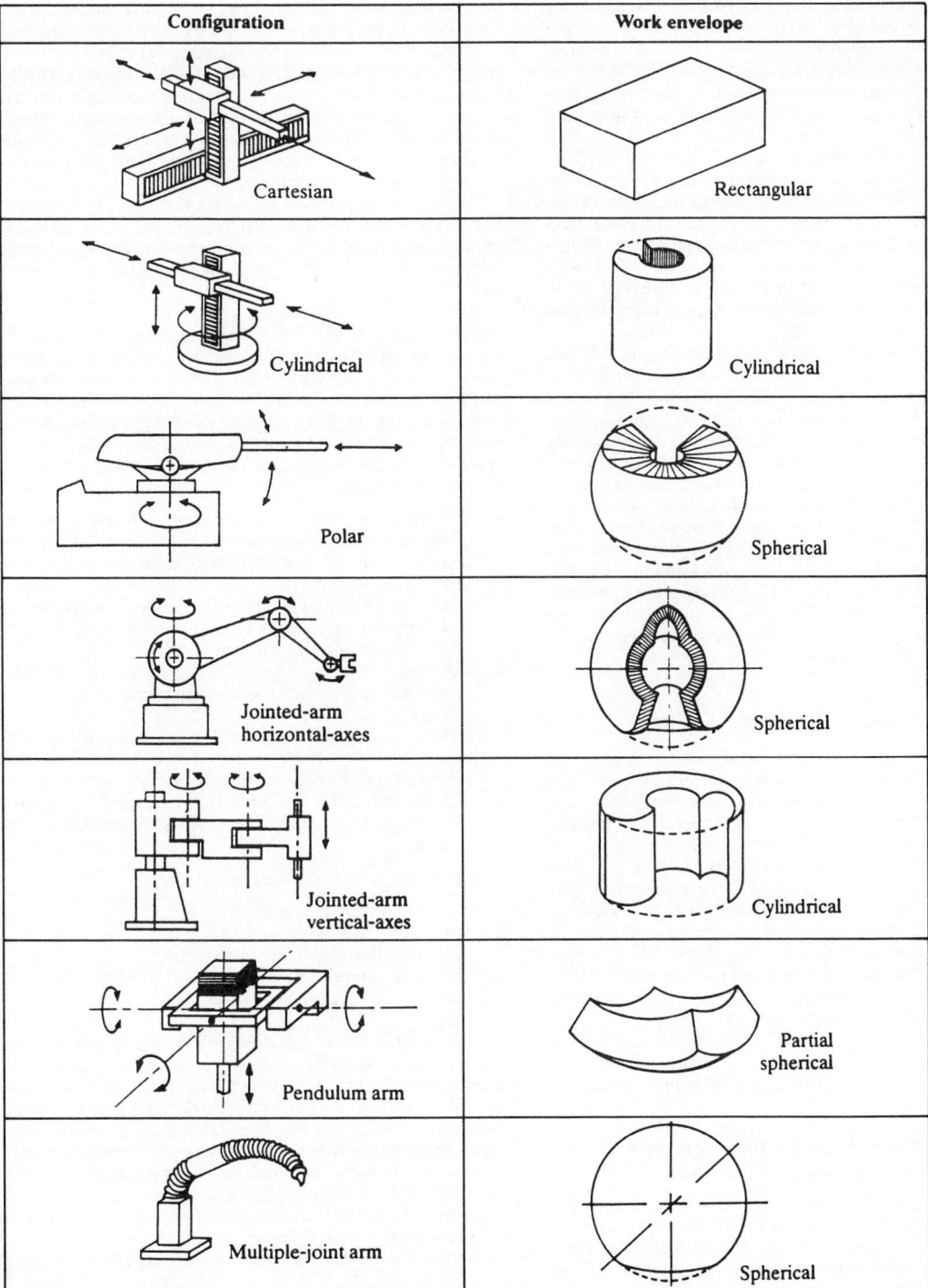

Configuration	Work envelope
Cartesian	Rectangular
Cylindrical	Cylindrical
Polar	Spherical
Jointed-arm horizontal-axes	Spherical
Jointed-arm vertical-axes	Cylindrical
Pendulum arm	Partial spherical
Multiple-joint arm	Spherical

Figure 5.13 Robot classification by geometric configuration and work envelope. (From Mair, G. M., *Industrial Robotics*, Prentice-Hall, 1988)

tasks. Point-to-point with coordinated path control is suitable for tasks such as arc welding, sealant application and spot welding of moving components (e.g. car bodies on a conveyor system). Continuous path control is used where the dexterous movement of a human operator has to be mimicked (e.g. in spray painting, or where complex contouring movements are necessary).

In point-to-point control the robot will move between defined points without regard to the path taken between them. In some robots an additional software facility allows the choice of movement between points in either the shortest travel time or in a straight line. In revolute robots movement in an 'elbow up' or 'elbow down' mode can also be selected. When using robots with simple PTP control particular care has to be exercised to ensure that collisions with obstacles will be avoided when the robot is running.

In point-to-point with coordinated path the control software allows the path the end effector will follow between points to be determined. Straight lines, circles, arcs and other curves can be defined. Two points are all that is necessary to define a line; circular movement can be programmed by specifying three points on a circumference or a centre point and a radius.

Full continuous path control is most often obtained by playing back the information recorded from physically leading the robot through a desired task. Every movement of the arm is recorded in real time by sampling joint positions at a high frequency, and this is used in on-line programming. Continuous path control is also employed in off-line programming in some systems where there exists the facility to insert the mathematical equations for desired curves. These curves will then be followed by the end effector.

5.2.1.5 Robot programming methods

Industrial robots may be programmed using a number of techniques. The most basic methods, employed in some very early hydraulic robots, used rotating drums on which pegs could be set to close microswitches. These switches operated solenoid valves, so controlling the flow of fluid to the actuators. Adjustable mechanical stops were fixed on the moving members to contact limit switches at appropriate points. These mechanical programming methods provided control of sequence and distance moved. More recently, the use of programmable logic controllers has been applied to sequence the movement of pneumatic robots and modular units. These are all non-servo control systems employing simple feedback from limit switches or proximity sensors.

Full servo controlled robots employ dedicated microprocessor-based controller units as described earlier. This allows sophisticated on- and off-line programming techniques to be employed.

On-line programming Here the robot arm itself is used during the direct programming operation. This method has the following advantages.

1. The robot can be observed as programming progresses; this increases confidence in the finished program since possible collisions and other robot 'no-go' areas will be easily identified.
2. In applications such as spray painting and welding where human experience is important, the direct programming of the arm by an expert effectively produces a transfer of skill from the human to the robot.
3. This type of programming is easy to learn and can often be accomplished by the personnel that the robot is replacing.
4. Less expensive computing hardware and software is involved than in on-line programming.

On-line programming also has one main disadvantage. Where the task being programmed is complex the programming time may be prohibitively long. For example, if a deburring operation is to be carried out and the robot has to be taught the shape of every hole and curve, the programmer will find the exercise extremely tedious and the cost benefits may be trivial. It is therefore a technique suited to tasks which are highly repetitive in nature, i.e. one sequence of movements can be repeated automatically a number of times.

There are basically two methods of on-line programming: teach by lead-through and teach by pendant. It is also possible to program directly from a computer terminal attached to the controller, which is similar in principle to the teach by pendant method.

Off-line programming Off-line programming involves creating the program for a robot task, without the need to be connected physically to the robot or even to be anywhere near its physical presence. In fact, when coupled with simulation techniques, off-line programming can be carried out before deciding on which robot to purchase for a specific application. Some of the advantages of off-line programming are as follows:

1. The robot for which the program is being made can continue working on its old task until ready for the new program. This obviously reduces robot down-time and increases productivity.
2. If the control system and language allows, it is possible to build into the program collision avoidance, error recovery and other contingency routines.
3. Compared to on-line programming, it is easier to make alterations to cope with variations in products and design changes.
4. Off-line programming is suited to full computer integration of a facility. For example, if a robotized computer-controlled machining cell is in operation, then with off-line programming the problems of downloading programmes to the robot at the appropriate times is greatly reduced.

There are also some disadvantages:

1. Real-world contact is lost.
2. More expensive hardware and software is required.
3. More programming skill is needed.
4. Fine adjustments under production conditions on the shopfloor are usually necessary.

Simulation Graphical simulation of the robot and its environment offers many benefits to the industrial engineer and programmer. Figure 5.14 shows an example of a low-cost simulation system display. This particular software, called 'Workspace', runs on a PC. Simulation packages provide the ability to model the robot kinematically and often dynamically to give an animated, real-time, visual representation of how the robot will work under programmed conditions. Three-dimensional wire frame – sometimes with colour shading or solid modelling – techniques are used.

There are a number of advantages associated with simulation:

1. Simulation allows a prospective robot purchaser to try out various models at relatively little cost before making a decision. Parameters such as work envelope, cycle times and joint configuration limitations can be compared for all the robots held in the library.
2. The immediate robot work area can be simulated and various permutations of machines and operation sequences experimented with before finalizing the layout.

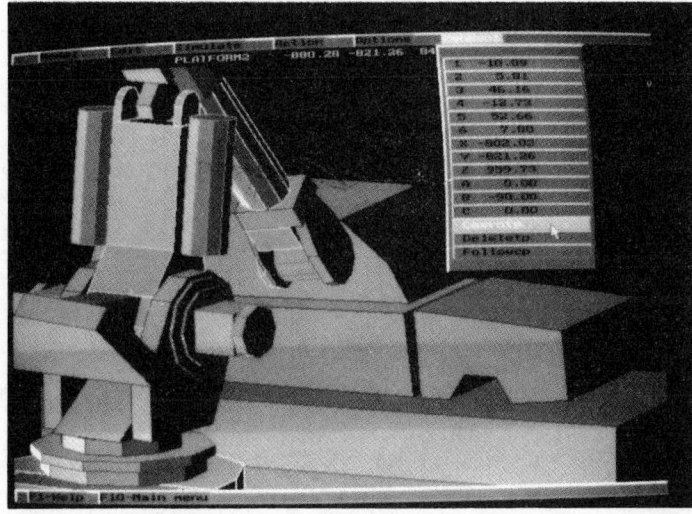

Figure 5.14 Robot simulation on a PC using 'Workspace' as an aid to programming (courtesy Robert Simulations)

3. Potential collisions can be detected at an early stage and programs and layouts modified to suit.
4. Simulation is very suitable for teaching and training purposes. Mistakes in programming can be observed and learned from without the hazards and potential costly damage that would be experienced in the real world.
5. At the robot design stage, simulation delays the need for physical prototypes to be built, thus reducing research and development costs.

These advantages, coupled with improving computing power-to-cost ratios, mean that simulation is becoming an increasingly popular robot-programming tool in industry, education and research.

5.2.1.6 Using industrial robots

Suitable applications The potential advantages of robotization can be maximized by making wise application selections.

Industrial robots realize their full economic potential in applications where product volume is large enough to recoup the expenditure on hardware, programming and engineering costs, yet is sufficiently low to prevent justification of dedicated special-purpose equipment. However, high-volume work with frequent model or option changes, such as is found in autombile assembly, is suitable for robotization. The effect of robots on the cost per unit, in relation to volume, is shown in Figure 5.15. The following are some further indicators as to applications that should provide suitable opportunities:

1. Tasks which are carried out in (or create) an unpleasant or hazardous environment. For example, toxic or flammable atmospheres are created by processes such as arc welding and spray painting, and removing human operators from these jobs can improve quality and increase production rates.
2. Jobs that are tiring or boring. Robotization of these eliminates absenteeism and labour turnover problems and usually improves quality.

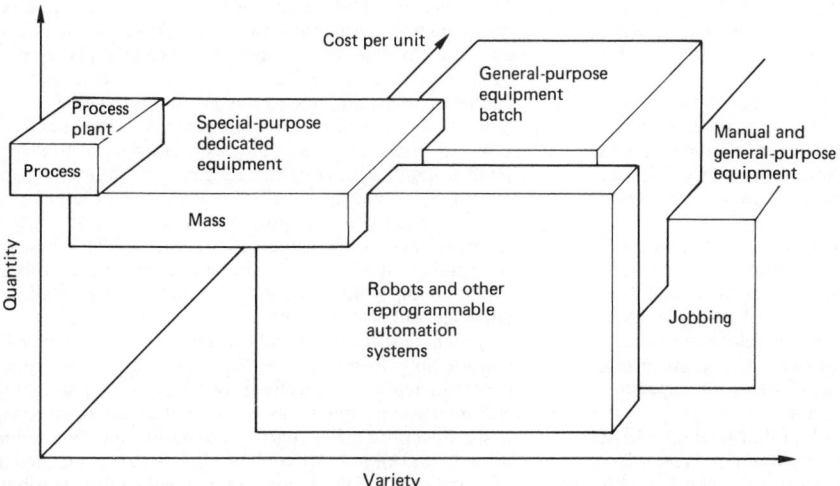

Figure 5.15 Effect of robots and other reprogrammable automation systems on cost per unit. (From Mair G. M., *Industrial Robotics*, Prentice-Hall, 1988)

3. Repetitive and simple operations requiring simple movements allow the least expensive robots to be used and minimize installation and programming problems.
4. Desired cycle times should not be too short. For example, if the cycle time is greater than, say, 3 seconds then the choice of robot is relatively wide. However, if very short cycle times are required, as in PCB component placement, then more specialized and expensive high-speed robots will be necessary.
5. The tolerances on the components and tools should allow robots of average precision to tackle the work.
6. The variety of products expected to be handled by the robot should not be large nor changes from one product to another too frequent. This will keep engineering and reprogramming costs to a minimum.

The following points are also relevant. If the task being considered has an integral inspection element, then additional costs will be incurred when vision or other sensing methods are added to carry out that inspection. In materials-handling applications very heavy loads will demand larger and more expensive robots. If an ordered environment exists (or can be made to exist) around the robot then robotization is simplified. If possible, work should be oriented and positioned at the previous operation before presentation to the robot. Most robots available commercially have limited reasoning ability, therefore tasks should demand little in the way of intelligence or judgement.

Once the task to be robotized has been selected the next stage is the selection of an appropriate robot.

Selecting the robot   Robot selection should be carried out after listing task demands such as cycle time, payload required, necessary precision, and cost. These demands will be compared against the specifications provided by the robot supplier or manufacturer, some of which are now listed.

1. *Speed.* Having decided the speed required of the robot from the work analysis, the detailed specification should now be examined. Some manufacturers may give maximum speeds for each axis of the robot, some the maximum speed of the end effector. These should be given for maximum load and at maximum reach as well as for under optimal conditions. It should be remembered, however, that maximum speed is not necessarily a very useful piece of information, because a robot arm must accelerate to and decelerate from this speed. Figure 5.16 shows a typical robot velocity curve. For some applications, particularly assembly, a 'goalpost' time is a more useful specification. This is the time, supplied by the robot manufacturer, that it should take the robot to complete a standard series of movements carrying a standard load. For example, the movement may be close gripper, move up 30 mm, move across 300 mm, move down 30 mm, open gripper. This will prove more useful for estimating than a maximum speed figure.
2. *Payload.* The maximum load expected to be encountered will have been determined, and a robot with sufficient strength to handle a considerably greater load should be selected. The specification should show whether the maximum load capacity is given with the arm close to the body or at full extension where the capacity will be much less due to leverage. Robots are available with capacities ranging from a few grams to 2 tonnes.
3. *Precision.* The overall precision of a robot is composed of three elements, i.e. resolution, repeatability, and accuracy. The resolution of a robot is normally a feature that is transparent to the user and is therefore not included in standard specification sheets. It refers to the smallest

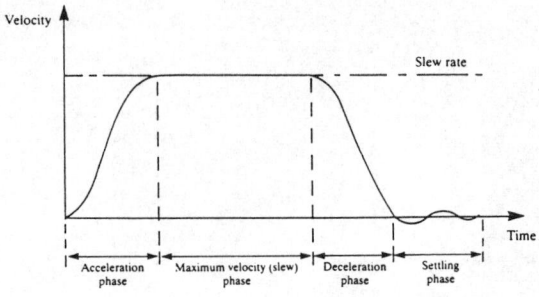

Figure 5.16   Robot velocity curve

controlled movement the end effector is capable of making. This is determined by (a) the resolution of the computer controller (i.e. the number of bits used to define a position over a given range), (b) the resolution of the drive system (e.g. the number of steps per revolution provided by a stepper motor and associated gearing) and (c) the resolution of the feedback elements such as shaft encoders. The repeatability of a robot is determined by its resolution plus clearances and wear on moving parts plus any other inaccuracies and errors in the total system. It is a statistical term describing how well the robot can return consistently to a taught point. This is the most common figure relating to precision to be included in robot specification sheets. Repeatabilities of ± 1 or 2 mm medium-duty work, ± 0.05 to ± 0.03 mm for medium assembly, and ± 0.01 mm for precision assembly are typical.

4. *Accuracy.* Assume that a computer-controlled robot is to move to a point in space. This point is defined by entering the coordinates into the control system. The difference between the taught target point and the actual position achieved by the robot, in the real world, is the 'accuracy'. This will be determined by the resolution, inaccuracies in the 'model' of the robot held in memory and other factors such as bending or thermal expansion of the robot arm. The relationship between accuracy and repeatability is shown in Figure 5.17.
5. *Configuration.* The supplier will provide information on the geometric configuration and dimensions for the effective work envelope of the robot. These can then be used to construct templates either on card or on computer to enable an appropriate work layout to be designed.
6. *Control system and programming method.* The specification will provide information on whether point-to-point, point-to-point with coordinated path or continuous path control is provided. It will also state the programming methods used. For CP programming by lead-through a slave arm may be available and for PTP or PTP-CP teach by pendant methods may be used. For many robots programming using a computer terminal and a high-level language will also be available.
7. *Cost.* The cost of a complete robot installation can vary considerably from that of the basic robot. The robot chosen can influence this total cost. Ease of programming, and interfacing capabilities will influence the engineering costs. Cost of fixturing, parts presentation and orientation devices, and end-of-arm tooling will have to be included in the total. Also, if working to a fixed budget for the robot, there will probably have to be a trade-off between precision, speed, strength and reach.

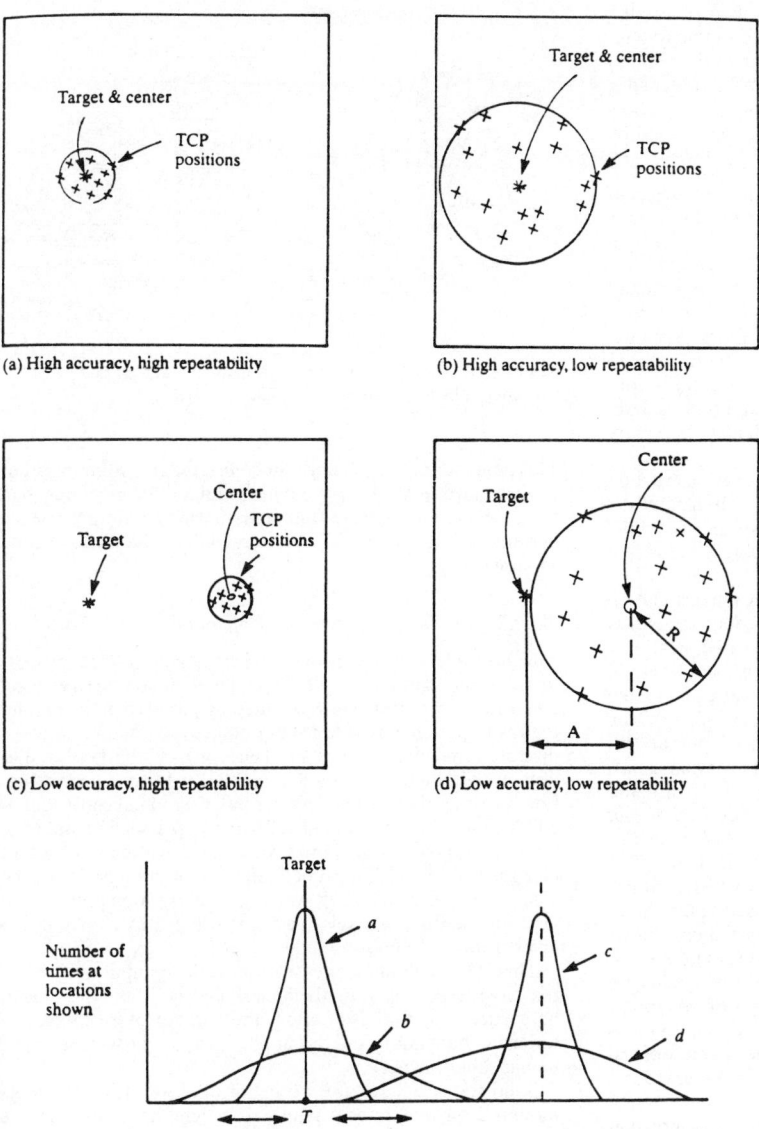

Figure 5.17 Schematic representation of accuracy and repeatability. (From Mair G. M., *Industrial Robotics*, Prentice-Hall, 1988)

Other specifications that should be considered include drive system, number of degrees of freedom, type and number of input and output ports, and memory size.

Robot safety As well as presenting the normal safety problems associated with moving equipment that is electrically, hydraulically or pneumatically powered, and machines that are under microprocessor control, industrial robots present some additional problems that are unique:

1. While executing a program the robot appears to the inexperienced observer to be moving spontaneously and unpredictably, each movement being difficult to anticipate. This applies particularly when the robot is at a 'dwell' point in its work cycle. It may appear to be deactivated but in fact it will spring into action as soon as it receives an appropriate command from the system controller.

2. Most robot arms sweep out a work volume much larger than that occupied by their base. With a six-degree-of-freedom robot the arm movements and positions are therefore difficult to visualize.

3. The integrity of the control system hardware and software is particularly crucial since faults will produce erratic and unpredictable behaviour.

4. Heavy-duty robots are built to be rugged and inelastic, and fast-moving arms are therefore extremely dangerous. Size is not necessarily important as was proved when one person was killed when struck on the back of the neck by a small teaching robot. Accidents can be caused by human

carelessness, insufficient training, poor robot or installation design, poor quality components used in the system, and software errors. Most accidents occur to those familiar with the robot such as programmers, maintenance engineers and operators. Those unfamiliar with robots tend to be more wary – it is the complacency caused by familiarity that is dangerous.

Industrial robot safety should be considered at the stages of robot design, supply, installation, programming and everyday usage. The designer should ensure all controls conform to good ergonomic practice. Controls and displays should obey standard conventions, mushroom-shaped stop buttons should protrude from surfaces and there should only be one start button, which should be recessed. All emergency stops should be hardwired into the power supply and not rely on software execution. Conventional good design practice should be observed, moving parts should not be exposed and there should be no trapping points for limbs or fingers, with no unnecessary protrusions capable of inflicting injury.

The robot supplier should ensure that proper instruction and training is given to appropriate personnel designated by the purchaser. The supplier should also make the user fully aware of the robot's limitations and any possible hazards that may be encountered.

When the user is planning and implementing the installation full consideration should be given to the robot's position within the factory, e.g. it should not be located near any trapping points such as roof pillars or stanchions, and it should not be possible for it to reach into passageways or manual work areas. Preparation of safety manuals, or safe working procedure documentation may also be carried out at this stage. Reference should always be made to appropriate rules, regulations and guidelines. In the UK there is the Health and Safety Executive (HSE) guidance booklet *Industrial Robot Safety*, and the MTTA booklets *Safeguarding Industrial Robots* Parts 1 and 2. There is also the British Standard 5304 Code of Practice – Safeguarding of Machinery which contains the basic principle of safeguarding, i.e. unless a danger point or area is safe by virtue of its position, the machinery should be provided with an appropriate safeguard which eliminates or reduces danger. Light curtains and pressure-sensitive mats are commonly used around the immediate vicinity of the robot. For maximum safety a 2 m high cage around the robot is recommended. This should have doors electrically interlocked to the power supply to ensure that unauthorized entry deactivates the robot.

During everyday operation management must ensure that only fully trained personnel operate the robot. Established safety procedures must be adhered to, appropriate warning signs given high visibility and, generally, a state of continual safety 'awareness' cultivated.

5.2.2 Industrial vision systems

5.2.2.1 Vision system components

The basic elements of an industrial vision system are shown in Figure 5.18. A camera is first necessary to acquire an image. This camera may be a vacuum-tube or a solid-state type, the latter now being the most popular. The signal from the camera is then processed in the vision system computer. The image observed by the camera and the digitized image used for computer processing are observed on a monitor which is switchable between them, or two monitors may be used, one for each image. A means of communicating with the system is necessary and this would take the form of a computer terminal and visual display unit. A means of allowing an automated physical reaction in response to the vision analysis is required.

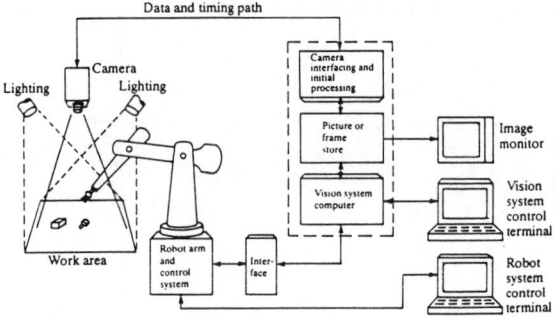

Figure 5.18 Robot vision system elements. (From Mair G. M., *Industrial Robotics*, Prentice-Hall, 1988)

This demands an interfacing unit connected to a robot or other device (say, a simple pneumatic cylinder for rejecting bad parts). Finally, to ensure optimum viewing conditions special lighting arrangements may be necessary to avoid distracting shadows or glare.

5.2.2.2 Vision system types and operation

Vacuum-tube cameras provide an analogue voltage proportional to the light intensity falling on a photoconductive target electrode. This electrode is scanned by an electron beam and the resulting signal is sampled periodically to obtain a series of discrete time analogue signals. These signals are then used to obtain digital approximations suitable for further processing. For example, if an A/D converter has a sampling capability of 100 ns, the image is scanned at 25 frames per second and each scan is composed of 625 lines, then there will be 640 picture elements, or 'pixels', per line.* Since some time is lost as the electron beam switches off when moving from one line to the next, the number of pixels in a frame will be 625×625, which gives a total of almost 400 000 pixels.

This is difficult and expensive to handle computationally in real time, especially if mathematical analysis of the image is to be carried out. For this reason, the number of pixels can be reduced, depending on the application, to provide a more manageable image.

Solid-state cameras use arrays of photosensitive elements mounted on integrated circuits. The light from the scene is focused by the camera lens onto the IC chip. Charge-coupled devices (CCDs) or photodiode arrays are scanned to provide a voltage signal from each light-sensitive element. These solid-state arrays are available in many densities, often from 32×32 to 1000×1000 pixel arrays. The larger pixel densities are too high for real-time vision analysis but they do provide high-quality video pictures. The voltage signals from the photosites are again digitized before further processing. In a 'line scan' camera linear arrays of photosites are used rather than the area type. Line scan cameras can be used where the object is moving steadily across the field of view. For example, an object passing under the camera on a conveyor belt can be scanned repeatedly and an image built up in the vision system memory.

Recently, solid-state cameras have become widely available and relatively inexpensive due to their large-volume production. They have a number of advantages over the vacuum-tube

---

* *Pixel* is derived from '*pic*ture *ele*ment' and usually represents the smallest possible display feature or point on the display device.

type, i.e. they are much smaller and lighter, more robust, more reliable, use less power, have a broader temperature operating range at lower temperatures, and are less likely to be damaged by high light intensities.

The pixel voltage signals from the camera are now assigned to a finite number of defined amplitude levels. The number of these 'quantization' levels is the number of 'grey levels' used by the system. An 8-bit converter will allow 256 grey levels to be defined. In practice, this number of levels is often unnecessary and processing time can be reduced by using only 16 grey levels. In some cases only two grey levels, i.e. black and white, are necessary, this is termed 'binary' vision.

Each grey level is next 'encoded', i.e. it is given a digital code, and the data stored in memory. In a computer vision system this is done for one picture 'frame' and the data stored in memory in what is termed a frame buffer or picture or frame store. Various algorithms are then used to minimize the data for analysis and organize them in such a way as to allow feature extraction and object recognition. Objects are usually recognized by the system by first showing it a sample of the object. The system 'remembers' the object by storing information on features such as object area, perimeter length, number of holes, and minimum and maximum radii from the centre of gravity. The sequence of all these processes is shown in Figure 5.19.

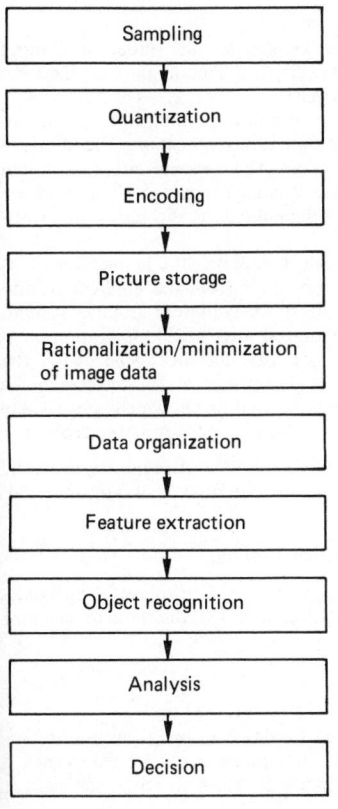

Figure 5.19 Typical operation sequence implemented in a vision system

5.2.2.3 Vision system applications

As the capabilities of vision systems increase so also does their popularity. Custom-designed hardware and developments in algorithms mean that the systems are becoming faster and more reliable. They are now found in a wide variety of industrial applications and are sometimes supplied as integral components or programmable electronic component placement machines and other robotic systems. Three main areas of application are listed below.

Identification Here the system is used to identify a product or individual component. For example, it may involve character recognition, as in reading alphanumeric data on a product label or recognizing a component on a workbench prior to assembly.

Inspection This is one of the major applications of vision systems as it is estimated that visual inspection accounts for around 10% of total manufacturing labour costs. This percentage can be very much higher in some industries (e.g. electronic product manufacturing such as PCBs, computers and other consumer goods). Sensible application choices can prove very cost-effective. Inspection is generally further divided into 'qualitive' and 'quantitive' inspection. In qualitive inspection it is attributes that are examined (e.g. glass bottles may be checked for flaws or castings checked for cracks, or the number of pins on an IC chip verified). In quantitive inspection dimensional or geometric features of a product are measured and checked (e.g. the diameter of a component turned on a lathe or the width of a steel strip coming from a rolling mill).

Decision making This is a general term which implies a number of applications. For example, a vision system could be used to guide the welding head of a robotic welder along the seam of a fabrication, or it could assist an automatic guided vehicle find its way around a factory shopfloor. In conjunction with artificial intelligence techniques vision can be used to provide the information input necessary to provide autonomous working of rotobic devices in unstructured environments.

In conclusion, vision system technology can be said to be rapidly improving and, in conjunction with advances in related technologies, it will continue to make a significant impact on factory automation for many years to come.

5.3 Computer graphics systems

5.3.1 Introduction

Before desk-top or home computers were readily available computer graphics was considered to be an expensive tool for industry, requiring large computing resources and custom interfaces. Some limited graphic representations used line printers or computer terminals but with very low resolution since images were constructed from printable characters.

Early microcomputers with a screen resolution of 256×256 pixels made simple line drawings possible. The computation required to process such an image is well within the capability of 8-bit microprocessors and the frame buffer and display hardware is simple and low-cost.

Microcomputers such as the Apple II and the BBC introduced colour and improved resolution to the graphics display so that most programmers could think in terms of presenting output in a graphical manner. Even if the output was a pure text display, then at least these data could be shown formatted with some overall layout in mind. These computers were usually programmed in BASIC which had graphics instructions such as MOVE, PLOT and DRAW added to the language. The graphics procedures were mostly unstructured in that objects were defined in terms of absolute coordinates using straight lines. Only later did graphics primitives appear, so that circles, ellipses and rectangles could be drawn in one instruction.

The basic display was a binary one where each point or pixel could be displayed in either black or white. If colour were used, then pixels would be displayed in black, red, green, blue or combinations of these three colours. Thus eight different colours are possible with such a system. This does not permit completely realistic displays to be constructed but it can allow useful representations of objects to be shown. If a monochrome display is used instead, then the colour information can be employed to generate eight different shades of grey, ranging from black to full white (referred to as the *grey-scale*).

5.3.1.1 Applications of low-resolution systems

There are many uses even for such simple systems. Waveforms, graphs and barcharts can present tabular data in a visual manner. Engineering drawings, architectural plans or printed-circuit board layouts can likewise be shown with sufficient accuracy. Early attempts at computer art successfully used these comparatively low-resolution, colour displays. Lack of a sufficient range of colours or an extended grey-scale was compensated for by the use of 'dithering' or 'hatching' patterns. By these means, a small group of adjacent pixels are integrated by the eye to form a single pixel of another hue. Thus spatial resolution is exchanged for increased colour resolution.

Computer-Aided Design (CAD) usually benefits from a graphics display and most early packages were of the line-drawing type (e.g. PCB design, wire-frame representations of solid objects and building plans). Animated drawings are possible but the computations required for even simple objects limits the maximum rate at which the display can be updated. The computer games repertoire includes aircraft flight simulators but the display cannot be realistic, as only a few wire-frame objects with a limited number of sides are possible.

5.3.1.2 High-resolution systems

Once the market had been established for computer graphics, the demand for higher resolution and more colours required an increase in performance which was even greater than the new 16- and 32-bit microprocessors could provide. The development of new graphics processing software could not completely close the gap in performance. The display hardware and frame buffer architecture had to be changed as well. New graphics processors and display devices began to appear. In this way, interactive and high-resolution graphics displays have been developed where resolutions of 1024×1024 are common and 2048×2048 displays (or greater) are readily available but expensive.

There is always a price to pay for increased resolution; 256×256 binary displays have 65 536 pixels or 8196 bytes for each frame, $1024 \times 1024 \times 8$-bit displays require 1Mbyte of storage. Each pixel in an image must potentially be calculated before the display frame is complete. As resolutions and the number of bits per pixel increase, the processing overhead and the requirements of the display hardware rise exponentially! Whereas 256×256 displays require a data rate of only 5 Mbps between the frame buffer and the monitor screen, high-resolution displays have a data rate of well over 100 Mbps so that the physical as well as the electrical design of such systems is no longer trivial.

5.3.1.3 Current applications

The largest usage of computer graphics is in the realm of CAD. Design and drafting tools are available for the production of working drawings. Modelling and simulation tools allow designs to be verified. Using shading and rendering of three-dimensional objects, the visual appearance of objects can be created. For example, stresses in mechanical systems can be calculated and the design altered to keep stresses within certain limits (Figures 5.20 and 5.21).

All the features described above can be obtained before any prototype is built, which saves time and money. The cycle of design, simulation, verification and rectification can be very short – far shorter than could be achieved by building physical prototypes. The number of faults in the new system is much reduced, many systems working first time!

Image Processing is a separate field where real images are processed so that features are enhanced, extracted or modified (Figures 5.22 and 5.23).

5.3.1.4 Interacting with the picture

Input devices are used to define or modify the image (see Section 5.3.3.3). These might directly change the attributes of the image (real-time update) or alternatively, the input device could mark or describe the desired changes which are then implemented when the display is next updated.

5.3.2 Creating a computer graphics image

The essential features required of a computer graphics system are as follows:

1. Hardware to output images to the screen and input devices to control the display. This might be called the *computer graphics system*.
2. A set of commands to generate the image. This is therefore a program and is called the *application*. The interface between the application and the computer graphics system is usually defined by a standard (e.g. GKS*) so that the application level is independent of the hardware which produces the image.
3. A processing device which supports the application. This converts high-level commands from the application into the low-level commands of the computer graphics system. This is usually incorporated into a desktop computer or *workstation* and the graphics may be displayed on the workstation's screen. Alternatively, a separate screen can be used for displaying the final image which is especially useful if very high-resolution images are being created.

These three modules, the *computer graphics system*, the *application* and the *workstation*, are now dealt with in detail.

5.3.3 Computer graphics systems

These are treated in three sections: (1) displays, (2) graphics generation and manipulation and (3) the human–machine interface.

5.3.3.1 Displays

CRT/LCD Most graphics displays are based on the *cathode ray tube* (CRT). A CRT contains an electron beam which is focused onto a phosphor-coated glass screen when light is

---

* Graphics Kernel System.

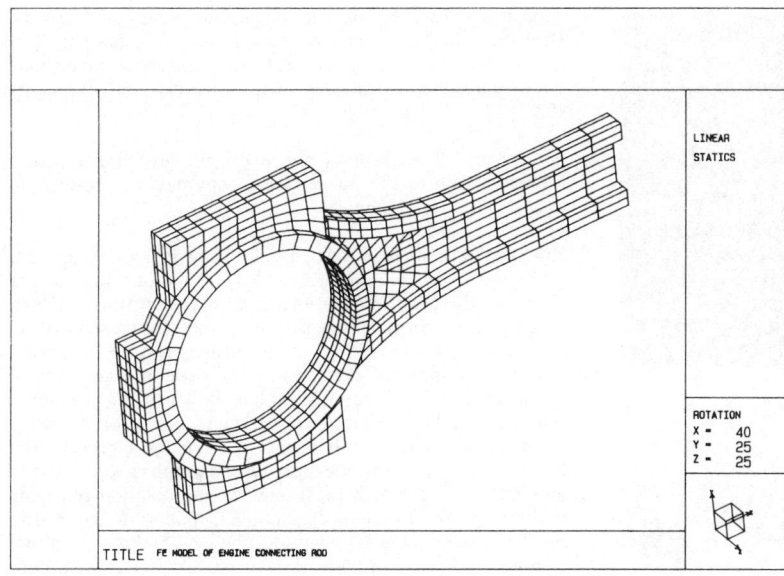

Figure 5.20 Finite-element mesh on connecting rod (Pafec)

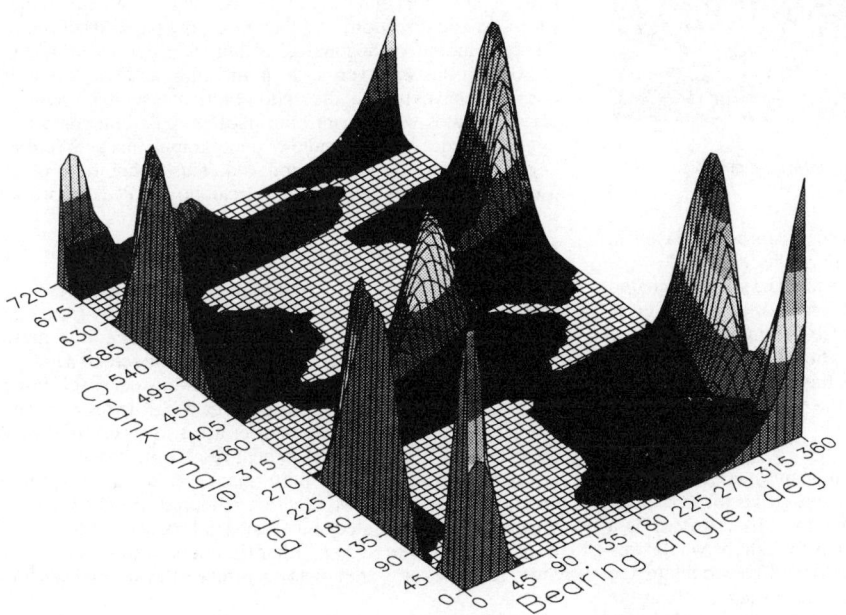

Figure 5.21 Pressure distribution on big-end bearing (Uniras)

emitted. High-quality CRTs allow the spot to be focused down to a very small area which improves the resolution of the device. The *resolution* is the maximum number of points which can be displayed per centimetre in the horizontal or vertical directions. The electron beam must be deflected in order to draw points anywhere on the whole screen surface. For high-resolution systems this requires a fast scan rate and complex deflection systems in order to maintain a linear display, and the video monitor is an expensive item in such systems. The electron beam may be modulated in intensity to give a range of grey tones in a monochrome system or to offer a range of hues in a colour system. The maximum rate of

Figure 5.22 Original image

Figure 5.23 Figure 5.22 processed using *threshold* operation

modulation determines the detail which can be displayed in the image and is the *video bandwidth* of the monitor.*

Most displays are now full colour where the screen is coated with phosphor dots or stripes in red, green or blue. However, the shadow mask required in such CRTs limits the ultimate resolution to the *dot pitch* of the phosphor stripes. Mono-chrome screens are not subject to this limitation, and thus offer improved resolution where the use of colour is not essential.

Vector and raster scan The electron beam in a CRT may be deflected by drawing the image one line or vector at a time. Since each line could be in any direction this is known as a random-scan display. The alternative is to scan the whole area of the screen repeatedly (raster-scan) and to modulate the intensity of the beam such that an image is shown only at the desired points on the screen. While vector scan displays are capable of higher resolution, the refresh rate is dependent on the number of vectors, and this becomes a limitation where there are a large number of vectors to draw. Raster-scan displays have a fixed screen refresh rate of between 25 and 80 frames per second and this is independent of the complexity of the image.

An alternative form of display uses plasma-panel or liquid-crystal (LCD) technologies. Here, transparent electrodes are

* This is the combination of the bandwidth limitation of the driving electronics and the CRT bandwidth.

deposited onto a glass screen. By placing a voltage across the electrodes the plasma may be made to glow (plasma-panel) or the cell may become opaque (LCD). In this way, pixels may be displayed to form the image. Resolutions exceeding 1024 × 1024 are possible.

Hard copy The displays above do not give a permanent representation of the image. Hard-copy devices produce an image on paper or film.

Printers Although printers usually print text they can generally be used to produce a low-resolution image. For example, the head of a dot-matrix printer can print dots in a regular array. Since this is similar to the arrangement of an image split into individual pixels, images can be produced. However, all dots are printed with the same intensity, and so this technique is only suitable for line drawings. An attempt at shading can be produced by grouping pixels or by using dithering techniques, although at the expense of resolution. Standard dot-matrix printers can offer a graphics resolution of about 60 dots per inch (dpi) or with interpolation requiring multiple passes this may be doubled. Twenty-four-pin dot-matrix printers allow a resolution of over 200 dpi with ink-jet printers and laser printers starting at 300 dpi as standard. These are usually monochrome displays. It is possible to use colour ink-jet or laser printers to produce colour displays.

Plotters Plotters produce an image by moving a pen over the paper and constructing the image as a series of vectors. Some plotters, for convenience, move the pen across the paper width (in one direction) and then move the paper itself under the pen in the orthogonal direction in order to draw the vectors. In this way, very large plots can be drawn. A plotter does not have to use ink; photoplotters use light-pens of varying widths which print onto photographic film contained in a light-proof box. This allows photographic masters (either negative or positive) to be produced. This technique is often used in the production of printed circuit board (PCB) artwork.

5.3.3.2 Graphics generation and manipulation

Frame buffer architecture The frame buffer for a vector-scan system is a display-list of the vectors required to form the image. A raster-scan frame buffer typically requires more memory since the intensity of each pixel on the screen needs to be defined and this could require up to 24 bits per pixel. For a 1024 × 1024 pixel display, this equates to 3 Mbytes of frame buffer! The data in the frame buffer are also accessed at very high speeds (10 ns per pixel in high-resolution systems). One technique for giving this performance is to use dual-port memory (or video RAM) where a random-access port allows the display to be updated and a fast serial access port is used to transfer data to the screen. Other techniques using FIFOs and Shift Registers are employed to achieve the desired performance.

Frame grabbers Not all images can be described by simple mathematical functions. Real images need to be displayed and processed sometimes with computer graphics superimposed. To acquire a real image, a camera is used. The camera signal is captured and digitized using a frame grabber. This allows single frames or a succession of frames to be transferred to the computer's memory as blocks of data. These can be manipulated and incorporated in the output image.

Scanners and digitizers Instead of using a camera, as in a frame grabber, documents may be scanned at high resolution and converted into a bit-mapped image suitable for further processing. A digitizing tablet is used to extract information from existing documents rather than merely to scan the whole image and so this requires a human operator. The pointer (usually a cross-hair device) selects the major features of the document and the coordinates of these points are transferred to the processing system where the image feature may be reconstructed.

Drawing The drawing operation takes the image parameters and converts them into a set of pixels in the frame buffer which define the display. The frame buffer contents are then a map of what is seen on the output device and this is therefore a bit-map or pixel-map of the image.

Converting the image parameters into pixels is not always simple. The change from a continuous function such as a straight line to a discretized version (pixels on a screen) can create unusual effects which are discussed in Section 5.3.4.2.

Graphics processors Processing graphical data requires considerable processing power. If this processing is performed in software then the range of processing operations is large, limited only by the ability of the programmer. The more computing-intensive the operation, the more the throughput suffers, in terms of frames processed per second. One way of alleviating the problem is to perform some processing operations using dedicated hardware. Such devices include convolvers for filtering and masking images and SIMD or MIMD devices for post-processing images. Parallel-processing techniques are used to increase the speed of these computing-intensive operations. In addition to the architectures mentioned above, the transputer is often used for graphics applications.

Colour look-up tables (palettes) If a display were to offer a realistic range of colours then the information that would need to be stored would require a very large frame buffer. Fortunately, not all colours need to be available at once in a given image. For example, a programmer may select 64 out of 4096 possible colours. This implies that while the system is capable of representing 4096 *physical* colours, only 64 *logical* colours are used. A means of mapping the logical colours to the physical colours is provided by the Colour Look-up Table (CLUT) or Palette. Thus the programmer writes the Palette once per image and can then refer to physical colours using one of the 64 logical colour numbers. These logical numbers may be re-used for another image to represent other colours by rewriting the Palette. In a similar way, monochrome images can be given a *false-colour* rendering by assigning colours (using the Palette) to each intensity level.

5.3.3 Human–machine interface

Input devices In order to define a graphical display, two main methods exist. The first describes the desired display using some form of 'language'. This is a *text-based* system where each element on the screen and its position is described by a set of alphanumeric commands entered using a *keyboard*. To modify the display, a text file is edited or special editing commands are issued and the screen is recompiled.

The second method uses *schematic entry*, where the user directly manipulates the screen interactively by using a *pointing device* to select the position on the screen where drawing or editing operations are to take place. This is more akin to drawing with pencil and paper and thus is preferred by most users. It is also essential for computer art, where the image cannot be easily described textually.

For formal graphics (e.g. electronic circuit diagrams) where the number of symbols to be drawn is limited and conventions are defined, either textual entry or schematic capture methods can be used. Experienced users will often prefer textual entry, being a faster method (especially for repetitive features) and in which error checking is simplified. This method only requires the use of a keyboard for data entry. Inexperienced users tend to prefer a direct representation where they can see what they are drawing. If a hard-copy output is required, this is often the only suitable technique. For schematic entry systems a pointer device is required to enter coordinates to the system and such devices are described below. Pointer devices are often used in conjunction with a keyboard in order to enter data by the most efficient means for greater productivity. However, they may sometimes be used alone.

Mouse, tracker ball, cursor key and joystick These are devices capable of passing orthogonally related coordinates to the application. All except the cursor keys are able to enter two coordinates simultaneously. Cursor keys are usually part of the keyboard assembly and are the slowest of the above devices to use. The amount that the coordinate is incremented for each depression of the key is usually variable to give coarse and fine positioning of the desired point.

A mouse device contains a small ball which is moved across the surface of a desk. The movement of the ball is detected optically or mechanically and is converted to digital pulses, the number and rate of which determine the distance to move and the rate of movement. The mouse often contains switches so that the terminal position can be marked. In this way, the mouse can be driven 'single-handed'. Using a mouse requires a free area of around 300×300 mm.

To overcome this restriction, the tracker ball inverts the mouse so that the ball is moved directly by hand. The body of the tracker ball does not move but the ball may be freely moved in any direction without limit. Again, switches may be fitted to make a self-contained input device.

A joystick operates in a similar way to the tracker ball except that movement of the joystick arm is limited to a few centimetres either side of a central position. The joystick may be biased to return to the central position when pressure is removed. Because of the limitation of movement of the joystick, it is more useful where absolute positioning is required, whereas the mouse or tracker ball indicate a relative position. However, using velocity sensing for the joystick, this limitation may be overcome.

Graphics tablet The graphics tablet represents a drawing area where information is transferred to the application. The tablet has sensors embedded in its surface which detect the position of a stylus. These sensors are often arranged in a matrix. When used with a stylus, data are entered free-hand in much the same way as a user would sketch a design using pencil and paper. The stylus may have a switch in the tip so that pressing the stylus indicates a selection. When existing drawings are to be digitized, these are attached to the tablet and reference points on the drawing are converted to coordinates using a cross-hair device and switch. The application can then use the reference points to recreate the drawing. When used in this way, the graphics tablet is more commonly known as a digitizer. The graphics tablet area may also have a reserved space around its perimeter which is not used for drawing, but which is divided into small areas used for the selection of parameters.

Light-pen and touch screen These operate in a similar way to the graphics tablet, except that the monitor screen is used. The light-pen detects the light generated when the CRT electron beam strikes the phosphor coating and the position of the pen is determined from the timing of the electrical pulse generated. A touch screen may have sensors arranged around the perimeter of the screen which detect when a light beam is broken by the pointing finger. Other forms of touch screen exist (for example, two transparent panels with electrically conducting surfaces will make contact when light pressure is applied at a point).

The disadvantage of these forms of input is that the screen is obscured. The chief advantage is that the choice of items to select is infinitely variable. However, the resolution of these systems is limited; a touch screen to the area of a finger tip and a light-pen by the problem of focusing or refraction since the CRT faceplate is quite thick and light from a number of adjacent pixels can trigger the light-pen.

5.3.4 Applications

At the applications level, graphics instructions from a display list or one of the input devices are interpreted so that an image is drawn on an output device.

5.3.4.1 World, normalized and device coordinates

Most graphics systems use the Cartesian coordinate system. A single coordinate system is not usually possible since the graphics representation and the image it represents differ in scale and reference frame. Thus three coordinate systems are commonly used. *World* coordinates are those specified by the user. If the image represents a real object, then the world coordinates might be a set of physical coordinates describing the real-world object. For convenience and ease of processing, world coordinates are usually converted into *normalized* coordinates which have a range of values from 0 to 1 and are real numbers. This system allows processing operations to proceed without having to worry about arithmetic overflows where numbers grow too large to be represented by 32 bits, for example. Physical devices require the normalized coordinates to be mapped to a set of *device* coordinates. In this way, a number of output devices can be driven from the same application but with a particular set of mappings from normalized to device coordinates for each device. If the output device has different resolutions along each axis, then the scaling factor will alter with the resolution.

The use of these coordinate systems is important when the image is transformed by rotation, zooming or clipping (see Sections 5.3.4.4 and 5.3.4.5).

5.3.4.2 Output primitives

Line drawing This requires converting each point along the line into a pixel coordinate which must then be written into the frame buffer for display. For example, a diagonal line is represented by a set of pixels which are in fixed positions on the pixel matrix. In most cases the result is adequate (Figure 5.24) using an algorithm which calculates the pixel position by simple integer division. However, for a line which is close to the vertical or horizontal axis, this algorithm does not give acceptable results. A better algorithm is required which calculates the *nearest* pixel to the ideal line so that even steps are produced (Figure 5.25). Such an algorithm was proposed by Bresenham (see Appendix) which has the advantage of only requiring addition operations in order to plot the line after a few initial calculations have been performed (Figure 5.26).

Circle drawing The advantage of circle drawing is its symmetry. Once one *x,y* pair has been calculated, then eight points on the circle can be defined (Figure 5.27). Calculating the plotted points using equal increments along the *x* axis is

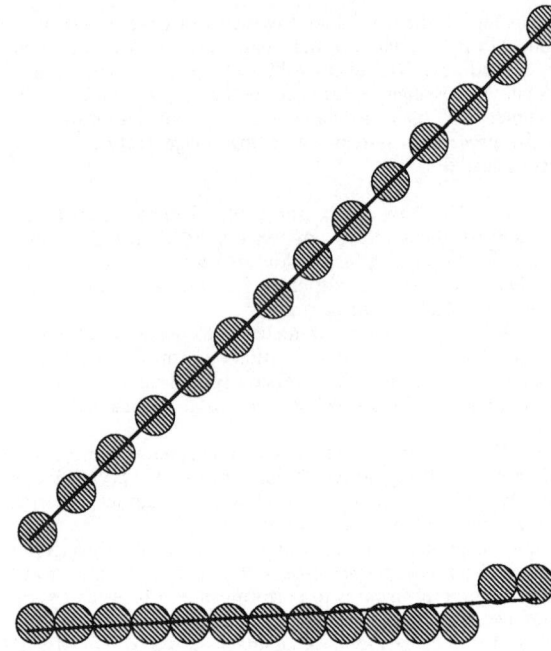

Figure 5.24 Pixels plotting using integer division

Figure 5.25 Pixels plotted using *nearest-pixel* algorithm

unsatisfactory as shown in Figure 5.28. Better results are obtained when points are plotted at equal angular rotations. However, the calculation involves evaluating a *sine* and *cosine* function; trigonometric functions use a lot of CPU time. Some means of reducing the number of trigonometric functions which need to be evaluated is desirable.

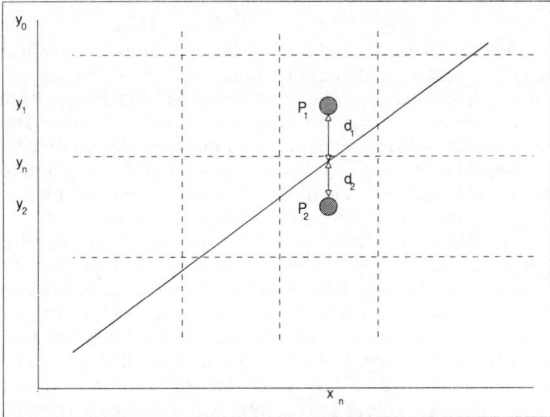

Figure 5.26   Bresenham's line-drawing algorithm

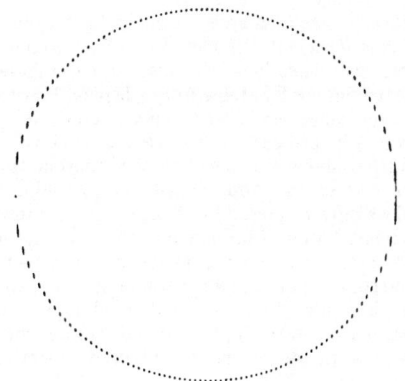

Circle drawn using constant x increments

according to: $y = \sqrt{1 - x^2}$

Figure 5.28   Circle plotted using equal increments along the x-axis

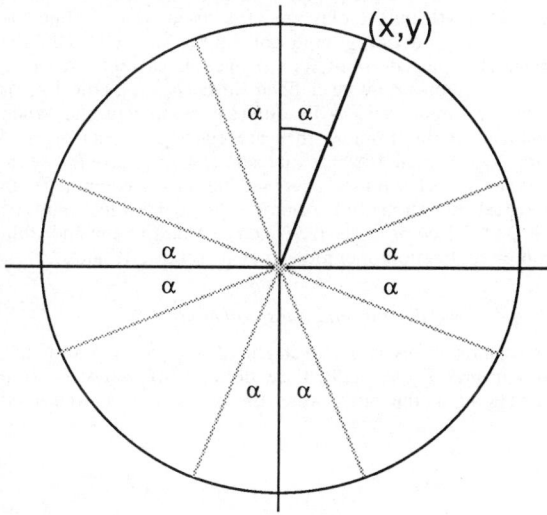

Figure 5.27   Using the circle's symmetry, eight points can be plotted for each (x,y) coordinate pair

Polygon method   If the circle is drawn as a polygon, then only a few calculations are required to determine the vertices after which a straight-line algorithm is used to join the vertices.

Use is made of the following relationships:

$$\sin(A + B) = \sin(A)\cos(B) + \cos(A)\sin(B)$$

$$\sin(A - B) = \sin(A)\cos(B) - \cos(A)\sin(B)$$

$$\cos(A + B) = \cos(A)\cos(B) - \sin(A)\sin(B)$$

$$\cos(A - B) = \cos(A)\cos(B) + \sin(A)\sin(B)$$

For this polygon, the 'radius' is incremented by $2\pi/n$ radians for each of the n vertices. If the angle A represents the current vertex, then the next vertex is found at an angle of $A + 2\pi/n$. Instead of calculating the sine and cosine of this new angle, the previous values of sine and cosine are incremented according to the expressions above. Thus:

$$\sin(A + 2\pi/n) = \sin(A)\cos(2\pi/n) + \cos(A)\sin(2\pi/n)$$

$$\cos(A + 2\pi/n) = \cos(A)\cos(2\pi/n) - \sin(A)\sin(2\pi/n)$$

Now $\sin(A)$ and $\cos(A)$ become the new sine and cosine values which will be updated for the next vertex. The two multiplication operations and one addition operation per function considerably reduce the computation required since $\cos(2\pi/n)$ and $\sin(2\pi/n)$ only need to be calculated once. The initial sine and cosine values can be selected to be '0' and '1' if a full circle is to be drawn. Only the first $2\pi/8$ radians need to be calculated as shown above if one takes advantage of the circle's symmetry. This method is prone to cumulative errors, but if these are less than half a pixel in total, then the method is satisfactory.

Other curves   Functions in which the gradient is predictable or always less than unity (e.g. a circle) can always be plotted by incrementing in unit steps along one axis and calculating the other coordinate. Complex curves may require the computation of the inverse function especially when the gradient is large, if gaps in the curve are to be avoided. This is computationally expensive. Curve-fitting techniques and straight-line approximations (e.g. polygon methods) considerably reduce the computation required if the resulting accuracy is acceptable.

Characters   Most applications require text to be displayed. The most common form of manipulating text is to hold *bit-mapped fonts* in memory, individual characters of which are copied to the screen at the desired position. These characters may be rotated in increments of 90° by manipulating the matrix to allow vertical or inverted text. Different font sizes may be produced by scaling the matrix although this only gives acceptable results for a small range of font sizes. A better solution is to hold each font in a variety of font sizes.

The above techniques only permit text to be aligned to one of the axes. For text to be produced at any angle or orientation, matrix transformations are possible but do not give good results. Using a *stroked font* where characters are represented by a small number of curves (or strokes) means that the character definitions are independent of angle and also the displayed size.

Most applications allow the user to define custom characters or symbols. In this way, fonts containing other than Roman characters may be used.

Move and copy   Defined areas of the image can be quickly and easily *copied* using BitBlt operations, thus avoiding repetition of previous calculations. This method is commonly used to enter text from the font table to the display. However, not all parts of the image can be so simply copied since overlapping blocks may be present. In this case the block will have to be recalculated and redrawn at the new coordinates. *Move* operations require the steps above and, in addition, the original block must be erased by recalculating and subtracting from the frame buffer. Alternatively, to erase the image, a rectangular area enclosing the image could be set to the background colour (thus erasing overlapping blocks within the area) and then any blocks partly defined in the area are redrawn with windowing applied to reconstitute the image. The options available in move and copy operations are discussed in Section 5.3.4.3.

Area-fill   If the shape of filled area is known, then the operation employs a polygon-drawing algorithm using a plot colour or pattern. When a pre-drawn area is to be filled, the shape of the area may not be known so a *flood-fill* algorithm is required. To fill such an area with a colour or pattern, a closed area is essential and a seed point within that area must be supplied from which the fill will be determined. The fill operation will set the pixels one row at a time within the desired area until a boundary is reached. For example, boundary may be defined as a foreground colour or the background colour. Fill operations on areas containing patterns give uncertain results if the pattern contains the boundary colour. Most fill algorithms are recursive so that complex areas may be filled. In such cases, the fill routine keeps a list of start points for each line of pixels which are to be filled. When it meets a boundary, it returns to the seed point and looks in other directions where the fill might proceed.

Narrow areas of one or two pixels in width might prematurely terminate a fill operation. Since the fill proceeds one row at a time, the narrow section might become blocked and appear to be a vertex of the enclosed area, thus terminating the fill. Section 5.3.4.3 describes some of the attributes of fill operations.

Aliasing   Since pixels can only be drawn on a finite matrix, continuous functions, when displayed, appear to have edges which do not exist. This artifact is called *aliasing* and its effect is to give diagonal lines a jagged appearance. In order to reduce this effect, various means of anti-aliasing are employed. If data in the frame buffer are processed to search for edges some will be found to be true edges (i.e. exist in the real image) and can be ignored. Where aliasing is found to occur the intensity of these and adjacent pixels can be modified to mask the edge. A form of Bresenham's line algorithm may be used to detect the relative position of a pixel from the true line and the intensity is then set in inverse proportion.

Hardware techniques exist to reduce the 'jaggies' which include pixel phasing and convolution operations.

Grids   A deliberate form of aliasing is used where the application demands that all points be plotted on a grid or in an orthogonal-only mode. For example, all pixel positions calculated by the application or entered by an input device which fall within predefined areas are converted to the same pixel position – that is, the centre of the defined area. The defined areas depend on the grid spacing, which may be altered.

In an orthogonal-only mode, one coordinate from the previously displayed point is fixed and only the other coordinate is free to change (usually constrained to a grid).

5.3.4.3   Attributes of output primitives

Attributes may be defined for drawing operations which affect line styles, colour and intensity. Line-style options include the line width and pattern. The pattern may be a hatching pattern in one colour or a pattern using a number of colours. The pattern will typically repeat every 8 or 16 pixels and may be considered to be 'tiled' across the whole display. Only where the line coincides with the tiled pattern are those pixels plotted as part of the image. The most common line style is *solid*.

Note that some attributes are not relevant or possible for certain display devices. While the intensity of a line can be varied for display on a CRT monitor, the same image will lose intensity information when plotted on a monochrome laser printer, for example. Referring to attributes individually, they are called *unbundled*. When used in this way, the application might require modification acccording to the display device used. Similarly, colour information will not remain constant when different displays are used, even for devices capable of using colour. As an example, a CRT display normally draws in white on a black background whereas a colour plotter would draw in black on white paper; both would display a red line in red. Thus attribute tables are often used which define the foreground and background colours to be used when the image is displayed on a CRT, to give one example. A whole set of attributes may be defined for each display device, or even for similar devices by different manufacturers. When arranged in this fashion, they are given the name *bundled atttributes*. Similar attributes are available to control fill styles.

When a block is moved or copied this may be combined with a logical operation. For example, the source block may be ANDed, ORed or Exclusive-ORed with destination and addition or subtraction operations may be set as attributes.

5.3.4.4   Two-dimensional transformations

Translation   This is a movement of a graphics object in a straight line (Figure 5.29). If the distance (dx, dy) is added to each point in the object then the object will be translated

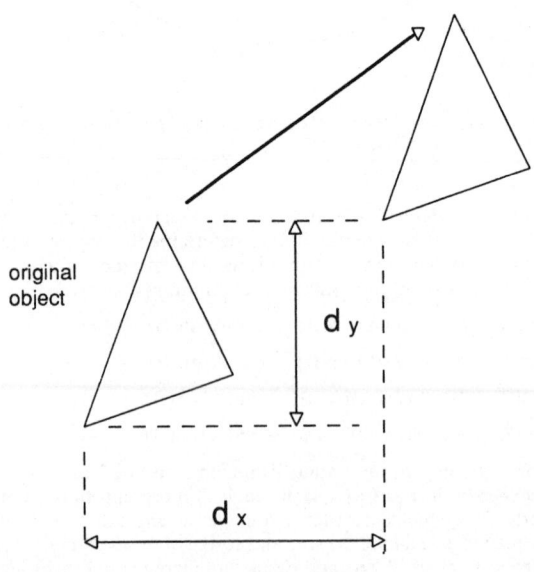

original
object

d y

d x

Figure 5.29   Linear translation of an object

linearly when redrawn. This is acceptable for lines or polygons (which can be represented as a set of lines). For circles and arbitrary curves, the offset is applied to the reference point (e.g. the centre of the circle) and the object redrawn.

Note that when an object is complex the redrawing of translated objects can be quite slow. In an interactive mode this can be a drawback. Hence some applications do not update the display completely except on request. This possibly leaves some extraneous pixels set in the display but which are cleared on the next display refresh operation. If BitBlt operations are not possible (due to overlapping objects, for example) then some applications calculate a bounding box and a few reference marks on its edge in order to temporarily describe the object. This *outline* image can be moved interactively at high speed and the object is only fully redrawn when the destination is fixed.

Scaling requires all relative distances of points within an object to be multiplied by a factor (Figure 5.30). This factor is usually the same for horizontal and vertical directions to retain the proportions of the original object. If the scaling factor differs in each direction, then the object will appear to be stretched or compressed.

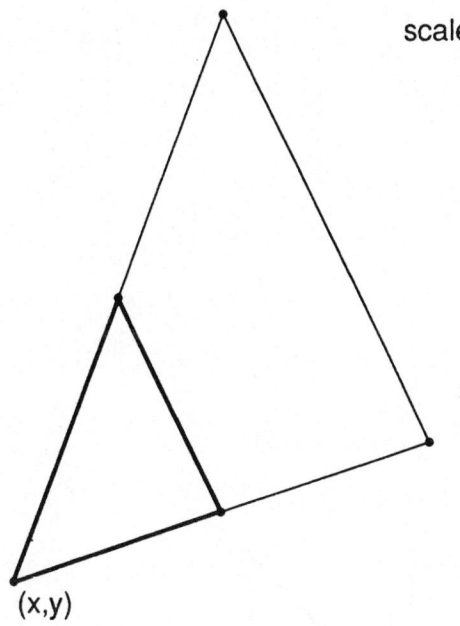

scale

Figure 5.30 Scaling operation. A scale factor of 2 is applied to the object relative to the point (x,y)

Rotation requires multiplication of coordinates by $\sin\theta$ and $\cos\theta$, where θ is determined from the pivotal point. The new coordinate is calculated from its position relative to the pivotal point (Figure 5.31).

Reflection produces an image which may be mirrored with respect to the x-axis, y-axis or a user-defined axis (Figure 5.32). Changing the sign of one or both sets of world coordinates will convert a point so that it is mirrored about one or both orthogonal axes.

Shear transformations can distort images (or correct for perspective distortions) by making the transformation factor a function of the coordinate values (Figure 5.33). Thus the transformation factor varies across an object.

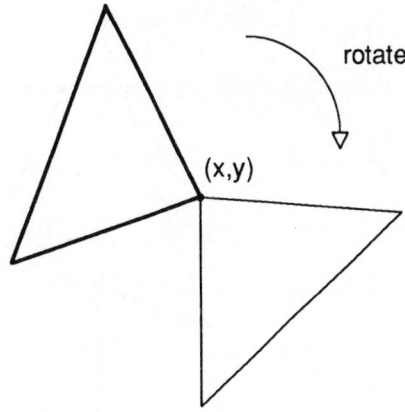

rotate

(x,y)

Figure 5.31 Rotation of an object about the pivotal point (x,y)

Matrix representations   All of the transformations above can be reduced to a sequence of basic operations, each of which can be represented as a 3×3 matrix for a two-dimensional display. For example, a *linear translation* of an object by a distance (dx, dy) requires the coordinate $[x\ y\ 1]$ to be multiplied by the matrix:

$$\begin{vmatrix} 1 & 0 & 0 \\ 0 & 1 & 0 \\ dx & dy & 1 \end{vmatrix}$$

Successive translations are additive such that two translations of (dx, dy) and (δx, δy) are equivalent to a translation of (dx + δx, dy + δy):

$$\begin{vmatrix} 1 & 0 & 0 \\ 0 & 1 & 0 \\ dx & dy & 1 \end{vmatrix} \cdot \begin{vmatrix} 1 & 0 & 0 \\ 0 & 1 & 0 \\ \delta x & \delta y & 1 \end{vmatrix} = \begin{vmatrix} 1 & 0 & 0 \\ 0 & 1 & 0 \\ dx + \delta x & dy + \delta y & 1 \end{vmatrix}$$

The *scaling* process requires more than one operation. The first *translates* the object to the graphics origin. Thus the second (*scaling*) operation can multiply all coordinates by the same factor (i.e. with respect to the origin). The final operation translates the object back to its original position. Thus one scalar and two translation operations are required in the following order:

$$\begin{vmatrix} 1 & 0 & 0 \\ 0 & 1 & 0 \\ -dx & -dy & 1 \end{vmatrix} \cdot \begin{vmatrix} mx & 0 & 0 \\ 0 & my & 0 \\ 0 & 0 & 1 \end{vmatrix} \cdot \begin{vmatrix} 1 & 0 & 0 \\ 0 & 1 & 0 \\ dx & dy & 1 \end{vmatrix}$$

where mx and my are the scaling factors and dx and dy are the distance of the object from the origin. These matrices may be combined to give the scaling matrix:

$$\begin{vmatrix} mx & 0 & 0 \\ 0 & my & 0 \\ (1 - mx)dx & (1 - my)dy & 1 \end{vmatrix}$$

The *rotation* process also requires translation to the origin before the rotate operator is applied and the inverse translation (as above). The *rotation* matrix is:

$$\begin{vmatrix} \cos\theta & \sin\theta & 0 \\ -\sin\theta & \cos\theta & 0 \\ 0 & 0 & 1 \end{vmatrix}$$

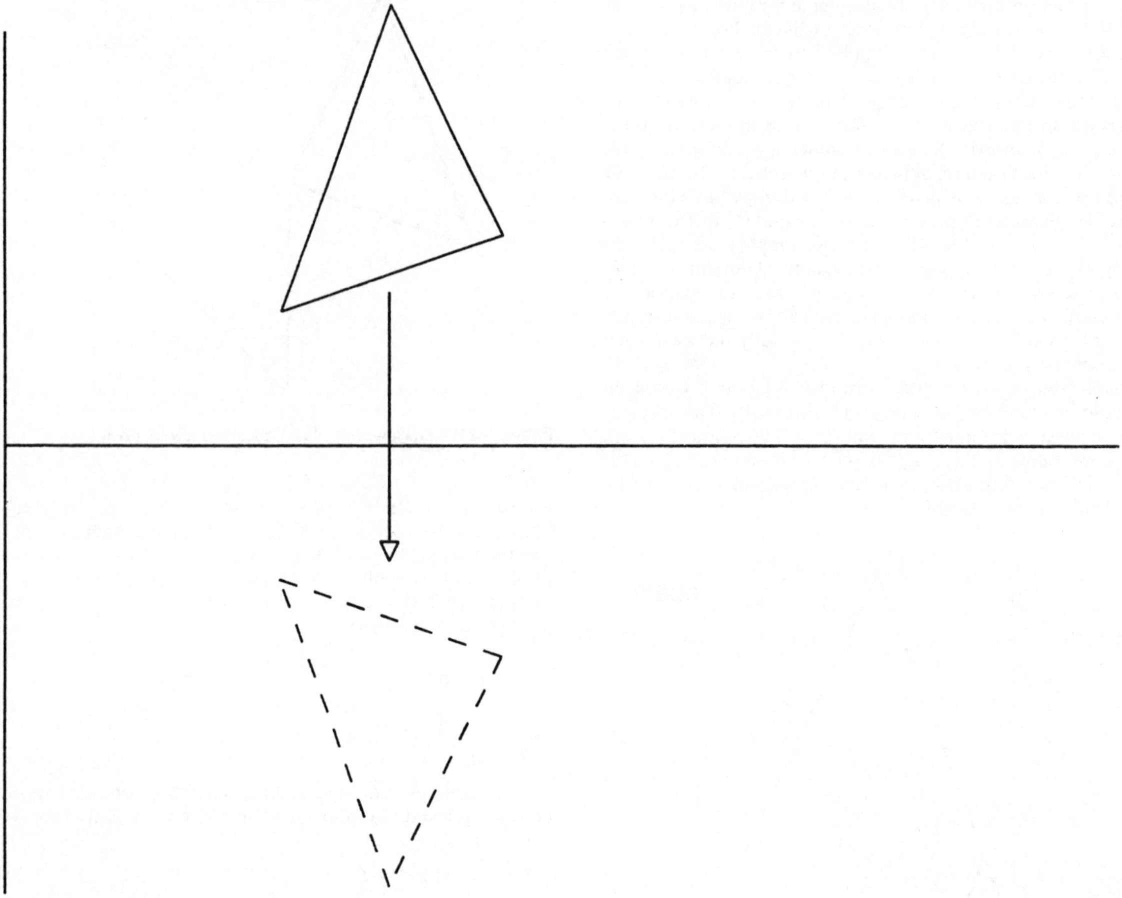

Figure 5.32   Reflection of an object about the *x*-axis

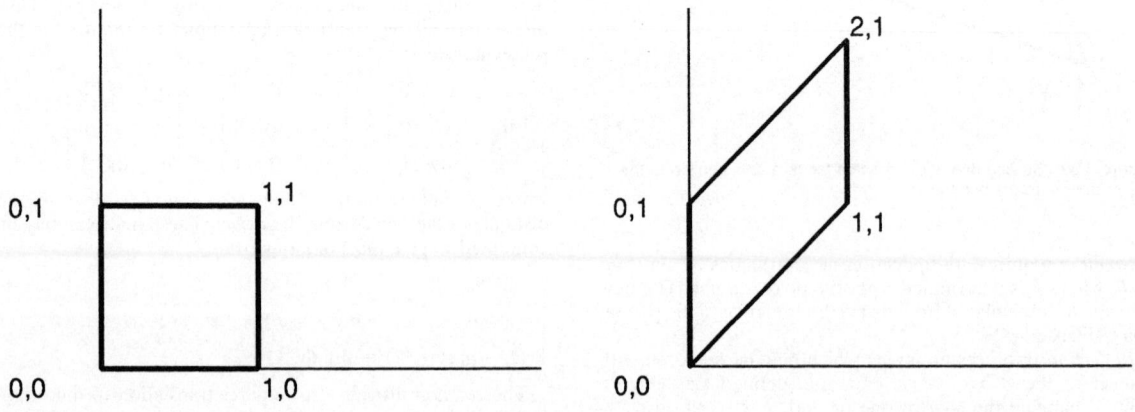

Figure 5.33   A *y*-direction shear transformation on a unit square using a shear factor of 1

where θ is the angle of rotation. With the two translation operations added, the overall matrix becomes:

$$\begin{vmatrix} \cos\theta & \sin\theta & 0 \\ -\sin\theta & \cos\theta & 0 \\ (1 - \cos\theta)dx + dy\sin\theta & (1 - \cos\theta)dy - dx\sin\theta & 1 \end{vmatrix}$$

Since all matrices can be multiplied together, then complex transformations can be constructed by applying the matrix operations in the desired order.

5.3.4.5 Windowing and clipping

Windowing A window is a rectangular display area. There is normally a single window displayed which occupies the whole screen. However, it is now common to find software which uses windows freely and there may be several windows displayed at once. An architecture which allows only a single process to run at any one time may display multiple windows, but only one can be an *active* window. Multi-tasking or multi-processor systems may have several windows which are active, i.e. each is controlled by a different *process* which is running.

Where multiple windows are displayed they will often overlap so that the window which has lower precedence (or is a *background* window) is partially or totally obscured (Figure 5.34). Hardware techniques are available to manage such overlaps, but more commonly this is performed in software. *Clipping* operations are performed when the contents of a window are being displayed so that only pixels within the permitted window limits are drawn; pixels outside the window area are *clipped* (Figure 5.35). The window boundaries and attributes are defined in a higher layer of the software – the *window manager*, which is conceptually part of the operating system. The window manager may draw a border around the window itself and label the border appropriately, but this is transparent to the process using the window. It is possible to define the windowing operation in terms of world or display coordinates (see Section 5.3.4.1) and 'window' is often used interchangeably when referring to either coordinate system. Where a distinction needs to be made between the two, the term *viewpoint* refers to the rectangular area on the display device.

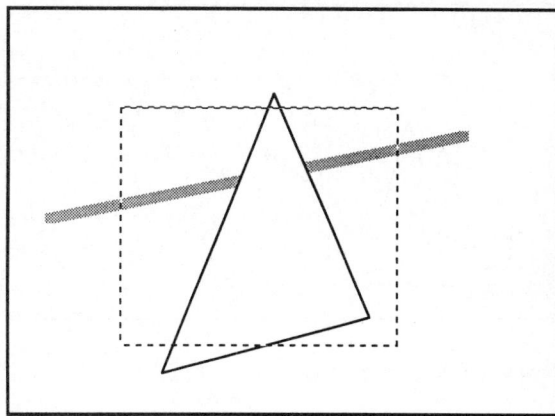

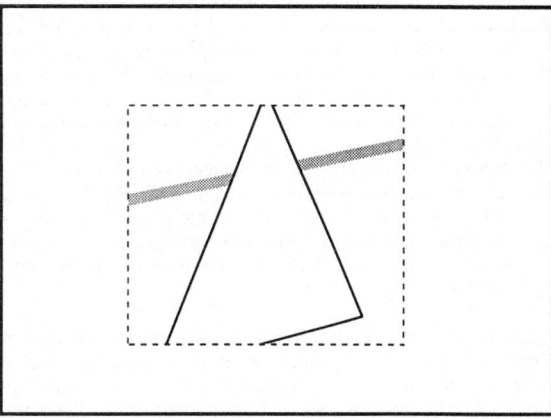

Figure 5.35 Clipped graphics

The most common operations to be performed on a window are described below and are implemented by calls to the window manager.

1. *Create* The dimensions and position of the new window are given and a *handle* is returned if the window is successfully created. This handle is used in future graphics calls to specify the window in which drawing operations are to take place. A newly created window will normally have the highest priority so that it may obscure parts of existing windows.
2. *Clear and delete (close)* The window handle is used to specify the window to be cleared or closed. It may not be possible to close a window if the process which owns it is still active.
3. *Drag (move)* The size and contents of the window are unchanged, but the position in the display is altered (Figure 5.36). A translation operation is used to perform this. The window position is normally constrained so that no part may be dragged off the display, otherwise further clipping may become necessary.
4. *Resize* The dimensions of the window are changed by altering the clipping parameters. The contents of the window which are visible before and after this operation remain unchanged (Figure 5.37).

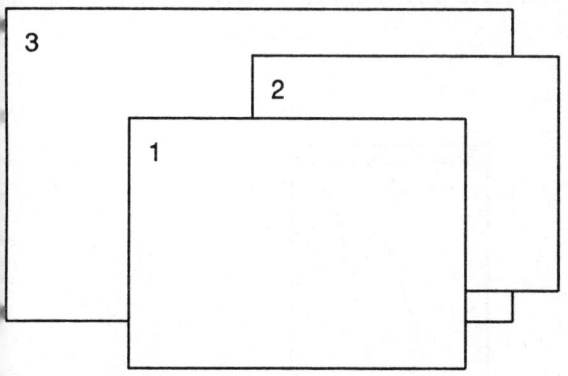

Figure 5.34 Multiple overlapping windows

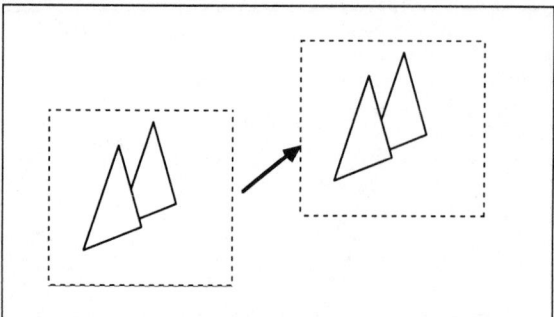

Figure 5.36 Dragging a window

5. *Zoom* The contents of the window are recalculated using a new scaling factor (Figure 5.38).
6. *Pan* Here, the *viewport* is unchanged in position and size, but the window moves 'behind' the viewpoint. This is a translation operation but the clipping attributes do not 'move' with the window (as for *drag*) but rather remain constant as far as the display coordinates are concerned (Figure 5.39).
7. *Priority* A window can be brough to the foreground or sent to the background by assigning it the highest or lowest priority attribute. If an intermediate priority is assigned, then the window may obscure parts of some windows and may itself be partly obscured by other windows (Figure 5.40).

Clipping text Where the clipped object comprises text, then clipping at the window boundary can leave partial characters visible in the same way as graphics objects are clipped at the pixel level. Sometimes this is visually undesirable. Thus text may be treated differently such that if any part of a character would be clipped, then that character is not displayed (Figure 5.41).

Updating the display When an operation takes place which disturbs the boundaries of the viewpoint then it is not only the window itself which needs to be redrawn; any part of the display which was partially obscured by the old viewport will also need to be redrawn (Figure 5.42).

If the background is now visible, the revealed areas are simply cleared to the background colour. If parts of other windows are revealed then two strategies exist. Either the whole window is redrawn and the window manager clips the pixels according to the window's priority, or an 'intelligent' process will only redraw those parts of the image that had previously been obscured. The first strategy is the simplest, but has the disadvantage of redrawing even those parts of the image that are correctly displayed – which means that overall system performance suffers. The second strategy is the most efficient in that only the area which requires redrawing is changed. This requires that the process itself can determine which objects or parts of objects were obscured and then require redrawing. This is not always easy to do or to calculate.

If the windowing is performed in hardware, then the display buffers do not become corrupted where windows overlap as each window has its unique, non-overlapping buffer. Thus when moving a window reveals another, no redrawing of the image buffer is required. The display hardware fetches data from the appropriate buffer as each window or part thereof is displayed on the output device.

5.3.4.6 Segments

Graphics objects may sometimes be repeated within an image; it is wasteful to store the same information several times so such objects may be stored as subpictures or *segments*. These objects are not restricted to being identically portrayed in the output image since variations of the same object can be produced by changing the attributes of the object.

A related hardware technique involves the use of *sprites*. In this way a graphics object can be predefined and held in memory. Whenever this object is required, it can be quickly copied into the frame buffer at the required position without requiring graphics processing operations to draw it. However, there is usually the restriction that attributes cannot be changed and so the sprite is fixed in size and colour.

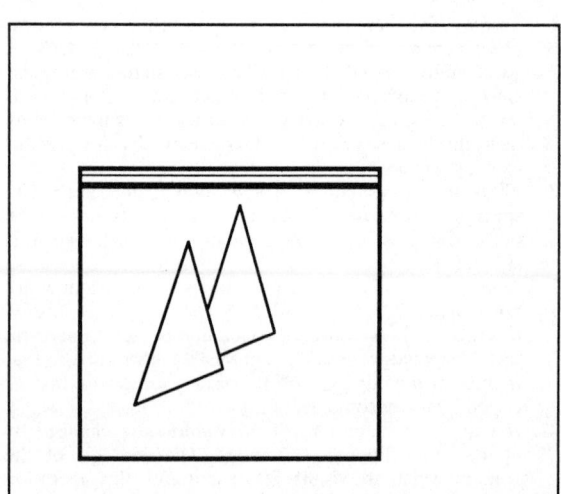

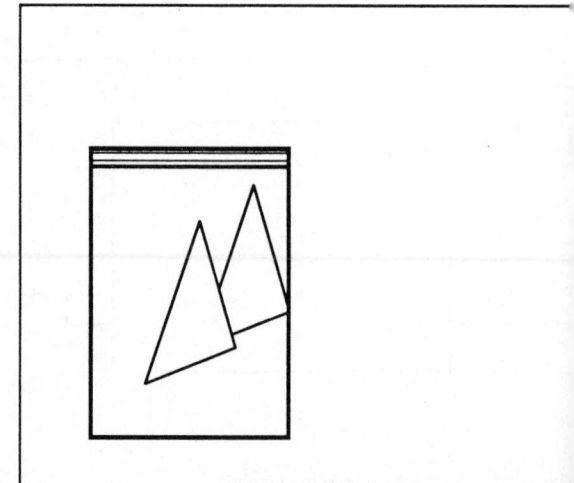

Figure 5.37 Resizing a window

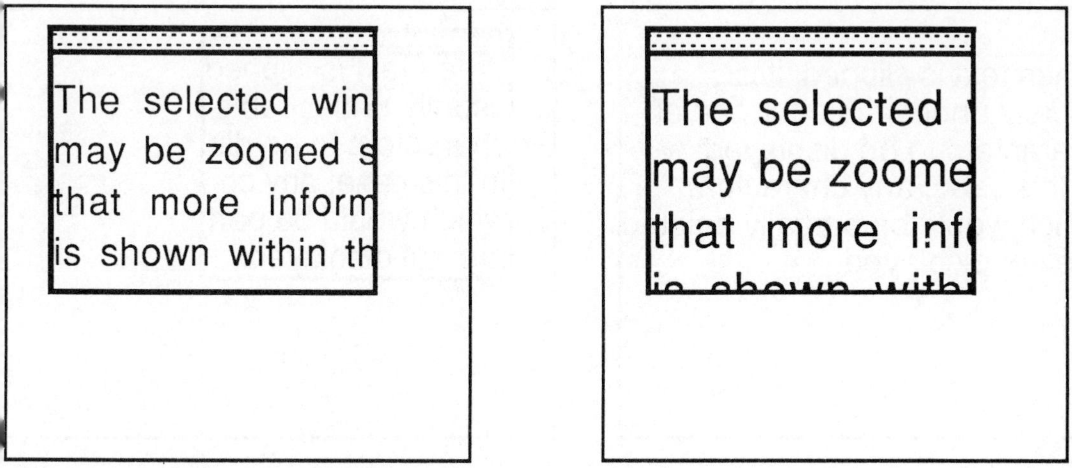

The selected win
may be zoomed s
that more inform
is shown within th

The selected
may be zoome
that more inf
is shown with

Figure 5.38 Zoom operation on a window

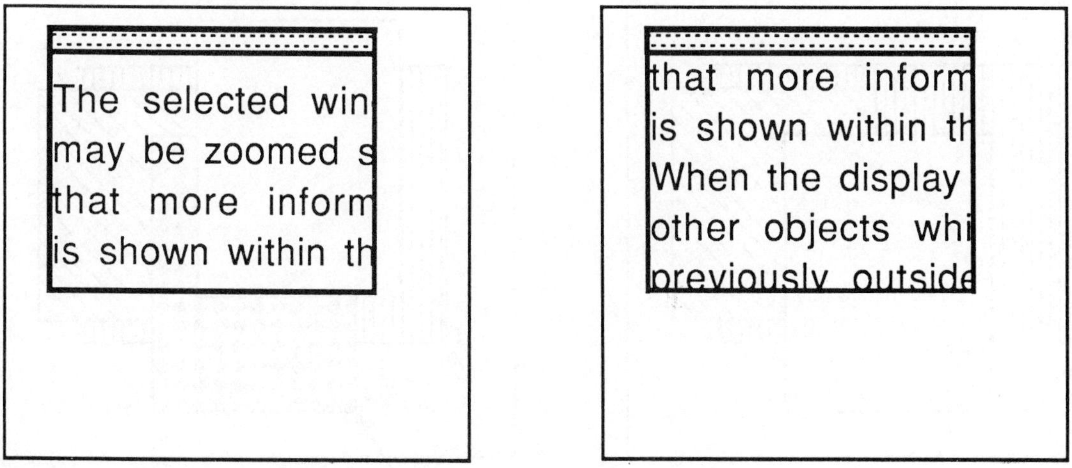

The selected win
may be zoomed s
that more inform
is shown within th

that more inform
is shown within th
When the display
other objects whi
previously outside

Figure 5.39 Pan operation on a window

Original display

Window 2 to foreground

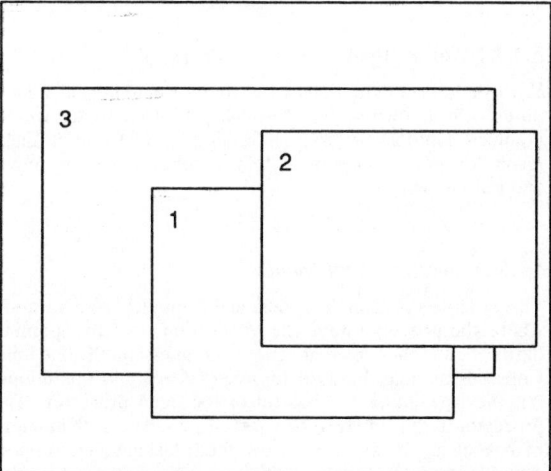

Figure 5.40 Changing the priority of window 2

When text is clipped, it is
usually undesirable for partial
characters to be displayed.
In this case, any characters
which would be partially deleted
are not displayed.

When text is clipped
usually undesirable
characters to be dis
In this case, any ch
which would be part
are not displayed.

Figure 5.41 Clipped text

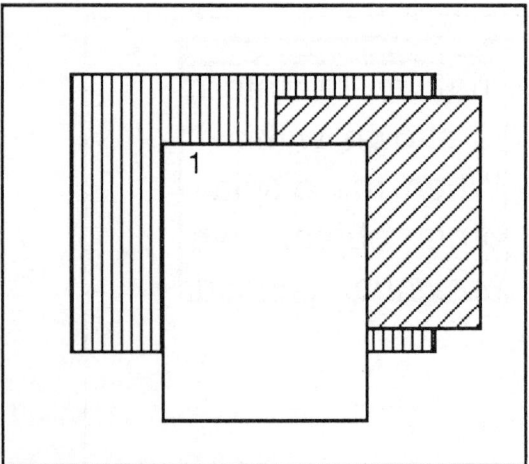

Figure 5.42 When a window is moved, the area exposed must be redrawn

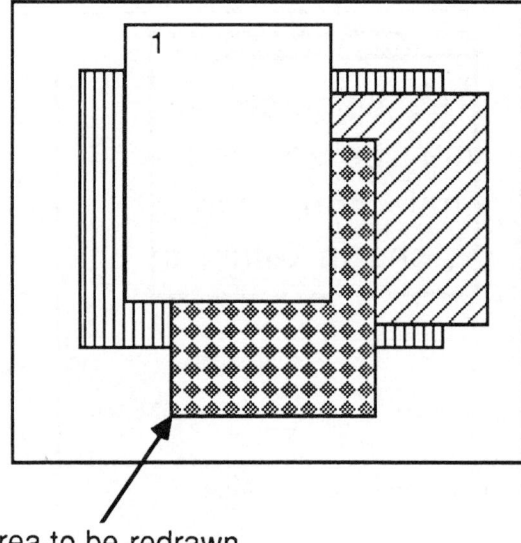

Area to be redrawn

5.3.5 Workstations

Workstation configurations vary from a single personal computer with its own screen, through a host computer with a graphics coprocessor using a separate screen, to a high-resolution multi-processor system incorporating many input and output devices.

5.3.5.1 *Integrated workstations*

This describes the standard 'personal computer' configuration where the processor runs the application, and the graphics hardware and the screen are contained in essentially one unit. Coprocessors may be used to accelerate certain operations, but they are under the control of the main processor. The processor itself is likely to be a fast 32-bit device with memory management. It will also allow multi-tasking and support multiple windows.

5.3.5.2 *Host/slave configuration*

In this case, the host computer provides mass storage, keyboard entry and interfaces for printers and plotters. The slave graphics processing unit contains the frame buffer, drawing and display hardware and is semi-autonomous. The two units are linked by a bus. In this way, the host runs the high-level application software and produces a display list to the slave. The slave is optimized to interpret the display list and perform the drawing operations at high speed.

The two parts of the system partition the process into high- and low-level operations and can work in parallel for much of the time.

5.3.5.3 *Operating systems*

Many PC-based systems use MS-DOS or some form of display manager. Only '386 systems and later allow true multi-tasking. The virtue of such a system is that it is multi-purpose and can provide a system for both word processing and graphics processing, it is therefore cost-effective.

UNIX*-based systems are multi-tasking and usually multi-user as well, although only a limited number of high-resolution terminals are allowed per node before the system performance suffers. UNIX allows several input and output devices to be added to the system and accessed by various users. Portability of applications between UNIX systems of different manufacture is a strong advantage. UNIX systems may range from a desk-top computer to a large main-frame installation.

VMS† systems based around the VAX architecture are not generally as portable to third-party hardware. However, VAX installations are fairly common for this not to be a severe drawback. A large resource of VMS-based graphics software is available.

An increasingly common feature is *networking*, where a number of high-performance graphics workstations are net-worked together to share central resources (or distributed resources). Since each workstation has its own processor (or processors) other users of the network are not disadvantaged when one user initiates some computing-intensive task. Only when simultaneous access is made by two workstations to a shared resource (e.g. a fileserver) is any drop in performance apparent.

5.3.6 Three-dimensional concepts

Three-dimensional concepts and operations are simply an extension of the two-dimensional concepts described in Section 5.3.4.4. For display purposes on two-dimensional devices further transformations must take place to give a three-dimensional *representation* in two dimensions.

5.3.6.1 Introduction

The coordinate system used is normally three orthogonally related axes: x, y and z. Translation, scaling, rotation reflection and shear operations are performed in a similar manner to the two-dimensional operations but for three dimensions. For example, the *linear translation* of an object by a distance (dx, dy, dz) requires the coordinate $[x\ y\ z\ 1]$ to be multiplied by the matrix:

$$\begin{vmatrix} 1 & 0 & 0 & 0 \\ 0 & 1 & 0 & 0 \\ 0 & 0 & 1 & 0 \\ dx & dy & dz & 1 \end{vmatrix}$$

Compare this with the two-dimensional linear translation matrix and observe the similarity:

$$\begin{vmatrix} 1 & 0 & 0 \\ 0 & 1 & 0 \\ dx & dy & 1 \end{vmatrix}$$

More three-dimensional transformations are given in Section 5.3.6.4.

5.3.6.2 Three-dimensional display techniques

A three-dimensional object will be projected onto a two-dimensional plane for display. If the z-axis is arranged to be normal to the plane of the display then no transformation of

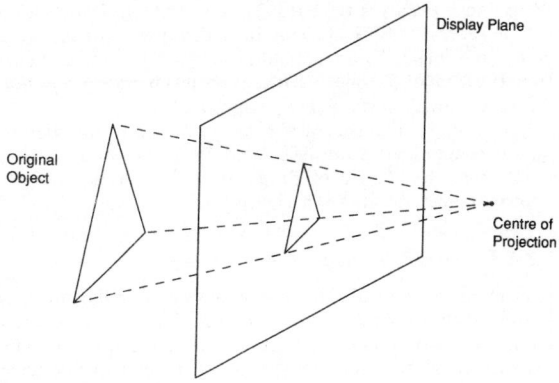

Figure 5.43 Perspective projection

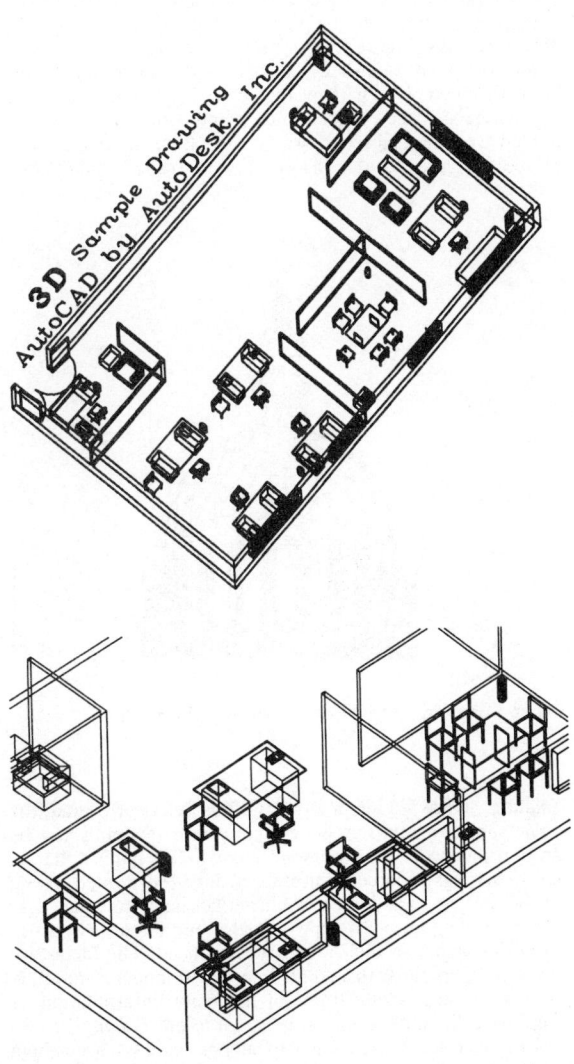

* UNIX is a Trademark of AT&T Bell Laboratories, Inc.
† VMS is a Trademark of the Digital Equipment Corporation.

Figure 5.44 Two views of an office from different viewpoints (courtesy of AutoCAD)

the x and y planes is required. However, depth information can be represented by allowing the z distance to offset the x and y coordinates by an amount proportional to the distance from the front of the object. This results in a *parallel projection* onto the viewing surface (flat perspective).

For a more realistic *perspective projection*, the depth-modified coordinates are calculated from the distance of a point from the Centre of Projection. Thus distant objects appear smaller, as shown in Figure 5.43.

5.3.6.3   *Three-dimensional representations*

It is not always desirable to view an object along the z-axis as described above. Any arbitrary *viewing point* could be chosen so that views from any point around an object and from any distance could be chosen (Figure 5.44). Indeed, the viewing point may be inside an object, giving an internal view. The projection onto the display plane requires the translation, scaling and rotation operations described in Section 5.3.6.4.

If all points in an object are translated from the three-dimensional object onto the viewing plane, then a *wire-frame* drawing results (Figure 5.45). This is acceptable as a representation, but is not realistic. In a solid object, many points are hidden from view by parts of the object itself. The image can be processed to determine which points would normally be hidden from the chosen viewpoint and these points are not plotted (Figure 5.46). This process is called *hidden-line removal*.

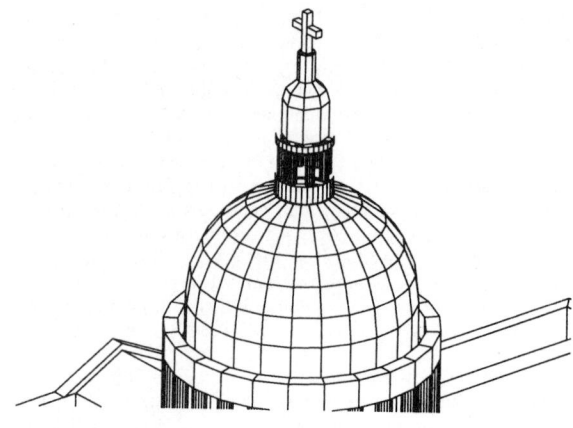

Figure 5.46   Figure 5.45 with hidden-line elimination (courtesy of AutoCAD)

Shading itself can be performed in a number of ways. *Dithering* retains a fixed pixel size and density but groups pixels into *superpixels* which may contain 2×2, 3×3 or larger arrays. This is also called *half-toning*. A 2×2 array may represent five grey levels according to the number of pixels set in the superpixel (0 to 4), but the resolution is halved in this example. *Continuous tone* is used to give 'photographic' quality since the density of a pixel may be varied over a continuous range from black to white. The final displayed resolution is not affected by this process.

5.3.6.4   *Three-dimensional transformations*

The transformation mentioned in Section 5.3.6.1 can be reduced to a sequence of basic operations, each of which can be represented as a 4×4 matrix. Only the basic operations are described here.

Figure 5.45   Wire-frame drawing (courtesy of AutoCAD)

Shading   Once hidden line removal has been performed, the object appears solid. A better impression of depth can be given if the facets or surfaces are *shaded*. This implies that an imaginary light source is introduced into the model. Those surfaces at an angle which would reflect light from the light source to the viewpoint appear bright, and as surfaces differ from this angle, so their intensity is reduced. The facets are still clearly visible at this point. In order to model a smoothly curved surface some form of intensity interpolation is employed, the most common being *Gouraud shading*.

This idea can be extended to model *shadows* for a high degree of realism and to perform *ray-tracing* so that the effect of transparent and refracting objects can be represented (Figure 5.47).

Figure 5.47   Ray-traced image (Acorn Archimedes computer, software by Beebug)

The *linear translation* of an object by a distance (dx, dy, dz) requires the coordinate $[x\ y\ z\ 1]$ to be multiplied by the matrix:

$$\begin{vmatrix} 1 & 0 & 0 & 0 \\ 0 & 1 & 0 & 0 \\ 0 & 0 & 1 & 0 \\ dx & dy & dz & 1 \end{vmatrix}$$

The *scaling* process requires more than one operation. The first *translates* the object to the graphics origin. The second *scaling* operation can scale all coordinates by the scaling factor with respect to the origin. The final operation translates the object back to its original position. The scaling matrix is:

$$\begin{vmatrix} mx & 0 & 0 & 0 \\ 0 & my & 0 & 0 \\ 0 & 0 & mz & 0 \\ 0 & 0 & 0 & 1 \end{vmatrix}$$

where mx, my and mz are the scaling factors in each dimension. If dx, dy and dz are the distances of the object from the origin in each dimension, then the combined scaling operation with translation becomes:

$$\begin{vmatrix} mx & 0 & 0 & 0 \\ 0 & my & 0 & 0 \\ 0 & 0 & mz & 0 \\ (1 - mx)dx & (1 - my)dy & (1 - mz)dz & 1 \end{vmatrix}$$

The *rotation* process also requires translation to the origin before the rotate operator and the inverse translation are applied (as above). In the three-dimensional case, the axis of rotation is arbitrary and is not necessarily aligned to any of the three axes. Taking the simple case where a z axis rotation is performed, the *rotation* matrix is:

$$\begin{vmatrix} \cos\theta & \sin\theta & 0 & 0 \\ -\sin\theta & \cos\theta & 0 & 0 \\ 0 & 0 & 1 & 0 \\ 0 & 0 & 0 & 1 \end{vmatrix}$$

where θ is the angle of rotation. It can be seen that this is similar to the two-dimensional case.

Acknowledgement

Dr D. N. Fenner, King's College, London (Figures 5.20 and 5.21).

Appendix: Bresenham's line algorithm

In Figure 5.26 the line is assumed to have a slope of less than 1. The straight line is plotted for constant x increments which are equal to the x pixel increment. Thus the position x_n is an integral pixel coordinate and the distance $x_n - x_{n+1}$ is equal to the x pixel increment. In order to plot the line, the equivalent y pixel positions must be found for each x pixel coordinate.

The real y coordinate is given by:

$$y_n = mx_n + b \tag{5.1}$$

This will normally not fall onto an integer pixel position so the *nearest y* pixel must be found.

In Figure 5.26 the distance of the point (x_n, y_n) from the neighbouring y pixel positions (P_1 and P_2) are shown to be d_1 and d_2. The smaller of d_1 or d_2 is used to select the pixel to be plotted. These distances are calculated as follows:

$$d_1 = y_1 - y_n$$
$$= y_1 - mx_n - b \tag{5.2}$$
$$d_2 = y_n - y_2$$
$$= mx_n + qb - y_2 \tag{5.3}$$

The difference $d_1 - d_2$ is calculated. If the result is positive, then the line is closer to P_2. If the result is negative, the line is closer to P_1. Now,

$$d_1 - d_2 = y_1 + y_2 - 2mx_n - 2b \tag{5.4}$$

Here, $y_2 = y_1 - 1$ since y_1 and y_2 are integer pixel coordinates which simplifies equation (5.4), as shown later.

Once the start of the line has been calculated and the nearest pixel found, it is not necessary to calculate $y = mx + b$ each time and then calculate the nearest y pixel position. Instead, a constant (Δy) is added to the previous real y value for each increment of Δx along the x axis, where Δx is the x pixel increment and $\Delta y = m\Delta x$.

Substituting for y_2 in equation (5.4) and multiplying by Δx gives:

$$(d_1 - d_2)\Delta x = 2\Delta xy_1 - 2\Delta yx_n - (2b + 1)\Delta x \tag{5.5}$$

The left-hand side of the expression is positive if the line is closer to P_2 or negative if closer to P_1. It is possible to calculate the new lhs of the expression from the previous one as shown below, where the lhs is denoted ω_n. The final term in equation (5.5) is a constant term, therefore:

$$\omega_n = 2\Delta xy_1 - 2\Delta yx_n - c$$

where $c = \Delta x(2b + 1)$ and

$$\omega_{n+1} = 2\Delta xy_0 - 2\Delta yx_{n+1} - c$$

Therefore ω_{n+1} may be derived from ω_n since

$$\omega_{n+1} - \omega_n = 2\Delta x(y_0 - y_1) - 2\Delta y$$

since $x_{n+1} = x_n + 1$.

For the first point in a line, ω_1 is calculated from $2\Delta y - \Delta x$. Thereafter, ω_{n+1} is evaluated using

$$\omega_{n+1} = \omega_n + 2\Delta x(y_0 - y_1) - 2\Delta y \tag{5.6}$$

The sign of this expression determines whether the upper or lower pixel is plotted.

It can be seen that multiplication by an integer and addition and subtraction operations are required, and these operations are easily performed by digital processors.

References

1 Tanenbaum, *Structured Computer Organisation*, Prentice-Hall, Englewood Cliffs, NJ (1984)
2 IGES/PDES Organization (IPO) – Committee: ISO TC 184/SC4/WG1 and WG2 ref.: NCGA, PO Box 3412, McClean, Virginia, USA

Further reading

Angel, E., *Computer Graphics*, Addison-Wesley, Reading, MA (1990)
Arthur (ed.), *CADCAM: Training and Education through the '80s (Proceedings of CAD ED '84)*, Kogan Page, London (1985)
Berk, *Computer Aided Design and Analysis for Engineers*, Blackwell Scientific, Oxford (1988)
Bertoline, *Fundamentals of CAD*, Delmar Publishing Inc. (1985)
Bielig-Schulz, G. and Shulz, C., *3D Graphics in Pascal*, Wiley, Chichester (1989)
Blauth and Machover, *The CAD/CAM Handbook*, Computervision Corporation, Bedford, MA (1980)
Bono, P., Encarnacao, J. L., Encarnacao, L. M. and Herzner, W. R., *PC Graphics with GKS*, Prentice-Hall, Englewood Cliffs, NJ (1990)

Bowman and Bowman, *Understanding CAD/CAM*, Howard Sams and Co., Indianapolis (1987)

Boyd, A., *Techniques of Interactive Computer Graphics*, Chartwell-Bratt, Bromley (1985)

Bresenham, J. E., 'Algorithm for computer control of digital plotter', *IBM Systems Journal*, **4**, 25–30 (1965)

Burger, P. and Gillies, D., *Interactive Computer Graphics*, Addison-Wesley, Reading, MA (1989)

Chang and Wysk, *An Introduction to Automated Process Planning Systems*, Prentice-Hall, Englewood Cliffs, NJ (1985)

Earnshaw, Parslow and Woodwark, *Geometric Modelling and Computer Graphics, techniques and applications*, Gower Technical Press, Aldershot (1987)

Farin, *Curves and Surfaces for Computer-Aided Geometric Design – A Practical Guide*, Academic Press, San Diego (1988)

Foley, J. D., van Dam, A., Feiner, S. K. and Hughes, J. F., *Computer Graphics*, 2nd edn, Addison-Wesley, Reading, MA (1990)

Gerlach, *Transition to CADD*, McGraw-Hill, New York (1987)

Groover and Zimmers, *CAD/CAM: Computer Aided Design and Manufacturing*, Prentice-Hall, Englewood Cliffs, NJ (1984)

Haigh, *An Introduction to Computer Aided Design and Manufacture*, Blackwell Scientific, Oxford (1985)

Hawkes, *The CADCAM Process*, Pitman, London (1988)

Hearn, D. and Baker, M. P., *Computer Graphics*, Prentice-Hall, Englewood Cliffs, NJ (1986)

Hewitt, T., Howard, T., Hubbold, R. and Wyrwas, K., *A Practical Introduction to PHIGS*, Addison-Wesley, Reading, MA (1990)

Hoffmann, *Geometric and Solid Modelling – An Introduction*, Morgan Kaufmann, San Mateo, CA (1989)

Ingham, *CAD Systems in Mechanical and Production Engineering*, Heinemann Newnes, Oxford (1989)

Kingslake, R., *An Introductory Course in Computer Graphics*, Chartwell-Bratt, Bromley (1986)

Laflin, S., *Two-Dimensional Computer Graphics*, Chartwell-Bratt, Bromley (1987)

Machover, *The C4 Handbook – CAD CAM CAE CIM*, Tab Books, PA (1989)

Mair, G. M., *Industrial Robotics*, Prentice-Hall International (UK) Ltd (1988)

Majchvzak, Chang *et al.*, *Human Aspects of Computer Aided Design*, Taylor & Francis, London (1987)

Port, *Computer Aided Design for Construction*, Collins, London (1984)

Rooney and Steadman, *Principles of Computer Aided Design*, Pitman/Open University (1987)

Voisinet, *Introduction to CAD*, McGraw-Hill, New York (1986)

6

Design standards

Martin Hodskinson
(Section 6.1)

Roy D. Cullum
(Sections 6.2–6.6)

Contents

6.1 Standardization in design 6/3
    6.1.1 Introduction 6/3
    6.1.2 Modular design 6/3
    6.1.3 Preferred sizes 6/3
    6.1.4 Levels of standardization 6/3
    6.1.5 Machine details 6/3
    6.1.6 Design procedure 6/3
    6.1.7 Codes of practice 6/3
    6.1.8 Abbreviations used 6/4

6.2 Drawing and graphic communications 6/4
    6.2.1 Drawing references 6/4
    6.2.2 Management and organization of engineering
         information 6/5
    6.2.3 Microfilm and computer technologies 6/5

6.3 Fits, tolerances and limits 6/7
    6.3.1 Conditions of fit 6/7
    6.3.2 Definition of terms 6/7
    6.3.3 Selecting fits 6/13
    6.3.4 Tolerance and dimensioning 6/13
    6.3.5 Surface condition and tolerance 6/14

6.4 Fasteners 6/14
    6.4.1 Automatic insertion of fasteners 6/15
    6.4.2 Joining by part-punching 6/17
    6.4.3 Threaded fasteners 6/17
    6.4.4 Load sensing in bolts 6/17
    6.4.5 Threadlocking 6/19
    6.4.6 Threaded inserts and studs for plastics 6/20
    6.4.7 Ultrasonic welding 6/20
    6.4.8 Adhesive assembly 6/20
    6.4.9 Self-tapping screws 6/20
    6.4.10 Stiff nuts 6/21
    6.4.11 Washers 6/21
    6.4.12 Spring-steel fasteners 6/21
    6.4.13 Plastics fasteners 6/21
    6.4.14 Self-sealing fasteners 6/21
    6.4.15 Rivets 6/21
    6.4.16 Suppliers of fasteners 6/21

6.5 Ergonomic and anthropometric data 6/23

6.6 Total quality – a company culture 6/29
    6.6.1 Useful addresses 6/33

References 6/34

6.1 Standardization in design

6.1.1 Introduction

Standardization in design is the activity of applying known technology and accepted techniques in the generation of new products. This may be interpreted as almost a contradiction in terms: if something is standard then there is little left to design. On the other hand, if standardization is seen as a means to promote communication in design and manufacture then its usefulness is clearer: time is not wasted in redesigning the wheel. Again, if standardization is seen as providing targets that products must attain then we have criteria for acceptance in performance and quality: you know what you are getting.

These points illustrate the difficulty in understanding the role of standardization within industry at large, and why it is too often ignored by designers and management. One does not want to be constrained. In fact, the converse is usually true, that by adopting standards in an appropriate manner, the designer is freed from many detail decisions that would otherwise hinder the overall scheme. It has to be acknowledged that some standards are inherently retrospective. That is, a standardized design method or procedure is bound to be based on past practice, and in some circumstances this may inhibit flexibility in adopting new methods. But to counteract this, the designer may well contribute to updating and development of new standards. Indeed, the standards organizations have development groups and committee structures for this very purpose.

6.1.2 Modular design

Standardization in design may be applied by the idea of modularization. That is, a range of products of related type or size may be designed (or redesigned) so that complex assemblies may be made up from a few simple elements or modules. For example, a vast range of electric motors and gearboxes can be made from a few each of motors, bases, flanges, gear cases, gears, shafts and extensions.

This approach applies particularly well to complex, low-volume production of products where each customer wants his or her own version or specification. The economics of individually designing for each customer would be prohibitive but the design of a few well-chosen modules would cover the majority of requirements. Other examples where modular design may be applied are: overhead travelling cranes, water turbines, hydraulic cylinders, machine tools (BS 3884: 1974).

6.1.3 Preferred sizes

Related to the modular design concept is the idea of preferred sizes. In any range of products there is usually some characteristic feature such as size, capacity, speed, power. It is observable that an appropriate series of numbers can cover a large proportion of requirements. For many practical situations the geometric series known as the Renard Series may be used (BS 2045: 1965 = ISO 3). The key point in application is the product-characteristic feature to which the Series is applied. This application of standardization is one of the best ways of promoting economy in manufacture by variety reduction. it also facilitates manufacture by sub-contracting and eases the problems of spares availability for maintenance. A further discussion is given in reference 3.

6.1.4 Levels of standardization

Standardization can operate at different levels: company, group, national, international, worldwide. It can be applied to many aspects of the activities of design and manufacture, e.g. terminology and communication, dimensions and sizes, testing and analysis, performance and quality. It is a broad-ranging subject and standardization will be exemplified below in further detail but only in so far as it serves the design function. The descriptions are drawn largely from experience with British Standards, but at the time of writing (1988) considerable efforts are continuing to harmonize BSI activities with European counterparts. Indeed, many British standards are entirely compatible with ISO standards and are listed as such in the BSI Catalogue which is published annually.

6.1.5 Machine details

Standard proportions are laid down and specified precisely with dimensions and preferred sizes. This is the most obvious use of standardization in design and is employed widely, almost without conscious effort.

6.1.5.1 Examples

BS 3692: 1967 Metric nuts bolts and screws (ISO 4759)
BS 1486: 1982 Lubricating nipples
BS 4235: 1986 Parallel and taper keys for shafts (ISO 774)
BS 6267: 1982 Rolling bearing boundary dimensions (ISO 15) (also BS 292: 1982 for more detailed dimensional specifications)
BS 1399: 1972 Seals, rotary shaft, lip. Shaft and housing dimensions
BS 3790: 1981 Belts – vee section. Dimensions and rating (ISO 155)
BS 11: Railway rails (ISO 5003)

6.1.6 Design procedure

A procedure, algorithm or method is described for application in circumstances where long experience has established satisfactory results. This is often very useful to the designer as a starting point but evolution beyond the standards is very probable in particular industries. Such standards are among those most likely to be influenced and developed by designers themselves. They may, however, be adopted as part of a commercial contract and thereby promote confidence in safety and reliability.

6.1.6.1 Examples

BS 436: 1986 Spur gears, power capacity (ISO 6336)
BS 545: 1982 Bevel gears, power capacity (ISO 677, 678)
BS 721: 1983 Worm gears, power capacity
(note the recent revisions of these very well-established standards)
BS 2573: 1983 Stresses in crane structures (ISO 4301)
BS 1726: 1964 Helical coil springs
BS 4687: 1984 Roller chain drives (ISO 1275)
BS 1134: 1972 Surface finish (ISO 468)
BS 5078: 1974 Jigs and fixtures
BS 5500: 1985 Unfired welded pressure vessels (ISO 2694)

6.1.7 Codes of practice

Methods for design, manufacture and testing are recommended in these types of standard, and, like 'design procedures' described above, they may form the basis of a commercial contract. Again, they are codes that may or may not be followed entirely but may be regarded as facilitating effective communication between vendor and purchaser or any other interested parties.

6.1.7.1 Design: Examples

BS 5070: 1974 Graphical symbols and diagrams
BS 308: Part 1: 1984 Engineering drawing practice (ISO 128):
BS 308: Part 2: 1985 Dimensioning (ISO 129) (see also reference 1)
BS 4500: 1969 Limits and fits (ISO 286)
BS 5000: Part 10: 1978 Induction motors (CENELEC HD231)
BS Au 154: 1989 Hydraulic trolley jacks
BS CP117: Part 1: 1965 Simply supported beams
BS 4618: 1970 Presentation of plastics design data

6.1.7.2 Manufacture: Examples

BS 1134: 1972 Surface texture assessment (ISO 468)
BS 5078: 1974 Jig and fixture components
BS 5750: 1987 Quality assurance system (ISO 9000) (see also reference 2)
BS 970: 1983 Steel material composition (ISO 683)
BS 6323: Parts 1–4 Steel tubes, seamless
BS 4656: Parts 1–34 Accuracy of machine tools (ISO various) (e.g. Part 34: 1985 Power presses = ISO 6899)
BS 4437: 1969 Hardenability test (Jominy) (ISO 642)
BS 6679: 1985 Injection moulding machine safety (EN 201)

6.1.8 Abbreviations used

BS   British Standard
BSI British Standards Institution
Au   Automobile Series (BS)
ISO International Standards Organization
EN   European Normalen
CP   Code of Practice

Notes: 1. Some standards are divided into separately published parts. Details are given in the BS Catalogue.
       2. Many colleges, polytechnics, universities and public libraries hold sets of standards. A list is given in the BS Catalogue.

6.2 Drawing and graphic communications

Since the days of cave dwellers, drawings have been a prime means of communications and today, with all the latest hi-tech innovations, drawings and graphics still hold a pre-eminent position. The needs are the same – all that has changed is the methods by which drawings are made, stored, retrieved and used. Never was a saying more true than 'One picture is worth a thousand words'.

In the world of engineering, drawings are absolutely essential. They may first be sketches arising from customers' needs or competitive designs but, from several sources, ideas can be given a form around which discussions can take place as to the idea's marketability, usefulness, etc. From this point, a component design starts to emerge. It will undergo many changes to satisfy the demands of various departments, e.g. production, servicing, installation, stress analysis, marketing, etc. All these changes will be documented, particularly if the company is adhering to the BS 5750 quality assurance code.

Because we see in a three-dimensional way, perspective and isometric three-dimensional drawings always provide a quicker appreciation of a design than the more familiar plans and elevations. Perspective and isometric drawings are, however, more difficult and time consuming to produce, require special skills and do not lend themselves to dimensioning. Thus the communication of an idea to the shopfloor for manufacture has generally been through the medium of plans and elevations of the item.

In preparing such drawings, dimensions should always be taken from datum lines which, ideally, should be coincident with the hardware itself rather than being a mere line in space. Using such a datum as a reference avoids a dimensional build-up of accumulated errors.

In the late nineteenth and early twentieth centuries, when life progressed at a more leisurely pace, engineering drawings and architectural plans were often works of art. Parts were shaded and coloured. This practice died out in the 1920s, although even today many architects' drawings not only use shading and colour but widely employ isometrics and perspectives. However, the audience for these drawings is rather different to that viewing engineering drawings. Architects' ideas and proposals are largely for public consumption and will be read by lay people as well as by builders and other planners. The public are often bemused by plans and elevations, particularly when some practices use third-angle projection and others first-angle projection.

Engineering drawings are mainly read by craftsmen who are, however, well trained in reading such drawings. A notable exception was during the Second World War, when many engineering companies found it necessary to include isometric or perspective drawings on the plans to give a visual appreciation of the part in question to the wartime workers on the shopfloor, many of whom had no knowledge or training of either mechanical engineering or drawings.

While a drawing will detail the geometry of a part and its method of assembly it cannot define graphically such things as finish, smoothness or material, although there are standard symbols which give a measure of surface roughness. Nevertheless, for these, annotations are required. Text and tables are often included on drawings and, together with drawn lines, complete a total engineering drawing.

Design is essentially a 'doing' activity and design skills develop through practice. Design learning needs a structured approach aimed at building up confidence and competence with definite educational objectives at each phase. Much of the first year at college is spent on learning drawing skills such as orthographic projection, sketching, tolerances and assembly drawings. These are essential for later work, and much emphasis is placed on the fact that drawings are the medium used by engineers to communicate ideas.

Ideally, a drawing should transmit from one person to another every aspect of the part – not only its shape but also what is is for and how it should be made and assembled.

6.2.1 Drawing references

Not the least important aspect is how the parts are referenced. It may seem a simple enough task to apply a number to a drawing or part and listing it in a register, but subsequent identification may prove more of a problem. Therefore specific numbering systems are introduced. Depending on the users' requirements, systems can be simple or complicated. For instance, every drawing of a part belonging to one complete assembly may be given a prefix, either character or number (for example, A/1234, where A defines the particular assembly and 1234 the individual part). This means that if a spare is required, the first sort (A) in the drawing register defines a particular assembly while the number will locate the individual part. This can be developed further by taking, say, the first two digits of the number 1234 (i.e. 12) and stating that all parts bearing the first two digits are made from flat metal sheet as distinct from, say, 13 for castings, 14 for plastics parts and so on. The original part A/1234 would thus be defined as belonging to a particular assembly (A) and would be seen as a part made from flat metal sheet (12) with the individual part number of 34. This can be further enhanced (or complicated) by the addition of a modification number or character which introduced the part or represents the year of manufacture and/or a part modified in shape or dimension. The whole purpose of the system is to locate the part number and/or its drawing from a part in service which may be damaged or overpainted, eliminating the part number.

Spares lists often carry exploded drawings in which a particular assembly is shown as a set of parts approximating in position to the assembled location. All these ploys help to identify any particular part whose title (for example, plate, angle, strut) is usually insufficient to positively identify the part. It is therefore obvious that, when preparing drawings, some thought should be given to the title. This should, as far as possible, within the limits of space, be brief but also describe the part such that its position and actual duty is defined.

Identification of parts is not only desirable for new designs but sometimes it is worthwhile being able to recall a part from a previously made assembly to see if it can be used in a new design. If this were possible, there would be several advantages:

1. The drawing would exist so that a redraw would be unnecessary;
2. There should be a history of the part showing the reason for its introduction and any modifications and why these were necessary, following perhaps, a malfunction in service;
3. A manufacturing routine would exist for the production of the part; and
4. Jigs and fixtures would also exist.

It may well be that in order to preserve the drawing-identification system the part may be given a new number and called a new part. But it is always an advantage to annotate the drawings so that their origins are preserved. It is more than possible that, although these parts were originally identical, subsequent modifications may have led to an entirely different part which could lead to confusion if they both bore the same part number.

Several systems have been devised to store and retrieve information on like parts. These often rely on comparative geometries (for example, cylindrical shapes where the aspect ratio of length to diameter is large). Another example would be flat plates where thickness or surface shape/area are a criterion. Another way of segregating the parts would be by use; for example, (a) rotating shafts, bearings and supports, (b) structural channels, etc.

Each company will have to work out its own philosophy as regards a drawing numbering system. Obviously, a company whose product range is small and whose subsequent models still have a strong family relation to earlier models will have a totally different system to one with a wide range of differing products.

6.2.2 Management and organization of engineering information

Engineering organization is concerned with the relationship between work and the people who do it. It varies greatly in complexity, depending on the range of products, whether, for example, they are aircraft or lawnmowers. However, each company must have facilities for conceiving a design and developing and then producing it. The prime task of engineering management is to create and sustain such an organization and direct it to the achievement of the declared objective.

Behavioural science focuses attention on the nature of individual and group interaction in an organization; that is, how people can work together in harmony. It emphasizes people rather than jobs, and underlines the simple truth that every organization must solve the problem of relationships among its staff if an efficient working environment is to be realized. Argysis[4] has stated:

> The individual and the organization are living organisms each with its own strategy for survival and growth. The individual's strategy for existence is, at crucial points, antagonistic to the strategy that guides the formal organization. This may lead to continual conflict between the individual and the organization. The conflict, however, can be a source for growth as well as a stimulant for disintegration.

Management has to solve problems within the context of the organization. The individual cannot usually work without a relevant organization structure, and the organization, in turn, depends almost entirely on the individual. However, their interests differ, and the manager's job is to develop a properly balanced interaction of individual needs and organizational demands.

Engineering design is a corporate activity, involving teams of people whose job it is to provide a prescription of what is to be produced. How people interact can be determined by the shape of the organization.

The resources of design are not like production facilities. In a machine shop it is possible to programme work for optimum utilization in accordance with machine capabilities. Designers, however, cannot be so programmed and their performance, for many valid reasons, is often unpredictable. Nevertheless, a design organization is expensive, and it is essential to study its function in the same way as machine functions.

In design and development organizations a careful balance between order and flexibility has to be obtained. Should, for example, a design office be centralized for several production groups or decentralized? Where there is little change, formal policies, procedures and rules to obviate communication difficulties and aid designers would be preferable. Where change is more or less continuous, a more dynamic grouping or task forces may work better in relation to specific jobs.

Over the years, considerable research has been undertaken (1) to find out what information designers want and how it should be best presented and (2) to develop convenient methods by which relevant information and data can be transferred to others. Several universities, polytechnics and colleges have been involved in this particular work and, since 1971, the Design Group at Southampton University have set up a series of seminars to discuss these various activities. During the 1960s a number of commercial systems were devised to provide designers with product data. Most fell by the wayside over the years although one, TI Index, not only survived but has been developed to take advantage of today's technology in microfilm and computer-aided storage and retrieval.

6.2.3 Microfilm and computer technologies

In addition to providing input data for a company, these advanced technologies have simplified the handling, storage and retrieval of engineering drawings and made possible the implementation of control systems for the issue of drawings.

Apart from microfilm leading to a reduction in the number of drawings being copied, it has contributed towards the continual process of identifying those areas which require further control and standardization.

The introduction of computers to the design office has considerably altered the role that microfilm can play. No longer need microfilm be looked upon as a stand-alone technology but can now be integrated into computer and manually controlled drawing-office information systems.

The introduction of CAD increased the speed with which design drawings could be completed and gave designers the freedom and ability to produce exploded and sectional views which previously were economically impossible. These extra drawings, which help communication with the shopfloor, also added to the problem of drawing storage. There was also the risk of drawing losses, not only through the normal risks of fire and theft but also through a computer crash or accidental erasure of storage tape and disk.

Through the years, a growing confidence in the permanence of microfilm records and widespread acceptance has not entirely overcome fears on how to modify a drawing held only on a 35 mm microfilm or how to produce true to scale drawings. However, the use of plain paper print technology and polyester film has overcome most of these problems. To meet the difficulties of controlling the sheer bulk in microfilm libraries a computerized drawing registry offers an engineer quick access to all the data associated with a drawing. Drawing number, issue date, levels of modification, source of drawing, application and schedule of materials can all be made available. The automatic marriage of a computer-held index of information and the physical microfilm image is now possible. The advantage of computer-held drawing graphics has been proven in CAD systems and now hardcopy and microfilm drawings can be incorporated into a CAD system. In essence, this allows conversion of hardcopy and microfilm into a raster digital format. Images can be stored on either magnetic or optical systems.

The days have gone when designers could remain isolated from manufacturers and confine their activities to creating what they considered the best possible designs without thought for the subsequent conversion of their designs into realizable products. Moreover, with the gradual replacement of paper drawings by digitized data, it is important that all concerned should have the means of exchanging the relevant data, which, in turn, means that all parties need compatible systems that intercommunicate.

Full-size CAD-produced designs can be transferred to microfilm either by conventional film camera or by using a microfilm plotter which reads the information from disk or tape outputting directly onto microfilm using laser technology. Alternatively, the cathode ray tube (CRT) may be used to imprint the image on film. For those wanting the ultimate quality, the microfilm plotter must be the first choice.

Microfilm originally came to be seen as an archival activity and as a positive insurance against loss of drawings. Microfilm cards can now be titled, indexed and have distribution data added by computer printing, optical character recognition (OCR) printing, bar codes or punched holes entered direct from the camera control panel. Information from the drawing can be fed directly from this keyboard into the design office data-processing system. As revised cards are issued, the computer can automatically update and re-issue all scheduling and listing. Today's microfilm camera can be upgraded to an electronic information management tool.

Microfilm, in its many forms, is a practical and economical technology to install within a drawing office system. Looking to the future, it is not without interest that an ultra-powerful electron microscope with a beam diameter of only 5 Å has been recently commissioned. Funded by SERC and housed at Liverpool University, the machine is capable of writing the entire contents of the 29 volumes of *Encyclopedia Britannica* on the head of a pin. Each of the dots making up a character is only 10 nm in diameter. Some exciting applications exist, one of which is the possibility of storing information by drilling an array of holes in a given pattern in a manner similar to that used for piano rolls which stored music to be read by a pianola. Information could be stored at densities at least one thousand times greater than in the latest computer storage device.

Another area for future developments can be seen from a recent contract won by SD-Scion to advise the European Space Agency (ESA) on the use of expert systems technology. The contract involves investigating how expert systems can help ESA's European Space Research Institute (ESRIN) in information retrieval. The study will also investigate the needs of ESA's new information systems to be used for project documentation handling, spacecraft payload data and other information. The study will involve several technologies in addition to expert systems, on-line data retrieval, intelligent access to databases, distributed knowledge bases and the use of hypertext to browse through text files.

It was, however, inevitable that with the proliferation of software and hardware, companies each would go their own way for various reasons and would finish up with systems that were incapable of communicating with each other without the use of some form of translator. Standards now coming into use should prevent the same degree of proliferation in the future, but for those systems which do exist (and many of them are highly efficient in their particular sphere), a translator is required. Possibly the best-known data exchange format is IGES (Initial Graphics Exchange Specification), which is a means of transferring data between two incompatible CAD systems by creating an intermediate neutral file that contains elements common to both systems. Data are fed from one system into IGES file and subsequently from IGES to the output translator. If a particular CAD system is to exchange data it needs to have software interfaces for reading (Pre-processor) and writing (Post-processor).

Studies show that designers spend an average of 30–50% of their time on documentation plus 50–70% on seeking relevant information, which leaves little time for creative design activities. These figures reveal an obvious need for tools to help manage product development by managing the documents that control it. Engineering processes in many industries involve interaction among many engineers and even among several divisions or even companies. Engineers must therefore cooperate in creating product concepts, proposals, requirements and specifications.

Tools to support these processes need to be readily available to engineers and be able to incorporate text, graphics and tabular data from an engineering database. To support evolution or change, these tools also need to be able to simplify the process while dealing with formal and informal controls.

An ideal documentation and document change control system would provide true-to-type pagination software with a full set of automatically maintained features as well as the tools to manage the evolution of the documents. The system should run on general-purpose engineering workstations and support very large team documents by demand, paging the latest version of reference text or graphics from anywhere on the network.

It is abundantly clear that the ability to manage change has become crucially important in developing a product and managing it throughout its life cycle. Providing the necessary support with accurate and reliable change-management tools paves the way to better products and product management within the company, which leads to increased competitiveness in the industrial marketplace.

6.3 Fits, tolerances and limits

The successful functioning of any assembly depends to a large degree on the interrelationship of the individual items. Whether machined or fabricated, the amounts by which the sizes of the items can deviate from the ideal or norm must be stated on the drawings so that the shopfloor can know to what tolerances they are expected to work.

In this respect the designer has a great responsibility, since tighter than necessary tolerances can escalate costs while too relaxed tolerances can mean difficulties in assembly and in the overall efficiency of the structure or machine. The degree of tolerance is also strongly influenced by surface finish which, like dimensional limits, can escalate the costs if very fine finishes are called for unnecessarily. Each design has to be considered on its merits. For instance, the bore of a cylinder or valve body may require a high degree of finish if it is to remain leakproof and not unduly wear a soft rubber or plastics seal.

With a shaft revolving in a bearing the clearances are more important than the surface finish, since there must be sufficient clearance to maintain an oil film but not so much that radial float is present. The surface finish is less important since, in most cases, the shaft is revolving on a film of oil and any small asperities can be catered for within the thickness of the oil film. While surface finish and dimensional tolerances are interrelated they can, in an operational sense, be considered independently.

6.3.1 Conditions of fit

Using a shaft and a housing as a practical example, there are five broad conditions of fit:

1. Running fit
2. Push fit
3. Driving fit
4. Force fit
5. Shrinkage fit

With the last three it is obvious that the diameter of the shaft exceeds that of the hole but by different amounts. With the first two, the shaft must be less than the hole diameter. In all cases the crucial question is, by how much?

For example, with a shaft of a nominal diameter of 50 mm, a machining tolerance may be given of +0.00 mm and − 0.05 mm. Its housing may be bored to a maximum diameter of 50.08 mm and a minimum diameter of 50.01 mm. Thus clearances could vary between 0.13 mm maximum and 0.01 mm minimum and consideration has to be given to the effect that these values could have on the efficient running of the machine. The clearances quoted may also be affected by what tolerances can be achieved in manufacture (e.g. the accuracy of a machine tool). It is no use specifying a tolerance of ±0.002 mm if the machine tool can only work to ±0.005 mm.

The ISO system (BS 4500) is designed to provide a comprehensive range of limits and fits for engineering purposes and is based on a series of tolerances to suit all classes of fits. The limits put on the shaft and the bore will determine these conditions and will vary according to the diameter of the shaft and bore. In selecting a fit for a given application, it is necessary to choose an appropriate tolerance for each member so that functional requirements are achieved.

While shafts and holes given in BS 4500 are referred to explicitly, the recommendation can apply equally well to other sections, and the term 'hole' or 'shaft' can be taken as referring to the space contained by or containing two parallel faces or tangent planes.

In the ISO system, 18 tolerance grades are provided and are designated as IT 01, IT 0, IT 1, IT 2, IT 3, . . . , IT 16 (see Table 6.1). The system provides 27 different fundamental deviations for sizes up to and including 500 mm and 14 for larger sizes to give different types of fit ranging from coarse clearance to heavy interference. The values associated with each of these deviations vary with size so as to maintain the same fit characteristics. In each case, the variation is determined by an empirical formula again based on extensive practical investigations.

Tables 6.2 and 6.3 contain the standardized values for the 27 deviations. Each deviation is designated by a letter. Letter (a) represents a large negative deviation (interference fit) while the letter (z) represents a positive deviation (clearance fit). The ISO system provides for the use of both hole-basis and shaft-basis fits. If the hole-basis is used, a series of deviations for shafts is required: if the shaft-basis is employed, a series of deviations for holes is needed. The 27 deviations of the system can be used for shafts or holes. The same letters are used for designations in each case but upper-case (capital) letters designate hole deviations and lower case (small) letters define shaft deviations. Shaft deviations are opposite in sign to those for holes.

6.3.2 Definition of terms

At this stage it is appropriate to define some of the terms used in BS 4500:

Deviation The algebraic difference between a size (actual, maximum, etc.) and the corresponding basic size.
Actual deviation: The algebraic difference between the actual size and the corresponding basic size.
Upper deviation: The algebraic difference between the maximum limit of size and the corresponding basic size. This is designated (ES) for a hole and (es) for a shaft.
Lower deviation: The algebraic difference between the minimum limit of size and the corresponding basic size. This is designated (EI) for a hole and (ei) for a shaft.
Zero line: In a graphical representation of limits and fits, the straight line to which deviations are referred. The zero line is the line of zero deviation and represents the basic size. By convention, when the zero line is drawn horizontally, positive deviations are shown above and negative deviation below it (Figure 6.1).
Tolerance: The difference between the maximum limit of size and the minimum limit of size (the algebraic difference between the upper and lower deviation). The tolerance is an absolute value without sign.
Tolerance zone: In a graphical representation of tolerance the zone between the two lines representing the limits of tolerance and defined by its magnitude (tolerance) and by its position in relation to the zero line.
Fundamental deviation: That one of the two deviations, being the one nearest to the zero line, which is conventionally chosen to define the position of the tolerance zone in relation to the zero line.
Grade of tolerance: In a standardized system of limits and fits a group of tolerances considered as corresponding to the same level of accuracy for all basic sizes.
Standard tolerance: In a standardized system of limits and fits, any tolerance belonging to the system.
Standard tolerance unit: In the ISO system of limits and fits a factor expressed only in terms of the basic size and used as a basis for the determination of the standard tolerances of the

Table 6.1 Standard tolerances of ISO system of limits and fits (courtesy of BSI)

Tolerance unit 0.001 mm

Nominal sizes		Tolerance grades																	
Over	To	IT 01	IT 0	IT 1	IT 2	IT 3	IT 4	IT 5	IT 6[a]	IT 7	IT 8	IT 9	IT 10	IT 11	IT 12	IT 13	IT 14[b]	IT 15[b]	IT 16[b]
mm	mm																		
—	3	0.3	0.5	0.8	1.2	2	3	4	6	10	14	25	40	60	100	140	250	400	600
3	6	0.4	0.6	1	1.5	2.5	4	5	8	12	18	30	48	75	120	180	300	480	750
6	10	0.4	0.6	1	1.5	2.5	4	6	9	15	22	36	58	90	150	220	360	580	900
10	18	0.5	0.8	1.2	2	3	5	8	11	18	27	43	70	110	180	270	430	700	1100
18	30	0.6	1	1.5	2.5	4	6	9	13	21	33	52	84	130	210	330	520	840	1300
30	50	0.6	1	1.5	2.5	4	7	11	16	25	39	62	100	160	250	390	620	1000	1600
50	80	0.8	1.2	2	3	5	8	13	19	30	46	74	120	190	300	460	740	1200	1900
80	120	1	1.5	2.5	4	6	10	15	22	35	54	87	140	220	350	540	870	1400	2200
120	180	1.2	2	3.5	5	8	12	18	25	40	63	100	160	250	400	630	1000	1600	2500
180	250	2	3	4.5	7	10	14	20	29	46	72	115	185	290	460	720	1150	1850	2900
250	315	2.5	4	6	8	12	16	23	32	52	81	130	210	320	520	810	1300	2100	3200
315	400	3	5	7	9	13	18	25	36	57	89	140	230	360	570	890	1400	2300	3600
400	500	4	6	8	10	15	20	27	40	63	97	155	250	400	630	970	1550	2500	4000
500	630	—	—	—	—	—	—	—	44	70	110	175	280	440	700	1100	1750	2800	4400
630	800	—	—	—	—	—	—	—	50	80	125	200	320	500	800	1250	2000	3200	5000
800	1000	—	—	—	—	—	—	—	56	90	140	230	360	560	900	1400	2300	3600	5600
1000	1250	—	—	—	—	—	—	—	66	105	165	260	420	660	1050	1650	2600	4200	6600
1250	1600	—	—	—	—	—	—	—	78	125	195	310	500	780	1250	1950	3100	5000	7800
1600	2000	—	—	—	—	—	—	—	92	150	230	370	600	920	1500	2300	3700	6000	9200
2000	2500	—	—	—	—	—	—	—	110	175	280	440	700	1100	1750	2800	4400	7000	11 000
2500	3150	—	—	—	—	—	—	—	135	210	330	540	860	1350	2100	3300	5400	8600	13 500

[a] Not recommended for fits in sizes above 500 mm.
[b] Not applicable to sizes below 1 mm.

system. (Each tolerance is equal to the product of the value of the standard tolerance unit for the basic size in question by a coefficient corresponding to each grade of tolerance.)

Clearance: The difference between the sizes of the hole and shaft, before assembly, when this difference is positive.

Interference: The magnitude of the difference between the sizes of the hole and the shaft, before assembly, when this difference is negative.

Clearance fit: A fit which always provides a clearance. (The tolerance zone is entirely above that of the shaft.)

Interference fit: A fit which always provides an interference. (The tolerance zone of the hole is entirely below that of the shaft.)

Transition fit: A fit which may provide either a clearance or interference. (Tolerance zones of the hole and shaft overlap.)

Minimum clearance: In a clearance fit the difference between the minimum size of the hole and the maximum size of the shaft.

Maximum clearance: In a clearance or transition fit the difference between the maximum size of the hole and the minimum size of the shaft.

Minimum interference: In an interference fit the magnitude of the (negative) difference between the maximum size of the hole and the minimum size of the hole before assembly.

Maximum interference: In an interference or transitional fit the magnitude of the (negative) difference between the minimum size of the hole and the maximum size of the shaft before assembly.

Table 6.2 Fundamental deviations for shafts (courtesy BSI)

Fundamental deviation	Upper deviation es												Lower deviation ei				
Letter	a^a	b^a	c	cd	d	e	ef	f	fg	g	h	js^b	j			k	
Grade	01 to 16												5–6	7	8	4–7	≤3 >7
Nominal sizes																	
Over / To (mm)																	
– / 3	−270	−140	−60	−34	−20	−14	−10	−6	−4	−2	0		−2	−4	−6	0	0
3 / 6	−270	−140	−70	−46	−30	−20	−14	−10	−6	−4	0		−2	−4	–	+1	0
6 / 10	−280	−150	−80	−56	−40	−25	−18	−13	−8	−5	0		−2	−5	–	+1	0
10 / 14	−290	−150	−95	–	−50	−32	–	−16	–	−6	0		−3	−6	–	+1	0
14 / 18																	
18 / 24	−300	−160	−110	–	−65	−40	–	−20	–	−7	0		−4	−8	–	+2	0
24 / 30																	
30 / 40	−310	−170	−120	–	−80	−50	–	−25	–	−9	0		−6	−10	–	+2	0
40 / 50	−320	−180	−130														
50 / 65	−340	−190	−140	–	−100	−60	–	−30	–	−10	0		−7	−12	–	+2	0
65 / 80	−360	−200	−150														
80 / 100	−380	−220	−170	–	−120	−72	–	−36	–	−12	0	±IT/2	−9	−15	–	+3	0
100 / 120	−410	−240	−180														
120 / 140	−460	−260	−200														
140 / 160	−520	−280	−210	–	−145	−85	–	−43	–	−14	0		−11	−18	–	+3	0
160 / 180	−580	−310	−230														
180 / 200	−660	−340	−240														
200 / 225	−740	−380	−260	–	−170	−100	–	−50	–	−15	0		−13	−21	–	+4	0
225 / 250	−820	−420	−280														
250 / 280	−920	−480	−300	–	−190	−110	–	−56	–	−17	0		−16	−26	–	+4	0
280 / 315	−1050	−540	−330														
315 / 355	−1200	−600	−360	–	−210	−125	–	−62	–	−18	0		−18	−28	–	+4	0
355 / 400	−1350	−680	−400														
400 / 450	−1500	−760	−440	–	−230	−135	–	−68	–	−20	0		−20	−32	–	+5	0
450 / 500	−1650	−840	−480														
Grade	6 to 16																
500 / 630	–	–	–	–	−260	−145	–	−76	–	−22	0		0				
630 / 800	–	–	–	–	−290	−160	–	−80	–	−24	0		0				
800 / 1000	–	–	–	–	−320	−170	–	−86	–	−26	0		0				
1000 / 1250	–	–	–	–	−350	−195	–	−98	–	−28	0	±IT/2	0				
1250 / 1600	–	–	–	–	−390	−220	–	−110	–	−30	0		-0				
2000 / 2500	–	–	–	–	−480	−260	–	−130	–	−34	0		0				
2500 / 3150	–	–	–	–	−520	−290	–	−145	–	−38	0		0				

[a]Not applicable to sizes up to 1 mm.
[b]In grades 7 to 11, the two symmetrical deviations ± IT/2 should be rounded if the IT value in micrometres is an odd value by replacing it by the even value immediately below.

Table 6.2

Fundamental deviation		Upper deviation ei													
Letter		m	n	p	r	s	t	u	v	x	y	z	za	zb	zc
Grade		01 to 16													
Nominal size															
Over	To														
mm	mm														
–	3	+2	+4	+6	+10	+14	–	+18	–	+20	–	+26	+32	+40	+60
3	6	+4	+8	+12	+15	+19	–	+23	–	+28	–	+35	+42	+50	+80
6	10	+6	+10	+15	+19	+23	–	+28	–	+34	–	+42	+52	+67	+97
10	14	+7	+12	+18	+23	+28	–	+33	–	+40	–	+50	+64	+90	+130
14	18								+39	+45	–	+60	+77	+108	+150
18	24	+8	+15	+22	+28	+35	–	+41	+47	+54	+63	+73	+98	+136	+188
24	30						+41	+48	+55	+64	+75	+88	+118	+160	+218
30	40	+9	+17	+26	+34	+43	+48	+60	+68	+80	+94	+112	+148	+200	+274
40	50						+54	+70	+81	+97	+114	+136	+180	+242	+325
50	65	+11	+20	+32	+ 41	+53	+66	+87	+102	+122	+144	+172	+226	+300	+405
65	80				+43	+59	+75	+102	+120	+146	+174	+210	+274	+360	+480
80	100	+13	–23	+37	+51	+71	+91	+124	+146	+178	+214	+258	+335	+445	+585
100	120				+54	+79	+104	+144	+172	+210	+254	+310	+400	+525	+690
120	140	+15	+27	+43	+63	+92	+122	+170	+202	+248	+300	+365	+470	+620	+800
140	160				+65	+100	+134	+190	+228	+280	+340	+415	+535	+700	+900
160	180				+68	+108	+146	+210	+252	+310	+380	+465	+600	+780	+1000
180	200	+17	+31	+50	+77	+122	+166	+236	+284	+350	+425	+520	+670	+880	+1150
200	225				+80	+130	+180	+258	+310	+385	+470	+575	+740	+960	+1250
225	250				+84	+140	+196	+284	+340	+425	+520	+640	+820	+1050	+1350
250	280	+20	+34	+56	+94	+158	+218	+315	+385	+475	+580	+710	+920	+1200	+1550
280	315				+98	+170	+240	+350	+425	+525	+650	+790	+1000	+1300	+1700
315	355	+21	+37	+62	+108	+190	+268	+390	+475	+590	+730	+900	+1150	+1500	+1900
355	400				+144	+208	+294	+435	+530	+660	+820	+1000	+1300	+1650	+2100
400	450	+23	+40	+68	+126	+232	+330	+490	+595	+740	+920	+1100	+1450	+1850	+2400
450	500				+132	+252	+360	+540	+660	+820	+1000	+1250	+1600	+2100	+2600
Grade		6 to 16													
500	560	+26	+44	+78	+150	+280	+400	+600							
560	630				+155	+310	+450	+660							
630	710	+30	+50	+88	+175	+340	+500	+740							
710	800				+185	+380	+560	+840							
800	900	+34	+56	+100	+210	+430	+620	+940							
900	1000				+220	+470	+680	+1050							
1000	1120	+40	+66	+120	+250	+520	+780	+1150							
1120	1250				+260	+580	+840	+1300							
1250	1400	+48	+78	+140	+300	+640	+960	+1450							
1400	1600				+330	+720	+1050	+1600							
1600	1800	+58	+92	+170	+370	+820	+1200	+1850							
1800	2000				+400	+920	+1350	+2000							
2000	2240	+68	+110	+195	+440	+1000	+1500	+2300							
2240	2500				+460	+1100	+1650	+2500							
2500	2800	+76	+135	+240	+550	+1250	+1900	+2900							
2800	3150				+580	+1400	+1200	+3200							

Table 6.3 Fundamental deviations for holes (courtesy BSI)

Fundamental Letter	Lower deviation EI A[a]	B[a]	c	CD	D	E	EF	F	FG	G	H	J[b]	Upper deviation ES J 6	7	8	K ≤8	>8	M ≤8[c]	>8	N ≤8	>8[d]
Grade	01 to 16												6	7	8	≤8	>8	≤8[c]	>8	≤8	>8[d]
Nominal sizes Over To	+	+	+	+	+	+	+	+	+	+	+		+	+	+						
– 3	270	140	60	34	20	14	10	6	4	2	0		2	4	6	0	0	−2	−2	−4	−4
3 6	270	140	70	46	30	20	14	10	6	4	0		5	6	10	−1+Δ	–	−4+Δ	−4	−8+Δ	0
6 10	280	150	80	56	40	25	18	13	8	5	0		5	8	12	−1+Δ	–	−6+Δ	−6	−10+Δ	0
10 14	290	150	95	–	50	32	–	16	–	6	0		6	10	15	−1+Δ	–	−7+Δ	−7	−12+Δ	0
14 18																					
18 24	300	160	110	–	65	40	–	20	–	7	0		8	12	20	−2+Δ	–	−8+Δ	−8	−15+Δ	0
24 30																					
30 40	310	170	120	–	80	50	–	25	–	9	0		10	14	24	−2+Δ	–	−9+Δ	−9	−17+Δ	0
40 50	320	180	130																		
50 60	340	190	140	–	100	60	–	30	–	10	0		13	18	28	−2+Δ	–	−11+Δ	−11	−20+Δ	0
65 80	360	200	150																		
80 100	380	220	170	–	120	72	–	36	–	12	0	±IT/2	16	22	34	−3+Δ	–	−13+Δ	−13	−23+Δ	0
100 120	410	240	180																		
120 140	460	260	200	–	145	85	–	43	–	14	0		18	26	41	−3+Δ	–	−15+Δ	−15	−27+Δ	0
140 160	520	280	210																		
160 180	580	310	230																		
180 200	660	340	240	–	170	100	–	50	–	15	0		22	30	47	−4+Δ	–	−17+Δ	−17	−31+Δ	0
200 225	740	380	260																		
225 250	820	420	280																		
250 280	920	480	300	–	190	110	–	56	–	17	0		25	36	55	−4+Δ	–	−20+Δ	−20	−34+Δ	0
280 315	1050	540	330																		
315 355	1200	600	360	–	210	125	–	62	–	18	0		29	39	60	−4+Δ	–	−21+Δ	−21	−37+Δ	0
335 400	1350	680	400																		
400 450	1500	760	440	–	230	135	–	68	–	20	0		33	43	66	−5+Δ	–	−23+Δ	−23	−40+Δ	0
450 500	1650	840	480																		
Grade	6 to 16																				
500 630	–	–	–	–	260	145	–	76	–	22	0					0		−26		−44	
630 800	–	–	–	–	290	160	–	80	–	24	0					0		−30		−50	
800 1000	–	–	–	–	320	170	–	86	–	26	0					0		−34		−56	
1000 1250	–	–	–	–	350	195	–	98	–	28	0	± IT/2				0		−40		−66	
1250 1600	–	–	–	–	390	220	–	110	–	30	0					0		−48		−78	
1600 2000	–	–	–	–	430	240	–	120	–	32	0					0		−58		−92	
2000 2500	–	–	–	–	480	260	–	130	–	34	0					0		−68		−110	
2500 3150	–	–	–	–	520	290	–	145	–	38	0					0		−76		−135	

[a]Not applicable to sizes up to 1 mm.
[b]In grades 7 to 11, the two symmetrical deviations ± IT/2 should be rounded if the IT value in micrometres in an odd value by replacing it by the even value immediately below.
[c]Special case: for M6, ES = −9 from 250 to 315 (instead of −11).
[d]Not applicable to sizes up to 1 mm.

Table 6.3

| Nominal sizes (mm) | | Fundamental deviation — Upper deviation ES | | | | | | | | | | | | | Values for Δ* (Grades) | | | | | |
Over	To	P to ZC (≤7)	P	R	S	T	U	V	X	Y	Z	ZA	ZB	ZC	3	4	5	6	7	8
–	3	Same deviation as for grades above 7 increased by Δ	6	10	14	–	18	–	20	–	26	32	40	60	0	0	0	0	0	0
3	6		12	15	19	–	23	–	28	–	35	42	50	80	1	1.5	1	3	4	6
6	10		15	19	23	–	28	–	34	–	42	52	67	97	1	1.5	2	3	6	7
10	14		18	23	28	–	33	–	40	–	50	64	90	130	1	2	3	3	7	9
14	18							39	45	–	60	77	108	150						
18	24		22	28	35	–	41	47	54	63	73	98	136	188	1.5	2	3	4	8	12
24	30					41	48	55	64	75	88	118	160	218						
30	40		26	34	43	48	60	68	80	94	112	148	200	274	1.5	3	4	5	9	14
40	50					54	70	81	97	114	136	180	242	325						
50	65		32	41	53	66	87	102	122	144	172	226	300	405	2	3	5	6	11	16
65	80			43	59	75	102	120	146	174	210	274	360	480						
80	100		37	51	71	91	124	146	178	214	258	335	445	585	2	4	5	7	13	19
100	120			54	79	104	144	172	210	254	310	400	525	690						
120	140		43	63	92	122	170	202	248	300	365	470	620	800	3	4	6	7	15	23
140	160			65	100	134	190	228	280	340	415	535	700	900						
160	180			68	108	146	210	252	310	380	465	600	780	1000						
180	200		50	77	122	166	266	284	350	425	520	670	880	1150	3	4	6	9	17	26
200	225			80	130	180	258	310	385	470	575	740	960	1250						
225	250			84	140	196	284	340	425	520	640	820	1050	1350						
250	280		56	94	158	218	315	385	475	580	710	920	1200	1550	4	4	7	9	20	29
280	315			98	170	240	350	425	525	650	790	1000	1300	1700						
315	355		62	108	190	268	390	475	590	730	900	1150	1500	1800	4	5	7	11	21	32
355	400			114	208	294	435	530	660	820	1000	1300	1650	2100						
400	450		68	126	232	330	490	595	740	920	1100	1450	1850	2400	5	5	7	13	23	34
450	500			132	252	360	540	660	820	1000	1250	1600	2100	2600						

Grade 6 to 16

Over	To	P	R	S	T	U
500	560	78	150	280	400	600
560	630		155	310	450	660
630	710	88	175	340	500	740
710	800		185	380	560	840
800	900	100	210	430	620	940
900	1000		220	470	680	1050
1000	1120	120	250	520	780	1150
1120	1250		260	580	840	1300
1250	1400	140	300	640	960	1450
1400	1600		330	720	1050	1600
1600	1800	170	370	820	1200	1850
1800	2000		400	920	1350	2000
2000	2240	195	440	1000	1500	2300
2240	2500		460	1100	1650	2500
2500	2800	240	550	1250	1900	2900
2800	3150		580	1400	2100	3200

* In determining K, M, N up to Grade 8 and P to ZC up to Grade 7, add the Δ value appropriate to the grade as indicated, e.g. for P7 from 18 to 30, Δ = 8 therefore ES = −14.

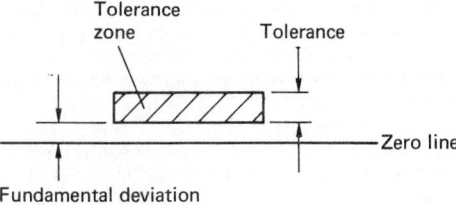

Figure 6.1 Zero line and its meaning

6.3.3 Selecting fits

In a national standard it is impossible to recommend selection of fits which are appropriate to all sections of industry. While the ranges shown will meet many requirements, there will be some sections of industry which will need coarser or finer qualities. The principles around which BS 4500 is founded provides for engineers to determine their own tolerances should this be necessary to suit the particular applications.

A typical fit is illustrated conventionally as in Figure 6.2, where the two rectangles represent the hole and the shaft tolerance zones in magnitude and position with respect to the basic size or zero line. The illustration shows a hole-basis clearance fit but the same convention can be used to represent any type of fit.

The characteristics which determine the nature of a hole-basis fit are the magnitude and sign of the fundamental deviation of the shaft and the magnitudes of the hole and shaft tolerances. When the shaft fundamental deviation is positive in sign, the fit becomes either a transition fit or an interference fit, depending on the magnitude of the fundamental deviation and the tolerances. The converse applies in the case of a shaft-basis fit, where the fundamental deviation of the hole and the magnitude of the hole and shaft tolerances determine its character.

For example, instead of H8 and f7, the qualities of H7-f6 or H9 and f8 could be used. Alternatively, instead of H7-h6, the qualities H8-h7 or H6-h5 could be employed. In all cases, however, the starting point should be the provision of the minimum number of tolerance grades for the basic member of the fit.

BS 4500A contains a range of hole-basis fits derived from the selected hole and shaft tolerances given in BS 4500: Part 1. When using a shaft-basis fit, the equivalents of hole-basis fits are given in BS 4500B. The ISO system is so comprehensive and flexible that it can provide a considerable range of fits based on an almost infinitely variable combination of deviations and tolerances. In practice, however, most normal engineering requirements can be met by a more limited selection of hole and shaft tolerances.

The advantages of standardizing on a limited range cannot be overemphasized. If a user selects three or four standard holes and perhaps 10 standard shafts (or vice versa) they can be combined in numerous ways to provide clearance, transition and interference fits but will be able to use the same range of tools and gauges for all of them.

'Force fit' is a term used when a pin or shaft is forced into a hole of slightly smaller size. As a rule, force fits are restricted to parts of small to medium size whereas shrinkage fits are specially applicable where maximum seizure or accurate knowledge of the stresses induced are required.

The most important point to consider when calculating shrinkage fits is the stress in the housing into which the shaft is shrunk; this stress will depend largely on the shrinkage allowance. If the allowance is excessive, the elastic limit of the

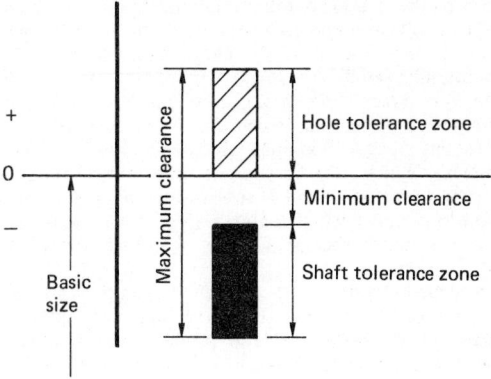

Figure 6.2 A typical fit

material will be exceeded and either permanent set will occur or the housing will fracture or burst. The intensity of the grip will depend on the wall thickness of the housing – the greater the thickness, the greater the grip.

6.3.4 Tolerance and dimensioning

Tolerance (see also Section 6.3.2) is the amount of variation permitted on dimensions or surfaces of parts. The tolerance is equal to the difference between the maximum and minimum limits of any specified dimension. As applied to the fitting of machined parts, the word 'tolerance' means the amount that duplicate parts are allowed to vary in size in connection with manufacturing operations. Tolerance may also be defined as the amount that duplicate parts are permitted to vary in order to secure sufficient accuracy without unnecessary refinement. The designer should always be aware that in manufacture, steps can be taken on the shopfloor to ensure that parts mate. However, in service, the replacement of parts in the field can only be easily effected if interchangeability is considered at the design stage. The degree of tolerance and the way in which parts are dimensioned can profoundly affect interchangeability.

In practice, tolerances should show the permissible amount of variation in the direction that gives the least cause for concern. When a variation in either direction is of equal importance a bilateral tolerance should be given. When a variation in one direction is more dangerous than a variation in the other, a unilateral tolerance should be given in the less dangerous direction.

Where tolerances are required between holes they are usually bilateral, as variations in either direction are usually of equal concern. With shafts for gears the tolerance on shaft centres should be unilateral to avoid the possibility of gears meshing too tightly. A small amount of backlash is likely to be of less consequence.

Only one dimension in the same straight line can be controlled within fixed limits. That is the distance between what would be the cutting surface of the tool and the locating or registering surface of the part being machined. It is therefore important not to locate any point or surface with tolerances from more than one point in the same straight line.

Every part of a mechanism must be located in each plane and every moving part must be given appropriate clearances. Dimensions should be given between those points or surfaces that are essential to hold them in correct relation to each other. This applies particularly to those surfaces in each plane which control the location of other component parts. It is good practice to establish a common locating point in each plane and give, as far as is possible, all dimensions from this common locating point.

Although ordinary limits of size, in themselves, only comprise limiting sizes between which the measured size of the features at any cross-section must fall, conventions have been established by which such limits exert some control on geometric form. In selecting limit of sizes for mating features it is important to consider whether this degree of control is adequate or whether a specific form of tolerance should be applied.

The recommendations in BS 4500 enable limits of size for mating features to be derived from a standard series of tolerances and deviations in order to provide various types of fit. However, the mating features of a fit are composed of surfaces, and the surface texture and geometric form of these surfaces frequently have a considerable bearing on satisfactory functioning. Surface texture is particularly important in the case of precise fits involving relative movement.

6.3.5 Surface condition and tolerance (see also Chapter 9)

All surfaces which have been finished by turning, milling, grinding, honing or other means consist of minute irregularities. These may be compounded by waviness attributable to vibration and machine deflections. There are a number of ways by which surface roughness can be measured and expressed numerically but the two best known are (1) root mean square (rms) value and (2) centreline average (cla). The latter has been brought into line with ISO recommendations and is now known as the R_a value (arithmetical mean deviation).

With the rms designation the heights and depth of the irregularities are measured from a central reference line which is at the mean of the heights and valleys. These measurements are taken at equal intervals and if the sum of such measurements is divided by the number of points at which the measurements are made, the resulting average can be used for comparative purposes:

$$\text{rms} = \sqrt{\frac{a + b + c}{m}} \ . \ . \ . \ .$$

Table 6.4 gives an indication of the sort of surface roughness values (R_a) to be expected from various manufacturing processes. These values should not be placed on the drawings without taking into account all other relevant factors such as load, lubrication, speed, temperature, types of materials, direction of movement and last, but not least, cost. To specify a surface texture value is virtually an instruction to the shopfloor with ultimate inspection routines. To specify an unnecessarily smooth surface can be economically unsound. Figure 6.3 shows the increase in production time needed to achieve smoother surfaces and this would, in cost terms, be increased also by the higher capital costs of equipment and higher running costs.

Control of surface texture is generally instituted to secure a surface texture of a known type and roughness value which experience has proved to be most suitable to give that particular part a long life, fatigue resistance, maximum efficiency and functional interchangeability at lowest cost together with attendant benefits such as reduction in vibration wear and power consumption.

BS 1134 deals with the assessment of surface texture and is in two parts. Part 1 covers method and instrumentation and Part 2 gives general information and guidance regarding the application of the method of Part 1.

Two schools of thought exist regarding control of surface roughness. One view is that all surfaces should be specified and the other is to apply control only to essential parts, leaving the rest to the standard and, hopefully, good practices of the company. It is not the function of BS 1134 to advocate either and to do so would be unmerited. However, it is important to bear in mind that quoting any form of surface control mut be done with the full understanding of the definition and attributes of a surface.

6.4 Fasteners

Fasteners embrace a wide range of common and specialized components, but the emphasis for the designer of today is on fastening systems largely because economy in production demands overall consideration of the assembly process. Fasteners inserted by hand are labour intensive and therefore make the assembly expensive.

Designers have to ask themselves which fastening system will serve the design purpose at the lowest overall cost. Does the design warrant fasteners and, if so, is it for assembly, maintenance, transport, etc.? The reason for a fastening device will lead to the choice of system, ranging from welding through adhesives, bolts, rivets, etc. to highly specialized fastening devices. Despite the growing popularity of structural adhesives, mechanical fasteners are still widely used but it must be emphasized again that the cost of inserting a fastener is likely to outweigh that of the individual fastener. Therefore the designer must be convinced that a mechanical fastener is really the best answer for the job in hand. At this point it would be convenient to enumerate questions such as:

1. Is a fastener really necessary?
2. Will the minimum number of fasteners be consistent with reliability and safety as well as economical in production?
3. Will the fastener specified perform its task efficiently?
4. Will it be simple to install and be capable of being placed by automatic methods?
5. Will the fastener have to be removed during service for maintenance and repair and will it be readily accessible with standard tools?
6. If the fastener is to be removed during service, is it of a type that can be safely used again?
7. Will the fastener material be compatible with that of the products?
8. Bearing in mind the additional cost involved, will it be necessary to specify a special fastener instead of a standard one?

There are other questions that could be asked and these depend on the type of assembly being considered, but the questions given above are basic to most designs.

The choice of fastener systems currently available is bewilderingly wide, but it is the designer's responsibility to select from the right category, to assess performance from the suppliers' data, and to consider in-place costs. Fasteners that fail in service are not only unreliable but uneconomic. The conditions under which the fastener operates must be known and the selected fastener must resist these conditions, which could include extremes of temperatures, corrosive environments, vibration and impact loads.

Table 6.4 Surface roughness values produced by common production process and materials (courtesy BSI)

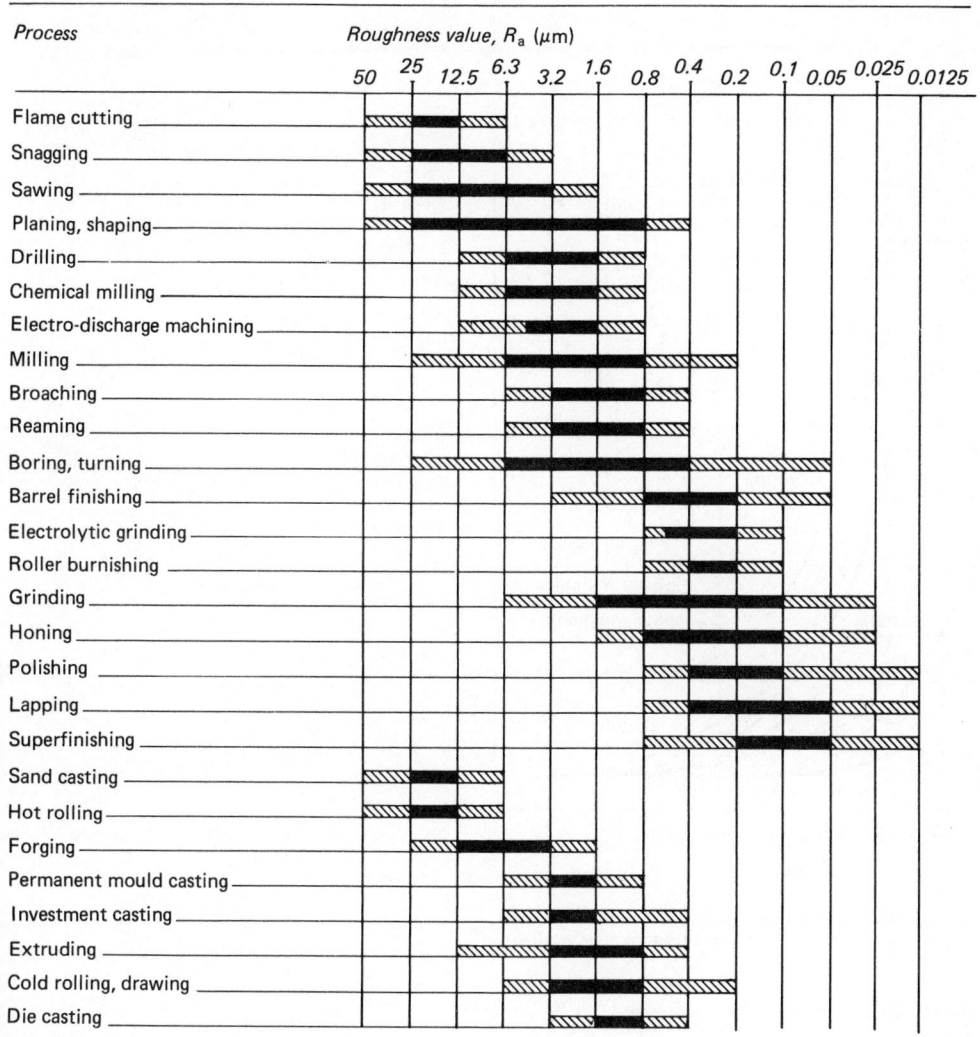

Key

■■■■ Average application

▧▧▧ Less frequent application

Note: The ranges shown above are typical of the processes listed. Higher or lower values may be obtained under special conditions.

6.4.1 Automatic insertion of fasteners

Depending on specific applications, most fasteners in use today can be installed automatically and since some 80% of the total cost of fastening is on-assembly cost, a mere saving of 10% in assembly costs is more significant than a 40% saving in piece part costs. Nevertheless, 80% of manufacturers assembling products with automated or robotic screwdriving equipment claim to have difficulties with their fasteners. This is the finding of a recent research survey carried out for European Industrial Services (EIS), who manufacture Nettlefolds screws. Faulty fasteners brought machinery to a halt and, in a significant number of instances, caused damage to machine or products.

For 23% of all users of automated equipment the problem is a continuing one which production engineers appear prepared to live with either because they are unaware that a solution exists or because they believe that any solution must be too expensive. It is against this background that European Industrial Services introduced, in 1987, a fault-free product known as Nettlefolds Gold Seal to the robot-user market. However, there appear to be just as many problems being experienced by people with automatic assembly equipment. Now, EIS are offering the Gold Seal products for use with automated screwdriving equipment.

Virtually all the screw products used in robotic and auto-mated screwdriving are hardened, mainly fastening metal to

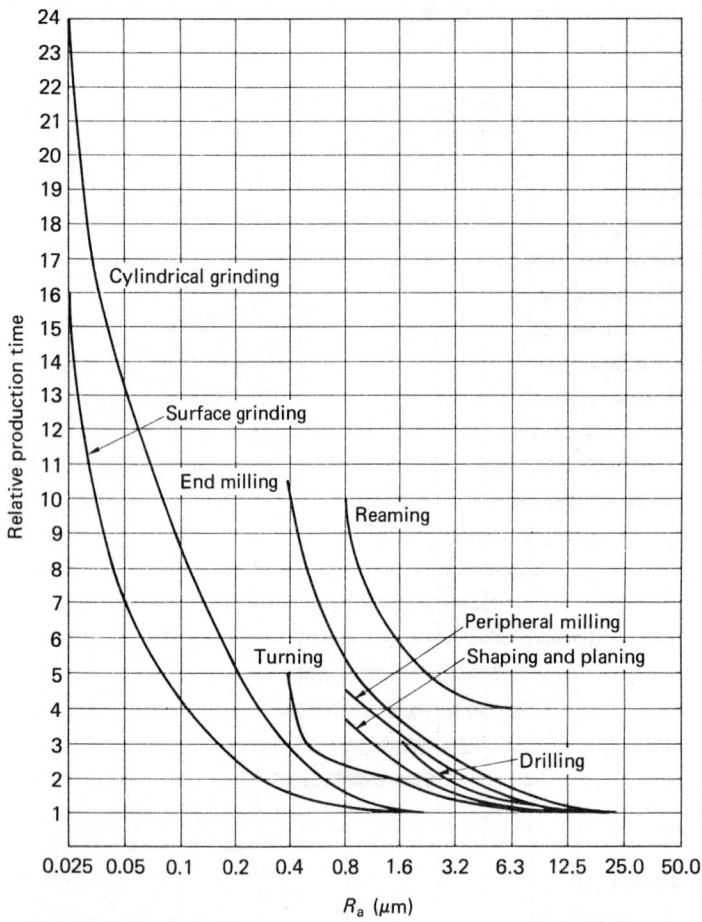

Figure 6.3 Relative production times for commonly used processes and materials (courtesy BSI)

Example: If a given area of surface is surface ground to 3.2 μm R_a taking approximately 1 minute, then to achieve 0.2 μm would take approximatly 2.5 minutes.

metal but with a growing proportion joining plastics or plastics to metal. The Gold Seal error-scanned products therefore cater for these requirements with a variety of self-tapping screws for metal applications and Polymate, a specially designed EIS patented screw for thermoplastics.

The Pemserter is a high-volume automated fastener insertion press from PEM International for inserting any of the range of bushes in the PEM range. The press has an intelligent main cylinder containing an optical encoder which transmits absolute ram position to a microprocessor-based logic control system. Operating and safety windows coupled with an end-of-ram sensor allow the machine to perform at its maximum rate while providing operator and tooling safety.

During a set-up procedure, these safety windows are automatically determined. Any variation from the initial parameters such as the operator removing the anvil, or the presence of a foreign object such as a finger, will cause the ram to immediately stop and return. The company claims that their press achieves new standards in performance, safety and ease of operation.

Another example of a modern assembly tool is the fully automated twin-headed Heatserter capable of installing up to 45 fasteners per minute in plastics mouldings (see Figure 6.4). The new CNC machine from PSM International plc can be pre-programmed in three axes – x, y and z – to cater for plastics mouldings of a wide range of shapes and sizes. Pre-heating circuitry is used to facilitate high-speed post-moulding fastener installation and the equipment features light guarding to provide easy tool changing. The Heatserter can be interfaced with moulding presses to form a link in an automated production line and is capable of installing the full range of PSM proprietary fasteners for thermoplastics.

The problems of automated assembly are not just handling and orientation but are often those of quality and consistency. Design for automated assembly will therefore embrace both component design and the manufacturing process.

Figure 6.4 PSM Automated Systems new CNC Heatserter (courtesy Fox Photography)

6.4.2 Joining by part-punching

A very effective way of fastening sheets together is by part-punching the top sheet into the bottom one. Material is partly cut from the top sheet and pushed through the lower sheet. The lower sheet is formed simultaneously so that an overlap between top and bottom sheet occurs. As the partial cut is connected to the upper sheet by webs, a join of considerable strength is produced.

The Trumpf TF300 power fastener has been developed to cut and form the sheets at the rate of two strokes per second and consists of a three-part tool set comprising punch, stripper and die. Developed for static or portable use, an optional coordinate table ensures the accurate positioning of joints. The power tool is particularly useful where box-shape enclosures are being assembled or where flat strips are to be joined together.

6.4.3 Threaded fasteners

Where threaded parts are concerned, designers have been bedevilled by the numerous thread forms that have been available in the past, and many are still with us today, in spite of the UK's declared policy to recommend the adoption of the ISO metric screw thread system. Existing thread systems include Whitworth, British Standard Fine (BSF), British Association (BA) as well as a range of pipe threads, gas threads, oil industry threads and fire service threads. However, there has been a part change typified by the use of ISO inch series of threads (Unified Coarse and Unified Fine). Nevertheless, wherever possible, new designs and updates should call for ISO metric thread systems.

The choice between Unified Coarse and Unified Fine threads depends on the complete understanding of the assembly in which screws and bolts are used. In general terms, the coarse thread (UNC) can fulfil most design requirements. One small practical advantage with the fine threads (UNF) is when they are used in conjunction with a slotted nut and split pin. To align the slot in the nut with the hole in the bolt, less torque is required with UNF for a given bolt tension.

Where bolts are used in tapped holes, a failure under tension loads should be by breakage at the core of the bolt thread rather than by stripping of the thread. Where bolts are in shear, designers should ensure that the loads are carried on the bolt shank rather than on the threaded portion. Another important aspect is the clearance of the bolt in the hole, particularly when several bolts appear in a row. The tolerance for dimensions must be carefully considered, otherwise undue strain will be put upon a single bolt rather than being equally distributed among them all.

With nuts and bolts there will always be a problem of making certain that, in field replacements, the bolt removed is replaced by one of similar strength. Members of the British Industrial Fasteners Federation (BIFF) all produce bolts of high quality and standards, and these can be readily identified from their head markings. Additional markings also indicate tensile strength and thread forms. Figures 6.5–6.7 show the most commonly used head forms for screws, bolts and nuts.

Threaded fasteners in general are used to provide a clamping force between two or more parts and where there is likely to be a need to dismantle at some time in the future, whether for access or for maintenance. Clamping force can be a major problem, particularly when clamping soft materials which may shrink in service and cause loosening of the bolts. In cases like this, locking the nuts and bolts together may not always give a satisfactory answer.

6.4.4 Load sensing in bolts

Much research has been carried out on problems relating to bolt tension and, in many critical applications, ranging from car engine cylinder heads to sophisticated chemical and nuclear plants, it is necessary to apply the right amount of torque, usually through the use of a torque spanner. A clever idea to indicate bolt tension is to be found in the new fastener, which is the brainchild of John Hirst, who heads Exotech, a design and development company based on the University of Warwick's Science Park. West Midlands-based T. W. Lench, one of the UK's leading nut and bolt manufacturers, is currently setting up a special manufacturing unit at its Warley factory to produce the new Hi-Bolt fastener as it is called (see Figure 6.8).

The system involves a structural bolt with an in-built load-sensing device. It is identical to a standard bolt apart from the actual sensor – a steel pin running through the shank which is gripped as the nut is tightened. Before tightening, the pin is free to move. As the nut is tightened, the bolt shank contracts and grips the pin. When this happens, the maximum pre-load in the bolt has been reached.

Using the same principle in a different way is the RotaBolt, in which the internal pin is secured at one end within the shank of the bolt. The free end holds a Rota washer and cap clear of the bolt head. As the bolt stretches under the tension imposed by tightening the nut, it pulls the Rota washer into contact with the bolt head and locks it. The gap between the Rota washer and bolt head has been set previously at the manufacturing stage, so when the Rota washer locks, the predetermined tension in the bolt has been achieved.

To determine the actual bolt tension required for a particular application is far more complicated and, as previously

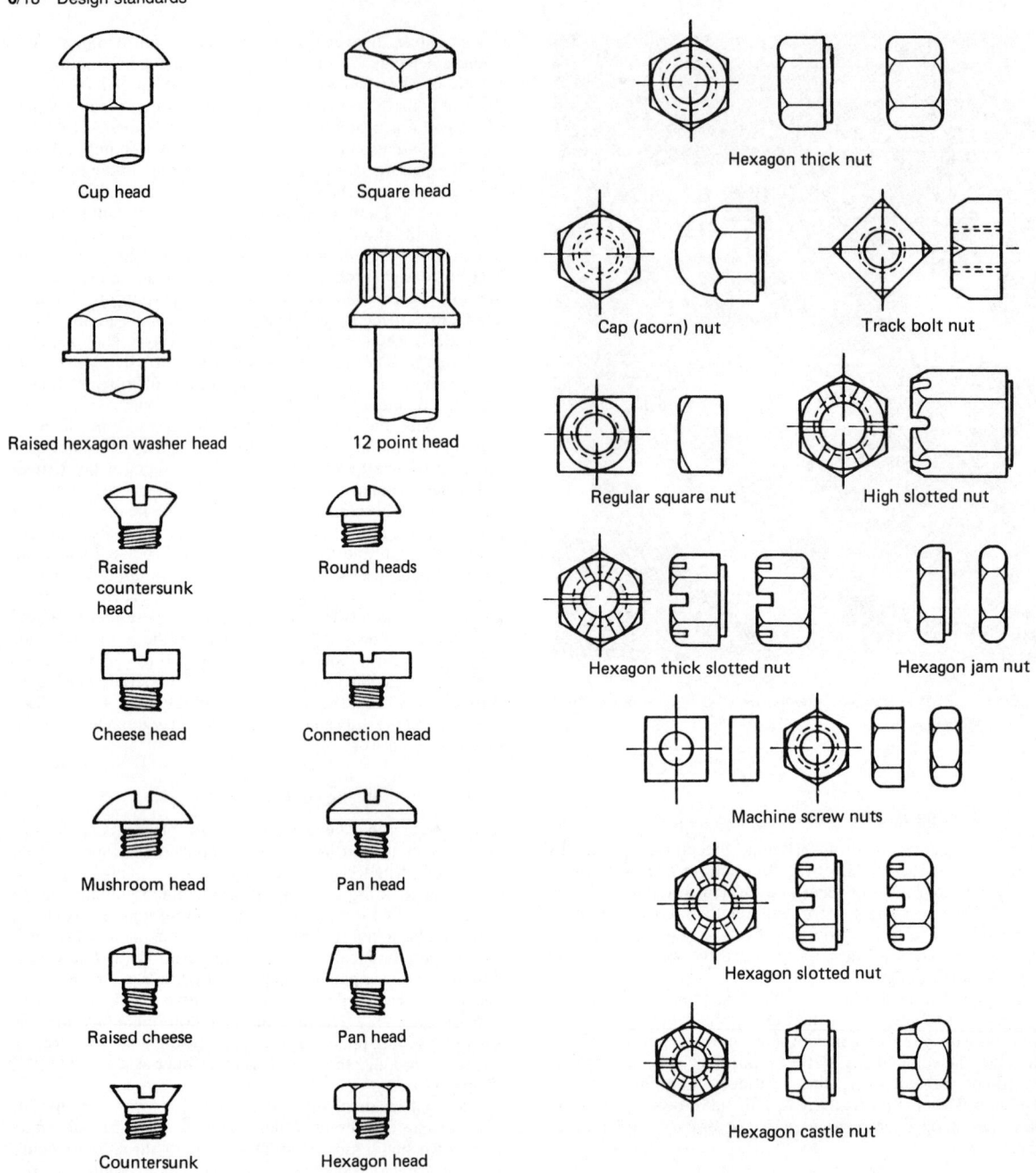

Cup head

Square head

Raised hexagon washer head

12 point head

Raised countersunk head

Round heads

Cheese head

Connection head

Mushroom head

Pan head

Raised cheese

Pan head

Countersunk head

Hexagon head

Figure 6.5   Head forms for bolts and screws

Hexagon thick nut

Cap (acorn) nut

Track bolt nut

Regular square nut

High slotted nut

Hexagon thick slotted nut

Hexagon jam nut

Machine screw nuts

Hexagon slotted nut

Hexagon castle nut

Figure 6.6   Free-spinning nuts

mentioned, particularly so when dealing with soft materials such as glass fibre-reinforced plastics (GRP). A research team at Imperial College London undertook an extensive test programme to determine the static strength of bolted joints in several forms of GRP. The experimental results were used to produce design charts for both single- and multi-bolt joints subjected to tensile/shear loading. While design methods established for structural joints in metals are applicable in a general way to GRP joints, the physical nature of the material does introduce problems not encountered with metals. The anisotropic stiffness and strength means that unexpected failure modes may be introduced, the low interlamina shear and through-thickness tensile strengths being a particular difficulty in this respect. Joint strengths were found to be strongly dependent on choice of reinforcing fabric.

Many technical devices have been used to prevent the unscrewing of nuts and bolts, and a selection is shown in Figures 6.7 and 6.11. Each has its own advantages and

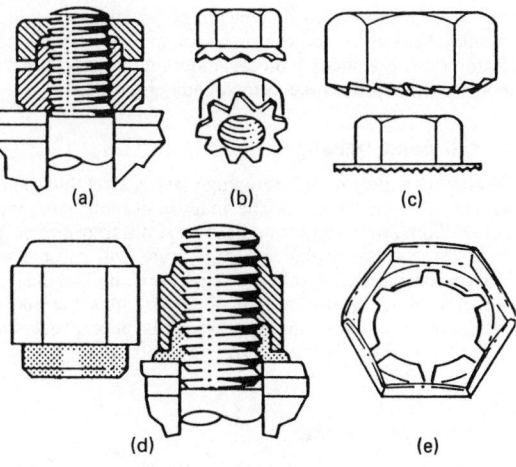

Figure 6.7 Free-spinning lock nuts. (a) Upper half of two-piece nut presses collar of lower half against bolt; (b) captive tooth washer provides locking with spring action; (c) ratchet teeth bite into bearing surface; (d) nylon insert flows around bolt to lock and seal; (e) arched prongs of single-thread lock unit grip screw

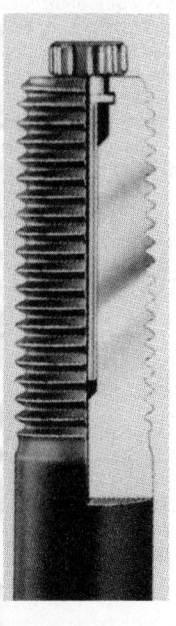

Figure 6.8 A close-up of the Hi-Bolt load-sensing system from Lench (courtesy Caters Photographic)

disadvantages. Some can only be effectively used once and, once removed, should be discarded. Serrations which effectively lock the bolt head or nut to a surface could initiate cracks in sensitive parts. In other cases, although the nut is locked, the bolt can turn. The most effective way of preventing relative rotation is by the use of liquid adhesives or anaerobic cements.

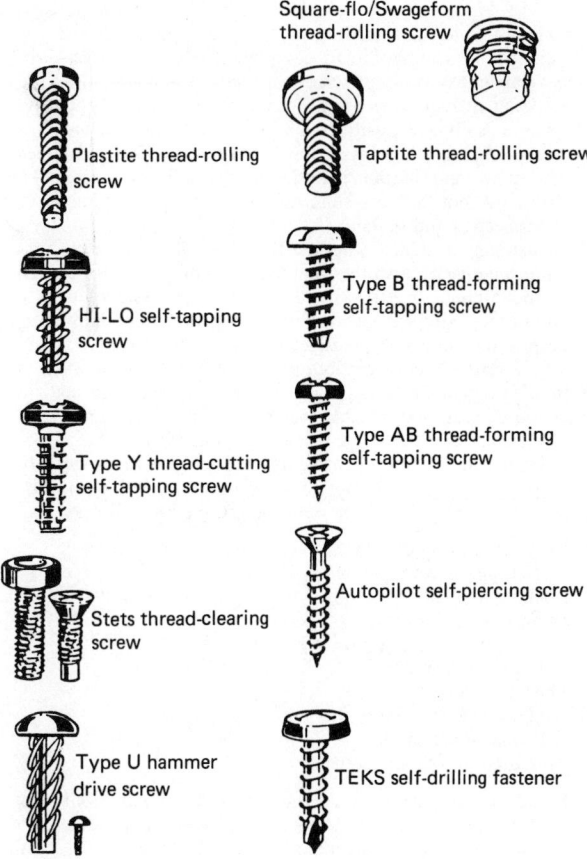

Figure 6.9 Self-tapping and thread-forming screws

6.4.5 Threadlocking

The selection of a chemical threadlocking compound will depend on the following criteria:

1. The ultimate shear strength of the fastener must not be exceeded, although this is usually only of importance for screw sizes of 5/16-inch diameter and less;
2. The severity of the loosening tendencies;
3. The size of the threads and therefore the viscosity of the compound to ensure thread filling;
4. The method of application and requirements for testing and putting into service;
5. The environmental requirements of temperature and chemical resistance.

The viscosity should be such that the compound can be easily applied, not run off, and will fill threads which have the maximum clearance. Pre-applied dry materials are available for application situations where liquids are not acceptable. They are easy to inspect and provide some re-use. The application is usually carried out by the bolt supplier and at least four formulations are available:

Low-strength locking and sealing
Medium strength – plated fasteners
Medium strength
High strength

PSM Fasteners, for example, provide a comprehensive range of bolts and studs with the pre-applied process for locking and sealing thread components. Their Scotch-Grip two-part epoxy system microencapsulates the resins while the hardener is freely carried in the coating substance. When the fastener is engaged with its mating thread, the capsules are crushed and the shearing action of the rotating fastener mixes the epoxy and hardener initiating the adhesive cure. An alternative but similar system uses microencapsulated anaerobic adhesives and sealants.

Fastening systems, whether they be adhesives, nuts and bolts, welding or any other method, are inherent in the design of the product or structure destined to be assembled as a one-off, by batch production or mass production. However, despite the popularity of adhesives, mechanical fasteners still have a prime role to play, but unless sufficient care is taken in their selection the installation of such fasteners can lead to increased costs and possibly to a decrease in the required level of mechanical efficiency or in-service reliability.

To provide a comprehensive list of all of the various types of fasteners available is impossible in this text, but it may be helpful to the reader if they can be grouped as follows.

Bolts, screws, studs, nuts and washers
Self-tapping screws
Locking and stiff nuts
Blind bolts/screws
Rivets (including blind rivets)
Spring-steel fasteners
Plastics fasteners
Quick-release fasteners
Self-sealing fasteners
Threaded inserts and studs

Some of these are illustrated for general guidance but with each manufacturer producing a wide range of products for similar applications, the reader would be wise in the final analysis to obtain specific data from these manufacturers. Several are listed in Section 6.4.16.

6.4.6 Threaded inserts and studs for plastics

A method of securing a threaded bore in sheet is by the use of bushes which are designed to bite into the sheet and prevented from rotating by its hexagon shape, serrated shoulder or spigot. A large variety of these types of bush is available.

Whenever components have to be fastened with screws to a plastics moulding, thread inserts which only require standard machine threads have distinct advantages:

1. The total re-usability of the assembly screw;
2. Unlike self-tapping or thread-forming screws which are applied directly into the plastics, there is no risk of thread stripping;
3. Often the assembly screw must remain tight and maintain its clamping effect. Threaded inserts allow this as they do not cause problems of stress relaxation in the material which is common when screws are driven directly into plastics.

The advantages of placing inserts after moulding instead of moulding-in are:

1. Reduced moulding time and therefore unit costs;
2. No expensive damage to tools due to misplaced inserts;
3. No metal swarf from inserts to contaminate a moulded surface; and
4. Availability of cost-effective assembly methods for post-moulded inserts.

Obviously, each job has to be considered on its merits, and post-moulded inserts tend to come into their own on long production runs. For small numbers, moulding-in can often be advantageous, in spite of the points made above.

6.4.7 Ultrasonic welding

Ultrasonics are widely used for joining plastics assemblies and for securing threaded inserts. The main limitation is joining parts made from different thermoplastics. A difference of only a few degrees in the softening temperature will cause one material to flow before the other. The pressure applied during ultrasonic welding is also a critical factor, and the most up-to-date ultrasonic welding equipment is microprocessor controlled to give the optimum pressure and heating conditions.

6.4.8 Adhesive assembly

Since its introduction, the Evo-Stik Thermaspray range of adhesives is finding increasing application in industry. The Thermaspray adhesives are pressure-sensitive, hot-melt spray-applied materials which offer distinct health and safety benefits, as well as an adhesion performance superior to conventional neoprene or polyurethane adhesives.

The Thermaspray products are high-tack adhesives based on a blend of synthetic rubbers, tackifying resins and liquid modifiers. The necessary investment in spray equipment and adhesive does mean that the Thermaspray system is not suitable for low-quantity production. However, given a sufficient production run, the Thermaspray system becomes one of the most efficient methods of applying adhesive evenly, economically and swiftly in the context of existing production lines. Several large manufacturing companies have invested in the system and are finding that the economics are working for them as efficiently as the adhesive.

Heat-sensitive substrates will not be damaged by the application of Thermaspray adhesives due to the low heat intensity of the spray deposit. The permanent tack of the adhesive provides a long open time, which means that large assemblies can be made with components being joined several minutes after the application of the spray deposit. Some low-stress assemblies have been tested successfully at up to 120°C. Higher-stressed component assembly may only be successful at lower temperatures.

The adhesives can be used to bond laminating panels, thin-gauge plastics and fibrous panels such as cork, fibreglass and rockwool. Other suitable materials for Thermaspray bonding include painted and unpainted metals, carpet materials, polyurethane, polythene and polypropylene sheet and foams.

6.4.9 Self-tapping screws

These screws fall into two main categories: (1) true self-tapping screws which, like a tap, cut their own threads in a prepared hole; and (2) thread-forming screws which form their own threads by a rolling or swaging action (Figures 6.9 and 6.10). Typical of the latter are the proprietary types Taptite and Swageform. Taptite incorporates machine screw threads formed on a trilobular shank and the Swageform has special lobes at 120° intervals along the tapered portion of the shank to impart a three-dimensional swaging action to form the threads.

Self-tapping screws are widely used in the assembly of sheet metal components in which the thread-cutting action results in a good fit and is resistant to vibration and shock loads. Thread-forming screws are more applicable to thicker ma-

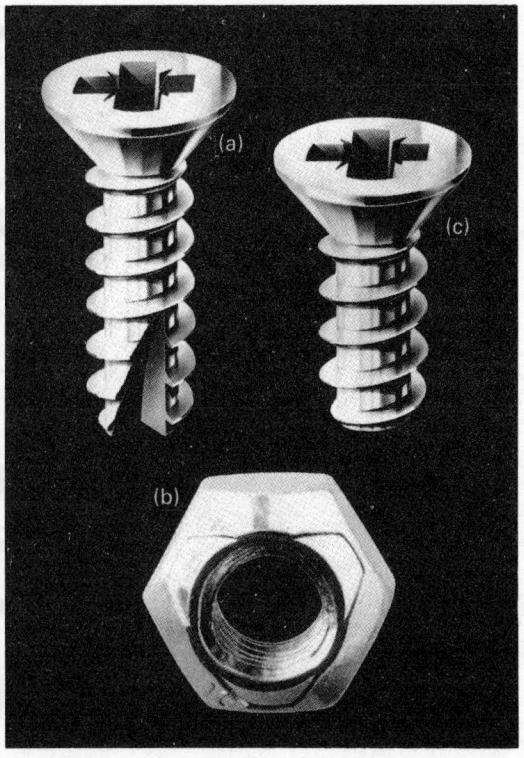

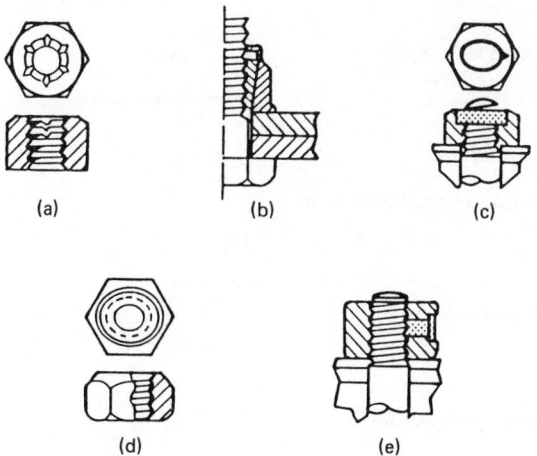

Figure 6.11 Stiff nuts. (a) Deformed thread has depressions in the face of the nut to distort a few threads; (b) slotted section forms beams that deflect inward and grip bolt; (c) threaded elliptical spring-steel insert produces locking; (d) non-metallic plug insert grips bolt threads; (e) out-of-round threads cause wedging action

Figure 6.10 *(Left to right)* Plastform thread-cutting screw; Toploc re-usable locking nut; Plastform screw for plastics

terials and blind holes. The self-drilling versions (for example, Teks and Drill-kwik) are designed with a true drill point to produce the necessary pilot hole for the threads.

6.4.10 Stiff nuts

Stiff nuts, as distinct from free-spinning nuts, generally incorporate either a locking element to generate an elastic contact between the nut and bolt threads or a differential pitch or deformed threads to create the required friction between the mating threads. Examples are shown in Figure 6.11.

6.4.11 Washers

Washers are used to provide a seating for nuts and bolts in order to distribute the load over an area greater than that provided by the bolt head or nut. They also prevent damage to the surfaces being joined due to rotation of the nut or bolt. Washers can also seal, cover up oversize holes and act as a spring take-up between fastener and workpiece. They also have the special task of preventing unwanted rotation of the nut or bolt. These include spring washers and serrated tooth types as shown in Figure 6.12.

6.4.12 Spring-steel fasteners

Although these include certain standard parts (for example, caged nuts), most spring-steel fasteners are designed for specific applications where speed of assembly is important and the cost of the fasteners is low. They are mass-produced from hardened and tempered steel strip and are used in non-critical

applications in cars and domestic electrical appliances and in the sheet metal industries. Some of the various shapes which can be produced are shown in Figure 6.13.

6.4.13 Plastics fasteners

These products fall into two categories: standard thread forms made from plastics such as nylon and the specially configured fasteners for specific applications. These can be self-coloured and are non-corrosive. Some examples are shown in Figure 6.14.

6.4.14 Self-sealing fasteners

Bolts, screws, washers and nuts can be made into self-sealing elements by the addition of a sealing element in applications where it is necessary to seal the fastener hole against the leakage of liquids or gases (see Figure 6.15). The choice of the fastener and the sealing material is dependent on the specific application.

6.4.15 Rivets

Rivets are low-cost permanent fasteners suitable for manual and automatic setting. They can be classified as solid, tubular, semi-tubular, bifurcated and blind, all of which are available in several different materials and head styles. Correct joint design and preparation together with the right fastener will ensure a strong, trouble-free and cost-effective assembly. A range of available types is shown in Figures 6.16 and 6.17.

6.4.16 Suppliers of fasteners

Armstrong Screws and Fixings, 72 Great Barr Street, Birmingham B9 4BJ

Bollhoff Fastenings Ltd, Midacre, The Willenhall Estate, Willenhall WV13 2JW

British Industrial Fasteners Federation, Queens House, Queens Road, Coventry CV1 3EG

Dzus Fasteners Europe Ltd, Farnham Trading Estate, Farnham, Surrey GU9 9PL

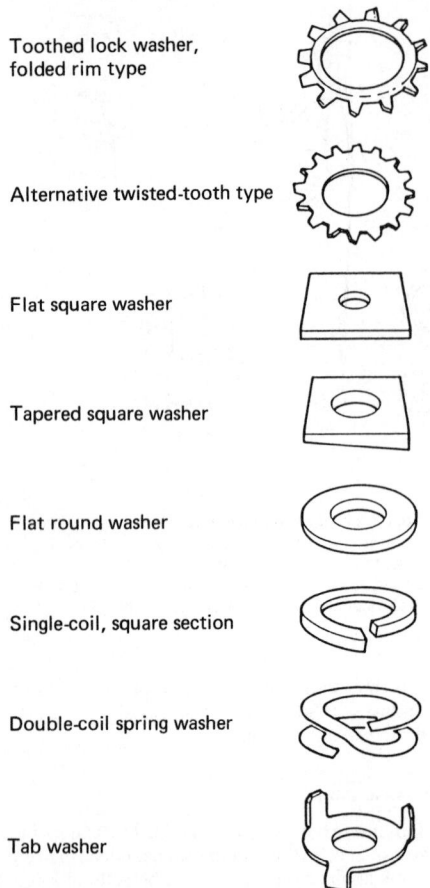

Toothed lock washer, folded rim type

Alternative twisted-tooth type

Flat square washer

Tapered square washer

Flat round washer

Single-coil, square section

Double-coil spring washer

Tab washer

Figure 6.12 Washers

Figure 6.13 Selection of spring-steel fasteners

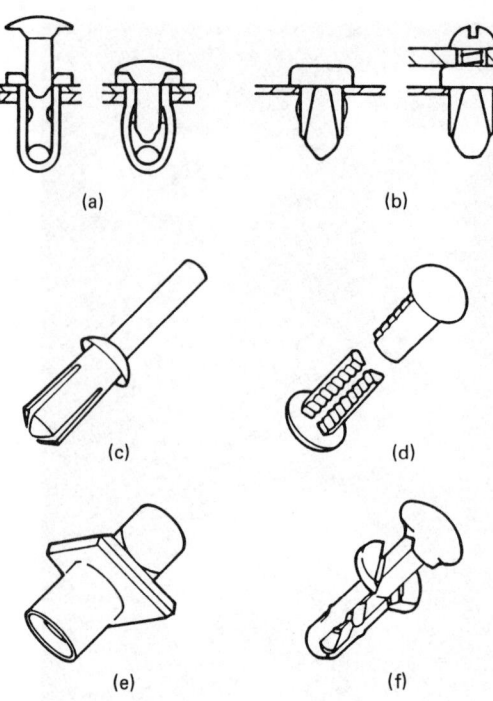

Figure 6.14 Various types of plastics fasteners. (a) Dart-type clip; (b) nylon expansion nut; (c) plasti-rivet; (d) rachet rivet; (e) plastics rivet; (f) Triploc rivet

European Industrial Services Ltd, Woden Road West, Kings Hill, Wednesbury, West Midlands WS10 7TT

Evode Ltd, Como Road, Stafford ST16 3EH

The Fasteners Centre Ltd, Hale House, Ghyll Industrial Estate, Heathfield, East Sussex TN21 8AW

ISC Fasteners, 180 Rooker Road, Waltham Abbey, Essex EN9 1JJ

Jetpress Sales Ltd, Nunn Brook Rise, Huthwaite, Notts NG17 2PP

T. W. Lench Ltd, PO Box 31, Excelsior Works, Rowley Regis, Warley, West Midlands B65 8BZ

Micro Plastics International Ltd, Unit 2 Healey Road, Coventry CV2 1SR

PEM International, Kirk Sandall Industrial Estate, Sandall Lane, Doncaster, South Yorkshire

PSM Fastener Systems Ltd, Longacres, Willenhall, West Midlands WV13 2JS

SEAC Ltd, 46 Chesterfield Road, Leicester LE5 5LP

Spiral Industries Ltd, Princewood Road, Corby, Northants NN17 2ET

Southco Fasteners Ltd, Unit E, Gregory's Bank Industrial Estate, Gregory's Bank, Worcester

TR Fastenings, Trifast House, Framfield Road, Uckfield, East Sussex TN22 5AR

Tappex Thread Inserts Ltd, Masons Road, Stratford-on-Avon, Warks CV37 9NT

Trumpf Machine Tools Ltd, Lyon Way, Hatfield Road, St Albans, Herts AL4 0LB

Unbrako, Gunns Lane, West Bromwich, West Midlands B70 9HF

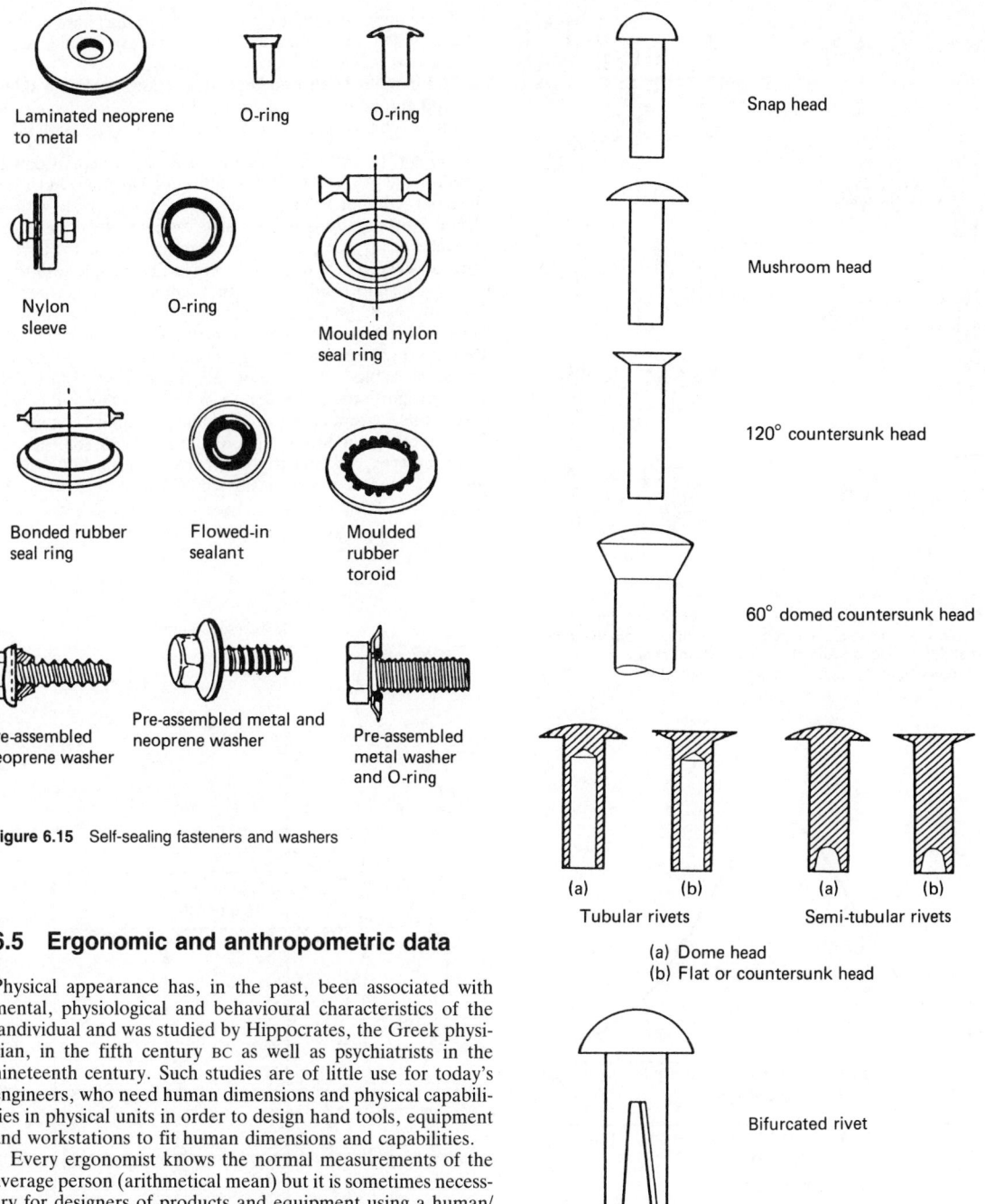

Figure 6.15 Self-sealing fasteners and washers

Laminated neoprene to metal

O-ring

O-ring

Nylon sleeve

O-ring

Moulded nylon seal ring

Bonded rubber seal ring

Flowed-in sealant

Moulded rubber toroid

Pre-assembled neoprene washer

Pre-assembled metal and neoprene washer

Pre-assembled metal washer and O-ring

Snap head

Mushroom head

120° countersunk head

60° domed countersunk head

| (a) | (b) | (a) | (b) |
| Tubular rivets | | Semi-tubular rivets | |

(a) Dome head
(b) Flat or countersunk head

Bifurcated rivet

Figure 6.16 Selection of rivets

6.5 Ergonomic and anthropometric data

Physical appearance has, in the past, been associated with mental, physiological and behavioural characteristics of the iandividual and was studied by Hippocrates, the Greek physician, in the fifth century BC as well as psychiatrists in the nineteenth century. Such studies are of little use for today's engineers, who need human dimensions and physical capabilities in physical units in order to design hand tools, equipment and workstations to fit human dimensions and capabilities.

Every ergonomist knows the normal measurements of the average person (arithmetical mean) but it is sometimes necessary for designers of products and equipment using a human/machine interface, to consider certain limits, for instance, the 5th and 95th percentile or, for safety reasons, the 1st and 99th percentile of the respective population.

Average people are more prevalent than extremes and it is possible to present this in a frequency distribution chart shown in Figure 6.18, where frequency is plotted vertically against man/woman height plotted horizontally. Since the curve is symmetrical, 50% of the population are shorter than average and 50% are larger. Thus the mean is equal to the 50th percentile. The 5th and 95th percentile, being the shortest distance along the horizontal axis, encloses 90% of the population and this, in the main, represents the normal distribution used in designing workstations.

A more scientific approach to the problem and supported by a mass of useful anthropometric data for those wishing to study the subject more deeply is given in the book *Bodyspace* by Stephen Pheasant,[16] which also provides a considerable number of relevant supporting references. It is also worth repeating the author's five fundamental fallacies.

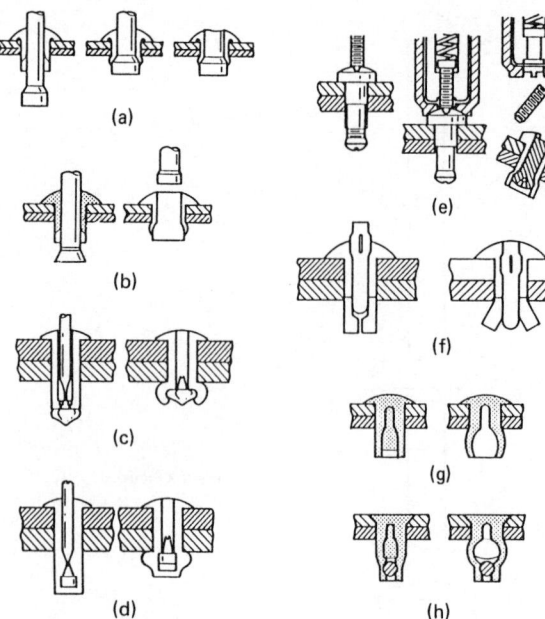

Figure 6.17 Blind rivets. (a) Break-stem mandrel; (b) pull-through mandrel; (c) break-stem mandrel (open end); (d) break-stem mandrel (closed end); (e) threaded mandrel; (f) drive pin; (g) chemical expanded (open end); (h) chemical expanded (closed end)

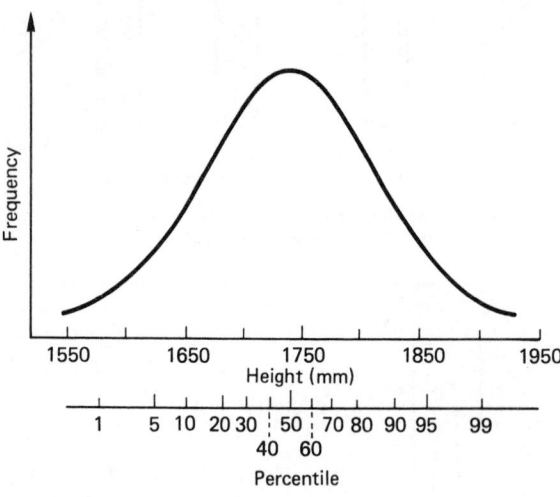

Figure 6.18 Size distribution in a population

4. Ergonomics is expensive and since products are actually purchased on appearance and styling, ergonomic considerations may conveniently be ignored.
5. Ergonomics is an excellent idea. I always design things with ergonomics in mind – but I do it intuitively and rely on my common sense so I do not need tables of data.

Figures 6.19 and 6.20 give the generally accepted dimensions for the 5th and 95th percentile female/male, but using these normal dimensions of the average female/male without careful consideration of the range of possible dimensional combinations is, for example, one of the reasons for uncomfortable seats. It is very often assumed that people with short legs have a short trunk. In fact they can have a long trunk. In the same vein, long-legged people can have long or short trunks. This is compounded by variations in arm length and distances between joints (e.g. ankle and knee, wrists and elbow; see Table 6.5). Therefore it is not always satisfactory to take a proportional dimension from a person's overall height. Large-scale surveys have provided an opportunity to compare the morphology of different populations and gain some idea of the differences. Another factor can arise in respect of posture. Different people, geometrically similar, sit differently in the same seat.

1. This design is satisfactory for me – it will, therefore, be satisfactory for everybody else.
2. This design is satisfactory for the average person – it will, therefore, be satisfactory for everybody else.
3. The variability of human beings is so great that it cannot possibly be catered for in any design – but since people are wonderfully adaptable, it does not matter anyway.

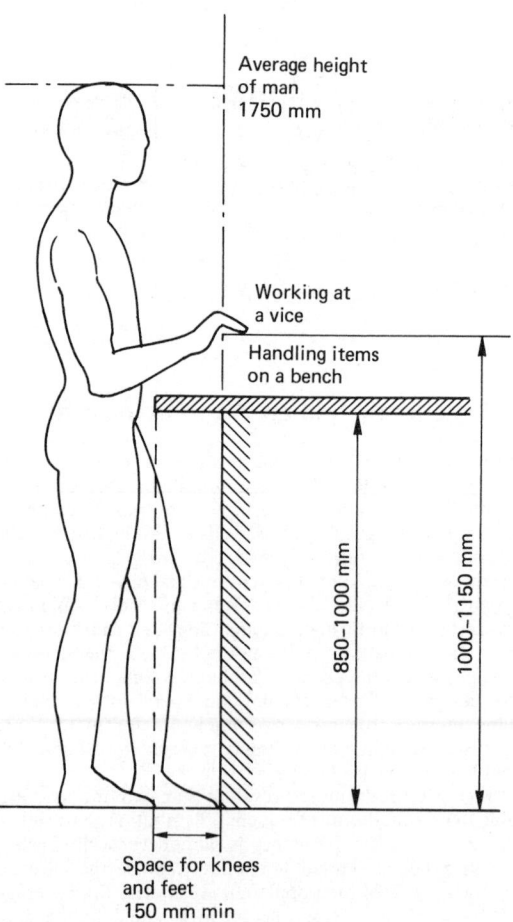

Figure 6.19 Worker standing at a bench

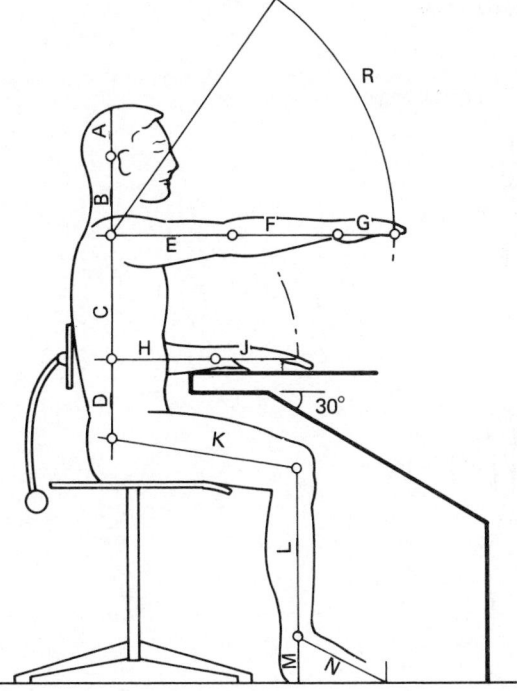

Figure 6.20 Sitting posture with limb dimensions (see Table 6.5)

Table 6.5 Dimensions between joints for woman/man (see also Figure 6.20)

Limb	Female (mm)	Male (mm)
A	105	125
B	170	185
C	290	295
D	170	185
E	290	295
F	200	255
G	160	195
H	200	255
J	160	195
K	430	460
L	370	395
M	110	105
N	220	240

In 1972 a set of three-dimensional data was collected on 281 subjects in a US Federal Aviation Administration-sponsored study of female flight attendant trainees. Each subject was asked to sit in the same hard seat as if she were in a commercial aircraft. The anthropometrist then attempted to align her mid-saggittal plane coincident with the mid-line of the seat. A multitude of mid-points were measured and posture was found to be not standardized. In fact, the variety of postures which an individual or a group of individuals may assume, is, for all practical purposes, infinite. Therefore the designer can only set limits and, within these limits, provide all the support that is needed, irrespective of posture.

Ergonomics or human engineering has four main disciplines:

- Anatomy and physiology (human structure and functioning)
- Anthropometry (body size)
- Physiological psychology (brain and nervous system)
- Experimental psychology (human behaviour)

It is not suggested that designers should be experts in these disciplines – these are largely the area for the scientists. Nevertheless, designers and engineers should be aware of the need to address the problems of a human/machine interface, which, of course, includes human comfort if the maximum output is to be obtained with minimum input.

Many factors are involved – not just the position, for example, of a person in relation to controls and instruments or working geometry of seats or workstations, but also to climate, comfort, noise, light and vision, body balances and the senses. Instrument and control knobs are a study in themselves. Should instruments be analogue or digital? Should they display up and down or from left to right or vice versa? Is colour important? Should control knobs be of a particular shape or texture? Should they be pulled or pushed? Rotated to left or right? And so on.

Considerable research has been carried out into these aspects and a number of conclusions have been formed upon which modern controls and instrumentation layout is based. Nowhere is this more critical than in aircraft cockpit layouts, PSVs and train controls, where speed and accuracy of human response is all-important, perhaps to the extent of a few milliseconds between safety and disaster.

Having decided upon a basic layout, the next problem is often a knowledge of forces that the human limbs can exert. Pedals and levers in particular must be designed to accommodate the appropriate human forces, bearing in mind that while a maximum force can be applied, to do this repeatedly will exhaust the operator in a very short time. As a guide, loads used in repeated movements should be no more than 10% of the maximum. Static loads such as those arising from a cramped work posture add considerably to fatigue and impair efficiency. Static loading cannot be entirely eliminated, as the stability of the human frame and the control of precise movements all necessitate certain static muscular loads. Wherever possible, the number of groups of muscles used at any one time to perform a given task should be limited and torsional loads around the vertical axis of the body avoided.

The amount of force that can be exerted by any limb depends largely on body posture and the direction of the force (see Figure 6.21). It must also be borne in mind that movements are less taxing when they are in alignment or sympathy with the natural movements of the human body. For example, in moving objects on a table it is less fatiguing to move them at 45° to the table edge than parallel and at 90° (see Figure 6.22).

The amount of information available on forces that can be generated during specific types of movement is considerable. Research has been carried out on sports such as golf, tennis, running, bowling, etc. where forces and reactions have been measured. Thus, there is no reason for manufacturers getting their designs wrong for lack of information.

The types of control should permit the efficient use of the limbs (see Figure 6.23). Hands, for example, should be used for quickness and accuracy; feet can exert considerable force. Controls in general should follow the movements of the objects being controlled, and where there are several controls in a group, sufficient space (certainly not less than 50 mm)

The amount of force that can be exerted by the limbs depends largely on the body posture and on the direction of the force.

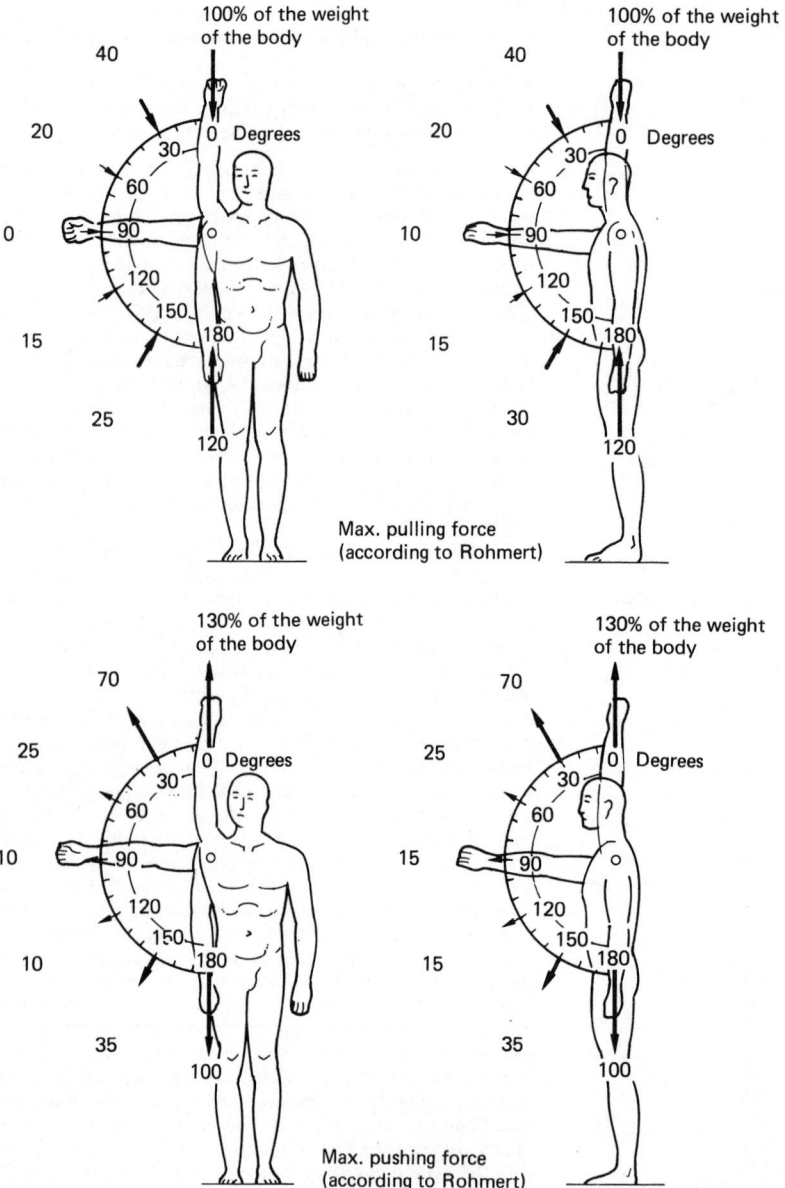

Max. pulling force
(according to Rohmert)

Max. pushing force
(according to Rohmert)

Maximum pulling and pushing forces as dependent on the position and direction
of the arm.
Sample of factory workers.
Men: mean weight 64.5 kg
Women: mean weight 57.9 kg
Age group from 15 to 57

Figure 6.21   Maximum pulling and pushing forces capable by the arm (courtesy Vademecum Ergonomics in Industry, Philips Technical Library)

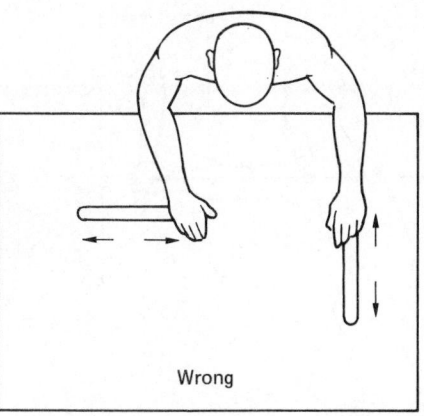

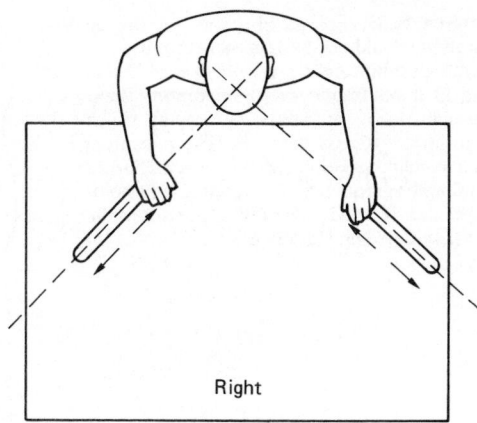

Figure 6.22 Relative simultaneous/consecutive movements

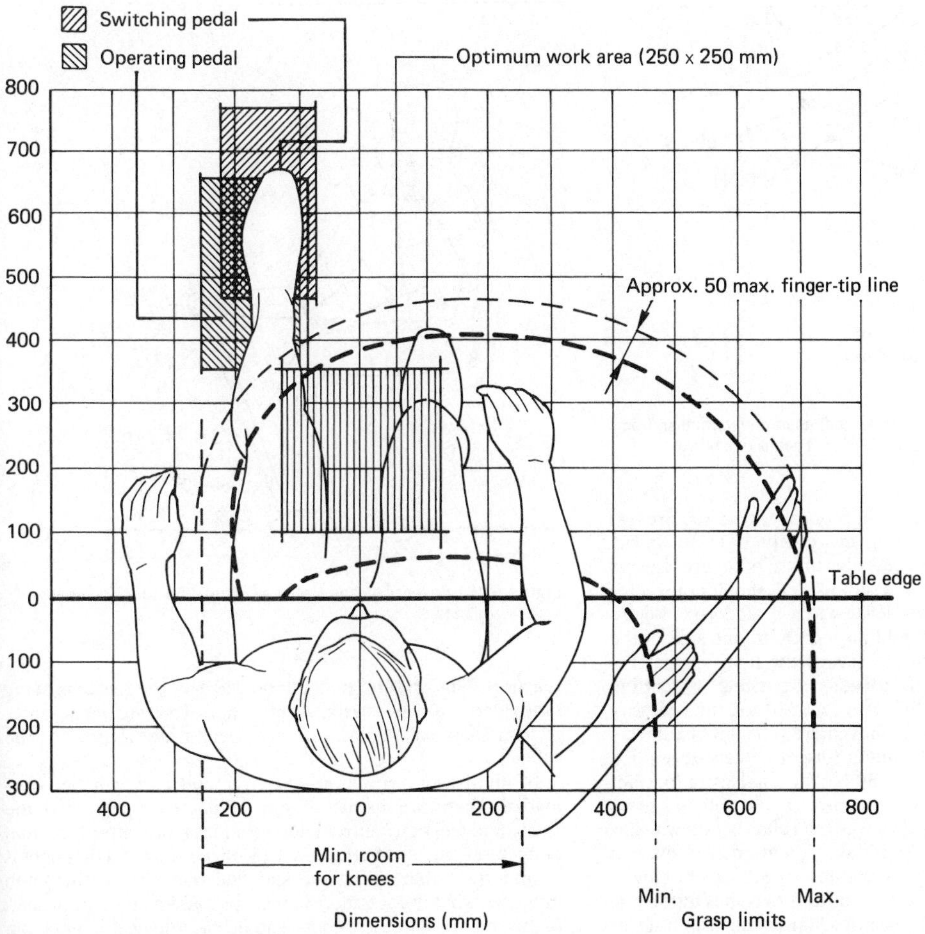

Figure 6.23 Working space for sitting operatives (courtesy Vademecum Ergonomics in Industry, Philips Technical Library)

should be left between them. For pushbuttons and toggle switches, this separation should not be less than 15 mm.

The force to operate pushbuttons can vary between 2.5 and 11 Newtons having in mind frequency of operation. Foot-operated pedals usually mean an ankle movement which should be limited to an arc of less than 25°. The maximum force applied from a standing position can be between 90 and 900 N, again depending on frequency of operation and holding time. Where the leg is used to apply a pressure, the diagram in Figure 6.24 offers some guidance.

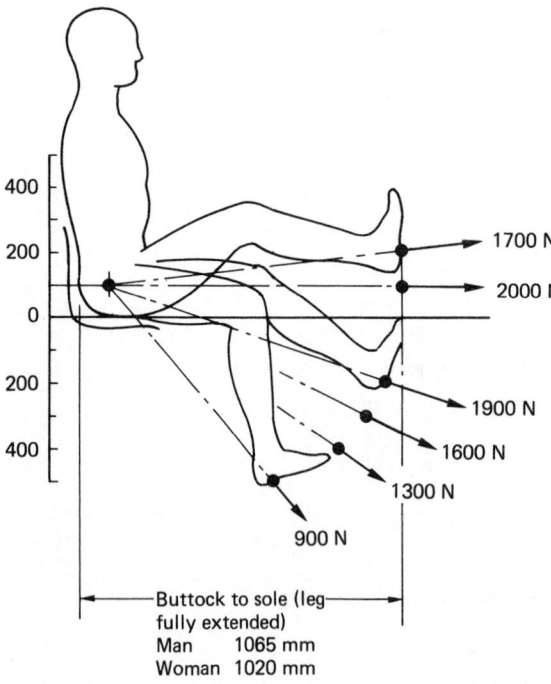

Figure 6.24 Forces from leg movements (based on information from Vademecum Ergonomics in Industry, Philips Technical Library)

Long levers are preferable to short ones, as they require less force in operation and the arm movement is more linear. Hand motion of short levers describes more of an arc. Levers necessitating considerable force should be at shoulder level for standing work and at elbow level when seated (see Figure 6.25). In both cases they should be slightly to one side of the operator and move towards the axis of the body so that the body is subjected to as little torsion as possible. Operating angles of levers should ideally not exceed 90° and the necessity perhaps of additional sideways movement (e.g. car gear lever) should be kept in mind. Maximum force that can be applied fore and aft should not exceed 130 N. Not the least important aspect is the feel of the lever or control knob, since in some applications the grasping of the lever is a reflex action which is an automatic reaction to a stimulus. Conditioned reflex is a reaction not natural to a specific stimulus but can be developed by association and repetition until it becomes automatic and appears to be a natural reaction. Many industrial tasks of a repetitive nature become conditioned reflex actions. An example is driving a car.

An important criterion is response initiate time, i.e. the time taken from the application of a stimulus to the start of a

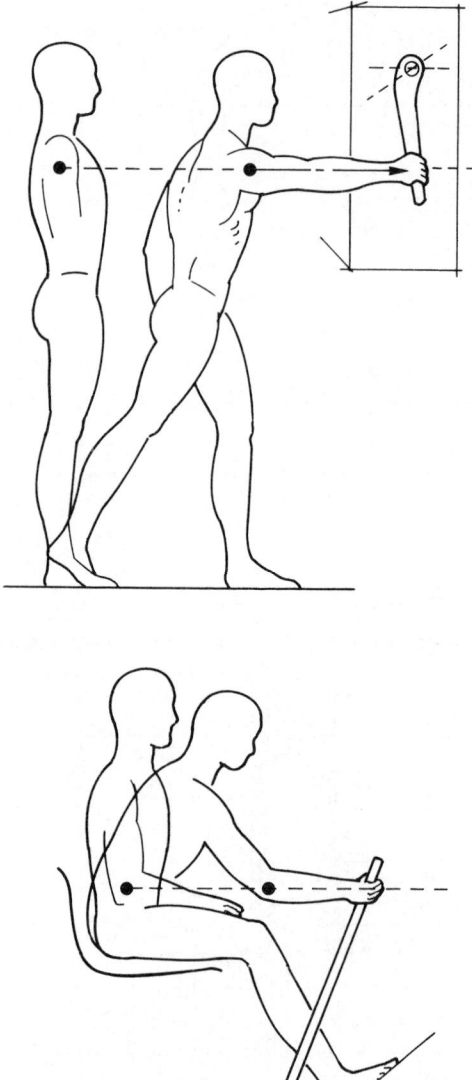

Figure 6.25 Lever positions (courtesy Human Factors Engineering, McGraw-Hill Ltd)

reaction. This, in turn, depends on which of the senses is being motivated – touch, hearing, sight, smell. The rate of response, in turn, depends on the sex, age, health and arousal of the subject.

In many work situations the eyes and ears can be overloaded, therefore tactile sensing comes into play, and the shape and feel of a control knob becomes more important than being just a device to effect control. In working in a dim light it is important that the shape and feel and size of the knob indicates what it does and whether it should be pulled, pushed or turned. In all cases, there should be a logical connection between the direction of movement of a control and the reaction of the object being controlled.

An important area in human/machine relationship is instrumentation, the ability to see what is happening to a process and to be able to interepret variations to detect a possible calamity and to be able to reach relevant alarms and other controls in an emergency (see Figure 6.26).

The cockput of a modern aircraft provides a typical example of the importance of instrument placing. While each application will demand its own particular answer, it is generally accepted that in an instrument control panel the most important instruments and the most often used controls should be placed in the lower right-hand area, while those requiring less attention are spaced in the remaining area. Each application, of course, may have over-riding considerations and these must be taken into account. Again, in general, the instruments should be within the normal field of vision, i.e. immediately in front of the observer and in the zone from eye-level to approximately 30° below.

There are very many types of visual indicators, ranging from the simple circular dial through linear or logarithmic displacement indicators to digital readouts. Each has its own particular advantages and disadvantages. Digital readouts are, for example, excellent for reading absolute values. Dial and linear indicators provide an indication of values within the context of limits (e.g. zero and a maximum). For example, it is more comforting to drive a car knowing the fuel tank is half-full of petrol than knowing there are 5 gallons in the tank, because the driver has to mentally calculate what proportion 5 gallons is to a full or empty tank. Where displays are calibrated, care should be taken that the legends are readable. Colour can play an important part in defining certain areas.

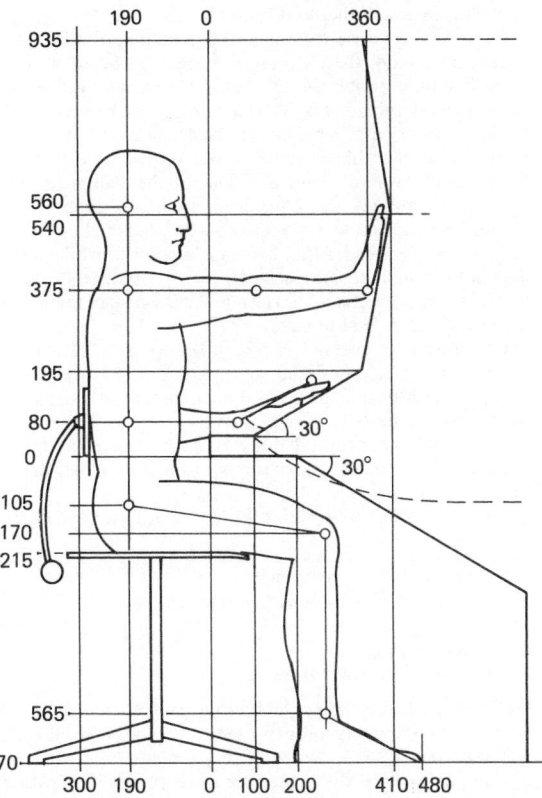

Figure 6.26 Layout of instrument panel in relation to sitting operative (based on information from Vademecum Ergonomics in Industry, Philips Technical Library)

In production engineering, flexible manufacturing systems (FMS) are now being widely employed to improve productivity and quality as well as the quality of life for the operator. The new techniques will have a major impact on economic factors. Research, however, has been mainly in the development of the technical aspects of software and hardware. Not so much has been directed to identify specific ergonomic problem areas, although universities worldwide have been studying problems of ergonomics and anthropometry. In the UK these include:

Loughborough University
University of Aston, Birmingham
City of London University
John Moores University
University of Surrey

More specifically directed research is being undertaken by

Central Electricity Generating Board, Barnwood, Glos
ITT Europe – ESC Research Centre, Harlow
RAE, Farnborough
National Institution of Agricultural Engineers, Silsoe
MRC Applied Psychology Unit, Cambridge

Much has been learned but there is much more to know – all of which will be invaluable for improving human/machine relationships.

6.6 Total quality – a company culture

Quality is and must be an emotional issue – largely a question of attitudes to be taken seriously. Failure to understand how to improve quality and value of goods and services on an ongoing basis means dissatisfied customers, declining markets, vanishing profits, reduced salaries and benefits and, ultimately, loss of jobs.

Customers expect quality. Once goods and services are delivered, reworking them to correct deficiencies is expensive both for the manufacturer and the customer. The purpose of quality management is to enhance the ability to improve products and services, to remove obstacles that prevent this and to use the collective talents and skills of the entire workforce. It means designing work processes so that each person has the right information, equipment, systems, procedures and training whenever and wherever it is needed.

Robert Gavin, the chairman of Motorola, regards people as the ultimate high technology. Companies that can best mobilize the collective insight and the talents of their employees have a great competitive advantage.

Quality is defined in the *Shorter Oxford English Dictionary* as 'Character, disposition, nature, capacity, ability or skill; excellence of disposition; an accomplishment or attainment; rank or position, nobility; an attribute, degree or grade of excellence'. As a company culture and a means of improving a product or service, quality is synergistic.

To achieve quality in products and service each company must lay its own stepping-stones to achieve the elimination of defects and to ensure consistency in the end product or service. It must first measure and review the current situation and then see where defects are occurring and, more importantly, why. This must include the setting up of feedback routines from the marketplace. People are of vital importance and training and retraining schemes must be envisaged.

Demand from companies wanting help in setting up quality management systems and meeting the requirements of BS 5750 are growing. Smaller companies now require this support as large firms increase their demands for improved quality standards from their suppliers. Today, major purchasers demand proof of a company's ability to produce quality products and provide a quality service and the provision of this proof in the form of independent third-party certification is possibly the most important factor in the successful marketing of a supply company. BS 5750 registration is going to become increasingly a condition of contract supply as currently practised by many large establishments such as British Gas, the CEGB and motor manufacturers such as Ford.

The BSI (British Standards Institution), through its certification and assessment services, provides industry with BS 5750, a first-class product certification and company quality-assessment scheme that will efficiently and effectively satisfy the customer. BS 5750 is divided into three distinct parts:

Part 1 – Specification for design, manufacture and installation
Part 2 – Specification for manufacture and installation
Part 3 – Specification for final inspection and test

Part 1 would cover a company engaged in the total activity of design and manufacture. Part 2 is for those companies only manufacturing and assembling to designs other than their own, while Part 3 is for companies engaged in final testing and distribution.

To be registered by BSI means that the company in question has been independently assessed by a team experienced in quality assurance and the technical conditions applied to that particular industry. The first essential for assessment is that a quality system is being operated which meets the requirements of BS 5750, and this must be documented. Registration commences when the assessment team is satisfied that the firm is operating its acceptable quality system and have agreed to the required level of audit tests.

From a realistic point of view, it is essential that the top person, whether he or she be Chairman, Managing Director or Chief Executive, is committed to quality assurance. Only he or she can inspire the workforce and generate the enthusiasm necessary to get the job right first time.

To do this it is necessary to have effective and continuing market research on the customers' quality needs. Unless these needs are fully understood, the designer will be unable to cater for them. Marketing must also ensure accurate promotion of the product and service, delivery on time as specified and efficient servicing and maintenance.

In the majority of cases a company will buy in raw materials and basic components. It is in the firm's interest that the quality of these parts is beyond doubt. Pressure should be exerted on supply companies to ensure that goods have zero defects. It is also important to bear in mind that true costs of poor materials, unreliable components and shoddy sub-assemblies far outweigh the savings made through the purchase of inferior stock. This becomes even more important when the JIT (Just in Time) technique is being used on the shopfloor. One of the implications of JIT is the requirement to issue parts from suppliers direct to the manufacturing lines without the more conventional goods-inward inspection operation to detect failing batches.

It is worth considering the conventional inspection process. At the end of the line, it only weeds out the faults that should never have happened in the first place. It does not place quality into the product. Inspection is a waste of effort if expensive equipment and skills are inspecting quality at the end of the manufacturing line. They should be part of the manufacturing process.

Because it is axiomatic that the customer is always right, complaints from customers are generally interpreted as being caused by a product or service failure. But research finds that the customer often contributes directly or indirectly to the cause of the problem. An example concerns the use of a product. If the customer fails to carry out specific directions for the product's use and/or maintenance then a defect may arise. The answer is to ensure that adequate instructions are provided and that there is no question of such instructions being misinterpreted. If manufacturers concentrate on the more traditional aspects of quality assurance then they may be forced to make extensive and expensive modifications to the product, whereas the provision of adequate instructions together with, perhaps, a HELP telephone number, would be just as effective and much less expensive.

Typical causes of customer dissatisfaction fall into four categories. Only those in one category (i.e. problems caused by production defects) can be totally fixed within the plant. In the other categories – (1) customer based, (2) marketing/delivery based and (3) product design – customers often constitute a part of the problem either because they have not adhered to the instructions or because the company has not fully allowed for customer behaviour.

Making a product 'idiotproof' is nearly always possible (at a price) but rarely practised. A manufacturer has to make a product 'pro idiotproof' in line with what constitutes a reasonable attitude on the part of an average customer. Sometimes they get it wrong, but too often it is because the company wants to keep down costs.

This is where designers can exercise their ingenuity by assessing the problem and approaching a solution by lateral thinking. Taking a simple example would be a hook on which fingers and clothes could be caught. Instead of redesigning the hook, would it not be possible to create a different design approach in which the hook could be eliminated?

Other defects arise from incorrect handling of components during installation because the designer had failed to appreciate handling problems. For example, heavy valves were being loaded/unloaded by threading a chain through the flow area, creating internal damage which caused the valve to malfunction. The blame must be laid on the lack of instructions on the way to load and unload the valves. A more meaningful ploy would have been to fit an eyebolt in an obvious and prominent position on the valve, thus drawing attention to the loader that here was a point to which a chain and hook could be attached. Not only would this prevent internal damage but it would also make the task of loading and unloading easier and quicker.

Humans, like rivers, will also take the path of least resistance. It is no good designing a path laid as two sides of a triangle to get from A to B when the hypotenuse is the shorter route. As mentioned earlier, quality is an emotional experience and it is important that this is at the back of the mind of every designer and that he or she is aware of human shortcomings.

There are five major routes for feedback from customers to manufacturers:

1. Direct customer complaints
2. Complaints from sales force/retailers
3. Warranty returns
4. Direct sampling
5. Customer-satisfaction survey

Each route has its own limitations and therefore all five sources of complaints must be used to get to the kernel of the problem.

Companies have discovered that pre-emptive education is cheaper than later repairs. One motor manufacturer is compensating its sales force for providing a 20-minute customer-education session on how to use the vehicle and to appreciate maintenance. This can have a significant impact on the

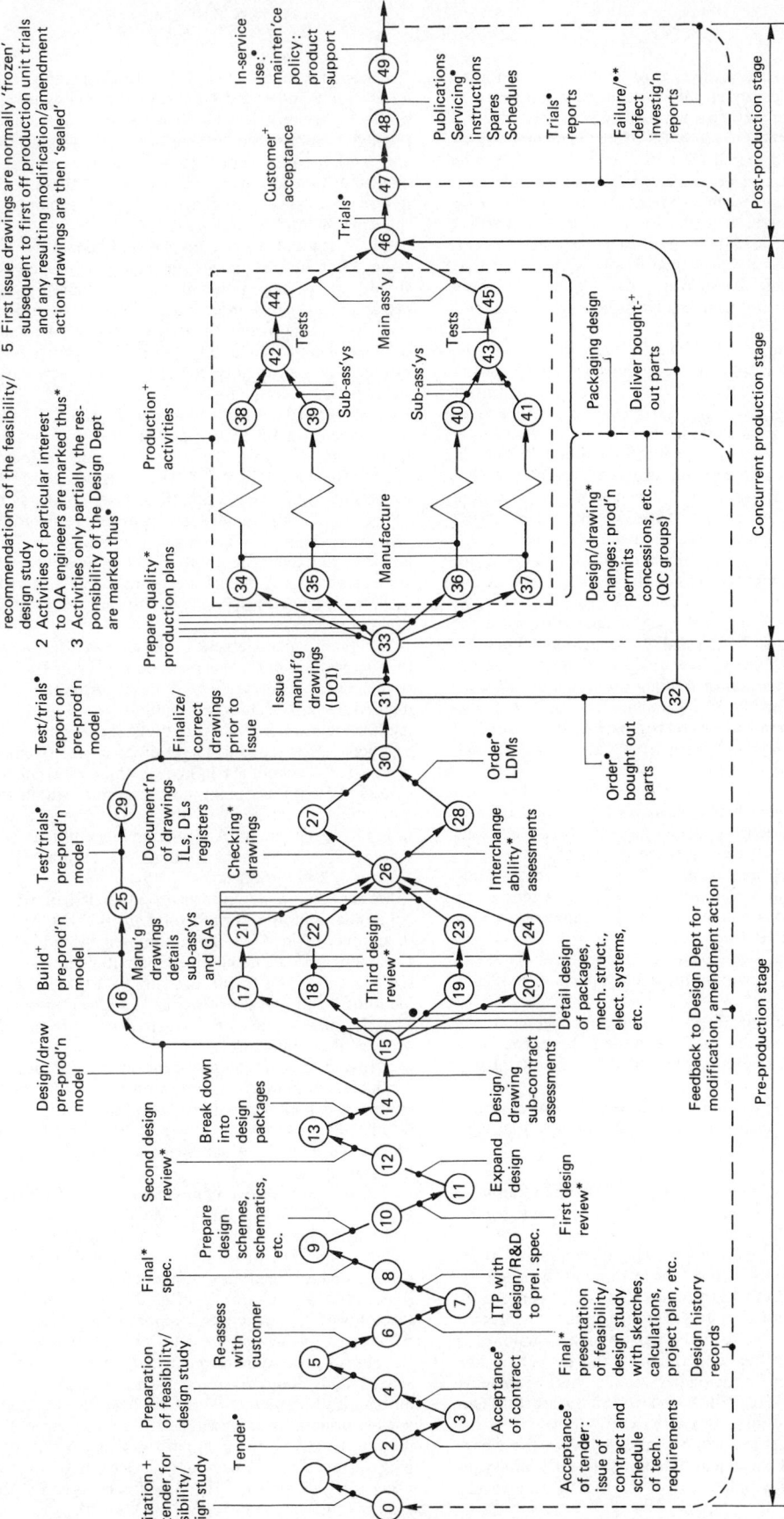

Figure 6.27 Flow chart of design activities for a typical project/product work programme fully or partially the responsibility of the Design Department (courtesy of the Institution of Engineering Designers)

Notes:
1 Project may be abandoned at event No. 7 depending upon the results/ recommendations of the feasibility/ design study

2 Activities of particular interest to QA engineers are marked thus*

3 Activities only partially the responsibility of the Design Dept are marked thus•

4 Activities that are not the direct concern of the Design Dept are marked thus+

5 First issue drawings are normally 'frozen' subsequent to first off production unit trials and any resulting modification/amendment action drawings are then 'sealed'

long-term satisfaction and loyalty of customers. Another manufacturer provides similar information via video tapes.

The first move for any company wishing to improve the quality of its products is to study BS 5750. First impressions may be that it is too generalized in its approach, but it has to apply to a wide range of very different industries and allow scope for its recommendations to be incorporated into existing management structures. It seeks to formalize the various stages in design and manufacture so that each step can be analysed and documented. As time goes by, the results so obtained can be used for comparisons, to seek out weak spots and where high costs are involved. With these data available, remedial measures can be adopted.

There is no magic in BS 5750. It really states the obvious – obvious, that is, to many companies who have had an efficient scheme to check the development of a product from concept to marketing. What BS 5750 does do is to bring some degree of standardization to these states.

Numerous bodies exist that are well qualified to help in setting up quality assurance. For example, there is EnTra (Engineering Training Authority), which took over from the original EITB (Engineering Industries Training Board) in the UK. Field training advisors are listed by the DTI (Department of Trade and Industry) as consultants, and their services are available to any company which seeks to introduce or improve its quality-assurance system.

More recently, the Danbury Park Management Centre has set up what is the first Total Quality Management Unit. Its overall purpose is to create a centre of excellence for the development and provision of training programmes and consultancy in Total Quality Management. The unit will also provide a forum for managers to compare quality initiative plans and performance and learn from each other through sharing experiences.

- To provide an open access facility which would offer a range of open-learning opportunities and tutorial help and support in all aspects of quality management.
- To assist with the identification of training and development needs of quality management staff in the engineering and associated industries and to provide a range of training programmes to meet those needs.
- To provide a focal point for the design and development of short courses and seminars for quality management and related staff.
- To provide a forum for directors, senior management and quality managers to share experiences, ideas and approaches to the resolution of problems related to quality management.
- To create a central location for the collection of information relevant to quality management – books, journals, articles, courses, etc.

It is obvious that a company would benefit from the services of a trained quality assurance manager or someone who can be a focal point for quality matters. Various training courses are available, either externally or within a company, from establishments such as PERA and Cranfield Institute of Technology. There is also an Institute of Quality Assurance as well as MSc courses for Quality Assurance so that there is no excuse for not being able to find a person highly qualified in quality matters. With the cooperation of the DTI there are also free consultancies for companies wishing to set up systems to conform to BS 5750. PERA and Cranfield are but two establishments geared up to this activity.

Not without interest is a self-appraisal kit from the Chartered Institute of Management Accountants for identifying and measuring quality related costs in the manufacturing industry.

A designer should always seek to reduce the complexity of an assembly and thereby improve the quality of the finished product. As an example, in an electronic assembly it was possible to reduce the number of fasteners from 120 to just 40 and, and in consequence, to decrease the number of adjustments in final test from 200 to 30. All these contribute to improved quality, reliability and serviceability as well as reducing manufacturing costs.

Quality should not be confused with reliability. The two are quite different. Establishing a quality assurance programme should ensure that reliability is consistent for each similar product. In other words, if a product has a designed life of three operating years, then every one of the same line of products should have this life span within narrow limits, whereas without quality control, the life may vary between wider limits such as 88–120%.

Measurement and documentation, which are part of the requirements of BS 5750 approval, are important for keeping track of modifications and improvements in a quantitative sense. The emphasis on measurement is comforting, as it focuses attention on the fact that better control is required.

The importance of selecting appropriate measurements in order to track improvements in any process, direct or indirect, is unquestionable. The ability to select those areas for improvement which gives the greatest return on investment in either equipment or skills depends in itself on the measurement of existing processes.

For quality management to be seen as effective and taking its true position in management planning it is essential that its performance be measured. Without an effective system for quality cost associated data gathering and analysis management are unaware of the true costs of not getting things right and hence are unable to target areas of improvement.

RAPRA Technology is investigating what quality costing systems (if any) are being used currently within the plastics and rubber industries. With this as background and in consultation with companies participating in a multi-client project, a system to complement companies' existing management systems will be produced.

More and more designing is being carried out with the aid of computers and appropriate software. CAD/CAM/CAE/CIM (Comptuer Aided Design/Computer Aided Manufacture/Computer Aided Engineering/Computer Integrated Manufacture) are essential to the designer and these techniques are ideal for keeping track of design modifications and to ensure that the maintenance of quality always prevails during the process of manufacture.

Software specific to the maintenance of quality is now available, and under the terms of a recent agreement, quality control and assurance software produced by the Valisys Corp. (a subsidiary of FMC Corp.) is being made available by McDonnell-Douglas to satisfy manufacturing industry requirements in the UK.

Valisys software complements McDonnell-Douglas' UNIGRAPHICS II CAD/CAM/CAE software, widely used by industry and available as nine fully integrated modules. In combination, they provide an unparalleled degree of automation in quality assurance inspection, evaluation and accept/reject tasks. A high degree of quality intelligence is built into the software to ensure that integrity of the original design can be realized throughout the entire production process.

Valisys verifies dimensioning and tolerancing specifications, speeds and simplifies tooling and fixturing, and executes automatically generated inspection paths. It allows for the correct orientation of machine tools with respect to part and feature location and compares finished parts against soft gauges (functional inspection gauges created in software form using UNIGRAPHICS II). Benefits claimed by McDonnell-Douglas for Valisys include:

- 70% reduction in inspection machine programming time
- 50% saving in cost/procurement of hard gauging
- 40% less time needed to set up critical machining operations
- 80% fewer hours spent reworking complex parts.

It is all too easy for a designer to get sidetracked. Starting with a concept and finding that other problems arise, be they manufacturing or costs, it is all too easy to concentrate on these new issues and alter the design, overlooking some of the earlier concepts and reasons. Every alternative design must be considered in the light of previous considerations.

In a paper to the Institution of Engineering Designers, G. J. Baker points out that:

> Quality points affecting design and design procedures come under the headings of Safety, Reliability, Life, Conformance, Interchangeability, Identification, Traceability, Communications, Verification, Unambiguity, Clarity etc. This means that most of the tangible design activities are of interest and concern to the QA engineer but in the main, the intangible activities are not. The intangible activities in general are the creative thought and artistic processes essential to the overall design exercise. They are a combination of natural flair, imagination and perception that enable the design engineer to arrive at the shape, colour and texture of a particular item, which when coupled with fundamental engineering principles, interest and effort in the work, form the basis of design solution conceptions.

It is stressed that the work carried out at the design stage lays the keystone of quality in the final product because the manufacturing drawings, as the final medium of expression of the design, form the basic work instructions for the production phases. Figure 6.27 presents the tangible design activities involved in a full project/product work programme and covers the pre-production, concurrent and post production phases.

Many companies are recognizing the value of ensuring quality design and manufacture and are producing various ideas for ensuring this. Sigma Ltd, for example, have announced a computerized data-collection system that, they say, will improve quality control and productivity. Their 769 DataMyte Attributes Collector is a complete system for the checking and analysing of attributes at any stage of manufacture or even in a service industry, where defects are not normally quantifiable. For example, it records whether there are dents, scratches, assembly errors, textile faults, etc. and the regularity of their occurrence. The basis of the DataMyte 769 system is a hand-held battery powered data-collector featuring a two-line character display. It is menu-driven with application-specific functions programmable through either the keyboard or a computer (see Figure 6.28).

Another company, Mari Advanced Microelectronics Ltd, has been awarded a certification of approval for the prestigious quality management system BS 5750: Part 1. Mari specializes in the provision of CAD/CAM equipment, support and consultancy with the emphasis on mechanical engineering. With BS 5750 being increasingly demanded by end users in the aerospace, automotive, defence and engineering industries, the approval will enable Mari to comply with the quality assurance standards required of suppliers to these and other industries.

Yet another example of a company's desire to provide quality products concerns the 1989 Ford Fiesta, which went on sale in the spring of that year. As part of a major quality campaign, Ford earlier set up a special fleet of pre-production cars which it placed with selected business and public users in European countries (Figure 6.29). These vehicles were moni-

Figure 6.28 The 769 DataMyte Attributes Collector (courtesy Sigma Ltd)

Figure 6.29 The pre-production Ford Fiesta (courtesy Ford Motor Company Ltd)

tored weekly for quality, reliability, durability and operating economy. The information was analysed by Ford in order to detect and correct any quality concerns before any units were actually sold to the public.

6.6.1 Useful addresses

British Standards Institution, 2 Park Street, London W1A 2BS

British Standards Institution, Certification and Assessment Dept, Maylands Avenue, Hemel Hempstead, Herts HP2 4SQ

Department of Trade and Industry, Directory of Quality Assurance Consultants, 29 Bressenden Place, London SW1E 5DT

Engineering Training Authority, Vector House, 41 Clarendon Road, Watford, Herts WD1 1HS

Hutchinson and Co. (Publishers) Ltd, *Manual of British Standards in Engineering Drawing and Design; Manual of British Standards in Engineering Metrology*, 17–21 Conway Street, London W1P 6JD

Institute of Quality Assurance, 54 Princes Gate, Exhibition Road, London SW7 2PG

Institution of Engineering Designers, Courtleigh, Westbury Leigh, Westbury, Wilts BA13 3TA

Mari Advanced Microelectronics Ltd, Mari House, Old Town Hall, Gateshead, Tyne and Wear NE8 1HE

McDonnell-Douglas Information Systems Ltd, Meirion House, Guildford Road, Woking, Surrey GU22 7QH

Quality and Reliability by Design, Engineering Design '86 Congress, The Institution of Mechanical Engineers, 1 Birdcage Walk, London SW1H 9JJ

Quality Assurance Training Services, School of Industrial Science, Cranfield Institute of Technology, Cranfield, Bedford MK43 0AL

RAPRA Technology, Shawbury, Shrewsbury SY4 4NR

Sigma Ltd, Avenue One, Business Park, Letchworth, Herts SG6 2HB

References

1. Parker, M., *Manual of British Standards in Engineering Drawing and Design*, BSI/Hutchinson, London (1984)
2. *Quality Assurance*, BSI Handbook No. 22 (1981)
3. Pahl, G. and Beitz, W., *Engineering Design*, Design Council (1984) (Springer-Verlag, 1977), Chapter 7
4. Argysis, C., 'Today's problems with tomorrow's organisations', *The Journal of Management Studies*, **4**, No. 1 (1957)
5. Cullum, R. D., *Handbook of Engineering Design*, Butterworths, London (1988)
6. Design Group, University of Southampton, *Conference Proceedings Information for Designers*
7. SERC, Polaris House, North Star Avenue, Swindon, Wilts SN2 1ET
8. SD-Scion, 127–147 Fleet Road, Fleet, Hants GU13 8PP
9. Brooker, K., *Manual of British Standards in Engineering Metrology*, BSI/Hutchinson (1984)
10. BS 1134 Method for the assessment of surface texture
11. BS 1916 Limits and fits for engineering
12. BS 4500 ISO Limits and fits; Part 1 General, tolerances and deviations; Part 2 Working limits on untoleranced dimensions
13. BS 4500A Data Sheet – Selected ISO fits, hole-basis; BS 4500B Data Sheet – Selected ISO fits, shaft-basis
14. *Institution of Engineering Designers Official Reference Book and Buyers Guide*, Sterling Publications (1988)
15. Anthropometry and Biomechanics. NATO conference series. Plenum Press, New York (1982)
16. Pheasant, S., *Bodyspace*, Taylor and Francis, London (1986)
17. McCormack, *Human Factors Engineering*, McGraw-Hill, New York (1970)
18. International Ergonomics Association, *Proceedings of Ninth Congress*, Taylor and Francis, London (1985)
19. Hammond, J., *Understanding Human Engineering*, David and Charles, Newton Abbot (1978)
20. *Vademecum Ergonomics in Industry*. Philips Technical Library, distributed in the UK by Cleaver-Hulme Press (1963)

7

Materials, properties and selection

Leslie M. Wyatt
(Sections 7.1–7.6, 7.8 and 7.9)

Vince Coveney and Carl Riddiford
(Section 7.7)

Roy Sharpe
(Section 7.10)

Contents

7.1 Engineering properties of materials 7/3
    7.1.1 Elastic properties 7/3
    7.1.2 Tensile-testing parameters 7/3
    7.1.3 Hardness 7/3
    7.1.4 Fracture toughness and impact testing 7/4
    7.1.5 Fatigue 7/4
    7.1.6 Creep and stress rupture 7/4
    7.1.7 Thermal properties 7/4
    7.1.8 Electrical properties 7/5
    7.1.9 Optical properties 7/5

7.2 The principles underlying materials selection 7/5
    7.2.1 Introduction 7/5
    7.2.2 Techniques of materials selection 7/6
    7.2.3 Preliminary examination of design 7/6
    7.2.4 The Classical Procedure 7/6
    7.2.5 Drawbacks of the Classical Procedure 7/7
    7.2.6 The Imitative Procedure 7/7
    7.2.7 The Comparative Procedure 7/8
    7.2.8 Information sources 7/8
    7.2.9 Computerization of materials selection 7/8
    7.2.10 An example of the application of the Classical Procedure 7/9
    7.2.11 An example of problems which may arise during information transfer 7/10

7.3 Ferrous metals 7/11
    7.3.1 Introduction and standards for wrought steels 7/11
    7.3.2 Types of wrought steel 7/12
    7.3.3 Steel making and ingot casting 7/12
    7.3.4 Mechanical working of metals 7/17
    7.3.5 Constitution and heat treatment of ferritic steels 7/21
    7.3.6 Carbon/carbon manganese steels 7/25
    7.3.7 High-strength low-alloy steels 7/27
    7.3.8 Electrical steels 7/28

7.3.9 Hardened and tempered steels 7/28
7.3.10 Free-cutting steels 7/30
7.3.11 Case-hardening steels 7/31
7.3.12 Stainless steels 7/35
7.3.13 Corrosion resistance of stainless steels 7/40
7.3.14 Heat-resisting steels 7/42
7.3.15 Toughness in steels 7/44
7.3.16 Maraging steels 7/45
7.3.17 Weldability of steels 7/45
7.3.18 Tool steels 7/48
7.3.19 Steels for springs 7/49
7.3.20 Cast steel 7/49
7.3.21 Cast iron: general 7/52
7.3.22 Grey cast iron 7/52
7.3.23 Nodular graphite ('SG') iron 7/52
7.3.24 Malleable iron 7/54
7.3.25 Austenitic cast irons 7/55
7.3.26 High-silicon cast iron 7/55
7.3.27 White cast iron (abrasion-resisting white iron) 7/55
7.3.28 High-chromium iron 7/55

7.4 Non-ferrous metals 7/55
    7.4.1 Copper and its alloys 7/55
    7.4.2 Aluminium and its alloys 7/62
    7.4.3 Titanium and its alloys 7/66
    7.4.4 Magnesium and its alloys 7/74
    7.4.5 Nickel and its alloys 7/75
    7.4.6 Zinc and its alloys 7/87
    7.4.7 Lead, tin and their alloys 7/88
    7.4.8 Cobalt and its alloys 7/90
    7.4.9 Other non-ferrous metals 7/91

7.5 Composites 7/92
    7.5.1 Introduction 7/92
    7.5.2 Reinforcing fibres 7/93
    7.5.3 Polymer matrices 7/93

7.5.4   Manufacturing processes for filamentary polymer composites 7/95

7.5.5   Properties of filamentary polymer composites 7/95

7.5.6   Applications of filamentary polymer composites 7/99

7.5.7   Discontinuous fibre-reinforced polymer composites 7/99

7.5.8   Carbon–carbon composites 7/101

7.5.9   Fibre-reinforced metals 7/101

7.5.10 Fibre-reinforced glasses and ceramics 7/103

7.5.11 Reinforced concrete 7/103

7.5.12 Particulate composites 7/105

7.5.13 Laminar composites 7/105

7.5.14 Wood and resin-impregnated wood 7/105

7.6   Polymers 7/106

7.6.1   Introduction 7/106

7.6.2   Classification of polymers 7/109

7.6.3   Polymer processing 7/111

7.6.4   Design of plastic components 7/113

7.6.5   Polymer characteristics, properties and applications 7/117

Appendix: Worked examples of design of plastic components 7/130

7.7   Elastomers 7/132

7.7.1   Introduction 7/132

7.7.2   Vulcanization, compounding and moulding 7/133

7.7.3   Basic elastomer properties 7/134

7.7.4   Thermoplastic elastomers 7/137

7.7.5   Design of elastomeric components 7/138

7.7.6   Elastomeric composites 7/138

7.7.7   Elastomers in use – further properties 7/139

7.7.8   Elastomers selection – some guidelines 7/143

7.7.9   Example applications 7/145

7.7.10 Conclusions 7/147

7.8   Engineering ceramics and glasses 7/147

7.8.1   Introduction 7/147

7.8.2   Standards 7/147

7.8.3   Clay-based ceramics 7/147

7.8.4   Oxide-based ceramics 7/148

7.8.5   Non-oxide ceramics 7/149

7.8.6   Carbons and graphites 7/150

7.8.7   Miscellaneous ceramics 7/150

7.8.8   Glasses 7/150

7.8.9   Glass ceramics 7/151

7.8.10 Mechanical properties 7/151

7.8.11 Manufacturing procedures 7/153

7.8.12 The future prospects of engineering ceramics 7/153

7.9   Corrosion 7/153

7.9.1   Introduction 7/153

7.9.2   Corrosion of metals in air or water 7/155

7.9.3   Resistance of metals to specific corrosion mechanisms 7/155

7.9.4   Biodeterioration 7/163

7.9.5   The prevention of corrosion 7/165

7.9.6   Procedure for identifying origin and mechanism of a corrosion failure 7/166

7.10 Non-destructive testing 7/166

7.10.1 Definition 7/166

7.10.2 Overall scope 7/166

7.10.3 Application areas 7/168

7.10.4 Methods of employing NDT in practice 7/169

7.10.5 Range of techniques available 7/171

7.10.6 Individual techniques 7/172

7.10.7 NDT method selection 7/177

7.10.8 Conclusions 7/178

Acknowledgements 7/178

References 7/178

Further reading 7/179

7.1 Engineering properties of materials

7.1.1 Elastic properties

Elastic or Young's modulus, E (units are GPa). The stress required to produce unit strain in the same direction, i, in the absence of restraint in the orthogonal directions:

$$E_i = \sigma_i \epsilon_i^{-1} \tag{7.1}$$

where σ is the stress and ϵ the strain which it produces. A standard testing method is described in ASTM E231.

Shear modulus, G (units are GPa). The shear stress required to produce unit angular rotation of a line perpendicular to the plane of shear:

$$G = T \cdot \phi^{-1} \tag{7.2}$$

where T is the shear stress and ϕ the angular rotation in radians.

Bulk modulus, K (units are GPa). The hydrostatic pressure p required to effect unit change in volume V:

$$K = p \cdot V \cdot (\Delta V)^{-1} \tag{7.3}$$

Poisson's ratio, ν. The ratio of the strain in a direction orthogonal to the direction of stress to the strain in the direction of stress:

$$\nu = \epsilon_{j,k} \cdot \epsilon_i^{-1} \tag{7.4}$$

These four basic elastic properties apply to homogeneous and isotropic materials and are related by the equations

$$E = 3K\,(1 - 2\nu) \tag{7.5}$$
$$ = 2G\,(1 + \nu) \tag{7.6}$$

In the case of a material which has anisotropic elastic properties the terms used may have different meanings and stresses and strains should be related using tensor analysis.

7.1.2 Tensile-testing parameters

When considering the properties obtained from the tensile test it should be realized that the results are always reported as though the load was applied to the initial cross section, A_o, of the test piece. Any reduction of this cross section is ignored. The test subjects a sample of material of circular or rectangular cross section, of a specific gauge length and equipped with end pieces of larger section which taper smoothly to the gauge length.

When subjected to uniaxial tension beyond the limit of proportionality the material within the gauge length elongates plastically, contracts uniformly or locally transversely and work hardens. The stress σ in the material increases but, because of the decrease in the cross-sectional area, A, the stress, S, calculated from the load the original cross-sectional area, A_o, increases more slowly, attains a maximum value S_u and (usually) declines before the specimen breaks.

Limit of proportionality. The stress at which elastic behaviour of a material is replaced by a combination of elastic and plastic behaviour. This is normally expressed either as

Yield stress S_y. MPa or as

Proof stress $S \begin{Bmatrix} 0.5 \\ 0.2 \\ 0.1 \end{Bmatrix} \%$. MPa

where

$$S = \frac{P \left(\begin{matrix} \text{yield or} \\ \text{strain offset} \begin{Bmatrix} 0.005 \\ 0.002 \\ 0.001 \end{Bmatrix} \end{matrix} \right)}{A_o} \tag{7.7}$$

where P is the load.

Ultimate tensile stress, S_u (MPa), is the maximum load at which a ductile material fractures in the tensile test divided by the original cross-sectional area. S_u is not to be confused with σ_u, the true stress:

$$\sigma_u = S_u \cdot A_o \cdot A_u^{-1} \tag{7.8}$$

where A_u is the cross-sectional area at the time of failure. S_u depends on the dimensions of the specimen (the length of the gauge length is normally $0.565 \sqrt{A_o}$ but it may be 50 mm or some other value), and the rate of application of stress. Both these parameters should be recorded.

Fatigue endurance is related to S_u rather than S_y. The difference between S_u and S_y is a measure of the safety margin against accidental overload.

Most modern design codes base the permissible stress in a material on a factor (say, 66%) of S_y. Some other codes use a factor of S_u as a design criterion. This is cost effective and safe when using a ductile material such as mild steel.

Tensile ductility is reported either as

Elongation e (%),

$$e = \frac{\delta}{L} = \frac{L - L_o}{L_o} \times 100 \tag{7.9}$$

where δ is the extension to fracture, or

Reduction in area, A_R(%):

$$A_R = \frac{A_o - A_u}{A_o} \times 100 \tag{7.10}$$

Ductility is the property that confers tolerance to flaws to a material, and is also an indication of material quality and correct heat treatment. Standards usually specify a minimum ductility. Standard procedures for tensile testing are given in BS 18 and ASTM E8, E345 and B557.

Flexural strength S (MPa) is the calculated maximum stress on the tensile side of a beam which failed when stressed in bending. It is used to measure the strengths of materials such as cast iron and ceramics which are too brittle to be tested by the standard tensile test. A beam stressed in three-point loading has the maximum stress applied only on one line on the surface.

Multiple testing is required to produce results which can be used in design and much higher safety factors (see Section 7.7) are required than are used for ductile materials tested by the standard tensile test. A standard testing method is described in ASTM C580.

7.1.3 Hardness

Hardness is the resistance of a material to permanent deformation by indentation or scratching. It is not a simple intrinsic property of a material but a complex response to a test. Vickers, Brinell and Knoop compare the load and the area of the impression produced by an indenter, Rockwell the load and the depth of the impression, Shore the rebound of an indenter and Mohs the ability of one material to scratch another.

Vickers Hardness *HV* (strictly, the dimensions are those of force per unit area but in practice Vickers and Brinell hardnesses are comparative numbers) is the quotient obtained by dividing the load *F* (kgf) by the sloping area of the indentation left in the surface of the material in mm^2 by a 136° pyramidal diamond indenter:

$$HV = \frac{2F \sin (136/2)}{d^2} = 1.854 \, Fd^{-2} \quad (7.11)$$

where *d* is the diagonal of the indentation.

Hardness is a measure of the wear resistance of a material. Used on metals the Brinell hardness value of a medium-carbon steel is directly related to the UTS, the Vickers hardness to the proof stress. Vickers, Brinell and Rockwell can be used to ensure that heat treatment has been correctly carried out.

Hardness testing of ceramics is performed with very light loads to avoid failure of the material.

Standards for hardness testing are

Vickers	BS 427, ASTM E92
Brinell	BS 240, ASTM E10
Rockwell	BS 891, ASTM E18
Schleroscope	ASTM 4448

7.1.4 Fracture toughness and impact testing

7.1.4.1 Fracture toughness testing

Plane strain fracture toughness K_{1c} ($Nm^{-3/2}$). The limiting stress intensity required to cause crack extension in plane strain at the tip of a crack when the stress is transverse to the crack. K_{2c} and K_{3c} are parameters corresponding to stresses in the plane of the crack. Standard testing methods are BS 5447, ASTM E399.

Elastic-plastic fracture toughness, J_{1c}. The limiting value of the *J* integral (which is a line or surface integral used to characterize the fracture toughness of a material having appreciable plasticity before fracture) required to initiate crack intension from a pre-existing crack.

Stress intensity to initiate stress corrosion, K_{1scc} ($Nm^{-3/2}$). The limiting stress intensity required to initiate propagation of a crack in a specific environment at a specific temperature.

7.1.4.2 Impact testing

In contradistinction to fracture toughness testing which quantifies a material property, Izod cantilever and Charpy beam type impact test results are a function of the method of testing. They particularly use a machined, rather than a fatigue-propagated notch. Results are expressed as the energy *J* (in joules) required to break the cross-sectional area behind the notch.

Testing a number of specimens of body-centred metals, ceramics and polymers over a range of temperature will reveal a transition temperature below which brittle behaviour is observed. This is reported as the fracture appearance transition temperature (f.a.t.t) (°C) at which half of the fracture surface is fibrous and half crystalline or as the fracture energy transition temperature (°C) at the inflection in the energy curve. This is a criterion of use for assessing material composition, treatment and behaviour. Standards for impact testing are BS 131, ASTM E23, E812 and E602 (sharp notch tension testing).

7.1.5 Fatigue

S–N curve. Graphical relationship between stress, *S*, and number of cycles, *N*, required to cause failure of a material in a fatigue test. This depends on mean stress, frequency and shape of stress cycle, temperature and environment, all of which should be specified. Note that this applies to high cycle fatigue.

High strain fatigue is strain, not stress related and plastic strain per cycle, ϵ_p, is inversely proportional to $N^{1/2}$ for almost all engineering materials.

Fatigue endurance limit, σ_e (MPa). The maximum stress below which a material is presumed to be able to endure an infinite number of cycles. This applies only to certain specific engineering materials such as steel and titanium.

Fatigue limit, σ_{10^7} (MPa). The maximum stress below which a material is presumed to be able to endure a specific number of cycles, usually 10^7 to 10^8 but may be lower for specific applications. Fatigue endurance limit and fatigue limit are both statistical quantities and depend on the same parameters as are detailed under the *S–N* curve.

Standard methods for fatigue testing are BS 3518, ASTM E513, E912, E206, E742, E466, E606, E468, E739. There are other ASTM standards in Vol. 03;01.

Fatigue life for p% survival (MPa). The maximum stress below which not less than *p%* of tested specimens will survive.

Fatigue notch factor, K_f. Ratio of the fatigue strength of a notched to that of an unnotched specimen.

Fatigue notch sensitivity g (ratio)

$$g = \frac{(K_f - 1)}{(K_t - 1)} \quad (7.12)$$

where K_t is the stress concentration factor. When *g* approaches 1 a material is fully sensitive. When *g* approaches 0 a material is insensitive.

7.1.6 Creep and stress rupture

Creep range. The temperature range, usually above half the melting point temperature in °K, at which the design stress computed from creep or stress rupture is lower than that calculated from yield or 0.2% proof.

Stress to rupture, σ_R (MPa). The tensile stress at which a material will fail if held at a specific temperature for a specific time, depending on the type of application.

Stress to a certain creep strain, σ_ϵ (MPa). The tensile stress at which a material will creep to a specific strain, ϵ (ignoring the initial strain on loading), if held at a specific temperature for a specific time. σ_ϵ and σ_R are related for a specific material.

Creep rupture elongation, *(%)*. The percentage of the original length by which a creep rupture specimen extends before failure.

Larson–Miller Parameter, P. A parameter used to extrapolate the results of creep rupture tests carried out at relatively short times to longer times. The rate equation is

$$P = T (\log t_R + C) \quad (7.13)$$

where *T* is the absolute temperature and *C* an empirically determined constant. Other rate equations have been derived by Sherby-Dorn and Manson-Haferd.

Standards for creep and stress rupture testing are BS 5447 and ASTM E1 329 for metals and BS 4618 for plastics.

7.1.7 Thermal properties

Specific heat per unit mass, C_p ($Jkg^{-1}K^{-1}$). The rate of change of heat content of 1 kg of the material with temperature. Specific heats are often quoted as $Jg^{-1}K^{-1}$ or in compilations of thermodynamic data as $cal.mole^{-1}K^{-1}$. They may also be quoted as mean specific heats over a range of temperature, usually 25°C to a specific elevated temperature.

Specific heat per unit volume, C_v ($Jm^{-3}K^{-1}$). Specific heat C_v of a gas at constant volume does not include the work required to expand the gas and is therefore lower than C_p.

Thermal expansion. Linear thermal expansion α (K^{-1}). The fractional increase in length, l, per degree rise in temperature at a specific temperature, T:

$$\alpha(T) = l^{-1}\frac{dl}{dT} \tag{7.14}$$

More commonly, a mean expansion between two temperatures

$$\bar{\alpha} = l_0^{-1}\frac{\Delta l}{\Delta T} \tag{7.15}$$

when Δl is the change in length from l_0 at temperature T_0 when the temperature is changed by $\Delta T = T - T_0$, is quoted. In data compilations T_0 is often 25°C.

In anisotropic materials (single crystals or materials with preferred orientation), the thermal expansion coefficient may differ between each of the three orthogonal directions x_i, x_j and x_k.

Thermal conductivity, λ ($Wm^{-1}K^{-1}$) is the heat flow per unit area generated by unit temperature gradient

$$\lambda = A^{-1}\frac{dQ}{dt} \cdot \frac{dl}{dT}$$

where dQ/dt is the rate of heat flow across area A and dT/dl is the temperature gradient. λ is normally a function of temperature and, in anisotropic materials, of direction.

Thermal diffusivity, D (m^2s^{-1}), is a measure of how fast a heat pulse is transmitted through a solid:

$$D = \frac{\lambda}{\rho C_p}$$

where λ is the thermal conductivity, ρ the density and C_p the specific heat. Thermal diffusivity varies with temperature but can be measured more quickly and accurately than thermal conductivity.

7.1.8 Electrical properties

Volume resistivity, ρ (Ωm). The resistance in ohms of a metre cube of the material:

$$R = \rho l A^{-1} \tag{7.16}$$

where R is the resistance of a body, l its length and A its uniform cross section.

Dielectric breakdown. No standard symbol ($KVmm^{-1}$ or KV). Measured according to IEC 672, BS 1598:1964, ASTM D116 or DIN 40685.

Relative permittivity. The ratio of the charge storage capacity of a material in an electric field which results from crystal realignment of its structure compared with the charge storage capacity of empty space.

Permittivity ($AsV^{-1}m^{-1}$), where A is current in amps, s time in seconds, V electric potential in volts and m metres.

Dielectric loss, tan δ. The phase angle introduced by the time taken for polarization to occur on application of a field. Tan δ is frequency and (usually) temperature dependent.

7.1.9 Optical properties

Spectral absorption coefficient, K (mm^{-1}). The log of the ratio of the incident to the transmitted light intensity through unit thickness:

$$K = (\log I_0 - \log I)\, x^{-1} \tag{7.17}$$

where I_0 is the incident intensity, I is the transmitted intensity and x the thickness in mm. K varies according to the wavelength of the incident light.

Refractive index, μ. The ratio of the velocity *in vacuo* to that in the medium:

$$\mu = V_{vacuo}. V_{medium}^{-1} \tag{7.18}$$
$$= Sin\, i \cdot sin^{-1}r$$

where i and r are the incident and refracted angles of the beam to the surface.

7.2 The principles underlying materials selection

7.2.1 Introduction

The requirement for selection of a material may arise from a number of circumstances, including:

1. An entirely new component is to be developed to perform functions not hitherto visualized;
2. A component is required to perform an increased duty which has rendered the performance of the material previously used unsatisfactory;
3. The incidence of failure in the material which has so far been specified is too high, or occurs at too early a stage in the life of the component;
4. Some material shortcoming not strictly related to operational performance has become apparent. A material which was acceptable initially may become unsatisfactory because;
   (a) it has become so expensive, relatively or absolutely, that the equipment, of which the component is a part, can no longer fulfil an economic function;
   (b) it is no longer available locally or globally (or might become unavailable in the event of an emergency);
   (c) it is no longer acceptable on grounds of health, safety, aesthetics or public sentiment.

Examples of materials which have been developed in answer to the listed circumstances are:

1. The 'magnox' can for the first-generation gas cooled power reactor;
2. Superalloy blades of progressively increasing creep resistance, culminating (so far) in the directionally solidified castings now used;
3. Notch ductile aluminium killed steels to replace the materials which failed by brittle fracture in the 'liberty' ships;
4. (a) Steel-cored aluminium instead of copper conductors for overhead power lines;
   (b) Nickel-based superalloys for military jet aircraft after it was realized that the sources of cobalt in cobalt-based superalloys were situated in volatile African or Iron Curtain countries;
   (c) Ceramic fibres to replace asbestos as a binder for heat insulation because of the hazard of 'Asbestosis'.

All of these examples of material choice were developed by means of the techniques which will be described later in this section. The materials selected have performed entirely satisfactorily, and in those cases where operational parameters were not the cause of replacement the substitute material has in fact performed better than the original.

7.2.2 Techniques of materials selection

There are at least three different techniques by which the optimum material for use in a specific component may be selected:

1. The 'Classical Procedure' using functional analysis and property specification;
2. The 'Imitative Procedure' which consists of finding out what material has been used for a similar component;
3. The 'Comparative Procedure' which consists of postulating that the component be made from some cheap and well-understood engineering material, assessing in what ways such a material's performance would be inadequate and from this arriving progressively at the right material.

The Classical Procedure is the only one that is universally applicable and it is essential, even when procedures 2 or 3 are followed, to check the findings by functional analysis and property specification. By itself, however, the Classical Procedure is expensive and time consuming and requires a considerable amount of prototype testing to ensure that no critical requirement or essential property has been overlooked.

The Imitative and Comparative Procedures, where applicable, will provide invaluable shortcuts, save a great expenditure of time and money and will help to ensure that no essential parameter has been overlooked. The materials engineer will be wise to employ all three techniques in parallel wherever practicable.

7.2.3 Preliminary examination of design

It is essential, whichever procedure is employed, to commence with the analysis of the function of a component, the critical examination of the design and the establishment of materials property requirements. Design affects materials selection procedure at all stages. A component may fulfil its function in more than one way using different designs which result in different materials property requirements and hence different optimum materials, and different manufacturing routes. For example, a box with a hinged lid may be made from two pieces of thin metal sheet and a pin or from one piece of polypropylene.

The effect of design on manufacturing process is particularly important when considering a materials change in an existing product (for example, from metal to plastic or ceramic). Design and materials selection constitutes an iterative process: design affects the optimum material which, in turn, affects the optimum design.

7.2.4 The Classical Procedure

7.2.4.1 Functional analysis

Functional analysis is a formal way of specifying material properties starting from the function of a component. This involves specifying

1. The functions of a component
2. The requirements for a component
3. The requirements for material properties

Function The overall function should be specified as broadly as possible to allow the greatest number of options in design. Where there are several functions all must be specified. This latter requirement is essential even when the choice of material has been necessitated by the failure of a material to perform one specific function, because a change in a material to make it capable of fulfilling one function may make it incapable of fulfilling another. Using a higher-tensile steel to carry an increased load may result in brittle fracture under shock, for example.

Component requirements When the functions have been established the component requirements can be identified. For example, the one-piece box mentioned above must be capable of being opened and closed an indefinite number of times.

In specifying component requirements it is important to remember that it must be possible to produce the article in the required form, and that the component must withstand the environment in which it is operating at least for its design life.

Materials property requirements From the component requirements the materials property requirements can be established. The material for the one-piece box must have an almost infinite resistance to high strain fatigue in air at room temperature. This is obtainable from a polypropylene component manufactured in a specific way.

The property requirements established by functional analysis may be quantitative or qualitative. For example, the material for an automotive exhaust *must* be sufficiently strong and rigid to withstand weight and gas pressure forces. Quantitative requirements must be established by analysis of the design and operating conditions. In comparison, the requirement to resist corrosion and oxidation is qualitative. Property requirements may also be classified as essential and desirable. The strength requirement in the material for the exhaust is essential. Environmental resistance is often sacrificed to minimize initial cost (even when, as in this example, a more resistant material may have superior economics considered over the total life of an automobile).

7.2.4.2 Materials requirement checklist

The next stage is the formulation of a materials requirement checklist, and the properties which will feature in this checklist are given in Section 7.1.

The reader should not be discouraged by the length and complexity of this list. It will, in many cases, become evident that whole ranges of properties (and materials) may safely be ignored at first glance. For example, if the component is required to transmit or refract light the choice of material is immediately limited to a glass, mineral or polymer and design and property specification is restricted. If electrical conductivity is significant, choice is limited to conducting metals, resistive or semiconducting materials or insulators.

Important characteristics The important characteristics requiring consideration for many engineering components are: mechanical properties; stiffness, strength and ductility; physical properties, thermal, electrical, magnetic and optical properties; environmental resistance and wear including applicability of corrosion protection; capacity for fabrication and cost which includes material, manufacturing, operating and replacement cost.

Mechanical properties Resistance to manufacturing and in-service loads is a requirement of all products. The material must not buckle or break when the component comes under load. It must also have an economic life in fatigue or under creep conditions. Where a number of materials meet the minimum strength and stiffness requirements a preliminary shortlist can be made on the basis of cost per unit strength or unit stiffness (or, in the case of space or transport applications, strength per unit weight).

Physical properties Physical properties such as specific gravity are important for most applications. For some, as noted above, optical or electrical properties may be paramount.

Environmental resistance (corrosion) Environmental resistance is a property whose universal importance has been obscured by the circumstance that it has been inherent in the choice of materials for most common applications.

Corrosion resistance requirements vary, from the absolute, where even a trace of contamination in a fine chemical food or cosmetic is unacceptable, to the barely adequate, where the cheapest material whose integrity will survive the minimum economic life should be chosen.

When assessing corrosion resistance, attention should be paid not only to the rate of general corrosion but also to the possibility of localized corrosion which, as described in Section 7.8, may destroy component integrity without significant dimensional changes. Corrosion mechanisms may cause the disintegration or deterioration of metals, polymers, ceramics, glasses and minerals.

Wear resistance Wear is the product of relative movement between one component and another or its environment. Its prevention depends principally on design and operation, but can be minimized or eliminated by the correct choice of material, material pair, or coatings.

Manufacture and cost Manufacturing routes are selected on the basis of lowest total cost to produce the desired performance. In the past, performance requirements have favoured certain processes such as forging instead of casting, but more recently attention to quality improvement techniques in casting have levelled up in-service properties and cost is emerging as the deciding factor.

It is difficult to assess the relative total costs of different material/manufacturing route combinations at the early stage of a design and, wherever possible, finalizing precise geometries should be delayed until possible materials and manufacturing routes have been identified, otherwise there will be an avoidable cost penalty.

Cost of a component includes:

In-position costs which comprise, material cost influenced by quality and quantity, manufacturing cost, quality control cost and administration cost and
Lifetime costs which comprise servicing, maintenance, warranty, outage and replacement cost.

Costs which accrue at different periods must be discounted to a common date. Differences in discounting rates between different countries or organizations can lead to the selection of different materials for applications which are, in all other respects, identical.

Comprehensive knowledge of the application is important in assessing the relative importance of cost and performance. Cost is paramount in the case of a widely marketed consumer item where small differences in reliability and life have little influence on saleability but performance is paramount for certain sporting or military applications.

There may, for example, be no advantage to be gained by incurring additional expense to prolong the life of a car exhaust system from 5 to 7 years when the purchaser intends to replace the car after 2 years. The sales department must always be consulted before the final material choice is made in cases such as this.

On the other hand, the material from which a racing-car spring is made must have the maximum possible specific rigidity, regardless of cost. The potential rewards for employ-ing the optimum material far outweigh any cost saving which might be obtained by choosing the second-best.

Material selection When the properties of candidate materials have been ascertained (by procedures which will be discussed later) a shortlist should be established. If it is immediately obvious that one material is outstandingly superior the choice is straightforward. Often there is one property requirement that outweighs all the others. When this is the case the choice is simplified. There may, however, be a number of possible materials, or none may meet all requirements.

A number of procedures have been proposed for eliminating all but one of a number of possible materials. These include an advantage/limitation table, an elimination grid[1] and ranking methods for properties, and number of properties, that meet requirements.

Often, local factors (i.e. using a material which is familiar locally, using a material which has a margin in one specific property that may be of value in a future marque of component or using a material that is suitable for a locally available fabricating or machining technique) will influence the final choice. When no material meets the requirement, a careful re-examination may reveal that a change in design, environment or operating conditions will enable satisfactory performance at minimum extra cost. As a last resort, it may be possible to arrange for easy replacement after a fixed time, and to hold a supply of spares.

7.2.5 Drawbacks of the Classical Procedure

The application of the procedures outlined above guarantees success if followed logically and completely and design, operating and material parameters are thoroughly understood. However, this is seldom the case in practice. Designs cannot always be evaluated precisely, material properties are seldom specified fully and it is impossible to predict exactly what an operator will do.

In most cases the Classical Procedure requires a considerable amount of mechanical property evaluation, possibly materials and process development and a substantial programme of prototype testing, before satisfactory performance can be guaranteed. Time may not be available to undertake this. The easiest way to a short cut is the Imitative Procedure.

7.2.6 The Imitative Procedure

Find out what material has been used for the same component or a component as similar as possible and use this, an improved material, or a material modified for the difference in conditions. Successful implementation of this procedure not only verifies design and reduces material property evaluation but also very substantially reduces prototype testing because the most likely causes of failure have already been experienced and cured.

The problem is to ensure that the information obtained is accurate, comprehensive and fully understood. Even within an organization, operators' reports are not completely reliable. A report of satisfactory performance may merely mean that operators know when the component is about to fail so that they can replace it without extra outage. One operator may have found out how to handle this component and a similar component may fail disastrously in the hands of another. These difficulties are compounded when information is obtained from an outside source, whether rival or friendly. Informants do not mean to mislead. The information they withhold is usually information that they cannot imagine that the recipient does not already possess.

The ability to obtain information when it is required depends on appropriate organization. There should be a materials engineer who combines knowledge of all the materials and requirements of the organization with an acquaintanceship or, ideally, friendship with all similar persons throughout the world. The right person is, when presented with a problem, able to contact someone who already has experience of the matter wherever he or she may be and obtain the benefit of that experience. His or her knowledge of the other organization would be sufficiently comprehensive to enable him or her to assess the effect of different procedures between the two organizations. The chemical industry (as described by Edeleanu[2]) operates a worldwide information system with personnel of this type and has found that information on what can be done and how to do it may most efficiently and quickly be obtained in this way.

7.2.7 The Comparative Procedure

The Comparative Procedure for materials selection operates by selecting a cheap, tolerant and well-understood material and investigating to what extent its properties fall short of those required for the component to operate satisfactorily. A typical example, and one for which this procedure is extremely suitable, is the specification of a material for chemical process plant.[3] A scheme design is produced using carbon steel which is cheap, readily produced, easily fabricated, ductile and therefore tolerant of flaws and geometrical irregularities and corrodes uniformly at a predictable rate. If carbon steel is shown not to be satisfactory the unsatisfactory property or properties can be modified. The necessary change may impair other properties but will do so in a predictable way. Thus:

> Improved corrosion resistance may be obtained by the use of a steel with a higher chromium and possibly a higher nickel content. This will increase cost and probably also delivery time, render design and fabrication more sensitive and may enhance sensitivity to localized corrosion.

Improved strength may be obtained by the use of a steel with increased carbon and alloy content with drawbacks similar to those that applied in the case of the improved corrosion-resistant material.

A higher temperature of operation may require the use of a creep-resisting steel, again with similar disadvantages. Operating at a lower temperature may require a steel with guaranteed low-temperature properties or may, in the limit, require an aluminium alloy.

Evidently, this procedure, with the exception of the case where a change is made to a completely different material, involves changes which are progressive, and whose effects can be foreseen. Therefore the chances of encountering some unexpected drawback are minimized and the requirement for component testing is minimized also.

7.2.8 Information sources

It has so far been assumed that staff charged with material selection have at their disposal a complete range of information on material properties. This may be the case when electronic databases[4] now being developed are perfected. In the meantime they should have available for reference, British, American and possibly German materials standards, and such volumes as the *ASM Metals Handbook*, the *Plastics Encyclopaedia* and as up-to-date a ceramics work as is then available. In addition, the *Fulmer Materials Optimizer*, which shows properties of all types of engineering material in the form of comparative diagrams, will prove an invaluable guide to materials selection.

When the field has been narrowed down to a few materials the materials manufacturer should be consulted. Organizations such as steelmakers or polymer manufacturers possess more information on their products than has been published and also experience in their application. They can provide valuable guidance on final selection, design and manufacture.

Furthermore, it should be remembered that a standard steel obtained from one manufacturer may differ in some relevant characteristic from the same steel purchased from a competitor. A reputable manufacturer is aware of this and should warn of problems which will have to be overcome in addition to extolling advantages.

7.2.9 Computerization of materials selection

Much effort is at present deployed on the production of databases for material properties. Three recent international conferences have been devoted to this subject[4] and a directory of databases for materials is available.[5] These databases are not necessarily material-selection systems and much interest has been directed to providing systems which will undertake material selection by the classical functional analysis and property-specification procedure.

It is not possible, legitimately, to computerize the Imitative Procedure because no organization can be expected knowingly to provide another (possibly competing) organization with access to programs intimately concerned with its own design philosophy and development programme. There is, on the other hand, no difficulty in computerizing the Comparative Procedure of materials selection. So long as (1) the materials involved form a very closely related family with very similar properties, (2) no novel and unforeseen failure mechanism takes over and (3) the properties of the candidate materials have been determined comprehensively, a computer program which will select the optimum material for a specific application can easily be produced.

Two such programs are known to exist. They are ICI (EPOS) for the selection of polymers and a Sandvik program for selection of cutting tools. These are knowledge-based systems dealing with families of essentially similar materials. A computer program for selection of process plant materials as described in Section 7.2.7 would be equally straightforward, provided the requirements could be met by a steel and no unforeseen failure mechanism took over.

The requirements for a computer program to undertake selection by the Classical Procedure are much more general and much less well defined. The starting point is a product design specification (PDS), which is a functional and formal statement of what is required from the product to be designed, not a description of the product. The PDS contains a material design specification (MDS) which, like the PDS, is incomplete and ill-defined. The computer must match this MDS to descriptions of existing materials and materials specifications (MS) which may be incomplete and reflect various levels of confidence. The result of the analysis may be a requirement to modify the PDS, to develop a new material or to acquire additional information concerning specific materials.

A computer system capable of selecting materials requires:

1. The ability to deal with simple and complex data structures;
2. Powerful structures for data acquisition and updating by augmentation and modification;
3. The ability to manage sparse data;
4. The ability to compare incomplete descriptions;
5. The ability to distinguish the relationships, and sometimes

lack of relationships between materials, or parameters nominally in the same classification;

6. The ability to be easily extensible.

It must take into consideration:

1. The duty or function of the component;
2. The materials properties;
3. The manufacturing route;
4. Shape, dimensions and failure mode;
5. The relative cost of the materials, manufacturing routes and designs considered.

It must, in addition, have the following user characteristics so that it can be operated by designers and engineers and free the materials engineer for long-term, difficult and strategic problems. It should:

1. Be rapid in use;
2. Require a minimum of learning;
3. Be accessible at different levels to suit different levels of user;
4. Have text and graphical output; and
5. Have recording facilities.

Various procedures for optimized decision making have been put forward, including linear programming methods[6] and numerical algorithms.[7] There is a tendency to rely on ranking methods which allocate a rank from 0 to 3 for each material property. This introduces an imprecision which should not be necessary in the application of a computer capable of relating property variation with overall cost.

There exists at least one system, which is claimed to be applicable at the innovation stage in design. A brief description of this system is given as an example of methods which could be employed.

PERITUS[8] is a knowledge-based system which comprises three main stages:

1. A Director stage which directs the non-specialist to data and knowledge modules. The structure of this is shown in Figure 7.1;
2. A Presort stage which produces a shortlist of candidates from the materials indicated by the Director stage utilizing ranking lists;
3. An Evaluation and Optimization stage. This can either display the shortlist with deviations compared with the ideal or, where the required modules exist, optimize according to failure modes, component dimensions and properties (presumably including overall cost).

The system is modular and additional modules may be added as required or when available (see Figure 7.2). Further information on the PERITUS system is to be found in references 9 and 10.

7.2.10 An example of the application of the Classical Procedure

In conclusion, the principles governing materials selection are illustrated by a description of the choice of materials for the core of the magnox reactors. Design and materials selection for the magnox reactors had to fulfil the following terms of reference:

The reactors must operate on natural uranium because enrichment would not be available within the required time scale. The reactors must produce electricity safely, reliably and economically and their construction must meet a time scale.

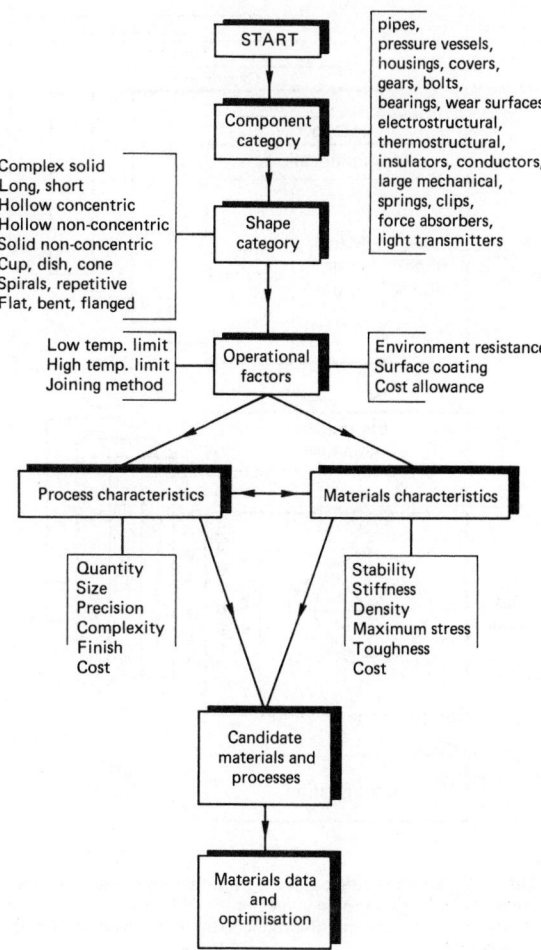

Figure 7.1 The structure and features of the Director stage of the PERITUS knowledge-based system for the selection of engineering materials. (Reproduced by permission of *Metals and Materials*)

The requirement to operate on natural uranium required, from nuclear considerations, that the fuel elements should be based on bars of almost pure uranium between 2.5 and 3 cm diameter and restricted the choice of moderator material to a dense form of carbon, beryllia or heavy water. Adequate supplies of heavy water would not become available within the time scale. Beryllia was hazardous and little was known about the effect on it of the helium that would be produced during irradiation. The core therefore had to be a graphite structure which must be self-supporting and provide channels for the coolant. A research programme demonstrated that this was practicable. Information available from abroad indicated that pure uranium distorted severely under irradiation. A research programme also showed that it also swelled severely.

An intensive programme resulted in the development of 'β quenched uranium', a dilute alloy with aluminium and iron. Distortion is minimized by means of a fine random-grain structure and swelling by a precipitate of uranium aluminium intermetallic which restricts diffusion of fission product gases.

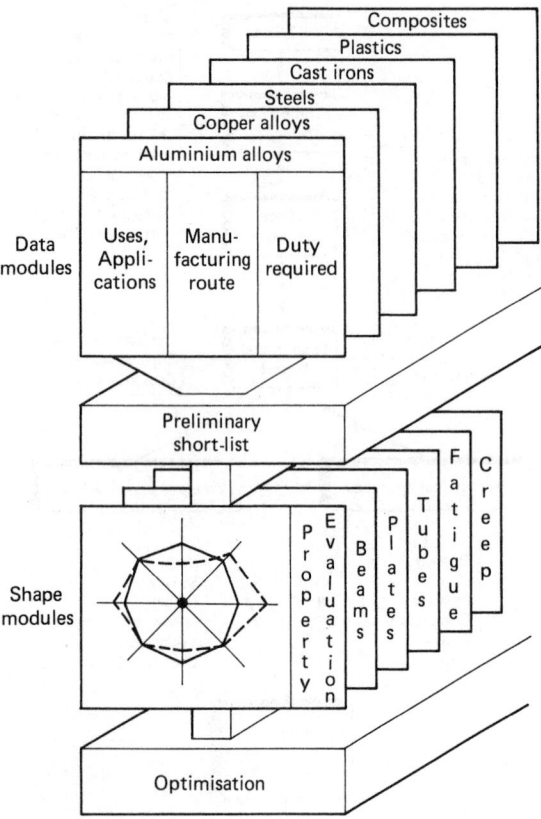

Figure 7.2 A representation of the organization of the pre-sort, evaluation and optimization stages of the PERITUS knowledge-based system. (Reproduced by permission of *Metals and Materials*)

The size of the gas bubbles is thereby minimized and their growth is restrained by surface tension. This keeps swelling to an acceptable value.

The development programme also showed that the uranium must not be heated long or often into the β phase. This limited the temperature of the centre of the fuel elements, and therefore their rating.

The next step was to choose a gaseous coolant and a canning material that was impermeable to uranium, fission products, plutonium, the chosen coolant, atmospheric air and water. The possible gaseous coolants were helium, hydrogen, oxygen, steam, air, nitrogen and CO_2/CO. Hydrogen was eliminated on grounds of safety, nitrogen on nuclear grounds (a loss of nitrogen coolant would greatly increase reactivity) and oxygen or steam would attack graphite. It was doubtful whether adequate supplies of helium could be made available. Carbon dioxide (which would equilibrate in the circuit with carbon monoxide) was cheap, readily available, had good heat-transfer properties and a low neutron capture cross section, and was relatively inert. Provided that a suitably compatible canning material could be found, carbon dioxide would serve.

The canning material had to have a low neutron absorption cross section. It had to be compatible with uranium, carbon dioxide and, unless a suitable separating material could be

incorporated, with graphite. The material had to be fabricated, and sealed into an impervious can in direct contact with uranium and the can had to be finned to increase heat transfer. It had to be strong enough for the fins to withstand coolant forces at an operating temperature up to, say, 500°C and to be ductile enough to deform rather than fracture when subjected to the permanent strains imposed by the distortion of uranium under irradiation (which had been minimized but not eliminated) and by the cyclic strains caused by differential expansion.

The materials permitted by nuclear considerations were carbon, beryllium, magnesium, aluminium and zirconium. Carbon lacked the necessary ductility for a can of the size required in direct contact with uranium. (Although it has since been used successfully in the form of small spheres containing uranium oxide or carbide with a helium coolant. This solution was ruled out because of the enrichment required.) Beryllium lacked the required ductility and alloyed with uranium at reactor operating temperatures. Aluminium had been shown to alloy with uranium in a way that destroyed the integrity of the can at reactor operating temperatures. Satisfactory cans could be made from either magnesium or zirconium.

Zirconium as mined and extracted contains hafnium, which has a high neutron capture cross section and would have to be removed before the zirconium could be used for cans. This process would have to be developed, the resulting zirconium would be very expensive and the fabrication cost of cans would also be high and require a long period of development.

Magnesium was therefore chosen, but because of its known tendency to oxidize in air (not in carbon dioxide) beryllium was added in the form of an aluminium beryllium alloy so that the cans would be immune from oxidation in the event of an incident in which air gained access to the reactor charge of cans at high temperature.

Economic methods of fabricating finned magnox cans for service with uranium and methods of sealing which guaranteed less than 1 in 10^4 defective cans were developed. Many millions of cans have been manufactured and have operated without fault, and the use of the classical functional analysis and property specification method of materials selection is fully vindicated.

7.2.11 An example of problems which may arise during information transfer

In the early days of operation of the magnox reactors discussions were held with an organization engaged in developing the same type of reactor. The specification and performance of the successful magnox cans were described, but the other organization decided to use instead of the magnox composition a magnesium–zirconium alloy whose mechanical properties were superior. It was very much later, when very large numbers of cans had been produced, that it was discovered that, while both materials were completely impervious to uranium, plutonium which was produced during irradiation diffused through the aluminium-free cans. Although the amount diffusing through was too small to have any effect on operation, or even, when the cans had been removed from the reactor, on health, it was sufficient to interfere with the operation of the burst-slug detection gear and if continued would eventually have made the reactors inoperable.

This information had not been withheld intentionally. It was appreciated that plutonium could and might diffuse through magnesium but the possible influence of this diffusion on burst-cartridge detection and the effectiveness of aluminium in blocking it had not been recognized at this early stage.

7.3 Ferrous metals

7.3.1 Introduction and standards for wrought steels

The development of alloys of iron (which include wrought iron, steel and cast iron) has been essential to a technological society. Iron ore is widespread, cheap and easily mined, and from it can be produced alloys with the widest range of properties of any material. Steels may be manufactured with properties which vary continuously from soft and ductile to strengths and hardnesses which, until the very recent development of the sialons, exceeded those of any relatively inexpensive material.

There are a very large number of wrought steels and numbering systems. German specifications alone include 1400 grades. Because of the rapid evolution of ferrous metallurgy none of these classifications and numbers are ever fully up to date. In engineering, standard steels are commonly referred to by their AISI/SAE number[11] based on composition (Table 7.1). The last two figures indicate the carbon content in the case of non-stainless steels; the corresponding general British specification[12] is BS 970: 1983 and earlier. In it steels are identified by a three-digit number denoting the type of steel followed (in the case of non-stainless steels) by a letter denoting the type of specification – A for analysis, H for hardenability or M for mechanical properties, followed by two numbers denoting carbon content. For stainless steels the first three figures are identical with the AISI figures; the letter is S, and the final two numbers are coded. Wherever possible, in the account of classes of steel which follows, the AISI/SAE and BS ranges will be given. There are also a number of specifications depending on product or application which do not always follow the BS 970 or AISI/SAE numbering system. In addition to the standard steels, there are well-known steels which are recognized by designations used by their originators and are in no way inferior to standard steels. Many steels of this kind are recognized and used worldwide in aircraft specifications and are made by many different steelmakers, although they have not yet been recognized by the several national bodies that govern steel specifications. Cast irons and cast steels have separate specification numbering systems.

Table 7.1 Basic numbering system for AISI/SAE steels

Numerals and digits	Type of steel and average chemical contents (%)
	Carbon steels
10XX	Plain carbon (Mn 1.00% max)
11XX	Resulphurized
12XX	Resulphurized and rephosphorized
15XX	Plain carbon (max Mn range – over 1.00–1.65%)
	Manganese steels
13XX	Mn 1.75
	Nickel steels
23XX	Ni 3.50
25XX	Ni 5.00
	Nickel–chromium steels
31XX	Ni 1.25; Cr 0.65 and 0.80
32XX	Ni 1.75; Cr 1.07
33XX	Ni 3.50; Cr 1.50 and 1.57
34XX	Ni 3.00; Cr 0.77

Numerals and digits	Type of steel and average chemical contents (%)
	Molybdenum steels
40XX	Mo 0.20 and 0.25
44XX	Mo 0.40 and 0.52
	Chromium–molybdenum steels
41XX	Cr 0.50, 0.80 and 0.95; Mo 0.12, 0.20, 0.25 and 0.30
	Nickel–chromium–molybdenum steels
43XX	Ni 1.82; Cr 0.50 and 0.80; Mo 0.25
43BVXX	Ni 1.82; Cr 0.50; Mo 0.12 and 0.25; V 0.03 minimum
47XX	Ni 1.05; Cr 0.45; Mo 0.20 and 0.35
81XX	Ni 0.30; Cr 0.40; Mo 0.12
86XX	Ni 0.55; Cr 0.50; Mo 0.20
87XX	Ni 0.55; Cr 0.50; Mo 0.25
88XX	Ni 0.55; Cr 0.50; Mo 0.35
93XX	Ni 3.25; Cr 1.20; Mo 0.12
94XX	Ni 0.45; Cr 0.40; Mo 0.12
97XX	Ni 0.55; Cr 0.20; Mo 0.20
98XX	Ni 1.00; Cr 0.80; Mo 0.25
	Nickel–molybdenum steels
46XX	Ni 0.85 and 1.82; Mo 0.20 and 0.25
48XX	Ni 3.50; Mo 0.25
	Chromium steels
50XX	Cr 0.27, 0.40, 0.50 and 0.65
51XX	Cr 0.80, 0.87, 0.92, 0.95, 1.00 and 1.05
501XX	Cr 0.50
511XX	Cr 1.02
521XX	Cr 1.45
	Chromium–vanadium steels
61XX	Cr 0.60, 0.80 and 0.95; V 0.10 and 0.15 minimum
	Tungsten–chromium steels
71XXX	W 13.50 and 16.50; Cr 3.50
72XX	W 1.75; Cr 0.75
	Silicon–manganese steels
92XX	Si 1.40 and 2.00; Mn 0.65, 0.82 and 0.85 Cr 0.00 and 0.65
	Low-alloy high tensile steels
9XX	Various
	Stainless steels AISI (*not* SAE)
2XX	Chromium–manganese nickel nitrogen austenitic steels
3XX	Chromium–nickel austenitic steels
4XX	Chromium–ferritic & martensitic steels
5XX	Silicon–chromium steels
	Boron-intensified steels
XXBXX	B denotes boron steel
	Leaded steels
XXLXX	L denotes leaded steel

7.3.2 Types of wrought steel

There are at least eleven separate classes of wrought steel most of which are further sub-divided.

Carbon steel (AISI 1006–1572–BS 970 000–119) is the basic type which far exceeds all other metals in tonnage produced. Low-carbon steels are sub-divided into hot-rolled (see Manufacturing Procedures below) and cold-rolled steel.

Hot-rolled steel has low strength (although the higher-carbon versions can be heat treated to high hardness in small sections) and low toughness but is readily available at low cost and is easily formed, welded and machined. Cold-rolled steel is harder, has good surface finish and dimensional tolerances, high strength at low cost and good machinability.

High-strength low-alloy steels are proprietary steels with low carbon made to SAE 950. They have significantly higher strength and are easily formed and welded.

Hardened and tempered steels (AISI/SAE 31–98 BS 970 500–599, including higher-carbon steels) are steels containing sufficient carbon and alloy to enable them to be heat treated to the desired strength and toughness at the design thickness. They may have high toughness and high strength at elevated temperature but they are more expensive than carbon steels and the higher-alloy steels have poor weldability and machinability.

Case-hardening steels are of relatively low carbon content (final BS specification Nos 12 to 25). They may be surface hardened by carburizing, carbonitriding or nitriding when heat treatment will produce a very hard surface and a softer (but, where necessary, strong) ductile core. They are used when wear resistance must be combined with core toughness.

Stainless steels (AISI 200–499, BS 970 300S–499S) contain chromium in amounts above 12% so that the magnetite layer formed on the surface of iron becomes, at lower chromium levels, a spinel and at higher levels chromic oxide. The introduction of chromium into the oxide layer greatly increases its stability and integrity and provides much-increased resistance to corrosion and oxidation, but also influences significantly the structure and properties of the underlying metal. The addition of further alloying elements results in five separate classes of stainless steel. These are, respectively:

1. *Ferritic stainless steels* are alloys of iron with up to 18% chromium and relatively small amounts of other alloy. They are ductile, the high-chromium versions have good corrosion resistance, and they are, relative to other stainless steels, inexpensive. They have a tendency to grain growth and are therefore difficult to weld. The recently developed 'Low Interstitial' ferritic stainless steels have chromium contents between 17% and 30% and very low carbon contents. These are claimed to have outstanding corrosion resistance but may be difficult to obtain.
2. *Martensitic stainless steels* are limited in chromium content to about 17% so that they may be hardened by quenching to give high hardness and strength.
3. *Austenitic stainless steels* avoid the problems which result from the addition of chromium to the ferrite matrix by the addition of nickel and other elements which change the structure to the high-temperature gamma form. The resultant alloys may have very high corrosion resistance, good ductility and/or high hot strength. Their very highly alloyed versions merge at iron contents below 50% into nickel alloys (see Section 7.4.5) and the very high creep strength versions are described as 'superalloys of iron'.
4. *Duplex stainless steels* have compositions which produce a mixed ferrite/austenite structure. They have excellent mechanical strength and corrosion resistance but may be difficult to obtain.

Precipitation-hardening stainless steels combine very high mechanical strength with excellent corrosion resistance. They require complex treatments.

Intermetallic strengthened (maraged) steels may be formed or machined in the soft condition and then aged. The maraged steels are very strong and very tough but very expensive.

The steel classes described so far have been based on composition and structure. Other classifications based mainly on application (which, to some extent, cut across the classification already described) include:

- *Electrical steels*: very low-carbon steels containing about 3% silicon supplied as strip, hot or cold rolled, with the surfaces insulated;
- *Spring steels*: in which a very high hardness can be produced by working, quenching or precipitation hardening. They may be carbon, alloy or stainless;
- *Tool steels*: used for forming or cutting materials. Their essential properties are high hardness, resistance to wear and abrasion, reasonable toughness and, in the case of high-speed steels, high hot hardness;
- *Creep-resisting steels*: with high creep and creep rupture strengths at high temperatures. They may range from bainitic through martensitic to austenitic steels and superalloys. The higher-temperature steels are also oxidation resistant because of chromium additions;
- *Valve steels*: with high-temperature tensile and creep strengths and good high-temperature oxidation and corrosion resistance. In the UK the title is restricted to certain martensitic and austenitic steels but in other countries it includes all steels which may be used for IC engine poppet valves.

7.3.3 Steel making and ingot casting

The raw materials for steelmaking (iron ore and coke) are converted in the blast furnace to molten iron containing about 3% carbon at a rate which varies up to about 8 million tons per annum. (This makes the blast furnace by far the most economic process for producing steel, but economies which have not the capacity to utilize steel production of this magnitude may utilize the direct reduction of iron ore to sponge and powder instead.) Figure 7.3 depicts a typical blast furnace.

The conversion of the iron from the blast furnace to flat products, billet bar and sections has, since World War II, undergone a revolutionary change due to the development of oxygen argon lancing, vacuum treatment, continuous casting and rapid in works analysis. Iron from the blast furnace is conveyed while still molten to the basic oxygen converter (BOC or BOF) (see Figure 7.4) where oxygen gas is passed through it reducing its carbon content to approximately that of the specified steel. It may then proceed direct to the pouring ladle, where alloying additions are made and slagging processes undertaken. The molten steel is poured into a lander which conveys it to a water-cooled mould whose base is formed by the previously poured solid metal. This metal is retracted through the mould which may oscillate in a vertical axis. The whole of the blast furnace output may pass through a single mould of this type (Figure 7.5(a)) and the cast metal may be fed directly into a rolling mill which reduces it to plate bar or section as required (Figure 7.5(b)). The process as described is applicable to very large throughputs but there is no reason successive ladle charges should not differ in composition, the different alloys being separated later in the mill train. Where higher-quality or special steels are required the liquid steel from the BOC may be transferred to a vacuum plant for further treatment prior to pouring[13] (see Figure 7.6).

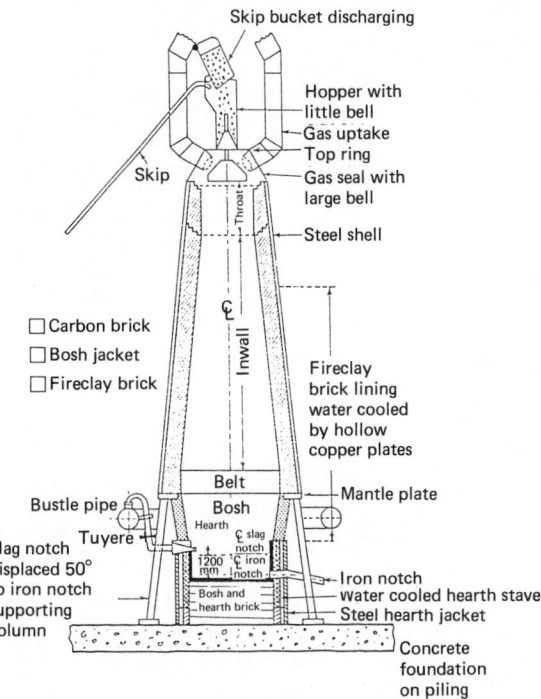

Figure 7.3 Blast furnace

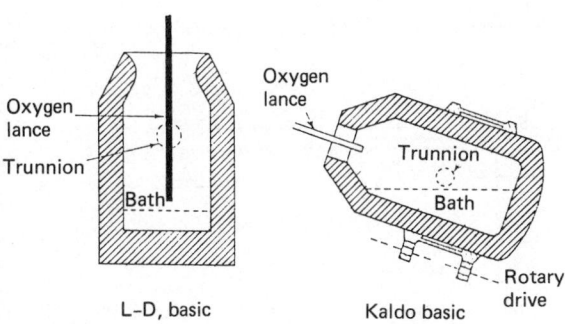

Figure 7.4 Types of BOC (Basic oxygen converter) or BOF (Basic oxygen furnace)

The Dortmund Harder (DH) and Ruhrstahl Heracus (RH) processes transfer the metal from a ladle into a superimposed vessel, the DH by sucking it up by vacuum, the RH by driving it up by pressure of argon. A further improvement has been to equip a vacuum argon treatment vessel with electric arc heating (Vacuum Arc Deoxidation (VAD))[14] and effectively to produce a secondary steel-making process (see Figure 7.7).

Most special steels, particularly stainless steels, cannot be decarburized by the BOF because the oxygen which reduces the carbon present also oxidizes chromium. Oxidation may be prevented by blowing oxygen in a vacuum but there are practical problems involved. These may be overcome by the AOD process, in which an artificial vacuum (so far as the partial pressure of carbon monoxide is concerned) is produced by diluting the blown oxygen with argon. This process is carried out in a tiltable vessel[15] (see Figure 7.8) with base tuyères which agitate the vessel contents. With this process removal of carbon, slag reduction, metal dioxidation, desulphuration and alloying are easily achieved.

Conversion of the molten iron from the blast furnace direct to the finished product in an integrated steelworks not only gives significant economies in energy and labour but can also provide steel of a quality equal or better than was previously available from electric arc melting.

In the first place, oxygen blown steel is naturally low in nitrogen so that the toughness problems associated with strain ageing are eliminated. However, where, as in austenitic stainless steels, nitrogen has a beneficial effect on tensile strength, controlled amounts may be introduced by replacing some or all of the argon in the AOD process.

The greatest improvement in liquid metal quality is, however, gained by vacuum treatment. Raising and lowering the steel in the DH or RH vessel very significantly reduces the partial pressures of hydrogen and carbon monoxide in the melt and the

$$C + Fe\,O \rightarrow CO + Fe$$

reaction proceeds until equilibrium between carbon and oxygen is established at a lower oxygen level. Line A in Figure 7.9 represents the oxygen carbon equilibrium at 1013 mbar (1 atm) over the carbon range 0.03–0.13%; the effect of reducing the pressure (e.g. by vacuum degassing) to 133.3 and 13.33 mbar appears in curves B and C, respectively. Since carbon is lost to the system as well as oxygen the theoretical effect on a steel initially at 0.05% C is shown in line 1 connecting the three curves. The actual effect on steels with various vacuum degassing techniques is shown in lines 2–5. Vacuum degassing thus effects reduction of the oxygen content to levels less than half those obtained in the best practice in steels air melted and refined at atmospheric pressure. Consequently, final residual deoxidation can be effected with much smaller amounts of aluminium or silicon and much cleaner steel can be produced with (if required) lower carbon levels. Vacuum carbon deoxidation is extremely beneficial in the production of plate with good through-thickness ductile properties because it virtually eliminates planar concentrations of non-metallic inclusions.

Vacuum degassing techniques, properly applied to permit carbon deoxidation, are the most economical way of upgrading steels and, in particular, low-carbon steels. It is further possible to make injections in a stream of inert gas of such elements as calcium or calcium carbide which can reduce sulphur and phosphorus and to replace the sometimes damaging inclusion of silicons and aluminium with other inclusions which improve transverse ductility, fatigue and machinability.

Additional advantages are gained by the substitution of continuous for ingot casting. Carbon, sulphur and phosphorus have been shown to segregate enough to give a concentration ratio of 3 or 4 to 1 between the top and bottom of a large conventionally poured ingot. Silicate inclusions segregate to different parts of the ingot. No longitudinal segregation can develop in a continuously cast ingot once the casting process has reached equilibrium. The only longitudinal variation of composition is that between one ladle charge and another, and this is revealed by analysis and may be corrected before pouring. There will be some compositional variation across the section, and possibly between dendrites, but at the outside this is unlikely to exceed 5% on either side of the mean. Continuous casting has the further advantage that the cross section of the ingot can be much better matched to that of the final rolled product than is possible with a conventionally cast

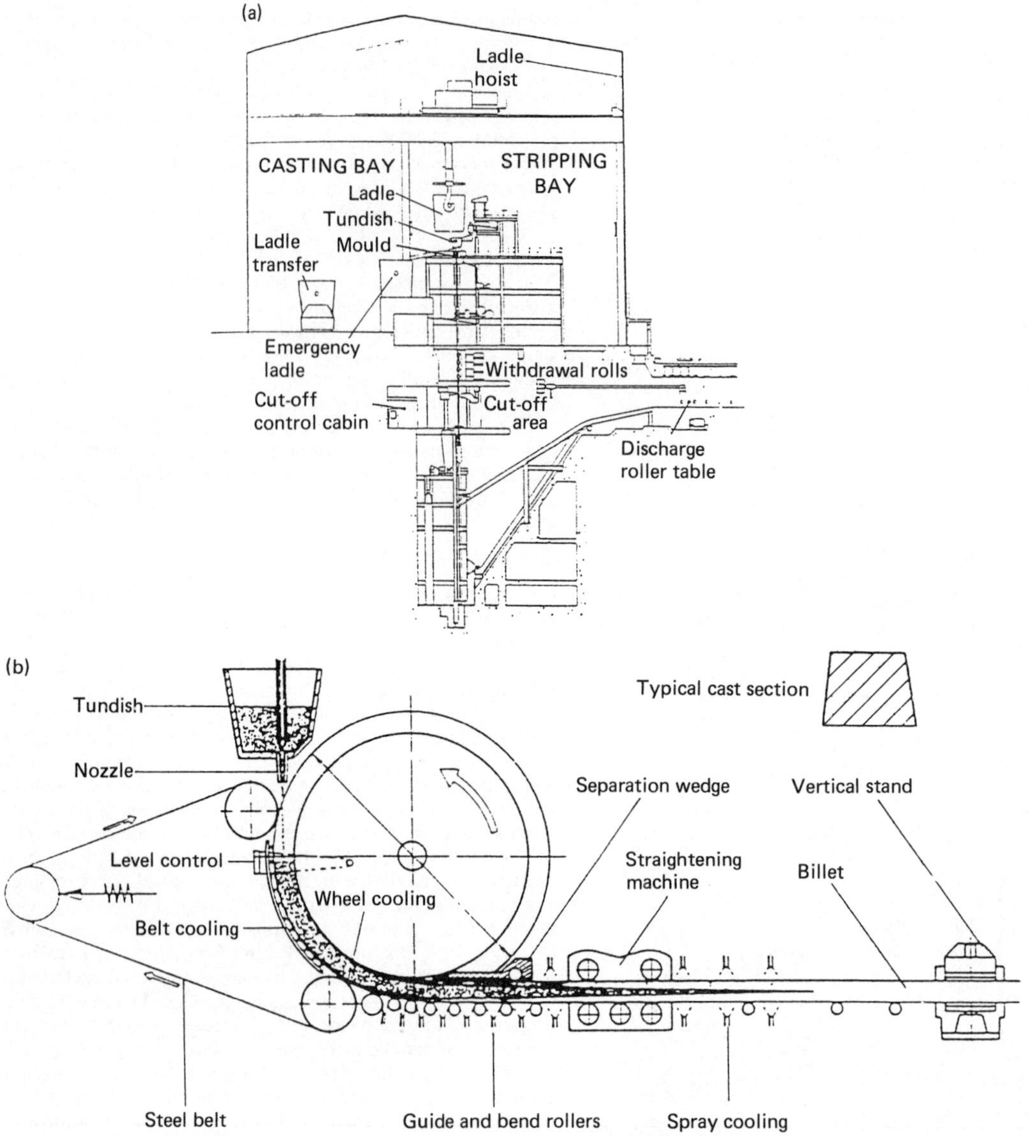

(a)

Ladle hoist

CASTING BAY   STRIPPING BAY

Ladle

Tundish

Ladle   Mould
transfer

Emergency
ladle

Withdrawal rolls

Cut-off
control cabin   Cut-off
area

Discharge
roller table

(b)

Tundish   Typical cast section

Nozzle   Separation wedge   Vertical stand

Level control   Straightening   Billet
machine

Wheel cooling

Belt cooling

Steel belt   Guide and bend rollers   Spray cooling

Figure 7.5 Continuous casters: (a) with cut-off; (b) with in-line roll stand. (Reproduced by permission of the Institute of Metals).

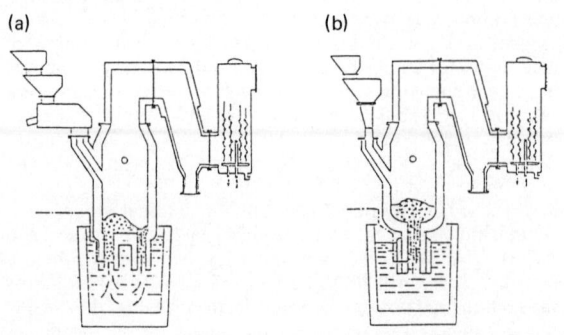

(a)   (b)

Figure 7.6 Vacuum degassers (deoxidizers, decarburizers): (a) RH; (b) DH. (Reproduced by permission of the Institute of Metals)

ingot. A large plate requires a large ingot, which in conventional casting must be of large cross section and may be too large for the available rolls. With continuous casting the size of plate is only limited by handling down the line and the metal quality is improved by the finer structure achieved by casting a smaller section.

However, it is seldom that the quantity of special and stainless steels warrants production from hot metal in proximity to the blast furnace. Such steels are generally produced from cold metal in the arc furnace. Here the quality of the steel is dependent on careful selection of the charge, so that quantities of tramp elements tin, arsenic, antimony, bismuth, copper, etc., which have serious effects upon ductility and cannot be reduced by steel making are minimized. Electric arc furnace steel is usually considered superior to steel made in the BOF, but the application of AOD and VAD can produce equivalent steel from either.

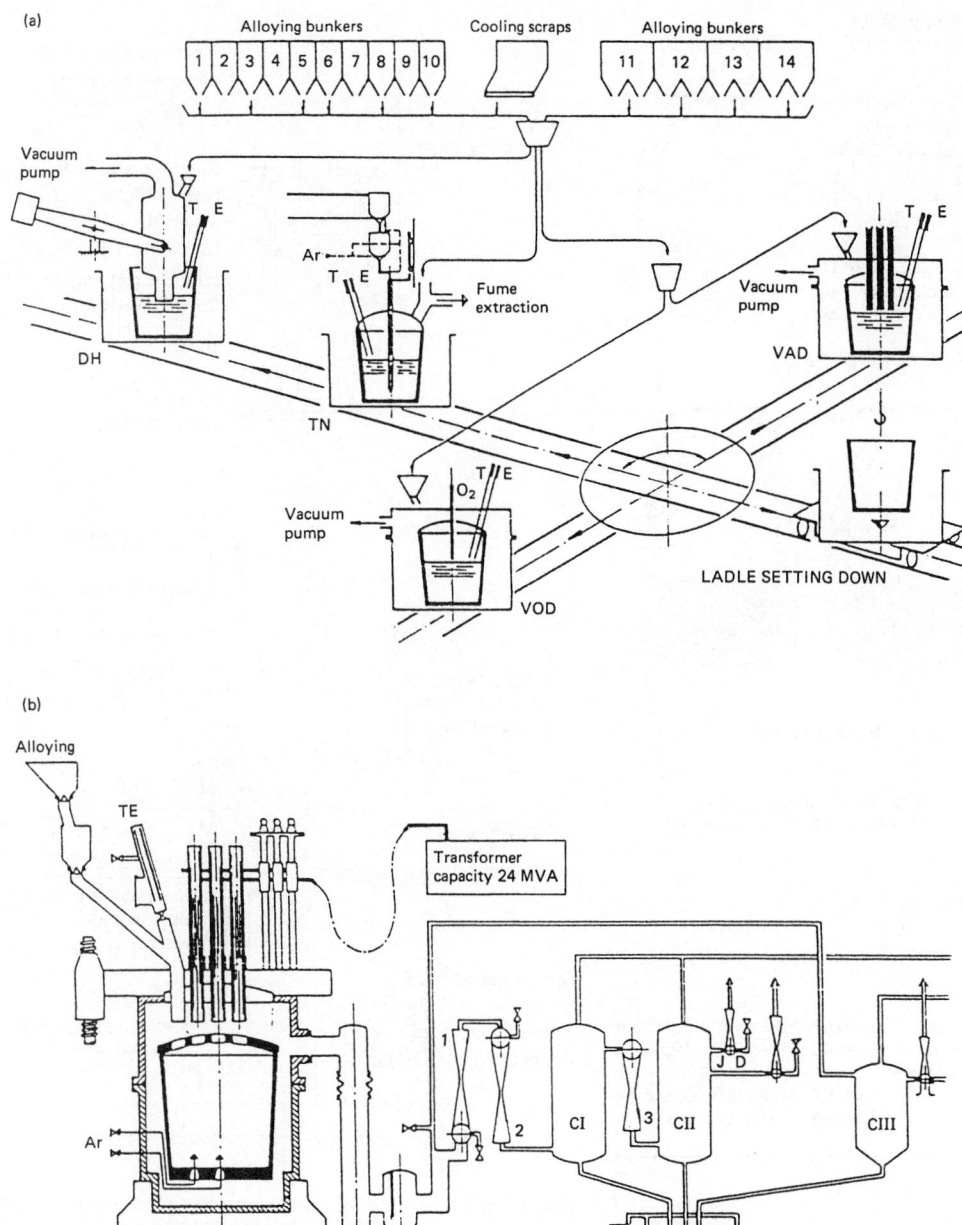

Figure 7.7 Secondary steelmaking plant: (a) General layout; (b) VAD unit.[13] (Reproduced by permission of the Institute of Metals)

There are other steel-making processes that will produce very high-quality metal, but all depend on the use of good-quality raw material. The vacuum high-frequency furnace will produce metal of low oxygen, nitrogen and hydrogen content, but is not well adapted to a continuous casting process. It is mainly used to produce small quantities of metal as small ingots (or castings). The highest-quality steel is produced by the consumable vacuum arc or the ESR processes. In both of these an electrode (or electrodes) is made of steel of the target composition and this electrode is progressively melted by striking an arc between it and a starting pad which is progressively withdrawn (or the electrode is withdrawn), thereby producing a semi-continuous casting (see Figure 7.10). The metal is refined on passing through the vacuum in the case of the consumable vacuum arc or through a low melting point slag in the case of the ESR. The solidifying metal is very well fed from the pool of molten metal and inclusions are dispersed and very substantially reduced in size. The consumable vacuum arc removes all gaseous or gasifiable inpurities but does not significantly influence the proportion of non-metallics.

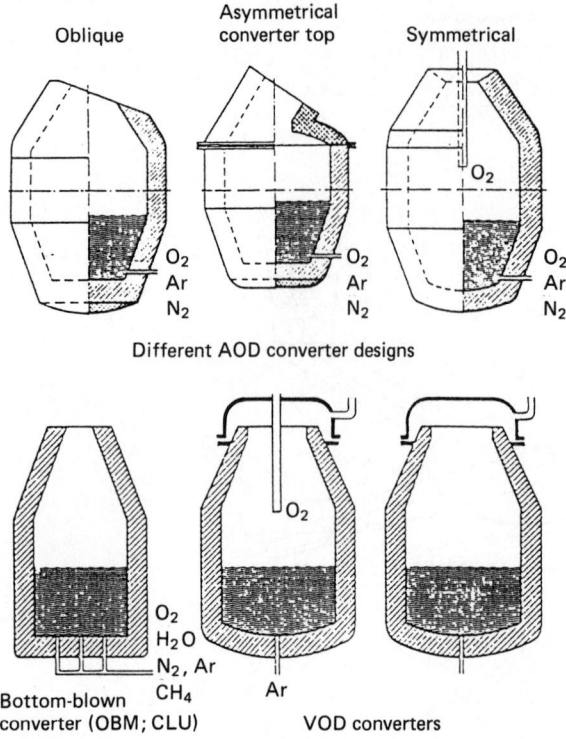

Different AOD converter designs

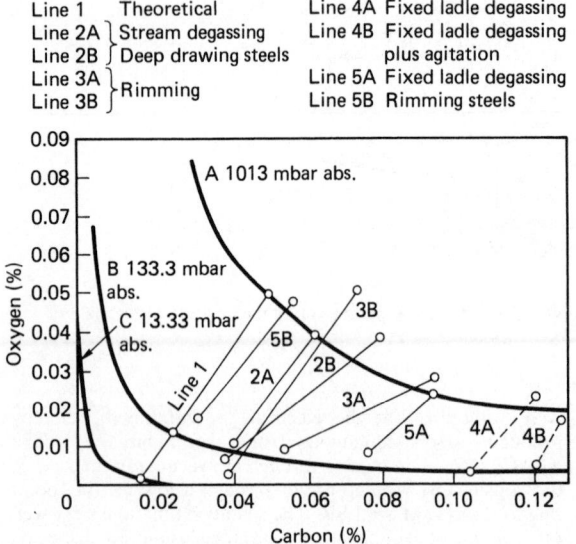

Bottom-blown
converter (OBM; CLU)

VOD converters

Figure 7.8 Converter processes for oxidizing and treating non-alloy and high-chromium alloy premelts. (Reproduced by permission of John Wiley)

Line 1	Theoretical	Line 4A	Fixed ladle degassing
Line 2A	Stream degassing	Line 4B	Fixed ladle degassing
Line 2B	Deep drawing steels		plus agitation
Line 3A	Rimming	Line 5A	Fixed ladle degassing
Line 3B		Line 5B	Rimming steels

Figure 7.9 Decrease in oxygen and carbon produced by degassing

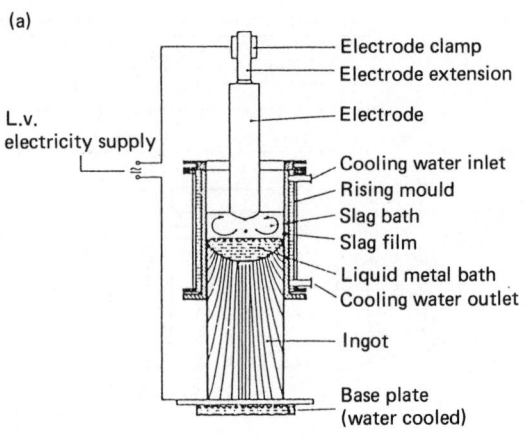

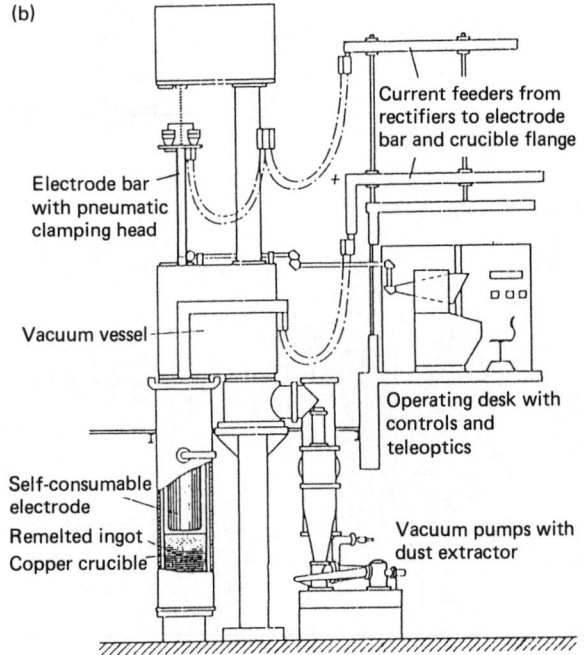

Figure 7.10 (a) ESR and (b) vacuum arc casting.[14] (Reproduced by permission of John Wiley). *Note*: For the convenience of the reader these figures also appear in Section 16.5.3.5

ESR can transfer non-metallic inclusions into the slag but does not remove hydrogen. ESR consumables must therefore be hydrogen-free.

The influence of these melting processes on such important properties as ductility or fatigue strength is very significant. The highest property values are not usually improved significantly compared with conventionally melted steel, but the proportion of values falling below a specific standard is very significantly reduced so that a much higher component performance may be guaranteed. A small proportion of steel is electron beam or plasma melted.

The proportion of steel that is continuously (or semi-continuously) cast has greatly increased.[15] The large integrated steelworks may have its entire throughput fed to a one stream continuous casting of 1 m × 2 m section[16], but batch metal from an electric arc melting shop is often semi-continuously cast in sections down to approximately 10 cm. Mould costs are reduced and throughputs and metal recovery considerably increased compared with the older type of mould. The proportion of wastage at the feeding head and the chances of segregation in the pipe are reduced.

Ingot is still employed where the quantity of steel required is insufficient to justify the use of continuous casting or for very large forgings, where the cross section of the ingot required to withstand the reduction needed is too large for it to be continuously cast.

7.3.4 Mechanical working of metals

A metal is worked mechanically either to generate a shape more economically than can otherwise be obtained or to provide improved properties in one or all of strength, ductility or fatigue (not creep, for which a cast structure is superior). The improvement in properties may be overall, directional to resist a directional stress system, or statistical to ensure that a lower percentage of components fall below a specified level of properties.

Metals can be mechanically worked because they differ from other crystalline solids in that the atoms in a crystal are not linked by valency bonds to adjacent atoms. The free electron metallic bond is non-specific, pulling equally hard in all directions. Metal atoms are therefore bound tightly in certain regular crystalline structures. There are seven crystal systems giving 14 possible space lattices and all engineering metals crystallize in one of three of these lattices (see Table 7.2 and Figure 7.11). Some metals are polymorphic: for example, in the case of iron, austenite is face-centred-cubic, ferrite is body-centred-cubic.

These crystalline structures are resistant to tensile stress but can be sheared along certain crystallographic planes. The planes of lowest resistance to shear have the closest packing of atoms in the plane. Each trio of atoms in a close-packed layer surrounds a space (or hollow) in which an atom in the next layer can rest. Slip occurs when the atoms in a plane move into the next hollow in the adjacent plane. The shear stress required to move a whole plane of atoms is very high and a perfect metal crystal (such as occurs in 'whiskers') would deform elastically between 3% and 10% before deforming plastically in shear.

The crystals in massive metals are not perfect. 'Dislocation' rows occur in the lattice where an atom is missing or an additional atom is present. These dislocations allow the planes to shear one row of atoms at a time, thus greatly reducing the critical shearing force. More dislocations come into play as the metal is deformed and, in most cases, two or more dislocations

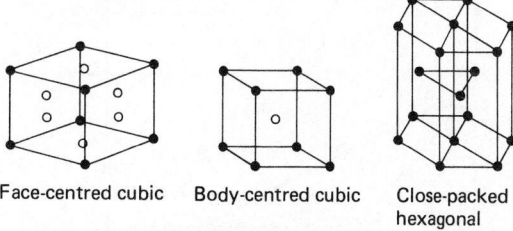

Face-centred cubic   Body-centred cubic   Close-packed hexagonal

○ Atoms at centres of faces or body of cube

Figure 7.11   Three simple space-lattice systems unit cells

may interact and stop each other. The critical shear stress therefore increases and the metal work hardens, increasing its tensile strength. Finally, no more shear is possible and the metal fractures. If temperature is increased, diffusion and recrystallization can release stopped dislocations. In hot working, diffusion occurs while the metal is worked, but in cold working and annealing the two processes occur separately. In either case the grain of the metal may, if the conditions are chosen correctly, be refined.

Mechanical working has other beneficial effects. The structure of cast metal may contain micro segregates, lumps and planes of constituents which separate out during casting. Mechanical working, correctly applied, may break up and disperse these to a size at which they can be diffused during heat treatment.

Castings may contain cracks, voids and non-metallic inclusions. Mechanical working, correctly applied, may close up and remove the cracks and voids (or aggravate them so that the part must obviously be rejected) and break up non-metallic inclusions. Alternatively, defects may be aligned and elongated in the direction of the tensile component of stress so that they do not impair resistance to fracture caused by tension or fatigue. Of the three space lattice systems shown in Figure 7.11 the face-centred-cubic has more planes on which slip can occur than the other two (which are themselves greatly superior to other lattice systems). Face-centred-cubic austenite is therefore more ductile than body-centred-cubic ferrite, and steel is usually hot worked in the austenite (γ) phase.

7.3.4.1 Hot working

Before working, the steel (unless already hot from continuous casting) is reheated into the γ-phase. Care must be taken to prevent excessive scaling and to avoid overheating (or burning) the steel. The limiting temperatures for carbon steels can be estimated from Figure 7.12 and are given in Table 7.3. Further forging temperature ranges are given in Table 7.4. Grain growth occurs when steel is held in the γ-phase, but subsequent working reduces grain size. Figure 7.13 illustrates these effects. The grain size of mild steel increases from ASTM 10 to ASTM 6 on heating to 900°C but six rolling passes reduce it to ASTM 10. Heating to 900°C has no effect on the grain size of two other steels but in both cases heating to higher temperatures increases grain size and rolling reduces it again. The finer austenite resulting from hot working transforms to a finer grain ferrite on cooling.

Table 7.2 Space-lattices of engineering metals

Lattice	Metals
Face-centred cubic	Aluminium, titanium,[a] lead, copper, iron,[a] cobalt,[a] nickel,[a] gold, silver
Body-centred cubic	Iron,[a] vanadium, chromium,[a] niobium, molybdenum, tungsten, zirconium
Hexagonal close packed	Titanium,[a] chromium,[a] cobalt,[a] nickel,[a] magnesium, zinc

[a] Metals having polymorphic habit.

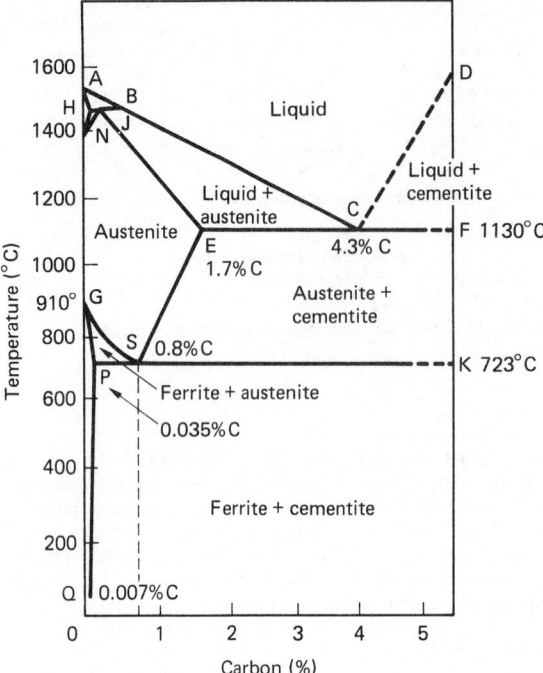

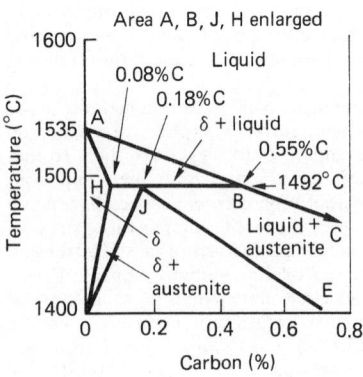

Figure 7.12 Iron-carbon constitution diagram

Table 7.3 Maximum forging temperatures for plain carbon[a] steels

Carbon (%)	Maximum forging temperature (°C)
0.1	1315
0.3	1286
0.5	1259
0.7	1215
0.9	1176
1.1	1133

[a] Maximum forging temperatures for alloy steels may be up to 50°C lower.

7.3.4.2 Rolling

Rolling is the most economic way of working steel if adequate quantities are produced. Steel for strip is cast into slabs, scarfed to ensure a defect-free slab and hot rolled using automatic gauge control. Carbon steels may be 'controlled rolled', that is, finish rolled at relatively low temperatures thereby inducing a fine ferritic grain structure with improved tensile strength and notch toughness.

Hot-rolled strip is cold rolled to reduce thickness and give the required surface finish and forming qualities. Finishing capabilities include rewinding to coil or cutting to length, side trimming and oiling. Roll trains for producing hot- and cold-rolled strip are shown in Figure 7.14.

By the use of a cluster mill, such as the Sendzimir (see Figure 7.15) strip down to 0.1 mm thickness can be produced. Grooved rolls are used to manufacture blooms, billets, bar and sections. Hollow sections can be produced by suitably designed mill trains and strip can be passed through forming rolls which turn the edges towards each other to form a seamed tube which is then welded continuously by inert metal arc or electric resistance. Seamless tube may be formed by cross rolling a billet in a Mannesman piercer in which the rolling action produces a tensile stress at the centre of the workpiece.

7.3.4.3 Forging

Forging is the process of working hot metal between dies either under successive blows or by continuous squeezing. It may be used to break down an ingot into a bloom or bar, to work down an ingot or billet to a rough finished shape before finishing or to make a forging. There are two essential differences compared with rolling. It is almost always possible (1) to design the dies and to arrange the sequence of forging to impose a higher ratio of compression to tension forces than is possible by rolling and (2) to ensure that the grain of the metal is in a preferred direction and not purely longitudinal as in rolling. For breaking down an ingot in, for example, tool, high-speed and some stainless and heat-resisting steels that have a two-phase structure, hot rolling would (at least before the development of ESR or vacuum arc casting) lead to ruptures due to the strong tension forces induced. Such ingots are usually broken down by hammer cogging. Forging is also used for making very large components such as turbine rotors which are usually 'open-die' forged in a press. Large forging ingots have a cross section whose circumference comprises a number of arcs meeting at cusps, because this shape minimizes surface cracking during casting. The first forging operation removes the cusps to form an approximately circular cross section and the forging is then drawn out through successive shape changes from octagon to square and back to octagon using dies of shapes shown in Figure 7.16. If the geometry of forging and press permits, the forging is upended and upset to produce some radial grain flow and it is then drawn out again.

Hollow forgings are made by punching a hole in the centre of a cylindrical workpiece and 'becking', working the die against a stiff bar passing through the forging and supported on 'V' blocks.

Small closed-die forgings are made in two dies attached, respectively, to the hammer ram and bed which have successive cavities to mould the stock progressively into a final shape in the last or finish cavity. For larger forgings a number of dies are made to perform one operation each. A wide variety of shapes can be made, depending only on ability to make and extract the component from two meeting dies with a parting line which may or may not be planar.

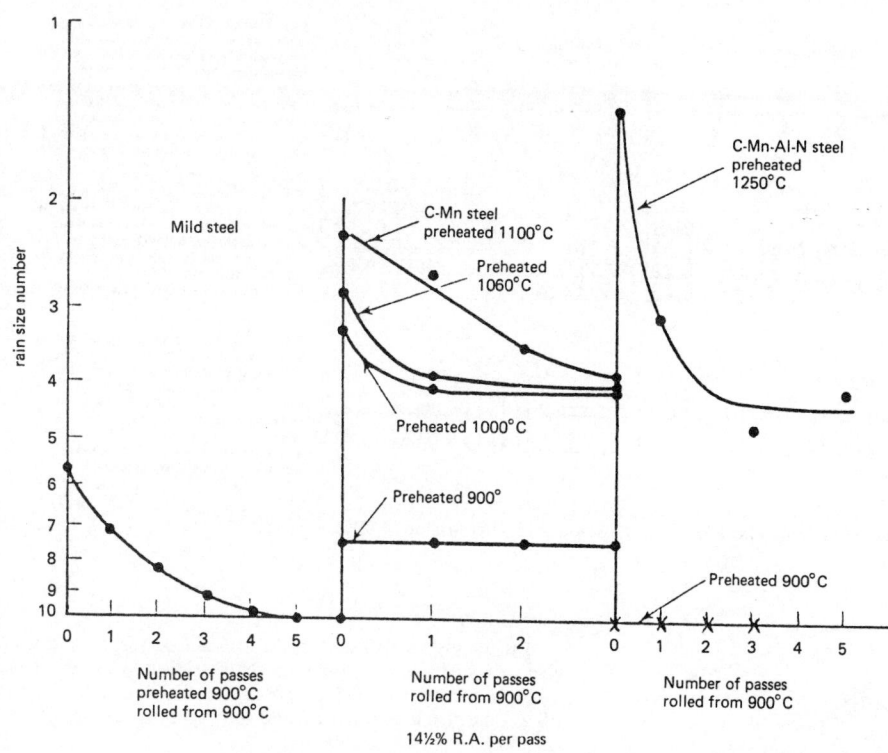

Figure 7.13 Austenite grain size related to heating temperature and number of roll passes

Table 7.4 Forging temperature ranges, tool and hot work die steels (American and British practice)

Type of steel	C	Cr	Mo	W	V	American		British		Remarks
						Start °C	Finish °C	Start °C	Finish °C	
Carbon tool (water hardening)	0.7/1.2 0.85/1.11	— —	— —	— —	— —	— 981–1093	— 815	950 —	750 —	
Tungsten (oil hardening)	0.9/1.15 1.25	0.5 0.5	— —	0.5/1.6 1.5	— —	981–1063 — 1008–1063	871 —	— 950	— 750 800	General tools and dies Drill and saw steel
Tungsten finishing steel (water hardening)	0.1/0.3	—	—	3.5/3.75	—		871	950–1010		Preheat slowly to 842°C. Slow cool in insulating material. For fine finishing cuts, cold dies, punches, gauges, etc.
Tungsten HSS. (oil or salt bath hardening)	0.8	4	—	18	1	1120–1176	926	1050–1100	900	Slow cool and anneal after forging
Hot work die steel Cr–Mo	0.3	5	1	1.25	—	1093–1149	898	1050–1090	850	Preheat 650–700°C, slow cool, anneal
Tungsten hot work steel	0.35	3	—	10	0.5	1120–1176	898	1100	900	Preheat 800–850°C, slow cool, anneal

American data from Sub-Committee on Tool Steels, *Metals Handbook* 1939 Edition, pp. 991–1031.

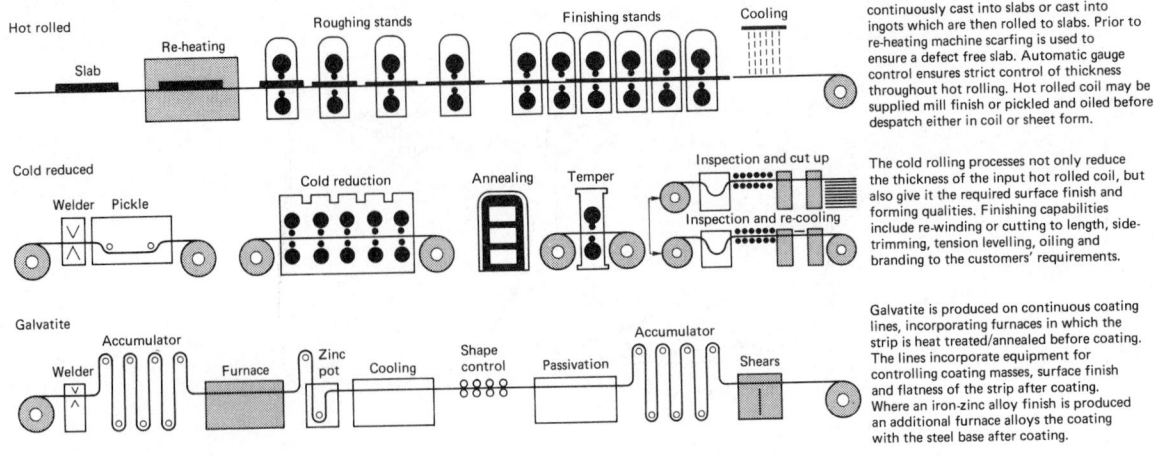

Hot rolled

Slab — Re-heating — Roughing stands — Finishing stands — Cooling

Steel from a basic oxygen furnace is either continuously cast into slabs or cast into ingots which are then rolled to slabs. Prior to re-heating machine scarfing is used to ensure a defect free slab. Automatic gauge control ensures strict control of thickness throughout hot rolling. Hot rolled coil may be supplied mill finish or pickled and oiled before despatch either in coil or sheet form.

Cold reduced

Welder — Pickle — Cold reduction — Annealing — Temper — Inspection and cut up — Inspection and re-cooling

The cold rolling processes not only reduce the thickness of the input hot rolled coil, but also give it the required surface finish and forming qualities. Finishing capabilities include re-winding or cutting to length, side-trimming, tension levelling, oiling and branding to the customers' requirements.

Galvatite

Welder — Accumulator — Furnace — Zinc pot — Cooling — Shape control — Passivation — Accumulator — Shears

Galvatite is produced on continuous coating lines, incorporating furnaces in which the strip is heat treated/annealed before coating. The lines incorporate equipment for controlling coating masses, surface finish and flatness of the strip after coating. Where an iron-zinc alloy finish is produced an additional furnace alloys the coating with the steel base after coating.

Figure 7.14 Mill trains for producing hot- and cold-rolled strip (Reproduced by permission of BSC)

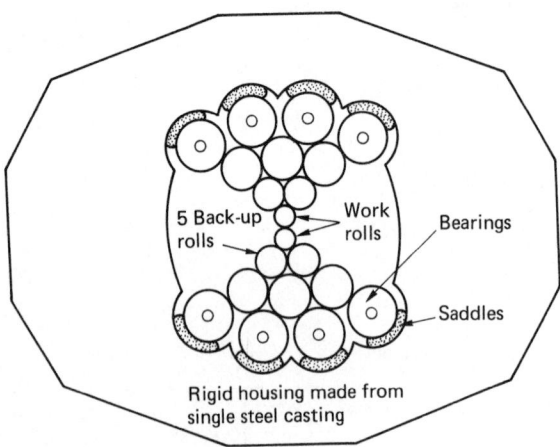

5 Back-up rolls — Work rolls — Bearings — Saddles — Rigid housing made from single steel casting

Figure 7.15 Sendzimir cluster reversing mill

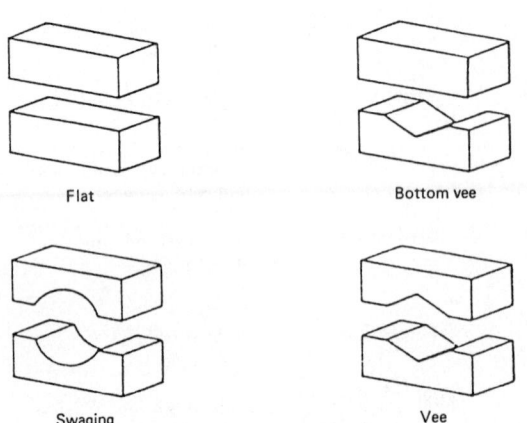

Flat   Bottom vee

Swaging   Vee

Figure 7.16 Die shapes for open-die forging

Most of the steels listed in BS 970 can be forged and will give properties appropriate to their section. Carbon steel forgings for engineering purposes are listed in BS 24 and BS 29 and forgings for fired and unfired pressure vessels in BS 1503.

The most demanding requirements for aircraft and similar requirements are met by ESR or consumable vacuum arc cast ingot usually made to manufacturers' own specifications (agreed, where appropriate, by official inspecting bodies). There are some 30 hot and cold metal-working processes which can be considered to be some version of forging.

7.3.4.4 Drawing and extrusion

A third important group of metal-working processes involves shaping metal by pushing ('extruding') or pulling through a die ('drawing'). Extrusion is usually performed hot, drawing, usually cold.

Extrusion Hot extrusion consists of placing a hot cylindrical billet into a container and either forcing a die with a centrally placed orifice onto one end (indirect extrusion) or applying pressure to a ram at the other end, the die being held stationary (direct extrusion). In either case metal is extruded though the die in the form of an elongated bar having the same cross section as the orifice.

Extrusion is applied widely to non-ferrous metals which soften at lower temperatures than the steel or hard metal dies. Bar and very complex sections can be economically produced and cut into a wide variety of shapes. Tubes and hollow sections can be made either by extruding metal through bridge-type dies in which the metal stream separates and rewelds at a later point or by extruding a hollow billet over a central mandril. Non-ferrous metals are extruded hot or cold. Only the softer steels are extruded cold.

Hot extrusion of steel requires a glass lubricant. Very rigid and powerful presses are required and their high cost plus the need for machined billets limits the process to high-cost steel, unusual shapes and tubes (which are mandril extruded). There are besides extrusion other methods of tube production

Drawing Usually after a 'semi' has been produced hot it is cold drawn to reduce diameter, wall thickness or both. Cold drawing through dies is used to produce wire, tube and light bar. This process requires considerable skill and attention to detail in die design, lubricants, wire rod cleansing and baking to remove hydrogen introduced during cleansing.

7.3.5 Constitution and heat treatment of ferritic steels

The versatility and adaptability of steel depends on the ability to vary and control the constitution, distribution and nature of the microstructure constituents by varying chemical composition, heat treatment and hot and cold working. This provides a wide variation in mechanical properties over a range extending from very low temperatures to around 1000°C. To indicate how this variation and control is achieved it is necessary to consider matters such as allotropy, solid solutions, constituents, phases, equilibrium, and the iron–carbon constitution diagram.

7.3.5.1 Allotropy of iron

Each of the allotropic modifications of pure iron (see Figure 7.12) is stable within certain ranges of temperature as shown in Table 7.5. The change from one modification to the other which occurs on heating or cooling through the critical temperatures 910°C and 1400°C is accompanied by recrystallization.

7.3.5.2 Solid solutions

Solid solutions are formed when a metallic solid dissolves one or more elements or compounds. The solute elements can diffuse as in liquid solutions. If the atomic size of the solvent and solute atoms is similar, the solute atoms tend to replace the solvent atoms in an unchanged space-lattice, but if they are substantially smaller they will most probably be situated interstitially between the solvent atoms.

Pure iron at room temperature is composed of grains of alpha iron (ferrite). Ferrite is capable of dissolving in limited quantity elements such as silicon, phosphorus, nickel, copper, arsenic, etc., so that commercial steel of ultra-low carbon content will have ferrite grains with these elements in solid solution. Alpha iron in commercial steel is referred to as ferrite. Gamma iron (austenite) also dissolves elements to form solid solutions. If, as is usually the case, gamma iron dissolves an element to a greater extent than alpha iron, when the change to alpha iron occurs, on cooling through the critical temperature, the alpha must accommodate more atoms of the element than its solubility allows and becomes supersaturated. Under suitable conditions, excess atoms will precipitate to arrive at a stable condition.

The high solubility of iron carbide (Fe_3C), 'cementite' in austenite and its relatively low solubility in ferrite is the basis for the heat treatment of hardened and tempered steel. Although steels containing cementite are strictly metastable, for all practical purposes they may be considered as stable because the precipitation of graphite (which is the stable form

Table 7.5 Stable temperature ranges of the allotropic modifications of pure iron

Modification	Crystal structure	Temperature range of stability (°C)
Delta, δ	Body-centred	1535–1400
Gamma, γ	Face-centred	1400–910
Alpha, α	Body-centred	910 and below

of carbon) occurs too slowly to be of interest except under special conditions.

7.3.5.3 Phases, equilibrium and the iron–carbon phase diagram

Constituents are the components of a metal alloy revealed by the microscope after suitable etching. Phases are physically and chemically homogeneous entities separated from the rest of the alloy by definite bounding surfaces. Phases in solid steel are austenite, ferrite, cementite, graphite, alloy carbides and other intermetallics. All phases are constituents but some constituents are not necessarily phases, e.g. pearlite is a constituent but it consists of the phases cementite and ferrite in a particular arrangement.

The phase diagram represents the phases present, at given temperatures, which are in equilibrium with each other. Equilibrium is attained by slow heating or cooling to allow constitutional changes to be completed.

The iron–carbide phase diagram (Figure 7.12) shows the effect of adding carbon to molten iron on the temperature ranges of stability of the delta, gamma and alpha phases in the solidified alloy and the effect of the varying solubility of iron carbide (cementite) on the phases present at temperatures from solidification downwards. The diagram represents nominally pure alloys; similar diagrams for other iron alloy systems (Fe–Cr, Fe–Mn, etc.) show the effect of other elements such as chromium, manganese, etc. on the range of stability of the phases. Since commercial plain carbon steels contain manganese and other elements in varying degree, Figure 7.12 does not accurately represent the critical temperatures, but it is a useful starting point for heat treatment.

7.3.5.4 Equilibrium decomposition of austenite: consideration of the iron–carbon constitution diagram

Point A in Figure 7.12 is at 1535°C, the highest temperature for the existence of solid delta-phase in carbon-free iron; line ABC shows the fall in the temperature, at which the first phase to separate from the solidifying liquid appears, as carbon content is increased. Line AHJEF (the solidus line) shows the temperatures below which all alloys are solid. As is shown in Figure 7.17, pure iron changes from face-centred-(austenite) to body-centred-cubic (ferrite) at 910°C; adding carbon lowers this temperature, as shown by line GS and, as the solubility of cementite in ferrite is less than in austenite, the change results in an increase in the carbon content of the remaining austenite, which, on further cooling, then forms more ferrite with again an increase in carbon content of the remaining austenite. This process continues until the last small amount of austenite contains 0.80% carbon, which changes at the fixed temperature of 723°C to a mixture of cementite and ferrite called Pearlite or Eutectoid, PSK being the Eutectoid line.

Alloys below and above 0.8% carbon (hypo- and hypereutectoid) both contain ferrite and cementite when cooled to room temperature but the mode of occurrence (the microstructural constituents of hypo-eutectoid steels being ferrite and pearlite, those of hyper-eutectoid being cementite and pearlite) of the phases is different with markedly different effect on mechanical properties. Descriptive sketches or actual photomicrographs may be used to illustrate the mode of occurrence of the phases. The constituents in the phase field and the changes in constituents on cooling for pure iron, 0.10%C and 0.80%C steel are shown schematically in Figure 7.17 when equilibrium conditions exist at extremely slow cooling or heating so that the diffusion of carbon is not inhibited.

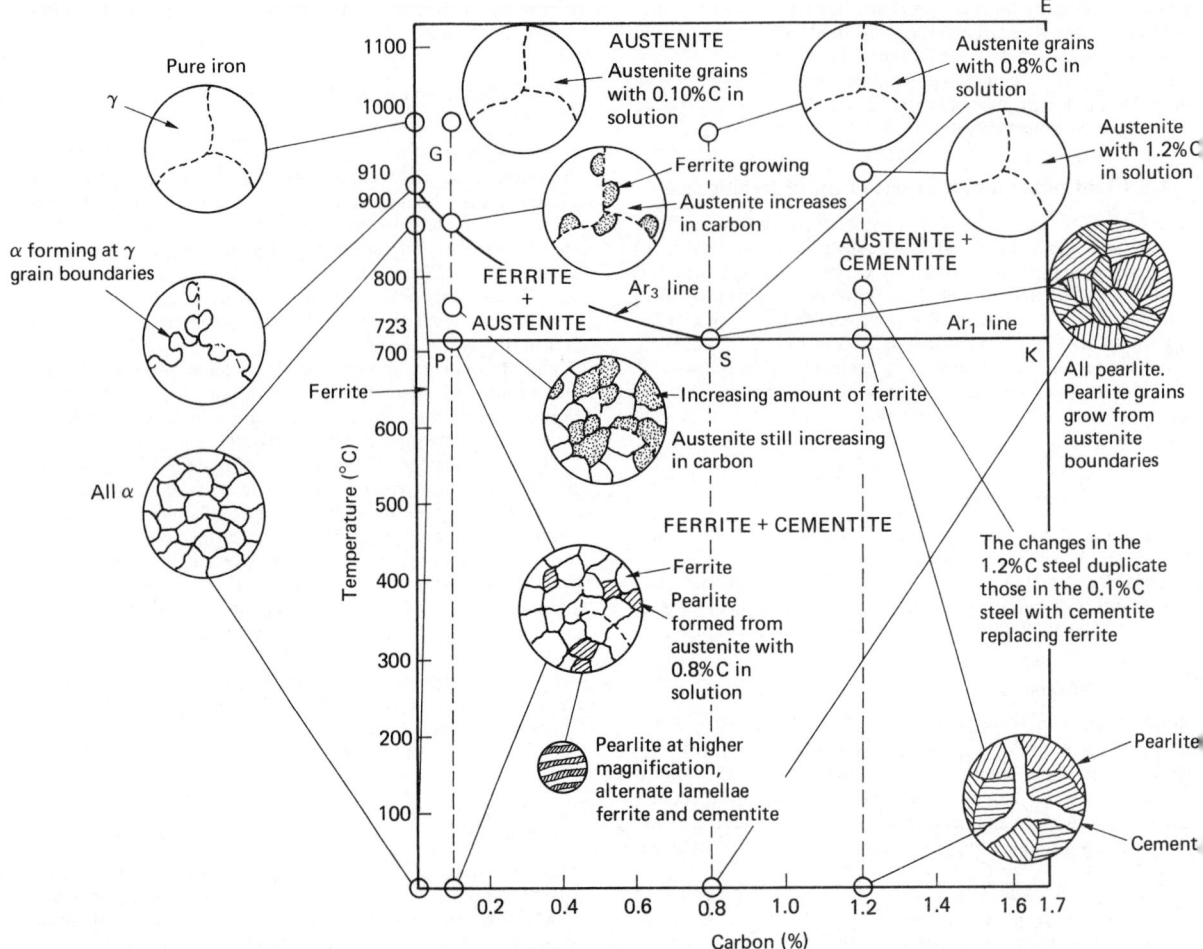

Figure 7.17 Steel portion of iron–carbon constitution diagram. Changes in grain structures and constituents with falling temperatures. The Ar$_3$, Ar$_1$ line GS, SK is the line at which austenite transforms on *cooling* during heat-treatment. The Ac$_3$ Ac$_1$ line (not shown) is the line at which ferrite/pearlite/cementite transforms to austenite on *heating* and is a few degrees higher, depending on the rates of heating and cooling, than the Ar$_3$ Ar$_1$ line

The solid solubility of carbon in gamma iron is 1.7% (point E). Hyper-eutectoid steels between 0.8% and 1.7% C precipitate cementite (6.8% C) when cooled to line SE, which reduces the carbon level of the remaining austenite thus requiring a further lowering of temperature before more cementite can precipitate. Continuous cooling down to the eutectoid line PSK results therefore in continuous precipitation of cementite until, at the eutectoid, the austenite contains 0.8% C and precipitates ferrite and cementite simultaneously to produce pearlite. Steels contain up to 1.7% C, cast irons more than 1.7% C.

It is common practice to use terms such as 'primary austenite' to differentiate austenite which freezes out of hypo-eutectoid alloys from austenite formed by heating up into the austenite phase field. Also, ferrite formed along line GS from steels with less than 0.8% C is called 'pro-eutectoid ferrite' and cementite formed along SE is called 'pro-eutectoid cementite'. The ferrite phase in the eutectoid constituent pearlite is called pearlitic ferrite and the cementite pearlitic cementite.

The microstructural constituents shown schematically in Figure 7.17 are the basis of most commercial steels but the proportions in which they occur and the mode of their occurrence (distribution and form) are profoundly influenced by rate of cooling from the austenite phase field. The equilibrium conditions necessary to the behaviour shown in Figure 7.17 and the approximate temperature indicated do not obtain in steel heat treatment practice except possibly when very large forging ingots are annealed.

Moreover, there is a thermal lag in the allotropic change so that the eutectoid line PSK is at an approximately 3°C higher temperature during heating up than on cooling down. With normal rates of heating and cooling of steel the eutectoid change is raised or lowered with faster heating or cooling. Consequently, in practice, a plain carbon steel has a temperature range in which the eutectoid temperature can vary according to heating or cooling rates. This also applies to the position of the austenite to ferrite change line GS.

However, Figure 7.17 illustrates an essential feature of steel heat treatment, namely, that austenite in transforming through the lines GS, SE or PSK develops a number of grains of the new constituents in each austenite grain, thereby refining the grain structure. The mechanical properties of steels consisting of ferrite and pearlite are strongly influenced by the average grain size of ferrite as well as the amount and type of pearlite (coarse lamellar, fine lamellar, etc.). The yield stress varies linearly with the reciprocal of the square root of the grain size.

On heating steel through the critical temperatures into the austenitic phase field the behaviour observed on cooling is reversed in the following manner:

Steel with 0.1%C. On passing through the lower critical temperature (Ac_1), which is higher than Ar_1, the pearlite areas first transform to austenite of 0.8%C content. This austenite grows by dissolving the surrounding ferrite grains as the temperature is raised and its carbon content is reduced. However, the austenite areas developing from the pearlite consist of numerous crystals so that just above line GS, when the structure is wholly austenitic containing 0.1%C, it consists of numerous small austenite grains. Heating to higher temperatures in the austenite phase field causes grain growth, some grains growing by absorbing smaller ones around them.

Eutectoid steel, 0.8%C. On heating above the lower critical temperature Ac_1 (which coincides with the upper critical temperature Ac_3 at the eutectoid composition), theoretically the pearlite should transform to austenite of 0.8%C content. In practice, it does so over a temperature range, the ferrite lamellae absorbing cementite to form a lower carbon austenite which then dissolves the remaining cementite. Grain growth follows on heating to higher temperatures in the austenite phase field.

Hyper-eutectoid steel, 1.2%C. At the eutectoid line the pearlite starts to transform to austenite of 0.8%C content. As the temperature is raised through the austenite plus cementite phase field, pro-eutectoid cementite is gradually dissolved by the austenite adjacent to it and eventually, by carbon diffusion, above the upper critical temperature the austenite attains a uniform carbon content at 1.2%. Grain growth follows on heating to higher temperatures in the austenite phase field.

The above simple behaviour of carbon steel relies on adequate time for diffusion of carbon being available.

When time at temperature is reduced, the diffusion is inhibited in varying degrees with a pronounced effect on the transformation changes. Thus, in plain carbon steels, increasing the rate of cooling through the critical temperature range Ar_3–Ar_1 lowers this range and alters the proportions of ferrite and pearlite. Steels with less than 0.25%C show refinement of ferrite grains, the growth of individual grains being suppressed, and the pearlitic constituent has finer cementite lamellae. Steels with more than 0.25%C show an increased amount of pearlite and decreased ferrite.

If the steel contains 0.35 or higher %C it is possible by a sufficient increase in cooling rate to produce a structure consisting entirely of pearlite. This pearlite will differ from equilibrium pearlite in having very thin cementite lamellae separated by wide ferrite lamellae. Since pearlite (with hard cementite lamellae) is the main contributor to tensile strength in ferrite–pearlite steels, its proportion and morphology in the structure are a prime consideration for heat-treatment practice. If the cooling rate through the critical range is increased still further, the austenite transformation may be entirely suppressed, and the steel remains as unstable austenite down to a lower temperature, when transformation begins with the formation of lower-temperature products (e.g. martensite,

bainite). When this happens the steel has been cooled at its 'critical rate'.

Martensite has special characteristics which are of great importance in the heat treatment of steel. The essential difference between the mode of formation of martensite and that of pearlite is that the change from the face-centred-cubic austenite lattice to the body-centred-cubic ferrite lattice occurs in martensite formation without carbon diffusion, whereas to form pearlite, carbon diffusion must take place, producing cementite and ferrite. The effect of this is that the carbon atoms strain the alpha martensite lattice producing micro-stresses and considerable hardness. The higher the carbon content, the greater the hardness of martensite (Figure 7.18), and the lower the temperature at which the change to martensite begins.

Furthermore, martensite, which is characterized by an acicular appearance, forms progressively over a temperature range as the temperature falls; if the temperature is held constant after the start no further action takes place. Martensite formation produces an expansion related to the carbon content. The mechanical properties of martensite depend on the carbon content; low-carbon martensites (less than 0.08%C) have reasonable ductility and toughness, high-carbon martensites have no ductility or toughness and extreme hardness and, because of the state of internal stress, are very liable to spontaneous cracking. Thus low-carbon martensite can be used for industrial purposes, e.g. welded 9% Ni steels for low-temperature applications have low-carbon martensitic heat-affected zones. High-carbon martensite must be tempered before it is allowed to cool to room temperature, e.g. carbon tool steels are water quenched to exceed the critical cooling rate but the tool is withdrawn from the bath while still hot and immediately tempered.

7.3.5.5 Isothermal decomposition of austenite

Reference was made in the previous section to the fact that if the gamma to alpha transformation is suppressed by fast

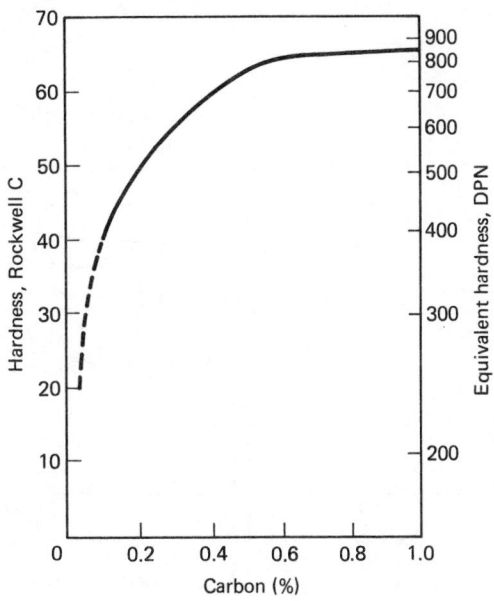

Figure 7.18 Hardness of martensite related to carbon content

cooling, the austenite is in an unstable condition. If, before reaching the temperature at which martensite begins to form, the cooling is arrested and the steel held at a constant temperature, the unstable austenite will transform over a period of time to a product which differs markedly from pearlite and has some visual resemblance to martensite in being acicular. This structure is called bainite; it is formed over a range of temperatures (about 550–250°C) and its properties depend to some degree on the transformation temperature. Bainite formed at a lower temperature is harder than bainite formed at a higher temperature. It is tougher than pearlite and not as hard as martensite. It differs fundamentally from the latter by being diffusion dependent, as is pearlite.

This type of transformation, at constant temperature, is important in the heat treatment of steel and is called isothermal transformation. It is characterized by an induction period, a start and then a gradual increase in speed of decomposition of the austenite which reaches a maximum at about 50% transformation and then a slow completion.

An isothermal transformation diagram which gives a summary of the progress of isothermal decomposition of austenite at all temperatures between Ac_3 and the start of martensitic transformation can be constructed. This is done by quenching small specimens of a steel (which have been held for the same time at a fixed temperature in the austenite field above Ar_3) to the temperature at which transformation is desired, holding for various times at this temperature and determining the proportion of transformed austenite. Such a diagram provides information on the possibilities of applying isothermal heat treatment to bring about complete decomposition of the austenite just below Ac_1 (isothermal annealing) or just above M_s (austempering), or of holding the steel at sub-critical temperatures for a suitable period to reduce temperature gradients set up in quenching without breakdown of the austenite, as in martempering or stepped quenching. Furthermore, if the steel is air hardening or semi-air hardening, the cooling rate during most welding processes exceeds the 'critical rate'. Therefore by using the isothermal diagram, the preheat temperatures and time necessary to hold a temperature to avoid martensite and obtain a bainitic structure can be assessed.

The principle of the isothermal diagram (also known as T–T–T diagrams) is illustrated schematically in Figure 7.19. The dotted lines showing estimated start and finish of transformation indicate the uncertainty of determining with accuracy the start and finish. The main feature of isothermal transformation – the considerable difference in time required to complete transformation at different temperatures within the

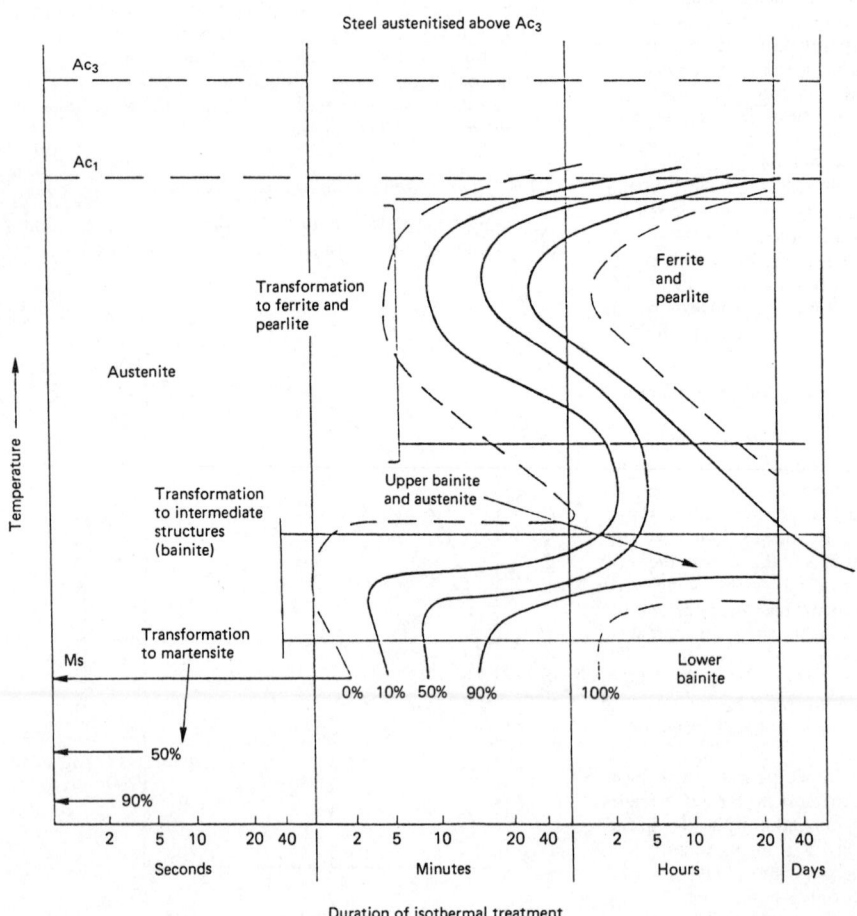

Figure 7.19 Schematic isothermal transformation diagram

pearlitic and bainitic temperature ranges – is to be noted. These diagrams vary in form for different steels. They also differ according to the austenitizing temperature (coarseness of gamma grains) and the extent to which carbides are dissolved in the austenite.

In alloy steels containing chromium, molybdenum or tungsten, segregation and carbide banding (size of carbides) varies and can affect the extent of carbide solution. In applying these diagrams it is usual to allow a considerably longer time for completion of transformation than that indicated on the diagram to cover the inherent uncertainties in individual consignments of steels.

7.3.5.6 *Effect of carbon and alloying elements on austenite decomposition rate*

As carbon content is increased the isothermal diagram is moved to the right, which indicates that austenite transformation is rendered more sluggish. Alloying elements increase the induction period, thus delaying the start, and they also increase the time necessary for completion. Furthermore, the effect of adding alloying elements is cumulative, but because they have different specific effects on transformation in the pearlitic or bainitic ranges it is not generally possible to predict the behaviour of multi-alloy steels.

7.3.5.7 *Decomposition of austenite under continuous cooling conditions*

It will be appreciated that while the isothermal transformation diagram provides the basic information about the characteristics of isothermal transformation for austenite of given composition, grain size and homogeneity, the common heat treatments used in steel manufacture such as annealing, normalizing or quenching are processes which subject the austenite to continuous cooling. This does not necessarily invalidate the use of isothermal diagram data for continuous cooling conditions because, as the steel passes through successively lower temperatures, the microstructures appropriate to transformation at the different temperatures are formed to a limited extent, depending on the time allowed, instead of proceeding to completion. The final structure consists of a mixture which is determined by the tendency to form specific structures on the way down, this tendency being indicated by the isothermal diagram.

The time allowed for transformation in the ferrite–pearlite and intermediate (bainite) regions obviously depends on cooling rate. A continuous transformation diagram will therefore have, as its essential features, means for indicating the amount of ferrite, pearlite, bainite and martensite which is obtained at various defined cooling rates; these are usually appropriate to heat treatment or selected welding cooling rates. Such a diagram is shown schematically in Figure 7.19.

The effect of continuous cooling is to lower the start temperatures and increase the incubation period so the transformation time tends to be below and to the right of the isothermal line for the same steel, these effects increasing with increasing cooling rate. As indicated in Figure 7.19, the time axis may be expressed in any suitable form; transformation time in Figure 7.20(a), bar diameter for bars as in Figure 7.20(b).

The positions of the lines defining the transformation products obviously vary according to the steel composition and austenitizing temperature. Diagrams for welding applications which have five cooling rates appropriate to the main fusion welding processes applied to various steel thicknesses have been produced by the Welding Institute, Cambridge, and by other welding research institutions in connection with the

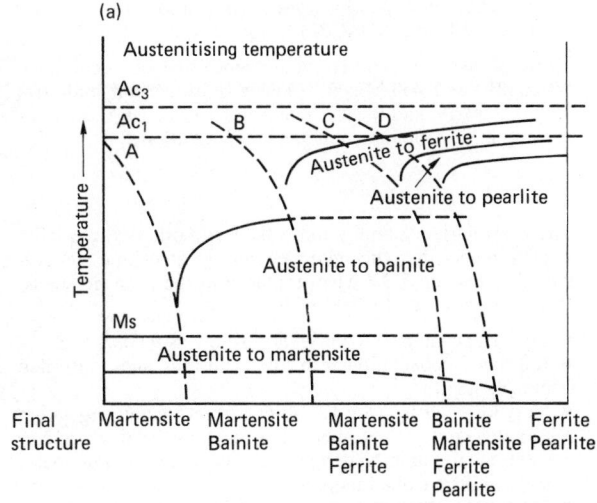

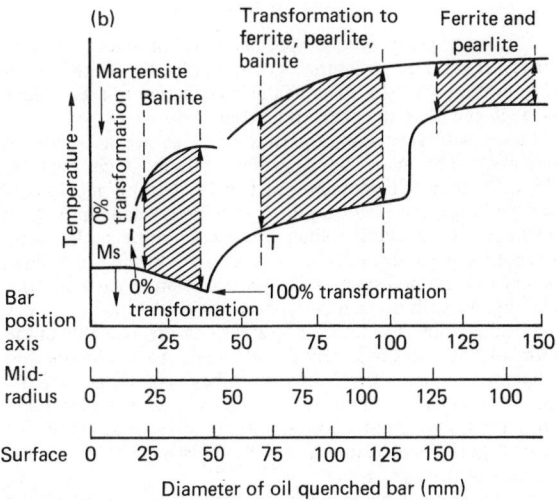

Figure 7.20 Continuous cooling time–temperature transformation diagram. (a) Applicable to forgings, plates and sections; (b) applicable to heat treatment of bars

development of weldable high-tensile steels. Manipulation of composition and heat treatment give rise to the several classes of steel already listed.

7.3.6 Carbon/carbon manganese steels

Rolled or hollow sections of carbon steels with carbon below about 0.36% constitute by far the greatest tonnage of steels used. In addition to the general specification of steels by analysis, they are sold by specifications depending on product form, and BS 970 is applied mainly to bar.

7.3.6.1 Weldable structural steels (Specification Nos BS 4360: 1970 and ISO R630)

These steels have yield strengths depending on section between 210 and 450 M Nm^{-2} achieved by carbon additions between 0.16% and 0.22%, manganese up to 1.6% and, for some qualities, niobium and vanadium additions.

7.3.6.2 Structural plates[17]

These products exemplify more than any others the quality improvements that the improvements in steel making described in Section 7.3.3 have produced in the tonnage steels. Plates can now be obtained with

- Lower maximum sulphur levels (as low as 0.008%);
- Improved deoxidation with low inclusions and controlled morphology;
- Very low hydrogen levels resulting from vacuum degassing;
- Greater control of composition resulting from secondary steel-making units and rapid in plant analysis, low inclusions and controlled morphology;
- Guaranteed high impact and elongation in the transverse direction;
- High impact values at low temperature in heat-affected zones.

There are many private specifications, primarily for material for offshore structures. For example, the British Steel Corporation's 'Hized' Plate will give reduction in area values through the plate thickness of around 25%.

Plates with superior properties (such as are used for oil pipelines) are made by controlled rolling steels such as BS 4360, grade 50E containing up to 0.1% Nb and/or 0.15% V and, although this is not explicitly specified, small amounts of nitrogen. Controlled rolling produces appreciably extra strength, e.g. yield and tensile values up to 340 and 620 MPa in a very fine-grained steel due to precipitation of carbonitrides and the low-carbon equivalent promotes weldability.

Besides plates, weldable structural steels are available in the form of flats, sections, round and square bars, blooms and billets for forging, sheet, strip and tubes. The range of flats, sections and bar is slightly restricted compared with plates and properties show minor variations. A very wide range of beams, guides and columns may be fabricated by automatic welding of plate steels.

Increased use is being made of hollow sections as they take up less space than angles or 'I' sections, decrease wind resistance and allow increased natural lighting and because, with care in design, they need not be protected on the inside and are cheaper to paint. Cold-forming sections increases strength and improves finish.

Forgings in weldable structural steels are included in BS 970. Tubes specified in BS 6323 may be hot or cold finished, seamless or welded in various ways. Yield strengths of hot-finished carbon steel tubes vary between 195 and 340 MPa and cold finished between 320 and 595 MPa. Cold-finished tubes are available in a variety of heat treatments.

The cheapest available steels to the specifications listed may, if purchased from a reputable steel maker, be used with confidence for most engineering purposes (with the exception of pressure vessels). If service conditions are known to be onerous, more demanding specifications and increased testing may be required.

7.3.6.3 Pressure vessel steels

The range of engineering plates, tubes, forgings (and, included here for convenience, castings) is matched by equivalent specifications for pressure vessel steels. Pressure vessel plate steels, specified in BS 1501: Part 1: 1980, are similar to structural steels but differ in that:

- Pressure vessel steels are supplied to positive dimensional tolerances, instead of the specified thickness being the mean. A batch of pressure vessel plates will therefore weigh more than the equivalent batch of structural plates (and cost more). A tensile test must be carried out on every plate (two for large plates) instead of one test per 40 tonne batch.
- Elevated temperature proof tests are specified for all pressure vessel plates.
- All pressure vessel plates have the nitrogen content specified and some the soluble aluminium content.
- All pressure vessel plates are supplied normalized.
- Pressure vessel tube steels are similar to those used for plates but, to facilitate cold bending, some of the grades are softer.
- The relevant specifications are, for seamless tube, BS 3601: 1974, for electric welded tube BS 3602: 1978 and for submerged arc welded tube BS 3603: 1977.
- Yield points lie between 195 and 340 MPa and Charpy V-notch impact must exceed 27 J at −50°C. (For lower-temperature service steels with up to 9% nickel, austenitic stainless or even martempered steels should be used.)

Carbon–manganese steel forgings for pressure vessels are specified in BS 1503: 1980. Materials are available with yield strengths varying (depending on section) between 215 and 340 MPa.

Carbon–manganese steel castings for pressure vessels are specified in BS 1504: 1976. These castings may contain up to about 0.25% of chromium, molybdenum, nickel and copper (total max. 0.8%) and 0.2% proof stresses range between 230 and 280 MPa.

7.3.6.4 Coil and sheet steel

BOF steel is continuously cast into slabs and rolled hot to coil or cut sheet. Hot-rolled strip is available in thicknesses above 1.6 mm up to 6.5 mm pickled and oiled and 12.7 mm as rolled in widths varying up to 1800 mm in:

- Forming and drawing quality aluminium killed
- Commercial quality
- Tensile qualities to BS 1449: Part 2 and BS 4360

in a variety of specified minimum yield strengths above 280 MPa. Weathering steel which develops an adherent coating of oxides and raised pattern floor plate is also available hot rolled.

Cold-reduced strip is available in thicknesses above 0.35 mm up to 3.175 mm and in widths varying up to 180 mm in:

- Forming and drawing qualities (typically, 180 MPa yield UTS 620–790 MPa to BS 1449: Part 1)
- Tensile qualities with yield points for low-carbon phosphorus-containing steels of 125 and 270 MPa and micro-alloyed with niobium of 300 and 350 MPa.

Cold-rolled narrow strip is available to BS 1449 and other more exacting specifications in thicknesses between 0.1 and 4.6 mm and widths up to 600 mm. Cold-rolled strips may be supplied in a variety of finishes, hot-dip galvanized to BS 2989, electro-galvanized, electro-zinc coated, ternplate (coated with a tin–lead alloy which facilitates forming and soldering) or coated with a zinc–aluminium alloy with exceptional corrosion resistance.

7.3.6.5 Steel wire

Wire with carbon contents ranging from 0.65% to 0.85% is specified in BS 1408. Carbon steel wire in tensile strengths 1400–12050 MPa (for coiled springs) and 1400–1870 MPa for zig-zag and square-form springs are listed, respectively, in BS 4367: 1970 and BS 4368: 1970.

The heat treatment of wires, including annealing and patenting, differs appreciably from other heat treatment processing. The increase in tensile strength as the amount of drawing increases for three carbon ranges is shown in Figure 7.21. Ductility falls as the tensile strength increases (Figure 7.22). When the limit of reduction has been reached the wire must be heat treated to remove the hard-drawn structure and replace it by a suitable structure for further reduction. For low-carbon steel this treatment is an anneal, just below the lower critical temperature, which recrystallizes the ferrite grains to an equiaxed form.

Medium- and high-carbon wires are generally patented (fairly fast cooling from above the upper critical point by air cooling or quenching in lead) to give a coarse pearlitic structure which will draw to very high tensile strengths. In addition to sub-critical annealing and patenting, the heat treatments used in wire production include normalizing, annealing, hardening and tempering and austempering, all designed to confer structures and properties which have particular relevance to the requirements of specific wire applications.

The tensile strength obtainable depends on carbon content and an approximate indication of the relationship for an-

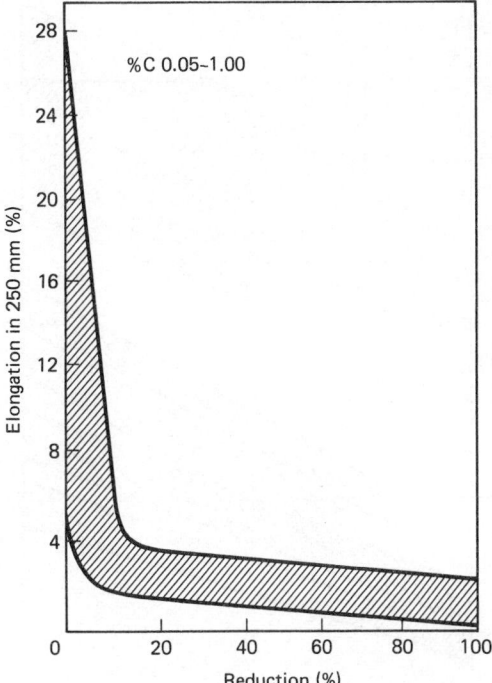

Figure 7.22 Decrease in ductility related to amount of reduction in wire drawing

nealed, patented and hardened and tempered wire is shown in Figure 7.23. Wire has a relatively large surface-to-volume ratio so that any decarburization due to heat treatment has a proportionately more significant effect than in heavier steel products. Consequently, wire heat treatment is conducted in specialized equipment (i.e. salt baths, atmosphere-controlled furnaces, etc.) aimed at minimizing any such difficulties.

Cold drawing through dies requires considerable skill and attention to detail in die design, lubricants, wire rod cleansing and baking to remove hydrogen introduced during cleaning.

7.3.7 High-strength low-alloy steels (HSLA steels)

High-strength low-alloy steels are proprietary steels manufactured to SAE 950 or ASTM 242 with carbon max. 0.22%, manganese max. 1.25% and such other alloying elements as will give the minimum yield point prescribed for various thicknesses ranging between 12 and 60 mm. Steels are available with yield points from 275 to 400 MPa, and the restriction on carbon and manganese content is intended to ensure weldability. Quenched and tempered weldable steels with significantly higher yields are also available.

Reduction in weight of steel gained by utilizing the higher yield stress in design is unlikely to reduce the cost of the material compared with that of the greater weight of a standard weldable structural steel purchased from BSC. Cost benefits arise, however, from handling the smaller quantity and welding the reduced thickness of the steel, and in transport applications from increased pay load, decreased fuel costs, freedom from weight restrictions and reduced duty imposed on other components of the vehicle.

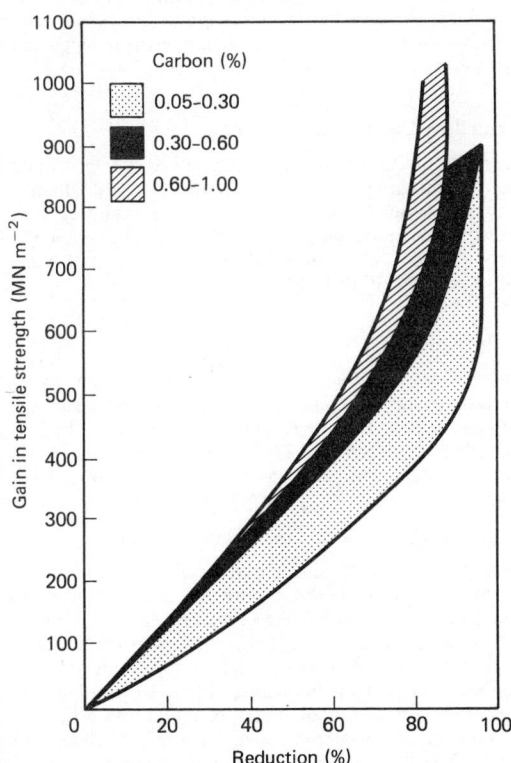

Figure 7.21 Increase in tensile strength related to amount of reduction in wire drawing for three levels of carbon

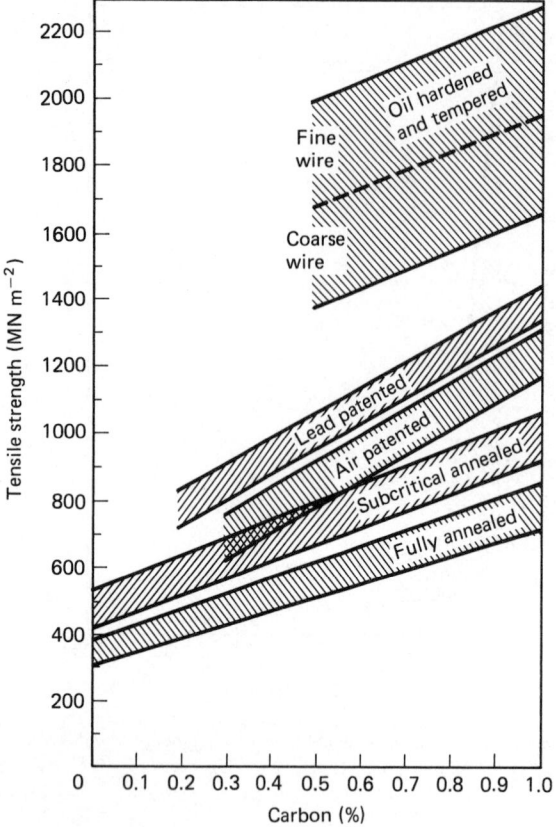

Figure 7.23 Tensile strength/carbon content relationship for wire

Attention must be paid to the following considerations:

1. The modulus of elasticity of an HSLA steel is the same as that of other ferritic steels. Any design which is buckling critical will require stresses and therefore sections identical to those of steels with lower strengths, and there will be no saving in quantity of steel.
2. Stress intensity is proportional to the second power of stress and fatigue growth rate per cycle is proportional to the fourth power of the range of stress per cycle. If brittle or fatigue fracture is a ruling parameter in design, a much more severe standard of non-destructive testing is needed for a component made from steel operating at a higher stress. In the limit, the critical defect size may fall below the limit of detection.
3. The notch ductility of an HSLA steel varies greatly according to the alloying elements used by the steel maker. If there is a risk of brittle fracture, values of Charpy V-notch energy and transition temperature should be specified by the designer.

Spectacular failures have resulted from ignoring these precepts.

7.3.8 Electrical steels

Electrical steels are a class of steel strip which is assembled and bolted together in stacks to form the magnetic cores of a.c. plant, alternators, transformers and rotors. Its essential properties are low losses during the magnetizing cycle arising from magnetic hysteresis and eddy currents, high magnetic permeability and saturation value, insulated surfaces, and a low level of noise generation arising from magnetostriction.

These parameters are promoted by maintaining the contents of carbon, sulphur and oxygen to the minimum obtainable and increasing grain size (which together minimize hysteresis loss) and incorporating a ferrite soluble element (usually silicon at a level of 3% to increase resistivity and therefore eddy current loss). The thickness of the steel must be optimized – reduction in thickness minimizes the path available for eddy currents but reduces the packing fraction and hence the proportion of iron available and increases handling problems. The surfaces are coated with a mineral insulant to prevent conduction of eddy currents from one lamination to the next. Accurate control of thickness and flatness minimizes stress when the laminations are bolted together and therefore reduces magnetostrictive noise which is promoted by stress.

There are two principal grades of electrical steel, differing essentially in loss characteristics. Hot-rolled strip is supplied to ASTM 840–85 in gauges of 0.47 and 0.64 mm with guaranteed losses of 13.2 and 16 W kg^{-1} at 15 Kilogauss induction and 60 Hz. Cold-rolled strip is supplied to ASTM 843–85 in gauges of 0.27, 0.3 and 0.35 mm with respective guaranteed losses of 1.10, 1.17 and 1.27 W kg^{-1} at 17 Kilogauss induction and 50 Hz.

Cold-rolled strip is manufactured by first rolling a sulphurized steel, followed by a programme of rolling and heat treatment which eliminates sulphur and produces a Goss or 'rooftop' texture. In this structure the [100] crystallographic direction which is most easily magnetized lies longitudinally in the strip. Cold-rolled strip is normally used for large alternators and transformers where the saving in lost power (and the problems of disposing of heat generated) outweigh the additional cost compared with hot rolled.

7.3.9 Hardened and tempered steels

At a content of carbon above about 0.35% (or less when alloying elements are present) useful increases in strength may be obtained by transformation. The most important class of steel to which this procedure is applied is the 'Hardened and Tempered Steels'. These will be chosen from AISI/SAE 1035–4310 and BS 970 080A32–945A40.

7.3.9.1 Heat treatment

The steel heat treatments, quenching and tempering, austempering, martempering, annealing and isothermal annealing can be described most simply by means of the isothermal diagram (Figure 7.20). (There are other heat-treatment procedures, notably ageing and controlled rolling.)

7.3.9.2 Quenching and tempering (Figure 7.24(a))

Steel quenched to martensite is hard and brittle due to the carbon being in unstable solid solution in a body-centred tetragonal lattice[15,16] and has high internal stresses. Heating (tempering) at 100°C causes separation of a transition phase, ϵ, iron carbide (Fe$_{2.2}$C) from the matrix, this being the first stage of tempering; slight hardening may occur initially. As the temperature is increased, relief of stress and softening occurs due to cementite formation and release of carbon from the matrix. The steel becomes significantly tougher.

Steels of suitable composition quenched fully to martensite and tempered at appropriate temperatures give the best combination of strength and toughness obtainable. There is a

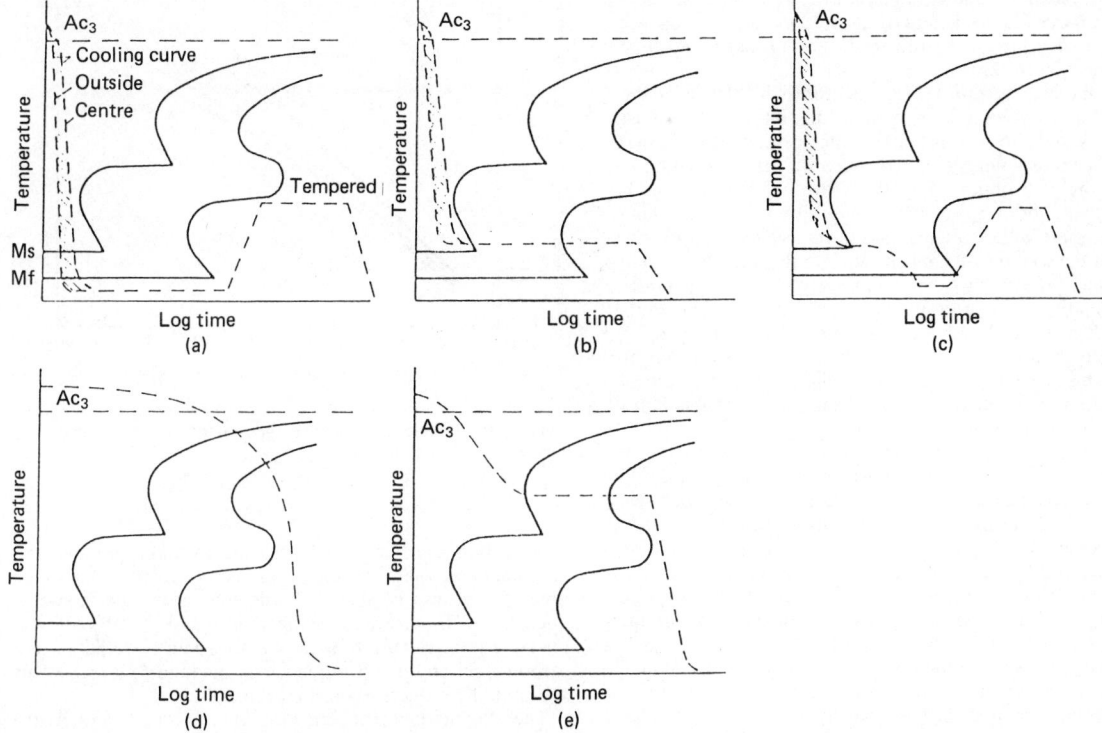

Figure 7.24 Isothermal diagrams showing heat treatment of steel. (a) Quenching and tempering. Product tempered martensite; (b) Austempering. Product lower bainite; (c) Martempering. Product tempered martensite; (d) Annealing. Product ferrite and pearlite; (e) Isothermal annealing. Product ferrite and pearlite

tendency, varying with different steels, for a degree of embrittlement to occur when tempering within the range 250–450°C, so steels are either tempered below 250°C for maximum tensile strength or above about 550°C for a combination of strength, ductility and toughness due to increasing coalescence of carbides.[17]

7.3.9.3 Austempering (Figure 7.24(b))

The purpose of this treatment is to produce bainite from isothermal treatment; lower bainite is generally more ductile than tempered martensite at the same tensile strength but lower in toughness. The main advantage of austempering is that the risk of cracking, present when quenching out to martensite, is eliminated and bainitic steels are therefore used for heavy-section pressure vessels.

7.3.9.4 Martempering (Figure 7.24(c))

The risk of cracking inherent in quenching to martensite can be reduced considerably while retaining transformation to martensite by quenching into a salt bath which is at a temperature slightly above that at which martensite starts to form, and then, after soaking, allowing the steel to air cool to room temperature. Distortion in quenching is a problem in pieces of non-uniform section and this is also considerably reduced by martempering.

7.3.9.5 Annealing (Figure 7.24(d))

Maximum softness is attained by annealing, involving slow cooling through the ferrite–pearlite field. The pearlitic structure developed provides optimum machinability in medium-carbon steels.

7.3.9.6 Isothermal annealing (Figure 7.24(e))

This treatment is used to produce a soft ferrite–pearlite structure. Its advantage over annealing is that, with appropriate steels and temperatures, it takes less total time because cooling down both to and from the isothermal treatment temperature may be done at any suitable rate, provided the material is not too bulky or being treated in large batches.

7.3.9.7 Hardenability of steel

Hardenability in this context refers to the depth of hardening, not the intensity. Hardening intensity in a quench is dependent on the carbon content. Plain-carbon steels show relatively shallow hardening; they are said to have low hardenability. Alloy steels show deep hardening characteristics to an extent depending primarily on the alloying elements and the austenitic grain size.

Hardenability is a significant factor in the application of steels for engineering purposes. Most engineering steels for bar or forgings are used in the oil-quenched and tempered condition to achieve optimum properties of strength and toughness based on tempered martensite. It is in this connection that hardenability is important; in general, forgings are required to develop the desired mechanical properties through the full section thickness.

Since the cooling rate in a quench must be slower at the centre of a section than at the surface, the alloy content must be such as to induce sluggishness in the austenite transformation sufficient to inhibit the ferrite–pearlite transformation at the cooling rate obtaining at the centre of the section. It follows that, for a given steel composition and quenching medium, there will be a maximum thickness above which the centre of the section will not cool sufficiently quickly except in those steels which have sufficient alloy content to induce transformation to martensite in air cooling (air-hardening steels). The practical usefulness of engineering steels, ignoring differences in toughness, can therefore be compared on the basis of this maximum thickness of cooling section which must be taken into account when considering selection of steel for any specific application.

A method for determining hardenability is to cool a bar of standard diameter and length by water jet applied to one end only. The cooling rate at any position along the bar will progressively decrease as the distance from the water sprayed end increases. The hardness is determined on flats ground on the bar surface located at 180°. The greater the hardenability, the further along the bar is a fully martensitic structure developed. This method of assessment is known as the Jominy end-quench test (for full details, see BS 3337).

Typical end quench (Jominy) curves for steels of medium and high hardenability are shown in Figure 7.25. A relationship between end-quench hardenability curves and the diameter of oil-quenched bars is shown in Figure 7.26. This can be used to choose a size of bar which will harden fully.

Jominy curves are provided by the SAE/AISI for steels to which the letter 'H' is added to the specification number and to BS 970 steels with the letter 'H' in the specification. Alternatively, a steel which will through-harden to the required yield stress at the design diameter may be selected from Table 7.6.

7.3.9.8 The function of alloying elements in engineering alloy steels

Apart from specialized functions – corrosion resistance, abrasion resistance, etc. – alloying elements are most widely used in engineering alloy steels with carbon in the range 0.25–0.55% or less than 0.15% for case hardening. Their function is to improve the mechanical properties compared

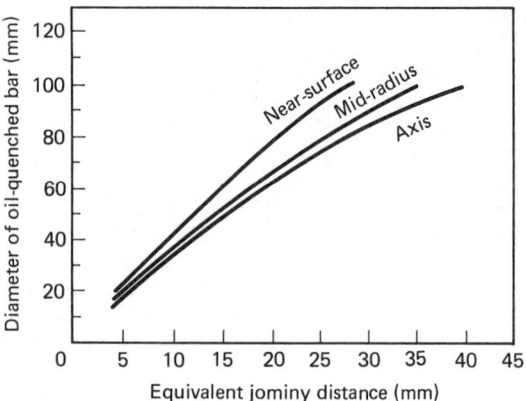

Figure 7.26 Relationship between end-quench hardenability curves and oil-quenched bars

with carbon steel and, in particular, to make possible the attainment of these properties at section thicknesses which preclude the use of shallow-hardening carbon steels, water quenched. They increase hardenability and thereby allow a lower carbon content to be used than would be required in a carbon steel and a softer quenching medium (e.g. oil). This substantially reduces quench cracking risks.

The alloying elements are Mn, Ni, Cr, Mo, V and Al (as grain-refining element). An important function of alloying elements, by rendering austenite transformations sluggish, is to make possible treatments which depend on an arrested quench followed by a timed hold at somewhat elevated temperature (austempering, martempering) which reduce internal stress and minimize distortion and cracking risks.

For full effectiveness in increasing hardenability, the alloy elements should be completely dissolved in the austenite before quenching. This is no problem with Mn and Ni but Cr, Mo and V form carbides which, in the annealed steel prior to quenching, may be of comparatively large size and, owing to a slower solution rate than cementite, are more difficult to dissolve. Solution temperatures may therefore be increased and/or times increased.

The effect of alloying elements when tempering is important.[18] In general, they retard the rate of softening during tempering compared with carbon steel but the effect, in this respect, of the carbide formers (Cr, Mo, V) is much greater than that of the other elements. They increase the tempering temperatures required for a given degree of softening, which is beneficial for ductility and toughness. Mo and V, at higher levels, confer an increase in hardness at higher tempering temperatures due to alloy carbide precipitation; this is 'secondary hardening' and is the basis of hardness in heat treatment of alloy tool steels. The effect of individual elements on the properties of steel is given in Table 7.7.

7.3.10 Free-cutting steels

Most free-cutting steels and those with the largest number of (and the most important) applications are carbon/carbon manganese steels. Some hardened and tempered and a few stainless steels are also free cutting. AISI/SAE free-cutting carbon/carbon manganese steels have 11 or 12 as the first two digits instead of 10 and the BS 970 designations have as the first digit a 2 while the second and third figures indicate the mean, or the maximum, sulphur content.

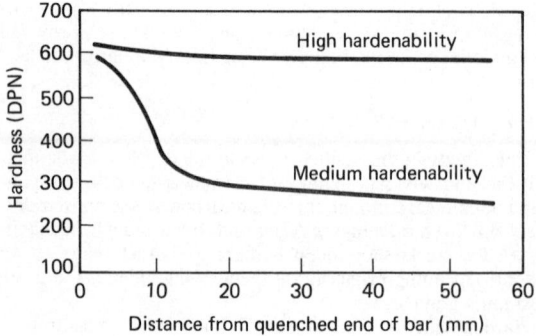

Figure 7.25 End-quench (Jominy Curves for steels of medium and high hardenability)

Free-cutting steels are really composites with additions which form a soft particulate second phase which acts as a 'chip breaker' during machining. This reduces tool wear, greatly diminishes the time and cost of machining and makes it easier to obtain a good finish.

The addition is usually sulphur in amounts between 0.1% and 0.33%. These steels were formerly manufactured by using a less effective sulphur-removing slag but present procedure is to resulphurize and the additional processing stage results in a slightly higher price for free-cutting steels. There is no systematic nomenclature for direct-hardening resulphurized alloy steels.

Additions of lead in amounts between 0.15% and 0.35% in addition to sulphur make steel even easier to machine. Specifications indicate leaded steels by inserting an L as an additional third letter in AISI/SAE grade numbers or adding Pb to BS 970 grade designations.

Free-cutting austenitic steels are limited to 303 or 303 Se which are standard 18/8 304 steels with sulphur or selenium additions. Free-cutting versions of 13% chromium steels are available to BS 970 416 S21, 416 S29 and 416 S37.

The particulate phase in free-cutting steels reduces their resistance to fatigue and may introduce other drawbacks. Free-cutting steels may be safely used in low-duty applications in non-aggressive environments for components which are not to be welded. It is essential, however, to ensure that components for severe duties are not made from them. This is of great importance when ordering components from a machining firm which will supply components made from free-cutting steel wherever possible to reduce costs. In particular, the designation 18/8 should not be used when ordering a steel as the supplier can supply 303 or 304. The AISI number should always be specified.

7.3.11 Case-hardening steels

Case hardening produces a very hard wear- and fatigue-resisting surface on a core which is usually softer but strong and tougher than that of a hardened and tempered steel. Besides its obvious advantages, case hardening usually improves fatigue endurance, partly because of the compressive stress induced at the surface. There are at least five different processes:

Surface hardening
Carburizing
Carbonitriding
Nitriding
Ion implantation

7.3.11.1 Surface hardening

Surface hardening is achieved by austenitizing only the surface of the steel by applying a high heat flux by electrical induction or by direct flame impingement, and then quenching in moving air, water or oil. Any steel of high enough carbon content may be surface hardened. Those most usually employed are carbon and free-cutting steels with 0.45–0.65% and hardened and tempered steels with 0.35–0.55% carbon.

The properties of the core are those to which the steel has originally been heat treated while hardnesses of from 50 to 65 Rockwell C are produced on the case. These hardnesses are lower than those available from other case-hardening processes but surface hardening is very versatile.

The depth of case produced by induction hardening may be varied by varying frequency from 0.64 mm at 600 kHz to 5 mm at 1 kHz. This is a much thicker case than can be

produced by any other method and very valuable for combating abrasive wear.

In flame hardening the surface is heated by one or more gas burners before quenching. The process can be applied to workpieces whose shape and size precludes other methods of case hardening.

7.3.11.2 Carburizing

Any carbon, free-cutting or direct-hardening alloy steel with 0.23% or less carbon is suitable for carburizing. The steel should be chosen according to the properties desired in the core. BS 960 and SAE publish lists of carburizing steels with hardenability data. Core strengths between 500 and 1310 MPa are available and Charpy impact toughness up to 55J (68 with 5 Ni 0.15 Mo steel). Case hardnesses of 64 Rockwell C for low-hardenability steels and 60 Rockwell C for high-hardenability steels can be obtained and the case, which contains a proportion of cementite, is hard wearing.

Carburizing is achieved by exposing the surface of the steel to a gas or liquid with a high carburizing potential at a temperature up to 925°C. Surfaces not required to be carburized should be masked, possibly by copper plating or, better, the carburized layer should be machined off before it has been hardened. There are three processes.

In pack carburizing the component(s) are placed in a heat-resisting box surrounded by a carburizing powder consisting basically of coke or charcoal particles and barium carbonate. The coke and barium carbonate react to produce carbon monoxide from which carbon diffuses into the steel. The process is simple, of low capital cost and produces low distortion, but it is wasteful of heat. It is also labour intensive because the boxes have to be packed and later emptied before heat treatment.

In liquid carburizing the component is suspended in a molten salt bath containing not less than 23% sodium cyanide with barium chloride, sodium chloride and accelerators. The case depth (which is proportional to time) is 0.3 mm in 1 hour at 815°C and 0.6 mm in 1 hour at 925°C. The process is efficient and the core can be refined in (and the component hardened from) the salt bath, but the process uses very poisonous salts, produces poisonous vapours and maintenance is required.

In gas carburizing, hydrocarbon gas is circulated around the workpiece at between 870° and 925°C. The relationship between case depth temperature and time is the same as for liquid carburizing. The process is clean, easy to control, suited to mass production and can be combined with heat treatment but the capital cost of the equipment is high.

7.3.11.3 Carbonitriding

Carbonitriding is achieved by heating the steel in a bath similar to a liquid carburizing bath but containing 30/40% sodium cyanide which has been allowed to react with air at 870°C (liquid carbonitriding) or in a mixture of ammonia and hydrocarbon (gas carbonitriding) at a lower temperature than is used for gas carburizing. The case produced is harder and more wear and temper resistant than a carburized case but is thinner. Case depths of 0.1–0.75 mm can be produced in 1 hour at 760°C and 6 hours at 840°C, respectively.

Steels which are carburized can also be carbonitrided, but because the case is thinner there is a tendency to use steels of slightly higher carbon and alloy content so that the harder core offers more support to the thinner case. A significant advantage of carbonitriding is that the nitrogen in the case significantly increases hardenability so that a hard case may be obtained by quenching in oil which can significantly reduce

Table 7.6 BS 970 and BS 4670 steels classified by tensile strength and maximum diameter, hardened and tempered*

Max. diameter (in.)	(mm)	Tensile strength 40–50 ton in.⁻² $630\text{–}790\ MN\ m^{-2}$	45–55 ton in.⁻² $690\text{–}850\ MN\ m^{-2}$	50–60 ton in.⁻² $750\text{–}930\ MN\ m^{-2}$	55–65 ton in.⁻² $85\text{–}1000\ MN\ m^{-2}$	60–70 ton in.⁻² $930\text{–}1080\ MN\ m^{-2}$
1/2	12	070M26 0.26C 215–430	080M36 0.36C 245–480 212M36 0.36C 310–495	080M46 0.46C 280–555 150M28 0.28C1.5Mn 325–570 212M44 0.44C 370–540	080M50 0.50C 280–595 150M36 0.36C1.5Mn 355–635 225M44 1.5Mn 415–600	
3/4	19	080M30 0.3C 230–450 216M28 0.28C 325–430	080M40 0.4C 345–510 120M19 0.19C1.2Mn 265–510	120M36 0.3C1.2Mn 340–570	070M55 0.55C 310–620	
7/8	22			503M40 1Ni 295–585		
1 1/8	28	120M19 1.2Mn 265–510	080M46 0.46C 280–555 150M19 1.5Mn 295–510 120M28 1.2Mn 310–510 120M36 1.2Mn 355–570 216M36 1.5Mn 310–510	080M50 0.50C 280–585 150M36 1.2Mn 355–635 225M44 1.5Mn 415–600	530M40 1Cr 510–680 606M36 1.5MnMo 510–680	*606M30 1.5MnMo* 510–850 605M36 1.5MnMo 480–850 640M40 1.25NiCr 510–755 708M40 1CrMo 510–755 945M31 1.5MnNiCrMo 480–850
2 1/2	64	216M36 0.36C 310–510 212M36 0.36C 310–495 225M36 0.36C 370–480 150M19 1.5Mn 295–510	503M40 1Ni 295–585 080M50 0.5C 280–585 150M28 1.5Mn 325–570 150M36 1.5Mn 355–635 212M44 0.44C 370–540	070M55 0.55C 310–620 530M40 1Cr 510–680 606M36 1.5MnMo 510–680 905M31 1.25CrAlMo 510–585	*605M30 1.5MnMo* 510–850 605M36 1.5MnMo 480–850 640M40 1.25NiCr 510–755 708M40 1CrMo 510–755 905M39 1.5CrAlMo 510–680 945M38 1.5MnNiCrMo 480–850	*608M38 1.5MnMo* 480–850 653M31 3NiCr 570–755 709M40 1CrMo 480–850 *818M40 1.5NiCrMo* 540–850
4	100	080M46 0.46C 280–555 120M28 1.2Mn 310–510 120M36 1.2Mn 340–570 503M40 1Ni 295–585	070M55 0.55C 310–620 212M44 0.44C 370–540 225M44 1.5Mn 415–600 530M40 1Cr 510–680 606M36 1.5MnMo 510–680 905M31 1.5CrAlMo 510–585	605M30 1.5MnMo 510–850 605M36 1.5MnMo 480–850 640M40 1.25NiCr 510–755 708M40 1CrMo 510–755 905M39 1.5CrAlMo 510–680 945M38 1.5MnNiCrMo 480–850	526M60 0.75Cr 620–740 *680M38 1.5MnMo* 480–850 653M31 3NiCr 570–755 709M40 1CrMo 480–850 *816M40 1.5NiCrMo* 540–850	917M40 1.5NiCrMo 635–1240
6	150	150M28 1.5Mn 325–530 150M36 1.5Mn 355–635 503M40 1Ni 295–585 785M19 1.5MnNiMo 430–465	605M30 1.5MnMo 510–850 605M36 1.5MnMo 480–850 608M38 1.5MnMo 480–850 640M40 1.25NiCr 510–755 708M40 1CrMo 510–755 905M39 1.5CrAlMo 510–680 945M38 1.5MnNiCrMo 480–850	608M38 1.5MnMo 480–850 653M31 3NiCr 570–755 709M40 1CrMo 480–850 816M40 1.5NiCrMo 540–850	722M24 3CrMo 635–755 817M40 1.5NiCrMo 635–1240 823M30 2NiCrMo 635–1235 826M31 2.5NiCrMo 635–1235 830M31 3NiCrMo 635–940	722M24 3CrMo 635–755 823M30 2NiCrMo 635–1235 826M31 2.5NiCrMo 635–1235 826M40 2.5NiCrMo 635–1235 830M31 3NiCrMo 635–940
10	250	503M40 1Ni 295–585 785M19 1.5MnNiMo 430–465 722M29 3.25CrMo 450–780	605M36 1.5MnMo 480–850 608M38 1.5MnMo 480–850 709M40 1CrMo 480–850 945M38 1.5MnNiCrMo 480–850 711M40 1CrMo 480–600 722M29 3.25CrMo 450–780	608M38 1.5MnMo 480–850 709M40 1CrMo 480–850 816M40 1.5NiCrMo 540–850 711M40 1CrMo 480–600 818M40 1.5NiCrMo 590–780 826M31 2.5NiCrMo 635–1235	722M24 3CrMo 635–755 817M40 1.5NiCrMo 635–1240 823M30 2NiCrMo 635–1235 826M31 2.5NiCrMo 635–1235 830M31 3NiCrMo 635–940 818M40 1.5NiCrMo 590–780 826M31 2.5NiCrMo 635–1235 826M40 2.5NiCrMo 640–900 897M39 3.25CrMoV 640–940 722M29 3.25CrMo 450–780 976M33 3.25NiCrMoV 700–980	823M30 2NiCrMo 635–1235 826M31 2.5NiCrMo 635–1235 826M40 2.5NiCrMo 635–1235 818M40 1.5NiCrMo 590–780 826M31 2.5NiCrMo 635–1235 976M33 3.25NiCrMoV 700–980 897M39 3.25CrMoV 640–940
20	500	785M19 1.5MnNiMo 430–465 722M29 3.25CrMo 450–780	711M40 1CrMo 480–600 722M29 32.5CrMo 450–780	818M40 1.5NiCrMo 590–780 826M31 2.5NiCrMo 635–1235 722M29 3.25CrMo 450–780	818M40 1.5NiCrMo 590–780 826M31 2.5NiCrMo 635–1235 826M40 2.5NiCrMo 640–900 722M29 3.25CrMo 450–780 976M33 3.25NiCrMoV 700–980 897M39 3.25CrMoV 640–940	818M40 1.5NiCrMo 590–780 976M33 3.25NiCrMoV 700–980 897M39 3.25CrMoV 640–940
39	1000	722M29 3.25CrMo 450–780	722M29 3.25CrMo 450–780	818M40 1.5NiCrMo 590–780 826M31 2.5NiCrMo 635–1235 722M29 3.25CrMo 450–780	826M40 2.5NiCrMo 640–900 976M33 3.5NiCrMoV 700–980 897M39 3.25CrMoV 640–940	976M33 3.25NiCrMoV 700–980 897M39 3.25CrMoV 640–940

1. In each block, the three sub-columns denote:
   1st Present designation of the steel in BS 970. The 'M' in the designation indicates that the steel is to be ordered to specific mechanical property requirements. However, many steels may be ordered to analysis only, when the letter becomes 'A', or to analysis and hardenability (end-quench) specifications, when the letter becomes 'H'. The first three figures indicate the broad analysis classification, the last two the carbon content.
   2nd Type of steel by broad analysis. These do not give full analyses, only the medium content of the leading element and a list of other alloying elements. For the straight carbon steels, the carbon content is repeated here, but for the alloy steels the carbon content can be inferred from the designation.
   3rd Yield stress range available for the steel at the specified equivalent diameter.

2. Steels marked with an asterisk are free-machining qualities.

3. Steels with first three digits 905 have high aluminium contents and are specifically intended for surface-hardening by nitriding. However, steels 722M24 and 897M39 are also suitable for nitriding, as well as for general purposes.

Table 7.6 (*continued*)

Tensile strength														
65–75 ton in.⁻² 1000–1160 MN m⁻²			70–80 ton in.⁻² 1080–1240 MN m⁻²			75–85 ton in.⁻² 1160–1310 MN m⁻²			80–90 ton in.⁻² 1240–1390 MN m⁻²			>100 ton in.⁻² >1540 MN m⁻²		
605M30	*1.5MnMo*	*510–850*												
605M36	1.5MnMo	480–850												
608M38	*1.5MnMo*	*480–850*	817M40	1.5NiCrMo	635–1240	817M40	1.5NiCrMo	635–1240				817M40	1.5NiCrMo	635–1240
709M40	1CrMo	480–950										897M39	3.25CrMoV	640–940
816M40	*1.5NiCrMo*	*540–850*												
945M38	1.5MnNiCrMo	480–850												
526M69	0.05Cr	620–740	830M31	3NiCrMo	635–940	823M30	2NiCrMo	635–1235				823M30	2NiCrMo	635–1235
817M40	1.5NiCrMo	635–1240				826M31	2.5NiCrMo	635–1235				826M31	2.5NiCrMo	635–1235
												897M39 (85min.)	3.75CrMoV	1110–1235
830M31	3NiCrMo	635–940	823M30	2NiCrMo	635–1235							826M40	2.5NiCrMo	640–900
			826M31	2.5NiCrMo	635–1235									
823M30	2NiCrMo	635–1235	826M40	2.5NiCrMo	640–900	826M40	2.5NiCrMo	640–900	826M40	2.5NiCrMo	640–900	835M30	4NiCrMo	1125–1235
826M31	2.5NiCrMo	635–1235												
826M40	2.5NiCrMo	640–900												
826M40	2.5NiCrMo	640–900	826M40	2.5NiCrMo	640–900									
897M39	3.25CrMoV	640–940	976M33	3.25NiCrMoV	700–980									
976M33	3.25NiCrMoV	700–980	897M39	3.25CrMoV	640–940									

4. As a general rule, steels quoted in any one block can be tempered down to the next lower tensile range. Equally, where a tensile range is quoted up to a certain maximum diameter, the properties can be attained on smaller diameters, but in practice this may be wasteful of the alloy content, and a cheaper steel may be satisfactory.

5. All steels below the dashed line in the 10″ diameter blocks refer to heavy forgings as specified in BS 4670, but all steels may be used as smaller section forgings as well as rolled bar or billet.

6. This table is based on the 1970 edition of BS 970 and the 1971 edition of BS 4670 except that steels in italics have been eliminated from the 1983 edition of BS 970.

Reproduced from *The Fulmer Optimizer*, by courtesy of Elsevier.

Table 7.7 Influence of added (and adventitious) elements in steel

Element	Dominant characteristic	Influence in ferritic steel	Influence in austenitic steel
Carbon	Strong austenite former	Strongly increases strength and hardability Decreases ductility	Causes weld decay unless stabilized Stabilizes austenite
Nitrogen	Strong austenite former	Increases strength Decreases toughness	Increases strength Stabilizes austenite
Manganese	Austenite former	Strongly increases strength Increases hardenability Increases tendency to quench cracking Neutralizes harmful effect of sulphur	Stabilizes austenite
Nickel	Austenite former	Refines grain Increases toughness Increases hardenability Slightly increases strength	Stabilizes austenite Stress corrosion cracking peaks at content of 17%
Chromium	Ferrite former Carbide former	Improves corrosion and scaling resistance Improves hardenability Slightly increases strength Retards softening in tempering	Improves corrosion and scaling resistance Destabilizes austenite In high concentration forms brittle 'sigma' phase with iron
Molybdenum	Ferrite former Carbide former	Strongly increases hardenability Moderately increases strength Retards softening on tempering Strongly increases strength at high temperature Alleviates temper embrittlement	Improves corrosion resistance Strongly increases strength at high temperatures
Vanadium	Ferrite former Carbide former	Strongly increases hardenability Moderately increases strength Strongly increases high-temperature strength Increases toughness Alleviates embrittlement by nitrogen	
Silicon	Deoxidizer	Improves scaling resistance Increases hardenability Reduces toughness Increases resistivity Promotes decarburization	Improves scaling resistance
Niobium	Strong carbide former	Increases strength of carbon steel by age hardening	Stabilizes against weld decay Increases strength at high temperature
Titanium	Strong carbide former (with aluminium)	 Strongly increases strength by age hardening	Stabilizes against weld decay Increases strength at high temperature Very strongly increases strength by age hardening
Aluminium	Deoxidizer	Increases toughness by combining with nitrogen. Increases sealing resistance. Renders steel suitable for gas nitriding	
Boron		In small amounts greatly increases hardenability. Improves strength at high temperatures	Greatly improves creep and rupture strength
Sulphur	Impurity except when added to improve machinability	Reduces cleanliness Reduces ductility ~0.3% added to improve machinability	Reduces cleanliness Reduces ductility ~0.3% improves machinability

Table 7.7 (continued)

Element	Dominant characteristic	Influence in ferritic steel	Influence in austenitic steel
Lead	Improves machinability	Added to improve machinability	
Selenium	Improves machinability		Added to improve surface finish on machining
Phosphorus	Impurity	Reduces ductility and cleanliness Can improve strength of carbon steel	Reduces ductility and cleanliness
Copper	Normally impurity	Improves corrosion resistance May improve strength but reduces ductility by ageing	Improves corrosion resistance Can increase strength at high temperature
Sn, Sb, As, Bi	Tramp element Impurities	Strongly reduces ductility Promotes temper embrittlement	Fortunately seldom encountered
Hydrogen	Impurity Decarburizer	Strongly promotes rupture and fracture	Fortunately seldom encountered

distortion in heat treatment. Case hardnesses of 65 Rockwell C may be produced with the same range of core strengths as by carburizing.

7.3.11.4 Nitriding

Nitriding may be achieved by heating steel in a cyanide bath or an atmosphere of gaseous nitrogen at 510–565°C. The steel component is heat treated and finish machined before nitriding.

Liquid nitriding uses a bath of sodium and potassium cyanides, or sodium cyanide and sodium carbonate. The bath is pre-aged for a week to convert about a third of the cyanide into cyanate. Two variants of the process are liquid pressure nitriding in which liquid anhydrous ammonia is piped into the bath under a pressure of 1–30 atm, and aerated bath nitriding in which measured amounts of air are pumped through the molten bath. All the processes provide excellent results, depth and hardness of case being the same as obtained from gas nitriding. Unlike gas nitriding, carbon steels can be liquid nitrided and the case produced on tool steels is tougher and lower in nitrogen than a gas-nitrided case. On the other hand, liquid nitriding uses a highly poisonous liquid bath at a high temperature and the process may take as long as 72 hours. It is really only suitable for small components.

Gas nitriding is achieved by introducing nitrogen into the surface of a steel by holding the metal at between 510°C and 565°C in contact with a nitrogenous gas, usually ammonia. A brittle nitrogen-rich surface layer known as the 'white nitride layer' which may have to be removed by grinding or lapping is produced. There are two processes: single- and double-stage nitriding.

In the single-stage process a temperature between 496°C and 524°C is used and about 22% of the ammonia dissociates. This process produces a brittle white layer at the surface. The first stage of the double-stage process is the same as the single stage but, following this, the ammonia is catalytically dissociated to about 80% and the temperature increased above 524°C. Less ammonia is used in the double-stage compared with the single-stage process and the brittle white layer is reduced in depth and is softer and more ductile. Process times are in the order of 72 hours.

Gas nitriding can only be used if the steel contains an alloying element such as aluminium, chromium, vanadium or molybdenum that forms a stable nitride at nitriding temperatures. The film produced by nitriding carbon steels is extremely brittle and spalls readily. In general, stainless steels, hot-work die steels containing 5% chromium and medium-carbon chromium containing low-alloy steels have been gas nitrided. High-speed steels have been liquid nitrided.

There are also a number of steels listed in AISI/SAE or BS 970 (or having the name 'Nitralloy') to which 1% aluminium has been added to make the steel suited for gas nitriding. AISI 7140 (BS 970 905 M39) is typical.

Nitriding can produce case hardnesses up to 75 Rockwell C depending on the steel. This hardness persists for about 0.125 mm but depths of case with hardness above 60 Rockwell of 0.8 mm may be produced.

The relatively thin case compared with other methods of case hardening make it customary to use fairly strong core material. For ferritic steels a UTS between 850 and 1400 MPa is usual. Typical components nitrided are gears, bushings, seals, camshaft journals and other bearings, and dies – in fact all components which are subject to wear. In spite of their relatively low hardness, austenitic stainless steel components are nitrided to prevent seizure and wear, particularly at high temperatures. Two considerations apply.

First, stainless steels must be depassivated by mechanical or chemical removal of the chromic oxide film before nitriding. Second, nitriding decreases corrosion resistance by replacing the chromic oxide film by a chromium nitride film and should not be employed when corrosion resistance is of paramount importance.

Ion implantation is achieved by bombarding the surface of a steel with charged ions, usually nitrogen when the object is to harden the surface. The cost is high, the quantity of nitrogen implanted small, and it can only be carried out by a laboratory which has an accelerator such as AERE. It is used for special applications which will probably increase in number.

7.3.12 Stainless steels

The addition of strong oxide-forming elements (aluminium silicon and chromium) replaces the oxide on the surface of iron by a tenacious film, which confers corrosion and oxidation resistance.[19] Alloys of iron with substantial proportions of aluminium and silicon have undesirable properties so that chromium additions which in progressively increasing quanti-

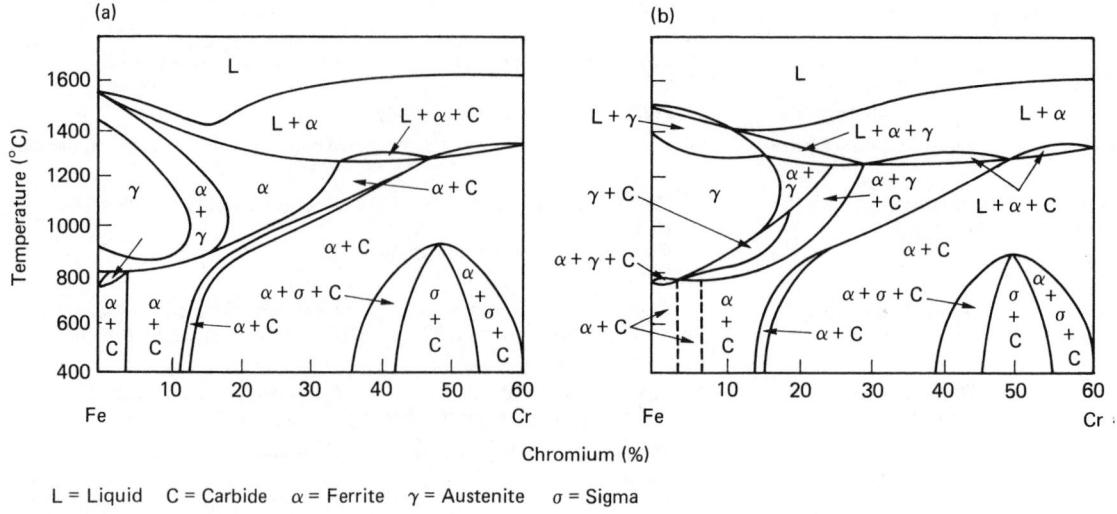

L = Liquid   C = Carbide   α = Ferrite   γ = Austenite   σ = Sigma

Figure 7.27 Iron–chromium–carbon phase diagrams (a) at 0.10% carbon, (b) at 0.50% carbon

ties change the oxide film first to a spinel and then to chromium tri-oxide must be employed. Stainless steels are alloys with a minimum of 50% iron and a minimum of 12% chromium.

7.3.12.1 Metallurgy of stainless steels

Iron forms a complete series of solid solutions with nickel and with chromium; the alpha or delta form (ferrite) will form solid solutions with chromium up to 100% of the alloying element but will dissolve only a limited amount of nickel and the gamma form (austenite) will dissolve up to 100% of nickel without a new phase appearing but can dissolve only limited amounts of chromium.

The above comments are reflected in the phase diagram for the Fe–Cr system (Figure 7.27). Of particular significance is the small austenite field known as the gamma loop; alloys to the right of this loop are ferritic and undergo no allotropic changes in heating or cooling, consequently grain refinement by such changes is not possible. The amount of chromium which closes this loop if no other element is present is 12.8%. Above this figure pure Fe–Cr alloys are ferritic and subject to grain growth as temperatures are raised to the liquidus. Addition of austenite formers enlarges the gamma loop so that, in the limit, the austenitic phase is stable over the entire range of temperature. Varying the proportions of chromium and nickel (and manganese and nitrogen) produces the several types of stainless steel.

Ferritic stainless steels contain between 11 and 30% of chromium, a minimum of austenite formers (see Table 7.7) such as carbon whose influence on the extent of the gamma loop is shown in Figure 7.27 and often some other ferrite formers so that they always retain a ferritic structure.

The standard ferritic (and martensitic) stainless steels have '400' series AISI and BS 970 numbers. These numbers increase with the chromium content, low numbers (e.g. 403) denoting 12% chromium. Other things being equal, therefore, a higher-numbered steel will have a better resistance to general corrosion than a lower-numbered one. The following numbers indicate a ferritic steel: 405, 409, 430, 434, 436. The non-standard steels include Carpenter 182 FM and four aluminium-containing steels Armco 18 SR, and BSC Sichromal 9, 10 and 12.

Ferritic stainless steels are marketed only in the form of plate and strip and all have similar mechanical properties: UTS 415–460 MPa, yield strength 275–550 MPa, elongation 10–25% depending on thickness of plate. They require no heat treatment beyond an anneal at about 800°C followed by air or furnace cooling. The steels are easily drawn and pressed and their machinability is good, 430 FSe being naturally the best. They are prone to grain growth particularly during welding and this impairs toughness and ductility.

The steels are virtually immune to chloride-induced stress corrosion cracking at the relatively low temperatures at which they are used and have good resistance to scaling at elevated temperatures, the aluminium containing varieties (e.g. the Sichromals being some of the best available materials in this respect). They are significantly cheaper than austenitic steels and are used for chemical plant components, domestic and catering equipment, automobile trim, domestic and industrial heater parts, exhaust systems and fasteners. The higher numbers, which have greater resistance to general corrosion, are used for the more demanding applications.

'*Low Interstitial*' grades characterized by carbon and nitrogen contents below 0.03%, chromium contents between 17 and 30% usually with molybdenum and other additions are recently developed ferritic stainless steels. These include one standard steel, 444 (with, in spite of its high number, only 18.5% Cr) and non-standard steels Alleghenny Ludlum 'E Brite 261' 'A129.4.4' and 'A294C', Nyby Uddeholm 'Monit', Crucible 'Seacure/SCI' and Thyssen 'Superferrit'.

These steels, particularly the versions which contain 28% Cr and 4% or more percentage Mo are claimed to have exceptional resistance to general, stress and pitting corrosions and to be suitable for the most aggressive environments obtaining in chemical plant and elsewhere.

Martensitic stainless steels contain 11–18% of chromium and some austenite formers (see Table 7.7) such as carbon (see Figure 7.27) so that they can be hardened by cooling through the gamma/alpha phase transformation.

The US martensitic stainless steels also have 400 series numbers, 403[B], 410[B], 414[B], 416[B], 420[B], 422, 431[B] and 440 (the affix B indicates a BS 970 version) with chromium contents increasing with specification number from 12% to 17% (the highest chromium content at which a steel can have a fully martensitic structure). They have therefore less general corrosion resistance than the ferritic stainless steels but have fair resistance to stress corrosion. The steels can be hardened by quenching from above 950°C to form a hard and brittle structure which must be tempered. Tempering at 150–370°C improves ductility with little loss of strength but above 500°C the strength falls off rapidly. Holding at temperatures between 370°C and 600°C causes temper embrittlement which reduces impact resistance and must be avoided.

The martensitic high carbon grades are difficult to form and weld. They are particularly suited for operations requiring resistance to wear and manufacture of a cutting edge and their applications include valves, tools, cutlery, scissors, turbine blades, coalmining equipment and surgical instruments. The most widely used (and therefore most easily available) martensitic and ferritic stainless steels are listed in Table 7.8.

Austenitic stainless steels contain 15–27% Cr and, in the case of the '300' series 8–35% Ni. In the '200' series for which there is no BS 970 equivalent some of the nickel is replaced by Mn and N, which cost less than nickel. These steels can be cold worked to higher strengths than the '300' series steels.

Table 7.8 Most readily available martensitic and ferritic stainless steels

AISI No.	Approximate composition	UTS (MPa)	Additional information
403	C 0.08 max. Cr 12.0/14.0 Ni 0.50 max.	420	A low-carbon stainless iron suitable for rivets, split pins, and lightly stressed engineering fittings. Nearest equivalent specifications BS 970 403S17 (1970) BS 1449 403S17 (1970) BS 1501: Part 3 403S17 (1973)
405	C 0.08 max. Cr 12.0/14.0 Ni 0.50 max. Al 0.10/0.30	420	Non-hardenable. Suitable for welded fabrications Nearest equivalent specifications BS 1449 405S17 (1970) BS 1501: Part 3 405S17 (1973)
409	C 0.09 max. Cr 11.0/13.0 Ni 0.70 max. Ti 5 × C/0.60	420	Non-hardenable. Suitable for welded fabrications Nearest equivalent specifications BS 1449 409S17 (1970)
410	C 0.09/0.15 Cr 11.5/13.5 Ni 1.00 max.	540/690	Martensitic stainless steel for general engineering applications Nearest equivalent specifications BS 970 410S21 (1970) BS 1449 410S21 (1970)
420	C 0.14/0.20 Cr 11.5/13.5 Ni 1.00 max.	690/850	Surgical instruments, Scissors, taper and hinge pins. General engineering purposes Nearest equivalent specifications BS 970 420S29 (1970)
420	C 0.20|0.28 Cr 12.0/14.0 Ni 1.00 max.	690/850	Valve and pump parts (which are not in contact with non-ferrous metals or graphite packing) Surgical instruments Nearest equivalent specifications BS 970 420S37 (1970)
420	C 0.28|0.36 Cr 12.0/14.0 Ni 1.00 max.	690/930	Cutlery and edge tools Nearest equivalent specifications BS 1449 420S45 (1970) BS 970 420S45 (1970)
430	C 0.10 max. Cr 16.0/18.0 Ni 0.50 max.	430	Ferritic stainless. Domestic and catering equipment, motor car trim, domestic and industrial heater parts Nearest equivalent specifications BS 970 430SI5 (1970) BS 1449 430SI5 (1970)
431	C 0.12/0.20 Cr 16.0/18.0 Ni 2.00/3.00	850/1000	General engineering. Pump and valve parts (in contact with non-ferrous metals or graphite packing) Nearest equivalent specifications BS 970 431S29
434	C 0.10 max. Cr 16.0/18.0 Mo 0.90/1.30 Ni 0.50 max.		Ferritic stainless. Motor car trim Nearest equivalent specifications BS 1449 434S19
SF67[a]	C 0.70 Cr 13.0		Razor blade strip

Free-machining versions of 13% Cr steels are available to BS 970 416S21, 416S29, 416S37.
[a] BSC trademark.

Table 7.9 Most readily available austenitic stainless steels

AISI No.	Approximate composition	UTS (MPa)	Additional information
202	C 0.07 max. Mn 7.00/10.0 Cr 16.5/18.5 Ni 4.00/6.50 N 0.15/0.25	630	Nearest equivalent specifications BS 1449 284S16 (1970)
301	C 0.15 max. Cr 16.0/18.0 Ni 6.00/8.00	540/1240[a]	Readily hardens by cold working. Structural steels for applications where high strength is required Nearest equivalent specifications BS 1449 301S21 (1970)
302	C 0.08 max. Cr 17.0/19.0 Ni 8.00/11.0	510/790[a]	For spoons and forks, holloware, architectural and shop fittings, domestic catering, food manufacturing, dairy and brewery equipment Nearest equivalent specifications BS 970 302S25 (1982) BS 1449 302S17, 302S25 (1970)
303	C 0.12 max. S 0.15/0.30 Cr 17.0/19.0 Ni 8.00/11.0	510/790[a]	A general-purpose austenitic free-cutting steel Nearest equivalent specifications BS 970 303S21 (1983)
304L	C 0.03 max. Cr 17.5/19.0 Ni 9.00/11.0	490	A low-carbon version of 304, fully resistant to weld decay. For chemical plant, food manufacturing, dairy and brewery equipment Nearest equivalent specifications BS 970 304S12 (1983) BS 1449 304S12 (1970) BS 1501: Part 3 304S12 (1973)
304LN	C 0.03 max. Cr 17.5/19.0 Ni 9.00/12.0 N 0.25 max.	590	A high proof stress version of 304L. For cryogenic, storage, and pressure vessels Nearest equivalent specification BS 1501: Part 3 304S62 (1973) (Hi-proof 304L)
304	C 0.06 max. Cr 17.5/19.0 Ni 8.00/11.00	510/790[a]	Holloware, domestic, catering, food manufacturing, dairy and brewery equipment. Recommended for stretch-forming applications. Readily weldable Nearest equivalent specifications BS 970 304S15 (1983) BS 1501: Part 3 304S15, 304S49 (1973) BS 1449 304S15 (1970)
304	C 0.06 max. Cr 17.5/19.0 Ni 9.00/11.0	510	As above. Preferable for deep-drawing applications Nearest equivalent specifications BS 1449 304S16 (1970)
304N	C 0.06 max. Cr 17.5/19.0 Ni 8.00/11.0 N 0.25 max.	590	A high proof stress version of 304. Cryogenic, storage and pressure vessels Nearest equivalent specification BS 1501: Part 3 304S65 (1973) (Hi-proof 304)
305	C 0.10 max. Cr 17.0/19.0 Ni 11.0/13.0	460	Dental fittings, thin-walled deep-drawn pressings. Low cold working factor and very low magnetic permeability Nearest equivalent specifications BS 1449 305S19 (1970)
316L	C 0.03 max. Cr 16.5/18.0 Ni 11.0/14.0 Mo 2.25/3.00	520	A low-carbon version of 316 fully resistant to weld decay. For chemical and textile plant, dairy and food equipment Nearest equivalent specifications BS 970 316S12 (1983) BS 1449 316S12 (1970) BS 1501: Part 3 316S12 (1973)
316LN	C 0.03 max. Cr 16.5/18.5 Ni 11.0/14.0 Mo 2.25/3.00 N 0.25 max.	620	A high proof stress version of 316L. Cryogenic storage and pressure vessels Nearest equivalent specification BS 1501: Part 3 316S62 (1973) (Hi-proof 316L)
316	C 0.07 max. Cr 16.5/18.0 Ni 10.0/13.0 Mo 2.25/3.00	540	Chemical and textile plant. Dairy and food equipment. A lower ferrite content version is for use in special applications (e.g. urea plant) Nearest equivalent specifications BS 970 316S16 (1983) BS 1449 316S16 (1970) BS 1501: Part 3 316S16 (1973)

Table 7.9 (continued)

AISI No.	Approximate composition	UTS (MPa)	Additional information
316N	C 0.07 max. Cr 16.5/18.5 Ni 10.0/13.0 Mo 2.25/3.00 N 0.25 max.	620	A high proof stress version of 316. For cryogenic storage and pressure vessels Nearest equivalent specification BS 1501: Part 3 316S66 (1973) (Hi-proof 316)
317L	C 0.03 max. Cr 17.5/19.5 Ni 14.5/17.0 Mo 3.00/4.00	490	A low-carbon version of 317 fully resistant to weld decay. For chemical plant Nearest equivalent specifications BS 970 317S12 (1983) BS 1449 317S12 (1970)
317	C 0.06 max. Cr 17.5/19.5 Ni 12.0/15.0 Mo 3.00/4.00	540	For chemical plant Nearest equivalent specifications BS 970 317S16 (1983) BS 1449 317S16 (1970)
320Ti	C 0.08 max. Cr 16.5/18.0 Ni 11.0/14.0 Mo 2.25/3.00 Ti 4 × C/0.60	520	Fully stabilized against weld decay Nearest equivalent specifications BS 970 320S17 (1983) BS 1449 320S17 (1970) BS 1501 Part 3 320S17 (1973)
321	C 0.08 max. Cr 17.0/19.0 Ni 9.00/11.0 Ti 5 × C/0.70	540	Fully stabilized against weld decay. Chemical, dairy and brewing plant, food manufacturing and textile equipment. Domestic and catering equipment Nearest equivalent specifications BS 970 321S12, 321S20 (1983) BS 1449 321S12 (1970) BS 1501: Part 3 321S12, 321S49 (1973)
Warm Worked[b] 321	C 0.08 max. Cr 17.0/19.0 Ni 9.00/11.0 Ti 5 × C/0.70	620	A high proof stress version of 321 obtained by controlled low-temperature hot working Nearest equivalent specification BS 1501: Part 3 321S87 (1973)
325[b]	C 0.12 max. Cr 17.0/19.0 Ni 8.00/11.0 Ti 5 × C/0.90 S 0.15/0.30	510/790	A free-cutting version of 321, fully stabilized against weld decay Nearest equivalent specification BS 970 325S21 (1983)
347	C 0.08 max. Cr 17.0/19.0 Ni 9.00/11.0 Nb 10 × C/1.00	510/540	Chemical, dairy and brewing plant. Food manufacturing and textile equipment. Domestic and catering equipment. Particularly suitable for use in welded plant in contact with nitric acid Nearest equivalent specifications BS 970 347S17 (1983) BS 1449 347S17 (1970) BS 1501: Part 3 347S17, 347S49 (1973)
347N	C 0.08 max. Cr 17.0/19.0 Ni 9.00/12.0 Nb 10 × C/1.00 N 0.15/0.25	650	A high proof stress version of 347 Nearest equivalent specification BS 1501: Part 3 347S67 (1973) (Hi-proof 347)

[a] Depending on size. [b] BSC Trade Name.

Austenitic materials with much more than 30% nickel are 'Nickel Alloys'. If they contain age hardening Al,Ti additions they are iron (or Nickel) Superalloys. The mechanical properties of austenitic steels range between UTS 490–680 MPa, Yield Strength 205–575 MPa, elongation 30–60%.

Some of the AISI specification numbers are followed by letters, and these letters (and where applicable to BS 970 numerical codes) are:

H: (BS Code 49). These steels contain 0.006B and 0.15Nb (except 347, which already has a higher Nb content) and have creep-resisting properties.

Se: This steel contains 0.15% Se and is free machining.

L (BS Code 11): These steels contain a maximum of 0.03% C.

N (BS Code 6X): These steels contain 0.2% nitrogen and therefore have proof stresses from 50 to 130 MPa higher than the non-nitrogen-containing steels.

Ti or Cb (BS Code 40): These steels contain Ti or Nb to combine with the carbon and thereby prevent weld decay.

There are over 50 standard AISI and slightly less BS 970 austenitic stainless steels. Table 7.9 lists those most commonly used and therefore most readily available. (Steels suitable for use at elevated temperatures are listed in Table 7.10.)

There are, in addition, a very large number of non-standard austenitic steels of which the following list is a small selection:

- Alleghhenny Ludlum 'A286': Really a superalloy but used also as a stainless steel because of its high yield strength.
- Armco 'Nitronic' high-nitrogen steel also with high yield strength.
- Avesta 254 and 654 'SMO': High-molybdenum-containing steel with exceptional resistance to pitting corrosion.
- Carpenter '20Cb3': Really a nickel alloy but generally known as a stainless steel, has a high resistance to sulphuric acid attack.
- BSC 'Esshete 1250': Steel with exceptional creep resistance and high yield.

Austenitic stainless steels are chosen on account of their resistance to general corrosion which is superior to that of a ferritic steel of similar chromium content and also because of the high ductility of the face-centred gamma structure which confers high hot and cold formability and high toughness down to cryogenic temperatures.

It is not possible to state exactly where the limits of stability of austenite steel lie at room temperature because transformation can be too sluggish to permit precise delineation of the phase fields and is influenced by further alloy addition such as Mo, Si and N. The austenite should ideally be 'Persistent', that is, it should not transform under the temperature or working conditions encountered in fabrication and service. The range of compositions with 'Persistent' austenite at room temperature is shown in Figure 7.28 (labelled 'A').

Austenite stability is increased by increasing nickel, manganese, carbon and nitrogen. Partial transformation will cause the steel to lose its non-magnetic character, impair its deep drawing characteristics and reduce notch toughness at cryogenic temperatures. There may be other drawbacks but service performance is not usually impaired.

Two substantial advantages are conferred by the presence of a proportion of ferrite; the prevention of fissuring on solidification and resistance to intergranular corrosion. Except in the case of welding (see Section 7.3.17.1) these advantages apply to cast rather than wrought austenitic steels.

Many austenitic stainless steels (including 304, typically 18.8 grade) are partially transformed by cold work and work harden appreciably. Steels such as these are air cooled in thin section but thicker sections are water quenched. Besides promoting stability this retains carbide in solution.

7.3.12.2 Duplex stainless steels

Duplex stainless steels contain 18–27% Cr, 4–7% Ni, 2–4% Mo with copper and nitrogen in proportions which ensure that they have a mixed ferritic austenite structure that is not heat treatable (see Figure 7.28). Their mechanical properties range between UTS 600–900 MPa, Yield Strength 410–850 MPa, Elongation 16–48%. The one standard duplex stainless steel is AISI 329, but there are, in addition, BSC 'SF22/5', Langley Alloys 'Ferralium 255', Sandvik '2RE60' and 'SAF2205/AF22' and Sumitomo 'DP3'.

The duplex stainless steels have outstanding properties. Their resistance to stress corrosion cracking is superior to that of comparable austenitic steels and they have good resistance to pitting corrosion. They have better toughness than ferritic steels and are easily welded. Those containing nitrogen can be cold worked to higher strengths than ferritic or austenitic steels, and are highly weldable provided that a welding consumable that will ensure the presence of ferrite in the weld metal is employed.

They have so far been used for tube plates, for marine applications, for sour gas pipeline and acetic acid production. When they are better known and more widely available they should become used in preference to austenitic steels for the more demanding applications.

Precipitation-hardening stainless steels contain 12–28% Cr, 4–7% Ni, Al and Ti to give a structure of austenite and martensite which can be precipitation hardened. The mechanical properties of these steels range between UTS 895–1100 MPa, Yield Strength 276–1000 MPa, Elongation 10–35%.

No precipitation-hardening stainless steels are standardized by AISI or in BS 970 but Firth Vickers 'FV 520' is covered by BS 1501 460552 for plate and BS Specification S143, 144 and 145 for bars, billets and forgings. Non-standard steels include Armco '15–5PH', '17–4 PH' and '17–7 PM' and Carpenter 'Custom 450' and 'Custom 435'.

Their excellent mechanical properties and corrosion resistance has caused precipitation-hardened stainless steels to be used for gears, fasteners, cutlery and aircraft and steam turbine parts. They can be machined to finished size in the soft condition and precipitation hardened later. Their most significant drawback is the complex heat treatment required which, if not properly carried out, may result in extreme brittleness.

7.3.13 Corrosion resistance of stainless steels

Corrosion resistance of stainless steels depends on surface passivity arising from the formation of a chromium-containing oxide film which is insoluble, non-porous and, under suitable conditions, self-healing if damaged. Passivity of stainless steel is not a constant condition but it prevails under certain environmental conditions. The environment should be oxidizing in character. Other factors affecting corrosion resistance include composition, heat treatment, initial surface condition, variation in corrosion conditions, stress, welding and service temperature.

7.3.13.1 Composition

Those ferritic and martensitic steels with roughly 13% Cr are rust resisting only and may be used for conditions where corrosion is relatively light (e.g. atmospheric, steam and oxidation resistance up to 500°C). Applications include cutlery, oil-cracking, turbine blades, surgical instruments, automobile exhausts, etc. 17% Cr (ferritic and martensitic) steels are corrosion and light acid resisting. They have improved general corrosion resistance compared with 13% Cr steels. Applications for the ferritic grade include domestic and catering equipment, automobile trim, and industrial heater parts. The martensitic grade is used in general engineering, for pump and valve parts in contact with non-ferrous metals or graphitic packings.

The addition of molybdenum significantly improves the integrity of the oxide film. The ferritic 434 and 436 grades can withstand more severe corrosive conditions and the martensitic, 440 grades are used where wear and acid resistance is required such as in valve seats.

Table 7.10 Steels suitable for use at elevated temperatures showing 0.2% proof and creep rupture strengths near the top of their useful temperature ranges.

BS 1501, 2 or 3 designation	AISI equivalent	Type of steel	Min. 0.2% PS at temperature (MPa)	10^5 h rupture strength at temperature (MPa)	Note
161 Grade 28	1025	Si killed carbon	147.5 at 450°C	133 at 450°C	
221 Grade 32	1527	Si killed carbon manganese	172 at 450°C	147 at 450°C	
223 Grade 32		Si killed carbon manganese Nb treated	173 at 450°C	142 at 450°C	
271		Mn Cr Mo V	292 at 450°C	309 at 450°C	Used for boiler drums in heavy sections
620		1% Cr $\frac{1}{2}$%Mo	136 at 550°C	49.4 at 550°C	
622		$2\frac{1}{4}$Cr 1%Mo	145 at 550°C	72.5 at 550°C	
625		5%Cr $\frac{1}{2}$%Mo	210 at 550°C	290 at 550°C	Used in refinery Not in power plant
660		9%Cr% 1%Mo	210 at 550°C	84 at 550°C	
		$\frac{1}{2}$%Cr $\frac{1}{2}$%Mo $\frac{1}{4}$V	199 at 550°C	74 at 550°C	
Jessups H46		12Cr 0.5Mo V NB N B	181 at 600°C		Gas turbine disk or steam turbine blade material
BS 4882 B16A (Durehete 1055)		1CR 1Mo $\frac{3}{4}$V 0.1Ti 0.005B	Stress relaxation specification		Bolting materials for temperature range 500–565°C
304 S49	304H	18Cr 8Ni 0.15Nb 0.005B	100 at 600°C	74 at 600°C	
316 S49	316H	18Cr 12Ni 0.15Nb 0.00B	100 at 600°C	118 at 600°C	
321 S49	321H	18Cr 10Ni 0.5% 0.45% 0.005B	111 at 600°C	105 at 600°C	
317 S49	347H	18Cr 12Ni 1Nb 0.005B	123 at 600°C	106 at 600°c	
	310	25Cr 20Ni	120 at 550°C	120 at 550°C	
BSC Esshete 1250		15Cr 6Mn 10Ni 1Mo 1Nb $\frac{1}{2}$V 0.006B	140[a] at 650°C	160 at 650°C	
Iron Superalloy		15Cr 25Ni Mo V 3Ti 0.3Al	150 at 700°C	79 at 700°C	Used in aircraft gas turbines

[a] 1% proof stress.

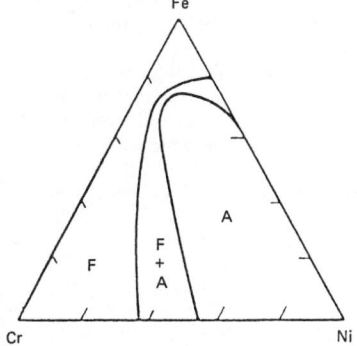

Figure 7.28 Iron–nickel–chromium phase diagram at room temperature showing persistent austenite (A) and duplex ferrite austenite (F+A) phase fields

Additional amounts of nickel above 8% to form an austenitic structure in the '300' steels further improve resistance to corrosion and acid attack. Applications include domestic, shop and office fittings, food, dairy, brewery, chemical and fertilizer industries.

The stainless steels with the highest corrosion resistance are those with even higher chromium contents such as 310 with 25% Cr and the low interstitial steels with up to 30% Cr. The addition of molybdenum up to 6% is also highly beneficial.

The resistance to sulphuric acid attack of Carpenter 20 Cb-3 which contains 3.5% Cu as well as 2.5% Mo has already been mentioned.

Note that to ensure an austenitic structure, the nickel content of the Mo-bearing steels increases above 8% as the content of Mo and other ferrite-stabilizing elements (Ti, Nb, etc.) increases. The 4.5% Mo alloy 317 LM is used in sodium chlorite bleaching baths and other very severe environments in the textile industry.

7.3.13.2 Heat treatment

Heat treatment has a significant influence on corrosion resistance. Maximum resistance is offered when the carbon is completely dissolved in a homogeneous single-phase structure. The 12–14% Cr steels are heat treated to desired combinations of strength, ductility and toughness and because of their low carbon content are generally satisfactory unless tempered in the range 500–600°C.

Austenitic steels (18/8) are most resistant when quenched from 1050°C to 1100°C, their normal condition of supply. A steel chosen for welded fabrications should be Ti or Nb stabilized (AISI, 'Ti' or 'Cb', BSI '40') or, better, an extra-low carbon (AISI 'L' or BSI 11) grade. Quenching after welding is usually impracticable.

7.3.13.3 Surface condition

For maximum resistance to corrosion the passive film must be properly formed; this is ensured by removing all scale, embedded grit, metal pick-up from tools and other surface contaminants. Polishing improves resistance. Passivating in oxidizing acid (10–20% NHO_3 by weight) solution at 25°C for 10–30 min confers maximum resistance to austenitic steels. The ferritic–martensitic grades are passivated in nitric acid–potassium dichromate solution (0.5% nitric acid + 0.5% potassium dichromate at 60°C for 30 min).

7.3.13.4 Variation in corrosion conditions

In the absence of experience, samples of the proposed steels should be tested in the condition in which they will be used (i.e. welded, if fabricated) in the intended environment taking full note of any possible variation in service conditions. The effect of welding on corrosion resistance is considered in Section 7.3.17.

7.3.13.5 Service temperature

Because stainless steels (other than those with very low carbon) which are unstabilized or partly stabilized with Ti or Nb may show chromium carbide precipitation when subjected to service temperatures above 350°C (see Section 7.3.17) this should be the upper limit for service in corrosive environments. Fully stabilized steels are not restricted in this manner.

7.3.13.6 Localized corrosion of stainless steels

The considerations discussed in Sections 7.3.13.1–5 apply principally to general corrosion which progressively reduces the thickness of a component until it is completely dissolved or its strength is so reduced that it can no longer withstand imposed stress. More insidious attack mechanisms on stainless steels are the five varieties of localized corrosion, galvanic, crevice, pitting, stress corrosion and intergranular penetration. These are confined to isolated areas or lines on the surface but penetrate through the thickness of a component to destroy its integrity without materially affecting its dimensions. Their incidence is less predictable and their onset more difficult to predict than general corrosion but their effect may be catastrophic.

Crevice and galvanic corrosion must be countered by designing to eliminate crevices and the juxtaposition of metals of different solution potential. Intergranular penetration will be discussed in the section dealing with welding. Pitting and stress corrosion are composition dependent.

Pitting occurs in conducting aqueous liquid environments (usually halide solutions) when local penetration of the oxide film creates stagnant locations in which diffusion generates strongly acid environments which rapidly penetrate a component. Figures 7.29(a) and (b) show, respectively, a pit in an early and in a very late stage. Resistance to pitting in low interstitial ferritic steels increases with a rise in chromium

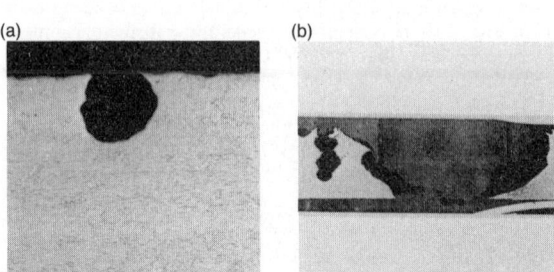

Figure 7.29 Pits, at (a) an early and (b) a late (penetrating) stage

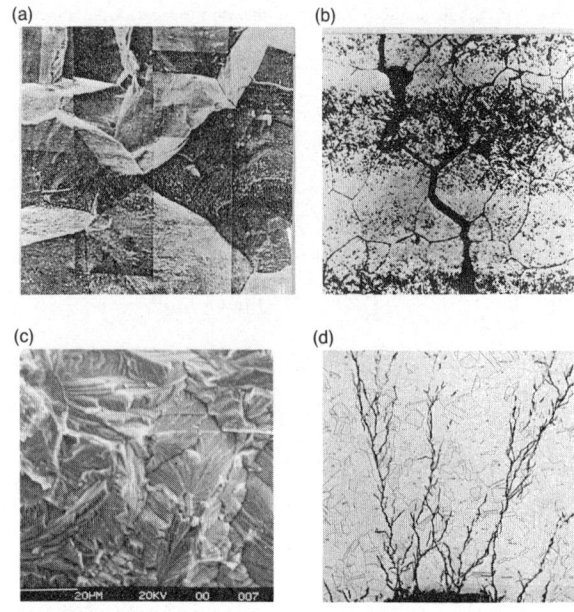

Figure 7.30 Stress corrosion cracks. (a) and (b) Intergranular; (c) and (d) transgranular

content from 18% to 29% and molybdenum from 1% to 4% while austenitic steels require at least 20% chromium and between 4.5% and 6% Mo. Typical austenitic steels with very high resistance are Alleghenny Ludlum 'A16 × M', Avesta '254 and 654 SMO' and Weir Material Services Zenon 100, all of which are claimed not to pit in stagnant seawater. These steels are also claimed to resist crevice corrosion should this not have been eliminated by design. The best standard austenitic stainless steel is 317 LM.

Stress corrosion cracking occurs when a material is stressed in tension in an aggressive aqueous environment, usually an alkali metal halide or hydroxide solution. Cracks may be transgranular as in Figure 7.30(a) and (b) or intergranular as in Figure 7.30(c) and (d). The tendency to stress corrosion cracking of a material is measured by its K_{1scc} value, which is the lowest value of stress intensity in $MNm^{-3/2}$ at which a crack will propagate in a specific medium at a specific temperature.

The growth rate of stress corrosion cracks is highly temperature dependent, increasing about 500 times with a rise in temperature from 20°C to 100°C. Most austenitic steels are resistant at ambient temperature but if temperature rises above about 40°C in a saline environment a change should be made to a high-molybdenum steel, a duplex steel, a nickel-free ferritic steel or a nickel alloy, depending on the severity of the conditions.

Materials are tested for resistance to stress corrosion by exposure to stress in a boiling aqueous solution of 42% $MgCl_2$. Very few materials will withstand this for long.

7.3.14 Heat-resisting steels

The range of operating temperature of carbon steels is limited to about 400°C by decrease in resistance to deformation, and in oxidizing atmospheres to about 500°C by diffusion of oxygen through the oxide film. Operation above these temperatures is achieved by the addition of alloying elements such as

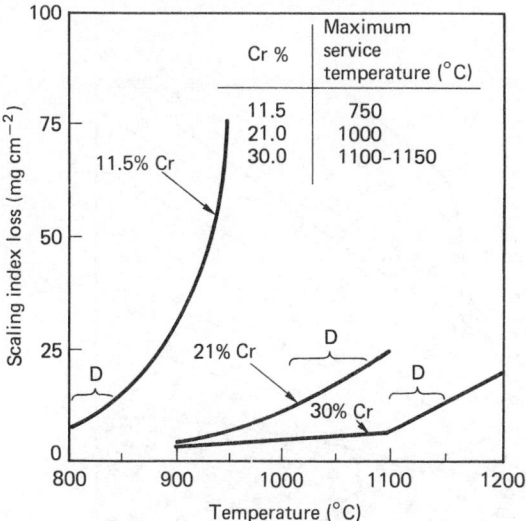

Figure 7.31 Relationship between scaling loss, temperature and chromium content of chromium steels

Cr %	Maximum service temperature (°C)
11.5	750
21.0	1000
30.0	1100–1150

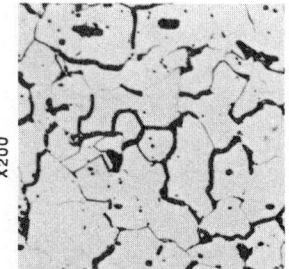

Figure 7.32 Micro fissures caused by hydrogen in steel

chromium, aluminium and silicon which render the oxide film more tenacious and limit diffusion (see Figure 7.31) and chromium, molybdenum, vanadium and niobium which, in solid solution or as carbides, impart stability and increase resistance to deformation.

The scope for providing high-temperature yield and creep strength in ferritic steels is limited, therefore for service above about 550°C, austenitic steels are used. Steels for use at high temperature must, in addition, be stable and capable of fabrication to the design shape. Material requirements vary with application.

Steam power plant materials, with minor exceptions, operate below 650°C and in atmospheres which are only a little more aggressive than air. But they must last for at least 10^5 hours and preferably two or three times that figure.

Chemical and refinery plant may be required to operate over a wider temperature range in very varied environments but they have shorter lives (usually about 2–4×10^4 hours). Aircraft propulsion turbines require materials to withstand high stresses at very high temperatures but component operating lives are seldom above 10^3 to 10^4 hours.

The materials used to meet these requirements are basically as follows. For components which must resist oxidation but are not stressed, ferritic chromium steels, preferably also containing silicon and aluminium, may be chosen. The choice will depend on cost, temperature, aggressiveness of environment and ability to fabricate from 405, 409 (the cheapest weldable stainless steel), 446 and BS Sichromal '9', '10' and '12'. For stressed components operating at high temperatures a choice may be made from the steels listed in Table 7.10. This table gives only a few out of many alternatives but provides at least one steel that may be selected with confidence to operate over any part of the temperature range.

At temperatures below 450°C silicon-killed carbon or carbon manganese steels are used except for heavy pressure vessels, where bainitic steels such as BS 1501 271 or 281 which have a high proof stress in the normalized condition are used.

Time-dependent deformation becomes more important than yield at temperatures above 400°C and the design criterion changes from a factor of the proof stress to a factor of the creep rupture stress at the design life of the component. As temperature increases above 400/450°C carbon steels start to give place to chromium–molybdenum bainitic or martensitic steels. In power plant, strength rather than corrosion resistance is the critical parameter and the lower chromium steels are preferred.

In chemical or refining plant the environment may be hydrogenous and higher chromium contents are essential to prevent hydrogen which diffuses into the steel, combining with carbon to cause internal ruptures (see Figure 7.32).

Increase in temperature beyond 550°C requires the higher creep resistance of an austenitic steel. One of the steels designated 'H' by AISI or coded 44 by BS 970 should be selected. These are stabilized by niobium and their creep rupture strengths and ductilities improved by the addition of 0.006% boron.

Standard steels are satisfactory up to about 600°C but non-standard steels such as BS 'Esshete 1250' have an increased temperature range and allow the use of thinner sections. In critical locations the very high scaling resistance of 310 may be used as the corrosion-resistant face of a laminated structure backed with Esshete 1250. For the higher temperatures and higher stresses in aircraft gas-turbine engine blades or disks recourse must be had to superalloys such as Alleghenny Ludlum 'A286' or 'Discoloy'. Where high temperature strengths of these are inadequate recourse must be had to nickel alloys (see Section 7.4.5). Scaling resistance of aircraft gas-turbine blade materials is provided by coating with aluminium.

7.3.14.1 Structural stability

Stainless steels heated to above 600°C in fabrication or operation are subject to embrittlement mechanisms which have, in the past, given rise to severe problems. These mechanisms are listed and the compositions over which they may occur are indicated in Figure 7.33. The better understanding of the problems has resulted in a method of avoidance or the appreciation that they are not so serious as was originally considered.

Embrittlement due to carbide precipitation is avoided by using a low-carbon or a stabilized steel (AISI 'L' or 'H' or BS 970 Code '11' or '44'). Straight chromium steels with chromium contents less than 27% Cr are not subject to sigma-phase embrittlement (precipitation of an intermetallic FeCr phase). Steels with more than 27% Cr should not be employed within the temperature range 520–700°C at which sigma-phase embrittlement occurs. The sigma phase is dissolved by heating to 820°C.

Straight chromium steels with more than 15% Cr suffer from '475°C' embrittlement if held in (or slowly cooled through) the range 525–425°C.

Austenitic steels, particularly 310 with silicon in excess of 1.5%, develop sigma phase when heated in the range

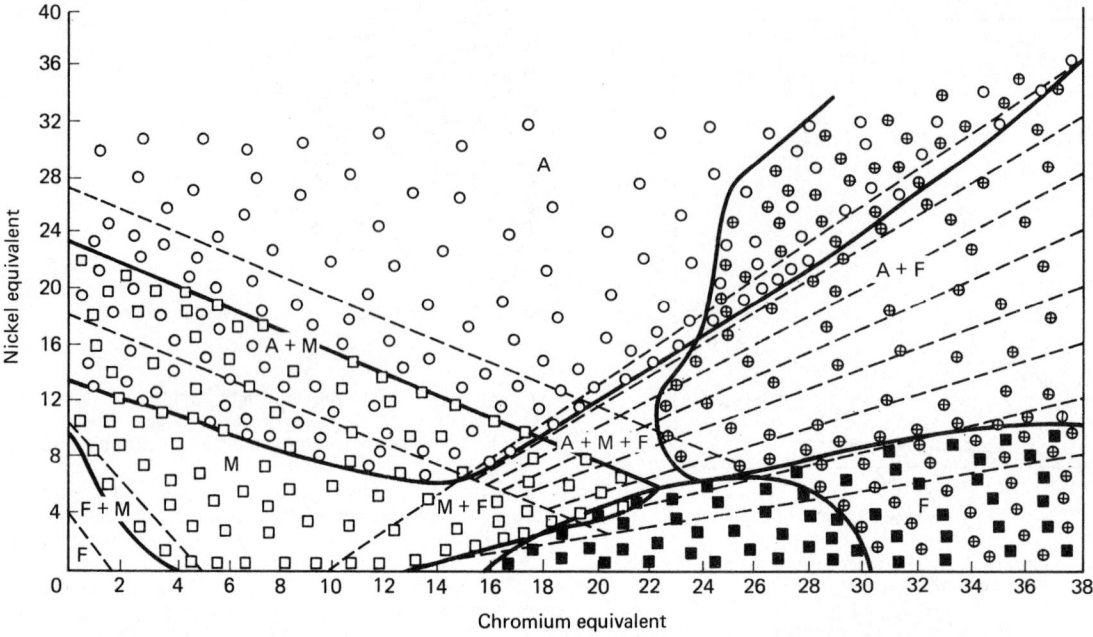

Figure 7.33 Embrittlement mechanisms in stainless steels related to composition (□) Martensitic cracking between 0 and 290°C; (○) hot cracking above 1250°C; (⊕) σ phase embrittlement after heat treatment or service at 500–900°C; (■) cold brittleness after gain growth due to high temperatures (>1150°C)—ductile above 400°C. Where symbols overlap, the material shows the characteristics of both mechanisms

590–925°C but, provided local stressing by differential expansion is prevented by design, the embrittlement has little effect on service performance even though the steel has zero room temperature ductility.

Austenite transformation to ferrite may be avoided by ensuring that the composition of the steel is such as to produce persistent austenite. There is little evidence of the transformation leading to problems in service even when this condition has not been met.

7.3.14.2 Valve steels

Internal combustion engine valves operate under severe conditions of fatigue, impact, high-temperature corrosion and wear. In the USA the SAE lists a special category which includes all types of steel which are used for valves (see the *SAE Handbook*[11]).

In the UK five steel types classified as stainless steels in BS 970 are described as Valve Steels. They are:

1. Grade 401 S45; 3% Si, 8% Cr (strictly, not a stainless steel), martensitic steel used for inlet valves in petrol engines and exhaust valves in medium-duty diesels. Limiting temperature 700°C
2. Grade 382 S34; 21% Cr, 12% Ni. Austenitic steel used for diesel exhaust valves, must be hard faced above 700°C.
3. Grade 443 S62; 2% Si, 20% Cr. Martensitic steel used for exhaust valves in petrol engines. Limiting temperature 750°C.
4. Grades 331 S40 and 331 S42 (KE965)*: 14% Cr, 14% Ni-Si-W. Austenitic steel suitable (with hard-faced seats) for temperatures up to 800°C.
5. Grades 349 S52, 349 S54. 352 S52 and 352 S54 (the S54 types are free cutting with sulphur additions). Scaling resistance to 900°C. Used for petrol engine exhaust valves.

*Common (or trade) name.

7.3.15 Toughness in steels

Toughness is the property that prevents failure of a material when a load is either rapidly applied or generates a high stress intensity at the root of a discontinuity. It is defined as the critical stress intensity resulting in fracture K_{Ic}, MNm$^{-3/2}$ or Charpy Impact Energy J. In the case of metals with a body-centred-cubic structure (ferritic steels) or a hexagonal structure (magnesium), which decline sharply in toughness over a narrow temperature range, Impact Transition Temperature (f.a.t.t.°C) is also used. These parameters are discussed in Chapter 8, Section 8.3. This section is confined to a description of materials which meet requirements for specific applications.

Other things being equal, fracture toughness bears an inverse relationship to tensile strength, grain size, and carbon content of a steel. Martensitic structures are tougher than bainitic which are themselves tougher than pearlitic structures with the same hardness.

Toughness is reduced by increase in the content of hydrogen, oxygen, nitrogen and sulphur and the so-called tramp elements, phosphorus, antimony, arsenic and tin, which cause 'temper embrittlement'. It is increased by a rise in content of nickel, manganese and appropriate amounts of aluminium, vanadium, niobium and molybdenum which specifically reduces temper embrittlement. The face-centred-cubic austenitic steels do not suffer from a ductile/brittle transition at low temperature.

Two examples of failures which were eliminated by a change to a tougher material are:

- Failures in the original welded ships which, in some cases, split in half. These failures all occurred at low temperature. In one specific case failure occurred at a weld start strake at 2°C in a steel with a ductile brittle transition temperature of 30°C and a Charpy energy at failure temperature of 11J. Failures were eliminated by deoxidizing steel with 0.15–0.3% Si and 0.02–0.05% Al which refines grain size and combines with nitrogen. Ship plate is now specified to have a Charpy V notch J value of 20 at 4°C, a figure which is well within the capability of modern steels low in hydrogen, oxygen, sulphur and phosphorus.

- Failures in heavy-section turbogenerator forgings, which operated at relatively low temperature, due to embrittlement by hydrogen combined with temper embrittlement. Hydrogen has been eliminated by vacuum treatment of the molten steel. 'Lower nose' temper embrittlement is associated with the migration of 'tramp' elements such as phosphorus, arsenic, antimony and tin, which are taken in solution at the tempering temperature and reprecipitate at grain boundaries at temperatures around 500°C. The migration is promoted by carbon, silicon, nickel and manganese but retarded by molybdenum. The embrittlement could be avoided by quenching the steel from its tempering temperature but the internal stresses so produced would be worse than the temper embrittlement. Temper embrittlement is minimized by reducing the content of tramp elements and by using carbon vacuum deoxidation which obviates the need for silicon.

Figure 7.34 and Table 7.11 show the properties available in modern large forgings in a 3.5 NiCr V steel.

Control of embrittlement is also important to avoid the risk of failure in light water pressure vessels made from ASTM 533B MnMoNi and 508 NiCrMo steels and is achieved by a specification with limits of: Cu, 0.10; P, 0.012; S, 0.015 and V, 0.05% which guarantees a K_{1c} value of 176 MNm$^{-3/2}$ at room temperature.

7.3.15.1 Cryogenic applications

Care must be taken when choosing steels for cryogenic applications for which, as a result of their ductile/brittle transition, the normal ferritic steels are unacceptably brittle. All the common standard austenitic stainless steels have excellent toughness at temperatures down to −240°C, measured by Charpy impact values usually between 140 and 150. Tensile and 0.2% proof stress increase as the temperature is lowered to around 1500 and 456 MPa, respectively, and elongations decrease slightly but remain adequate at 40–50%. The 0.2% nitrogen grades (typically, 316N, H (316 S66)) have higher proof stresses and are particularly suited to cryogenic applications, because the nitrogen ensures that the austenite is persistent. There is some evidence that the endurance limit of austenitic steels increases as temperature decreases.

If a non-stainless steel is preferred there are the French 'Afnor' specification steels, 3.5% Ni, 5% Ni and 9% Ni whose low-temperature properties improve with increasing nickel content. If high strength combined with high toughness at cryogenic temperatures is required, a maraging steel should be specified.

7.3.16 Maraging steels

Maraging steels are supplied to ASTM A579. They are high-nickel steels which are hardened by precipitation of an aluminium titanium compound on ageing at 500°C. They have a number of advantages including high strengths, normally ranging between 1100 and 1930 MPa (but a steel with a proof stress of 2400 and a UTS of 2450 MPa is available), excellent toughness even at −196°C and good resistance to stress corrosion cracking.

Their greatest advantage is, however, ease of fabrication. They can be machined at their low-solution-treated hardness of 300 VPN and then aged to their optimum hardness at 500°C with minimal distortion and no risk of cracking. They have good weldability, needing no preheat and their properties may be restored after welding by ageing.

Their main disadvantage is their high cost and the fact that, to obtain optimum toughness, they should be made by ESR or vacuum arc melting. Also, because of the absence of hard carbides, they are inferior in wear properties to hardened and tempered steels.

7.3.17 Weldability of steels

Steels may be welded by almost all varieties of electric arc welding methods, including gas-shielded MIG and TIG with and without filler, flux shielded manual metal arc, submerged arc, electro slag, spot, projection and flash butt. Other fusion methods include the more recently developed electron beam and laser, and relatively old-fashioned gas welding. Solid-phase methods, forge, diffusion, friction and explosive welding may also be used. Many of these procedures are concerned with relatively thin sections or special design and applications.

7.3.17.1 Weldability of non-stainless steels

'Weldability' of steels usually implies the ability to make long runs in fairly large sections either by manual metal or submerged arc and is governed in ferritic steels by the 'Carbon Equivalent':

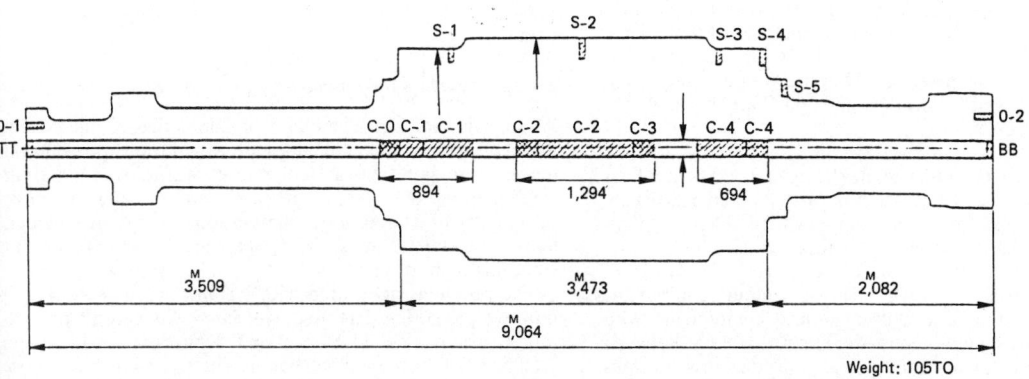

Figure 7.34 Dimensions of typical large rotor forging showing location of test specimens in Table 7.11

Table 7.11 Mechanical properties of the rotor shown in Figure 7.34 after quality heat treatment*

Sample No.	Position and orientation of sample		Tensile test				Charpy impact test (notch: 2mm V)							
			$S_{0.2}$ (kg mm^{-2})	S_u (kg mm^{-2})	e (%)	A_R (%)	As received		De-embrittled†		Embrittled‡		$\Delta f.a.t.t._1$ (°C) AsRe-WQ	$\Delta f.a.t.t._2$ (°C) STC-WQ
							Energy (J)	f.a.t.t. (°C)	Energy (J)	f.a.t.t. (°C)	Energy (J)	f.a.t.t. (°C)		
Desired properties														
—	Surface	R	82.4	84.5	15	45	4.1	≤+15.5						
—	Centre of core bar	L	81.0	84.5	15	45	—	—						
		T	—	—	—	—	4.1	≤+15.5						
Actual properties														
S—1	Surface radial	O	84.0	95.3	19.1	69.6	out. 11.0	<−75						
		I					in. 12.1	−65						
S—2		O	83.8	96.0	19.1	65.6	out. 12.7	−63						
		I					in. 12.1	−29						
S—3		O	83.7	94.8	20.5	68.3	out. 12.7	−75						
		I					in. 12.5	−37						
S—4		O	92.1	102.8	18.0	62.8	out. 9.3	<−75						
		I					in. 11.0	<−75						
S—5		O	90.9	101.6	19.1	68.8	out. 10.8	<−75						
		I					in. 11.5	<−75						
O-1	Ends	L	90.9	101.9	20.5	67.4								
O-2			92.8	103.6	20.5	67.9								
C-1	Centre of core bar	L	84.3	98.4	19.8	57.9	4.6	+15	10.5	+1	4.1	+38	+14	+37
		T	83.8	98.7	17.7	51.9	5.9	+12	6.9	+13	4.1	+43	−1	+30
C-2		L	85.0	98.5	18.3	60.5	6.7	0	12.5	−18	7.7	+6	+18	+24
		T	85.0	99.0	16.9	53.9	7.3	0	13.0	−20	9.2	+5	+20	+25
C-3		L	84.0	97.1	19.9	59.5	8.7	−8						
		T	83.8	96.4	18.2	57.8	8.7	−10						
C-4		L	83.4	96.0	20.8	62.6	8.1	+11	14.0	−4.5	7.1	+20	+15.5	+24.5
		T	83.4	95.7	18.7	59.9	10.6	+10	9.0	+8	5.8	+22	+2	+14

R, radical; T, transverse; L, longitudinal; I, inner; O, outer.
△f.a.t.t., increase in f.a.t.t.; AsRe-WQ, s re-water quenched; STC-WQ, step cooled and water quenched.
* Reproduced by courtesy of Japan Steelworks.
† De-embrittled: 590°C × 1 h − WQ.
‡ Embrittled: step cooled.

$$CE = C\% + N\% + \frac{Cr\% + Mo\%}{5} + \frac{V\% + Ni\% + Cu\%}{15}$$

(Boron is not taken account of in this equation but has a great influence on hardenability and therefore on weldability.) Steels with carbon equivalent below 0.14% are readily welded without special precautions in a wide range of thicknesses.

Steels with carbon equivalent between 0.14% and 0.45% require the following precautions, depending on the value of carbon equivalent and section size to prevent the formation of austempered martensite cracking aggravated by hydrogen.

Specification of low-hydrogen electrodes This is always desirable but requires operator skill to compensate for the more sluggish metal and slag flows compared with other electrodes.

Use of preheat before welding The preheat temperature required depends on the CE and the metal thickness; for a carbon equivalent of 0.2, 40°C and 110°C are advisable for respective metal thicknesses of 25 and 225 mm, while for a carbon equivalent of 0.45, 170°C and 260°C are advisable for the same thicknesses.

Control of heat input Other things being equal, a higher heat input gives less risk of formation of austempered martensite than a lower heat input but care must be taken to limit distortion and the introduction of stress.

Use of post-heat after welding This is seldom required for CEs below about 0.35%, but high-duty components with restrained welds should be post-weld heat treated at between 600°C and 650°C for one hour per 25 mm thickness. Besides preventing immediate cracking (or making it obvious during inspection) post-weld heating improves dimensional stability.

It is essential when welding thick and complex structures to post-heat-treat one weld before commencing to weld a cross seam.

Steels with carbon equivalent above 0.45 These present very severe problems in welding. Very high preheats ranging up to 340°C for carbon equivalent 0.6 and 225 mm thickness, low-hydrogen electrodes (preferably lower in carbon equivalent than the parent material) and immediate post-heat-treatment at temperatures around 800°C are essential. Sample test welds are advisable.

Maraging steels With carbon contents around 0.03% these have a soft martensite matrix and are highly weldable with no risk of decarburization, distortion or cracking. They should be used where very high strength combined with weldability is required.

7.3.17.2 Welding of stainless steels

Welding is the normal method of fabricating stainless steel vessels. The heat-affected zones are raised to incipient fusion temperature but time at temperature varies with different welding processes. Argon arc and spot welding are most satisfactory in heating for minimum time; metal arc welding, inert-gas metal arc and submerged arc are less so in that order from this point of view.

The problems associated with welding stainless steels fall into two categories. The first, associated with carbide precipitation, includes 'Weld Decay' and 'Knife Line Attack' and affects mainly corrosion behaviour. The second includes those phenomena which may be assessed by means of the Schaeffler Diagram (see Figure 7.33).

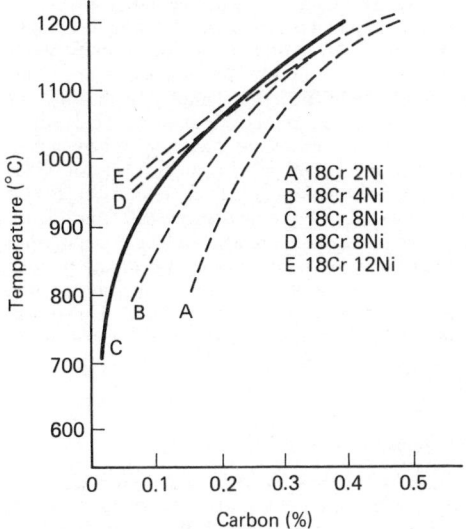

Figure 7.35 Variation of solubility of carbon in 18% chromium steels with temperature and nickel content

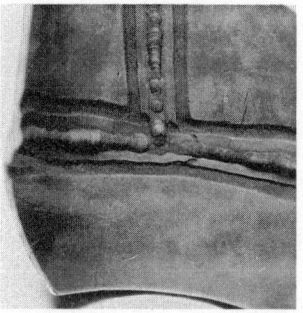

Figure 7.36 Example of 'weld decay' in an austenitic steel

7.3.17.3 Carbide solution and precipitation

The solubility of chromium carbide in austenite decreases with decreasing temperature and increasing nickel content (Figure 7.35). At room temperature the solubility in 18% Cr 8% austenite (solid line) is approximately 0.03%. If an 18% Cr 8% Ni alloy containing, say, 0.06% C is annealed at 1050–1100°C all chromium carbide is in solution and remains in unstable solution after quenching to room temperature. If the alloy is heated to an intermediate temperature excess carbide is precipitated.

The mode of precipitation of carbide is dependent upon whether or not the austenite has been worked. If the quenched but unworked alloy is heated in the temperature range 450–750°C chromium carbide is precipitated at the grain boundaries; the lower the temperature, the longer the time required. Thus at around 450°C the time taken for precipitation can be about two years whereas at 700°C it is a matter of minutes.

Precipitation is effected by diffusion of carbon atoms to the grain boundaries where they each combine with approximately four times the number of chromium atoms. Diffusion of carbon is relatively fast at these temperatures but that of chromium extremely slow. Consequently, the chromium atoms are almost entirely supplied by the grain boundaries so that the grain boundary chromium content is substantially lowered. This local depletion of chromium causes loss of passivity in acid corrodants with consequent attack along grain boundaries.

7.3.17.4 Weld decay

The resultant 'Intergranular Penetration' in a casting or, if the heating has been caused by welding, 'Weld Decay' (see Figure 7.36) can completely disintegrate the material. Precipitation in cold-worked material takes place along slip planes as well as grain boundaries, consequently the distance that the chromium atoms must diffuse is small. Hence, although the same amount of chromium is removed as carbide, the depletion is more uniformly distributed with a consequential lowering of general corrosion resistance but a lower tendency to intergranular failure.

There are two alternative approaches to the problem of preventing intergranular corrosion. Either the carbon content of the steel is limited, by using an AISI 'L' or BS 970 Code '11' steel, to 0.03% at which precipitation of carbide in sufficient quantity to cause trouble is impossible, or an element such as titanium or niobium, which has a stronger affinity for carbon than chromium, is added to form the appropriate carbide by using an AISI 'Nb' or 'Ti' or BS 970 Code '40' steel. The theoretical amounts required to ensure that all carbon in excess of 0.02% is combined are titanium = 4 × excess carbon, niobium = 8 × excess carbon. In practice, allowance must be made for nitrogen combining with the added element (particularly Ti) and for the efficiency of combination – carbon levels below 0.06% requiring a higher titanium or niobium to carbon ratio for complete combination than those above 0.08%.

7.3.17.5 Knife-line attack

Furthermore, when a stabilized (Ti or Nb treated) steel is heated to successively higher temperatures above 950°C up to 1250°C the carbide enters solution and is broken down into its constituent elements to an increasing extent so that above 1100°C a relatively small amount of carbon remains combined. The free carbon is then available to form chromium carbide on subsequent re-heating in the 450–750°C range. Combination of carbon with titanium occurs in the 850–950°C range given adequate time. The whole question of stabilization is concerned with time, temperature and amount of free carbon.

Time at temperature affects the extent of re-solution of the titanium and niobium carbides present in stabilized steels; titanium carbide dissolves more rapidly than niobium carbide. Re-solution takes place at temperatures in excess of approximately 1200°C under welding conditions. The extent depends on carbide particle size as well as time at temperature. If reheated within the sensitization temperature range (around 650°C) this narrow zone immediately adjacent to the weld metal precipitates intergranular chromium carbide, because combination of Ti or Nb with C cannot occur at this temperature.

Thus, although the stabilized steels will not precipitate chromium carbide in the region of the heat-affected zone raised to 650°C by welding, there is the possibility, in conditions where the edge of the weld metal is reheated to 650°C, that intergranular attack can occur. The existence of such conditions depends on the welding practice but in most fabricated articles, as distinct from samples with single-run

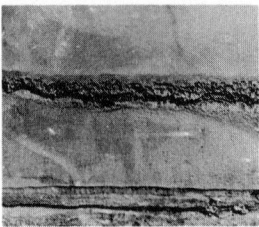

Figure 7.37 Example of 'knife-line' attack on stainless steel weld

welds, positions must arise at weld junctions where these conditions will obtain; welded samples should therefore always have crossed welds.

This particular type of intergranular attack at the weld metal edges is known as 'knife-line' attack (see Figure 7.37). It is most likely to be seen in boiling dilute nitric acid solutions. The composition of the steel affects its incidence; steels with lower nickel-to-chromium ratio, which produce a greater amount of delta ferrite in the knife-line zone, are less susceptible. Fully austenitic Ti stabilized grades appear to be more susceptible than fully austenitic Nb stabilized.

Where corrosion conditions are known to offer a knife-line hazard, treatment of the fabrication at 870°C will promote precipitation of the carbon as titanium of niobium carbide with consequent resistance to attack. The unstabilized 431 and 434 grades are susceptible to intergranular attack after welding. This can be prevented by heat treating for 2 hours at 600–800°C which coalesces the carbide films.

7.3.17.6 Weld problems which may be assessed by means of the Schaeffler Diagram

Fully austenitic weld metal tends to crack on solidification because of inherent weaknesses at the boundaries of columnar grains. The composition of the filler metal is therefore adjusted to ensure that all the molten zone contains a small proportion of ferrite (that is, it lies on the A + F side of the A/A + F line in Figure 7.33) and also to ensure that its

strength is not excessive compared with the parent metal.

As long as the weld metal composition is maintained within the zone in which ferrite and austenite co-exist austenitic steels have excellent weldability. Ferritic stainless steels are weldable but suffer from the brittleness and grain growth problems described in Section 7.3.12. Martensitic stainless steels suffer from the same brittleness problems as carbon and quenched and tempered steels unless the carbon content is below 0.12%.

Both precipitation hardening and duplex stainless steels (the compositions of which can be roughly estimated from Figure 7.28) are fully weldable without preheat, and the precipitation-hardening steels may be hardened by precipitation after welding. Brief summary notes on corrosion and welding aspects of stainless steels are given in Table 7.12. This is also referred to in Section 7.9 where for the convenience of the reader some of the figures in Sections 7.3.13 and 7.3.17 have been repeated.

7.3.18 Tool steels

The name tool steels (BS 4659: 1971 and AISI/SAE 'Tool Steels') covers a wide variety of steels used for forming and cutting materials which have as essential properties high hardness, resistance to wear and abrasion and adequate toughness. There are (or have been) some 82 AISI standard steels, 25 BS steels and many non-standard steels, but it should be possible to meet almost all requirements from the nine steels listed here.

Carbon steels are used for hand tools and other applications where high levels of toughness are required and where some distortion in heat treatment can be tolerated. Recommended steels are:

AISI 109; BS 4659 BWIA; 0.9%C steel with good combination of hardness and toughness, good general-purpose steel
AISI 210; BS 4659, BW2; 1%C, 0.25%V. Retains a sharp edge and withstands shock better than BWIA.

A carbon tool steel should be quenched in water or brine and tempered as soon as its temperature has been quenched to 'hand warm'. Carbon tool steels will soften and lose their edge if appreciable heat is generated by the cutting action.

Table 7.12 Corrosion resistance and weldability of stainless steels

| AISI No. | Nominal composition | | | | | Corrosion notes |
	C	Cr	Ni	Mo	Other	
410	0.08–0.4	12–14	—	—	—	Rust resisting. Higher-carbon grades for engineering applications, turbine blades, cutlery, etc.
465	0.08 max.	13	—	—	Al	Weldable grade.
430	0.1 max.	16.5	—	—	—	Resists mild acids. Special feature is resistance to nitric acid. May require heat treatment after welding (600–800°C) to avoid intergranular attack. Forming of sheets up to 3 mm at room temperature; greater thickness at 200–350°C.
430Ti	0.1 max.	17.5	—	—	Ti	Weldable grade not requiring heat treatment. Argon arc (gives minimum grain growth) preferred. Both grades, if welded, should not be applied under conditions of shock loading or vibration.
304	0.08 max.	18	10	—	—	Rust and acid resistant. Suitable for welding in certain applications.
304L	0.03 max.	18	10	—	—	Extra low carbon. Very resistant to intergranular corrosion. Weldable for practically all applications.
309Cb } 321	0.1 max.	18	10.5	—	Nb or Ti	Not susceptible to intergranular attack (but see reference to knife-line attack in text). Applicable above 300°C. Weldable.
316	0.07 max.	17.5	11	2.2	—	Resistance to chemical attack better than 18/8 (e.g. severe acid attack). Resists intergranular attack up to 6 mm thickness. Applicable below 300°C. Weldable for most applications.
316L	0.03 max.	17.5	11	2.2	—	Superior resistance to intergranular corrosion, suitable for thicknesses greater than 6 mm.
316Cb } 316Ti	0.1 max.	17.5	11.5	2.2	Nb or Ti	Not susceptible to intergranular attack (but see knife-line attack). Applicable above 300°C. Suitable for strong acids at elevated temperatures. Weldable.
317	0.07 max.	17.5	12	2.8	—	Resists intergranular attack up to 6 mm thickness. Applicable below 300°C. Corrosion resistance superior to 2% Mo alloys. Weldable for most applications.
317Cb } 317Ti	0.1 max	17.5	12.5	2.8	Nb or Ti	For strong acids at high temperatures. Applicable above 300°C. Weldable.
317LM	0.03 max.	17	13.5	4.5	—	Resistance to strong organic acids at elevated temperatures. Increased resistance to pitting. Applicable below 300°C. Resists intergranular attack. Weldable for most applications.

High-speed steels have a high content of carbide forming elements W, V and Cr and therefore retain their hardness at high temperatures (i.e. they have good 'red hardness'). Recommended steels are:

AISI M2; BS 4659 BM2 for normal duty;
AISI T4; BS 4659 BT4 for faster cutting and increased output;
AISI M42; BS 4659 BM42 for cutting hard materials.

'T' steels are tungsten steels and 'M' steels molybdenum steels which are cheaper but slightly more difficult to heat treat. Heating must be carried out in atmosphere-controlled furnaces to prevent decarburization; slowly to 825°C then quickly to the manufacturer's recommended temperature around 1300°C, followed by quenching in air blast, oil or salt bath at 525°C and air cooling. After an optional refrigeration treatment the steel must be tempered (secondary hardened) two or three times at about 545°C, again in controlled atmosphere.

For many purposes high-speed steels are being replaced by sintered carbides or ceramics such as sialons (see Section 7.5) which have exceptional wear and heat resistance even though they may not be as tough as high-speed steels.

Hot-work steels are used for forming (not cutting) hot materials. They must not soften at temperature and must have good wear resistance. They must also be able to resist thermal fatigue when heated and cooled (sometimes by water jets). Their metallurgy is similar to that of high-speed steels. Recommended steel is:

AISI H13; BS 4659 BH13. This steel has the highest and deepest hardness of the hot work steels.

Cold-work steels are used for forming cold materials and resistance to abrasive wear is of highest importance. In addition, they may have to be machined to very complex shapes and must therefore have very high dimensional stability during heat treatment.

Recommended steels are:

AISI 01; BS 4659 B01; 0.95CW,W.V. Steel for light duties, simple to heat treat.
AISI D2; BS 4659 BD2; 1.5C, 12Cr, MoV. Martensitic stainless steel with very high hardness and wear resistance for general application.

Shock-resisting steels are used for tools which are subject to heavy vibration or hammering; they must be hard but also have reasonable toughness to avoid failure by brittle fracture. Recommended steel is:

AISI, S1, BS 4659 BS1; 0.5C SiMnCrW. Metallurgy is relatively uncomplicated and heat treatment straightforward.

7.3.19 Steels for springs

There are three different types of spring steel. 'Patented' and cold-drawn carbon steel wire is used for small coil springs. 'Patenting' consists of heating the billet to roughly 1000°C to develop a coarse grain size so that after slow cooling the steel has a coarse pearlite/bainite structure which is readily drawn into wire.

The steels used and the properties of the wire are covered by BS 5216 and ASTM A227 and 228 specifications. They have carbon contents varying between:

- 0.65% for 'Hard drawn spring wire' which has the largest diameter (up to 9 mm), the poorest surface finish and the lowest tensile strength (less than 940 MPa); and
- 0.85% for 'Piano or Music wire', which has the smallest diameter (0.1 mm minimum), the best surface finish and the highest tensile strength (up to 3780 MPa).

Many ranges of tensile strength are available. Springs are cold coiled from the wire. Carbon and alloy spring steels are made to specifications in BS 970: Part 5.1972 and corresponding AISI/SAE grades.

Coil springs are usually made from hot-rolled and ground bar of the diameter required for the final spring. The bar is heated to a temperature within the hardening temperature range, coiled on a mandrel, slipped off the mandrel, quenched and tempered to a tensile strength around 1650 MPa. Carbon steels are used for springs up to 13 mm diameter, more highly alloyed steels for higher diameters, the maximum around 80 mm diameter being made from BS 925 A60 SiMnMo steel.

The purchase specification must strictly limit decarburization of the surface (to which silico manganese steel, which is popular for springs, is particularly prone) because fatigue cracking, which will propagate across the spring, may start in a soft decarburized surface layer. The surfaces of all but the smallest springs are conditioned by shot peening which induces a compressive surface stress and increases fatigue strength by 25–30%. 'Scragging', which overloads the spring in the direction it will be used in service, produces residual stresses which oppose service stresses in the surface layers and therefore improves endurance.

Rust is harmful to spring performance and, to prevent it, a spring should be protected immediately after peening.

Corrosion-resistant steel springs are covered by an old British Standard (BS 2056: 1953) which uses the EN designations. In practice, stainless steel wire for springs is usually supplied to AISI number.

Martensitic steels are usually supplied softened and lightly cold drawn to a UTS between 620 and 850 MPa. They are hardened and tempered after forming. Austenitic steels are cold drawn to UTS between 1800 and 2000 MPa for diameters below 2 mm and 1000 MPa for diameters up to 10 mm. One precipitation-hardening stainless steel DTD5086 can be supplied for forming in the softened condition and can then be precipitation hardened to 1800 MPa.

7.3.20 Cast steel

All the types of steel described earlier in this section can, in principle, be produced as castings. In practice, the steel grades listed in BS and the several US standards authorities are confined to a limited number of types given in Table 7.13. This includes:

- BS specifications
- ASTM grades for carbon steels and for steels with alloy content up to 8% and UTS between 482 and 827 MPa
- ACI (Alloy Castings Institute of the USA) grades for Heat Resistant and Corrosion Resistant Steel Castings.

While each grade in a Steel Castings Specification is the equivalent of a grade in BS 970 or SAE/AISI they differ in important aspects. For example, a foundry is less likely to be equipped for carbon vacuum deoxidation than is a large steelworks. To allow for this, the silicon content of steel castings is usually set at a higher level than for the corresponding wrought steel and the very low carbon grades are not included. This may require the content of other alloying additions to be adjusted. Also, an austenitic steel casting often contains more ferrite than the corresponding wrought steel to prevent fissuring during solidification (both in casting and welding) and to resist intergranular penetration. A designer is advised therefore when ordering a casting to specify the BS 1504, BS 3100 or ACI grade rather than the BS 970 or AISI grade number for the corresponding wrought steel.

If a compelling reason exists for specifying a steel not listed in a standard casting specification the casting will almost

Table 7.13 Standards for steel castings

(a) Non-stainless steels

BS 3100: 1976 Grade	BS 1504: 1976 Grade	ASTM[a]	UTS (MPa)	Steel type	Special requirements
A1, 2 + 3	430, 480, 546	A27 + A148 A356	430, 490, 540 430, 480, 540	Carbon steel for general purposes Carbon steel for pressure vessels	0.2% PS specified at temperature
AL1			430	Carbon steel for low temperature	Charpy 20J at −40°C
A4, 5 + 6			540, 620, 690	Carbon–manganese steel for general purposes	
B1	26	A27 A356	460 460	Carbon–molybdenum steel for elevated temperature Carbon–molybdenum steel for pressure vessels	0.2% PS specified at temperature
	27		460	3.5% Nickel steel for pressure vessels	Charpy 20J at −60°C
BL1 BL2			460 460	0.5% molybdenum steel at low temperatures 3% nickel 0.5% molybdenum steel	Charpy 20J at −50°C Charpy 20J at −60°C.
B2	28		480 480	1.25% chromium molybdenum steel 1.25% chromium molybdenum steel	0.2% PS specified at temperature
B3	29	A389	540 540	2.25% chromium molybdenum steel 2.25% chromium molybdenum steel	0.2% PS specified at temperature
B4	30		620 620	3% chromium molybdenum steel 3% chromium molybdenum steel	0.2% PS specified at temperature
B5	31		620 620	5% chromium molybdenum steel 5% chromium molybdenum steel	0.2% PS specified at temperature
B6	32		620 620	9% chromium molybdenum steel 9% chromium molybdenum steel	0.2% PS specified at temperature
B7	33		510 510	0.5% chromium 0.5% molybdenum 0.25% vanadium 0.5% chromium 0.5% molybdenum 0.25% vanadium	0.2% PS specified at temperature
BS 3146 1A, B + C 2 – 12		A732 A732		Carbon steel investment castings Alloy steel investment castings	

(b) Stainless and heat-resisting steels

BS 3100: 1976 Grade	BS 1504: 1976	ACI numbers	UTS (MPa)	
302C25		CF20[b]	480	
302C35				
304C12		CF-3	430	
304C15		CF-8	480	Low carbon 18/8 type
309C30		CH-20	560	
309C32			510	
309C35			450	
309C40		CK-20		
311C11				
315C15		CF-16F	480	115% molybdenum 18/8
315C16			480	
316C12	316C12	CF-3M	430	2.5% molybdenum 18/10 low carbon
316C16	316C16	CF-8M	480	2.5% molybdenum 18/8
316C71	316C71		510	2.5% molybdenum 18/8
	317C12		430	3.5% molybdenum 18/10 low carbon
317C16	317C16	GG-8M	480	3.5% molybdenum 18/10
318C17		CF-12M	480	2.5% molybdenum niobium 18/10
347C17	347C17	CF-8C	480	Niobium stabilized 18/12
364C11		CN-7M	430	Chromium–nickel–copper
410C21		CA-15	540	13% chromium martensitic steel
420C29	34	CA-40	690	13% chromium martensitic steel
425C11	35	CA-6NM		13% chromium 4 nickel
452C11				28% chromium 1.5 molybdenum ferritic steel
452C12				28% chromium 0.5 molybdenum ferritic steel

BS 3146 Pt 2

Corrosion and heat-resisting investment castings.

Most grades are covered by ASTM A743 and A744.

[a] US alloying practice does not correspond with British, so ASTM designations do not correspond exactly with BS. Other ASTM specifications include: A567, A128, A487, A216, A217, A757, A352, and A747.

[b] The figures indicate carbon content.

certainly be more expensive because the foundry may have to make experimental castings and will not be able to recycle scrap directly. Also, it may be more difficult to obtain a guarantee of quality.

Design of castings is too complex a subject for detailed consideration here. The essential criterion is to make absolutely sure that nowhere within the casting exists a point where, through a local increase in section, metal is left to solidify surrounded by metal that has already solidified. Walls should be of as uniform thickness as possible, corners radiused, multiple junctions eliminated, changes of section tapered and where large sections are inevitable they should be so placed that they solidify progressively towards a feeding head. Where isolated large increases in section are unavoidable chills may be used. (See *The ASM Metals Handbook*, 1961 edition, pp. 122–146.)

In principle, the properties of a casting should be identical to those of a forging of the similar composition, and in practice castings are available with UTS to match any forgings up to 827 MPa UTS and corresponding yield strengths are available. There are, however, significant differences in the cast and wrought structure, particularly in alloys with more than one phase present.

Castings have a 'cast structure' which is effectively a skeleton of intermetallics that tend to limit and restrict slip. In a correctly worked wrought alloy this skeleton is broken up so that it becomes effectively a dispersion of fine particles rather than a network. Working refines the grain and renders the alloy more susceptible to heat treatment, and has two effects which are significant in design. Ductility is increased and creep strength decreased.

The reduction in ductility of castings compared with wrought steel has a negligible effect on design with steels with UTS around 500 MPa. However, with strengths of 800 MPa and above, the fracture toughness of cast material is lower and more variable and the fatigue endurance limit about 20% lower than that of wrought material. In addition, the continuous-casting process for manufacturing ingots from which wrought material is forged has much superior feeding and segregation characteristics than is possible in a large sand casting so that the material is inherently superior. Furthermore, forging, correctly programmed, can be made to align the grain (and any discontinuities) in the principal stress direction and thus make the component more resistant both to brittle fracture and fatigue. On the other hand, the transverse properties of a casting should be superior, and thorough inspection will reduce or eliminate dangerous defects.

The improved creep resistance of the cast structure is of considerable value in the case of large turbine castings, while the creep properties of the small lost wax castings (listed in BS 3146) could not be obtained in forgings. Even higher creep properties could be obtained by directional solidification but this is used for the more highly creep-resistant nickel alloys rather than steels.

7.3.21 Cast iron: general

Cast iron is an alloy of iron with carbon in the range between 1.7% (the eutectic composition) and 4.5%. There are two basic types, one of which is a composite of steel and graphite while the other, white cast iron, consists of cementite in a matrix of steel.

White Cast Iron, Low Alloy White Cast Iron, Martensitic White Cast Iron and High-Chromium White Cast Iron have special wear and environmental resistant properties. The graphite-containing cast irons which include the Flake Graphite, Nodular Graphite and Malleable grades have been regarded as a cheap and brittle substitute for other engineering

materials, but have in addition to their relatively low cost very definite technological advantages. These are particularly evident in the case of the newly developed Austempered Ductile Irons.

The several grades of cast iron, with the permission of BCIRA, are given according to BS specification in Table 7.14 which lists all (with the possible exception of damping capacity) relevant physical and mechanical properties.

7.3.22 Grey cast iron

Grey cast iron (Flake Graphite Iron) can be 'non-alloyed' 'low alloy' or acicular. Design stresses etc. given in Table 7.1 (BS 1452 and ASTM A48 Class 20–60) are for non-alloyed grey cast iron with carbon contents varying from 3.65% to 2.7%, silicon from 2.5% to 1.35%, phosphorus from 0.5% to 0.09% and manganese around 0.6%.

This is the cheapest engineering metal, not only because the raw materials – pig iron, cast iron and steel scrap, limestone, coke and air – are all relatively inexpensive but also because melting costs in a cupola are relatively low. Casting is very easy because cast iron is more fluid, has a narrower solidification range and a lower in-mould shrinkage than steel. Machinability is excellent because graphite acts as both a chip breaker and a tool lubricant.

Grey cast iron has good dry-bearing qualities and its freedom from scuffing makes it a good material for automobile cylinder walls. Its wear resistance is assisted by slight chilling and a hard network of phosphide eutectic. It has also excellent damping capacity, particularly in the lower (higher-carbon) grades and is particularly suitable for machine-tool bases and frames. On the other hand, it is brittle because the graphite flakes reduce strength, the maximum recommended tensile design stress is only 25% and the fatigue loading limit between 11% and 16% of the tensile strength. (It should be remembered that tensile stress is measured by bend – see Section 7.1.)

There are (or were) two variants with better fatigue properties. Compacted graphite iron or meehanite is made by inoculating an iron which would otherwise solidify white.

Haematite high-carbon low-phosphorus iron was originally made from haematite pig iron. Its low phosphorus content improves its fatigue properties (while reducing fluidity).

Low-alloy and Acicular Cast Irons made by adding Ni, Cu, Cr, Mo, V or Sn (and in the case of acicular cast iron reducing the phosphorus content) enable grey cast iron to be used in higher-duty applications without redesign or technological change.

7.3.23 Nodular graphite ('SG') iron

Nodular or spheroidal graphite, 'SG' cast irons are available in grades corresponding to those of grey cast iron but are produced by inoculation of the melt with nickel, magnesium and caesium compounds which change the form of the graphite to near-spheroidal nodules (see Figure 7.38). This produces material which has strength, ductility and thermal shock resistance more typical of steel but castability, damping capacity and machinability more typical of cast iron.

The recommended, design, tensile and fatigue stresses are a much higher proportion of the UTS than is the case with cast iron. Steel castings, fabrications and sometimes forgings may be replaced with considerable economic advantages. Matrix structures can be varied by changing cooling rate or alloying between ferrite, pearlitic carbide and acicular structures for higher-duty applications.

The development of nodular and other higher-duty irons detailed here has accelerated the trend to modern melting

Table 7.14 Cast irons: classes, grades and properties

Safe design stresses and other properties ↓ / Cast irons →	Engineering cast irons																									Cast irons resistant to heat, abrasion, corrosion, thermal shock															
	Flake graphite iron, BS 1452: 1977								Nodular graphite (SG) iron, BS 2789: 1973							BS 309: 1972		BS 310: 1972			BS 3333: 1972					BS 1591: 1975			BS 4844: 1972-1974			Austenitic cast iron, BS 3468: 1974						BCIRA Broadsheet 63				
																																Flake		Nodular								
	Grade 150	Grade 180	Grade 220	Grade 260	Grade 300	Grade 350	Grade 400	Compacted graphite iron	Grade 370/17	Grade 420/12	Grade 500/7	Grade 600/3	Grade 700/2	Grade 800/2	Austempered ductile iron	Grade W340/3 Whiteheart	Grade W410/4 malleable iron	Grade B290/6 Blackheart	Grade B310/10 malleable	Grade B340/12 iron	Grade P440/7	Grade P510/4 Pearlitic	Grade P540/5 malleable	Grade P570/3 iron	Grade P690/2	Si 10 to 14 High-silicon	Si 16 iron	Si Mo 4 0.5	Grades 1A to 1C Abrasion	Grades 2A to 2E resisting	Grades 3A to 3E white iron	Ni Cu Cr 15 6 2	Ni Cr 20 2	Ni Mn 13 7	Ni Cr 20 2	Ni Mn 23 4	Ni Cr 35 3	Cr Mo 15 3 High-	Cr 23-28 chromium	Cr 30-35 iron	'Hematite' high-C low-P iron	
Tensile stress (N/mm²)	38	45	55	65	75	88	100	92–103	129	138	145	156	173–236	198–270	270–620	81–63	88–83	102	110	121	120–144	130–154	134–158	162–210	257			140				54	49–57	90	86	77	90				35	
Elongation (%)	<1	<1	<1	<1	<1	<1	<1	4–2	25–17	20–12	15–7	15–3	5–2	5–2	10–1	>3	>4	>6	>10	>12	>7	>4	>5	>3	>2			15–10						15	7	25						
Compressive stress (N/mm²)	156	187	229	270	312	364	416	150	152	173	204	216	238–322	271–367	350–850	93–116	120–140	130	138	149	172–270	184–230	188–233	238–279	353																150	
Unnotched-fatigue stress (N/mm²)	23	27	33	39	45	50	51	60	63	67	75	83	93	101	130–170	41–51	53–62	58	62	68	79	85	88	71–91	93								35		50							
Hardness (HB)	130–160	150–180	160–200	180–220	200–250	225–275	250–305	140–150	115–140	140–170	170–215	215–250	250–265	265–305	250–500	120–180	120–180	125–140	130–140	130–140	140–170	170–180	180–190	190–240	240–270	450–520	450–520	200–220	250–400	500–600	400–650	140–200	120–215	130–170	140–200	150–180	140–190	550–750	550–750	250–340	90–120	
Young's modulus (GN/m²)	100	109	120	128	135	140	145	155–165	169	169	169	174	172–176	172–176	155–170	176	176	169	169	169	172	172	172	172–176	176	124	124	140–200	180–200	180–200	180–200	85–105	85–105	140–150	112–130	120–140	112–123		217	224	80–100	
Notched-impact value, J, at 20 °C								7–3	15–13	10–5	5–2	5–2	5–2	5–2	8–4			17–13	13–12	17–13	10–5	5–2	5–2	5–2	5–2			11–2	12–6					27–15	27–13	24	7					
Notched-impact transition temperature (°C)									−10–+30	−10–+30	40–100	80–120	100–120	100–120				−10–+30	−10–+30	−10–+30	40–100	40–100	40–100	0*–100	0*–100																	*Hardened & tempered
Fracture toughness K_{Ic}, (MN/m$^{3/2}$)		12	20	24–20					46	46	45–37	45–37	45–25	45–25	100*–70*			55–35	55–35	55–35	50	50–40	50–40	50–30	50–30				30–25	25–20	20–25											*Equivalent K_{Ic}
Service temperature (°C), max./min.	500–−50	500–−50	500–−50	500–−50	500–−50	500–−50	500–−50	500	500–−40	500–0	500–20					500–0	500–0	500–−40	500–−40	500–0	500–0	500–0	500–0	500–0	500–0	250–0	250–0	800–−40	50–0	50–0	50–0	700–−40	700–−80	500–−80	700–−80	700–−200	800–−80		900	1050	900*–500*	*Cycling *Steady
Design stress (N/mm²) + 450 °C/ − 50 °C	38	45	17–55	18–65	18–75	30–88	30–100		22–129	22–138	23–145	27–156	32–173*	32–198*		19–72	19–85	19–102	19–110	19–121	23–120*	23–130*	23–134*	28–162	28–257							35–54	39–57		62–86	59–77		70	70	70		*Static loading
Density (g/cm³)	7.0	7.1	7.2	7.2	7.2	7.3	7.3	7.2	7.1	7.1	7.1–7.2	7.1–7.2	7.2	7.2	7.4	7.4	7.4	7.3	7.3	7.3	7.3	7.3	7.3	7.3	7.3	7.0	7.0	6.85	7.7	7.7	7.7	7.3	7.3	7.3	7.4	7.4	7.6	7.67	7.69	7.43	7.0	
Thermal expansion (10^{-6}/K) 20 °C/20 – 400 °C	10–12.5	10–12.5	10–12.5	10–12.5	10–12.5	10–12.5	10–12.5	10–12.5	10–12.5	10–12.5	10–12.5	10–12.5	10–12.5	10–12.5		11–13	11–13	11–13	11–13	11–13	11–13	11–13	11–13	11–13	11–13	12–15.9	12–15.9	12–13.5				18	18	18	18	18	18					
Thermal conductivity (W/m.K) 100 °C/400 °C	53–50	52–49	50–47	49–46	47–44	46–43	44–41	41–38	37–36	37–36	36–35	33–32	31	31		36–35	36–35	38–36	38–36	38–36	36–35	35–34	35–34	34–33	33–32			27–25													59–36	
Electrical resistivity (μΩ.m)	0.85	0.78	0.76	0.73	0.70	0.67	0.64		0.50	0.50	0.51	0.53	0.54	0.54		0.24–0.26	0.24–0.26	0.37	0.37	0.37	0.40	0.40	0.40	0.40	0.40	0.50	0.50															
Maximum magnetic permeability (μH/m)	310–380	310–380	310–380	310–380	310–380	310–380	310–380		2140	2140	1600	870	500	500		910–1820	910–1820	1900	1900	1900	1280–900	900–750	750–650	650–400	400–380									Non-mag.								
Hysteresis loss (B = 1T), (J/m³)	2500–3000	2500–3000	2500–3000	2500–3000	2500–3000	2500–3000	2500–3000		600	600	1300	2200	2700	2700		1500–850	1500–850	450	450	450	1300–1900	1900–2200	2200–2400	2400–3800	3800–4500			*				*	*	*	*	*	*	*	*	*		Heat
																													*	*	*							*	*	*		Abrasion
Poisson's ratio	← 0.26 →								← 0.275 →							← 0.26 →										*	*	*				*	*	*	*	*	*	*	*	*		Corrosion
																																									*	Thermal shock

(a)

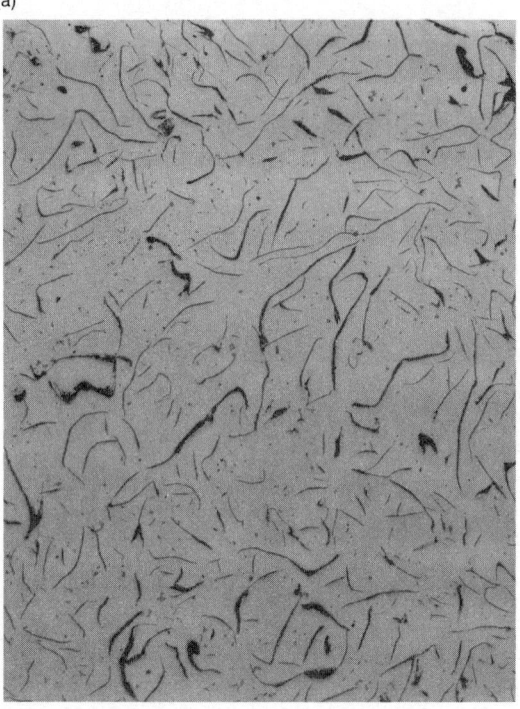

(b)

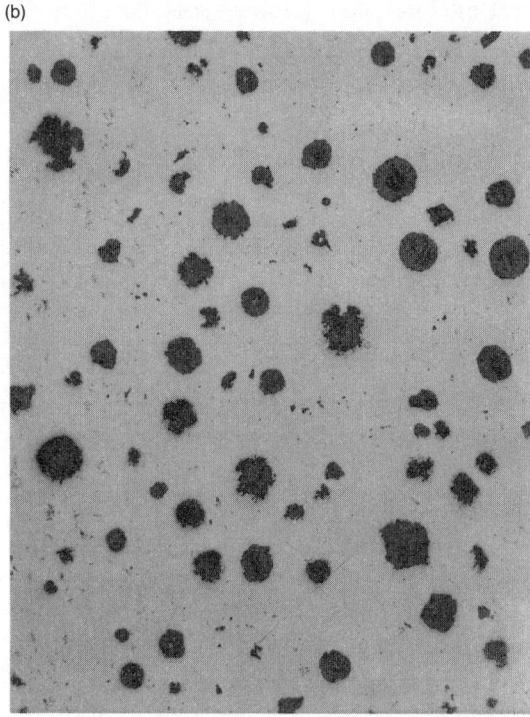

Figure 7.38 Contrast in graphite morphology (a) in grey cast iron, (b) in 'SG' iron (courtesy of Roger Davies, Fulmer Technical Services)

practice. Casting from a cupola is not amenable to the precise composition control which is possible with an electric or gas furnace. Even where a cupola is used for the actual melting, final control of composition requires a holding furnace.

7.3.23.1 Austempered ductile iron

Austempered ductile iron is SG iron with added alloying elements, (usually Mo, Ni and/or Cu) sufficient for a bainitic structure, usually with retained austenite, to be produced in the section size by austempering. Such material can have yield strength and UTS up to 1150 and 1400 MPa with elongations of 6% and fatigue limit up to 33% of the UTS. Wear resistance because of the graphite and retained austenite is superior to steel of the same hardness and components such as gears are quieter in operation. The potential of austempered ductile iron, which is substantially cheaper than forged steel and can be cast closer to shape than steel is usually forged, exceeds that of any other recently developed material.

Obtaining the required properties requires dedication to process control in foundries and heat-treatment departments. The most economically rewarding application for austempered ductile iron is as a material for gears which can be made quieter, lighter and cheaper than the equivalent steel gears. One disadvantage, for the highest-rated gears, is the lower fatigue strength of austempered ductile iron compared with that of steel, but this is being overcome by shot peening the teeth of the gears. Austempered ductile iron has been used successfully for tracks for off-the-road vehicles, pump bodies, agricultural equipment, friction blocks and drive shafts.

7.3.24 Malleable iron

Malleable iron is cast with a white cementite structure which is converted to a steel–graphite composite by annealing. The requirement for the as-cast structure to be graphite-free limits the maximum section to about 38 mm and the general run of castings weigh under 5 kg and have a maximum section of 25 mm. There are three varieties:

1. Whiteheart Malleable has a carbon content of about 3.5% which improves castability compared with the other varieties. Other alloying elements are Si 0.6%, Mn 0.25%, $S < 0.3$, $P < 0.1$. It is heat treated for 5–6 days at 875°C packed in an oxidizing medium to produce spidery graphite aggregates in a pearlite/ferrite matrix. This long heat treatment increases cost, limits rate of production and decarburizes the surface layer.

2. Blackheart Malleable has compositions varying between C 2–2.65%, Si 0.9–1.65%, Mn 0.25–0.55%, $S < 0.05$–0.18 and $P < 0.18$. It is heat treated in a neutral atmosphere for 40–60 hours at 860°C, cooled to 690°C, held for 4–5°C per hour and air cooled. It has graphite aggregates in a ferrite matrix (no decarburization) and, although not so easy to cast as whiteheart has rather better properties and the best combination of machinability and strength of any ferrous material. There are two ASTM A47 grades, 32510 and 35018 and one A197 'Cupola' grade of lower quality.

3. Pearlitic Malleable has usually a higher manganese content, varying from 0.25% to 1.25% and may be cooled rapidly after annealing. It has higher strength than the other malleables and, unlike them, has good wear resistance and is difficult to weld.

Because of their low cost and excellent shock resistance the malleable irons have been used extensively in the power train, frame, suspension and wheels of motor vehicles, rail, agricultural and electrical equipment but the market for them has contracted except for galvanized pipe fittings.

7.3.25 Austenitic cast irons

Austenitic cast irons have an austenitic matrix containing either flake or nodular graphite. They are non-magnetic, have thermal expansion coefficients similar to low-expansion aluminium alloys (with which they can be used as wear- and thermal fatigue-resistant inserts for pistons) and are available in a wide range of grades including:

- Ni-resist 14–32% Ni, 20% Cr for resistance to medium concentration acids and
- Nirosilal; Ni + Si for resistance to high-temperature oxidation and growth up to 950°C.

High-nickel nodular irons have excellent ductility and are suitable for cryogenic applications.

7.3.26 High-silicon cast irons

High-silicon cast irons, composition 10–17% Si, Mo < 3.5%, have a silico ferritic solid solution matrix with dispersed graphite, exceptional corrosion resistance to mineral-oxidizing acids and, although extremely brittle, are used as pipes, stills and vats where strength is not needed. The 4% Si, 0.5% Mo grade has good resistance to oxidation and acids and better strength than the high-silicon grades.

7.3.27 White cast iron (abrasion-resisting white iron)

There are four types of White Cast Iron; unalloyed, low-alloy, martensitic and high-chromium.

Unalloyed White Cast Iron has a reduced content of silicon so that on fairly rapid cooling after casting no graphite is formed and the carbon is in the form of cementite or pearlite. Chill in white cast iron is increased by raising the content of carbon and manganese but reduced by increasing sulphur and phosphorus.

White irons with carbon content above 3.5% can have Brinell hardness of up to 600. However increased carbon decreases transverse breaking strength and causes brittleness. Low-alloy white irons have added elements (usually Cr and Ni) that increase chill and improve toughness and wear resistance but are insufficient to produce a martensitic structure).

Martensitic white cast iron (e.g. Nihard) has sufficient alloying elements (usually Cr and Ni) to produce a cementite/martensite structure with higher hardness (up to 90 schleroscope) and toughness than other cast irons and is also stable at temperatures up to 550°C. Martensitic white cast iron should preferably be stress relieved.

The white cast irons can be machined only with difficulty using carbide tools and should be cast as nearly to size as possible. They have higher solidification shrinkage than other cast irons and require careful running and feeding. They are used for grinding and ore crushing equipment, mill liners, tables, rollers and balls and other applications requiring wear resistance.

The selection of the correct wear-resistant material for any application depends on relative life and relative cost. Martensitic white irons cost more than low-alloy which cost more than unalloyed but, depending on the application, the life in wear of the most expensive material may be between 50% and 100% longer than the cheapest. Also, the more ductile, more expensive material should be less prone to fracture, but this

may depend more on the events in the mill or on the technique of the supplying foundry. It is advisable to carry out comparative trials on different materials including, where appropriate, forged martensitic steel and deposited carbides and to standardize on that material which proves to be most economical for the specific application.

7.3.28 High-chromium iron

Irons of chromium content 15–35% have a partially austenitic structure with higher toughness and strength than Ni-based and high corrosion and oxidation resistance. They have a higher resistance to strong acids than silicon cast irons and can be used for heat-treatment equipment, melting pots for lead, zinc and aluminium, other parts exposed to corrosion at high temperature and for wet-grinding operations.

7.4 Non-ferrous metals

7.4.1 Copper and its alloys

7.4.1.1 General

Copper is basically more expensive than iron but has important advantages for special applications, the most significant of which is conduction.[20] Its electrical conductivity when pure is superior on a volume basis to all metals other than silver, and on a weight (and specific cost) basis to all metals other than aluminium. The same relationships apply to its thermal conductivity. Both properties are reduced by alloying, but the conductivities of copper alloys are superior to those of steels.

Copper's second most important characteristic is its resistance to natural environments. Where iron rusts, copper remains bright or develops an attractive patina, and this characteristic is improved by appropriate alloying. In marine environments the toxicity of copper prevents fouling. There is, however, a temperature limitation on the use of copper alloys compared with steels.

Copper and a high proportion of its alloys are highly ductile so that they are eminently suited to forming operations. Some, particularly the leaded brasses, are also highly machinable so that the finished cost of a brass component may well be competitive with that of any other material when allowance is made for the value and easy recovery of scrap.

The mechanical properties of copper, tensile and fatigue strength and creep resistance can be improved by alloying, without, however, achieving the strengths of steels or approaching the specific strengths of the light metals. The good mechanical properties are retained at cryogenic temperatures but are inferior to steels at elevated temperatures.[21,22] Other properties also benefit from alloying. The influence of specific additions is indicated in Table 7.15.

Copper alloys are by no means the easiest to cast or weld and their toxicity, although having useful biocidal applications, prohibits contact with foodstuffs. They are divided into classes, the main classes having traditional names. The main classifications together with their BS designations are listed in Table 7.16. The British Standards for product forms are given in Table 7.17 and the material condition codes in Table 7.18.

7.4.1.2 Copper

'Copper' is an alloy of copper and oxygen. The oxygen content of the conductivity grades is not such as to affect their electrical conductivity, but unless an 'Oxygen-Free' grade is used, would cause problems in welding[23] and also when heated in a reducing atmosphere. Non-conductivity grades are deoxi-

Table 7.15 Influence on copper of alloying additions

Improved property	Alloying addition
Strength	Aluminium, beryllium, chromium, zirconium, zinc, tin, phosphorus, silicon, nickel, manganese, iron
Corrosion resistance	Nickel, aluminium, silicon, tin, arsenic, manganese, iron
Wear resistance	Tin, cadmium, silicon, aluminium, silver
Bearing properties	Lead
Colour	Zinc, tin, nickel
Cost	Zinc
Machinability	Lead, tellurium, sulphur, zinc
Castability	Zinc
Resistance to annealing (conductivity not impaired)	Silver

dized, usually with phosphorus which reduces electrical conductivity.

The suitability of copper for electrical conductors depends on its high conductivity combined with a high resistance to atmospheric corrosion and ease of drawing and fabrication. Material for conductors should be selected from grades C100–104 or 110, all of which have electrical conductivities of at least 101 IACS annealed, 97 IACS cold drawn (100 IACS = 0.019 $\mu\Omega$m). If the conductor is to be heated in a reducing atmosphere the oxygen-free grade C103 should be used. The high-conductivity grades have UTS 385 and proof stress 325 MPa hard and 220 and 60 MPa annealed.

Additions of silver between 0.02% and 0.14% improve creep strength and resistance to annealing without impairing conductivity and should be used for rotating machinery or where a component must be heated during manufacture.

For general engineering and building operations where conductivity is not significant any of the grades C101–107 may be used. Arsenical grades have slightly better strength at high temperature, phosphorus deoxidized grades are better to braze or weld.

7.4.1.3 High-conductivity copper alloys

There are a number of alloys containing a high percentage of copper which balance the minimum possible reduction in conductivity against some other desirable property, machinability (C109, 110) strength (CB101) wear resistance (C108) or strength at high temperature (CC101, 102). Strengths range from 495 to 1346 MPa and conductivities from 90 to 20 IACS. Several of these alloys are available in cast form.

7.4.1.4 Brass

The range of composition of copper–zinc alloys is illustrated in Figure 7.39. There are two classes, α brasses containing between 24% and 37% zinc and duplex brasses with 40–47% zinc.

All brasses can be hot worked; α brasses are readily cold worked and cast but duplex brasses are significantly more workable at elevated temperatures and can be extruded and forged into complex sections and shapes. This formability has the result that brasses are available in a very wide range of shapes. They are intrinsically easy to machine and machinability is improved even more by the addition of low percentage of lead. Brasses have a very useful strength range (330–810 MPa), are resistant to atmospheric and natural water corrosion and the incorporation of zinc lowers their cost appreciably compared with copper.[23]

The most suitable alloy for high-speed machining is the leaded CZ1214 PB and for hot stamping CZ122. These alloys

Table 7.16 Classes of copper alloy

Common name	Description	British Standard Wrought	British Standard and ASTM Cast	ASTM designation Wrought
Copper	Alloy of copper and oxygen. Sometimes accompanied by deoxidant. Sometimes contains silver	C100–107, 110	HCC1 (BS 4577 A1/1)	OF ETP OLP
High-conductivity copper alloy	Alloy of copper with additions which improve strength at a minimum loss in conductivity	C108, 110, 111–113 CB101, CC101/2	A2/1–4/2	
Brass	Alloy of copper and zinc to which other elements (usually with their name as a prefix) may have been added.	CZ101–137	SCB1–5 DCB1–3 PCB1 HTB1–3	2XX 3XX (headed)
Bronze Phosphor-bronze Leaded bronze Leaded phosphor-bronze Copper.tin	Traditionally alloys of copper and tin Copper, tin and phosphorus Copper, tin and lead Copper, tin, phosphorus and lead Copper and tin	PB101–104	PB1–4 LB1–5 LPB1 CT	5XX
Silicon bronze	Copper, silicon and manganese	CS101		
Gunmetal Nickel gunmetal Leaded gunmetal	Copper, tin and zinc Copper, tin, zinc and nickel Copper, tin zinc and lead		G1 & 2 G3 LG1–3	
Nickel silver Leaded nickel silver	Copper, nickel and zinc Copper, nickel, zinc and lead	NS103–109 NS101, 103, 111		7XX
Cupronickel	Copper, nickel, manganese (sometimes iron)	CN102–108	CN1, 2	7XX
Aluminium bronze	Copper aluminium (iron, nickel and/or manganese silicon)	CA101–107	AB1–3, CMA1	6XX
Copper–lead	Bearing alloy. Copper 20–40% lead		(ISO Cu-Pb20–40)	
Constant-resistivity alloys		(ISO Cu Mu13, Ni3 Cu Mu13, Al2)		
Memory alloys	Copper, zinc and aluminium		No standard	

Table 7.17 British and some ASTM standards for copper alloys in product form

British Standard	ASTM (where equivalent[a])	
BS 6017	Depend on material	Copper refinery shapes
BS 1400	B176, 584, 148	Cast copper alloys
BS 2870	B248	Copper and copper alloys – sheet, strip and foil
BS 2871	B251	Copper and copper alloys – tubes (3 Parts)
BS 2872	B124	Copper and copper alloys – forging stock and forgings
BS 2873	B250	Copper and copper alloys – wire
BS 2874	B249	Copper and copper alloys – rods and sections (other than forging stock)
BS 2875	B248	Copper and copper alloys – plate
BS 1432		Copper for electrical purposes – strip with drawn or rolled edges
BS 1433		Copper for electrical purposes – rod and bar
BS 1434		Copper for electrical purposes – commutator bars
BS 1977		High-conductivity copper tubes for electrical purposes
BS 4109		Wire for general electrical purposes and for insulated cables and flexible cords
BS 4608		Copper for electrical purposes – rolled sheet, strip and foil

[a] These are a selection of ASTM standards, many of which refer to particular material applications.

Table 7.18 Material condition coding for copper alloys

0	Annealed
$\frac{1}{4}$H, $\frac{1}{2}$H, H, EH	Harder tempers produced by cold working (or part annealing)
SH, ESH	'Spring hard' tempers produced by cold rolling thin material
M	'As manufactured'
W	Solution treated. Will precipitation harden
W($\frac{1}{4}$H), W($\frac{1}{2}$H), W(H)	Material solution treated and then cold worked to progressively harder tempers
WP	Solution and precipitation treated
W($\frac{1}{4}$H)P, W($\frac{1}{2}$H)P, W(H)P	Material solution treated, cold worked to progressively harder tempers and then precipitation treated

have ultimate tensile strengths of 450 MPa and they are, because of their high zinc content, the least expensive of the wrought brasses.

Where a higher tensile strength is required there are a number of high-tensile brasses, with additions which include aluminium, iron, manganese and silicon. The wrought alloys CZ114–116 and 135 have tensile strengths between 430 and 770 MPa, the cast alloys between 470 and 810 MPa. Corrosion resistance may be improved by adding tin, aluminium, arsenic and nickel.

Two important considerations must be observed when introducing brass components into service in aqueous environments:

1. In some potable, marine or industrial waters containing sulphur compounds many brasses, particularly duplex, are prone to 'dezincification' – the preferential dissolution of zinc leaving a weak copper sponge. This can usually be prevented by choosing an alpha brass inhibited with arse-

nic, and recently a duplex alloy CZ132 has been developed which, when correctly heat treated, is claimed to resist potable waters (see 'dezincification-resistant brass, Figure 7.39(a)).

2. Stress corrosion may be caused by ammonia derived from organic refuse, by chlorine or by mercury. Brass which may be exposed to these environments should be given a stress-relief heat treatment.

Where there is a possibility that either of these phenomena will occur, trials should be made of the behaviour of the brass in the chosen environment. Alternatively, a bronze or aluminium bronze (but *not* manganese bronze) should be used.

7.4.1.5 Bronze and gunmetal

Brasses all contain zinc but the parallel does not apply to bronzes, which traditionally are copper–tin alloys. 'Aluminium Bronze', 'Silicon Bronze' and 'Lead Bronze' contain no tin and 'Manganese Bronze' is a brass.

The true wrought bronzes contain up to 8% tin and are deoxidized with phosphorus. They have a single-phase solid solution structure at the annealing temperature but the solubility of tin falls with decreasing temperature (see Figure 7.39(b)). Cold working the resultant supersaturated solution gives them excellent elastic properties to which, coupled with a high resistance to corrosion, they owe the majority of their applications.

The phosphor bronze most generally used is PB102 containing 5% tin but higher and lower tin contents with UTS ranging between 590 and 680 MPa hard and 320 and 380 MPa annealed are available. They are used for springs, instrument components and bearings.

Cast phosphor bronzes PB1, 2 and 4 contain 10% tin, more than the wrought alloys and are also used for bearings with hardened steel where load-carrying capacity at low speed is required. Gunmetals (single-phase alloys of copper, tin and zinc) G1, also with lead LG1, 2 and 4, which combine modest strength (UTS ranging between 180 and 440 MPa) with good corrosion resistance are more widely used because of their better castability. Applications include pumps, valves, bearings and statuary.

Also used for bearings are the leaded bronzes LB1–LB5 which have better plasticity and are particularly suitable for mating with soft-steel journals which would be scored by

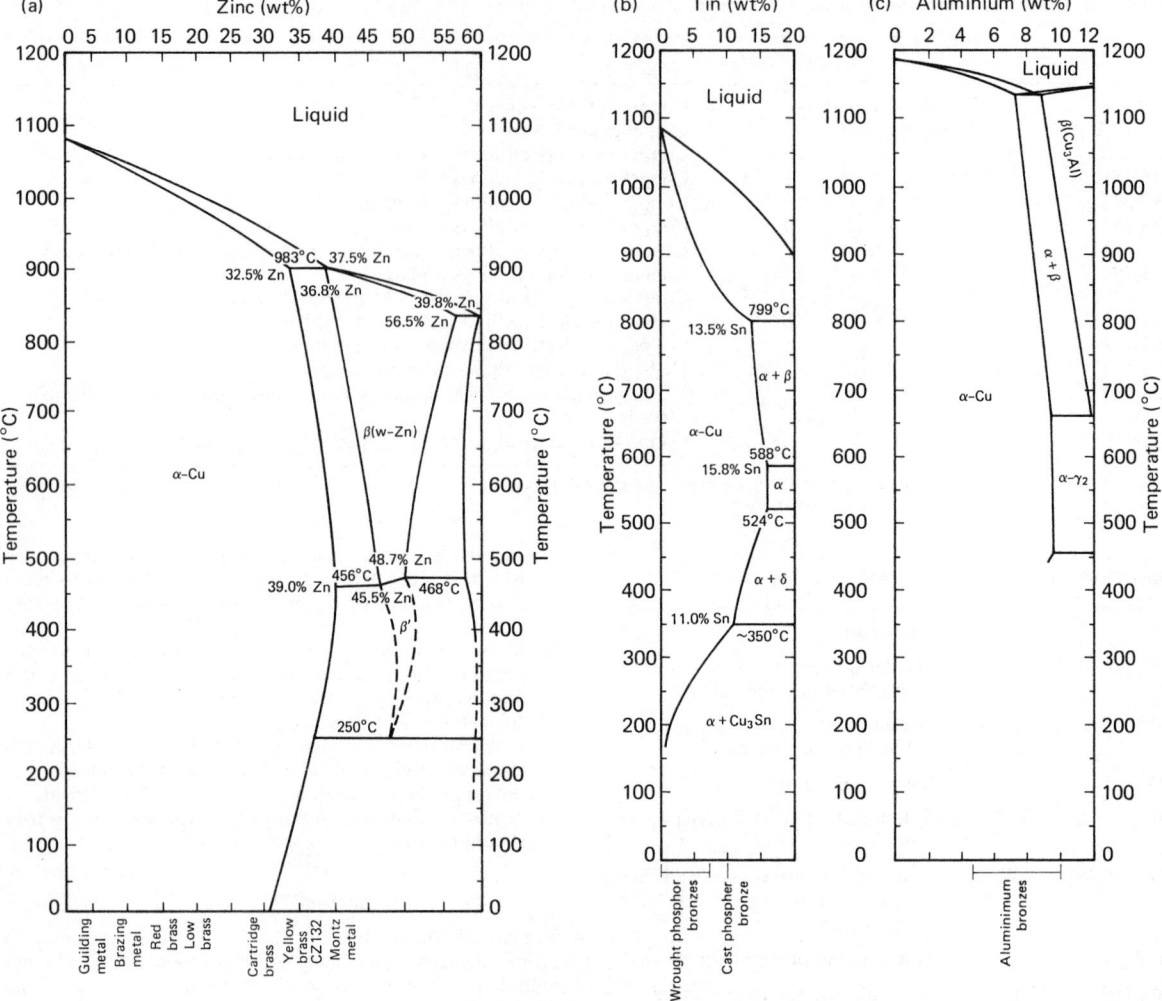

Figure 7.39 Partial phase diagrams. (a) The copper–zinc alloys; (b) The copper–tin alloys; (c) The copper–aluminium alloys

harder materials. The addition of lead (which does not dissolve in solid or liquid copper but forms a composite) confers machinability and very considerably improves bearing characteristics.

7.4.1.6 Aluminium bronze

Aluminium bronzes CA102 and 104–7 and AB1–3 contain between 6% and 10% aluminium together with iron, nickel or silicon (see Figure 7.39(c)). They all have excellent corrosion-resistant properties, due basically to the combination of the electrochemical properties of copper with the tenacious oxide film of aluminium, and good mechanical properties (ultimate tensile strengths ranging from 430 to 820 MPa wrought and from 460 to 700 MPa cast).

Aluminium bronzes can be used in a very wide range of environments including dilute acids with the exception of nitric. They are particularly suited to marine environments where, unlike most steels, they resist pitting and crevice corrosion and because of the biocidal nature of copper are not subject to biodeterioration. The alloys with higher aluminium contents (see Figure 7.39(c)) have a duplex structure of α solid

solution and β Cu_3Al and develop high strengths when worked.

Although aluminium bronzes have quite good wear resistance they should not be used in applications which involve relative movement in contact with other metals because the oxide film causes galling, fretting and seizure. This restriction apart, aluminium bronze has a very high potential as an engineering material which has yet to be realized fully.

7.4.1.7 Silicon bronze

There is one silicon bronze CS101 (CuSi3Mn) used mainly as wire for marine fasteners.

7.4.1.8 Copper–nickel alloys

The copper–nickel alloys CN101, 2, 4, 5 and 7 wrought and CN1 and 2 cast have even better corrosion resistance than the aluminium bronzes. They have moderate strengths, UTS ranging from 300 to 390 MPa annealed and from 360 to 650 MPa hard. They may be used with confidence in the most severe conditions of marine pollution and corrosion. However, no copper alloy is completely resistant to corrosion/erosion. Where, as is the case with those marine condenser tubing installations from which sand or silt cannot be excluded, corrosion erosion is the principal cause of failure, titanium or a pitting-resistant stainless steel performs better.

7.4.1.9 Nickel–silvers

Nickel–silvers NS102–6 are copper alloys with 10–25% Ni and 18–28% Zn. They have good corrosion resistance and attractive colour and are available as strip and wire with proof stress between 600 and 620 MPa, which makes them very suitable for relay springs at ambient and moderately elevated temperatures. Leaded nickel–silvers NS1OI and NS111, which have improved machinability, are available also as bar.

7.4.1.10 Miscellaneous alloys

Copper alloys with between 20% and 45% lead are used for bearings whose loadings are too high for white metals. They were originally cast into a steel backing but are now usually made by powder metallurgy techniques to improve consistency and produce a more favourable structure.

Alloys of copper with manganese and aluminium or manganese and nickel can have very small temperature coefficients of electrical resistance. They are used in instruments and are usually supplied as annealed wire. Other copper manganese alloys have high damping capacities.

Typical compositions and mechanical properties of these alloys (which are not covered by British Standards) are listed in Table 7.19.

Alloys of copper with zinc 12–40% and aluminium 2–8% are capable of existing in two distinct configurations above and below a critical temperature which may be selected to lie between −100°C and +100°C. This 'Shape Memory' characteristic enables the manufacture of a component that will generate a force 200 times that which can be obtained from a bimetallic strip of similar size and is used for controlling temperature.

7.4.1.11 Selection of copper alloys

Choice of the alloy of copper that will most satisfactorily fulfil most appropriate engineering applications may be made by consulting Tables 7.20–7.23.

Table 7.19 Some miscellaneous copper alloys

(a) Copper–lead alloys for bearings (lead bronzes)

Type	Composition (wt %)					Tensile strength (MN/m²)	Characteristics
	Cu	Pb	Sn	Ag	Others		
80/20 Cu–Pb	78	20	—	—	1.2–5 Ni	140	Bonded to steel shells for bearings. Alloys up to 45% lead are in use
74/24 Cu–Pb	74	24	2	—	—	140	Good conductivity
70/30 Cu–Pb	69	30	—	0.6	—	140	Withstands 'pounding'
60/40 Cu–Pb	59	40	—	1	—	—	Lead distribution improved by powder metallurgical processing
55/45 Cu–Pb	55	45	1			120	

(b) Copper–manganese alloys

Type	Composition (%)				Condition	Tensile properties		Characteristics
	Cu	Mn	Al	Ni		Strength (MN/m²)	Elongation (%)	
Cu-Mn-Ni ('Manganin')	Balance	12	—	2–4	Drawn – Soft	420	30	High specific resistance alloys with low temperature coefficient of electrical resistance
Cu-Mn-Al	Balance	13	2	—	Drawn – Soft	430	30	
					– Hard	680	10	
General-purpose high-damping alloy ('Incramute')[a]	58	40	2	—	Hot and cold worked			High capacity for vibration damping and good corrosion resistance. Good casting and hot and cold forming properties, but difficult to machine. Costly
Marine high-damping alloy ('Sonoston')[b]	63.5	30	4	2.5	Cast			High capacity for vibration damping and good marine corrosion resistance. Casting requires skill. Poor machinability. Costly. High strength suitable for springs
	60	20	—	20	Heat treated			

[a] International Copper Research Association Inc.
[b] Stone Manganese Marine Ltd.

Reproduced by permission of the heirs of the compiler, Dr E.A. West.

Table 7.20 Initial guide to selection of copper and copper alloys for electrical purposes

Applications	Wrought	Cast
Power cables, overhead lines, industrial and domestic wiring	C101, C102	
Telecommunication and coaxial cables	C101, C102	
Overhead lines and electric-traction catenaries	C101, C102, C108	
Bus bars	C101, C102	
Flexible cables	C101, C102, C108	
Generator windings and transformers at low and normal temperatures	C101, C102	
Generator windings at raised temperatures	CC101, CC102, Cu + 0.1Ag	
Commutators	C101, C102, A2/1 Cu + 0.1Ag	
Slip rings	CZ108	PB4, HTB1, SCB4, SCB6
Electronic components in vacuum	C103, C110	
Switch blades, etc.	C101, C102, CB101, Cu + 0.1Ag	HCC1, CC1
Machined components for fittings (terminals, etc.)	C111, C109	
Cast components for conductors,		HCC1
spring contacts, etc.	PB101, PB102, PB103, C108, CB101, A3/2 NS103, NS105, NS106, NS107	

Reproduced by permission of the heirs of the compiler, Dr E.A. West.

Table 7.21 Initial guide to selection of copper and copper alloys for applications in mechanical and chemical engineering

Applications	Wrought	Cast
Bellows and diaphragms	PB101, PB102	
Valve bodies for high duties	CA104, CA105	AB1, AB2
Pressure vessels	C106, C107, CS1C1	
Wear-resistant cams, guides, etc.	CB101, CA105	AB1, AB2
Pump components	CA106, PB103, PB104	AB1, AB2, CMA1, G1, LG2, LG4
Springs	CB101, PB102, PB103, NS104, NS106, NS107 CZ106, CZ107, CZ108, C101	
Bearings – heavy duty, rolling mills	CA106, CA105, Copper lead	AB2
– marine		G1
– non-critical, low loads		LG2, LG4, LB4
– light duties	CZ120, CZ121, CZ122	DCB1, DCB2, DCB3
– average duties, good lubrication	PB104, CZ124	LG2, LG3
– average duties, poor lubrication		LB2, LB3, LB4, LPB1
– for hard shafts	PB104, CZ124	PB1, PB2, PB4, G1
– for soft shafts, low loads	PB103, PB104, CZ121, CZ124	LB1, LB4, LB5
– plates for bridges	CA104, PB104	PB1, PB2, PB4
– ball and roller cages	A/3/2	
Automobile radiators – tubes	C101, C102, C106, CZ105, CZ106	
– strip	Copper + 0.1Ag	
– tanks	CZ105, CZ106, CZ107	
Gears – light duty		G1, LG2, LG3, LG4, DCB1
– moderate duty	PB103, PB104, CA103, CA104, CZ105	DCB2, DCB3, AB1
– moderate duty	PB103, PB104, CA103, CA104, CA105	PB1, PB2, PB4
– very heavy duty, low speed		PB2, AB2
– high loading	CA103, CA104	PB1, PB2, PB4, AB2
– high abrasive loading	CA104	AB1
– pinions	PB103, PB104	PB1
– clocks and similar	CZ118, CZ120, CZ122	
– instruments (high precision)	CZ120, CZ122	
Bushings for sleeves	PB103, PB104, CZ120, CZ121, CZ122	
Deep drawn and pressed items	CZ105, CZ106, CN104, NS104, NS105, C104	
Repetition-machined items	C109, C111, CZ118, CZ119, CZ120, CZ121, CZ122	
Brazed assemblies	C106, C107, CZ103	
Non-sparking tools, etc.	CB101, CA103, CA104, CA105	AB1, AB2
Chains	CB101, CA104	AB1
Hot forgings and stampings	CZ109, CZ120, CZ122, CZ123	

Reproduced by permission of the heirs of the compiler, Dr E.A. West.

Table 7.22 Initial guide to selection of copper and copper alloys for corrosion resistance

Environment and applications	Wrought	Cast
Rural and industrial		
Roofing sheet and cladding, flashing and gutters	C101, C102, C104	
External decorative items, formed strip and sections	C102, C104, CZ107, CZ121, CZ125, NS101, NS102, NS111, NS112	
Lightning conductors	C101, C102, C104	
Wall ties and masonry fittings	PB102, CA103	PB1, AB1, AB2
Window frames and statuary	CZ121	
Statuary	C101, C102, C104, C106	Special tin bronzes
Solar heating panels	C106, C107	
Damp-proof course and weather strip	C104, C107, CZ101, CZ102	
Tubes for water and gas	C106	
Water cylinders, calorifiers	C106, CS101	
Tubes for soil and waste systems	C106	
Hinges and butts	CZ108, CZ121	
Water fittings, taps, etc.		SCB1, SCB3, SCB6, DCB1, DCB3, PCB1
Valves for water		SCB1, SCB3, SCB6, DCB3, LG2, LG4, G1
Nails and screws	C102, CS101, CZ106, CZ108, CZ121, CZ124	
Marine		
Condenser tubes	CN101, CN107, CZ109, CZ110, CZ112, CA105	
Tubes for sea water	CN101, CN107, CZ110, CS101, CA104, CA105	
Oil tank heaters	CZ110	
Fittings for sea water	CZ112, HTB1, CN107, CS101, CA104, CA105	PB1, PB4, G1, LG2, LG4
Valves and pump components	CA104, CA105	G1, LG2, LG4, PB1, PB4, AB2
Boiler feedwater fittings	GT1, PB1, PB3	AB2, G1, LG2, LG4, PB1, PB4
Propellers		CMA1, (CMA2), AB2, HTB1
Portholes, deadlights, windows	CA104	G1, AB2
Non-magnetic fittings	CZ112, CA103, CA104	G1, LG2, LG4, AB1, AB2
Small fittings and cleats	CZ110, CZ111, CZ112, CS101, CA104	HTB1, LG2, LG4
Nails, clouts and screws	CS101, CZ112, CA103, CA105	

Reproduced by permission of the heirs of the compiler, Dr E.A. West.

Table 7.23 Initial guide to selection of copper and copper alloys for applications at raised temperatures

Applications	Wrought	Cast
Superheated steam valves and fittings	CA103, CA104, CA105	LG2, LG4, AB2, CMA1
Oxygen lance heads	CC101, CC102	
Arc furnace electrode holders	C108	CC1-TF
Disk brakes	CC101, CC102	
Aircraft brakes	CC101, CC102	
Spot, seam and flash butt welding electrodes and dies	C108, CB101, CC101	

Reproduced by permission of the heirs of the compiler, Dr E.A. West.

7.4.2 Aluminium and its alloys

7.4.2.1 General

Although aluminium has been used commercially for little more than a hundred years it has become, on a volume basis, the most widely used non-ferrous metal. It has the essential characteristic of high corrosion resistance in normal atmospheres and conditions and its cost per unit volume is usually lower than that of any other metal except steel.[24] One of the most important characteristics of pure aluminium is its high electrical conductivity, 61 IACS on a volume basis, but because of its low specific gravity, aluminium conductors are usually cheaper than copper conductors.

A most important point to consider in the selection of aluminium is the cyclic price variation. The supply of aluminium is inelastic because furnace capacity cannot be expanded quickly. While therefore aluminium components cost less than copper or magnesium components when supply exceeds demand, magnesium components and occasionally even copper conductors may become superior economically when aluminium is in short supply. In extreme cases a correct choice made when a component is designed may prove to be incorrect when it comes to be manufactured.

Aluminium is ductile and easily fabricated and, although its mechanical properties are not outstanding, the specific strength and modulus of its alloys are excelled only by titanium and magnesium among common engineering metals because of its low specific gravity (2.7).

The tensile properties of aluminium improve with reduction in temperature and, because it has a face-centred-cubic structure, it does not embrittle at cryogenic temperatures. It is therefore suitable for components for operation at cryogenic temperatures.

The casting and fabrication of aluminium are easy due to its low melting point and high ductility. Because of this, and because of its wide usage, aluminium is readily available in all wrought forms, shapes and finishes, and all varieties of castings. BS specifications covering aluminium alloys are listed in Table 7.24.

Most wrought alloys can be hardened by cold working and are available in a variety of tempers. Some alloys, both cast and wrought, are amenable to heat treatment by solution treatment and ageing. The several heat treatments and conditions are denoted by letter and number codes which are listed in Table 7.25. All except some heat-treated alloys are very easy to weld.

Aluminium is easily machined and, in spite of its ductility, capable of a good finish. It is non-sparking and non-magnetic.

Drawbacks are poor fatigue and elevated temperature properties. Aluminium has no endurance limit, the fatigue strength of the pure metal at 10^8 cycles is only about 60% of its tensile strength and this ratio is lower, the higher the tensile strength of the alloy.[25] Because of its low melting point, high-temperature capability is very limited and the tensile strength of wrought alloys falls very sharply above 220°C and that of cast alloys above 270°C.

Although corrosion resistance in normal atmospheres is good, aluminium is prone to galvanic corrosion when coupled to most engineering metals (except magnesium and zinc) and certain alloys in high heat-treatment conditions are prone to localized corrosion.

7.2.4.2 Wrought aluminium alloys (see Table 7.26)

All aluminium alloys are collectively known as 'aluminium'.[26] 'Pure aluminium' is an alloy of aluminium, iron and silicon derived from the reduction process, in contents varying according to grade up to 1% with minor quantities of other metals. 'Bright Trim' grades have slightly higher impurity contents and small additions of magnesium. In 'Super Pure' aluminium, which is double refined, the impurities Cu, Si and Fe total less than 0.01%.

The tensile strengths of the pure aluminium grades vary from about 55 MPa (temper designation 'O') annealed to about 160 MPa Strain Hardened Full Hard (temper designation 'H.8'). Increases in strength are obtained by alloy additions. Manganese and magnesium in amounts up to 1.25% Mn and 4.5% Mg increase the UTS up to about 300 MPa annealed and 375 strain hardened half-hard.

None of these alloys owes its tensile properties to heat treatment, and corrosion resistance is reduced only marginally compared with that of pure aluminium. All of the alloys are readily welded.

Table 7.24 British Standard specifications covering aluminium alloys

BS 1490:1970	*Aluminium and aluminium alloy ingots and castings*
	Applications
BS 1470:1985	Wrought aluminium and aluminium alloys for general engineering purposes – plate, sheet and strip
1471	Drawn tube
1472	Forging stock and forgings
1473	Rivet, bolt and screw stock
1474	Bars, extruded round tubes and sections
1475	Wire
BS 3087:1974	Anodic oxide coatings on wrought aluminium for external architectural applications
BS 1615:1972	Anodic oxidation coatings on aluminium
BS 5762:1979	Methods for crack opening displacement (COD) testing
BS 8118:1985	Code of practice for the design of aluminium structures

See also BS L series and DTD specifications for defence materials.

Table 7.25 Codes for aluminium product forms and conditions used by BS

Wrought material
Prefixes indicating product form

B = bolt and screw stock
C = clad
J = longitudinally welded tube
R = rivet stock
E = bars, extruded round tube and sections
F = forging stock and forgings
G = wire
S = plate, sheet and strip
T = drawn tube

Suffixes indicating condition

F = as fabricated
O = annealed
H = strain hardened (i.e. strengthened by cold
      working – not heat-treatable materials)
H2 = approximately equivalent to previous BS
      designation 1/4H
H6 = approximately equivalent to 1/2H
H8 = approximately equivalent to 3/4H

Heat treatable:

T4 = solution treated and naturally aged (formerly TB or
      W)
T3 = solution treated, cold worked and naturally aged
      (formerly TD or WD)
T5 = precipitation treated (formerly P or TE)
T8 = solution treated, cold worked and precipitation
      treated (formerly TH or WDP)
T6 = solution treated and precipitation treated (formerly
      TF or WP)

A special suffix E indicates an electrical grade of material.

Cast material
Code indicating heat treatment

M = as cast
TB = solution heat treated and naturally aged (formerly
      W)
TB7 = solution heat treated and stabilized (W special)
TE = precipitation treated (P)
TF = solution heat treated and precipitation treated (WP)
TF7 = solution heat treated, precipitation treated and
      stabilized (WP special)
TS = thermally stress relieved

Ingot materials have the same designations as castings but no code letters for condition.

The lower-strength alloys of this group are used for domestic and culinary equipment and automobile trim; the medium-strength alloys for architectural, marine and commercial road vehicles; and the higher-strength alloys for marine, welded structural applications and aircraft tubing.

Higher strengths are obtained by quenching suitable alloys from elevated temperatures and then ageing by reheating at lower temperatures. There are several groups:

1. Alloys containing amounts of Mg and Si in the ratio of the compound Mg_2Si with Mn or Cr. These have UTS up to 330 MPa without any real deterioration in corrosion resistance. They are used for applications similar to those of the non-heat-treated alloys but requiring slightly higher strength.

2. Alloys containing Cu as well as Mg_2Si with one or more of Mn, Fe, Ni, Cr, Ti, Zr and Nb may age harden at room temperature after quenching or may require precipitation heat treatment to give UTS up to 450 MPa. The corrosion resistance of these alloys is impaired so that they must be protected from weathering. These alloys may be used for aircraft structures, other miscellaneous structural applications, and in some cases forged aircraft engine parts, including forged pistons.

3. Alloys containing Zn + Mg + sometimes Cr, some of which give UTS around 450 MPa while others range up to 600 MPa. Alloys heat treated to the higher levels may have their fracture toughness as well as their resistance to stress corrosion impaired. These are mainly used for aircraft structural applications but other structural applications are becoming important.

4. The addition of lithium in amounts around 2.5% to CuMgZr alloys gives an increase in modulus and a decrease in specific gravity of 10% in alloys with 0.2% PS and UTS of 500 and 420 MPa, elongations around 7%, and plane strain fracture toughness values around 35, which are to be compared with the properties of 2324 or 2014 for which the lithium alloys might be substituted. Fatigue crack growth rates also appear comparable or better than the standard alloys, as does the range of general corrosion. These alloys are, however, prone to stress corrosion. Samples of these alloys have been flying for several years and qualifying tests for major usage are proceeding.[27-29]

7.4.2.3 Aluminium casting alloys (see Table 7.27)

The alloying additions for casting alloys differ significantly from those for wrought because of the role of silicon in improving castability and reducing thermal expansion. The coefficient of thermal expansion is reduced by 33% for a silicon content of 23% and proportionately for other contents – an effect which can be important in components such as pistons. All aluminium casting alloys contain approximately 0.2% titanium as a grain refiner. In addition, aluminium silicon alloys must be 'modified', that is, the acicular silicon phase in the aluminium silicon eutectic must be converted to a fine spheroidal particle by the addition of sodium, either as a metal or as a fluoride compound, if the alloy is to have a reasonable ductility. The eutectic composition of modified silicon aluminium is approximately 12.6%. If, to reduce the coefficient of thermal expansion, the silicon content is raised above this value, the elongation will become very low because of the presence of hyper eutectoid silicon.

Five major casting alloy groups can be identified:

1. In castings additions of manganese or up to 12% silicon increase the UTS up to 170 MPa without heat treatment. The straight aluminium silicon alloys are used where castability is essential but only moderate strength is required.

2. Moderate increases in tensile strength without heat treatment are obtained by additions of copper sometimes accompanied by iron, magnesium or zinc to the basic silicon alloys. These alloys are used where good casting characteristics, weldability and pressure tightness are required with moderate strength. Applications include mani-

Table 7.26 US and UK wrought aluminium alloy designations

Composition type	USA AA	UK BS (where different)	Whether heat treatable	UTS (MPa)	Weldability		Notes
					Fusion	Resistance	
Al 99.99	1099, 1199	–	No	50–100	Very good	Good	
Al 99.8	1060		No	50–150	Very good	Good	
Al 99.5	1050	IE	No			Very good	
Al 99.5	EC	IE	No			Very good	
Al 99.0	1200		No	100–150	Very good	Very good	
Al-4CuSiMg	2014		Yes	400–500	No	Very good	
Al-4Cu-Mg	2024	L97, L98	Yes	400–500	No	Good	
Al-2Cu-1.5Mg-1Fe-1Ni	2618		Yes	400–500	No	No	
Al-2Cu1Ni1MgFeSi	–				No	No	
Al-5.5CuPbSi	–	FC1			No	No	Free-cutting
Al-1Mn	3103		No	200	Good	Good	
Al-MnMg	3105		No	200	Good		
Al-5Si	4043/4543	–	No				
Al-12Si	4047		No				
	4032	DTD324B					
Al-1Mg	5005				Excellent	Excellent	
Al-2Mg	5052		No	150	Excellent	Excellent	
Al-3.5Mg	51.54A	5154	No	200	Excellent	Excellent	
Al-5Mg	5056A	5056	No	200–300	Excellent	Excellent	
Al-3MgMn	5454		No	250–300	Excellent	Excellent	
Al-4.5MgMn	5083		No	300–400	Excellent	Excellent	
Al-MgSi	6063		Yes	150–200	Good	Good	
Al-MgSi	6463		Yes	150–200	Good	Good	
Al-1MgSiCu	6061		Yes	200–300	Difficult	Good	
Al-MgSiMn	6082		Yes	200–300	No	Good	
Al-4.5Zn1Mg	{ 7005 { 7075	DTD5074A	Yes	350	No	Good	
Al-2.5Li 1.3Cu 0.8Mg 0.1Zr	8090		Yes	400–500	Not stated	Not stated	Low density
Al-2.3Li 2.7Cu 0.12Zr	2090		Yes	400–500	Not stated	Not stated	High modulus

Table 7.27 UK and US aluminium casting alloy designations

Alloy type	Composition type	UK BS	USA AA	UTS[a] (MPa)	Notes
Pure Al	99.5	LMO		80	
AlSi	Al-5Si	LM18	443	120–150	
	Al-12Si	LM6	–	170–200	Linear exp. coeff. 20×10^{-6}
AlSiMg	Al-15Si1CuNg	LM16	355	210–310[TF]	
	Al-7SiMg	LM25	356	140–310[TF]	
	Al-11SiMgCu	LM13	A332	150–290[TF]	
	Al-12SiMg	LM9	A360	180–310[TF]	
	Al-19SiCuMgNi	LM28		130–200[TF]	
	Al-23SiCuMgNi	LM29		130–210[TF]	Linear exp. coeff. 16×10^{-6} Low ductility
AlSiCu	Al-10Cu-2SiMg	LM12	222	180	
	Al-5Si1Cu	L78		20–31[TF]	
	Al-5Si-3Cu	LM4	319	150–310[TF]	
	Al-5Si-3CuMn	LM22	–	260	
	Al-6Si-4CuZn	LM21	319	180–200	
	Al-7Si-2Cu	LM27	319	150–180	
	Al-8Si-3CuFe	LM24	A380	200	
	Al-9Si-3CuMg	LM26	F331	230	
	Al-10Si2CuFe	LM2	–	180	
	Al-12SiCuFe	LM20	413	220	
	Al-17Si-4CuMg	LM30	390	160	
AlMg	Al-5Mg	LM5	314	170–230	
	Al-10Mg	LM10	520	310–360	
AlCuX	Al-2Cu-1.3Si-1.3Ni-1Mg-1Fe	L51	–	215[TF]	Max. imp. 250
	Al-4Cu	L91/92		280–355[TF]	Max. imp. 250
	Al-4Cu-2Ni-2Mg	L35		220–310[TF]	Max. imp. 270
	Al-4.5Cu-0.7Ag-0.34Mg 0.10Si-0.14Fe-0.4Mn-0.4Cr	Not in specification		450[TF]	High-technology sand casting
AlZnMg	Al-5ZnMg	DTD5008		215	
	Al-7.7Reg.1.2Zn	DTD5018		275–305	

[a] Weakest sand casting – strongest die casting.
TF Solution heat treated and precipitation treated.

folds, valve bodies, ornamental grills and general-purpose castings.

3. Heat-treated alloys, either AlSiMg or AlSiCu, may have UTS ranging up to 310 MPa but ductility is low with elongations values ranging around 1% or less, particularly with the higher silicon-content materials. They are used for automobile and diesel pistons and pulleys.

4. UTS ranging up to 300 MPa with elongation between 4 and 10 may be obtained by heat treating aluminium magnesium, aluminium copper, aluminium zinc magnesium or more complex alloys of low silicon content. These are used for more highly stressed components such as aircraft fittings, internal combustion engine pistons and cylinder heads where higher strength and hot strength are important but castability is not so critical.

5. There is one, silver containing, casting alloy Al4.5 Cu, 0.7 Ag, 0.34 Mg, 0.10 Si, 0.14 Fe, 0.4 Mn, 0.4 Cr not in any standard which, when manufactured by a high-technology casting process, may be heat treated to UTS 450 MPa elongation 10%, equivalent to many forging alloys. It is expensive and unlikely to have a high resistance to corrosion but it offers great advantages for complex highly stressed components such as may be used in fighter aircraft.

7.4.2.4 Fabrication of aluminium and its alloys

Aluminium has a face-centred-cubic structure and its consequent ductility combined with its relatively low melting point make it easy to cast and work. It is available in the form of sand, gravity and pressure die castings, a very wide range of shapes and sizes of extruded sections, forgings, plate, sheet and strip, many sheet metal products, rod, wire and tube. Most of its alloys can be welded by inert gas shielded metal arc or electric resistance processes. Powder processes have been developed, principally for producing composites usually of an alumina or other ceramic in an aluminium matrix.

High-quality aluminium components depend on the production of a fine-grained defect-free casting. Aluminium has no allotropic modification and cannot be refined by heat treatment. While aluminium alloys can be recrystallized after working, their fatigue characteristics depend on the production of an inherently fine-grained casting with a very fine dispersion of intermetallic particles. Further, because of the high solidification contraction, aluminium alloys depend for their soundness on very efficient feeding during solidification. These problems are compounded by the tendency of molten aluminium to absorb hydrogen which is evolved to form cavities in the metal on solidification and by the tenacious oxide film which forms on the surface of the molten (as of the solid) metal, and may be carried under the surface of the casting by turbulence to form discontinuities and interrupt feeding.

Hydrogen must be minimized by melting practice and eliminated by fluxing shortly before pouring. A fine cast structure and satisfactory feeding of ingots for mechanical working can be achieved by continuous casting and this process is amenable to smooth liquid metal transfers which minimize or eliminate oxide inclusions.

The production of high-quality conventional sand and gravity die castings has, in the past, depended to a very large extent on the experience and skill of the foundry. Recent developments in the technique, both of sand and die casting, have gone far to ensure rapid and efficiently fed solidification of a melt free from oxide inclusions. The most modern techniques of sand and die casting consist of causing molten alloy to rise from the centre of the melt into a superimposed die or mould. The liquid metal (which has therefore no oxide

skin) rises with an unbroken surface during all stages of filling with metal, transferred in non-turbulent conditions, from a large quiescent enclosed bath of molten alloy. The pressure head developed to raise liquid smoothly into the mould is maintained to feed the casting during solidification. The more even flow of metal itself increases the chilling effect of the mould on the solidifying metal but this effect is augmented by using mould materials with greater heat capacity.

Cosworth Foundry, whose casting unit is represented diagrammatically in Figure 7.40, uses zircon sand which, besides producing a greater chill, has a lower thermal expansion than silica sand and significantly improves accuracy. The newly developed casting processes improve tensile and fatigue strength by about 30% and double elongation compared with conventional sand casting.[30]

7.4.2.5 Identification and standards of aluminium alloys

Wrought products The BSI, the US Government and the standards organizations of many countries use the four-digit system of the Aluminum Association of the US to identify

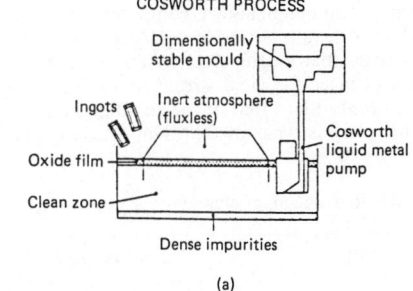

COSWORTH PROCESS

(a)

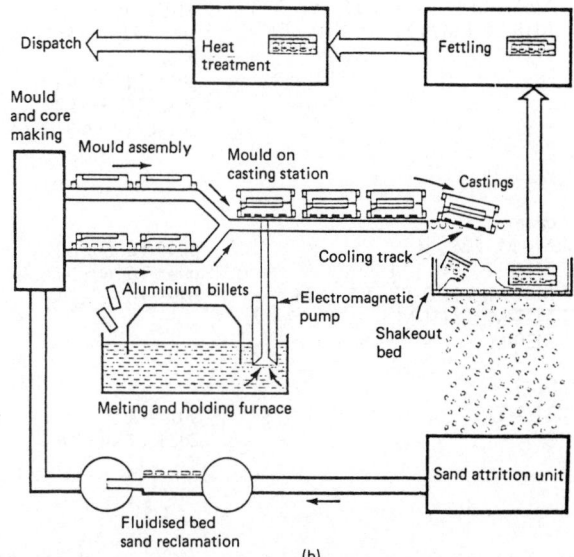

(b)

Figure 7.40 The Cosworth casting process. (a) The casting unit; (b) flow diagram of the Cosworth process. (Reproduced by permission of *Metals and Materials*) Note: For the convenience of the reader this figure is also reproduced in section 16.5.44

wrought aluminium alloys as listed in Table 7.26. In most cases the BS numbers are identical with those of the USA, but a few British alloys, for which there is not always a US equivalent, retain the old L system of numbering and a few newer alloys are so far covered only by DTD specifications.

Castings Casting alloy numbers have not been homologized. Most BS alloys have 'LM' numbers but a few 'L' numbers survive and some alloys are covered by DTD specification. At least three distinct number classifications are used in the USA. Aluminum Association casting alloy numbers and BS Product Specifications are listed in Table 7.24 and condition codes in Table 7.25.

7.4.2.6 Design using aluminium

The attention of the design engineer is called to three matters which are of great importance in the design and operation of aluminium components. Creep buckling in compressed struts and sheets can occur under loads far smaller than those calculated from the normal Euler formula using the modulus of elasticity.[31]

Creep buckling is not unique to aluminium alloys but the loading conditions in aircraft struts are often such that it is critical.

When a typical aluminium alloy is compared with a steel the fatigue crack growth rate per cycle da/dN for the aluminium alloy is forty times that for the steel. When compared with a titanium alloy of twice the yield stress, dA/dn for the aluminium alloy is twenty times that of the titanium alloy. K_{Ic} for the aluminium alloy is typically 20 MNm$^{-3/2}$ compared with 160 MNm$^{-3/2}$ for the steel and 60 MNm$^{-3/2}$ for the titanium alloy (see Chapter 8, Section 8.3). This is, to some extent, offset by the lower stresses used in design with aluminium. Even so, fatigue crack growth calculations play a much more significant part in the design process for aluminium alloys than for medium-strength steels.

The fatigue crack growth versus stress intensity curve shows less tendency to turn down to the vertical with aluminium alloys than with steel. Therefore aluminium alloys do not have a well-defined endurance limit. This is not so serious as it might seem, because aluminium alloy components are seldom designed to withstand cyclic loading in the region of 10^8 cycles. It should be noted that the endurance limits normally quoted for sand castings refer to those produced by conventional methods and not the high-technology methods described in Section 7.4.2.4.

The choice of alloy to meet specific requirements will be assisted by reference to Table 7.28 for resistance to environmental corrosion, Table 7.29 for resistance to aqueous corrosion, Table 7.30 for resistance of castings, Table 7.31 for wrought products to stress at room temperature, and Table 7.32 for resistance to stress at elevated temperatures.

7.4.3 Titanium and its alloys

7.4.3.1 General

The importance of titanium as an engineering material depends on specific strengths which (except beryllium, which is costly and toxic) are greater than those of any other metal

Table 7.28 Guide to exposure of aluminium alloys to rural, marine and industrial environments

Castings (BS 1490)	Wrought (BS 1470–75)	Notes
Group 1 LM5-M LM10-TB LM6M, LM9-TE and TF LM18-M, LM20-M	1060, 1050, 1200 3103, 3105 5005, 5251, 5154, 5454 5083 6060-T4, 6061-T4, 6082-T4, 6463-T4. 2014C-T4, 2024C-T4	Weathers to pleasant grey colour, deepens to black in industrial atmospheres Superficial pitting occurs initially and gradually ceases Seldom needs painting except for decoration May be anodized for appearance but some alloys (e.g. LM6) give dark films
Group 2 LM4-M, LM2-M LM16-TF, LM13-M LM21-M, LM22-TB, LM24-M LM27-M, LM31-M LM20-M	6060-T6, 6061-T6, 6065-T6, 6082-T6, 6463-T6	Weathers as above Is normally painted in severe industrial environments and for marine service May be anodized as above
Group 3 LM12-TF	2014-T4 2024-T4 and T6	Painting needed in marine and industrial atmospheres, but coatings need only infrequent renewals Sprayed aluminium coatings, or cladding, give excellent protection Seldom anodized

Aluminium retains its initial appearance if it is washed periodically at 3- to 12-month intervals, depending on the severity of the pollution in the air. For many applications the weathered surface is satisfactory but anodizing is often undertaken, particularly for architectural items, to preserve a smooth appearance but periodic washing is desirable, especially when the anodic film is coloured. Anodizing should be in accordance with the relevant British Standards BS 1615:1972 and BS 3987:1974.

Reproduced by permission of the heirs of the compiler, Dr E.A. West.

Table 7.29 Initial guide to corrosion resistance of aluminium alloys at normal temperatures

Exposure media/applications	Wrought products (BS 1470–77, 4300)		Castings (BS 1490)	
Inland atmospheres/ Building components, roofs	1st choice:	1050, 1090, 1200, 3103, 3105, 5005, 5251	1st choice:	LMO, 5, 6, 9, 10, 18, 25
	2nd choice:	5083, 5154, 5454, 6061, 6063, 6082, 7020	2nd choice:	LM2, 41, 13, 16, 20, 21, 22, 24, 26, 27, 28, 29, 30
Marine – boats and ships, fittings	1st choice:	5005, 5083, 51540, 5251, 54540, 6061TB, 6063TB, 6082TB	1st choice:	LM5, 10
	2nd choice:	5154H, 5454H, 6061TF, 6062TF, 6082TF, 7020TB and TF	2nd choice:	LMO, 9, 18, 25
Chemical and food plant	1st choice:	1080, 1050, 3103, 3105, 51040	1st choice:	LMO, 5, 10
Structural items	1st choice:	6061TB, 6063TB, 6082TB	1st choice:	LM5, 10
	2nd choice:	6061TF, 6063TF, 7020 6082TF, 2014A (clad)	2nd choice:	LM6, 9

Reproduced by permission of the heirs of the compiler, Dr E.A. West.

Table 7.30 Initial guide to selecting cast aluminium alloys according to static strength[a]

UTS (MPa)	Elongation (%)	Alloy reference (BS 1490)	Form of casting	Condition
120–140	1–5	LM4, LM5, LM18, LM25 LM27	Sand	M
		LM28, LM29	Sand	TF
		LM28	Chill	TE
150–180	1–5	LM2, LM4, LM5, LM12, LM18, LM21, LM24, LM25, LM27, LM28, LM30	Chill	M
		LM16	Sand	TB
		LM6	Sand	M
		LM25	Sand	TE
190–210	1–5	LM6, LM9, LM20	Chill	M
		LM13, LM25, LM26, LM29	Chill	TE
		LM28, LM29	Chill	TF
210–245	1–8	LM4, LM9, LM6, LM25	Sand	TF
		LM9	Chill	TE
		LM16, LM22, LM25	Chill	TB
250–310	1–2	LM4, LM9, LM13, LM16, LM25	Chill	TF
	8–12	LM10	Chill	TB
450	10	Special silver alloy listed in Table 7.27	High-technology sand	TF

[a] Figures are those obtained on standard test bars.
Reproduced by permission of the heirs of the compiler, Dr E.A. West

alloys. It has excellent creep resistance up to 600°C and exceptional corrosion resistance to oxidizing media. The high specific strengths derive from alloys with tensile and yield strengths up to 1380 and 1230 MPa, respectively, and a specific gravity pure of 4.51 rising to a maximum of 4.85 alloyed, roughly half that of steel. The corrosion and oxidation resistance is due to a tenacious oxide film on a metal above silver and nickel and below only Hastelloy and Monel in the

electrochemical series for structural metals in sea water. One research programme on coupled specimens in sea water indicated titanium to be the most noble of structural metals. Titanium alloys are very resistant to cavitation erosion.

These advantages are reinforced by a high fatigue resistance, (endurance limits are roughly half of UTS) and a thermal expansion (8–10×10^{-6}°C^{-1} between 0°C and 500°C), roughly half that of austenitic stainless steels and aluminium

Table 7.31 Initial guide to selecting wrought aluminium base materials according to static strength[a]

UTS (MPa)	Elongation (%)	Alloy reference (BS 1470–75, 4300)	Condition range
55–95	35–20	1080A, 1050A, 1200, 1350	0
90–155	25–16	3103, 3105, 5005	0
	18–16	6082	0
	10–3	1080A, 1050A, 1200, 1350	H4–H8
145–185	8–1	3103	H2–H4
	15	3103, 5005	H4–H8
	12	6082, 6463	TB/TF
		6061, 6063	TF
160–200	20–18	5251	0
215–285	18–12	5154A, 5454	0
	8–3	5251	H3–H6
	12	7020	TB
	9	6061, 6063	TF
275–350	16–12	5083	0
	8–3	5154A, 5454	H2–H4
	8	6082	TF
	10	7020	TF
	12	2031	TB
375–400	14–8	2014A, 2014 (clad), 2618A	TF
	10–4	5083	H2–H4
400–480	8–5	2014A, 2014 (clad), 2618A	TF
480–580	4	7075	T6

[a] As there are variations depending on form and thickness, the details must be obtained direct from the relevant British Standards.
Reproduced by permission of the heirs of the compiler, Dr E.A. West.

Table 7.32 Initial guide to selecting aluminium alloys for use at elevated temperatures[a]

UTS (MPa)	Temperature (°C)	BS references	
		Castings	Wrought forms
100–155	100	LM4, LM5, LM6, LM10 TB, LM12 TF, LM13 TF, LM25 TF, L35, 2L92 TF	1200 H8, 2014 T6, 2618 T6, 3103 H4–H8, 5083 0, 5251 0 H4–H8, 5454 0/H2–H4, 6063 T6, 6082 T6, 7075 T6.
	150	LM4, LM5, LM6, LM10 TB, LM12 TF, LM13 TF, LM25 TF, L35, 2L92.	1200 H8, 2014 T6, 2618 T6, 3103 H4–H8, 5083 0, 5251 0/H4–H8, 5454 0/H2–H4, 6063 T6, 6082 T6, 7075 T6.
	200	LM4, LM5, LM6, LM10 TB, LM12 TF, LM13 TF, L35, 2L92	2014 T6, 2618 T6, 5083 0, 5251 0/H4–H8, 5454 0/H2–H4, 7075 T6.
	250	LM4, LM5, LM10 TB, LM12 TF, LM13 TF, L35, 2L92.	5083 0, 5454 0/H2–H4, 6082 T6, 7075 T6.
	300	LM10 TB, LM12 TF, L35, 2L92.	– –
150–200	100	LM4, LM5, LM6, LM10 TB, LM12 TF, LM13 TF, L25 TF, L35, 2L92.	1200 H8, 2014 T6, 2618 T6, 3103 H8, 5083 0, 5251 0/H4–H8, 5454 0/H2–H8, 6063 T6, 6082 T6, 7075 T6.
	150	LM5, LM10 TB, LM12 TF, LM13 TF, LM25 TF, L35, 2L92.	2014 T6, 2618 T6, 3103 H8, 5083 0, 5251 0/H4–H8, 5454 0/H2–H4, 6082 T6, 7075 T6.
	200	LM10 TB, LM12 TF, LM13 TF, L35, 2L92.	2618 T6, 5083 0, 5251 H4–H8, 5454 0/H2–H4.
200–250	100	LM10 TB, LM12 TF, LM13 TF, LM25 TF, L35, 2L92.	2014 T6, 2618 T6, 5083 0, 5251 H4–H8, 5454 0/H2–H4, 6063 T6, 6082 T6, 7075 T6
	150	LM10 TB, LM12 TF, LM13 TF, L35, 2L92	2014 T6, 2618 T6, 5083 0, 5251 H4–H8, 5454 H2–H8, 7075 T6.
	200	LM10 TB, LM12 TF, LM13 TF, L35, 2L92.	2618 T6
	250	LM12 TF, L35.	– –

[a] Based, in general, on static tests undertaken at temperature often soaking at the same temperature.
Reproduced by permission of the heirs of the compiler, Dr E.A. West.

and two thirds that of ferritic steels. Most titanium alloys may be welded using gas-shielded or diffusion methods, and some are capable of superplastic forming.

The main disadvantage of titanium is its cost.[32] Ore separation and reduction to metal is expensive and the metal reacts readily when hot with oxygen and nitrogen so that melting and casting must be done *in vacuo* and welding under a protective atmosphere. Further features are a modulus 105–125 GN m^{-2}, compared with 180 GN m^{-2} for steel and 65 GN m^{-2} for aluminium alloys, and low electrical and thermal conductivities (only about 4% of that of copper). (The higher modulus compared with aluminium can have advantages for springs, ultrasonic devices, surgical implants, etc.) Also, in spite of its generally excellent performance, the corrosion reactions of titanium with very strongly oxidizing media can be catastrophic and it is attacked by uninhibited reducing media.

There are essentially six major groups of engineering applications of titanium:

- Roofing and cladding
- Process plant and flue lining
- Marine condenser tubing
- Electrochemical fitments
- Aerospace frames
- Aero engines

The most exciting applications lie in the compressors (and other components) of aircraft turbine engines. The development of these engines has depended on the parallel development of alloys, with progressively improved creep resistance without sacrificing fatigue resistance. The development of titanium alloys has kept pace with these requirements as is illustrated diagrammatically in Figure 7.41.

7.4.3.2 Titanium metal and alloy structure

The structure of titanium metal is transformed from the low-temperature close-packed hexagonal α-phase to the higher-temperature body-centred β-phase on heating above 850°C. Alloying elements influence the properties of one or both phases, and their relative stability.

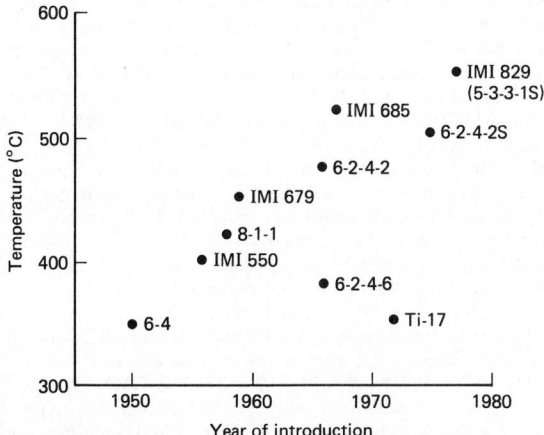

Figure 7.41 High-temperature alloys of titanium at present in production aero engines: year of development and temperature capability under optimum stress conditions. (Reproduced by permission of IMI Titanium Ltd)

Titanium with α-stabilizing and strengthening elements Zirconium and tin are two additions that strengthen the α-phase considerably, particularly at high temperature. Unfortunately, they also raise the density. Oxygen and nitrogen which may be introduced during reduction and processing stabilize the α-phase and increase strength at the expense of toughness. Their content must therefore be limited. The addition of small quantities of palladium (IMI Grade 260, 262) markedly improves resistance to reducing environments and the addition of Mo and Ni (ASTM Grade 12) is a cheaper but less effective attempt to obtain a similar result.

Aluminium stabilizes the α-phase and also reduces density. The alloy Ti 5 Al-2.5 Sn (ASTM Grade 6) has useful tensile properties up to 300°C and is weldable but difficult to cold form. Unfortunately, attempts to increase strength still further by increasing the alloy content resulted in embrittlement due to ordering and formation of Ti$_3$ Al which led to the rule

$$Al + \frac{Sn}{3} + \frac{Zr}{6} + 10 \times O_2 < 9 \text{ w\%}$$

Titanium with β-stabilizing and strengthening elements Mn, Fe, Cu, Cr, V and Mo additions tend to stabilize the β-phase. The most favourable elements are those that do not have a peritectic reaction (see Figure 7.42).

Very high strengths (up to 1500 MPa) can be achieved in metastable β-alloys but although alloys such as Ti 8 Mn (which was processed in the $\alpha + \beta$-phase region) and Ti 13V-11Cr 3Al (which was strengthened by precipitation of the α-phase) have ben developed, the alloys can embrittle on heating above about 300°C and none have found high-temperature applications in aero engines.

Titanium with both α- and β-stabilizing and strengthening elements Titanium alloys with added elements that stabilize both the α- and β-phases simultaneously provide significantly increased room and high-temperature strengths. Ti-6Al-4V is rated up to 350°C and has found widespread application. The addition of Si significantly increases creep strength and Ti-4Al-2Sn-4Mo-0.5Si (IMI550) not only is 10% stronger than Ti-6Al-4V but, in addition, can be operated up to 400°C. (It has been later found that the optimum benefit of silicon additions occurs at a lower level of addition to near-α-titanium alloys (see Figure 7.43).

$\alpha + \beta$ alloys have also been developed to have high strength for moderate-temperature applications (up to about 400°C). These include Ti-2.1/4 Al–11 Sn-4 Mo-0.2 Si (IMI680) and Ti-6Al-5Zr-4 Mo-1Cu-0.2 Si (IMI700) and in the USA Ti-6Al-2 Sn-4 Zr-6 Mo and Ti-5Al-2Sn-2Zr-4 Mo-4Cr.

Near-α-titanium alloys The near-α-titanium alloys have the maximum amount of α-stabilizers with β-stabilizing elements high enough to give medium-strength levels but low enough to avoid problems in welding and embrittlement in creep at high temperature. Ti-11Sn-2.1/4 Al-5 Zr-1 Mo-0.2 Si (IMI679) can be operated up to 450°C and the American alloy Ti-8Al-1V-1Mo up to 400°C, although it suffers from ordering embrittlement and needs careful control in processing. Ti-6Al-2Sn-4Zr-2Mo ('6.2.4.2') was developed to have a temperature capability of about 470°C and has been widely used in the USA.

All these alloys had previously been worked and heat treated in the $\alpha + \beta$-phase region. β-heat treatment significantly improved creep strength but reduced tensile ductility to an unacceptably low level. This problem was overcome by the development of near-α Ti-6Al-5Zr-0.5 Mo-0.25 Si (IMI685) which is β heat treated to produce an acicular microstructure and in this temper has creep capability up to 520°C (see Figure 7.44).

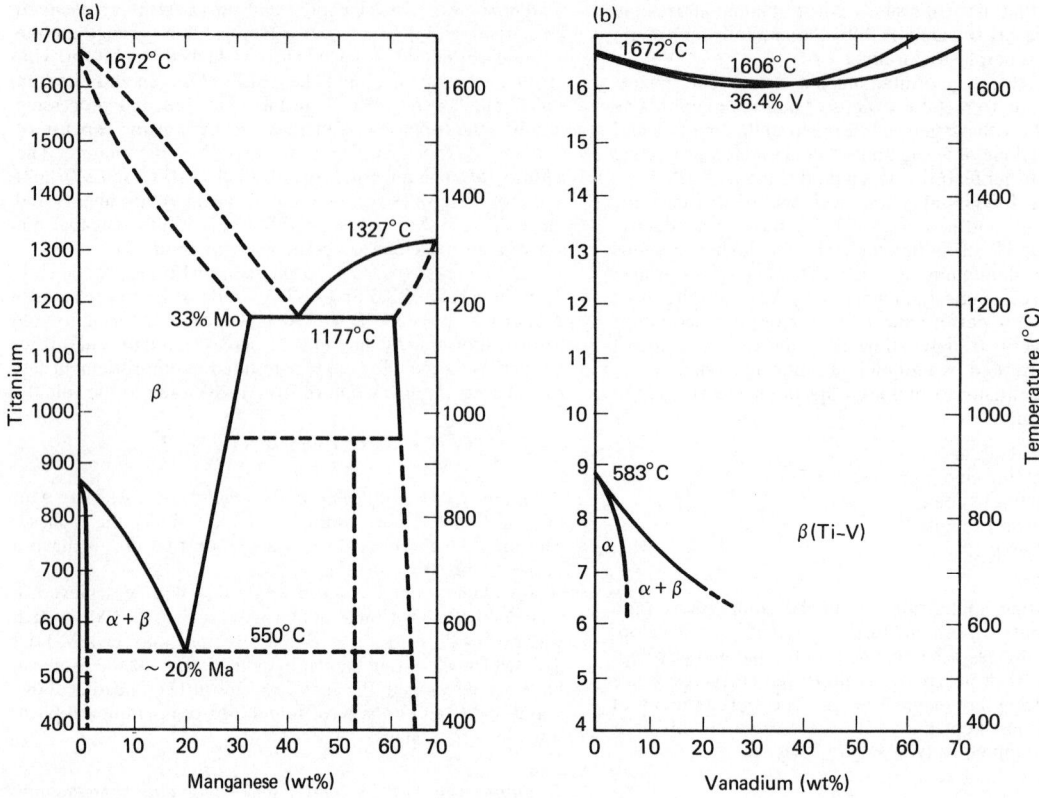

Figure 7.42 Partial phase diagrams of β-stabilizing additions to titanium. (a) Unfavourable structure of manganese alloys showing peritectic; (b) favourable simpler structure of vanadium alloys

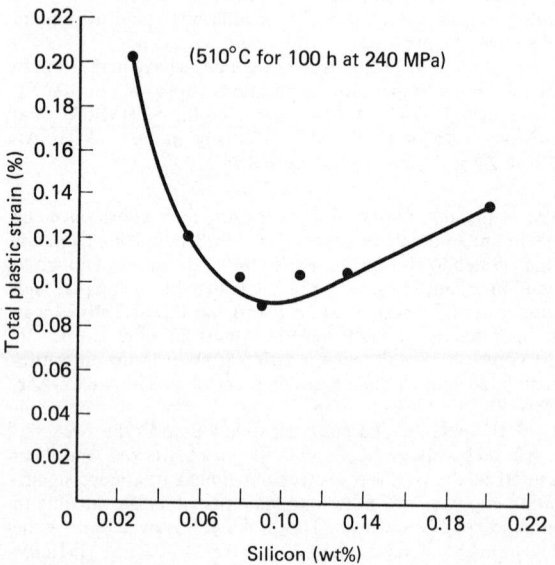

Figure 7.43 Effect of silicon on the creep strain in 6-2-4-2. (Reproduced by permission of IMI Titanium Ltd)

β heat treatment by itself may not be sufficient to maximize creep resistance. Cooling rate is also critical and an example of this is shown in Figure 7.45.

Fracture toughness and fatigue crack growth rate are improved by changing from α + β to β heat treatment. This is illustrated in Figure 7.46. Unfortunately, room-temperature tensile ductility and fatigue crack initiation are worsened. The effective initiation unit is the prior β grain or the α colony which, in a β-treated structure, can be from 0.5 to 2 mm in size compared with 10–15 μm in an α + β structure.

Reduction in size of structural features while retaining an acicular structure can be achieved by utilizing thermomechanical processing using multi-recrystallization stages, and by alloy development. This procedure has resulted in β grain sizes between 0.5 and 0.75 mm in an alloy Ti-5.6Al-3.5Sn-3Zr-Nb 0.3Mo-0.3Si (IMI829). Improvements have also been carried out in the USA on Ti6Al-2Sn-4Zr-2Mo.

The best creep properties (see Figure 7.44) together with an acceptable fatigue crack growth rate are available in Ti-5.8 Al, 4.4 Sn-3.5 Zr -0.7 Nb-0.5 Mo-0.35 Si-0.06C (IMI834). This alloy is characterized by a more gradual change in α-phase content with heat-treatment temperature as in the comparison of β transus approach curves for this alloy and 829 shown in Figure 7.47. This allows the alloy to be heat treated in the high α field for optimum creep and crack propagation resistance but with β grain size controlled to about 0.19 mm for improved fatigue performance.

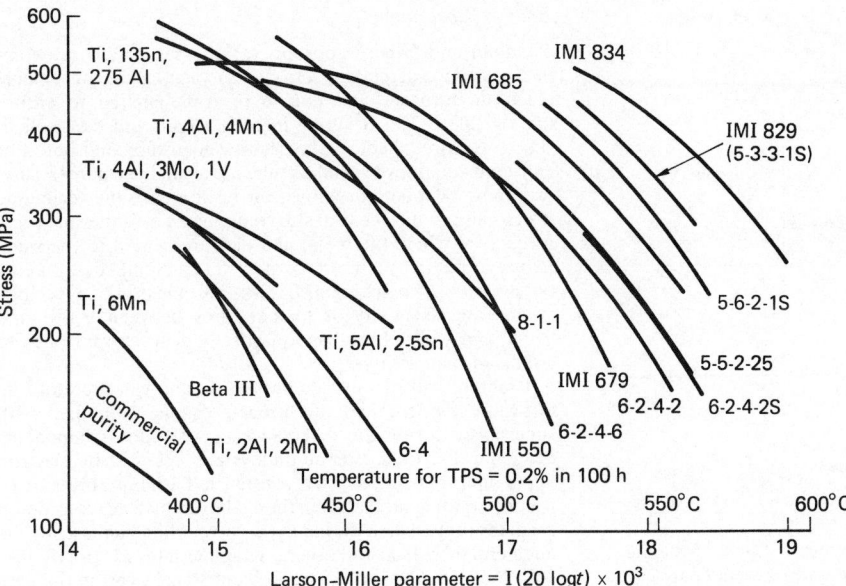

Figure 7.44 Creep-resistant alloy development. Larson-Miller plot showing improvements over the last 30 years. (Reproduced by permission of IMI Titanium Ltd)

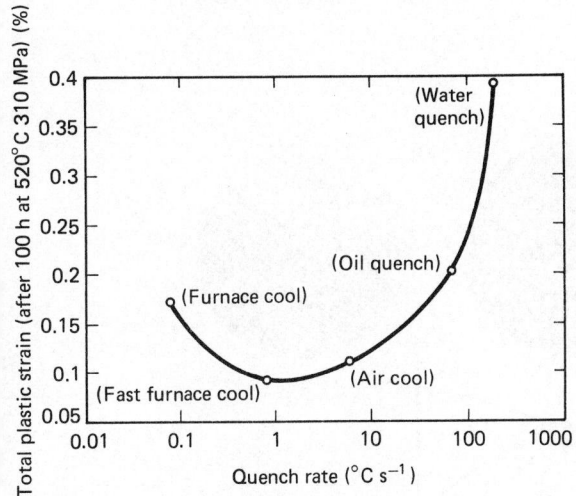

Figure 7.45 Effect of cooling rate on the creep strain in IMI 685. (Reproduced by permission of IMI Titanium Ltd)

The relationship between heat-treatment temperature and the balance of creep and fatigue properties is illustrated schematically in Figure 7.48. The optimum balance can be adjusted for a particular application.

7.4.3.3 Nomenclature and standards

The chemical composition, nomenclature (where available), UK and US specification numbers, tensile properties and weldability are listed in Table 7.33. In Europe titanium alloys

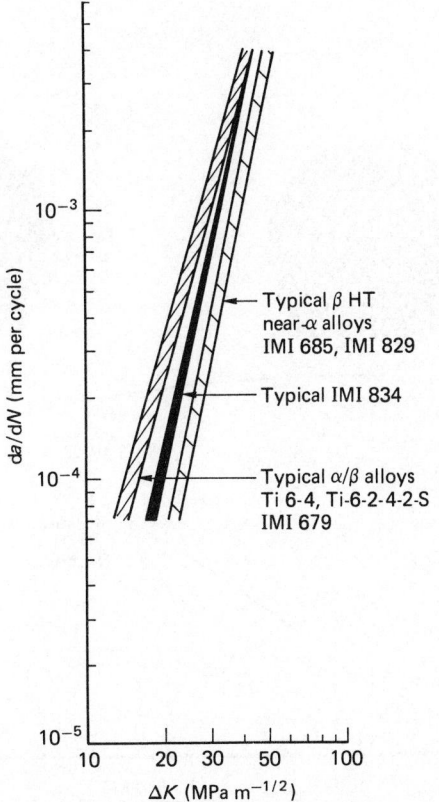

Figure 7.46 Crack-propagation data of high-temperature alloys. (Reproduced by permission of IMI Titanium Ltd)

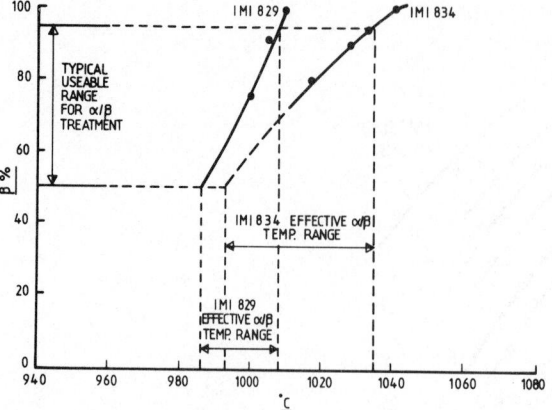

Figure 7.47 β transus approach curves

are often referred to by their IMI number. Some of the more recent titanium alloys had not, at the time of writing, appeared in national material specifications. The properties of the more attractive of these (IMI 829 and 834) are therefore listed in Table 7.34.

Such is the pace of titanium alloy development that designers are recommended to approach the manufacturer for information on materials available at the time before finalizing their designs.

7.4.3.4 Fabrication

Titanium and (where appropriate) alloying metals are compacted and welded into electrodes for vacuum arc consumable electrode furnaces and double or triple melted to ensure homogeneity. The resulting ingot is forged and then rolled. Bar, rod, wire, sheet, plate, seam-welded tube and bored or extruded and drawn seamless tube are available commercially.

Almost all titanium alloys can be forged. One technique which can produce near to size components is isothermal press forging, in which the metal and dies are heated to the same temperature between 900°C and 950°C. Some fine-grained duplex alloys, including IMI 318, 550 and 6242, have high strain rate sensitivity at temperatures between 900°C and 950°C and will deform superplastically at low strain rates to strains of over 1000%.

Titanium will dissolve its own oxide and can therefore be diffusion bonded, and this process can be combined with superplastic forming to make a variety of hollow components economically. Most titanium alloys are weldable by electron beam, inert gas-shielded arc, resistance, flash butt, pressure or friction welding and, even in those alloys which are considered to deteriorate when welded, the properties can usually be augmented by heat treatment. Titanium alloys IMI 110–160, 260, 262, 318, 685 and 829 can be centrifugally cast in rammed graphite or investment moulds using a consumable electrode or skull melting furnace.

Tensile and yield strengths are about 5% below those of forgings and dimensional tolerances vary from 0.6% to 2%, depending on process and size. It is becoming increasingly common for castings to be hot isostatically pressed to give properties comparable to those of forgings.

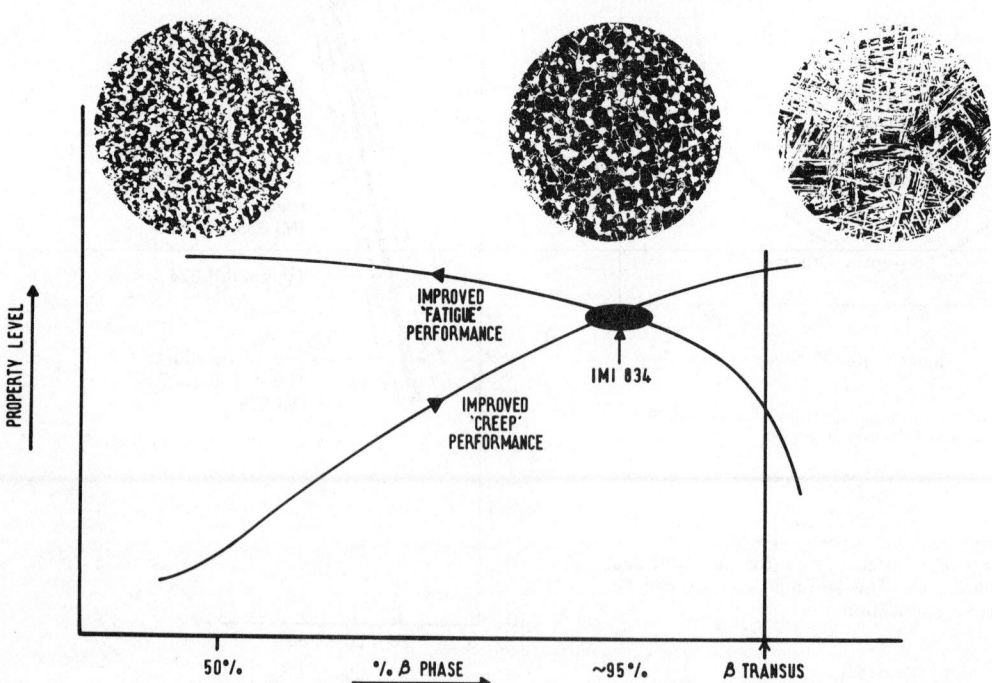

Figure 7.48 Effect of α/β phase proportions on creep and fatigue properties (schematic). (Reproduced by permission of IMI Titanium Ltd)

Table 7.33

Nominal composition	IMI No.	BS	ASTM	Typical mechanical properties		Weldability
				0.2% PS (MPa)	UTS (MPa)	
(a) α-Phase alloys of titanium						
Ti-0.05 O_2/N_2	110		Grade 1 (B265, 337, 338, 348, 367, 381)[a]	190	310	
Ti-0.07 O_2/N_2	115	2TA.1		250	375	
Ti-0.13 O_2/N_2	125	2TA.2, 3, 4, 5	Grade 2 (ditto)	330	465	
Ti-0.2 O_2/N_2	130		Grade 3 (ditto)	410	540	
Ti-0.28 O_2N_2	155	2TA6	Grade 4 (B265)	540	650	
Ti-0.30 O_2/N_2	160	2TA7, 8, 9	Grade 4 (B348, 367, 381)	500	670	Excellent
Ti-0.15 Pd 0.07 O_2/N_2	260		Grade 11 (B265, 337, 338, 348, 381)	220	330	
Ti-0.15 Pd 0.13 O_2/N_2	262		Grade 7 (B265, 337, 338, 348, 381)	350	465	
Ti-2.5 Cu	230	2TA 21–24 TA 52–55, 58		530[b] 667(STA)[b]	620[b] 720(STA)[b]	
Ti-0.3 Mo 0.8 Ni	Grade 12		Grade 12 (265, 337, 338, 348, 367, 381)	400 400	500 500	
Ti-5 Al 2.5 Sn	317	TA14–17	Grade 6 (265, 337, 338, 348, 367, 381)	850 850	950 950	Very good/ Excellent
(b) β-Phase alloys of titanium						
Ti-6 Al, 4V	318	2TA11–13, 28 TA 56, 59	Grade 5 (B265, 337 338, 348, 367, 381)	930(a) 1070(STA)	1030(a) 1130(STA)	Good
Ti-4 Al, 3Mn, 2Sn 0.5Si	550	TA45–51, 57		1040(STA)	1170(STA)	Poor, not normally welded except by diffusion
Ti-4Al 4Sn 0.5Si	551	TA38–42		1200(STA)	1310(STA)	
Ti-6Al, 6V, 2Sn	(662)			960(a)	1020(a)	
Ti6Al, 2Sn, 4Zr, 6Mo	(6246)			1080(STA) 1145(STA)	1150(STA) 1260(STA)	
Ti-5Al, 2Sn, 2Zr, 4Mo, 4Cr	(T.17)			1090(STA)	1170(STA)	Poor, not normally welded
Ti-4.5Al, 5Mo, 15Cr, 0.13 O	(Corona 5)			890(a) 1172(STA)	956(a) 1255(STA)	
(c) Near-α phase titanium alloys						
Ti-2.25Al, 11Sn, 5Zr1Mo 0.25Si	679	TA18.20, 25, 26, 27		1025	1230	Poor
Ti-6Al, 2Sn, 4Zr, 2Mo, 0.1Si	(62425)			950	1050	Poor
Ti-8Al, 1Mo,-1V	(811)			975	1095	Fair
Ti-6Al, 5Zr, 0.6Mo, 0.25Si	685	TA 43, 44		900	1020	Very good/ excellent
Ti 5.6Al, 3.5Sn, 3Zr, 1Nb, 0.25Mo, 0.3Si	829			850	1000	Very good/ excellent
Ti-5.8Al-4.4Sn, 3.5Zr, 0.7Nb, 0.5Mo, 0.35Si, 0.06C	834			950	1090	Very good/ excellent
(d) Metastable β-phase alloys of titanium						
Ti-15V, 3Cr, 3Sn, 3Al				1450	1500	Fair
Ti-10V, 2Fe, 3Al, 0 < 0.16				1115	1170	–

Designations in parentheses are not IMI numbers.

[a] In US specifications
B265 Plate Sheet and Strip
B337 Seamless and Welded Pipe
B338 Seamless and Welded Tube
B348 Bar and Billet
B367 Castings
B381 Forgings

[b] Annealed: (STA) Solution Treated and Aged

Table 7.34 Properties of IMI 829 and 834

Alloy No.	Temperature (°C)	0.2% PS min (MPa)	UTS min (MPa)	Elongation (min)	Stress for 0.1% total plastic strain in 100 h (MPa)	Modulus of elasticity (GPa)	Endurance limit (% of UTS)	Fracture toughness K_{Ic} (MNm$^{-3/2}$)	Specific gravity	Coefficient of thermal expansion (20–1000°C K^{-1})
829	20	820	950	10		~120	50	78	4.54	10.4×10^{-6}
	540	450	590	12	200	~93				
834	20	910	1030	6		~120	50	37	4.59	11.3×10^{-6}
	600	450	585	9	150	~93				

7.4.3.5 Applications

So-called 'commercially pure' titanium (IMI 110–160) is primarily a chemical plant vessel and piping material, resisting sea water, halogen compounds, oxidizing and organic acids and many gaseous environments. It is also used for non-consumable anodes, sometimes coated with precious metal, and fitments for metal finishing and other electrochemical operations. IMI 260 or 262 may be substituted if there is any possibility that the environment may become reducing in nature. A coarse-grain faceted structure can be produced which can be selectively anodized and used for decorative and jewellery products. The ultimate tensile strengths of these alloys range up to 650 MPa so that they can be used for less highly stressed aircraft components. Aircraft engine casings and bypass ducts require stronger and more creep-resistant material, and for these the higher-strength more creep-resistant readily formed and welded alpha-strengthened IMI 230 or, if there is no severe forming operation involved ASTM Grade 5 (IMI 318) may be used. $\alpha + \beta$ alloys (for example, IMI 550) have much higher tensile and fatigue strengths (which are exceeded only by the less stable β-strengthened alloys) and are used for discs, blades and highly stressed airframe components.

The creep properties of the near-α alloys have already been described. They are used for the higher-temperature compressor discs, rings, blades and impellers at temperatures approaching 600°C. Above this temperature the limitation for titanium appears to be in the field of oxidation rather than strength. Oxidation-resistant noble metal ion plated coatings and Ti-Al and Ti-Nb-Al alloys or compounds are showing promise.

7.4.4 Magnesium and its alloys

7.4.4.1 Introduction

Magnesium is the least dense of the engineering alloys and, because of this, the specific strengths of its alloys are superior to those of aluminium and medium-strength steels.[33,34] More important, although the specific rigidities $E\rho^{-1}$ of the alloys differ little from those of aluminium and steel, the stiffness in bending of a section of equal weight (which is related to $E\rho^{-3}$) is far superior.

Substantial savings in weight are possible for designs for which rigidity is a significant criterion. A further advantage of magnesium is its high damping capacity which helps to minimize vibration fatigue. Further scope for weight reduction in comparison with aluminium arises from the excellent castability of some alloys for components where section thickness is determined by casting considerations. Die casting benefits from the fact that dies are not attacked by molten magnesium as they are by molten aluminium. Magnesium is readily extruded and forged at elevated temperatures. It is highly machinable but suffers from a risk of fire and even explosion unless precautions are taken to prevent accumulation of swarf.

These advantages are (or have in the past been) offset by three major drawbacks. The major one compared with aluminium is the less protective nature of the oxide. The early magnesium alloys had very poor corrosion resistance, and although this disability has to a large extent been overcome in more recently developed alloys, the liability to galvanic corrosion has not been lessened and care must be taken to prevent it. A second drawback arises from the hexagonal structure of the magnesium crystal, which prevents fabrication by cold working. A third was the higher price of magnesium compared with aluminium which, even allowing for the lower specific gravity, confined its use to applications in which weight reduction is at a premium. These include aerospace, components for high-performance automobiles, materials handling, portable tools, high-speed machinery and high-value portable consumer goods. Recently, however, the relative costs of the two metals have reversed and magnesium components may cost less than aluminium.

7.4.4.2 Alloy designation and standards

Magnesium alloys are usually referred to by their ASTM designations which, together with their temper designations, are given in a four-part code. The alloy designation consists, first, of *letters* which identify the major alloying elements, given in order of decreasing concentration. The letters are followed by *numbers* which give, to the nearest percentage, the weight percentage of the alloying elements. The code which relates the letters to alloying elements is as follows:

A	aluminium	N	nickel
B	bismuth	P	lead
C	copper	Q	silver
D	cadmium	R	chromium
E	rare earth	S	silicon
F	iron	T	tin
H	thorium	W	yttrium
K	zirconium	Y	antimony
L	lithium	Z	zinc
M	manganese		

In the ASTM system only the *two most concentrated* alloying elements are designated. Thus MEL alloy ZCM711, which contains 6.5 wt% Zn–1.2 wt% Cu–0.7 Mn has the ASTM designation ZC71. The *third* part of the alloy designation consists of a letter which refers to a standard alloy within the broader composition range specified by the first two parts of the designation. The *temper* designation, which is separated from the first two or three parts of the code by a hyphen, is the same as that adopted for aluminium alloys. This is given in full in ASTM specification B296-67. Thus a complete designation might read as

QH21A-T6

The alloy contains approximately 2 wt% silver and 1 wt% thorium; its exact composition is covered by the specification for the 'A' version of the alloy; and it is in the solution-treated and artificially aged condition. The alloy designation is given in full in ASTM specification B275. ASTM specification numbers comprise:

- B92 magnesium ingot
- B93 magnesium alloy ingot
- B94 die castings
- B50 sand castings
- B199 permanent mould castings
- B403 investment castings
- B90 sheet and plate
- B91 forgings
- B107 extruded bar shapes and tubes

BS 2970 lists alloys, cast and wrought and has a code for conditions similar to ASTM with the addition of M for as-cast. The designations of the major British supplier, Magnesium Electron, are roughly similar to those of the ASTM. Tables 7.35 and 7.36, respectively, list the cast and wrought alloys with their magnesium electron designations, British and ASTM specifications, principal mechanical properties and characteristics.

7.4.4.3 Development

The first magnesium alloys contained aluminium up to 10% and sometimes also up to 6%. These alloying additions increase strength and give precipitation-hardening properties. The addition of manganese reduced iron pick-up and so improved corrosion resistance. Further development was aimed at improving creep and corrosion resistance, both of which were very poor in the earlier alloys.

Addition of zirconium refines the grain structure and improves the strength and ductility, both hot and cold. About 0.5% of this alloying addition together with additions of thorium, rare-earth metals, silver and/or copper provide a combination of strength, castability and high-temperature tensile and creep strengths comparable with the high-temperature aluminium alloys.

Improvements in corrosion resistance have resulted from an understanding of the effect of heavy-metal impurities. Aluminium- and zinc-containing casting alloys are now produced with the following impurity limits: $Fe < 0.005\%$, $Ni < 0.001\%$, $Cu < 0.015\%$ with $0.15\% < Mn < 0.25\%$ to suppress iron pick-up. As a result, the high-purity versions, ASTM designations AZ1D and AM60B, have corrosion rates in salt solution only about 2% of those of the original alloys. Alloy WE54 is unique among the creep-resisting alloys in that its corrosion resistance is excellent and on a par with that of the high-temperature aluminium casting alloys.

7.4.4.4 Casting of magnesium alloys

A major problem in the manufacture of magnesium castings has been the prevention of oxidation of the molten metal which has, in the past, been prevented by melting under a layer of flux. It has always been difficult to prevent entrainment of flux, and the possibility of flux inclusions in castings has worsened the problems arising from corrosion. New techniques of fluxless melting and stirring have been developed and the introduction of low-pressure casting as used for aluminium alloys should bring further improvements.[35]

7.4.4.5 High-temperature strength of magnesium alloys

The stress to produce 0.2% total strain at 1000 hours is shown in Table 7.37 and the variation of ultimate tensile and proof stress of the creep resisting cast alloys of magnesium in Figure 7.49. These properties are adequate for many aerospace and automobile engine components.[36] A creep-resistant forging alloy ZT is available but its application is restricted compared with that of castings.

Table 7.37 Stress to produce 0.2% total strain in 1000 hours in creep-resisting cast magnesium alloys

Temperature (%)	150	204	260	315
Alloy		Stress	MPa	
WE54	88	40	23	
HK31		55	24	6.7
EZ33		38	18	5.5
HZ32		44	28	12
QE22		31	12	
ZCM630	61	41		

Courtesy of Mr W Unsworth, Magnesium Elektron Ltd.

7.4.4.6 Corrosion

The improved corrosion-resistant alloys of magnesium have been referred to in Section 7.4.4.3. The high-purity casting alloys have performed at least as well as high-temperature aluminium alloys in salt fog corrosion tests. Painted diecast high-purity AZ91D has shown negligible scribe corrosion creepages and excellent stone chip resistance.

Unfortunately, the galvanic corrosion of magnesium is not affected, and great care must be taken in design not to place magnesium in electrical contact with a metal with a significantly more positive electrode potential. Where juxtaposition is unavoidable for service in a corrosive atmosphere, reliable insulation should be inserted between the two components, or the magnesium component should be protected in some other way. This applies particularly to riveted joints. Because of the limitation on ductility imposed by the hexagonal crystal structure, magnesium cannot be used for rivets. Rivets of aluminium which has a relatively small positive electrode potential to magnesium are used. Steel or copper rivets must not be employed.

7.4.4.7 Applications

Improvements in high-temperature and corrosion resistance have increased the possible applications of cast magnesium alloys in automobile, aerospace and other applications where reduction in weight is significant. These include helicopter gearboxes, aircraft engine casings, high-performance car and motorcycle components (including wheels), computer parts, military equipment and video and conventional cameras. The applications of wrought alloys are more restricted because of the problems of cold forming and riveting, but include railings, ladders, brackets and cans for 'magnox' reactors.[37,38]

7.4.5 Nickel and its alloys

7.4.5.1 General

Although the major proportion of nickel mined is used as an alloying agent for ferrous metals, nickel as a major constituent

Table 7.35 Casting alloys of magnesium

Typical chemical composition – Major alloying elements (%)	Elektron alloy	Condition	Tensile properties[a]			Compressive properties		Fatigue endurance values[c]		Hardness
			0.2% Proof stress (MPa)	Tensile strength (MPa)	Elongation[b] (%)	0.2% Proof stress (MPa)	Ultimate strength (MPa)	Unnotched (MPa)	Notched (MPa)	Brinell
Y 5.25 Nd and other heavy rare earth metals 3.5 Zr 0.5	WE54	Solution and precipitation treated								
		Sand cast	185	255	2	—	—	95–100	—	80–90
		Chill cast	185	255	2	—	—	—	—	—
Rare earth metals 3.0, Zn 2.5, Zr 0.6	ZRE1	Precipitation treated								
		Sand cast	95	140	3	85–120	275–340	65–75	50–55	50–60
		Chill cast	100	155	3					
Zn 4.2, rare earth metals 1.3, Zr 0.7	RZ5	Precipitation treated								
		Sand cast	135	200	3	130–150	330–365	90–105	75–90	55–70
		Chill cast	135	215	4					
Zn 5.8, rare earth metals 2.5, Zr 0.7	ZE63	Solution and precipitation treated								
		Sand cast	170	275	5	190–200	430–465	115–125	70–75	60–85
Th 3.0, Zn 2.2, Zr 0.7	ZT1	Precipitation treated								
		Sand cast	(85)	185	5	85–100	310–325	65–75	55–70	50–60
		Chill cast	(85)	185	5					
Zn 5.5, Th 1.8, Zr 0.7	TZ6	Precipitation treated								
		Sand cast	155	255	5	150–180	325–370	75–80	70–80	65–75
		Chill cast	155	255	5					
Ag 1.5 Nd rich rare earth metals 2.0 Zr 0.6 Cu 0.07	EQ21A	Solution and precipitation treated								
		Sand cast	175	240	2	165–200	310–385	100–110	60–70	70–90
		Chill cast	175	240	2					
Ag 2.5, Nd rich rare earth metals 2.5, Zr 0.6	MSR-B	Solution and precipitation treated								
		Sand cast	185	240	2	165–200	310–385	100–110	60–70	70–90
		Chill cast	185	240	2					
Ag 2.5, Nd rich rare earth metals 2.0, Zr 0.6	QE22 (MSR)	Solution and precipitation treated								
		Sand cast	175	240	2	165–200	310–385	100–100	60–70	70–90
		Chill cast	175	240	2					
Al 8.0, Zn 0.5, Mn 0.3	A8	As cast								
		Sand cast	(85)	140	2	75–90	280–340	75–85	58–65	50–60
		Chill cast	(85)	185	4					
		Solution treated								
		Sand cast	80	200	7	75–90	325–415	75–90	60–70	50–60
		Chill cast	80	230	10					
Al 9.0, Zn 0.5, Mn 0.3, Be 0.0015	AZ91	plus Be Die cast	(150)	(200)	(1)					
Al 9.5, Zn 0.5, Mn 0.3	AZ91	As cast								
		Sand cast	(95)	125	—	85–110	280–340	77–85	58–65	55–65
		Chill cast	(100)	170	2					
		Solution treated								
		Sand cast	80	200	4	75–110	185–432	77–92	65–77	55–65
		Chill cast	80	215	5					
		Solution and precipitation treated								
		Sand cast	120	200	—	110–140	385–465	70–77	58–62	75–85
		Chill cast	120	215	2					
Al 7.5–9.5, Zn 0.3–1.5, Mn 0.15 min.	C	As cast								
		Sand cast	(85)	125	—	65–90	278–340	73–80	58–65	50–60
		Chill cast	(85)	170	2					
		Solution treated								
		Sand cast	(80)	185	4	75–90	330–415	77–85	62–73	50–60
		Chill cast	(80)	215	5					
		Solution and precipitation treated								
		Sand cast	(110)	185	—	90–115	340–432	62–73	58–62	70–80
		Chill cast	(110)	215	2					
Zn 5.5–6.5 Cu 2.4–3.0 Mn 0.25–0.75	ZCM 630-T6	As sand cast	125	210	2			94	57	55–65

The tensile properties quoted are the specification minima for the first specification listed for that alloy and condition. The ranges given are the specified minima; values in parentheses are for information only.

[a] The values quoted are for separately cast test bars and may not be realized in certain portions of castings.
[b] Elongation values are based on a gauge length of $5.65\sqrt{A}$ except in the case of thin material where a gauge length of 50 mm may be used (see BS 2 L.500, 3370 and 3373). With the latter gauge length, elongation requirements for sheet and plate depend on thickness and a range of minima is quoted.
[c] Endurance values for 50×10^6 reversals in rotating bending-type tests; semi-circular notch, radius 1.2 mm; SCF approx. 2.0. Reversed bending for sheet.

Courtesy of Magnesium Elektron Ltd.

| Description | Min. of Def. Procurement Executive (DTD Series) | Specifications | | | ASTM | |
| | | British Standards | | Alloy designation and temper | Specification |
		Aircraft	General Engineering		
Excellent retention of strength after long exposure at 250°C Good castability, weldable Good corrosion resistance	— —	— —		WE54A-T6 —	
Creep-resistant up to 250°C. Excellent castability. Pressure tight and weldable	— —	2 L.126 2 L.126	2970 MAG6-TE 2970 MAG6-TE	EZ33A-T5	B80-76
Easily cast, weldable, pressure tight, with useful strength at elevated temperatures	— —	2 L.128 2 L.128	2970 MAG5-TE 2970 MAG5-TE	ZE41A-T5	B80-76
Excellent castability, pressure tight and weldable with high developed properties in thin wall castings	5045	—	—	ZE63A-T5	
Creep-resistant up to 350°C. Pressure tight and weldable	5005A 5005A	— —	2970 MAG8-TE 2970 MAG8-TE	HZ32A-T5	B80-76
Stronger than, but as castable as RZ5, weldable, pressure tight	5015A 5015A	— —	2970 MAG9-TE 2970 MAG9-TE	ZA62A-T5	B80-76
	— —	— —	— —		
Heat-treated alloys with high yield strength up to 200°C. Pressure tight and weldable	5035A 5035A	— —	— —		
	5055 5055	— —	— —	QE22A-T6	B80-76
General-purpose alloy. Good founding properties. Good ductility, strength and shock resistance. Also available as a high-purity grade	— — — —	— — 3 L.122 3 L.122	2970 MAG1-M 2970 MAG1-M 2970 MAG1-TB 2970 MAG1-TB	AZ81A-F AZ81A-T4	B80-76
General-purpose pressure diecasting alloy. Draft ISO specification	—	—	—	AZ91B-F	B94-76
	— —	— —	2970 MAG3-M 2970 MAG3-M	AZ91C-F	
General-purpose alloy. Good founding properties. Suitable for pressure die castings	— —	3 L.124 3 L.124	2970 MAG3-TB 2970 MAG3-TB	AZ91C-T6	
	— —	3 L.125 3 L.125	2970 MAG3-TF 2970 MAG3-TF		
	— —	— —	2970 MAG7-M 2970 MAG7-M		
General-purpose alloy with good average properties	— —	— —	2970 MAG7-TB 2970 MAG7-TB		
	— —	— —	2970 MAG7-TF 2970 MAG7-TF		
Good founding properties Good creep resistance	—	—	—	—	—

Table 7.36 Wrought magnesium alloys

Typical chemical composition – Major alloying elements (%)	Elektron alloy	Condition	Tensile properties[a]			Compressive properties 0.2% Proof stress (MPa)	Compressive strength (MPa)	Fatigue properties[c]	
			0.2% Proof stress (MPa)	Tensile strength (MPa)	Elongation[b] (%)			Unnotched (MPa)	Notched (MPa)
	ZCM711								
		Extruded bars and sections 0–13 mm diameter	160	240	7	—	—	—	—
Zn 6.5, Cu 1.2, Mn 0.7		As extruded							
		Precipitation treated	200	250	5	—	—	—	—
		Fully heat treated	300	325	3	—	—	—	—
		Forgings[d]	—	—	—	—	—	—	—
	ZW3								
		Extruded bars and sections 0–10 mm	200	280	8	—	—	—	—
Zn 3.0, Zr 0.6		10–100 mm	225	305	8	200–250	385–365	110–135	85–95
		Extruded forging stock 0–10 mm	195	280	8	—	—	—	—
		10–100 mm	205	290	8	—	—	—	—
		Forgings[d]	205	290	7	165–215	370–340	—	—
	AZM								
		Extruded bars and sections and extruded forging stock							
Al 6.0, Zn 1.0, Mn 0.3		0–75 mm	180	270	8	130–180	370–420	125–135	90–95
		75–150 mm	160	250	7	115–165	340–400	—	—
		Extruded tube	150	260	7	130–180	—	—	—
		Forgings[d]	160	275	7	130–165	340–400	115–125	80–90
Al 8.5, Zn 0.5, Mn 0.12 min	AZ80	Forgings – precipitation treated	200	290	6	—	—	—	—
	AZ31								
Al 3.0, Zn 1.0, Mn 0.3		Sheet – soft 0.5–6.0 mm	(120)	220–265	10–12	—	—	—	—
		Extruded bars and sections 0–10 mm	150	230	8	—	—	—	—
		10–75 mm	160	245	10	—	—	—	—
	ZTY								
Th 0.8, Zn 0.6, Zr 0.6		Extruded forging stock 0–25 mm	130	230	6	—	—	75	45
		25–50 mm	110	200	8	—	—	—	—
		above 50 mm	95	200	8	—	—	—	—
		Forgings[d]	130	230	6	—	—	—	—

Larger sizes than those shown above are available: when required, property levels will be by agreement.

[a] The tensile properties quoted are the specification minima for the first specification listed for that alloy and condition. Where a range is quoted the specification requirements depend on sheet thickness. Bracketed values are for information only.

[b] Elongation values are based on a gauge length of 5.65 $\sqrt{A}$ except in the case of thin material where a gauge length of 50 mm may be used (see BS 2 L.500, 3370 and 3373). With the latter gauge length, elongation requirements for sheet and plate depend on thickness and a range of minima is quoted.

Impact value		Hardness			Specifications				
				Min. of Def. Procurement Executive (DTD Series)	British Standards		ASTM		
Unnotched (J)	Notched (J)	(v.p.n.)	Description		Aircraft	General Engineering	Alloy designation	Standard No.	
—	—	—	The highest strength magnesium wrought alloy when fully heat treated. Weldable	—	—	—	ZC71A-T6		
—	—	—		—	—	—			
—	—	—		—	—	—			
—	—	—		—	—	—			
—	—	65–75		—	2 L.505	3373 MAG-E-151M			
23–31	9.5–1.2	65–75	High strength extrusion, and forging alloy. Weldable under good conditions	—	2 L.505	3373 MAG-E-151M	ZK30A		
—	—	65–75		—	L.514	3372 MAG-E-151M			
—	—	65–75		—	L.514	3372 MAG-E-151M			
6.27	4.7–9	60–80		—	L.514	3372 MAG-F-151M			
34–43	6.7–9.5	60–70		—	L.512 and 3	3373 MAG-E-121M		}B107-76	
—	—	55–65	General-purpose alloy. Gas and arc weldable	—	L.512 and 3	3373 MAG-E-121M	AZ61A-F	}B91-72	
—	—	60–70		—	2 L.503 and	3373 MAG-E-121M			
16–23	3.4–4	60–70		—	L.513	3373 MAG-F-121M			
—	—	60	High-strength alloy for forgings of simple design	—	—	—	AZ80A	B91-72	
—	—	50–65	Medium-strength sheet and extrusion alloy.	—	—	3370 MAG-S-1110	AZ31B-O	B70-90	
—	—	50–65	Good formability.	—	—	3373 MAG-E-111M	}AZ31B-F	B107-76	
—	—	50–60	Weldable	—	—	3373 MAG-E-111M			
—	—	50–60		5111	—	—			
—	—	50–60	Creep resistant up to 350°C	5111	—	—	HZ11A		
—	—	50–60		5111	—	—			
—	—	50–65	Fully weldable	5111	—	—			

[c] Endurance values for 5×10^6 reversals in rotating bending-type tests; semi-circular notch, radius 1.2 mm; S.C.F. approx. 2.0. Reversed bending for sheet.
[d] Forging properties quoted are those in the most favourable direction of flow; the manufacturer should be consulted on directionality.

Courtesy of Magnesium Elektron Ltd.

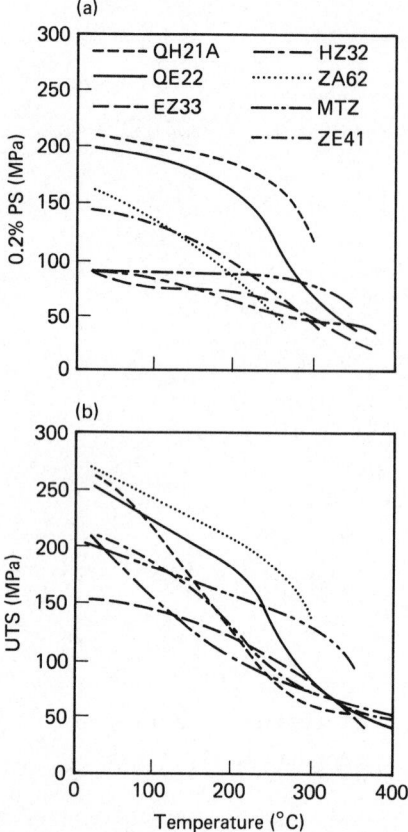

Figure 7.49 Effect of temperature on (top) the tensile 0.2% proof stress and (bottom) the ultimate tensile strength of various magnesium cast alloys. QH21 is QH22 with part of the neodymium replaced by thorium: MT2 is HZ32 with no zinc. (Reproduced by permission of Mr W. Unsworth, Magnesium Elektron Ltd)

forms many alloys which have a very wide range of outstanding properties.

The face-centred-cubic lattice of nickel persists without allotropic change from very low temperatures to its melting point. This confers ductility down to cryogenic temperatures and strength up to 70% of the melting point, which makes nickel an outstanding base for creep-resistant materials.

Pure nickel has excellent corrosion resistance to non-oxidizing media, and this property is enhanced by the addition of copper, chromium and molybdenum. Oxidation and scaling resistance is conferred by the addition of chromium and adherence of the chromic oxide surface layer is improved by rare-earth additions.

The range of properties available in nickel alloys includes very low thermal expansion coefficients and almost constant elastic moduli over limited ranges of temperature. Other alloys have a wide range of electrical resistance and in some this is almost constant over a range of temperature. Excellent magnetic properties are available but only at low magnetic induction.

The major drawback of nickel alloys is their cost. Nickel is not widely distributed and some of the more abundant depo-

sits are expensive to process. The cost of the metal is augmented by that of some of the alloying additions. The embrittling action of some impurities requires their content to be kept very low. This complicates the refining and scrap-segregation procedures and further increases cost.

The manufacturing cost of components from nickel alloys is high, partly because of their poor castability, partly because of their high resistance to deformation at elevated temperatures and partly because of their high rate of work hardening. Very powerful equipment operating at high temperatures is required to work nickel alloys and they are hard to machine. The resistance of nickel to oxidizing gases is poor and the action of sulphidizing gases catastrophic.

There are six main groups of nickel alloys. These comprise: corrosion-resistant, high-temperature, electric, magnetic, controlled physical property and hard-facing alloys. There are also a few miscellaneous types.

7.4.5.2 Nomenclature and standards

Some of the more important alloys are covered by national specifications, but among engineers they are normally referred to by the designation given by their supplier, usually a subsidiary of the International Nickel Corporation. Sometimes the identical material is available from another supplier, and in this case it is usually referred to by the same number but without the trade name (e.g. 'Alloy 800' instead of 'Incoloy 800').

Most wrought alloys are provided in all the semi-finished forms and may also be available as castings. Cast alloys are usually available only in that form. Such British and ASTM standards as exist are listed in Table 7.38.

Table 7.38 British and ASTM standards for nickel and nickel alloys

---

Refined nickel, principally cathodes, briquettes and pellets.
BS 375: five grades (99.5–99.95) of refined nickel
ASTM B39: 99.8% refined nickel
Depending on grade, nickel may contain from 0.005% to 1.5% Co, 0.05% to 0.1% C and 0.002% to 0.15% Cu.
Minor impurities are restricted to values between 0.00002% and 0.002%.

Nickel alloy castings:
BS 3071 Nickel–copper (~30%) castings
ASTM 494 Chemical composition and tensile requirements for 11 casting alloys

Wrought nickel alloys:

BS 3072	Nickel and nickel alloy sheet and plate
BS 3073	Nickel and nickel alloy strip
BS 3074	Nickel and nickel alloy seamless tube
BS 3075	Nickel and nickel alloy wire
BS 3076	Nickel and nickel alloy bar

ASTM B564	Nickel alloy forgings
ASTM B161	Nickel seamless pipe and tube
ASTM B162	Nickel plate sheet and strip
ASTM B168	Nickel chromium iron alloys plate sheet and strip
ASTM B69	Welded nickel alloy pipe
ASTM B670	Precipitation hardening nickel alloy plate sheet and strip
ASTM B335	Nickel molybdenum alloy rod

7.4.5.3 Corrosion-resistant alloys

The corrosion-resistant alloys include:

- 'Nickel', a name which covers a range of general-purpose corrosion-resistant materials.
- 'Monels', essentially nickel alloys with 30% copper which have exceptional resistance to aqueous environments.
- 'Hastelloys' (a Cabot Corporation trade name), nickel–molybdenum alloys which have excellent resistance to acids at high temperature.
- 'Iliums', nickel–chromium–molybdenum copper alloys which resist sulphuric acid.
- 'Inconels', nickel–chromium–iron alloys and 'Incoloys', nickel–chromium–molybdenum alloys, which are really high-temperature alloys but inconel 625 and incoloy 800 have exceptional resistance to stress corrosion and incoloy 825, with additional copper resists strong mineral acid.

The corrosion-resistant alloys are listed, with their standard designations, tensile strengths and dominant characteristics in Table 7.39.

7.4.5.4 High-temperature alloys

The high-temperature alloys of nickel (with the exception of the dispersion-strengthened composites 'SDO') can be regarded either as based on Nimonic 75 (80/20 nickel chrome) or as an extension of the austenitic steels with progressively reducing contents of iron and progressively increasing strengthening additions. There are two classes:

(1) *High-temperature corrosion-resistant alloys*: these have relatively low hot strength (for a nickel alloy) but good scaling resistance (see Table 7.39). They include:
   (a) Nimonic 75 (originally developed as Brightray wire) but available in other forms;
   (b) Inconel 600 and 601 with higher scaling resistance are similar to Brightray but contain some iron;
   (c) 50/50 nickel chrome available as castings or as a cladding material (Incoclad) and IN657 (with niobium) available as castings are the materials with the highest resistance to fuel ash corrosion (except for precious metals).
(2) *Creep and corrosion-resistant alloys (nickel superalloys)*: these are alloys of nickel and chromium, or nickel, chromium and iron with strengthening additions. They are employed anywhere that strength and oxidation resistance are required at high temperature but their development as materials to withstand progressively more severe conditions in gas turbine engine components has stimulated research both in alloy composition and in manufacturing technique. The progress and results of this development are summarized in the following section. The relative rupture strengths of the several alloys produced in different ways are shown in Figure 7.54.

7.4.5.5 Nickel alloys for gas turbines[39]

Materials for forged blades[40–42] The original nickel alloy turbine blades were forged and their creep and creep rupture strengths were improved by increasing the proportion of ordered precipitating phase γ' ($NiAl_3$) and solid solution strengthening additions, chromium, cobalt, molybdenum, tungsten and tantalum. These elements also dissolve in the γ' phase and may have a twofold hardening effect. The compositions must be optimized to obtain the most favourable balance of creep and thermal fatigue strength and oxidation resistance. Unfortunately, increasing content of γ' has to be balanced by a reduction in chromium (which reduces the solubility of aluminium and titanium) but the effect of this on scaling resistance is offset to some extent by the increase in aluminium. At the highest temperatures oxidation resistance is obtained by coating with aluminium.

The increased temperature capacity of the improving forged nimonic alloys is apparent from Figure 7.50 and from the following list:

Nimonic alloys Basic grade(s)	Improved corrosion resistant grade(s)	Operating temperature °C
80 A	81	815
90. Inconel 750	91	920
105		940
115		1010

Increase in γ' content reduces the range of temperature available for hot working between softening and incipient melting. This can be offset to some extent by the addition of cobalt which increases the solvus temperature. Even so, the small margin between maximum preheating temperature (1100°C) and minimum working temperature (1050°C) of Nimonic 115 indicates that this is likely to remain the forged alloy with the best high-temperature properties.

Cast blades Vacuum casting provided an alternative manufacturing route. Problems met in the earlier alloys included cracking due to the separation of massive carbides on solidification and the formation of σ-like intermetallic compounds. However, in one specific alloy, IN100 up to 70% γ' fraction can be achieved. This had the immediate effect of eliminating the need for solid solution strengthening so that a low-density high creep-resistant material became available. Early casts of IN100 embrittled after prolonged exposure at 850°C due to the formation of intermetallic compounds σ, μ, X, π or Laves phases had to be avoided by computerized phase control (Phacomp) techniques.[43,44]

The achievement of yet higher-temperature capability required the addition of a high melting point solid solution strengthener such as tungsten. The further addition of 2% hafnium improved ductility and minimized cracking by increasing the amount of low melting point eutectic in (Martin Marietta) MarM002. This and the alloy IN138 which, although it lacks the best creep rupture properties, is extremely well suited to coating with aluminium and is therefore used where the maximum creep resistance is not essential, were standardized.

The structure of the grain boundaries is important. Too little strengthening impairs creep resistance, too much impairs creep ductility and therefore thermal fatigue. The grain boundary structure is controlled through precipitation of carbides, usually chromium carbides (but substituted with other carbides) and by boron and zirconium additions. Compositions of wrought and cast superalloys are listed in Table 7.40.

Directional solidification[45–48] The next improvement in temperature capability came with directionally solidified blades made in a plant of the type shown in Figure 7.51. The mould is enclosed within a hot zone and the heat extracted through a chill plate. The blade consists effectively of a bundle of crystals each with the (100) direction longitudinal and having random rotational orientations transverse to this direction.

This structure has little effect on creep but, because there are no transverse grain boundaries, very significantly increases rupture ductility, particularly in the highly creep-resistant

Table 7.39 Corrosion-resistant nickel alloys

Alloy	Ni	C	Mn	Fe	S	Si	Cu	Cr	Co	Mo	Al	Ti	Others	BS No.	Developer or proprietor	Form	Tensile strength (MPa)	Corrosion resistance
Low/medium temperature																		
Nickel 200	99.5	0.08	0.18	0.2	0.005	0.18	0.13	—	—	—	—	—	—	3072/6 NA11	International Nickel	Annealed→cold worked	380→550	Aqueous and general
Nickel 201	99.5	0.01	0.18	0.2	0.005	0.18	0.13	—	—	—	—	—	—	3072/6 NA12	International Nickel	Annealed→cold worked	340–410	Stress corrosion and general
Nickel cast	98	0.1	1.0	0.2	—	1.0	—	—	—	—	—	—	Mg 0.1		International Nickel	Cast bar	360–420	Aqueous and general
Monel alloy 400	63.0 min.	0.15	1.0	2.5 max.	0.024 max.	0.5 max.	31.0	—	—	—	—	—	—	3072/6 NA13	International Nickel	Annealed bar	480–620	Seawater, polluted water and acids
Monel alloy 410	66.0	0.2	0.8	1.0	0.008 max.	1.6	30.5	—	—	—	—	—	—	3071 NA1	International Nickel	Cast bar	450–580	Seawater, polluted water and acids
Monel alloy K-500	63.0 min.	0.15	1.5	2.0	0.010 max.	0.5	30.0	—	—	—	2.9	0.6	—	3072/6 NA18	International Nickel	Heat-treated bar	620–760	Higher strength alloy resists seawater
Hastelloy alloy B	bal.	0.09 max.	1.0 max.	5.0 max.	0.03	1.0	—	1.0	2.5	28.0	—	—	V0.3		Cabot Corporation	Solution-treated bar	880	Hydrochloric acid
Hastelloy alloy C276	bal.	0.10	1.0	5.5	0.03	1.0	—	16.0	2.5	16.5	—	—	V0.3		Cabot Corporation	Solution-treated plate	780	Chlorides and hypochlorites
Hastelloy alloy D	bal.	0.12	1.0	2.0	—	10.0	3.0	1.0	1.5	—	—	—	—		Cabot Corporation	Solution-treated cast bar	790	Sulphuric acid
Hastelloy alloy G	bal.	0.05 max.	1.5	19.5	—	1.0 max.	2.0 max.	22.0	2.5 max.	6.5	—	—	W1 max. Nb+Ta2		Cabot Corporation	Annealed sheet	710	Hot sulphuric and phosphoric acids
Hastelloy alloy N	bal.	0.06 max.	0.8 max.	5.0 max.	—	1.0 max.	0.35 max.	7.0	0.2 max.	16.5	0.5 max.	—	B0.01 max.		International Nickel	Solution-treated sheet	700	Molten fluorides
Illium alloy B	bal.	0.05 max.	1.0 max.	1.5 max.	—	3.5 max.	5.5 max.	28.0	—	8.0	—	—	—		International Nickel		415–485	All concentrations sulphuric acid to BP
Illium alloy G	56.0	—	—	—	—	—	6.5	22.5	—	6.5	—	—	—		International Nickel		470	
Illium alloy R	68.0	—	—	—	—	—	3.0	21.0	—	5.0	—	—	—		International Nickel		780	60% sulphuric acid to 50°C
Illium alloy 98	bal.	0.07 max.	1.5	1.5	—	1.25	5.0	28.0	—	8.5	—	—	—		International Nickel		370–540	60% sulphuric acid to BP
Inconel 625	60.5	0.10 max.	0.25 max.	5.0 max.	0.015 max.	0.5 max.	—	21.5	—	9.0	0.25	0.25	Nb+Ta3.65	3072/6 NA16	International Nickel	Annealed cold	830→1040 worked	Seawater
Incoloy alloy 825	42.0	0.05 max.	1.0 max.	Bal.	0.03 max.	0.5 max.	2.25 max.	2.15	—	3.0	0.20	0.9	—		International Nickel	Annealed bar	590–730	Strong mineral acids
High temperature																		
80 Ni-20 Cr	bal.	0.13 max.	1.0 max.	5.0 max.	0.02 max.	1.0 max.	0.5 max.	19.5	—	—	—	0.40	—		International Nickel	Cast and extruded clad	~300	Resist fuel ash corrosion
50 Ni-50 Cr	50	0.1 max.	0.3 max.	1.0 max.	—	0.5 max.	—	50.0	—	—	—	—	—		International Nickel	Cast	600	
IN 657	bal.	—	—	—	—	—	—	48.52	—	—	—	—	Nb1.5		International Nickel	Cast		
Inconel alloy 600	bal.	0.15 max.	1.0 max.	8.0 max.	0.015 max.	0.5 max.	0.5 max.	15.5	—	—	—	—	—	3072/6 NA14	International Nickel	Annealed bar	550–690	
Inconel alloy 601	60.5	0.05 max.	0.5 max.	14.1	0.007 max.	0.25	0.25	23.0	—	—	1.35	—	—		International Nickel	Annealed bar	740	
Incoloy alloy 800	32.5	0.10 max.	1.5 max.	bal.	0.015 max.	1.00 max.	0.75 max.	21.0	—	—	0.38	0.38	—	3072/6 NA15	International Nickel	Annealed bar	590	Originally developed for electric kettle elements
Incoloy alloy DS	37.0	0.1 max.	1.2 max.	bal.	—	2.3 max.	0.5 max.	18.0 max.	—	—	—	—	—		International Nickel	Annealed bar	730	

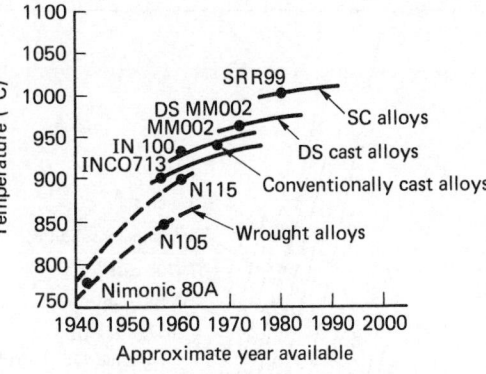

Figure 7.50 Increases in temperature capability for turbine blade alloys, based on creep rupture in 1000 h at 150 MPa. (Reproduced by permission of *Metals and Materials*)

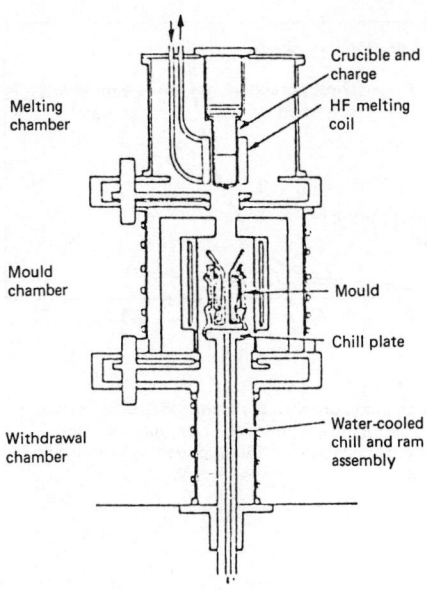

Figure 7.51 Schematic of Rolls-Royce directional solidification plant. (Reproduced by permission of *Metals and Materials*)

alloys such as Mar M200 (see Figure 7.52). This produces greatly increased thermal fatigue life, which is further enhanced by a low elasticity modulus in the longitudinal (100) direction. This allows a greater elastic extension in a thermal stress cycle and reduces $\Delta\epsilon_p$ (see Chapter 8, Section 8.4).

Single crystals Even greater improvements accrue from casting blades as single crystals either by restricting the cross section of the solidifying metal in advance of the moulded blade (see Figure 7.53) or by using a seeding crystal. There are

Table 7.40 Typical compositions of some wrought cast and powder nickel base superalloys

		Co	Cr	Al	Ti	C	Ta	Mo	W	Nb	Fe	Zr	B	Others	Developer/user
Turbine blade alloys															
Wrought	Inconel 600	—	15.5	—	—	0.08	—	—	—	—	8.0	—	—	—	
	Inconel 722	—	15.0	0.6	2.5	0.04	—	—	—	—	7.0	—	—	—	
	Inconel X-750	0.5 max.	15.0	0.9	2.5	0.04	—	—	—	1.0	7.0	0.03	—	—	
	Nimonic 75	—	19.5	—	0.4	0.10	—	—	—	—	5.0 max.	0.05 max.	0.001 max.	—	
	Nimonic 80A	2.0 max.	19.5	1.4	2.5	0.10 max.	—	—	—	—	1.0 max.	0.10 max.	0.005 max.	—	
	Nimonic 90	18.0	19.5	1.5	2.4	0.13 max.	—	—	—	—	1.0 max.	0.10 max.	0.005 max.	—	International Nickel
	Nimonic 105	20.0	14.8	4.7	1.2	0.16 max.	—	5.0	—	—	1.0 max.	0.14 max.	0.008 max.	—	
	Nimonic 115	14.8	15.0	5.0	4.0	0.15	—	4.0	—	—	1.0 max.	0.20 max.	0.06 max.	—	
	Nimonic 120	10.0	12.5	4.5	3.5	0.08 max.	—	5.7	—	—	1.0 max.	0.05 max.	0.025	—	
	Nimonic 81	2.0 max.	30.0	0.9	1.8	0.05	—	—	—	—	1.0 max.	0.06	0.002	—	
	EPK 55	20.0	28.5	1.2	2.3	0.05	—	—	—	0.75	0.5 max.	0.07	0.006	—	
	EPK 57	19.7	24.3	1.4	3.0	0.05	—	1.5	—	0.95	0.5 max.	0.05	0.012	—	
	Udimet 500	16.5	17.5	2.9	2.9	0.15 max.	—	4.0	—	—	4.0 max.	—	0.010 max.	—	
	Udimet 700	17.0	15.0	4.3	3.4	0.15 max.	—	5.3	—	—	4.0 max.	—	0.05 max.	—	
Cast	IN 713 LC	—	12.5	6.1	0.8	0.12 max.	—	4.2	—	2.2†	0.5 max.	0.10	0.012	—	International Nickel
	IN 100	15.0	10.0	5.5	4.7	0.18	—	3.0	—	—	0.5 max.	0.06	0.014	1.0 V	Pratt and Whitney
	B 1900	10.0	8.0	6.0	1.0	0.025	4.0	6.0	—	—	—	0.04	0.015	—	
	MarM200	10.0	9.0	5.0	2.0	0.15	—	—	12.5	1.0	—	0.05	0.15	—	
	Mar M002	10.0	9.0	5.5	1.5	0.15	2.5	—	10.0	—	—	0.05	0.15	1.5 Hf	Martin Marietta
	Mar-M-247	10.0	8.3	5.5	1.0	0.15	3.0	0.67	10.0	—	—	0.05	0.015	Hf	
	IN 738	8.5	16.0	3.4	3.4	0.17 max.	1.8	1.8	2.6	0.9	0.5 max.	0.10	0.010	—	International Nickel
	René 77	15.0	14.5	4.3	3.3	0.07	—	4.2	—	—	—	0.04	0.016	—	General Electric
	René 80	9.5	14.0	3.0	5.0	0.17	—	4.0	4.0	—	—	0.03	0.015		
Mechanically alloyed powder	MA6000	—	12.0	4.5	2.5		2.0	2.0	4.0					1.1Y₂O₃	
Rolled from cast billet	*Sheet alloys*														
	PK16	18.0	16.5	1.2	1.2	0.06	—	3.3	—	—	2.2	0.06 max.	0.05 max.		
	PK33	14.0	18.0	2.9	2.2	0.07	—	7.0	—	—	1.0	—	—		
Forged from cast billet	*Turbine disk alloys*														
	Nimonic 901	13.5	12.5	0.3	2.9	0.04	—	5.7	—	—	3.0	—	—		
	Waspaloy	13.5	19.5	1.3	3.0	0.08	—	—	—	—	—	0.06	0.006		
	Astroloy	17.0	15.0	4.0	3.5	0.06	—	5.25	—	—	—	—	0.03		
Forged from powder	AP1	17.0	15.0	4.0	3.5	0.025	—	5.0				0.04	0.025		

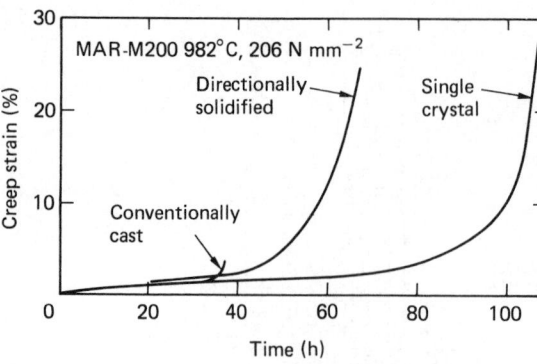

Figure 7.52 Comparison of creep curves for Mar-M200 tested at 1255 K and 206.8=MPa in conventionally cast, directionally solidified and single-crystal forms. (Reproduced by permission of The Metals Society)

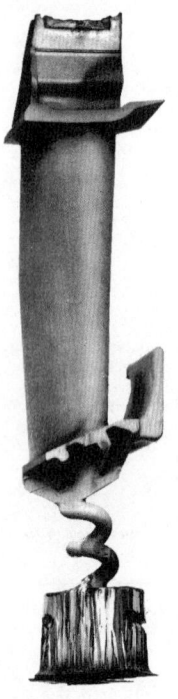

Figure 7.53 Single-crystal blade with spiral constriction. (Reproduced by permission of *Metals and Materials*)

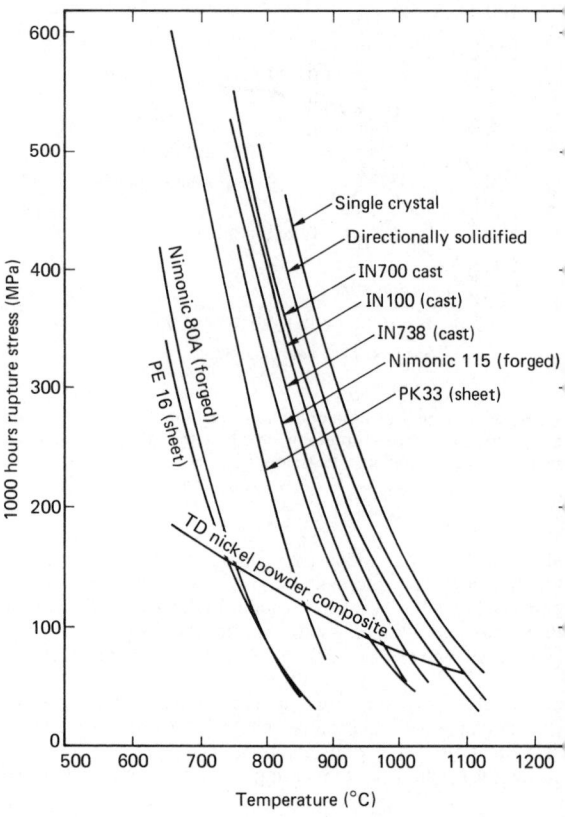

Figure 7.54 Comparative 1000 h to rupture strengths of nickel alloy gas turbine blading materials produced by a variety of techniques

no grain boundaries in a single crystal, consequently elements introduced to strengthen grain boundaries such as carbon, boron, zirconium and hafnium may be omitted. Higher-solution treatment temperatures can therefore be used and stronger more uniform γ′ precipitates produced. Thus, in the UK SRR99 will replace MarM002, low-density RR2,000 will replace IN100 and RR 2060 will be used for nozzle guide vanes.

The kind of performance that may be expected from directionally solidified and single-crystal materials is indicated

in Figure 7.54. The compositions of superalloys that are suitable for directional solidification are listed in Table 7.41. Any significant further advance in gas-turbine blade materials must come from the use of composites.

Composite blades Composites can be produced by casting eutectic alloys in directionally solidifying furnaces or by powder techniques. Directionally cast eutectics normally have a laminar structure but can, theoretically, be seeded to produce fibre reinforcement in a matrix. Typical compositions are listed in Table 7.42. Unfortunately, some directionally solidified eutectics have limited oxidation resistance, some delaminate on thermal cycling and some have uneconomically long production times. None have so far been exploited commercially.

Dispersion-strengthened (ODS) alloys resist creep at very high temperatures (see TD Nickel, Figure 7.54) but it is extremely difficult to produce a material with high strength at lower temperatures. One such alloy MA6000 (see Table 7.40) is said to be strong enough for stubby tapered unshrouded blades which can be run at higher temperatures than other nickel base-alloy blades.

Turbine disk materials Turbine disk materials require good tensile strength (to prevent bursting in the event of an overspeed) and good low-cycle fatigue life (to withstand the changes of stress undergone in the flight cycle) rather than

Table 7.41 Compositions of high-temperature alloys that have been directionally solidified: superalloys

Alloy	Ni	Co	Cr	Al	Ti	C	Ta	Mo	W	Nb	Zr	B	Others
IN713LC	Bal.	—	12.0	5.9	0.6	0.05	—	4.5	—	2.0	0.1	0.01	—
IN713C	Bal.	—	12.5	6.1	0.8	0.12	—	4.2	—	2.0	0.1	0.012	—
IN100	Bal.	15.0	10.0	5.5	4.7	0.18	—	3.0	—	—	0.06	0.014	1.0 V
MarM200	Bal.	10.0	9.0	5.0	2.0	0.15	—	—	12.5	1.0	0.05	0.015	—
MarM002	Bal.	10.0	9.0	5.5	1.5	0.15	2.5	—	10.0	—	0.05	0.015	1.5 Hf
MarM246	Bal.	10.0	9.0	5.5	1.5	0.15	1.5	2.5	10.0	—	0.05	0.015	—
B1900	Bal.	10.0	8.0	6.0	1.0	0.10	4.0	6.0	—	—	0.10	0.015	—
IN738LC	Bal.	8.5	16.0	3.5	3.5	0.11	1.6	1.75	2.5	0.7	0.08	0.008	—
IN939	Bal.	19.0	22.5	1.9	3.7	0.15	1.4	—	2.0	1.0	0.09	0.009	—
MarM247	Bal.	10	8.4	5.5	1.05	0.15	3.3	0.65	10.0	—	0.05	0.015	1.4 Hf
UTRCMMT 143	73.9	—	—	5.8	—	—	6.0	14.3	—	—	—	—	—
X 40	10	Bal.	25.0	—	—	0.5	—	—	7.5	—	—	—	—
MarM509	10	Bal.	23.5	—	0.2	0.6	—	—	7.0	—	0.5	—	—
P & W 444	Bal.	—	8.6	5.1	1.98	—	—	—	11.1	—	—	—	—
P & W 454	Bal.	5.0	10.0	5.0	1.5	—	12.0	—	4.0	—	—	—	—
Single-crystal alloys — NASAIR 100	Bal.	—	9.0	5.75	1.2	<0.01	3.3	1.5	10.5	—	—	—	—
RR SR99	Bal.	5.0	8.5	5.5	2.2	<0.015	2.8	—	9.5	—	—	—	—
RR 2000	Bal.	15.0	10.0	5.5	4.0	0.02	—	3.0	—	—	—	—	1.0 V
RR 2060	Bal.	5.0	15.0	5.0	2.0	0.02	5.0	2.0	2.0	—	—	—	—

Reproduced (with additions) by permission of the Metals Society.

Table 7.42 Compositions of directionally solidified eutectic composites

	Structure		Composition (wt%)												
Alloy	Matrix	Reinforcement	Ni	Co	Cr	Al	Nb	Ta	C	W	Re	V	Mo	Y	Fe
NITAC3116A	γNi+γ'Ni$_3$Al	TaC	Bal.	3.7	1.9	6.5	—	8.2	0.24	—	6.3	4.2	—	—	—
COTAC 744	γNi+γ'Ni$_3$Al	NbC	Bal.	2.0	10.0	4.0	4.9	—	0.6	10	—	—	—	—	—
γ–γ'–Cr$_3$C$_2$	γNi+γ'Ni$_3$Al	Cr$_3$C$_2$	Bal.	—	12.3	6.9	—	—	1.8	—	—	—	—	0.5	—
(Co, Cr)–Cr$_7$C$_3$	Co, Cr solid soln	Cr$_7$C$_3$	—	Bal.	41.0	—	—	—	2.4	—	—	—	—	—	—
γ–γ'–δ	γNi+γ'Ni$_3$Al	δ Ni$_3$Nb	Bal.	—	6.0	2.5	20.1	—	0.06	—	—	—	—	—	—
γ–γ'–α	γNi+γ'Ni$_3$Al	α Mo	Bal.	—	—	6.0	—	—	—	—	—	—	32	—	—

Reproduced by permission of the Metals Society.

creep resistance at high temperatures. The alloys adopted, Astroloy, Waspaloy and Nimonic 901 (see Table 7.40) are similar in composition to turbine blade alloys but are given low-temperature thermomechanical treatments which generate a homogeneous dislocation structure, improve tensile strength by up to 40% and produce a fine-grained microstructure which has good fatigue properties.

Ingots cast from more highly alloyed disk materials are prone to segregation and difficult to forge. It should be possible to overcome this by powder processing but, so far, the presence of defects too small to be detected by NDT has prevented the application of powder techniques to turbine disks. The mechanically alloyed powder API (Table 7.40) is an example of this type of alloy.

Casing materials Sheet materials have compositions similar to blade materials but their creep resistance is usually lower, partly because it is not needed, partly because the high-technology casting procedures are not applicable to sheet and partly because they usually have to be welded. The creep rupture properties of two sheet materials, PE16 and PK33, which span the available range are indicated in Figure 7.54.

The compositions of a selection of available (and potential) nickel gas-turbine alloys are listed in Tables 7.40–7.42.

This account has been based on experience at Rolls-Royce but similar developments, using alloys of marginally different compositions tailored to meet specific design requirements, have been achieved at General Electric and United Technologies.

7.4.5.6 Electrical alloys

Nickel and nickel alloys are used, usually in the form of wire or strip, because of their individual characteristics of resistivity, electron emission, thermoelectric properties or corrosion resistance. Table 7.43 lists these materials, giving composition, supplier, tensile strength, and property of major interest. They are usually supplied in the form of annealed bar for drawing into wire or strip.

7.4.5.7 Magnetic alloys

The magnetic alloys of nickel are principally those with high magnetic permeability in low- or medium-strength magnetic

Table 7.43 Electrical alloys of nickel

Alloys	Ni	C	Mn	Fe	S	Si	Cu	Cr	Co	Mo	Al	Ti	Others	BS No.	Supplier	Form	Tensile strength	
Nickel 205	99.5	0.08	0.18	0.10	0.004	0.08	0.08	—	—	—	—	0.03	Mg0.05	—	International Nickel	Annealed rod	460	High damping. Electronic valve electrodes
Nickel 212	97.7	0.10	2.0	0.05	0.005	0.05	0.03	—	—	—	—	—	—	—	International Nickel	—	476	Resists sulphur embrittlement. Support wires
Nickel 222	99.5	0.01	0.02	0.04	0.0025	0.01	0.01	0.01	0.06	—	0.01	0.01	Mg0.08	3504	International Nickel	—	340	Emits electrons. Valve cathode sleeves
Nickel 240	95.0	—	2.0	—	—	0.45	—	1.7	—	—	—	0.3	Zr0.15		International Nickel	—	—	Resists lead and sulphur. Spark plug electrodes
Nickel 270	99.98	0.01	0.003	<0.001	<0.001	<0.001	<0.001	<0.001	<0.001	—	—	<0.001	Mg0.001		International Nickel	Annealed strip	340	Powder product, highly deformable. Valves, etc.
Constantan (Ferry)	45	—	—	—	—	—	55	—	—	—	—	—	—		ITT (Harlow)	Annealed bar	415	Low temp. coeff. of resistivity. Resistors. Thermocouples
Brightray B	59.0	0.1	1.0 max.	Bal.	—	0.35	—	16.0	—	—	—	0.17	—		Wiggin Alloys	Annealed bar	686	Heating elements up to 950°C
Brightray C	Bal.	0.1 max.	0.25	1.0 max.	—	1.5	—	19.2	—	—	0.26	—	Rare-earth metals 0.05		Wiggin Alloys	Annealed bar	735	Heating elements up to 1150°C
Brightray S	Bal.	0.1 max.	0.4	1.0 max.	—	1.0 max.	—	20.0	—	—	—	—	—		Wiggin Alloys	Annealed bar	734	Strip heating elements up to 1150°C
Brightray 35	37.7	0.1 max.	1.2 max.	Bal.	—	2.2	—	18.0	—	—	—	—	—		Wiggin Alloys	Annealed bar	740	Resists carburization. Heating elements up to 1050°C
Chromel alloy P	Bal.	—	—	0.2	—	0.4	—	10.0	—	—	—	—	—		British Driver Harris			Thermocouples up to 1100°C
Alumel	Bal.	—	1.75	0.1	—	1.2	—	—	—	—	1.6	—	—		British Driver Harris			Thermocouples up to 1100°C
Nicrosil	Bal.	—	—	—	—	1.5	—	14.3	—	—	—	—	—		British Driver Harris			Thermocouples up to 1100°C
Nisil	Bal.	—	—	—	—	4.5	—	—	—	—	—	—	Mg0.1		British Driver Harris			

fields or some special form of magnetic hysteresis loop. They are used mainly in the form of tapes or sheet, or as powder for cores for electronic equipment. Very careful control is needed in production. Pure nickel or nickel-rich cobalt alloys are used as magnetostriction transducers. Compositions, tensile strengths and an indication of magnetic characteristics are listed in Table 7.44. They may be obtained from Telcon Metals, ITT (Harlow) and International Nickel.

7.4.5.8 Alloys with special dimensional and elastic properties

Certain alloys of nickel have minimal or controlled coefficients of expansion over certain ranges of temperature. Alloys with minimal expansion coefficients are used in instruments and those with controlled coefficients for sealing to glass or ceramics in vacuum devices. Alloys with small positive temperature coefficients are used for temperature-insensitive vibrating instrument devices. There is also a low-temperature shape memory alloy Nitinol. The compositions, ultimate tensile strengths and an indication of the characteristic properties of nickel alloys with dimensional property applications are listed in Table 7.45.

7.4.5.9 Hard-facing materials[57,58]

A large number of commercial hard-facing alloys depend on a nickel matrix, usually containing a dispersion of chromium and chromium carbide, boron and sometimes molybdenum and silicon. Hardness of the deposited material varies according to composition between 300 and 720 VPN, and wear resistances can be obtained superior to other materials at room temperature but inferior to cobalt alloys at high temperature or in resistance to aqueous environments.

7.4.6 Zinc and its alloys

7.4.6.1 Introduction and standards

Zinc owes its commercial applications to three characteristics:

1. Zinc components are essentially low priced. The metal requires less energy to produce and to cast than any of its competitors and is cheap to machine.
2. It has quite exceptional corrosion properties. Only aluminium, magnesium and the alkali metals are anodic to it so that it will protect all other metals sacrificially and can be

Table 7.44 Magnetic alloys of nickel

Alloy	Ni	C	Mn	Fe	S	Si	Cu	Cr	Co	Mo	Al	Ti	Others	BS No.	Supplier	Form	Tensile strength (MPa)	Comment
Nickel 205	9.95	0.08	0.18	0.10	0.004	0.08	—	—	—	—	—	0.03	Mg0.05					High magnetostriction. Ultrasonic transducers
Nickel-4% Co	Bal.	—	—	—	—	—	—	—	4.0	—	—	—			Int. Nickel			High magnetostriction. Ultrasonic transducers
Nickel-18% Co	Bal.	—	—	—	—	—	—	—	18.0	—	—	—						
75 Ni-25 Fe	70–80	—	—	Bal.	—	x	x	—	—	x	—	—	Small additions of Si, Mo or Cu		Telcon Metals	Annealed strip	540	
50 Ni-50 Fe	50	—	—	Bal.	—	x	x	—	—	x	—	—		2875A	ITT Harlow	Annealed strip	430	
36 Ni-74 Fe	36	—	—	Bal.	—	x	x	—	—	x	—	—			Wiggin Alloys	Annealed strip	530	
JAE metal	70	—	—	—	—	30	—	—	—	—	—	—	—		Wiggin Alloys	Annealed bar	430	
30 Ni-70 Fe	30	—	—	70	—	—	—	—	—	—	—	—	—			Annealed strip	430	High-temp. coeff. of permeability. Magnetic temp. compensation

Trade names and suppliers of magnetic alloys are Mumetal and Radiometal (Telcon Metals Ltd). Permalloy (ITT Harlow) Nilomag and SAE metal (Wiggin Alloys Ltd).

Table 7.45 Controlled expansion and constant modulus alloys of nickel

Alloy[a]	Ni	C	Mn	Fe	S	Si	Cu	Cr	Co	Mo	Al	Ti	Others	Form	Tensile strength (MPa)	Comment
Nilo alloy 36[1]	36.0	0.5 max.	0.5	Bal.	—	0.5 max.	0.5 max.	—	—	—	—	—	—	Annealed bar	460	Very low thermal expansion 20–100°C. Metrology. Chronometry
Nilo alloy 42[2]	42.0	0.15 max.	0.5	Bal.	—	0.5 max.	0.5 max.	—	—	—	—	—	—	Annealed bar	525	Thermal expansion $\sim 5.5 \times 10^{-6}\,K^{-1}$ 20–300°C. Thermostats and sealing into glass
Nilo alloy 48[2]	48.0	0.15 max.	0.5	Bal.	—	0.5 max.	0.5 max.	—	—	—	—	—	—	Annealed bar	494	Thermal expansion $\sim 9 \times 10^{-6}\,K^{-1}$ 20–400°C. Thermostats and sealing into soft glass
Nilo alloy K[2]	29.5	0.05 max.	0.3	Bal.	—	0.5 max.	0.5 max.	—	17.0	—	—	—	—	Annealed bar	525	Thermal expansion $\sim 6 \times 10^{-6}\,K^{-1}$. Sealing into borosilicate glass
Inconel alloy 903	38.0	—	—	Bal.	—	—	—	—	15.0	—	0.7	1.4	Nb3.0	Warm-worked bar	—	Thermal expansion $\sim 8 \times 10^{-6}\,K^{-1}$. Components stressed at variable temps
Ni-Span alloy C-902[3]	42.25	0.1 max.	0.5	Bal.	—	0.6	—	5.3	—	—	0.55	2.5	—	Heat-treated bar	1240	Low positive coeff. of elastic modulus. For accurate vibrating devices
55 Nitinol[6]	55	—	—	—	—	—	—	—	—	—	—	45	—	Annealed bar	860	
60 Nitinol	60	—	—	—	—	—	—	—	—	—	—	40	—	Annealed bar	940	
														Heat-treated bar	1070	

[a] These are Wiggin designations. Other designations are.
[1]Telcon Metals "Invan" (British Driver Harris Therlo), [2]Telcon Metals Telcoseal (British Driver Harris Therlo) [3]Telcon Metals Elinrar.
[b] Nitinol is the designation of the Naval Ordnance Laboratory, Washington, DC, USA.

used as the anode in an electrolytic cell, but in normal atmospheres it forms a protective film which is penetrated only about 0.01 mm/yr in the worst industrial atmospheres.

3. Its low melting point makes it exceptionally suited as a material for die casting. Its drawbacks are that it is relatively soft and weak, has poor creep resistance and its hexagonal structure makes it brittle at room temperature.

Zinc's major uses are:

- As a coating to protect steel
- As a base for casting alloys
- As a sheet material for battery cases and roofing.

Standards covering zinc metal include:

BS 1004 and ASTM B669. Zinc alloys for casting
BS 3436 Ingot zinc and ASTM B6 Zinc
BS 5338 Code of practice for zinc alloy pressure die casting
ASTM B69 Rolled zinc
ASTM B 418 Cast and wrought galvanic zinc anodes for use in saline electrolytics

7.4.6.2 Zinc-based casting alloys[49,50]

Zinc-base alloys may be used for all types of sand, investment, gravity and pressure die casting. Melting is clean and easy, casting is less sensitive to problems of oxide inclusions, misruns and voids, finish is superior, wall thicknesses can be thinner, detail sharper, and shapes more complex than with such competitive materials as aluminium and copper alloys and cast irons. Zinc alloys have excellent machinability and less machining is required than with castings in other metals. Complex shapes can therefore be produced more cheaply in them than in any other metallic material. A further advantage of the zinc alloys is that for those low-speed bearing applications for which copper alloys have traditionally been used there are indications that zinc casting alloys perform better than leaded gunmetal.

Casting alloys, which usually have aluminium as their main alloying element, are commonly known by their 'ZA' numbers.[51,52] Their composition and properties are listed together with those of competitive materials in Table 7.46.

7.4.6.3 Alloy development

Zinc-base alloys early developed a poor reputation because some castings disintegrated in service and electroplate sometimes showed poor adherence. The cause of disintegration was traced to the presence of heavy metals, iron, lead, cadmium and tin, which are held strictly to acceptable limits in modern high-purity zinc and research on plating eliminated the problems of adherence.

Alloys ZA3 and ZA5 are very suitable for hot-chamber high-pressure die casting because the molten metal has a very low affinity for iron and there is very little wear on the dies. Castings in these alloys are therefore inexpensive to produce and very suitable for complex lowly stressed components.

Increase in aluminium content very significantly increases strength but also increases aggressiveness to iron so that above 8% (the aluminium content of ZA8) the alloys cannot be hot-chamber die cast. Although more expensive in the form of components, the higher aluminium casting alloys are more than competitive with cast iron, copper and aluminium alloys in situations where their low hot strength is not detrimental. ZA12 should not be considered for stressed applications at or above 120°C but the ASME boiler code design stress for ZA27 at 150°C (which is the stress required to produce a secondary

creep of 1% in 100 000 hours) is 69 MPa. As long as these limitations are observed, ZA12 and ZA27 have a potential market for components for mass-produced cars equivalent to that of magnesium alloys for high-performance sports cars.[53]

7.4.6.4 Wrought zinc alloys

Zinc is available in the form of sheet, extrusions (and forgings) and wire. Sheet zinc is used mainly for roofing and as anode cans in the common dry cell. Roofing zinc can be pure, or the more creep-resistant zinc copper or zinc titanium alloys which are an economic replacement for copper or lead. Zinc strip for battery cases contains approximately 0.5% lead and zinc plate for photo-engraving (an important application), 0.2% lead and 0.2% cadmium. Seventy-eight per cent zinc, 22% aluminium sheet can be made superplastic and drawn and stretched into shapes that require a very high degree of deformation.

Extruded zinc is available as the copper–titanium alloy. These alloys are used for architectural purposes in conjunction with roofing or walling sheet, but their applications have been largely superseded by plastics. A major remaining application for pure extruded (or sheet) zinc is as sacrificial anodes to protect steel ships or buried pipes.

Zinc wire is largely used for spray metallizing but is available for nails, hooks, gauze and similar wire products and for solder.

7.4.6.5 Zinc coating for corrosion protection

The major use for zinc is as a coating on steel for corrosion protection. Zinc operates in two ways. Its room-temperature oxidation product is adherent (as opposed to that of steel in moist conditions) but after it has eventually been penetrated it protects the underlying metal electrochemically. It makes an excellent base for paint. There are three types of zinc coating:

- 'Hot dip galvanizing' is achieved by immersing the steel in a bath of molten zinc. The zinc and steel form an alloy at the interface.
- 'Electrogalvanizing' is achieved by electroplating zinc through an aqueous electrolyte. The thickness of coating is more uniform but usually thinner than that produced by other methods.
- 'Sherardizing' (heating the component to 370°C in zinc powder) produces a layer of uniform thickness.

'Zalutite' (a trademark of BSC), 55% aluminium, 43.5% zinc and 1.5% silicon coated steel, combines the protection of zinc and aluminium and, in many environments, is superior to galvanized.

7.4.7 Lead, tin and their alloys

7.4.7.1 Introduction

Lead and tin are soft metals with low melting points. Lead is used pure or strengthened by alloying with antimony for roofing, cable sheathing, radiation shielding, battery electrode and chemical plant materials. Both lead and tin alloys are used for bearings.

Alloys of lead and tin have melting points lower than those of either metal and may be further alloyed with antimony and bismuth to reduce the melting point still more or to confer other properties. Tin is used for coating steel and may be alloyed with lead for this purpose.[54]

Table 7.46 Composition and properties of zinc casting alloys and competitive materials

	ZA3	ZA5	ZA8			ZA12			ZA27				Brass	Aluminium	Cast iron	
Alloy															Blackheart malleable	Grey
Specification	BS 1004	BS 1004	ASTM B-669-82										BS 1004 SCB3	LM6M		
Aluminium	3.9–4.3	3.9–4.3	8–8.8			10.5–11.5			25.0–28.0							
Copper	0.10	0.75/1.25	0.8–1.3			0.5–1.25			2.0–2.5							
Magnesium	0.025/0.05	0.03/0.06	0.015–0.030			0.015–0.030			0.010–0.020							
Iron	<0.075		<0.010			<0.075			<0.10							
Lead						<0.004										
Cadmium						<0.003										
Tin						<0.002										
Zinc						Balance										
Casting condition	Pressure die	Pressure die	Sand cast	Gravity die	Pressure die	Sand cast	Gravity die	Pressure die	Sand cast	Sand H.T.	Gravity die[a]	Pressure die				
Mechanical properties (20°C)																
UTS (MPa^2)	283	324	248–276	221–255	365–386	275–317	310–345	392–414	400–440	310–324	424 / 27.5	407–441	185–240	160–185	290–345	160–345
Elongation (% in 2 inches)	15	9	1–2	1–2	6–10	1–2	1–2	4–7	3–6	8–11	1	1	15–30	5–7	6–12	<0.5
Young's modulus (GPa)	83	92	88	85	—	83	—	—	78	79	—	—	83	71	169	75–145
Hardness (BHN)	83	92	82–89	85–90	99–107	90–110	85–95	95–105	110–120	90–100	110–120	116–122	45–65	55–60	110–149	200–250
Physical properties (20°C)																
Density (gm/cm^3)	6.7	6.7	6.3			6.0			5.0				8.5	2.6	7.3	7.3
Electrical conductivity (%IACS)	26	26	27.7			28.3			29.7				20	37	—	—
Thermal conductivity (W/m°C)	113	110	115			116			125.5				90	142	49	42–50
Melting range (°C)	382–387	379–388	375–404			380–430			380–490				920–1000	580–640	1450–1550	1090–1260

[a] Preliminary data.

7.4.7.2 Nomenclature and standards

Chemical lead: BS 334:   Type A: Pure lead
                      Type B1: Copper lead
                      Type B2: Copper tellurium lead
                      Type C: Antimonial lead
         ASTM B 29: Pig lead

White metal-bearing alloys (Babbitt metal): BS 3332 and ASTM B23
Solder metal: ASTM B 32
Tinplate: BS 2920

7.4.7.3 Lead in corrosion service

Lead forms adherent corrosion products which may be sulphate, oxide, carbonate, chromate or more complex compounds. These coatings protect the base alloy and have led to its use for roofing, underground pipes and contact with sulphuric, sulphurous, chromic and phosphoric acids. The two main drawbacks of lead are its toxicity, which has caused its withdrawal from all applications associated with potable liquids and its low creep strength.

Antimony additions increase both the UTS and the creep strength of lead. The tensile strength of pure lead is about 15 MPa and the stress to cause 1% creep per year at 30°C about 2.1 MPa. Lead containing 8% antimony has a tensile strength of 60 MPa and a stress to cause 1% creep per year at 30°C of 2.7 MPa.

7.4.7.4 Bearing metals (babbitts)

'Babbitt'-bearing metals contain up to 90% tin with antimony, copper and in some cases lead, or up to 90% lead with antimony and tin. Their important characteristics are 'anti-seizure' properties and 'fatigue resistance'.

Anti-seizure properties imply that the material is readily wetted with oil, but that, should the oil film break down, the material will not adhere to a steel journal but flow out of the way locally and embed and cover any hard particle which may have gained access. Both lead and tin have these characteristics.

Fatigue resistance in a bearing implies that the material will, when the bearing is subjected to an alternating load, resist the formation of cracks that initiate at right angles to the bearing surface, propagate almost to the backing and then turn at right angles to form 'loose tiles' that will erode away. Traditional tin-base babbitts were considered to have greater fatigue resistance than lead base but it is now appreciated that lead is superior if it is made thin enough, preferably down to a thickness of 0.125 mm.

Both tin- and lead-base babbitts are easy to cast, easy to bond to the backing and easy to machine. They are useful for low-duty bearings made in short production runs. They have, however, been supplemented by other types of bearing for high-duty service and mass production. Often a very thin layer of pure lead bonded to the surface of a stronger support material is used.

7.4.7.5 Low melting-point alloys

Solders are alloys of tin and lead in varying proportions (for example, 38 Pb 62 Sn for tinman's solder and 66 Pb 34 Sn for plumber's solder). There are a number of lead–tin alloys with bismuth or antimony whose applications depend on their low melting points. The most important are the type metals which contain from 12% to 30% antimony. Antimony expands on solidification and an alloy containing between 20% and 30%

of this metal has a negligible contraction and produces a clear typeface.

Tin and terneplate Electrolytic or immersion tinplate has been used to protect steel sheet used mainly in the food industry. For other than food products, terneplate (which contains 10–25% tin, remainder lead) is supplanting tinplate, largely on a cost basis but also because the lubricity of lead assists drawing and forming operations.

7.4.8 Cobalt and its alloys

7.4.8.1 Introduction[55,56]

Cobalt is a soft, silvery metal readily corroded by aggressive environments but properties developed by alloying include a high coercive force, exceptional resistance to corroding and oxidizing environments at ambient and elevated temperatures, exceptional hardness and wear resistance and high hot strength. It can also be transmuted by irradiation with neutrons to Co-60 which emits 7×10^6 eV γ-rays and is used for high-penetration radiography.

Its major drawbacks are its high cost and the distribution of its ores which are found in politically unstable countries. Also, its alloys compare unfavourably in high-temperature strength to those of nickel (see below).

7.4.8.2 Applications of cobalt alloys

The major application of cobalt is as a permanent magnet material. Alnico 5, 24% Co, 14% Ni, 8% Ai, 3% La, 51% Fe has, for example, retentivity B_r 12 500 Gauss and coercive force H_c 550 Oersteds. Cobalt is also a constituent of high-permeability magnets. Permendur has saturation induction B_s 24 500 Gauss and coercive force H_c 2 Oersteds.

The next most important group of applications are those which depend on strength, hardness and corrosion resistance at room and elevated temperatures. These include gas-turbine materials, furnace hardware and wear-resistant and spring alloys. Almost all of these materials are based on an alloy containing between 20% and 30% chromium to which are added other constituents conferring specific mechanical properties.

Because cobalt is a high melting-point metal with no allotropic modifications its prospects as a high creep-strength material for aircraft turbines would appear highly favourable. Unfortunately, no strengthening mechanism comparable with the γ'-precipitation process in nickel has been found. This deficiency might be overcome by the development of a creep-resisting composite but there is little incentive to develop such a material on a base whose supply might fail in an emergency. Nickel has therefore supplanted cobalt as a gas-turbine material. The deformation, corrosion and oxidation resistant characteristics of the cobalt superalloys, LG05, HS188, UMCo50 and stellite 250 have led to their extensive application as furnace hardware.

There are a number of dental and prosthetic alloys specified in ANSI/ASTM F75–76 for cast alloys and ANSI/ASTM F90–76, F562–78, and ASTM F563–78 for wrought alloys. BS 3561: Part 2: 1980 includes both cast and wrought.

Cobalt alloys are superior for dental purposes to gold because of their higher strength and lower specific gravity. The decision as to which cobalt alloy to use for prosthetic purposes, or whether to use surgical stainless steel, titanium or tantalum instead depends on the preference of the surgeon.

Cobalt-based wear-resistant products are, in effect, alloys of cobalt/20% chromium with tungsten carbide. They are available to a number of designations depending on the wear and other characteristics required and in forms which include castings, forgings, powder for compaction and hard-facing consumables (see References). The most comprehensive classification is that of the Australian Welding Research Association (AWRA).

Cobalt is also the basis for a number of alloys for springs ('Elgiloy' or 'Cobenium'), low-expansion corrosion-resistant alloy ('Stainless Invar') and alloys with low-temperature coefficients of modulus of elasticity.

7.4.9 Other non-ferrous metals

There are some 28 non-ferrous metals not considered individually. Many of them are used as alloying additions to the metals described in Sections 7.3.1–7.3.7. Only those which form the major constituents of alloys will be considered here.[57–59]

7.4.9.1 Metals used for alloying steels

Tungsten has the highest melting point (3410°C) of any metal and the highest strength at elevated temperatures. The UTS (of 1 mm dia. wire) is approximately 2600 MPa at room temperature and 30 MPa at 2800°C. It has good corrosion resistance but oxidizes in air at temperatures over 500°C.

Tungsten is normally fabricated by powder metallurgical techniques but can be cast by vacuum arc or electron beam. Its applications depend on its high-temperature properties in protective atmospheres and include:

- Electric light filaments. These are made by a powder metallurgy technique incorporating an oxide which restricts grain growth so that the filament consists of a bundle of single crystals each of which is continuous along the length of the filament.
- Electrodes in electron tubes.
- Electrodes for inert gas welding. For this and the previous application, thorium oxide may be incorporated to promote ionization and the smooth striking of an arc.
- Electrical contact materials for highly repetitive and continuous arcing applications. Tungsten has outstanding resistance to arcing, welding or sticking. It has, however, a tendency for the positive terminal to oxidize, which can be prevented by substituting palladium or platinum for this terminal only.

Molybdenum[60] has many of the high-temperature characteristics of tungsten but its melting point (2610°C) is lower and its high-temperature strength, although good, does not compare with that of tungsten. Like tungsten, it oxidizes in air above 500°C unless protected. It is, however, more ductile and more easily fabricated. It is highly resistant to liquid media including glasses and molten metals, but is attacked by oxidizing agents. It is fabricated by consumable arc melting. Molybdenum's applications include:

- Electrical and electronic parts.
- High-temperature furnace parts (particularly for vacuum furnaces).
- Glass-melting furnaces.
- Hot-working tools.
- Dies and cores for die casting.

Chromium[61] is a light, silvery metal which is highly resistant to oxidation and to many corroding media, but not to hydrochloric acid. It is, however, extremely brittle (the production of ductile chromium has been reported but not followed up) and this restricts its use. Applications are:

- Mirrors.
- X-ray targets.
- Decorative and wear resistant electroplate.

The hardness of electroplated chromium depends on its hydrogen content and may be up to 1200 VPH as-plated reducing to 70 VPH annealed. Chromium has a low coefficient of friction; 0.12 sliding, 0.14 static and can be plated in a porous form to retain oil.

Tantalum and *niobium* have excellent fabricability, high hot strength and a low ductile brittle transition but oxidize in air above 300°C. Tantalum is used for chemical plant liners, surgical implants and, because of the electrical properties of its oxide film, for rectifiers and capacitors.

7.4.9.2 Precious metals

The precious metals have very high resistance to oxidation and corrosion. They include:

- The platinum group metals: platinum, palladium, iridium, rhodium, osmium and ruthenium, which have high melting points, (1760–3050°C), high strengths and excellent dimensional stability at elevated temperatures; and
- Gold and silver, which have low melting points (1063°C and 961°C, respectively) and are soft and ductile.

Precious metals are specified for applications which demand extreme reliability or absolute freedom from corrosion and those in which high recovery value and long trouble-free service offset a high original cost. Applications of specific precious metals include:

- *Silver*
  Electrical conductors which, unlike copper, do not oxidize at elevated temperatures and contacts.
  Corrosion-resistant containers for food processing (now largely superseded by stainless steels).
  Bearings.
  Ornaments and jewellery (sometimes as plates).
  High melting-point solder.
- *Gold*
  Instruments requiring corrosion resistance.
  Conductors for transistor circuitry.
  Dentistry.
  High melting-point solder.
  Coinage, jewellery and ornamental uses.
- *Platinum* (and, to a lesser extent, *palladium*)
  Catalyst, particularly for automobile exhaust systems and chemical synthesis.
  Furnace windings.
  Laboratory crucibles and containers.
  Resistance thermometers and thermocouples.
  Linings for optical glass and fluoride process plant.
  Electrical contacts.

Iridium and rhodium are used mainly as alloying agents for platinum, but pure iridium has been used for the manufacture of high-quality glass and rhodium as an electroplate having even more resistance than platinum. Osmium and its alloys have very high hardness (approximately 800 VPH), resistance to wear and corrosion and moduli. They are used for fountain-pen nibs, record-player needles, instrument pivots and electrical contacts.

7.4.9.3 Nuclear metals

Nuclear metals are divided into 'fissile', fertile', 'canning' and 'control' materials. Fissile metals undergo fission when irradiated by neutrons, disintegrating into two major fission products, a number of neutrons which serve to carry on the chain reaction, other particles and energy including γ-radiation. Fissile materials include U-235 (a constituent of natural uranium), U-233 (a product of neutron capture by thorium) and *plutonium* (a product of neutron capture by U-238, the major constituent of natural uranium). They constitute the fuel in nuclear reactors. 'Fertile' metals include U-238 and *thorium*. They are incorporated into nuclear reactor fuel or used separately in 'blankets' to absorb neutrons and produce additional fissile material. 'Canning' metals are used to contain nuclear fuel in a reactor, maintain its integrity and dimensions, protect it from attack by the coolant, retain fission products so that they do not contaminate the coolant (and, through it, the environment), transfer the heat produced efficiently and absorb a minimum proportion of neutrons. Canning and core structural materials now in use include *stainless steel* for sodium-cooled and high-temperature gas-cooled reactors, *magnesium alloy* for the original 'magnox' reactors, *zirconium* for pressurized water and boiling water-cooled power reactors and *aluminium* for water-cooled research reactors.

Zirconium[62] occurs naturally together with *hafnium*, which has high neutron-absorbing properties. These must be separated by a complex chemical process before they can be used in water reactors: zirconium as a core structural material, hafnium as a control rod material. Both have excellent resistance to pressurized water attack if they are suitably alloyed and satisfactorily pure.

Beryllium combines a very low nuclear capture cross section with good strength and hardness at moderately high temperatures. It appeared to have great promise as a canning and core structural material but the promise has not been fulfilled mainly because of its lack of ductility and resistance to environmental attack and partly because of doubts concerning the effect of helium, which is produced when beryllium is irradiated by neutrons. The applications of alloys based on beryllium are confined to those such as spacecraft, where its high specific strength outweighs its high cost and hazard to health. (Its oxide causes 'berylliosis', similar to silicosis but more virulent when ingested by breathing.)

7.4.9.4 Metals used in integrated circuits

Silicon and *germanium*, which when pure are very poor electronic conductors of electricity, can be transformed by 'doping'. Introducing into the lattice pentavalent elements, phosphorus, arsenic or antimony creates free electrons and gives rise to negative or *n*-type conductivity. Introducing trivalent elements boron or aluminium reduces the number of electrons to form 'holes' and gives rise to positive or *p*-type conductivity.

Junctions between regions of these two conductivity types are called *p-n* junctions. These are at the heart of most semiconductor devices: diodes, transistors, solar cells, thyristors, light-emitting diodes, semiconducting, lasers, etc. By taking a slice of highly pure single-crystal silicon, diffusing into it *p*- and *n*-type atoms in a geometrical pattern controlled photographically and then insulating or interconnecting regions by metallization, circuits with millions of components can be formed on one silicon chip.

Highly pure, zone-refined single-crystal silicon has completely superseded germanium for the manufacture of transistors and silicon integrated circuits. The quantity used is small, amounting only to tens of tons per annum, but its technological importance is enormous.

7.5 Composites

7.5.1 Introduction

A composite is a combination of two or more constituents to form a material with one or more significant properties superior to those of its components. Combination is on a *macroscopic* scale in distinction to alloys or compounds which are *microscopic* combinations of metals, polymers or ceramics. Those properties that may be improved include:

Specific gravity
Elasticity and/or rigidity modulus
Yield and ultimate strength and, in the cases of ceramics and concrete, toughness
Fatigue strength
Creep strength
Environmental resistance
Hardness and wear resistance
Thermal conductivity or thermal insulation
Damping capacity and acoustical insulation
Electrical conductivity
Aesthetics (attractiveness to sight, touch or hearing)
Cost

Not all these properties can (or should) be improved at the same time, but the consideration which governs the choice of a composite is that a critical property has been adequately improved, while deterioration in other properties has not been significant.

Usually (but not invariably) a composite consists of a matrix which is relatively soft and ductile containing a filler which is harder but may have low tensile ductility. The use of composites has persisted ever since tools of wood or bone (which are naturally occurring composites) were used by primitive humans. The earliest human-made composite was probably straw-reinforced mud for building. The Egyptians invented plywood, an early example of the improvement (which continues to the present day) of the natural composite, wood.

There are two ways of classifying composites: according to either the material of the matrix or the geometrical distribution of the components. Composites classified according to geometry include:

Particulate composites which are distributions of powder in a matrix;
Laminar composites which comprise layers of two or more materials;
Fibrous composites which comprise a matrix that is usually relatively soft and ductile surrounding a network of fibres which are usually stronger but may be brittle relative to the matrix. The fibres may be short and their orientation effectively random or they may be long and carefully aligned. Composites with long fibre reinforcement are known as 'Filamentary Composites' and the fibres may be aligned on one, two or three directions.

Composites classified according to matrix include:

Fibre-reinforced or powder-filled polymers
Concrete, reinforced concrete and prestressed concrete
Wood and resin-impregnated wood
Metal matrix composites
Fibre-reinforced ceramics and glasses
Carbon fibre-reinforced carbon

Of these, reinforced concrete probably has the greatest industrial importance but fibre-reinforced polymers have the greatest technological and engineering interest and the major part of this section will be devoted to them. Some other classes of composite will be described briefly and their properties and applications outlined.

7.5.2 Reinforcing fibres

Reinforcing fibres may be long (even continuous) and carefully aligned, or short with only the limited degree of alignment that is produced by flow during fabrication. The ability to control directional properties by the use of long fibres has opened up new fields in design, particularly for aerospace. Fibre material may be glass, ceramic, high-modulus polymer, carbon, boron or metal. Most types of fibre are manufactured and used in long lengths but the cheaper materials may be used in short lengths. In the case of glass the fibres are manufactured long but may be chopped. Some ceramic fibres, notably alumina, can be produced short or long depending on the technique of manufacture.

The most widely used glass fibres are 'E glass', a non-alkaline alumino borosilicate of average composition SiO_2, 54; Al_2O_3, 14; B_2O_5, 8.5; CaO, 18.5; MgO, 4; rare earth oxide, 0.8. This material was originally developed for electrical insulation and represents a compromise between mechanical property considerations and ability to form fibres from the melt. Fibres are made by feeding the liquid glass through a multitude of orifices to form up to 4000 filaments approx. 20 μm dia. in one strand of roving. E-glass fibres have a specific gravity of 2.5 and a breaking strength of 2.6 GPa, higher than that of massive glass probably because discontinuities are less severe. However, their modulus of 86 GPa is low and this makes it difficult to utilize the high tensile strength in design because of the danger of buckling and the high elastic deformation.

Glasses with higher strength and slightly higher modulus are available but it is more difficult to manufacture fibre from them. Quartz fibres have excellent high-temperature strength but must be drawn individually from heated rod, and are therefore expensive.

E-glass cannot be used to reinforce concrete because it is attacked by alkali. Glass in which the alumina is replaced by zirconia is more difficult to convert to fibre, but retains its strength much better in concrete and is marketed by Pilkington under the name of 'Cemfil', as a substitute for asbestos, which may no longer be used. The specific gravity of Cemfil is 2.5, its UTS 3.6 GPa and its elastic modulus 75 GPa.

The most extensively used oxide fibre is alumina prepared by spinning or extruding an aqueous solution of a precursor, aluminium hydroxide/chloride, with a slurry of alumina in suspension. Spinning produces relatively short staple fibre 2–4 cm long in blanket form; extrusion produces continuous filament. Small amounts of silica may be present. The spun or extruded fibre is converted into alumina by heating and may have UTS up to 1.8 GPa and elastic modulus up to 380 GPa. Alumina fibres have no application for resin matrix composites but their high-temperature strength makes them an attractive constituent of metal matrix composites and they have been used both in the staple form and as whiskers.

Carbon and graphite fibres are made by spinning a precursor filament, usually polyacrylonitride (PAN), which is then heat treated while being subjected to tensile stress. The heat treatments include stabilization, oxidation, pyrolysis and graphitization at temperatures increasing from around 200 to 3000°C. Fibres with tensile properties ranging between 3.4 GPa UTS, 235 GPa elastic modulus and a breaking strain of 1.5% (for Courtauld's Graphil AXS) and 2.5 GPa UTS, 340 GPa elastic modulus with a breaking strain of 0.6% (for Graphil HMS) may be obtained in the more expensive grades by varying the temperatures and stress applied. The production of carbon fibres is illustrated schematically in Figure 7.55. Cheaper grades are available with inferior properties. Cheaper precursors for carbon fibre include cellulose and pitch of high-aromatic low-impurity content.

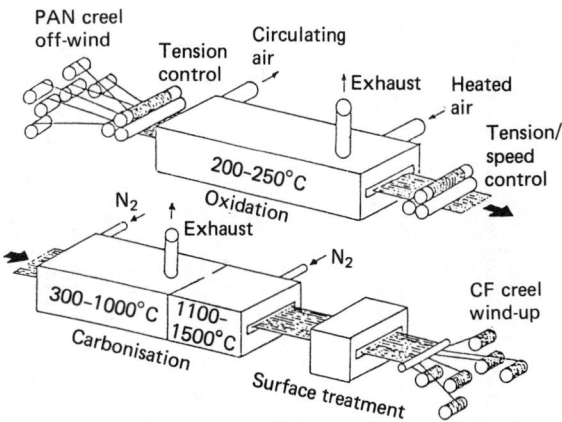

Figure 7.55 Schematic production of carbon fibre. (Reproduced by courtesy of North-Holland Publishing Company)

Boron filaments B/W Fibres 95–200 μm diameter are manufactured by the chemical deposition of boron from a gaseous mixture of boron trichloride and hydrogen at betwen 1000 and 1300°C on a 12.5 μm diameter tungsten wire. B/C fibres have been made experimentally by deposition on carbon filaments. The tensile strength of B/W fibres range from 3.4 to 3.9 GPa, their elastic modulus is 390 GPa and their specific gravity 2.5. Although the ductile properties of boron fibres are no better than those of a ceramic they do not suffer long-time loss of strength and their creep properties are superior to those of tungsten. They are the strongest and stiffest fibres at present used for composite reinforcement.

Silicon carbide fibres are manufactured by spinning high molecular weight polycarbon silane fibres at between 200°C and 300°C, curing in air at 200°C and heat treating at 1250°. Depending on the grade of precursor, fibres 10 μm diameter may have UTS between 2 and 3 GPa and elastic moduli between 150 and 200 GPa. They have excellent high-temperature strength and useful electrical properties.

Para-orientated aromatic polyamide (Aramid) fibres are manufactured by spinning a liquid crystalline solution (or 'dope') of their sulphuric acid ester to form highly ordered domains of extended polymer chains. This structure confers a much higher modulus than is common in polymers and is the basis of the commercial exploitation of polyparaphenylene teraphthalamide (Kevlar) fibres. After spinning, the 'dope' is spun into fibres, coagulated in an aqueous bath at low temperature, stretched, washed, dried and heat treated at temperatures between 250°C and 550°C. The resultant fibres have specific gravities around 1.45, UTS 2.64 GPa, moduli between 59 GPa and 127 GPa and breaking strains between 2.8 and 4.0%.

Fibres with moduli up to 60 GPa can be made from polyolefins and polyoxymethylene. Metal-reinforcing fibres include carbon, alloy and stainless steels, tungsten, niobium and tantalum. The characteristics and properties of these materials are described in Section 7.3. The properties of fibres and whiskers appropriate for use in composites are listed in Table 7.47.

7.5.3 Polymer matrices

The essential qualities of a matrix material are the ability to infiltrate among and bond strongly to reinforcing fibres. After

Table 7.47 Properties of fibres and whiskers appropriate for use in composites

Fibre	Type of filament	Density (kgm^{-3})	Elasticity modulus E (GPa)	Tensile strength, σ_u (MPa)	Specific strength, σ_u/ρ (km)	Specific stiffness, E/ρ (mm)	Elongation, ϵ (%)	Type of composites used in:
	Whisker							
	Alumina	3880	380	18 000	4.5	9.8		Metal matrix
Asbestos	Chrysotile	2550	164	1 000	0.4	6.5	2.5	Previously cement
	Crocidolite	3370	196	3 500	1.1	6	2.5	
Boron		2520	390	3 400	1.35	15		Polymer matrix
	Boron carbide	2470	450	6 700	2.70	18		Metal matrix
Carbon	High modulus	1900	340	3500	1.8	18	0.5	Polymer
	Low modulus	1900	235	2350	1.37	12	1.0	Polymer, metal and ceramic matrix
	Carbon	1630	980	21 000	13.00	60		
Cellulose		1200	10	400	0.33	0.8	15	Concrete matrix
	Copper	8740	124	3 000	0.34	1.4		Polymer matrix
Glass	E	2500	86	3 200	1.37	3.4	4.8	Polymer matrix
	Alkali resistant	2700	75	2 500	0.95	3	3.6	Polymer and concrete matrix
	Iron	7680	200	13 000	1.70	2.6		Polymer matrix
Kevlar	High modulus	1440	133	2 900	2.01	9	2.1	Polymer matrix
	Low modulus	1140	69	2 900	2.63	6	4	Polymer matrix
	Nickel	8790	215	3 900	0.44	2.4		Polymer matrix
Nylon		1140	>4	850	0.75	0.35	13.5	Concrete matrix
Polypropylene		900	>8	400	6.44	0.9	18	Concrete matrix
	Silicon carbide	3120	840	11 000	3.50	27		Polymer and metal matrix
Steel	High tensile	7860	200	2 000	0.25	3.1	3	Concrete matrix
	Stainless	7860	160	1 700	0.22	2.5	3.5	Polymer matrix

these criteria have been met it should set as quickly as possible to a strong and heat- and environment-resisting solid. Polymers share with concretes the advantage over other possible matrix materials that they fulfil these requirements at a relatively low processing temperature. There are two classes: thermosetting and thermoplastic polymers (see Section 7.4). Thermosetting resins compounded with a hardener may be infiltrated between fibres while liquid and allowed to harden at room or elevated temperature. They include unsaturated polyesters, which are relatively cheap and easy to work but do not bond well to fibres and have a relatively high shrinkage. These are used for large and comparatively low-duty composites, usually with glass reinforcement.

Epoxide resins are the most extensively used matrix materials for high-duty carbon, boron and aramid fibres. They perform excellently at temperatures up to the region of 160–200°C.

Thermosetting resins which have been used as matrices operating at higher temperatures include phenolics, phenol arakyls and the recently developed polyphenylene quinoxialine. Resins which are beginning to replace epoxies for high-temperature service with carbon reinforcement are bismaleides (BMI) and polyimides (PI) which have continuous service capabilities of 200°C and 300°C respectively. (Some polyimides have survived short exposures to 760°C.) These polymers are, however, difficult to handle and polyimides in particular are expensive and require high cure temperatures.

Thermoplastic matrix materials are tougher than thermosets, have an indefinite shelf life, the semi-finished composite can be hot formed and in some cases have better high-temperature and solvent resistance. However, the molten polymer has a higher viscosity than an uncured thermoset, fabrication temperatures are high and some are expensive.

Many thermoplastics have been used, ranging from the cheapest (nylon) to the highly expensive polyamide imide (PAI) and polyether-ether ketone PEEK. PEEK composites have a maximum service temperature of 250°C, a work of fracture up to thirteen times that of epoxide composites and significantly better fatigue resistance, but are expensive.

7.5.4 Manufacturing procedures for filamentary polymer composites

Filamentary composites are manufactured by 'lay-up', a term used for positioning the fibres and matrix to form the shape of the final component. Lay-up may be accomplished by 'pultrusion', 'winding' or 'laying', 'tow', 'tape', 'cloth' or 'mat'. In none of these forms are the fibres twisted to form a yarn. All forms of sub-assemblies can be obtained as 'prepregs' saturated with the resin which is later to form the matrix.

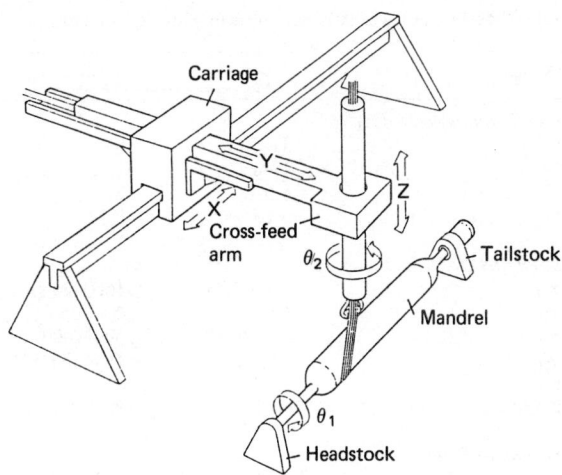

Figure 7.57 Gantry type five-axis filament winding machine. (Reproduced by permission of *Metals and Materials*)

In *pultrusion* (see Figure 7.56) the reinforcing fibres are used to pull the material through a die.

In *winding*, impregnated single filaments, rovings or tapes are wound onto a former or mandrel. Figure 7.57 shows a winding machine which may be computer controlled to produce any convex shape from which the mandrel can be removed. Filaments may be orientated according to the pattern of stresses that are to be withstood.

Cloth winding or laying utilizes pre-impregnated cloth which is deposited in the desired form and orientation. The bidirectionality and convolutions of the fibres in cloth make for lower precision in strength and stiffness. Cloth laying is therefore often used for filling where strength and stiffness are not critical. *Moulding* can start with a deposition of pre-cut layers of prepreg fibres which are compressed at elevated temperature to form the final laminate. *Continuous lamination* is the application of pressure by rolling to bond layers of prepreg cloth or mats.

7.5.5 Properties of filamentary polymer composites

Filamentary polymer composites consist, in principle, of 'laminae' which are assembled into 'laminates'. A 'lamina' is a flat or curved assembly of unidirectional fibres in a matrix. It is

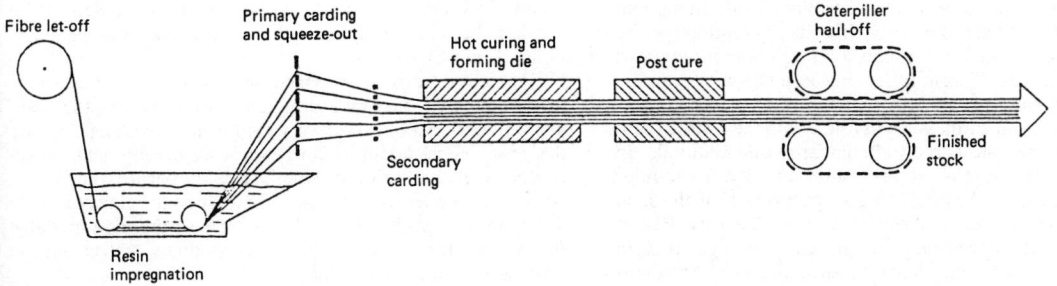

Figure 7.56 Pultrusion process in which the reinforcing fibres are used to pull the material through the die. (Reproduced by permission of *Metals and Materials*)

Table 7.48 Properties of 60% fibre plies in epoxide laminates

Property	E glass	S glass	Kevlar 49	HT-CFRP	HM-CFRP	Boron
Elastic modulus (GPa)						
E_{11}	37–50	55	77–82	140–207	220–324	210
E_{22}	12–16	16	5.1–5.5	9.8–10.0	6.2–6.9	19
G_{12}	4.5–6	7.6	1.8–2.1	5–5.4	4.8	4.8
ν_{12}	0.20	0.26	0.31	0.25–0.34	0.20–0.25	0.25
Strengths (MPa)						
σ_{1T}	1100–1200	1600–2000	1300–2000	1240–2300	783–1435	1240
σ_{2T}	40	40	20–40	41–59	21	70
σ_{1C}	620–1000	690–1000	235–280	1200–1580	620	3300
σ_{2C}	140–220	140–220	140	170	170	280
τ_{12}	50–70	80	40	80	60–70	90
ILSS	60	80	60	90–100	60–90	90
Strains to failure						
ϵ_{11}	2–3	2.9	1.8	1.1–1.3	0.5–0.6	0.6
ϵ_{22}	0.4	0.3	0.5	0.5–0.6	<0.7	0.4
ϵ_{11C}	1.4	1.3	2.0	0.9–1.3	—	1.6
ϵ_{22C}	1.1	1.9	2.5	1.6	2.8	1.5
Thermophysical						
SG	1.9–2.1	2.0	1.35–1.38	1.5–1.6	1.63	2.2
$\alpha_1, \times 10^{-6}\,K^{-1}$	6.3	3.5	−4 to −4.7	+0.4	−0.43 to −0.8	4.5
$\alpha_2, \times 10^{-6}\,K^{-1}$	30	29	60–87	25	27–32	23
$k_1, Wm^{-1}\,K^{-1}$	1.26	1.58	1.7–3.2	10–17	48–130	—
$k_2, Wm^{-1}\,K^{-1}$	0.59	0.57	0.15–0.35	0.7	0.8–1.0	—
Specific heat, $J\,kg^{-1}\,K^{-1}$	840	840	1260	840	840	1260

Reproduced by courtesy of *Metals and Materials*, from a paper by R. Davidson.

highly anisotropic, having low stiffness and strength transversely (see Table 7.48).

Laminae of varying orientations are therefore superimposed in a stack to form a 'laminate' with directional properties tailored to match the stress. Laminates are therefore essentially two-dimensional structures (the 'dimensions' may be curved when the component is a cylinder or sphere) and the mechanical properties in any of the principal directions of a laminate are inferior to those in the principal direction of one of the constituent laminae. Additionally, the thermal stresses which arise on cooling from the curing temperature may impair strength.

Three-dimensional reinforcement such as is employed in carbon/carbon composites (see Sections 7.5.8 and 7.5.11) is not normally applied to laminated plastics and shear and transverse tensile stresses can result in delamination.

The matrix supports, protects, distributes load among and transmits load between the fibres. If a fibre should break the matrix, stressed in shear, transmits load from one broken end to the other and to adjacent fibres. Because boron or graphite fibres in a polymer matrix provide by far the greater proportion of strength and stiffness, composites with these fibres can, in most cases, be considered to be linear elastic materials. In composites with glass or aramid fibres the lower modulus results in the matrix bearing a higher proportion of the load and the stress strain relation may depart from linearity. Elastic and physical properties may, in the case of high-strength composites, be calculated from classical theory.[63] Strengths are more difficult to calculate because the secondary stresses induced in a composite may exceed the transverse shear strengths and may themselves cause failure.

The parameters which must be taken into account in design include:

1. *Elastic properties*: Longitudinal Stiffness E_{11}, Transverse Stiffness E_{22}, In-Plane Shear Modulus G_{12}, Poisson's ratio ν_{12}.
2. *Strength properties*: Longitudinal Tensile Strength $\sigma_{1,T}$, Transverse Tensile Strength σ_{xr}, Longitudinal Compressive Strength $\sigma_{1,c}$, Transverse Compressive Strength $\sigma_{2,c}$, Yield Strength σ_y, In-plane Shear Strength $T_{1,2}$.
3. *Physical properties*: Specific Gravity SG, Longitudinal Thermal Expansion Coefficient α_1, Transverse Thermal Expansion Coefficient α_2, Longitudinal Thermal Conductivity k_1, Transverse Thermal Conductivity k_2.

The Specific Strengths and Moduli of Fibrous Composites and other engineering materials are illustrated diagrammatically in Figure 7.58. (In this figure specific properties are derived by dividing the modulus or strength by the density and a gravitational term of 9.81.)

The fatigue processes which occur in composites differ fundamentally from those in metals, and, providing that they are well understood, offer very significant advantages to the designer. High-modulus fibres such as carbon and boron confer excellent tension/tension fatigue properties, the fatigue stress at 10^7 cycles of longitudinal boron epoxy being only 15% less than the tensile stress. This is because the high-modulus fibres limit the stress in the lower-modulus matrix and so protect it from fatigue damage.

However, in those plies in which fibres are orientated transverse to the principal cycle stress the matrix is subjected to transverse tensile and shear stresses which cause cracking

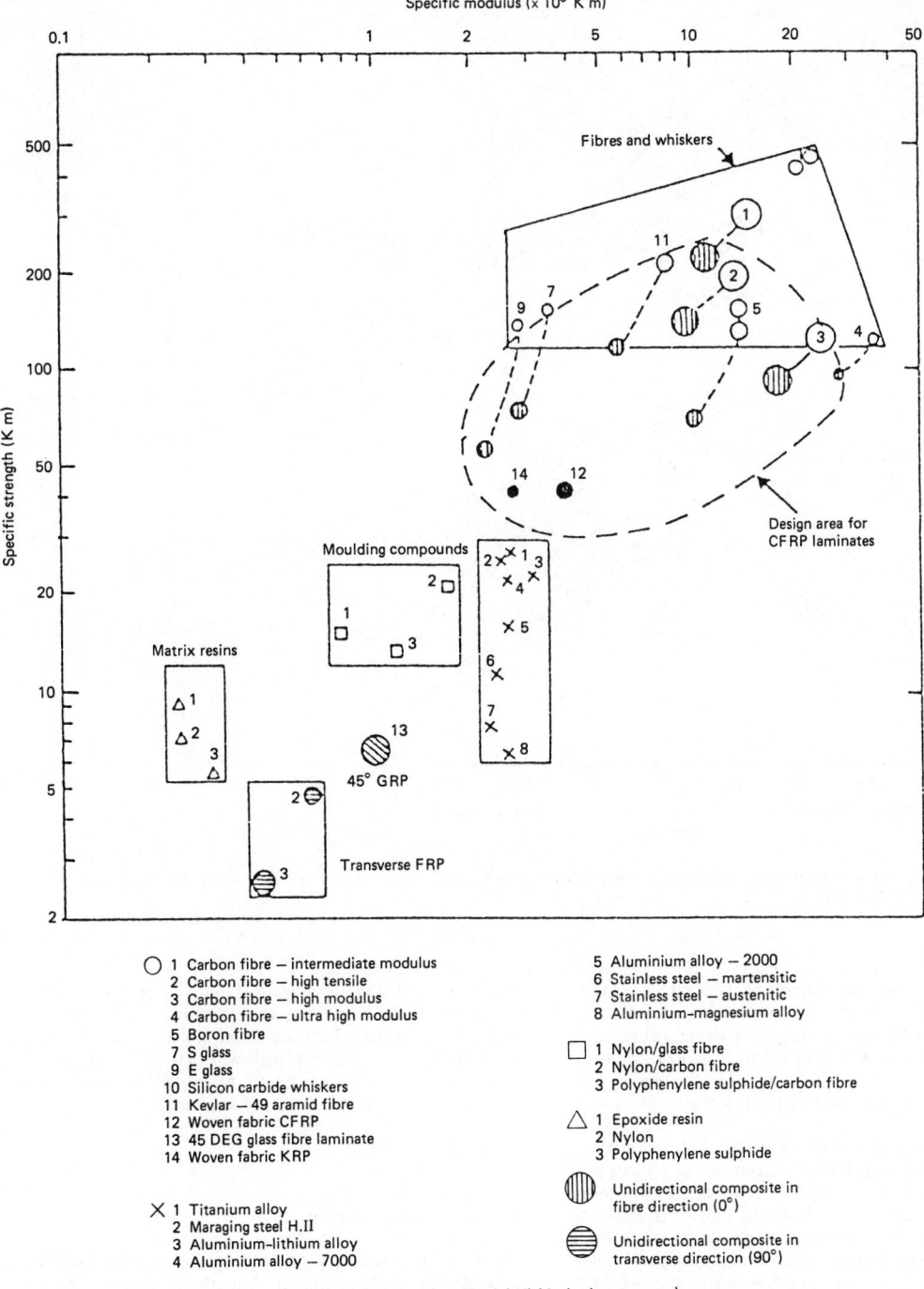

Figure 7.58 Specific strengths and moduli of composites and competing materials. (Reproduced by permission of *Metals and Materials*)

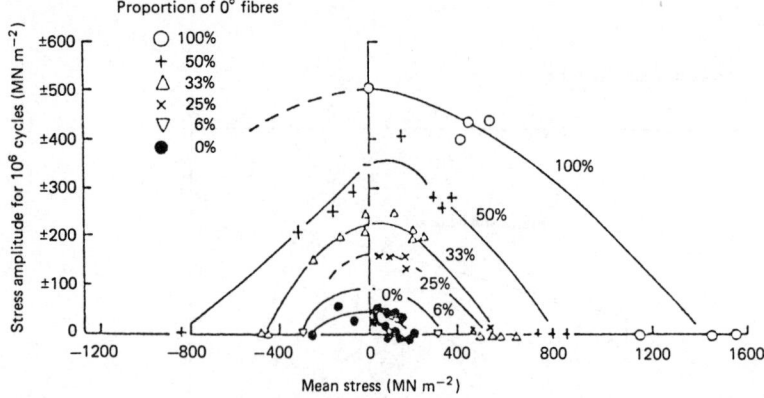

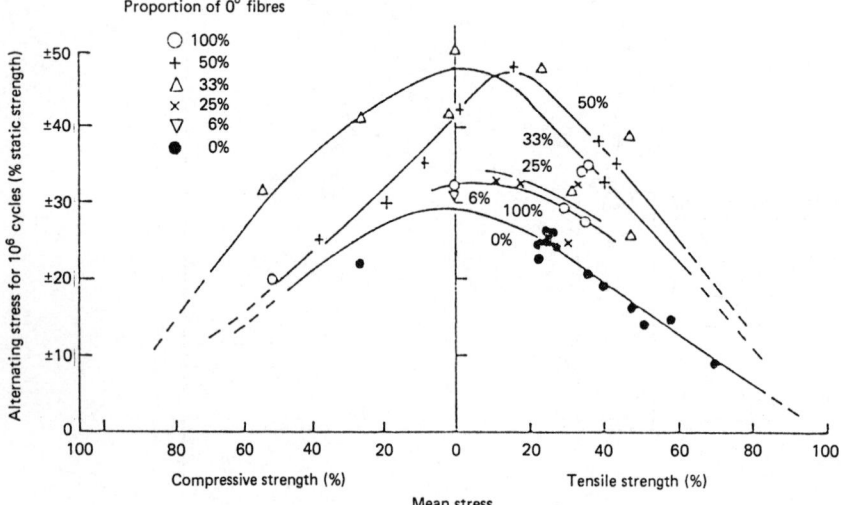

Figure 7.59 Maximum and minimum stress in fatigue cycling causing failure at 10^6 cycles in various CFRP laminates. (Reproduced by permission of *Metals and Materials*)

parallel to the plies and delamination. The effect of fibre orientation on CFRP laminates is shown in Figure 7.59.

Only glass composites have a steep *S-N* slope, presumably caused by the diffusion of moisture which causes cracks to initiate in the glass fibres. Even so, the *specific* fatigue resistance of longitudinal fibreglass is far superior to that of any metal.

A further advantage of composites subjected to fatigue is that, whereas in metal fatigue there is, during the greater part of the life of a component, no superficial evidence of deterioration, there is in filamentary reinforced plastics a slow and progressive deterioration revealed at an early stage by a decrease in modulus or an increase in cracking in specific plies which is easily detectable by non-destructive examination. This reduction in modulus could, if allowed to continue, lead to failure by buckling, but both because of the higher specific fatigue strength and because of the more obvious incidence of failure, catastrophic fatigue failures in filamentary composites are much less likely than in metals.

The assessment of the influences of impact on filamentary composites is more complex than metals because of their anisotropy and large number of failure mechanisms. Where, for example, in a jet engine a titanium blade will shear undamaged through the body of an intruding bird, a composite blade of equivalent strength will shatter. It can be stated that, in terms of impact strength, for composites the common fibres may be ranked in order of superiority:

1. Kevlar 29, Glass
2. Kevlar 49, boron
3. High-tensile carbon
4. High-modulus carbon

The resistance to attack of polymers depends on the specific polymer and its environment. Traditional matrices based on polyesters, vinyl esters and epoxies perform very successfully in atmosphere, soil and many items of chemical plant. Protection may, however, be needed against degradation by ultraviolet radiation from sunlight. Some polymers, including fluoroplastics PTFE and PDF and polyether ether ketone PEEK have exceptional resistance to radiation damage and may be used as matrices and as coatings.

7.5.6 Applications of filamentary polymer composites

The cost of GFRP is of a similar order to steel and aluminium or timber and where its lightness and corrosion resistance are advantageous, and its fabrication methods suitable for the specific component, it is used. Applications include small boats (and not so small minesweepers), roofing and cladding for buildings and many components for road and rail transport.

Other uses of GFRP are promoted by one or more specific property parameters. It is, for example, displacing steel for vehicle leaf springs on account of its lightness and fatigue resistance. It is replacing porcelain and glass for electrically insulating components on account of its strength and insulating properties. It is replacing steel for aqueous liquid vats, tanks and pipes because of its lightness, strength and corrosion resistance.

High-performance composites are used in aerospace or sport, where the requirement for the specific stiffness and/or specific strength justifies the increased cost. The aerospace applications of CFRP include the basic structures of spacecraft and commenced with ancillary fittings, floors and furniture of aircraft, but is now extending to major structural items such as stabilizers, tailplanes and fins. Future fighter aircraft will probably contain a high proportion of CFRP and will benefit from a reduced sensitivity to radar.

High-performance sports goods are also increasingly made of CFRP because the reward of coming first in a race (or a fishing contest) far outweighs the additional cost of a CFRP racing-car skin or a CFRP fishing rod compared with any conceivable alternative material, except possibly boron-fibre (BF) reinforced composites.

The combination of a specific tensile strength around 0.8 and modulus around 105 GPa m^3 kg^{-1} can only be obtained from BF-reinforced plastics. Boron fibres may be used by themselves or as a hybrid composite, part BF, part CF for horizontal and vertical stabilizers, control surfaces, wing skins, flaps, slats, tail surfaces, spars, stringers, fuselage-reinforcement tubes, spoilers, airhole flaps, doors, hatches, landing-gear struts, helicopter rotor shafts and blades for military and civil airplanes and space shuttles. The use of such materials (including aluminium matrix composites) can reduce weight by from 12% to 45%, almost double service life, and decrease fuel usage and maintenance by about 10%. BF-reinforced composites are also used for the pickup arms for high-fi record-playing decks where specific stiffness is paramount.

The relative cost of glass, carbon, hybrid and boron-reinforced plastics is 1, 10, 20 and 30, but the cost of the high-strength high-modulus fibres is reducing with time.

The use of 'aramid' para-orientated aromatic polyamids fibres has been restricted because their relatively low moduli (58.9–127.5 GPa) makes it difficult to take advantage of their high UTS, (up to 2.64 GPa) in designs which may be buckling critical. They have been used for golf shafts, tennis racquets and boat hulls, Kevlar T950 for tyres and Kevlar T956 for other rubber components.

7.5.7 Discontinuous fibre-reinforced polymer composites

7.5.7.1 General

Discontinuous fibres of an average length in the region of 380 μm may be incorporated in proportions up to about 25% by volume in mouldable polymers to enhance their stiffness, strength, dimensional stability and elevated temperature performance. Reinforced thermoplastic materials (RTP) may be shaped by melt fabrication techniques, injection moulding, extrusion, blow moulding and thermoforming. The material is melted or plasticized by heating, shaped in the plasticized condition and cooled to resolidify. Reinforced thermosets may be made to flow in the pre-cured state and cured or cross linked to an infusible mass in the hot mould.

'Commodity' thermoplastics, polyolefins, polystyrene, polyvinyl chloride, etc. are utilized mainly in the non-reinforced form but are marketed in the fibre-reinforced form. A much higher proportion of engineering thermoplastics, polyamides, polyacetyls and thermoplastic polyesters are reinforced, usually with short glass fibre and the specialized high-performance thermoplastics such as polysulphones are also reinforced, often with short carbon fibre. Short glass-fibre reinforcement is used for thermosets such as phenolic, amino and melamine formaldehyde resins which may be injection moulded, although the curing time lengthens the manufacturing cycle.

An important class of composite are the long fibre-reinforced sheet-moulding compounds (SMC) and the dough-moulding compounds (DMC) based on unsaturated polyester, vinyl ester and epoxide resins. These materials are normally compression moulded (see Figure 7.60) and have to compete with steel pressings. Similar composites are based on the thermoplastics which are produced as sheets that are heated and then pressed between cold dies.

Two materials are used for discontinuous fibre reinforcement: short and long staple glass fibre, and short staple carbon fibre. Aramid fibres have the required properties but polymers compounded with them are not yet obtainable commercially. Discontinuous fibre-reinforced plastics cost less to fabricate than the corresponding filamentary reinforced materials but their mechanical properties are significantly inferior. This is because the rule of mixture that is obeyed precisely so far as modulus is concerned, and approximately so far as yield strength and UTS is concerned, for high-modulus continuous fibres is no longer obeyed for discontinuous fibres. The strength of short fibre-reinforced polymers is controlled by a complex series of interactions between the fibres and the matrix.

The fibre/matrix interface is usually the weakest link. In aligned fibres the end becomes debonded at quite low loads and the debonding spreads along the fibre as the load increases. Debonding reduces the stiffening efficiency of the fibre and constitutes a microcrack which may extend into the matrix.

The mechanical strengths of typical short and woven fibre-reinforced thermosets are listed in Table 7.49. Table 7.50 details the mechanical properties of short fibre-reinforced

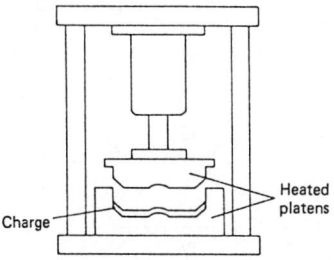

Figure 7.60 Press moulding arrangement for discontinuous fibre reinforced plastics. (Reproduced by permission of *Metals and Materials*)

Table 7.49 Properties of short fibre and woven fibre reinforced thermosets

Property	BMC[a]	SMC[b]	Glass fibre polyester	Glass fibre epoxide	Woven CF epoxide	Woven Kevlar epoxide
Stiffness (GPa)						
E_{11}, E_{22}	11	12–13	17–21	23–26	70	31
ν_{12}	0.11	0.11	0.11–0.12	0.12–0.16	0.08	—
Strengths (MPa)						
σ_{1T}, σ_{2T}	60–69	75–120	303	379–517	586–620	517
σ_{1C}, σ_{2C}	138	179–193	276	345–413	690	83
σ_f	103	138–172	214	517–624	841–1034	345
τ_{1LSS}	13.8	17–28	24	28	55–67	55
Izod impact, J m^{-1}	430–640	640–850	750–960	1600	—	—
SG	1.65–1.80	1.7	1.7–1.8	1.8–1.9	1.59	1.33
α_1, $\alpha_2 \times 10^{-6}$ K^{-1}	18–31	22–36	10–16	10.6	3	0
k_1, k_2 Wm^{-1} K^{-1}	0.1–0.23	0.6–0.22	0.16–0.20	0.16–0.33	—	—
Specific heat, J kg^{-1} °C^{-1}	850	850	850	850	—	1260

[a] 15–25% glass.
[b] 30–40% glass.
Reproduced by courtesy of *Metals and Materials*, from a paper by R. Davidson.

Table 7.50 Mechanical properties of short fibre reinforced thermoplastics

1B Polymer	2B Glass fibre content (W%)	(V%)	9 Water absorption (max) (%)	10 Flexural modulus (GPa)	11 UTS (MPa)	12 Tensile elongation (%)	13 Notched Izod impact (J m^{-1})
1. Polyethylene (HD)	20	9	0.1	4.0	55	2.5	50
2. Polyethylene (HD)	40	20	0.3	7.5	80	2.5	70
3. Polypropylene	20	8	0.02	4.0	63	2.5	75
4. Polypropylene (chemically coupled)	20	8	0.02	4.0	79	4	90
5. Polypropylene (chemically coupled)	40	19	0.09	7.0	103	4	100
6. Nylon 6	40	23	4.6	10.5	180	2.5	150
7. Nylon 6.6	20	10	5.6	9.0	130	3.5	100
8. Nylon 6.6	40	23	3.0	15	210	2.5	136
9. Nylon 6.10	40	22	1.8	9.0	210	2.5	170
10. Nylon 11	30	15	0.4	3.2	95	5	—
11. Acetal homopolymer	20	12	1.0	4.3	60	7	40
12. Acetal co-polymer	30	19	1.8	9.0	90	2	40
13. Acetal (chemically coupled)	30	19	0.9	9.7	135	4	95
14. Polystyrene	40	22	0.1	11.3	103	2.5	60
15. SAN	40	22	0.28	13.4	128	2.5	60
16. ABS	40	22	0.5	7.6	110	3.5	70
17. Modified PPO	40	22	0.09	8.6	135	3.5	80
18. PETP	30	18	0.24	8.3	130	4	85
19. PBTP	40	26	0.4	9.6	150	4	155
20. Polysulphone	40	26	0.6	11.0	138	2	100
21. Polyethersulphone	40	26	—	11.0	205	—	80
22. PPS	40	26	0.06	12.5	160	3	80
23. Polycarbonate	20	10	0.19	5.8	110	6	180
24. Polycarbonate	30	17	0.18	8.2	127	5	190
25. Polycarbonate	40	24	0.16	10.3	145	4	200
Carbon fibre filled materials							
26. Nylon 6.6	30	21	2.4	20.0	240	3.5	75
27. PETP	30	24	0.3	13.8	138	2.5	60
28. Polysulphone	30	24	0.4	14.0	158	2.5	60
29. PPS	30	24	0.1	16.9	186	2.5	55

Reproduced by permission of North-Holland Publishing Co.

thermoplastics which includes some carbon fibre-reinforced materials. The superiority of filamentary reinforcement is evident. Short fibre reinforcement shows to even less advantage in fatigue, creep and impact loading and is not to be recommended for highly stressed parts. Short fibre reinforcement is, however, much cheaper than filamentary reinforcement and is used extensively for a great variety of domestic, architectural, engineering, electrical and automotive components.

7.5.8 Carbon–carbon composites

Carbon–carbon composites retain their strength to a higher temperature than any competitive material (see Figure 7.61). They are unique in that the matrix is identical in composition to that of the reinforcing fibres. They differ from the polymer composites already described in that the matrix which can exist in any number of quasi-crystalline forms from 'glassy' or amorphous carbon to graphite has low strength and negligible ductility. While, therefore, single and bidirectionally reinforced carbon–carbon composites are manufactured the need to avoid delamination has promoted three-directional reinforcement.

Complex weaving equipment has been developed to achieve multilayer locking by means of structures such as are shown in Figure 7.62. Even more complex patterns are employed.

As an alternative to three- (or eleven-) directional weaving, the directional reinforcement may be produced by fabric piercing. Arrays of layers of two-directional fabric are pierced with metal rods or needles. The metal needles are withdrawn and replaced by yarns or by pre-cured resin yarn rods. Fabric piercing is versatile and can produce a higher overall fibre volume and a higher preform density than weaving.

Other techniques for producing multi-directional structures involve the assembly of rods consisting of yarns pre-rigidized with phenolic resins by pultrusion. These can be used to form '4D' tetrahedral structures or by incorporating a filament winding operation into a cylindrical structure. Densification of the structure with carbon is achieved by impregnation with

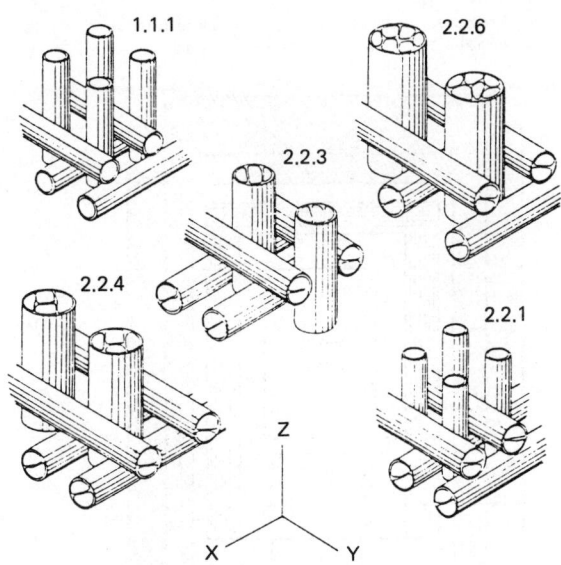

Figure 7.62 Three-dimensional orthogonal weaves for carbon carbon composites. (Reproduced by courtesy of North-Holland Publishing Company)

pitch, a thermosetting phenol or furfural type resin or by depositing carbon from a hydrocarbon (CVD process).

The preform may be impregnated with liquid by a vacuum process, carbonized at 650–1100°C at low pressure and then graphitized within the range 2000–2750°C. The cycle is repeated until the desired density is achieved. Alternatively, the preform may be impregnated with pitch, carbonized and then graphitized at high pressure in a HIPIC furnace; and the cycle repeated as required. In this process the workpiece must be isolated from the pressure vessel in a furnace of the type shown in Figure 7.63.

Impregnation by carbon by the CVD process is carried out by feeding hydrocarbon gas through and into the pores of the preform, isothermally, under a thermal gradient or under differential pressure. Carbon is deposited at 1100°C and in this case, as in impregnation with a thermosetting resin, a carbon rather than a graphite matrix is formed. The tensile properties of carbon–carbon composites with various matrices are listed in Table 7.51.

The application of carbon–carbon composites has so far been restricted by high cost and their susceptibility to oxidation at temperatures above 400°C. Coatings to protect against oxidation are under development. Their most important application has been as rocket nozzles, thrust chambers, ramjet combustion lines and heat shields for space vehicles. They are used commercially for aircraft brake systems for Concorde and military aircraft as well as for hot pressing moulds. They can also be employed for very high-temperature heat shields and elements for vacuum furnaces. Their high-temperature strength will favour a large number of uses if their cost is reduced.

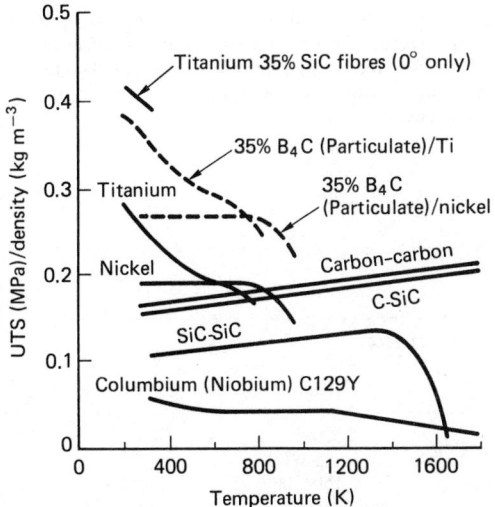

Figure 7.61 Strength-to-density ratio for several classes of high temperature materials with respect to temperature. (Reproduced by permission of *Metals and Materials*)

7.5.9 Fibre-reinforced metals

The potential of fibre-reinforced metals is so great that they have been declared a strategic material in the United States.

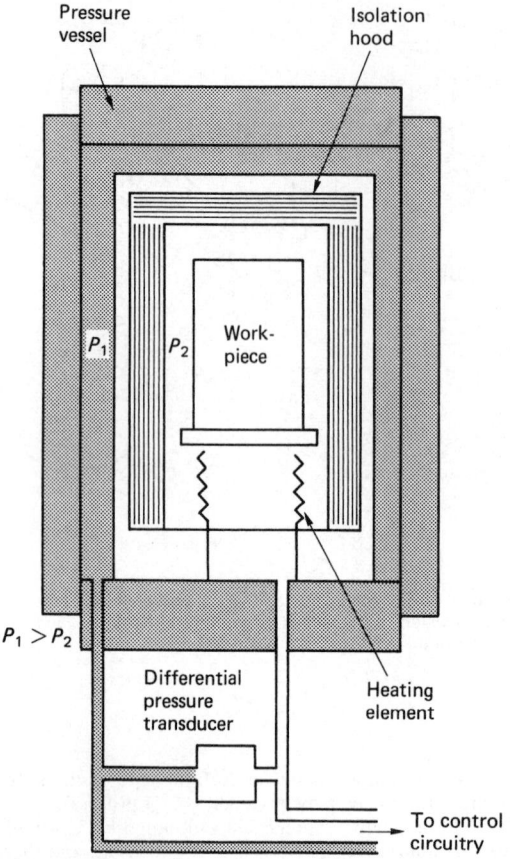

Pressure vessel

Isolation hood

P_1

P_2

Work-piece

$P_1 > P_2$

Differential pressure transducer

Heating element

To control circuitry

Figure 7.63 Isolation hood and differential pressure equipment for HIPIC processing. (Reproduced by permission of *Metals and Materials*)

The fibres may be:

- Whiskers, usually silicon carbide made by pyrolysis of rice hulls;
- Discontinuous fibres, alumina and alumina silica often felted together as insulation blankets;
- Continuous fibres, boron, silicon, carbide, alumina, graphite, tungsten, niobium zinc and niobium titanium.

Most engineering metals could be used as a matrix for a composite. Matrices of titanium, magnesium, copper and superalloys are the subject of investigation but almost all the applications so far recorded have used an aluminium matrix.

Typical properties of metal-matrix composites so far investigated are shown in Table 7.52. The fabrication techniques used include conventional and squeeze casting, powder methods including hot moulding and isostatic pressing, diffusion bonding and vapour deposition. A major problem is to prevent damage or dissolution of the fibres during the manufacturing process while producing a good metallurgical bond with the matrix. Surface treatments are sometimes employed to promote one or both of these objectives.

Silicon carbide particles and whiskers are given a special surface treatment which promotes wetting by aluminium. After casting, the composite may be worked by any of the conventional methods.

Powder metallurgy techniques are best suited to the manufacture of particle-filled composites but, provided the pressing operation is designed to avoid damage to the fibres either by carefully controlled direction of the pressure or by hot hydrostatic pressing, it can be used to make fibrous metal-matrix composites.

In 'Squeeze casting' pressure is used to force molten metal into the interstices of fibre preforms which have preferably been evacuated. Production of a piston by this process is illustrated in Figure 7.64.

Defusion bonding has been used to fabricate boron fibre-reinforced aluminium by the process illustrated in Figure 7.65 but it can also be used for magnesium or titanium. Fibres of silicon carbide or aluminium may be coated with aluminium by vapour diffusion or by passing through molten aluminium, and bundles of coated files may then be compacted by rolling, swaging or hot hydrostatic pressing.

Almost all the manufacturing processes for metal-matrix filament composites are expensive and most of the applications have so far been limited to space technology. The aluminium boron composite described earlier has been used for tubular spars for the space shuttle. An antenna boom is being built for the NASA space telescope from graphite fibre-reinforced aluminium which has the advantage of a very small thermal expansion coefficient and a good thermal conductivity.

A commercial application is the manufacture of the pistons illustrated in Figure 7.64 with fibrous alumina reinforcement in the region of the top ring groove and the crown. The piston outperforms the previous marque of piston which had niresist inserts and is cheaper.

Table 7.51 Typical mechanical properties of carbon–carbon composites

Property	Matrix			
	CVD pyrocarbon	*Pitch*	*Pitch*	*Phenolic resin*
Heat treatment temp.(°C)	2000–2500	1700	2500	1000
Bulk density (g m^{-3})	1.6	1.6–1.65	1.60–1.9	1.73
Flexural strength (M Nm^{-2})	250	200–240	150–200	220
Tensile strength M Nm^{-2})	180		240	300
Young's modulus (G Nm^{-2})	90	70–90	75–95	65
ILSS (M Nm^{-2})	8	15–12	12–15	12

Reproduced by courtesy of *Metals and Materials*, from a paper by Dr G. Savage.

Table 7.52 Representative properties of metal-matrix composites

Matrix	Reinforcement	Reinforcement (vol. %)	Modulus (GPa)		Tensile strength (MPa)	
			Longitudinal	Transverse	Longitudinal	Transverse
Aluminium	None	0	70	70	280–490	280–490
Epoxy	High-strength graphite fibres	60	147	10.5	1260	35
Aluminium	Alumina fibres	50	203	154	1050	175
Aluminium	Boron fibres	50	203	126	1530	105
Aluminium	Ultrahigh modulus graphite fibres	45	350	35	630	35
Aluminium	Silicon carbide particles	40	147	147	560	560
Titanium	Silicon carbide monofilament fibres	35	217	188	1750	420

Reproduced by courtesy of *Metals Engineering*.

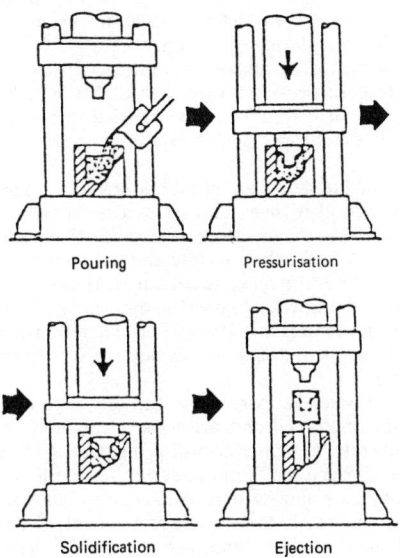

Pouring   Pressurisation

Solidification   Ejection

Figure 7.64 Production of a piston with fibrous inserts by squeeze casting

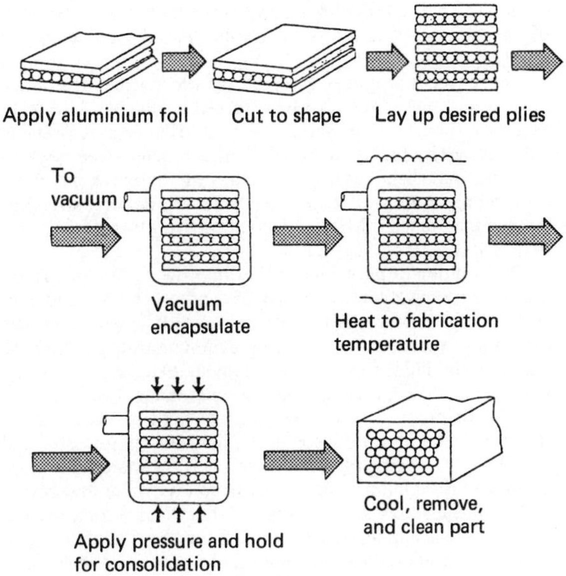

Apply aluminium foil   Cut to shape   Lay up desired plies

To vacuum

Vacuum encapsulate   Heat to fabrication temperature

Apply pressure and hold for consolidation   Cool, remove, and clean part

Figure 7.65 Aluminium–boron fibre composite fabrication by diffusion bonding

7.5.10 Fibre-reinforced glasses and ceramics

The advantages which would accrue from conferring increased toughness to a material such as a ceramic which has high hardness, high temperature strength and chemical inertness has stimulated investigations into the manufacture and properties of fibre-reinforced ceramics and glasses. Enhanced toughness properties have been achieved in composites which include:

- Graphite fibre/lithium alumino silicate matrix. These had fracture strengths ranging up to about 10^3 MPA and work of fracture γ_F* about 10^4 Jm^{-2} compared with 10 Jm^{-2} for the majority of ceramics.
- Silicon carbide fibre/silica glass matrix. These had fracture strengths ranging up to 600 MPa and work of fracture γ_F up to 600 Jm^{-2}.

*The work of fracture γ_F is the integrated area below the load-deflection curve divided by twice the specimen cross-sectional area.

- Silicon carbide fibre/silicon nitride matrix. These had fracture strengths between 55 and 127 MPa and work of fracture γ_F between 200 and 900 Jm^{-2}.

These materials are very costly to produce and, in the case of glass matrix composites, the glass tends to attack the fibres at temperatures above 400°C. Also, for reasons too complex to be detailed here, low fibre/matrix bond strengths produce higher toughness values (but lower strengths) than high bond strength composites. For these reasons, the only composite of this class to have achieved significant commercial application has been glass reinforced with steel wire mesh which continues to present a barrier after the glass has been shattered. Further information on the principles and technique of fibre reinforcement of ceramics is to be found in references 64 and 65.

7.5.11 Reinforced concrete

Concrete and mortar constitute, in terms of volume, the most important matrix materials for human-made composites. They are particulate composites. 'Concrete' comprises a matrix of hydrated Portland cement (other cements have been used) surrounding mineral particles, usually silica in the form of sand and aggregates. 'Fine' aggregates are limited in size to 5 mm, 'coarse' aggregates to between 5 and 20 mm. 'Mortar' is hydrated Portland cement and sand.

Portland cement should conform to BS 12. The sampling and testing of aggregates is described in BS 812 in conjunction with BS 882.

The maximum compressive strength of concrete structures has over 30 years increased from 40 to over 100 MPa provided that Norwegian codes are worked to. This has been achieved by: reducing the water required (and therefore the voidage) to provide workability during pouring and compaction by adding super plasticizers and water reducing agents; adding silica fume which forms an extremely dense matrix with Portland cement; and partially replacing cement by p.f.a and g.g.b.s. which react more slowly than cement and achieve the same strength without a damaging temperature rise. This development has been retarded in the EC by the limitation in BS 8100 of the design shear stress to a value based on strength of 40 MPa.

The tensile strength of concrete is affected by slow crack growth and must therefore be assessed by the Weibull techniques referred to in Section 7.6: 5 MPa is a reasonable working estimate. A material, 'macro defect-free (MDF) cement', has been made by removing macroscopic flaws during preparation of cement paste having flexural strengths between 60 and 70 MPa and compressive strengths greater than 200 MPa.[66]

The addition of polymers further improves concrete. There are two types: polymer-impregnated and polymer-added concrete. Polymer-impregnated concretes (PICs) are made by drying and vacuum/pressure impregnating hardened concrete with a liquid monomer such as methylmethacrylate to fill the voids and then polymerizing the monomer by radiation, thermal or promoter catalysis. The strength of PIC is about four times that of normal concrete, 200 MPa compressive and 20 MPa tensile being obtained. However, the material is more prone to brittle failure. Water permeability, water absorption and chemical attack are also reduced, but PIC is expensive and its commercial application is therefore limited. Polymer-added concrete is prepared by the addition of a polymer or monomer during the mixing stage. The increases in strength are not as great as those of PIC, compressive strengths being limited to about 100 MPa while tensile strengths up to 18 MPa are reported. Polymer-added concretes have increased resistance to abrasion and chemical attack and bond well to existing concrete. Since the increase in cost is only the cost of the materials they have extensive applications for items such as floors subject to heavy wear.

Concrete, like carbon, is a brittle material and benefits from three-dimensional reinforcement by both short and long fibres, as is shown (for both concrete and mortar) in Table 7.53. Chopped steel, glass and polypropylene have been used for precast concrete sections and steel for *in-situ* concrete. Flexural reinforcement can be achieved by adding 2% of random steel fibre to concrete.

The main problem is to obtain an adequate dispersion of this concentration. The mix must contain 50% of fines and, if vibration compacting is used, care must be taken to avoid unacceptable fibre alignment. 0.44% polypropylene has been used successfully for pile sections.

Glass fibre and steel bar and wire are used for filamentary reinforcement. Concrete pipes are manufactured with 1% wound glass fibre concentrated at the inner and outer sections. Reinforced concrete contains a three-dimensional network of steel bar and/or wire aligned to resist tensile stress. Compression stresses are resisted by the concrete, but those regions of the concrete near to a steel bar which is stressed in tension are subjected to tensile stress during operation.

The versatility of reinforced concrete is illustrated by its use in the Thorpe railway suspension bridge. This bridge actually contained more steel than would have been used in a steel girder bridge, but it was cheaper steel, and does not require to be painted.

In prestressed concrete steel tendons may be stressed, while the concrete is poured over them, and released to compress the concrete after it has cured. More frequently they pass through channels or holes in the concrete and are stretched, and the ends secured after the concrete has cured. In correctly designed structures of prestressed concrete the whole of the concrete should be in compression. Prestressed concrete structures can be designed to have much lighter sections than reinforced concrete.

Corrosion of the steel reinforcement can be a serious problem with both reinforced and prestressed concrete. A sound layer of concrete 25 mm thick will protect steel from corrosion but cracks which may form under either tensile or compressive loading may allow water, which is very likely to contain salt, to gain access to the steel reinforcement. If the steel rusts it will increase in volume and eventually cause disintegration of the concrete.

The techniques for improving the fracture strength of concrete discussed earlier should prove extremely beneficial, not only in preventing failure due to corrosion but also, by alloying reinforced and prestressed concrete structures to be designed to higher stresses, in opening up new applications in the field of mechanical as well as civil engineering. Evidence is accumulating as a result of studies on biomechanics[67] that comparatively high values of modulus and toughness can be

Table 7.53 Fracture strengths and work of fracture of cement/mortar composites

Material	Flexural cracking strength (MN m^{-2})	Ultimate flexural strength (MN m^{-2})	Work of fracture, γ_F (KJ m^{-2})
Carbon fibre/cement	30–50	130–185	2–8
Steel fibre/cement	6–12	6–17	2–4
Asbestos/cement	—	17–38	—
Glass/mortar	5–8	7–17	0.6–1.0
Polypropylene/mortar	~5	~5	—
Plain concrete	~5	~5	~0.03

Reproduced by courtesy of Applied Science Publishers Ltd.

achieved in such materials as nacre (mother of pearl) and antler by geometrical arrangements of calcium carbonate and small quantities of organic material. It is conceivable that high modulus, high strength and high toughness might be achieved at relatively low cost by combining high-strength high-modulus fibres, an MDF cement matrix and a thin tough, flexible polymer or elastomer interlayer which would bond to both the fibres and the matrix.

7.5.12 Particulate composites

The most important metal-matrix particulate composites are cemented carbides or hard metals. These are cermets, consisting of finely divided hard particles of carbide of tungsten, usually accompanied by carbides of titanium or tantalum, in a matrix, usually cobalt, but occasionally nickel and iron.

Hard metals have the high elastic moduli, low thermal expansions and low specific heats of ceramics combined with the high electrical and thermal conductivities of metals. They have high abrasive wear and corrosion resistance, good resistance to galling and good friction properties and they are, compared with ceramics, ductile, having fracture strengths about 1000 MPa and work of fracture γ_F about 250 Jm^{-2}.

Hard metal dies and tools are manufactured by a powder route. A 'green' or partially sintered compact is machined into shape and then sintered in a hydrogen atmosphere at a temperature approaching (or even reaching) the melting point of the matrix metal. After sintering (or solidification) cobalt occupies the interstices between the grains as an almost pure metal with its original ductility. If nickel or iron are used as a binder they tend to dissolve more tungsten carbide than does cobalt and the ductility of the resulting composite is impaired. Increasing the percentage of cobalt increases ductility but decreases hardness, modulus, resistance to wear, galling and crater formation.

Cemented carbides with 3% cobalt have a hardness (HV 500 g) above 1900, a flexural strength of 2200 MPa and an elasticity modulus of 675 GPa. Increasing the cobalt percentage to 25 decreases hardness to 950 and the modulus to 462 GPa but increases flexural strength to 3200 MPa.

In general, high-carbide versions, particularly those with added titanium carbide, are used for finishing cuts. Medium-carbide content materials are used for roughing cuts and low carbide content materials for high-impact die applications. Tantalum carbides are used for applications involving heat.

Tungsten carbides may be used in oxidizing conditions up to about 550°C and in non-oxidizing conditions up to about 850°C. Titanium carbides can be employed at temperatures up to about 1100°C.

The application of hard metals for wear-resistant cutting tools is now being challenged (in the absence of shock) by ceramics such as alumina and sialon. British standards for hard metals include:

BS 3821: 1974 Hard metal dies and associated hard metal tools
BS 4193: 1980 Hard metal insert tooling
BS 4276: 1968 Hard metal for wire, bar and tube drawing dies

A new and important development in metal-matrix particulate composites is the reinforcement of aluminium by silicon carbide particles. Forty volume per cent of silicon carbide in aluminium doubles the modulus, halves the coefficient of thermal expansion, increases tensile strength and greatly reduces frictional wear. This composite is available as sheet, extrusions, forgings and castings.

The field of metal in metal particulate composites is, by comparison, restricted. Examples are the additions of lead to steel to promote machinability and to copper to produce a bearing material. These applications are both giving place to other materials. In particular, bearings are made of porous sintered bronze impregnated with PTFE.

7.5.13 Laminar composites

The number of laminar composites is vast and defies classification. Metal/metal laminates usually comprise a substrate that provides strength but reduces cost with a surface material that resists environment or improves marketability. Examples include rolled gold, Sheffield and electroplate, tinplate, galvanized iron or titanium-sheathed steel. Alternatively, the core may be ductile and the surface hard enough to provide a cutting edge (e.g. a damascus sword blade).

Wrought iron consists of layers of iron and slag which confers corrosion resistance (and solid-phase weldability). Glass is laminated with transparent plastic for automobile windscreens.

A most important class of laminar composite is the sandwich (lightweight) structure which comprises two high-strength skins which may be metal, wood, plastic or cardboard separated by a core that may be basically lightweight such as balsawood or foam or of honeycomb construction.

7.5.14 Wood and resin-impregnated wood

Wood is a natural composite and is one of (if not the) oldest composite used by humans. It is reinforced by a system of parallel tubes constructed of cellulose fibres which confer longitudinal properties such as those shown in Table 7.54. This structure has developed by natural selection in such a way as to ensure that failure of one element does not interact with an adjacent element is such a way as to lead to progressive failure. The low specific gravity of wood gives it specific strengths comparable with steel. Its transverse properties are very poor and it very easily splits longitudinally. Provided, however, that this is allowed for in design, timber structures, such as the Lantern Tower at Ely Cathedral, may be designed to support heavy loads for many centuries.

Table 7.54 Mechanical properties of raw materials used for wood resin composites

		Douglas Fir	Dry Yellow Birch	West System Epoxy
Specific gravity		0.52	0.62	1.14
Compressive strength	(MPa)	48.2[a]	56.8[a]	96.5
UTS	(MPa)	103.4[a]	138[a]	62
Elastic modulus	(GPa)	14	14.3	2.06

[a] Parallel to grain.

Plywood was developed by the ancient Egyptians to provide strength in two directions and to prevent warping. The next advance was to impregnate laminated wood with resin. This can be achieved in two ways:

1. Veneers of softwoods such as Douglas Fir are impregnated with epoxy resin, laid up in a female mould and cured under a vacuum bag. Figure 7.66 shows a section near the root end of an aerogenerator blade made by joining two half-sections made in this way. Design allowables (based on wooden propellor blade experience) are given in Table 7.55. Tests on prototype aerogenerator blades have indicated that the mechanical properties are adequate and in series production the blades would be cheaper than any other material (with the possible exception of prestressed concrete).
2. Compressed impregnated wood (as manufactured by Permali, Gloucester) involves laying up birch or beech veneers interleaved with phenolic resin and bonding at

high temperature and pressure. This product is a highly weather-resistant electrical insulating material with mechanical properties as listed in Table 7.56 and a substantially flat S-N curve with fatigue strength better than 90 MPa at 10^9 cycles.

7.6 Polymers

7.6.1 Introduction

A 'polymer' is, strictly, any molecule which consists of two or more molecules of another chemical compound which is known as its 'monomer'. Polystyrene $(C_5H_8)_n$, which is a glassy solid, is formed by the polymerization of its monomer styrene or phenylethylene into a long-chain compound (see Figure 7.67). This occurs on standing but is accelerated by

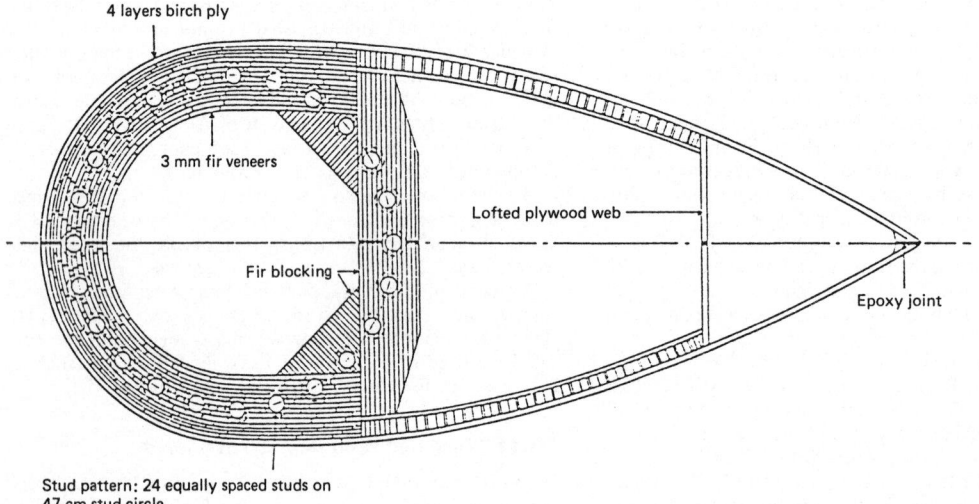

Figure 7.66 Section near the root end of an impregnated wood aero generator blade

Table 7.55 Design allowable for wood laminates

		Static allowables		4×10^8 cycles Fatigue allowables	
		'One time'	Working	$R = -1$	$R = 1$
Parallel to grain					
Flexural	(MPa)	39.25	32.87	16.69	39.25
Tensile	(MPa)	34.43	28.84	16.69	34.43
Compressive	(MPa)	30.21	25.30	16.69	30.21
Shear	(MPa)	4.71	3.94	2.11	4.71
Perpendicular to grain					
Tensile	(MPa)	1.15	0.97	0.47	1.15
Compressive	(MPa)	9.92	8.25	3.81	9.92
Rolling shear	(MPa)	0.70	0.59	0.32	0.70

Douglas Fir laminae, values corrected to 8% moisture, 49°C except $R = -1$ (room temperature).

Table 7.56 Mechanical properties of Permali impregnated compressed wood

		Longitudinal grain	*75% longitudinal grain*	*Equal longitudinal cross grain*
Crossbreaking strength	(MPa)	190	150	85
Compressive strength				
Parallel to laminae	(MPa)	170	170	140
Perpendicular to laminae	(MPa)	105	190	205
Shear strength				
Endwise	(MPa)	20	34	48
Flatwise	(MPa)	62	55	48
Elastic modulus	(GPa)	17.2	15.2	13.8

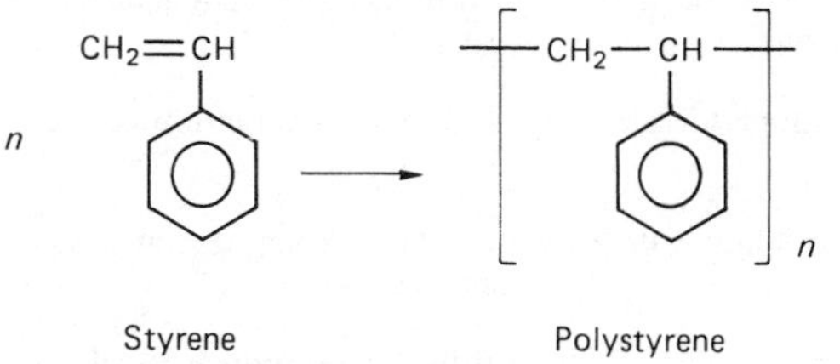

Figure 7.67 Styrene becomes polystyrene by the conversion of a double bond into two links

heat or the additions of an accelerator (for example, benzoil peroxide).

True polymerization requires the presence in the monomer of a double bond as in styrene or of a molecular structure which can be broken to provide links to other molecules. The name 'polymer' has, however, been extended, first, to include products of high molecular weight formed by 'condensation polymerization', which occurs by the elimination of a small molecule (usually water at each polymerization step), and, second, the formation of a large molecule from equal numbers of two different molecules. A typical example of these processes is the formation of polyamide 'Nylon' by the action of heat on adipic acid and hexamethylene diamine to form a chain which contains a repeating unit (see Figure 7.68).

The term 'polymer' (or 'elastomer') is now used for all synthetic resins, compounds of carbon or silicon with one or more of hydrogen, oxygen, nitrogen, chlorine, fluorine and sulphur, which contain large numbers of repeating units grouped in long chains or in cross-linked structures. An

elastomer differs from a polymer in having a comparatively low modulus and in being capable of elastic extension of several hundred per cent. Elastomers are more closely related to natural products because a large number of them are rubbers and the compositions of many synthetic elastomers are similar in composition to rubber.

There are natural polymers. Amber or the adhesive mentioned in the Bible as the binder for straw-reinforced bricks are among the oldest materials used by humans.

The great technological importance of polymers and elastomers arises because they can be compounded with additives, fillers and reinforcing agents (see Table 7.57) to become plastics, fibres, coatings, fluids, rubbers or composites from which useful artefacts or engineering structures may be manufactured. Plastics are generally less stiff and weaker than metals but their ease of processing, low density, attractive appearance, resistance to environment, usually lower cost and potential for recycling often render them the preferred raw material.

Polymers are usually purchased from the manufacturer or supplier in the form of plastics. It often happens that manufacturers have given names to the plastics which they supply that differ from those of the polymers which are their most important constituents. To avoid confusion, this section refers to the materials which it describes by the name of the polymer or elastomer. Where the manufacturer's name has obtained very wide currency it is also mentioned, together with the name of the manufacturer. All commercially available polymers, their US supplier and trade names are listed in *The Plastics Encyclopaedia*[68] published by Modern Plastics and the UK suppliers are listed in *Kompas UK*[69] published by Reed Information Services in association with the CBI. Both these are published annually and, in case of difficulty in obtaining a specific plastic, the reader is recommended to consult them.

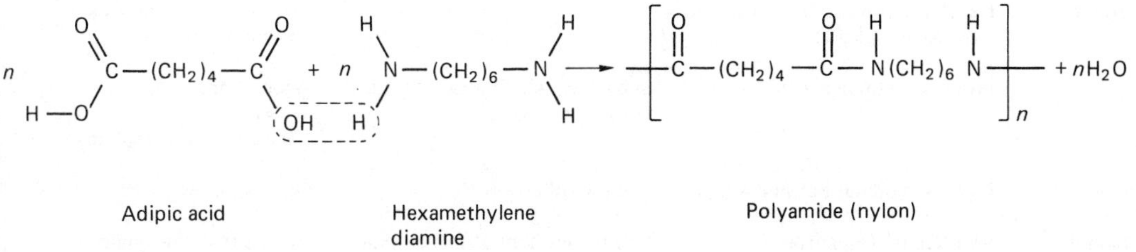

Figure 7.68 Formation of polyamide (nylon 66) from molecules of adipic acid and hexamethylene diamine with elimination of water

Table 7.57 Fillers and reinforcement for polymers

Type	Function	Type of material	Example of material
(a) Additive fillers			
Inert filler	Reduce cost. Reduce tackiness Improve electrical insulation Reduce distortion on moulding Reduce die swell on extrusion Reduce wear	Inert and insoluble mineral	Calcium carbonate China clay Talc Barium sulphate Carbon black
Coupling agent	Promote adhesion between polymer and filler	Adhesive	γ Mercaptopropyl-tri methoxysilane
Plasticizer	Lower melt viscosity (therefore facilitating processing)	Non-volatile solvent (high molecular weight)	Di-n-butyl phthalate metallic soap
Extender	Cheap replacement for properties of plasticizer	Compatible with plasticizer	Chlorinated paraffin wax
Lubricant	Prevent sticking of compound to processing equipment	Limited compatibility with compound	Lead or calcium stearate or stearic acid
Impact improver	Improve toughness of rigid amorphous thermoplastics	Synthetic rubber	Ethylene-propropylene rubber in polypropylene ABS and polystyrene
Stabilizer	Prevent ageing due to (a) Oxidation (b) Ultraviolet light	Chain breaking antioxidant Converts UV radiation to heat Shielding material	Substituted phenol (Benzo triazole) Carbon black
Colorant	Mass coloration	Pigment or die	
Cross linking agent	Produce cross links in linear polymers	Vulcanizing agent in rubber Accelerators in thermosets	Sulphur in rubber Aminos in epoxide resins Peroxides
Friction reducer	Reduce friction between plastic used as a bearing and (usually metal) shaft or plate	Material with low inherent friction	Graphite PTFE PTFE fibre Molybdenum disulphide
Nucleating agent	Promote rapid freezing, refine crystallization, improve clarity and reduce voids	Polymer of similar cohesive energy density but higher melting point	Nylon 66 in Nylon 6 Fine silica
Antistatic additive	Prevent generation of static electricity during service	Conducting compound which will migrate to the surface	Quaternary ammonium compound Glycol alkyl ester
Conducting additive	Confer some conductivity to the plastic	Conducting powder or flake that forms a network through the plastic	Silver, carbon black, graphite
Flame retardant	Resist the propagation of fire in and along a plastic		Trichloroethyl phosphate Antimony trioxide
Blowing agent	Produce cellular plastic	Substance which generates gas during processing	Azocarbonamide Water Fluorocarbon (obsolete)
Density reducer	Increase volume but not weight	Hollow spheres	Glass spheres
Photo-degradant	Renders polymer from photo-degradable by ultraviolet light	Ultraviolet light absorber which generates reactive chemical intermediates	Iron di-thio-carbonate.

Table 7.57 *(continued)*

Type	Function	Type of material	Example of material
Biodegradant	Render plastic degradable by biological action	Constituent which can be acted upon by naturally occurring enzyme	Cellulose or urea derivative
(b) Reinforcing fillers			
Reinforcing particle	Increase hardness, wear resistance and tensile strength mainly in the case of elastomers	Inert particle, to which the polymer will bond sometimes assisted by coupling agent	Carbon black Also silica, aluminium hydroxide, zinc oxide and calcium silicate
Reinforcing fibres	Increase modulus and tensile and impact strength usually in specific direction	(a) Generally used fibrous material	Wood flour and shaped wood particles, cotton flock, macerated fabric and synthetic organic fibre (nylon)
		(b) Cut fibre, usually of uniform length oriented to reinforce in specific direction	Glass, aramid and carbon Recently whiskers
		(c) Fibres running the full component length. (See section 7.5 Composites)	Glass, aramid and carbon

7.6.2 Classification of polymers

Polymers are classified into three categories: 'thermoplastics', 'elastomers' and 'thermosets'.

7.6.2.1 Structure of thermoplastics

The structures of thermoplastics (and elastomers) are essentially long single-chain molecules but chain branching occurs either with the production of relatively short and well-defined spur groups or (sometimes unintentionally) the chain may bifurcate into two long branches. The backbone of the chain consists, in the simpler cases, of carbon atoms but other atoms such as nitrogen and oxygen may substitute for carbon atoms at regular (or intentionally irregular) intervals; and in the case of silicones, the backbone consists of alternating silicon and oxygen atoms.

Each carbon atom can, in principle, link with four atoms, two to continue the chain and two to link with other atoms. The natural angle between two carbon bonds is 109°28', therefore a carbon chain would naturally assume a zig-zag pattern, but these angles are flexible and can, under certain conditions, permit rotation of one carbon atom with its attached groups relative to the next atom. The chains can therefore, in principle, bend to make way for other chains and to accommodate bulky side groups attached to the chain.

In the case of some polymers the link between two adjacent carbon atoms is made by two bonds (see Figure 7.69). Double bonds prevent rotation (although they may facilitate rotation round the next single bond in the chain), restrict bending and leave only one link free for hydrogen or a side group to be attached to each of the two carbon atoms on either side of the double bond.

An important characteristic of double bonds both in the main chain or in a side group is that they may be broken to leave two carbon atoms free to link with a side group or to

(a)   (b)

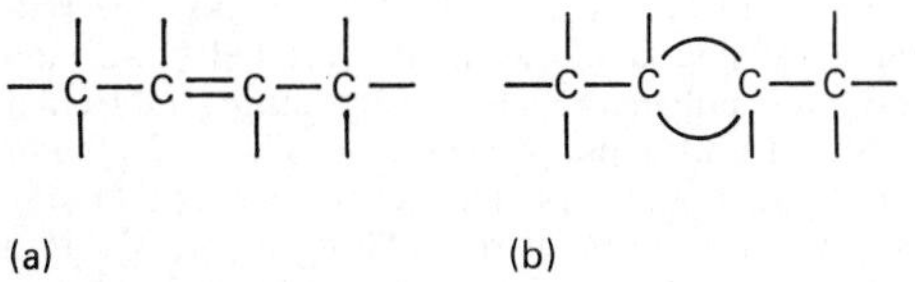

Figure 7.69 Double bond in carbon chain. (a) Chemical representation; (b) representation illustrating strain in bond

cross link to another chain. This is important in the case of elastomers and thermosets where chains are linked together. Some polymer chains contain cyclic groups and have therefore two strands at intervals along their length. A chain is inflexible (and bulky) at these points. Flexibility is further reduced when two adjacent cyclic groups are linked at two points (see Figure 7.70). The structure of the chain controls the characteristics of the polymer and therefore the plastic derived from it.

7.6.2.2 Influence of chain structure and length on properties of thermoplastics

Phase diagrams of thermoplastics There are two types of long-chain thermoplastic: amorphous and crystalline, depend-

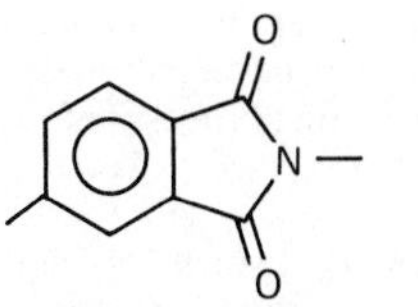

Figure 7.70 Phenyl group linked at two points to imide group

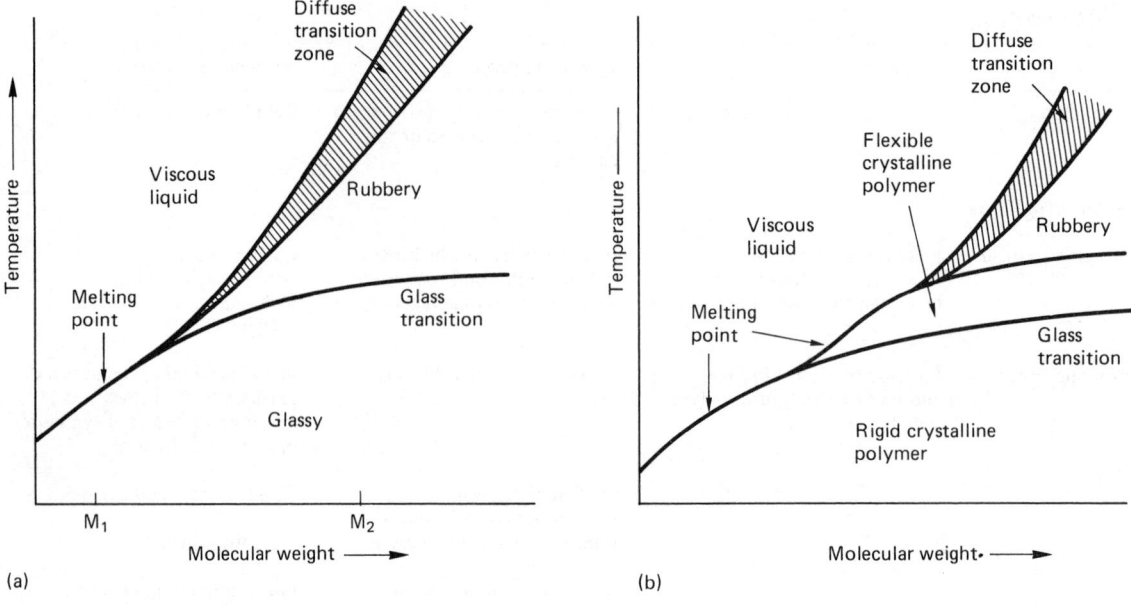

Figure 7.71 Temperature molecular weight diagram (a) of amorphous thermoplastics, (b) of crystalline thermoplastics

ing on the structure of the chain. Simple regular configurations promote packing and favour crystallization. Bulky side radicals or irregularly spaced blocks inhibit crystallization and favour amorphous polymers. The temperature, molecular weight diagrams of both types are shown in Figure 7.71. For high molecular weight polymers there is usually a liquid phase which, in cooling, transforms at the melting point, T_m, to a rubbery phase. On further cooling the rubbery phase transforms in the case of an amorphous polymer to a glassy phase at the glass transition temperature. In the case of a crystalline polymer the rubbery phase transforms to a flexible crystalline phase which itself transforms to a rigid crystalline phase at the glass transition temperature.

Properties of the liquid phase In the liquid phase the molecules of a thermoplastic polymer have sufficient energy to move independently of each other and the polymer is capable of viscous flow. The melting-point temperature depends on the stiffness of the chain – the stiffer the chain, the higher the melting point. With very stiff-chain polymers the liquid phase may be absent. The viscosity of the liquid depends on the length of the chain – the longer the chain, the higher the viscosity.

The preferred fabrication process for thermoplastics ('melt processing') is carried out in the liquid phase. A high melting point and viscosity require high pressures and fabricating equipment strong enough to withstand these pressures at high temperatures. In the limit the fabricating temperature may exceed that at which the polymer boils or decomposes so that the polymer cannot be melt processed. Certain very high molecular weight thermoplastics require to have the chains broken up into smaller lengths by a masticating process before they can be melt fabricated.

Properties of the 'rubbery' state Below T_m (which for high melting-point polymers is not sharp but is a diffuse transition) a thermoplastic enters a rubbery state. A simple (but not

universally accepted) concept is that in the rubbery state molecular rotation about single bonds in the chain can occur within a short time scale. The molecules are highly coiled, and uncoil on the application of a tensile stress and will accommodate reversible elongations as high as 1200%.

Influence of the glass transition (T_g) of amorphous thermoplastics Below the glass transition temperature T_g an amorphous polymer changes from a rubbery to a glass-like state. If this occurs below ambient temperature the polymer is an 'elastomer' and is capable of large extensions when stressed in tension. In natural and synthetic rubbers the highly flexible polymer molecules must be lightly cross linked to prevent them slipping past one another when stressed and in the rubber industry diene rubbers are 'vulcanized' with sulphur. When the glass transition temperature is above ambient the polymer is a transparent glass-like thermoplastic.

Properties of crystalline thermoplastics Thermoplastics with highly regular structures such as PTFE, polyethylene, nylon and aliphatic polyester are able to crystallize and this makes the glass transition temperature less definite. In high molecular weight polymers a flexible crystalline phase appears between the rubbery phase and the glass transition temperature. The rigid crystalline phase is less transparent (highly crystalline polymers may be opaque) and generally less ductile than the glass phase of amorphous thermoplastics. Polymers may also show shrinkage for long periods while they crystallize, but this may be avoided by annealing at a temperature where crystallization takes place rapidly.

The ability of a polymer to crystallize may be controlled or eliminated by introducing irregularities into the chain. One very effective way is by copolymerization, which can introduce two quite different groups at random (or controlled) intervals along the chain:

1. *Fibre-forming thermoplastics*: these include polyamides (nylon 66) and polyethylene terephthalate (Terylene). Most amorphous polymers have yield strengths around 55 MPa. Crystalline polymers may vary between 14 MPa (polyethylene) and 83 MPa (nylon). When, however, a crystalline polymer is stretched in the flexible crystalline phase (room temperature for nylon 66 and above 67°C for Terylene) it can develop strengths approaching 700 MPa, many times greater than unorientated polymers. High strengths can also be induced in films and sheet by two-way stretching. Besides high strength, biaxially stretched films of Melinex (polyethylene terephthalate), Saran (polyvinylidene chloride) and polypropylene have excellent clarity because the biaxial orientation of the crystallites does not affect the transmission of light waves.
2. *Liquid crystal polymers*: Even higher tensile strengths can be achieved in the so-called 'liquid crystal polymers' which consist of rod-like molecules that tend to orientate in the direction of shear when stretched (see Section 7.6.5). Tensile strengths up to 3.8 GPa, which (except for a lower modulus) is equivalent to steel or carbon for composite reinforcement, are available in poly-*p*-phenylenediamine ('Kevlar': Dupont).

Other properties of thermoplastics dependent on structure

1. *Creep strength*: Creep of a thermoplastic is thermally activated sliding of the chains relative to each other. This is hindered by a number of factors – spatial interference, chain stiffness and electrical and chemical interactions between the constituents of chains. A special case is hydrogen bonding, where the hydrogen of one chain is shared with oxygen or nitrogen at a nearby location on an adjacent chain. Most creep-resisting thermoplastics contain phenyl groups with sulphone, ether, ketone or imide links or a combination of all three. A high melting point does not by any means imply high resistance to creep.
2. *Density*: Density both of thermoplastics and thermosets depends mainly on the individual atoms present. Hydrocarbons do not contain heavy atoms and have specific gravities of 0.86–1.05. Where chlorine or fluorine are present specific gravities can range up to 2.2. Crystallization implies closer packing and therefore increases density.
3. *Electrical properties*: Electrical properties of thermoplastics and thermosets depend on the presence within the molecule of dipoles, pairs of atoms with contrasting values of electronegativity (see Table 7.58). The greater the difference in the electronegativity of the atoms bonded together, the greater the polarity of the bond. Where this difference is greater than 2, electrovalent bonds are commonly formed. Where it is less than 2, the bond is usually

covalent but it may also be polar. Thus a carbon fluorine bond is more polar than a carbon hydrogen bond. The influence of dipoles is only felt if they are unsymmetric. In the case of symmetrical molecules or groups such as carbon tetrachloride, phenyl groups, polyethylene and polyisobutylene the polar effects neutralize each other, therefore any electrical property is controlled solely by electron displacement. Such compounds and non-polar polymers have high-volume resistivities of the order of 10^{19}–10^{20} Ωm, high dielectric strengths (180–320 kV cm^{-1}), low dielectric constants (2–2.5) and low power factors (loss tangents less than 0.0003). On the other hand, polymers containing unbalanced dipoles, such as cellulose nitrate, have volume resistivities down to 10^{13} Ωm, dielectric strengths down to 120 kV cm^{-1}, dielectric constants as high as 7 and loss tangents as high as 0.06. Loss factor and dielectric constant are frequency related. At low frequencies the polar molecules are able to vibrate in-phase with the electric field and losses are low. They are unable to do this at higher frequencies and losses reach a maximum but become smaller at higher frequencies still because the dipoles cannot respond quickly enough. The effects are complex and the reader is recommended to consult reference 88.

7.6.2.3 Structure and properties of thermosets

Thermosetting polymers are, compared to thermoplastics, relatively small molecules, either branched or straight chains which each contain at least three reactive groups that can link with reactive groups in other molecules. These groups can be caused to interact by heat, by the addition of a hardener or accelerator, or by the action of all three. The result is a rigid cross-linked polymer which is chemically stable, does not melt and cannot be subjected to significant deformation. Such a polymer cannot be fabricated by any conventional process and must be processed in the low molecular weight form known as the 'A'-stage resin. The fabricated shape is then cross linked to form the 'C'-stage resin.

7.6.3 Polymer processing

7.6.3.1 Melt processing

Melt processing is the most important, most economical way of shaping plastics. It can be used for those thermoplastics with melting point and viscosity low enough not to subject die materials to excessive temperatures and stresses.

Melt processing of thermoplastics The raw material to be processed is usually supplied in the form of regular shape and even-size granules, because this makes quantity metering easier and quicker and leads to more uniform and predictable heating. One technique which has been developed recently is to pultrude impregnated continuous-fibre rovings to produce a lace which is chopped into 10 mm lengths (Verton: ICI). This produces a 10 mm fibre reinforcement which has greatly improved properties compared with normal short-fibre reinforcement. Whatever form of raw material is used, it is essential that it should be dried thoroughly before use.

The most important processing procedures involve the use of a screw pump.[70] The equipment used for 'injection moulding' is shown in Figure 7.72(a). The granules are fed into a heated cylinder by a screw which first recedes to provide space for the material and then advances to inject it into a relatively cool mould in which it sets. When the plastic has set the mould is opened and the moulding removed. Figure 7.72(b) shows an 'extruder' which will produce rod, tube or filament that can be

Table 7.58 Electronegativity values of some common elements

Element	Electronegativity	Element	Electronegativity
Caesium	0.7	Iodine	2.4
Potassium	0.8	Carbon	2.5
Sodium	0.9	Sulphur	2.5
Lithium	1.0	Bromine	2.8
Aluminium	1.5	Nitrogen	3.0
Silicon	1.8	Chlorine	3.0
Hydrogen	2.1	Oxygen	3.5
Phosphorus	2.1	Fluorine	4.0

(a)

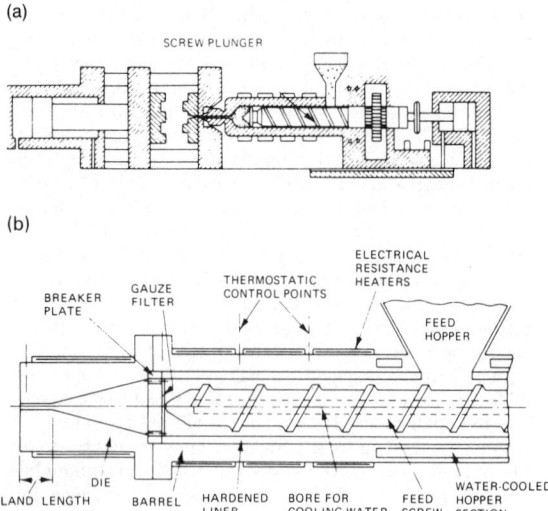

(b)

(c)

(d)

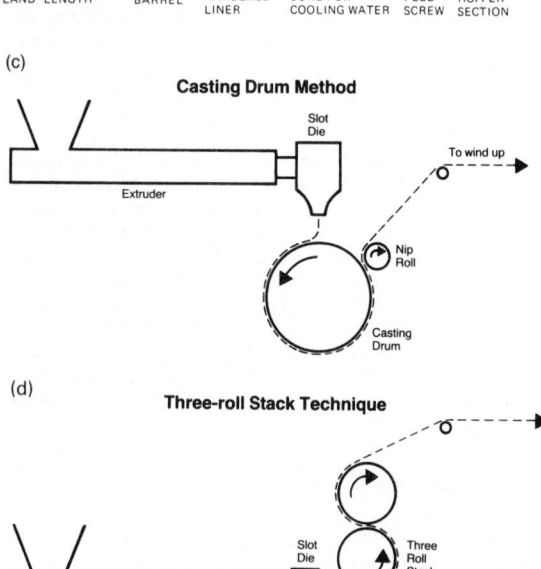

Figure 7.72 (a) Screw injection moulder; (b) screw extruder; (c) and (d) techniques used to produce film

(a)

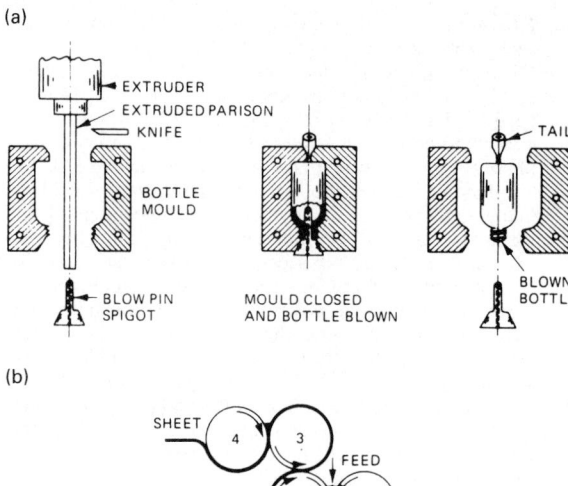

(b)

Figure 7.73 (a) Extrusion blow moulding; (b) calendering; flattening out molten material into sheet between moulds

quenched and drawn, sheet or strip in conjunction with a casting drum or three-roll stack (see Figure 7.72(c)) or coated wire.

Bottles may be produced by 'extrusion blow moulding' (see Figure 7.73(a)) in which a rod is extruded against a 'blow pin spigot'. After extrusion a mould is closed on the extruded parison which is still above softening point and the bottle is blown.

Procedures which do not involve extrusion are:

1. *Calenderizing*: flattening material out to sheet between rolls (Figure 7.73(b)). (This process is used extensively for plasticized PVC because problems have arisen in extrusion.)

2. *Centrifugal casting*: this involves filling a mould with (preferably) powder, spinning, heating to melt the powder, cooling slowly and removing the hollow casting.
3. *Compression moulding*: this process is more usually employed for thermosetting plastics but it can be used for thermoplastics, preferably in the form of powder. The compression moulding cycle involves compacting the polymer in the mould, heating above melting point and then applying a heating compression/relaxation technique; typically, 3.5 MPa for one minute, release, then 7 MPa for one minute, release, then 14 MPa for 10 minutes and cool at 40°/minute to removed included air.

Melt processing of thermosets Because a fully cross-linked thermoset will not melt, melt processing must be carried out on an 'A'-stage resin, a low molecular weight cross-linkable polymer which may be compounded with a hardener and/or accelerator. Cross linking is normally initiated by the heat of the mould and shaping must be completed before cross linking has occurred to such an extent that it prevents flow. This clearly raises problems in extrusion or injection moulding, because any material remaining in the extruder will set and become difficult to remove. Extrusion of thermosets must therefore be a discontinuous process in which a high-pressure reciprocating ram forces the thermoset through the die.

The most commonly used process is compression moulding, illustrated in Figure 7.74. The mould is heated to approximately 170°C and the material softens and flows to fill the mould before casting. The moulding material may be powder, a dough-moulding compound (DMC), a sheet-moulding compound (SMC) or a preform moulding.

Alternative processing methods There are a variety of alternatives to melt processing, some of which are used as a matter of convenience while others may be employed for materials which cannot be melt processed. PTFE, which cannot be melt processed, is formed by powder compression and sintering. Solution processing – dissolving the thermoplastic in a solution which is later allowed to diffuse out – may be used for

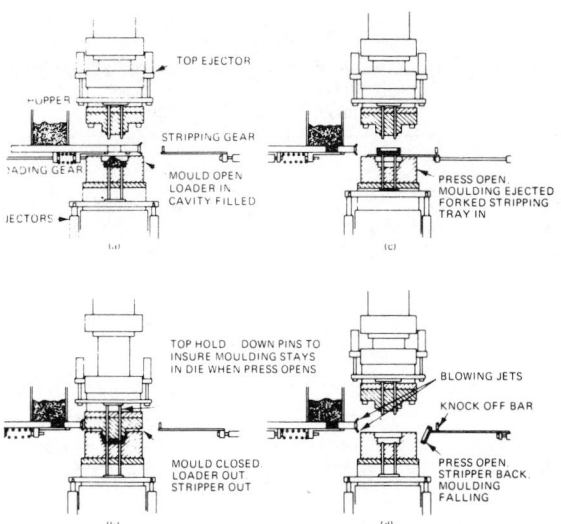

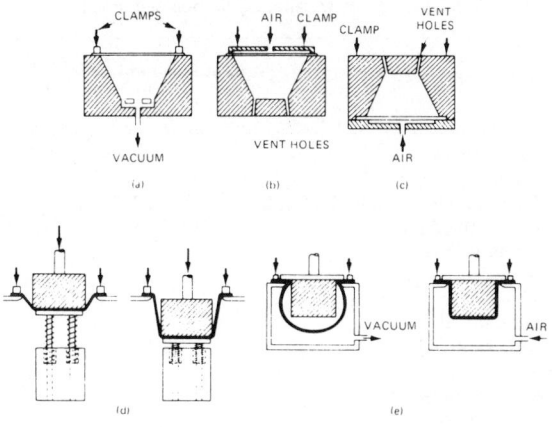

Figure 7.75 Shaping of sheet in the rubbery phase. (a) Application of vacuum; (b) and (c) air pressure; (d) mechanical pressure; (e) combination of methods (vacuum snapback)

Figure 7.74 Typical process for the compression moulding of thermosetting plastics. (a) Load; (b) mould; (c) eject, stripper in; (d) strip

other difficult thermoplastics. This process may also be employed for casting, film casting and fibre spinning.

Suspension processing is important because many polymers occur in latex form, i.e. as polymer particles of diameter of the order of 1 μm suspended in a liquid, usually an aqueous medium. The suspension is formed to shape by dipping, spraying or some other means of deposition and then coagulated by means of an acid. Other suspensions include particles of PVC suspended in a plasticizer.

Polymerization casting can be used for both thermoplastics and thermosets. A liquid monomer or the low molecular weight polymer is poured into a mould and polymerizes *in situ*. This process is mainly used for small sections to avoid problems associated with the heat involved during polymerization. A most important application is the encapsulation of small electrical or electronic components by epoxide resins. Other materials suitable for casting are acrylics, nylon and polyester resin.

Reaction injection moulding (RIM) is a form of polymerization casting which can produce components with foam cores and a non-porous skin as well as solid components. The components of the polymer are mixed with a liquid blowing agent in a reaction vessel and transferred to a mould. The heat of reaction volatilizes the blowing agent (which has in the past been a fluorocarbon but can be water) in the core but not at the surface of the component, where it is cooled by the mould. RIM is used for thermoplastic polyurethane elastomers.

Processing in the rubbery state is used where (as is the case with PTFE, high molecular weight polyethylene and cast polymethyl methacrylate) the polymer cannot be melt processed or where processing from sheet is the most convenient and economical means of manufacture. PTFE and polymethyl methacrylate are warm coined.

The forming of sheet is usually carried out by a vacuum-forming process (Figure 7.75) but polymers which are more difficult to deform such as polymethyl methacrylate and unplasticized PVC may require mechanical pressure or positive air pressure (Figures 7.75(d) and (e)).

There are, in addition to the processes described in this section, a variety of procedures for the fabrication of composites. These are described in Section 7.6.5.

7.6.4 Design of plastic components

When choosing a plastic material for a new or existing component the designer must be satisfied that the material can be fabricated to the required shape at an economic cost; will withstand environmental and stressing conditions which it will undergo during service and any emergencies that may arise; will satisfy any electrical, optical or aesthetic conditions that may be inherent in its operation; and will under no circumstances give rise to any safety or toxicity hazard. The tables associated with each class of polymer in Section 7.6.5 will provide the information required to satisfy many of these requirements. Answers to other questions may be obtained from fabricators and material suppliers or impartial authorities such as RAPRA[71] and the BPF.[72]

There exist, for all plastics, limiting minimum wall thicknesses that it is unwise to venture below. These are shown in Table 7.59. Wall thicknesses should be as even as possible because sudden changes in thickness in thermoplastics can lead to residual stresses and distortion and moulding faults such as sinks and voids. Corners should be radiused to reduce stress concentration. The effect of the radius of a fillet upon the stress concentration is shown in Figure 7.76. There is little benefit in increasing the fillet radius to thickness ratio above about 0.6.

The rigidity of plates or walls can be increased by the use of ribs which may be used as part of the runner system in injection moulding. Rib dimensions may be calculated by the standard strength of materials formulae provided that the plastic stiffness data used are appropriate to the temperature and loading conditions of service. Sinks or voids at the junction of the rib and wall are minimized by making the section of the rib less than that of the wall.

Detailed descriptions of the design of ribs, bosses, inserts, undercuts and general mould design are available in the literature.[73–76] Information on mould design for specific plastics may be obtained from the supplier.

Table 7.59 Suggested wall thicknesses of moulded articles

	Minimum for any article (mm)	For small articles (mm)	Average for most articles (mm)	Large to maximum articles (mm)
Thermosetting				
Phenolics				
General-purpose and flock-filled	1.27	1.58	3.17	4.75 to 25.4
Fabric-filled	1.58	3.18	4.75	4.75 to 9.53
Mineral-filled	3.18	3.18	4.75	5.08 to 25.4
Alkyd				
Glass-filled	1.02	2.36	3.18	4.75 to 12.7
Mineral-filled	1.02	3.18	4.75	4.75 to 9.53
Ureas and melamines				
Cellulose-filled	0.89	1.58	2.54	3.18 to 4.75
Fabric-filled	1.27	3.18	3.17	3.18 to 4.75
Mineral-filled	1.02	2.36	4.75	4.75 to 9.53
Thermoplastic				
Acrylics	0.64	0.89	2.36	3.18 to 6.35
Cellulose acetate	0.64	1.27	1.91	3.18 to 4.75
Cellulose acetate butyrate	0.64	1.27	1.91	3.18 to 4.75
Ethyl cellulose	0.89	1.27	1.58	2.36 to 3.18
Polyamide	0.38	0.64	1.53	2.36 to 3.18
Polyethylene	0.89	1.27	1.58	2.36 to 3.18
Polystyrene	0.76	1.27	1.58	3.18 to 6.35
Polyvinyls	1.36	1.58	2.36	3.18 to 6.35

Reproduced from *Plastics Engineering Handbook* by courtesy of the Society of the Plastics Industry Inc.

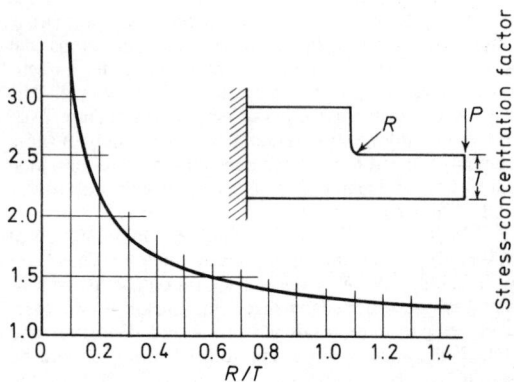

Figure 7.76   Effect of fillet radius on stress concentration factor.
P = applied load, R = fillet radius, T = thickness

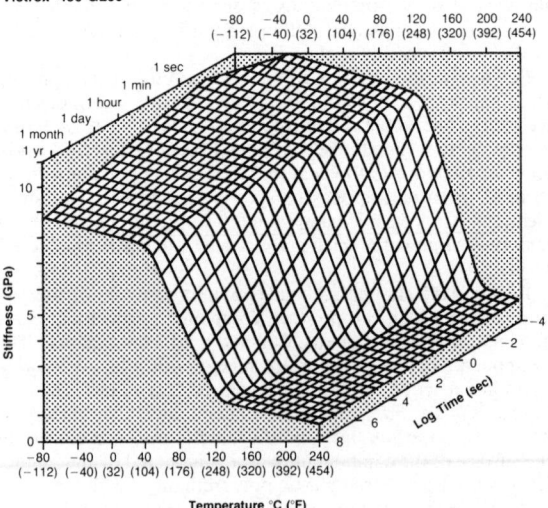

Figure 7.77   Stiffness versus temperature versus time for Victrex PEEK 450 GL30 (three-dimensional plot)

7.6.4.1 Designing for stiffness

The properties of a plastic depend on time and temperature, on any fibre reinforcement which may have been incorporated and on the fibre orientation. A typical three-dimensional plot of stiffness versus temperature versus time is shown in Figure 7.77. The influence of the glass transition temperature of 143° is evident.

The data to be used are supplied by the supplier of the plastic (in this case, ICI) as plots of tensile creep modulus against time at a series of temperatures (see Figure 7.78). The influence of anisotropy is taken into consideration by providing data measured along and transverse to the flow direction. Every attempt should be made to align the flow in the direction of the principal stress.

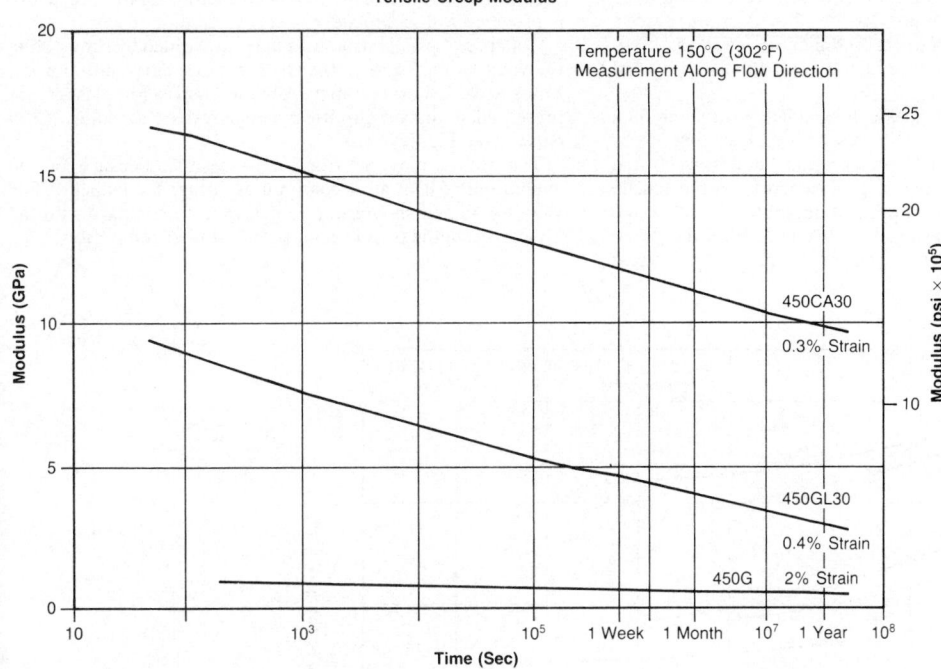

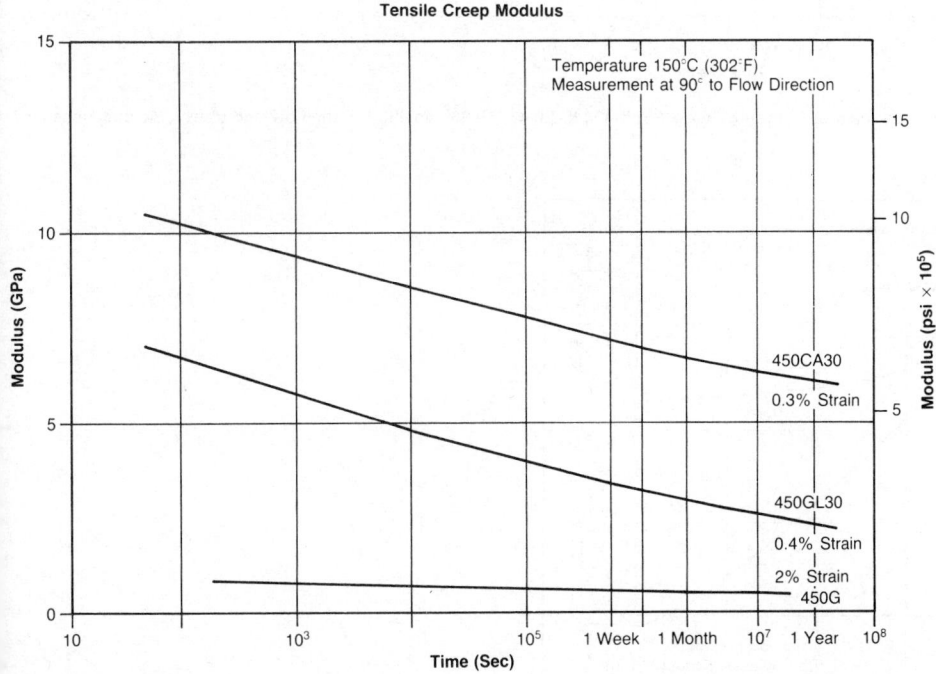

Figure 7.78 Plots of tensile creep modulus against time for three grades of Victrex PEEK in flow direction and transverse to flow direction (courtesy of ICI)

It is reasonable to assume that creep moduli in compression and tension are similar. The ratio between creep modulus in tension and in shear is approximately 2.9. A more precise value can be obtained from the supplier for each material. Tensile creep rupture data (see Figure 7.79) are also provided by the supplier.

It is first necessary to decide how much distortion of a component is allowable over its design life. The dimensions (thickness) of the component are then calculated from normal elastic formula using a value of E obtained from the tensile creep modulus at the appropriate temperature and design lifetime. When the dimensions have been decided the maximum stress can be calculated by elastic formulae. If this stress

is not more than 60% of the creep rupture stress at the design lifetime the design is safe. A worked example of this procedure is provided in the appendix to this section.

Collapse of plastic structures may be avoided by substituting the value of the stress in the stress rupture curve into the von Mises yield criteria. Fatigue characteristics for plastics are presented as maximum stress versus cycles to failure (SN) curves (see Figure 7.80).

The values may be frequency sensitive because of the generation of heat at discontinuities at high frequencies. The values obtained should not be expected to compare with the fatigue strengths of continuously reinforced composites.

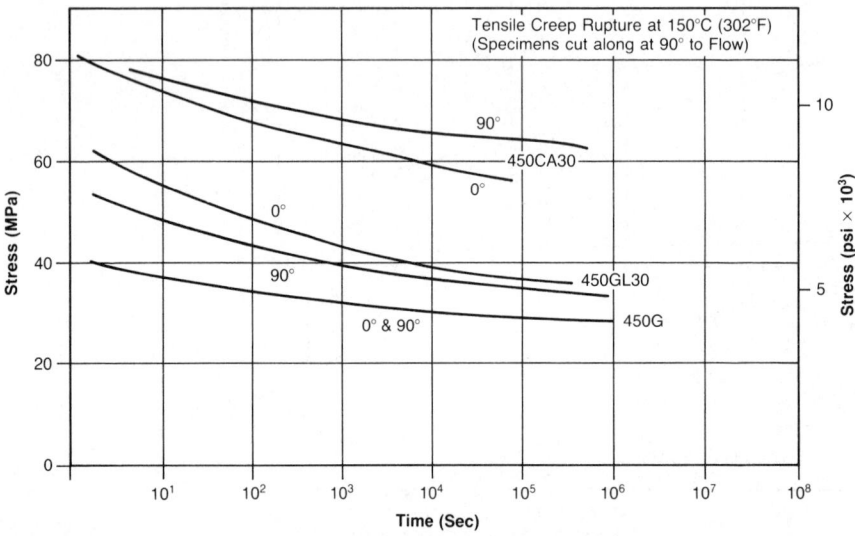

Figure 7.79 Tensile creep rupture data at 150°C on Victrex PEEK 400 CA30 450 GL30 and 450G in the flow and transverse directions

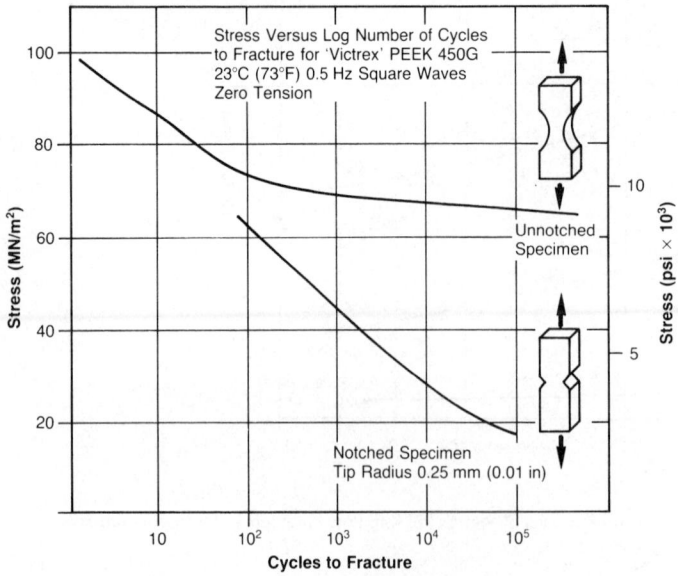

Figure 7.80 Example of fatigue data provided by manufacturer (ICI)

7.6.5 Polymer characteristics, properties and applications

7.6.5.1 Thermoplastics

Thermoplastics fall naturally into three classes: commodity thermoplastics, which are inexpensive and have relatively low strength and temperature resistance; engineering thermoplastics, which are stronger but rather more expensive; and high-temperature thermoplastics, which can have properties equal to or better than thermosets and compete for some applications with metals.

Commodity thermoplastics Properties of typical commodity thermoplastics are listed in Table 7.60.

- *Polyethylene*: This is the simplest polymer, essentially a long-chain aliphatic hydrocarbon (Figure 7.81). Its advantages are low cost and easy processability, it is an excellent electrical insulator and is tough and flexible. Its disadvantages are a low softening point and tensile strength and very poor creep resistance. The low-density form (LDPE) is used mainly for packaging and piping. The high-density form (HDPE) is fabricated by injection into domestic articles and by blow moulding into household bottles.
- *Polypropylene*: This is essentially polyethylene with an R(CH₃) group replacing one hydrogen molecule attached to every other carbon atom in the chain (Figure 7.82). Commercial polypropylene is essentially 'isotactic', that is, the methyl groups are all disposed on one side of the chain. Compared with polyethylene, polypropylene has a lower density, a higher softening point and therefore a higher service temperature and appears to be free from environmental stress cracking. On the other hand, it has a higher brittle point and is more susceptible to oxidation. It is used in preference to polyethylene, where its slightly higher stiffness and higher service temperature are advantages and also where its great resistance to fatigue in flexure (such as is required in integral hinges) is an advantage.
- *Polyvinyl chloride*: This is essentially polyethylene with a chlorine atom replacing a hydrogen atom bonded to every other carbon atom in the chain (Figure 7.83). The polymer itself is unstable, and in order to allow it to be processed a stabilizer (white lead for non-food applications) or a metal-

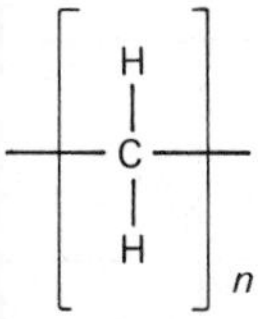

Figure 7.81 Structure of polyethylene

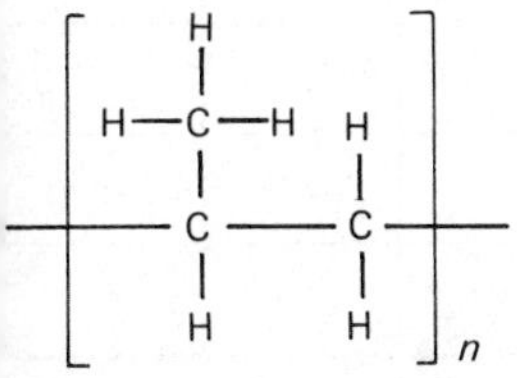

Figure 7.82 Structure of polypropylene

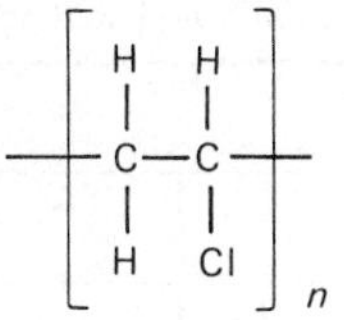

Figure 7.83 Structure of PVC

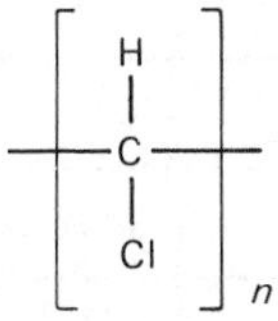

Figure 7.84 Structure of chlorinated polyvinyl chloride

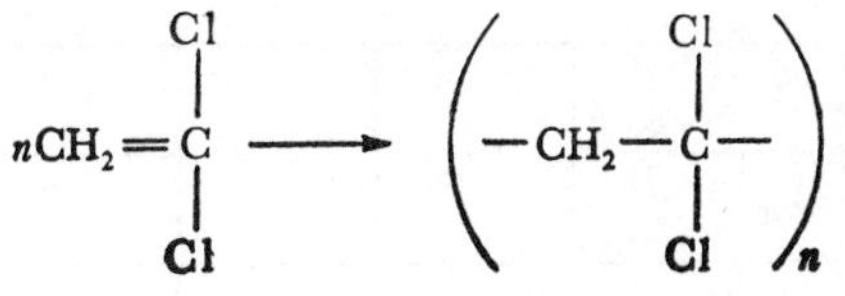

Figure 7.85 Structure of polyvinylidine chloride

lic soap must be added. Polyvinyl chloride is a rigid plastic but the addition of plasticizers such as di-isoethyl phthalate imparts flexibility and extensibility to produce materials with a wide range of properties. This, assisted by a very reasonable price, has resulted in very extensive application. PVC will also form pastes and latexes with plasticizers which extend its applications to coatings and rotationally moulded hollow articles. Further chlorination of vinyl chloride results in chlorinated polyvinyl chloride (Figures 7.84 and 7.85).

- *Polystyrene*: This is essentially polyethylene with a phenyl group replacing a hydrogen group bonded to every other carbon atom in the chain (see Figure 7.67). The effect of the benzene ring which, in contrast to the methyl group in polyethylene, can be attached in any direction, is to make the polymer stiff and brittle but amorphous, and it has poor chemical resistance. It is used for applications mainly promoted by its low cost. The impact strength of polystyrene can be improved by alloying, progressively with a rubber (styrene butadiene or polybutadiene) or with acrylonitrile to form SAN and with both butadiene rubber and acronitrile to produce ABS. The improved toughness greatly increases the applications and ABS competes propertywise with engineering plastics at a significantly lower cost.
- *Polymethyl pentene (TPX)*: This is an ethylene chain thermoplastic with a rather bulky and knobbly side chain. TPX has a very low density, very high transparency and a continuous service temperature of 134°C. This promotes applications for lamp covers, electrical equipment which has to withstand soldering and encapsulation and for transparent sterilizable equipment in spite of relatively high cost, low impact strength and poor ageing properties. TPX was developed by ICI but is now marketed by Mitsui.

Table 7.60 Properties and an indication of price for typical commodity thermoplastics

Group	Property		Test method			Polyethylene				Polypropylene	
	Specific		ASTM	Other	Units	Low density LDPE	High density HDPE	HDPE 30% glass fibre	Ultra high molecular weight		30% glass fibre filled
Mechanical	Tensile strength 23 °C		D638	ISO R527	MNm⁻²	0.4 - 16	21 - 38	70	17 - 24	27 - 40	47 - 103
	Flexural strength 23 °C		D638	ISO R178	MNm⁻²	–	35 - 50	80	28	40 - 50	63 - 137
	Elongation at fracture 23 °C		D790	ISO R527	%	90 - 650	600 - 1500	2 - 3	300 - 500	30 - >200	2 - 4
	Tensile elastic modulus 23 °C		D638	ISO R527	MNm⁻²	120 - 240	420 - 1400	–	140 - 750	500 - 1900	4000 - 6900
	Flexural elastic modulus 23 °C		D790	ISO R178	MNm⁻²	55 - 410	690 - 1800	6300	900 - 960	1200 - 1700	4200 - 6550
	Impact Izod 3.5mm. Strength notch 23 °C		D256	ISO R180A	J/cm of notch	No break	0.6 - no break	0.55	No break	0.2 - 1.2	0.3 - 1.6
	Hardness ᵃ Rockwell or Shore		D785		Shore	D41 - 505	D60 - 70	R85	D60 - 70	R80 - 105	R90 - 115
Thermal	Max. continuous service temp. (no load)			UL746B	°C	85 - 100	120 - 130	–	90	110 - 120	90 - 120
	Deflection 0.45 MNm⁻²		D648	ISO R75	°C	38 - 50	60 - 95	132	95	95 - 135	146 - 163
	Temp. 1.81 MNm⁻²				°C	32 - 50	43 - 55	127	65 - 82	46 - 67	130 - 150
	Thermal conductivity		C177		Wm⁻¹ K⁻¹	0.33	0.46 - 0.02	0.37	0.35 - 0.44	0.17 - 0.20	0.33
	Thermal expansion		D696		10⁻⁶ K⁻¹	100 - 200	110 - 130	50	200	68 - 104	31 - 38
Electrical	Volume resistivity			IEC 93	ohmm	1.2×10¹³->10¹⁴	> 10¹⁴	-	> 10¹⁴	> 10¹⁴	10¹³-1.5×10¹⁴
	Dielectric strength		D149	IEC 243	kVm⁻¹	180 - 400	180 - 200	-	280	240 - 280	150 - 205
	Dielectric constant or Relative permittivity	60 Hz	D150	IEC 250		2.25 - 2.35	2.30 - 2.35	-	2.3	2.2 - 2.6	2.4
		10⁶Hz				2.25 - 2.35	2.30 - 2.35	-	2.3	2.1 - 2.6	2.2 - 2.9
	Power factor or Loss tangent	60 Hz	D150	IEC 250		< 0.0005	< 0.0005	-	0.0002	0.0005 - 0.0007	0.001
		10⁶Hz				< 0.0005 - 0.0009	< 0.0005	-	0.0002	0.0002 - 0.012	0.001-0.003
	Arc resistance		D495		S	135 - 160	-	-	-	125 - 136	70 - 80
Other	Specific gravity		D792			0.918 - 0.948	0.950 - 0.065	1.17	0.937 - 0.060	0.90 - 0.91	1.12 - 1.13
	Melting point			DSC	°C	102			132	180	
	Mould shrinkage				%	1.5 - 5	2.5	0.3 - 0.4	8 - 10	1.0 - 2.5	0.4 - 0.6
	Water absorption 24 h		D570	ISO R62A	%	< 0.01 - 0.03	< 0.01	0.2	< 0.01	0.01 - 0.3	0.03 - 0.04
	Refractive index		D542			1.5 - 1.51	1.5 - 1.51			1.50	
	Flammability			UL 94		V - 2				V0/HB	
Econ-omic	Cost ᵇ		Best estimate of mid-1991	£ per kilo for 1 tonne quantity		0.80	1.20	2.00	1.10	0.75	1.60

ᵃ R. Rockwell M or R scla – S. Shore D or A scale

ᵇ Approximate costs at mid-1991. Prices vary according to manufactured quantity supplied and also generally with inflation and relative to each other with time. The prices quoted are for guidance in preliminary selection and should be verified before a final choice is made.

Polystyrene		ABS		Poly Methyl pentene (TPX)		Poly Methyl methacrilate		Cellulose acetate	Cellulose nitrate	PVC Flexible (plasticized)
General-purpose	20/30% glass filled	General-purpose	20/40% glass filled	Unreinforced LC	LC glass filled	Cast sheet and mouldings	Low/high impact			
35 - 84	63 - 105	17 - 62	60 - 133	22 - 25	56	55 - 85	40 - 60	29 - 77	35 - 70	7 - 26
83 - 118	74 - 140	68	112 - 190		96	80 - 140	60 - 100	20 - 76	56 - 70	
1.0 - 4.5	0.70 - 1.3	10 - 140	2.5 - 3.0	10 - 55		5 - 6	20 - 100	6 - 70	10 - 40	200 - 450
2800 - 3500	5900 - 9000	9000 - 2500	4200 - 7000	1200 - 1400		2400 - 3300	1400 - 2800	650 - 4000	1400 - 2800	3 - 20
2800 - 3200	5600 - 7000	2100 - 3100	5.6 - 9.8	800 - 1800		2400 - 3800	1500 - 2600	800 - 2480		
–	0.2 - 2.25	0.95 - 2.00		0.2 - 2.8		0.2	1.0	0.2 - below 20 °C / 3.2 - above 20 °C		Varies with plasticizer
M65 - M90	M70 - M95	R100	0.5 - 1.2	R70 - 90		M100	M70 - 100	R54 - 125	M25 - 50	S.A - 75
105	90 - 105	75 - 85	95 - 110	134		80 - 90	60 - 80	55 - 95	60	Cable insulation grade 105 .
			100 - 105			100 - 105	85	52 - 105		–
65 - 113	100 - 110	83 - 87	98 - 116			90	80	48 - 86		
	0.12	0.18 - 0.32		0.17		0.16 - 0.25	0.16 - 0.2	0.26 - 0.32	0.13 - 0.21	0.125 - 0.16
60 - 80	36 - 38	60 - 130	29 - 36	11.7		70	75	80 - 190	130 - 160	50 - 180
$10^{11} - 10^{15}$	3.2×10^{14}	1.2×10^{14}		$> 10^{14}$		10^{13}	10^{13}	$10^{12} - 10^{13}$	$10^{12} - 10^{14}$	$10^{9} - 10^{13}$
200 - 280	140 - 170	140 - 185		276		200	200	100 - 240	120 - 240	120 - 160
2.45 - 3.1	–	2.4 - 5.0		2.1 - 2.12		3.3 - 3.9	3.3 - 4.5	3.5 - 7.5	6.7 - 7.3	5 - 9
2.4 - 2.7	–	2.4 - 3.8		2.12		2.2 - 3.2	2.1 - 3.2	3.5 - 5.0	6.2	3.3 - 3.45
0.0001 - 0.0006	0.004-0.014	0.003-0.008		0.00007		0.05	0.05	0.01 - 0.06	0.06 - 0.15	
0.0001 - 0.0004	0.001-0.003	0.007-0.015		0.000025		0.15 - 0.3	0.15 - 0.03	0.01 - 0.10	0.07 - 0.10	Varies according to formulation
60 - 140	25 - 40	50 - 85				No tracking		50 - 310	60 - 80	
1.04 - 1.11	1.2 - 1.33	1.04 - 1.07	1.10 - 1.38	0.835 - 0.845		1.19	1.15 - 1.2	1.22 - 1.34	1.35 - 1.40	1.2 - 1.55
230				245		160 - 200		300		1.2 - 1.45
0.1 - 0.6	0.1 - 0.2	0.4 - 0.9	0.1 - 0.2	1.5 - 3.0		0.1 - 0.8	0.4 - 0.8			
0.03 - 0.4	0.05 - 0.10	0.35 - 0.4	0.18 - 0.40	0.01		0.3 - 0.4	0.4			
1.59 - 1.60	Translucent/opaque	1.536	Translucent/opaque	1.463		1.49	1.49	1.46 - 1.50	1.50	Opaque
V0/HB		V0/HB				HB		HB	Inflammable	Slow / Self ext
0.85	1.30	1.80	2.10				2.20	2.60		0.75

- *Acrylics*: Acrylic plastics are a further example of an ethylene chain. They are produced by the polymerization of ethyl methacrylate or, more usually, methyl methacrylate. They are hard, transparent materials. The applications of polymethyl methacrylate stem from its good light transmission and outdoor weathering properties. It is used mainly for covers of light signs.
- *Cellulosic polymers*: 'Celluloid' (cellulose nitrate plasticized with camphor and cellulose acetate, made respectively by the nitration and acetylation of cotton linters or wood pulp) was among the earliest plastics but most of the applications have been discontinued in favour of vinyl plastics. Cellulose nitrate is still used for knife handles, spectacle frames and table-tennis balls. Cellulose acetate, because of its clarity, is used for films and sheeting. Its application for mouldings and extrusion is being displaced by styrene polymers and polyolefins.

Engineering thermoplastics Engineering thermoplastics combine reasonable strength, stiffness and toughness. Their raw material cost varies from at least twice to several times that of the commodity thermoplastics. The properties of typical engineering thermoplastics are listed in Table 7.61.

- *Polyamides (nylons)*: (Aliphatic) polyamides differ from polyethylene-type polymers by having polar CONH groups spaced out at regular intervals along the aliphatic chain. The polar groups cause the polymers to crystallize with a high molecular interaction while the intervening aliphatic chain segments provide flexibility in the amorphous region. In principle, polyamides are formed by the condensation of a diamine with a dicarboxylic acid (Figure 7.86). Nylons are classified by numbers according to the number of carbon atoms ($m + 2$, p) in the material or materials from which they are formed. Thus nylon 6 which is formed from caprolactum, which contains six carbon atoms, has the formula shown in Figure 7.87. The number(s) provide a measure of the spacing of the CONH groups in the chain. The closer the CONH groups, the higher the melting point and heat deflection temperature and the higher the tensile strength, rigidity, hardness, resistance to creep and hydrocarbons and the water absorption.

  Nylon 6 and nylon 66, which have relatively closely spaced CONH groups (Figures 7.87 and 7.88), are very suited to fibre production which takes betwen 80% and 85% of their production. This gives these two grades an economy of scale advantage so that the other grades are only used where nylon 6 and nylon 66 are unsuitable, usually because of their high water absorption. Nylon 6 and nylon 66 plastics are used mainly for engineering applications where their good bearing qualities are advantageous, but recently acetal resins, which are superior in everything except toughness (the toughness of nylon is enhanced by water absorption), are giving nylons increasing competition. Nylon 11 and nylon 12 are used for electrical insulation where water absorption may be critical. Glass-filled nylons such as ICI Verton, which has aligned long fibres, are taking an increasing share of the market.

$$\left[C-O-N-H-C_2H_5 \right]_n$$

Figure 7.87 Structure of nylon 6

$$\left[OC-C_4H_8-CONH-C_6H_{12}NH \right]$$

Figure 7.88 Structure of nylon 66

- *Aromatic polyamides*: These, derived from aliphatic amines and terephthalic acid, are, not surprisingly, amorphous, more rigid, harder, have lower water absorption, lower expansion coefficient, better heat and moisture resistance, better insulation properties than nylon and are transparent. They compete with polymethyl methacrylate, polycarbonate, polysulphone and ABS in toughness but have not the heat resistance of polysulphone and ABS (Figure 7.89).
- *Aramid fibres*: There is a special class of aromatic polyamide fibre defined by the US Federal Trade Commission as 'aramid fibres' for which the fibre-forming substance is a long-chain synthetic polyamide in which at least 85% of the amide linkages are attached directly to two aromatic rings. Two such liquid-crystal fibres, poly-*p*-phenylene terephthalimide (Kevlar), and poly-*m*-phenylene isophthalimide (Nomex), are marketed by Dupont (Figure 7.90). Kevlar has exceptional strength (up to 3.8 GPa) and is used for reinforcement. It is equivalent in strength but inferior in modulus to steel or carbon.

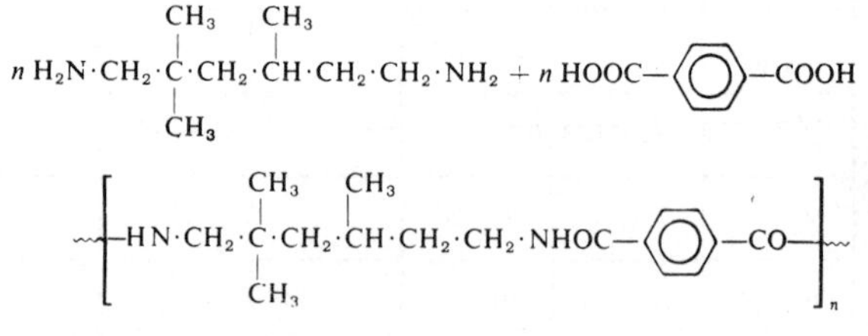

Figure 7.89 Structure of typical aromatic polyamide

$$n\ HOOC-C_mH_{2m}-CO\,OH + m\ H_2N-C_pH_{2p}-NH_2 \longrightarrow$$

dicarboxylic acid   diamine

$$\left[OC-C_mH_{2m}-CONH-C_pH_{2p}NH \right]_n + nH_2O$$

nylon m + 2, p

Figure 7.86 Formation reaction for polyamides

Table 7.61 Properties and an indication of price of typical engineering thermoplastics

	Polyamide						Acetal resin		Polycarbonate		Polyester resins			Modified PPO		Polyphenylene sulphide	
	Nylon 6.6 Dry	Nylon 6.6 50%RH	Nylon 6 Dry	Nylon 6 50%RH	Nylon 6.6 30% glass fibre	Nylon 6 30% glass fibre	Homo polymer	Teflon fibre filled (Delrinsaf)	Unfilled	40% glass	PETP glass reinforced	PBT unreinforced	PBT glass reinforced	Cheapest unfilled (110)	Dearest 30% glass filled GPN3SE1	Unfilled	40% glass
Tensile yield strength	65	50	75	40	150	160 - 140	66	48	64	120	167	34	150	45	78	75	135
Flexural strength	110	40		35	260	260	100		102	190	230		170	88	142	140	141
Elongation at fracture	60	400	20	190	3	3 - 5	45	5	80 - 120	2 - 3	3	300	2.5	60	4 - 6	1 - 6	1 - 6
Tensile elastic modulus	3200	2600	2800	1400	9600	9600/5600	3500		2500	10 000	9700		8400	2500	8400	3300	3800
Flexural elastic modulus	3000	1000	3000	950	9100	7600	2700	2800	2500	10 000		2400	8200	2.5	7700	3800	3800
Impact strength	0.5	1	0.5	3	8.0	1.4	1	29	6 - 9	1	1.01	0.5	1.1	2.1	0.9	0.02	0.31
Hardness	RM 118	RM 108	RM 119	RM 109	E 60	RM 21	RM 94	RM 118	RM 93	RM 93	RM 100	RM 71	RM 71	R 115	L 109	R 124	R 121
Max. continuous service temp. (no load)	100	95	95	90	120	100			125	130			170	-	105	-	185
Deflection 0.45 MNm^{-2}	230		170	170	260	212	170	168					217		142		
Temp. 1.81 MNm^{-2}	85		80	80	200	200	122	168	140	145	224	50	208	110	137	137	243
Thermal conductivity	0.25	0.25	0.25		0.2				0.21	0.22			0.24	0.23	0.16		
Thermal expansion	90		95		27	25	81		60 - 70	20	32	89	44 - 80	67	25	49	28
Volume resistivity	10^{13}	10^{11}	10^{12}		5×10^{13}	5×10^{8}	$>10^{13}$	5×10^{15}	10^{15}	10^{15}	10^{13}	1.2×10^{12}	7.5×10^{12}	10^{15}	10^{15}	10^{16}	2×10^{15}
Dielectric strength	200		200		210	160	150	5	155	200	224	50	170	195	212	150	177
Dielectric constant 60 Hz	4	6	4.2	4.2	4.5	4.2	3.7	3.1	3.1	3.4	3.6	3	3.9	2.65	3.15		
Dielectric constant 10^6 Hz	3.5	4.7	3.4	3.4	3.7	3.7	3.7	3.1	3.0	3.4	3.5	3	3.6	2.64	3.11		3.8
Power factor 60 Hz	0.014		0.04		0.018	0.018	0.0048	0.009	0.0009	0.0013	0.005	0.002	0.0025	0.0004	0.0020	0.0005	0.0013
Power factor 10^6 Hz	0.04	0.03	0.03		0.017	0.017	0.0048		0.01	0.0067	0.012	0.021	0.015	0.0009	0.0021		
Arc resistance	120				110	100	129		10 - 11	5			150	75	120	34	
Specific gravity	1.14		1.13		1.37	1.34	1.4	1.54	1.2	1.52	1.56	1.31	1.52	1.06	1.29	1.34	1.60
Melting point	264		213		264	213	175	175	220/230	220/230	250/255	224	224	250	290	285 - 295	285 - 295
Mould shrinkage	1.2 - 1.8		0.5 - 1.5				2.3	2.0	0.4 - 0.7	0.1 - 0.3			0.2 - 0.8	0.5 - 0.7	0.1 - 0.3	1.0	0.2 - 0.6
Water absorption	1.5		1.8		0.9		0.25	0.18	0.12 - 019	-			0.06	0.07	0.07	<0.02	<0.05
Refractive index	Translucent		Translucent		Opaque	Opaque	Opaque	Opaque	1.54 - 1.6			1.54 - 1.57	Opaque				
Flammability	VO	Self-extinguishing	Self-extinguishing		Slow-burning	Slow-burning	Slow-burning		V - 2	V - 1			94HB/V0	Self-extinguishing non-dripping	Self-extinguishing non-dripping	Non-burning non-dripping	V - 0/5V
Cost [a]	2.9		3.0		3.0	3.1	2.4		3.0	3.5	2.4	2.65	2.75	2.65	3.7	5.2	6.75

[a] Approximate mid-1991 prices, £ per kilo per tonne.

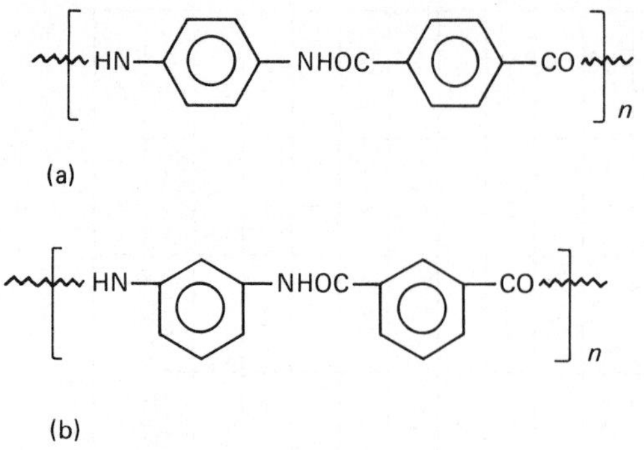

(a)

(b)

Figure 7.90 Structure of (a) Kevlar, (b) Nomex

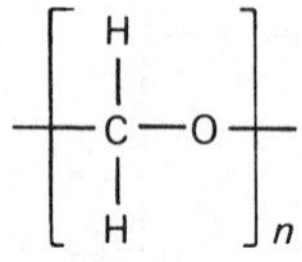

Figure 7.91 Structure of acetal chain

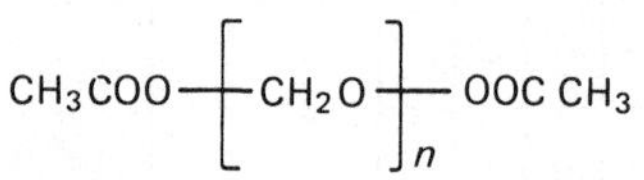

$$CH_3COO \left[CH_2O \right]_n OOC\,CH_3$$

Figure 7.92 End sealing of acetal chain

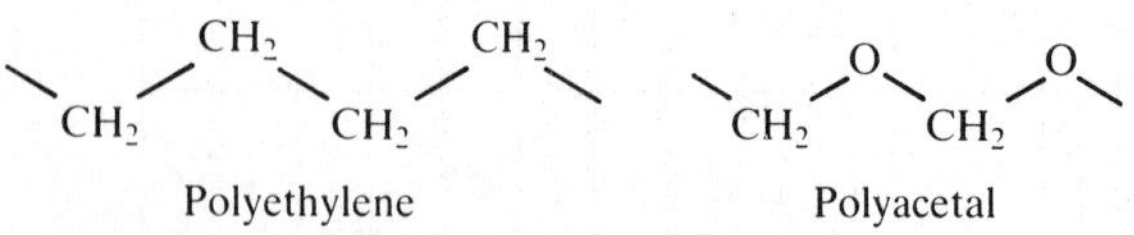

Polyethylene   Polyacetal

Figure 7.93 Polyethylene and acetal chains compared

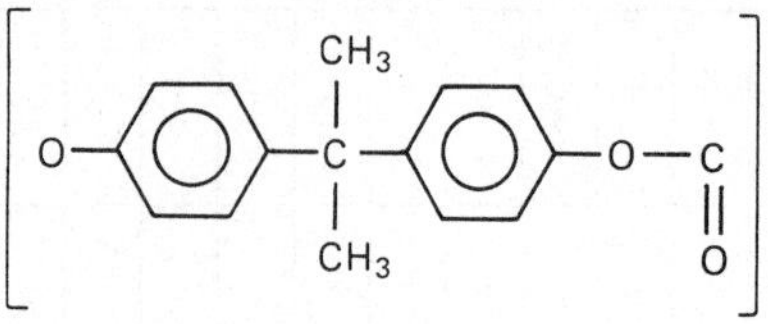

Figure 7.94 Structure of bisphenol carbonate

insulating properties and is self-extinguishing. It has a very wide application where two or more of these attributes are required and its production capacity has doubled in seven years. Its major application is in electronic and electrical engineering. Its disadvantages are special care needed in processing, limitations in chemical and ultraviolet light resistance, moderate electrical tracking resistance and notch sensitivity.

- *Polyesters*: Linear polymers with very varied structure and properties may be obtained from esters of dihydroxy alcohols and dicarboxylic acids. The possibilities of variation of the polyesters are almost infinite. The principal types are: polyethylene terephthalate (PETP), well known as a fibre (Terylene), as a film (which is stabilized and crystallized by two-way stretching and annealing), as a reinforced and nucleated moulding material and again by two-way stretching as a bottle material in this case gas permeability is reduced by coating with vinylidene chloride base copolymers and polybutylene terephthalate (*PBT*): This has a lower processing temperature but also lower values of T_m and T_g. PBT has better processability than PETP and has therefore a wider range of application for moulded equipment. Other polyesters marketed include polyarylates with high mechanical strength and low notch sensitivity and oxybenzoyl polyesters which have continuous-use temperatures up to 300°C. They are both highly expensive, as are the liquid-crystal polyesters which have excellent mechanical properties and ductility.

- *Modified polyphenylene oxide* (*Noryl: GEC*): Polyphenylene ether, whose structure consists of benzene rings linked by oxygen atoms, would be expected to have high thermal and dimensional stability. Such a material with 2.6 methyl side groups was found to form and polymerize very easily and is marketed by GE under the (incorrect) name polyphenylene oxide. It is a rigid, heat resistant, more or less self-extinguishing polymer with good electrical and chemical resistance, low water absorption and very good dimensional stability. Unfortunately, its price is too high to justify more than very limited application and GE therefore introduced a series of blends with polystyrene-type plastics. They are available in combinations of self-extinguishing and non-self-extinguishing and glass-reinforced and non-reinforced grades. They go some way to bridging the gap between the engineering thermoplastics and the more recently developed high-temperature thermoplastics.

- *Polyphenylene sulphides*: These present a problem in classification. They have rather more heat resistance than standard engineering polymers such as polycarbonates but not enough to bring them into the classification of high-temperature polymers. They are, in principle, thermoplastic but can be cross linked by air ageing. Their advantages are:

Heat resistance better than polycarbonate
Flame resistance UL temperature index 240, Oxygen Index 53%

- *Acetal resins*: Very pure formaldehyde can be persuaded to polymerize into a linear polymer as shown in Figure 7.91. In contrast to the polymers already described, however, the chain is unstable unless the ends are esterified, as shown in Figure 7.92. The –C–O– bond is shorter than the CH_2–CH_2 bond in the polyethylenes and the polyacetals pack more closely. The resultant polymer is therefore harder, more dense and has a higher melting point and a higher crystallinity (Figure 7.93). Compared with the nylons, acetal resins have better stiffness, fatigue endurance, creep resistance and water resistance and a lower coefficient of friction against steel but are less ductile. Acetals are available blended with PTFE to give a very high operating value of pressure–velocity product above 15 000 against a steel shaft, and blended with polyurethane to give high toughness.

- *Polycarbonates*: These are esters of dibasic alcohols with carbonic acid H_2CO_3. In practice, the market is dominated by bisphenol carbonate, but other polymers are being developed (Figure 7.94). This polymer has rigidity, extremely good toughness up to 140°C, transparency, good

Chemical resistance next to PTFE
Good electrical insulant but not as good as PTFE and polyethylene.

Their disadvantages are:

Brittleness
High mould temperature
High cost (compared with polycarbonate).

High-temperature thermoplastics The properties of typical high-temperature thermoplastics are listed in Table 7.62. High-temperature thermoplastics combine good oxidative stability with high-temperature strength. They compare in resistance to environments at high temperature with fluorocarbons but in most cases cost less and are easier to process. They can be melt processed as thermoplastics and compete for most applications with thermosets and for many applications with metals.

Commercially available high-temperature thermoplastics include polysulphone (Udel: Union Carbide), polyether sulphones (Victrex and Astrel: ICI and Carborundum, and Radel: Union Carbide); polyether ketones (PEEK and PEK: ICI), polyetherimide (Ultem: GE), polyimide (Kapton: Dupont) and polyamide-imide (Torlon: Amoco).

- *Polysulphone*: The simplest polysulphone (polyphenylene sulphone) is not thermoplastic and decomposes as it melts. The material marketed as a sulphone, Udel (Union Carbide), supplements the sulphone linkage with other linkages and an isopropyledene group which provides flexibility, imparting toughness and improving processability. The glass-transformation temperature T_g is 150°C.
- *Polyether sulphones*: In polyether sulphone Victrex (ICI) the ether group allows mobility of the chain in the melt phase while the sulphone group gives high-temperature performance. The glass transformation temperature is 230°C. So-called polyarylsulphone Astrel (Minnesota Mining and Manufacturing) has a glass transformation temperature of 285°C. Polyether sulphone Radel (Union Carbide, now marketed by Amoco) is tougher. The polyether sulphones have good tensile and creep strength and impact resistance at both high and ambient temperatures. They have good resistance to attack by petrol, oil and acids, resist burning, produce very little smoke and can be sterilized.

They are excellent insulators with a high dielectric constant. They can be melt processed. Their application can be recommended where their superior performance at high temperature as compared with polycarbonate justifies their higher cost. Blends of polysulphone with ABS such as Mindel (marketed by Amoco) are cheaper, easier to process and have higher impact strengths but lower glass-transformation temperature than the unblended homopolymer. They may be used at lower temperatures.

- *Polyether ketones*: Victrex PEEK (ICI), whose structure is illustrated in Figure 7.95, is an outstanding heat-resisting thermoplastic because it has a very high oxidative stability (Underwriters Laboratory temperature index of 240°C), very low flammability without the need for additives, the lowest toxic gas and smoke emission of any thermoplastic (see Figure 7.96) and a high softening point. In addition, it is tough, has excellent fatigue resistance, outstanding hydrolytic stability, low moisture absorption and good radiation resistance. It is easily processed but at high melt temperatures. It can be recommended for use where its outstanding high-temperature properties (its glass-transformation temperature is 144°C) justify its high cost. Its most important application is as an insulator for electrical wiring in locations where its resistance to fire makes it pre-eminent. Polyether ketone (PEK: BASF Hoescht and ICI) has the structure illustrated in Figure 7.97. PEK has

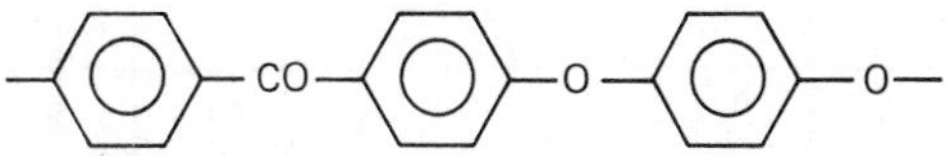

Figure 7.95 Structure of polyether ether ketone PEEK

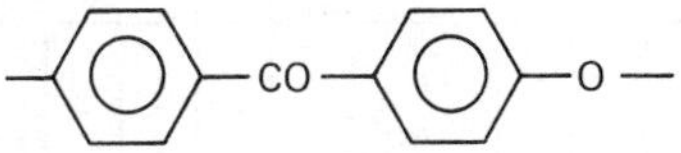

Figure 7.97 Structure of polyether ketone

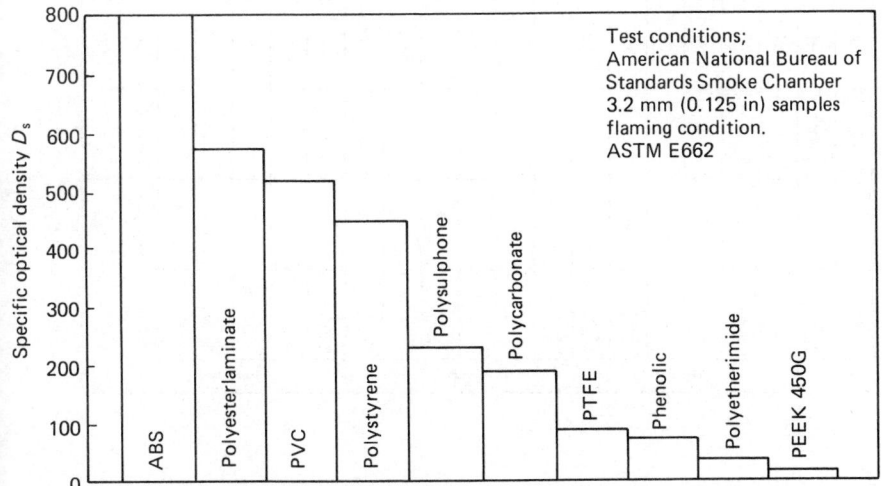

Figure 7.96 Smoke emitted on burning some plastics compared

Test conditions;
American National Bureau of Standards Smoke Chamber
3.2 mm (0.125 in) samples flaming condition.
ASTM E662

Table 7.62 Properties and an indication of price of typical high-temperature thermoplastics

Property	Polysulphone Unfilled	Polysulphone 30% glass filled	Polyether sulphone Unfilled	Polyether sulphone 30% glass filled	"Polyaryl" sulphone Unfilled	Poly-etheretherketone Unreinforced	Poly-etheretherketone 30% glass fibre reinforced	Poly-etheretherketone 30% carbon fibre reinforced	Polyimide Thermoplastic	Polyimide Thermoset	Polyimide Thermoset 30% glass filled	Polyamide-imide 20% graphite 3% PTFE	Polyamide-imide 30% glass fibre 1% PTFE	Polyetherimide Unfilled	Polyetherimide 30% glass
Mechanical															
Tensile yield strength	60 - 74	117	84	124 - 140	90	100	170	226	120	90	190	125	195	105	160
Flexural strength	106	172	129	180	129	170	233	355	205	120	345	175	318	145	230
Elongation at fracture	50 - 100	2 - 3	40 - 80	3	13	50	2.2	1.3	10	8	<1	6	5	60	3
Tensile elastic modulus	2840	7515	2440	7800	2550	3600	9700	13 000	1330	1400	19 600			3000	9000
Flexural elastic modulus	2690	8720	2600	7800	2780	4100	10 000	20 000	3360	3000	22 000	7170	11 090	3300	83000
Impact strength	0.7	0.96	0.78	0.8	0.8	0.83	0.96	0.85	0.35	0.5	8.5	0.6	1.08	0.5	1.0
Hardness	R120	M84	M88	M98	M110	M99	M103	107				–			
Thermal															
Max. continuous service temp. (no load)	150	177	190	–	260	250	250	250	290	260	>260	–		170	170
Deflection Temp. 0.45 MNm⁻²	187	198		216										210	212
Deflection Temp. 1.81 MNm⁻²	174	185	203	213	270	160	315	315	353	357	350	266	274	200	200
Thermal conductivity	0.26	0.32	0.17	0.23	0.19	0.25	0.43	0.02	0.1	0.35	0.48			0.22	
Thermal expansion	56	25	55	24	47	47	22	15	50	27	15	23	10	62	20
Electrical															
Volume resistivity	5×10^{14}	10^{15}	10^{15}-10^{16}	$>10^{14}$	3.2×10^{16}	4.9×10^{16}		1.4×10^{5}	$>10^{14}$	10^{14}-10^{15}	5×10^{13}			6.7×10^{15}	3×10^{14}
Dielectric strength	150 - 170	190	160	160	138	190			224	224	200			330	300
Dielectric constant (60 Hz)	3.07	3.55	3.5	3.8	3.9	3.25			3.34		4.84			3.15	
Dielectric constant (10⁶ Hz)	3.03	–	3.5	3.76											
Power factor (60 Hz)	0.0008	0.0019	0.001		0.003	0.003			0.0055		0.0034			0.0013	0.0015
Power factor (10⁶ Hz)	0.0034	–	0.0035	–											
Arc resistance	60 - 122	114	20 - 120	–	67				230					128	85
Miscellaneous															
Specific gravity	1.24	1.41	1.37	1.56	1.36	1.32 - 1.26	1.49	1.44	1.43	1.43	1.9	1.45	1.57	1.27	1.51
Melting point	190	190	230	230	285	340	340	340	Very high	300	300			300	
Mould shrinkage	0.7		0.6	0.2 - 0.3	0.7 - 0.9	1.2 - 0.7	1.1 - 0.2	0.3 - 4			0.1 - 0.2	–		0.5 - 0.7	0.2
Water absorption	0.3	0.22	0.43	0.31	1.1	0.5	0.11	0.06	0.3	0.32	0.2	0.19	0.22	0.25	0.18
Refractive index	1.57 - 1.59		1.65		1.67	Opaque	Opaque	Opaque	Opaque	Opaque	Opaque	Opaque	Opaque	Opaque	Opaque
Flammability	Non-burning	Non-burning	Non-burning	Non-burning	Self extinguishing non-dripping	V - 0	V - 0	V - 0						V - 0	V - 0
Economic															
Cost [a]	7.9	6.9	7.9	7.4	11.0	11.0			5.0			52.0 - 56.0		10.3	9.2

[a] Approximate mid-1991 prices, £ per kilo per tonne.

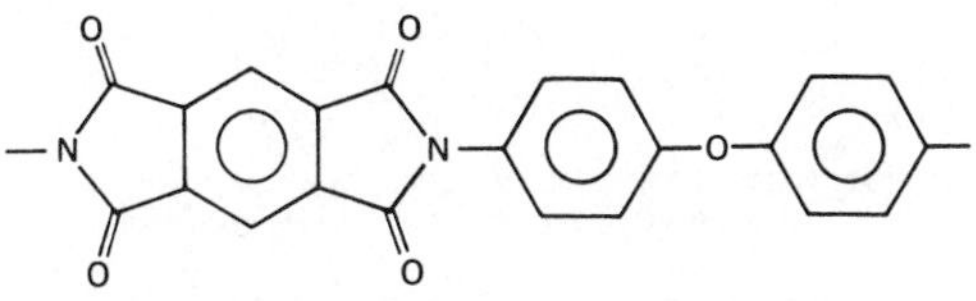

Figure 7.98 Structure of thermoplastic polyimide Kapton

the excellent attributes of PEEK but a superior heat-deformation resistance.

- *Polyimide*: Polyimides such as Kapton (Dupont), whose structure is illustrated in Figure 7.98, have excellent electrical properties, solvent resistance, flame resistance, outstanding abrasion resistance and exceptional heat resistance. The limited tractability of the polymer makes it almost impossible to process by conventional thermoplastic methods and it is very expensive. It may, however, be made into laminates by dissolving a precursor in acetone, impregnating the glass or carbon-fibre reinforcement and curing. The main application is for seals for jet engines and similar rubbing applications where the very high cost is justified. Polyimides may, however, crack in water or steam at 100°C.

- *Polyamide-imide*: Introducing an amide link NHOC in series with the imide group produces a polyamide-imide such as Torlon (Amoco), whose structure is illustrated in Figure 7.99. The polyamide grouping provides sufficient flexibility to allow compression and injection moulding but still retain many of the high-temperature characteristics of polyimides. Polyamide-imides may be used where high-temperature stiffness and creep resistance is required but their main use is for high-temperature bearings. For this purpose they are usually blended with PTFE and graphite.

- *Polyetherimide Ultem* (GE): Ultem, a polyetherimide introduced in 1982, has ether linkages in series with the polyimide linkages. The ether linkages give sufficient flexibility to permit melt processing albeit at high temperatures and the material retains many of the high-temperature properties of the polyimides (Figure 7.100). The material is claimed to have high temperature strength, high softening point, high UL temperature index, high flame resistance, low smoke emission and excellent hydrolytic stability at a price competitive with that of polycarbonates.

- *Fluoroplastics*: The fluoroplastics may be regarded as derived from the polyethylenes—polypropylenes by the substitution of fluorine for hydrogen. The carbon fluorine bond is very stable, having a bond strength up to 504 kJ while fluorine oxygen has a very low bond strength and the polymers are therefore very stable and resistant to attack by a wide range of chemicals. The properties of typical fluoroplastics are listed in Table 7.63.

Polytetrafluoroethylene (PTFE): PTFE lacks flexibility and has a high melting point. It cannot therefore be melt processed and must be fabricated by powder techniques, pressing and sintering, free sintering or a form of very slow 'extrusion', in which the powder is fed cold into a long heated sintering die. However, the intermolecular attraction between PTFE chains is very small and, as a result, the bulk polymer lacks rigidity and tensile strength and its creep resistance is very low (Figure 7.101). PTFE is also relatively expensive. It has, however, the advantages of inertness, is even more resistant than PEEK, is very resistant to weathering, resists ignition and, if burned by sustained flame from other sources, produces very little smoke. It has excellent electrical characteristics, is non-adhesive and has the lowest coefficient of friction of any material. It is used mainly for coatings and anti-friction additions to other plastics.

Other fluorine-containing polymers: The desirable properties of PTFE have promoted the development of fluorine-containing polymers that can be melt processed. The properties of PTFE are very closely approached by tetra fluoroethylene-hexa fluoropropylene, Teflon FEP (Dupont), which has a lower maximum service temperature and tetra fluoroethylene–ethylene copolymer Tefzel ETFE (Dupont), whose maximum service temperature is still lower. The best processable alternative to PTFE is probably a copolymer of tetrafluoroethylene and perfluoropropyl vinyl ether, Teflon PFA (Dupont). This material has a higher melting point than Teflon FEP, easier processability and improved high-temperature mechanical properties. All these polymers are costly even compared with PTFE and the only fluoroplastics to have gained a significant market share are polyvinyl fluoride, PVF Tedlar (Dupont) which is supplied as a weather-resistant glazing film and polyvinylidine fluoride which has piezo and pyroelastic properties and forms the basis for many fluoro rubbers.

- *Silicones (thermoplastics)* The silicon atom is more electropositive than carbon. Therefore although silicon will form compounds analogous to the lower molecular weight hydrocarbons the mutual repulsion between the silicon atoms limits their maximum number in a chain to six. Silicon has, however, a great affinity for oxygen. Silicone polymer chains have therefore a backbone structure of alternate silicon and oxygen atoms as shown in Figure 7.102, where R is normally hydrocarbon ($CH_3C_2H_5$) but may be halide, hydroxyl group or another oxygen-linked silicon radical (siloxane). The most important thermoplastic silicones are silicone fluids, most of which are dimethyl siloxane chains as in Figure 7.103. (Some specialized fluids have branched chains.) They are colourless, odourless, of low volatility, non-toxic and stable at temperatures below 150°C. Their physical properties are summarized in Table 7.64. For linear silicones at 25°C the viscosity (η) in centistokes and the number n of dimethyl siloxy groups are connected by the relationship

$$\log \eta = 0.1\sqrt{n} + 1.1$$

Silicones are used as polish additives, release agents for rubber moulding, lubricants, greases and anti-foaming agents.

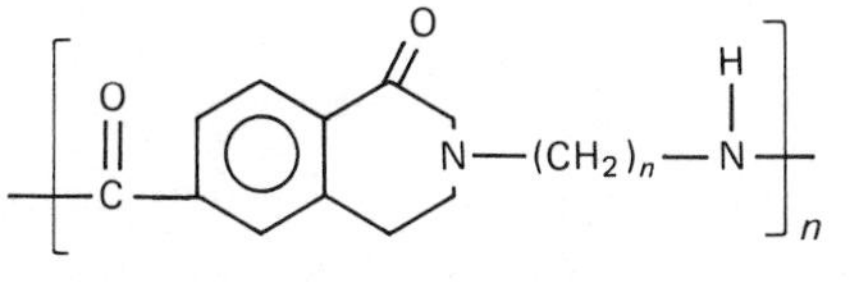

Figure 7.99 Structure of polyamide-imide Torlon

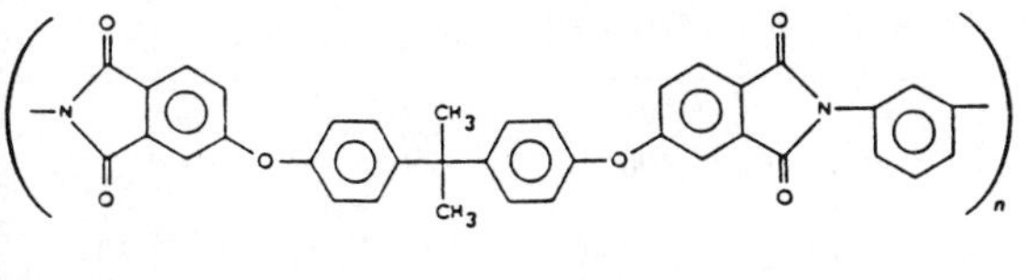

Figure 7.100 Structure of polyetherimide Ulten

Table 7.63 Properties and an indication of price of typical fluoroplastics

	Property	PTFE		FEP
		Unfilled	30% glass filled	Unfilled
Mechanical	Tensile strength	10–44	12–20	17–24
	Flexural strength			
	Elongation at fracture	250–450	200–300	250–330
	Tensile elastic modulus	410–550		350–480
	Flexural elastic modulus	350–700	1,650	630–670
	Impact strength	1.3–2.1	2.2	No break
	Hardness	SD^a 50–55	SD^a 60–75	SD^a 58
Thermal	Max. continuous service temp. no load	260	260	204
	Deflection 0.45 MNm^{-2}	120	—	70
	temp. 1.8 MNm^{-2}		—	
	Thermal conductivity	0.24	0.36	0.20
	Thermal expansion	99	60–80	85–189
Electrical	Volume resistivity	>10^{16}	>10^{13}	2 × 10^{16}
	Dielectric strength	155–200		120–160
	Dielectric 60 Hz	2.1		2.1
	constant 10^6 Hz	2.05	2.35	2.1
	Power 60 Hz	0.0002		0.0002
	Factor 10^6 Hz	0.0002–0.00004		0.0007
	Arc resistance	>300		165–300
Miscellaneous	Specific gravity	2.1–2.2	2.25	2.14–2.17
	Melting point	327^b		290
	Mould shrinkage	0.5–3.0	2.0	3–6
	Water absorption	0.01		<0.01
	Refractive index	Opaque		1.34
	Flammability	Self-extinguishing	Self-extinguishing	Self-extinguishing
	Economic Cost	10.60		21.30

^{a} SD = Shore D scale.
^{b} Not a true liquid above this temperature.

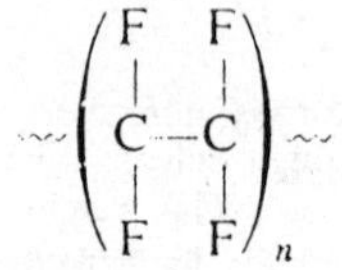

Figure 7.101 Structure of polytetrafluorethylene

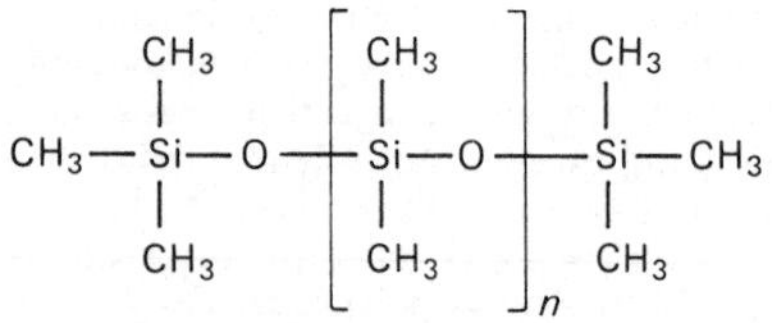

Figure 7.103 Structure of dimethyl silicone fluid

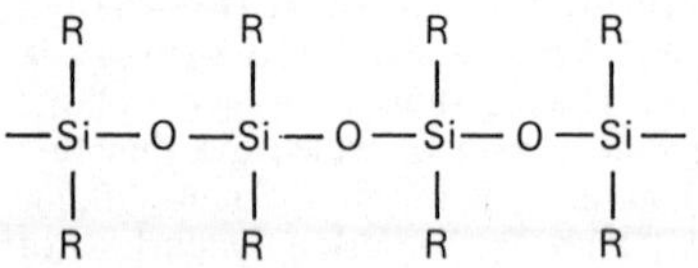

Figure 7.102 Basic structure of silicone chain

Table 7.64 Some physical properties of dimethylsiloxane polymers of the type $(CH_3)_3SiO[Si(CH_3)_2O]_nSi(CH_3)_3$

Value of n	1	3	6	14	90	210	350
Viscosity (centistokes)	1.04	2.06	3.88	10	100	350	1000
Specific gravity d_{25}	0.818	0.871	0.908	0.937	0.965	0.969	0.970
Refractive index n_D	1.382	1.390	1.395	1.399	1.403	1.403	1.404

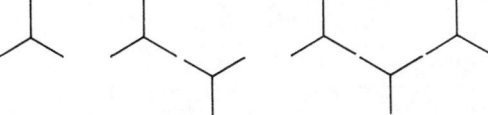

Figure 7.104 Proliferation of reactive groups as thermosets cross link

7.6.5.2 Thermosets

Consider a monomer with at least three reactive points where linkage can occur. A dimer formed by linkage of two of these molecules will have four reactive groups and a trimer will have five. In general, an *n*-mer will have $n + 2$ reactive groups (Figure 7.104).

Linkage from an 'A'-stage resin may therefore start fairly slowly but it accelerates rapidly in the later stages, forming large cross-linked molecules of 'C'-stage resin. Polymers which form these large cross-linked molecules are known as thermosets. They can be made to polymerize by heating, by the addition of a hardener and/or an accelerator or by a combination of both. Polymers cannot be melt processed in the 'C' stage and processing must be carried out before or in the early stages (the 'A' stage) of polymerization.

Thermosets are rigid at room temperature and have markedly greater resistance to distortion or creep at elevated temperature and higher resistance to abrasion than thermoplastics. A consideration which is growing in importance is that, unlike thermoplastics, they cannot be recycled. Although there is no commonly accepted division into commodity and engineering thermosets such a distinction will, for convenience, be made in this section.

'Commodity' thermosets The properties of 'commodity' thermosets are listed in Table 7.65.

- *Phenolics*: The phenolic resins were the first to be produced commercially from compounds of low atomic weight. They are produced by the action of formaldehyde (in the form of formalin, a roughly 37% aqueous solution) on phenol. Novolaks are prepared by reaction of phenol with formaldehyde in the ratio 1:08 under acid conditions (Figure 7.105). They cannot be hardened by heating alone but require the addition of a hardener, usually hexamine tetramine prepared by the action of gaseous amonia on, a formaldehyde solution. Resols are prepared by reacting phenol with an excess of formaldehyde under basic conditions. They will cross link without a hardener if the pH is adjusted (Figure 7.106). A phenol formaldehyde moulding powder will con-

Table 7.65 Properties and an indication of price of typical 'commodity' thermosets

		Phenolics		Aminos — Melamine formaldehyde		Aminos — Urea formaldehyde	
	Property	Woodflour filled	Glass filled	Cellulose filled	Glass filled	Cellulose filled	Woodflour filled
Mechanical	Tensile strength	30–60	30–120	41–70	35–138	40–90	40–55
	Flexural strength	38–42	103	70–115	105–150	70–120	75–115
	Elongation at fracture	Not measurable	Not measurable	0.6–0.9	Not measurable	0.5–1	—
	Tensile elastic modulus	5000–12 000		7000–11 000	17 000	7000–10 000	7000–10 000
	Flexural elastic modulus	7000–9000		6900–12 400	13 000–18 000	9000–11 000	—
	Impact strength	230	100–9200	0.1–0.3	0.3–0.9	0.12–0.28	0.12–0.15
	Hardness			RM 120		RM 115	RM 115
Thermal	Max. continuous service temp. no load	100–150	175–290	100	150–205	75	75
	Deflection 0.45 MNm^{-2} temp. 1.8 MNm^{-2}	130–190	150–315	180	205	130–145	—
	Thermal conductivity	0.16–0.3	0.32–0.88	0.29–0.42	0.56–1.6	0.28–0.14	0.25–0.38
	Thermal expansion	30–45	8–20	20–50	15–17	22–36	3–6
Electrical	Volume resistivity			10^9–10^{11}	2×10^9	10^{10}–10^{14}	10^9–10^{12}
	Dielectric strength	100–160	56–160	110–120	70–120	120–160	120–180
	Dielectric 60 Hz	5–13	5–7.1	6–9	10–11	7–9.5	7–9
	constant 10^6 Hz	4–6	4.5–6.6	7–9	7	6.5–7	6–8.5
	Power 60 Hz			0.04–0.08	0.2	0.25–0.43	0.25–0.45
	factor 10^6 Hz			0.025–0.045	0.014	0.2–0.3	0.25–0.4
	Arc resistance			110–114	180	80–150	80–130
Miscellaneous	Specific gravity	1.3–1.5	1.7–2	1.47–1.55	1.9	1.5	1.45
	Mould shrinkage	0.4–0.9	0.15–0.3	0.4–1.5	0.1–0.4	0.6–1.4	0.6–1.0
	Water absorption	1	0.03–1.20	0.1–0.6	0.1–0.2	0.4–0.8	0.5–1.2
	Refractive index	Opaque	Opaque	Translucent	Opaque	Transparent opaque	Opaque
	Flammability	Self-extinguishing	V-O-HB	V-O-HB	VIO	V-O-HB	V-O-HB
Economic	Cost[a]	1.0	1.62	1.44		1.0	0.73

[a] Approximate mid-1991 prices, £ per kilo per tonne.

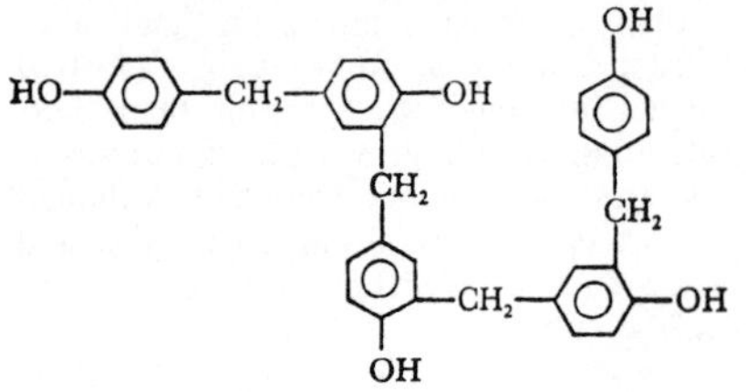

Figure 7.105 Structure of phenolic novolak

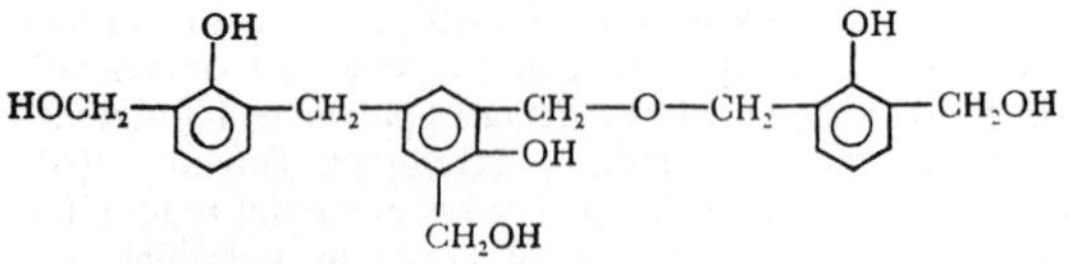

Figure 7.106 Structure of phenolic resol

tain: resin, hardener, (if novolak), accelerator (lime or magnesium oxide), filler, (woodflour, mica, cotton flock, chopped fabric, or twisted cord), lubricant, pigment and (optionally) plasticizers. The filler is essential to combat shrinkage because of the high heat evolution during curing. Phenolics can be moulded at 100–110°C and are cured at 170°C. They are used for domestic and industrial hardware. A large range of phenolic laminates are manufactured. Although they are being replaced by thermoplastics for many applications, a process that is being accelerated by environmental considerations, the general increase in living standards has maintained the overall demand for them.

- *Aminos*: The most widely produced aminos are the urea and melamine formaldehyde resins. Both depend, in a somewhat complicated way, on reactions between methylene groups, between methylene and amino groups and on the elimination of ether linkages, as shown in Figure 7.107. The advantages of urea formaldehyde over phenol-formaldehyde are low cost, high mechanical strength and wear resistance, freedom from taste and odours, the ability to self-extinguish, good electrical properties and a wide colour range. On the other hand, aminos must be filled for successful moulding, have a poor long-term oxidation resistance, poor impact properties, cannot be used above 77°C and have a high moulding shrinkage. The moulding powders are used for bottle caps and electrical fittings and urea formaldehyde adhesives have a large application in the manufacture of chipboard. Melamine reacts with formaldehyde to form trimethylol melamine and hexamethyl melamine, as shown in Figure 7.108. The melamine derivatives have, respectively, three and six reactive groups and will condense to form very hard plastics which have a good

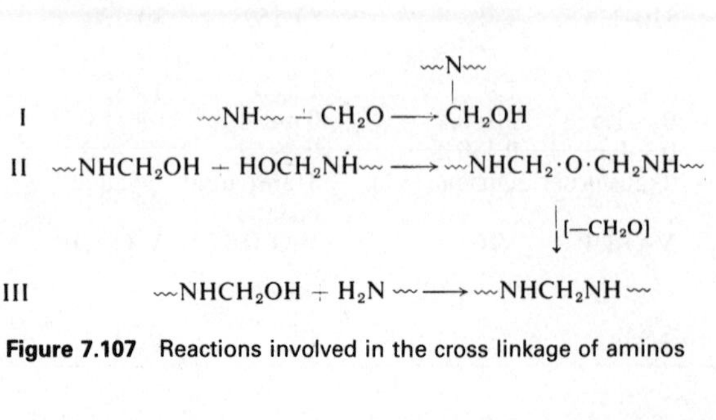

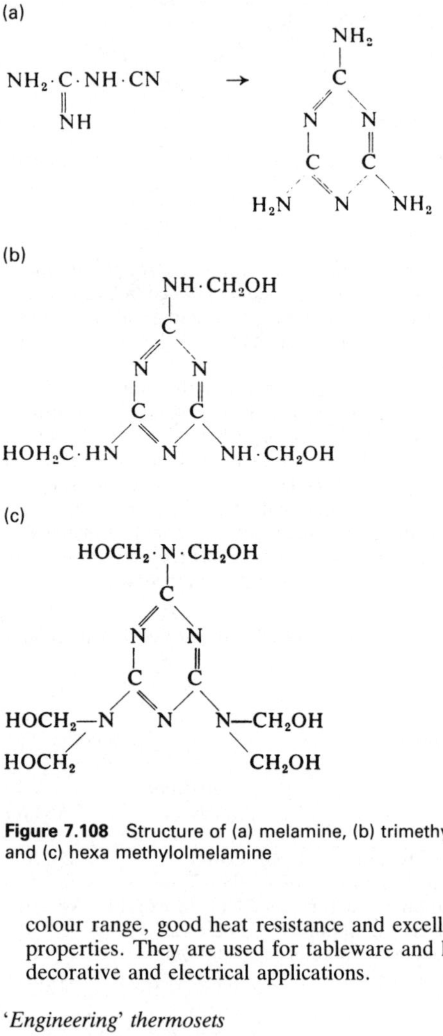

Figure 7.108 Structure of (a) melamine, (b) trimethylolmelamine and (c) hexa methylolmelamine

colour range, good heat resistance and excellent electrical properties. They are used for tableware and laminates for decorative and electrical applications.

'Engineering' thermosets

- Unsaturated polyester resins:
  Unsaturated polyesters are esters, usually of polyhydric alcohols with unsaturated acids. The resins considered here include: laminating resins, moulding compounds, allylics and alkyds.

(a) Moulding and laminating polyester resins
Moulding compounds are usually maleic acid (or its trans-isomer fumaric acid) esters of 1, 2 propylene glycol as in Figure 7.109. Since any COOH group can react with any OH group it is clear that there is scope for ester molecules with several points where cross linking can occur. In practice, a proportion of saturated acid is added to space out the double bonds and reduce the density of cross linking. Phthalic acid is used in the form of its anhydride because it provides an inflexible link which maintains rigidity in the cured resin and is cheap. The use of isophthalic acid instead of phthalic anhydride gives higher heat-distortion temperatures and flexural modulus and better craze resistance. Styrene is added as a reactive copolymer diluent and because more can be added to isophthalate resins the water and alkali resistance of these can be made better than that of the terephthalates. These resins are used, cured with benzoylperoxides, for high-temperature curing or methyl ethyl ketone peroxide or cyclohexanone

Figure 7.107 Reactions involved in the cross linkage of aminos

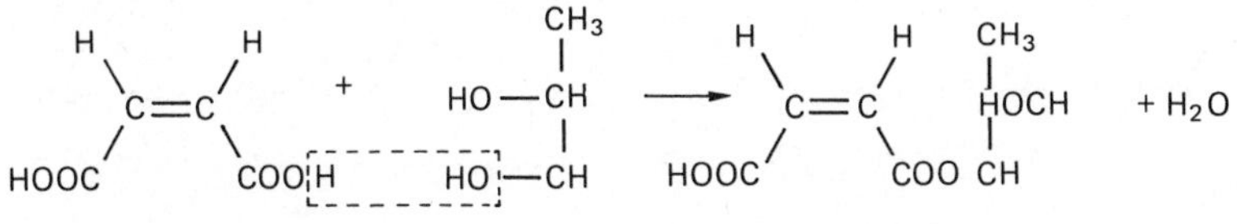

Maleic acid   Propylene glycol   Unsaturated polyester

Figure 7.109 Formation of unsaturated polyester from maleic acid and propylene glycol

peroxide with a cobalt compound as an accelerator for room-temperature curing. They are used to make laminates.

(b) Polyester moulding compounds
Dough moulding compounds (DMC) are blends of polyester resin reinforcing fibre, mineral filler, pigment, catalyst (usually benzoyl peroxide) and lubricant. They are usually press moulded. Problems have in the past been experienced with DMC but most of these are being overcome by incorporating thermoplastic acrylic/styrene polymers. The problem of damage to fibres is overcome by substituting a sheet moulding compound (SMC) which is made via a system of rolls which produce sheet sandwiched between polythene films. SMC can be moulded and cured to produce many designs for which DMC otherwise have to be used. The polyesters are inferior in strength, adhesiveness and resistance to environments to the epoxide resins, but are much less expensive, and this has ensured their application for all but the most demanding engineering purposes. Developments in thermoplastic resins and their fabrication techniques, and, even more important, the environmental benefit which accrues from their recycling capacity (which the thermosets lack) is enhancing the competition which polyesters have to face.

(c) Alkyds
Alkyd moulding compounds are polyesters blended with such fillers as cellulose pulp, mineral fillers, lubricants, pigments and peroxide curing agents mixed on hot rolls and powdered. The name 'Alkyd' is also employed for polyester-type surface coatings. In this case the unsaturated acids are derived from natural oils such as linseed oil, soya bean oil and tung oil and they form glycerides with phthalic acid. The coating resins are effectively thermoplastic when applied as a thin coating but air oxidation causes them to cross link so that they harden and set.

(d) Allylics
Allylic resins are polyesters of vinyl alcohol, which is unsaturated, with phthalic acid. There are two versions, diallyl phthalate (DAP) and diallyl isophthalate (DAIP), as shown in Figure 7.110. The DAP resins are superior for use as electrical insulators to phenolic resins and alkyds in tracking resistance at elevated temperature. The DAIP resins are more expensive and are claimed to withstand 220°C for long periods. Allylics are available in a wide range of colours. They have the drawback of a high shrinkage on cure and, where this is important, epoxide moulding compounds should be used instead. Diallyl glycol carbonate, also an unsaturated glycol ester of dibasic acids, is transparent, hard, has excellent optical properties and is used for spectacle lenses and coatings.
● *Epoxide resins*: These are characterized by the possession of at least two highly strained and therefore highly reactive epoxide groups (see Figure 7.111). These groups react particularly with substances which provide hydrogen ions (Figure 7.112). Since there are at least two epoxide groups per molecule cross linking occurs very easily. As there is no

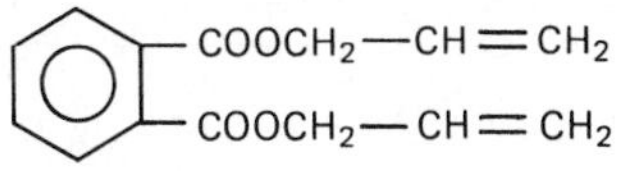

(a)

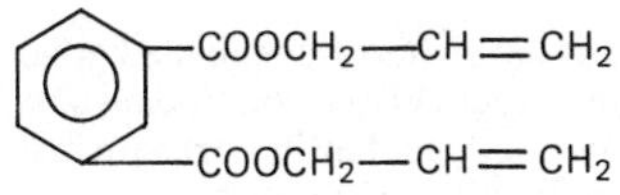

(b)

Figure 7.110 Structure of (a) diallyl phthalate DAP and (b) diallyl isophthalate DAIP

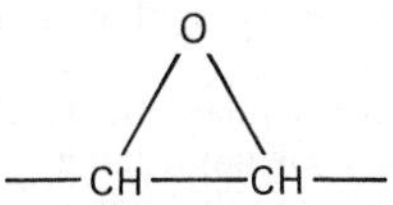

Figure 7.111 Structure of epoxide group

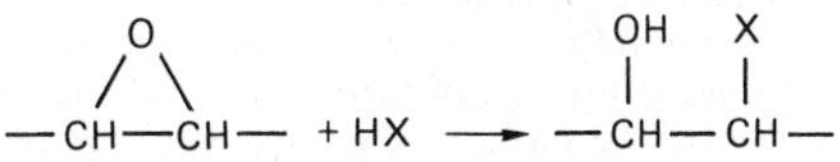

Figure 7.112 Cross linking reaction of epoxide group

elimination of water, epoxies exhibit a lower curing shrinkage than many other thermosets. Eighty to ninety per cent of epoxide resins are produced by reacting bisphenol A and epichlorhydrin to produce a compound or polymer with the structure shown in Figure 7.113. These will link not only by the epoxy groups but also by the hydroxyl group(s) but other forms include cyclic aliphatic, acrylic, aliphatic and other glycidyl ether resins. There is not only a variety of monomers but also a choice of hardening systems. These include amines of which primary and secondary amines enter themselves into the cross linking while tertiary amines are catalytic, acids, and acid anhydrides. Anhydrides are preferred to acids because they release less water on cure, and are in other respects equivalent. Amines give a more rapid cure and the polymer has greater resistance to alkalis but precautions must be taken to protect the skin of the operator from attack. Acids are less skin sensitive and give higher heat-distortion temperatures. That of amine-cured resin seldom exceeds 100°C, that of phthalic anhydride is about 120°C while methylnadic anhydride gives 202°C. Other additions include diluents and flexibilizers.

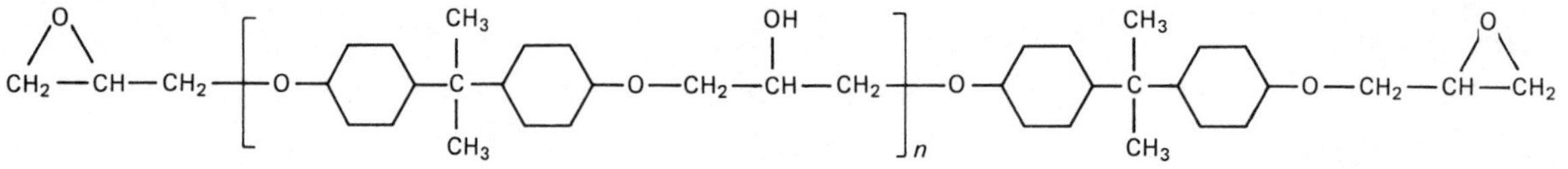

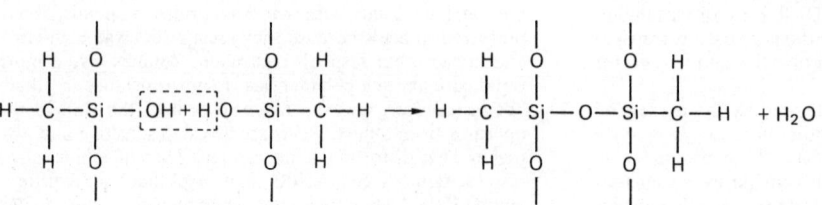

Figure 7.113 Glycidal ether epoxy resin

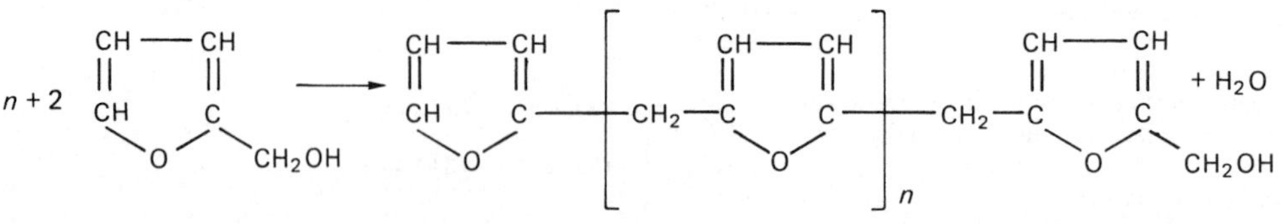

Figure 7.114 Suggested mechanism of furan resin formation

In general, epoxide resins have almost unequalled adhesion, low shrinkage on cure, freedom from volatile solvents, very high chemical resistance, high tensile strength and creep resistance. This combination of properties has ensured their application in spite of their relatively high cost. Their major application is for surface coatings. Another important use is for adhesives.

Liquid resins are used to encapsulate electrical and electronic circuit components. There is a small production of moulding powders. The most significant technological application of epoxide resins is as a matrix for composites. These have outstanding value where a high strength/weight ratio is paramount.

- *Thermosetting polyimides*: These have similar structural features and properties to thermoplastic polyimide but have the advantage that they can be processed in the monomer form and then hardened. They are used mainly for high-temperature bearings reinforced and filled with graphite or PTFE or both. Their properties are listed in Table 7.66 along with thermoplastic polyimides.
- *Furan resin, Quacorr RP100A (Quaker Oats Co)*: Furan resin is a condensation product of furfural alcohol (which itself is prepared from oat husks by digesting and hydrogenating – see Figure 7.114). When hardened by, for example, the addition of 4% of toluene *p*-sulphonic acid the chains cross link as shown in Figure 7.115. The cross linked molecules are rigid and inflexible and therefore have a high deformation temperature, and chemical and water res-

istance better than phenolics, polyesters and epoxides. Furan resins are used for lining of tanks, vats and piping in chemical plants and as laminates to resist organic fluids, particularly at high temperature.
- *Silicone resins (thermosetting)*: Silicone resins which contain OH groups (usually produced by the action of water on chlorosilanes) cross link by condensation (Figure 7.116). Clearly, an increase in the proportion of hydroxyl (–OH) groups to methyl groups (CH_3) attached to the silicon atoms in the chain would provide additional cross linkage and a 1:1 ratio of methyl to silicon would be brittle when hardened. Methyl/silicon ratios between 1.2 and 1.6 are used in practice. Phenyl groups C_6H_5 can be substituted for a proportion of the methyl groups. Methyl-phenylsilicone resins are used mainly for glass cloth laminates for electrical applications at high temperature. They are cured, initially by heating and finally by a catalyst such as triothanolamine zinc octoate or dibutylin octoate. Compression-moulding compounds may be used for high-temperature electrical applications but are even more expensive than PTFE.

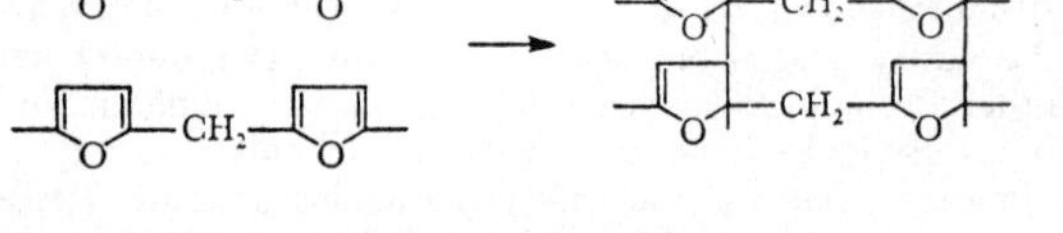

Appendix: Worked examples of design of plastic components

Cantilever beam

A rectangular cantilever beam of length 20 cm is required to support an end load of 5 N at a temperature of 150°C. At the end of its useful life of 1 year the end deflection should be no more than 3 mm. Select an appropriate grade and beam dimension.

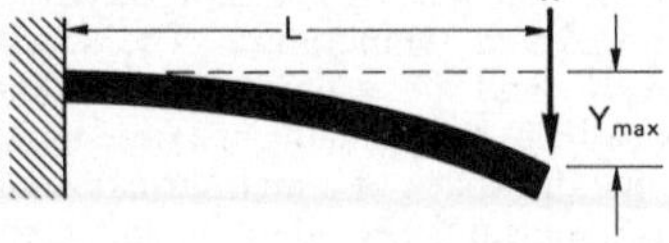

Figure 7.115 Cross-linking reaction of furan resin

Figure 7.116 A cross-linking mechanism for silicone resins

Table 7.66 Properties and an indication of price of typical 'engineering' thermosets

Category	Property	Unsaturated polyester — DMC	SMC	Rigid cast	Alkyds — Mineral filled	Glass filled	Allylics — DAP glass filled	DAIP glass filled	Allyl dicyclol carbonate cast	Epoxide resins — Cold cured	Hot cured	Phenol Novolak	Glass filled moulding compound	Furan — Chopped strand laminate	Silicone — Cast resin	Glass fibre filled
Mechanical	Tensile strength	30 - 70	50 - 90	4 - 21	20 - 60	25 - 65	40 - 80	25 - 35	35 - 40	50 - 70	70 - 90	35 - 105	70 - 140	>140	2 - 7	28 - 45
	Flexural strength	49 - 140	140 - 210	60 - 160	60 - 180	60 - 180	75 - 175	60 - 65	40 - 100	84 - 105	110 - 125		70 - 240			70 - 100
	Elongation at fracture	0.5	9	<5	Not measurable	Not measurable	Not measurable	Not measurable	Not measurable	3 - 6	3 - 6	2 - 6	4		100 - 10 000	
	Tensile elastic modulus	12 000 - 14 000	9 000	—	3000 - 21 000	14 000 - 20 000	10 000 - 15 000	—	2000	2800 - 4200	2100	1000 - 4000	21 000		630	
	Flexural elastic modulus	7000 - 14 000	7000 - 16 000	—	14 000		8000 - 11 000	—	1.8 - 2.4	2800 - 4200	2800 - 4200		18 000 - 32 000			7000 - 18 000
	Impact strength	0.8	3.5 - 11	0.1 - 0.2	0.2 - 0.3	0.25 - 8	0.16 - 7.5	0.15	0.1 - 0.2	0.2 - 0.25	0.2 - 0.35	0.2 - 0.4	11 - 15			0.15 - 5
	Hardness									RM 99 - 108	RM 105 - 115		RM 105		SA15 - 65	RM80 - 90
Thermal	Max. continuous service temp. no load	150	120 - 205	120	150 - 230	235	150 - 230	230 - 260	100			260	150 - 260	218	260	>335
	Deflection Temp. (0.45 MNm², 1.8 MNm²)	200 - 230	190 - 230	110 - 200	180 - 260	205 - 260	150 - 260	160	60 - 90	70 - 110	120 - 160	150 - 260	120 - 260			>500
	Thermal conductivity	0.2 - 0.25	0.2 - 0.25	0.16	—	8.8	0.2 - 0.6		0.2	—	0.14 - 0.20	0.14 - 0.6	0.16 - 0.4		0.14 - 0.6	0.28 - 0.36
	Thermal expansion	20 - 33	18 - 33	55 - 100		15 - 33	10 - 36		81 - 143	45 - 65	45 - 65		11 - 35		50 - 300	20 - 50
Electrical	Volume resistivity	10^{11} - 10^{13}	10^{12} - 10^{15}	10^{15}		10^{10} - 10^{13}	10^{11} - $>10^{14}$		$>4 \times 10^{12}$	10^{11} - 10^{14}	$2\text{-}3 \times 10^{14}$	10^{13}	$>10^{12}$		10^{12} - 10^{13}	10^{12}
	Dielectric strength	120 - 170	150 - 160	100 - 160	140 - 180	100 - 210	130 - 220	130	150	140 - 180	140 - 180	120 - 240	120 - 160		220	100
	Dielectric constant 60 Hz	5.3 - 7.3	4.4 - 6.3	4.4 - 8.1	5.1 - 7.5	5.9 - 7.3	4.3 - 4.6	4.7	4.4	4 - 5	3 - 6	3.7	3.5 - 5		2.75 - 4.2	3.5 - 5
	Dielectric constant 10^{6} Hz	5.2 - 6.4	4.2 - 5.8	4.1 - 5.9		5.2 - 6.8	3.5 - 3.9		3.5 - 3.9	3.3 - 4	3 - 6	3.7	3.5 - 5		2.6 - 2.7	3.2 - 4.3
	Power factor 60 Hz	0.011 - 0.041	0.007 - 0.021	0.026 - 0.31		0.007 - 0.041	0.006 - 0.019	0.017	0.01	0.002 - 0.010	0.01 - 0.04	0.003	0.01		0.001 - 0.025	0.004 - 0.03
	Power factor 10^{6} Hz	0.007 - 0.02	0.016 - 0.024	0.023 - 0.06		0.008 - 0.023	0.04 - 0.06		0.04 - 0.06	0.03 - 0.05	0.018 - 0.09	0.024	0.01		0.001 - 0.002	0.002 - 0.02
	Arc resistance	50 - 240	120 - 200			130 - 420	115 - 250	115	120 - 250	45 - 120	45 - 120		120 - 180		115 - 130	175 - 250
Miscellaneous	Specific gravity	1.7 - 2.3	1.6 - 2.4	1.01 - 1.2	1.6 - 2.3	2.1 - 2.2	1.5 - 2.08	1.61 - 1.79	1.3 - 1.4	1.1 - 1.4	1.1 - 1.4	1.16 - 1.21	1.6 - 2	1.8	0.99 - 1.5	1.8 - 1.9
	Mould shrinkage	0.1 - 1.2	0.1 - 0.4	0.5 - 2.5	0.1 - 0.4	0.1 - 1.5	<0.1 - 1.5	0.3 - 0.5	0.2	0.1 - 1	0.1 - 1		0.1 - 0.5		0 - 0.06	1.8 - 1.9
	Water absorption	0.06 - 0.28	0.1 - 0.15				0.12 - 0.6	0.50	0.2	0.08 - 0.15	0.08 - 0.15		0.05 - 0.2		0.08	0 - 0.06
	Refractive index	Opaque	Opaque	Transparent to opaque	Opaque	Opaque	Opaque	Opaque	1.536	1.54 - 1.60	1.54 - 1.60	1.54 - 1.60	Opaque	Opaque	Opaque	Opaque
	Flammability	Flame retardant	—		V - 0		Self-extinguishing non-burning	Self-extinguishing	—	Slow-burning	Slow-burning	Slow-burning	Self-extinguishing	Resists burning		
	Cost [a]	1.37	1.25	1.25	1.8	1.8	5.65	8.75		2.5			4.6	3.0		15.0

[a] Approximate mid-1991 prices, £ per kilo per tonne.

From Strength of Materials formula

(1) $Y_{max} = \dfrac{WL^3}{3EI}$

Hence $\dfrac{WL^3}{3EI} = 3 \times 10^{-3}$m

(2) $l = \dfrac{bd^3}{12}$

Assume the bar is moulded from the end and hence along-the-flow data can be used.
Try the carbon reinforced grade 450 CA30.
From the graph of Tensile Creep Modulus versus Time at 150°C

$E \sim= 10$ GPa

From (1)

$3 \times 10^{-3} = \dfrac{5 \times 0.2^3}{3 \times 10 \times 10^9 \times 1}$

Therefore $I = 4.44 \times 10^{-10}$ m^4

From (2)

$\dfrac{bd^3}{12} = 4.44 \times 10^{-10}$ m^4

The most economical use of material will be with b as thin as feasible, say 2 mm. Therefore

$444.4 \times 10^{-12} = \dfrac{2 \times 10^3 \times d^3}{12}$

and $d = 13.9$ mm

In 450 CA30 the beam dimensions will be

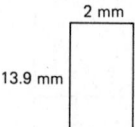

2 mm

13.9 mm

Check stress level

$\dfrac{M}{I} = \dfrac{\sigma}{y}$

$M = 5 \times 0.2 = 1$ Nm

$I = 4.44 \times 10^{-10}$

$y = \dfrac{d}{2}$

$= 6.95 \times 10^{-3}$m

Therefore $\sigma = \dfrac{1 \times 6.95 \times 10^{-3}}{4.44 \times 10^{-10}}$

$= 15.6$ MPa

From the tensile creep rupture curve this is seen to be well on the safe side.

Key to symbols:
Y_{max} = maximum deflection I = moment of inertia
y = distance from neutral axis L = length of bean
W = load b = breadth of beam

d = depth of beam E = modulus
M = bending moment σ = stress

Thin-walled tube

A thin-walled tube of 10 mm internal diameter is required to carry hot gas under a pressure of 30 bar. The tube will operate at 100°C. Calculate the wall thickness required in PEEK 450G.
Thin cylinder assumption is valid if $t/d \leqslant 0.1$.

Circumferential (hoop) stress $\sigma_c = \dfrac{pd}{2t}$

Longitudinal stress $\sigma_L = \dfrac{pd}{4t}$

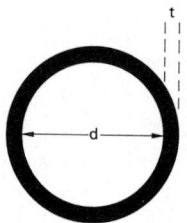

From the graph of Tensile Creep Modulus versus Time at 100°C E for 450G at long times and 100°C = 2.5 GPa. This is valid up to a 2% strain.

$\sigma_c = 2.5 \times 10^9 \times 0.015$
$\quad = 37.5$ MPa

Assume a design strain level of 1.5%.

$\dfrac{pd}{2t} = \dfrac{30 \times 10^5 \times 0.01}{2 \times t}$

Therefore, wall thickness $t = \dfrac{30 \times 10^5 \times 0.01}{2 \times 37.5 \times 10^6}$ metres

$= 0.4$ mm

Check thin wall assumption $t/d = \dfrac{0.4}{10}$

$= 0.04$

This is less than 0.1 and therefore the assumption is valid.

Key to symbols:
t = wall thickness σ_L = longitudinal stress
d = internal diameter p = pressure
σ_c = circumferential (hoop) stress E = modulus

7.7 Elastomers

7.7.1 Introduction

Elastomers, or rubbers, have been used in engineering for a little over one hundred years. For the first half of this time there was no practical alternative to the natural material; the fact that examples have survived to present times shows that components made with elastomers can perform useful functions after long periods. Today elastomers are a large and growing class of materials. At one end of the spectrum, natural rubber (NR) and synthetics such as chloroprene rubber (CR) are soft materials which have the ability to undergo extensions of many hundreds of per cent yet return rapidly to their original shape when the deforming force is removed; at the other extreme lie some of the thermoplastic elastomers (TPEs) which are harder and less elastic. The compliance of rubbers in shear and extension combined with their near-incompressibility means that particular techniques are often necessary for stress analysis with these materials.

Natural rubber (NR) was used in the ancient civilizations of Central and South America. It was not possible to exploit fully its elastic characteristics for mechanical applications until Charles Goodyear discovered in 1839 that the addition of sulphur, followed by heating, changed the rubber into a truly elastic material. This process, known as vulcanization (or curing), results in cross links being formed between the long molecular chains of the base polymer.

In the 1920s synthetic elastomers began to appear. Today there are over 20 types of synthetic elastomer produced commercially, each with its own characteristics. Advances have also been made in the vulcanizing systems, anti-degradants, fillers and oils used with the elastomer in the formulation of an engineering material. With modern formulation methods, elastomers are available to cater for a wide range of mechanical requirements, service temperatures and environmental conditions, singly or as blends.

The mechanical properties of NR epitomize those of a strong, fully rubbery material, and it is these attributes, together with its competitive price, which account for the continued widespread use of NR in vehicle suspension springs, for example. Where excellent resistance to weathering and high temperatures is required with a good degree of elasticity, synthetic elastomers such as ethylene propylene rubber (EPM/EPDM) are used. A relatively new set of additions to the family of elastomers are the thermoplastic elastomers (TPEs). These are materials with properties intermediate between elastomers and thermoplastics. It is sometimes possible to redesign rubber-based components, replacing the elastomer with a TPE and achieving cost and/or weight reductions.

7.7.2 Vulcanization, compounding and moulding

During vulcanization or 'curing' the polymer chains of a base elastomer are chemically cross linked (Figure 7.117). Unvulcanized natural rubber was used for rubber balls, erasers and macintoshes, but vulcanization resulted in an altogether stronger, more environmentally resistant material and paved the way to the use of rubber in mechanically demanding applications.

The original curing process used sulphur and high temperatures (hence the name 'vulcanization'). The majority of cross-linking processes for natural and other rubbers still involve the use of sulphur and elevated temperatures. However, vulcanization conditions can vary a great deal from high-temperature (up to 200°C) cures lasting only a few seconds, through 'normal' cures typically lasting about 15 min at 150°C, to room-temperature vulcanization taking place over days or longer. The properties of the vulcanized rubber (or 'vulcanizate') are influenced by the number of cross links and their type: the properties can also be affected by other molecular modifications to the elastomer which occur during the vulcanization process. Vulcanizing systems with relatively high loads of sulphur are termed 'conventionally accelerated', those with low sulphur levels are called 'EV' (efficient vulcanization) systems. EV systems produce NR and styrene–butadiene (SBR) vulcanizates with more stable properties at high temperature and lower creep (see Section 7.7.3.2). For NR, EV leads to a reduction in fatigue resistance but in SBR fatigue resistance is maintained.

The term 'compounding' in rubber technology refers to the mixing of various ingredients into rubber. Mixing of solid rubber is performed on open two-roll mills and in closed mixers. The high shear forces produced are used to disperse ingredients and can also be used to break down the molecular weight or 'masticate' certain rubbers. In addition to the vulcanizing chemicals, the ingredients usually include particulate fillers. These can be non-reinforcing (clay, chalk and some carbon blacks, finely divided carbon powders fall into this category) or reinforcing (carbon black or fine silica powder). Non-reinforcing fillers, added to harden or cheapen a rubber, generally have a neutral or negative effect on strength properties. Reinforcing fillers, in contrast, tend to have a positive effect on strength properties. This is particularly true for some of the synthetics, a number of which are weak without reinforcing fillers. Silicone rubber is an example of an elastomer the strength properties of which benefit greatly from the presence of an appropriate silica filler. Fillers can also markedly increase the damping and the stiffness of the elastomer – particularly at small strains – and can lead to complex history-dependent behaviour.

Rubber compounds routinely include some oil or other diluent added to aid processing of the rubber prior to vulcanization and/or to influence the properties of the vulcan-

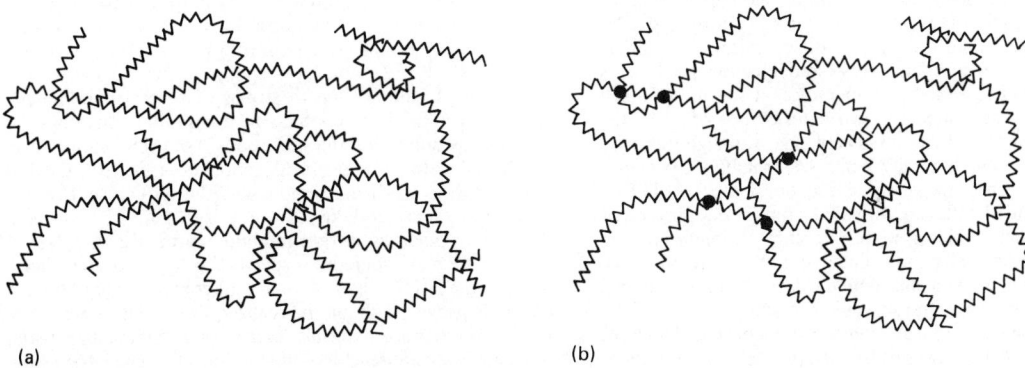

(a) (b)

Figure 7.117 Cross linking of elastomer chains. (a) Elastomer chains kinked, coiled and entangled prior to vulcanization; (b) elastomer chains cross linked (●) together

izate – notably stiffness or low-temperature resistance. Many elastomers have anti-degradants of various kinds added during compounding to improve the resistance of the final vulcanizate (to attack, for example, by oxygen or ozone). Natural, butadiene and styrene–butadiene rubbers (NR, BR and SBR) are examples of elastomers whose environmental resistance is greatly improved by the presence of appropriate anti-degradants.

Vulcanized elastomeric products can be given their final shape in a number of ways. Machining of vulcanized rubber is relatively rare, except for trimming; but calendering of unvulcanized sheet, coating of fabric from rubber solution, extrusion and moulding are all common.

Broadly, there are three types of moulding. In compression moulding a quantity of unvulcanized material is placed in a single-cavity mould (or mould area) with a closing lid or plunger. The mould is then pressed between heated platens of a press whereupon the elastomer is shaped and vulcanized. Transfer moulding is also widely used and here the plunger is located in the first cavity; when the moulding is under the press the rubber is forced through transfer ports into the second, shaping, cavity. Compression or transfer moulding time is normally not less than 5 min although there is considerable variation depending on size of component, rubber formulation and temperature. It should be realized that conventional thermosetting elastomers such as NR, EPM/EPDM, BR, SBR, chloroprene rubber (CR) and nitrile (NBR) behave as complex non-linear viscoelastic materials during the moulding process, and that it is only under high pressures that they can be made to flow at a satisfactory rate. Injection moulding of a thin-walled elastomeric component typically takes less than 1 min for conventional elastomers. With conventionally moulded elastomers it is not unusual for products to be heated in an oven after removal from a hot mould; this process, termed 'post-cure', is particularly common for certain speciality rubbers (e.g. fluorinated types). Post-cure can improve elastic recovery in the vulcanizate and reduce the tendency to creep without the need for long and expensive moulding cycles.

Steam autoclaves and molten-salt baths are two further methods for the cure of rubber articles, although previous shaping is necessary in such low-pressure curing. Millable polyurethanes are processed as conventional elastomers but castable polyurethanes flow more easily so lighter moulds can be used. Although castable elastomers are not processed/vulcanized in the same way as conventional elastomers, a form of chemical cross-linking does normally take place during and after moulding.

For thermoplastic elastomers (TPE), however, chemical cross-linking does not take place during moulding. Instead, within the TPE, thermoplastic areas, molten during and solid after moulding, take the place of chemical cross-links. Because there is no cure time, very rapid injection moulding is possible for TPEs, and the temperature of the mould itself needs to be lower. There are two main classes of TPE; the block co-polymers and the physical blends, although there are also materials with thermally labile (reformable) cross-links (e.g. ionomeric polymers, which can be classed as TPEs). Physically blended TPEs are prepared from thermoplastic and rubbery polymers which generally remain distinguishable in the final material. The molecular chain structure of block polymer TPEs consists of alternate blocks of hard, amorphous or crystalline, polymer and soft rubbery polymer.

Block co-polymer TPEs are composed of alternate blocks of hard polymer (high glass-transition temperature or crystalline) and soft rubbery polymer (low glass-transition temperature) on a molecular scale. At high temperatures the 'hard' segments are sufficiently soft to allow flow into shape to take place. However, at low and moderate operating temperatures the hard segments link the soft segments together and have a similar function to chemical cross-links and reinforcing filler in a conventional vulcanized and filled rubber. Styrene–butadiene–styrene (SBS) block co-polymers are an early example: the glassy polystyrene blocks link the rubbery polybutadiene blocks. Polyether–polyester block copolymer rubbers first became commercially available a decade after SBS, but now, together with polyurethane and polyether–polyamide materials, epitomize the high, mechanical performance TPEs. In polyether–polyester TPEs most of the polyester (hard) material is crystalline at operating temperature but some is in the amorphous phase with the polyether. This interlinked morphology makes for a strong material. Polyurethane TPEs consist of urethane-sparse (soft) and urethane-rich (hard) segments; their morphology is broadly similar to that of polyether–polyester TPEs. Polyether–polyamide block copolymer TPEs consist of soft (polyether) and hard (polyamide) segments.

TPOs, TPVs and TPEAs are all essentially physical blends between rubbers such as EPM/EPDM, NR or NBR and polyolefin thermoplastics – typically polypropylene but sometimes polyethylene. Depending on the proportions present, method of preparation and other factors, the elastomer and thermoplastic or the thermoplastic alone form continuous phases. Materials with more rubbery properties can be obtained by vulcanization of the rubber phase during blending with the thermoplastic; these dynamically vulcanized materials are sometimes called 'thermoplastic elastomer alloys'. The addition of fillers and liquid diluents is normally made at the production stage of TPE materials.

7.7.3 Basic elastomer properties

7.7.3.1 Elasticity

Natural rubber (NR) was the first elastomer and its mechanical properties set a standard by which other elastomers are judged. NR when compounded without filler (gum NR) can have a very low shear modulus (~ 0.3 MNm^{-2}) depending on the degree of cross-linking, and can undergo deformation of several hundred per cent without failure or significant permanent deformation (Table 7.67). Some insight into the mechanical properties of NR can be gained by considering its molecular structure (Figure 7.118). NR consists principally of chains of cis-1,4 polyisoprene units. These polymer chains consist typically of 20 000 repeat units. The IV–I carbon–carbon bonds permit easy rotation between adjacent isoprene units. Thus it is possible for a polymer chain to have a relatively straight configuration (low entropy) or a very twisted configuration (high entropy). It is these *entropic* changes which are primarily responsible for the mechanical behaviour of gum NR up to strains of several hundred per cent. This is in sharp contrast to many other materials for which distortion of interatomic bonds is of prime importance; the fact that the tension in a section of stretched gum NR *increases* when it is heated emphasizes this difference. In crude terms the shear modulus of vulcanized gum NR is about 0.4 MNm^{-2} whereas the bulk modulus is around 2000 MNm^{-2}.

Elastomers are sometimes specified in terms of their hardness, on the IRHD (international rubber hardness) or Shore A scales. However, it should be realized that a given percentage hardness tolerance typically corresponds to a much wider percentage tolerance of modulus. Filled elastomers have non-linear stress–strain characteristics, particularly at small strains, and hardness indentation produces non-uniform strain within the elastomer; this combination leads to uncertainty in the

Table 7.67 Properties of natural and chloroprene rubber and other materials

Property	Units	Natural rubber		Chloroprene rubber		Polyethylene (high density)	Nylon 66	Aluminium	Mild steel	Water
		Unfilled[a]	Filled[b]	Unfilled	Filled[c]					
Density	Mgm^{-3}	0.95	1.12	1.32	1.4	0.94–0.97	1.14	2.70	7.75	1.00
Thermal properties										
Specific heat	Jg^{-1}°C^{-1}	1.83	1.50	2.2	1.7	2.3	1.7	0.95	0.48	4.19
Conductivity	Wm^{-1}°C^{-1}	0.15	0.28	0.19	0.21	0.45–0.52	0.25	230	46	0.65
Coeff. of lin. expansion ($\times 10^5$)	°C^{-1}	22	18	15	11[d]	11–13	8	2.3	1.1	7
Mechanical properties										
Brinell hardness	kgf mm^{-2}	≪1	≪1	≪1	≪1	1	—	22	140	0
International Rubber Scale hardness[e]	IRHD	45	65	40[f] 46[g]	71	98	100	100	100	0
Tensile strength	MNm^{-2}	24[h]	24[h]	12 24	29	20–35	60–80	70–100	400–500	0
Elongation at break	%	700	550	--- 800	550	20–600	60–100	5	30	0
Young's modulus, E[i]	MNm^{-2}	1.9	5.9	1.6	—	~1000	1000–3000	70 000	2.1×10^5	0
Bulk modulus, K	MNm^{-2}	2000	~2000	~2000	~2000	3300	—	70 000	1.8×10^5	2050
Poisson's ratio, ν	—	0.499	0.499	0.499	0.499	0.38	0.38	0.345	0.291	0.5
Acoustic properties										
Velocity of sound[j]	ms^{-1}	1600	1700[d]	1400	1500	—	—	6400	6000	1500
Electrical properties										
Volume resistivity	Ωm	10^{14}	$10^{10,k}$	10^9	k	10^{15}	10^{12}	10^{-8}	10^{-6}	10^2–10^5
Dissipation factor	—	0.002–0.04	0.006–0.14	0.03	—	<0.005	0.01–0.02	—	—	—

[a] Gum vulcanizate containing cross-linking ingredients only.
[b] Vulcanizate containing approx. 50 parts of a reinforcing carbon black per 100 parts, by weight, of natural rubber.
[c] GN (sulphur modified) type plus 44 parts of a reinforcing carbon black per 100 parts, by weight of chloroprene rubber.
[d] Estimated value.
[e] Hardness range is zero (e.g. liquid) to 100 (e.g. glass).
[f] W type.
[g] GN type.
[h] Calculated using original unstrained cross-sectional area; when the cross section just before break is used the value can be up to 200 MNm^{-2}.
[i] For isotropic materials, at small strains, E is related to G, K and ν by $E = 2G (1 + \nu) = 3K (1 - 2\nu)$.
[j] Plane compression waves in medium of large extent.
[k] Depends on filler type and concentration, etc. High values obtainable by replacing carbon black by silica. Values under 10 Ωm achievable by using specially conductive carbon blacks.

relationship between hardness and modulus, with hardness often corresponding best with shear moduli measured at about 5% strain.[77]

When strains are large the equations of classical elasticity are no longer strictly applicable; so deformations in gum elastomer are often described in terms of extension ratios (λ_1, λ_2, λ_3) of final length (l) to original length (l_0) in the three principal directions:

$$\lambda_1 = l/l_0$$

For elastomers it is usually appropriate to consider true stresses, based on current areas, rather than engineering stresses, based on original areas. In rubber elasticity theory, also known as hyperelasticity and finite strain theory, widespread use is made of the three invariants:

$$I_1 = \lambda_1^2 + \lambda_2^2 + \lambda_3^2$$
$$I_2 = \lambda_1^2\lambda_2^2 + \lambda_2^2\lambda_3^2 + \lambda_3^2\lambda_1^2$$
$$I_3 = \lambda_1^2\lambda_2^2\lambda_3^2$$

which are independent of the directions of the axes chosen. Provided that energy dissipation processes can be ignored, the mechanical behaviour of the elastomer can be predicted from the strain energy density, or stored free energy, W. (In

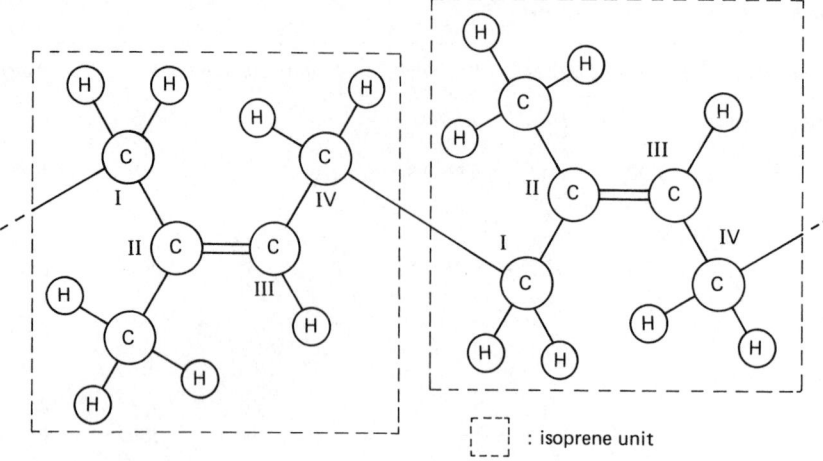

Figure 7.118   Molecular structure of *cis*-1,4-polyisoprene

practice, this is applicable to a well-vulcanized gum, unfilled, elastomer operating well above its glass-transition temperature, T_g.) The principal (true) stress in direction 1, for example, is given by

$$\sigma_1 = \frac{1}{\lambda_2\lambda_3} \frac{\partial W}{\partial \lambda_1} (I_1, I_2, I_3)$$

The simplest hyperelastic strain energy density function is given by the statistical (mechanics) theory, which assumes incompressibility and therefore independence of W from I_3:[78]

$$W = \tfrac{1}{2}G(I_1 - 3)$$

The statistical theory is capable of giving approximate values in stress calculations and provides some useful insights into the way that elastomers behave. If an element of elastomer is subjected to a simple shear, γ (Figure 7.119), it can be shown that $I_1 - 3$ is γ^2, so that statistical theory predicts

$$W = \tfrac{1}{2}G\gamma^2$$

and the shear stress

$$\sigma_{xy} = \frac{dW}{d\gamma} = G\gamma$$

indicating that shear stress is proportioned to shear. This behaviour is observed for *unfilled* vulcanization up to quite

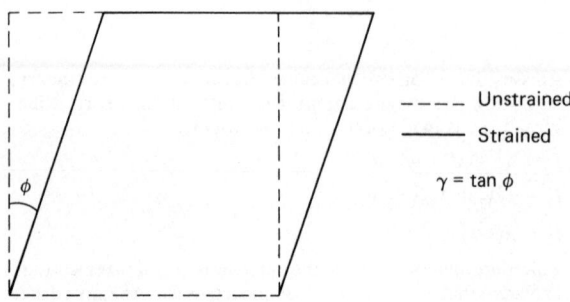

Figure 7.119   Simple shear

high deformations ($\gamma \sim 1$). However, such simple linear relationships are not predicted for other modes of deformation. For homogeneous compression as in a tensile test, the principal extension ratios are:

$$\lambda_1 = \lambda \text{ and } \lambda_2 = \lambda_3 = \lambda^{-1/2}$$

The extension or compression stress given by statistical theory is

$$\sigma_1 = G\left(\lambda^2 - \frac{1}{\lambda}\right)$$

where it has been assumed that $\sigma_2 = \sigma_3 = 0$. Note, too, that the stress given is the *true* stress – based on *final* cross-sectional area; the corresponding engineering stress based on the original cross section is

$$\sigma_E = G\left(\lambda - \frac{1}{\lambda^2}\right)$$

This result can be seen to coincide with the classical result for infinitesimal longitudinal strains ($\epsilon = \lambda - 1$) if Poisson's ratio is taken as $\tfrac{1}{2}$:

$$\sigma_E = E\epsilon = 2(1 + \nu)G\epsilon = 3G\epsilon$$

The Mooney–Rivlin equation for W is sometimes used as an improvement on the statistical theory:

$$W = C_1(I_1 - 3) + C_2(I_2 - 3)$$

This equation gives the true stress for homogeneous elongation or compression as:

$$\sigma_1 = 2C_1\left(\lambda^2 - \frac{1}{\lambda}\right) + 2C_2\left(\lambda - \frac{1}{\lambda^2}\right)$$

Once again the side stresses σ_2 and σ_3 have been taken to be zero. The Mooney–Rivlin equation is the first two terms of the more general expression for the strain energy density function:[79]

$$W = \sum_{i,j=1}^{N} C_{ij}(I_i - 3)^i (I_j - 3)^j$$

In principle, by taking a sufficient number of terms in this or other series, strain energy density functions can be represented to a high degree of accuracy. However, there are problems which remain. When elastomer is confined it cannot be assumed incompressible. Compressibility is sometimes allowed for by adding terms in (powers of) $(I_3 - 1)$ or $(I_3^{1/2} - 1)^2$ to the strain energy power series.

The stress–strain curves for filled elastomers often exhibit sharp changes of slope – especially for small strain amplitude deformations. This can lead to computational difficulties with the power series representation of the strain energy density function. A related and more fundamental problem is that when filled elastomer is deformed not all the work done is stored – some is lost from the mechanical system in the form of heat. The stress is therefore no longer simply a function of the state of strain: it also depends on how that state of strain was arrived at.

7.7.3.2 Creep and stress relaxation

A well-vulcanized gum elastomer, such as polybutadiene (BR) or natural rubber (NR), operating well above its T_g has low creep. Stress relaxation ($\dot{S}$) and creep ($\dot{C}$) rates are defined:[80]

$$\dot{S} = -\frac{1}{\sigma}\frac{\partial \sigma}{\partial t}\bigg|_{\epsilon}, \quad \dot{C} = \frac{1}{\epsilon}\frac{\partial \epsilon}{\partial t}\bigg|_{\sigma}$$

So their interrelationship is:

$$\dot{C} = \frac{\sigma}{\epsilon}\frac{\dot{S}}{(\partial\sigma/\partial\epsilon)_t}$$

showing that for a given stress relaxation rate and secant modulus, creep is lowest for stiffening spring and highest for softening spring behaviour. The theory of linear viscoelasticity holds reasonably well for gum elastomers, especially in shear, so that time, physical stress relaxation, and the mechanical loss tangent (tan δ) are interrelated:

$$t\dot{S} \approx \frac{2}{\pi}\tan\delta$$

Therefore unfilled elastomers with high mechanical damping such as nitrile (NBR) have relatively high creep and stress relaxation rates. For filled materials based on low-damping elastomers, damping is largely non-viscoelastic and the previous equation substantially overestimates $\dot{S}$.

Plots of deformation during creep or stress during relaxation against log t are approximately linear. The rate of stress relaxation is generally reported as a percentage (of the stress 1 min after the displacement was applied) per decade (tenfold increase in time); creep is similarly reported in percentage per decade. As an indication, stress relaxation for a normal gum NR vulcanizate is about 1.5% per decade, while for a heavily filled NR vulcanization about 5% per decade can be expected. After long periods, stress relaxation may largely be caused by chemical changes in the elastomer rather than by viscoelastic or other physical processes. The 'chemical' relaxation rate is approximately independent of time and has a different dependence on temperature than does physical relaxation. For a typical NR-based material at 10°C, the anaerobic chemical relaxation rate is about 0.5% of initial stress per annum. In sustained cold conditions, some elastomers crystallize, also resulting in stress relaxation; these effects, in contrast to other physical and chemical relaxation processes, can be reversed by sufficient warming.

7.7.3.3 Effects of filler

It can be seen in Figure 7.118 that each isoprene unit in cis-1, 4-polyisoprene contains a double bond between carbon atoms II and III: the polymer chain's backbone is said to be 'unsaturated'. The prevalence of these double bonds in NR renders it vulnerable to oxidative attack by diatomic oxygen, ozone or ultraviolet radiation, for example. There are similar potential problems for other unsaturated elastomers such as styrene–butadiene (SBR). However, replacement of carbon atoms II or III by an atom of high electropotential and/or large radius can provide some protection by stabilization of the double bond by electron sharing and/or by the blocking effect of the large atom; the chlorine atom in chloroprene rubber performs these roles. Vulnerable elastomers can also be protected from diatomic oxygen and ozone by the inclusion into the rubber compound of anti-degradant chemicals which impede the oxidation process. Wax included in the elastomer formulation diffuses to the surface after moulding and can form an effective barrier against ozone in some applications. Addition of even a small amount of carbon black powder gives protection against ultraviolet radiation.

When NR is subjected to very large strains, highly ordered regions of polyisoprene are formed. (A corresponding process occurs in other elastomers – notably chloroprene rubber.) This process of strain crystallization, which is quickly reversed once the strains are reduced, gives NR a good measure of resistance to fatigue crack propagation – even when the elastomer contains little or no filler. In contrast, many other elastomers, notably EPM/EPDM and styrene–butadiene (SBR), normally require substantial loadings of reinforcing filler to give adequate fatigue life.

Inclusion of filler in a rubber compound can also modify processing properties and harden the vulcanized material. However, the presence of significant amounts of reinforcing filler in a rubber compound leads to appreciable levels of mechanical damping, particularly at strain amplitudes around 3%. The filled material can be much stiffer at small strains than at large strains and the mechanical properties depend on strain history to a marked extent. Overall, the effect of a substantial filler loading on an elastomer is to create a material which is of great practical use but whose detailed mechanical behaviour is difficult to model with accuracy. Filled elastomer mechanical behaviour includes:

1. Non-elliptical force deformation loops;
2. High damping without strong frequency dependence of modulus;
3. Strain and history dependence of properties.

The 'triboelastic' model for filled elastomers has recently been developed to incorporate some aspects of (1)–(3) above; notable for its relative simplicity it is based on combinations of spring and frictional elements.[81] Fractional calculus offers another approach to modelling the weak frequency dependence of modulus in high-damping filled elastomers.[82] However, modelling the dynamic behaviour of filled elastomers remains a research area.

7.7.4 Thermoplastic elastomers

Thermoplastics elastomers (TPEs) are materials which have mechanical properties intermediate between conventional elastomers and normal thermoplastics. TPEs share the advantages of normal thermoplastics: rapid moulding, re-usable scrap and the potential to combine a number of parts in a single moulded unit. Where a metal or thermoplastic component is replaced by a TPE, machining costs can sometimes be reduced because dimensional tolerances are less critical for the TPE as it is more deformable. Environmental and fluid resistance is often higher for TPEs than for the rubber which they seek to replace. The high ozone resistance of TPEs can lead to very high fatigue lives at low strain amplitudes. (Often it is ozone cracking which initiates failure in vulnerable elastomers operating at low strain amplitudes.)

In general, TPEs show less complete recovery immediately after deformation than do many conventional elastomers. Chemical creep and relaxation rates, dominant at long times, tend to be lower for TPEs although physical, shorter-term, relaxation is generally higher. In components moulded from TPE, residual stress can be significant and further study of the effects of this on long-term performance would be beneficial.

Higher modulus materials, with Young's moduli up to hundreds of MNm^{-2} can be obtained with TPEs than is normal for conventional elastomers; however, the yield strains of these high-modulus TPEs is correspondingly small (typically 10%) when judged by the standards of other elastomers. The relatively high moduli of some TPEs means that weight savings can be achieved compared with conventional elastomers where a particular stiffness is a prime requirement in the product. However, the lower extensibility of hard and moderately hard TPEs means that redesign is often necessary.

7.7.5 Design of elastomeric components

Frequently, an elastomeric component is required to have a particular stiffness and to be able to support a given load safely. For dynamic applications a damping level over a particular frequency range may also be specified. (Dynamic stiffness and damping may also interact significantly with temperature effects (see Section 7.7.7.2).)

Approximate design of elastomeric components to a stiffness is relatively straightforward for simple geometries. However, some account must generally be made of the history-dependent non-linear behaviour of the material and the state of cure of various parts of the component. The qualifying geometries are limited to the high aspect-ratio bonded pad and a few others. For these, analytical small-strain formulae are available.[83-86] In the case of a circular bonded pad of diameter, d, and thickness, t, the shape factor, S, is given by

$$S = \frac{d}{4t}$$

If $2.5 < S < 7$ and G is the shear modulus of the elastomer, the translational shear stiffness (k_s), the torsional shear stiffness (T_s), the compressive stiffness (k_c) and the tilting stiffness (T_c) are given (Figure 7.120), to a reasonable approximation, by:

$$k_s = \pi d^2 G/4t, \ T_s = \pi d^4 G/32t$$
$$k_c = 3\pi d^2 GS^2/2t, \ T_c = \pi d^4 GS^2/32t$$

Also, if the pad is compressed by an amount d, the shear strains produced in the elastomers are typically $\sqrt{6} \ Sd/t$ rising to a maximum of $6Sd/t$ at the edges of the bonds. Note also that the horizontal stiffness of a multiple layer pad may be significantly reduced because of tilting of the layers.

For more complex geometries, in the past a combination of approximate formulae, experience, intuition, trial and error and rubber compound adjustments have been needed to obtain a component of a given stiffness. Today, finite-element analysis (FEA) is increasingly being used to calculate the force–deformation behaviour of (and stresses within) elastomeric components. For unfilled elastomers, there is little difficulty in describing the material properties needed for either small- or large-strain programs (see Section 7.7.3). Small-strain linear programs may be adequate in many cases where the strains are approximately 30% or less since the relative error will be of the order of the prevailing strain. The near-incompressibility of elastomers does, however, present a difficulty for many FE packages. The small-strain non-

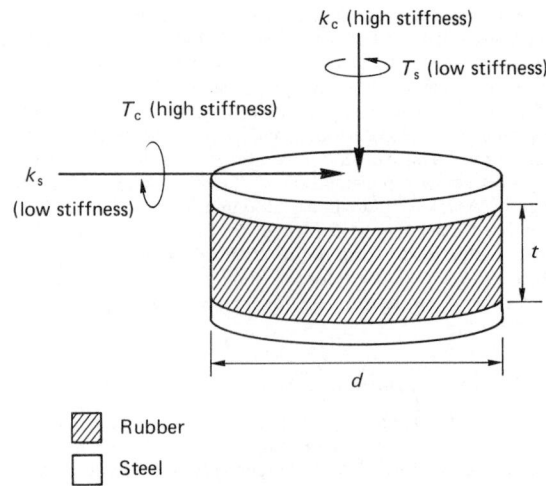

Figure 7.120 Elastomer/steel pad

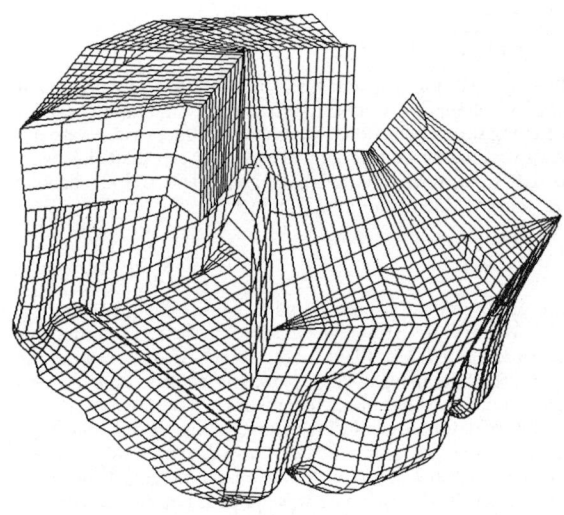

Figure 7.121 Finite-element mesh for an engine mount. (Courtesy of Clevite, Chippenham)

linearity of filled elastomers can present a serious difficulty for small- or large-strain FE programs and prediction of the dynamic behaviour of components made with highly filled elastomer remains a research area. Finite element analysis is, however, very useful for gaining insight into the behaviour of components of complex shape (Figure 7.121).

7.7.6 Elastomeric composites

A widely used type of elastomeric composite consists of layers which are virtually rigid alternating with (and bonded to) layers of elastomer. The rigid layers are often made from steel and are commonly referred to as 'reinforcements'. As shown by the formulae in Section 7.7.5, the stiffness in a direction perpendicular to the planes of the reinforcements can be made much greater than the in-plane stiffness. Many components, such as laminated earthquake isolation bearings and the suspension springs for some lorries, do have simple planar reinforcements. However, in other units the reinforcement layers can be V-shaped, as in suspension springs for some trains, or two-dimensional curves, as in certain helicopter rotor bearings.

Elastomeric composites are also made in which the bonded reinforcement layers are woven fibres of glass. These reinforcements are relatively inextensible and such pads have been used widely as expansion bearings in bridges in California. Another subtly different type of composite pad consists of alternate layers of nylon fabric with some extensibility and cork-filled elastomer. Such bearings are used widely in Britain for vibration isolation. Although the shape factors are high in many of the types of composite listed above, in most cases their compliance derives principally from the filled rubber layers; in a further commercially available type of composite a significant amount of compliance is provided, initially at least, by the non-elastomer layers. In this case the elastomer layers are hundreds of times thinner than the plan dimensions of the pad and the non-elastomer layers are woven textile fibre.

Most elastomeric composite springs in use are of compact design. Their forte is the support of large loads – especially those acting perpendicular to the plane of reinforcements; furthermore, the translational stiffnesses in all three directions can be adjusted independently of one another by judicious choice of reinforcement, shape and spacing. Such 'conventional' elastomeric springs do have their limitations, however: one is that the maximum repeated deflection is limited to about the minimum dimension of the spring. Larger deflections can be achieved by combining the rubber unit with a lever. In one novel rubber spring under development the elastomer is sandwiched between relatively flexible but inextensible layers; the spring operates in flexure so that large deflections can be achieved but, unlike a normal beam, the ends remain parallel.[87] Other types of elastomer-based composites under development include short-fibre-reinforced elastomers.

In general, reinforcement of elastomers by short discontinuous fibres gives a composite having a performance lying between high-structure carbon black reinforced rubber and continuous fibre rubber composites. The aim of short-fibre elastomer composites is to maintain flexibility while exploiting the high moduli of reinforcing materials such as fibres of carbon, glass or polymers. Detailed properties depend on the fibre and matrix materials; adhesion between the two; and on dispersion, orientation and aspect ratio of the fibres. Aspect, length to width, ratios of 1000:1 are usual; dimensions are at the micrometre level.

The properties that short-fibre reinforcement produce are highly anisotropic unless special procedures are adopted in processing. Appreciable increases in modulus are achievable, relative to the matrix elastomer, and high compressive loads can be applied to the composite material, which retains some recoverable flexibility and extensibility. Abrasion resistance and high-temperature properties are frequently improved. These results can often be achieved with short-fibre reinforced elastomers by processes which are less laborious than those, such as laminating, required for continuous fibres.

Short-fibre reinforcement has been shown to be particularly effective in improving dimensional stability for blended, unvulcanized, TPEs because of cross-interface location of the fibres. Aramid fibres increase the heat-distortion temperature of polyester TPEs from 80° to 200°C and wear life is improved four hundredfold. Silk fibre has been shown to raise the longitudinal tensile strength of TPEs by a factor of two.

High-modulus short fibres can be effective only if forces are transferred to them. However, force transfer is limited as fibre levels must be kept relatively low if some degree of elastomeric behaviour is to be retained; this constrains their use in elastomers as compared to thermosetting plastics. In order to be fully effective, and to avoid premature failure in moulded components, the reinforcing fibres must be fully dispersed and wetted by the matrix elastomer; attainment of these objectives requires continuing development of process methods.

For continuous fibre reinforced materials it is often assumed that the strains in the matrix and reinforcement materials are equal. This assumption is not entirely valid for elastomer composites with relatively low levels of short fibres when stressed in the longitudinal direction.

'Stress quenching' mechanisms of fibre pull-out can occur during impact and improve impact resistance for plastic-fibre composites; these mechanisms are not helpful in fibre-reinforced elastomers, for which good bonding is essential.

With particulate filler in the elastomer matrix, fibre dispersion is improved, and the strains in the matrix and fibre are less dissimilar when the composite is under load. Under these circumstances the minimum length for the reinforcements to be effective (L_c) can be estimated from the fibre radius (R), the interfacial shear strength (τ_i) and the tensile breaking stress of the fibres (σ_c):

$$L_c \approx R\sigma_c/\tau_i$$

However, in practice, processing does not give ideal alignment of the fibres and this makes analysis difficult.

7.7.7 Elastomers in use – further properties

7.7.7.1 Crack growth and fatigue

The fact that the stiffness of an elastomer originates largely from configurational changes in the polymer chains rather than from distortions in interatomic bonds accounts for the low elastic (shear) moduli that can be obtained with these materials – down to about 0.3 MNm^{-2}. Their strength, however, does depend on interatomic bonds; thus the failure stress even for soft vulcanizates, such as unfilled or lightly filled natural rubber, is about 25 MNm^{-2} in tension – comparable to that for thermoplastics such as high-density polyethylene and nylon 66 (Table 7.67). For practical purposes, failure and the elastic limit are coincident in conventional elastomers; similar fracture mechanics theory is applied to elastomers as to other materials but with the proviso that the strains are orders of magnitude larger.

According to fracture mechanics, the propagation rate of a crack into an elastomer is governed by T:

$$T = - \left.\frac{\partial U}{\partial A}\right|_l$$

where U is the total elastic energy and A is the area of one fracture surface; the partial derivative is evaluated with boundaries of the elastomer a fixed distance l apart so that the external forces do no work. T (also written as G) is known as the 'strain energy release rate', although 'rate' here refers to changes in strain energy as a function of crack size rather than of time.

If T is less than a threshold value T_0, the crack will propagate only by ozone attack for susceptible elastomeric materials (Tables 7.68 and 7.69). T_0 is ~50 Jm^{-2}, but there are significant differences between elastomers, while compounding details such as cross-link density and filler loading also have an effect.

The existence of a threshold value for T implies indefinite life for components (in the absence of attack by ozone, etc.) if the deformation is kept below a threshold level. Deformation

Table 7.68 Selection guide: common elastomers

Name	Symbol[a]	Operational temperature range[b] (°C)	Tolerance[c]			Comments
			Ozone	Water (20°C)	Mineral oil (20°C)	
Natural	NR	−60/+80	P	G	P	Excellent tear strength properties unfilled or filled. Can give low hardness and low mechanical damping. Properties relatively insensitive to temperature and frequency. Good tolerance of gamma radiation. Low-temperature crystallization stiffening most rapid at about −25°C. $T_g \approx -70°C$.
Synthetic polyisoprene	IR	−60/+80	P	G	P	Strength properties best for highest *cis* types. Properties can be similar to NR's; creep rate and strength generally slightly lower for IR.
Butadiene	BR	−70/+80	P	G	P	As is the case for many elastomers in this and Table 7.69, a generic title covering many structural variations. Damping and temperature/frequency dependence of modulus can be even lower than for NR because of low T_g (about −90°C).
Butyl Halobutyl	IIR, CIIR BIIR	−60/+120	F	G	P	Very good strength properties when filled, although unfilled generally inferior to NR. High damping over a wide temperature range but frequency sensitive. Lower gas permeability than most elastomers ($T_g \approx -65°C$) but some flexibility retained even at cryogenic temperatures. Cure of IIR is sluggish: halogenated butyls are faster, thus more suitable for covulcanization with other elastomers.
Chloroprene	CR	−30/+100	F	F	F	Excellent tear strength properties unfilled or filled. Self-extinguishing because of halogen in chemical structure but evolved halogen compounds may cause problems. Excellent fire resistance with appropriate additives. Particular formulations may be required for prolonged contact with water. $T_g \approx -40°C$.
Ethylene-propylene Copolymers Terpolymers	EPM EPDM	−40/+130	G	G	P	Standard high-temperature non-oil-resistant elastomer. Large amounts of filler and oil can be added to modify properties. Acceptable strength properties when filled but no match for NR when strength/fatigue properties and low heat build-up due to mechanical working are paramount.
Nitrile	NBR	−30/+100	P	G	G	NBR is a standard oil-resistant elastomer. Properties depend on acrylonitrile content of copolymer. Brittleness temperature, oil and high temperature resistance all increase and rubberiness decreases with increasing acrylonitrile. $T_g \approx -45$ to $-15°C$ depending on acrylonitrile content. Acceptable strength properties only when filled. High mechanical damping and low gas permeability. XNBR has better strength and abrasion properties than NBR. HNBR has better mechanical properties and higher resistance to heat and to oxidized fuels (sour gasolines) than NBR.
Hydrogenated	HNBR[d]	−30/+140	F	G	G	
Carboxylated	XNBR[d]	−30/+120	P	F	G	
Styrene butadiene	SBR	−40/+80	P	G	P	A generic title covering many structural variations. The styrene content greatly influences final properties. Good strength properties and abrasion resistance when filled.
Urethane						Wide range of hardness available – hardest have high strength but low deformability before yield. Range of damping available through the hardness range. Creep relaxation generally higher than for NR. Very good resistance to gamma radiation. Continuous contact with hot water should often be avoided.
Polyester	AU	−30/+80	G	F	G	
Polyether	EU		G	G	G	

[a] Unless otherwise indicated ASTM D1418 or ISO 1629 symbols are used.
[b] Indications only of typical long-term continuous operating temperatures: formulation details are often important. See also ASTM D2000.
[c] Intrinsic level: G, good; F, fair; P, poor. Additives can improve tolerance. F means either intrinsic resistance moderate or special care needed in formulation. See also ASTM D2000 and ISO TR 7620.
[d] Not an ASTM/ISO symbol.

Table 7.69 Selection guide: special-purpose elastomers

Name	Symbol[a]	Operational temperature range[b] (°C)	Tolerance[c] Ozone	Tolerance[c] Water (20°C)	Tolerance[c] Mineral oil (20°C)	Comments
Chlorinated polyethylene	CM	−20/+140	G	G	G	Good environmental resistance but not a very 'rubbery' elastomer.
Chlorosulphonated polyethylene	CSM	−20/+130	G	G	G	Temperature range and other properties depend on degree of chlorosulphonation. Elasticity and strength properties do not match those of NR. Brilliant white vulcanizates available.
Epichlorhydrin Homopolymers	CO	−10/+140	G	P	G	Fair strength properties. ECO can give relatively low mechanical damping. Gas permeability for CO can be significantly lower than for IIR.
Copolymers	ECO	−20/+140	G	F	G	
Epoxidized natural	ENR[d]	−20/+80	P	G	F–G	Degree of oil resistance and other properties depends on degree of epoxidation. Strain crystallizes to give excellent tear strength filled or unfilled. High damping and low gas permeability obtainable.
Ethylene-methylacrylate	AEM (EACM[d])	−40/+150	G	G	F	Strength properties fair. High damping over a wide temperature range. Can accept substantial quantities of filler – e.g. fire retardant.
Ethylene-vinylacetate copolymers	EVA[d]	−40/+120	G	G	F	Properties depend on proportion of vinylacetate-when above about 1/3, essentially elastomeric. Substantial quantities of filler accepted.
Fluorinated hydrocarbon (fluoroelastomers)	FKM (FPM)	−20/+230	G	G	G	Vulcanizates quite hard. Relatively low extensibility.
Fluorosilicone	FVMQ	−90/+230	G	G	F	See silicone.
Perfluorinated	FFKM	−20/+260	G	G	G	As FKM but tolerant of a wider range of fluids.
Polyacrylics	ACM	−10/+140	G	P	G	Vulcanizates relatively hard. Moderate strength properties.
Polynorbornene	PNR[d]	−40/+80	P	G	P	These elastomers are highly plasticized polynorbornene – properties can depend on the plasticizing oil. Very soft vulcanizates available with acceptable strength and elasticity.
Polypropylene oxide	GPO	−50/+130	G	F	F	Lower mechanical damping possible than for most. Moderate strength properties.
Polysulphide	TM	−40/+80	G	G	G	Excellent tolerance to wide range of organic fluids at moderate temperatures. Rather weak.
Silicone	MQ, VMQ, PMQ, PVMQ	−90/+240	G	G	P	Poor strength properties at room temperature can be improved by silica filler – to a point. Strength properties decline only gradually with increasing temperature. High permeability to gases.
Tetrafluoroethylene –propylene copolymers	TFE/P[d]	−20/+200	G	G	G	As FKM but resistant to a wider range of fluids.

[a] Unless otherwise indicated ASTM D1418 or ISO 1629 symbols are used.
[b] Indications only of typical long-term continuous operating temperatures: formulation details are often important. See also ASTM D2000.
[c] Intrinsic levels: G, good; F, fair; P, poor. Additives can improve tolerance. F means either intrinsic resistance moderate or special care needed in formulation. See also ASTM D2000 and ISO TR 7620.
[d] Not an ASTM/ISO symbol.

in simple extension provides an illustration. For an edge crack in a tensile test piece of an elastomer with Young's modulus E, small-strain theory gives

$$T \approx 3E\epsilon^2 c$$

where ϵ is the tensile strain and c is the crack length. For a high-quality well-mixed elastomeric material $c = 25~\mu m$ is representative of flaw size so, for typical values of $E = 3~MNm^{-2}$ and $T_0 = 50~Jm^{-2}$, indefinite life is predicted if the strain is kept below about 50%. This does occur in the absence of ozone cracking.

If T exceeds T_0, mechanical crack growth proceeds at a rate which depends on T. The progress of a crack may be slowed by rate-dependent energy-dissipation processes, particularly in high glass-transition temperature elastomers such as nitrile rubber (NBR). In strain-crystallizing elastomers such as NR or chloroprene rubber (CR) even at quite high values of T crack growth can be disrupted by the formation of crystallites at the crack tip. In many circumstances, then, when a constant stress is applied to a piece of NR or CR a crack will not grow; the crack advances only each time the stress is released and restored and by an amount which depends on the maximum T during the stress cycle; if the stress is only partially released, the crack growth per cycle is reduced by an amount depending on the minimum stress level. For a strain-crystallizing elastomer it is only when T exceeds the threshold value by orders of magnitude that a component of any bulk will fail during a single application of stress. Elastomers such as nitrile (NBR) which do not strain crystallize suffer mechanical crack growth even under non-cyclic conditions; this is termed 'time-dependent' crack growth. For susceptible elastomeric materials time-dependent crack growth is again a function of T; under cyclic conditions there is an extra contribution to crack growth. The presence of reinforcing fillers in the formulation can impede time-dependent crack growth in susceptible elastomers but, at high concentrations, they may produce side-effects which are not required. Thus NR or CR are frequently chosen for components which support substantial static loads. Nevertheless, under some conditions even strain-crystallizing elastomers may require significant amounts of reinforcing filler to improve their fatigue resistance.

It should be emphasized that fatigue in elastomers is a complex area. Elastomers vary in their fatigue resistance ranking according to the mechanical severity of the application and the mode of deformation.

7.7.7.2 Temperature, frequency and damping

For sinusoidal strain cycling of amplitude ϵ_0 and radian frequency ω

$$\epsilon = \text{real or imaginary part of } [\epsilon_0 e^{j\omega t}]$$

In general, the steady-state, cyclic stress response (σ) of an elastomer component is non-sinusoidal but can be described by a Fourier series:

$$\sigma = \text{Re or Im}\left[\epsilon_0 \sum_{n=0}^{\infty} A_n^* e^{jn\omega t}\right]$$

$$= \text{Re or Im}\left[\epsilon_0 \sum_{n=0}^{\infty} |A_n^*| e^{j(n\omega t + \phi_n)}\right]$$

where $A_n^* = A_n' + jA_n''$. If the component and material are symmetrical about $\epsilon = 0$, all even terms will be zero; when only the $i = 1$ term is non-zero this corresponds to the linear viscoelastic case:

$$\sigma = \text{Re or Im}[\epsilon_0 A_1^* e^{j\omega t}]$$

$$= \text{Re or Im}[\epsilon_0 |A_1^*| e^{j(\omega t + \phi_1)}]$$

ϕ_1 is the mechanical energy loss angle and is usually written as δ. Because of the orthogonality of terms in the Fourier series, δ is an accurate measure of mechanical energy dissipation even when there are higher harmonics in the stress response. For simple shear:

$$A_1^* = G^* = |G^*| e^{j\delta} = G' + jG''$$

where G^*, G', and G'' are the complex storage and loss shear moduli, respectively. (Even for heavily filled NR the third harmonic contribution to the stress response is unlikely to be more than ~3% of the fundamental up to shear-strain amplitudes of 100%.)

An elastomer's damping is often quoted in terms of rebound resilience (R). This is the fraction (or percentage) of mechanical energy remaining after a collision between two very dissimilar masses. The fractional resilience is related to the loss factor (tan δ):

$$R \approx e^{-\pi \tan \delta}$$

However, caution needs to be exercised as tan δ can depend both on strain amplitude and on frequency – hence on the duration of the collision. The damping ratio (ζ or $u\%$), relative to critical, of a damped spring-mass system is given by

$$\zeta \approx \tfrac{1}{2} \tan \delta$$

at the natural frequency. The approximation is very good for $\delta \leqslant 10°$.

According to linear viscoelasticity theory, the loss tangent of elastomers implies frequency dependence in the storage modulus:[88]

$$\frac{d\ln[G'(\omega)]}{d\ln(\omega)} \approx \frac{2}{\pi} \tan \delta$$

The equation is mainly applicable to vulcanizates which obtain their damping from the base elastomer rather than filler effects; it indicates that high-damping elastomers will have moduli which increase strongly with increasing frequency. The low intrinsic damping at ambient temperatures of elastomers such as butadiene rubber (BR), natural and synthetic poly-isoprene (NR, IR), and polypropylene oxide (GPO) is associated with a weak dependence of modulus on frequency. Frequency-insensitivity of modulus can be an advantage in some circumstances: increases in $|G^*|$ of $\leqslant 10\%$ per hundred-fold increase in frequency are typical for lightly filled NR, while for heavily filled NR the corresponding increase can be about 25% – depending on strain amplitude. Quite high levels of damping can nevertheless be obtained from materials such as filled NR – δ up to ~20° at low strain amplitudes even at frequencies around 1 Hz. (δ peaks at about 3% strain.) In contrast, the loss angle of unfilled or lightly filled NR is ~1.5° under similar conditions.

As in other glass-forming materials, the mechanical properties of elastomers exhibit a particularly strong temperature dependence as the temperature is reduced towards the glass-transition temperature T_g. G' rises towards the glassy value (about a thousand times the rubbery value); G'' also rises, but passes through a maximum just above T_g.

In the range T_g to $T_g + 100°C$ the effects of temperature and frequency are interrelated, inversely, in a manner which holds for a wide variety of polymers. For a given elastomer, the frequency dependence of the viscoelastic properties can be estimated from their temperature dependence over a restricted frequency range and vice versa. The formula most widely used for time–temperature superposition in elastomers is the WLF (Williams, Landel, Ferry) equation. According to the WLF transform, the storage moduli and loss angles measured at temperature and frequency (T_1, f_1) and at (T_2, f_2) are the same if:

$$\frac{1}{A}\log_{10}\frac{f_2}{f_1} = \frac{T_2 - T_s}{B + T_2 - T_s} - \frac{T_1 - T_s}{B + T_1 - T_s}$$

For a given material, constants A, B and the reference temperature T_s should be determined from the data; however, they may be estimated from the following:

$A \simeq 8.86$, $B \simeq 101.6$, $T_s \simeq T_g + 50°C$

Sometimes, before the transform operation, storage moduli, G', measured at an absolute temperature, T, are 'reduced' or 'normalized' relative to a chosen temperature, T_r, by an equation of the form

$$G'_n = \frac{\rho_r T_r G'}{\rho T}$$

where ρ and ρ_r are the densities at T and T_r, respectively.

The WLF and other transform equations are particularly useful because it can be difficult to obtain accurate dynamic data for elastomers at frequencies approaching 1 kHz and above. However, such transform equations should be used with caution as they can sometimes give misleading results. One conclusion is inescapable, though: choice of a high T_g elastomer, such as many nitriles (NBR), to achieve high damping, results in strong temperature and frequency sensitivity.

It should be noted that the effect of T_g-related damping in elastomers is to produce a decrease in modulus with increase in temperature whereas for a low-damping elastomer the statistical theory, supported by experiment, predicts the opposite effect.

It is not only the modulus of an elastomer which changes with temperature. Short-term effects of elevated temperature can include reduction in tear strength to unacceptably low values.

Long-term effects of high temperature include progressive breakdown of molecular bonds and/or progressive cross-linking reactions which soften and harden the vulcanizate respectively; an appropriately chosen cure system for high-temperature service can balance hardening and softening processes – producing little change in modulus with time. Strength properties show a more rapid deterioration at high temperatures.

7.7.7.3 Effects of liquids

Both equilibrium and time-dependent effects can be important when an elastomer is exposed to a liquid. Solubility is an equilibrium parameter of key importance when the elastomeric component is small in section or when immersion is continuous. In the absence of chemical cross-links susceptible elastomers can be dissolved in appropriate solvents. (Crepe rubber soles and diesel fuel is an example.) Well-vulcanized elastomers do not dissolve but can still be swollen and weakened considerably. The degree of swelling depends on the degree of chemical compatibility between the elastomeric material and the fluid concerned. The mineral (petroleum-based) oil compatibility of some hydrocarbon based elastomers such as natural (NR) and styrene–butadiene (SBR) is high and they can absorb large amounts. Other elastomers such as acrylonitrile–butadiene ('nitrile' NBR) have much lower mutual solubility with mineral oil and are therefore classed as oil resistant. Elastomers containing large numbers of chlorine or fluorine atoms have low compatibility with mineral oil. However, the presence of such strongly electronegative atoms can increase compatibility with water. The degree of compatibility of a liquid can also depend on non-rubbers in the vulcanized material. Polyisoprene has low compatibility with water but absorption can be increased by hydrophilic substances (e.g. protein) present in natural rubber (NR) but removed in deproteinized natural rubber.

When an elastomeric component is immersed in a soluble fluid the fluid does not saturate the elastomer instantaneously; in many cases the non-equilibrium concentration (c) is governed by the diffusion equation:

$$\frac{\partial c}{\partial t} = \mathrm{div}(D\,\mathrm{grad}\,c)$$

If D, the diffusion coefficient or diffusivity, is independent of the concentration the following results:

$$\frac{\partial c}{\partial t} = D\nabla^2 c$$

Analysis is complicated by the large amount of swelling that can occur, but the above equation indicates that the time required for a liquid to penetrate to a particular depth is roughly proportional to the square of the depth.

The diffusivity of a fluid in an elastomer depends, inversely, on the mechanical damping of the unfilled elastomer and strongly on the viscosity of the fluid (η). The diffusion coefficient of mineral oil is considerably lower in chloroprene rubber (CR) than in NR; in butyl rubber (IIR) and most nitriles (NBR) the diffusion coefficients are lower still. Whereas the penetration into NR of low- and moderate-viscosity petroleum-based oils is rapid it can take months for a viscous oil ($\eta \sim 0.3$ Nsm^{-2}) to penetrate much more than about 1 mm. It is this type of time-dependence which enables elastomers which are classified as non-oil-resistant to be used in circumstances where they are exposed to oil periodically.

7.7.7.4 Heat transfer

The diffusion equations given in Section 7.7.7.3 are also applicable to non-equilibrium thermal conditions. In this case temperature (T) is the relevant 'concentration'; the thermal 'diffusivity' is calculated from the thermal conductivity (k), the density (ρ) and the specific heat (s) of the elastomer.

$D = k/\rho s$

The thermal conductivity of many elastomeric materials is low in the absence of high carbon black loadings (~ 0.2 Wm^{-1} °C^{-1}) and the specific heat is relatively high (~ 2000 Jkg^{-1}°C^{-1}) while the density is ~ 1000 kgm^{-3} giving a value of $D \sim 10^{-7}$m^{2}s^{-1}. The low thermal diffusivity of elastomers means that although these materials can be vulnerable to sustained high temperatures, thick sections are resistant to high-temperature exposure of short duration. The relatively low thermal conductivity of elastomers can, however, be a disadvantage where there is significant mechanical energy dissipation in large components as heat build-up can result in damaging temperature levels.

7.7.8 Elastomer selection – some guidelines

In elastomer selection it is usual to categorize the properties of elastomeric materials into mechanical aspects such as modulus, damping or fatigue behaviour and into the elastomer's tolerance to environmental factors such as high or low temperature, exposure to oils, ozone, or ionizing radiation. The balance between the importance of mechanical properties and the degree of hostility of the environment, considered with the design of the component, will help determine whether an elastomer is suitable for a particular function. Simplified tables and guides to the performance of different elastomers may sometimes be misleading, as environmental factors can interact with each other and with mechanical conditions.

Although indications are given here of some of the complications, the main purpose is to give broad indications of the behaviour of various elastomer types (Tables 7.68 and 7.69).

The fact that some elastomers such as natural rubber (NR) and certain grades of chloroprene rubber (CR) and synthetic polyisoprene (IR) crystallize significantly when subjected to high strain gives these materials inherent advantages in strength and resistance to fatigue. Non-strain-crystallizing elastomers such as silicone rubber and most styrene–butadiene rubbers (SBR) require reinforcing fillers such as finely divided carbon black or silica powder to give them the strength properties required for many applications. Some elastomers, such as ethylene–propylene rubber (EPM/EPDM) and silicon rubber, have high innate resistance to degradation by ozone and diatomic oxygen whereas others, notably nitrile (NBR) and fluoroelastomers (fluorinated hydrocarbon rubbers), are relatively unaffected by many mineral (petroleum-based) oils.

Component geometry can have a profound effect on performance with regard to fatigue or environmental attack: fatigue life can be greatly extended by the avoidance of stress concentrations, while thick sections of 'non-resistant' rubber can often function adequately despite oil splashes or ozone attack.

A number of elastomers capable of withstanding high temperatures for long periods without deterioration are significantly weaker at the elevated temperature. Other elastomers can sometimes be used successfully above their normal operating limit; what constitutes an acceptably long life varies according to the application from hours, or less, to tens of years. It should also be remembered that the properties for a given elastomer type can vary widely depending on 'compounding' ingredients, and processing details (Section 7.7.2). Furthermore, elastomers are frequently blended with each other to influence performance or processing behaviour.

When high extensibility combined with strength and elasticity are the main requirements an elastomeric material rich in NR is often selected – especially so when the component is bulky. One reason that NR is favoured in these circumstances is that a low surface/volume ratio tends to protect the component against failure produced by surface degradation. A further advantage of NR for bulky components is that lower mechanical damping is obtainable with the natural material than with most other elastomers; it follows that problems of heat build-up are less when NR is used in components subjected to repeated deformation. (Similar comments apply to certain grades of synthetic polyisoprene, IR.) NR is, however, unsuitable for heavy exposure to mineral oils of anything but the highest viscosity; because of this and its inability to tolerate sustained temperatures much above 100°C NR is classified as a 'general-purpose' elastomer. Natural rubber is, however, resistant to a surprisingly wide range of liquids. The vulnerability of NR, and other elastomers such as SBR, to oxidative ageing and ozone attack can be overcome in many circumstances by incorporating anti-degradants into the formulation.

A typical temperature range for components made from NR is −60°C to +80°C for continuous operation. However, time-dependent stiffening due to crystallization can occur at temperatures well above −60°C. The maximum crystallization rate occurs in the region of −25°C. Although such stiffening is reversed by moderate strains or warming, the use of crystallization-resistant compounds (conventional sulphur vulcanization) is advisable in some applications.

The general-purpose (GP) base elastomers are competitive in price – natural rubber being least expensive. Although the price of ethylene–propylene base elastomers is currently four to five times that of NR this is offset by the fact that EPM/EPDM can accept large quantities of diluent oil of similar price to NR. The base elastomers of chloroprene rubber materials (CR) are also several (about five) times more expensive than NR. However, CR possesses a useful balance of mechanical and environmental characteristics which ensures it is widely used – although not in such quantities as the GP materials (Table 7.68). Chlorine compounds evolved on combustion give CR a level of fire retardancy but, in some circumstances, can pose problems. (Similar comments apply to other elastomers which contain chlorine or fluorine.) Improved fire resistance can be obtained with appropriate additives. Special formulations may be required for prolonged contact with water. Crystallization-resistant grades of CR are available, but in other grades quite rapid low-temperature crystallization-induced stiffening can take place – especially at temperatures around −10°C. Plasticizers can be used to depress the T_g of −40°C but care needs to be exercised, as this can sometimes increase the rate of stiffening caused by low-temperature crystallization.

Acrylonitrile–butadiene elastomers (NBR), commonly known as 'nitriles', are widely used because of their oil-resistant properties and the moderate price of the base elastomers (about four times that of NR). Typically, nitriles are significantly less rubbery than NR or CR. Increasing acrylonitrile content increases T_g, minimum usable temperature and oil resistance while reducing elasticity. (For 50%, 40% and 28% acrylonitrile content in NBR the approximate T_g are −15°C, −20°C and −45°C, respectively.) Nitrile is attacked by ozone, and polyvinylchloride (PVC) is sometimes blended in to improve resistance.

Isobutyl–isoprene elastomers (IIR), referred to as 'butyls', are similar in price to nitriles. In spite of their low T_g (about −65°C) unfilled gum butyl elastomers can give quite high levels of damping over a fairly wide range of temperatures. Other useful attributes of IIRs include low gas permeability and quite good tolerance to ozone.

Urethane elastomers (polyurethanes) can be formulated to give a particularly wide range of properties. It is this versatility coupled with convenient processing characteristics which accounts for the widespread use of urethane elastomers. Among this class of elastomers are materials with higher strengths and with moduli two orders of magnitude greater than conventional elastomers. High and low damping grades of urethane elastomers are available over a wide hardness range – although low damping grades have loss angles several times the lowest available for NR vulcanizates. Creep rates are generally significantly higher for urethane rubber than for NR vulcanizates. Some types of urethane elastomers have very good abrasion resistance. Urethane rubber is generally tolerant of oxygen and mineral oils, but polyester types (AU) are vulnerable to degradation by moisture (hydrolysis) in hot conditions, and to subsequent microbial/fungal attack. Polyether type urethanes (EU) are much more resistant to aqueous fluids but continuous use in water at over 80°C is not recommended and contact with strongly acidic or alkaline solutions should be avoided. On combustion, polyurethanes can produce highly toxic fumes so flame retardancy can be particularly necessary; this can be provided by additives or by polymer design. Urethane elastomers have good resistance to gamma radiation compared to most other elastomers. (The tolerance of NR is also good but halogen-containing elastomers have poor resistance to radiation.)

A number of special-purpose elastomers are available which can withstand chemical environments at temperatures around 150°C higher than those tolerated by general-purpose and other widely used elastomers such as chloroprene, nitriles and urethanes (Table 7.69). However, a high price must be paid for these remarkable elastomers. Fluorinated hydrocarbon elastomers (fluoroelastomers FKM/FPM) are currently about

Table 7.70 Typical properties of thermoplastic elastomers

	Styrenic (SBS and SEBS)	Block polyamide	Block polyester (YPBO)	Polyurethane	Ionic	Soft rubber blends[a] (TPV)
Hardness (Shore A & D)[b]	30A–90D	70A–80D	40A–75D	70A–70D	40A–90A+	50A–75D
Density (Mg^{-3})	0.9–1.3	1.0–1.15	1.1–1.4	1.1–1.3	0.95–1.95	0.90–1.30
Tensile strength (MNm^{-2})[c]	5–35	25–50	20–50	25–60	5–30	4–25
Elongation at break (%)[c]	250–1200	300–700	350–800	300–700	300–800	250–600
Operational temperature range (°C)[d]	−60/+110	−40/+140	−50/+140	−40/+110	−40/−[g]	−50/+120
Tolerance[e] Ozone	F–G[f]	G	G	G	P–G	G
Mineral oil	P–F	G	G	F–G	G	P–G

[a] Properties are for materials with a cross-linked component.
[b] The Shore A and the IRHD scales approximate to each other.
[c] Harder grades yield before failure.
[d] Indications only of typical long-term continuous operating temperatures.
[e] Indications only: G, good; F, fair; P, poor.
[f] Use of saturated blocks enhances environmental resistance.
[g] Temperature of dissociation for essentially monovalent neutralized aggregates is around 100°C.

50 times as expensive as general-purpose elastomers. Moreover, elastomers such as FKM can be more difficult to process and they produce vulcanizates which are harder and less elastic than the general-purpose materials. A useful attribute of silicone elastomers is that their strength properties decrease only gradually with increasing temperature.

Thermoplastic elastomers (TPEs) are available with a broad range of properties (Table 7.70). Styrenic (styrene–butadiene–styrene block copolymer) TPEs alone can have shear moduli as low as 0.3 MNm^{-2} or as high as several hundred MNm^{-2}. Although for harder types of TPE the elastic limit may be only about 10%, many of them yield before failure, making them tough materials. Hard, strong materials are available: the polyether–polyester block copolymers or block polyesters (YPBO); the polyurethane, or urethane TPEs; and the polyether–polyamide block copolymers or block polyamides. The chemical/environmental resistance of the YPBOs is also generally excellent – although at high temperatures additives may be required to give protection against attack by aqueous media. Environmental/chemical tolerance of urethane TPEs is generally good although there is some susceptibility to hydrolysis (attack by aqueous fluids) – especially for polyester-based materials at high temperatures. Resistance to water is better for polyether-based urethane TPEs and for harder grades.

The properties of thermoplastic elastomer blends vary widely according to the type of elastomer and thermoplastic and according to their composition and method of preparation. Many have higher resistance to ozone and diatomic oxygen than their rubbery component alone possesses, but their tolerance to mineral oils is generally not good unless an oil-resistant elastomer, such as nitrile, is used. Thermoplastic elastomer blends cannot withstand such high stresses as the urethane block polyamide and block polyester types. Short-term creep rates at ambient temperature for blends based on comparable contents of NR, EPDM and IIR are around 2.5%, 3% and 4% per decade, respectively (see also Section 7.7.3). However, at low (around −40°C) and high temperatures

(about 100°C) EPDM-based TPEs can have lower creep rates than those based on the natural materials. The fact that short-term, physical, creep and stress-relaxation rates are relatively high for TPEs is offset by their long-term, chemical, creep rates being relatively low.

7.7.9 Example applications

By definition, elastomers are soft materials able to undergo large strains and remain within their elastic limit; often these characteristics make them natural materials for springs for vibration 'isolation' or energy absorption. Elastomeric springs are used for the reduction of dynamic force transmission (isolation) as major components in the suspension systems of trucks and rail vehicles. Often, these are elastomer–steel laminates giving different force–deflection characteristics in different directions. Usually an elastomer with good mechanical strength and low damping is used, such as natural rubber, to avoid excessive heat build-up in these large components. Damping, if required, is supplied by a separate damper. Although damping is detrimental to isolation performance well away from resonance, in many practical vibration sources such as bumpy roads or piston engines, a very wide range of frequencies occur: some near-resonant behaviour then becomes inevitable and a measure of damping can be beneficial. Where heat build-up is not a problem a high-damping elastomeric material, such as one based on butyl rubber, is therefore sometimes used. However, natural rubber continues to be used very extensively in engine mounts in spite of the high ambient temperatures under the bonnets of many contemporary cars.

One difficulty with plain elastomer engine mounts is that they tend to have low damping and stiffness at low frequencies and high displacement amplitudes and high damping and stiffness at high frequencies and low displacement amplitudes. The ideal characteristics are the reverse: high stiffness and damping to restrain engine movement at idling speed and low stiffness and damping to provide vibration isolation at running speed. These ideal requirements are now met to a large degree by fluid-filled mounts encased in elastomer. A subtle but important aspect of vibration isolation which is sometimes overlooked arises from the fact that any real 'isolation' spring has mass: vibration waves can therefore travel through it, greatly reducing the effectiveness of isolation at audio frequencies. Springs based on elastomers are much less susceptible to these 'wave effects' than are those made from materials with very low damping such as steel. Moreover, the high-frequency transmissibility of steel springs can be reduced by using elastomeric elements in series. Elastomers perform these and similar force reduction functions in many parts of an automobile; it should not be forgotten that the tyres themselves are an important part of the suspension system (Figure 7.122).

The mechanical properties of elastomers, particularly natural and chloroprene, are widely exploited in other areas for vibration isolation. Notable examples include the suspension of underground rail track on elastomeric springs or matting to reduce the transmission of groundborne noise to nearby buildings; conversely, the buildings themselves, or parts of them, can be supported on isolation springs. The same principles of groundborne noise/vibration/shock isolation are applicable to situations ranging from the operation of a large steam-hammer to that of an electron microscope; frequently isolation of the source or the receiver is accomplished with the assistance of elastomers. Recently, important buildings in earthquake-prone areas have begun to be built on elastomeric-steel mounts for earthquake protection;[89] even in areas of relatively low seismicity potentially hazardous plant must now be earthquake resistant – seismic isolation mounts can be used towards this end.

Elastomers are capable of absorbing large amounts of mechanical energy per unit volume and returning to their original shape: their capacity is orders of magnitude greater than that of steel, for example. This property has led to the widespread application of elastomers in devices to reduce forces occurring in collisions which are a regular occurrence; particular examples are: dock and other types of fenders, bump-stops and snubbers, and playground surfaces. In accordance with Newton's second law, collision impulse reduced if rebound does not occur, thus high damping can be an advantage. However, other factors such as (fatigue) strength also need to be taken into consideration. For a given travel, forces are minimized if they are near the average value throughout; this type of characteristic is frequently obtained for rubber springs by designing them to buckle under design loads. Such buckling constructions include matched pairs of columns, tubes and also cellular elastomers.

Low-damping elastomers, such as unfilled or lightly filled natural rubber, can be used effectively to store mechanical energy. Many devices to exploit this attribute have been considered over the years but, apart from toys, the only widespread application is in small harpoon guns. However, this is a position that could change in a future dominated by rising energy costs.

The low moduli of elastomers enable them to conform well to surface roughness and accounts for their good grip. The high friction of rubbery materials is used in a large number of devices: belt drives, conveyor belts, solid and pneumatic tyres are just a few examples – many are multifunction devices. The highest friction is obtained with very soft vulcanizates; often these are highly oil-extended – as in racing-car 'slicks', for example. Contrary to popular belief, friction coefficients in excess of 1 are possible. Conversely, much lower friction can be obtained by means of formulation additives or by using fluoroelastomers.

The use of elastomers in joints and bearings can pay rich dividends in terms of: reduction of complexity, avoidance of stick-slip behaviour and of backlash problems, low mainte-

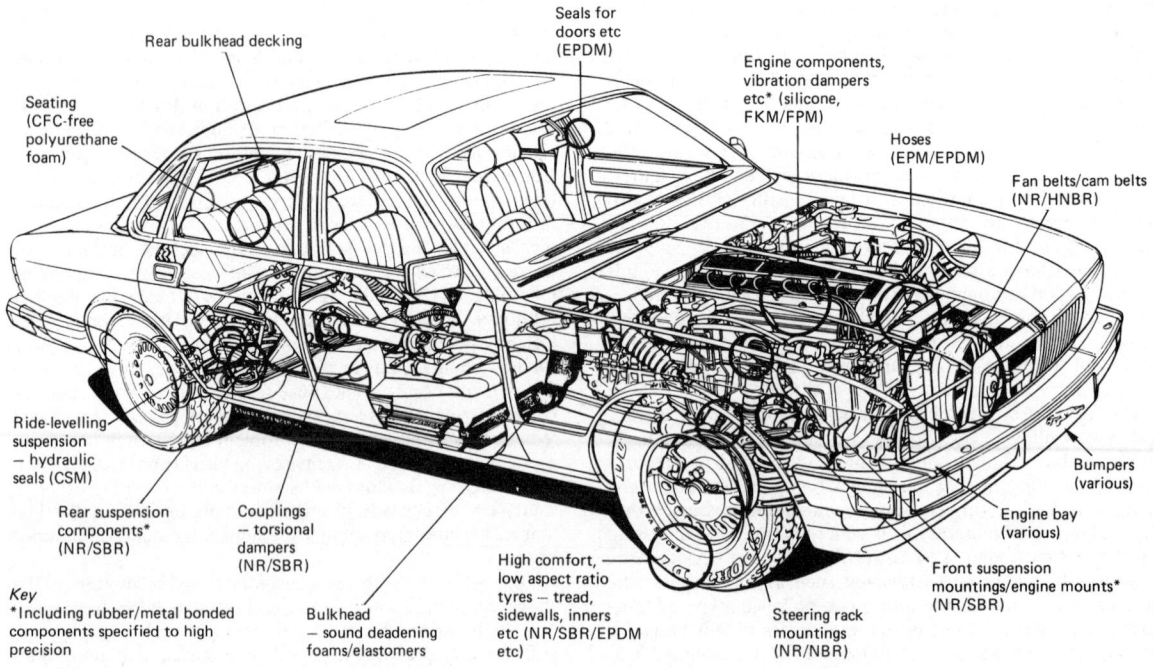

Figure 7.122 Elastomers in the automobile: the Jaguar XJ40 (Courtesy of Jaguar Cars, Coventry)

stick-slip behaviour and of backlash problems, low maintenance, and weight saving. Elastomer-based bearings can be used to allow for thermal expansion of structures. They are widely employed to accommodate misalignment and relative movement in shaft couplings where they also give force reduction for shock loadings. Other related applications include helicopter rotor bearings and track pads in tracked vehicles.

Elastomers find another range of applications in the containment, piping or sealing of fluids where flexibility is required. In many cases less than full rubbery behaviour is tolerable. This is fortunate, since some of the most widely used mineral-oil-resistant elastomers are not very rubbery. Ethylene–propylene is generally used for radiator hose and for seals in these and other hot water systems. There are potential uses for some of the thermoplastic elastomers in these applications. One aspect of fluid sealing where fully elastic behaviour is beneficial is where reciprocating and rotary sliding occurs; such seals are often designed to allow slight leakage to lubricate the seal. In addition to being extensively employed in fluid containment – especially where flexibility is a requirement – elastomers are widely used for corrosion and abrasion-resistant surfaces. For abrasion-resistant coatings the choice between elastomer types and between elastomers and other materials such as ceramics depends on the details of the application.

Applications of short-fibre reinforced elastomer and TPE composites include: diaphragms, hose, static seals, conveyor belting and belt drives. These composites are also used as 'bead' and 'chafer' materials in off-the-road types.

The promise of TPEs in mainstream mechanical engineering design has been demonstrated in a number of applications including bicycle wheels and tyres, hose, car bumpers and hinges. However, the potential of TPEs remains to be fully exploited by industry.

7.7.10 Conclusions

Many aspects of contemporary life such as road transport would be much more difficult and noisy and less comfortable without elastomers. Nevertheless, elastomers remain under-exploited. This under-use can be partly explained by difficulties in modelling the mechanical behaviour of elastomers – particularly when filled. Also, frequently, elastomers are inadequately or inappropriately characterized. Too often, elastomers are regarded as fringe materials. In reality, elastomers are natural materials to consider in many situations where a spring, joint or bearing is required. The expanding range of mechanical and environmental properties available from elastomers offers designers a valuable resource and a worthy challenge to their ingenuity.

7.8 Engineering ceramics and glasses

7.8.1 Introduction

Ceramics are inorganic crystalline materials, low in ductility and high in melting point which are usually fabricated not by melting but by processes involving powder compacting and sintering at very high temperatures. They are usually compounds of metals with non-metals and they owe their high temperature resistance to deformation to directed covalent or electronic bonds between the constituent atoms which do not permit plastic flow at operating temperatures too low for diffusion to occur.

Glasses have analogous compositions to ceramics but have lower melting points and retain their amorphous liquid structure on cooling. They have, instead of a melting point, a glass-transformation temperature above which their viscosity is low so that they can be melt formed, and below which the viscosity is high so that they have effective shape retention equivalent to a solid. *Glass ceramics* or melt-formed ceramics can be melt formed as glasses but crystallize on a micron or sub-micron scale at a lower temperature. *Single-crystal materials* are special products with a uniform structure.

Machinable ceramics fall into two types. One type has platey structures which can be readily cleaved and the other consists of particles that are only weakly bonded together so that they can be readily chipped apart. These, unlike other ceramics that have to be ground with an abrasive, can be machined with conventional tools.

7.8.2 Standards

Most national standards are highly oriented to testing or application. Only in the field of electrical insulation has any systematic classification been attempted in IEC 672 1980 which is to be the basis for BS 6045. So far as is possible, the IEC classification will be followed in this section.

7.8.3 Clay-based ceramics

Technical porcelains are ceramics based on clay and other silicates with added alumina. Fine-grain versions IEC IXX, 2XX and 4XX are used primarily for electrical purposes but coarser grain size chemical stoneware and kiln furniture are also included. These are well vitrified and therefore non-porous.

IEC Class C110, Porcelains with <30% alumina, consist of quartz grains and massive mullite surrounded by a matrix of glassy feldspar, and the shrinkage of the quartz on transformation at 530°C leads to microcracking. They have low strengths (50–100 MPa) and the lowest thermal conductivities (1.7–2.1 Wm^{-1} K^{-1}). They are good insulators at ambient temperature but resistivity falls off as temperature increases. They are cheap and used for internal and external insulators.

IEC Class C111, Pressed porcelains with <30% alumina, are similar in composition to Class 110 but because the pressing process, which is carried out with a clay containing less water, allows trapping of air between the granules, pressed porcelains are more porous and have lower strengths. They must be glazed for outdoor applications and cannot be used at high voltages.

IEC Class C112, Crystobalite porcelains, contain crystobalite instead of quartz but are otherwise similar to Class 110.

IEC Class C120, Aluminous porcelains with 30–50% alumina: most or all of the quartz in Class 110 is replaced with alumina giving improved strength (*ca* 100 MPa).

IEC Class C130, Aluminous porcelains with 75% alumina: feldspar porcelains in which all the quartz has been replaced by alumina improving modulus and strength (up to 200 MPa).

IEC Class C2XX, Steatite and forsterite ceramics replace the quartz or alumina in class 1XX with enstatite or forsterite talc. These materials comprise:

- IEC Class C210, Low-voltage steatites. These are porcelains with 80–90% talc, 5–10% plastic clay and 5–10% feldspar. They have low strength (50–100 MPa) and because they have about 0.5% open porosity are limited to low-voltage insulation.
- IEC Class C220, Normal steatites. In these the electric losses of the glassy phase are reduced by the substitution of barium or calcium carbonate for the feldspar and the

addition of magnesia and zirconia. Class C220 has higher strength, resistivity and lower dielectric dissipation than Class C210.

- IEC Class C221, Low-loss steatites. Alkali levels are lower than those of Class C220 to reduce dielectric loss by minimizing the glassy phase.
- IEC Class C230, Porous steatites. These are low-loss steatites with *ca* 30% open porosity which improves machinability and thermal shock resistance but limits the material to low-voltage applications.
- IEC Classes C0240, Porous forsterites, and C0250, Dense forsterites. In forsterites the magnesia level is raised so that forsterite is formed instead of steatite to give a coefficient of expansion *ca* $10 \times 10^{-6} \text{ K}^{-1}$ which matches titanium and some nickel–iron alloys for ceramic to metal assemblies in high-frequency power devices and microwave tubes.
- IEC Class C410, Dense cordierite. Cordierite is a magnesium aluminium silicate which is made synthetically from clay, talc and sillimanite and has a low coefficient of thermal expansion. Ceramics consisting of this material with 5% vitrifying feldspar are readily fired and can be made dense. Although their strength is only 50–100 MPa their low thermal expansions *ca* $3 \times 10^{-6} \text{ K}^{-1}$ confer shock resistance.
- IEC Class C420, Dense Celsian and Zircon Porcelains (unclassified) are little used.

Porous ceramics (IEC Classification C5XX) may be based on clay, cordierite or alumina up to about 80%. (Porous ceramics with higher alumina contents are included under high aluminas, Class C7XX.) They are used primarily as electrical heating and thermocouple insulations where their porosity confers resistance to thermal shock and machinability. These ceramics are compounded from fine aggregates for electrical applications, but coarse aggregates may be used for crucibles and tubing for high-temperature processing.

IEC Class C510, Porous aluminosilicates: these are manufactured primarily from refractory clays with additions of quartz, sillimanite or alumina to increase refractoriness. They contain very little fluxing material and their 30% porosity gives them the thermal shock resistance required for such applications as gas fire radiants.

IEC Classes C511, 512, 520, Porous magnesium aluminosilicates: these are developed from Class C510 by the addition of magnesia-containing minerals to produce cordierite which greatly enhances shock resistance and the distinctions between the classes are based on performance. The highest-performance C520 Class will stand severe thermal cycling and are used for experimental heat exchangers and automobile exhaust catalyst supports.

IEC Class C530, Porous aluminous materials: these ceramics have alumina contents up to 80% and are more refractory but less shock resistant than the other classes of porous ceramic.

Impermeable aluminosilicate and mullite ceramics, IEC Class 6XX, are impermeable fine grain materials with high strength (100–200 MPa) and high dielectric strength. They are used for gastight tubing and high-temperature electrical insulation and are classed according to alumina content.

IEC Class C610, 50/60% alumina ceramics: these are made from clay and alumina (or aluminosilicates) and consist of mullite in a silicous glassy phase.

IEC Class C620, 60–80% Alumina ceramics: this specification covers materials with a wide range of properties and crystalline compositions because mullite, which is 71.8% Al_2O_3 and 28.2% SiO, is highly refractory and compositions containing more Al_2O_3 are also highly refractory (and require a high firing temperature). Compositions with more silica than

mullite are less refractory and the addition of other oxides such as CaO reduces refractoriness considerably. If a highly refractory material is required this must be specified closely. However, all materials in this class are fine grained with no open porosity and have fair shock resistance.

7.8.4 Oxide-based ceramics

Oxide-based ceramics, IEC classification C3XX, C7XX and C8XX and non-classified materials consist of synthetic oxides together with bonding materials which sometimes form a glassy phase. The amount and composition of the glass has a significant influence on the engineering properties. In general, a low glass content gives a high refractoriness, a large grain size, high electrical resistance and low dielectric loss. On the other hand, strength and wear resistance are favoured by a low grain size which requires more glass to reduce the firing temperature.

IEC Classification 7XX, High alumina ceramics: in these ceramics the second two digits denote the minimum percentage of alumina. This degree of classification is presumably adequate for electrical purposes but not for all applications.

IEC Class C799, High-purity aluminas (>99.7% Al_2O_3): these are single-phase ceramics which may be sintered to give a large grain size or hot pressed to give low porosity and a small grain size. If sintered in a reducing atmosphere they can be made translucent or transparent and may be used, for example, for sodium vapour lamp envelopes.

Aluminas with over 99.7% Al_2O_3 may be made from relatively lower-purity aluminas or may have added magnesium oxide which restricts grain growth. The magnesia-free products are more refractory but weaker (100–200 MPa tensile strength) while those containing magnesia have tensile strengths of 2–400 MPa and are more suitable for thin-walled tubing or small diameter rods. Aluminas with 99–99.7% Al_2O_3 have usually small amounts of glassy phase which facilitates sintering and controls electrical properties.

These are the most widely used of the alumina ceramics for engineering and electrical purposes, those with fine grains having strengths up to 400 MPa while those with coarse grains are used for electrical insulation.

IEC Class C795: These aluminas all have a deliberately formulated glassy phase which facilitates metallizing by manganese/molybdenum.

IEC Class C786: this is a somewhat heterogeneous class, some having a glassy phase while others have additions of manganese and titanium dioxides which impair electrical properties and refractoriness. Those with lower alumina contents are non-refractory and are used for low-temperature electrical insulators and mechanical components.

IEC Class C780: there are two types of material in this class, one is refractory consisting of mullite and silica, the other non-refractory and limited to low temperatures and non-critical uses.

There are also aluminas which do not fall within the IEC classification. Porous alumina ceramics are available in the same composition ranges with high resistance to corrosion or slag attack. Unlike the non-porous materials, these can be machined (or rather chipped) by hard metal tools. The addition of 10–20% of zirconia to alumina ceramics enhances strength to over 400 MPa, making them suitable for cutting tool tips. The addition of 20–40% TiC also increases strength and stiffness for similar applications. The addition of TiO (a high-permittivity material) significantly increases the dielectric constant of alumina ceramics where this is desirable.

Synthetic alumina crystals such as sapphire and ruby are used for high temperatures and scratch-resistant windows and for bearings and laser elements.

IEC Class C830, Zirconia ceramics: zirconia is more refractory than alumina, it can be machined readily to give a low coefficient of friction against metals and it is a useful oxygen ion conductor above about 700°C. It is, however, more expensive than alumina, which is therefore used preferentially where either will suffice and its use is complicated by phase instability. Pure zirconia is monoclinic at room temperature, transforms to tetragonal form at about 1200°C and to a cubic form at about 2370°C. Its use as a ceramic depends on stabilization of the cubic phase, of a fine tetragonal phase within cubic crystals or of an all-tetragonal material.

Truly stable zirconia would be very strong (above 400 MPa) and highly refractory. Development is proceeding and, if manufacturers' claims are fully justified, zirconia ceramics have wide applications. Meanwhile they are used for oxygen probes, small crucibles, dishes, etc. and heating elements for operation in air at high temperature.

IEC Class C3XX, Titania ceramics: ceramics based on TiO_2 have been used for applications requiring high permittivity and clay-bonded sub-stoichiometric materials are used for thread guides in the textile industry because they are semiconducting and discharge static electricity.

IEC Class C820, Magnesia ceramics: magnesia possesses few advantages over alumina as a ceramic but, considered as a refractory, has a higher thermal conductivity, greater refractoriness and greater resistance to basic slags. Magnesia ceramics are therefore used in applications requiring resistance to corrosion at high temperatures and, specifically, for electrical purposes as an insulator in mineral insulated cabling.

IEC Class C810, Beryllia ceramics: beryllia is much more expensive than alumina and highly toxic. However, it has a thermal conductivity at room temperature of 500 $Wm^{-1} K^{-1}$, an order of magnitude greater than other electrical insulating material, and it is therefore used for some electronic devices. Its widely heralded future as a nuclear material has not yet been realized.

Other oxide ceramics not in the IEC classification include lime, thoria, uranium dioxide (used as a nuclear fuel), ferrites (which have very extensive magnetic applications) and titanates.

Ferrites, compounds of Fe_2O_3 with other oxides such as ZnO, are available with very high magnetic permeability and can be made in the powder form to have very low losses so that they may be used for cores for high-frequency inductors. Certain ferrites will form strong permanent magnets.

Titanates are used for their electrical properties. Some (usually barium titanates) have permittivities up to 5500 and are used as dielectrics for electronic circuitry. Magnesium titanates have permittivity-compensating properties. Lead zirconate–titanates have strongly piezoelectric properties.

7.8.5 Non-oxide ceramics

Non-oxide ceramics differ essentially from oxide ceramics in that they oxidize at high temperatures in air and cannot therefore be fabricated by a conventional oxide-sintering process. They must therefore be fabricated by the alternative routes listed in Section 7.8.11 and their compositions and microstructures are adjusted accordingly. The most important are carbides, silicides, borides and carbons.

7.8.5.1 Silicon carbide (carborundum)

This owes its application to its high hardness, very high refractoriness, high thermal conductivity and its semiconducting properties. Its electrical conductivity is low at low temperature and low currents, but increases by orders of magnitude at high temperatures and high currents. Although silicon

carbide is oxidized in air at high temperature many of the commercial materials form a protective layer of silica.

The name 'silicon carbide' comprises a number of products manufactured in different ways. The conventional method of manufacture is the reduction of silica sand by carbon in the arc furnace which produces the high-temperature stable α-phase. The phase is produced by low-temperature gas reactions and is converted to the β-phase by heating above 2000°C. Commercial products include:

1. Clay-bonded silicon carbide containing 10–15% clay which bonds chemically to the surface film of silica on sintering;
2. Other additives which enhance sintering, but permit a higher proportion of silicon carbide than is possible with clay bonding, have been used with reported success;
3. Very high-temperature sintering of prime silicon carbide produces a coarse-grained, porous and highly refractory product;
4. Hot-pressed silicon carbide requires the addition of a secondary material, usually about 2% alumina. It is very strong (above 400 MPa), fine-grained and non-porous but very expensive;
5. Reaction-bonded silicon carbide is made by two distinct processes. In one type a compact of silicon carbide, carbon and an organic binder is infiltrated by liquid silicon which bonds the particles by the formation of additional silicon carbide. The free silicon remaining may, if required, be leached out with acid to leave an open porous refractory. The other process achieves the same object by firing a compact of silicon carbide and silicon in nitrogen to produce a silicon nitride bond.
6. Coatings or thin-walled components of pyrolytic silicon carbide can be produced by decomposing a gas mixture containing both silicon and a carbon on a heated substrate. This process found an important application in the manufacture of fuel elements for high-temperature gas-cooled reactors. In these fuel elements a silicon carbide shell contains fission products produced in small spheres of fissile material;
7. Graphite can be 'case hardened' with silicon carbide produced from solid (or gaseous) silicon.

The original application of silicon carbide was as an abrasive for grinding steel, but it has been replaced for this purpose by fused alumina and its use as an abrasive is now restricted to shaping materials of lower tensile strength. It is used extensively for refractories, particularly where the shock resistance conferred by its high thermal conductivity is an asset both in vacuum or reducing conditions and in air. A major use is in electrical resistance heating elements and it is also employed for susceptors in high-frequency induction fields. Its high hardness makes it suitable for abrasion-resistant components, particularly for non-lubricated bearings.

7.8.5.2 Silicon nitride-based materials

Silicon nitride and its derivatives have properties which augur a great engineering potential because of its high hardness, high strength, refractoriness, low coefficient of thermal expansion which confers thermal shock resistance and its resistance to oxidation which, like that of silicon carbide, is due to the formation on its surface of a silica film.

The applications of silicon nitride ceramics has, however, been hindered by fabrication difficulties. Like silicon carbide, it cannot be readily sintered, tending to dissociate at above 1850°C and alternative methods of manufacture are difficult to control or expensive. Two products are available:

1. *Reaction-bonded silicon nitride*: This can be produced by converting a silicon powder compact to nitride by sintering in a nitrogen atmosphere. The resulting compact is porous and the product variable.

2. *Hot-pressed silicon nitride*: This requires the addition of an oxide (typically 1–5% MgO) which provides a liquid phase at the sintering temperature. This produces a fine, strong product but limits the temperature at which the high strength is retained. Hot pressing is expensive and the shapes which can be produced are limited.

The applications of silicon nitride have so far been limited to high-temperature industrial applications. It is extremely valuable for such components and operations as furnace supports, heating tubes, jigs and printer's saggars, silver soldering and vacuum and copper brazing.

7.8.5.3 Sialons

The difficulties experienced with silicon nitride have, to a large extent, been overcome by the partial substitution in the molecule of aluminium for silicon and oxygen for nitrogen. The resulting sialons can be hot pressed from mixtures of silicon nitride, alumina, silica and aluminium nitride or, if additional densifying constituents are added, they can be sintered. These materials merit and have received considerable study and, as a result, porosities as low as those in hot pressings and strengths above 400 MPa may be achieved with sintered materials.

An alternative route to sintering is the nitriding (and subsequent high-temperature sintering) of a mixture of oxides and silicon. Sialons and related products are finding wide and increasing applications in manufacturing processes where resistance to heat and abrasion are required.

7.8.5.4 Boron carbide

Boron carbide is one of the hardest ceramic materials available and can be hot pressed to achieve tensile strengths above 400 MPa. It is used for very abrasive conditions (e.g. shot-blast nozzles and for ballistic armour). Its main drawback is its cost, which is high. The sintered product is more porous and has a coarser grain size.

7.8.5.5 Boron nitrides

Boron nitride in its most commonly available form differs from the ceramics already considered in that it has a laminar structure and is therefore soft and machinable. The properties of the boron nitride crystal are highly anisotropic and shapes deposited pyrolytically on a substrate have thermal and electrical conductivities and coefficients of thermal expansion which may be up to a hundred times greater parallel to the substrate than perpendicular to it. This property and the fact that many molten metals do not wet boron nitride has led to its use for crucibles for metal melting. There is also a cubic form made under high pressure which is used as an abrasive.

7.8.6 Carbons and graphites

Three allotropic modifications give rise to a very wide range of carbons and graphites with a correspondingly large range of applications:

1. Diamond is a cubic crystalline material with a very stable structure. It is the hardest substance known, highly refractive and highly transparent and very expensive when pure. It can, with difficulty, be manufactured but the manufac-

tured product, which can be used as an abrasive, is barely competitive with the mined material.

2. Graphite has a laminar hexagonal ring structure. Each ring is strongly bonded to six other rings in approximately the same plane but the bonding betwen planes is weak. Graphite has therefore excellent lubricating properties, is highly refractory and, if the layers of hexagonal rings are continuous and aligned, very high specific strength and specific elasticity modulus in a direction in the plane of the rings.

3. Amorphous carbon is, as its name implies, amorphous to X-rays but this structure is probably a large-scale statistical phenomenon rather than a true liquid type structure. The available amorphous carbons are probably random assemblies of graphite platelets, the size and distribution of which govern the properties.

Solid carbon and graphite products are usually made from a mixture of graphite (which occurs naturally) or carbon particles (usually derived from coal or oil cokes) pressed with a carbonaceous binder to form a solid block. The mass is converted to carbon on firing. Very high-temperature firing graphitizes both the amorphous carbon and the carbon derived from the binder.

Both industrial carbons and graphites are highly refractory materials which must be protected from oxidation at high temperatures with strengths around 45 MPa and electrical resistivities around 30 Ωm. The thermal conductivity of graphite is higher than that of carbon. Carbons may be impregnated with resin to improve strength and soundnes or with metals.

Carbons and graphites manufactured by polymer carbonization are assuming great technical importance. They include:

1. Carbon fibres, made by carbonizing polymer fibres which are subjected to tension during carbonization and graphitization. They are highly oriented graphite fibres with high specific tensile strengths and specific moduli exceeded only by boron fibre. They are used for manufacture of composites described in Section 7.6. Carbon/carbon fibre combines the refractory and electrical conducting properties of graphite with high specific strength.

2. Vitreous carbon, produced by controlled carbonization of a cross-linked polymer, is glassy in appearance, and has no open porosity. It is available in plate, dish and crucible form and, according to grade, will withstand temperatures of 1000–2500°C in inert atmospheres.

3. Graphite foam and graphite or carbon felts are produced by carbonization or graphitization of the appropriate polymer foam or mat.

The refractoriness, electrical conductivity, lubricity and (in the form of fibre) high mechanical properties of carbon and graphites together with their reasonable cost render them suitable for many and varied engineering and consumer applications.

7.8.7 Miscellaneous ceramics

A number of other ceramics have specific applications. They include refractory metal, usually tungsten or titanium, carbides which are used extensively for cutting or forming tools. These are usually bonded with cobalt, nickel or nickel molybdenum and are described in Section 7.6.

7.8.8 Glasses

All commercial glasses are based on silica (SiO_2) to which may be added other oxides which progressively reduce softening point, increase thermal expansion and impact other character-istics. 99.8% silica glasses include:

- Fused silica, which contains small bubbles, and is used for applications that require good thermal shock and corrosion resistance;
- Fused quartz which is transparent and used for high-quality tubing; and
- Vitreous silica which is used for high-equality optical com-ponents.

Aluminosilicate glasses have service temperatures next to silica glasses and are resistant to alkalis. They are used for high-performance electronic applications and other industrial applications. Borosilicate glasses have lower softening tempe-ratures than aluminosilicate glasses and, because of the lower cost of fabrication, have more widespread applications. Soda lime silicate glasses have still lower softening temperatures and higher coefficients of expansion. They are widely used for windows and other domestic applications because of their relatively low cost. Alkali-lead silicate glasses have, besides the lowest softening points commonly available, refractive indices and dispersive power which makes them useful for optical and similar applications. Other glasses include pho-tochromic glasses, semiconducting glasses, solder glasses for glass–metal seals and ophthalmic glasses.

7.8.9 Glass ceramics

The properties of glass ceramics can be tailored to have specific properties to fit the desired application. These in-clude:

- Thermal expansion, which can be made to match that of the material to which the glass ceramic is to be fused, or possibly to approach zero;
- Refractoriness;
- Transmission of light;
- Colour, or
- Machinability.

Where an application is foreseen, manufacturers or specialists should be consulted for advice on manufacturing procedure and practicability.

7.8.10 Mechanical properties

The applications of ceramics are strongly influenced by diff-erences in mechanical behaviour compared with metals.

7.8.10.1 Elastic properties

The elastic properties of ceramics are influenced by the level of porosity and also by the proportion of different phases present. The effect of small, closed, randomly distributed spherical pores on the modulus of a material of Poisson's ratio $v = 0.3$ has been shown to be approximately:

$$E = E_0(1 - 1.9v + 0.9v^2)$$

where E_0 is the modulus of the non-porous material.

Each phase in a multiphase material contributes, to a first approximation, according to its volume fraction. Where, as is common in engineering ceramics, a high-modulus major phase is bonded by a low-modulus minor phase, the modulus reduces rapidly with increase in the content of the minor phase. Usually modulus decreases with increasing temperature but glasses and complex ceramics are available with moduli which increase with increasing temperature (over a limited range). The range of moduli for ceramics is indicated in Table 7.71.

Table 7.71 Typical room temperature elastic moduli of glasses and ceramics

Material	Young's modulus (GNm^{-2})	Note
Glass	63–73	
Glass ceramics	78–120	
Carbons and graphites	9–13	
Porous ceramics	54	
Porcelains	69–138	
Oxides	132–155	
Zirconias	230–580	
Aluminas ⪦ 99%	280–350	
Silicon carbide	385–470	
Boron nitride	44–104	Anisotropic
Silicon nitride	160–310	
Sialon	280	
Boron carbide	440	

7.8.10.2 Strength and ductility

The tensile strength values for ceramics quoted in this section and elsewhere are not comparable with those quoted for metals. In a metal the strength of the material obtained from a simple mechanical test adjusted by a safety factor chosen on the basis of previous experience may be used to determine a safe loading stress. In ceramics and other brittle materials measurements on nominally identical materials will show scatter in strength results with standard deviations around 20% of the mean strengths. Catastrophic failure occurs by rapid propagation of a crack as soon as the stress at the most severe flaw reaches a critical value. (For more detailed discussion of this see references 90 and 91.) The consequences of this are:

1. Since specimens contain critical flaws of varying severity nominally identical specimens exhibit scatter in strength.
2. Large components and components with larger surface areas are, on average, weaker than smaller ones because there is a higher chance of finding a flaw of given severity. (It is also likely that a manufacturing method will produce a larger flaw in a larger component, but this is secondary to the argument.)
3. Because a severe flaw at a stress less than the maximum stress in a component may reach a critical value for fracture before a less severe flaw subjected to the maxi-mum stress, a component will not necessarily fail at the point of maximum stress. A thorough stress analysis of a component must therefore be carried out.

There is therefore no single measure of strength applicable both to a test specimen and a component, and no absolute guarantee can be given that a component will not fail under its design load. Design must be based on a probabilistic approach which will keep the frequency of failure below an acceptable figure, depending on the seriousness of the consequences.

7.8.10.3 Methods of testing

The standard uniaxial tensile and compression tests used for metals are not applicable to ceramics because of the difficul-ties of gripping the specimen and of aligning it axially. The tests used apply specific parts of the specimen to tensile stress by loading other parts in compression. A number of tests have been devised[92] and the most straightforward method is the

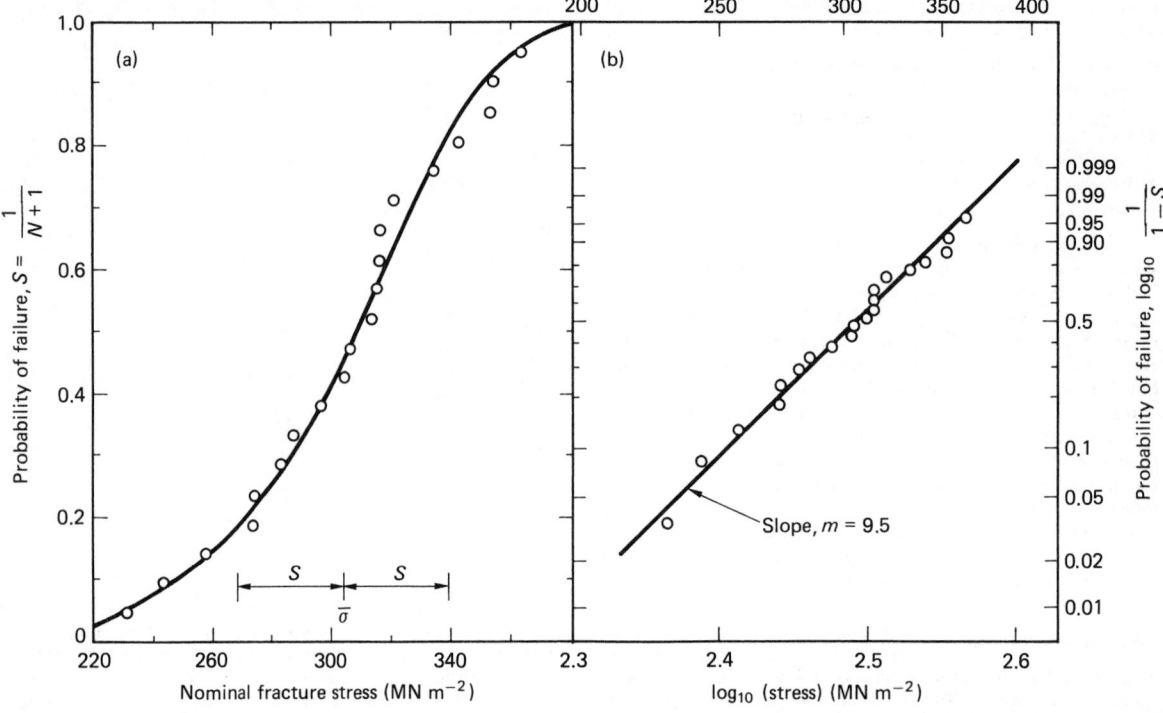

Figure 7.123 The fracture strengths of a batch of 95% alumina specimens tested in four-point bending at room temperature with a 320 grit ground surface finish are plotted (a) as a cumulative distribution and (b) as a Weibull plot to get a straight-line fit to the points. Using the Weibull parameters calculated the solid line in (a) represents the Weibull distribution calculated. The slope m of the straight line is the Weibull modulus. (Reproduced by courtesy of Dr R. Morrel[90])

three-point bend test, in which the tensile stress rises to a maximum on a line on one surface of the specimen.

There is no recognized international standard for the testing technique, specimen size, surface finish and shaping procedure, therefore small differences in strength quoted by a number of manufacturers should not be regarded as sigificant where it is desired to select the strongest. Procedures have been developed for determining the load which may be applied to a component to give a specific probability of failure in the short term.

It is necessary to undertake a programme of tests on at least 20 specimens from which a statistical distribution known as the 'Weibull modulus', m, which gives an estimate of the degree of scatter of the data (see Figure 7.123) may be determined. From the value of m and the effective volume or surface of the component the probability of failure at a given load may be calculated. Working stresses for a 10^{-4} failure probability may be of the order of 25–50% of the mean bend stress. The procedures are described in detail in references 90 and 93.

7.8.10.4 Long-term strength

The long-term strengths of many ceramics, particularly silicate-bonded materials, are lower than the short-term strength. This phenomenon has been described (usually referring to the behaviour of glasses) as 'static fatigue' but is more correctly termed 'slow crack growth'. The crack-like defects present in all ceramics (except whiskers) extend by a thermally activated atomic process, driven by the applied stress, influenced by the environment at the crack tip and

strongly dependent on temperature: In oxide ceramics OH ions can break metal-oxide bonds, therefore clay bonded materials may have lower long-term strengths in high-humidity atmospheres and particularly in alkaline environments. With most non-oxide ceramics, water vapour has little influence, so that little effect may be observed at stresses well below the fracture stress.

Most ceramic components are subjected to low values of tensile stress in service but where a guarantee of survival for a specific period at stresses in excess of 10% of the nominal breaking stress is required the only solution is a short-term overload proof stress which rejects components in the low-strength tail of the distribution.[92]

7.8.10.5 Fatigue under cyclic loading

Little systematic work has been carried out on the effect of cyclic loading on ceramics. If cyclic loading cannot be avoided the effect may best be estimated by integrating the varying stress intensity factor raised to a power n (obtained from the reference to the previous paragraph) with time.

7.8.10.6 Strength at elevated temperatures

As described in Section 7.8.10.4, slow crack growth increases with increasing temperature, consequently increasing temperature may initially reduce long-term strengths. However, further increase in temperatures may result in flaw rounding, which increases strength to a maximum. Strength is reduced by creep and creep rupture at still higher temperatures. Few

Table 7.72 Approximate temperatures at which ceramics and glasses are suitable for long-term operation under load

Type of ceramic	Temperature (°C)	Notes
Soda-lime glass	300	
Silicon glass	800	
High-purity alumina	1400	
Silicon nitride	1600	Oxidizes in air above 900°C
Graphite	>1500	Oxidizes in air above 400°C

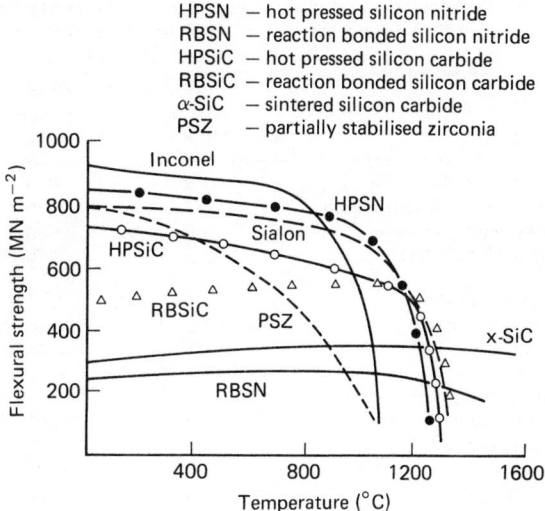

HPSN — hot pressed silicon nitride
RBSN — reaction bonded silicon nitride
HPSiC — hot pressed silicon carbide
RBSiC — reaction bonded silicon carbide
α-SiC — sintered silicon carbide
PSZ — partially stabilised zirconia

Figure 7.124 Flexural strength variations with temperature of several materials. (Reproduced by permission of *Metals and Materials*)

data are available beyond statements on 'upper use temperature' usually under no-load conditions, which implies suitability for short-term use without disastrous changes in microstructure. Temperatures at which ceramics are suitable for long-term use under load are indicated in Table 7.72.

7.8.11 Manufacturing procedures

The manufacture of ceramics includes shaping with or without the application of pressure and consolidation by firing. The characteristics of the ceramic and its cost depend on the process which must be used. The simplest (and cheapest) is compaction and sintering.

Compaction may be carried out by a variety of processes which may involve the addition of a binder or lubricant or may include the application of pressure. 'Sintering', in its pure sense, implies the densification of an assembly of particles at high temperatures by solid-state diffusion processes. This process is relatively slow, needs very high temperatures and a dense impervious body is achieved only with difficulty. Ceramics other than oxides may require a protective atmosphere.

Vitrification (or liquid-phase sintering) is achieved by incorporating a component which melts to a viscous liquid during sintering and solidifies (in the case of a clay-based ceramic) to a glassy phase. The term 'well vitrified' indicates that a dense impervious body has been achieved.

'Reaction bonding' is achieved by arranging for two or more components of a compacted body to react to form both a desired phase and a bond between particles. The reacting components may all be solid or one of them may be gaseous or liquid and introduced during the sintering process. 'Self-bonding' is bonding the major constituent of a ceramic with its constituents formed by reaction during sintering.

'Hot pressing' is achieved by the application of pressure during sintering. There are two variants. Hot pressing as commonly understood consists in applying pressure unidirectionally by a plunger acting on material in a rigid container. Hot hydrostatic pressing consists in submitting the material contained in a membrane to hydrostatic pressure during heating. Both processes produce material of a higher density and superior quality more quickly than can be achieved by sintering but they are substantially more expensive and component shape is restricted.

'Pyrolysis' is the process whereby a ceramic is deposited from a gaseous environment onto a (usually) heated substrate. This process is usually employed to deposit coatings or produce thin-walled components of graphite or silicon carbide from carbon or silicon-containing gas.

7.8.12 The future prospects of engineering ceramics

Ceramics and glasses have a wide range of applications in manufacturing technology and even wider applications in domestic use. If they are to achieve the really widespread engineering application that their properties appear to merit they must find this in some field such as the automotive industry.

Some components, including oxygen sensors, catalyst supports, diesel glow plugs, rocker arm pads, pre-combustion chambers and turbocharger rotors have already been standardized, usually for high efficiency and military engines. Tables 7.73 and 7.74 and Figure 7.124 list the properties and applications of the ceramics at present available for the required breakthrough into the mass market.

The major problem is that, depending on fabrication procedure, the cost of a ceramic component is from five to fifteen times that of the presently used steel component so that a substantial improvement in performance or a marked reduction in cost is required to justify the use of the ceramic. The potential advantage of materials such as sialons, which combine excellent properties and a relatively inexpensive manufacturing route (sintering) is evident.

7.9 Corrosion

7.9.1 Introduction

Corrosion is the term used to describe any effect which results in the impairment of the service performance of a material component by the action of another entity, usually its environment. It is usually envisaged to be the reduction in size or weakening of a component by removal of material, but deterioration in performance can occur in a number of ways which include increase in volume of the component or contamination of contained material.

Corrosive effects can be classified according to the substrate (the material subjected to corrosion), the corrodant (the

Table 7.73 Properties of selected metals and technical ceramics (properties are at room temperature unless otherwise stated). The values should be used as a guide only, since various ceramic grades exist under one generic name

	Density $(10^3 kgm^{-3})$	Bend strength (MNm^{-2})	Young's modulus (GNm^{-2})	Fracture toughness $(MNm^{-3/2})$	Thermal[a] expansion $(10^{-6}K^{-1})$	Thermal[a] conductivity $(Wm^{-1}K^{-1})$
Alumina 99%	3.9	400	400	~3.0	9.0	24
Zirconia toughened alumina	4.1	450	340	~8	8.1	23
Aluminium titanate	3.0	~40	20	—	0	1.5
Cordierite (MAS)	2.5	120	110	~2.5	2.0	1.5
Silicon carbide						
Reaction bonded	3.1	~500	410	~4.5	3.8	100
Sintered	3.1	460	400	~4.5	4.0	90
Silicon nitride						
Hot pressed	3.2	800	310	6	3.2	~20
Reaction bonded	~2.5	200	170	~3	3.0	12
Sintered	3.2	~400–700	250	5	3.4	~16
Sialon	~3.2	~950	~290	~8	~3.1	~21
Zirconia						
Plasma sprayed	5.2	6–80	48	~2	8.0	1.0
Partially stabilized zirconia	5.6	500	205	8	9.5	1.7
TZP (Y)	6.05	1000	210	15	9.0	2.0
Fully stabilized	5.8	180–250	160	~4	10.0	2.0
Cast iron	7.2	—	117	—	12.0	54
Nimonic[b] (80A)	8.2	—	200	—	13.0	12
Aluminium alloy IM27	2.75	—	71	—	21.0	155

[a] Values for temperature range 300–600 K.
[b] Inconel 751 has a $K_{1c} = 80$–100 $MNm^{-3/2}$.
Reproduced by permission of *Metals and Materials*.

Table 7.74 Automotive ceramic components

Component	Material	Car maker	Ceramic manufacturer	Year of introduction
Engine				
Diesel glow plug	Silicon nitride	Isuzu	Kyocera	1981
		Mitsubishi	Kyocera	1983
		Toyota	Toyota	1984
		Caterpillar		1986
			Kyocera	
IDI precombustion chamber	Silicon nitride	Isuzu	Kyocera	1981
		Mazda		1986
		Toyota		1986
Exhaust port liner	Aluminium titanate	Porsche	Hoechst	1986
Rocker arm pad	Silicon nitride	Mitsubishi		1984
Turbocharger rotor	Silicon nitride	Nissan	NTK	1985
		Buick		1987
Knock sensor	PZT Pb(ZrTi)O₃	Toyota		1980
Other				
Oxygen sensor	Zirconia/titania	Vehicles with 3-way catalytic converters		
Catalyst support	Cordierite	Vehicles with 3-way catalytic converters		
Warning alarm	PZT	Toyota plus other		
Sensors	NTC/PTC[a] thermistors			

[a] Negative/positive temperature coefficient ceramics (Al_2O_3–Cr_2O_3/$BaTiO_3$).
Reproduced by permission of *Metals and Materials*.

entity, usually a material the presence of which causes the attack), the mechanism of the attacking process or the resultant effects. An understanding of corrosion mechanisms enables the correct choice of material and conditions of operation to be made and therefore minimizes cost and enhances safety.

Cost includes not only the manufactured cost of a material component but also the cost of inspecting, replacing and servicing and the cost of outages arising from failure. A component must not fail in a manner which generates a hazard to operating personnel or to the public. Corrosion science and technology involves not only the correct choice of a material, its working environment and any protective techniques adopted to eliminate or postpone failure, but also the elucidation of the cause of failure.

7.9.2 Corrosion of metals in air or water

7.9.2.1 Oxidation

Thermodynamics The corrosion process is usually one of several reactions often forming a metal oxide (or hydroxide).[129] This reaction involves the oxidation (loss of electrons) of the metal to form an ion:

$$M - ne^- = M^{n+}$$

and the reduction of the oxygen to form an oxygen ion:

$$O_2 + 4e^- = 2O^{--}$$

The driving force for this reaction to occur is a reduction in free energy, G. For all metals, with the exception of gold, the change in free energy of the reaction

$$M + O_n = = MO_n - n\Delta G$$

is negative. The greater the decrease in free energy, the stronger the tendency to react.

Kinetics Reactions are normally activation energy controlled, and therefore, provided that the change of temperature does not change, the mechanism

$$d \log K/dt = -E/RT^2$$

where K is the rate of corrosion, E the heat of activation and T the absolute temperature. With the values of E which occur in most aqueous and oxidation types of corrosion, it can usually be taken with sufficient accuracy for most practical purposes that an increase in temperature of 10°C will double the rate of reaction.

Effect of film formation A metal surface reacts with oxygen extremely rapidly but, if an adherent film of oxide forms, subsequent oxidation depends upon oxygen or metal ions diffusing through the film. While the process of diffusion proceeds at an equilibrium rate the rate of increase of thickness is governed by the equation

$$y^2 = K_0 \exp(-Q/RT)t$$

where y is the thickness of the oxide film,
  K_0 is a constant,
  Q is the activation energy of diffusion through the film,
  R the gas constant,
  T is the temperature (K), and
  t is the time.

With metals forming strong stable films the parabolic rate is never approached and the rate decreases asymptotically. With metals where diffusion through the film is more rapid the parabolic rate may, after a period, give place to a quasi-

straight-line relationship. Corrosion after this event is known as 'catastrophic' or 'breakaway' corrosion and usually results in failure of the component.

Electrochemical corrosion Metals can be oxidized by aqueous environments but the dominant mechanism in conducting electrolytes is electrochemical. This requires both anodic and cathodic sites to exist on the surface of the metal exposed to the electrolyte. The metal is oxidized at an anodic site forming an ion in solution:

$$M - e = M^+$$

and liberating hydrogen gas or reducing dissolved oxygen

$$2H^+ + 2e^- = H_2$$

$$O_2 + 4e^- + 2H_2O = 4OH^-$$

at the cathodes. There must clearly be a complete electrical circuit between the anode(s) and the cathode(s) to permit electrons to flow through the electrolyte and the metal.

If the reaction products are soluble, the metal dissolves progressively and this process is termed 'general corrosion'. The reaction may be slowed or terminated by an insoluble layer. When this occurs, circumstances may arise in which the protective layer fails at points or lines so that the component is weakened or fractured, although its overall dimensions remain substantially unchanged. This form of corrosion is termed 'localized' corrosion.

The potential differences between two areas of substrate which lead to electrochemical corrosion may arise in a number of ways. These include externally applied potential (sometimes arising from stray currents), differences in temperature, local differences in concentration of the electrolyte (such as differences in oxygen concentration) and, most important, differences in composition of the substrate. The last may arise from the structure of the metal or from the presence of two different metals in electrical contact and exposed to the corroding solution. In this latter case the anodic or least-noble metal is corroded and the cathodic or most-noble metal protected. The effect is enhanced as difference in reactivity of the metals increases and depends on the relative area of the two metals. A smaller area of one metal compared with a larger area of the other suffers more intense corrosion or receives enhanced protection.

The order of reactivity in which the metals can be placed differs from one corroding solution to another. A typical order is shown in Table 7.75.

As a general rule, resistance to corrosion may be obtained by the choice of a metal or alloy which is high in the galvanic series or which forms a protective coating under the expected corroding conditions, or in which these two characteristics are combined. Resistance to the various types of corrosion classified according to specific metals and alloys is described in other sections. The following sub-section describes the specific types of corrosion in air and aqueous electrolytes and indicates how metals react to them.

7.9.3 Resistance of metals to specific corrosion mechanisms

7.9.3.1 General corrosion in air

All metals from zinc upwards in the galvanic series in Table 7.75 can be used unprotected at low temperatures. Care must be taken in marine and industrial atmospheres. At high temperatures stainless steel and nickel alloys in which diffusion through the oxide film is reduced (or eliminated) must be used.

Table 7.75 A typical galvanic series for some commonly used metals and alloys

Protected (cathodic or most-used noble metal)

Order of increasing reactivity ↓

Platinum
Gold
Graphite
Silver
Stainless steel (high molybdenum)
Stainless steel 304[a]
Stainless steel (ferritic)
Inconel
Nickel
Silver solder
Monel
Aluminium bronze
Tin bronze
Copper
Brass
Tin lead
Lead tin solder
Carbon steel
Pure aluminium
Cadmium
Aluminium alloys
Zinc
Magnesium alloys
Magnesium

Corroded (anodic or least-noble metal)

[a] Where it is feared that there is a danger of localized pitting or stress corrosion the common austenitic stainless steels should be placed much lower in the table, perhaps adjacent to brass.

7.9.3.2 General corrosion in aqueous solutions

All metals from aluminium upwards in the galvanic series can be used in pure water. More aggressive electrolytes in chemical process plant (and pure chemicals that must not be contaminated) require progressively more resistant stainless steels, nickel alloys, hastelloys and titanium.

Both water (including steam) and air oxidize steel but provided that the section is adequate and moderate increase in dimensions can be tolerated (see below) carbon steels give satisfactory service up to 350°C. Above this temperature progressively increasing proportions of chromium must be added.

Corrosion jacking　This is a consequence of general corrosion in which a restraining or enveloping component is distorted or ruptured by an increase in volume of a metal, usually iron or steel. The phenomenon received its greatest publicity when bolts in the Magnox reactors were stretched and broken by oxide growing on the faces of washers (see Figure 7.125(a)) which had been made from low-silicon rimming steel. The problem was cured by reducing the operating temperatures of existing reactors and by using a steel of higher resistance to oxidation in the later ones. It has been a very serious problem in reinforced concrete road bridges where sometimes brackish water permeated through the concrete (Figure 7.125(b)). The cure is to maintain an adequately thick covering (*ca* 1 inch) of impervious concrete.

The problem was understood by the original builders of the Parthenon in Athens, who cast lead round their wrought-iron reinforcing bars. Later repairers (e.g. Belano) sometimes ignored the problem, usually with disastrous results (Figure

(a)

(b)

(c)

Figure 7.125 Typical examples of corrosion jacking. (a) Bolts, typical of those in magnet reactors before and after stretching because of washers corroding; (b) spalling of reinforced concrete; (c) damage caused by Belano's repairs to Propylaea using mild steel supports.

7.125(c)). Belano's steel is now being removed and titanium substituted.

7.9.3.3 Galvanic corrosion

'Galvanic corrosion' is homogeneous corrosion accelerated locally by an electric potential difference. It usually occurs when two metals fairly far apart in the galvanic series are in electrical contact and exposed to the same corroding solution. Figure 7.126 shows typical instances of galvanic corrosion. Other examples have occurred when copper and iron water-pipe fittings are used, when aluminium and stainless steel were employed together in an aircraft fuselage and when uninsulated steel bolts were used to assemble magnesium alloy castings. One way of producing galvanic corrosion (and, incidentally, protection) is to connect two pieces of identical metal in the same solution in series with a source of potential (this may be caused accidentally by a stray current). The effect is exaggerated when the areas of the two metals differ. The corrosive (or protective) action is enhanced at the surface which has the smaller area.

The influence of the galvanic effect is not uniformly disadvantageous. It may be used to protect a component by connecting to it one or more electrodes (or a coating) of a more electronegative metal (for example zinc) immersed in the corroding solution. Alternatively, the same 'cathodic protection' may be afforded by means of a resistant (for example, platinized titanium) electrode which is maintained at an appropriate negative potential by an EMF connected in series.

7.9.3.4 Galvanic corrosion: selective leaching

'Selective leaching' is a form of galvanic corrosion in which the two metals form part of the structure of the alloy corroded.

Two examples are typical. A cast iron sea-water condenser has been known to have all the iron leached away in certain parts, leaving a friable graphite skeleton. Brass exposed to industrial atmospheres has the whole of the zinc leached away, leaving a sponge of copper having the original dimensions of the component but negligible strength.

The most common examples of selective leaching of zinc arise from the use of brass where bronze has (or should have) been specified. The result can be disastrous when (as has occurred) the component is a steeplejack's attachment ferrule. The situation is complicated by the fact that some materials with the trade name 'bronze' are in fact brasses prone to selective leaching. Brasses which are claimed to resist selective leaching have, however, been developed recently. Where there is any possible doubt a qualified corrosion expert must be consulted.

7.9.3.5 Homogeneous corrosion erosion

When corrosion and erosion occur simultaneously their effects are mutually reinforced because erosion removes any protective film that might otherwise impede corrosion. Steam/water mixtures may corrode/erode plant, as is shown in Figure 7.127. The cure here is to eliminate crevices through which the expanding steam/water mixture can flow. Cavitation (an example of which is shown in Figure 7.128) can also initiate corrosion. In cases where erosion and pitting both occur pitting-resistant stainless steels are replacing titanium for condenser tubing.

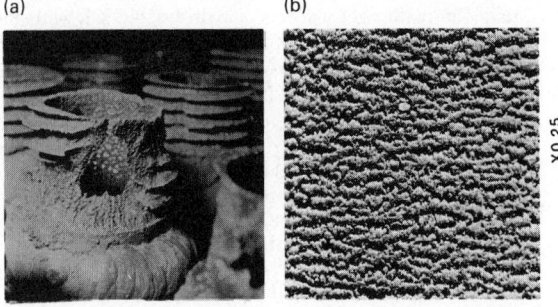

Figure 7.127 Typical instances of homogeneous corrosion/erosion, (a) Hot water seeping along threads and evaporating erodes/corrodes carbon steel bolts and sockets; (b) very high velocity steam at evaporation zone of once through boiler causes rippled magnetite which impedes flow and therefore reduces power

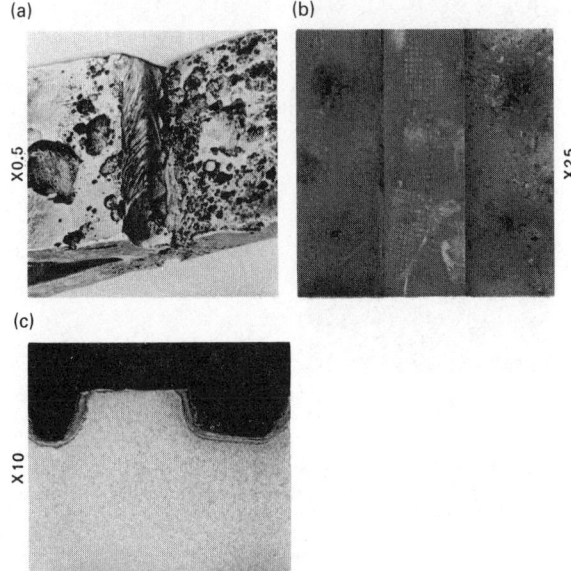

Figure 7.126 Typical instances of galvanic corrosion. (a) Characteristic appearance of galvanic corrosion caused by welding an austenitic steel gauze onto a 403 steel plate; (b) local pitting caused by deposit of tin on steel surface; (c) micro section through pits in (b)

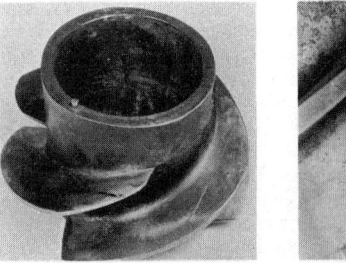

Figure 7.128 Cavitation damage on a disposable stainless steel feed pump inducer. (a) Inducer; (b) cavitation damage

(a)   (b)

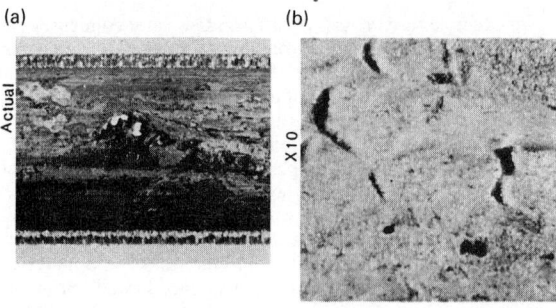

Figure 7.129 Heterogeneous erosion/corrosion of a condenser tube. (a) Actual size; (b) magnified ten times

7.9.3.6 Heterogeneous erosion corrosion

Where solid materials are transported by a corrosive solution or gas or where materials handling is carried out in the presence of corrosive liquids the impingement of solid particles causes erosion which mutually reinforces corrosion. Typical examples are shown in Figure 7.129. The best way of reducing this effect is to design to minimize impingement. If this cannot be achieved the material at the points of maximum impingement should be coated or replaced by a hard, corrosion-resistant material such as stellite (see Section 7.1).

7.9.3.7 Localized corrosion

Ionic concentration This occurs where a corrosive solution contains an anion which forms a hydrolysable salt with the substrate or another cation. It can lead to the local concentration of quite strong acids, which attack what are normally protective films. The typical processes of attack are:

1. *Crevice corrosion*: This type of corrosion frequently occurs with stainless steel. Strong acids form in crevices particularly where oxygen is reduced compared with the main body of the solution. The cure is to eliminate crevices by design. If this is not possible and the corroded material is a stainless steel a more resistant stainless steel (see Section 7.1) may be tried. Figure 7.130 shows the effects of crevice corrosion on a titanium tube plate.

2. *Pitting corrosion*: This is crevice corrosion in which the corrosion mechanism has formed its own crevice. Figure 7.131 illustrates the way development of a pit can lead to failure by perforation. Any material which depends on an oxide film for resistance to corrosion can suffer from pitting corrosion but it occurs mainly in austenitic steels in aggressive aqueous environments, including stagnant sea water and marine condenser tubes. It is easier to cure than is crevice corrosion. Filtration to remove particulate matter and avoidance of organic growths is sometimes successful. Austenitic stainless steels have been developed which are claimed not to pit in stagnant sea water.

3. *'On-load' pitting corrosion*: This is a form of corrosion limited to steam generator tubes. Pressure, temperature, and heat rating of steam plant advanced more quickly than technological capability. As a result, pits of the type illustrated in Figure 7.132 developed in boiler tubes in ships and power stations. The pits occurred where debris collected in the tubes at positions of high heat flow or where the flow of water was reduced by faults in design or manufacture. Water containing chlorides and oxygen

*For the convenience of the reader some of the figures in sections 7.3.13 and 7.3.17 are repeated here.

(a)   (b)

(c)

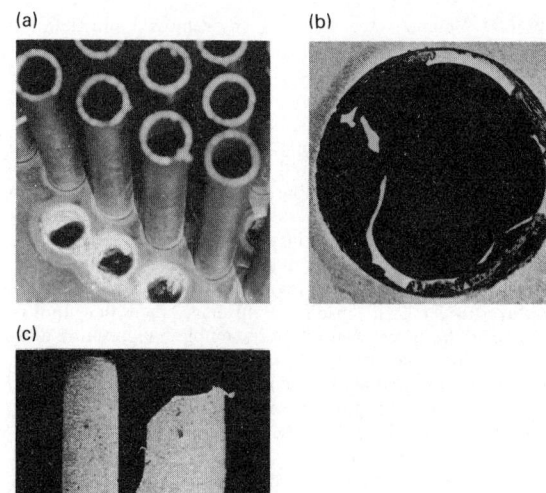

Figure 7.130 Three views of crevice corrosion of a titanium tube plate. (a) Heavy corrosion confined to crevices; (b) end view of corroded weld; (c) side view of corroded weld

(a)   (b)

(c)

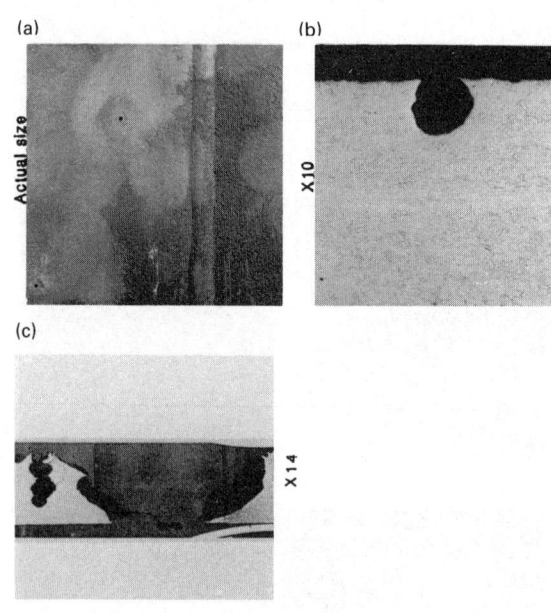

Figure 7.131 Failure develops from pitting corrosion. (a) Appearance of pit on surface; (b) section through pit; (c) section through pit which has penetrated wall

flowed towards the tube wall and steam flowed away from the wall, leaving behind impurities, and strong acids developed. 'On-load' pitting corrosion has been almost completely eliminated by improving:

● Overall boiler design to eliminate hot spots and points of reduced flow;

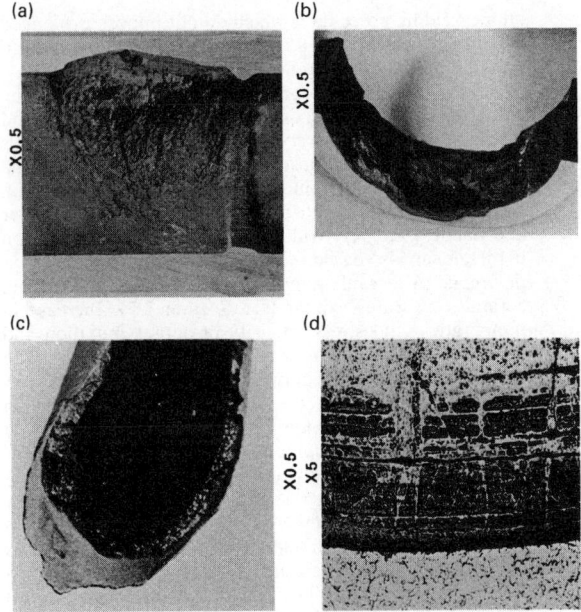

Figure 7.132 Characteristic appearance of 'on load' pitting corrosion. (a) Early stage of pit; (b) pit about to penetrate wall; (c) pit has caused hydrogen degenerated metal behind it to fracture; (d) structure of layered magnetite in pit and hydrogen degeneration in metal

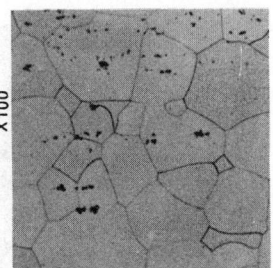

Figure 7.133 Structure of sensitized stainless steel showing chromium carbide precipitation

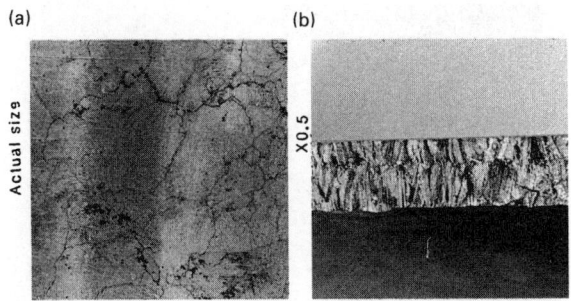

Figure 7.134 Characteristics of penetration cracking of sensitized austenitic steel. (a) Multiple crack system oriented at random (i.e. not aligned transverse to tensile stress); (b) appearance of fracture

- Construction to eliminate irregularities (e.g. weld protrusions) which caused local stagnant areas;
- Feedwater purity to eliminate inpurities such as oxygen and chlorides.

Penetration cracking corrosion processes There are a number of corrosion mechanisms which leave a substantial part of the substrate material intact but destroy its integrity by penetrating specific localities either as trenches or cracks. Some mechanisms depend for propagation on a combination of applied tensile stress and an aggressive environment, but in other cases the environment attacks specific structures in the absence of stress.

1. *Intergranular penetration.* This is corrosive attack by aqueous solutions, usually chlorides, along the grain boundaries of austenitic steels containing above 0.02% carbon which have been 'sensitized' by heating within the range 425–500°C. This precipitates chromium carbide at the grain boundaries (see Figure 7.133) and locally reduces the concentration of chromium and therefore the corrosion resistance. Intergranular penetration may be recognized by the presence of a multiple system of cracks which are orientated at random (see Figure 7.134). It can be prevented by:

   - Using a steel with a carbon content below 0.02%;
   - Incorporating into the steel stabilizing elements Nb, Ti, Mo, which are strong carbide formers;
   - Subjecting the steel to a desensitizing heat treatment above 870°C followed by rapid cooling.

2. *Intergranular corrosion in a gaseous environment.* Gaseous intergranular corrosion can arise in a number of ways. Typical is attack on nickel alloys by flue gas containing

reducing sulphur which forms a nickel sulphide eutectic at the grain boundaries. The only cure is to avoid using this combination of materials in environments where this type of attack is known to occur.

3. *Weld decay.* This is a form of intergranular penetration of austenitic steel specific to the heat-affected zones of welds. An example of this (which can proceed until the weld drops out of the component) is shown in Figure 7.136. The cure is identical to that for intergranular penetration.

4. *Knife-line attack.* If the heating during welding is sufficiently prolonged and intense, stabilization of austentic stainless steels by additions of Nb and Ti will not prevent attack. Under these conditions corrosive attack is concentrated at the edges of welds (Figure 7.137). The cure is to

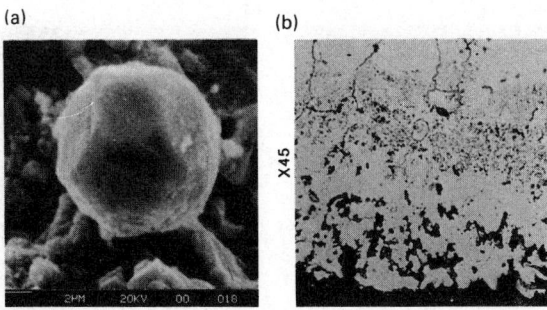

Figure 7.135 Intergranular corrosion of nickel alloy by furnace gas containing sulphur. (a) Nickel sulphide in scale; (b) sub-surface penetration by nickel sulphide

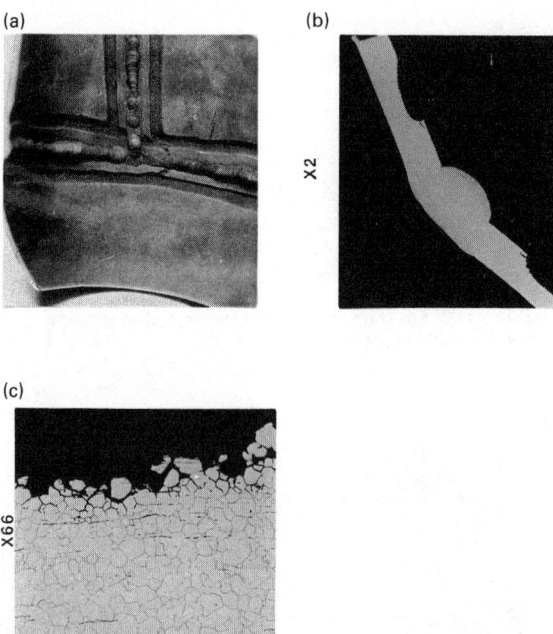

Figure 7.136 Characteristic appearance of weld decay. (a) Surface appearance showing trenches at side of welds; (b) section showing form of trenches; (c) intergranular attack at sides and base of trenches

adopt welding procedures which do not impose prolonged and intense heating. If this cannot be avoided the weld must be post-solution treated.

5. *Stress corrosion.* Stress corrosion cracking depends on the co-existence at a surface of an aggressive environment and a tensile stress. It has very great importance in materials technology because almost every known material is prone to stress corrosion by some environment and the two main constituents which lead to stress corrosion in metals (water and halides) are very widely distributed. Stress corrosion in metals can also be caused by liquid metals (see below) in polymers by organic solvents (see Section 7.5) and in ceramics by water vapour (see Section 7.7). Increase in temperature and stress intensity promotes initiation and accelerates propagation of stress corrosive cracks. The numerical relationships between these parameters are now being evaluated (see Section 7.1) but they are statistical in nature and one specimen may survive long periods in apparently arduous conditions while an apparently similar specimen may fail rapidly in relatively easy conditions. A component may last for a long period apparently undamaged then fail completely in a short time. Stress corrosion cracks in metals are seldom discrete but usually form a network of branching cracks which propagate in a direction normal to stress. The crack starts at a small defect which generates, with the stress, a stress intensity above a threshold value.

Stress corrosion of austenitic stainless steels by chloride is often *transgranular* and the cracks are transverse to the tensile stress but change direction where they intersect grain boundaries (see Figure 7.138(a)). The appearance of the fracture as studied by SEM (Figure 7.138(b) and (c)) and the microstructure (Figure 7.138(d)) are very charac-

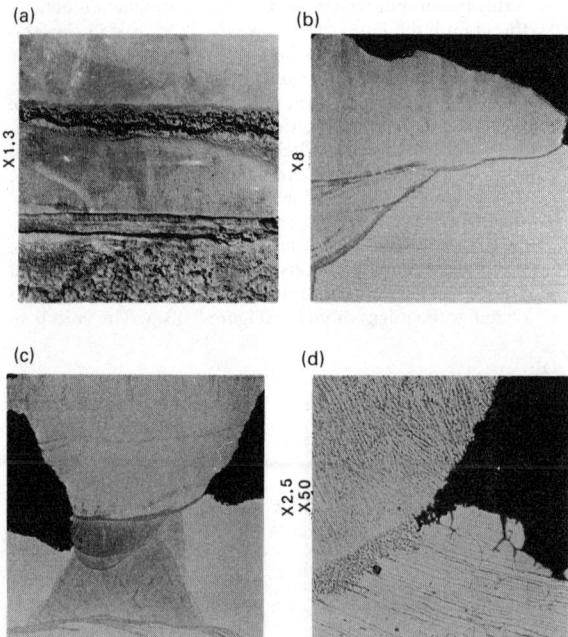

Figure 7.137 Characteristic appearance of knife-line attack. (a) Appearance of surface; (b) section at early stage; (c) section at late stage; (d) intergranular nature of attack

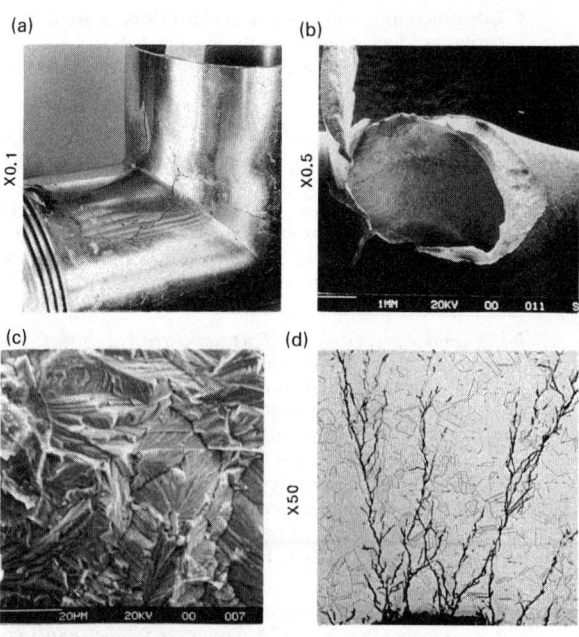

Figure 7.138 Characteristic appearance of transgranular stress corrosion. (a) Surface at low magnification. Multiple crack system roughly transverse to tensile stress but changing direction where they intersect grains. No deformation of metal. (b) Fracture at low magnification. Stress corrosion crack gives way to ductile tear. (c) Fracture at high magnification. Very characteristic appearance. (d) Micro section. Multiple branching thin transgranular cracks may be filled with corrosion products

teristic. However, fracture has been known to change from trans- to intergranular and back in a short distance along the crack.

Most forms of stress corrosion cracking are *intergranular* and, in the case of metals, require the presence of liquid water before they will propagate (chloride stress corrosion also requires the presence of oxygen). Stress corrosion can therefore be prevented by elimination of water and aggressive agents such as hydroxides, chlorides and thyonic acids. Component finish should be improved, stress reduced and (for austenitic steels) sensitization avoided. The appearance of intergranular stress corrosion can be seen in Figure 7.139. Data on stress corrosion of stainless steels are provided in Section 7.3.

6. *Hydrogen cracking (usually of steel).* Hydrogen can be introduced into metals in a variety of ways, including the

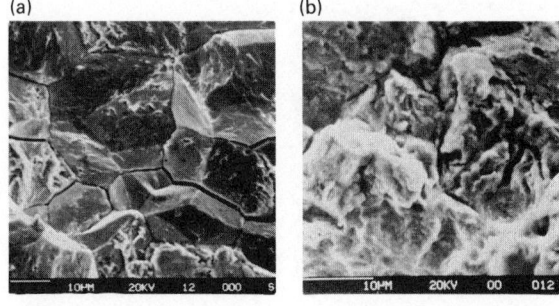

Figure 7.140 Two manifestations of hydrogen cracking. (a) Intergranular. Resembles SCC but hydrogen crack will not contain corrosion product. (b) Quasi-cleavage. Can only be hydrogen cracking

Figure 7.139 Characteristic appearance of intergranular stress corrosion. (a) Surface at low magnification. Multiple crack system roughly transverse to tensile stress but changing direction at corners of grains. No transverse strain. (b) Fracture at low magnification. SCC crack darker and covered with corrosion product gives way to intergranular brittle fracture. (c) Higher magnification at interface with brittle fracture shows continuous wall. (d) Network of cracks at grain boundaries may be filled with corrosion product

extraction process, pickling, plating, welding, exposure to an environment such as H_2S or holding in a hydrogen environment and will promote cracking where, for example, a steel of UTS greater than $930\ MNm^{-1}$ is stressed in tension. Hydrogen cracking may be intergranular (see Figure 7.140(a)) when it resembles and may indeed co-exist with stress corrosion. It can be distinguished from stress corrosion by a characteristic quasi-cleavage appearance (Figure 7.140(b)) and, unlike stress corrosion, it does not necessarily start from a surface. Hydrogen cracking may be avoided by preventing ingress of hydrogen into the metal at all stages of manufacture, processing and operation.

7. *Hydrogen degradation.* When the quantity of hydrogen absorbed by a metal such as steel, zirconium, titanium or aluminium greatly exceeds the amount associated with hydrogen cracking it may lead to 'hydrogen degradation'. There are several manifestations. Hydrogen produced in steel making or by low-temperature aqueous corrosion migrates to inclusions or discontinuities, collects as molecular hydrogen or methane and, because neither can be re-absorbed, generates very high pressures which cause 'fish eyes' (see Figure 7.141) or blisters which disrupt the steel and may cause brittle fracture. Hydrogen absorbed at high temperature from 'on-load corrosion' or from hydrocarbon in a refinery first decarburizes the steel and then forms a grain boundary network (see Figure 7.142) which leads to crumbling of the steel. The remedies recommended for hydrogen penetration will also cure hydrogen degradation. In refinery service additions of chromium to the steel, increasing with increasing service temperature,

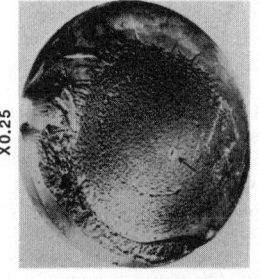

Figure 7.141 Fish-eye appearance typical of hydrogen degradation

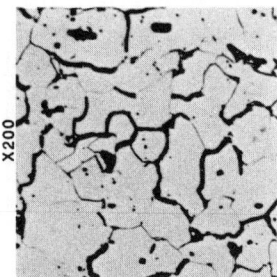

Figure 7.142 Microstructure of hydrogen-degraded steel showing microfissures where hydrogen (or methane) was trapped and decombines atom

(a)   (b)

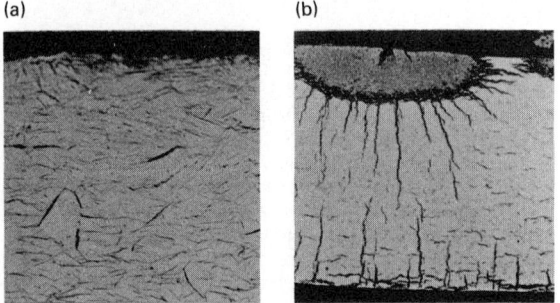

Figure 7.143 Hydride degradation in titanium and zirconium. (a) Hydride platelets in a titanium alloy; (b) 'sunburst' fracture aggravated by hydride in a leaking zirconium canned fuel element

(a)   (b)

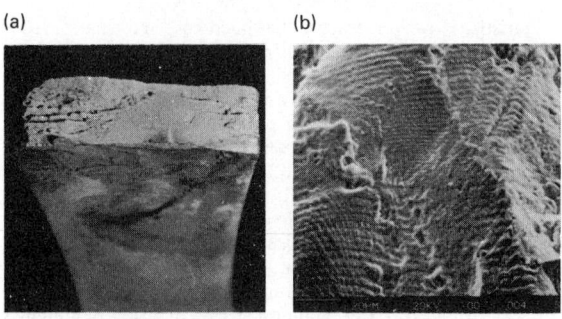

Figure 7.144 Characteristic features of corrosion-initiated fatigue. (a) In this instance fatigue initiated not at the point of highest loading, but at a point where crevice corrosion had occurred; (b) The crack showed striations typical of fatigue, with no blurring or corrosion product

(a)   (b)

(c)   (d)

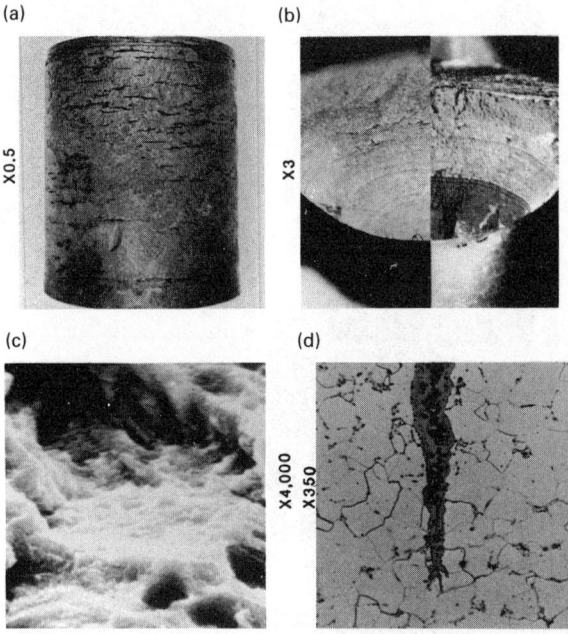

Figure 7.145 Characteristics of corrosion-accelerated fatigue. (a) Surface covered with corrosion product. Many parallel fatigue cracks; (b) fracture covered with corrosion product. Before and after cleaning; (c) striations blurred; (d) crack filled with corrosion product

prevent hydrogen degradation. Zirconium alloys in pressurized water reactors have suffered hydrogen degradation when the incorrect alloy has been used or when the alloy has been contaminated by nitrogen during fabrication. Hydride platelets are found in specific metallographic directions in both zirconium and titanium (see Figure 7.143(a)) and the failures may be spectacular (Figure 7.143(b)). Modern alloys and fabrication techniques have eliminated this problem.

8. *Corrosion fatigue*. Corrosion may initiate, accelerate, retard or prevent fatigue cracking. Corrosion *initiates* fatigue when a fatigue crack is initiated by a cyclic stress below the endurance limit of the material because the environment has attacked it at that point. Often, as in the case shown in Figure 7.144, the fatigue cracking starts away from the most highly stressed area of the specimen, and the main body of the fracture shows typical fatigue striations.

Corrosion may *accelerate* fatigue when the corrosive environment diffuses along the fatigue crack and increases the rate of propagation. The appearance of the fracture is similar to that of fatigue but the surfaces are corroded (see Figure 7.145(a) and (b)), the striations blurred (Figure 7.145(c)) and the crack may be filled with corrosion product (Figure 7.145(d)). Instances have occurred when an apparently aggressive environment such as oxygenated seawater has *prevented* the initiation or *retarded* or *stopped* the propagation of a fatigue crack at a stress at which fatigue failure would have occurred in air. Corrosion fatigue is avoided by removing the aggressive agent in the environment or affording protection from it and/or by reducing cyclic and tensile stress.

7.9.3.8 Corrodants of metals other than water or air

Metals may be corroded by many agents besides aqueous and atmospheric environments. These include solid and liquid metals, furnace slags, glasses and gases.

Corrosion by solid metals

- *Fretting corrosion*. The most important solid/solid corrosion of metals is 'fretting corrosion', which occurs when vibration causes low-displacement slippage between two surfaces pressed together. It is caused either by repeated welding and separating of high spots or by abrasion by oxide layers.

(a)   (b)

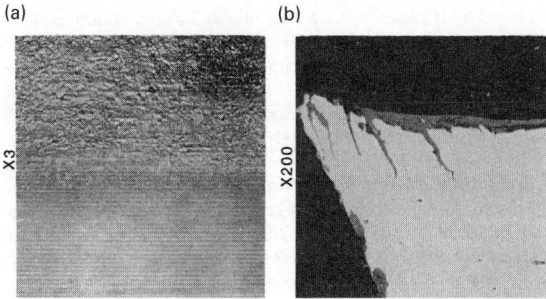

X3

X200

Figure 7.146 Characteristics of fretting corrosion. (a) Roughened pitted surface; (b) fatigue cracks developed from fretted surface

Most material combinations are prone to fretting corrosion but the presence of oxide films aggravates it, especially when oxygen is present. Failure occurs by accumulation of oxide debris, seizing, galling or loss of tolerance. Fatigue may initiate at fretted parts (see Figure 7.146) caused by small displacements but large displacements tend to 'rub out' the initiating fatigue cracks. Fretting corrosion is eliminated by stopping relative motion and alleviated by lubrication, using harder materials, fitting a gasket or by increasing relative motion.

● *Solid-state diffusion.* Almost any two metals that will form an alloy can, in theory, corrode each other. A typical example was the corrosion penetration of the prototype aluminium cans of the first uranium metal reactors. On heating at temperatures simulating reactor conditions the uranium alloyed locally with the aluminium can to form UAl_3 pyramids which penetrated the can, allowing ingress of oxygen. This destroyed the fuel element by oxide jacking. Inserting an uranium oxide/graphite layer between uranium and can prevented failure by this mechanism in the reactor.

7.9.3.9 Liquid metal corrosion

General liquid metal corrosion Any liquid metal that comes into contact with a solid metal with which it will form an alloy (for example, molten aluminium in a steel melting pot) is liable to corrode it away. The effect is enhanced if the liquid metal forms a circuit operating between a range of temperature. For example, molten bismuth corrodes steel slowly (if at all) but if it is used in a heat-transfer circuit the steel is dissolved at the higher temperature and deposited at the lower.

Liquid metal penetrations Far more potentially dangerous is liquid metal penetration, a form of stress corrosion in which the aggressive agent is a liquid metal which can penetrate rapidly along the grain boundaries of a metal stressed in tension and cause cracking. Many metals, including lead, bismuth, tin and zinc (cracking from which caused the Foxborough disaster), will cause penetration cracking in steel. One example was the failure of an overheated shaft (see Figure 7.147) caused by penetration of copper from brazing metal into the steel of the shaft which resulted in cracking.

7.9.4 Biodeterioration

Biodeterioration is, with few exceptions, corrosion promoted and aggravated by biological action. It affects all classes of

(a)   (b)

(c)

(d)

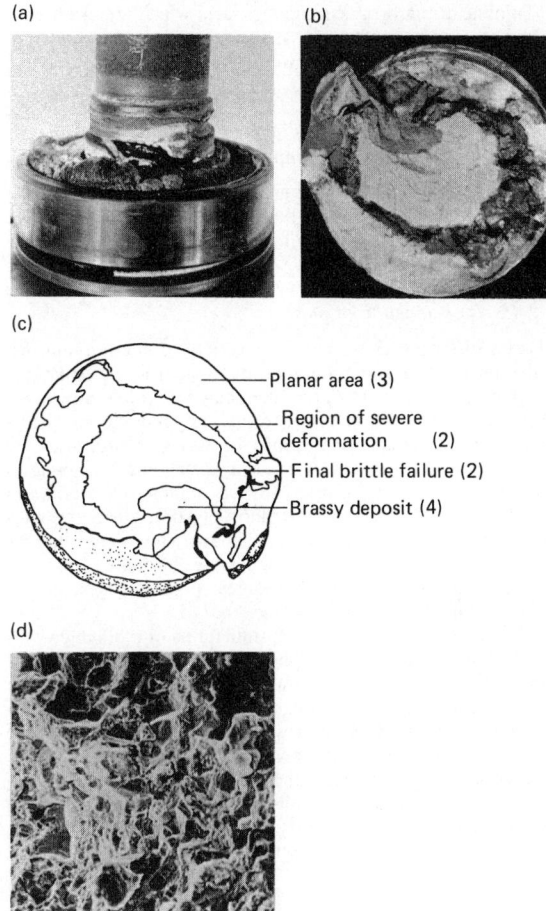

Planar area (3)

Region of severe deformation (2)

Final brittle failure (2)

Brassy deposit (4)

Figure 7.147 An example of penetration of liquid copper from brazing metal into the steel of an overheated shaft which resulted in failure by stress corrosion. (a) Appearance of failed shaft; (b) appearance of fracture showing copper deposit; (c) key to (b); (d) intergranular cracking and copper deposit

materials but the corrosive effects always require the presence of moisture.

7.9.4.1 Biodeterioration of metals

The most important biological effect in metal corrosion is pitting caused by the formulation of a differential aeration cell by a colony of bacteria or plants on the metal surface. Cast iron, carbon and low-alloy steel pit in fresh water and some stainless steels in seawater. Frequent changes of water help, but infected systems must be disinfected by a biocide such as chlorine.

Some organisms can produce organic acids (for example, *Cladisporum Rejinae*, which attacks aluminium at kerosene/water interfaces in aircraft fuel tanks). Sulphate-reducing bacteria can produce sulphides which corrode iron pipes, piles and the hulls of ships berthed in estuaries. This is very difficult to cure completely but pipelines may be surrounded by sand or chalk (with a biocide added) and piles and ships' hulls painted or cathodically protected.

Thiobacilli can produce sulphuric acid which attacks metals in mines or sewers. The cure is to use lime as an inhibitor in buried metals, to aerate sewers and to use acid resisting stainless steel in mines.

7.9.4.2 Biodeteriation of building materials

The same thiobacilli can also greatly accelerate the corrosion of concrete and limestone buildings. Where possible, the source of sulphur should be eliminated.

7.9.4.3 Biodeteriation of plastics

Plastics suffer attack from a great variety of microorganisms. Polymers derived from natural products incorporate structures to which enzymes can key or which may be broken down into products upon which microorganisms can feed and are therefore prone to attack. Resistance to attack is promoted by increased halide content, increased chain length, increased cross linking, toxicity of breakdown products to bacteria and the presence of sulphur. Thus fluoroplastics, vinyls, epoxies and polystyrenes are (as far as is known) immune, whereas animal and vegetable glues, melamine formaldehyde, cellulose derivatives, polyvinyl acetates, polyester-type polyurethanes and natural rubbers are attacked.

Most plastics are, however, formulations of a polymer with a filler reinforcer and/or a plasticizer or stabilizer. Natural fillers or reinforcers such as sawdust, starch paper or cellulose may be attacked. The major cause of breakdown in properties of plastic materials is, however, attack of the plasticizer (usually a long-chain organic acid or ester) which can embrittle and cause changes in shape and texture of individual vinyls.

Complete exclusion of moisture inhibits microbiological attack. Where this is not practicable an internally plasticized thermoplastic resin should be used.

7.9.4.4 Biodeterioration of natural products

Natural products such as wool, cotton and wood are all subject to attack by bacteria or moulds. The golden rule in every case is to keep the material clean and dry, in the case of wood by good design and regular painting, in the case of textiles by correct conditions of storage. Alternatively (or in addition) the material should be impregnated with a biocide, such as a chlorinated phenol or an organic wood preservative. (Further information on biodeterioration is available from the Department of Biodeterioration, University of Aston.)

7.9.4.5 Corrosion of ceramics and glasses

Ceramics are, in general, highly corrosion resistant. They will resist strongly acid aqueous solutions much better and more economically than most metals and, when used for this purpose, have glazed surfaces. Glasses are also available that will resist strong alkalis. Borosilicate glasses can contain phosphoric acid. All glasses, however, dissolve slowly in hydrofluoric acid.

Corrosion of ceramics at elevated temperatures Oxide ceramics resist oxygen at elevated temperatures but may be attacked by gases containing sulphur. In air, carbide ceramics oxidize at high temperatures, boron carbides and graphite oxidize rapidly above 500°C while silicon carbide is limited to 1400°C. Silicon nitride can withstand temperatures up to 1400°C whereas boron nitride oxidizes rapidly above 900°C.

Corrosion of ceramics by liquid metals Many ceramics resist attack by liquid metals and silica and aluminium-based refractories are widely used to contain them. Sometimes corrosive attack is advantageous. In basic steel furnaces attack by sulphur and phosphorus on a dolomite lining play an essential role in removal of these metals from steel.

Corrosion of ceramics by fused salts and slags Fused salts and slags are perhaps the most aggressive corroding media. Ceramic materials probably form the most suitable containers for these materials but the correct choice of refractory is important, particularly with silicate slags.

Stress corrosion of oxide ceramics Stress corrosion caused by the moisture in air exerts a most important influence on the performance of oxide ceramics stressed in tension. Cracks propagate through the material from the surface and when the stress intensity reaches a critical value the material fractures. (Since this is a statistical phenomenon it follows that a large component is weaker than a small one and great caution must be exercised in subjecting ceramics to tensile stress.) This cracking process (which has been incorrectly termed 'Static Fatigue') may not necessarily occur with carbide or nitride ceramics.

7.9.4.6 Corrosion of plastics

An attempt to provide a brief summary dealing with the corrosion of plastics is made difficult by the very large number of polymers whose behaviour needs to be considered and complicated by the presence of fillers and reinforcing agents. Plastics are called on to withstand the action of four main classes of environment: air often in conjunction with heat, water, organic solvents and ionizing radiation.

The most important requirement for many plastics is their resistance to combined heat and air. There are two criteria by which this may be assessed: the Oxygen Index ASTM 02863-70 test and the Underwriter's Laboratory UL 94 Burning Rating Code, which also assesses the performance under electrical stress.

The most resistant materials are the fluoroplastics followed by high-temperature thermoplastics and thermosets, most of which char gradually when heated. Many commodity thermoplastics burn freely and some (typically, celluloid) almost explosively. This danger can, however, be reduced by the addition of suitable anti-flash additives. All suppliers will provide suitable guidance.

Second in performance to flame resistance is the volume and toxicity of the smoke which may be generated when a plastic catches fire. ABS, polyester, PVC and polystyrene are particularly poor in this respect. PEEK and polyetherimide are outstandingly good.

Water, whose effect may be aggravated by acids, alkalis, ultraviolet light and general weathering, can hydrolyse certain plastics, causing general deterioration in many thermoplastics, cellulose esters and some polyesters. Fluoroplastics and epoxide resins are highly resistant. The effect may be critical with glass reinforcement which may retain negligible strength in a polyester matrix under marine conditions whereas carbon fibre in epoxide is resistant. Nylons will absorb water and this reduces tensile strength but greatly increases elongation and notch toughness.

Resistance to water and air at and above 100°C is essential for polymers used in medicine which have to withstand sterilizing. Polypropylene is good, polysulphone, polyether sulphone and PEEK are excellent.

Resistance to acids, alkalis and organic solvents varies with the type of acids and polymer. Generally, thermosetting polymers such as epoxide resins, polyimides and fluoroplastics have good resistance to most agents. Some thermoplastics such as acrylics, cellulosics and nylons have very variable resistance. Manufacturers should be consulted.

The resistance of polymers to sun (ultraviolet) light is variable. Fluoroplastics, polyimides, polyacetals and silicones are excellent, the rest rank from 'good' to 'poor'. Where a plastic with inferior resistance to sunlight has to be specified, an outer layer of resistant polymer should be provided. Alternatively, a filler which absorbs radiation may be incorporated.

Ionizing radiation has a beneficial effect on some plastics, converting, in some cases, a thermoplastic into a resistant thermoset by promoting cross branching.

7.9.5 The prevention of corrosion

7.9.5.1 Factors influencing corrosion

The chance that a component may fail by corrosion should, whenever possible, be eliminated by attention to:

Material choice
Control of environment
Design
Operation

The material may be a metal, ceramic, mineral, plastic or natural product. If a metal, it should be sufficiently high in the electropositive series or should form an adequately resistant oxide film. Plastics are usually resistant to the environment in which they normally operate but certain plastics (for example, epoxide resins and fluorocarbons) are exceptionally resistant. Ceramics are normally highly resistant but some (e.g. alumina silica and zirconia) may have superior resistance. Some natural products are better than others (e.g. oak and reed thatch outlast soft wood and wheat straw).

Control of environment may involve the exclusion of water and industrial pollution from air or the exclusion of halides, sodium hydroxides, sulphur compounds or other industrial waste from water. At high temperatures alkali chlorides, sulphates and vanadates should, whenever possible, be eliminated from gaseous environments.

Design to prevent corrosion may include temperature limitation (e.g. by eliminating hot spots and flame impingement in furnaces). It must eliminate crevices and stagnant areas and prevent electrical contact between metals wide apart in the galvanic series in aqueous systems. Where appropriate, operating lives must be limited. Operating procedures should avoid temperature excursions, prevent stagnation in aqueous systems, eliminate stray currents, provide regular maintenance (cleaning and greasing) and should arrange to replace components which suffer corrosion at regular intervals. Requirements of cost and mechanical strength may, however, prevent the use of an adequately resistant single material and it may be necessary to provide additional protection.

7.9.5.2 Corrosion protection of metals

Metals may be protected against corrosion in three ways (which may be combined).

Provision of a surface barrier layer A surface layer may be provided that excludes contact with the corrosive environment. This barrier layer may be paint, polymer, vitreous enamel, conversion coating (anodizing, phosphating or chromating), diffusion coating (aluminizing, chromizing or siliconizing) or a metallic coating. A metallic coating may be rhodium, platinum, gold, silver, chromium, nickel, cadmium, aluminium, zinc or a zinc/aluminium alloy.

Provision of a protective surface layer A surface layer that affords cathodic protection may be provided. This consists of a metal, cadmium, aluminium, zinc or zinc/aluminium alloy lower in the electro positive series than the substrate. Such a layer will continue to protect even if incomplete and may, with advantage, underlie a barrier layer.

Provision of cathodic protection When the environment is a corrosive liquid protection may be afforded by immersing electrodes in the liquid connected electrically with the substrate. The electrodes may be sacrificial (e.g. zinc which dissolves preferentially and requires periodic renewal) or resistant, such as platinized titanium which requires a source of potential to be connected in series and electrical energy provided.

Coating processes Paints are easy to apply and normally cheap and readily renewable.

Polymer coatings are more expensive but more resistant than paint.

Vitreous enamelling is expensive, gives complete protection, is heat and acid resistant, electrically insulating, easy to clean and has an excellent appearance. It requires a hard and heat-resistant substrate.

Conversion coatings are cheap and form an excellent basis for paint. Anodizing, which can be applied only to certain aluminium magnesium and zinc alloys, is very adherent, wear resistant and durable. Chromating and phosphating, which can be carried out in a bath or continuously, can be applied to steel, zinc or cadmium.

Electroplating may be used for zinc, cadmium, chromium, nickel, copper, tin, silver, gold, platinum and rhodium. This process can be used for exterior and marine environments but its bonding is not as strong as hot-dip and diffusion coatings and its use in corrosive media is limited.

Electroless deposition may be used for nickel (and nickel phosphide), copper, gold and cobalt. It has good throwing power and is used for aggressive environments subject to wear.

Hot-dip coating may be used for zinc, tin, aluminium and zinc–aluminium alloy. It forms a good bond with the substrate and applies thicker coats than electroplating. Its cost is low, it is suitable for exterior use and it forms an excellent basis for paint.

Metal spraying is applicable to most metals. It is, as a method of depositing zinc, competitive with galvanizing for one-off applications and for applying to specific areas which require lengthy protection.

Diffusion coatings include aluminizing, chromizing, siliconizing and sheradizing (zinc). The process is slow and relatively expensive but will produce very resistant coatings for high duties.

Vacuum evaporation and sputtering can be used to deposit thin but continuous films of aluminium or gold. For some purposes the economy in deposited material outweighs the high cost of the plant.

Plasma coating is used for depositing wear- and corrosion-resistant coatings of materials such as stellite or tungsten carbide-based material.

Choice of coating material and process is governed by engineering, economic and environmental considerations. A steel structure protected by galvanizing or zinc spraying normally lasts between 10 and 15 years (which usually covers its design life) without further treatment other than for decoration purposes in most atmospheric environments. Paint is much cheaper and easier to apply but requires maintenance at periods between 3 and 12 years. The aggregate costs over 25 years of painting could well be much greater than galvanizing.

An additional consideration is that a barrier layer is only effective if it is complete. If a coating is defective at one point the effect of a corroding liquid may be much greater at that point (because of the concentration of the electric potential) than it would have been if it had access to the whole surface, and the structure will be damaged more quickly than if it had not been coated. If there is any risk of this, a protecting coat (e.g. zinc) should be applied beneath the barrier layer (e.g. paint).

7.9.5.3 Protection of plastics

Plastics are painted for a great many reasons, among them protection from corrosion, light and weathering. Great care must be taken with the process and choice of material to obtain good adherence and avoid damaging the material.[130]

7.9.5.4 Protection of concrete and masonry

The only protection that can reasonably be afforded to concrete and masonry is to prevent, wherever possible, atmospheric pollution and to take extreme precautions to avoid the possibility of 'jacking corrosion'.

7.9.5.5 Protection of wood

There are two golden rules to follow to ensure the long life of wooden architectural and other components:

1. Keep the wood dry by design and operation and apply and maintain a good coat of paint.
2. Impregnate the wood with one or more of a number of agents toxic to insects, moulds and bacteria.

7.9.6 Procedure for identifying origin and mechanism of a corrosion failure

Corrosion failures occur in plant and the procedure for identifying the cause must start at the plant and will probably continue in a laboratory. The procedures are complex and their detailed description would occupy many pages. Figure 7.148 lays down a logical basis of procedure which, if followed, will ensure that the investigation acquires all relevant information.

7.10 Non-destructive testing

7.10.1 Definition

Non-destructive testing (NDT) forms an integral part of quality control, a term used to describe the procedures which contribute to total quality assurance. A formal definition of the subject, agreed by the International Committee for Non-destructive Testing (ICNDT) and accepted later by the International Standards Organization (ISO) states:

Non-destructive testing is a procedure which covers the inspection and/or testing of any material, component or assembly by means that do not affect its ultimate serviceability.

In practice, the scope and importance of NDT can tend to be confused by the diffuseness of its boundaries as set by this definition, and by differing interpretations of how it should best be used with economy and effect to achieve its objectives.

It is difficult to quantify the savings that can be achieved by the effective application of modern techniques of NDT to control the quality and reliability of manufactured products; or by adopting a change of philosophy from using NDT merely for post-production inspection to one of incorporating suitable techniques in management planning to ensure an adequate level of quality assurance and general fitness-for-purpose of a product. However, from evidence that is available from many industries, it is apparent that savings of both a direct and consequential nature can be very substantial and worthwhile.

On the other hand, there is no lack of awareness of the expense and loss of commercial credibility which can so easily accrue from improperly planned and managed NDT, or from unfortunate errors of judgement in defect interpretation, leading to a poor-quality product, or plant failure, with consequent loss of life or environmental pollution.

7.10.2 Overall scope

There are four main ways in which non-destructive testing can be incorporated into manufacturing practice:

1. To provide control of quality at product manufacture or during plant fabrication;
2. To ensure that an item conforms to specification;
3. To examine plant, equipment or components during service, in order to meet statutory requirements or as an insurance against premature breakdown or failure;
4. As a diagnostic tool in research and development.

There is a tendency in some quarters to associate non-destructive testing merely with 'flaw detection'. This narrow interpretation has unfortunately tended to identify the subject with 'testing for scrap', so that it has come to be regarded rather disparagingly by some industrial managements as an unavoidable but costly overhead charge on production. However, non-destructive testing, if judiciously used, has a far more positive role to play; not least, in significantly lowering total manufacturing costs. This is particularly true if one can reject potentially defective material at an early stage of processing, especially in industries where significant scrap can occur in the manufacture of products with a high added value.

With engineering and constructional materials, properties of prime concern, such as strength, fracture toughness, fatigue or corrosion resistance cannot generally be measured non-destructively and, as a consequence, it is necessary to approach the problem indirectly and look for secondary features likely to be significant. For example, shrinkage and porosity in cast metals, defective welding, lamination in sheet and cracks in forgings are obvious suspect features, for which efficient non-destructive testing techniques have been developed. As materials and conditions of service get more complex, less obvious features such as microstructure, composition, internal stress and homogeneity become important. This means that they too may need to be carefully controlled and, as a consequence, monitored non-destructively.

Non-destructive testing is not confined to the factory and foundry. On-site testing of pressure vessels, pipelines and bridges, and in-service maintenance of airframes, aero-engines and refinery installations all present special problems to both instrument designer and operator. Then again, automated inspection, computer-aided manufacture and in-line process control raise quite different problems.

Although the non-destructive testing techniques required may be similar, different situations may call for very different levels of sensitivity, and working in this way to a specification requires experience and considerable interpretative skill, whichever NDT technique is used. Indeed, many of the techniques of non-destructive testing are now so well developed and advanced as regards sensitivity that what they are capable of revealing in the limit is often an embarrassment, and of little practical relevance to performance. Figure 7.149 shows a range of surface and internal variables that may need

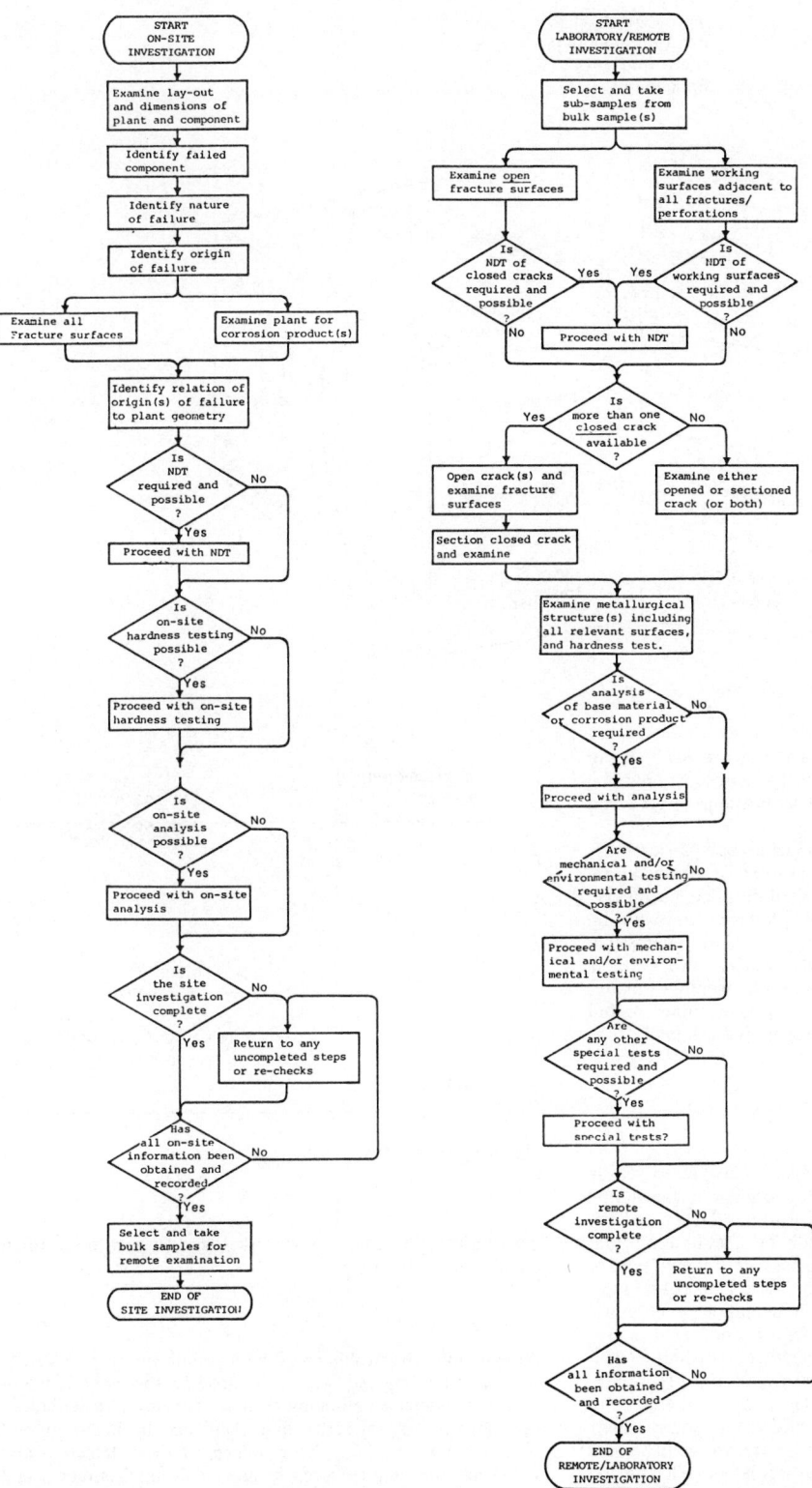

Figure 7.148   Procedure for identifying origin and mechanism of a corrosion failure

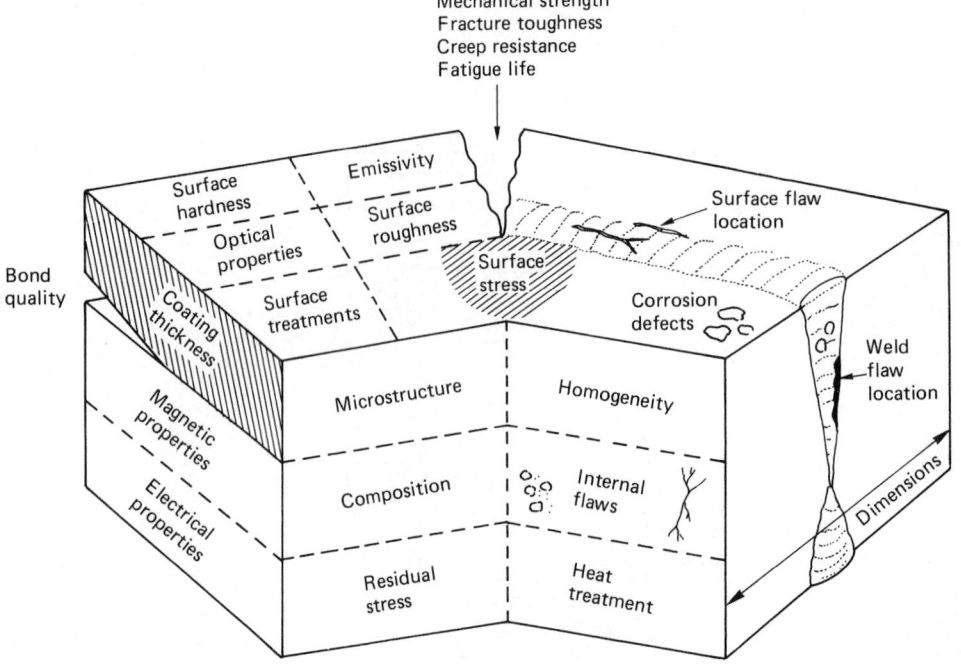

Figure 7.149 The scope of non-destructive testing

to be controlled and hence monitored non-destructively, again emphasizing that flaw location and sizing represents only one of the many facets that may need to be considered when assessing overall quality.

Non-destructive testing during production and fabrication is closely allied with 'condition' monitoring of plant during service and regular 'health' monitoring of installed machinery. Many of the techniques can also be adapted for manufacturing control, enabling flaws to be eliminated as early as possible during the manufacturing process. NDT is also now an important element in the 'fitness-for-purpose' philosophy of manufacture in which design, materials selection, manufacture and quality control are integrated and properly coordinated.

7.10.3 Application areas

7.10.3.1 Materials control

Control of materials 'quality', which can be so easily influenced and modified by casting, forging or machining aberrations, represents the traditional use of NDT, and the variability to be monitored can broadly be classified into that associated with surfaces and that hidden within the volume of the material (Figure 7.150). Cracks and visible discontinuities have always tended to capture the imagination as potential and often catastrophic sources of failure and, as a consequence, have developed an almost emotive association with the 'folklore' surrounding non-destructive testing. However, the overall 'quality' of a material and its ability to match up to a performance specification can be affected by many other, perhaps more insidious, pockets of structural variability which are often difficult to define and categorize, let alone locate and identify. These might be locked-in stresses, grain variability, distributed porosity, depth of surface treatments, inclusion distribution or constituent diffusion. Many of these still present a challenge when it comes to specifying reliable non-

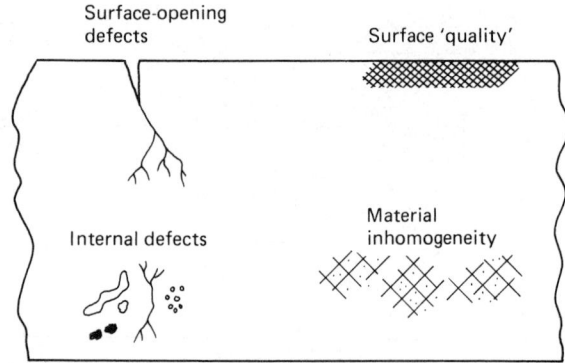

Figure 7.150 Material 'quality'

destructive tests that can be used and simply interpreted outside of a laboratory.

7.10.3.2 Assembly

At assembly, the problems of the test multiply (Figure 7.151) Welding, bonding and bolting all introduce their own brand of specialized defects and many of these, because of orientation, geometry or lack of accessibility, still tax the ingenuity and skill of those to whom the problems of their detection and sizing are presented – often in unfriendly environments and inconvenient situations. It is here that close links between designer, tester and operator are so important.

Every effort must be made by the designer to build in inspectability, to understand the problems and frustrations of

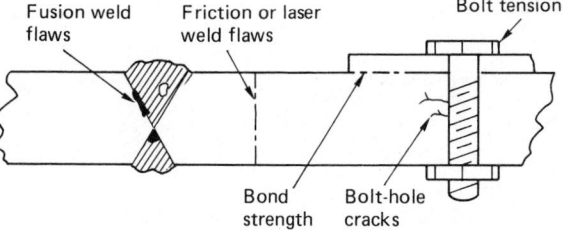

Figure 7.151 Joint quality

the tester and to realize the limitations of NDT technology on which the tester's judgement – and hence reputation – are based. Every effort must also be made to set up a dialogue between design and inspection teams as early as possible so that when design demands special inspection problems, sufficient time is available to develop, evaluate and calibrate suitable inspection procedures and train and validate the operators.

7.10.3.3 Automated metrology

The physical principles on which the more conventional flaw-detection NDT techniques are based can also be adapted for automated metrology and so, in a sense, this is a subject which is appropriately included under the NDT 'banner'. Ultrasonic techniques have been developed for accurate and high-speed monitoring of tube-wall, or plate, thickness; capacitance gauges have been designed for tube-bore measurement; and laser beam techniques have been used to obtain quantitative data on surface profiles and surface smoothness, where such factors need to be precisely controlled for reasons of heat transfer, assembly precision or to optimize performance. Most techniques can be adapted to give digital signals and in measurement-type tests on production components, where very large numbers of individual readings are inevitably made, data reduction and analysis can then be readily performed to simplify interpretation and provide archival data.

7.10.3.4 Materials sorting

Another area of application of non-destructive testing methods is in product or materials sorting to ensure uniformity of size, heat treatment or composition (Figure 7.152). Composition control, in particular, is a problem of growing importance as nuances in alloy composition become subtler, as specifica-

tions reflect the importance of trace-element constituents, or as property tolerances need to be controlled in alloys of the same nominal composition. The glib answer is that the problem of manufacturing variability can be solved by good 'housekeeping', but accidents and oversights occur in the most carefully organized factory. There is no 'panacea' instrument and each problem has to be carefully analysed around existing NDT technique possibilities. Portable spectroscopes, electrical or magnetic property analysis, thermoelectric (hot-probe voltages) or triboelectric (friction-generated voltages) techniques all have (or have had) a part to play, but this is still a fertile field for further technological exploration.

7.10.3.5 Plant surveillance

A major call on NDT expertise is to satisfy surveillance requirements for installed plant to ensure that an appropriate level of structural integrity is being maintained during service. Much of the requirement is covered by legislation and is closely specified by regulatory Codes of Practice. The requirement, of course, particularly applies to installed pressure vessels and pipework, ships operating to Classification Society rules, operating aircraft and aerospace structures, nuclear power plant installations and offshore platforms and their associated pumping plant and pipelines.

It is here that traditional NDT technology is pushed to its limits. Problems of access and geometry, site hazards, difficulties of positive interpretation and quantitative evaluation of defect dimensions, pressures to get plant back 'onstream' and interference with other maintenance work if radiation sources are required, all add up to a challenge of considerable magnitude and one where the inspector needs all the support and cooperation possible. The concept of 'fingerprinting' a structure before it is put into operation, so that changes in defect content or growth of cracks during service can be more positively monitored, is becoming a favoured approach to this particular problem. This raises attendant problems of long-term control of test sensitivity, data recording, archival storage and the need for automated test procedures which are reliable and repeatable over what might be 20–40 years of periodic application.

7.10.4 Methods of employing NDT in practice

Bearing in mind that non-destructive testing is an integral part of the wider management function of quality assurance, there are different ways in which NDT tests can be incorporated (as illustrated in Figure 7.153).

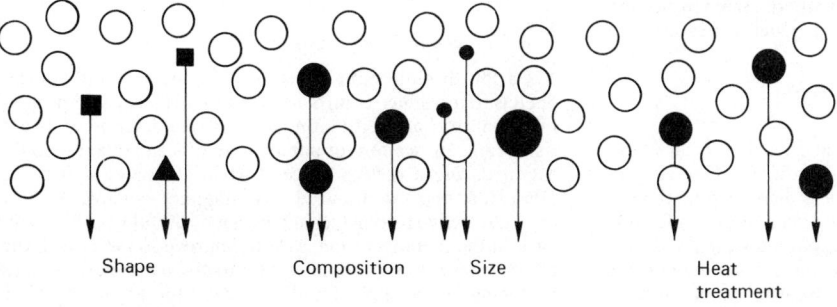

Shape   Composition   Size   Heat treatment

Figure 7.152 Materials sorting

Inspection

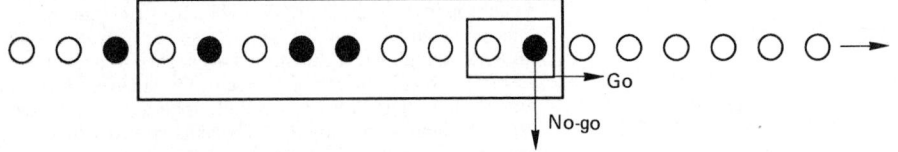

Quality control

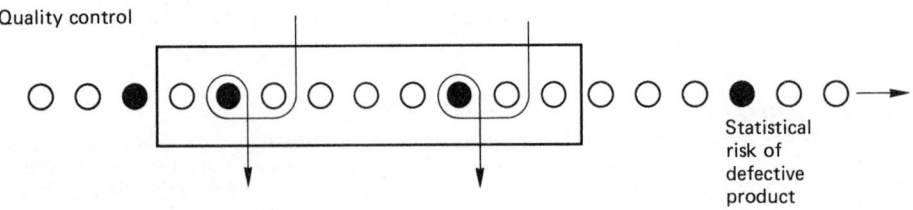

Process control

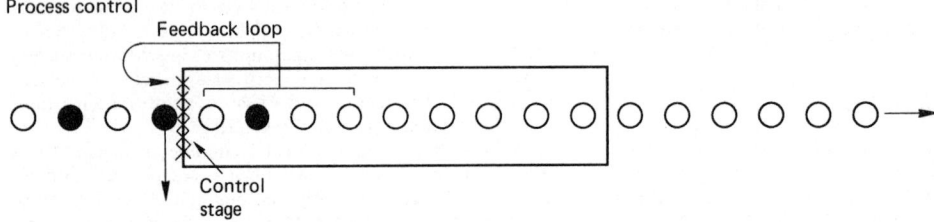

Pre-production control

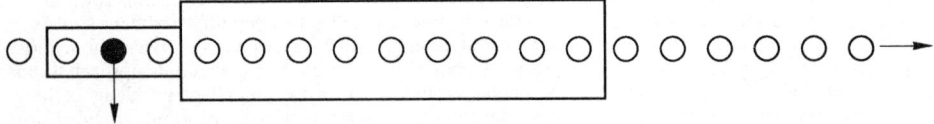

Figure 7.153   Methods of incorporating NDT into a production line

7.10.4.1 Inspection

At various points appropriate to the production (often only at the end of the line), some or all of the product is inspected and the 'sheep' separated from the 'goats'. This can be costly, time consuming and wasteful, although, in many situations, necessary. Such, unfortunately, is the limited measure of confidence that some managements place on the technology and in the way it is sometimes applied that this process is often used when a product leaves a supplier firm and the same inspection repeated by the customer when the product comes into the input stores!

7.10.4.2 Quality control

By invoking the concept of statistical checking at all stages along a production line to identify out-of-specification products as they arise, a smoother and less disruptive method of control results, and a measured risk is taken of the uniformity and correctness of the final product. Non-destructive testing techniques naturally have a role in this product-monitoring function and complement the more usual metrological measurements from which statistical quality-control procedures have traditionally evolved.

7.10.4.3 Process control

The third principle of operation is to have continuous monitoring of a product line with a feedback signal to the mechanics controlling the process. Thickness control by feedback adjustment of roll pressure, property control by feedback adjustment of furnace temperature, coating control and spot welding control are all manifestations of this positive approach to product quality uniformity.

7.10.4.4 Pre-production control

Yet a fourth philosophy which has not yet been fully developed is to introduce control at such an early stage in a process that quality is assured before added manufacturing costs are significant. Inspection for potential defects and quality rectification *during* a casting process, while a spot weld is being fabricated, or as an arc weld is cooling, are examples of how non-destructive testing techniques can be usefully pushed right back in the manufacturing cycle to improve to the utmost the effectiveness and economics of quality assurance. Similar reasoning can be applied to the incentive for developing NDT methods of monitoring NC machining by positive metrological control before the component is removed from the machine.

7.10.5 Range of techniques available

7.10.5.1 Surface flaws

The surface of a component has always attracted considerable attention from the standpoint of non-destructive testing. This is partly, of course, because the accessibility of surfaces makes inspection easier and interpretation more direct. It is, however, primarily because so many of the variables associated with surfaces have a significant effect on either the serviceability or saleability of a product. From economic considerations, inspection for saleability can often be as important as inspection for serviceability, and to satisfy both requirements, surfaces may be protectively or decoratively treated. These surface-treatment processes themselves often require special non-destructive tests to ensure adequacy of protectiveness and uniformity of coverage.

Surface discontinuities can act as stress raisers; they can reduce mechanical strength, especially bend strength and fatigue strength; and they can act as initiating points for brittle failure. Not only do surface discontinuities need to be located, but the depth, shape, nature, orientation and position are usually significant. Non-destructive methods for detecting surface discontinuities are well established and widely practised and are summarized in Figure 7.154.

7.10.5.2 Internal flaws

An internal flaw is one that cannot be detected by visual inspection, or one whose depth or extent cannot be accurately gauged by a surface-inspection technique. It may be an original casting defect or a defect introduced subsequently by a deformation process such as forging, extrusion, heat treatment or a joining process such as welding or brazing. The detection of internal defects is an area of non-destructive testing which has received considerable attention over the years and one which has resulted in major technological advances, particularly in the fields of radiology and ultrasonics. The range of currently available techniques is illustrated in Figure 7.155.

7.10.5.3 Structural variability

Discrete flaws are not the only cause of product failure. General microstructural variability (both at the surface and internal), preferred orientation, residual stress levels, heat-treatment variations, anisotropy, compositional non-uniformity, variations in electrical or magnetic properties, moisture content, and dislocation density are just some that may need to be controlled. A wide range of techniques is

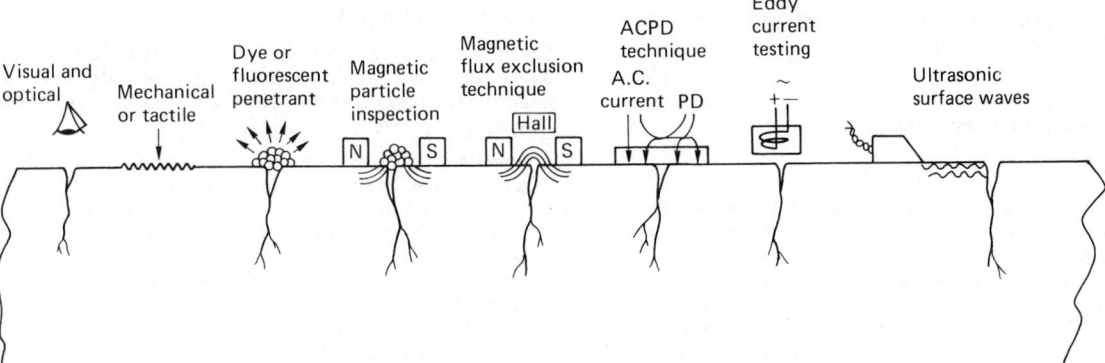

Figure 7.154 NDT techniques for locating surface flaws

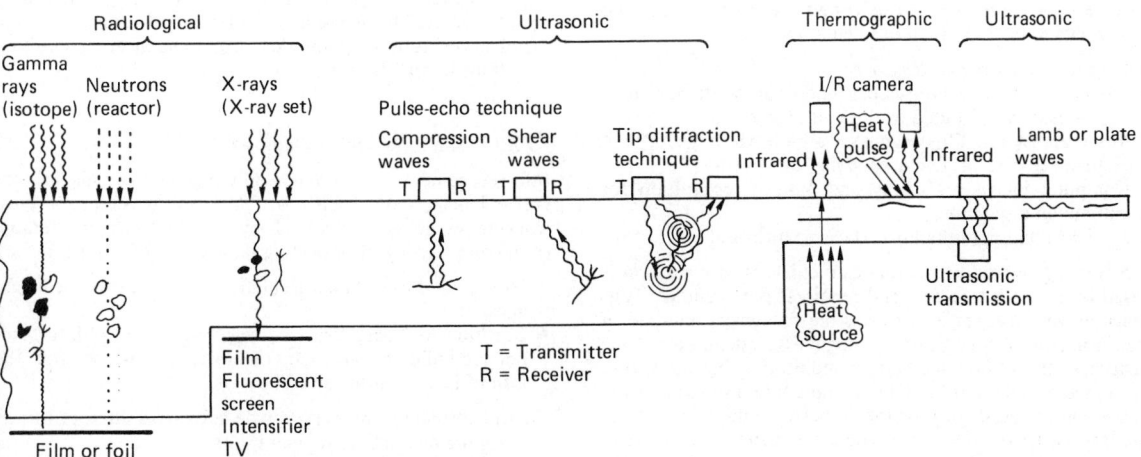

Figure 7.155 NDT techniques for locating internal flaws

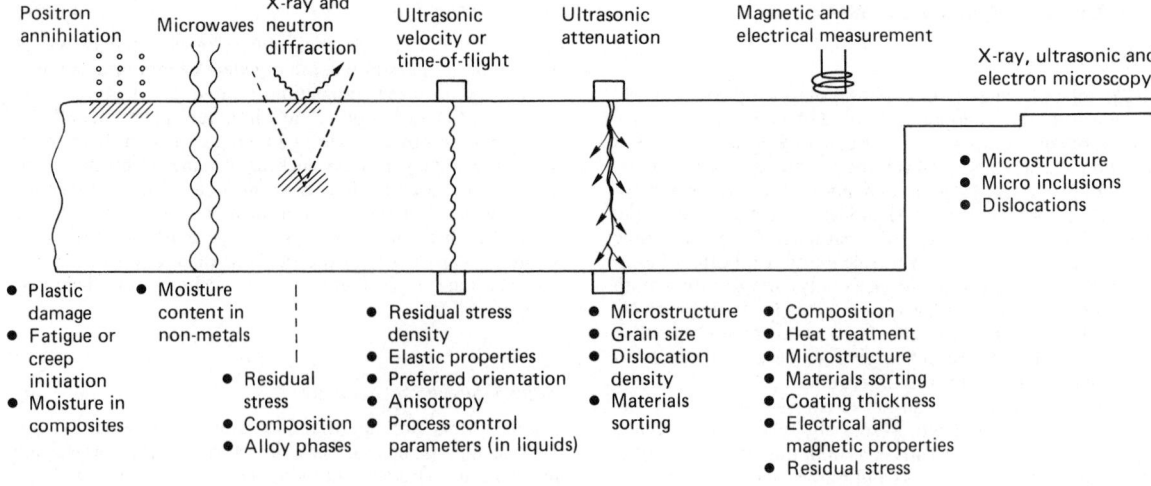

Figure 7.156 NDT techniques for monitoring structural variability

currently available to handle the associated monitoring requirements (Figure 7.156).

7.10.6 Individual techniques

7.10.6.1 Direct surface inspection

Visual inspection is the oldest, simplest and most widely used of all non-destructive testing techniques. All visual inspections are subjective and so may be influenced by outside factors. They should therefore be carried out under the best possible conditions. There must be adequate lighting with no dazzle or glare, and the inspector should be comfortable and protected from disturbing factors such as noise, draughts, extremes of temperature and inclement weather.

A systematic approach is essential; if it is not possible to cover a large area in a natural viewing sequence then a grid system should be used and one square examined at a time. Beware of eye fatigue and plan the work so that there are opportunities for the eyes to be rested. Technical aids to the humen eye to extend visual inspection include:

- Lenses, microscopes, telescopes
- Fibre-optic devices, boroscopes, endoscopes, etc. for viewing surfaces with limited or difficult access
- Mono and stereo TV systems for remote viewing
- High-speed cine or TV for studying fast events
- Computer image processing and pattern recognition for automated inspection
- Schlieren photography for surface texture studies

Another group of direct non-destructive tests for surfaces is based on tactile interaction and requires some contact. Thus stylus movement over a surface is the recognized method of measuring and defining surface roughness; indentation of the surface is the classic method of monitoring hardness (i.e. resistance to indentation). This group could also contain the classic metrological tools of the inspector – the micrometers and the callipers – for dimensional measurement, although conventional metrology is normally considered outside the accepted boundaries of non-destructive testing.

7.10.6.2 Optical metrology

Lasers and other optical devices are now available to extend direct visual and tactile techniques and provide accurate, non-contacting means of measuring size, position, component spacing, profile, etc.:

- Laser interferometers measure displacements to the highest accuracy, e.g. in linear measuring machines. They can also measure vibration amplitudes in machinery, bearings, loudspeakers, ultrasonic transducers, etc. and display these in three-dimensional form.
- Optical diode arrays can provide passive means of monitoring displacement, automated size measurement, etc.
- Reflectometers and laser-scatter devices can provide ways of monitoring and mapping surface smoothness of metals and other reflective materials.

Optical holography is a method of storing three-dimensional images on film for detecting strain regions and defects, particularly in composite structures.

Microwaves with wavelengths of $1–10^3$ mm (radar range) can also be used for non-contact detection of surface cracks in metal as well as to pinpoint moisture or chemical composition variations in non-metals.

7.10.6.3 Thermographic techniques

Infrared cameras are commercially available which produce thermal images on a CRT or TV monitor screen. Typical working range is 0–800°C. They provide a rapid, remote (non-contact) way of inspecting large areas of plant for

- Hot spots due to failed insulation, fouling, coolant blockages, etc.
- Location of leaks, corroded regions, hidden objects, embedded building construction features, etc. which affect the rate of inward diffusion of heat.

Thermochromic paints can provide a low-cost alternative way of mapping out surface temperatures.

SPATE (Stress Pattern Analysis by Thermal Emission) is a non-contact thermographic way of measuring local stresses by observing the minute surface temperature rises that occur in cyclically loaded structures. Pulsed Video Thermography (PVT) is a developing technique using transient heat sources to show up sub-surface defects on a video monitor (e.g. voids, delaminations in composites, poor adhesion of coated materials, etc.).

7.10.6.4 Liquid penetrant inspection

This is a simple, widely used, low-cost way of detecting surface-opening cracks, porosity, etc. in non-porous, clean objects (e.g. metal castings, forgings, weldments and ceramics). A liquid (containing a coloured dye or UV-sensitive substance) is sprayed onto the object. The liquid is drawn into any surface-opening crack by capillary action, thereby highlighting its presence after subsequently cleaning the surface and applying a developing agent. The technique can detect minute defects and be automated for continuous production application, but is generally unsuitable for rough, or dirty surfaces. Penetrants are formulated in three groups:

- Solvent-removal penetrants;
- Post-emulsion penetrants where an emulsifier is subsequently applied for a predetermined time and the whole removed by water;
- Water-washable penetrants.

Developers are also of three basic types:

- A hard-drying developer paint which cannot be accidentally removed;
- A soft-drying developer paint which is easily removable;
- A dry-powder developer which is applied by blowing or dusting.

Stages in liquid penetrant testing

- Precleaning and degreasing; a vapour-phase degreasing bath is ideal;
- Preliminary inspection;
- Application of penetrant;
- Soaking to allow the penetrant to enter the cracks;
- Spraying of post-emulsifier (where applicable);
- Removal of penetrant: (a) excess by tissues or rags: (b) remainder by solvent or water;
- Dry, hot air is very suitable (except where dry powder developer is used);
- Application of developer;
- Soaking of developer: this may take anything from 10 minutes to 24 hours, depending on size of crack and sensitivity required;
- Final visual inspection.

Safety in using penetrant materials   Solvents used in penetrant materials and cleaning agents may be both flammable and toxic. When carrying out liquid penetrant testing great care should be taken to ensure good ventilation and that precautions are made when using flammable liquids. Aerosols should not be used in confined spaces and indeed are now regarded as environmentally 'unfriendly' and to be avoided.

7.10.6.5 Magnetic inspection

Magnetic Particle Inspection (MPI) is a technique for locating surface and near-surface defects (in ferromagnetic materials only). The principle is that locally magnetized and magnetic discontinuities around defects are shown up by patterns formed by magnetic powder particles applied to the object's surface, usually in fluid form. It is a rapid, low-cost and sensitive way of detecting surface and near-surface cracks, inclusions, porosity and seams in castings, forgings, bars, etc. The methods is used both during manufacture and for in-service inspection of plant. It is excellent for detecting very small, tight service-induced cracks and can be used in either portable or on-line automated systems.

There are five principal ways of applying the magnetic field:

- Permanent magnet
- Electromagnet, solenoid coil or magnetic yoke
- Current flow, using prods or contacts
- By means of a threading bar
- Induction, or induced current flow

The selection of the most suitable magnetizing method depends on the size and shape of the object and also the orientation of the defect. It may be necessary to apply the magnetizing field in more than one direction, or indeed to apply more than one technique, in order to achieve a complete examination.

Sensitivity levels vary, but a field strength of between 3 and 6 kA/m (38 and 75 oersteds) is generally acceptable. The field strength (and direction) may be measured with a tangential field strength meter. The magnetic particles are coloured black, red or yellow and may also be fluorescent in order to achieve maximum contrast against the background colour of the item being inspected. Specially formulated white paints may also be applied before the test in order to obtain better contrast.

Magnetic particle testing requires skill and training in interpreting indications. In addition to discontinuities, particles may be attracted to sharp changes of section, surface scratches and boundaries of dissimilar metals. Sub-surface defects are recognized by their slightly blurred outline. A suspected crack may be confirmed or otherwise by removing a film of metal with emery paper or a smooth file and retesting. Many engineering components need to be demagnetized after testing. This is achieved by passing the components two or three times through a coil carrying alternating current.

An extension of MPI is the magnetic flux leakage technique. If ferromagnetic components are magnetized (by a permanent or electro magnet) to near-saturation, local flaws or metal thinning can divert some of the magnetic flux outside the component and this can be detected by Hall-type search probes. This provides a relatively fast, 'first-look' method for detecting corrosion thinning and pits in large-area steel plates up to 1 cm thick such as the 'bottom' plates of oil storage tanks, reinforcing bars in concrete, broken strands in cables, etc.

7.10.6.6 Electrical techniques

Eddy currents are produced in any electrically conducting material when influenced by a coil placed close to it in which an a.c. is flowing. For the purpose of non-destructive testing, two coils are generally used, a small a.c. being generated in the primary coil and this induces an alternating magnetic field in the specimen; this in turn generates an a.c. of similar frequency in the secondary receiver coil which induces an opposing alternating magnetic field in the component.

Eddy currents provide a sensitive, versatile method of inspection using either portable instruments or in-line automated systems, and are used for inspecting electrically conducting materials. These are generally below 1 cm in thickness since the a.c. probe coil (1 kHz–5 MHz) only excites eddy currents within the object's surface layers. The flow of these currents is affected by many surface or near-surface defects or regions of varying conductivity or permeability. The currents picked up in the search coil are then amplified, analysed and displayed in terms both of amplitude and phase on either a meter or oscilloscope or as a digital signal. Calibration with reference blocks is usually required. Applications include:

- Detection of surface-opening or near-surface cracks, seams, pits, etc.
- Sorting metals of dissimilar composition, heat treatment or microstructure
- Measuring thickness of coatings on metals

Heat exchanger tubes in non-ferrous or austenitic materials can be inspected by using internal coils pulled through the tubes.

Eddy currents are not as penetrating as either ultrasonics or radiography, but have three outstanding advantages:

- Physical contact is not required.
- It is a very rapid form of inspection.
- It is easily automated.

Multifrequency and pulsed eddy current techniques are now available which increase the sensitivity and selectivity of the testing.

An alternative electrical technique is based on potential difference measurements. An electric current (a.c. or d.c.) is induced to flow between two contact prods pressed onto the surface of the object. Surface cracks in the current path between these contact points increase the apparent electrical resistance between the prods. These resistance changes, measured by the potential difference across a second pair of contact probes, can be related to crack depth. The technique forms the basis of low-cost portable instruments for measuring depths of surface-opening cracks in conducting materials, but results can be variable if cracks are 'tight' or provide an electrical 'bridge' for the current.

7.10.6.7 Radiography

X-rays and gamma-rays at the short wavelength end of the electromagnetic spectrum have the important characteristic of being able to penetrate considerable thicknesses of materials that are opaque to light. They have, therefore, been widely used to provide non-destructive techniques to locate internal flaws. Fortunately, they produce, like light, an image in a photographic emulsion (X-ray film) and so radiographs can be obtained directly to provide a pictorial and permanent inspection record.

Although a powerful non-destructive testing tool, radiography does require considerable skill and training to achieve the correct sensitivity and image definition, and considerable experience and judgement in interpreting the images correctly in relation to the flaws contributing to the 'image' pattern. It is essentially a 'shadow' technique and, as image contrast is caused by differential absorption, the technique is more suited to finding 'volumetric' defects (i.e. pores and cavities) than cracks and laminations. Indeed, these latter can be virtually invisible if the radiation is directed across the tight interfaces of such defects. Penetration depends on the energy of the radiation and a range of X-ray sources from 50 kV to 400 kV potential (in hot-cathode Coolidge tubes) are widely used in engineering practice and cover section thicknesses up to around 10 cm of steel or 25 cm of aluminium alloy.

Linear accelerator ('Linac') X-ray sources, up to 8–9 MeV, allow radiography of very thick sections (e.g. 1.5 m concrete or 0.5 m steel). They are used for inspecting reinforced concrete bridge sections, heavy castings or large assemblies such as gas turbines. Linacs are expensive, heavy sources posing a significant potential radiation hazard.

Microfocal X-ray sources are now available which produce sharp, enlarged images (up to 15 ×) projected onto film

placed a metre or so from the object. Main applications are in the detection of tiny cracks or microporosity (down to 20 μm in size) in high-performance materials (e.g. ceramics and Nimonics) or resolving fine detail in small mechanisms, fibre composites, etc.

Gamma-rays provide an alternative to X-radiography for thick sections or in-field applications where access is difficult or power for an X-ray set is not available. Gamma-rays from a portable isotope source pass through an object and, like X-rays, form a direct image on film. Gamma radiography is also used for detecting cracks, corrosion defects, inclusions and for resolving internal structures. Isotope sources are selected for the penetration required (e.g. cobalt-60 can inspect up to 250 mm of steel). Image resolution is generally inferior to X-radiography and some isotopes have relatively short useful lifetimes. Panoramic (360°) imaging is possible, using small isotopes inside hollow objects, tubes, etc.

Historically, gamma-rays from radium, mesothorium and radon sources were used for materials inspection around 1925 and the pioneer work on the systematic investigation and evaluation of gamma radiography was carried out around 1930 when artificial radioisotopes were first produced by cyclotron bombardment. Satisfactory source activities for gamma radiography were not practically possible until preparation by reactor irradiation became feasible. There is now a wide choice of sources and their characteristics relevant to radiography have been widely evaluated and documented. The three most commonly used are given in Table 7.76.

Because of their different characteristics, X-ray and gamma-ray sources complement each other in their application:

- For light-alloy and low density specimens there is no suitable high-intensity radioisotope source. The preferred radiation source for such applications will, therefore, practically always be X-rays. Generally, for steel or copper alloys in thicknesses less than about 20 mm, X-rays will be preferred.
- For steel thicknesses less than about 50 mm the highest sensitivities can always be obtained with X-rays, but the differences between X- and gamma-radiography are not so large at the upper thickness end of this range.
- If large amounts of radiography are contemplated there will be a preference for using X-rays because of the shorter exposure times which are possible.
- If the radiation source has to be taken to the specimen, gamma-ray sources in their shielded containers are more portable than large X-ray sets, and the relative importance of convenience and sensitivity must be assessed for the particular application. X-rays will usually give the better sensitivity, but it may not be possible to handle an X-ray set under 'site' conditions. X-ray sources can be 'switched off', gamma-ray sources cannot.
- If only a small amount of radiography is required and the highest attainable sensitivities are not essential, a gamma-

Table 7.76 Properties of the three most common isotopes used in industrial radiography

Isotope	Half-life	Optimum thickness range for steel (mm)
Iridium-192	74.4 days	12-60
Caesium-137	30 years	50-100
Cobalt-60	5.26 years	50-200

ray source would initially be cheaper to buy. Again, choice of source must finally depend on the specimen thicknesses involved: gamma-rays, for example, should not be used for the radiography of very thin specimens unless there are very strong reasons for not employing an X-ray set.

Other radiation-based techniques include the following.

Radiometry This is used for detecting voids and density variations, and measuring wall thickness or fill-levels in concrete or metal components. The attenuation of gamma- or X-rays passing through the object is measured by a radiation detector, and can be related to density, thickness, etc. Lightweight isotope sources are used for in-field applications and backscatter methods for single-sided access. It is fast, accurate and easily automated and is sometimes referred to as the 'Source-and-counter' method.

'Real-time' radiography Replacing radiographic film by a fluorescent imaging screen linked to a TV system allows test objects to be inspected immediately (i.e. in real time), with the object either static or in motion. Resulting images can be stored on video tape and replayed, and, if required, enhanced or analysed automatically by computer-image processing. Real-time techniques can be used with neutrons and gamma-rays as well as X-rays. Main uses include on-line NDT of components, foreign-body detection and studies of internal workings of engines and other machinery.

Neutron radiography This is a technique which uses a beam of neutrons as the sensing radiation. Neutrons of various energies can be used (usually referred to, in order of decreasing energy as fast, epithermal, thermal and cold) and the source can be a nuclear reactor, an accelerator or certain isotopes. Neutrons do not produce a photographic image directly on film and some intermediate foil is necessary to give fluorescence, or an intermediate radioactive image which can subsequently be transferred to a film by autoradiography. Neutron radiography is particularly useful for inspecting brazed and adhesively bonded structures, monitoring the filling of ordnance devices and for locating hydrogen-rich areas and certain structural components with a hydrogen content (e.g. paper and rubber) which cannot be revealed by X- or gamma-rays. It is also widely employed for detecting residual refractory core material in the cooling channels of precision-cast turbine components. Neutron diffraction and scattering can be used for accurate measurement of residual stresses in thick components or evaluation of microstructures.

Electron radiography – beta-radiography This is a radiographic technique using beta particles (electrons). Since electrons are so readily absorbed this form of radiography can only be used effectively with foils, papers and paper products for studying fibre structures and watermarks, etc. The electrons can be excited in a front foil of metal or in a metal backing plate. Alternatively, the source can be a beta-emitting radioactive foil placed on one face of the specimen with the film on the other.

Positron annihilation This is a technique which uses a beam of positrons from a small, low-intensity isotope source for detecting early (pre-crack) fatigue or creep damage in metals and for monitoring moisture levels in composites.

Low-voltage radiography This is a radiographic technique for inspecting thin samples (e.g. papers) or low-density samples (e.g. fibre-reinforced plastics or beryllium). Low-voltage (5–10 kV) X-ray sources are used to give as high an attenuation as possible within the sample. It is often necessary to have a vacuum or helium path between the X-ray source and sample to cut down air absorption of the low-energy X-rays. It is usually necessary to expose on bare film in a darkened room to cut out the image of the paper of the film envelope.

Panoramic radiography This is a 360° degree radiographic arrangement using a central isotope source or a rod-anode X-ray tube. It can also be applied to a radiograph built up by moving sample and film on opposite sides of a narrow slit in an opaque screen during the radiographic exposure.

Dynamic radiography – in-motion radiography This refers to radiographic techniques which 'freeze' a situation in a sample in which changes are occurring with time. It is usually applied to the study of moving machinery or assemblies during operational functioning (e.g. aero engines or pumps), or to structural changes occurring more slowly due to metallurgical failure processes (e.g. creep cavitation), or to heat treatments and corrosion.

Flash radiography – high-speed radiography This is a technique of dynamic radiography in which motion is 'frozen' by using an extremely short pulse (tens of nanoseconds duration) of high-intensity radiation. Field-emission X-ray tubes have been constructed which can give sufficient output in 50 ns to produce a radiograph during very high speed motion, such as in studies of high-speed liquid flow and molten metal pouring during casting.

Xeroradiography Here a radiographic image is formed on a semiconductor (usually a selenium coating supported on a metal plate) instead of on film. The semiconductor is charged initially and the loss of charge when the plate is irradiated is proportional to the radiation dose received at every point. The residual charge pattern is revealed by dusting the plate with powder, which is preferentially attracted to the charge. On the radiograph there is a characteristic pattern of image enhancement at density steps such as edges, which can often help to improve image visibility. The technique is cheaper to operate than film radiography but is not widely used at present.

Tomography By suitably rotating, rocking or translating the specimen in relation to the source of radiation and the film, or vice versa, it is possible to produce a sharp radiograph of a particular plane within a solid sample. The rest of the image detail is blurred because of the geometric unsharpness caused by the relative motion purposely introduced between the rest of the sample and the film.

Computerized tomography (or CAT scanning), originally developed for medical use, is now finding wider application in industry for observing detail in a particular plane at right angles to the X-ray beam.

Legislation The use of ionizing radiations for the purpose of industrial radiography is subject to special legislation in most countries. In the UK there is:

- *The Radioactive Substances Act 1960.* This requires all users of radioactive material to register with (or apply for exemption from) the appropriate government department. It is also concerned with registration of mobile radioactive apparatus and the right of entry and inspection by the government department Inspectors.

- *The Ionizing Radiations (Sealed Sources) Regulations 1969.* These deal principally with industrial radiographic processes and the protection of both users of ionizing radiations and the general public. The Factory Inspector for the district must be notified in writing before any work with ionizing radiations takes place for the first time in any factory. Workers are 'classified', which means they are under medical supervision and detailed records are kept of their individual radiation dosages. The regulations also deal with administration, notification and records, basic principles of protection, radiological supervision, organization of work, monitoring and the schedule of maximum permissible radiation doses.
- *The Radioactive Substances (Carriage by Road) GB Regulations 1940.* These impose requirements on packaging, labelling and procedures involved in the transportation of radioactive substances on roads to which the public has access.
- *The Radioactive Substances (Road Transport Workers) GB Regulations 1970.* These are concerned with the protection of workers engaged in the transportation of radioactive substances. The Code of Practice for the Carriage of Radioactive Materials by Road is intended to assist all concerned to discharge their obligations under the law. Notification containing prescribed information on vehicles employed in regulated transport operations must be given to the local licensing authority.

Radiation protection The basic principles of radiation protection are distance, shielding and time and, in practice, combinations of all three factors are used. The most efficient form of protection is, wherever possible, to use the radiation in a directional beam pointing vertically into the ground. Shielding is either by the use of dense materials like steel or lead or greater thicknesses of cheaper, less dense materials such as concrete and brick. Shielding is usually used in conjunction with distance as the most effective and economical form of protection.

7.10.6.8 Ultrasonic and acoustic testing

Moving out of the electromagnetic spectrum we can look into the frequency spectrum of elastic waves. This is another very fruitful source of NDT techniques since, again, elastic waves can move fairly freely through solid materials and can carry with them information about any imperfections or reflecting interfaces that lie in their path:

<table>
<tr><td></td><td colspan="2">*Acoustic testing*</td><td></td></tr>
<tr><td>*Audible (sonic)*</td><td></td><td>*Ultrasonic testing*</td><td></td></tr>
<tr><td>1 10 100</td><td>1 10 100</td><td>1 10 100</td><td>1 10</td></tr>
<tr><td>Hz</td><td>kHz</td><td>MHz</td><td>GHz</td></tr>
</table>

The techniques either use fairly well-defined beams which can readily be generated in the megahertz ultrasonic range (generally 0.2–25 MHz) or rely on an analysis of vibrations in a lower frequency range generated by setting up forced resonances in a sample, or by analysing the 'white acoustic noise' introduced by striking a sample.

Ultrasonic techniques are now widely used and are proving an extremely versatile complement to radiography since they are particularly sensitive for detecting lamellar and planar flaws (such as forging and lack-of-fusion welding defects) which can be quite invisible to X-rays. However, beam orientation is important and significant lack-of-fusion flaws can also be missed by ultrasonics if they are not orientated

suitably for reflection or if the transducers are not positioned in such a way that the reflected energy can be received.

Ultrasonic energy is 'transduced' to and from electrical voltage signals by ceramic piezoelectric plates built into 'probes' and the resonant frequency of the plate governs the frequency of the ultrasound that is emitted from the probe. The response characteristics, size, damping, pulse length and attachment of these piezoelectric plates in the ultrasonic probe are important and often insufficient attention is paid to this vital aspect of the whole test. Within the metal, the ultrasound pulses can be transmitted in a number of modes and an understanding of their velocity, beam direction and attenuation is essential if the test is to be carried out in an optimum way. 'Compression' and 'shear' waves are used as probing beams for detecting internal flaws, the latter being particularly useful for weld inspection. Rayleigh or 'surface' waves can be generated for locating near-surface flaws and Lamb or 'plate' waves for inspecting sheet, plate and tubing.

The ultrasound pulses travel through materials by transferring energy from atom to atom and thus their velocity depends on the elastic properties and density of the medium. The wavelength of the ultrasound (typically in the range 0.3–3 mm) roughly defines the minimum size of defect that can be detected; this varies quite significantly in different materials.

Flaw *location* is now well established by ultrasonic techniques, by accurately timing the round trip of the pulses. However, flaw *sizing* is less well established but is currently receiving considerable attention in order to back up the quantitative predictions being made by fracture mechanics analyses of critical crack dimensions for brittle fracture. Ultrasonic techniques are also widely used to monitor thickness of sections in order to check for corrosion or erosion. Pulse transit timing is used for thicknesses over 0.5 mm, although for the 0.1–2.5 mm range detecting the frequency at thickness resonance usually gives a more precise measurement.

Much sophistication and improvement has gone into ultrasonic instrumentation over the years and at one end of the range reliable, rugged, portable equipment is now widely available both for flaw location and thickness monitoring. At the other end of the scale automated, modular, instrumentation with computerized digital data processing is available for on-line product quality monitoring and flaw analysis.

Determination of position and size of flaws Before use, a flaw detector is calibrated over a suitable range using an approved calibration block. The position of a flaw is then ascertained by a direct reading in the case of a compression-wave probe, and by simple trigonometry or a specially designed slide rule when using a shear-wave probe. The size of flaws may be estimated by using one or more of the following methods:

- The maximum amplitude technique in which the defect is scanned from as many angles and orientations as possible. At the point of maximum amplitude the beam path and surface distance readings are noted and plotted. The plots give a facsimile outline of the flaw.
- In the 20 dB technique, the beam profile of each probe is plotted out 20 dB each side of the centre of the beam. This beam is then logged on a slider of transparent plastic. A cross-section of the weld or detail of the item under examination is drawn full size onto another piece of plastic. The slider is superimposed onto the latter, duplicating the position of the probe on the item being examined. The origins of any indications are simply identified and the flaws may be sized by noting the position of probe and reflection when the defect reflection falls 20 dB either side of the maximum.

- Ultrasonic velocity measurements are in use for in-field measurement of residual (retained) stresses in structures, and for monitoring process control parameters (e.g. flow rate and fluid concentration).
- In the 'acoustic' range of frequencies there is growing interest in developing techniques for fairly coarse quality assessment based on small changes in resonant frequency when the structure contains a flaw; or by changes in spectral content in a pulse of energy induced by a short sharp hammer blow (acoustic impact testing). At these frequencies there are no directional beam characteristics as such so that large structures can be examined 'globally' as a prelude to a more detailed examination should an anomaly be indicated.

In all the above techniques acoustic or ultrasonic energy is introduced into the sample as a necessary precursor to the inspection. However, the initiation or growth of cracks and relative movement of crack faces can themselves produce elastic waves at sonic or ultrasonic frequencies. A network of piezoelectric detector probes clamped to a structure can locate the position of cracks and continuously monitor their growth over long periods by monitoring this in-built acoustic emission. The main applications of this technique are in pressure vessels and other plant. Problems can arise from background noise and in the interpretation of complex signals.

Another parallel technique for 'health' monitoring of plant is based on analysing the noises and vibrations (over a broad frequency spectrum) emanating from moving machinery and engines and interpreting the information in terms of bearing wear and component alignment. Indeed, plant surveillance and machinery condition monitoring are important potential areas for further development, and new forms of 'health' monitor are continually being sought to meet improved safety and integrity requirements.

7.10.7 NDT method selection

Because of the range and complexity of NDT techniques now available or the subject of current research, it is becoming increasingly important to ensure that an optimum approach is made to any particular inspection requirement. The NDT Method Selector (Table 7.77) relates the techniques referred

Table 7.77 NDT method selector

Main NDT methods worth considering
● Good prospects
○ Some prospects

Test object material	Inspection task	1	2	3	4	5	6	7	8	9	10	11	12	13	14	15	16	17	18	19	20	Other techniques
Metals	Surface opening cracks	●			●	●	●	●			○	○	●	●	●	●		○	○	○		21, 22, 23, 24, 30
	Surface corrosion, pits etc.	●					○	●		○		○	●	●	●		○				○	21, 23, 27
	Severe corrosion thinning	●	○	○			○	●		●	○	○	●	●	●		●				○	23, 28
	Internal cracks							○		○		○	○	●	●	●		○	○			27, 30
	Porosity					○						○	○	●	●	●	●			○		23, 27
	Lack-of-fusion defects									○			○	○	○	●						30
	Internal voids, inclusions					○				○	●	○	●	●	●	○		○				23, 27, 30
	Defect sizing		○		○			●		●			●	●		●			●			27, 30
	Thickness measurement			○	●		○	○		●						●	●			○		23, 29
	Microstructure variation		○					●								○						25, 26, 28
	Stress/strain measurement			○																		21, 22, 25, 29
Coated metals	Coating thickness measurement				○		●			●		○	○	○	○		●					21, 23
	Coating delamination	○					○						○	○	●	●	○					21, 23
	Coating 'pin holes'	○				●																
Composite materials	Delaminations and disbonds	○					○						●	●	●	●	●	●		○		21, 23
	Fibre/matrix ratio evaluation						○						○	○	○							
	Incomplete cure of resin															○						29
	Internal porosity												○	●	○							23
Concrete	Concrete thickness measurement								●	○							○					
	Reinforcing-bar corrosion	○											●		○				○		○	24, 27
Ceramics	Surface cracks	○				●							○	●	○	○		○	○			
	Internal cracks, porosity												○	●	○	●		○	○	○		23

Key to techniques:

1 Visual inspection	11 Neutron radiography	21 Optical holography
2 Thermography	12 X-radiography	22 SPATE
3 Optical metrology	13 Microfocal radiography	23 Pulsed video thermography
4 Liquid penetrant inspection	14 'Real-time' radiography	24 Microwaves
5 Magnetic particle inspection	15 Ultrasonic flaw detection	25 Neutron diffraction and scattering
6 Eddy current testing	16 Ultrasonic thickness gauging	26 Positron annihilation
7 Magnetic flux leakage	17 Acoustic impact testing	27 X-ray tomography
8 Potential drop crack sizing	18 Acoustic emission monitoring	28 Ultrasonic attenuation spectroscopy
9 Radiometry	19 Leak testing	29 Ultrasonic velocity measurement
10 Gamma radiography	20 Plant-condition monitoring	30 Ultrasonic time-of-flight diffraction

From 'Quality Technology' wallchart published by the National NDT Centre in collaboration with the British Institute of NDT.

to in this section to some of the more common inspection problems likely to be encountered in practice. However, professional guidance on the current state-of-art of the techniques and available instrumentation should be sought in the first instance. Two ready sources of such guidance are the Trade Group of the British Institute of NDT in Northampton and the National NDT Centre, operated by AEA Technology at Harwell Laboratory. Advice on the status of appropriate National, European and International Standards which set down the methods of applying many of the techniques are available from the British Standards Institution in London.

Information on the nationally approved and internationally accepted NDT operator approval scheme PCN (Personnel Certification in Non-destructive Testing) is available from the PCN secretariat at the British Institute of NDT in Northampton.

Further up-to-date information on all aspects of NDT is listed in the annual *NDT Yearbook* published by the British Institute of NDT.

7.10.8 Conclusions

NDT has a vital role to play in monitoring and improving the quality of manufactured products and ensuring integrity of fabricated plant. The techniques available are many and varied and there is a range of new methods and equipment likely to emerge from the active research and development now being directed to this subject, anticipating the requirements of new materials, new manufacturing processes and new legislation. NDT is part of a wider management function in mechanical engineering and should be considered and developed with full appreciation of its proper context and wide potential.

Acknowledgements

The author has incorporated and updated some material from the Non-destructive Testing section of the previous 11th edition of this book prepared by Mr J. G. Rees. Some of the information on NDT techniques and the NDT Method Selector have been drawn from the wallchart 'Quality Technology: A Guide to Non-destructive Testing Methods, Services and Information Sources' in the preparation of which the author was associated. Due acknowledgement is given to the National NDT Centre and the British Institute of NDT who jointly distributed the wallchart. The author has also drawn freely from specialist papers on NDT subjects that he personally authored or presented at conferences when employed by AEA Technology.

References

1 Gillam, E., *Metallurgist and Material Technologist*, **9** 521–525 (1979): Crane, F. A. A. and Charles, J. A., *The Selection and Use of Engineering Materials*, Butterworths, London (1984)
2 Edeleanu, G., 'Information requirements and the chemical industry', *Metals and Materials*, **3**, No. 3, 43–44 (March 1987)
3 Turner, M. E. D., 'Materials for the process industry', *Metals and Materials*, **3**, No. 3, 136–139 (March 1987)
4 *Proceedings* of a Workshop held at Fairfield Glade, Tennessee, November 1982: Steering Committee of the Computerized Materials Data Workshop (1983): *Proceedings* of the 9th International CODATA Conference Jerusalem, June 1984, North-Holland, Amsterdam (to be published): *Proceedings* of CEC Workshop of Factual Materials Data Banks, Petten, The Netherlands (to be published)
5 Hampel, V. E., Bollinger, W. A., Gayner, C. A. and Oldani, J. J., UCRL Report No. 90942 (June 1984)
6 Breen, D. H., Walker, G. H. and Sponzill, J. T., *Metal Progress*, **103**, 83–88 (1973)
7 Farag, M. M., *Materials and Process Selection in Engineering*, Applied Science Publishers, London (1979)
8 Matsel Systems Ltd, 14 Mere Farm Road, Birkenhead, Merseyside, L43 9TT, UK
9 Swindells, N. and Swindells, R. S., 'System for engineering materials selection', *Metals and Materials*, **1**, No. 5, 301–304 (May 1985)
10 Dimaid, A. and Zucker, J. J., 'A conceptual model for materials selection', *Metals and Materials*, **4**, No. 5, 291–296 (May 1988)
11 *The SAE Handbook* (current edition)
12 *Iron and Steel Specifications*, 5th edn, British Steel
13 The Institute of Metals, *Secondary Steelmaking for Product Improvement*, London (1985)
14 Plockinger, M. E. and Etterich, O., *Electric Furnace Steel Production*, John Wiley, Chichester (1985)
15 The Institute of Metals, *Continuous Casting*, London (1982)
16 The Institute of Metals, *Continuous Casting*, London (1985)
17 BSC Plates Steel Specification Comparisons: Part 1, Structured Steels; Part 2, Pressure Vessel Steels
18 The Mond Nickel Co., *Transformation Characteristics of Direct Hardening Nickel Alloy Steels*, 3rd edn
19 The Institute of Metals, *Stainless Steels '84* (1985)
20 West, E. G., *Copper and its Alloys*, Ellis Horwood, Chichester (1982)
21 Uphegrove, C. and Burghoff, H. L., *Elevated Temperature Properties of Coppers and Copper-Base Alloys*, ASTM Special Publication No. 181 (1956)
22 Smith, C. S., 'Mechanical properties of copper and its alloys at low temperatures – a review', *Proc. ASTM*, **39**, 642–648 (1939)
23 Dawson, R. J. C., *Fusion Welding and Brazing of Copper and Copper Alloys*, Butterworths, London (1973)
24 *The Properties of Aluminium and its Alloys*, Aluminium Federation (1983)
25 Woodward, A. R., *The Use of Aluminium for Stressed Components*, *Selection of Materials in Machine Design*, I. Mech. E. Conference Publication 22 (1973)
26 *Key to Aluminium Alloys – Designations, Compositions and Trade Names of Aluminium Materials*, compiled by W. Hufnagel, Aluminium-Zentrale (1982)
27 Evans, B., McDarmaid, D. S. and Peel, C. J., The evaluation of the properties of improved 'aluminium–lithium alloys for aerospace applications', Paper to SAMPE Conference, Montreux, June 1984, from RAE Farnborough.
28 Grimes, R., Cornish, A.J., Miller, W. S. and Reynolds, M. A., 'Aluminium–lithium based alloys for aerospace applications', *Metals and Materials*, **1**, No. 6, June, 357–363 (1985)
29 Peel, C. J., Evans, B. and McDarmaid, D. S., 'Development of aluminium–lithium alloys in the U.K.', *Metals and Materials*, **3**, No. 8, August, 449–455 (1987)
30 Lavington, M. H., 'The Cosworth Process – a new concept in aluminium alloy casting production', *Metals and Materials*, **2**, No. 11, November, 713–719 (1986). The Cosworth Process may be licensed through the International Mechanite Co.
31 Cottrell, A. H., *Creep Buckling. The Mechanical Properties of Matter*, John Wiley, Chichester (1964)
32 Lutjering, G., Zwicher, U. and Burrk, W., *Titanium Science and Technology*, Deutsche Gesellschaft für Metallkunde EV (1984)
33 Beck, A., *The Technology of Magnesium and its Alloys*, Magnesium Elektron (1940)
34 Alico, J., *Introduction to Magnesium and its Alloys* (1945)
35 Unsworth, W., 'Developments in magnesium alloys for casting applications', *Metals and Materials*, **4**, No. 2, February (1988)
36 Unsworth, W., 'Meeting the high temperature aerospace challenge', *Light Metal Age*, August (1986)
37 Grenfield, P., *Engineering Applications of Magnesium*, Mills and Boon, London (1972)
38 Ray, M. S., *The Technology and Applications of Engineering Materials*, Prentice-Hall International, Hemel Hempstead, UK (1987)
39 Driver, D., 'Developments in aero engine materials, *Metals Materials*, **1**, No. 6, June (1985)

40 White, C. H., *The Development of Gas Turbine Materials*, Applied Science, London (1981)
41 Bettridge, W. and Heslop, J., *The Nimonic Alloys*, 2nd edn, Edward Arnold, London (1974)
42 Beeley, P. R. and Driver, D., *Metals Forum*, **7**, 146 (1984)
43 Wallace, W., *Metal Science*, **9**, 547 (1975)
44 Ashdown, C. P. and Grey, D. A., *Metal Science*, **13**, 627 (1979)
45 Meetham, G. W., *Metallurgist and Mater. Tech.*, **9**, 387 (1982)
46 Beeley, P. R. and Driver, D., *Metals Forum*, **7**, 146 (1984)
47 Sims, C. T. and Hagel, W. E. (eds), *The Superalloys*, John Wiley, New York (1972)
48 McLean, M., *Directionally Solidified Materials for High Temperature Service*, The Institute of Metals
49 Engineering Properties of Zinc Alloys, International Lead Zinc Research Organization, Inc. (ILZRO) (April 1981)
50 Morgan, S. W. K., *Zinc and its Alloys*, Macdonald and Evans, Plymouth (1972)
51 Barber, N. I. and Jones, P. E. 'A new family of foundry alloys', *Foundry Trade Journal*, 17 January (1980)
52 Gervais, E., Levent, H. and Bess, M., 'Development of a family of zinc base foundry alloys', *84th Casting Congress of the American Foundrymen's Society*, St Louis, Missouri, April 1950
53 Lyon, R., 'High strength zinc alloys for engineering applications in the motor car', *Metals and Materials*, January, 55–57 (1985)
54 Carvis, J. H. and Gilbert, P. T., *The Technology of Heavy Non-Ferrous Metals and Alloys: Copper, Zinc, Tin and Lead*, Newnes, London (1967)
55 A general account of cobalt alloys can be found in Cobalt Monograph 1960, Centre d'Information du Cobalt, Brussels
56 A full account of the cobalt superalloys is to be found in reference 47 and the Further Reading at the end of this chapter
57 Simon, E. W., *Guide to Uncommon Metals*, Frederick Muller, London (1967)
58 McGachie, R. O. and Bradley, A. G., *Precious Metals*, Pergamon Press, Oxford (1981)
59 Wickers, R. R., *Newer Engineering Materials*, Macmillan, London (1969)
60 Harwood, J. S., 'The metal molybdenum', Symposium *Proceedings* of the ASM, Cleveland, Ohio (1958)
61 Sully, A. H. and Brandes, E. A., *Chromium*, Butterworths, London (1967)
62 Miller, G. L., *Zirconium*, Butterworths, London (1957)
63 Jones, A., *Mechanics of Composite Materials*, Scripto Book Co. (1975)
64 Kelly, A. and Mileiko, S. T., *Fabrication of Composites*, North-Holland, Amsterdam (1983). This book is part of a series intended to cover all aspects of composites
65 Jayatilaka, A. de S., *Fracture of Engineering Brittle Materials*, Applied Science Publishers, London (1979). This book deals with the mechanisms whereby fibre reinforcement improves strength and, sometimes, toughness
66 Birchall, J. D., Howard, A. J. and Kendall, K., 'Flexural strength and porosity of cements', *Nature*, **289**, 288–289 (1981)
67 Jackson, A. P., Vincent, J. R. F. and Turner, R. M., 'The mechanical design of nacre', *Proc. Roy. Soc. Lond.*, **B234**, 415–440 (1988)
68 The *Modern Plastics Encyclopaedia* is distributed free to subscribers to *Modern Plastics*, a McGraw-Hill publication
69 'Kompas' is a classified list of companies and suppliers for each country published by Reed Information Services in association with the CBI
70 Brydson, J. A., *Plastics Materials*, 5th edn, Butterworth-Heinemann, Oxford (1989)
71 The Rubber and Plastics Research Association of Great Britain, Shawbury, Shrewsbury SY4 4NR
72 British Plastics Federation, 47 Piccadilly, London W1V 0DN
73 Hardy, D. V. N. and Megson, N. J. L., *Quant Rev.* (*London*), 25 (1948)
74 Beck, R. D., *Plastic Product Design*, Butterworths, London (1970)
75 Page, R. G. W., *Injection Mould Design*, Butterworths, London (1968)
76 Bebb, R. H., *Plastics Mould Design*, Vol. 1 *Compression and transfer moulds*, Butterworths, London
77 Muhr, A. H. and Thomas, A. G., 'Allowing for non-linear stress–strain relationships of rubber in force deformation calculations, Part II: Relationship of hardness to modulus', *NR Technology*, **20**, No. 2, 27–32 (1989)
78 Treloar, L. R. G., *The Physics of Rubber Elasticity*, 3rd edn, Oxford University Press, Oxford (1975)
79 Oden, J. T., *Finite Elements of Nonlinear Continua*, McGraw-Hill, New York (1972)
80 Gent, A. N., 'Relaxation processes in vulcanized rubber, Part I', *J. Appl. Polym. Sci.*, **6**, 433–441 (1962)
81 Turner, D. M., 'A trioelastic model for the mechanical behaviour of rubber', *Plastics and Rubber Processing and Applications*, **9**, 197–201 (1988)
82 Koh, C. G. and Kelly, J. M., 'Application of fractional derivatives to seismic analysis of base-isolated models', *Earthquake Engineering and Structural Dynamics*, **19**, 229–241 (1990)
83 Adkins, J. E. and Gent, A. N., 'Load-deflection relations of rubber bush mountings', *British J. Appl. Phys.*, **5**, 354–358 (1954)
84 Göbel, E. F., *Rubber Springs Design*, Newnes-Butterworths, London (1974)
85 Hill, J. M., 'Radial deflections of rubber bush mountings of finite lengths' *Int. J. Engineering Science*, **13**, 407–423 (1975)
86 Thomas, A. G., 'Design of laminated bearings – I', *Proceedings* of Conference on NR for Earthquake Protection of Buildings, Malaysian Rubber Research and Development Board, Kuala Lumpur, 1983
87 Thomas, A. G., 'A novel design of rubber spring', International Rubber Conference, Rubber Research Institute of Malaysia, Kuala Lumpur, 1985
88 Schwarzl, F. R. and Struick, L. C. E., 'Analysis of relaxation measurements', *Advances in Molecular Relaxation Processes*, **1**, 201–255 (1967/1968)
89 Coveney, V. A., 'Earthquake base isolation – past, present and future', *Progress in Rubber and Plastics Technology*, **7**, No. 4, 298–307 (1991)
90 Morrel, R., *Handbook of Properties of Technical and Engineering Ceramics*, Parts 1 and 2, National Physical Laboratory
91 Davidge, R. W., *Mechanical Behaviour of Ceramics*, Cambridge University Press, Cambridge (1979)
92 Shook, W. B., Critical Survey of Mechanical Property Test Methods for Brittle Materials, Technical Report ASD-TDR-63-491, AD 417620 (1963)
93 *The Fulmer Optimizer*, The Fulmer Institute

Further reading

The Fulmer Optimizer (current edition available from the Fulmer Institute; new edition to be published by Elsevier) – materials
The ASM Metals Handbook (current edition) – metals and material properties
The Metals Reference Book (Butterworth-Heinemann, current edition) – metals

Section 7.3

The following organizations publish information on all aspects of their subjects: The Welding Institute, Abington, Cambs; The Steel Castings Research and Trade Association, 5 East Bank Road, Sheffield; The Drop Forging Research Association; The British Cast Iron Research Association. Some (but not all) of this information is restricted to their members

Section 7.4

More than 20 brochures, giving technical data, properties, manufacturing procedures, applications and suppliers of copper alloys are available without charge from The Copper Development Association, Orchard House, Mutton Lane, Potters Bar, Herts EN6 7AP

Magnesium Elektron Ltd, Royal House, London Road, Twickenham TW1 3QA, publishes brochures detailing the properties of the principal magnesium alloys

IMI, PO Box 704, Witton, Birmingham B6 7UR, publishes a series of brochures listing the properties, fabrication procedures and applications of their standard range of titanium alloys

Inco Alloys International, Holmer Road, Hereford, publish free of charge a series of brochures on the properties and applications of nickel alloys

The Wear Technology Division of the Cabot Corporation (Deloro Stellite), Stratton St Margaret, Swindon, Wilts, issue publications on their 'Deloro' series of wear-resistant cobalt-base alloys and 'Triballoy' intermetallic compounds for wear-resistant coatings

Wall Colmonoy Ltd, Pontardawe, West Glamorgan, publish brochures listing their nickel alloys for wear-resistant coatings

The Lead Development Association, 3 Berkeley Square, London W1X 6AS, publishes a number of brochures on lead and its alloys

The International Tin Research Institute, Fraser Road, Perivale UB6 7AQ, will provide information on the properties of tin and its alloys

Section 7.5

Metals and Materials, Volume 2, Nos 4, 6, 7, 9, 10 and 12, contain a series of articles, contributed by members of the Materials Development Division of the AERE, dealing with all aspects of filamentary and short-fibre composites (with the exception of cement-based composites). Volume 4, No. 5 contains additional information on filament winding techniques.

Volume 3, No. 11 deals with some aspects of automotive applications of composites and Volume 4, No. 7 with aircraft applications.

Volume 4, No. 9 contains two articles on carbon–carbon composites. Volume 2, No. 3 has an article on metal matrix composites

Lubin, G., *Handbook of Composites*, Van Nostrand Reinhold, New York (1982)

Section 7.7

American Society for Testing and Materials, Standard Classification Systems for Rubber Products in Automotive Applications, ASTM D2000

Bhowmick, A. K. and Stephens, H. L. (eds), *Handbook of Elastomers – New Developments and Technology*, Marcel Dekker, New York (1988)

Blow, C. M. and Hepburn, C. (eds), *Rubber Technology and Manufacture*, Butterworth-Heinemann, Guildford (1987)

Brydson, J. A., *Rubbery Materials and their Compounds*, Elsevier, London (1988)

Ferry, J. D., *Viscoelastic Properties of Polymers*, 3rd edn, John Wiley, New York (1980)

Freakley, P. K. and Payne, A. R., *Theory and Practice of Engineering with Rubber*, Applied Science, London (1978)

Fuller, K. N. G., Gregory, M. J., Harris, J. A., Muhr, A. H., Roberts, A. D. and Stevenson, A., 'Engineering use of natural rubber', in Roberts, A. D. (ed.), *Natural Rubber Science and Technology*, Oxford University Press, Oxford (1988)

Hepburn, C. and Reynolds, R. J. W. (eds), *Elastomers: Criteria for Engineering Design*, Applied Science, London (1979)

International Organization for Standardization, *Rubber Materials – Chemical Resistance*, ISO TR 7620 (1986)

Lindley, P. B., *Engineering Design with Natural Rubber*, Malaysian Rubber Producers Research Association (MRPRA), Hertford, UK

Malaysian Rubber Producers Association, *Natural Rubber Engineering Data Sheets*, MRPRA, Hertford, UK (1979)

Morton, M. (ed.), *Rubber Technology*, 3rd edn, Van Nostrand Reinhold, New York (1987)

Murray, R. M. and Thompson, D. C., *The Neoprenes* (International edition), E. I. DuPont de Nemours & Co., Wilmington, Delaware (1963)

Rader, C. P. and Stemper, J., 'Thermoplastic elastomers – A major innovation in rubber', *Prog. Rubber Plast. Technol.*, **6**, No. 1, 50–99 (1990)

Snowdon, J. C., *Vibration and Shock in Damped Mechanical Systems*, John Wiley, New York (1968)

Section 7.8

Harris, J. E., 'Oxidation-induced deformation and fracture', *Proc. 6th International Conference on Fracture* (ICF6), New Delhi, India (1984)

Kirk, J. N., 'Ceramic components in automotive applications', *Metals and Materials*, **3**, No. 11, 647–652 (1987)

Mass, M., 'Paint finishing of plastics', *SITEV 81*

National Engineering Laboratory, *Engineering Ceramics as Applied to Reciprocating Engines* (1987)

Section 7.10

Current papers on the application of NDT in engineering practice appear in the regular issues of:

The British Journal of NDT (British Institute of NDT)
The European Journal of NDT (British Institute of NDT)
NDT & E International (Butterworth-Heinemann)
Ultrasonics (Butterworth-Heinemann)
Materials Evaluation (American Society for NDT)

The International Institute of Welding publishes on a variety of NDT subjects (Radiography, Ultrasonic testing, Magnetic testing, Residual stress monitoring, Offshore NDT). Also Guidance Document SST-1157-90, Assessment of the Fitness-for-purpose of Welded Structures

American Society for Metals, *Non-destructive Inspection and Quality Control*, Volume 11 in the *Metals Handbook* series

American Society for NDT, *NDT Handbook* (7 volumes) (1985–1991)

British Institute of NDT, *The Capabilities and Limitations of NDT* (8 parts)

British Institute of NDT, *NDT Annual Year Book*

British Standards Year Book, BSI, Milton Keynes

Halmshaw, R., *Industrial Radiology Techniques*, Wykeham Publications, London (1982)

Höller, P. *et al.* (eds), *Non-destructive Characterisation of Materials*, Springer-Verlag, New York

Krautkramer, J. and Krautkramer, H., *Ultrasonic Testing of Materials*, 4th edn, Springer-Verlag, New York (1990)

Rao, R. *et al.* (eds), *Condition Monitoring and Diagnostic Engineering Management*, Chapman and Hall, London (1990)

Sharpe, R. S. (ed.), *Research Techniques in NDT* (8 volumes) Academic Press, New York (1970–1985)

8

Mechanics of solids

Peter Myler
(Sections 8.1 and 8.2)

Leslie M. Wyatt
(Sections 8.3–8.5)

Contents

8.1 Stress and strain 8/3
    8.1.1 Fundamental definitions 8/3
    8.1.2 Linear elasticity 8/4
    8.1.3 Stress systems for isotropic materials 8/4
    8.1.4 Plane stress system 8/4
    8.1.5 Compliance relationship 8/7
    8.1.6 Stress concentrations 8/7
    8.1.7 Impact stresses in bars and beams 8/8
    8.1.8 Orthotropic material 8/8
    8.1.9 Plasticity 8/10

8.2 Experimental techniques 8/12
    8.2.1 Strain gauges 8/12
    8.2.2 Basic principles 8/13
    8.2.3 Gauge factor 8/13
    8.2.4 Strain gauge arrangements 8/14
    8.2.5 Photoelasticity 8/16
    8.2.6 Holography 8/18
    8.2.7 Thermo-elastic analysis 8/18
    8.2.8 Brittle-coating 8/19
    8.2.9 X-ray analysis 8/19

8.3 Fracture mechanics 8/19
    8.3.1 Introduction 8/19
    8.3.2 Linear elastic fracture mechanics 8/19
    8.3.3 Fracture toughness testing 8/19
    8.3.4 Influence of shape of defect 8/19
    8.3.5 Typical ranges of fracture toughness values 8/21
    8.3.6 Post-yield fracture mechanics 8/21

8.4 Creep of materials 8/23
    8.4.1 Introduction 8/23
    8.4.2 Creep and stress rupture testing 8/23
    8.4.3 Deformation mechanisms and laws 8/23
    8.4.4 The interpretation of creep and stress rupture
        data 8/25
8.4.5 Parameters favouring high creep and rupture
    strengths 8/26
    8.4.6 Design in the creep range 8/26
    8.4.7 Creep of ferritic pressure vessel steels 8/26
    8.4.8 Remanent life 8/30

8.5 Fatigue 8/31
    8.5.1 Introduction 8/31
    8.5.2 Fatigue crack initiation 8/31
    8.5.3 Fatigue crack propagation 8/32
    8.5.4 Unstable crack growth 8/32
    8.5.5 Final specimen failure 8/32
    8.5.6 Parameters that influence fatigue
        performance 8/32
    8.5.7 Fatigue testing 8/37
    8.5.8 Design for fatigue 8/37
    8.5.9 Parameters and symbols employed 8/40

References 8/40

Further reading 8/42

8.1 Stress and strain

8.1.1 Fundamental definitions

8.1.1.1 Direct stress σ

The level of direct stress at a point within a loaded body may be considered analogous to the pressure acting in a fluid. It is the measure of the level to which the bonds that hold a structure are being pushed closer (compression – defined negative) or pulled apart (tension – defined positive) (see Figure 8.1). It is defined as the normal force per unit area, acting at a point within a material and not associated with any specific area.

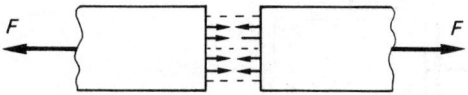

Figure 8.1   Direct stress acting on a section in a bar

8.1.1.2 Shear stress τ

This is again a point quantity and is measured in units of force per unit area. It measures the level by which the bonds are translated with respect to each other. In reality, there is no sign convention for shear stress since there is no physical difference between shearing from right to left or left to right (see Figure 8.2).

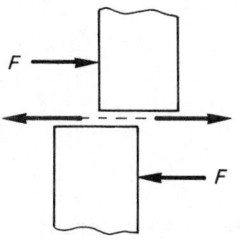

Figure 8.2   Shearing action at a section

8.1.1.3 Strain ε

Strain is a measure of how far apart the bonds are pulled, pushed or translated with respect to each other. Strains are usually quoted as dimensionless quantities with respect to some length dimension.

Engineering strain is that strain which is dimensionalized with respect to the original shape before any strain is induced into the material (see Figure 8.3), i.e.

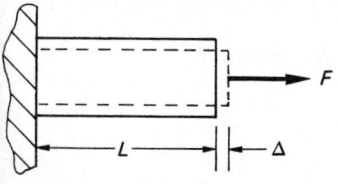

Figure 8.3   Deformation of a loaded bar

$$\epsilon = \frac{\Delta}{L} \tag{8.1}$$

Strain values for materials such as steel and aluminium are very low in the working environment (usually $\pm 1 \times 10^{-3}$).

8.1.1.4 Poisson's Ratio ν

This ratio is that of lateral strain to longitudinal strain under conditions of uniaxial longitudinal stress.

8.1.1.5 Elasticity

A material may be considered to behave elastically if, when the force/stress is applied to it, the resulting deformation/strain appears instantaneously; and if, when the forces are removed, the material returns to its initial shape.

8.1.1.6 Modulus of elasticity – Young's Modulus E

If under elastic conditions there is a linear relationship between stress and strain, the ratio of the induced stress to its corresponding strain is defined as Young's Modulus. Materials with a linear relationship between stress and strain are termed Hookean.

8.1.1.7 Plasticity

This is the property of sustaining appreciable permanent deformation without rupture. Materials such as steel and cast iron when stressed beyond the elastic limit become partially plastic, the degree of plasticity growing with increased stress.

8.1.1.8 Yield point

This is the lowest stress at which strain increases without an increase in stress. Only a few materials exhibit a true yield point; for other materials the point is used as a transition point.[1]

8.1.1.9 True stress

When dealing with stresses in the plastic condition the stresses must be calculated using the current deformed state of the material and not on the original geometry as for elastic behaviour.

8.1.1.10 True strain – natural strain or logarithmic strain e

As with true stresses, plastic strains are based on the current deformed geometry and not on the original geometry. It is equal to $\log_e(1 + \epsilon)$, where ϵ is the engineering strain.

8.1.1.11 Dynamic stresses

These are associated with any system where loads applied to a component are time dependent. They include creep, fatigue, impact and relaxation stresses.

8.1.1.12 Creep stresses

When the strain varies progressively with time under constant or decreasing stress the material is under a state of creep. This phenomenon usually occurs at elevated temperatures.

8.1.1.13 Fatigue stresses

Stresses that vary periodically with time (cyclic) are classed as being under a fatiguing environment. Fracture under fatigue can occur at stresses much lower than the levels achieved under steadily increasing load conditions.

8.1.1.14 Impact stresses

Components which are subjected to transient loading have shock or stress waves induced into them, provided the duration of the load is of the same order of magnitude as the natural period of vibration of the component.

8.1.2 Linear elasticity

The most common form of stress analysis used in engineering deals with material behaviour that is said to be linear elastic, hence conforming to Hooke's law, which states that strain is linearly proportional to stress.

The constant of proportionality is termed Young's Modulus or the Modulus of Elasticity. When the strength and elastic properties are the same in any direction the material is said to be isotropic. If the strength and elasticity are different for different directions the material is said to be anisotropic.

Hookean behaviour can be assumed for most common engineering materials up to a specific strain level, and mathematical theories can be applied with an acceptable level of accuracy for engineering approximations. A level of caution must be used when detailing certain materials such as wood and cement where the material's response is dependent upon the rate of loading. The apparent Young's Modulus must also be carefully chosen for such applications.

thickness direction. In other words, the thickness dimension is an order of magnitude smaller than any other dimension.

The state of stress acting in a body at a point can be described by the system acting in the xy plane shown in Figure 8.5(a). By taking moments about a corner of the element, it can be shown, ignoring secondary terms, that $\tau_{xy} = \tau_{yx}$. A state of plane stress can then be expressed by a combination of three stresses σ_{xx}, σ_{yy} and τ_{xy}. The above system describes the general state of stress at a point. By taking a plane at an angle θ relative to the xy system (Figure 8.5(b)) another system

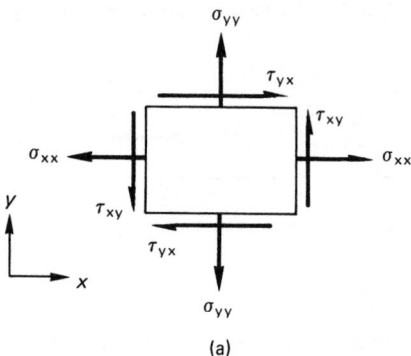

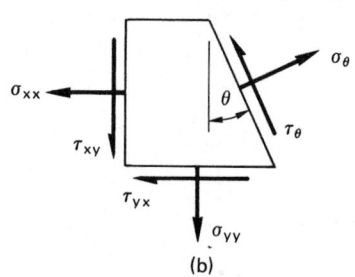

Figure 8.5 (a) Plane stress representation; (b) stresses acting on a plane inclined at $\theta°$ to the xy plane

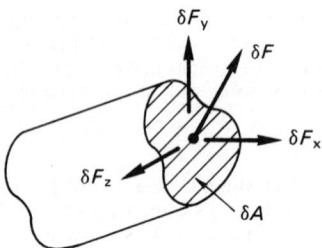

Figure 8.4 Resolution of force δF along cartesian axes

8.1.3 Stress systems for isotropic materials

Consider a point in a material subjected to a force δF acting on a plane (area) δA. If the force is resolved into three mutually perpendicular planes giving three forces δF_x, δF_y and δF_z and using the definition of stress in Section 8.1.1, three stresses can be identified:

$$\lim_{\delta A \to 0} \frac{\delta F_x}{\delta A} = \sigma_{xx}, \quad \lim_{\delta A \to 0} \frac{\delta F_y}{\delta A} = \sigma_{xy} \tag{8.2}$$

$$\lim_{\delta A \to 0} \frac{\delta F_z}{\delta A} = \sigma_{xz}$$

The first suffix gives the direction of the normal to the area on which the stress acts and the second the direction of the stress with respect to the plane (Figure 8.4)

8.1.4 Plane stress system

Many engineering systems can be regarded as or approximated to a plane stress system (e.g. plates and shell). A plane stress system is where one of the three mutually perpendicular stresses σ_{xx}, σ_{yy} and σ_{zz} is taken to be constant or zero. This state of stress occurs when a component is 'thin' in the

acting on the θ plane can be expressed in terms of the three stresses acting in the xy plane. From equilibrium,

$$\sigma_\theta = \frac{\sigma_{xx} + \sigma_{yy}}{2} + \frac{\sigma_x - \sigma_y}{2} \sin2\theta + \tau_{xy} \sin2\theta \tag{8.3}$$

$$\tau_\theta = \tau_{xy} \cos2\theta - \frac{\sigma_x - \sigma_y}{2} \sin2\theta \tag{8.4}$$

Equations (8.3) and (8.4) describe the stresses as a function of θ. The maximum and minimum stresses can be obtained by differentiating equation (8.3) and equating this to zero, i.e.

$$\frac{\mathrm{d}\sigma_\theta}{\mathrm{d}\theta} = -\sigma_x - \sigma_y \sin2\theta + 2\tau_{xy} \cos2\theta = 0$$

Hence

$$\tan2\theta = \frac{2\tau_{xy}}{\sigma_x - \sigma_y} \tag{8.5}$$

Substituting this angle into equation (8.3) gives the maximum and minimum stress as:

$$\sigma_1 = \frac{\sigma_{xx} + \sigma_{yy}}{2} + \sqrt{\left[\left(\frac{\sigma_{xx} - \sigma_{yy}}{2}\right)^2 + \tau_{xy}^2\right]}$$

(8.6)

$$\sigma_2 = \frac{\sigma_{xx} + \sigma_{yy}}{2} - \sqrt{\left[\left(\frac{\sigma_{xx} - \sigma_{yy}}{2}\right)^2 + \tau_{xy}^2\right]}$$

Occurring at an angle

$$\theta = \frac{1}{2}\tan^{-1}\left(\frac{2\tau_{xy}}{\sigma_{xx} - \sigma_{yy}}\right)$$

(8.7)

Equations (8.3) and (8.4) can be used in a very useful form graphically known as the Mohr's stress circle. If a graph is plotted whose ordinate is the shear stress and the abscissa the normal stress the locus of a stresses can be plotted with the circle centre located on the horizontal axis at $(\sigma_{xx} + \sigma_{yy})/2$ and whose radius is

$$\sqrt{\left[\left(\frac{\sigma_{xx} - \sigma_{yy}}{2}\right)^2 + \tau_{xy}^2\right]}$$

All angles associated with the circle are double those acting on the actual elements.

A positive shear pair produces a shear couple acting on an element (Figure 8.6). If the couple tends to rotate the element clockwise this is defined as a positive shear pair and is plotted vertically on the circle.

Figure 8.6 Shear couple notation

As an example, consider the state of stress shown in Figure 8.7. By inspection: $\sigma_{xx} = 45$ N/mm², $\sigma_{yy} = -15$ N/mm² and $\tau_{xy} = 40$ N/mm². Using these values the circle can be constructed in stages as outlined below.

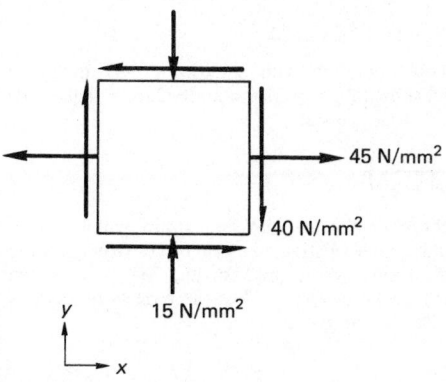

Figure 8.7 Stressed element example

Two points are plotted in $\sigma = \tau$ space using the shear notation (Figure 8.8). Then the centre of the circle is obtained and, using Pythagoras, the circle radius calculated (Figure 8.9). The circle can now be constructed, and where the circle intercepts the σ axis the maximum and minimum stresses are given. These stresses are termed principal stresses. The direction of these stresses can be obtained from the circle by following the direction given on the circle. For the above case σ_1 is acting at angle θ clockwise relative to the x direction as in Figures 8.10 and 8.11. The maximum shear stress is given by the radius of the circle, and this occurs on a plane at $\pm 45°$ from the principal planes, i.e. $\pm 90°$ on the circle. In this particular example, $\sigma_1 = 65$ N/mm², $\sigma_2 = -35$ N/mm² and $\tau_{max} = 50$ N/mm²; $\theta = 35°$ cw from the x axis.

In a two-dimensional stress field several sets of lines describing the state of stress are used in producing information about the stress distribution within a component. The most common terms are:

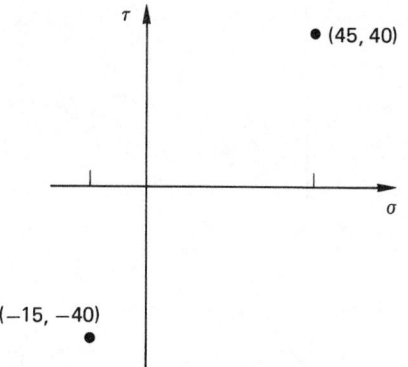

Figure 8.8 Mohr's circle coordinates

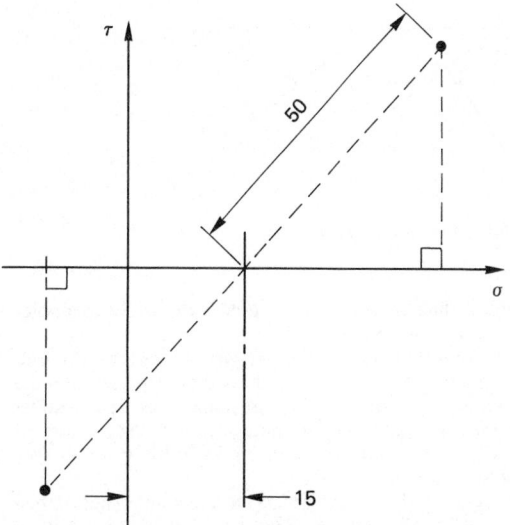

Figure 8.9 Mohr's circle centre and radius

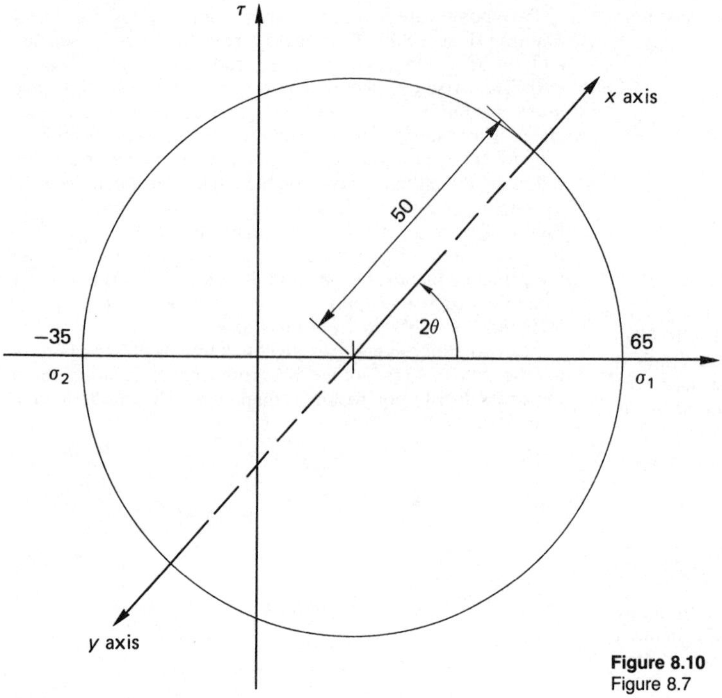

τ

x axis

50

-35

σ_2

2θ

65

σ_1

y axis

Figure 8.10 Completed Mohr's circle for the stressed element in Figure 8.7

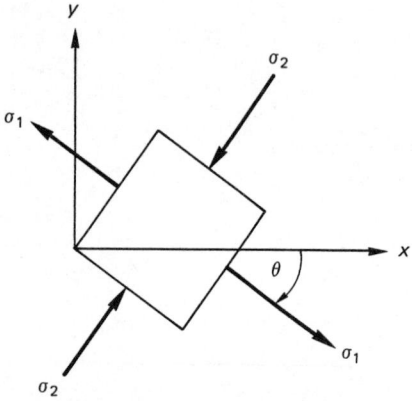

Figure 8.11 Principal stress directions

Isoclinic: A line of constant, θ, used extensively in photoelastic analysis (see Section 8.2.2);
Isostatic: An orthogonal network of curves, one set representing maximum principal stress, the other representing the minimum principal stress. These are sometimes termed stress trajectories and are used in finite element stress plots or observed in crack patterns occurring in brittle coatings (see Section 8.2.8);
Isochromatic: A line along which the maximum shear stress $(\alpha[\sigma_1 - \sigma_2])$ is constant, so called because this represents a constant colour band in photoelasticity. This is also useful in estimating ductile failure;
Isopachic: A line along which the sum of the principal stresses, $\sigma_1 + \sigma_2$, is constant. These are usually associated with field problems.

While stress is an engineering concept and not measurable, engineering strains which are associated with body deformations can be measured. In most applications elastic strains are small in magnitude (between $\pm 1 \times 10^{-3}$) and are measured by techniques discussed in Section 8.2.1.

Using the notation in Section 8.1.2, the representation of strain in any plane can be expressed similarly to those obtained for stresses. Consider Figure 8.12, which shows a small element within a component subjected to a direct stress, shearing stress and the induced strains. The normal and shear strain acting on any plane θ are given by

$$\epsilon_\theta = \frac{\epsilon_x + \epsilon_y}{2} + \frac{\epsilon_x - \epsilon_y}{2}\cos2\theta + \frac{\gamma_{xy}}{2}\sin2\theta \tag{8.8}$$

$$\gamma_\theta = \gamma_{xy}\cos2\theta - (\epsilon_x - \epsilon_y)\sin2\theta$$

Comparing these equations with equations (8.3) and (8.4) there is a direct substitution of ϵ for σ and $\gamma/2$ for τ. This leads to the principal strain axes at

$$\theta = \frac{1}{2}\tan^{-1}\left(\frac{\gamma_{xy}}{\epsilon_x - \epsilon_y}\right) \tag{8.9}$$

Relative to these planes the shearing strain will be zero. Further, a Mohr's circle of strain can also be constructed using ϵ (the normal strain) as the abscissa and $\gamma/2$ (half of the shearing strain) as the ordinate. Thus the circle will have its centre on the horizontal axis at

$$\frac{\epsilon_{xx} + \epsilon_{yy}}{2}$$

and a radius of

$$\sqrt{\left[\left(\frac{\epsilon_x - \epsilon_y}{2}\right)^2 + \left(\frac{\gamma_{xy}}{2}\right)^2\right]}$$

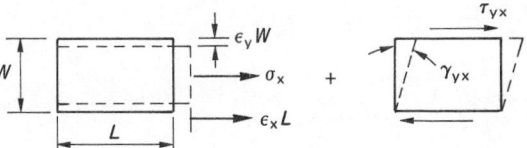

Figure 8.12 Strain notation

In determining principal strains at a point in a component, equation (8.8) is often used with three strain-measuring elements (see Section 8.2.1) attached at angles θ_1, θ_2, θ_3 such that

$$\epsilon_{\theta i} = \frac{\epsilon_x + \epsilon_y}{2} + \frac{\epsilon_x + \epsilon_y}{2} \cos 2\theta i + \frac{\gamma_{xy}}{2} \sin 2\theta i \qquad (8.10)$$

where $i = 1, \ldots, 3$

Choices of angles are usually 0, 45, 90 or 0, 60, 120 for most commercial applications. Using the three measured strains in a known direction the Mohr's circle can be constructed and hence the principal strains evaluated. If knowledge of the principal directions is available a two-element system may be used. A 0, 60, 120 unit can be represented by the Mohr's circle shown in Figure 8.13.

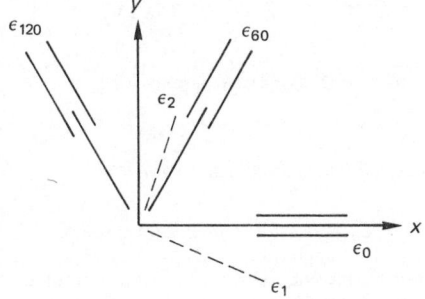

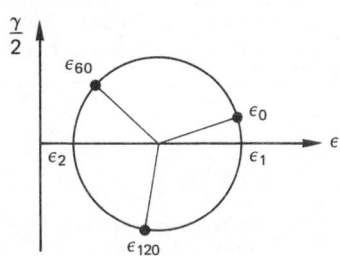

Figure 8.13 Mohr's circle of strain for a 0, 60, 120 rosette

8.1.5 Compliance relationship

Two important phenomena are associated with homogeneous, isotropic materials in the relationship between stress and strain at a point:

1. Hooke's law in one dimension, which states the proportionality between uniaxial stress and strain in the same direction, i.e. $\sigma_x = E \epsilon_x$, the constant of proportionality E termed Young's Modulus or elastic modulus.
2. The Poisson effect, i.e. the observation that a stress in one direction induces not only strain in that direction but also strains in the other two orthogonal directions, $\epsilon_y = \epsilon_z = -\nu \epsilon_x$, where the constant of proportionality, ν, is known as Poisson's ratio.

Using the two effects, a generalized expression of Hooke's law in three dimensions can be obtained by simple superposition, e.g.:

$$E \epsilon_x = \sigma_x - \nu \sigma_y - \nu \sigma_z$$
$$E \epsilon_y = \sigma_y - \nu \sigma_z - \nu \sigma_x \qquad (8.11)$$
$$E \epsilon_z = \sigma_z - \nu \sigma_x - \nu \sigma_y$$

Expression (8.11) can be rearranged in terms of stress in the following format:

$$\sigma_z = \frac{1 - \nu}{1 + \nu} \frac{E}{1 - 2\nu} \left\{ \epsilon_z + \frac{\nu}{1 - \nu} (\epsilon_y + \epsilon_x) \right\} \qquad (8.12)$$

These expressions may be reduced for plane stress problems to

$$\sigma_x = \frac{E}{1 - \nu^2} (\epsilon_x + \nu \epsilon_y)$$

$$\sigma_y = \frac{E}{1 - \nu^2} (\epsilon_y + \nu \epsilon_x)$$

$$\tau_{xy} = G \gamma_{xy} = \frac{E}{2(1 + \nu)} \gamma_{xy}$$

or in matrix form

$$\begin{Bmatrix} \sigma_x \\ \sigma_y \\ \tau_{xy} \end{Bmatrix} = \frac{E}{1 - \nu^2} \begin{bmatrix} 1 & \nu & 0 \\ \nu & 1 & 0 \\ 0 & 0 & \dfrac{1 - \nu}{2} \end{bmatrix} \begin{Bmatrix} \epsilon_x \\ \epsilon_y \\ \tau_{xy} \end{Bmatrix} \qquad (8.13)$$

or

$$\{\sigma\} = [D] \{\epsilon\}$$

Conditions of stress or strain within a material can be difficult to calculate mathematically for most shapes, with the exception of a few geometries such as bars, cylinders and plates. To obtain the stresses in other shapes it is necessary to use experimental techniques or computational procedures such as finite or boundary element analyses.

8.1.6 Stress concentrations

When designing load-bearing components the engineer uses the basic strength of materials equations in order to obtain stresses in, for example, beams and plates. In carrying out these calculations the engineer must proceed with the utmost caution, since these equations assume no discontinuities within the material. The presence of cracks and holes is very common. These discontinuities introduce stress concentrations which, in some instances, are the critical stress levels and lead to localized yielding or failure, depending upon the material property behaviour. As a result, it is necessary to consider a stress-concentration factor K_t, which is defined by the following relationship:

$$K_t = \frac{\text{Maximum stress}}{\text{Background stress}} = \frac{\sigma_{max}}{\sigma_n} \qquad (8.14)$$

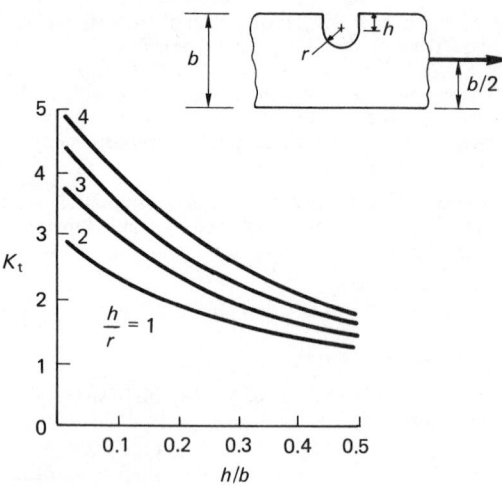

Figure 8.14 Stress concentration factor K_t for a notched sample

In general, σ_{max} will be determined by computational or experimental methods, and σ_n by simple theory such as $\sigma_n = P/A$, $\sigma_n = (My)/I$, $\tau = (Tr)/J$. For ductile materials stress concentrations may not be critical when under static loading but critical under dynamic loading. For brittle materials such as carbon-fibre composites, stress concentrations can be the governing conditions because the material cannot relieve the stresses by yielding. A full list of stress concentration factors for components under such loads as bending and torsion are given in references 2–4.

8.1.7 Impact stresses in bars and beams

8.1.7.1 Stress waves or pulses

A stress wave passes through a material when the different sections are not in equilibrium, as in the case of colliding bodies. Due to the material properties of a body, a finite time is required for this disequilibrium to be felt by other parts of the body. The lack of load equilibrium is observed by the presence of stress waves moving through a particular section.

The two most common wave forms are longitudinal and torsional. Longitudinal waves are of the form of tensile waves or compression waves. In the case of tensile pulses, sections of a body move in the opposite direction to the travelling wave, whereas compression pulses travel in the same direction. Torsional waves travel or oscillate in a plane which is transverse to the direction of the wave motion.

The conditions of one-dimensional impact stress waves can be devised from basic energy principles. Consider the prismatic section in Figure 8.15. Defining:

C_L = the longitudinal wavefront speed
V_o = the velocity of the prismatic section
A_o = the plane area of the prismatic section
σ_o = the stress acting on area A_o
ρ = the density of the section
and equating the change in momentum to the impulse load gives

$$(A_o C_L t \rho)V_o = (\sigma_o A_o)t$$

Thus

$$\sigma_o = \rho C_L V_o \qquad (8.15)$$

where

$$C_L = \sqrt{\frac{E}{\rho}} \qquad (8.16)$$

A similar expression can be derived for torsional loading on a 'thin-walled' tube, outside diameter d_o:

$$\tau = \rho C_T d_o \frac{\omega}{2} \qquad (8.17)$$

where

$$C_T = \sqrt{\frac{G}{\rho}} \qquad (8.18)$$

and ω = angular velocity.

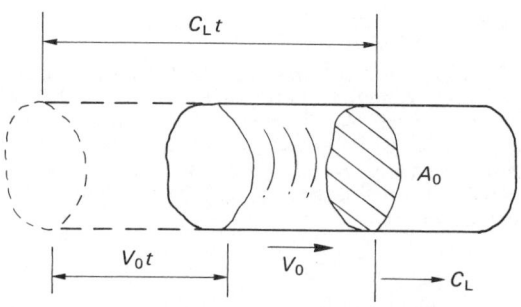

Figure 8.15 Uniform bar under the action of a stress wave

8.1.7.2 Typical elastic longitudinal and torsional wave speeds

In Table 8.1 C_L and C_T vary with temperature because E is especially sensitive to temperature. Also, C_L varies with direction for non-isotropic materials such as fibre-reinforced composites. With material such as concrete, Young's Modulus is different between tension and compression, producing unusual behaviour.

The mathematical and experimental behaviour of materials under impact loading can be very complex and is described in more detail in references 5 and 6. The analysis outlined in these references is the foundation upon which explosive forming or welding is based.

Table 8.1

	C_L (m/s)	C_T (m/s)
Mild steel	5150	3230
Aluminium	5090	3110
Glass	5300	3260
Epoxy resin	1540	950
Cast iron	3960	2470

8.1.8 Orthotropic material

An important classification of material used more frequently in recent years is the fibre-reinforced composite. This material has different properties in different directions, the high-stiffness properties being in the direction of the reinforcing

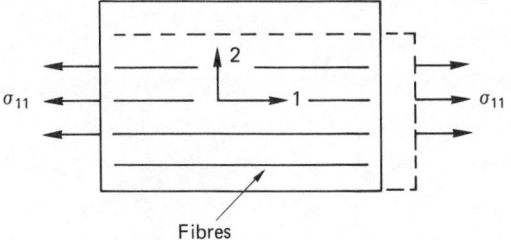

Figure 8.16 Principal directions for a laminate

fibres. A set of equations similar to those used for isotropic materials can be derived and used with care for certain design considerations.

These types of materials are constructed from thin sheets, known as laminates. They are fabricated by stacking together sets of unidirectional or bidirectional layers (also termed plies or lamina) in predetermined directions and thickness to give the desired performance characteristics.

Consider a single lamina with the axis set shown in Figure 8.16 and assuming plane stress conditions, the relationship between stress and strain is as follows:

$$\epsilon_{11} = \frac{\sigma_{11}}{E_{11}} - \nu_{21}\frac{\sigma_{22}}{E_{22}}$$

$$\epsilon_{22} = \frac{\sigma_{22}}{E_{22}} - \nu_{12}\frac{\sigma_{11}}{E_{11}} \qquad (8.15)$$

$$\epsilon_{12} = \frac{\tau_{12}}{G_{12}}$$

From symmetry,

$$\frac{\nu_{12}}{E_{11}} = \frac{\nu_{21}}{E_{22}}$$

By manipulating equations (8.15) a stress–strain stiffness matrix can be formed:

$$\begin{Bmatrix} \sigma_{11} \\ \sigma_{22} \\ \sigma_{12} \end{Bmatrix} = \begin{bmatrix} Q_{11} & Q_{12} & 0 \\ Q_{21} & Q_{22} & 0 \\ 0 & 0 & Q_{66} \end{bmatrix} = \begin{Bmatrix} \epsilon_{11} \\ \epsilon_{22} \\ \epsilon_{12} \end{Bmatrix} \qquad (8.16)$$

where

$$Q_{11} = \frac{E_{11}}{(1 - \nu_{12}\nu_{21})}$$

$$Q_{12} = Q_{21} = \frac{\nu_{12}E_{22}}{(1 - \nu_{12}\nu_{21})} = \frac{\nu_{21}E_{11}}{(1 - \nu_{21}\nu_{12})}$$

$$Q_{22} = \frac{E_{22}}{(1 - \nu_{12}\nu_{21})}$$

$$Q_{66} = G_{12}$$

Equation (8.15) reduces to equation (8.11) if $E_{11} = E_{22}$ and $\nu_{12} = \nu_{21}$. If the stiffness matrix $[Q]$ is inverted by matrix manipulation the compliance matrix $[S] = [Q]^{-1}$ is obtained. The corresponding strain–stress relationship can then be written

$$\begin{Bmatrix} \epsilon_{11} \\ \epsilon_{22} \\ \epsilon_{12} \end{Bmatrix} = \begin{bmatrix} S_{11} & S_{12} & 0 \\ S_{21} & S_{22} & 0 \\ 0 & 0 & S_{66} \end{bmatrix} \begin{Bmatrix} \sigma_{11} \\ 12 \\ \sigma_{12} \end{Bmatrix} \sigma_{22} \qquad (8.17)$$

where

$$S_{11} = 1/E_{11} \quad S_{22} = 1/E_{22} \quad S_{21}$$
$$S_{12} = -\nu_{12}/E_{11} \quad S_{66} = 1/G_{12}$$

When a unidirectional reinforced orthotropic lamina is loaded only in the principal material directions the deformation is independent of any shearing effects and there is no coupling between in-plane and shear strain. This is not the case when the loading is applied at some arbitrary angle θ to the principal axis (Figure 8.17).

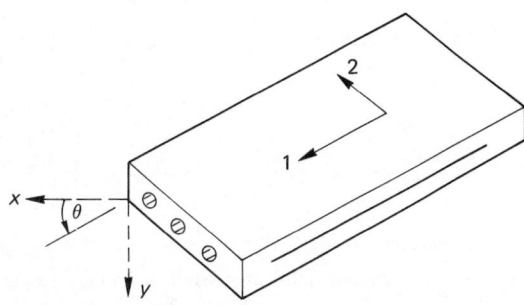

Figure 8.17 Orientated

It is possible to derive a transformation relating strains and stresses in two axes by a process similar to that used to derive Mohr's circle. As with Mohr's circle, half the 'engineering' shear strain must be used in the transformation.

Expressing stress in the x,y coordinate system in terms of stresses in the 1,2 coordinate system in the following way:

$$\begin{Bmatrix} \sigma_x \\ \sigma_y \\ \sigma_{xy} \end{Bmatrix} = \begin{bmatrix} m^2 & n^2 & -2mn \\ n^2 & m^2 & 2mn \\ mn & -mn & m^2 - n^2 \end{bmatrix} \begin{Bmatrix} \sigma_1 \\ \sigma_2 \\ \tau_{12} \end{Bmatrix} \qquad (8.18)$$

where $m = \cos \theta$ and $n = \sin \theta$,

$$\{\sigma_{xy}\} = [T] \{\sigma_{12}\}$$
$$\{\sigma_{xy} = [T]Q\{\epsilon_{12}\} = [T] [Q] [T]^{-1}\{\epsilon_{xy}\} = [\overline{Q}] \{\epsilon_{xy}\}$$

where

$$\overline{Q}_{11} = Q_{11}m^4 + 2m^2n^2(Q_{12} + 2Q_{66}) + Q_{22}n^4$$
$$\overline{Q}_{22} = Q_{11}n^4 + 2m^2n^2(Q_{12} + 2Q_{66}) + Q_{22}m^4$$
$$\overline{Q}_{12} = m^2n^2(Q_{11} + Q_{22} - 4Q_{66}) + Q_{12}(m^4 + n^4)$$
$$\overline{Q}_{66} = m^2n^2(Q_{11} + Q_{22} - 2Q_{12} - 2Q_{66}) + Q_{66}(m^4 + n^4)$$
$$\overline{Q}_{16} = m^3n(Q_{11} - Q_{12} - 2Q_{66}) + (Q_{12} - Q_{22} + 2Q_{66})n^3m$$
$$\overline{Q}_{26} = n^3m(Q_{11} - Q_{12} - 2Q_{66}) + (Q_{12} - Q_{22} + 2Q_{66})m^3n$$

and $[T]$ is called the transformation matrix.

Using the same matrix manipulation, the reduced compliance matrix can be generated as follows:

$$\begin{Bmatrix} \epsilon_x \\ \epsilon_y \\ \epsilon_{xy} \end{Bmatrix} = \begin{bmatrix} \overline{S}_{11} & \overline{S}_{12} & \overline{S}_{16} \\ \overline{S}_{12} & \overline{S}_{22} & \overline{S}_{26} \\ \overline{S}_{16} & \overline{S}_{26} & \overline{S}_{66} \end{bmatrix} \begin{Bmatrix} \sigma_x \\ \sigma_y \\ \sigma_{xy} \end{Bmatrix} \qquad (8.19)$$

where

$$\overline{S}_{11} = S_{11}m^4 + m^2n^2(2S_{12} + S_{66}) + S_{22}n^4$$

$$\overline{S}_{22} = S_{11}n^4 + m^2n^2(2S_{12} + S_{66}) + S_{22}m^4$$

$$\overline{S}_{12} = m^2n^2(S_{11} + S_{22} - S_{66}) + S_{12}(m^4 + n^4)$$

$$\overline{S}_{66} = 2m^2n^2(2S_{11} + 2S_{22} - 4S_{12} - S_{66}) + S_{66}(m^4 + n^4)$$

$$\overline{S}_{16} = m^3n(2S_{11} - 2S_{12} - S_{66}) - (2S_{22} - 2S_{12} - S_{66})n^3m$$

$$\overline{S}_{26} = n^2m(2S_{11} - 2S_{12} - S_{66}) - (2S_{22} - 2S_{12} - S_{66})m^3n$$

Using equations (8.19) it can be shown that the elastic properties can also be defined in the x,y direction relative to the principal planes, so that

$$\frac{1}{E_x} = \frac{m^2}{E_1} + \left(\frac{1}{G_{12}} - \frac{2\nu_{12}}{E_1}\right)m^2n^2 + \frac{n^2}{E_2}$$

$$\frac{1}{E_y} = \frac{n^2}{E_1} + \left(\frac{1}{G_{12}} - \frac{2\nu_{12}}{E_1}\right)m^2n^2 + \frac{m^2}{E_2} \qquad (8.20)$$

$$\frac{1}{G_{xy}} = 2\left(\frac{2}{E_1} + \frac{2}{E_2} + \frac{4\nu_{12}}{E_1} - \frac{1}{G_{12}}\right)m^2n^2 + \left(\frac{m^4n^4}{G_{12}}\right)$$

$$\nu_{xy} = E_x\left\{-\frac{\nu_{12}}{E_{11}}(m^4 + n^4) - m^2n^2\left(\frac{1}{E_1} + \frac{1}{E_2} - \frac{1}{G_{12}}\right)\right\}$$

8.1.8.1 Laminates

In the cases of laminates fabricated from individual lamina equations (8.15) to (8.19) are used to evaluate the properties of each layer. The layers are then summed to obtain the in-plane and bending stiffnesses of the laminate. This stiffness is then used in conjunction with the known applied loads in order to obtain the laminates in-plane strains and curvature based on plate or beam theory.

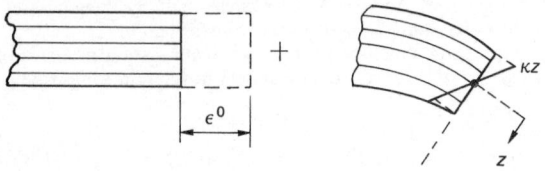

Figure 8.18 Laminate strains

Using the notation in Figures 8.18 and 8.19, the relationship between strain and load is given by

$$\begin{Bmatrix} N \\ M \end{Bmatrix} = \begin{bmatrix} A & B \\ B & D \end{bmatrix} \begin{Bmatrix} \epsilon \\ \kappa \end{Bmatrix} \qquad (8.21)$$

where

$$A_{ij} = \sum_{k=1}^{n} [\overline{Q}_{ij}]_k \{Z_k - Z_{k-1}\}$$

$$B_{ij} = \frac{1}{2}\sum_{k=1}^{n} [\overline{Q}_{ij}]_k \{Z^2_k - Z^2_{k-1}\}$$

$$D_{ij} = \frac{1}{3}\sum_{k=1}^{n} [\overline{Q}_{ij}]_k \{Z^3_k - Z^3_{k-1}\}$$

ϵ = strain to normal extension
κ = curvature

Thus the stress at any layer is given by

$$\sigma_x = \overline{Q}_j\epsilon_x + z\overline{Q}_j\kappa_x$$

$$\sigma_y = \overline{Q}_j\epsilon_y + z\overline{Q}_j\kappa_y \qquad (8.22)$$

$$\sigma_{xy} = \overline{Q}_j\epsilon_{xy} + z\overline{Q}_j\kappa_{xy}$$

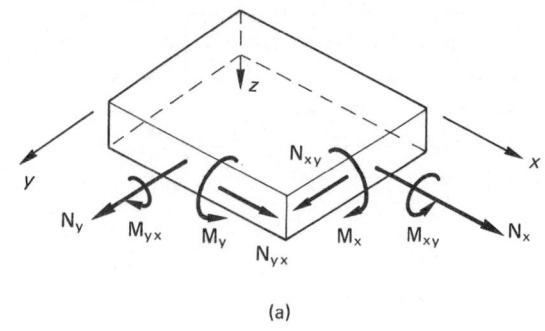

(a)

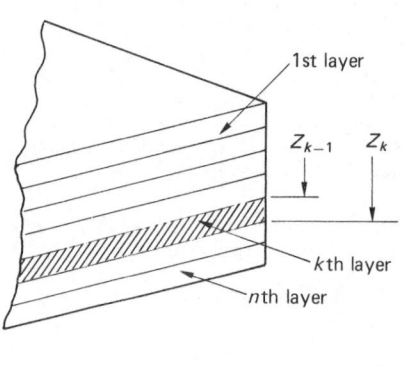

(b)

Figure 8.19 (a) Notation for a full laminate; (b) notation for a single laminate

These calculations can be carried out on almost any commercial microcomputer that has matrix inversion routines. This type of analysis can also be adapted for finite-element methods in order to obtain the relative stiffness of a particular section in a component. The above analysis must be used with caution if the laminate becomes quite thick and plane stress conditions no longer occur, i.e. through-thickness stresses may become significant (see references 5–10 for more information).

8.1.9 Plasticity

The behaviour of materials beyond the level of strain whereby there is no longer a linear relationship between stress and strain is called plasticity. In this type of behaviour the material will no longer return to its original shape once the load is removed; in fact a permanent deformation will occur in the material (Figure 8.20).

In dealing with stress systems beyond the elastic limit, similar to those associated with metal forming, the 'engineer's' definition of stress and strain becomes obsolete and stress and strain are defined with respect to the current deformed states. These stresses and strains are usually termed 'true stress and true strain'.

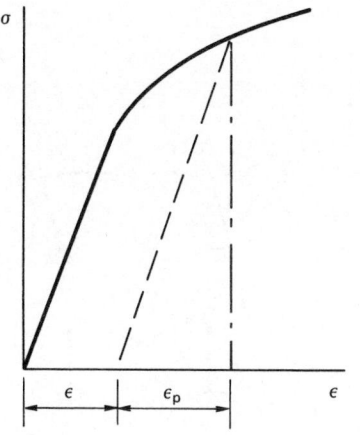

Figure 8.20 Typical stress–strain curve

True stress, σ, is defined as $\dfrac{\text{Load}}{\text{Current cross-sectional area}}$

$$(8.23)$$

True strain, ϵ_p, is defined as $\dfrac{\text{Change in length}}{\text{Current length}}$

$$\epsilon_p = \int_{L_0}^{L} \frac{dL}{L} = \ln\left(\frac{L}{L_0}\right) = \ln(1 + \epsilon) \qquad (8.24)$$

where L is current length and L_0 is the original length.

True strain is often quoted as natural or logarithmic strain. The advantage of natural strains over engineering strains is that the strains are additive and for most large strain (plastic) processes the change in volume during the processes is approximately zero. Hence, the sum of the three principal strains can be approximated to zero, i.e. $\epsilon_{p1} + \epsilon_{p2} + \epsilon_{p3} = 0$, a condition exploited in the analysis of metal forming.

Problems associated with plasticity can sometimes be those in which a section has partly yielded but other connecting sections remaining elastic, the plastic strains being of similar magnitude to the elastic strains. The compatibility equations and the stress–strain relationships become difficult to handle and complete solutions are very rare.

In cases where the plastic strains are large compared to the elastic strains it is sometimes permissible to neglect the elastic strains, hence greatly simplifying the solution process. The idealization of the material behaviour within this range of applicability is used by engineers for design work in the fields of structural design, dynamic behaviour and metal forming.

Material idealization usually falls into five categories:

1. Perfectly elastic (Figure 8.21(a));
2. Rigid, perfectly plastic (Figure 8.21(b));
3. Rigid, linear work hardening (Figure 8.21(c));
4. Elastic, perfectly plastic (Figure 8.21(d)); and
5. Elastic, linear work hardening (Figure 8.21(e)).

Models can be more complicated to take structural damping or load rating into account but analyses become extremely complex.

In the field of structural engineering the use of rigid-perfectly plastic idealization can be a useful design tool for selecting a specific structural member. The philosophy is based on that a frame or beam cannot deflect indefinitely or

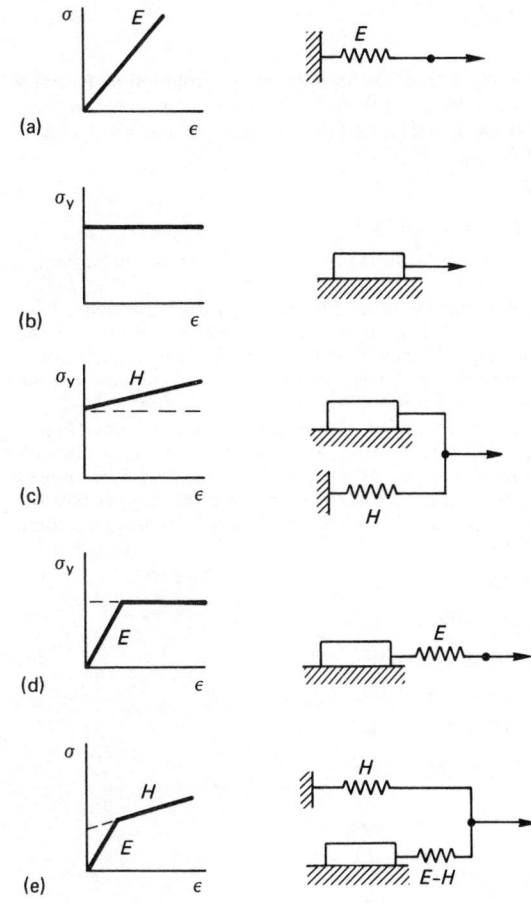

Figure 8.21 Schematics of material behaviour

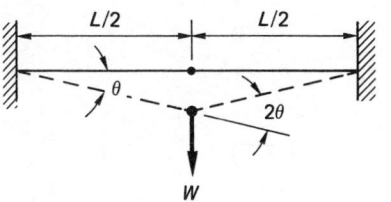

Figure 8.22 Plastic collapse mechanism for a built-in beam

collapse until the full plastic moment, M_p, has been developed at critical sections to form a collapsed structure.

A simple illustration of the procedure is shown in the following example of a uniform beam built in at both ends carrying a central concentrated load as in Figure 8.22. Under the action of increasing load, W, plastic hinges form once the section becomes fully plastic at three points. No collapse will occur unless three hinges are formed. Between the region of the plastic hinge the beam is considered to be rigid. At the instant of collapse the work done by the load W is $W(L/2)\phi$. The rate of energy loss by the plastic hinges is $M_p\phi + M_p\cdot 2\phi + M_p\phi$. Thus equating the two gives

$$W = \frac{8 \cdot M_p}{L}$$

This method produces an upper-bound solution or an over-estimate to the collapse load. The stress–moment relationship is the same as that for simple bending providing that the plastic modulus (z_p) is used, i.e.

$$\sigma = \frac{M_p}{z_p}$$

Values of plastic modulus are given in most beam-design tables.

In determining loads for more complex problems (e.g. portal frames) it is the practice to 'guess' the position of the plastic hinges so that the structure will become a mechanism. The procedure becomes a trial process, the optimum solution being the mechanism that gives the lowest collapse load.

Consider the design of a rectangular portal frame (Figure 8.23) made from column and beam sections. Since the columns have different collapse moments from the beams, hinges can occur in a variety of combinations producing six possible modes of collapse. Denoting beam collapse by M_b and column by M_c we have:

(a) $4M_b = Wl$

(b) $2M_c = Wl$

(c) $2M_c + 4M_b = 3Wl$

(d) $2M_b + 2M_c = Wl$

(e) $2M_b + 2M_c = 2Wl$

(f) $4M_c + 2M_b = 3Wl$

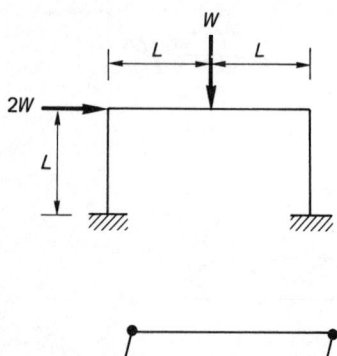

Figure 8.23   Rectangular portal frame

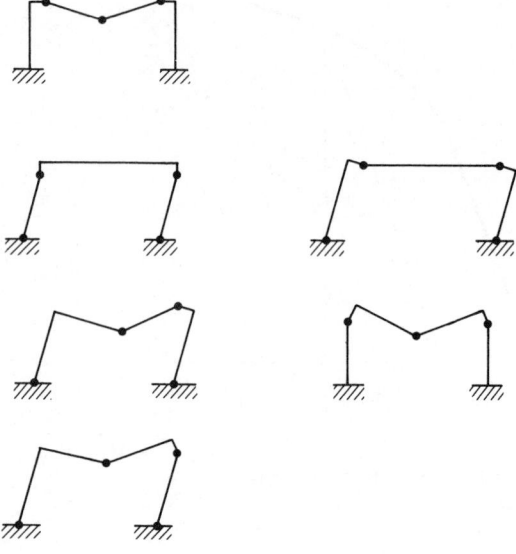

Figure 8.24   Collapse mechanism for the portal frame in Figure 8.23

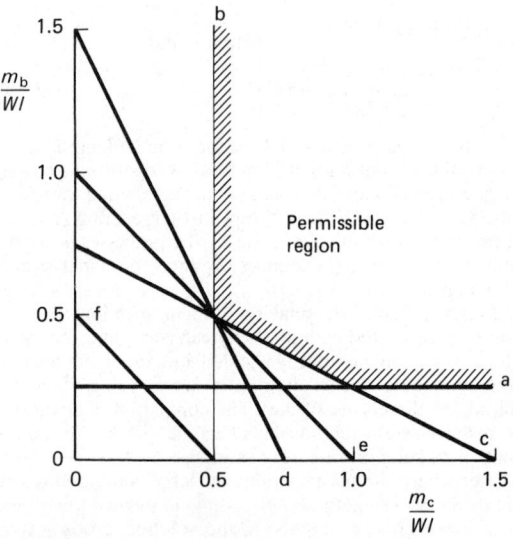

Figure 8.25   Geometrical analog for the plastic design of a portal frame

These sets of equations can be represented by coordinate axes in Figure 8.25. It follows that all possible combinations of M_b and M_c required by the frame under the given loads are represented by the line segments a–f which are convex towards the origin. The shaded region in the figure is called the permissible region, since plastic collapse does not occur for any design represented by a point lying in this region. Points to the origin side of the boundary a,b,c,d,e,f represent designs which cannot support the given loads.

8.2   Experimental techniques

8.2.1   Strain gauges

Electrical resistance strain gauges are the most frequently used devices in experimental stress analysis today. This type of gauge is frequently employed as a sensor in transducers to measure load, torque, pressure and acceleration.

The basic principles go back to 1856, when Lord Kelvin used copper and iron rods to observe the characteristic of the modern strain gauge. This has led to the following developments and observations:

1. The resistance of wire changes as a function of strain.
2. Different materials have different sensitivities.
3. The Wheatstone bridge can be used to measure the relevant quantities more accurately than single gauges.

Today bonded foil gauges monitored with a Wheatstone bridge has become a high perfected (accurate) measuring system. Precise results can be obtained quickly using relatively simple methods with inexpensive gauges and instrumentation systems.

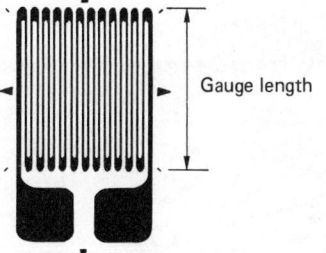

Figure 8.26 Typical strain gauge unit

8.2.2 Basic principles

The resistance, R, of a uniform conductor length, L, cross-sectional area, A, specific resistance, ρ, is related by

$$R = \frac{\rho L}{A} \tag{8.25}$$

Differentiation with respect to R gives

$$\frac{dR}{R} = \frac{d\rho}{\rho} + \frac{dL}{L} + \frac{dA}{A}$$

$$= \frac{d\rho}{\rho} + (1 + 2\nu)\frac{dL}{L} \tag{8.26}$$

Defining sensitivity by

$$\frac{(\Delta R/R)}{\epsilon} = 1 + 2\nu + \frac{(d\rho/\rho)}{\epsilon} = F \tag{8.27}$$

The sensitivity is not constant but is dependent upon the amount of coldwork, impurities and strain rate. Most commercial gauges are based upon a copper–nickel alloy because of the linearity over a wide range and excellent thermal stability.

Modern gauges are of the foil type, that is, a thinly etched pattern on a metal foil (Figure 8.26). Common foil gauges have a lower limit on resistance of about 90 Ω, ranging in length from 2 mm to 100 mm. Standard resistances are 120 Ω and 350 Ω, although some gauges are available in 1000 Ω resistance.

The gauges are commercially supplied on a thin plastic (epoxy or polyamide resins) base because of the fragile nature of the metal foil. The carrier base also acts as an electrical insulation. Gauges are also commonly supplied in the form of rosettes (two or three gauges) as in Figure 8.29 or stack configurations or specific arrangements for diaphragm transducers used in pressure measurement (see Figure 8.28). Occasionally, special-purpose gauges are available for which bonded wire, weldable strain gauges or semiconductor gauges are the most suitable solution to the problem, but they are usually expensive.

A major concern with the application of the strain gauges is the surface adhesion between the gauge unit and the surface of the component. The adhesive serves as a vital function: it must transmit the strain from the component's surface to the gauge. Adhesives can influence the gauge factor, hysteresis and temperature performance. Modern adhesives are relatively cheap but incorrect application can prove to be very costly. Surfaces for strain gauge application must be clean, degreased and surface treated if metals are involved. Environment effects attack most adhesives, so if hazardous conditions prevail, the gauge unit and adhesive are usually covered with a polymeric coating.

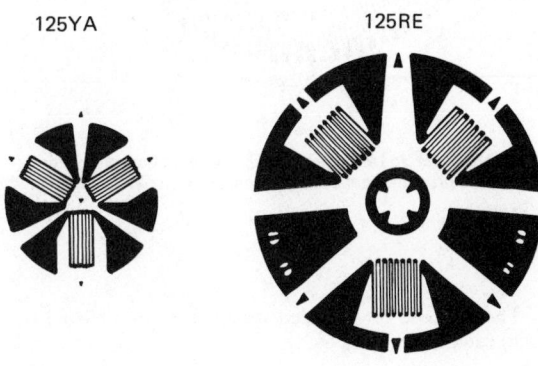

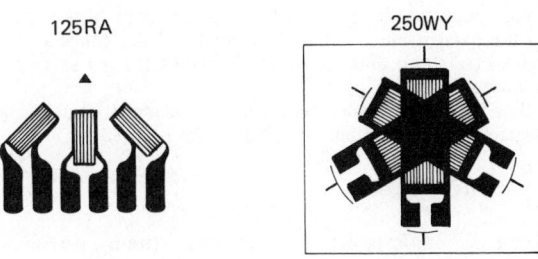

Figure 8.27 A selection of strain gauge rosettes (by permission of Welywn Strain Ltd, UK)

8.2.3 Gauge factor

The end loops of a foil gauge (see Figure 8.26) are enlarged to desensitize the effects of cross (transverse) sensitivity. The axial segments of the grid pattern have a large width-to-thickness ratio, hence some amount of transverse strain will be transmitted through the adhesive and the carrier will produce some response in addition to the axial-strain response. The magnitude of transverse strain is a function of all three materials, namely, the gauge, the adhesive coating and the component.

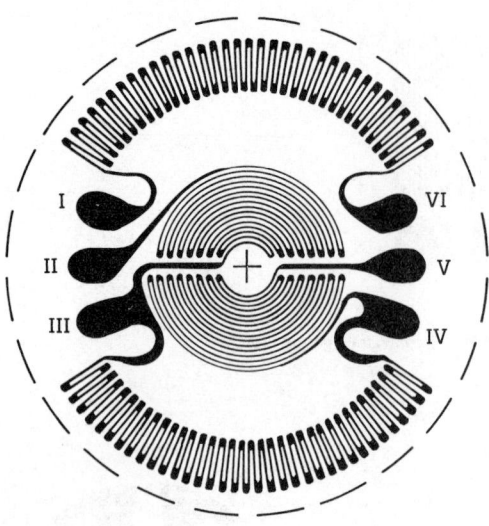

Figure 8.28 Micro-Measurements 'JB' pattern strain gauge for diaphragm pressure transducers (by permission of Welwyn Strain Ltd, UK)

The response of a bonded strain gauge to a biaxial strain field can be expressed as

$$\frac{dR}{R} = F_a\epsilon_a + F_t\epsilon_t + F_s\gamma_{at} \tag{8.28}$$

where ϵ_a is the normal strain in the axial direction of the gauge, ϵ_t is the normal strain in the transverse direction of the gauge, γ_{at} is the shear strain, F_a is the sensitivity of the gauge to the axial strain, F_t is the sensitivity of the gauge to the transverse strain and F_s the sensitivity of the gauge to shearing strain.

For most gauges the shear sensitivity is small and can be neglected. Hence equation (8.28) can be reduced to

$$\frac{dR}{R} = F_a(\epsilon_a + K_b\epsilon_t) \tag{8.29}$$

where $K_t = F_t/F_a$ is defined as the cross- (transverse) sensitivity factor for the gauge.

Strain gauge manufacturers provide a calibration constant known as the gauge factor F for each gauge supplied, defined as $dR/R = F\epsilon$. The calibration test is usually carried out by a uniaxial tensile test on a piece of material with a Poisson's ratio, ν_c; thus $\epsilon_t = -\nu_c\epsilon_a$. Equation (8.29) now can be written as

$$\frac{dR}{R} = F_a(1 - \nu_cK_c)\epsilon_a \tag{8.30}$$

Comparing equations (8.28) and (8.30) gives

$$F = F_a(1 - \nu_cK_t) \tag{8.31}$$

It is important to realize that for any strain field except that corresponding to a uniaxial stress field on a material with the same Poisson ratio as the calibration material there will always be an error in the indicated strain if the cross sensitivity is not zero. In many instances, the error is small enough to be neglected, but in some cases it is not. The error due to cross sensitivity for a strain gauge oriented at any angle, in any strain field, on any material, can be expressed as

$$ne = \left\{ \frac{K_t(\epsilon_t/\epsilon_a) + \nu_c}{1 - \nu_cK_t} \right\} \cdot 100 \tag{8.32}$$

where ne is the error as a percentage of the actual strain along the gauge axis. By inspecting equation (8.32) it can be observed that the cross sensitivity increases with the absolute values of K_t and ϵ_t/ϵ_a. Equation (8.28) can be approximated to

$$ne \sim K_t\frac{\epsilon_t}{\epsilon_a} \cdot 100 \tag{8.33}$$

provided the strain ratio ϵ_t/ϵ_a is not close to ν_c. A plot of equation (8.32) is shown in Figure 8.29 for convenience in estimating errors indicating strain readings. The approximated equation can also be checked against these curves.

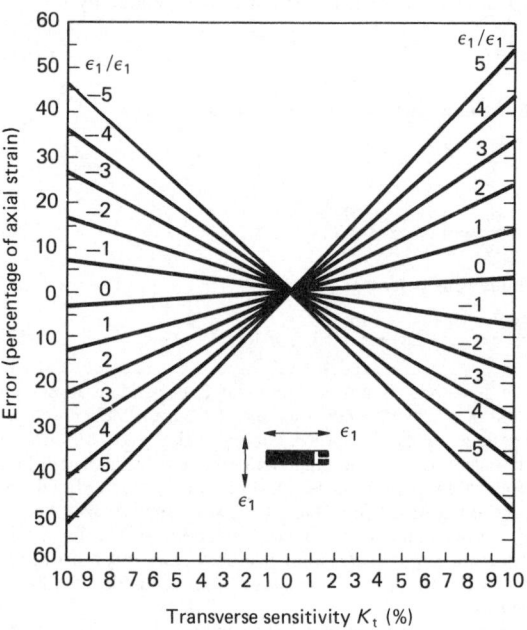

Figure 8.29 Errors due to transverse sensitivity of a strain gauge

8.2.4 Strain gauge arrangements

8.2.4.1 The Wheatstone bridge

Wheatstone bridges are common circuits frequently used in strain gauge systems in order to improve the sensitivity and cancel out temperature effects.

In the circuit shown in Figure 8.30 the change in output voltage (ΔV_o) is proportional to the strain reading, and can be shown to be

$$\Delta V_o = \frac{V_iR_2}{R_1[1 + (R_2/R_1)]^2} \left(\frac{\Delta R_1}{R_1} - \frac{\Delta R_2}{R_2} + \frac{\Delta R_3}{R_3} - \frac{\Delta R_4}{R_4} \right) \tag{8.34}$$

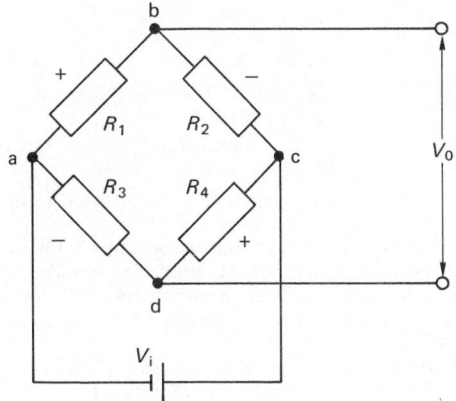

Figure 8.30   Constant-voltage Wheatstone bridge

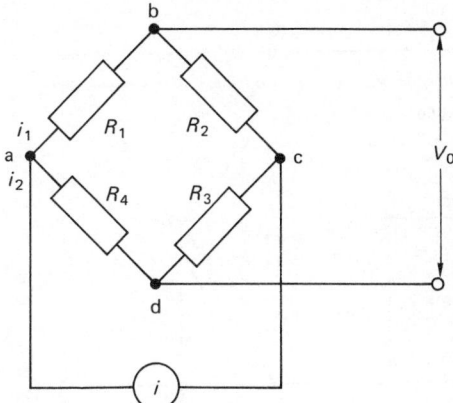

Figure 8.31   Constant-current Wheatstone bridge

Defining sensitivity of the circuit by

$$S_o = \frac{\Delta V_o}{\epsilon} = \frac{V_i R_2}{R_1[1 + (R_2/R_1)]^2}\left(\frac{\Delta R_1}{R_1} - \frac{\Delta R_2}{R_2} + \frac{\Delta R_3}{R_3} - \frac{\Delta R_4}{R_4}\right)$$
$$(8.35)$$

it can be seen that the circuit sensitivity is a function of the number of active arms, the gauge factor, the input voltage and the ratio of resistances R_1/R_2. The maximum sensitivity of the circuit is when $R_1 = R_2$. With four active arms in the bridge a circuit sensitivity of $F_g V_i$ can be achieved, whereas with one active arm a circuit sensitivity of only $F_g V_i/4$ can be obtained. When the bridge supply voltage V_i is selected to drive the gauges in the bridge so that they dissipate the maximum allowable power, a different sensitivity equation must be used. This is fully described in reference 13. Most experimental tests use single-active-arm circuits with the signal from a bridge amplified by factors of 10–1000 before records are taken. Multi-active-arm circuits are normally employed for transducer application where care must be taken as to which arm is used for either temperature or the Poisson gauge to give maximum sensitivity (see reference 13).

Constant-current systems are also used, though these systems exhibit non-linear output whenever the changes in resistance are large; this therefore limits their usage. Recent advances in the area of electronics have led to constant-current devices being common in modern industry. These are based on a high-impedance system which changes the output voltage with respect to the resistive load in order to maintain a constant-current set-up.

Consider the arrangement in Figure 8.31 for a Wheatstone bridge with a constant current. It can be shown that the change in output voltage ΔV_0 is given by

$$\Delta V_o = \frac{I_i R_1 R_3}{\Sigma(R + \Delta R)}\left(\frac{\Delta R_1}{R_1} - \frac{\Delta R_2}{R_2} + \frac{\Delta R_3}{R_3} - \frac{\Delta R_4}{R_4} + \frac{\Delta R_1 \Delta R_3}{R_1 R_3}\right.$$
$$\left. - \frac{\Delta R_2 \Delta R_4}{R_2 R_4}\right)$$
$$(8.36)$$

where

$$\Sigma(R + \Delta R) = R_1 + \Delta R_1 + R_2 + \Delta R_2 + R_3 + \Delta R_3$$
$$+ R_4 + \Delta R_4$$

Equation (8.36) shows that there are non-linear terms in the equation. However, the non-linear effects in this circuit are less than those associated with constant-voltage systems. Good circuit design can lead to the non-linear terms being negligible, even for large changes in resistance.

The resistance change for a metallic foil strain gauge is quite small. Consequently, any item which produces resistance changes within the Wheatstone bridge is extremely important. Since all foil gauges have lead wires, soldered joints and binding posts connected to them, their effects on the resistance to a particular gauge can be significant. Lead wires which, in general, are long compared to the strain gauges can cause significant errors due to the temperature effects associated with these leads. These effects are usually minimized by employing the three- or five-wire system similar to that shown schematically in Figure 8.32. In this circuit, both the active and dummy gauges are placed at a remote location. One of the three wires is used to connect terminal a of the bridge to a remote location. This wire is not a lead wire since it is not connected to R_1 or R_4. The active and dummy gauges have one long wire each with a resistance R_i and one short wire with negligible resistance. The bridge is still initially balanced since both arms R_1 and R_4 are increased by R_l. The change in output voltage is now given by

$$\Delta V_o = \frac{V_o R_1}{R_4[1 + (R_1/R_4)]^2}\left\{\left(\frac{\Delta R_1}{R_1 + R_l}\right)_\epsilon + \left(\frac{\Delta R_1}{R_1 + R_l}\right)_{\Delta t}\right.$$
$$(8.37)$$
$$\left. + \left(\frac{\Delta R_l}{R_1 + R_l}\right)_{\Delta t} - \left(\frac{\Delta R_1}{R_1 + R_l}\right)_{\Delta t} - \left(\frac{\Delta R_l}{R_1 + R_l}\right)_{\Delta t}\right\}$$

where ϵ represents induced strain and ΔT temperature changes. Temperature compensation is thus achieved since all the temperature-related terms (ΔT) cancel each other out.

Other effects such as switches, electrical noise and slip rings are beyond the scope of this chapter but are described in detail in reference 13.

8.2.4.2   Load cells

Strain gauge circuits are used frequently in load cell design because of their relative ease and cheapness. To produce a simple load-measuring transducer, a simple tension bar is used

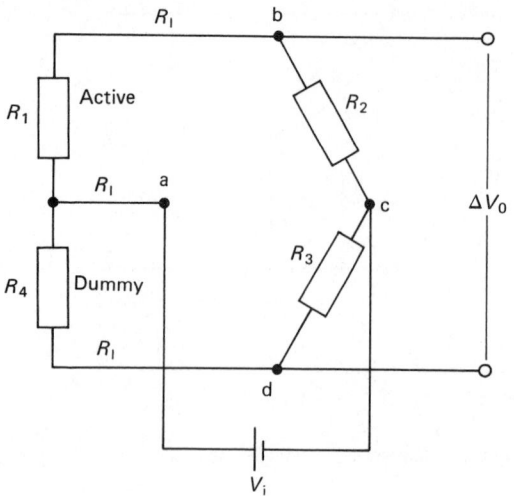

Figure 8.32 Three-wire compensation arrangement

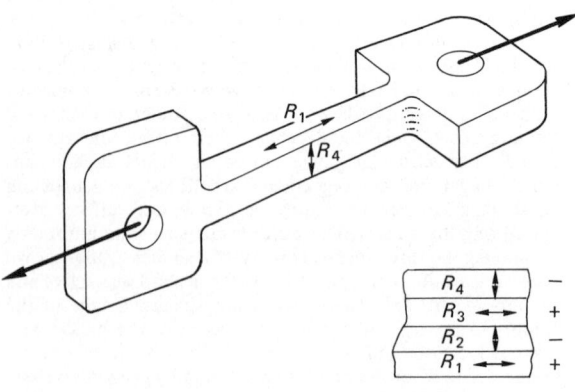

Surface development

Figure 8.33 Strain gauge set-up for a load cell

with four or eight gauges attached to the surface at the central region of a simple bar. Two gauges are mounted in the axial direction and two opposite gauges in the transverse direction as shown in Figure 8.33. If the tension bar is subjected to a load P the following strains are produced:

$$\epsilon_a = \frac{P}{AE}, \quad \epsilon_t = \frac{\nu P}{AE} \tag{8.38}$$

where A is the cross-sectional area,
   E is the Young's modulus, and
   ν is the Poisson ratio,

If the four gauges are positioned in the Wheatstone bridge as shown in Figure 8.30, the ratio of output voltage to supply voltage $\Delta V_o/V_i$ is given by

$$\frac{\Delta V_o}{V_i} = \frac{1}{4}\left(\frac{\Delta R_1}{R_1} - \frac{\Delta R_2}{R_2} + \frac{\Delta R_3}{R_3} - \frac{\Delta R_4}{R_4}\right) \tag{8.39}$$

since

$$\frac{\Delta R_1}{R_1} = \frac{\Delta R_3}{R_3} = F\epsilon_a = \frac{FP}{AE}$$

and

$$\frac{\Delta R_2}{R_2} = \frac{\Delta R_4}{R_4} = F\epsilon_t = -\frac{\nu FP}{AE}$$

$$\frac{\Delta V_o}{V_i} = \frac{FP}{2AE}(1 + \nu) \approx \frac{P}{AE}(1 + \nu) \text{ for } F \approx 2.0 \tag{8.40}$$

Equation (8.36) shows that the output is proportional to the applied load P.

The advantage of placing the four gauges allows for the nullifying of an eccentricity of load, producing bending or torsion loads since tensile and compressive strains will cancel out each other. Temperature compensation is also achieved with four active gauges in the bridge.

A special type of strain gauge is used as part of a pressure transducer. The strain gauge pattern is called the 'JB' pattern and fits centrally on one side of a thin diaphragm (see Figure 8.28). The diaphragm pressure transducer is small, easy to fabricate and inexpensive. Maximum strains occur at the centre and edges of the diaphragm, hence the requirement for the design pattern shown in Figure 8.28. Note that the solder tabs are positioned at regions of low strain. Averaging the strains over the region covered by each sensing element and averaging the output, the total gauge output ϵ_o can be expressed as

$$\epsilon_o = 820\frac{PR_0^2}{t^2 E}(1 - \nu^2)\text{mV/V} \tag{8.41}$$

where P is the pressure, R_0 the diaphragm radius, t the diaphragm thickness and the gauge factor is assumed to be 2.0.

Strains induced by the pressure are non-linear if the central deflection of the diaphragm is greater than a quarter of the thickness. Therefore this type of transducer is usually calibrated for a specific pressure range in order to keep non-linear effects small.

Diaphragm pressure transducers are also used to measure pressure response, therefore it is important that, when designing the transducer, the diaphragm should have a natural frequency greater than five times the highest applied frequency. The natural frequency, F_n, can be expressed as

$$\frac{0.471t}{R_0^2}\sqrt{\left[\frac{E}{\rho(1 - \nu^2)}\right]} \tag{8.42}$$

where ρ is the density of the diaphragm material (kg/mm^3). The natural frequency of an existing unit can easily be determined experimentally by tapping the transducer at the centre of the diaphragm and noting the response on an oscilloscope.

8.2.5 Photoelasticity

Photoelasticity is a useful tool for estimating stress distributions in components with complicated geometries with or without complex load cases for which mathematical techniques are difficult or almost impossible. It provides quantitive evidence of highly stressed regions and peak stresses. Equally important, it highlights areas of low stress, leading to designs whereby materials can be utilized efficiently. It is widely used for problems in which stress or strain information extends into regions of the structure.

The photoelastic phenomenon occurs in particular plastics when they are subjected to a strain, and under this strain the light becomes polarized. When a particular plastic model (early work was carried out on glass) is stressed a ray of light enters along one of the principal stress directions (see Figure 8.34) and is divided into two component waves each with its plane of vibration (plane of polarization) parallel to one of the remaining principal planes. Furthermore, the relative retardation of the light along these paths is dependent upon the magnitude of the two remaining stresses. This phenomenon is termed birefringence or double refraction. The two pheonomena are brought together in the photoelastic polariscope (see Figure 8.35).

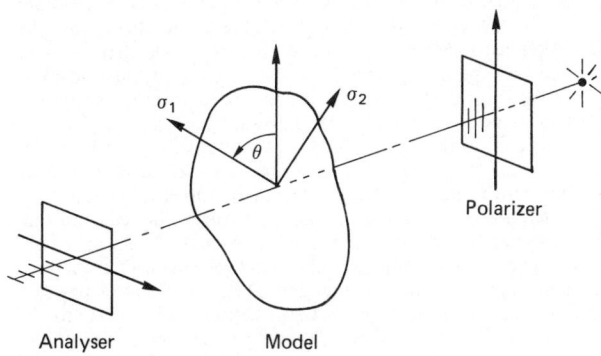

Figure 8.34 A stressed photoelastic model in a plane polariscope

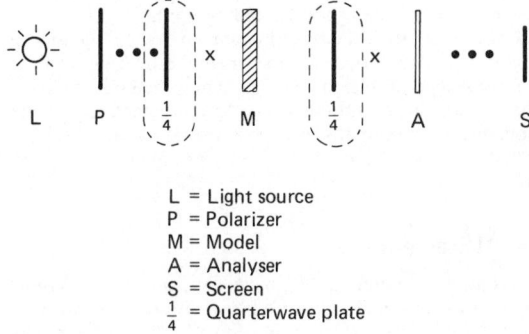

L = Light source
P = Polarizer
M = Model
A = Analyser
S = Screen
$\frac{1}{4}$ = Quarterwave plate

Figure 8.35 Polariscope arrangement with circular polarized light

If the relative retardation N is 0,1,2,3, etc. cycles, the waves of the light intensity diminish to zero (extinction) in a dark-field polariscope. Extinction may also occur in 1/2, 3/2, 5/2, 5/2, etc. cycles in a light field polariscope. That is, in a darkfield polariscope a locus of dark bands corresponding to the level of maximum shear stress are observed in the model, like the pressure potentials on a weather map. A succession of light and dark bands in a monochromatic light source or filter (usually sodium) occurs. These bands are called fringes and their value is defined as the fringe order (see Figure 8.36).

8.2.5.1 Two-dimensional analysis – plane stress

Models used for plane stress analysis are those in which the thickness of the material through which the light travels is small relative to the other dimensions. If the principal stresses and their associated angles are required for the model they can be quickly obtained at the model's boundary and relatively easily internally by the equations of elasticity.

In order to determine these values a model is first placed in a plane polariscope and then loaded. Dark lines are seen on

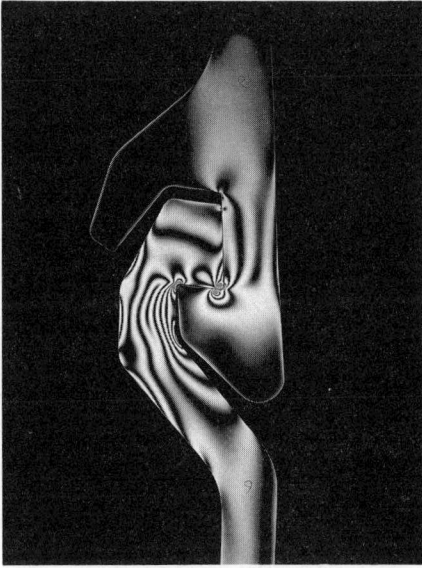

Figure 8.36 Fringe pattern for a tension coupling in a sodium light source (by permission of Sharples Stress Consultants)

the model. These lines correspond to the principal planes or planes of symmetry and are termed isoclinics. Under load they define the planes but not stress magnitude. When the model is loaded a series of fringes occurs in monochromatic light which are proportional to the stress levels. Unfortunately, the fringe patterns are undistinguishable from the isoclinics, therefore the plane polariscope is only used to determine the principal planes. In order to determine the stress levels in a model it is inserted in what is termed circular polarized light. This is achieved by inserting quarter-wave plates which, due to their nature, extinguish the isoclinics, leaving only the isochromatics (fringe patterns).

In a monochromatic light source a pattern of dark and light bands is formed under load which are proportional to the model loading. Care must be taken not to exceed the elastic limit of the material.

The fringe pattern is related to the stress system by the stress optical law:

$$2\tau_{max} = (\sigma_1 - \sigma_2) = \frac{FN}{t} \tag{8.43}$$

where t is material thickness (mm), F is the fringe value (N/mm/fringe) and N is the measured fringe value. In terms of the fringe pattern this is defined as the number of fringes (N) that pass through these points during the application of external loading and is given as the fringe order at that point.

By removing the quarter-wave plates either manually or optically the direction as well as the magnitude at this point can instantly be obtained. Measurements of fractional fringe values as well as integer values can also be obtained by a variety of compensation techniques described in detail in references 13–16.

The F value for the model is material-dependent and is usually calibrated from the same batch of material as the model. This is obtained from testing on a piece of material with a known stress distribution, i.e. a simple tension bar, a beam under four-point bending (pure flexure) or a disk in diametrical compression.

In general, the photoelastic model is much smaller than the actual one. Thus, the prototype stress, σ_p, can be calculated from the model stress, σ_m, from

$$\sigma_p = \sigma_m \frac{t_m}{t_p} \frac{P_p}{P_M} \frac{L_M}{L_p} \qquad (8.44)$$

where the subscript m represents the model and p the prototype, P is the loading, t the thickness and L the relative length. Displacements, δ, can also be calculated by

$$\delta_p = \delta_m \frac{P_p}{P_M} \frac{E_M}{E_p} \frac{t_M}{t_p} \qquad (8.45)$$

where E is the relative Young's modulus.

It is clear from the above equations that component stresses can be obtained for most two-dimensional problems. The modulus of elasticity is never a problem in determining the stress distributions unless the loading is such that contact stresses occur. Also, Poisson's ratio need not be considered when the model is simply connected and the body force (gravity) is negligible.

8.2.5.2 Two-dimensional model

Materials used to produce the models must fulfil the following criteria if successful results are to be obtained:

- Linear properties
- Mechanical and optical isotropy and homogeneity
- Low creep
- Low moisture absorption
- Inexpensive
- Easily machined

Epoxy resins are the most common material for static stress conditions because of their high photoelastic sensitivity and increased freedom from mechanical and optical creep. Urethane rubber is also used, especially for dynamic problems since the material has a low modulus and the stress waves (see Section 8.1.7) induced are less than 90 ms^{-1}. Machining problems do occur with this material since the material has to be below its transition temperature before it can be machined.

An extremely useful extension of two-dimensional photoelasticity is the surface-coating method, whereby the photoelastic material is applied as a thin layer on the surface of an actual component. The cement bonding the coating to the component is usually a small-volume fraction of powdered aluminium in order to produce a reflective surface so that light travels through the coating and reflects back for a second traversal of the coating. This method is extremely useful for obtaining stresses on free surfaces of three-dimensional coatings. Measurements are usually carried out by hand-held instruments and are not limited to the linear ranges of the component, although the accuracy and resolution is much reduced.

8.2.5.3 Three-dimensional analysis

Photoelastic techniques can be applied to three-dimensional problems but with a great deal extra of work and analysis compared to two-dimensional ones. Accuracy is also reduced and sensitivity can be a problem due to large amounts of careful machining that must be carried out before any readings can be obtained.

Three-dimensional analysis is performed by using 'frozen stress' or scattered-light methods. With stress freezing techniques the stresses are locked into the model due to heat and are retrieved by slicing the model in order to produce a quasi-two-dimensional model. Scattered-light techniques do not use sliced models.

Stress freezing is the most common method adopted for stress analysis. The stresses are locked into the model by loading the model and applying a uniform heat up to a temperature above the critical temperature which unlocks the secondary bonds. The load is then kept constant for the remainder of the procedure. The temperature is set constant for a period of time and then allowed to drop with the secondary bonds permanently set into the model. The primary bonds are permanently locked into the model by the reforming of the secondary bond, producing a permanent stress pattern locked into the model. At the same time as the model is subjected to the heat cycle a calibration piece must be under the same heat cycle, since the material's fringe value will be different from that obtained by room-temperature tests.

After stress freezing the model it is carefully cut into thin slices ready for inspection in a polariscope. The stresses are then obtained by similar methods to two-dimensional analysis. Unfortunately, unless the model can be cut on a principal axis, the secondary principal stress method will have to be employed (see references 18 and 19). However, sometimes stresses are assumed not to change through the thickness of the slice and two-dimensional technique can be used. In order to check this, models are sometimes shaved by taking small cuts off the thickness and the stress pattern checked to see if it is independent of the thickness. Since all epoxy materials are abrasive, carbon-tipped tools should be used.

The use of photoelasticity and strain gauges still tend to be the most common forms of experimental stress analysis. In recent years there has been an influx of new methods available to the stress analyst, although, due to the complexity of the equipment, they are relatively expensive and require specialist knowledge. A selection of the new developments is given below.

8.2.6 Holography

By recording a holographic image of an object's visible surface, any changes in shape can be monitored by comparing the new shape with the recorded image. This is commonly known as holographic interferometry. The technique uses a fine laser to produce full-field quantitative strain maps. When used with computer-image processing, general-purpose plots can be produced for a wide range of engineering products with complex geometries, such as turbine blades and piston rings. Holography can be used for both dynamic and static loading.

Since holography is an optical technique it is a non-contacting method and problems associated with strain gauge technology do not exist. Future development in the equipment may yield cheaper and more compact apparatus, thus giving the prospect of portable systems. A detailed discussion on the usage of holography is given in references 20–23.

8.2.7 Thermo-elastic analysis

When a material is compressed adiabatically it has work done on it, and this work is evidenced by a rise in temperature. This temperature change, however, is very small – about 0.2°C at the elastic limit. By inducing a cyclic strain on the component surface, and measuring the temperature changes by infrared thermography, in conjunction with computer enhancement a sensitivity of 1×10^{-3}°C can be obtained in the measurement of temperature. This means that a typical scan of a component section can take several hours of cyclic loading, but the only specimen preparation needed is to apply a coat of matt-black paint to the surface.

Infrared radiation is emitted from a surface in all directions and therefore measurement can be made on surfaces which are not normal to the reading lens. This enables cyclindrical components to be scanned and, in some instances, irregular curved shapes. Sensitivity of stress is around 0.5–1 N mm^{-2} for common metals, although coarser readings can be used for general stress analysis. It is also possible to monitor cracks and their development under cyclic loading.

One of the present drawbacks of this technique is that it cannot tolerate large movements acting normal to the line of sight of the scanning unit.

Present thermo-elastic analysis in the UK often comes under the heading of stress-pattern analysis by thermal emission (SPATE) and is detailed in references 24 and 25.

8.2.8 Brittle-coating[13,26,27]

This technique provides a simple and direct solution to the stress distribution in a component when extreme accuracy is not required. It assumes perfect adhesion of a coating to the component. The coating is designed to have brittle characteristics.

Coatings are selected such that they fracture (crack) at relatively low levels of stress. Thus the technique is classed as non-destructive.

Modern coatings tend to be room-temperature cure resins which are pigmented.

8.2.9 X-ray analysis[28,29]

This method is particularly useful in determining residual stresses occurring in a component. It is perhaps the only non-destructive technique available to the stress analyst at present which can carry out this task.

A disadvantage of this technique is the limitation on the test sample size imposed at the X-ray diffractometer stage in the process, although this may be remedied by equipment developments in the future.

8.3 Fracture mechanics

8.3.1 Introduction

If the stresses in a cracked structure close to the tip of an infinitely fine crack were computed by elastic stress concentration factors the stress would approach infinity and a material with a crack of any size would fail under any tensile stress. The linear elastic fracture mechanics (LEFM) approach avoids this difficulty by analysing the stress field that surrounds the crack tip rather than the infinite stress at the crack tip and then identifying certain features of the field which are common to all elastic bodies.

Only a summary of the results of the analysis can be given here. For a full analysis the reader is referred to texts in the Further Reading at the end of the chapter. (Note: To facilitate comparison with the references throughout this section σ, the resolved or calculated stress, has been used for stress. In Section 8.4 S, the applied load, has been used. σ and S are sometimes (but not always) interchangeable. Some references use P for S.)

8.3.2 Linear elastic fracture mechanics

For a brittle material, A. H. Griffiths showed in 1920 that the fracture stress for an elliptical crack is

$$\sigma = (4\gamma_s E \pi^{-1} a^{-1})^{1/2} \tag{8.46}$$

where γ_s is the surface energy of work of fracture per unit area de/da, E is the modulus of elasticity and a is the critical crack length for equilibrium.

Equation (8.46) satisfactorily predicts the behaviour of a brittle material but, for a material in which plastic flow will occur at the crack tip, a plastic work term must be added to the γ_s term. $4\gamma_s$ then becomes G_c, the strain energy release rate and a stress intensity factor K is defined as $K = \sigma\sqrt{\pi a}$ for an infinite plate. K_c, the fracture toughness, is the critical value of K for crack propagation (K_{Ic} is K_c in tension). Then

$$K_c = \sqrt{EG_c} \tag{8.47}$$

$$= A\sigma_{crit}\sqrt{\pi a} \tag{8.48}$$

If, for a given material, the value of K_c is determined for one value each of σ_{crit} and a, the result can be used to determine the value of σ_{crit} for any other value of a (or size of flaw) *provided the stress–strain system is identical to that in an infinite plate (plane strain conditions)*.

8.3.3 Fracture toughness testing

Test procedures for carrying out such a test are laid down in ASTM 399-83 Standard Test Method for Plane Strain Fracture Toughness of Metallic Materials and the nomenclature used is defined in ASTM E616-82. BS 5447: 1977 stipulates similar methods.

There are a number of specimen designs but the most straightforward compact tension (CT) specimen 1 is shown in Figure 8.37 (Figure A4.1 from ASTM 399). Plane strain conditions are maintained by ensuring that both specimen thickness B and the crack length, a, exceed 2.5 $(K_{Ic}/\sigma_y)^2$. K_{Ic} should be overestimated. Testing under these conditions measures plane strain fracture toughness K_{1c} in tension σ_x. Fracture toughness K_{2c} and K_{3c} can also be measured in xy shear and yz shear respectively. The customary coordinate system is illustrated in Figure 8.38 (Figure 2 from ASTM E 616). If the specimen is too small, shear will initiate at the ends of the crack and the test is then stated to be invalid.

8.3.4 Influence of shape of defect

Fracture toughness specimens contain (so far as can be ensured) cracks which are straight and orthogonal in the specimens. Crack-like defects which occur in practice are often irregular and are usually considered to have the form of an enveloping ellipse transverse to the applied stress.

For an elliptical surface flaw (see reference 35)

$$K = \sigma\sqrt{\left(2\frac{\pi}{Q}a\right)} \tag{8.49}$$

and for an elliptical embedded flaw

$$K = \sigma\sqrt{\left(\frac{\pi}{Q}a\right)} \tag{8.50}$$

where

$$Q \text{ (the shape parameter)} = (\phi^2 - 0.212\sigma^2 \cdot \sigma_y^{-2}) \tag{8.51}$$

and

$$\phi = \int_0^{\pi/2} \left[\sin^2\theta + \left(\frac{a}{e}\right)^2 \cos^2\theta\right] d\theta \tag{8.52}$$

ϕ is the complete elliptical integral of the second kind, σ the working stress and σ_y the 0.2% proof stress.

Q is plotted in Figure 8.39 for elliptical flows having crack depth and semi-minor axis a, crack length and major axis $2c$

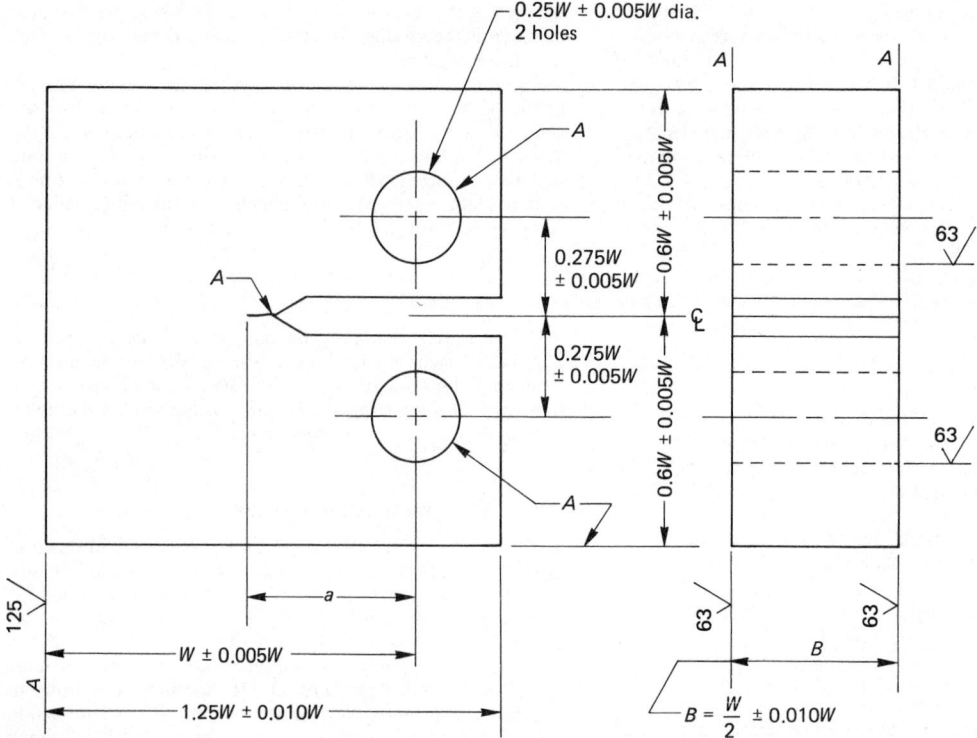

Figure 8.37 Compact specimen C (T) standard proportions and tolerances (reproduced by permission of ASTM). *A* surfaces shall be perpendicular and parallel as applicable to within 0.002*W* TIR. The intersection of the crack starter notch tips with the two specimen surfaces shall be equally distant from the top and bottom edges of the specimen within 0.005*W*. Integral or attachable knife edges for clip gauge attachment to the crack mouth may be used. *W* is a parameter defined to accommodate hollow cylinders. For this type of specimen *W* = 2*B*

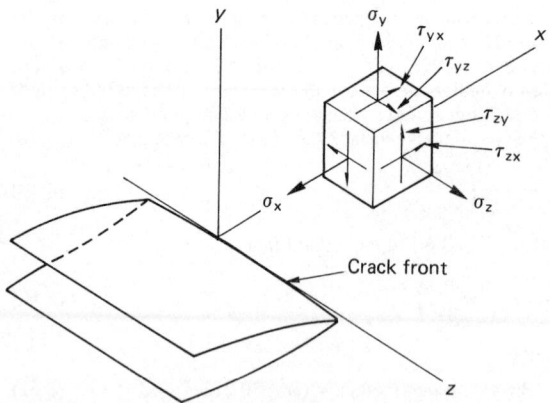

Figure 8.38 Customary coordinate system and stress on a small-volume element located on the *x*-axis just ahead of the crack front (reproduced by permission of ASTM)

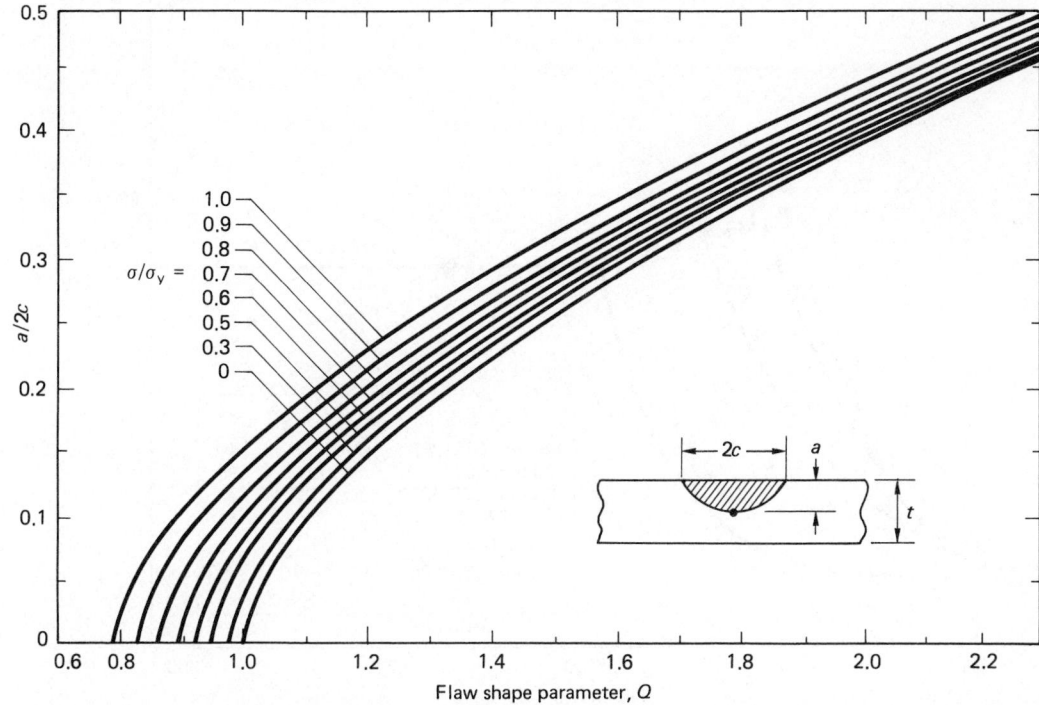

Figure 8.39 Flaw-shape parameter for surface and internal flaws

and a number of values of $\sigma \cdot \sigma_y^{-1}$. The 'corrected' value of a (derived from equations (8.49) or (8.50)) is designated a_p.

8.3.5 Typical ranges of fracture toughness values

The properties of high-tensile steels, aluminium and titanium alloys are such that plane strain conditions can be obtained from specimens that can readily be produced. Typical values are listed in Table 8.2. Linear elastic fracture calculations can be readily carried out on structures made from these materials provided that the size and shape of defect can be determined or estimated.

Many structural steels have lower yield strengths and superior fracture toughness, and rigorous fracture toughness tests at room temperature for such materials would require specimen thicknesses of at least 300 mm. Material of this section and apparatus for carrying out such a test are by no means

Table 8.2 Typical values of fracture toughness and yield strength for steels, titanium and aluminium alloys and pressure vessel steels

	Yield strength σ_y (MPa)	Fracture toughness K_{Ic} (Mn$^{-3/2}$)
Hardened steels	1500–1100	30–160
Titanium alloys	1300– 700	30–150
Aluminium alloys	600– 400	18– 35
A533 Grade B 12-inch pressure vessel steel plate	350	>160

universally available, nor, if they were, would the results necessarily reproduce the behaviour of a component of smaller section size.

When tests are carried out over a range of temperatures on ferritic steels the results correspond to those shown in Figure 8.40 (see reference 36). The high-temperature shelf (the upper shelf) of the fracture toughness value is characteristic of high toughness ductile fracture behaviour.

8.3.6 Post-yield fracture mechanics

When materials fail under ductile fracture conditions, LEFM is not applicable. The extra plasticity associated with the ductile fracture mechanism requires a rigorous elastic plastic fracture analysis.

There are many methods for performing elastic plastic fracture analysis. Each of them is tailored to a different type of problem, but they all reduce to one of two basic failure concepts, the J integral and the crack opening tip displacement δ. The different methods centre around the methodology of calculating and using these two parameters. Moreover, there is an identity between the two parameters, i.e.

$$J = m\sigma_y\delta \tag{8.53}$$

where m is a constant factor, hence there is a common link between all modern methods of fracture mechanics analysis.

8.3.6.1 The J integral method

The J integral was first proposed by Rice:[30,31]

$$J = \int_P \left(u\mathrm{d}_y - T_i\left(\frac{\delta u_i}{\delta x}\right)\right)\mathrm{d}s \tag{8.54}$$

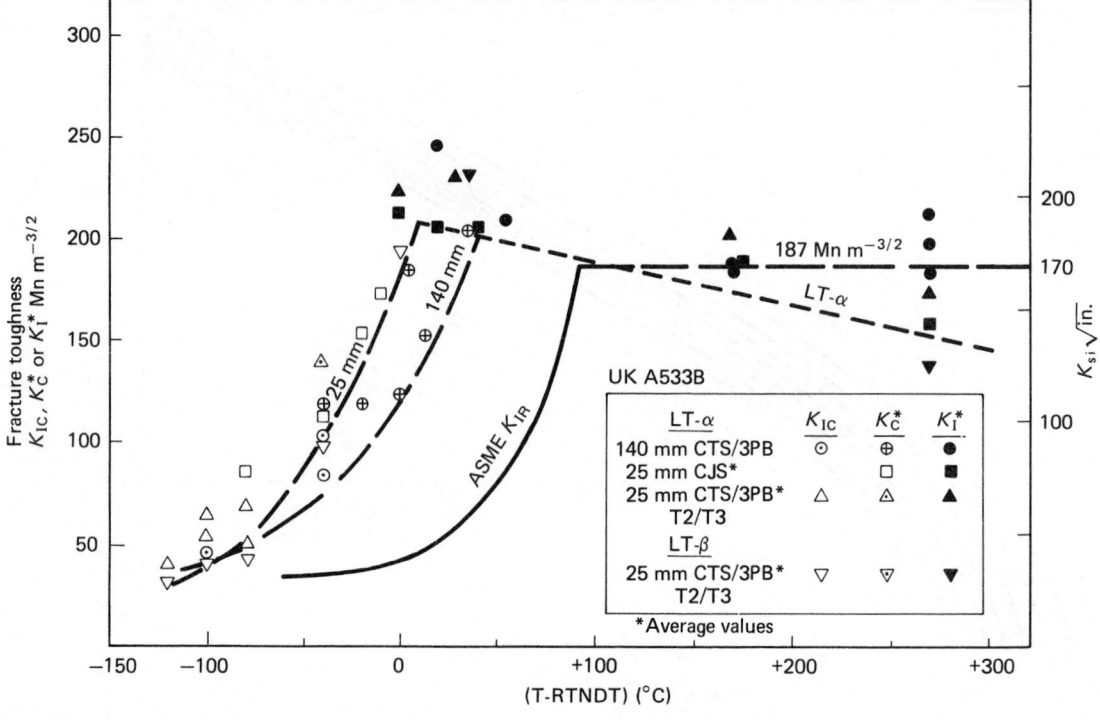

Figure 8.40 Comparison of fracture toughness data for A533 Grade B Class 1 steel of UK origin with the ASME K_{IR} curve and a constant upper shelf plateau of 170 ksi $\sqrt{\text{in}}$ (reproduced by permission of Applied Science Publishers)

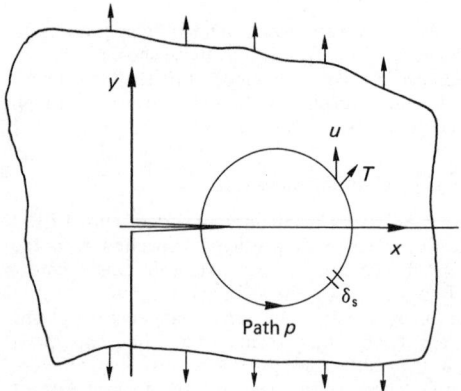

Figure 8.41 Integration path (anti-clockwise direction) about the crack tip for Rice integral (reproduced by permission of Applied Science Publishers)

where J is the path-independent integral (where $J = G$ in an LEFM case), ds is an element of the path P (see Figure 8.41), U is the strain energy density, T_i is the tension vector and $T_i du$ is the work done when components of T move through a distance du.

To perform an analysis, it is necessary to calculate the J applied to a structure under the loading conditions pertinent and to compare this with J_c, the calculated material fracture toughness. Calculation of J applied is often a complicated business, frequently requiring finite element analysis. This is time consuming and expensive and makes the J method difficult to apply. Compendia have been developed[37-39] which use standardized routines to calculate J for a given set of geometries, and these may be used when the geometry of interest is closely simulated to the geometry contained within the compendium. However, for most practical cases, this situation does not apply.

8.3.6.2 The crack-tip opening displacement (COD) method

This method was the first practical method of performing an elastic plastic analysis.[33,40,41] The method depends upon the critical crack opening displacement on a specimen of the material and comparing this with the applied crack-tip opening displacement. The applied COD value is calculated from a curve developed to describe pessimistically the failure condition of a number of tests on cracked wide plates loaded to failure. The method is therefore empirical, but has a large body of supporting data to validate it. Consequently, this method was the main one contained in the British Standard Institute's published document PD6493, 1980.

8.3.6.3 The CEGB's R6 method

The R6 method uses a concept of a failure assessment diagram (FAD) to link LEFM with fully plastic fracture mechanics. It was written as a standardized method which is revised periodically and is now in its third revision, where it is capable of performing a full J analysis for any type of structure. The concept of the failure-assessment diagram simplifies an otherwise very complex analysis by requiring only two basic calculations to be performed. The first is in the LEFM regime and requires the calculation of K_r:

$$K_r = K_I/K_{Ic} \qquad (8.55)$$

and the second is in the fully plastic regime and requires a calculation of L_r:

$$L_r = \sigma/\sigma_u^{(\text{load applied/plastic yield load})} \qquad (8.56)$$

The R6 (and J analysis) failure-assessment diagrams for a carbon manganese steel are shown in Figure 8.42 together with data obtained from centre cracked panel test specimens.

R6 is at the time of writing the most universally applied method of analysis. It is being incorporated into BSI document PD6493, along with a revised version of COD analysis based upon the R6 FAD, and it is being adapted for use in many other countries, most notably in the USA, where a new ASME X1 Appendix will be based upon R6.[33,42,43]

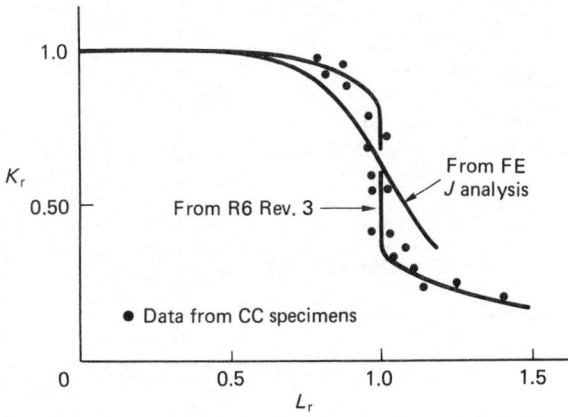

Figure 8.42 FADs for C–Mn steel and tensile test results

8.4 Creep of materials

8.4.1 Introduction

This section describes the behaviour of materials under the combined effects of (constant) elevated temperature and stress.

Creep is the strain developed in a material over a period of time by the application of a (constant) uniaxial or biaxial stress. Creep is sensibly permanent in a metal but may, to some extent, be reversible in a plastic.

Stress relaxation is the progressive reduction in stress in a material which is held for a period of time with one or two of its orthogonal dimensions greater than its rest dimensions.

The creep stress, σ_ϵ, is the stress at which a test piece made from a material strains by a specified proportion (often 1%) when held for a specific time at a specific temperature.

The creep rupture stress, σ_R, is the stress at which a test piece made from a material fractures when held for a specified time t_R at a specified temperature.

The creep rupture elongation is the proportion, expressed in percentage of the original length, by which a test piece made from a material extends before fracture.

The creep range is a range of temperature within which the criterion for design in a material is σ_ϵ or σ_R, *not* σ_y or S_n. The lower limit of the creep range approximates to $0.5T_m$. (T_m is the melting point of a material in °K.)

Static fatigue is a term frequently used to describe the failure of non-metallic materials after a sustained period of loading. In some cases static fatigue is an alternative and misleading designation for creep rupture. In others it refers to a separate phenomenon (see Section 7.6).

8.4.2 Creep and stress rupture testing

Most creep and stress rupture testing is carried out in tension at a fixed elevated temperature on cylindrical specimens 5 mm or more in diameter. Where rupture data only are required the specimens are identical in shape to the corresponding tensile test specimens.

As many as ten specimens may be loaded as a longitudinally connected string in one testing machine. Strain measurements are normally carried out cold on the specimens at infrequent intervals after the furnace holding them has cooled and also following the failure of one of the specimens.

Where precise creep data are also required circumferential vee-shaped ridges are machined on the specimen at the ends of the gauge length. Extensometer systems are clamped on the ridges, and connected to precise measuring equipment. At low temperatures strain gauges may be used. Usually, only one specimen is tested in a machine.

Tests may be carried out in special or protective environments, but normally for components operating in steam or flue gas, air is used. Conditions of testing, including accuracy and reliability of temperature control, stress and strain measurement are laid down in BS 3500 Methods for creep and stress rupture testing of metals and ASTM E139 Standard practice for conducting creep, creep rupture and stress rupture testing.

Creep, in common with most metal properties, is a statistical phenomenon and many tests must be undertaken before data that may be used with confidence in design are obtained. Figure 8.43 gives some idea of the variability in tests carried out by a number of laboratories on one grade of steel of a similar section thickness from a number of sources.

A number of other test procedures have been employed. Rectangular section specimens are sometimes used for sheet or strip materials. Testing with stress systems other than uniaxial tension have included the internal pressure testing of hollow cylinders. In this case, longitudinal tension is sometimes also applied to equalize stress in two orthogonal directions.

In one series of tests torsional shear and longitudinal tension have been applied simultaneously. Uniaxial compression may be used for brittle materials. Bend testing has been employed.

The creep testing of plastics is discussed in BS 4618: Part 1 and ASTM D2990.

8.4.3 Deformation mechanisms and laws

Creep may occur by a number of atomic mechanisms which include: thermally activated glide, viscous glide, dislocation climb and diffusional mass transport through grain interiors and along grain boundaries. The relative contribution of a particular mechanism depends on the value of stress and temperature and the relationships between strain, stress and temperature vary according to which mechanism is operative. Deformation mechanism maps[44] have been produced for a number of materials.

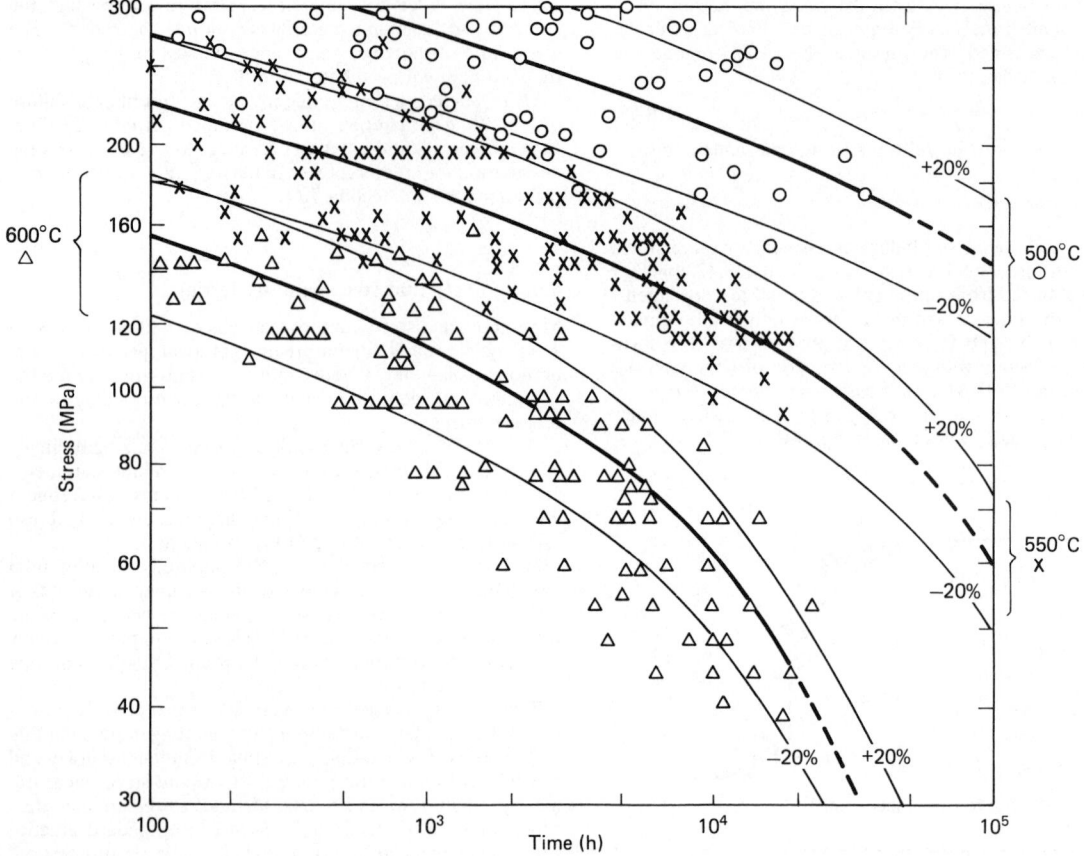

Figure 8.43 Comparison of derived curves with data points, ½Cr ½Mo ½V steel, 1.0% creep strain (reproduced by permission of the I. Mech. E.)

A simplistic mechanism, which forms the basis for most dislocation creep theories put forward by Bailey and Orowan considers that creep results from the competing processes of work hardening and thermal recovery.

'Creep' in metals is the summation of five processes. These include first, an initial, recoverable, elastic extension. This is followed by 'primary creep' during which, in the case of most alloys, the rate of deformation, $\dot{\epsilon}$, decreases and deformation follows approximately the relation

$$\dot{\epsilon} = A't^{1/3} \tag{8.57}$$

where A' is a constant and t is time.

At low temperatures (below 0.3–$0.5T_m$) the creep rate approaches zero at shorter or longer times and the three later stages do not exist. At temperatures above $0.3T_m$ some alloys which have been designated class 1 (for example, Al-3% Mg) exhibit a continuously increasing creep rate in the early stages of creep. At temperatures above $0.5T_m$ primary creep is succeeded by 'secondary creep' during which the rate is sensibly constant with time and at intermediate stress levels, follows the relationship for stress:

$$\dot{\epsilon}_s = A''\sigma^n \tag{8.58}$$

and for temperature and stress

$$\dot{\epsilon}_s = A'''\sigma^n \exp - (Q_c/RT) \tag{8.59}$$

where Q_c is the activation energy for creep.

The exponent n is an important parameter for design under creep conditions because it controls the distribution of stress in complex structures. For pure metals at intermediate stresses it generally varies between 4 and 5, for solid solution alloys it is approximately 3 and it can range as high as 30 for dispersion-strengthened alloys. At very high stresses creep rate varies exponentially with applied stress.

The constant A'' is related to the activation energy of the mechanism that controls the creep process. The activation energy is relatively low at low temperatures but increases to a value equal to that for self-diffusion at $0.6T_m$ where dislocation climb is the dominant deformation mechanism. In this temperature regime and where stress dependence follows a power law, steady-state creep follows the Dorn equation[45]

$$\dot{\epsilon}_s = AD_L \frac{Gb}{kT} \left(\frac{\sigma}{G}\right)^n \tag{8.60}$$

Figure 8.44 Cavitation in a material at a late stage of tertiary creep (produced by replica technique) (reproduced by permission of ERA Technology)

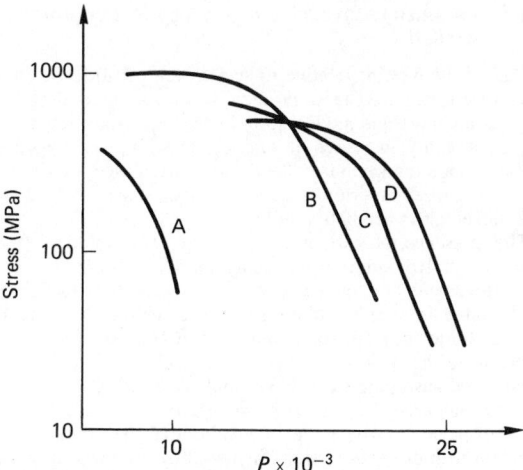

Figure 8.45 Larson–Miller master creep curves (reproduced by permission of Prentice-Hall). A: heat-treated aluminium alloy; B: titanium alloy; C: cobalt-molybdenum steel; D: 18–8 austenitic stainless steel

where A and n are constants that must be measured empirically, b is the burgers vector representative of the discontinuity caused by the strain, D_L is the self-diffusion coefficient, k is Boltzman's constant, T the absolute temperature, σ the applied stress and G the shear modulus.

Secondary creep eventually gives place to 'tertiary creep' which follows a power law, usually approximately cubic $\epsilon = ct^3$

$$\epsilon_t = ct^3 \qquad (8.61)$$

The early stages of tertiary creep may be marked initially by metallographic changes through which the microscopic and sub-microscopic constituents that restrain dislocation movement lose their effectiveness. Precipitates coarsen and carbides spheroidize. Later, cavities develop along grain boundaries (and elsewhere) and cracks originate at triple points where grain boundaries meet. These defects decrease cross-sectional area locally, increase the strain rate, and therefore the size and number of defects. Figure 8.44 shows the microstructural appearance of a material at a late stage of tertiary creep.

Finally, the defects and cavities unite and the material ruptures.

8.4.4 The interpretation of creep and stress rupture data

Creep tests suffer from scatter of data arising from material and testing variables. The replication required to allow for the scatter and the extended life of components, particularly in power plant, make the complete experimental characterization of the creep and stress rupture properties of a material prohibitively expensive. Moreover, the power plant would be built and operating long before such a test programme could be completed.

There is therefore a great incentive to develop a formula which relates stress, temperature and strain and allows long-term behaviour to be predicted from relatively short-term data. Such methods or parameters have been put forward by

Larson-Miller, Manson-Haferd, Goldhoff-Sherby, Manson-Succop, Or-Sherby-Dorm and Monkman Grant.

Most of these parameters derive from:

$$P = \frac{\log t/\sigma \cdot Q - \log t_A}{(T - T_A)^R} \qquad (8.62)$$

where t is the rupture time (in hours), σ is stress, T is test temperature and T_A, t_A, Q and R are constants derived from the experimental test data. They facilitate interpolation and may be used with reasonable success to extrapolate the creep behaviour of relatively simple materials. None of them, however, can deal with a curve with an inflection such as is shown in Figure 8.46 because they all have an increasing downward-slope characteristic. Most creep-resisting steels exhibit an inflection.

It is clear that extrapolation from data up to time A to time B will give an optimistic value, while extrapolation from time B to time C will give a pessimistic value of rupture strength. The design strengths of many creep-resisting steels have been determined from the second type of extrapolation. The results of this are considered in Section 8.4.8.

It is not possible to place the techniques in order to merit. Some fit some data best while others are more suitable for other data. That technique which best fits the data under consideration should be chosen. The Larson–Miller method, which has been most extensively used, assumes that the steady-state creep rate is equivalent to the reciprocal of the time for complete failure to occur. The rate equation is

$$P = T(\log t_R + C) \qquad (8.63)$$

where P is the Larson–Miller parameter, C is an empirically determined constant (approx. 20), t_R is the time to rupture (in hours) and T is the absolute temperature. Values of P are determined over a range of stresses, usually from short-term results at high temperatures.

Figure 8.45 shows a series of master Larson–Miller Curves for different materials from which creep rupture performance may be derived. Too much confidence should not be placed on the results of extrapolation to conditions diverging widely from those of the tests actually carried out.

8.4.5 Parameters favouring high creep and rupture strengths

A high melting point relative to service temperature favours resistance to creep. This trend may, however, be modified if there is an allotropic modification in the solid state, because atomic mobility increases at temperatures approaching the phase change temperature. Therefore, austenitic steels have creep properties superior to those of ferritic steels, and cobalt and nickel alloys are better still.

The presence of certain elements in solid solution (for example, molybdenum in austenitic steels) restrains dislocation movement and increases creep resistance. Precipitates precipitated from solid solution increase resistance to creep provided that they are stable and do not coarsen. The most effective is the γ' precipitate in nickel alloys. Carbides in ferritic and austenitic steels have similar effects.

Very high creep resistance is imparted by the presence of hard dispersed particles such as alumina in metals. The problem with dispersion-strengthened alloys is that it has not yet proved possible to confer strengthening at lower temperatures comparable with the improvement in creep resistance at the higher temperatures.

Other things being equal, a material with a coarse grain structure has a higher creep resistance than one with a fine grain structure. In particular, castings which usually have a coarser grain structure (and often also a cast intermetallic network) have substantially higher resistance to creep than forgings. This more than compensates for their lower ductility. Single crystals are best of all.

8.4.6 Design in the creep range

For any material the lower temperature limit of the creep range depends on the design lifetime, which may vary from seconds for a missile to 10^5 hours for power plant and on the code to which a component is designed. For a short-lived component control by σ_Y extends to a higher temperature than for a long-lived one. If a design code lays down a relatively high safety factor on σ_R compared with that on σ_Y, σ_R takes over at a relatively lower temperature than when this situation is reversed.

The lower limit of the creep range varies therefore from industry to industry and from one country to another, and a higher proof stress material may have advantages economically in some locations and for some applications and a higher rupture stress in other cases.

In the creep range at least two out of three parameters are required for design. For rotating and other machinery where minimum clearances must be maintained it is essential to know the maximum total strain for given values of stress, time and temperature.

For components where strain is not critical the value of the creep rupture stress, σ_R, must be known. σ_R varies statistically and design codes specify a factor, usually $0.6 \times$ mean or/and $0.8 \times$ minimum for the design stress.

Whether the design is strain or rupture controlled, it is essential to know the stress distribution, and for this a knowledge of the stress/strain exponent n is required. When the value of this parameter differs significantly from unity, stress analysis becomes very much more complex than under linear elastic conditions.

The stress distribution can be calculated reasonably accurately for straightforward geometries such as a hollow cylinder, but component geometries such as a cylinder/cylinder intersection require finite-element analysis and have, in the past, been designed using empirical thickness compensation rules which depend on experience for safety in design.

Provided that data such as are described in Section 8.4.7 are available, computer programs can analyse designs as complex as the cylinder/sphere intersection. 'Bearsafe', produced by the CEGB Berkeley Nuclear Laboratory, is a typical program. In principle, these programs are capable of analysing non-axisymmetric designs such as the cylinder/cylinder intersection, but the cost of computer time would be prohibitive. The cylinder/cylinder intersection has to be stressed by a procedure involving analogy with the cylinder/sphere.

A further parameter derived from creep rupture testing is rupture ductility, but this relates more to performance under high strain fatigue loading (see Section 8.3) rather than to creep.

8.4.7 Creep of ferritic pressure vessel steels

The creep properties of the ferritic pressure vessel steels have received very intensive study. The results have been analysed and reported in reference 49. Examples relating to one material are taken from this publication to illustrate characteristics and methods of presentation of creep and stress rupture data. The material selected is $\frac{1}{2}$Cr $\frac{1}{2}$Mo $\frac{1}{4}$V steel of light section.

Elevated temperature proof stress values which are design criteria below the creep range are listed in Table 8.3 and hot tensile curves in Figure 8.47.

Table 8.3 Elevated temperature proof stress values of a ½Cr ½Mo ¼V steel (MPa)

Proof strain (%)	Temperature (°C)						
	20	100	200	300	400	500	600
0.1	265	254	240	170	150	130	115
0.2	270	259	235	192	177	155	135
0.5	272	261	235	240	220	185	158
1.0	286	269	271	275	260	205	167

The 0.2% proof stress is that given by ISO 2604/IV steel – Figure 8.47 gives the averaged experimental stress–strain curves to 5% strain, the post-yield stresses having been scaled down to comply with this minimum 0.2% proof stress. Other proof stresses in the table have then been taken from Figure 8.47. (Reproduced by permission of the IMechE.)

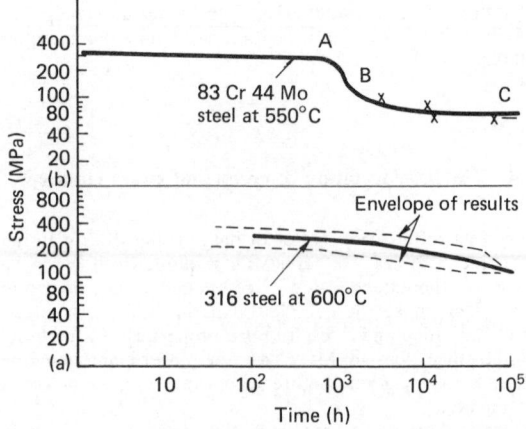

Figure 8.46 Comparison of rupture behaviour of 316 and a Cr 0.5 Mo steel showing an inflected curve

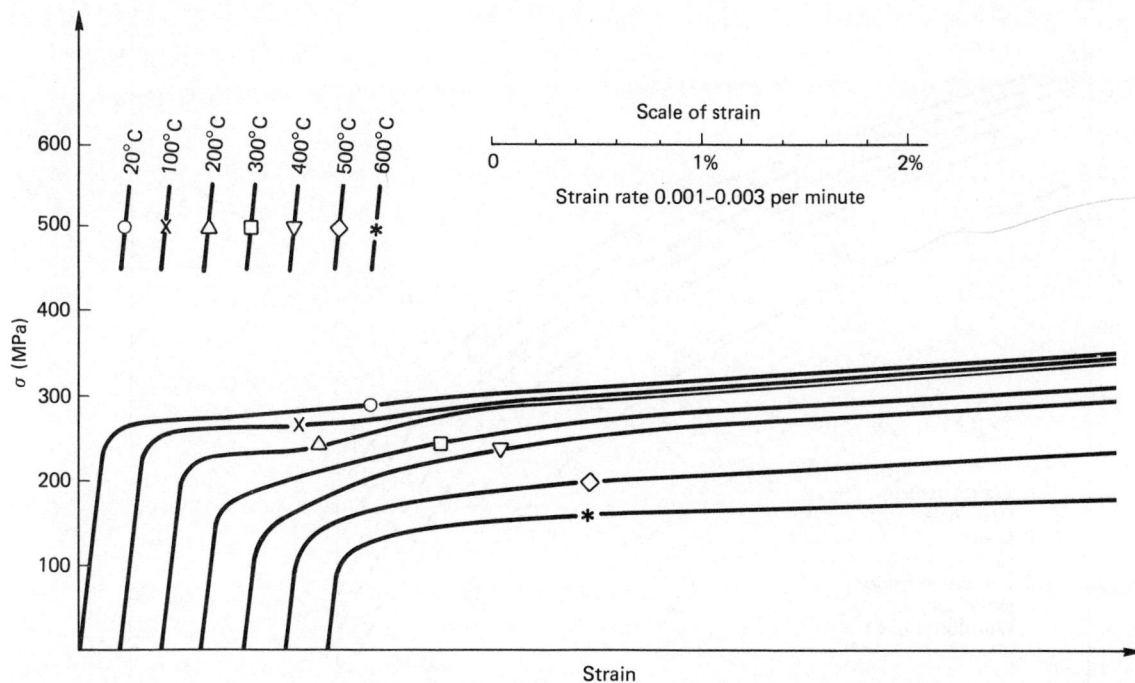

Figure 8.47 Hot tensile curves for ½Cr ½Mo ¼V steel (light sections) referred to minimum 0.2% proof stress (see Table 8.3) (reproduced by permission of the I. Mech. E.)

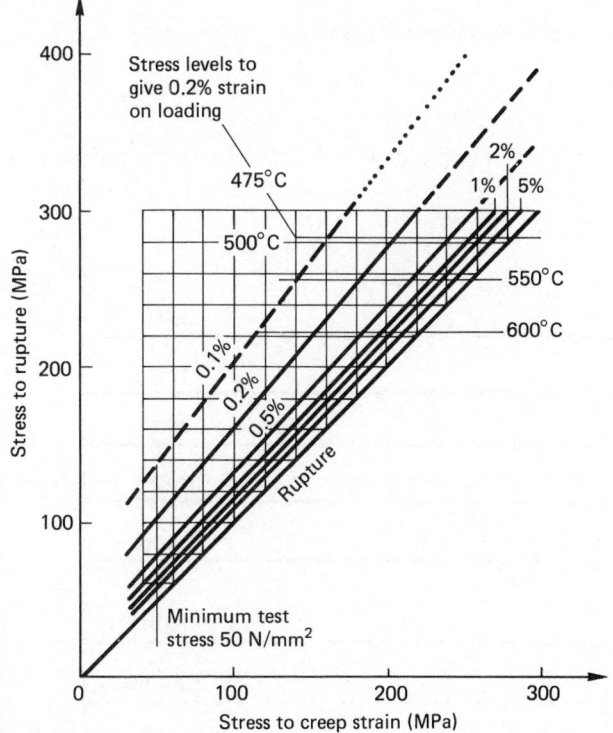

Figure 8.48 Parametric lines relating stress to rupture and stress to a certain creep strain in the same time and temperature, ½Cr ½Mo ¼V steel, light sections (0.1% line obtained by extrapolation) (reproduced by permission of the I. Mech. E.)

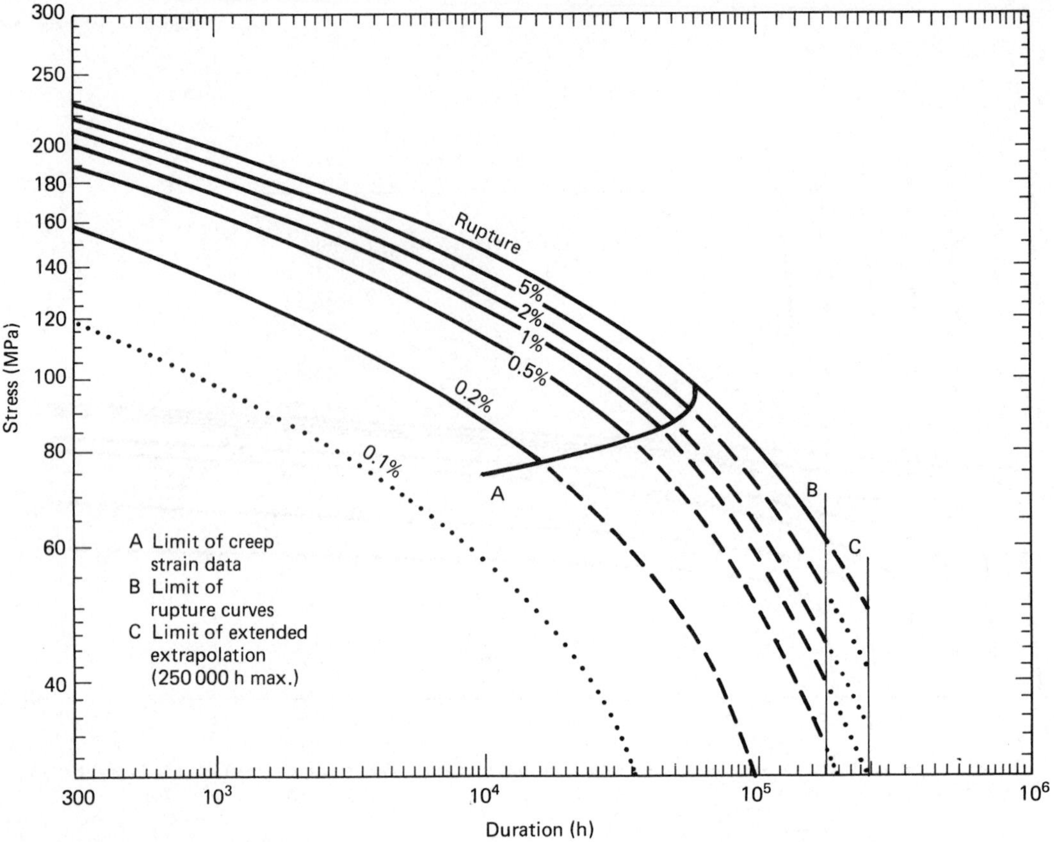

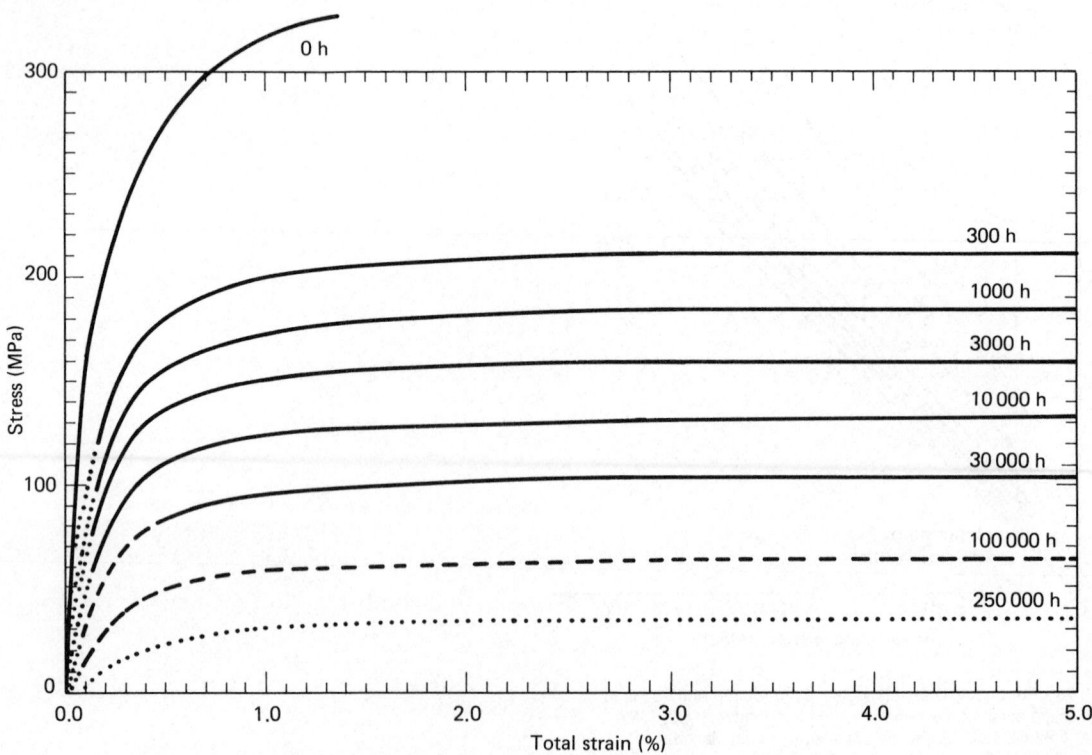

Figure 8.49 Creep strain and rupture contours, ½Cr ½Mo ¼V steel, light sections (reproduced by permission of the I. Mech. E.)

Figure 8.50 Isochronous curves of stress against total strain, ½Cr ½Mo ¼V steel, 550°C, light sections (reproduced by permission of the I. Mech. E.)

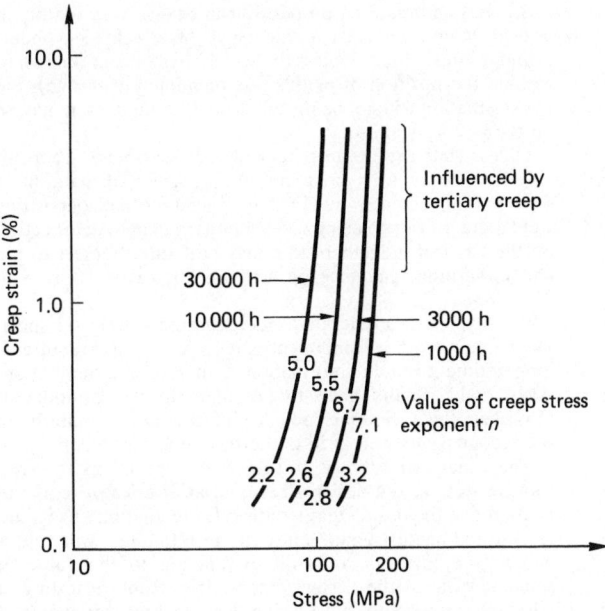

Figure 8.51 Isochronous curves of creep strain against stress, ½Cr ½Mo ¼V steel, light sections, 550°C (reproduced by permission of the I. Mech. E.)

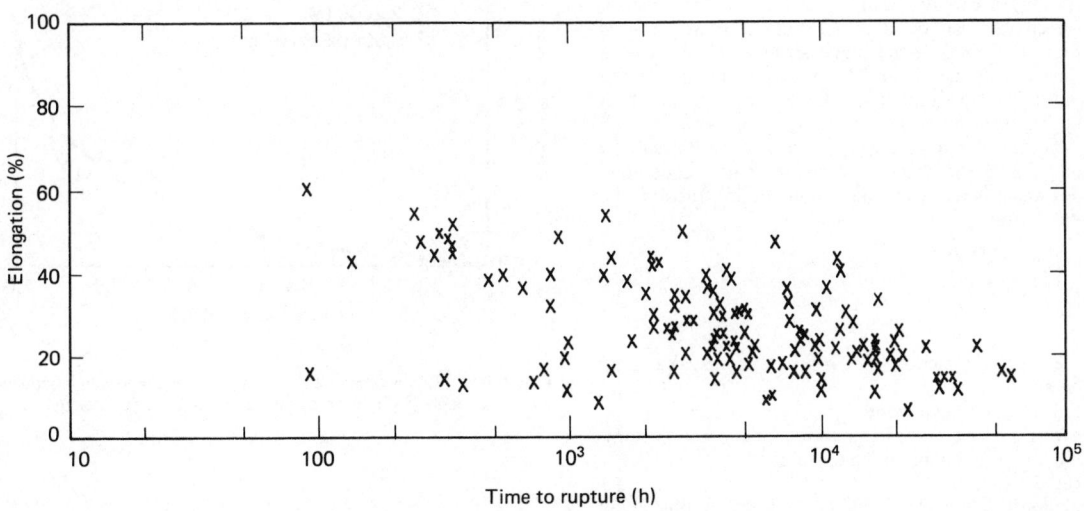

Figure 8.52 Creep rupture elongation for ½Cr ½Mo ¼V steel, 526–575°C, light sections (reproduced by permission of the I. Mech. E.)

A typical parametric equation relating the stress to rupture, σ_R (MPa), to the stress, σ_ϵ, to produce a given strain ϵ% in the same time and at the same temperature is

$$\sigma_R = \left(1 + \frac{0.0287}{\epsilon}\right)\sigma_\epsilon + 11.58 + \frac{7.714}{\epsilon} - \frac{0.1621}{\epsilon^2} - 0.2\epsilon^2 \tag{8.64}$$

between 30 and 300 MPa stress, 0.1–5% creep strain and at temperatures between 475° and 600°C.

Parametric lines relating stress to rupture and stress to a certain creep strain are shown in Figure 8.48 and typical creep strain and rupture contours in Figures 8.49 and 8.50. Typical isochronous curves of creep against total strain are illustrated in Figure 8.49 and isochronous curves of creep strain against stress showing a range of values of the creep stress exponent n are given in Figure 8.51. Figure 8.52 shows creep rupture elongations.

8.4.8   Remanent life

It is often necessary, when an item of plant approaches its design life or has been operated for a period under more severe conditions than were allowed for in design, to assess whether and for how long operation can safely continue. Life-fraction rules have been devised which allow for the effects of out-of-design operation provided that the times, stresses and temperatures are known. The simplest empirical rule defines the life fraction as

$$[(t/t_R)(\epsilon/\epsilon_R)]^{1/2} \qquad (8.65)$$

Often, the design conditions for plant impose very little creep damage from service exposures up to 200 000 hours. This applies to carbon, carbon molybdenum and chromium molybdenum steels operated at ASME boiler and pressure vessel code allowable stress. Some of the reasons for this have been discussed in Section 8.4.4.

A further reason is that the effect of oxidation on small-sized rupture specimens tested for long times in air is more severe than it is on relatively large section plant items and σ_R values are therefore pessimistic. An accurate determination of the remanent life of plant which, for these reasons, has been overdesigned will, by deferring replacement, provide substantial cost savings.

There are two experimental procedures for examining a plant item whereby its remanent creep rupture life may be determined; metallographic examination and measurement of rupture properties after service. Metallographic examination can be used to detect structural changes or creep cracking and cavitation damage.

Structural changes in creep-resisting steels consist mainly of spheroidization of carbides. This usually marks the onset of tertiary creep but can be detected only if there is good documentation of the original metal structure. This type of creep damage may be removed completely by giving the component a heat treatment similar to the original heat treatment. The rupture properties are fully restored.

Creep cavitation damage can be assessed quantitatively by determining the number fraction of cavitated or cracked grain boundaries which is related to the remanent life fraction, t/t_r, by the relationship:

$$A = 1 - \left(1 - \frac{t}{t_r}\right)^{(\lambda-1)/n\lambda} \qquad (8.66)$$

where

A	= the number fraction of cavitated or cracked grain boundaries,
n	= the creep rate exponent,
λ	= ϵ_r/ϵ_s (ratio of rupture strain to secondary creep strain),
e_s	= $\dot{e}_{min}t_r$ (e_{min} is minimum creep rate), and
t_r	= rupture life.

The relationship between A and t/t_r is shown graphically in Figure 8.53 and the time that is expected to be taken for an observed defect to grow is illustrated in Figure 8.54.

The most effective non-destructive technique for monitoring microstructural degradation and crack initiation is surface replication. The area to be examined is prepared as for conventional metallography and a cellulose acetate film softened by immersion in acetone is applied to the surface. The film is peeled off when dry and examined by optical and scanning electron microscopy. A typical replicated microstructure is shown in Figure 8.55. If the results of examination indicate that cavities or cracks are present, but that a substantial time interval is likely to elapse before failure, this is not necessarily a reason for complacency.

Stresses and material properties can change very sharply in critical geometries such as the angle of a welded cylinder/cylinder intersection. Unless the surface examined is known to include the position of peak stress or minimum strength the worst situation will not be revealed and the component may be on the point of failure.

Unless staff experts in remanent life assessment are available it is better to remove cracked or cavitated material as soon as conveniently possible by machining and depositing weld metal or by replacing the component than to defer action on the basis of judgement that may be faulty. Subject to the above warning, guidance on action to be taken is given in Table 8.4.

Post-exposure accelerated creep rupture testing on specimens taken from a component is more costly and requires a longer time to yield information than replica examination. Also, weld repairs have to be made to the component. However, this technique provides information on material in the secondary as well as in the tertiary stages of creep.

The usual procedure is to subject the specimens to creep rupture tests at the design stress and at a series of temperatures above the design temperature. If the log times to failure are plotted against temperature the resulting line will approximate the linearity and will extrapolate to the value of remanent life at the design temperature. Rupture testing at constant temperature but varying stress and measurements of creep rate on material taken from service have also been used to determine remanent life.

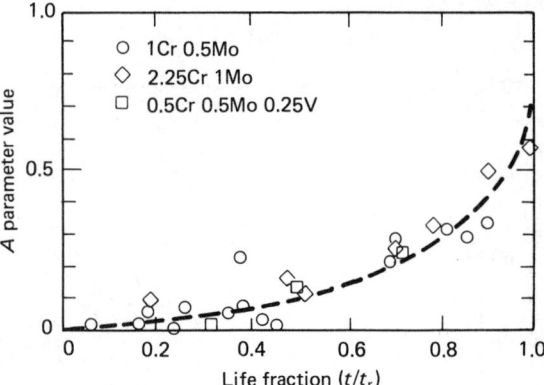

Figure 8.53   Development of creep cavitation damage with creep life for ferritic steels (reproduced by permission of ERA Technology)

Table 8.4   Suggested course of action arising from observation on replica (after ERA Technology)

Observed phenomena	Action
Isolated cavities	Observe
Oriented cavities	Observe. Inspect again after appropriate interval
Cavity linkage microcracks	Repair or replace at next convenient outage
Macrocracks	Repair or replace immediately

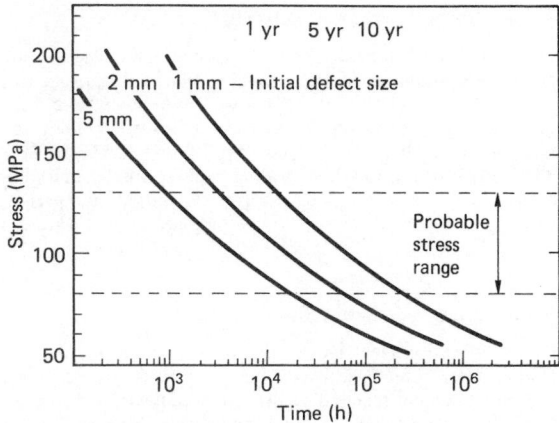

Figure 8.54 Prediction of time for defect to grow to 10 mm depth (reproduced by permission of ERA Technology)

Figure 8.55 Appearance of cavities revealed by replica technique (see Figure 8.44) (reproduced by permission of ERA Technology)

8.5 Fatigue

8.5.1 Introduction

Fatigue is the failure of a metal under repeated (or cyclic) loading under stress(es) lower than would lead to failure if applied singly. Failure occurs by the propagation, under repeated loading, of a crack which may have initiated at a surface or started at a defect at the surface or internally. This section is confined to fatigue of metals but analogous phenomena occur in polymers (see Section 7.5) and composites (see Section 7.7).

There are two basic types of fatigue; *high-strain fatigue* in which the maximum tensile stress in the cycle is larger in magnitude than the yield stress of the material and leads to failure in from 2 to 10^5 cycles; and *high-cycle fatigue* in which the stress may be smaller in magnitude than the yield stress and failure occurs at more than 10^5 cycles. Stresses are usually applied mechanically. *Thermal fatigue* failure is caused by stresses arising from cyclic temperature gradients. Historically, the term has also been applied to the distortion and

weakening of tin (and, presumably, other anisotropic materials such as uranium) which derives from the thermal cycling of adjacent crystals of differing orientations (and therefore differing thermal expansion coefficients).

The term 'fatigue' is sometimes applied to the failure of ceramics (see Section 7.7) or polymers (see Section 7.5) under constant loads of long duration. *Corrosion fatigue* refers to the situation which arises when the influence of the environment is such as to accelerate failure significantly relative to fatigue in air. *Creep fatigue* is a term loosely applied to describe the influence of interactions between creep and fatigue.

The study of fatigue has recognized a number of parameters which, insofar as they have not been defined in Sections 7.1 and 8.3, are listed and defined in Section 8.5.9 (see Table 8.5).

Failure by fatigue takes place in three stages: initiation, propagation and sudden failure of the remaining cross-section by rupture or buckling. (See note to Section 8.3.1.)

Table 8.5 ASTM standards and recommended practices for fatigue testing

ASTM E4 Practices for Load Verification of Testing Machines
E6 Definitions of Terms Relating to Methods of Mechanical Testing
E8 Methods of Tension Testing of Metallic Materials.
ASTM E206 Standard Definitions of Terms Relating to Fatigue Testing and the Statistical Analysis of Fatigue Data
ASTM E337 Test Method of Measuring Humidity with a Psychrometer (The Measurements Wet-Bulb and Dry-Bulb Temperatures)
ASTM E466 Standard Practice for Conducting Constant Amplitude Axial Fatigue Tests of Metallic Materials
ASTM E467 Standard Practice for Verification of Constant Amplitude Dynamic Loads in an Axial Load Fatigue Testing Machine
ASTM E468 Standard Practice for Presentation of Constant Amplitude Fatigue Test Results for Metallic Materials
ASTM E513 Standard Definitions of Terms Relating to Constant Amplitude Low-Cycle Fatigue Testing
ASTM E561 Practice for R-Curve Determination
ASTM E606 Standard Practice for Constant Amplitude Low-Cycle Fatigue Testing
ASTM E739 Standard Practice for Statistical Analysis of Linear or Linearized Stress-Life (S-N) and Strain-Life (ϵ-N) Fatigue Data
ASTM E742 Standard Definitions of Terms Relating to Fluid, Aqueous and Chemical Environmentally Affected Fatigue Testing

8.5.2 Fatigue crack initiation

A fatigue crack may initiate at the surface of a metal when it is subjected to cyclical shear in the plane of the surface. The presence of a tensile stress is generally considered essential. The commonest (and most easily understood) mechanism results from slip bands ratcheting on two planes at separate parts of the cycle, generating tongues of metal and intrusions (see Figure 8.56 and reference 51) which may be observed microscopically. Other observed phenomena are tubular holes which are associated with nucleation on persistent slip bands

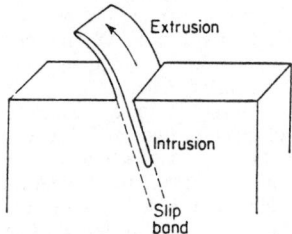

Figure 8.56 Fatigue initiation caused by extrusion and intrusion on a slip band (reproduced by permission of John Wiley)

and twin boundaries. Grain boundaries are also nucleation sites – the nucleation process in this instance being the creation of voids and their coalescence to form a crack. This nucleation process is particularly prevalent when creep, and consequential grain boundary cavitation, play a part in accelerating failure by fatigue. Increasing hardness reduces shear strain and therefore increases the stress required to initiate fatigue. Reducing grain size, particularly of a softer phase, also increases resistance to nucleation.

Fatigue cracks grow very slowly during the nucleation stage and under the lower stresses at which high-cycle fatigue occurs this stage may last 75% of the fatigue life of a component. Cracks may stop growing altogether because of (1) some stress relief effect (if they have started from a very sharp defect), (2) some form of strain ageing, (3) some inherent saturation mechanism or (4) environmental action.

If the fatigue crack growth curve (see Figure 8.57) is asymptotic to a value of stress intensity ΔK_{th} there will be a value of load range ΔS below which failure will not result, however much of the number of cycles, N, is increased. Half this value of ΔS is called the endurance limit S_f. This is the case with steels and titanium alloys.

On the other hand, an initiated intrusion may extend along the slip plane until the stress intensity factor range ΔK reaches a value at which a crack will propagate.

When the fatigue crack growth curve shows no sign of becoming asymptotic to any value of stress intensity there is no endurance limit S_f and an increase in the value of N will lead to failure at a lower value of ΔP. This has generally been considered to be the case with aluminium alloys but the fatigue crack growth curve may become asymptotic to very low values of ΔK even with these (see Section 8.5.5).

8.5.3 Fatigue crack propagation[55,56]

When the fissure produced by crack initiation reaches the size at which the stress intensity range in the cycle is sufficient the crack propagates in a direction normal to the tensile stress. The characteristics of fatigue crack propagation are summarized in Figure 8.57. This shows three regions of fatigue crack growth. The first region, slow crack growth, may exhibit a fatigue crack growth threshold stress intensity range, ΔK_{th} below which cracks do not propagate. The magnitude of the threshold value is sensitive to both stress ratio R and environment, and in some materials is believed not to exist.

During the second stage the log-log plot of ΔK versus da_p/dN usually follows a straight line corresponding to the relationship

$$\frac{da_p}{dN} = C(\Delta K)^n \qquad (8.67)$$

where C and n are constant for a given material, stress ratio R and environment.

8.5.4 Unstable crack growth

During the third stage crack growth rate increases rapidly and becomes unstable as failure of the specimen is approached. This instability is due (in the case of high-strength low-toughness metals) to the peak stress intensity equalling the fracture toughness K_{Ic}. In the case of ductile metals, the instability is due to the peak loading equalling that required to cause (in the size of specimen generally used to measure fatigue crack growth) plastic flow of the reduced metal section behind the crack.

8.5.5 Final specimen failure

Unless vibration, leakage or the results of non-destructive testing have already caused the component to be taken off-load, a stage is reached, as the length of the crack increases and the thickness of residual material behind the crack decreases, when the component fails either by brittle fracture or by elastic collapse.

The appearance of fatigue fracture is shown in Figure 8.58. Illustrated in Figure 8.58(a) is a low-magnification view showing origin, leading lines and bench marks where fatigue has been interrupted.

8.5.6 Parameters that influence fatigue performance

For simplicity, this account has so far assumed that the component or specimen is smooth and uniform, the load constant is sinusoidal in form with constant amplitude, and the environment is dry air at room temperature. The following parameters must be considered in design and testing.

1. The form of the cycle may be sinusoidal, triangular, square or trapezoidal; the frequency of the cycle may be constant or variable; and the load amplitude may be constant or vary in a random or regular manner. Variations in frequency, cycle shape or hold time under load have significant influence (a) when the environment is aggressive, (b) in high-strain fatigue and (c) at elevated temperature (when creep plays a part in the process). They have little influence on region two high-cycle fatigue crack growth rates or on endurance value in benign environments. Increase in R value has usually little effect on the component n of creep crack growth (although it sometimes causes n to reduce). It does cause an increase in the constant C in equation (8.67) and a decrease in K_{th} and, therefore, a decrease in endurance limit. Typical experimental values obtained from fatigue crack growth tests in steel and aluminium are shown in Figure 8.59. The influence of R value and the tendency of growth rates in steel to approach a threshold limit at much higher values of ΔK and da_p/dN than aluminium are apparent.
2. Notches and other stress raisers affect fatigue life. In particular, inclusions or defects may increase stress intensity to a value at which the initiation stage is bypassed and life calculations need only consider propagation and final failure.
3. Environment has a very significant effect on fatigue initiation and propagation – usually adverse, but a more aggressive environment does not always impair fatigue performance.
4. Different materials behave in different ways. Higher-strength and finer-grained materials usually have higher endurance limits, but the effect of increasing hardness on fatigue properties usually reaches a limiting value at a certain hardness. The value of n in equation (8.67) varies

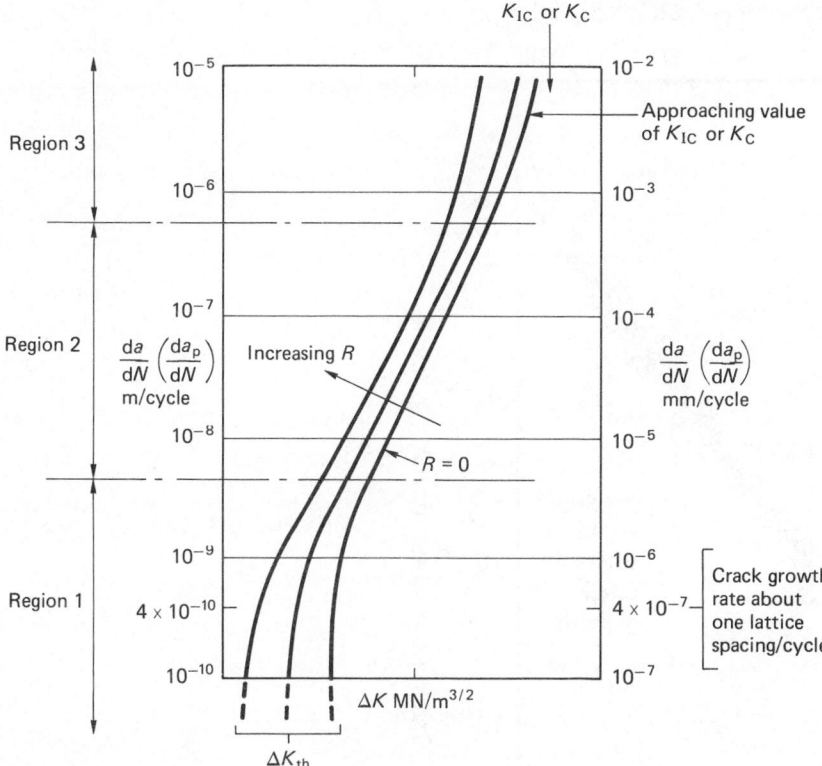

Figure 8.57 Fatigue crack-propagation curves (reproduced by permission of ESDU International from ESDU 80036)

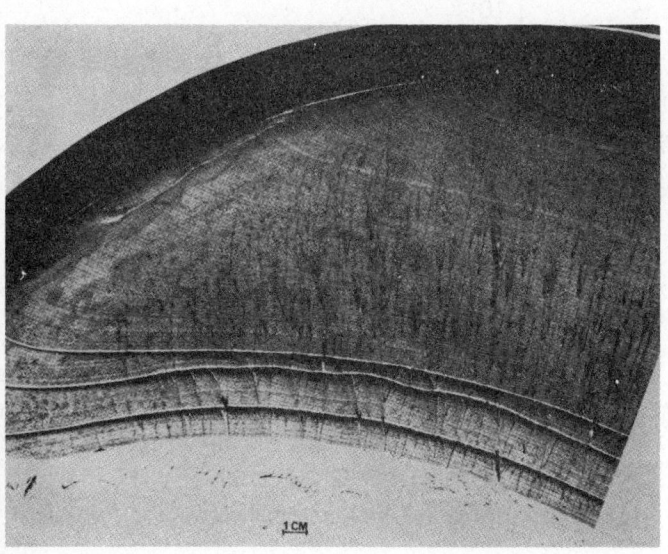

Figure 8.58 Characteristic appearance of fatigue failure at low magnification

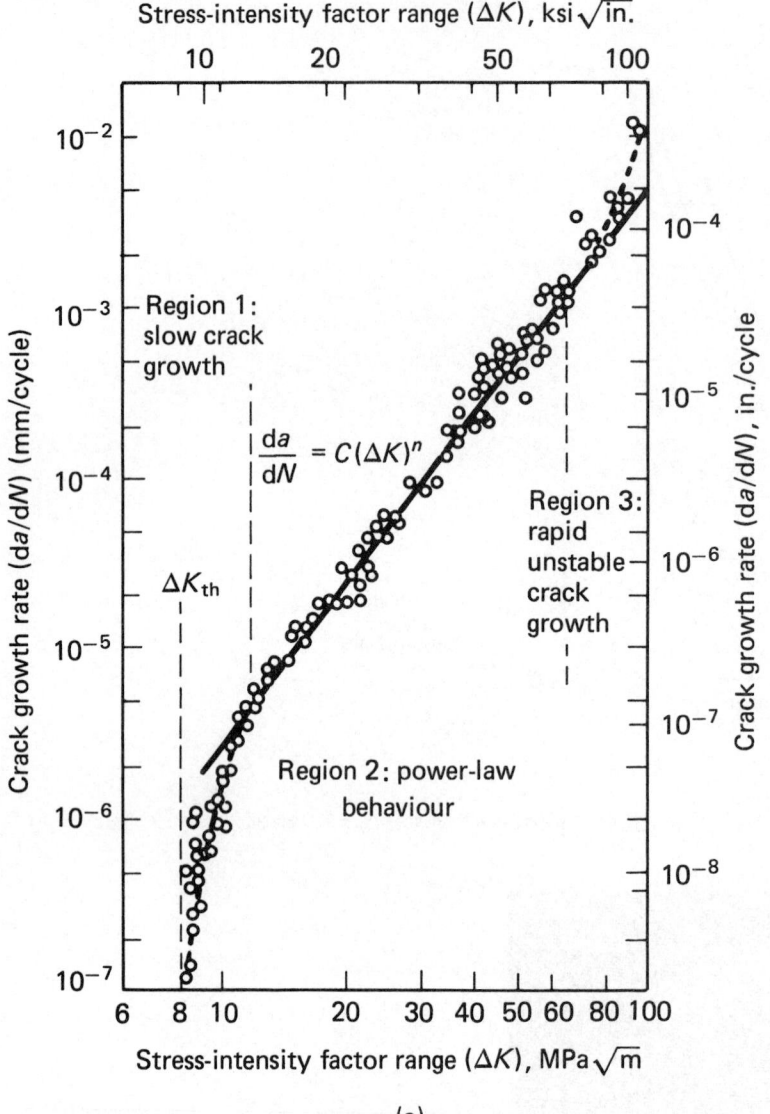

Figure 8.59 (a) Fatigue crack growth behaviour of ASTM A533 B1 steel (reproduced by permission of the ASM *Metals Handbook*); (b) and (c) fatigue crack growth of 7075-T6 solution-treated and artificially aged Al-Zn alloy at high and low values of *ΔK* (reproduced by permission of ESDU International from ESDU 83007)

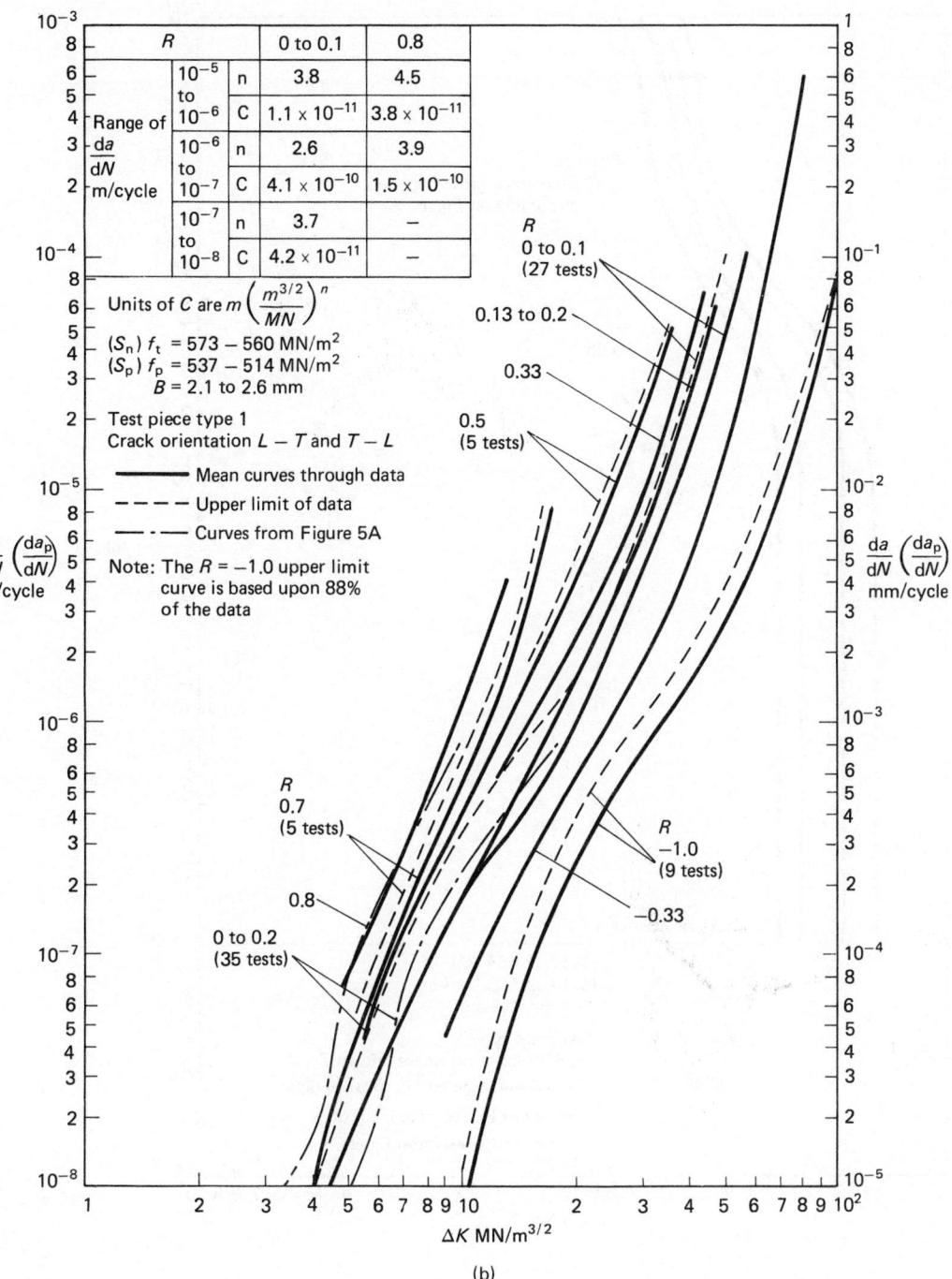

R			0 to 0.1	0.8
10^{-5} to 10^{-6}		n	3.8	4.5
		C	1.1×10^{-11}	3.8×10^{-11}
10^{-6} to 10^{-7}		n	2.6	3.9
		C	4.1×10^{-10}	1.5×10^{-10}
10^{-7} to 10^{-8}		n	3.7	–
		C	4.2×10^{-11}	–

Range of $\dfrac{da}{dN}$ m/cycle

Units of C are $m \left(\dfrac{m^{3/2}}{MN} \right)^n$

$(S_n)\ f_t = 573 - 560\ \text{MN/m}^2$
$(S_p)\ f_p = 537 - 514\ \text{MN/m}^2$
$B = 2.1$ to 2.6 mm

Test piece type 1
Crack orientation $L - T$ and $T - L$

——— Mean curves through data
– – – Upper limit of data
–·–·– Curves from Figure 5A

Note: The $R = -1.0$ upper limit
curve is based upon 88%
of the data

R
0 to 0.1
(27 tests)

0.13 to 0.2

0.33

0.5
(5 tests)

R
0.7
(5 tests)

R
−1.0
(9 tests)

0.8

−0.33

0 to 0.2
(35 tests)

$\dfrac{da}{dN} \left(\dfrac{da_p}{dN} \right)$ m/cycle

$\dfrac{da}{dN} \left(\dfrac{da_p}{dN} \right)$ mm/cycle

$\Delta K\ \text{MN/m}^{3/2}$

(b)

Figure 8.59 (continued)

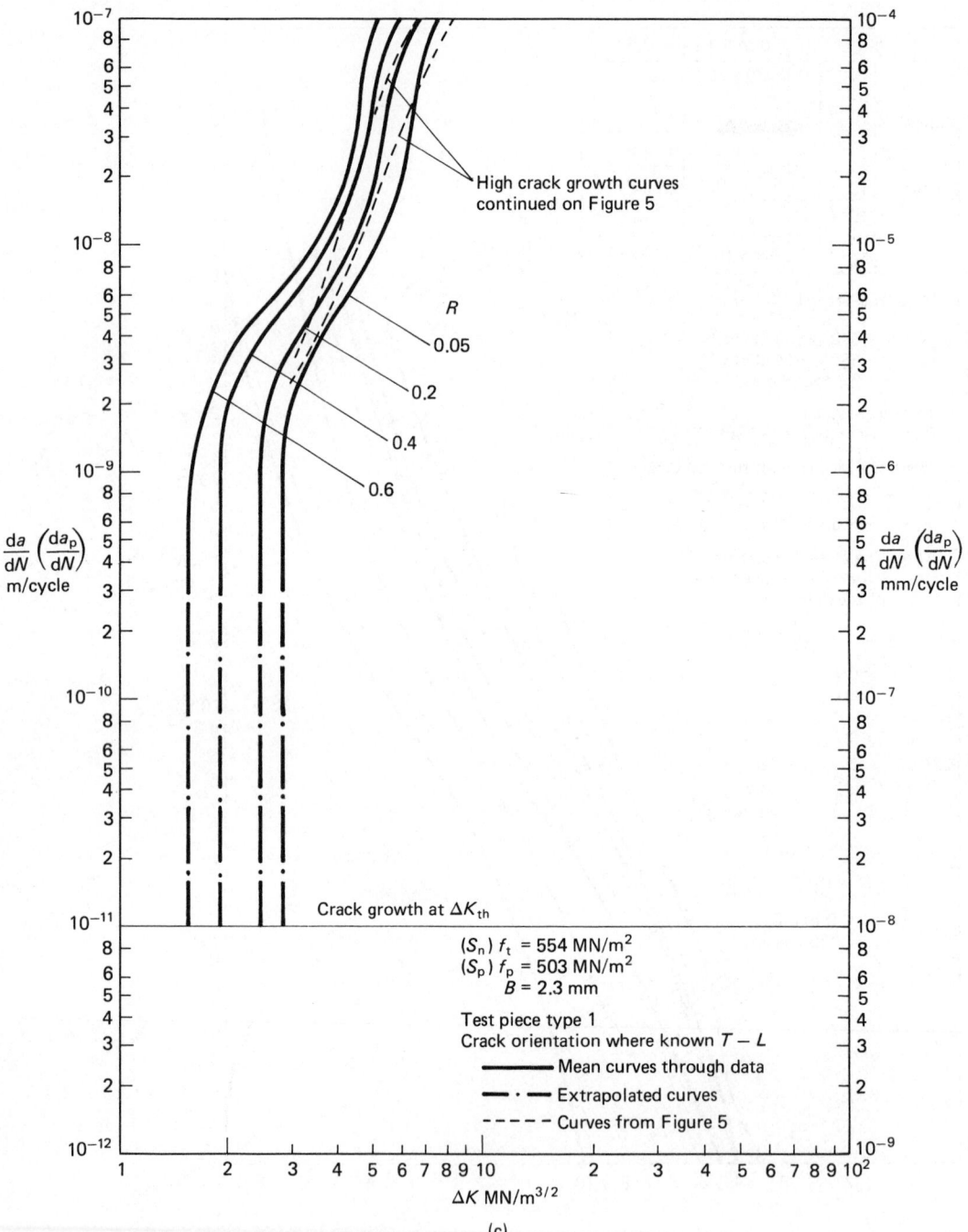

Figure 8.59 (continued)

little from 2.25 for most engineering materials but the value of the constant C is lowest for steels, intermediate for titanium and highest for aluminium alloys. Fatigue properties can be improved by suitable heat treatment, but a heat treatment that increases ΔK_{th} may also increase C and vice versa. This is exemplified in the heat treatment of titanium alloy IMI 834, which can, according to the requirements of the design, be heat treated either from the $\alpha\beta$ phase to optimize fatigue endurance limit or from the β phase to optimize fatigue crack growth (see Section 7.4.3).

8.5.7 Fatigue testing

Design data in fatigue may be collected by three distinct procedures:

1. Rotating beam-fatigue testing machines are cheap and inexpensive to build, simple to operate and represent the least expensive method of determining endurance limits or fatigue strengths. They are, however, limited to high cycle fatigue, a sinusoidal wave form and one value of R ($= -1$).
2. Uniaxial fatigue testing machines are free from these limitations. Stress-controlled tension/tension testing with sinusoidal loading and an R value of 0.1 is standard procedure. Negative R values require very rigid machines and very accurate alignment and expense is augmented if capacity to provide non-sinusoidal wave forms, strain measurement and control and/or special environments are required.
3. Fatigue crack growth testing uses similar test pieces and equipment to that used for measurement of fracture toughness (see Section 8.3). As in (2) above, an R value of 0.1 is standard practice and negative R values, non-sinusoidal waveforms and special environments increase complexity and expense.

Standards issued by the ASTM (listed in Table 8.5) cover uniaxial fatigue testing procedures. The materials engineer is recommended to study the relevant standard before embarking on a programme of fatigue testing.

8.5.8 Design for fatigue

8.5.8.1 High-strain fatigue

The Coffin–Mansion relationship (see Figure 8.60) for failure in high-strain fatigue

$$\Delta\epsilon_p \propto N^{-1/2} \qquad (8.68)$$

(where $\Delta\epsilon_p$ is the plastic strain range and N the number of cycles to failure) has been derived theoretically and shows little variation experimentally for metals and alloys (and indeed for all materials having measurable ductility). The relationship may be used with confidence in design provided:

1. The material does not undergo a structural change during service;
2. The environment does not significantly influence crack initiation or propagation;
3. There is no significant creep at any stage of the cycle.

The influence of environmental attack, which can seriously reduce fatigue resistance, may be gauged from a study of references 70–72. The influence of creep which can, at cycles intermediate between 2 and 10^5, reduce fatigue resistance by as much as 50% is illustrated in Figure 8.61.

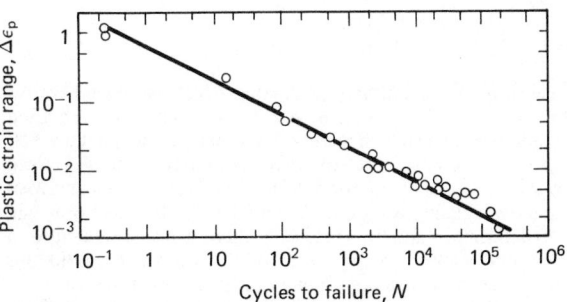

Figure 8.60 Low-cycle fatigue curve ($\Delta\epsilon_p$ versus N) for type 347 stainless steel (reproduced by permission of the ASM *Metals Handbook*)

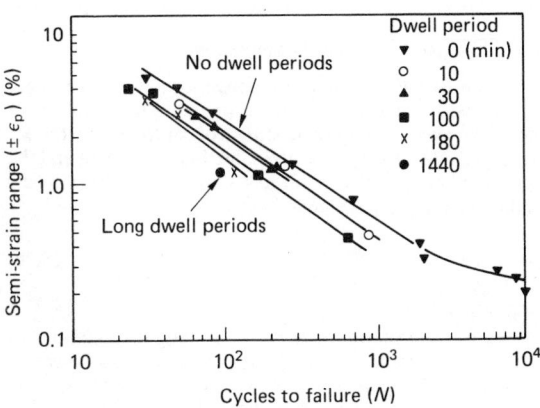

Figure 8.61 High strain fatigue resistance of IP rotor materials

8.5.8.2 High-cycle fatigue[52,59,62,63]

The behaviour of a material in high-cycle fatigue (which is stress related) up to 10^5 cycles may be derived by combining the Coffin–Manson relationship (which is strain related) and the stress–strain relationship in the standard tensile test. Above 10^5 cycles the log-log SN curve continues with a decreasing downward slope. With some materials, the curve levels off at the endurance limit which, for titanium alloys and medium-strength steels, is roughly half the UTS. Conventionally manufactured steels with UTS above 500 MPa show a considerable scatter, the best values of endurance limit continuing along the 50% of UTS line, while the worst are roughly 50% of 500 MPa, which is therefore a limiting value for design. (High-tensile steels refined by vacuum arc or ESR have superior fatigue properties.)

With other materials the log-log SN curve continues its decreasing downward slope. Design with such materials should be based on the fatigue strength at a specified number of cycles (10^8, for example) set by the designer.

8.5.8.3 Influence of mean stress

The effect of mean stress on the fatigue limit for a given number of cycles may be calculated from the empirical relationship[76]

$$S_a = S_f\left[1 - \left(\frac{S_m}{S_u}\right)^m\right] \qquad (8.69)$$

where S_a is the limiting alternating stress associated with a mean stress S_m. (When $m = 1$ this becomes the modified Goodman Law and when $m = 2$, Gerber's Parabolic Law.) If, as in Figure 8.61, a set of curves is plotted relating S_a/S_f to S_m/S_u, and results appropriate to some value of mean stress are available in addition to the value of S_f, the value of m can be determined and the value of S_a thereby also obtained for a required value of S_m/S_u. If S_f only is known it should, for purposes of provisional estimation, be assumed that $m = 1$.

This procedure has the effect that the yield point of the material is exceeded at the peak of the cycle for certain values of S_m and S_a. If a straight line (shown interrupted in Figure 8.62) is drawn between the values of the yield point S_p on the two axes (points A and B) yielding can be avoided by designing to ensure that stress conditions always lie to the left of this line.

8.5.8.4 Influence of stress concentration

The presence of a notch in a component or a test piece might be expected to reduce the fatigue or endurance limit by a factor equivalent to the elastic stress concentration factor K_t. In fact, yielding is likely to occur at the tip of the notch and the actual stress cycle at that point has a lower value than elastic calculations using the formula

$$S_n = \frac{S}{K_t}\left[1 - \left(K_t^1\left(\frac{S_m}{S_u}\right)^m\right)\right] \qquad (8.70)$$

would imply.[75]

A sharp notch therefore reduces the fatigue strength of a strong but less ductile material more than that of a softer more ductile material. Figure 8.62[53] shows the influence of variation in notch radius of a steel heat treated to a range of UTS values.[75]

Because initiation occurs at the surface, fatigue properties are generally very sensitive to surface conditions, including variations in finish, composition and residual stress. Rough finishes introduce notches and stress raisers and the fatigue properties of surfaces improve roughly in the order: sandcast, forged, hot rolled, die cast, cold rolled, machined, ground and polished. A hard surface improves fatigue resistance whereas cracks may develop in a soft surface layer and propagate into the core. Case-hardening steel or work hardening its surface therefore improves fatigue resistance, whereas decarburizing a steel or bonding a surface layer of soft aluminium onto Duralumin worsens resistance. The greatest improvement in fatigue resistance is gained by nitriding the surface of a stress-raising notch.

Introducing a compressive stress by case hardening[67] or shot peening increases resistance to fatigue as does quenching a uniform-shaped component. If, however, a component core in which tensile stresses have been introduced by quenching is bared locally by machining its fatigue resistance is impaired.

8.5.8.5 Influence of variable-amplitude loading

Miner[68] proposed the hypothesis that each load cycle of any amplitude damages the material and that the damage is the same per cycle as would lead to failure under constant amplitude. This hypothesis leads to the equation

$$p_f = D_f\left\{\sum_{b=1}^{b=B}\left(\frac{n_b}{N_b}\right) = \frac{D_f}{D_1}\right. \qquad (8.71)$$

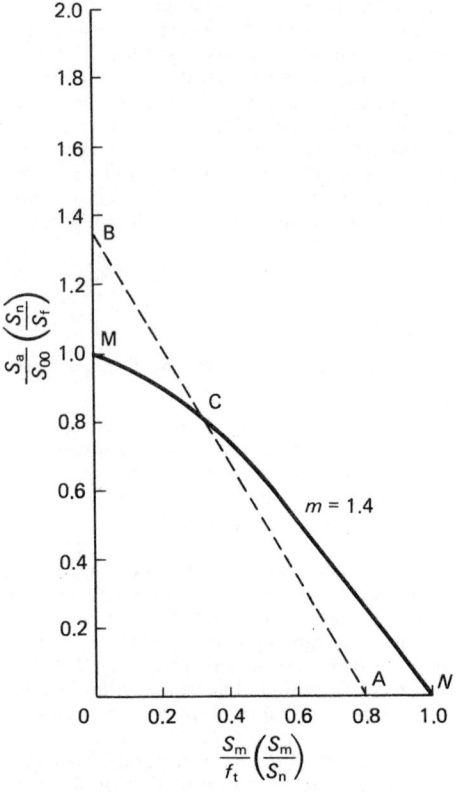

Figure 8.62 Curves obtained by assigning values to m in equation (8.70)

where p_f is the number of programmes to failure,
D is the summation of damage increments,
D_f is the value of D at failure,
B is the number of cycles in cycle by cycle programme or a number of blocks in block programme,
b is the cycle number in any one cycle by cycle programme or block number of blocks in any one block programme,
N_b is the endurance, in cycles under constant amplitude loading at same stress cycle as block or cycle b, and
n_b is the number of stress cycles in programme block b (equals unity for cycle by cycle programme).

This approach has some application when the peaks of all the stress cycles remain below the yield point. However, the stress–strain conditions at the root of a notch where yielding occurs differ significantly between constant- and variable-amplitude loading. In cases where this applies the designer is advised to consult reference 74.

8.5.8.6 Differing stress configurations

It has been stated that the presence of a tensile stress at some part of the cycle is a necessary condition for fatigue failure. However, drive shafts and torsion bars, which, in principle, are loaded in shear, are known to fail by fatigue. Fatigue cannot be caused by compressive loading alone.

Where, as is the case with cylinders containing gas under pressure, a material is subject to fatigue under multiaxial tensile loading it has been stated that the effective fatigue behaviour can be calculated by using von Mises yield criterion.[51]

8.5.8.7 *Influence of environment*

Figure 8.63 shows how atmospheres with varying capacity to introduce hydrogen have very significant influence on ΔK_{th} and on rate of creep crack growth but little (if any) influence on the exponent n.

Etching and electroplating components which will later be subject to fatigue must be very carefully controlled and any hydrogen introduced should be removed by annealing. An environment which favours stress corrosion may also have a catastrophic effect on fatigue failure. These considerations apart, accurate prediction of the effect of a specific environment on fatigue is extremely difficult, although it may well be very significant.

It cannot always be assumed that fatigue performance will be worse in a more aggressive environment. The fatigue resistance, for example, of magnesium in helium is worse than that in carbon dioxide. Oxidizing seawater, by a blunting mechanism, may stop the growth of fatigue cracks which would propagate under the same stress conditions in air.

The designer who wishes to subject a material to fatigue in an environment other than air at room temperature is advised to consult references 52 and 70–72.

8.5.8.8 *Fatigue crack growth calculation*

Fatigue crack-growth calculation techniques[53,55–58,71] are employed to determine the number of cycles required for a crack to grow from the size of an existing defect to one at which the component will fail. The size of the existing defect is the crack size a_{pf} (calculated by procedures described in Section 8.3 from the size and shape of the largest defect) that will *not* be detected by the system of inspection employed, oriented in the most damaging direction possible. The component will fail when the crack attains a size a_{pf}, at which the component will leak, buckle or crack in plane strain K_{Ic} or plane stress K_c under the action of the peak stress of the cycle. Failure in plane strain is usually the most sensitive and, because of its relative simplicity, is the criterion that will be considered here, but the possibility of a soft material buckling should not be forgotten.

Fatigue cracks propagate according to a law of the form

$$\frac{da_p}{dN} = f(R\Delta K) \tag{8.72}$$

and can be represented graphically by curves similar to those shown in Figure 8.57. In region 1 the curves approach a threshold value K_{th} which is very sensitive to R and is of interest in the design of lightly stressed components that are subject to a large number of service loadings.

Designs that lead to fatigue crack growth in region 3 must be avoided. Most cases of fatigue crack growth that are of interest in design occur in region 2, where it can be represented by:

$$\frac{da_p}{dN} = C(\Delta K)^n \tag{8.73}$$

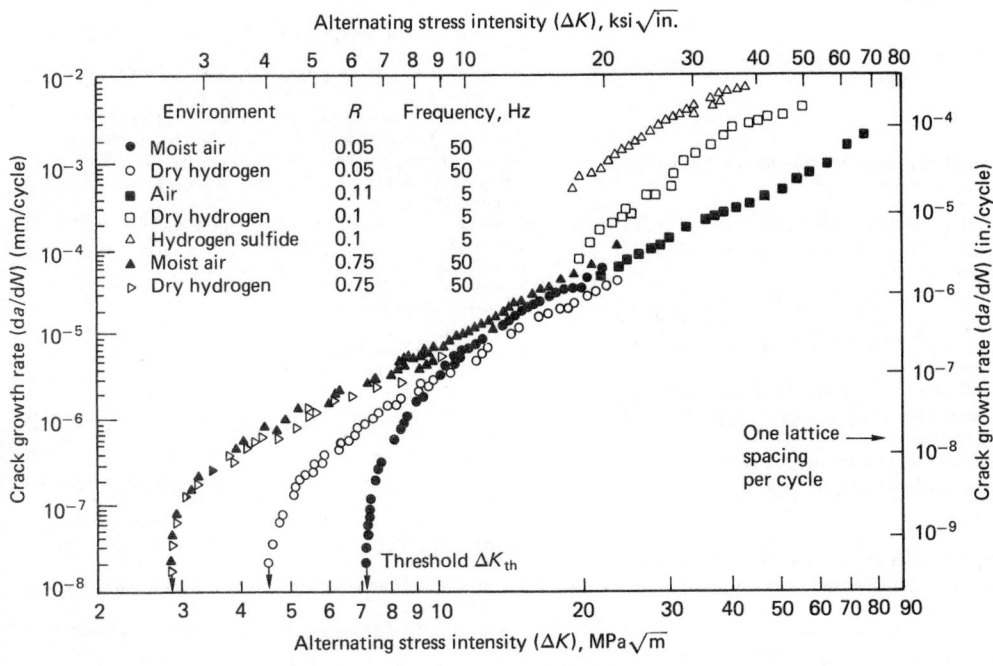

Figure 8.63 Fatigue crack propagation data in a martensitic 2¼Cr IMo steel (SA542-C12) in atmospheres with differing hydrogen potential and at varying values of *R* (reproduced by permission of the ASM *Metals Handbook*)

Many attempts have been made to derive laws that describe the complete crack-propagation curves.[55,56] One example which takes account of high crack growths in region C but not the low crack growth regimes is

$$\frac{da_p}{dN} = \frac{C(\Delta K)^n}{(1 - R)K_c - \Delta K} \tag{8.74}$$

It is essential, if the simpler equation (8.67) is used, to ensure that the value of R in the component is the same as the value from which the test data are derived. The designer can estimate the life of a structure or of a component by analytical or graphical methods.

For plane strain the integration of equation (8.67) yields the solution

$$N\,_{N_i}^{N_f} = \frac{1}{C[S_{max} - S_{min}]\pi^{1/2}]^n} \left[\frac{a_p^{(1-n/2)}}{1 - n/2} \right]_{a_{pi}}^{a_{pf}} \tag{8.75}$$

If N_f, the number of cycles at which a component will fail, is calculated from equation (8.75) to be smaller than will be imposed during the design life of a component the following options may be available:

1. A higher standard of inspection possibly coupled with an improved manufacturing procedure may reduce the value of a_{pi}.
2. A design modification may have the effect of reducing $S_{max} - S_{min}$ or of reducing C by means of reducing the value of R.
3. A material or heat-treatment change may have the effect of reducing the value of C.
4. A change in operating procedure may have the effect of reducing the number of cycles imposed, or of reducing $S_{max} - S_{min}$ or of reducing the value of C.
5. Reducing the service life of the component before repair or replacement will decrease the number of cycles imposed.

The option, or combination of options which are least damaging economically, should be adopted.

8.5.9 Parameters and symbols employed (that have not already been defined in Sections 7.1 or 8.3)

Cycle (fatigue cycle, stress cycle, load cycle or strain cycle): the smallest segment of the stress-time or strain-time cycle which is repeated periodically.

N (dimensionless number): the number of cycles.

S_{max} [F] maximum load: the load having the highest algebraic value in the cycle, tensile load being considered positive and compressive load negative.

S_{min} [F] minimum load: the load having the lowest algebraic value in the cycle, tensile load being considered positive and compressive load negative.

ΔS [F] range of load: the algebraic difference between the maximum and minimum loads in one cycle:

$$\Delta = S_{max} - S_{min}$$

R (dimensionless ratio) load ratio (stress ratio): the algebraic ratio of the minimum to the maximum load in a cycle:

$$R = P_{min}/P_{max}$$

$\Delta\epsilon_p$[L] range of plastic strain: the algebraic difference between the maximum and minimum plastic strains in one cycle.

K_{max} [$FL^{-3/2}$]: maximum stress intensity factor. The highest algebraic value of the stress intensity factor in a cycle.

K_{min}[$FL^{-3/2}$]: maximum stress intensity factor. The lowest algebraic value of the stress intensity factor in a cycle. (This is set equal to zero when R is less than or equal to zero.)

ΔK [$FL^{-3/2}$]: range of stress intensity factor. The algebraic differences between the maximum and the minimum stress intensity factor in a cycle.

ΔK_{th} (orΔK_{TH}) [$FL^{-3/2}$]: threshold value of ΔK. The highest value of ΔK that does not cause crack extension.

da/dN (L) (in fatigue, a always implies a_p, see Section 8.3): the rate of crack extension caused by constant-amplitude loading expressed in terms of crack extension per cycle.

S_a [F]: half the stress range in a fatigue cycle:

$$S_a = S_r/2 = (S_{max} - S_{min})/2$$

S_m [F]: the static or mean stress in a fatigue cycle.

SN curve: the results of fatigue tests plotted as S_m or S_a to number of cycles to failure using a logarithmic scale for the number of cycles and a linear or logarithmic scale for stress.

S_f [F] fatigue (endurance) limit: the value of S_a corresponding to the horizontal portion of the SN curve with a 50% probability of failure. For many non-ferrous metals the 'fatigue strength' at 10^8 cycles is quoted. This is *not* the endurance limit.

K_t (dimensionless ratio): stress concentration factor. The ratio of the area test stress in the vicinity of a notch (or other stress concentration) to the corresponding nominal stress.

K_f (dimensionless ratio): fatigue notch factor. The ratio of the fatigue strength of a plain unnotched specimen to that of a notched specimen at the same number of cycles.

Note: throughout, S has been used for stress based on cross-sectional area of a component (not to be confused with σ, the calculated local stress). Calculations based on ASTM or ASME procedures will use P instead of S.

References

1 Johnson, W. and Mellor, P. B., *Engineering Plasticity*, pp. 2–3, Van Nostrand Reinhold, New York (1980)
2 Peterson, R. E., *Stress Concentration Factors*, Wiley, Chichester (1974)
3 *British Engineering Science Data*, 65004, ESDU (1972)
4 Roark, R. J. and Young, W. C., *Formulas for Stress and Strain*, 5th edn, pp. 560–606, McGraw-Hill, New York (1985)
5 Johnson, W., *Impact Strength of Materials*, Edward Arnold, London (1972)
6 Goldsmith, W., *Impact*, Edward Arnold, London (1960)
7 Morley, J. G., *High-performance Fibre Composites*, pp. 89–113, Academic Press, New York (1987)
8 Tsai, S. W. and Hahn, H. T., *Introduction to Composite Materials*, Tech. Pub. AG (1980)
9 Timoshenko, S. and Goddier, S., *Theory of Elasticity*, McGraw-Hill, New York (1951)
10 Jones, R. M., *Mechanics of Composite Materials*, Scripta Book Co. (1975)
11 Chakrabarky, J., *Theory of Plasticity*, 1st edn, McGraw-Hill, New York (1987)
12 Heyman, J., *Struct. Eng.*, **31**, 125 (1953)
13 Dally, J. W. and Riley, W. F., *Experimental Stress Analysis*, 2nd edn, McGraw-Hill, New York (1978)
14 Post, D., 'Photoelasticity', in *Manual of Engineering Stress Analysis*, pp. 30–46, Prentice-Hall, Englewood Cliffs, NJ (1982)
15 Sharples, K., 'Photoelasticity', *Chartered Mechanical Engineer*, October (1981)
16 Frocht, M. M., *Photoelasticity*, Vol. 2, Wiley, Chichester (1982)
17 Hetenyi, M., 'The fundamentals of three-dimensional photoelasticity', *J. App. Mech.*, **5**, No. 4, 149–155 (1938)
18 Frocht, M. M. and Guermsey, R. Jr, 'Studies in three-dimensional photoelasticity', *Proc 1st US Nat. Cong., App. Mech.*, 301–307 (1951)

19 Frocht, M. M. and Gremsey, R. Jr, 'Further work on the general three-dimensional photoelastic problem', *J. App. Mech.*, **22**, 183–189 (1955)

20 Taylor, G. E., 'Holography', in *Manual of Engineering Stress Analysis*, 3rd edn, pp. 70–79, Prentice-Hall, Englewood Cliffs, NJ (1982)

21 Jones, R. and Wykes, C., *Holography and Speckle Interferometry*, Cambridge University Press, Cambridge (1983)

22 Wykes, C., 'Holography in engineering design', *Engineering*, April (1983)

23 Robinson, D. W. and Williams, D. C., 'Automatic fringe analysis in double exposive and live fringe holographic interferometry', *Proc. SPIE*, 599, Cannes, December (1985)

24 Webber, J. M. B., 'Principles of spate technique for full-field stress analysis', *1st Int. Conf. Stress Analysis by Thermoelastic Techniques*, SIRA, London, November (1984)

25 Thompson, D., 'In full spate design assurance', *Engineers Digest*, December, pp. 1cw Version 3.10 (1982)

26 Hickson, V. M., 'Some new techniques in strain measurement', *Stress Analysis*, Wiley, New York (1965)

27 Ellis, G., 'Practical strain analysis by the use of brittle coating', *Proc. SESA*, Vol. I, No. 1, pp. 46–53 (1963)

28 Macherauch, E., 'X-ray stress analysis', *Proc. 2nd SESA Int. Cong. on Experimental Mechanics*, Washington, September (1965)

29 Macherauch, E., 'Acta', *Physics, Austr.*, **18**, 364 (1964)

30 Rice, J. R., *J. Appl. Mech. Trans. ASME*, **35** (1986)

31 Rice, J. R., in Liebowitz, H. (ed.), *Fracture – Advanced Treatise*, Academic Press, New York (1968)

32 Witt, F. J., 'Equivalent energy procedures for predicting gross plastic fracture', *Fourth National Syposium on Fracture Mechanics*, Carnegie-Mellon University (1970)

33 Harrison, R. P., Milne, I. and Loosemore, K., 'Assessment of the integrity of structures containing defects', *CEGB Report R/H/R6-Rev 1* (1977)

34. Jayatilaka, A. de S., *Fracture of Engineering Brittle Materials*, Applied Science Publishers, Barking (1979)

35 Rorke, D. P. and Cartwright, D. J., *Compendium of Stress Intensity Factors*, HMSO, London (1976)

36 Ingham, T. and Sumter, J. D. G., 'Design against fast fracture in thick walled pressure vessels', *Proc. Conf. Tolerance of Flaws in Pressurized Components*, May 1978, Inst. Mech. Engrs, London

37 Kumar, V., German, M. D. and Shih, C. F., 'Estimation technique for the prediction of elastic plastic fracture of structural components of nuclear systems', *General Electric Co., report SRD-80-094*, Schenectady, New York (1980)

38 Kumar, V., German, M. D. and Shih, C. F., 'An engineering approach to elastic plastic fracture analysis', *EPRI report NP1931* (198?)

39 Kumar, V., German, M. D., Wilkening, W. W., Andrews, W. R., de Lorenzi, H. G. and Mowbray, D. F., 'Advances in elastic plastic fracture analysis', *EPRI report NP3607* (1984)

40 Burdekin, F. M. and Dawes, M. G., *Proc. of Conf. on Application of Fracture Mechanics to Pressure Vessel Technology*, I. Mech. E. London (1971)

41 Kamath, M. S., 'The C.O.D. design curve: an assessment of validity using wide plate tests', *Welding Institute report 71/1878/E* (1979)

42 Burdekin, F. M., Milne, I. and Garwood, S. J., 'The background to BS PD6493 fracture section revisions', Paper 37, *International Conf. on Weld Failures* (1988)

43 Milne, I., 'The state of the art in assessing structural integrity', *CEGB Report NO. TPRD/L/3184/R87* (1987)

44 Ashby, M. F., 'A first report on deformation mechanism maps', *Acta. Met.*, **20**, 887–888 (1972)

45 Mukherjee, A. H., Bind, J. E. and Donu, J. E., 'Experimental correlations for high temperature creep', *ASM Trans. Quart.*, **62**, 155–179 (1969)

46 *High Temperature Properties of Steel*, BISRA/ISI Conference, Eastbourne (1966)

47 *Proceedings of the Joint International Conference on Creep*, New York and London, Institute of Mechanical Engineers, London (1963)

48 Ray, M. S., *The Technology and Applications of Engineering Materials*, Prentice-Hall, Englewood Cliffs, NJ (1979)

49 The Creep of Steels Working Party, *High Pressure Design Data for Ferritic Pressure Vessel Steels*, Mechanical Engineering Publications Ltd, London (1983)

50 Cane, E. S. and Bell, K., 'Remanent life assessment of process plant', Seminar on High Temperature Crack Growth, I. Mech. E. 10 November 1987.

51 Cottrel, A. H., *The Mechanical Properties of Matter*, Wiley, Chichester (1964)

52 Coffin, L. F., 'Fatigue at high temperatures – prediction and interpretation', James Clayton Lecture, Proc. Inst. Mech. Engineers, Sheffield

53 *The Metals Handbook*, Edition 9, Vol. 1, The American Society for Metals

54 *The Fulmer Optimizer*, Fulmer Research Institute and Elsevier, New York

55 Paris, P. and Erdogan, F. A., 'Critical analysis of crack propagation laws', *Trans. Am. Soc. Mech. Engrs J. Bas. Engng*, **85**, Series D, No. 4. December 528–534 (1963)

56 Paris, P. C., 'The fracture mechanics approach to fatigue', in *Fatigue: An Interdisciplinary Approach*, pp. 107–132, Syracuse University Press, New York (1964)

57 Rice, J. R., 'Mechanics of crack tip deformation and extension' 242–311 (1967)

58 Shijic, H., 'Significance of fatigue cracks in micro- and macro-range', *ASTM STP 415*, 415–419 (1967)

59 Coffin, L. F., Jr, *Nat. Eng. Quart.*, **3**, 22 (1963)

60 Manson, S. S., *NASA Report TN D-1574*, NASA, April (1963)

61 Manson, S. S. and Hirschberg, M. H. (eds), *Fatigue: An Interdisciplinary Approach*, p. 133, Syracuse University Press, New York (1964)

62 Coffin, L. F., Jr, *Trans. ASME*, **76**, 931 (1954)

63 Tavernelli, J. F. and Coffin, L. F., Jr, *Trans. ASM*, **51**, 438 (1959)

64 Sines, G. and Weisman, J. R. E., *Metal Fatigue*, McGraw-Hill, New York (1959)

65 Horger, O. J., 'Fatigue characteristics of large sections', in *Fatigue*, American Society for Metals (1953)

66 Phillips, C. E. and Heywood, R. B., *Proc. Inst. Mech. Eng. (London)*, **165**, 113–124 (1951)

67 *Fatigue Durability of Carburized Steel*, American Society for Metals (1957)

68 Miner, M. A., 'Cumulative damage in fatigue', *Trans. ASME*, **67**, A159 (1945)

69 Grover, H. J., 'Fatigue of aircraft structures', *NAVAIR 01-1A-13*, Naval Air Systems Command, US Department of the Navy (1966)

70 Devereux, O., McEvily, A. J. and Staehle, R. W., (eds), *Corrosion Fatigue: Chemistry, Mechanics, and Microstructure*, Conference Proceedings, University of Connecticut, 14–18 June 1971, National Association of Corrosion Engineers, Houston

71 Craig, H. L., Crooker, T. W. and Hoeppner, P. W. (eds), *Corrosion – Fatigue Technology*, Symposium Proceedings, Denver, 14–19, November 1976, ASTM, Philadelphia

72 'Corrosion – fatigue failures', in *Metals Handbook*, Vol. 10, 8th edn, pp. 240–249, American Society for Metals (1975)

73 Use of Linear Elastic Fracture Mechanisms in Estimating Fatigue Crack Growth Rates and Residual Strength of Components, ESDU 80036, ESDU International Ltd, 251/a Regent St, London W1R 7AD

74 Fatigue Life Estimation under Variable Amplitude Loading, ESDU 76014, 76016 and 77004 (see reference 73).

75 Estimation of Endurance and Construction of Constant Amplitude SN Curves for Related Data Corrected for Notch and Mean Stress Effects, ESDU 76014 (see reference 73)

76 The Effect of Mean Stress on Fatigue Strength (Plain Test Pice), ESDU fat 00.01. ESDU also publish information on fatigue data for steels, and aluminium and titanium alloys and components (and other engineering topics)

Further reading

ASME III, Appendix G and ASME XI, Appendix A describes fracture mechanics methods of assessing the significance of defects in thick-walled pressure vessels for nuclear reactor systems.

Nichols, R. W., *Developments in Pressure Vessel Technology*, Applied Science Publishers, Barking (1979), reviews the state of assessment of the effect of flaws in pressure vessels at the date of publishing

The *ASTM Metals Handbook* deals with the whole field of fracture mechanics

The Fulmer Optimizer describes the method of calculation of the effect of flaws, both in LEFM and all the available methods of calculation for post-yield fracture mechanics. Fracture toughness data on steels, titanium and aluminium alloys are provided

ESDU 80036 introduces the concepts of linear elastic fracture mechanics and shows how the concepts may be used for analysing the behaviour of cracked structures under fatigue or static loadings (ESDU International Ltd, 251/a Regent Street, London W1R 7AD)

ERA Technology Ltd, Cleeve Road, Leatherhead, Surrey, publish brochures relating to remanent life assessment

The American Society of Mechanical Engineers, United Engineering Centre, 345 East Street, New York, NY10017, USA, publishes the *ASME boiler and pressure vessel code* in a series of numbered volumes which cover all aspects of design, manufacture and testing and are widely used as standards for pressure vessels

9 Tribology

Edward H. Smith (Sections 9.1 and 9.3)

John Weston-Hays (Section 9.2)

Bert Middlebrook (Section 9.4)

Dennis R. Hatton (Section 9.5)

Tony G. Herraty (Section 9.6)

Philip Eliades (Section 9.7)

Keith T. Stevens and A. Davies (Section 9.8)

Michael W. J. Lewis (Section 9.9)

Ian Sherrington (Section 9.10)

Contents

9.1 Basic principles 9/3
    9.1.1 Introduction 9/3
    9.1.2 Lubrication regimes 9/3
    9.1.3 Friction, wear and lubrication 9/4

9.2 Lubricants (oils and greases) 9/5
    9.2.1 Physical characteristics 9/5
    9.2.2 Additives 9/6
    9.2.3 Lubricating-oil applications 9/7
    9.2.4 General machinery oils 9/7
    9.2.5 Engine lubricants 9/7
    9.2.6 Hydraulic fluids 9/10
    9.2.7 Machine tools 9/13
    9.2.8 Compressors 9/18
    9.2.9 Turbines 9/20
    9.2.10 Transformers and switchgear 9/20
    9.2.11 Greases 9/21
    9.2.12 Corrosion prevention 9/22
    9.2.13 Spray lubricants 9/23
    9.2.14 Degreasants 9/23
    9.2.15 Filtration 9/23
    9.2.16 Centrifuging 9/24
    9.2.17 Centralized lubrication 9/25
    9.2.18 Storage of lubricants 9/25
    9.2.19 Reconditioning of oil 9/26
    9.2.20 Planned lubrication and maintenance
        management 9/26

9.2.21 Condition monitoring 9/26
9.2.22 Health, safety and the environment 9/26

9.3 Bearing selection 9/27
    9.3.1 Characteristics of bearings with continuous
        motion 9/27
    9.3.2 Bearing selection charts 9/28

9.4 Principles and design of hydrodynamic bearings 9/30
    9.4.1 Introduction 9/30
    9.4.2 Principles of hydrodynamic lubrication 9/30
    9.4.3 Viscosity 9/31
    9.4.4 Journal bearing design 9/32
    9.4.5 Self-contained bearings 9/39
    9.4.6 Thrust bearings 9/39

9.5 Lubrication of industrial gears 9/41
    9.5.1 Methods of lubrication 9/42
    9.5.2 Types of gear oils 9/44
    9.5.3 Heat dissipation 9/45
    9.5.4 Selection of gear lubricants 9/47
    9.5.5 Service life of gear lubricants 9/49

9.6 Rolling element bearings 9/49
    9.6.1 Introduction 9/49
    9.6.2 Types 9/50
    9.6.3 Selection 9/52

9.6.4 Bearing life 9/52
9.6.5 Bearing friction and speed limits 9/54
9.6.6 Application 9/55
9.6.7 Lubrication 9/56
9.6.8 Bearing mounting and maintenance 9/58
9.6.9 Bearing failures 9/59

9.7 Materials for unlubricated sliding 9/60
9.7.1 Introduction 9/60
9.7.2 Performance rating 9/60
9.7.3 Counterface materials 9/62
9.7.4 Polymers and polymer composites 9/62
9.7.5 Carbon graphites 9/67
9.7.6 Solid lubricants 9/68
9.7.7 Metallic alloys and composites 9/69
9.7.8 Surface treatments and coatings 9/70
9.7.9 Conclusions 9/70

9.8 Wear and surface treatment 9/70
9.8.1 Introduction 9/70
9.8.2 Wear types 9/71
9.8.3 Surface treatments and coatings 9/73
9.8.4 Tribological data 9/83

9.8.5 Selection philosophy 9/85
9.8.6 Quality control 9/87
9.8.7 Closure 9/88

9.9 Fretting 9/88
9.9.1 Introduction 9/88
9.9.2 Source of relative movement 9/89
9.9.3 Characteristics 9/90
9.9.4 Parameters influencing fretting 9/91
9.9.5 Theoretical considerations 9/95
9.9.6 Fretting wear evaluation 9/98
9.9.7 Preventative measures – some palliatives 9/98
9.9.8 Summary of palliatives 9/101

9.10 Surface topography 9/102
9.10.1 Effects of surface topography 9/102
9.10.2 Measurement 9/109
9.10.3 Characterization of surface measurements 9/11
9.10.4 Summary, conclusions and future developments 9/124

References 9/125
Further reading 9/131

9.1 Basic principles

9.1.1 Introduction

Machines transmit power between their moving components which gives rise to friction and wear of the components. Tribology is, essentially, concerned with minimizing this friction and wear, and is formally described as *'the science of interacting surfaces in relative motion'*. The tribologist has to understand about the roughnesses of surfaces, the properties of liquid and solid lubricants, the principles of hydrodynamic and hydrostatic lubrication, the design of rolling element bearings, the behaviour of surfaces under stress, the nature of wear processes, and many other items which demand an interdisciplinary approach to design and problem solving.

9.1.2 Lubrication regimes

The most usual way to reduce friction and wear is by lubrication, employing gases, liquids or solids. Four basic lubrication regimes can be considered, characterized by the ratio, M, of the combined roughness, R, of both surfaces (in R_a) to the mean film thickness, h, i.e. $M = R/h$. Three of these regimes are illustrated in Figure 9.1, where the roughness of the surfaces is greatly exaggerated for presentation reasons. In reality, most surfaces more resemble gentle rollings hills on the earth's surface than the craggy, mountainous features illustrated. Figure 9.2 shows the more realistic dimensions and structure of a real surface, and a more detailed discussion of surface topography is presented in Section 9.10.

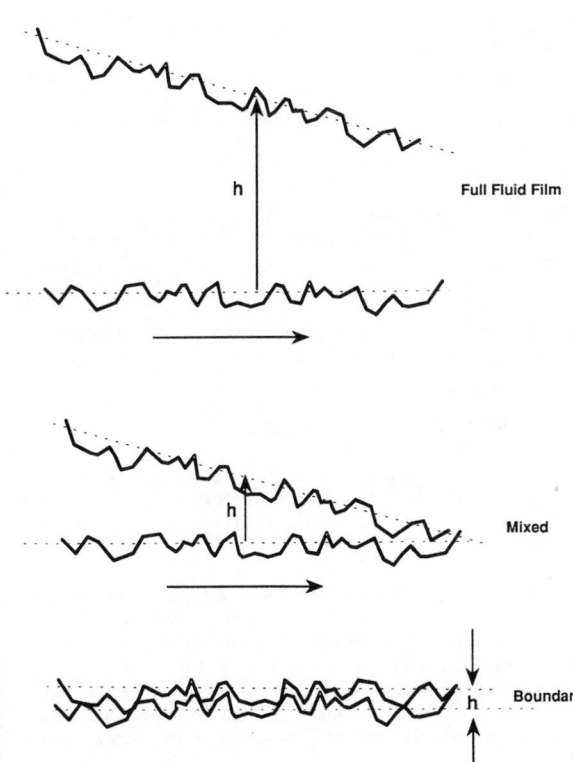

Figure 9.1 Lubrication regimes

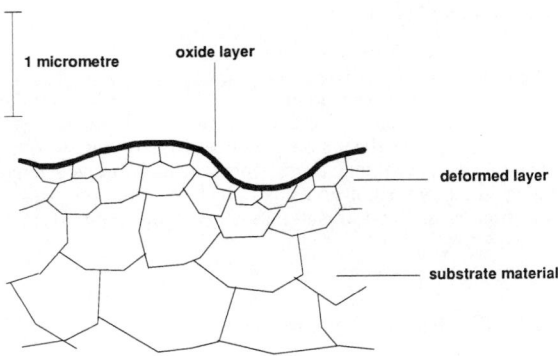

Figure 9.2 A typical engineering surface

9.1.2.1 Rigid, full fluid-film (5 < M < 10)

In this situation the surfaces are kept apart by a pressurized fluid such as oil or air. The clearance space is much larger than the average surface roughness, and therefore the surfaces can be considered smooth. The pressurization of the fluid is achieved by external means in hydrostatic bearings, but is accomplished in hydrodynamic contacts by the relative motion and geometry of the surfaces. Examples are crankshaft main bearings in internal combustion engines (oil-lubricated hydrodynamic journal bearings), shaft support bearings in power-generating turbines (often a combination of oil-lubricated hydrostatic and hydrodynamic journal bearings), piston rings (oil-lubricated hydrodynamic slider bearings), dentists' drills (air-lubricated hydrostatic bearings for shaft support) and magnetic disk heads (air-lubricated hydroynamic slider bearings). The friction coefficient in liquid-lubricated contacts falls typically within a range of 0.004 to 0.01. An outline of the principles of rigid, hydrodynamic lubrication is presented in Section 9.4.

9.1.2.2 Elastic full fluid-film (1 < M < 10)

There are situations in liquid lubricated arrangements where the loads are so high that the pressure in the lubricant causes local elastic distortions of the surfaces. This form of hydrodynamic lubrication is called elastohydrodynamic lubrication. The nominal clearance can still be such that surface roughness effects can be ignored, although mixed elastohydrodynamic lubrication can occur.

Elastohydrodynamic lubrication is most commonly found in gears and rolling element bearings, which are discussed in Sections 9.5 and 9.6, respectively.

9.1.2.3 Mixed (1 < M < 5)

Loads or speeds can be such that the opposing surfaces occasionally come into contact. In these contacts, part of the load is carried by the liquid or gaseous lubricant, and part by the interacting asperities. The rubbing together of asperities will increase friction, and this can be minimized by attention to the compounds which can be made to adhere to the surfaces. Hypoid gears in automobile rear axles at times exhibit this regime. Increased wear occurs, and friction coefficients in liquid-lubricated contacts of between 0.01 and 0.06 are found.

9.1.2.4 Boundary (M < 1)

In this regime, all the applied load is carried by the surface asperities, and the friction and wear which arises depends upon the lubrication properties of the molecules on the surfaces. Coefficients of friction between 0.06 and 0.1 are typical when a low shear strength surface film is present. If no such film is present, then coefficients ranging between 0.2 and 0.4 can be exhibited, even rising as high as 1.0 in some cases. The friction and wear of materials in dry sliding is discussed in Section 9.7.

9.1.3 Friction, wear and lubrication

9.1.3.1 Friction

There are three general laws of friction:

1. Friction is independent of apparent contact area
2. Limiting friction is proportional to the normal load
3. Limiting friction is independent of sliding speed

Limiting, or breakout, friction is the friction force obtaining at the point of slippage. The first two laws, proposed by Amonton in 1699, are generally applicable. The third law, proposed by Coulomb in 1785, has a reduced area of applicability.

The second law is more commonly written as:

$$\mu = F/W$$

where

μ = static, or breakout, coefficient of friction,
F = limiting friction force,
W = load applied normal to the surfaces.

In the light of law 3, a dynamic coefficient of friction can also be defined as the ratio of the sliding friction force to the normal load. It can be significantly lower than its static counterpart.

The friction occurring at asperity contacts is usually attributable to one of two causes: abrasion, where the harder surface (or a harder particle) ploughs through the softer counterface; or adhesion, where the intimate contact between asperities generates local adhesive bonds which break when motion is initiated. Friction can also arise due to hysteresis, as in the case of rolling friction.

The simple adhesive theory of friction deduces that:

μ = shear strength of surface/hardness of surface

This is a gross oversimplification[1] but serves to indicate that adhesive friction can be reduced by reducing the shear strength of the bonds (by using dissimilar materials in opposition), introducing low shear strength surface layers (often as boundary lubricants such as stearates, sulphides, etc.), and/or increasing the hardness of the surface. There is evidence that a soft, thin layer of low shear strength material on a harder substrate can produce a component whose surface exhibits both low shear strength and high hardness – the latter being acquired from the parent material if the surface film is sufficiently thin.

The friction effects of abrasion can be minimized by suitable choice of counterface materials so that ploughing is reduced or prevented, and/or the addition of boundary lubricants which can ease the ploughing process. Sometimes boundary lubricants occur naturally as, for example, oxide layers; sometimes a boundary lubricant can be added to a liquid lubricant, e.g. stearates which have polar properties permitting them to attach to a surface like bristles on a brush (see Section 9.5.2). Arnell et al.[1] and Rabinowicz[2] provide more detailed discussions.

9.1.3.2 Wear

Wear occurs through adhesion, abrasion, fretting, corrosion or fatigue. The first two modes are sometimes called sliding wear and are discussed below. Further detailed discussion of wear is presented in Sections 9.7–9.9. Suh[3] has suggested that sliding wear can also be caused by delamination of surfaces. In addition to the discussion in this chapter, the reader is referred to the *Wear Control Handbook*[4] for further details.

Adhesive wear arises from the breaking of the adhesive bonds formed when asperities contact and deform. The resulting wear particles may be smeared on one of the surfaces, form work-hardened abrasive particles, or be ejected from the contact. Abrasive wear occurs when harder asperities plough through a softer counterface (two-body abrasion) or when a harder third particle gouges the surface of a softer material (three-body abrasion).

Wear equations Wear by sliding (adhesive, abrasive or delamination) is usually measured in terms of the specific wear rate, k, defined as the volume of material worn away for a unit load and unit sliding distance. Thus:

$$k = v/(W.L) \tag{9.1}$$

where

v = volume of worn material (m^3),
W = normal applied load (N),
L = total sliding distance (m).

The standard units are m^2 N^{-1}. Values of 10^{-16} or lower are indicative of mild wear, and values greater then 10^{-14} represent severe wear. Most material wear rates lie within the range 10^{-17} and 10^{-12}.

PV factors in wear Rearrangement of equation (9.1) can be effected to produce:

$$v = k.W.L \tag{9.2}$$

Dividing both sides of this equation by the apparent area of contact, a, yields:

$$h = (k.W.L)/a \tag{9.3}$$

where h is the wear depth.

Since W/A is the bearing pressure, P, equation (9.3) can be written as:

$$h = k.P.l \tag{9.4}$$

Dividing both sides of the equation by time, t, and rearranging gives

$$k = h/(P.V.t) \tag{9.5}$$

where the sliding speed, V, is equal to L/T.

Sometimes the constant, k, which is numerically identical to that in equation (9.1), is called the Wear Factor.

Equation (9.5) can be rearranged to yield:

$$k.t/h = 1/(P.V) \tag{9.6}$$

If the parameter k is a constant for a particular configuration, then an acceptable value for the term $(k.t/h)$ can be determined since the required life, t, and the acceptable maximum wear depth, h, can both be specified. Equation (9.6) then shows that the parameter $(k.t/h)$ is inversely proportional to the product of P and V. This is often quoted by manufacturers of dry or partially lubricated bearings as the bearing's 'PV factor'.

This same product, PV, also influences the temperature rise at the sliding contact. This can be deduced from the fact that:

- Power loss is proportional to the product of friction force and speed.
- Heat loss by conduction is proportional to temperature difference, T, between the sliding surface and its surroundings, the apparent area of the contact, A, and the conductivity of the bearing and its surroundings.

Thus the temperature rise, T, at the sliding interface is given by:

$$T = c.\mu P.V$$

where c is a constant which characterizes the resistance of the bearing configuration to heat dissipation. This is discussed further in Section 9.7.

9.1.3.3 Fluid film lubrication

This form of lubrication can be achieved using liquids or gases. The aim is to keep the surfaces separated during normal operation. The most common liquid lubricants are mineral oils, which are discussed in Section 9.2. Air is probably the most common gaseous lubricant.

Liquid film bearings usually exhibit laminar flow, and can be designed using the Reynolds equation as discussed in Section 9.3. A modified form of the familiar Reynolds number, Re, is used in fluid film lubrication studies, defined as:

$$Re = (\rho.u.l./\eta).(h/l)$$

where

ρ = density of the lubricant,
u = a representative velocity of the lubricant,
l = characteristic dimension of the bearing (e.g. length or width),
η = dynamic viscosity of the lubricant,
h = characteristic film thickness.

This ratio assesses the relative importance of viscous and so-called inertia forces. If $Re \ll 1$, the former dominate the latter and laminar flow pertains. When $Re \gg 1$, the latter dominate the former and turbulent flow is present. Some higher speed bearings operate in the turbulent regime. Further reading on turbulent flow can be found in reference 5.

9.2 Lubricants (oils and greases)

9.2.1 Physical characteristics

Reference will be made to the physical characteristics of lubricants as they affect their selection for various applications. These terms are well known to the lubricant supplier but are not always fully understood by the user. Brief descriptions of these characteristics are therefore given so that their significance may be appreciated.

9.2.1.1 Viscosity

This is the most important physical property of a lubricating oil; it is a measure of its internal friction or resistance to flow. In simple terms, it provides a measure of the thickness (not density) of a lubricating oil at a given temperature; the higher the viscosity, the thicker the oil. Accurate determination of viscosity involves measuring the rate of flow in capillary tubes, the unit of measurement being the centistoke (cSt). As oils become thinner on heating and thicker on cooling a viscosity figure must always be accompanied by the temperature at which it was determined.

The number of commercial viscosity systems can be confusing, and as kinematic viscometers are much more sensitive and consistent, there is a growing tendency to quote kinematic viscosities. The International Standards Organization (ISO) uses kinematic viscosity in its viscosity grade classification (Table 9.1). These ISO grade numbers are used by most oil companies in their industrial lubricant nomenclature. This provides the user with a simple verification of conformity regarding viscosity between plant manufacturer and oil supplier recommendations and also in the monitoring of correct oil usage on his plant.

9.2.1.2 Viscosity Index (VI)

This is a way of expressing the rate of change of viscosity with temperature. All oils become less viscous as the temperature increases. The rate of change of viscosity varies with different oils and is mainly dependent on the type of crude from which the oil is derived and the refining method. The higher the VI figure, the lower is the variation in viscosity relative to temperature. The VI of an oil is an important property in applications where the operating temperature is subject to considerable change.

9.2.1.3 Pour point

This is a rough measure of a limiting viscosity. It is the temperature 2.5°C above that at which the oil ceases to flow when the vessel in which it has been cooled is held horizontally for 5 s. The pour point is a guide to behaviour and care should always be taken that operating temperatures are above the figure specified by the oil manufacturer as the pour point of a given oil.

9.2.1.4 Flash point

The flash point of an oil is the temperature at which it gives off, under specified conditions, sufficient vapour to form a flammable mixture with air. This is very different from the

Table 9.1 ISO viscosity grade chart

ISO viscosity grade	Mid-point kinematic viscosity	Kinematic viscosity limits cSt at 40°C (104°F)	
		min.	max.
2	2.2	1.98	2.42
3	3.2	2.88	3.52
5	4.6	4.14	5.06
7	6.8	6.12	7.48
10	10	9.00	11.0
15	15	13.5	16.5
22	22	19.8	24.2
32	32	28.8	35.2
46	46	41.4	50.6
68	68	61.2	74.8
100	100	90.0	110
150	150	135	165
220	220	198	242
320	320	288	352
460	460	414	506
680	680	612	748
1000	1000	900	1100
1500	1500	1350	1650

temperature of spontaneous combustion. The test is an empirical one and the result depends upon the instrument used and the prescribed conditions. For example, the flash point may be 'closed' or 'open', depending on whether the test apparatus has a lid or not. As far as lubricating oils are concerned, the test is of limited significance, although it can be indicative of contamination (for example, the dilution of crankcase oil by fuel).

9.2.1.5 Penetration of grease

The most important physical property of a lubricating grease is its consistency, which is analogous to the viscosity of a liquid. This is determined by an indentation test in which a weighted metal cone is allowed to sink into the grease for a specified time. The depth to which the cone penetrates, in tenths of a millimetre, is a measure of the consistency. There is a widely accepted scale, that of the American National Lubricating Grease Institute (NLGI), that relates penetration to a consistency number.

The penetration test is used mainly to control manufacture and to classify greases and is, within limits, a guide to selection. Penetrations are often qualified by the terms 'worked' and 'unworked'. As greases are thixotropic, that is, they soften as a result of shear but harden again after shearing has stopped, the worked penetration for a particular grease may be appreciably greater than the unworked penetration. The difference between these two figures may be a useful guide to the selection of greases for operating conditions that involve much churning – as small a difference as possible being desirable (see Table 9.2).

9.2.1.6 Drop point of grease

The drop point of a grease is an indication of change from a soft solid to a viscous fluid; its value depends completely on the conditions of test, particularly the rate of heating. The grease sample, which is held in a small metal cup with an orifice, is heated at a predetermined rate. The drop point is the temperature at which a drop of the sample falls from the cup.

The drop point is of limited significance as far as the user is concerned, for it gives no indication of the condition of the grease at lower temperatures, or of change in consistency or structure with heat. It is a very rough indication of a grease's resistance to heat and a guide to manufacture. The difference between the highest temperature at which a grease can be used and the drop point varies very much between types. It is at its maximum with some soda greases and much smaller with multi-purpose lithium products and modern complex greases.

Table 9.2 NLGI consistency classification for greases

NLGI number	ASTM worked penetration at 77°F
000	455–475
00	400–435
0	355–385
1	310–340
2	265–295
3	220–250
4	175–205
5	130–160
6	85–115

9.2.2 Additives

Much highly stressed modern machinery runs under conditions in which a straight mineral oil is not adequate. Even the highest quality mineral oil can be unsatisfactory in response of its resistance to oxidation and its behaviour under pure boundary conditions, but it is possible to improve these characteristics by the addition of relatively small amounts of complex chemicals. This use of additives resembles in many ways the modification of the properties of steel by the addition of small amounts of other chemicals. It will be of value to have some knowledge of the effect of each type of additive.

9.2.2.1 Anti-oxidants

When mixed with oxygen, lubricating oil undergoes chemical degradation resulting in the formulation of acidic products and sludge. This reaction, which is affected by temperature, the presence of catalysts such as copper and the composition of the oil, can be delayed by the inclusion of suitable additives.

Anti-oxidants are the most extensively used additives and will be found in oils and greases which are expected to operate for considerable periods or under conditions which would promote oxidation. Typical examples are crankcase oils and bearing greases.

9.2.2.2 Anti-foam

The entrainment of air in lubricating oil can be brought about by operating conditions (for example, churning) and by bad design such as a return pipe which is not submerged. The air bubbles naturally rise to the surface, and if they do not burst quickly, a blanket of foam will form on the oil surface. Further air escape is thus prevented and the oil becomes aerated. Oil in this condition can have an adverse effect on the system which, in extreme cases, could lead to machine failure. The function of an anti-foam additive is to assist in the burst of air bubbles when they reach the surface of the oil.

9.2.2.3 Anti-corrosion

The products of oil oxidation will attack metals, and this can be prevented by keeping the system free from pro-oxidative impurities and by the use of anti-oxidants. These additives will not, however, prevent rusting of ferrous surfaces when air and water are present in the mineral oil. The presence of absorbed air and moisture is inevitable in lubricating systems and therefore the oil must be inhibited against rusting. These additives, which are homogeneously mixed with the oil, have an affinity for metal, and a strongly absorbed oil film is formed on the metal surface which prevents the access of air and moisture.

9.2.2.4 Anti-wear

The increasing demands being made on equipment by the requirement for increased output from smaller units create problems of lubrication, even in systems where full-fluid film conditions generally exist. For instance, at start-up, after a period of rest, boundary lubrication conditions can exist and the mechanical wear that takes place could lead to equipment failure. Anti-wear additives, by their polar nature, help the oil to form a strongly absorbed layer on the metal surface which resists displacement under pressure, thereby reducing friction under boundary conditions.

9.2.2.5 Extreme pressure

Where high loading and severe sliding speeds exist between two metal surfaces, any oil film present is likely to be squeezed out. Under these conditions very high instantaneous pressures and temperatures are generated. Without the presence of extreme pressure additives the asperities would be welded together and then torn apart. Extreme pressure additives react at these high temperatures with the metal or another oil component to form compounds which are more easily deformed and sheared than the metal itself, and so prevent welding. Oils containing extreme pressure additives are generally used in heavily loaded gearboxes which may also be subjected to shock loading.

9.2.2.6 Detergent/dispersant

The products of combustion formed in internal combustion engines, combined with water and unburnt fuel, will form undesirable sludge which can be deposited in the engine and so reduce its operation life and efficiency. Detergent/dispersant additives prevent the agglomeration of these products and their deposition in oilways by keeping the finely divided particles in suspension in the oil. They are used in engine-lubricating oils where, when combined with anti-oxidants, they prevent piston-ring sticking. They are essential for high-speed diesels, and also desirable for petrol engines.

9.2.2.7 Viscosity Index improvers

When mineral oils are used over an extended temperature range it is frequently found that the natural viscosity/temperature relationship results in excessive thinning out in the higher-temperature region if the desired fluidity is to be maintained at the lower region. The addition of certain polymers will, within limits, correct this situation. They are of particular value in the preparation of lubricating oils for systems sensitive to changes in viscosity such as hydraulic controls. They are also used in multigrade engine oils.

9.2.3 Lubricating-oil applications

There is a constant effort by both the supplier and consumer of lubricants to reduce the number of grades in use. The various lubricant requirements of plant not only limit the extent of this rationalization but also create the continuing need for a large number of grades with different characteristics.

It is not possible to make lubricants directly from crude oil that will meet all these demands. Instead, the refinery produces a few basic oils and these are then blended in varying proportions, together with additives when necessary, to produce an oil with the particular characteristics required. In some instances the continued increase in plant performance is creating demands on the lubricant which are at the limit of the inherent physical characterisics of mineral oil. Where the operational benefit justifies the cost, the use of synthetic base stocks is being developed.

Where these are considered for existing plant, seal and paint compatibility needs to be reviewed before such products are introduced. The problems which face the lubricant supplier can best be illustrated by looking at the requirements of certain important applications.

9.2.4 General machinery oils

These are lubricants for the bearings of most plant, where circulating systems are not involved. These are hand, ring, bottle or bath lubricated bearings of a very wide range of equipment; line shafting, electric motors, many gear sets and general oil-can duties. The viscosity of these oils will vary to suit the variations in speed, load and temperature.

While extreme or arduous usage conditions are not met within this category, the straight mineral oils which are prescribed must possess certain properties. The viscosity level should be chosen to provide an adequate lubricant film without undue fluid friction, though this may also be influenced by the method of application. For instance, a slightly higher viscosity might be advisable if intermittent hand oiling has to be relied upon. Although anti-oxidants are not generally required, such oils must have a reasonable degree of chemical stability (Figure 9.3).

9.2.5 Engine lubricants

The type of power or fuel supply available will influence the decision on prime mover to be used. This is often electric power, but many items of plant such as compressors, generators or works locomotives, will be powered by diesel engines, as will most of the heavy goods vehicles used in and outside the works.

The oils for these engines have several functions to perform while in use. They must provide a lubricant film between moving parts to reduce friction and wear, hold products of combustion in suspension, prevent the formation of sludges and assist in cooling the engine. Unless the lubricant chosen fulfils these conditions successfully, deposits and sludge will form with a consequent undesirable increase in wear rate and decrease in engine life.

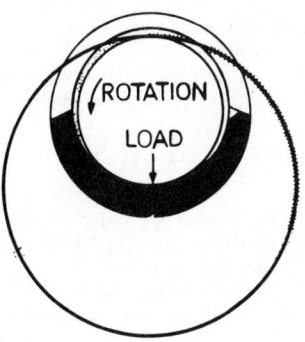
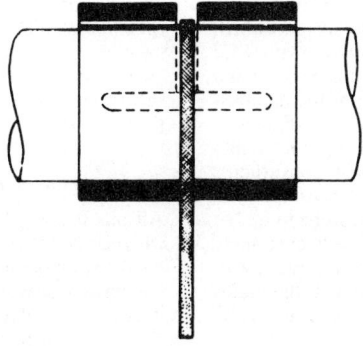
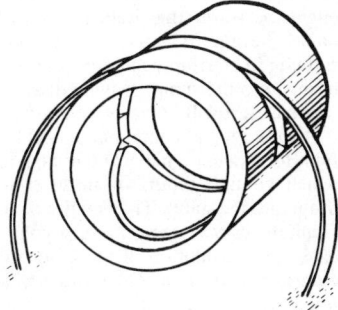

Figure 9.3 Ring oiled bearings

9.2.5.1 Frictional wear

If the effects of friction are to be minimized, a lubricant film must be maintained continuously betwen the moving surfaces. Two types of motion are encountered in engines, rotary and linear. A full fluid-film between moving parts is the ideal form of lubrication, but in practice, even with rotary motion, this is not always achievable. At low engine speeds, for instance, bearing lubrication can be under boundary conditions.

The linear sliding motion between pistons, piston rings and cylinder walls creates lubrication problems which are some of the most difficult to overcome in an engine. The ring is exerting a force against the cylinder wall while at the same time the ring and piston are moving in the cylinder with a sliding action. Also, the direction of piston movement is reversed on each stroke. To maintain a full fluid oil film on the cylinder walls under these conditions is difficult and boundary lubrication can exist. Frictional wear will occur if a lubricant film is either absent or unable to withstand the pressures being exerted. The lubricant will then be contaminated with metal wear particles which will cause wear in other engine parts as they are carried round by the lubricant.

9.2.5.2 Chemical wear

Another major cause of wear is the chemical action associated with the inevitable acidic products of fuel combustion. This chemical wear of cylinder bores can be prevented by having an oil film which is strongly adherent to the metal surfaces involved, and which will rapidly heal when a tiny rupture occurs. This is achieved by the use of a chemical additive known as a corrosion inhibitor.

9.2.5.3 Products of combustion and fuel dilution

As it is not possible to maintain perfect combustion conditions at all times, contamination of the oil by the products of combustion is inevitable. These contaminants can be either solid or liquid.

When an engine idles or runs with an over-rich mixture the combustion process is imperfect and soot will be formed. A quantity of this soot will pass harmlessly out with the exhaust but some will contaminate the oil film on the pistons and cylinders and drain down into the crankcase. If there is any water present these solids will emulsify to form sludges which could then block the oilways. Filters are incorporated into the oil-circulation system to remove the solid contaminants together with any atmospheric dust which bypasses the air filters.

One of the liquid contaminants is water, the presence of which is brought about by the fact that when fuel is burnt it produces approximately its own weight in water. When the engine is warm this water is converted into steam, which passes harmlessly out of the exhaust. However, with cold running or start-up conditions this water is not converted and drains into the sump. Having dissolved some of the combustion gases, it will be acidic in nature and will form sludges.

Another liquid contaminant is unburnt fuel. A poor-quality fuel, for example, may contain high boiling point constituents which will not all burn off in the combustion process and will drain into the sump. The practice of adding kerosene to fuel to facilitate easy starting in very cold weather will eventually cause severe dilution of the lubricating oil. Excessive use of over-rich mixture in cold weather will mean that all the fuel is not burnt because of the lack of oxygen and again, some remains to drain into the sump.

Poor vaporization of the fuel will also produce oil dilution. Generally, this fuel will be driven off when the engine becomes warm and is running at optimum conditions. However, severe dilution of the oil by fuel could have serious results as the viscosity of the oil will be reduced to an unacceptable level.

9.2.5.4 Oxidation

The conditions of operation in an engine are conducive to oil oxidation, and this is another problem to be overcome by the lubricant. In the crankcase, the oil is sprayed from various components in the form of an oil mist which is in contact with a large quantity of air and at a fairly high temperature. Oxidation produces complex carbonaceous products and acidic material and these, combined with fuel contaminants, will form stable sludges. In the combustion chamber, where the temperatures are very much higher, the oil is scraped up the cylinder walls by the piston ascending at very high speeds and is again present in the form of an oil mist. A form of carbon deposit is produced by a combination of heat decomposition and oxidation. Some of this deposit will remain, but some will pass into the sump. The effect of oxidation adds to the problem of oil contamination by the products of combustion, resulting in the formation of a resin-like material on the pistons and hot metal parts known as 'lacquer' and acidic material which will attack bearing metals such as copper-lead.

These problems of engine lubrication can be overcome by using a highly refined oil. The resistance to oxidation is further enhanced by the use of anti-oxidants. The addition of corrosion-inhibitors counters acidic materials produced by combustion at low engine temperatures.

Detergent–dispersant additives are incorporated so that the carbonaceous matter produced by imperfect combustion is retained in suspension in the oil, preventing it from being deposited on the engine surfaces. Such an oil is known as a fully detergent-type lubricant. All these additives are gradually consumed during operation and the rate of decline in their usefulness will determine the oil-change period. This rate is, in turn, influenced by the conditions of operation.

9.2.5.5 The SAE viscosity system

This classification was devised by the Society of Automotive Engineers (SAE) in America by dividing the viscosity span into four and giving each of the divisions a number – SAE 20, 30, 40 and 50. The thinnest (SAE 20), for example, covered the range 5.7–9.6 cSt specified at 210°F, which was considered to be a temperature typical of a hot engine. (The SAE originally specified temperatures in °F, because that was the convention. Today, temperatures are quoted in °C.)

Later, the SAE series was extended to include much lower-viscosity oils because of the growing demand for easier winter starting. The viscosities of the three new grades were specified at 0°F (typical of cold morning temperatures) and each was given the suffix W for Winter – SAE 5W, 10W and 20W. Later still, grades of 0W, 15W and 25W were added to satisfy the more precise requirements of modern engines (Table 9.3).

9.2.5.6 Multigrades

All oils become less viscous when heated and more viscous when cooled, but some are less sensitive than others to these viscosity/temperature effects. The degree of sensitivity is known as Viscosity Index (VI). An oil is said to have high VI if it displays a relatively small change of viscosity for a given change of temperature.

In the 1950s, developments in additive technology led to the production of engine oils with unusually high VIs, known as

Table 9.3 Viscosity chart

	Maximum viscosity cP at °C[a]	Maximum borderline pumping temperature (°C)[b]	Viscosity (cSt) at 100°C min.	max.
0W	3250 at −30	−35	3.8	—
5W	3500 at −25	−30	3.8	—
10W	3500 at −20	−25	4.1	—
15W	3500 at −15	−20	5.6	—
20W	4500 at −10	−15	5.6	—
25W	6000 at −5	−10	9.3	—
20	—	—	5.6	9.3
30	—	—	9.3	12.5
40	—	—	12.5	16.3
50	—	—	16.3	21.9

[a] As measured in the Cold Cranking Simulator (CCS).
[b] As measured in the Mini Rotary Viscometer (MRV).

multigrade oils. A multigrade oil's high resistance to temperature change is sufficient to give it the combined virtues of a low-viscosity grade at low (starting) temperatures and a high-viscosity one at running temperatures. An SAE 20W-40 multigrade, for example, is as thin at −20°C as a 20W oil, but as thick at 100°C as an SAE 40 oil. Thus the multigrade combines full lubrication protection at working temperatures with satisfactorily easy starting on frosty mornings. Figure 9.4 is a viscosity–temperature graph for six monograde oils and a 10W-40 multigrade, showing how the multigrade has the high-temperature properties of an SAE 40 oil and the low-temperature properties of an SAE 10W. Thus the multigrade is suitable for all-year-round use.

9.2.5.7 Performance ratings

The SAE numbering system refers purely to the viscosity of the oil, and is not intended to reflect lubricating performance (there is no such thing as an 'SAE quality' oil, for example).

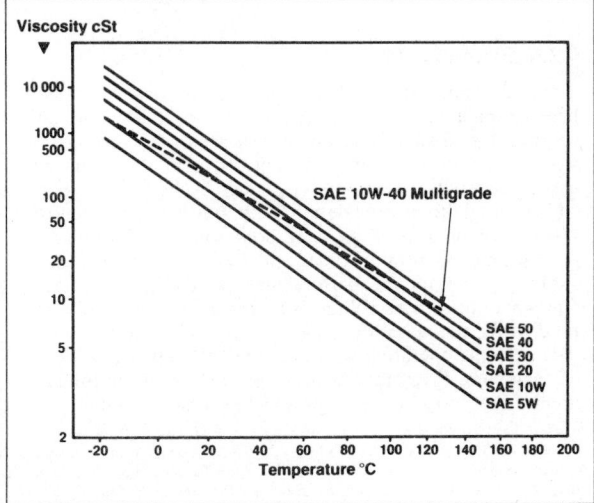

Figure 9.4 Multigrade chart

Engine oils are marketed in a range of performance levels, and need to be classified according to the severity of service conditions in which they are designed to operate. Accordingly, the American Petroleum Institute (API) has drawn up a coding system in which oils are subjected to a series of classifying bench-tests known as the 'Sequence' tests.

9.2.5.8 The API service classifications

In the API system the least demanding classification for a petrol engine was originally designated SA. The most demanding is, at present, SG. (The S stands for Service Station.) Constant development of both engines and oils means that from time to time the highest ratings are superseded by even higher ratings. The API system also classifies diesel engine oils by their severity of service. Here the categories have the prefix C, which stands for Commercial.

Petrol engines
SA   Service typical of engines operated under mild conditions. This classification has no performance requirements.
SB   Service typical of engines operating in conditions such that only minimum protection of the type afforded by additives is desired. Oils designed for this service have been used since the 1930s; they provide only anti-scuff capability and resistance to oil oxidation and bearing corrosion.
SC   Service typical of petrol engines in 1964–1967 cars and trucks. Oils designed for this service provide control of high- and low-temperature deposits, wear, rust and corrosion.
SD   Service typical of 1967–1970 petrol engines in cars and some trucks; but it may apply to later models. Oils designed for this service provide more protection than SC against high- and low-temperature deposits, wear, rust and corrosion; and may be used where SC is recommended.
SE   Service typical of petrol engines in cars and some trucks in 1972–1979. Oils designed for this service provide more protection against oxidation, high-temperature deposits, rust and corrosion than SD or SC, and may be used where those classifications are recommended.
SF   Service typical of petrol engines in cars and some trucks from 1980. Oils developed for this service provide better oxidation stability and anti-wear performance than SE oils. They also provide protection against engine deposits, rust and corrosion. Oils meeting SF may be used wherever SE, SD or SC is recommended.
SG   Service typical of petrol engines in present cars, vans and light trucks. Oils developed for this service provide improved control of engine deposits, oil oxidation and engine wear relative to oils developed for previous categories. Oils meeting SG may be used wherever SF, SE, SF/CC or SE/CC are recommended.

Diesel engines
CA   Service typical of diesel engines operated in mild to moderate duty with high-quality fuels. Occasionally this category has included petrol engines in mild service. Oils designed for this service were widely used in the late 1940s and 1950s; they provided protection from bearing corrosion and light-temperature deposits.
CB   This category is basically the same as CA, but improved to cope with low-quality fuels. Oils designed

for this service were introduced in 1949.

CC   Service typical of lightly supercharged diesel engines operated in moderate to severe duty. Has included certain heavy-duty petrol engines. Oils designed for this service are used in many trucks and in industrial and construction equipment and farm tractors. These oils provide protection from high-temperature deposits in lightly supercharged diesels and also from rust, corrosion and low-temperature deposits in petrol engines.

CD   Service typical of supercharged diesel engines in high-speed high-output duty requiring highly effective control of wear and deposits. Oils designed for this service provide protection from bearing corrosion and high-temperature deposits in supercharged diesel engines running on fuels of a wide quality range.

CDII Service typical of two-stroke cycle diesel engines requiring highly effective control over wear and deposits. Oils designed for this service also meet all the requirements of CD.

CE   Service typical of certain turbocharged or supercharged heavy-duty diesel engines operating under both low speed–high load and high speed–low load conditions. Oils designed for this service must also meet the requirements specified for CC and CD classifications.

Before an oil can be allocated any given API performance level it must satisfy requirements laid down for various engine tests. In the SG category, for example, the engine tests are as follows:

Service IID measures the tendency of the oil to rust or corrode the valve train and to influence the value lifter operation.

Sequence IIIE measures high-temperature oil oxidation, sludge and varnish deposits, cam-and-tappet wear, cam and lifter scuffing and valve lifter sticking.

Sequence VE evaluates sludge deposits, varnish deposits, oil-ring clogging and sticking, oil-screen plugging and cam wear.

Caterpillar IH2 determines the lubricant effect on ring sticking, ring and cylinder wear, and accumulation of piston deposits.

CRC L-38: the characteristics assessed are resistance to oxidation, bearing corrosion, tendency to formation of sludge and varnish, and change of viscosity.

In the CE category the tests are:

Caterpillar IG2: the lubricant characteristics determined are ring sticking, ring and cylinder wear, and accumulation deposits under more severe test conditions than those for Caterpillar IH2.

Cummins NTC-400 measures crownland and piston deposits, camshaft roller follower pin wear and oil consumption.

Mack T6 assesses oil oxidation, piston deposits, oil consumption and ring wear.

Mack T7 evaluates oil thickening.

CRC L-38 (as above).

Other specifications Various authorities and military bodies issue specifications relating to the service performance of engine oils. In some instances the ratings are almost identical with those of the API, but most of them are not precisely parallel because they cover performance factors encountered in particular engines and particular categories of service.

The most common of the other specifications are those with the prefix MIL, issued by the US military authorities. MIL-L-2104E approximates to the API CE rating for diesel lubricants, although it also relates to petrol engines that require

API SE performance. MIL-L-46152D covers oils for both diesel and petrol engines, and approximates to API SG/CC (Figure 9.5).

CCMC ratings Another important set of performance specifications is produced by the European Vehicle Manufacturers' Association, known by its initials CCMC*. The CCMC rating G-1 corresponds roughly to API SE, and G-2 to API SF. G-3 (comparable to MIL-L-46152B, for petrol engines only) covers fuel-efficient and light-viscosity lubricants blended from special high-quality base oils. CCMC also issues specifications for diesel lubricants: D-1 approximates to API CC, D-2 to API CD and MIL-L-2104D, and D-3 to API CE and MIL-L-2104E.

To qualify for the CCMC categories G and D, an oil must meet the requirements of the following tests in addition to the relevant API classification tests.

For the G category

Ford Kent, which evaluates cold ring sticking, piston skirt varnish, oil thickening and consumption.

Fiat 132, to evaluate the tendency of the oil to cause pre-ignition.

Daimler Benz OM 616 to evaluate wear of cylinders and cams.

Bosch Injector Rig, measuring the mechanical stability of the oil to assess its shear stability.

Noack Test, to measure the weight loss due to evaporation of the oil.

High shear/high temperature viscosity test, to assess the oil's capability for resisting shear, and so retaining its viscosity, at high temperatures.

Tests for oil/seal compatibility and oil consumption are still to be established.

For the D category

Bosch Injector Rig, *Noack*, and *D-B OM* 616 tests as above together with:

For D1 and D2 only, *MWM-B* evaluating varnish, carbon deposits, and ring-sticking;

For D3 only, *D-B OM 352A* bore polishing and piston cleanliness;

For PD1 only, *VW 1.6L* to evaluate ring sticking and piston cleanliness.

9.2.6 Hydraulic fluids

The wide application of hydraulic systems has undoubtedly been stimulated by the increasing use of fully automatic controls for sequences of operations where the response to signals must be rapid and the controls themselves light and easily operated. These needs are met by hydraulic circuits which, in addition, provide infinitely variable speed control, reversal of high-speed parts without shock, full protection against damage from overhead and automatic lubrication.

Over the years the performance standards of hydraulic equipment have risen. Whereas a pressure of about 7000 kPa used to be adequate for industrial hydraulic systems, nowadays systems operating with pressures of 15 000–25 000 kPa are common. Pressures above 35 000 kPa are to be found in applications such as large presses for which suitable high-

---

*Comité des Constructeurs d'Automobiles du Marché Commun represents joint industry opinion on factors such as lubricant specifications, emissions, vehicle design and safety standards. With regard to crankcase lubricants, CCMC defines sequences of engine tests, and the tests themselves are defined by CEC (Coordinating European Committee for the Development of Performance Tests for Lubricants and Engine Fuels: a joint body of the oil and motor industries).

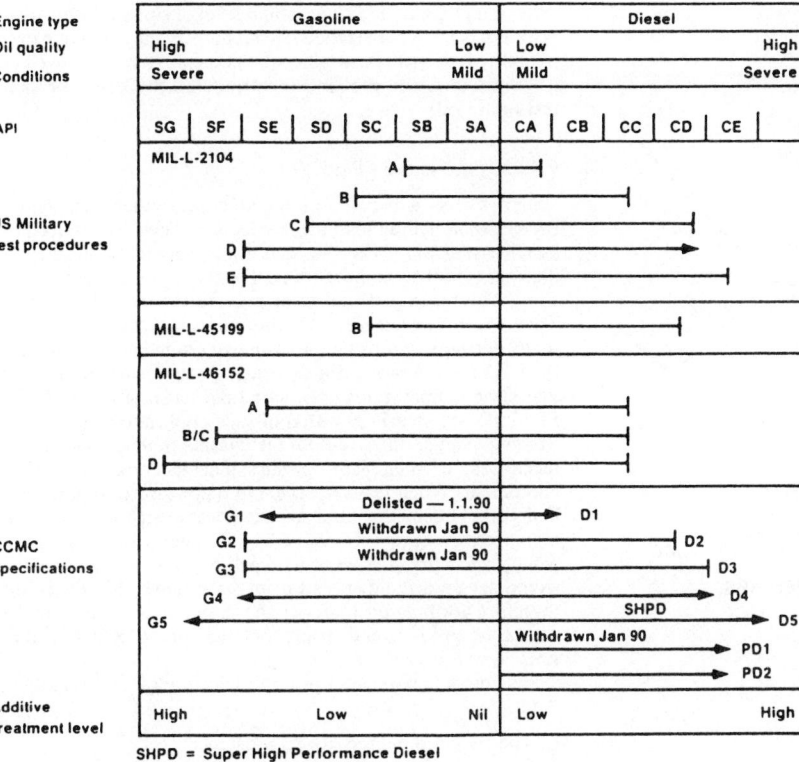

Engine type	Gasoline							Diesel					
Oil quality	High						Low	Low					High
Conditions	Severe						Mild	Mild					Severe
API	SG	SF	SE	SD	SC	SB	SA	CA	CB	CC	CD	CE	

US Military test procedures — MIL-L-2104 (A, B, C, D, E); MIL-L-45199 (B); MIL-L-46152 (A, B/C, D)

CCMC specifications — G1 (Delisted — 1.1.90), G2 (Withdrawn Jan 90), G3 (Withdrawn Jan 90), G4, G5 / D1, D2, D3, D4, D5 (SHPD, Withdrawn Jan 90), PD1, PD2

Additive treatment level	High	Low	Nil	Low	High

SHPD = Super High Performance Diesel
PD = Passenger (Car) Diesel

Figure 9.5 Approximate relationship between classifications and test procedures

pressure pumps have been developed. Additionally, systems have to provide increased power densities, more accurate response, better reliability and increased safety. Their use in numerically controlled machine tools and other advanced control systems creates the need for enhanced filtration. Full flow filters as fine as 1–10 μm retention capability are now to be found in many hydraulic systems.

With the trend toward higher pressures in hydraulic systems the loads on unbalanced pump and motor components become greater and this, coupled with the need for closer fits to contain the higher pressures, can introduce acute lubrication problems. Pumps, one of the main centres of wear, can be made smaller if they can run at higher speeds or higher pressures, but this is only possible with adequate lubrication. For this reason, a fluid with good lubrication properties is used so that 'hydraulics' is now almost synonymous with 'oil hydraulics' in general industrial applications. Mineral oils are inexpensive and readily obtainable while their viscosity can be matched to a particular job.

The hydraulic oil must provide adequate lubrication in the diverse operating conditions associated with the components of the various systems. It must function over an extended temperature range and sometimes under boundary conditions. It will be expected to provide a long, trouble-free service life; its chemical stability must therefore be high. Its wear-resisting properties must be capable of handling the high loads in hydraulic pumps. Additionally, the oil must protect metal surfaces from corrosion and it must both resist emulsification and rapidly release entrained air that, on circulation, would produce foam.

Mineral oil alone, no matter how high its quality, cannot adequately carry out all the duties outlined above and hence the majority of hydraulic oils have their natural properties enhanced by the incorporation of four different types of additives. These are: an anti-oxidant, an anti-wear agent, a foam-inhibitor and an anti-corrosion additive. For machines in which accurate control is paramount, or where the range of operating temperatures is wide – or both – oils will be formulated to include a VI improving additive as well.

9.2.6.1 Viscosity

Probably the most important single property of a hydraulic oil is its viscosity. The most suitable viscosity for a hydraulic system is determined by the needs of the pump and the circuit; too low a viscosity induces back-leakage and lowers the pumping efficiency, while too high a viscosity can cause overheating, pump starvation and possibly cavitation.

9.2.6.2 Viscosity Index

It is desirable that a fluid's viscosity stays within the pump manufacturer's stipulated viscosity limits, in order to accommodate the normal variations of operating temperature. An oil's viscosity falls as temperature rises; certain oils, however, are less sensitive than others to changes of temperatures, and these are said to have a higher VI. Hydraulic oils are formulated from base oils of inherently high VI, to minimize changes of viscosity in the period from start-up to steady running and while circulating between the cold and hot parts of a system.

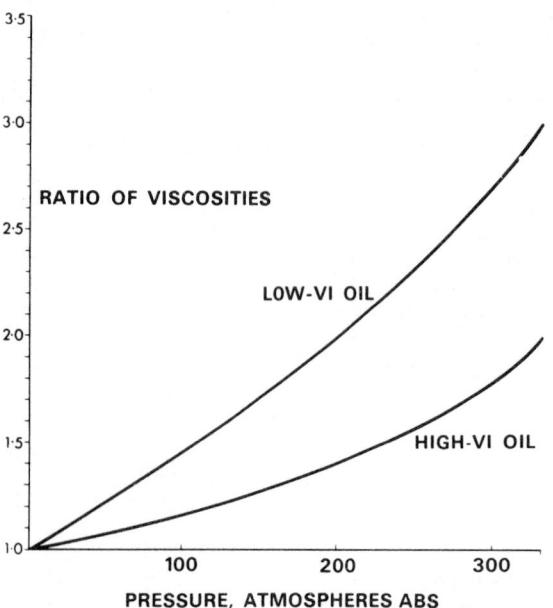

RATIO OF VISCOSITIES

LOW-VI OIL

HIGH-VI OIL

PRESSURE, ATMOSPHERES ABS

Figure 9.6

9.2.6.3 Effects of pressure

Pressure has the effect of increasing an oil's viscosity. While in many industrial systems the working pressures are not high enough to cause problems in this respect, the trend towards higher pressures in equipment is requiring the effect to be accommodated at the design stage. Reactions to pressure are much the same as reactions to temperature, in that an oil of high VI is less affected than one of low VI. A typical hydraulic oil's viscosity is doubled when its pressure is raised from atmospheric to 35 000 kPa (Figure 9.6).

9.2.6.4 Air in the oil

In a system that is poorly designed or badly operated, air may become entrained in the oil and thus cause spongy and noisy operation. The reservoir provides an opportunity for air to be released from the oil instead of accumulating within the hydraulic system. Air comes to the surface as bubbles, and if the resultant foam were to become excessive it could escape through vents and cause loss of oil. In hydraulic oils, foaming is minimized by the incorporation of foam-breaking additives. The type and dosage of such agents must be carefully selected, because although they promote the collapse of surface foam they may tend to retard the rate of air release from the body of the oil.

9.2.6.5 Oxidation stability

Hydraulic oils need to be of the highest oxidation stability, particularly for high-temperature operations, because oxidation causes sludges and lacquer formation. In hydraylic oils, a high level of oxidation stability is ensured by the use of base oils of excellent quality, augmented by a very effective combination of oxidation inhibitors.

A very approximate guide to an oil's compatibility with rubbers commonly used for seals and hoses is given by the Aniline Point, which indicates the degree of swelling likely to

arise; a high figure indicates a high level of compatibility. This system has been superseded by the more accurate Seal Compatibility Index (SCI), in which the percentage volume swell of a 'standard' nitrile rubber is determined after an immersion test in hot oil.

9.2.6.6 Fire-resistant fluids

Where fire is a hazard, or could be extremely damaging, fire-resistant hydraulic fluids are needed. They are referred to as 'fire resistant' (FR) so that users should be under no illusions about their properties. FR fluids do not extinguish fires: they resist combustion or prevent the spread of flame. They are not necessarily fireproof, since any fluid will eventually decompose if its temperature rises high enough. Nor are they high-temperature fluids, since in some instances their operating temperatures are lower than those of mineral oils. FR fluids are clearly essential in such applications as electric welding plants, furnace-door actuators, mining machinery, diecasters, forging plant, plastics machinery and theatrical equipment. When leakage occurs in the pressurized parts of a hydraulic system the fluid usually escapes in the form of a high-pressure spray. In the case of mineral oils this spray would catch fire if it were to reach a source of ignition, or would set up a rapid spread of existing flame. FR fluids are therefore formulated to resist the creation of flame from a source of ignition, and to prevent the spread of an existing fire.

Four main factors enter into the selection of a fire-resistant fluid:

1. The required degree of fire-resistance
2. Operational behaviour in hydraulic systems (lubrication performance, temperature range and seal compatibility, for example)
3. Consideration of hygiene (toxicological, dermatological and respiratory effects)
4. Cost

9.2.6.7 Types of fluid

The fluids available cover a range of chemical constituents, physical characteristics and costs, so the user is able to choose the medium that offers the best compromise for operational satisfaction, fire-resistance and cost effectiveness. Four basic types of fluid are available and are shown in Table 9.4.

In a fully synthetic FR fluid the fire resistance is due to the chemical nature of the fluid; in the others it is afforded by the

Table 9.4 CETOP classifications of fire-resistant hydraulic fluids

Class	Description
HF-A	Oil-in-water emulsions containing a maximum of 20% combustible material. These usually contain 95% water
HF-B	Water-in-oil emulsions containing a maximum of 60% combustible material. These usually contain 40–45% water
HF-C	Water–glycol solutions. These usually contain at least 35% water
HF-D	Water-free fluids. These usually refer to fluids containing phosphate esters, other organic esters or synthesized hydrocarbon fluids

CETOP: Comité European des Transmissions Oleohydrauliques et Pneumatiques.

presence of water. The other main distinction between the two groups is that the fully synthetic fluids are generally better lubricants and are available for use at operating temperatures up to 150°C, but are less likely to be compatible with the conventional sealing materials and paints than are water-based products.

When a water-based fluid makes contact with a flame or aaa hot surface its water component evaporates and forms a steam blanket which displaces oxygen from around the hot area, and this obviates the risk of fire. Water-based products all contain at least 35% water. Because water can be lost by evaporation, they should not be subjected to operating temperatures above about 60°C. Table 9.5 shows a comparison of oil and FR fluids.

9.2.6.8 High water-based hydraulic fluids

For a number of years HF-A oil-in-water emulsions have been used as a fire-resistant hydraulic medium for pit props. Concern over maintenance costs and operational life has created interest in a better anti-wear type fluid. Micro-emulsions are known to give better wear protection than the normal oil-in-water emulsions. At the same time the car industry, in attempts to reduce costs especially from leakages on production machinery, has evaluated the potential for using HWBHF in hydraulic systems. As a result, in many parts of industry, not only those where fire-resistant hydraulic fluids are needed, there is a increasing interest in the use of HWBHF.

Such fluids, often referred to as 5/95 fluid (that being the ratio of oil to water), have essentially the same properties as water with the exception of the corrosion characteristics and the boundary lubrication properties which are improved by the oil and other additives. The advantages of this type of fluid are fire resistance, lower fluid cost, no warm-up time, lower power consumption and operating temperatures, reduced spoilage of coolant, less dependence on oil together with reduced transport, storage, handling and disposal costs, and environmental benefits.

In considering these benefits the the user should not overlook the constraints in using such fluids. They can be summarized as limited wear and corrosion protection (especially with certain metals), increased leakage due to its low viscosity, limited operating temperature range and the need for additional mixing and in-service monitoring facilities.

Because systems are normally *not* designed for use with this type of fluid, certain aspects should be reviewed with the equipment and fluid suppliers before a decision to use such fluids can be taken. These are compatibility with filters, seals, gaskets, hoses, paints and any non-ferrous metals used in the

Table 9.5 Comparison of oil and FR fluids

	Mineral oil	Water-in-oil emulsion	Water–glycol	Phosphate ester
Fire resistance	Poor	Fair	Excellent	Good
Relative density	0.87	0.94	1.08	1.14
Viscosity Index	High	High	High	Low
Vapour pressure	Low	High	High	Low
Special seals	No	Partly	Partly	Yes
Special paints	No	No	Yes	Yes
Rust protection	Very good	Good	Fair	Fair

equipment. Condensation corrosion effect on ferrous metals, fluid-mixing equipment needed, control of microbial infection together with overall maintaining and control of fluid dilution and the disposal of waste fluid must also be considered. Provided such attention is paid to these design and operating features, the cost reductions have proved very beneficial to the overall plant cost effectiveness.

9.2.6.9 Care of hydraulic oils and systems

Modern additive-treated oils are so stable that deposits and sludge formation in normal conditions have been almost eliminated. Consequently, the service life of the oils which is affected by oxidation, thermal degradation and moisture is extended.

Solid impurities must be continuously removed because hydraulic systems are self-contaminating due to wear of hoses, seals and metal parts. Efforts should be made to exclude all solid contaminants from the system altogether. Dirt is introduced with air, the amount of airborne impurities varying with the environment. The air breather must filter to at least the same degree as the oil filters.

It is impossible to generalize about types of filter to be used. Selection depends on the system, the rate of contamination build-up and the space available. However, a common arrangement is to have a full-flow filter unit before the pump with a bypass filter at some other convenient part of the system. Many industrial systems working below 13 500 kPa can tolerate particles in the order of 25–50 μm with no serious effects on either valves or pumps.

Provided that the system is initially clean and fitted with efficient air filters, metal edge-strainers of 0.127 mm spacing appear to be adequate, although clearances of vane pumps may be below 0.025 mm. It should be remembered that an excessive pressure drop, due to a clogged full-flow fine filter, can do more harm to pumps by cavitation than dirty oil.

If flushing is used to clean a new system or after overhaul it should be done with the hydraulic oil itself or one of lighter viscosity and the same quality. As the flushing charge circulates it should pass through an edge-type paper filter of large capacity. It is generally preferable to use a special pump rather than the hydraulic pump system, and the temperature of the oil should be maintained at about 40°C without local overheating.

9.2.7 Machine tools

Lubricants are the lifeblood of a machine tool. Without adequate lubrication, spindles would seize, slides could not slide and gears would rapidly distintegrate. However, the reduction of bearing friction, vital though it is, is by no means the only purpose of machine-tool lubrication. Many machines are operated by hydraulic power, and one oil may be required to serve as both lubricant and hydraulic fluid. The lubricant must be of correct viscosity for its application, must protect bearings, gears and other moving parts against corrosion, and, where appropriate, must remove heat to preserve working accuracies and aligments. It may additionally serve to seal the bearings against moisture and contaminating particles. In some machine tools the lubricant also serves the function of a cutting oil, or perhaps needs to be compatible with the cutting oil. In other tools an important property of the lubricant is its ability to separate rapidly and completely from the cutting fluid. Compatibility with the metals, plastics, sealing elements and tube connections used in the machine construction is an important consideration.

In machine-tool operations, as in all others, the wisest course for the user is to employ reputable lubricants in the manner recommended by the machine-tool manufacturer and the oil company supplying the product. This policy simplifies the selection and application of machine-tool lubricants. The user can rest assured that all the considerations outlined above have been taken into account by both authorities.

The important factors from the point of view of lubrication are the type of component and the conditions under which it operates, rather than the type of machine into which it is incorporated. This explains the essential similarity of lubricating systems in widely differing machines.

9.2.7.1 Bearings

As in almost every type of machine, bearings play an important role in the efficient functioning of machine tools.

9.2.7.2 Roller bearings

There is friction even in the most highly finished ball or roller bearing. This is due to the slight deformation under load of both the raceway and the rolling components, the presence of the restraining cage, and the 'slip' caused by trying to make parts of different diameter rotate at the same speed. In machine tools the majority of rolling bearings are grease-packed for life, or for very long periods, but other means of lubrication are also used (the bearings may be connected to a centralized pressure-oil-feed system for instance). In other cases, oil-mist lubrication may be employed both for spindle bearings and for quill movement. In headstocks and gear-boxes, ball and roller bearings may be lubricated by splash or oil jets.

9.2.7.3 Plain journal bearings

Plain bearings are often preferred for relatively low-speed spindles operating under fairly constant loads, and for the spindles of high-speed grinding wheels. These bearings ride on a dynamic 'wedge' of lubricating oil. Precision plain bearings are generally operated with very low clearances and therefore require low-viscosity oil to control the rise of temperature. Efficient lubrication is vital if the oil temperature is to be kept within reasonable limits, and some form of automatic circulation system is almost always employed.

9.2.7.4 Multi-wedge bearings

The main drawback of the traditional plain bearing is its reliance on a single hydrodynamic wedge of oil, which under certain conditions tends to be unstable. Multi-wedge bearings make use of a number of fixed or rocking pads, spaced at intervals around the journal to create a series of opposed oil wedges. These produce strong radial, stabilizing forces that hold the spindle centrally within the bearing. With the best of these, developed especially for machine tools, deviation of the spindle under maximum load can be held within a few millionths of a centimetre.

9.2.7.5 Hydrostatic bearings

To avoid the instabilities of wedge-shaped oils films, a lubricating film can be maintained by the application of pressurized oil (or, occasionally, air) to the bearing. The hydrostatic bearing maintains a continuous film of oil even at zero speed, and induces a strong stabilizing force towards the centre which counteracts any displacement of the shaft or spindle. Disadvantages include the power required to pressurize the oil and the necessary increase in the size of the filter and circulatory system.

9.2.7.6 Slideways

Spindles may be the most difficult machine-tool components to design, but slideways are frequently the most troublesome to lubricate. In a slideway the wedge-type of film lubrication cannot form since, to achieve this, the slideway would need to be tilted.

9.2.7.7 Plain slideways

Plain slideways are preferred in the majority of applications. Only a thin film of lubricant is present, so its properties – especially its viscosity, adhesion and extreme-pressure characteristics – are of vital importance. If lubrication breaks down intermittently, a condition is created known a 'stick-slip' which affects surface finish, causes vibration and chatter and makes close limits difficult to hold. Special adhesive additives are incorporated into the lubricant to provide good bonding of the oil film to the sliding surfaces which helps to overcome the problems of table and slideway lubrication. On long traverses, oil may be fed through grooves in the underside of the slideway.

9.2.7.8 Hydrostatic slideways

The use of hydrostatic slideways – in which pressurized oil or air is employed – completely eliminates stick-slip and reduces friction to very low values; but there are disadvantages in the form of higher costs and greater complication.

9.2.7.9 Ball and roller slideways

These are expensive but, in precision applications, they offer the low friction and lack of play that are characteristic of the more usual rolling journal bearings. Lubrication is usually effected by grease or an adhesive oil.

9.2.7.10 Leadscrews and nuts

The lubrication of leadscrews is similar in essence to that of slideways, but in some instances may be more critical. This is especially so when pre-load is applied to eliminate play and improve machining accuracy, since it also tends to squeeze out the lubricant. Leadscrews and slideways often utilize the same lubricants. If the screw is to operate under high unit stresses – due to pre-load or actual working loads – an extreme-pressure oil should be used.

9.2.7.11 Recirculating-ball leadscrews

This type was developed to avoid stick-up in heavily loaded leadscrews. It employs a screw and nut of special form, with bearing balls running between them. When the balls run off one end of the nut they return through an external channel to the other end. Such bearings are usually grease-packed for life.

9.2.7.12 Gears

The meshing teeth of spur, bevel, helical and similar involute gears are separated by a relatively thick hydrodynamic wedge of lubricating oil, provided that the rotational speed is high enough and the load light enough so as not to squeeze out the lubricant. With high loads or at low speeds, wear takes place if the oil is not able to maintain a lubricating film under extreme conditions.

Machine-tool gears can be lubricated by oil-spray, mist, splash or cascade. Sealed oil baths are commonly used, or the gears may be lubricated by part of a larger circulatory system.

9.2.7.13 Hydraulics

The use of hydraulic systems for the setting, operation and control of machine tools has increased significantly. Hydraulic mechanisms being interlinked with electronic controls and/or feedbacks control systems. In machine tools, hydraulic systems have the advantage of providing stepless and vibrationless transfer of power. They are particularly suitable for the linear movement of tables and slideways, to which a hydraulic piston may be directly coupled.

One of the most important features for hydraulic oil is a viscosity/temperature relationship that gives the best compromise of low viscosity (for easy cold starting) and minimum loss of viscosity at high temperatures (to avoid back-leakage and pumping losses). A high degree of oxidation stability is required to withstand high temperatures and aeration in hydraulic systems. An oil needs excellent anti-wear characteristics to combat the effects of high rubbing speeds and loads that occur in hydraulic pumps, especially in those of the vane type. In the reservoir, the oil must release entrained air readily without causing excessive foaming, which can lead to oil starvation.

9.2.7.14 Tramp oil

'Tramp oil' is caused when neat slideway, gear, hydraulic and spindle lubricants leak into water-based cutting fluids and can cause problems such as:

- Machine deposits
- Reduced bacterial resistance of cutting fluids and subsequent reduction in the fluid life
- Reduced surface finish quality of work pieces
- Corrosion of machine surfaces

All these problems directly affect production efficiency. Recent developments have led to the introduction of synthetic lubricants that are fully compatible with all types of water-based cutting fluids, so helping the user to achieve maximum machine output.

9.2.7.15 Lubrication and lubricants

The components of a hydraulic system are continuously lubricated by the hydraulic fluid, which must, of course, be suitable for this purpose. Many ball and roller bearings are grease-packed for life, or need attention at lengthy intervals. Most lubrication points, however, need regular replenishment if the machine is to function satisfactorily. This is particularly true of parts subjected to high temperatures.

With the large machines, the number of lubricating points or the quantities of lubricants involved make any manual lubrication system impracticable or completely uneconomic. Consequently, automatic lubrication systems are often employed.

Automatic lubrication systems may be divided broadly into two types: circulatory and 'one-shot' total-loss. These cover, respectively, those components using relatively large amounts of oil, which can be cooled, purified and recirculated, and those in which oil or grease is used once only and then lost. Both arrangements may be used for different parts of the same machine or installation.

9.2.7.16 Circulatory lubrication systems

The circulatory systems used in association with machine tools are generally conventional in nature, although occasionally their exceptional size creates special problems. The normal installation comprises a storage tank or reservoir, a pump and

filter, suitable sprays, jets or other distribution devices, and return piping. The most recent designs tend to eliminate wick feeds and siphon lubrication.

Although filtration is sometimes omitted with non-critical ball and roller bearings, it is essential for most gears and for precision bearings of every kind. Magnetic and gauze filters are often used together. To prevent wear of highly finished bearings surfaces the lubricant must contain no particle as large as the bearing clearance.

Circulatory systems are generally interlocked electrically or mechanically with the machine drive, so that the machine cannot be started until oil is flowing to the gears and main bearings. Interlocks also ensure that lubrication is maintained as long as the machine is running. Oil sight-glasses at key points in the system permit visual observations of oil flow.

9.2.7.17 Loss-lubrication systems

There are many kinds of loss-lubrication systems. Most types of linear bearings are necessarily lubricated by this means. An increasingly popular method of lubrication is by automatic or manually operated one-shot lubricators. With these devices a metered quantity of oil or grease is delivered to any number of points from a single reservoir. The operation may be carried out manually, using a hand-pump, or automatically, by means of an electric or hydraulic pump. Mechanical pumps are usually controlled by an electric timer, feeding lubricant at preset intervals, or are linked to a constantly moving part of the machine.

On some machines both hand-operated and electrically timed one-shot systems may be in use, the manual system being reserved for those components needing infrequent attention (once a day, for example) while the automatic systems feeds those parts that require lubrication at relatively brief intervals.

9.2.7.18 Manual lubrication

Many thousands of smaller or older machines are lubricated by hand, and even the largest need regular refills or topping up to lubricant reservoirs. In some shops the operator may be fully responsible for the lubrication of his own machine, but it is nearly always safer and more economical to make one individual responsible for all lubrication.

9.2.7.19 Rationalizing lubricants

To meet the requirements of each of the various components of a machine the manufacturer may need to recommend a number of lubricating oils and greases. It follows that, where there are many machines of varying origins, a large number of lubricants may seem to be needed. However, the needs of different machines are rarely so different that slight modification cannot be made to the specified lubricant schedule. It is this approach which forms the basis for BS 5063, from which the data in Table 9.6 have been extracted. This classification implies no quality evaluation of lubricants, but merely gives information as to the categories of lubricants likely to be suitable for particular applications.

A survey of the lubrication requirements, usually carried out by the lubricant supplier, can often be the means of significantly reducing the number of oils and greases in a workshop or factory. The efficiency of lubrication may well be increased, and the economies effected are likely to be substantial.

Table 9.6 Classification of lubricants

Class	Type of lubricant	Viscosity grade no. (BS 4231)	Typical application	Detailed application	Remarks
AN	Refined mineral oils	68	General lubrication	Total-loss lubrication	May be replaced by CB 68
CB	Highly refined mineral oils (straight or inhibited) with good anti-oxidation performance	32 68	Enclosed gears – general lubrication	Pressure and bath lubrication of enclosed gears and allied bearings of headstocks, feed boxes, carriages, etc. when loads are moderate; gears can be of any type, other than worm and hypoid	CB 32 and CB 68 may be used for flood-lubricated mechanically controlled clutches; CB 32 and CB 68 may be replaced by HM 32 and HM 68
CC	Highly refined mineral oils with improved loading-carrying ability	150 320	Heavily loaded gears and worm gears	Pressure and bath lubrication of enclosed gears of any type, other than hypoid gears, and allied bearings when loads are high, provided that operating temperature is not above 70°C	May also be used for manual or centralized lubrication of lead and feed screws
FX	Heavily refined mineral oils with superior anti-corrosion anti-oxidation performance	10 22	Spindles	Pressure and bath lubrication of plain or rolling bearings rotating at high speed	May also be used for applications requiring particularly low-viscosity oils, such as fine mechanisms, hydraulic or hydro-pneumatic mechanisms electro-magnetic clutches, air line lubricators and hydrostatic bearings
G	Mineral oils with improved lubricity and tackiness performance, and which prevent stick-slip	68 220	Slideways	Lubrication of all types of machine tool plain-bearing slideways; particularly required at low traverse speeds to prevent a discontinuous or intermittent sliding of the table (stick-slip)	May also be used for the lubrication of all sliding parts – lead and feed screws, cams, ratchets and lightly loaded worm gears with intermittent service; if a lower viscosity is required HG 32 may be used.

Class	Type of lubricant		Typical application	Detailed application
HM	Highly refined mineral oils with superior anti-corrosion, anti-oxidation, and anti-wear performance	32 68	Hydraulic systems	Operation of general hydraulic systems	May also be used for the lubrication of plain or rolling bearings and all types of gears, normally loaded worm and hypoid gears excepted, HM 3X and HM 68 may replace CB 32 and CB 68, respectively
HG	Refined mineral oils of HM type with anti-stick-slip properties	32	Combined hydraulic and slideways systems	Specific application for machines with combined hydraulic and plain bearings, and lubrication systems where discontinuous or intermittent sliding (stick-slip) at low speed is to be prevented	May also be used for the lubrication of slideways, when an oil of this viscosity is required

Class	Type of lubricant	Consistency number	Typical application	Detailed application
XM	Premium quality multi-purpose greases with superior anti-oxidation and anti-corrosion properties	1 2 3	Plain and rolling bearings and general greasing of miscellaneous parts	XM 1: Centralized systems XM 2: Dispensed by cup or hand gun or in centralized systems XM 3: Normally used in prepacked applications such as electric motor bearings

Note: It is essential that lubricants are compatible with the materials used in the construction of machine tools, and particularly with sealing devices.
The grease X is sub-divided into consistency numbers, in accordance with the system proposed by the National Lubricating Grease Institute (NLGI) of the USA. These consistency numbers are related to the worked penetration ranges of the greases as follows:

Consistency number	Worked penetration range
1	310–340
2	265–295
3	220–250

Worked penetration is determined by the cone-penetration method described in BS 5296.

9.2.8 Compressors

Compressors fall into two basic categories: positive-displacement types, in which air is compressed by the 'squashing' effect of moving components; and dynamic (turbo)-compressors, in which the high velocity of the moving air is converted into pressure. In some compressors the oil lubricates only the bearings, and does not come into contact with the air; in some it serves an important cooling function; in some it is in intimate contact with the oxidizing influence of hot air and with moisture condensed from the air. Clearly, there is no such thing as a typical all-purpose compressor oil: each type subjects the lubricant to a particular set of conditions. In some cases a good engine oil or a turbine-quality oil is suitable, but in others the lubricant must be special compressor oil (Figure 9.7).

9.2.8.1 Quality and safety

Over the years the progressive improvements in compressor lubricants have kept pace with developments in compressor technology, and modern oils make an impressive contribution to the performance and longevity of industrial compressors. More recently a high proportion of research has been directed towards greater safety, most notably in respect of fires and explosions within compressors. For a long time the causes of such accidents were a matter of surmise, but it was noticed that the trouble was almost invariably associated with high delivery temperatures and heavy carbon deposits in delivery pipes. Ignition is now thought to be caused by an exothermic (heat-releasing) oxidation reaction with the carbon deposit, which creates temperatures higher than the spontaneous ignition temperature of the absorbed oil.

Experience indicates that such deposits are considerably reduced by careful selection of base oils and antioxidation additives. Nevertheless, the use of a top-class oil is no guarantee against trouble if maintenance is neglected. For complete safety, both the oil and the compressor system must enjoy high standards of care.

9.2.8.2 Specifications

The recommendations of the International Standards Organization (ISO) covering mineral-oil lubricants for reciprocating compressors are set out in ISO DP 6521, under the ISO-L-DAA and ISO-L-DAB classifications. These cover applications wherever air-discharge temperatures are, respectively, below and above 160°C For mineral-oil lubricants used in oil-flooded rotary-screw compressors the classifications ISO-L-DAG and DAH cover applications where temperatures are, respectively, below 100°C and in the 100–110°C range. For more severe applications, where synthetic lubricants might be used, the ISO-L-DAC and DAJ specifications cover both reciprocating and oil-flooded rotary-screw requirements.

For the general performance of compressor oils there is DIN 51506. This specification defines several levels of performance, of which the most severe – carrying the code letters VD-L – relates to oils for use at air-discharge temperatures of up to 220°C.

The stringent requirements covering oxidation stability are defined by the test method DIN 51352, Part 2, known as the Pneurop Oxidation Test (POT). This test simulates the oxidizing effects of high temperature, intimate exposure to air, and the presence of iron oxide which acts as catalyst – all factors highly conducive to the chemical breakdown of oil, and the consequent formation of deposits that can lead to fire and explosion.

Rotary-screw compressor mineral oils oxidation resistance is assessed in a modified Pneurop oxidation test using iron naphthenate catalyst at 120°C for 1000 h. This is known as the rotary-compressor oxidation test (ROCOT).

9.2.8.3 Oil characteristics

Reciprocating compressors In piston-type compressors the oil serves three functions in addition to the main one of lubricating the bearings and cylinders. It helps to seal the fine clearances around piston rings, piston rods and valves, and thus minimizes blow-by of air (which reduces efficiency and can cause overheating). It contributes to cooling by dissipating heat to the walls of the crankcase and it prevents corrosion that would otherwise be caused by moisture condensing from the compressed air.

In small single-acting compressors the oil to bearings and cylinders is splash-fed by flingers, dippers or rings, but the larger and more complex machines have force-feed lubrication systems, some of them augmented by splash-feed. The cylinders of a double-acting compressor cannot be splash-lubricated, of course, because they are not open to the crankcase. Two lubricating systems are therefore necessary – one for the bearings and cross-head slides and one feeding oil directly into the cylinders. In some cases the same oil is used for both purposes, but the feed to the cylinders has to be carefully controlled, because under-lubrication leads to rapid wear and over-lubrication leads to a build-up of carbon deposits in cylinders and on valves. The number and position of cylinder-lubrication points varies according to the size and type of the compressor. Small cylinders may have a single point in the cylinder head, near the inlet valve; larger ones may have two or more. In each case the oil is spread by the sliding of the piston and the turbulence of the air.

In the piston-type compressor the very thin oil film has to lubricate the cylinder while it is exposed to the heat of the

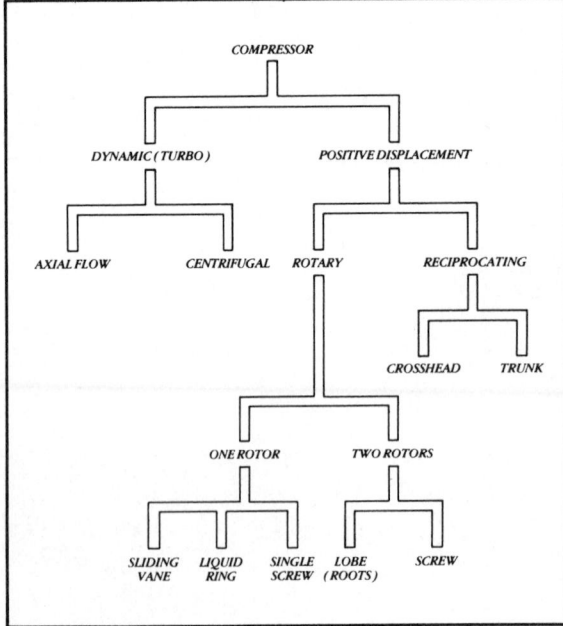

Figure 9.7 Compressor types

compressed air. Such conditions are highly conducive to oxidation in poor-quality oils, and may result in the formation of gummy deposits that settle in and around the piston-ring grooves and cause the rings to stick, thereby allowing blow-by to develop.

Rotary compressors – vane type The lubrication system of vane-type compressors varies according to the size and output of the unit. Compressors in the small and 'portable' group have neither external cooling nor intercooling, because to effect all the necessary cooling the oil is injected copiously into the incoming air stream or directly into the compressor chamber. This method is known as flood lubrication, and the oil is usually cooled before being recirculated. The oil is carried out of the compression chamber by the air, so it has to be separated from the air; the receiver contains baffles that 'knock out' the droplets of oil, and they fall to the bottom of the receiver. Condensed water is subsequently separated from the oil in a strainer before the oil goes back into circulation.

Vane-type pumps of higher-output are water-jacketed and intercooled: the lubricant has virtually no cooling function so it is employed in far smaller quantities. In some units the oil is fed only to the bearings, and the normal leakage lubricates the vanes and the casing. In others, it is fed through drillings in the rotor and perhaps directly into the casing. This, of course, is a total-loss lubrication technique, because the oil passes out with the discharged air.

As in reciprocating units, the oil has to lubricate while being subjected to the adverse influence of high temperature. The vanes impose severe demands on the oil's lubricating powers. At their tips, for example, high rubbing speeds are combined with heavy end-pressure against the casing.

Each time a vane is in the extended position (once per revolution) a severe bending load is being applied between it and the side of its slot. The oil must continue to lubricate between them, to allow the vane to slide freely. It must also resist formation of sticky deposits and varnish, which lead to restricted movement of the vanes and hence to blow-by and, in severe cases, to broken vanes.

Rotary compressors – screw type The lubrication requirements for single-screw type compressors are not severe, but in oil-flooded rotary units the oxidizing conditions are extremely severe because fine droplets of oil are mixed intimately with hot compressed air. In some screw-type air compressors the rotors are gear driven and do not make contact. In others, one rotor drives the other. The heaviest contact loads occur where power is transmitted from the female to the male rotor: here the lubricant encounters physical conditions similar to those between mating gear teeth. This arduous combination of circumstances places a great demand on the chemical stability, and lubricating power, of the oil.

Other types Of the remaining designs, only the liquid-piston type delivers pressures of the same order as those just mentioned. The lobe, centrifugal and axial-flow types, are more accurately termed 'blowers', since they deliver air in large volumes at lower pressures. In all four cases only the 'external' parts – bearings, gears or both – require lubrication. Therefore the oil is not called upon to withstand the severe service experienced in reciprocating and vane-type compressors. Where the compressor is coupled to a steam or gas turbine a common circulating oil system is employed. High standards of system cleanliness are necessary to avoid deposit formation in the compressor bearings.

Refrigeration compressors The functions of a refrigerator compressor lubricant are the same as those of compressor lubricants in general. However, the close association between refrigerant and lubricant does impose certain additional demands on the oil. Oil is unavoidably carried into the circuit with refrigerant discharging from the compressor. In many installations provision is made for removal of this oil. However, several refrigerants, including most of the halogen refrigerants, are miscible with oil and it is difficult to separate the oil which enters the system which therefore circulates with the refrigerant. In either case the behaviour of the oil in cold parts of the systems is important, and suitable lubricants have to have low pour point and low wax-forming characteristics.

Effects of contamination The conditions imposed on oils by compressors – particularly by the piston type – are remarkably similar to those imposed by internal combustion engines. One major difference is, of course, that in a compressor no fuel or products of combustion are present to find their way into the oil. Other contaminants are broadly similar. Among these are moisture, airborne dirt, carbon and the products of the oil's oxidation. Unless steps are taken to combat them, all these pollutants have the effect of shortening the life of both the oil and the compressor, and may even lead to fires and explosions.

Oxidation High temperature and exposure to hot air are two influences that favour the oxidation and carbonization of mineral oil. In a compressor, the oil presents a large surface area to hot air because it is churned and sprayed in a fine mist, so the oxidizing influences are very strong – especially in the high temperatures of the compressor chamber. The degree of oxidation is dependent mainly on temperature and the ability of the oil to resist, so the problem can be minimized by the correct selection of lubricant and by controlling operating factors.

In oxidizing, an oil becomes thicker and it deposits carbon and gummy, resinous substances. These accumulate in the piston-ring grooves of reciprocating compressors and in the slots of vane-type units, and as a result they restrict free movement of components and allow air leakages to develop. The deposits also settle in and around the valves of piston-type compressors, and prevent proper sealing.

When leakage develops, the output of compressed air is reduced, and overheating occurs due to the recompression of hot air and the inefficient operation of the compressor. This leads to abnormally high discharge temperatures. Higher temperature leads to increased oxidation and hence increased formation of deposits, so adequate cooling of compressors is very important.

Airborne dirt In the context of industrial compressors, dust is a major consideration. Such compressors have a very high throughput of air, and even in apparently 'clean' atmospheres, the quantity of airborne dirt is sufficient to cause trouble if the compressor is not fitted with an air-intake filter. Many of the airborne particles in an industrial atmosphere are abrasive, and they cause accelerated rates of wear in any compressor with sliding components in the compressor chamber. The dirt passes into the oil, where it may accumulate and contribute very seriously to the carbon deposits in valves and outlet pipes. Another consideration is that dirt in an oil is likely to act as a catalyst, thus encouraging oxidation.

Moisture Condensation occurs in all compressors, and the effects are most prominent where cooling takes place – in intercoolers and air-receivers, which therefore have to be drained at frequent intervals. Normally the amount of moisture present in a compression chamber is not sufficient to affect lubrication, but relatively large quantities can have a serious effect on the lubrication of a compressor. Very wet conditions are likely to occur when the atmosphere is excessively humid, or compression pressures are high, or the compressor is being overcooled.

During periods when the compressor is standing idle the moisture condenses on cylinders walls and casings, and if the oil does not provide adequate protection this leads to rusting. Rust may not be serious at first sight, and it is quickly removed by wiping action when the compressor is started, but the rust particles act as abrasives, and if they enter the crankcase oil they may have a catalytic effect and promote oxidation. In single-acting piston-type compressors, the crankcase oil is contaminated by the moisture.

9.2.9 Turbines

9.2.9.1 Steam

Although the properties required of a steam-turbine lubricant are not extreme it is the very long periods of continuous operation that creates the need for high-grade oils to be used. The lubricating oil has to provide adequate and reliable lubrication, act as a coolant, protect against corrosion, as a hydraulic medium when used in governor and control systems, and if used in a geared turbine provide satisfactory lubrication of the gearing. The lubricant will therefore need the following characteristics.

Viscosity For a directly coupled turbine for power generation a typical viscosity would be in the range of 32–46 cSt at 40°C. Geared units require a higher viscosity to withstand tooth loadings typically within the range of 68–100 cSt at 40°C.

Oxidation resistance The careful blending of turbine oils, using components which, by selective refining, have a reduced tendency to oxidize, produces the required long-term stability. The high temperatures and pressures of modern designs add to these demands, which are combatted by the incorporation of suitable anti-oxidant additives.

Demulsibility The ability of the lubricant to separate readily and completely from water, in either a centrifuge or a settling tank, is important in a turbine lubricant. Otherwise the retained water will react with products of oxidation and particle contaminants to form stable emulsions. These will increase the viscosity of the oil and form sludges which can result in a failure. Careful and selective refining ensures a good demulsibility characteristic. Inadequate storage and handling can seriously reduce this property.

Corrosion resistance Although the equipment is designed to keep the water content at a minimum level, it is virtually impossible to eliminate it entirely. The problem of rusting is therefore overcome by using corrosion inhibitors in the lubricant formulation.

Foaming resistance Turbine oils must be resistant to foaming, since oil-foam reduces the rate of heat transfer from the bearings, promotes oxidation by greatly extending the area of contact between air and oil. It is also an unsatisfactory medium for the hydraulic governor controls. Careful refining is the primary means of achieving good resistance to foaming. Use of an anti-foam additive may seem desirable but this should be approached with caution. If it is used in quantities higher than the optimum it can in fact assist air entrainment in the oil by retarding the release of air bubbles.

9.2.9.2 Gas

The lubricants generally specified for conventional gas turbines invariably fall within the same classification as those used for steam turbines and are often categorized as 'turbine oils'. In those cases where an aircraft type gas turbine has been adapted for industrial use the lubricant is vitally important to their correct operation. Specifications have been rigidly laid down after the most exhaustive tests, and it would be unwise, even foolhardy, to depart from the manufacturers' recommendations. No economic gain would result from the use of cheaper, but less efficient, lubricants.

9.2.9.3 Performance standards

In the UK there is BS 489:1983. In Europe there is DIN 51515 together with manufacturers' standards such as those set by Brown Boverie and Alsthom Atlantique. In the USA there are the ASTM standards and the well-known General Electric requirements.

The total useful life of a turbine oil is its most important characteristic. ASTM method D943 (IP 157) measures the life indirectly by assessing the useful life of the oxidation inhibitor contained in the formulation and is often referred to as the TOST 'life' of the oil. Rust prevention is generally assessed by the ASTM D665 (IP 135) method.

There are many other specifications designed by equipment builders, military and professional societies, as well as users. Care always needs to be taken when purchasing turbine oil to specification. The cheapest oil, albeit conforming to the specification, may not necessarily be the best within that specification for the particular purpose. For instance, the additive package is rarely (if ever) defined, so that unexpected reactions can occur between oils which could affect overall performance.

9.2.10 Transformers and switchgear

The main requirement for a power-transmission equipment oil is that it should have good dielectric properties. Oil used in transformers acts as a coolant for the windings; as an insulant to prevent arcing between parts of the transformer circuits; and prevents the ionization of minute bubbles of air and gas in the wire insulation by absorbing them and filling the voids between cable and wrapping. In switchgear and circuit breakers it has the added function of quenching sparks from any arc formed during equipment operation. Oils for use in power transmission equipment should have the following properties; high electric strength, low viscosity, high chemical stability and low carbon-forming characteristics under the conditions of electric arc.

9.2.10.1 Performance standards

The efficiency of transformer oils as dielectrics is measured by 'electric strength' tests. These give an indication of the voltage at which, under the test conditions, the oil will break down. Various national standards exist that all measure the same basic property of the oil. In the UK it is BS 148: 1984. There is an international specification, IEC 296/1982, which may be quoted by equipment manufacturers in their oil recommendations.

9.2.10.2 Testing

How frequently the oil condition should be tested depends on operating and atmospheric conditions; after the commissioning sample, further samples should be taken at three months and one year after the unit is first energized. After this, under normal conditions, testing should be carried out annually. In unfavourable operating conditions (damp or dust-laden atmospheres, or where space limitations reduce air circulation and heat transfer) testing should be carried out every six months.

Testing should include a dielectric strength test to confirm the oil's insulation capability and an acidity test, which indicates oil oxidation. While acid formation does not usually develop until the oil has been in service for some time, when it does occur the process can be rapid. If acidity is below 0.5 mg KOH/g no action would seem necessary. Between 0.5 and 1 mg KOH/g, increased care and testing is essential. Above 1 the oil should be removed and either reconditioned or discarded. Before the unit is filled with a fresh charge of oil it should be flushed. These suggestions are contained in a British Standards Code of Practice.

Sludge observations will show if arcing is causing carbon deposits which, if allowed to build up will affect heat transfer and could influence the oil insulation. There is also a flash point test, in which any lowering of flash point is an indication that the oil has been subjected to excessive local heating or submerged arcing (due to overload or an internal electrical fault). A fall in flash point exceeding 16°C implies a fault, and the unit should be shut down for investigation of the cause. Lesser drops may be observed in the later stages of oil life, due to oxidation effects, but are not usually serious. A 'crackle' test is a simple way of detecting moisture in the oil. Where water is present the oil should be centrifuged.

9.2.11 Greases

Grease is a very important and useful lubricant when used correctly, its main advantage being that it tends to remain where it is applied. It is more likely to stay in contact with rubbing surfaces than oil, and is less affected by the forces of gravity, pressure and centrifugal action. Economical and effective lubrication is the natural result of this property and a reduction in the overall cost of lubrication, particularly in all-loss systems, is made possible.

Apart from this, grease has other advantages. It acts both as a lubricant and as a seal and is thus able, at the same time as it lubricates, to prevent the entry of contaminants such as water and abrasive dirt. Grease lubrication by eliminating the need for elaborate oil seals can simplify plant design.

Because a film of grease remains where it is applied for much longer than a film of oil, it provides better protection to bearing and other surfaces that are exposed to shock loads or sudden changes of direction. A film of grease also helps to prevent the corrosion of machine parts that are idle for lengthy periods.

Bearings pre-packed with grease will function for extended periods without attention. Another advantage is the almost complete elimination of drip or splash, which can be a problem in certain applications. Grease is also able to operate effectively over a wider range of temperatures than any single oil.

There are certain disadvantages as well as advantages in using grease as a lubricant. Greases do not dissipate heat as well as fluid lubricants, and for low-torque operation tend to offer more resistance than oil.

9.2.11.1 Types of grease

The general method of classifying greases is by reference to the type of soap that is mixed with mineral oil to produce the grease, although this has rather less practical significance nowadays than it had in the past. One example of this is the multi-purpose grease that may replace two or three different types previously thought necessary to cover a particular field of application. Nevertheless, there are unique differences in behaviour between greases made with different metal soaps, and these differences are still important in many industrial uses, for technical and economic reasons.

Calcium-soap greases The line-soap (calcium) greases have been known for many years but are still probably the most widely used. They have a characteristic smooth texture, thermal stability, good water resistance and are relatively inexpensive. The softer grades are easily applied, pump well and give low starting torque. Their application is limited by their relatively low drop points, which are around 100°C. This means that, in practice, the highest operating temperature is about 50°C.

Nevertheless, they are used widely for the lubrication of medium-duty rolling and plain bearings, centralized greasing systems, wheel bearings and general duties. The stiffer varieties are used in the form of blocks on the older-type brasses. Modifications of lime-base grease include the graphited varieties and those containing an extreme pressure additive. The latter are suitable for heavily loaded roller bearings such as in steel-mill applications.

Sodium-soap greases The soda-soap (sodium) greases were, for some considerable time, the only high-melting point greases available to industry. They have drop points in the region of 150°C and their operating maximum is about 80°C. These greases can be 'buttery', fibrous or spongy, are not particularly resistant to moisture and are not suitable for use in wet conditions. Plain bearings are very frequently lubricated with soda-based greases.

For rolling-contact bearings, a much smoother texture is required, and this is obtained by suitable manufacturing techniques. Modified grades may be used over the same temperature range as that of the unmodified grade and, when they are correctly formulated, have a good shear resistance and a slightly better resistance to water than the unmodified grades.

Lithium-soap greases These products, unknown before the Second World War, were developed first as aircraft lubricants. Since then the field in which they have been used has been greatly extended and they are now used in industry as multi-purpose greases. They combine the smooth texture of the calcium-based greases with higher melting points than soda-soap greases, and are almost wholly manufactured in the medium and soft ranges. Combined with suitable additives, they are the first choice for all rolling-contact bearings, as they operate satisfactorily up to a temperature of 120°C and at even higher for intermittent use. Their water resistance is satisfactory and they may be applied by all conventional means, including centralized pressure systems.

Other metal-soap greases Greases are also made from soaps of strontium, barium and aluminium. Of these, aluminium-based grease is the most widely used. It is insoluble in water and very adhesive to metal. Its widest application is in the lubrication of vehicle chassis. In industry it is used for rolling-mill applications and for the lubrication of cams and other equipment subject to violent oscillation and vibration, where its adhesiveness is an asset.

Non-soap thickened greases These are generally reserved for specialist applications, and are in the main more costly than conventional soap-based greases. The most common substances used as non-soap thickeners are silicas and clays prepared in such a way that they form gels with mineral and synthetic oils. Other materials that have been used are carbon black, metal oxides and various organic compounds.

The characteristic of these non-soap greases which distinguishes them from conventional greases is that many of them have very high melting points; they will remain as greases up to temperatures in the region of 260°C. For this reason, the limiting upper usage temperature is determined by the thermal stability of the mineral oil or synthetic fluid of which they are composed. Applications such as those found in cement manufacturing, where high-temperature conditions have to be met, require a grease suitable for continuous use at, say, 204°C. Although it is difficult to generalize, the non-soap products have, on the whole, been found to be somewhat less effective than the soap-thickened greases as regards lubricating properties and protection against corrosion, particularly rusting. Additive treatment can improve non-soap grades in both these respects, but their unique structures renders them more susceptible to secondary and unwanted effects than is the case with the more conventional greases.

Filled greases The crude types of axle and mill grease made in the early days frequently contained large amounts of chemically inert, inorganic powders. These additions gave 'body' to the grease and, possibly, helped to improve the adherence of the lubricating film. Greases are still 'filled' but in a selective manner with much-improved materials and under controlled conditions. Two materials often used for this purpose are graphite and molybdenum disulphide.

Small amounts (approximately 5%) of filler have little or no effect on grease structure, but large amounts increase the consistency. However, the materials mentioned are lubricants in themselves and are sometimes used as such. Consequently it is often claimed that when they are incorporated into the structure of the grease the lubricating properties of the grease are automatically improved. A difference of opinion exists as to the validity of this assumption, but it is true that both molybdenum disulphide and graphite are effective where shock loading or boundary conditions exist, or when the presence of chemicals would tend to remove conventional greases.

Mixing greases The above comments on the properties of the various types of grease have shown that very real differences exist. Each one has its own particular type of structure, calls for individual manufacturing processes and has its own advantages and disadvantages. It is because of these distinct differences that the mixing of greases should never be encouraged. If greases of different types are mixed indiscriminately there is a risk that one or other of them will suffer, the resulting blend being less stable than either of the original components and the blend may even liquefy.

9.2.11.2 Selecting a grease

A few brief notes on the fundamental factors that influence a choice of grease may be helpful. The first essential is to be absolutely clear about the limitations of the different types, and to compare them with the conditions they are to meet. Table 9.7 gives the characteristics of high-quality greases.

Greases with a mixed base are not shown in the table because, in general, they are characterized by the predominant base; for example, a soda-lime grease behaves like a soda grease. Temperature limits may be modified by the required length of service. Thus, if a soda grease requires to have only a short life, it could be used at temperatures up to 120°C.

When the type most suitable for a particular application has been chosen, the question of consistency must be considered.

Table 9.7 Characteristics of high-quality greases

Grease (type of soap)	Recommended maximum operating temperature (°C)	Water resistance	Mechanical stability
Lime	50	Good	Good
Soda	80	Poor	Good
Lithium	120	Good	Good
Aluminium	50	Fair	Moderate

The general tendency over the last two decades has been towards a softer grease than formerly used. Two factors have probably contributed to this trend; the growth of automatic grease dispensing and the use of more viscous oils in grease making.

In practice, the range of grease consistency is quite limited. For most general industrial applications, a No. 2 consistency is satisfactory. Where suitability for pumping is concerned, a No. 1; for low temperatures, a No. 0; and for water pumps and similar equipment, a No. 3.

9.2.11.3 Grease application

In applying lubricating grease the most important aspect is how much to use. Naturally, the amount varies with the component being serviced, but some general rules can be laid down. All manufacturers agree that anti-friction bearings should never be over-greased. This is particularly true of high-speed bearings, in which the churning of excess lubricant leads to overheating. The rise in temperature of a bearing as the amount of grease increases has been recorded. With the bearing housing one-third full, the temperature was 39°C; at two-thirds full the temperature rose to 42°C; and with a full charge of grease it went up to 58°C.

The general recommendations for grease packing are:

1. Fully charge the bearing itself with grease ensuring that it is worked around and between the rolling elements.
2. Charge the bearing housing one-half to two-thirds full of grease.

Churning, and its attendant high temperature, may change the structure of the grease permanently, in which event softening may result in leakage and stiffening in lubricant starvation. There is no fixed rule for the period between re-greasings, since this depends on the operating conditions. Most recommendations suggest inspection and possible replenishment every six or twelve months, though the general tendency as grease quality improves has been to extend this period. The higher the temperature of a machine, the more frequently it must be greased because of possible losses of softened lubricant or changes in its structure.

It is not always incorrect to over-grease. With a sleeve bearing, for instance, gun pressure may be maintained until old grease exudes from the ends of the bearing, and the same is true of spring shackles. For the sake of economy and cleanliness, however, this should never be overdone.

9.2.12 Corrosion prevention

Most plant has to work under adverse conditions, in all sorts of weather, and subject to contamination by various agents. However, as long as it is in use it can be reasonably sure of receiving at least a minimum amount of regular maintenance and attention, and this will reduce the likelihood of working parts being attacked by corrosion when plant is in service. However, when plant has to be laid up until required, no matter how carefully matters have been planned, corrosion is always a serious possibility. Modern machinery, with highly finished surfaces, is especially susceptible to atmospheric attack. The surfaces of components also require protection during transport and storage.

Even today, rusting of industrial plant and material is accepted by some as an inevitable operating expense. There is no necessity for this attitude, however, as the petroleum industry has evolved effective, easily applied temporary protectives against corrosion, which are well suited to the conditions met in practice.

9.2.12.1 Categories of temporary corrosion preventives

Temporary corrosion preventives are products designed for the short-term protection of metal surfaces. They are easily removable, if necessary, by petroleum solvents or by other means such as wiping or alkaline stripping. Some products for use in internal machine parts are miscible and compatible with the eventual service lubricant, and do not, therefore, need to be removed.

The major categories of temporary corrosion preventives are:

Soft-film protectives
Dewatering fluids giving soft/medium films
Non-dewatering fluids giving soft films
Hot-dip compounds
Greases

Hard-film protectives
Oil-type protectives
General-purpose
Engine protectives

The development of products in these categories has been guided by known market demands and many manufacturers have made use of established specifications for temporary protectives. In the UK, for example, British Standard 1133, Section 6 (covering all categories) and British Government Specifications CS 2060C (PX10 dewatering fluid) are frequently followed.

9.2.12.2 Selection of a corrosion preventive

Temporary corrosion preventives are in some cases required to give protection against rusting for periods of only a few days for inter-process waiting in factories. Where the protected components are not exposed to the weather, protection can be given for up to a year or more for stored components in internal storage conditions. On the other hand, components may require protection for a few days or even weeks under the most adverse weather conditions. Some components may have to be handled frequently during transit or storage. In general, therefore, the more adverse the conditions of storage, the longer the protective periods, and the more frequent the handling, the thicker or more durable the protective film must be.

Because of the wide variation in conditions of exposure it is not possible to define the length of protection period except in general terms. Solvent-deposited soft films will give protection from a few days to months indoors and some weeks outdoors; a solvent-deposited medium film will give long-term protection indoors and medium-term protection outdoors. Hot-dip compounds and cold-applied greases give films that can withstand considerable handling and will give medium to long protection. Solvent-deposited hard-film protectives will give

long-term protection but are fairly difficult to remove. Oil protectives give short- to medium-term protection of parts not subjected to handling and are also much used for the preservation of internal working parts; they need not be removed and can in some instances serve as lubricating oils.

'Short term', 'medium term' and 'long term' are expressions that are not rigorously defined but are generally accepted as meaning of the order of up to 6 months, 12 months and 18 months, respectively, in temperate climates. Where local conditions are more severe (in hot, humid climates, for example) the protection periods are less. These protection periods are related to the preventive film alone, but where transit or storage conditions call for wrapping or packaging then longer protection periods can be obtained.

The distinction between a simple part and a complex assembly is an important factor in selecting a temporary protective. The solvent-containing protectives may not be suited to treating assemblies, because:

1. Assemblies may contain non-metallic parts (rubber, for example) that could be attacked by the solvent;
2. The solvent cannot evaporate from enclosed or shielded spaces and the intended film thickness will not be obtained;
3. Evaporated solvent could be trapped and could then leach away the protective film.

Hence the hot-dip compounds, or greases smeared cold, are better for assemblies with non-metallic parts masked if necessary. Solvent-containing protectives therefore find greater application in the protection of simple parts or components. The available means of application, the nature of any additional packaging and the economics and scale of the protective treatment are further factors that influence the choice of type of temporary corrosion preventive.

9.2.13 Spray lubricants

There are several applications where the lubrication requirement is specialized and very small, needing precise applications where access is limited because of equipment design or location. In these instances lubricant application by aerosol is the most suitable method. Extreme-pressure cutting fluid for reaming and tapping, etc., conveyor and chain lubricant, anti-seize and weld anti-spatter agents, release agents, electrical component cleaner and degreasants are examples of the ever-widening range of products available in aerosol packs.

9.2.14 Degreasants

Often, before any maintenance work starts it is necessary (and desirable) to remove any oil, grease and dirt from the equipment concerned. It may also be necessary to clean replacement components before their installation. Solvents, emulsions and chemical solutions are three broad types of degreasants. The method of degreasing (direct onto the surface, by submersion, through degreasing equipment or by steam cleaners), component complexity and the degree of contamination will all have to be taken into account when selecting the type of product to be used.

9.2.15 Filtration

Some 70–85% of failures and wear problems in lubricated machines are caused by oil contamination. Clean oil extends machine and oil life and gives greater reliability, higher productivity and lower maintenance cost. Hence some type of filter is an essential part of virtually all lubrication systems.

Cleaning of oil in service may be accomplished quite simply or with relatively complex units, depending on the application and the design of the system. Thus for some operations it is enough to remove particles of ferrous metal from the oil with a magnetic system. In a closed circulatory system, such as that of a steam turbine, the nature of the solids and other contaminants is far more complex, and the treatment has therefore to be more elaborate. In an internal-combustion engine both air and fuel are filtered as well as crankcase oil.

The efficiency of filtration must be matched to the needs of the particular application, and this is true both quantitatively (in relation the anticipated build-up of solids in the filters) and qualitatively (in relation to the composition of the contaminants and their size). Dirt build-up varies considerably, but it is probably at its maximum with civil engineering equipment. In this field, diesel engines in trucks will steadily accumulate something like 0.3 kg of solids in the crankcase oil within a month.

Particle size is naturally important. It is generally assumed that particles of less than 3 μm in diameter are relatively harmless. However, this is on the assumption that the oil film is itself of this, or greater, thickness; in other words, that full fluid-film hydrodynamic lubrication persists during the whole working cycle of the machine. This is seldom the case, for there are either critical areas or critical phases at or during which mixed or even wholly boundary conditions prevail – when, in fact, the oil film is less than 3 μm thick. The tendency of modern industrial equipment to operate at higher speeds and under greater pressures leads to higher wear rates. Increased pump capacity, as in hydraulic circuits, coupled with a decreased oil volume means a relatively greater amount of contamination. All in all, much more is demanded of the filter today, whatever the application, than at any time in the past.

9.2.15.1 Types of filter

The terms 'filter' and 'strainer' are in common use and many lubricant systems contain both. The word 'strainer' is often associated with the removal of large particles, and though it is true that in the majority of cases a strainer is in fact employed to remove coarse particles, the fundamental difference between it and a filter is not one of porosity but purely one of geometry. In a strainer the liquid passes through in a straight line, but in a filter a far more devious route is followed.

Strainers are usually made from woven wire gauze, like a sieve, and though today the pre-size can be made very small indeed (BSI 300 mesh gauze separates particles of roughly 50 μm) they are mainly included for the exclusion of large particles. Filters deal with the removal of very much smaller particles.

Naturally from the above definition there is some unavoidable overlapping, and a really fine strainer of, say, stainless steel 'cloth' is regarded as a filter. There are five main types of filtering units as follows.

Surface films   These are usually constructed of woven metal gauze, paper or cloth. The paper filter may have the working surface enlarged by pleating and the paper impregnated and strengthened. As an example, one proprietary pleated model gives, from an element 11.5 cm long and 8.5 cm in external diameter, a filtering surface of some 3250 cm^2. This type, sometimes described as a radial-fin unit, has a good throughput and is easy to clean or replace. Filters in this class generally have porosities from 100 μm down to 10 or, in extreme cases, even down to 2μm.

Edge filters   A typical unit comprises a pack of metal or paper discs with a washer between each, the gauge of the latter governing the degree of filtration. The oil flows from the outside and is discharged through a central channel. Some designs can be cleaned without dismantling or interrupting the flow.

An alternative method of manufacturing is to employ a coil of flat metal ribbon as the element, each turn spaced from the next by small lateral protuberances. The principle of filtration is the same. Porosities of both types are identical and cover a wide range, usually from 100 μm down to 0.5 μm.

Depth filters (*absorption-type filters*)

1. *Chemically inactive*: There are made from a variety of materials that include wound yarn, felt, flannel, cotton waste, wood pump, mineral wool, asbestos and diatomaceous earths. The solid particles are trapped and retained within the medium. Certain types will remove water, as well as large and small particles of solids in a range down to 10 μm. Ceramics are sometimes employed for depth filtration, as also are special sintered metals.

2. *Chemically active*: These filters are similar in design to the non-active depth units but the filtering media used are so chosen that contaminants adhere by chemical attraction. Thus there is a dual action, mechanical and chemical. The materials used include various activated clays, Fuller's earth, charcoal and chemically treated paper. Their cleansing action is much more thorough than that of the purely mechanical devices, for they are capable of removing matter actually in solution in the oil.

Magnetic and combined magnetic filters   In its simplest form the magnetic filter comprises a non-magnetic outer casing with an inner permanent magnetic core round which the liquid flows. Because of the magnetic anisotropy of the field the ferrous particles are continuously diverted to the area of strongest attraction coinciding with the direction of flow. A more elaborate design of magnetic clarifier has its elements mounted in a rotating disc. The dirty fluid flows through the chamber in which the disc dips, and ferrous particles adhering to the magnetized areas are removed by the action of scrapers and collected in containers. The capacity of one such disc has been given as 2250 l/h with a range of sludge removal as high as 30 kg/h. Combined units may have the magnet located within a coil of wire that forms the permeable, mechanical filter.

For its specialized application (cleaning the coolants used for metal-machining operations such as grinding and honing) the magnetic filter is easily maintained and cleaned. It has a high throughput and will remove ferrous particles as small as 1 μm. Some of the non-magnetic material is associated with the ferrous particles suspended in the fluids and this is also removed with them.

The centrifugal filter   This is a specialized design and is, in effect, a true centrifuge of small size that operates on the reaction turbine principle, an oil-circulating pump providing the necessary power. One advantage claimed for this type is that it operates at a steady flow rate, whereas the flow rate through a felt or paper element diminishes as the bed of dirt is built up. The centrifugal filter has been successfully applied to diesel engines where the greater part of the dirt particles are under 2 μm in diameter.

9.2.16 Centrifuging

The centrifugal separation of solid impurities is adopted either as an alternative to filtration or combined with it. For example, a lubricant circulating system can be cleaned by having fixed-element filters that arrest larger particles, and a centrifuge system that removes the finer solids in suspension together with any water contained in the oil.

The centrifuge is a powerful tool. The magnitude of the available centrifugal force – the product of the mass of the particle and its acceleration – is easily appreciated when the speeds and dimensions of a commercial unit are considered. A vessel with a diameter of 25.4 cm spinning at 1700 rev/min gives an acceleration at the centrifuge wall of some 400 g. In terms of settling this means that centrifuging a crude oil for 30 s is at least equivalent to simple gravitational settling over a 24 h period.

The advantage of the modern continuous centrifuge is the rapidity with which it will separate both solids and immiscible liquids. Another stems from the larger volume of oil it can handle in a given time.

9.2.17 Centralized lubrication

Manual application of lubricants has the inherent risk of failure due to omission. With the increasing complexity of plant, the costs of lost production and of manpower to try to prevent such omissions are becoming unacceptable.

Mechanized methods of pumping oil and grease to bearings and other components are becoming increasingly utilized. Some of these systems are fundamentally suited to either oil or grease, but others, including all those where continuous circulation is involved, are suitable only for oil.

Built-in mechanized grease lubrication is nearly always of the centralized 'one-shot' variety, in which a single pump stroke supplies grease simultaneously to a number of bearings. The amount supplied to each station is regulated by suitable valves or adjustable metering orifices. The pump may be manually operated or connected to a suitable machine component, whereby grease is fed only when the machine is actually running and at controlled temperatures. Pneumatic or electric pumps are also used, set in operation at regular intervals by an automatic timing device.

One-shot metered lubrication is eminently suited to oiling systems and can be employed either in an 'all-loss' arrangement or as part of a circulatory system. Sight-glasses or other indicators should be incorporated, since such lubricating mechanisms are nowadays so reliable that a blockage or other failure might not be suspected until too late.

Circulatory systems often use an intermediate header tank, from which the bearings are supplied by gravity. The complete system may comprise, in addition and according to the size of the installation, heat exchangers or coolers, filters, strainers, settling tanks, centrifuges and other purifying equipment.

Oil mist feeds are used less for plain bearings than for lubricating some other types of machine parts, but applications are increasing in number. A stream of dry compressed air is used both to generate the mist and to carry it to the bearing. The atomized oil droplets are released from air suspension at points of turbulence around bearings, gears and other moving components or in a special re-classifying fitting at the end of the supply line. Reclassifiers are generally employed when plain bearings are to be lubricated by oil mist, but the method is fundamentally unsuited for bearings requiring hydrodynamic thick-film lubrication.

Special precautions must be taken with oil-mist feeds to ensure that the compressed air, which greatly enhances the rate of heat dissipation, can escape from the housing. If vents or other outlets become blocked, the back pressure may stop the flow of lubricant.

9.2.18 Storage of lubricants

It cannot be emphasized too strongly that dirt and correct lubrication are incompatible. The lubricant manufacturer has a comprehensive system of classification, filtration and inspection of packages which ensures that all oils and greases leaving his plant are free from liquid and solid contaminants. It is in his own interests that the user should take the same care to ensure that the lubricant enters his machinery in as clean a condition as that in the bulk tank or barrel. The entry of abrasive dust, water and other undesirable matter into bearings and oilways may result if lubricants are handled carelessly.

The conditions in a plant are often far from ideal and usually storage facilities are limited. This, however, should serve as a constant reminder of the need for continual care, the adoption of suitable dispensing equipment, organized storekeeping and efficient distribution methods. Furthermore, the arrangements on any particular site will be governed by local organization and facilities. Technical personnel from lubricant suppliers are available to assist and advise plant management on the best methods for a particular site. The general recommendations given about the care of lubricants consist of elementary precautions which are mainly self-evident and yet, unfortunately, are often ignored.

The modern steel barrel is reasonably weatherproof in its original condition, but if stored out of doors and water is allowed to collect in the head, there may, in time, be seepage past the bung due to the breathing of the package. Exposure may also completely obliterate the grade name and identification numbers, as is evidenced by the frequent requests made to sample and test lubricants from full packages that have been neglected on-site because no other method of identification is possible. Unless it is absolutely unavoidable, packages should never be stored in the open and exposed to all weather. Even an elementary cover such as a sheet of corrugated iron or a tarpaulin may provide valuable protection.

However rudimentary the oil stores, the first essential is cleanliness; the second is orderliness. These two essentials will be easily achieved if maximum possible use is made of bulk storage tanks. In the case of bulk storage of soluble oils the need for moderate temperatures is vital, and the tanks should be housed indoors to protect their contents against frost. There are several other benefits to be derived from the use of tanks, i.e. reduction in storage area, handling of packages and, possibly, bulk-buying economics. All barrels should be mounted on a stillage frame of suitable height, fitted with taps and the grade name clearly visible. The exterior surfaces of both tanks and barrels should be kept scrupulously clean and each container provided with its own drip tray or can.

The storage and handling of grease presents more problems than are encountered with fluid lubricants, as the nature of the material and design of the conventional packages make contamination easier. Lids of grease kegs must be kept completely free from dust and dirt, and should be replaced immediately after use. The most common way in which solids enter a grease package is by the user carelessly placing a lid either on the ground or on some other unsuitable surface. Fortunately, there are available today a number of simple dispensing units which can entirely obviate this danger and which can be adapted to all types of packages.

Wherever manual distribution has to be adopted, containers should be reserved for the exclusive use of specific units and their operators and, as far as possible, for a particular grade. When not in use they must be stored away from all possible sources of contamination. To promote economy and reduce waste due to spillage, their shape and proportions must be suited to the application.

While it is impossible to describe a system of storekeeping and distribution suitable for every site there are certain essential principles which should be adhered to if cleanliness, order and economy are to be maintained. How these principles should be applied is for individual managements to decide. The keynote, however, should be simplicity. Distribution should be controlled by a storekeeper familiar with both grades and needs. While the lubrication schedule for any particular unit is generally the concern of the operator, the storekeeper must equally be aware of it and have a comprehensive list of the different grades, their applications, quantities, daily and other periodic needs. On such a basis he will be able to requisition and store the necessary lubricants in the most convenient and economic quantities and packages, and ensure that supplies are used on a 'first in, first out' basis.

Care and good housekeeping at every stage from handling, stacking and storage, right through to dispensing and application will:

- Ensure that the correct product reaches the point of application and is free from contamination;
- Help towards maximum efficiency in the use of lubricants and the equipment in which they are employed;
- Avert accidents and fire hazards arising from mishandling;
- Prevent any adverse effects on people, equipment and the environment.

9.2.19 Reconditioning of oil

Reconditioning is the removal of contaminants and oxidation products (at least in part) but not previously incorporated additives. It may also involve the addition of new oil and/or additives to adjust the viscosity and/or performance level. This process is sometimes referred to as 'laundering' or 'reclamation'. The method treats used lubricating oil to render it suitable for further service, either in the original or a downgraded application. Two types of treatment are generally employed.

1. Filtration to remove contaminants, followed by the addition of new oil and/or additives to correct performance level;
2. A simple filtration process to remove contaminants.

In practice, treatment (1) usually involves a contractor collecting a segregated batch of oil, reconditioning and returning it for re-use. The simple filtration process can be carried out by a contractor, but is more usually done on-site. Re-refining is the removal of contaminants and oxidation products and previously incorporated additives to recover the lube base stock for new lubricant or other applications.

9.2.20 Planned lubrication and maintenance management

Having the correct lubricant in each application will only give the maximum benefit if and when it is applied at the correct frequency and quantity. With the increasing complexity of plant this is becoming more vital and, at the same time, more difficult to achieve. The solution to this problem is planned lubrication maintenance, which, in essence, is having the right lubricant in the right place at the right time in the right amount.

Most oil companies offer a planned lubrication maintenance (PLM) service that will meet these requirements with the minimum of effort on the part of the customer. These schemes provide logical routing for the lubrication operative, balanced work loads and clear instructions to those responsible for specific tasks associated with lubrication and fault-reporting facilities. Many schemes are now designed for computer operation which also accommodate plant and grade changes, operation costings and manpower planning. It is essential that any such scheme should be adaptable to individual requirements.

There are a few computerized PLM schemes which are dynamic systems and can be integrated into an overall maintenance management information system. These contain maintenance, inventory and purchase order modules and go far beyond 'just another work order system'. They provide the necessary information to control complex maintenance environments, thereby improving productivity and reducing operational costs.

9.2.21 Condition monitoring

Condition monitoring is an established technique which has been used by capital-intensive or high-risk industries to protect their investment. The concept has developed radically in recent years largely due to advances in computerizations which offer greater scope for sophisticated techniques. These fall into three types of monitoring: vibration, performance and wear debris. The last monitors particulate debris in a fluid such as lubricating oil, caused by the deterioration of a component.

Oil-related analysis encompasses a variety of physical and chemical tests such as viscosity, total acid number and particulate contamination. This is often extended to include the identification of wear debris, as an early warning of component failure, by either spectrographic analysis or ferrography or both. The former is commonly used in automotive and industrial application for debris up to 10 μm and the latter mainly for industry users covering wear particles over 10 μm. Ferrography is relatively expensive compared with many other techniques, but is justified in capital-intensive areas where the cost is readily offset by quantifiable benefits such as longer machinery life, reduced loss of production, less downtime, etc.

9.2.22 Health, safety and the environment

There are a wide variety of petroleum products for a large number of applications. The potential hazards and the recommended methods of handling differ from product to product. Consequently, advice on such hazards and on the appropriate precautions, use of protective clothing, first aid and other relevant information must be provided by the supplier.

Where there is risk of repeated contact with petroleum products (as with cutting fluids and some process oils) special working precautions are obviously necessary. The aim is to minimize skin contact, not only because most petroleum products are natural skin-degreasing agents but also because with some of them prolonged and repeated contact in poor conditions of personal hygiene may result in various skin disorders.

9.2.22.1 Health

It is important that health factors are kept in proper perspective. What hazards there may be in the case of oil products are avoided or minimized by simple precautions. For work involving lubricants (including cutting fluids and process oils) the following general precautions are recommended:

- Employ working methods and equipment that minimize skin contact with oil;
- Fit effective and properly positioned splash guards;
- Avoid unnecessary handling of oily components;
- Use only *disposable* 'wipes';
- Use soluble oils or synthetic fluids at their recommended dilutions only, and avoid skin contact with their 'concentrates'.

In addition to overalls, adequate protective clothing should be provided. For example, a PVC apron may be appropriate for some machining operations. A cleaning service for overalls should be provided and overalls should be cleaned regularly and frequently. Normal laundering may not always be sufficient to remove all traces of oil residues from contaminated clothing. In some instances dry cleaning may be necessary. Where this applies to cotton overalls they should first be dry cleaned and then laundered and preferably starched, in order to restore the fabric's oil repellancy and comfort. As a general rule, dry cleaning followed by laundering is always preferable to minimize the risk of residual contamination wherever heavy and frequent contamination occurs and when the type of fabric permits such cleaning.

Overalls or personal clothing that become contaminated with lubricants should be removed as soon as possible – immediately if oil soaked or at the end of the shift if contaminated to a lesser degree. They should then be washed thoroughly or dry cleaned before re-use.

Good washing facilities should be provided, together with hot and cold running water, soap, medically approved skin-cleansers, clean towels and, ideally, showers. In addition, reconditioning creams should be available. The provision of changing rooms, with lockers for working clothes, is recommended.

Workers in contact with lubricants should be kept fully informed by their management of the health aspects and the preventive measures outlined above. Any available government leaflets and/or posters should be prominently displayed and distributed to appropriate workers.

It should be made clear to people exposed to lubricants that good standards of personal hygiene are a most effective protection against potential health hazards. However, those individuals with a history of (or thought to be particularly predisposed to) eczema or industrial dermatitis should be excluded from work where, as in machine-tool operation, contact with lubricants is virtually unavoidable.

Some industrial machining operations generate a fine spray or mist of oil, which forms an aerosol – a suspension of colloidal (ultra-microscopic) particles of oil in air. Oil mist may accumulate in the workshop atmosphere, and discomfort may result if ventilation is inadequate. Inhalation of high concentrations of oil mist over prolonged periods may give rise to irritation of the respiratory tract, and in extreme cases to a condition resembling pneumonia. It is recommended that the concentration of oil mist in the working environment (as averaged over an 8-h shift) be kept below the generally accepted hygiene standard of 5 mg/m^3. This standard does, however, vary in some countries.

9.2.22.2 Safety

In the event of accident or gross misuse of products, various health hazards could arise. The data provided by the supplier should outline these potential hazards and the simple precautions that can be taken to minimize them. Guidance should be included on the remedial action that should be taken to deal with medical conditions that might arise. Advice should be obtained from the supplier before petroleum products are used in any way other than as directed.

9.2.22.3 Environment

Neat oils and water-based coolants eventually reach the end of their working lives, and then the user is faced with the problem of their correct disposal. Under *no* circumstances should neat oils and emulsions be discharged into streams or sewers. Some solutions can, however, be fed into the sewage system after further dilution – but only where permitted.

There are many companies offering a collection service for the disposal of waste lubricating oil. The three main methods employed are:

1. Collection in segregated batches of suitable quality for use by non-refiners
2. Blending into fuel oil
3. Dumping or incineration

If method (3) is used due regard must be paid to the statutory requirements that must be met when disposing of waste material. These are covered in two main items of legislation; namely, the Deposit of Poisonous Waste Act 1972 and the Control of Pollution Act 1974. It is the responsibility of the producer of waste oil to ensure that the waste is disposed of in the correct manner, to ensure that no offence is committed and that the contractor is properly qualified to execute the service.

Acknowledgements

The editor is grateful to BP Oil UK Ltd for their help in writing this chapter and for their permission to reproduce the figures and tables.

9.3 Bearing selection

Neale[6] suggests that bearings can be classified according to the type of relative movement which they permit between the opposing surfaces. Four categories are proposed, namely, movement about a point, about a line, along a line, and in a plane. Each of these categories can be subdivided into oscillatory or continuous motion. Probably the most common bearings are those which exhibit continuous motion either about a line (such as journal bearings) or in a plane (such as thrust bearings). In turn, these bearings can be classified according to their load-carrying capacity/speed characteristics. The selection of an appropriate bearing for an application will entail matching the required characteristics to those provided by a particular bearing type. This matching of characteristics is only one step, and the designer must also consider geometric and environmental constraints, cost and predicted bearing life. Reference 6 provides useful further reading, along with appropriate ESDU Design Guides.[7]

9.3.1 Characteristics of bearings with continuous motion

In all the figures in this subsection the acceptable operating range for the bearing is below the solid line and within the maximum speed limit.

9.3.1.1 Liquid-lubricated, hydrodynamic journal bearings (Figure 9.8)

At lower speeds, the operating limit is determined by the minimum operating film thickness allowed. This in turn will be determined by the roughness of the opposing surfaces (asperity contact must be avoided) and the filtration level of the lubricant (particles of the same order as the minimum film thickness may cause surface damage). As speed increases, the lubricating liquid gets hotter and its viscosity reduces. This in turn reduces the maximum load which can be carried for an acceptable minimum film thickness. The upper speed limit is determined by the bursting speed of the shaft. A similar

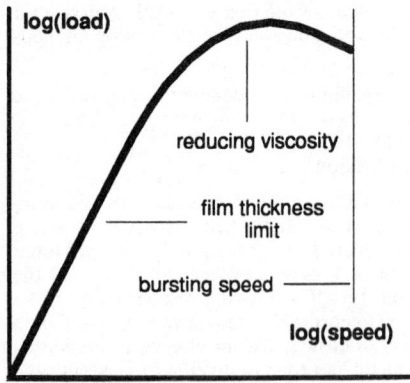

Figure 9.8

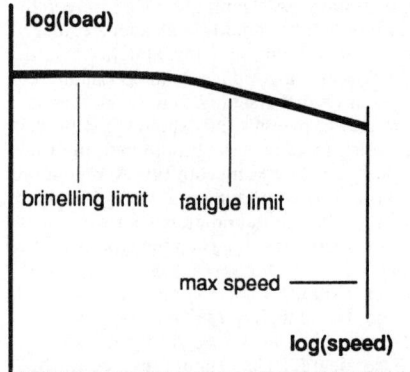

Figure 9.10

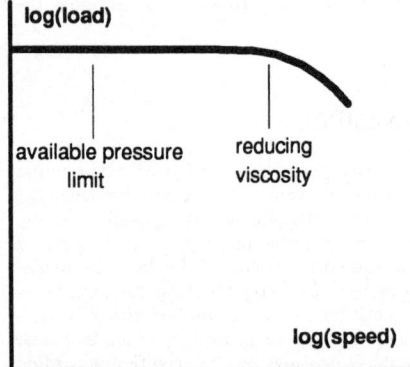

Figure 9.9

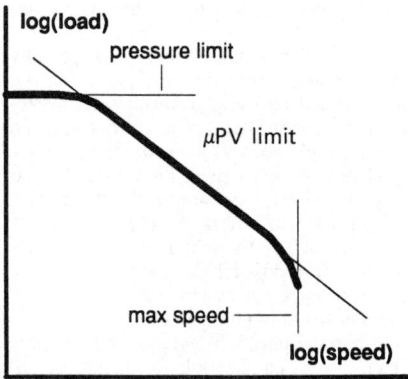

Figure 9.11

similar diagram can be drawn for liquid-lubricated hydrodynamic thrust bearings.

9.3.1.2 Liquid-lubricated, hydrostatic bearings (Figure 9.9)

These bearings have a sizeable load-carrying capacity at zero surface speed because this parameter is determined by the pressure of the supply liquid. The magnitude of this pressure is limited by the capabilities of the pressurizing apparatus and the associated equipment. At higher speeds, viscosity effects due to sliding become more pronounced, as in the case of hydrodynamic contacts.

9.3.1.3 Rolling element bearings (Figure 9.10)

The load limit at zero or low speeds arises from the tendency of the rolling elements to deform the races because of the high contact pressures. Since this is similar to the effect produced by the Brinell hardness test, the term 'brinelling limit' is employed. At higher speeds the races tend to fail through fatigue caused by the cyclical stress patterns induced as the elements pass repeatedly over the same points. For cylindrical rollers, the slope of this line is (−10/3), and for ball bearings it is (−3). At the highest speeds, failure may be due to excessive forces on the cage, or unwanted skidding of the rolling elements giving rise to severe wear.

9.3.1.4 Partially lubricated bearings (Figure 9.11)

These bearings have a lubricant embedded in the solid material. The former slowly escapes into the contact thus providing a partial level of lubrication. At low speeds, the maximum load is dictated by the structural strength of the bearing material. As speed increases, the load is limited by the temperature rise at the sliding interface, and the bearing life which are controlled by the product PV (see Section 9.1). An upper limit on speed is determined from temperature limitations.

9.3.1.5 Dry bearings (Figure 9.12)

Similar characteristics apply to these bearings as to partially lubricated contacts, but poorer load/speed characteristics are exhibited because of the absence of a lubricant.

9.3.2 Bearing selection charts

Figures 9.13 and 9.14 are taken from reference 6 and indicate the operating characteristics of the bearing types in Section 9.3.1. Figure 9.13 gives guidance on the type of bearing which has the maximum load capacity at a given speed and shaft size. It is based on a life of 10 000 h for rubbing, rolling and porous metal bearings. Longer lives may be obtained at reduced loads

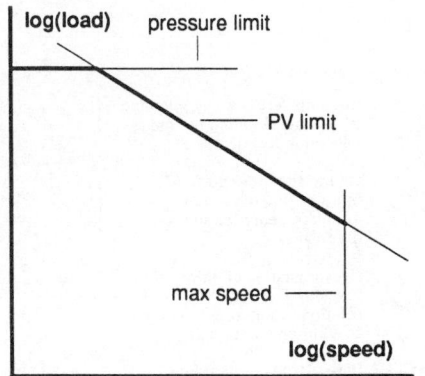

Figure 9.12

and speeds. For the various plain bearings, the width is assumed to be equal to the diameter, and the lubricant is assumed to be a medium-viscosity mineral oil. In many cases the operating environment or various special performance requirements, other than load capacity, may be of overriding importance in the selection of an appropriate type of bearing. See the tables in Section A2 of reference 6 in these cases.

Figure 9.14 gives guidance on the maximum load capacity for different types of bearing for given speed and shaft size. In many cases the operating environment or various special performance requirements, other than load capacity, may be of overriding importance in the selection of an appropriate type of bearing. See the table in Section A3 of reference 6 in these cases. Further details on design with these bearings can be found in reference 7, where advice is given on the selection and design of an appropriate bearing for a particular duty.

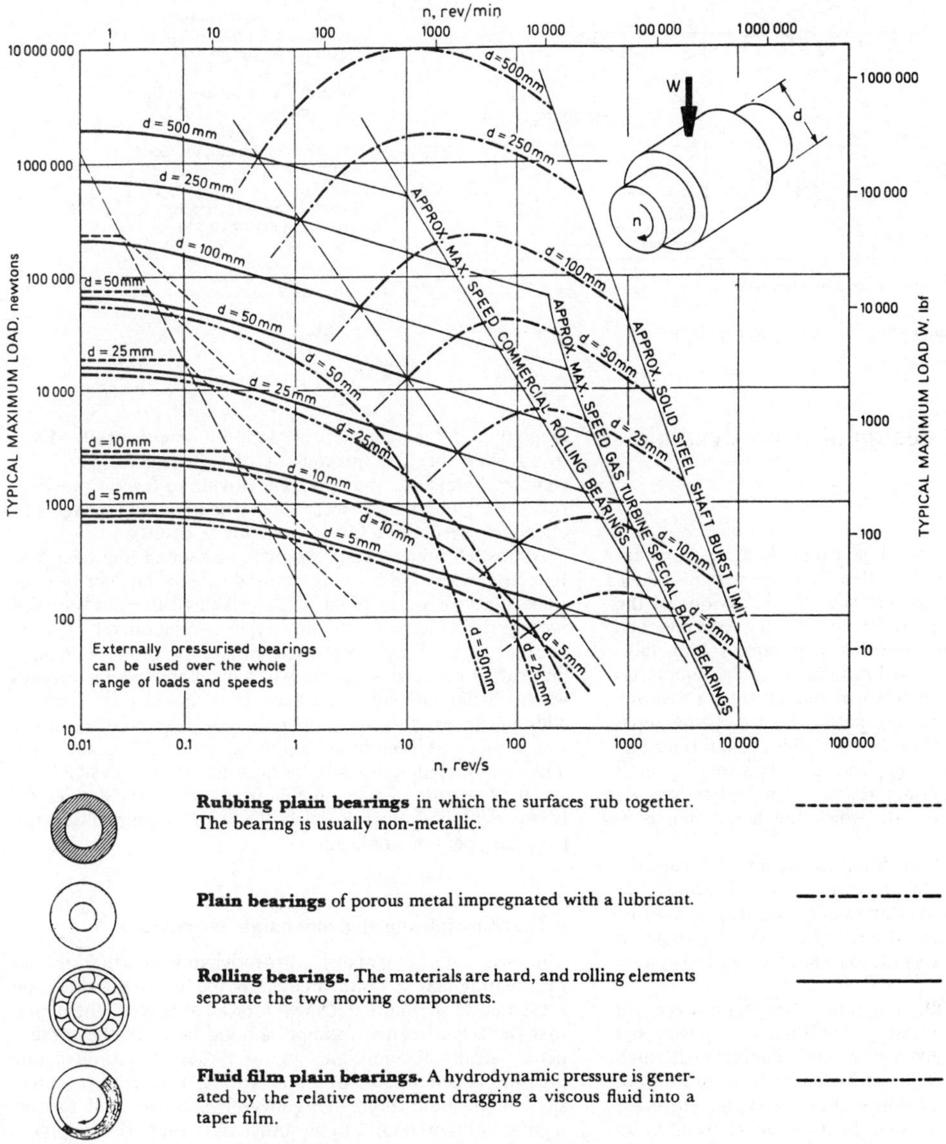

Rubbing plain bearings in which the surfaces rub together. The bearing is usually non-metallic.

Plain bearings of porous metal impregnated with a lubricant.

Rolling bearings. The materials are hard, and rolling elements separate the two moving components.

Fluid film plain bearings. A hydrodynamic pressure is generated by the relative movement dragging a viscous fluid into a taper film.

Figure 9.13   Selection by load capacity of bearings with continuous rotation

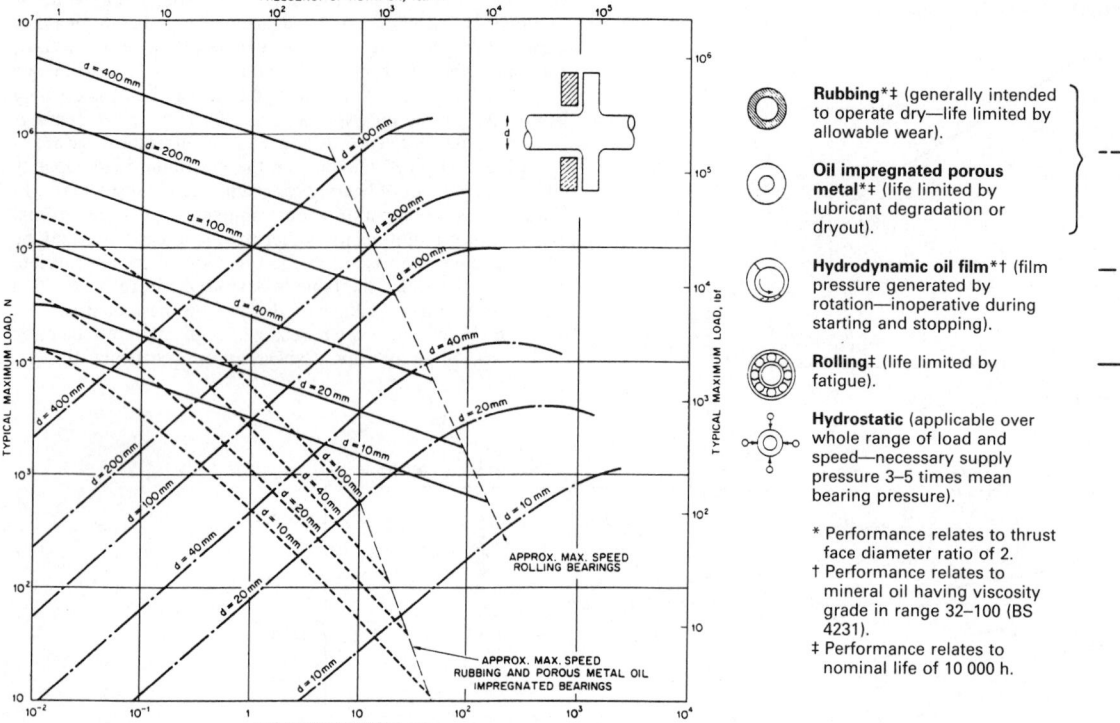

Figure 9.14 Guide to thrust bearing load-carrying capability

9.4 Principles and design of hydrodynamic bearings

9.4.1 Introduction

The subject of hydrodynamic (liquid film) bearings is essentially the subject of lubrication, therefore the design of such bearings is concerned principally with the behaviour of the liquid film separating the relatively moving components. Engines, turbines, motors, gearboxes, pumps, rolling mills and many of the machines used in industry (for example, in packaging, printing and production manufacture) are basically made of stationary and moving parts, the two being separated – at least in the ideal case – by a film of liquid, usually oil but not always. The moving part is usually a rotating shaft carrying a gear, impeller, armature, etc., and the bearings are the stationary components with which the liquid film is in immediate contact.

Typically, film bearings are fitted to accurately locate the rotating system within the machine. Two bearings are normally required for radial location of the shaft, plus a thrust bearing (usually two) mounted one either side of a collar or disk fixed to one end of the shaft, to locate the rotating system axially.

The major reasons for adopting film bearings, however, are to optimize load-carrying capacity, film thickness, power loss and heat generation for a given speed and diameter. Although the bearing diameter is usually set by the shaft or rotor of the machine, the length can be adjusted at the design stage for optimum performance. Note that the frictional resistance of a typical liquid film is extremely low, being about two orders of

magnitude lower than that for metal-to-metal contact. Moreover, when the film thickness is of sufficient magnitude to completely separate the relatively moving surfaces – the ideal case – then the rate of wear of the bearing/shaft surfaces is effectively zero and a long service life is ensured.

Almost all fluids, even gases, can be used in film bearings. Indeed, process fluids are often used for convenience (e.g. in some types of water pump). Nevertheless, the fluid usually preferred for bearings of any type is mineral oil. This is because it is cheap, possesses inherently good boundary-lubricating properties (useful when inevitable contact occurs when starting and stopping) and can be dosed with chemical additives to enhance its properties (e.g. improved oxidation resistance, rust inhibition, anti-wear etc.). Also, and very importantly, mineral-based lubricating oils are available in about 18 viscosity grades ranging from 2 to 1500 cSt at 40° C. It is possible therefore to select the most suitable oil for any particular bearing application.

9.4.2 Principles of hydrodynamic lubrication

The basic requirements of a hydrodynamic bearing are that the bearing has a finite area; that the bearing surface be presented to the fluid at a slight attack (or wedge) angle; and that there is a relative 'sliding' motion between the components. If these conditions are met then a hydrodynamic pressure is generated along the bearing surface by compression of the fluid along the converging wedge, and this integrated film pressure can be sufficient to support the applied bearing load on the fluid film.

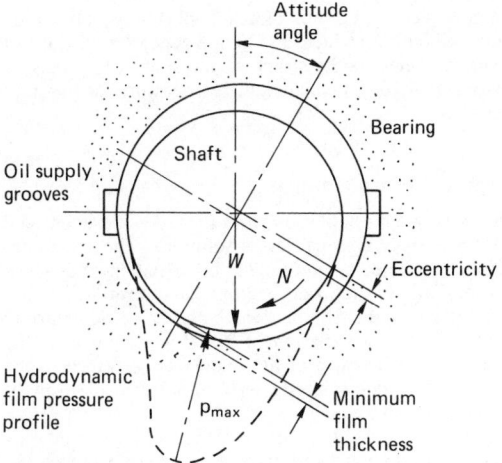

Figure 9.15 Hydrodynamic journal bearing

A converging wedge fluid film is generated automatically in a lubricated journal bearing by virtue of the necessary running clearance between the journal and the bearing bore, combined with the effect of load and rotation which produces a displaced, eccentric disposition of the journal (Figure 9.15).

The principle of hydrodynamic film pressure lubrication in a journal bearing was first observed experimentally by Towers[8] in 1883. Sponsored by the Institution of Mechanical Engineers, his 'First report on friction experiments (friction of lubricated bearings)' describes how a cork, then a wooden plug, fitted in the loaded zone of the bearing crown to stop up the oil hole, was 'forced out by the oil in a way which showed that it was acted on by a considerable pressure'.

Reynolds's paper[9] to the Royal Society in 1886 explained the phenomenon by analysis showing that a converging wedge-shaped film was necessary to generate pressure within the film. This classic paper is the basis of all hydrodynamic bearing theory.

A simplified version of Reynolds's equation is

$$\frac{\partial}{\partial x}(h^3.\partial p/\partial x) + \frac{\partial}{\partial y}(h^3.\partial p/\partial y) = 6.U.\eta.dh/dx$$

where

p = pressure in lubricant (Pa),
h = film thickness (m),
x, y = coordinates within the plane of the film,
U = velocity in the x-direction (m s^{-1}),
η = the lubricant dynamic viscosity (Pa.s).

However, in 1854 Hirn[10] in France had established some important factors from friction tests on oils and other fluids. He found that bearing lubrication was a function of: lubricant viscosity; rotating speed; and applied load. Hirn's results were analysed in 1883 by a Russian scientist, Nikolai Pavlovich Petrov, who used Newton's hypothesis of 1668 regarding fluid shear friction or viscosity, and showed that bearing friction could be explained by the behaviour of the fluid film. Petrov's Law is:

$$\mu = F/W = \eta.U.A/(W.c) = 4.\pi^2.r^2.b.(\eta.N)/(W.c)$$

where

μ = friction coefficient,
F = friction force (N),
W = applied load (N),

A = bearing wetted area $(2.\pi.r.b)$,
r = bearing radius (m),
b = bearing length (m),
c = radial clearance (m),
N = rotational speed (s^{-1}).

As the average pressure (p^{av}) exerted by the load on the relevant area of a journal bearing is W/A (load divided by projected area $2.r.b.$), then the basic equation for hydrodynamic film friction for a constant value of clearance ratio c/r (commonly 1/1000) becomes:

$$\mu = f(\eta.N/p^{av})$$

The basic equation for the dimensionless minimum film thickness ratio in the loaded zone of a bearing is similar:

$$h_{min}/c = f(\eta.N/p^{av})$$

where h_{min} = minimum film thickness (m).

Therefore in all studies of hydrodynamic bearings the essential factors are those which determine the behaviour of the separating film, which, for our purposes, we will refer to as the oil film. These factors are:

Oil viscosity
Oil flow
Bearing dimensions
Bearing geometry
Applied load
Rotating speed.

For practical use in bearing design, however, Reynolds's equation was too difficult to solve and Petrov's Law could only be applied to a non-representative case – that of a concentric or nearly concentric bearing. It was not until some 20 years later in 1904 that Sommerfeld[11] in Germany derived from Reynolds's differential equation a simple and usable set of equations for load capacity, friction moment and friction.

Sommerfeld's work showed that, neglecting cavitation in the unloaded portion, and assuming no end leakage of fluid (i.e. an infinitely long bearing), the load-carrying capacity of a journal bearing per unit length could be described using all the physical parameters normally available to the designer.

Michell[12] in 1905 proposed a method of integrating Reynolds's equation for application to plane surfaces whereby the $\delta/\delta x$ term was dropped. Twenty-five years later the method was applied to journal bearings by other workers, and by 1952 Ocvirk[13] produced the following usable equation for short bearings which has been shown to correlate well with experimental results:

$$p = \pi^2.\eta.N.(b/d)^2.(d/C_d)^2.(\epsilon/(1 - \epsilon^2)^2).(1 + 0.62.\epsilon^2)^{0.5}$$

where

b = bearing length (m),
d = bearing diameter (m),
C_d = total diametral clearance (= $2.c$),
ϵ = eccentricity ratio (= ratio journal centre displacement to the radial clearance).

Note minimum film thickness $h_{min} = c.(1 - \epsilon)$ or $C_d/2.(1 - \epsilon)$.

There are several other milestones in the development of our understanding of film bearings, but one in particular should be mentioned for background.

9.4.3 Viscosity

Sir Isaac Newton's hypothesis of 1668 is still recognized as the basis for understanding viscous or laminar flow in a fluid.

Newton explained the internal friction property of fluids as resembling the friction between two solid sliding surfaces. He demonstrated from experiments with two concentric cylinders, submerged in water, that a force was required to rotate one cylinder with respect to the other. Newton showed that the required force was a measure of the internal frictional shear resistance (or viscosity) of the fluid, and that it was associated with the shear area, the rotational speed and the film thickness in the following manner:

$$F = \eta.A.U/h \qquad \text{therefore: } \eta = F.h/(A.U)$$

This equation defines Absolute (or Dynamic) Viscosity which, in appropriate units, is required for bearing analysis and design.

It has become standard practice to specify lubricating oils by their kinematic viscosity, which is a convenient method of measuring viscosity using gravity flow. Multiplying by the fluid density is necessary to convert to absolute viscosity for use in bearing calculations.

A lubricating oil may have many chemical and physical properties which affect its behaviour, but for hydrodynamic bearings it is clear that the characteristic of viscosity is the most important. For a given bearing, such as is used in typical engineering applications, and for given operating conditions of load, speed, oil flow and supply temperature, it is the viscosity of the lubricating oil in the bearing separating film that finally determines the power loss, the heat generation, the system temperature and the load-carrying capacity.

If we regard the basic parameter $\eta N/p^{av}$, or, as it is frequently referred to, ZN/P, as an index of bearing performance then clearly the correct oil viscosity can be chosen to match the speed, the applied loading and the size of bearing. Viscosity is a measure of the physical ability of the oil to maintain a separating film under the specified bearing conditions.

However, viscosity is also a measure of the internal frictional shear resistance of the fluid, and so the process of shearing the oil film in a bearing has the effect of generating frictional heat within the film. Inevitably, the work done in shearing the film raises the film temperature, and in many applications a flow of oil in and out of the bearing is necessary to remove the generated heat and to maintain a reasonable system temperature.

The business of designing hydrodynamic bearings is therefore also associated with the selection of the lubricant and the bearing material, and specifying the oil feed system details such that:

1. The applied load will be carried on an adequate separating oil film at the operating speed.
2. The heat generation will be reasonably low commensurate with maintaining acceptable oil and bearing temperatures.
3. The bearing material fatigue strength will be adequate to tolerate the imposed pressure and the generated temperature, and will operate safely without serious surface damage when inevitable contact occurs at starting and stopping.

A major difficulty in analysing the performance of oil film bearings is the marked variation of viscosity with temperature. A typical bearing oil may show at least an order and possibly two orders of magnitude viscosity variation between the full range of operating conditions from cold start to maximum film temperature.

The viscosity within the film will vary between inlet and maximum temperature conditions. Estimating the effective temperature to obtain the effective film viscosity therefore requires iteration, and this is where modern computer methods are useful.

Some of the heat generated will be lost via the structure, but this proportion is usually small in pressure-fed applications

and can be neglected. For non-critical applications, lubrication is by static oil bath, in which case all the generated heat is lost to the surroundings via the structure and shaft, and a reasonable estimate of the effective film temperature is therefore required.

9.4.4 Journal bearing design

Methods have been established from theory, experiment and practice to produce bearing design solutions. The basic non-dimensional parameter required to be specified and which incorporates all the relevant factors is a term which has become known as the Sommerfeld Number or Sommerfeld Reciprocal. This is a variation of ZN/P which includes the 'clearance ratio' (the ratio of the diametral clearance to the diameter) and is conveniently used in reciprocal form as 'dimensionless load'.

Much work has been done over the years in analysing bearings of various length-to-diameter ratio to establish the variation of several parameters against dimensionless load. These are heat generation, oil flow, eccentricity ratio (i.e. the eccentricity of the shaft within the bearing ($\epsilon = 0$ for concentric operation and $\epsilon = 1$ for fully eccentric, i.e. touching)) and the attitude angle, i.e. the angle of disposition of the shaft/bearing centres to the load line which is always beyond the load line in the direction of rotation (Figure 9.15). Typical values are shown in Figure 9.16.

Another variable which may have to be considered is the angle of bearing arc (Pinkus and Sternlicht[14]). Few bearings have a full 360° bore, because of the need for oil supply which is usually pressure fed to longitudinal grooves cut within the bore and along the bearing length, and normally terminating short of the ends. Typically, bearings are made in two halves to allow assembly into the machine (e.g. engine main bearings), and this allows the oil grooves to be conveniently cut at the joint faces, thus reducing the bearing to two plain halves of, say, 150° each.

It is evident that with longitudinal oil supply grooves the direction of load application must be fairly constant and towards the centre plane portion as in turbines or engine main bearings. A bearing has a much-reduced load-carrying capacity if the load is applied towards the oil groove, because the hydrodynamic wedge length is significantly shortened. In cases where the load direction is indeterminate or variable (e.g. engine big-ends) then the answer can be to use a circumferential oil groove at the centre of the bearing length.

Charts of dimensionless parameters are useful for setting out the basic design of bearings operating under constant speed and load. A well-developed chart method by the Engineering Sciences Data Unit[15] for typical split bearings enables the relevant parameters to be determined, and computer programs are also available for standard designs. Programs also exist which allow evaluation of designs incorporating complicated multi-arc geometry, and which will solve for thermal variations, distortion and oscillating load.

Assuming a typical bearing, the first step for a preliminary design evaluation is to collect the physical data for the bearing system and to calculate the dimensionless load:

$$\text{Dimensionless load} = \frac{W.(C_d/d)^2}{\eta.N.d.b}$$

The value of the viscosity term must be assumed initially because of its dependence on temperature, heat generation, etc. The other parameters are usually known (e.g. load, speed, diameter).

The bearing eccentricity ratio (ϵ) is then obtained from charts of eccentricity ratio against dimensionless load such as

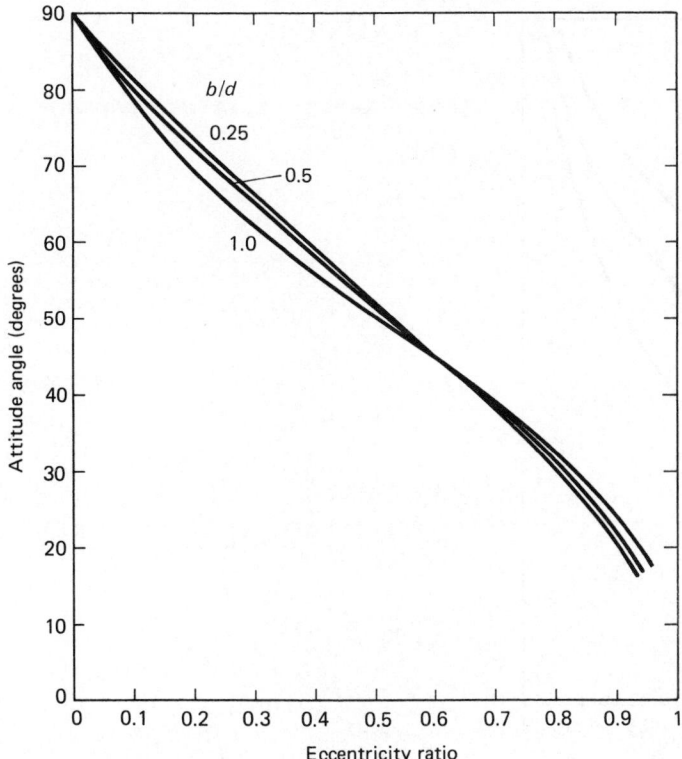

Figure 9.16   Journal bearing: attitude angle

Figure 9.17 and the associated minimum film thickness (*h*) determined where:

$$h_{min} = (C_d/2).(1 - \epsilon)$$

The oil flow, heat generation, etc. are also determined from charts. The process is repeated until reasonable coincidence is obtained depending on whether the requirement is to design a bearing for a particular duty or to evaluate an existing design.

For an initial assessment of a typical bearing the assumptions to make for a safe and reliable design are generally length-to-diameter ratio (*b/d* ratio) of, say, 0.7, and a minimum clearance ratio (C_d/d) of about 0.001 (remember to take tolerance limits into account). Long bearings (of *b/d* ratio greater than unity) are prone to misalignment problems. Small values of clearance ratio less than 0.001 can lead to reduced oil flow and high temperatures, particularly in high-speed bearings where a larger clearance is required (see below).

The dimensionless load should be in the range 10–60. Values approaching 100 indicate very high eccentricity and small film thickness and are only acceptable in large bearings. However, values exceeding 100 are usual in very large, heavily loaded, slow-speed bearings where the shaft surface finish dimension is very small in relation to the diameter. Low values (say, below 10 (high-speed light-loads)) increase the risk of instability. Eccentricity ratio should generally be in the range 0.7–0.95.

There are other limiting factors which define the safe operating zone and which must be properly considered in producing a design to operate reliably in service. These are discussed below.

9.4.4.1 Diametral clearance

The bearing clearance can be adjusted at the design stage to optimize the film thickness, the heat generation, the oil flow and the temperature. Note that as the clearance dimension is very small and may vary on manufacturing tolerances by a factor of two, the calculations should therefore cover the extreme limits of tolerance. As a general rule, the minimum clearance should be adequate to allow sufficient oil flow so as to limit the temperature rise within the bearing to an acceptable value.

High-speed bearings are generally lightly loaded but need large clearances to reduce heat generation and to promote stability against film whirl. Heavily loaded slow-speed bearings have marginal film separation and need small clearances to improve the hydrodynamic performance and to allow greater film thicknesses to be generated. Empirical selection is adequate for the initial design, and guidance for typical bearings is given:
Minimum clearance ratio = 0.0005. (shaft speed in rev/sec)$^{0.24}$
i.e. C_d/d = 0.0005.(rps)$^{0.24}$ (for diameters of 0.1 m or greater)
Clearance ratio should be increased for bearings of diameter less than 0.1 m.

9.4.4.2 Surface roughness

The roughness of engineering surfaces is usually measured by traversing a stylus and recording the undulations as a root mean square (RMS) or centre line average (*Ra*) value.
Typical *Ra* values for shafts and bearings are:

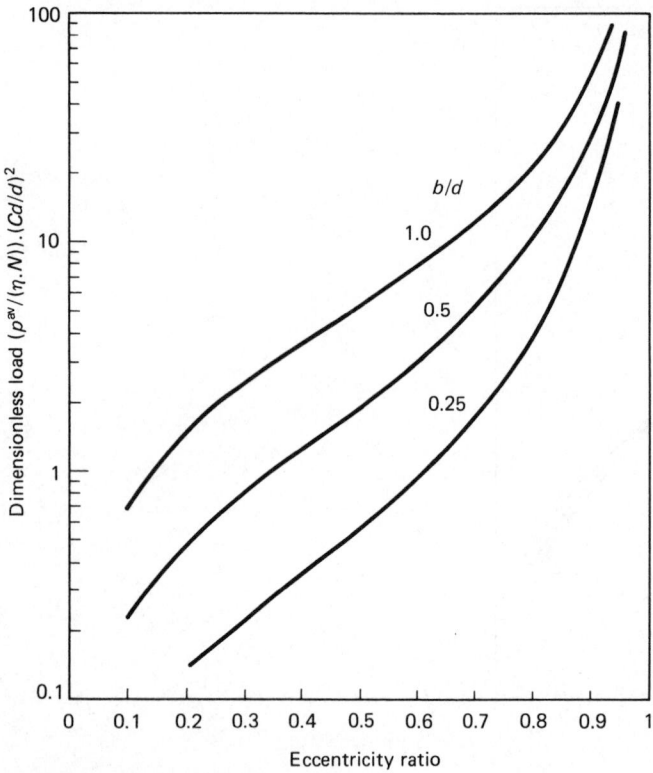

Figure 9.17 Journal bearing: dimensionless load

Turned surfaces:	3.2 to 12 μm
Fine turned:	1.6 to 0.8 μm
Ground surfaces:	1.6 to 0.4 μm
Fine ground:	0.4 to 0.1 μm
Lapped:	0.2 to 0.05 μm

The peak-to-valley dimension is, however, much greater than indicated by these averaging methods, and can be from about four to ten times greater, dependent on the method of machining and the magnitude of the average dimension. It is the peak-to-peak contact that represents the ultimate 'touch-down' condition of bearing operation and the design must take it into account.

9.4.4.3 Minimum allowable film thickness

The basic calculation of minimum film thickness inherently assumes the bearing and journal surfaces to be smooth and parallel to one another. This happy condition is, of course, seldom true, and while parallelism may be achieved on assembly, thermal distortion inevitably introduces a degree of misalignment when in operation, and this should be considered in the design.

In a precisely aligned system the ultimate minimum allowable film thickness is set by the combined roughness of the surfaces at the point which will just allow the surface asperities to come into contact, thus increasing friction and heat generation. (see Figure 9.18) At best, the result will be light burnishing or polishing of the surfaces, at worst, failure due to wear, local melting or seizure dependent on the sliding speed

and the materials. 'Safe' minimum film thickness values are therefore specified in design which take account of bearing size, rotating speed, materials, application and method of surface finishing.

Note from Figure 9.18 that the 'knee' of the curve represents the point of surface asperity contact, and its position in relation to ZN/P for a precisely aligned system is only dependent on the combined surface finish. 'Running-in' of new bearings has the effect of reducing the point at which the 'knee' occurs.

For normal use, empirical data for typical bearing and shaft surfaces are adequate to produce recommended safe values of minimum film thickness. The values given by the following equation allow a factor of >1.5 on the peak-to-valley dimension of typical journal surfaces. This minimum margin is generally safe for correctly aligned and clean systems:

Minimum allowable film thickness = dia(mm)$^{0.43}$ (μm)

i.e. $h_{\min} = d^{0.43}$ (μm)   (d in millimetres)

In many heavily loaded slow-speed applications of large bronze bearings with grease lubrication the operating minimum film thickness falls well below the safe recommended value as given by the above equation, and wear inevitably takes place. This is usually unavoidable, but nevertheless acceptable, as the wear rate can be minimized by adopting self-aligning features; very fine surface finishes, particularly of the shaft; by maximizing the lubrication using a high-viscosity oil component in the grease; by careful detail design of the bearing and grease grooves; and by using effective grease-feeding arrangements.

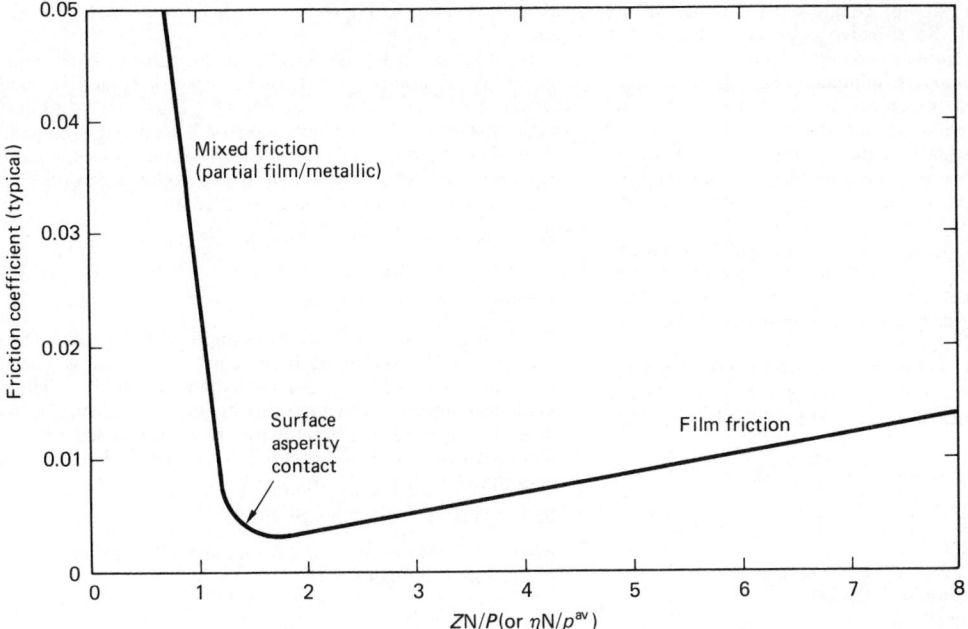

Figure 9.18   Journal bearing: friction coefficient

9.4.4.4 Lubrication

The preferred lubricant for most bearings is oil. Lubricating oils are designed to operate satisfactorily in terms of the oil itself to give a long life in properly maintained and clean systems, but, as previously stated, the important property for hydrodynamic bearings is the oil viscosity. Oils are formulated to be suitable for given applications (e.g. turbines, compressors, gears, internal combustion engines), therefore turbine oils are solvent refined and usually incorporate only anti-oxidant and anti-rust additives. They are particularly efficient at separating out from water, which is an inevitable contaminant in a steam turbine system. Gear oils, on the other hand, may contain additives to protect against gear tooth scuffing and other surface damage. Such additives are not generally required for film bearings.

The life of the oil is usually limited by oxidation, which produces acidic residues. Oxidation life is very sensitive to temperature and by rule of thumb can be assumed to halve for every 10°C. A maximum safe upper limit temperature for most mineral oils is 120°C, which coincides with the safe upper operating limit for white metals, but the oil would have a short life at this temperature. Typical bulk temperatures in a large system would be 40–60°C with a maximum bearing metal temperature of, say, 80–90°C. In automobile engines oil temperatures frequently reach 150°C but then these systems are comparatively small and are required to be changed frequently due not only to oxidation but also to contamination by sludge, carbon, fuel, water, etc. For very high temperatures up to about 250°C, synthetic oils are available.

While the correct viscosity grade of oil can be identified from rigorous bearing calculations to optimize the film thickness, heat generation and temperature rise, it is quite adequate and certainly more convenient to use empirical selection for a preliminary assessment.

The general guidance in Table 9.8 applies, where VG represents the viscosity grade of the oil according to the ISO definition: 'kinematic viscosity in centistokes at 40°C.' Data are given in oil company literature which usually specifies the viscosities at 40°C and 100°C to enable the variation over the working temperature range to be estimated.

9.4.4.5 Pressure-fed lubrication systems

Bearings in large machines (e.g. turbines, industrial and marine gears, compressors, automotive engines) are usually supplied with oil from a pressurized positive-feed system to ensure safety, reliability, uniform operating conditions and effective cooling. The system requires a storage tank, pump and return line, and can be usefully enhanced by the inclusion of filtration equipment, and instrumentation to measure and control pressure and temperature.

Table 9.8

Diameter × speed (mm × rps)	'Viscosity grade' of oil required at the operating temperature of:			
	40°C	60°C	80°C	100°C
3400–6000	15 VG	32 VG	68 VG	150 VG
1700–3400	22	46	100	220
850–1700	32	68	150	320
500–850	46	100	220	460
250–500	68	150	320	680
140–250	100	220	460	1000
80–140	150	320	680	1000
30–80	220	460	1000	1500

The total bearing design procedure must consider all the factors which control the effective operating bearing film temperature and the effective viscosity, which in turn controls all other relevant parameters: minimum film thickness; bearing attitude angle; heat generation or power loss; rate of oil flow; maximum oil/bearing temperature; and the outlet oil temperature. All these factors are interrelated and therefore the equilibrium operating condition for the total bearing system is determined by simultaneous solution of all the variables.

Oil flow through a journal bearing is principally from two components: the pressurized local flow through the bearing clearance via the feed grooves, and the hydrodynamic flow, which depends on the disposition or eccentricity of the rotating shaft. Pressurized film flow occurs whether the system is rotating or stationary, and is the flow from the feed grooves out through the clearance space. Assessment is based on a modification of the general expression for fluid flow through thin slots:

Pressurized flow rate = $dp.R_s.t^3/(12.\eta)$ (m^3 s^{-1})

where

dp = pressure drop (Pa),
R_s = width-to-length ratio of the rectangular slot,
t = thickness of the thin film (m), and
η = fluid film viscosity (Pa.s).

Modification of the equation is necessary for a journal bearing, because the slot width is not constant but widens out from the groove ends as the flow progresses out through the clearance spaces to the ends of the bearing. A factor is therefore used based on eccentricity and length/diameter ratio.

Hydrodynamic flow is controlled by the bearing dimensions, geometry, clearance, the journal's eccentricity and running speed. It is zero at zero eccentricity (concentric, no-load operation). For normal operation of a loaded bearing the hydrodynamic oil flow is calculated from the following basic equation, and factors are used to take account of eccentricity, length/diameter ratio and oil groove length.

Hydrodynamic flow rate = f($C_d.N.d.b$)

9.4.4.6 Heat generation

The heat generation or power consumption in a bearing film is produced by the work done in shearing the film. This is related to Petrov's Law which gives a good approximation for lightly loaded (therefore near-concentric) bearings. For typical bearings carrying load, the basic Petrov relation is modified by a dimensionless number[15] appropriate to the design and the operating eccentricity:

Power loss: $H = H^1.(\eta.N^2.d^3.b/C_d)$

where H = film power loss (Watts) and H^1 = dimensionless power loss (see Figure 9.19).

9.4.4.7 Bearing materials

The bearing material must be of adequate hardness and strength to support the load, particularly if there is a fatigue element in the load cycle as would be the case, for example, in

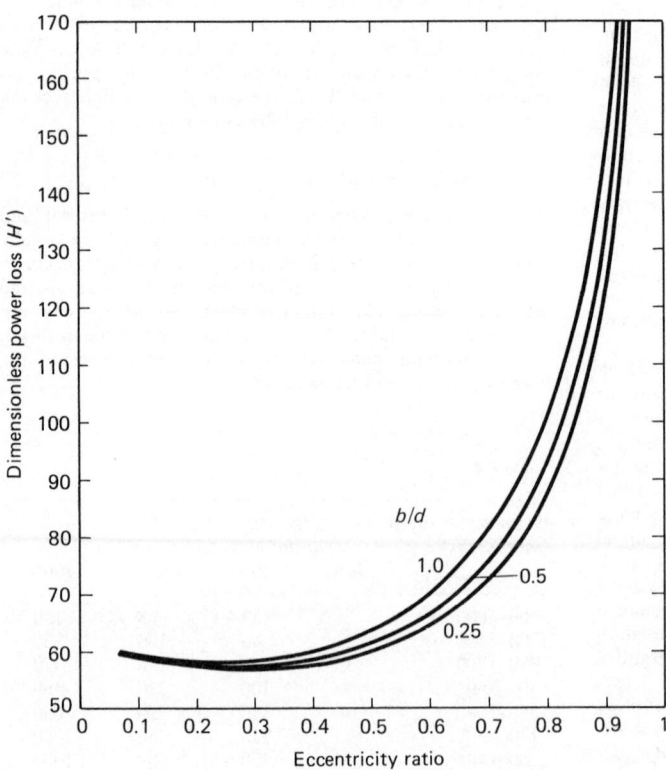

Figure 9.19 Journal bearing: dimensionless power loss

an internal combustion engine. Conversely, the material should be soft enough to embed particulate debris so as not to score the shaft. Good compatibility is important, that is, the ability to 'self-heal' or to withstand an occasional 'touch' by the shaft, particularly if lubrication becomes marginal. Corrosion resistance is necessary to prevent the acidic products of oxidation in the lubricating oil from corroding the bearing surfaces. The bearing material must satisfy in measure all these requirements.

Some of the more common bearing materials used with oil or grease lubrication and their general properties are given in Table 9.9. The fatigue limit values given in the table relate to the average bearing pressure (p^{av}), i.e. the load divided by the 'projected' area ($b.d$). The localized peak film pressures (p_{max}) are several times the average (Figure 9.20).

Within each group of bearing material there are available many variations of the proportions of constituent elements to give selected properties. Also, the high strength capacity of the harder materials can be used to support thin surface layers of softer material to allow improved contact compatibility and to enhance the fatigue life of the softer material. Production-manufactured engine bearings are normally tin plated to protect against corrosion from acidic residues in the oil.

In general, the softer bearing materials (e.g. the white-metals) are best for high-speed lightly loaded applications such as turbine bearings and will withstand an occasional 'touch' without serious damage. Comparatively, the harder bronzes generate higher local temperature on contact and are best for low-speed heavy-load applications such as gearboxes and general engineering equipment. Hard phosphor–bronze is used for highly loaded small-end bearings in some internal combustion engines because of its excellent fatigue capacity.

Table 9.9

Material	Hardness at room temp. (HV)	Fatigue limit (MPa)	Operating temperature limit (°C)	Min. shaft hardness (HV)
White-metal (Pb base)	16–20	7	120	160
White-metal (Sn base)	27–32	7–14	120	160
Aluminium–tin	30–60	16	160	200
Copper–lead	45–50	10	190	300
Lead–bronze	60–80	25	200	400
Phosphor–bronze	100–150	70	230	400

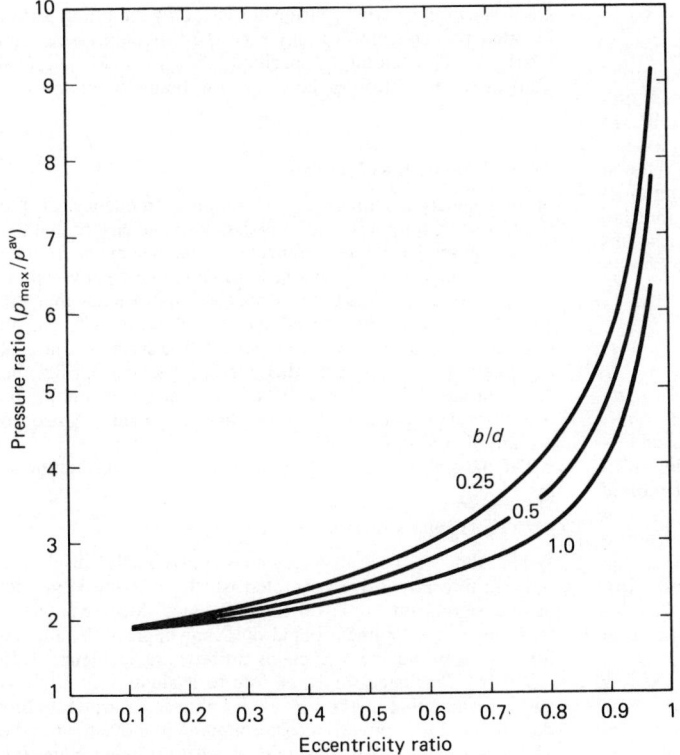

Figure 9.20 Journal bearing: film pressure ratio

9.4.4.8 Oscillating load

In some applications the bearing load is not steadily applied, but oscillates with rotation. These include bearings in engine crankshafts (mains and big-ends) and piston-pins, as well as forging and coining presses. In this situation the bearing load capacity is enhanced over and above the limit for a steadily applied load. The reason is that, although the film thickness begins to reduce as the peak cyclic load is applied, it takes a finite time for the oil to be squeezed out from the small clearance. During this period of time and before an equilibrium minimum film condition is reached, the peak load is relieved which then allows the high eccentricity to reduce and the film to recover. This is known as the 'squeeze-film' effect.

Several theories and empirical methods have been suggested to consider this effect. Unfortunately, the results appear to vary markedly. The general principle adopted is to relate the change in shaft eccentricity with the time interval required to squeeze the oil in the small reducing clearance space from one shaft position to another. The direction of flow has to be assumed.

A simple approximation for hand calculation was proposed in 1954 by Archibald[16] based on the general equation for flow through thin slots (previously discussed). Archibald's equation for a full (360°) bearing is:

$$dt = (6.\pi.\eta.b.d^3/[C_d^2.W]).(\epsilon_2/[1 - \epsilon_2^2]^{0.5} - \epsilon_1/[1 - \epsilon_1^2]^{0.5})$$

where

dt = time interval from ϵ_1 to ϵ_2 (s),
η = viscosity (Pa.s),
b = bearing length (m),
d = bearing diameter (m),
C_d = diametral clearance (m),
W = applied peak load (N),
ϵ_1 = initial eccentricity ratio, and
ϵ_2 = final eccentricity ratio.

Generally, the usefulness of the above equation is as a check on the likely change in eccentricity for a given time interval based on rotating speed to ensure that the reduced film thickness is adequate. The equation is only applicable to bearings which are long in relation to their diameter, because the flow is assumed to be circumferential, therefore the end flow negligible. The method can only give an approximation of the bearing behaviour, but, nevertheless, will usually indicate whether or not the design is adequate. Computer programs have also been devised based on finite-element and finite-difference analyses.

9.4.4.9 Film instability

While it is obviously desirable to operate with an adequate separating film to give a high degree of reliability, care must be taken in design to ensure that the bearing will behave in a stable manner and will not induce self-sustaining oil film whirl, a common problem with lightly loaded, high-speed journal bearings. The problem is associated with the operation of circular bearings at low eccentricity where the journal tends to operate in an almost concentric disposition relative to the bearing and with a large attitude angle (Figures 9.15 and 9.16). In this condition the oil film which rotates at about 0.48 times shaft speed tends to push the journal around bodily, also at about half shaft speed.

Oil film whirl is particularly worrying when the rotating shaft speed is about twice its first critical frequency. The resultant vibration which can occur at operating speeds between, say, 1.8 and 2.4 times the first critical can be very difficult to remedy.

Changes in bearing clearance generally have only a small effect on the running eccentricity of an oil-fed journal bearing because oil flow, heat generation and film temperature also change. Adjusting the clearance, however, can have a useful effect on whirl frequency and can be used to change the frequency if film whirl is present, as the following approximation[17] shows:

$$f_{oil} = ([3 - \epsilon]/4\pi).(g/c)^{0.5} \quad \text{for } b/d = 1$$

where

f_{oil} = whirl frequency (Hz),
ϵ = eccentricity ratio,
c = radial clearance (m), and
g = gravitational acceleration (m s^{-2})

Other techniques for shifting the operating mode to a more stable condition involve either adjusting the operating conditions in the case of a running bearing (e.g. by allowing a higher oil film temperature) or, for a new machine, by using a shorter bearing or a different geometrical concept. The objective in all cases is to de-rate the load-carrying capacity of the bearing such that the minimum oil film thickness is reduced and the shaft disposition moved to a condition of greater eccentricity and smaller attitude angle, thus producing a more stable operating mode, as is the case with a heavily loaded bearing.

Several designs have been produced (Figure 9.21), including:

Elliptical bore ('Lemon' bearing) ⎫ in which the bearing lobe
Three-lobe and   ⎬ radii are greater than the
Four-lobe bore   ⎭ bearing contact radius
Four longitudinal grooves
Offset-halves
Tilted offset-halves
Tilting pads/pivoting pads

Of all these alternative bearing designs, only the tilting pad or pivoting pad concept is really successful in suppressing film whirl. This type should be specified for lightly loaded vertical shaft units for critical applications (e.g. boiler circulators).

9.4.4.10 Shaft critical frequency

In high-speed machines the vibrational frequency of the rotating shaft system is determined in design to ensure that the operating speeds do not coincide or even come near to a critical frequency at which a self-sustaining destructive vibration can develop. However, it must be appreciated that the critical frequency value calculated for the rigid shaft can, in practice, be reduced to a lower value by the spring stiffness of the bearing oil film. For this reason, bearing spring and damping coefficients have to be taken into account in the overall design, particularly where lightly loaded, high-speed oil film bearings are used.

The natural frequency of the undamped system is defined[17] by:

$$f = (1/2\pi).(\text{film stiffness/journal mass})^{0.5} \text{ (Hz)}$$

When the shaft centre becomes positionally displaced, squeeze-film forces are generated which are quite large and have a significant effect on the frequency. Moreover, shaft movement is never linear but always at an angle to the applied force, thus producing a changing attitude angle (Figures 9.15 and 9.16). It is necessary therefore to evaluate four stiffness and four damping terms YY, YX, XX and XY which define the force in the Y direction corresponding to movement in the Y direction and so on, the units of stiffness being N/m, and damping N.s/m.

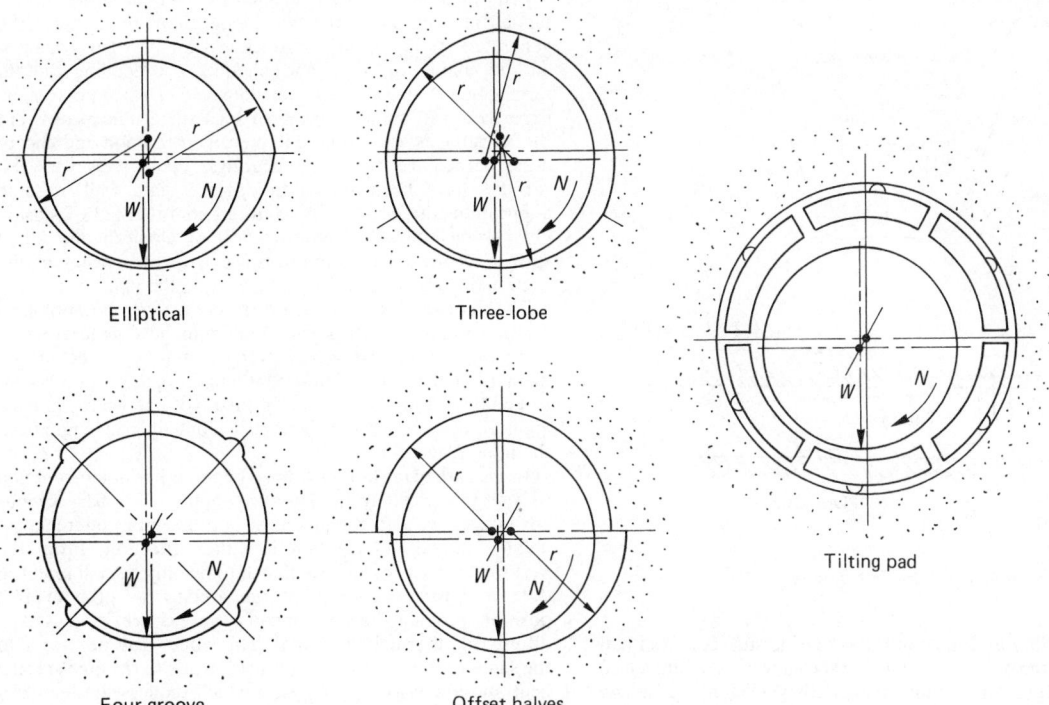

Figure 9.21 Alternative journal bearing designs

For a simple assessment the effect of bearing whirl on the shaft critical speed may be approximated by adding the reciprocals squared of the frequencies of the rigid rotor and the oil whirl value (previous section) to produce the reciprocal squared of the effective rotor critical frequency:

$$(1/f_{\text{eff}})^2 = (1/f_{\text{oil}})^2 + (1/f_{\text{rotor}})^2$$

9.4.4.11 Process fluid lubrication

It is sometimes convenient to operate film bearings in the process fluid. Typical examples are boiler circulators, long-shaft river-water pumps and acid pumps. Unlike oil, water has no boundary lubricating properties and has a low viscosity (0.001 Pa.s = 1 cP at 20°C) therefore the unit loading has to be kept low and the materials have to be suitable.

Fluted rubber bearings are generally very effective for lightly loaded long-shaft river-water pumps and small-boat propeller shafts, as they can accommodate grit and sand without damage. The bearing works hydrodynamically by a wedge-shaped surface contour which develops by elastic deformation of the rubber, thus allowing separating films to be generated on the several lands. Rapid failure occurs if the rubber is allowed to run dry.

Glandless boiler circulators frequently employ bearings made from fabric-reinforced thermosetting resins of which there are many types available. Some designs incorporate helical debris wash-out grooves which inevitably reduce the load-carrying capacity. Nevertheless, using hardened and finely ground shaft sleeves a long operational life is obtained even with calculated film thicknesses as low as 0.005 mm in small bearings.

Water-lubricated bearings tend to be long in relation to diameter to give greater bearing area in view of the low viscosity of water (which reduces with increasing temperature). A high standard of surface finish is therefore required on the mating shaft (say, 0.1–0.4 μm R_{a}). Good alignment is also essential. Carbon bearings are used for very high temperature water.

For bearings operating in corrosive fluids an important property is their resistance to corrosive attack. Hard materials such as austenitic high-nickel cast iron are used but have poor bearing properties relating to the accommodation of debris, which then tends to score and damage the shaft. Thermoplastic polymers have solved these problems and many filled and unfilled types are quite suitable in acids and alkalis, the filled thermoplastics being generally preferred because of their low wear rate.

9.4.5 Self-contained bearings

Assemblies of self-contained bearing systems are available for incorporation into machines such as turbines, centrifugal compressors, pumps and motors. The action of the rotating film is utilized to pump oil from the bearing to an associated cooler then back to the bearing. One end bearing would be journal only, the other combined with a thrust bearing for axial location.

9.4.6 Thrust bearings

Multi-pad thrust bearings operate on hydrodynamic principles. However, unlike journal bearings which automatically adjust to allow the converging or tapered film wedge, the taper

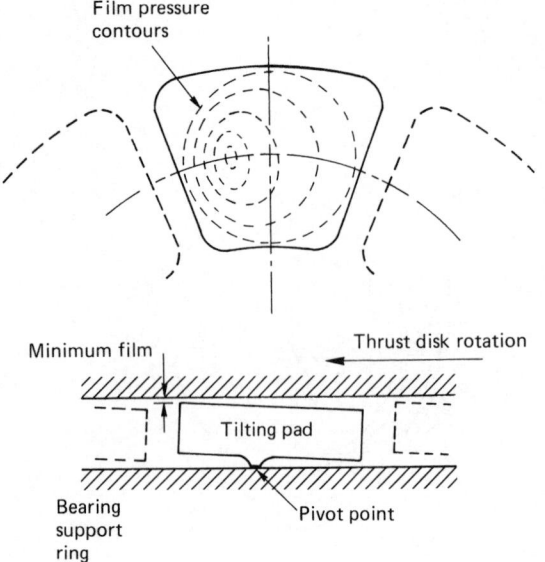

Figure 9.22 Hydrodynamic thrust bearing

has to be either machined into the bearing surface (fixed pad bearing) or the bearing arranged in such a way as to allow it to tilt when in operation (tilting pad bearing) (Figure 9.22). The shaft incorporates a collar or thrust disk which is normally located by two thrust bearings, one either side. Each bearing comprises several pads (Figure 9.23). The preferred pad geometry is nearly 'square', that is the radial breadth equal to the circumferential length, or: $b/L = 1$.

Fixed pads have to be machined to a very accurate surface taper, the pad rise dimension being about three times the generated minimum oil film thickness for best performance, and to allow for wear and unequal pad loading due to misalignment and varying pad thickness. Difficulties can be experienced in manufacturing to such small dimensions, but the advantages are simple one-piece construction and cheapness, particularly for small bearings. Fixed pads with flat surfaces have been shown to operate quite well in some applications due principally to differential thermal expansion effects which allow a contour to develop along the metal pad surface as the oil film temperature rises as a result of the frictional work done in its passage along the pad.

Theoretically, tilting pad thrust bearings can be optimized for maximum film thickness, minimum heat generation or minimum temperature. For a given geometry of inside/outside diameter and pad circumferential length, optimization is controlled by setting the position of the pivot line which, in most applications, is at a point about 60% along the pad length from the inlet edge.

In the case of oil-lubricated pad bearings it is not uncommon to site the pivot at the half-way point. This has obvious advantages in both manufacture and assembly interchangeability, and allows rotation in either direction. From the performance aspect there is shown to be only a small practical difference using a centre pivot compared to the optimum pivot position, probably because of pad surface convexity caused by differential thermal expansion. This effect may not apply to the same degree with water lubrication due to the high specific heat and low viscosity of water, which reduces the temperature rise along the pad to very small values.

The tilting pad automatically adjusts when in operation to the optimum tilt angle, which will vary with changes in operating conditions: load, speed and temperature/viscosity. The leading-edge film thickness for a 0.6 pivot position is

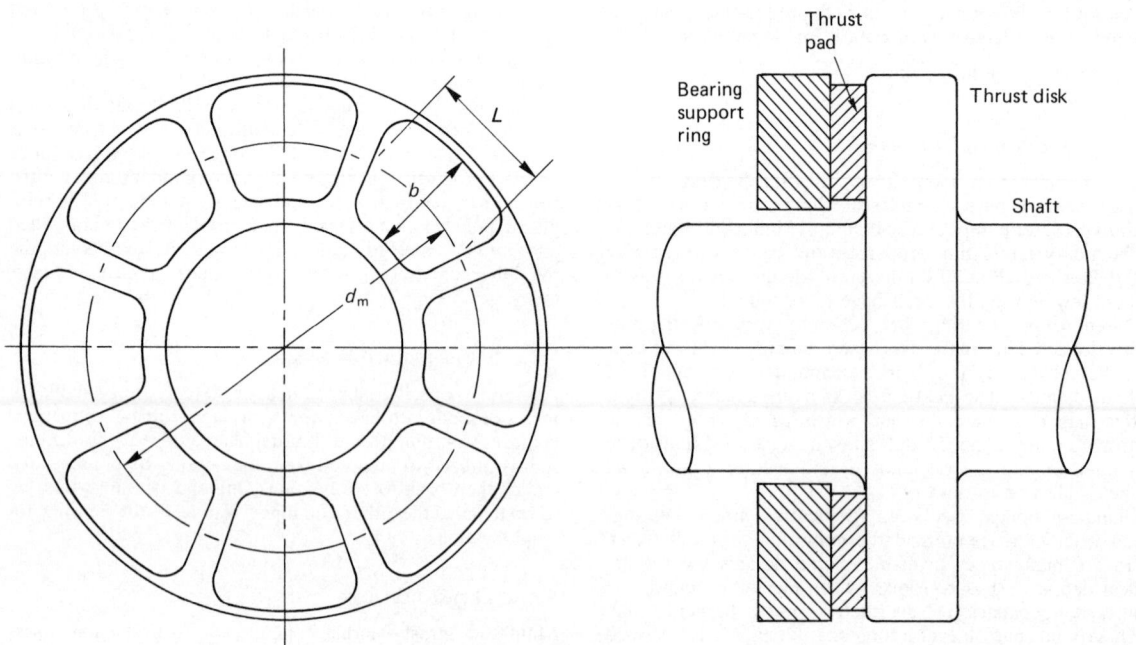

Figure 9.23 Eight-pad hydrodynamic thrust bearing

approximately equal to 2.4 times the trailing-edge film thickness under all conditions of film operation.

For a preliminary bearing design assessment the effective film viscosity is required, therefore the film temperature is assumed, and should be at least of the same order as the expected drain temperature. The minimum or trailing-edge film thickness can then be calculated from the following equation, which is suitable for 'square' or nearly square pads (i.e. $b/L = 1$):

$$h_0 = 0.46(\eta.N.d_{\mathrm{m}}.b/p^{\mathrm{av}})^{0.5}$$

where

h_0 = minimum film thickness at the trailing edge (m),
η = effective film dynamic viscosity (Pa.s or cP/1000),
N = speed (Hz),
d_{m} = mean diameter of disk (m),
b = radial width of pad (m),
L = circumferential length of pad (m)
p^{av} = mean pad pressure = total load/($b.L.z$) (Pa), and
z = number of pads.

The safe minimum allowable film thickness for design purposes is generally based on the pad dimension rather than the diameter. The following equation is for a precisely aligned high-quality bearing:

Minimum allowable film thickness = 1.7.pad width$(\mathrm{mm})^{0.44}$ (μm)

i.e. $h_0 = 1.7.b^{0.44}$ (μm) (b in millimetres)

However, some degree of misalignment is unavoidable in most machines, as are variations in pad thickness within the manufacturing tolerance band. These factors must therefore be taken into account at the design stage and the design film thickness adjusted accordingly.

An important difference between pad-type thrust bearings and journal bearings is the behaviour of the oil flow through and out of the bearing. In pressure-fed journal bearings, fresh cool oil is pumped into the grooves incorporated into the bearing surface, and most of it – apart from the groove end leakage – is carried by the rotating journal into the loaded area. In bath-lubricated pad thrust bearings, however, the hot oil film emerging from the trailing edge of one pad tends to be dragged by the rotating disk into the entry gap of the next pad, thus giving a hot oil carryover effect.[18] In some cases this can raise the effective film temperature to a value significantly higher than the oil drain temperature, considering that most of the oil flowing between the pads to drain plays little or no part in film cooling. Correct assessment of the oil film temperature is therefore very important in the calculations.

In a typical multi-pad bearing, however, most of the generated film heat is transferred by convection and conduction to the rotating disk, which usually acts as a very good heat sink.[19] Nevertheless, film temperatures can be quite high in some cases. For such applications the pad film temperature can be reduced significantly by adopting directed lubrication from spray nozzles positioned between each pad, which direct discrete jets of oil onto the rotating disk surface. The jets tend to displace the adhering hot film and allow cool oil to enter the pad surface.

The heat generated at each pad by oil film friction only is:

$$H_{\mathrm{pad}} = 13.8.L.(N.d_{\mathrm{m}})^{1.5}.(\eta.b.p)^{0.5} \text{ (Watts)}$$

and the total bearing heat from oil film friction only is:

$$H_{\mathrm{film}} = z.H_{\mathrm{pad}}$$

Heat is also generated in conventional oil bath bearings due to churning of the oil by the rotating disk, and this effect is also eliminated by using directed lubrication. The heat generated by oil churning can be the significant component, particularly in large, high-speed units.

Modern methods for the design of multi-pad thrust bearings are available.[20,21] These take account of all the relevant factors needed to determine operating performance.

9.5 Lubrication of industrial gears

In order to ensure the effective lubrication of industrial gears it is first necessary to have a basic knowledge of the tribological implications in their design, i.e. to study the conditions which can arise when interacting surfaces are in relative motion. All types of gear teeth transfer power and motion through relatively small areas of contact in the form of very narrow bands or ellipses which are known as the 'lines of contact'. Since these contact areas are so small they are subjected to very high stresses and the gear teeth therefore have to be made from strong, hard materials such as steel and the harder bronzes.

Figure 9.24 shows a pair of contacting involute gears with their respective 'pitch circles' and the 'line-of-action' which is determined by the pressure angle selected during design. Contact of the gear teeth begins as the trailing edge or the top of the driven gear tooth crosses the line-of-action and engages with the root of the leading face of the pinion gear tooth. Contact continues until the leading edge of the tip of the pinion tooth crosses the line-of-action. Lines drawn on the gear tooth edge for equal angular displacement as the contacting faces cross the line-of-action show that sliding takes place during tooth contact. It can also be seen that the sliding velocity is at a maximum at the start of contact but decreases continuously as the line of contact approaches the pitch circle. At this point sliding becomes zero and with continued movement the direction of sliding reverses and accelerates until the teeth disengage.

During contact it will be noted that on the pinion, sliding is always away from the pitch line, whereas on the driven gear, sliding is always toward the pitch line. Also, tooth contact only occurs on the line-of-action.

Pure rolling motion, which only occurs momentarily at the pitch point, does not promote rapid wear of the surfaces, but sliding, which occurs over the remainder of the contacting surfaces, does. Also, the faster the rate of sliding, the more severe the rate of wear and consequently the more difficult the conditions of lubrication become. Since sliding speed increases with increased pitch-line speed of the gears and the distance of the point of contact from the pitch point, high-speed gears are made with a large number of small teeth to reduce their sliding speed.

When studying tooth action, consideration must also be given to the important differences in gear tooth contact between spur and helical gears and between straight-bevel and spiral-bevel gears. In spur and straight-bevel gears there are alternately one pair and then two pairs of teeth in contact. Any wear that occurs during single-pair contact will not result in relief by transferring the load to another pair of teeth, so that, once started, wear will continue. In contrast, in helical and spiral-bevel gears there are always at least two pairs of teeth in contact so that should wear occur between one pair of teeth, more load is transferred to the other teeth in contact, thus reducing the load on the teeth subjected to wear. For this reason helical and spiral bevel gears are easier to lubricate than spur or straight-bevel gears. Figures 9.25 and 9.26 show the lines of tooth contact for spur and helical gears.

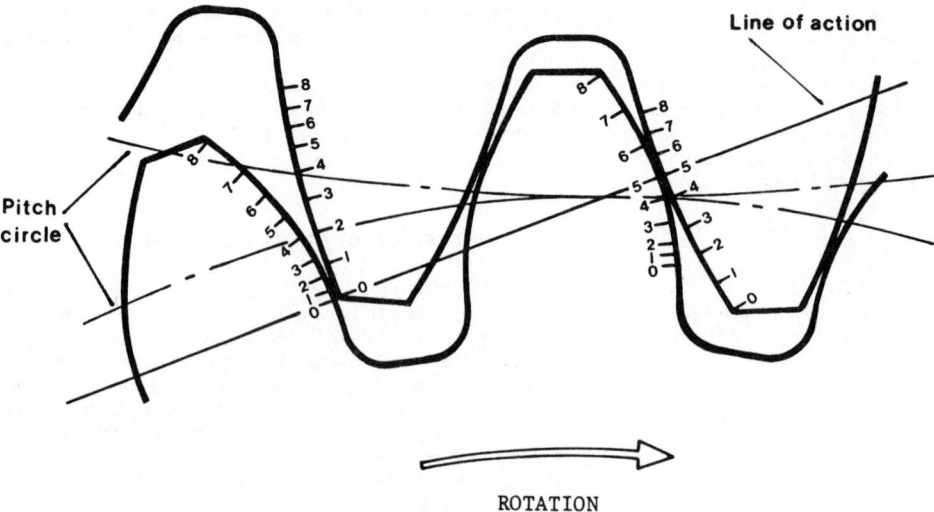

Figure 9.24 Relative sliding in tooth contact area

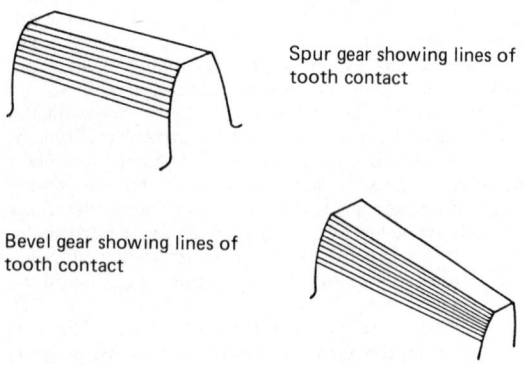

Spur gear showing lines of tooth contact

Bevel gear showing lines of tooth contact

Figure 9.25 Lines of contact – spur and bevel gear

As shown in Table 9.10, the various types of gears may be divided into groups whose conditions of tooth operation, i.e. sliding and rolling, are similar.

9.5.1 Methods of lubrication

There are three principal methods of applying gear lubricants: by spraying, by bath and by hand. With enclosed spur, helical and bevel gear units, at pitch line speeds above about 12.5 m/s, the oil is forced by a pump through special nozzles to spray onto the gear teeth at a pressure of about 0.7–6 bar, depending on the viscosity of the oil and the speed of the gear. Usually special nozzles with deflector surfaces are adopted so that the oil is projected in a fan-shaped spray over the gear teeth. These nozzles are generally pitched at intervals of 75–125 mm across the width of the gear. They are sensitive to

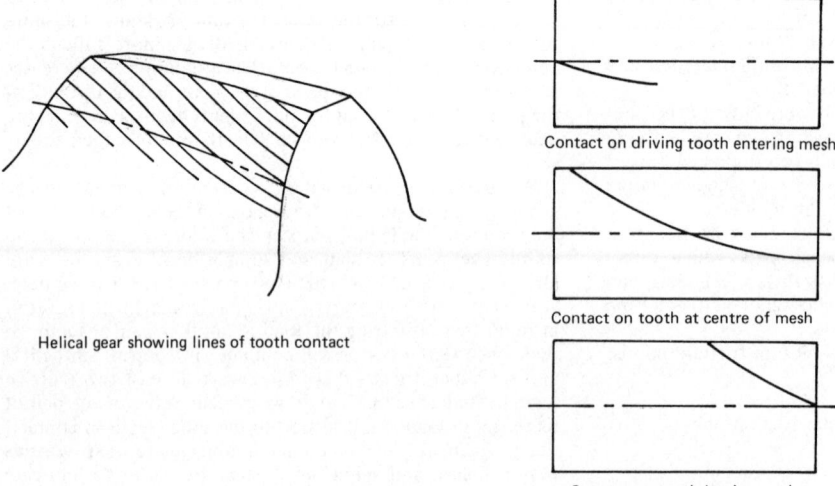

Helical gear showing lines of tooth contact

Contact on driving tooth entering mesh

Contact on tooth at centre of mesh

Contact on tooth leaving mesh

Figure 9.26 Lines of contact – helical gear

Table 9.10 Gear types

Shaft position	Type of gear	Usual materials	Maximum pitch line speed (ms^{-1})	Theoretical tooth contact	Type of sliding	Recommended type of lubricant	Remarks
Parallel	Spur Helical	} Steel/ Steel	} 5 100	} Line	Sliding in transverse plane only, i.e. at right angles or nearly so to line of contact.	Straight, mineral or EP	Full EP only where teeth of case-hardened steel and heavily loaded,
Intersecting	Straight bevel Spiral bevel		2.5 2.5		Relative sliding between two surfaces except at pitch line, where rolling contact only.		EP oils, particularly for heavily or shock loaded gears, e.g. mill gears or where good protection against rustling required.
	Hypoid	Case-hardened steel/case-hardened steel	—	Line	Considerable sliding along line of contact. Surfaces move in diverging directions. Sliding at all positions of contact.	EP	May require special break-in oils for severe service.
Perpendicular and non-intersecting	Worm	Case-hardened steel/phosphor–bronze	20[a]	Line	Considerable sliding along line of contact. Parts of surfaces move in opposite directions. Sliding at all positions of contact.	HVI straight, mineral or compound	EP oil may be used if temperature not over 60°C. Compounded oils preferable for high torque/low speed conditions.
Non-parallel and non-intersecting	Crossed helical[b] (spiral)	Steel/bronze	20[a]	Point	Sliding at all positions of contact.	Ditto	Ditto

[a] Rubbing speed.
[b] Light load only.

blocking by dirt, paint flakes, etc., and must be regularly inspected and cleaned. To minimize blocking, the spray holes are generally not less than approximately 2.5 mm diameter, or if the orifice is a slit, a width of not less than 0.75 mm. To obtain a suitable spray from such a hole, a minimum flow rate of about 0.05 litres per second is necessary. Large and high-speed gears may need a greater rate of oil flow than this minimum and approximately 3×10^{-4} litres per second per kW transmitted may be required for each gear train, or otherwise 1.5×10^{-4} litres per second per kW for both gears and bearings.

The oil is usually directed onto the teeth as they go into mesh, but sometimes for high-speed gears the oil is sprayed onto the teeth as they come out of mesh in order to reduce power loss due to excessive churning in the gear mesh while supplying sufficient oil to the gears to keep them cool. The round jet of oil delivered from the hole in the nozzle is often directed straight onto the gear teeth, where it spreads over the surface. It is possible with high-speed gears that the peripheral speed is greater than the speed of the oil leaving the nozzle, with the result that the tips of the teeth overtake the jet of oil. At very high speeds this has been known to cause tip erosion of the teeth. This can be avoided by directing the oil onto a deflecting surface attached to the nozzle so that the oil jet is spread out into a fan-shaped and less concentrated spray. Worm gears having a surface speed of the worm greater than about 10 m/s are best lubricated by two solid high-pressure jets as distinct from sprays, at right angles, and parallel, to the worm axis.

For slow and medium-speed enclosed gear units below about 12.5 m/s pitch line speed for spur, helical and bevel units and below about 10 m/s for worm surface speed for worm gears, bath lubrication is commonly adopted. At such speeds, provided the extent of dip is not excessive, the churning loss in the bath amounts to only about 0.75% of power transmitted per train compared with a total loss of about 1.5%. The dipping wheel need not be immersed in the oil bath to more than three times the height of the tooth and this should not be exceeded by very much in high-speed units if high temperature rise and power loss is to be avoided. Slow-speed units may have a greater depth of immersion without these disadvantages and therefore can tolerate greater variations in oil level.

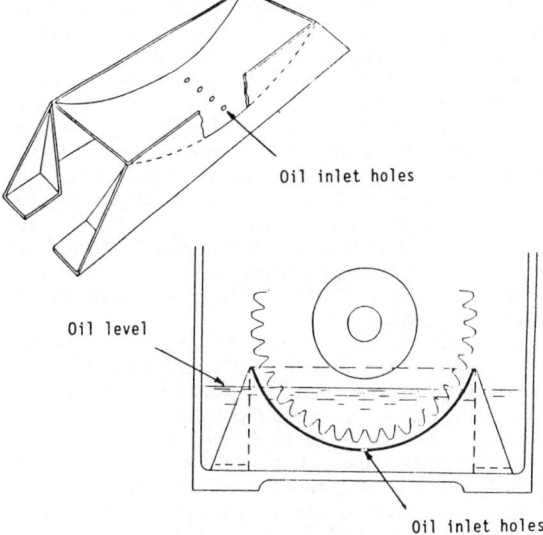

Figure 9.27 Typical trough or shroud to provide restricted dip lubrication

For some applications of helical gears it is feasible to use bath lubrication at speeds in excess of 12.5 m/s by partially enclosing the lower part of the gear in a trough or shroud, as shown in Figure 9.27. Oil is fed from the bath to the gear through a row of holes in the bottom of the casing for lubrication and cooling while minimizing churning.

Where sealing is difficult, low-power gear units are sometimes lubricated with grease. Grease, however, has a number of disadvantages when compared with oil. If the grease is too stiff, it will 'channel', i.e. the gears will cut a channel through it which will not be refilled quickly enough to prevent lubricant starvation and gear failure. Tests have shown that the optimum consistency of the grease to avoid this happening is

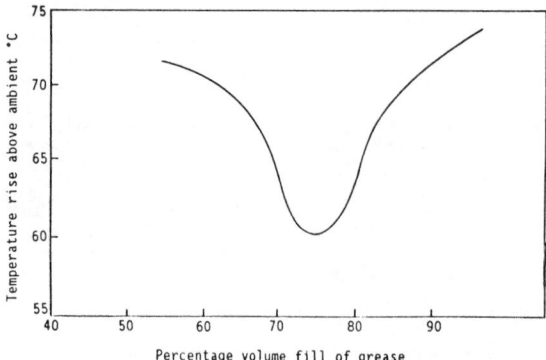

Figure 9.28 Variation of temperature rise with the amount of lubricant used

approximately an NLGI grade 00 or 0. Even greases of this soft consistency will still have a tendency to channel and therefore the grease fill quantity is important. The optimum quantity should be sufficient to ensure adequate lubrication but not too much to increase churning losses and temperature rise. This amount will depend on the design of the gearbox and the orientation in which it is run and can only be determined by carrying out heat-rise tests. A typical graph showing variation of temperature rise with the amount of grease used is given in Figure 9.28.

An important function of a gear lubricant is to carry away the heat generated in the contact region. Greases have poor heat-transmitting properties and are therefore not very efficient in this respect. They also have the disadvantage of retaining abrasive wear debris in the vicinity of the gear teeth. The development of greases based on polyglycol fluids has overcome many of these problems and are now used in small worm and helical gearboxes where a 'lubricated for life' system is required.

Below pitch line speeds of about 2.5 m/s gears are normally not enclosed and are referred to as 'open'. Very slow-speed gear units may be lubricated from small open baths or 'slush pans' by heavy oils or by the more liquid of the bitumen compounds. Otherwise, up to 2.5 m/s, the stiffer bitumen compounds either heated or cut back with solvent may be applied intermittently by hand at intervals of up to about two weeks. It is generally an advantage to apply the lubricant by hand at the end of a working shift when the gears are hot and the lubricant has time in which to set to a tough, durable film.

Intermittent mechanical spray systems are now frequently used, particularly for the higher-speed or less accessible open gears, and are capable of spraying both heavy oils and residual compounds.

9.5.2 Types of gear oils

The principal function of a gear lubricant is to provide a constant film which will effectively reduce the metallic contact between the opposing surfaces, thereby reducing the amount of wear. Normalized and through-hardened gears are satisfactorily lubricated by straight mineral oils. These oils are petroleum products which do not contain additives to enhance their properties, and their ability to keep surfaces separated and so reduce friction and wear depends largely on their viscosity.

In certain applications although straight mineral oils will give adequate wear protection, operating conditions can be

such that additional protection is necessary against corrosion of the metal and oxidation of the oil. In these applications turbine-type oils should be used which are inhibited against rust and oxidation. These oils are often referred to as R&O type oils.

Probably the most commonly used lubricants in industrial gear units are those containing anti-wear or extreme-pressure additives. These oils are necessary where operating conditions are severe or where adequate protection cannot be given by straight mineral oils. With these oils protection is achieved by incorporating so-called boundary lubricants into the oil to produce, by physical absorption or chemical reaction, a film which will be soft and easily sheared but difficult to penetrate or remove from the surfaces.

One such boundary lubricant is long-chain fatty acids, a typical example of which is stearic acid. This forms a closely packed film either by absorption of its acid end groups onto the surface oxide or by reaction with the oxide to form a soap of the gear metal. This type of film is illustrated schematically in Figure 9.29. Unfortunately, the life of all metal soaps is limited by their melting point or desorption temperature, above which the films become ineffective and provide little more protection than a straight mineral oil. For long-chain fatty acids the desorption temperature would be about 100°C and within this limit there will be a very low level of friction. In gear lubrication, this type of boundary additive is used mainly in worm gear lubricants, where the bronze wheel forms a chemically reactive partner and where low friction is especially desirable.

These long-chain fatty acids are not suitable, however, for the more severe conditions which are encountered in some steel–steel gears where loads and sliding speeds are high. For these applications additives are required which form films with higher melting points and have greater adherence to the metal surface. These additives are generally known as 'Extreme Pressure' or EP additives, although a more appropriate description would be 'Extreme Temperature'. These types of

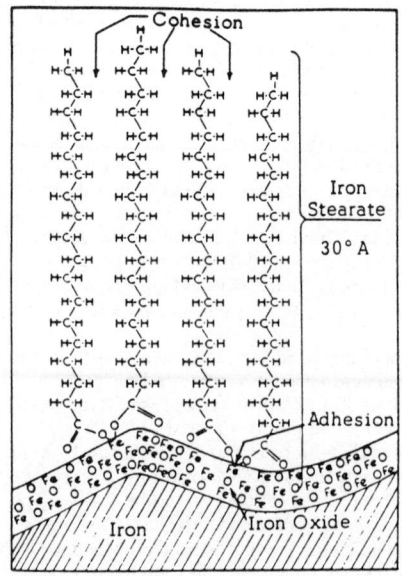

Figure 9.29 Schematic diagram representing chemisorption of stearic acid on an iron surface to form a monolayer of iron stearate with an essentially organic outer surface

additives are usually the more chemically reactive oil-soluble materials containing one or more of the elements chlorine, phosphorus, sulphur, lead and zinc.

The detailed mechanism involved with these additives in practice is seldom fully understood, but it would appear that under mild operating conditions they produce a mainly hydrocarbon film which is firmly attached to the metal surface by their reactive groups. Under more severe conditions, however, these additives decompose and react with the metal surfaces to produce a film which is softer, less brittle, more adherent and therefore more effective than the oxide layer. The films found are largely inorganic (for example, iron phosphates, chlorides and sulphides). This type of film is illustrated schematically in Figure 9.30. Phosphorus-containing additives are generally effective under relatively mild conditions, whereas additives containing chlorine and especially those containing sulphur are required for the more severe operating conditions. Since these additives are not effective until the temperature of reaction is reached, a combination of additives such as sulphur and phosphorus is used to give good anti-wear protection for mild conditions and protection against scuffing when operating under more severe conditions.

To be effective the additive should absorb onto the metal surface and highly polar substances are therefore advantageous. To achieve this economically, the oil should not contain components or other additives which are more polar than the EP additive, since this may give insufficient EP activity or require the EP additive to be used at a higher concentration than would otherwise be necessary.

These additives are soluble in the base oil but additives which are a dispersion in the base oil are also used. In the past, the large particle sizes of these types of additives have given problems of storage instability. However, newly developed additives such as the finely dispersed borate materials have shown very good EP and anti-wear performance combined with very good thermal and oxidation stability. Molybdenum disulfide dispersed in colloidal form in mineral oil has also been shown to give improved efficiency in worm gear speed reducers, but non-colloidal molybdenum disulfide did not give the same improvement. Results obtained with the FZG Gear Tester have shown that in the presence of certain EP additives, the effectiveness of molybdenum disulfide can be inhibited.

9.5.2.1 Synthetic lubricants

Increasing interest is developing in the use of synthetic and semi-synthetic lubricants for gears, particularly where operating conditions are severe or where temperatures are extreme. Where operating temperatures are constantly above 100°C the life of mineral oil-based lubricants is limited due to the increased rate of oxidation and the formation of deposits, which may cause operating problems. At very low operating temperatures the increase in the viscosity of mineral oil-based lubricants may impair its flow properties, resulting in lubricant starvation or channelling. Under these extreme temperatures the viscometric properties of some synthetic lubricants have a distinct advantage. Figure 9.31 gives an indication of the comparative temperature limits for mineral oils and some synthetic fluids. The actual temperature limits will depend on a number of factors, such as viscosity, composition, operating conditions and system design.

Not all the synthetic fluids give improved lubrication performance in gears when compared with mineral oil-based lubricants. An indication of their comparative performance in respect of anti-wear and friction is given in Table 9.11. The use of polyglycols, particularly in worm gears, has many advantages. Its very low frictional properties with the material combination of steel and phosphor-bronze results in lower operating temperatures.

Figure 9.32 shows that considerably higher critical torques at the tooth/flanks can be achieved with the use of polyglycols. The figure also illustrates their potential for energy saving. When considering the use of synthetic fluids to replace mineral oil-based lubricants, compatibility with paints and sealants and miscibility with mineral oil should not be overlooked. Poly-alphaolefins have an advantage in this respect since they are miscible in any proportion with mineral oils.

9.5.3 Heat dissipation

The importance of the surface temperature of gear teeth, as discussed above, underlines the significance of the cooling function of the oil. The quantity of oil necessary to dissipate the heat generated by a pair of gears is usually greater than the quantity required for adequate lubrication.

For splash-lubricated gears the level of oil necessary is determined in the gear design stage and is specified by the gear manufacturer. It is important, however, that this oil level should not be exceeded, since this could increase the churning losses in the oil and thereby the temperature of the gears.

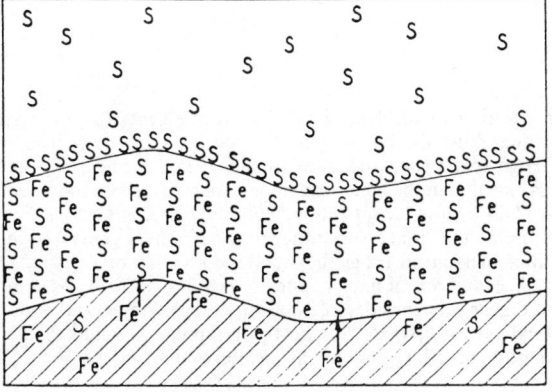

Figure 9.30 Schematic diagram representing inorganic film of iron sulphide on iron formed by sulphur in the oil

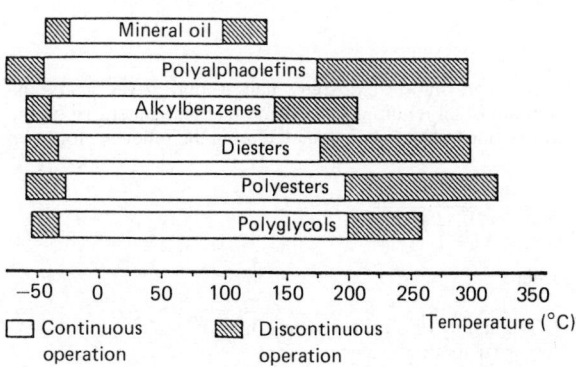

Figure 9.31 Approximate operating temperature limits of various lubricants (according to J. J. Obrzut)

Table 9.11 Comparison of synthetic fluids

	V/T^a behaviour	Anti-wear performance	Friction	Compatibility with paints	Compatibility with sealants	Miscibility with mineral oil	Low-temperature behaviour	High-temperature behaviour
Mineral oil	0	0	+	+++	+++		0	0
Polyalphaolefin	+	0	+	+++	+++	+++	+	++
Alkylbenzene	0	0	+	+++	+++	+++	+	0
Diester	++	0	+	−	0	+	+	+
Polyester	++	0	++	−	0	0	++	+++
Polyglycol	++	+++	+++	+	+	−	+	+++
Phosphoricaciol ester	−	++	++	−	0	−	0	+
Silicone oil	+++	−	−	++	+++	−	+	+

+++ = excellent; ++ = very good; + = good; 0 = sufficient; − = poor.
^{a} V/T = viscosity/temperature.

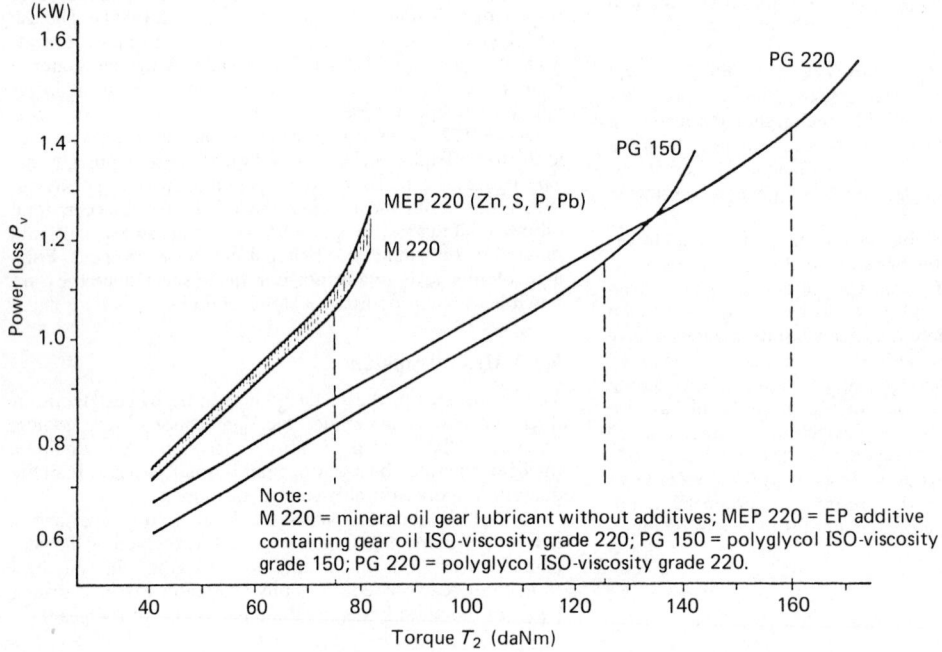

Figure 9.32

For spray-lubricated spur, helical and bevel gears the amount of oil required to dissipate the heat generated by the contacting teeth can be calculated using the following formula:

$$Q = \frac{0.588\,(1 - E)k}{t_2 - t_1}$$

where

Q = quantity of oil (litres/second),
E = gear tooth efficiency,
t_1 = temperature of inlet oil (°C),
t_2 = temperature of outlet oil (°C), and
k = kilowatts transmitted.

In addition to the heat generated by the gears themselves, account must also be taken of that generated by the bearings.

High-speed gears mounted on rolling element bearings are less exacting in their lubrication demands than those mounted on plain bearings, and consequently higher viscosity oils can be used. Also, since very little frictional heat is generated in rolling element bearings they required no additional quantity of oil. However, in plain bearings which are normally used in large gear sets considerable heat can be generated and additional oil to dissipate this heat is required. A guide to the quantity of oil required may be derived from the Petroff equation:

$$Qb = \frac{3.3\,ln^2d^3\eta}{(t_2 - t_1)\,c \times 10^5}\text{litres/second/bearing}$$

where

l = length of bearing (m),
d = diameter of journal (m),
c = clearance (m),
n = journal speed (Hz),
η = oil viscosity (cS at inlet temperature),
t_1 = temperature of inlet oil (°C), and
t_2 = temperature of outlet oil (°C).

To allow for adequate heat disposal and to give an acceptable life for the oil, the quantity in the system should be equal to a minimum of 3 minutes' supply.

The quantity of oil required to dissipate the heat generated in forced lubricated worm gears can be calculated from the following:

$Q = 5C$ litres/second

where C = centre distance of gears (m). Again, the quantity of the oil in the system should not be less than 3 minutes' supply.

9.5.4 Selection of gear lubricants

An oil selected for the lubrication of a gear set should fulfil certain basic requirements which, although simple, are often overlooked. These basic requirements can be summarized as follows:

- To minimize friction and reduce wear between sliding surfaces by the provision of a thin film of lubricant which will prevent metal-to-metal contact between the elements of bearings and the faces of gear teeth at the point of contact;
- To protect the gear set against rust and corrosion;
- To transfer the heat generated in enclosed gears to the gear case walls or to other cooling surfaces.

The ability of the lubricant to satisfy these together with any other special requirements depends on its fundamental characteristics, some of which may be inherent properties of the straight lubricants while others stem from the use of suitable additives. The most important of these properties are reviewed below.

9.5.4.1 Viscosity

Undoubtedly, the most important single property of any gear lubricant is its viscosity or, to be more accurate, its viscosity/temperature/pressure relationship. Irrespective of the types of additive that may be used, the efficient lubrication of a gear train under hydrodynamic conditions is dependent on viscosity.

One of the major considerations in the lubrication of gears is to prevent scuffing of the gear teeth, and it is possible, given sufficient data, to calculate approximately the minimum oil viscosity required to prevent this type of damage. One method is the use of the following equation:

$\nu = 352 \, [F(R + 1)/(bd \, RV)]^{0.415}$

where

ν = viscosity at 38°C (cSt),
F = tooth force (Newtons),
b = gear face width (mm),
d = pinion pitch diameter (mm)
R = gear ratio, and
V = pitch velocity (m/s).

Elasto-hydrodynamic and flash temperature theory can be applied in certain cases but the calculations are somewhat involved and are often outweighed by practical considerations.

Use is therefore often made of the various empirical methods and viscosity recommendations supplied by gear manufacturers. For situations where a viscosity recommendation is not available, use can be made of alternative methods such as the Deutsche Industrie Normen (DIN) system, DIN 51 509 or the American Gear Manufacturers Association, AGMA standard 251.02 for open gears and AGMA standard 250.04 for enclosed gears.

It is common knowledge that, as the viscosity increases, the thicker will be the film of oil that it provides between the mating surfaces, and consequently the greater the protection against scuffing. There are also indications that an increase in viscosity of the oil is accompanied by greater protection against other forms of tooth surface failures such as deformation, peening and, possibly, pitting (see Figure 9.33).

It has been recognized for many years that pitting fatigue life increases with the viscosity of the lubricating oil, but it was assumed that viscosity was important only in determining whether the oil could penetrate and propagate cracks in the manner described above. Recent developments in elasto-hydrodynamic lubrication have shown, however, that viscosity is important in other respects. In contrast with earlier beliefs, it is now accepted that full oil films can exist in heavily loaded contacts and that viscosity is the most important property of the oil in determining the thickness of the films. In any given contact, film thickness is approximately proportionate to (oil viscosity × rolling speed of the surfaces)$^{0.7}$. The significance of this in relation to pitting is that an increase in film thickness might be expected to reduce asperity contact and so raise the fatigue limit and extend fatigue life in the same way as a reduction in surface roughness. Experiments conducted with disk machines at a constant stress greater than the fatigue limit of the materials used show that a very close correlation between tendency to pitting, and the ratio, D, can be derived as follows:

$$D = \frac{\text{Sum of initial peak to valley roughness}}{\text{Theoretical elasto-hydrodynamic film thickness}}$$

The practical implication of these observations is that if a machine is operated at contact stresses greater than the fatigue limit of the softer surface, the onset of pitting may be delayed by using a higher viscosity oil. However, since the loaded

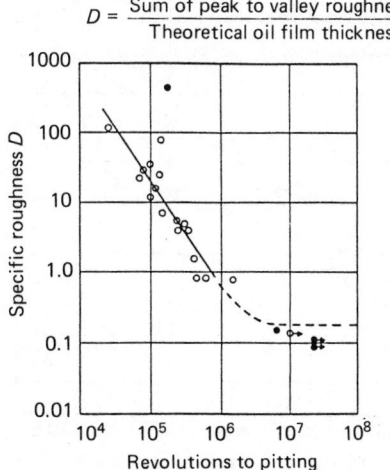

Figure 9.33 Influence of specific surface roughness D on fatigue pitting

surfaces will then tend to run hotter than with a less viscous oil, the potential benefit of using an oil of higher viscosity will not be fully realized.

From these considerations it would appear that the higher the viscosity of the oil, the more effective it should be as a gear lubricant. Unfortunately, this is not so, and to understand why, the disadvantages of using such an oil must be considered. The main disadvantage is that an increase in oil viscosity gives rise to increased power losses and a higher temperature rise of the gear unit and of the oil. At very high temperatures, bearings may fail and the oil may rapidly oxidize and thicken, thereby reducing the life of the oil. In addition, lubrication starvation may occur through channelling or by blockage of oil holes, oil galleries, etc. by insoluble oxidation products.

An extremely important factor in the lubrication of gears is that the viscosity of all gear lubricants is markedly affected by changes in temperature, the viscosity decreasing sharply with increasing temperature and vice versa. As a result, a lubricant may become too thin to give satisfactory service at high temperatures or so thick in cold weather that it may be difficult to start a gear set. In practice, some lubricants are more resistant than others to changes in viscosity with temperature, as shown in Figure 9.34, which compares the effect of temperature on the viscosity of a typically highly refined mineral oil and on a synthetic oil. This resistance to change to viscosity with temperature can be conveniently indicated by the viscosity index, the higher viscosity index number signifying the greater resistance to change. All oils have a natural viscosity index which can be improved by incorporating an additive suitable for this purpose.

From these considerations it is obvious that in cases where gears are required to operate over a wide temperature range it is essential to use lubricants with a high viscosity index to ensure satisfactory performance and the avoidance of difficulties arising from channelling, excessive pressure drops in circulating systems and excessive starting torque.

Another point in connection with viscosity is that it is now becoming increasingly important for the life of an oil charge to be as long as possible, because of not only the cost of the lubricant but also the down-time involved in carrying out oil changes. To a large extent, the selection of the correct viscosity of the oil for a gear unit can be a strong contributory factor in determining the life of the oil. This is because low-viscosity oils are able to shed contaminants more quickly or easily than high-viscosity oils. The usual contaminants encountered in enclosed gear units are dust, scale, abrasive metal particles and water. With solid contaminants, settling is quicker in thinner oils, which assists in reducing the chances of abrasive wear of teeth and bearings. Furthermore, filtration is easier with thin oils as a fine-mesh can be used.

It should also be noted that, as a general rule, low-viscosity oils have less tendency to foam than high-viscosity oils, the rate of release of air entrained by the oil being inversely proportional to the viscosity. At the same time, the demulsibility of an oil, i.e. its ability to separate from water, is better with low-viscosity oils than with high-viscosity oils.

It can be seen, therefore, that the use of an oil of correct viscosity is very important and that it can, in certain gear units, be critical. For this reason, much time and effort is still being spent on investigating means of calculating the correct viscosity for any gear unit.

9.5.4.2 Anti-foaming properties

With medium- and high-speed gears some foaming of the oil is inevitable, and this condition is particularly aggravated in circulating oil systems. Although, as already mentioned, low-viscosity oils can normally release entrained air rapidly, occasions can arise when the rate of release is insufficient. In such cases, oils containing anti-foaming additives may be used.

9.5.4.3 Anti-corrosion properties

The prevention of rust and corrosion in the gear unit must always be considered and it may, in fact, be the determining factor in the choice of an oil. Gears operating in a moist or wet atmosphere are easily recognizable and can be catered for by the selection of a suitable lubricant, but moisture can also be present in a gearbox when operating in apparently dry conditions. This moisture is due to condensation in the box as a result of cooling of the unit when not in use, and it can be particularly troublesome when the gears are inoperative for long periods.

If the oil is to prevent corrosion it must be able to deny the access of moisture to the metal surfaces. In other words, it must have good metal-wetting properties. New straight mineral oils are limited in this respect but their metal-wetting properties generally improve soon after use. This is due to the small amounts of products caused by early deterioration, including the polar compounds which promote metal-wetting. Where a high degree of resistance to rust and corrosion is required, it is common practice to add to the oil special rust inhibitors, the molecules of which are strongly polar and have the property of attaching themselves firmly to the metal, thereby promoting the displacement of moisture at the surface.

9.5.4.4 Oxidation resistance

Oils exposed to high temperatures and contact with air will inevitably oxidize the rate and extent of oxidation, depending on the chemical composition of the oil and the conditions of exposure, i.e. the higher the temperature and the greater the exposure to air, the more rapid will be the reaction. Contact with certain metals and metallic compounds, particularly copper and oxides of iron, will also speed up the reaction.

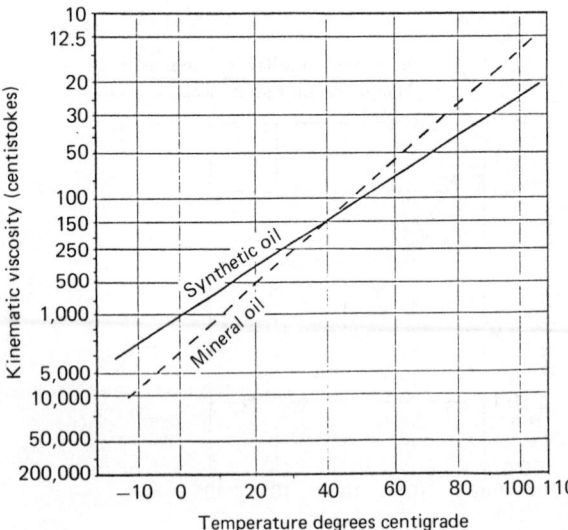

Figure 9.34 Viscosity as a function of temperature for a typical highly refined straight mineral oil and a typical synthetic gear oil

Oxidation results in a formation of oil-soluble organic acids and insoluble compounds which appear as lacquer, sludge and carbonaceous deposits, with a consequent reduction in the life of the lubricant. In some gear systems, the operating conditions may increase the rate of oxidation appreciably, in which case it is necessary to use oils containing oxidation inhibitors.

In gears lubricated by oil spray, the mixture of air and oil passing through the mesh of high-speed gears can be considerably heated by compression. Temperatures as high as 160°C have been measured in gears running at 150 m/s, although tooth temperatures were only around 110°C.

9.5.4.5 Anti-wear properties

Since the highest viscosity oil that can be used is limited by the considerations previously reviewed, the margin of safety against wear with straight mineral oils is less in some gears than in others. Straight mineral oils may therefore prove inadequate under certain operating conditions (for example, during the running-in of some gears or in subsequent service with heavily loaded gears). To meet these requirements, oils containing anti-wear additives are used. The action of these additives has been discussed above.

9.5.4.6 Extreme-pressure properties

Under severe conditions of loading, anti-wear additives are ineffective and it is therefore necessary to utilize the more active substances designated extreme pressure (EP) additives. Depending on the degree of reactivity conferred, lubricants embodying these additives are known as mild extreme-pressure oils and full extreme-pressure oils, the latter being intended for very severe conditions. Extreme-pressure additives are stable at normal ambient temperatures but decompose at the high local temperatures which are likely to occur at the point of contact between heavily loaded gear teeth. The additives then react with the metal to produce compounds which form a film on the gear teeth which is more readily sheared, thereby preventing damage to the metal.

9.5.5 Service life of gear lubricants

In terms of the physical and chemical properties of gear lubricants, it can be seen that, apart from giving the desired performance characteristics, the careful selection of viscosity and appropriate additives can materially increase their service life. There are, however, other factors which can have a marked effect on service life and which are mainly related to the conditions to which the lubricant is subjected in use. Of these factors, the most important are oxidation and contamination. It has been pointed out that the rate of oxidation is governed by the operating temperature and the extent of aeration of the oil. Systems should therefore be designed to reduce the influence of these factors to a minimum.

For example, insofar as an operating temperature is concerned, it should be noted that some systems, particularly large gear sets in steel works, utilize heavy gear oils which must be heated to facilitate pumping. In some systems it is important that the heaters used for this purpose are carefully selected to avoid the possibility of local overheating and cracking of the oil. The greatest danger arises when the heaters are functioning but the oil is not circulating, in which case the rate of flow of the oil over the heated surfaces will be low, as it will depend on natural convection. To combat the risk of overheating, the ratings of heaters should therefore be in accordance with the type of oil concerned. Thus, with straight mineral oils a heater rating of approximately 520 kW m^{-2} of heater surface is usually recommended. Much

lower ratings are necessary with additive-type oils, particularly the extreme-pressure leaded oils, and these ratings can be as low as 160 kW m^{-2} with the heaviest grades.

A major requirement in regard to excessive aeration of the oil is to maintain the ingress of air into the system at a minimum. Probably the most common cause of turbulence in an oil tank is unsuitable design of oil-return pipes. These should be carried to a point well below the surface of the oil to eliminate the possibility of the returning oil splashing into the tank and creating excessive turbulence and entrainment of air. In addition, the tank should be designed to permit air bubbles to rise freely to the surface, and this can be accomplished by means of baffles which divide the tank into inlet and outlet sections and rise vertically from the bottom of the tank to a point approximately one-third of the depth of the oil below the surface.

Entrainment of air will also be reduced by locating the suction connection of the tank well below the surface of the oil and as far as possible from the return pipe. However, to minimize contamination, care must be taken not to locate the suction connection so low in the tank that water and other impurities can be collected from the tank bottom. It is also advisable to fit a drain cock at the lowest point in the tank to enable impurities to be drained off at regular intervals.

The service life of an oil can also be affected by the presence of certain materials in the system. For instance, although copper is widely used owing to its strong catalytic oxidation effect on the oil it is preferable that some alternative material be used. Also, components with galvanized and other zinc-coated surfaces should not be allowed to come into contact with the oil, as the zinc may react with the additives it contains, particularly those present in many of the extreme-pressure oils.

9.6 Rolling element bearings

9.6.1 Introduction

The use of rolling elements, balls or some form of cylinder to separate the two moving components within a mechanical system has been universally accepted by design engineers since the turn of the twentieth century. The development of superior bearing steels over the past 40 years, constant improvement in manufacturing techniques and the detailed development of many different bearing types and variants has resulted in the universal availability of standard bearing ranges of outstanding performance at relatively low cost.

In terms of their speed capability, rolling element bearings are ideally applied in the mid-ground of bearing usage between rubbing plain bearings and fluid film plain bearings, providing good load capacity for their projected area. They have excellent friction characteristics and very low system power consumption. The rolling contact is inherently low friction and the majority of assemblies are self-contained requiring no external oil circulation system.

Rolling element bearings, correctly applied, are zero-wear systems, hence high location accuracy of a shaft can be achieved and maintained over a long period of time. The great majority of rolling bearings throughout the world are metric sized; typically, bore sizes in 5 mm steps, i.e. 25 mm, 30 mm, 35 mm, etc. Inch-sized bearings are still available for replacement purposes, particularly deep-groove ball bearings and taper roller bearings, but their use for new equipment is not recommended. Figure 9.35 shows the terminology generally used. The cage is not universally fitted as some low-speed, high-load capability bearings are completely filled with rolling elements. This is known as a crowded assembly.

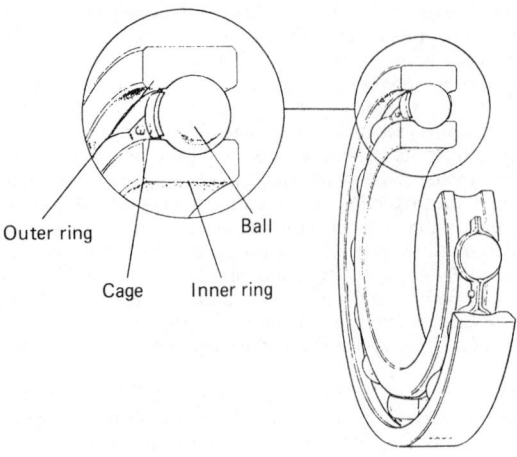

Figure 9.35 Bearing terminology

9.6.2 Types

All rolling element bearings can be divided into two main categories, i.e. those using balls as the rolling element and those using a roller of some form as the rolling element. Ball bearings basically have point contact between rolling elements and rings, implying a lower load capacity for the same sized bearing than roller bearings, which all have a line contact between rolling elements and rings.

9.6.2.1 Deep-groove ball bearings

Deep-groove ball bearings are the most common type and are particularly used in low to moderate load applications such as domestic appliance electric motors. Ball bearings are non-separable, usually with two-piece pressed steel cages riveted or welded together. They are designed for high radial load capability, and the high conformity between balls and raceway grooves enables them to carry considerable axial loads in either direction.

Ball bearings have been particularly developed in recent years for their inherent low noise characteristics, high accuracy and the availability of sealed and greased-for-life variants. Contacting seals and non-contacting shields are usually available (Figure 9.36), the choice depending upon the friction characteristics required and the severity of the external environment.

Ball bearings have a high maximum speed capability (e.g. a 608 bearing, 8 mm bore, has a nominal speed limit of 32 000 rpm using grease lubrication). Larger bearings have lower speed limits (e.g. a 6207 bearing, 35 mm bore, has a

nominal speed limit of 9000 rpm for grease lubrication). In general, the bearing manufacturer's stated limiting speed values can be exceeded with careful attention to lubrication and heat extraction.

9.6.2.2 Angular-contact ball bearings

Angular-contact ball bearings are similar to deep-groove ball bearings except that the line of action of the load acting through the contacts between balls and raceways forms an angle with the bearing axis. The bearings are therefore particularly suitable for carrying combined (axial and radial) loads. Several different contact angles are available but the 40° version is a generally available variant, giving excellent axial and moment load capability while retaining sufficient radial load capacity. Angular-contact bearings are often used in pairs mounted side by side; back-to-back or face-to-face to carry axial loads in both directions, or tandem to carry very high axial loads (see Figure 9.37). It is essential that the bearings are specified as suitable for pairing by the manufacturer to ensure correct sharing of the loads. A typical application of angular-contact bearings is in centrifugal pumps, where a back-to-back arrangement is often used to give moment stiffness to an overhung shaft load.

Cages are available in pressed steel, reinforced polyamide, solid brass, and, less commonly, pressed brass and solid steel. There is a general trend towards reinforced polyamide giving excellent dynamic features. A steady-state temperature limit of 110–120°C can be problematic in high-temperature environments. The pressed steel or solid brass cages may then be preferred.

Angular-contact ball bearings have excellent high-speed capability, low-contact angle variants for the aerospace industry being the fastest bearings currently in use.

9.6.2.3 Self-aligning ball bearings

These bearings have two rows of balls and a common sphered raceway in the outer ring (Figure 9.38). The latter feature gives the bearing its self-aligning property, permitting minor angular misalignment of the shaft relative to the housing – both static and dynamic. The bearings are particularly suitable for applications where misalignment can arise from errors in mounting, due to structural inaccuracy or from shaft deflection. The self-aligning ball bearing has the high-speed capability of all ball bearings but only a moderate load capability, which limits the possible uses of these bearings. They are, however, widely used in applications such as

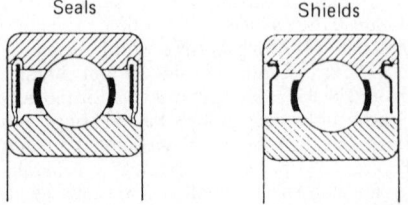

Figure 9.36 Contacting seals, non-contacting shields

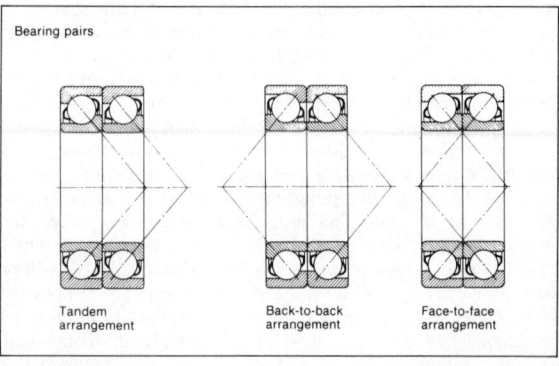

Figure 9.37 Angular-contact ball bearing paired arrangements

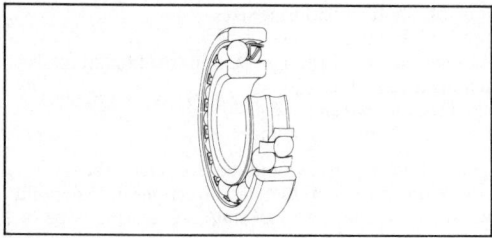

Figure 9.38 Self-aligning ball bearing

high-speed fan assemblies which are required to run with long relubrication intervals.

The self-aligning ball bearing is commonly available with cylindrical or tapered bores. The tapered bore variant can be fitted, using an adaptor sleeve, onto a parallel shaft with no machining or further fastening required. This feature is available on some other bearing types.

9.6.2.4 Cylindrical roller bearings

A line contact bearing is one in which the cylindrical rollers are axially guided between integral flanges on one bearing ring (Figure 9.39). The flanged ring with the roller and cage assembly forms a unit which can be removed from the other ring. This separable feature of the bearing design facilitates mounting, particularly where interference fits for both rings are necessary.

Cylindrical roller bearings have exceptionally low friction characteristics, allowing high-speed operation and a high radial load capability. The variant with thrust flanges only on one ring, type N or NU, is free to float axially and can be used in a shaft system to allow for axial displacement, perhaps due to temperature effects. These are known as non-locating bearings. Type NJ with one thrust flange on the inner ring as well as two flanges on the outer ring is used as a locating bearing and can be used to carry axial load. Type NUP, having

a full set of flanges, may be used to locate the shaft in both directions or carry axial loading in both directions. Recent developments in cylindrical roller bearings have significantly increased the axial load capability but actual capacity values vary from one manufacturer to another. Reinforced polyamide cages are widely used. These suffer from the temperature limitation of 110–120°C indicated above but have otherwise been found to be excellent in service. Pressed steel cages would be the normal substitute for the polyamide in high-temperature environments. Full-complement cylindrical roller bearings have exceptionally high radial load capability and are particularly suitable for high-load, low-speed applications. The high friction inherent in the crowded assembly considerably reduces the maximum speed capability.

9.6.2.5 Taper roller bearings

The single-row taper roller bearing (Figure 9.40) has the ability to carry combinations of large radial and thrust loads or to carry thrust loads only. In the taper roller bearing the line of action of the resultant load forms an angle with the bearing axis. There is a difference in the inner and outer raceway contact angles, resulting in a force component which drives the tapered rollers against the guide flange. Because of the relatively large sliding friction generated at this flange, the bearing is not generally suitable for high-speed operations. The bearings are of separable design. The outer ring (cup) and the inner ring with roller and cage assembly (cone) may be mounted separately.

Single-row taper roller bearings can carry axial loads in one direction only. A radial load imposed on the bearing gives rise to an induced axial load which must be counteracted. The bearing is therefore generally adjusted against a second taper roller bearing. The setting of axial end float or preload of the bearing pair is critical to the success of the application. A taper bearing arrangement has high moment stiffness and is therefore frequently applied in areas, such as gearboxes, where correct alignment is important.

Double and four-row taper bearings are also available for specific, very high load applications. These usually have a built-in clearance requiring no adjustment.

9.6.2.6 Spherical roller bearings

Spherical roller bearings have two rows of rollers which run on a common sphered raceway in the outer ring (Figure 9.41). The two inner ring raceways are inclined at an angle to the bearing axis. The bearings are self-aligning and insensitive to minor misalignments of the shaft. In addition to radial loads, the bearing can also accommodate axial loads in both directions but cannot support moment loading. Spherical roller bearings are most commonly applied in heavy engineering applications where the misalignment capability, high load

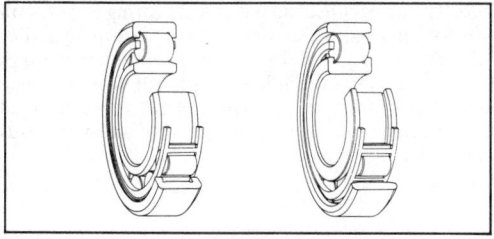

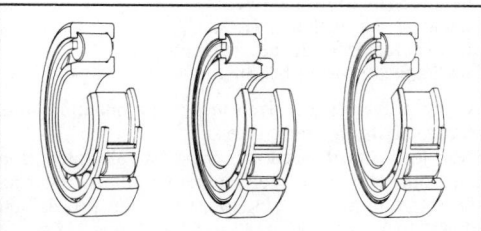

Figure 9.39 Cylindrical roller bearing flange variants

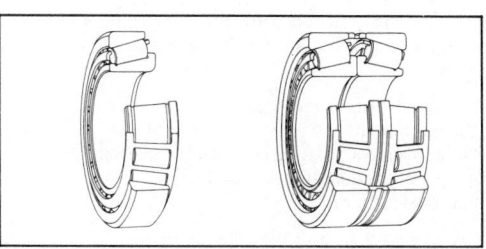

Figure 9.40 Taper roller bearing. Single row and paired

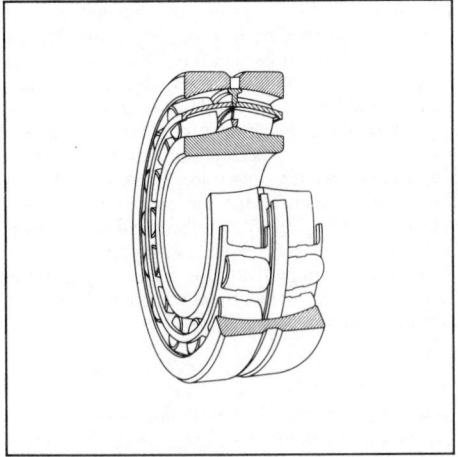

Figure 9.41   Spherical roller bearing

capacity and general robustness are most appreciated. They are manufactured in a very wide size range (e.g. 20 mm bore to 1180 mm as standard, or larger to special order from one major manufacturer).

9.6.2.7 Needle roller bearings

A needle roller bearing is a cylindrical roller bearing having rollers of considerably greater length than diameter. They are designed to fit into applications where radial space is at a premium. In some applications the needles run directly on a hardened shaft, saving the space of the inner ring.

Generally, needle roller bearings are less accurate than cylindrical roller bearings and have considerably higher friction. They are most suited to applications where the loading is light or intermittent.

9.6.2.8 Thrust bearings

The bearing types described in Sections 9.6.2.1 to 9.6.2.7 are basically designed to sustain radial loads, albeit with a considerable thrust capacity in many cases. In a situation where the thrust (axial) load is greater than the radial, then one of the thrust bearings is often preferable. Both ball and roller versions are available. Generally, the speed capability of the bearings is much reduced compared to their radial equivalents and the friction increased.

9.6.3 Selection

The selection of the correct bearing for a particular application requires consideration of several main parameters for the bearing type:

- Misalignment/stiffness requirements
- Speed requirement
- Available space and existing geometric constraints
- Mounting and dismounting considerations
- Friction consideration.

Table 9.12 gives an overall assessment of the main bearing types. Additionally, to decide on the precise bearing size and designation we must consider:

- The possible shaft and housing sizes
- The service life requirement
- The lubrication and, if applicable, relubrication intervals
- The sealing arrangements
- The axial location arrangements
- Static and shock/impact loading.

The general procedure would be to select the type of bearing most appropriate to the load directions in the application, bearing in mind that the simplest bearing type will usually be the easiest to adjust and be most economical. The bore size of the bearing is usually given by the shaft diameter. Consideration of the service life requirement will indicate the minimum basic dynamic load rating of a suitable bearing, and a nomogram (Figure 9.42) will be of use. Knowing the speed of the bearing and the required service life in hours (or millions of revolutions), the ratio C/P can be found from the chart, where C = basic dynamic load rating of the bearing (N) as defined in ISO 281 and P = equivalent dynamic load acting on the bearing (N).

Typical service life requirements for types of machines are given in Table 9.13. Knowing the working load, P, the minimum dynamic capacity, C, necessary to do the job can be easily determined. Inspection of the bearing tables from the manufacturers will indicate which bearing section will be appropriate.

Further consideration must then be given to manufactured clearance for a one-piece bearing, grease type and seal execution for self-contained bearings and other optional features to obtain the exact bearing designation. More information is given in subsequent sections.

9.6.4 Bearing life

Bearing life calculations carried out for design and selection purposes have been based upon a prediction of a rolling contact, subsurface initiated, spall or pit in the raceway. The method was first described by Lundberg and Palmgren (1952) and is specified in ISO 281.

The basic rating life of a rolling bearing is the life in millions of revolutions associated with 90% reliability of a bearing population, i.e. an L_{10} life is the life to failure of 10% of the population. The median life (L_{50}) is normally about five times the L_{10} life. The actual life of an individual bearing is therefore meaningless when used to describe bearing performance. For radial ball bearings $L_{10} = (C/P)^3$ and for radial roller bearings $L_{10} = (C/P)^{10/3}$. P is the radial load which will have the same effect upon the bearing as the actual radial and axial load conditions. It can be calculated from the radial and axial loads using X and Y factors:

$$P = XF_r + YF_a$$

where

P = equivalent dynamic bearing load (N),
F_r = actual radial bearing load (N),
F_a = actual axial bearing load (N),
X = radial load factor for the bearing, and
Y = axial load factor for the bearing.

Actual X and Y factors are given in bearing manufacturers' literature and vary from type to type.

Often the loads on the bearings must be calculated from gear loads, out of balance, etc. external to the bearing. Computer programs are available to assist in this process and can be applied to most situations.

The basic rating life is a conservative estimate of the service life of good-quality bearings used with good-quality, clean

Table 9.12 Selection guide for rolling bearings

Bearing type	No. of rows	Load capability Radial	Axial	Allowable misalignment (degrees)	Comments
Deep-groove ball bearing	1	Medium	Medium	0.01–0.05	General-purpose. Widely used, particularly smaller sizes.
Self-aligning ball bearing	2	Medium	Light	2.0–3.0	High speed with misalignment capability.
Angular-contact ball bearing	1	Medium	Medium-heavy	0	Frequently used in matched pairs for combined loads.
Cylindrical roller bearing Cage	1	Heavy	Light[a]	0.03	Heavy radial loads, high speed. Locating and non-locating variants.
Full-complement	1	Very heavy	Light	0.03	Very heavy radial loads at low speed.
Taper rolling bearing	1	Heavy	Medium-heavy	—	Mounted as opposing pair. Very stiff assembly.
Spherical roller bearing	2	Very heavy	Light-medium	1.5–3.0	High radial loads with misalignment, reasonable speed capability.
Needle roller bearing	1	Light-medium	Zero	0	Excellent radial load capability for section height.
Thrust ball bearing	1	Zero	Light-medium	0	Single direction thrust, no radial capability.
Spherical roller thrust bearing	1	Medium[b]	Heavy	1.5–3.0	Single direction thrust with some radial, takes misalignment.

[a] Suitable flanged designs variant required.
[b] Radial load must not exceed 55% of simultaneous axial load.

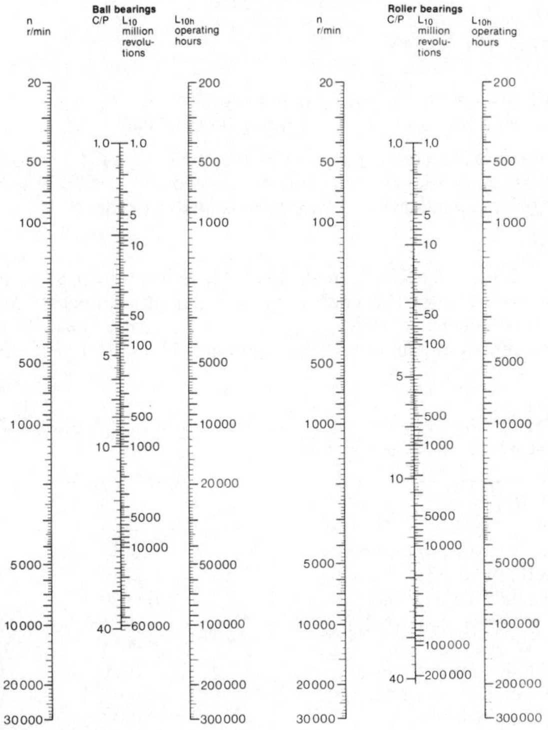

Figure 9.42 Life calculation chart

Table 9.13 Typical service life requirements (L_{10} hours)

Machine	Operating hours L_{10}
Domestic appliances Farm equipment	300–3000
Intermittent use Electric hand tools Lifting tackle Construction machinery	3000–8000
Machines used 8 h/day but not fully utilized General-purpose gearboxes Electric motors	10 000–25 000
Machines used 8 h/day and fully utilized Machine tools Conveyors, printing equipment	20 000–30 000
Machines for 24 h/day use Rolling mill gears Compressors Pumps	40 000–50 000
Water works machinery Rotary furnaces	60 000–100 000
Pulp and papermaking industry Power station plant, mine pumps and mine ventilator fans	100 000

Table 9.14 Life adjustment factor for reliability, a_1

Reliability (%)	a_1
50	5
90	1
95	0.62
96	0.53
97	0.44
98	0.33
99	0.21

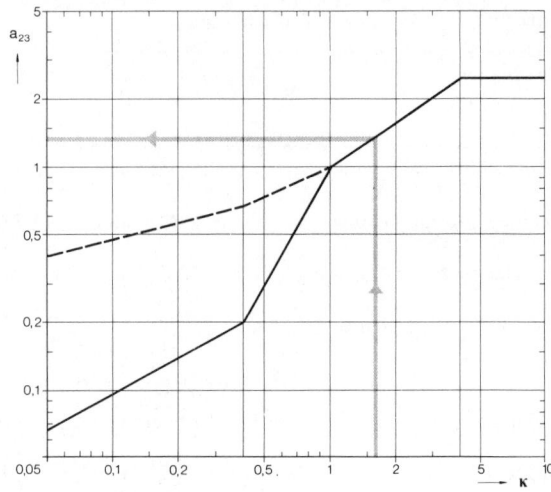

Figure 9.43 Life adjustment factor a_{23} derived from viscosity ratio κ

lubricant. Where the real bearing operating conditions can be more exactly described an accurate modified life equation may be used:

$$L_{\mathrm{na}} = a_1 a_2 a_3 \left(\frac{C}{P}\right)^p$$

where

L_{na} = adjusted rating life (M revs),
a_1 = life adjustment factor for reliability (Table 9.14),
a_2 = life adjustment factor for material, and
a_3 = life adjustment factor for operating conditions.

The a_2 material factor is accepted as 1 for modern bearings accurately manufactured from clean chromium steel. Values higher than 1 may be appropriate for special steels or heat treatments. The a_3 factor is essentially determined by bearing lubrication quality provided that temperatures are not greater than 150°C. The lubricant provides an oil film which ideally keeps the rolling contact surfaces of the bearing apart. The fatigue life of the bearing lengthens as the separation of the surfaces increases. Under the conditions prevailing in an adequately sealed rolling bearing the a_3 factor is based on the viscosity ratio, κ. This is defined as the ratio of the actual viscosity of the lubricant in the bearing, V, to the viscosity V_1 required for adequate lubrication. Both viscosity values are at the operating temperature of the bearing, which is usually the highest temperature within the bearing system. The determination of V_1 is described in Section 9.6.7. Both factors a_2 and a_3 depend, to some extent, on the bearing manufacturer's processes – quality, steel cleanliness, heat treatment, raceway surface roughness specifications. Some manufacturers prefer to combine a_2 and a_3 and give an a_{23} factor, others use only the a_3 factor. A sample plot of the effect of the viscosity ratio κ on the life adjustment factor a_{23} is shown in Figure 9.43.

In practice, the calculated life is often exceeded by bearings running lightly loaded in clean conditions with good lubrication (e.g. a sealed and greased-for-life bearing in an electric motor). However, bearings running in contaminated conditions, such that solid or liquid contamination works its way into them, will frequently achieve a much inferior life which may be reduced to only 10% of the calculated value. Heavily loaded bearings running in clean conditions probably achieve their calculated life. A calculation method taking the effect of contamination into account and introducing a fatigue limit stress is now available and, provided sufficient detailed information of the application is available, will give a more exact service life prediction.

9.6.5 Bearing friction and speed limits

9.6.5.1 Friction

The friction within rolling element bearings is dependent upon several factors, of which the bearing type and specific design features, the load, the lubricant properties and the rotational speed are the most important. The coefficient of friction varies for different bearing types (see Table 9.15). As a guide, the frictional moment, M, can be calculated as

$$M = uP \; \frac{d}{2} \quad \text{(Nm)}$$

where u = coefficient of friction for the bearing type, P = bearing load (N) and d = bearing bore diameter (m).

For more accurate calculations the total frictional moment must be considered as having two components: M_0 (which is load independent) and M_1 (which is load dependent):

$$M = M_0 + M_1$$

M_0 predominates in lightly loaded bearings at high speed as it results from the energy losses within the lubricant. M_1 predominates in slowly rotating bearings under load as it results from the elastic deformation and local slip in the

Table 9.15 Coefficient of friction

Bearing type	u
Deep-groove ball bearings	0.0015
Self-aligning ball bearings	0.0010
Angular-contact ball bearings	0.0020
Cylindrical roller bearings	
With cage	0.0011
Full-complement	0.0020
Spherical roller bearings	0.0018
Thrust ball bearings	0.0013
Spherical roller thrust bearings	0.0018

raceway contacts. While accurate calculation of M_0 and M_1 is possible, knowing the precise application conditions and the internal design of the bearing, it is best left to the bearing manufacturers.

It should be noted that rubbing seals can often produce more friction than the bearings they are protecting. This is particularly true of small deep-groove ball bearings where non-contacting shields may be the best low-friction seal solution, even though lubricant loss or contaminant ingress may then be problematical.

9.6.5.2 Speed limits

The limiting factor for the maximum speed of operation of ball and roller bearings is primarily the permissible operating temperature of the lubricant. The lubricant is heated by the bearing friction, hence higher speeds can generally be attained from a lower friction bearing or, indeed, if the heat can be removed by cooling, so that the lubricant does not reach its permissible temperature limit.

The limiting speeds for oil and grease lubrication are given in manufacturers' tables. These usually apply to a so-called 'normal' application situation and can be exceeded provided attention is given to cooling or use of a lubricant with a higher temperature limit. Only very rarely is the maximum speed for a bearing the result of a structural strength limitation.

9.6.6 Application

9.6.6.1 Axial location

The bearing arrangement of a rotating shaft usually requires at least two bearings to support and locate the shaft relative to the stationary part of the machine. One bearing should be locating, providing radial support and axial guidance in both directions, and the other non-locating, giving only radial support. Only bearings capable of taking both radial and axial loads are suitable for the location position. The non-locating bearing must permit axial displacement either by movement of the whole bearing in the housing or by movement within the bearing as for NU- or N-type cylindrical roller bearings. Examples of locating and non-locating arrangements are shown in Figure 9.44. The arrangement where each of the two bearings supports the shaft axially in one direction is known as cross-location and is used on short shafts with, principally, taper roller bearings to provide a very stiff arrangement with very precise location for a gear mesh.

The ring with the tightest fit (usually the inner one) must always be axially secured. The other ring on locating bearings must also be axially secured. On non-locating bearings the loose ring must be axially free to attain its own equilibrium position. Inner ring axial location is usually by means of a shoulder on the shaft on one side and a locknut and washer on the other. Lightly loaded bearings can be located using circlips and larger bearings using an end plate on the shaft end. Location onto a parallel shaft is carried out using an adaptor sleeve pulling into a taper bore bearing. This system does have an axial load limitation, dependent on the friction between shaft and sleeve.

Outer ring axial location is usually by means of the housing end cover but a threaded ring screwed into the housing bore or an external circlip can also be used.

9.6.6.2 Radial location

The inner and outer rings of rolling bearings must be correctly secured on the shaft or in the housing to prevent the 'progression' of the ring and the onset of fretting damage and

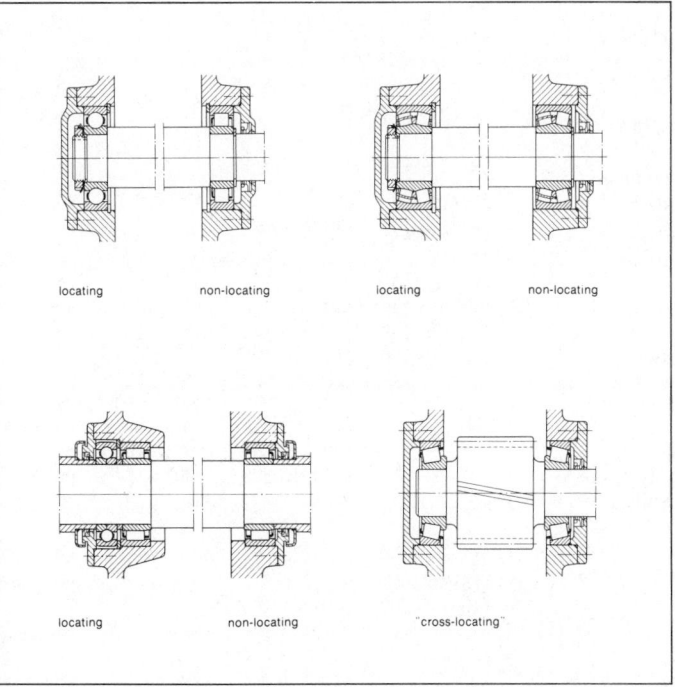

Figure 9.44 Examples of locating and non-locating bearing arrangements

possible overheating. Generally, a machine arrangement with a rotating shaft and steady radial (and axial) loads on the bearings will require an interference fit on the inner ring/shaft position and a clearance fit on the outer ring. If the direction of load is indeterminate or if heavy loads are involved, it is desirable that both rings have an interference fit. This requires the use of either separable bearings or non-separable bearings with tapered bores and adaptor sleeves to be used. Recommended fits for solid shafts and housings are given in Tables 9.16 and 9.17. Rolling bearings of the same size from different manufacturers made to ISO 281 standards will have common external dimensions and tolerances, and should therefore be interchangeable without the need to modify shaft or housing tolerances.

An interference fit on one or both of the bearing rings means that the manufactured internal clearance is reduced. Clearance must therefore be selected with the reduction in mind so that an internal clearance will remain after assembly. This clearance will in turn generally be further reduced during operation by thermal expansion of the bearing inner ring. The extent of expansions can be accurately predicted using computer programs.

A bearing fitted onto a hollow shaft will, in principle, require a heavier interference fit than that used if the shaft were solid to achieve the same contact pressure between inner ring and shaft. There is generally no correction necessary to the fit until the shaft wall thickness is less than 0.25 of the shaft outside diameter.

9.6.7 Lubrication

Effective lubrication is essential for rolling bearings to achieve their performance potential. Lubrication prevents the metal-to-metal contact between rolling elements, raceways and cage which would otherwise result in wear, friction and poor service life. The lubricating properties of an oil or grease deteriorate with time as a result of mechanical working and ageing, and will therefore require replenishment at some time.

Grease lubrication is generally used when rolling bearings operate under moderate conditions of load, speed and environmental temperature. Grease is particularly convenient to handle in service and is less likely to leak – all reasons for its use for the majority of bearings. Housings should not be fully packed with grease, and a 30–50% fill of the free volume is recommended. When selecting a grease, the consistency, temperature range and the rust-inhibiting factors must be considered. Consistency is quoted in terms of the National Lubricating Grease Institute (NLGI) scale. Metallic soap

Table 9.16 Recommended fits for solid steel shafts: radial bearings with cylindrical bore

Conditions	Examples	Shaft diameter (mm)			Tolerance
		Ball bearings	Cylindrical and taper roller bearings	Spherical roller bearings	
Stationary inner ring load					
Easy axial displacement of inner ring on shaft desirable	Wheels on non-rotating axles				g6
Easy axial displacement of inner ring on shaft unnecessary	Tension pulleys, rope sheaves				h6
Rotating inner ring load or direction of loading indeterminate					
Light and variable loads ($P \leqslant 0.06$ C)	Conveyors, lightly loaded gearbox bearings	(18) to 100 (100) to 140	40 (40 to 100	— —	j6 k6
Normal loads and heavy loads ($P > 0.06$ C)	Bearings applications generally electric motors, turbines, pumps, internal combustion engines, gearing woodworking machines	18 (18) to 100 (100) to 140 (140) to 200 (200) to 280 — — —	— 40 (40) to 100 (100) to 140 (140) to 200 (200) to 400 — —	— 40 (40) to 65 (65) to 100 (100) to 140 (140) to 280 (280) to 500 500	j5 k5 m5 m6 n6 p6 r6 r7
Very heavy loads and shock loads with difficult working conditions ($P > 0.12$ C)	Axleboxes for heavy railway vehicles, traction motors, rolling mills	— — —	(50) to 140 (140) to 200 200	(50) to 100 (100) to 140 140	n6 p6 r6
High demands on running accuracy with light loads ($P \leqslant 0.06$ C)	Machine tools	18 (18) to 100 (100) to 200 —	— 40 (40) to 140 (140) to 200	— — — —	h5 j5 k5 m5
Axial loads only					
	Bearing applications of all kinds	250 250	250 250	250 250	j6 js6

Table 9.17 Recommended fits for cast iron and steel housings: radial bearings – solid housings

Conditions	Examples	Tolerance	Displacement of outer ring
Rotating outer ring load			
Heavy loads on bearings in thin-walled housings, heavy shock loads ($P > 0.12$ C)	Roller bearing wheel hubs, big-end bearings	P7	Cannot be displaced
Normal loads and heavy loads ($P > 0.06$ C)	Ball bearing wheel hubs, big-end bearings, crane travelling wheels	N7	Cannot be displaced
Light and variable loads ($P \leqslant 0.06$ C)	Conveyor rollers, rope sheaves, belt tension pulleys	M7	Cannot be displaced
Direction of load indeterminate			
Heavy shock loads	Electric traction motors	M7	Cannot be displaced
Normal loads and heavy loads ($P > 0.06$ C), axial displacement of outer ring unnecessary	Electric motors, pumps, crankshaft bearings	K7	Cannot be displaced as a rule
Accurate or silent running			
	Roller bearings for machine tool work spindles	K6	Cannot be displaced as a rule
	Ball bearings for grinding spindles, small electric motors	J6	Can be displaced
	Small electric motors	H6	Can easily be displaced

thickened greases of consistency 1, 2 or 3 are those normally used for rolling bearings. The consistency should not change greatly with mechanical working or the elevated temperatures encountered in service. Vibrating applications will require mechanically stable greases. Calcium-, sodium- and lithium-based greases are commonly used with differing operating temperature ranges. The lithium-based greases are now becoming more common due to generally hotter environments. Synthetic greases are also becoming more used in special applications where their extreme temperature ranges are required.

In heavily loaded applications, and particularly at low speeds, greases containing EP additives will give superior bearing performance, but such greases will not give any benefit on small bearings. Mixing of different grades and brands of grease is not recommended. Greases having the same thickener and a similar base oil should be compatible but, although lubricating properties may not be impaired, the resultant mixture may have a softer consistency than either of the component greases and leakage may occur.

Oil lubrication is used when high speeds or temperatures rule out greases or when adjacent machine parts are oil lubricated. Bearing limiting speeds are higher for oil lubrication than for grease lubrication.

Oil lubrication can most simply be carried out by an oil bath in which the oil is picked up by the rolling elements and returns to the sump by gravity. The oil level, at standstill, should be no higher than the mid-point of the lowest ball or roller in the bearing to avoid excessive oil churning. An oil-circulation system is recommended for arduous conditions.

The oil within the bearing is then at a minimum but is constantly replaced by cool oil from a large reservoir. Such a system, while more expensive to install, results in extended oil life and optimum bearing performance.

For extremely high speeds the extraction of heat from the bearing by the oil is of paramount importance and oil jet systems may be necessary. Solvent refined mineral oils should be used for the lubrication of rolling bearings. The minimum kinematic viscosity of the mineral oil required to ensure adequate lubrication at the operating temperature is calculated using Figure 9.45. The mean diameter of the bearing, d_m ($= 0.5$ $(d + D)$) is required and the operating speed. The required viscosity, V_1 (mm^2 s^{-1}), can then be read off. Knowing, or predicting, the operating steady state temperature of the oil in the vicinity of the bearing the viscosity V at 40°C can be obtained from Figure 9.46 (40°C is the internationally accepted standardized reference temperature). This will define the minimum grade of oil required. Bearing life can, however, be extended by selecting an oil with viscosity greater than V_1 (see Section 9.6.4).

Example A bearing having a bore diameter, d, of 340 mm and outside diameter, D, 420 mm, is required to operate at a speed, n, of 500 rpm. From Figure 9.45 the minimum viscosity, V_1, to give adequate lubrication for a speed of 500 rpm is 13 mm^2 s^{-1}. Assume the operating oil temperature to be 70°C (a reasonable first estimate for normal conditions).

From Figure 9.46 an oil having a viscosity of 30 mm^2 s^{-1} at 40°C will have the required 13 mm^2 s^{-1} viscosity at 70°C. The correct oil will be one having a viscosity greater than

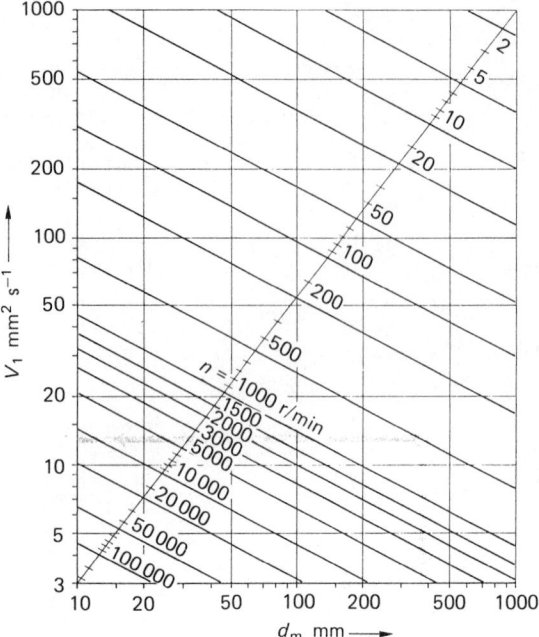

Figure 9.45 Calculation of minimum required viscosity, V_1

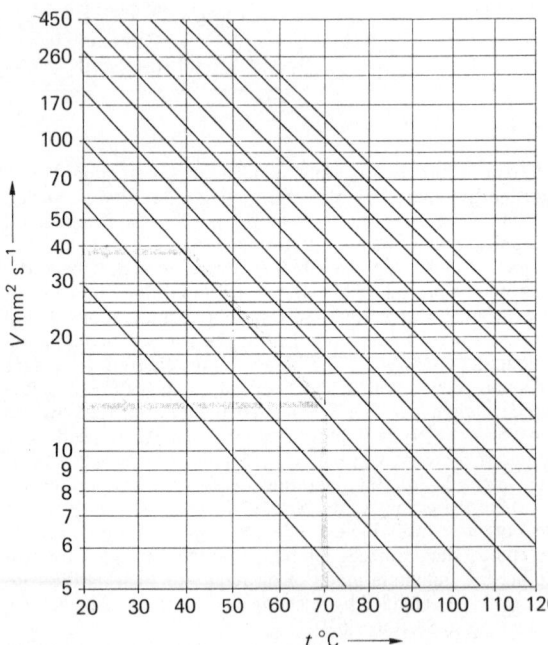

Figure 9.46 Determination of oil viscosity at 40°C, V

does not exceed 50°C and there is no contamination. Circulating-oil systems contain a higher volume of oil and may therefore run longer. Oil changes should be determined by regular oil examination or reference to the equipment manufacturer's handbook, particularly when gears or other equipment are lubricated by the same oil supply.

9.6.8 Bearing mounting and maintenance

Rolling bearings are high-precision engineering components and must be treated as such during their entire working life. Large-scale original equipment manufacturers who build in large numbers of rolling bearings take great care in the design and execution of their assembly equipment. It is rare that bearing damage will occur in such circumstances. The major areas of concern are where the assembly of bearings is infrequent. It must be done in a field environment, or where the correct tools are not available.

The study of a suitable mounting procedures guide will repay itself handsomely and attendance at a course on bearing maintenance is advised. In service, repairs are difficult because of the possibility of airborne contamination getting into the bearing assembly. Bearings should be taken out of their protective wrapping at the last possible moment and covered (for example, with a plastic bag) if left exposed on a shaft. Grease tins should not be left open and the grease should be applied with clean implements.

Inexpensive tooling such as drifts and pullers will ease assembly and disassembly on small to medium-sized bearings. Larger bearings will require heavier-duty equipment such as induction heaters (which have largely replaced oil baths or hot plates on safety grounds), hydraulic nuts and oil injection equipment. These items can often be hired if their use does not justify purchase.

It is very important that the bearing rings, cages or rolling elements do not receive direct blows during mounting as this will cause damage. Under no circumstances should pressure be applied to one ring in order to mount the other ring. Before mounting, the seating surfaces should be lightly smeared with oil.

One-piece bearings with cylindrical bores are usually mounted on the ring having the tighter fit first. Small bearings normally having a light interference fit on the inner ring may be positioned using light hammer blows to a soft metal drift or a length of tubing applied to the inner ring face.

If a non-separable bearing is to be pressed into position on the shaft and in the housing at the same time a mounting ring should be placed between the bearing and the mounting dolly so that the mounting force can be applied evenly to both inner and outer rings.

Separable bearings can be mounted independently on the inner and outer rings. When the shaft, with the bearing inner ring already mounted, is fitted into the housing containing the outer ring, great care must be taken to avoid scoring the raceways and rolling elements. Even light scores will rapidly develop into failure sites during rotation.

Larger bearings cannot always be pressed into position due to the large forces required. The larger of the two mating surfaces are therefore heated before mounting. The expansion of a bearing inner ring or housing when heated to 100°C is sufficient to allow troublefree fitting. The inner rings of bearings with a tapered bore are always mounted with an interference fit. The degree of interference is determined by the amount the bearing is driven up the taper of the shaft or adaptor sleeve. During mounting the radial internal clearance of the bearing is reduced. The clearance reduction is a measure of the degree of the interference fit. Self-aligning ball bearings are driven up until the outer ring can still be easily

$39 \text{ mm}^2 \text{ s}^{-1}$. The nearest thickness grade from a quality supplier would be preferred.

Oil change intervals are mainly dependent on the operating conditions and on the quantity of oil used. Oil bath systems should be changed once a year, provided the oil temperature

rotated, but a slight resistance is felt when the outer ring is swivelled. Spherical roller bearings are driven up until a specified internal radial clearance is attained. The resultant clearance is specified in bearing manufacturers' tables. Alternatively, the axial displacement of the bearing from an initial position of contact can be used. This method is, however, more liable to error.

Maintenance requirements vary for the type of bearing, its duty cycle and its working environment. The smaller ball and roller bearing used in domestic appliances and private cars generally do not require any attention during the working life of the machine. Larger industrial plant, generally using roller bearings, running with grease or oil lubrication will require regreasing or an oil change at regular intervals to ensure that the quality of the lubricant remains adequate.

Regreasing intervals for rolling element bearings are shown in Figure 9.47. The figure is based upon the use of an age-resistant average quality grease and is valid for bearing temperatures of +70°C measured on the outer ring. The relubrication interval should be halved for every 15°C increase above 70°C. The maximum operating temperature of the grease should obviously not be exceeded. Conversely, if operating temperatures are lower than +70°C the intervals can be lengthened to about twice for operating temperatures of 50°C and below.

Where there is a risk of the grease becoming contaminated the relubrication intervals should be reduced. The amount of grease needed for relubrication can be obtained from

$$G = 0.005\ D.B$$

where G = grease quantity (g), D = bearing outside diameter (mm) and B = bearing total width (mm). Where infrequent lubrication is indicated it is sufficient to expose the bearing in its housing, remove the used grease and pack the fresh grease between the rolling elements.

Where more frequent relubrication is required proper provision should be made for regreasing. Preferably a grease nipple should be fitted to the housing, positioned so that the fresh grease actually reaches the bearing and replaces the old. Provision should be made for the escape of excess used grease so that the bearing assembly does not become fully packed with grease. This will result in hot running.

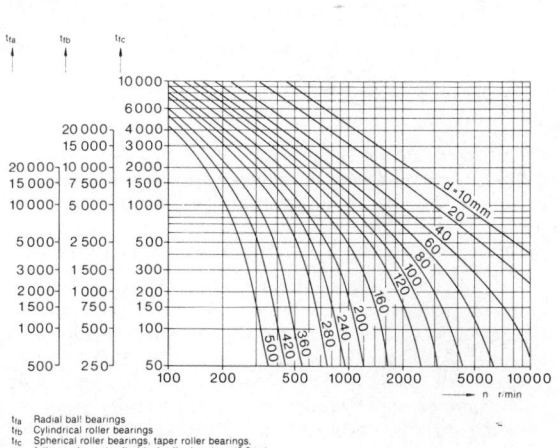

Figure 9.47 Relubricating interval determination

9.6.9 Bearing failures

Repeated bearing failures on a particular machine can be indicative of a number of problems (e.g. a misapplied bearing, poor maintenance or incorrect mounting practice). It is therefore important to examine the failed bearing and attempt to interpret the cause of the failure so that effective remedial action can be applied. A systematic approach must be adopted. Samples of the grease or oil lubricant should be examined for their quality and possible contamination by solids or liquids. The distribution of lubricant should be noted. Bearings should be cleaned and disassembled for a thorough analysis. Careful observations should include the following points.

9.6.9.1 Load paths

The burnished areas on inner and outer ring raceway which show where the balls or rollers have actually been working repay close observation. Very often the actual load path differs from what it was designed to be. The difference then accounts for the failure. Failures under this heading could be caused by:

- Radial preload in a bearing, leading to a load path all around the inner and outer rings. This could be caused by too tight a fit on the inner ring, an expansion of the inner ring due to temperature rise on a bearing that was selected with too small a manufactured clearance.
- Axial loading on a bearing that was intended only to be a radial bearing. Often this is a result of axially locating both bearings on a shaft not giving a facility for axial expansion.
- Misalignment of a shaft causing excessive internal loading of a bearing.
- A bent shaft or misfitment of the bearing in the housing or on the shaft. This will result in a load pattern running circumferentially from one side of the raceway to the other in ball bearings.
- Out-of-balance loads on rotating equipment. This is particularly devastating on heavy machinery running at high speeds where the design specification assumes good balance. The telltale markings on the bearing rings will be a path completely around the outer ring and a localized load area on the inner ring (coincident with the radial direction of the out-of-balance).

9.6.9.2 The raceway surfaces

The classical rolling element bearing failure is a localized, subsurface-initiated spall or pitting, rolling contact fatigue. Due to improvements in bearing steel quality and manufacturing accuracy over recent years, this failure mode is now comparatively rare. It represents the best life that can be attained from the bearing. Further endurance from a new bearing will only be attained by improving the lubrication quality or reducing the applied loads.

The more common failure mode is a surface-initiated distress over the entire surface of the failed ring caused by metal-to-metal contact of the rings and rolling elements – two-body abrasive/wear. This failure is indicative of inadequate lubrication quality, i.e. the lubricant viscosity at the operating temperature is inadequate to create a sufficiently thick lubricant film between rings and rolling elements to escape the dangerous metal-to-metal contacts. In the early stage raceways will become glazed in appearance, followed by a dull frosty surface, eventually leading to flaking.

A general surface distress failure can also be caused by contaminants within the bearing assembly resulting in three-body abrasive wear. Solid and liquid contaminants have the same effect. Both can usually be identified by examination of the grease (oil) or of the bearing cavities.

Poor lubrication can often be distinguished from other surface-initiated failures by examination of the cage pockets. The rolling element/cage contact is a sliding mode and, as such, is susceptible to wear if the lubricant is suspect.

9.6.9.3 Raceway indentations

Regular raceway indentations at ball or roller pitch indicate overload damage under non-rotating conditions. Poor bearing mounting techniques, pushing the interference fit ring into place through the rolling elements and loose ring, can very easily indent one or both rings. Failure will occur very quickly once the bearing is in service.

Indentations can also be caused by false brinelling, i.e. the internal oscillation of the bearing due to vibration of the whole machine. This is common on standby equipment situated in a vibrating environment. The remedy is either to regularly rotate the machine to spread the load and relubricate the bearings or to preload the bearing in some manner. Brinelling can only occur where there is movement within the bearing.

9.6.9.4 Electrical discharge damage

Electrical current leakage across a rolling element bearing will frequently result in severe arcing pits or valleys on the raceways. These will typically have a regular washboard appearance and will cause severe vibration. The use of inadequately earthed electric arc welding, perhaps during repairs on a structure containing bearings, is a common originator of such damage. The remedy is either better earthing using an earth strap or, in extreme cases, insulation of the bearing by coating the bearing outer diameter with a suitable non-conducting layer. Such bearings, obtainable from the major manufacturers, are becoming common in traction motors and other bearings in electrical environments.

9.7 Materials for unlubricated sliding

9.7.1 Introduction

There are literally hundreds of materials which are suitable for unlubricated sliding. The main selection criterion is the maximum acceptable level of friction and wear which can be tolerated in the bearing contact. In general, this level will be higher than in lubricated contacts and therefore dry rubbing bearings find application where the use of a fluid lubricant is not possible. In practice, this is where:

- Fluids are adversely affected by the environment (e.g. high or low temperature, vacuum, radiation or solvents);
- Product contamination is undesirable (e.g. food or optics);
- Maintenance and relubrication is not possible (e.g. in remote locations);
- A combination of low speed and high load breaks down the fluid film (e.g. oscillatory, offshore buoy moorings).

In addition to satisfying tribological requirements, the material must support the applied load without significant distortion, under the operating conditions, to prevent severe surface damage during sliding. Provided these criteria can be met, the use of dry bearings can offer advantages in terms of cost, simplicity of design and weight.

Five groups of materials can be identified which exhibit relatively low friction and wear characteristics under unlubricated sliding conditions:

Polymers and polymer composites
Solid lubricants
Self-lubricating bronzes
Carbons and graphites
Hard-facing alloys, ceramics and cermets.

The different mechanical, thermal and tribological properties of these groups of materials satisfy a wide range of operating specifications. They include materials capable of continuous operation in extremes of temperature, from $-250°C$ to in excess of $500°C$, at bearing pressures in excess of 100 MPa and at sliding speeds over 10 m/s. When sliding is intermittent, even higher pressures and speeds can be tolerated.

The tribological performance and operating limits of materials within these groups is presented in the following sections. The best known and most frequently used dry bearings are polymers and polymer composites. While there is an abundance of information available on these materials, data on some of the other groups are extremely limited. Wherever possible, a list of material suppliers is therefore included from whom further technical advice can be obtained.

In addition to these five groups of dry bearing materials, there are certain porous metals which can be impregnated with fluid lubricants and operate without external lubrication for limited periods. Since a requirement for regular, wet lubrication can prevent the use of many metallic materials in sliding applications, consideration will be given to these prelubricated materials which can exhibit extremely low friction and wear under relatively high load and sliding speed conditions.

Before presenting specific data on materials, the parameters used to rate performance, their limitations and the factors which can influence them must be considered. The interpretation of manufacturers' data, when predicting bearing performance, without an appreciation of these factors can be extremely misleading.

9.7.2 Performance rating

9.7.2.1 Wear

As discussed in Section 9.1, wear data are normally expressed in terms of a specific wear rate, k. This is defined as the volume of material worn, per unit load, per unit sliding distance. Hence

$$k = \frac{v}{WL} \tag{9.7}$$

where

v = worn volume of material,
W = load between the bearing surfaces (m^3), and
L = total sliding distance (m).

Standard units are $m^2 \, N^{-1}$. Values of $10^{-16} m^2 \, N^{-1}$ or lower are indicative of mild wear and values higher than $10^{-14} m^2 \, N^{-1}$ represent severe wear. Most material wear rates are within the range 10^{-17} to $10^{-12} m^2 \, N^{-1}$.

Two common features of dry bearing wear should be noted:

1. The wear rate is often time dependent. A typical plot of wear volume versus time for polymer-based materials sliding against a smooth, hard counterface is shown in Figure 9.48. From this graph it can be seen that linear and steady-state wear is only observed after a more rapid running-in period. During this period, transfer from the bearing material to the counterface takes place and the polymer surface conforms to the harder surface geometry. Subsequent contact is between the bearing material and the transferred film which gives the low friction and wear

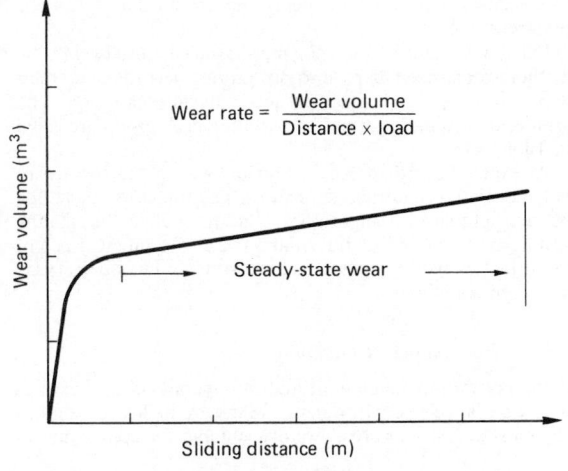

Figure 9.48 Typical plot of wear versus sliding distance for PTFE-based materials under dry conditions

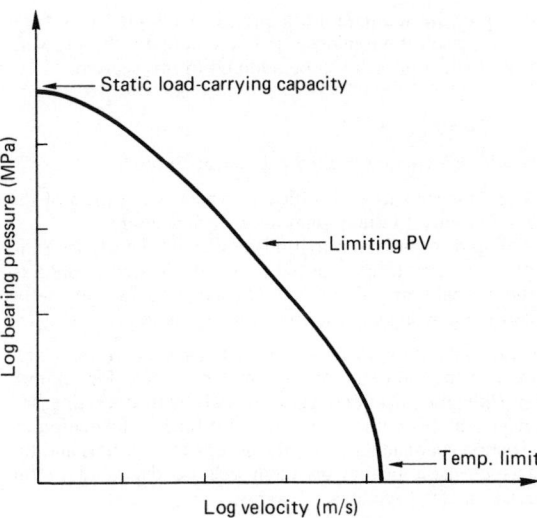

Figure 9.49 Typical 'PV' plot for dry bearings

characteristics generally quoted in manufacturers' literature. Wear rates during running-in can be an order of magnitude higher than the steady-state value. Allowance for this wear in terms of fits and tolerance may be necessary during the design stage.

2. The concept of a material-specific wear rate is based on the assumption that changes in the applied load do not cause changes in the wear characteristics of the material. This is generally the case under low loads but the wear performance of many materials, especially polymers, is influenced by the bearing pressure and operating temperature once they exceed a critical limit. Since the temperature rise at the surface due to frictional heating is proportional to the sliding speed, it is appropriate to incorporate a pressure velocity term into this wear equation, as discussed in Section 9.1, equation (9.5).

9.7.2.2 PV factors

Limiting pressure velocity (PV) values are normally quoted in manufacturers' data. Alternatively, a PV factor may be quoted in terms of a maximum wear rate. Above these limits, either excessive frictional heat will be generated at the bearing surface, which will cause the wear rate and friction to increase, or the stresses in the material will exceed the elastic limit.

There are, however, several limitations on the use of PV limits:

1. Quoted PV values normally refer to operation at ambient temperatures of about 20°C. For higher-temperature applications a reduced PV limit will apply since the ambient temperature will be superimposed on the frictional heat generated at the surface. Guidance on PV temperature correction factors is given in an ESDU polymer materials selection guide.[22]
2. If the in-service PV value is within the limit only because one of the two parameters is very low, the wear rate may again be adversely affected, due either to frictional heat in the case of excessive speed or overloading. It is therefore useful if a complete PV curve is presented, as shown in Figure 9.49.

3. If the heat generated at the rubbing surface can be readily dissipated from the contact area, either by forced air cooling or through the bearing housing, a higher PV limit can be applied.

Since frictional heat is often the limiting factor, especially with polymer sliding, the concept of a pressure temperature limit is, at first sight, attractive. However, whereas the calculation of bearing pressure and sliding speed is normally straightforward, the bearing surface temperature is the sum of three components:

- The ambient temperature;
- The mean surface temperature rise during sliding;
- The flash temperature rise at the localized asperity contacts on the surface.

The temperature rise over the apparent surface area due to frictional heating is proportional to the friction coefficient between the mating surfaces, the applied load and the sliding speed. However, an estimate of the heat dissipated from the bearing or through the housing is also required for an accurate prediction of this temperature rise. Hence:

$$T = R.\mu.W.V \qquad (9.8)$$

where

T = mean surface temperature rise,
μ = friction coefficient,
W = normal applied load, and
V = sliding velocity.

This equation can, alternatively be expressed as in Section 9.1:

$$T = c\mu.PV$$

R and c are constants which characterize the resistance to heat dissipation from the rubbing surfaces. Details of this calculation are given in several design guides on polymers.[22,23]

The calculation of flash temperature is fairly complex.[24,25] However, if the sliding speed is well below its limiting value, as given by the pressure velocity curves, this component can normally be ignored in the estimation of the bearing surface temperature.

This pressure temperature approach is adopted in several dry bearing selection guides.[23,26] For simplicity, the pressure velocity limit approach will be adopted in this section.

9.7.2.3 Friction

Two different friction coefficients can be quoted:

1. The dynamic friction coefficient, which is the ratio of the friction force to the normal force during sliding;
2. The static or breakout friction coefficient. This is the ratio of the friction force required to initiate relative sliding, to the normal force. This breakout friction coefficient can be considerably higher than the dynamic value.

It is also important to recognize that the friction characteristics, like the wear rate of most materials used for unlubricated sliding applications, vary not only with operating and environmental conditions but also with time. Information on the friction coefficients of specific groups of materials and the effects of these factors on them will be discussed in the relevant material sections.

9.7.3 Counterface materials

In any dry contact there are two mating surfaces. The counterface material and its surface finish greatly influence the friction and wear characteristics of most dry bearing materials. In general, lower wear rates will be obtained with hard, smooth counterfaces but the costs of preparation often dictate the choice of material and finish within acceptable tribological performance limits. Surface treatments and coatings are available which can produce a hard bearing surface on a relatively cheap substrate (Section 9.8). Considerable development has been made in these fields during the past ten years and their application as a technique for controlling wear is becoming more widespread.[27]

From this brief discussion of material performance data presentation and its limitation, it should be apparent that a universally applicable material wear rate and friction coefficient does not exist for a given material. It is important, therefore, to ascertain whether quoted values can be justly applied to a specific application when predicting bearing performance. Whenever possible, short-term, accelerated laboratory tests should be carried out under representative in-service conditions to obtain relevant tribological data.

9.7.4 Polymers and polymer composites

9.7.4.1 Material classification

The main advantage of polymers over metals is their inherently low friction and wear characteristics in the absence of external lubrication. In comparison to mild steel, their tensile strength is a factor of ten lower, their elastic modulus a factor of one hundred lower and their coefficient of thermal expansion a factor of ten times higher. These properties, which would impose severe limitations on their use, are greatly improved by the use of fillers and there are over 70 commercially available polymers and polymer composites in the UK today.

Polymers fall into two distinct classes: thermoplastics and thermosets. Thermoplastics are characterized by softening on heating and retaining their plastic properties on cooling. Thermosets undergo extensive cross linking of their molecular chains on heating which causes irreversible hardening of the material. Most thermoplastics therefore have the advantage over thermosets that they can be injection moulded. This facilitates cheap mass production. Thermosets are generally

produced as bars, tube or sheets and are machined to their required shape.

For convenience, the polymer-based materials will be further categorized here into six groups according to their appearance, performance or load-carrying capacity. The groups and examples of materials within each group are listed in Table 9.18.

As explained earlier, the friction and wear characteristics of polymer-based materials depend on many factors, including bearing pressure, temperature, sliding speed, the mating surface material and its finish, and the environment. Each of these factors will be considered individually but their effects are often combined.

9.7.4.2 Environmental suitability

Polymers are extremely inert and can operate as dry bearings in a wide range of environments. Many are highly resistant to chemical attack and are therefore suitable for application in

Table 9.18 Polymers and polymer composite classification

	Suppliers
1. *Unfilled thermoplastics*	
Polyamide (nylon)	Du Pont, ICI
High-density polyethylene	BP
Polytetrafluorethylene (PTFE)	Du Pont, ICI, Hoechst
Ultra-high molecular weight polyethylene (UHMWP)	Hoechst
Polyacetal	
2. *Filled thermoplastics*	
Polyacetal with oil	Railko
Polyamide with oil	Courtaulds
Carbon-fibre reinforced polyamide	Courtaulds
Polyurethane/elastomer blend	Thordon
Polyacetal with PTFE	Du Pont
Nylon with glass fibre	DSM Rim Nylon
3. *Thin-layer materials*	
PTFE/glass fibre and resin	Ampep
PTFE Flock/Nomex cloth/resin	Rose
PTFE/lead, bronze sinter	Glacier
Injected resin/fillers	Barden
4. *Filled PTFE*	
PTFE/bronze	Polypenco, Shamban, Fothergill
PTFE/glass fibre	Fothergill
PTFE/carbon/graphite	Morganite
PTFE/mineral	Henry Crossley
PTFE/bronze/graphite	Glacier
5. *High-temperature polymers*	
Polyimide (Vespel)	Du Pont
Polyimide and graphite	Du Pont
Poly-ether-ether ketone (PEEK)	ICI
Polyamide/imide	Polypenco
6. *Reinforced thermosets*	
Polyester resin/graphite	Orkot
Phenolic resin/asbestos	Railco
Epoxide resin/graphite	Ciba Geigy
Phenolic resin/cotton	Tufnol

Table 9.19 Environmental suitability

Material/group	Acids	Alkalis	Solvent	Vacuum	Inert gas	Radiation	Water
PTFE	✓	✓	✓	✓	✓	0.1M Rad	R
Nylon	W	W	✓	X	✓	1	S
HDP	W	✓	✓	✓	✓	10	✓
UHMWP	✓	✓	✓	✓	✓	100	✓
Polyacetal	W	✓	✓	✓	✓	1	✓
Polyimide	✓	✓	✓	✓	✓	1000	✓
PEEK	✓	✓	✓	✓	✓	1000	✓
Fillers							
Carbon fibre/graphite	✓	✓		X	R	>1000	C
Glass fibre	✓	✓	✓	✓	✓		✓
Bronze	X	X	✓	✓	✓	>1000	C
Thin-layer materials							
PTFE/glass/resin	W	W	✓	✓	✓	100	R
PTFE/Nomex/resin	W	W	✓	✓	✓	10	R
PTFE/Pb/bronze	X	X	✓	✓	✓	100	✓
Resin/fillers	W	W	✓	✓	✓	10	R

Key: ✓ – suitable
  X – unsuitable
  S – swelling
  R – reduction in properties
  C – corrosion
  W – weak acid or alkali only

acids, alkalis and solvents. The effect of chemicals on any fillers present in the material must, however, be considered. Polymers have a lower resistance to ionizing radiation than metals, but doses as high as 1000 MRads can be tolerated by polyimides with only a slight reduction in mechanical properties and tribological performance. Table 9.19 summarizes the suitability of the polymers and fillers for use in different environments and can be used to eliminate materials on an environmental suitability basis. This is often the simplest and first stage of the materials selection process.

Several general observations can be made from Table 9.19:

1. Composites containing graphite are unsuitable for use in a vacuum or dry, inert gas. This is because graphite requires the presence of a condensible vapour, such as water, to maintain its natural lubricity. Further information on the properties of carbon graphites is given in Section 9.7.5.
2. Immersion in water has an adverse effect on the performance of many of these dry bearing materials since it removes or prevents the establishment of a transfer film, on which their low friction and wear characteristics depend. Since many of the polymers are, to some extent, porous, swelling can occur in water. This is particularly notable with polyamide materials, which can increase in volume by up to 3% in water. This can result in a reduction in clearance between bushes and shafts and an increase in bearing pressure. Where this may present a problem, only those materials ticked as suitable in Table 9.19 should be considered.
3. Unfilled polyimide and PEEK are extremely inert and can operate in most environments. They are slightly porous but swelling in water is limited to about 0.1% by volume. The main limitation on the use of these two materials is their high cost, which is about £30/kg for PEEK.
4. Although a radiation dose of 0.1 MRads causes a severe reduction in the mechanical properties of bulk PTFE-based materials, tests have shown that thin-layer materials containing PTFE can withstand doses as high as 100 MRads

without any major increase in wear rate. This is because the reinforcement carries the load and the reduction in properties of the PTFE has little effect on load-bearing capacity.

9.7.4.3 Pressure velocity limits

The pressure velocity limiting curves presented here (Figures 9.50 to 9.52) are based on a depth wear rate of 25 μm per hundred hours sliding in air at an ambient temperature of 20°C. Where the ambient temperature is hotter, a lower pressure velocity limit will apply. These indicate a range of bearing pressures of up to 100 MPa at low speeds and a maximum sliding speed of about 10 m/s at low loads within this group of materials.

The filled polyimide and PEEK high-temperature materials are well suited to high-load, high-speed applications (Figure 9.50). This is to be expected, since heat generated at the rubbing surface due to friction is often the limiting factor in material performance. At lower speeds, the thin-layer materials offer extremely good load-carrying capacity (Figure 9.51), the limitation on speed being imposed by softening of the resin due to frictional heating. In contrast, the pressure velocity limits of unfilled PTFE are low but increase by a factor of two with the addition of most fillers (Figure 9.51).

The performance of reinforced thermosets is directional with optimum conditions when sliding is normal to the laminate. In this orientation, they are suitable for high-load, low-speed sliding applications.

The pressure velocity limits of unfilled thermoplastics are lower than those of the reinforced grades (Figure 9.52). The use of fillers can increase their maximum sliding speed from 1 m/s to 3 m/s and their load-carrying capacity from 4 MPa to in excess of 10 MPa.

9.7.4.4 Temperature limits

With the exception of PTFE which does not mechanically degrade at temperatures below 200°C, the unfilled and filled

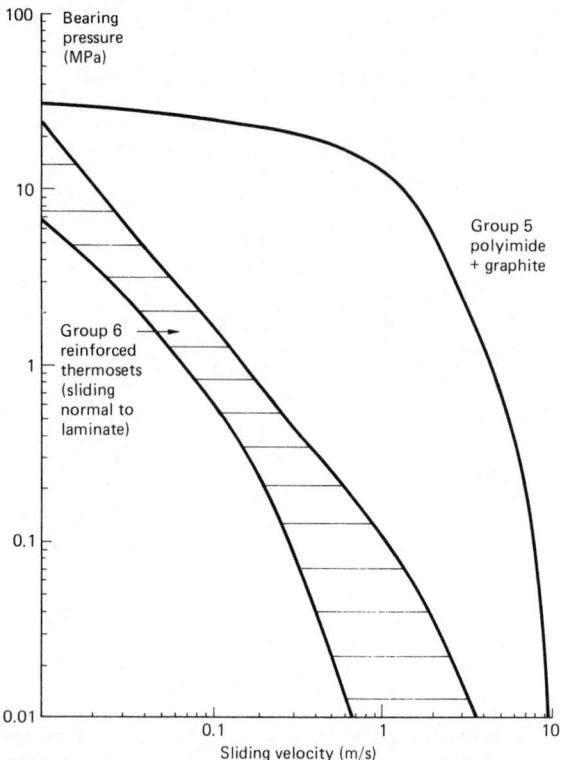

Figure 9.50 PV limiting curves for reinforced thermosets and polyimides

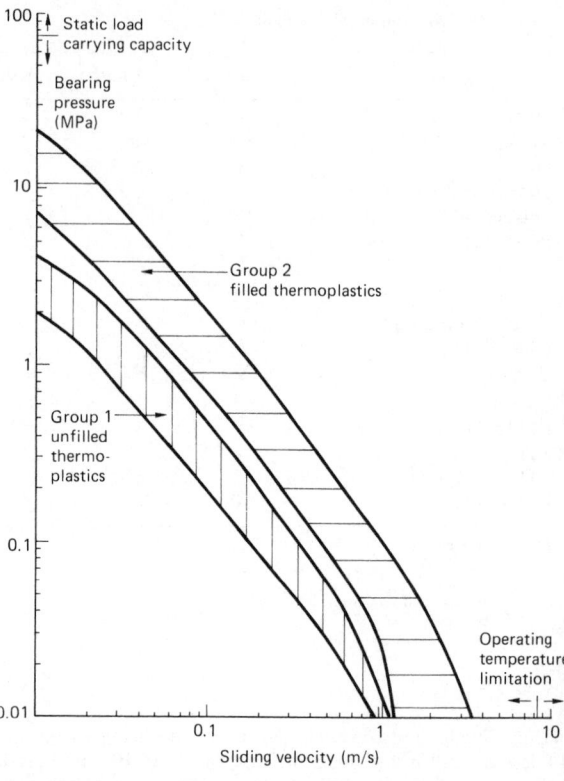

Figure 9.52 PV limiting curves for unfilled and filled thermoplastics

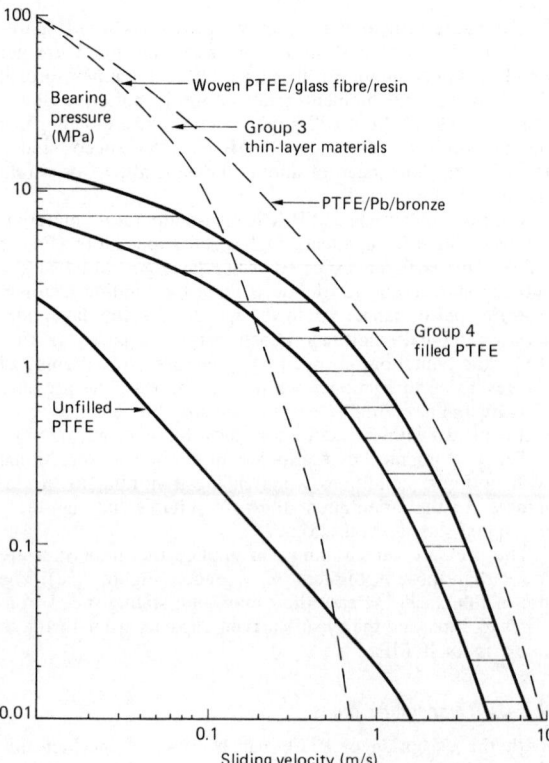

Figure 9.51 PV limiting curves for PTFE-based materials

grades of thermoplastic materials start to degrade when their surface temperature exceeds 100°C. For higher-temperature applications, most of the thin-layer materials (Table 9.18) are largely unaffected by temperatures of up to 150°C and the filled PTFE materials can be used at temperatures of up to 250°C. When the bearing surface temperature is expected to reach up to 300°C, polyimide or PEEK may be suitable.

9.7.4.5 Friction and wear data

The effect of pressure and temperature on the wear rate of polymer-based materials is illustrated in Figure 9.53. From this graph it is apparent that there is a low-pressure, low-temperature region where the wear rate is essentially independent of these factors. This wear rate is referred to as the K_0 wear rate. The K_0 values presented in Figure 9.54 were measured using an annular thrust washer test geometry. The polymer washer was loaded against a static mild steel counterface, ground to a surface finish of 0.2 μm R_a, under a bearing pressure of 1 MPa and continuously rotated at a sliding speed of 0.03 m/s. Under these conditions, all the materials tested were within their K_0 limits.

Although renowned for its low friction, the wear rate of unfilled PTFE is very high. The addition of fillers reduces the wear rate but increases friction. Its use in the unfilled form is not therefore recommended where high rates of wear are unacceptable (e.g. rotary seals).

The thin-layer materials all exhibit low wear rates but their wear depth is limited to between 0.05 mm and 0.3 mm depending on the material. Beyond this limiting depth, their tribological properties deteriorate rapidly. Figure 9.55 shows a typical wear volume versus sliding distance plot for these materials. It is recommended that, where possible, the pro-

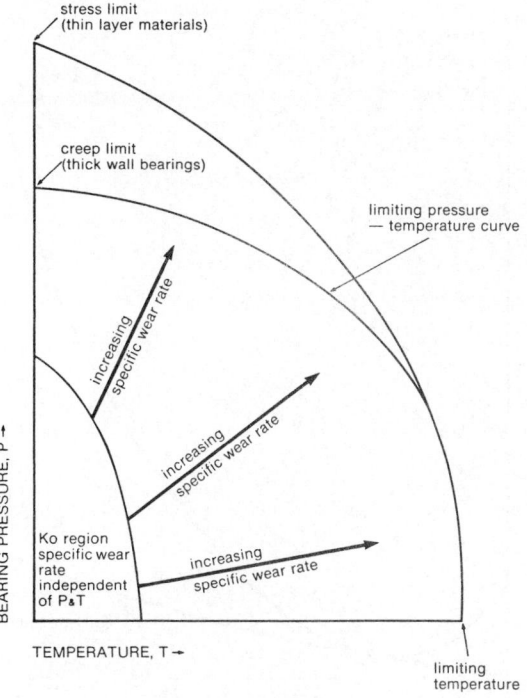

Figure 9.53 Bearing pressure and temperature limits for polymer-based dry bearing materials

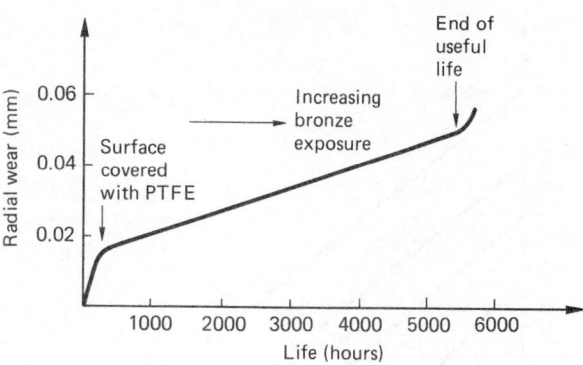

Figure 9.55 Typical plot of wear versus time for thin-layer bronze/PTFE material

gressive wear of these materials should be monitored to predict useful operating lives.

Pressure effects on wear rate are often closely linked to temperature rise.[28] This is not surprising, since frictional heating is a function of the applied load. Anderson has derived pressure temperature wear correction factors for a wide range of polymers which permit a prediction of the wear rate under different operating conditions.[23,29] This wear correction factor is multiplied by the K_0 value to predict the in-service wear rate. For most polymer-based materials, this factor increases from one to twenty as the temperature increases (Figure 9.56) to the softening point of the material. This increase in wear

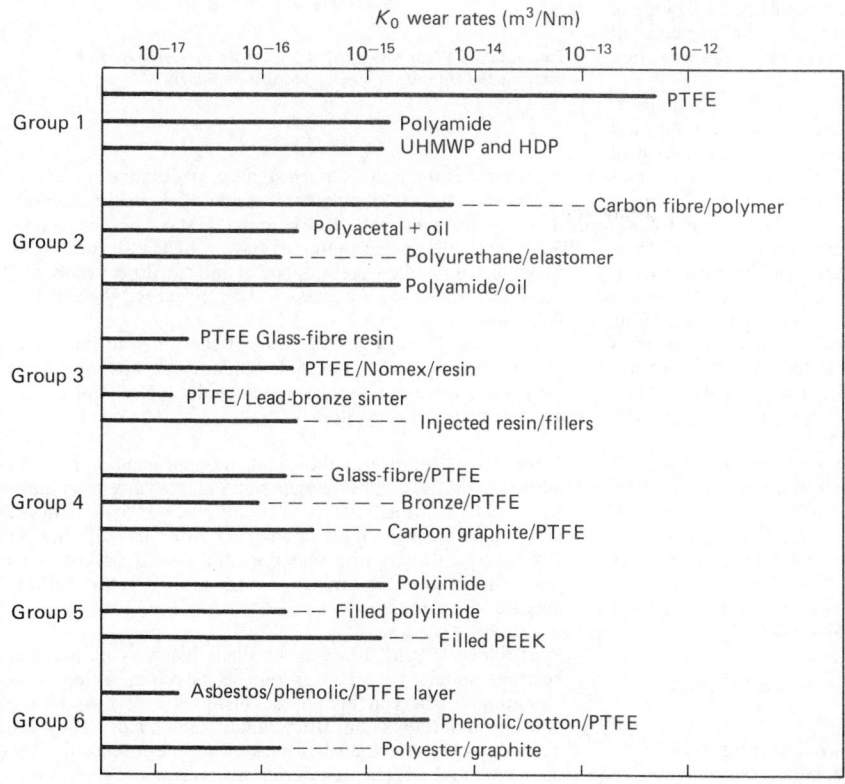

Figure 9.54 K_0 wear rates

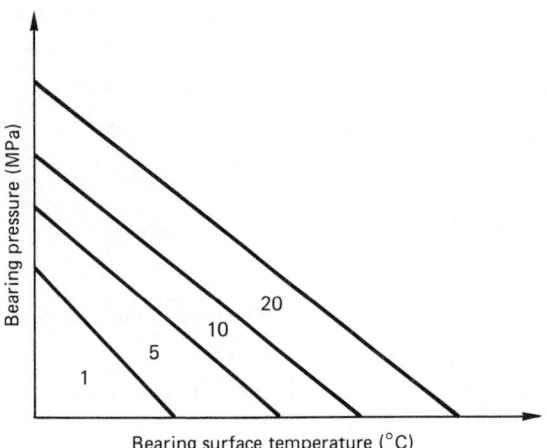

Figure 9.56 Typical wear correction factor plot from reference 23

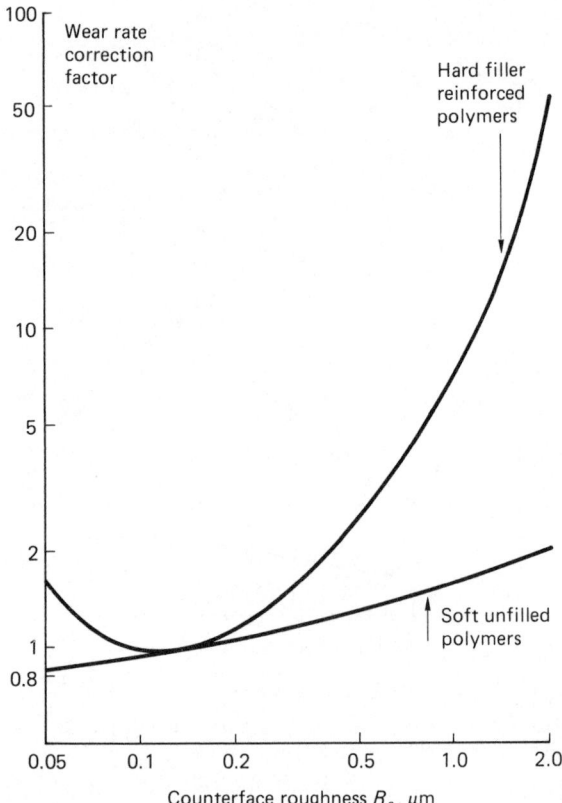

Figure 9.57 Wear rate correction factors for polymer and composites sliding on steel (hardness < 300 HV)

correction factor is offset against a reduction in value with decreasing pressure. In general, to maximize wear life, the bearing surface temperature should be maintained in the range 0–40°C and the bearing pressure kept as low as possible.

These K_0 wear rate values were measured during sliding against a smooth, polished counterface. If the counterface material corrodes or the surface becomes contaminated with relatively hard particles such as sand, the wear rate of the materials will increase significantly. Figure 9.57 shows the effect of counterface roughness on the wear rate of a range of polymers. It should be remembered that the presence of lubricants, even in trace quantities, will reduce the wear rate of most polymers.

The range of dynamic friction coefficients exhibited by polymer materials within their pressure velocity limits are presented in Figure 9.58. The dynamic friction coefficient of both unfilled and filled grades of PTFE increases from about 0.1 to 0.3 with decreasing pressure and increasing temperature. At high bearing pressures of about 100 MPa the friction coefficient of the thin-layer PTFE materials can be as low as 0.03 provided the surface temperature remains below about 150°C. This value increases to 0.2 mainly with decreasing pressure. The trend of friction increasing with decreasing pressure and increasing temperature is also true for most thermoplastic materials, in both unfilled and filled forms. With reinforced thermosets the friction actually reduces slightly with increasing temperature but still increases with decreasing pressure.

The static friction coefficient of most thermoplastics is similar to their dynamic value. Certain thermosets, however, exhibit static friction coefficients almost 50% higher than their dynamic value. Where this is the case, stick slip or jerky motion often ocurs. An indication of materials prone to stick slip is given in Figure 9.58.

Material suppliers from whom additional technical information can be obtained are listed in Table 9.20.

9.7.4.6 Design considerations

Aspect ratio The aspect ratio of journal bearings is the ratio of the bearing length to its internal diameter. Normal practice is to design with a ratio of between 0.5 and 1.5. If this range is greatly exceeded, problems are encountered with bearing alignment. If the ratio is much smaller, the establishment of a coherent low-friction polymer transfer film on the counterface, as observed with PTFE and UHMWP, is less likely. Reinforced thermoset materials have a higher flexural modulus than most filled thermoplastics and the aspect ratio with these materials can, if necessary, slightly exceed these guideline values.

With thrust bearings, the ratio of the outer to inner diameter is important. A ratio of less than two is advisable since the outside edge can become overheated if high sliding speeds are attained at this position.

Clearances Because of their high thermal expansion, polymeric materials require a larger running clearance than lubricated metal bearings. A typical running clearance is 5 μm per mm diameter with a minimum clearance of 125 μm. For thin-layer materials, the manufacturer should be consulted since the required clearance is influenced by the backing material (Table 9.20).

Wall thickness and installation Wall thickness of polymer bearings should be made as thin as possible, within wear limitations. This will aid in the dissipation of heat to the housing and reduce distortion under high loads. They are generally located in their housings by an interference fit. The recommended degree of interference varies between 5 μm/mm and 10 μm/mm for the relatively rigid thermoplastics (e.g. polyamides, polyacetals, UHMWP) to between 10 μm/mm

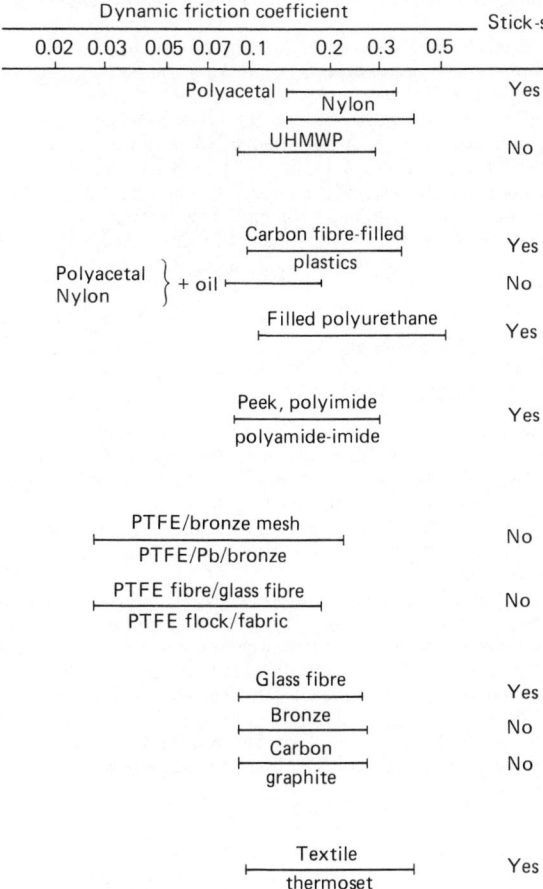

Figure 9.58 Range of friction coefficients for polymers and polymer composites

Table 9.20 Material suppliers

Company	Address
Ampep	Clevedon, Avon BS21 6QQ
Barden Corp	Western Road, Bracknell, Berks
BP Chemicals	Sully, Penarth, S. Glamorgan
Courtaulds	Carbon Fibres Unit, PO Box 161, Coventry
Du Pont	Maylands Avenue, Hemel Hempstead
Fothergill	Bridgwater, Somerset
Glacier Ltd	Alperton, Wembley
Henry Crossley	Hill Mill, Astley Bridge, Bolton
Hoechst	Salisbury Road, Hounslow
ICI	Welwyn Garden City, Herts
Morganite	Grove Road, Romford, Essex
Orkot	Orgreave, Sheffield
Pampus	Loomer Road, Newcastle, Staffs
Polypenko	83 Bridge Road East, Welwyn Garden City, Herts
Railco	Loudwater, High Wycombe
Rose	Saxilby, Lincoln
Shamban	St James Terrace, Nottingham
Tufnol	Perry Barr, Birmingham

and 20 μm/mm for the less rigid material (e.g. PTFE). Thin-layer metal-backed bearings should be designed using the properties of the backing material. Further advice can be obtained from the suppliers listed in Table 9.20.

Counterface materials The choice of counterface material is influenced by both the fillers present in the polymer and the environmental conditions. Where potential abrasives are present, such as grit or sand, hard fillers can improve the abrasion resistance of polymers. In general, hard particles can become embedded in the softer polymers and scoring of the counterface is often more severe than polymer wear under these conditions. When hard fillers are present in the polymer (for example, glass) the use of a counterface material with a hardness value exceeding 600 HV is recommended to prevent counterface damage.

In water, the possibility of electrolytic corrosion must be considered when selecting counterface materials for use with graphite-reinforced polymers. Laboratory tests are recommended to ascertain the potential for corrosion which will lead to significant increases in friction and wear.

9.7.5 Carbon graphites

9.7.5.1 Material properties

Carbon graphite materials possess several attractive properties which make them suitable for applications where polymers and lubricants cannot be used. They can operate at temperatures of up to 500°C in air, this temperature being limited by the oxidation resistance of the rubbing surface. They are therefore able to operate at relatively high sliding speeds compared to other unlubricated materials. Their high thermal conductivity also promotes dissipation of the frictional heat, generated at the bearing surface, to the housing. Although their porosity can be as high as 15%, they exhibit good dimensional stability and this can be further improved by impregnation of the pores with fillers, as discussed below. Their high electrical conductivity makes them ideal dry bearing materials for applications such as electric motor contact brushes and slip rings. Since these are brittle materials, care must be taken to minimize impact loads in service.

Carbon graphites are only suitable for unlubricated sliding in the presence of a condensible vapour, such as water. If the water vapour content of the atmosphere surrounding the sliding surface falls below 600 ppm, high wear or dusting will occur.[30] This dusting is so called on account of the dust generated from the carbon graphite surface which can wear at a rate measured in centimetres per hour under these drier conditions. Consequently, these materials are unsuitable for use as bearings or sliding contacts in a vacuum or dry, inert gas. However, in these environments, there is little change in their mechanical properties at temperatures of up to 1000°C.

Three graphitic grades are commonly used for unlubricated sliding: carbon graphite, metal-impregnated carbon graphite and electrographite.

Furnace-fired carbon graphites are porous and can operate at temperatures up to 350°C. Their compressive strength varies between 80 MPa and 160 MPa,[31] increasing with graphite content. Their tensile strength is considerably lower. Impregnation with thermoset resins, such as epoxide resin, limits operating temperatures to about 200°C but reduces permeability and increases their strength and wear resistance.

Metals impregnated into carbon-graphites include copper and copper–lead mixes.[32] These increase load-carrying capacity without reducing temperature limits.

Electrographite is produced by heat treatment of hard carbons at temperatures of up to 2800°C.[31] These resist oxidation in air better than the ungraphitized materials. The more graphitized materials readily transfer to the counterface and their friction and wear characteristics are generally lower than other grades. Since water can disrupt this transfer layer its presence is detrimental to performance.

The wear rate of most grades of graphite is within the range 10^{-15}m^3/Nm to 10^{-14}m^3/Nm. Dynamic friction coefficients of these materials vary between 0.1 to 0.3; the higher value is exhibited by metal-filled grades in high-load, low-speed situations and the lower value by electrographite (bearing grade) under low-load, high-speed conditions.

The wear rate of most grades of graphite is within the range 10^{-15}m^3/Nm to 10^{-14}m^3/Nm. Dynamic friction coefficients of these materials vary between 0.1 to 0.3; the higher value is exhibited by metal-filled grades in high-load, low-speed situations and the lower value by electrographite (bearing grade) under low-load, high-speed conditions.

9.7.5.2 Limiting pressure velocity

Limiting PV curves for these materials are presented in Figure 9.59 for a wear rate of 25 μm per 100 h.[31] These indicate a maximum possible sliding speed of 10 m/s. However, other sources claim that the sliding speed of all carbon graphite materials should be limited to 1.25 m/s.[32,33] It is therefore recommended that manufacturers be consulted for information on specific grades of these materials.

9.7.5.3 Design considerations

A ground, hard counterface (400–600 HV) is recommended for use with carbon graphites to promote material transfer and minimize abrasion. In the presence of an electrolyte, such as

water, the shaft material must be selected to avoid electrolytic corrosion.

Typical wall thickness for carbon graphite bushes is about 4 mm for a 25 mm bore and 12 mm for a 100 mm bore.[33] This relatively thick wall is necessary on account of the low impact and tensile strengths of these materials. Bushes of up to 25 mm bore can be press fitted in their housings with an interference of up to 2 μm/mm. Larger bushes are normally shrink fitted (for example, by freezing in liquid nitrogen) with an interference of about 4 μm/mm. This interference should be increased if the housing material is expected to expand considerably more than the bush during operation.

9.7.6 Solid lubricants

9.7.6.1 Applications

The inclusion of solid lubricants in a section on materials for unlubricated sliding may appear inappropriate. However, in many applications where the presence of a lubricant (or, more specifically, a wet lubricant) is considered undesirable or impossible, the use of a dry, solid lubricant can reduce friction and wear without adversely affecting the materials or the environment.

At extremes of temperature where oils and greases cease to lubricate due to changes in viscosity, dry lubricants can maintain a low-friction film between mating surfaces in relative motion. Many solid lubricants are also extremely inert and can withstand radiation levels and chemicals which cause wet lubricants to degrade. They are also usually very clean and adhesive to the bearing surface. This reduces the likelihood of product contamination which is essential in many industries.

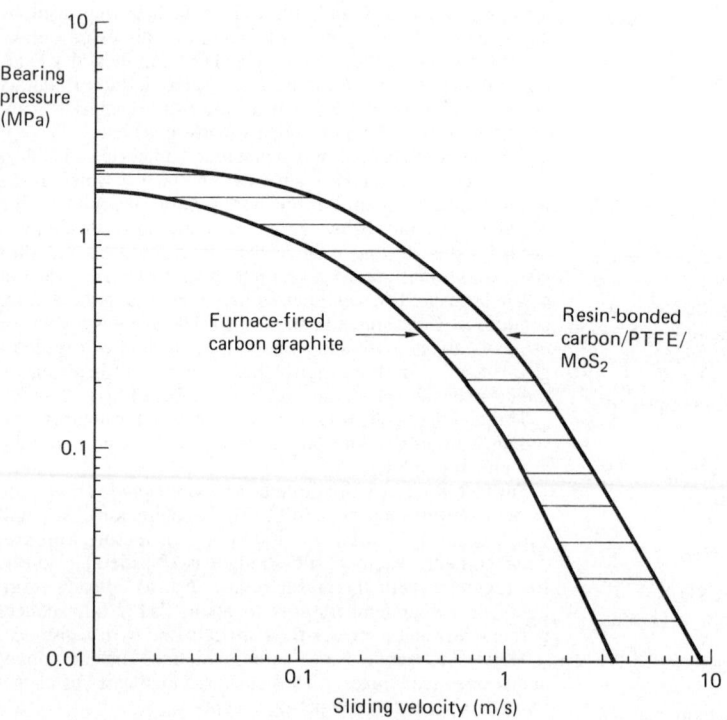

Figure 9.59 PV limiting curves for carbon graphites

The three most popular solid lubricating materials are PTFE (which has already been discussed in the section on polymers), graphite and molybdenum disulfide. As lubricants, they are normally contained in a resin binder, such as phenolic, epoxide or cellulose resins. In general, a high ratio of lubricant to resin will lower friction whereas lowering the ratio minimizes wear. They are available as spray or paste, both of which are easily applied to bearing surfaces.

The low shear strength of all these materials ensures rapid transfer to the counterface and subsequent low friction. Adhesion to the surfaces becomes important after the initial transfer and the presence of fluids such as water is often detrimental to performance.

9.7.6.2 Molybdenum disulfide

Molybdenum disulfide is a lamellar solid, particularly suited to vacuum applications. The simplest method of applying lamellar solids is burnishing of the dry powder into the surface with a cloth, taking care not to scratch the surface. Sprays are also available containing bonding agents which cure at room temperature. For applications where thin films of very high quality are required this lubricant can be applied by a sputtering process.[34] Tests in vacuum indicate that sputtered coatings of this material, only 0.5 μm in thickness, have better tribological properties than bonded films over ten times thicker. The maximum operating temperature of molybdenum disulfide is 350°C.

The wear rate of MoS_2 coatings is influenced by environmental factors and the type of binder.[35,36] The substrate has little effect on wear rate provided the coating is not penetrated. Specific wear rates in the range of $1 \times 10^{-15} m^3/Nm$ to $1 \times 10^{-14} m^3/Nm$ are typical for all but the hardest and softest coatings. Dynamic friction coefficients as low as 0.03 are not uncommon with sputtered coatings of MoS_2 at high loads in air. This rises to within the range 0.15–0.2 as film thickness increases and the load decreases.

9.7.6.3 PTFE and graphite

The properties of these two materials in bulk form are discussed in Sections 9.7.4 and 9.7.5. Both are available in spray form with binders which cure at room temperature.

The most common dry lubricant sprays are based on PTFE or molybdenum disulfide. Typically, the coating thickness is between 1 μm and 15 μm, although thicker coatings are possible. It is difficult to predict the life of the coating and when no other lubricant is present their main use is in assembly of components and where only a small amount of sliding will occur. Friction coefficients can be as low as 0.03. They are also used as a surface coating with the pocket lubricated materials described in Section 9.7.7.2. The coating protects the pockets during installation and promotes rapid formation of a transfer film on the mating surface. When sprayed on metal, the maximum bearing pressure limit is considerably higher than those quoted for their bulk properties.

Lead, gold and silver are also used as dry lubricants for space applications but the deposition process is expensive.[34]

9.7.7 Metallic alloys and composites

9.7.7.1 Copper-based alloys

The main group of metallic alloys used for bearing applications are bronzes. These copper-based alloys offer advantages over most polymeric materials in terms of higher load-carrying capacity, lower thermal expansion and greater resistance to high temperatures. In contrast, their friction and wear characteristics are higher than most polymers and polymer composites. The use of dry lubricants is therefore becoming increasingly popular with these materials to reduce friction and wear.

Bronzes range in hardness from the relatively soft leaded bronze, through the harder phosphor bronze to the aluminium bronzes.[37] The compressive strength of the harder materials can be as high as 300 MPa. Specific data on their tribological properties are limited but wear rates of about $1 \times 10^{-13} m^3/Nm$ and dynamic friction coefficients of between 0.5 and 0.6 are typical under low-load, low-speed conditions. The addition of graphite powder (Devametal by Glacier Ltd) or lead during sintering of the bronze can reduce the material wear rate by an order of magnitude and also reduce friction. Bearing pressures of up to 400 MPa are possible. Graphite in nickel–copper alloys has a similar effect on wear rate and this material can operate in temperatures up to 500°C. Both these materials exhibit friction coefficients in the range 0.25–0.4.

9.7.7.2 Dry lubricated bronzes

Bronzes with discrete pockets of dry lubricant in the bearing surface are commercially available. The pockets normally contain a combination of thermoset resin, PTFE, molybdenum disulfide, graphite and wax in various quantities. If graphite is present, the possibility of electrolytic corrosion in service must be considered. Pressure velocity limits with the pocket-lubricated materials are in the range 0.5–4 MPa × m/s and the maximum bearing pressure (at low speeds) is about 100 MPa. These materials are marketed by several firms including FW Birketts Ltd, Merriman Inc. and Johnson Metals Ltd. Their operating temperature is limited, by the softening point of the resin, to between 150°C and 250°C. Wear rates of as low as $1 \times 10^{-15} m^3/Nm$ have been measured at The National Centre of Tribology (NCT) when sliding against hard, smooth counterfaces and friction values are reduced to between 0.1 and 0.2.

In general, the leaded bronzes should be used in conjunction with an unhardened steel shaft whereas the harder phosphor and aluminium bronzes perform better with a hardened steel counterface. The strength of all these alloys can be increased by applying them as a thin lining to a steel backing.

9.7.7.3 Self-lubricating porous metals

Although dry bearings are normally selected for applications where the use of an oil is undesirable or impossible, this can be due to the problems associated with maintenance or the supply of lubricant to the rubbing surface. Oil-impregnated porous metal bearings do not require an external lubricant supply and offer low friction and wear characteristics over a range of relatively high loads and speeds compared with polymers.

Porous metals are produced by the sintering of loosely compacted metal powders. The degree of porosity can be increased or decreased by varying the compacting loads during the sintering process but is normally in the region of 20–30%[37] of the total volume. Oil is then impregnated into the pores. To ensure maximum penetration of the oil into the material, this is normally carried out under a vacuum.

The two main self-lubricating porous metals are porous bronze and iron. These can be impregnated with a range of oils. In general, higher-viscosity oils, in the range 75–100 centipoise at 600°C should be used for high-load, low-speed applications.[37] Lower-viscosity oils should be selected for the low-load but high sliding velocity conditions. Most porous metal bearings are supplied with a medium viscosity oil such as SAE30.

The life of these bearings is difficult to predict. As the oil in the pores becomes depleted, the absorption ability of the metal increases. Oil loss from the bearing is therefore more rapid during the early stages of operation, reducing to a lower rate with decreasing oil content. The rate of loss is also governed by the operating temperature of the bearing. Since oil viscosity reduces with increasing temperature, the rate of oil usage increases under high speeds as the frictional heat generated at the surface is high. A pressure velocity limit of about 3 MPa × m/s is recommended to minimize the wear rate of these materials when used in bulk form.

A composite material of porous bronze, filled and overlaid with an acetal resin co-polymer is available in strip form which can be formed into bushes and thrust washers (Glacier DX). The friction and wear characteristics of this material are inferior to those of the thin-layer polymer based materials described in Section 9.7.4. However, the material can operate under conditions which would not normally sustain a film of lubricant between the sliding surfaces.

The bearing surface is indented and can be packed with grease. The indentations retain the lubricant and sustain a film on the surface. This enables the material to operate under oscillating conditions where conventional lubricated materials would not sustain a film of lubricant between the surfaces.

Friction coefficients during sliding against steel or cast iron mating surfaces are claimed to be in the range 0.01–0.1. Since the lubricant remains on the surface, the static friction coefficient is also low in the range 0.02–0.15. The maximum operating temperature is limited by the softening point of the acetal resin to 130°C. A new, similar material with PEEK in place of acetal is now available which can operate at 250°C (Glacier Hi-Ex). Above temperatures of 80°C, the use of silicone grease is recommended. With grease lubrication, a pressure velocity limit of 0.7 MPa × m/s is specified but this can be increased to almost 3 MPa × m/s if oil lubrication is possible.

As with polymers, a counterface surface finish of 0.4 μm R_a or less is recommended. The dimensional stability of both these materials is extremely good. The recommended operating clearance for journal bearings manufactured in this material are influenced by the operating pressure and velocity. For a journal diameter of 10 mm the clearance varies between 0.02 mm and 0.06 mm as the pressure velocity term increases from 0.1 MPa × m/s to 2.8 MPa × m/s. For larger journals of about 100 mm diameter, a clearance of between 0.06 mm and 0.2 mm is recommended over a similar range of pressure velocity limits.

9.7.8 Surface treatments and coatings

Surface coatings and treatments offer an ideal solution to many tribological problems. The substrate material can be chosen for strength or corrosion resistance and then treated or coated to give a hard wear-resistant bearing surface.

9.7.9 Conclusions

The prediction of the friction and wear characteristics of materials in sliding contact is not straightforward. Many factors influence performance, including pressure, sliding speed and surface temperature and their effects are often combined. Selection of materials from generalized catalogue data can therefore lead to problems in service. Where an accurate prediction of these characteristics is required, laboratory tests under conditions representative of those found in service should be conducted. With the wide range of dry bearing materials available, the designer should be able to identify one that suits her or his particular requirements.

9.8 Wear and surface treatment

9.8.1 Introduction

Surface engineering is an old concept. A classic example is that of the wooden stagecoach wheels fitted wth a cast-iron hoop to give improved strength and wear resistance. This example embodies all the principles of surface engineering; the wooden wheel is light, tough, flexible, cheap and easy to manufacture. The iron hoop, shrunk fitted, gives increased surface wear resistance and strengthens the whole structure by placing the body of the wheel in compression. Moreover, a worn hoop can be removed and replaced by a new one.

The principles arising in the example above are no less relevant to modern engineering. In all engineering applications it is the surface material of any component which has to co-exist with the external environment, such as contacting surfaces and process fluids. It is, therefore, not surprising that the majority of engineering failures arise from deficiencies in the surface material in dealing with fatigue, friction, wear, corrosion and so on. It follows that the designer, like the wheelwright, should choose the bulk material from the standpoint of structural and economic criteria, and the surface material to deal with the external conditions. The required surface properties can be achieved either by the application of specialized coatings or by modification of the bulk material.

In many applications the concept described above is not correctly applied. All too often it is the engineer, when making and using a component, who discovers the unexpected problem of wear, fatigue or corrosion and then returns to the designer to seek a solution. This usually turns out to be a compromise, in the form of a surface treatment or coating which does not affect the existing design or structural requirements of the component but gives only a partial solution to the tribological problem. It is better if the tribological demands are considered at the design stage and that surface engineering is employed alongside the requirements of the bulk material.

It is the designer, therefore, who has the job of engineering a component while it is still on the drawing board. He or she needs to assess the operational and environmental demands before deciding on the bulk material. Also, an understanding of the tribology of the application is required before an appropriate surface can be selected.

It would be a rare service situation that exhibited a single environmental feature that needed consideration, and most engineering components are subject to a complex range of factors, each affecting material selection. It is widely known, for instance, that complex demands are placed on turbine blade and ring materials for jet engines, where resistance to creep, fatigue, high-temperature oxidation, erosion, fretting, thermal shock, thermal cycling and thermal stability are all required in a single component. Similarly, complex demands are placed on many components in highly sophisticated modern engineering plant, typically in the petrochemical, nuclear process and manufacturing industries. However, even when the environmental demands are not as exacting as those mentioned, there is still a need to consider the effects of service and environmental factors on performance.

To be in a position to make a choice of a particular wear-resistant surface for an engineering application, the designer must first be able to predict the mechanism of wear that is likely to take place in service. He or she then should have some knowledge of what surface treatments and coatings are available. It is not necessary to know the precise details of how the processes are applied, but an awareness of the basic properties such as coating or case-depth, hardness and the likely effect of distortion or tempering on the substrate should suffice. It has to be remembered that surface treatments cover a very wide scale, from several centimetres thick to only fractions of a micrometre, and from comparatively soft materials to extremely hard ceramics. The task would also be made easier if data on friction and wear, preferably obtained under wear conditions appropriate to the particular application, were available.

Accordingly, this section is aimed at the designer; providing some insight into different wear mechanisms and how to identify them, listing and describing the main types of surface treatments and coatings available and then discussing tribological data and their use and interpretation. Finally, it presents a selection philosophy and discusses the vital area of quality control; an important aspect when applying any new process to critical or expensive components.

9.8.2 Wear types

Before prescribing a solution, it is important that the designer or engineer should understand which wear process will be in operation in a particular application. The main wear processes occurring in industry are listed below. For a further description of the fundamental aspects of friction and wear the reader is referred to Section 9.1.

9.8.2.1 Abrasive wear

This form of wear, which is said to account for more than 50% of wear problems in industry, occurs when hard particles such as rocks or metal fragments slide or roll under pressure across a surface. This action tends to cut grooves in much the same way as a cutting tool, displacing elongated chips or slivers of metal.

'Two-body' abrasive wear occurs when particles such as coal or metal ores are transported across a surface. A similar wear process, on a much finer scale, is found in areas such as printing or textiles. This time, the abrasion is by the small particles of pigment carried in the product.

'Three-body' abrasion results from high-stress situations where the particles are trapped between the loaded surfaces in relative motion. The particles are often reduced in size as they would be in ore-crushing machines or when foreign particles ingress into bearings. Generally, only a small proportion of the particles cause wear but, as the attack angle is raised, high wear rates are experienced. With abrasive wear, the load on the particles, their shape and hardness are all important; as, together with component hardness, they will determine whether surface penetration or particle crushing occurs.

9.8.2.2 Adhesive wear

When metal surfaces slide over one another, and the contact pressure is sufficiently high to cause local plastic deformation and welding, adhesive wear occurs. Clean surfaces, non-oxidizing conditions and chemical similarities promote this type of wear, which accounts for about 15% of industrial wear problems. Initially, it occurs between a few particles, which increase in size as motion continues and eventually the junctions rupture to give material transfer. The adhesion and transfer process can lead to debris trapping which may add a three-body abrasive wear mechanism to the process.

Wear under adhesive conditions is often subject to sharp transitions in behaviour and two regimes, 'severe wear' and 'mild wear', can be encountered. These are caused when changes in load or sliding speed produce thermal changes at the rubbing surface. In 'mild wear', the surface oxide protects the rubbing parts and damage is slight. In 'severe wear', the oxide is disrupted, metal-to-metal adhesion results and wear

rates can rise by two orders of magnitude. This form of severe wear is common with metals having only thin, brittle oxide films and results in 'galling' of materials such as stainless steel and aluminium and titanium alloys.

Another form of adhesive wear is termed 'scuffing'. It occurs under partially lubricated conditions when the oil film thickness is insufficient to give complete protection to the rubbing surfaces. Metal-to-metal contact leads to considerable surface deformation and heating, and with additional carbon being supplied by degradation of the lubricant, metallurgical changes are produced at the surface. It is quite common to find hard and brittle layers of carbide produced at the surface and for subsequent cracking and spalling to occur.

9.8.2.3 Erosive wear

Erosion is a form of abrasive wear resulting from impact and cutting by hard particles carried in a gas or liquid. It occurs in components such as turbines, pipework, pumps and valves. Erosion requires a different treatment because the contact stresses arise from the kinetic energy of the particles as they encounter the surface. Particle velocity, abrasive size and impacting angle are all important in determining the amount of erosion. Generally, the wear volume is proportional to the cube of the velocity. When the impacting angle is small, a cutting action takes place and an increase in the material hardness reduces the wear. At large impacting angles, the solution can be quite complex; the wear is due to surface deformation and soft materials may perform better than hard ones.

A particular form of erosion can occur without any abrasive content in the fluids. High-velocity impact of fluid droplets can cause plastic deformation of local areas of a surface and the repeated action results in a fatigue-type pitting failure. Additionally, 'cavitation' erosion damage can be caused by the impact produced from the collapse of vapour or gas bubbles formed in contact with a rapidly moving or vibrating surface. Again, the resultant damage is typical of a fatigue-type process.

9.8.2.4 Fretting wear

Fretting is produced by small oscillatory displacements at interfaces; it occurs in heat exchangers, gas turbines, bearings and where there is a source of continual vibration. There are a number of stages, starting with adhesion and metal transfer between the rubbing parts. This produces wear debris which then oxidizes to form an abrasive. Because the motion is small, the abrasive is retained in the contact area and itself becomes a source of further wear. Eventually, a steady-state wear rate is established.

When dealing with ferrous materials, operating in air, it is normal to get reddish-brown oxide particles or a 'red mud'-like substance surrounding the damaged area; this is the result of 'fretting corrosion'. The high-cycle oscillatory displacements can also lead to surface fatigue and thus 'fretting fatigue' can apply.

Fretting, therefore, involves a complex interaction of adhesion, corrosion, abrasion and fatigue. Further details are given in Section 9.9.

9.8.2.5 Corrosive wear

Oxidative or corrosive wear occurs when the sliding surfaces are attacked by a highly oxidizing or corrosive environment. This results in a build-up of loosely adherent corrosion products which are quickly removed by the wear process. Thus, a fresh metal surface is exposed to further rapid corrosion. Since no passivating layer is maintained, the rate of material loss can be very rapid and can far exceed the sum of the two processes (dry wear or simple corrosion) acting separately.

9.8.2.6 Fatigue wear

Fatigue wear is characteristic of surfaces contacting under high Hertzian contact stresses; typically, the rolling/sliding conditions in ball bearings and gears. The surfaces are usually lubricated and, for long periods, no significant wear takes place. However, the cyclic compressive and tensile stresses produce metallurgical changes in the surface and cracks are initiated. The cracks propagate steadily into the material at an acute angle to the rolling direction and then parallel to the surface to detach material and form a pit. Once such a situation is reached, the rate of material loss becomes rapid and catastrophic failure results.

9.8.2.7 Identifying the wear process

There are two approaches to wear identification. The first is to anticipate it at the design stage; the second usually takes the form of failure analysis; examining worn parts or wear debris with a view to prescribing a solution.

With careful thought at the design stage, it is usually possible to anticipate likely wear problems. If there is metal-to-metal contact in a clean, marginally lubricated machine, wear problems are likely to be of an adhesive type. With unhardened surfaces and no lubrication, the wear may be in the severe regime. If materials such as stainless steel, aluminium or titanium are involved in high load, dry contact, then severe adhesive wear and galling are almost inevitable. When abrasives are present, one needs to obtain data on the size and hardness of the particles, together with information on likely contact loads. This will enable one to characterize the probable stress. If a corrodent is present, then one can anticipate aggressive chemical wear. When vibrations are present, then fretting is likely and with cyclic loading, particularly under Hertzian contact conditions, fatigue wear could occur. Stress calculations, particularly those occurring sub-surface, are a vital part of any selection procedure.

Examination of worn parts, even with a simple low-powered magnifying glass, can yield a useful insight into the wear process. Grooving and cutting (Figure 9.60(a)) indicates abrasive wear, while a torn and rough bright surface (Figure 9.60(b)) suggests severe adhesive wear. Mild adhesive wear is characterized by a smooth oxidized rubbing surface (Figure 9.60(c)). Significant quantities of oxidized debris (Figure 9.60(d)) indicates fretting, and fatigue wear is, of course, characterized by pitting (Figure 9.60(e)).

Although it is possible for the engineer to make some characterization of the wear process, the analysis is best left to the expert in tribology. An alternative approach is to monitor the wear debris since wear particles are unique, having individual characteristics which bear evidence of the conditions under which they were formed. Careful examination of the morphology and determination of the composition of wear particles can thus yield specific information concerning the surface from which they were produced, the mechanism of their formation and the operative wear mode in the system from which they were extracted. Ferrography[38] is a technique developed to separate wear debris and contaminant particles conveniently from a lubricant for examination and analysis. It consists of decanting a sample of diluted machine lubricant down a glass slide held over a magnetic field. Particles are then separated according to size and composition. Particles generated by different wear mechanisms have characteristics which may be identified with a specific wear mechanism. For

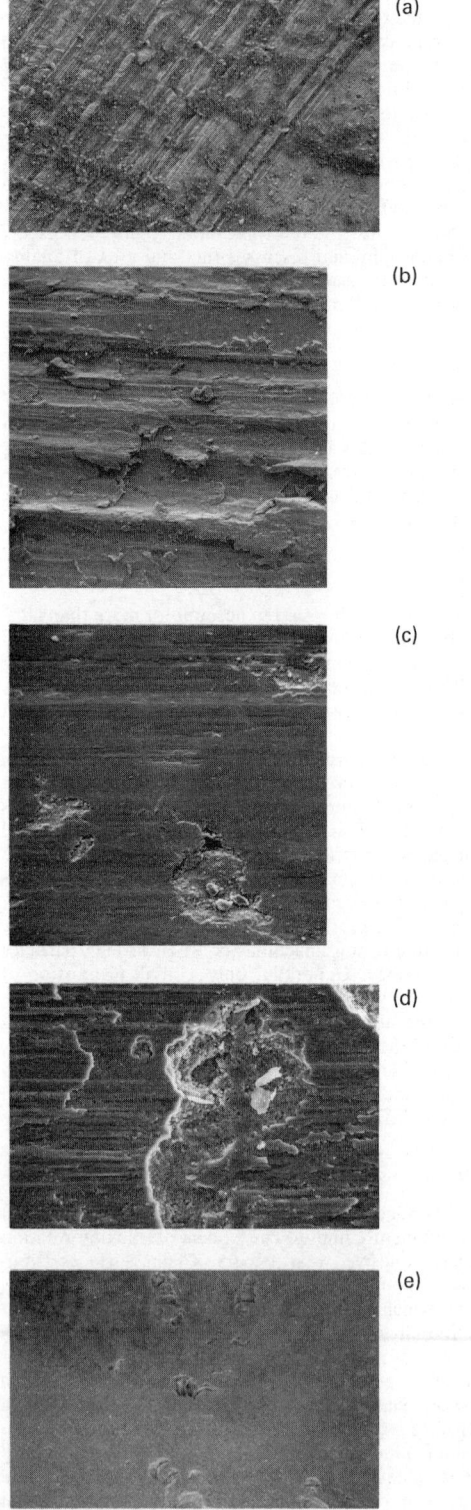

Figure 9.60 Appearance of worn surfaces. (a) Abrasive wear; (b) severe adhesive wear; (c) mild oxidative wear; (d) fretting wear; (e) fatigue pitting

example, wear particles found in most machines have the form of platelets and indicate normal permissible mild adhesive wear. Abrasive wear gives a debris in the form of miniature spirals, loops and bent wires similar to cutting swarf from a machining operation. A concentration of such particles is indicative of severe abrasive wear. Severe adhesive wear produces large metallic particles and fatigue gives laminar, spherical and then chunky particles as the failure progresses. A wear particle atlas[39] has been produced to assist with particle characterization.

It is possible, therefore, to identify wear processes, but by this time costly failures have occurred and there is no substitute for addressing the problems at the design stage and employing the correct form of surface engineering.

9.8.3 Surface treatments and coatings

There are many ways of treating metal surfaces to improve their wear resistance or to reduce friction. They can be grouped into three broad categories:

1. Modifying the surface without altering the chemistry;
2. Changing the chemistry of the surface;
3. Adding a layer of new material to the surface.

Techniques from any of these categories may be used in combination with others; for example, electroplating may be followed by a diffusion treatment, a sprayed coating may be sealed with a polymer or lubricant or two different coating processes may be used to lay down duplex layers. The various processes are described in detail below and the main properties are summarized in Table 9.21.

9.8.3.1 Modifying the surface without altering the chemistry

The most common method of modifying the surface is to transformation-harden the outermost region of the component by localized heating. The method of heating is usually by induction or by flame, but more specialized treatments using a laser or electron beam as the heat source are also available. First, the desired core properties are developed by standard heat treatment practices. The surface is then hardened by rapid heating to transform the surface regions to austenite, and quenched to produce martensite. In most cases, it is usual to follow the hardening by a low-temperature tempering treatment to relieve the internal stresses. The process has the advantage that it can be applied to selected areas of a component without affecting the bulk properties of the material.

Transformation-hardening is used to combat wear under high loads, where there is sufficient case depth generated to support the contact stresses. Hardening the steel surface results in increased support to the oxide layer and, in adhesive wear situations, gives mild wear. It is also effective against high-stress abrasive wear (unless the abrasive is exceptionally hard, such as alumina or silicon carbide). However, transformation-hardening produces little change in the corrosion properties and components can suffer both corrosive wear or fretting damage.

Induction hardening[40] When an electric current is passed through a conducting coil a magnetic field is produced inside it. If the magnetic flux is created by a high-frequency alternating current it gives rise to eddy currents in the surface of any metal object placed within the coil and surface heating results. In this way, the surface of steel parts can be heated above the austenitic transformation temperature. Quenching of the surface is usually achieved by using a water- or oil-emulsion spray attached to (or below) the heating coil, with the component being passed through at a predetermined rate.

Steel grades suitable for induction hardening are those having a carbon content in the range 0.30–0.50%. If steels with a higher carbon content than 0.5% are to be treated, there is a risk of cracking and the heat treatment must be carefully controlled. The steel, prior to induction hardening, should be in the hardened and tempered or normalized state. Fully annealed steel is unsuitable, because the time needed to dissolve the carbides is longer than the heating-up time.

The depth of hardening and actual hardness depend on a number of factors. The carbon content of the steel has a decisive influence on the hardness, and the depth of hardening is determined by the temperature, the holding time, the frequency of the inductive current and the quench rate. For

Table 9.21 Important properties of wear-resistant surface treatments and coatings

Treatment type	Substrates	Treatment temperature (°C)	Typical case or coating depth	Hardness (HV)
Induction hardening	0.3–0.5%C steels	Over 800 (surface)	Several millimetres	600–800
Flame hardening	0.3–0.5%C steels	Over 800 (surface)	Several millimetres	600–800
Laser hardening	0.3–0.5%C steels	Over 800 (surface)	Up to 0.75 mm	600–800
Cold working	Steels	Ambient	Up to 0.25 mm	Slight increase
Carburizing	Steels	Over 800 (bulk)	Several millimetres	600–800
Carbonitriding	Steels	Over 800	Several millimetres	600–800
Boriding	Steels	800–1000	0.1 mm	Up to 1500
Carbide diffusion	High-carbon steels	1000	10 μm	Up to 3500
Nitriding	Alloy steels	500–600	0.2 mm	700–1200
Nitrocarburizing	(i) Alloy steels	500–600	0.2 mm	700–1200
	(ii) Carbon steels	500–600	20 μm	500–700
Oxidation	Steels	300–600	5 μm	approx 350
Plating/diffusion	Steels	500–600	20 μm	200–950
Anodizing	Aluminium alloys (low Cu+Si)	60	20 μm	500–600
Electrolytic/diffusion	(i) Aluminium alloys	150	20 μm	400–450
	(ii) Copper alloys	400	20 μm	400–450
Ion implantation	All metals	100	0.1 μm	Unknown
Physical vapour deposition	Steels	200–500	3 μm	Up to 4000
Chemical vapour deposition	Tungsten carbide steels	Over 800	5 μm	Up to 2000
Chromium plating	Most metals	60	20–100 μm	900
Electroless nickel	Most metals	60	10–50 μm	550–1000
Flame spraying	Most metals	100–200	Up to 0.5 mm	Depends on material
Spray/fuse	Steels	700–800	Up to 0.5 mm	Up to 900
Plasma spraying	Most metals	100	Up to 0.2 mm	800–1600
High-velocity spraying	Most metals	100	Up to 0.2 mm	800–1600
Welding/cladding	Steels	Over 800	Several centimetres	Depends on materials

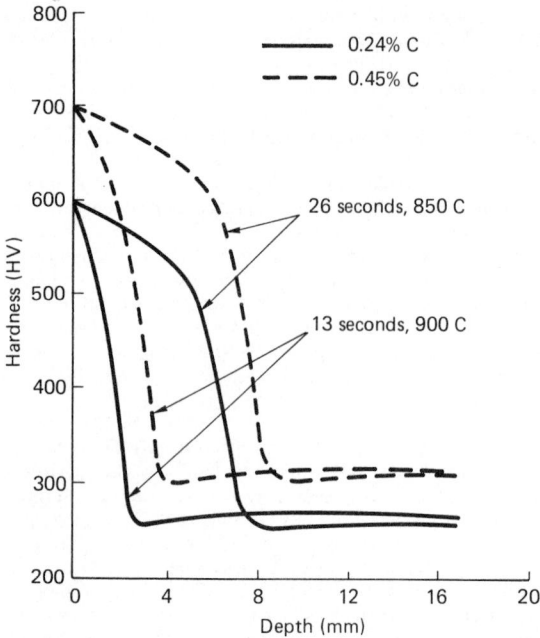

Figure 9.61 Hardness profiles for 0.24%C and 0.45%C steels after induction hardening (after Thelning[40])

instance, Figure 9.61 shows the typical behaviour of two steels with 0.24% and 0.45% carbon with different holding times. In fact, the temperature, heating time and holding time during induction hardening can all be closely controlled, so it is possible to select in advance the required hardness and depth.

Induction hardening is ideal for large objects and has the advantages of low cost, short heating periods with minimum surface decarburization and oxidation. Dimensional changes are minimal and case depths are of the order of several millimetres.

Flame hardening[40]   Flame hardening and induction hardening give broadly equivalent results. Flame hardening is achieved by heating the steel to the hardening temperature using a welding torch and then quenching in oil or water. It is less easy to control and surfaces can sometimes be overheated, oxidized and decarburized. Best results are obtained on bodies having rotational symmetry by spinning them on a rotating table under the flame.

The grades of steel recommended for induction hardening are also suitable for flame hardening and the resulting hardness and case depth is again determined by carbon content, hardening temperature and quench rate. Flame hardening has the advantage that it can be directed to selected surface areas.

Laser and electron beam hardening   Lasers are traditionally used for welding and cutting but they can now be employed as a versatile heat treatment tool.[41] The laser is focused onto the surface and scanned over the specific area where increased hardness is required. Because of the massive amount of steel around the narrow heated area, the process is self-quenching. As with induction and flame hardening, the ideal steels are those with a medium carbon content but the achievable case depth is limited by the onset of surface melting and is usually 0.5–0.75 mm.

There are many advantages in using laser processes; they provide fast production, low distortion and avoid the necessity for post-treatment operations, including machining or grinding. However, lasers are expensive and there is some indication that electron beam processes may be more economical than laser techniques since both capital and operating costs are lower. In electron beam hardening the beam operates in a vacuum but the workpiece needs only to be at 0.06 m bar pressure. Area hardening is obtained by scanning the area in a raster; the dwell time, beam energy and focus all being controlled by computer.

Cold working   An alternative to transformation hardening the surface is to increase the hardness by cold working. The most common method is to use shot peening[42] where compressive stresses are induced in the exposed surface layers of metallic parts by the impingement of a stream of shot. The shot, made of iron, steel or glass, is directed at the metal surface at high velocity with careful control of angle, scanning rate and shot recycling.

The increase in hardness produced by this treatment is modest. The most important effect is the production of a surface compressive stress, extending to a depth of up to 0.25 mm. If a bar is treated on one side only, the effect is dramatically illustrated by significant bending of the component. The process is, therefore, used more to improve fatigue properties and to increase resistance to stress corrosion cracking rather than to give better abrasive or adhesive wear resistance.

9.8.3.2 Changing the chemistry of the surface

The surface chemistry can be altered in many ways. Most commonly, it involves the introduction of interstitial elements[43] such as carbon, nitrogen or boron or a combination of carbon and nitrogen into the metal (usually ferrous) surface at elevated temperature. Other processes, such as aluminizing, chromizing and siliconizing, use metallic or metalloid elements to alter the chemistry of the surface. Equally, the interstitial element can be implanted into the surface (at ambient temperature) in the form of a high-energy ion beam (ion implantation). There are also processes which combine electroplating and diffusion treatments. These involve the deposition of materials such as tin, sulphur or bronze onto a ferrous component followed by a diffusion treatment to form compounds at the surface. Similar types of treatment are also available for non-ferrous materials such as bronze–titanium alloys and aluminium alloys. Oxide layers can produce significant tribological advantages and these can be produced on both ferrous materials (by thermal treatment) and on some non-ferrous materials (by anodizing).

By far the most important of these techniques are the thermochemical diffusion treatments for ferrous metals. These come under two headings: those carried out at relatively low temperatures in the ferritic range, and those performed at high temperature in the austenitic range. The high-temperature processes are discussed first.

Carburizing   Carburizing[40] (also called case hardening) is the most commonly used of the surface-modification processes. Ferrous alloys are heated to a temperature above the ferrite/austenite phase transformation temperature (825–925°C) while in contact with a solid, liquid or carbonaceous medium. This results in the diffusion of carbon into the steel to a depth controlled by the diffusion time and temperature, so that the surface can then contain up to 0.7–1.0% carbon. The steel is then quenched to produce a martensitic structure with a hardness of up to 1150 HV. Tempering, to reduce residual stresses, is usually performed at between 150°C and 180°C.

There are four different techniques for carburizing. In pack carburizing, the parts are placed in contact with charcoal in a sealed box and carburizing takes place via a reaction at the surface to carbon dioxide and carbon monoxide. In salt bath carburizing, the active agent is sodium cyanide and the reaction again proceeds by the gaseous phase. Nowadays, the most popular method is by gas curburizing, when hydrocarbons, as well as carbon monoxide and carbon dioxide, are employed. Finally, there is also the possibility of plasma carburizing,[44] a technique using a low-pressure glow discharge. It is still in its infancy but promises to be more energy-efficient and can produce deeper cases than conventional carburizing.

Carburizing, like induction hardening, produces hardened layers to a depth of several millimetres. The depth of case hardening is defined as the distance from the surface to a plane at which the hardness is 550 HV. Figure 9.62 shows, schematically, the hardness profile across a section of a typical case-hardened steel. The actual hardness and depth of hardening are dependent on several factors; for instance, the carbon potential of the carburizing medium, the quenching temperature, the quenching medium (water quenching produces a greater hardened depth than oil quenching) and the steel composition.

The dimensional changes that occur during case hardening are critical, since they determine the extent of any post-grinding operations. Changes are usually of the order of 0.1% but they can be both expansion or contraction, depending on the type of steel, the depth and method of hardening and material dimensions. The user is referred to reference 40 for detailed examples on how different-shaped components are distorted during treatment.

Carburizing is used for a wide range of components, specifically to improve abrasive or adhesive wear resistance and to reduce fatigue. A high hardness and good case depth are essential to combat high-stress abrasive wear (for instance, on pump parts). Fatigue properties are improved by the increased hardness and also by the production of a compressive stress. Many gears are carburized. However, carburizing will have little benefit to corrosion resistance so that corrosive wear or fretting situations may not be suitable applications.

Carbonitriding[40] In the case of carbonitriding, the name is misleading. It is not a form of nitriding, which is performed at a relatively low temperature with the steel in the ferritic condition, but it is a form of carburizing. The treatment involves the introduction into the steel of nitrogen as well as carbon at temperatures between 700°C and 900°C, either from a molten salt bath or in a gas-carburizing atmosphere into which ammonia gas has been introduced. It might have been more logically named nitrocarburizing but this term has been used to describe processes at lower temperature (discussed later) and confusion is bound to continue.

The addition of nitrogen to the case of a steel considerably increases its hardenability and it leads to improved wear resistance compared to ordinary carburizing. Consequently, engineering components made from plain carbon steel, which on carburizing would have to be quenched in water, can be quenched in oil with less danger of distortions.

There are several proprietary variations on this type of process designed to give surface layers with specialized properties. The Noskuff[45] treatment consists of an initial carburizing treatment at 800–900°C before the hot parts are transferred to a Noskuff bath. This contains cyanide and 10–20% sodium cyanate at a temperature of 700–760°C and parts are treated for 5–15 min. The parts are then quenched into oil or water, in order to harden the carbon-rich case which would have remained in the austenitic conditions during the time of the Noskuff treatment. As the name implies, this treatment gives good resistance to scuffing by producing a hard martensitic layer near the surface (Figure 9.63).

Boriding[40,46,47] This heat treatment is based on the diffusion of boron into plain carbon and low-alloy steels, converting the metallic surface of components into iron boride. It is a pack process in which engineering components are placed into a muffle furnace and entirely surrounded by a boron-rich granulate. Heat treatment is carried out at 800–1000°C for 1–8 h, depending on the depth of the required layer. The surface hardness of treated steels is high, typically 1500 HV, and the case depth to a hardness value of 500 HV is typically 0.05 mm.

All steels can be borided but the thickness of the borided layer decreases as the alloying element content increases.

Chromizing, aluminizing, siliconizing[48] There are a number of high-temperature pack processes aimed at surface diffusion

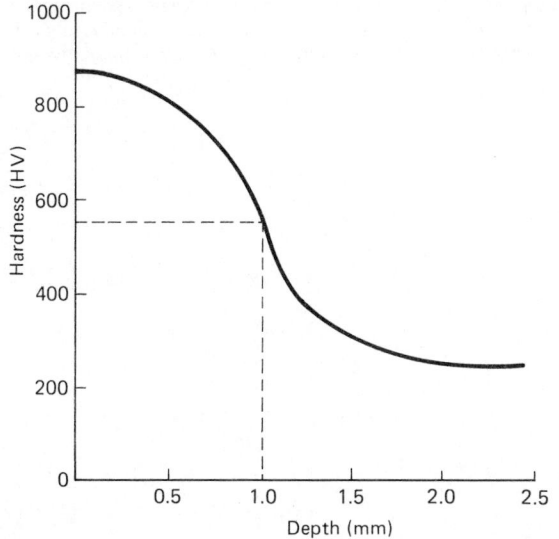

Figure 9.62 Schematic representation showing depth of case-hardening

Figure 9.63 Polished microsection through a Noskuff layer

of metals such as chromium, aluminium and silicon into ferrous surfaces. These are very specialized techniques for improving wear and corrosion at elevated temperatures in hostile environments. For instance, they are used in plant in nuclear reactors and for containing agressive chemicals in electrical batteries.

Carbide diffusion treatment Carbide diffusion coating is a relatively new addition to the range of high-temperature processes. By using very specialized salt baths at closely controlled temperatures (typically, 1020–1050°C), compound layers of alloy carbides, such as vanadium carbide, niobium carbide and chromium carbide, can be produced on the steel surface. The mechanism of carbide layer formation relies on carbon from the substrate migrating to the surface to combine with the metallic elements in the salt. The process produces a dense layer with a clear substrate boundary (for example, the Toyota Diffusion Process[49] – Figure 9.64). The hardness varies with the type of carbide, but can be as high as 3500 HV.

The process is used extensively to treat tools used for steel forming, cutting, forging or press tools and greatly reduces adhesive wear and metal pick-up. The high surface hardness also gives good resistance to abrasive wear. However, the layers are relatively thin (5–15 μm) and loads should not be excessive.

Component growth is equivalent to the thickness of the layer.

Nitriding[40] All the processes described above are performed at a high temperature, with the steel in the austenitic condition. They involve the diffusion of interstitial alloying elements followed by a quench and temper and, therefore, significant distortion. Nitriding is carried out with the steel at a lower temperature, in a ferritic condition and there is no phase transformation on cooling and, consequently, less distortion. Because the temperature is lower, diffusion is slower and the case depths are more shallow than those produced by austenitic thermochemical diffusion treatments.

Nitriding is a case-hardening process in which nitrogen is diffused into the steel surface. The process is carried out at 500–525°C, which is below the eutectoid temperature, and involves the formation of hard, wear-resistant iron-alloy nitrogen compounds. No phase transformation can occur on cooling of the steel and no quenching is required to develop surface hardness. Only steels which contain stable nitride-forming elements such as aluminium, chromium, molybdenum and vanadium are suitable for nitriding. Prior to nitriding, the steel must be hardened and tempered at a temperature of about 650°C to produce a fully stabilized tempered martensitic structure.

There are three methods of nitriding – gas nitriding, plasma (or ion) nitriding and powder nitriding. These can each involve the absorption of nitrogen only, but a fourth technique, the salt bath process, always employs the diffusion of both nitrogen and carbon. This is often referred to as salt bath nitriding but it is more properly called 'nitrocarburizing'. This process is discussed below and is not to be confused with the austenitic treatment; 'carbonitriding' (Section 9.8.3.2). Certainly, any engineer or designer can be excused if they become confused. The gas or plasma nitriding processes produce only nitrides but if hydrocarbons are added to the gas mixtures, both nitrogen and carbon are diffused into the surface. Hence, both 'gas nitrocarburizing' and 'plasma carburizing' can be added to the bewildering list of processes.

Gas nitriding[40] takes place at about 520°C for periods between 10 and 100 h, giving case depths between 0.2 and 0.7 mm. (The case depth for nitrided layers is usually defined as the depth at which the hardness falls to 400 HV.) The process uses ammonia flowing over the components in an electrically heated furnace. A typical nitrided case is shown in Figure 9.65 for a 3%CR 0.5%Mo alloyed steel. There is a surface compound layer of iron nitrides (called 'which' layer because it is non-etching) which can be up to 50 μm thick. Below that there is a hard diffusion zone with alloy nitride precipitates.

Plasma nitriding[40,50] uses an ionized low-pressure gas, usually a mixture of nitrogen and hydrogen, which serves as both a source of heat and a supply of nitrogen. Ionization is created by applying a potential (up to 1000 V) between the workpieces and the chamber. The characteristics of the nitrided cases are similar to those produced by gas nitriding but the plasma process is flexible and can be more closely regulated. The temperature can be varied between 420°C and 700°C and the composition of the surface iron nitride compound layer can be closely controlled. The process can effectively treat the internal surface of holes as small as 2 mm diameter and chambers are available that can accommodate components up to 5 m long. Powder nitriding[40] is probably the least used of the nitriding processes. It is similar to pack carburizing, the parts being packed in contact with a nitrogen-

Carbide layer

Substrate

20 μm

Figure 9.64 Toyota diffusion treatment

Gas-nitrided layer

Substrate

50 μm

Figure 9.65 Gas-nitrided Cr/Mo steel

rich powder. It gives equivalent results to gas nitriding but is limited in processing time (and, therefore, case depth) to about 10 h because of the possibility of pitting.

As stated above, any steel with a nitride-forming alloying element can be treated. However, the response to nitriding differs greatly according to the composition of the steel. Figure 9.66 shows typical hardness profiles for four plasma nitrided steels; a 3%Cr 0.5%Mo steel, a high-speed steel, a hot-work steel and an austenitic stainless steel. Broadly, the case hardness increases as the content of alloying element increases (e.g. chromium) but the case becomes more shallow and abrupt. Steels that are very rich in chromium can only be effectively treated by the plasma process, ion bombardment being the only method of breaking down the chromium oxide surface layer to allow diffusion of the nitrogen. There is also a difference in the corrosion properties. A nitrided low-alloy steel can form a surface compound layer (iron nitrides), which improves corrosion resistance over that of the substrate. In contrast, plasma nitrided stainless steel forms no compound layer and corrosion resistance is seriously impaired. In fact, nitriding a stainless steel which has been selected for its corrosion properties is generally a pointless exercise.

Compared to the austenitic processes, the dimensional changes associated with nitriding are small. Typically, a component will swell by 0.02 mm per side.

Although nitriding produces high values of surface hardness, the treatment is not necessarily suitable for abrasive wear situations. Under high point loadings, the case is liable to chip or flake and parts with sharp corners or cutting edges can be particularly vulnerable. In fact, nitriding finds most of its uses in reducing adhesive wear; particularly eliminating severe wear and scuffing. The compound layer tends to give reduced friction and, because of its improved corrosion resistance, can reduce chemically assisted wear. It must be remembered, however, that if the compound layer is removed (for instance, by grinding to a required dimension) or if the substrate is a stainless steel, no such improvement will be gained.

Nitriding also improves the fatigue properties and, because of its good hot hardness and resistance to tempering, it can be applied to parts that might operate at temperatures up to 600°C.

Nitrocarburizing[40,51] The longest-established method of nitrocarburizing is, as stated above, the salt bath process. It uses a mixture of salts based on sodium and potassium cyanide

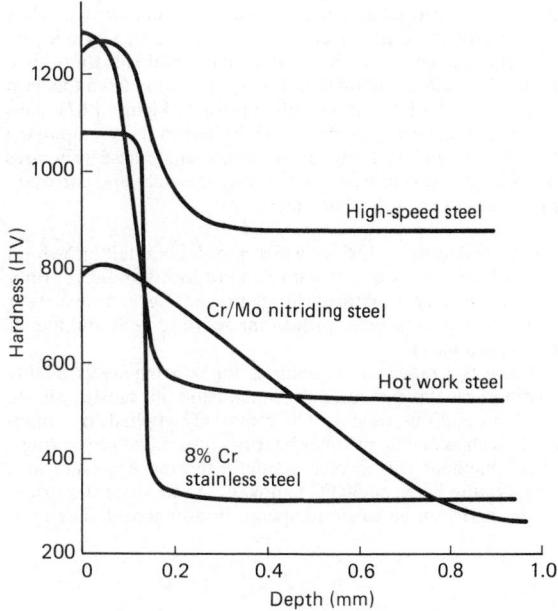

Figure 9.66 Typical hardness profiles for four different plasma-nitrided steels

which react to form cyanates and supply nitrogen and carbon to the surface being treated. Treatment times are usually 2–4 h. There are a number of proprietary variations on the process, each bringing a new development. The Tufftride process uses a titanium rather than an iron crucible to minimize decomposition of the salt and air is injected into the bath to promote the formation of the cyanate. In the Sulfinuz process, sulphides are added to the salt and these help to catalyse the conversion of the cyanides to cyanates and also lead to diffusion of sulphur into the surface. This gives excellent anti-scuffing properties and parts which have been Sulfinuz treated should not be finish ground. A second variation of the sulphur-rich treatments is the Sursulf process, where lithium compounds are used instead of cyanides to reduce environmental problems.

It is these environmental problems with salt baths that have led to the development of the gas nitrocarburizing and plasma nitrocarburizing processes. The gas process involves treatment at about 570°C in an atmosphere of ammonia and endothermic gas for 2–3 h, followed by a quench in oil (to give maximum improvement in fatigue properties) or cooled under recirculating gas. As with the salt bath processes, there are several proprietary variations and some of these (for instance, the Nitrotec[52] process) are finished with an oxidation stage, to give a slightly porous surface, and then waxed or oiled. This reduces friction and improves corrosion resistance.

The plasma process is similar to that for nitriding but methane gas is added to the gas mixture to contribute carbon as well as nitrogen.

Broadly, nitrocarburizing acts in a similar way to nitriding and, for alloy steels, gives equivalent case properties and dimensional changes. However, the nitrocarburizing processes are now used increasingly to treat (and uprate) plain carbon steels. The improved control over the chemistry now allows the processes to produce hard, dense surface layers of iron nitrocarbides; the same 'white' or 'compound' layers which used to be regarded as a nuisance during plain nitriding. They are usually 25–50 μm thick. In the absence of any alloying elements in the steel, there is no hardened diffusion zone under the surface and there is a sharp transition from the hard layer (over 600 HV) to the soft substrate (Figure 9.67). This means that applications should be limited to those employing light loads but there can be no doubt that these processes provide an excellent way of improving the wear and corrosion properties of cheap carbon steels.

Oxide treatments[53]   Oxides which are deliberately grown on the surface of steel components can produce significant tribological advantages particularly in preventing adhesive wear and, when oil is present, preventing scuffing wear and metal-to-metal transfer.

There is a range of possibilities for treating steels. Oxide coatings can be produced by immersion in caustic nitrate solutions and this treatment is commonly applied to components such as needle or roller bearings, gears and piston rings. Other methods can involve a simple thermal exposure at a temperature between 300°C and 600°C to produce the oxide, or the parts can be steam tempered or autoclaved. The most striking example of the use of steam tempering is on high-speed steel cutting tools where the oxide reduces pick-up and extends the life by about a factor of 2 over that given by bright components. The oxide film is mainly Fe_2O_3 and is about 3 μm thick.

Iron oxides are not particularly hard or tough and cannot be expected to be effective in reducing abrasive wear.

Plating/diffusion treatments   Although these processes involve the addition of a new material to the surface in the form of a coating, the post-diffusion treatment brings them under the category of 'altering the surface chemistry'. There are a number of proprietary treatments under this heading, each aimed at giving a ferrous surface the desirable tribological properties associated with some non-ferrous materials.

The Stanal[54,55] process is a two-stage treatment involving electrodeposition on ferrous metals of a tin–antimony alloy, followed by heat treatment at 500°C in a nitrogen atmosphere. The alloy coating becomes integrated with the metal base and tin, being the principal element added by diffusion, forms intermetallic Fe–Sn–C such as Fe_3SnC.

The hardness of the diffused layer can vary between 600 and 950 HV, depending on the carbon content of the steel. In addition, the presence of carbon assists the mutual diffusion of iron and tin, and increases the depth of the diffused layer. Carbon steel treated by this process has exceptional corrosion resistance and, when tested in a salt spray, is claimed to be only marginally inferior to stainless steel. Treatment thickness is about 20 μm.

The Forez[54,55] process was designed to apply to ferrous components (principally, mild steels and cast iron) a bronze coating which was capable of deforming plastically during running-in wear. The process is based on the electrolytic deposition of copper and tin followed by heat treatment at between 550°C and 600°C. This provides complete mutual diffusion of the elements within the coating and some diffusion of these metals into the steel. The hardness of the bronze coating can vary between 170 and 550 HV but is usually of the order of 300 HV. Thickness is typically 20 μm.

In the case of the Sulf-BT[55] process, a coating of iron sulphide (FeS) is applied to the surface of steel components. This electrolytic process uses a molten salt bath as the electrolyte at a temperature of only 190°C; the components to be treated are made the anode and the bath is made the cathode. The treatment time is nominally 10 min for which an FeS conver-

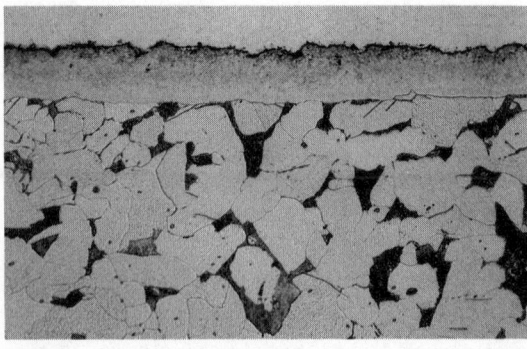

Ni plate

Gas-nitrocarburized layer

Substrate

$\vdash\!\!-\!\!-\!\!\dashv$
50 μm

Figure 9.67   Gas-nitrided 0.4%C steel

sion coating of 7.5 μm may be expected. The treatment may be applied to all iron and steel components including those containing up to 13% chromium, whether or not they have been previously heat treated. The low operating temperature allows hardened or case-hardened parts to be treated without tempering-back as would occur after treatment by the nitro-carburizing processes.

Treatments for non-ferrous materials Although the majority of tribological applications involve steel components, there will be some occasions when the designer will wish to use non-ferrous materials. This may, for instance, be for reasons of weight saving, for increased specific strength or to satisfy a need for good thermal conductivity. These requirements might be met by titanium, aluminium or copper alloys but their use will inevitably bring tribological problems. In particular, aluminium and titanium alloys are very prone to severe wear, seizure and galling, and some sort of surface treatment will be required.

Anodizing[56,57] can be applied to a range of metals, including aluminium and titanium. It is an electrolytic process with the metal as the anode so that an oxide coating is developed on the surface (Figure 9.68). For wear resistance, a thickness of at least 20 μm is required, giving a growth in the surface of half that value. Oxide layers produced by anodizing have a hardness around 500–600 HV and, provided the loads are low, do give protection from adhesive wear. For aluminium in particular, the situation is very much one of 'ice-on-mud' with a hard, brittle layer on a very soft substrate. Also for aluminium, some anodizing processes are available with the addition of a solid lubricant such as PTFE. This further reduces friction and gives good 'non-stick' properties.

Other treatments are available for aluminium and titanium alloys. The Zinal[56] treatment for aluminium alloys involves a two-stage process; electrolytic deposition of 1 μm of zinc, 2 or 3 μm of copper and 12–15 μm of indium, then a diffusion treatment at 150°C for 3 h. The polymetallic coating becomes integrated within the aluminium alloy and intermetallic compounds of indium and copper are formed which gives rise to a surface layer hardness of 400–450 HV. The process improves the resistance to adhesive wear and galling (cold welding) and has no significant detriment to electrical or thermal conductivity or to corrosion resistance. Titanium alloys can be plasma nitrided, giving a thin layer of titanium nitride, but the process requires a temperature near 900°C and may give unacceptable

distortion. Alternatively, they can be salt bath treated (the Tiduram[56] process) giving a compound layer of nitrides and carbides with good resistance to adhesive wear.

The Delsun[55] process has been developed specifically for copper-based alloys, brasses and bronzes. The process is similar in principle to the Zinal process for aluminium alloys, and involves the electrolytic deposition of an alloy of tin, antimony and cadmium followed by heat treatment at a temperature of 400°C. Diffusion occurs and intermetallic compounds are formed giving a surface layer with a hardness of 400–450 HV. It increases resistance to adhesive wear and seizure and improves corrosion resistance of both brass and bronze. Treatment thickness is of the order of 2 μm.

Ion implantation[58,59] Ion implantation involves the bombardment of the metal component surface with a high-energy beam of ions. Any species can be implanted but the usual material is nitrogen. The process takes place in a vacuum chamber and the accelerating voltage is of the order of 100 kV, giving a depth of penetration of about 0.1 or 0.2 μm. Thus, the affected zone is extremely shallow but the local effect on the surface properties is significant. A high-volume fraction of ions is forced into the metal (irrespective of the mutual solubility), creating a high compressive stress and increasing the hardness. The actual effects, because they act over such a shallow depth, are difficult to quantify but the process does bring benefits to both wear resistance and corrosion protection. The process appears to be most effective in situations where component wear is already low but further improvement is required. For instance, ion implantation can reduce adhesive wear on tungsten carbide forming tools and it is also effective against low-stress abrasive wear in plastic moulding tools.

Ion implantation is a low-temperature process (about 100°C) giving insignificant dimensional changes to the component. It is a 'line-of-sight' process so the parts to be treated have to be manipulated under the beam, inside the vacuum chamber. It is not, therefore, a cheap process but, for high-value components, the benefits can be well worth while.

9.8.3.3 Adding a layer of material to the surface

There is an enormous list of processes which involve coating a component with an extra layer of material. They range in thickness from just 1 or 2 μm to several centimetres and can

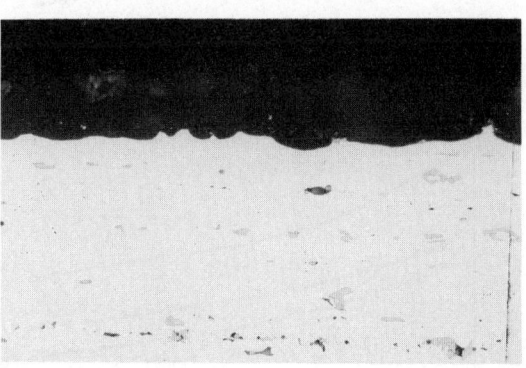

'Black' anodized layer

Substrate

20 μm

Figure 9.68 Anodized aluminium alloy

be ceramics, metals, alloys, plastics, solid lubricants or composites of more than one of these materials. They are applied for a variety of reasons; to reduce wear or friction, to give corrosion protection or to reclaim a component which has already suffered material loss by corrosion or wear. The main processes are described below in approximate order of thickness, starting with the thinnest coatings.

Ion plating, physical vapour deposition Ion plating[60] involves the deposition of thin coatings on components placed within a vacuum chamber. The system is evacuated and then refilled to a low pressure of argon gas. A high voltage is applied and a glow discharge is struck between the components (the cathode) and the chamber wall (the anode). When the material to be deposited is introduced into the discharge (say, by evaporation using a crucible with resistive heating) some of the atoms become ionized and are attracted towards the components. The ions transfer energy to the remaining neutral atoms by collisions and these, too, deposit on the components. Thus, a coating is steadily built up.

The wider terminology for this type of process is physical vapour deposition (PVD). Many materials can be deposited and there are many variations on the main principle, the differences occurring in the way material is introduced into the discharge. Metals such as aluminium or lead can be evaporated by resistive heating; aluminium coating being used for corrosion protection while lead is employed as a solid lubricant for applications in vacuum. To evaporate the high melting point refractory metals such as titanium or zirconium it is necessary to use either an electron beam (E-beam evaporative PVD) or an electric arc discharge (arc-evaporative PVD). Such systems are used to deposit coatings such as titanium or zirconium nitride, the nitrogen being added to the argon to create a reaction with the evaporating metal within the glow discharge. The process is also termed reactive ion plating. An alternative way of introducing the metal is to use the argon ion-bombardment from the glow discharge itself to 'knock off' atoms from the target by momentum transfer. This is called sputtering[61] and is a slower process than evaporative PVD. The sputtering process can be enhanced by the use of magnetic fields; this is termed 'magnetron sputtering'.[62]

PVD has two basic advantages. The first is that the thermal energy that would normally be required is replaced by ionization energy so that the temperature of deposition is relatively low (450°C or below). This minimizes distortion or tempering of components. The second advantage is that, prior to deposition, the components can be negatively biased so that they are atomically cleaned by ion-bombardment. This gives good adhesion to the coating.

The main coating applied by these PVD techniques is titanium nitride[63] (TiN). It has a gold colour, is extremely hard (more than 3000 HV) and significantly reduces adhesive wear. The coating thickness is usually 2–5 μm (Figure 9.69). TiN is applied widely to high-speed steel cutting tools and can increase the life of twist drills by up to a factor of 6 or, more usefully, can allow tools to be operated at greatly increased production rates (higher feeds and speeds – Figure 9.70). This is also used on tool-steels, improving the life of tools in cold forming, warm forming, blanking, rolling and shearing; typically by a factor of 3.

Because of its resistance to oxidation and adhesive wear, TiN can be used to resist fretting wear. It can be applied to shafts or housings for contact with plain or ball-bearings. However, like any thin coating, TiN is best suited to hard substrates which give it proper support.

Because of its very high hardness, TiN is excellent at resisting low-stress abrasive wear. It is particularly useful for protecting components that handle materials such as paper,

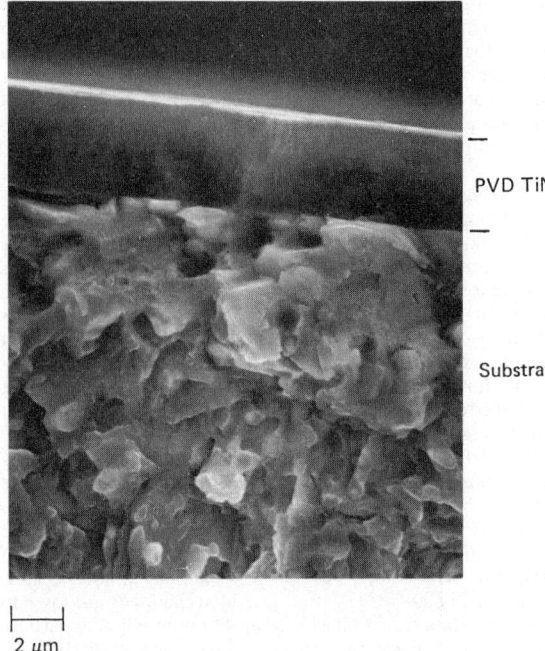

PVD TiN

Substrate

$\vdash\!\!-\!\!\dashv$
2 μm

Figure 9.69 Fracture section through PVD TiN coated high-speed steel

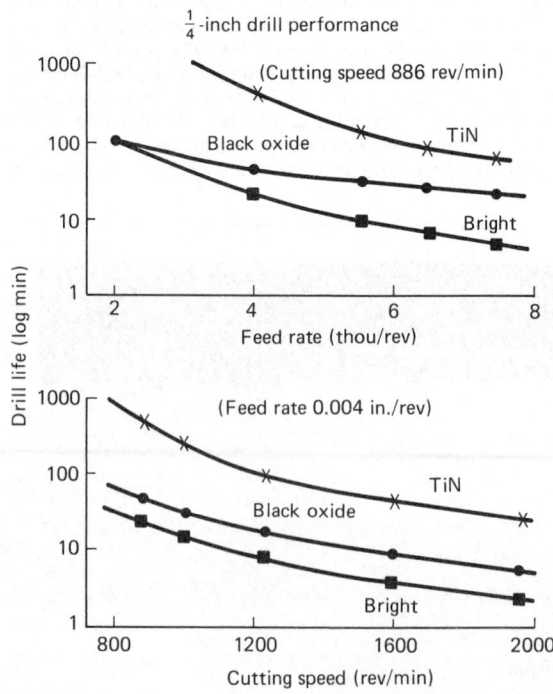

Figure 9.70 Improved life of PVD TiN coated twist drills

textiles and filled plastics. It can also be applied to shafts, gears or plain bearings in equipment handling abrasive slurries. If there is an element of corrosion then the substrate itself needs to be resistant. Alternatively, TiN can be applied over the top of another coating (for instance, electroless nickel). It can be applied to carburized surfaces with only minimal tempering, but if it is to be applied to a nitrided or nitrocarburized material, the compound layer must be removed first.

There are now PVD chambers that can coat components up to 3 m long and processing time and costs are being continuously reduced. PVD is a semi-'line-of-sight' process and so holes and re-entrant areas will receive less coating. Typically, at the equivalent of one diameter deep down a hole, the coating thickness will be 30% of that outside the hole. The coatings replicate the underlying surface texture without build-up on edges or corners.

Finally, PVD is a high-technology process. It is best suited for high-quality components; bright, clean, free from oxide, burrs and contaminants.

Chemical vapour deposition[64,65] Chemical vapour deposition (CVD) is, in many ways, a competitor of physical vapour deposition. The process is used to deposit metals and ceramics by the decomposition of a reactive gas at the surface of components placed within the chamber. For instance, titanium carbide is produced by a reaction between titanium tetrachloride and methane, or titanium nitride can be produced by replacing the methane with nitrogen or by using hydrogen and ammonia.

Deposition temperatures are very high, usually above 800°C. Thus there are potential problems with softening and distortion of substrates. In fact, the CVD process is most commonly used to coat tungsten carbide tooling, particularly the indexable inserts used for high-speed turning. The cobalt-cemented carbide is an ideal substrate because it has a similar coefficient of thermal expansion to the TiC or TiN coatings. Also, it does not suffer any volume change during cooling and gives good support.

There can be problems of decarburization of substrates during deposition and it is common to apply duplex or multiple coatings (for instance, TiN on TiC or Al_2O_3 on TiC) to produce graded properties. The coating thickness is typically 1–5 μm. It is possible to apply the process to high-speed or tool steel components but these must be reheat-treated to retrieve the substrate hardness. This will produce distortion and, in many cases, will bring an unacceptable loss in precision. However, for tools with non-critical dimensions, CVD coatings can bring the same benefits to life as those applied by PVD. Equally, because the coatings have high hardness, they can be used to resist low-stress abrasive wear. The system is well suited to handling large numbers of small items, the parts being simply jigged on trays in the furnace.

CVD coatings are also used to protect against corrosion and corrosive wear. For instance, chromizing (referred to in Section 9.8.3.2), even though it is a pack process, can be considered as CVD because it occurs by the decomposition of a gas.

Finally, it is now possible to combine the PVD and CVD principles in the form of plasma-assisted chemical vapour deposition (PACVD) bringing the flexibility of the chemical process at a much lower temperature (less than 300°C). However, as yet, deposition rates are low and applications are mainly in thin (sub-micron) coatings for electronics applications.

Electrolytic and electroless coatings Over 30 metals can readily be deposited from aqueous solution[66] but they do not include alkali or alkali-earth metals and refractory metals such as tungsten, molybdenum and vanadium. While there is a tendency to think that electrolytic deposits are mainly for corrosion resistance, decorative purposes or electrical/electronics uses, there are many engineering/tribological applications for electroplates. Hard and soft plates are used, depending on the particular function required. To resist abrasive wear, adhesive pick-up and corrosion, hard chromium plates[67] are ideal. Porous or intentionally cracked chromium plates are used for oil retention, as in automotive cylinder liners. The hardness is 850–900 HV and coatings thickness is typically 20–100 μm. There are several proprietary variations on the plating techniques; some baths giving ultra-hard (1110 HV) deposits and others giving a dense, crack-free, but rather softer (700 HV) coating. Electrolytic coatings tend to build up on outside corners and sharp edges (because of a concentration in the current density) and to give reduced thickness in holes and on inside corners.

Soft plates of tin are used to facilitate 'running in', prevent fretting and confer corrosion resistance. Plates of silver, lead, cadmium, tin and antimony are used in heavy-duty sleeve bearings, particularly in aircraft power units. Nickel plate can be deposited from a wide range of solutions and is used to minimize abrasive wear in cases such as sliding contacts on hydraulic rams. Some care should be taken in the selection of electrolytic nickel, particularly with respect to the counter-face, because of its tendencies to gall. Nickel is a good undercoat plate for hard chromium. However, because the shock resistance of chromium is poor, it is prudent to make the bulk of the layer nickel, and give a relatively thin top-coat plate of chromium.

Electroless nickel plates,[68] autocatalytically depositing nickel-phosphorus (Ni–P) or nickel–boron (Ni–B), have many useful tribological applications. In the case of the Ni–P deposits, a hardness of about 500 HV is obtained but can be thermally aged to a hardness in excess of 1000 HV. This is achieved after one hour at 400°C by the precipitation of nickel phosphides. Such hardness is not retained at high temperatures but the Ni–B deposits are superior in this respect. The range of applications of nickel plates can be increased by incorporating fine dispersions of wear-resistant particles in the plating solution (NiC, SiC or Al_2O_3). Such coatings are particularly effective in high-temperature wear situations. Electroless nickel is also available with the addition of PTFE. This duplex coating is less hard but has excellent non-stick properties.

In general, electroless nickel is not as abrasion resistant as hard chromium plate but, because it is not an electrolytic process, it does perfectly replicate the component surface without build-up on edges or corners. It is normally applied in thickness between 10 and 50 μm and it has excellent corrosion resistance, particularly in the 'as-plated' condition, i.e. straight from the plating bath.

Both electroless nickel and hard chromium plate are commonly applied to ferrous materials but they are now also used increasingly for components made in aluminium. Plating baths can usually accommodate components up to a metre or more in size.

Phosphating[69] These coatings are mainly used as a base for paint but they are effective in the presence of a lubricant to ease the deep drawing of steel, and to decrease the wear and fretting of sliding parts, particularly during 'running-in' processes. The identity of the phosphating process is often concealed under a proprietary treatment. In general, they are based on dilute phosphoric acid solutions of iron, manganese and zinc phosphates either separately or in a combination. Accelerators are added to shorten the process time to just a few minutes (in the temperature range 43–71°C). The simplest phosphate coating consists of grey to black crystals of $Fe_3(PO_4)_2$ and $FeH PO_4$. Zinc and manganese phosphates produce more complex layers which absorb lubricant more readily, and are effective in reducing adhesive processes such as galling, pick-up and scuffing. In addition to phosphating, there are many chemical conversion coatings which involve dipping components in solutions to develop specific compounds. Treatments such as chromating, used on non-ferrous alloys to prevent corrosion, will hold lubricants, and provide a base for bonded lubricant coatings.

Sprayed coatings[70] In spraying techniques, powders are heated to a semi-molten state and deposited at high velocities onto the component surface. Coating thicknesses vary from about 0.05 to 1.00 mm. The techniques can be divided broadly into flame gun, arc, plasma-arc and detonation gun processes. The merits of any one technique over another need to be assessed with reference to the particular job in hand. Obviously the selection must be on a cost-effective basis taking due account of the integrity required for a particular duty. One of the difficulties with these processes is to assess the substrate bond integrity, porosity and general coating qualities on a production basis. Suppliers are well aware of this, and the usual approach is to design their coating methods with care, so that tight control of the process variables is maintained by following the set procedure at every stage.

Electric arc spraying is used for metal deposition for wear resistance, corrosion resistance or reclamation. The coating material is fed as two wires and an electric arc is struck between them to cause melting. The molten metal is then propelled onto the substrate by compressed air.

In flame spraying, the source of heat is a burning gas, such as acetylene, and the coating material is fed into the gun either as a wire or a powder. It is a relatively cheap process and gives a high deposition rate but, in general, the bond strength is lower and the porosity is higher than that achieved by the electric arc process. The spray-fuse process takes the technique a step further by first spraying and then fusing with a second heat source, such as flame, torch or by induction. This is the basic technique used for the nickel- or cobalt-based Stellite-type alloys and gives excellent resistance to corrosion, erosion, abrasion and fretting. Alloys can have hardnesses up to 900 HV with near-zero porosity. There will, of course, be substrate distortion and the parts will require finish grinding. Coating thicknesses are typically up to 0.5 mm.

The plasma spray process makes use of an ionized gas (usually argon or nitrogen) to produce much higher temperatures than those created during flame spraying. This allows deposition of higher melting point materials such as metal oxides or metal/ceramic mixtures. Such coatings have relatively good substrate adhesion and porosity levels are usually in the range 2–4% (Figure 9.71). A typical coating thickness would be 0.1 mm and the process finds a wide range of applications in resisting abrasive wear. Substrate heating is minimal.

There are several variations on the plasma arc technique. The process can be conducted in a partially evacuated chamber (Low Pressure Plasma Spraying – LPPS), giving reduced porosity, reduced oxidation of the coating material and, because the substrate reaches a higher temperature, better adhesion. The same advantages can be gained by shrouding the arc in a non-oxidizing gas (Inert Atmosphere Plasma Spraying) and both these processes are now used increasingly to deposit the nickel–cobalt–chromium–aluminium–yttrium (MCrAlY)-type alloys which are used extensively in the aircraft industry to resist high-temperature oxidation erosion and fretting. A third variation is the Transferred Plasma-Arc process in which a secondary electric current is established between the arc and the workpiece. This promotes substrate heating and surface melting and gives more dense and more adherent coatings. Deposition rates can be very high and the technique is typically used to deposit thick (up to 10 mm) abrasion- and erosion-resistant coatings for use in applications such as mining and agriculture. The substrate must be electrically conducting and be able to withstand some thermal distortion.

The ultimate spray techniques are those producing the highest particle temperatures and velocities. These are achieved in the proprietary techniques such as Detonation Gun, Jet Coat and Mach Stream. As their names suggest, they are high-velocity techniques based on combustion of high-octane fuels. Porosity is very low (less than half of 1%) and substrate adhesion is excellent. They are used to deposit tough, abrasion- and erosion-resistant coatings such as chromium carbide or tungsten carbide/cobalt cermets. Thickness is typically 0.1 mm and substrate heating is minimal.

For all spray techniques, a correct substrate preparation is essential. The surface should be clean, free from scale, flash and burrs and should be pre-roughened by a grit-blasting procedure. Attention should be paid to the required coating

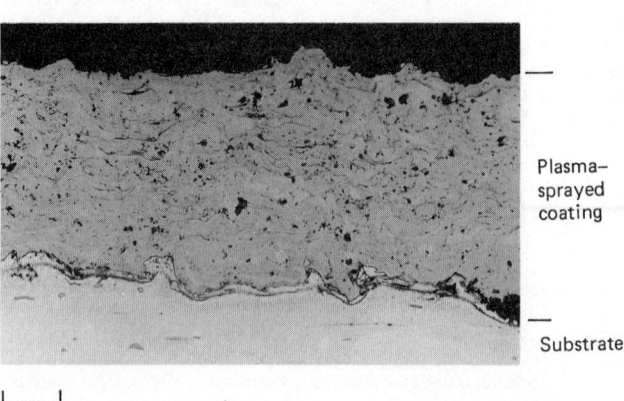

Plasma–sprayed coating

Substrate

|———|
20 μm

Figure 9.71 Plasma-sprayed chromium oxide layer

distribution so that the spraying procedure can be optimized and, for mass production, automated and computer controlled.

Laser alloying and cladding[41,71] These are relatively new processes. For cladding, the powdered material is blown directly into the laser-generated melt pool. Laser alloying is a similar process, except that the energy is increased to produce more substrate melting and complete surface alloying with the powdered material. The reaction area is shielded with an inert gas. A particular advantage of such techniques is that a specific area can be treated, thus minimizing component distortion.

Corrosion-resistant surfaces and metallic glasses are being produced but more development is required before the processes have wide industrial application. One of the problems is that of controlling the depth of heating, particularly when thin-layer fusion zones are required.

Welding and roll-cladding[72] These processes involve relatively thick layers, typically up to 2 or 3 cm. Welding can be used to good effect in tribological situations where high-stress abrasive wear is the problem, such as coating digger teeth, tank tracks and on ore-handling equipment. For instance, some of the Stellite-type coatings are applied by weld deposition.

Cladding is usually associated with corrosion or mild wear problems that are encountered in the chemical, food processing or printing industries. The two processes, roll-cladding and weld cladding, are complementary; hard abrasion-resistant materials are difficult to fabricate and are best deposited by welding, while the more corrosion-resistant materials are based on ductile austenites and are amenable to roll-cladding and forming.

9.8.4 Tribological data

Friction and wear data are available for surface treatments and coatings. Designers can sometimes find them in the open literature or in the advertising information which accompanies the various products. However, before they can make use of such data, they need to consider how the wear tests were performed, i.e. the type of laboratory wear machine that was used, and relate the conditions to those present in their own applications. The object of this sub-section is to acquaint the designer with the wear data that are available and to assist in interpretation and application.

As described in Section 9.1, wear is expressed in terms of the rate of material volume removal, usually as a function of applied loading and rubbing distance. For metal-on-metal rubbing under dry conditions, the rate of wear is often proportional to load and distance, and it is just possible that data could be related to a designer's specific application. However, it is more likely that there will be some further influence on the rubbing situation (for instance, a process fluid or lubricant) and that absolute wear data cannot be obtained.

The situation in abrasive wear is even more complex. Wear rate is a function of so many factors (for example, hardness, impact velocity and particle shape and size) that there is no prospect of users finding data which have any absolute meaning to their applications. Relative wear data that rank a range of treatments in order of wear resistance can usually be found but, even then, the range of wear will depend greatly on the aggressiveness of the abrasive. For instance, two sets of wear tests, one using SiC (2500 HV) as the abrasive, the other using SiO_2 (800 HV), may both produce the same ranking for a number of coatings or treatments but the relative wear rates will be very different. The harder abrasive will tend to

compress the apparent differences between coatings and it will be difficult for users to predict the effects in their own applications, where completely different abrasives are likely to be present. In addition, a designer or user has to be sure that wear data were obtained under relevant conditions. If high loads are present, it is no use relying on data produced under low-load conditions. Thin coatings or treatments, even though they might be very hard, can be crushed into the substrate and torn away without any benefit.

For other wear situations the idea of wear rates is equally doubtful. The designer will rely on data or experience which shows, for instance, that certain coatings are better in fretting situations than others, or that some treatments are effective in corrosive applications. Equally, in situations involving severe wear problems such as seizure, galling and scuffing, wear rate has no meaning. Material is transferred, torn and deformed; surfaces weld and seize. The objective is to eliminate the problem, not merely to reduce it. Wear data, therefore, are most commonly published for relatively simple situations of adhesive or abrasive wear. The range of wear machines is extensive and some, together with examples of wear data, are described below.

Adhesive wear data are produced in many test geometries. In most cases, the geometry is chosen because it allows a quick and easy measure of wear, rather than in any attempt to simulate a particular application. The most common wear machine is probably a 'pin-on-disk' arrangement, with the wear on the pin being measured from its weight loss or reduction in length and that of the disk from profilometry of the worn groove. Figure 9.72 illustrates the typical mild/severe wear transition that takes place with a normalized medium-carbon steel rubbing dry against itself. If the disk is through-hardened, the severe wear regime is suppressed and a mild

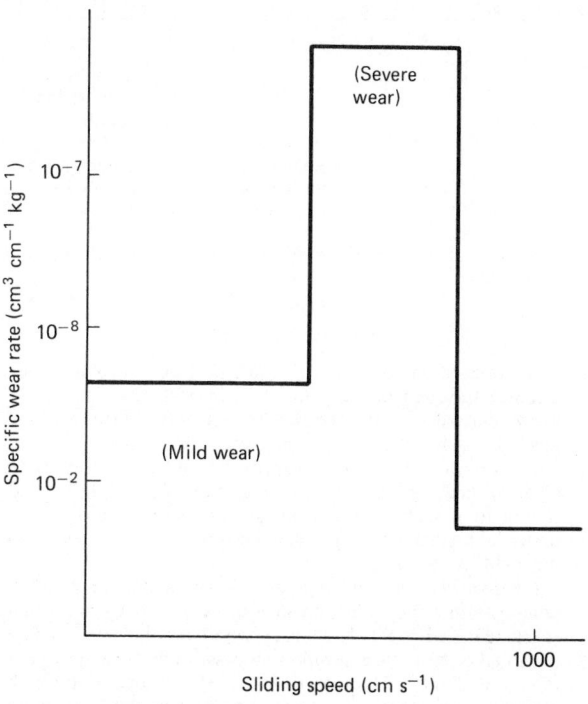

Figure 9.72 Pin-on-disk tests for normalized 0.4%C steel, showing transition in wear

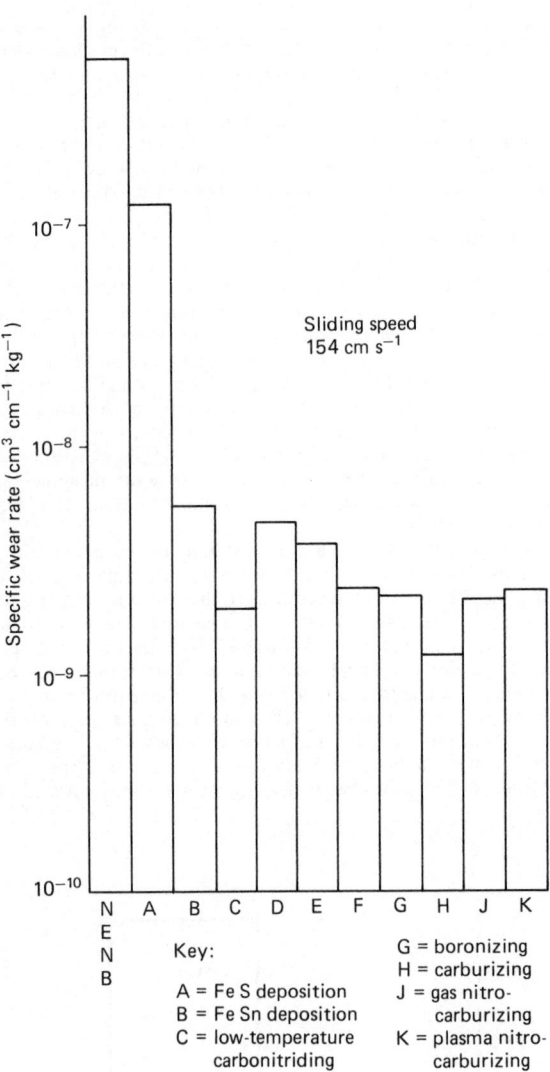

Figure 9.73 Pin-on-disk tests on various surface treatments on 0.4%C steel

Key:

A = Fe S deposition
B = Fe Sn deposition
C = low-temperature carbonitriding

G = boronizing
H = carburizing
J = gas nitro-carburizing
K = plasma nitro-carburizing

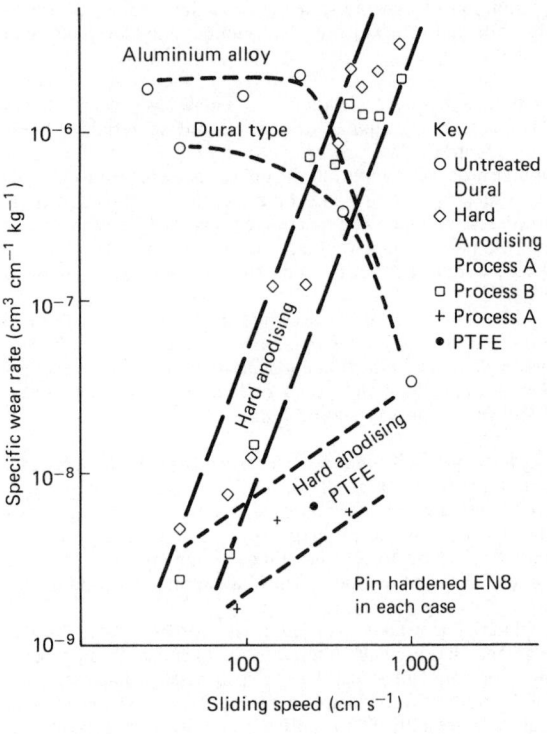

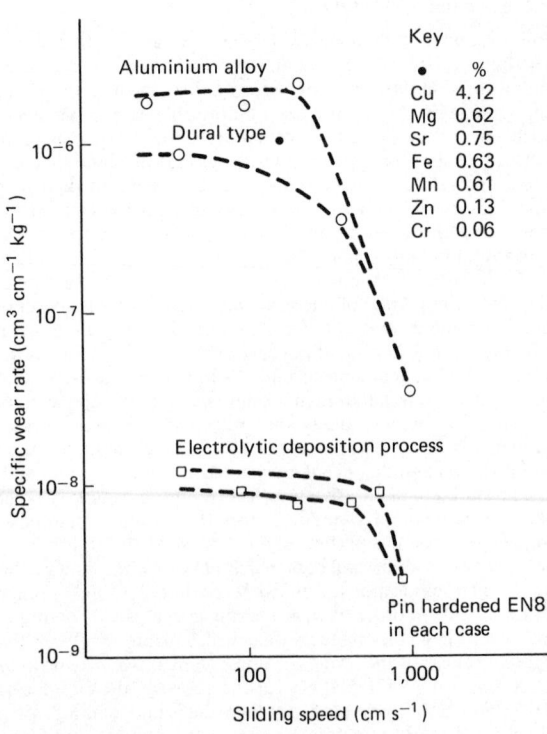

Figure 9.74 Pin-on-disk tests on treated aluminium alloy

wear rate of about 1×10^{-9} cm³ cm⁻¹ kg⁻¹ (volume per distance travelled per unit load) is maintained. Figure 9.73 shows that most of the thermochemical diffusion treatments have the same effect, with only the sulphur-based process failing to give mild wear. This provides an excellent example of the difficulties inherent in wear testing. The sulphur-based treatment is aimed specifically at eliminating scuffing, i.e. under lubricated conditions, and a different test is required to highlight its advantages.

Figure 9.74 shows similar wear data for an aluminium alloy, demonstrating the reduction in wear achieved by anodizing (particularly when PTFE is incorporated into the layer) and by the Zinal electrolytic deposition process. In this case, the wear rates are about a factor of 10 higher than those found with medium-carbon steel. Another test geometry used for producing adhesive wear data is 'crossed-cylinders'. Here, the contact stress is high and certainly unsuitable for evaluating thin coatings. Tests may also be quoted from Falex 'pin-and-jaw', or 'four-ball' machines. These evaluate scuffing or seizure resistance and a 'failure-load' rather than a wear rate is the usual quoted result. A 'disk-on-disk' geometry (sometimes tested with a combination of rolling and sliding) gives a line contact and tests can lead to fatigue pitting as well as adhesive wear.

The range of machines used for performing abrasive wear tests is more limited. Data are produced by rubbing samples against abrasive paper disks (sometimes in a spiral to ensure that fresh abrasive is acting throughout the test). It may also be produced from a grit-blasting machine (erosion) or from a rubber wheel tester. In each case the result is likely to be expressed as a weight loss or penetration depth per unit time; i.e. simple relative data. An example of rubber wheel test data for abrasion by SiO_2 is shown in Table 9.22. It demonstrates

Table 9.22 Abrasive wear rates for several coatings and substrates: rubber wheel test

Coating/material	Hardness (HV 100 g)	Worn volume after 100 revs at 130 N load (mm^3) ($\times 10^{-3}$)
PVD TiN	3500	0.5
CVD TiN	1800	1.0
CVD TiC	3200	5.1
Sprayed WC	1500	9.8
Sprayed Al_2O_3	1200	19.3
Sprayed Cr_2O_3	1300	23.2
High-speed steel	850	5.7
Mild steel	180	142.0

the excellent abrasion resistance of PVD titanium nitride under the light load that was used. However, if high loads or large abrasive particles are employed the results would be completely changed, with only the thicker surface treatments surviving.

The lesson on wear data is, therefore, very clear. The designer should treat it with caution, making sure that it involves the appropriate wear mechanism and that the test conditions have some relevance to the particular application.

9.8.5 Selection philosophy

As stated earlier, the process of selecting the wear-resistant surface should be started at the design stage. If due consideration is given to the environmental factors, and if the user processes are considered at the onset, then preliminary selection will be relatively easy (Figure 9.75).

The first stage will be to satisfy the mechanical engineering demands for the component. Ideally, the lowest cost component which will meet their demands is required. This usually means that, when manufacturing a component, it must be made with the cheapest material compatible with its design requirements, and be fabricated with the minimum number of low-cost operations. During this stage, the designer will be considering all the mechanical and environmental requirements; for example, there may be a need for corrosion resistance, good fatigue properties or resistance against creep or impact damage.

The next stage is to consider the wear. If the environment or engineering demands dictate that a material, or the condition in which it must be used, are not compatible with the wear processes likely to be encountered in the system, then wear-resistant surface treatments must be employed. It is important that the type of wear is carefully identified; abrasive (two-body, three-body, high or low stress), adhesive, erosive,

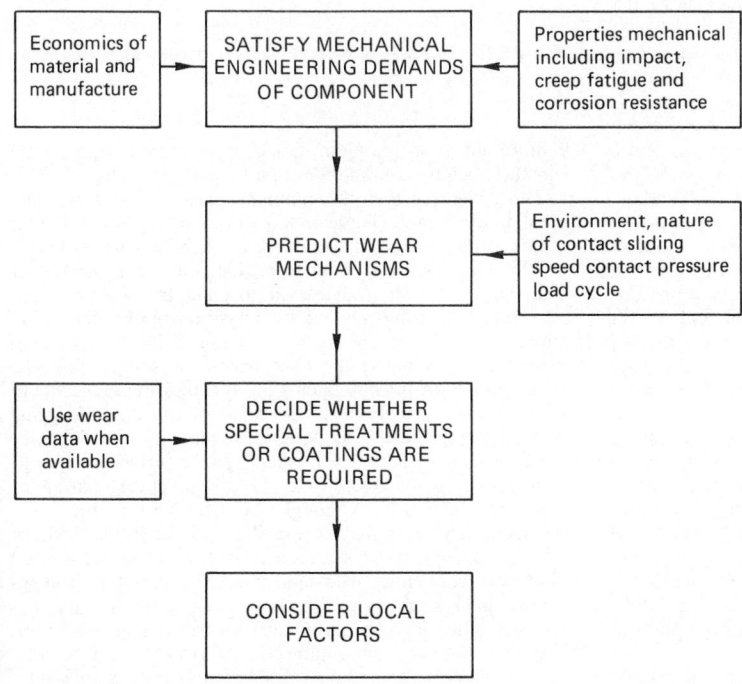

Figure 9.75 Selection of wear-resistant surface – initial procedure

fretting, chemical or fatigue. Also, it is necessary to obtain data on likely rubbing conditions, such as sliding speed, contact pressures, load cycles, hardness and type of any abrasives and the presence of any corrosive medium.

If wear data are available and those data are relevant to the load, sliding speed, environment and counterface material, then this will assist in the selection process. However, the designer should look carefully at any wear data found in the trade literature supplied by the surface treatment specialist, or in published papers and journals, and ensure that they were obtained under a relevant wear regime. It is totally inappropriate to assume that wear data obtained under, say, abrasive conditions, would have any meaning to an application involving another wear process such as adhesion, fatigue or corrosion. Obviously, each design and component cannot be subjected to tribological testing, but there are wear and performance characteristics of classes of materials and surface treatments which allow a primary selection.

By this stage of the selection process, the engineer should have a reasonably short list of materials and surface treatments that could be used, and should have considered possible manufacturing routes and worked out some detailed requirements; for instance, depth of surface treatment (determined by the loading conditions), hardness, core properties, whether the total surface area of the component needs treating, the required wear life, number of components to be produced, etc. However, a number of options for surface treatment will remain and should be reconsidered when assessing the 'local' factors. The final selection must be based on what is practicable, available and economic (Figure 9.76).

First, there will be the question of availability. Usually, the user will like to sub-contract any surface treatment fairly locally. The designer will have to ensure that the selected process is available and that the processor has the necessary equipment, working skill and quality control to provide the treatment reliably and reproducibly at the volume of production anticipated.

Second, the designer must consider the geometry of the component. Complex geometries, incorporating a number of section thicknesses, are likely to distort when treated at high temperature. This may limit the choice to low-temperature treatments. The heat-treatment history of the component is important. Specific core properties may have been produced by hardening and tempering so that any subsequent surface treatment must be applied at a temperature below that of the final temper. If this is not so, the core may be softened and the coating/substrate combination may then have insufficient load-carrying capacity to withstand the service contact stresses. Depending on the areas to be treated, the geometry may also exclude all 'line-of-sight' plating or coating processes. Obvious examples of this are the ceramic and cermet spray processes such as plasma arc and detonation gun. Ion implantation is also a line-of-sight process and physical vapour deposition techniques have only limited ability to penetrate re-entrant surfaces or holes.

In some applications, the replication of the surface shape may be important. The electrolytic plating processes tend to deposit thicker coatings on sharp corners and peaks while leaving valleys with little or no coverage. In contrast, electroless coatings tend to replicate the shape perfectly and may be preferred. On a finer scale the designer may need to preserve a particular surface finish, perhaps applied to the component for reasons of controlling friction or to provide specific optical properties, and so may be limited to ion implantation, physical vapour deposition or electroless coatings.

In all cases, the designer will be considering the dimensional requirements and making allowances for any final machining or grinding after treatment. Low-temperature processes that

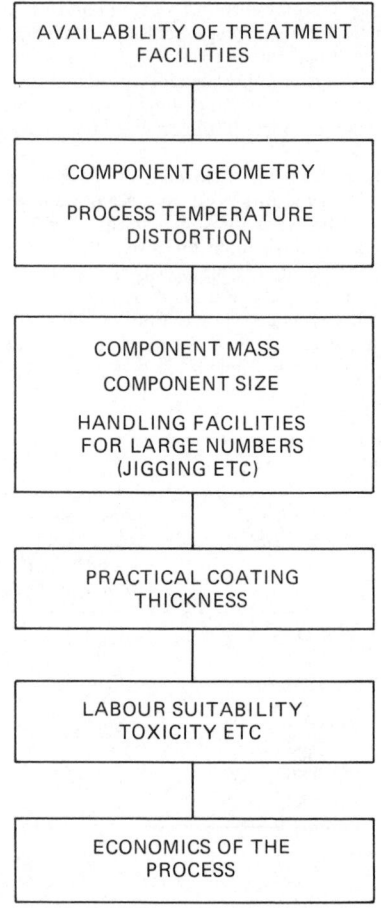

Figure 9.76 Selection of wear-resistant surfaces – local factors

require no post-treatment finishing operations and retain dimensions after treatment are particularly attractive.

Third, the question of component mass, size and numbers must be considered. These are obvious points to look for but it is surprising how easy it is to overlook such facts as: the length of the component you want to nitride is a few inches longer than any salt bath available in the country, a component requiring a coating applied by physical vapour deposition turns out to be too heavy to be handled inside a vacuum chamber, or a particular heat treatment service does an excellent job on 10 components but is totally impracticable to deal with 100 000, because the logistics and cost of jigging would be prohibitive. The designer should always seek advice from the contractor before finalizing on the surface treatment.

A fourth point to consider is the practical coating or treatment thickness. Although the wear performance of a particular treatment may be excellent it is advisable to check whether the particular process is capable of giving the desired depth of treatment to withstand the design pressure throughout its anticipated service life. Some of the processes, particularly ion plate deposits, have excellent hardness, low friction, good corrosion and anti-galling characteristics but it is impractical to deposit more than a few microns onto a substrate. Similarly, some thermochemical diffusion treatments, electrolytic/electroless plates and sprayed surfaces cannot be developed to adequate depths for highly stressed surfaces. For this reason, the design engineer must consider not only the surface but also sub-surface stressing on a particular component.

If the coating or surface treatment technique is to be brought 'in-house', i.e. out of the hands of sub-contractors, it is important to consider all the implications. When introducing new skills, new processes, high technology, processes requiring the use of chemicals, salt baths, toxic materials or radiation, all the technical and practical aspects must be appreciated. It is equally important to consider how they will be received at shopfloor level.

Finally, it is important to consider the economics. Surface treatments can provide tribological and environment-compatible surfaces on relatively cheap substrates, and unless the designer has a knowledge of the range of treatments available, it will be difficult to select the most cost-effective solution.

9.8.6 Quality control

A designer or engineer who is proposing the use of a surface coating or treatment on a critical component requires confidence, not only that he or she has chosen the correct solution to the problem, but confidence in the integrity and reliability of the treatment process. Certification of a product should, of course, be the duty of the supplier, but it is up to both the user and the surface treater to agree what surface properties are important and how to specify them. This should lead to a formal quality-control procedure with full documentation.

The surface properties to be specified depend on the specific application. However, it is likely that, for modification techniques such as case hardening, nitriding, induction hardening, etc., the two key properties will be hardness and case depth. For coatings, the thickness and hardness will again be important but the bond strength to the substrate will be critical. No matter how hard or thick it is, if the coating falls off, it cannot do its job. For sprayed coatings, porosity may be an issue (for instance, allowing a corrosive medium access to the substrate) and, in some applications, surface finish may need to be specified.

A further question to be answered is whether the quality-control tests should be carried out on the actual component (in which case they may need to be non-destructive) or whether they should be performed on a test coupon (a small, flat plate sample) that is treated at the same time. The danger in using a coupon is that it may not experience the same treatment conditions as the component. For instance, it is nearly always easier to effectively treat a flat plate than to deal with a sample with complex geometry. The ideal solution is to include with the batch some components (or representative sections) that are expendable and can be used for destructive quality-control tests. The number of samples and the frequency of checks would, again, be a matter of agreement between user and supplier. It should be decided on the basis of batch sizes and on the likely consequences to the end product of faulty surface treatment being undetected.

The detection of mistakes is, of course, the very purpose behind quality-control tests and some ideas for test procedures are given below. However, the real key to control is in quality *assurance*, i.e. the close definition and monitoring of the treatment procedure itself. The key treatment parameters should be identified and the user should demand full documentation of each treatment cycle. This way, the product is reliable and there is less need for costly and time-consuming monitoring of the final items. There should always be some quality-control testing, the unknown can sometimes occur

and, with good documentation, faults can be traced back through the process and quickly rectified.

9.8.6.1 Hardness and case-depth of treated surfaces

The only reliable way of determining the hardness and depth of hardening is to prepare a polished cross-section through the treated sample and perform a series of hardness tests from the surface into the bulk. Obviously, this is destructive and must be performed either on a tab sample or on sacrificed components.

For case-hardened components the depth of hardening is defined as the distance from the surface to a plane at which the hardness is 550 HV. The measurements must be made using a load of 1 kg. The same method is applied to induction and flame-hardened surfaces but this time the convention is to define the limiting hardness at 400 HV. This definition also applies to nitrided surfaces, but only for conventional nitriding steels. For other nitrided steels (for instance, a high-alloy stainless steel) the specification would have to be agreed between the parties concerned. All these conventions apply to steels with a case depth of more than 0.3 mm and are covered by ISO Standards 2639, 3754 (1976) and 4970 (1979). For cases less than that value it will be necessary for parties to draw up their specification. In particular, for very shallow treatments (for instance, the surface 'compound' layers produced by nitrocarburizing processes) it may be necessary to use very light hardness loads, perhaps as low as 0.1 kg. This demands expert metallographic preparation and careful measurements, taking account of the wide statistical spread in low-load hardness values. For ultra-thin diffusion layers, such as those produced by ion-implantation, detection of the hardened surface is possible only by an ultra micro-hardness technique.[73] However, this is really an academic tool and not suitable for quality control.

9.8.6.2 Hardness of coatings

In the case of coatings, the whole layer usually has a constant hardness, so the idea of a 'case-depth' is inappropriate. For dense coatings such as electrolytic chromium plate or electroless nickel the measurement of hardness on a polished section is straightforward, usually being performed at a load of 1 kg. For sprayed coatings the results may be affected by porosity, the material collapsing into sub-surface voids and giving low hardness values. In this case, hardness would be quoted alongside porosity (discussed below).

With very thick coatings, the hardness can be measured directly on the surface, provided that the surface finish is good and the load is not so high that the coating collapses into the substrate. For very thin coatings, only 1 or 2 μm thick (typical of coatings applied by physical vapour deposition), no direct method is practicable. The hardness can be predicted by performing a series of tests into the surface at different loads and using the analysis described by Thomas.[74]

9.8.6.3 Coating thickness

Obviously, a micrograph prepared at a known magnification from a polished section provides a positive record of coating thickness. However, this is a measure at one single point and it is often important to map the coating thickness over the contours of the components.

For coatings in the thickness range up to 30 μm, X-ray fluorescence[75] gives excellent results (provided there is atomic number contrast between the coating and substrate). It now supersedes the traditional method using electrons – Beta-backscattering. It is particularly effective for measuring the thickness of coatings such as titanium nitride or zirconium nitride applied by either physical or chemical vapour deposition.

For thicker coatings, other techniques are available. These can be based on the use of eddy currents, ultrasonics, thermal waves, etc.

9.8.6.4 Coating porosity

Porosity is important in most sprayed coatings and the best method of measurement is to prepare micrographs of polished sections. Porosity can then be assessed by a visual comparison with 'standard' photographs. The major coating companies have established such standards, usually covering a range from 0.25% to 10% porosity. An absolute measure of porosity can be made from a micrograph by making an area of line tracing and determining either the area or length of the voids as a ratio of the total (for a homogeneous structure, volume, area and line porosity are equal). It is important that the polished section is prepared with care, so that the polishing procedure itself does not pluck out material and create 'false' porosity. This is a particular problem with the harder plasma-sprayed ceramics and causes frequent disputes between users and coating contractors. The polishing procedure should have been worked out previously and it should then be included in the overall quality-control specification.

A second possibility is to measure porosity by vacuum impregnating the coatings with a low-viscosity fluid and measuring the weight taken up. This relies on accurate knowledge of the volume of coating (i.e. the area and the average thickness), on the pores being interconnected and on all the pores being filled. (For very fine pores, capillary forces may prohibit complete filling.) These are significant points but, if the process can be perfected, it does have the merit of determining a volume porosity. In contrast, a polished section gives only a single point value.

9.8.6.5 Coating adhesion

The adhesion is the key property and the most difficult to quantify. In fact, measurements are usually qualitative rather than quantitative.

To produce an absolute measure of bond strength (in units of force/unit area required to detach the coating) it is possible to use a tensile-type test. This relies on gluing a peg to the coating with a high-strength epoxy and measuring the tensile force required to pull off the coating. In practice, the glue usually has a lower bond strength than the coating/interface or the coating fails cohesively. Additionally, the difficulties in setting up and pulling the pins accurately at right angles to the surface mean that a statistical approach, with multiple tests, is required.

For thin coatings, such as those applied by physical or chemical vapour deposition, a scratch test technique[77] is available. This uses a spherical diamond which is dragged across the surface at a steadily increasing load. The point of coating detachment is detected by measuring the friction or by monitoring the acoustic emission, and a 'critical' load is assigned as a measure of adhesion. Such a test usually requires a flat plate tab sample.

For thicker coatings there are a number of alternatives. Samples can be bent in a closely defined way and the bent area inspected for flaking. Alternatively, they can be thermally shocked, thermally cycled, impacted or subjected to high-load indentations (for instance, using a Vickers hardness tester with a pyramidal diamond indentor). In such tests, it is usual for coatings to crack, but without flaking from the substrate. The procedure should be worked out between the user and con-

tractor and designed so that an unsatisfactory coating is highlighted. It may actually be necessary for the coater to deliberately produce coatings with a poor bond so that a relevant procedure can be developed. It is also preferable for the test to bear some relation to the actual application so that, for instance, components subjected to bending in service would be given a bend-type adhesion test.

9.8.6.6 Surface finish

If there is a need to preserve a particular surface finish, the quality control is a matter of a 'before' and 'after' texture measurement using a stylus profilometer. In most cases, it will be sufficient to determine the most common texture parameter, R_a, the 'centre-line-average'. However, some coatings, even though they may generally replicate the underlying surface, have a subtle texture of their own. This is true of some of the evaporated physical vapour deposition coatings and also some electrolytic coatings. In that case, it may also be necessary to monitor other surface finish parameters, particularly those relating to the 'shape' of the surface (skewness) and the sharpness of peaks and valleys (rms slope or average wavelength).

9.8.7 Closure

Surface engineering is the obvious solution to many of today's engineering and wear problems. It allows optimization of the surface and the substrate in a cost-effective way. To be most effective, the principle of surface engineering must be adressed at the design stage of a component, with the designer making full use of available data on the coating or treatment properties, including friction and wear. Then, having made a selection, the designer should be entitled to a good quality and reliable treatment service and it is the duty of the coating contractor to provide a certified product.

The science of surface engineering is expanding quickly. There are increasing possibilities of combining treatments and coatings to produce even more specialized properties. Coatings might be ion-implanted, either after or during their deposition. They might be laser-glazed to further increase hardness or diffusion. One coating might be applied to another to produce a combination of properties (e.g. abrasion resistance and corrosion protection). The time may come when expensive alloying elements can be added exclusively to the surface of a cheap component (perhaps by ion-plating) so that it can then be effectively nitrided. Designers and engineers should be kept abreast of developments and be continuously aware of possibilities for improving component life, reliability and economics.

Acknowledgements

The authors would like to thank Mr M. Farrow of the Surface Science Division, Northern Research Laboratories, UKAEA, for his invaluable technical assistance in assembling this information.

9.9 Fretting

9.9.1 Introduction

Tomlinson[78] first investigated the phenomenon of fretting in 1927 after observations of red rusting of the grips of fatigue-testing machines. He coined the term 'fretting corrosion', by which name it is commonly known, and carried out the first quantitative study.[79] He considered that damage of the surfaces by fretting was initiated by mechanical wear produced by sliding of one surface on another and that the corrosion observed with base metals in air was a consequence of the wear. He stated that 'although the presence of oxidation products shows that chemical action accompanies fretting, the process is nevertheless certainly not one of corrosion as ordinarily understood'.

The distinction between fretting wear and ordinary wear is that fretting generally occurs at contact surfaces that are intended to be fixed in relation to one another but which actually undergo minute alternating relative movement. A classical example is the damage to the races of wheel bearings of automobiles during shipment by rail or ship, initially ascribed to brinelling caused by impact as a result of vibration but now known to be caused by slip between ball and race (and called 'false brinelling').

When fretting occurs on base metal in air, the wear debris always consists of oxides of the metal so that the most common symptom of fretting is the red-brown mud (comprising essentially red iron oxide mixed with oil or grease) shown as a patchwork over contacting steel surfaces.

Fretting damage occurs when two loaded surfaces in contact undergo relative oscillatory tangential movement (known as 'slip') as a result of vibration or stressing. Amplitudes of relative moment are small and often difficult to measure or even to predict by analysis. It is a deterioration process often ignored or not understood by designers. As a consequence, many fretting problems often come to light only late in the product development or, even worse, when the product is out in service. In such cases it is often too late to allow anything but minor redesign, and the only option then available is to employ some type of palliative (which is rarely a long-lasting solution), when fundamental changes in design concept can be the only successful solution. Possible situations in which relative movement and hence fretting can occur, either intentionally or otherwise, are legion. More common ones are flexible couplings, press fits of bearing raceways and hubs on axles, riveted aircraft structures, screws in surgical implants, steel ropes, heat exchangers, business machines and electrical contacts.

Reports of the occurrence of fretting have increased over the years, in part due to the increased demands placed on materials by virtue of higher power densities and stresses in modern machinery. Fretting and fretting fatigue have become increasing problems in the aerospace industry, particularly with the use of exotic alloys based on titanium and aluminium. A number of excellent review papers are available in the literature which provide summaries of the state of knowledge at various times.[80-84]

Although fretting damage by itself is seldom sufficient to cause failure directly, the irregularities produced in the surface cause loss of dimensional accuracy and wear products can cause excessive friction and seizure of closely fitting parts.

A number of different terms have been used in the literature, including fretting, fretting wear, fretting corrosion, fretting fatigue, false brinelling, rubbing fretting, impact fretting and impact-slide fretting. It is worth providing some definitions to distinguish between these various terms. Although fretting was initially described as fretting corrosion, it is better to use the term 'fretting' as a general title to cover a number of aspects of the phenomenon. In the first instance, fretting was observed with materials such as low-carbon steels, which at room temperature and in the presence of oxygen and water vapour, produce copious amounts of finely divided red oxide identified as haematite (Fe_2O_3) and fretting was understood to be closely associated with corrosion on the material surface. The oxidized particles are extremely hard and become em-

bedded in the contact surfaces producing a red staining which even mild abrasive cleaning will not remove. This staining can often be used to diagnose the occurrence of fretting. It has often been stated that the oxidized debris was responsible for abrasive wear of the surfaces or even that it alleviated the damage by acting as a miniature ball bearing. However, fretting is known to occur in non-oxidizing environments such as high-purity inert gases where there is no oxidized debris and corrosion effects are minimal.

It is generally accepted that fretting processes are caused by high-frequency relative movement of contacting surfaces even with slip amplitudes of less than 1 μm. The upper transition between fretting and oscillatory wear is not well defined, but somewhere between 100 and 200 μm is the upper limit at which the wear process assumes the characteristics of unidirectional or reciprocating sliding wear.

Fretting wear is best used to describe processes in which ultimate failure of the component occurs by loss of the material surface leading to fracture, loss of function or pressure boundary rupture (applies to heat exchanger tubes, pressure vessels). In such instances, the component geometry is such as to allow escape of the wear debris with continuing penetration of the component thickness. This is to be contrasted with the more common fretting processes in which the geometry traps the debris, restricts access of the environment and produces little loss of section. Seizure of closely fitting parts such as bearing raceways, press fits, machine tool slideways and gear couplings is a characteristic mode of failure for fretting wear.

Fretting fatigue is a consequence of fretting damage to components subjected to cyclic stressing in which the fretting scar acts as a fatigue crack initiator. It usually occurs where fretting wear is minimal since either the production of copious amounts of wear debris interfere with the fretting process or the wear front progresses at such a rate so as to obliterate the propagating crack. The elimination of excessive wear of a component can often lead to fretting fatigue problems, much to the consternation of the machine designer, developer or operator. A number of papers and publications provide reviews of this extensive topic.[85,86]

False brinelling is a specific term used to describe the damage to rolling element bearings particularly where the bearing has been subjected to vibration from neighbouring machinery during a period of inoperation. The damage does resemble a brinelling indentation and can lead to rough running and premature fatigue failure during subsequent operation of the bearing.[87]

Although the classical forms of fretting involve loaded contacting surfaces, it is now recognized that fretting can occur between surfaces that undergo separation and repeated contact. Thus, the term *rubbing fretting* should be used to differentiate between loaded non-separating surfaces and those which experience periodic impact and slide, for which the terms *impact fretting* or *impact-slide fretting* are used. It should be noted that oblique collision of elastic bodies yields small-scale sliding under the Hertzian contact forces and milliseconds contact duration which can cause significant fretting wear damage. It is largely due to such occurrences of impact fretting problems in nuclear power plant that this topic has received much attention in recent years.[88-93]

9.9.2 Source of relative movement

9.9.2.1 Force/stress excited

Since it can be stated that wear cannot arise from surfaces which are not subject to relative slip and that the most successful method of overcoming fretting is to eliminate the fretting movement, it is worth examining how the relative movement between surfaces is generated. An important distinction can be made between systems in which the relative movement comes from an alternating force or stress applied to elements of a machine, and ones in which the movement is caused by defined displacements. This distinction must be made, as the nature of any palliative is critically dependent upon the source of relative movement.

In force- or stress-excited systems the degree of slip at a surface is a non-linear function of the applied force or stress. It is influenced mainly by damping, which is a function of normal load, static and dynamic friction coefficients and a number of other parameters. The success of a palliative such as a soft interfacial layer, which changes not only the fretting wear resistance but also the friction coefficient, cannot be guaranteed if perhaps the vibration amplitudes increase as a result of the use of such a layer.

Forces or stresses can arise from a large number of sources, including acoustic noise, eddy or vortex shedding (flow-induced vibration), pressure pulsations, mechanically transmitted vibrations, aerodynamic loading and electrical noise. As part of any solution to a fretting problem, consideration must be given to the vibration source and, if possible, some attempt made to reduce it.

In many mechanical systems subject to force or stress excitation, the fretting wear rate has a maximum at some level of load between the contacting surfaces (Figure 9.77). For low-contact forces, the wear rate is low because the surfaces are lightly loaded. Suppose one decided that an increase in load was necessary to attempt to reduce the fretting damage. It is possible for the wear rate to increase as the increased load has little effect upon the amplitude of slip. However, as one passes the maximum, a further increase in load results in a reduction in wear rate as the increased load damps out vibrations and significantly reduces the relative slip. The effectiveness of changing load in this type of system critically depends therefore on where the machine is currently operating on this curve and in part explains the contradictory results obtained with some palliatives. Force- or stress-excited systems probably comprise 90% of the systems in which fretting is observed and, in general, the movements are not intentional.

9.9.2.2 Amplitude driven

There are some mechanical components which are subjected to a fixed amplitude of movement irrespective of loading (for example, misaligned gear couplings where the amplitude of movement is related to the degree of angular misalignment and not to the transmitted torque). Other examples are in business machines such as impact printers where the print head is designed to move through a defined travel. A characteristic of these systems is that movements are generally intentional and little can be done to reduce or limit the relative slip at the surface without affecting the function of the machine.

9.9.3 Characteristics

9.9.3.1 Nature of damage

The debris produced by fretting in oxidizing environments is largely the oxide of the metals involved and therefore occupies a greater volume than the metal destroyed. Conventional static oxidation kinetics are not necessarily obeyed with oxidation of all constituents of an alloy, including those generally conferring corrosion resistance. In a confined contact zone with little possibility of debris escape, considerable pressure can build up, leading to seizure (for example, in closely fitting parts, shrink or press fits) or the formation of corresponding debris-filled pits in both surfaces as a result of wear from the mechanical action of the sheared debris. In some instances, surfaces can be jacked apart by these oxide-filled pits. Some examples of components in which debris entrapment can occur are machine governors, vehicle leaf springs and locked coil steel ropes.

Where debris can escape, loss of fit can result, with continuing and perhaps accelerating fretting wear (particularly if slip amplitudes increase as loads reduce and wear progresses). The presence of oxide can affect the performance of electrical contacts or create safety hazards such as the pyrophoric oxide produced by the fretting of stacked aluminium ammunition boxes subject to vibration in ships' holds.

Under some circumstances the production of debris can lead to a significant reduction in fretting wear rate if, for example, the oxide can be compacted into a wear-resistant layer, such as occurs in steels used in high-temperature carbon dioxide for nuclear reactor heat exchangers; while in other situations, corrosion is accelerated by the continual removal of oxide (e.g. the passivating chromium oxide on austenitic stainless steels), particularly in aqueous environments.

9.9.3.2 Nature of debris

On steel surfaces fretted in air at temperatures less than 150°C, the debris is largely hexagonal alpha-Fe_2O_3, a reddish-coloured oxide which is non-magnetic and often contains metallic iron. On aluminium, the debris is black alumina containing about 25% metallic aluminium. There is increasing evidence to suggest that the fretting debris is platelike in nature and these observations have led to proposed theories of a delamination process along the lines suggested by Suh.[94] In some instances spherical particles have been observed in the fretting of silver[95] and mild steel in argon at room temperature to 500°C.[96]

9.9.3.3 Mechanism of fretting wear

It was originally thought that three processes were involved in fretting:

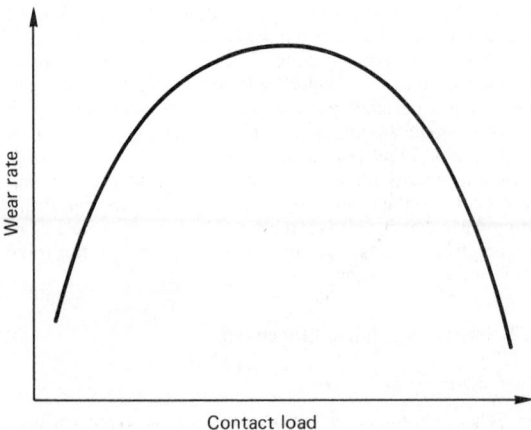

Figure 9.77 Variation of wear rate with load for force-induced vibration systems

1. The production of loose metallic particles by local welding and adhesion which were subsequently oxidized;
2. Abrasion of the surfaces by hard oxide debris;
3. Continual scraping off and regrowth of oxide films.

On this basis, an accelerating wear process would be expected as more debris is produced, which is contrary to most observations. Indeed, debris may have a beneficial effect. As theories of fretting have developed, a review by Hurricks[82] considered three stages to be involved:

1. Adhesion and metal transfer;
2. Production of wear debris by mechanicochemical action;
3. Steady-state production of debris by fatigue action rather than abrasion,

and this is now largely accepted to be the most plausible explanation of fretting processes. Adhesion and welding in the early stages, with steel and noble metals causing material to be raised above the level of the original surface, followed by smearing and, finally, metal removal by a delamination process. Suh's[94] delamination theory of wear envisages the coalescence of subsurface voids produced by dislocation pile-ups at obstacles such as inclusions under the action of the alternating shear stress. It is possible that slip amplitude effects (discussed later) may be explained by a critical amplitude above which dislocations do not return to their original position, thus accelerating fatigue crack initiation and growth. Delamination theory explains the platelike appearance of the debris and sub-surface cracks have been observed in sections through fretted surfaces. Observations made by Sproles and Duquette,[97] Jahanmir[98] and Waterhouse[99] also suggest that a form of delamination wear also takes place in fretting. The delamination model proposes that fatigue cracks nucleate beneath the surface. The cracks then propagate parallel to the surface, until instability or a material flaw forces the crack to the surface to produce a flake wear particle. Sproles and Duquette describe how delamination occurs in multiple layers over the fretting contact, resulting in a flaky metallic scale-like covering over the surface. The observations were made under fretting fatigue conditions, i.e. in the presence of a bulk alternating stress; the surface stress level was high.

The current theory therefore suggests that initial damage arises from adhesion and welding at points of real contact, resulting in material being raised above the original surface. The extent and severity of this stage depends upon the reactivity of the metal and the corrosivity of the environment. Raised material is then smeared out and the surface is removed by a delamination process to produce plate-like debris which are essentially metallic but covered with oxide. Comminution of debris by grinding between the surfaces may take place with further oxidation as the particles are reduced in size. Complete oxidation of very fine debris may ultimately occur. The compaction of debris in the contact zone allows continued transmission of the alternating shear stresses. Work hardening and work softening can have adverse effects upon the fatigue properties and accelerate the process.

For cyclically stressed surfaces, propagating fatigue cracks may be initiated in the early stages from boundaries between slip and no-slip regions. Although wear rates in the initial stage are comparable with adhesive wear, in most cases the wear rate falls significantly as steady-state conditions are established. For further discussions on fretting theories, see references 100–103.

9.9.3.4 Concept of specific wear rate

Archard[104] has expressed the wear behaviour of materials under unidirectional or reciprocating sliding with the term 'specific wear rate', which relates the volume of material worn to the load and total sliding distance. The specific wear rate, k, can be calculated as follows:

$$k = \frac{\text{Volume of material removed}}{\text{Load} \times \text{sliding distance}} \quad (9.7)$$

The normal units for specific wear rate are $m^3 N^{-1} m^{-1}$. It is useful to use a similar concept for fretting in order to express the specific wear rate as a function of the normal load, frequency and slip amplitude. The relationship becomes:

$$k = \frac{\text{Volume of material removed}}{2 \times \text{load} \times \text{frequency} \times \text{cycles} \times \text{p-t-p slip}} \quad (9.8)$$

where the slip amplitude is the peak-to-peak value. The specific wear rate can therefore be used to express results of tests to evaluate fretting wear, from which calculations of wear behaviour under different conditions of time, load and slip amplitude may be made.

9.9.4 Parameters influencing fretting

There are a significant number of parameters which can affect the fretting damage sustained by a sliding interface. Among these are:

- Time or number of cycles
- Slip amplitude
- Normal load
- Slip frequency and surface velocity
- Contact geometry
- Materials properties (hardness, work hardening)
- Environment (humidity, temperature, medium, oxygen potential)
- Lubrication

9.9.4.1 Time or number of cycles

Figure 9.78 presents a schematic representation of the progression of wear with time or number of cycles. An initial

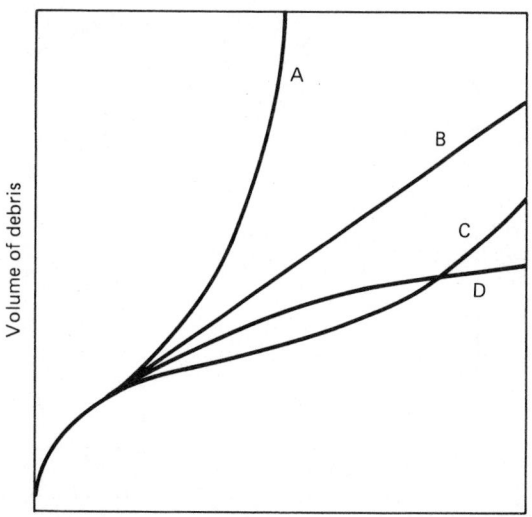

Figure 9.78 Schematic representation of debris production as a function of time

high wear is followed by a decrease in the volumetric removal rate. As oxidized debris is formed, the pattern for mild steel falls to that of curves B or D, with the latter typical of low slip amplitude. Curves A or C typify soft materials with a hard abrasive oxide which can dominate the material wear behaviour in the second stages of wear.

When expressed in terms of change in specific wear rate with time, the wear rate shows a decrease with time (Figure 9.79) for mild steel fretted in air at 20°C using the results of Ming-Feng and Uhlig,[102] Ohmae and Tsukizoe[105] and Ming-Feng and Rightmire.[106] The specific wear rate reduces by more than two orders of magnitude when tests are taken from a few hours to a few hundred hours. This demonstrates the necessity to extend any laboratory to realistic timescales in order to reach an equilibrium value of specific wear rate.

9.9.4.2 Slip amplitude

There is general agreement that above an amplitude of 100 μm the volume of material removed is directly proportional to the slip amplitude. In other words, the wear volume is directly proportional to the load and the sliding distance and specific wear rates correspond typically to those of sliding wear. At low slip amplitudes, the evidence suggests that damage is much lower. Some investigators claim that there is no measurable damage below 100 μm and elastic movement takes up all the displacement betwen the surfaces. Stowers and Rabinowicz[107] claim difficulty in measuring the actual movement and comment upon the possibilities of lost motion in the test apparatus. However, Tomlinson[78] found damage down to 2 nm, and data on mild steel at 20°C, compared by Lewis and Didsbury[108] to other workers[102,105,109] have shown an increase in specific wear rate of over two orders of magnitude over the slip amplitude range 50–100 μm (Figure 9.80). This observation has an important effect in any attempt to cure a fretting problem since a twofold reduction of slip amplitude from 100 to 50 μm can lead to over a hundredfold reduction in wear volume.

The occurrence of fretting damage even at very low levels of slip means therefore that any machine element at which relative movement can occur will see some level of damage. A lower limit of damage of 0.75 μm has been suggested by Kennedy et al.[110] so that it very difficult to reduce slip to a level at which no damage will occur. The exact threshold is, to some extent, dependent upon the properties of the materials in use. Microslip can occur in many supposedly fixed or rigid connections (e.g. press or shrink fits of hubs or bearing raceways). At very low levels of slip, fretting fatigue becomes a potential problem at the boundary between slip and no-slip regions.

It should be noted that slip amplitude effects are not necessarily a feature of all materials and are dependent upon the environmental conditions. Slip amplitude effects for mild steel disappear with increasing temperature, which is probably more a reflection of the increasing influence of protective oxides (to be discussed later).

9.9.4.3 Normal load and surface pressure

Most workers[102,103] report a constant specific wear rate as a function of load, providing the movement is forced and takes place over the entire contact area (Figure 9.81). If, however,

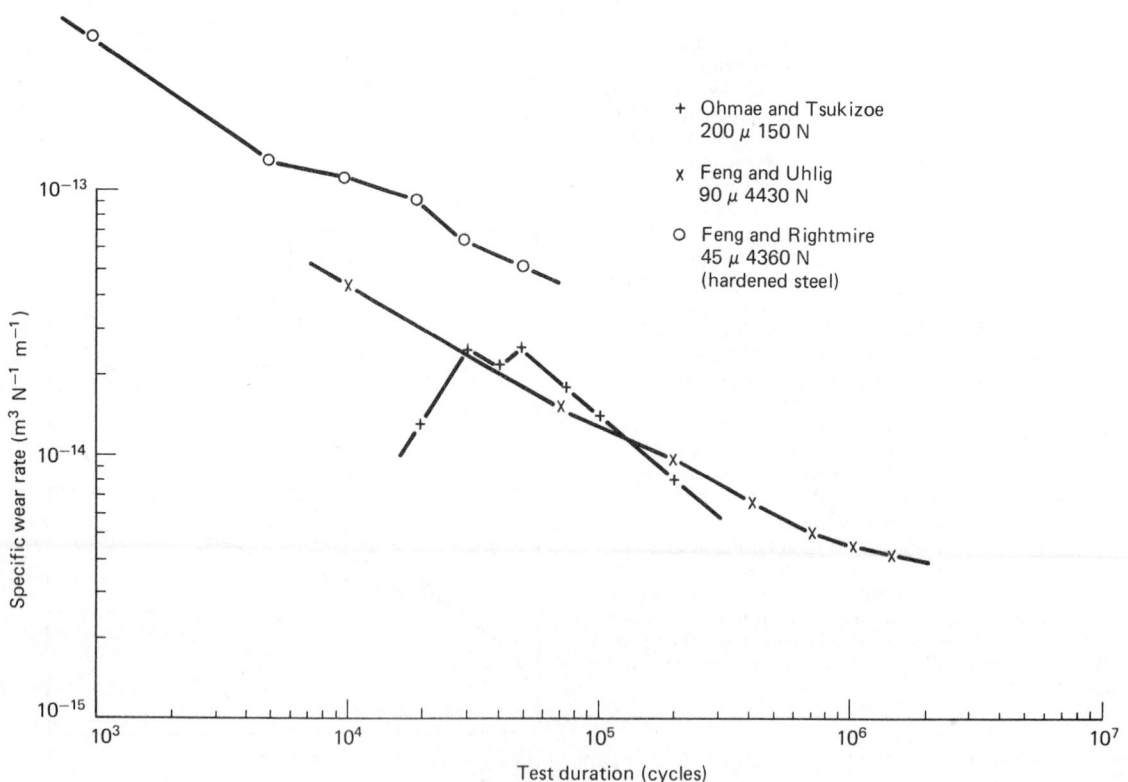

Figure 9.79 Effect of test duration on specific wear rate for mild steel under rubbing fretting conditions

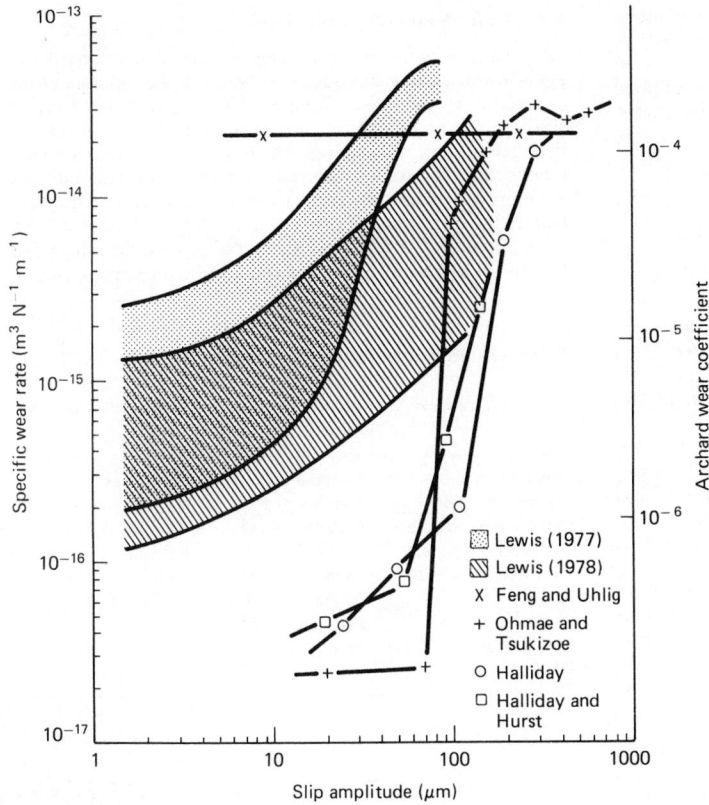

Figure 9.80 Effect of slip amplitude on specific wear rate for mild steel under rubbing fretting conditions: comparison with other workers

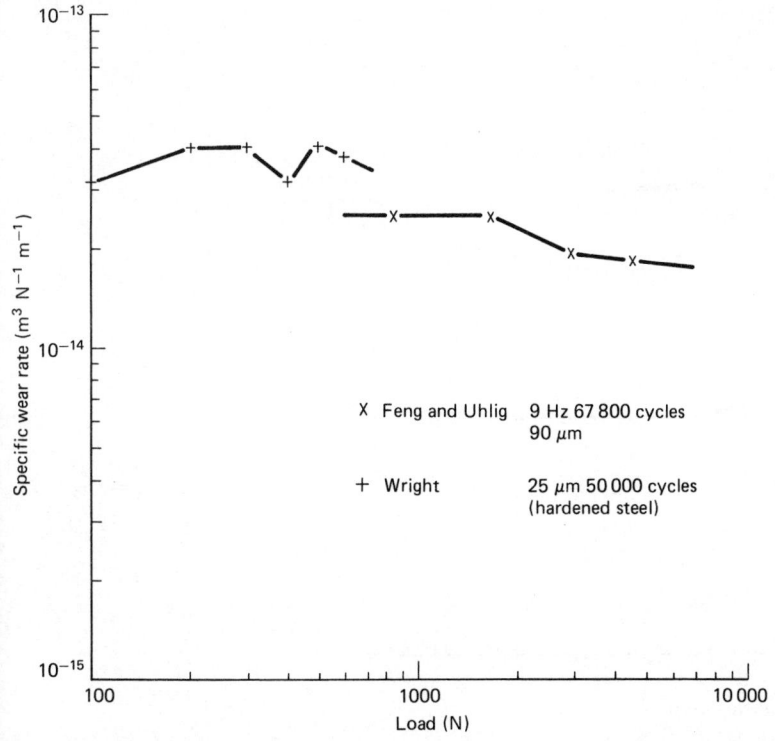

Figure 9.81 Effect of load on specific wear rate for mild steel under rubbing fretting conditions: results of Feng and Uhlig and Wright

the increased load results in a decrease in the slip amplitude then the rate at which fretting occurs can be reduced. A reduction in contact area to increase the pressure has often been used to overcome fretting problems, but one should be careful that the increased pressure does not exacerbate fretting fatigue problems since the increased pressure increases the stress concentration at the slip boundary.

9.9.4.4 Slip frequency and surface velocity

Slip frequency may affect the rate of fretting if the process is removal and regrowth of oxide. If the oxide grows according to a logarithmic law, then frequency effects are probably negligible above 17 Hz, because only that portion of the oxidation curve at short exposure times is in effect.

Figure 9.82 shows results from Ming-Feng and Uhlig[102] for mild steel fretted in air over the frequency range 1–60 Hz. There is little influence of frequency indicated, while Soderberg et al.[111] suggest little effect at high slip amplitudes for a frequency range of 10–1000 Hz and recommend frequency as a useful parameter for accelerating fretting wear tests.

It is interesting to note that greater surface damage has been reported for very low frequencies (in which surface velocity is typically 1 mm/day) and oxide growth and removal effects dominate. This has been particularly important in electrical contacts which can be subject to low slip at very low frequency.

9.9.4.5 Hardness and surface finish

Observations would generally suggest that increased hardness gives lower fretting damage. It is believed that combinations of similar metals of differing hardness give good fretting resistance, though wear in general is thought to be alleviated by combinations of materials which have low mutual alloying tendencies. This view is partially supported by Sakman and Rightmire,[101] who produced a ranking of fretting wear resistance (given in Table 9.23).

Opinions on the effect of surface finish are quite contradictory in the literature. Some investigators believe that a smooth

Table 9.23 Fretting resistance of various materials in dry air

Low	Medium	High
Steel on steel	Cadmium on steel	Pb on steel
Nickel on steel	Zinc on steel	Ag plate on steel
Aluminium on steel	Cu alloy on steel	Ag plate on Al plate
Al-Si alloy on steel	Zn on Al	Parcolubrited steel on
Sb plate on steel	Cu plate on Al	steel
Tin on steel	Ni plate on Al	
Al on Al	Ag plate on Al	
Zn plated steel on Al	Fe plate on Al	
Fe plated steel on Al		

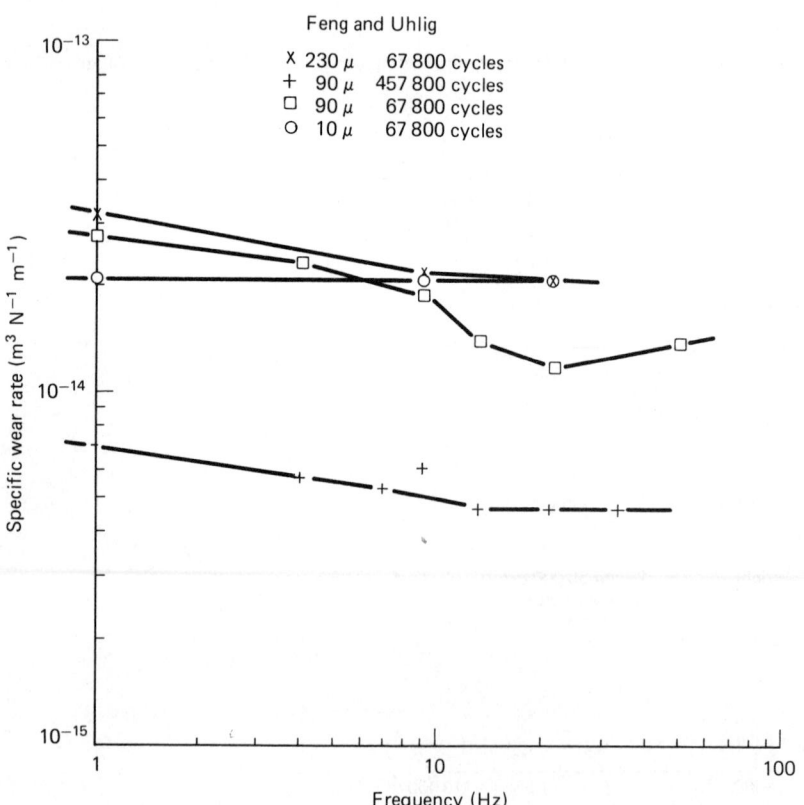

Figure 9.82 Effect of frequency on specific wear rate for mild steel under rubbing fretting conditions: results of Feng and Uhlig

surface is more susceptible to fretting damage while others consider that super-finishing can prevent excessive damage. The weight of evidence points to the superiority of a rough surface for lubricated contacts, the surface irregularities providing a reservoir for lubricant.

9.9.4.6 Contact geometry

The geometry of contact is an important variable in fretting. In general, large-area nominally flat or conforming contacts tend to trap the debris produced and give an over-deepened pitted surface during fretting. Conforming contacts are poor from the point of view of lubricant access since the sliding movement is not sufficient to entrain lubricant. Some surface treatments which are used to alleviate fretting problems are most effective with conforming contacts because of their low thickness. Conforming contacts are most prone to seizure. In contrast, non-conforming contacts allow debris to escape.

9.9.4.7 Environment

In an inert atmosphere, such as nitrogen or vacuum, the fretting wear of steel reduces but the incidence of material transfer between the surfaces increases. The initial phase of adhesion and welding is preserved rather than the transition to oxidative wear appearing. Little loose wear debris tends to be produced. Humidity also reduces the amount of fretting damage as the hydrated oxides of iron are less hard than their dehydrated counterparts.[102]

Since fretting involves both the mechanical properties of a material and its reactivity with the environment, it is to be expected that changes in temperature would have considerable effect on fretting wear. The fretting behaviour of mild steel has been investigated by Hurricks[112,113] and his results are given in Figure 9.83. A significant reduction in specific wear rate is observed at 150–200°C due to the greater thickness and adherence of the oxide magnetite (Fe_3O_4) formed. This behaviour has also been found by Lewis[114] with mild steel in air up to 500°C, in which the specific wear rate for high-amplitude (100 μm) fretting shows a considerable reduction with increasing temperature with a lesser reduction for low slip (10 μm).

It would appear that the formation of a compacted oxide layer on mild steel increasingly reduces the fretting wear rate by providing a protective wear-resistant film. These so-called 'glazes' are observed at high temperatures on corrosion-resistant materials such as austenitic stainless steels and nickel-based alloys similar to those formed in reciprocating sliding.[115,116]

In aqueous environments a passivating film is relied upon to limit corrosion of the metal surface. Disruption of the film by fretting can produce large changes in electrode potential for the baser metals. Fretting can therefore show its effects more by continuous disruption of the passivating film. Experiments in which the potential of the fretting surfaces is kept constant show a linear relationship between corrosion current and slip amplitude. Calculations show that the bulk of the material removed is the result of mechanical action rather than chemical dissolution.[84]

Corrosion reactions can be controlled by imposing a cathodic potential on the system (or on metals which display passivation, an anodic potential). It has been observed that cathodic protection can give a significant improvement in fretting fatigue behaviour.[81]

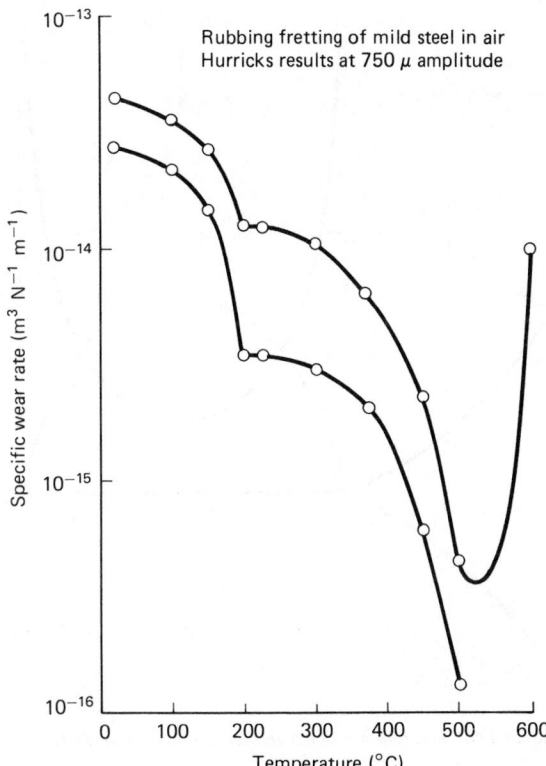

Figure 9.83 Rubbing fretting of mild steel in air (after Hurricks)

9.9.4.8 Lubrication

The low relative velocity and high contact pressures of fretting surfaces are not conducive to effective lubrication. In addition, the escaping debris provides a barrier to lubricant access, as well as thickening the oil or grease by mixing with it (e.g. the thick red 'cocoa' observed in many lubricated fretted contacts). Shear-susceptible greases (i.e. those whose viscosity falls with shearing) appear to be the most effective and the literature generally recommends greases of low base viscosity to improve the flow into the fretting zone. The selection of greases for lock coil steel ropes is quite different, however, with high viscosity being most suitable.

Lubricants containing molybdenum disulfide have frequently have recommended for the alleviation of fretting. However, in line with other work for unidirectional sliding, the effectiveness of MoS_2 is much reduced in the presence of a liquid because it cannot effectively attach to the metal surface when there are other surface-active agents competing (e.g. anti-wear additives). Considerable promise, however, has been shown in dry-bonded MoS_2 films for the control of fretting.[117]

9.9.5 Theoretical considerations

A classic example of microslip in fretting contacts is given in the elastic contact of two spheres or a sphere on flat where the surfaces are subject to an alternating tangential force. Mindlin[118] showed that for a sphere pressed against a plane by a normal force, N, and subjected to a tangential force, T, the shear traction is unbounded towards the edge of the contact (see Figure 9.84). In reality, this distribution is not possible

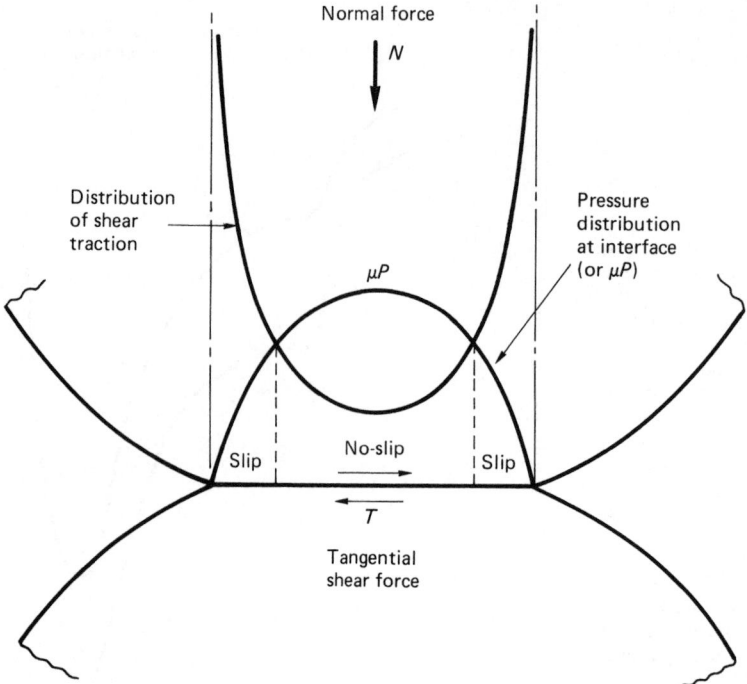

Figure 9.84 Pressure and shear stress distributions at elastic contact of two spheres

and a relief mechanism must operate. This mechanism is provided by micro-slip, which occurs over an annular area at the edge of the contact. As the tangential force is increased, the annular area of micro-slip grows until:

$$T = \mu N$$

where μ is the coefficient of friction. At this point the whole of the two surfaces are in relative motion, i.e. gross sliding takes place.

Mindlin's analysis illustrates how the localized concentration of shear traction gives rise to fretting. Elimination of the shear concentration is therefore a fundamental means of avoiding fretting. Furthermore, the designer has the potential to predict fretting early in the design process. By carrying out a stress analysis of the contact interface, areas of potential fretting can be identified where the following inequality holds:

$$\tau > \mu \sigma_n$$

where τ is the shear stress at the surface and σ_n is the in-plane stress normal to the surface.

The shear traction also has an important effect on the generated contact stress field which is now considered important in the mechanism of fretting fatigue. Without a shear traction (zero friction coefficient) only a small tensile stress is generated at the edge of a hemispherical contact. In such cases, failure is likely beneath the surface where the material is subjected to the maximum yield stress. As the friction coefficient increases the point of maximum yield stress moves towards the surface, reaching it at a friction coefficient of 0.33.

In the fretting of steels, friction coefficients approaching unity are not uncommon. Figure 9.85 illustrates the rapid increase in the tensile stress at the trailing edge of contact as the friction coefficient is increased. Figure 9.86 shows the damage developed between a steel sphere and a flat as a result

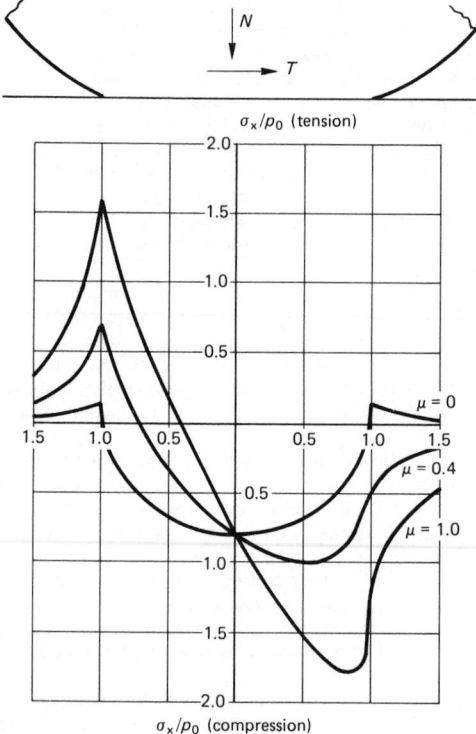

Figure 9.85 Ratio of in-plane stress σ_x to maximum contact pressure p_0

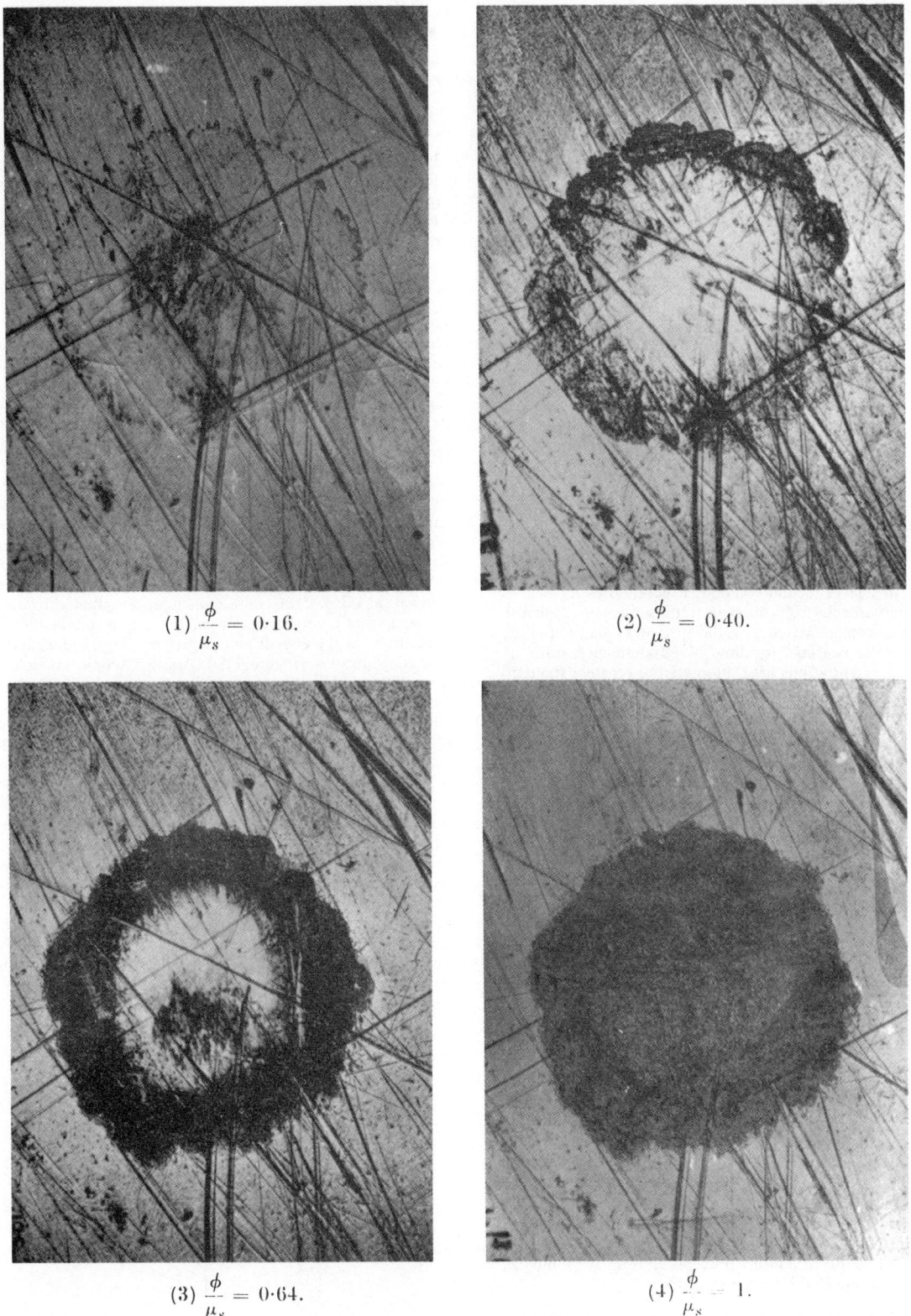

(1) $\dfrac{\phi}{\mu_s} = 0\cdot16$.

(2) $\dfrac{\phi}{\mu_s} = 0\cdot40$.

(3) $\dfrac{\phi}{\mu_s} = 0\cdot64$.

(4) $\dfrac{\phi}{\mu_s} = 1$.

Figure 9.86   Fretting damage for sphere-on-flat for various applied shear forces

of fretting under increasing values of tangential force.[119] Figure 9.86(1) shows a very slight outer ring for a value of $\phi/\mu_s = 0.16$ where $\phi = T/N$ (the shear coefficient) and μ_s = static friction coefficient. The damage extends inwards as the area of microslip grows until the entire surface is subjected to gross slip when the value of ϕ/μ_s reaches 1 (Figure 9.86(2)).

It is at the boundary between the regions of slip and no-slip that fretting fatigue cracks can initiate and grow, though fretting wear in the outer area of slip causes a redistribution of the contact pressure and often fatigue effects are halted as the wear zone progresses inwards.

9.9.6 Fretting wear evaluation

There are a plethora of testing facilities described in the literature, each developed to study a particular aspect of fretting or fretting fatigue. Confusion over the effects of slip amplitude has arisen in the past, largely because some investigators have ignored or underestimated the lost motion that inevitably occurs. Modern fretting apparatus are generally designed with very stiff drive mechanisms in order to ensure that relative slip is occurring at the contact interface and often take great care to measure the actual slip close to the interface.

There are no standard fretting test machines, each industry developing a design to suit its needs. There is no standard test geometry, with test specimen designs varying from hemisphere-on-flat to crossed-cylinders, cylinder-on-flat and flat-on-flat. In many instances, actual component geometries have been used. The particular difficulties in fretting machine design to ensure area contact with flat-on-flat specimens lead to the use of hydrostatic bearings to allow a self-aligning feature. A number of organizations have attempted to create a standard for fretting evaluation, including the American Society for the Testing of Materials.[120]

9.9.7 Preventative measures – some palliatives

9.9.7.1 Some lines of attack

There are a number of methods which may be used to reduce or eliminate fretting damage:

- Prevent the relative movement by a change in design or modifying the source of vibration.
- Provide a treatment to the surfaces to reduce or eliminate welding and adhesion processes, remembering that all materials are subject to fretting to a greater or lesser extent.
- Encourage movement to allow access of lubricants, usually a difficult line of attack in view of the high contact pressures, the generation of wear debris and low relative velocity of sliding.

A number of observations can also be made which can assist in solving a fretting problem:

- No wear takes place between surfaces subject to vibration unless its amplitude is sufficient to produce slip.
- Damage is produced on any solid surface, whether metallic or non-metallic, if slip takes place.
- Fretting wear of base metals in the absence of oxygen gives less rapid accumulation of debris and less intense damage than in air or oxidizing atmospheres.
- No liquid lubricant entirely prevents fretting but some can produce a considerable improvement.

Additionally, one must make the distinction between the fretting of components (such as shrink fits, press fits, bolted flanges, etc.) which are not intended to undergo relative motion and those (such as flexible couplings, bearings, universal joints) which undergo intentional oscillatory movement.

9.9.7.2 Prevention of gross slip

In order to overcome fretting damage in components not intended to slip, gross slip can be reduced or eliminated by:

- Increasing the load or closeness of fit;
- Increasing the coefficient of friction (e.g. with grit blasting followed by lead plating[121]);
- Interposing a layer having high elastic strain limit (e.g. rubber or polymers).

9.9.7.3 Design analysis to prevent or reduce fretting damage

Fundamentally, the best solution to a fretting problem is to 'design it out' if at all possible by using all-welded or unit construction. Design modifications may be necessary to reduce or eliminate the source of vibration (e.g. isolation from local excitation sources). Bearing in mind that slip occurs at an interface because the shear force is greater than the opposing friction force, one can reduce the shear stress or increase the friction force. This reduces to the problem of eliminating the concentration of shear stress over the surface. The friction force may be increased by reducing the apparent area of contact while keeping the load constant, thus increasing the pressure. Similarly, some platings such as cadmium can increase the friction coefficient by promoting seizure.

In situations where the concentration of shear traction occurs at the edge of the contact, a simple design modification may be all that is required. For example, on a clamped joint, undercutting at the edge is an effective means of reducing the shear concentration (Figure 9.87). Another simple example is a press-fitted hub on a shaft in which stepping down the shaft diameter significantly reduces the shear concentration (Figure 9.88). Unfortunately, simple geometries such as these are the exception. In the majority of real engineering assemblies, while geometric modifications are usually possible, it is not often clear how effective the modifications will be in eliminating the shear concentration. In these cases the answer is to carry out investigations of the joint by a numerical stress analysis technique such as finite-element analysis.

Several commercially available finite element codes have special elements for modelling Coulomb friction which will provide detailed information about the behaviour of a potential fretting interface. However, since sliding is a non-linear element in the problem, these programs are usually expensive in computing time. Fortunately, in the majority of cases the main interest is in predicting whether fretting will occur or not. This simplifies the analysis and allows the interface to be

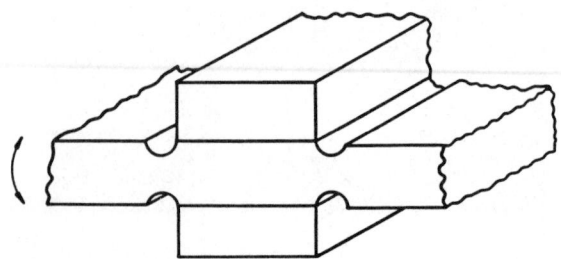

Figure 9.87 Undercutting at clamped joints to avoid stress concentration

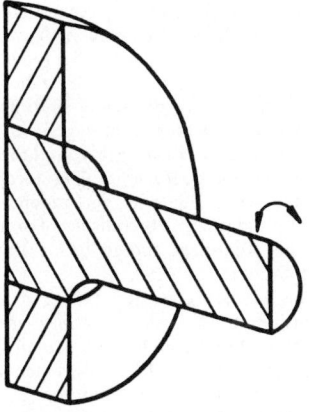

Figure 9.88 Stepped shaft to avoid stress concentration

modelled by simpler linear elastic elements. Provided a sufficient refinement of elements is used in modelling the joint, the design can be established as satisfactory, if the surface shear stress is less than the product of contact pressure and friction coefficient (i.e. $\tau < \mu\sigma_n$) everywhere across the joint. If the condition fails at any element node, then the extent of fretting cannot be predicted, and the design should be considered as prone to fretting.

In certain cases, optimizing the design of an assembly on the basis of other essential design criteria may mean that fretting becomes inevitable. For this situation, a palliative may be the only option available to minimize the fretting damage. In such instances a detailed analysis of the fretting interface can be useful to categorize the type of fretting problem. This, in turn, can assist with the selection of a suitable palliative.

9.9.7.4 Selection of fretting palliatives

To aid with the selection of a palliative, fretting problems can be classified into one of two categories determined by whether the movement is force controlled or amplitude controlled. To explain these categories, it is useful to consider again the Mindlin problem. In the partial slip regime according to Johnson,[122] the ratio $T/\mu N$ determines the level of micro-slip. The slip amplitude is therefore a function of the forces involved, and it is proposed that the problem can defined in the category 'force controlled'. Under gross sliding, i.e. $T = \mu N$, clearly the forces involved play no part in determining the slip amplitude. Fretting in this case is put in the category 'displacement controlled' since the slip amplitude is essentially fixed.

In reality, defining in which category the problem falls is not, in the majority of cases, straightforward. Fretting involving wear between components which are intended to move (e.g. oscillating bearings, ball joints, slideways, etc.) is, in general, a displacement controlled problem and there is little ambiguity. However, in clamped assemblies where movement occurs unintentionally, fretting allows strains of different magnitudes in the component mating surfaces. If this 'differential strain' is largely unaffected by the applied loads, then the problem effectively is one of displacement controlled fretting. If, however, the differential strain is substantially modified by the applied loads, then the problem should be classed as one of force controlled fretting. A finite-element analysis can be used to determine in which category the problem falls.

For a displacement controlled problem the approach to alleviate fretting wear or fretting fatigue is to try to reduce the normal load and the friction coefficient. With fretting fatigue this palliative is clearly aimed at reducing the contact stress levels. In a fretting wear problem the palliative should also be effective since it has been found that the wear volume is proportional to the applied load.[104]

In a force controlled fretting wear problem the requirements of a palliative are completely the reverse. Increasing normal load and friction coefficient can act as effective palliatives by lowering slip. Slip has been shown to be one of the most important fretting parameters. Work by Ohmae and Tsukizoe,[105] Halliday,[123] Halliday and Hirst[109] and Lewis and Didsbury[108] all suggest that the specific wear rate decreases with decreasing slip amplitude. Figure 9.80 summarizes the results of these investigations into the effects of slip amplitude. The data from the individual experiments suggest that the specific wear rate (volume lost per unit load per unit sliding distance) varies as a function of the slip amplitude raised to a power ranging between 2 and 4. Clearly, if fretting cannot be entirely prevented, even a modest reduction in slip may be sufficient to reduce the level of wear down to acceptable proportions.

In the alleviation of force controlled fretting fatigue, reducing the shear traction is the primary objective. Thus, reducing the friction coefficient and normal load are desirable palliatives providing the increase in wear can be tolerated.

9.9.7.5 Lubricants

The detailed choice of a palliative treatment is still largely empirical. However, with a basic understanding of fretting and a detailed analysis of the problem, a rational decision about the selection of the most appropriate category of palliative can be made. For instance, in the general category of lubricants, while it may not be possible to identify immediately the most suitable individual lubricant, one can say in what circumstances a lubricant should be tried.

Lubricants can obviously produce a large decrease in the coefficient of friction. In force controlled situations this effect is likely to result in unsatisfactorily high slip levels. Also, with clamped assemblies the load transmission across the joint may be modified and fasteners may become overloaded. However, lubricants can prove very useful in displacement controlled fretting fatigue. Application of a lubricant will lower the level of shear traction which will considerably reduce the contact stress level. Also, in displacement controlled fretting wear, boundary lubrication can significantly improve performance. However, unless continuously replenished, the self-cleaning action of fretting very quickly removes the boundary-lubricating film. Gaining access to the inner regions of the fretting contact without separating the surfaces is difficult for oils of high viscosity and virtually impossible with greases. A thin, penetrating oil is therefore the best choice. Vapour blasting or phosphating may also assist in providing a reservoir for the lubricant.

9.9.7.6 Solid lubricants

Often the difficulty of containment precludes the use of an oil and an alternative in this situation is a solid lubricant such as molybdenum disulfide or zinc oxide. Solid lubricants are very effective at reducing fretting damage over a limited life.[117] If the problem involves high numbers of fretting cycles ($>5 \times 10^5$), then solid lubricants such as described are unlikely to be satisfactory without re-application.

Phosphate treatment prior to application of the MoS_2 coating can produce a significant improvement in fretting wear resistance.

9.9.7.7 Thermochemical treatments

These treatments involve the diffusion of carbon, nitrogen and, less usually, chromium or boron. The most common thermochemical treatments are carburizing, carbonitriding, nitrocarburizing and gas nitriding. The factors involved in the selection of a thermochemical treatment are too numerous to go into in detail here, but it is possible to make generalizations about this form of palliative.

This category of treatments is designed to increase surface hardness. In general, the adhesive and abrasive resistance improve with increased surface hardness. The most effective treatment against fretting wear is therefore likely to be the one producing the largest increase in hardness. However, the effect of the treatment on the friction coefficient must not be ignored. For instance, on En32 steel, liquid nitrocarburizing using the Cassel Sulfinuz process reduced the friction coefficient in slow linear sliding from 0.8 down to 0.2.[124] Clearly, for displacement controlled fretting, reduced frictional traction will enhance the effectiveness of the palliative. However, in a force controlled situation the reduction in friction may negate the benefits.

By virtue of the compressive stresses developed in the outer surface layers the fatigue strength of steel is normally enhanced by thermochemical treatments. As a general principle, any process which increases the normal fatigue strength of the steel, providing it is not accompanied by a significant increase in friction coefficient, will improve the fretting fatigue performance. The process giving rise to the greatest increase in normal fatigue strength is likely to provide the largest improvement in fretting performance.

9.9.7.8 Surface coatings

In previous investigations of palliatives, both hard and soft coatings have been employed to mitigate both fretting wear and fretting fatigue. However, it is probably true to say that there are more conflicting opinions in the literature on the performance of individual coatings than for any other class of palliative. The reasons for the conflicting opinions on coating performance are not always clear. However, sometimes a successful coating performance can be dependent on the characteristics of the test rig.

For example, in a rig producing low-amplitude force controlled fretting, such as that described by Budinski,[125] coatings having high coefficients of adhesion may bring about seizure of the contacts. A palliative which is given a favourable assessment by such a rig may in fact only be effective in force controlled situations where seizure can be brought about. Sikorskii[126] has correlated high coefficients of adhesion with soft coatings such as silver, indium and lead. It is perhaps therefore no surprise to find that these are precisely the coatings which are said to be beneficial in fretting.[127] It has also been suggested that soft metal coatings work by absorbing the fretting movement.[84] However, the amount of movement which can be absorbed is likely to be so low that it does not seem to the author a credible explanation.

With thin soft coatings applied on a hard substrate, a good palliative performance may be the result of a low coefficient of friction which can occur. Halling[128] has proposed an explanation for this low-friction phenomenon. In general, however, the poor durability of soft coatings leads the author to believe that, with one or two special exceptions, soft coatings are likely to be of little benefit as a fretting palliative.

With electroplated chromium, Alyab'ev et al.[129] found that the conditions of deposition and the coating thickness can influence the wear resistance. In this work it was also found that the chromium deposition with the highest hardness had the best fretting wear resistance. However, under fretting fatigue the high levels of residual stress present in the coating reduce the normal fatigue performance considerably; and the coating has little effect on the friction coefficient. Therefore, although improvements in the fretting fatigue performance have been claimed for chromium plating,[130] there is no reason why under dry conditions this treatment should be of any significant benefit in fretting fatigue.

In general, because of the adverse effect on the normal fatigue limit, it is not considered that hard coatings are effective palliatives for fretting fatigue unless the coating significantly reduces the friction coefficient. Electroless nickel impregnated with PTFE can reduce friction and therefore has the potential to give improved fretting fatigue performance. While improved wear performance using electroless nickel has been reported by Gould et al.[131] there is no reported work to test the effectiveness on fretting fatigue of a nickel coating impregnated with PTFE.

Hard coatings deposited by spraying or welding can give high wear-resistant surfaces. There are numerous coatings commercially available, and deposition can be by several methods. Little is known about the effectiveness of the majority of these coatings as palliatives in fretting wear situations. However, in work carried out at the National Centre of Tribology (NCT), various coatings applied by the Union Carbide detonation-gun process based upon bonded carbides have been tested. At temperatures below 200°C in a carbon dioxide environment, the application of these coatings to mild steel or stainless steel reduced the wear rate by up to an order of magnitude.

9.9.7.9 Polymeric materials

Replacing one of the fretting surfaces with a polymer material can be effective in reducing fretting damage to a steel surface. However, the choice of polymer is all-important. This is because hard abrasive oxide debris produced from a steel surface can become embedded in the polymer. The polymer then becomes an effective carrier of the abrasive debris and the wear process becomes two-body abrasion. One of the best polymer materials to use in a fretting wear problem is PTFE. This is probably because PTFE works by forming a transfer film on the steel counterface. Any abrasive debris is then encapsulated by the PTFE transfer film.

Unfortunately, PTFE is a rather soft polymer that will move under load, and it is therefore not generally suitable for the high-load situations of fretting fatigue. However, in commercial bearings additional support has been achieved by incorporating PTFE into a sintered phosphor-bronze matrix.

Tests at NCT have also shown the benefits to be gained by the use of PTFE-based materials (see Table 9.24). The fretting wear of mild steel was considerably reduced by the use of a fabric-based PTFE weave counterface, the steel exhibiting an almost imperceptible level of damage and wear, while wear of the polymer was also low. Clearly, consideration should be given to these types of material for fretting problems in flat-on-flat geometries.

PTFE can also be constrained and supported by incorporating it into a hard electroless nickel coating. The fretting resistance of the hard nickel is enhanced and certainly for low-cycle fretting wear this coating is a very effective palliative.[131] In fretting fatigue the benefits are likely to come from the low friction, therefore the palliative may not be effective in force controlled situations. Also, the reduction in fatigue

Table 9.24 Rubbing fretting wear of various materials versus mild steel at 125 μm slip amplitude

Counterface	Specific wear rate $(m^3N^{-1}m^{-1})$ ($\times 10^{-16}$)
Mild steel	300–600
Nylon 6.6	80–160
Carbon-graphite filled PTFE	10–20
Amorphous carbon-graphite	9
Woven PTFE fibre/glass fibre + resin	6–8
PTFE flock/resin/Nomex cloth	4–5

For additional information on materials and manufacturers, see reference 132.

strength associated with the hard nickel plating may be detrimental to the fretting fatigue performance.

9.9.7.10 Interfacial layer

It has been shown that micro-slip can be prevented when a thin layer of flexible material (for example, rubber or Terylene) is interposed between the fretting surfaces.[133] With a layer of the right compliance and thickness, the shear stress concentration at the edge of the contact giving rise to the slip can be eliminated from flat contacting surfaces. With Hertzian contacts because the contact pressure goes to zero at the edge, to achieve the same results it is necessary to bond the layer to both the surfaces.

Although this palliative can be effective at preventing relative sliding there are design limitations to its applicability. Clearly, the palliative can, in practice, only be used with relatively small slip amplitudes, therefore it is not likely to be suitable in large-amplitude displacement controlled situations. The increase in compliance produced by the palliative may also be unacceptable to the overall performance of the design. The durability of the interfacial layer under the alternating loading may also be a limiting factor.

Notwithstanding these limitations, the ability of interfacial layers to eliminate the shear concentration can prove very effective in many fretting fatigue problems involving bolted or riveted joints. Sandifer[134] has carried out practical testing of interfacial layers for use in the aircraft industry and found the palliative to be a success.

9.9.7.11 De-stressing notches

This has been suggested as a palliative only to fretting fatigue problems. This palliative was investigated by Kreitner,[135] who tested a flat fatigue specimen machined with closely spaced de-stressing notches (0.4 mm deep) running laterally and longitudinally over the contact area. The tests showed that with the de-stressing notches fretting was prevented from influencing the fatigue strength of the specimen. Moreover, although there was a reduction in the fatigue limit caused by the notches, the improvement in fretting fatigue strength was still in excess of 100%. The effectiveness of this palliative has been confirmed by Bramhall,[136] who also obtained a significant increase in fretting fatigue strength using de-stressing grooves cut in a lateral direction only.

The advantage of this palliative is that the fatigue strength is governed purely by the notch effect of the de-stressing notches; it is therefore more predictable. Also, the palliative is applicable to both displacement and force controlled problems.

9.9.7.12 Surface cold working by shot peening

Shot peening is one of a number of treatments which can be used to produce surface cold working. This type of treatment increases the hardness and induces a compressive stress in the surface layers. Both of these features, as previously explained, can be advantageous in reducing fretting damage.

In addition, shot peening produces surface roughening which may well have benefit in both fretting fatigue and fretting wear problems. It has been suggested that rough surfaces are more resistant to fretting fatigue.[137] This may be because a rough surface has, to a limited extent, the same de-stressing effect as the notches described previously.

A shot-peened surface will also retain lubricant better, each indentation acting like a small oil reservoir. Therefore, in both displacement controlled fretting fatigue and wear, shot peening may be expected to enhance the effectiveness of an oil lubricant.

9.9.8 Summary of palliatives

9.9.8.1 Palliatives for fretting fatigue

Palliatives for fretting fatigue problems have been reviewed by Chivers and Gordelier[86,138,139] in an attempt to provide a rational approach to their selection. They observed that a useful starting point is provided by the work of Nishioka and Hirakawa[140] in which an equation is developed for the initiation of fretting fatigue cracks in flat fatigue specimens which are fretted by cyclindrical fretting pads. The fretting pads were orientated such that the direction of slip was orthogonal to the cylinder axes. They derived the expression:

$$\sigma_{fwl} = \sigma_{wl} - 2\mu P_0 \left\{ 1 - \exp\left(-\frac{s}{k} \right) \right\} \tag{9.9}$$

where σ_{fwl} is the alternating stress necessary to initiate fretting fatigue cracks, σ_{wl} the alternating stress necessary to initiate cracks in the absence of fretting, μ the coefficient of friction, P_0 the peak Hertzian stress, s the relative slip and k a constant depending on the material and surface condition. This equation relates to the condition of partial slip in which relative movement takes place over only part of the contact area. When the full slip condition is reached and s becomes large, equation (9.9) reduces to:

$$\sigma_{fwl} = \sigma_{wl} - 2\mu P_0 \tag{9.10}$$

An improvement in fretting fatigue behaviour can therefore be achieved by raising the fatigue strength, σ_{wl}, of the base material, reducing the relative slip, s, either to achieve partial slip or to reduce the degree of partial slip present, lowering the contact pressure such that P_0 is reduced or lowering the coefficient of friction μ.

Excluding a change in base material, the problem is that the remaining options are interactive since a reduction either in contact pressure or in friction coefficient may result in an increase in the slip amplitude. An important consideration is what form the driving force to produce the movement takes, that is, whether conditions are controlled by displacement or force.

With displacement control, the amplitude of movement will be constant regardless of the force required. Considering equations (9.9) and (9.10), it is apparent that for either partial or full slip conditions, and with constant slip amplitude, reductions in either friction coefficient or contact force should result in an improvement in fretting fatigue performance.

With force control, the peak oscillating force to produce the movement is fixed in magnitude. In some circumstances therefore a modest increase in either contact force or friction coefficient could have a large effect in inhibiting the slip that occurs. If, however, these increases should fail to achieve a partial slip condition, equation (9.10) shows that a deterioration in fretting fatigue performance would be expected and hence the result of a modification to the conditions of contact is dependent on the slip regime. These results are summarized in Table 9.25. Additionally, Chivers and Gordelier reviewed possible mechanisms of alleviation and these are summarized in Table 9.26.

9.9.8.2 Palliatives for fretting wear under force excitation

A similar distinction based upon the source of excitation may be made for fretting wear. There are a wide range of possible palliatives for control of fretting wear caused by force or stress excitation and these are summarized in Table 9.27.

9.9.8.3 Palliatives for fretting wear under amplitude excitation

Palliatives for control of fretting wear under amplitude-driven movements are summarized in Table 9.28.

9.10 Surface topography

9.10.1 Effects of surface topography

All machining processes leave characteristic topographic features on the surfaces of components. These features vary in amplitude, spacing and shape and can exert a significant effect on the component's function or aesthetic appearance.

The British Standard, BS 1134: 1988,[141] identifies two components of surface topography which are generated by most common machining processes and classified according to their cause:

1. Roughness: 'The irregularities in the surface texture that are inherent in the production process but excluding waviness and errors of form.'
2. Waviness: 'That component of surface texture upon which roughness is superimposed. This may result from machine or work deflection, vibrations or release in machinery.

Three surface characteristics are illustrated by Figure 9.89. In addition to the above definitions a third surface shape component, that of 'form', is generally recognized as being the

---

Table 9.25 Changes in contact parameters to improve fretting fatigue behaviour

Slip	Depth wear rate ≪ 1 Crack growth rate		Depth wear rate ≤ 1 Crack growth rate	
	Displacement driven	Force driven	Displacement driven	Force driven
Full	Lower μ Lower P	Lower μ Lower P	Increase P	Increase P
Partial	Lower μ Lower P	Increase μ Increase P	Increase P	Decrease P

Table 9.26 Possible mechanisms of alleviation for various palliatives

Palliative	Probable mechanism	Comments
Chemical treatment Sulphidizing	Induces compressive stress	Should always work
Mechanical treatment Shot peening	Induces compressive stress	Should always work
Metal coatings Hard hard particles in metallic binder	Reduces μ or stops cracks in lamellar structure	Useful if μ decrease required
chromium	Increases μ	Useful if μ increase required
Soft cadmium silver copper	Reduces μ	
Non-metallic coatings Solid lubricant (e.g. MoS_2 or PTFE)	Reduces μ (may act as elastic interface)	Limited life with wear
rubber insert	Allows relative movement without generation of additional stress	Energy dissipation may cause degradation

Table 9.27 Palliatives for fretting wear under force excitation

Palliative	Mode of action
Design changes	Reduce source of excitation
Increase load	Reduces amplitude of slip
Increase pressure	Reduces amplitude of slip (beware of fretting fatigue)
Increase μ	Reduces amplitude of slip
Elastic interface polymer rubber	Movement taken up in elastic shear of interfacial layer
Surface treatments soft, e.g. Ag Au Pb	Prevention of adhesion
hard e.g. anodizing sulphidizing phosphating hard Cr electroless Ni TiC, WC, TiN	Improved fretting wear resistance (beware lower fretting fatigue limit) (beware lower fretting fatigue limit)
Surface coatings detonation gun plasma spray	Improved fretting wear resistance
Solid lubricants MoS$_2$, PTFE filled-PTFE	Improved fretting wear resistance

Table 9.28 Palliatives for fretting wear under amplitude excitation

Palliative	Mode of action
Design changes	Change design to reduce amplitude of slip Use different design concept to avoid slip
Reduce load	Reduces wear rate
Reduce pressure	Reduces wear rate
Increase surface area	Lower depth loss (beware decreased lubricant access)
Decrease and lubricate	Increased amplitude of slip to allow better access of lubricant
Surface treatments soft, e.g. Ag Au Pb	Prevention of adhesion
hard	Improved fretting wear resistance
e.g. anodizing sulphidizing phosphating electroless Ni TiC, WC, TiN	(acts as good lubricant reservoir)
Surface coatings detonation gun plasma spray	Improved fretting wear resistance
Solid lubricants MoS$_2$, PTFE filled-PTFE	Improved fretting wear resistance

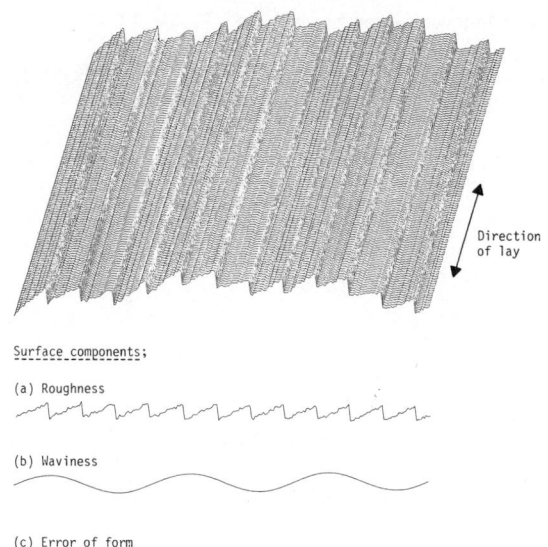

Surface components:

(a) Roughness

(b) Waviness

(c) Error of form

Figure 9.89 Diagrammatic representation of a machined surface illustrating the classification of topography by the separation of profile components

nominal shape of the component. i.e. the surface shape when roughness and waviness are neglected. Normally, deviations from ideal form are referred to as 'errors of form'. Figure 9.89 also shows that machining marks in surfaces may predominantly fall along one direction. This is frequently called the 'lay' of the surface.

These definitions of surface features are arbitrary and, therefore, also ambiguous. It is possible for surface features of entirely different scales, produced by different machining processes, to have the same classification. The term 'surface roughness' is also frequently regarded to be synonomous with 'surface texture' and 'surface topography' and, as a result, could potentially be confused with the 'roughness component' of a surface structure. Despite such difficulties, these terms are well established.

The sizes of machining marks are usually discussed in units of millionths of a metre, i.e. micrometres (abbreviation μm). In mechanical engineering metrology, the size of features of interest generally ranges from a few hundred micrometres down to hundredths of a micrometre and the scale of these features is not easy to visualize. Figure 9.90 attempts to place objects of this scale into perspective by relating their dimensions to physically important parameters of interest in the study of surface topography and the size of some items familiar from daily life.

It appears that no surface can be regarded as being perfectly flat. (Even though its imperfections may arise only as a result of misalignment in the positions of atoms in its surface.) As a consequence, considerations relating to surface finish stretch across a broad band of subjects which relate to the functional performance and manufacture of engineering components. The influence of surface topography on physical phenomena varies widely in scope and includes lubrication, electrical contact resistance, heat transfer, fluid flow, noise generation, the performance of optical components and the fidelity of thin-film coating processes. The influence of surface topogra-

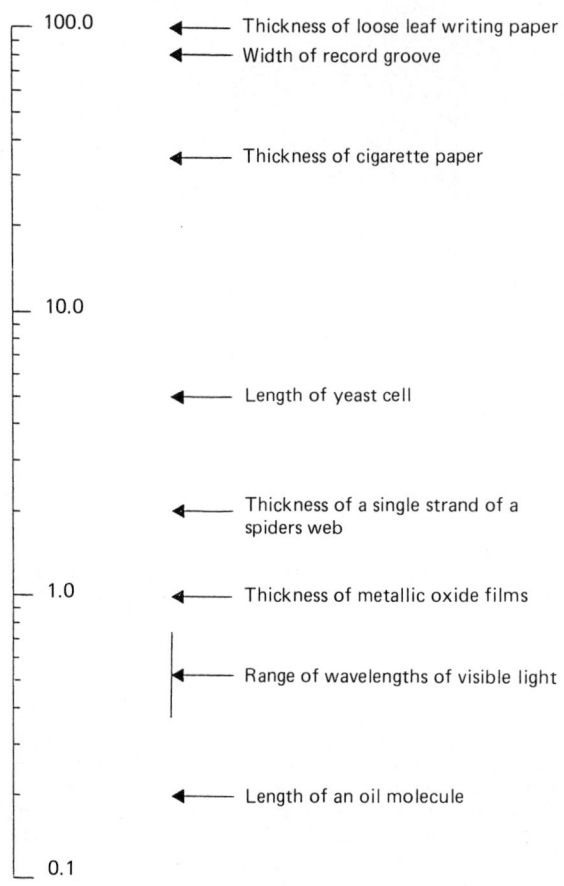

Length (micrometres)

— 100.0 ◄—— Thickness of loose leaf writing paper
    ◄—— Width of record groove

    ◄—— Thickness of cigarette paper

— 10.0

    ◄—— Length of yeast cell

    ◄—— Thickness of a single strand of a spiders web

— 1.0 ◄—— Thickness of metallic oxide films

    ◄—— Range of wavelengths of visible light

    ◄—— Length of an oil molecule

— 0.1

Figure 9.90 List of typical dimensions to demonstrate the perspective of features of significance in the surface topography of engineering components

phy on manufacture is principally one in which 'quality' is traded off against cost and a number of these influences are discussed below.

9.10.1.1 Liquid-film lubrication

Theoretical investigations into the influence of surface roughness in lubrication has been a topic of considerable interest over recent years. Surface roughness is of importance because fluid flow rate, the fundamental principle on which the Reynolds equation is formulated, depends on the cube of film thickness. Flow rates will suffer perturbation as a result of local changes in film thickness due to surface roughness.

Several analyses of lubrication which involve surface roughness have been described and the simplest of these studies investigate the influence of longitudinal striated roughness on hydrodynamic lubrication. Michell[142] examined the effect of sinusoidal roughness in slider bearings while Dowson and Whomes[143] investigated the lubrication of cylindrical rollers containing vee-shaped, square, and sinusoidal roughness running along a circumferential direction. A study which examined the effects of non-dimensional roughness was conducted by Christensen.[144]

The investigation of transverse roughness is a more difficult problem because an additional term arises in the Reynolds equation when the direction of motion of a surface does not coincide with the direction of its tangent. Dyson[145] points out that some analyses of transverse roughness (e.g. Christensen and Tonder[146]) did not appreciate this and consequently their results are only applicable in situations where a plane surface moves over a stationary rough one. An accurate analysis of the effects of transverse roughness was given in a study of the lubrication of rough disks by Berthe and Godet.[147]

The inclusion of surface roughness effects in the solution of the Reynolds equation in EHL problems is uncommon, as the situation is very complicated even in the treatment of ideally smooth surfaces. However, a perturbation analysis, applicable in situations where the rms surface roughness is less than one third of the film thickness has been developed[148] and applied in a study of finite bearing lubrication.[149]

The effects of surface roughness in hydrodynamic lubrication and EHL can be summarized. In general, for an infinitely long bearing the effect of transverse roughness is to increase load capacity while longitudinal roughness decreases load capacity. This occurs because transverse roughness ridges inhibit lubricant flow, generating an increase in pressure in the load-bearing zone. In contrast, longitudinal ridges provide flow channels for the lubricant to escape from the load-bearing zone, reducing its load capacity. In a finite bearing the situation is more complex. Narrow bearings have the greatest side leakage flow and, therefore, support the lowest loads. Longitudinal roughness tends to inhibit side leakage and, as a consequence, will tend to improve the load-carrying capacity of 'short' bearings. However, it should be noted that the net effect of roughness depends on the length-to-width ratio of the bearing as well as the nature of its roughness. Load-carrying capacity is reduced when the bearing is wide and increased for very narrow configurations. The friction force also depends on the bearing length-to-width ratio, with friction for longitudinal roughness being greater than that for transverse roughness.

The influence of surface roughness in mixed lubrication has also been considered. Patir and Cheng[151,152] have derived a model which allows any three-dimensional roughness structure to be analysed. Mixed lubrication is a complex situation but, in general, this model shows that as 'roughness' is increased, friction and load-carrying capacity of a surface is also increased. Additionally, it has been of value in studying the power loss at the piston-ring/cylinder interface in internal combustion engines.[153]

The majority of lubrication analyses employ the Reynolds equation, which assumes that the cross-film lubricant flow is negligible compared with the flow in other directions. It should be noted that as surface roughness is increased this assumption becomes less valid. Under such conditions the more complex Navier–Stokes equations should be used in principle. The lubrication of a rigid slider bearing has been considered using the Navier–Stokes equations and it has been demonstrated that the Reynolds equation progressively underestimates the load capacity of such an arrangement as roughness is increased.[154]

Since surface roughness has an influence on friction it is not surprising that it also affects wear. Dawson[155] found that the ratio, D, given by

$$D = 1/M$$

was inversely proportional to the tendency for a bearing to begin pitting; a standard mode of fatigue failure experienced by roller bearings and gear teeth. The stresses generated by the repeated elastic deformation caused by the passage of a ball or roller result in this failure which manifests itself as a small 'pit' in a component where a fragment of metal has fallen out. (The mechanism of pitting in relation to surface roughness is described in detail in reference 156.)

Scuffing is a second form of failure which is influenced by surface roughness. When brief inter-asperity contact or asperity-particle contact occurs it causes friction which generates heat. Scuffing is thought to arise when the speed and pressure at the contact are increased so much that, at a critical point, the heat generated cannot be dissipated in the surrounding bulk material without melting at the contact. An expression to describe friction in gears which includes a surface roughness term has been developed[157] and it appears that over a limited range of roughness friction can be reduced by selecting an appropriate surface roughness for gears. As a consequence, resistance to scuffing is increased.

The phase of wear which occurs in the early stages of sliding is usually referred to as running-in. The influence of surface topography in running-in has been reported under both lubricated conditions[158] and dry sliding conditions.[159] During the running-in phase rough surfaces tend to become smoother, the rate of wear depending on both the initial surface topography and the applied load. Other studies have found that once running-in is complete the wear rate becomes independent of the initial surface roughness[160] and can be described by a mechanism known as delamination.[161] According to delamination theory, a material fatigued under sliding develops subsurface dislocations and voids which run parallel to the surface. When these faults coalesce, sheet-like wear particles are released. Under low loads the initial wear rate of a smooth surface is higher than that of a rough one because delamination commences relatively quickly compared to rough surfaces, where delamination is delayed until the asperities are worn smooth. The opposite is true under high loads as rough surface asperities are easily and quickly removed by adhesive and abrasive processes under such conditions (see Halling[162] for descriptions of adhesive and abrasive wear mechanisms).

Bayer and Sirico[163] have conducted tests into the influence of anisotropy of surface roughness on wear. Spherical specimens with isotropic surface topography were positioned against flat planes with a well-defined lay and caused to slide under a fixed load in the boundary lubrication regime. Using sliding parallel and perpendicular to the lay of the anisotropic plane Bayer and Sirico showed that as the coarseness of surface increased, the sensitivity of wear to orientation became greater. However, below a limiting roughness the influence exerted on wear by changes in roughness was a more significant effect than the directional dependence.

9.10.1.2 Contact mechanics

Surface topography exerts an influence on the processes which occur at static junctions as well as ones where surfaces are in relative motion. Theoretical and experimental studies of static contact are diverse and range from such subjects as the analysis of the stiffness of machine tool joints[164] to the investigation of heat transfer and electrical resistance.[165]

The main aim of theoretical studies of static contact is to predict the real area of contact between the components which form the interface so that related physical characteristics can be derived. During the last three decades several models of various levels of geometric and statistical complexity have been proposed to describe contact at interfaces.

One of the earliest contact models was of a geometric nature and was proposed by Archard.[166] Archard considered the deformation of spherical asperities as they were loaded against a non-deformable plane. He showed that by progressively covering each asperity with similar ones of smaller radius the relationship between the true area of contact and the load which it supported elastically approached direct proportion-

ality. This would seem to support the validity of Archard's model, as it is usually assumed that the force of friction is proportional to the true area of contact between two bodies. (Since friction is proportional to load, according to Amonton's law, it is reasonable to expect that the true area of contact is proportional to load.)

Subsequent models of surface contact, based mainly on statistical approaches, have introduced descriptions of surface topography which have varying levels of realism. Many models assume surface asperities to have a Gaussian height distribution. Experimental investigations imply that at least some surfaces produced by engineering processes do have near Gaussian height distributions, so such an assumption can be justified.[167]

There are several versions. Greenwood and Williamson[168] constructed a model which used Gaussian and exponential functions to describe the height distribution of parabolic asperities prior to contact with a non-deformable plane. Greenwood and Tripp[169] constructed a similar model which employed spherical asperities on both surfaces. Whitehouse and Archard[170] conducted an analysis which avoided stringent assumptions regarding the geometry of individual asperities by assuming instead that the surfaces had an exponentially decaying autocorrelation function. Nayak[171] applied the technique of random process theory to describe engineering surfaces in a modification of the more complex analysis of random, moving, ocean surfaces conducted by Longuet-Higgins.[172,173] This work enabled useful relationships between the statistics of surfaces and profiles to be derived and applied later in an analysis of plastic contact.[174]

An academic debate exists regarding the nature of the deformation at asperity contacts. Some of the authors of the surface models outlined above derive a dimensionless parameter, the plasticity index, in an attempt to establish the proportion of plastic/elastic deformation which ocurs at asperity contacts. These plasticity indices relate the elastic modulus, the material hardness and a measure of surface roughness in order to achieve this.

The Greenwood and Williamson[168] model considers deformation of a rough surface of hardness, H, which has a Gaussian distribution of asperity peak heights with standard deviation, σ, and constant tip radius β, in contact with a plane. If the joint, elastic modulus of the contact is E the plasticity index, ψ, of the rough surface given by this model is

$$\psi = \frac{E^1}{H}\left(\frac{\sigma}{\beta}\right)^{1/2}$$

where $\dfrac{1}{E^1} = \dfrac{1 - v_1^2}{E_1} + \dfrac{1 - v_2^2}{E_2}$. Note that if $E_1 > E_2$ then

$E^1 = \dfrac{E}{(1 - v^2)}$ with E_1, E_2 and v_1, v_2 being Young's Modulus

and Poisson's Ratio respectively for the interface materials 1 and 2.

The index indicates if a surface is likely to suffer a significant proportion of plastic deformation of asperities at a nominal pressure, P. Surfaces which have a plasticity index greater than unity will undergo plastic deformation at very low loads and can be expected to deform plastically in most practical applications. If ψ is less than 0.6, significant plastic deformation will only arise under very high loads, which would not be encountered in routine applications. Greenwood and Williamson[168] give an example in which a 'significant' proportion of plastic deformation is defined as plastic deformation over an area $\geq 2\%$ of the real area of contact. They state that a nominal pressure of 2.0×10^{-2} is required to cause significant plastic deformation if $\psi = 1$ and that a nominal pressure of 4.3×10^5 kg m^{-2} is required to generate the same proportion of deformation for a surface where $\psi = 0.6$. They also point out that, in practice, the index ψ can vary between 0.1 and more than 100.

Whitehouse and Archard[170] derived a more general version of the plasticity index from their surface model which does not rely on the assumption that surface asperities have a constant radius. It has the form

$$\psi^* = K\left(\frac{E'}{H}\right)\left(\frac{\sigma}{\beta^*}\right)$$

where K is a numerical constant, H is the hardness of the material, β^* is a parameter (defined in Section 9.10.3.8, called the 'correlation distance', and derived from the autocorrelation function of the surface shape).

Onions and Archard[175] found that if the constant, K, took the value unity then the plasticity index, ψ^*, could coincidently, be interpreted in the same way as the index proposed by Greenwood and Williamson. This suggestion appears to be generally accepted.

The plasticity index is potentially a very valuable parameter because it provides an assessment of the physical and geometrical properties of a surface. It has been suggested that it could provide a useful guide to the condition of a surface during the running-in process since the variation of the index with time will show how quickly the surfaces involved approach an elastic contact condition.[176]

Differences between the predictions of these surface models are discussed by Thomas.[177] These differences arise principally from variations in the form of the asperity models adopted, illustrating that the character of a surface, as well as its scale of roughness, has an effect on the nature of the contact between the surfaces.

9.10.1.3 Fluid flow

The topography of a surface influences the flow of fluids in other situations in addition to lubrication. This influence is economically significant in a number of engineering applications, including sealing and the propulsion of ships and aircraft.

Many different types of seal are used in engineering applications. Establishing the desirable surface characteristics of an efficient seal is not a trivial problem. One approach has been outlined in an investigation of the performance of lip seals.[178,179] This approach involved dividing the members of a collection of lip seals into two categories: those which sealed effectively and those which leaked. The topography of all the seals was characterized by measuring a number of commonly used surface parameters, and some of these parameters showed a loose correlation with the tendency of a seal to leak. By constructing a suitable algebraic function it was possible to use all the measured parameters together to differentiate between sealing and leaking topography much more effectively, making it possible to identify the characteristics of an ideally good and an ideally bad lip seal surface.[180]

Fundamental work on fluid flow conducted by Nikuradse in the 1930s has been recently reviewed.[181] Nikuradse investigated the flow of fluids through pipes roughened by sand grain coatings and demonstrated that the skin friction factor was a function of the ratio of the grain size to the pipe diameter. He showed that in the turbulent flow regime of a rough pipe, the logarithm of the friction factor is not inversely proportional to the logarithm of the Reynolds number as it is in the case of an ideal smooth pipe. Instead, the linear dependence ends at some point and the friction factor tends to a constant value which is related to the surface roughness/pipe diameter ratio as the Reynolds number increases.

The work of Nikaradse has been criticized on the basis that it does not simulate the topography generated by machining processes very effectively and is, therefore, of limited value. As a consequence, alternative methods of analysing fluid flow over rough surfaces in the form of model cracks have been investigated.[182]

Fluid flow over rough surfaces is of considerable importance in ship hull design, and it is reported that between 80% and 90% of the total resistance of a ship to motion is due to skin friction.[183] In addition, relatively small changes in roughness exert a significant effect. Conn et al.[184] describe an early investigation into ship hull roughness effects. For one particular ship they observed that different paint finishes could vary its frictional resistance by up to 5% while allowing the ship hull to foul could increase it up to nearly 50%.

A discussion of ship trial results collected over many years indicates that, typically, a 1% increase in power is required to maintain ship speed for every 10 μm increase of a roughness parameter called the mean apparent amplitude (MAA).[185] It is also stated that typical hull deterioration rates are between 10 μm and 40 μm MAA per annum, with new ships having a hull roughness of approximately 130 μm MAA while that of old ships may exceed 1000 μm MAA.

The efficiency of aircraft is also strongly influenced by skin friction. However, methods being considered to reduce skin friction seem to be directed along lines which are not associated with studies into the effect of the surface topography of the wings/fuselage.[186]

9.10.1.4 Vibration

Machines with moving parts inevitably produce noise when in operation. Noise can be generated as a result of the misalignment of components and the occurrence of mechanical resonances at particular speeds. However, a contribution to the noise spectrum can also arise as a result of surface roughness. Thomas[187] discusses three separate sources of noise arising as a consequence of surface roughness:

1. Noise generated by the elastic deformation and release of form and waviness features during rolling. This type of noise can be generated in EHL. At low speeds it contributes to the low-frequency end of the noise spectrum and it is unlikely to cause problems. However, at higher speeds both the frequency of the noise and its energy are increased and its effect may be significant.

2. Shock noise caused by the elastic deformation and release of asperities within the Hertzian contact zone. This form of noise arises in both rolling and sliding contacts. It is apparently the most dominant form of surface-generated noise and it may affect a broad range of frequencies, although its most severe influence is exerted between 300 Hz and 10 000 Hz.

3. Shock noise arising from asperity collisions and debris collisions. This form of noise arises in EHL where very thin films separate moving surfaces. Under such conditions inconsistencies in surface roughness can cause inter-asperity contact or collision with entrained debris to generate transient noise.

Two principal conclusions may be drawn with regard to noise generated by surface roughness:

1. By identifying the operating conditions of a component it should be possible to determine a surface finish which maintains low levels of acoustic noise.

2. Surface-generated noise may be valuable as a form of 'non-invasive' surface monitoring. It has already been

shown that surface-generated noise can be used to detect wear.[188] It is therefore potentially useful for the diagnosis of potential failure in rolling and sliding contacts.

9.10.1.5 Coating technology

The appearance and service performance of coated surfaces depends strongly on the surface topography of the overlaid substrate. As a result, materials which are to be coated are generally required to have a surface finish which is uniform and of a high quality, i.e. free from pits and scratches.

Cold-rolled steel is a typical example of a material for which a good-quality surface finish is essential to maintain the aesthetic appearance of objects constructed from it. It is widely used for the body panels of cars where small blemishes are visible even when sprayed. Avoiding visible defects is exceedingly difficult. The human eye can detect pits only 0.05 μm deep in a 'smooth' surface, and such features are only just within the range of routine surface-measurement procedures. Eliminating them by improvements in the manufacturing method can be difficult and expensive. Other applications in which control of surface roughness plays an important part in maintaining an acceptable aesthetic appearance include the manufacture of card, paper and photographic film.

As well as influencing aesthetic appearance, substrate roughness can also affect the service performance of painted materials. Substrate roughness causes variability in the film thickness of painted materials, which has an observable effect on corrosion resistance and ease of cleaning.[189] In general, it seems that the effect of painting a 'rough' surface is to make it 'smoother' whilst the consequence of painting a 'smooth' surface is to make it 'rougher'.[190]

9.10.1.6 Optics

Optical components such as lenses, prisms and mirrors are used in many different types of equipment. This equipment is used in a wide range of applications for both conventional purposes (e.g. microscopy, photography, astronomy) and more 'exotic' ones (e.g. laser beam collimation, the control of synchrotron radiation and infrared surveillance). Common applications also include the rapidly increasing use of optical components in communication devices,[191] transducers[192] and data-storage equipment.[193]

In general, in order to achieve satisfactory performance, exceedingly stringent tolerances on the roughness of optical components must be attained. Typically the standard deviation of their roughness is in the range 0.001–0.01 μm.[194] It is important to be able to characterize the surface topography of optical components so that limitations in performance can be estimated. Beckmann and Spizzichino[195] explain how a statistical description of a component such as a lens or mirror can be used to predict its scattering properties. In addition, Mie theory which requires a knowledge of the size, shape and optical constants of the feature can be used to predict scattering from small isolated features on a surface[196]. The effects of scratches, digs and other surface defects which have dimensions much greater than the wavelength of light can be determined only by relating their dimensions to those of standard scratches whose scattering characteristics are known.

9.10.1.7 Electronics and computer hardware

In recent years the size of integrated circuits has been reduced considerably while the number of electronic elements in a given area has approximately doubled each year. Ultimately, fundamental physical criteria will limit this progress.[197] However, for the moment, such factors are not constraining ones and the difficulties which limit the miniaturization of electronic circuits are engineering problems.

The surface roughness of component substrates appears to be one consideration in this respect. Photolithographic processes are used in the manufacture of integrated circuits. Deviation from flatness in the substrate surface causes gaps between the substrate and the emulsion of the photoplate which allows stray light to enter, causing defective circuit geometry.[198] Similar problems are encountered when circuit films are deposited through an evaporation mask. Deviation in flatness between the deposition mask and the substrate permit evaporated material to enter and thus destroy line geometry.[199] In addition to these effects, substrate roughness has been found to influence the resistance of thin-film resistors[200] and the charge-storage capabilities of thin-film capacitors.[201] Roughness effects appear to become significant when the thickness of the film layer – typically, 0.1–0.01 μm – is of the same order as the average roughness of the substrate.[199]

The design of magnetic recording media is another branch of modern technology in which surface topography is of importance. The quality of a magnetically recorded signal depends on head-to-media spacing and, therefore, on surface roughness. Computer disk systems are one important application of magnetic recording technology. The read/write heads which transfer data to and from hard disks must maintain close proximity to the disk without actually coming into contact and causing wear. They operate as small air slider bearings located between 0.25 μm and 0.4 μm from the disk surface, whose roughness amplitude is generally an order of magnitude less than the air/film thickness. White[202] has analysed the performance of thin-film air slider bearings with two-sided roughness. He suggests that, in practice, the character of the topography of the read/write head and the disk will not have an important influence on its operation, and shows that only the amplitude of the surface roughness therefore needs to be considered in the design of such components.

9.10.1.8 Bio-engineering, pharmacy and hygiene

There are several naturally occurring circumstances in biology where surface topography is important (e.g. the lubrication of synovial joints as described by Tandon and Rakesh[203]). However, such examples cannot strictly be regarded as engineering applications! More valid examples are to be found in the discipline of bio-engineering.

The performance of temporary artificial blood vessels is one aspect of this subject. The range of roughness in tubes used to bypass blood during heart operations varies from about 2 μm up to 15 μm.[204] One problem associated with these tubes is that they can damage the red blood cells flowing through them. Stewart[204] found that twenty times more red blood cells were destroyed by some tubes in comparison with others. His investigation established a correlation between the scale of roughness of the bypass vessel and the damage to red blood cells passing through them. In addition, surface character appeared to exert an effect: 'smooth' tubes were found to cause less damage than 'rough' ones.

Another aspect of the influence of surface topography on performance is evident in the pharmaceuticals industry. The manufacture of tablets is conducted in shaped dies and the efficiency and reliability of the tableting process is influenced by the surface roughness of these dies. Dies which have an unsuitable topography result in the incorrect formation of tablets or their fracture on removal.

Surface topography also has an important bearing on hygiene. Many surgical, catering and household implements are manufactured with high-quality surfaces because smooth, polished surfaces possess relatively few sites which harbour bacteria during cleaning.

9.10.1.9 Considerations in production

As long ago as 1933 a pioneering paper published by Abbott and Firestone[205] recognized some of the inadequacies of simplistic schemes of surface characterization. The fact that it is not possible to fully specify the character and scale of a surface with a small set of arithmetic indicators still remains a serious problem for production/design engineers. Often a description of a surface is made by specifying a particular machining process to identify the principal surface character and accompanying this by statistical parameters which give further detail.

The topography of an engineering component is largely determined by two factors. The type of machining method used controls its character, while the 'coarseness' of the operation, and to a point the time spent on manufacture, determines the roughness scale.

BS 1134[206] indicates that in a number of finishing processes there is a range of finish over which the cost of production changes only minimally with reductions in surface roughness. It also points out that there is normally a transitional finish beyond which the cost of producing a surface with further reductions in roughness increases rapidly. It is, therefore sensible to manufacture components with the scale roughness which still maintains acceptable performance. In addition the least costly process which produces a surface character with acceptable functional performance should also be selected.

Selection of an appropriate manufacturing process is not a straightforward decision because it involves a detailed consideration of how the entire component is to be made. Many components are made by multi-pass operations in which a coarse maching process is used to obtain the nominal component shape while a fine is used to establish the desired surface finish. This might involve applying the same type of operation more than once using different tool or machine settings or it may mean that a combination of entirely different processes are used to obtain the final surface. The combination of operations which gives rise to target specification most economically may not be obvious.

Establishing the machining conditions for a finishing process to obtain a specified topography is also not straightforward, as many interacting factors are involved. Under ideal circumstances the factors to be considered would involve only the operational setting of the machine (e.g. the geometric characteristics of the cutting tool, the work speed, the tool feed rate and the type of cutting fluid used). Even under ideal conditions it is only possible to calculate the theoretical roughness developed in a machining operation for the simplest types of process, i.e. single-point tool cutting. Relationships of this type for turned surfaces generated by tools with a range of tip profiles have been reported by Dickinson.[207] Relationships

for less traditional metal-shaping processes such as electro-discharge machining (EDM), electro-chemical machining (ECM) and ultra-sonic machining (USM) have also been deduced.[208] However, they cannot be applied directly as they involve a parameter called a 'surface finish factor', which depends on the material of the workpiece/tool and the ultimate surface finish of the component. In order to apply the equations it is necessary to conduct trial tests to determine the surface finish factor for specific situations. (Some values of the surface finish factor for specific situations are given by Ghabriel et al.[208].)

Expressions which describe the relationships between operating conditions and surface roughness for complex cutting operations such as grinding and milling are apparently not available. Companies employ their own heuristic guidelines to estimate surface finish in these cases.

In practice, it is not usually feasible to obtain the surface finish predicted by theoretical conditions. Several factors combine to prevent this. Probably the most significant factor is the accretion of cutting debris on the tip of the tool. This accretion is in a dynamic state, continually breaking down and being replaced with new debris, causing the tool to have variable cutting characteristics. In general, the roughness of the surface produced by the built-up edge of the tool increases as the built-up edge grows larger. Changes in machining conditions which tend to reduce chip-tool friction/adhesion result in improved surface finish. These include increased workpiece speed, use of different tool materials (e.g. carbide tool tips) and the application of a good cutting lubricant. A detailed study of the effect of built-up edge on the surface roughness of a turned component is presented by Selvam and Radhakrishnan.[209]

Many other factors contribute to non-ideal surface finish arising under practical machining conditions. The most common of these are: chatter vibration of the machine tool, inaccurate tool movement, defects in the composition of the workpiece and discontinuous chip formation in the machining of brittle materials. In addition to these problems it appears that where several types of machining are applied in the manufacture of a component the 'shadows' of earlier operations sometimes remain evident in the topography of the finishing process.[210]

From these studies it is apparent that many problems are associated with the production aspects of surface roughness. Some are concerned with achieving good functional performance but just as many are of an economic nature. BS 1134[206] points out that 'seeking too good a finish' is a common error leading to waste in production time. It also outlines two simple philosophies to guide designers in controlling surface texture, but it concedes that the practice to be adopted depends very strongly on the area of engineering concerned.

The optimization of manufacturing operations is a highly complex subject, even when only the problems outlined above are considered. However, in the real manufacturing environment many other important variables also become involved (e.g. the power consumed by the various machining processes, the effect of the type of material being worked, and the quantities of raw material wasted in manufacture). Surface finish is just one consideration that must be reconciled with many others.

9.10.2 Measurement

Surface structure exists in two directions: the vertical direction, where it is characterized by height (or amplitude) parameters, and the horizontal direction where it is characterized by spatial (or wavelength) parameters. Any measurement method should be able to record roughness variations in both directions.

In the vertical direction roughness amplitudes vary widely. Coarse machining operations can produce features several hundred micrometres high. However, in contrast, some surfaces manufactured for special applications may contain perturbations of only molecular dimensions. In the horizontal plane, roughness variations arise on a scale which varies from the dimensions of the specimen down to atomic diameters. It is clear, therefore, that wide extremes of range need to be encompassed by any technique used to measure surface topography.

The relevance of surface topography to the functional performance and production cost of a component over a very wide range of applications has precipitated the development of a plethora of instruments to record and parameterize the structure of surfaces, and each of these devices has different limitations and advantages. In recent years development has centred largely on instruments which measure surface topography without contact with the specimen, i.e. mainly optical methods. However, these devices are, as yet, not particularly well established in terms of the number of applications in which they are used routinely. This situation may change in the future but, for the moment, more traditional techniques (i.e. stylus instruments and 'mechanical' comparators) share the bulk of the workload.

9.10.2.1 Stylus instruments

For many years the most popular device used to measure surface topography has been the stylus instrument. Despite some disadvantages, it has proved versatile and reliable in both the manufacturing and research environments. It is also the instrument in terms of which all national standards are defined.[211]

Construction and operation A typical commercial stylus instrument comprises five basic components: a stylus transducer, an amplifier, a chart recorder, a traverse unit and a meter system, as illustrated by the block diagram in Figure 9.91(a).

A common form of stylus transducer, which employs a linear variable differential transformer (LVDT), is manufactured by Rank Taylor Hobson. It consists of a beam which is pivoted on two knife edges and carries the stylus at one end with a ferrite block at the other (Figure 9.91(b)). The ferrite block is located between two coils. As the stylus is drawn over the irregularities of a surface by the traverse unit the stylus is displaced, causing the ferrite block to move between the coils. The coils form part of an inductance bridge circuit which is balanced when the stylus is in a neutral position. When the stylus changes position it causes a change in the mutual inductance in the coils modulating a high-frequency carrier signal in proportion to the displacement of the stylus. The relative change in the phase of the carrier signal indicates the

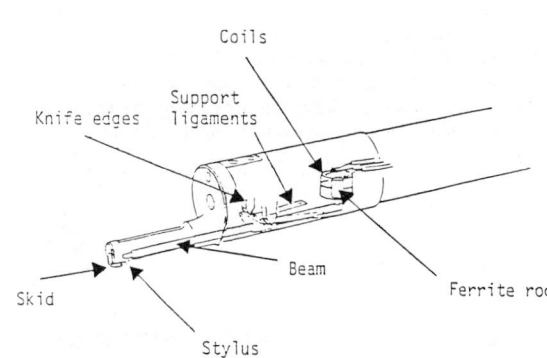

(a)

(b)

Figure 9.91 (a) Principal components of a commercial stylus instrument; (b) construction of a linear variable differential transformer stylus transducer (based on Figure 66 of *Exploring Surface Texture* by H. Dagnall and reproduced with the permission of Rank Taylor Hobson)

direction of displacement of the stylus. The carrier signal is amplified and demodulated to yield a signal representing a surface profile which may be output to a chart recorder or used to evaluate some parametric assessment of surface roughness given on a meter or visual display unit.

In addition to LVDTs, several other forms of displacement transducer are in use (e.g. optical interferometers, variable capacitors[213]).

Range and resolution LVDT transducers usually have a maximum vertical range of between 0.5 mm and 1.0 mm. The vertical resolution of stylus instruments depends mainly on the level of ambient vibration in the vicinity of the instrument but it is also influenced by electrical noise inherent in the amplifier. Divisions of 0.01 μm are common on the output of chart recorders, suggesting that this level of accuracy can be approached under controlled conditions.

The horizontal range of a stylus instrument is determined by the distance which the stylus is able to traverse. The horizontal resolution depends on the shape and dimensions of the stylus tip, and it is a limit which cannot be defined exactly.[214]

Datum arrangements The output of a stylus transducer depends on the difference in height between the stylus and a reference datum. To exclude the form of a component from measurements of topography it is necessary to generate this datum by causing the transducer to follow a path parallel to the nominal shape of the component. This can be achieved in two ways:

1. Using an independent datum as illustrated in Figure 9.92(a). Here an accurately flat or curved shape which corresponds to the nominal shape of the specimen is used to constrain the vertical position of the transducer as it traverses the specimen.
2. Using a skid datum which is attached to the transducer. The skid rests on the specimen and follows its form as the transducer is drawn along by the traverse unit in the manner illustrated by Figure 9.92(b).

When the independent datum is used, all departures from the nominal shape are recorded. However, it is a time-consuming method of measurement as an extensive setting-up procedure is often required to align the specimen and datum. Use of the skid datum circumvents this difficulty but surface profiles recorded using this datum may not be accurate. A specimen surface containing features with wavelengths greater than the length of the skid surface profile will be subjected to a high-pass filtering operating as the skid changes its vertical position when crossing these features. Isolated peaks in a surface can also generate spurious artefact valleys.[212]

A number of experimental and theoretical investigations have been conducted to examine the extent of profile distortion by the skid datum. The main conclusions are:

1. Use of the skid datum, in appropriate circumstances, only has a small influence on the parametric characteristics of a profile.[215]
2. When parametric influences are observed the peak parameters of a profile are those most significantly affected.[216]
3. Inaccuracies caused by use of a skid are greater for surfaces with prominent directional characteristics.[217]

Stylus effects Early publications report that gramophone styli were used as the transducer probe.[218] These have since been replaced by specially designed styli which are smaller and tipped with diamond to improve resistance to wear. Two types of stylus shape are commonly used: a conical stylus with a spherical tip and a pyramidal stylus with a truncated, nominally flat tip. The spherical tip of a conical stylus usually has a radius of less than 10 μm. The flat area of a pyramidal stylus is typically a rectangle of dimension 8 μm $\times$ 4 μm, the longer side being normal to the direction of traverse.

The profile recorded by the stylus tip is never a perfect assessment of the shape of the specimen. This arises as a consequence of two factors; the finite size/geometry of the stylus and the load it applies to the specimen surface:

- *Geometric/size considerations*
  The shape recorded by a stylus as it traverses a surface is the locus of a fixed point on the stylus and is known as the 'effective profile'.[141] As a consequence of its shape and finite size the profile recorded by the stylus can differ from the true surface shape; i.e. its shape can be distorted and, in some circumstances, its amplitude may be attenuated. In general, the shape and finite dimensions of the stylus cause the radius of curvature of peaks to be slightly enhanced and the width of valleys to be slightly reduced (Figure 9.93). Attenuation of high-frequency contributions to the profile shape also occurs. This arises because the stylus is unable to enter valleys which are narrower than its own length. This is a progressive effect which begins when the radius of curvature of a profile valley is smaller than the effective radius of curvature of the stylus.
- *Physical interactions*
  In most stylus instruments a spring is used to apply a small force, approximately 0.001 N, to the stylus to ensure that it always remains in contact with the specimen. As this load is supported by a very small area at the stylus tip, high pressures are generated at the contact with the specimen and temporary or permanent deformation can ensue. A number of investigators have estimated the magnitude and form of deformation caused by styli. However, theoretical estimates of deformation due to stylus pressure tend to vary widely depending on the assumptions made in the calculation. Many physical and chemical factors such as variation in hardness, oxidation level and fatigue influence the result as well as parameters which relate to the stylus geometry. Additional problems can also arise as the bulk properties of a material do not necessarily give an acceptable model of its microscopic behaviour (e.g. microscopic hardness has been found to be a function of load[219]).

It appears to be accepted that the errors caused by the effects described above are negligible under most circumstances. However, care should be taken when features of interest have the same order of magnitude of width as the stylus tip and, additionally, when the specimen has a low bulk hardness value. Methods for estimating errors due to the shape and load of styli are given in reference 214.

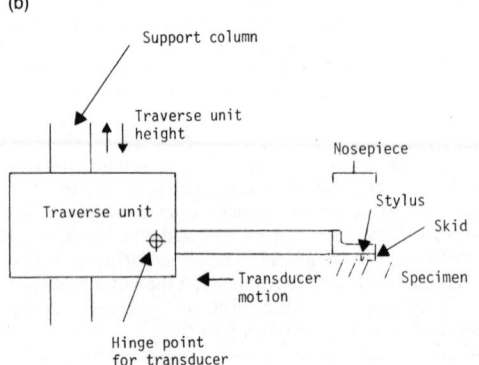

Figure 9.92 Datum arrangements for nominally flat specimens. (a) Use of the independent datum; (b) use of the skid datum

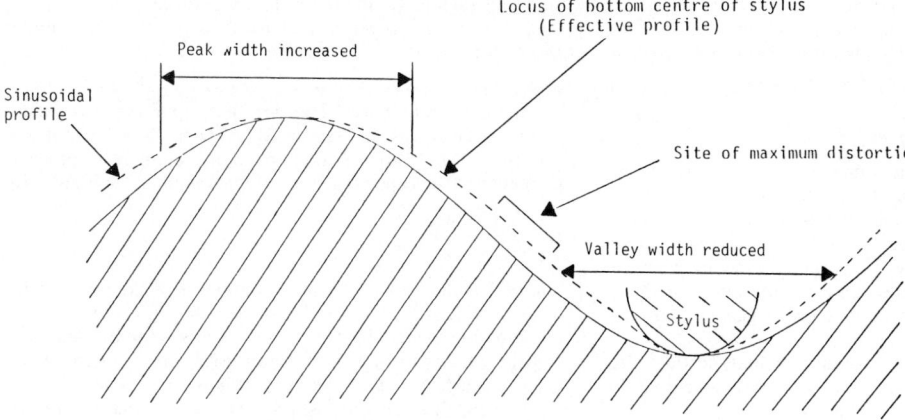

Figure 9.93 Distortion of a sinusoidal profile when traversed by a spherical stylus

Digital data-acquisition systems The development of flexible digital systems has permitted the value of stylus instruments in experimental studies in surface topography to be extended considerably. Digitized data are commonly acquired by sampling the amplifier output from stylus instruments. In the earliest systems, analogue data were converted into digital form and stored on punched paper or magnetic tape and then taken to a remote computer for subsequent analysis. Such an approach was adopted by Williamson.[220]

More recently, there has been an increase in the use of on-line systems. Equipment typically used in on-line data acquisition is illustrated in Figure 9.94. The principal components are a commercial stylus instrument which is linked to a computer through an analogue-to-digital converter (ADC) and an interface which usually includes a filter to set a high-frequency cut-off point in the data. To generate the output signal the transducer gearbox may be operated manually or automatically by the computer via a series of electric relays. Collected data are normally stored on-disk or in memory for subsequent processing rather than being analysed in real time. A number of on-line data systems have been described.[221–224]

One common criticism of stylus instruments is that they are normally restricted to supplying only profile data. However, stylus instruments can be adapted to allow an areal record of a surface to be obtained, rather than a simple profile. This is achieved by recording several parallel profiles, each displaced laterally from the previous one by a short distance. If all the profiles are referenced to a common origin they form a raster-scan record of the surface. A raster scan recorded using

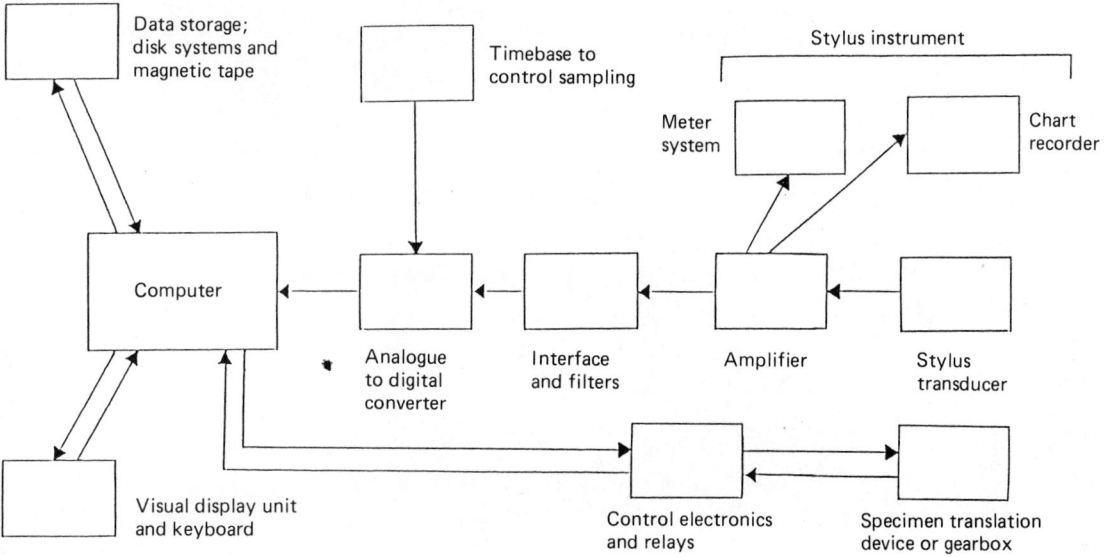

Figure 9.94 Equipment for the on-line acquisition of surface data

the author's equipment is illustrated in Figure 9.95. Areal measurements are attracting increasing interest and devices which perform raster measurements have been described in several publications.[225–227]

9.10.2.2 Comparative techniques

A wide range of instruments which use comparative techniques have been developed. Generally, their principle of operation is based upon a physical phenomenon which shows some dependence on surface roughness. Sometimes these devices are designed to give a parametric assessment of surface roughness, although the reliable operation of this type of instrument is often limited to a given class of surface topography.

Comparative devices have several advantages; they are relatively inexpensive, they are quick to apply, and they do not require extensive special training to operate.

Friction tests Probably the most well-known comparative measurement technique is the tactile test. In this test the finger is drawn across a specimen and its 'feel' (i.e. the frictional resistance) is then compared to that produced by a set of calibrated samples manufactured by the same process. The sample whose 'feel' most closely resembles that of the specimen is noted.

It has been found that the tactile test is significantly more reliable than visual inspection as a method for assessing surface roughness. However, the range of accurate comparison appears to depend on the character of the surface under inspection.[228]

Electrostatic techniques The distance separating two plane metal surfaces can be estimated by using them to form a capacitor. If they are separated by an insulating layer of air of thickness t_a, the capacitance, C, of the combination is given by

$$C = \frac{K\epsilon_a A}{t_a}$$

where A is the area of the smaller plane, ϵ_a is the permittivity of air and K is a constant. Measuring the capacitance of the combination using a bridge circuit permits t_a to be calculated. In principle, by forming a capacitor using a free electrode and a rough surface it is possible to obtain some assessment of surface roughness by measuring the capacitance of the arrangement and calculating t_a, which will then be some average measure of separation. (Note that it is not the distance of the mean plane of the surface from the probe.)

A number of transducers of this form have been described in publications[229,230] and the most reliable one has a flexible electrode coated with a dielectric material.[230] The flexible electrode allows the transducer to conform to the surface shape to remove the effect of waviness and form errors. The dielectric layer serves to make the transducer less sensitive to contact pressure. Capacitance measured roughness, t_a, shows

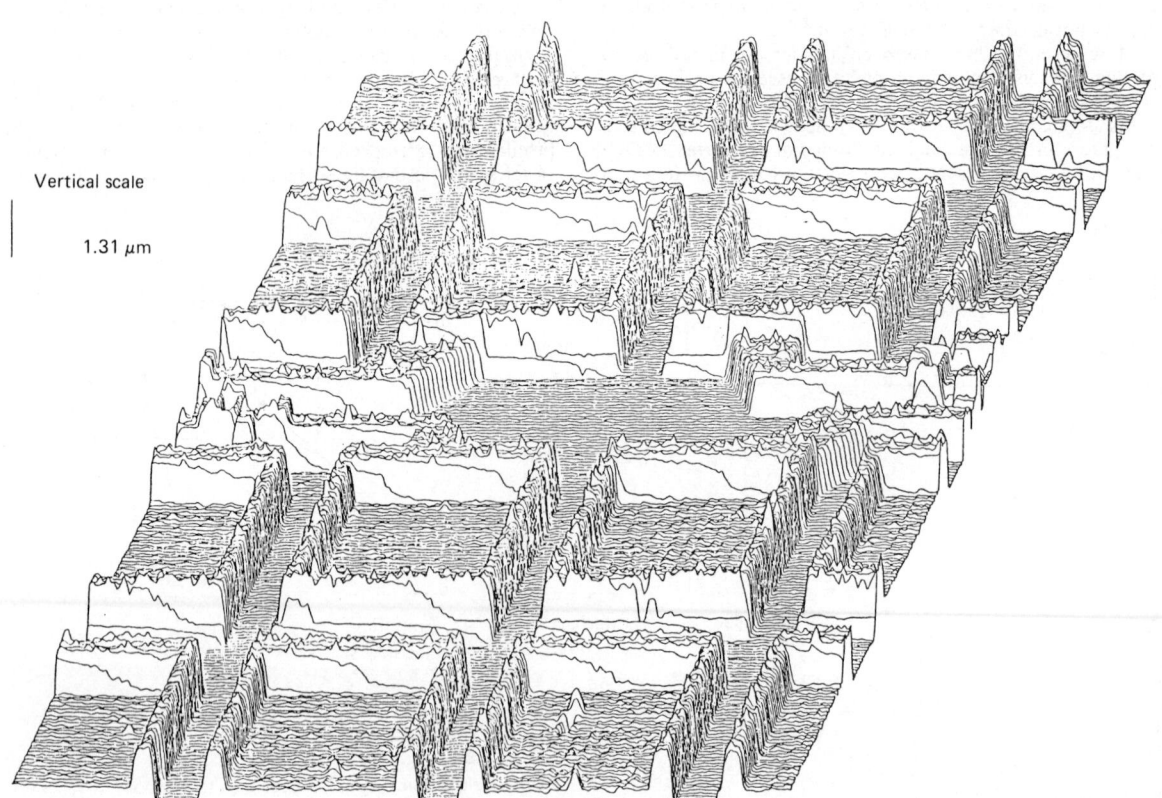

Vertical scale

1.31 μm

Figure 9.95 A raster-scan of the surface of an integrated circuit

some correlation with a parameter called 'depth of surface smoothness', defined by reference 231.

Pneumatic methods Surface roughness can be measured by a method called air gauging, which assesses the flow of air through a gap between a specimen surface and an open-ended nozzle placed facing downwards onto it. The nozzle is normally connected to a source of air at a constant pressure, P, via an intermediate chamber. The flow of air into this chamber is regulated by a valve whose open area is a; its flow out is controlled by the area, A, of the gap between the nozzle and the surface asperities which support it. Escape of air from the nozzle causes the pressure in the intermediate chamber to fall to p. It can be shown that, over a certain range, the relationship between p/P and a/A is linear, allowing the method to be used to assess surface roughness.[232]

Experimental measurements made by air gauging have been found to exhibit a linear relationship with an average roughness of $R_a = 0.1$ μm to $R_a = 5$ μm.[233] The extent of correlation of back pressure with other roughness parameters has also been investigated by experiment.[234]

A novel approach to air-gauging techniques has been described by Tanner.[235] Rather than measuring pressure differences, a null method based on a pneumatic analogue of a Wheatstone bridge was used to measure pressure in the nozzle. Several stages of development resulted in a compact device able to give an electrical signal/readout proportional to the average surface roughness.[236–238]

Air-gauging methods provide a simple, inexpensive, portable, quick and robust way of assessing surface roughness well suited to use in the quality control of surfaces on the shopfloor and the technique appears to find particular favour in the paper industry. British Standards have been compiled in an attempt to develop a common approach in its use in this application (e.g. reference 239). Table 9.29 summarizes the performance of three comparative techniques used to assess surface roughness.

9.10.2.3 Optical methods

Although the stylus instrument is currently very widely used in the measurement of surface roughness, it suffers several significant disadvantages:

1. It normally provides information only for a profile section of a surface.

2. The finite size of the stylus and the load it applies to the specimen can give rise to measurement errors and specimen damage.
3. Its operation is relatively slow. It cannot be used as an 'in-process' measurement technique (e.g. to monitor the performance of numerically controlled tools).

The advent of sophisticated digital electronic systems has, in recent years, allowed optical data to be processed more easily, and this has encouraged the development of instrument systems which measure surface roughness through optical phenomena. These methods have several attributes which no other single technique can claim:

1. Normally, areal data can be obtained easily.
2. There is no contact between the specimen and the instrument, so no surface damage is caused during measurement.
3. Measurements can be performed quickly.
4. Measurements can be made on any type of material (i.e. they are not restricted to observations of electrically conductive materials as required by electron microscopy).
5. The specimen does not need to be in a vacuum.
6. Complex specimen preparation is not required.

Techniques which measure surface roughness using optical phenomena can be divided into two broad categories: non-parametric techniques, which are able to record the actual topographic structure of a specimen, and parametric techniques, which are only able to assess the general characteristics of an area of surface (e.g. its rms roughness).

Non-parametric instruments A number of non-parametric techniques are available and can be divided into three categories by their principle of operation:

1. Light sectioning
2. Interference microscopy
3. Focus feedback methods.

Light sectioning is a non-destructive analogue of the process of taper sectioning whereby an optical microscope is used to examine a section cut through a specimen surface at a shallow angle to magnify height variations.[240] Light (or optical) sectioning is sometimes referred to as the Schmaltz technique, after its inventor. Figure 9.96 illustrates the principle of light sectioning. The surface is illuminated by a light beam collimated by a narrow slit. This is then viewed from the side, usually by means of a microscope. Topographic features in the surface, illuminated by the beam, appear as profile sections. Illumination and viewing angles of 45° are normally adopted, leading to a magnification of the height of features in the profile by a factor of ($\sqrt{2}$).

The practice of illuminating the specimen from an angle introduces distortion into the observed image. The features visible in it do not represent a true cross section of the surface because the valleys are displaced laterally from the peaks by a small distance. This can be overcome by illuminating the specimen from a normal angle, although this reduces the vertical magnification.

Optical sectioning is suitable for examining surfaces whose roughness range is between 2 μm and 200 μm. The vertical resolution of the technique is about 0.5 μm.[241]

Until recently, because of the difficulty of extracting quantitative information, optical sectioning has only been used to assess the profile range, R_t, and for quantitative examination of surface features. This difficulty has been overcome by digitally recording an image delineated by a narrow laser beam using a television camera. This image is then transferred to a computer for storage, processing and display.[242,243]

Table 9.29 A summary of the performance of parametric techniques of surface roughness measurement

Technique	Usable R_a range	Notes
Friction tests	0.025–3.2 μm Ref. 228	Accuracy depends on surface type
Electrostatic measurements	1.5–10 μm Ref. 230	Data not available on wider range measurements
Pneumatic methods	0.16–5.69 μm Ref. 233	Range apparently extendable if required Ref. 235

Note: R_a range is given as a measure of performance. This is not intended to imply that the raw instrument output correlates best with this parameter.

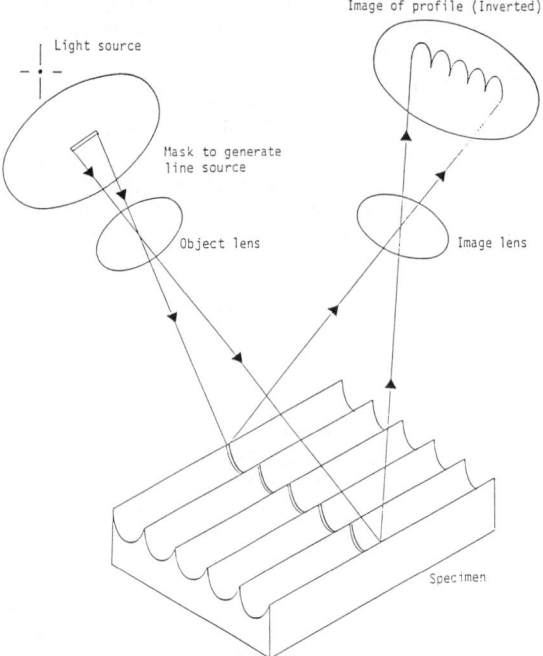

Figure 9.96 Principle of the light-sectioning method. (Based on Figure 128, Dagnall, H. *Exploring Surface Texture* and reproduced by permission of Rank Taylor Hobson)

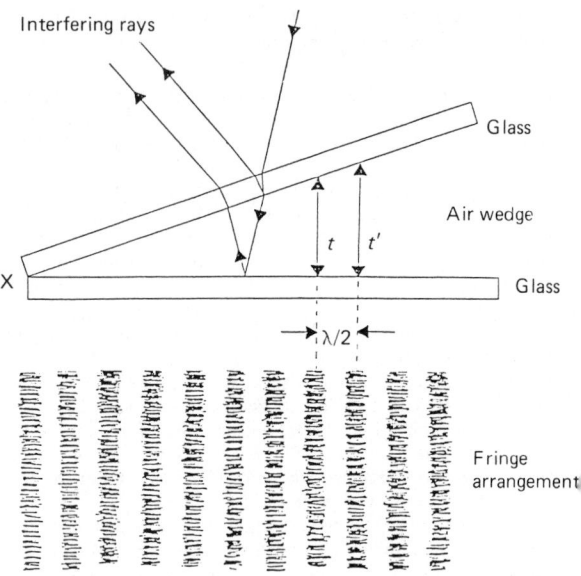

Figure 9.97 Interference fringes caused by an air wedge (wedge angle exaggerated). At X, the path difference, PD, is zero but a dark fringe is observed. This is caused by the 180-degree phase change which occurs when light is reflected at a dense medium. The phase change is equivalent to a PD of $\lambda/2$. At the mth dark fringe where the wedge thickness is t, the effective PD, $m\lambda$, is therefore

$$m\lambda = 2t + \lambda/2 \quad m - \text{integer}$$

At the next fringe where the thickness is t' the PD is

$$(m + 1) = 2t' + \lambda/2$$

Hence the fringe separation is

$$t - t' = \lambda/2$$

In *interference microscopy*, if two slightly inclined glass plates are illuminated by a coherent monochromatic light source, a series of parallel light and dark bands will be visible when the arrangement is viewed from above (see Figure 9.97). The dark bands arise as a result of the destructive interference of light wave fronts. The distance between neighbouring fringes is $\lambda/2$, where λ is the wavelength of the illumination. In principle, interference techniques can be used to examine surface topography by replacing the lower glass plate with a reflective specimen and the upper one with some form of reference plane nominally aligned with the specimen. Interference between light beams from these surfaces will generate a contour pattern of the surface irregularities of the specimen. Interference microscopy is suitable for measuring surface irregularities with low slopes and roughness ranges less than 1 μm. (In areas of high slope, surface contours blend together and cannot be interpreted.) Several types of interference microscopes exist and their design and capabilities are reviewed by references 244–246. The two main categories of interferometer are the double- and the multiple-beam types.

Typically, the double-beam interferometer has a horizontal resolution of 1.0 μm and a vertical resolution of 0.01 μm. The corresponding figures for the multiple-beam device are 2.5 μm and 0.001 μm.[247] (In multiple-beam interferometry the beam is displayed sideways at each of the 50–100 reflections, causing a reduction in horizontal resolution.)

Until recent years interferometric measurements have been restricted in their use to the examination of specific features, such as the height of steps and the depth of grooves, because interferograms could not be readily transformed into an electronic signal which could be submitted for further analysis.

Developments in the field of electronics in the last fifteen years have made the automatic collection of qualitative details in interferograms more straightforward, permitting them to be used to obtain parametric assessments of surface topography.

The use of television cameras, charged coupled devices (CCDs) and phototransistors now allows fringe patterns to be recorded relatively easily for storage and processing by computers. With this modification, 'standard' interferometers have been used to examine the surfaces of semiconductor wafers,[248] magnetic recording media,[249] gauge blocks[250] and optical components.[251] Digitized images are also of value because they can be signal averaged to allow the fidelity of fringe patterns to be enhanced. This technique has been used to obtain accurate surface statistics for use in verifying light-scattering theory.[246]

Interferometric measurements of even higher accuracy can be obtained using phase detecting systems in which two polarized beams of slightly different frequency are focused onto a point on the specimen surface and a stationary reference. The beat frequency of interfering return beams is directly proportional to any change in height of the specimen surface. The 'interrogating' beam can then be swept around the surface to obtain height measurements at specific points. Circular[252] and linear[253,254] scan patterns have been used in surface examination. These systems are able to examine small areas about $0.5 \text{ mm} \times 0.5 \text{ mm}$ and have vertical and horizontal resolutions of 0.5 nm and 1.25 nm, respectively.

An ingenious variation on the use of interferograms involves holography. The approach involves the use of standard holographic techniques to form two holographic images of an object on the same plate. Each image is generated using illumination of slightly differing wavelengths. When the plate is developed and re-illuminated with monochromatic light, interference between wavefronts produced by the two slightly different images generates a fringe contour map. The difference in the height represented by these contours can be varied between 1 μm and several millimetres, allowing surface form as well as roughness to be examined. Holographic techniques have been applied to study wear[255] as well as surface shape.[256,257]

A number of research instruments have been developed and tested which can be classified as *focus feedback methods*. They all operate on similar principles although the detailed design of these instruments may differ significantly.

The operation of all these devices relies on a narrow beam of light 0.5–4 μm in diameter which is accurately focused on the specimen surface. If the spot is caused to traverse the specimen surface, undulations will cause the vertical location of the spot to be modified, thus destroying its exact focus. A number of methods of assessing the extent of this defocus are available. They are used to evaluate an error signal which refocuses the spot by adjusting the position of the specimen,[258] or a lens,[259–261] or to calculate the vertical displacement of the spot directly by monitoring parameters related to its size.[262] Several optical configurations can be used to monitor defect of focus.[263] Sometimes the best aspects of two designs can be combined to give better performance.[264,265]

In general, the performance of focus feedback-based instruments can be as good as that of stylus instruments. Their horizontal resolution falls in the range of 5–0.01 μm and their vertical resolution typically is 0.1 μm, although vertical resolution of 0.001 μm has been claimed.[261] The output of these devices is rather similar to the output of stylus instruments, i.e. a voltage proportional to the height of the surface relative to a fixed point. However, their vertical range is usually larger than that of a stylus instrument of comparable resolution and, in addition, no contact is involved in the measurement of height, and sometimes surfaces which have variable reflectance create problems. Despite this, instruments based on focus feedback appear to have the potential to be a serious commercial rival to the stylus instrument in the future.

Parametric methods A wide range of measurement methods use the reflection of light to quantify surface roughness in parametric terms. Many of these techniques have considerable potential for 'in-process' use as they offer a 'real-time' assessment. In addition, they examine an area of a surface rather than a profile. This is often considered to be a desirable attribute for any method which quantifies surface topography.

Three basic optical phenomena are employed in attempts to measure surface topography:

1. Scattering
2. Speckle
3. Polarization.

Scattering measurements Theories which reliably predict the distribution of light scattered from a rough surface are only available for cases where the surface structure takes specific forms. The domain of validity for those theories available has been evaluated for the case when the target surface is a sine wave.[266]

However, it was found that these theories only provided a good approximation for the distribution of light reflected from 'general' engineering surfaces when the rms roughness of the surface, R_q, was much less than the wavelength of illumination, λ. Nevertheless, a number of studies have addressed measuring surface roughness in this range using scattered light. These measurements are classified according to the component of the reflected beam which is recorded. These are:

1. The specular beam
2. The diffuse beam
3. Total scatter
4. The angular distribution.

When a collimated beam of light is reflected from a plane, conducting surface, all the light remains in a collimated beam known as the *specular beam*. If the reflecting surface is rough, light is scattered into a distribution of angles. In general, as 'roughness' increases, less light is scattered along the specular direction.

Measurement of the intensity of the specular beam produced by a narrow incident beam reflected from a rough surface is the most straightforward type of reflectance assessment. Measurements of this type have been used for many years, strictly for comparative purposes, using an instrument called a glossometer.[267]

Measurements of the intensity of this beam can also be used to predict the R_q roughness of surfaces with a Gaussian height distribution provided $R_q < \lambda/10$.[268] Beyond this limit the specular beam disappears. However, it appears that some form of correlation between specular intensity and roughness can still be identified, even when the surface height distribution is not Gaussian.[269]

As surface roughness increases, the pattern of the scattered radiation tends to become more diffuse. Investigators who use the *diffuse beam* to assess surface roughness attempt to identify a parameter which characterizes the 'diffusiveness' of the scattered radiation and then try to relate this parameter to surface roughness by empirical studies.

Attempts to measure diffuseness have involved comparing the intensity of the specular beam with the intensity of rays at a fixed angle away from it[270] and measuring the 'width' of the specular beam itself.[271] The techniques only appear to show limited correlation with surface roughness parameters but may provide the basis of useful 'in-process' comparators.

Total scatter measurements involve assessing the total intensity of light diffusely scattered from a surface, i.e. it is the total integrated scatter (TIS). The arrangements required to make TIS measurements are more elaborate than those necessary to assess specular reflectance. The main component required to record TIS is a reflective sphere, called a Coblentz sphere, which surrounds the specimen and collects all the light reflected from it. (Except for that projected along the specular direction. Light collected by the sphere is thrown onto a detector.) The proportion of diffuse light reflected by the specimen is deduced by directing the incident beam alternately onto the specimen and directly onto a second detector, and comparing the output of the two detectors.[272]

TIS measurements give the same information as specular reflectance measurements about surface roughness. However, it appears that TIS measurements are particularly valuable in the optics industry as they provide a fundamental method of appraising the performance of optical components which can be automated relatively easily.[272]

Under the usual condition that $R_q \ll \lambda$ it can be shown, using scalar scattering theory, that the angular distribution of light scattered from a rough surface directly is a map of the power spectral density of the surface topography.[273] For light normally incident on a specimen surface it has been reported that the scattering angle, θ_s, is related to the roughness component of frequency, r, by

$$\sin \theta_s = (\lambda/2\pi)^r$$

and the *angular distribution* of the scattered radiation can be related to the power spectrum.[266]

The bandwidth of the roughness spectrum which can be measured using the angular distribution can also be predicted. The long wavelength limit is set by $\theta_s = 90°$ with the short wavelength limit being determined by the minimum scattering angle which can be measured with respect to the specular beam, typically 0.5°. HeNe and CO_2 lasers are commonly used to provide illumination. The long and short wavelength limits spanned by these sources are approximately 0.8–80 μm and 10–1000 μm, respectively.[273]

The great advantage of angular distribution measurements over other forms of scatter measurement is that they supply spatial data as well as height information about surface roughness. Several investigators have constructed apparatus to measure the roughness of optical components and other 'high-finish' surfaces using angular distribution measurements.[274–279]

Speckle pattern measurements When a rough surface is illuminated with partially coherent light, random patterns of bright and dark regions can be observed in the reflected beam. This phenomenon is known as speckle. The spatial distribution and contrast of speckle patterns has been found to depend on surface roughness. As a result, the use of such patterns for the measurement of surface roughness has been investigated. Three main forms of speckle measurement are under development:

1. Monochromatic speckle contrast
2. Polychromatic speckle contrast
3. Speckle pattern correlation

1. *Monochromatic speckle contrast*: In speckle contrast investigations changes in intensity of the speckle pattern are commonly assessed by a parameter called average contrast, V. The contrast in speckle patterns produced by monochromatic light has been found to correlate with the average roughness, R_a, of surfaces up to a limit of about 0.13 μm.[280] Beyond this point a saturation effect is evident. Further increases in roughness produce no change in the speckle contrast, V. The use of radiation of longer wavelength will, theoretically, permit surfaces of roughness up to about $R_a = 1$ μm to be assessed. Further increases in the measurement range of speckle techniques can be achieved by using polychromatic illumination.

2. *Polychromatic speckle contrast*: Speckle patterns can also be produced by reflecting light which covers a band of frequencies, $\Delta\nu$, from rough surface specimens.[281] Speckle patterns produced using polychromatic illumination display contrast variations which increase initially and then decrease near to the point where monochromatic speckle begins to saturate. This measurement technique has been used to obtain an assessment of rms roughness, R_q, on a range of manufactured surfaces.[282] The range of validity for measurement of R_q is

$$\frac{\lambda}{2} < R_q < \frac{2c}{\Delta\nu}$$

where c is the velocity of light. With the laser sources currently available this allows roughness measurements up to $R_q = 5$ μm to be made.[273]

3. *Speckle pattern correlation*: In this technique two speckle patterns are produced from a specimen using illumination either from two different angles or from two sources of different wavelength. Discrimination of these patterns can be achieved either by double exposure of a photographic plate or by the use of frequency-sensitive photodetectors. This permits their degree of correlation to be established by means of the intensity cross-correlation function. Good agreement between roughness assessments by speckle contrast measurements and stylus instrument measurements has been observed over wide ranges ($R_q = 0.13$–6 μm,[283] $R_q = 8$–32 μm[284]).

Polarization measurements The science of measuring the change in the state of polarization of a beam of light when it is reflected from a surface is known as ellipsometry, which is conventionally used to measure the thickness of surface films. However, the polarization of a reflected beam is also influenced by surface roughness, so the possibility that it may be used to measure roughness arises.

The value of ellipsometry in surface studies has not been firmly established. There is some evidence that suggests an almost linear relationship between the R_a roughness of specimen surfaces and parameters which describe the state of polarization of light reflected from them.[285,286] However, other studies have encountered difficulty because it has not been possible to isolate changes in polarization due to oxide film effects from those due to surface roughness.[287,288] This problem appears to be a significant drawback to this method of roughness measurement.

9.10.2.4 Microscopy

The use of conventional optical microscopes is limited by a compromise which exists between the resolution and depth of field obtained in the image of a specimen. An inspection of resolution/depth of field data for optical microscopes reveals that they are not suitable for general use in the study of engineering surface topography because they cannot offer a depth of field which maintains all parts of the specimen in focus simultaneously while providing enough resolution to expose the relevant details of machined surfaces.[289] (This excludes interference microscopes, which operate on a different principle and are discussed in Section 9.10.2.3.)

Electron microscopes offer significantly better resolution and depth of field than optical counterparts.[290] The improvement is chiefly due to the very short wavelength of electron microscope beams in comparison with visible light.[291] As a consequence, electron microscopes tend to be more useful in the study of surface topography than conventional optical light microscopes.

Electron microscopy There are two basic types of electron microscopy:

1. Transmission electron microscopy (TEM)
2. Reflection electron microscopy (REM)

In TEM an image is produced when electrons are incident on a thin specimen (<1 μm thick) which deflects and scatters the electrons as they pass through it. A lens system magnifies and focuses scattered electrons to form an image on a screen which can be viewed directly or photographed.

In order to examine the surface of a metal component using TEM it is usually necessary to fabricate a replica of the specimen surface. This arises because it is difficult to remove a thin layer from a component for inspection without distorting it and/or the investigator may wish to examine the specimen without destroying it. The process of manufacturing surface replicas is not without problems.[289] TEM is typically able to resolve features up to a separation of around 0.3 nm[291] and has been employed to study the changes in surface structure during wear.[292]

In REM, electrons scattered from the surface of the specimen are focused to form an image. (This removes the requirement for specimens to be thin.) Commonly, this image is produced by a beam of finely focused electrons which scans the specimen in a raster pattern, a process called scanning electron microscopy (SEM). Scanning electron microscopes have a typical optical resolution of about 10 nm, a little less than that available in TEM.[291] However, this disadvantage is compensated by the fact that specimen preparation is considerably easier. Conductive specimens can be inspected directly and replicas can have a deep construction, which makes them more robust. SEM is also a popular technique used in the study of wear mechanisms.[293]

Several special problems arise when examining specimens with any form of electron microscopy. First, specimens must be small enough to fit into the specimen cell of the microscope (typically, no more than 20 mm by 20 mm). As a result, the examination of the surfaces of large components can only be achieved by destructive means (i.e. cutting the specimen) or by fabricating replicas which may be unreliable. Specimens must preferably be electrically conductive, but non-conductors can be examined successfully by coating them in a thin layer of metal by a vacuum evaporation process. Special techniques to observe non-conductors for long periods without this preparation are also available.[294] The examination of surfaces using electron microscopy is never a quick operation. The preparation of the specimen takes some time, as does the formation of the vacuum in the microscope – required so that electrons can travel through the microscope without being scattered by air molecules.

Electron microscopy is a popular technique used in the investigation of surface structures and wear particles. However, the interpretation of the images it forms is not necessarily straightforward. Additionally, these images do not readily yield quantitative data about the height of surface features. Recently some attempts have been made to derive height data from this process.[295] This was achieved by digitizing micrograph images using a microdensitometer to produce an array of grey scale points. As it has been shown that the slopes of surface elements are proportional to the micrograph density, integrating the density data allows a surface profile to be generated.[296] Assembling an ensemble of parallel profiles calculated in this way allows three-dimensional views of the surface to be drawn and parameters to characterize the surface shape to be calculated.

Photogrammetric methods have been used to obtain height data from both TEM and SEM images.[297,298] In this technique two micrographs of the surface are formed by viewing the specimen at different inclinations to the main beam. Surface contours are then calculated by viewing the images as a stereoscopic pair and evaluating changes in relative height from parallax differences using a photogrammetric plotter.

An alternative method of deriving topographic information from SEM images using feedback control has also been described.[299] This approach relies on feedback adjustment of the objective lens of the microscope through computations based on the intensity gradient of the image. The signal required to keep the image in an exact in-focus condition reflects height changes at the specimen surface. The method has a vertical resolution of about 1 μm.

Vacuum tunnelling microscopy The development of an instrument called a 'topografiner' was reported by a member of the US Bureau of Standards in 1966.[300] This device uses piezoelectric translators to position the specimen and probe. (The length of piezoelectrical crystals can be controlled very accurately by applying a potential difference across them.) Two piezoelectric translators are used to position a pointed emitter probe in the x-y plane and a third to adjust the position of the specimen in the z (vertical) plane. The emitter probe, located 50 nm above the specimen surface, is connected to a constant-current supply, establishing a potential difference between the emitter probe and the specimen. This potential difference is dependent on the spacing between the specimen and the probe, which is controlled by a servo system which applies a signal to the z-axis piezo crystal. As the emitter probe is translated in a raster pattern by the x and y axis piezo crystals the specimen is raised and lowered to compensate for changes in the specimen-emitter potential due to the specimen roughness. The signal to the z-axis piezo crystal corresponds to the specimen surface shape and is used to fabricate areal representations of its topography.

Following further developments, the topografiner ultimately had a vertical resolution of 3 nm and a horizontal resolution of 0.4 μm.[301,302] This performance is in about the same range as that of an electron microscope and suffers similar limitations of application (e.g. specimens must be small, conductive and contained in a vacuum).

The topografiner was a forerunner of a device called the scanning tunnelling microscope, which was similar in construction, but operated on a different physical principle. Development of this instrument by Binnig and Rohrer has led to substantial improvements in resolution.* It is reported[303] that one of their early effective microscopes had a horizontal resolution of 0.1 nm and a vertical resolution of 0.001 nm. This high resolution made the device suitable for the study of surfaces on the atomic level, including investigations of the shape of catalytic molecules, viruses and semiconductors.[304] Further development of the device was undertaken by Binnig to produce a variant instrument called the 'force microscope', which is able to examine the surface shape of insulating materials. It has a vertical resolution of less than 0.1 nm and a horizontal resolution of 3 nm.[305]

For convenient comparison, Table 9.30 summarizes the typical performance of several types of microscope along with the specification of a commercially available stylus instrument.

9.10.3 Characterization of surface measurements

Parameters which are used to characterize measurements of surface topography are defined almost entirely in terms of the profile signal generated by stylus-type measuring instruments. This arises for historical reasons. For several decades, stylus instruments have provided the only reliable means of obtaining a qualitative record of the shape of a specimen surface. The design and form of this instrument has consequently influenced the development of parameters used to assess surface topography. Until recently, any method adopted as a means of parameterizing the transducer signal needed to be relatively simple and based on mathematical expressions which could easily be evaluated by means of analogue circuits. Several surface parameters evolved within this constraint and

---

*In recognition of their outstanding contribution to the development of the tunnelling microscope Binnig and Rohrer were jointly awarded the 1986 Nobel Physics Prize, sharing the award with Ernst Ruska, acclaimed as the inventor of the electron microscope.

Table 9.30 A summary of the typical specifications of instruments used in the laboratory measurement of surface topography

Device	Resolution				Vertical measurement range or depth of field		Measurable area
	Lateral		Vertical		At lowest resolution	At highest resolution	
	Lowest	Highest	Lowest	Highest			
Stylus instrument (Talysurf VI; Rank Taylor Hobson, Leicester)	Not easily defined Ref. 214		0.5 μm	0.00025 μm	500 μm	0.25 μm	Depends on traverse length. Typically, a few millimetres
Optical light microscope (OLM)	2.5 μm Ref. 290	0.1 μm	Not applicable		42 μm Ref. 290	0.04 μm	Depends on magnification
Transmission electron microscope (TEM)	2.5 nm Ref. 290	0.5 nm	Approximately the same as the lateral resolution Ref. 297		400 nm Ref. 290	80 nm	Depends on magnification
Scanning electron microscope (SEM)	5 μm Ref. 290	10 nm	Not available		1 mm Ref. 290	2 μm	Depends on magnification
Tunnelling microscope	0.4 μm Ref. 298	0.1 μm Ref. 302	3 nm Ref. 298	0.001 nm Ref. 302	Depends on range of piezoelectric elements. Typically, 1 μm or less Ref. 298		Depends on range of piezoelectric elements. Typically less than 10 μm^2 Refs 298, 301

they were mainly related to the statistical properties of the amplitude distribution of an arbitrarily filtered version of the transducer output. Their suitability for assessing surface topography was limited although some remain in common use in the manufacturing environment (e.g. R_a). The advent of digital electronics has removed most computational restrictions so that parameters of virtually unlimited complexity can now be defined and evaluated for surface data. The availability and flexibility of this new computational power beckons investigations which develop more useful and relevant ways of characterizing surface topography.

This section presents several parameters and methods used to characterize surface topography in both industry and research. The inherent advantages and weaknesses of these techniques are discussed.

9.10.3.1 The influence of measurement bandwidth

The length of a profile determines a long wavelength limit for the features which may be observed within it. If the profiles measured from the surface of a specimen with a periodic structure are not sufficiently long to admit all the periodic content, parameters assessed using profiles recorded from different sites on the specimen will be inconsistent. Parameters assessed from profiles of differing lengths will also yield variable results. Even more intractable difficulties arise if the specimen surface contains an appreciable random component in its structure. A non-stationarity has been identified in manufactured surfaces which have a random structure.[306] Consequently, the statistical properties of a profile depend on its length. Its power spectral density function, $G(\omega)$, is described by the relationship

$$G(\omega) = \frac{2\pi K}{\omega^2} \tag{9.11}$$

where $\omega = (2\pi/\text{wavelength})$ and K is a constant known as the 'topothesy' of the surface. As power spectral density is a function which describes the variance of a signal as a function of wavelength and the variance of a sine wave is half the square of its amplitude, it can be seen that equation (9.11) effectively states that the amplitude of the sine wave components in the spectrum of a profile are proportional to wavelength. As a consequence, it was proposed that the total variance of a profile, σ_T, could be written as the sum of the periodic component, σ_P, and the random component, σ_R, thus:

$$\sigma_T = \sigma_P + \sigma_R = \sigma_P + (KL)^{1/2}$$

where L is the length of the profile.

Manufacturers of surface-measurement instruments attempt to overcome the non-stationarity of surface profile measurements by filtering away frequency components of the transducer signal which lie below a given limit called the 'cut-off'. Standardized limits are defined by National Standards documents such as BS 1134.[307] Surface roughness parameters are usually evaluated from a profile called the 'assessment length', which is normally taken to be five times longer than the cut-off limit of the amplifier filter. (The cut-off limit is referred to as the 'sampling length'.) Such filtering is effective because it removes the variance contribution of surface features with wavelengths longer than the 'cut-off' distance. However, resulting roughness measurements are arbitrary, as the selection of the cut-off point does not seem to be based on any physically significant criterion.

The short-wavelength limit of surface structures presents similar difficulties. Irregularities smaller than atomic dimensions are believed to exist, indeed it would be surprising if this were not so. Effectively, this means that a clear, natural cut-off point is absent. The problems that this situation causes are eloquently outlined by Thomas and Sayles.[308] They define the moments, m_n, of power spectra, $G(\omega)$, of profiles to be of the form

$$m_n = \int_0^\infty \omega^n G(\omega) \, d\omega \qquad (9.12)$$

The density and curvature of asperities and the mean slope of a surface are functions of the first three even spectral moments: m_0, m_2 and m_4.[309] As equation (9.12) is not integrable, due to the infinite upper limit, the existence of a fixed value for these important and widely used parameters is excluded.[308,309]

The solution to this problem is again a pragmatic one. National Standards documents define standard styli in an attempt to make the high-frequency cut-off points of all measurement devices similar. Although the choice of a standard dimension for the stylus tip is an arbitrary one, depending largely on a compromise between useful lifetime and ease of manufacture, at least this arrangement means that measurements made by different instruments are likely to agree to within an acceptable tolerance of each other. (Note that the 'frequency response' of a stylus depends strongly on its shape and size. This dependence is a complex one which is discussed in detail elsewhere.[214])

Thus it can be seen that in the absence of natural physical constraints on the bandwidth of measurements from surfaces, an artificial limit is imposed by National Standards which define standard high- and low-frequency 'cut-off' points. The choice of these limits is arbitrary, defining a measurement bandwidth which is more suitable for quality control rather than fundamental study of phenomena associated with surface topography.

9.10.3.2 The effect of sampling conditions

Surface parameters are frequently calculated from a set of discrete samples $f(y_k)$ drawn from a continuous profile $f(y)$. These discrete samples are separated by a short horizontal interval Δy. It has been shown that the choice of sampling interval, Δy, influences the assessment of some parameters, such as the mean slope and mean curvature of a profile.[310] Subsequently, such parameters cannot be considered to be intrinsic properties of a surface profile. For surfaces with a Gaussian height distribution it has been shown that mean slope is proportional to $(\Delta y^{-1/2})$.[310]

The effect of sample interval size on the evaluation of other parameters has also been examined.[311] It has been found that sample interval size 'has no appreciable effect' on pure height parameters such as average roughness. However, shortening the sample interval is found to increase the value of extreme range parameters and to reduce the 'width' of those parameters which had some dependence on the spatial characteristics of the profile, such as average wavelength. Further discussion of the influence of sample interval size on profile parameters can be found in reference 312.

9.10.3.3 Reference lines

Many statistical parameters assess surface roughness by measuring the departure of the profile from a reference (mean) line. Two systems which define mean lines exist: the M-system and the E-system. Of these, the M-system is by far the most widely used.

The M-system This system is specified by several National Standards.[313,314] In this system the mean line is defined by the electrical filter through which the profile is passed to remove its long-wavelength components.

The E-system The E (or envelope) system mean line is the locus described by the centre of a circle with a radius of a few millimetres as it rolls across a profile and is then displaced downwards by a distance such that the areas enclosed by the effective profile above and below it are equal.[315]

Neither of these systems has any particular physical importance but they do provide an effective practical reference for statistical parameterization of profiles. A number of these parameters are discussed below.

9.10.3.4 Amplitude parameters

A wide range of parameters are used to assess the height deviations of the profile from the mean line.

Average roughness (R_a) Average roughness, R_a, is also known as 'centre line average', or 'arithmetic average' being denoted by 'cla' or 'AA'. However, average roughness, R_a, is the preferred designation. It is defined by

$$R_a = \frac{1}{L} \int_0^L |f(y)| \, dy$$

where $f(y)$ is the departure of the profile from the mean line and L is the length of the profile. In digital implementations R_a is given by

$$R_a = \frac{1}{M} \sum_{k=1}^M |f(y_k)|$$

where M discrete samples $f(y_k)$ are drawn from $f(y)$.

One of the main criticisms of R_a is that it cannot distinguish between profiles of different shape. Profiles having the same R_a value may be entirely different in character. Despite this drawback, R_a is in almost universal use for quality control, being defined by standards authorities in many countries, including Britain, Germany, France and the USA.[316] To obtain a reasonably reliable assessment of R_a an average over several sampling lengths, L, is normally evaluated.

Root mean square roughness (R_q) The root mean square roughness is identical to the standard deviation, σ, of the profile height distribution. It is defined by

$$R_q = \left(\frac{1}{L} \int_0^L f^2(y) \, dy \right)^{1/2}$$

and its digital implementation is

$$R_q = \left(\frac{1}{M} \sum_{k=1}^M f^2(y_k) \right)^{1/2}$$

For surfaces with a Gaussian height distribution the R_q parameter can be used to calculate R_a via the relationship

$$R_a = \left(\frac{2}{\pi} \right)^{1/2} R_q \approx 0.8 R_q$$

This relationship is a good approximation for most surface types provided their height distribution is almost symmetric, even if it is not Gaussian.[317]

One advantage which the R_q parameter offers over R_a is that it appears to be more sensitive to changes in shape of the profile when material is conserved than R_a.[318] This arises as a result of the form of the definition of R_a. R_a is assessed by reference to a mean line positioned so that the areas enclosed by the profile above it and below it are equal. Redistribution of material from one side of the mean line to the other will cause a shift in the position of the mean line and may leave the R_a value unchanged. The R_q value, however, is more likely to be modified by the operation. This lower sensitivity to changes in profile shape when material is conserved means that R_a is less suitable than R_q for monitoring certain surface processes e.g. running-in under pure rolling.[319] (It should be noted, however, that R_q itself is not certain to identify changes in material redistribution. Neither R_a or R_q have values which are unique to specific profile shapes.)

Although the R_q parameter suffers fewer drawbacks than R_a, it is less widely used than R_a. This is particularly surprising when many theoretical treatments of surface topography yield results which are expressed in terms of R_q rather than R_a.

Depth of smoothness (R_p)   Depth of smoothness is a roughness parameter which can be measured by friction tests.[320] It has also been found to correlate well with capacitance measurements of surface roughness.[321] Depth of smoothness is defined by

$$R_p = \frac{1}{L} \int_0^L (f(y)_{max} - f(y))\, dy$$

where $f(y)_{max}$ is the highest peak in a profile. This parameter is not commonly evaluated by stylus methods.

Extreme-value parameters   It is sometimes useful to have a measure of the extremes of departure of a profile and many different forms of these parameters have been defined. The most commonly used of these are: maximum peak-to-valley height, R_t, a parameter known as ten-point height, R_z, and the maximum departures of the profile above and below the mean line referred to as R_p and R_v, respectively. Sometimes these parameters are assessed for individual sampling lengths. The maximum peak-to-valley height within a sampling length is then denoted by R_{ti}, and the maximum R_{ti} in a whole profile is designated R_y (formerly R_{max}). Note that R_y is distinct from the maximum peak-to-valley height of the profile, which is given by $(R_y + R_p)$.

Ten-point height, R_z, is the mean separation of the five highest peaks and the five lowest valleys in the profile and is defined by

$$R_z = \frac{1}{5} \left(\sum_{i=1}^{5} P_i - \sum_{i=1}^{5} V_i \right)$$

where P_i and V_i denote maxima and minima of $f(y)$.

Extreme-value parameters are sensitive indicators of high peaks or deep scratches in a surface. As a consequence, their principal use is in quality-control applications. They are also of use in lubrication studies.

Moments of amplitude distribution   None of the parameters described above provide information regarding the shape of the profile. However, such details can be derived from moments of the amplitude distribution $P(f(y))$ of the points in the profile.

The third moment of the amplitude distribution is known as skewness and denoted by R_{sk}, where

$$R_{sk} = \frac{1}{R_q^3} \int_{-\infty}^{\infty} f^3(y)\, P(f(y))df(y)$$

or, for discrete data,

$$R_{sk} = \frac{1}{MR_q^3} \sum_{k=1}^{M} f^3(y_k)$$

The fourth moment is known as kurtosis, R_k, and given by

$$R_k = \frac{1}{R_q^4} \int_{-\infty}^{\infty} f^4(y)P(f(y))\, df(y)$$

and in digital implementation by

$$R_k = \frac{1}{MR_q^4} \sum_{k=1}^{M} f^4(y_k)$$

Symmetric profiles with equal numbers of peaks and valleys of the same height/density have zero skewness. Profiles in which peaks dominate have positive skewness and profiles dominated by valleys have negative skewness.

Kurtosis can be used as an assessment of the 'spikiness' of a profile. If $R_k < 3$ the profile has relatively few tall, narrow spikes. As R_k increases above 3 the shape of its peaks and valleys become less rounded. When $R_k = 3$ the profile heights have Gaussian distribution.[321a]

Skewness and kurtosis have been shown to be of value in studies of running-in wear[322] and analyses of surfaces which attempt to relate profile character and the functional requirements of surfaces.[323]

Bearing ratio (t_p)   In applications where two surfaces are in loaded contact, a parameter known as the bearing ratio has been found useful for defining the characteristics of surfaces

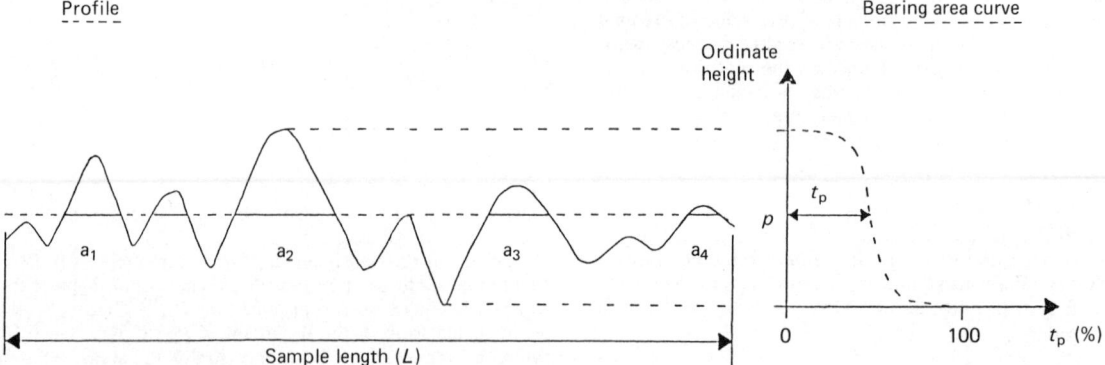

Figure 9.98   Definition of the bearing ratio parameters

which will perform a given function reliably. It is frequently used to assess surfaces in finishing processes such as honing and lapping. Figure 9.98 illustrates its method of evaluation. The bearing ratio is given by

$$t_p = \frac{\Sigma a_i}{L} \times 100 \ (\%)$$

In practical situations a surface profile is recorded prior to finishing. If experience shows that a bearing ratio of 69% is required for effective and reliable operation, the height, p, to which the surface crests must be reduced, can be determined.

An alternative method of measuring the bearing ratio has been suggested.[324] The cumulative height distribution of the profile, also known as the Abbott bearing area curve, is illustrated in Figure 9.98. The curve describes the percentage of solid material lying above a certain height. Abbott and Firestone[324] have suggested that 'bearing area fraction' was identical to the bearing (length) ratio, and this has since been verified for random surfaces.[325]

9.10.3.5 Spatial parameters

Several parameters have been devised to describe spatial characteristics of surfaces. They have not been included as standard instrument readings until recent years and this may account for their use being less widespread than parameters such as R_a and R_q.

Average wavelength (λ_a) The average wavelength, λ_a, of the spatial features in a surface profile, $f(y)$, is given by

$$\lambda_a = 2\pi \frac{R_a}{\Delta_a}$$

where Δ_a is the average slope of the profile as defined in reference 326 (see Section 9.10.3.6). This parameter is apparently useful for assessing the surface quality of sheet steel. It is able to measure the 'openness' of a surface and is found to correlate well with its visual appearance. It is sometimes used to measure the quality of sheet steel for car bodies.[326a]

Another valuable use of the average wavelength parameter has been found to be in machine condition monitoring.[327] Average wavelength is a sensitive detector of the feed marks of a machine tool. If machine settings are incorrect, λ_a will change significantly even though changes in R_a or other height parameters may be relatively small.

Root mean square wavelength (λ_q) Root mean square wavelength is defined by

$$\lambda_q = 2\pi \frac{R_q}{\Delta_q}$$

where Δ_q is the root mean square slope of a profile and has similar attributes to λ_a discussed above.

High spot count (HSC) Peaks in a surface profile are frequently important from a functional viewpoint. It has been reported that the spacing of roughness peaks exerts an important influence on the manufacture of sheet steel.[327] Careful control of the roughness of the steel is necessary when pressing the sheet to obtain consistent lubrication and thus avoid scoring.

Normally, high spot count is assessed by determining the number of peaks within a sampling length. However, several different definitions of a 'peak' are encountered.[328] Most commonly, high spot count is determined by counting the number of excursions per unit length above a profile mean line.[316]

A further spatial parameter which is sometimes used instead of HSC is the mean high spot spacing, S_m. This parameter is the reciprocal of HSC.

9.10.3.6 Hybrid parameters

A small number of parameters are in use which do not measure solely amplitude or spatial characteristics of profiles but serve to assess other familiar geometric characteristics.

Mean slope (Δ_a) The average slope, Δ_a, of a profile $f(y)$ is given by

$$\Delta_a = \frac{1}{L} \int_0^L \left| \frac{df(y)}{dy} \right| dy$$

In digital implementations Δ_a is calculated using

$$\Delta_a = \frac{1}{M-1} \sum_{k=1}^{M-1} \left| \frac{f(y_{k+1}) - f(y_k)}{y_{k+1} - y_k} \right|$$

The average slope of a profile can be used to estimate the actual length of a profile L', so that it may be compared with its nominal length, L, using the relationship

$$\frac{L'}{L} \approx 1 + \frac{\Delta_a^2}{2}$$

as illustrated by Figure 9.99.

It has been indicated that the ratio (L/L') is useful in painting applications, where it can be used as a guide to the surface available for 'keying'.[328a] In addition, average slope itself is a valuable parameter in several applications such as the

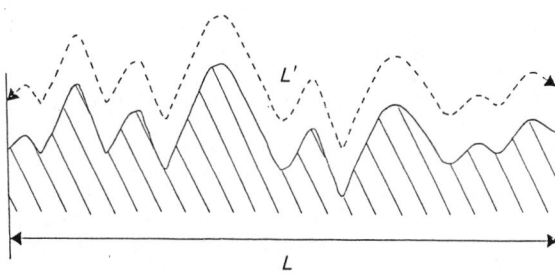

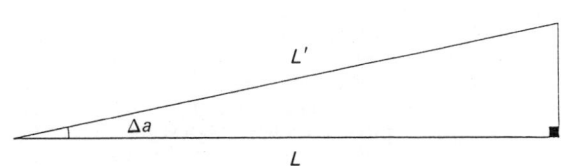

Figure 9.99 Estimation of the true length (L') of a profile:

$$\frac{L}{L'} = \cos \Delta a$$

$$\frac{L'}{L} = \frac{1}{\cos \Delta a} = \frac{1}{1 - \frac{1}{2}(\Delta a)^2}$$

$$[(1 - (\tfrac{1}{2}\Delta a^2)]^{-1} = 1 + (-1)(-\tfrac{1}{2}\Delta a^2) + \ldots$$

$$= 1 + \tfrac{1}{2}\Delta a^2 + \ldots$$

Hence $\dfrac{L'}{L} \approx 1 + \tfrac{1}{2}\Delta a^2$ provided $-1 < (-\tfrac{1}{2}\Delta a^2) < 1$

estimation of sliding friction[329] and the study of the reflectance of light from surfaces.[330]

Mean peak radius (ρ_m) The numerical value of ρ_m depends on the definition of a peak. Thomas[316] defines a peak for a digitized profile $f(y_k)$ as an ordinate which is greater than the two adjacent to it, so that its curvature, C_p, is given by

$$C_k = \frac{(d^2 f(y_k))/dy^2}{\{1 + (df(y_k)/dy)^2\}^{3/2}} = \frac{d^2 f(y_k)}{dy^2}$$

As $df(y_k) = 0$ at a turning point,

$$C_k = \frac{2f(y_k) - f(y_{k-1}) - f(y_{k+1})}{(y_{k+1} - y_k)^2}$$

Peak radius, ρ, is the inverse of curvature so that mean peak radius, ρ_m, is given by

$$\rho_m = \frac{1}{M-2} \sum_{k=1}^{M-2} (1/C_k)$$

The parameter ρ_m has been found useful in theoretical and experimental studies of the friction and adhesion of elastomers.[331]

9.10.3.7 Relationships between topographic parameters

When the distribution of ordinate heights in a profile is Gaussian a relationship between R_a and R_q can be obtained and an estimate of one parameter calculated from a knowledge of the other. Such relationships can be derived for many theoretical height distributions. For example, the ratio between R_a and R_q for a sinusoidal profile which has a 'U-shaped' height distribution is $R_a/R_q = 2^{2/3}/\pi = 0.90$.[332]

The relationships between a wide range of profile parameters have been investigated. It has been shown that a knowledge of R_a, R_q, R_{sk} and R_k defines the shape of a certain class of height distribution curves of the ordinates (i.e. Pearson distributions and Johnson distributions) sufficiently well to allow a number of well-known parameters to be estimated reliably from them.[332] For example, a knowledge of the skew and kurtosis of a profile permits the ratio R_a/R_q to be determined. The skew and kurtosis parameters have also been used to estimate parameters of the bearing area curve.[332]

Extreme parameters such as R_y, R_t, R_p, etc. cannot be estimated from a knowledge of the height distribution alone as they depend on spatial considerations for a given length of profile. However, a method of estimating a variable called the 'equivalent uncorrelated sample size', M_{eq}, using time-series models of surface profiles has been suggested.[332] This parameter can then be used with reduced (i.e. uncorrelated) profile data in equations which estimate the expected values of R_t and R_y for examples of uncorrelated points of size M (where $M = M_{eq}$). Thus the parameters R_t and R_y are obtained from

$$E(\text{Range}) = \int_{-\infty}^{\infty} 1 - (1 - P(f(y)))^M - (P(f(y)))^M \, dy$$

where $P(f(y))$ is the probability density function of the ordinate heights. For $M \geq 1000$ the relationship

$$E(\text{Range}) = 2\left(\frac{0.577}{(2 \log_e M)^{1/2}} + (2 \log_e M)^{1/2}\right)$$

is proposed to be sufficiently accurate.

It is possible to use a statistical approach of the above type to estimate unknown parameters of a profile given a suitable set of variables which describe the height distribution of its ordinates. However, estimating parameters which describe the surface as a whole from details of profiles is more difficult. This arises for two principal reasons: first, the statistical parameters that characterize a profile from a surface do not characterize the surface as a whole; second, large variations can arise in the same parameter when it is measured on different parts of a specimen.

Attempts have been made to study the link between surface and profile parameters and some limited success has been achieved using an idealized surface model and carefully manufactured surface specimens.[333]

An experimental study of the parametric variation in profiles recorded from different parts of the same surface has also been undertaken.[311] Calibration specimens as well as surfaces produced by standard machining processes were examined. On calibration specimens it was found that the least variable parameters were R_a, R_q and Δ_a. However, even on standards for R_a, variations of R_a of nearly 9% were observed. On machined surfaces variations of 50% in some parameters such as R_z and HSC were not uncommon. (A cursory statistical discussion indicated that such variations were within the bounds expected under statistical fluctuation.)

9.10.3.8 Autocorrelation analysis

Autocorrelation analysis reveals details of the spatial structure of a profile. Bendat and Piersol[334] define the autocorrelation function (ACF), $R(\tau)$, as

$$R(\tau) = \lim_{L \to \infty} \frac{1}{L} \int_0^L f(y)f(y + \tau) \, dy \qquad (9.13)$$

The equivalent digital form of this equation is

$$R(\tau) = \frac{1}{M-i} \sum_{k=1}^{M-i} f(y_k) \, f(y_{k+i})$$

$$i = 0, 1, 2 \ldots M - 1$$

From equation (9.13) it can be seen that $R(\tau)$ is the average product over the profile length, L, of the profile $f(y)$, and a version of this profile displaced in space by a distance τ, i.e. $f(y + \tau)$. The ACF is often normalized by dividing $R(\tau)$ by the variance of the profile R_q^2. To distinguish between the normalized and non-normalized ACF, some authors refer to equation (9.13) as the autocovariance function (ACVF), with its normalized form being the ACF.[335]

Certain characteristics of profiles are clearly reflected in the form of their autocorrelation functions. Profiles with a periodic structure possess periodic ACFs while those with a random structure possess ACFs which slowly decay from unit to zero. Penklenik[336] recognized this and suggested a scheme for grouping surface types according to the form of the ACF. The approach provided a qualitative method for profile description but not a qualitative method of characterization.

Whitehouse and Archard[310] recognized that many types of surfaces were random and had Gaussian height probability density functions. Consequently, their ACFs took on an exponential form, which could be approximated by

$$R(\tau) = \exp(-\tau/\beta^*)$$

where β^* was called the correlation distance. At $\tau = 2.3 \, \beta^*$, $R(\tau)$ has decayed to 10% of its initial ($\tau = 0$) value. β^* was regarded as the spacing at which two points on a profile could be regarded as independent events. By applying a statistical analysis known as a Markov process to profiles it was found that many statistical properties of the surface could be derived as functions of only R_q and β^*. This work was later applied to obtain tribological parameters such as the elastic contact area and the plastic force per unit nominal area in terms of R_q and β^* for surfaces with parabolic asperities in contact with a non-deformable plane.[337]

One serious problem associated with the Whitehouse and Archard model is that the parameter is strongly influenced by the frequency bandwidth of the profile measurement from which the ACF is calculated.[308] In order to obtain consistent data it is necessary to standardize the measurement bandwidth of the profile, particularly in the light of the non-stationarity of random surfaces discussed in Section 9.10.3.1. Nayak[309] discusses some consequences of this position in formal mathematical terms.

Although the ACF is valuable for obtaining statistical and tribological details of surfaces in some circumstances, it is less suitable for studying changes in topography during wear, deformation or similar processes. King and Stout[338] calculated the ACFs of profiles recorded from a phosphor-bronze pin at various stages of wear, and although the shape of the profile was modified considerably at each stage the ACF exhibited only small changes. (Smith and Walmsley[339] discuss why this arises.) In order to enhance the value of ACF data in such studies the use of an alternative relation the 'structure function', $S(\tau)$, given by

$$S(\tau) = 2R_q(1 - R(\tau))$$

has been proposed.[340]

9.10.3.9 Spectral analysis

As an alternative to autocorrelation, profile spatial information can also be presented as a spectrum; i.e. a function which decomposes a profile into a number of frequency components of varying amplitudes. The power spectral density function (PSDF), $G(p)$, a tool borrowed from communication theory, is often used for this purpose. The PSDF is given by

$$G(p) = \frac{1}{L} \left| \int_0^1 F(y) \exp(-2\pi py) \, dy \right|^2$$

where p is the frequency of a given component. (Special techniques exist for evaluating this function digitally.[341]) The PSDF is related to the ACF by

$$G(p) = 4 \int_0^{L-\tau} R(\tau) \cos 2\pi p\tau \, d\tau$$

where $R(\tau)$ is defined by equation (9.13).

The nth moment, M_n, of the power spectral density function is defined by

$$M_n \int_0^\infty G(p) \, \omega^n \, d\omega$$

where $\omega = 2\pi p$.

In the 1950s the moments of power spectra were used to develop theoretical models to predict the statistics of sea surfaces with random, Gaussian height distributions.[342] More recently, this work has been successfully adapted to describe engineering surfaces with both isotropic (i.e. non-directional[309]) and anisotropic (i.e. directional[343]) characteristics. Attempts have also been made to estimate plasticity indices from these models to predict the predominant mode of contact between a surface of this type and an inelastic flat plane.[344]

Further forms of modelling in spectral analysis involve the use of the Fourier coefficients rather than those of the power spectrum. Fourier series retain details of the phase relationships between frequency components of the original signal and

can, therefore, be used to reconstruct the original signal. These series have been used to describe the frequency content of surface profiles in this way. It has been shown that many terms in the Fourier series of profiles from some surfaces are negligibly small, permitting the original profile to be reconstructed by using only a fraction of the Fourier coefficients.[345] The error in the reconstructed profile depends on the number of Fourier coefficients used to synthesize it. It has been demonstrated that 99.5% of the power in profiles recorded from shaped, turned and ground specimens is contained by 12, 20, and 50 of the most significant terms of the Fourier series, respectively. This effect has also been observed in the power spectra of areal data.[346]

In addition to surface modelling, the spectral analysis of surface profiles recorded from manufactured parts has been proposed as a method of machine-tool surveillance.[347] Normally such surveillance is conducted by the analysis of machine noise and vibration which does not readily yield information regarding the total performance of the machine. However, such problems as bearing faults and spindle assembly errors are likely to be manifest in the workpiece as characteristic profile shapes, permitting the fault to be identifiable by spectral analysis of the surface profile. Although work in this area as yet does not extend to such classifications, it has been shown that specific indicators of the condition of certain parts of numerically controlled machines can be identified.

Spectral analysis can be performed by decomposing a signal using elementary functions other than the traditional trigonometric ones of sine and cosine. It has been suggested that the two-state Walsh functions may be appropriate for analysing surface profiles as the discontinuous state of a digitized surface may be more satisfactorily represented by discontinuous basis functions than continuous ones.[339] Additional conveniences of computational efficiency are also available. Spectral analyses of surface profiles using Walsh functions have subsequently been compared with conventional spectral analysis in the investigation of wear processes.[348]

9.10.3.10 Time-series modelling

Analytical models of surface profiles are normally highly idealized, and this tends to limit their range of practical applicability. The use of computers extends the possibility of using simulation techniques which describe actual profile shapes in a more representative way. Time-series methods provide an appropriate tool for this approach and have been proposed by some investigators as a means of modelling surface profiles.

Three basic types of time-series models – autoregressive (AR), moving average (MA) and mixed autoregressive/moving average (ARMA) – are available. An AR model of order p of a digitized profile $f(y_k)$ is given by

$$f(y_k) = \phi_1 f(y_{k-1}) + \phi_2 f(y_{k-2}) \ldots + \phi_p f(y_{k-p}) + a_{y_k}$$

where a_{y_k} is residual white noise of zero mean and, usually, a Gaussian distribution. An MA model of order q is given by

$$f(y_k) = \theta_0 a_{y_k} + \theta_1 a_{y_{k-1}} \ldots + \theta_q a_{y_{k-q}} (\theta_0 = 1)$$

An ARMA model is given by

$$f(y_k) = \phi_1 f(y_{k-1}) + \phi_2 f(y_{k-2}) \ldots + \phi_p f(y_{k-p})$$
$$+ \theta_0 a_{y_k} + \theta_1 a_{y_{k-1}} + \ldots + \theta_q a_{y_{k-q}}$$

The parameters ϕ_0 and θ_q can be determined from the ACF of the process being modelled using methods described by Box and Jenkins.[349]

AR models have been employed by several investigators to model surface profiles.[350-352] The theory of AR models is formally applicable only to random data and it has been confirmed that only random profiles and not periodic ones can be analysed successfully by this approach.[351] AR models have been used successfully for profiles with random, Gaussian and non-Gaussian height distributions.[352] However, the presence of long-wavelength components, such as those found in real random surfaces, was not taken into account by low- (second-) order models.

ARMA models have also been used for modelling profiles with both Gaussian and non-Gaussian height distributions.[353,354] One of these models includes an account of a method for modelling profiles which contain periodic as well as random components in their structure.[354] Modelling is achieved by separating the random and periodic parts of the signal using a time-domain technique described by Hannan.[355] (Another method for separating the random and periodic parts of the profile by manipulating its PSD and ACF has also been suggested.[350])

It is clearly possible to employ time-series models as methods for accurately characterizing the shape of measured profiles. Such models would be of much greater value if they could then be used to predict tribological parameters or to describe the details of wear processes. However, no work appears to have been conducted in this area up to the present.

9.10.4 Summary, conclusions and future developments

Surface topography has importance in engineering from both a functional and a production standpoint. There are many applications in which an appropriate choice of finish for a component will optimize some aspect of its performance (e.g. efficiency or lifetime). It is also desirable to minimize the production cost of a component. This can be achieved at the design stage by recognizing the correlation between the functional suitability of the finish of a component.

It is particularly important to avoid overstringent specification of topography. A greater freedom of choice in the character of a component surface permits more flexibility in the selection of the machining method used to produce it and, in addition, a larger scale of permitted roughness normally reduces the machining time required to produce it. Consideration of both these factors can lead to more economic production.

Appreciation of the importance of surface topography in many areas of technological endeavour has led to the development of a wide range of instruments which are able to measure and record surface roughness. Of these, the stylus instrument is the most widely used. Although several forms of inaccuracy are inherent in the data produced by these instruments, the sources of these errors are well known as they are widely discussed in research publications.

Other methods of assessing surface topography are also in use. Electron microscopy is commonly employed to study qualitative changes in topography during wear experiments. Interference microscopes are proving to be a valuable tool for inspecting the high-quality surfaces of the substrates of electronic components and magnetic recording media. The potential of both these forms of microscopy has been extended considerably in the last decade as the capability to interface them with computers has allowed the generation of quantitative output in a much more rapid, varied and convenient form.

The influence of digital electronics is also evident as a factor which has encouraged the development of other optical methods of measuring surface topography. Focus feedback techniques can provide measurement accuracy as good as that of stylus instruments. As they offer several operational advantages over the latter device (e.g. they are non-contact

devices). It is likely that commercial development will be considered and it is possible that these instruments will become a serious rival to stylus instruments in the market at some future date.

Many instruments are designed to directly record the topography of a surface in terms of the variation in its height as a function of position. However, it is possible to design instruments which use changes in a physical phenomenon which depends on surface structure to provide a parametric assessment of its height/spatial variations over a defined area. These include measurements based on gas flow, electrical capacitance and friction. These measurement methods provide a useful means of checking the quality of surface finish on the production line. However, they require contact with the specimen and cannot really be used to provide an 'in-process' assessment of large areas of a surface being produced. Again recent developments in the field of electronics mean that optical methods which provide a parametric assessment of surface roughness can be implemented with more convenience and cost effectiveness.

One limitation of most optical techniques which supply a parametric assessment of surface roughness is they can only be applied reliably in the measurement of 'high-quality' surfaces such as those found on some optical components. However, it appears that one of these methods (i.e. assessment of speckle pattern contrast) has great potential as a technique which can be applied to inspect components with a wide range of roughness. It is one alternative which may allow 100% real-time inspection for quality control of the surface of components produced by numerically controlled systems. This is an industrial requirement which is likely to become more pressing in the next few years.

The assessment of surface topography is a two-part process involving both measurement and characterization of the data it produces. Methods which characterize surface roughness ideally need to parameterize both height and spatial characteristics.

The most common method of characterizing surface topography is by statistical assessment of profiles of the type recorded by stylus instruments. A wide range of parameters which assess the amplitude and spatial characteristics of profiles, either independently or simultaneously, have been discussed. The main drawback in the interpretation of these parameters is that their numerical value depends on the bandwidth of the profile measurement and, for digital data, on the sampling interval employed during acquisition of the profile. This effect arises because specimen surfaces have no natural bandwidth limit, so an arbitrary limit is unavoidably imposed by the measurement operation.

The value of profile parameters is also restricted because the statistical properties of profiles do not necessarily reflect the overall statistical properties of the surface from which they were recorded. Under some idealized circumstances it is possible to derive relationships between surface parameters and profile parameters. However, in many practical circumstances material quality, tool wear/damage, machine characteristics and the presence of cutting debris can cause inhomogeneity in the surface structure of the specimen and render such estimates invalid.

In spite of these drawbacks, profile parameters remain widely used and commonly form the basis of theoretical models of surface topography. Indeed, it seems likely that the future will see an extension of such models to include both profile and areal parameters along with material data in attempts to predict phenomenological parameters of many types. Clearly, the development of these models, even if they are of limited applicability, is of great potential benefit in many areas of technology.

Acknowledgements

This section has been compiled using material from the following papers, reproduced by permission of the publishers or their patron.

Section 9.9.1

Sherrington, I. and Smith, E. H., 'The significance of surface topography in engineering', *Prec. Eng.*, **8(2)**, April, 79–89 (1986). Published by Butterworth Scientific

Section 9.9.2

Sherrington, I. and Smith, E. H., 'Modern techniques in surface metrology. Part 1 – Stylus instruments, electron microscopy, and non-optical comparators', *Wear*, **125(3)**, 271–288 (1988). Published by Elsevier Sequoia
Sherrington, I. and Smith, E. H., 'Modern techniques in surface metrology. Part 2 – Optical instruments', *Wear*, **125(3)**, 289–308 (1988). Published by Elsevier Sequoia

Section 9.9.3

Sherrington, I. and Smith, E. H., 'Parameters for characterising the surface topography of engineering components', *Proc. I. Mech. E.*, **201(C4)**, 297–306 (1987), Published by MEP
This contribution is reproduced from the *Proceedings of the Institution of Mechanical Engineers*, Part C, by permission of the Council of the Institution.
© Institution of Mechanical Engineers 1987.

References

1 Arnell, R. D., Davies, P. B., Halling, J. and Whomes, T. L., *Tribology: principles and design applications*, Macmillan, London (1991)

2 Rabinowicz, E., *Friction and Wear of Materials*, Wiley, Chichester (1965)

3 Suh, N. P. in *Fundamentals of Tribology*, MIT Press, Cambridge, MA (1978), pp. 443–453

4 Winer, W. (ed.), *Wear Control Handbook*, ASME

5 Taylor, C. M. *Turbulent Flow Bearings: Design and Energy Losses*, ASME Centennial Conference, San Francisco, (1980) pp. 161–185

6 Neale, M. J., *Tribology Handbook*, Newnes-Butterworths, London (1976)

7 Engineering Science Data Units on Tribology, London (various dates)

8 Towers, B., 'First report on friction experiments', *Proc. Inst. Mech. Engrs*, 632–666, November (1883): 'Second report', 58–70 (1885)

9 Reynolds, O., 'On the theory of lubrication and its application to Mr Beauchamp Towers' experiments', *Phil. Trans. Roy. Soc.*, **177**, 157–234 (1886)

10 Hirn, G. A., 'Sur les principaux phenomènes que presentent les frottements mediats', *Bull. Soc. ind. Mulhouse*, **26**, 188–277 (1854)

11 Sommerfeld, A. J. W., 'Zur Hydrodynamischen Theorie der Schmiermittelreibung', *Zeits. f. Math. u. Phys.*, **40**, 97–155 (1904)

12 Michell, A. G. M., 'The lubrication of plane surfaces', *Zeits f. Mathe u. Phys.* **52**, Pt 2, 123–137 (1905)

13 Ocvirk, F. W., 'Short bearing approximation for full journal bearings', NACA Technical Note 2808 (1952): also with G. B. DuBois, 'Analytical derivation and short bearing approximation for full journal bearings', NACA Report 1157 (1953)

14 Pinkus, O. and Sternlicht, B., *Theory of Hydrodynamic Lubrication*, McGraw-Hill, New York (1961)

15 ESDU Data Item 84031, 'Calculation methods for steadily loaded axial groove hydrodynamic journal bearings', Engineering Sciences Data Unit, London (1984)

16 Archibald, F. R., 'Load capacity and time relations for squeeze films; American Soc. Mech. Engrs, Paper No. 54-A-50, November (1954)

17 Cameron, A. *The Principles of Lubrication*, Longmans Green, London (1966)

18 Ettles, C., 'Hot oil carry-over in thrust bearings', *Proc. Inst. Mech. Engrs*, **184**, Pt 3L, 75–81 (1970)

19 Neal, P. B., 'Heat transfer in pad thrust bearings', *Proc. Inst. Mech. Engrs*, **196**, No. 20, 217–228 (1982)

20 ESDU Data Item 82029, 'Calculation methods for steadily loaded fixed-inclined-pad thrust bearings', Engineering Sciences Data Unit, London (1982)

21 ESDU Data Item 83004, 'Calculation methods for steadily loaded, off-set pivot, tilting-pad thrust bearings', Engineering Sciences Data Unit, London (1983)

22 ESDU Data Item 76029, 'A guide on the design and selection of dry rubbing bearings', Engineering Sciences Data Unit, London (1976)

23 'Polymer materials for bearing surfaces. Selection and performance guide', National Centre of Tribology, Risley, Warrington (1983)

24 Archard, J. F., 'The temperature of rubbing surfaces', *Wear*, **2**, 438 (1958)

25 Jaeger, J. C., 'Moving sources of heat and the temperature at sliding contacts', *Proc. R. Soc., NSW*, **76**, No. 56, 203 (1942)

26 ESDU Data Item 87007, 'A guide on the design and selection of dry rubbing bearings', Engineering Sciences Data Unit, London

27 Farrow, M. and Gleave, C., 'Wear resistant coatings', *Transactions of the Institute of Metal Finishing*, **62**, Pt 2 (1984)

28 Crease, A. B., 'Design data for the wear performance of rubbing bearing surfaces', *Tribology*, **6**, No. 1, 15 (1973)

29 Anderson, J. C. and Robbins, E. J., 'The influence of temperature generation on the wear of some polymers', *The Wear of Non-Metallic Materials, Proceedings of 3rd Leeds-Lyon symposium* on Tribology (1976)

30 Amateau, M. F. and Glaeser, W. A., 'Survey of materials for high temperature bearing sliding applications', *Wear*, **7**, No. 5, 385 (1964)

31 Norbrac Carbon, Technical data sheets

32 Pratt, G. C., 'Graphite/metal composites for dry and sparsely lubricated bearing applications', *Tribology*, **6**, No. 6, 259 (1973)

33 Lancaster, J. K., 'Dry bearings: a survey of materials and factors affecting their performance', *Tribology*, **6**, No. 6, 219 (1973)

34 National Centre of Tribology Course Notes, 'Tribology and design of spacecraft mechanisms' (1985)

35 Salomon, G., De Gee, A. W. J. and Zaat, J. H., 'Mechano-chemical factors in MoS_2 film lubrication', *Wear*, **7**, No. 1, 87 (1964)

36 Halling, J. (ed.), *Principles of Tribology*, Macmillan, London (1973)

37 Morgan, V., 'Porous metal bearings', *Engineering*, August (1980)

38 Seifert, W. W. and Westcott, V. C., 'A method for the study of wear particles in lubricating oil', *Wear*, **21**, No. 1, 27–42 (1972)

39 Bowen, E. R. and Westcott, V. C., *Wear Particle Atlas*, Foxboro/Trans-Sonics, MA, July (1976)

40 Thelning, K. E., *Steel and its Heat Treatment*, 2nd edn, Butterworths, London (1984)

41 Steen, W. M. and Weerarasinge, V. M., 'The laser's other role', in *Recent Developments in Surface Coating and Modification Processes*, Mechanical Engineering Publications, London (1985), pp. 49–52

42 Mehelich, C. S. and Woelfel, M. W., 'Shot peening', in *Metals Handbook*, 9th edn, Vol. 5 (1982)

43 Bell, T., *Survey of the Heat Treatment of Engineering Components*, Metals Society, published by the Iron and Steel Institute (1976)

44 Stains, A. M. and Bell, T., 'Technological importance of plasma-induced, nitrided and carburised layers on steel', *Thin Solid Films*, **86**, 201–211 (1981)

45 Internal document, National Centre of Tribology, UKAEA, Risley, Warrington, Cheshire (1982)

46 Habig, K. H., 'Wear protection of steels by boroding, vanadizing, nitriding, carburising and hardening', *Materials in Engineering*, **2**, 83–93 (1980)

47 Child, H. C., 'Improving tool steels by thermochemical treatments', *The Metallurgist and Materials Technologist*, 303–309 (1981)

48 Shrier, L. D. (ed.), 'Principles of applying coatings by diffusion', in *Corrosion 2; Corrosion Control*, Newnes-Butterworths, London (1976)

49 Arai, T. and Komatsu, N., Internal document, Toyota Central Research and Development Laboratory, Nagoya, 468 Japan

50 Edenhofer, B., 'Physical and metallurgical aspects of ion nitriding', *Heat Treatment of Metals*, 23–28 (1974)

51 Clark, D. S. and Cherry, F. K., 'A new low temperature surface treatment – nitrocarburising', *Machine Tool Research*, **XXIII**, No. 1 (1984)

52 'Nitrotech, surface treatment process shows its mettle', *Metallurgia*, February (1985)

53 Peller, S., *Corrosion Causes and Prevention*, McGraw-Hill, New York (1951)

54 Gregory, J. C., 'Low temperature metal diffusion treatments for the improvement of scuffing and wear resistance of ferrous and non-ferrous metal parts', *Heat Treatment of Metals*, **2**, 33–38 (1978)

55 Gregory, J. C., 'Chemical conversion coatings to reduce wear', *Tribology International*, 105–113 (1978)

56 Gregory, J. C. and Caubet, J. J., 'Thermal and chemico-thermal treatments of non-ferrous materials to reduce wear', *Tribology*, 8–13 (1971)

57 ASM Committee on Finishing of Aluminium, 'Cleaning and finishing of aluminium and aluminium alloys', in *Metals Handbook*, 9th edn, Vol. 5 (1982)

58 Hochman, R. F., 'Ion implantation', in *Metals Handbook*, 9th edn, Vol. 5 (1982)

59 Dearnaley, G., 'Practical applications of ion implantation', *Journal of Metals*, **34**, No. 9, September, 18–28 (1982)

60 Teer, D. G. and Arnell, R. D., *Recent Developments in Surface Coating and Modification Processes, Principles of Ion Plating*, Mechanical Engineering Publications, London (1985), pp. 21–27

61 Thornton, J. A., 'Sputtering', *Metals Handbook*, 9th edn, Vol. 5 (1982)

62 Thornton, J. A., 'Recent advances in sputter deposition', *Surface Engineering*, **2**, 283–293 (1986)

63 Stevens, K. T. and Douglas, A., 'PVD coatings – their properties and potential applications', *Tribology 50 years on*, International Conference organized by the Institute of Mechanical Engineers, 1–3 July 1987

64 Blocher, J. M., 'Chemical vapour deposition', in *Metals Handbook*, 9th edn, Vol. 5 (1982)

65 Hitchman, M. L., *Recent developments in surface coatings and modification processes, Chemical Vapour Deposition for Surface Modification*, Mechanical Engineering Publications, London (1985), pp. 3–19

66 Bidmead, G. F. and Davies, G. R., *Trans. Institute of Metal Finishers*, **56**, 97 (1978)

67 Chessin, H. and Fernald, E. Jr, 'Hard chromium plating', in *Metals Handbook*, 9th edn, Vol. 5 (1982)

68 Fields, W. D., Duncan, R. N., Zickgraf, J. R., Baudrand, D. W. and Russell, H. Jr, 'Electroless nickel plating', in *Metals Handbook*, 9th edn, Vol. 5 (1982)

69 Shrier, L. D. (ed.), 'Phosphate coatings', in *Corrosion 2; Corrosion Control*, Newnes-Butterworth, London (1976)

70 Clare, J. H. and Crawmer, D. E., 'Thermal spray coatings', in *Metals Handbook*, 9th edn, Vol. 5 (1982)

71 Draper, C. W., 'Laser surface alloying – state of the art', *Journal of Metals*, 25–33 (1982)

72 Bell, T., Institute of Metallurgists, Spring Review, Course series, No. 10 (1978)

73 Pollock, H. M., Maugis, and Barquis, M., 'Characterisation of sub-micron surface layers by indentation', *Micro-indentation Hardness Testing*, ASTM special publication (1986)

74 Thomas, A., 'Microhardness measurement as a quality control technique for thin, hard coatings', *Surface Engineering*, **3**, No. 2 (1987)

75 Latter, T. D. T., 'X-ray fluorescence for coating thickness measurements', *Circuit World*, **10**, No. 1, Institute of Circuit Technology

76 Laeng, P. and Steinmann, P. A., Internal document, Laboratoire Suisse de Recherches Horlogères, CH-2000 Neuchatel 7, Switzerland

77 'A microprocesser-controlled instrument for coating adhesion assessment', Technical Literature, Ion Coat Limited, Hull

78 Tomlinson, G. A., 'The rusting of steel surfaces in contact', *Proc Roy Soc (London) Series A*, **115**, 472–483 (1927)

79 Tomlinson, G. A., Thorpe, P. L. and Gough, H. J., 'Investigation of the fretting corrosion of closely fitting surfaces', *Proc I. Mech E.*, **141**, 223 (1939)

80 Campbell, W. E., 'Fretting', in *Boundary Lubrication – A Review of World Literature*, ASME (1967)

81 Waterhouse, R. B., 'Fretting', in Scott, D. (ed.), *Treatise on Materials Science and Technology*: Vol. 13, *Wear*, Academic Press, London (1969)

82 Hurricks, P. L., 'The mechanism of fretting – a review', *Wear*, **15**, 389–409 (1970)

83 Waterhouse, R. B., 'Fretting wear', *Wear*, **100**, 107–118 (1984)

84 Waterhouse, R. B., 'Fretting corrosion', in Hopkins, D. W. (ed.), *International Series of Monographs on Materials Science and Technology*, Vol. 10, Pergamon Press, Oxford (1972)

85 Waterhouse, R. B., *Fretting Fatigue*, Applied Science, London (1981)

86 Chivers, T. C. and Gordelier, S. C., 'Fretting fatigue palliatives: some comparative experiments', *Wear*, **96**, 153–175 (1984)

87 Almen, J. O., 'Lubricants and false brinelling of ball and roller bearings', *Mech. Eng.*, **59**, 415 (1937)

88 Chivers, T. C., 'Nuclear tribology – a personal perspective', *Tribology International*, **19**, No. 5, October (1986)

89 Lewis, M. W. J. and Campbell, C. S., 'Fretting and wear of stainless and ferritic steels in liquid metal fast breeder reactors', *Proc. 2nd Conf. on Materials Performance in Nuclear Steam Generators*, St Petersburg, Florida (1980): published in *Nucl. Tech.*, **55**, No. 2, 460–469 (1981)

90 Lewis, M. W. J. and Campbell, C. S., 'Aluminised surfaces for the prevention of galling and fretting wear in fast reactor steam generators', *Proc. 3rd BNES Int. Conf. on Liquid Metal Engineering and Technology*, Oxford (1984), pp. 91–98

91 Chivers, T. C. et al., 'Vibration data and its employment for component life projections in AGR circulators', *BNES Conf. on Vibration in Nuclear Plant*, Keswick, UK, p. 685

92 Chivers, T. C., 'Aspects of fretting wear of sprayed cermet coatings', I. Mech. Eng. Fretting Wear Seminar, Nottingham, UK (1985): published in *Wear*, **106**, 63–76 (1985)

93 Low, M. B. J., 'Fretting problems and some solutions in power plant machinery', I. Mech. Eng. Fretting Wear Seminar, Nottingham, UK, 1985: published in *Wear*, **106** 315–335 (1985)

94 Suh, N. P., 'The delamination theory of wear', *Wear*, **25**, 111–124 (1973)

95 Stowers, I. F. and Rabinowicz, E., *Appl. Phys*, **43**, 2485 (1954)

96 Hurricks, P. L., *Wear*, **27**, 319 (1974)

97 Sproles, E. S. and Duquette, D. J., 'The mechanism of material removal in fretting', *Wear*, **49**, 339–352 (1978)

98 Uhlig, H. H., 'Mechanics of fretting corrosion of mild steel in air and nitrogen', *Trans. ASME J. Appl. Mech.*, **21**, 401–407 (1954)

99 Jahanmir, S., Suh, N. P. and Abrahamson, E. P., 'Microscopic observations of the wear sheet formation by delamination', *Wear*, **28**, 235 (1974)

100 Waterhouse, R. B. and Taylor, D. E., 'Fretting debris and the delamination theory of wear', *Wear*, **29**, 337 (1974)

101 Sakman, B. W. and Rightmire, B. G., 'An investigation of fretting corrosion under several combinations of oxidation', NACA TN 1492 (1948)

102 Ming-Feng, I. and Uhlig, H. H., 'Fretting corrosion of mild steel in air and nitrogen', *Trans. ASME J. Appl. Mech.*, **76**, 395–400 (1954)

103 Wright, K. H. R., 'An investigation of fretting corrosion', *Proc. Inst. Mech. Engrs (London)*, Series B, 56–574 (1952–1953)

104 Archard, J. P., 'The contact and rubbing of flat surfaces', *Applied Physics*, **24**, 981 (1953)

105 Ohmae, N. and Tsukizoe, T., 'The effect of slip amplitude on fretting', *Wear*, **27**, 281–294 (1974)

106 Ming-Feng, I. and Rightmire, B. G., 'An experimental study of fretting wear', *Proc. Inst. Mech. Eng.*, **170**, 1055 (1956)

107 Stowers, I. F. and Rabinowicz, E., 'The mechanism of fretting wear', *Jour. Lub. Tech.*, **95**, 65 (1973)

108 Lewis, M. W. J. and Didsbury, P. B., UKAEA Internal Document (1977)

109 Hallliday, J. S. and Hirst, W., 'The fretting corrosion of mild steel', *Proc. Roy. Soc. (London)*, Series A, 236, 411–425 (1956)

110 Kennedy, P. J., Petersen, M. B. and Stallings, L., 'An evaluation of fretting at small amplitudes', Symposium sponsored by ASTM Committee G-2 on Erosion and Wear, Warminster, PA, June 1981.ASTM Special Technical Publ. 780

111 Soderberg, S., Bryggman, U. and McCullough, T., 'Frequency effects in fretting wear', *Wear*, **110**, 19–34 (1986)

112 Hurricks, P. L., 'The fretting wear of mild steel from room temperature to 200C', *Wear*, **30**, 217–229 (1974)

113 Hurricks, P. L., 'The fretting of mild steel from 200–500C', *Wear*, **30**, 189–212 (1974)

114 Lewis, M. W. J. and Strong, N. A., UKAEA Internal Document (1986)

115 Stott, F. H., Lin, D. S. and Wood, G. C., 'The structure and mechanism of formation of the "glaze" oxide layers produced on nickel-based alloys during wear at high temperatures', *Corros. Sci.*, **13**, 419 (1972)

116 Stott, F. H., Lin, D. S., Wood, G. C. and Stevenson, C. W., 'The tribological behaviour of nickel and nickel–chromium alloys at temperatures from 20 to 800C', *Wear*, **36**, 147 (1976)

117 Godfrey, D. and Bisson, E. E., 'Effectiveness of molybdenum disulfide as a fretting corrosion inhibitor', NACA Tech. Note No. 2180 (1950)

118 Mindlin, R. D., 'Compliance of elastic bodies in contact', *Trans. ASME J. App. Mech.*, **16**, 259–268 (1949)

119 Bowden, F. P. and Tabor, D., *The Friction and Lubrication of Solids*, Vol. II, Oxford University Press, Oxford (1964)

120 'Materials evaluation under fretting conditions', Symposium sponsored by ASTM Committee G-2 on Erosion and Wear, Warminster, PA, June 1981. ASTM Special Technical Publ. 780

121 Gray, H. C. and Jenny, R. W., 'An investigation of chafing on aircraft engine parts', *Trans. SAE*, **52**, 511 (1944)

122 Johnson, K. L., 'Energy dissipation of spherical surfaces in contact transmitting oscillating forces', *J. Mech. Eng. Sci.*, **3**, 362–368 (1961)

123 Halliday, J. S., 'Experimental investigation of some processes involved in fretting corrosion', Conference on Lubrication and Wear (Inst. of Mech. Engrs), 1957, pp. 640–646

124 *The Cassel 'Sulfinuz' Process*, ICI Mond Division, p. 4

125 Budinski, K. G., 'Control of fretting corrosion', *Thin Solid Films*, **64**, 359–363 (1979)

126 Sikorskii, M. E., 'Correlation of the coefficient of adhesion with various physical and mechanical properties of metals', *Trans. ASME J. Basic Eng.*, **85**, 279–285 (1963)

127 Waterhouse, R. B., Brook, P. A. and Lee, M. C., *Wear*, **5**, 235–244 (1962)

128 Halling, J., 'Surface coatings; material conservation and optimum tribological performance', *Tribology*, October, 203–208 (1979)

129 Alyab'ev, A., Shevelya, V. V., Gladchenko, A. N., Venediktov, V. A., Ryaboi, A. and Dolzhanskii, Y. M., 'The resistance of thick-layer chromium coatings to fretting corrosion', *Fiz. Khim. Mekm Mater.*, **15**, No. 4, 26 (1970)

130 Wise, S. and Burdon, E. S., *Inst. Loco. Eng.*, **54**, 298 (1964–1965)

131 Gould, A. J., Boden, P. J. and Harris, S. J., 'Fretting wear and fretting fatigue properties of electroless nickel–phosphorus coatings', *Trans. Inst. of Metal Finishing*, **61**, 97–104 (1983)

132 National Centre of Tribology, *Polymer Materials for Bearing Surfaces – Selection and Performance Guide* (1983)

133 Johnson, K. L. and O'Connor, J. J., 'Mechanics of fretting', *Proc. Inst. Mech. Engrs*, **178**, Pt 3, Paper 11 (1963–1964)

134 Sandifer, J. P., 'Evaluation of methods of reducing fretting fatigue in 2024–T3 aluminium lap joints', *Wear*, **26**, (3) 405–412 (19??)

135 Kreitner, L., 'The effect of brinelling and fretting fatigue on the fatigue life of assembled machine components', *Royal Aircraft Establishment* (1979 trans.), p. 1998

136 Bramhall, R., *Studies in Fretting Fatigue*, DPhil dissertation University of Oxford (1973)

137 Leadbeater, G., Noble, B. and Waterhouse, R. B., 'The fatigue of an aluminium alloy produced by fretting on a shot peened surface', *Advances in Fracture Research*, Proc. 6th Inter. Conf. on Fracture, Vol. 3, Pergamon Press, Oxford (1983) pp. 2125–2132

138 Chivers, T. C. and Gordelier, S. C., 'A literature review of palliatives for fretting fatigue', *Wear*, **56**, 177–190 (1979)

139 Chivers, T. C. and Gordelier, S. C., 'Fretting fatigue and contact conditions: a rational explanation of palliative behaviour', *Proc. I. Mech. Eng.*, **199**, No. C4 (1985)

140 Nishioka, K. and Hirakawa, K., 'Fundamental investigations of fretting fatigue – part 5', *Bull. JSME*, **12**, 692–697 (1969)

141 BS 1134: Part 1: 1988 Method for the assessment of surface texture

142 Michell, A. G. M., *Lubrication – its principles and practice*, Blackie, London (1950), pp. 281–291

143 Dowson, D. and Whomes, T. L., 'The effect of surface roughness upon the lubrication of rigid cylindrical rollers. I. Theoretical, II Experimental', *Wear*, **18** 129–140 and 141–151 (1971)

144 Christensen, H., 'Stochastic models for hydrodynamic lubrication of rough surfaces', *Proc. I. Mech. E.*, **184**(1), 1013–1026 (1969–1970)

145 Dyson, A., 'Hydrodynamic lubrication of rough surfaces – A review of theoretical work', 4th Leeds-Lyon Symposium, September (1977) pp. 61–70

146 Christensen, H. and Tonder, D., 'The hydrodynamic lubrication of rough bearing surfaces of finite width', *Trans. ASME. J. Lub. Tech.*, **95**(2), 166–172 (1973)

147 Berthe, D. and Godet, M., 'A more general form of Reynolds' equation. Application to rough surfaces', *Wear*, **27**, 345–355 (1973)

148 Bush, A. W., Gibson, R. D. and Skinner, P. H., 'The effect of surface roughness in elasto-hydrodynamic lubrication', *Wear*, **96**(2), 177–202 (1984)

149 Bush, A. W., Skinner, P. H., and Gibson, R. D. 'Surface roughness effects in point contact elastohydrodynamic lubrication', *Wear*, **83**, 285–301 (1982)

150 Tonder, K. 'The lubrication of surfaces having a cross-striated roughness pattern', 4th Leeds-Lyon Symposium. September (1977) pp. 80–87

151 Patir, N. and Cheng, H. S., 'An average flow model for determining effects of three-dimensional roughness on partial hydrodynamic lubrication', *Trans. ASME. J. Lub. Tech.*, **100**, January, 12–17 (1978)

152 Patir, N. and Cheng, H. S., 'Application of average flow model to lubrication between rough sliding surfaces', *Trans. ASME, J. Lub. Tech.*, **101**, 220–230 (1979)

153 Rohde, S. M., 'A mixed friction model for dynamically loaded contacts with application to piston ring lubrication', Seventh Leeds-Lyon Symposium (1980), Paper IX (v), pp. 262–278

154 Sun, D. C. and Cheng, K. K., 'First effects of Stokes roughness on hydrodynamic lubrication technology', *Trans. ASME, J. Lub. Tech.*, **99**, 2–9 (1977)

155 Dawson, P. H., 'Effect of metallic contact on the pitting of lubricated surfaces', *J. Mech. Eng. Sci.*, **4**, 16–21 (1962)

156 Berthe, D., Michau, B., Flamand, L. and Godet, M., 'Effect of roughness ratio and hertz pressure on micro-pits and spalls in concentrated contacts: Theory and experiments', 4th Leeds-Lyon Symposium, September (1977), pp. 233–242

157 Kelley, B. W. and Lemanski, A. J., 'Lubrication of involute gearing', *Proc. I. Mech. E.*, **182** (Part 3A, paper 11), 173–184 (1967–1968)

158 Queener, C. A., Smith, T. C. and Mitchell, W. L., 'Implications of wear theories on surface quality', *Ann. CIRP*, **25**(2), 513–519 (1976)

159 Rowe, G. W., Kaliszer, H., Trmal, G. and Cotter, A., 'Running-in of plain bearings', *Wear*, **34**, 1–14 (1975)

160 Jahanmir, S. and Suh, N. P., 'Surface topography and integrity effects on sliding wear', *Wear*, **44**, 87–99 (1977)

161 Suh, N. P., 'The delamination theory of wear', *Wear*, **25**, 111–124 (1973)

162 Halling, J. (ed.), *Principles of Tribology*, Macmillan, London (1975), Chapter 5, pp. 94–127

163 Bayer, R. G. and Sirico, J. L., 'The influence of surface roughness on wear', *Wear*, **35**, 251–260 (1975)

164 Thomas, T. R. and Sayles, R. S., 'Random process approach to the prediction of joint stiffness', *Trans. ASME, J. Lub. Tech.*, **99B**, 250–256 (1977)

165 Yip, F. C. and Venart, J. E. S., 'Surface topography effects in the estimation of thermal and electrical contact resistance', *Proc. I. Mech. E.*, **182** (3K) 81–91 (1967–1968)

166 Archard, J. F., 'Elastic deformation and the laws of friction', *Proc. Roy. Soc. (London)*, **A243**, 190–205 (1957)

167 Sayles, R. S. and Thomas, T. R., 'Measurements of the statistical microgeometry of engineering surfaces', *Trans. ASME J. Lub. Tech.*, **101**, October, 409–418 (1979)

168 Greenwood, J. A. and Williamson, J. B. P., 'Contact of nominally flat surfaces', *Proc. Roy Soc. (London)*, **A295**, 300–319 (1966)

169 Greenwood, J. A. and Tripp, J. H., 'The contact of two nominally flat rough surfaces', *Proc. I. Mech. E.*, **185** (48/71), 25–633 (1970–1971)

170 Whitehouse, D. J. and Archard, J. F., 'The properties of random surfaces', *Proc. Roy. Soc. (London)*, **A316**, 97–121 (1970)

171 Nayak, P. R., 'Random process model of rough surfaces', *Trans. ASME, J. Lub. Tech.*, **93**, 398–407 (1971)

172 Longuet-Higgins, M. S., 'The statistical analysis of a random, moving surface', *Phil. Trans. Roy. Soc.*, **A249**, 321–387 (1957)

173 Longuet-Higgins, M. S., 'Statistical properties of an isotropic random surface', *Phil. Trans. Roy. Soc.*, **A249**, 157–174 (1957)

174 Nayak, P. R., 'Random process model of rough surfaces in plastic contact', *Wear*, **26**, 305–333 (1973)

175 Onions, R. A. and Archard, J. F., 'The contact of surfaces having a random structure', *J. Phys. D. (Appl. Phys.)*, **6**, 289–304 (1973)

176 Halling, J. (ed.), *Principles of Tribology*, Macmillan, London (1975), Chapter 3, p. 66

177 Thomas, T. R., *Rough Surfaces*, Longman, New York (1982), Chapter 8, pp. 168–188

178 Thomas, T. R., Holmes, C. F., McAdams, H. T. and Bernard, J. C., 'Surface features influencing the effectiveness of lip seals: A pattern recognition approach', SME paper IQ75–128 (1975)

179 Thomas, T. R., Holmes, C. F., McAdams, H. T. and Bernard, J. C., 'Surface microgeometry of lip seals related to their performance', *Proc. 7th International Conf. on Fluid Sealing*, Paper J2 (BHRA, Cranfield) (1975)

180 Thomas, T. R., Holmes, C. F., McAdams, H. T. and Bernard, J. C., 'Discussion to "Surface features influencing the effectiveness of lip seals: A pattern recognition approach"', SME paper IQ75–128 (1975)

181 Thomas, T. R., 'Some applications of statistical surface measurements to engineering problems', *Mechanique, matériaux, electricité*, **337**, January, 7–16 (1978)

182 Manning, P. T., 'The influence of surface roughness on gas flow through cracks', *Wear*, **57**, 365–376 (1979)

183 Lackenby, H., 'Resistance of ships, with special reference to skin friction and hull surface condition', *Proc. I. Mech. E.*, **176**, 981–1014 (1962)

184 Conn, J. F. C., Lackenby, H. and Walker, W. P., 'BSRA resistance experiments on the Lucy Ashton (Part II)', *Trans. Inst. Naval Arch.*, **95**, 350–388 (1953)

185 King, M. J., 'The measurement of ship hull roughness', *Wear*, **83**, 385–397 (1982)

186 Postlethwaite, A., 'Wear and tear float away', *Technology*, 2 April, 19 (1984)

187 Thomas, T. R., *Rough Surfaces*, Longman, New York (1982), pp. 223–226

188 Ananthapadmanaban, T. and Radhakrishnan, V., 'An investigation of the role of surface irregularities in the noise spectrum of rolling and sliding contacts', *Wear*, **83**, 399–409 (1982)

189 Hansen, C. M., 'Surface roughness profiles and coatings performance', *J. Paint Technol.*, **44**(570), July, 61–66 (1972)

190 King, M. J. and Thomas, T. R., 'Stylus measurement of the microgeometry of a coated surface', *J. Coating Tech.*, **50**, 56–61 (1978)

191 Onians, S., 'Telecoms boosts optics market', *Technology*, 16 April, 9 (1984)

192 Place, J. D., 'Fibre optics', *Measurement and Control*, **19**, September, 35–37 (1986)

193 Bennion, I., Bowman, R., Crawford, M., Hurditch, R. I. and Reid, D. C. J., 'A high speed optical data recorder', *New Electronics*, April, 22–24 (1980)

194 Bennett, J. N. and Dancy, J. H., 'Stylus profiling instrument for measuring statistical properties of optical surfaces', *Appl. Opt.*, **20**(10) May, 1785–1802 (1981)

195 Beckmann, P. and Spizzichino, A., *A Scattering of Electromagnetic Waves from Rough Surfaces*, Pergamon, New York (1963)

196 Van de Hulst, H. C., *Light Scattering by Small Particles*, Wiley, New York (1957)

197 Wallmark, J. T., 'Physical limitations on the design of microcircuits', *Physics Bulletin*, **33**(10), 362–365 (1982)

198 Glang, R. and Schaible, P. M., 'Tolerance limits of etched film resistors', *Thin Solid Films*, **1**, 309–322 (1967/1968)

199 Anderson, R. M. and Neudeck, G. W., 'Flatness and surface roughness of some common thin film substrate materials', *J. Vac. Sci. and Tech.*, **8**(2), 454–457 (1971)

200 Coffman, B. and Thurnauer, H., 'Effect of ceramic substrates on the resistance of vacuum deposited thin metal films', *Trans. 9th Nat. Vac. Symp.*, Macmillan, New York (1962), pp. 89–95

201 Schwartz, N. and Brown, R., A stylus method for evaluating the thickness of thin films and substrate surface roughness', *Trans. 8th Nat. Vac. Symp.*, Macmillan, New York (1962), pp. 836–845

202 White, J. W., 'The effect of two sided surface roughness on ultra thin gas films', *Trans. ASME, J. Lub. Tech.*, **105**, 131–137 (1983)

203 Tandon, P. N. and Rakesh, L., 'Effects of cartilage roughness on the lubrication of human joints', *Wear*, **70**, 29–36 (1981)

204 Stewart, J. W., 'Haemolysis caused by tubing in extracorporal circulation', *The Lancet*, No. 7068, 340–342 (1959)

205 Abbott, E. J. and Firestone, F. A., 'Specifying surface quality', *Mech. Eng.*, **55**, 569–572 (1933)

206 BS 1134: Part 2 1990 Method for the assessment of surface texture

207 Dickinson, G. R., 'Survey of factors affecting surface finish', *Proc. I. Mech. E.*, **182**(3K), 135–147 (1967–1968)

208 Ghabriel, S. R., Saleh, S. M., Kohail, A. and Moisan, A., 'Problems associated with electro-discharge machined, electrochemically machined and ultrasonically machined surfaces', *Wear*, **83** 275–283 (1982)

209 Selvan, M. S. and Radhakrishnan, V., 'Groove wear, built-up edge, and surface roughness in turning', *Wear*, **30**, 179–188 (1974)

210 Chetwynd, D. G., McKee, F. A. and Rakels, J. H., 'Machined surfaces: Final texture and underlying structure', *Wear*, **83**, 233–240 (1982)

211 Thomas, T. R., 'Measurement techniques. Notes from a short course on surface topography in engineering', Teesside Polytechnic, September (1980)

212 Dagnall, H., *Exploring Surface Texture*, Rank Taylor Hobson, Leicester (1980), pp. 75–127

213 Garrat, J. D., 'Survey of displacement transducers below 50 mm', *J. Phys. E.*, **12**, 563–573 (1979)

214 Sherrington, I. and Smith, E. H., 'A quantitative study of the influence of stylus shape and load on the fidelity of data recorded by stylus instruments', Presented at the Second National Conference on Production Research, Edinburgh (September 1986)

215 Nara, J., 'On CLA value obtained with direct reading, surface roughness testers – effects of skid and high pass filter', *Bull. Japan Soc. Proc. Engng.*, **1**, 263–273 (1966)

216 Whitehouse, D. J., 'Surface texture assessment errors of finishing processes caused by skid distortion', *J. Phys. E.*, **15**, 1337–1340 (1982)

217 Ishagaki, H. and Kawaguchi, I., 'Effect of a skid on the accuracy of measuring surface roughness', *Wear*, **68**, 203–211 (1981)

218 Abbott, E. J. and Goldschmidt, E., 'Surface quality – A review of *Technische Oberflachenkunde* by G. Schmaltz', *Mech. Engng.*, **59**, 813–825 (1937)

219 Gane, N. and Cox, J., 'The microhardness of metals at very

low loads', *Philos Mag.*, **22**, 881–889 (1970)

220 Williamson, J. B. P., 'Microtopography of surfaces', *Proc. I. Mech. E.*, **182(3K)**, 21–30 (1967–1968)

221 Kinsey, D. and Chetwynd, D. G., 'Some aspects of the application of digital computers to the on-line measurement of surfaces', *Conf. Proc. of IMEKO VI*, Dresden, paper B523, June (1973)

222 Edmonds, M. J., Jones, A. M., O'Callaghan, P. W. and Probert, S. D., 'A three-dimensional relocation profilometer stage', *Wear*, **43**, 329–340 (1977)

223 Thomas, T. R. and Walker, J., 'Roughness measurement with a micro-computer', *Microcomputer Applications*, **3**(4), 40–47 (1979)

224 Bennett, J. M. and Dancy, J. H., 'Stylus profiling instrument for measuring statistical properties of smooth optical surfaces', *Appl. Opt.*, **20**(10), 1785–1802 (1981)

225 Sayles, R. S. and Thomas, T. R., 'Mapping a small area of surface', *J. Phys. E.*, **9**, 885–861 (1976)

226 Tsukada, T. and Sasjima, K., 'A three-dimensional measuring technique for surface asperities', *Wear*, **71**, 1–14 (1981)

227 Teague, E. C., Scire, F. E., Baker, S. M. and Jensen, S. W., 'Three-dimensional stylus profilometry', *Wear*, **83**, 1–12 (1982)

228 Rubert, M. P., 'Functional assessment of surface roughness', *Proc. I. Mech. E.*, **182(3K)**, 350–359 (1967–1968)

229 Sherwood, K. F. and Crookall, J. R., 'Surface finish assessment by an electrical capacitance technique', *Proc. I. Mech. E.*, **182(3K)**, 344–349 (1967–1968)

230 Brecker, H. N., Fromson, R. E. and Shum, L. Y., 'A capacitance based surface texture measuring system', *Ann. CIRP*, **25**(1), 375–377 (1977)

231 Thomas, T. R., *Rough Surfaces*, Longman, New York (1982), p. 87

232 Graneck, M. and Wunsch, H. L., 'Application of pneumatic gauging to the measurement of surface finish', *Machinery*, **81**, 701–707 (1952)

233 Wagner, J. C., 'Surface effects in pneumatic gauging', *Int. J. Mach. Tool Des. Res.*, **7**, 1–14 (1967)

234 Radhakrishnan, V. and Sagar, V., 'Surface roughness assessment by means of pneumatic measurement', *Proc. IV AIMTDR Conf. ITT-Madras* (1970), pp. 487–494

235 Tanner, L. H., 'A pneumatic Wheatstone bridge for surface roughness measurement', *J. Phys. E.*, **12**, 957–960 (1979)

236 Tanner, L. H., 'An improved pneumatic Wheatstone bridge for roughness measurement', *J. Phys. E.*, **13**, 593–594 (1980)

237 Tanner, L. H., 'A self balancing pneumatic potentiometer and Wheatstone bridge with electrical readout', *Precis. Eng.*, **3**(4), 201–207 (1981)

238 Tanner, L. H., 'A self balancing Wheatstone bridge for surface roughness measurement', *Wear*, **83**, 37–47 (1982)

239 BS 6563: 1985 Method for the determination of the roughness of paper and board by the Parker Print-Surf apparatus

240 Nelson, H. R., 'Taper sectioning as a means of describing the surface contour of metals', 2nd edition of *Proceedings* of the special summer conference on Friction and Surface Finish held at Cambridge, Mass. (5–7 June 1940), MIT Press, Cambridge, Mass. (1969)

241 Thomas, T. R., *Rough Surfaces*, Longman, New York (1982), p. 45.

242 Uchida, S. N., Sato, H. and O-Hori, M., 'Two dimensional measurement of surface roughness by the light sectioning method', *Ann. CIRP*, **28**(1), 419–423 (1979)

243 Sato, H. and O-Hori, M., 'Characteristics of two dimensional surface roughness. Taking self excited chatter marks as objective', *Ann. CIRP*, **30**(1), 481–486 (1981)

244 Trumpold, H., 'Limits of application of the interference methods for surface measurements', *Proc. I. Mech. E.*, **182(3K)**, 241–254 (1967–1968)

245 King, R. J., Downs, M. J., Clapham, P. B., Raine, K. W. and Talim, S. P., 'A comparison of methods for accurate film thickness measurement', *J. Phys. E.*, **5**, 445–449 (1972)

246 Bennett, J. M., 'Measurement of the rms roughness, autocovariance function and other statistical properties of optical surfaces using a FECO scanning interferometer', *Appl. Opt.*, **15**(11), November, 2705–2721 (1976)

247 American National Standards Institute, 'Surface texture: surface roughness, waviness and lay', Document B 46.1 (1972)

248 Jarosz, M., Kocsanyi, L. and Giber, J., '*In situ* deformation measurement on the surface of silicon wafers', *J. Phys. E.*, 746–748 (1982)

249 Wahl, F., So, S. and Wong, K., 'A hybrid optical-digital image processing method for surface inspection', *IBM, J. Res. Develop.*, **27**(4), July, 376–385 (1983)

250 Pugh, D. J. and Jackson, K., 'Automatic gauge block measurement', Paper 1.2, NELEX '82 Conference at NEL, Glasgow (September 1982)

251 Bruning, J. H., Herriott, D. R., Gallagher, D. P., Rosenfeld, D. P., White, A. D. and Brangaccio, D. J., 'Digital wavefront measuring interferometer for testing optical surfaces and lenses', *Appl. Opt.*, **13**(ii), November, 2693–2703 (1974)

252 Sommargen, G. E., 'Optical heterodyne profilometry', *Appl. Opt.*, **20**(4), February, 610–618 (1981)

253 Wyant, J. C., Koliopoulos, C. I., Bhushan, B. and Basila, D., 'Development of a three dimensional non-contact digital optical profiler', *J. Tril.*, **108**, January, 1–18 (1986)

254 Perry, D. M., Moran, P. J. and Robinson, G. M., 'Three dimensional surface metrology of magnetic recording materials through direct phase detecting microscopic interferometry', *Proc. Fifth Conf. on Video and Data Recording*, University of Southampton (April 1974)

255 Groves, D., Lalor, M. J., Cohen, N. and Atkinson, J. T., 'A holographic technique with computer aided analysis for the measurement of wear', *J. Phys. E. (Sci. Instrum.)*, **13**, 741–746 (1980)

256 Tsurota, T. and Shiotake, N., 'Holographic generation of contour map of diffusely reflecting surface by using immersion method', *Japan. J. Appl. Phys.*, **6**, 661–662 (1967)

257 Lech, M., Mruk, I. and Strupnicki, J., 'Holographic contouring of the surface topography and study of the contact rigidity of rough surfaces', *Wear*, **57**, 236–268 (1979)

258 Fairman, Y., Lenz, E. and Shamir, J., 'Optical profilometer: a new method for high sensitivity and wide dynamic range', *Appl. Opt.*, **21**(7), September, 3200–3208 (1982)

259 Arecchi, F. T., Bertani, D. and Ciliberto, S., 'A fast versatile optical profilometer', *Opt. Comm.*, **31**(3), 263–266 (1979)

260 Dupuy, O., 'High precision optical profilometer for the study of micro-geometrical surface defects', *Proc. I. Mech. E.*, **182(3K)**, 255–259 (1967–1968)

261 Thwaite, E. G., 'The roughness of surfaces', *Australian Physicist*, November, 170–174 (1977)

262 Mignot, J. and Gorecki, C., 'Measurement of surface roughness comparison between a defect of focus optical technique and the classical stylus technique', *Wear*, **87**, 39–49 (1983)

263 Simon, J., 'New contact devices for measuring small microdisplacements', *Appl. Opt.*, **9**(10), October, 2337–2340 (1970)

264 Dobosz, M., 'Optical profilometer: a practical approximate method of analysis', *Appl. Opt.*, **22**(24), December, 3983–3987 (1983)

265 Dobosz, M., 'Accuracy of profile measurements by means of a focused laser beam', *Wear*, **98**, 117–126 (1984)

266 Teague, E. C., Vorburger, T. V. and Maystre, D., 'Light scattering from manufactured surfaces'. *Ann. CIRP*, **30**(2), 563–569 (1981)

267 Stout, K. J., 'Optical assessment of surface roughness', *Precis. Eng.*, **6**(1), 35–39 (1984)

268 Beckmann, P. and Spizzichino, A., *The Scattering of Electromagnetic Waves from Rough Surfaces*, Pergamon Press, New York (1963)

269 Tanner, L. H. and Fahoum, M., 'A study of the surface parameters of ground and lapped metal surfaces using specular and diffuse reflection of laser light', *Wear*, **36**, 299–316 (1976)

270 Takeyama, H., Sekiguchi, H., Murata, R. and Matsuzaki, H., 'In process detection of surface roughness in machining' *Ann, CIRP.*, **25**, 467–471 (1976)

271 Clarke, G. N. and Thomas, T. R., 'Roughness measurement with a laser scanning analyser', *Wear*, 107–116 (1979)

272 Bennett, J. M., Burge, D. K., Rahn, J. P. and Bennett, H. E., 'Standards for optical surface quality using total integrated scattering', *Proc. SPIE*, **181**, 124–132 (1979)

273 Vorburger, T. V. and Teague, E. C., 'Optical techniques for on-line measurement of surface topography', *Precis. Eng.*, **3**,

61–83 (1981)

274 Church, E. L. 'The measurement of surface texture and topography by differential light scattering', *Wear*, **57**, 93–105 (1979)

275 Thwaite, E. G., 'Power spectra of rough surfaces obtained by optical Fourier transform', *Ann. CIRP*, **29**(1), 419–422 (1980)

276 Thwaite, E. G., 'A quantitative comparison of the wavelength spectrum of a surface obtained by optical Fourier transformation from profile measurements', *Wear*, **83**, 181–187 (1982)

277 Hingle, H. T. and Rakels, J. H., 'The practical application of diffraction techniques to assess surface finish of turned parts', *Ann. CIRP*, **32**(1), 499–501 (1983)

278 Rakels, J. H., 'Diffraction, an old optical phenomenon, used as an advanced metrology tool', *Proc. of 'Modern production and production metrology'*, 2–4 April 1986, Technische Universität, Vienna

279 Rakels, J. H., 'The use of Bessel functions to extend the range of optical diffraction techniques for in-process surface finish measurements of high precision turned parts', *J. Phys. E.*, **19**, 76–79 (1986)

280 Fujii, H. and Asakura, T., 'Roughness measurement of metal surfaces using laser speckle', *J. Opt. Soc. Am.*, **67**, 1171–1176 (1977)

281 Sprague, R. A., 'Surface roughness measurement using white light speckle', *Appl. Opt.*, **11**, 2811–2816 (1972)

282 Fujii, H. and Asakura, T., 'A contract variation of image speckle intensity under illumination of partially coherent light', *Opt. Comm.*, **11**, 35–38 (1974)

283 Fujii, H. and Lit, J. W. Y., 'Surface roughness measurement using dichromatic speckle pattern: An experimental study', *Appl. Opt.*, **17**, 2690–2694 (1978)

284 Leger, P. and Perrin, J. C., 'Real-time measurement of surface roughness by correlation of speckle patterns', *J. Opt. Soc. Am.*, **66**, 1210–1216 (1976)

285 Lonardo, P. M., 'Measurement of smooth surface roughness by means of a photometric method', *Ann. CIRP*, 189–190 (1974)

286 Lonardo, P. M., 'Testing a new optical sensor for in-process detection of surface roughness', *Ann. CIRP*, **27** 531–534 (1978)

287 Smith, T., 'Effect of surface roughness on ellipsometry of aluminium', *Surface Sci.*, **56**, 252–271 (1976)

288 Vorburger, T. V. and Ludema, K. C., 'Ellipsometry of rough surfaces', *Appl. Opt.*, **19**, 561–573 (1980)

289 Rochow, T. G. and Rochow, E. G., *An Introduction to Microscopy by Means of Light, Electrons, X-rays or Ultrasound*, Plenum Press, New York (1979)

290 Bowen, D. K. and Hall, C. R., *Microscopy of Materials*, Macmillan, London (1975)

291 Goodhew, P. J., *Electron Microscopy and Analysis*, Wykeham, London (1975), pp. 9–13

292 Halliday, J. S., 'Application of reflection electron microscopy to the study of wear', *Proc. 1st Int. Conf. on Lubrication and Wear*, Paper 40 (1957), pp. 647–665

293 Quinn, T. F. J., 'Dry wear of steel as revealed by electron microscopy and X-ray diffraction', *Proc. I. Mech. E.*, **182**(3N), 201–213 (1967–1968)

294 Crawford, C. K., 'Charge neutralisation using very low energy electrons', *Scanning Electron Microscopy*, **2**(31), 31–46 (1979)

295 Rasigni, M., Rasigni, G., Palmori, J. P. and Llebaria, A., 'Validity of surface roughness study using microdensitometer analysis of electron micrographs of surface replicas', *Opt. Soc. Am.*, **71**(12), 1549–1550 (1981)

296 Rasigni, F., Varnier, M., Palmari, J. P. and Llebaria, A., 'Spectral density function of the surface roughness for polished optical surfaces', *J. Opt. Soc. Am.*, **73**(10), 1235–1239 (1983)

297 Butler, D. W., 'A stereo electron microscope technique for microtopographic measurements', *Micron*, **4**, 410–424 (1973)

298 Matsuno, Y., Yamada, H., Harada, M. and Kobayashi, A., 'The microtopography of the grinding wheel surface with SEM', *Ann. CIRP*, **24**(1), 237–242 (1975)

299 Holburn, D. M. and Smith, K. C. A., 'On-line topographic analysis in the SEM', *Scanning Electron Microscopy*, **2**(31), 47 (1979)

300 Young, R. D., 'Field emission ultramicrometer', *Rev. Sci.*

Inst., **37**(3), March, 275–278 (1966)

301 Young, R. D., 'Surface microtopography', *Physics Today*, **24**, November, 42–49 (1971)

302 Young, R., Ward, J. and Scire, F., 'The Topografiner: An instrument for measuring surface microtopography', *Rev. Sci. Inst.*, **43**(7), 999–1011 (1972)

303 Binnig, G., Rohrer, H., Gerber, Ch. and Weibel, E., 'Surface studies by scanning tunnelling microscopy', *Phys. Rev. Lett.*, **49**(1), 57–60 (1982)

304 Quate, C. F., 'Vacuum tunneling: A new technique for microscopy', *Physics Today*, **39**(8), August, 26–33 (1968)

305 Newmark, P. and Garwin, L., 'Electron microscopy acclaimed', *Nature*, **323**(23), October, 663 (1986)

306 Sayles, R. S. and Thomas, T. R., 'Surface topography as a non-stationary random process', *Nature*, **271** 431–434 (1978)

307 BS 1134: 1972 Assessment of surface texture, Part 2: General information and guidance, p. 12

308 Thomas, T. R. and Sayles, R. S., 'Some problems in the tribology of rough surfaces', *Tribology Int.*, **11**, 163–168 (1978)

309 Nayak, P. R., 'Random process model of rough surfaces', *ASME, J. Lub. Tech.*, **93**(F), 398–407 (1971)

310 Whitehouse, D. J. and Archard, J. F., 'The properties of random surfaces of significance in their contact', *Proc. Roy. Soc. (London)*, **A316**, 97–121 (1970)

311 Thomas, T. R. and Charlton, G., 'Variation of roughness parameters on some typical manufactured surfaces', *Precis. Eng.*, **3**, 91–96 (1981)

312 Whitehouse, D. J., 'The digital measurement of peak parameters on surface profiles', *J. Mech. Eng. Sci.*, **20**(4), 221–227 (1978)

313 BS 1134:1972. Assessment of surface texture, Part 1, p. 10

314 American National Standard, Surface texture-surface roughness waviness and lay, ANSI document B 46.1 (1978)

315 Radhakrishnan, V., 'Analysis of some of the reference lines used for measuring surface roughness', *Proc. I. Mech. E.*, **187**, 575–582

316 Thomas, T. R., 'Characterisation of surface roughness', *Precis. Eng.*, **3**(2), 97–104

317 Thomas, T. R., *Rough Surfaces*, Longman, New York (1982), p. 92.

318 Leaver, R. H., Sayles, R. S. and Thomas, T. R., 'Mixed lubrication and surface topography of rolling contacts', *I. Mech. E.*, **188**, 461–469 (1974)

319 Thomas, T. R., *Rough Surfaces*, Longman, New York (1982), p. 94

320 Bikermann, J. J., *Physical surfaces*, Vol. 20, *Physical chemistry*, Academic Press, London and New York (1970)

321 Brecker, J. N., Fromson, R. E. and Shum, L. Y., 'A capacitance based surface texture measuring system', *Ann. CIRP*, **25**(1), 375–377 (1977)

321a Thomas, T. R., 'Characterisation of surface roughness', *Precis. Eng.* **3** (2), 97–104 (1981)

322 King, T. G., Watson, W. and Stout, K. J., 'Modelling the micro geometry of lubricated wear', *Proc. 4th Leeds-Lyon Symp.*, Paper XII (iv) (1978), pp. 333–343

323 Stout, K. J. and Davis, E. J., 'Surface topography of cylinder bores – the relationship between manufacture, characterisation and function', *Wear*, **95**, 111–123 (1984)

324 Abbott, E. J. and Firestone, F. A., 'Specifying surface quality', *Mech. Eng.*, **55**, 569–572 (1933)

325 Finkin, E. F., 'The bearing area of surfaces', *Trans. ASME, J. Lub. Tech.*, **90F**, 329–330 (1968)

326 Spragg, R. C. and Whitehouse, D. J., 'A new unified approach to surface metrology', *Proc. I. Mech. E.*, **185**(47/71), 697–707 (1970–1971)

326a Dagnall, H. *Exploring Surface Texture*, Rank Taylor Hobson (1980), p. 62

327 Dagnall, H., *Exploring Surface Texture*, Rank Taylor Hobson, Leicester (1980), p. 57

328 Russel, A., 'Peak counter for surface texture measurement', *Proc. I. Mech. E.*, **182**(3K), 327–329 (1967–1968)

328a Dagnall, H. *Exploring Surface Texture*, Rank Taylor Hobson (1980), p. 60

329 Woo, K. L. and Thomas, T. R., 'Roughness, friction and wear: The effect of contact planform', *Wear*, **57**, 357–363 (1979)

330 Tanner, L. H., 'A comparison between talysurf 10 and optical measurements of roughness and surface slope', *Wear*, **57**, 81–91 (1979)

331 Briggs, G. A. D. and Briscoe, B. J., 'Surface roughness and the friction and adhesion of elastomers', *Wear*, **57**, 269–280 (1979)

332 King, T. G. and Spedding, T. A., 'On the relationships between surface profile height parameters', *Wear*, **83**, 91–108 (1982)

333 Ismailov, V. V. and Kourova, M. S., 'Correlation between surface topography and profile statistical parameters', *Wear*, **59**, 401–421 (1980)

334 Bendat, J. S. and Piersol, A. G., *Random Data: Analysis and measurement procedures*, Wiley-Interscience, New York (1971)

335 Thomas, T. R., 'Recent advances in the measurement and analysis of surface microgeometry', *Wear*, **33**, 205–233 (1975)

336 Penklenik, J., 'New development in surface characterisation and measurements by means of random process analysis', *Proc. I. Mech. E.*, **182**(3K), 108–125 (1967–1968)

337 Onions, R. A. and Archard, J. F., 'The contact of surfaces having a random structure', *J. Phys. D.*, **6**, 289–304 (1973)

338 King, T. G. and Stout, K. J., 'Some topographic features of the wear process', *Ann. CIRP*, **25**(1), 351–356 (1977)

339 Smith, E. H. and Walmsley, W. M., 'Walsh functions and their use in the assessment of surface texture', *Wear*, **57**, 157–166 (1979)

340 Sayles, R. S. and Thomas, T. R., 'The spatial representation of surface roughness by means of the structure function; a practical alternative to correlation', *Wear*, **42**, 263–276 (1977)

341 Cooley, J. W., Lewis, P. A. W. and Welch, P. D., 'The Fast Fourier Transform and its applications', *Trans. IEEE (Education)*, **12**(1), 27–34 (1969)

342 Longuet-Higgins, M. S., 'The statistical analysis of a random moving surface', *Phil. Trans. Roy. Soc.*, **A249**, 321–387 (1957)

343 Bush, A. W., Gibson, R. D. and Keogh, G. P., 'Strongly anisotropic rough surfaces', ASME paper, 78-Lub-16 (1978), pp. 1–6

344 Nayak, P. R., 'Random process model of rough surfaces in plastic contact', *Wear*, **26**, 305–333 (1973)

345 Raja, J. and Radhakrishnan, V., 'Filtering of surface profiles using fast Fourier transform', *Int. J. Mach. Tool Des. and Res.*, **19**(3), 133–141 (1979)

346 Sherrington, I., 'Fourier models of the surface topography of engineering components', *Surface Topography*, **1**(1), March, 11–25 (1988)

347 Raja, J. and Whitehouse, D. J., 'An investigation into the possibility of using surface profiles for machine tool surveillance', *Int. J. Prod. Res.*, **22**(3), 453–466 (1984)

348 Yolles, M. I., Smith, E. H. and Walmsley, W. M., 'Walsh theory and spectral analysis of engineering surfaces', *Wear*, **83**, 151–164 (1982)

349 Box, G. E. P. and Jenkins, G. M., *Time series analysis: Forecasting and Control*, Holden-Day, San Francisco (1970)

350 Staufert, G., 'Description of roughness profiles by separating the random and periodic components', *Wear*, **57**, 185–194 (1979)

351 DeVries, W. R., 'Autoregressive time series models for surface profiles characterisation', *Ann. CIRP*, **28**(1), 437–440 (1979)

352 Watson, W., King, T. G., Spedding, T. A. and Stout, K. J., 'The machined surface – time series modelling', *Wear*, **57**, 195–205 (1979)

353 Staufert, G., 'Characterisation of random roughness profiles – A comparison of AR modelling technique and profile description by means of commonly used parameters', *Ann. CIRP*, **28**(1), 431–435 (1979)

354 Watson, W. and Spedding, T. A., 'The time series modelling of non-Gaussian engineering processes', *Wear*, **83**, 215–231 (1982)

355 Hannan, E. J., *Time Series Analysis*, Methuen, London, (1960)

Further reading

Section 9.2

BP publications:
Industrial lubrication
Machine tools and metal cutting
Lubricants for heavy industry
Gear lubrication
Hydraulic fluids
Machine shop lubricants
Cutting fluids
Compressor lubrication
Greases
Temporary corrosion preventives
Degreasants
Storage and handling of lubricants
Aerosols
Health, safety and environmental data sheets

Further information on lubrication can be obtained from:

Booklets and leaflets published by most oil suppliers
Libraries of Institute of Plant Engineers
       Institution of Production Engineers
       Institute of Petroleum
       Institution of Electrical Engineers
       Institution of Mechanical Engineers
Literature published by additive companies
Literature published by the American Society of Lubrication Engineers
Libraries of universities
National Centre of Tribology

With such a wide and important subject it is not possible to provide a full list but the above will indicate some initial contact points. The author hopes that no offence is caused by any omission.

Section 9.5

Adams, H. J., 'Borate – A new generation EP gear lubricant', *Lubrication Engineering*, **33**, 5, 241–246 (19??)

Auswahl von Schmierstoffen für Zahnradgetriebe Schmieröle, DIN 51 509, Teil 1

Bartz, W. J., 'Selection of lubricating oils for gears – an introduction to Standard DIN 51 509, *I. Mech. E. Tribology Group Convention* (1976)

Bartz, W. J. (NLGI spokesman). 'Interrelations between molybdenum disulfide and oil soluble additives', December (1989)

Bathgate, J. and Lucas, A. G., 'Lubricated-for-life industrial gear units', *Proc. I. Mech. E.*, **184**, Pt 30 (1969–1970)

Dawson, P. H., 'Further experiments on the effect of metallic contact on the pitting of lubricated rolling surfaces', *Proc. I. Mech. E.*, **180**, Pt 3B (1965–1966)

Fowle, T. I. 'Gear lubrication: relating theory to practice', ASLE 75AM-5A-2 (1975)

Gear Lubrication, British Gear Manufacturers Association, Technical Memorandum No. 11

Lauktotka, E. M., 'Lubrication of gears with synthetic lubricants', *J. Syn. Lubrication*, **2**, No. 1 (1985)

Pacholke, P. J. (NLGI spokesman), 'Effects of stable molybdenum disulfide lubricant additives on worm gear efficiency', **1**, L111, No. 3

Shell International Petroleum Company Ltd, *The Lubrication of Industrial Gears*, London (1964)

Watson, H. J., *The lubrication of gears operating under unusual conditions*, May (1964)

Section 9.6

Harris, T. A., *Rolling Bearing Analysis*, 2nd edn, Wiley-Interscience, New York (1984)

International Standard ISO 281/1, Rolling Bearings – Dynamic Load Ratings and Rating Life (1977–03–15)

Lundberg, G. and Palmgren, A., *Dynamic Capacity of Rolling Bearings*, Acta Polytechnica, Mech. Eng. Ser. 2, R.S.A.E.E. No. 4, 96 (1952)

Nisbet, T. S. and Mullett, G. W., *Rolling Bearings in Service*, Hutchinson Benham, London

Palmgren, A., *Ball and Roller Bearing Engineering*, 3rd edn, Burbank, Philadelphia (1959)

Section 9.9

Bryggman, U. and Soderberg, S., 'Contact conditions in fretting', *Wear*, **110**, 1–18 (1986)

Collins, J. A. and Torey, F. M., 'Fretting fatigue mechanisms and the effect of direction of fretting motion on fatigue strength', *J. Mat.*, **7**, 460–464 (1974)

Klaffke, D., 'Fretting wear of ceramic-steel; the importance of wear ranking criteria', *Wear*, **104**, 337 (1985)

Klint, R. V., 'Oscillating tangential forces on cylindrical specimens in contact at displacements within the regions of no gross slip', *ASLE Trans.*, **3**, 437–445 (1968)

10

Power units and transmission

James Carvill
(Section 10.1)

Roy D. Cullum
(Section 10.2)

Contents

10.1 Power units 10/3
    10.1.1 Heat engines 10/3
    10.1.2 Steam turbines and steam plant 10/9
    10.1.3 Gas turbines 10/13

10.2 Power transmissions 10/15
    10.2.1 Belt drives 10/15
    10.2.2 Gears and gearing 10/26
    10.2.3 Chains 10/35
    10.2.4 Shaft couplings 10/36
    10.2.5 Clutches, freewheels and brakes 10/43

Further reading 10/49

10.1 Power units

10.1.1 Heat engines

10.1.1.1 Heat engine cycles

Carnot cycle

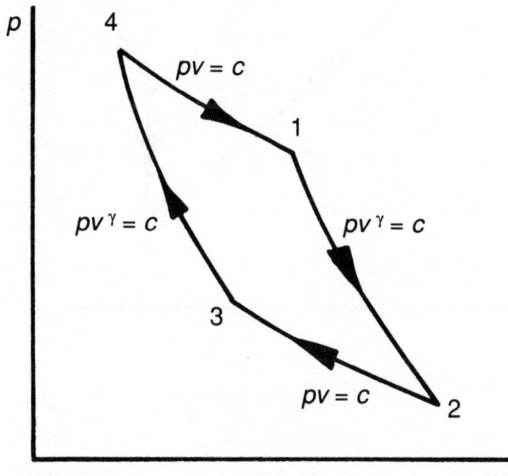

Figure 10.1 The Carnot cycle

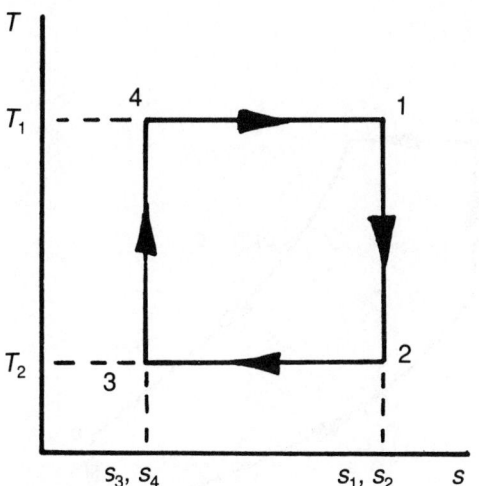

Figure 10.2

Heat supplied per kg $Q = T_1(s_1 - s_4)$

The idea gas cycle is the Carnot cycle, and, in practice, only about half of the Carnot cycle efficiency is realized between the same temperature limits (see Figures 10.1 and 10.2).

Constant-pressure cycle

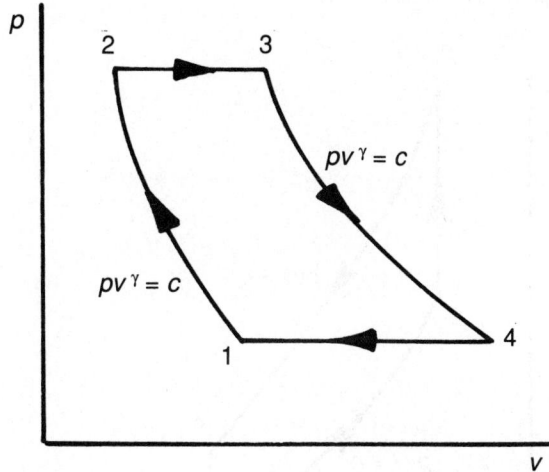

Figure 10.3 Constant-pressure cycle

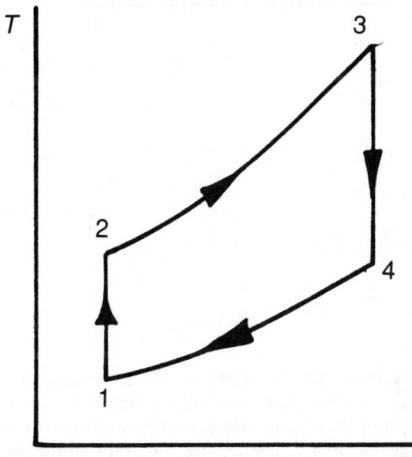

Figure 10.4

Efficiency $\eta = 1 - \dfrac{T_2}{T_1}$

$(s_1 - s_4) = R \ln \dfrac{p_4}{p_2} - c_p \ln \dfrac{T_1}{T_2}$

Work done per kg $W = (T_1 - T_2)(s_1 - s_4)$

Work ratio $= \dfrac{W}{W_{\text{gross}}} = \dfrac{(s_1 - s_4)(T_1 - T_2)}{T_1(s_1 - s_4) + c_v(T_1 - T_2)}$

Efficiency $\eta = 1 - \dfrac{1}{r^{\frac{\gamma-1}{\gamma}}}$, where $r = \dfrac{p_2}{p_1}$

$W = c_p(T_3 - T_4) - c_p(T_2 - T_1)$,

Work ratio $= 1 - \dfrac{T_1}{T_3} \, r^{\frac{\gamma-1}{\gamma}}$

$$Q = c_p(T_3 - T_2)$$

In this cycle heat is supplied and rejected at constant pressure; expansion and compression are assumed to take place at constant entropy. The cycle was once known as the Joule or Brayton cycle and was used for hot-air engines. It is now the ideal cycle for the closed gas-turbine unit (see Figures 10.3 and 10.4).

Otto (constant-volume) cycle

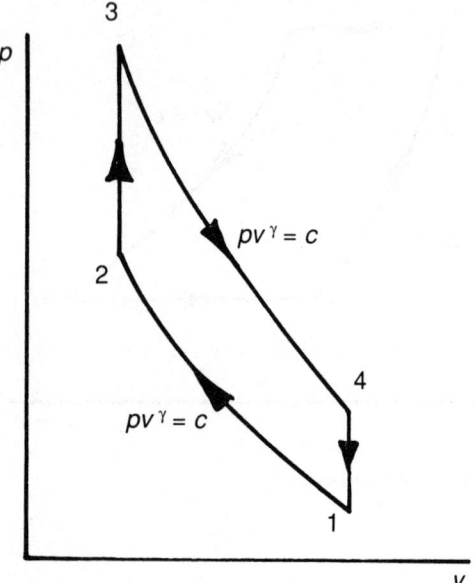

Figure 10.5 Otto cycle

Efficiency $\eta = 1 - \dfrac{1}{r^{\gamma-1}}$

where $r = \dfrac{v_1}{v_2}$ and $\dfrac{T_2}{T_1} = \dfrac{T_3}{T_4} = r^{\gamma-1}$

$W = c_v(T_3 - T_2 - T_4 + T_1)$

$Q = c_v(T_3 - T_2)$

This is the basic cycle for the petrol engine, the gas engine and the high-speed oil engine. Heat is supplied and rejected at constant volume, and expansion and compression take place isentropically. The thermal efficiency depends only upon the compression ratio (see Figure 10.5).

Diesel (constant-pressure) cycle

Efficiency $\eta = 1 - \dfrac{(\beta^{\gamma} - 1)}{(\beta - 1)\gamma r^{\gamma-1}}$

where $r = \dfrac{v_1}{v_2}$ and $\beta = \dfrac{v_3}{v_2}$

$W = c_p(T_3 - T_2) - c_v(T_4 - T_1)$

$Q = c_p(T_3 - T_2)$

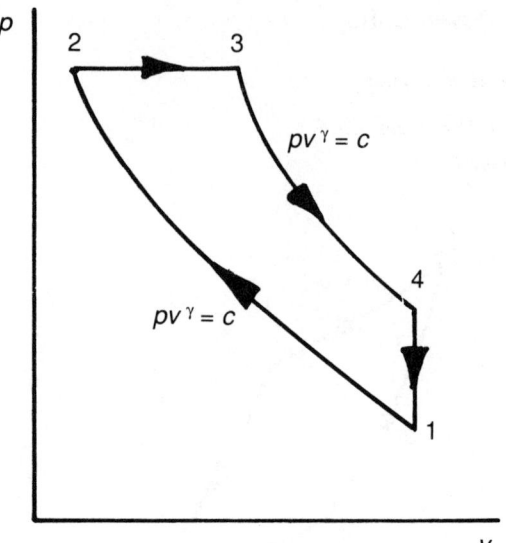

Figure 10.6 Diesel cycle

Although this is called the Diesel cycle, practical diesel engines do not follow it very closely. In this case heat is added at constant pressure, otherwise the cycle is the same as the Otto cycle (see Figure 10.6).

Dual-combustion cycle

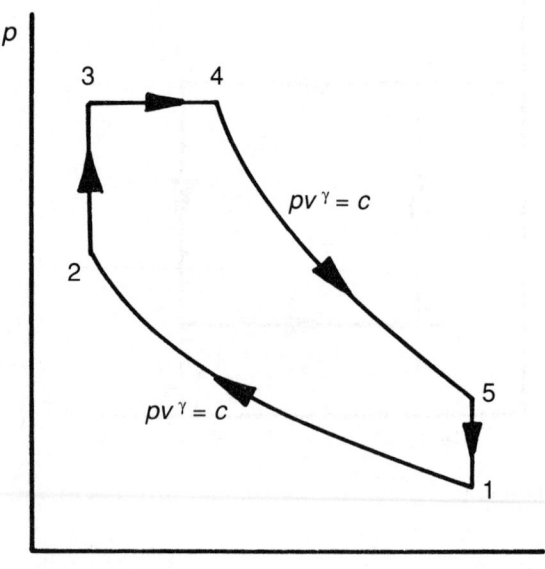

Figure 10.7 Dual-combustion cycle

Efficiency $\eta = 1 - \dfrac{(k\,\beta^{\gamma} - 1)}{[(k - 1) + (\beta - 1)\gamma k]r^{\gamma-1}}$

where $r = \dfrac{v_1}{v_2}$, $k = \dfrac{p_3}{p_2}$, $\beta = \dfrac{v_4}{v_3}$

$W = c_v(T_3 - T_2) + c_p(T_4 - T_3) - c_v(T_5 - T_1)$

$Q = c_v(T_3 - T_2) + c_p(T_4 - T_3)$

Modern diesel engines follow a cycle similar to this ideal one. In this case combustion takes place partly at constant volume and partly at constant pressure (see Figure 10.7).

Practical engine cycles   In actual engines the working substance is air only in the induction and compression strokes. During expansion and exhaust the working substance consists of the products of combustion with different properties to air. In addition, the wide variation in temperature and pressure results in variations in the thermal properties. Another factor is 'dissociation', which results in a lower maximum temperature than is assumed by elementary treatment of the combustion process.

10.1.1.2   Reciprocating spark-ignition internal-combustion engines

Four-stroke engine   The charge of air and fuel is induced into the engine cylinder as the piston moves from top-dead-centre (TDC) to bottom-dead-centre (BDC). The charge is then compressed and ignited by the sparking plug before TDC, producing a high pressure and temperature at about TDC. The gas expands and work is produced as the piston moves to BDC. A little before BDC the exhaust valve opens and the gases exhaust. The process is completed during the next stroke. A typical pv diagram is shown in Figure 10.8 and formulae are given for power, mean effective pressure, efficiency and specific fuel consumption (see also Figures 10.9–10.11).

Pressure–volume (pv) diagram
A = area of power loop (mm^2) L_d = length of diagram (mm)

B = area of pumping loop (mm^2) K = indicator constant (Nm^{-2} mm^{-1})

Indicated mean effective pressure
$$p_i = (A - B)\dfrac{K}{L_d} \ N\,m^{-2}$$

Indicated power
N = rev. s^{-1}
n = number of cylinders
A_p = piston area (m^2)
L = stroke (m)

Indicated power $P_i = p_i A_p L N \dfrac{n}{2}$ watts

Torque
$T = F R$ (Nm) where F = force on brake arm (N)
  R = brake radius (m)

Brake power
$P_b = 2\pi N T$, watts

Friction power
$P_f = P_i - P_b$

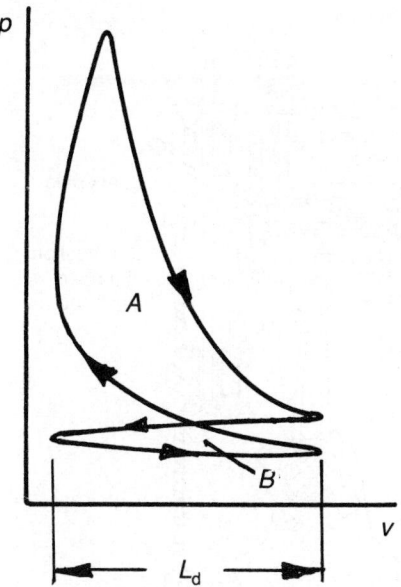

Figure 10.8 PV diagram: four-stroke engine

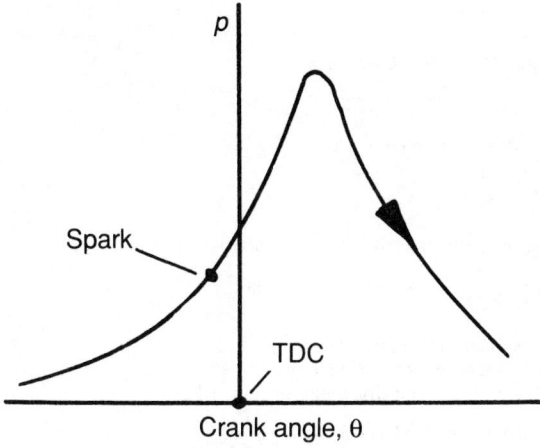

Figure 10.9 Pressure crank angle diagram

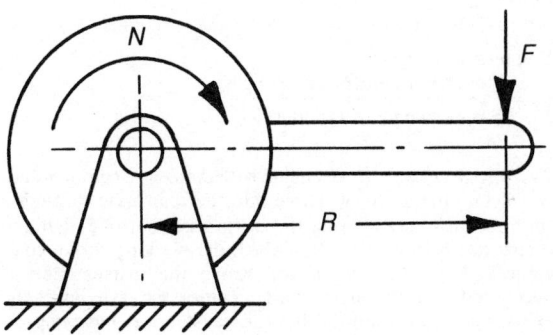

Figure 10.10 Brake

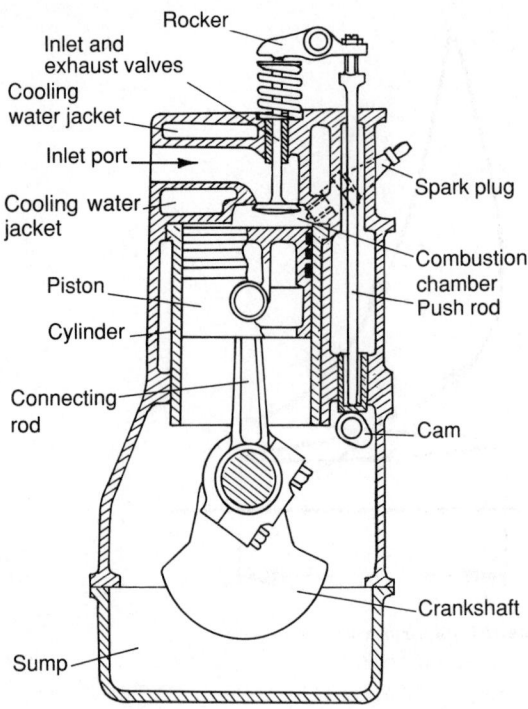

Figure 10.11 Four-stroke petrol engine

Mechanical efficiency
$$\eta_m = \frac{P_b}{P_i}$$

Brake mean effective pressure (bmep)
$$p_b = \frac{4\pi T}{A_p \, L \, n} = \text{constant} \times T \ N \ m^{-2}$$

Brake thermal efficiency
$\dot{m}$ = mass flow rate of fuel kg.s^{-1}
LCV = lower calorific value of fuel, J.kg^{-1}

Brake thermal efficiency $\eta_b = \dfrac{P_b}{\dot{m} \, LCV}$

Specific fuel consumption
$$\text{SFC} = \frac{\dot{m}}{P_b}, \ \text{kg.s}^{-1}.\text{W}^{-1}$$

Volumetric efficiency
$$\eta_v = \frac{\text{Volume of induced air at NTP}}{\text{Swept volume of cylinder}}$$

Two-stroke engine   In an engine with crankcase compression, the piston draws a new charge into the crankcase through a spring-loaded valve during the compression stroke. Ignition occurs just before TDC after which the working stroke commences. Near the end of the stroke the exhaust port is uncovered and the next charge enters the cylinder. The exhaust port closes shortly after the transfer port and compression begins. The piston is shaped to minimize mixing of the new charge with the exhaust.

Compression-ignition engines Both four- and two-stroke engines may have compression ignition instead of spark ignition. The air is compressed to a high pressure and temperature and the fuel injected. The high air temperature causes combustion (see *Figures 10.12–10.14*).

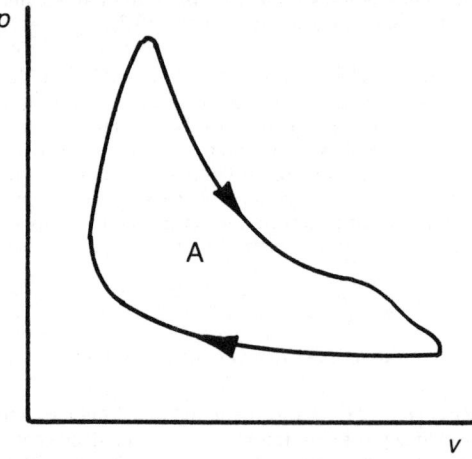

Figure 10.12 Cylinder diagram

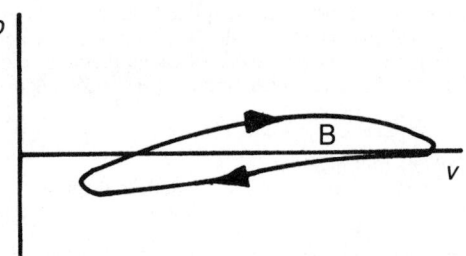

Figure 10.13 Crankcase diagram

PV diagram for two-stroke engine
A = area of power loop
B = area of pumping loop

Indicated mean effective pressure (imep)
$$p_i = (A - B) \, \frac{K}{L_d}, \ K = \text{indicator constant}$$

Indicated power
$P_i = p_i \, A_p \, L \, N \, n$

Brake mean effective pressure (bmep)
$$p_b = \frac{2\pi T}{A_p \, L \, n}$$

Other quantities are as for four-stroke engines.

Timing diagrams
Four-stroke engine (Figure 10.15)
IO = inlet valve opens
IC = inlet valve closes

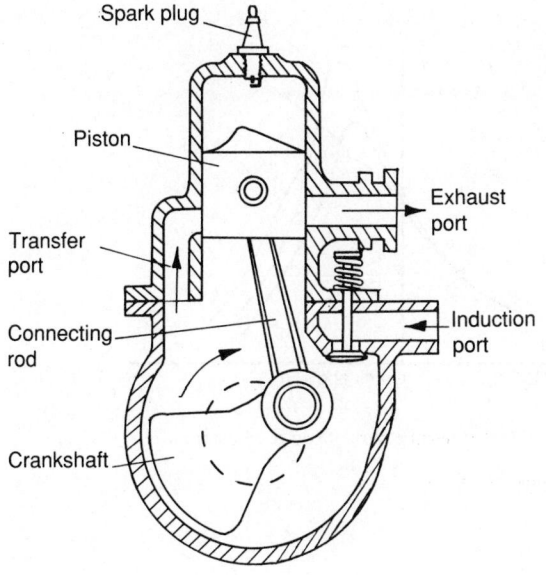

Figure 10.14 Two-stroke petrol engine

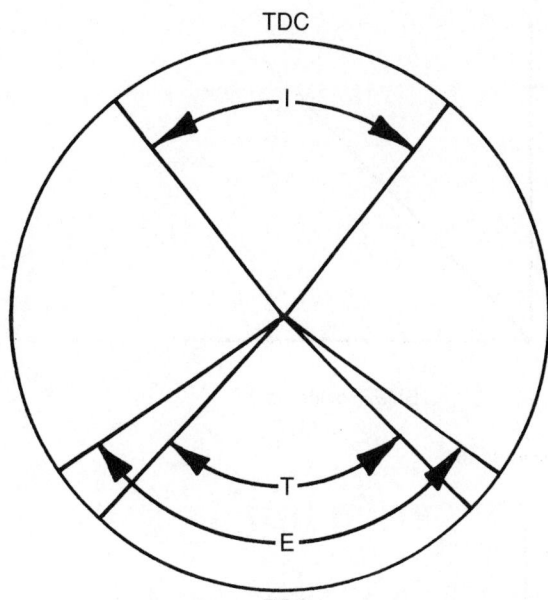

Figure 10.16 Timing diagram: two-stroke engine

E = exhaust angle (approx. 120°)
TR = transfer angle (approx. 100°)

Performance curves for internal-combustion engines Typical curves are shown in Figures 10.17–10.23 for mechanical efficiency versus brake power, bmep versus torque, and volumetric efficiency versus speed. The effect of mixture strength on the *pv* and *p*θ diagrams is shown and curves of power and mep against speed given. The curve of specific fuel consumption versus brake power, known as the Consumption Loop, shows the effect of mixture strength on fuel consumption.

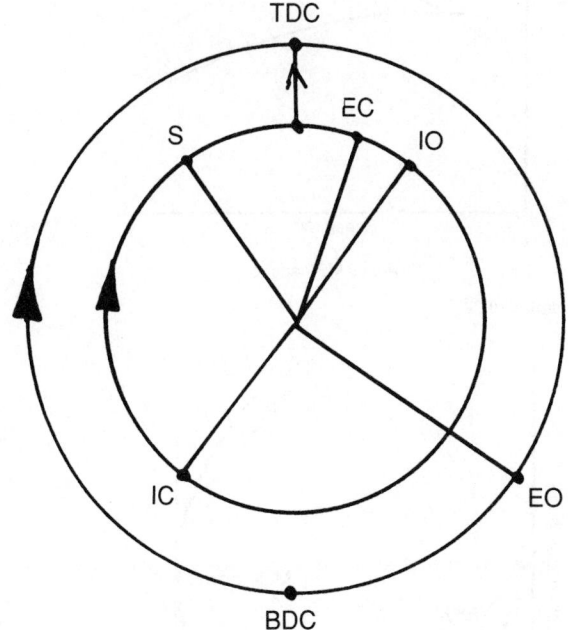

Figure 10.15 Timing diagram: four-stroke engine

S = spark occurs
EO = exhaust valve opens
EC = exhaust valve closes

Two-stroke engine (Figure 10.16)
I = inlet angle (approx. 80°)

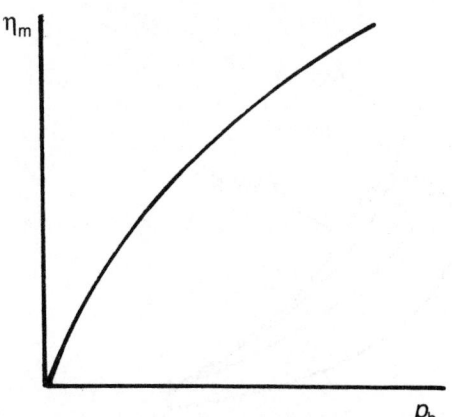

Mechanical efficiency vs brake power

Figure 10.17 Performance curve for internal-combustion engines (see also Figures 10.18–10.23)

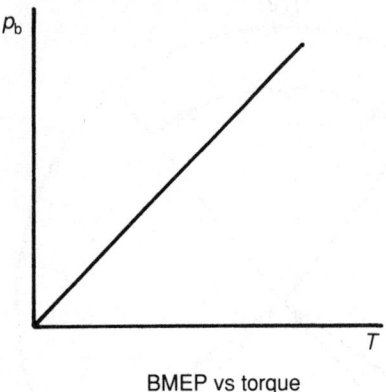

BMEP vs torque

Figure 10.18

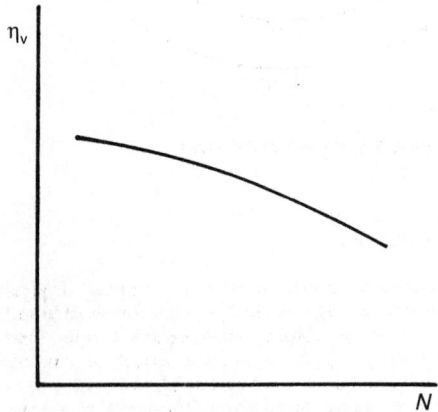

Volumetric efficiency vs speed

Figure 10.19

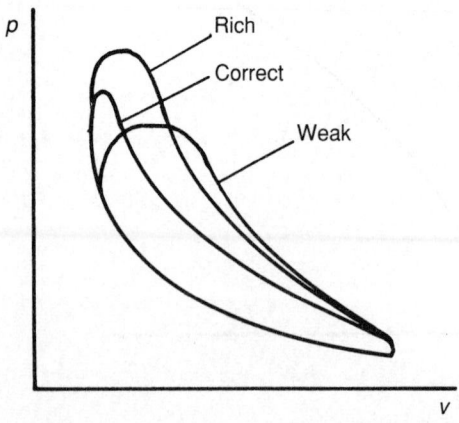

Effect of mixture strength on $p - v$ diagram

Figure 10.20

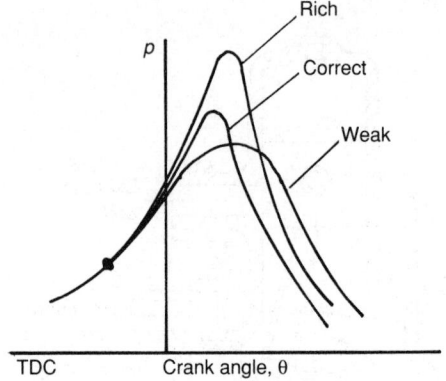

Effect of mixture strength on $p - \theta$ diagram

Figure 10.21

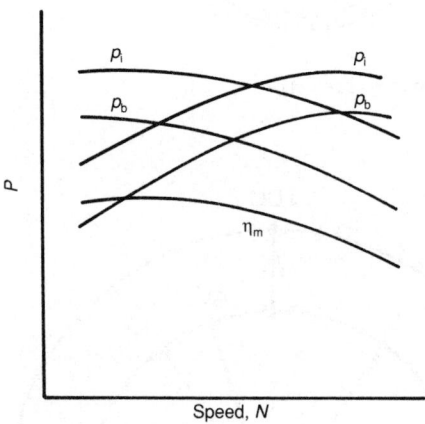

Power vs MEP speed

Figure 10.22

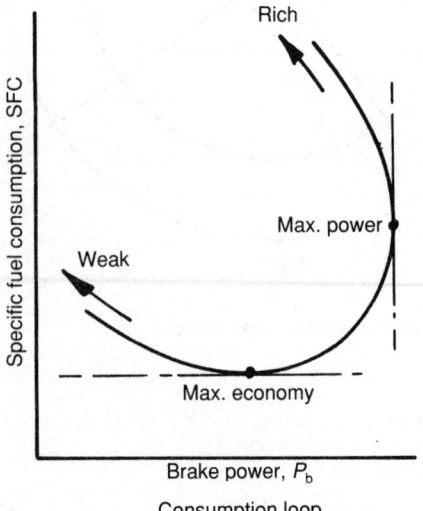

Consumption loop

Figure 10.23

10.1.2 Steam turbines and steam plant

10.1.2.1 Steam plant

The simplest steam cycle of practical value is the Rankine cycle with dry saturated steam supplied by a boiler to a power unit (e.g. a turbine) which exhausts to a condenser where the condensed steam is pumped back into the boiler. A higher efficiency is obtained if the steam is initially superheated and also a fall in specific steam consumption which means smaller plant. If the steam is 'reheated' and passed through a second turbine the final dryness fraction is increased with beneficial effects (e.g. reduced erosion of turbine blades due to water droplets). In addition, there is a further reduction in specific steam consumption.

In the 'regenerative cycle' the efficiency is improved by bleeding off a proportion of the steam at an intermediate pressure and mixing it with feedwater pumped to the same pressure in a 'feed heater'. Several feed heaters may be used but these are of the 'closed' variety to avoid the necessity for expensive pumps.

Rankine cycle – dry saturated steam at turbine inlet (Figures 10.26–10.27). From the Ts diagram:

$$s_2 = s_1, \quad x_2 = \frac{(s_2 - s_{f2})}{s_{fg2}}$$

$$h_2 = h_{f2} + x_2 h_{fg2}$$

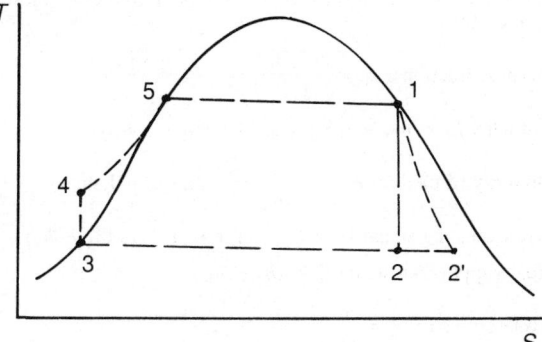

Figure 10.24 Rankine cycle

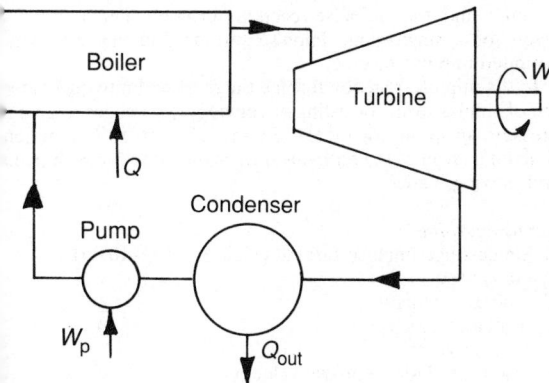

Figure 10.25 Rankine cycle – steam boiler plant

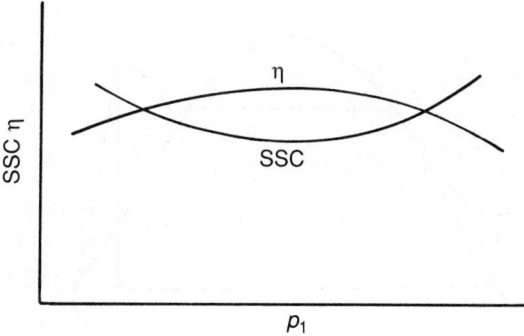

Figure 10.26 Rankine cycle – effect of inlet pressure

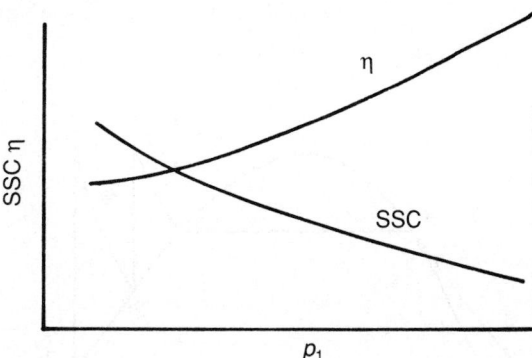

Figure 10.27 Rankine cycle – effect of superheat temperature

Work output $W = (h_1 - h_2)$

Heat supplied $Q = (h_1 - h_{f3})$

Cycle efficiency $\eta = W/Q$ (neglecting pump work)

Specific steam consumption (SSC) = $3600/W$ kg/kWh

Note: If the turbine isentropic efficiency η_i is allowed for:

$W = (h_1 - h_2)\eta_i$ and expansion is to 2'.

Rankine cycle – with superheat (Figure 10.28) h_1 is the enthalpy for superheated steam. The method is the same as for dry saturated steam. The graph shows the effect of superheat temperature on efficiency and specific steam consumption.

Rankine cycle with reheat (Figure 10.29) At 2 the steam is reheated to 6 and passed through a second turbine:

$W = (h_1 - h_2) + (h_6 - h_7)$
$Q = (h_1 - h_3) + (h_6 - h_2)$

p_6 is found from $T_6 = T_1$ (usually) and $s_{g2} = s_1$, from which h_6 is found. h_7 is found from $s_7 = s_6$.

Regenerative cycle (Figures 10.30 and 10.31)
Turbine inlet conditions p_1, t_1, h_1

Turbine outlet conditions p_2, t_2, h_2

Bleed steam conditions p_b, t_b, h_b

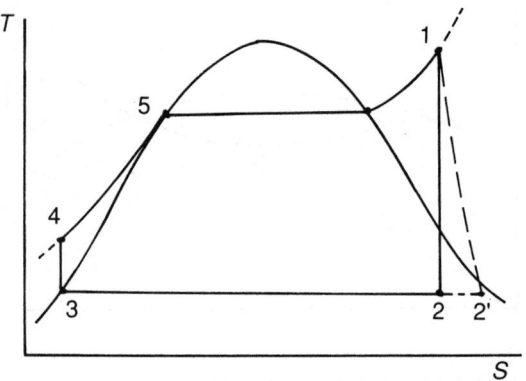

Figure 10.28 Rankine cycle with superheat

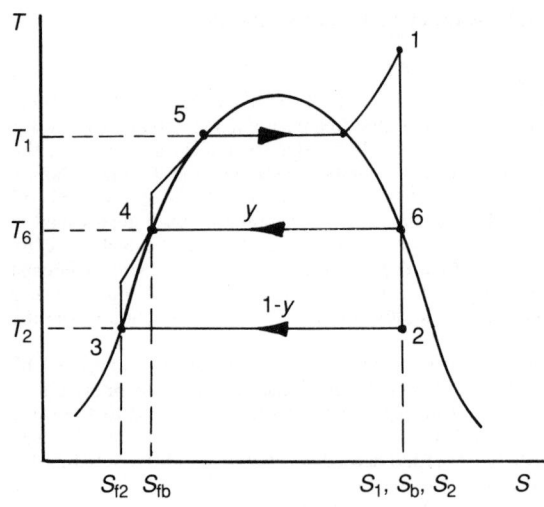

Figure 10.31 Regenerative cycle on Ts diagram

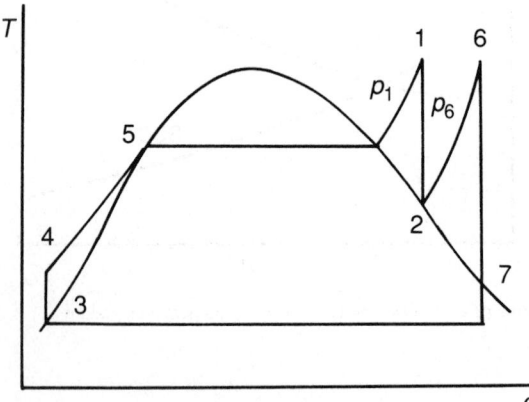

Figure 10.29 Rankine cycle with reheat

For maximum efficiency − $t_b = \dfrac{(t_1 + t_2)}{2}$

A bleed pressure p_b is selected to correspond to the saturation temperature t_b.

Dryness fractions: $x_b = \dfrac{s_1 - s_{fb}}{s_{fgb}}$, $x_2 = \dfrac{s_1 - s_{f2}}{s_{fg2}}$

Enthalpy $h_b = h_{fb} + x_b h_{fgb}$ and $h_2 = h_{f2} + x_2 h_{fg2}$

Quantity of bled steam $y = \dfrac{h_{fb} - h_{f2}}{h_b - h_2}$ kg/kg total steam

Work done/kg steam $W = (h_1 - h_b) + (1 - y)(h_b - h_2)$

Heat supplied/kg steam $Q = (h_1 - h_{fb})$

Cycle efficiency $\eta = \dfrac{W}{Q}$

Specific steam consumption (SSC) $= \dfrac{3600}{W}$ kg/kW h

10.1.2.2 Steam turbines

There are two main types of steam turbine: the 'impulse turbine' and the 'impulse-reaction turbine'. The theory is given for a single-stage impulse turbine and velocity compounded impulse turbine.

In the impulse-reaction turbine the fixed and moving blades are of similar form consisting of converging passages to give a pressure drop in each case. In the case of 50% reaction (Parson's turbine) the enthalpy drop is the same for both fixed and moving blades.

Impulse turbine
1. Single-stage impulse turbine (Figures 10.32–10.34)
Symbols used:
C = nozzle velocity
C_b = blade velocity
C_a = axial velocity
ρ = ratio of blade to nozzle velocity
β_1 = blade inlet angle
β_2 = blade outlet angle (in this case $\beta_1 = \beta_2$)

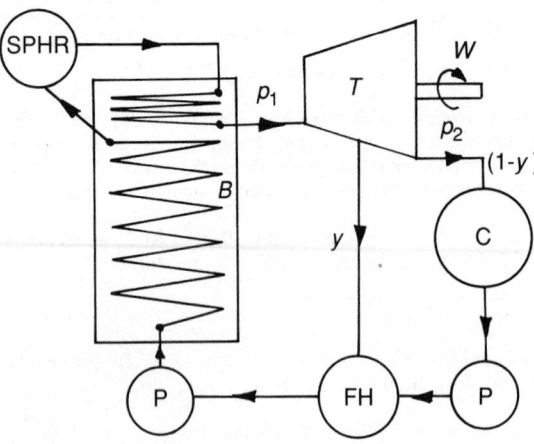

Figure 10.30 Steam plant for regenerative cycle. T = turbine, C = condenser, P = pump, FH = feed heater, B = boiler, SP = superheater

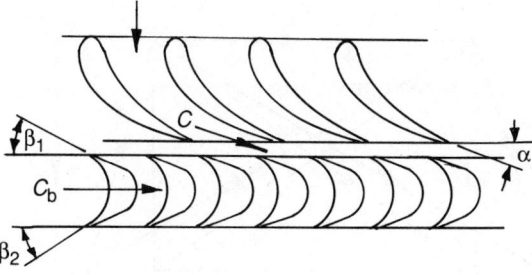

Figure 10.32 Impulse turbine

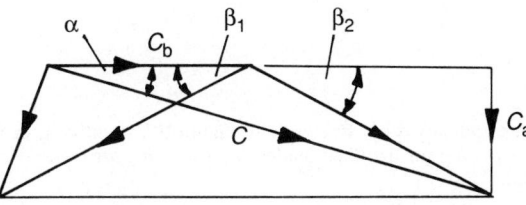

Figure 10.33 Single-stage impulse turbine velocity triangles

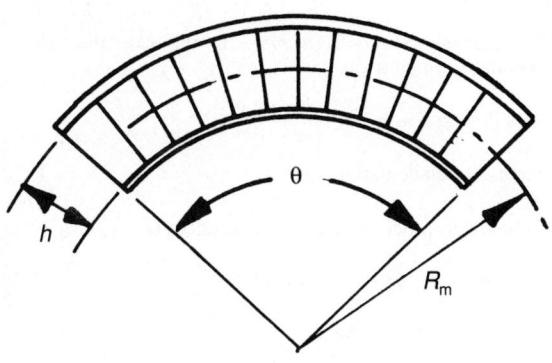

Figure 10.34 Nozzles

α = nozzle angle
$\dot{m}$ = mass flow rate of steam
k = blade friction coefficient = $\dfrac{\text{outlet relative velocity}}{\text{inlet relative velocity}}$
P = stage power
η = stage diagram efficiency
T_a = axial thrust on blades
R_m = mean radius of nozzle arc
v = specific volume of steam at nozzle outlet
θ = nozzle arc angle (degrees)
N = speed of rotation
h = nozzle height
A = nozzle area

Power $P = \dot{m}C^2\rho(\cos \alpha - \rho)(1 + k)$

where $\rho = \dfrac{C_b}{C}$ and $C_b = 2\pi R_m N$

Stage efficiency $\eta = 2\rho(\cos \alpha - \rho)(1 + k)$

Maximum efficiency $\eta_{max} = (1 + k)\cos^2 \dfrac{\alpha}{2}$, at

$\rho = \cos \alpha \dfrac{(1 + k)}{2}$

Axial thrust $T_a = \dot{m}C(1 - k)\sin \alpha$

Mass flow rate $\dot{m} = \dfrac{C_a A}{v}$

Nozzle area $A = \dfrac{\pi R_m \theta h}{180}$

2. **Pressure-compounded impulse turbine (Figure 10.35)**
The steam pressure is broken down in two or more stages.
Each stage may be analysed in the same manner as before.
3. **Velocity-compounded impulse turbine (Figures 10.36–10.38)**
One row of nozzles is followed by two or more rows of moving blades with intervening rows of fixed blades of the same type which alter the direction of flow.
Two-row wheel: Assume $\beta_1 = \beta_2$ and $k = 1$ and that all blades are symmetrical.

Maximum efficiency $\eta_{max} = \cos^2 \alpha$ at $\rho = \dfrac{\cos \alpha}{4}$

In which case the steam leaves the last row axially.

Impulse-reaction turbine (Figures 10.39 and 10.40) In this case there is 'full admission', i.e. $\theta = 360°$. Both nozzles and moving blades are similar in shape and have approximately the

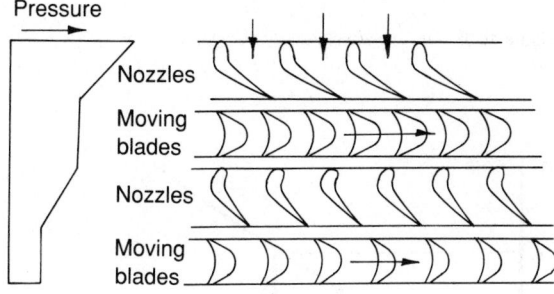

Figure 10.35 Pressure-compounded impulse turbine

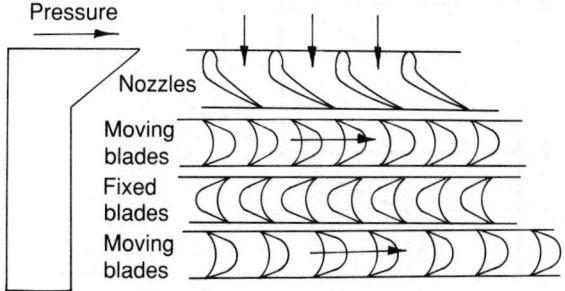

Figure 10.36 Velocity-compounded impulse turbine

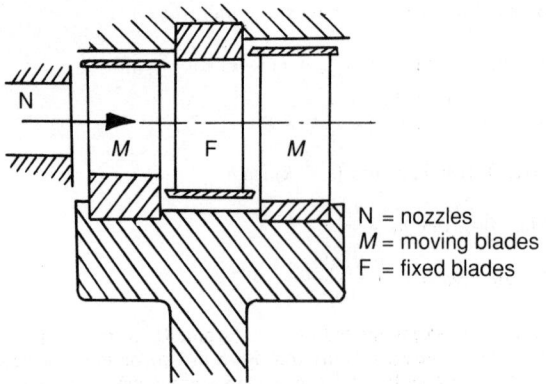

N = nozzles
M = moving blades
F = fixed blades

Figure 10.37 Two-row wheel, M = moving blade row, F = fixed blade row

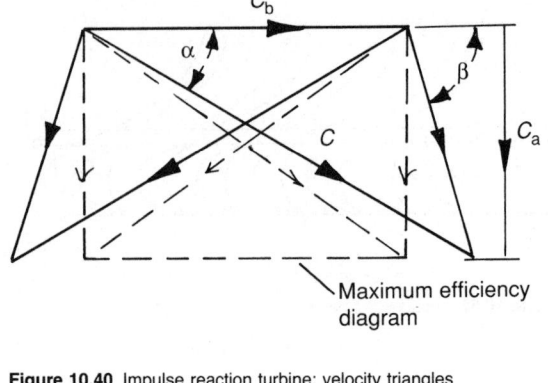

Maximum efficiency diagram

Figure 10.40 Impulse reaction turbine: velocity triangles

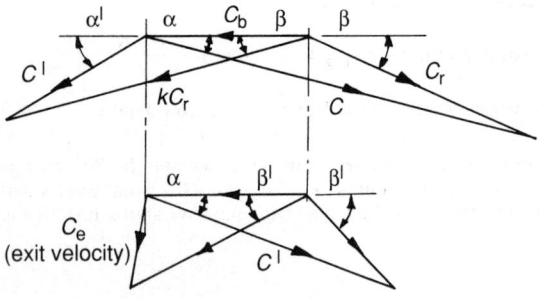

Figure 10.38 Two-row wheel velocity triangles

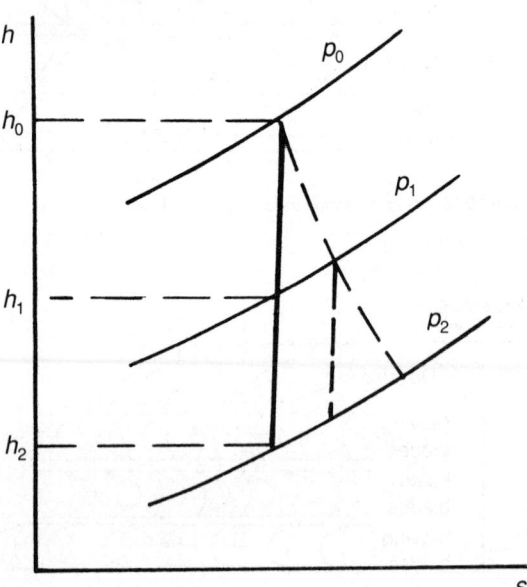

Figure 10.39 Stage heat drops

same enthalpy drop. Referring to Figure 10.39, enthalpy drop is $(h_0 - h_1)$ for the fixed blades and $(h_1 - h_2)$ for the moving blades.

Mass flow rate $\dot{m} = \dfrac{C_a A}{v}$

Area of flow $A = 2\pi R_m h$

Degree of reaction $R = \dfrac{(h_1 - h_2)}{(h_0 - h_1) + (h_1 - h_2)} = \dfrac{h_1 - h_2}{h_0 - h_2}$

50% reaction (Parson's) turbine In this case the velocity diagram is symmetrical:

Mass flow rate $\dot{m} = \dfrac{2\pi R_m h C \sin \alpha}{v}$

where α = blade outlet angle

Enthalpy drop per stage $\Delta h_s = C^2 \rho \,(2 \cos \alpha - \rho)$

where $\rho = \dfrac{C_b}{C}$ and $C_b = 2\pi R_m N$

Stage power $P_s = \dot{m}\,\Delta h_s$

Stage efficiency $\eta_s = \dfrac{2\,\rho(2 \cos \alpha - \rho)}{1 + \rho\,(2 \cos \alpha - \rho)}$

Maximum efficiency $\eta_{max} = \dfrac{2 \cos^2 \alpha}{(1 + \cos^2 \alpha)}$ when $\rho = \cos \alpha$

Reheat factor and overall efficiency (Figures 10.41 and 10.42) Referring to the 'condition curve' in Figure 10.42:

Δh_A = available stage enthalpy drop
Δh_I = isentropic stage enthalpy drop
Δh_{OA} = available overall enthalpy drop
Δh_{OI} = isentropic overall enthalpy drop

Stage efficiency $\eta_s = \dfrac{\Delta h_A}{\Delta h_I}$

Overall efficiency $\eta_o = \dfrac{\Delta h_{OA}}{\Delta h_{OI}}$

Reheat factor (RF) = $\dfrac{\eta_o}{\eta_s}$

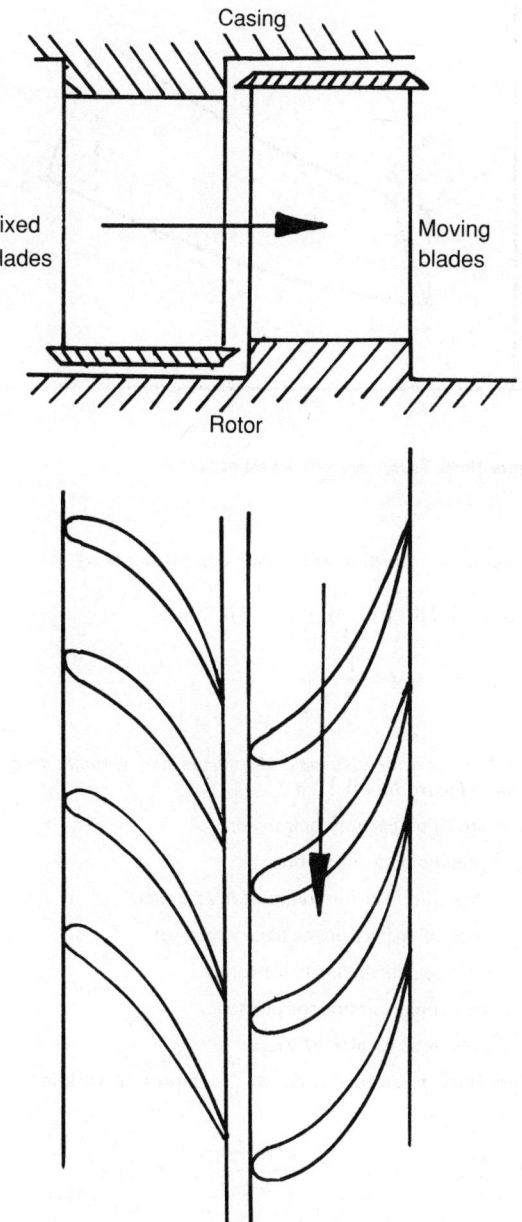

Figure 10.41 Reaction blading. F = fixed blades, M = moving blades

10.1.3 Gas turbines

The gas turbine unit operates basically on the constant-pressure cycle, particularly in the case of the 'closed cycle'. In the 'open cycle' air is drawn in from the atmosphere, compressed and supplied to a combustion chamber where fuel is burnt with a large amount of 'excess air'. The hot gases drive a turbine which drives the compressor and also provides useful work. The efficiency increases with compression ratio and the output power increases with both compression ratio and turbine inlet temperature.

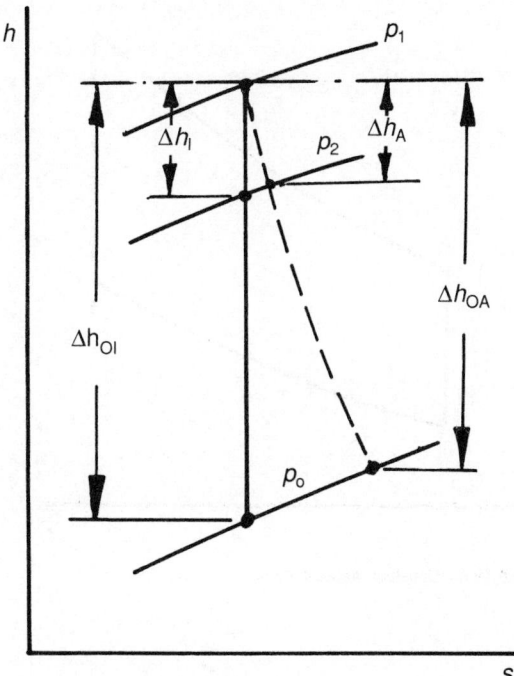

Figure 10.42 Reheat factor and overall efficiency

The effect of losses and variation in fluid properties is shown on the basic cycle. The efficiency of the basic cycle can be greatly increased by incorporating a heat exchanger between the compressor outlet and the combustion chamber inlet. It uses the exhaust gases from the turbine to preheat the incoming air.

10.1.3.1 Simple cycle (Figures 10.42–10.45)

Compression ratio $r = \dfrac{p_2}{p_1} = \dfrac{p_3}{p_4}$,

Let $c = r^{(\gamma-1)/\gamma} = \dfrac{T_2}{T_1} = \dfrac{T_3}{T_4}$, and $t = \dfrac{T_3}{T_1}$

Heat supplied $Q = c_p T_1(t - c)$ per kg of air

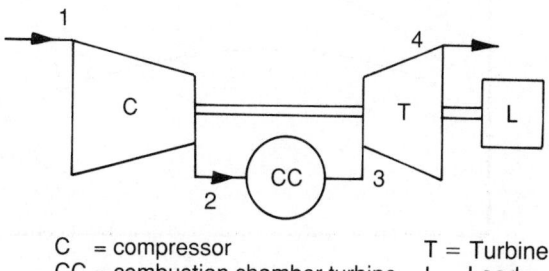

C   = compressor   T = Turbine
CC = combustion chamber turbine   L = Load

Figure 10.43 gas turbine – simple cycle

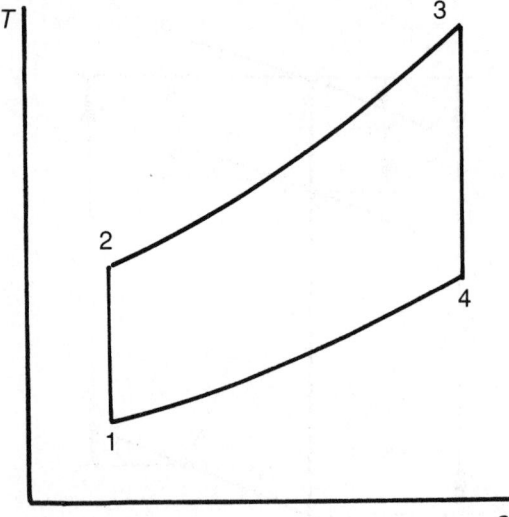

Figure 10.44 Constant-pressure cycle

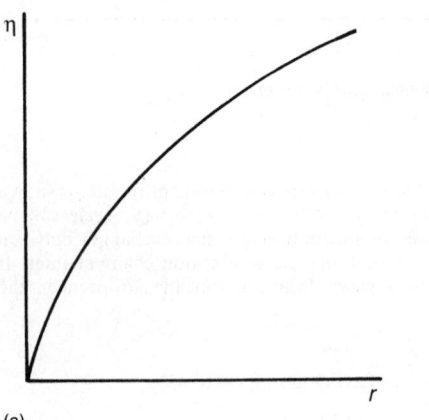

(a)

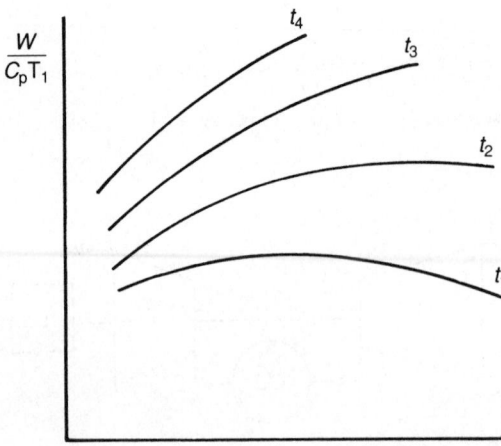

(b)

Figure 10.45 (a) Efficiency of simple gas turbine; (b) work function versus compression ratio characteristics

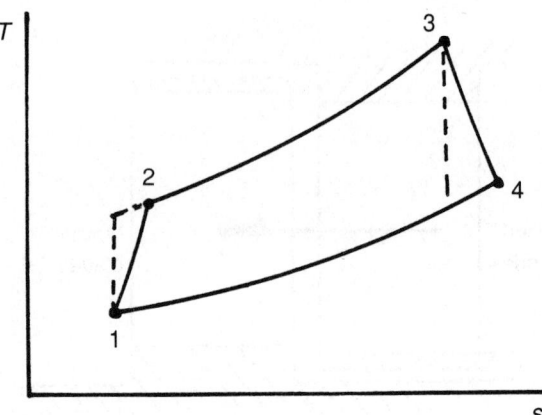

Figure 10.46 Ts diagram for a simple gas turbine

Work done = turbine work out − compressor work in

$$= W = c_p\, T_1 \left[t\left(1 - \frac{1}{c} \right) - (c - 1) \right]$$

Efficiency $\eta = 1 - \dfrac{1}{c}$

Simple cycle with isentropic efficiencies and variable specific heats   (*Figure 10.46*)   Let

$_c c_p$ = specific heat for compressor

$_t c_p$ = specific heat for turbine

$_{cc} c_p$ = specific heat for combustion chamber

γ_c = ratio of specific heats for compressor

γ_t = ratio of specific heats for turbine

η_c = isentropic compressor efficiency

η_t = isentropic turbine efficiency

Work done = turbine work out − compressor work in

or

$$W = {}_t c_p (T_3 - T_4)\eta_t - {}_c c_p \frac{(T_2 - T_1)}{\eta_c} \quad \text{per kg of air}$$

Heat supplied $Q = {}_{cc} c_p \left(T_3 - T_1 - \dfrac{(T_2 - T_1)}{\eta_c} \right)$ per kg of air

Work ratio $= \dfrac{\text{Net work out}}{\text{Gross work}} = \dfrac{W}{{}_t c_p (T_3 - T_4)\eta_t}$

Efficiency $\eta = \dfrac{W}{Q}$

Note that $r = \dfrac{p_2}{p_1}, \dfrac{T_2}{T_1} = r^{(\gamma_c - 1)/\gamma_c}, \dfrac{T_3}{T_4} = r^{(\gamma_t - 1)/\gamma_t}$

10.1.3.2 Simple cycle with heat exchanger   (Figures 10.47–10.49)

Heat supplied $Q = c_p T_1 \, t\left(1 - \dfrac{1}{c} \right)$

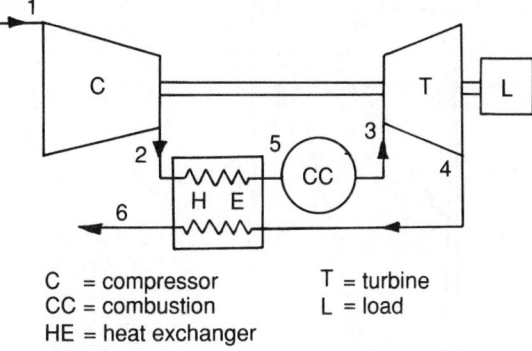

C = compressor   T = turbine
CC = combustion   L = load
HE = heat exchanger

Figure 10.47 Gas turbine cycle with heat exchanger

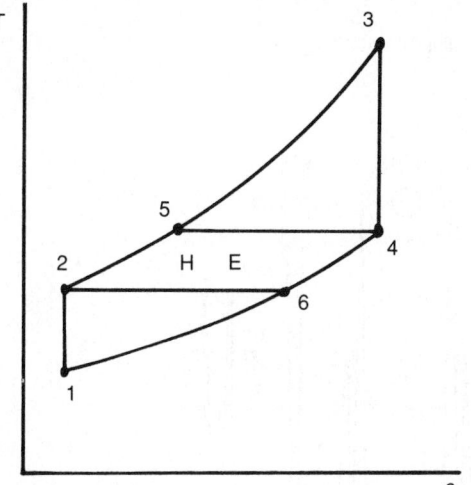

Figure 10.48 Ts diagram for gas turbine with heat exchanger

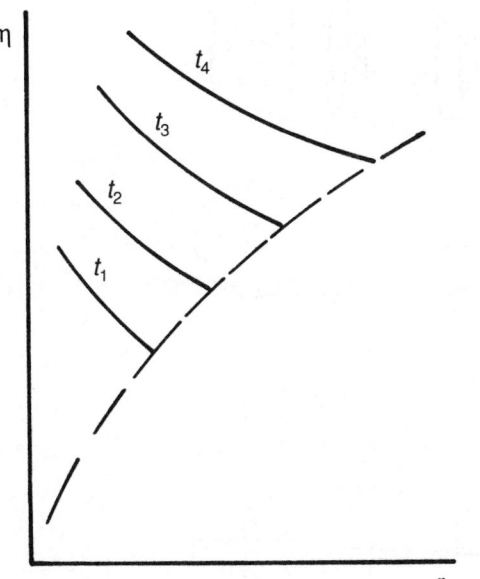

Figure 10.49 Efficiency curves for gas turbine with heat exchanger

Work done $W = c_p T_1 \left[t \left(1 - \dfrac{1}{c} \right) - (c - 1) \right]$

Efficiency $\eta = 1 - \dfrac{c}{t}$

10.2 Power transmissions

Good design demands that the power source is placed as close as possible to the driven unit and is one reason why the unit gearbox integral with electric motor is so popular. Unfortunately, in many instances, circumstances such as geometry, environmental considerations or lack of electricity supply dictate placing the prime mover at a distance from the driven unit and probably not in the same plane. Thus the designer must adopt a system of power transmission which may be shafts, belts, chains, gears, fluid power or all five. Inevitably, these will introduce complications and the successful system is one that is fundamentally simple, with the least number of parts that are able to transmit the required power. Modern machinery and equipment frequently also demand the installation of clutches, brakes and couplings, etc. and the selection of the right components for a specific task is usually a compromise. Nevertheless, engineers accept these limitations and successfully bring together many component parts to operate at a high degree of efficiency backed with established reliability.

10.2.1 Belt drives

The simplest form of power transmission is probably the belt drive. Progress in power belts has been considerable within the last decade and much of this effort has been concerned with finding improved materials more suited to specific types of transmissions. Belts can be round, flat, V-form or toothed and some designs incorporate features of more than one type. The original form of round belting for power transmission was the rope drive using cotton ropes and, while this form of drive is still used, the modern round belt is usually elastomeric or woven from synthetic fibres.

For light loads extensible round belts ranging from 3 mm (or smaller) to 12 mm diameter are available with elastomeric cores. These are useful as, within limits, they can be stretched to different lengths. They can also damp out vibration and, after fitting, tension can be maintained without the need for adjustment or idler pulleys. A further advantage is that they can be fitted to pulleys without the need to first reduce the centre distance. Round belts are usually intended to run in U-grooved pulleys which are essential for cotton-based ropes, but other materials will usually operate efficiently in V-grooves of 40–45°. With cotton ropes, the pulley grooves should be highly polished to minimize the wear of the cotton.

A modern hard-wearing material with a high coefficient of friction is polyurethane, while various other polymer formulations (sometimes with a mix of suitable elastomers) can provide a flexible, strong and long-lasting belt.

10.2.1.1 Flat belts

For light loads, and where small-diameter pulleys are in use, a flat belt made from woven fibres provides a highly flexible drive. However, historically, the flat belt was usually made of leather and enabled overhead shafts to be driven over considerable distances with power take-off at convenient points along their length. These drives were very tolerant of misalignments and were also highly flexible since they could be crossed

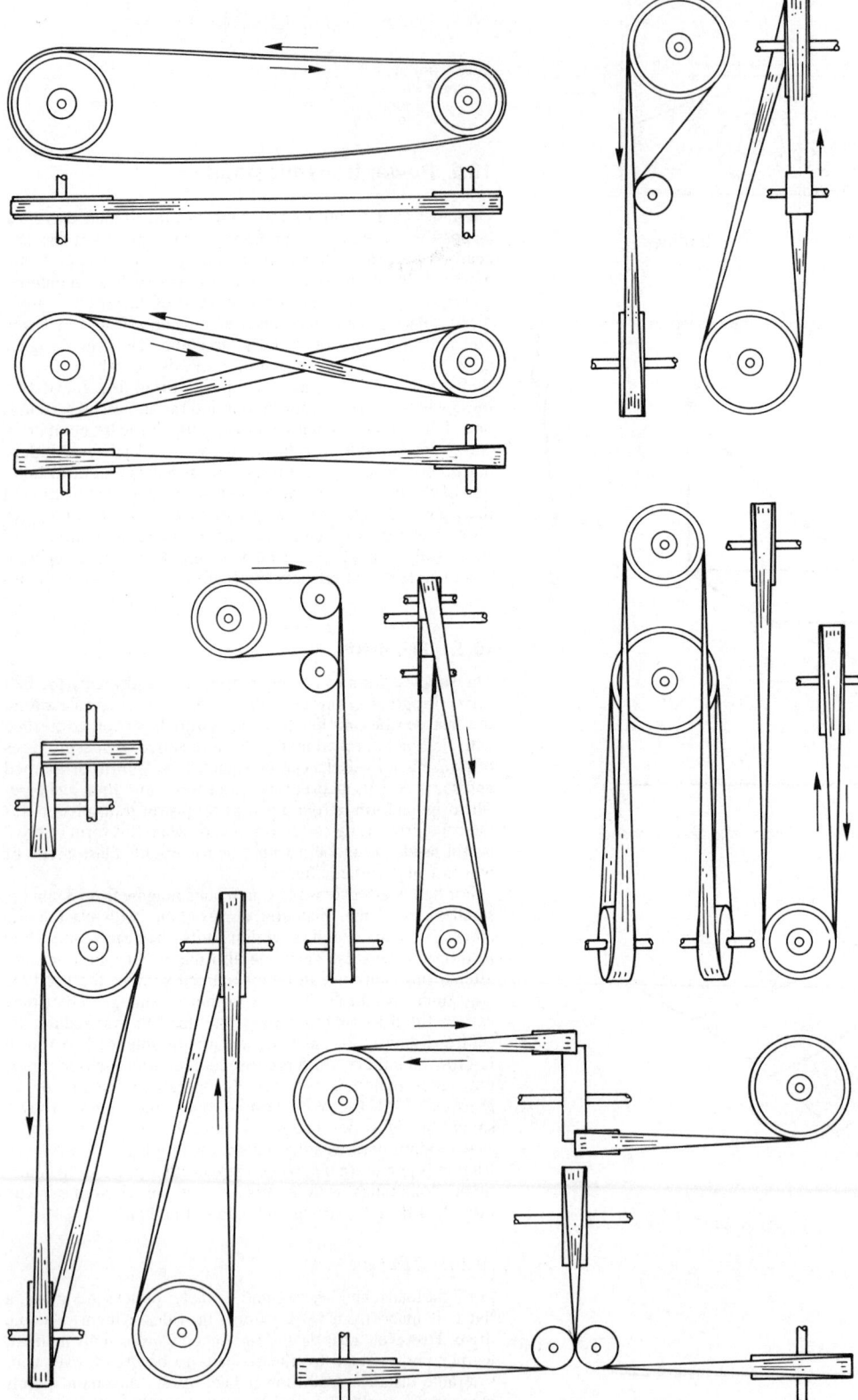

Figure 10.50 Various ways in which a belt drive may be used

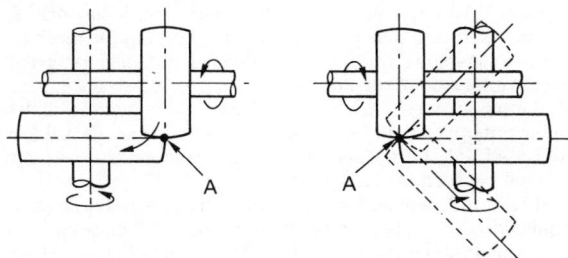

Figure 10.51 A general rule for aligning belt pulleys connecting non-parallel shafts

to give opposite direction of rotation and could be used to drive shafts which were not necessarily in the same plane (see Figure 10.50). In this connection, a general rule for aligning belt pulleys connecting shafts which are not parallel is for the centre of the face of the driven pulley to be aligned with the centre of that face of the driving pulley from which the belt leaves, as at A in Figure 10.51. The manner in which the belt passes over the pulleys is indicated by the arrows showing the direction of rotation. The driven pulley can be set at any angle in relation to the driving pulley provided it is turned about point A, as indicated by the dotted lines. The direction of rotation should not be reversed unless the relative positions of the pulleys are changed in accordance with the above rule.

The modern flat belt is usually of composite construction, power being transmitted by a high-tensile core often made from nylon or other high-tensile fibres (metal is also used) with the friction grip provided by leather or suitable polymer bonded to the core. Nevertheless, belts have several disadvantages, some of which arise from exposure to atmospheric and environmental changes that affect their stability, particularly those employing leather. Such belts often need dressing to avoid high slippage, while over-tensioning places heavy loads on the shaft bearings. Improvements in materials with the introduction of rubber and rubber/cotton composite belts

followed by plastics belts ameliorates some of these disadvantages. However, the death knell of large-scale flat belt drives was sounded with the introduction of unit construction for machines and equipment. Electric motors became smaller and more efficient and their price was low enough for them to be economically used on the basis of one motor to each machine, or indeed several motors to provide the power for different tasks in a single machine. There was still the need for drives, but these were considerably shorter while the power requirements were still rising. The V-Belt, in its different forms, became the generally accepted method of efficiently transmitting power.

Nevertheless, there are still some applications for flat belt drives and the following information will help the engineer to arrive at a belt size for specific applications. As in gears, the ratio of speed of driven to drive shaft is a matter of pulley diameter and is derived from the following formulae (see also Figure 10.52):

$$\text{Rev/min of driven pulley} = \frac{\text{Dia. of driver} \times \text{rev/min of driver}}{\text{Dia. of driven}}$$

The length of a belt is derived from the formula:

Open belt: $L = \pi(R + r) + 2\sqrt{[c^2 + (R - r)^2]}$ (approx.)

Equal pulleys: $L = \pi(R + r) + 2c$

Crossed belt: $L = \pi(R + r) + 2\sqrt{[c^2 + (R + r)^2]}$ (approx.)

where

R = radius of larger pulley,
r = radius of smaller pulley,
c = centre distance,
L = total length of belt.

Allowable power that can be transmitted by a particular belt is best derived in conjunction with manufacturers' literature, as it will be dependent on the choice of belt material.

Tension relationship between the slack side of the belt (which should be above the pulleys) is given by:

$$T_1/T_2 = 2.7^{\mu\theta}$$

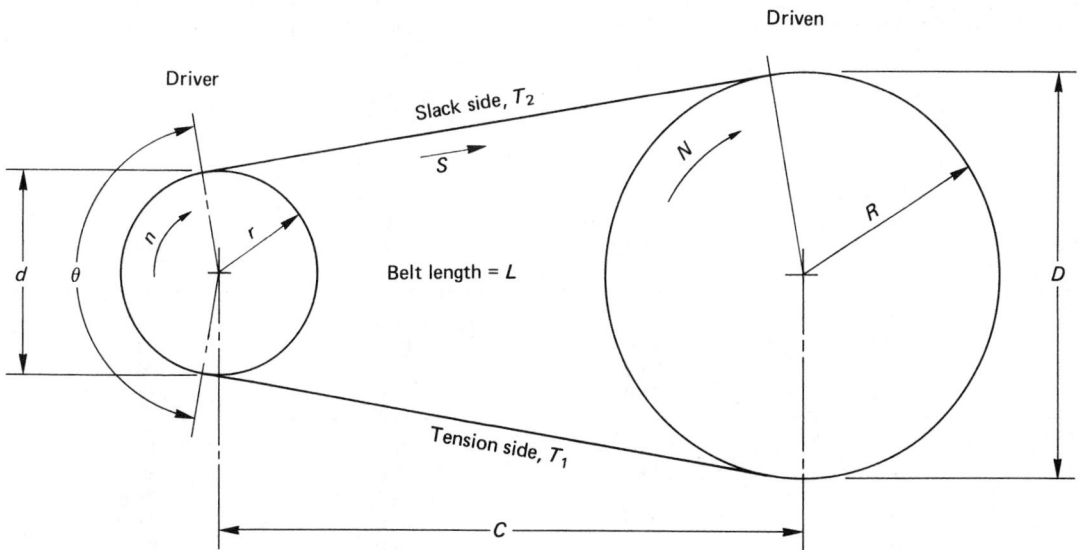

Figure 10.52 Drive geometry

and

$$T_1 - T_2 = \frac{kW \times 1000}{S} \text{ Newtons}$$

where

T_1 = tension in tight side of belt,
T_2 = tension in slack side of belt,
θ = angle of wrap of belt around the smaller pulley (in radians)

$$= 2\text{Cos}^{-1}\frac{(D - d)}{2c}$$

μ = coefficient of friction,
S = belt speed (m/s)

$$= \frac{d \times n}{19\ 100} \text{ m/s}$$

Pulleys should slightly exceed the belt width and should be slightly crowned. Ideal belt speed is around 18 m/s. At belt speeds above 18 m/s there may be a loss of power due to centrifugal force, and it is common for belts, apparently tight when stationary, to fail to exert a grip when centrifugal effects predominate.

10.2.1.2   V-Belts

The endless V-Belt drive, used either singly or in multiples, has a very wide range of uses and has applications in almost all types of machinery. There are two families of belts: classical, sometimes called industrial, and wedge, sometimes called narrow. The wedge belt differs from the classical belt, from which it was developed, in having a greater height-to-top width ratio. The benefit to the user of the wedge belt system is that generally a more compact (and possibly cheaper) drive can be designed on account of the greater individual belt load-carrying capacity. Both belt types are the subject of a Standard Specification, BS 3790: 1981, and this document embodies, where appropriate, the work of the International Standards Organization (ISO). This conformity with ISO permits a high level of international interchangeability.

Applicationally, V-Belt drives provide a high degree of flexibility, with a wide choice of speed ratios and shaft centre distances. The drives are quiet in operation and the use of multiple belts removes the likelihood of a sudden breakdown.

Special drives for shafts at right angles and V-flat drives, where the driven pulley is flat and can be part of the machine (such as a tanning drum), can be designed, although particular criteria must be observed.

Automotive belt drives are selected and operated under quite different conditions to industrial belts and are the subject of BS AU150: 1984. Their design and performance are usually agreed between manufacturers.

The V-Belt is manufactured with a core of high-tensile cord embedded in rubber or synthetic rubber and encased in a fabric and rubber reinforcement. The core supplies the tensile strength and the casing forms the friction surface in contact with the pulleys. Different belt constructions are available which provide anti-static capacity, oil tolerance and flame-resistance properties. All the belts have a standard angle of 40°, and the angle of the pulley groove varies to accommodate the slight bulging of the sides of the belt as it flexes around the pulley.

BS 3790: 1981 specifies six sizes of classical belt, designated Y, Z, A, B, C and D, and four sizes of wedge belt, SPZ, SPA, SPB and SPC. The cross-sectional dimensions of the sections are shown in Table 10.1. It should be noted that while the overall dimensions are given for recognition purposes, the pitch dimensions are used for the calculation of power, centre distance, etc. Pitch dimensions must be regarded as absolute and used for the derivation of other dimensions, which can be nominal or toleranced.

V-Belt pulleys: Figure 10.52 and Table 10.2 show the standard profiles, which in four cases are equally suitable for either classical or wedge belts. These profiles, which are defined in BS 3790, also conform to ISO 4183. (See also Figure 10.54.) In Table 10.2, dimensions b and h are minima, although in the case of the four dual profiles (e.g. Z-SPZ) dimension h can be reduced by 20% if it is intended to use only classical belts. The tolerances for dimensions e, f and a can be found in BS 3790: 1981.

Pulleys must be produced from materials with adequate strength for the power and speed requirements of the drives. Light alloys are not usually employed for continuous industrial duty because groove wear would cause damage to the belts. Cast iron pulleys are the most common: BS 1452, Grade 12 for belt speeds up to 25 m/s, and Grade 14 up to 40 m/s. For belt speeds above 40 m/s special pulleys may be necessary.

Table 10.1

Belt section	Pitch width (mm)	Nominal top width (mm)	Nominal height (mm)	Minimum recommended diameter (mm)
SPZ	8.5	9.5	8.0	67
SPA	11.0	13.0	10.0	100
SPB	14.0	16.0	14.0	160
SPC	19.0	22.0	18.0	224
Y	5.3	6.5	4.0	20
Z	8.5	10.0	6.0	50
A	11.0	13.0	8.0	75
B	14.0	17.0	11.0	125
C	19.0	22.0	14.0	200
D	27.0	32.0	19.0	355

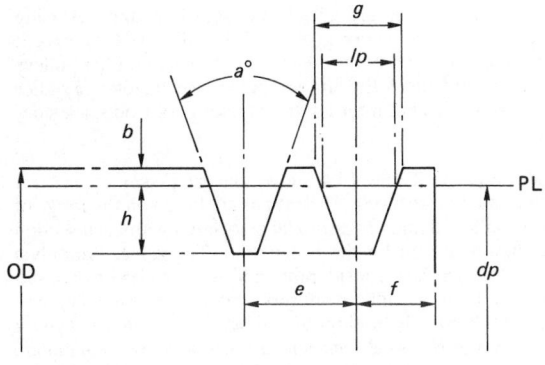

Figure 10.53 Multi-groove pulley cross section. PL = pitch line, OD = outside diameter, h = groove depth from PL, b = distance between OD and PL, a° = groove angle, e = centre-to-centre of grooves, f = edge of pulley to centre of first groove, g = top width of groove, dp = pulley pitch diameter, lp = pitch width of V-Belt

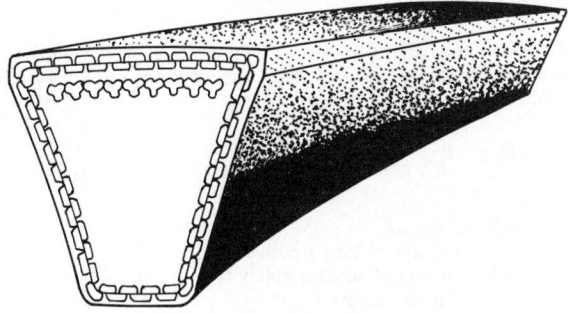

Figure 10.54 Narrow V-Belts (better known as wedge belts) have a narrower profile than classical V-Belts, with a relative height of approximately 0.9. There is a more even distribution of the tension forces between the reinforcing cords and thereby a higher power rating compared to classical V-Belts of the same top width

The pulley surfaces and especially the groove sidewalls should be machined such that the surface finish is a maximum of 3-2 μm when determined by the method described in BS 1134. Pulleys must be balanced either statically or dynamically, depending on the rim speed and ratio of face width to diameter. BS 3790 contains comprehensive balancing information. In general, for all pulleys operating below 10 m/s rim speed, and for pulleys with face widths half or less of the diameter operating below 20 m/s rim speed, static balancing is adequate.

Pulleys employing split-taper bushings are most convenient for installation and removal in that they avoid the need for interference fits, keys, etc. A Taper-Lock pulley manufactured by J. H. Fenner & Co. Ltd is shown in Figure 10.55.

Figure 10.55 A Fenner Taper-Lock bushing for securing a large pulley to a shaft

Table 10.2

Belt section	lp (mm)	b (mm)	h (mm)	e (mm)	f (mm)	a (°) (mm)	dp	g (mm)
Y	5.3	1.6	4.7	8.0	7.0	32	≤63	6.2
						36	>63	6.3
Z-SPZ	8.5	2.5	9.0	12.0	8.0	34	≤80	10.0
						38	>80	10.2
A-SPA	11.0	3.3	11.0	15.0	10.0	34	≤118	13.1
						38	>118	13.3
B-SPB	14.0	4.2	14.0	19.0	12.5	34	≤190	16.6
						38	>190	16.9
C-SPC	19.0	5.7	19.0	25.5	17.0	34	≤315	22.5
						38	>315	22.9
D	27.0	8.1	19.9	37.0	24.0	36	≤500	32.3
						38	>500	32.6

Selection of belt drives   BS 3790: 1981 contains all the information necessary to design a drive; power ratings, standard pulley diameters, service factors, correction factors for belt length, arc of contact and speed ratio. Similar details are usually given in the catalogues of manufacturers, some of whom operate a technical advisory service.

The number of belts required for a particular drive can be obtained using the power table for the selected type and size of belt. The power rating given in the table for the particular pulley diameter and shaft speed is multiplied by the correction factors for belt length, arc of contact, etc., and then divided into the design power (actual power × service factor) of the drive. If the result of the division contains a fraction, the next whole number of belts is used.

Power-correction factors for industrial service   These are based on prime movers classified into two separate groups, with reference to Driven Machinery classified into four separate groups as detailed below. Table 10.3 gives the factors for periods of up to 10 hours, 10 to 16 hours and over 16 operational hours per day. The four separate groups of driving machines are defined as follows:

Light duty – Agitators for liquids, blowers and exhausters. Centrifugal pumps and compressors. Fans up to 7.5 kW. Light-duty conveyors.
Medium duty – Belt conveyors for sand, grain, etc. Dough mixers. Fans over 7.5 kW. Generators. Line shafts. Laundry machinery. Machine tools. Punches, presses and shears. Printing machinery. Positive-displacement rotary pumps. Revolving and vibrating screens.
Heavy duty – Brick machinery. Bucket elevators. Exciters. Piston compressors. Conveyors (drag-panscrew). Hammer mills. Papermill beaters. Piston pumps. Positive-displacement blowers. Pulverizers. Sawmill and wood-working machinery. Textile machinery.
Extra heavy duty – Crushers (gyratory-jaw-roll). Mills (ball-rod-tube). Rubber calenders, extruders, mills.

For the above four groups (1) for speed-up and reversing drives multiply the factor given in Table 10.3 by 1.25, except where high torque is not present on starting. (2) If idler pulleys are used, add the following to the service factors: (a) idler pulley on slack side, internal, 0; (b) idler pulley on slack side, external, 0.1.

Power ratings   Table 10.4 shows typical power ratings for each of the belt sections. The ratings are based on the range of motor pulley diameters normally associated with each section and the speeds are for the faster shaft. The values are only a guide and can vary considerably and it is prudent to consult the Standards or manufacturers' catalogues for a precise selection. Normally, pulleys should be chosen which will give a belt speed in the 15–20 m/s speed range and are of adequate diameter in relation to the motor bearings (see Table 10.4).

Minimum motor pulley diameter   Table 10.5 shows the minimum pulley diameter suitable for British metric electric motors, to BS 5000: Part 10: 1978. The diameters were calculated to give a minimum bearing life (B_{10}) of 12 000 hours, and a tight to slack side tension ratio of 5 was assumed (180° arc of contact). All dimensions are in millimetres. Smaller diameters can be used but the drive end load should be calculated and referred to the motor manufacturer.

Arc of contact correction factor   The arc of contact x on the small pulley can be calculated from the following formula (see Table 10.6):

$$x = 180 - \frac{(D - d) \times 60}{C}$$

where
x = angle of contact,
D = pitch diameter of larger pulley (mm),
d = pitch diameter of smaller pulley (mm),
C = centre distance (mm),
F = arc of contact factor.

Table 10.3 Service factors for V-Belt drives

Driven machines group	Electric motors A.C. – star delta start D.C. – shunt wound Internal combustion engines with four or more cylinders. All prime movers fitted with centrifugal clutches, dry or fluid couplings or electronic soft-start devices			Electric motors A.C. – direct-on-line start D.C. – series and compound wound Internal combustion engines with less than four cylinders Prime movers not fitted with soft-start devices		
	10 and under	*Over 10 to 16 incl.*	*Over 16 and continuous service*	*10 and under*	*Over 10 to 16 incl.*	*Over 16 and continuous service*
Light duty	1.0	1.1	1.2	1.1	1.2	1.3
Medium duty	1.1	1.2	1.3	1.2	1.3	1.4
Heavy duty	1.2	1.3	1.4	1.4	1.5	1.6
Extra heavy duty	1.3	1.4	1.5	1.5	1.6	1.8

Table 10.4

Belt section	Pulley diameter (mm)	Power (kW) at motor speeds		
		2880	1440	960
SPZ	67	2.11	1.24	0.89
	140	7.72	4.40	3.09
SPA	100	5.03	3.06	2.22
	200	16.93	10.31	7.32
SPB	160	14.35	9.00	6.50
	315	–	23.75	17.37
SPC	224	18.70	16.60	12.70
	560	–	–	53.30
Y	20	0.04	0.02	0.01
	50	0.19	0.11	0.08
Z	50	0.26	0.16	0.12
	90	0.60	0.36	0.28
A	75	1.42	0.91	0.66
	125	3.63	2.24	1.61
B	125	3.96	2.72	2.02
	200	7.82	5.90	4.35
C	200	7.80	7.49	5.76
	400	–	18.20	15.50
D	355	–	21.20	19.30
	560	–	–	35.30

Note: The values are for 180° arc of contact on the small pulley. Interpolation can be used for speeds between those shown. The presence of a dash indicates that the pulley rim speed is above 40 m/s and therefore not recommended for cast iron pulleys.

Table 10.5

Nominal motor speed (rev/min)	Motor power (kW)										
	3.0	4.0	5.5	7.5	11.0	15.0	18.5	22	30	37	45
	(mm)	(mm)	(mm)	(mm)	(mm)	(mm)	(mm)	(mm)	(mm)	(mm)	(mm)
2880	67	67	67	67	71	80	95	90	90	106	112
1440	67	85	75	100	95	118	118	140	140	160	190
960	67	71	95	80	118	125	112	140	160	160	190

Table 10.6

$x°$	F	$x°$	F	$x°$	F
180	1.000	160	0.95	139	0.89
177	0.99	157	0.94	136	0.88
174	0.99	154	0.93	133	0.87
171	0.98	151	0.93	130	0.86
169	0.97	148	0.92	127	0.85
166	0.97	145	0.91	123	0.83
163	0.96	142	0.90	120	0.82

Note: Arcs of contact below 120° should not be used without confirmation of the drive details by the belt manufacturers.

V-Belt formulae Let

R = speed ratio
C = centre distance (mm)
L = pitch length of belt (mm)
d = pitch diameter of small pulley (mm)
D = pitch diameter of large pulley (mm)
V = velocity or speed of belt (m/s)
F = arc of contact correction factor
K = service factor
E = belt length factor
N = number of belts required
R = prime mover speed ÷ driven machine speed
$V = d \times$ rev/min of small pulley ÷ 19.100
$L = 2C + (D - d)^2/4C + 1.57 (D + d)$
$C = A + \sqrt{(A^2 - B)}$ where $A = L/4 - 0.3925 (D + d)$ and
$B = (D - d)^2/8$

$$N = \frac{\text{Actual power} \times K}{\text{Basic power per belt} \times F \times E}$$

Note: Although contemporary practice uses pitch dimensions for all calculations, in the past it was common to define classical belts by inside length. In the event of only the inside length of a belt being known, a conversion to pitch length can be made by adding the following constants (dimensions in millimetre units):

A	B	C	D
35	43	56	79

Example Determine the basic drive equipment for a piston pump running at 1150 rev/min and driven by a 1440 rev/min, 22 kW electric motor, star delta starting 12-hour day duty, approximate centre distance 730 mm.

1. Service factor = 1.3
2. Minimum motor pulley = 140 mm
3. Speed ratio = 1440 ÷ 1150 = 1.25:1
4. Choose standard pulleys 160 and 200 mm
5. By observation it can be seen that a 160 SPB pulley running at 1440 rev/min transmits 9.00 kW
6. Belt length required = $2 \times 730 + (200 - 160)^2$ ÷ $4 \times 730 + 1.57 (200 + 160) = 2026$ mm (SPB 2020 will suffice)
7. Arc of contact = $180 - \dfrac{(200 - 160) \times 60}{730} = 177°$

   ∴ Factor = 0.99
8. From BS 3790, belt length factor for SPB 2020 = 0.93
9. $N = \dfrac{22 \times 1.3}{9.00 \times 0.99 \times 0.93} = 3.45$ (say, 4 belts)

Installation of V-Belts When fitting it is necessary to move the motor towards the driven pulley so that the belts may be placed in their grooves by hand. The use of a lever of any kind to force the belts onto the pulley can damage the load-bearing cords leading to premature failure.

The accepted method of belt tensioning is by the application of a force normal to the belt spans, at the span centre, to achieve a stated deflection. This method is fully described in both BS 1440 and BS 3790, and also in manufacturers' catalogues and installation instructions. The high performance of modern belts, especially wedge, can only be realized by proper tensioning and this is particularly important in the early life of the drive when bedding-in and initial stretch are taking place; nothing damages belts more rapidly than the heat generated by slip.

Where an adjustable centre distance cannot be arranged it is necessary to use a jockey pulley tensioning device. With classical belts this may be either a flat-faced pulley running on the outside of the belts or a grooved pulley running on the inside. For wedge belts only the latter should be used. In either case, it should be positioned so as to preserve the arc of contact on the powered small pulley and any adjustment to the service and arc of contact factors, occasioned by its use, made to the design calculations.

When multi-belt drives are installed, matched sets of belts, coded for length, must be used to ensure correct load sharing. When replacing belts always order a matched set and do not mix old and new belts. Finally, pulleys should be properly aligned by normal workshop methods and the drive fitted with a ventilated guard for safety and to allow heat dissipation and air calculation.

Raw-edge V-Belts Recent years have seen the development of the raw-edge V-Belt. These are available with a smooth flat underside or a cogged underside and are manufactured by accurately cutting cured sleeves to the required section dimensions. Raw-edge V-Belts have no textile case, and this, together with a cogged underside, reduces resistance to bending and allows them to operate on smaller pulley diameters than the conventional V-Belt. However, when cogged belts are used in larger pulleys the contact area and therefore the power-transmission capability are somewhat reduced.

Raw-edge V-Belts are normally manufactured in the wedge belt sections but they are also available from some manufacturers in the classical sections. They are commonly used as fan belts for cars but have become of growing importance in the industrial market.

10.2.1.3 Synchronous belt drives

Both flat belts and V-Belts lose a very small amount of speed (less than 1%) due to belt 'creep' (a condition not to be confused with slip) which is due to the change in belt section and tension as it moves around the pulley. If absolute synchronization is required then some type of geared drive is called for.

The idea of cogged, rubber driving belt for synchronous power transmission originated with the Singer Sewing Machine Company in America. The aim was to maintain register of the different moving parts of the machines without the possibility of oil contamination. The idea became a reality in 1940 and the use of synchronous belts spread to other small machines and instruments. This concept was developed and applied to other machinery and became more generally accepted during the 1950s.

As with chain, the tooth pitching became standardized and the early types were based on the inch system of units. There are five pitches generally available: XL, L, H, XH and XXH. XL is generally restricted to small business machines such as electric typewriters and photocopiers and XXH tends to be uneconomical for the power capacity, leaving L, H and XH in general industrial use. The teeth have an involute shape the same as gears to ensure smooth, rolling contact as the belt enters and leaves the pulley. Tooth form and size are covered by BS 4548. Figures 10.56 and 10.57 show the tooth profile and dimensions for L and H pitch.

Because stable length is essential for synchronous belts they were originally reinforced with steel. Today glass-fibre reinforcement is common and aramid is used if maximum capacity is required. The load-carrying tension numbers are moulded into a very thin layer of neoprene (synthetic rubber). To this are moulded the uniformly spaced and pitched neoprene teeth. The facing material is a layer of nylon fabric, providing

L pitch

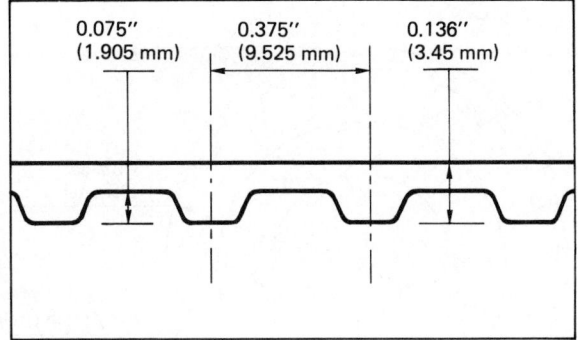

0.075″ (1.905 mm) 0.375″ (9.525 mm) 0.136″ (3.45 mm)

Figure 10.56

H pitch

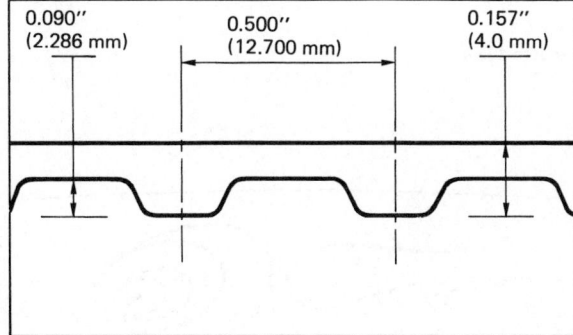

0.090″ (2.286 mm) 0.500″ (12.700 mm) 0.157″ (4.0 mm)

Figure 10.57

a wear-resisting surface for the entire running surface of the belt and its teeth. Figure 10.58 illustrates the typical components of a timing belt.

Depending on the application, timing belts have considerable advantages over chains, gears, V-Belts and flat belts due to one or more of the following features:

1 Owing to the use of a fibre-glass cord the timing belt has no stretch whatsoever in service, and this in itself eliminates the necessity of expense on automatic take-up devices and/or periodic maintenance. It also permits installation in otherwise inaccessible locations. Fixed-centre drives become possible. Except gears, all other forms of indirect transmission require periodic take-up.
2 The timing belt drive requires no lubrication and this allows for very substantial economies in initial drive design since oiltight housing and gear cases, seals, lubrication lines and accessories are all completely eliminated, while, at the same time, maintenance costs are also drastically reduced. In many industries such as food handling, strict process restrictions do not permit the use of lubricants in close proximity to the products being processed.
3. The timing belt drive allows for positive synchronization and this feature is daily becoming of more importance with the greatly increased use of automation, computerization and the necessity for very accurate, synchronized industrial drives.
4. Because of the very thin cross section, timing belts are extremely flexible and will operate efficiently over smaller pulleys than those used with comparable V-Belt or flat belt drives. Since arc of contact is not as critical a feature in timing belt drive design, larger ratios and shorter centre distances can be easily accommodated, ensuring considerable saving in space and weight. While arc of contact is not a critical design feature, it is most important, in order to gain the full advantage of belt width, to note that the belt teeth in mesh with the pulley grooves must not be less than 6. When the belt teeth in mesh are 5 or less the shear strength of the tooth becomes the critical factor in design, and this invariably results in an increase in belt width.

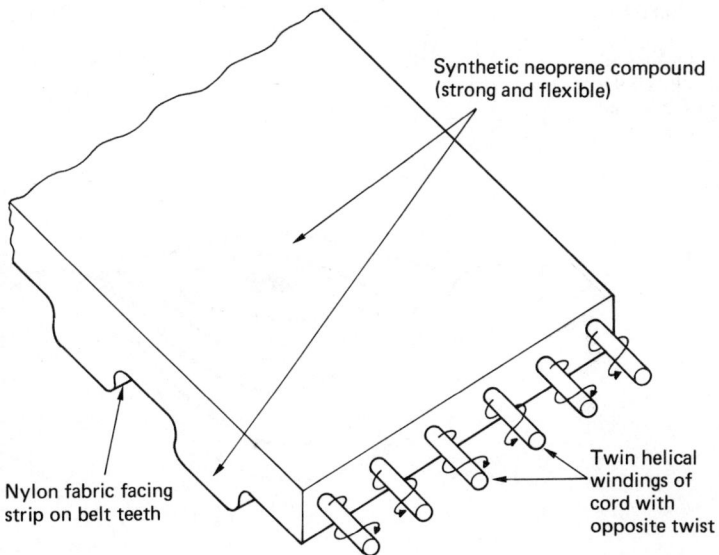

Synthetic neoprene compound (strong and flexible)

Nylon fabric facing strip on belt teeth

Twin helical windings of cord with opposite twist

Figure 10.58

5. The very thin section ensures that the heat generation when the belt is flexing around the pulley is kept to an absolute minimum; furthermore, there is no creep or slip to generate heat. The belt profile also allows for very high belt speeds of up to 60 m/s, although drives of above 30 m/s must be carefully considered because of pulley material.

Timing pulleys Standard timing pulleys are normally produced from steel and cast iron, and most manufacturers follow a similar coding system consisting of numbers and letters. The first numbers indicate the number of grooves in the pulley, the letter represents the pitch of the grooves and the final number the belt width that the pulley accepts.

Therefore, the code symbol 24H200 represents a timing pulley with 24 grooves, $\frac{1}{2}$-inch pitch and accepts a 2-inch wide belt. Pulleys are also recognizable by 'type', which refers to the particular design of pulley. All timing pulleys up to and including 48 grooves in L and H pitch are supplied with flanges. Even on perfectly aligned pulleys, a standard construction timing belt will 'track', and it is for this reason that one pulley (generally, the smaller of the two) is flanged to prevent the timing belt 'walking off' the drive. Figure 10.59 shows a typical flanged pulley.

Unlike any other type of drive, the pitch diameter of the timing pulleys is so arranged that it is actually in the centre of the flexing part of the timing belt where the load-carrying cords are situated. As mentioned previously, because of this, the pitch diameter of the timing pulley is always greater than its o.d. Figure 10.60 shows the basic dimensional details.

HTD drives Recent modifications of traditional trapezoidal tooth profiles to more circular forms offer a more uniform load distribution, increased capacity and smoother, quieter action. These newer synchronous belts with rounded curvilinear tooth design are known as HTD, which stands for High Torque Drive. Figure 10.61 shows a comparison between the standard involute belt and the newer HTD curvilinear design and illustrates the different stress patterns.

The HTD belt was developed to handle the higher torque capabilities normally associated with chain. The new design allowed, for the first time, metric pitched drives, and the standard pitch dimensions are 3 mm, 5 mm, 8 mm and 14 mm. Figure 10.62 shows dimensional details of 5, 8 and 14 mm pitch belts.

Both the belts and pulleys are manufactured in similar materials to the standard timing belt range. As the belt is fully metric the designation is straightforward. For example, in 1610-14M-85 mm, the first figure indicates the pitch length in

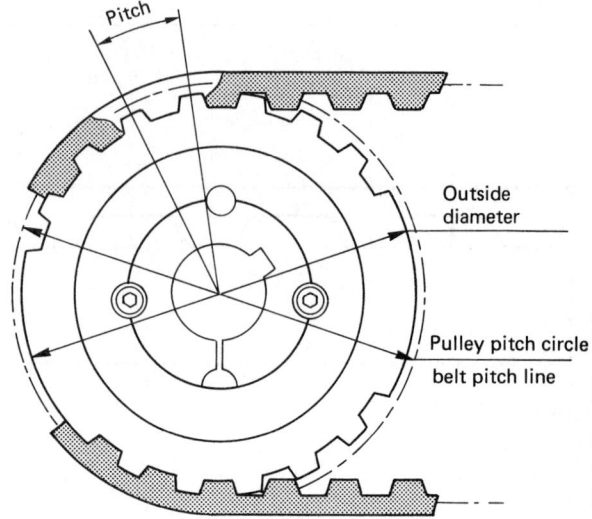

Figure 10.60

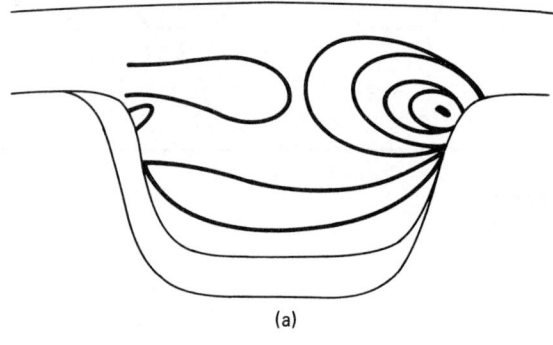

(a)

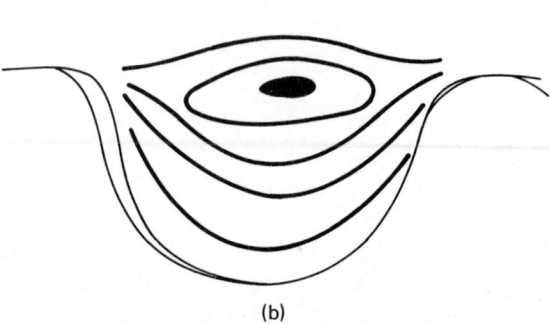

(b)

Figure 10.61 (a) Standard involute belt tooth profile showing loading concentrated at the point of contact; (b) HTD curvilinear design showing improved contact and substantial root stress distribution

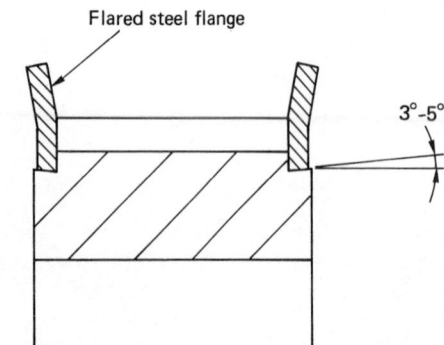

Figure 10.59

5 mm HTD

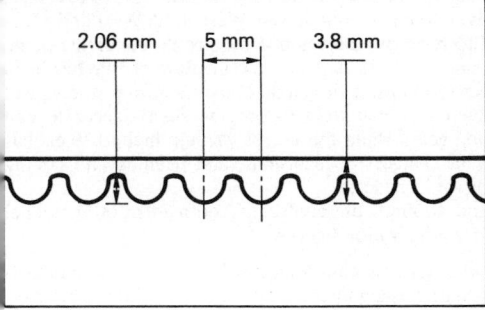

8 mm pitch

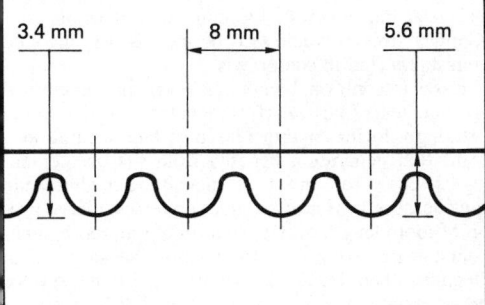

14 mm pitch

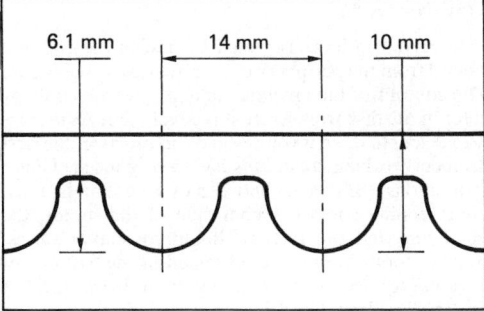

Figure 10.62

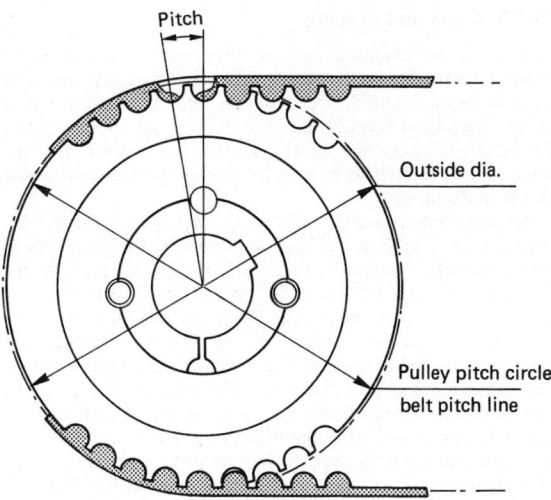

Figure 10.63

millimetres, followed by the metric pitch and lastly the belt width. The belt pitch length is the total length (circumference in millimetres as measured along the pitch line). The theoretical pitch line of a HTD belt lies within the tensile member (see Figure 10.63). The belts are available in a range of standard lengths up to 4500 mm pitch length and a range of widths from 9 mm to 125 mm.

10.2.1.4 Miscellaneous belt drives

In addition to the flat belts and V-Belts described above, there are also V-link belts made up from a number of separate links fastened together to form an endless belt. With these, access to pulley areas necessitated by the use of endless belts does not apply. Second, since belt length can be adjusted by increasing

or decreasing the number of links, fixed pulley centres can be used.

In general, V-link belts are more expensive than the endless belt but some economic advantages may be gained by holding a stock of links rather than a variety of endless belts. When flexing round a pulley, V-link belting does not suffer distortion as does the solid V-Belt; adjacent links slide over one another and there is little or no internal stress generated and in consequence, lower heat. Thus smaller-diameter pulleys can be used.

While most of these belts are made from various polymers in combination with fibre reinforcements, there are also all-metal belts. These are made from thin metal strips ranging from carbon steel through beryllium copper to stainless steel, titanium and, in the case of high temperatures, Inconel. The belt is perforated with holes and the pulleys can have teeth of various shapes, ranging from round or rectangular pegs to formed teeth. These belts are not in common use but offer potential in new projects.

10.2.1.5 Manufacturers

Graton and Knight Ltd, Warwick Road, Borehamwood, Herts WD6 1LX	Extruded polyester V-belting Nycor flat belting Round section polyurethane belting ($\frac{3}{16}-\frac{1}{2}$in dia.)
J. H. Fenner & Co Ltd, Marfleet, Hull HU9 5RA	V-belting Wedge belting Synchronous belts
BTL Ltd, Hudson Road, Leeds LS9 7DF	Link belts and special section belts
Pirelli Transmissions (UK) Ltd, Arthur Drive, Moor-Farm Industrial Estate, Kidderminster, Worcs	Timing belts
The Gates Rubber Co. Ltd, Heathhall, Dumfries, Scotland DG1 1QA	Synchronous belts

10.2.2 Gears and gearing

Gearing is an essential part of most power transmission systems and, wherever possible, the use of ready-prepared units is recommended. Gear design and manufacture is a highly specialized venture, and success can only be bought at the price of experience. Noise, vibration and short life are some of the penalties to be paid for gears imperfectly designed and manufactured.

An ambitious research programme involving a further investment of £8 million has been approved by the government without which it is believed that much of the UK gear industry would decline significantly. The programme is the result of several years of planning by the BGA (British Gear Association) Gear Research Council which has determined and prioritized the industry's research needs and established where the research might be carried out. The programme relates to four main technological themes: gear materials, gear design, gear lubrication and gear manufacturing and metrology. It is expected that it will develop to include projects in other areas of mechanical power transmission technology such as clutches and flexible couplings. The programme will be flexible to cater for the changing needs of the industry and as such, indicates the prudency of buying-in ready-made gears.

This programme is timely, as a deal of confusion exists in the mind of many engineers regarding gear design and selection. It appears to be centred first, on the change from imperial to metric working and second, the introduction of new geometry considerations. It must also be recognized that the majority of manufacturers' literature and technical data is still given in imperial dimensions. This is primarily to cater for spares and replacements, although most companies cover metric gears which are not direct replacements for imperial-dimensioned gears.

As part of the engineering commitments of the BGA, new teaching modules are being developed in conjunction with the University of Sheffield. In the following, formulae have been given using descriptive terms together with the new symbols from the teaching modules (where these are known) in parentheses.

It is, of course, recognized that the use of standard gear units may not always be possible but the guiding principle is that, wherever possible, use standard bought-out manufactured gears of gear units. The cost of cutting, grinding and finishing is likely to be expensive with any new in-house operation.

However, it is important that basic aspects of gear design are understood so that the limitations are recognized. Other matters of significance include methods of securing gears to their shafts, their lubrication, their size in relation to their duties and the selection of appropriate materials.

10.2.2.1 Tooth profile

The profile of a gear tooth must be chosen bearing in mind the following:

1. All the gears must mate and mesh with a smooth uniform action.
2. The tooth must have a section sufficiently strong for the applied loads.
3. The tooth must be free from weakening undercuts.
4. The tooth will mesh at the correct shaft centre distance.
5. The profile of the teeth offers no manufacturing difficulties.
6. The geometry provides an adequate tooth overlap.

The involute curve provides the most widely used profile for gear teeth although there are other profiles such as the cycloid

and a variety of profiles found in horological designs. There has also been a revival of the basic Russian Novokov gear, which never found favour in the West until Westland Helicopters Ltd recently redeveloped the profile under the name of conformal gears. In industry, the involute profile has been the subject of intensive design and manufacturing studies and had enabled manufacturers to provide silent, accurate and long-lasting gears while the use of vacuum-melted steels has removed the dangers of inclusions, and peening and honing have improved surfaces.

Westland adopted the conformal tooth form in a parallel shaft gear configuration because:[*]

1. They are more tolerant than involutes to the large out-of-plane shaft misalignments experienced in high power-to-weight ratio aircraft transmissions. This is explained by the differences in contact geometry (see Figure 10.64) misalignment resulting in localized concentration of the narrow line contact of the involute form compared with an inconsequential axial movement of the elliptical conformal contact. Contact stresses would thus be increased in involute teeth but unaffected in conformals.
2. Power losses in conformal teeth are lower than in equivalent involute gears (particularly a planetary set) due to the lower sliding velocities and increased surface separation.
3. Lubricant film generation benefits from the greater entraining speeds – an order of magnitude higher than involutes because conformal contact traverses a large proportion of tooth length during rotation of one tooth pitch.
4. Conformal gears have proved to be more tolerant to tooth imperfections than involutes, whether these be surface damage or variations in long-wave surface finish characteristics within manufacturing tolerances.

10.2.2.2 Involute profile

An involute curve can be constructed by tracing the end of a cord unwound from the periphery of a circular disk (see Figure 10.65). The contour of the involute curve is governed only by the diameter of the disk from which it is developed. As there is no limit to the length of an involute curve, in practice, the best portion to meet working conditions has to be chosen. Under working conditions, the contact between two teeth at the pitch point is pure rolling contact. Either side of that point, the contact is sliding and the rate of sliding constantly varies. Standard gear tooth forms are obtained using cutters of standard geometry and corresponding to a basic rack as defined in BS 436: Parts 1 and 2.

Gear teeth are sometimes crowned (see Figure 10.67(b)), which is a progressive reduction of the tooth thickness from the middle part towards each end face, in order to ensure the transmittance of the stresses of a flank to its mating flank under the best conditions.

The choice of a suitable pressure angle for the basic rack (see BS 436: Part 2) is important, for it governs the thickness of the tooth at the root, the length of contact made by teeth on the flanks of the mating gear and the number of teeth in a small-diameter pinion before tip interference commences. Although pressure angles used in the past varied from $14\frac{1}{2}°$ to $20°$, experience has shown that the generally accepted pressure angle is the British Standard value of $20°$. As the number of teeth in a gear diminishes, a point is reached where good

---

[*]According to a paper presented by Cox and Rees of Westland Helicopters Ltd at a Seminar on 'Transmission technology for propfan and geared fan engines', IMechE Aerospace Division, 1985.

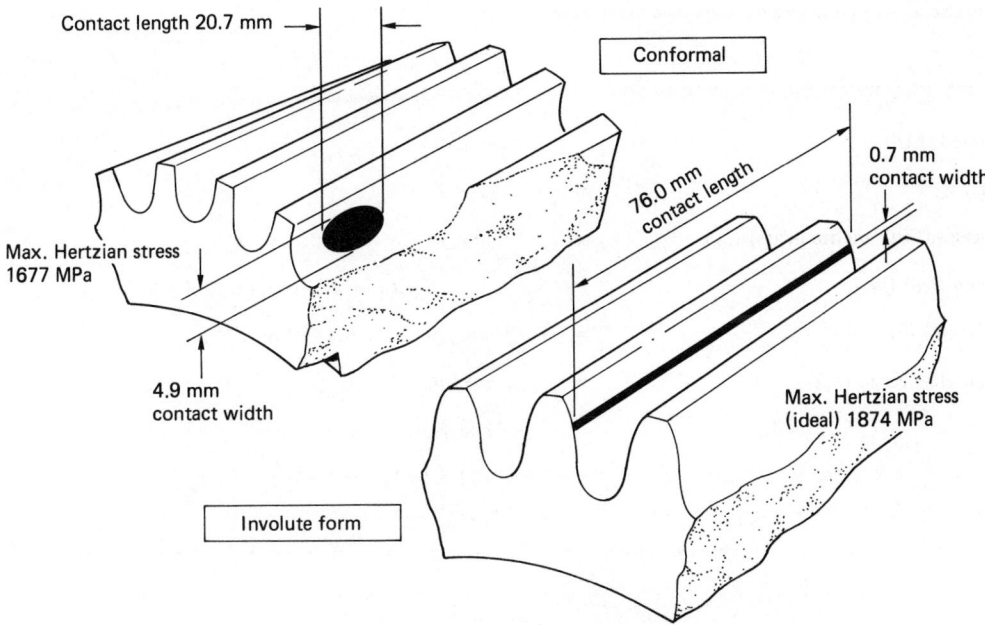

Figure 10.64 Comparison of contact areas and stresses for involute and conformal gears of similar pitch circle diameters and tangential load

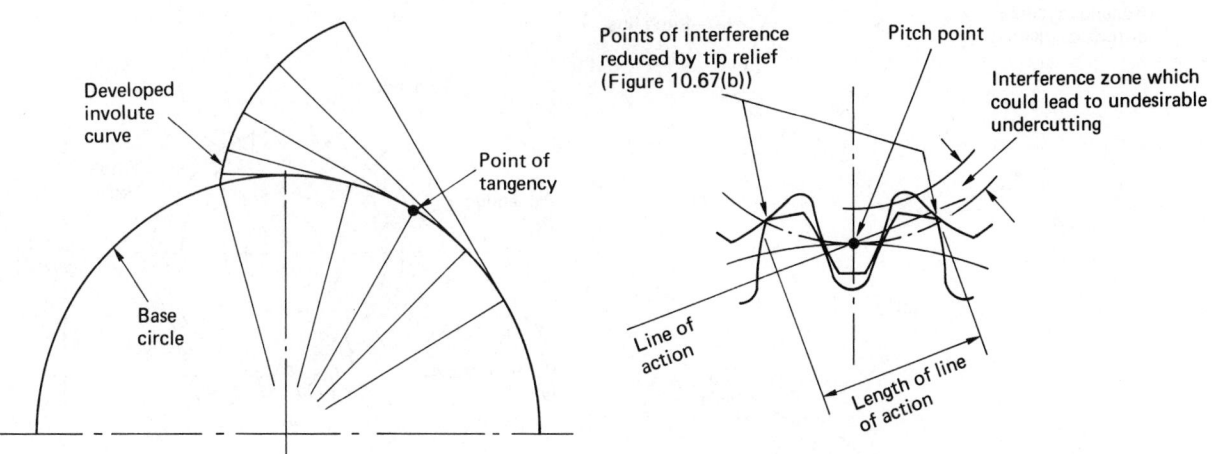

Figure 10.65 Developing an involute curve

Figure 10.66 Tooth interference

contact between the mating gears cannot be maintained. For a full-depth involute tooth form, the minimum number of teeth is given by the expression:

$\dfrac{2}{\text{Sin}^2\,\alpha}$ where α is the pressure angle (20°)

∴ Minimum number of teeth = $2/\text{Sin}^2\,20 = 2/0.342^2 = 17.09$

In practice this would mean, say, 17 teeth, but with adequate radius at the tip of the tooth the minimum could be reduced to 14 without undercutting the roots of the teeth (see Figure 10.66).

Table 10.7, used in conjunction with Figures 10.67 and 10.68, indicates some of the general terms and formulae used in connection with the design of gears and gearing. For efficient running it is important that correct meshing of teeth is ensured, and with bought-out gears this comes down primarily to establishing the correct centre distances for the shafts. Tolerances will depend on size and duty, and values are given in BS 436: Parts 1 and 2. The addendum modification consists of shifting the profile of the gear teeth to compensate for deflection under load and for manufacturing errors, and this involves certain limiting values which are summarized in British Standards PD 6457.

Table 10.7 General formulae for spur gears (without addendum modification) (dimensions in mm)

Pitch circle diameter (reference circle diameter) (d)	=	$z \times m_n$
Overall diameter (d_a)	=	$(z + 2) \times m_n$
Diametral pitch (not used with metric gears) (p_n)	=	$1/m_n$ or z/d (reciprocal of m_n)
Module (denotes tooth size) (m_n)	=	p_n/π or d/z (reciprocal of p_n)
(number of teeth) (z)	=	d/m_n or $(d_a \times p_n) - 2$ or $d \times p_n$
Circular pitch on reference circle (p_n)	=	$\pi \times m_n$
Addendum (h_a)	=	$1 \times m_n$
Tooth thickness	=	$p_n/2$ or $\pi/2p_n$ or $1.57/p_n$
Tip/root clearance	=	$0.25m_n$

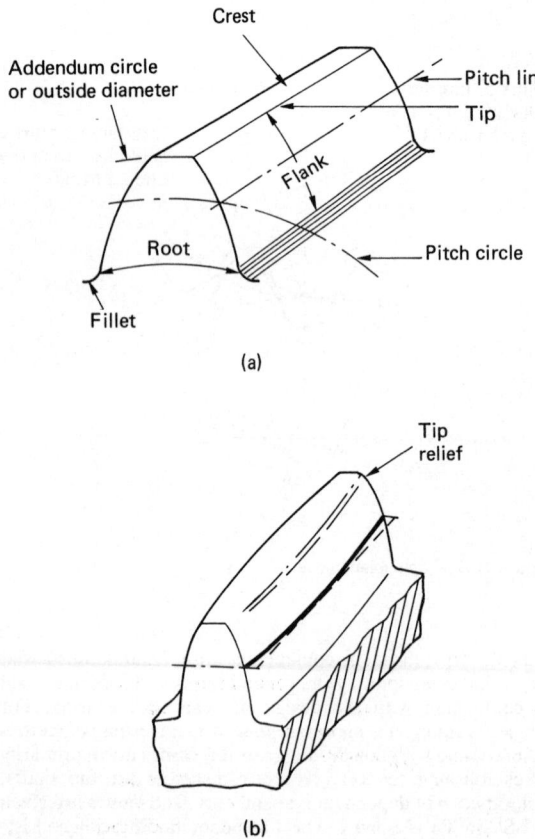

Figure 10.67 (a) Nomenclature for gear tooth; (b) crowning of gear teeth

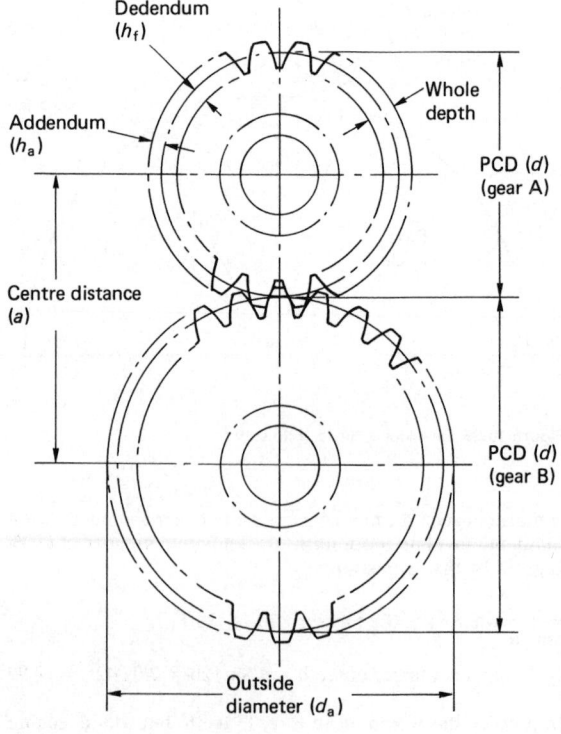

Figure 10.68 Nomenclature for meshing gears

For fine-pitch gears ($m_n < 1.0$ mm) some manufacturers tend to increase the tip/root clearance by reducing the diameter of the dedendum circle. For machine-cut gears these can be stated as follows.

Standard dedendum = $1.4 \times m_n$ for fine-pitch gears with modules below 1.00

Dedundum = $1.25 \times m_n$ for fine-pitch gears with modules 1.00 and above

For general considerations the dedendum can be $1.25 \times m_n$, which is taken from the British Standard rack. Variations in these values for the tooth profile in the past gave rise to some confusion. The reason is that experience and method of manufacture dictated to individual manufacturers the best values for them to accept and, not unnaturally, different manufacturers took different values.

Obviously, when choosing gears of different overall diameters it is important that the tooth sizes are identical, and this can be expressed by the module which is the reference circle diameter (d) divided by the number of teeth (z), i.e.

$$m_n = d/z$$

This can be seen from the illustration in Figure 10.66 of two meshing gears of differing diameters. As an example we can give each gear a physical size:

Gear A		Gear B	
(z) = 35	Ratio 2:1	(z) = 70	
(d) = 31.5 mm		(d) = 63 mm	
(m_n) = 31.5/35 = 0.9		(m_n) = 63/70 = 0.9	
(p_n) 0.9 = 2.83		(p_n) = 0.9 = 2.83	

Theoretical tooth thickness (s) = 2.83/2　　　　= 1.42
　　　　　　Addendum (h_a)　= 1×0.9　　= 0.9
　　　　　　Dedendum (h_f)　= 1.4×0.9　= 1.26
　　　　　　Working depth　　= $h_a + h_f$　　= 2.16
　　　　　　Whole depth　　　= $h_a + h_f + 0.4$ = 2.302

It will be seen that, by cross checks with the individual formulae given in Table 10.7, irrespective of gear diameter, providing the module and basic generating rack are identical, the gears will mesh satisfactorily (see Figure 10.69).

Gears used in clocks have already been briefly mentioned and there is currently a growing interest in small mechanisms

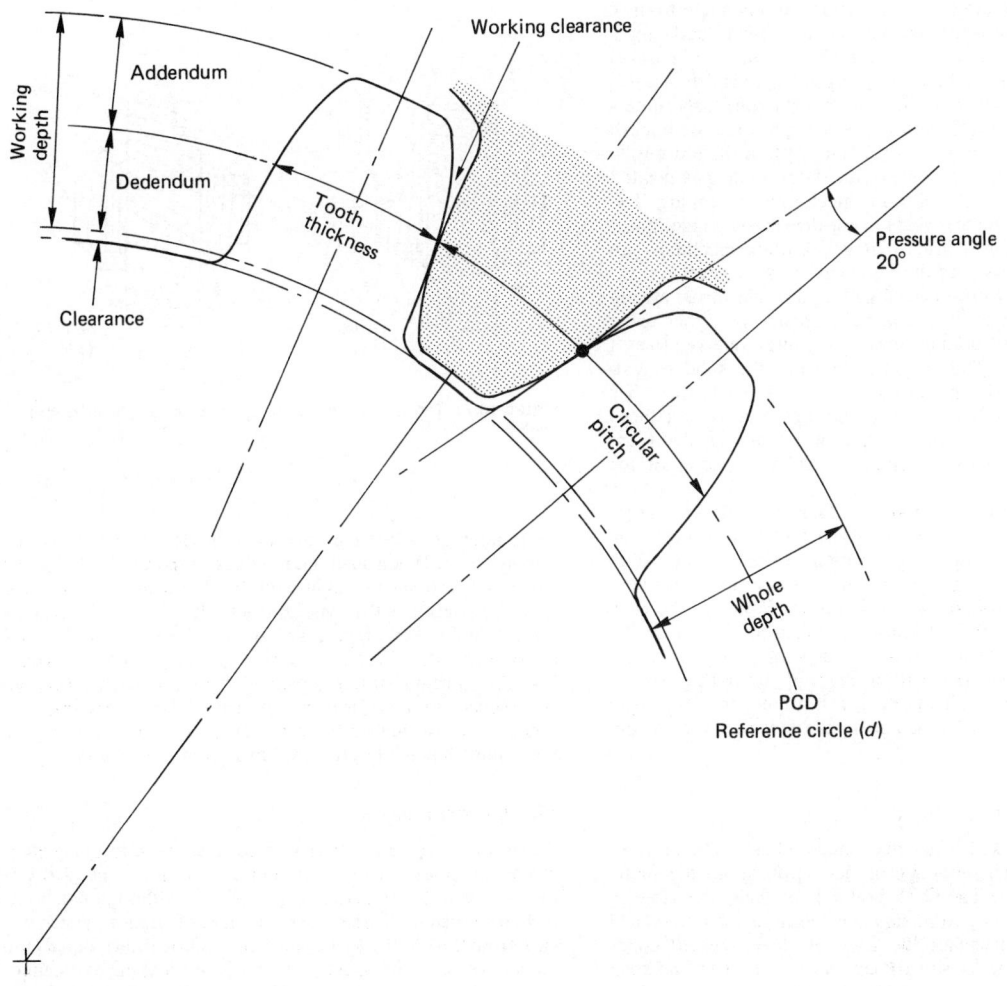

Figure 10.69 Tooth engagement

such as may be used in instrumentation, etc. These interests have focused attention on the need for a standard tooth configuration for small gears. While these mechanisms cannot be regarded as power transmissions, they may be of interest to readers, since the British Standards Institution have put forward recommendations in BS 978 (imperial) and BS 4582 (metric module) which discuss the use of involute tooth profiles together with a double-arc profile for very small pinions with between six and 12 teeth. The scope of these British Standards relates to four accuracy grades for involute spur, helical and crossed helical gears having modules of 1 or finer (diametrical pitches finer than 20 in imperial units). Examples are:

Class A   Scientific instruments and control systems
Class B   Navigational instruments and high-speed components of control systems where quietness and smooth running are essential and machine tool speed control
Class C   General-purpose instruments, counters, clockwork mechanisms

10.2.2.3 Helical gears

Helical gears have several advantages over straight-cut spur gears. One is that shafts can be inclined at any angle from 0° (parallel) to 90° (crossed axis helicals with small load capacity), the helix being adjusted to suit. One advantage of parallel axis helicals is that, unlike the spur gear in which the load is taken over the width of the next tooth instantaneously (unless the tooth has tip relief), with the helical gear, the teeth mesh gradually so that at no time is the full width of the tooth fully engaged. This eliminates some of the shock loading associated with straight teeth and makes for much quieter running. The cost of producing helical gears is not the disadvantage that it used to be. Modern designs of gear-cutting machines can handle helical gears with the same ease as spur gears.

One drawback to the helical gear is the side thrust arising from the helix angle. This can be overcome by either using thrust bearings or a double helical gear, often referred to as a herringbone gear. This may be cut from the solid or two separate gears used, one with a left-hand helix and the other with a right-hand helix. Some authorities suggest that these gears should be avoided, as spur gears are as good for low-quality drives with the single helical being superior for precision drives.

For shafts lying parallel to each other one gear will have a right-hand helix and the matching wheel a left-hand helix. On both gears the helix angle of generation will be the same. With shafts at 90° to each other both gears will have the same hand (either left- or right-hand helix). Thrust reactions are shown in Figures 10.70 and 10.71. With shafts arranged at less than 90° to each other, if the sum of the helix angle of both gears equals the shaft angle, the hand will be the same on both gears. If, however, the helix angle of one gear is greater than the angle between the shafts, then each gear will be handed (see Figure 10.70).

10.2.2.4 Bevel gears

Bevel gears are used to connect shafts whose axes lie at an angle to each other, although in most applications the shafts are at right angles. The tooth profile is basically the same as used for spur gears except that the tooth gets progressively smaller as it approaches the apex of the projected cone. Normally the teeth are straight cut and radiate from the apex of the pitch cone, but it is possible to give them curved, skew or spiroid form. Generally, the shafts of conventional bevel

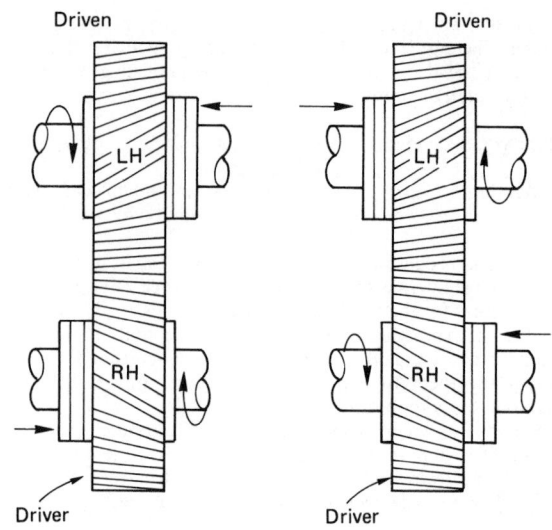

Figure 10.70 Thrust reactions using helical gears on parallel shafts

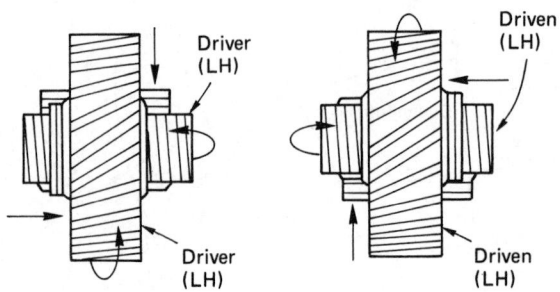

Figure 10.71 Thrust reactions using helical gears on shafts at right angles

gears intersect, although bevels can be designed to have the pinion offset. When such a pinion has radial teeth, the crown wheel will also have straight teeth but offset in relation to the axis. A variation is the hypoid, where the teeth on both gears are cut on the skew (Figure 10.73), in which situation they will act similarly to helical gears with consequent smoother running. The spiroid gear has curved teeth and, in many cases, can be likened to an offset worm drive. These systems do, however, cause higher tooth pressures and, as a result, it is important that really efficient lubrication is provided.

10.2.2.5 Worm drives

Worm drives have a number of advantages, one being that, given the helix angle is around 20° or less, the drive is considered to be unsatisfactory in reverse although not positively irreversible. Where the coefficient of friction, which can vary from 0.01 to 0.1 (with indifferent lubrication), equals the tangent of the helix angle, the gear is self-locking and cannot be turned by the gearwheel. However, the coefficient is not accurately predictable, as it can be affected by vibration, the

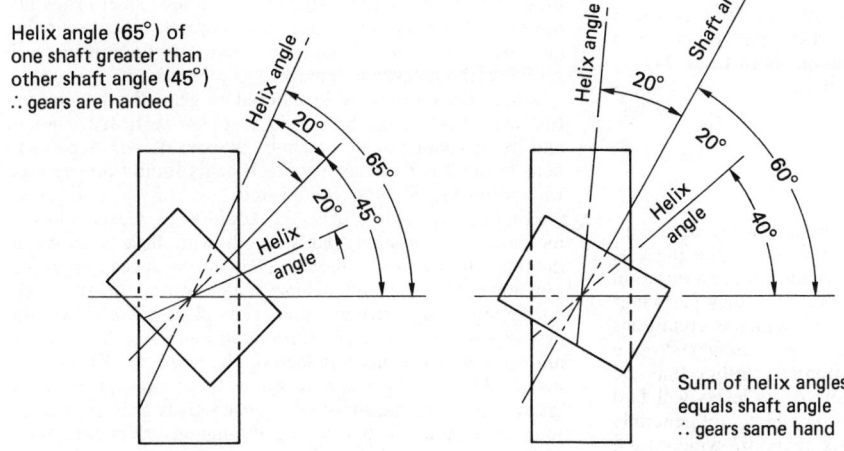

Helix angle (65°) of one shaft greater than other shaft angle (45°)
∴ gears are handed

Sum of helix angles equals shaft angle
∴ gears same hand

Figure 10.72 Handling of helix with shafts at an angle of less than 90°

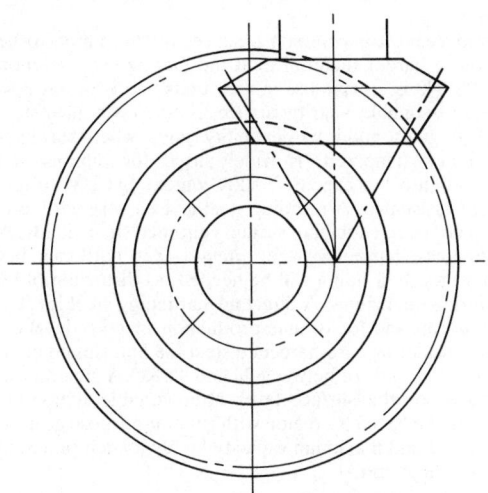

Figure 10.73 Hypoid gears allowing pinion offset

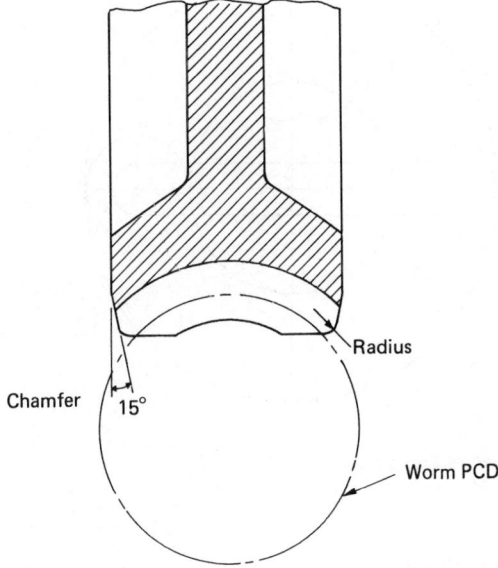

Figure 10.74 Typical worm wheel rim section

finish of the tooth surface and, above all, the degree and efficiency of the lubrication. For this reason, if a truly irreversible drive is required it is prudent to fit a brake in the system. Experiments have shown that the efficiency of a worm can be quite high; the best figures being when the helix angle is about 45° although the rate of increase in efficiency is markedly slower between 25° and 45° helix angle.

Shafts normally lie at right angles with a worm drive but other angles can be accommodated by adjusting the helix angle on the worm. While the worm can work in conjunction with a spur gear, the contact area between the teeth is limited and full power cannot be transmitted. It is usual for the worm wheel to fit closely to the diameter of the worm itself, thus providing the maximum surface on which to transfer the load (see Figure 10.74).

Worms can be single-start or multi-start. With a single-start worm the velocity ratio between worm and wheel is equal to the number of teeth on the wheel. With multi-start worms the velocity ratio can be expressed thus:

$$\text{Velocity ratio} = \frac{\text{Number of teeth on wheel}}{\text{Number of teeth on worm}}$$

The worm should be kept as small as possible consistent with stiffness needed to avoid deflection, as this will keep down surface speeds and friction. BSI recommendations for the design of worms and worm wheels are given in BS 721.

10.2.2.6 Gear trains

Internal gears are used in a variety of designs such as reduction gears and epicyclic units. In general, such units use standard pinions or bevels and have the involute tooth form. Figure 10.75 shows a typical epicyclic gear train and Figure 10.76 a differential unit as used in an automobile for drive to the wheels.

10.2.2.7 Gear materials

Materials are normally selected according to the duties intended for the gears. For power transmissions, the physical properties are critical and selection becomes more a question of choosing the right steel rather than any other particular material. The exception would be if environmental conditions dominated the specification. For example, in a corrosive atmosphere, stainless steel or reinforced plastics may be considered. Slow-moving gears at low stress levels will find cast iron a suitable material while worm wheels are generally made from phosphor-bronze. Plastics gears are widely used

for various mechanisms, but for reliable power transmissions they could be more expensive than a suitable steel. They do, however, have the advantage that they are quieter running and have the ability to take up small deformation in tooth profile without causing damage.

When choosing a suitable material for gears the questions of first cost, ease of machining, its response to heat treatment and its behaviour in service must be considered. A point to bear in mind is that case-hardened gears should not be used with softer metals such as mild steel.

For heat-treatable steels, the addition of *nickel* tends to increase the hardness and strength with little sacrifice in ductility. Its use as an alloying element produces less distortion due to lower quenching temperature. In the case-hardening group, carburization takes place more slowly but grain growth is restricted. *Chromium* increases the hardness and strength over that obtained by the addition of nickel but loss of ductility is greater. It refines the grain and imparts a greater depth of hardness. *Manganese* gives greater strength than nickel and a higher degree of toughness than chromium. *Vanadium* has a similar effect to manganese but the loss of ductility is greater: hardness penetration is also greater but machining is difficult. *Molybdenum* has the property of increasing the strength without affecting ductility. For the same hardness, steels containing molybdenum are more ductile than any other alloy steel and, having nearly the same strength, are tougher without increasing the difficulty of machining.

For spur gears it is common practice for the pinion to be made from a harder material than the mating gear. Carbon steel to BS 970 is in wide use in gear units and is of low cost and offers reasonable wear resistance. A Ni/Cr/Mo alloy steel to BS 970 gives good hardenability and, when through-hardened and tempered, is widely used for pinions and wheels. A slightly less expensive alloy containing Cr/Mo offers good wear resistance. A nitriding steel gives a very hard case but this is not deep enough to sustain continual shock loads. A carburizing case-hardening steel offers a deep hard case but subsequent tooth grinding will be needed, as distortion often occurs during quenching. A direct air-hardening steel is often used for worms and for volume-production speed reducers.

When considering case-hardened steel the aim should be to have a core hardness of between 30 and 40 Rc. A general run of hardness for the surface layer after suitable tempering should be in the 55–65 Rc region with general-purpose gears at the lower end and maximum capacity highly loaded precision gears at the high end.

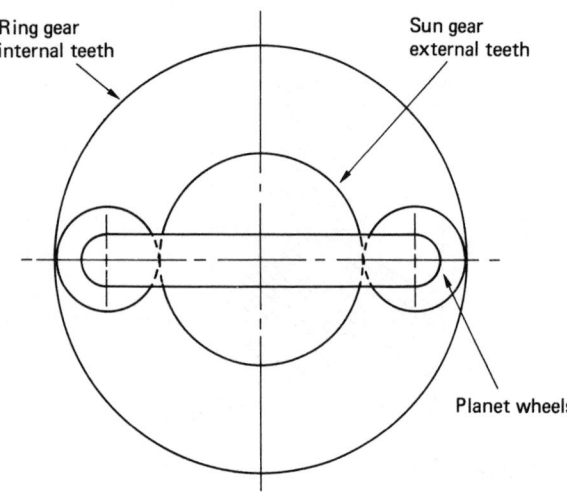

Figure 10.75 Epicyclic gear train

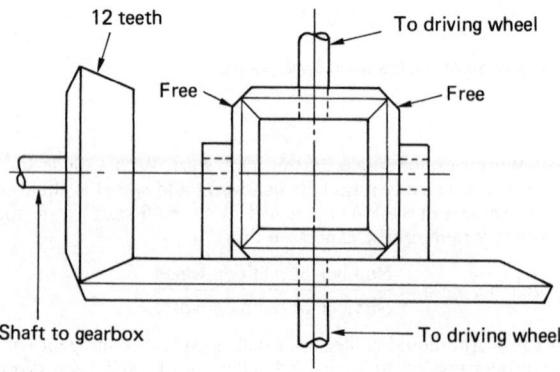

Figure 10.76 Differential gear as used in automobiles

10.2.2.8 Securing gear wheels to shafts

Securing the gear wheel to the shaft can be done in a variety of ways. The criterion should be that the gear remains square with the shaft, that the means of securing can meet the imposed loading without loosening and that it does not unduly weaken the shaft or gear. Figure 10.77 shows a number of different ways of securing the gear in small or lightly loaded drives, but in many cases the arrangement for securing may be part of an assembly and therefore a special integral arrangement may be devised. Friction holds such as a set screw (A in Figure 10.77) must be carefully considered. It would be inexcusable to use this method in, say, an automobile gearbox and for positive engineering applications, keyways and splined shafts are to be preferred.

A standard square key (B in Figure 10.77) is ideal for a positive drive in one direction only. For very large gears undergoing shock loads, two square keys – a tangential key (Figure 10.78) or a Kennedy key (Figure 10.77) – can be used.

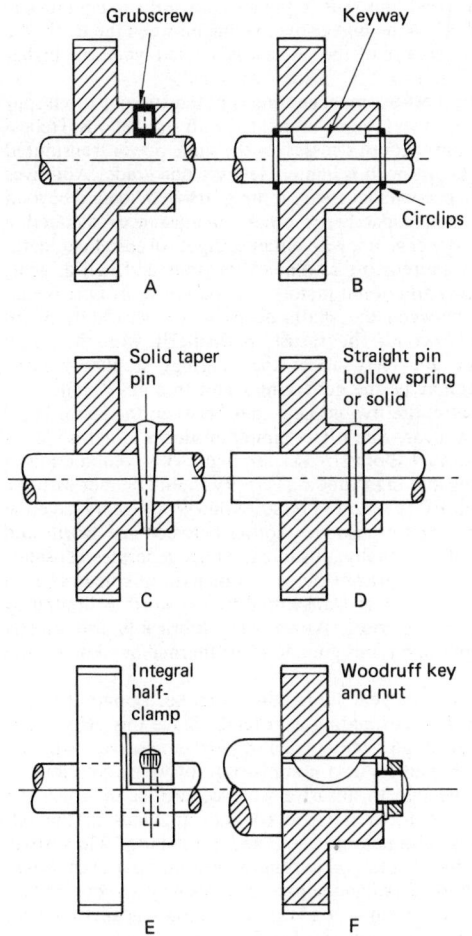

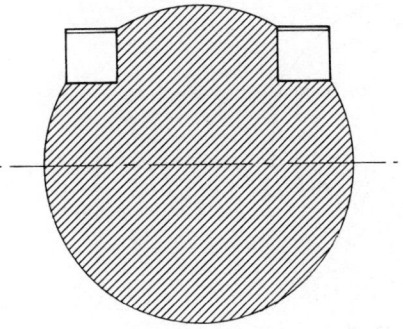

Figure 10.79 Kennedy keys

For smaller high-speed gears, splined or serrated shafts, though more expensive to produce, provide a positive drive and can allow, where necessary, some axial movement (see Figures 10.80 and 10.81). There are also a number of proprietary devices on the market in the form of a bushing which can be expanded to grip both gear wheel and shaft.

10.2.2.9 Gear units

During the last decade there has been a marked increase in the ranges of standard gearboxes employing spur, helical, bevel and worm gears suitable for a wide range of powers and with an extensive selection of fixed ratios in single- and multiple-reduction types, with a choice of parallel shafts, right-angled output shafts, co-axial shafts and offset parallel axis shafts. A high degree of standardization has been achieved and interchangeable components have been developed by many manufacturers. There are now British Standards for many external dimensions, particularly in respect to methods of mounting the units.

It is usual to consider the use of motorized units so that a complete package of motor/reduction unit is available requiring only the choice of coupling to the driven machine. This gives a single form of transmission and provides an economical and efficient package.

Figure 10.77 Methods of securing gear wheel to shaft (B and F are the preferred engineering solutions)

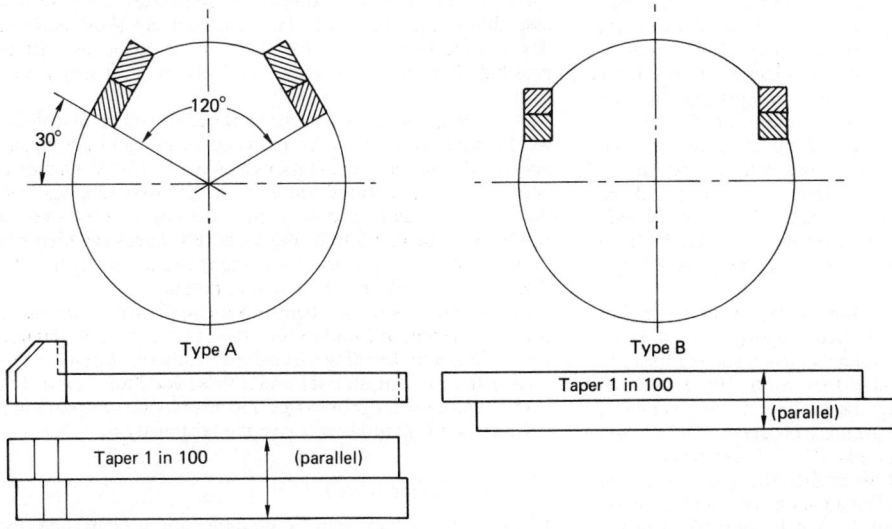

Figure 10.78 Tangential keys

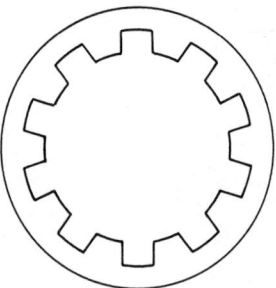

Figure 10.80 Square spline

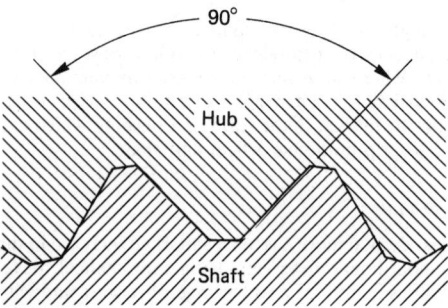

Figure 10.81 Serrated shaft

10.2.2.10 Lubrication

An essential factor in the lubrication of gearing is the maintenance of a fluid film between the surfaces of the gear teeth. By separating the surfaces in this way, there is a reduction in the degree of contact between the asperities on the mating parts. Inadequate separation leads to scuffing and possibly the localized welding together of the mating surfaces. In practice, gearing will most often be operating under combined conditions of boundary and hydrodynamic lubrication, which means that some of the loading will be taken directly by the opposing asperities and some by the fluid film. Choosing a lubricant for a given duty thus becomes a matter of ensuring that an adequate film thickness is always present at the mating surfaces. This film thickness should be in excess of the combined heights of the asperities, and will depend on the relative velocity of those surfaces, their dimensions and the viscosity of the oil. The major oil companies have considerable documentation which they will generally make available to engineers concerned with the selection of the correct lubricant.

Surface finish is dependent, within limits, on the method of manufacture and the material used. Lightly loaded gears hobbed from steel banks will probably have a surface finish of 1.0–2.5 μm (cla) and require no further finish. Those intended for heavier duties will possibly be hardened and tempered, and the teeth may then need grinding or shaving, which will produce a surface finish of around 0.13–1.0 μm (cla).

While it may be seen that the higher the viscosity of the lubricant, the better for providing an adequate film thickness, a high viscosity also means oil drag or frictional loss in the transmission system. This in itself will raise the operating temperature and thus reduce the viscosity until an equilibrium is attained. Thus the higher the loading between the teeth, the higher the viscosity of the oil that is needed while the higher the speed, the lower the viscosity needed.

In practical terms, most gearing is of the straight-tooth spur or helical types and can be lubricated with straight oils. Helical gears with lower tooth stresses for the same power transmitted would be happy with a lighter, less viscous grade. Additives would only become necessary if the gears were loaded beyond their designed capacity or other circumstances dictated a lighter oil. Bevel gears, with either straight- or spiral-cut teeth, will have requirements similar to the spur and helical gear, although two additional factors may be relevant. One is that the angle between the shafts needs to be accurately maintained and second, the thrust, particularly with the spiral bevel, may adversely affect the bearings so that bearing lubrication may be the governing factor in a bevel unit.

Generically, the hypoid gear falls between the spiral bevel and a worm gear. The offset pinion produces a high slide-to-roll ratio so that tooth stresses are high. This combination of high loading and high rubbing velocity should be met with the use of a full EP (Extreme Pressure) oil. Worm gears have the distinction that the relative motion between the worm and worm wheel is virtually all sliding, which generates considerable heat. The lubricant helps to dissipate this heat and, in most cases, the power transmitted by the worm is limited by this temperature rise. Worm gear lubricants are mostly straight mineral oils designed to resist thermal breakdown and oxidation.

The essence of gear lubrication is to keep apart the conforming surfaces of mating gear teeth. Since this objective is never achieved, gear teeth will always wear in service, and the best to be hoped for is a nominal rate of wear. Accelerated wear may be due to abrasive wear conditions in which the surface of the teeth is removed and circulates in the oil, producing additional abrasion and scratching. Thus some method of filtering the oil to remove any hard particles should be considered, although the most obvious way is to ensure that the teeth have as smooth a surface as economically possible. Further details are presented in Chapter 9.

10.2.2.11 Transmission shafts

Shafts used in power transmissions will invariably be either solid or thick-walled tubes. In gearboxes and similar assemblies, the shafts will be comparatively short and the design objective will be for these to be made as stiff as possible. Torsional stresses are unlikely to be of major concern.

In many applications a transmission shaft will be machined, will incorporate many different diameters along its length and may contain splines or serrations as well as through-diametric holes. All these features add up to stress discontinuities, and therefore adequate attention must be given to avoiding a sudden change of section and to ensure adequate blending radii. If the shaft is particularly complicated it is likely that the chief mode of failure will be one of fatigue.

For relatively short straightforward shafts where torque is transferred from one end to the other, the stresses are limited to torsion and bending. Standard textbook formulae for solving the working stresses under these conditions exist. For more complex shaft geometries, recourse to the computer and finite-element methods will give the best analysis.

10.2.2.12 Bearing reactions

When a single-spur gear is mounted on a shaft and the bearings are assumed to provide simple support, the bearing loads can be assessed by taking moments in terms of the nominal tooth load. This load in all but the most detailed of calculations can be taken as the resultant of the tangential load at the PCD and what can be termed the separation load tending to force the teeth out of mesh. In the simple arrangements in Figures 10.82 and 10.83 it can be seen that the major bearing reaction will be to the tooth load; reaction to the separation force, S, will be at right angles to the tooth load. With a helical gear, part of the tooth load will be translated into axial thrust.

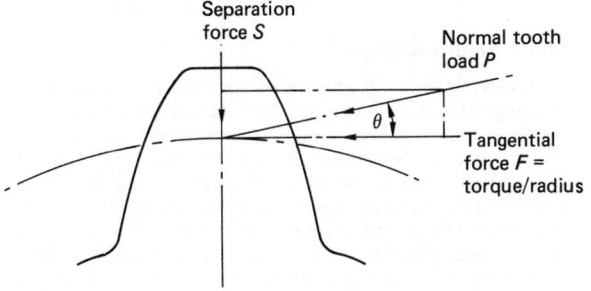

Figure 10.82 Resolution of tooth loads

With straight-cut bevels there will be an end thrust in both shafts tending to force the gears out of mesh. The magnitude of these thrusts will alter proportionally when curved teeth are used in the bevels.

In a worm drive the major load is end thrust in the worm shaft, its magnitude depending on its helix angle. There is also a side thrust on both worm and worm wheel arising from the frictional force between worm and worm wheel teeth.

10.2.3 Chains

Through the industrial ages, chains have provided a positive means of transferring power from one shaft to another, parallel, shaft. Today, bushed roller chains provide a simple, robust drive with the minimum of friction, their manufacture being laid down in BS 228: 1984 (ISO 60–1982). Until the introduction of flexible toothed belts, chain drives were almost universally used in automobile engines to drive the camshaft. A disadvantage of the chain drive is that centre distance for sprockets has to be either adjustable or worked on the basis of standard chain pitches. Alternatively, a jockey wheel or similar device can be used on the slack side of the chain, although this will generally preclude the drive being reversed.

Chains need to be adequately lubricated and must work in a clean environment. Dust and dirt will quickly wear chains and

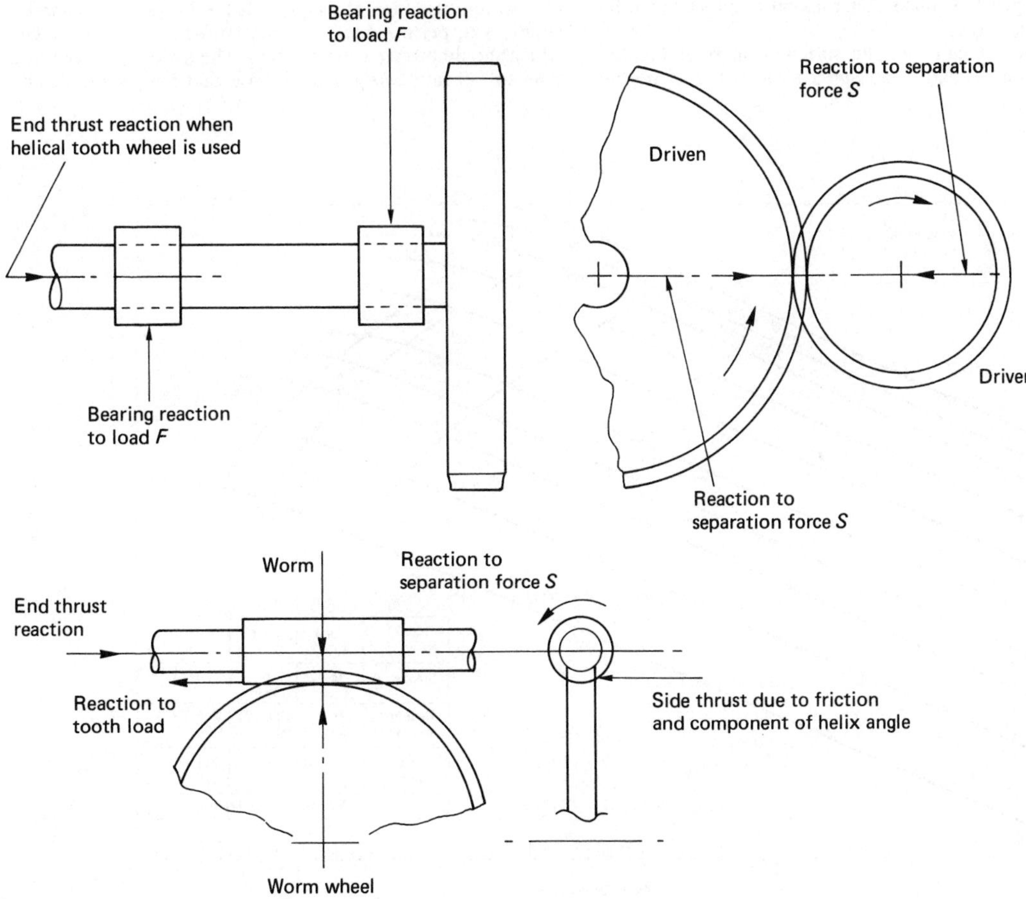

Figure 10.83 Bearing reactions arising from gear loads

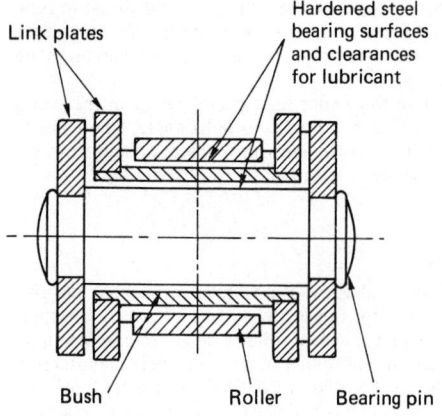

Link plates

Hardened steel
bearing surfaces
and clearances
for lubricant

Bush   Roller   Bearing pin

Figure 10.84 Typical section through roller chain (with acknowledgements to *Tribology Handbook*, Butterworths)

they will become noisy. A section through a typical roller chain is shown in Figure 10.84.

In addition to the standard roller chains, attempts have been made to design chains to operate even more quietly. These are based on a shaped link plate that rolls on a sprocket tooth rather than impacting.

For 15 000-hour chain life, the selection chart in Figure 10.85 is based on a steady load application with a 19-tooth sprocket. For different numbers of teeth, the relevant selection factor should be included. Where impulsive loads are encountered the selection factor should also be applied, irrespective of the sprocket size. Tables 10.8 and 10.9 indicate the types of loading likely to be encountered.

In general, the smallest pitch should be used, even if it means going to duplex or triplex chains. The centre distance should normally be kept to within 30–80 times the chain pitch. For large ratio drives the angle of lap on the sprocket should be not less than 120°. Manufacturers' literature usually contains the necessary data to establish which chain should be used for a specific task.

10.2.4 Shaft couplings

An important element in power transmission systems is the coupling whereby two relating shafts can be joined together. For large-diameter shafts revolving at low speeds, the simplest device is a solid flanged coupling but this does not allow for any misalignments in the shafts. For smaller installations it would be prudent to introduce a coupling that could cater for small shaft offsets and deviations in parallelism. A coupling becomes even more important if one of the shafts is the output from a prime mover where there are inherent vibrations. Selecting the right coupling will avoid transmitting these vibrations into the second shaft and its assembly.

There are numerous designs of couplings to cater for almost every conceivable condition, and the task of the engineer is to choose the right sort of coupling at the lowest cost commensurate with performance. Torsion stiffness is another factor which should receive consideration. The majority of couplings tend to fall into two groups – those that have some flexible

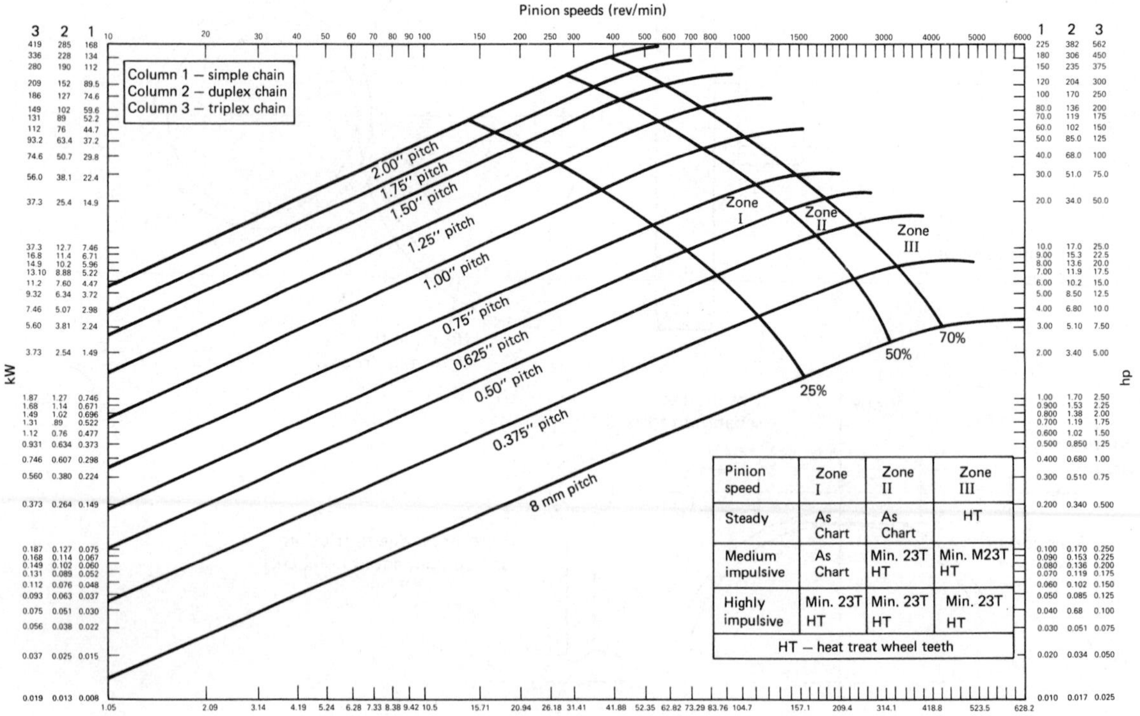

Column 1 – simple chain
Column 2 – duplex chain
Column 3 – triplex chain

Pinion speeds (rev/min)

2.00" pitch
1.75" pitch
1.50" pitch
1.25" pitch
1.00" pitch
0.75" pitch
0.625" pitch
0.50" pitch
0.375" pitch
8 mm pitch

Zone I
Zone II
Zone III

25%   50%   70%

kW
hp

Pinion speeds, rad/s

Pinion speed	Zone I	Zone II	Zone III
Steady	As Chart	As Chart	HT
Medium impulsive	As Chart	Min. 23T HT	Min. M23T HT
Highly impulsive	Min. 23T HT	Min. 23T HT	Min. 23T HT
HT — heat treat wheel teeth			

Figure 10.85 Performance curves for roller chain drives to BS 228: 1984 (with acknowledgements to *Tribology Handbook*, Butterworths)

Table 10.8 Selection factors

Machinery characteristics	No. of sprocket teeth				
	17	19	21	23	25
Steady	1.1	1.0	0.9	0.85	0.75
Medium impulsive	1.7	1.5	1.4	1.3	1.2
Highly impulsive	2.2	2.0	1.8	1.7	1.6

Table 10.9 Machinery characteristics

Steady	Medium impulsive	Highly impulsive
Alternators	Reciprocating pumps and compressors	Welding generator
Generators (excluding welding)	Cable reels	Excavators
Drilling machine	Cranes and hoists	Trenchers
Rotary pumps and compressors	Ball mills	Hammer mills
Conveyors and elevators – uniform feed	Propeller drives	Oilwell machinery
	Screens	Pulverizers
	Winches	Rubber machinery
	Wire drawing	

medium interposed between two halves each of which is carried by the shafts to be connected, and those that are mechanically flexible, examples being the internal gear coupling and Hooke joint. Some of the many variations are given below. Most manufacturers carry a wide range of variations in each half of the coupling to allow the use of different methods by which these can be secured to the shafts (e.g. using standard keys or proprietary shaft locking bushes).

The optimum choice of a flexible coupling for any application is the result of a compromise between many factors, and while performance at minimum cost is important, subsequent maintenance should also be considered. Specifically, the following points should be investigated:

1. Decide if the coupling should be torsionally soft or rigid. Soft types are generally less expensive.
2. Consider whether a small amount of backlash is acceptable. Backlash-free couplings are usually more expensive.
3. Calculate the required torque and add in any appropriate service factors. For example:

Torque = $9550 \times (P/N) \times k \times S$
where P = power transmitted (kw); N = rev/min, k = starting frequency based on maximum number of starts:

$k = 1$ for   0–30 starts/h
   1.2 for   30–60
   1.5 for   60–120
   2.0 for 120–180

and S = service factor which, for continuous rating, ranges from

1.0 to 1.75 for light even loads
1.25 to 2.25 for irregular shock loading
1.5 to 2.5 for arduous drive conditions

10.2.4.1 Types of couplings

In addition to the standard types given below there are special couplings made to meet specific requirements and may incorporate the features of one or more of the standard types: for example, telescopic couplings to allow considerable axial movements; quick-disconnect couplings; spacer-type couplings to take up any space between the ends of the two shafts; couplings with shear pins which free one half of the coupling when a severe overload is transmitted, etc.

Pin couplings   These couplings can cater for a wide range of power ratings by the optimum use of pins ranging in numbers usually from three to 16. The pins are rigidly fixed to one half of the coupling with the free end terminating in a flexible bush of rubber or plastics material. An example would be the Renold Pinflex (see Figure 10.86).

Flexible disk couplings   This type uses steel pins fixed in the metal half bodies of the coupling to transmit the torque through a flexible disk interposed between the two halves. The disk can be of staggered layers of rubberized fabric or a suitable solid polymer such as polyurethane (see Figure 10.87).

Flexible spider coupling   The flexible spider coupling transmits the torque through an oil-resistant rubber spider assembled between two metal half bodies. In some designs the spider is replaced by separate rubber blocks, manufacturers having their own particular designs to give a positive drive and to take up angular and linear displacements as well as to absorb any shock loads (see Figure 10.88).

Tyre-type coupling   This coupling consists of two half bodies connected by an external polymer tyre and is available in a range of shaft sizes to cater for torques from 65 to 1690 Nm,

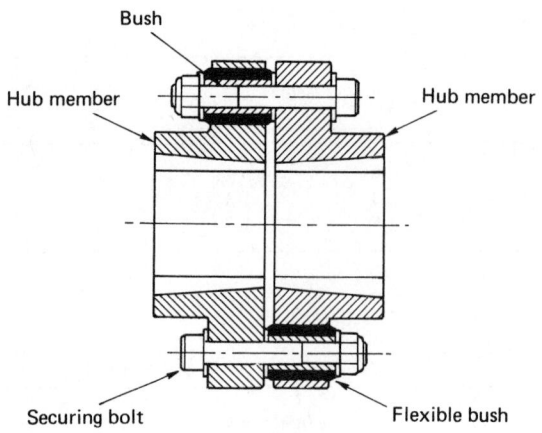

Figure 10.86 Section through Pinflex coupling (with acknowledgements to Renold Gears)

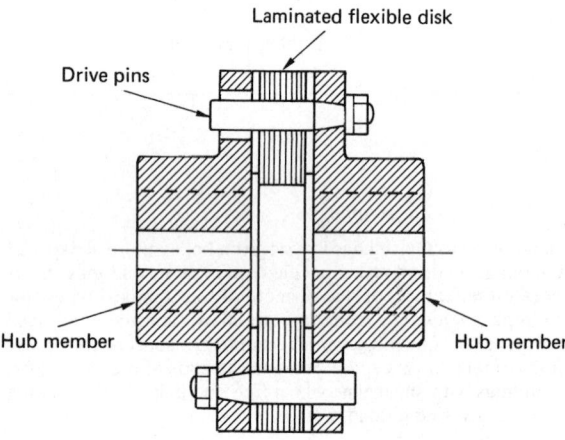

Figure 10.87 Section through disk flexible coupling (with acknowledgements to Renold Gears)

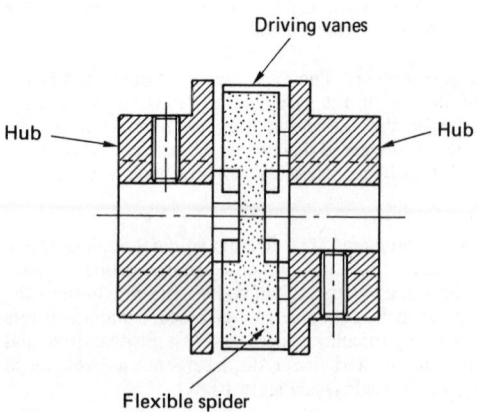

Figure 10.88 Section through spider flexible coupling (with acknowledgements to Renold Gears)

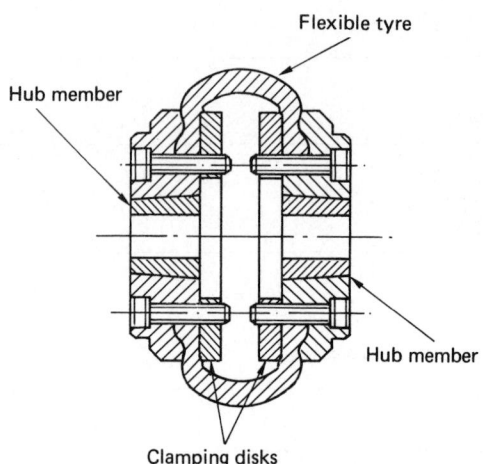

Figure 10.89 Section through a Renold Uratyre coupling (with acknowledgements to Renold Gears)

each size being able to handle 4° angular misalignment as well as end-float and axial displacement. A typical example is the Renold Uratyre (see Figure 10.89).

Chain coupling A chain coupling comprises two chain sprockets encircled by a duplex chain and contained within a housing. Disconnecting the chain provides a quick and easy means of disconnecting the shafts. This type of coupling is not designed to cater for anything other than minimal misalignment (see Figure 10.90)

Internal gear coupling Two basic types of internal gear couplings are available, the operating principle using a pair of externally cut gear hubs engaging the teeth of an internally cut gear in the housing. The teeth on the hubs are radiused so that the coupling can accommodate limited angular deflections (see Figure 10.91(a)). An alternative is to use one gear hub and to fit the engaging ring on the other shaft (see Figure 10.91(b)).

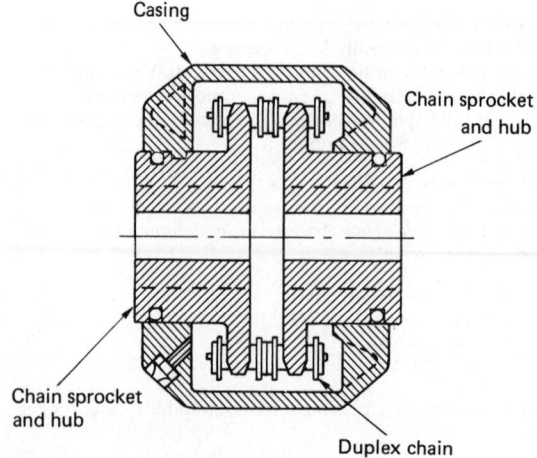

Figure 10.90 Section through a chain type coupling (with acknowledgements to Renold Gears)

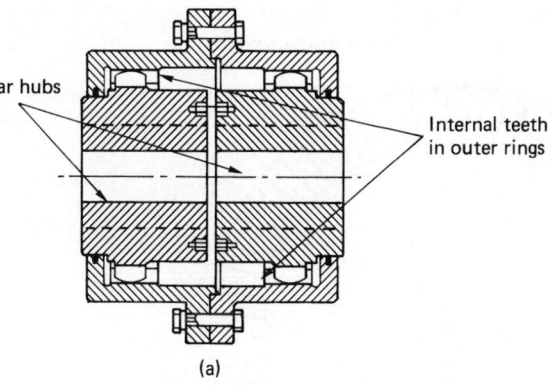

(a)

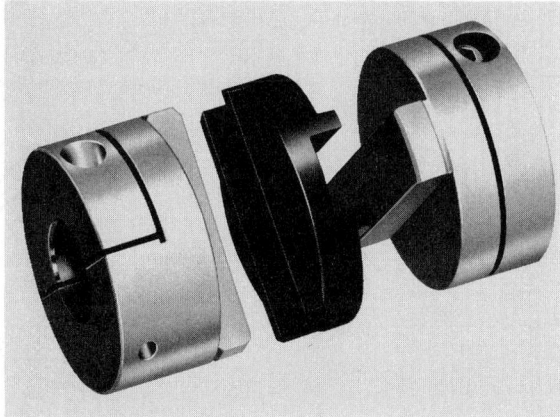

Figure 10.92 Exploded view of a Huco Oldham coupling for light power transmission

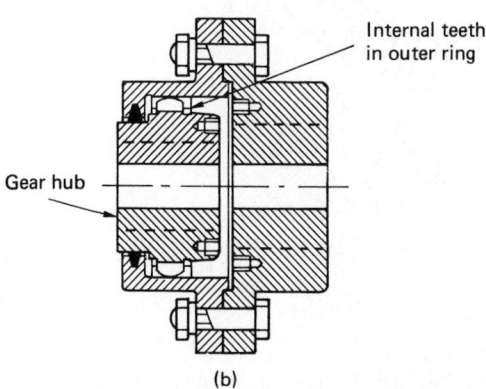

(b)

Figure 10.91 Sections through internal gear coupling. (a) Renold standard double-engagement type); (b) Renold single-engagement type

Oldham coupling This coupling consists of two halves each containing a diametric tenon placed at 90° to each other and mating with a centre-floating disk with two mating grooves. This is historically one of the earliest designs of couplings and will accommodate both angular and axial displacements. A typical example of the small size of the Huco Oldham coupling is shown in Figure 10.92.

Face tooth coupling Face tooth coupling rings may be used wherever precise indexing or positioning of one shaft to another is required. A pair of rings constitute a coupling for accurate location capable of transmitting high torque. The mating faces of each ring are machined to produce straight-sided radial V-teeth which, when meshed together, form a rigid angular and radial location. The rings can be bolted direct to a flanged member fitted to the ends of each shaft (see Figure 10.93).

Hooke coupling or universal joint This form of coupling has long been used in automobiles to accommodate the angular movement of the carden shaft connecting the gearbox to the rear axle; it is similarly used in front-wheel drive cars. Basically, the joint consists of two fork members attached to the ends of each shaft, the fork ends being secured to a centre

cruciform member. Because there are speed fluctuations proportional to the angle of the two shafts, as given in the following example, it is usual to employ a universal joint at each end of a power transmission shaft, the combination cancelling out any speed variations (see Figure 10.94).

Thus, if the driver rotates at a constant speed of 100 rev/min and the shaft angle is 25°, the maximum speed of the driven shaft is at a rate equal to the secant of the angle times 100, which equals $1.1034 \times 100 = 110.34$ rev/min. The minimum rate speed equals the cosine of the shaft angle multiplied by the revolutions per minute which, in the example, is $0.9063 \times 100 = 90.63$.

Spring coupling These consist of two hubs connected by single or multiple torsion spring elements. They can be wound different hands to cater for reversible drives. Standard couplings can be used for torques up to 900 Nm and different hub fittings are widely available (see Figure 10.95).

Bellow coupling These are usually regarded as more suitable for low-power transmission with standard couplings from, for example, Simplatroll, available to take torque up to 1 Nm. They are torsionally rigid, free from backlash and extremely light (see Figure 10.96).

All-metal dry flexible coupling The all-metal construction of this type of coupling avoids the need for any form of lubrication or extensive maintenance. High-torsion rigidity and good inherent balance makes the coupling ideal for high-speed applications. In essence, it is similar to the disk coupling shown in Figure 10.87, except that the centre disk is usually made of layers of thin flexible metal disks.

The future for dry membrane couplings is excellent as industry looks more and more for increased machinery life between shutdowns and reduced maintenance services. Another trend is towards gas turbine drives for non-generating equipment such as pumps and compressors. Gas turbine makers have developed aero-derivatives for industrial applications. Here the needs of the couplings will be lighter weight, reduced overhung moments and higher speed and power capabilities. Performance monitoring is another area that is developing as rotating equipment operators seek improved plant efficiencies, preventative maintenance programmes and

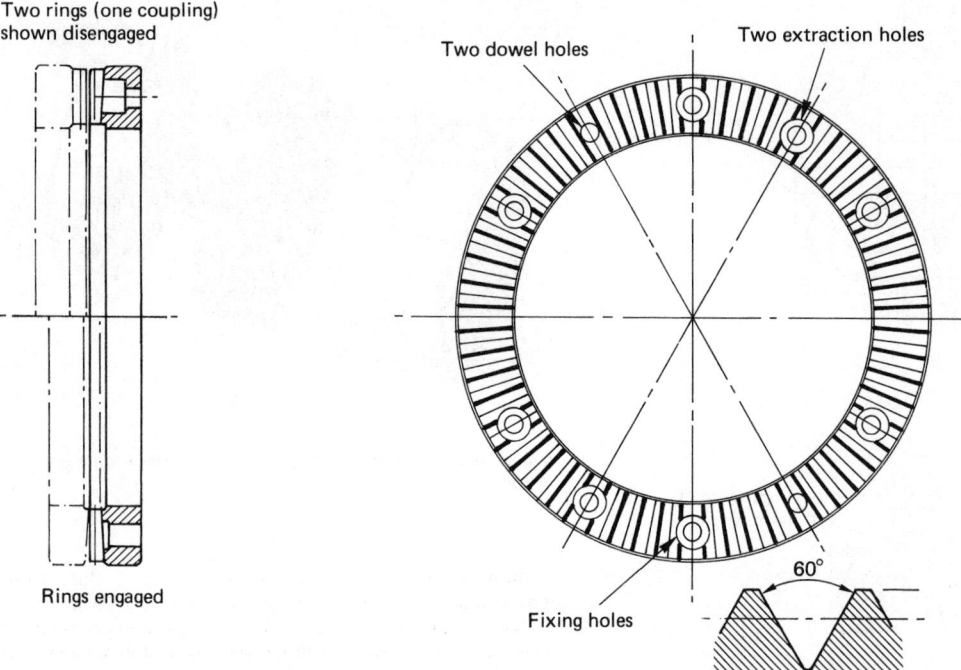

Two rings (one coupling) shown disengaged

Two dowel holes

Two extraction holes

Rings engaged

Fixing holes

60°

Figure 10.93 A TI Matrix Engineering face tooth coupling ring

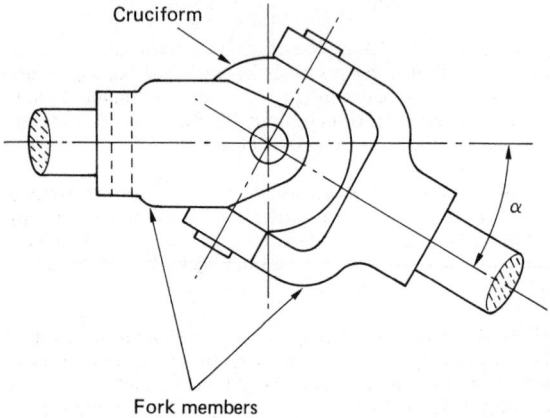

Cruciform

α

Fork members

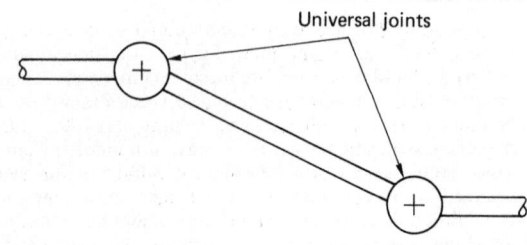

Universal joints

Figure 10.94 Hooke coupling or universal joint

longer operating life. To this end, Flexibox have developed a torque measuring system as an integral part of their Metastream flexible couplings.

The following (contributed by Flexibox Ltd) will give users and designers an appreciation of the capability and design of all-metal multiple membrane type power transmission couplings and, by the application of simple disciplines, obtain the practically infinite life for which the couplings are designed. While reference is made extensively to Flexibox Metastream ranges of power transmission couplings, many of the principles and most of the practices can be applied to power transmission couplings in general. Flexibox Metastream power transmission couplings have proved their effectiveness in transmitting torque under shaft misalignment in a wide variety of driver/driven machinery combinations.

10.2.4.2 All-metal mebrane couplings

The M, LS and T series couplings are all-metal non-lubricated membrane couplings which require no additional services, have no wearing parts and are easily installed without special tools. These couplings accommodate shaft misalignment in deflection of thin metal membranes. These are normally stainless steel, although non-ferrous membranes in Monel and Inconel are used for special applications where safety or corrosion merit particular attention. A number of these membranes are assembled into a pack (membrane bank) according to torque rating of the coupling. The membrane bank is (usually permanently) built into a membrane unit. In the spacer type coupling a membrane unit is fitted to either end of a spacer piece to produce a spacer unit (transmission unit). The M series range includes a double-bank membrane

(a)

(b)

Figure 10.95 (a) Examples of torsion spring type couplings; (b) typical application of torsion spring coupling (with acknowledgements to Simplatroll Ltd)

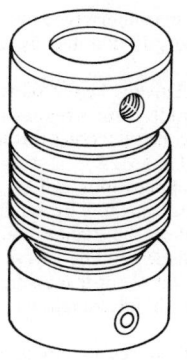

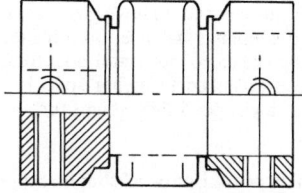

Figure 10.96 Examples of bellow type shaft couplings (with acknowledgements to Simplatroll Ltd)

unit (MODO) where two membranes banks are permanently fixed to a central ring, which is effectively a short spacer. The transmission unit is fitted between the driving and driven flanges of hubs or adaptors on the respective machines. The membrane units and transmission units include spacer-retention features that prevent parts of the couplings being thrown if the membranes shear.

In the ring form LS and T series ranges a spacer-retention feature is provided by bushes on the drive bolts. These bushes are shrouded by clearance holes in the coupling flanges and retain the spacer assembly if the membranes are damaged. This shrouded bush arrangement also provides an emergency/overload drive facility.

Speed and dynamic balance All-metal membrane couplings are the ideal choice for high-speed rotating machinery because

they can be dynamically balanced to high qualities and, as there are no wearing parts, this quality will not deteriorate during the life of the coupling, provided a few basic disciplines are observed.

Couplings are dynamically balanced to reduce the dynamic loads generated by mass eccentricity and rotation of non-symmetric masses.

Concentricity between coupling components and machinery shafts is achieved by close control of spigot/recess locations and is maintained by rigid assembly at the interfaces.

The influences of non-symmetrical masses are minimized by either removing material or adding counterbalance masses. Dynamic balancing equipment is used to indicate the magnitude and position of such corrections.

Match-marking of corresponding flanges and match-weighing of replaceable fasteners enable duplication of the quality achieved on balancing machines to be ensured.

The high-quality manufacturing procedures used by Flexi-box produce couplings that are symmetrical and will rotate concentric with machine shafts. The couplings have no wearing parts and are torsionally rigid so the balance quality will not change over the coupling life. Standard membrane couplings are used on most low- and medium-speed applications without dynamic balancing.

Coupling ranges such as MHS, TSK and the high-performance ranges for high-speed operation are dynamically balanced according to the needs of the equipment. The ultimate speed limit of balanced couplings is dictated by the material strength under centrifugal forces. A change of material (for example, to a high-grade steel such as EN24T or high-strength alloy) can increase the coupling speed capability.

Dynamic balancing ISO 1940, 'Balancing quality of rotating rigid bodies', specifies permissible residual unbalance of rotating components as a function of machinery type and speed of operation. The type of machinery is denoted by a Quality Grade 'G'.

Thus, a component for a relatively heavy diesel engine would be balanced to quality grade G16, whereas a component for a comparatively lightly constructed gas turbine would require a balance quality grade G2.5.

Although the majority of applications where dynamic balancing is necessary would be satisfied by quality grade G6.3, Flexibox has standardized on the higher-quality grade G2.5 for normal commercial balancing. The acceptable residual unbalance and couple per unit of rotor mass in 8 mm/kg on centre of gravity displacement in micrometres is shown in Figure 10.97.

Normal commercial practice produces a coupling with only the transmission unit dynamically balanced. The hubs or adaptors are balanced after fitting to their appropriate shafts as part of the machine rotor assembly by the machine builder.

For very high-speed applications and lightweight equipment, Flexibox have supplied couplings balanced to a higher quality as well as individually balanced hubs and adaptors. The high-performance range is balanced to higher specifications in line with the needs of high-speed lightweight turbine drives.

Torque and misalignment Power transmission coupling elements are subjected to various stresses which may initially be considered separately as steady and cyclic stresses:

Steady
Torque
Centrifugal
Axial misalignment
Design and manufacturing technique stresses
Cyclic
Axial shuttle
Torsional fluctuations
Angular and lateral misalignment

Steady stress factors are accommodated in the basic coupling design, giving an adequate design margin over ultimate stress capabilities of the membrane material used. Maximum torque capacity is usually expressed as a power-to-speed ratio, i.e. the coupling rating:

$$\text{Selection: Coupling rating} = \frac{\text{Power}}{\text{Speed}}$$

This is usually expressed in kW per 1000 rev/min or HP per 100 rev/min.

Maximum axial misalignment capacities of couplings are very generous and alignment within 10% of the coupling's limit is easily achieved. Allowance for the thermal growth of shafts can normally be made without exceeding the coupling's capacity in the cold and hot dynamic states.

Because cyclic stresses have a great effect on coupling life, these must be given more attention.

Axial shuttle is not normally a problem on machines where the shaft positions are axially located within the bearing arrangements. Moreover, stress levels caused by axial shuttle are low, and Flexibox membrane couplings have a non-linear axial stiffness characteristic which tends to damp out axial exciting vibrations.

Occasionally, however (for example, on sleeve bearing motor applications), it may be necessary to move the operating position up the stiffness curve by deliberately adding an axial displacement, thereby inducing a resisting force against the axial excitation. Cyclic stresses due to torsional fluctuations are usually accommodated by the use of a service factor in the coupling selection procedure. These service factors have been derived from a wealth of experience and knowledge of the torque characteristics of driving and driven machinery:

$$\text{Max. torque} = \frac{\text{Power}}{\text{Speed}} = \frac{\text{Coupling rating}}{\text{Service factor}}$$

Therefore:

$$\text{Required rating} = \frac{\text{Power} \times \text{service factor}}{\text{Speed}}$$

While the service factor effectively reduces the coupling rating by increasing the design margin, the axial and lateral misalignment capacities are unaffected. However, because higher speeds mean higher cyclic frequencies, it becomes necessary at very high speeds to reduce the angular (and, consequently, lateral) misalignment limit of flexible couplings according to design and speed.

Lateral (or angular) misalignment leads to many more coupling failures than all other causes combined. Lateral misalignment is accommodated in an angular deflection of each of the membrane banks in a spacer coupling configuration. The effects of angular and lateral shaft misalignment are therefore additive in producing cyclic stresses in the coupling.

To reduce these stresses within the membrane material capacity, therefore, accurate shaft alignment is crucial. A relatively small improvement in angular/lateral alignment greatly reduces cyclic stress levels and consequently extends coupling life expectancy. Coupling misalignment capacities and ratings given in suppliers' technical literature should allow for the anticipated stresses due to misalignment and torque simultaneously.

10.2.4.3 Flexible coupling ranges

Metastream Flexible couplings are designed to accommodate the inevitable displacement which occurs between the centre lines of two rotating shafts. Flexible elements are in the form

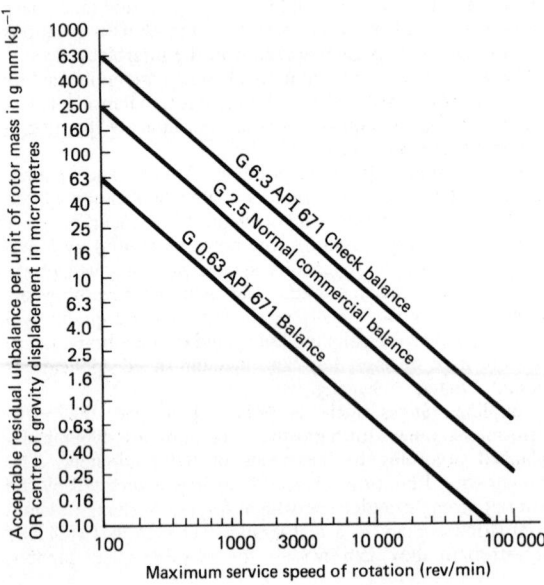

Figure 10.97 Unbalance versus speed (with acknowledgements to Flexibox Ltd)

of membrane banks, usually made of stainless steel or other corrosion-resistant materials. As all flexing occurs within the membranes, there are no wearing parts and therefore no necessity for lubrication, adjustment or any other form of maintenance. All Flexibox M, LS and T series membrane couplings incorporate spacer-retention anti-fly devices.

M series 'spoke form' membrane couplings The flexible element of the M series is a bank of spoked membranes secured rigidly at their inner and outer diameters. Designed for low-, medium- and high-speed operation, the M series design meets the AP1 610 specification and is particularly suitable for process pump applications. The axial stiffness of the M series membrane unit is strongly non-linear, making it inherently self-damping and therefore ideal for unlocated rotor motor applications without recourse to end stops.

In the event of a seizure or mal-operation of the driven or driving machinery, the coupling will spin freely after membrane failure and can therefore be used to protect major equipment. The MHSO Single bank non-spacer coupling accommodates axial and angular misalignment only. Its use is limited to such applications as three-bearing systems and cardan shaft configurations, using one coupling at each end of long spacer shaft or tube. The MHSS single-bank spacer coupling accommodates axial, angular and lateral misalignments. Lateral misalignment capacity is increased by longer DBSEs (distance between shaft ends). The coupling is used in most process and industrial machinery, particularly where there is a need for a shaft gap when changing machine bearings, seals, etc., without disturbing either machine.

The MODO non-spacer double-bank coupling accepts axial, angular and limited lateral misalignment and is used on close coupled machinery, where some lateral shaft misalignment has to be anticipated.

Ring-form tangential transmission link designs Metastream 'T' series flexible couplings employ ring-form banks of stainless steel membranes to combine a high power transmission/weight ratio with maximum flexibility and high torsional stiffness. The membranes are arranged to transmit the driving torque as a pure tensile load. The 'waisted' form ensures that the bending and fatigue stresses arising from misalignment are at a minimum in the critical areas around the driving bolts. This form permits high torque ratings with relatively small diameters so that the coupling can be used at high speeds without exceeding acceptable levels of stress.

The T series is sub-divided into the several ranges of couplings which, between them, offer a wide variety of capabilities and features required by rotating equipment designers. The LS spacer coupling design is an inexpensive simple arrangement for general industrial applications at low and medium speeds, with ratings from 2 up to 24 000 kW per 1000 rev/min.

High-performance couplings The high-performance range of Metastream couplings has been specifically designed to meet the requirements of manufacturers and users of high-performance rotating equipment and comply with API standard 671. High-performance coupling speeds can be as high as 30 000 rev/min. Coupling sub-assemblies are dynamically balanced to a limit of G1.25 ISO 1940, and assemblies check balanced to G6.3. The designs are specified for unspaced turbine and compressor applications at ratings up to 38 MW per 1000 rev/min. The inverted hub design allows the coupling effective centre of gravity to be moved close to the bearing for reduced overhung moment and minimized bearing loads (see Table 10.10).

Condition-monitoring couplings Torsionally stiff membrane couplings have no wearing parts; they need no lubrication or adjustment; they have a predictable high torsional stiffness that does not alter over a period of time and they have accessible low-stressed spacer tubes. These features make such couplings ideal for both the train-gauge and phase-displacement torque-measuring systems. The coupling is supplied with a factory-assembled transmission unit ensuring dynamic balance integrity and measuring system accuracy throughout the virtually unlimited life of the coupling.

Axially split couplings The TSEW coupling is a precision built, six-link, non-spacer T series coupling with two membrane units connected by an axially split spacer. The coupling is designed to allow maintenance of the membrane assemblies without moving either of the rotating machines. Shaft separations of only 3 mm can be achieved while allowing a practical misalignment capacity.

Optional features Membrane couplings can be supplied in spark-resistant designs for hazardous areas. Normal atmospheric corrosion protection is provided by stainless steel membranes and a phosphated finish on other steel parts. For more severe environments, other materials such as titanium, aluminium alloys and other types of finishes can be specified. For applications that need continuity of drive under overload or after membrane failure, modifications are available to meet these requirements. There are also torque-limiting devices and overload shear devices as well as an *in-situ* line-balancing correction.

10.2.5 Clutches, freewheels and brakes

Clutches, freewheels and brake units are important components in transmission systems and can be included as separate items or integrated with other transmission units. In the following the various types of clutches and brakes will be described as separate items.

In most machinery, the clutch or brake is remote from the operator and consequently they are provided with the means of remote control. These may be electric, pneumatic or hydraulic, and most manufacturers have a range of components catering for all these alternative means of control. For example, a typical air-operated clutch from Wichita is shown in Figure 10.98.

10.2.5.1 Dog clutches

These are positive-drive components and are normally operated only when they are stationary. Various tooth forms are used (see Figure 10.99); (1) straight-cut square teeth, (2) sawtooth formation and (3) gear type radial teeth. Because these components are more often regarded as couplings, mention has been made of ring-face tooth couplings in Section 10.2.4. An example of an electromagnetically operated tooth clutch is shown in Figure 10.100.

10.2.5.2 Freewheel clutches

These are more often referred to as freewheel, or over-running, clutches or even jamming roller clutches. There are two types. The first uses either balls or rollers spaced in inclined wedge-shaped spaces around the periphery of the hub (see Figure 10.101). If the speed of the driven shaft overtakes that of the driver, the balls or rollers tend to roll back out of contact with the driven member and a positive drive is disconnected. If the speed of the driver increases beyond the driven member, the balls or rollers are dragged into contact

Table 10.10 Gear coupling versus dry membrane coupling

	Gear	Membrane
Initial cost	Low on low-duty applications	Competitive
Operating and maintenance costs	High-lubrication and cooling circulation and filtration. Oil and oil seals frequently renewed	Negligible
Spares costs	Up to full replacement and regular oil seals	Membrane assemblies
Refurbishing costs	New coupling. Frequent new oil seals	Low, even if re-aligning and balancing
Inspectability	Dismantle coupling	Membranes are visible
Lubrication considerations	Lube oils can be tailored to suit specific application for best service, but 'compromised' by economic expedient. Must be clean and cool. Oil must be chemically compatible. The oil seal can limit misalignment capacities	None
Speeds	High	Very high
Temperature range	Limited by oil and oil seals (100°C typical)	Not usually a problem up to 200°C
Angular misalignment	Comparable	Comparable
Axial misalignment	High	Adequate even with thermal growth
Power:weight ratio	Very efficient use of materials	Very efficient use of materials
Close-coupled	Ideal	Standard modification
Dirty environment	Lube oil contamination	Generally no problem
Wear	Progressive. Rate is sensitive to the efficiency of the lube system. 'Wear band' can reduce misalignment capability unless designed with 'full tooth engagement'	None
Corrosion	Plating and oil seals must resist environmental and internal (oil) attack. Teeth are exposed to oil	Stainless St membrane and phosphated steel parts. Painting/special materials available for hostile areas
Mechanical properties	Change with wear.	Accurately predictable and consistent
Balance	Changes with wear, lube oil path, centrifuging	High qualities achievable Consistent over coupling life.
Torque overload effects	Excessive overload will do permanent damage	Taken by collars to protect against permanent damage
Anti-fly	By nature	By guard ring designs
Failure analysis	Chemical and metallurgical analyses can give clear indication of causes	Membrane fracture pattern point to possible causes
Backlash	Some, initially increasing with wear	Virtually zero
Sleeve-bearing rotors	Needs end stop	Non-linear axial stiffness gives inherent damping. No end stop needed
Vertical	Needs modification	Usually no modification
Misalignment loads	Negligible in ideal conditions. But 'torque lock' is common and imposes high loads on bearings, gears, seals, etc.	Very low generally, but excessive axial displacement produces high thrust reaction forces
Quality design and manufacture	Manufacturing accuracy is crucial to evenly stressed teeth and shared torque load sharing. Staggeringly complex design calculations lead to arbitrary and empirical formulae	Relatively easy to get right
Failure mode	Disintegration and loss of drive function	M series – drive disconnected LS & T series – drive maintained

with the outer member and the positive drive is established. Balls are only used in very light power applications as they have only a point contact; rollers, on the other hand, have a line contact and can be used for substantial torque loadings.

Generally, the larger the angle of the wedge, the greater the roller diameter and hence torque capacity while a small angle provides a more positive engagement. Response is virtually instantaneous since the rollers are always in contact with the inner and outer races. Thus, taking up the drive is a matter of breaking the intervening oil film and the natural deformation of the material under load. Rollers are normally energized by spring and plunger assemblies acting on them in the direction

of the trapping angle. In a phased roller clutch, the rollers are precisely located and guided by a cage which is spring energized so that all rollers engage in unison. Compared with the individual roller clutch, it can offer more uniform loading of the rollers and a greater torque capacity for a given size.

A variation of this type of clutch or freewheel is the sprag clutch in which the space between an inner and outer revolving race is filled with a series of cams or sprags whose major diameter is slightly greater than the radial space between the races (see Figure 10.102). Rotation of one race in the driving direction causes the sprags to tilt, thus transmitting the torque in full from one race to the other. Conversely, rotation of the

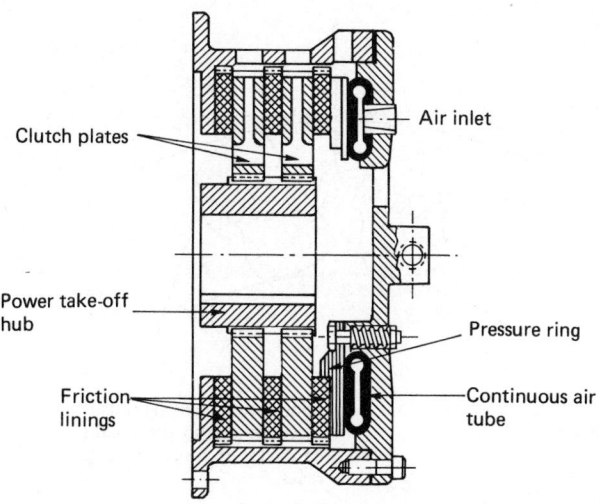

Figure 10.98 Typical air-operated clutch (with acknowledgements to Wichita Co. Ltd)

Figure 10.100 Suredrive electromagnetic tooth clutch (with acknowledgements to TI Matrix Engineering)

race in the other direction frees the sprags and permits over-running between the races. A tilting force keeps the sprags in light contact with both inner and outer races and this can be done using various spring arrangements. There is thus no loss of motion, the driving torque being instantaneously transmitted between race. In general, sprag clutches are able to transmit greater torques for a given overall size than other types of freewheel devices.

10.2.5.3 Cone friction clutch

The cone clutch (see Figure 10.103) embodies the mechanical advantage of the wedge which reduces the axial force required to transmit a given torque. In general engineering its use is restricted to the more rugged applications such as contractors' plant. In a smaller form it is often used in machine tools.

10.2.5.4 Plate friction clutch

This can be of single-plate type (see Figure 10.104) or multi-plate construction (Figure 10.105). Basically, the clutch consists of friction lining(s) sandwiched between driving and driven plate(s). Springs usually provide the clamping pressures. With

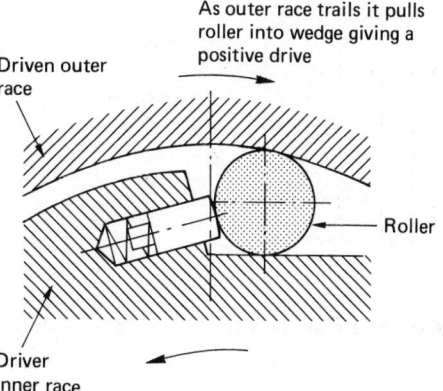

Figure 10.101 The wedging of rollers between inner and outer races to provide power transmission

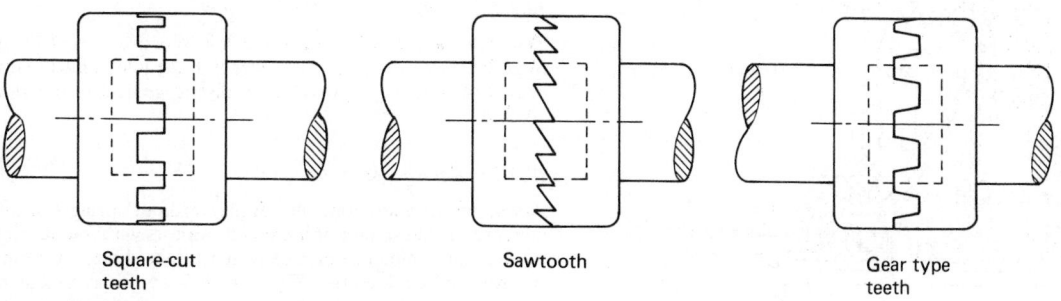

Square-cut teeth

Sawtooth

Gear type teeth

Figure 10.99 Various forms of teeth in a dog clutch

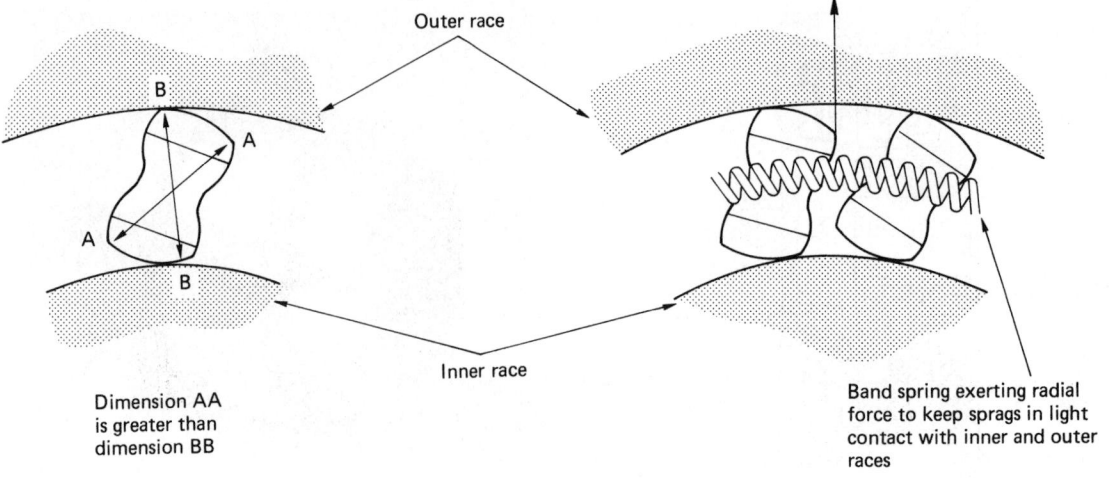

Figure 10.102 Elements in a sprag clutch

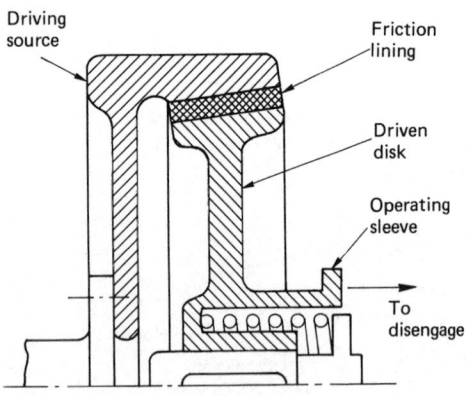

Figure 10.103 Section through a typical cone type friction clutch

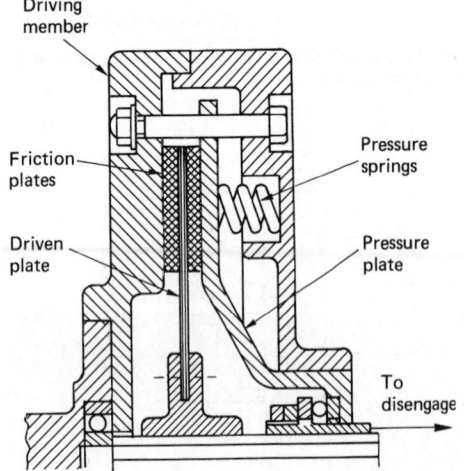

Figure 10.104 Section through a typical single-plate friction clutch

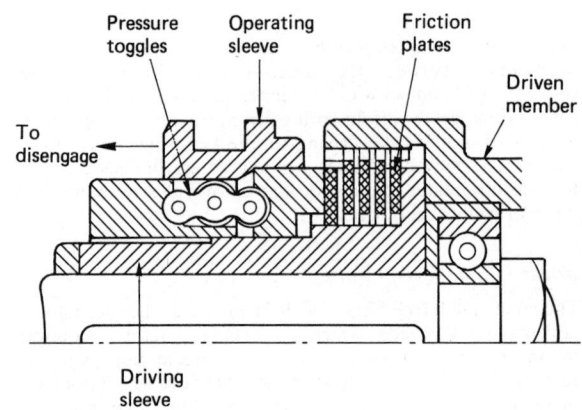

Figure 10.105 Section through a typical multi-plate friction clutch

multi-plate clutches the diameter can generally be reduced for a given torque as against a single-plate clutch. Many multi-plate clutches run in oil which helps to conduct away the generated heat.

10.2.5.5 Expanding ring friction clutch

This will transmit high torque at low speed and centrifugal force increases the gripping power but adequate clutch withdrawal force must be provided for disengagement (see Figure 10.106).

10.2.5.6 Centrifugal friction clutch

Automatic in operation, the torque, without spring control, increases as the square of the speed. Motors with low starting torque can commence engagement without shock. A spring control can be added (see Figure 10.107) so that engagement does not take place until the motor has reached a predefined speed (often 75% of full speed).

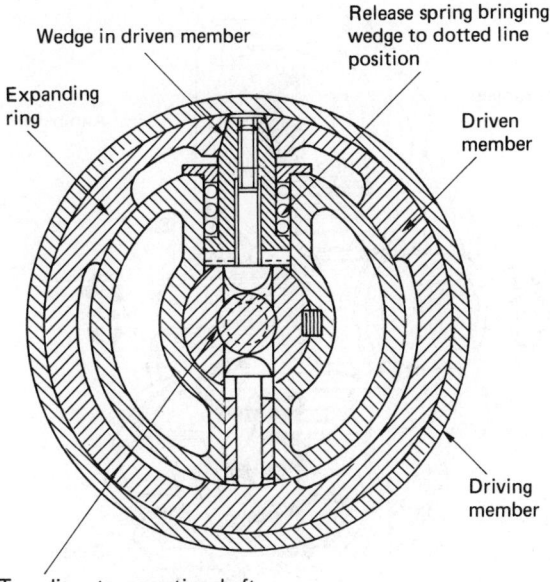

Figure 10.106 Expanding ring type of friction clutch

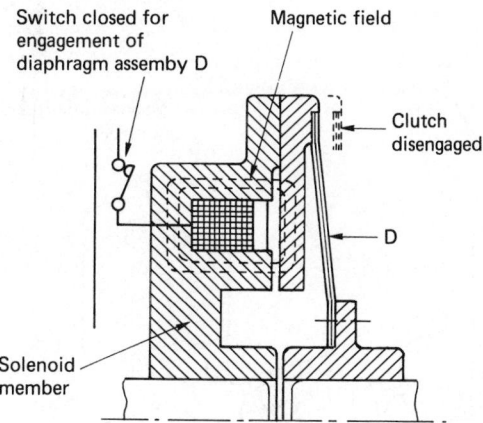

Figure 10.108 Typical single-plate electromagnetic clutch

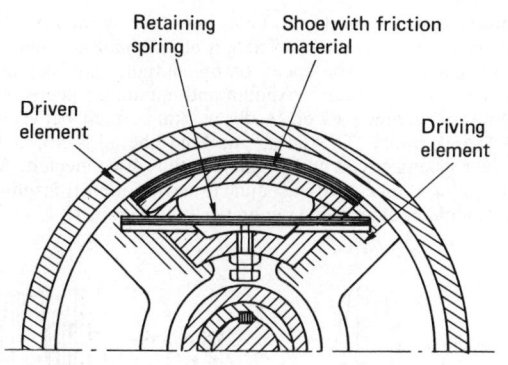

Figure 10.107 Typical mechanism of a centrifugal clutch with spring control

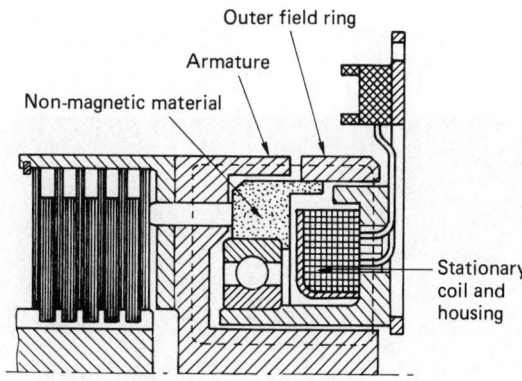

Figure 10.109 Typical multi-plate electromagnetic clutch

10.2.5.7 Magnetic friction clutches

These are compact units and operated by a direct magnet pull with no end thrust on the shafts (see Figures 10.108 and 10.109). It is ideal for remote control.

10.2.5.8 Particle clutches

These consist of inner and outer races with the annular space between being filled with magnetic particles. When a suitable current is applied the particles lock together with the races and form a drive. They can be used when constant slip is required and are suited to repetitive starts and stops or controlled accelerations. The same principle can be applied to braking.

For example, the Magne range of magnetic particle clutches and brakes from R. A. Rodriguez consist of only two parts, the inner race called the rotor and the outer race or drive cylinder. The space between the members is filled with a fine magnetic powder, and when a magnetic field is created through a stationary d.c. coil the powder forms a link between the two members and torque is thus transmitted. With the Magneclutch, both members are rotating to provide transmitted torque; with the Magnebrake, the outer member is held stationary, resulting in braking torque. Transmitted torque is proportional to the strength of the magnetic field. A typical application is shown in Figure 10.110.

10.2.5.9 Wrap spring clutch

These are generally used for low-torque applications and low speeds. They consist of a helical spring arranged to wrap against a drum surface. As the grip of the spring increases in proportion to the transmitted torque, the helical spring locks the driving and driven drum together. When the sleeve is released, the clutch is engaged; holding the sleeve unwinds the spring and this engages the clutch.

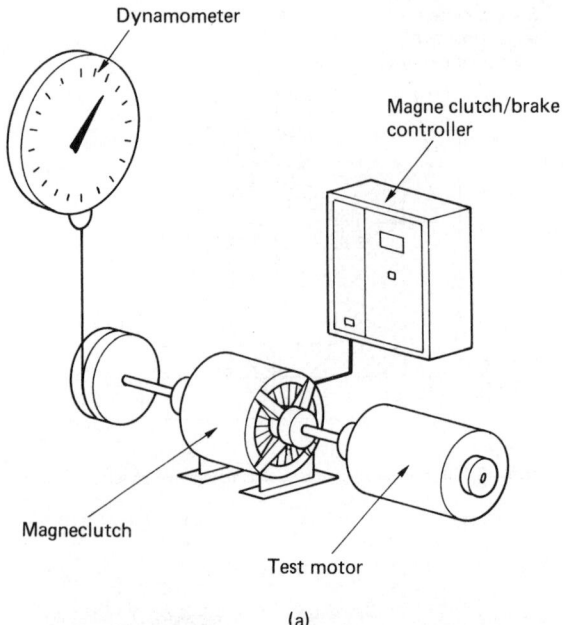

(a)

(b)

Figure 10.110 (a) Application of Magne particle clutch (with acknowledgements to R. A. Rodriguez); (b) typical magnetic particle clutches (with acknowledgements to Huco Engineering Industries Ltd)

10.2.5.10 Fluid coupling

A very important type of coupling or clutch is that employing a fluid drive. These couplings give the engineer an efficient, simple and reliably mechanical means of controlling the speed of the driven machinery at the same time, allowing the use of comparatively low-cost constant-speed squirrel-cage motors. In addition, they offer the advantage of a no-load start, smooth and progressive acceleration and protection from shock loadings.

The fluid drive is situated in the drive line between the motor and driven machine. As in all fluid couplings operating on the hydrodynamic principle, there are only two basic elements – the impeller and the runner (see Figure 10.111).

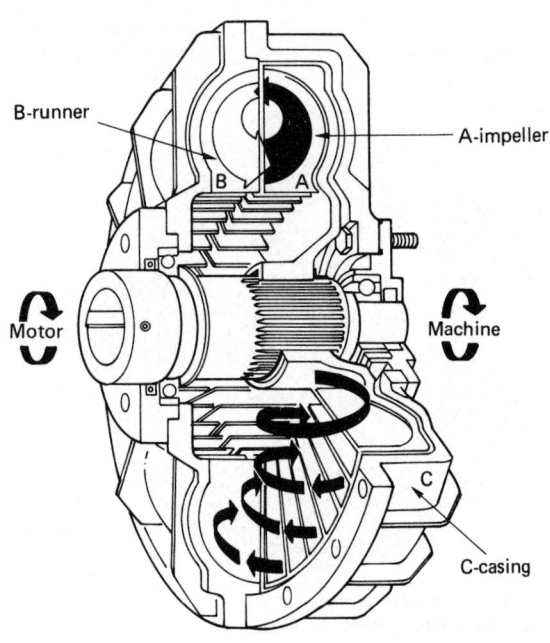

Figure 10.111 The principle of fluid coupling (with acknowledgements to Hansen Transmissions Ltd)

The power is transmitted from input to output by the flow of oil between the two elements. There is no mechanical connection between them. The speed of the output shaft can be varied steplessly between maximum and minimum speeds by adjusting the quantity of oil in the working circuit between impeller and runner. To stop the machine, the oil is emptied from the working circuit and the drive is thus disconnected. A fluid coupling can be used in conjunction with other transmission elements as shown diagrammatically in Figure 10.112.

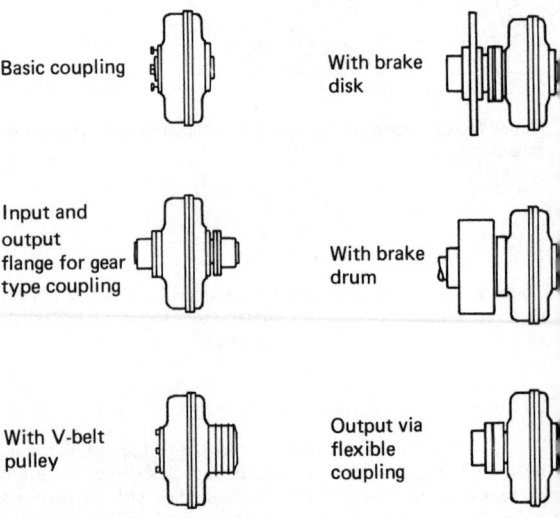

Figure 10.112 Examples of using a fluid coupling in conjunction with other transmission elements

10.2.5.11 Brakes

Many of the principles used in friction clutches can be applied to brakes. Large brake units of the type used in contractors' equipment can be band, caliper disk or drum types (Figures 10.113–10.115). Smaller versions than those used in contractors' machinery are available and an example of a caliper brake is shown in Figure 10.116. It can be used for dynamic braking to bring equipment to rest or as a holding brake to prevent motion.

10.2.5.12 Suppliers of couplings, clutches and brakes

Alanco-Alamatic Ltd, Wilton Street, Denton, Manchester M34 3WD	Disk brakes

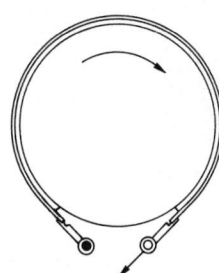

Simple band brake

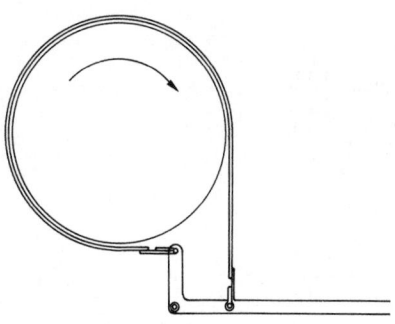

Reversible band brake

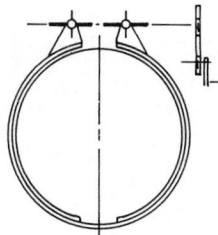

Screw-operated reversible band brake

Figure 10.113 Examples of band brakes

Flexibox Ltd, Nash Road, Trafford Park, Manchester M17 1SS	Couplings
Fluidrive Engineering Co. Ltd, Broad Lane, Bracknell, Berks RG12 3BH	Fluid couplings
Hansen Transmissions, Beeston Royds Industrial Estate, Geldern Road, Leeds LS12 6EY	Fluid couplings, gear units, disk and drum brakes, flexible couplings
Huco Engineering Industries Ltd, Peerglow Centre, Marsh Lane, Ware, Herts SG12 9QL	Couplings
Renold Gears, PO Box 224, Wentloog Corporate Park, Wentloog Road, Cardiff CF3 8YT	Flexible couplings
R. A. Rodriguez (UK) Ltd, Icknield House, Eastcheap, Letchworth, Herts SG6 3DF	Electromagnetic particle brakes and clutches
Simplatroll Ltd, Caxton Road, Bedford MK41 0HT	Torque limiters, electromagnetic clutches and brakes
Stieber Ltd, Stieber House, Works Road, Letchworth, Herts SG6 1PF	Disk caliper brakes, drum brakes, freewheels, couplings
TI Matrix Engineering Ltd, Brechin, Angus, Scotland DD9 7EP	Clutches, brakes, couplings, caliper disk brakes
Twiflex Ltd, The Green, Twickenham, Middlesex TW2 5AQ	Disk brakes, flexible couplings, clutches, fluid couplings
Voith Engineering Ltd, 6 Beddington Farm Road, Croydon, Surrey CR0 4XB	Fluid couplings, mechanical couplings, hydrodynamic caliper and drum brakes
Warner Electric Ltd, St Helen Auckland, Bishops Auckland, Co. Durham DL14 9AA	Electromagnetic couplings and brakes, wrap spring brakes, freewheels
Wellman Bibby Co. Ltd, Cannon Way, Mill Street West, Dewsbury, West Yorkshire WF13 1EH	Flexible couplings, torque limiters, brakes, clutches
Wichita Co. Ltd, Ampthill Road, Bedford MK42 9RD	Air-operated clutches and brakes, flexible couplings

Acknowledgements

Sections 10.2.1.2 (V-Belts) and 10.2.1.3 (Synchronous belt drives) were provided by J. M. Woodcock, Group Product Manager, Indirect Drives at Fenner Power Transmissions, and Section 10.2.4.2 on the Metastream ranges of power transmission couplings was supplied by Flexibox Ltd. These contributions are gratefully acknowledged.

Further reading

Dudley, D. W., *Handbook of Practical Gear Design*, McGraw-Hill, New York
Dyson, Evans and Snidle, 'Wildhaber–Novokov circular arc gears: Some properties of relevance to their design', *Proc. Royal Society* (1989)
Gear Lubrication, BGA Technical Memorandum No. 11
Merritt, H. E., *Gear Engineering*, Wiley, Chichester

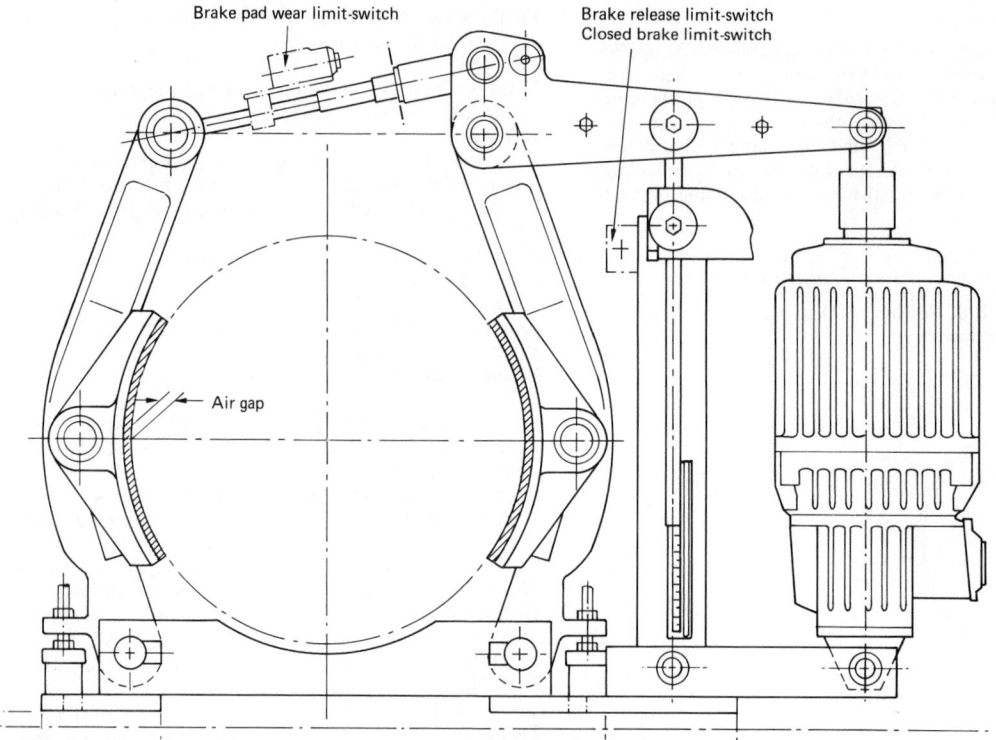

Brake pad wear limit-switch

Brake release limit-switch
Closed brake limit-switch

Air gap

Figure 10.114 Example of industrial drum brake (with acknowledgements to Stieber Ltd)

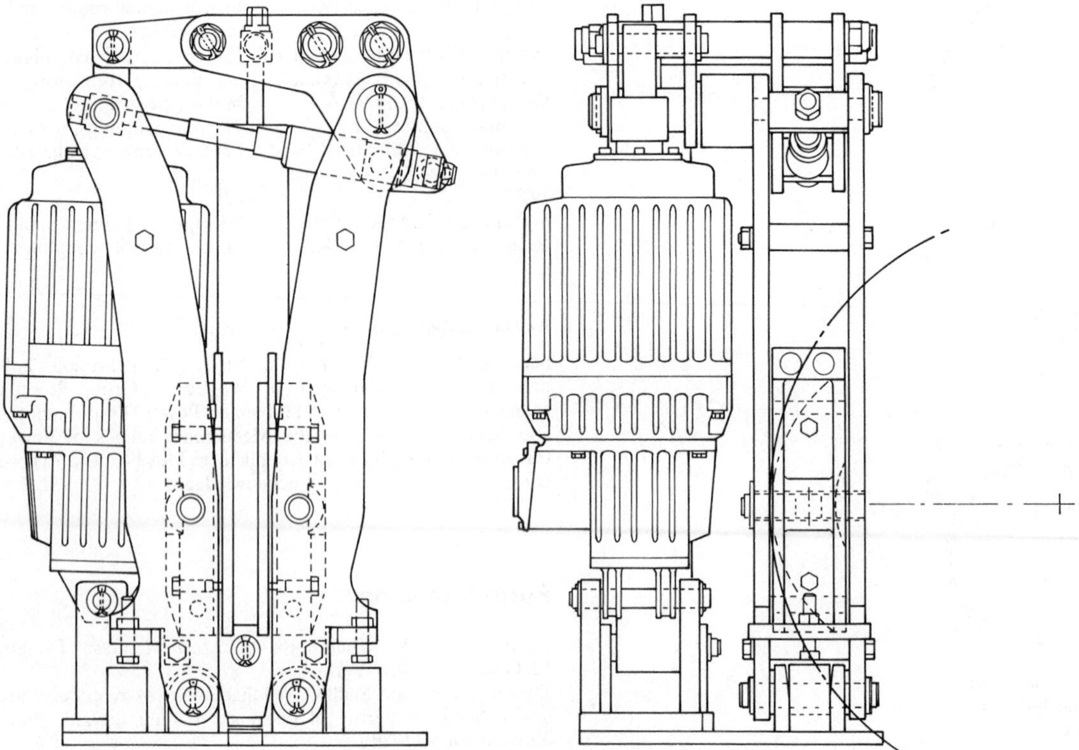

Figure 10.115 Example of a caliper disk brake (with acknowledgements to Stieber Ltd)

Figure 10.116 Surestop electromagnetically released caliper brake system (with acknowledgements to TI Matrix Engineering)

Mott, R. L., *Machine Elements in Mechanical Design*, Merrill, New York
Neale, M. J., *Tribology Handbook*, Butterworths, London
Watson, H. J., *Modern Gear Production*, Pergamon, Oxford

BS 228: 1984 (ISO 60–1982) Gears for electric traction (includes guidance for tooth profile modification)

BS 436: 1986: Parts 1, 2 and 3 Spur and helical gears

BS 545: 1982 Bevel gears (machine cut)

BS 721: Specification for worm gearing
Part 1: 1984 Imperial units
Part 2: 1983 Metric units

BS 978: 1968
Part 1 Fine pitch gears: involute spur and helical
Part 2 Gears for instruments and clockwork mechanisms: cycloid type gears plus Addendum No. 1 (1959) on double circular arc type gears
Part 3 Gears for instruments and clockwork mechanisms: bevel gears
Part 4 Gears for instruments and clockwork mechanisms: worm gears
Part 5 Fine pitch gears: hobs and cutters

BS 2519: 1976 Glossary of gears
Part 1 Geometrical definitions
Part 2 Notation

BS 4582 Fine pitch gears (metric module)
Part 1: 1984 Involute spur and helical gears
Part 2: 1978 Hobs and cutters

BS 3027: 1968 Dimensions of worm gear units

BS 3696 Specification for master gears
Part 1: 1984 Spur and helical gears (metric module)

BS 4185 Machine tool components

BS 5265: Part 1: 1979 Mechanical balancing of rotating bodies
API 671 Special purpose couplings for refinery services (American Petroleum Institute)

PD 3376 (1984) Addendum 1 to BS 978: Part 2 Double circular arc type gears

PD 6457 (1984) Guide to the application of addendum modification to involute spur and helical gears

Further information may be obtained from the British Gear Association, St James's House, Frederick Road, Edgbaston, Birmingham B15 1JJ

11

Fuels and combustion

Eric M. Goodger

Contents

11.1   Introduction   11/3

11.2   General fuel types   11/3

11.3   Major property overview   11/3
    11.3.1   Fuel density   11/3
    11.3.2   Calorific value   11/5

11.4   Major fuel groupings   11/5
    11.4.1   Gaseous fuels   11/5
    11.4.2   Liquid fuels   11/8
    11.4.3   Solid fuels   11/12

11.5   Combustion   11/14
    11.5.1   Fundamentals   11/14
    11.5.2   Applications   11/16

11.6   Conclusions   11/19

References   11/20

11.1 Introduction

The conventional chemical (i.e. non-nuclear) fuels are essentially materials containing stored energy that can be released as heat by the oxidation reaction of combustion. The released heat may be utilized as such for space heating, cooking, the generation of steam or treatment of industrial products. Alternatively, the heat may be converted partially into mechanical work through some type of heat engine, the extent of conversion being limited fundamentally by the second law of thermodynamics, and practically by the nature of the engine and its conditions of operation. Combustion performance is thus a prime parameter of any fuel, but the handling characteristics involved in storage, distribution and fire safety are also all of practical importance. Fuels are therefore required to burn readily and completely when required within the combustor but not at all outside it, and to resist the disturbing effects of extreme conditions of storage, with no damage to the materials of the fuel and combustion systems, and all at realistic costs. As a result, several property requirements may conflict, and specifications are customarily drawn up by authoritative bodies to establish the compromises necessary, and set the standards for fuel quality.

In this chapter, properties, specifications, handling, applications and, finally, combustion performance are considered for typical members of the conventional gaseous, liquid and solid fuels in use today, with brief reference to the likely candidates for the future.

11.2 General fuel types

Most conventional fuels are derived from sources of fossil origin, mainly by direct processing into the required commercial products but, in some cases, manufactured as by-products from other industrial processes. There is also a small but growing input from industrial and municipal wastes, used either directly or as derivatives.

The demise of living organisms results in deposits of hydrogen–carbon–oxygen compounds (carbohydrates) and water which, through the long passage of time, tend to lose their oxygen and water components. The main constituents of naturally occurring gases and petroleum liquids therefore comprise hydrogen–carbon compounds (hydrocarbons) only, whereas the solids still retain some oxygen and water, depending on the extent of coalification that has taken place. Industrially manufactured gases may also contain free hydrogen, carbon monoxide, carbon dioxide, nitrogen and hydrogen sulphide. Traces of sulphur, lead, vanadium, etc. exist in most liquid and solid fuel sources, and may need to be either reduced or removed to meet specification limits.

Atoms of carbon and hydrogen combine in very precise patterns dependent upon their bond-forming capacity (valency). With a valency of four for carbon and unity for hydrogen, and with single and multiple bonds possible between carbon atoms in both straight- and ring-chain structures, several million different molecular structures are possible, ranging from the simplest (CH_4, methane) upwards, each with its individual name, and physical and chemical properties.[1] With a few exceptions such as the naturally occurring gases, individual hydrocarbons are not used as such in large-scale practice since their extraction and preparation would be inordinately expensive. A conventional liquid fuel, for example, consists of several hundred different types of hydrocarbon that have been separated as a group from the parent crude oil, the overall properties of the group being determined by the combined properties of the individual components.

The physical nature of a fuel is determined largely by its component molecular size. The chemical bonding referred to above is mainly electrostatic in nature, some influence of which is also felt beyond the outer boundaries of the molecule. These external forces around small molecules, e.g. from C_1 (methane) to C_4 (butanes and butenes), are so weak that the molecules are able to move freely relative to each other, the bulk materials thus existing as gases. Since the external fields of larger molecules are correspondingly stronger, these materials take the form of liquids, any imposed relative movement giving rise to viscous resistance forces. Very large molecules, on the other hand, are so strongly bound together that they comprise solids (Table 11.1).

The chemical nature of a fuel is determined largely by molecular shape and by the nature of the carbon–carbon bond. As discussed later, compact shapes of molecules imply greater resistance to thermal agitation and spontaneous ignition. Single bonding between carbon atoms results from hydrogen saturation of the molecule and indicates storage stability, whereas multiple bonding arising from unsaturation leads to instability due to the tendency of these bonds to open out and form new bonds with like molecules (polymerization), so increasing the overall size of individual molecules, resulting in self-contamination with solid particles.

11.3 Major property overview

Whereas certain properties are specific to a given fuel type and its application, the following major properties apply throughout the chemical fuel range, and are therefore presented here on an overview basis.

11.3.1 Fuel density

Density is defined as mass per unit volume. For fluid fuels, this is qualified with some standard conditions of test in order to allow for volume change with temperature and in, the case of gases, with pressure. For solid fuels, allowances may be made for the air voids between individual lumps. This gives rise to the following definitions, shown with units recommended by the Institute of Petroleum:

Density of gaseous fuel at t_1 and p_1 =

$$\frac{\text{Mass of sample}}{\text{Volume of sample at } t_1 \text{ and } p_1}, \text{kg/m}^3 \text{ at 15°C and 1 atm}$$

Density of liquid fuel at t_1 =

$$\frac{\text{Mass of sample}}{\text{Volume of sample at } t_1}, \text{kg/l at 15°C}$$

Lump density of solid fuel =

$$\frac{\text{Mass of single lump of sample}}{\text{Volume of single lump of sample}}, \text{kg/l}$$

Bulk density of solid fuel =

$$\frac{\text{Mass of sample lumps}}{\text{Volume of sample lumps and voids}}, \text{kg/l}$$

These density terms are largely replacing the original 'specific gravity' (SG) and 'relative density' (rel. d), the numerical values themselves being equal in all three cases when applied

Table 11.1 Typical fossil-derived fuels (in increasing level of density)

Group	Specific name	Formula	Typical density at 15°C G (kg m^{-3})	L or S (kg l^{-1})	Specific energy (MJ kg^{-1} fuel)	(MJ kg^{-1} st.mix)	Energy density G (MJ m^{-3})	L or S (MJ l^{-1})
GASES	*Manufactured*	H$_2$	0.085	0.070^a	120.24	3.41	10.22	
	Hydrogen		0.47		37.66	2.92	17.70	
	Coke oven gas		0.67		15.21	3.01	10.19	
	Blue water gas		0.77		22.48	2.89	17.31	
	Carburetted water gas		1.09		4.34	2.05	4.77	
	Coke producer gas	C$_2$H$_2$	1.10		48.26	3.41	53.13	29.44
	Acetylene	CO	1.185		10.10	2.92	11.97	
	Carbon monoxide		1.27		2.46	1.53	3.13	
	Blast furnace gas							
	Naturally occuring	CH$_4$	0.679	0.424^a	50.00	2.75	33.95	21.20
	Methane	C$_3$H$_8$	1.90	0.505	45.48	2.78	86.42	22.97
	Propane	C$_4$H$_{10}$	2.53	0.580	44.43	2.78	112.41	25.77
	Butane							
LIQUIDS	*Petroleum-derived blends*	(C$_8$H$_{14.6}$)	5.0	0.75	43.0	2.77		32.3
	Motor gasoline	(C$_{12.5}$H$_{24.4}$)	7.8	0.80	43.4	2.76		34.7
	Aviation kerosine	(C$_{15}$H$_{27.3}$)	9.3	0.84	42.9	2.77		36.0
	Gas oil			0.87	41.8	2.75		36.4
	Diesel fuel			0.97	40.3	2.70		39.1
	Residual fuel oil (Class G)			1.02	38.2	2.92		39.0
	Coal tar fuel 50			1.25	36.0	2.67		45.0
	Coal tar fuel 400							
SOLIDS	*Naturally occuring*			0.5–(0.8)–1.1	17.30	2.84		(13.84)
	Wood			1.15–(1.2)–1.25	19.7	3.08		(23.64)
	Peat			1.25	25.7	3.18		32.13
	Lignite/brown coal			1.28	32.62	3.52		41.75
	Coal, general-purpose			1.35	34.30	3.31		46.31
	Coal, high volatility			1.45	35.34	3.20		51.34
	Coal, medium volatility			1.4–(1.55)–1.7	33.33	3.21		51.66
	Anthracite							
	Manafactured	C(gr)		1.75–(1.88)–2.0	33.08	3.23		62.19
	Coke			2.21	32.76	2.63		72.40
	Carbon (graphite)							

G = Gas   L or S = Liquid or solid   ( ) = Average ^{a} At boiling point and 1 atm.

to liquids and solids. Relative density is still used in connection with API gravity (see below) and Wobbe number (Section 11.4.1.1), and is given by expressions of the following type:

$$\text{rel. d at } t_1/t_2 = \frac{\text{Mass of given volume of sample at } t_1}{\text{Mass of equal volume of reference fluid at } t_2}$$

For gases, the reference fluid is atmospheric air, and for liquids and solids is pure water.

The test method used for the density of gaseous fuels can be the direct measurement of the mass of the sample in a Chancel flask of known volume. Alternatively, the relative density can be determined using a single orifice by comparing the flow rates of equal volumes of sample and air at common upstream conditions, since flow rate is inversely proportional to (density)$^{1/2}$. Hence

$$\text{rel. d of sample} = \left(\frac{\text{Sample flow time}}{\text{Air flow time}}\right)^2$$

For a liquid fuel, as before, the most direct method of determining density involves measuring the mass of the sample contained in a density bottle of known volume, but the commonly used indirect method for the less viscous liquids is based on the buoyancy of a hydrometer floating in the sample,

with a correction for every degree of displacement from the standard temperature level. For liquid petroleum fuels in the United States, relative density is used in terms of the gravity system of the American Petroleum Institute, as follows:

$$\text{Degrees API} = \frac{141.5}{\text{rel. d at } 60/60°C} - 131.5 \quad \text{(see Figure 11.1)}$$

For solid fuels, the more precise lump density is determined by the displacement method in which the sample lump is weighed both in air and in water, the difference indicating the weight of the volume of water displaced by the lump, and therefore the volume of the lump itself. Bulk density can be determined by using measured quantities of water, mercury or other immiscible liquid to fill the voids between a representative number of lumps of sample in the normal packing density.

For the conventional fuels, typical densities range from 0.679 to 2.53 kg/m^3 at 15°C and 1 atm for gases, 0.72 to 1.25 kg/l at 15°C for liquids and 0.5 to 2.21 kg/l for solids and, as shown in Figure 11.2, the progressive rise in density from gaseous methane to solid graphite is associated with a reduction in hydrogen content, and corresponding increase in carbon content. The density of conventional fuel therefore gives an instant approximate indication of the general type of

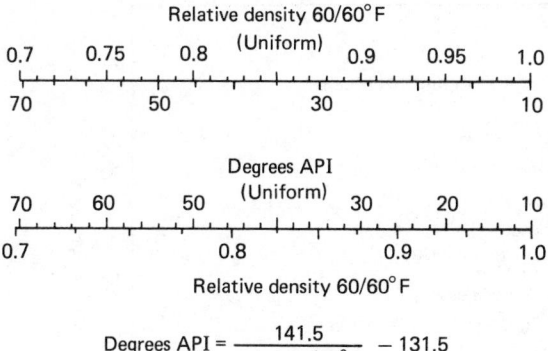

Degrees API = $\dfrac{141.5}{\text{rel. d } 60/60°F}$ − 131.5

Note: rel. d 15/4°C = 0.99904 (rel. d 60/60°F)

Figure 11.1 Relationship between relative density and degrees API[2]

the fuel (e.g. natural gas, kerosene or coal) in terms of a simply determined single value. In the case of liquid fuels, it also relates the more fundamental unit of mass to the customary handling quantity of volume, and therefore permits the determination of the storage tank size for a given mission or energy requirement.

With the blending of a number of fluids of different densities, and on the assumption of no chemical or physical interactions between them, blend density bears a linear relationship with concentration on a component volume basis, but not on a component mass basis, as shown in the following expressions:

Blend density = $0.01 \, (v_A \, d_A + v_B \, d_B + \ldots)$

Blend density = $\dfrac{100}{(m_A/d_A + m_B/d_B + \ldots)}$

where v_A, v_B, etc. are percentage volume concentrations of components A, B, etc., m_A, m_B, etc. are percentage mass concentrations of components A, B, etc. and d_A, d_B, etc. are densities of components A, B, etc.

11.3.2 Calorific value

Since a fuel is, in essence, an energy store, the quantity of energy stored within a given fuel is a property of fundamental importance. For gaseous fuels, the volumetric basis is invariably used, and calorific value is then defined as Energy Density, with recommended unit MJ/m³. For liquid and solid fuels, on the other hand, the gravimetric basis is generally employed, together with the term Specific Energy, and recommended unit MJ/kg; when Energy Density is required in these cases, the recommended unit is MJ/l.

Laboratory determination of these quantities is based on complete combustion at conditions of constant-pressure steady-flow for the gaseous fuels, and constant-volume non-flow for the liquids and solids, the released heat being absorbed in cooling water. This is not only convenient experimentally, but in each case meets the thermodynamic requirements for the energy to be released entirely as heat transfer rather than partly as work transfer, the latter being far more difficult to measure. In the flow test, corrections are made for the energy remaining in the flue gases, and for the expansion of the cooling water, whereas in the non-flow test allowance is made for the energy contributions of the ignition wire (and of

the gelatine capsule used for containing the more volatile samples), together with the heat of formation of the acids produced. Gaseous fuels burn completely at atmospheric pressure, but liquids and solids require higher pressures and an oxygen atmosphere. All the experimental results are 'gross' values since the final temperature is so slightly above ambient that the combustion-generated water condenses and so adds its latent heat of vaporization to the combustion energy. In practice, however, combustion products invariably leave the combustor at high temperature, either to provide draught in the case of boilers/furnaces, or as constrained by the heat–work conversion limit imposed by the second law of thermodynamics in the case of heat engines. Hence, 'net' values are more meaningful, and are obtained by suitable corrections for latent heat, as indicated in the latest issue of reference 4.

As shown in Figure 11.2 for the hydrocarbon fuels, the level of hydrogen content falls with increase in fuel density, and since the specific energy for hydrogen is far greater than that for carbon, the level of specific energy also tends to fall, as shown in Figure 11.3(a). Using the recommended units, values of energy density in MJ/l for the liquids and solids are determined directly by multiplying specific energy with density (MK/kg × kg/l). As seen from Figure 11.3(b), the small fall in specific energy is more than compensated by the relatively large rise in fuel density, consequently energy density rises with fuel density. Thus the highly attractive specific energy of hydrogen, for example, is offset to some extent by its very low energy density necessitating large storage volumes.

The above considerations on calorific value are important in terms of storage capacity and combustor fuel consumption for a given power output. One additional point of interest follows from the fact that combustors necessarily operate with mixtures of fuel and air rather than fuel alone, hence the key parameters here are the calorific values of these mixtures. Using stoichiometric mixtures (see Section 11.5.1) as a basis for comparison, the air/fuel mass ratios are found generally to fall with increasing fuel density. Hence, although the level of specific energy is lower with the heavier fuels, less air is required for their combustion and thus less mass in which to distribute the resulting heat. Broadly, therefore, little change is expected in the specific energies of the stoichiometric-air mixtures for all the fuels within the hydrocarbon range. This conclusion can be seen to hold also with fuels that already contain some components of air (e.g. carbon monoxide, CO; methanol, CH_3OH; and ammonia, NH_3). As shown in Figure 11.4, a mean value of about 2.9 MJ/kg of stoichiometric fuel–air mixture applies to a wide range of hydrocarbons and related fuels, and similar remarks apply to energy densities of these mixtures. Thus a changeover from a high- to a low-energy fuel with a constant *mixture* mass flow rate and comparable mixture strength would not be expected to entail a marked change in power output, but would, nevertheless, give rise to a higher rate of fuel consumption.

11.4 Major fuel groupings

Having reviewed those properties applying throughout the range of conventional chemical fuels, attention now turns to the main fuel groups in which certain production procedures and properties are specific.

11.4.1 Gaseous fuels

This group subdivides into those that occur naturally throughout the world, with broadly comparable characteristics, and

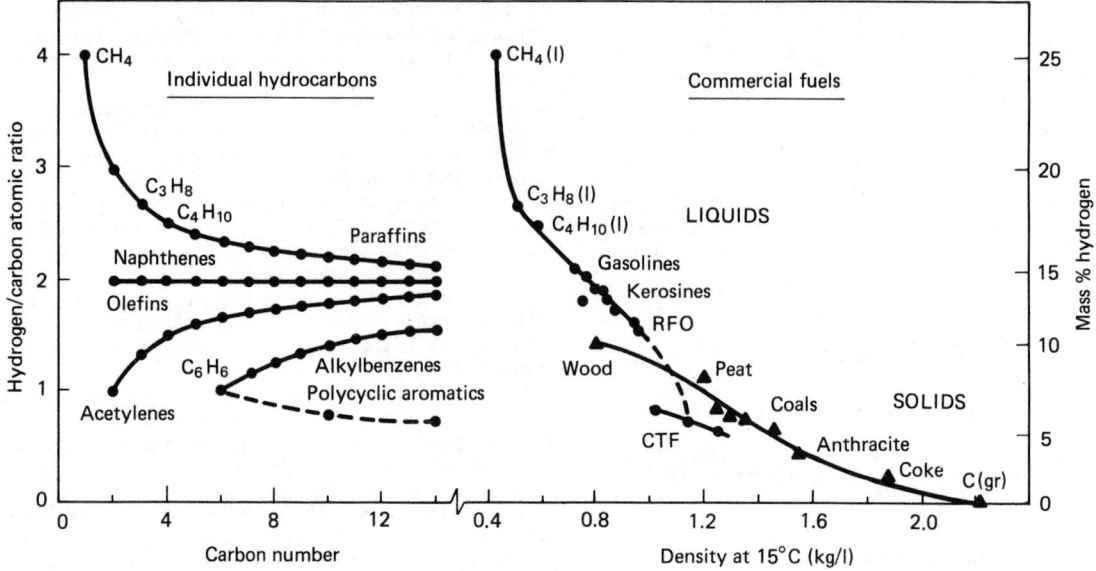

Figure 11.2 Hydrogen content of hydrocarbons and commercial fuels[2]

those that are manufactured within industry, frequently as by-products, and utilized wherever possible on-site.

11.4.1.1 Naturally occurring gases

These comprise the 'natural gases' based largely on methane, and the 'petroleum gases' based on propane and butanes. Worldwide deposits of natural gas exist either freely or in association with petroleum. In addition to methane, sources of natural gas may contain some condensable higher hydrocarbons (from ethane to octanes), and are hence referred to as 'wet' gases, together with traces of such inerts as nitrogen, carbon dioxide, water vapour and hydrogen sulphide. Expanding the gas at the well head reduces the temperature and freezes the water and heavier hydrocarbons which are separated from the gas stream and subsequently melted. At the gathering station, further stripping of the gas takes place by means of compression and cooling, together with sweetening by absorption or adsorption. In some instances, methane-containing gases are collected from coalmines, landfill garbage sites and sewage farms, and then used to power diesel-electric generating sets.

Natural gas has now become the standard commercially supplied gas for industrial and domestic space heating, hot water supplies and cooking, and also for some applications to boilers, furnaces, kilns and industrial gas turbine engines.

The condensate gases extracted from wet natural gas consist mainly of propane and butane together with traces of lighter and heavier hydrocarbons, as do the gases released from petroleum itself. The commercial grades of propane and butane each contain a minimum of 95% by volume of the named gas, with concentrations of other materials limited by such specifications as BS 4250.[5] Commercial mixtures of propane and butane are available under the name of petroleum gases and, when liquefied by compression and cooling for convenient storage, of LPG. The principal treatment of petroleum gases comprises either the removal or sweetening of mercaptan sulphur compounds in order to combat odour

and corrosion. Petroleum gases are used for industrial, commercial, agricultural and domestic heating, drying and cooking, and one particular use of propane is in hot-air balloons. The major properties of the naturally occurring gases are shown in Table 11.2.

In some heating applications where different fuel gases become available, interchangeability requires that the thermal throughput remains constant. Using the same upstream gas pressure and fixed orifice size, the volumetric flow rate is inversely proportional to (fluid density)$^{1/2}$, as discussed in Section 11.3.1. Hence the fuel interchangeability for constant heating rate requires near-constancy of the following parameter:

$$\text{Wobbe number} = W_0 = \frac{\text{Energy density}}{(\text{rel. d})^{1/2}} \text{ MJ/m}^3$$

If a different gas pressure is used with the different fuel, interchangeability is based on

$$\text{Extended Wobbe number} = W_0(\Delta p)^{1/2} \text{ MJN}^{1/2}/\text{m}^4$$

This leads to the concept of a 'simulated' fuel, as, for example, with a volumetric blend of 61 propane/39 air to simulate natural gas by matching the net Wobbe number of 45.61 MJ/m^3. A difference in W_0 of 5% is usually acceptable.

11.4.1.2 Industrially manufactured gases

Members of this group range in increasing density from hydrogen to blast-furnace gas, and are mostly by-products of other manufacturing processes. The source material is either coal or coke, and production is by carbonization (heating in the absence of air), by gasification through partial oxidation (heating with limited air), or by catalytic hydrogenation of coal directly to gas. More recent processes involve nitrogen-free gases for blast purposes, higher pressures, fluidized solid beds and/or removal of ash in the form of slag.[6]

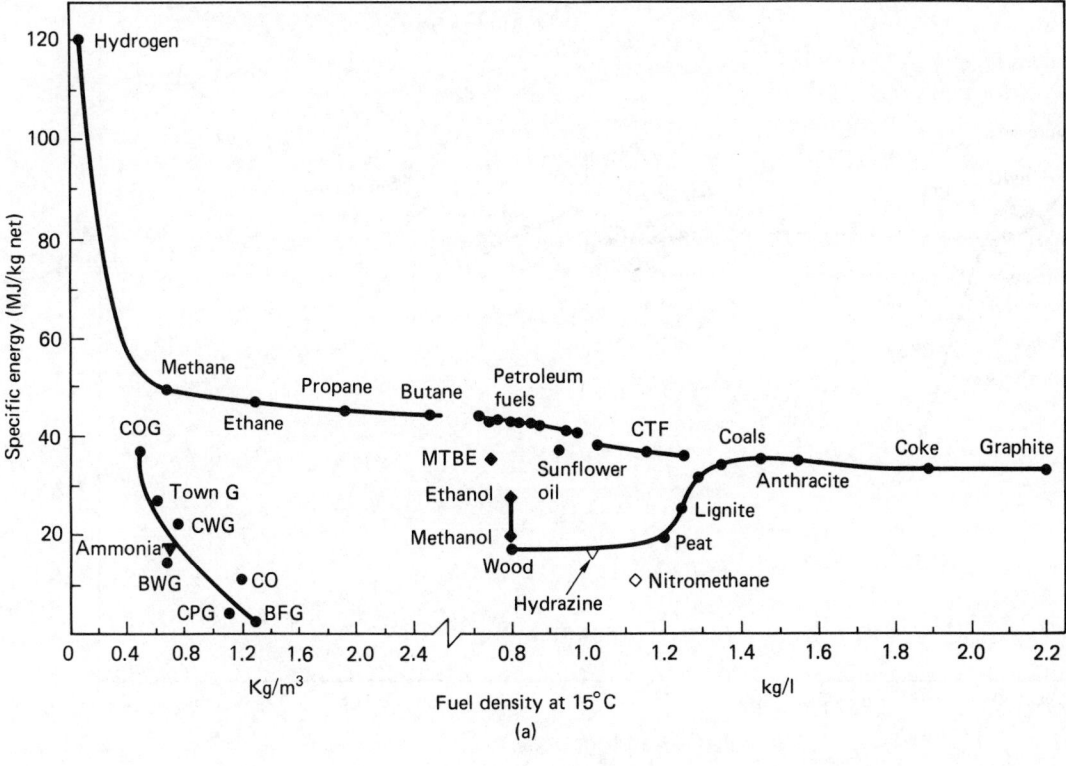

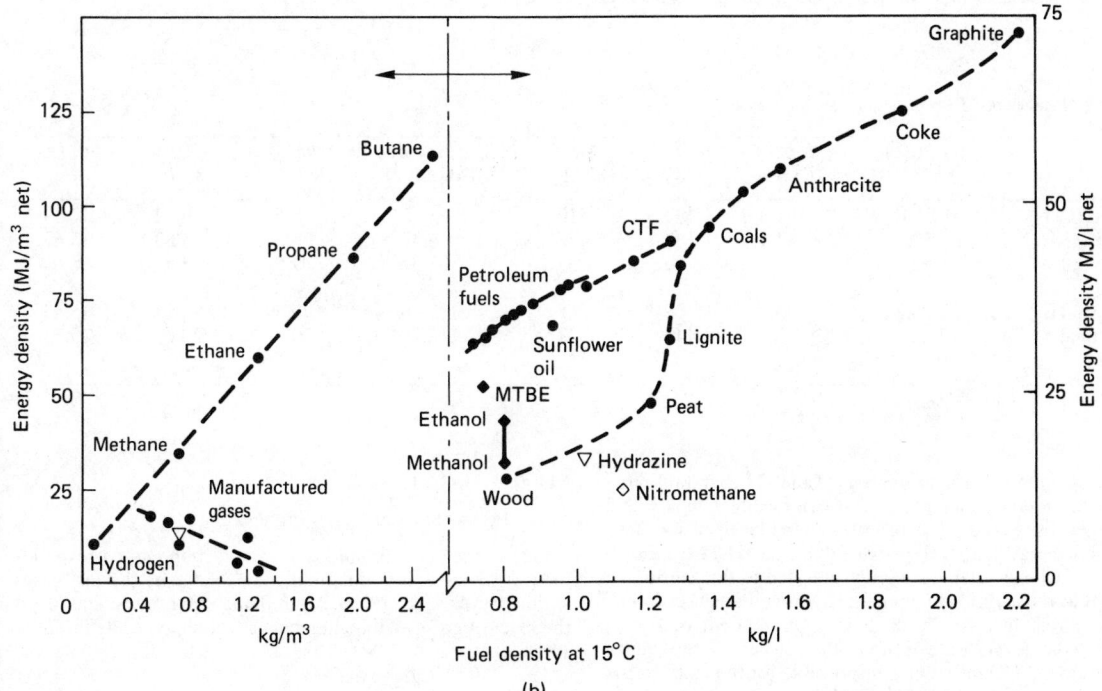

Figure 11.3 (a) Variation of net specific energy with fuel density;[2] (b) Variation of net energy density with fuel density[2]

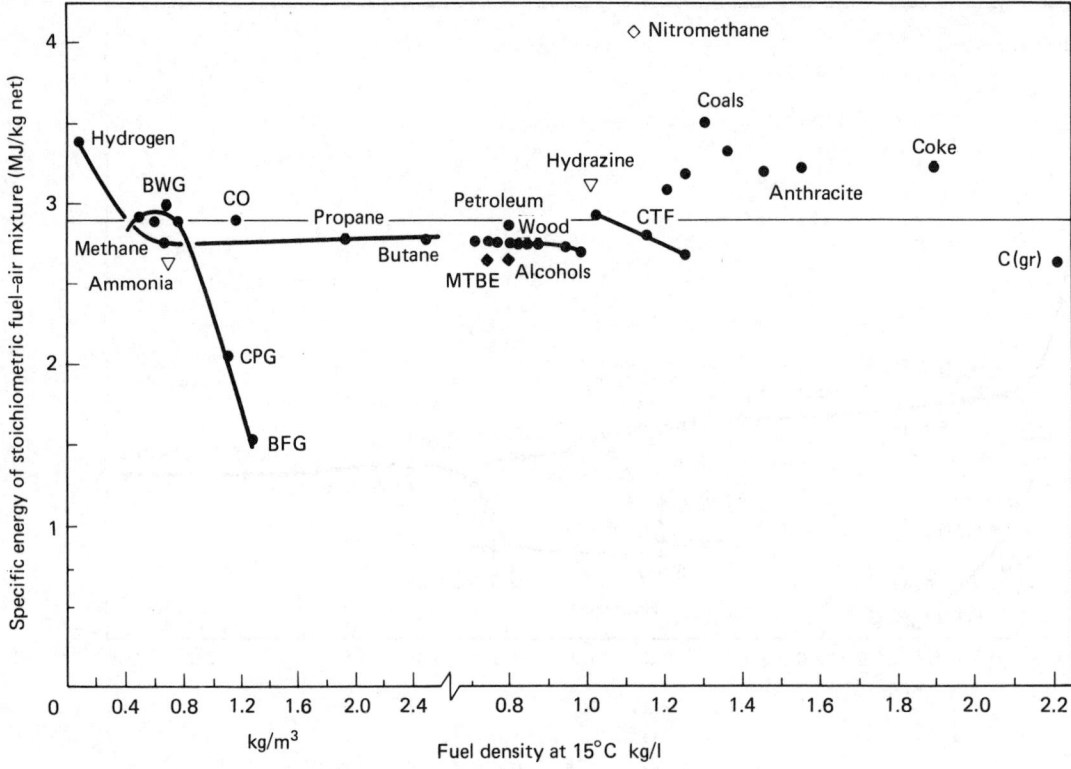

Figure 11.4 Illustration of minor variation in specific energy of stoichiometric fuel–air mixtures[2]

Table 11.2 Naturally occurring gaseous fuels (properties)

Fuel	rel. d (g) at 15°C, 1 atm	b.p. (°C)	m.p. (°C)	ΔH_{vap} (kJ kg^{-1})	Net energy density (MJ m^{-3})	Net W_o (MJ m^{-3})	$(A/F)_s$ Molar	$(a/f)_s$ Mass
Methane (CH$_4$)	0.554	-161.5	-182.5	510	33.95	45.61	9.52	17.19
Propane (C$_3$H$_8$)	1.522	-42.1	-187.7	426	86.47	70.09	23.80	15.63
Commercial propane	1.52	-41	-186	358	85.7	69.5	23.5	15.6
Butane (C$_4$H$_{10}$)	2.006	-0.5	-138.4	385	112.46	79.40	30.94	15.42
Commercial butane	2.01	-3	-140	372	112.2	79.1	30.0	15.3

ΔH_{vap} = enthalpy of vaporization.

The data for the gaseous fuels in Table 11.3 show that, broadly, the total volume content of combustibles falls with increasing density, and this is reflected in the levels of specific energy and energy density shown in Table 11.1. The exception is CWG, which has relatively high values due to the 'carburation', that is, the addition of energetic hydrocarbon gases from cracked petroleum oils. The high nitrogen content of BFG constrains the levels of calorific value, whereas the high concentration of CO in BWG is responsible for the characteristic blue coloration of the flame. The general term 'water gas' follows from the water–gas reaction

$$C(gr) + H_2O(g) \rightleftharpoons CO(g) + H_2(g)$$

and the water–gas shift reaction

$$CO(g) + H_2O(g) \rightleftharpoons CO_2(g) + H_2(g)$$

where (g) = gas, and (gr) = graphite

Due to the wide variations in composition and corresponding properties, it is common practice to classify the fuel gases within families, sometimes containing sub-groups, based on the level of gross Wobbe number, as shown typically in Table 11.4.

11.4.2 Liquid fuels

This group sub-divides into the liquefied fuel gases, and the wide range of products derived from refining the various types of crude oil found worldwide, and the tars derived from coals.

Table 11.3 Manufactured gaseous fuels (production, composition and properties)

Fuel	Production method	Composition % volume						rel.d 15°C 1 atm	NET W_0 (MJm^{-3})	$(A/F)_s$ Molar	$(a/f)_s$ Mass
		Combustible			Non-combustible						
		C_xH_y	H_2	CO	CO_2	O_2	N_2				
Hydrogen (H_2)	Decomposition of hydrocarbons and water	0	100	0	0	0	0	0.070	38.62	2.380	34.190
Coke oven Gas (COG)	Carbonization of coal optimized for coke	30.6	54.0	7.4	2.0	0.4	5.6	0.38	28.72	4.572	11.916
Blue water Gas (BWG)	High-quality BFG from coke and steam	0.8	49.0	41.0	4.7	0	4.5	0.55	13.74	2.218	4.055
Carburetted water gas (CWG)	BWG enriched with hydrocarbons	21.0	37.0	30.5	5.6	0.4	5.5	0.63	21.81	4.270	6.793
Coke producer gas (CPG)	Complete gasification of coal by partial oxidation	0.5	11.0	29.0	5.0	0	54.5	0.89	5.06	1.000	1.114
Acetylene (C_2H_2)	Controlled decomposition of methane	100	0	0	0	0	0	0.899	56.03	11.900	13.255
Carbon monoxide (CO)		0	0	100	0	0	0	0.967	12.17	2.380	2.461
Blast furnace gas (BFG)	Low-quality CPG from coke and air	0	2.5	24.0	17.5	0	56.0	1.04	3.07	0.631	0.609

Table 11.4 Representative families of fuel gases

Family	Gross Wobbe number (MJm^{-3})	Examples
Low energy	Up to 16	BFG, CPG, CO, BWG
1	24.4–28.8	CWG, COG
2	48.2–53.2	NG
3	72.6–87.8	PG, C_2H_2

11.4.2.1 Liquefied fuel gases

After sweetening at the gathering station, natural gases destined for liquefaction have the remaining water vapour removed by refrigeration, absorption in concentrated glycol solution, or adsorption in such solid desiccants as silica gel, alumina or molecular sieves. Liquefaction itself follows by compression and cooling below the dew point. Since its critical temperature of −82.6°C is below ambient, methane-based natural gas is a true gas rather than a vapour, and needs to be stored at its boiling point (−161.5°C) if required in the liquid phase at atmospheric pressure. Loss of refrigeration by inward heat leakage during storage is restricted by the use of double-skinned cryogenic tankage of metal, or sometimes of pre-stressed concrete or underground caverns. Such vaporization as does occur (e.g. 0.1% per day) is handled by reliquefaction on site. Rapid revaporization for immediate usage is achieved by means of heat exchangers fed with combustion gases, steam, or water.

The petroleum gases propane and butane are readily lique-fied, either individually or as LPG mixture, by compression and cooling. Since their critical temperatures are above ambient, and their boiling points not far below ambient, their storage as liquids at ambient temperature requires only moderate pressures, and hence reasonably lightweight containers.

11.4.2.2 Natural gasoline

The vapours of the heavier hydrocarbons ranging from C_5 (pentanes) to C_8 (octanes) obtained by stripping natural gases comprise 'natural gasoline' which is usually stabilized by removal of traces of petroleum gas components through more precise distillation at pressures above atmospheric. Such stabilized natural gasoline is commonly blended with conventional petroleum-derived gasoline.

11.4.2.3 Petroleum fractions

Most crude petroleums are yellowish-brown to black, relatively free-flowing liquids of densities ranging from about 0.78 to 1.0 kg/l, consisting essentially of hydrogen-saturated hydrocarbon compounds. The precise composition, and hence properties, depend upon the location, age and nature of the organic source materials, and crudes are classified on a semi-empirical basis, with increase in density showing a change from paraffinic to naphthenic.

The primary step in refining is the thermal separation by distillation into a number of fractions ranging from gasolines to residual fuel oils (Tables 11.1 and 11.5), each of which contains several hundred different hydrocarbons boiling within certain predetermined ranges of temperature. The relative quantities of these 'straight-run' products are then adjusted to meet the demand pattern by chemical methods of molecular-

Table 11.5 Liquid fuels derived from petroleum and coal tar (typical values of main properties)

Fuel	Boiling range (°C)[a]	Freezing or pour point (°C)	Kinematic viscosity (cSt at 15°C)	Flash point (°C)	SIT (°C)	H_{vap} at bp (kJ kg^{-1})	(a/f) mass
Aviation gasoline	46–145	− 65	0.7	− 40	427	292	14.9
Motor gasoline	33–190	− 65	0.6	− 43	400 (4*)	279	14.5
Kerosene	144–252	− 48	1.8	55	255	207	14.7
Gas oil	180–360	− 7	6.0	68	247	177	14.5
Diesel fuel	200 +	− 20(P)	25	80	245		14.2
RFO (Class F)		0(P)	800	93	300		13.9
RFO (Class G)		10(P)	6000	116			13.9
CTF 50			11	82			12.1
CTF 400			23	120			12.5

[a] Not measurable above 370°C due to cracking.
(P) = Pour point.
H_{vap} = enthalpy of vaporization

bond breaking and joining, followed by various forms of finishing treatment ready for the energy market. For convenience, the following UK Joint Services Designations are used to identify the lower-density fuels:[7]

AVGAS – Aviation gasoline
MOGAS – Motor gasoline
AVTAG – Aviation wide-cut turbine fuel (a blend of gasoline and kerosene components used widely until recently for military aircraft)
AVTUR – Aviation turbine fuel (aviation kerosine)
AVCAT – Aviation high-flash turbine fuel (lower-volatility aviation turbine fuel for naval aircraft)

A number of properties relevant to petroleum fuel blends are grouped in the following two sections.

Volatility This concerns the readiness of a liquid to convert to vapour at the prevailing conditions of temperature and pressure. Vaporization takes place relatively slowly at ambient conditions, as with the humidity of the atmosphere, but becomes vigorous with the rapid evolution of bubbles when the level of vapour pressure reaches that of the atmosphere itself, i.e. at the boiling point. Broad comparisons of fuel volatility are therefore made using a standard test method by measuring the boiling point or, in the more usual case of fuel blends, the boiling range. A sample of the fuel is vaporized at a controlled rate by heating in a flask, condensed in an ice-cooled side tube, and collected in a graduated cylinder. Corresponding readings of vapour temperature and condensate recovery are plotted as a distillation curve showing the boiling levels of the various components (Figure 11.5 and Table 6). With petroleum fuels, a maximum test temperature limit of 370°C is set in order to avoid cracking of the heavier hydrocarbon molecules. More precise results for research purposes are obtained by measuring the vapour/liquid volume ratios on either side of the initial boiling point.

For the lower-density fuels at the more volatile end of the range, volatility is also assessed by measuring the absolute vapour pressure. In the standard Reid test, the liquid fuel sample is heated in a container to a moderate temperature (37.8°C), the vapour rising through an air chamber fitted with a pressure gauge. A modified method is used for liquefied gases, and a micro method is also available.

For the range of petroleum fuels from gasolines to fuel oils, the weak *mixture* limit of flammability is common at about 1%

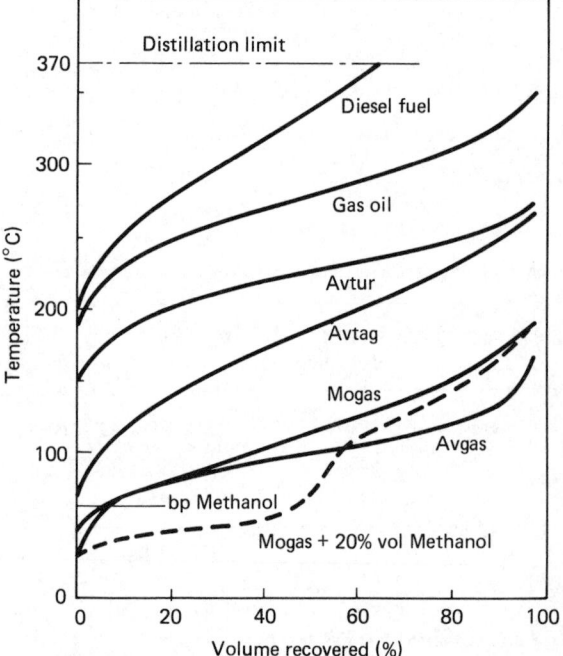

Figure 11.5 Distillation curves for representative petroleum fuels, showing effect of blending methanol with motor gasoline[2]

by volume of fuel vapour in the mixture with air, as outlined in Section 11.5.2. However, due to a progressive reduction in volatility over this range, the weak *temperature* limit of flammability at which this 1% fuel concentration is reached rises with fuel density. This temperature limit is known as the 'flash point', and is determined by warming the liquid sample in a standard design of container with successive introductions of a standard-sized flame into the vapour space until the temperature is reached when the flame flashes across the space and then extinguishes. Since flash points lie below ambient for the light fuels, the method is generally used for

Table 11.6 Influence of volatility changes on gasoline performance

Fractions changed	Volatility raised (lower temperature level)	Volatility reduced (higher temperature level)
Heavy	Mixture distribution more uniform Inlet system deposits reduced Combustion deposits reduced Oil dilution reduced	Cruise economy improved
Medium	Warm-up more rapid Acceleration smoother Short-trip economy improved	Carburettor icing reduced
Light	Cold starting improved	Hot starting improved Vapour loss reduced Vapour lock reduced

kerosines and heavier fuels. Flash point is mainly of concern with regard to fire safety in handling, and materials with flash points lying within the ranges 0°C and below, 0–20°C, and 21–55°C are classified respectively as extremely flammable, highly flammable, and flammable.[8]

Viscosity and freezing At the lower end of the temperature scale, fuel pumping, filtering and spraying are influenced by the ability of the fuel to flow. As discussed in Section 11.2, the attractive force fields between individual molecules give rise to resistance to internal displacement and flow. This resistance constitutes 'viscosity'.

The dynamic viscosity (symbol η) of a sample may be defined as the tangential force on unit area of either of two parallel planes at unit distance apart when the space between them is filled with the sample, and one of the planes moves with unit velocity in its own plane relative to the other. Hence, it is the force per unit area to produce a unit velocity gradient, as shown in Figure 11.6. The units are g cm^{-1} s^{-1}, or poise, P. Since this is inconveniently large in practice, the centipoise, cP, is adopted, where 1 cP = 0.01 P = 1 mN s/m^2. The viscosity of heavy fuel is determined on a dynamic basis from the resistance incurred by motor-driven rotation of one cylinder inside another when located in the sample.

The kinematic viscosity (symbol v) may be defined as the quotient of the dynamic viscosity and the density (ρ) of the sample. Hence, $v = \eta/\rho$, and the units are cm^2/s, or stokes, St. Again, the smaller unit is more convenient, where 1 cSt = 0.01 St = 1 mm^2/s. Kinematic viscosity is of general interest with the light and medium-heavy fuels in connection with pumping and spraying, bearing a direct relationship with the mean droplet size of the spray. Kinematic viscosity may be determined from the efflux time of a given volume of sample flowing under its own pressure head through a capillary forming one leg of a glass U-tube viscometer suspended in a thermostatically controlled water bath. This follows because the force *resisting* the laminar (low-speed) flow of fluid through a restriction is directly proportional to the dynamic viscosity, whereas the force *promoting* the flow is that due to gravity, and thus directly proportional to density. The resultant flow time is thus a function of the quotient of these two properties, and an equation converts the measured time to kinematic viscosity in cSt units.

Since the electrostatic forces between the fuel molecules vary with molecular spacing on an inverse square basis, fuel expansion on heating results in a similar relationship between kinematic viscosity and temperature. However, the greater convenience of straight-line plotting is achieved by modifying both scales on a logarithmic basis, as in the ASTM and 'Refutas' methods, examples of which are shown plotted in Figure 11.7.

On cooling, individual materials freeze at fixed temperatures, commonly described as 'melting points'. With fuel

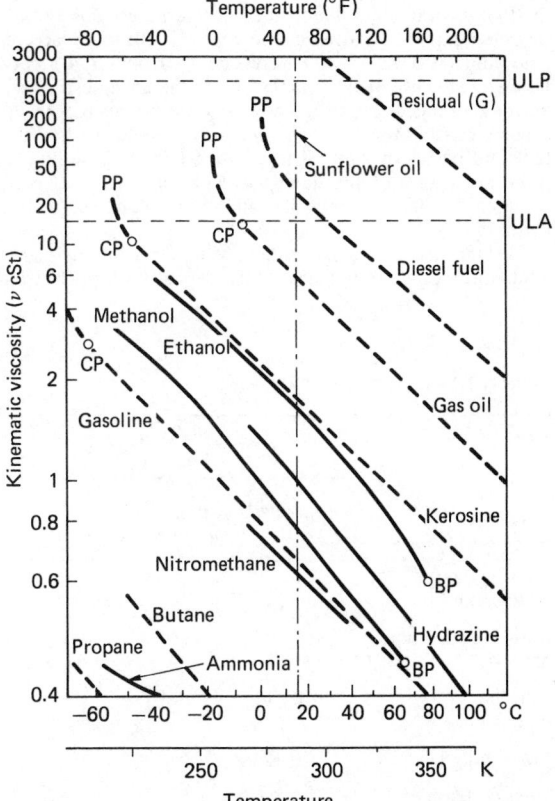

Figure 11.7 Variation of fuel kinematic viscosity with temperature[2] ULP = upper limits on pumping; ULA = upper limits on atomization; CP = cloud point; PP = pour point

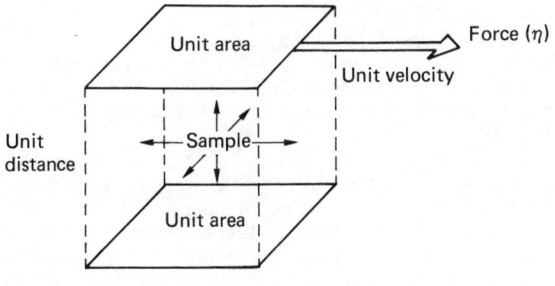

Figure 11.6 Schematic for definition of dynamic viscosity[2]

blends, however, the process of solidification is progressive, comparable to the variation in boiling temperatures throughout the distillation process. With continuous reduction in temperature, the heavy fractions of the blend will start to solidify and promote a cloud of wax crystals within the body of the liquid, the temperature at which these first appear being defined as the cloud point. In the case of aviation fuels, a comparable condition is noted as the freezing point when the wax crystals just disappear while the chilled sample is warmed and stirred.

With further cooling of a sample, the crystals increase in size and number, and eventually coalesce to form a rigid structure which shows no movement on tilting the test jar to the horizontal for a period of 5 s. A temperature 3°C above this test temperature is defined as the pour point. Clearly, wax crystals within a liquid will create interference in the capillary tube of a viscometer, and give an apparent increase in viscosity over and above that due to the falling temperature. For this reason, the modified plot of viscosity versus temperature loses its linearity, and curves upwards as the temperature falls from the cloud to the pour point, as shown in Figure 11.7.

11.4.2.4 Coal tar fuels

The tars produced by high-temperature carbonization of coal may be distilled to yield light oils, carbolic oil, creosote oil and anthracene oil, leaving residual pitch.[9] Re-blending of these liquid distillates gives rise to a range of aromatic-rich coal tar fuels (CTF). The original classification ranges from CTF 50 through CTF 100, 200, 250, 300 and 400 to pulverizable CTF, the figures in the middle members of the range representing the temperature (°F) at which the viscosity falls to the acceptable spraying level of 100 Redwood I seconds (24.1 cSt). However, the use of these fuels outside the tar industry itself has now dropped to a very low level, and no changes in the original classification are envisaged at this time. The first two fuels are blends of light oil fractions, whereas the following three are blends of the liquids with the medium soft pitch residue CTF 400. The densities range from about 0.95 to 1.31 kg/l.

Coal tar fuels are not compatible with petroleum fuels since admixture leads to unmanageable deposits of pitch from solution but, for use in oil burners, the following pairs are regarded as roughly equivalent:

CTF 100 – light fuel oil, class E
CTF 200 – medium fuel oil, class F
CTF 250 – heavy fuel oil, class G

Although the specific energies are lower, and the difficulties of spraying greater, coal tar fuels can be used on occasions in preference to petroleum fuels owing to their relative freedom from sulphur and vanadium, and their higher luminosity.

11.4.3 Solid fuels

The majority of solid fuels stem from vegetable matter, either as sources that are continuously renewed within the span of human life or as fossil material derived during periods of formation up to several million years. The following examples are presented in increasing order of age.

11.4.3.1 Fuel wood and straw

In view of its wide availability, rapid reproduction, free-burning quality and relative lack of ash (0.6% only), fuel wood still represents a major source for domestic heating and cooking in areas where other fuels are not available. The water content of woods ranges from about 50% by mass in the freshly cut 'green' condition, to about 10% when air dried, and the oxygen content is also high. The modern approach to fuel wood preparation is to reduce both bulk timber and forestry wastes into small chips which are then compressed into dry briquettes for furnace combustion with almost complete absence of smoke and ash.

Interest in straw as a fuel has been heightened by the adverse public opinion of on-field burning of stubble, and the major handling problem of bulk due to low density is being tackled by high-intensity compression, either cold or with added heat to melt the waxes (see Table 11.7). Projects are in hand whereby large rural buildings can be heated by using straw grown on part of their own estates (for example, Woburn Abbey).

11.4.3.2 The peat-to-anthracite series

The solid fuels derived from vegetable matter deposited in water constitute the 'peat-to-anthracite' series, and arise through a process of coalification initiated by anaerobic bacteria, and continued under the action of temperatures and pressures within the earth's crust over periods of several million years. This process is associated with a progressive loss in the volatile matter (VM) containing the water vapour, and an increase in the relative proportion of carbon. Since variable amounts of water and ash will be present, solid fuel samples are usually air-dried under specified conditions before testing, with the results expressed on a dry mineral-matter-free (dmmf) basis (Table 11.8).

The first stage in the overall process is peat, a partly decomposed material with a growth rate of about 1 mm per year. Raw peat is associated with over 90% of water, and so requires draining before harvesting and subsequent drying to about 25% of water.

The second member of this series is lignite, or brown coal, which is dark brown in colour with a pronounced earthy or woody texture, the moisture content ranging from about 60% as mined to about 20% when air-dried. The third member is black lignite, or high-volatility sub-bituminous coal, which is pitch-like, being sufficiently mature to show signs of fibrous structure. Medium-volatility bituminous coal is black, hard and usually banded parallel to the bedding plane. Two signifi-

Table 11.7 Wheat straw data in UK, 1986

| | Straw | | |
	Untreated	Compressed and heated	Coal
Specific energy Net MJ kg^{-1}	16.4	15.5[a]	32.6
Mass consumption Mt a^{-1}	7	7	110
Energy equivalence EJ a^{-1}	0.114	0.019	3.60
Density kg l^{-1}	0.11	1.10	1.28
Energy density Net MJ l^{-1}	1.8	17.1[a]	41.7

[a] Allowing for energy input at 1000 bar and 80°C in Cranfield rig.
EJ = J × 10^{18}.

Table 11.8 Solid fuels (composition and property)

| Fuel | Composition % mass dmmf | | | | | | $(a/f)_s$ |
	C	H	N+S	VM	O	Ash	mass
Wood	50	6	1	85	43	0.6	5.1
Peat	57.5	5.5	2	68	32	5	5.4
Lignite (brown coal)	70	5	2	53	21	5	5.5
High-volatility coal	85	5.3	2.6	35	16	5	9.2
Medium-volatility coal	89.9	5	2.2	27	8	5	10.1
Low-volatility coal	91.8	4.4	2.1	16	5	4	10.2
Anthracite	93	3.5	1.9	7	4	3	10.0
Coke	96.5	0.5	2	1	0	7	9.6
Carbon (graphite)	100	0	0	0	0	0	11.48

cant properties of coal fuels are their ability for individual particles to adhere to one another on heating (caking), and to soften, swell and resolidify with shrinkage to form coke (coking) rather than remaining as a soft, easily crushed charred residue. All coking coals are therefore caking, but only some caking coals are able to form hard coke of sufficient strength to maintain a cellular structure on the furnace bed, permitting gaseous through-flow. The ability to cake is determined by the Gray–King assay test involving slow all-directional heating for $1\frac{1}{2}$ h, and increases with level of rank up to a maximum at about 90% carbon. The swelling necessary for coking is determined by the crucible test involving violent unidirectional heating for $2\frac{1}{2}$ min, followed by visual assessment of the shape of the residual coke button.[10]

Low-volatility semi-bituminous coal, or semi-anthracite, represents a hard intermediate stage between bituminous coal and true anthracite which itself is the ultimate form of metamorphosis of the original vegetable matter, and is produced by the intense pressures exerted during geological movement. Anthracite has a sub-metallic lustre, with zero caking power.

More precise methods for classifying the various types of coal include Seyler's coal chart in which percentage carbon is plotted against percentage hydrogen, the location of the point within a broad band representing the types of coal in the given coalfield, and being related to many other properties superimposed on the chart. More recent systems include the Coal Rank Number of British Coal, based jointly on the percentage content of volatile matter and the caking quality. Some coals exist outside the above peat-to-anthracite series owing to their derivation from spores, algae and other small types of plant life. Cannel coal, for example, contains a relatively high proportion of hydrogen and volatile matter generally, and is so named from the candle-like nature of its flame. It is hard and tough, with a dull black lustre, whereas boghead coal or torbanite appears dull brownish.

11.4.3.3 Cokes

The production of coke is achieved by thermal decomposition of suitable types of coal. On heating in the absence of air, an initial release of hydrocarbon gases occurs at the decomposition temperature of about 350°C. At the same time, a plastic layer forms on the outer surface of the coal lump, and begins to advance towards the centre with the penetration of the heat. Since the plastic layer is impervious to gas, the internally released gases cause the lump to swell. The plastic layer subsequently resolidifies into a soft, dull black semi-coke which, in turn, releases hydrogen and carbon monoxide,

shrinking and cracking into the much harder coke at about 1000°C, with a conversion mass ratio of about 0.7 of the original coal.

11.4.3.4 Graphite

Although graphite is not used as a fuel, being far more valuable for other applications, it is of interest chemically as one form of elemental carbon and since it represents the heavier end of the hydrogen–carbon range within which lie all the hydrocarbon fuels. Because of its plate-like structure of benzene-type hexagon rings, the space between the plates is sufficient to accommodate gaseous molecules which then act as bearing balls, permitting the plates to slide over each other.[1] Graphite is therefore a valuable solid lubricant, and being unaffected by radiation is suitable for use in nuclear plants. It is of interest to note that, in another allotropic form, this time based on the methane structure, carbon exists as diamond. In one of its third, amorphous, forms it constitutes soot.

11.4.3.5 Solid wastes

In industry, combustible materials arising as waste frequently lend themselves to useful consumption on-site, particularly with the more tolerant furnaces as distinct from engines, since fuel transfer, and probably preparation, costs are minimal, and the overall efficiency of the industry is improved accordingly.

In addition, everyday human life generates appreciable quantities of waste on a steady basis, the traditional method of disposal being by dumping in landfill sites. Eventually, the shortages of suitable sites led to the introduction of incineration – without heat recovery – in order to reduce the bulk by about 90%, and render the residue sterile and odourless. About 40% by mass of solid municipal waste (SMW) is combustible and, with the development of incineration towards the useful application of the heat, SMW becomes classified as a fuel, with a net specific energy of about 8 MJ/kg, i.e. about one fifth that of coal. The waste can be pulverized and compressed into briquettes, then burnt under a boiler either alone or as a 50/50 mixture with coal. Alternatively, the waste may be processed physically, or by fermentation, and then upgraded by removal of some impurities and/or by the addition of refinery wastes or sewage sludge to give solid refuse-derived fuels (RDF) of about 15 MJ/kg net specific energy.

11.5 Combustion

The term 'combustion' is generally accepted to mean oxidation with the evolution of light and heat, the most common oxidant being atmospheric air except in the case of certain metallurgical furnaces and rocket engine applications. In broad terms, the ultimate requirements for fuels on reaching the combustor are ease of ignition, rapidity and completeness of combustion, with minimal production of unburnt products and associated pollutants. For boilers and furnaces, the additional requirement is for radiant heat transfer to be a maximum, whereas for engines this should be a minimum, since work transfer is the desired objective.

11.5.1 Fundamentals

Certain basic aspects of the combustion process that are relevant to all practical applications concern the proportions of the initial fuel–oxidant reactants and their resulting products, the maximum temperature reached in the flame, the behaviour of flame after its initiation, and the initiation itself, as discussed below.

11.5.1.1 Stoichiometry

Stoichiometry is the study of the proportions in which materials react chemically, and a 'stoichiometric' combustible mixture is one in which no excess or deficiency exists in either fuel or oxidant, being thus described as 'chemically correct'. The standard oxidant considered is atmospheric air, with a nitrogen/oxygen molar (i.e. volumetric) ratio of 3.76 and a molar mass of 28.96. (Note: 1 mole of a material is the quantity having a mass, in grams, equal to the molecular weight – now described as the 'molar mass' – of the material concerned. Hence the masses of one mole of oxygen *gas* and nitrogen *gas* are 32 g and 28 g, respectively, and a mole of any gas occupies a volume of 22.414 l approximately at 1 atm and 0°C.)

For a hydrocarbon fuel of general formula C_xH_y, the various molar quantities in the stoichiometric combustion equation are readily determined by balances of carbon, hydrogen, oxygen and accompanying atmospheric nitrogen, as follows:

$$C_xH_y + m_s(O_2 + 3.76 N_2) = x\,CO_2 + \frac{y}{2}H_2O + 3.76\,m_sN_2 \tag{11.1}$$

where m_s = stoichiometric moles of O_2 per mole of fuel. The molar balance for oxygen gives the following values:

$m_s = x + y/4$

Stoic. air/fuel molar ratio $= (A/F)_s = \dfrac{4.76}{1}m_s = 4.76(x + y/4)$ (11.2)

Stoic. air/fuel mass ratio $= (a/f)_s = (A/F)_s \dfrac{28.96}{\text{Molar mass of fuel}}$

$$= \frac{4.76(x + y/4)\,28.96}{12x + y}$$

$$= \frac{137.85(x + y/4)}{12x + y} \tag{11.3}$$

Based on the approximate mean formulae for conventional hydrocarbon fuels, values of stoichiometric air/fuel ratios are as listed in Tables 11.2, 11.3, 11.5 and 11.6.

The hydrogen component of hydrocarbon fuel burns fairly directly to H_2O, whereas the released carbon goes through an intermediate stage of CO before completion to CO_2. In practice, heat evolved from the combustion is of sufficient magnitude for the oxidation products to become so agitated thermally as to start dissociating, that is, breaking down and reverting back to their H_2, CO and O_2 reactant forms. For this reason, the individual elemental combustion reactions are written with reversible arrows, as shown:

$$H_2 + (1/2)O_2 \rightleftharpoons H_2O \quad \text{and} \quad CO + (1/2)O_2 \rightleftharpoons CO_2$$

The extent of this dissociation is determined by the level of temperature obtaining, and recourse must be made to the literature for the values of the equilibrium constant, which can be expressed in terms of the ratio of the moles of products and reactants concerned, indicating the dissociated quantities at any given temperature.[11]

At the moderately high levels of combustion temperature for hydrocarbons (> 2000 K), therefore, the stoichiometric equation would appear in the following form:

$$C_xH_y + m_s(O_2 + 3.76 N_2)$$
$$= n_1CO_2 + n_2H_2O + n_3CO + n_4H_2 + n_5O_2 + 3.76m_sN_2 \tag{11.4}$$

with all n values calculable using the values of equilibrium constants appropriate to the level of combustion temperature. These effects also occur, on a smaller scale and particularly at the higher temperatures, in reactions between atoms of H, O and N, giving a more comprehensive and complex stoichiometric equation of the following form:

$$C_xH_y + m_s(O_2 + 3.76 N_2) = n_1CO_2 + n_2H_2O + n_3CO$$
$$+ n_4H_2 + n_5O_2 + n_6H + n_7O + n_8OH + n_9NO + n_{10}NO_2$$
$$+ n_{11}N_2 \tag{11.5}$$

Subsequent cooling of the combustion products should, theoretically, bring an end to this dissociation and a return to the completely burnt quantities shown in equation (11.1). In practice, however, due to local chilling of the reactions by turbulence and/or contact with the cooler walls of the container, parts of the mixture endure in the dissociated form, hence equations (11.4) or (11.5) still apply. The resulting product concentrations may then be determined experimentally using such techniques as chemical absorption or adsorption, infrared radiation, or paramagnetism.

As discussed in the following section, fuels can burn effectively at mixtures other than stoichiometric, in which case the non-dissociated combustion equations may be expressed as follows:

Fuel-weak mixtures:

$$C_xH_y + m(O_2 + 3.76_2) = xCO_2 + (y/2)H_2O + n_5O_2$$
$$+ 3.76mN_2 \tag{11.6}$$

where the moles of excess oxygen $= n_5 = m - (x + y/4)$

Fuel-rich mixtures:

$$C_xH_y + m(O_2 + 3.76N_2) = n_1CO_2 + n_2H_2O + n_3CO$$
$$+ n_4H_2 + 3.76mN_2 \tag{11.7}$$

where the insufficiency of oxygen results in a sharing between the CO and H_2, and the moles of incompletely burnt products, n_3 and n_4, may be determined using the equilibrium constant from the water–gas reaction (Section 11.4.1.2).

When additional dissociation into molecular and/or atomic species is considered, as in the stoichiometric case above, the more comprehensive equations (11.4) or (11.5) also apply to these non-stoichiometric cases, and the complexity of solution generally warrants computation.

11.5.1.2 Combustion temperature

The maximum temperature reached in the flame occurs when combustion is just complete, and no energy is permitted to escape from the products (i.e. adiabatic conditions with no work transfer). For a fuel of known composition, this combustion temperature may be calculated from a knowledge of the mixture proportions and the calorific value of the fuel (usually expressed as enthalpy of reaction), since a thermal balance is envisaged between the energy released by the reactants and that absorbed by the products. Even in the non-dissociated case, the calculation must be iterative, since a temperature must first be assumed in order to select the appropriate values for energy absorbed by each product. The energy released is then compared with the total energy absorbed, and the final non-dissociated temperature found by trial and error to give equality to these two energy quantities.

A similar procedure is followed in the dissociated case, except that the assumed temperature must first be used by trial and error to determine the quantities of products in the combustion equation before proceeding to the energy comparison, which is unlikely to balance at the first attempt. Derivation of the correct temperature is usually possible by linear interpolation with the result of a second attempt.[12]

In the event, little difference is found between combustion temperatures over a wide range of fuels. This follows because although those fuels with higher concentrations of hydrogen exhibit high levels of calorific value, these same fuels generate more H_2O as a product, and this has a specific heat capacity nearly three times that of CO_2. Consequently, this additional heat capacity absorbs the additional heat released, and most theoretical adiabatic combustion temperatures for stoichiometric mixtures of hydrocarbon fuels in air at 1 atm lie within the range 2300 – 2500 K, although acetylene reaches about 2600 K. In practice, of course, variable heat losses are inevitable, and temperatures do not exceed about 2200 K.

Combustion temperatures are reduced on the weak side of stoichiometric due to the additional heat capacity of the excess air, and on the rich side due to the unavoidable incompleteness of combustion. Values for both dissociated and non-dissociated mixtures are shown in Figure 11.8 for methane and iso-octane, indicating the extent of the effects of both mixture strength and dissociation on combustion temperature.

11.5.1.3 Flammability

Although the stoichiometric mixture represents the ideal case with maximum release of heat energy and no excess reactants to limit the temperature by absorbing part of the heat, fuels are able to burn at mixtures both weaker and richer than stoichiometric, as outlined above. However, mixture limits are reached when the flame is unable to propagate owing to the local insufficiency of molecules of either fuel or oxygen. These weak and rich mixture limits of flammability in air lie at 5 and 15 volumetric per cent of fuel vapour (34.3 to 10.2 air/fuel mass ratio) for methane, and reduce progressively through the hydrocarbon gases to almost constant values of 1 and 5 (22 to 4 air/fuel ratio) for the petroleum liquids ranging widely from aviation gasoline to residual fuel oil.[2]

As indicated in Section 11.4.2.3, the flash point, that is, the liquid temperature required to derive the weak mixture limit above the free surface, is not common throughout this range of fuels owing to the differing volatilities. An upper flash point can be envisaged, corresponding to the reasonably constant rich mixture limit of flammability, but this is not normally measured, being accepted as lying approximately 30°C above the conventional flash point. Plotting curves for both flash points in Figure 11.9 indicate the temperature range of

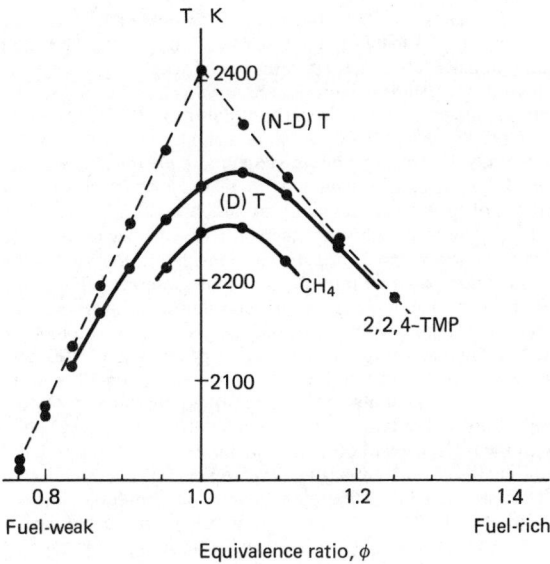

Figure 11.8 Mixture effect on combustion temperature for methane and iso-octane (2,2,4-trimethylpentane)[12]

$$\text{Equivalence ratio} = \frac{\text{Stoichiometric a/f}}{\text{Actual a/f}}$$

(D) = dissociated; (N-D) = non-dissociated

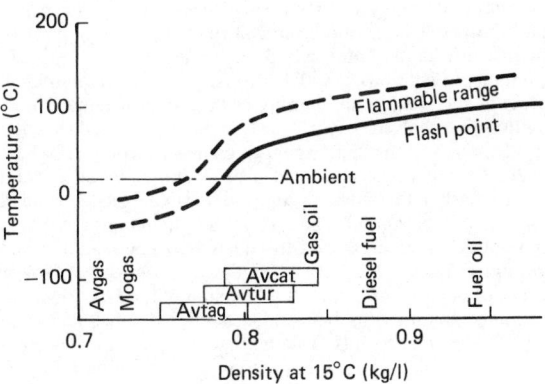

Figure 11.9 Variation of flash point with density for representative commercial hydrocarbon fuels[1]

flammability, and it is worth noting that, under ambient storage, the vapour–air mixtures in a kerosene tank lie within the flammable range, whereas those in a gasoline tank do not. Nevertheless, fire danger from the gasoline tank is still considered greater since high volatility demands tank venting, and the vented over-rich mixture rapidly weakens in the presence of atmospheric air, generating flammable mixtures *outside* the tank where vulnerability to ignition may be greater.

In addition to fire safety in handling, flammable mixture limits together with speed of flame propagation determine the behaviour of flames within practical combustors, and a certain extent of control lies in the design of the combustor and the fuel–air feed systems. The flame speeds of fuel–air mixtures can be measured, usually on a laminar basis, by igniting the mixtures in glass tubes or soap bubbles and recording the movement of the resulting flame. Laminar flame speeds of gaseous and vaporized liquid hydrocarbon fuels in air at atmospheric pressure are generally fairly common at the relatively low level of 0.5 m/s. Hydrogen exhibits a comparatively high level of 3.5 m/s. In practice, these values are increased considerably by the effects of turbulence, temperature and pressure.

Flame control can be exercised either by propagation across a confined space, as in spark-ignition piston engines, or by stabilization, as in an intermittent spray-generated flame in a compression-ignition engine, or a continuous spray or vapour-generated flame within a furnace, gas turbine, ramjet or rocket engine. In the former case, propagation rates and patterns are determined by chamber geometry, levels of turbulence, wall temperatures, stratified mixture strengths, and so on. In the latter case, matching of the low flame speed with the high speed of entry air is achieved by decelerating the air in a diffuser, and diverting (even reversing) the direction of the airstream so that the flame is able to maintain a constant location. At all points in the stabilized flame, a balance is achieved between the opposing speeds of the approaching mixture and the flame. High flame speeds permit shorter lengths of combustor and/or higher engine speeds and, together with wide ranges of flammability, give greater flexibility for changing combustion conditions without the hazard of flame-out.

11.5.1.4 Ignitability

The ignition of a fuel–air mixture may be either forced or spontaneous. In the former case, a localized addition of concentrated energy creates a small nucleus of flame which then grows outwards under the control of the flammability of the mixture. In the latter, a bulk addition of energy results in spontaneous ignitions at widely diffuse points, with an overall effect of the whole bulk igniting at the same instant. Forced ignition is involved in the flash point apparatus (Section 11.4.2.3) and in the spark-ignition engine (Section 11.5.2.1), whereas spontaneous ignition can be examined by means of a heated flask into which is injected a small predetermined quantity of fuel sample. The test starts at high temperature so that ignition is assured, and the delay between injection and ignition is noted. The test continues at progressively lower levels of temperature until the minimum level is found to promote ignition, this being associated with the maximum delay of several seconds. This minimum temperature is commonly referred to as the spontaneous ignition temperature, or SIT.

The variation with fuel density of SIT can be appreciated, in simple terms, by noting the size and shape of the fuel molecules, and their behaviour under the action of the thermal agitation within the flask. Small, compact molecules can be imagined as resisting this agitation quite effectively, consequently exhibiting high levels of SIT. The larger and more complex molecules, on the other hand, can be visualized as vulnerable to the energetic disturbances within the heated air, and tending to be cracked, i.e. broken into smaller sub-groups of atoms. The free bonds exposed by cracking of the molecules of fuel and oxygen lead to the possibilities of partial oxidation to certain C–H–O compounds which may be unstable and proceed to further reactions, many of which are exothermic (heat producing). These localized energy releases, repeated in many other points within the mixture, can be sufficient to cause ignition, consequently levels of SIT could be expected to be lower than those for the lower-density fuels. The curve in Figure 11.10 shows this to be the case, consequently the

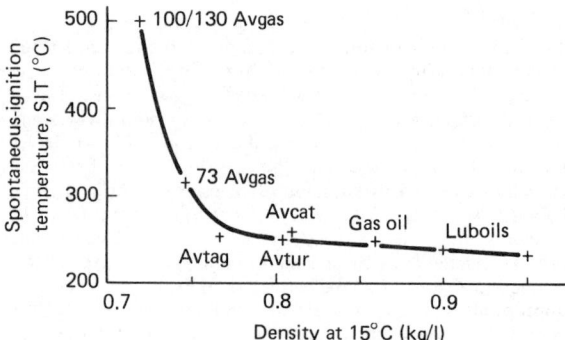

Figure 11.10 Variation of spontaneous-ignition temperature with density for representative commercial hydrocarbon fuels[1]

heavier fuels are more ignitable than their lighter counterparts, whereas the latter were seen to be more flammable (Section 11.5.1.3).

Research at Cranfield has shown these clear-cut differences in ignitability to be lost at atmospheric pressure when higher temperatures are used to reduce the delay periods to milliseconds, but to be regained when these short-delay tests are conducted at the higher pressures used in practical combustors.[13]

11.5.2 Applications

All practical applications of combustion involve mixtures of fuel and oxidant which lie within the limits of flammability and are subjected to some form of ignition. However, each application is associated with a particular form of mixture preparation and of flame generation, and has its own specific requirements for optimal performance, as discussed below.

11.5.2.1 Discontinuous combustion engines

The long-established internal-combustion engines operating on repeated mechanical cycles comprise the reciprocating-piston type engines, recently joined by the rotary Wankel type, in each case sub-divided by the mode of ignition into spark (forced) and compression (spontaneous).

The spark energy required for ignition of a stagnant fuel–air mixture has been found to vary from about 0.2 to 0.3 mJ, but considerably more energy is supplied in practice to cater for mixture turbulence and a wide range of operating conditions. Under normal circumstances in the spark-ignition engine, the flame, once ignited, propagates across the chamber until all the mixture is consumed, the resulting rise in pressure serving to drive the piston towards bottom dead-centre and so provide the required power.

Under adverse circumstances where the operating pressure and temperature become excessive, as in low-speed hill climbing in high gear, or due to inadequate quality of fuel, the rising pressure and temperature in the chamber are sufficient for the remaining end gases to ignite instantly and spontaneously, giving rise to an uncontrolled explosion rather than a progressive burn under the control of the advancing flame. This phenomenon is termed 'spark knock', and is manifested by a high-frequency noise as the gaseous vibrations excite the engine structure. Although light knock may be tolerated for short periods, sustained heavy knock at high speed can result in damage to the cylinder-heat gasket on the side opposite the plug, plus erosion of the piston crown and top land, with eventual break-up of the top piston ring.

The thermal efficiency of any piston-type engine improves with rise in compression ratio (volume at bottom dead-centre/ volume at top dead-centre), but spark knock sets a limit of about 10/1 on the compression ratio of automobile engines and, should a change have to be made to lower-quality fuels, the ignition timing may have to be retarded by several degrees of crank angle to avoid knock. To be suitable for spark-ignition applications, therefore, a fuel must be low in radiation and also *poorly ignitable*, hence, as shown in Figure 11.10, gasoline (which happens to be *highly flammable*) is optimal. Anti-knock quality of a gasoline is determined by means of a specially designed engine that permits variation in compression ratio while the engine is running. Using the test sample, the compression ratio is increased until the onset of a standard intensity of knock. At the same compression ratio, a blend of two reference fuels is found that matches the knock intensity of the test sample. The reference fuels used are iso-octane, i-C_8H_{18} (with a compact molecule giving high resistance to spontaneous ignition) and normal heptane, n-C_7H_{16} (with a non-compact straight-chain molecule). The anti-knock quality of the sample is then specified in terms of the Octane Number, which is the volume percentage of iso-octane in the matching reference blend. The 'Research' octane number is appropriate to moderate-speed cruising, and the 'Motor' octane number, determined under more severe conditions of test, to high-power high-speed driving (Table 11.9). Standard aviation gasoline (designated 100LL, i.e. Low Lead), of which comparatively little is now in use, has an octane number of 100 at weak-mixture cruising conditions, together with a performance number of 130 (30% more power than iso-octane) under rich-mixture take-off conditions.

Spark-ignition engines can also suffer from pre-ignition when combustion deposits within the chamber are caused to glow. This is equivalent to multiple ignition, which is advantageous in principle but, in this case, is dangerous since it is uncontrolled and can eventually lead to melting of the piston crown. Some fuel molecules, such as alcohols and aromatics, although highly resistant to spark knock, are prone to pre-ignition at the lower pressures and temperatures of operation. Pre-ignition, which is now virtually eliminated through improved design and materials, sometimes gave rise to continuing operation of the engine (run-on) on a regular basis after the electrical ignition was switched off. With modern engines, low-speed run-on may occur erratically owing to the spontaneous ignition arising from the high temperature level of the charge and the time available for reaction.

Under normal conditions in the compression-ignition engine, air alone is compressed in the cylinder and the fuel then introduced by means of an injector, the initial droplets in the spray igniting spontaneously after a short delay within the compression-heated air. The flame then propagates upstream towards the injector, burning the remaining fuel charge as it enters. The compression ratio must be high, typically from 13.5/1 to over 20/1, in order to achieve the high temperatures and pressures required for ignition, consequently the engine needs to be rugged and tends to be heavier than its spark-ignition counterparts. Furthermore, the fuel-injection system must operate at very high pressures in order to provide adequate pressure drop across the injector for good-quality spray into the chamber.

Under adverse conditions where the temperatures and pressures are insufficient, as in cold starting, or due to inadequate quality of fuel, the ignition delay of the initial droplets is so extensive that most of the fuel charge has entered the chamber by the time ignition does occur. The resulting combustion is so rapid and uncontrolled as to be explosive, giving rise to rough running and the characteristic heavy noise of 'diesel knock'.

To be suitable for compression-ignition applications, therefore, a fuel must give low radiation and be *readily ignitable*, hence, as shown in Figure 11.10, gas oil and diesel fuel (which happen to be *poorly flammable*) are optimal. Anti-knock quality of a compression-ignition engine fuel is also determined in a variable-compression engine, and a blend of two reference fuels is found which matches the ignition delay of the test sample. These reference fuels are normal cetane, n-$C_{16}H_{34}$ (a non-compact straight-chain molecule), and, currently, a compact isomer of cetane, the anti-knock quality of the sample being specified in terms of Cetane Number, which is the volume percentage of normal cetane in the matching reference blend. The minimum anti-knock limits for UK diesel fuels are specified as 50 CN for automotive use and 45 CN for general purposes.[14]

Since the ignition requirements are diametrically opposed in these two types of piston engine, it follows that an inverse relationship holds with octane and cetane number (Figure 11.11), and that fuels suited to one type of engine are unsuited to the other.

11.5.2.2 Continuous-combustion engines

Clearly, the power-density of an engine, in either watts/m^3 or watts/kg, would be increased if the combustion could be maintained on a continuous basis. This is achieved by continuous spraying of the fuel into an open chamber where the supply of air, and rejection of products, can also be continuous. In the gas-turbine engine, some of the energy in the

Table 11.9 Specified minimum octane ratings for motor gasoline

Leaded (BS 4040) Max. lead 0.15 g l⁻¹			*Unleaded* (BS 7070) Max. lead 0.013 g l⁻¹		
Grade	RON	MON	Grade	RON	MON
4 star	97	86	–	–	–
–	–	–	Premium	95	85
3 star	94	82	–	–	–
2 star	90	80	Regular	90	80

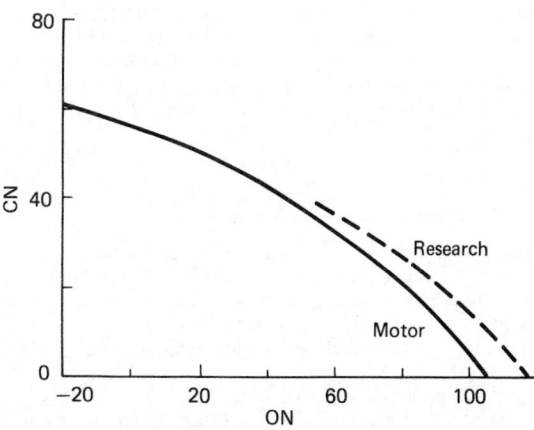

Figure 11.11 Representative relationships between octane and cetane numbers[2]

product gases is extracted by means of a rotary turbine, which, in turn, drives the shafting to an upstream compressor, the remaining energy being extracted either by an additional turbine stage or by a rearward propelling nozzle. For hypersonic flight (greater than Mach 5), the compression effect within the entry air duct is sufficient for combustion purposes, so that the rotating elements of compressor, shaft and turbine may be discarded, and the engine is then described as a ramjet. For still higher velocities, and for flight outside the earth's atmosphere, the vehicle must carry its oxidant as well as its fuel, consequently the entry air duct is not necessary, and the engine is described as a rocket. In all these cases, the flame is established in a location constant within the chamber, and mixtures of fuel and oxidant are fed to it. Combustion requirements therefore include ready vaporization of the fuel and mixing of the vapour with oxidant, together with adequate flame speed and flammable mixture range, and clean complete combustion with minimal emissions.

In continuous-flow combustors, the air pattern is designed to assist flame stabilization by means of swirlers and secondary air side entries, the primary requirement being high relative velocity between fuel and air. In gas-turbine engines, the turbine blades not only experience mechanical stress through high-speed rotation but are exposed constantly to the temperature of the working gas stream in contrast to piston engines, where such metal heating occurs during only part of the cycle.

Consequently, a large proportion of the incoming air is used for cooling the combustion gases immediately downstream of the flame zone. For gas-turbine engine fuels, the best compromises are found in kerosine (aviation) and gas oil (industrial), although in the latter case adaptations can be made to suit gaseous fuels and the heavier fuel oils. Hypersonic ramjets call for specially designed fuels of high energy density in order to minimize storage volume and thus aircraft drag, and high heat capacity to be able to withstand the kinetic heating effects of high-speed flight. The candidate fuels include kerosines of high thermal stability, specially selected heavy hydrocarbons, and, eventually, endothermic fuels that react to absorb heat during flight. Supersonic combustors call for particularly high flame speeds, as with hydrogen. Space rockets, on the other hand, require fuel-oxidant pairs of maximum specific energy, and these include hydrogen–oxygen, kerosine–oxygen, hydrazine–hydrogen peroxide, and a range of solid materials that contain both fuel and oxygen components. In view of the absence of diluent nitrogen, combustion temperatures are particularly high in rocket chambers (e.g. 3000 K), consequently the chemical dissociation and re-association reactions within the propelling nozzle are complex, but they can be handled by computation. The performance of such high-speed engines is usually assessed on a basis of the thrust produced per unit mass flow rate of propellants (i.e. fuel + oxidant), and this parameter is termed the 'specific impulse', with unit N s/kg.

11.5.2.3 Continuous-combustion heaters

Applications of heat to manufactured stock for processing, or to water for raising steam, usually take place in a continuous-flow combustion chamber with the flame stabilized in the stream of air and fuel, as in Section 11.5.2.2.

Fuel may be introduced by the following methods;

- As gas into a mixing zone with air
- As low-pressure liquid for vaporization within a flame-heated tube
- As a centrifugal spray from a rotary-cup burner fed at low pressure

- As a low-pressure spray subjected to locally high velocities of air or steam
- As a high-pressure-jet spray into relatively low-speed air.
- As a bed-supported layer of solids, with streams of air supplied below and above
- As solids fluidized with air in a vertical reactor
- As pulverized solids suspended in a large-scale chamber

Since the output of a pressure-jet burner is proportional to the square root of the pressure drops across the orifice, a turn-down ratio (maximum/minimum flow rate) of, say, 10/1 requires a pressure ratio of 100/1. Furthermore, since low pressures tend to degrade spray quality, and high pressures incur excessive mass and expense of fuel systems, high turn-down ratios are achieved by such means as concentric, multiple nozzles designed for different entry pressures, by over-supplying the nozzle and spilling the excess back to the tank, or by variable opening of the entry ports to the nozzle.

The transfer of combustion heat from the flame takes place by a combination of forced convection and radiation. The former process is used in the low-temperature furnaces, augmented by an air fan, whereas at temperatures above about 750°C, radiation, which varies with the fourth power of absolute temperature, becomes dominant, and the heat transfer may take place either directly from the flame or indirectly from heated refractory surfaces.

The gases are the most convenient fuels for furnace use because of their control flexibility and clean burning, and pre-heating of air and fuel is used with the low-energy gases in order to achieve the required combustion temperatures. Liquid fuels in furnaces range from gas oil through to the heaviest residual fuel oil and coal tar fuels, the presence of particles of solid carbon in the flame increasing radiation. Solid fuels tend to be used more for the larger power-station boiler furnaces, where large-scale provision can be made for mechanical stoking, ash removal and flue-gas treatment. Ash can be removed in part by prior fuel washing, or handled subsequently by controlling the temperature below the level at which the ash becomes sticky. Improved overall control of solid fuel appears to result from the use of a pulverized fuel in a fluidized bed giving better heat conduction and greater tolerance to coarse particle size with high ash content.

11.5.2.4 Emissions control

In Section 11.5.1.1, the major products of combustion were seen to comprise carbon dioxide and water plus, of course, unreacted atmospheric nitrogen, together with variable quantities of carbon monoxide, oxygen and, possibly, unreacted hydrogen, depending upon the strength of the mixture. Since these volumetric concentrations within the total products lie above about 1%, they are described as macro. In addition, there may well be micro concentrations of nitrogen oxide and dioxide (together represented as NO_x), the hydroxyl radical, and atomic oxygen and hydrogen. A method of calculating all these concentrations was outlined in Section 11.5.1.1, and practical methods of measurement include chemical absorption in liquid or solid reagents, or physical methods based on thermal conductivity, infrared absorption, chromatography, paramagnetism or sonic velocity.

Although the water vapour produced by combustion eventually condenses and returns to the global water cycle, such carbon dioxide that is not absorbed by the world's vegetation and oceans has been identified as a major cause of extending the greenhouse effect. Carbon monoxide is a recognized poison, and the nitrogen oxides, together with unburnt hydrocarbons, give rise to smog. The polynuclear aromatic hydrocarbons are also suspected of carcinogenic effects, whereas any sulphur in the fuel will emerge as corrosive sulphur dioxide. Carbonaceous soot particles are probably also formed in fairly copious proportions in most flames as a result of imperfect mixing, cracking and/or slow reaction rates, but they tend to be oxidized in later stages of combustion. Luminous non-sooting flames are required for heat transfer in furnaces, and non-luminous non-sooting flames for work transfer in engines. Unburnt soot gives rise to smoke as a further pollutant of the atmosphere, and also to the possibilities of carbon deposits within the engine and exhaust systems. Such additives as lead have also become unacceptable to the public, and unleaded fuels are now becoming increasingly available.

Typical volumetric concentrations of the combustion products from a spark-ignition engine are shown in Figure 11.12. These indicate, as discussed in Section 11.5.1.1, that the normally expected weak and rich products do, in fact, extend into each other's domains. Operation on the weak side not only improves fuel economy but reduces most of these products, the exception being NO_x, which is generated by direct reaction of atmospheric nitrogen and oxygen at high temperatures. Methods of controlling these emissions include exhaust air injection, afterburners and catalytic reactors in the exhaust system to handle CO and hydrocarbons, and exhaust recirculation into the cylinder, and/or catalytic reactors, for the NO_x. Exhaust recirculation has the two effects of lowering the availability of oxygen, and the level of the peak flame temperature, but also reduces the power. Similar effects result from retarded ignition timing, but fuel economy is then impaired. Overall weak-mixture operation is feasible using a stratified charge in the chamber, and is achieved using fuel injection and suitable chamber design.

Similar problems of emissions arise with compression-ignition engines, where high utilization of air is the target, and where soot reaches objectionable levels if the fuel–air mixing and/or the maintenance of the pumps and injectors is poor. Control can be effected by the use of pre-combustion chamber, and of fuel additives based on barium.

The continuous-flow combustors tend to protect the flame region so that the mixture strength is near-stoichiometric, and then dilute the combustion products with sufficient air to reduce the mixed temperature to a level that downstream surfaces can withstand. Emission concentrations are therefore diluted, but their overall quantities are still significant, and careful combustor design for good mixing and flame control is always important.

11.6 Conclusions

Over many decades of use, the optimal fuel-combustor pair has emerged for each application, but developments continue into improving efficiency, startability, smooth running, clean burning and general overall performance. However, these targets never remain stationary for very long, and two of the current concerns are the continuing availability of conventional fuels well into the next century, and the sinister extension of the essential greenhouse effect owing to increased concentrations of combustion-generated carbon dioxide, and some hydrocarbons themselves.

The subject of alternative fuels has comprised a special study at Cranfield since the first warnings in 1973, and brief conclusions appear as follows. The remaining lifetimes of conventional fossil sources are, of course, dependent on many – as yet unknown – factors of an economic and political nature, but a broad assessment is gained from the simple (reserves/production) ratios. In effect, this assumes no changes either in the global reserves of a particular resource or in the annual production rate over the period concerned (whereas in fact both may rise slightly initially), but the results are useful on a comparative basis and, for 1991, appear as follows;

Natural gas: 58.7 years
Petroleum: 43.4 years
Coal: 239 years

Several alternative fuels could be made available, and are classified under the two headings of Supplemental and Substitute in Table 11.10. From the above lifetime figures, coal is seen as not only likely to endure as a fuel in its own right well into the future, but also to serve as a potential source for supplementing the conventional oil- and gas-derived fuels.

Applications for all the substitute fuels can be found, and some actually offer improved performance in certain instances, their higher prices being offset to some extent eventually by bulk production. However, only hydrogen offers relief from CO_2 pollution and even with this fuel oxides of nitrogen would still be produced. Methane is also a greenhouse gas and, although it, in turn, produces CO_2, it could be used to advantage as a substitute fuel rather than allowed to escape to the atmosphere.

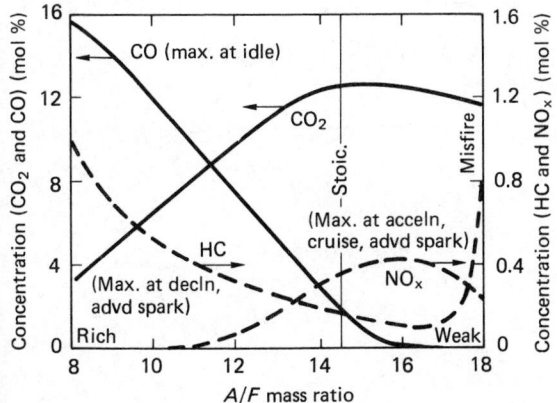

Figure 11.12 Volumetric concentrations of products of hydrocarbon combustion[1]

Table 11.10 Alternative fuels (suggested nomenclature)

1. Conventional	C_xH_y
2. Supplemental Similar to conventional fuels but alternative sources	Petroleum-type products from: NG Coal Shale oil Tar Peat Biomatter
3. Substitute Unrelated to conventional fuels or sources	Alcohols, C_xH_yOH Nitroparaffins, $C_xH_yNO_2$ Nitrohydrides, N_xH_y Liquid methane, LCH_4 Liquid hydrogen, LH_2 Biofuels, $C_xH_yO_z$

Certain engine developments may bring success to the hybrid-type piston engine, which is so closely allied to both the spark- and compression-ignition engines that octane and cetane rating will have no significance, and a wide range of fuels be acceptable. Many furnaces are already omnivorous, tolerating a wide range of gaseous, liquid and solid fuels, including wastes. Electric propulsion is already in small-scale use for delivery vehicles, etc., the parent chemical fuel being burnt under more controlled conditions within power stations, and further development of the fuel cell will permit higher efficiencies by direct conversion from chemical to electrical energy without the thermodynamic constraints imposed by the intermediate form of heat energy.

Should the greenhouse effect threaten to expand to disastrous proportions, consideration would have to be given not only to further reduction of energy demands but also to sources other than those generated by oxidation. Nuclear power is one such, and environmental objections may be allayed by success in the development of cleaner-reacting fusion. The ultimate sources may be geothermal, or more direct forms of solar power received either as such or through the media of winds, waves or tides, in each case converted to conveniently handled electricity.

Conventional fuels and combustion engines therefore occupy a unique position in the technical world by offering a very valuable and versatile service, but with disadvantages that may eventually be resolved only by abandonment and replacement before irreparable damage is done to the environment. As ever in such situations, research is the key, and adequate support for its continuation is nothing less than vital.

References

1 Goodger, E. M., *Hydrocarbon Fuels*, Macmillan, London (1975)
2 Goodger, E. M., Alternative Fuel Technology Series, Vol. 2, *Comparative Properties of Conventional and Alternative Fuels*, Cranfield Press, Bedford (1982)
3 Anon., *Recommended SI and other metric units for the petroleum and petrochemical industries*, Institute of Petroleum, London (1974)
4 Anon., Definitions of the calorific value of fuels, BS 526, latest issue
5 Anon., Specifications for commercial butane and propane, BS 4250, latest issue
6 Goodger, E. M., *Alternative Fuels*, Macmillan, London (1980)
7 Goodger, E. M. and Vere, R., *Aviation Fuels Technology*, Macmillan, London (1985)
8 Anon., *Classification and labelling of substances dangerous for supply and/or conveyance by road*, Health & Safety Commission, HMSO, London (1986)
9 Rose, J. W. and Cooper, J. R., *Technical Data on Fuel*, Scottish Academic Press, Edinburgh, 7th edition (1977)
10 Macrae, J. C., *An Introduction to the Study of Fuel*, Elsevier, Amsterdam (1966)
11 Goodger, E. M., *Combustion Calculations*, Macmillan, London (1977)
12 Goodger, E. M., *Calculated adiabatic combustion temperatures of hydrocarbon–air mixtures*, Cranfield Report No. SME 6, June 1974
13 Goodger, E. M. and Eissa, A. F. M. 'Spontaneous ignition of falling droplets in the Cranfield pressure rig', *Journal of Institute of Energy*, December, 199–208 (1987)
14 Anon., British Standard Specification for fuel oils for oil engines and burners for non-marine use, BS 2869, latest issue
15 Goodger, E. M., 'Alternative fuel and power options', Cranfield/MOD first international seminar on aviation turbine fuel specifications, Bath, September 1988, Cranfield Press, Bedford 1989)

12 Alternative energy sources

J. Cleland McVeigh

Contents

12.1 Introduction 12/3
    12.1.1 The Earth's energy flows 12/3
    12.1.2 Energy from solar radiation 12/3
    12.1.3 Energy from the Earth's interior 12/3
    12.1.4 Energy from the tides 12/3
    12.1.5 The energy flow system 12/3

12.2 Solar radiation 12/4
    12.2.1 Flat-plate collectors 12/6
    12.2.2 Space heating 12/8

12.3 Passive solar design in the UK 12/8
    12.3.1 The main technical issues in passive design 12/9
    12.3.2 The energy-performance assessments 12/9
    12.3.3 Technology transfer 12/9
    12.3.4 Some UK applications 12/9
    12.3.5 Atria 12/10

12.4 Thermal power and other thermal applications 12/10
    12.4.1 Solar ponds 12/11
    12.4.2 Heliohydroelectric power generation 12/11
    12.4.3 Distillation 12/12
    12.4.4 Ocean thermal energy 12/12
    12.4.5 Refrigeration and cooling 12/12

12.5 Photovoltaic energy conversion 12/14
    12.5.1 Introduction 12/14
    12.5.2 Basic principles 12/14
    12.5.3 Developments 12/14
    12.5.4 Systems 12/15
    12.5.5 Applications 12/15

12.6 Solar chemistry 12/15

    12.6.1 Introduction 12/15
    12.6.2 Photochemical processes 12/16
    12.6.3 Chemical conversion with photoelectrochemical cells 12/16
    12.6.4 Fuels 12/17
    12.6.5 Chromogenic materials 12/17
    12.6.6 Transparent insulation materials 12/18
    12.6.7 Solar detoxification 12/18
    12.6.8 Chemical heat storage 12/18
    12.6.9 Other applications 12/18

12.7 Hydropower 12/19
    12.7.1 Introduction 12/19
    12.7.2 The basic hydropower plant 12/19
    12.7.3 Types of turbine 12/19
    12.7.4 Hydropower potential 12/20
    12.7.5 Pumped storage 12/21
    12.7.6 Small-scale hydropower 12/21
    12.7.7 Economic, social and environmental issues 12/22
    12.7.8 Summary 12/23

12.8 Wind power 12/23
    12.8.1 Introduction 12/23
    12.8.2 Wind-energy potential 12/23
    12.8.3 Small to medium-range windmills 12/24
    12.8.4 The vertical-axis windmill 12/25
    12.8.5 The development of large horizontal-axis wind turbines and some national programmes 12/25
    12.8.6 Some environmental issues 12/26
    12.8.7 Summary 12/27

12.9 Geothermal energy 12/27
    12.9.1 Geothermal resources 12/27
    12.9.2 Geothermal areas, fields and aquifers 12/28
    12.9.3 Thermal applications 12/28
    12.9.4 Electricity generation 12/29
    12.9.5 Hot dry rocks 12/29
    12.9.6 Some factors influencing developments 12/29
    12.9.7 Summary 12/30

12.10 Tidal power 12/30
    12.10.1 Introduction 12/30
    12.10.2 Tidal power principles 12/30
    12.10.3 Tidal power schemes 12/31
    12.10.4 Tidal power sites 12/31
    12.10.5 Possible impacts of tidal schemes 12/32
    12.10.6 The economics of UK tidal power (and of many of the renewables) 12/33

12.11 Wave power 12/33
    12.11.1 Introduction 12/33
    12.11.2 Wave theory and characteristics 12/33
    12.11.3 Types of wave energy convertor 12/34
    12.11.4 Shore-based systems 12/35

12.11.5 Summary   12/35

12.12 Biomass and energy from wastes   12/36
    12.12.1 Introduction   12/36
    12.12.2 Photosynthesis   12/36
    12.12.3 Energy resources   12/36
    12.12.4 Conversion of biomass to fuels and other
           products   12/36
    12.12.5 Cooking – the major application of
           biomass   12/37
    12.12.6 Energy from waste materials   12/37

12.13 Energy crops   12/39
    12.13.1 Short-rotation forestry   12/39
    12.13.2 A fuel-alcohol plant   12/39
    12.13.3 Marine and aqueous applications   12/39
    12.13.4 Choice of system   12/40
    12.13.5 Summary   12/40

References   12/40

12.1 Introduction

Renewable energy sources have been recognized as major alternatives for many years, but it is only since 1973 that substantial research, development and demonstration projects have been undertaken. Two of the main applications are to supply heat and electricity. Many examples can be found throughout the world and the opportunities to develop export markets in the supply of materials, equipment and services are considerable.

The analysis commences with an overview of the Earth's energy flows. It can be seen that solar energy is by far the greatest resource and effectively 'drives' most of the other renewable sources. The one major alternative which cannot strictly be regarded as renewable is geothermal energy, but it will be shown that new research work could release very significant amounts of energy from this resource.

The availability of solar radiation is then discussed and it can be seen that the main drawback in high-latitude countries such as the UK is that there are long periods in the winter months when the total radiation received is probably too small to make a useful contribution.

Solar energy can be easily converted into heat and could provide a significant proportion of the domestic hot water and space heating demand in many countries. The main applications are discussed in the following sections. Active solar systems absorb solar radiation on a collecting surface. This converts the radiation into heat, which can be transferred by a circulating fluid (often water or air) to the point of use or to a heat-storage system. Solar heat can also be used for cooling.

Thermal systems for power generation have a potential in the lower latitudes and details of some of the main systems are given in this chapter.

Passive systems use solar energy naturally by involving conventional building elements for its collection, storage and distribution. Although passive solar and associated building design features involve extra design effort, they do not necessarily lead to extra construction costs in new buildings. It has been found that passive features in the refurbishment of older buildings can also be cost-effective.

Photovoltaic systems are well established in the growing consumer product market. Other applications in the UK are limited to a few small-scale systems (e.g. marine light beacons). Applications in warmer countries range from small-scale power supplies for rural villages to power generation in the megawatt range.

The one renewable resource which has a long-established record for electric power generation is hydropower. It is still the fourth-ranking major energy resource (after oil, coal and natural gas) and the potential for further development is still considerable.

Biomass is often completely overlooked as a potential energy source, but estimates of its present position suggest that it supplies up to 15% of world energy demand, mainly in the form of fuelwood use in the developing countries.

Windpower has had a long history of development in many countries, with mechanical power mentioned for many years in the literature. Modern developments are discussed, together with an assessment of wind energy potential and some national programmes. The UK has, perhaps surprisingly, one of the best wind-energy regimes in the world.

Geothermal energy is often considered to be of minor importance and limited to the major fault lines in the Earth's surface in countries such as Iceland or New Zealand. However, the UK are leaders in research into the development of 'hot rock' technology and this is examined.

Tidal power on a world scale is insignificant, but a few countries, including the UK, have suitable tidal sites and their potential is assessed.

Wave power is, in theory, a massive alternative energy resource, but the development problems have been very considerable. Here the UK, as an island community, probably leads in the development of land-based wave energy conversion systems.

12.1.1 The Earth's energy flows

There is a continuous flow of energy through the Earth's atmosphere and surface. By far the greatest energy source is the sun. The other two sources, heat from the Earth's interior and tidal energy caused by the gravitational forces of the earth–moon–sun system, are almost negligible in comparison. A world energy flow diagram is shown in Figure 12.1.[1] The main flow paths were originally suggested by Hubbert[2] in 1962, who subsequently quantified all of them in 1971.[3] Some of these values have been revised to include later work, but it must be appreciated that all the values are approximate.

12.1.2 Energy from solar radiation

The rate at which the incoming solar radiation is intercepted at the edge of the Earth's atmosphere is 1373Wm^{-2} with a probable error of 1–2%.[4] This is known as the 'solar constant', defined as the energy received by a unit surface perpendicular to the solar beam at the Earth's mean distance from the sun. During the year the solar constant can vary by ±3.4%, partly due to variations in the Earth–sun distance. The total solar power intercepted by the Earth is given by $1373\pi r^2$, where r is the mean radius of the earth (6.371×10^6 m). This gives a figure of 1.75×10^{17} W or an annual total of 5.52×10^{24} J. For comparison, the world's total commercial energy consumption in 1988 was 3.61×10^{20} J[5] or some 15 000 times smaller than the solar energy intercepted by the Earth.

12.1.3 Energy from the Earth's interior

Heat flow by conduction through the Earth's solid crust has been estimated to be approximately 0.063 Wm^{-2}.[3] Although there are a few areas surrounding active volcanoes and hot springs where the heat flow can be very much greater than this value, these sources are estimated to contribute at a rate of perhaps 1% of the conduction rate. The conduction rate is given by $0.063 \, 4\pi r^2$, where $4\pi r^2$ is the Earth's surface area. This gives a figure of 3.21×10^{13} W for conduction and a total of approximately 3.24×10^{13} W for the terrestrial energy flow.

12.1.4 Energy from the tides

Tidal energy has been estimated from studies of the rates of change of the periods of rotation of the Earth and moon to be 3×10^{12} W.[3]

12.1.5 The energy flow system

Some 30% (5.25×10^{16} W) of the incoming solar radiation is directly reflected and scattered back into space as short-wave radiation. The Earth's atmosphere, the oceans and the land masses absorb about 47% (8.17×10^{16} W). This is converted directly into heat at the ambient surface temperature and is reradiated as long-wave radiation. The hydrological cycle uses about 23% (4.04×10^{16} W). As water evaporates and changes into water vapour, heat is absorbed. This heat is released when the water is precipitated. The water vapour is often convected high up in the atmosphere. When precipitation occurs, water from these high levels provides all the potential energy in lakes and the kinetic energy of rivers. Most of this energy is dissipated as low-temperature heat as it descends to sea level. A very small proportion of the incoming

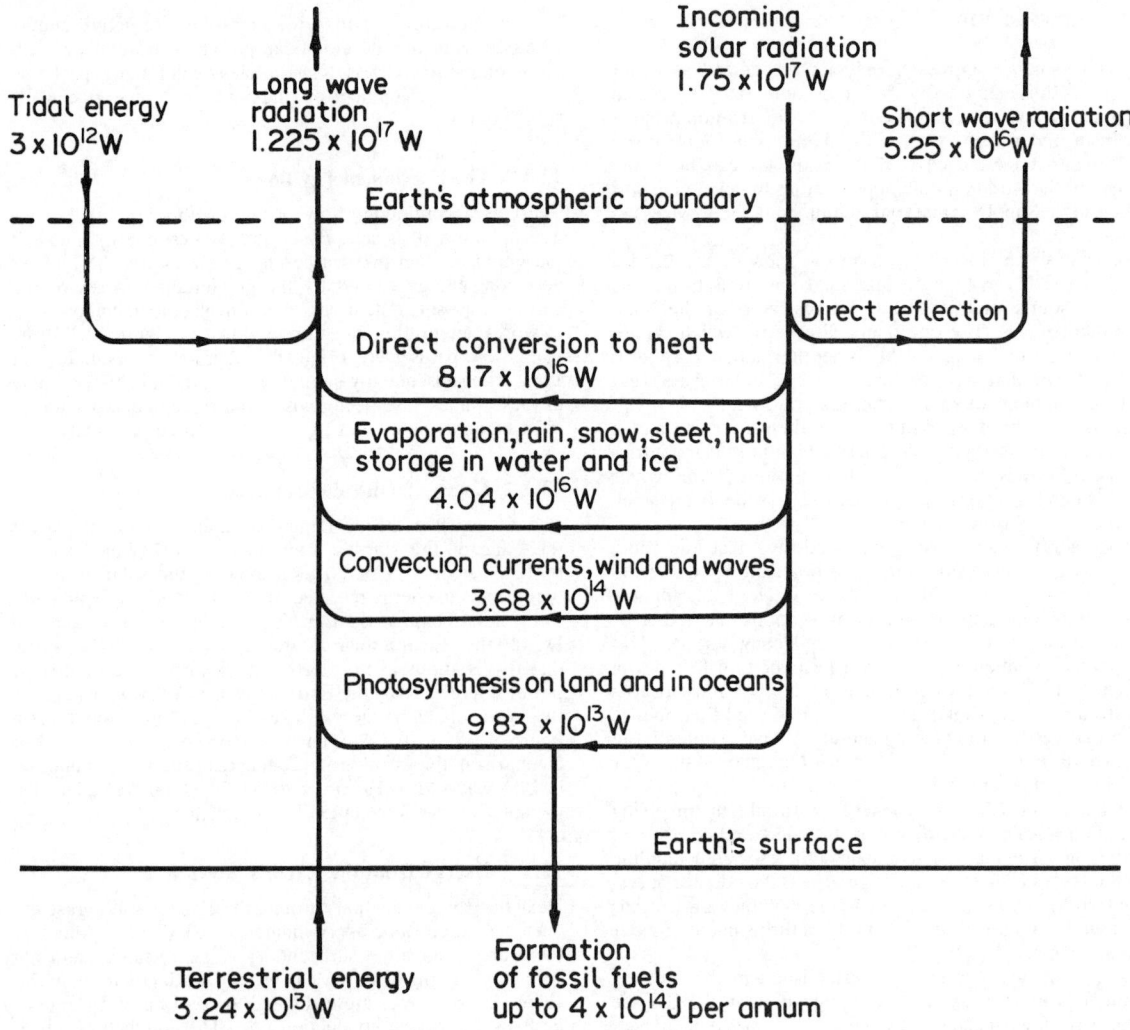

Figure 12.1 World energy flow diagram (after McVeigh[1])

solar radiation, about 0.21% (3.68×10^{14} W), creates the movements in the Earth's atmosphere which cause the winds and ocean waves and currents. These are also dissipated as low-temperature heat.

The amount of organic matter synthesized annually on land and in the oceans by photosynthesis is equivalent to 3.1×10^{21} J of stored energy.[6] This represents less than 0.06% (9.83×10^{13} W) of the incoming solar radiation, but is nevertheless about ten times the world's total commercial energy consumption in 1982. Photosynthesis is the conversion of solar energy into the fixed energy of carbohydrates through the reaction:

Solar energy + CO_2 + H_2O → carbohydrates + O_2

About 0.5% of this photosynthetically fixed carbon is consumed annually as food[7] and up to 1.5% is directly consumed throughout the world as fuelwood. During the past 600 million years the fossil fuels, coal, oil and natural gas, were formed from a very small fraction of the synthesized organic matter. This matter is laid down each year in places such as swamps and peat bogs where its rate of decay is greatly retarded. The annual rate of formation can be calculated by dividing the estimated total fossil fuel resources, approximately 2×10^{24} J, by the total formation period. This gives a figure of up to 4×10^{14} J. In these calculations the various potential energy resources from nuclear fuels have been omitted.

12.2 Solar radiation

The two major problems associated with all applications of solar energy are its relatively low intensity at the Earth's surface and its intermittent nature, both from day to night and through the seasons. The maximum intensity of solar radiation received at the Earth's surface occurs within an hour or two of mid-day on clear days and is limited to about 1 kW m^{-2}.

A detailed description of solar radiation and its availability throughout the world is beyond the scope of this chapter, but can be found in many of the standard texts (for example, references 8–10). For the European Community countries a *Solar Radiation Atlas* was published in 1984[11] giving full details of global radiation on horizontal surfaces and both global and diffuse radiation on vertical and inclined surfaces on a monthly averaged basis. For most practical purposes the simple classification of solar radiation into direct, diffuse or global is adequate.

Direct solar radiation, I, is the solar radiation flux associated with the direct solar beam from the direction of the sun's disk, which may be assumed to be a point source, and is measured normal to the beam (that is, on a plane which is perpendicular to the direction of the sun). Diffuse radiation, D, reaches the ground from the rest of the whole sky hemisphere from which it has been scattered in passing through the atmosphere. Global solar radiation, G, includes all the radiation, direct and diffuse, incident on a horizontal plane. The distribution of diffuse radiation is not uniform over the whole sky hemisphere and is more intense from a zone of about 5° radius surrounding the sun. This is known as circumsolar radiation. Radiation may also be reflected from the ground onto any inclined surface, though this is very difficult to assess. The relationship between direct radiation, I, the diffuse radiation, D, and the global radiation, G, is given by

$$G = D + I \sin \gamma \qquad (12.1)$$

where γ is the solar altitude above the horizon.

Solar radiation, or irradiance, received on the Earth's surface averaged over 24 hours on each day throughout the year, is shown in Figure 12.2. The average in the UK is close to 100 Wm^{-2} while in the Arab countries, such as Egypt and the Sudan, it is close to 300 Wm^{-2}. A detailed look at the mean daily totals of solar radiation, averaged for each month, in London and Khartoum is shown in Figure 12.3.

In the UK and other Northern European countries there is about a fourfold variation between the worst three-month winter period and the best three-month summer one. About half the radiation in Northern Europe is diffuse and this puts a limit on applications which require focusing. In many of the Arab countries there is only a small variation from one month to the next and there is a greater proportion of direct radiation. The worst month in the Sudan has a considerably greater total than the best month in the UK.

In the UK it can be seen from data derived from Meteorological Office records[12] that there are two main trends – higher levels of global radiation in the west of the country, where the skies are, in general, clearer, and lower levels towards the north, due to the higher latitudes. These data are for the period 1965–1970, with the exception of Aldergrove, which is for 1969 and 1970 only. The comparatively low radiation levels in the winter period are combined with an increased proportion of diffuse radiation, which greatly reduces the effectiveness of potential solar applications (Table 12.1).

Many solar applications require a knowledge of the total radiation on an inclined surface facing in any direction while the only available datum is the total global radiation on a horizontal surface in the same location or within a reasonable distance. Sometimes the only records consist of daily sunshine hours. There are numerous examples in the literature giving various approaches to modelling solar radiation data. For example, reference 13 cites over 50 other references. Another approach can be adopted where a country has an established radiation network. This is to select an 'example year' which most closely represents the statistical average (e.g. reference 14).

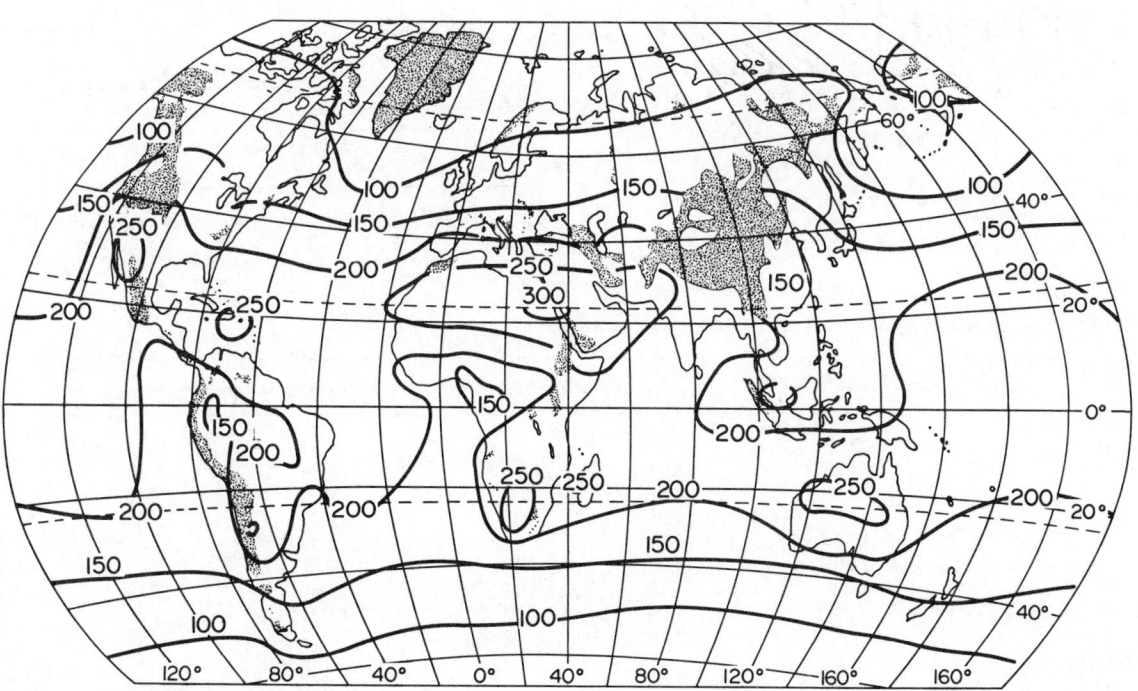

Figure 12.2 Solar radiation received on the Earth's surface averaged over 24 hours each day (W m^{-2}) (after McVeigh[1])

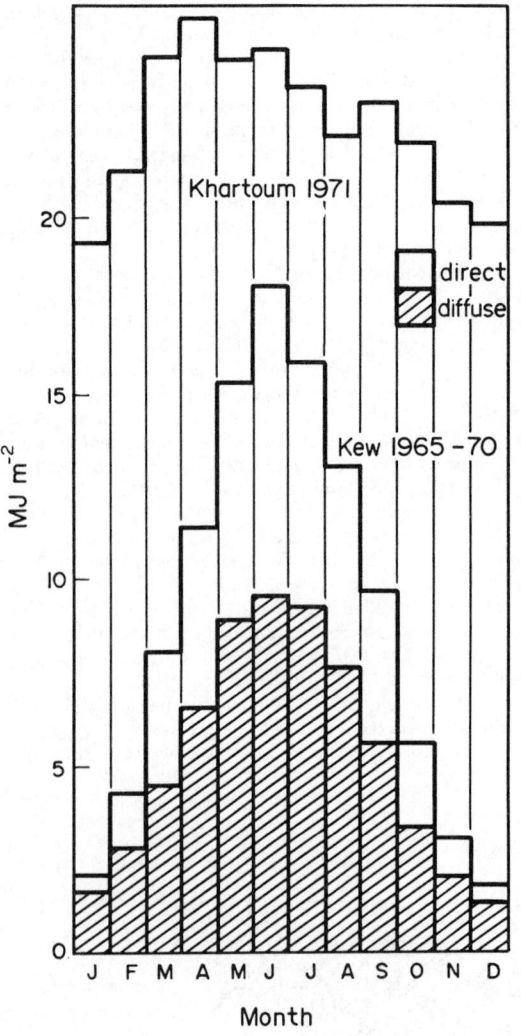

Figure 12.3 Mean daily totals of solar radiation in London and Khartoum (after McVeigh[1])

12.2.1 Flat-plate collectors

Solar energy can be easily converted into useful heat which could provide a significant proportion of the domestic hot water and space heating demand in many countries. One of the drawbacks in high-latitude countries, such as the UK, is that there are many days in the winter months when the total radiation received will be too small to make any useful contribution. The most widely known and understood method for converting solar energy into heat is by the use of a flat-plate collector for heating water, air or some other fluid. The term 'flat plate' is slightly misleading and is used to describe a variety of different collectors which have combinations of flat, grooved and corrugated shapes as the absorbing surface, as well as various methods for transferring the absorbed solar radiation from the surface of the collector to the heated fluid. Many different types of collector have been built and tested by independent investigators over the past 90 years, the early work being carried out mainly in the USA, Australia, Israel, the UK and South Africa.[10] Tests were made in specific locations with wide variations in test procedures and availability of solar radiation. The main objective of these tests has been to convert as much solar radiation as possible into heat, at the highest attainable temperature, for the lowest possible investment in materials and labour.[15]

The majority of flat-plate collectors have five main components, as shown in Figure 12.4:

1. A transparent cover which may be one or more sheets of glass or a radiation-transmitting plastic film or sheet;
2. Tubes, fins, passages or channels integral with the collector absorber plate or connected to it, which carry the water, air or other fluid;

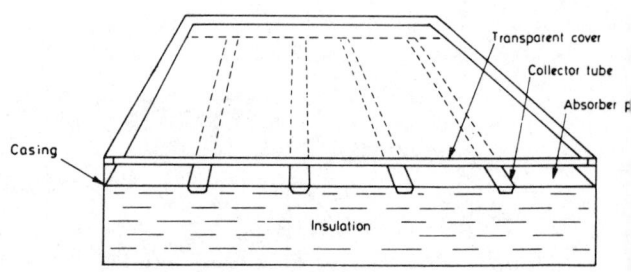

Figure 12.4 The components of a flat-plate collector (after McVeigh[10])

Table 12.1 Annual variation of the mean daily totals of global solar radiation on a horizontal plane (MJ m^{-2})

	Kew	Aberporth	Aldergrove	Eskdale Muir	Lerwick
January	2.13	2.39	1.67	1.54	0.82
February	4.13	5.03	4.50	4.36	2.97
March	8.06	9.51	7.29	7.30	6.41
April	11.62	14.25	12.22	11.47	12.21
May	15.54	16.89	14.65	12.99	13.60
June	18.06	20.09	19.48	16.57	16.90
July	16.03	18.13	15.35	13.64	15.42
August	13.29	15.08	13.45	12.24	11.93
September	9.73	10.58	8.86	7.79	6.88
October	5.79	6.16	4.39	4.52	3.54
November	3.00	2.97	2.66	2.41	1.37
December	1.72	1.94	1.46	1.37	0.55

3. The absorber plate, normally metallic and with a black surface, although a wide variety of other materials can be used, particularly with air heaters;
4. Insulation, which should be provided at the back and sides to minimize the heat losses;
5. The casing or container which encloses the other components and protects them from the weather.

Components (1) and (4) may be omitted for low-temperature-rise applications, such as the heating of swimming pools.

Flat-plate collectors can also be classified into three groups according to their main applications:

1. Applications with a very small rise in temperature, such as in swimming pools where the collector needs no cover or insulation at the back or sides. A high rate of flow is maintained to limit the temperature rise to less than 2°C;
2. Domestic heating and other applications where the maximum temperature required is not more than 60°C. Insulation at the back and at least one transparent cover are necessary;
3. Applications such as process heating or the provision of small-scale power, where temperatures considerably above 60°C are necessary. A more sophisticated design approach is needed to reduce heat losses from the collector to the surroundings.

One basic equation, the Hottel–Whillier–Bliss equation,[16] expresses the useful heat collected, Q, per unit area, in terms of two operating variables, the incident solar radiation normal to the collector plate, G_c, and the temperature difference between the mean temperature of the heat-removal fluid in the collector, T_m, and the surrounding air temperature, T_a, as follows:

$$Q = F\left[(\tau\alpha)G_c - U(T_m - T_a)\right] \qquad (12.2)$$

where F is a factor related to the design of the collector plate and the effectiveness of heat transfer from the collector plate to the heat-removal fluid.

The transmittance–absorptance product $(\tau\alpha)$ takes account of the complex interaction of optical properties in the solar radiation wavelengths.[16] It is actually some 5% greater than the direct product of the transmittance through the covers, τ, and the collector plate absorptance, α, because some of the radiation originally reflected from the collector plate is reflected back again from the cover. The heat loss coefficient, U, rises very rapidly with increasing wind velocity if there are no covers, but is less dependent when the collector has at least one cover. The number and spacing of the covers and the conditions within the spaces can be significant; for example, an evacuated space greatly reduces heat losses.

The three design factors, F, $(\tau\alpha)$ and U, define the thermal performance of the collector. The overall efficiency of the collector, $\eta = Q/G_c$, can then be expressed in terms of the temperature difference, $(T_m - T_a)$, and the incident solar radiation, G_c:

$$\eta = \frac{Q}{G_c} = F\left((\tau\alpha) - \frac{U}{G_c}(T_m - T_a)\right) \qquad (12.3)$$

The longwave radiation emitted from the collector plate surface can be considerably reduced by treating the collector surface. The treatment reduces its emissivity in the longwave spectrum without greatly reducing the absorptivity for short-wave radiation. Any collector surface with this treatment is known as a selective surface. Many methods have been used to achieve these surfaces, including chemically applied coatings.

Equation (12.3) can be used to compare the performance characteristics of different types of collector and to explore the

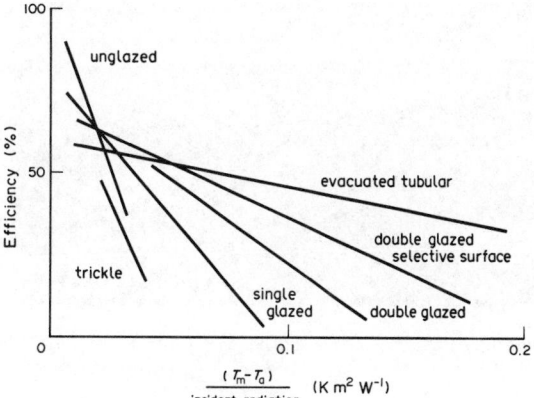

Figure 12.5 Performance of six different types of collector (after McVeigh[1])

effects of altering the various design parameters. This is illustrated in Figure 12.5, which describes the performance of six different types of collector.

The unglazed collector performs well for small values of $(T_m - T_a)$ and with good radiation conditions. A similar performance is obtained from the trickle collector, based on the use of black corrugated sheets with water flowing down the channels from a perforated pipe. For higher temperatures one or two transparent covers are needed and the use of a good selective surface also enhances the performance with lower incident radiation values. The best all-round performance to provide temperatures above 60°C is given by the evacuated tubular collector described later in Section 12.4.

By the end of the 1980s a solar water heating industry had been established in many of the Western industrial countries. Its size depended on many factors, the main influences being the annual solar radiation level and the extent of government subsidies and support. Countries such as the United States, France, Germany, Italy and Japan had a substantial number of the larger multinational companies involved in solar water heating, while solar industries were also being established in many of the Arab countries. In the UK a possible potential market of some 20 million houses was identified, apart from commercial, industrial and government buildings. By 1985, after a research and development programme on active solar heating, the Department of Energy[17] pointed out that although the technical potential was as much as 45 Mtce (million tonnes of coal equivalent) annually, the high cost of the installations could reduce this to as little as one million tonnes per annum. Even this would require an installation rate substantially above that attained in recent years. Despite the publishing of Codes or Guides to good practice[18–22] a number of adverse factors had combined to inhibit the development of a satisfactory market in the UK for active solar water and space heating systems.[23] The main factor was the collapse in oil prices and the correspondingly small rises in the price of coal and natural gas. Nevertheless, one application in the UK has been successful – the heating of swimming pools in the summer season, where the demand for water at about 25°C can be met.

An example of one of the many urban solar water heating projects to be found in countries with lower latitudes than the UK is illustrated in Figure 12.6, where several systems with several thousands of square metres of flat-plate collector are shown on the roofs of blocks of flats in Bucharest.

Figure 12.6 Flat-plate collectors on the roofs of blocks of flats in Bucharest

12.2.2 Space heating

The possibility of at least partially heating a building by solar energy has been demonstrated on many occasions over the past 40 years. The criteria originally suggested by Telkes[24] in 1949 for solving the problems of collection, storage and distribution of solar energy have been somewhat modified since then as experience has been gained with an increasing number of installations. In the early work, emphasis was placed on the collection of winter sunshine. This has been broadened to include the very substantial contribution that can be obtained from diffuse radiation. The three terms 'efficient', 'economic' and 'simple' used by Telkes in relation to the collector have always been the goal of sound engineering practice and an analysis of some of the earlier solar installations indicated that very few can satisfy all three.

The associated problem of storing the solar energy collected in the summer for use the following winter has attracted a considerable research effort. The principle of using a very large, heavily insulated water storage tank buried underneath the building was tried some 50 years ago at the first Massachusetts Institute of Technology solar house.[25] The comment that this approach was highly uneconomical had a considerable influence on the direction of storage system research over the next two decades. The effect of latitude and local radiation characteristics are now widely appreciated. It was originally thought that only a few days' storage could be economically viable so that the solar energy received during clear winter days would be made available for successive overcast periods and that this could only occur to any extent in parts of the world where there are appreciable amounts of winter sunshine. However, longer storage periods of up to several months were achieved in several solar houses during the mid-1970s and by the end of the decade interseasonal storage has been shown to be not only possible but also cost-effective. For shorter-term storage considerable reductions in the total volume of the store are possible by the use of chemical storage methods pioneered by Telkes.

The term 'solar house' first became familiar in the USA during the 1930s, when architects began to use large, south-facing windows to let the lower-slanting rays of the winter sun penetrate into the back of the room.[26] At that time there was no real distinction made between the active systems, which use pumps or fans to circulate heat, and passive systems, which use solar energy naturally, involving the conventional building elements for solar energy collection, storage and distribution. The large windows saved fuel during the day, but it was not possible to store the solar energy. At night and during cloudy days the heat loss was so great that there was a relatively small saving on fuel during the entire heating season. This approach was later known as the 'direct gain' system.

Prior to 1973 there were probably less than 50 well-known solar houses or buildings in the world. Among these, Europe's largest, and for many years the world's largest passive building was the annexe to St George's School, Wallasey,[10] completed in 1962. There are now many thousands of solar houses and buildings throughout the world. In parts of the USA and Northern Europe the earlier designs in the 1970s were mostly active systems and the need for long-term interseasonal heat storage began to attract considerable attention. By the late 1980s over 30 operating solar heating systems using interseasonal heat storage in some 12 countries had been identified and characterized.[27] Proposals to use large plants with over 100 000 m² of collector arrays were planned. Most of these systems involved some type of *in-situ* construction using aquifers, earth coils, drilled rock, water pits or rock caverns.[27] One common feature noted was the economy of scale as storage volume increased.

The great majority of all these systems used some form of active solar collector system, with pumped circulation of the heat transfer fluid (usually water). Results from the early monitored tests also indicated that large interseasonal stores are not suitable for smaller individual houses and the emphasis for this category, including commercial buildings, changed to passive applications. Passive systems use solar energy naturally by involving conventional building elements for the collection, storage and distribution of solar energy.

In addition to the direct gain system, mentioned above, four other passive systems have been categorized[28] as follows:

1. *The Trombe wall* or thermal storage wall in which the heat is stored in a wall which also absorbs the solar energy when it comes through the glazing;
2. *The solar greenhouse*, in which the features of the solar thermal storage wall and the direct gain approach can be combined by building a greenhouse on the south side of a building (in the northern hemisphere). This approach attracted very considerable attention in the 1980s;
3. *The roof pond*. In this system a shallow pond or tank of water sits on a flat roof with its surface generally contained by a transparent plastic sheet. Movable insulation is essential because solar input during the summer months can be very large and it is needed to protect the building from heat losses in the winter. In the winter, the movable insulation covers the top of the pond at night and the heat from the pond can radiate down to the house. During the summer, the insulation covers the top of the pond in the day and is removed at night when the heated water can radiate out to the colder night sky;
4. *The natural convective loop*. In this system air is usually used as the heat-transport medium and works in the same way as the classic thermosyphon water heating system.[10]

Many countries now have their own national passive solar programmes. Within the Commission of the European Communities a major research, development and demonstration programme has already resulted in the publication of a *European Passive Solar Handbook*[29] which covers basic principles and concepts. Also included are case histories, references and suggestions for further reading.

12.3 Passive solar design in the UK

Passive solar design is now regarded by the government as one of the most promising renewable energy technologies in the UK.[30] The combination of current energy efficiency measures with passive solar design techniques could reduce energy costs in buildings by as much as 40% from today's levels with existing standards. The initial passive solar programme funded by the Department of Energy concentrated on establishing the worth of the resource in energy terms and it found that a contribution of up to 14 Mtce (million tonnes of coal equivalent) annually could be expected.[31].

Their first two groups of studies were carried out on housing and non-domestic buildings. Design studies were selected rather than field tests so that comparative tests could be made with one particular thermal simulation model. In the second phase of the programme several design practices have been invited to prepare building designs incorporating passive solar features. An additional feature of the second phase has been the appointment of a 'solar expert' to each project. His or her role will be to comment on all the proposals in the design. The client or customer will also be represented at this stage by a person who is already a builder or property developer. These two additions to the design process (the 'expert' and 'client') will provide extra technical knowledge and an appreciation of real market forces.

The main objective of the Department of Energy's programme is the establishment of a Design Advice Scheme.[32] One of the anticipated outcomes from the preliminary programme is the preparation of a solar design guide for UK conditions. The Department of Energy's Passive Solar Programme is closely coordinated with other government building research and development programmes, e.g. the Department of the Environment and the Energy Efficiency Office. The Building Research Establishment also takes an active part in providing advice and participating in the programme.

12.3.1 The main technical issues in passive design

A number of issues have been identified in the literature and the UK government programme will broadly follow the guidelines already laid down elsewhere, i.e.:

1. The influence of the local climate on the optimum design;
2. How the patterns of occupancy affect the various criteria which are being evaluated;
3. The perceived advantages and disadvantages of the various passive design elements in the selected project, measured against energy use and cost, the feeling of comfort and general well-being, the amenity value and whether or not people are likely to pay for the passive features – the marketability of the design;
4. The integration of energy use into the overall design process.

12.3.2 The energy-performance assessments

As well as these design studies, a number of both domestic and non-domestic buildings have been monitored for a year each to determine the energy flows and to evaluate the quality of the environment as perceived by the users. Each study is carried out according to a clear and well-defined methodology. Standard profiles are produced at the end of each assessment and the performance is assessed alongside a similar building design without such passive features.

12.3.3 Technology transfer

This has been identified as perhaps the most important and significant part of the whole project. However, until the pattern of designs and assessments begins to emerge, it will not be possible to do more than hold some pilot or trial dissemination studies, based on the present results. Commercial clients, the householder or house purchaser, local and public authorities and institutions, and the decision makers will all need to be informed, as well as the energy professionals, especially those in the design practices not yet involved in this work. Details of the Department of Energy's pilot Design Advice Scheme were announced in 1989.[32]

12.3.4 Some UK applications

12.3.4.1 Parklands Estate, Bourneville[33]

The Parklands Estate forms part of the Bourneville Solar Village. The Solar Village covers seven sites, each containing a variety of low-energy or passive houses and believed to be the largest single solar housing project in Europe. The entire project has been supported by the Bourneville Village Trust, the Housing Corporation and the Commission of the European Communities.

The project was started in 1984 following detailed consultations with local residents and various national and government organizations. There are a number of different house types, including direct-gain passive houses for single families and two blocks of flats for elderly people. The aim of Parklands was to provide 90 single-family houses for first-time buyers and the houses had to be built within strict cost limits. Particular attention was paid to the orientation of the houses, with the majority facing south to maximize solar gain. Monotony of appearance was avoided by variety in design and by placing some houses off due south. It has been found that, on average, there is a direct-gain solar contribution of some 33% to the gross space heating load. A number of improvements have been suggested for future designs.[33] North-facing windows were thought to be too small to allow enough daylight to penetrate. The remedy would be to have larger, perhaps triple-glazed windows. There were component problems with double glazing, cavity-wall insulation and the thermal blinds (about half the houses reported blind failures). The value of the energy savings were compared with the additional cost of the energy saving measures and gave an average payback period of just under 15 years – a figure which the Bourneville Village Trust considered acceptable.

One of the 90 houses was specially adapted to have other solar and passive features, and is separately monitored. In addition to heavy insulation and large south-facing direct-gain windows, it has a large attached greenhouse (or sun space), a variable flowrate active solar water heating system linked to variable volume storage, another active solar space-heating system with phase-change energy storage, a heat-recovery system and some photovoltaic power generation.[34] This house is illustrated in Figure 12.7.

12.3.4.2 The JEL Building, Stockport[33]

This is a low energy production and office building for a company heavily involved in the manufacture of energy control and monitoring systems. It has a fully glazed south facade, in the centre of which is a two-storey atrium. The top of this atrium serves as a collection point for the solar-heated air which can be transferred directly to the production area or used as preheated air as the input to conventional heaters. Blinds on the south wall are set automatically to admit or reflect solar radiation, and other measures include a zoned building energy management system. The energy savings are approximately 50% annually for an additional 3% spent on building costs, giving an estimated simple payback period of 6 years.

Figure 12.7 One of the houses from the Parklands Estate, Bournville

12.3.5 Atria

Modern interest in the glazed courtyard or atrium has been said to date from 1967, when two innovative buildings were completed in the USA. Since then there have been many developments in North America and Scandinavia. In the UK the majority of the 500 retail and leisure developments at present under construction are reported to feature atria.[35] Atria provide buffer spaces which protect the occupied parts of the building from the full impact of the external environment. Care is needed during the design stage in the UK to ensure that the provision of passive solar gains for heating during the heating season do not result in overheating in the summer. Another advantage of atria is that the glazed space has relatively high daylight levels. Thus window sizes can be increased as the energy penalties are reduced and daylight can be used more consciously as an alternative to artificial lighting.[36] One of the best-known examples of this design approach is the Wiggins Teape's headquarters at Basingstoke, designed by Arup Associates. This consists of a hollow rectangle built around a full-height atrium 20 m wide by 40 m long. Half the offices look outward and the rest face into the atrium, which makes daylighting and natural ventilation possible without the associated energy penalties. The economics appear to be very encouraging as the building cost 30% less than its deep-plan, artificially lit, air-conditioned predecessor (also designed by Arup Associates) and is planned to use 25% less energy.[37]

12.4 Thermal power and other thermal applications

The earliest recorded example of a solar-powered engine was reported in France in 1860 and a solar thermal water distillation plant was built in Chile with a total area of some 4700 m^2 of glass in 1872.[10] There has always been a major need for water pumping in arid regions where clear sky conditions often give long periods of direct solar radiation. The early experiments led to the most spectacular solar engine development of the time – the Shuman–Boys Sun-Heat Absorber at Meadi in Egypt in 1913. The absorber consisted of five large parabolic mirror sections each 62.5 m long and 4.1 m wide between the edges of the mirrors, giving a total collecting area of 1277 m^2. Although the maximum recorded pumping horsepower was only 19.1, this was attributed to the poor performance of the steam engine and pumping plant.[38] Using the same steam conditions a modern low-pressure steam turbine could produce about 100 hp. After these technically promising early results, very little more work was carried out on solar thermal applications for the next 40 years. Power generation from coal and oil became widespread and relatively inexpensive.

System efficiency is proportional to the product of the collector efficiency, the ideal Carnot efficiency and a factor relating actual engine efficiency to the Carnot efficiency. There is clearly a conflict between the desired collector temperature for maximum collector efficiency and the desired temperature for high Carnot efficiencies for any given ambient temperature. High collector efficiencies require the maximum temperature to be as low as possible, while high Carnot efficiencies need as high a maximum cycle temperature as possible.

It can be shown that unless a large area of relatively inexpensive collectors can be used, collectors capable of working at temperatures up to at least 250°C are essential for thermal power applications. These can be classified into two main classes – tracking and non-tracking focusing collectors – with a number of sub-groups in each class.

The two most important non-tracking focusing collectors are the compound parabolic concentrator (CPC) and the evacuated tubular collector. The CPC was developed in 1974 by Winston.[39] In cross section it appears to take the form of a truncated parabola with a vertical axis. Winston's innovation was to place the focus of the parabola forming the right-hand section at the base of the left-hand section and vice versa. With careful choice of dimensions all the incoming direct radiation was concentrated on the absorber near the base. Concentration factors up to ten can be achieved without diurnal tracking. Evacuated tubular collectors use vacuum technology to surround the absorbing collector surface, often a selectively coated pipe, with a double-walled evacuated glass tube. The focusing element plays a relatively minor role with this type. These collectors are commercially available, with typical values of between 0.84 and 0.86 for $(\tau\alpha)$ and 1.5 W m^{-2} K^{-1} for U.[40]

Tracking focusing systems can be classified into three groups. Point focusing collectors usually have a tracking parabolic reflecting dish, in which the direct solar beam is reflected to a central collecting point fixed to the dish. Line focusing collectors have an absorbing pipe placed at the focus. They are either parabolic in cross section or focused through a linear Fresnel lens, which consists of a series of wedge-shaped sections in a single flat unit, with each segment designed to concentrate the incident radiation onto the absorbing pipe. This type is often known as a trough collector. Concentration ratios of up to 40:1 can be achieved.

For high temperatures and large-scale power generation, the central 'Power Tower' system was considered. A large field of steered mirrors, or heliostats, reflect solar radiation to a single central receiver mounted on a tall tower. Concentration ratios approaching 3000:1 have been achieved with this type. The best example in the 1970s was the 1 MW solar furnace at Odeillo, built by the French Centre National de la Recherche Scientifique.[41] The parabolic mirror, which is 39.6 m high and 53.3 m wide, contains 9500 individual mirrors with a total reflecting area of 1920 m². It faces a field of 63 heliostats, with a total mirror area of 2839 m². During the 1970s studies in the United States[42] considered the design of individual units of 100 MWe capacity. The next major step towards this concept was the commissioning of a 10 MWe plant, solar one, at Barstow, California, in January 1982.

In Europe there were a number of smaller-scale complementary programmes supported by the EEC and the IEA. Their largest, known as Eurelios, was completed in Sicily for the EEC in 1981 and had two types of heliostat with a total area of 7800 m² and a designed output steam temperature of 450°C. Europe's largest solar power tower system was the French Themis, rated at 2.5 MW and situated a few kilometres from Odeillo at Targasone at an altitude of 1700 m. The main system consisted of a group of 201 heliostats with a total area of 10 740 m², facing a receiver on the top of a 101.5-m tower. There was also an auxiliary system of eleven parabolic mirror concentrators, each with a surface area of 75 m², to maintain the salt solution used as the heat transfer fluid above its solidification point of about 150°C. Problems were experienced with the corrosive salt solution at high temperatures and with gusty wind conditions shattering some mirrors. By July 1984 it was producing between 7000 and 8000 kWh daily, but further work on the system was terminated.

Writing about this phase of the developments in 1987, Kesselring[43] commented that there had been a number of important achievements. The amount of material (and hence overall costs) needed to build plants had been reduced, systems for higher temperatures and heat fluxes had been developed and significant progress had been made in both heliostat and receiver design.

Progress in the development of commercial line focusing solar thermal collector systems had been far more successful.[44] By 1990 over 200 MW of thermal-electric capacity had been installed in the USA, with the latest system, the Luz SEGS VIII, rated at 80 MW and having a total collector area of nearly 500 000 m². Estimated costs in the region of 8 cents kWh⁻¹ were forecast.[45]

12.4.1 Solar ponds

A solar pond is a body of saline water in which the concentration increases with depth. In a natural pond when solar radiation heats the water below the surface the action of convection currents causes the heated water to rise to the surface and the pond temperature normally follows the mean temperature of the surroundings. A solar pond contains concentrations of dissolved salts which gradually increase with depth, causing the density of the water to increase towards the base of the pond, which is often black. Solar radiation penetrates to the base, heating the water at this lower level, but any convection currents are suppressed by the density gradient. Heat losses from the surface are reduced, compared with a natural pond, and the temperature at the bottom of the pond rises.

While there are daily fluctuations in both ambient air temperature and in the upper water layers, the temperature at the bottom of the pond, where heat would be extracted, remains fairly uniform.[46] A solar pond is both a massive heat collector and heat storage system and, compared with a

conventional collector and heat store, is relatively inexpensive. Other advantages are:

1. Problems which could occur with dirt settling on the surface and reducing collection efficiency are eliminated;
2. Extracting energy from the pond is very straightforward as the lowest hot layer can be pumped to the power station and returned to the pond for reheating;
3. Solar ponds can operate continuously throughout the year provided the storage capacity is chosen to match the demand;
4. A solar pond power system can – like a hydroelectric plant – provide peaks of power on demand.

The system is illustrated in Figure 12.8.

The development of the solar pond concept has been in a series of order-of-magnitude steps, with the few kilowatts of the early 1960s being followed by the 150 kWe solar pond at Ein Bokek in the Dead Sea area, developed by Dr Harry Tabor of the Scientific Research Foundation in Israel. This was switched on at midnight on 16 December 1979 and has subsequently given outputs approaching 300 kWe. The third stage was the 5 MWe solar pond at Beth Ha'arava, also in the Dead Sea area. During the commissioning stage in 1983 temperatures above 86°C were achieved in March.[47]

Following these successful trials, in 1984 plans were announced in Israel to develop up to 2000 MW from solar ponds over a 10- to 20-year period. A disadvantage of the system is that overall conversion efficiencies from solar radiation to electricity are relatively low – in the order about 1%. However, this is offset by the equally low relative costs of the 'collector'.

Solar ponds could be particularly interesting for developing countries as they are likely to have good solar radiation conditions and about half the total costs could be found locally (in the simple materials used for the construction of the pond). The range of possible applications continues to increase, with space heating for buildings and greenhouses, process heating, space cooling, desalination and salt production already either in operation or under consideration.[10]

12.4.2 Heliohydroelectric power generation

The concept of heliohydroelectric power generation is to convert solar energy into electricity by first transforming it into hydraulic energy. If a closed reservoir is completely sealed off from the sea, the level of the reservoir will tend to decrease as a result of evaporation. Hydroelectric generators

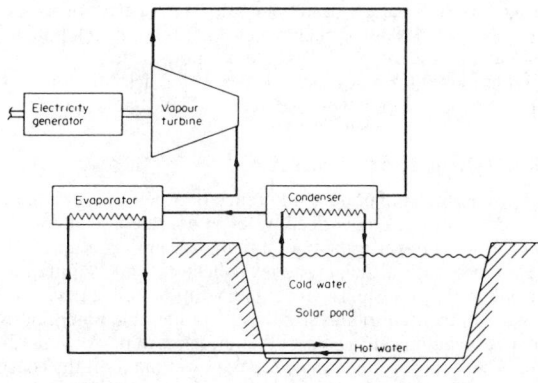

Figure 12.8 A solar pond system (after McVeigh[10])

could be placed at the reservoir end of pipes connecting the reservoir to the sea. The fall in the level of the reservoir induces a flow of water from the sea, and the potential energy caused by the difference in water levels could be transformed into electrical power. By choosing suitable water levels and power systems, it would be possible to have a continuous process. This topic has been extensively studied in Saudi Arabia by Kettani and Gonsalves,[48] who have measured evaporation rates and compared these with meteorological data. Kettani has also explored the possibility of building a dam across the Gulf of Bahrain, using the whole sealed-off Bay to create a hydraulic head from the open sea in the Gulf.

Connecting the Mediterranean Sea with the Qattara Depression for a hydroelectric scheme was first considered in 1916. This project was revived during the 1970s by a joint Egyption–Federal Republic of Germany feasibility study. The results of this study, which was completed in December 1980, were reported in 1982 but a firm decision on whether or not to proceed with the scheme was not taken.

12.4.3 Distillation

One of the major problems in many parts of the world is the scarcity of fresh water, and the development of inexpensive large solar distillation units capable of easy transportation and handling is increasingly important.[10] Solar distillation is another application which dates back to the nineteenth century and the simplest form still now in use is basically unchanged from the early designs, which consisted of a shallow tray, filled with salt or brackish water, and covered by a sloping glass cover plate. The solar radiation heats the water in the tray and evaporates it. When the vapour comes in contact with the colder surface of the glass it condenses, forming fresh water which runs down the inner surface in the form of droplets and can be collected in a trough at the lower edge. Under good radiation conditions an output of about 4 litres of fresh water can be obtained daily per square metre of cover.

Australia has gained considerable practical experience over many years, and a guide to the design, construction and installation of a solar still developed by the Commonwealth Scientific and Industrial Research Organization was published in 1965.[49] One of the best-known large installations in recent years, with an evaporating surface area of 8667 m^2, was completed on the island of Patmos in the Aegean in 1967.[50] The average distillation rate was 3.0 kg m^{-2} per day, with a maximum of 6.2 kg m^{-2} at midsummer. The first large installation designed and manufactured in the UK was a 185 m^2 unit for Aldabra in the Indian Ocean[51] in 1970. The development of various methods of improving simple still designs, such as the use of stepped salt water shelves or wicks, up to the early 1980s has been reviewed.[10] Later developments include the use of an 8 kW (peak) photovoltaic array to provide power for a reverse osmosis water desalination unit for a community of 250 people near Jeddah, Saudi Arabia.[52]

12.4.4 Ocean thermal energy

The use of the temperature difference between the surface of the ocean and the colder deep water to operate a heat engine was proposed towards the end of the nineteenth century. The oceans are natural solar energy collectors and require no special storage systems or manufactured collectors and, because of their enormous size, have considerable potential to compete economically against other methods of power generation. The earliest system was a 22 kW power plant off the coast of Cuba developed by Claude[53] in the late 1920s. It had an overall efficiency of less than 1% and operated with an open

Rankine cycle in which the higher-temperature sea water was passed directly to a low-pressure evaporator to provide steam to power the turbine. It was uneconomic at that time, as was a subsequent larger project by the French some 20 years later, and further work was discontinued.

Renewed interest in the concept came in the 1960s when the possibility of using a closed Rankine cycle was suggested[54] in the USA. This work formed the basis for several major large-scale theoretical investigations which showed very similar overall net cycle efficiencies, ranging from 2.1% to 2.4%. Following these systems feasibility studies, a joint project called Mini-OTEC was developed by the State of Hawaii, Lockheed Missiles and Space Company and the Dillingham Corporation. Mini-OTEC was a 50 kW (18 kW net) plant which operated off the coast of Hawaii during 1979 to establish that the problems raised in the feasibility studies were surmountable. It was also intended to draw public and political attention to the OTEC system in contrast to the better-known solar technologies. The project was described by Henry J. White,[55] who concluded that all the goals had been achieved and that a combination of private industry with enlightened state support could design, construct, test and operate a sophisticated energy system at a cost, and in a time scale, which a more formalized procedure might be unable to attain.

Among the advantages listed for OTEC by Avery and Dugger of the Ocean Energy Program at the Johns Hopkins University[56] were

1. It can deliver energy to regions bordered by warm ocean waters via direct electric-power transmission;
2. It can also deliver energy to all other regions via an energy-intensive material such as ammonia, produced on OTEC plant ships in the tropics.

Ammonia was considered to be an outstanding choice as it could save the natural gas already used in conventional ammonia plants, it could serve as a synthetic fuel or it could provide an easily transported and storable source of hydrogen for fuel cells to generate electric power. The costs of OTEC electricity delivered directly or via fuel cells were projected to be competitive with costs from coal or nuclear plants if rapid OTEC development were to be pursued.

In the 1980s, several other countries apart from the USA had active OTEC programmes. These included France, Japan and Eurocean, a European consortium.[57] However, it was found that the main obstacles to the rapid development of OTEC systems were those facing a number of other interesting large-scale renewable energy systems. There is clearly a concern that some of the environmental problems may not have been fully appreciated. The enormous cost of the development of full-size systems in the order of 400 MW could only be provided by government funding, and there were no signs that this was likely in the early 1990s.

12.4.5 Refrigeration and cooling

The great advantage of using solar energy in refrigeration and cooling applications is that the maximum amount of solar energy is available at the point of maximum demand. There are two quite different major applications, the first in the cooling of buildings and the second in refrigeration for food preservation or for storing vaccines for medical purposes.[1]

In building applications, solar collectors can be used to provide a high-temperature input to cooling systems using solar-powered engines or absorption systems. This part of the system could be used to provide heating outside the hot midsummer period and the cost of the system could be shared between the two functions. The cooling demand is also at a maximum during the early afternoon, depending on the orientation of the building and its thermal mass, so that the storage capacity for cooling is only a few hours, in contrast to the very much greater periods required for heating systems.

Solar-powered refrigeration was first demonstrated over a hundred years ago[10] and several different systems based on the familiar absorption refrigerator were commercially available by the early 1980s. The four main methods used for refrigeration and cooling systems are as follows:

1. The vapour compression refrigeration cycle
2. Absorption systems
3. Evaporative cooling
4. Radiative cooling

In its simplest theoretical form the well-known domestic refrigerator can be thought of as a reversed heat engine. In practice, nearly all vapour-compression refrigerators are driven by electric motors. Although small solar-powered engines have been used to replace the electric motor in laboratory trials, it has proved to be too expensive to proceed to commercial applications. For future small-scale applications it is likely that conventional electrically powered refrigerators will be driven by photovoltaic cells through a d.c. to a.c. inverter, a method which is already proving to be economic in some applications.

There are two main types of absorption cooling systems. For continuous operation, the cycle is illustrated in Figure 12.9. The working fluid is a solution of refrigerant and absorbent, the most common being ammonia–water or water–lithium–bromide. When solar heating is supplied to the generator some refrigerant is vaporized and a weak mixture is left behind. The vapour is then condensed and expands to the lower-pressure evaporator, where it is vaporized and refrigerates the external working fluid, which would be air for an air-conditioning application. The cycle is completed in the absorber when the refrigerant recombines with the original solution and is pumped back to the generator.

The technology was well established by the end of 1983 when one of the largest systems in the world, the solar air-conditioning system at the Chaim Sheba Medical Centre in Israel, had been operating successfully for three years.[58] The system is powered by approximately 3000 m^2 of flat-plate solar collectors and annual savings of electricity are about 120 kWh per square metre of collector.

Intermittent ammonia–water absorption units are relatively simple to operate and have been successfully developed at the Asian Institute of Technology (AIT), Bangkok, by Professor R.H.B. Exell.[59] During the day, heat from solar collectors vaporizes the ammonia and it condenses in the condenser. At night the ammonia is passed through an expansion valve into an evaporator, where the refrigeration effect produces ice.

During the early 1980s the principal unit at the AIT could produce up to 30 kg of ice from water at 28°C, with a collector area of 5 m^2 linked to the ammonia–water system. The solar radiation was enhanced by the use of plane reflecting mirrors. Subsequently a much larger unit was developed for use in rural village communities.

Evaporative systems achieve a cooling effect through the evaporation of water. Evaporative cooling can be achieved without making the air within a building more humid by combining a vertical water heating panel with a horizontal shaded roof pond.[60] Heat is absorbed from the room, as shown in Figure 12.10. The heated water rises to the roof pond where it is cooled by evaporative cooling. An earlier simple

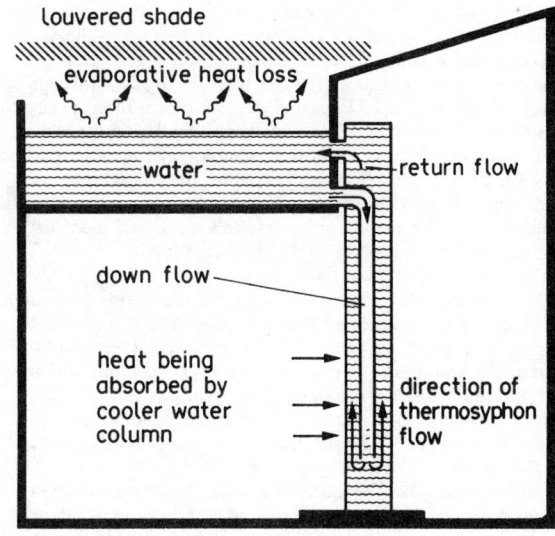

Figure 12.10 A room-cooling system (after McVeigh[1])

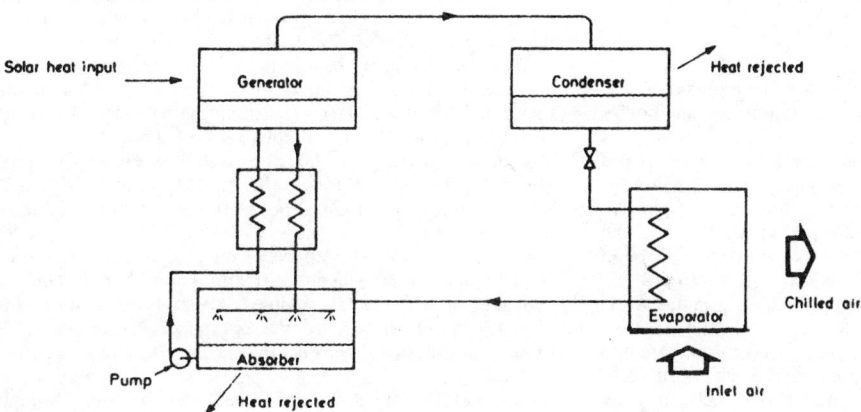

Figure 12.9 An absorption cooling system for continuous operation (after McVeigh[10])

method described by Thomason[61] takes water from a house storage tank and trickles it down an unglazed north-facing (in the northern hemisphere) roof.

Radiative cooling to the sky is possible at night with clear atmospheric conditions. Several examples of the use of the technique of pumping water into various roof-mounted collectors have been described.[10] Further examples of various applications, together with a detailed examination of the theoretical aspects of solar cooling and refrigeration, are given in a report to UNESCO by McVeigh.[62]

12.5 Photovoltaic energy conversion

12.5.1 Introduction

The direct conversion of solar energy into electrical energy has been studied since the end of the nineteenth century.[10] The early work was concerned with thermocouples of various different alloys and efficiencies were very low, usually less than 1%. Modern developments date from 1954, when the Bell Telephone Laboratories discovered that thin slices of silicon, when doped with certain traces of impurities, achieved efficiencies some ten times greater than the traditional light-sensitive materials used in earlier devices. Since then, photovoltaic cells, modules and systems have developed rapidly.[63] Improvements in their cost-effectiveness, reliability and life have resulted in photovoltaics becoming the first choice in a wide range of applications on both engineering and economic grounds, with the more recent emphasis on 'clean' electricity likely to become an increasingly important factor.

The cost of the photovoltaic module, described as 'the basic building block of the system', has fallen in real terms to about one tenth of its value in 1980 and is expected to continue to fall, unlike almost all other competitive electricity-generation systems.[64] Among the advantages listed for the modern solar cell during the 1970s were that it had no moving parts and an indefinitely long life, required little or no maintenance and was non-polluting.[65] The lifetime should now be modified to 'probably more than 20 years' and there are the normal environmental hazards associated with the manufacture of semiconductors. Unlike other types of electric generator, it is suitable for a very wide range of power applications from a few milliwatts to several thousand megawatts. Photovoltaics is now widely recognized as a mature technology, ready to move into a new phase in an expanded product market which already includes telecommunications, cathodic protection, remote power and utility demonstration projects.[64]

12.5.2 Basic principles

The basic principle of the solar cell is that the doping of a very pure semiconductor with small traces of impurities can modify its electrical properties, producing the p-type, having fixed negative and free positive charges, and the n-type (fixed positive, free negative). A typical p-type (positive) is boron-doped silicon and a typical n-type (negative) is phosphorus-doped silicon. If these two types are placed together and the surface is exposed to sunlight, electrons will diffuse through the p–n junction in opposite directions, giving rise to an electric current. The positive contact is usually a metal layer, or contact plate at the back of the cell, while the negative contact at the top (or front) of the cell collects the current and also allows as much light as possible to enter the cell. Manufacturers often quote the output of their cells in 'peak watts' or Wp. This is measured at a standard temperature of 25°C with a sunlight intensity of 1 kW m^{-2}, which is some-

times reached at noon in many Middle East countries. The power output of the solar cell varies almost directly with the intensity of the light falling on it. Halving the light intensity would also halve the current output, but the voltage would drop by only a few per cent. However, the voltage output is very temperature dependent, and decreases by about 0.5% for every degree Celsius rise above 25°C.

12.5.3 Developments

Three main types of photovoltaic module design have shown particular promise in both performance and cost. These are the crystalline flat plate, the thin-film flat plate and concentrator modules. Currently available commercial flat-plate modules are based on either crystalline silicon or thin-film amorphous silicon cells. Concentrator module installations all use single-crystal silicon cells. Crystalline silicon flat-plate modules dominated the power market, defined as arrays larger than 50 Wp, by the early 1990s, but informed opinion agreed that amorphous silicon thin-film flat-plate modules were getting an increasing share and could be dominating the market by the end of the decade.

Amorphous silicon cells have had considerable development for the consumer market. This includes all products in which the power supply is built in, such as calculators or watches. The expanding consumer market accounted for over 25% of total world production by 1990, mostly in Japan.[64]

For power applications the commonest module type is the flat-plate crystalline silicon module, using cells made from single crystal or polycrystalline material. Individual cells are series strung and then encapsulated behind low-iron-content tempered glass. By 1990, module efficiencies of 11% for single crystal cells and 10% for polycrystalline cells had been reported,[64] with the possibility that module efficiencies greater than 15% could be achieved using cells with 18% efficiency.

A European Standard exists for the testing of photovoltaic modules. UK suppliers can have equipment tested for electrical characteristics at the Royal Aircraft Establishment, Farnborough, and for mechanical integrity and environmental tests at the Solar Energy Unit at Cardiff University.[64]

It is expected that the design features leading to this improved efficiency will soon be introduced into commercial modules and that module efficiencies of 17–18% could be anticipated with single-crystal cells. One of the advantages of the crystalline silicon flat-plate module is its excellent reliability. A lifetime of between 5 and 10 years is normally guaranteed at present and operation for periods greater than 20 years is expected.

The cost of crystalline silicon modules in the early 1990s was in the order of $4–$5 per Wp. At this level they were becoming increasingly competitive with other electricity supplies.

Thin-film modules have much lower fabrication costs, because a smaller amount of the expensive semiconductor material is used and large-scale automated fabrication processes are possible. Hence the costs of amorphous silicon modules are lower than those of crystalline silicon modules. However, they have a lower efficiency, typically 6–7%, and production costs less than $1 per watt were anticipated during the early 1990s.[64]

Efficiencies of 10–11% have been reported in laboratory tests of amorphous silicon cells and these should soon be available in commercial modules. Other types of thin-film modules based on both copper indium diselenide and cadmium telluride will also be commercially available in the near future.[64]

Considerably more expensive cells can be used through optical concentration and high-efficiency cells based on both gallium arsenide and silicon as well as multi-junction cells are under development. This work aims to achieve cells with over 30% efficiency, and already an efficiency of 37% has been reported in laboratory tests[63] with a tandem of GaAs/GaSb cells. Should cells of this efficiency potential be incorporated into commercial modules, then efficiencies greater than 20% are technically feasible.

12.5.4 Systems

In addition to the basic photovoltaic module, both a supporting structure and some form of electrical conditioning are required. In space applications the supporting structure and associated electrical conditioning are highly specialized to provide both protection from the space environment and to withstand severe launching and operational conditions.

A typical terrestrial module produces about 14 V d.c. and about 3 A under the peak watt intensity of 1 kW m^{-2}. Series and/or parallel connection can be used to obtain higher power levels. The load must be matched to the array and the module must be mechanically supported to withstand wind loads. Correct orientation is essential and higher outputs can be obtained if the array tracks the sun. The power output must be delivered in an acceptable form, such as a constant d.c. voltage or 240 V 50 Hz a.c. Normally this requires power-conditioning equipment.

Electrical power must be used as efficiently as possible in any stand-alone system to minimize array size and cost for any specific application. Items of equipment other than the photovoltaic modules are known as the Balance of Systems (BOS). Their cost is the BOS cost. Some BOS costs are directly related to array area and can be reduced through increased module efficiency. Area-related BOS costs include the cost of the land, preparation of the site, the array structure and associated wiring and maintenance. Small arrays can often be fitted into otherwise unused land or onto an existing supporting structure, but for arrays above 1 kWp this may not be possible. However, the area-related BOS costs can show considerable benefits of scale if large numbers of identical sub-arrays can be installed at the same time on the same site.

Power-related BOS costs include the relatively inexpensive (for all except the smallest systems) electronic control equipment. The major costs are for batteries in d.c. systems and inverters for a.c. systems. Battery characteristics (e.g. storage density and charging efficiency for low currents) are being continually improved. Battery costs represent about 50% of the module costs for most typical applications.

Inverters have also been developed specifically to match photovoltaic arrays, but the scope for significant cost reduction is limited unless a completely new inverter technology can be discovered. Total BOS costs for large arrays on low-cost easy-access land could be in the order of $5 per Wp, but can often be well above $10 per Wp. However, this might reduce to about $3 per Wp by the end of the decade, with the introduction of the improvements discussed above.

12.5.5 Applications

In addition to battery charging for many different applications in the leisure and military markets, photovoltaic modules are used in applications where a grid supply cannot reach. Photovoltaic power systems are already the first choice for some telecommunications applications and for cathodic protection because of their reliability and low maintenance requirements. Photovoltaics is also economically competitive in certain situations in developing countries (e.g. for water pumping, lighting and providing small-scale power). When problems of limited distribution and a maintenance network for modules and systems are removed, these markets could grow rapidly.

It is not yet economically feasible to connect photovoltaics into a utility grid, although there have been exceptions. For example, in the south-west of the USA, where the utility peak load coincides with the period around noon, photovoltaics are competitive with the utility peak load costs, mainly for air-conditioning loads. Several countries in Southern Europe are examining the possibility of using photovoltaics to supply mountain or island communities, especially those with large summer tourist populations.

An alternative approach to the identification of the markets for terrestrial photovoltaics has been given by Hill,[66] who identified the following four categories:

1 *Professional*. This includes cathodic protection, telecommunications, utility demonstration projects and remote power, including hybrid systems, for a variety of loads (e.g. desalination plants or navigation lights).
2 *Developing countries*. These are usually quite small-scale applications such as lighting, refrigeration, radio and TV, water pumping and normal village electrification.
3 *Consumer power*. Applications where an individual is both the purchaser and the user (e.g. in caravans, boats, in the leisure industry, or in a remote location).
4 *Consumer products*. This market is mainly for the solar-powered calculator, but includes other small items.

A major role for electricity generation within the next 25 years, using photovoltaic cells as cladding on commercial buildings, has also been envisaged by Hill.[67] This could well be the first major role for photovoltaics in the UK, as these buildings are normally only occupied during the daylight hours. To make maximum use of winter sunshine, the photovoltaic cells would be placed on south-facing vertical panels. On completion of the first stage of feasibility studies[68] the next stage should be government-supported installations of appropriate hardware to gain 20 years of real-time experience. This would give architects and clients the confidence to specify and install photovoltaic cells when they become cost-effective.

The next stage could be 500 kW (0.5 MW) plants in selected areas to provide grid-support for local utilities (e.g. for the more isolated communities). Photovoltaics could then be seen to be cost-effective on buildings other than commercial office blocks.

The UN Conference on the Environment and Development in Rio, June 1992, was asked to endorse a major role for photovoltaics, particularly for basic electrical services to the 2000 million rural poor people. Approximately 1 million basic photovoltaic systems, ranging in power from some 30 to 100 W, could be in place by 1995, rising to some 10 million by the end of the century. This would give a total installed capacity of some 300 MW. The present annual world output is estimated to be some of 60 MW (1991 estimate) and growing at about 20% per annum.

12.6 Solar chemistry

12.6.1 Introduction

The direct conversion of solar energy into stored chemical free energy has attracted research workers for many years. For example, Tsubomura[68] cites Ciamician,[69] who, in 1912, described his dream of new industry based on solar photochemical conversion which is self-sustaining and pollution-free. Ciamician also pointed out that even if in a distant future the supply of coal became completely exhausted, life and civilization would continue as long as the sun shines. An excellent review of the work in the field up to the early 1960s by Farrington Daniels[70] included a description of basic processes and suggested that detailed studies of the reaction of photosynthesis (the conversion of carbon dioxide and water into carbohydrate and oxygen) could help to point the way towards research in the quest for new photochemical reactions.

By the early 1990s, reaction temperatures between 300 K and 2500 K could be achieved by several solar collector/receiver systems. This meant that, in principle, a very large number of different thermal and/or catalysed chemical processes could be carried out.[71] The objectives of solar chemistry could be defined as the use of solar radiation to produce fuels and chemicals, or to drive chemical processes such as detoxification.[72] The three main paths which can be followed in the conversion of solar radiation are:

1. Pure thermally driven processes;
2. Totally photodriven processes; and
3. Electrochemical process;

with the end uses to thermochemistry, photochemistry and electrochemistry and the pairwise hybrids.

12.6.2 Photochemical processes

Photochemical processes have been defined by Bolton and Archer[73] as those in which the absorption of solar photons in a molecule produces excited states, or alternatively in a semiconductor raises electrons from the valence band to the conduction band. As a result of the chemical reactions which may then occur, some of the excitation energy may be stored as chemical energy or a useful chemical reaction may be catalysed. Tsubomura[69] defines photochemical conversion as a technology to synthetize valuable chemical materials or fuels by the use of solar energy. Approximately half of the total solar radiation which reaches the Earth arrives in the ultraviolet and visible range (300–700 nm) and can be used in various photochemical reactions. This often leads to systems containing highly coloured substances. The other half, which occurs in the infrared region, cannot make a useful contribution as its energy concentration is too low. The maximum overall efficiency of any photochemical energy conversion is limited to about 30% because some of the higher-energy photons of shorter wavelengths have some energy degraded as heat during the reaction. The majority of photochemical reactions are exothermic and are not suitable for converting solar radiation into stored chemical energy. The known endothermic (energy-storing) reactions which occur with visible light are, in theory, capable of producing valuable chemical fuels. A major problem has been that most of these endothermic reactions reverse too quickly to store the energy of the absorbed light.

It can be seen that two types of chemical reaction driven by sunlight are outlined above: those in which there is a net storage of solar energy and the product of the reaction could be either a fuel, such as methane or hydrogen, or electricity; and those where the reaction may be photocatalytic and the sunlight would be used to catalyse the synthesis of useful chemicals. Bolton and Archer[73] further divided the subject into the following two major areas:

1. Direct processes, in which the direct absorption of solar photons by dye molecules or a semiconductor leads to useful photochemistry or photoelectrochemistry. Work in this area is normally on a laboratory scale.
2. Thermal processes, in which concentrated sunlight generates the high temperatures used to drive useful chemical reactions. There are only a few large-scale facilities where high-temperature concentrated sunlight is available.

The general criteria for the selection of useful photochemical conversion and storage systems have been evolving since a 1957 Symposium on photochemistry in the liquid and solid states[74] (e.g. Porter and Archer,[75] Bolton,[76] Bolton and Hall[77]) and have been summarized by Bolton and Archer[73] as follows:

1. The overall photochemical process must be endergonic, i.e. result in a net storage of solar energy.
2. Side reactions leading to the irreversible degradation of the photochemical reactants and/or sensitizers must be almost totally absent.
3. The photochemical reaction should be capable of operating over a wide bandwidth of the solar spectrum, with a threshold wavelength in the red or near infrared region.
4. The quantum yield for the photochemical reaction should be very near unity.
5. The back reaction must be extremely slow under ambient conditions, but should proceed rapidly under controlled catalytic conditions, or at elevated temperatures to release the stored energy when needed.
6. The product(s) of the photochemical reaction should be readily separated from the reaction mixture and easily stored or transported.
7. The reagents and any container materials should be cheap and non-toxic, and the reaction should not be inhibited by the presence of oxygen.

Most of these requirements were not essential in photocatalytic processes where energy storage was not a consideration.[73] However, all the various systems which had been investigated up to 1990 were considered by Tsubomura[78] to be unsatisfactory because they failed to meet some of these criteria. He strongly believed that chemical conversion by means of semiconductor photoelectrochemical methods seemed to be more promising. He also pointed out that the technique of electrochemical synthesis combined with solar photoelectric conversion (by the use of solar cells) was more feasible and readily applicable, if not yet economic.

12.6.3 Chemical conversion with photoelectrochemical cells

The distinctive feature of the photoelectrochemical cell is that it can transform solar energy directly into stable chemical energy.[78] Research in this field expanded very rapidly since the work of Fujishima and Honda on water splitting with TiO reported in 1972.[79] The principle of the photoelectrochemical cell (PEC) is that either an n-type or a p-type semiconductor electrode is immersed in an electrolyte solution together with a counter-electrode. With an n-type semiconductor electrode an electric field is formed inside the semiconductor electrode which drives the light-induced conductive electrons into the outer circuit and drives the light-induced positive holes out into the electrolyte solutions. Where the solution contains redox agents R/Ox, the reduced species R will transform to the oxidized form Ox. If this is stable it drifts towards the counter-electrode, where the reverse reaction Ox + e → R occurs and no ultimate chemical change has taken place in the solution. As both a photovoltage and a photocurrent are generated in the outer circuit, the PEC is behaving like a photovoltaic cell. Memming[80] lists 14 PECs with different semiconductors and redox couples, four of which had relatively high (greater than 10%) conversion efficiencies. The advantage of using PECs for producing electrical energy is that they can be fabricated very easily without needing any diffusion or evaporation techniques for making the junction. Their disadvantage is that many of the known semiconductor electrodes with suitable band gaps are corrosive in aqueous solutions, while those which are stable in water have too wide a band gap.

12.6.4 Fuels

Environmental considerations are increasingly becoming more important in the utilization of solar energy to obtain new fuels at a competitive cost. One of the ideal cycles is water splitting to produce hydrogen. Hydrogen can be made by following any of the paths outlined above, e.g. thermochemical, photochemical, electrochemical or in a pairwise hybrid process.

The main features of the use of hydrogen as an energy carrier, the 'Hydrogen Energy' concept, were first outlined by Bockris and Triner[81] in 1970 and are shown in Figure 12.11.

Hydrogen forms the intermediate link between the primary energy sources and the energy-consuming sectors.[1] It is also independent of the primary energy sources used for its production. Even if these change, the intermediary energy systems of transmission, storage and conversion can remain unaltered. The hydrogen economy system is completely cyclic. Water from lakes, rivers or oceans is converted into hydrogen and oxygen. Its combustion product is water vapour, which is returned to the biosphere, and it is the least polluting of all the synthetic fuels. The well-known solar furnace at Odeillo[82] was first used to produce hydrogen and oxygen in the early 1980s, but the challenge is to make the process cost-effective, something which could come with the increasing emphasis on the true costs of fossil-fuel-induced environmental pollution. Hydrogen can also be applied in the traditional reactions of the chemical industry, or can be reacted with carbonaceous material, such as carbon dioxide, carbonates, or biomass to produce organic fuels and chemicals.[72]

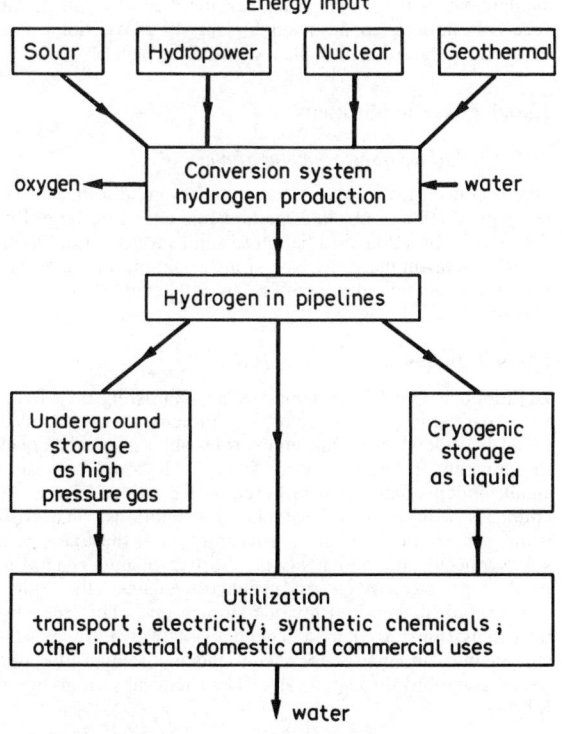

Figure 12.11 The main features of the use of hydrogen as an energy carrier (after McVeigh[1])

Another route for the thermochemical transport of solar energy is through the CO_2 reforming of methane. The endothermic reaction of CO_2 with CH_4 to yield CO and H_2 was suggested over ten years ago.[83] Methane and carbon dioxide are reacted in a solar receiver (reformer) at high temperature to produce energy-rich products. These can be stored at ambient temperatures and can be transported over distances of some hundreds of kilometres to the consumer site. Here the reverse reaction takes place in a thermal reactor (methanator), releasing the chemical energy.[84] The process could take place on one site, with the reformer–storage–methanator linked by pipework.

12.6.5 Chromogenic materials

Chromogenic materials offer the possibility of developing advanced glazings which combine variable control of solar gain with efficient thermal insulation.[85] In a major review of the technical properties and merits of known electrochromic phenomena in 1984, Lampert[86] pointed out that most of the wealth of technical literature and patents dealing with electrochromic materials and devices was primarily for electronic information display or other small-scale applications. Consequently, only minor attention had been paid to electrochromic devices as transmissive devices. Since then, transparent apertures employing photochromic, thermochromic or electrochromic materials have been the focus of intensive worldwide research by many groups[87,88] concerned with the efficient use of energy in buildings.

The electrochromic window is the most advanced example of these efforts. It is basically a multilayer thin-film device which performs as an electric cell, and consists of an electrochromic layer and a counter-electrode, or ion-storage layer, separated by an ion conductor. For window applications these layers are commonly sandwiched between two transparent electronic conductors which are deposited onto transparent substrates (e.g. glass or polymeric materials). In operation a d.c. electric field is applied across the transparent conductors and ions are driven either into or out of the electrochromic layer, causing reflectance and/or absorptance modulation of visible and near-infrared electromagnetic radiation and hence changes in the optical properties of the device. The electrochromic layer may be caused to colour or bleach in a reversible way under the influence of the external electric field. The principal aim of current research is the development of a stable, durable, all solid-state electrochromic device – the 'smart window'.[87]

Liquid-crystal-based chromogenic materials have also been successfully used as electrically activated devices.[89] Two transparent electrodes provide an electric field to change the orientation of liquid-crystal molecules interspersed between the electrodes. The orientation of the liquid crystals alters the optical properties of the device. Two main types of liquid crystal systems, the guest–host and polymer-dispersed or encapsulated devices, have been identified for large areas. Their disadvantages are that their unpowered state is diffuse, haze remains in the activated (transparent) state and ultraviolet stability is poor. A third approach uses suspended particle devices, but various technical problems such as long-term stability and cyclic durability have slowed their development.

Two non-electrically activated devices use photochromic or thermochromic materials. Their research histories date back at least 100 years.[87] When photochromic materials are exposed to light they change their optical properties, only reverting to their original properties in the dark. Photochromic plastic has been developed for ophthalmic use and could become useful for regulating solar glazings.[87] Thermochromic materials display a large optical property change when a particular temperature is exceeded. Above this critical temperature, transmittance is reduced and if this temperature is close to a comfort temperature thermochromism could be used for automatic temperature control in buildings.[89]

12.6.6 Transparent insulation materials

A new technology is emerging from European initiatives which will bring about revolutionary changes in the building industry.[90] Transparent (or translucent) insulation materials are a relatively new class of materials which combine the uses of glazing and insulation in the traditional design of any solar thermal system. While the primary use of glazing in buildings has also been to allow light to enter, its ability to transmit radiation gives it the subsidiary function of providing solar heat.

Insulation suppresses conduction and convection losses from buildings, but the traditional opaque materials such as polystyrene granules or foam which have been developed for this purpose are equally effective in suppressing solar gain from the outside of the building. For large energy gains both high irradiation levels and high values for the product of the solar transmittance and absorptance of the absorber are essential. The influence of insulation (U-value) depends on the temperature level of the system and the desired heat storage period. For low U-values very good absorption is needed in the thermal wavelengths, although infrared selective coatings can also be used on the front cover or the absorber plate to reduce infrared radiation losses.[91] Convection losses can be greatly reduced by the use of structured materials such as capillaries and honeycombs or low-pressure systems. Until recently no natural or man-made product could offer both high-transmission, low-conduction and strong convection suppressant characteristics, but by the mid-1960s it was possible to conceive the potential benefits of such a material.[92] Following rapid advances made in Germany during the 1980s the term 'transparent insulation' was accepted as best describing the goal of the technology.[93]

The first experimental results were presented in 1985 by Wittwer et al.[94] and the following four generic types which display different physical properties were first proposed in 1986:[95]

Type	Examples
Absorber-parallel	Multiple glazing, plastic films
Absorber-perpendicular	Honeycombs, capillaries
Cavity structure	Duct plates, foam
(Quasi)-homogeneous	Glass fibres, aerogels

By 1990 several systems had become commercially available and considerable potential for more scientific research work had been identified.

12.6.7 Solar detoxification

The ability of sunlight to detoxify waterborne chemicals is well known through the cleansing of polluted streams as they flow through areas open to direct solar radiation. Solar detoxification uses this natural process of degradation to produce non-hazardous substances from hazardous organic chemicals. As conventional detoxification methods do not always deal adequately with chemical wastes, the solar route is attracting attention.[96] Among the advantages are that, by using the sun as the energy source, there is no added airborne pollution or use of conventional fossil fuels. Solar detoxification breaks down the hazardous chemicals into an environmentally benign or easily treated end-product in one step. However, conventional processes often remove the wastes from the water and take them elsewhere for treatment, thus increasing the possibility of further contamination.

The detoxification of water is a photochemical process which destroys contaminants by the chemical action of light and a semiconductor catalyst.[97] When exposed to sunlight, the catalyst absorbs the high-energy photons and reactive chemicals, the hydroxyl radicals, are formed. These radicals are powerful oxidizers and break down the contaminant molecules, typically forming carbon dioxide, water and dilute mineral acids (e.g. hydrochloric acid), which can be neutralized in a post-treatment process before the treated water is discharged.

Among the common toxic chemicals which could be solar treated are trichloroethylene and other chlorinated solvents,[96] pesticides, wood preservatives, dyes typically found in textile mill effluent and leakages or spills of various liquid fuels.[96] Dioxins in PCBs can now be destroyed by solar energy and these laboratory techniques should reach the market by 1995.[98]

12.6.8 Chemical heat storage

Thermal energy storage is essential for many solar thermal applications. The properties of suitable salt hydrates were first discussed by Telkes[99] in 1974. Sodium sulphate decahydrate, mixed with 3–4% borax as a nucleating agent if complete crystallization is to be obtained, was the most-tried material, with a transition temperature close to 30°C. The problem of a barrier being formed between the liquid and solid phases proved very difficult to solve and numerous polymeric stabilizers were tried. Many different salt-phase change materials have been tried in the past two decades, including calcium chloride hexahydrate and sodium acetate trihydrate, and modified varieties covering a range of transition temperatures from about 8°C to 58°C, several of which were commercially available in 1990.[97] For high temperatures, molten nitrate salt receivers have been designed for the 10 MWe Solar One electricity generating pilot plant discussed in Section 12.4.

12.6.9 Other applications

12.6.9.1 Surface transformation of materials

Highly concentrated solar energy, typically greater than 1 MW m^{-2}, provides a controlled method for delivering large flux densities of broadband radiation to solid surfaces, thus creating the solar-induced surface transformation of materials. Candidate technologies identified by Pitts et al.[101] are shown in Table 12.2.

12.6.9.2 Thermochemical heat pump

A thermochemical heat-pump system, consisting of a fixed-focus parabolic solar collector, a stationary thermochemical metal hydride storage unit and various sub-systems has been investigated at the laboratory stage.[102] It could provide a small, independent solar-powered home-energy centre for countries with good solar radiation conditions. The basic principles are that sunlight is concentrated in the fixed-focus solar concentrator and heats the Stirling engine generator, producing electricity, and dehydrogenates the high-temperature magnesium hydride storage unit. The freed hydrogen is transferred to a second low-temperature hydride storage unit, and may be recycled to the high-temperature unit (hydrogenation) through a valve. The chemical systems are as follows:

Storage: $MgH \rightarrow Mg + H - 75$ kJ mol^{-1}
Release of stored heat: $Mg + H \rightarrow MgH + 75$ kJ mol^{-1}

Table 12.2

Technology	Present uses
(1) Chemical vapour deposition	Electronics, hard-facing corrosion
(2) Diffusion coatings	Pack cementation
(3) Layered thin films	Electronics, photovoltaics
(4) Melted powdered coatings	Corrosion, hard-facing, ceramics
(5) Rapid thermal annealing	Electronics, photovoltaics
(6) Self-propagating high temperature	Ceramics, refractory powders synthesis
(7) Transformation hardening	Steel industry
(8) Zone-melting recrystallization	Electronics

As yet, a Stirling engine for remote, maintenance-free applications has not been developed, but this remains one of the main development goals of the project team.[102] Main parameters for an aperture area of 3 m^2 are 4 kWh of high-temperature energy (heat) for cooking + 3.4 kWh of electricity + 3 kWh of heat for domestic hot water + 3 kWh cooling energy for a refrigerator.

12.7 Hydropower

12.7.1 Introduction

Hydroelectric power is the world's largest commercially available renewable energy source, accounting for about 6.7% of the total primary energy consumption.[103] Water has been used as an energy source for thousands of years, but the various traditional designs of watermill used until the nineteenth century could only lead to a technical dead end.[104] None of them were capable of using a head of water much greater than their own diameter. Further progress followed with the development of the water turbine which was subsequently linked to an electric generator. Although credit for the world's first hydroelectric plant is often attributed to the US plant which started in the autumn of 1882 at Appleton, Wisconsin, two plants were already operating in the UK at that time.[105] The earliest was Sir William Armstrong's small hydroelectric plant, rated at just under 5 kW, which was constructed in 1880 to light his picture gallery at Cragside, Northumberland, some 1.5 km away. The first public supply of electricity was reported from Surrey in 1881, when electric current generated from the waters of the River Wey was used to light the streets of Godalming. The cables had to be laid in the gutters as there was no legal authority to dig up the streets.

The world's first large hydroelectric plant was built in 1895 at the Niagara Falls in the United States, with two turbines each rated at 4100 kW.[106] The subsequent development of alternating current by George Westinghouse in 1901 allowed electric power to be transmitted over long distances.[107] By 1903 Canada had a 9.3 MW plant, also at Niagara Falls, and the era of modern hydropower had commenced. The first reliable survey of water turbines manufactured and installed throughout the world in the late 1920s[108] suggested that about 40% of the world's electricity was generated by hydropower, with the United States and Canada having a combined operating potential capacity of over 13 000 MW and five other countries (France, Japan, Norway, Sweden and Switzerland) with operating potential capacities greater than 1000 MW. A few of the earlier hydropower plants, known as run-of-the-river plants, could not generate any power when the river was low during the dry season, but by the 1930s the use of large dams had been established in the United States. The creation of the Tennessee Valley Authority in 1933 with their comprehensive approach to the planning and development of river basins set a pattern which has been widely followed in other countries.[107] Since then there has been a steady growth in hydropower throughout the world although the percentage share of hydropower in meeting world electricity demand had fallen to about 25% by the early 1990s.

During the 1980s, the total output from North America and Europe remained unchanged, but their share of the total world output dropped from just under 60% to 45%. Among the developing countries, Brazil, Ghana, Mozambique, Zaire and Zambia obtained over 85% of their electricity from hydropower.[103,109]

The potential for development of hydropower over the next 40 years is so great that it could provide an output equivalent to the total electricity generated in the world from all sources in the early 1980s. Most of this potential is in the developing countries, some of whom could, in theory, increase their present use of hydropower by a factor of ten or more.

12.7.2 The basic hydropower plant

The basic principles of hydroelectric power generation are shown in Figure 12.12. Water at a high level, often stored behind a dam, falls through a head z. Its gravitational potential energy is converted to kinetic energy and the flowing water drives a water turbine. The rotating turbine shaft drives the electric generator to produce electricity.

The theoretical maximum velocity is obtained by equating the gravitational and potential energies as follows:

$$gz = 1/2 \, v^z$$

If the volumetric flowrate is Q (m^3 s^{-1}), density ρ (kg m^{-3}), then the power output in watts is given by

$$\rho \, Q \, g \, z$$

12.7.3 Types of turbine

Turbines can be classified according to the direction of the water flow through the blades, e.g. radial, axial or combined-flow turbines, or as reaction, impulse or mixed-flow turbines. In reaction turbines there is a change of pressure across the turbine rotor, while impulse turbines use a high velocity jet impinging on hemispherical buckets to cause rotation. There are three basic types of turbine broadly related to low, medium or high heads.[1]

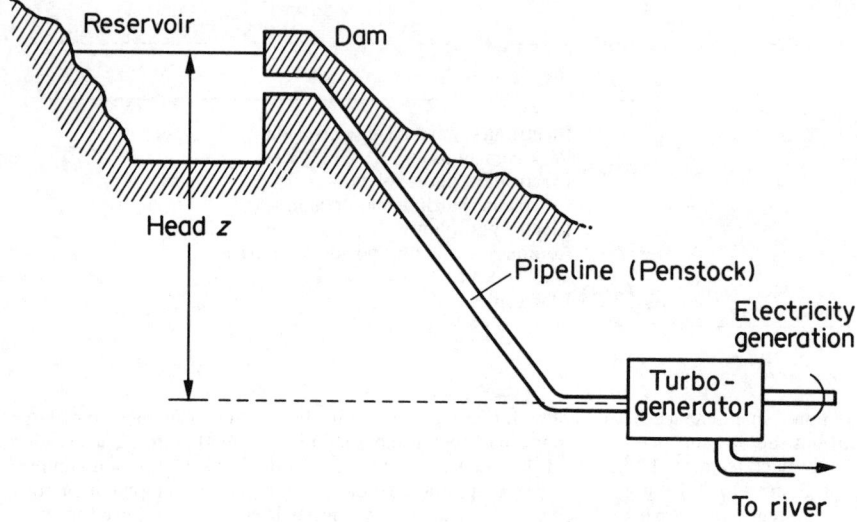

Figure 12.12 The basic principles of hydroelectric power generation (after McVeigh[1])

Propeller or axial flow turbines are used for low heads in the range from 3 to 30 metres. They can have relatively inexpensive fixed blades, which have a high conversion efficiency at the rated design conditions but a poorer part-load efficiency, typically 50%, at one third of full rated output. Alternatively, the more expensive Kaplan turbine has variable-pitch blades which can be altered to give much better part-load efficiency, perhaps 90% at one third of full rated output.

The Francis turbine is a mixed-flow radial turbine and is used for medium heads in the range from 5 to 400 m. It has broadly similar performance characteristics to the fixed-blade propeller type and its speed is controlled by adjusting the guide vane angle.

The best-known impulse turbine is the Pelton wheel. Each bucket on the wheel has a centrally placed divider to deflect half the flow to each side of the wheel. It is normally used for heads greater than 50 m and has good performance characteristics over the whole range, very similar to the Kaplan turbine, reaching 60% efficiency at one-tenth of full rated output.[1] The speed is controlled by a variable inlet nozzle, so that with a constant head, the delivered torque to the generator is proportional to the flowrate and the turbine speed can be held at that required for synchronous generation at the particular grid frequency. This type of installation is known as a constant-speed/constant-frequency system and optimization of the power output is relatively easy.[110] In smaller installations, optimum power cannot be obtained at constant speed where the hydraulic head is both relatively low and variable over a wide range.

A detailed description of methods which can be used for optimizing electric power from small-scale plant has been given by Levy.[110] He points out that small hydroelectric systems will become more financially attractive through developments of low-cost power converters (from 100 W upwards), special variable-speed/constant-frequency generators and cheap computing units for on-line power measurement and optimizing control. This means that many run-of-the-river sites that were considered in the past to be unsuitable for electricity generation can now be used.

12.7.4 Hydropower potential

The Earth's energy flow diagram (Figure 12.1) shows that just over 4×10^{16} W flows in the hydrological cycle of evaporation, rain, other precipitation and storage in water and ice. A very small proportion of this hydrological energy flow, probably between 0.01 and 0.015%, is considered to be theoretically available for conversion into hydropower.[109,111] This theoretical world hydropower potential is calculated as the total energy potential of river discharges relative to a datum of sea level or the base level of erosion for closed basins and is widely quoted as 44.28×10^{12} kWh per annum.[109,111] However, this figure does not seem to include the 3.94×10^{12} kWh for the former USSR, which was separately listed by the 1980 World Energy Conference,[109] and there is also some doubt as to whether the 6×10^{12} kWh estimated for the People's Republic of China has been included.[111] A better assessment is the 'technically usable hydropower potential', which allows for the unavailability of certain river reaches, mainly those near estuaries. This is less than half the theoretical value. The 'economic potential' includes all hydropower resources which are regarded as economic compared with alternative sources of electric power at the time of the assessment. These can be classified into three categories: operating, under construction and planned, as shown in Table 12.3. The economic operating hydropower potential of 372.1 GW represented just under 16% of the technically usable potential.

The world operating potential of some 372.1 GW could, in theory, have provided $372.1 \times 10^9 \times 365 \times 24$ watt-hours or 3.26×10^{12} kWh.[7] The actual energy generated was 1.65×10^{12} kWh.[7] This represents 50.6% of the potential, a typical figure for most hydroelectric plant. Not only are there seasonal fluctuations in water availability, but the demand for electricity fluctuates and plants need to close for maintenance. In the United States and Canada the figure of (actual energy generated) divided by (theoretically available potential) was 47.7% in 1979. This ratio is known as the load factor. As the electrical power from a hydroelectric plant can be used directly without the conversion losses and wasted heat associated with conventional fossil fuel power plant, the primary energy equivalent of

Table 12.3 Hydropower potential (GW) (after reference 109)

Region	Technically usable	Economic		
		Operating	Under construction	Planned
Asia[a]	609.6	53.1	9.1	42.0
Latin America	431.5	34.1	40.5	92.4
Africa	358.4	17.2	5.4	22.9
USA and Canada	356.2	128.9	34.6	39.0
Former USSR	250.0	30.3	21.8	19.4[c]
China[b]	216.9	5.7	5.9	Unknown
Europe	163.2	96.1	10.7	22.5
Rest of world	44.5	6.7	2.3	3.6
Total	2430.3	372.1	130.3	241.8

[a] Figures from Asia probably do not include data from the People's Republic of China.[111]
[b] Figures may not include all small hydropower plant.
[c] Estimated.

hydroelectricity is usually taken as about three times its actual output.

A common conversion is that 4000 kWh of 'electricity generated' is considered to have the primary energy equivalent of one tonne of oil.[103]

By 1980 there had been a steady growth in hydropower for many years at about 3.5% per annum, representing a doubling period every 20 years. This figure was used by the 1980 World Energy Conference to estimate that hydropower could be quadrupled by 2020, reaching a total of over 1600 Mtoe. A more realistic figure was suggested by McVeigh[1] with a logistic equation approach giving a growth rate of just under 3%. By the early 1990s, however, it could be seen that growth in the decade of the 1980s had only averaged 2.0%, and that the doubling period had stretched to about 35 years. Reasons for this could include the gradual reduction in performance of some of the older hydro schemes due to silting, and the changing patterns of rainfall, which, in turn, could be due to global warming.

12.7.5 Pumped storage

Pumped storage systems are used at times of peak demand for electricity. The water can be pumped to an upper storage reservoir usually at night when the demand is low, and then allowed to flow down through the turbines, generating electricity when it is required.

Although small pumped storage schemes were first built in the 1890s, the first large system in the UK was built at Ffestiniog, Wales, in 1963, with four 90 MW generators each coupled to separate pumps and turbines on the same vertical shaft.[112] This was soon followed by a 4 × 100 MW system at Cruachan in Scotland.

The largest pumped storage system in Europe was completed in 1984 at Dinorwig, near Llanberis in North Wales. During its construction 3 million tonnes of rock were excavated from the heart of the mountain between two reservoirs and 16 km of shafts and tunnels were created.[112] The upper reservoir is 568 m above the underground power station and the horizontal distance between the upper and lower reservoirs is 3200 m. There are six turbogenerator units, each rated at a nominal 300 MW. It can generate at full output for about 5 hours. The overall efficiency of any pumped storage system

is less than the 'once-through' conventional plant, as the pumping efficiency during the return flow to the upper reservoir must be included. This pumping efficiency, typically about 90%, reduces the overall efficiency to about 70–75%. However, the economics are quite different. Pumping to the upper reservoir only occurs when the electricity tariffs are low. Electricity is supplied to meet peak demands when tariffs are usually at their highest. Further, the use of a pumped storage system reduces the need for additional conventional plant which would only be needed for very short periods each year. The pumped storage plant at Dinorwig can be generating electricity within 10 seconds of requirement.

Other benefits, apart from the reduction in utilization of both high-cost oil-fired and low-efficiency coal-fired plant during the peak demand periods, include a reduction of both the extent and the duration of frequency excursions arising from large losses of generation output or the sudden increase in consumer demand experienced at the end of many popular television programmes.[112] This is often up to 2000 MW in a few minutes. Dinorwig has given the system the ability to pump prior to the impact of the television 'pick-up', thus creating an artificial demand. As the real demand increases, the pumps can be reversed to generate within 90 seconds.[112]

12.7.6 Small-scale hydropower

One of the needs in many parts of the world is for electrical power in remote regions far from a conventional transmission system. Small-scale hydropower is again becoming considered for an increasing number of these applications.

Recently, the energy policies adopted by the different Member States of the European Economic Community to reduce their dependence on third countries as suppliers of energy, together with the technical improvements outlined above, have made it possible for small hydropower plants to become competitive in many parts of Europe.[113]

The early history of hydropower up to the 1930s was largely dominated by small plants, less than 1 MW in capacity, but then the economies of scale began to favour large-scale development. Until fairly recently it was necessary to match the turbine design very carefully to the particular site. This resulted in an expensive special 'one-off' hydropower generator. The smaller the application, the greater the installed cost per kilowatt of capacity. The need for these specially designed systems has been largely overcome by the use of standardized turbines and associated equipment, with the acceptance of some loss in overall plant efficiency and performance.

As outlined above, one of the major factors which could favourably influence the economics of small-scale hydropower is the development of microprocessor-based electronic load governors. These can overcome problems of instability in matching waterflow to a variable demand and can also reduce costs as expensive mechanical controls are no longer necessary.

Definitions of the size of any hydro scheme into Large, Small, Mini and Micro appear in the literature, but there seems to be no agreement on what these sizes represent, as Table 12.4 shows.

Bazaga[113] points out that the definition of small hydropower plant is 'not exactly the same in the different Member States of the EEC' and bases his analyses on a power capacity less than or equal to 10 MW. Among the distinctive features of small hydropower plants which he identifies are:

1. They are usually run-of-the-river and the energy produced depends on the available flow.
2. They rarely contaminate the environment and do not give off heat.
3. They can be built in a short period of time, with standard equipment and well-known construction processes.
4. Projects can be developed which combine electricity generation with other uses.
5. The power source is reliable, within its hydrological limitations. The equipment and facilities involved have a long life, require little maintenance and seldom break down.
6. The technology involved is well developed and overall efficiency is over 80%.
7. Operating systems are often automatic, leading to low operation and maintenance costs.

By the 1980s the country with the greatest experience in small-scale hydropower development was the People's Republic of China, where nearly 100 000 plants have been constructed in the past 20 years.[111] Most of their recent plants have a rated output of some 300 kW and much of their projected increase in hydropower over the next 20 years will also be small-scale.

12.7.7 Economic, social and environmental issues

The costs and benefits of hydropower plant are usually evaluated by an economic comparison with conventional thermal or nuclear power stations. The main factors which must be considered, in addition to increases in construction costs, are changes in the cost of fossil fuels and in environmental protection regulations. Although there has been a steady growth in power station construction costs in all countries over the past two decades, thermal and nuclear

power station costs have risen at a greater rate than those of hydropower plants. There are two reasons for this. The technology and management of the construction of hydropower plants has improved relative to conventional power station construction and new environmental protection and safety regulations have adversely affected the cost of nuclear and coal-fired power stations.[111] These new regulations have resulted in greatly increased expenditure for the control of air and water pollution with coal-fired stations, and for radiation monitoring and control together with improved safety standards in nuclear installations. Some of the adverse effects of hydropower schemes, such as the essential reinforcement of river banks or compensation for moving and resettling whole communities from flooded land, have always been included in the overall construction costs.

An economic advantage when considering the later stages in any hydropower scheme is that dams with existing hydropower schemes can be raised to provide both additional storage capacity and a potentially increased output. Turbine generators can be added to some existing storage reservoirs to create new generating capacity.

The economics of any hydropower system are absolutely site-specific, depending critically on the topology, geology and hydrology of the site.[64] These factors influence the power capacity and developments costs, which, in turn, depend on what is required from the system (e.g. a high or low load factor) or whether storage is required or not. Hydropower, like tidal power, is highly capital-intensive and can have a very long life, often over a hundred years for the basic civil engineering work. With the low operation and maintenance costs, together with the other advantages outlined by Bazaga above, the main economic problems arise from the financial requirements of high interest rates and the demand for short 'payback' periods.[64] Again, as with tidal power, there is a zero fuel cost, and the UK Watt Committee also commented[64] that it is paradoxical that investment in hydro schemes looks extremely favourable in retrospect.

Rivers and streams are regarded in the great majority of countries throughout the world as a public resource. Their use in potential hydropower schemes is subject to government control. Hydropower development may be socially acceptable to some sectors of the community and have quite disastrous effects on others. For example, the construction of the Aswan High Dam in Egypt resulted in the destruction of the sardine fishing industry in the Eastern Mediterranean, but this was balanced by the development of a new fishing industry on the newly created Lake Nasser.[107] There have been many studies on the adverse impacts on health which can result from the large dams associated with hydropower projects[107] and it would appear that there is still a need for major health-education programmes to be associated with these projects, so that diseases such as bilharzia and malaria could be eliminated. Other associated environmental problems include the need for extensive drainage systems on newly irrigated land and the threats to new dams caused by widespread deforesta-

Table 12.4 Power output ranges

Source	UK Watt Committee (1990)[64]	Hurst and Barnett (1990)[114]	Bazaga (1988)[113]
Large	50 MW	Greater than 1 MW	
Small	5 – 50 MW	0.5 – 1 MW	10 MW or less
Mini	0.5 – 5 MW	100 – 500 kW	
Micro	Less than 500 kW	Less than 100 kW	

tion and soil erosion many kilometres upstream. Some existing aquatic and terrestial ecosystems have been disrupted and there may have been a loss of visual amenities in scenic areas.

On the other hand, the United Nations Hydropower Panel[111] has also drawn attention to the positive effects of hydropower reservoirs on the environment. The creation of regulating reservoirs has been shown to make a substantial improvement in the water supply for domestic, industrial and agricultural purposes in many cases. The danger of catastrophic floods has often been eliminated. The overall effects of hydropower schemes throughout the world have been beneficial, although there have been some largely unanticipated adverse reactions with the environment. These could either be reduced or eliminated through careful resource planning.

12.7.8 Summary

Hydropower is the only renewable energy resource with a fully developed technological base and a relatively predictable growth rate over the next few decades. Its industrial infrastructure is well established in many countries, and it provides very substantial proportions of the electricity demand in a number of countries. Although it accounted for only 6.7% of the world's primary energy consumption in 1990, this figure could easily rise to over 10% by the middle of the next century. It is particularly suitable for the needs of remote communities in the developing countries.

12.8 Wind power

12.8.1 Introduction

Energy from the wind is derived from solar energy, as a small proportion of the total solar radiation reaching the Earth causes movement in the atmosphere which appears as wind on the Earth's surface.[1] The wind has been used as a source of power for thousands of years and the traditional horizontal axis tower mill for grinding corn, with sails supported by a large tower, rather than a single post, had been developed by the beginning of the fourteenth century in several parts of Europe. Its use continued to expand until the middle of the nineteenth century, when the spread of the steam engine as an alternative, cheaper, source of power started its decline. Nevertheless, before the end of the nineteenth century several countries used the windmill as one of their main sources of power. In the Netherlands[115] there were about 10 000 windmills giving power outputs of up to 50 kW. In Denmark housemills were often mounted on the roofs of barns and, together with industrial mills, were estimated to be producing about 200 MW from over 30 000 units.[116] In the United States[115] an estimated 6 million small multi-bladed windmills for water pumping were manufactured between 1850 and 1940.

Work on the development of wind-generated electricity started in Denmark in 1890 when Professor P. La Cour obtained substantial support from the Danish government, which not only enabled him to erect a windmill at Ashov but provided a fully instrumented wind tunnel and laboratory. Between 1890 and his death in 1908, Professor La Cour developed a more efficient, faster-running windwheel, incorporating a simplified means of speed control, and pioneered the generation of electricity. The Ashov windmill had four blades 22.85 m in diameter, mounted on a steel tower 24.38 m high. Power was transmitted, through a bevel gearing, to a vertical shaft which extended to a further set of bevels at

ground level, and the drive was connected to two 9 kW generators – the first recorded instance of wind-generated electricity. By 1910 several hundred windmills of up to 25 kW capacity were supplying villages with electricity. The use of wind-generated electricity continued to increase in Denmark and a peak of 481 785 kWh was obtained from 88 windmills in January 1944.[117]

Large-scale modern windpower dates from the designs of an American engineer, Palmer C. Putnam[118] in the 1930s. He was responsible for the Smith–Putnam windmill which was erected at Grandpa's Knob in central Vermont in 1941. It had two blades with a diameter of 53.34 m, and at that time it was the world's largest ever windmill, a record it was to hold for the next 35 years. The synchronous electric generator and rotor blades were mounted on a 33.54 m tower and electricity was fed directly into the Central Vermont Public Service Corporation network. The windmill was rated at 1.25 MW and worked well for about 18 months until a main bearing failed in the generator, a failure unconnected with the basic windmill design. It proved impossible to replace the bearing for over two years because of the war and during this period the blades were fixed in position and exposed to the full force of the wind. Also, in 1942, cracks had been noticed around some rivet holes, but these were considered to be so small that they could be ignored. On 26 March 1945, less than a month after the bearing had been replaced, the cracks widened suddenly and a spar failed, causing one of the blades to fly off. The S. Morgan Smith Company, who had undertaken the project, decided that they could not justify any further expenditure on it, apart from a feasibility study on the installation of other units in Vermont. This indicated that the capital cost per installed kilowatt would be some 60% greater than conventional systems.

Although sceptics have tended to regard this experiment as an expensive failure, it was the most significant advance in the history of windpower. For the first time, synchronous generation of electricity had taken place and been delivered to a transmission grid. Both mechanical failures were due to a lack of knowledge of the mechanical properties of the materials at that time. Bearing design and the problems of fatigue in metals have been studied extensively since then and similar failures are less likely to occur in modern windmills.[119] Their research programme included an extensive series of on-site measurements, which proved that the actual site at Grandpa's Knob had a mean wind velocity of only 70% of the original estimated velocity and that many other sites should have been selected. The technical problems of converting wind energy into electricity had been largely overcome and the possibility of developing wind power as a national energy resource in any country with an appropriate wind climate has been established. However, very few wind turbines were to be built over the next 30 years.[1]

12.8.2 Wind-energy potential

Wind has a dependable annual statistical energy distribution but a complete analysis of how much energy is available from the wind in any particular location is rather complicated. It depends, for example, on the shape of the local landscape, the height of the windmill above ground level and the climatic cycle. Somewhat surprisingly, the British Isles have been studied more extensively than practically any other country in the world[120,121] and the west coast of Ireland, together with some of the western islands of Scotland, have the best wind conditions with mean average wind speeds approaching 9 ms^{-1}. The kinetic energy of a moving air stream per unit mass is $\frac{1}{2}V^2$ and the mass flow rate through a given cross-sectional area A is $\rho A V$, where ρ is the density. The theoretical power available in the air stream is the product of these two terms:

$$\tfrac{1}{2}\rho A V^3$$

If the area A is circular, typically traced by rotor blades of diameter D, then $\pi/4 D^2 = A$, and the power available becomes

$$\frac{\pi}{8}\rho D^2 V^3$$

The actual power available can be conveniently expressed as

$$C\tfrac{1}{2}\rho A V^3$$

where C is the coefficient of performance or power coefficient.

The maximum amount of energy which could be extracted from a moving airstream was first shown by the German engineer Betz, in 1927, to be 16/27 or 0.59259 of the theoretical available power. This efficiency can only be approached by careful blade design, with blade-tip speeds a factor of six times the wind velocity, and is known as the Betz limit. Modern designs of windmills for electricity generation operate with power coefficient values (C) of about 0.4, with the major losses caused by drag on the blades and the swirl imported to the air flow by the rotor.[122] Any aerogenerator will only operate between a certain minimum wind velocity, the starting velocity V_S, and its rated velocity V_R. Typically, V_R/V_S lies between 2 and 3. If the pitch of the blades can be altered at velocities greater than V_R, the system should continue to operate at its rated output, the upper limit depending only on the design. In some systems the whole rotor is turned out of the wind to avoid damage at high wind speeds. An annual velocity duration curve for a continuously generating windmill is shown in Figure 12.13.

The effect of the height of the windmill tower on the performance can be significant and empirical power law indices have been established[123] relating the mean wind velocity V to the height H, in the equation $V = H^a$. A value of $a = 0.17$ is the accepted value in the UK for open, level ground, but this rises to 0.25 for an urban site and 0.33 for a city site. An ideal site is a long, gently sloping hill.

The mean annual wind velocity is normally used to describe the wind regime at any particular location, but the output from a windmill is proportional to V^3. Since a transient arithmetic increase in wind velocity will contribute much more energy to the rotor than an equal arithmetic decrease will deduct, the mean of V^3, which is always much greater than the cube of the mean annual wind velocity, should be used. For example, if the mean wind velocity is 8 ms^{-1} the most common variation in wind velocity occurs at frequent short intervals between 6 ms^{-1} and 10 ms^{-1}[116] and $8^3 = 512$, whereas $\tfrac{1}{2}(6^3 + 10^3) = 608$. A useful concept is the velocity exceeded for 50% of the year (4380 hours), shown in Figure 12.13 as V_{50}. This is quite close to the mean annual wind speed and has been used to give the annual extractable energy E_a if the rotor shaft is attached to an electrical generator as[124]

$$E_a = 3.2289\, D^2\, V_{50}^3\ \text{kWh}$$

The Betz limit, outlined above, is purely theoretical, and in practice the power extraction efficiency will be reduced if either:[125]

1. The blades are so close together or rotating so rapidly that a following blade moves into the turbulent air created by a preceding blade; or
2. The blades are so far apart or rotating so slowly that much of the air passes through the cross section of the device without interfering with a blade.

The rotational frequency of the wind turbine must be matched to particular wind speeds to obtain the optimum efficiency. The power extraction is therefore a function of the time taken by a following blade to reach the position occupied by the preceding blade, and the time taken for the normal airflow to become re-established once the disturbed air has left that position. This has resulted in a very important parameter – the tip speed ratio – defined as the speed of the turbine blade tip divided by the speed of the normal airstream, or oncoming wind. A more detailed analysis can be found in the standard literature.[125]

For the great majority of wind-power applications, however, it is more important to know the probability that a minimum site wind velocity will be exceeded. Long periods of no wind or only light winds are obviously unacceptable. Matching the wind turbine to the characteristics of any particular site has needed the use of probability functions, the best known being the Weibull function.

12.8.3 Small to medium-range windmills

Multi-bladed windmills for water pumping are still being manufactured in several countries and an estimated one million were in use in the early 1980s.[115] These windmills have a high solidity, or area of blade relative to total swept area. This gives a high starting torque but a relative low power coefficient, typically about 0.2. Wind energy was considered to have a significant role in pumping water in the developing countries by the United Nations Technical Panel,[115] but they also identified three problems with existing designs: they were too complicated for local manufacture, too expensive and too difficult to maintain and repair. Several new designs appeared in the late 1970s and early 1980s. These could be made locally and were relatively inexpensive, but a wider educational programme was still needed before the technology could be disseminated.[115]

Small low-solidity wind turbines for generating electricity in the range up to 10 kW are widely available in many countries. Windmills in Sri Lanka, for example, locally developed in the early 1980s, could give an output of up to 400 W and would cost no more than $200 to build.[126] Prototypes are used to charge locally manufactured lead-acid batteries which power low-energy consumption fluorescent tubes. This provides an electric lighting system at about half the cost of conventional kerosene lamps.

Isolated communities in good wind areas, especially in mountain regions, on islands or in coastal areas, can meet their power needs in the 10–1000 kW range by a combination of wind power and a suitable back-up system.

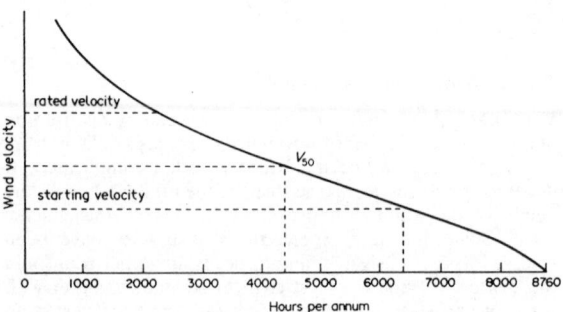

Figure 12.13 Annual velocity duration curve for a continuously generating windmill (after McVeigh[1])

By the mid-1980s the combination of wind and diesel generators was attracting very considerable international research and development activity. The results of much of this work were summarized by Lipman in 1990,[127] who pointed out that a wind power system may be fully meeting an autonomous load at one moment and be in considerable power deficit a few seconds later. Strategies which were being tried included various types of load control, both long- and short-term energy storage, hybrid systems using flywheels and multiple diesels[127] or a pumped hydroelectric system.[115]

Among the smaller UK companies the Northumbrian Energy Workshop (NEW) have helped the government of the Seychelles with wind-resource assessment using data loggers they have installed under a United Nations Development Programme (UNDP).[128] NEW is also continuing to support a UNDP project in the very different climatic conditions of Mongolia for which they supplied 27 Marlec WG910 50 W windchargers and four Dyna Technology 200 W windchargers. NEW, together with the National Centre for Alternative Technology in Wales, and Marlec, have also supplied some small solar-photovoltaic-wind hybrid systems for projects in Tanzania and Kenya.

Most of the Marlec WG910 windchargers are exported to remote parts of both developing and developed countries. A particularly interesting user is the 'Footsteps of Scott Expedition', which reported using their Marlec aerogenerator at temperatures down to $-40°C$ and windspeeds exceeding force 12 and averaging 40 mph over 12 h. They reported 'faultless' operation under these extreme conditions.[128,129]

12.8.4 The vertical-axis windmill

The modern vertical-axis windmill is a synthesis of two earlier inventions. These are the Darrieus[130] windmill with blades of symmetrical aerofoil cross section bowed outward at their mid-point to form a catenary curve and attached at each end to a vertical rotational axis perpendicular to the wind direction and the Savonius[131] windmill or S-rotor, in which the two arcs of the 'S' are separated and overlap, allowing air to flow through the passage. The Darrieus windmill is the primary power-producing device, but, like other fixed-pitch high-performance systems, is not self-starting. The blades rotate as a result of the high lift from the aerofoil sections, the S-rotor being used primarily to start the action of the Darrieus blades. The wind-energy conversion efficiency of the Darrieus rotor is approximately the same as any good horizontal system[132] but its potential advantages are claimed to be lower fabrication costs and functional simplicity.[133] In 1981 the largest American Darrieus machine, with three blades, had developed 500 kW. A 4 MW machine jointly funded by the Canadian National Research Council and the Institut de Recherche d'Energie du Quebec was completed on a site in the St Lawrence river valley, Quebec, in 1985. An earlier feasibility study concluded that Darrieus machines up to 8 MW in size could be built.

In the UK, an analysis of the Darrieus rotor suggested to Musgrove[134] that straight-bladed H-shaped rotors, with the central horizontal shaft supporting two hinged vertical blades, could be a more effective system. A variety of designs based on Musgrove's work in the UK during the 1970s and early 1980s have been studied and a small, 6 m diameter, three-bladed version was commercially available by 1980.

This work was followed by a 25 m diameter 130 kW machine at Carmarthen Bay, which started a test and monitoring programme in November 1986. Full details of the development of this design are available.[135] Following the highly successful trials, a larger version, known as VAWT 850, was inaugurated in August 1990. The '850' refers to the swept area of the blades. Its rated capacity is 500 kW, with a cut-in windspeed of 6 m s^{-1} and a shutdown windspeed of 23 m s^{-1}.[136]

Musgrove also considered the possibility of siting groups or clusters of windmills in shallow offshore locations in the UK such as the Wash. Two advantages of this proposal are the higher mean windspeeds and the greatly reduced environmental objections.

12.8.5 The development of large horizontal-axis wind turbines and some national programmes

Details of the largest horizontal-axis wind turbines built or planned in Europe during the period from the late 1970s to the mid-1980s showed that four countries, Denmark, Germany, Sweden and the UK, had major programmes.[1] In 1979, the Danish machine at Twind, rated at 2 MW with a blade diameter (three blades) of 54 m, became the largest in the world since the Smith–Putnam machine.[137] This was a private venture and it never achieved the full rated power.

The official Danish programme for large electricity producing wind energy systems started in 1977 with a joint programme directed by the Energy Ministry and the Electricity Utilities. Their major project was the design, construction and operation of two machines, Nibe A and B, which were erected in 1979. These turbines are sited close to each other and are identical, apart from their rotor blades. Those for the A machine are supported by stays while the blades of the B machine are self-supporting.

Construction of a 2 MW wind turbine near Esbjerg, in Western Jutland, was completed in 1988, with grants from the EC.[138] The main parameters were a blade diameter (three blades) of 61 m, a hub height of 60 m and a rated windspeed of 15 m s^{-1}. The estimated annual output was 3.5 GWh y^{-1} and the estimated capacity of Danish windfarms was approaching 100 MW at the same time.[139]

The German wind programme, known as the Growian programme, had some 25 projects in operation during the early 1980s, ranging from some small, low-cost units rated at 15 kW for production in developing countries and a medium-sized 25 m diameter twin-bladed 265 kW machine, the Voith-Hutter commissioned in 1981, to the large Growian I machine, rated at 3 MW, with the world's largest blade diameter of 100 m. The rated capacity of the German Research, Development and Demonstration programme was 8 MW towards the end of the 1980s.

The main feature of the Swedish programme was related to the design, construction and operation of two large-scale prototypes, located at Maglarp in the province of Skane in southern Sweden, and Nasuden on the island of Gotland. These projects, with rated capacities of 3 MW and 2 MW, respectively, formed the main basis of Swedish work during the decade.

In the United States the first major project in the official wind energy programme was the ERDA Model Zero (MOD-0) 100 kW windmill which consisted of a two-bladed, 38.10 m diameter, variable-pitch propeller system driving a synchronous alternator through a gearbox, mounted on a 30.48 m high steel tower.[140] The blades were located downstream from the tower and a powered gear-control system replaced the traditional tail fin of earlier designs. This initial test programme was designed to establish a database concerning the fabrication, performance, operating and economic characteristics of propeller-type wind turbine systems for providing electrical power into an existing power grid.

The next in the series, the MOD-1 windmill, became the world's largest machine in May 1979, when it was commissioned. This was also a twin-bladed downwind horizontal axis machine with a blade diameter of 60.96 m and rated at 2 MW. Problems of interference with television signals were overcome but a low-level, low-frequency noise could only be reduced by lowering the speed of rotation and output. This resulted in design changes in the later machines in the series. The MOD-1 machine was dismantled in 1983.[141]

Subsequent machines in the programme were planned to reach the MOD-5, rated at 7.3 MW with a blade diameter of 122 m, but the overall economics of windpower meant that efforts concentrated on designs in the 300–500 kW range. As early as 1983 a spokesman for the General Electric Company said that the future market for large wind turbines was very doubtful as forecasts of electricity load growths were lower than expected and subsidies for the use of renewable energy systems in the United States were planned to end.[142]

The UK programme could be regarded as dating from the early 1950s when two 100 kW machines were built, the John Brown machine which was erected in the Orkneys, and the Enfield-Andreau machine, which was eventually built in Algeria in 1957.[10] A wind database was also established in the 1950s by the Electrical Research Association. Preliminary work with a design feasibility and cost study of large wind turbine generators suitable for network connection was carried out in 1976 and 1977 by a group comprising British Aerospace Dynamics Group, Cleveland Bridge and Engineering Co. Ltd, Electrical Research Association Ltd, North of Scotland Hydro-Electric Board, South of Scotland Electricity Board and Taylor Woodrow Construction Ltd.[143] A reference design was evolved for a 60 m diameter turbine in 1977. This became the WEG (Wind Energy Group) 3 MW design for the machine which was eventually inaugurated at Burgar Hill, Orkney, in November 1987. The main design features of the twin-bladed horizontal axis machine included a rated output of 3 MW at 17 m s^{-1}, a blade diameter of 60 m and a hub height of 46 m. A smaller prototype, a 20-m diameter 250 kW machine, was commissioned in the summer of 1983.

A 1 MW wind turbine at Richborough, Kent, began generating at the end of 1989. It is also a project in the European Commission's large wind-energy machines programme. The site was selected as typical of an average mainland UK location, and has a mean annual windspeed of 6.8 m s^{-1} at a hub height of 45 m, compared with the 10.5 m s^{-1} at Burgar Hill. It has three blades, each 26.5 m long and is an 'extended' version of the James Howden 300–750 kW horizontal-axis range.[144]

In parallel with the development of the range of prototype machines outlined above, plans for the first commercial windfarm in the UK were being finalized at Delabole, Cornwall, in 1991.[145]

During 1985 the National Engineering Laboratory (NEL) at East Kilbride, Glasgow, established a National Wind Turbine Centre (NWTC) at Myres Hill, some 8 km to the south-west of the NEL, near the village of Eaglesham. The 7-ha site is on a moorland ridge 350 m above sea level with an open outlook and clear view of the Irish coast.[129] It has three 'universal' concrete foundation pads, each complete with a monitoring hut and capable of taking a wide range of machines of both the conventional (horizontal-axis) type and vertical-axis machines. A further test pad of the same design is located at the NEL. A high mast fitted with a wide range of anemometry and other metereological instrumentation will provide information of the wind profile and climate over the 50 m height.

The NWTC has been created to assist companies by:

1. Providing independent accreditation of machine performance and quality;
2. Supplying engineering and technical expertise to improve the design and cost-effectiveness of machines;

3. Seeking to promote reliability through assuring high standards of materials and quality of manufacture.

The main data system collects signals at a central building at Myres Hill, and transmits these data by microwave link to the NEL Power Systems Engineering Division for analysis in a real-time facility. The microwave link allows two-way communication and some control activities may be undertaken remotely in the future.[129] Facilities for loading the different types of machine are as follows:

1. Machines with induction generators will supply electricity to the National Grid;
2. Machines with synchronous generators will be connected to individual resistive load units with programmable controllers;
3. Water pumping and direct heat production machines will utilize a 70 000-litre water reservoir on the test site.

A major review of renewable energy in the UK carried out in 1988[31] concluded that the onshore technical potential of windpower was 45 TWh yr^{-1}, and that, in principle, some 30 TWh yr^{-1} could be provided by the year 2025. Making the assumption that a broadly similar ratio of installed capacity to annual output experienced in California[119] could be applied to these projected UK figures, the equivalent UK installed capacity in 2025 would be about 20 GW. This methodology, based on California data, but remembering that several UK systems have been installed in California, was probably used as the basis for the 1991 ETSU report to the CEC,[146] which gives a potential of 17 GW for land-based windpower in the UK. This is quoted as being 'technically available' in 1990, with a further 12 GW for offshore wind by the year 2000.

There now appears to be no technical barrier to this potential. For example, the Watt Committee[64] considered that the main strategic reason for introducing wind power or any other renewable energy source into the UK system would be to increase the security of the system by adding to the diversity of the plant. At least 20% of the system peak load could be accepted from variable supply sources, such as windpower, without significant cost penalties on the operation of existing plant.

12.8.6 Some environmental issues

The most important effect of wind energy is that it is relatively non-polluting during its working lifetime when it has a zero fossil fuel input. A life-cycle analysis would show that very minor pollutants were emitted during the construction and site works period.

Several negative effects are often quoted in the literature.[1,64] The first is the visual impact, but visual impact is difficult to quantify and depends on subjective judgements. The blades could also cause a rotating shadow pattern which might present visual problems. A major area of concern has been the danger of birds colliding with the blades, but this was considered to be a negligible hazard.[141]

Electromagnetic interference can be caused by a windmill and the considerable amount of research suggests that where significant TV interference appears possible, remedial action should be taken before the problem arises. The area around the Burgar Hill site for the WEG 3 MW machine on mainland Orkney had poor reception prior to 1983. A new repeater station was installed so that better signal reception was established before the wind turbines were operating. Noise is also a problem, particularly at low frequencies, and the amount of data is sparse.[141] A large amount of low-frequency noise appears to be common to all machines and the only solution is to site them sufficiently far away from objectors.

This could severely limit potential wind sites. Safety is an obvious problem and blades have been known to become detached and fly off. Some towers have suffered structural failure in high winds.

12.8.7 Summary

By the end of the 1980s approximately 100 manufacturers had been identified who had supplied well over 100 000 wind turbines and pumps of various sizes throughout the world.[139] Of these, some 20 000 were connected to the grid with a total installed capacity of over 2000 MW. The largest single group of installations is in California, where they are situated mainly in the Altamont Pass region, some 60 miles east of San Francisco near Livermore, the Tehachapi region, about 100 miles from Los Angeles near Mojave, and the San Gorgonio Pass region, again about 100 miles from Los Angeles and near Palm Springs.[119] In Europe the installed capacity in 1991 was as shown in Table 12.5.

Of the 100 manufacturers identified above, only about 30 were regarded as 'well established' by the European Wind Energy Association. 'All but a handful' were based within the European Community and a series of goals for the exploitation of Europe's wind energy was also established in 1991[147] as follows:

4 000   MW by the year 2000
11 500   MW by the year 2005
25 000   MW by the year 2010
100 000   MW by the year 2030

The final figure is equivalent to 10% of the Community's total electricity demand in 1990.

The three leading countries (Table 12.5) had plans to enhance their installed capacities during the 1990s. Both Denmark and the Netherlands were aiming at 1000 MW each, with 250 MW in the Netherlands by 1995. Germany was planning to install at least 100 MW over a five-year period.[147] Italy's projections were to provide an additional 300 MW to 600 MW over the decade.

Table 12.5 Installed European capacity, 1991 (in MW)[119]

Denmark	360
Netherlands	55
Germany	55
Spain	15
UK	10
Greece	5
Italy	5
Belgium	2
Portugal	2
Total	509

12.9 Geothermal energy

Geothermal energy is thermal energy stored in the Earth. Although the Earth's heat can be regarded as an infinite source of energy, prolonged exploitation can exhaust a geothermal field. Geothermal energy is, therefore, not strictly a renewable source of energy compared with, for example, solar energy or hydro power.

In the nineteenth century it was believed that the residual heat in the centre of the Earth was the source of the natural geothermal phenomena such as hot springs with jets of steam that could be seen on the Earth's surface.[148] It is now widely accepted that there are two heat sources. The first arises from radioactive decay and the geological evidence points strongly to potassium, uranium and thorium contained in the rocks that form the Earth's crust.[149] The second comes from the mantle which lies below the crust and which may also contain small concentrations of radioactive elements. The crust is some 30–35 km thick below the continental land masses and the boundary between the crust and the mantle is known as the Mohorovicic seismic discontinuity, or Moho.[148,149] According to plate tectonic theory the crust is not a solid shell but consists of rigid segments or plates which can move relative to each other over the mantle. Pressure builds up at plate boundaries and the resulting sudden movement results in earthquakes and promotes the movement of large masses of molten rock or magma upwards into the Earth's crust, causing volcanic activity.[149] The thermal effects of the interaction between plates can extend several hundred kilometres from the boundaries. Major plate boundaries are well known and indicate the areas where exploitation of Earth's heat would be most likely to be successful.

Measurements of temperatures taken in mines and boreholes penetrating into the crust show that, with the exception of a very shallow zone near the Earth's surface, the temperature rises as the depth increases. The rate of increase, or thermal gradient, is between 20 and 30 K per kilometre for non-volcanic regions, with a smaller increase for older rock[149] and a much higher increase near magma penetration. Most of the heat which reaches the Earth's surface does so by conduction, but some is transferred by convection to the free water in the outer few kilometres of the crust. This can occur by the simple process of groundwater sinking through permeable rock such as the sandstones or limestones, or where the rock has been fractured, into the hotter regions and then circulating back towards the surface, or by the heating of the groundwater by igneous activity.[149] Temperatures of these hydrothermal fluids in the range from 100° to 200°C are common and in places have reached 400°C, conditions which can result in some of the water flashing (changing state from liquid to vapour with rapid reduction in pressure) into steam and appearing as hot springs or geysers. The average value for terrestrial heat flow on the continental land masses is about 0.06 W m^{-2}.[150,151] A slightly higher value, 0.063 W m^{-2}, is also widely quoted in the literature,[152,153] but in the exploited geothermal fields the heat flow carried to the surface by the fluids can be from 200 to 1700 times this value.[153]

12.9.1 Geothermal resources

Compared with the proved reserves and ultimate resources for the fossil fuels, which are published at regular intervals, estimates for both national and world geothermal heat resources must be regarded with very considerable caution. Armstead[154] points out that it is of more immediate interest to have an approximate idea of the amount of geothermal energy that could be obtained under existing economic and operating conditions, but that any attempt to estimate this must be highly speculative. Many regions in the world have had no geothermal exploration and it is very difficult to place any confidence in much of the published work. However, as a starting point the World Energy Conference[155] produced an assessment which is given in Table 12.6.

In the early 1990s total world primary energy consumption was approximately 3.6×10^{20} J and the primary energy equivalent of electricity generated was about a quarter of this

Table 12.6 Estimates of geothermal resources for electricity generation (10^{20} J)

Resource base, taking into consideration the continental land masses to a depth of 3 km and a datum of 15°C	410 000
Of this resource base only 2% is assumed to be of adequate temperature for electricity generation	8 200
Assume that the overall recovery and conversion efficiency is about 2.2%	180
One fifth is convertible by existing technology	36

Data derived from references 154 and 155.

figure. Table 12.6 suggests that the convertible geothermal resource is some forty times greater than the world annual production of electricity. This figure may be of the right order, but cannot be regarded with confidence.

The World Energy Conference also gave a figure of 2.9 × 10^{24} J for the estimated recoverable thermal energy which was theoretically available for direct applications at lower temperatures. Again, this figure must be fairly speculative as it amounts to over 7% of the total estimated resource base for electricity generation. When estimates of geothermal resources are made for individual countries the same reservations must be applied. The wide range of estimates which can be given emphasize the need for caution. For example, a series of estimates for Japan ranged from 40 000 MWe for the next thousand years, representing some 35% of the total world potential, to 8650 MWe.[154]

12.9.2 Geothermal areas, fields and aquifers

The surface of the Earth can be classified into three main areas as follows:[154]

1. Non-thermal areas with temperature gradients between 10 and 40 K per kilometre of depth;
2. Semi-thermal areas with temperature gradients approaching 80 K per kilometre of depth;
3. Hyperthermal areas with considerably larger temperature gradients.

An important distinction must be made between a geothermal area and a geothermal field. Many thermal areas are associated with rock of low or zero permeability and cannot be exploited under existing economic and operating conditions. Geothermal fields contain the hot water or steam in permeable rock formations and a number of these are operating commercially. They can also be classified into three main types:[154]

1. *Semi-thermal fields* which can produce hot water at temperatures up to 100°C from depths up to 2 km;
2. *Wet fields* which produce water under pressure at temperatures greater than 100°C. When this water reaches the surface, its pressure falls and some flashes into steam, the remainder being boiling water at atmospheric pressure;
3. *Dry fields* which produce dry saturated or superheated steam at pressures above atmospheric pressure.

Another source of useful hot water is the low-grade aquifer, which can produce water up to a temperature of about 75°C by drilling to depths of between 1.5 and 2 km, corresponding to a

temperature gradient of 30 K per kilometre. Low-grade aquifers can be found in non-thermal areas but are only worth exploiting if they are located fairly close to an appropriate application, such as space heating in a town or city.

12.9.3 Thermal applications

The earliest application of geothermal energy was the use of natural hot springs for bathing or medical purposes. One hot spring near Xian, the capital city of ancient China, has been used for over a thousand years and still attracts many bathers. The history of the use of geothermal energy for industrial applications probably started in the Larderello area of Tuscany, in central Italy, in 1827. Thermal energy from the hot wells was used in the crystallization of boric acid, which was also obtained from the natural pools formed from condensed steam and rainwater[153] and a flourishing chemical industry developed there over the next hundred years.

In Europe, one of the richest geothermal sources is in Hungary, where geothermal baths have been used for hundreds of years. Geothermal heating was first introduced in the 1960s and is used in greenhouses covering a total area of 2 million m^2. Uses in other agricultural applications include corn drying and poultry farms.[156]

An increasing number of applications for space heating have also been reported since the 1960s. Examples of two different types are the semi-thermal fields in Iceland and the low-grade aquifer found in the Paris basin. In Iceland the Reykjavik Municipal District Heating Services were able to sell hot water at less than one fifth of the cost of heating with oil.[157]

The capital cost of the geothermal plant installed in the Paris basin per housing unit was US$ 2000 (at 1980 prices) which was said to be comparable with any conventional system.[157] District heating using geothermal energy was first introduced in Hungary in 1985, and by 1990 over 6000 dwellings in six cities were being supplied during a heating period which lasts about six months.[156] The main development in the UK is in the Wessex basin. The first borehole at Marchwood in 1979/1980 produced a flow of 30 l s^{-1} at 72°C from a depth of 1700 m. This was originally intended to pre-heat feedwater for the power station, but this was closed before the project could be taken further.[158] The second was the Southampton borehole. The pumping rate was too low for an extensive heating scheme, but a later, smaller, scheme proposed by the Southampton City Council in 1987 now provides about one MW of geothermal heat as part of a larger 12 MW system. The maximum pumping rate is limited to 12 l s^{-1} to ensure an operating life of 20 years.[158] A very small warm water scheme (*ca* 22°C) is operating from a 250 m experimental borehole in Cornwall for a horticultural application.[64]

The main features of a typical geothermal district heating system are shown in Figure 12.14, based on information from reference 159. The first borehole establishes the chararacteristics of the aquifer and then becomes the production well, out of which the hot water is pumped. The second borehole is the reinjection well, which is used to dispose of the saline water after the heat has been extracted in the heat exchanger. This well is approximately one kilometre from the production well to delay the return of cold water for 25–30 years. The auxiliary boiler can provide additional heat at periods of high demand and the whole system is connected to the housing units and buildings by a pipeline system. This must be no further than one kilometre from the geothermal wells, both for reasons of cost and prevention of heat loss. Among other thermal applications widely quoted in the literature are greenhouse heating including soil warming, drying of organic products, salt extraction and industrial process heating.

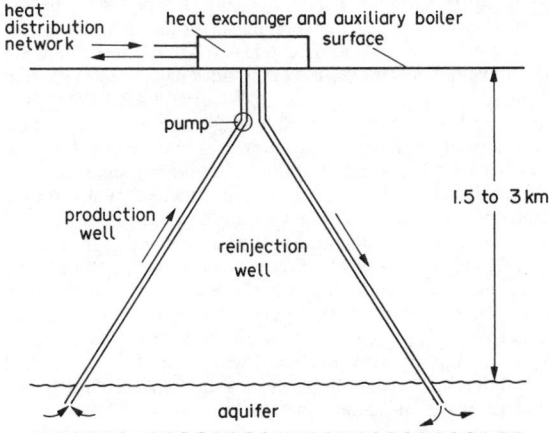

Figure 12.14 A typical geothermal district heating system (after McVeigh[1])

12.9.4 Electricity generation

The history of electricity generation also started in the Larderello area, when a simple steam engine coupled to a d.c. generator was driven by steam from the geothermal field. This provided some electric lighting for the town of Larderello.[152,153] The steam engine was replaced by a 250 kW turbo-alternator in 1913. Until 1958 Italy was the only country where natural steam was used for power generation on an industrial scale. Production commenced in New Zealand in that year, followed by the Geysers field in the United States in 1960, when the total installed capacity in the world was 369 MW.[150] Development over the next decade to 697 MW by 1970 represented an annual growth rate of 6.5%, but the 1970s saw a considerable increase in growth. Table 12.7 shows the installed and projected geothermal electrical generating capacity for 1980, 1987, 1989, 1990 and 2000, taken from data presented at the United Nations Conference on New and Renewable Sources of Energy in 1981[157] and from references 64 and 160 in 1990.

Data published by Shaw and Robinson[150], also in 1981, put installed capacity in 1980 as 2082 MW. This gives a growth rate of 11.6% per annum during the 1970s. By examining the known orders for new plant they concluded that a realistic assessment of installed capacity in 1990 would be between 3786 MW and 5645 MW, less than half the figure suggested by the United Nations Conference.[157] Their projected figures of

3786 MW and 5645 MW for 1990 would represent average annual growth rates over the decade of 6.1% and 10.5%, respectively. McVeigh's comment[1] in 1984 that the Shaw and Robinson projection seemed reasonable compared with the nearly 20% per annum growth rate to achieve the figures suggested by the UN Conference was fully justified with the publication of the 1989 data.

The risks and problems associated with geothermal projects are not unlike those in searching for oil. The success rate of geothermal drilling, when measured by the proportion of wells which strike exploitable hot water or steam, is probably greater than that of oil drilling. However, the rewards are much smaller and there are risks, as the first UK experience outlined above indicates.

12.9.5 Hot dry rocks

Geophysicists have suggested that rock at temperatures of 200°C can be found at drillable depths, less than 10 km, over large regions of the Earth's surface.[150] This has resulted in a number of major research projects in which deep holes are drilled into these hot rocks and a system of cracks is propagated between them.[150,154,157,159] Most of the research is aimed at establishing an optimum method for generating these fracture patterns. The basic technique uses hydraulic fracturing, and the first successful tests were carried out at Los Alamos in the United States during the early 1970s.[159] A fracture system some 600 m in diameter was created between wells 3 km deep and up to 4.5 MW was removed as heat during the initial test period of 2000 hours. This showed that the concept was valid. In the UK the Camborne School of Mines have extended the work in the United States by a more sophisticated approach to fracturing. They initiate the fracture system by explosives and then follow up with hydraulic fracturing.

Formal reviews of progress were undertaken in 1984, 1987 and 1990.[161] By 1984, two 2-km deep wells had been drilled, but the reservoir, or heat exchanger, between the two wells had relatively poor hydraulic properties. A third well was drilled and the second review concluded that while considerable progress had been made, there were several specific problems. The main one was still with the reservoir. This was a hundred times smaller than the size calculated to be necessary for a commercial reservoir and a reliable reservoir design process had not been validated. Three more years of experiments and studies still revealed that a satisfactory procedure for creating a commercial-scale reservoir had not been demonstrated. There was no reliable information about the properties of the rock likely to be encountered at the 6–7 km depths necessary for commercial exploitation. In a technical analysis of the work up to 1990, Parker[162] stated that until holes have been drilled to these depths the uncertainty will never be removed. Hot dry rock projects were unlikely to attract any private sector income in the short term.[161] Nevertheless, the potentially exploitable granites in south-west England alone contain the equivalent of 8000 million tonnes of coal.[163]

12.9.6 Some factors influencing developments

The economics of the applications of geothermal energy depend on the costs of competitive fuels. Where there are active geothermal fields and scarce indigenous resources, such as in Iceland or Hungary, geothermal power is already the economic choice. Financial constraints and the lack of a suitable technical infrastructure can inhibit development in some of the poorer developing countries who would appear to have considerable hyperthermal field potential.

Table 12.7 Installed and projected geothermal electrical generating capacity (MW)

1980[a]	1987[b]	1989[c]	1990[d]	2000[d]
2462	4707	5155	12 122	17 644

[a] UN installed figure.[157]
[b] UN Watt Committee, derived installed figure.[64]
[c] Dickson and Fanelli installed figure.[160]
[d] UN projected figures. (These UN projections were qualified by the comment that they were 'minimum' figures.)

Several other possibilities for using geothermal energy have been discussed, including the direct exploitation of the heat from active volcanoes.[149,154,157] This would have very considerable practical difficulties as it would involve tapping the magma at a depth of several thousand metres below the volcano, a technology which has not yet been developed. Among the ideas put forward for exploiting this source, Armstead[154] has suggested injecting water into the hot basaltic magma to produce hydrogen by dissociation.

The possible environmental problems which can arise from geothermal exploitation have been identified[154,159] and can include:

1. The use of land for initial drilling operations and possible noise and damage;
2. The long-term visual impact and use of land for the power or heat-extraction plant;
3. The development of a suitable heat-distribution and pipe-line system;
4. The release of gases, fluids and various chemicals during operation;
5. The physical effects on the geological structure of the area.

The earliest geothermal operations were carried out at a time when environmental issues were not taken into consideration. These early steam plants were reported to have unsightly tangles of steam-transmission pipes, clouds of waste steam accompanied by a strong smell of hydrogen sulphide and, eventually, significant surface subsidence. However, in recent years these adverse effects have been minimized. For example, air pollution standards at the world's largest field, the Geysers in the United States, have resulted in 'cleaner' air than before the field was exploited. For the low-temperature aquifer systems the environmental impact should be negligible.[159] The main problem is the safe disposal of the warm chemically laden water after it has passed through the heat exchangers. It has become normal practice to reinject the brine back into the other end of the aquifer as shown in Figure 12.14.

12.9.7 Summary

Unlike the other alternative energy sources, geothermal energy is capable of providing continuous heat and power. With electricity generation, the plant is particularly suitable for base-load operation. The use of low-grade aquifers can provide space heating at costs comparable with or below conventional systems. However, the long term future is still quite unpredictable. Several authorities believe that heat mining, the exploitation of hot dry rocks, could become a commercial reality within the next two decades.[153,154] Should this happen, a major new energy resource, comparable in size to the ultimate oil and gas resources, would be available, but the UK experience at the Camborne School of Mines serves as a warning against over-optimism.

12.10 Tidal power

12.10.1 Introduction

Tides are caused by the interaction of the gravitational and kinematic forces of the Earth, the moon and the sun. The gravitational force at any point on the surface of the ocean depends on the position of the moon and the sun and on their distance from the point. The period of the tides depends upon the 29.53-day period of rotation of the moon about the Earth, the Earth's daily rotation and upon the orientation of the Earth during its path round the sun.[164] The difference in length between the 24-hour solar day and the 24.813-hour tidal day causes the spring and neap tides. When the sun and moon are almost in line with the Earth the tides have their maximum amplitude and are known as the spring tides. When the moon–Earth–sun angle is a right angle the tides have their minimum amplitude and are known as the neap tides. The ratio between the greatest spring tide and the smallest neap tide can be up to 3:1.[122] The overall effect of tidal forces is surprisingly small. In the open ocean the tidal range, defined as the difference in amplitude between low and high tides, is typically about 1 metre.[122] Over the continental shelves the tidal range increases to about 2 metres and in some estuaries or deep narrow bays it can be up to 16 metres.

These increased tidal ranges in estuaries or bays are caused by the interaction of two types of wave. The first is the tidal wave advancing from the open sea and the second is the reflected waves from the sides of the estuary.[165] These two waves can reinforce each other at certain times, depending on the shape of the estuary and the period of tide, causing an amplification. Peak amplification occurs at resonance. Theoretically, a channel of uniform cross section would be resonant if its length were equal to one-quarter of the wavelength of the tidal movement. In practice, this length is modified by actual variations in depth and width. The rise and fall of the tide is also limited by the frictional losses caused by the action of the water over the sea bed.

Scientific publications on tidal schemes date from the early eighteenth century and various designs for dams and associated turbines appeared from the end of nineteenth century.[166] Modern proposals for exploiting tidal power are based on the use of the stored potential energy in a dam.[164] The use of the kinetic energy of the tidal current has been limited to a few very small-scale developments.[167]

12.10.2 Tidal power principles

Tidal power can be obtained from the flow of water caused by the rise and fall of the tides in partially enclosed coastal basins. This energy can be converted into potential energy by enclosing the basins with dams. This creates a difference in water level between the ocean and the basin. The resulting flow of water as the basin is filling or emptying can be used to drive turbo-generators. Electricity conversion removes the geographical restrictions placed on the earlier uses of tidal energy.[165]

The potential energy of a body of mass, m, at a height, z, above the datum line is mgz. If the surface area of a tidal basin is A m^2 and the mean tidal range is r m, then the maximum potential energy available during the emptying or filling of the basin is given by

$$\tfrac{1}{2} \rho g A r^2$$

The tide rises and falls twice during the tidal day of 24.814 hours, so the theoretical average power is four times the maximum potential energy divided by the total time in the tidal day or

$$\frac{4 \times \tfrac{1}{2} \rho g A r^2}{24.813 \times 3600}$$

Taking ρ as 1000 kg m^{-3} and $g = 9.81$ m s^{-2} the theoretical average power becomes

$$0.220 \, A r^2$$

If generation is only on the ebb tide the figure is halved. The actual power output is up to 25% of the theoretical average.

Some locations are particularly favourable for large tidal schemes because of the focusing and concentrating effect

Table 12.8 Mean tidal range in selected locations (m)[165]

Location	Range
Bay of Fundy, Canada	10.8
Severn Estuary, UK	8.8
Rance Estuary, France	8.45
Passamaquoddy Bay, USA	5.46
Solway Firth, UK	5.1

which can be obtained from the shape of their bays or estuaries. Typical ranges are shown in Table 12.8, which includes the world's largest tidal range in the Bay of Fundy and Europe's largest, the Severn Estuary.

In an ebb generation system the use of pumps to increase the level of water contained in the basin at high tide appears to be attractive. The principle is illustrated by a simple example. The additional energy required to raise the water level z m at high tide is

$$\tfrac{1}{2}\rho g A z^2$$

The maximum potential energy now available during the emptying of the basin becomes

$$\tfrac{1}{2}\rho g A (r + z)^2$$

giving a net gain of

$$\tfrac{1}{2}\rho g A (r^2 + 2rz + z^2 - z^2 - r^2)$$
$$= \rho g A r z$$

In practice there are two problems. Pumping involves some loss of overall efficiency and turbines capable of pumping as well as generating are more expensive. The power needed for pumping may be required when demand on the whole electricity system is high and could involve the use of an expensive form of generation in another section of the network.[122] The net gains in revenue from flood pumping in the proposed Mersey barrage (UK) have been examined[168] for various ratios of imported energy cost against exported energy value.

12.10.3 Tidal power schemes

There are a number of different schemes which can be grouped into two main combinations, depending on whether one or two basins are used:

1. Single basin, generation only on the ebb tide;
2. Single basin, generation only on the flood tide;
3. Single basin, generation with both the ebb and the flood tides;
4. Single basin, generation with both tides and pumped storage;
5. Double basin;
6. Double basin with a pumped storage system.

Any particular scheme could be optimized against any one of a number of different and distinct parameters. These include maximum net energy output; constant power output; constant-head operation; maximum pumped storage capacity or lowest initial capital requirements.[165]

Single-basin ebb generation allows the incoming tide to flow through sluice gates and the turbine passageways. These are closed at high tide and the water is retained until the sea has ebbed sufficiently for the turbines to operate. This is normally at about half the tidal range. Initially, the flow is restricted to maintain a high head and to operate the turbines at maximum efficiency. Later in the cycle the turbines are usually operated at maximum power.[122]

Single-basin flood generation is the reverse of ebb generation. It has a number of potential disadvantages, the main one being the prolonged periods of low tide experienced above the dam. A second disadvantage is that the amount of energy would be less than with an ebb generation scheme, as the surface area of the estuary decreases with depth.

Two-way generation with a single-basin system generates electricity from both the flood and ebb tides. This does not result in a greatly increased power output. Neither phase of the cycle can be taken to completion because of the need to reduce or increase levels in the basin for the next phase. There are also economic disadvantages. The turbines are more complex and less efficient if they are required to operate in both directions and the turbine water passages must be longer. An advantage is that power is available four times in the tidal day, rather than for two longer periods.

Double-basin schemes often include provision for pumped storage, but in their simplest form they could operate as two independent two-way generation schemes. In another form water would always flow from the higher-level basin to the second lower-level basin. The second basin could only be emptied at low tide.

A detailed discussion of the relative merits of the different schemes has been given by Taylor,[122] who points out that it is difficult to generalize, as a large number of variables, which vary from one site to another, need to be considered.

12.10.4 Tidal power sites

Many of the various design studies carried out until the early 1970s suggested that while tidal power systems were technically possible they would be unable to generate electricity at a competitive price. A notable exception was the first major report on the Severn Barrage, published in 1933,[169] but the recommendations were ignored and over half a century later further feasibility studies were still being carried out.[168]

Only three modern tidal power schemes were operating in the 1980s. The largest and oldest is the Rance Barrage near St Malo on the Brittany coast of France. Two much smaller schemes are in the former USSR and China. All three schemes have been built primarily to gain operating experience for the possible development of much larger systems.[122]

Work on the Rance site commenced in June 1960, the final closure of the estuary against the sea took place in July 1963 and the last of the 24 10 MW turbo-generators was commissioned in November 1967.[164] The overall width of the barrage is 750 m. The tides follow a fairly constant two-week cycle throughout the year. During the first week of the cycle the tidal range is between 9 m and 12 m and in the following week between 5 m and 9 m.[164] The mean tidal range is 8.45 m. For the lower tidal ranges the barrage operates only on the ebb tide with the basin level increased by pumping. For mean and spring tides two way generation is used, sometimes augmented by pumping. Electricité de France have shown[164] that the output/input ratio for pumping can be as high as 2.8:1. The operation of La Rance is computer controlled and optimized to match the period when it would be most expensive to generate electricity for the French national grid from conventional power stations. The nominal average output of between 50 and 65 MW is therefore not the maximum which could be obtained.

Nevertheless, it has been pointed out[168] that while La Rance tidal power is the cheapest electricity on the French system, Electricité de France comment that it would be too expensive to build any further tidal power systems.

In the former USSR a small 400 kW pilot scheme was completed in 1968 at Kislogubsk on the Barents Sea. The main objective was to try out a new construction method, the use of floated-in prefabricated caissons to form both the main power-house and spillway structures.[170] The overall dimensions of this structure were 36 m × 18.3 m by 15.35 m high and the single reversible turbine was purchased from the French company that supplied turbines for the Rance Barrage.[164] The People's Republic of China have a broadly similar pilot scheme rated at 500 kW at Jangxia Creek in the East China Sea.[122]

In the UK one site has been considered to be outstanding for nearly 70 years, the River Severn. Government interest in the Severn commenced in 1925 when the House of Commons established a Sub-Committee, which later became the Severn Barrage Committee of the Economic Advisory Council. The main conclusion of its report in 1933 was that the cost of power generated by a Severn Barrage with secondary storage some 20 km away would be only two thirds of the cost of that generated at equivalent coal-fired stations.[169] The scheme included road and rail crossings and the rated output was 804 MW. A further report in 1944 suggested doubling the output of the turbines to 25 MW while maintaining the rated output at some 800 MW. A later report in the early 1950s drew attention to the potentially high capital cost of any scheme.

Several further reports which appeared up to 1975 have been summarized by McVeigh,[1] who noted that in a comment on the various proposals that had been made over the past 50 years, the authors of an Institution of Chemical Engineers report[171] stated in 1976: 'A curious feature has been the regular conclusion that the scheme would have been economic if embarked upon on earlier occasions, but never on the current one . . .'

In 1977 the Department of Energy summarized the various proposals,[165] including the results of their own specially commissioned studies.[172,173] This was followed in 1978 by the establishment of a further pre-feasibility Severn Barrage Committee which reported in 1981[174] that it is 'technically feasible to enclose the estuary by a barrage located in any position east of a line drawn from Porlock due north to the Welsh coast'.

The most cost-effective of three schemes considered in detail was a single-basin, ebb-generation scheme with a 13 km barrage from Brean Down, a few kilometres south of Weston-super-Mare, to Lavernock Point, with an estimated annual output of 13 TWh at a cost of some 2.4 pence per kWh, at that time close to the official cost of nuclear electricity. This particular scheme was the subject of a two-year study jointly funded by government and industry which commenced in June 1983.

By 1990 all the recent studies concluded that the Severn Barrage was technically feasible and the main design parameters had been agreed for the Brean Down–Lavernock Point line outlined above.[168] These parameters included a nominal design life of 120 years, which was selected on the basis of it being a multiple of 30 and 60 years, time spans regarded as periods for major refurbishments. In practice, it was felt that the barrage could have an indefinite life. Other main parameters included an estimated total installed capacity of 8640 MW, obtained from 216 turbine generators, each 9 m in diameter and rated at 40 MW, and an annual output now estimated to be 17.0 TWh. This represented some 7% of the electricity consumption of England and Wales in 1989. But the possibilities for promoting and financing the scheme had to be delayed until after the UK electricity supply industry had been privatized.

The second largest potential tidal project in the UK is the Mersey Barrage.[168] Although the potential installed capacity is only in the order of 600 MW, it could have a major impact on the local economy. Overall in the UK, the theoretical tidal barrage capacity is approximately 25 GW.[168] Other possible smaller sites in the UK include Morecambe Bay, the Humber, the Wash and the Solway Firth.

In Canada tidal ranges of up to 16 m have been recorded in the upper regions of the Bay of Fundy in north-east Canada. This has been the subject of several investigations, including the working of the Atlantic Tidal Power Board[175] in 1969 (development not economically justified) and the Bay of Fundy Tidal Power Review Board[176] in 1977, who assessed the potential of 30 possible sites. The three with the best prospects, Cobequid Bay, Shepody Bay and Cumberland Basin, had an estimated potential output of 6.4 GWe.

Estimates of some 500 possible sites in the People's Republic of China have suggested a potential of over 110 GW.[122] Those for the former Soviet Union have concentrated on the White Sea between Murmansk and Archangel, where the potential could range from 16 to over 50 GW.

Other parts of the world with potential large-scale sites include the Kimberley Region of Western Australia, and South America, India and South Korea.

A series of estimates reviewed by Hubbert[177] suggested that the total tidal energy dissipated in the world's shallow seas was no more than 10^{12} W although these data appeared to omit any estimates from the People's Republic of China. The average maximum potential power which could be recoverable from these sites was estimated to be 64 GWe in 1969. When it is appreciated that some of the sites are very long distances away from any potential main user, this figure does not seem to be unduly pessimistic.

12.10.5 Possible impacts of tidal schemes

It is difficult to quantify the social, industrial and environmental impacts which any proposed scheme in the UK or elsewhere could have. These have been widely reviewed in the literature[122,178] and some of the main points are discussed briefly below:

1. *Water levels* both in the basin upstream of the barrage and to seaward could be changed.
2. *Tidal flows* reduce the strength of the currents upstream of the barrage. Downstream and to sea the effects could extend over 50 km.
3. *Sedimentation* may occur in the basin and could lead to a slow and possibly small reduction in basin volume. To seaward the sediments previously swept out by tidal flows may stay deposited.
4. *Mixing* will occur less in the water above the basin because of reduced currents and tidal excursions.
5. *Navigation*. Ships could be slowed by passing through locks; on the other hand, predictable periods of deeper water could be an advantage.
6. *Industry* could benefit during construction but may have to adopt higher standards in dealing with possible polluting liquid effluents.
7. *Land drainage* could be affected inside the barrage because of higher low-water levels.
8. *Sea defences* will be less liable to storm damage after the construction of a barrage.
9. *Ecosystem*. The aquatic ecosystem will always be affected by any changes in turbidity and salinity.
10. *Migratory fish* will face the obstacle of the barrage, but the

inward journey will probably be straightforward through the sluices in ebb-generation schemes.

11. *Recreational* opportunities could be enhanced in suitable locations, with less turbid water above the barrage.

There was unanimous agreement among members of the discussion panel at the third UK Tidal Power Conference[168] that several more years of environmental assessment would be needed before specific plans could be brought forward for the Severn Barrage. There were still many poorly understood or completely unknown oceanographic, sedimentary, engineering or environmental issues which had not yet been pursued, mainly because they were not 'make-or-break' issues in respect of the viability of the Severn project. The effect of global warming on mean sea levels, however, was not thought to impose any problems over the next few decades.

Some ideas of the scope and range of the environmental work on the Severn Barrage Project can be found by examining the full list of the 59 Environmental Impact Studies (and their associated contractors) during the two-year period from 1987 to 1989. It is difficult to see how all this work could be replicated for the many smaller UK sites.

12.10.6 The economics of UK tidal power (and of many of the renewables)

The discussion panel at the third UK Tidal Power Conference[168] were asked if they could assure the assembled delegates that there was common ground between, on the one hand, the opportunity to generate power without damaging emissions or radiation risks and, on the other, the need to provide power to the consumer on a commercially attractive basis. Dr John Chesshire[179] responded that he had stressed in his own publications and research work that there were inconsistencies in approach to capital expenditure and investment appraisal in the UK energy sector. A fourfold disaggregation was the best way to make this point briefly, i.e. between public and private sectors, and between energy efficiency and new energy supply projects.

Thus the financial return and depreciation period required from tidal barrage schemes would be similar to those expected by the UK financial sector for nuclear power and would lead to barrage generating costs in excess of 10 p kWh^{-1} (in 1989), as compared with combined cycle gas turbine generating costs of 2.2–2.5 p kWh^{-1}. It was therefore unlikely that, without regulatory or fiscal adjustment, there can be any common ground between the opportunity to generate power from nuclear and most renewable sources of energy and the need to provide electricity to consumers on a commercially attractive basis.

12.11 Wave power

12.11.1 Introduction

The history of wave energy conversion probably dates from the last few months of the eighteenth century, when the first patent for a wave energy device was filed in Paris.[180] Since then there have been hundreds of patents filed throughout the world with well over 300 in the UK alone between 1856 and 1973.[181]

Only a few proposals appear to have had model tests and these showed poor efficiency. Modern developments started in 1945 when Yoshio Masuda commenced privately funded research in Japan on a wide range of devices.[182] In 1960 his work received government support and he concentrated on wave-activated air turbines, one of which was installed in a

lighthouse in Tokyo Bay. This had a maximum output of 130 W. By the early 1970s over 200 small units were operating in Japan, mostly on buoys.

The UK Department of Energy wave energy programme commenced in 1974. This early work showed that the waves arriving at Britain's Atlantic coastline delivered a surprisingly large amount of energy[183,184] and the national programme quickly established that it was physically possible to generate useful power from ocean waves with reasonable efficiency.[185] However, after a decade of development, always exciting and sometimes frustrating, none of the devices investigated seemed to be capable of generating power at a cost which would be comparable to more conventional sources.[185] Some of the more promising concepts are discussed after the introduction to simple wave theory and characteristics, and the prospects for wave power in the 1990s are assessed.

12.11.2 Wave theory and characteristics

The basic characteristics of wave power can be studied from standard linear wave theory.[186] This considers a simple progressive sinusoidal wave of amplitude, a, wave length, λ, and period, T, progressing in deep water. Sinusoidal waves of a single wavelength are known as monochromatic waves and 'deep water' is defined as having a depth greater than half a wavelength. The velocity with which the wave propagates, the phase velocity, can be written as

$$\frac{\lambda}{T} \text{ or } \frac{gT}{2\pi}$$

By considering the rate of change of potential energy as the water in a wave above sea level falls into the troughs in front of the wave, the power in a wave front of width, W, is given by[187]

$$\frac{W\rho g^2\, a^2 T}{8\pi}$$

Real ocean waves are quite different from the theoretical ideal wave described above. Ocean waves are generated by the wind, so that wave energy is an indirect form of solar energy. The ocean acts effectively as an extremely large integrator for wind energy[122,188] and, in addition, the inertia of the water can provide a limited amount of short-term energy storage which can compensate for variations in wind velocity with time and place.[183] The waves arriving at a point can have originated from storms hundreds of kilometres away, the 'swell' sea, or from local winds, the 'wind' sea.[188]

The distance from the origin of swell waves is known as the fetch. Swell waves can appear to be substantially plane and monochromatic so that the longer the period between waves, the faster they travel. However, the local wind sea which is superimposed on it can be more complicated and random in wavelength, phase and direction. Any record of sea waves is therefore very complex and is best described as 'the linear sum of many monochromatic waves of random relative phase distributed both in direction and across the frequency spectrum'.[188] A detailed discussion of the spectral density function has been given by Pierson and Moskowitz.[189] However, for most practical purposes a very simplified relationship has been derived which is probably accurate to within ± 30%. This gives the power of a wind-generated wave system for any location in terms of the significant wave height, H_s, defined as the average height of the highest third of the waves, and the zero crossing period, T_z, defined as the time interval between successive upward movements of the water level past the mean position. For the Ocean Weather Ship *India* (50°N, 19°W) the power per metre of wave front has been shown[190] to be approximately

$$0.55 \, H_s^2 T_z \, \text{kW}$$

The estimates of power availability for wave energy systems in the mid-1970s were based on *India* data which suggested that an average power of 91 kW per metre of wave front was available.[184] This ranged from periods with very little power to severe storm conditions when the power level could exceed several megawatts.[191] More recent measurements near South Uist reported by the CEGB in 1983 showed that at inshore sites more suitable for the deployment of wave energy systems, power levels between 40 and 50 kW m^{-1} could be expected in water about 50 m deep. Levels of 25 kW m^{-1} could be expected off the north-east coasts of England or south-west Wales.

By the early 1980s a number of overall estimates had been made for the UK, giving a total resource capacity estimated to be 120 GW, based on a mean potential of 80 kW m^{-1} along a 1500-km coastline. When simple geographical limits were imposed, the potential dropped to 48 GW, and with further limitations such as device configuration, station design, capture and power train limitations, the achievable resource would reduce to about 6 GW, according to an ETSU assessment.[192] The 1988 UK review of renewable energy in the UK gave a technical potential of 30 GW, capable of providing some 50 TWh yr^{-1}, mainly off the Western Isles of Scotland and the coast of Cornwall,[31] but in 1991, a later report to the CEC[193] showed only 0.23 GW theoretically available by the year 2000.

12.11.3 Types of wave energy convertor

There have been a number of different classifications in the past decade, starting with Salter's flaps, floats, ramps, converging channels and liquid pistons or air bells.[187] The concept of 'active' and 'passive' systems has also been suggested,[194] active systems having parts which respond to the waves with power generated from the relative motion of these components while passive systems absorb wave energy by virtue of a fixed structure. One of the simplest methods for using wave energy would be to construct an immovable structure to capture large volumes of water which can subsequently be used to drive a water turbine. This system could also be classified as a ramp or passive system. The best-known proposal was initiated in Mauritius in the 1950s and has been the subject of a series of official reports, summarized by Bott[195] in 1979. All the reports agreed that the project would be technically feasible, but the economic viability has proved to be the stumbling block. By the early 1980s the more generally accepted scientific classification was that of terminators, attenuators and point absorbers.[191]

A terminator is defined as a wide structure which is aligned perpendicular to the incident wave direction. Much of the experimental work on this type has been carried out in narrow wave tanks and thus has resulted in good agreement with extensive theoretical studies, so that the performance of terminators is generally more fully understood than other types of device.

The best known is probably the 'Salter Duck', a system originally proposed by Salter in 1974.[187] The floating 'duck' section is an asymmetric cam-shaped device designed to extract energy through semi-rotary motion induced by the incident waves.[191] The system would consist of a long central core, or spine, upon which the duck sections are mounted. Large gyroscopes would be placed inside the nose of each duck, which rotates along its principal axis. The precessing motion of the gyroscopes could be used to drive hydraulic

pumps. The concept is revolutionary, as it would be designed to be maintenance-free over a 25-year lifetime.

The Cockerell raft was based on a simpler concept, that of a series of pontoons or rafts connected by hinges, with power generated from the relative movement of the rafts. Both the Duck and the Raft were taken to 1/10 scale model tests.[196,197] Several systems known as oscillating water column devices and which could also be considered as variations of Masuda's original air bell concept have been studied. One which was considered very promising and capable of proceeding to full-scale testing was the National Engineering Laboratory's breakwater system, which consists of a concrete structure mounted on the seabed. The motion of the waves causes a column of water inside the structure to rise and fall, inducing air flow through turbines. Proposals were suggested in 1982 by a private consortium and the National Engineering Laboratory for a 4 MW prototype to be built off the island of Lewis in the Outer Hebrides,[1] but this was not followed up. Almost a decade later a 30 MW array of oscillating water columns was being considered in Plymouth Sound. The estimated capital cost in 1991 was £35 million and the cost of the electricity generated would be 6p kWh^{-1}.[193] Other terminators have been studied in the UK, including the Russel Rectifier and the Sea Clam, and detailed descriptions are widely available.[122]

An attenuator is a long, thin structure which is aligned parallel to the incident wave direction. It was originally thought that energy could be progressively extracted along its entire length, but this has proved impractical because the rear element would need to extract as much energy as the front one for optimal operation.[191] Two devices considered in the UK were the Vickers Attenuator and the Lancaster flexible bag. Neither device was able to show a satisfactory performance in model testing.

The point absorber is an axially symmetric device, constrained to move vertically, which can absorb wave energy from any direction. Theoretically, it can absorb wave energy from an effective wave frontage of $\lambda/2\pi$.[191] This means that a number of interconnected but widely distributed point absorbers could produce as much power as a continuous line absorber having the same total length.

Research into the development of composite materials at Queen's University, Belfast, resulted in the development of a glass-fibre reinforced polyester resin and its utilization in the prototype of a new type of wave energy convertor, the Belfast Buoy. This can be considered as an oscillating water column device with an important difference; the vertical axis air turbine rotates in the same direction, irrespective of the direction of air flow. The turbine system has been named after its inventor, Professor Alan Wells.[198] Several different configurations have been suggested, but a good hydrodynamic performance can only be obtained from a very limited bandwidth.[191]

In 1982 a major review of renewable energy in the UK resulted in a recommendation that no new development work should be supported on large-scale offshore wave energy devices[192] mainly on economic grounds, as other renewables (e.g. wind and tidal power) were considered to be more economically attractive. Since then, wave energy work has concentrated on three systems:[199]

1. Point absorber devices, mainly at Lancaster University (the Flounder, Frog and PS Frog);
2. The Circular SEA Clam;
3. The shoreline wave energy resource, through the generation of electricity from oscillating columns located in shoreline rock gullies. This has included studies of the potential shoreline resource and the building by Queen's University, Belfast, of a device on Islay, which is described below.

12.11.4 Shore-based systems

Small shore-based wave power systems have been used by the Japanese to power lighthouses for nearly two decades.[200] During 1984–1985 the Wells turbine was used by the Kraener Brug company in a Norwegian 500 kW wave power station built on a cliff edge at Tostestallen, north of Bergen. This system was partially destroyed later in a severe storm. Another Norwegian device is the Norwave Tapchan, or tapering channel, invented by a mathematician, Even Mehlum.[200] This uses a funnel-shaped channel blasted out of rock to a predetermined profile which causes resonance with the local wave spectra. At the inner end of the channel is a wall, about 3 m above sea level. Waves enter the channel and propagate. As the walls narrow, the wave height increases until the water reaches the top of the wall and flows over it into a storage lagoon with a surface area of 8000 m^2. Water can then flow out of the lagoon and back to the sea through a 350 kW Kaplan turbine.[200]

In the mid-1980s, the Queen's University, Belfast, team followed their earlier work with a proposal to harness wave energy in relatively shallow natural rock gullies. The island of Islay in the Inner Hebrides was selected for the first prototype shore-based wave power system in the UK. The principle is illustrated in Figure 12.15. The device spans a natural rock gully in relatively shallow water and is being used as a test bed for a new two-stage Wells turbine.[201] The wave entering the gully oscillates a column of water inside the box, causing air to pass through the turbine in either direction. The official inauguration ceremony was held in 1991.

12.11.5 Summary

The future of wave power is particularly uncertain. None of the systems tested in the past decade have been able to demonstrate that they could generate electricity at a cost which would be comparable with other, more conventional, sources, except for a few specific locations in remote islands. The economics would alter in favour of wave power if conventional methods become more expensive, which many authorities believe is inevitable.[1]

However, a major reassessment of wave power was being carried out in the UK in the early 1990s, and the publication of the interim report in October 1991[199] showed a new approach, with all the active members of what is described as 'the wave energy community' participating, to produce a 'forward-looking review . . . based on best current knowledge'.

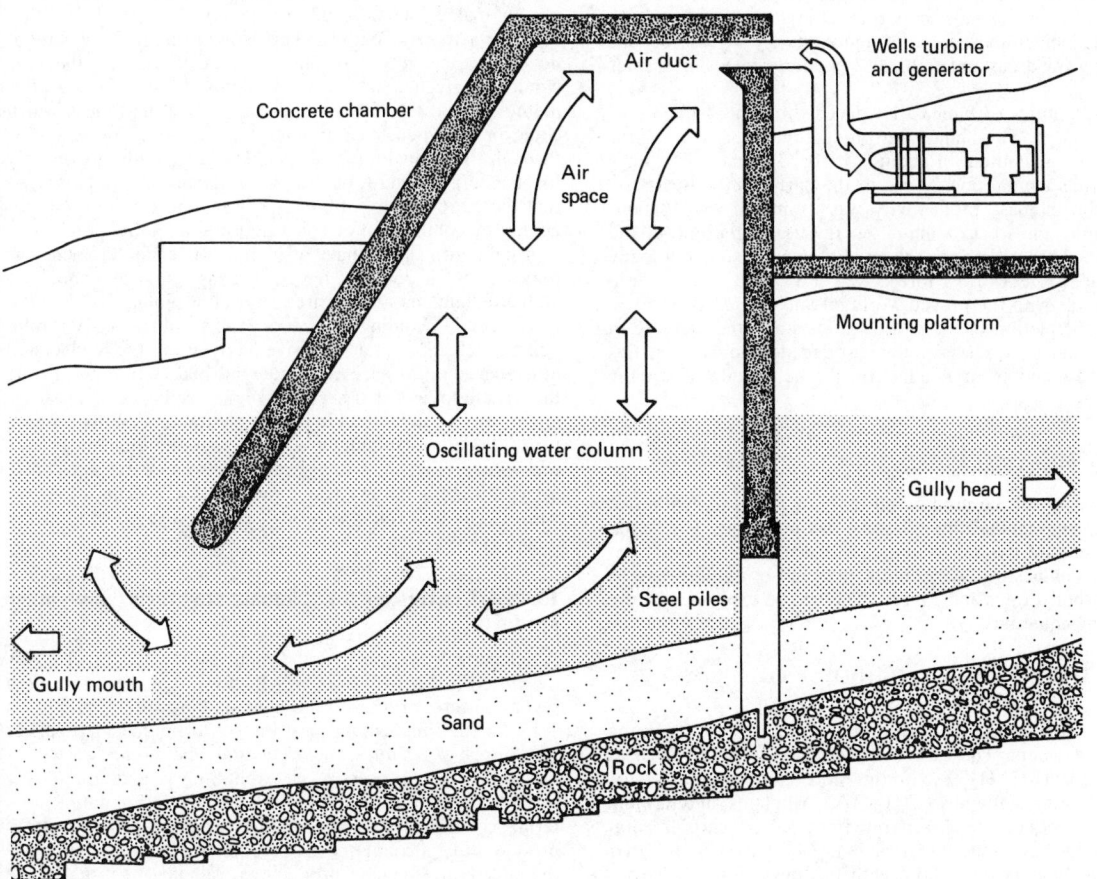

Figure 12.15 Schematic diagram of an oscillating water column device (after Review 7[201])

12.12 Biomass and energy from wastes

12.12.1 Introduction

The development of human life can be directly traced through biological conversion systems, initially through the provision of food, then food for animals, the materials for housing and energy for cooking and heating. The commencement of industrial activities was followed by the development of agriculture and forestry to their present levels. The renewed emphasis on biological conversion systems arises from the fact that solar energy can be converted directly into a storable fuel and other methods of utilizing solar energy require a separate energy-storage system. The carbohydrates can be reduced to very desirable fuels such as alcohol, hydrogen or methane, a process which can also be applied directly to organic waste materials which result from food or wood production.

Biomass can be defined as all types of animal and plant material which can be converted into energy. It includes trees and shrubs, grasses, algae, aquatic plants, agricultural and forest residues, energy crops and all forms of wastes. Estimates of how much of the world's energy demand is met by biomass range from 6% to 13%.[202,203] Among developing countries, biomass is the single most important source of energy, especially within the domestic sector, although some local industries such as bakeries, brick firing or steam production are also dependent on fuelwood. Nine tenths of the population in the poverty belts rely on wood as their chief source of fuel, and although in cooler regions some is used for heating, by far the most important energy need is for cooking.[204] Cooking fuel represents approximately 50% of fuel use in many rural areas and up to 90% of energy needs in warmer regions.[205]

The quantities of biomass produced throughout the world are very large. The annual net production of organic matter has an energy content of about 3×10^{21} J, some eight times the world's annual energy use in the early 1990s. In forests alone the biomass productivity was estimated to be about three times the world's annual energy use at the end of the 1970s.[203] At that time the potential biomass resource already standing in the world's forests was 1.8×10^{22} J, a figure comparable with the proven world oil and gas reserves. Even with the depletion of the world's forests and the increase in both oil and gas reserves in the past decade, biomass remains the largest and most familiar of all the renewable energy resources.

12.12.2 Photosynthesis

Solar energy can be used by all types of plants to synthesize organic compounds from inorganic raw materials. This is the process of photosynthesis. In the process, carbon dioxide from the air combines with water in the presence of a chloroplast to form carbohydrates and oxygen. This can be expressed by the following equation:

$$CO_2 + H_2O \xrightarrow[\text{Chloroplast}]{\text{Sunlight}} C_x(H_2O)_y + O_2$$

A chloroplast contains chlorophyll, the green colouring matter of plants. The carbohydrates may be sugars such as cane or beet, $C_{12}H_{22}O_{11}$, or the more complex starches or cellulose, represented by $(C_6H_{10}O_5)_x$. All plants, animals and bacteria produce usable energy from stored carbon compounds by reversing this reaction. Compared with other methods, biological or photosynthetic conversion efficiencies are much lower, but are potentially far less expensive. Photosynthetic efficiency is based on the amount of fixed carbon

energy produced by the plant compared with the total incident solar radiation. Plants can only use radiation in the visible part of the solar spectrum between wavelengths of 400–700 nm, known as the photosynthetically active radiation (PAR) region. This represents about 43% of the potentially available total radiation. At the plant some of the PAR is reflected and with other losses due to internal chemical processes the maximum attainable efficiency lies between 5% and 6%.

Under very favourable conditions, conversion efficiencies of between 2% and 5% have been recorded in the field for growth periods of a few weeks, but considerably lower efficiencies are achieved over longer periods of growth. Irish grasslands or forests with Sitka spruce are capable of dry matter yields greater than 16 tonnes ha^{-1} which represents an efficiency of about 0.7%. The main reasons for these relatively low efficiencies are environmental constraints, nutritional limitations and the incidence of pests and diseases.[206] Typical environmental constraints would include a drought or daily variations in ambient temperature. Nutritional limitations depend on the soil quality which, in turn, relies on the output of fertilizers.

12.12.3 Energy resources

There are five routes which can be followed to obtain the organic material or biomass which is the starting point for the energy conversion process. The first, and by far the simplest, is to harvest the natural vegetation. There are fertile regions in many parts of the world where the topography or some other reason makes the land unsuitable for agriculture or other valuable activities. With the harvesting of natural vegetation, no costs are involved in planting or clearing and the land would be given a new use. A major disadvantage of this method is that the yields are, at best, about half those which could be obtained from an energy plantation. The second is through the cultivation of a specific energy crop, grown only for its energy content, or the use of agricultural surpluses, so that the stored chemical energy can be converted into useful energy by combustion or converted into a storable fuel.

A land crop should have as high a conversion efficiency as possible, but it does not have to be digestible by animals or edible by humans. The entire material or biomass of the crop can be used, including the leaves, stalks and roots. By careful genetic selection and intensive cultivation the conversion efficiency should reach 3% under normal conditions. In the third route, trees and other types of lignocellulose material are grown specifically as fuel in energy plantations. Short-rotation forestry (described later) is a good example. The fourth uses the wastes from agro-industrial processes or residues from agriculture, animal wastes, straw, and all forms of urban wastes. The fifth route is through algae in the sea or grown in inland ponds.

12.12.4 Conversion of biomass to fuels and other products

A selection from some of the main conversion processes is illustrated in Figure 12.16, which shows that there are often several different routes to the same end product. Combustion is by far the simplest and best-known technique, particularly with forestry residues and industrial and urban wastes. A number of the processes are well known and are ideally suitable for producing fuels. With aerobic fermentation, materials containing starches and simple sugars can be used to produce ethyl alcohol or ethanol. Anaerobic fermentation has the added advantage of producing a valuable by-product, the nutrient-rich fertilizer from the digested slurry, when used to treat domestic sewage or animal wastes and produce biogas. In

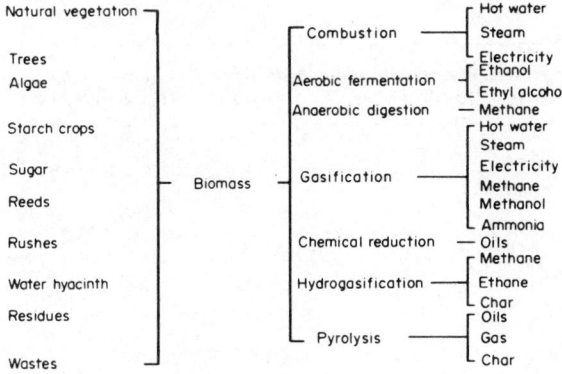

Figure 12.16 Some of the main conversion processes (after McVeigh[10])

Table 12.9

Product	Energy content[64]
Wood chips	18.6–20.9 MJ per kg dry weight
Biogas (two thirds CH_4, one third CO_2)	22–28 MJ m^{-3}
Ethanol	19 MJ l^{-1}
Methane	38 MJ m^{-3}
Methanol	16.9 MJ l^{-1}
Oils	30–40 MJ kg^{-1}
Pyrolytic oils	23–30 MJ kg^{-1}
Pyrolytic gas	8–15 MJ m^{-3}
Char	19–31.5 MJ kg^{-1}

the pyrolysis process the organic material is heated to temperatures between 500° and 900°C at ordinary pressures in the absence of oxygen, producing methanol, which was a by-product of charcoal in the last century. Methanol was first used as a fuel for high-performance racing cars and was subsequently studied as an additive in many laboratories. It is now considered to be an essential part of the future automobile fuel mixture.[207] The typical energy content of some of the products in Figure 12.16 is shown in Table 12.9.

12.12.5 Cooking – the major application of biomass

The United Nations have warned for many years that over 90% of wood cut in Africa is burnt as fuel.[208] Deforestation and desertification are widespread and increasing, with the southern edge of the Sahara extending by over 5 km yr^{-1}.

The scarcity of wood in some areas has meant that local inhabitants have had to move on or turn to substitutes. The World Bank[203] reported that between a half to one billion people use agricultural or animal wastes to fuel their fires. In India, cattle dung represents three quarters of the Indian domestic fuel consumption, robbing the land of valuable nutrients.[209] In parts of Africa crop residues and stubble are uprooted and used for fuel.

Cooking is a very cultural-specific activity. However, the most common means of cooking throughout the developing world is on an open fire. The most basic stove is simply three stones arranged on the ground in a triangle. The pan rests on the stones, between which three or more pieces of wood are placed. Efficiencies are low, between 2% and 10%, although much depends on the rate of burning, the air convection and other factors such as the height of the pan above the fire. Compared with the three-stones method, traditional woodfuel stoves are more efficient and the technology for improving their use of fuel exists. With a potential 30% fuel saving through the adoption of improved designs, fuel demand would be considerably reduced. Another possibility is the adoption of solar cookers, but these are too expensive and often local cooking habits or other sociological factors inhibit their acceptance, and they are unlikely to provide an adequate substitute for fuelwood, although they could provide a complementary energy source.

Adoption of alternative cooking methods also depends on such factors as aesthetic appeal and even attributes of social status. Each community is unique with regard to cooking modes and each will pertain to different values with consequent perceptions of suitability. It is therefore difficult to generalize from one community to another as to the most appropriate and acceptable fuelwood stove or solar cooker.[1]

12.12.6 Energy from waste materials

In the UK over 70 Mt of waste materials are generated annually[210] in homes, agriculture, commerce and industry. Because of changes in living patterns, especially in central heating and in the consumption of packaged goods, up to two thirds of collected domestic waste can be combustible. A breakdown of waste production and the distribution of combustible content are given in Table 12.10 (after Jackson and Tron[210]).

The calorific value of combustible waste can vary from 5.0 to 40 MJ kg^{-1} so that even at the lowest level (which predominates) a potential primary fuel content of 364 × 10^6 GJ per year or an installed generating capacity of 12 GW is available, of which some 60% is in domestic waste. A recent (1991) assessment of the overall gross calorific value of UK municipal waste was about 11 MJ kg^{-1}, a figure which was said to underline the advantage for energy-recovery purposes of burning the total combustible waste, rather than transporting the bulky waste and landfilling.[211]

The recent history of the use of waste as a fuel is not particularly encouraging. The variability of content of waste materials, the cost of acquisition and clean air legislation generally precludes all but the largest schemes, which inevitably mean district heating schemes run by local authorities. There are such installations in the UK at Edmonton and Nottingham, both of which are combined heat and power schemes.[23]

The late 1970s and early 1980s saw the development of pelleted Waste Derived Fuel (or Refuse Derived Fuel – RDF). Some types of refuse were dried, shredded and refined to concentrate the combustible and compressible fraction and produce a hard pellet-type fuel which could be used as a direct replacement for small coal. Only a few plants were built and operated in the UK. The completion of the plants coincided with the fall of conventional energy prices in the mid-1980s and new electricity tariffs made direct combustion of wastes a more economic proposition.[211]

Estimates for the annual average production of straw as a by-product of cereal crops in the UK range between 12 and 13.7 Mt.[23,64] Probably over half (50–60%) has been burnt in the fields to recycle minerals, but this was being phased out in the early 1990s. A total of some 166 000 t is used directly for farmhouse and animal house heating annually in over 7000

Table 12.10 Estimated UK waste and its energy content

	Gross weight $(t \times 10^{-6})$	Weight of combustible content $(t \times 10^{-6})$	Energy content $(GJ \times 10^{-6})$	(%)
Domestic refuse (including vegetable and putrescibles)	18.0	12.0		
Commercial refuse	3.8	1.6	166	46
Industrial refuse	19.0	4.1	50	14
Industrial process wastes	23.0[a]	2.3	30	1
Total: Domestic, Commercial and Industrial solid waste	63.8[a]	20.0	246	68
Agricultural (surplus) straw, plastics only	8.0	8.0	105	29
Forestry	1.0	1.0	13	3
Total: all sectors	72.8	29.0	364	100

[a]Excludes mining, quarrying, construction site wastes and power-station ash.

boilers.[64] Proposals have been made for a number of applications from heat production to producer-gas fuelled vehicles, particularly tractors.[212] The maximum UK potential for on-farm straw combustion is 1.9 Mt yr^{-1}, but 0.9 Mt yr^{-1} by the end of the century is thought to be more realistic.[213] Industrial heating in small industries such as food and drink, cement and brickmaking, and in light engineering could use a further 5 Mt, but this was not considered to be likely in the short term.[64]

Domestic and commercial refuse contains large quantities of organic matter. In the UK, over half this refuse is carbohydrate in origin, and each of the estimated 3300 active landfill sites can be considered as large 'bio-reactors' for the decomposition of this organic matter. In the first stage, microbial activity is high and the rapid depletion of the available oxygen results in anaerobic conditions. Anaerobic digestion is a complex process, involving the degradation of large organic compounds, such as vegetable matter or paper, to simpler substances such as sugars.[214] This is followed by the production of hydrogen, carbon dioxide and fatty acids prior to the generation of biogas in the final stage. This is, by definition, 'landfill gas', with a chemical composition and calorific value indicated for biogas in Table 12.9.

It has a fairly recent history, being first noticed in the United States and Germany in the 1960s,[214] and its first use as a gaseous fuel in the UK followed in the 1970s with the firing of brick kilns.[214] By the end of 1987, some 14 sites in the UK were using landfill gas directly[214] and the first five small electricity-generation stations had commenced operation. The total capacity of those sites fitted with generating equipment had reached 30 MW towards the end of 1991.[215]

In a survey of world trends at the end of the 1980s, the UK came second only to the United States in the commercial exploitation of landfill gas, with the main UK use being in gas kilns, furnaces and boilers. World use was projected to reach some 3.5 Mtce by 1992, with electricity generation at 440 MW. The UK potential for landfill gas is considered to be about 3 Mtce.

Anaerobic digestion occurs naturally in organic swamps, producing marsh gas. Man-made digesters can provide the best conditions for the controlled continuous production of biogas. Air is excluded and the digester container is held at about 35°C. The People's Republic of China has been recognized as the world's major user of biogas systems for many years and biogas production has become a comprehensive controlled method of waste disposal, supplying fertilizer and improving rural health in addition to providing a renewable energy source. By 1983 there were between 5 and 6 million operating biogas units and about 60 special biogas institutes in the provinces.[216] Their sole function is to carry out research and development work on biogas units and also to develop a trained group of technical teachers who would go out into the countryside to instruct others in the construction techniques, operation and maintenance.[1]

A basic Chinese family biogas unit is shown in Figure 12.17. This would use the wastes from the smallholding activities of a number of families. Most units of this type could produce about 6–7 m^3 gas daily during the summer months.[1]

It is formed from a horizontal concrete cylinder, buried about 1 m underground. Square-sectioned vertical entrance and exit chambers have tight-fitting concrete lids. The gas is generated in the upper section of the cylinder and the delivery pipe to the family kitchen branches to a large vertical water manometer, mounted on the kitchen wall, so that a careful check can be kept on the gas pressure, normally about 250 mm of water above atmospheric pressure. Both human and animal wastes are used as raw material for the units, as well as various types of vegetable waste matter. Basic loading and clearing the processed waste for use as fertilizer takes between one and two hours per week in the summer months and slightly longer in the spring and autumn, as more care has to be taken with the quality of the wastes in colder conditions.[216]

India has also had considerable experience with the development of biogas systems and some countries are now basing their designs on the established Chinese and Indian systems. For example, over 1000 *gobar* (cow-dung) plants have been built in Nepal, biogas has been used to replace diesel fuel in Botswana and plans to develop some 300 000 biogas systems in rural Thailand have been studied.[217] Among the industrial countries Romania, with extensive pig farms, was using biogas in an experimental bus during the 1980s.[1]

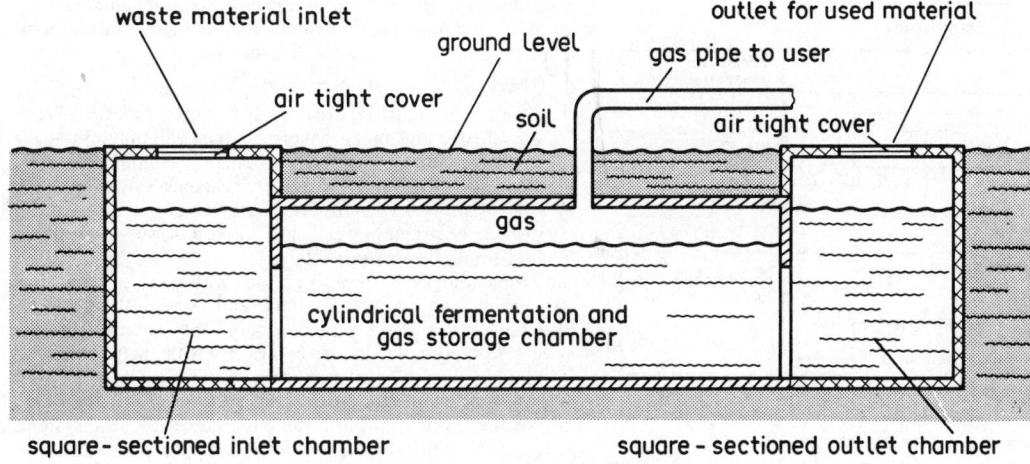

Figure 12.17 A Chinese family biogas unit (after McVeigh[1])

In the UK, the temperate climate is less encouraging for biogas, but careful overall system design can overcome this problem. A prototype unit for a dairy herd of 320 cows was completed in Kent in 1979. Electrical power was generated from a Ford diesel generator modified for gas combustion with spark ignition, with a continuous maximum power calculated to be about 25 kW.[218] A number of smaller units were also operating in other parts of the country.[219]

A leading commercial organization in the UK, Farm Gas, developed a range of small digesters mainly for farm use, from the mid-1970s. In 1991 they were examining the use of small digesters for municipal solid waste for the UK Department of Energy.[220]

12.13 Energy crops

12.13.1 Short-rotation forestry

The use of trees as energy crops has been proposed in several countries since the early 1970s. Detailed feasibility studies in the USA have shown that biofuels can be produced at competitive costs, by choosing the appropriate plant species, planting density and harvest schedule for each plantation site, thus minimizing the overall cost of the plant material.[221] In Ireland about 6% of the land area consists of bogland and less than a fifth of this area is being harvested for peat, which is either used directly as fuel in the home or for generating electricity.

Until recently it had been thought that bogland was unproductive, but grass, shrubs and trees have all been successfully grown. Even with a conversion efficiency of 0.5% for Sitka spruce, the same bogland area at present used for turf could produce exactly half the quantity of electricity through the combustion of the trees. The Irish government has demonstrated that woodchips obtained from short-rotation forestry can provide an economic alternative to oil.[222]

Briefly, the short-rotation forestry concept follows the sequence of selecting, planting, harvesting and utilizing as fuel the woodchips obtained from coppicing hardwood trees. The chips will be left to dry naturally in the fields, then collected, transported and burnt directly in specially modified power stations. Alternatively, they can be bagged and sold directly

for burning in central heating plants or for gasification. Harvesting would occur every three or four years. The trees are expected to regrow up to eight times from the existing root structure before replanting is necessary. A major advantage of the system is that the fuel can be stored indefinitely.[222]

An interesting statistic from the UK in 1991 pointed out that there were well over 200 woodfuelled combined heat and power plants.[223] While most of these were using by-products from sawmills or joinery works, the possibility of an alternative use of farmland for fuelwood production was being seriously considered, with support from the European Social Fund.

12.13.2 A fuel-alcohol plant

An important factor in considering energy crop conversion is the energy needed for harvesting and for fertilizers to increase the crop yields. Net energy analysis is used to assess the energy cost/benefit ratio of any proposed fuel conversion process. The energy inputs and outputs of the system can be measured and the net energy ratio (NER) can be defined as the ratio of the energy outputs to the energy inputs. Any application of this concept requires a careful definition of the system boundaries. The NER concept has been used in the world's first cassava (mandioca) fuel-alcohol commercial plant in Brazil.[224]

The system boundaries and energy flows are shown in Figure 12.18. The system consists of the cassava plantation, the fuel-alcohol distillery and the forest from which the fuelwood is obtained to provide process steam for the distillery. Energy optimization of cassava distilleries could lead to the development of varieties of cassava with larger stalk-to-root ratios, so that the cassava stalks could replace the fuelwood requirement.

Another self-sufficient process is the sugarcane fuel-alcohol system. The bagasse or by-product can generate all the necessary process steam. The NER of both systems is shown in Table 12.11 and based on 1 m^3 of anhydrous ethanol and total on-site generation of electric power.[224]

12.13.3 Marine and aqueous applications

In oceans the production of organic matter by photosynthesis is generally limited by the availability of nutrients and they have been compared to deserts because of their low productiv-

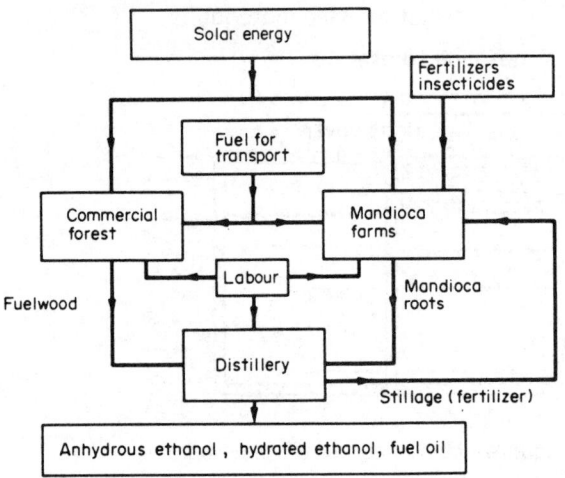

Figure 12.18 System boundaries and energy flows (after McVeigh[1])

ity. However, there are a few areas where natural flows bring the nutrients from the bottom of the ocean to the surface so that photosynthesis can take place. Particular interest has been shown in the cultivation of giant kelp (*Macrocystis pyrifera*), a large brown seaweed found off the west coast of the USA.

An early estimate examined the yield from an area of 600 000 km² and concluded that the equivalent of some 2% of the US energy supply could be provided.[225] One of the disadvantages of harvesting natural kelp beds would be the relatively low output caused by the lack of nutrients. Artificial kelp 'farms' have been suggested[226] and a 1000 m² system, with nutrient-rich deep water being pumped from deep water to the surface kelp, has been developed, as the first stage of a project which could lead to a 40 000 ha system.

Aquatic weeds can easily be converted to biogas. The water hyacinth (*Eichhornia crassipes*) has been extensively studied as a biogas source, particularly at the US National Aeronautics and Space Administration (NASA).[227] On a dry weight basis, one kilogram of water hyacinth can produce 0.4 m³ of biogas with a calorific value of 22 MJ m⁻³. Aquatic weeds grow prolifically in many tropical regions and are costly to harvest. However, as weed clearance is essential to keep waterways clear, biogas production could be regarded as a valuable by-product in these applications.

12.13.4 Choice of system

The factors affecting the choice of a particular biological conversion system identified by Hall and Coombs[228] include

considerations of agricultural capacity, environmental factors, population density, labour intensity in the agricultural sector and the energy demand per capita. Four regimes were distinguished in their simple classification:

1. Temperate industrial areas such as North America, Western Europe and Japan, where biomass will only produce a small fraction of the energy demand. Emphasis will be placed on the production of scarce chemicals from biomass. The by-products from certain industries may be used to provide heat and power, while the use of wood as a direct fuel source is also possible.
2. Tropical and sub-tropical regions with good soil and high rainfall such as parts of India and Africa, Brazil, Indo-China and north Australia. Energy from biomass has the greatest potential in these regions, with many examples already competing economically, e.g. biogas and the more efficient use of fast-growing wood species.
3. Northern polar and arid regions where biological systems are only possible in an artificial environment, e.g. use of the nutrient film technique.
4. Marine and aqueous regions through the use of fast-growing water weeds, seaweeds or micro-algae.

The development of photobiological energy-conversion systems can take place more readily in the temperate Western countries with their high technological background. However, these systems can function more effectively in the developing tropical and sub-tropical countries and could make a very significant contribution towards reducing their dependence on increasingly scarce and expensive oil.

12.13.5 Summary

Biomass is already the most important energy resource for between a quarter and half of the world's population. Its use on a commercial scale is increasing through technical development. Further rises in the costs of competitive energy will bring many more applications into widespread use. The resource base is very large and capable of at least a fourfold expansion from today's level, which is estimated to lie between 6% and 13% of total world energy use. There would appear to be no major technical obstacle to a fourfold expansion.[1]

References

1 McVeigh, J. C., *Energy Around the World*, Pergamon Press, Oxford (1984)
2 Hubbert, M. King, *Energy Resources: A Report to the Committee on Natural Resources*, National Academy of Sciences – National Research Council. Publ. 1000-D, Washington, DC (1962)

Table 12.11 Net energy analysis[224]

| Raw material | Output | Energy (10⁶ kcal) | | | | NER |
| | | Input | | | | |
		Agriculture	Distillery	Transport	Total	
Sugarcane	5.59	0.42	0.017	0.26	0.70	8.0
Cassava	5.59	0.30	0.045	0.27	0.62	9.0

3 Hubbert, M. K., 'The energy resources of the Earth', *Scientific American*, **225**, 61–70, September (1971)

4 Frolich, C., 'Contemporary measures of the solar constant', in *The Solar Output and its Variation*, Colorado Associated University Press, Boulder (1977)

5 *BP Statistical Review of World Energy*, The British Petroleum Company plc, London, June (1989)

6 Hall, D. O., in *Solar Energy: A UK Assessment*, UK Section, International Solar Energy Society, London, p. 266 (1976)

7 Hall, D. O. and Rao, K. K., *Photosynthesis*, Edward Arnold, London (1976)

8 Duffie, J. A. and Beckman, W. A., *Solar Engineering of Thermal Processes*, Wiley, New York (1980)

9 Jesch, L. F., *Solar Energy Today*, UK Section ISES, London (1981)

10 McVeigh, J. C., *Sun Power*, 2nd edn, Pergamon Press, Oxford (1983)

11 Palz, W. (ed.), *European Solar Radiation Atlas*, Commission of the European Communities, Brussels (1984)

12 Page, J. K., in *Solar Energy*, Memorandum by UK Section, International Solar Energy Society, Select Committee on Science and Technology, House of Commons Paper 156-i, HMSO, London (1975)

13 Saluja, G. S., Muneer, T. and Smith, M. E., 'Methods for estimating solar irradiation on a horizontal surface', *International Journal of Ambient Energy*, **9**, 2 (1988)

14 Holmes, M. J. and Hitchin, E. R. 'An "example year" for the calculation of energy demand in buildings', *Building Services Engineer*, **43**, 186–189 (1978)

15 Yellott, J. I., 'Solar energy utilization for heating and cooling', *ASHRAE Guide and Data Book series*, 59 (1974)

16 Smith, C. T. and Weiss, T. A., 'Design applications of the Hottel–Whillier–Bliss equation', *Solar Energy*, **19**, 109–113 (1977)

17 Energy Technology Support Unit, *Prospects for the exploitation of the renewable technologies in the United Kingdom*, Report R30, HMSO, London (1985)

18 Solar Trade Association, *Code of Practice for the Solar Water Heating Industry*, London (1979)

19 Heating and Ventilating Contractors Association, *Guide to Good Practice for Solar Heating for Domestic Hot Water*, London (1979)

20 British Standards Institution, Code of Practice for solar water heating systems for domestic hot water, BS 5918: 1980 (revised edition 1988)

21 British Standards Institution, Methods of test for the thermal performance of solar collectors, BS 6757: 1986

22 British Standards Institution, Code of Practice for solar heating systems for swimming pools, BS 6785: 1986

23 Sherratt, A. F. C. and McVeigh, J. C., 'Ambient energy in the UK', *International Journal of Ambient Energy*, **8**, 1 (1987)

24 Telkes, M., 'A review of solar house heating', *Heating and Ventilating*, **46**, 68–74, September (1949)

25 Hottel, H. C., and Woertz, B. B., 'The performance of flat-plate solar-heat collectors, *Trans ASME*, **64**, 91–104 (1942)

26 Nemethy, A., 'Heated by the sun', *American Artisan*, Residential Air Conditioning Section, August (1949)

27 Bankston, C. A., 'A summary of central solar heating plants with seasonal storage', *Advances in Solar Energy Technology*, 2, Pergamon Press, Oxford (1987)

28 Stromberg, R. F. and Woodall, S. O., *Passive Solar Buildings: A Compilation of Data and Results*, SAND 77–1204, Sandia Laboratories, Albuquerque, New Mexico (1977)

29 Commission of the European Communities, *European Passive Solar Handbook*, Brussels (1989)

30 Department of Energy, 'Gains all round through passive solar design', *Review*, 2, London (1988)

31 Department of Energy, *Renewable Energy in the UK: The Way Forward*, Energy Paper 55, HMSO, London (1988)

32 Department of Energy, 'EDAS – Energy Design Advice Scheme', *Review*, 9, London (1989).

33 Project Monitor Series, Commission of the European Communities, Brussels (1987–1989)

34 *The making of a solar village and a demonstration house in Bourneville*, Franklin Consultants, Birmingham (1985)

35 Saxon, R., 'Introduction to atria today', UK Section of the International Solar Energy Society, Conference, London (1987)

36 Aschehoug, O., 'Glazed spaces – research and development today', UK Section of the International Solar Energy Society, Conference, London (1987)

37 'The passive solar design of buildings', *Renewable Energy News*, 12, Department of Energy, London (1985)

38 Ackerman, A. S. E., 'The utilisation of solar energy', *Trans. Soc. Engrs*, 81–165 (1914)

39 Winston, R., 'Principles of solar collectors of a novel design', *Solar Energy*, 16, 89–95 (1974)

40 Horster, H. and Kersten, R., 'Evacuated solar collectors', UK ISES Conference (C25) on *Recent Developments in Solar Collector Design*, London (1981)

41 Trombe, F. *et al.*, 'First results obtained with the 1000kW solar furnace', *Solar Energy 15*, 63–66 (1973)

42 Blake, F. A., *100 MWe solar power plant design configuration and performance*, NSF-RANN Grant No. AER-74-07570, Martin Marietta Aerospace, Denver (1975)

43 Kesselring, R., 'Solar thermal power: advances in research and development', *Advances in Solar Energy Technology*, 1, Pergamon Press, Oxford (1987)

44 Jaffe, D., Friedlander, S. and Kearney, D., 'The LUZ Solar Electric Generating Systems in California', *Ibid.*

45 'SEGS VIII Project closes', LUZ *International Newsletter*, **4**, 2 (1990)

46 Saulnier, B. *et al.*, 'Field testing of a solar pond', *ISES Congress, Los Angeles, Extended Abstracts*, Paper 35/1, July (1975)

47 Jesch, L. F., Report on Solar World Congress, Perth, *Sun at Work in Britain*, **17**, November (1983)

48 Kettani, M. A. and Gonsalves, L. M., 'Heliohydroelectric power generation', *Solar Energy*, **14** (1972)

49 Read, W. R. W., 'A solar still for water desalination', *Report ED 9*, CSIRO, Melbourne (1965)

50 'Aegean Island installs world's largest solar distillation plant', *Civil Engineering and Public Works Review*, 1005, September (1967)

51 Porteous, A., 'Fresh water for Aldabra', *Engineering*, 490, 15 May (1970)

52 Crutcher, J. L. *et al.*, 'A stand-alone seawater desalting system powered by an 8 kW ribbon photovoltaic array', *Solar World Forum*, 2, 1110–1119, Pergamon Press, Oxford (1982)

53 Claude, G., 'Power from the tropical sea', *Mechanical Engineering*, **52**, 1039–1044, December (1930)

54 Anderson, J. H. and Anderson, J. H. Jr, 'Large-scale sea thermal power', ASME Paper No. 65–WA/Sol-6, December (1965)

55 White, H. J., 'Mini-OTEC', *International Journal of Ambient Energy*, **1**, 2, 75–88 (1980)

56 Avery, W. H. and Dugger, G. L., 'Contribution of ocean thermal energy conversion to world energy needs', *International Journal of Ambient Energy*, **1**, 3, 177–190 (1980)

57 *Expanded Abstracts of the 7th Ocean Energy Conference*, Washington, DC, June (1980)

58 *Solar absorption chiller and total climate control system*, Publicity Unit, Israel Export Institute, Tel Aviv, Israel (1984)

59 Excell, R. H. B. and Kornsakoo, S., 'The development of a solar-powered refrigerator for remote villages', *Solar World Forum*, **2**, 1049–1053, Pergamon Press, Oxford (1982)

60 Crowther, K. and Melzer, B., 'The thermosyphoning cool pool', *Proc. Third National Passive Solar Conference*, San José, California, Vol. 3, pp. 448–451 (1979)

61 Thomason, H. E. and Thomason, H. J. L., 'Solar houses/heating and cooling progress report', *Solar Energy*, **15**, 27–40 (1973)

62 McVeigh, J. C., *Solar Cooling and Refrigeration*, a report for UNESCO (1984)

63 Luque, A. and Araujo, G. L. (eds), *Physical Limitations to Photovoltaic Energy Conversion*, Adam Hilger, Bristol (1990)

64 Laughton, M. A. (ed.), *Renewable Energy Sources*, Report No. 22, The Watt Committee on Energy, Elsevier Applied Science, London and New York (1990)

65 *Solar Energy: A UK Assessment*, UK Section of the International Solar Energy Society, London (1976)

66 Personal communication, Professor R. Hill (1989)

67 Personal communication, Professor R. Hill, University of Northumbria at Newcastle (PV Applications Centre, Contract E/5A/1365/2716 with AEA Technology) (1991)

68 Tsubomura, H., 'Photochemical conversion of solar energy', *Proc. 1989 Congress ISES*, Volume 3, p. 2193, Pergamon Press, Oxford (1990)

69 Ciamician, G., 'The photochemistry of the future', *Science*, **36**, 385 (1912)

70 Daniels, F., *Direct Use of the Sun's Energy*, Yale University Press, Newhaven and London (1964)

71 Reller, A., Maurer, H. and Padeste, C., 'The use of solar energy for the generation of organic carbon compounds', in Gupta, B. P. and Traugote W.H. (eds), *Solar Thermal Technology – Research, Developments and Applications*, Hemisphere, New York, pp. 407–413 (1990)

72 Sizmann, R. and Nix, G., 'High temperature, high flux density solar chemistry', *Ibid.*, pp. 351–378.

73 Bolton, J. R. and Archer, M. D., 'Chemical conversion and storage of solar energy – an overview', *Proc. 1985 Congress of ISES*, Volume 3, p. 1843, Pergamon Press, Oxford (1986)

74 Heidt, L. J. *et al.*, *Photochemistry in the Liquid and Solid States*, Wiley, New York (1960)

75 Porter, G. and Archer, M. D., '*In vitro* photosynthesis', *Interdisc. Sci. Rev.*, **1**, 119 (1976)

76 Bolton, J. R., 'Solar fuels – the production of energy-rich compounds by the photochemical conversion and storage of solar energy', *Science*, **202**, 705 (1978)

77 Bolton, J. R. and Hall, D. O., 'Photochemical conversion and storage of solar energy', *Annual Rev. Energy 4*, 353 (1979)

78 Tsubomura, H. 'Chemical conversion and storage of solar energy', *Energy and the Environment into the 1990s*, Volume 3, p. 1430, Pergamon Press, Oxford (1990)

79 Fujishima, A. and Honda, K., 'Electrochemical photolysis of water at a semi-conductor electrode', *Nature*, **238**, 37 (1972)

80 Memming, R., 'Photoelectrochemical utilization of solar energy', *Energy and the Environment into the 1990s*, Volume 3, p. 1436, Pergamon Press, Oxford (1990)

81 Bockris, J. O'M., *Energy Options*, Taylor and Francis, London (1980)

82 Lede, J. *et al.*, 'Production of hydrogen by direct thermal decomposition of water', *International Journal of Hydrogen Energy*, **8**, No. 9, 675–679 (1983)

83 Chubb, T. A., 'Characteristics of CO_2–CH_4 reforming-methanation cycle relevant to the Solchem thermochemical power system', *Solar Energy*, **24**, 341–345 (1980)

84 Diver, R. B. *et al.*, 'Solar test of an integrated sodium reflux heat-pipe receiver/reactor for thermochemical energy transport', in Gupta, B. P. and Traugote, W. H. (eds), *Solar Thermal Technology – Research, Development and Applications*, pp. 517–526, Hemisphere, New York (1990)

85 Hutchins, M. G. *et al.*, 'Electrochromic properties of rf sputtered tungsten oxide films', *Applied Optics in Solar Energy III*, p. 34, Czechoslovak National Academy of Sciences (1989).

86 Lampert, C. M., 'Electrochromic materials and devices for energy efficient windows', *Solar Energy Materials*, **11**, 1 (1984)

87 Lampert, C. M., 'Advances in optical switching technology for smart windows', *Proc. 1989 Congress ISES*, Volume 3, p. 2183, Pergamon Press, Oxford (1990)

88 Lampert, C. M. and Granqvist, C. G. (eds), *Large-area Chromogenics: Materials and Devices for Transmittance Control*, SPIE Opt. Engr. Press, Bellingham (1989)

89 Granqvist, C. G., 'Solar energy materials: the role of research and development', *Energy and the Environment into the 1990s*, Volume 3, p. 1465, Pergamon Press, Oxford (1990)

90 Jesch, L. F., 'Using transparent insulation in solar energy applications', *Int. J. Ambient Energy*, **9**, 4, 203 (1988)

91 Wittwer, V., 'Transparent insulation materials', *Energy and the Environment into the 1990s*, Volume 3, p. 1344, Pergamon Press, Oxford (1990)

92 Hollands, K. G. T., 'Honeycomb devices in flat-plate collectors', *Solar Energy*, **9**, 3, 159 (1965)

93 Jesch, L. F., 'Conclusions', *Proc. Int. Workshop on Transparent Insulation Materials for Passive Solar Energy Utilisation*, p. 52, Freiburg (1986)

94 Wittwer, V. *et al.*, 'Translucent insulation materials', *Proc. 1985 Congress ISES*, Volume 2, p. 1333, Pergamon Press, Oxford (1986)

95 Platzer W. J. and Wittwer, V., 'Total energy transmission of transparent insulation material', *Proc. Workshop on Optical Measurement Techniques*, Ispra, Italy (1987)

96 Shedding a new light on hazardous waste', Solar Energy Research Institute, Golden, Colorado, February (1991)

97 Beard, J. T. and Ebadian M. A. (eds), *Solar Engineering – 1990*, Proc. 12th ASME Conf., ASME, New York (1990)

98 Moore, H., Plenary Session Address, ISES World Congress, Denver, Colorado, August (1991)

99 Telkes, M., 'Solar energy storage', *ASHRAE Journal*, 38, September (1974)

100 Brandstetter, A. and Kaneff, S., 'Materials and systems for phase change thermal storage', *Energy and the Environment into the 1990s*, Volume 3, p. 1460, Pergamon Press, Oxford (1990)

101 Pitts, J. R., Stanley, J. T. and Fields, C. L., 'Solar induced surface transformation of materials', in Gupta, B. P. and Traugote, W. H. (eds), *Solar Thermal Technology – Research, Development and Applications*, pp. 459–470, Hemisphere, New York (1990)

102 Kleinwachter, J. *et al.*, 'Solar power station with thermochemical storage', *Ibid.*, pp. 545–555

103 *BP Statistical Review of World Energy*, The British Petroleum Company plc, London, June (1991)

104 Foley, G., *The Energy Question*, 2nd edn, Penguin, Harmondsworth (1981)

105 *Electricity Supply in Great Britain*, The Electricity Council, London (1973)

106 *A History of Technology*, V, pp. 528–533, Oxford University Press, Oxford (1962)

107 Deudney, D., *Rivers of Energy: The Hydropower Potential*, Worldwatch Paper 44, Worldwatch Institute, Washington (1981)

108 *Power Resources of the World* (1929)

109 *Survey of Energy Resources*, World Energy Conference, London (1980)

110 Levy, D., 'Optimising electric power from a small-scale hydro-electric plant', *Journal of the Institute of Energy*, 109–118, September (1990)

111 Report of the Technical Panel on Hydropower, *United Nations Conference on New and Renewable Sources of Energy*, Nairobi (1981)

112 Mandle, K. T., 'Dinorwig pumped-storage system', *Power Engineering Journal*, 259–262, September (1988)

113 Bazaga, A., 'Small Hydropower', Euroforum New Energies, Volume 2, pp. 97–102, Commission of the European Communities, published by H. S. Stephens and Associates, Bedford (1988)

114 Hurst, C. and Barnett, A., *The Energy Dimension*, Intermediate Technology Publications, London (1990)

115 *Report of the Technical Panel on Wind Energy (second session)*, United Nations, Conference on New and Renewable Sources of Energy, Nairobi (1981)

116 Juul, J., 'Wind machines', *Wind and Solar Energy Conference*, New Delhi, UNESCO (1956)

117 Golding, E. W. and Stodhart, A. H., *The use of wind power in Denmark*, ERA Technical Report C/T 112 (1954)

118 Putnam, P. C., *Power from the Wind*, Van Nostrand, New York (1948); 2nd edn with G. W. Koeppl (1982)

119 *Assessment of Research Needs for Wind Turbine Rotor Materials Technology*, US National Academy of Sciences, National Research Council, Department of Energy, National Academy Press, Washington, DC (1990)

120 Golding, E. W., *The Generation of Electricity by Wind Power*, E. & F. Spon (1955); reprinted CTT (1976)

121 Golding, E. W. and Stodhart, A. H., *The potentialities of windpower for electricity generation*, British Electrical and Allied Industries Research Association, Tech. Rep. W/T16 (1949)

122 Taylor, R. H., *Alternative Energy Sources*, Adam Hilger, Bristol (1983)

123 Davenport, A. G., *Proceedings of the (1963) Conference on Wind Effects on Building and Structure Vol. 1*, HMSO (1965)

124 Caton, P. G., 'Standardised maps of hourly mean wind speed over the United Kingdom and some implications regarding wind speed profiles', *Fourth International Conference on Wind Effects on Building and Structures*, London (1975)

125 Twidell, J. W. and Weir, A. D., *Renewable Energy Resources*, E. & F. N. Spon, London (1986)

126 McVeigh, J. C., 'When nuclear power is not the answer', *Electrical Review International*, **1**, 1, February (1984)

127 Lipman, N. H. 'Overview of wind/diesel systems', in *Energy and the Environment into the 1990s*, Volume 3, pp. 1547–1569, Pergamon Press, Oxford (1990)

128 *Windirections*, **5**, No. 2, October (1985)

129 *National Engineering Laboratory Newsletter*, Issue 14, May (1986)

130 Darrieus, G. J. M., 'Turbine having its rotating shaft transverse to the flow of the current', *US Patent 1,835,018*, 8 December 1931

131 Klemin, A., 'The Savonius wing rotor', *Mechanical Engineering*, **47**, No. 11, November (1925)

132 South, P. and Rangi, R. S., *A wind-tunnel investigation of a 14 ft diameter vertical-axis windmill*, National Research Council of Canada, LTR-LA-105, September (1972)

133 South, P. and Rangi, R. S., 'The performance and economics of the vertical-axis wind turbine developed at the National Research Council, Ottawa, Canada', *Agricultural Engineer*, February (1974)

134 Musgrove, P. J. and Mays, I. D., 'The variable geometry vertical axis windmill', *Proc. 2nd Int. Symposium on wind energy systems*, BHRA Fluid Engineering, Cranfield, E4, 39–60 (1978)

135 VAWT Ltd, *Testing and monitoring of a 25m diameter vertical axis wind turbine*, ETSU Report WN 5025, Harwell, Oxfordshire (1990)

136 VAWT Ltd, Eaton Court, Maylands Avenue, Hemel Hempstead, Herts HP2 7DR

137 Hinrichsen, D. and Cawood, P., 'Fresh breeze for Denmark's windmills', *New Scientist*, 567–570, 10 June (1976)

138 Friss, P., 'Tjaereborgmollen at Esbjerg. The Danish 2MW wind turbine', *Euroforum New Energies*, Volume 3, pp. 688–690, H. S. Stephens and Associates, Bedford (1988)

139 Lindley, D., 'The commercialisation of wind energy', *op. cit.*, Volume 2, pp. 47–70.

140 Reed, J. W., Maydew, R. C. and Blackwell, B. F., Wind energy potential in New Mexico, *SAND-74-0077, Sandia Laboratories Energy Report*, July (1974)

141 Bedford, L. A. W. and Tolland, H. G., 'Wind energy and the environment', in *Energy and our Future Environment*, Institute of Energy, London (1983)

142 *World Solar Markets*, Financial Times Business Information Ltd, London (1983 and 1984)

143 Lindley, D. and Stevenson, W., 'The horizontal axis wind turbine project on Orkney', *Proc. Cranfield, Third BWEA Wind Energy Conference*, 16–23 (1981)

144 Bedford, L., 'The Richborough story', *Review 11*, 8–10 (1990)

145 Edwards, P., 'Viewpoint', *Review 16*, 12–13 (1991)

146 'The new ETSU report', *Renew, NATTA Newsletter 74*, 24 (1991)

147 *Wind Energy in Europe*, European Wind Energy Association, Via Bormida 2, 1-00198 Rome, Italy (1991)

148 Holland, M. B., 'Power from the Earth', *Chartered Mechanical Engineer*, 40–45, November (1978)

149 Dunham, K., 'Geothermal energy and heat pumps', Keynote Address, *Int. Conf. Future Energy Concepts*, Institution of Electrical Engineers, London (1979)

150 Shaw, J. R. and Robinson, P. E., 'Prospects for the future exploitation of geothermal energy', *Int. Conf. Future Energy Concepts*, Institution of Electrical Engineers, London (1981)

151 Pearson, C. M., 'Exploitation of hot dry rock geothermal energy in the United Kingdom', *Int. J. Ambient Energy*, **3**, 1, 19–26 (1982)

152 Hubbert, M. K., 'The energy resources of the Earth', *Scientific American*, **225**, 61–70, September (1971)

153 Leardine, T., 'Geothermal power', *Phil. Trans. R. Soc. Lond. A*, **276**, 507–526 (1974)

154 Armstead, H. C. H., *Geothermal Energy*, 2nd edn, E. and F. N. Spon Ltd, London (1983)

155 'World energy resources: 1985–2020', *World Energy Conference*, London (1980)

156 Lengyl, L. and Bohoczky, F., 'Geothermal energy resources in Hungary', *International Journal of Ambient Energy*, **12**, No. 3, 127–130 (1991)

157 *Report of the Technical Panel on Geothermal Energy (second session)*, United Nations Conference on New and Renewable Sources of Energy, Nairobi (1981)

158 'The Southampton story', *Review 15*, 8–9 (1991)

159 Garnish, J. D., 'Geothermal energy and the UK environment', in *Energy and our Future Environment*, Institute of Energy, London (1983)

160 Dickson, M. H. and Fanelli, M., 'Geothermal Energy and its utilization', in *Small Geothermal Resources*, published by UNITAR/UNDP Centre on Small Energy Resources, Rome (1990)

161 Wright, M., 'The Geothermal Programme – where does it stand?', *Review 13*, 8–9 (1990)

162 Parker, R. H., 'Progress with the development of hot dry rock geothermal energy at Camborne School of Mines', *Energy and the Environment into the 1990s*, Volume 5, pp. 2963–2968, Pergamon Press, Oxford (1990)

163 Garnish, J. D., 'Prospects for geothermal energy', Keynote Address, *Int. Conf. Future Energy Concepts*, Institution of Electrical Engineers, London (1981)

164 Holland, M. B., 'Power from the tides', *Chartered Mechanical Engineer*, 33–39, July (1978)

165 *Tidal power barrages in the Severn Estuary*, Energy Paper Number 23, Department of Energy, HMSO, London (1977)

166 Charlier, R. H., 'Tidal power plants: sites, history and geographical distribution', *Proc. 1st Int. Symp. on Wave and Tidal Energy*, BHRA, Cranfield (1979)

167 Wyman, P. R. and Peachey, C. J., 'Tidal current energy conversion', *Int. Conf. Future Energy Concepts*, 164–169, Institution of Electrical Engineers, London (1979)

168 'Developments in tidal energy', *Proceedings of the Third Conference on Tidal Power*, Institution of Civil Engineers, London (1990)

169 *Severn Barrage Committee Report, 1933*, quoted in *The Exploitation of Tidal Power in the Severn Estuary*, Fourth Report from the Select Committee on Science and Technology, HMSO, London (1977)

170 Tanner, R., Murphy, D. and Warnock, J. G., 'Influence of technological advances on potential tidal power developments', *Int. Conf. Future Energy Concepts*, 130–142, Institution of Electrical Engineers, London (1979)

171 *Materials and Energy Resources*, The Institution of Chemical Engineers, London (1976)

172 *Severn Barrage Study*, Ex 753, Hydraulics Research Station, Wallingford, November (1976)

173 *Pre-feasibility study on the closure of the Estuary*, Report for the UK Department of Energy, Netherlands Engineering Consultants, March (1977)

174 *Tidal power from the Severn Estuary*, Energy Paper Number 46 (two volumes), Department of Energy, HMSO, London (1981)

175 *Feasibility of Tidal Power Development in the Bay of Fundy*, Atlantic Tidal Power Programming Board, Ottawa (1969)

176 *Reassessment of Fundy Tidal Power*, Bay of Fundy Tidal Power Review Board, Ottawa (1977)

177 Hubbert, M. K., 'Energy Resources', in *Resources and Man*, National Academy of Sciences, W. H. Freeman and Company, San Francisco (1969)

178 Ruxton, T. D., 'Tidal power from the Severn Estuary: an initial environmental impact assessment', in *Energy and our Future Environment*, Institute of Energy, London (1983)

179 Dr John Chesshire, Science Policy Research Unit, Sussex University, Falmer, Sussex, UK

180 Ross, D., *Energy from the Waves*, 2nd edn, Pergamon Press, Oxford (1982)

181 Holland, M. B., 'Power from the waves', *Chartered Mechanical Engineer*, 41–44, September (1978)

182 Masuda, Y., Study of wave activated generator and future view as island power source', *Second Int. Ocean Development Conference*, 2074–2090 (1973)

183 Denton, J. D. *et al.*, 'The potential of natural energy

resources', *CEGB Research*, **2**, 36–39, May (1975)

184 Mollinon, D., Buneman, O.P. and Salter, S. H., 'Wave power availability in the NE Atlantic, *Nature*, **263**, 5574, 223–226, 16 September (1976)

185 Andrews, S. A. and Platts, M. J., 'Wave energy and the environment', in *Energy and our Future Environment*, Institute of Energy, London (1983)

186 Milne-Thompson, L. M., *Theoretical Hydrodynamics*, 4th edn, Macmillan, London (1962) or in Coulson, C. A., *Waves*, Oliver and Boyd, Edinburgh and London (1955)

187 Salter, S. H., 'Wave power', *Nature*, **249**, 5459, 720–724, 21 June (1974)

188 Glendenning, I., The potential of wave power', *Applied Energy*, **3**, 197–222 (1977)

189 Pierson, W. K. and Moskowitz, L., 'A proposed spectral form for fully developed wind seas', *J. Geophysical Res.*, **69**, 24 (1964)

190 Count, B. M. and Robinson, A. C., *On the estimation of power from limited data*, CEGB Report R/M/N 895, London (1976)

191 Count, B. M., Fry, R. and Haskell, J. H., 'Wave power: the story so far', *CEGB Research*, **15**, 13–24, November (1983)

192 Davies, P. G. (eds), 'Wave Energy – The Department of Energy's R & D Programme 1974–1983', ETSU Report R26, Hanwell, Oxfordshire, March (1985)

193 'Private wave power', *Renew, NATTA Newsletter*, **74**, 10 (1991)

194 Glendenning, I. and Count, B. M., 'Wave power', *Proc. Symp. Renewable Sources of Energy*, 50–81, Royal Society of Arts, London, June (1976)

195 Bott, A. N. W., 'Electro/mechanical aspects of the Mauritius "passive" type wave energy project', *Int. Conf. Future Energy Concepts*, 81–87, Institution of Electrical Engineers, London (1979)

196 Platts, M. J., 'The development of the wave contouring raft', *Ibid.*, 160–166

197 Bellamy, N. W., 'Wave power experiments at Loch Ness', *Ibid.*, 167–176

198 McIlhagger, D. S. and Long, A. E., 'Harnessing energy from the waves', *Queen's University Association Annual Review*, 48–52 (1979)

199 *Wave Energy Review; Interim report*, ETSU Report R60 for the Department of Energy, Harwell, Oxfordshire, October (1991)

200 Salter, S. H., 'World progress in wave energy – 1988', *International Journal of Ambient Energy*, **10**, No. 1, 3–24 (1989)

201 'Wave energy: shore-based devices', *Review 7*, 20 (1989)

202 Hall, D. O., 'Solar energy use through biology – past, present and future', *Solar Energy*, **22**, 307–329 (1979)

203 *Energy in the Developing Countries*, World Bank, Washington, DC (1980)

204 Brandt, Willy (Chairman), *North–South: A programme for survival*, The Report of the Independent Commission on International Development Issues, Pan Books, London and Sydney (1980)

205 Makhijani, A. and Poole, A., *Energy and Agriculture in the Third World*, Ballinger, Cambridge, Massachusetts (1975)

206 Jesch, L. F., *Solar Energy Today*, UK Section, International Solar Energy Society, London (1981)

107 Ward, R. F., 'Alcohols as fuels – the global picture', *Solar Energy*, **26**, 169–173 (1981)

208 Hancock, G., 'Premiers to discuss use of renewable energy', *The Guardian*, 10 August (1981)

209 Harrison, P., *Inside the Third World – the Anatomy of Poverty*, 2nd edn, Penguin, Harmondsworth (1981)

210 Jackson, D. V. and Tron, A. R., 'Energy from wastes', *International Journal of Ambient Energy*, **6**, No. 1, 31–44 (1985)

211 Loram, R. G., 'Energy from waste', *Energy World*, **191**, 12–14, September (1991)

212 Thring, M. W., 'Steam generation using renewable energy', *International Journal of Ambient Energy*, **4**, No. 3, 153–160 (1983)

213 Martindale, L. P., 'Straw as fuel and the Department of Energy's R & D programme', *Energy Management Focus*, No. 3, 17 (1985)

214 'How fuel gas is produced from refuse tips', *Review 1*, 20 (1987)

215 *Landfill Gas Trends*, Energy Technology Support Unit, Harwell, Issue 3, p. 1, October (1991)

216 McVeigh, J. C., Personal discussions in Hangzhou, People's Republic of China, March 1983.

217 *RERIC News*, Renewable Energy Resources Information Centre, PO Box 2754, Bangkok, Thailand, **4**, 3, December (1981)

218 Keable, J. and Dodson, C. A., 'Modular system for biogas production using farm waste', *Sun II*, 1, 83–87, Pergamon Press, Oxford (1979)

219 Chesshire, M., 'Anaerobic digestion of farm wastes', in *Solar Energy in Agriculture*, Conf. C.33, UK-ISES, London (1983)

220 McAngus, J., 'Reducing the goods', *Review 16*, 19–20 (1991)

221 Szego, G. C. and Kemp, G. C., 'Energy forests and fuel plantations', *Chemtech*, 275–284, May (1973)

222 Neenon, M., Lyons, G. and O'Brien, T. C., 'Short rotation forestry as a source of energy', *Solar World Forum*, Volume 2, pp. 1258–1262, Pergamon Press, Oxford (1982)

223 'Making farm woods pay', *Review 15*, 20–21 (1991)

224 *Centro de Tecnologia Promon Newsletter*, **3**, 1, Rio de Janeiro, February (1978)

225 Wolf, M., *Utilization of solar energy by bioconversion – an overview*, Testimony before the US House of Representatives Science and Astronautics Committee, 13 June 1974

226 White, L. P., and Plaskett, L. G., *Biomass as Fuel*, Academic Press, London (1981)

227 *Making Aquatic Weeds Useful: Some perspectives for developing countries*, National Academy of Sciences, Washington, DC (1976)

228 Hall, D. O. and Coombs, J., 'The prospect of a biological–photochemical approach for the utilization of solar energy, in *Energy from the Biomass*, pp. 2–14, The Watt Committee on Energy, Report No. 5, London (1979)

13

Nuclear engineering

Jeffery D. Lewins

Contents

13.1 Introduction 13/3

13.2 Nuclear radiations and energy 13/3

13.3 Mechanical engineering aspects of nuclear power
stations and associated plant 13/4
13.3.1 Fuel provision 13/4
13.3.2 Reactor types 13/5
13.3.3 The coolant challenge 13/5
13.3.4 Post-reactor fuel processes 13/8
13.3.5 Quality assurance and control 13/9

13.4 Other applications of nuclear radiation 13/9
13.4.1 Gauging and detection 13/9
13.4.2 Compact power sources 13/9
13.4.3 Materials and sterilization effects 13/9
13.4.4 Medical applications 13/9
13.4.5 Dating and tracing 13/10

13.5 Elements of health physics and shielding 13/10
13.5.1 Biological effects 13/10
13.5.2 Radiation limits 13/11
13.5.3 Elementary shielding 13/11

Further reading 13/12

13.1 Introduction

Advances in physics that occurred towards the end of the nineteenth century in understanding the make-up of atoms and their nuclei led in the twentieth century to engineering applications of nuclear energy. The two major applications have been to nuclear weapons and to the production of electricity. There are a number of peripheral uses of the energy released in nuclear reactions.

It is the task of the nuclear engineer to exploit the energy of the nucleus to the benefit of humankind. Nuclear engineering shares with other branches of engineering, metallurgy, chemistry, etc. the history of having been bred in war and only subsequently applied in peace. It is perhaps the most stark in its contrasts of the benefit it might bring to humankind and the dangers of misusing the power our command over the nucleus of matter can give.

The energy coming from nuclear reactions may be specific in terms of the major radiations (alpha-particles, beta-particles and gamma-rays and neutrons in particular) or may be exploited for the general energy associated with them as an intense source of heat. Table 13.1 lists some of the major facts about the energy of nuclear radiations. It is the intensity of the nuclear energy source that is the underlying phenomenon for both nuclear weapons and power stations.

Nuclear engineering calls for many skills and, arguably, is not a basic engineering discipline. It certainly has a major element of mechanical engineering involved in the economic and safe development of nuclear power. But it is a discipline that must synthesize abilities in physics and chemistry, mechanical and electrical engineering, metallurgy, process control, economics, biology and health physics and a broad view of the social responsibilities of engineers. The present survey seeks to cover two points only: an illustration of the role of the mechanical engineer in major aspects of the use of nuclear power for the production of electricity in land-based plants; and to summarize some of the restrictions and regulations involved in using ionizing radiations in the workplace.

It is necessary first to survey briefly the nature of the energy arising from nuclear reactions and the radiations produced. We also list some of the other applications involving mechanical engineers in nuclear engineering.

13.2 Nuclear radiations and energy

A number of naturally occurring elements are radio-active, particularly the higher mass elements immediately starting with uranium and thorium. The transformation of the nucleus is accompanied by the emission of energy (Table 13.2).

Table 13.2 Nuclear radiation examples

Natural chain, alpha-emission

$^{238}U \rightarrow {}^{234}Th + He (\alpha) + \gamma$

Beta decay – cobalt source

$^{60}Co \rightarrow {}^{60}Ni + \beta + \gamma + neutrino$

Fusion

$D + T = {}^{2}H + {}^{3}H \rightarrow {}^{4}He + {}^{1}n + Q: 17.6 \text{ MeV}$

Fission

$^{235}U + {}^{1}n \rightarrow {}^{236}U^*$ (typically)

$\rightarrow {}^{147}La + {}^{87}Br + 2 {}^{1}n + Q: \text{approx. } 200 \text{ MeV}$

Fission fragments

$^{135}I \xrightarrow{\beta} {}^{135}Xe \xrightarrow{\beta} {}^{135}Cs$ long-lived: half-lives (h)

6.7   9.2

Common forms for this energy to take are:

- Alpha-particles (the nucleus of helium) – recoil energy;
- Beta-particles (an electron; more rarely its complement, the positron) – recoil energy;
- Gamma-radiation; in nature identical to X-rays and characterized by a wave-frequency value proportional to the energy of the gamma photon.

Three chains exist in nature for the decay of the heavy elements to lead; the fourth has already decayed. One or two other naturally occurring radioactive elements include the potassium-40 isotope (present in fertilizers) and carbon-14 produced in the atmosphere by interaction with cosmic or solar radiation. Each radioactive species is characterized by its half-life (i.e. the mean time for half the nuclei to decay). For these chains, the shorter half-lives go with increasing kinetic energy of the radiation.

Neutrons, which with protons (the nucleus of hydrogen) form the building blocks or nucleons of the nuclei, can be obtained by stripping the proton out of deuterium, the naturally occurring heavy isotope of hydrogen, in a charged particle accelerator. However, neutrons are formed in more copious quantities in fusion and fission reactions.

In fusion, light elements have their nuclei brought together to fuse into a heavier nucleus that is more stable, thus releasing energy. In particular, the deuterium–deuterium fusion can proceed either to a nucleus of helium-3 and a neutron or to a nucleus of hydrogen-3 (tritium) and a proton.

Table 13.1

Radiation	Symbol	Rest mass (u)	Charge(e)	Ionizing	LET(a)
Alpha	He	4.003	+2	Direct	High
Beta	β	1/1873	−1	Direct	Low
(Positron)	β^+	1/1873	+1	Direct	Low
Gamma (X-ray)	γ	Zero	0	Direct	Low
Neutron	n	1.000	0	Indirect	High

u: one unified atomic mass unit, 1.66057×10^{-27} kg   a: linear energy transfer
e: one atomic charge, 1.60219×10^{-19} C

Such a process occurs in the stars and is being investigated (so-called cold fusion) in the laboratory. The competing fusion reaction being studied in magnetic and inertial confinement systems is the more favourable deuterium–tritium (d–t) reaction. Such a reaction yields 17 MeV (megaelectron-volts) of energy. The energy taken off by the neutron is 14 MeV, which therefore forms the principal source of energy to be exploited in making electricity, a formidable engineering task that has not yet been demonstrated. The favoured d–t route to fusion requires tritium. This can be bred locally from lithium and the excess neutrons leaving the fusion plasma. Tritium, a further isotope of hydrogen, is notoriously difficult to contain and is radioactive with a half-life of some 12 years. It therefore forms a hazard in such fusion reactors.

Fission, discovered in late 1938 and exploited in reactors and weapons in World War II, is the process of disrupting a heavy element, typically uranium-235 or plutonium-239, with a neutron to give neutron-rich fission products. In the process two or three neutrons are released making a chain reaction feasible. The energy release in fission is about 200 MeV of which 160 MeV are found in the kinetic energy of the fission products. These can be expected to be slowed down in the solid fuel which thus becomes hot. Coolants are essential and the hot coolant in turn can be made to raise steam to produce electricity. The fragments are unstable and decay largely by intense beta emission, thus heating the fuel even after the neutron reaction is shut down. This decay or 'after-heat' is a major safety implication of fission reactors calling for highly reliable design to ensure safety during an accident.

The excess neutrons that are not absorbed in a further fission will be captured by other elements in the fuel or structure or leak into the surrounding shield. This material, in turn, has radioactivity induced in it which may last, depending on the half-lives of the isotopes produced, for hundreds of years, a major factor in decommissioning fusion or fission reactors.

Capture in uranium-238 (the common isotope) leads to the production of plutonium-239. The former is called a 'fertile' material and is converted to the latter, a 'fissile' material, i.e. one that is readily fissioned by a slow or thermal neutron. (Essentially, any heavy element would fission if struck by a sufficiently energetic or fast neutron but there might then be no energy gain.) Breeder reactors (FBR) exploit this conversion of fertile to fissile material particularly using neutrons around 1 MeV to cause fission at an energy that produces about three neutrons per fission.

13.3 Mechanical engineering aspects of nuclear power stations and associated plant

In this section the broad outline of the nuclear electricity industry is described with reference to mechanical engineering aspects that embrace: pressure vessel design, heat transfer and fluid flow, remote manipulators and robotics, automatic control, etc.

13.3.1 Fuel provision

The only significant fuel currently used is natural uranium. Uranium is widely distributed throughout the world (and in sea water) and economic supplies are found particularly in Canada, the USA, Southern Africa, the former USSR and Australia. That these areas are in part distinct from oil-producing regions is a strategic attraction. Commercial ores are exploited to around 1% uranium so that mining, often

open-cast, is accompanied by an on-site concentration process.

Thorium, more widely distributed than uranium, is a potential fertile fuel since it can, in a reactor, breed the fissile uranium-233 isotope. It is not yet exploited commercially.

Natural uranium undergoes considerable further processes from the imported yellow cake (oxide) before final fabrication into fuel elements. Natural uranium has little risk of leading to a spontaneous critical reaction (except in some liquid processes) but has a natural radioactivity that may be an industrial hazard, particularly in the mining stage, where it is associated with radon gas as well as other decay daughters. In general, however, the metal needs more protection from the sweat of its handlers than vice versa. Impurities in the fuel and in structural or moderating materials used in reactors must be kept to exceptionally low levels to avoid competition for neutrons in the chain reactions.

Most commercial reactors now use slightly enriched uranium where the natural concentration of uranium-235 (around 0.7%) is raised to between 2.5% and 3%. Such enrichment is needed to offset the neutron capture in fuel clad with stainless steel or zirconium and to provide for long fuel life before the fissile material is burnt up.

The technology of uranium enrichment is well advanced. The major route in World War II and immediately after was through gaseous diffusion (more properly, gaseous fusion) with preferential separation through membranes of the low-pressure gas uranium hexafluoride or 'hex'. Major mechanical problems with these plants are the large runs of pipes at low pressures, the corrosive nature of fluorine compounds and the design of pumps. The high pumping costs of this process fostered the development of the rival ultra-centrifuge process for enrichment in which hex is separated by a thermally enhanced centrifuge. Major mechanical problems here are the reliability of the centrifuges, operating at a rim speed of some 300 m/s in vast cascade halls where the mechanical failure of one unit risks knock-on damage to the remainder. Gas and magnetic bearings have been developed to meet this challenge (Figures 13.1 and 13.2).

The UK's enrichment takes place at Capenhurst, where the centrifuge process has superceded the diffusion process since 1982. These plants are operated in conjunction with plants in Germany and Holland by URENCO in which British Nuclear Fuels plc have a third interest.

The enrichment plants, of necessity, produce depleted uranium. There is some prospect of this being used in fast breeder reactors and thus, like thorium, offering a greatly increased availability of fission fuel. Large stocks of depleted uranium are available from store in the UK. Its exploitation rests on rising fuel prices until the extra capital expense of the fast breeder reactors is offset by their cheap fuel. At least the availability of depleted uranium and the demonstrated ability of fast breeders to produce electricity puts a ceiling on competing fuels.

Enrichment also plays a part in preparing supplies of other items for the nuclear industry, particularly heavy water, and the isotopes of boron and lithium.

Plutonium bred in a nuclear reactor by the capture of neutrons in uranium-238 is an artificial source of nuclear fuel available from thermal and fast reactors. The immediate product of such breeding is the fissile isotope plutonium-239. It should be known that, like highly enriched uranium-235, this isotope of plutonium is a potential weapons material. Fuel removed from a reactor after a short exposure contains principally this weapons-grade plutonium. However, fuel left longer – with economic advantage in peaceful use – allows further neutron capture to the successive 240, 241 and perhaps 242 isotopes of plutonium. Plutonium-241 is fissile but the

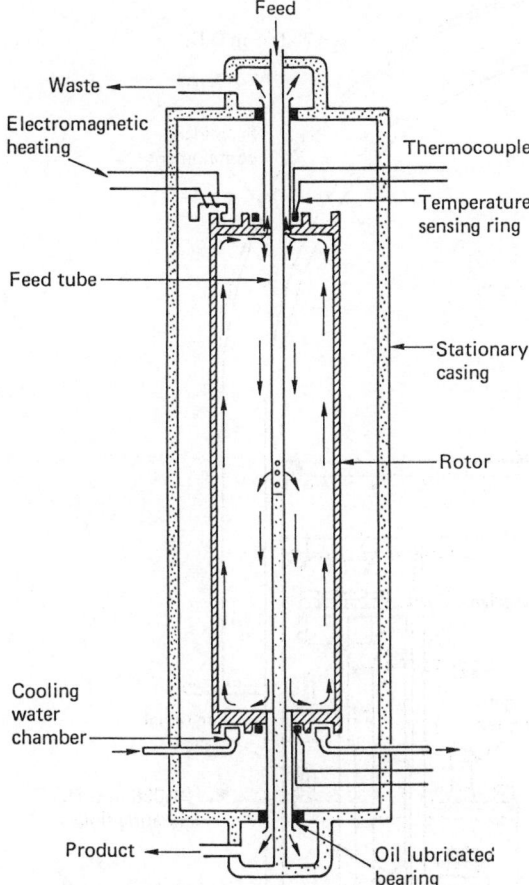

Figure 13.1 Centrifuge diagram

Figure 13.2 Centrifuge hall (courtesy BNFL and URENCO)

intermediate plutonium-240 is not. Thus such fuel is far less suitable for weapons and is known as commercial fuel.

Plutonium is separated in the reprocessing stages. If it is to be incorporated into fuel elements, either for the core of fast reactors or as a substitute for uranium-235 in thermal reactors, account must be taken of the greater radio-toxicity (some ten times more than uranium) and the risks of criticality accidents in the concentrated fissile plutonium. It is essential therefore for production to be undertaken in a controlled atmosphere, usually at lower than ambient pressures inside glove boxes to direct the leakage of air safely. The major risk is of inhalation with deposit in the lungs or of entry to the bloodstream via cuts. Ingestion is appreciably less risky as plutonium is not easily absorbed by the gut.

The fissile material is used in the dioxide form in most reactors since such ceramics can operate at considerably higher temperatures than the metal. Final preparation of fuel, usually after enrichment in uranium-235 or as a mixed oxide with plutonium (MOX), involves fabrication and cladding. Cladding is generally in a zirconium alloy (or in stainless steel for higher operating temperatures) with the major purpose of retaining fission products whose radioactivity would otherwise contaminate the plant. Extremely high standards of manufacturing integrity are required of fuel elements so that quality control and assurance, including non-destructive testing and the documentation of individual fuel elements, is essential.

Major mechanical implications of the special nature of producing nuclear fuel include therefore: the radiation and criticality hazard, the reliability of high-speed centrifuges, the leaktightness of many miles of vacuum pressure piping in enrichment plant, corrosion of hexafluoride material, the use of glove boxes and the control of quality in the final fuel element assembly.

13.3.2 Reactor types

Reactors may be classified in several ways, first by the predominant energy of the neutrons causing fission. Fast reactors produce neutrons that directly induce fission. They generally require highly enriched uranium or plutonium cores. Alternatively, in thermal reactors, the fast neutrons produced in fission are slowed down or moderated to thermal energies. The better yield of neutrons in fission allows natural or only slight enrichment of fuel.

Hydrogen is an effective moderating material and finds a particular role in light (ordinary) water-cooled reactors (LWR) which may be pressurized (PWR) or boiling (BWR). Alternatively, heavy water (deuterium oxide) is used. The other major category of thermal reactor has gas cooling and graphite moderation, as in the UK Magnox and Advanced Gas-Cooled Reactors (AGR).

Thus a second categorization of reactors is in their coolant and/or moderator: Light Water Reactors (LWR), Heavy Water Reactors, Gas-Cooled Reactors and Metal-Cooled Reactors where sodium or sodium potassium alloy (NaK) are used for good heat transfer in fast reactors in which moderation is not wanted.

13.3.3 The coolant challenge

Reactors are operated at high energy density to make them economic despite high capital costs. The energy density in a PWR is comparable to the boiler of an oil-fired station; in a FBR it is comparable to a chemical rocket. It is essential that coolant capacity is maintained throughout operation. In this respect, there is competition between a gaseous coolant with poorer heat transfer capability and a liquid coolant with greater heat transfer but the risk of two-phase loss of heat

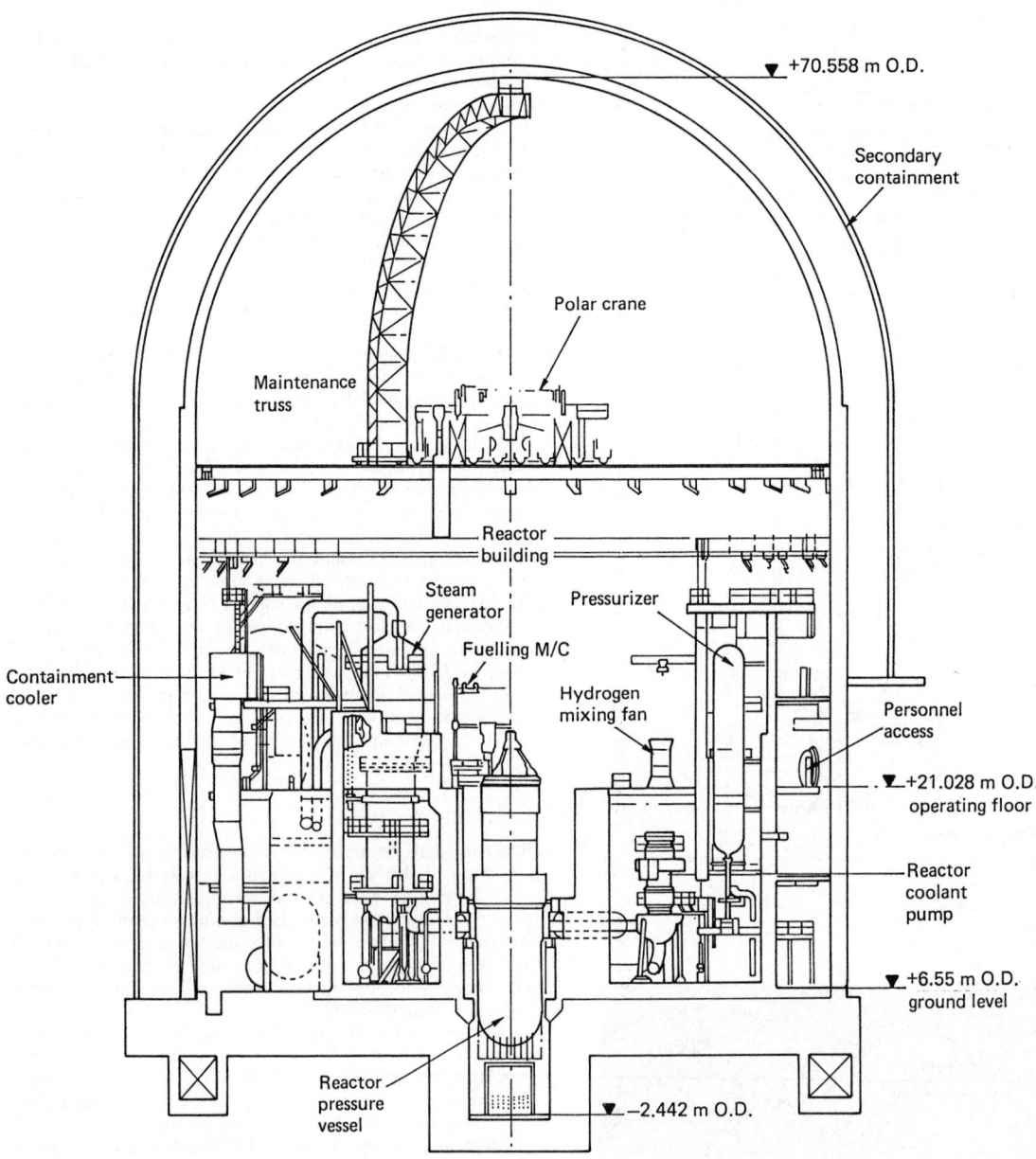

▼ +70.558 m O.D.

Secondary
containment

Polar crane

Maintenance
truss

Reactor
building

Steam
generator

Pressurizer

Fuelling M/C

Containment
cooler

Hydrogen
mixing fan

Personnel
access

▼ +21.028 m O.D.
operating floor

Reactor
coolant
pump

▼ +6.55 m O.D.
ground level

Reactor
pressure
vessel

▼ −2.442 m O.D.

Figure 13.3 Cross-section of PWR building

transfer on depressurization. Nuclear power has made considerable demands on the prediction of heat transfer in one- and two-phase flow.

Coolant capacity must be provided not only during normal operation but also on shutdown, particularly in accident conditions. Whereas the neutron chain reaction can be terminated rapidly with neutron-absorbing control rods, etc., the after-heat from the decay of previously formed fission products cannot be turned off. Roughly, this after-heat is at 10% of previous steady operating power for a few minutes from shutdown, decaying slowly until, say, 1% after a week, 0.1% after a month. Thus a reactor that had been designed for operation at 3000 MW (thermal) must be provided with a reliable shutdown and emergency cooling circuit(s) to cope with 300 MW after-heat.

PWRs operate with water as coolant-moderator at some 170 bar and with a core of some 3 m diameter. There is therefore a considerable challenge in providing a reliable pressure vessel. The challenge is eased in systems having multiple pressure tubes, one for each fuel bundle and therefore of smaller diameter (Figures 13.3 and 13.4).

A major failure of the vessel of a PWR would be catastrophic. It would imply no availability of coolant water and the overheating and probably meltdown of the fuel with the release of its residual radioactivity. Suffice it to say that 29 of the initial 31 casualties at Chernobyl (1986) died from radia-

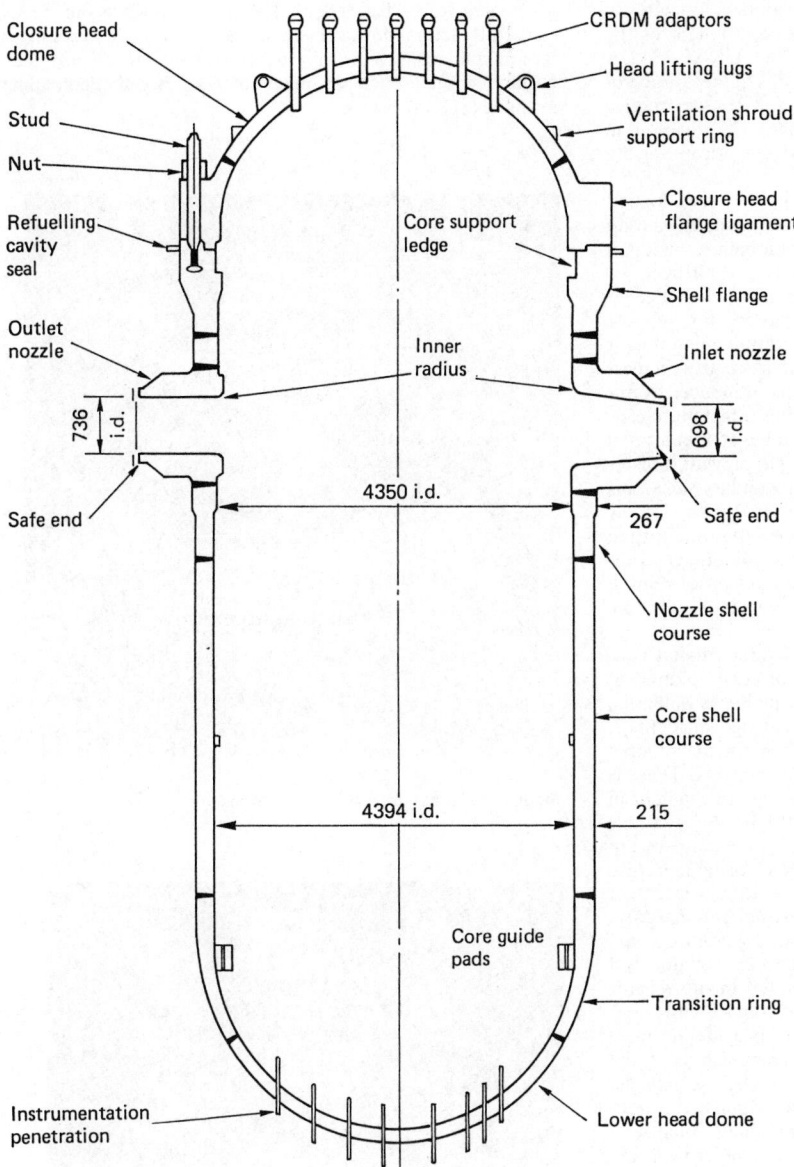

Figure 13.4 PWR pressure vessel

tion burns from the exposed fuel and fission fragments, following failures in a pressure-tube reactor.

The sodium coolant of a fast reactor has some disadvantages. The sodium is a fire hazard on contact with water in the steam-raising plant and sodium passed through the reactor core is activated by neutron capture. It is therefore seen to be essential to have an intermediate sodium loop between the primary core sodium and the boilers to prevent the reaction of active sodium and water. However, the metal coolant offers the advantages of good heat transfer at ambient pressures; no elaborate pressure envelope is called for and a pool design is possible.

Gas-cooled reactors have a pressure envelope between the high pressure of PWRs and the low pressure of FBRs. Carbon dioxide pressures were increased in successive designs, from around 10 bar in the early Magnox stations through 28 bar at Wylfa, the last Magnox, to around 35 bar for the AGRs.

The most challenging pressures are therefore found in PWRs, and here the loss of coolant capacity if the primary water were allowed a change of phase to steam might be catastrophic. It follows that pressure vessel design, fabrication and testing are a major goal in the PWR plant. It is inevitable that there will be some cracks in the forging and welding of a vessel using steel of more than 250 mm thickness. It is desired than any crack should allow leakage and hence detection before catastrophic (Griffiths) failure. Early comparisons under the Plate Inspection Steering Committee (PISC) cast doubt upon the guarantees of pressure vessel safety but non-destructive testing, chiefly ultrasonic, has been greatly developed under this challenge, including formal systems of validating testing methods and testers. Major developments in the UK in this area have included the establishment of the Inspection Validation Centre (IVC) at the UKAEA/AEA Technology site, Risley. This centre operates in the context of two competing systems of certification for non-destructive testing in the UK: the PCN or Personal Certification in Non-Destructive Testing and the ASNT or American Society for Non-Destructive Testing scheme.

Control of reactors is exercised using control rods loaded with neutron-absorbing materials such as hafnium, backed up with boron-loaded particles for use in emergencies and, in some cases, neutron-absorbing gases such as nitrogen or helium-3. There is a considerable mechanical challenge in designing the control rods and their motors for reliable operation in the fierce radiation and temperature environment of the reactor core. High-speed flow, particularly in gas-cooled reactors, has led to major problems of induced vibration resulting in fretting and cracking. Boiler tube failure can, in part, be met by overprovision of boiler tubes so that a small proportion can be lost from use by plugging to prevent transfer between the radioactive primary and the secondary sides; this also allows some margin for fouling.

However, where parts of the reactor core deteriorate or fail in operation, a major problem of repair in a radiation environment is faced. This has called for the development of remote sensing and manipulation tools able to penetrate into the reactor, often through tortuous paths.

Overall control is now the function of direct digital controllers which provide for the processing of great volumes of data needed for a historical verification of quality as well as an assessment of past accidents and near-accidents (incidents).

Special problems of fast reactors arise from the high energy density (for economy) and the liquid metal coolants. There is both a high temperature and a high-temperature gradient in Liquid-metal Fast Breeder Reactors (LMFBR). Fluctuations in the coolant flow lead to thermal striping or temperature cycles of around 1 Hz which imply problems of temperature creep. The fast neutron flux also leads to lattice displacement with problems of swelling and loss of structural integrity. Furthermore, there has been generally poor experience in the provision of sound heat exchangers between sodium and water. Since the primary sodium is activated in the reactor core and must not be allowed to come into contact with water in the steam generators, it is customary to provide an intermediate sodium–sodium loop. A further complication of the mechanical design has been, at least in practice so far, the provision of additional core supports since, unlike the well-moderated thermal reactor, it is conceivable that a collapse of the core of a fast reactor might lead to a supercritical and unacceptably dangerous condition. It follows that the capital costs of the FBR tend to be higher than for, say, PWR, and hence FBRs are unlikely to be economic until such times as a rise in the cost of natural uranium outweighs their greater capital cost.

13.3.4 Post-reactor fuel processes

Once exposed in a reactor, the fuel, which may have yielded 30–80 MWdays/kg, is highly radioactive and must be both cooled and shielded. Pressurized water reactors are shut down, flooded externally and the bolted vessel heads removed before fuel can be moved. (The design of the stud tensioners used in replacing a PWR vessel head is a major mechanical challenge; see Figure 13.5.) Fuel may be shuffled from one region of the reactor core to another as well as having a quarter or a third removed and replaced with fresh fuel. The exposed fuel is kept on-site for a period to allow immediate decay of fission products, either under water or in a dry,

gas-cooled store. The older UK Magnox and Advanced Gas Cooled Reactors, however, are designed on the philosophy of on-line fuel changing, with its economic advantages; they require a major commitment to fuel loading/unloading machines (Figure 13.6).

Figure 13.5 Pressure vessel stud tensioner

Figure 13.6 AGR refuelling machine at Heysham II Power Station (courtesy Strachan & Henshaw)

Whether the used fuel is reprocessed – to recover the unused uranium (for re-use after possible further enrichment), to obtain the plutonium bred in the reactor but not yet consumed *in situ* and to concentrate the fission products (about 1% of the used fuel mass) – is a decision that has been made on political as well as technical grounds. It is the practice in the UK and France to reprocess fuel. This has been done routinely for Magnox (metal-fuelled) reactors. It is proposed (but not yet practised in the UK) to reprocess the oxide fuel from AGRs and PWRs. In the USA, however, commercial-grade fuel is not reprocessed, in accordance with national policy associated with the non-proliferation of weapons.

Where reprocessing is undertaken, highly concentrated and radioactive fission products are obtained. Currently these are stored in double-lined stainless steel cans, perhaps for 50 years to allow the activity to decay. If the contents are then to be disposed of, i.e. without expectation of recovery, this must be done in a way that will prevent the escape of the residual active products to the biosphere over many thousands of years. A favoured route is for the fixing of the fission products in a glass followed by encapsulation in bitumen and metal containers for disposal in a deep vault on land or under the sea. There will be major mechanical engineering challenges when this stage is undertaken.

The reprocessing plant is largely a chemical engineering industry. It has major mechanical engineering aspects in: provision of reaction vessels, criticality control and shielding; the provision of heat transfer/coolants for used fuel elements and concentrated fission products; remote manipulation (Figure 13.7); and design of waste storage and disposal vaults.

13.3.5 Quality assurance and control

From the above it will be seen that quality is an essential element in nuclear engineering (as in much else). Its importance lies in two mutually supporting aspects: a necessity in the provision of safety and in the protection of commercial investment. The accident at Chernobyl in 1986 has led to massive direct costs associated with the contamination by radioactivity of the surrounding area. Yet the accident in 1978 at Three Mile Island in the USA, while leading to negligible activity release off-site, has in similar fashion enormous direct and indirect economic consequences. In both cases the cost of clean-up is measured in billions (gigapounds); the indirect costs have similar magnitudes, providing for substitute electricity or in the halting and perhaps denial of public acceptance of a nuclear power programme.

The emphasis on quality assurance and control is therefore understandable. The principles of quality assurance were indeed laid down in the US nuclear submarine programme by H. G. Rickover in 1942 and were re-expressed in 1970 as the eighteen criteria of quality assurance (US Code of Federal Regulation 10CFR50 Appendix B), much as the nuclear weapons programme was the driving force in the development of digital computers in the 1950s.

13.4 Other applications of nuclear radiation

13.4.1 Gauging and detection

Use is made of the specific nature of radiation in such instruments as beta gauges for determining the thickness of paper on rolling mills; to trace leakages and blockages in underground pipes from a gamma-radiation 'pig'; and in the testing of welds by gamma-radiation. In this respect, gamma-radiation and X-rays are indistinguishable in application and thus gamma-rays show up contrast in heavy elements. Neutrons can be used, however, to contrast good and poor scattering elements and even neutron radiographs and holographs of hydrogenous material are feasible. Ionizing radiation is used in fire (smoke) detectors and is cleared for domestic use in the UK.

13.4.2 Compact power sources

Medium-sized pressurized water reactors (PWR) are widely used in major navies for submarines (where their underwater endurance is unsurpassed) and capital ships, aircraft carriers and ice breakers. Only demonstrations have been made of commercial marine applications. In space, small fission reactors have been employed but simpler and more reliable long-life power plants can be made using the decay heat and charge release from isotopes of plutonium, etc. Some of these devices have been adapted to remote sea-navigation beacons and to heart pacemakers.

13.4.3 Materials and sterilization effects

Gamma-radiation promotes polymerization of plastics and is routinely used. Gamma-radiation, from, say, activated cobalt-60 is used to sterilize pre-packed medical instruments and is proposed for the treatment of food such as vegetables to delay decay. Such treatment does some damage (e.g. to vitamins) but does not induce radioactivity in the food which has been cleared for human consumption in a number of countries.

13.4.4 Medical applications

Ionizing radiations (X-ray and gamma) are also used for diagnostic visualization including tomography and for treatment, notably for cancer. Some experimental use has been made of neutron beams to treat cancer of the brain after ingestion of a suitable neutron-capture agent such as boron. The behaviour of the lung can be visualized by inhaling artificially produced technetium.

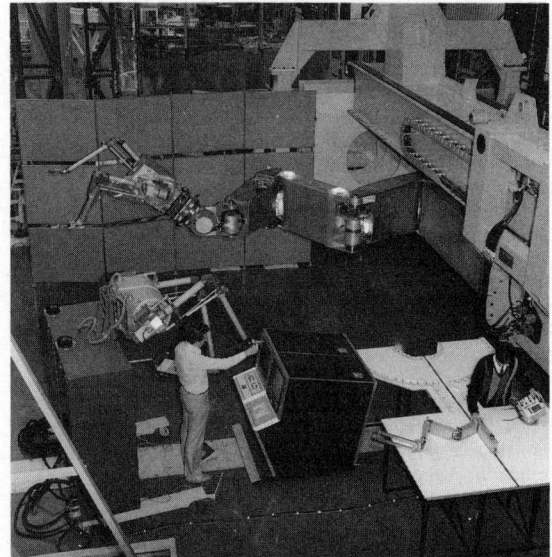

Figure 13.7 Remote manipulator for JET fusion experiment (courtesy JET Joint Undertaking)

13.4.5 Dating and tracing

Activated isotopes may be incorporated into materials and are particularly helpful in tracing metabolic reactions in biological substances. Dating of artefacts by measuring carbon-14 activity (or the modern method of isotopic analysis) has enabled wood and similar materials to be dated back some 8000 years.

13.5 Elements of health physics and shielding

Any engineer wishing to make use of nuclear energy in any form needs to understand the effect of the associated radiations on matter, inert and biological materials. Alpha- and beta-particles, and gamma-rays are directly ionizing; that is, their passage through matter is accompanied by the disruption of the neutral molecule to form charged radicals or ions. Neutrons are not directly ionizing but the results of their reactions are ionizing radiations. The damage done may also include displacement of the nucleus, particularly by neutrons, alpha-particles and other heavy ions. Thus the steel cladding of a nuclear fuel element may need to be designed for the swelling and atomic displacement in which an average atom is struck over one hundred times in its lifetime. The Windscale fire of 1957 is associated with the sudden release of the Wigner displacement energy imparted by neutrons to the graphite moderator.

In the modern approach to the measurement of the effects of ionizing radiation the physical nature is described in terms of the energy deposited per unit mass. The SI unit of one joule per kilogram is given the special name of the gray (symbol Gy). It replaces the older, obsolete and obsolescent units of the roentgen and the rad (Table 13.3).

The biological effects of ionizing radiations spring from their energy but vary with the nature of the radiation. To account for this, quality factors are employed to multiply the physical dose in gray to express an effective dose equivalent in sievert (symbol Sv), also in J/kg. These quality factors are thus non-dimensional numbers. Values of quality factors for some radiations are listed in Table 13.4. The primary reason for the variation in biological effectiveness is associated with the energy density along the track of the particle, described as LET (linear energy transfer). Thus alpha-particles, the nuclei of helium, being relatively massive and doubly charged have a high LET and correspondingly cause such severe damage to individual cells that they are less liable to recover. Gamma-rays, however, have a low LET (if they can be said to have a track at all), and even though the energy deposited per kg may be the same, the energy is spread over more cells which may therefore recover from the insult, leading to noticeably smaller biological damage.

On the other hand, the high LET radiations will be stopped more readily than low LET radiations. A sheet of paper might serve to stop most alpha-particles; beta-particles can penetrate a few centimetres of flesh; but gamma-rays have a mean-free path in air of several metres, depending on their energy. Thus alpha-particles may be seriously damaging if ingested but shielding suitable for beta and gamma radiation will be more than adequate against external alphas. Protection against radiation involves therefore shielding against external sources and the prevention of ingestion leading to internal doses.

Table 13.4 Quality factors for ionizing radiation

Radiation	Q
Photons (gamma- or X-rays)	1
Beta-particles	1
Protons	10
Alpha-particles and heavy multiply charged particles	20
Neutrons	
thermal	2.3
fast	10

13.5.1 Biological effects

The effects of radiation on living matter depend on the energy deposition (in gray), the quality factor for the radiation and thus the equivalent dose in sievert but also upon the rate at which the dose is applied. The slower dose rate provides for some opportunity for the body's defence mechanisms to counter its gross effects (Figure 13.8).

Another important distinction is between stochastic and non-stochastic (deterministic) consequences. Some consequences of disease can be correlated with the effective dose (e.g. clouding of the retina under neutron irradiation). The response is not necessarily linear but such effects can have an associated lower dose limit below which the effect is essentially not seen.

In other cases, however, only the incidence of the disease, not the consequence, is a function of the effective dose. In particular, cancers and genetic defects are initiated by radiation with a certain probability; the result, however, is not itself a function of the radiation received.

It follows that whereas safety limits could be stipulated such that injury of a deterministic nature is prevented, no such limit can be set for conditions of a stochastic type. Thus a lower limit can only be set comparatively; is the probability of injury acceptable in comparison with other, non-radiation, occupations?

Table 13.3 Radiation units: current and obsolescent

Activity		
Bequerel	Bq	1 disintegration/s
Curie	Ci or ci	37×10^9 Bq
Absorbed dose		
Gray	Gy	1 joule/kilogram
Rad	rad	100 Gy
Roentgen	R	see below
Dose equivalent		
Sievert	Sv	1 joule/kilogram
Rem	rem	100 Sv

Notes

Sv = Gy time-weighting factors. The most important weighting factor is the *quality factor*, Q (see Table 13.4).

The roentgen, R, has been defined in different ways at different times. One definition is 'that quantity of X- or gamma-radiation which produces in air at s.t.p. ions carrying a total charge of one electrostatic unit of either sign'. It is approximately 88 Gy.

The SI units of Bq, Gy and Sv are preferred.

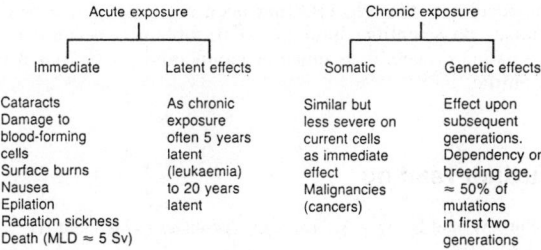

Figure 13.8 Biological effects of ionizing radiation

It does not follow that the incidence of stochastic symptoms is linearly proportional to the dose received, although this assumption of linearity is common in all administrative regulation of dose limits. On the one hand, there might be a threshold under which no incidence occurs; indeed, there are even arguments for homesis, where a little radiation (as perhaps the natural background we evolved in) does you good. On the other hand, there are arguments to support the quadratic relationship of stochastic incidence to effective dose. The linear hypothesis is simple and allows a ready assessment of radiation from combined sources.

A consequence of the assumption of linearity is the role of the population exposure, measured in (say) person sieverts. If the population dose is (say) one person-Sv, then the incidence of cancer can be expected to be the same, whether this is spread over a population of ten or of 100. It follows that the idea of enlarging the workforce to lower the average exposure per worker is not good practice.

13.5.2 Radiation limits

Formal statutory limits are imposed by national legislation and, in the UK, their observance is a matter for the Health and Safety Commission and its Executive (HSE). A distinction is made between normal practice and emergencies and between industrial limits and limits for the public.

To appreciate these limits, it should be said that the average exposure of the population in Britain is some 2 mSv a year, the major part coming from radon gas (and thus varying markedly over the country) and less than 1 μSv from the nuclear power industry. Acute exposure, on the other hand, of 5 Sv (some 2000 times natural background) may be fatal.

The limits are based on recommendations of the ICRP (International Commission for Radiation Protection). In arriving at these limits, the ICRP have made clear that they expect good practice to keep the average exposure below the limits (approximately one tenth of the limit itself). Industrial limits are arrived at by comparing current estimates of genetic damage and incidence of cancer with risks currently accepted in other, non-ionizing, industries. They are therefore subject to review over the years, generally downwards.

The following are based on the recommendation of the ICRP No. 60 (1990).

13.5.2.1 Industrial limits

Deterministic limit 500 mSv/annum (a) (eye 150 mSv/a)
Stochastic limits 20 mSv/a

In practice therefore, limits based on stochastic effects apply. The overall limit is arrived at by a suitable weighting of different organs.

In UK practice, an intermediate level of 30% of the limit is introduced as a reference level, i.e. at 6 mSv/a. If the working practice is such that it can be demonstrated that exposure to this intermediate level is unlikely, then the workers and the workplace need not be specially designated. If this demonstration cannot be made, then special precautions and administration is required of the plant, including long-term records of actual exposure, etc. It is obviously attractive to design the work involving small quantities of radiation to avoid 'designation' and a useful figure of merit is to convert the 6 mSv on the basis of a 2000-hour working year to give 3 μSv/h. If the dose equivalent rate is below this figure, designation is not required.

Obviously, this escape is not available in major installations. Such industries must have, by law, a Radiation Protection Adviser and a formal system of monitoring exposure. Special national records are maintained for workers with uranium and plutonium.

13.5.2.2 Public exposure

On the philosophy that the public have less choice of working conditions, the exposure limits for the public are set at one tenth of the industrial limit.

13.5.2.3 Emergencies

Somewhat higher than annual average limits can be accepted for special operations, such as planned maintenance, if the workers' dose over a number of years averages out satisfactorily. In an emergency where lives are at risk, larger doses again can be countenanced. There is obviously no easy way to provide for a legislative limit in these circumstances.

13.5.3 Elementary shielding

Protection of personnel, perhaps animals, and equipment can be provided by suitable application of the following guidelines:

1. Reduce the exposure to the source to a minimum time.
2. Use distance for protection; the radiation from a point source goes down about as rapidly as the square of the distance.
3. Employ a material shield suitable to interrupt the passage of the radiation.

A too-simple approach to shielding can be dangerous. Thus thermal neutrons can be captured by cadmium. But a cadmium suit would be fatal to the wearer in view of the gamma-rays emitted upon neutron capture. The high-energy gamma-rays go through several competing reactions before eventual absorption in material (where they ionize and ultimately heat matter). Material that is opaque to gammas at one energy may be transparent at another – there is a dangerous window in iron. Also, the travel of the gammas in these processes can lead to radiation appearing from unexpected directions – sky shine over a shielding wall.

There are other aphorisms arising in elementary shielding such as 'water has no cracks'; solid shields invariably have weaknesses at joints which may lead to unacceptable radiation streaming for which 'dogleg' design is appropriate.

Small sources can be shielded against in elementary ways by over-designing the shield. The cost, in size and weight, may be acceptable. But major nuclear sources require elaborate and expensive shielding. Correspondingly, major resources involving much computing time and perhaps experimental verification are devoted to the design of the shield.

These elements of shielding and the restrictions of radiological control lend themselves to an approach generally known as ALARA:

As Low As Reasonably Achievable, economic and social factors taken into account.

It is implied in ALARA that no radiation shall be employed for its own sake, but only if there is a benefit. Where there is a benefit, comparisons are due with non-radiation methods to the same end. Finally, statutory limits shall not be exceeded.

ALARA is internationally recognized, although in the UK the regulating authorities have made modifications such as ALARP (reasonably practical), etc.

In principle, the optimization of the design process and the operating process are conducted until the user is able to satisfy the regulators that ALARA has been achieved. There is, of course, no scientific optimum to be obtained because the regulating process is as much economic and political as it is scientific.

Further reading

Cumo, M. and Navigli, A., *Thermal Hydraulics* (Vols I and II), CRC (1988)

Marshall, W. (ed.), *Nuclear Power Technology*: Vol. 1 *Reactor Technology;* Vol. 2 *Fuel Cycles*; Vol. 3 *Nuclear Radiations*, Clarendon, Oxford (1983)

Martin, A. and Harbison, S.A., *Introduction to Radiation Protection*, 2nd edn, Chapman and Hall, London (1983)

The protection of persons against ionising radiation arising from any work activity, Health and Safety Executive, HMSO, London

14 Offshore engineering

Minoo H. Patel

Contents

14.1 Historical review 14/3

14.2 Types of fixed and floating structures 14/3

14.3 Future development 14/11

14.4 Hydrodynamic loading 14/13
14.4.1 Wave loads on slender structures 14/17
14.4.2 Wave loads on large bodies – diffraction theory 14/18

14.5 Structural strength and fatigue 14/19
14.5.1 Structural analysis method 14/19
14.5.2 Structure failure and fatigue life calculations 14/22

14.6 Dynamics of floating systems 14/23
14.6.1 Heave motion of cylindrical buoys 14/23
14.6.2 Heave motion of a semi-submersible 14/25
14.6.3 Heave and surge motions of tensioned buoyant platforms 14/26
14.6.4 Analysis with multiple degrees of freedom 14/29
14.6.5 Morison-equation approach for slender structures 14/30
14.6.6 Diffraction theory 14/31

14.7 Design considerations and certification 14/32

References 14/34

14.1 Historical review

The oceans of planet Earth exercise a dominant influence on its weather systems and ecology. They are believed to have been the medium in which life first appeared on the planet and gradually evolved into land-based plant and animal species. The continued existence of land-based life is still dependent on the oceans through their effect on the planet's weather and the crucial role that the oceans play in global ecology. Since prehistoric times and up to the early part of this century, however, our use of the oceans has been largely restricted to near-shore transportation and fisheries, although there is evidence to suggest migrations of people across very large oceanic distances. Since the fifteenth century, the pace of ocean transportation and deep-water fishing has gradually increased, but our utilization of the oceans has still been restricted to these two activities.

Over the last five decades, however, traditional uses of the oceans have expanded to include the exploitation of hydrocarbons below the sea bed and the potential of large-scale mineral gathering and energy extraction. The development of offshore oil and gas has, however, led the way, with land-based exploration and production for crude oil being extended to the sea bed as early as 1887, when the first exploration drilling was carried out, from over a few feet of water, in California.

Since then, the pace of oil exploration and production in shallow water has gradually increased, with the next major development occurring in 1910, when wells were sunk in Ferry Lake, Louisiana. This was followed in 1929 by the exploration and production of a large oil reservoir on Lake Maracaibo in Venezuela. The development of the Gulf of Mexico as an offshore area started in the 1930s, with oil first being produced in 1938 from a timber platform in 4.3 m (14 ft) of water, on the Creole Field, which was located 1.6 km (1 mile) off the coast of Louisiana. Following the Second World War, offshore activity in the Gulf of Mexico rapidly evolved with oil exploration and production in deeper water until by 1959, an oil production platform had been installed in 30.5 m (100 ft) of water by Shell on a block in Grand Isle. At the same time, British Petroleum were developing shallow-water oil finds in the Persian Gulf.

Such oil and gas exploration and development spread to most of the continental shelf areas of the world from the mid-1950s, starting first in the Gulf of Mexico, then spreading to the waters off Mexico, Brazil and Brunei in the Far East. The map of Figure 14.1(a) presents an overview of this worldwide offshore exploration and production activity, whereas Figure 14.1(b) uses sea-bed contours to show the extent of the continental shelves around all the major land masses. The continental shelves cover only a small fraction of the total ocean areas, but they are potential sites for exploration or are already known to have large reserves.

The offshore industry began a technically more challenging phase when the North Sea was first explored as a potential offshore area in the early 1960s. Figure 14.2 presents an overview of the fields developed in the North Sea since the mid-1960s. These fields range from large gas accumulations in the relatively calm waters off East Anglia in the south to harsh northerly environments in the vicinity of 60° North, where fields such as Magnus, Thistle, Murchison and Cormorant are exposed to some of the most hostile wind and wave conditions of any maritime area in the world. The North Sea is typical of many offshore areas where the development of technology and structures to exploit oil and gas fields has had a significant positive impact on the economies of surrounding countries.

The development of land-based and offshore oil and gas deposits has played an essential role in laying the foundations of the modern world with its high living standards. Figure 14.3 displays the relative volumes of onshore and total oil production over the last two decades and shows that offshore oil production is now approaching 30% of total worldwide production.

14.2 Types of fixed and floating structures

The design of offshore structures used for oil and gas production offers some indication of the following problems. A typical steel-framed jacket structure used for offshore oil drilling and production will encounter largely steady environmental forces from wind and current flows, and from the structure's self-weight. The jacket will also be subjected to high levels of cyclic load due to gravity waves (see Section 14.4), its sea-bed equipment will be working in the presence of very high sea-water pressure, the steel making up its structure will have to survive sea-water corrosion for long periods since large parts of the jacket structure are difficult to access and maintain. At the same time, the physical processes that govern interactions between the atmosphere and the ocean surface, the effect of the structure on the fluid around it and on the behaviour of the sea-bed foundation are not completely understood in scientific terms.

These problems are compounded by the uncertainties of predicting the most extreme environment likely to be encountered by the structure over its lifetime, which is measured in decades, and the uncertainty of categorizing the geology and recoverable resources of the hydrocarbon reservoir that the steel structure is designed to develop. All of these interacting problems offer unique challenges for advanced scientific analysis and engineering design.

The technical evolution of the modern offshore industry can be measured by the depth at which it has been able to carry out exploration drilling and by the structures or vessels that have made such drilling possible. Initially, exploration drilling was carried out from shallow water fixed platforms which were made of timber and piled to the sea bed. This was later followed by drilling from barges or drilling tenders mounted in sheltered locations. The water-depth capability of drilling equipment has gradually increased, first by the use of jack-up rigs – a typical modern jack-up rig is shown in Figure 14.4. These rigs are generally of triangular construction, containing steel-framed legs at each corner which can be jacked up or down by electric or hydraulic machinery. The jack-up hull is transported to the drilling site either by loading onto a barge or by towing the self-floating hull. The legs are jacked down at the drill site, and the platform hull is raised a sufficient distance above mean water level to prevent waves hitting the underside of the platform. Drilling then commences, usually for a work programme of between 2 to 4 months. When the drilling programme has been completed, the hull is lowered down to the sea surface, or onto a transportation barge, and the rig is moved to the next drilling location. Some jack-ups have also been used for oil production in shallow water, with modern jack-ups able to operate in severe weather at water depths of up to 107 m (350 ft).

Exploration drilling in waters of up to 18 m (60 ft) depth was also carried out from floating vessels with one or more pontoons or caissons supporting a deck with vertical columns. Such platforms were floated onto the drilling site and then ballasted down to the sea bed to carry out a drilling programme. During the flotation phase of deployment it was noticed that such column-stabilized platforms had exceptionally low motion response to waves. This feature was utilized to develop so-called semi-submersible floating platforms which did not have to be ballasted down to the sea bed

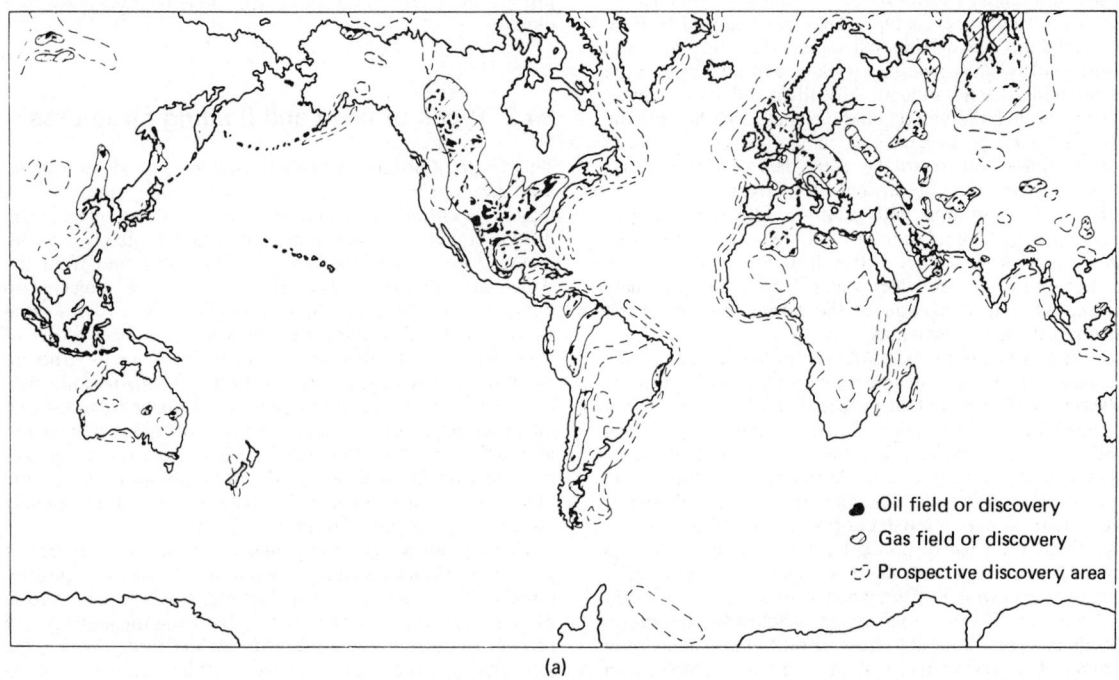

(a)

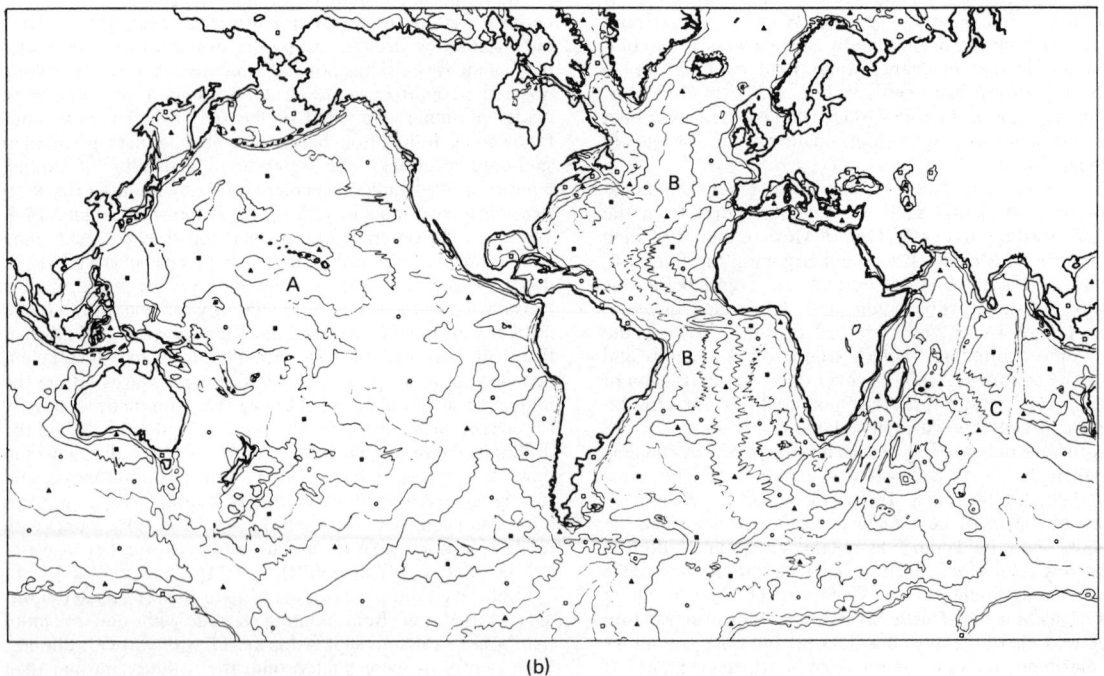

(b)

Figure 14.1 (a) Worldwide offshore development areas; (b) distribution of water depths in the world's oceans. □ 500 m, ▲ 1000 m, ○ 3000 m, ■ 5000 m. A – Pacific Ocean, B – Atlantic Ocean, C – Indian Ocean

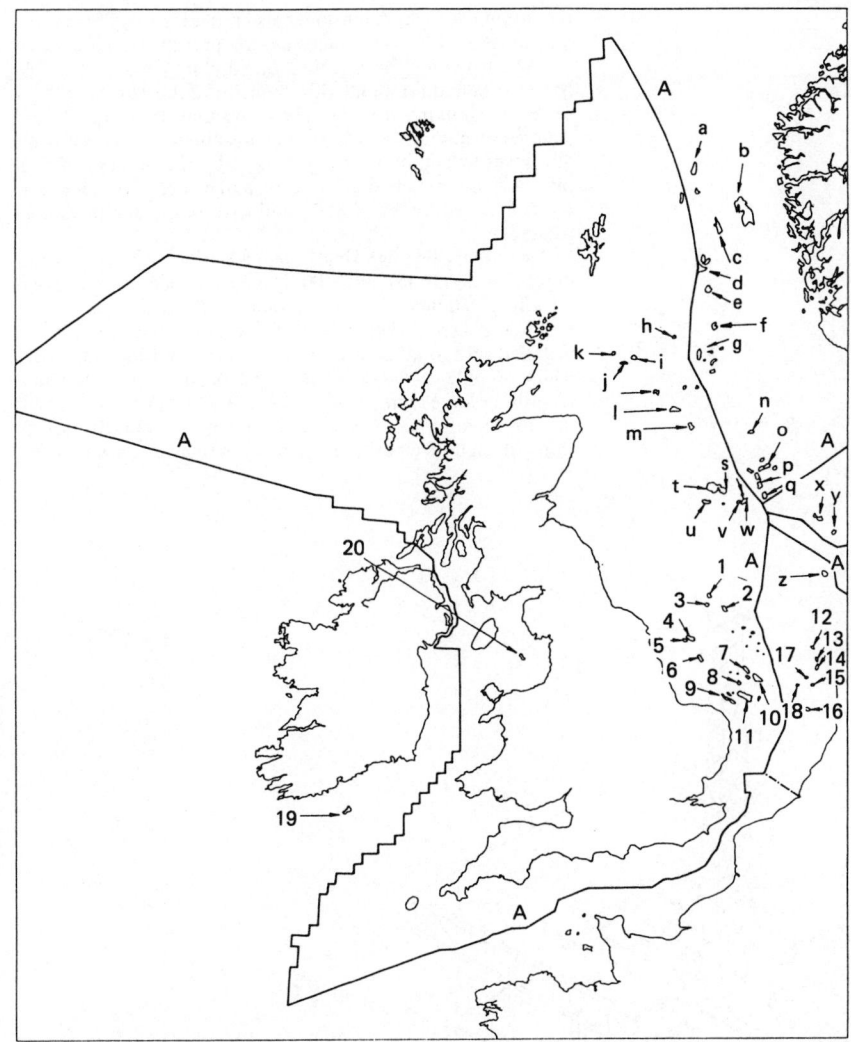

Figure 14.2 Distribution of oil fields in a typical offshore location – the North Sea. For clarity, only a representative selection of fields are illustrated. A – Median lines marking limits of neighbouring country's offshore sectors. Names of oil and gas fields: a – Snorr, b – Troll, c – Oseberg, d – Frigg, e – Heimdal, f – Balder, g – Sleipner, h – South Brae, i – Piper, j – Tartan, k – Claymore, l – Forties, m – Montrose, n – Ula, o – Ekofisk, p – Eldfisk, q – Valhall, r – Innes, s – Clyde, t – Fulmar, u – Auk, v – Dumcan, w – Argyll, x – Tyra, y – Dan, z – F2-3, 1 – Forbes, 2 – Gordon, 3 – Esmond, 4 – Ravenspurn, 5 – Cleeton, 6 – West Sole, 7 – Viking, 8 – Vulcan, 9 – Hewett, 10 – Indefatigable, 11 – Leman, 12 – L4-A, 13 – L7-C, 15 – K15-FA, 16 – P6, 17 – K14, 18 – K18, 19 – Kinsale Head, 20 – Morecambe (from Department of Energy[1])

in order to carry out drilling. Figure 14.5 shows a perspective view of one such modern vessel with two parallel submerged pontoons at keel level supporting a large deck area using eight vertical surface-piercing columns. The columns, pontoons and deck are braced to form a space frame structure which can be moored on location by a spread of catenary mooring ropes or chains. The deck contains a central opening called the moonpool through which drilling activities take place. Most semi-submersible vessels are equipped with accommodation, helicopter pads for crew transport, propulsion engines, pipe-handling equipment, cranes and a variety of marine services to support crew and drilling operations.

Drilling into the sea bed is carried out through a vertical pipe called the marine riser which connects the area under the

moonpool to a subsea wellhead. A drill bit on the end of the drill pipe is passed through the riser to penetrate the sea bed and carry out the actual drilling. A lubricating fluid called drilling mud is pumped to the drill face at high pressure through the hollow drill pipe and returned through the annulus between the drill pipe and riser. The drilling mud pressure is used to drive rotating cones on the drill bit with the drill bit and pipe itself rotating at up to 250 rev/min. The drilling mud serves several purposes. It lubricates the drill bit, drives the drill bit cones, sweeps out drill cuttings to the surface and provides a hydrostatic pressure head to contain high-pressure oil or gas that may be encountered. The small, but significant, heave motions of the floating drilling rig due to ocean waves are compensated for by a slip joint at the base of

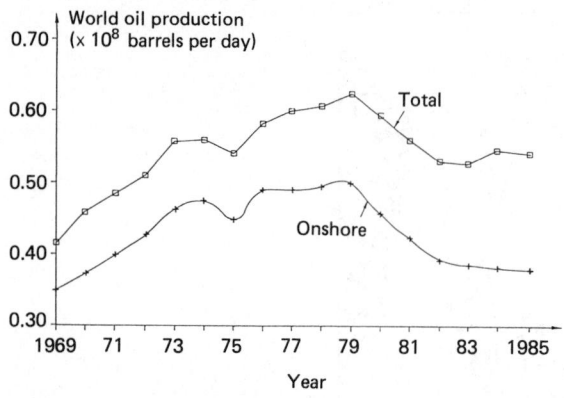

Figure 14.3 Growth of onshore and total oil production (from American Petroleum Institute[2])

the drilling derrick. A tensioner also provides an upward force on the riser to keep it in tension and prevent buckling of its slender structure. The moorings of a drilling semi-submersible have to maintain station to less than 7% of the water depth, in order to continue drilling. The maximum water depth and weather conditions in which a semi-submersible vessel drills are governed by its mooring system, by the amount of drill pipe and riser pipe that it can carry, and by heave compensator limits imposed by the vessel's motion response and prevailing weather.

A caternary line mooring system is impractical above water depths of about 457 m (1500 ft), and in such cases a dynamically positioned vessel is required. Dynamic positioning requires a vessel, whether a ship or a semi-submersible, to be installed with rotating thruster pods and propulsion machinery which is able to work against and balance environmental disturbing forces due to wind, current and wave drift. Dynamic positioning has been most commonly employed on drill ships which utilize position signals from sensors, such as a taut

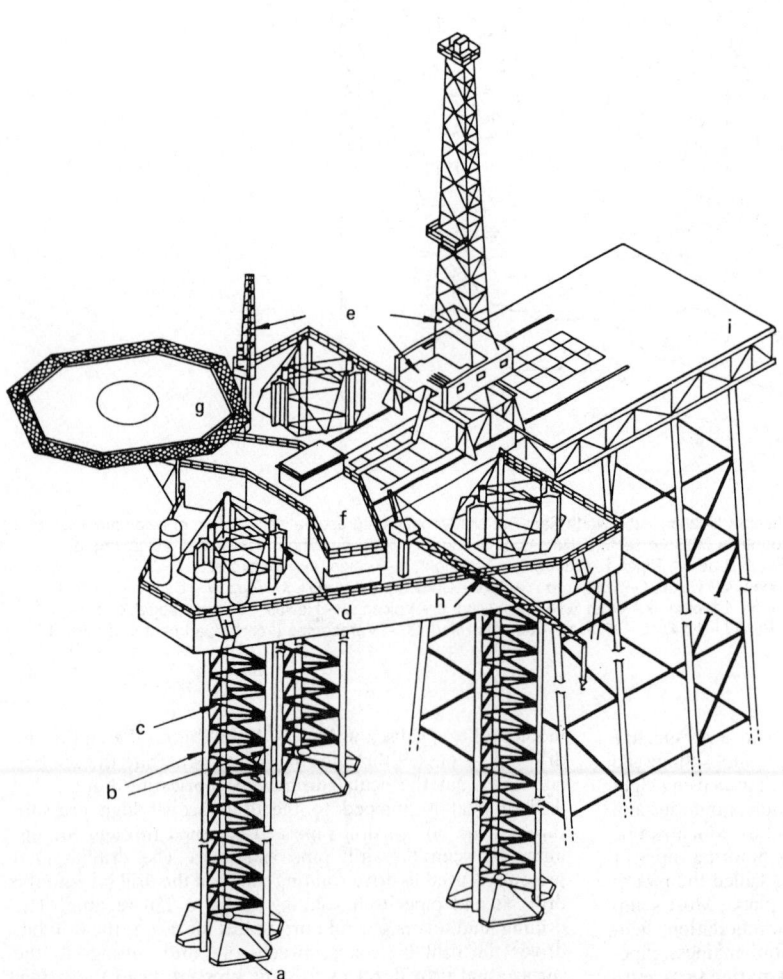

Figure 14.4 Typical jack-up rig: a – spud cans, b – elevating racks, c – legs, d – gear units, e – drilling derrick and equipment, f – accommodation, g – helicopter pad, h – cranes, i – nearby jacket platform

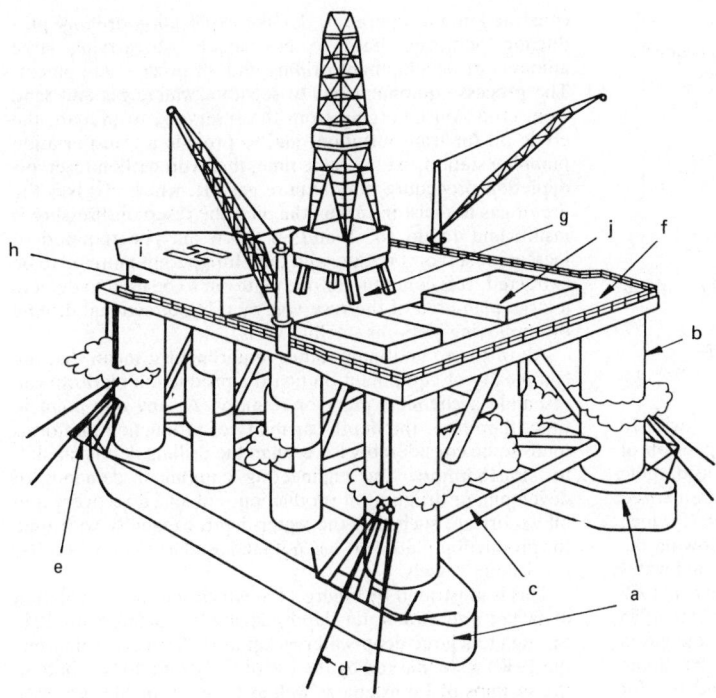

Figure 14.5 Typical semi-submersible vessel: a – submerged pontoons, b – surface-piercing deck-support columns, c – bracing members, d – mooring lines, e – anchor racks, f – deck structure, g – moonpool, h – accommodation, i – helicopter pad, j – drill pipe racks

vertical wire to the sea bed or from acoustic beacons on the sea bed around the wellhead, to obtain an error signal representing horizontal offset from above a subsea wellhead. The vessel is fitted with thrusters controlled from an automated computer-based algorithm which is programmed to reduce this offset and maintain the vessel as close to directly above the wellhead as is possible.

Figure 14.6 shows a schematic view of this arrangement with the drill ship connected by a riser to the blow-out preventer (BOP) stack, prior to that being connected to the wellhead. Acoustic beacons provide a position reference to the drill ship – the position data are then used by a computer to control the main propulsion system and lateral thrusters to resist wind, current and wave drift forces on the vessel. It is necessary to maintain position within a 'watch' circle that has a radius of about 7% of the water depth to continue drilling.

Several modern semi-submersibles have also been equipped with dynamic positioning systems. The maximum water depth to which a dynamically positioned drilling vessel can work is theoretically unlimited, although, in practice, limitations are imposed by the length of marine riser that can be carried on the vessel and by tensioner and heave compensator performance. Dynamically positioned vessels, nevertheless, have to disconnect the riser in extreme weather conditions, to ride out storms and then to reconnect in calmer weather.

There are approximately 500 offshore drilling rigs worldwide made up of about 250 jack-up rigs, 130 semi-submersible vessels, 90 drill ships and 30 bottom-standing submersible rigs – the last being of the type which can be floated out onto a shallow-water location and then ballasted down to rest on the sea bed prior to commencement of drilling. The worldwide active jack-up rig fleet began in 1954 and developed to 27 rigs by 1960 and 103 by 1970; whereas the first semi-submersible drilling vessel started operation in 1962, with 25 active vessels

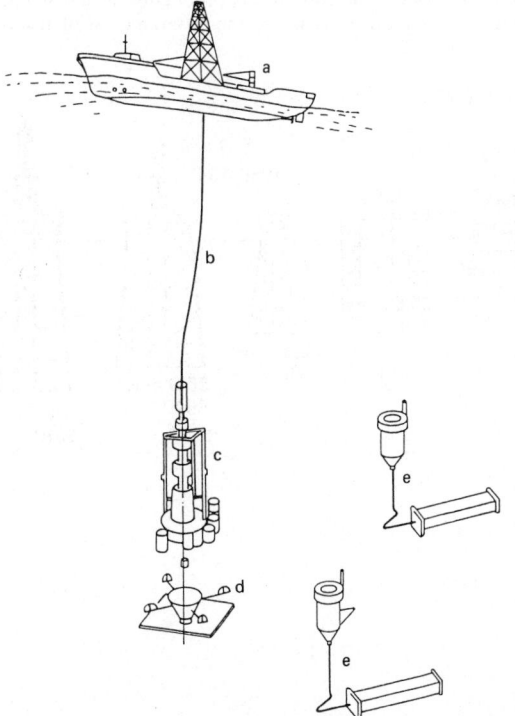

Figure 14.6 Dynamically positioned drill ship: a – drill ship, b – riser, c – blow-out preventer, d – wellhead on sea bed, e – acoustic positioning beacons

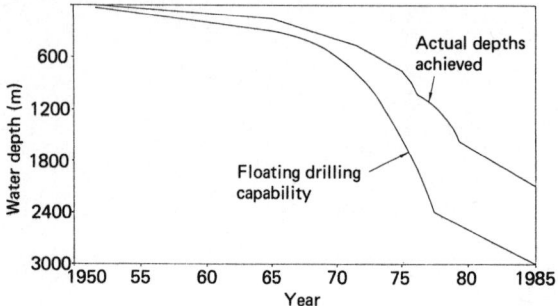

Figure 14.7 Deep-water drilling capability and achievements

by 1970. The first drill ship started operation in 1953, with 16 being developed by 1960, and 39 by 1970. The expansion of the worldwide drilling fleet has been matched by considerable increase in the water depths at which drilling can be carried out. This is illustrated most vividly by Figure 14.7, which presents water depth against year with two lines showing the maximum drilling depth capability, as well as the actual water depth records that were achieved in the years between 1952 and 1985. It can be seen that in the period from 1970 to 1985 the water depth at which drilling was carried out increased from around 1366 m (1200 ft) up to 2128 m (6981 ft). Some drilling in very high water depths has been carried out for geological research into the structure of the land mass below the oceans.

The water depths at which exploration drilling is carried out is an indicator of future requirements for oil production. In drilling programmes where significant discoveries of hydrocarbons are made, a decision on oil production is dependent on the prevailing price of oil and the economics of platform construction and operation. Unlike exploration drilling, production platforms have to be capable of carrying large amounts of development drilling and oil process equipment. The process equipment has to separate water, gas and sand from crude oil emerging from the reservoir, to prepare the crude oil for transportation, and to provide a transportation pumping station. At the same time, the hydrocarbon reservoir depletion procedure may require gas lift, which involves the use of gas to assist in raising the oil if the reservoir pressure is insufficient to do so. Water injection may be required to increase reservoir pressure and therefore permit more oil to be extracted. It is common for one platform to be used to deplete a large plan area of the reservoir by using directional drilling or employing step-out satellite wells.

All of these, and many other requirements, mean that the complexity of equipment on board a production platform can resemble a chemical plant or refinery. At any one point in time, therefore, the depth capability of production platforms tends to be considerably lower than the drilling depth capability. Furthermore, the engineering equipment demands of development drilling, oil production, enhanced recovery and oil export are such that the water-depth capability with time for production platforms has not increased as steeply as it has for drilling vessels.

This is illustrated by Figure 14.8, which shows the evolution of jacket size and water-depth capability starting with 1947 through to a large deep-water jacket in 1978. In general terms, the 1930s were marked by the use of timber platforms, both in the swamps of Louisiana as well as the Gulf of Mexico, with 1947 being the first year in which a steel platform was installed in 6 m (20 ft) water depth off the Louisiana coast and out of sight of land. By the mid-1950s, jacket platforms were being installed in water depths of about 30 m (98 ft) with the framed structures becoming more open. The 1950s to the early 1970s saw platform water depths extend to approximately 120 m (394 ft). Between 1970 and 1987, fixed-platform water-depth

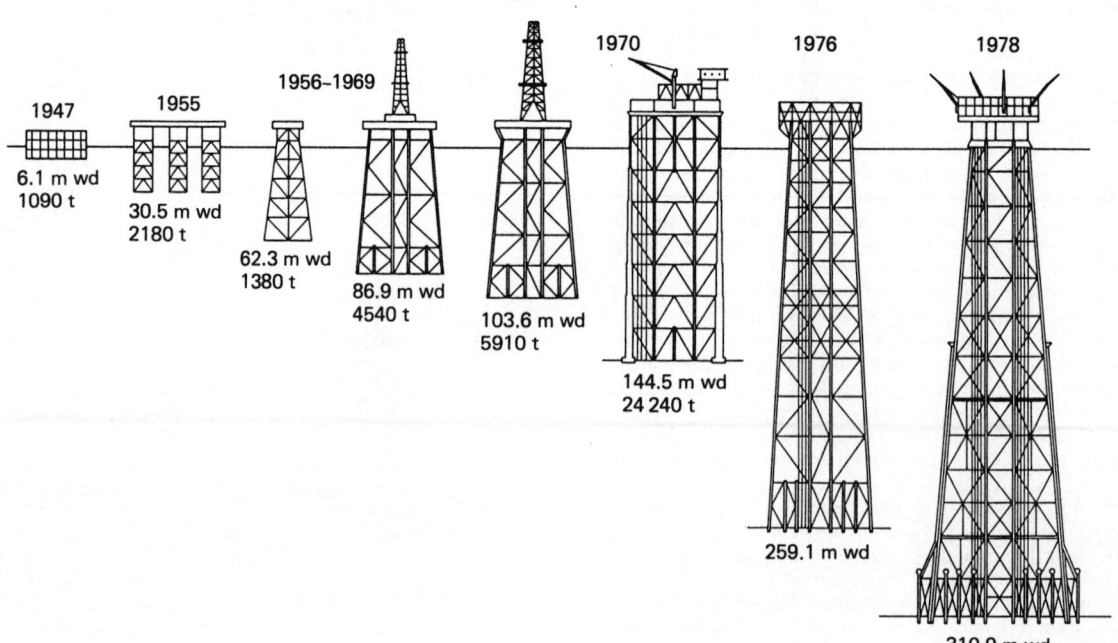

Figure 14.8 Evolution of deep-water production capability (wd = water depth) (from Lee[3])

capability increased substantially from around 120 m (394 ft) through to the largest depth to date for a fixed jacket structure – the Cognac platform in the Gulf of Mexico which stands in 311 m (1020 ft) of water.

It is instructive to examine the structure of a typical jacket for deployment at moderate water depth in the North Sea as illustrated by Figure 14.9. The jacket consists of a steel-framed tubular structure, connected to the sea bed by piles which are driven through pile guides on the outer members of the jacket. The jacket topsides consist of a series of modules which house drilling eqipment, production equipment, living quarters, gas flare stack and revolving cranes. The topside also has facilities for living quarters, survival craft, hotel and catering facilities, and a helicopter pad for transfer of crews from and to a shore

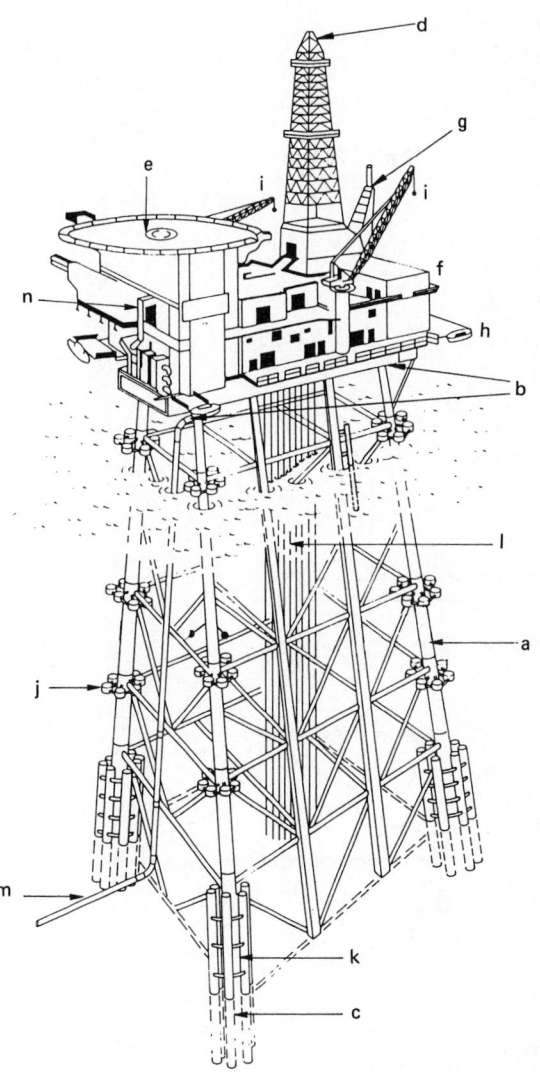

Figure 14.9 Typical offshore drilling and production platform: a – jacket, b – module-support frame, c – piles, d – drilling derrick, e – helicopter pad, f – drilling and production equipment, g – flare stack, h – survival craft, i – revolving cranes, j – pile guides, k – pile sleeves, l – drilling and production risers, m – export pipeline, n – accommodation

base. The topside modules are supported on a module support frame, which is mounted on the water surface-piercing jacket structure. The drilling and production tubulars are brought up from the sea bed to the topsides through conductor guides located within the jacket framing. Crude oil and gas are brought up to the surface from the reservoir for processing and then pumped back down through an export pipeline, either to a tanker loading buoy or by subsea pipeline to a shore terminal.

The detailed design of the framing for a jacket structure can vary considerably, depending on requirements of strength, fatigue and launch procedure. The main structural members and bracing can be connected by 'X' or 'K' shaped joints with sizes that can range from an extensive space frame of small-diameter 'X' braced or 'K' braced members to an alternative sparser framing with larger diameter members.

A jacket designer has to cope with an extensive list of constraints imposed on a jacket structure through its life. The life cycle consists of design, construction, load-out, launch, installation, piling and hook-up phases of the platform before it comes into oil production service. This is followed by a 10–25-year operational life through which the platform has to be maintained followed by the ecologically desirable requirement to remove and dispose of the platform after the reservoir has been depleted.

Jacket topside weights in the Central and Northern North Sea vary between 17 400 tonnes for the North Cormorant jacket up to 35 000 tonnes for the Statfjord and Brae platforms. This topside weight has to be broken down into a number of modules which can be installed offshore onto the module support frame on top of the jacket by crane barges. The design of the jacket and module arrangement is greatly influenced by the availability of crane barges and by their maximum lift capacities. Recent construction of several semi-submersible crane vessels with large lift load capacities is beginning to have a significant influence on offshore installation practice. The semi-submersible configuration offers a low downtime to weather, and the very large size of the vessel means that lift loads of up to 12 000 tonnes can currently be lifted. It is, therefore, feasible to construct and lift fewer modules each of larger tonnage and, therefore, reduce the complexity and cost of module installation and hook-up operations. The availability of larger module sizes also reduces hook-up times and enables earlier oil or gas production to be achieved.

Offshore oil developments in the North Sea have also pioneered a completely different design of fixed platform based on the use of concrete to create a large, heavy structure which can rest on the sea bed and remain stable under its own weight. The first such platform was installed in the North Sea on the Ekofisk field in 1973, and since then over 17 platforms have been installed. Although concrete gravity structures are considered relatively expensive compared to steel frame jackets, the structures do offer an attractive alternative to jackets in hostile waters like the North Sea and in situations where the uncertainty of a tanker export system demands a certain amount of oil storage on the platform. Other advantages of concrete gravity platforms over jackets are that the structures can be constructed onshore or in sheltered waters, with all the topsides installed, hooked up and tested prior to floating out and towing the structure to its offshore location. Installation then only requires ballasting the platform down onto the sea bed and consolidating the foundation below the platform by pumping grout into the spaces between the platform base and foundation.

The elimination of steel piling and of having a concrete structure tolerant to overloading and to degradation due to exposure to sea water offers other advantages over the use of steel. It has been demonstrated that concrete used in coastal installations in the 1930s and 1940s has survived, essentially unaffected by exposure to sea water up to the present day, whereas conventional steel structures would have been susceptible to sea-water corrosion and require substantial levels of maintenance and protection. These advantages have to be set against the fact that concrete gravity structures are relatively expensive. They often actually employ a greater mass of steel in their reinforcing members than would be required by an equivalent steel-frame jacket structure. Concrete gravity structures are likely to suffer from foundation settlement during their working lives which can reduce the air gap between the mean water level and the underside of the structure. Another disadvantage with concrete gravity structures is that no feasible means of removing the structure has been defined at present. Figure 14.10 shows a typical design for a Condeep concrete gravity structure.

The total cost for developing an offshore oil or gas field clearly plays a major part in the selection of the production platform. In the early growth years of the offshore industry, when very large oil deposits were being found, the economics of each development permitted large, expensive structures to be built. However, as more of the larger reservoirs were discovered and developed, the remaining reservoirs have a higher probability of containing smaller oil deposits and being in deeper waters and harsher environments. At the same time, the cost of fixed steel-frame jackets or gravity structures tends to increase in an approximately exponential fashion with water depth, primarily due to the larger platform structure and the greater volume of material required for structural integrity in deeper water.

Figure 14.11 illustrates a typical variation of jacket weight with water depth for two locations – the Gulf of Mexico and offshore California – as well as for the North Sea. The jacket

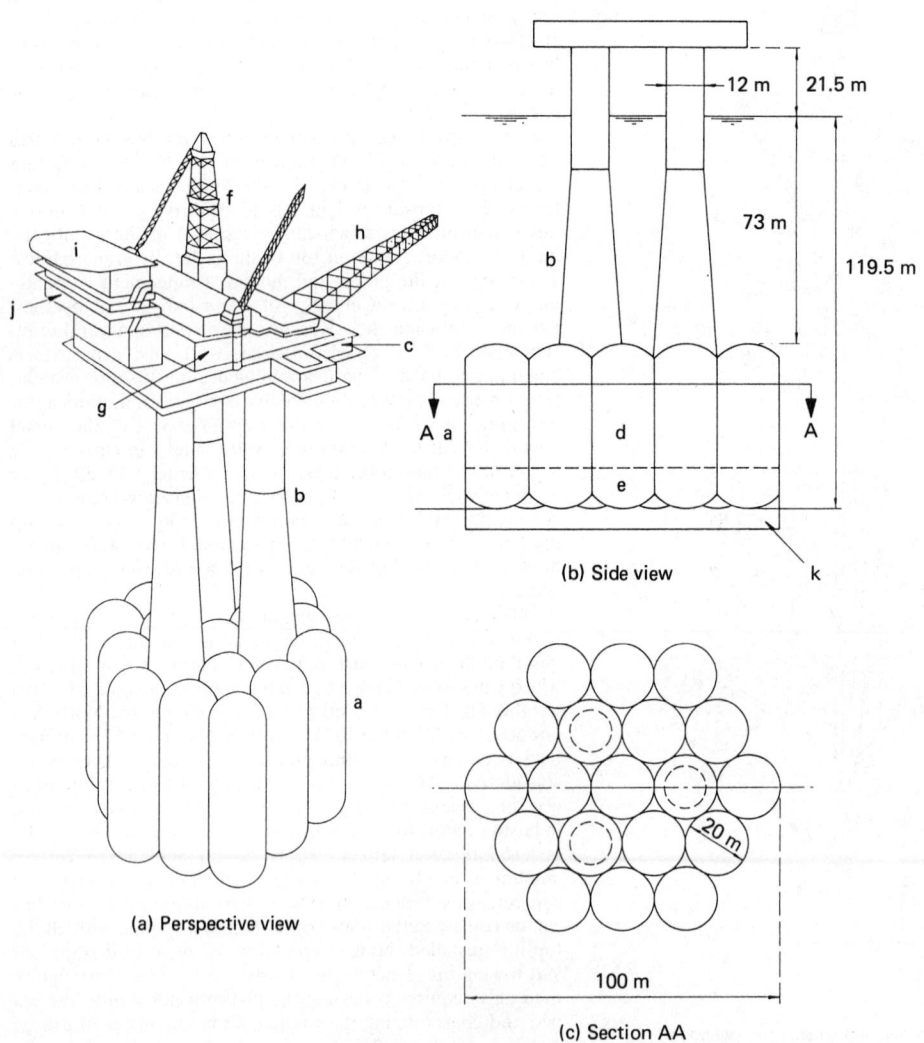

(a) Perspective view

(b) Side view

(c) Section AA

Figure 14.10 Typical Condeep concrete gravity structure: a – sea-bed caisson, b – surface-piercing columns, c – module-support structure, d – oil storage, e – sand ballast, f – drilling derrick, g – drilling and production equipment, h – flare stack, i – helicopter pad, j – accommodation, k – steel skirt (from Furnes and Loset[4])

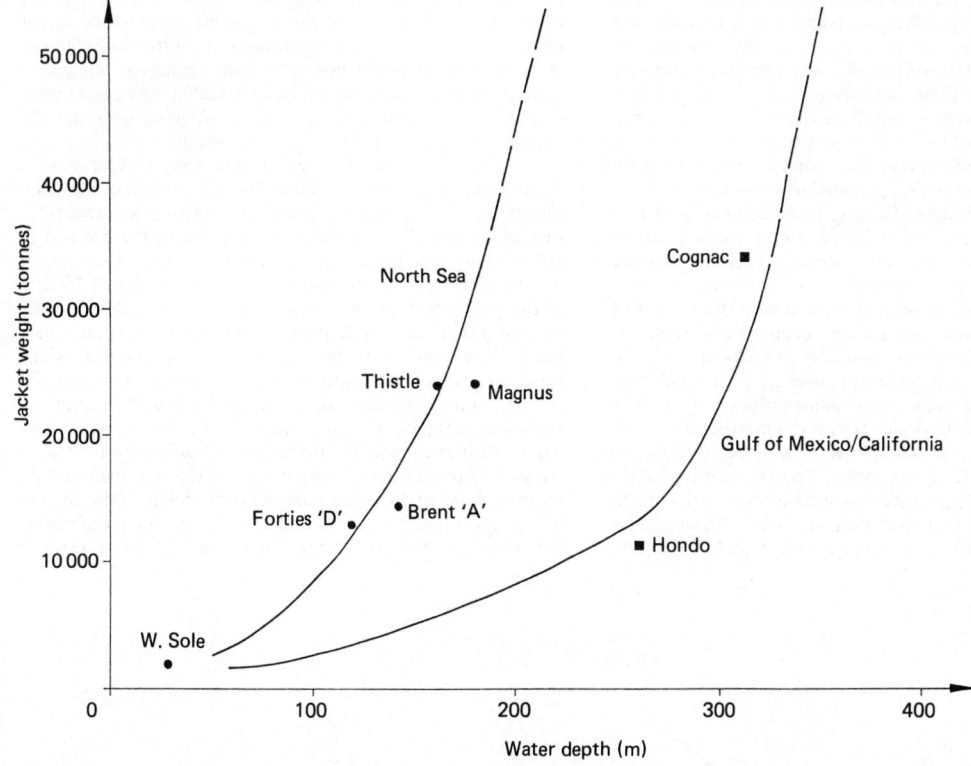

Figure 14.11 Variation of platform jacket weight with water depth (from Thornton[5])

weight on the vertical axis can be considered to be representative of the platform cost whereas the lines denoting North Sea and Gulf of Mexico and California are illustrative of the influence of environmental conditions on jacket weight. The full lines denote current developments and the dashed lines are an estimate of future projections using the same kind of technology. Figure 14.11 shows that the effect of more extreme environmental conditions for the North Sea in the 100–200 m (328–656 ft) water-depth range increases jacket weight and, therefore, cost by a factor of between 3–6. The figure also illustrates that a jacket weighing 30 000 tonnes in the Gulf of Mexico or offshore California could be used in up to 320 m (1050 ft) water depth whereas that jacket weight would only suffice for approximately 175 m (574 ft) water depth in the North Sea.

14.3 Future development

It is clear that the approximately exponential increase of jacket weight and therefore cost with increasing water depth is unsustainable for the North Sea sector since new discoveries are tending to have smaller recoverable reserves and are, therefore, of economically marginal value. This has set the scene for the development of new technology in oil production to try to reduce the rate of increase of cost with water depth. This need for reducing costs has prompted feasibility studies on a wide variety of new types of production platforms based on compliant structures which are fixed to the sea bed but respond with some degree of flexibility to the ocean environment. The consequent reduction in the volume of steel or

concrete material that connects the surface facility to the sea bed can lead to a potential reduction in costs. This is illustrated by Figure 14.12, which presents the cost of design, fabrication and installation for different platform types plotted against water depth.

It can be seen that new platform concepts such as the tensioned buoyant platform and semi-submersible or tanker-based floating production systems – all of which are described later – offer significant reductions in the rate of increase of

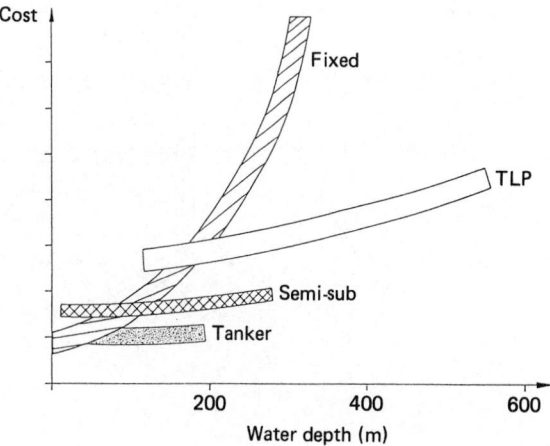

Figure 14.12 Cost comparisons for deep-water structures (from Hamilton and Perrett[6])

cost with water depth. It is now becoming increasingly obvious in the industry that oil production in deep water up to the end of this century is likely to be carried out by compliant or floating production platforms in all cases, except for the very largest of fields or for fields in shallow water. In deep waters above 457 m (1500 ft), a compliant or floating production system may in fact be the only means of producing a field. There are a number of existing and proposed alternatives for cheaper compliant or floating production systems.

One of the most popular floating production platforms is shown in Figure 14.13, and is based round a converted or new-build semi-submersible vessel which is catenary moored to the sea bed and is connected to a subsea manifold or wellhead either by rigid tensioned vertical multi-tube risers or by flexible risers. Such systems are technically suitable for depth ranges of between 70 and 250 m (230 and 820 ft) approximately. They are, however, limited by a number of factors and technical risks. Semi-submersibles only have a relatively small deck payload capacity and, therefore, the range of available oil process options is restricted compared to what is possible on fixed platforms. The compliancy of the platform and the catenary mooring system poses a risk of the vessel losing station and damaging its risers during severe environmental conditions or damage to a mooring line. The

platform cannot directly offer oil-storage facilities and the heave, roll and pitch motions of the platform to waves can limit or degrade processing operations. Nevertheless, catenary moored semi-submersibles offer an economic means of exploiting small reservoirs which have sufficiently good hydrocarbon field characteristics so as to require only a small amount of processing on the surface facility.

A technically more advanced form of compliant structure is offered by the tension or tethered buoyant platform, shown in Figure 14.14. The surface platform configuration resembles that of a semi-submersible but is tethered to the sea bed by vertical legs, which are kept in tension by excess buoyancy in the platform. The excess buoyancy is of the order of 15–25% of the platform displacement and tensions the tethers to such an extent that heave, roll and pitch motions of the platform to ocean waves are virtually eliminated. The platform does, however, experience surge, sway and yaw due to wave action but the vertical tethers are a reliable means of keeping the platform permanently on station above the subsea wells. These wells are connected to the surface facility by an array of risers. Oil processing is carried out on the platform and the hydrocarbon products are pumped back down to the sea bed to an export pipeline. Conoco Ltd in 1984 installed the world's first tethered buoyant platform in 148 m (485 ft) of water on

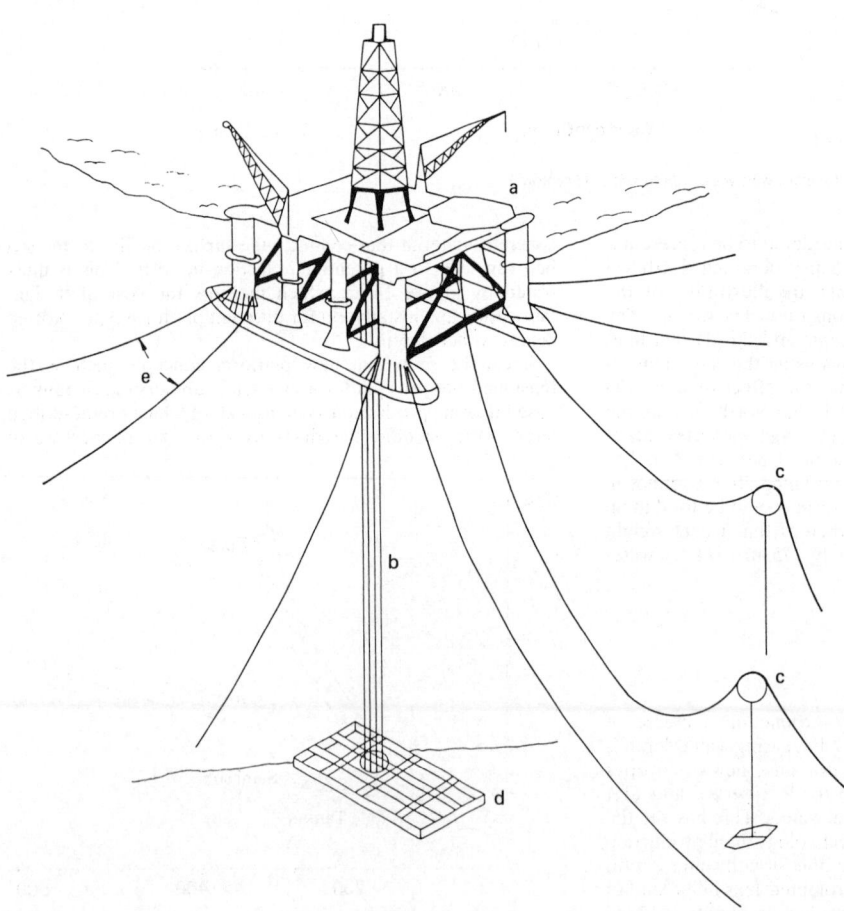

Figure 14.13 Typical semi-submersible-based floating production system: a – surface platform, b – multi-tube vertical drilling and production riser, c – flexible production risers, d – sea-bed template, e – catenary moorings

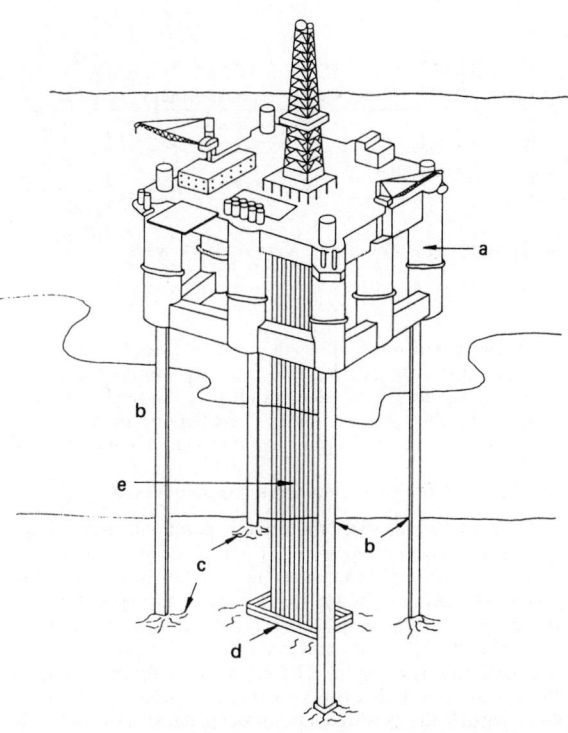

Figure 14.14 Typical tensioned buoyant platform: a – surface platform, b – tensioned tethers, c – tether foundations, d – template on sea bed, e – marine risers

the Hutton field in the North Sea. Due to the success of this development, it is likely that tethered buoyant platforms will be considered for a number of future field developments. The probable depth range of tensioned buoyant platforms is from approximately 120 m (394 ft) up to as high as 1500 m (4921 ft).

It is also possible to develop small isolated fields in water depths of up to 170 m (558 ft) in calm-weather areas by the use

of articulated column or single-anchor leg storage and tanker systems. Another form of economically viable compliant structure is offered by the guyed tower. These advanced compliant production platforms are described further by Thornton.[5]

14.4 Hydrodynamic loading

The design and operation of offshore structures requires a thorough understanding of the physical character and behaviour of the ocean environment and also of the atmosphere above it. Planet Earth is a globe with equatorial and polar diameters of 12 756 km (7926 mile) and 12 714 km (7900 mile), respectively, and a total surface area of 509.859×10^6 km^2 (19.686×10^7 mile^2). Over 70% of this surface area is covered by three large ocean areas which, in their order of size, are the Pacific, Atlantic and Indian Oceans. These oceanic areas also extend into the waters of the Antarctic, although this region can be regarded as a distinct southernmost ocean. The Arctic ocean is another body of water of significant size but is not comparable to the three major oceans of the globe.

Figure 14.1(b) gives a map of the surface of planet Earth and illustrates the extent of all the major and minor sea areas and continents. The average depths of all the oceans, including adjacent seas, is 3795 m (12 451 ft) and, therefore, on a global length scale, the oceans are very widely spread but thin sheets of water with an average vertical dimension of about 4 km (2.49 mile) compared to the Earth's equatorial circumference of 40 074 km (24 901 mile). Table 14.1 gives a summary of the physical properties of sea water, pure water and air.

Offshore structures in the oceans are subjected to environmental forces due to wind, current, tides and gravity waves. Atmospheric winds are induced by the latitude imbalance in the solar radiation absorbed by Earth. The general worldwide circulation pattern of atmospheric winds is caused by the consequent redistribution of energy that must take place. Atmospheric mean wind speeds can vary from still calm up to 52 m/s (101 knots) or more, with wind gust components having greater transient velocities. There also exists a corresponding current circulation in the oceans induced by a combination of

Table 14.1 Physical properties of pure water, sea water and air

Quantity and units	Pure water		Sea water (salinity 35°/oo)		Air at standard temperature and pressure
	0°C	20°C	0°C	20°C	0°C and 1.013 bar
Dynamic viscosity (kgm^{-1}s^{-1})	1.793×10^{-3}	1.002×10^{-3}	1.877×10^{-3}	1.072×10^{-3}	1.709×10^{-5}
Thermal conductivity (Wm^{-1}K^{-1}	0.561	0.599	0.566	0.600	0.0241
Kinematic viscosity (m^{2}s^{-1})	1.793×10^{-6}	1.004×10^{-6}	1.826×10^{-6}	1.046×10^{-6}	1.322×10^{-5}
Prandtl number	13.2	6.94	13.2	7.13	0.71
Bulk modulus (GNm^{-2})	1.993	2.190	2.172	2.388	0.0001013
Specific heat capacity (kJkg^{-1}K^{-1})	4.218	4.182	3.992	3.993	1.004[a]
Density (kgm^{-3})	999.839	998.205	1028.1	1024.5	1.293

[a] At constant pressure.

wind effects and the rotation of the planet. Typical current speeds vary between 0.25 m/s (0.49 knots) and 0.75 m/s (1.46 knots) at the ocean surface, although in some areas current speeds can be as high as 2.5 m/s (4.9 knots). A Meteorological Office publication[7] gives further details of global wind and current phenomena. Ocean waters are also subjected to the oscillatory astronomical phenomenon of tidal action caused by the gravitational acceleration of the Earth to the Moon and Sun – see Strahler and Strahler[8] for details.

However, by far the largest environmental loads on offshore structures are due to gravity waves. These can exist at the interface between any two fluids of different densities – air and water, for example – and are driven by an interaction between inertia of the fluid at the interface and its potential energy under gravitational action. The physical mechanisms underlying gravity wave formation by wind action and decay by internal losses are complex. Winds blowing over a water surface generate waves, with the length over which this occurs being called a 'fetch' and the resultant 'forced' waves classified as 'wind waves' or 'a sea'. Gravity waves are progressive (or travelling) in nature and can, therefore, move out of their area of formation and, when running free of the winds that generated them, are called 'a swell'. Such swell waves lose energy very gradually due to air resistance, internal friction and by friction with the sea bed in shallow water.

Thus swell waves can travel substantial distances across the Earth's surface with little reduction in intensity. Gravity waves are, however, dispersive in nature– their celerity (wave or phase speed) is a function of wave period or wave length so that they spread out (disperse) along their direction of wave propagation. Non-progressive (that is, standing) gravity waves also occur in confined waters such as harbours, inland waterways and in wave-testing tanks. These waves are not of interest for the design of offshore structures and so are not considered here.

In nature, many components of periodic waves with different wave heights, periods and directions of travel occur at the same time in a given area. The superposition of all of these wave components coupled with their dispersive behaviour leads to a randomly varying sea-surface elevation which can be treated by statistical methods. The simplest mathematical representation of gravity waves is known as linear wave theory. This theory permits the determination of irregular ocean waves by superposition of regular wave motions at different periods and from different directions. Gravity waves have periods ranging from as low as 0.8 s up to 25 s, with corresponding wave lengths of from 1 m (3.3 ft) up to 1 km (3290 ft), and with wave heights as high as 33 m (108 ft) from trough to crest.

Initially, some basic parameters describing the characteristics of a progressive wave are defined with reference to Figure 14.15. The wave number, k, is defined as

$$k = \frac{2\pi}{\lambda} \tag{14.1}$$

Wave frequency can be defined as f in Hz, ω in radians per second or period T in seconds. These parameters are related by

$$\omega = 2\pi f, \ T = \frac{2\pi}{\omega}, \ T = \frac{1}{f} \tag{14.2}$$

This theory, originally developed by Airy,[9] offers a powerful analytical solution for gravity waves. The following assumptions have to be invoked:

1. The water is assumed to be of uniform density with constant water depth.

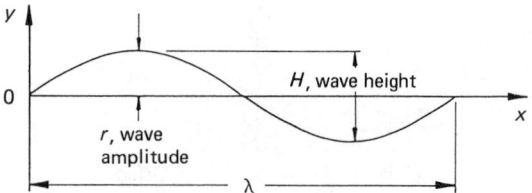

Figure 14.15 Wave elevation for a unit amplitude wave

2. Viscosity and surface-tension effects are neglected.
3. The square of particle velocities are considered to be negligible but vertical accelerations are not neglected. The wave heights are accordingly assumed to be very low, hence the theory is often called small-amplitude wave theory.
4. The wave motion is assumed to be irrotational.

Table 14.2 summarizes the velocity potential, stream function and derived properties for linear deep-water gravity waves. Equivalent expressions for shallow water are also listed. The expressions for shallow-water theory differ from the deep-water equations in having hyperbolic variations of vertical and horizontal particle displacements, velocities and accelerations with depth. The resultant difference in wave fluid behaviour is illustrated in Figure 14.16, which shows wave particle orbits with depth for deep and shallow water. In shallow water, particle motions describe ellipses instead of circles. The ellipses become shallower with decreasing depth and at the bottom the motion is purely horizontal. In terms of numerical values, the shallow-water equations tend to their deep-water simplifications for water depth $d > \lambda/2$.

Offshore structures are subjected to both steady and time-dependent forces due to the action of winds, currents and waves. Winds exert predominantly steady forces on the exposed parts of offshore structures, although there are significant gust or turbulence components in winds which induce high unsteady local forces on structural components as well as a low-frequency total force on the whole structure. Ocean currents also exert predominantly steady forces on submerged structures together with the localized effects of vortex shedding which can induce substantial unsteady forces on structural members. However, gravity waves induce by far the largest force on most offshore structures. The applied force is periodic in nature, although non-linear wave properties give rise to mean and low-frequency drift force. Non-linearities in the wave-loading mechanism also give rise to superharmonic force components. Both of these secondary forces can be significant if they excite resonance in a compliant structure. Only gravity wave loading is, therefore, considered here for brevity – see Patel[10] for further details on wind and current loading.

The procedure for calculating wave forces on offshore structures can be divided into fundamentally different approaches depending on the size of the structural member and the height and wavelength of incident waves. These parameters can be written in the form of two ratios: structural member diameter to wavelength (D/λ) and wave height to structural member diameter (H/D). Table 14.3 uses approximate values for the ratio K (which is a measure of (H/D)) to separate the wave regimes into its constituent parts.

For small structural members where $D/\lambda < 0.2$, Morison's equation is used to estimate forces due to wave action, with the implicit assumption that the diameter of the member is small enough in relation to the wave length so as not to alter incident wave characteristics to any significant extent. Mori-

Table 14.2 Summary of water wave properties

Quantity	Deep water	Shallow water ($d/\lambda < 1/2$)
Surface profile	$\eta = a\sin(kx-\omega t)$	$\eta = a\sin(kx-\omega t)$
Velocity potential	$\phi = -ace^{ky}\cos(kx-\omega t)$	$\phi = -ac\dfrac{\cosh k(y+d)}{\sinh kd}\cos(kx-\omega t)$
Stream function	$\psi = +ace^{ky}\sin(kx-\omega t)$	$\psi = +ac\dfrac{\sinh k(y+d)}{\sinh kd}\sin(kx-\omega t)$
Horizontal water velocity	$u = \partial\phi/\partial x$ $= +kcae^{ky}\cos(kx-\omega t)$	$u = +ack\dfrac{\cosh k(y+d)}{\sinh kd}\sin(kx-\omega t)$
Vertical water velocity	$v = \partial\phi/\partial y$ $= -kcae^{ky}\cos(kx-\omega t)$	$v = -ack\dfrac{\sinh k(y+d)}{\sinh kd}\cos(kx-\omega t)$
Horizontal acceleration	$= -k^2c^2ae^{ky}\cos(kx-\omega t)$	$= -k^2c^2a\dfrac{\cosh k(y+d)}{\sinh kd}\cos(kx-\omega t)$
Vertical acceleration	$= -k^2c^2ae^{ky}\sin(kx-\omega t)$	$= -k^2c^2a\dfrac{\sinh k(y+d)}{\sinh kd}\sin(kx-\omega t)$
Wave celerity	$c = \dfrac{\lambda}{T} = \dfrac{g}{\omega} = \left(\dfrac{g\lambda}{2\pi}\right)^{1/2} = \dfrac{gT}{2\pi}$	$c^2 = \dfrac{g\lambda}{2\pi}\tanh kd^{\,\text{a}}$
Wave length	$\lambda = \dfrac{2\pi c^2}{g} = \dfrac{2\pi g}{\omega^2} = \dfrac{gT^2}{2\pi}$	
Wave number	$k = \dfrac{2\pi}{\lambda} = \dfrac{\omega^2}{g} = \dfrac{g}{c^2} = \dfrac{4\pi^2}{gT^2}$	
Pressure[b] $P' - P_a$	$p = +\rho gae^{ky}\sin(kx-\omega t)$	$p = +\rho ga\dfrac{\cosh k(y+d)}{\cosh kd}\sin(kx-\omega t)$
Horizontal pressure gradient[b] $\partial p/\partial x$	$+\rho gake^{ky}\cos(kx-\omega t)$	$+\rho gak\dfrac{\cosh k(y+d)}{\cosh kd}\cos(kx-\omega t)$
Vertical pressure gradient[b] $\partial p/\partial y$	$+\rho gake^{ky}\sin(kx-\omega t)$	$\rho gak\dfrac{\sinh k(y+d)}{\cosh kd}\sin(kx-\omega t)$
Maximum wave slope	$\theta = ka = \dfrac{2\pi a}{\lambda}$	

[a] For very shallow water ($d \leqslant \lambda/20$), this expression becomes

$$c^2 = gd$$

with $c^2 = 0.968\,gd$ at $d = \lambda/20$.

For deeper water ($d > \lambda/2$), the expression becomes

$$c^2 = \frac{g\lambda}{2\pi}$$

with $c^2 = 0.996\,g\lambda/(2\pi)$ for $d = \lambda/2$.

[b] These terms exclude hydrostatic pressure.

son's equation, which was first put forward by Morison *et al.*,[11] is based on the assumption that wave forces can be expressed as the sum of a drag force due to wave fluid velocity and an inertia force due to wave acceleration. On the other hand, for larger structural members (with $D/\lambda > 0.2$), diffraction theories are necessary to account for the reflection and radiation of waves from the structural member. These potential flow methods, however, cannot account for viscous drag forces.

The second parameter of interest is the ratio H/D. Its importance is based on the fact that drag forces on structures (say, a pipe of circular cross section) in an oscillatory wave flow are dominated by the separation of flow behind the

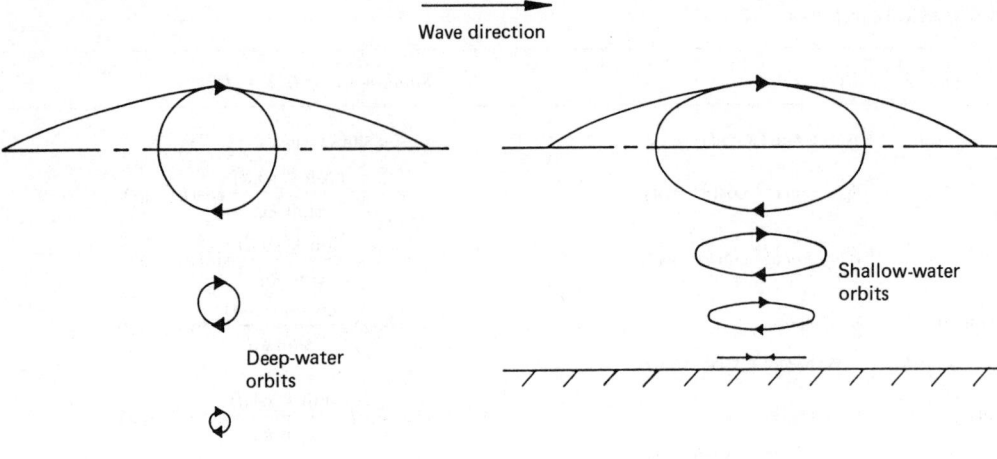

Wave direction

Deep-water orbits

Shallow-water orbits

Figure 14.16

Table 14.3 Wave loading regimes

D/λ	<0.20	>0.20
$K > 25$	Drag-dominated flow regime. Morison's equation with C_M and C_D values required for computing wave forces. Drag coefficient is function of Reynolds number.	
	For $R>1.5 \times 10^6$, $C_M = 1.8$, $C_D = 0.62$	
	For $10^5 < R < 1.5 \times 10^6$, $C_M = 1.8$, C_D varies from 1.0 to 0.6	
$5 < K < 25$	Intermediate regime between drag and inertia domination. Morison's equation applicable but published C_M and C_D values exhibit wide scatter. Flow behaviour and consequent loading complex and uncertain. For $R<1.5 \times 10^6$, $C_M = 1.8$, $C_D = 0.62$	
$K < 5$	Inertia-dominated regime. Morison's equation of diffraction theory for computing wave forces. $C_M = 2.0$ Effect of C_D is negligible	Morison's equation unsuitable for computing wave forces. Diffraction theory required

C_M = inertia coefficient, C_D = drag coefficient

$$R = \frac{U_m D}{\nu} \quad K = \frac{U_m T}{D}$$

U_m = peak velocity, T = wave period, ν = kinematic viscosity, R = Reynolds number K = Keulegan–Carpenter number.

cylinder and the formation of large vortices. For a small H/D ratio (<1.5), the wave height (and thus orbital diameter) are not unidirectional long enough for the flow to initiate separation and develop or shed vortices. In this case, drag forces are very small, acceleration-dependent inertia forces dominate and potential flow diffraction theory can be used to predict wave forces with confidence. At the other extreme for $H/D > 8$, approximately, the wave flow will have been unidirectional long enough for a substantial vortex flow to develop. Drag forces will then be large and a Morison formulation, which accounts for these, must be used. An intermediate region, where $1.5 < H/D < 8$, also exists where the flow regime can be

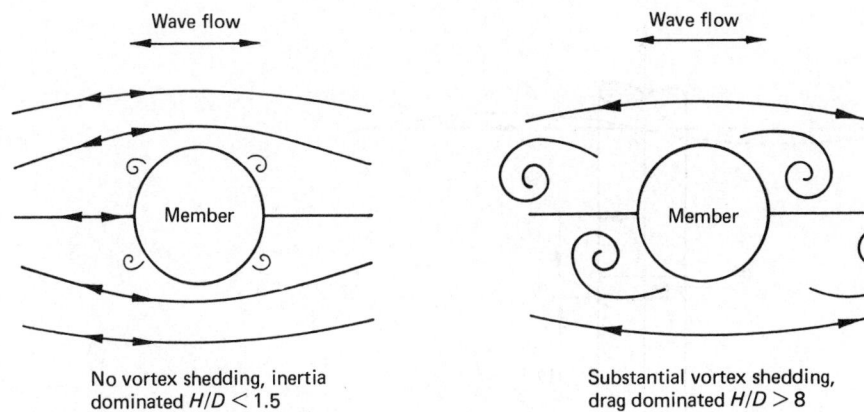

Wave flow

Member

No vortex shedding, inertia
dominated $H/D < 1.5$

Wave flow

Member

Substantial vortex shedding,
drag dominated $H/D > 8$

Figure 14.17 Inertia- and drag-dominated flows

highly complicated and wave forces are difficult to compute. The two flow regimes corresponding to H/D of < 1.5 and > 8 are illustrated in Figure 14.17. A non-dimensional parameter called the Keulegan–Carpenter number is a more rational measure of the ratio of water particle motion double amplitude to cylinder diameter than H/D, which is only valid close to the water surface. The Keulegan–Carpenter number is defined as

$$K = \frac{U_m T}{D} \qquad (14.3)$$

where U_m ms^{-1} is the maximum normal velocity in the oscillatory flow of period T seconds about the cylinder of diameter D metres.

The approximate range of numerical values encountered is from 0 up to 30 m for H and from 60 m to 620 m for λ. Thus a subsea pipe or the space frame of a tower jacket structure (with typical D from 0.4 to 3 m) would give large H/D and small D/λ – indicating a drag-dominated wave loading with use of Morison's equation being appropriate for computing the loading. Alternatively, a large monolithic gravity structure (with typical $D = 80$ m) would yield small H/D and large D/λ; thus pointing to an inertia-dominated loading regime and requiring potential flow diffraction theory for computing wave loads. The combination of both H/D and D/λ being small can occur but the other extreme of both H/D and D/λ being large cannot, because of the height limitation for stable waves which constrains the ratio H/λ to a maximum value of approximately 1/7.

Table 14.3 illustrates the wave-loading regimes described above. The Keulegan–Carpenter number (K) is used as a measure of ratio H/D.

14.4.1 Wave loads on slender structures

The mechanics for wave loading on slender tubular subsea pipelines or members of offshore jacket structures is described here. Such structures (with $D/\lambda < 0.2$) will be subjected to so-called Morison wave loading, which is based on the assumption that wave properties are not affected by the presence of the structure and that the total wave force can be expressed as a sum of inertia forces (due to wave fluid acceleration) and drag forces (due to wave fluid velocity).

An equation to represent this loading can be written as

$$dF = C_M \rho \, dV \, \dot{U}_n + C_D \frac{1}{2} \rho \, dS |U_n| U_n \qquad (14.4)$$

$\qquad$ inertia force $\qquad\qquad$ drag force

where dF is the total wave force on a pipeline element of volume dV and frontal area, dS, U_n, $\dot{U}_n$, are instantaneous wave fluid velocities and accelerations normal to the pipe axis, ρ is the fluid density and C_M and C_D are inertia and drag coefficients. The inertia coefficient is conventionally taken as the added mass coefficient plus one to account for the Froude–Krylov or undisturbed pressure force due to wave fluid acceleration – see Patel[8] for further details. The drag force term has a modulus sign to ensure that the drag force remains in the same direction as the wave velocity.

Although the Morison equation was originally developed for vertical piles in shallow water, for engineering purposes, it is assumed to be valid for cylinders of arbitrary orientation in deep and shallow water with the proviso that the coefficients C_M and C_D are chosen experimentally for the appropriate condition. Application of the Morison equation can be illustrated by developing expressions for the wave force on vertical and horizontal pipe segments.

Consider an element, dy, of a vertical circular cylinder of radius, r, as shown in Figure 14.18. The total force, dF, acting on this element in the direction of wave propagation is the sum of inertia and drag force components. This force can be integrated to yield the total force, F, and moment about the sea bed as

$$F = C_M \rho \pi r^2 \int_{-d}^{0} \dot{u} \, dy + C_D \rho r \int_{-d}^{0} |u| u \, dy$$

and

$$M = C_M \rho \pi r^2 \int_{-d}^{0} (d + y)\dot{u} \, dy + C_D \rho r$$

$$\int_{-d}^{0} (d + y)|u| u \, dy \qquad (14.5)$$

where u and $\dot{u}$ are horizontal components of wave velocity and accelerations. Using expressions for u and $\dot{u}$ for linear wave

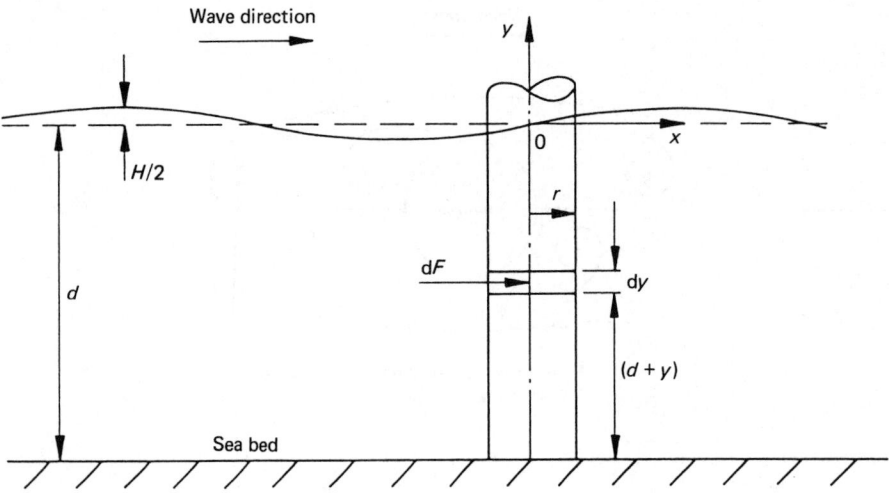

Figure 14.18 Wave force on vertical cylinder

theory from Table 14.2 with $x = 0$ and $\theta = \omega t$, equations (14.5) can be readily integrated to yield

$$F = -\frac{2\pi\rho r H^2 \lambda}{T^2} \{K_2\, C_M \cos\theta + K_1\, C_D|\sin\theta|\sin\theta\}$$

and

$$M = -\frac{2\rho r H^2 \lambda^2}{T^2} \{K_4\, C_M \cos\theta + K_3\, C_D|\sin\theta|\sin\theta\}$$

where

$$K_1 = \frac{1}{16\,\sinh^2 kd}\,[2kd + \sinh 2kd] \tag{14.6}$$

$$K_2 = \frac{\pi r}{2H}$$

$$K_3 = -\frac{1}{64\,\sinh^2 kd}\,\{2k^2 d^2 + 2kd \sinh 2kd + 1 - \cosh 2kd\}$$

$$K_4 = \frac{\pi r}{4H\,\sinh kd}\,\{1 + kd\,\sinh kd - \cosh kd\}$$

14.4.1.1 Drag and inertia coefficients

The value of drag coefficients, C_D, to be inserted in Morison's equation can only be obtained experimentally. In theory, the value of the inertia coefficient, C_M, can be calculated (it is, for example, 2.0 for a smooth cylinder in an ideal fluid). However, measured values are used in practice, particularly when drag is the dominant force. For irregular shapes, the inertia coefficient can be calculated from diffraction theory.

Many experiments have been designed to measure values of drag and inertia coefficients in steady or planar oscillatory flows but their results can only be used for wave force prediction with caution. The most useful experimental measurements have been made in circumstances which model full-scale conditions within the ocean environment. Measurements made offshore on large test structures in real seaways are especially valuable.

One problem facing the user of Morison's equation is the large scatter in values of the inertia and drag coefficients. However, there is a useful degree of correlation between the coefficients and two flow parameters – Keulegan–Carpenter number, K, and Reynolds number, R. Nevertheless, the scatter and hence some uncertainty remain. Table 14.3 gives a summary of generally accepted added inertia and drag coefficients for flow around circular cylinders in unconfined flow remote from solid boundaries such as the sea bed for pipelines. The effects of such sea-bed proximity on coefficient values are examined in more detail by Patel.[10]

Sarpkaya[12,13] has carried out many systematic studies of the variations of inertia and drag coefficients for circular cylinders in planar oscillatory flow.

14.4.2 Wave loads on large bodies – diffraction theory

The calculation of wave forces on bodies with dimensions that span a significant proportion of the wave length ($> 0.2\lambda$) must account for the fact that the wave flow will be influenced by the presence of the body. The solution of the Laplace governing equation with the usual sea-bed and free surface boundary condition together with the additional condition of no flow through the body surface will give rise to additional component waves in the solution which are called scattered or diffracted waves. The wave force on the body is then due to the incident wave as well as the scattered wave field arising from the presence of the body. Furthermore, the Froude–Krylov force in such a flow field with a large body can no longer be written as $\rho V \dot{U}$ (see Patel[10]) because the wave-induced pressure gradient around the body is no longer constant. At the same time, the conditions that require diffraction theory to be used ($D/\lambda > 0.2$) also imply that the ratio of wave height to body dimension, H/D, is less than unity since $H/\lambda < 1/7$. For such small values of H/D, drag forces will be small since flow amplitudes will be less than the body dimension and flow separation will not occur. These fortuitous circumstances mean that a potential flow solution of the wave diffraction problem will represent a physically realistic situation.

A statement of the full-wave diffraction problem in three dimensions can be written as follows using an axes system with

Oxy in the still water surface and Oz pointing vertically upwards. The governing equation is

$$\frac{\partial^2\phi}{\partial x^2} + \frac{\partial^2\phi}{\partial y^2} + \frac{\partial^2\phi}{\partial z^2} = 0 \tag{14.7}$$

with the linearized free surface boundary condition at $z = 0$ of

$$\frac{\partial^2\phi}{\partial t^2} + g\frac{\partial\phi}{\partial z} = 0 \tag{14.8}$$

and the sea-bed boundary condition at $z = -d$ of

$$\frac{\partial\phi}{\partial z} = 0 \tag{14.9}$$

together with the no-flow through the immersed body boundary condition of

$$\frac{\partial\phi}{\partial n} = 0 \tag{14.10}$$

at the body surface, where n denotes a direction normal to this surface. All the assumptions inherent in the governing equations and the linearized free surface boundary condition, therefore, apply to this linear diffraction problem, that is, irrotational and inviscid flow of small wave amplitude.

The solution to this wave-diffraction problem is obtained by writing the velocity potential as the sum of the incident and scattered potential, ϕ_i and ϕ_s, respectively. Thus

$$\phi = \phi_i + \phi_s \tag{14.11}$$

Furthermore, the scattered waves generated by the presence of the body boundary have to be restricted in the mathematical formulation to outgoing waves only. This requires that the scattered wave potential satisfies the condition

$$\frac{\partial\phi_s}{\partial r} + \frac{1}{c}\frac{\partial\phi_s}{\partial t} = 0 \tag{14.12}$$

where r is radial distance from a point on the body surface and c is wave celerity. This can be written[14,15] as

$$\lim_{r\to\infty} r^{1/2}\left[\frac{\partial\phi_s}{\partial r} - ik\,\phi_s\right] = 0 \tag{14.13}$$

where the factor $r^{1/2}$ takes account of the directional spreading of waves.

Since the incident wave potential, ϕ_i, is known, the boundary condition at the body surface can also be written as

$$\frac{\partial\phi_s}{\partial n} = -\frac{\partial\phi_i}{\partial n} \tag{14.14}$$

Once ϕ_s is obtained as a solution to this problem, wave-induced pressures can be obtained from the linearized Bernoulli equation and integrated to obtain forces and moments. Note that equations (14.11) and (14.14) only apply to a stationary body. A moving body will induce additional potentials due to its motions and the above equations will need to be extended to account for this as is described subsequently in Section 14.6.

Patel[8] gives a review of solution schemes for a diffraction analysis. Typical methods in the research literature are presented by MacCamy and Fuchs,[16] Garrison and Chow,[17] Eatock-Taylor and Waite[18] and Zienkiewicz et al.[19]

14.5 Structural strength and fatigue

This section describes the final stages of the design process for an offshore structure where static and dynamic environmental forces calculated from the methods outlined earlier are converted to structural stresses and stress load cycles. These are then used to deduce design safety factors and estimate fatigue lives. Static and dynamic structural analyses of offshore structures are complicated by non-linear effects, due predominantly to soil structure interaction in the former case and non-linearities arising from gravity wave mechanics and drag force loading in the latter. These methods of analysis and the effects of non-linear behaviour are considered in detail in this section.

14.5.1 Structural analysis method

Steel-framed jackets in shallow water can be analysed for structural stresses with acceptable accuracy by using a quasi-static environmental loading coupled with an elastic structural analysis modified for the non-linear behaviour in the soil foundation. In this approach, winds and currents are assumed to apply static loads with wave action applying a dynamic loading which is translated into dynamic structural stresses through a quasi-static stress analysis. The technique assumes that resonant frequencies of structural vibrations are sufficiently separated from wave frequencies so that dynamic magnification has a negligible effect on calculated stresses.

In the early days (before the 1960s), structural analysis of jackets was carried out using hand calculations with simple frame theory. However, the advent of computers and the development of finite-element structural analysis methods has revolutionized structural calculations in the offshore industry. Finite-element analysis program packages such as NASTRAN and STRUDL are used in conjunction with wave-loading and foundation-analysis programs to carry out complex computer-aided design calculations for three-dimensional steel-framed structures as well as for monolithic concrete ones.

The basic theory underlying one such finite-element analysis called the direct stiffness method is described here. The basis of the method is that the structure to be analysed is discretized into a number of small elements – with a framed jacket structure represented by an assembly of beam elements where as a concrete gravity structure could also be described by beam elements or, alternatively, by cylindrical shell elements. The displacements of nodal points on the elements are taken as unknowns and the finite-element method is implemented using the following steps:

1. Initially, physical data for the structure configuration, member properties and support constraints are identified. The structure is then discretized into finite elements and the unknown nodal displacements are identified and numbered using a systematic indexing procedure.
2. This is followed by evaluation of member-stiffness matrices using local member axes. These matrices are then transformed into global structure axes and assembled into a global stiffness matrix.
3. The loads acting on the members and nodes are also assembled into a global load vector. This requires that forces acting on each member other than at the nodes are transformed into equivalent joint loads before being added onto the known loads at the nodes such that the structure is considered as being loaded at the nodes only.
4. It is then necessary to rearrange the equation relating the nodal displacement vector and stiffness matrix to the load vector in order to substructure out degrees of freedom which are constrained (that is, their nodal displacements

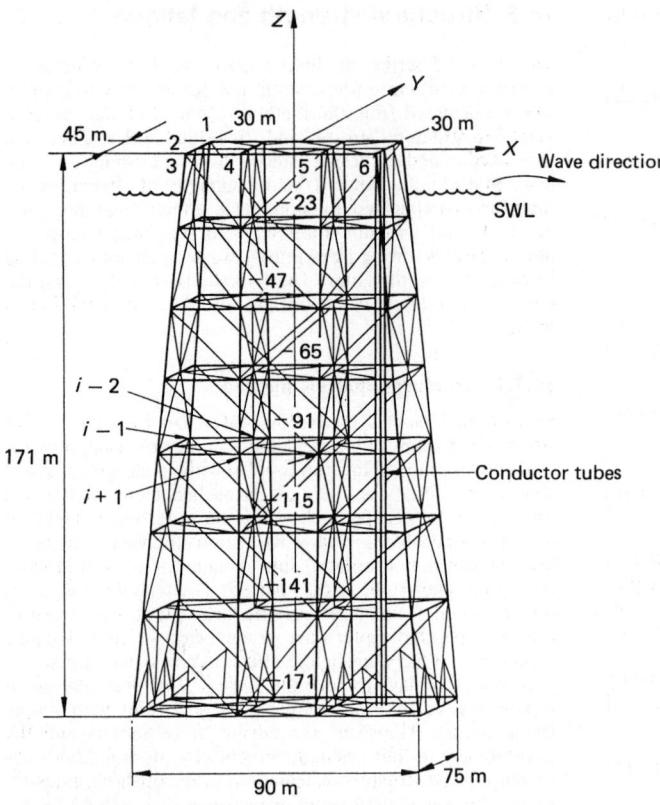

Figure 14.19 Typical idealization of a jacket structure

are known or are zero) in order that a matrix equation for the unknown displacements only can be formed.

5. The penultimate step in the procedure is to use the stiffness matrix and load vector to arrive at a solution for the displacements at each of the nodes.

6. In the final step, the displacements are used in combination with the equivalent joint loads to compute member forces and resultant internal stresses.

The mathematical formulation of the above procedures is described in detail for item (1) above as an example.

A three-dimensional (space frame) beam element idealization of a typical jacket structure is shown in Figure 14.19. The jacket structure consists of up to 600 beam elements to represent each of the tubular members making up the frame. A global axes system Oxyz is first defined with respect to the whole structure and all of the structural nodes are numbered in a systematic manner around the structure. A typical numbering system could, as shown in Figure 14.19, start from a horizontal level at the origin with increasing node numbering in a prescribed direction around the nodes at that level, followed by a similar procedure at the next level down, and so on. Once this node definition is complete, structure data on node coordinates relative to global axes, member lengths, areas of cross section and second moments of area can be systematically identified in terms of the node numbering. This is followed by definition of six degrees of freedom for each node – three in translation and three in rotation parallel to and about the global axes directions Ox, Oy and Oz. Thus the number of degrees of freedom will be six times the number of

nodes, although a small proportion of these freedoms (at the base of the jacket of Figure 14.19, for example) will be constrained to be zero or prescribed in other ways to account for the effect of foundations.

The next step in the analysis is to define the stiffness of each member in the framework as described in item (2) above. This and the subsequent procedures are described in more detail by Patel.[10]

Application of the above technique to gravity structures differs in a number of ways. The structural analysis can be carried out through a simple beam element model or, alternatively, more representative cylindrical shell elements can be used to model the caisson structure and vertical walls. In both cases, it is necessary to include some idealization of the foundation elasticity either by equivalent springs or by continuing the finite-element idealization into the foundation using solid elements as described by Penzien and Tseng.[20]

Furthermore, wave forces are calculated using diffraction theory only or a combined Morison/diffraction approach. It is usual to calculate the forces separately and to define them as a distribution of equivalent nodal loads for input to the analysis. Since the structural deflections of these relatively rigid structures are very small compared to water-particle displacements, the non-linear drag force in the Morison equation can be explicitly computed and does not present analysis problems. More flexible structures in deep-water locations and other compliant structures need to be analysed using the methods of structural dynamics. Patel[10] gives an overview of these with details presented by Bathe and Wilson[21] and Malhotra and Penzien,[22] among others.

14.5.1.1 Foundation effects

For piled jackets, the structural calculations described above have to be supplemented by a lateral pile-loading analysis which quantifies the coupled response of an embedded pile and non-linear material behaviour of the surrounding soil. When a single pile is loaded by a horizontal force due to environmental loading of the jacket structure of which it is a part, the pile will deflect against the surrounding soil and take up a typical deflection and bending moment profile shown in Figure 14.20. The magnitude of deflection, rotation and bending moment at the end of the pipe will depend on a balance of forces between the pile and surrounding soil. Determination of this behaviour requires data on the lateral soil reaction on the pile as a function of depth together with a means of structural analysis of the pile itself due to a varying soil loading. Taking the former first, the soil reaction is a function of soil type, pile properties, its loading, deflection and the soil depth, and may be expressed as

$$p = - E(x,y) \, y \qquad (14.15)$$

where p is the distributed soil reaction force, y is the lateral pile deflection and E is the modulus of elasticity of the soil which will vary with soil deflection, y, and depth, x. From basic beam-bending theory, the structural behaviour of the pile can be written as

$$E_p I_p \frac{d^4 y}{dx^4} = w \qquad (14.16)$$

where $E_p I_p$ is the elastic modulus of the pile and w is the distributed lateral load on the pile. Combining equations (14.15) and (14.16) yields the basic differential equation for a laterally loaded pile as

$$E_p I_p \frac{d^4 y}{dx^4} + E(x,y) \, y = 0 \qquad (14.17)$$

One of the greatest difficulties in solving this equation arises from the fact that the soil modulus, E, varies both with pile

deflection, y, and with depth, x. For many soil types, E increases with depth and can be modelled by the equation

$$E = k \, x \qquad (14.18)$$

where k is a constant. A simple but very approximate method for calculating pile behaviour is obtained by assuming that the pile is completely fixed against rotation at some point below the mud line (such as A in Figure 14.20). The position of this point is selected from experience and the forces exerted on the pile by the soil are neglected above the point. Despite the desirable simplicity of this method, it does not reflect the physical behaviour of the pile and soil and is not recommended for use.

An alternative method for computing pile/soil interaction is to solve equation (14.18) by a finite-difference numerical scheme, although analytical solution can be obtained for the special case of constant E (see Poulos and Davis[23] and Reese[24]).

However, this numerical analysis can only be carried out if the soil modulus of elasticity function $E(x,y)$ is known. One form of E that is commonly used is

$$E = E_L \left[\frac{x}{L} \right]^N$$

where E_L is the value of the modulus at the pile bottom and N is an empirical index equal to or greater than zero. $N = 0$ corresponds to a constant modulus whereas $N = 1$ gives a linear variation as in equation (14.18). In general, N is taken between 0 and 0.15 for clay soils and N equals 1 for granular soils. Analytic solutions for equation (14.17) are available for $N = 0$.

For real soils, however, the relationship between soil reaction and deflection is non-linear with the reaction reaching a limiting value when the deflection is sufficiently large. Figure 14.21 presents typical soil reaction against deflection curves for a consolidated clay soil and demonstrates the flattening out of the reaction/deflection curves at large deflections. These curves are conventionally referred to as p-y curves.

The non-linear soil reaction curves are accommodated in analysis by rewriting equation (14.17) as

$$\frac{d^2 M}{dx^2} + S(x) \frac{d^2 y}{dx^2} + E(x,y) \, y = 0 \qquad (14.20)$$

where M is the pile moment at depth x and the additional effect of a variation of axial force S with depth x is accounted

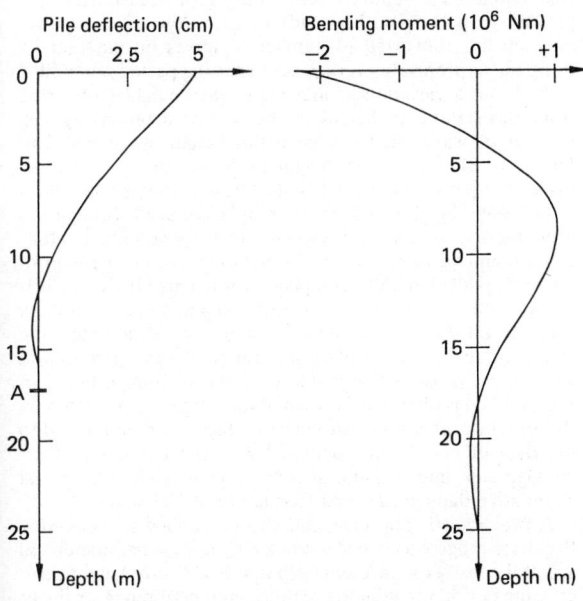

Figure 14.20 Deflection and moment of an embedded pile

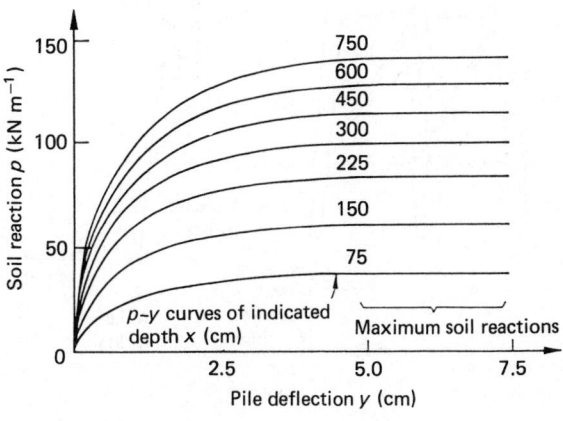

Figure 14.21 Typical p-y curves for a consolidated clay

for (see Reese[24] for further details). This equation can be solved by a finite-difference technique similar to that described above. The solution requires inputs of *p-y* curves at various depths (see Figure 14.21), although the method implicitly assumes that the soil behaviour at a particular depth is independent of that in adjacent layers. Equation (14.20) and the *p-y* curves are solved using an iterative finite-difference technique. The resultant shear forces and bending moments at the pile surface are derived as functions of pile top displacement and rotation, respectively, and combined with a jacket finite element analysis as in Figure 14.22 to complete definition of the pile and structure interaction.

The *p-y* data necessary for the analysis can be obtained in three different ways – full-size or model-scale lateral loading tests on instrumented piles are one of the most reliable methods, although practical considerations often restrict *p-y* data sources to laboratory tests or empirical correlations based on soil survey data.

Other pile and foundation analyses also use an elastic continuous mathematical model of the surrounding soil together with a finite-beam element model of the pile to quantify pile deflections, and soil behaviour. Poulos and Davis[23] give further details. Analysis of piles for jacket platforms often needs to account for pile groups and their interaction with the soil. Focht and Kock[25] and O'Neill et al.[26] describe extensions of the techniques described above for pile groups.

14.5.2 Structure failure and fatigue life calculations

The safe structural design of a jacket structure ultimately depends on the detailed design of individual tubular joints. The analyses described above are readily able to ensure that members between joints do not fail from buckling or over-stressing in tension or bending. There are, however, several failure mechanisms that have to be designed against in the definition of a tubular joint. A tubular joint brace loading the main member in compression can initiate failure by local buckling or by punching shear of the main member wall. Tensioned braces can cause failure due to crack growth or lamellar tearing of the main member wall.

However, a large amount of structural design studies, finite-element analyses and model tests have ensured that these failure mechanisms can be almost completely avoided. The majority of useful recommendations made by such work

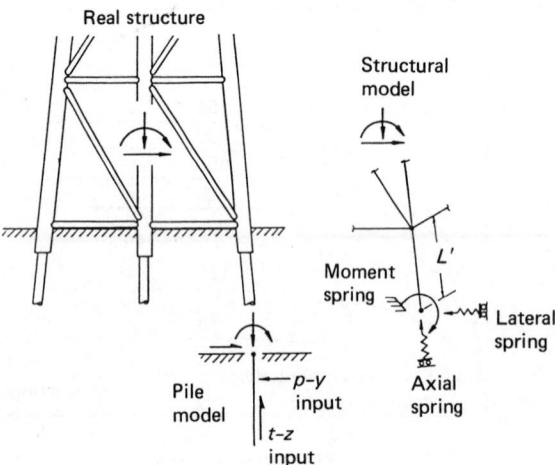

Figure 14.22 Spring model to represent pile

have been incorporated into codes of practice, of which those issued by the American, Petroleum Institute[27] and the British Standards Institution[28] are good examples.

At present, the largest source of structural failures in operation are due to fatigue damage compounded by the effects of corrosion. A metal will fail under repeated cycles of relatively low stress levels due to the growth of internal cracks that can be initiated from local construction defects or at stress-concentration sites. Such metal fatigue failures are quantified from experimental data by the use of *S-N* curves which are plots of stress range, *S*, against number of cycles to failure, *N*, with both axes plotted in logarithmic form. The fatigue behaviour of a variety of steels can, therefore, be characterized by *S-N* curves of the type shown in Figure 14.23 obtained from BS 6235: 1982. These curves show that steel specimens in air do exhibit a low stress range level (called the fatigue limit) below which the metal would have a theoretically infinite life, although this property disappears if the effects of combined stress range cycles and corrosion are considered.

The Palmgren–Miner cumulative fatigue damage rule is used to quantify fatigue damage caused by a large number of cycles at low stress ranges. The rules give the resultant cumulative damage ratio, *D*, as

$$D = \sum_{i=1}^{N} \frac{n_i}{N_i} \leqslant 1 \tag{14.21}$$

where *N* denotes the total number of stress range intervals being considered, n_i is the number of cycles experienced by the structure within the *i*th stress range and N_i is the number of cycles to failure given by the *S-N* curve for this *i*th stress range. Failure will occur when parameter *D* reaches unity. If the number of stress range cycles used in the calculation correspond to one year's operation then the fatigue life of the structure in years can be given by the reciprocal of the cumulative damage.

In design practice, the fatigue life is calculated using a deterministic approach illustrated by the left-hand column of the flow diagram of Figure 14.24. The starting point for such an analysis is the significant wave height against average zero crossing period scatter diagram, typically averaged for one year's weather conditions. The number of occurrences (expressed in parts per thousand) in each row of the scatter diagram are converted to number of waves in one year by using the appropriate average zero-crossing period and added to build up a variation of maximum wave height (converted from the significant height in the scatter diagram) against number of waves that exceeded this height, as in box 2 of Figure 14.24. Each wave height range in this figure has an associated period range that is also presented with the figure.

The wave heights and period ranges are used through the wave theory, loading and response analyses described earlier to yield an equivalent curve of stress range against number of loading cycles that will be applied at this range in box 3. It is usual practice in the calculation to select a wave period for any wave height that will produce the largest stress, although the choice of an average between the range of wave periods will yield a less conservative result. A curve such as in box 3 of Figure 14.24 is obtained for each of the fatigue-prone zones on the structure. A stress–concentration factor is then applied to the data of box 3 and combined with the Palmgren–Miner damage rule and a material *S-N* curve to yield the annual cumulative damage, *D*, and fatigue life of 1/*D* years.

A probabilistic approach can also be applied to estimating the stress range exceedence of box 8 by using a parameterized spectral density to represent each significant wave height/zero-crossing period range in the scatter diagram of box 1. A linear structural response analysis is then carried out to yield the

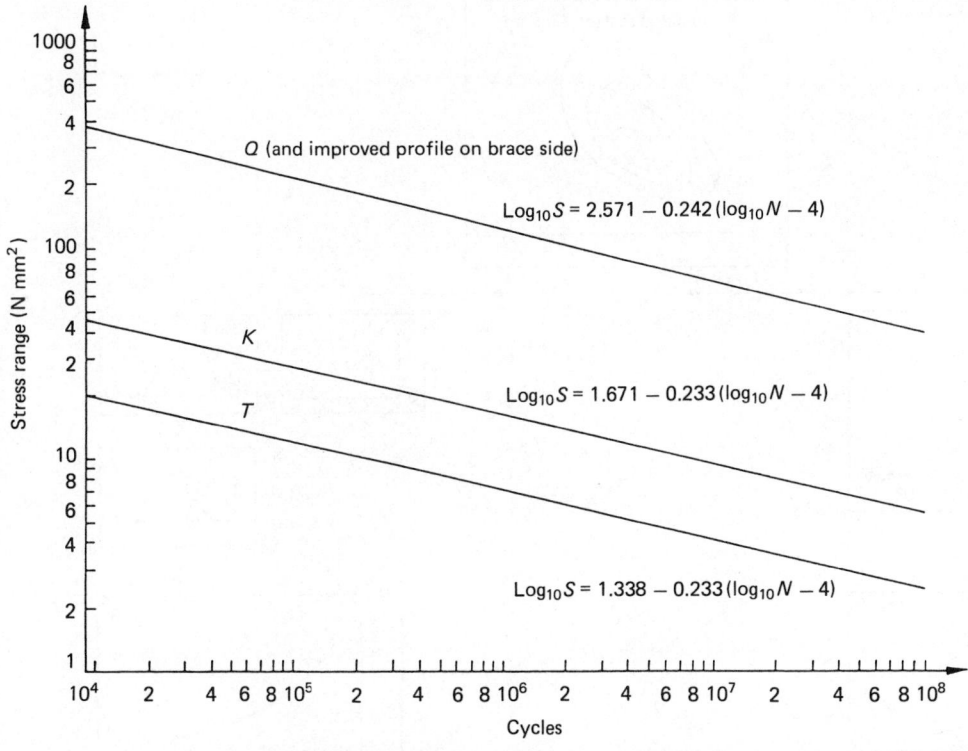

Figure 14.23 Typical *S-N* lines for welded tubular joints. *Q* line is used for general fatigue-life calculations and the *K* and *T* lines for an appropriate 'punching shear' method (from British Standards Institution[28])

stress spectrum (as in box 6) at all points of interest. A Rayleigh probability distribution of stress peaks (box 7) is then assumed to estimate the probability of occurrence of each stress range from which the figure of box 8 can be derived and the fatigue analysis completed. It should be noted that the average zero-crossing period of the stress spectrum in box 6 will be different from that of the wave spectrum of box 5, and will need to be calculated in order to determine the numbers of stress cycles applied in a year.

The effect of sea-water corrosion and the value of stress concentration factors to be used are two of the most important sources of uncertainty in fatigue analysis of an offshore structure. The American Petroleum Institute[27] and the British Standards Institution[28] give some guidance on these matters.

Much recent progress has been made on quantifying fatigue crack growth in tubular joints by the application of the theory of fracture mechanics coupled with more accurate in-service crack-detection techniques – see, for example, the work of Dover and Connolly[29] and Dover and Wilson.[30] Their combined use is leading to more reliable monitoring and rectification of fatigue crack growth during the life of an offshore structure.

14.6 Dynamics of floating systems

14.6.1 Heave motion of cylindrical buoys

The transformation of a problem concerned with fluid structure interaction of a floating body into the governing equation of a second-order dynamic system is illustrated by considering heave motion of the small buoy shown in Figure 14.25. Coordinate y denotes vertical motion of the water surface and z the resultant buoy vertical motion. Now the total force acting on the buoy will be due to added mass, damping and hydrostatic stiffness induced forces due to differential motion between the buoy and surrounding fluid. The net force due to these effects will serve to accelerate the buoy. Thus the equation of motion can be written as

$$m\,\ddot{z} = A\,(\ddot{y} - \ddot{z}) + B(\dot{y} - \dot{z}) + C\,(y - z) \qquad (14.22)$$

where m is the buoy mass and A, B, C are coefficients expressing the added mass, damping and hydrostatic stiffness induced forces on the buoy. Here the damping force is taken to be linear and the Froude–Krylov force – applying through dynamic wave pressure on the buoy base – is given by the stiffness term Cy for a buoy.

Rearranging equation (14.22) to bring unknown buoy motion terms to the left-hand side gives

$$(m + A)\,\ddot{z} + B\,\dot{z} + C\,z = F(t) \qquad (14.23)$$

where

$$F(t) = A\,\ddot{y} + B\,\dot{y} + C\,y \qquad (14.24)$$

Thus, there are two equations governing the behaviour of the buoy in waves. Equation (14.24) is concerned with the mechanism of wave elevation motion, y, exerting an exciting force $F(t)$ on the buoy whereas equation (14.23) is of similar form to the governing equation for a second-order dynamic system. The form of the above equations is typical of that obtained for all floating bodies in that two transfer functions

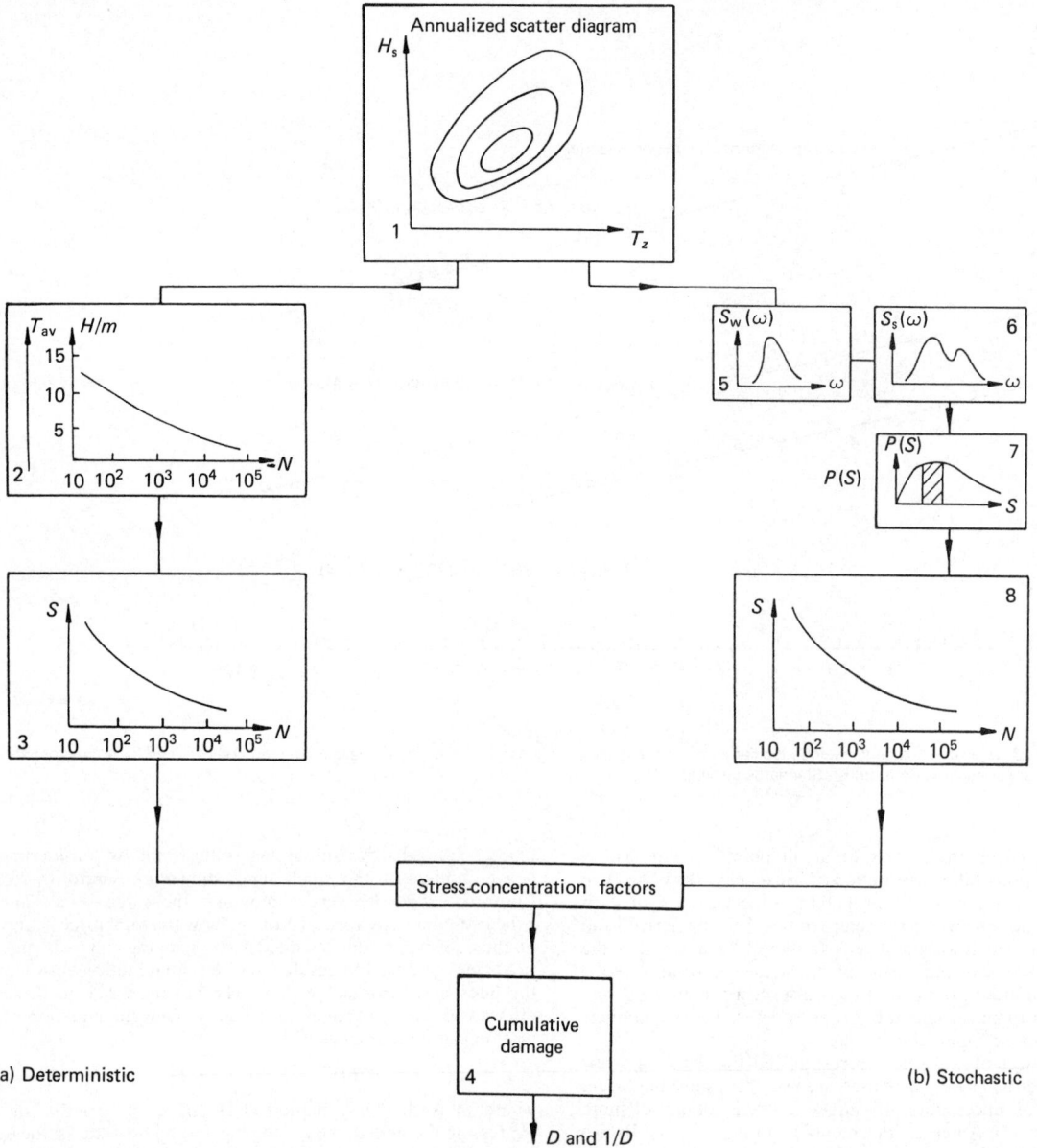

Figure 14.24 Fatigue-life calculation procedures

(from wave elevation to force and from wave force to motion) are involved in deriving a body-motion response from wave elevation.

A typical but very simplified solution procedure for the above equations is presented here to illustrate these ideas further. Taking buoy height, h, as very small compared to wave length, a wave of amplitude a with elevation

$$y = a \cos \omega t \tag{14.25}$$

can be used with equation (14.24) to give

$$F(t) = a \left[(C - A \omega^2) \cos \omega t - B \omega \sin \omega t\right] \tag{14.26}$$

Thus the wave force is a function of A, B, C, frequency, ω,

and wave amplitude, a, and can be expressed as an amplitude and phase transfer function of the form

$$F(t) = F_a \cos (\omega t + \epsilon)$$

where

$$\frac{F_a}{a} = \left[(C - A\omega^2)^2 + (B\omega)^2\right]^{\frac{1}{2}} \tag{14.27}$$

and

$$\epsilon = \tan^{-1} \left[\frac{-B\omega}{C - A\omega^2}\right]$$

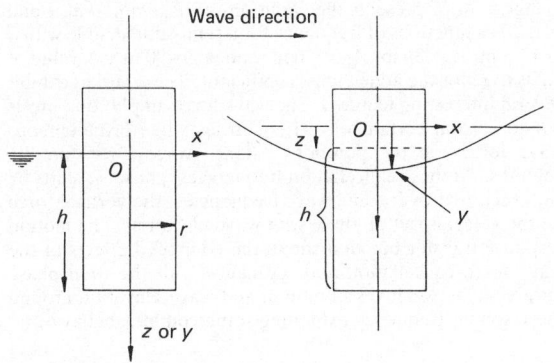

Figure 14.25 Heave motion of a buoy

Equation (14.27) possesses all the features of a second-order dynamic system transfer function (see Warburton[31]) with its amplitude and phase varying with frequency of excitation. Equation (14.23) then becomes

$$(m + a)\,\ddot{z} + B\dot{z} + Cz = F_a \cos(\omega t + \epsilon) \tag{14.28}$$

and has a solution of the form

$$
\left.
\begin{aligned}
& z = z_a \cos(\omega t + \epsilon + \emptyset) \\
& \text{where} \\
& \frac{z_a}{F_a} = \frac{1}{\{[C - (m + A)\omega^2]^2 + (B\omega)^2\}^{1/2}} \\
& \text{and} \\
& \phi = \tan^{-1}\left[\frac{B\omega}{C - (m + A)\omega^2}\right]
\end{aligned}
\right\} \tag{14.29}
$$

Note that the wave force leads the wave elevation by ϵ and the buoy motion lags the wave force by ϕ. When $m \to 0$, $\epsilon + \phi = 0$ and suggests that a mass-less buoy will follow the wave profile exactly without phase lag. The amplitude and phase transfer functions from wave force to buoy motion will be like that of a second-order dynamic system, but the overall wave elevation to buoy motion amplitude and phase transfer functions will be distorted by the presence of the wave elevations to wave force transfer functions. It is recommended that both of these transfer functions be evaluated independently during the motions analysis of a floating body to obtain a clearer physical understanding of its behaviour. (Figures 14.27, 14.29 and 14.30 present phase transfer for typical floating offshore structures.)

For a cylindrical buoy of radius, r, and height, h, floating in water of density, ρ, the terms m, c and A can be given by

$$
\left.
\begin{aligned}
& m = \rho \pi r^2 h \\
& C = \rho \pi g r^2 \\
& A = \frac{1}{3}\,\pi \rho r^3
\end{aligned}
\right\} \tag{14.30}
$$

where the added mass, A, in vertical motions of the cylinder end is approximated by that of a hemisphere of radius, r, and added mass coefficient 0.5 on the cylinder end. Other approximations for A involve treating the cylinder end as a flat disk moving along its axis of symmetry.

14.6.2 Heave motion of a semi-submersible

The heave motion to waves of a semi-submersible of simple geometry can be evaluated by treating the vessel as a collection of slender members on which the loading can be calculated using the Morison equation with deep-water linear wave theory. In this particular derivation, the damping and wave-induced drag forces are neglected for simplicity. Thus the equation of vertical motion for a four-column, twin-pontooned semi-submersible of the configuration shown in Figure 14.26 can be written as

$$(m + A)\,\ddot{z} + C\,z = F(t) \tag{14.31}$$

For a simple four-column, twin-pontooned semi-submersible of the type shown in Figure 14.26 the coefficients of this equation are

$$
\begin{aligned}
& m = 4\rho[A_v d + A_h L] \\
& A = 4\rho C_m A_h L \\
& C = 4\rho g A_v
\end{aligned}
$$

where A_v and A_h are the areas of cross section of the vertical columns and rectangular pontoons, respectively, and C_m the added mass coefficient for the pontoons in vertical motions.

The vertical wave-exciting force on the vessel can be taken as the sum of wave forces on the pontoons and columns to yield

$$
\left.
\begin{aligned}
& F(t) = \int_{-L}^{L} 2\rho A_h\,(1 + C_m)\,\dot{w} \cdot \mathrm{d}x \\
& \qquad\quad + 2A_v\,(p_f + p_r)
\end{aligned}
\right\} \tag{14.32}
$$

where $\dot{w}$ is the vertical wave particle acceleration at $z = -d$ and p_f, p_r are wave pressures at $z = -d$ and $x = +L$ and $-L$, respectively. Note that the pontoon/column intersections have been approximated by lengths d and L to simplify the above algebra.

The Froude–Krylov force normal to the pontoon axis is calculated using an inertia coefficient $(1 + C_m)$ whereas the axial Froude–Krylov force on the vertical columns is calculated as a product of dynamic wave pressure and column area of cross section. The former approach is valid for calculating normal forces on a slender member about which wave properties can be considered constant, whereas the wave-pressure formulation is necessary for axial forces on members whose length is comparable to incident wave length.

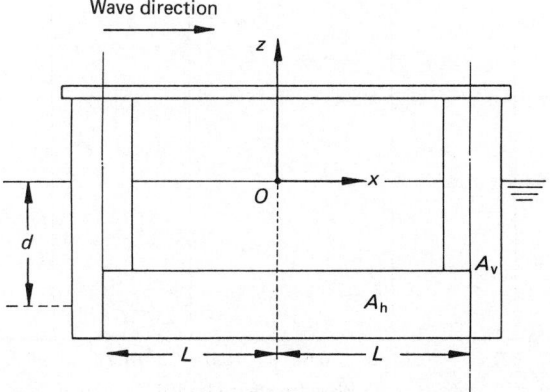

Figure 14.26 Side view of a four-column semi-submersible

Using the Oxz axes of Figure 14.26, the wave particle vertical acceleration and pressure can be written as

$$\dot{w} = -a\omega^2 \, e^{-kd} \sin(kx - \omega t)$$

and

$$p = +a \frac{\rho\omega^2}{k} e^{-kd} \sin(kx - \omega t) \qquad (14.33)$$

and substituted into equation (14.32) to yield the wave force relationship

$$F(t) = 4a\rho g e^{-kd}[-A_v \cos kL$$
$$+ A_h(1 + C_m) \sin kL] \sin \omega t \qquad (14.34)$$

Evaluation of equation (14.34) followed by the solution of equation (14.32) yields the wave-induced force and motion response, which can be written as

$$F(t) = F_a \sin(\omega t + \epsilon)$$

and

$$z(t) = z_a \sin(\omega t + \epsilon + \phi) \qquad (14.35)$$

Figure 14.27 presents the variations of $|F_a|/|a|, |z_a|/|a|$, ϵ and $(\epsilon + \phi)$ as functions of frequency for a semi-submersible with $d = 13.5$ m, $L = 36$ m, $A_h = 70$ m^2 and $A_v = 90$ m^2. A value of 1.2 is used for the added mass coefficient. These curves exhibit several interesting features. The wave force amplitude cancels out to zero at certain wave frequencies when instantaneous wave forces on the pontoons balance wave forces on the columns. At these cancellation frequencies, phase, ϵ, shifts by 180° such that over some wave frequencies, the vertical force on the vessel is out of phase with wave elevation. The motion response transfer function shows the combined effects of the wave force cancellation and resonance with the total phase difference between vessel motion and wave elevation around the resonant frequency exhibiting some complex behaviour.

14.6.3 Heave and surge motions of tensioned buoyant platforms

A tensioned buoyant platform (TBP) is a floating structure connected to a sea-bed foundation by vertical mooring tethers which are kept in tension by excess buoyancy over weight in the platform structure. The use of TBPs as alternatives to oil

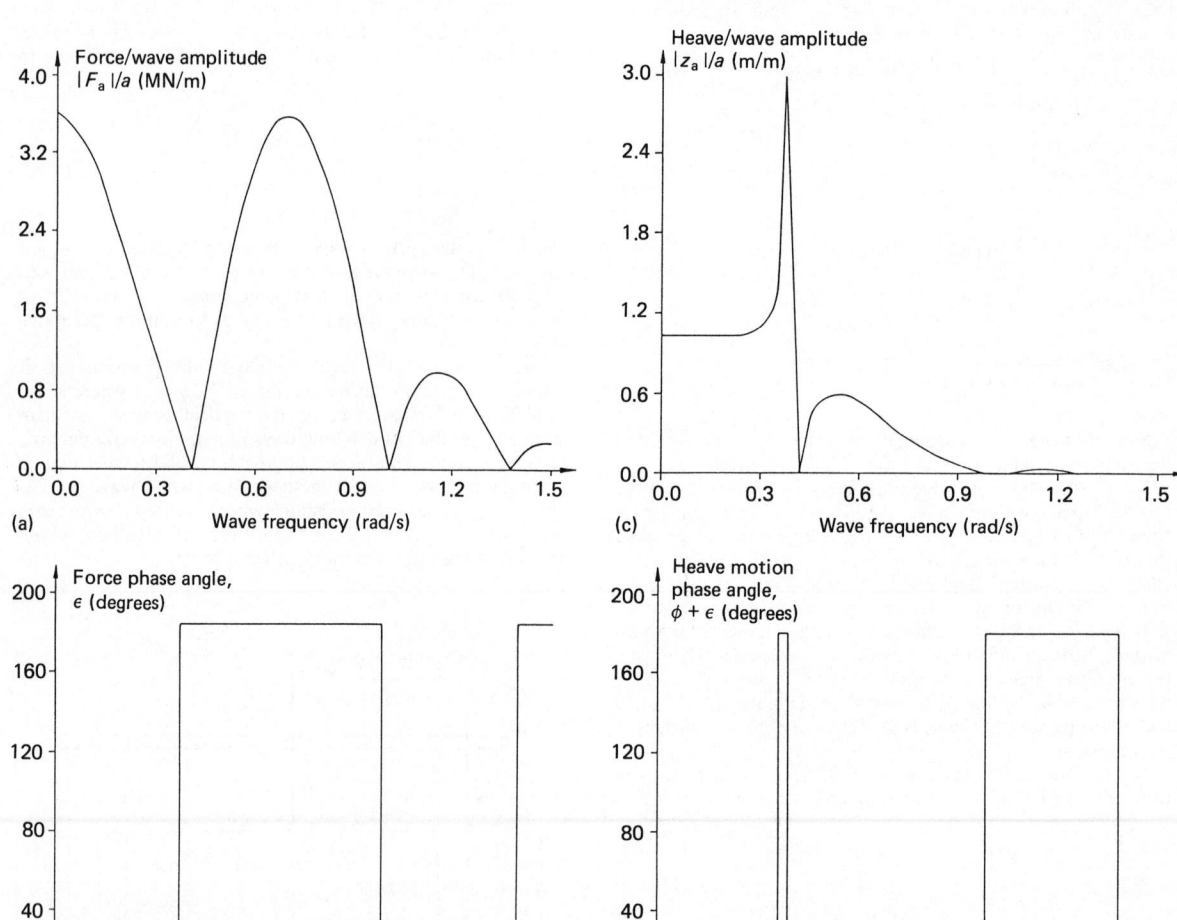

Figure 14.27 Heave force and motion response of a four-column semi-submersible

production from fixed or catenary moored semi-submersibles or other floating structures is now well established. A TBP possesses a combination of desirable characteristics as an offshore work platform. The absence of a fixed rigid structure from sea surface to sea bed coupled with the excellent station-keeping characteristics of tensioned vertical tethers satisfy the two important demands of economy and safe operation. Additionally, a TBP exhibits very much smaller wave-induced heave, roll and pitch motions due to the presence of the vertical mooring tethers. Large surge and sway offsets due to wind and current steady force components will, however, cause the platform to be drawn down lower in the water and exhibit a so-called 'set-down'. Nevertheless, the wave frequency surge and sway motions of a TBP are desirable, since horizontal wave loads are not fully reacted by the surface platform structure.

Conventional TBPs have structural features which are similar to those of semi-submersibles. These include the presence of deeply submerged main buoyancy chambers which give rise to a significant reduction in wave-induced forces due to the exponential decay of wave particle velocities and accelerations with depth below the surface. The combination of horizontal pontoons and vertical columns supporting the working deck produces wave force cancellation at certain frequencies. This occurs because the upward wave force on vertical columns due to buoyancy changes can be almost exactly cancelled out by the downward wave force on horizontal pontoons. Additionally, the stiffness of the tethers and their axial tensions will radically influence the dynamic response of the platform to waves.

Methods of analysis for a TBP are very similar to those outlined earlier for a semi-submersible structure, except for the insertion of a stiffness matrix induced by the presence of the relatively stiff platform tethers. Thus the equation of motion, the evaluation of the coefficient matrices and the solution of the equation are the same as above with the tether stiffness matrix C_t included.

The governing equation for heave motion of a tensioned buoyant platform of the configuration shown in Figure 14.28 is derived below. The platform is square in plan view with transverse pontoons (marked A in Figure 14.28) of length $2L$ and cross-sectional area, A_h. The equation for heave motions is

$$\left.\begin{aligned}
(m + A)\,\ddot{z} + (C + C_t)\,z &= F_z(t) \\[4pt]
\text{where} \\[4pt]
m = 4\rho\,[A_v d + 2A_h L] &- \frac{4T}{g}
\end{aligned}\right\} \quad (14.36)$$

with T being the tension in each of the tethers on the four corners of the platform,

$$\left.\begin{aligned}
A &= 8\rho\,C_{mz} A_h\,L \\[4pt]
C &= 4\rho\,g\,A_v \\[4pt]
\text{and} \\[4pt]
C_t &= 4\lambda'
\end{aligned}\right\} \quad (14.37)$$

where C_{mz} is the added mass coefficient for the pontoon cross section in vertical motion and λ' is the axial stiffness of each of the tethers. In this analysis, the drag damping coefficient and

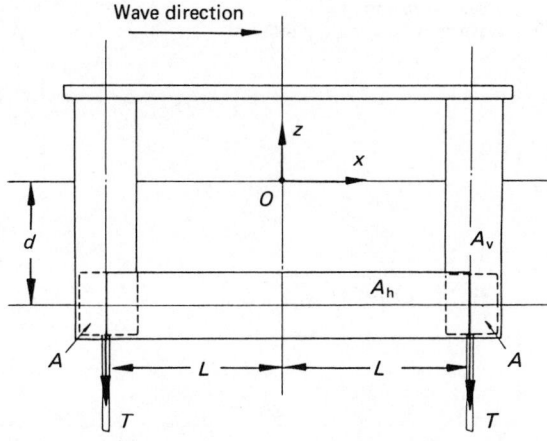

Figure 14.28 Side view of a four-column tensioned buoyant platform. A are transverse pontoons of length $2L$ and area of cross section A_h

wave-induced drag forces are neglected for simplicity. The wave force in heave is

$$\left.\begin{aligned}
F_z(t) &= \int_{-L}^{L} 2\rho\,A_h\,(1 + C_m)\,\dot{w}\,dx + 2A_v\,(p_f + p_r) \\[4pt]
&\quad + 2\,\rho\,A_h\,L\,(1 + C_m)\,[\dot{w}_f + \dot{w}_r]
\end{aligned}\right\} \quad (14.38)$$

where $\dot{w}_f$ and $\dot{w}_r$ are the vertical wave particle accelerations at $z = -d$ and $x = +L$ and $-L$, respectively, with the remaining notation from earlier in this section. Then, using equation (14.33), we get

$$F_z(t) = 4a\,\rho\,g\,e^{-kd}\,[-A_v \cos kL + A_h\,(1 + C_m)\sin kL$$

$$+ A_h.kL.(1 + C_m)\cos kL]\sin \omega t \quad (14.39)$$

These equations have most of the characteristics of the governing equation for semi-submersibles except for the presence of stiffness term, C_t, which is large and tends to substantially reduce the heave natural periods of such platforms to the range of 3 to 5 s from the usual semi-submersible heave reasonance period of from 18 to 25 s. Using the definition of equation (14.35), Figure 14.29 presents variations of the wave force and motion transfer functions with frequency for a TBP with $d = 35$ m, $L = 50$ m, $A_h = 100$ m^2 and $A_v = 150$ m^2. The total tether pre-tension and stiffness are taken to be 3.75 MN and 56.53 MN/m, respectively. The added mass coefficient for vertical motion of the rectangular pontoons is taken to be 1.2. The wave force is slightly modified due to the additional transverse pontoons, but the motion response exhibits a substantial reduction in magnitude due to the shift in resonant frequency.

A similar analysis to that above can be carried out for surge motion of a TBP with a governing equation of the form

$$(m + A')\,\ddot{x} + (C' + C'_t)\,x = F_x(t) \quad (14.40)$$

The additional terms in the above equation are

$$\left.\begin{aligned}
A' &= 4\rho\,C_{mx}\,A_h\,L + 4\rho\,A_v d \\[4pt]
C' &= 0 \qquad C'_t = \frac{4T}{s}
\end{aligned}\right\} \quad (14.41)$$

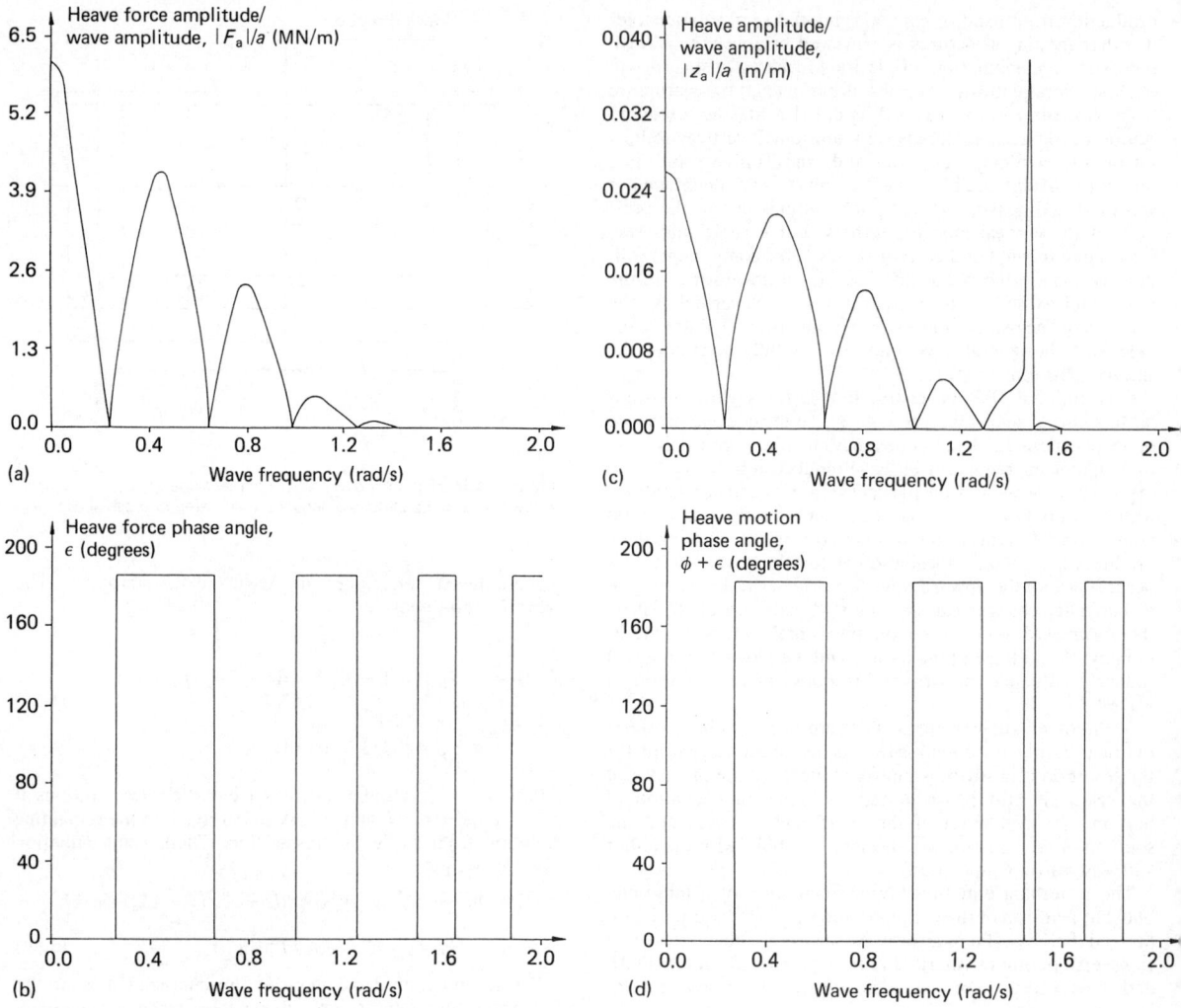

Figure 14.29 Heave force and motion response of a four-column tensioned buoyant platform

where C_{mx} and 1.0 are the added mass coefficients for horizontal flow normal to the pontoons and vertical columns, respectively. The hydrostatic stiffness term, C', is zero and the tether stiffness is expressed in terms of tension and tether lengths. An expression for $F_x(t)$ can be derived in the same way as for heave motion to be

$$F_x(t) = \int_{-d}^{0} 4\rho \, A_v \, [\dot{u}_f + \dot{u}_r] \, dz$$

$$+ \rho \, A_h \, L \, (1 + C_{mx}) \, [\dot{u}_f + \dot{u}_r]_{z=-d} \qquad (14.42)$$

$$- 2A_h \, [p_f - p_r]_{z=-d}$$

where $\dot{u}_f$, p_f and $\dot{u}_r$, p_r denote horizontal wave particle accelerations and wave pressures at $x = +L$ and $x = -L$, respectively, and are given by

$$\dot{u} = -a\,\omega^2 \, e^{+kz} \cos(kx - \omega t) \qquad (14.43)$$

and equation (14.33) with C_{mx} and 1.0 taken as the added mass coefficients for horizontal flow normal to the pontoons and vertical columns, respectively. Then, the wave force $F_x(t)$ can be written as

$$F_x(t) = -\left\{ \frac{8\rho a\omega^2 A_v}{k} [1 - e^{-kd}] + 2\rho a\omega^2 A_h L \right.$$

$$\left. (1 + C_{mx}) e^{-kd} \right\} \cos kL \cos \omega t \qquad (14.44)$$

$$- 4\rho g a \, A_h \, e^{-kd} \sin kL \cos \omega t$$

Note that the only force cancellations occurring in equation (14.44) are due to zeroes of the combined $\cos kL$ and $\sin kL$ terms.

Figures 14.30 present the surge force and motion response transfer function calculated from these equations. The tether length, s, is taken to be 450 m and $C_{mx} = 1.2$ with all other parameters being as in the calculation of heave response

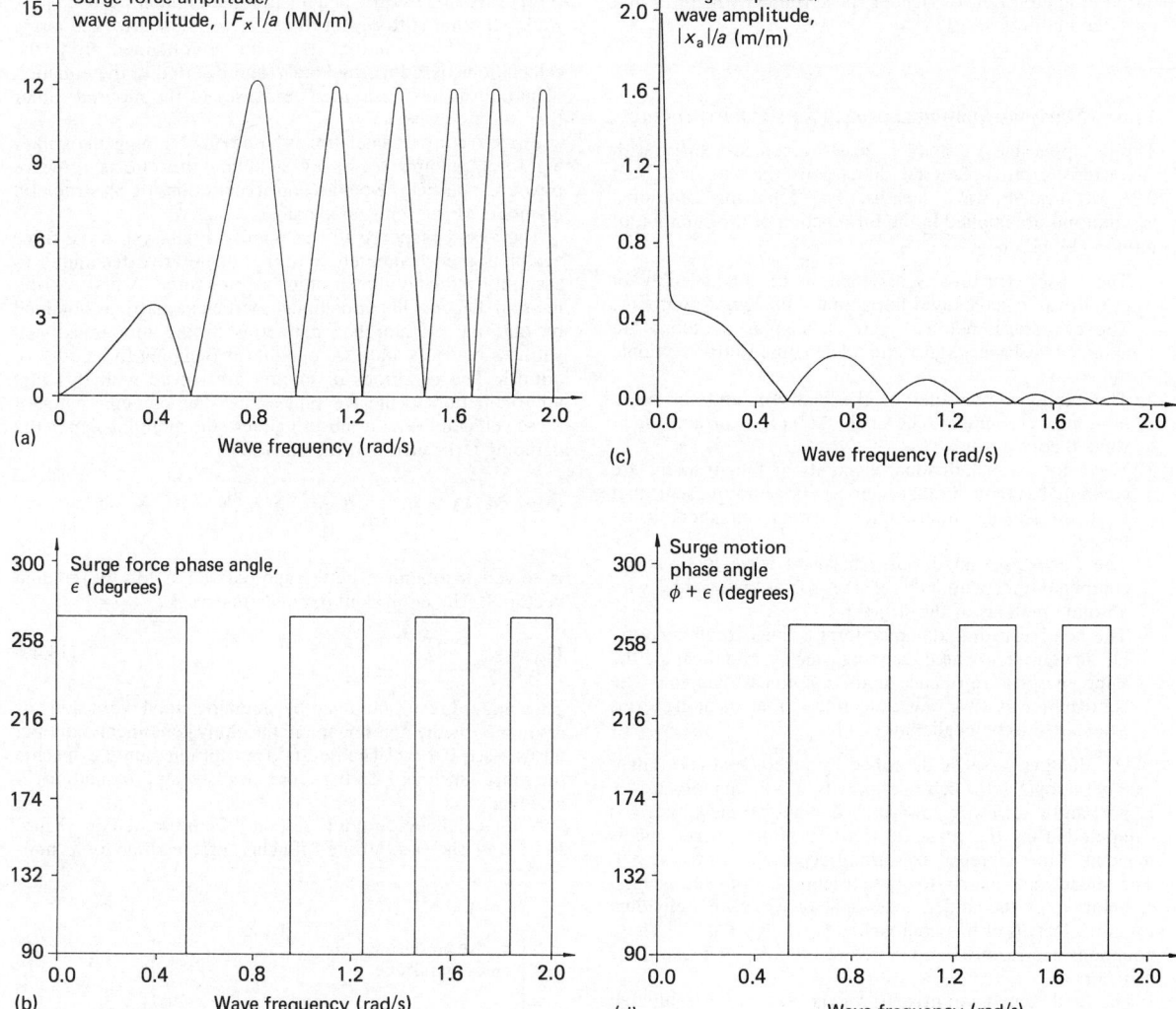

Figure 14.30 Surge force and motion response of a four-column tensioned buoyant platform

described above. Again, the effect of tether stiffness dominates the motion transfer function with a low natural period and relatively large surge motions at predominant wave periods.

14.6.4 Analysis with multiple degrees of freedom

The general multiple-degree-of-freedom equations of motions for a stationary floating body in gravity waves can be formulated using an approach similar to that used for single degrees of freedom earlier. It is usual to formulate these matrix equations in the six rigid-body degrees of surge, sway, heave, roll, pitch and yaw described by a column vector, $\mathbf{X}$, although additional degrees of freedom (such as structural deformations) can be incorporated, if required. The matrix equation of motion is

$$(\mathbf{M} + \mathbf{M}_A)\,\ddot{\mathbf{X}} + \mathbf{B}_V\,|\dot{\mathbf{X}}|\dot{\mathbf{X}} + (\mathbf{K} + \mathbf{K}_m)\,\mathbf{X} = F(t) \qquad (14.45)$$

where $\mathbf{M}$ and $\mathbf{M}_A$ are the (6×6) coefficient matrices quantifying structure physical mass and added mass, respectively, $\mathbf{B}_V$ is a (6×6) matrix representing a non-linear drag force while $\mathbf{K}$ and $\mathbf{K}_m$ are (6×6) stiffness matrices describing the hydrostatic and mooring restoring forces acting on the vessel.

The formulation of the equation of motion for a monolithic ship-shape hull form differs from that used for a semi-submersible since for ship shapes the diffraction of incident waves due to the large size of the hull form has to be accounted for. For these hull forms, the basic equation of motion can be written simply as

$$[\mathbf{M} + \mathbf{M}_A]\ddot{\mathbf{X}} + \mathbf{B}_p\dot{\mathbf{X}} + [\mathbf{K} + \mathbf{K}_m]\,\mathbf{X} = \mathbf{F}\,e^{i\omega t} \qquad (14.46)$$

where $\mathbf{M}$, $\mathbf{M}_A$, $\mathbf{K}$ and $\mathbf{K}_m$ are defined above, $\mathbf{B}_p$ is the matrix of potential damping coefficients and $\mathbf{F}\,e^{i\omega t}$ is a complex vector of wave forces obtained by summing hull surface pressures from a diffraction analysis.

The above basic equations of motion are solved using different approaches for slender space-framed structures and for large hull forms.

14.6.5 Morison-equation approach for slender structures

This is applicable to slender space-framed structures with characteristic cross-sectional dimensions that are less than 20% of incident wave lengths. The following additional assumptions are implied in the formulation of the equation of motion (14.44) above:

1. The vessel structure is assumed to be an assembly of cylindrical elements and horizontal rectangular pontoons. The cylindrical members and rectangular pontoons are assumed to have large ratios of length to cross-sectional dimension.
2. The motion amplitudes of the platform and the wave motions are assumed to be small. As a consequence, linear wave theory is used.
3. Wave forces on individual elements of the structure are computed as though other members were not present, that is, hydrodynamic interference between members is ignored.
4. The forces associated with sinusoidal wave motions are computed independently of the forces associated with absolute motions of the structure.
5. The non-linear drag damping term is linearized by assuming an equivalent linear damping which would dissipate the same energy at resonance as the non-linear damping. The contribution of wave radiation effects to the damping terms is assumed to be negligibly small.

The floating vessel is described by orthogonal coordinate axes with origin at the centre of gravity, x pointing forwards, y to port and z vertically upwards. The physical mass matrix is computed using the mass distribution of the vessel and is diagonal if the reference axes are principal axes of the vessel. The added mass matrix for the structure is obtained using a summation of the added mass matrix for each individual cylinder. Details of this approach are given by Patel.[10] These equations can readily be modified to account for rectangular cylinders.

The fluid damping matrix, **B**, for the structure is evaluated in a similar manner to the added mass matrix – as a sum of the contributions from each individual member. The derivation of the generalized damping matrix for an arbitrarily orientated circular cylinder can be derived using the method described by Patel[10] with the assumption that only drag forces normal to the cylinder axis are significant. Unlike the added mass matrix, the non-linear velocity square proportionality generates an asymmetric damping matrix. The corresponding matrices for the rectangular cylinder and non-elongated body are evaluated as special cases of the circular cylinder.

Contributions to the hydrostatic stiffness matrix, **K**, will only arise in the heave, roll and pitch degrees of freedom due to buoyancy forces in the water plane cutting members of the hull. If, for a member number, n, A_{wn} denotes the water plane area and (x_{wn}, y_{wn}) are the coordinates of the centroid of this water plane area, then the hydrostatic stiffness matrix elements k_{ij} can be written as the summations

$$\left. \begin{array}{l} k_{33} = \rho g \Sigma A_{wn} \\ k_{43} = \rho g \Sigma y_{wn} A_{wn} \\ k_{53} = -\rho g \Sigma x_{wn} A_{wn} \\ k_{54} = -\rho g \Sigma A_{wn} x_{wn} y_{wn} \\ k_{44} = \rho g V (GM)_p \\ k_{55} = \rho g V (GM)_R \end{array} \right\} \qquad (14.47)$$

where V is the vessel displacement by volume and $(GM)_r$, $(GM)_p$ are metacentric heights in roll and pitch, respectively, while all other stiffness terms are zero. A linearized mooring systems stiffness matrix, $\mathbf{K}_m$, can be obtained from the calculations described by Patel[10] and inserted in the equation of motion to obtain motion responses of the moored rather than the free vessel.

The wave force calculation is based on obtaining an oscillating force column vector by summing the effects of wave pressures, particle velocities and accelerations on all structural members of the semi-submersible.

The vessel response to this applied oscillating wave force vector is then computed by using an iterative technique to account for the non-linear drag damping force. A first approximation diagonal linear damping coefficient matrix is obtained by ignoring all non-diagonal terms in the total mass and stiffness matrices and assuming linear damping to be 10% of critical. The equations of motion are solved with this first approximation to the damping value. The column vector of these velocities is then substituted into the modulus sign in the damping term such that the equation

$$(\mathbf{M} + \mathbf{M}_A)\ddot{\mathbf{X}} + \mathbf{B}_v \frac{8}{3\pi} |\dot{\mathbf{X}}_{\text{last}} \;\; \dot{\mathbf{X}} + (\mathbf{K} + \mathbf{K}_m)\mathbf{X} = \mathbf{F}(t)$$
$$\text{approx} \qquad (14.48)$$

is solved to obtain a better approximation for the column vector **X**. The equivalent damping matrix $\mathbf{B}_{eq}$

$$\mathbf{B}_{eq} = \mathbf{B}_v \frac{8}{3\pi} |\dot{\mathbf{X}}| \qquad (14.49)$$

is a standard result obtained by assuming equal work done at resonance by the non-linear and the equivalent linear damping terms – see Patel.[10] The iteration is continued until a specific tolerance (such as 1%) between successive approximations is achieved.

A typical heave motion response is presented in Figure 14.31 and shows a strong influence of the drag force non-

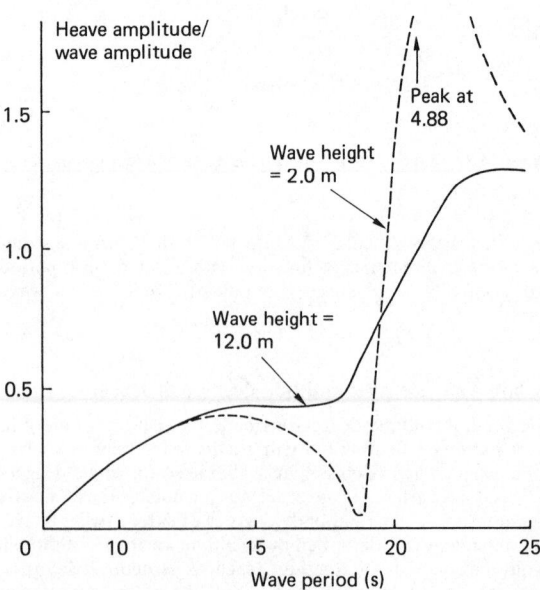

Figure 14.31 Heave motion response of a semi-submersible vessel in head seas

linearity around resonance with the heave response amplitude per unit wave amplitude reducing from 4.88 m/m at 1 m wave amplitude to 1.26 m/m at 6 m wave amplitude. The vessel motion response away from resonance is not significantly affected, although there is some increase in response around 16–19 s due to the corresponding increase in wave force amplitude at these periods. The large change in the unit heave response at and around resonance is to be expected, since the damping force in a vibratory system is dominant at resonance.

14.6.6 Diffraction theory

Calculations of wave-induced motions of a large non-space frame structure in gravity waves requires a solution of the wave problem with no flow boundary conditions at the moving body surface in addition to the free surface and sea-bed boundary conditions. The solution can be split into two related problems – the scattering wave problem defines wave forces on a floating body when fixed in space and with waves incident on it in an identical manner to the technique for computing wave forces on a fixed body described in Section 14.4. The radiation wave problem is concerned with defining forces on the body (added mass and damping) due to its oscillation in otherwise still water. These oscillations will induce wave potentials such that the total wave potential in the fluid is the sum of the incident, ϕ_w, scattered, ϕ_s, and forced wave potentials, ϕ_f, so that

$$\phi = \phi_w + \phi_s + \phi_f \tag{14.50}$$

and these must satisfy the boundary conditions at the body surface given by

$$\frac{\partial \phi_w}{\partial n} + \frac{\partial \phi_s}{\partial n} + \frac{\partial \phi_f}{\partial n} = V_n \tag{14.51}$$

where V_n is the velocity of the body surface in the direction n normal to the surface. This boundary condition can be applied at the mean body surface since the theory is applied for small motions. ϕ, together with its three components. It must also satisfy the Laplace equation and the free surface and sea-bed boundary conditions. Furthermore, ϕ_s and ϕ_f must satisfy the radiation conditions.

Boundary conditions for the scattering and radiation wave problems can be split up from equation (14.51) as

$$\left.\begin{array}{l} \dfrac{\partial \phi_w}{\partial n} + \dfrac{\partial \phi_s}{\partial n} = 0 \\[2mm] \text{and} \\[2mm] \dfrac{\partial \phi_f}{\partial n} = V_n \end{array}\right\} \tag{14.52}$$

respectively, both being applied on the body surface. The scattering problem is identical to the application of diffraction theory on fixed structures as described in Section 14.4. The radiation problem can also be solved by using either boundary-integral or boundary-element techniques. For brevity, only the solution using boundary-integral techniques is described here. As in Section 14.4, the analysis assumes inviscid, irrotational flow and that wave amplitudes are small. The unsteady flow around the floating vessel is calculated by introducing oscillating sources of unknown velocity potential on the vessel's submerged surface that is discretized by a mesh of facets with an oscillating source on the surface of each facet.

A Green's function is used to represent the velocity potential of each source which, because of the form of the Green's function, satisfies Laplace's equation, zero flow at the horizontal sea bed, the free surface and radiation boundary

conditions. Now, the solution for scattered wave potential due to the stationary floating body, subjected to incident waves of potential, ϕ_2, is identical to that described in Section 14.5 for fixed structures. A set of linear simultaneous equations are obtained by equating the flow due to the local source plus the additional flow due to all other sources to the negative of the flow due to the undisturbed wave for each facet on the body surface. Solutions of these equations yields the unknown source strengths and, therefore, the velocity potential, ϕ_s, which is used to derive pressures and wave forces by integration over the body surface. Thus the wave force vector, $\mathbf{F}$, of equation (14.46) may be obtained for an incident wave of specified frequency and direction.

The velocity potentials, ϕ_f, are obtained in a way similar to that above except for the use of a different boundary condition which reflects the fact that ϕ_f arises from body motions in otherwise still water. Thus, at all facets, the source strengths, ϕ_{fi}, are such that the flow due to the local source plus the flow due to all other sources equals the velocity component of the body along the facet normal. This velocity component will depend on the mode of motion (surge, sway, heave and so on) in which the body is moving. All of this can be represented by equating the normal velocity of the fluid and of the jth facet for the vessel moving in its kth mode of motion. This yields the equation

$$\sum_{\text{All } i} \frac{\partial \phi_{fi}}{\partial n_j} \sigma_{ik} = V_{jk} \tag{14.53}$$

where V_{jk} is the normal velocity of the jth facet with the vessel moving in its kth mode of motion. Furthermore, n_j is the normal to the jth facet, $\partial \phi_i / \partial n_j$ is the normal fluid velocity at the jth facet due to a unit source at the ith facet, and σ_{ik} are the unknown source strengths required in the kth mode. Application of equation (14.53) for all facets produces a system of complex equations to be solved for the source strengths. Once these are known, the pressures at the facets are evaluated and their effects integrated over the vessel surface to yield forces in each mode of motion to unit motion in the kth mode.

These forces may be written as a complex square matrix, $G(\omega)$ which can be decomposed into its real and imaginary parts through the equation

$$G(\omega) = \omega^2 \mathbf{M}_A (\omega) - i\omega \mathbf{B}_p(\omega) \tag{14.54}$$

to yield frequency-dependent added mass and damping matrices $\mathbf{M}_A(\omega)$ and $\mathbf{B}_p(\omega)$ which are required for equation (14.46).

The inclusion of physical mass, hydrostatic and mooring stiffness matrices, $\mathbf{M}$, $\mathbf{K}$ and $\mathbf{K}_m$ completes derivation of all of the coefficient matrices of equation (14.46). The hydrodynamic coefficient matrices are, however, frequency dependent and require carrying out a diffraction analysis at all frequencies at which motions are required. Equation (14.46) is linear and can readily be solved to yield the displacement vector $\mathbf{X}$. The exciting force vector $F(\omega)$ and the coefficient matrices $\mathbf{M}_A(\omega)$ and $\mathbf{B}_p(\omega)$ can also be derived using finite-element methods in a way analogous to that for the boundary-integral approach described above.

There is one further point of interest regarding the relationship between the scattered and forced wave potentials (ϕ_s and ϕ_f) for a floating vessel problem. The use of equations called Haskind relations (see Newman[32]) enables the scattered wave potential, ϕ_s, to be expressed in terms of the incident and forced wave potentials, ϕ_w and ϕ_f. Thus, once ϕ_f is calculated, ϕ_s need not be computed by diffraction analysis but can instead be derived using the Haskind relations.

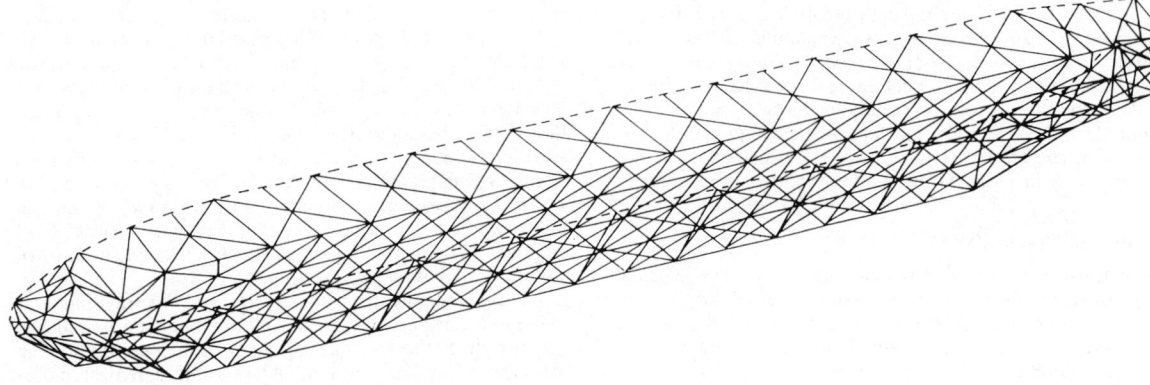

Figure 14.32 Facet discretization of a submerged ship hull for diffraction theory

Typical results of a boundary integral diffraction analysis for a ship-shaped hull are shown in Figure 14.33. The discretization of the submerged hull geometry is shown in Figure 14.32 using 277 triangular facets on the ship half-hull. The vessel is of 263.7 m overall length, 40.8 m beam and 145 937 t displacement with 14.80 m draught floating in deep water. Figure 14.33(a) presents the variation of added mass and radiation damping coefficients with frequency for heave and pitch motions. Note that the variation in added mass is relatively small but the radiation damping shows large changes with very small values at some wave periods. Wave-induced heave force and pitching moments and the resultant motion responses for head seas are presented in Figures 14.33(b) and 14.33(c).

14.7 Design considerations and certification

It is important to appreciate that the design procedures for jacket structures outlined in the previous three sections are

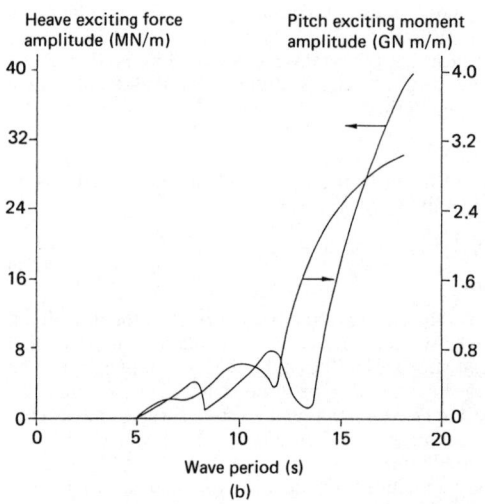

(b)

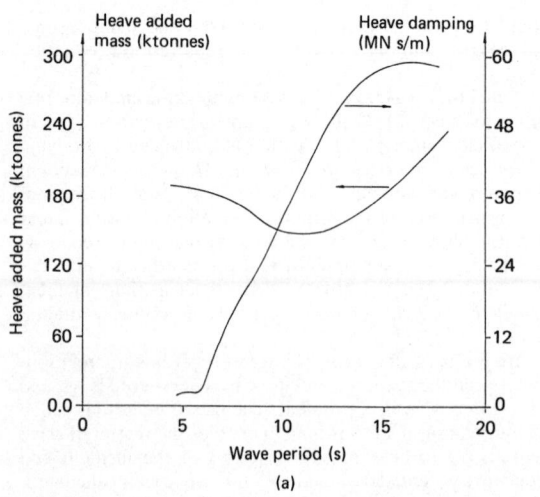

(a)

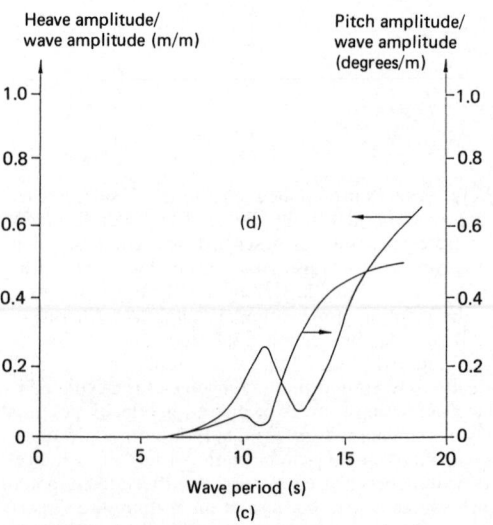

(c)

Figure 14.33 Variations of heave and pitch added masses, wave-excitation forces and motion response with wave period for ship hull

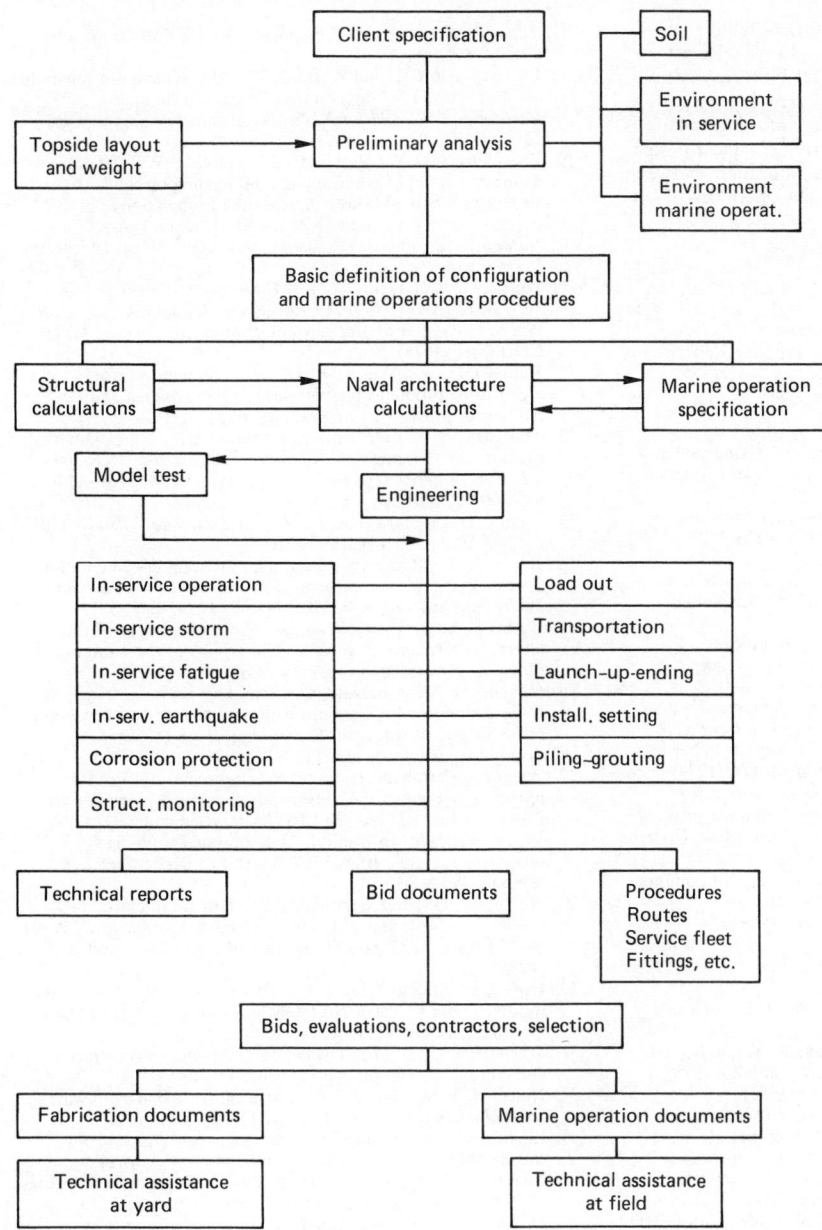

Figure 14.34 Design procedure for jacket structure

only a small part of the total design process. In order to illustrate this point, Figure 14.34 presents a flow chart showing the design procedures that need to be followed, from the initial specification through to commencing operation of a typical offshore structure. The jacket has to have sufficient strength, as it is assembled during the fabrication stage and loaded out of the yard. It has also to meet the naval architectural and structural requirements of tow-out, up-ending and installation as well as surviving for a 20–40-year life. Some of the supplementary design tasks not covered in this chapter include the response of the structure to earthquakes, the

provision of corrosion protection and in-service structural monitoring. The design procedure for large jackets invariably contains a model test phase for critical operations such as up-ending during installation. The documentation of the material, structural and welding details of the design during its certification, fabrication and service life pose an engineering management problem.

Certifying authorities play a key role in the design procedure for an offshore structure. The major certifying authorities in the United Kingdom, Norway and the United States have built up extensive codes of practice which reflect research work, in-service experience and the results of failure investigations over many years of operation (see Lloyd's Register of Shipping,[33] Department of Energy,[34] Det Norske Veritas,[35] and American Bureau of Shipping[36]). Certifying authorities also provide an independent check of many of the calculations and decisions that need to be made during a typical design. There tends to be close technical collaboration between research establishments, designers and the operators of offshore structures.

References

1. Department of Energy, *Offshore Installations, Guidance on design and construction*, Part II, Section 4.3, HMSO, London (1986)
2. American Petroleum Institute, *Basic Petroleum Databook*, Volume VI, No. 3, September, API, 1220 L Street NW, Washington, DC 20005, USA (1986)
3. Lee, G. C., 'Recent advances in design and construction of deep water platforms, Part 1', *Ocean Industry*, November, 71–80 (1980)
4. Furnes, O. and Loset, O., 'Shell structures in offshore platforms: design and application', *Engineering Structures*, **3**, July, 140–152 (1980)
5. Thornton, D., 'A general review of future problems and their solution', *Proceedings of the Second International Conference on Behaviour of Offshore Structures*, 28–31 August, Paper 88, BHRA Fluid Engineering, Cranfield, Bedford, UK (1979)
6. Hamilton, J. and Perrett, G. R., 'Deep water tension leg platform designs', *Proceedings of the Royal Institution of Naval Architects International Symposium on Developments in Deeper Waters*, 6–7 October, Paper no. 10 (1986)
7. Meteorological Office, *Meteorology for mariners*, 3rd edition, HMSO, London (1986)
8. Strahler, A. N. and Strahler, A.H., *Modern Physical Geography*, Wiley, New York (1978)
9. Airy, Sir G. B., 'Tides and waves', *Encyc. Metrop.*, Art. 192, pp. 241–396 (1845)
10. Patel, M. H., *Dynamics of Offshore Structures*, Butterworth Scientific, Guildford (1989)
11. Morrison, J. R., O'Brien, M. P., Johnson, J. W. and Schaaf, S. A., 'The forces exerted by surface waves on piles', *Petroleum Transactions*, **189**, TP 2846, 149 (1950)
12. Sarpkaya, T., 'In line and transverse forces on smooth and sand roughened cylinders in oscillatory flow at high Reynolds numbers', *Report No. NPS-69SL76062, Naval Postgraduate School*, Monterey, California (1976)
13. Sarpkaya, T. and Isaacson, M., *Mechanics of Wave Forces on Offshore Structures*, Van Nostrand Reinhold, New York (1981)
14. Sommerfield, A., *Partial Differential Equations in Physics*, Academic Press, New York (1949)
15. Stoker, J. J., *Water Waves*, Interscience, New York (1957)
16. MacCamy, R. C. and Fuchs, R. A., 'Wave forces on piles, a diffraction theory', *US Army Corps of Engineers, Beach Erosion Board*, Tech. Memo. No. 69 (1954)
17. Garrison, C. J. and Chow, P. Y., 'Wave forces on submerged bodies', *Journal of Waterways, Harbours and Coastal Division,*

American Society of Civil Engineers, **98**, No. WW3, 375–392 (1972)
18. Eatock-Taylor, R. and Waite, J. B., 'The dynamics of offshore structures evaluated by boundary integral techniques', *International Journal for Numerical methods in Engineering*, **13**, 73–92 (1978)
19. Zienkiewicz, O. C., Bettes, P. and Kelly, D. W., 'The finite element method of determining fluid loading on rigid structures – two and three dimensional formulations', in Zienkiewicz, O. C., Lewis, P. and Stass, K. G. (eds), *Numerical Methods in Offshore Engineering*, Wiley, Chichester (1978)
20. Penzien, J. and Tseng, W. S., 'Three dimensional dynamic analysis of fixed offshore platforms', in Zienkiewicz, O. C. *et al.* (eds), *Numerical Methods in Offshore Engineering*, Wiley, Chichester (1978)
21. Bathe, K. J. and Wilson, E. L., 'Solution methods for eigen-value problems in engineering', *International Journal for Numerical Methods in Engineering*, **6**, 213–216
22. Malhotra, A. K. and Penzien, J., 'Nondeterministic analysis of offshore tower structures', *Journal of Engineering Mechanics Division, American Society of Civil Engineers*, **96**, No. EM6, 985-1003 (1970)
23. Poulos, H. G. and Davis, E. H., *Pile Foundation Analysis and Design*, Wiley, New York (1980)
24. Reese, L. C., 'Laterally loaded pile; program documentation', *Journal of the Geotechnical Engineering Division, American Society of Civil Engineers*, **103**, No. GT4, 287–305 (1977)
25. Focht, J. A., Jr and Kock, K. J., 'Rational analysis of the lateral performance of offshore pile groups', *Proceedings of the Offshore Technology Conference*, OTC 1896 (1973)
26. O'Neill, M. W., Ghazzaly, O. I. and Ho, Boo Ha, 'Analysis of three-dimensional pile groups with nonlinear soil response and pile–soil–pile interaction', *Proceedings of the Offshore Technology Conference*, OTC 2838 (1977)
27. American Petroleum Institute, *Recommended practice for planning, designing and constructing fixed offshore platforms*, Dallas, Texas, Rpt No. API-RP-2A (revised annually) (1987)
28. British Standards Institution, Code of practice for fixed offshore structures, BS 6235: 1982, BSI, 2 Park Street, London, W1A 2BS
29. Dover, W. D. and Connolly, M. P., 'Fatigue fracture mechanics assessment of tubular welded Y and K joints', Paper No. C141/86, *Institution of Mechanical Engineers*, London (1986)
30. Dover, W. D. and Wilson, T. J., 'Corrosion fatigue of tubular welded T-joints', Paper No C136/86, *Institution of Mechanical Engineers*, London (1986)
31. Warburton, G. B., *The Dynamical Behaviour of Structures*, 2nd edition, Pergamon Press, Oxford (1976)
32. Newman, J. N., 'The exciting forces on fixed bodies in waves', *Journal of Ship Research*, **6**, 10–17 (1962)
33. Lloyd's Register of Shipping, *Rules and regulations for the classification of mobile offshore units*, January, Part IV, Chapter 1, Sections 2, 3, 4 and 5, Lloyd's Register of Shipping, 71 Fenchurch Street, London EC3 4BS (1986)
34. Department of Energy, *Development of the oil and gas resources of the United Kingdom*, Appendix 15, Department of Energy, HMSO (1986)
35. Det Norske Veritas, *Rules for classification of mobile offshore units*, Det Norske Veritas, PO Box 300, N-1322, Hovik, Oslo, Norway (1987)
36. American Bureau of Shipping, *Rules for building and classing mobile offshore drilling units*, ABS, 45 Eisenhower Drive, PO Box 910, Paramus, New Jersey, USA (1987)

15

Plant engineering

Robert K. Turton
(Section 15.1)

Raymond J. H. Eason and Neal Barnes
(Section 15.2)

Ernie Walker and Ronald J. Blaen
(Section 15.3)

George E. Pritchard
(Section 15.4.1)

Paul Compton
(Section 15.4.2)

John S. Bevan
(Section 15.4.3)

James Carvill
(Section 15.5)

Dennis H. Bacon
(Section 15.6)

Roger C. Webster
(Section 15.7–15.9)

Contents

15.1 Compressors, fans and pumps 15/3
    15.1.1 Design principles 15/3
    15.1.2 Machine selection 15/13
    15.1.3 Performance monitoring and prediction 15/14

15.2 Seals and sealing 15/18
    15.2.1 Compression packing 15/18
    15.2.2 Reciprocating hydraulic seals 15/32
    15.2.3 Rotary shaft lip seals 15/50
    15.2.4 Mechanical seals 15/59
    15.2.5 Clearance seals 15/68

15.3 Boilers and waste-heat recovery 15/75
    15.3.1 Types of boilers 15/75
    15.3.2 Application and selection 15/78

15.3.3 Heat transfer in industrial boilers 15/80
15.3.4 Terminology 15/83
15.3.5 Waste-heat boilers 15/84
15.3.6 Economizers 15/84
15.3.7 Superheaters 15/85
15.3.8 Steam storage 15/86
15.3.9 Water-level control 15/86
15.3.10 Automatic control 15/87
15.3.11 Feed-water requirements 15/88
15.3.12 Blowdown requirements control and tanks 15/89
15.3.13 Clean Air Act requirement for chimneys and flue designs 15/89
15.3.14 Energy conservation 15/90
15.3.15 Design Standards for pressure vessels, pipes and flanges 15/90

15.4 Heating, ventilation and air conditioning   15/91
    15.4.1 Heating   15/91
    15.4.2 Ventilation   15/97
    15.4.3 Air conditioning   15/106

15.5 Refrigeration   15/114
    15.5.1 Vapour compression cycle   15/115
    15.5.2 Pressure–enthalpy chart   15/115
    15.5.3 Gas refrigeration cycle   15/115

15.6 Energy management   15/116
    15.6.1 The energy manager   15/116
    15.6.2 Energy surveys and audits   15/116
    15.6.3 Applications   15/118
    15.6.4 Waste-heat recovery   15/122
    15.6.5 Control systems   15/123
    15.6.6 Summary   15/124

15.7 Condition monitoring   15/124
    15.7.1 Preventive maintenance   15/124
    15.7.2 Predictive preventive maintenance   15/124
    15.7.3 Condition monitoring   15/125
    15.7.4 The parameters   15/125
    15.7.5 Vibration monitoring for machine condition   15/125
    15.7.6 Vibration analysis techniques   15/126

15.8 Vibration isolation and limits   15/129
    15.8.1 Introduction   15/129
    15.8.2 Damping   15/130
    15.8.3 Multi-degree of freedom systems   15/130
    15.8.4 Vibration isolation   15/130
    15.8.5 Shock isolation   15/131
    15.8.6 Vibration attenuation   15/132
    15.8.7 Measurement of vibration   15/133
    15.8.8 Vibration limits   15/136

15.9 Acoustic noise   15/138
    15.9.1 Introduction – basic acoustics   15/138
    15.9.2 Sound intensity   15/139

15.9.3 Sound power   15/139
15.9.4 Addition and subtraction of decibels   15/139
15.9.5 Addition of decibels: graph method   15/139
15.9.6 The relationship between SPL, SIL and SWL   15/139
15.9.7 Frequency weighting and the human response to sound   15/140
15.9.8 Noise indices   15/140
15.9.9 Noise-rating curves   15/141
15.9.10 Community noise units   15/141
15.9.11 Road traffic   15/141
15.9.12 Air traffic   15/142
15.9.13 Railway noise   15/142
15.9.14 Noise from demolition and construction sites   15/142
15.9.15 Noise from industrial premises   15/142
15.9.16 Measurement of noise   15/142
15.9.17 Microphones   15/142
15.9.18 The sound-level meter   15/142
15.9.19 Digital signal analysis   15/143
15.9.20 Noise control   15/143
15.9.21 Noise nuisance   15/143
15.9.22 Health effects   15/144
15.9.23 Damage to plant/machinery/building structures   15/144
15.9.24 Legislation concerning the control of noise   15/144
15.9.25 British Standard 4142: 1990   15/145
15.9.26 Noise-abatement zones   15/146
15.9.27 Planning application conditions   15/146
15.9.28 The Health and Safety at Work etc. Act 1974   15/146
15.9.29 The Noise at Work Regulations 1989   15/146
15.9.30 Noise control engineering   15/147
15.9.31 Noise-reduction principles   15/147
15.9.32 Absorbers   15/148
15.9.33 Vibration isolation   15/148
15.9.34 Practical applications   15/149

References   15/150

15.1 Compressors, fans and pumps

15.1.1 Design principles

15.1.1.1 General

Compressors, fans and pumps are all devices for increasing the pressure energy of the fluid involved. Two basic types are used: rotodynamic, where flow is continuous, and positive displacement, where fluid is worked on in discrete packages defined by machine geometry. Compressors, fans and pumps may be rotodynamic, and compressors and pumps positive displacement. In general, the positive displacement machines give low mass flow and high pressure rise.

15.1.1.2 Rotodynamic machine principles

These can be discussed together as the Euler equation applies to all types, differences being due to the fluid involved and the flow path. Figure 15.1 illustrates flow path differences.

15.1.1.3 Forms of the Euler equation

Standard turbomachinery textbooks (see Turton[1]) derive this equation, so it will be applied here to centrifugal and axial machines. Considering Figure 15.2 (a simple centrifugal pump) the specific energy increase is given by the Euler equation

$$gH = u_2 V u_2 - u_1 V u_1 \qquad (15.1)$$

where u_1, u_2 are peripheral velocities $(=\omega r)$ $V u_2$, $V u_1$ are the peripheral components of the absolute velocities V_2 and V_1, respectively (see Figure 15.3).

$V u_1$ is usually considered as zero in design flow conditions, so

$$gH_{\text{IDEAL}} = u_2 \, V u_2 \qquad (15.2)$$

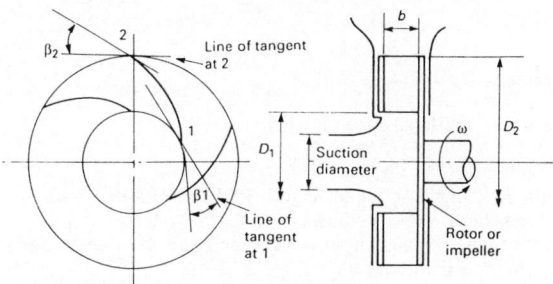

Figure 15.2   A simple radial outflow machine

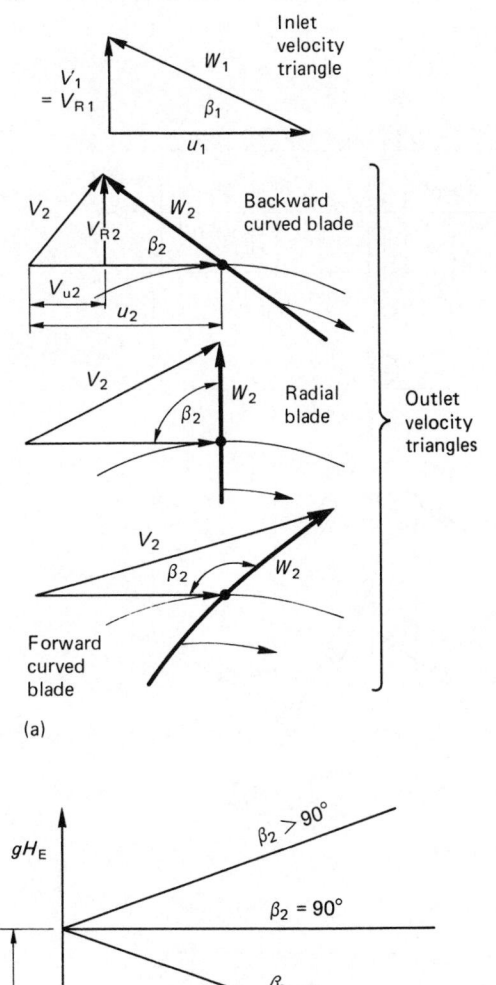

Figure 15.3   The effect of outlet angle on machine performance

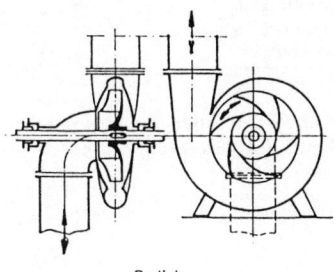

Radial

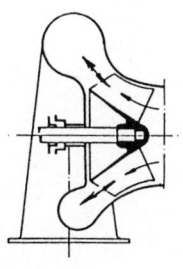

Mixed

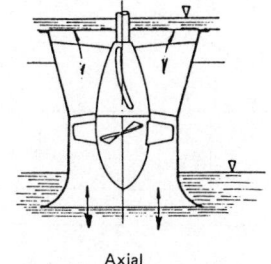

Axial

Figure 15.1   Flow paths used in rotodynamic machines

or

$$gH_{IDEAL} = u_2^2 - \frac{Qu_2}{A_2} \cot\beta_2 \qquad (15.3)$$

or when rotational speed is constant,

$$gH_{IDEAL} = K_1 - K_2 Q \qquad (15.4)$$

with K_2 depending on β_2. Figure 15.3 shows how varying β_2 affects both velocity diagrams and the gH to Q plot of performance plots, compressors being affected at lower flows by surge as discussed later.

A simple axial machine is shown in Figure 15.4, with typical general velocity diagrams, which define the geometry and terms used:

$$gH_{IDEAL} = u[Vu_2 - Vu_1] \qquad (15.5)$$

or if $Vu_1 = 0$ (zero inlet whirl) as assumed for pumps of fans:

$$gH_{IDEAL} = uVu_2 \qquad (15.6)$$

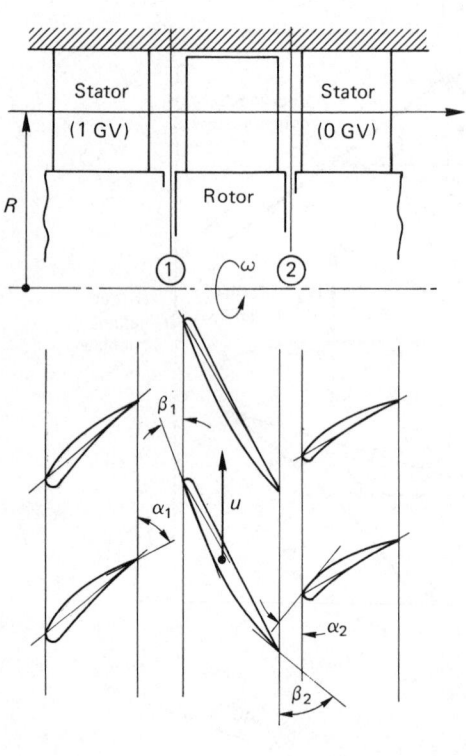

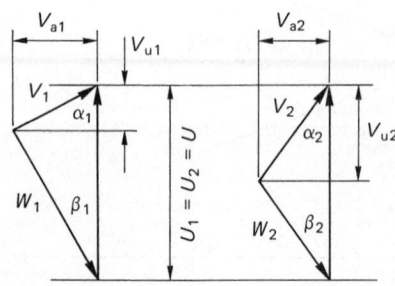

Figure 15.4 Axial flow pump or compressor stage and the 'ideal' velocity triangles

or

$$gH_{IDEAL} = uV_{A2} \cot\beta_2 \qquad (15.7)$$

V_{A2} is a function of Q and flow area and β_2 is related to blade angles.

For compressors, as Horlock[2] and Turton[1] show,

$$gH = \frac{\Delta p}{\rho} = Cp\ \Delta T = (u_2\ Vu_2 - u_1\ Vu_1) \qquad (15.8)$$

and for axial machines, this is usually written

$$\frac{\Delta p}{\rho} = Cp\ \Delta T = u\ (\Delta Vu) \qquad (15.9)$$

and the velocity diagrams combine, as shown in Figure 15.5, on a common base.

15.1.1.4 Definitions of efficiency

In all these machines efficiency statements are used:

$$\text{Overall efficiency } \eta_0 = \frac{\text{Power to fluid}}{\text{Power to shaft}} \qquad (15.10)$$

$$\text{Hydraulic efficiency } \eta_H = \frac{\text{Actual energy rise}}{\text{Euler energy rise}} \qquad (15.11)$$

$$\text{Volumetric efficiency } \eta_V = \frac{\text{Delivered flow}}{\text{Flow passing through rotor}} \qquad (15.12)$$

$$\text{Mechanical efficiency } \eta_M = \frac{\text{Fluid power}}{\text{Input shaft power}} \qquad (15.13)$$

$$\text{Thus } \eta_0 = \eta_M \times \eta_V \times \eta_H \qquad (15.14)$$

15.1.1.5 Reaction

This is defined for a compressor as:

$$R = \frac{\begin{array}{c}\text{Energy change due to or}\\\text{resulting from static pressure}\\\text{change in the rotor}\end{array}}{\text{Total change in the stage}} \qquad (5.15)$$

For an axial compressor 50% reaction means a symmetrical velocity diagram as shown in Figure 15.5.

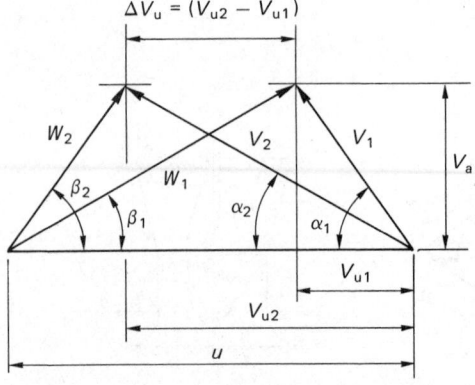

Figure 15.5 Axial velocity triangles based on a common base for an axial stage with 50% reaction ($V_1 = W_2$; $V_2 = W_1$)

If a simple pump is considered, it is possible to state that there must be a working relation between the power input P, the flow rate Q, energy rise gH, fluid properties ρ and μ, and size of the machine D. If a dimensional analysis is performed it can be shown that a working relation may exist between a group of non-dimensional quantities in the following equation:

$$\frac{P}{\rho\omega^3 D^5} = f\left[\frac{Q}{\omega D^3} \cdot \frac{gH}{\omega^2 D^2} \cdot \frac{\rho\omega^2 D^2}{\mu} \cdot \frac{\rho\omega^2 D^2}{K}\right] \quad (15.16)$$
$$\quad (1) \qquad (2) \qquad (3) \qquad (4) \qquad (5)$$

Term (1) is a power coefficient which does not carry any conventional symbol. Term (2) can easily be shown to have the shape V/U and is called a flow coefficient, the usual symbol being θ. Term (3) similarly can be shown to be gH/U^2 and is usually known as a head coefficient (or specific coefficient) ψ. Term (4) is effectively a Reynolds number with the velocity the peripheral speed ωD and the characteristic dimension being usually the maximum impeller diameter. Term (5) is effectively a Mach number, since K is the fluid modulus.

Since these groups in the SI system are non-dimensional they can be used to present the results of tests of pumps in a family of pumps that are geometrically similar and dynamically similar. This may be done as shown in Figures 15.6 and 15.7 and Figure 15.8 shows how the effect of changing speed or diameter of a pump impeller may be predicted, using the scaling laws:

$$\left.\begin{aligned}
\frac{P}{\rho\omega^3 D^5} &= \text{Const} \\[2mm]
\frac{Q}{\omega D^3} &= \text{Const} \\[2mm]
\frac{gH}{\omega^2 D^2} &= \text{Const}
\end{aligned}\right\} \quad (15.17)$$

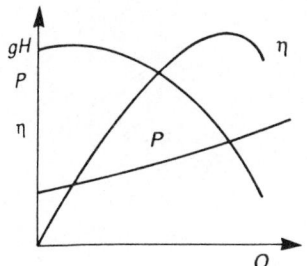

Figure 15.6 A pump characteristic for constant rotational speed

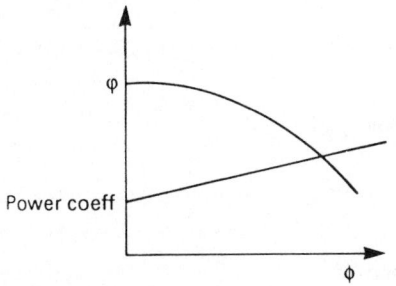

Figure 15.7 A non-dimensional plot for a pump

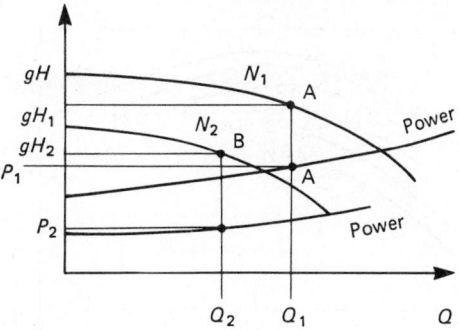

Figure 15.8 Prediction of speed change effect using equations (15.17)

In Figure 15.8 points A define the energy rise gH and power P_1 at a flow rate Q_1, when the pump is driven at speed ω_1. If equations (15.17) are applied, D and ρ being the same,

$Q_1/\omega_1 D^3 = Q_2/\omega_2 D^3$; hence Q_2

$gH_1/\omega_1^2 D^2 = gH_2/\omega_2^2 D^2$; hence gH_2

$P_1/\rho\omega_1^3 D^5 = P_2/\rho\omega_2^3 D^5$; hence P_2

This approximate approach needs to be modified in practice to give accurate results, for using model tests to predict full size power, as discussed by codes such as the American Hydraulic Institute Standards.[13]

The classical approach to the problem of characterizing the performance of a pump without including its dimensions was discussed by Addison,[3] who proposed that a pump of standardized size will deliver energy at the rate of one horsepower when generating a head of one foot when it is driven at a speed called the Specific Speed:

$$N_s = K\frac{N\sqrt{Q}}{H^{3/4}} \quad (15.18)$$

The constant K contains fluid density and a correction factor, and it has been customary to suppress K and use the equation:

$$N_s = \frac{N\sqrt{Q}}{H^{3/4}} \quad (15.19)$$

Caution is needed in using data as the units depend on the system of dimensions used, variations being litres/minute, cubic metres/second, gallons per minute or US gallons per minute as well a metres or feet. Plots of efficiency against specific speed are in all textbooks based upon the classic Worthington plot, and Figure 15.9, based on this information, has been prepared using a non-dimensional statement known as the characteristic number

$$k_s = \frac{\omega\sqrt{Q}}{(gH)^{3/4}} \quad (15.20)$$

This is based on the flow and specific energy produced by the pump at its best efficiency point of performance following the approach stated by Wislicenus:[4] 'Any fixed value of the specific speed describes a combination of operating conditions that permits similar flow conditions in geometrically similar hydrodynamic machines.'

Figure 15.10 presents, on the basis of the Characteristic number, the typical impeller profiles, velocity triangle shapes and characteristic curves to be expected from the machine flow paths shown. In the figure the characteristic ordinates are

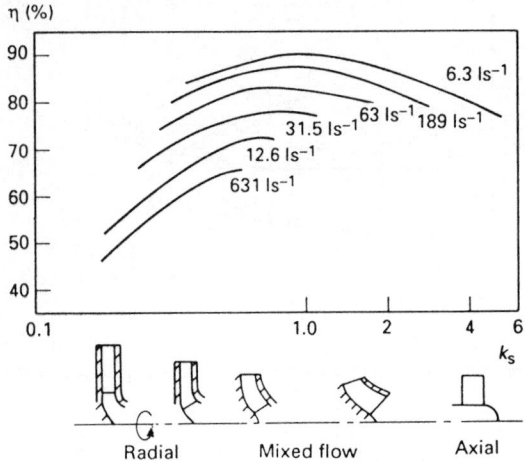

Figure 15.9 The variation of overall efficiency with non-dimensional characteristic number k_s for pumps (Turton[1])

are the ratios of actual head/design head and actual flow/design flow. This indicates the use of the number as a design tool for the pump engineer.

The scaling laws (equation (15.17)) may be used to predict the performance from change of speed as indicated in Figure 15.8. In many cases the pump engineer may wish to modify the performance of the pump by a small amount and Figure 15.11 illustrates how small changes in impeller diameter can affect the performance. The diagram in its original form appeared in the handbook by Karrasik et al.[5] and has been modified to

appear in metric form, as can be seen. The rules used are often called the Scaling Laws, written in the form:

$$D_2/D_1 = \sqrt{g_2 H_2/g_1 H_1}$$
$$Q_2/Q_1 = \sqrt{(g_2 H_2/g_1 H_1)^{3/2}} \qquad (15.21)$$
$$P_2/P_1 = (g_2 H_2/g_1 H_1)^{5/2}$$

Other methods of adjusting the output while keeping the speed constant consist of modifying the profiles of the blades at the maximum diameter of the impeller. This technique has been used for a long time and is often used to obtain a small energy rise when the pump is down in performance when tested. (The reader is referred to Karassik et al.[5])

For compressors equation (15.16) could be employed but convention generally uses:

$$\frac{P_{02}}{P_{01}} = f\left[\frac{\omega D}{\sqrt{RT_{01}}} \cdot \frac{m\sqrt{RP_{01}}}{D^2 P_{01}} \cdot R_E \cdot M_n\right] \qquad (15.22)$$

The temperature and pressure statements are conventionally stagnation values. Most compressor manufacturers use a dimensional form, and state the gas involved, so that equation (15.22) becomes:

$$\frac{P_{02}}{P_{01}} = f\left[\frac{\omega D}{\sqrt{T_{01}}} \cdot \frac{m\sqrt{T_{01}}}{D^2 p_{01}} \cdot R_E \cdot M_n\right] \qquad (15.23)$$

Figure 15.12 presents a typical compressor plot.

15.1.1.6 Positive displacement machine principles

Whether the machine is of reciprocating or rotary design, fluid is transferred from inlet to outlet in discrete quantities defined by the geometry of the machine. For example, in a single-acting piston design (Figure 15.13) the swept volume created by piston movement is the quantity delivered by the pump for each piston stroke, and the total flow is related to the number

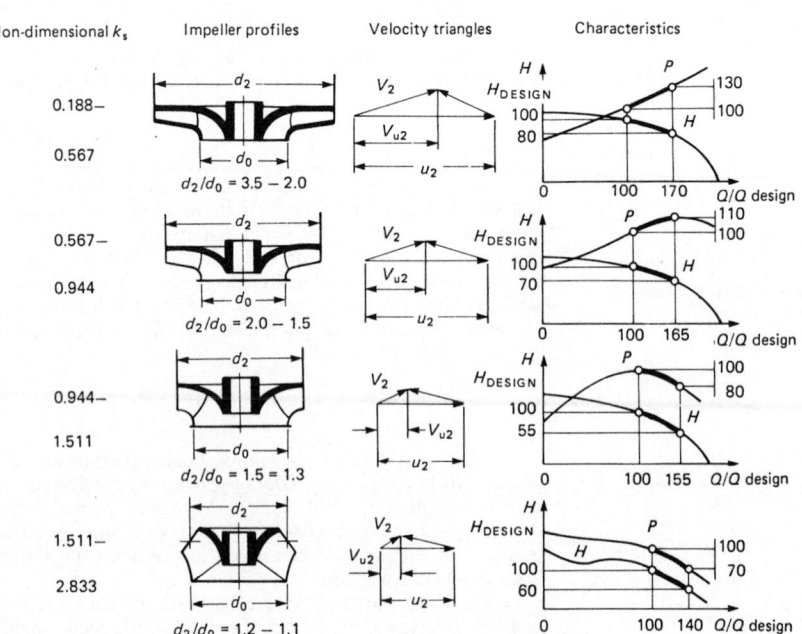

Figure 15.10 Impeller profiles, velocity triangles and typical characteristics as a function of k_s (Turton[1])

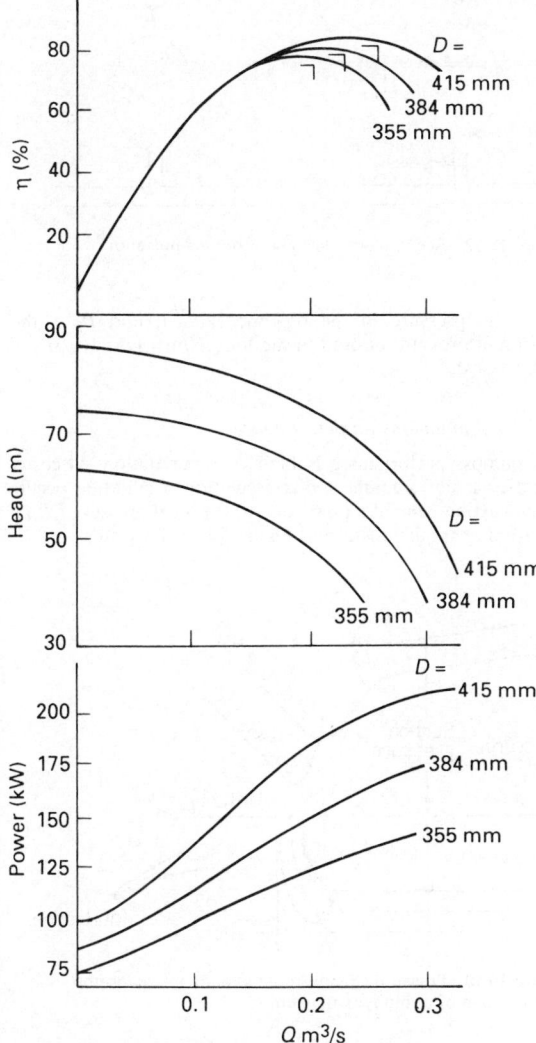

Figure 15.11 Pump scaling laws applied to diameter change (adapted from Karrasik *et al.*[5])

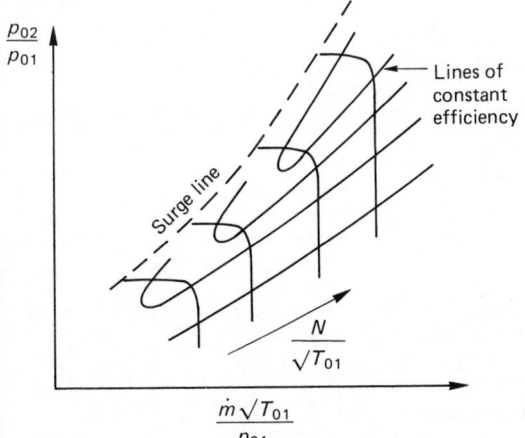

Figure 15.12 A typical compressor plot

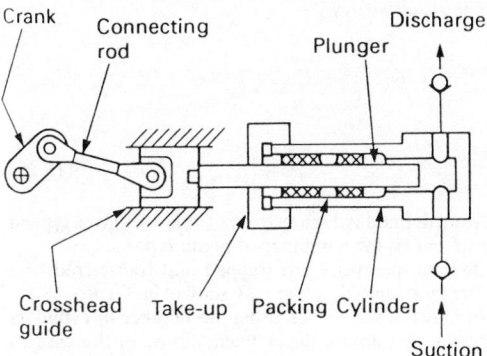

Figure 15.13 A plunger pump (or piston pump)

of strokes per unit time. Similarly, the spur-gear device (Figure 15.14) traps a fixed quantity in the space between adjacent teeth and the casing, and total flow rate is related to the rotational speed of the gear wheels.

The maximum possible flow rate

$$Q_0 = \text{displacement} \times \text{speed}$$

as shown in Figure 15.15. The actual flow is reduced by leakage, flow Q_L:

$$Q = Q_O - Q_L$$

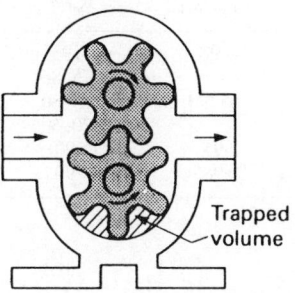

Figure 15.14 An external gear pump

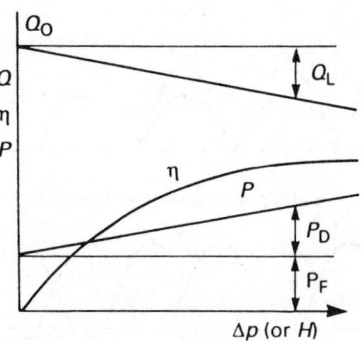

Figure 15.15 The typical characteristic of a positive displacement pump driven at constant speed

The volumetric efficiency

$$\eta_v = \frac{Q}{Q_O} = 1 - \frac{Q_L}{Q_O} \qquad (15.24)$$

and

$$\eta_0 = \frac{\rho Q g H}{P_D + P_L} \qquad (15.25)$$

P_D and P_L are defined in Figure 15.15. Table 15.1 gives typical values of η_v and η_0 for a number of pump types.

Since discrete quantities are trapped and transferred, the delivery pressure and flow vary as shown in Figure 15.16, which also illustrates how increasing the number of cylinders in a reciprocating pump reduces fluctuations. In the case of lobe and gear pumps the fluctuations are minimized by speed of rotation and increasing tooth number, but where, for control or process reasons, the ripple in pressure is still excessive a means of damping pulsations must be fitted. Often a damper to cope with this and pressure pulses due to valve closure is fitted, two types being shown in Figure 15.17. The capacity of the accumulator is important, and one formula based on experience for sudden valve closure is

$$Q_A = \frac{Q P_2 (0.016\ L - T)}{(P_2 - P_1)} \times 0.25 \qquad (15.26)$$

Here Q_A is the accumulator volume (m³); Q is flow rate (m³/s); L is pipe length (m); T is valve closure time (seconds);

Table 15.1 Some values of η_v and η_0 for positive displacement pumps

Pump	$\eta_V(\%)$	$\eta_0(\%)$
Precision gear	→98	→95
Screw	—	75–85
Vane	85–90	75–80
External gear	—	20–60
Radial – multi-piston	>95	>90
Axial – multi-piston	>98	>90

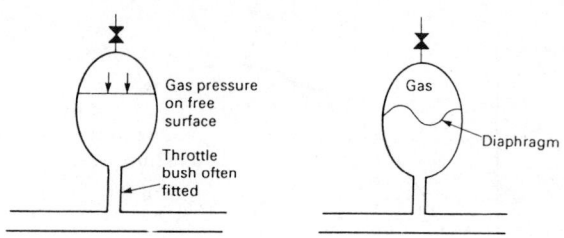

Figure 15.17 Accumulator designs to reduce pulsation

P_1 is the pressure in the pipeline (N/m²); and P_2 is the maximum pressure desired in the line (N/m²) ($P_2 = 1.5P_1$ in many cases).

15.1.1.7 Limitations on performance

For pumps, performance is limited by cavitation, viscosity effects, gas entrainment and recirculation. Cavitation occurs in the suction zone of a pump due to the local pressure falling to around vapour pressure as Figure 15.18 illustrates.

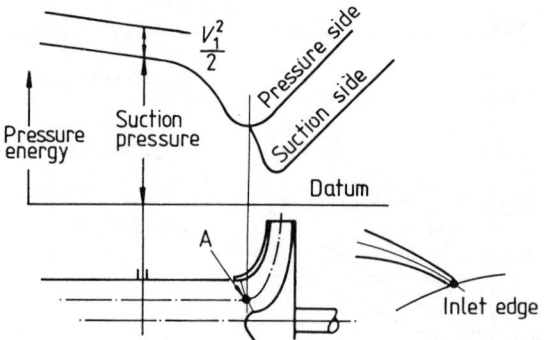

Figure 15.18 Pressure changes on a stream surface in the suction zone of a rotodynamic pump

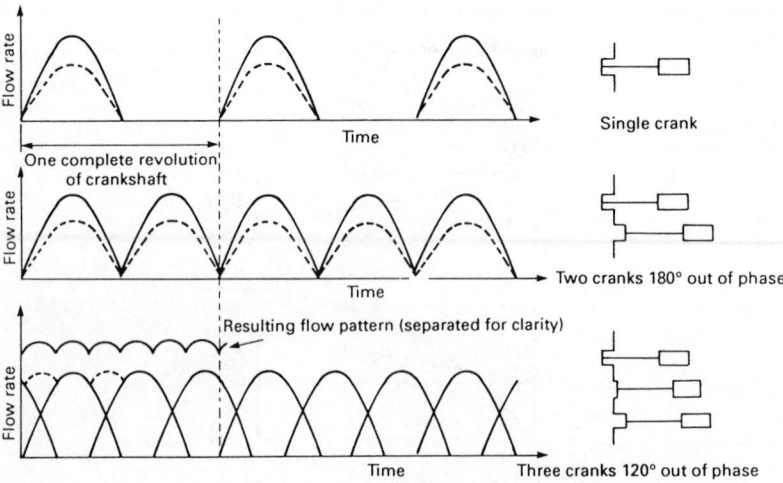

Figure 15.16 The variation in flow rate with numbers of cylinders caused by a reciprocating pump

The pump flow range is reduced as suction pressure reduces. Cavitation also causes considerable damage as bubbles of gas form and then collapse. Two criteria are used to judge whether a pump is in trouble from cavitation or not: one is the concept of NPSH (net positive suction head) and the other is the noise generated.

Net positive suction head is the margin of head at a point above the vapour pressure head. Two statements are used: NPSH available and NPSH required:

$$\text{NPSH}_\text{A} = \text{Total head at suction flange} - \text{vapour pressure head}$$

Figure 15.19 illustrates how system NPSH or NPSH$_\text{available}$ is calculated for the usual suction systems shown.

For a centrifugal pump, the basic NPSH is calculated from

$$\text{NPSH}_\text{A} = h_\text{s} - h_\text{f} \frac{10.2}{\rho} \left[\frac{B}{1000} + P_\text{i} - P_\text{v} \right] \qquad (15.27)$$

where

h_s = static suction head at the pump suction (m)
h_f = flow losses in suction system (m)
B = minimum barometric pressure (mbar)
  (use 0.94 of mean barometer reading)
P_i = minimum pressure on free surface (bar gauge)
P_v = vapour pressure at maximum working temperature
  (bar absolute)

In the process industries h_f is calculated for the maximum flow rate and the NPSH at normal flow allowed for by using the formula

$$\text{NPSH}_\text{A} = 0.8 \, [\text{NPSH}_\text{basic} - 1] \qquad (15.28)$$

This gives a 'target' value to the pump supplier that is 'worst' condition. In general, for cold-water duties equation (15.28) can be used for the duty flow required. Equation (15.27) is used for reciprocating and rotary positive displacement machines, but allowance is made for acceleration effects.

In reciprocators h_f is calculated at peak instantaneous flow, including maximum loss through a dirty filter, and an additional head 'loss' to allow for pulsation acceleration is used:

$$h_\text{A} = \frac{700 \, NQ}{2} \, \Sigma \, \frac{L}{d^2} \qquad (15.29)$$

where

N = crankshaft rotational speed (rpm)
Q = flow rate, $(1 \, \text{s}^{-1})$
L = length of line (m)
d = diameter of line (mm)

and

$$\text{NPSH} = \text{NPSH}_\text{A} - h_\text{A} \qquad (15.30)$$

For metering pumps,

$$\text{NPSH}_\text{basic} = h_\text{s} + \frac{10.2}{\rho} \left[\frac{B}{1000} + P_\text{i} - P_\text{v} \right] \qquad (15.31)$$

h_f is as for the reciprocating pump based on peak instantaneous flow and

$$h_\text{A} = \frac{6\delta P_\text{i}}{\rho} \qquad (15.32)$$

15.1.1.8 NPSH required (NPSH$_R$)

This is a statement of the NPSH that the pump can sustain by its own operation, so that the operating requirement is that NPSH$_\text{R}$ < NPSH$_\text{A}$ and Figure 15.20 indicates how the critical operating flow is related to NPSH$_\text{A}$ and NPSH$_\text{R}$. The usual operating criterion is based on a cavitation test (Figure 15.21). The critical NPSH$_\text{R}$ is defined as the point at which the pump head falls by $x\%$ (3% is often used).

For the centrifugal pump two terms are in common use: the Thoma cavitation number σ and the suction specific speed S_N:

$$\sigma = \frac{\text{NPSH}_\text{R}}{\text{Pump head rise}} \qquad (15.33)$$

NPSH$_\text{R}$ is defined as in Figure 15.22. This figure gives a typical plot of σ against k_s that may be used as a first 'design' estimate of NPSH$_\text{R}$, but in many applications test data are required:

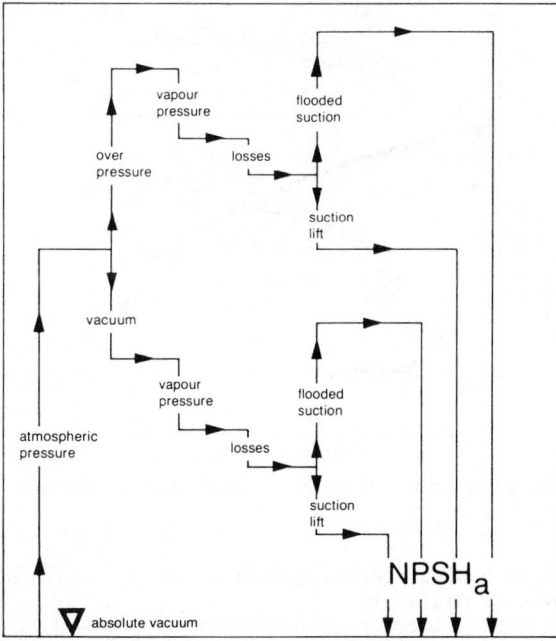

Figure 15.19 A visualization of the way NPSH $_\text{available}$ is calculated for a pump suction system. (Courtesy of Girdlestone Pumps Ltd)

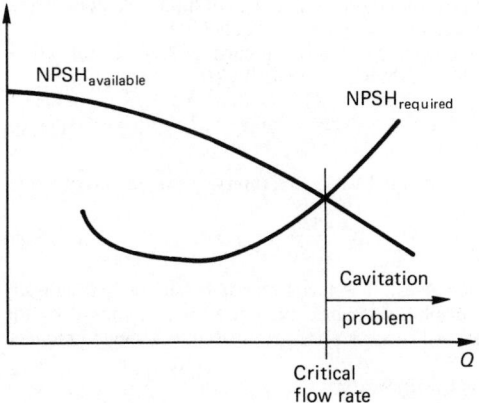

Figure 15.20 Critical flow rate determined by cavitation considerations

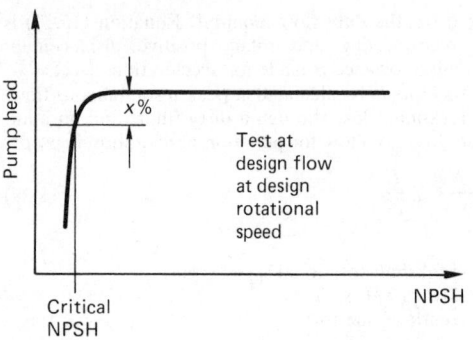

Figure 15.21   A conventional presentation of pump cavitation behaviour

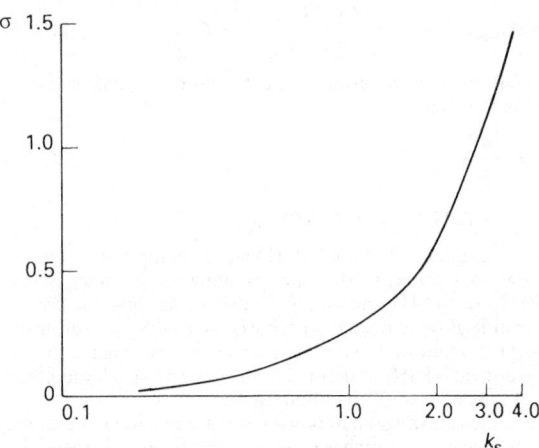

Figure 15.22   Variation of σ with k_s for rotodynamic pumps (Turton[1])

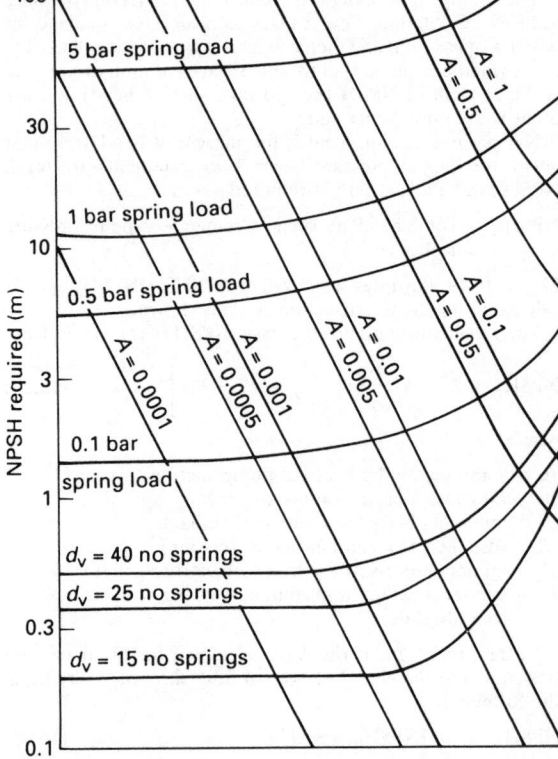

Figure 15.23   NPSH $_{required}$ for reciprocating metering pumps related to valve spring loading (equations (15.35) and (15.36))

$$S_N = \frac{N\sqrt{Q}}{K(\text{NPSH}_R)^{3/4}} \qquad (15.34)$$

where K is a constant = 175 if $g = 9.81$ m s^{-2}, Q is in l/s, N in revolutions/second, and NPSH$_R$ is m of liquid. A 'good' value of S_N for a centrifugal pump is around 10 000.

For reciprocating metering pumps NPSH$_R$ is related to valve loading as shown in Figure 15.23:

$$A = \frac{24\nu Q\rho}{Zd_V^{3}} + 5 \times 10^5 \frac{\rho Q^2}{Z^2 d_V^4} \qquad (15.35)$$

where d_V = nominal valve size (mm) for single valves, and

$$A = \frac{80\nu Q\rho}{Zd_V^{3}} + 15 \times 10^5 \frac{\rho Q^2}{Z^2 d_V^4} \qquad (15.36)$$

for double valves. It is recommended that for hydraulically operated diaphragm pumps the extra losses imposed by the diaphragm and support plate are treated as a single unloaded valve.

For other reciprocators

$$\text{NPSH}_R = 5U^2 + \frac{0.12(P_d)^{0.75}}{\rho} \qquad (15.37)$$

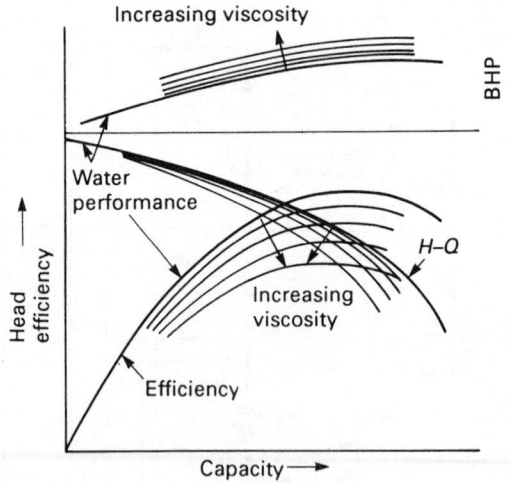

Figure 15.24   Effect of viscosity increase on centrifugal pump performance

where U = mean plunger speed (m s^{-1}) and P_d = discharge pressure bar absolute.

Viscosity affects pump performance by increasing flow losses. Figure 15.24 illustrates the deterioration as viscosity increases. If the kinematic viscosity is greater then 100 centistokes, water performance must be corrected as shown later in

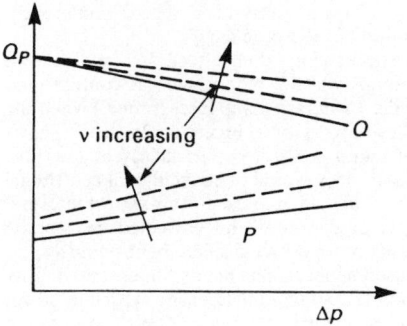

Figure 15.25 Effect of viscosity increase on positive displacement pump performance

Figure 15.27. Figure 15.25 indicates that in a positive displacement pump the volumetric efficiency improves and power requirement increases (with increasing viscosity).

Table 15.2 summarizes the effects of liquid changes (effectively, viscosity and density changes) on pump performance and Figure 15.26 presents material by Sterling[6] which illustrates how efficiency falls away with viscosity for two pumps working at the same duty point, graphically illustrating the rapid decay of efficiency as μ increases in a centrifugal pump.

Figure 15.27 demonstrates a well-known method of correcting for fluid change from water for a centrifugal pump. This allows an engineer to predict change in performance if the kinematic viscosity of the liquid to be pumped is known and the water test data are available.

Recirculation effects at low flow rates are now well documented, and can cause vibration and, in some cases, severe

Table 15.2 The effect of viscosity – a comparison

Type of pump	Significant viscosity levels[a]	Effect of viscosity level	Treatment and/or notes
Centrifugal	20	—	Performance maintained similar to water performance up to this level
	20–100	Lowering of H–Q curve increase in input hp	General lowering of efficiency but may be acceptable
	Above 100	Marked loss of head	Considerable reduction in efficiency, but high efficiencies may still be attainable from large pumps
Regenerative	Above 100	Marked loss of performance	Pumps of this type would not normally be considered for handling fluids with a viscosity greater than 100 centistokes
Reciprocating	Up to 100	Little	Performance generally maintained. Some reduction in speed may be advisable to reduce power input required
	Above 100	Performance maintained but power input increased	Speed is generally reduced to avoid excessive power inputs and fluid heating
	Above 1000	Flow through valves may become critical factor	Larger pump size selection run at reduced speed – e.g. 3 × size at 1000 centistokes running at one-third speed. Modification of valve design may be desirable for higher viscosities
Plunger	—	—	For very high-pressure deliveries only
Sliding vane	Above 100	Sliding action impaired: slip increased	Not generally suitable for use with other than light viscosity fluids
External gear	None	Power input and heat generated increases with increasing viscosity	May be suitable for handling viscosities up to 25 000 centistokes without modification. For high viscosities: (a) Clearances may be increased (b) Speed reduced (c) Number of gear teeth reduced
Internal gear	None	Power input and heat generated increases with increasing viscosity	For higher viscosities: (a) Speed may be reduced (b) Number of gear teeth reduced (c) Lobe-shaped gears employed
Lobe rotor	250	None	(a) Speed may have to be reduced
	Above 250	Cavitation may occur	(b) Modified rotor form may be preferred
Single-screw	None	—	Nitrile rubber stator used with oil fluids
Twin- or multiple-screw	Up to 500	Little or none	—
	Above 500	Increasing power input required	Speed may be reduced to improve efficiency

[a] Viscosity in centistokes.

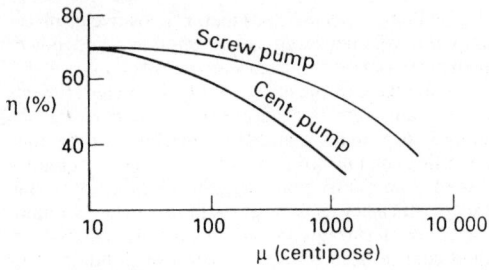

Figure 15.26 Comparison of efficiency reduction with viscosity increase for a screw pump and a centrifugal pump of similar duty

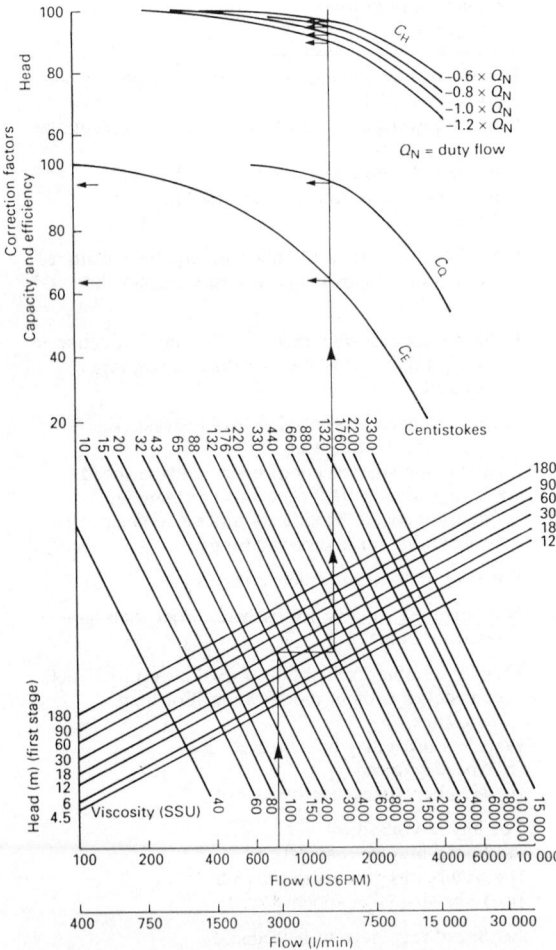

Figure 15.27 A method of correction for viscosity (adapted from American Hydraulic Standards[13]). Example: The pump is to handle 750 USGPM of 1000 SSU Liquid against a head of 30 m. From the diagram, $C_E = 0.64$; $C_Q = 0.95$; $C_M = 0.92$ at duty point ($1.0 \times Q_N$). To test on water needs tests at a flow rate of 789.5 USGPM and 32.6 m: if the test efficiency η is 75%, oil efficiency $\simeq 0.75 \times 0.64 = 48\%$

cavitation damage. Papers given at a recent conference[7] indicate the magnitude of the problem.

Gas content is another important effect. It is well known that centrifugal pumps will not pump high gas content mixtures, as flow breaks down (the pump loses 'prime') when the gas/liquid ratio rises beyond 15%. Figure 15.28 clearly shows how a centrifugal pump is affected particularly at low flow rates, and the behaviour is typical of conventional centrifugal pumps. Figures 15.29 and 15.30 present well-known information on the effects of dissolved and entrained gas on the volumetric efficiency of a positive displacement pump.

Fans are often used in near-ambient conditions, and density change is not significant, so that inlet density is used in power calculations. Care is needed in air-conditioning systems to correct for the temperature at the fan inlet. Axial fan performance is affected by blade stall as in compressors.

A compressor characteristic is shown in Figure 15.31. Flow is limited at the high mass flow end of the curve at any speed when local velocity in a passage (usually the last stage outlet guide vanes in an axial machine and the diffuser vane ring in a radial compressor) reaches sonic velocity and thus mass flow cannot increase further. The phenomenon of surge is more complicated as it is caused by flow instability. Its effects can be limited by reducing the pressure rise in an axial stage but not eliminated. Rotating stall occurs in both radial and axial machines and its action is shown in Figure 15.32. A vane stalls and affects flow round an adjacent vane which in turn stalls. This effect thus propagates round the blade row, in the opposite direction to rotation, at about half the rotational speed. Reference 2 gives more detailed discussion.

Figure 15.31 shows the total limitations on the compressor surge line and mass flow rate of stall and choking. For detailed discussion, textbooks such as those by Horlock[2] and Balje[10] may be consulted.

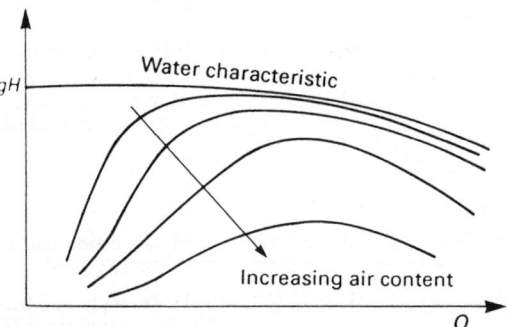

Figure 15.28 Effect of gas content on centrifugal pump performance

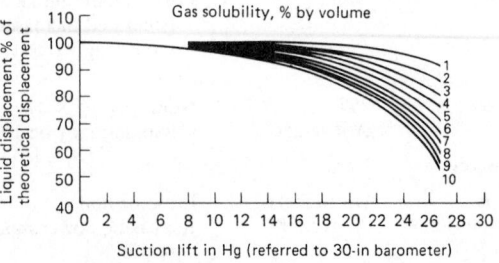

Figure 15.29 Gas solubility in water

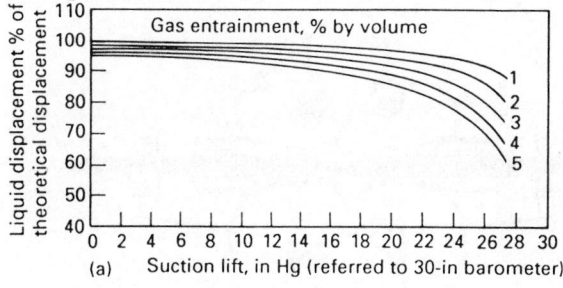

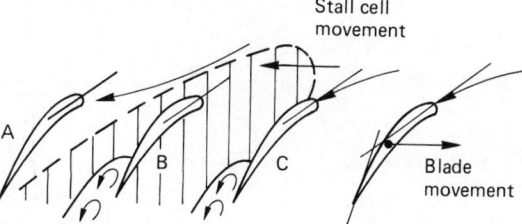

Figure 15.32 Rotating stall in an axial blade row

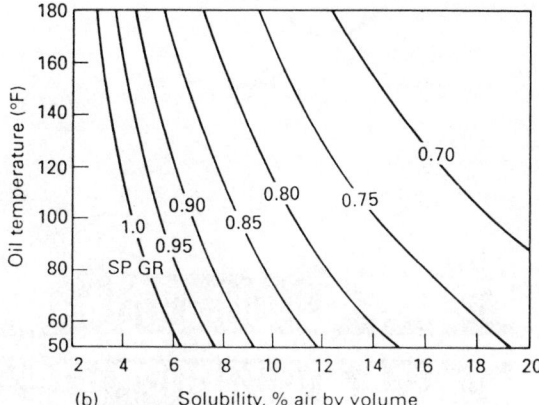

Figure 15.30 (a) Effect of entrained gas on liquid displacement for a positive displacement pump; (b) solubility of air in oil. *Example*: At a pressure of 5 inches Hg with 3% gas entrainment by volume pump capacity is reduced to 84% of theoretical displacement

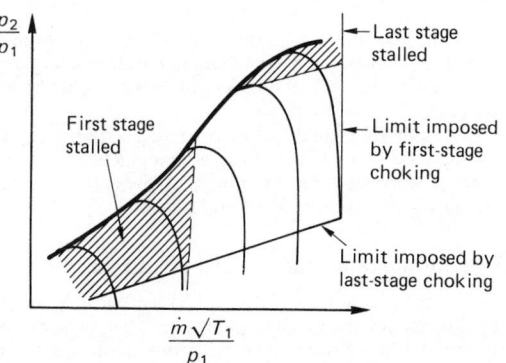

Figure 15.31 Limits placed on compressor performance

15.1.2 Machine selection

Although Balje[8] and Csanady[9] have proposed a common basis of performance presentations using a non-dimensionalized number resembling specific speed, each type of machine will be discussed separately. Engineers employed in water supply, the process industries and other spheres of activity have a formidable task when selecting equipment. If the equipment

they select does not come up to specification the maximum claim on the supplier is the price paid. The cost to their company is that of plant downtime and lost production which is likely to exceed equipment costs by many times. 'Buyer beware' is thus a normal rule. To assist the buyer there are BS and ISO specifications and codes of practice such as the American Petroleum Industry (API) standards, but in many areas there are no such aids, and the buyer has to rely on advice, experience and, ultimately, engineering common sense.

Any pump, fan or compressor selected must fulfil the specified duty (or duties) and be capable of operating safely and economically with a minimum of maintenance and down-time. The selector has therefore a challenging task. The first essential task is to prepare the technical brief which will become the tender document. This brief must state the entire operating envelope of the machine, with complete details of temperature, humidity, fluid properties and site variations, and detail the standards and codes which will apply, e.g. API 610[10] for refinery and petrochemical centrifugal pumps. This covers materials, bearing and seal systems, pressure testing of casings, vibration and noise limits, hydraulic performance, draft documents, shipping and installation in over 100 pages. In short it is a comprehensive document of mutual understanding between customer and supplier.

The project engineer needs data to decide which type of machine, likely size, rotational speed and drive system before submitting a detailed tender document. Some basic charts will therefore be discussed.

The principles of the two groups of pumps (rotodynamic and positive displacement) have been discussed, and Figures 15.33 and 15.34 illustrate the main types. A universal index of flow path and size for centrifugal pumps is the specific speed referred to above (Figure 15.10) which indicates the flow path shapes and probable characteristics. A useful pressure to flow rate envelope is shown in Figure 15.35. Once a type is decided, manufacturers' data may be consulted. Usually these are test data when pumping water for rotodynamic machines, and an approximate idea of performance can be obtained by converting water data using a conversion chart as described earlier.

When considering the selection of positive displacement pumps, Figure 15.36 is a useful range guide. Fan selection devolves into the choice of an axial or a centrifugal machine, and whether a single- or double-stage machine is required, but choice is usually determined by flow rate and pressure rise needed, and in some cases by the space available in which a machine will need to be installed. Table 15.3 gives a working basis for fan selection.

Figure 15.37 outlines the main types of compressor, and it must be said that when selecting turbo compressors the choice of machines is a function of delivery pressure and flow rate (as Figure 15.38 indicates). To extend consideration further, Figure 15.39, based on an article in a Sulzer Technical Review,

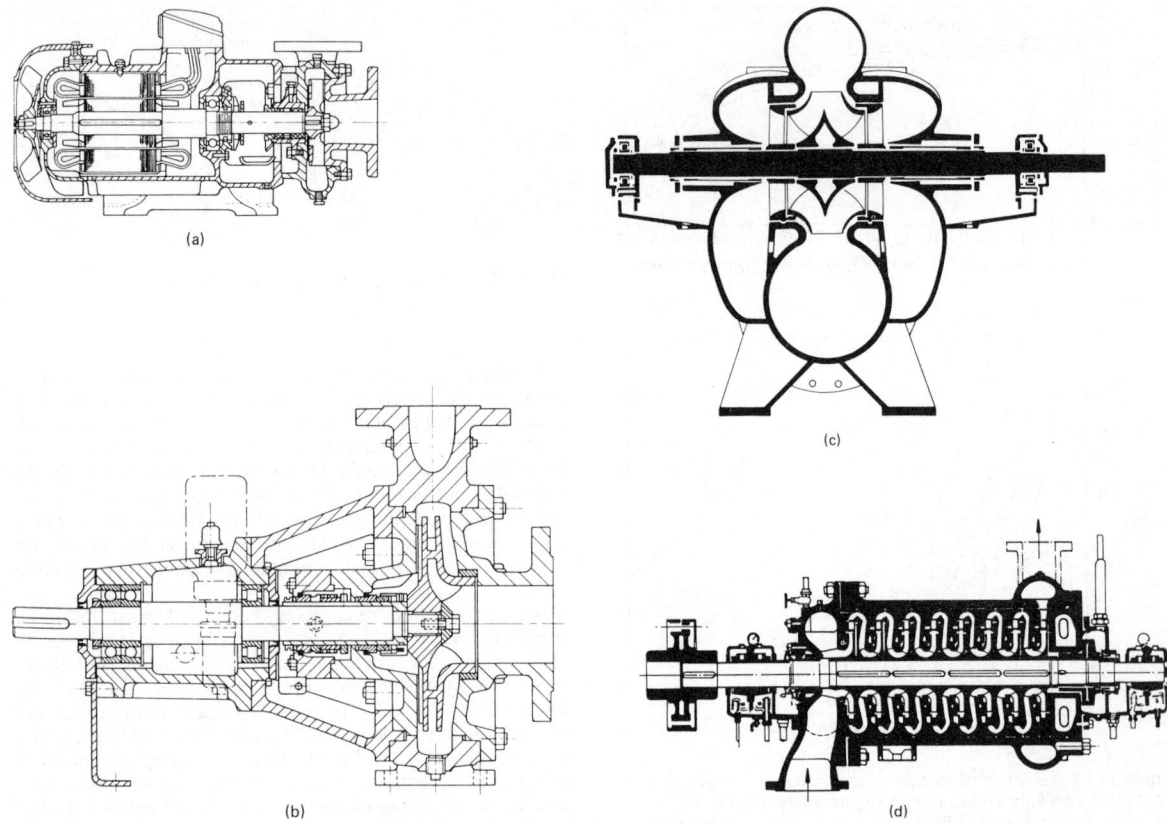

Figure 15.33   Some typical pump layouts. (a) A monobloc design with the impeller fixed on the motor shaft; (b) a modern back pull-out design; (c) a double-entry pump; (d) a multistage pump design

covers plant supplied by the company of both turbomachine and positive displacement, screw, vane, or diaphragm types. Most makers offer oil-lubricated and non-lubricated machines.

As Figures 15.38 and 15.39 indicate, the selection of compressor type depends on the pressure rise and flow rate required. Large compressors are supplied for a number of duties. About 20% are used for air compression, for factory services where usage is typically around 87 bar for energy storage, for other industrial duties, or in bottles. Industrial usage in the field of oxygen, nitrogen and medical gases accounts for a large sector. Natural gas transmission is also a substantial field of application. The industrial and process processes work on pressure up to about 400 bar. Polyethylene processes demand pressures up to 3500 bar. As the discussion on pumps indicated, when the duty could be met by a number of types, choice is often determined by experience in service, complexity or cost. The only positive attitude is probably to choose a turbomachine unless company policy dictates a positive displacement one. One factor with compressors is the temperature rise (over 150°C with a reciprocator). Cooling causes water and water vapour to accumulate with the consequent need for careful after cooling, intercooling between stages in multi-stage machines and water collection to prevent tools, instruments or equipment being damaged.

Oil injection is often used in screw and rotary vane machines to cool and to help eliminate water. Wear is also reduced. Oil injection at the rate of up to 20 mg m^{-3} of gas is used, and then filtered well below the contamination limit for factory air (5 mg m^{-3}). (An efficiency of recovery of 99.9999% has been claimed.)

Selection methods should reflect operating experience as well as being based on intelligent use of manufacturers' data, satisfactory performance results from rigorous adherence to company specifications as well as good selection.

15.1.3 Performance monitoring and prediction

Any pump, fan or compressor is supplied against a contract duty. In the case of many small pumps and fans which are quantity or batch produced makers will often quote against a typical performance which they check by routine testing, and will only do a full works test if a customer requires this. With larger pumps, fans and compressors, a full works test (usually witnessed) is required, and often check tests when installed in the systems will be needed; this latter point will be discussed together with routine monitoring.

15.1.3.1 Works tests

For back-pull out pumps ISO 5199[11] covers all aspects, including testing, seals, bearings, noise and vibration, and lists all the relevant ISO and related BS 5316 standards, among which Part 1 (for general-duty class C pumps) and Part 2 (for class B

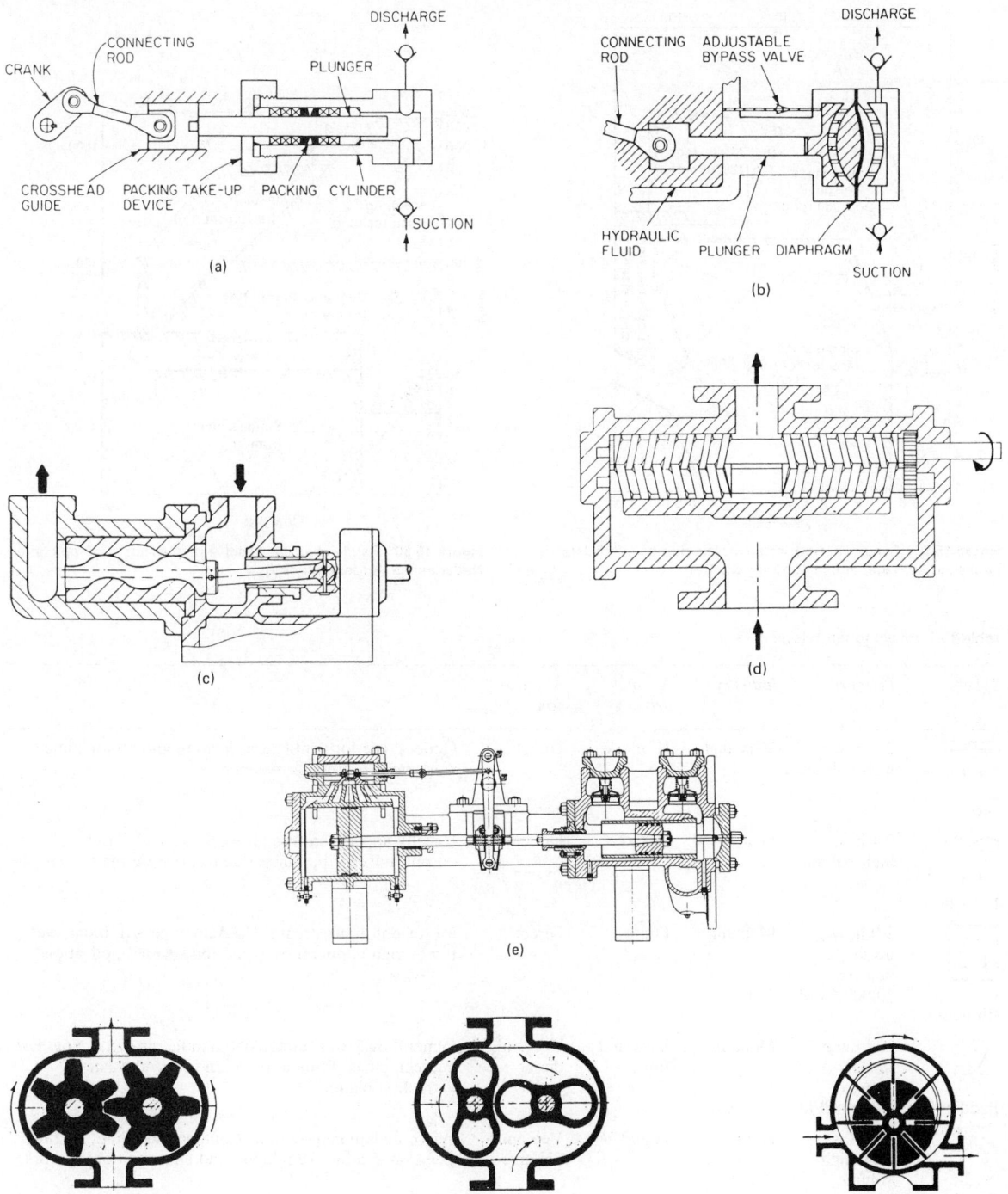

Figure 15.34 Some positive displacement pump designs. (a) Single plunger pump; (b) simple diaphragm pump; (c) mono pump; (d) twin-screw pump; (e) steam reciprocating pump; (f) gear pump; (g) lobe pump; (h) vane pump

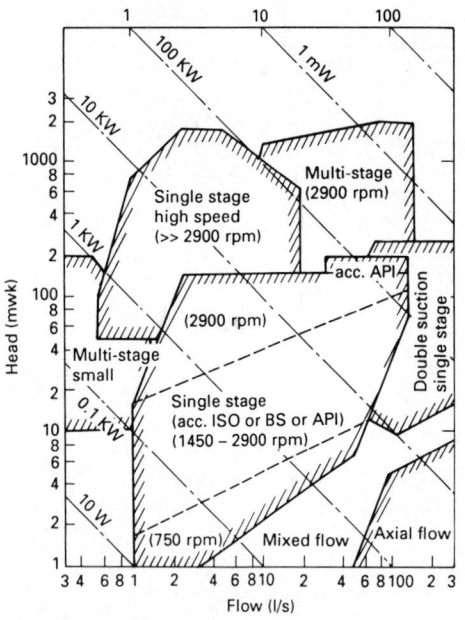

Figure 15.35 Range chart for rotodynamic pumps (after data published by Nederlandse Aardolie MIJ BV)

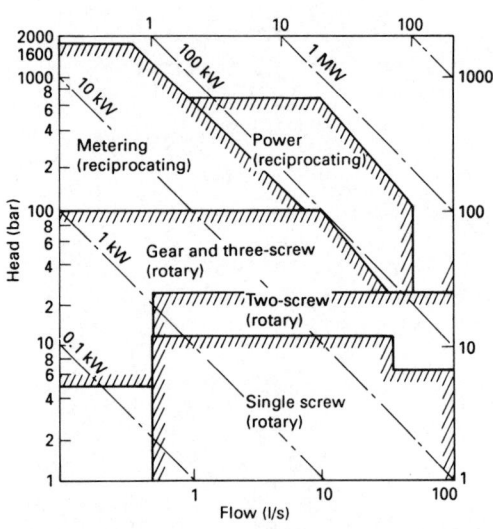

Figure 15.36 Range chart for positive displacement pumps (after Nederlandse Aardolie MIJ BV)

Table 15.3 An aid to fan selection

Type	Pressure volume	Industry	Normal drive	Application	
Axial	2.5 in w.g. high volume	Very high	H and V	Direct	General use for ventilation, heating and minor fume work on low-pressure systems
Propeller	0.4 in w.g. high volume	Low	H and V	Direct	Usually applied on free air work, such as input and output units for buildings due to pressure limitations
Bifurcated	1.0 in w.g. up to approx. 10 000 CFM	Medium	Fume	Direct	Motor not in air-stream. Used on explosive fume, wet fume, high-temperature work and severe applications
Paddle	12 in w.g. up to approx. 30 000 CFM	Medium	Dust and fume	Vee and direct	General dust and frame. Will handle air containing dust and chippings. Wide application in wood-waste extraction plants
Forward	6.0 in w.g. very high volume	High	H and V	Vee and direct	Will only handle clean air. Compact and quiet running. Used on heating, ventilation and air-conditioning work
Backward	20 in w.g. high volume	High	Dust and fume	Vee and direct	General dust and fume. High-pressure systems and on dust-collector plants. Will handle some dusty air
Blowers	42 in w.g. usually low volume	Medium	General	Direct	Furnace blowing, cooling, conveying and where there is a need for high pressures

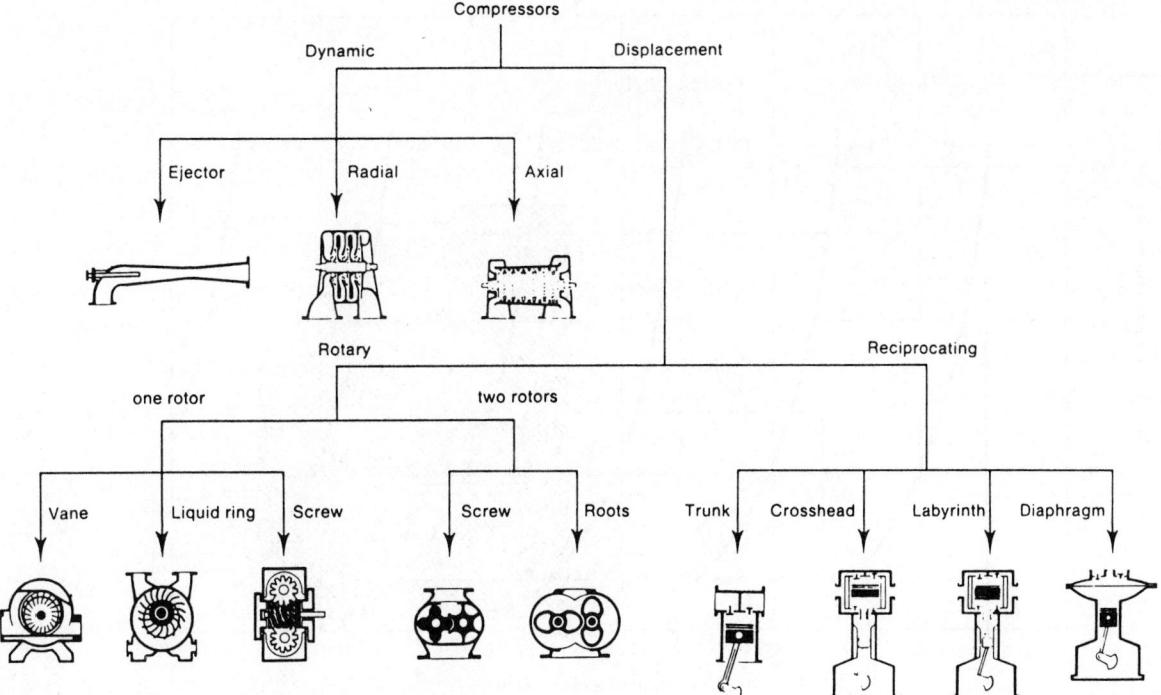

Figure 15.37 Basic compressor types

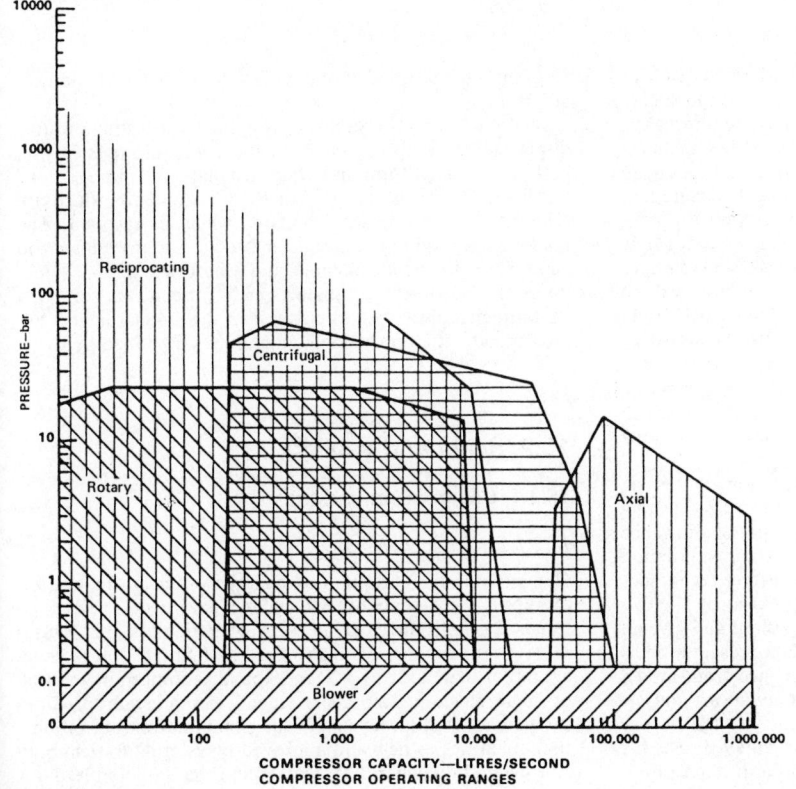

Figure 15.38 An approximate range chart for compressors

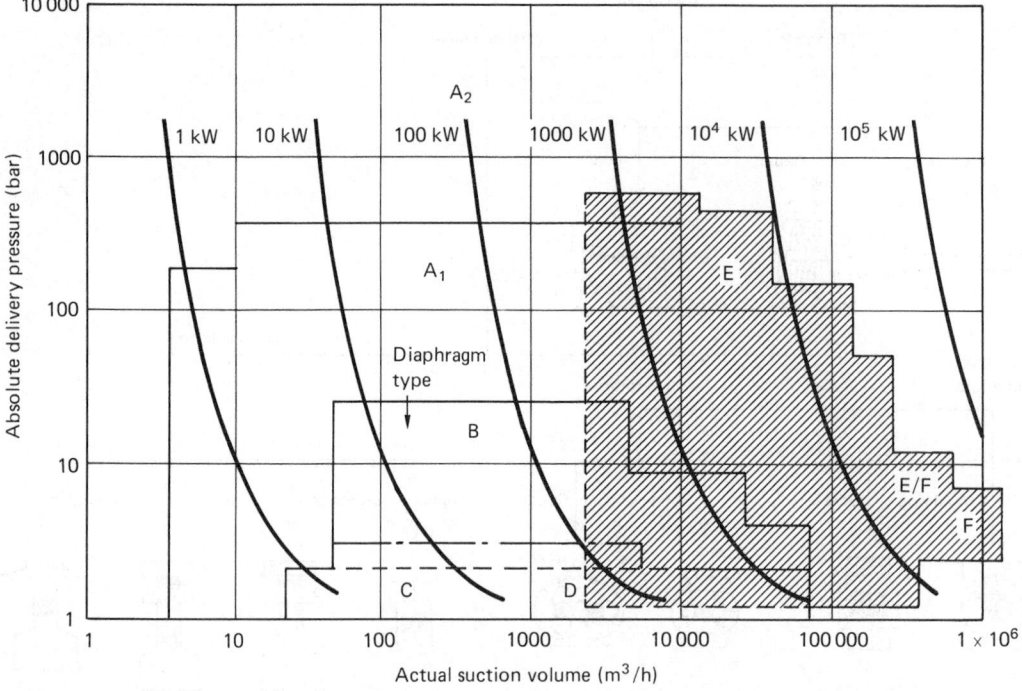

Figure 15.39 An example of a manufacturer's range chart (based on a *Sulzer Review* article[17]). A_1 – reciprocating compressor, lubricated and non-lubricated cylinders; A_2 – reciprocating compressors, lubricating compressors; B – screw compressors, dry or oil-flooded rotors; C – liquid ring compressor; D – rotary (Roots type); E – centrifugal compressors; F – axial compressors

pumps) detail test arrangements and procedures as well as instrumentation for pressure, flow, torque power and speed. Permissible bands of readings are specified as are alternative cavitation tests. In the case of BS 5316 it is stated in an annex that for mass-produced pumps the manufacturers, if they state that the standard is being satisfied, must be able to ensure that performance for any pump does not diverge from the published curve by more than ±6% for total head, ±8% for flow rate and ±8% for input power. This allows customers to have confidence in the published curves. Similar provisions will be found in the American Hydraulic Institute Standards.[13] If the pump is to follow API 610 these standards must be satisfied.

Where the liquid to be pumped is not water it is common practice to test on cold water and to predict the performance to be expected by using a chart such as Figure 15.36, which gives an example of how water test duty may be obtained if the duty is known.

For fans, standards also specify instrumentation and test rig layout. BS 848: Part 1[14] gives methods of standardized testing and also of prediction when models are used and of allowance for compressibility. Since fan noise is important in ventilation systems BS 848: Part 2[14] lays down noise-testing techniques and gives details of test chambers and site provisions. The two parts form an essential item of fan test provision, and give all the necessary equations required for test data presentation as well as for prediction of probable performance from model tests, and for correction for non-standard situations and air conditions.

A similar standard, BS 2009,[15] covers acceptance tests for turbo-type compressors and exhausters. This also states provisions for standardized rig layout and instrumentation and methods of presenting data in a standardized way. Corrections for compressibility and methods of performance prediction are all given.

BS 1571: Part 1[16] lays down provisions for testing positive displacement compressors of all the common types in use, both in packaged form and other installations.

All the standards give lists of British Standards which are relevant and quote ISO Standards which correspond. The reader is referred to the literature listed if test procedures and equipment are being planned and where standardized methods of performance are being sought for contract purposes. Performance prediction is covered in the standards and follows broadly the dimensionless quantities described here.

15.2 Seals and sealing

15.2.1 Compression packing

15.2.1.1 Introduction

Compared to the deterministic qualities of ferrous metals, for example, the essentially deformable nature of sealing materials has introduced a measure of variability that causes many commentators to regard fluid sealing technology as an art rather than a science. Seen as an anachronism in a period of high technological achievement, compression packings show no signs of losing significant ground in terms of production quantities as new and improved types proliferate in both Europe and elsewhere. To understand this situation requires some appreciation of the fundamental mode of operation of the adjustable gland or stuffing box shown in Figure 15.40.

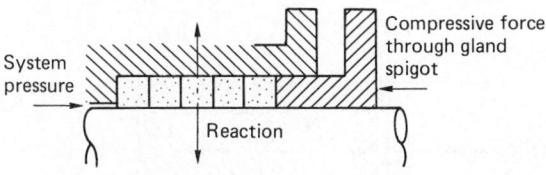

Figure 15.40 Compression packing

This may be filled with split packing rings chosen from a variety of materials and constructions, described elsewhere, which are persuaded to react against a shaft, whether rotary or reciprocating, to the extent that the radial force developed exceeds the pressure to be sealed. Packings in this category used for rotating or reciprocating equipment rely on a controlled leakage for long-term lubrication purposes if they are to survive for an adequate period. The continued justification for the compression packing might appear obscure against such a background but there can be no doubt that certain areas of application exist where no reasonable substitute is available.

Pumps Many reasoned and well-researched papers have been published to support mechanical seals against soft packing, and vice versa. There is no doubt that the former have supplanted packed glands as original equipment on the majority of rotodynamic pumps for a variety of process and service fluids, but there are operating parameters and cost considerations which will frequently dictate the choice of soft packing. Table 15.4 compares the relative attributes of the two contenders in basic terms.

In general, it may be said that, unless zero leakage is an absolute priority, compression packings will retain an important position wherever regular maintenance is available and the following considerations apply:

- Simplicity in gland design and ancillary equipment
- Ease of fitting
- Flexibility of supply and spares for plant utilizing many different types and sizes of pump handling a wide variety of fluids.

- Frequent ability to cater for adverse conditions without elaborate precautions

Valves If any doubt exists regarding selection on pumps then a much more obvious choice of soft packing applies to the valve scene. The relative lack of movement, ease of fitting and, in this case, lack of leakage requirement for lubrication purposes (plus the most decisive advantage of low cost) are factors which ideally relate to compression packings.

There are areas where moulded elastomeric seals present a reasonable alternative but even the most exotic compounds would seldom be used above 250°C – unless reinforced by asbestos fabric.

15.2.1.2 Operating principles

By comparison to the seal types described in the literature – particularly elastomeric lip and squeeze seals – compression packings respond to applied pressure in inverse proportion to the hardness of their construction and rely on an external force to produce the radial pressure required for effective sealing. The method of generating that force can vary but usually (and preferably) involves a bolted gland spigot as shown in Figure 15.40 where controlled axial movement is easily achieved by adjustment of the retaining nuts or studs. Spring loading is sometimes used in inaccessible situations but such a provision lacks the fine control demanded by some packing types and has a limited range of load capability.

While the sealing force can be adjusted to cater for service wear, care must be taken to avoid overcompression which will lead to excessive friction, shaft wear and premature packing failure.

To increase density and dissipate heat, soft packings invariably contain lubricants, loss of which, through excessive compression or overheating in service, will result in packing volume loss with subsequent reduction in the effective sealing reaction and correspondingly increasing leakage rates. By limiting compression to a point where slight controlled leakage is obtained, adequate lubrication of the dynamic surfaces is ensured and overcompression of the packing avoided. However, where lubrication is a problem – or a degree of gland cooling is required – a lantern ring can be incorporated into the gland area for the distribution of additional lubri-

Table 15.4

Comparison	Soft packing	Mechanical seal
Initial cost	Of the order of 10:1 in favour of soft packing depending on size and application	
	APPROXIMATELY EQUAL	
Reliability	Ample warning of impending failure with possibilities for correction	Little or no warning of end of useful life with possibility of sudden complete failure
Installation	Essentially simple – requiring no special skills if correct procedure adopted	Skilled fitting required – precisely defined environment and assembly
Maintenance	Regular and requiring experience	Zero
Spares	Facility for stocking length form material or complete pre-formed sets at relatively low cost	Spare seal components must be available – cost can be substantial
Shaft wear	Can be considerable; shaft sleeves reduce replacement costs Friction losses slightly higher with soft packing	Nil
Operating costs	Leakage losses zero with mechanical seals but positive with soft packing as lubrication of sealing rings is essential	

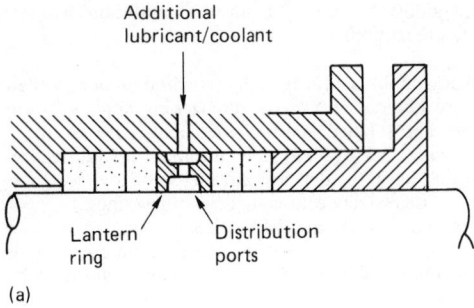

(a)

Additional lubricant/coolant

Lantern ring

Distribution ports

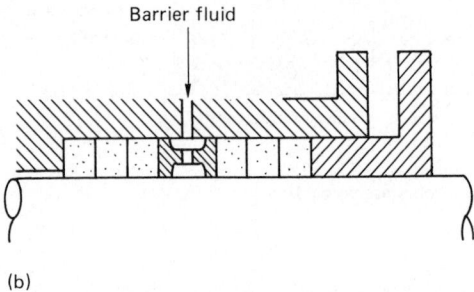

(b)

Barrier fluid

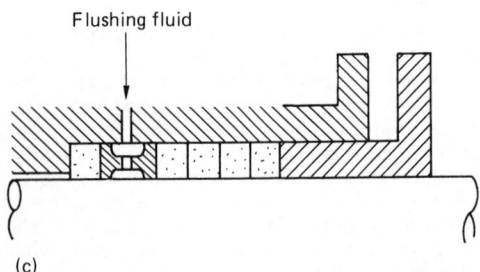

(c)

Flushing fluid

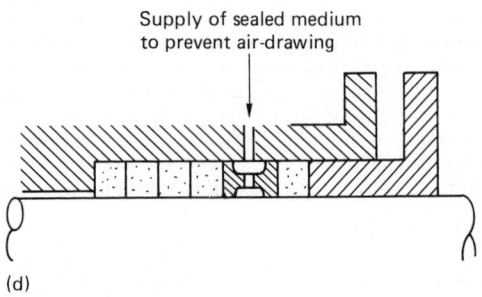

(d)

Supply of sealed medium to prevent air-drawing

Figure 15.41

cant/coolant (Figure 15.41(a)). The position of a lantern ring will depend on the nature of the application but, since the packing rings nearest to the gland spigot do most of the work, the additional fluid should usually be introduced near to that area.

If it is essential that the fluid being pumped does not escape to atmosphere (e.g. a toxic medium), the lantern ring may serve to introduce a barrier fluid at a pressure of 0.5–1 bar above that to be sealed (Figure 15.41(b)). Similarly, where there is a risk of severe abrasive wear to the packing, a flushing fluid may be introduced through the lantern ring (Figure 15.41(c)). For application with negative pump pressures (i.e. suction) a supply of the medium being sealed can be made through the lantern ring to prevent air-drawing (Figure 15.41(d)).

If extreme temperatures are to be encountered it is unlikely that cooling through the lantern ring will be sufficient and recourse must be made to internal cooling of the gland housing and shaft to reduce the temperature at the gland to a value

within the packing's capabilities. Conversely, when dealing with media which crystallize or congeal when cool (e.g. sugars, tars, etc.), the packing will face rapid destruction unless gland heaters or a steam-jacketed arrangement are employed to restore the fluid state before starting up.

It should always be remembered that the inclusion of a lantern ring into the gland area invariably complicates assembly and can provide a possible source of shaft scoring. They should, therefore, only be considered when the nature of the application absolutely demands their presence.

15.2.1.3 Gland design

At this juncture, few international standards exist to define housing design for soft packings but the dimensions shown in Table 15.5 should be satisfactory for most applications. Housing depths will vary with individual circumstances, such as the inclusion of a lantern ring, but five rings of square-section

Table 15.5 Suggested housing widths in relation to shaft diameters (all dimensions in millimetres)

All packings except expanded graphite		*Expanded graphite*	
Shaft diameter	Housing width	Shaft diameter	Housing width
Up to 12	3	Up to 18	3
Above 12–18	5	Above 18–75	5
18–25	6.5	75–150	7.5
25–50	8	150 and above	10
50–90	10		
90–150	12.5		
150	15		

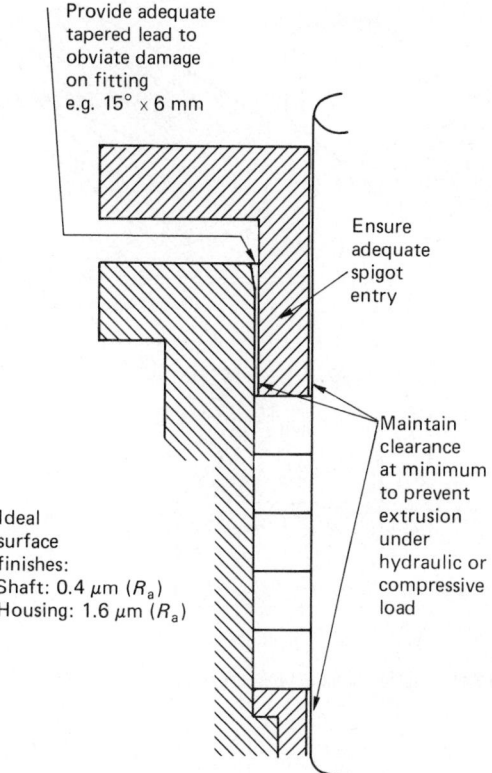

Provide adequate
tapered lead to
obviate damage
on fitting
e.g. 15° × 6 mm

Ensure
adequate
spigot
entry

Ideal
surface
finishes:
Shaft: 0.4 μm (R_a)
Housing: 1.6 μm (R_a)

Maintain
clearance
at minimum
to prevent
extrusion
under
hydraulic or
compressive
load

Figure 15.42

packings are usually recommended for the average, uncomplicated duty.

Other design considerations worthy of note, but often overlooked, are summarized as follows (see Figure 15.42):

1. The provision of an adequate tapered 'lead in' at the mouth of the gland to facilitate entry of the packing and to obviate the risk of damage in the assembly operation. A minimum of 15° × 6.5 mm usually represents good practice.
2. The provision of a reasonable surface finish on adjacent metal parts – particularly the dynamic surface. The better the finish, the less wear will occur; 0.4 μm (16 μ in) R_a on the shaft and 1.6 μm (64 μ in) R_a on the stuffing box bore should be ideal for most applications. The use of shaft sleeves can give considerable maintenance advantage when considering the question of surface finish.
3. The danger of extreme running clearances at the gland – particularly on the spigot side. In those exceptional cases where excessive clearance is unavoidable the packing should be protected by an independent ring of suitably robust material or construction which reduces the clearance to a minimum.
4. An allowance for entry of the gland spigot well into the gland area; certainly to an extent that exceeds substantially the depth of the tapered lead-in. The length of spigot selected must also cater for packing compression, resulting from gland adjustment. Typical entry lengths should be at least two times packing section. For packings of softer construction, maximum length should be provided. With

modern packing materials, bevelled glands are seldom an advantage and can actually promote movement of the sealing ring on the spigot side into the live clearance.
5. The need to avoid excessive shaft misalignment or whip.
6. The provision of adequate shaft support. The packing must not be used as a bearing.

15.2.1.4 Packing construction and materials

Fibres:	
Mineral	asbestos
Vegetable	cotton
	flax
	jute
	ramie
	sisal
Synthetic	aramid
	glass
	graphite filament
	graphite foil
	nylon
	polytetrafluoroethylene (PTFE)
	rayon

Lubricants:	
Dry	graphite
	mica
	talc
	molybdenum disulphide
Wet	tallow
	castor oil
	straight mineral lubricating oil
	petrolatum
	solid fractions
	paraffin wax
	soaps
	silicone grease
	PTFE dispersions

Metals:	
	lead foil and wire
	aluminium foil
	copper foil and wire
	brass wire
	monel wire
	inconel wire
	stainless steel wire
Elastomers:	natural and synthetic

The principal forms of constructions for fibrous compression packings are (see Figure 15.43):

- *Braided* Individual yarns are braided tube over tube and squared off. The density of this type of construction is high and ideal for many valve applications.
- *Plaited* Multiple yarns are interwoven in plaited bundles in such a way that the direction of fibre follows the periphery of the packing ring. The natural characteristic of this construction is more suited to centrifugal pump applications than valve service although the inherent flexibility of the form is popular with some users.
- *Cross-plait* All the yarns are interlocking and pass diagonally through the packing to provide a firm construction of consistent density and shape. Used extensively for synthetic yarn packings for valves and pumps.
- *Composite asbestos plastic* This packing category is of fairly recent origin and includes those types based upon a braided and reinforced asbestos jacket enclosing a 'plastic' core. Although inaccurate in the scientific sense, the term 'plastic' conveniently describes those many mixtures of

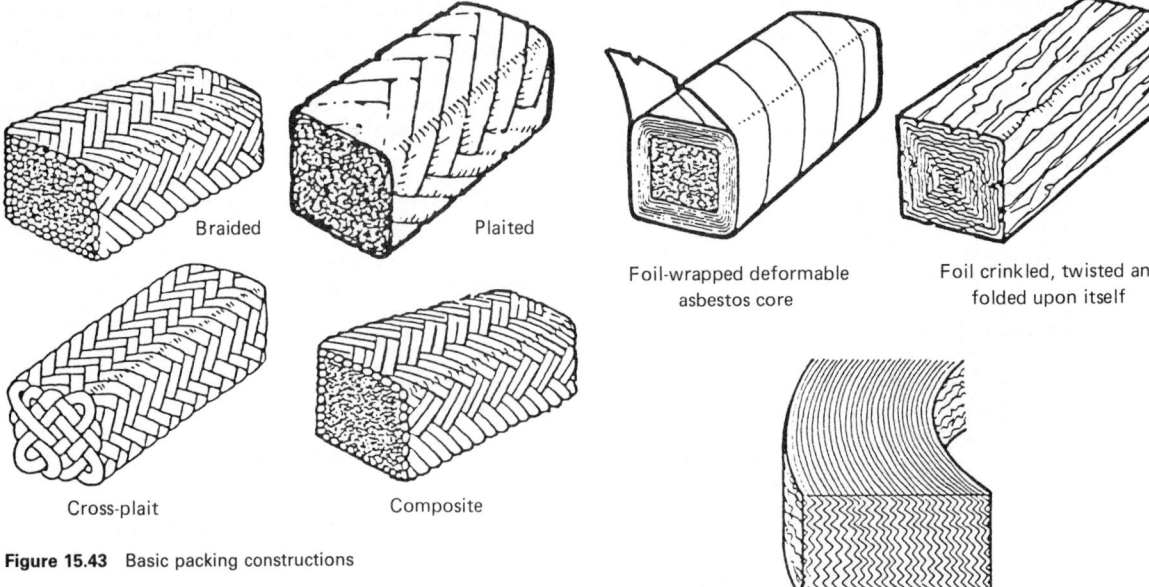

Figure 15.43 Basic packing constructions

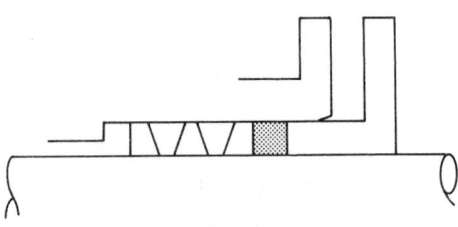

Figure 15.44 Typical metal foil-based packing construction

asbestos fibre and lubricant, both mineral and solid, from which readily deformable packing materials may be made. This packing is widely accepted for difficult valve-sealing duties.

All the fibre-based constructions described here are frequently reinforced with metal. This applies particularly to asbestos based products where the use of metal wire in the yarn can extend the service capability of the packing to 800°C and beyond. But for this feature, even the best quality non-metallic asbestos yarn packing would be restricted to temperatures of about 315°C maximum.

All the lubricants described above are used in conjunction with fibre packings of different sorts and are applied by dipping, coating, soaking, vacuum impregnation, dusting, etc. The prime object is maximum lubricant retention. Frequently, several treatments and repeat processes are employed to achieve this end.

In the field of metallic packings there are many constructions available the three most popular being foil-wrapped resilient asbestos core, foil crinkled and folded upon itself, and corrugated foil, concertina wound (see Figure 15.44). These useful types are suitable for both reciprocating and rotary shafts and are widely used on pumps, valves, turbines, compressors and refrigeration plant. All are normally lubricated with mineral oil and graphite. Lead and aluminium alloys are the most popular foil materials.

In the field of compression packings, elastomers are not widely used although some braided packings do employ yarns that are treated with a rubber proofing to render them more suited to difficult wet applications such as condensate duty. Rings of square or rectangular section compression packing, manufactured from folded, rolled or laminated elastomer proofed cloth, are still popular for relatively slow-moving, lower-pressure reciprocating pumps handling water or LP steam. One particular design, with a moulded, double-bevelled section, made from semi-metallic rubberized yarn, is particularly effective on rotary applications dealing with viscous media which solidify when the pump is idle and cause damage to conventional plaited packings on restarting from cold. This moulded packing is also suitable for duties involving solids and abrasives (see Figure 15.45).

Figure 15.45 Double-bevelled, elastomer-proofed fabric packing for abrasive duties

15.2.1.5 Type of wear

Although a typical set will be five rings, experience has shown that there is not a linear pressure drop through/across the five rings. The majority of the effective sealing is done by the rings adjacent to the gland spigot, i.e. on the atmosphere side of the set – as indicated in Figure 15.46.

The two rings on the pressure side will often be virtually uncompressed and still contain ample lubricant. In contrast, the rings at the gland/atmosphere side will invariably have lost all lubricant, be very hard and probably have suffered around 50% compression – it is these rings which have been doing most of the sealing. In desperate situations replacement of these two – or possibly three – rings can often restore performance and leakage to an acceptable level.

The quantity of packing to use and its size for a given application relies largely on the experience of the user/manufacturer in the type of duty being performed, or in liaison with a packing supplier at the design stage. The latter course of action is always to be favoured if any doubt exists, since an exact knowledge of the capabilities and limitations of the material employed can be found only with those specialists responsible for compounding and production.

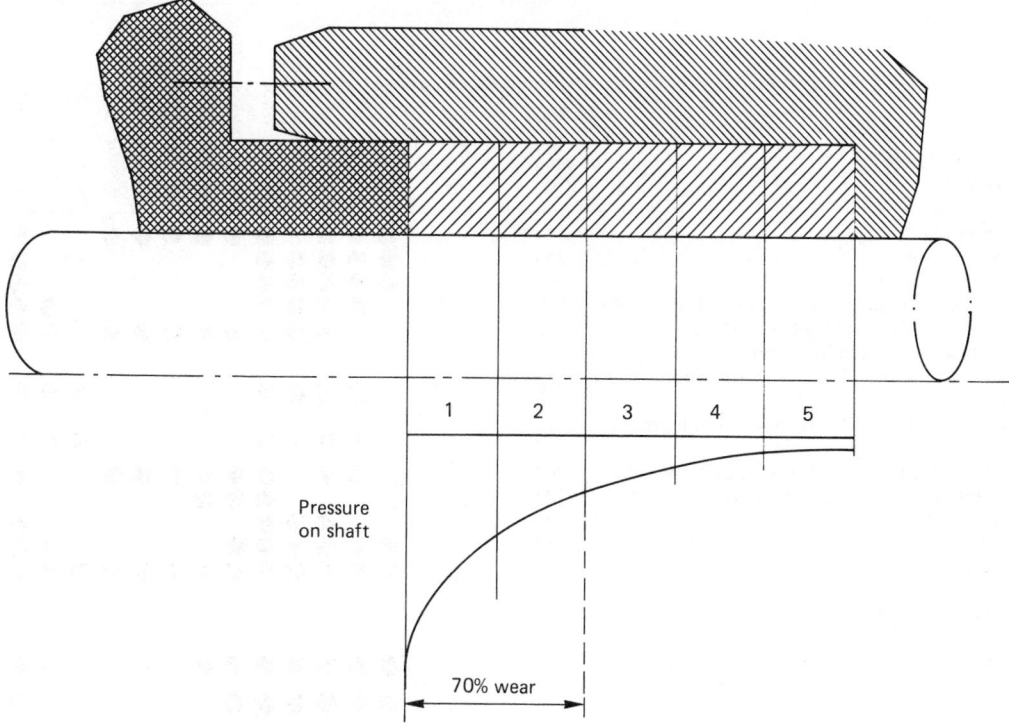

1   2   3   4   5

Pressure
on shaft

70% wear

Figure 15.46

Five rings of square section packing are often accepted as a sufficient number for the average uncomplicated duty but there are many pump applications where the presence of a lantern ring or similar consideration may dictate a greater quantity.

The appropriate packing section to use in relation to diameter is open to a degree of individual preference but broad recommendations are shown in Table 15.5. To give an idea of the capabilities of the various materials and constructions of soft packings which are readily available, reference may be made to Table 15.6 (suitability in different media/speed and temperature limits), Table 15.7 (comparative speed performance), Table 15.8 (comparative temperature performance) and Table 15.9 (comparative cost indication). (*Note*: The statement of speed and temperature limits for a given material should not be construed as meaning that a packing will be suitable for duties where such maxima are jointly encountered.)

15.2.1.6 Reciprocating pumps' duties

Much of the above information also relates to reciprocating pumps. While many years ago compression packings were used on such pumps, many engineers today would automatically think of using a more modern multi-lip type seal. However, for the more difficult applications compression packings are now being used again.

The most important difference from rotary applications is that for these reciprocating duties the packing is generally subjected to far higher operating pressures, and so requires a proportionately greater degree of gland spigot loading/compression in order to develop sufficient sealing force.

This extra compressive force combined with the frictional drag caused by the reciprocating movement means that extrusion of the packing into the gland bush clearance is often the main cause of seal failure. While the new synthetic yarns are extremely strong they are generally very small and so tend to extrude more easily. Once extrusion has occurred this will invariably stop any leakage completely and excessive temperatures are developed very quickly.

Often the operating clearance between the gland spigot and the ram is too wide to prevent extrusion and so some form of anti-extrusion element must be incorporated into the sealing arrangement. One very simple and very effective method is to use a hard fabric ring as is normally employed as the support ring of a typical set of multi-lip seals – as shown in Figure 15.47(a). Because the lip makes intimate contact with the ram and the wall of the stuffing box this contains the packing perfectly and prevents extrusion.

Many pumps are still fitted with multi-lip seals of the chevron type. Generally, these will operate well providing the conditions are within their capability. On many of the larger high-speed pumps, however, the frictional heat generated by the pressure and rubbing speed causes the rubber proofing of the seals to carbonize and soon the flexibility is completely lost and the set becomes virtually a solid mass which will not then respond to further gland adjustment. It is for these more arduous pump duties that a change back to compression packing is now occurring and is proving very successful.

On modern pumps, particularly the smaller sizes, there is a trend towards non-adjustable glands. Compression packings cannot be considered for such applications unless some form of spring loading is incorporated. In this respect the arrangement as shown in Figure 15.47(b) has proved very successful.

Table 15.6

Solvents columns span HYDROCARBON, OXYGENATED, HALOGENATED.

RECIPROCATING, ROTARY PUMPS AND VALVES	Max. rec. temp (°C)	Rotary Speed (m/s)	STEAM	CONDENSATE	WATER	AIR	MINERAL OILS	LIGHT HYDROCARBONS	HYDROCARBON	OXYGENATED	HALOGENATED	REFRIGERANTS	ACIDS	ALKALIS	NON-CORROSIVE SOLUTIONS
Lubricated aluminium foil	540	7.5	●	●	●		●	●	●	●	●	●			●
Lubricated braided asbestos	350	○	●	●	●	●	●								●
Lubricated plaited asbestos	315	20	●	●	●	●	●								●
Plaited, lubricated asbestos impregnated with PTFE dispersion	290	10	●	●	●	●	●							○	●
Plaited, lubricated asbestos impregnated with PTFE dispersion but with no additional lubricant	290	8	●	●	●	●	●	●	●	●	●	●		○	●
PTFE impregnated asbestos and glass fibre yarns with suitable lubricant	290	7.5	●	●	●		●						●	●	●
PTFE yarn impregnated with PTFE dispersion and inert lubricant	250	8	●	●	●		●						●	●	●
Soft lead-based foil wrapped round lubricated asbestos core	260	12	●	●				●	●	●	●				●
Hydrocarbon-resistant lubricated plated asbestos	200	7						●	●	●					
Lubricated plaited cotton	90	7			●	●	●								●
Cross-plait aramid fibre yarns	250	15	●	●	●	●	●	●	○	○	○	○	○	●	●
Pure graphite foil with no volatile additives	○	○	●	●	●	●	●	●	●	●	●	●	●	●	●
RECIPROCATING PUMPS AND VALVES															
Monel wire reinforced asbestos cover with plastic core *	480		●	●	●	●	●	●					○	●	●
Synthetic rubber bonded braided asbestos with brass wire reinforcement	310		●	●	●	●	●	●	○						●
Lubricated plaited flax	90				●	●	●								●
VALVES ONLY															
Constructed from a jacket of asbestos reinforced with inconel wire braided over a resilient asbestos core *	650		●	●	●	●	●	●	●	●	●	●	●	○	●
Lubricated braided asbestos with monel wire reinforcement *	600		●	●	●	●	●	●	●	●	●	●		●	●
Self-lubricating fibrous asbestos with flake graphite or mica	540		●	●	●	●	●	●	●	●					●
Lubricated braided asbestos and brass wire reinforced *	510		●	●	●	●	●	●		●		●			●
PTFE yarn impregnated with PTFE dispersion and inert lubricant but with no additional lubricant	250		●	●	●		●	●	●	●	●	●	●	●	●
Unsintered PTFE cord gland seal for rapid valve packing	250		●	●	●	●	●	●	●	●	●	●	●	●	●

SYMBOLS
● Recommended
○ Consult
* Corrosion inhibitor included

Table 15.7 Maximum rotary speeds for pump packings

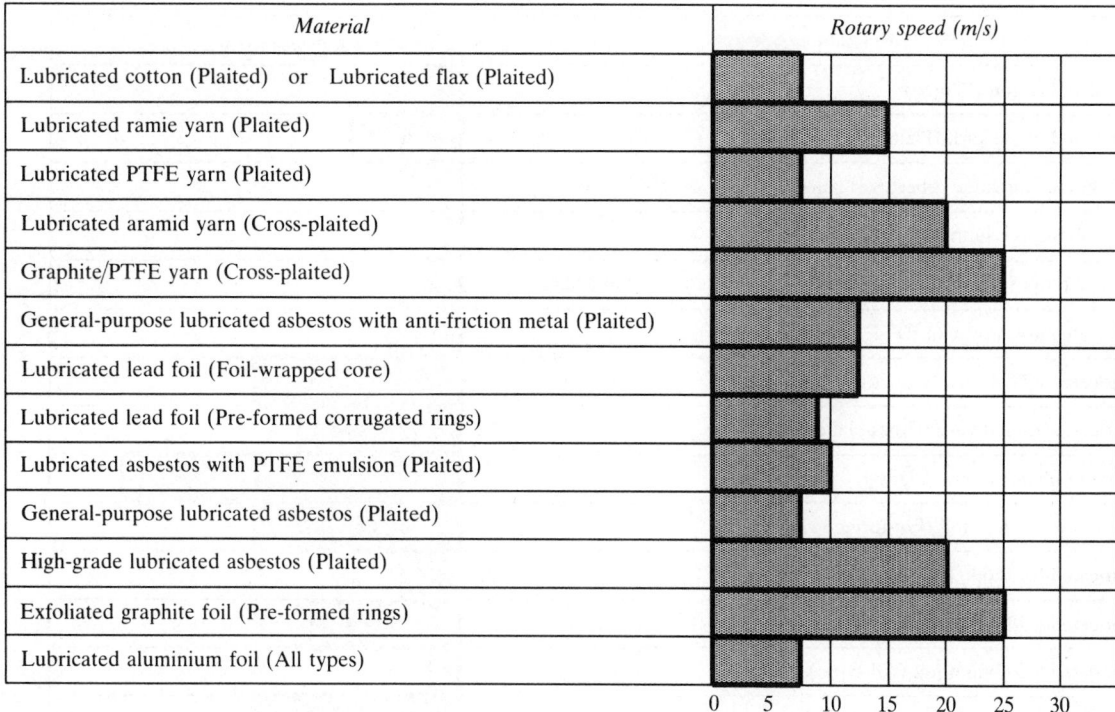

Table 15.8 Maximum service temperatures of pump packings

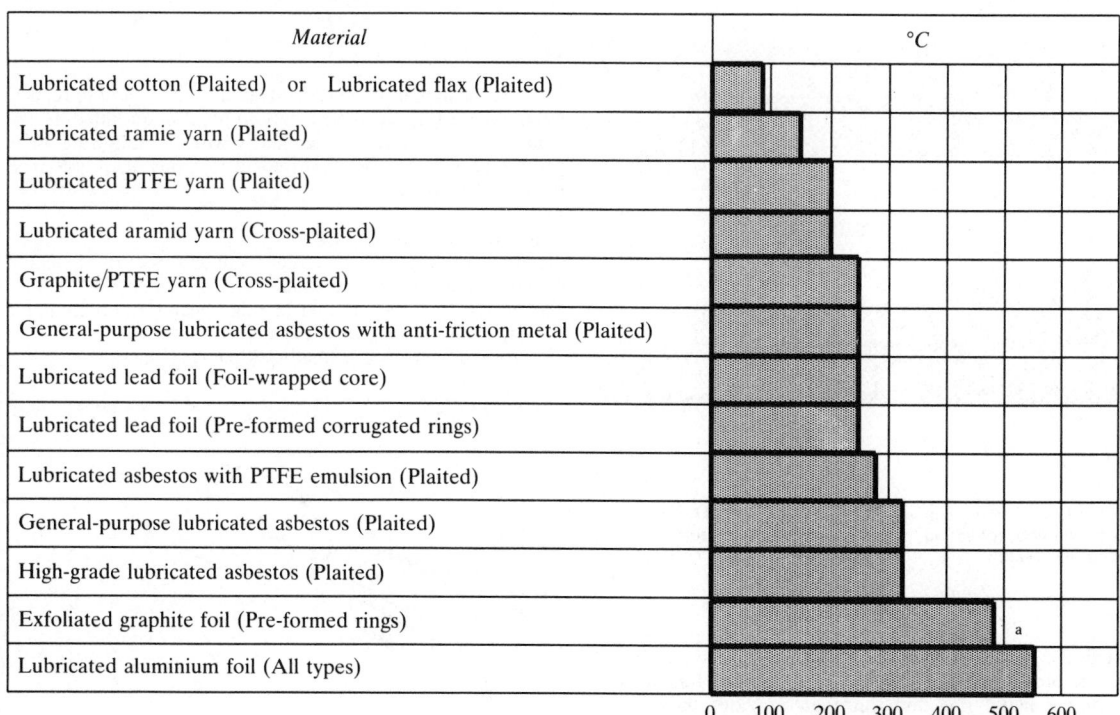

[a] In air, for oxygen-free duties, max. temperature 2500°C

Table 15.9 Typical relative costs of pump packings

Material	Relative cost
Lubricated cotton (Plaited)	
Lubricated ramie yarn (Plaited)	
High-grade lubricated asbestos (Plaited)	
General-purpose lubricated asbestos (Plaited)	
General-purpose lubricated asbestos with anti-friction metal (Plaited)	
Lubricated asbestos with PTFE emulsion (Plaited)	
Lubricated PTFE yarn (Plaited)	
Lubricated aramid yarn (Cross-plaited)	
Graphite-filament yarn (Plaited)	
Exfoliated graphite foil (Pre-formed rings)	
Lubricated lead foil (Foil-wrapped core)	
Lubricated lead foil (Pre-formed corrugated rings)	
Lubricated aluminium foil (All types)	
Hybrid – graphite/PTFE cross-plait	

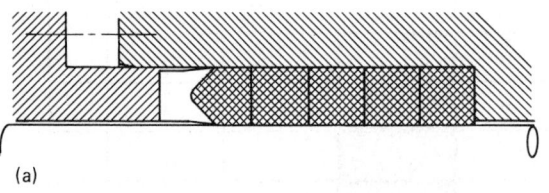

(a)

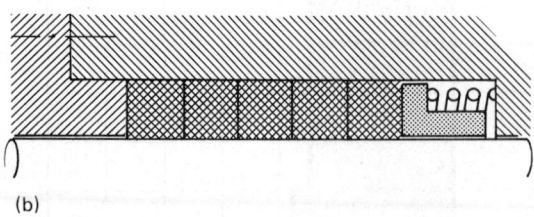

(b)

Figure 15.47   (a) Packing configuration with lip-profiled fabric anti-extrusion ring; (b) spring-loaded packing configuration for non-adjustable glands

Although the springs provide the basic force on the packing, during the pressure stroke the fluid acts over the full radial width of the metal ring and this then provides the extra 'hydraulic' force with which the conventional adjustable gland would have to be developed by appropriate gland loading.

15.2.1.7 Fitting

It is often assumed that unskilled labour can be used to repack pump glands but this is true only so long as 'unskilled' is not equated with 'unaware'. Familiarity with the following ideal procedure will be more than repaid in terms of trouble-free packing performance:

Where length form is used:
1. Spirally wrap the material around a rod of diameter equivalent to the pump shaft.
2. Cut the required number of rings cleanly to obtain good butt-joins (see *Figure 15.48*).
3. Proceed as for pre-formed split packing rings.

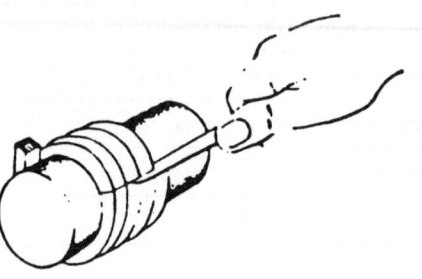

Figure 15.48

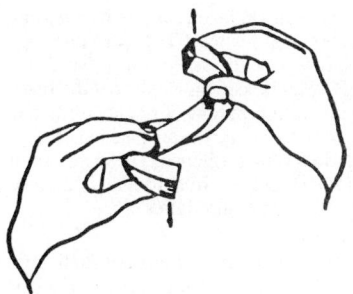

Figure 15.49

Where pre-formed split packing rings are used:

1. Carefully remove old packing (including, where appropriate, the packing on the far side of a lantern ring).
2. Thoroughly clean all surfaces that will contact the packing and, where permitted, smear with oil. Gland and neck bushes, shaft surface and bearings should also be checked for signs of wear and rectified as necessary.
3. Place first ring over the shaft by opening to an 'S' configuration to ensure that bending effects are spread over the whole ring (see *Figure 15.49*).
4. Insert first ring into stuffing box and lightly bed in with a split (wooden) distance piece and gland spigot. With plaited packing the 'V' formation on the outside diameter of the ring should be pointing in the direction of shaft rotation (see *Figure 15.50*).
5. Repeat (3) and (4) with remainder of rings ensuring that each ring is firmly seated and that the butt joins are staggered by at least 90°. (*Note:* The rings must be fitted individually and under no circumstances should complete sets be fitted as a unit.)
6. When the requisite number of rings have been fitted, tighten gland nuts until the shaft or spindle torque increases. Then slack off gland and pull up to finger tightness only. (If pump is to be stored before use leave gland slack so that packing resilience is not impaired.)
7. *Running-in pumps*

   Prime casing and run pump up to operating speed for 10–15 minutes. If pump is not fitted with gland cooling, a cold-water spray over the gland housing will avoid excessive heat build-up during this stage. If no leakage occurs, stop pump, vent casing pressure and slacken gland further. Repeat until leakage starts.

   The controlled leakage, essential for lubrication purposes, can then be obtained by running the pump and evenly tightening the gland nuts in increments of two flats until approximately one drop every few seconds is ob-

tained. Approximately 15 minutes should be left between successive adjustments. *Do not overtighten.*

8. Where loose-form material is used for valves proceed as (1) and (2) and tamp packing into a dense homogeneous mass, progressively filling the housing to the required degree.
9. Because of the danger of corrosion through electrolytic action, packings containing graphite should be avoided on valves or pumps with stainless or chrome steel stems. This risk is most acute when the packing remains in the gland during storage and is particularly aggravated by the presence of moisture.

15.2.1.8 Fault finding

A major advantage of compression packings is that breakdown is rarely sudden or catastrophic but rather a matter of a gradual build-up of leakage until an unacceptable level is reached. Normally, considerable life can be achieved by controlling leakage with further tightening of the gland nuts. (*Note:* the seepage of fluid which acts as a lubricant for the packing on rotary applications should not be confused with leakage and the rate of one drop every few seconds should be maintained.) However, if other than routine maintenance or just plain 'fair wear and tear' are suspected as the cause of leakage and the need for repacking, then the following hints could well prove useful:

1. Confirm that the packing is rated as suitable for the application.
2. If one or more rings are missing from the set, check for excessive neck bush clearance allowing extrusion of rings into the system. If the top ring has extruded between the gland follower, anti-extrusion rings could avoid replacement of metal parts.
3. If the packing's radial thickness appears diminished in one or more places, check for an undersize shaft or badly worn bearings which could cause shaft whip or spindle wobble.
4. If radial section of packing directly beneath the shaft is reduced or premature leakage occurs along the top of the shaft, check for misalignment of shaft centre to stuffing box bore.
5. If the packing is worn on the outer diameter, check for loose rings or rings rotating with the shaft due to insufficient gland load.
6. If the packing rings have bulges on their radial faces, the adjacent ring was probably cut too short, causing packing under pressure to be forced into the gap at the joint.
7. If the packing nearest the gland spigot shows excessive deformation while other rings are in fair condition, the set was probably incorrectly installed and subjected to excessive gland tightening. (*Note:* Overtightening is usually the greatest single cause of premature packing failure.)
8. If the cause of your particular problem is still not apparent, give equal attention to i.d. and o.d. leakage and check for a rough stuffing box bore before seeking specialist advice.

15.2.1.9 Standardization

In the interests of stock control it is clearly an advantage to rationalize the variety of packings used in any plant to that minimum number which will effectively cater for all the conditions likely to be encountered. If cost is no object, then there are single, sophisticated materials and constructions that will go some way towards satisfying most demands, but it is doubtful if cost effectiveness could be justified. Far better to compromise on a small number of reputable products developed for the areas in question, e.g. pumps, valves, etc.

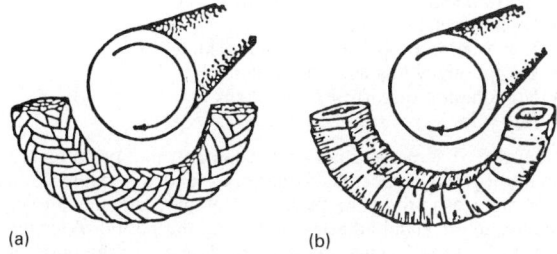

(a)   (b)

Figure 15.50 (a) Plaited and (b) foil-wrapped packing

There are few standards applying to compression packings on a national or international basis, although many individual companies and organizations have domestic standards which have, in many cases, been the subject of collaboration between user and packing manufacturer. BS 4371: 1968 specifies minimum standards for lubricated plaited cotton, lubricated plaited flax, lubricated plaited or braided asbestos, dry white non-metallic plaited or braided asbestos, plaited or braided asbestos, metallic wire reinforced, indurated asbestos, and lubricated fibrous asbestos and gives guidance on limiting operating parameters for these constructions. There are other pump and valve British Standards which specify common stuffing box dimensions such as BS 1414: 1975, BS 1873: 1975 and BS 3808: 1964.

Where packings are required for service with potable water in the water authority distribution system (which covers reservoir to tap), only those materials which have gained a National Water Council Approval may be used. Such products have been tested to establish that they produce no colour, taste or turbidity, are non-toxic and will not support microbial growth.

Statutory Instruments 1978, No. 1927, The Materials and Articles in Contact with Food Regulations 1978 required that compression packing materials, for example,

'......do not transfer their constituents to foods with which they are, or likely to be, in contact, in quantities which could –

(1) endanger human health or
(2) bring about a deterioration in the organoleptic [sensory] quality of such food or an unacceptable change in its nature, substance or quality.'

Such regulations inevitably restrict the range of available materials and lubricants. Consultation with the supplier is recommended to establish preferred grades.

15.2.1.10 Compression packings: material developments

Introduction The route to improve performance in most areas of sealing is generally in the use of new/improved materials – this is without doubt the case with compression packings. In recent years what can loosely be termed 'synthetic materials' have become available which have changed the overall attitude of the engineer regarding the use of the packed gland. These have three distinct advantages over the old-established materials:

1. The performance capabilities have been extended in terms of pressures, speed and temperature limitations.
2. Combined with these the resultant life has also been significantly increased.
3. The age-old problem of repeated gland adjustments has been reduced to the extent that on many applications it has been virtually eliminated – the 'non-adjustable' packed gland is now a reality.

Four materials have achieved these improvements and although not exactly new, PTFE is included, since by comparison with the natural fibres of asbestos and cotton, etc. it is relatively new and does occupy an important place in today's range of compression packings.

PTFE Polytetrafluoroethylene yarns provide soft packings for services where corrosive media are being handled or freedom from contamination is an essential requirement. A semi-rigid fluorocarbon plastic, PTFE is unique in possessing almost complete chemical resistance within its temperature range which, in this field of application, spans the cryogenic

area to 250°C. Another major advantage refers to its very low coefficient of friction. Lubricated plaited PTFE yarn packings are suitable for rotary surface speeds up to 8 m s^{-1} and are also finding increasing acceptance on high-speed, high-pressure, multi-ram reciprocating pumps. In solid form this material is not acceptable as a compression packing due to poor creep properties and lack of resilience. However, solid junk rings or spacers in PTFE are often used to enhance packing performance on arduous pump duties.

Aramid fibre Packings made from aramid fibre, usually of a distinctive yellow colour, are becoming increasingly popular for a variety of pump and valve services hitherto satisfied by PTFE-lubricated asbestos packings. It has a high tensile strength, excellent resilience and thermal stability up to 250°C and is resistant to a wide range of chemicals. Experience has shown that while it is extremely resistant to abrasives, it is also extremely tough such that it can sometimes severely damage/wear the surface of a rotating pump shaft. However, the latest quality yarns combined with an improved lubricant have reduced this problem and speeds of 20 m s^{-1} have now been successfully achieved. It has also proved to be successful in high-pressure reciprocating pump applications, being used typically with arrangements as shown in Figures 15.47(a) and (b).

Graphite Without doubt, this material more than any other is responsible for extending the performance of the packed gland. It is available in various forms, each having advantages and, hence, specific areas of use:

Graphite yarn/filaments
Yarn packings in this material are a development for rotary pump applications and provide possibilities for extending the range of the packed gland beyond boundaries hitherto established. A high coefficient of thermal conductivity, low friction and resistance to chemical attack are the useful characteristics of this material. Temperatures up to 400°C may be considered. If a good performance is to be obtained, then close attention must be paid to mechanical conditions such as shaft run-out and finish. Care in fitting and running-in is also mandatory.
Expanded graphite foil is the most recent and significant application of graphite, particularly in the context of valve applications. Expanded graphite materials combine the well-established thermal and friction characteristics, long associated with the correctly developed use of carbon-based products, with a unique flexibility and resilience. The attributes of this exfoliated form of graphite bear recording.

- Excellent resistance to compression set resulting in little loss of radial gland force or flange seating stress over long periods (see Table 15.10)
- No loss of volatiles even at high temperature, thus minimizing frequency of gland adjustment
- Resistance to a wide range of chemicals
- High-temperature capability, particularly in non-oxidizing environments
- High thermal conductivity
- Low friction properties – self-lubricating
- Exceptionally low chloride content
- No adhesion or corrosion problems
- Fire-safe

Figure 15.51 shows an interesting comparison of performance on a test gland between half-rings of expanded graphite and a lubricated asbestos yarn packing. Not only did the former require fewer gland adjustments during the period of testing but the average leakage rate was much less – to the point of running virtually dry for protracted periods.

Table 15.10

Compressibility/recovery ASTM F36-66 Procedure H (Major load 7 N mm^{-2}) (1000 lbf in^{-2})	Expanded graphite 1 mm thick 1.0 g cm^{-3}	Expanded graphite 2 mm thick 1.0 g cm^{-3}
Compressibility	39%	33%
Recovery	20%	25%
Stress relaxation BS 1832: 1972 Temperature 300°C for 16 h Initial stress of 40 N mm^{-2} Residual stress NB: 10 N = 1 kgf 1 N mm^{-2} = 10 kgf cm^{-2} approx.	40 N mm^{-2}	40 N mm^{-2}

Rather than use the tape form of expanded graphite which is primarily a useful maintenance expedient, moulded rings to a selected and controlled density should be the first choice. Although more costly than conventional packing materials, economies of radial width and number of rings used are feasible, quite apart from the performance advantage likely to be derived from the use of expanded graphite.

Hybrid graphite/PTFE yarn This latest material is a yarn combining PTFE and graphite; the PTFE is not added to the yarn as was the case with many other materials in the past. This proprietary blended yarn based on an exclusive process has achieved much success on high-speed rotating pump duties at speeds in excess of 25 ms^{-1}. Many manufacturers are now offering products based on this material, which is particularly insensitive to variations in fitting techniques, etc. Such packings bed-in easily and have characteristically lower leakage rates than many conventional variants.

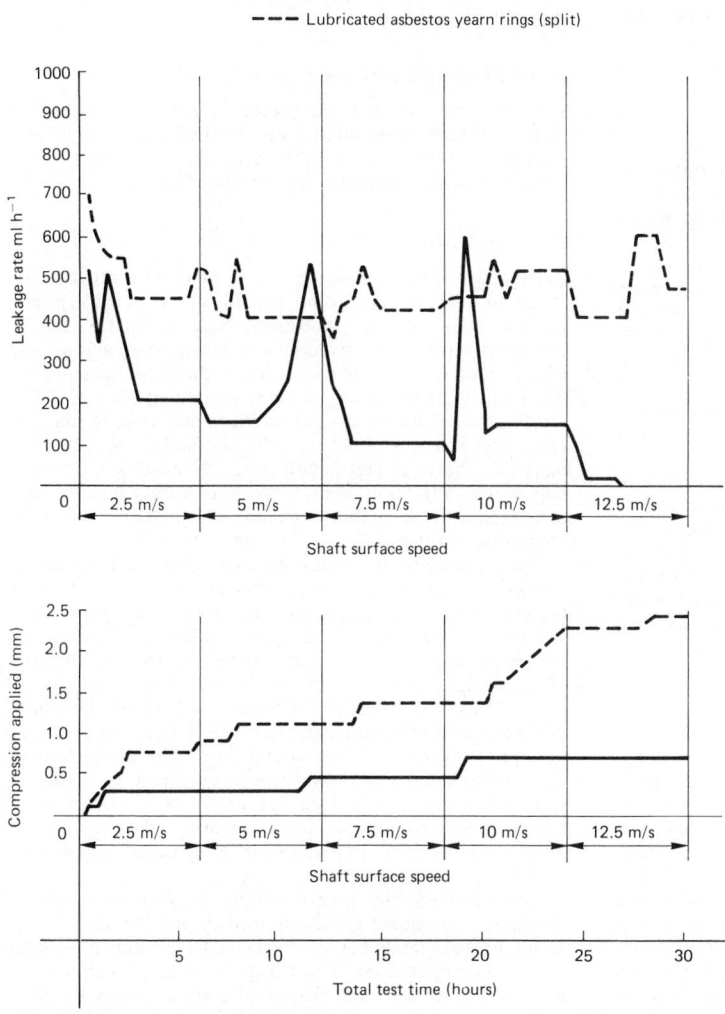

Test conditions: Distilled water   – 90°C
temperature of medium
Pressure of medium – 3.45 bar

—— Expanded graphite rings (in halves)

- - - Lubricated asbestos yearn rings (split)

Figure 15.51

Since these yarns are extremely small in size, extrusion can occasionally be a problem. Hence when using these new products, extra attention should be given to the extrusion clearance of the gland and neck bushes.

Due to its PTFE content, service temperatures are limited to 260°C. However, this is well below the operational gland temperature of most rotary shaft applications.

One cannot leave materials without special reference to the vital role played by *asbestos* – a much-denigrated mineral fibre without which economic and practical solutions to many sealing problems would not be feasible. Although understandable, in an age of correct awareness of health and safety matters, the over-reaction against asbestos has revealed many inconsistencies. Motor manufacturers may prohibit its use as a plant-maintenance material but continue to use asbestos in a brake-lining and clutch-facing role where residual dust is evident. Some users may seek to limit its application in a safe form as a valve packing or gasket but perpetuate its specification for fireproof positions and roofing where the mineral is cut in a dry form.

Hazards exist but adherence to basic advice on handling asbestos will result in a sensible balance between prohibition and practicality. In this context, users of asbestos-based pump and valve packings, gaskets or allied components might heed, to advantage, the statement issued by the Asbestos Information Committee to the British Valve Manufacturers' Association.[18–22]

15.2.1.11 Selection

Selection is the most vexing question as, for many duties, so many reasonable alternatives exist. Much will depend on personal experience, frequency of maintenance, original cost level, contamination considerations, size, etc.

In the case of a manufacturer producing large quantities of valves or pumps to standard dimensions there is much to be said for purchasing sets or rings rather than length-form packing. With the techniques available, packing can be supplied ready for immediate fitting with substantial reduction in that overall cost represented by receiving length-form that must be cut to size by skilled personnel. This economy is not confined to the large manufacturers but it is they who will enjoy the greater advantage.

On the other hand, in many instances, the problem of stocking rings or sets tailormade for an assortment of valves varying in origin, type and dimensions can prove intolerable. For these cases, there is a clear need for the versatility of packing in length-form. Comparable with this solution is the expedient provided by those packings of plastic nature that are available in loose form but this advantage must be weighed against the labour cost in the careful fitting required.

15.2.1.12 Cross-plait construction/mixed yarns

Virtually all these new synthetic yarn/filament materials are manufactured into length-form packings using the 'cross-plait' construction as shown in Figure 15.43, which results in a far better and more uniform construction and it does have the distinct advantage of enabling two different yarns to be used in the manufacture of a packing. This technique has resulted in the development of a new generation of cross-plait mixed-yarn products, which can maximize the benefits of the individual constituents and minimize their shortfalls. Typical examples are packings deploying tough aramid (yellow) yarn at the corners and the high-speed thermally superior hybrid yarn just described to most of their rubbing face – these offer excellent extrusion resistance and a higher-speed capability than a plain

aramid product. Also for chemical compatibility PTFE filaments can, of course, be combined with others as necessary.

15.2.1.13 Gland adjustment

The one disadvantage of the compression packing arrangement is that after installation subsequent periodic adjustment of the gland is normally required. The reason for this is that there has been a small loss of volume, either material or lubricant, which means that the density has been changed and so a reduction in the radial sealing force occurs, resulting in an increasing leakage. Gland adjustment, i.e. compression, then restores the density and the sealing force.

The new materials, particularly the graphite types, are self-lubricating and so very little lubricant is added to the packing. In consequence, because there is relatively no material or lubricant to be lost or forced out of the packing, the amount of subsequent gland adjustments are very much reduced, giving a longer life and reduced maintenance requirements. This aspect is demonstrated perfectly with the graphite foil used for valve sealing. The material is virtually pure graphite and is perfectly stable in size and density within the temperature range of −200 to +550°C, and once fitted and correctly adjusted it will usually provide years of trouble-free service.

15.2.1.14 Applications

Basically, the fundamental choice is generally that for a dynamic/rotary application the length-form cross-plaited yarn/filament product should be used and for valve spindles rings of expanded graphite are the first choice.

15.2.1.15 Valves

Expanded graphite is now the most common form of valve stem seal. It is the standard for virtually all valves in the power generation and nuclear industries, both in the UK and in Europe. Density of the material is very important and usually a density of 1.6 g/cm³ is used. This is the value specified by CEGB in their specification for this material (No. 155701). For this reason, it is always preferable to use moulded rings for critical applications rather than the alternative 'tape' form of material. The rings are manufactured by winding tape coaxially around a rod until the required o.d. is obtained and then compressing these within a mould. The material is usually compressed in the order of approximately 60%.

Once installed and the optimum gland adjustment is applied, no further adjustments should be necessary. The only reason subsequent adjustments are necessary is because the sealing force has reduced – this can only happen if the density has changed, which in turn can only be the result of some loss of material, i.e. extrusion of the top or bottom rings.

It is therefore extremely important to have the minimum possible extrusion clearances. However, if extrusion proves to be a problem, then the use of end rings of graphite filament (i.e. cross-plaited type) have proved extremely successful in solving this problem. Indeed, the use of these 'combination sets' is now becoming far more common and in the power-generation industry in France it is virtually a standard approach.

Compared with the older materials, these new sets require negligible subsequent gland adjustments and the use of 'live' spring-loaded glands has now achieved the ultimate in long trouble-free operations. The reason 'live' glands can now be used is due to the small degree of compressibility of these graphite materials. Once fitted, these sets will probably only compress by about 8–10% at most, whereas the older sets in lubricated asbestos materials could well compress ultimately by perhaps 35% or 40%, rendering spring-loaded glands impractical.

15.2.1.16 Problems with expanded graphite

Occasionally a problem can occur which initially causes a significant increase in friction, then is usually followed by leakage. It is often referred to as graphite 'pick-up' and, for some reason, graphite from the inner surface of the rings adheres to the valve stem and is dragged through the set. Apart from the very noticeable increase in friction, depending upon the length of movement, the graphite deposit is often seen on the emerging valve stem.

At present there appears no clear explanation of why or when this will happen, although the number of instances are relatively low. However, although the mechanism of how it occurs is not fully understood, it has been found that the use of end rings of graphite filament – as used primarily to prevent extrusion – also seems to prevent this problem. A typical set is shown in Figure 15.52.

15.2.1.17 Corrosion

Corrosion is an aspect often associated with the use of graphite when used on stainless steel valve stems. However, this is normally not caused by the graphite itself but by other impurities used in the old-type seals – in particular, moisture

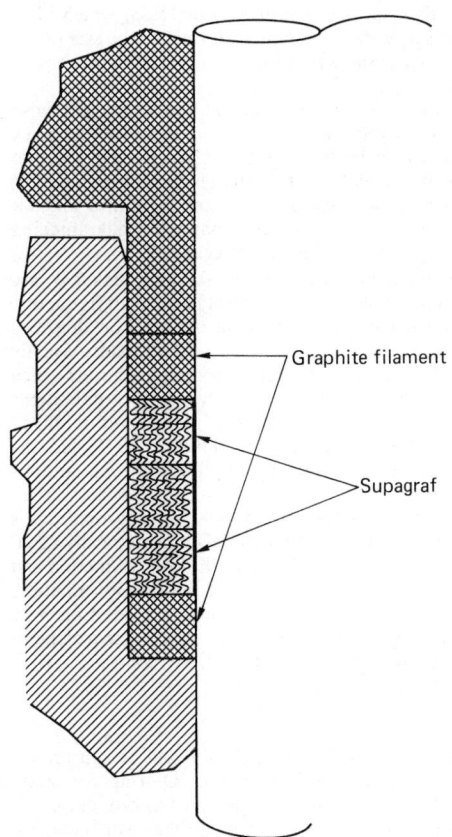

Figure 15.52

that enters the seal when a valve is initially hydraulically tested which then remains there until such time that the valve is put into service. By using a different seal for this hydraulic testing, drying out the stuffing box and then fitting the graphite seals, no corrosion normally occurs. For those users still concerned, these materials can be supplied containing a corrosion inhibitor.

15.2.1.18 Length-form packing

If length-form packing is preferred (possibly for maintenance reasons) then within its operational limitations plaited-type packings can be used in valves, although their efficiency will rarely be equal to that of expanded graphite. However, for easier operating conditions, particularly those valves handling fluids, they can prove very successful. Non-asbestos length-form packings based on reinforced braided glass yarns combined with a graphite based core are now available for temperatures up to at least 500°C.

15.2.1.19 Rotary shafts (pumps, etc.)

Expanded graphite should not really be considered for dynamic rotary applications, since it does not have the flexibility to cater for the inevitable minute radial movements of the shaft that are usually present. Having said this, it has operated on equipment providing only the minimum of leakage, but the units in question did have double-ended bearing arrangements to the shaft, etc. – providing almost perfect mechanical conditions.

The cross-plaited types are therefore the first choice for any rotary application and the new materials discussed will normally give excellent results in terms of reduced leakage, extended life and infrequent adjustments. On boiler feed pumps the 'hybrid' PTFE/graphite yarn type is now giving exceptional results and after initial bedding-in periods of over 18 months without further adjustments are quite common – and then the packing is replaced purely as a scheduled maintenance requirement, not because it has failed.

15.2.1.20 Rubber core

A recent trend which is proving extremely successful is to incorporate an elastomeric core into the cross-plaited type packings. This then gives a degree of recovery to the rings and assists in reducing the subsequent gland adjustments. It is also extremely successful when there is radial movement of the shaft such as on mixing vessels and agitators, etc. Depending upon the operating conditions, the elastomer used is frequently fluorocarbon (Viton) or ethylene propylene. This approach is also being used for length-form valve packing – typically in the water industry and, obviously, for potable-water applications only approved materials must be used.

15.2.1.21 A new natural fibre

In contrast to the synthetic materials, packings manufactured from ramie yarns are now available and proving very successful. This is a natural material and is extremely tough and water resistant. Although it has a temperature limitation of only 120°C, it is a far more economical alternative for basically water applications, including rotary and reciprocating pumps and valves.

15.2.1.22 The future

The recently developed materials such as aramid, the hybrid graphite/PTFE and the natural ramie yarns will form the bulk of compression packings for both reciprocating and rotary applications. Not only do these materials overcome the health problems associated with using asbestos they have extended the performance capability of the packed gland. While they are initially more expensive than the asbestos products, the improvement in performance, reduced maintenance and extended life invariably make them cost effective.

Although these new materials can replace asbestos products for most dynamic applications, much development is still taking place with glass/PTFE/graphite or ceramic constructions mainly to satisfy the higher temperatures that are encountered in valve applications. However, many problems remain to be solved in producing a general-purpose product that can compete with asbestos in terms of lubricant retention, absorbency or durability at comparable costs.

There can be no doubt that graphite-based materials will see even greater acceptance despite the higher initial costs, and with the current rate of development, asbestos as a compression packing material will soon be a thing of the past. Despite its simple concept, there really is no alternative for the valve stem seal. Indeed, its performance now is such that the question must be asked, 'Is there a need for an alternative?'

For rotary shafts the improved performance in terms of less leakage, less shaft wear and longer life with fewer adjustments, must make the choice between a compression packing and the mechanical seal an even more difficult decision than it was in the past.

Finally, not including the spring-loaded lip seal, since this is really only suitable for very low pressures, the only choice for a rotating shaft seal has been either the mechanical seal or a packed gland. If a steady leakage is acceptable – as is normally experienced from a packed gland – there is now a third option of a rotalion, an automatic throttle device giving controlled leakage, irrespective of pressure and rubbing speed.

15.2.1.23 Guidelines for removal of asbestos from plant

1. Assess the amount of work involved.
2. Decide when the work is to be carried out.
3. Supervision to notify Safety Department Supervision, who will then notify the Health & Safety Executive (Asbestos Licensing Regulations 1983). Twenty-eight days' notice required except in cases of extreme emergency.
4. Permit to Work Certificate to be issued by Supervision with reference to protective clothing, respiratory protective equipment and electrical isolation.
5. The area is to be enclosed with polythene sheets where possible.
6. Asbestos warning notices are to be suitably placed in the area.
7. Disposable overalls must be worn, a new pair each day. If leaving the site is necessary, overalls must be vacuumed before removal and fitted again on-site when returning. On removal at the end of the day, place carefully into the plastic bag provided, having first vacuumed down as stated above. The bag should be sealed and placed in the notifiable waste container.
8. Rubber boots should be worn with the overall trouser outside the boots. The soles and outside of the boots are to be completely vacuumed clean and washed after use.
9. Dust masks *must* be worn throughout the asbestos-removal operation and *must* be the last piece of protective wear removed on completion of work. Masks should then be vacuumed, placed in the container provided and returned to source.
10. Thoroughly wet asbestos with water before removal.
11. Remove wetted asbestos with care and place into red notifiable waste bags and seal.
12. The whole area and screens must be carefully vacuumed with asbestos approved cleaner. Remove screens, bag, seal and place in notifiable waste container.
13. Vacuum overalls on-site before removal then place into plastic bag, seal and place in notifiable waste container with removed bagged asbestos.
14. Safety Department Supervision should inspect the area and, if found to be satisfactory and suitable for normal working, will endorse the Permit to Work Certificate to this effect.
15. Supervision will then complete the Permit to Work Certificate.
16. *The HSE Factory Inspector may call at any time to check that the correct procedure is being carried out.*

15.2.2 Reciprocating hydraulic seals

15.2.2.1 Introduction

Sealing devices, particularly in the field of hydraulics, are usually of an essentially deformable nature, and permit fluid power designers to work within economically sensible conditions of fit, surface finish and fluid cleanliness across the range of temperatures and pressures which they are likely to encounter. In the context of those seals specified for linear dynamic motion, a much axially shorter seal has evolved which allows hydraulic cylinders, for example, to be produced to more compact proportions with the consequent saving in raw material costs and in the overall bulk of fluid power sub-assemblies. This is clearly seen in Figure 15.53, which illustrates how the seal design can directly influence the size of the unit and still give the same performance in terms of power and stroke length.

A proliferation of seal designs and materials exists and those responsible for selection are required to make their choice with discretion to ensure that the most cost-effective solution is found. Selecting a seal which facilitates the achievement of the cheapest production costs for the component into which it fits is seldom the most reasonable basis for guaranteeing minimum leakage and maximum serviceability.

It is an unfortunate fact that most fluid leakages are attributed to a faulty seal whereas investigation will frequently show that the number of occasions of malfunction due to a faulty product are few. The real source of difficulty usually stems from incorrect seal selection, working conditions which differ from those considered to apply, seal containments outside specified tolerances in terms of dimension or finish, faulty fitting, contaminants in the fluid to be sealed or, importantly, personal interpretation of 'leakage' in quantitative terms.

To understand the limitations of fluid seal performance it is necesary to know a little of the materials most typically used, the origins of fundamental seal designs, how they function and their refinement to the present state of the art. Fortunately, the seal is no longer an afterthought but more often receives the attention it properly deserves as a vital interactive element in any complete fluid power system.

15.2.2.2 Materials

Within the scope of this section, attention will be centred on solid elastomeric seals, typified by the 'O' ring for static connections, elastomer-containing seals such as cylinder packings based on proofed textile and those designs which employ plastic components for various functions. Tables 15.11 and 15.12 and Figure 15.54 are taken from the British Rubber

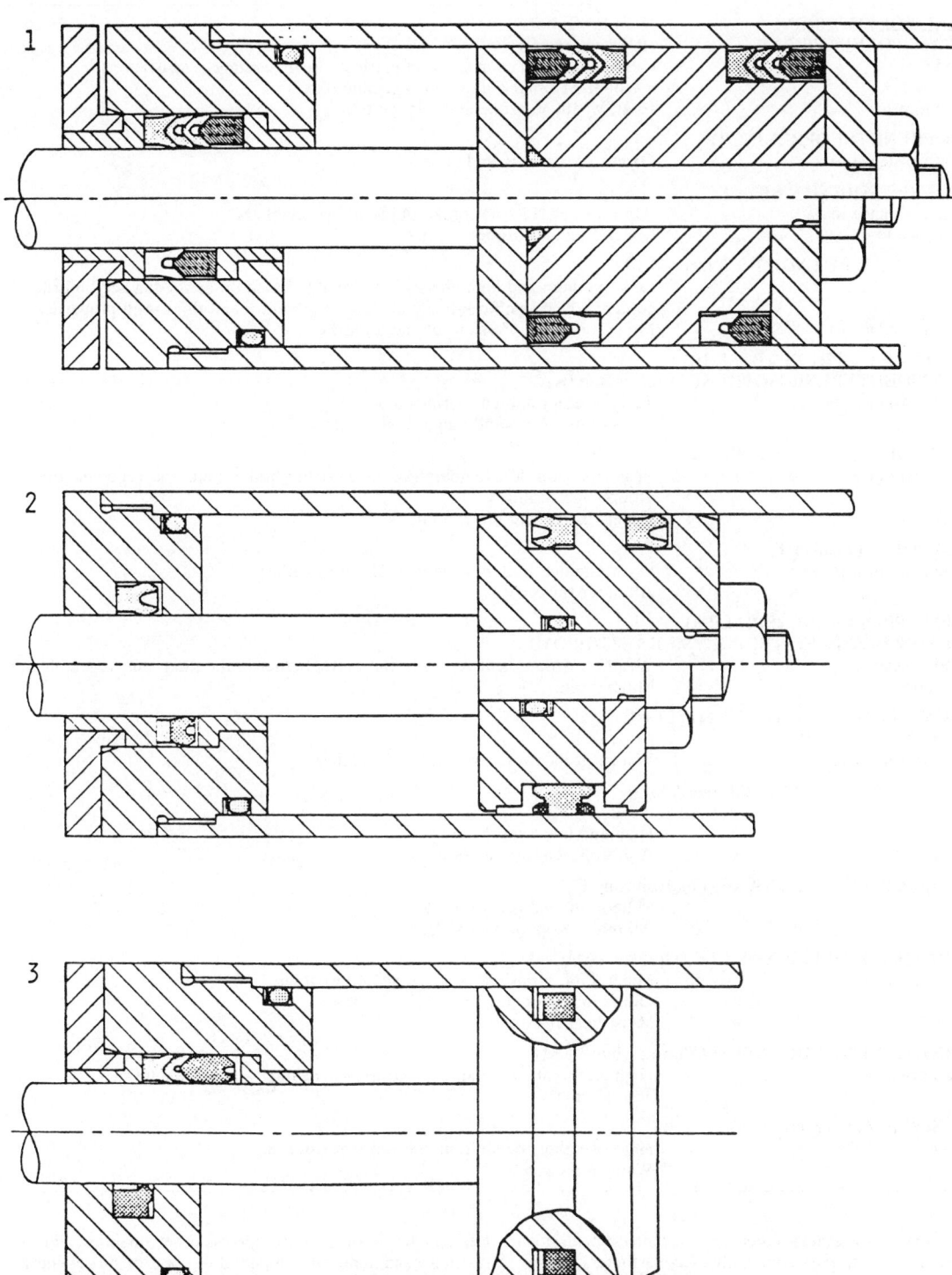

Figure 15.53

Table 15.11 Polymers – types and general properties

NATURAL POLYISOPRENE	(Natural Rubber-NR)
SYNTHETIC POLYISOPRENE	(Polyisoprene-IR)
Very good	General physical properties, in particular tear resistance, resistance to glycol-ether brake fluids and vegetable oils.
Poor resistance to	Heat, weather, ozone and mineral oil.
CHLOROPRENE (NEOPRENE-CR)	
Very good resistance to	Ozone, weathering and abrasion.
STYRENE-BUTADIENE (SBR)	
Very good resistance to	Abrasion, glycol-ether brake fluids and vegetable oil
Poor resistance to	Mineral oil and ozone.
ACRYLONITRILE-BUTADIENE (Nitrile-NBR)	
	Low, medium and high nitriles are available based on increasing acrylonitrile content which significantly affects low temperature and fluid swell properties.
Very good resistance to	Mineral oil, compression set and abrasion.
ISOBUTYLENE-ISOPRENE (Butyl-IIR)	
CHLORO-ISOBUTYLENE-ISOPRENE (Chloro-Butyl)	
Very good resistance to	Tear, weather and gas permeation.
Poor	Tensile properties and mineral oil resistance.
POLYBUTADIENE (Butadiene-BR)	
Very good resistance to	Tear, abrasion, low temperature, glycol-ether brake fluids and vegetable oils. Offers high resilience.
Poor resistance to	Water, ozone and mineral oil.
POLYSULPHIDE (Thiokol-T)	
Very good resistance to	Ozone, mineral oil, petroleum fuels and weather.
Poor	All other properties.
ETHYLENE PROPYLENE (EPR-EPM)	
ETHYLENE PROPYLENE TERPOLYMER (EPT-EPDM)	
Very good resistance to	Weather, ozone, heat, water, steam, glycol-ether brake fluids and vegetable oils.
Poor resistance to	Mineral oil.
CHLOROSULPHONATED POLYETHYLENE	
(Hypalon-CSM)	
Very good resistance to	Water, ozone, abrasion, acid and weather.
METHYL-VINYL SILOXANE (Silicone-VMQ)	
PHENYL-METHYL-VINYL SILOXANE (Silicone-PVMQ)	
Very good resistance to	High and low temperatures.
Poor resistance to	Abrasion, tear and tension.
TRIFLUOROPROPYL SILOXANE (Fluorosilicone-FMQ)	
Very good	Mineral oil and fuel resistance. All other properties as VMQ.
POLYURETHANE DI-ISOCYANATE (Urethane-AU)	
Very good resistance to	Mineral oil, abrasion, tear, ozone and weather. Offers high modulus and tensile properties.
Poor	Moist heat resistance.
FLUORINATED HYDROCARBON (Fluorocarbon-FKM)	
Very good resistance to	High temperature (in air and most oils), weather and petroleum fuel.
Poor	Tear strength.
POLYACRYLATE (Acrylic-ACM)	
Very good	Heat, weather, mineral oil and ozone resistance.
Poor	Water resistance.

Manufacturers' Association's *Guide to Elastomeric Seals* to give broad reference to most commonly used elastomers, their properties, fluid compatibility and temperature resistance. Of major interest to the fluid power industry are the following.

Acrylonitrile-butadiene (*nitrile – NBR*) Probably upwards of 80% of seals supplied to the fluid power industry are based on nitrile compositions. The balance of properties available from

NBR in terms of good compression set qualities, abrasion resistance, mineral oil compatibility, ease of processing and low initial cost favour this choice. Nitrile compositions are not, however, compatible with fire-resistant fluids of the phosphate ester type.

Isobutylene-isoprene (*butyl – IIR*) Satisfactory in service with phosphate ester fluids but rather lifeless from a physical

Table 15.12

	Natural Rubber	Neoprene	Nitrile	Butyl	EPDM	Fluorocarbon	Silicone	Acrylic	Urethane	Fluoro-silicone	SBR	Thiokol	Butadiene
Air or oxygen	N	A	A	R	A	R	R	R	R	R	N	R	A
Dilute acid	R	R	A	R	R	R	N	N	N	A	R	R	A
Dilute alkali	R	R	A	R	R	A	N	N	N	A	R	R	A
Water	R	A	R	R	R	A	A	N	N	R	R	R	R
Lower alcohols	R	R	R	R	R	R	N	N	N	R	R	A	A
Commercial petrols	N	N	A	N	N	R	N	A	A	A	N	R	N
Fuel and diesel oils	N	N	R	N	N	R	N	R	A	A	N	R	N
LUBRICANT OILS:													
(a) Mineral based	N	R	R	N	N	R	A	R	R	R	N	A	N
(b) Synthetic based	N	N	A	N	N	N	N	N	N	A	N	A	N
HYDRAULIC OILS:													
(a) Mineral based	N	N	R	N	N	R	N	R	R	A	N	A	N
(b) Ester based (non-flamm.)	N	N	N	A	A	R	A	N	N	N	N	A	N
(c) Water glycol based	N	N	R	R	R	A	A	N	N	N	N	A	N
(d) Chlorinated	N	N	N	N	N	A	N	N	N	N	N	N	N
(e) Silicone based	N	N	A	N	N	A	N	N	N	N	N	N	N
(f) Glucol-ether brake fluids and vegetable oils	R	A	N	A	R	N	N	N	N	N	R	N	R

R-recommended; A-acceptable; N-not recommended

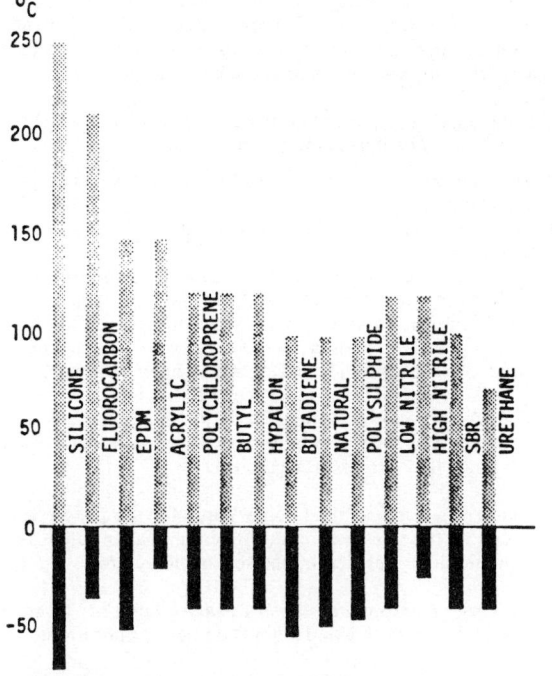

Figure 15.54 Temperature ranges for polymers

point of view. Not suitable for mineral oil service nor should butyl seals be smeared with mineral oil or grease on fitting.

Ethylene propylene (EPR – EPM) or ethylene propylene terpolymer (EPT – EPDM) Same service function and limitations as butyl but generally better physical properties.

Fluorinated hydrocarbon (fluorocarbon – FKM) The only usual selection where compatibility with mineral oil and phosphate ester is required.

Polyurethane (urethane – Au) Widely favoured as a dynamic seal material in the USA and in Germany, polyurethane exhibits excellent abrasion resistance and is very durable in situations which might adversely affect nitrile compositions or even proofed fabrics, e.g. passage of seal lips over ports in a cylinder wall. Poor low-temperature flexibility, hydrolysis in hot water and a lower operational temperature ceiling are limiting factors.

Other seal component materials worth mentioning are:

Textiles – woven cloth such as cotton, asbestos, terylene and nylon are used as the substratum in elastomer-proofed fabric packings for medium- and high-pressure duties. Non-woven materials such as polyester and polyamide are also available. The proofing elastomer can be varied to suit but will frequently be based on nitrile or fluorocarbon.

PTFE – employed as a back-up ring material for 'O' rings. Poor creep properties restrict its use as a seal material in its own right unless energized by an elastomeric component or spring.

Nylon and acetal – main function is heel support for dynamic seals or as component-bearing ring material.

Hytrel – the registered trade name of a Du Pont range of thermoplastic polyester materials which form a bridge between between elastomers and thermoplastics. Has similar properties to polyurethane but exhibits much better low-temperature flexibility. Used currently as an anti-extrusion element in certain spool type piston head seals.

It is important that seals are protected and stored in such a way that they will not deteriorate or become damaged before they are used. Recommendations for storage of seals, etc. are given in Section 15.2.2.19.

15.2.2.3 Seal designs

Within the boundaries set by fluid power systems the majority of seals employed can be categorized as 'lip' or 'squeeze' types, or indeed a combination of both. Figure 15.55 illustrates the simplest forms utilized for static or dynamic reciprocating situations – the 'U' ring and the 'O' seal – and indicates how both rely on interference stress for at least a component of their functional operation. Both are responsive to system pressure and will generate a radial stress greater than the pressure to be sealed. Neither rely on externally applied compression to any real degree as in the case of the soft-packed gland. This is particularly relevant to squeeze seals, which are usually intended to float axially in their housings.

The quality of the sealing contact area and the interference stress in the same zone will control low-pressure sealing efficiency. These factors plus overall seal geometry and composition will set the point at which time hydraulic response takes over from the manufactured interference condition. The hydraulic component may become significant at pressures of 40 bar for an elastomeric 'O' ring or as high as 200 bar for a rigid proofed fabric packing.

Sealing slack oil and low pressures is generally a function of seal integrity and is the more difficult condition. Preventing leakage at high pressures is a feature of seal containment, i.e. preventing extrusion or rapid wear if in a dynamic duty.

Figures 15.56 and 15.56 list typical lip and squeeze types in common and traditional usage. Some particular points to note on each category are as follows:

1. *Lip seals* are single-acting to applied pressure and must be separated by a fixed component in double-acting duties so that the hydraulic load from the element under pressure is not passed on to the trailing element. For maximum service life multi-ring packing sets are often preferred where the succession of sealing edges ensures that breakdown is not sudden. The use of split rings is also entirely feasible with most designs of this sort without sacrificing sealing performance to any critical degree. Savings in downtime will be obvious. Having said this, it will be shown later that for very arduous operating conditions multi-ring sets are not always the correct choice as they can in fact be self-destroying.

2. *Squeeze seals* have the advantage of being double-acting and are usually housed more economically than their corresponding lip brethren. These seals all operate on the basic principle of the simple 'O' seal. For piston head or gland sealing, the 'O' seal should not ideally be used as a dynamic seal, since, due to the linear motion, it can suffer from rolling/spiral twist, which in turn leads to leakage. If, because of space restrictions a seal of 'O' seal proportions must be used, then either a 'D' section or 'quad' shape seal should be used – as seen in Figure 15.57. Remember that, as for static 'O' seals, if the pressures are in excess of, say, 100 bar these seals would also require some form of back-up/anti-extrusion washers. Most seals in this category use a single sealing zone and damage in service will be followed by more immediate breakdown than in the case of multi-lip packings.

Some of these seal types are discussed in more detail later. However, regardless of the seal design, it can only function correctly if the mechanical aspects of the housing are correct, therefore some fundamental requirements on this subject are now outlined.

15.2.2.4 Seal housings

The design of both gland and piston head arrangements will vary from one manufacturer to another and, particularly in the case of the piston head, this will be influenced by the actual seal, i.e. whether or not it incorporates bearing elements. Obviously, operating pressures, degree of knuckling/offset forces, etc. will also be taken into consideration.

Gland design A typical basic gland arrangement is as seen in Figure 15.58. The important aspects are that:

- The supporting bearing bush for the rod/ram is of sufficient axial length.
- This neck bush is positioned inboard of the seal and is therefore lubricated.
- While this bush requires to be a close fit to the rod, such bushes can cause fluid starvation to the seal or in certain conditions an intensification of fluid pressure at the seal. Such potential hazards can be easily avoided by providing a pressure relief through or across this inner bearing bush. (This is discussed later under 'Drag Flow'.)
- The gland follower should have the minimum possible clearance with the rod to prevent seal extrusion, but a wider clearance than the inner neck bush.

Piston heads Similar basic points apply for the piston head although, if the seal incorporates bearing elements, the fits, etc. should be as detailed by the seal manufacturer.

Tolerances Appropriate tolerances and/or limits are given in Figure 15.59 for both gland and piston head applications.

Extrusion clearances Table 15.13 shows extrusion clearances appropriate to an H9/f8 guidance fit of the neck bush.

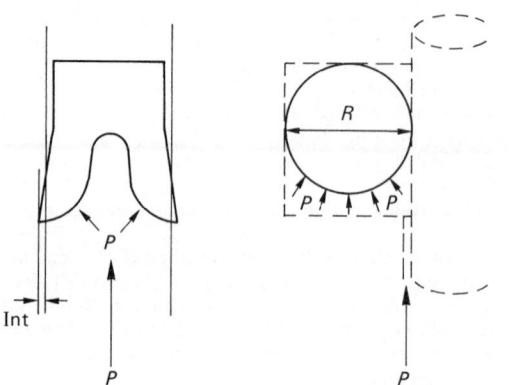

Figure 15.55 Lip and squeeze seals

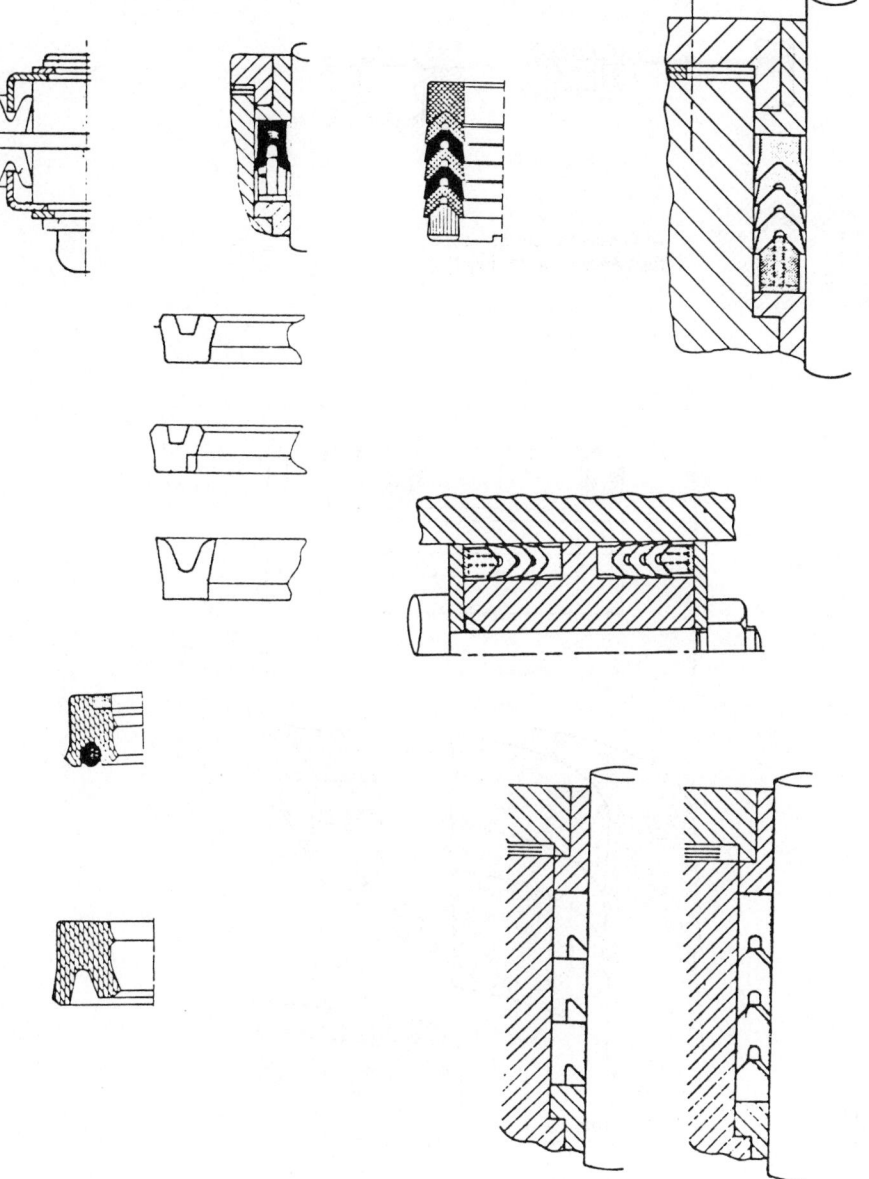

Figure 15.56 Lip seals

Lead-in chamfers To facilitate fitting of the seals both cylinder and piston rods should have an adequate lead-in/on. Suggested minimum values are given in Table 15.14 – these relate to an angle of 20°/30°.

Surface finish Piston rods and cylinder bores should be typically ground or honed and have a surface finish of 0.4 μm R_a or better, i.e. in accordance with BS 1134: Part 1. For critical applications this should be improved to 0.15/0.25 μm R_a. There is no advantage in trying to achieve a near-perfect finish, i.e. in the region of 0.05 μm R_a, as this will often prevent the establishment of a coherent fluid film under the seal. For the static seal contact surfaces, these should be better than 1.6 μm R_a.

Housing sizes Generally housing sizes will be taken from the seal suppliers' literature. However, for possible seal interchangeability housing sizes to an accepted standard should be used, i.e. ISO 5597/1. If there is a choice of seal sizes, the largest radial width/section seal should be chosen.

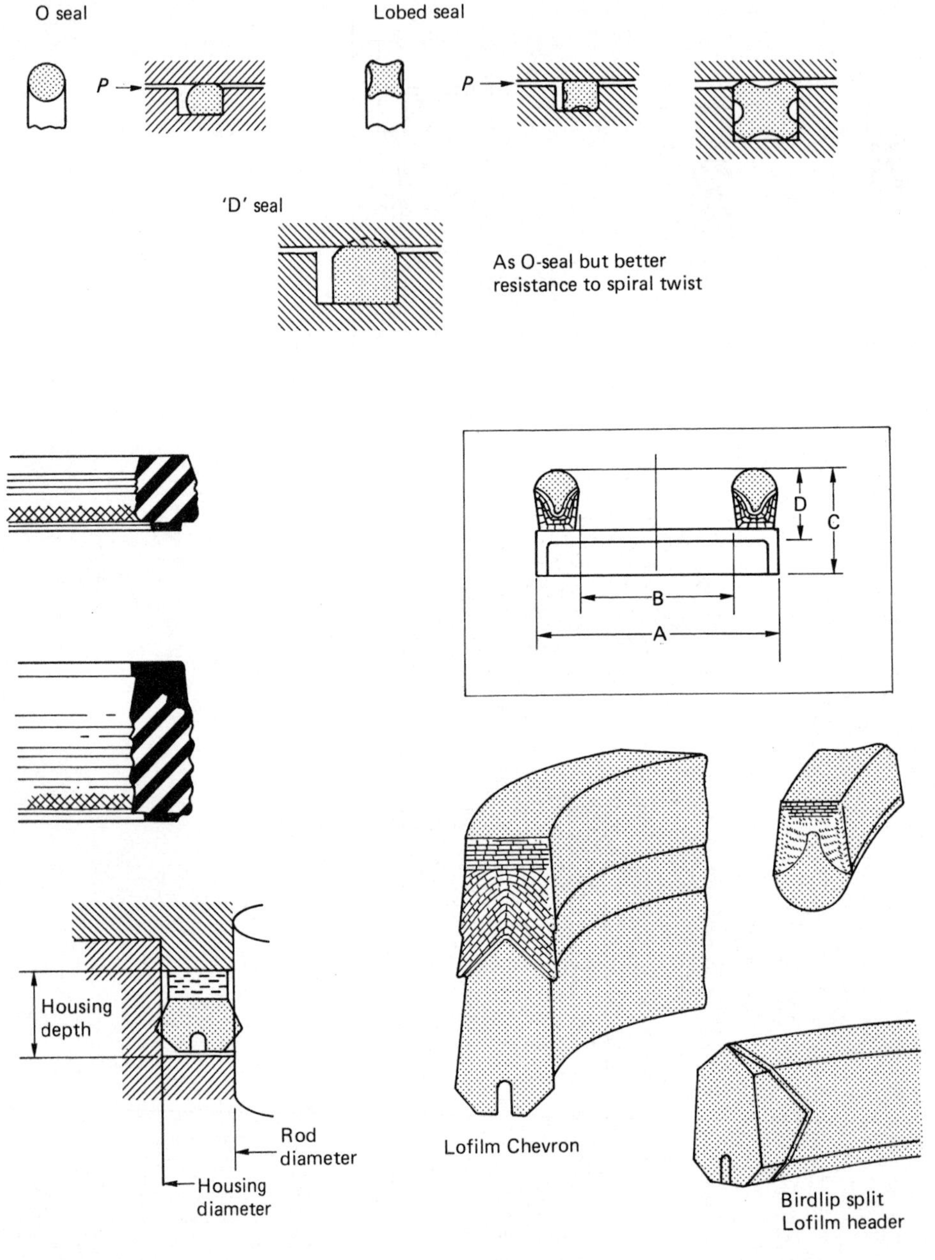

O seal

Lobed seal

$P \rightarrow$

$P \rightarrow$

'D' seal

As O-seal but better
resistance to spiral twist

D

C

B

A

Housing
depth

Rod
diameter

Housing
diameter

Lofilm Chevron

Birdlip split
Lofilm header

Figure 15.57   Single-acting squeeze seals

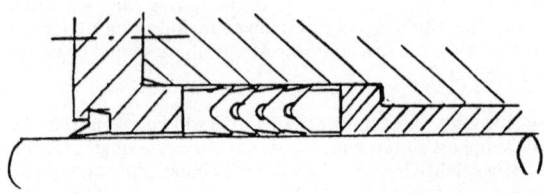

Figure 15.58

15.2.2.5 Type of fluid

As far as dynamic seals are concerned the great majority of mineral-based or phosphate ester fluids do not present lubricity problems. However, there is considerable development taking place to use water/soluble oil emulsions where seal performance can be directly influenced by the percentage of soluble oil.

One of the most demanding types of duty for any dynamic seal involves plain water at high pressure whether by design or by total loss of soluble oil content in a nominally lubricated

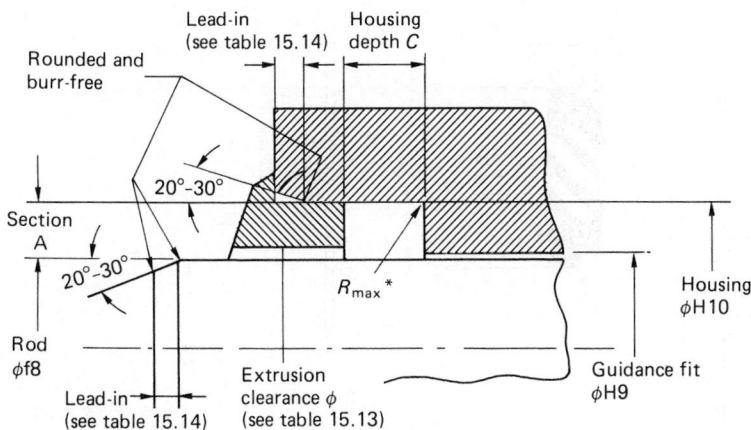

*See table 15.14

(a)

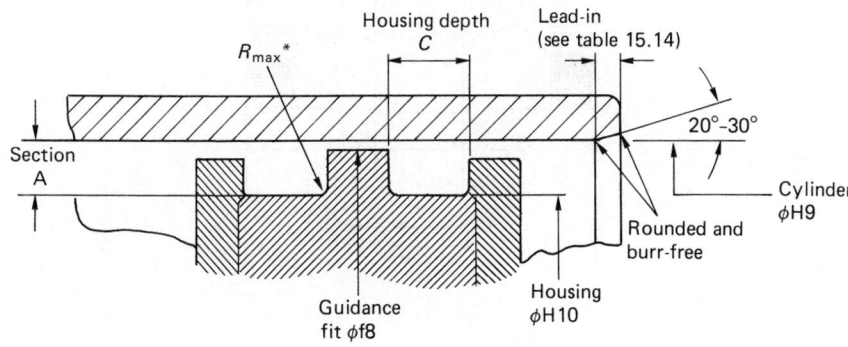

(b)

Figure 15.59 Examples of housing for (a) rod and (b) piston seals

system. High speeds are particularly difficult to sustain unless a minimum soluble oil level of 2% is guaranteed.

Tests conducted on three-throw mining pumps operating at 250 bar, 0.6 m/s, have shown that for even a 2% soluble oil content, average seal life will be extended by a factor of four by comparison with untreated water. Equally, a separate lubricant feed will produce similarly notable improvements.

The temperature of hydraulic fluid should not exceed 60°C if at all possible, as significantly faster swell and softening of proofed fabrics and straight polymers will occur above this value. To illustrate the point, the following data are based on immersion testing for 7 days in Shell Tellus 27 mineral oil:

Nitrile proofed fabric –
room temperature + 0.7% volume increase

60°C	1.2%
90°C	2.6%
120°C	3.3%

Filtration For the bulk of elastomer proofed fabric and solid elastomer seals fluid filtration of 25 μm should be perfectly adequate from the sealing performance aspect. Special seals such as those embodying PTFE wear faces will benefit from 10 μm filtration or better. In either event, filtration equipment offering a finer cut-off will probably be specified to suit control valve functions on a given press or hydraulic component.

The majority of conventional hydraulic seals which are tested in-house work in conjunction with no more than a coarse wire strainer in any hydraulic system. However, modern filtration aids must be considered an advantage particularly if the operation of the plant involved produces aggressive residues.

Air entrainment As cycle speeds become faster due to increasing work demands, system pressures must fluctuate more quickly. In many situations, full working pressure must be exhausted in milliseconds (e.g. die-casting machines, plastics injection moulding presses, etc.). If air is entrained in the hydraulic fluid such rapid decompressions can be exceedingly dangerous if no automatic venting is available, and can cause rapid seal failure – as discussed later.

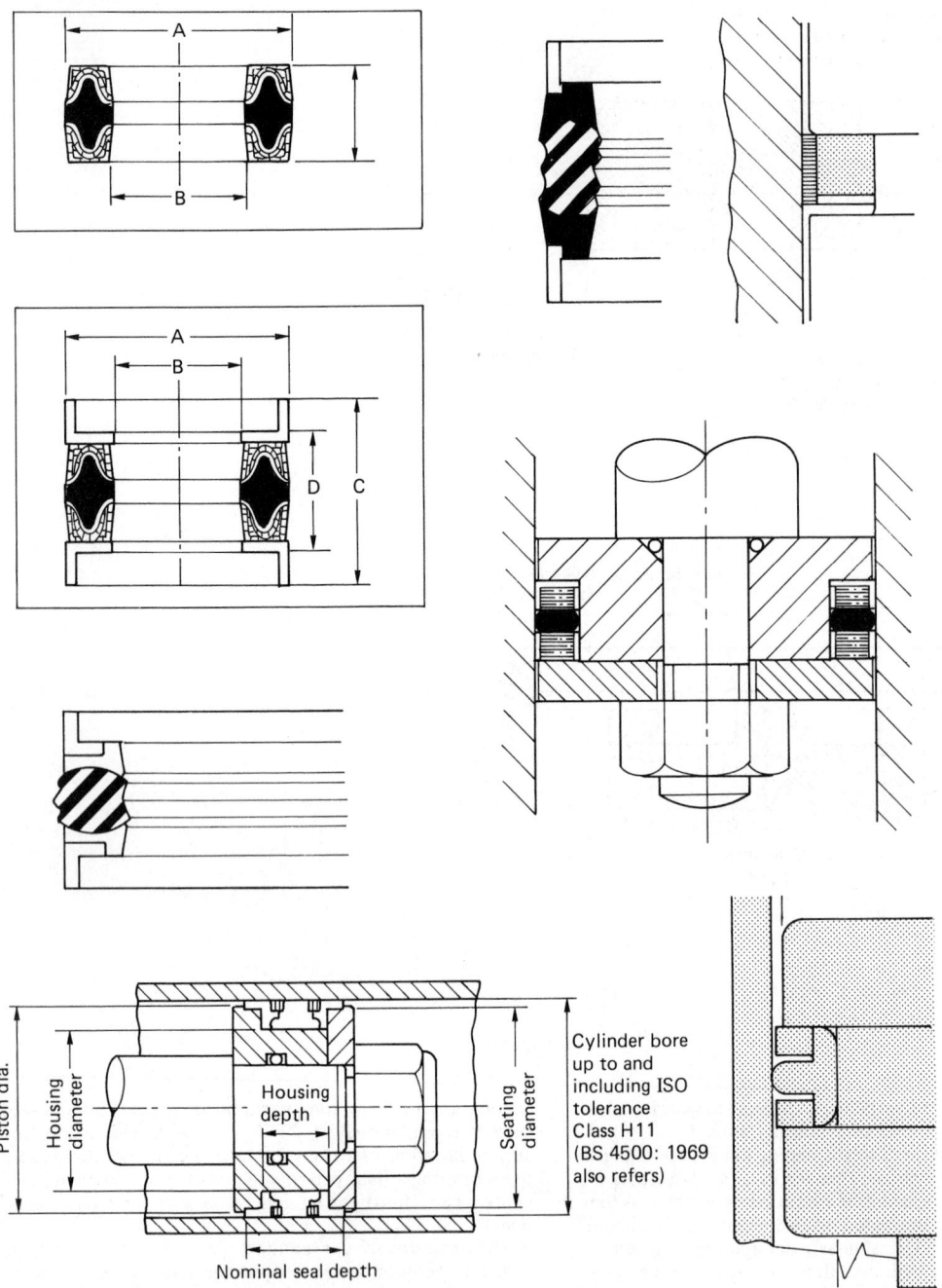

Figure 15.60 Double-acting squeeze seals

15.2.2.6 Tribological considerations

It is heartening to find the study of seal behaviour an essential part of tribology seminars, receiving as much attention, indeed, as bearings, lubricants and surface topography. In recent years much company research and independent study by organizations such as BHR Group, The Fluid Engineering Centre, has been expended on defining sealing mechanisms – particularly in relationship to reciprocating motion.

Film conditions All dynamic seals rely on a coherent fluid film under their contact area if they are to function consistently and predictably. Such films may stem from boundary

Table 15.13 Extrusion clearance (mm) (assuming H9/f8 guidance fit at neck bush)

Rod diameter		Gland follower bore limits
Above	Up to and including	
—	10	+0.072 +0.036
10	18	+0.086 +0.043
18	30	+0.104 +0.052
30	50	+0.124 +0.062
50	80	+0.148 +0.074
80	120	+0.174 +0.087
120	180	+0.200 +0.100
180	250	+0.230 +0.115
250	315	+0.260 +0.130
315	400	+0.280 +0.140
400	500	+0.310 +0.155

Values equal max. H9 plus H9 tolerance.

Table 15.14 Lead-in chamfer

Nominal radial section	4	5	7.5	10	12.5	15	20	25
Minimum length of chamfer	2	2.5	4	5	6.5	7.5	10	10
R_{max}		0.3	0.3	0.4	0.6	0.8	0.8	1.0	1.0

To aid assembly and to prevent damage to sealing lips, housings should be provided with a chamfer of 20–30° and the length as indicated above

All dimensions in mm

lubrication in some modes and be truly hydrodynamic in others. The film will vary in thickness according to seal profile, interference stress, pressure, speed, surface finish, type of fluid employed and its temperature. On reciprocating duties it might typically vary from 0.25 to 3.0 μm in thickness. On rotary shafts, values of 0.6–1.0μm would usually apply.

Film conditions – rod/gland seals Leakage from rotary shaft lip seals is seldom evident unless seal weardown or under-lip cracking has taken place – all other features being equal. However, in reciprocating applications passage of the rod emerging out through the gland will carry/drag the oil film to the atmosphere side of the seal as indicated in Figure 15.61. Under normal conditions, unless the seal is severely physically damaged, leakage will only occur on the emerging stroke – no leakage will occur when the rod retracts although it is during this retracting stroke that leakage 'appears' to develop.

Figure 15.62 is a diagrammatic representation of the worst film transport situation involving an emerging cylinder rod with slack pressure to the gland which retracts with the sealing element under load. In this mode the heel of the seal will be energized and can prevent the return of the total quantity of film carried by the rod. The fitting of an effective wiper can aggravate the condition.

To combat the emergent film, careful attention is required to pressure side seal geometry. Designs such as that shown in Figure 15.63(a) with knife-cut sealing edges and a specific relationship between contact edge and groove heights have proved very successful despite their short axial length.

For greater security, the revised Chevron seal, incorporating a squeeze-type sealing header, as shown in Figure 15.63(b) has proved extremely successful.

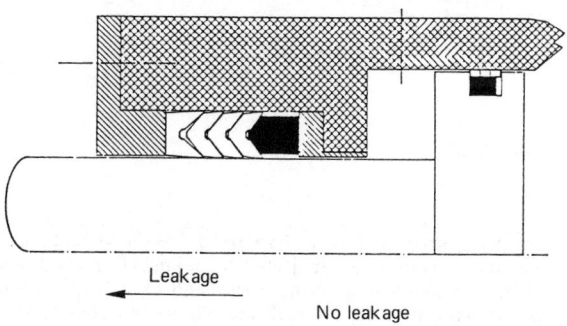

Leakage ← No leakage →

Figure 15.61

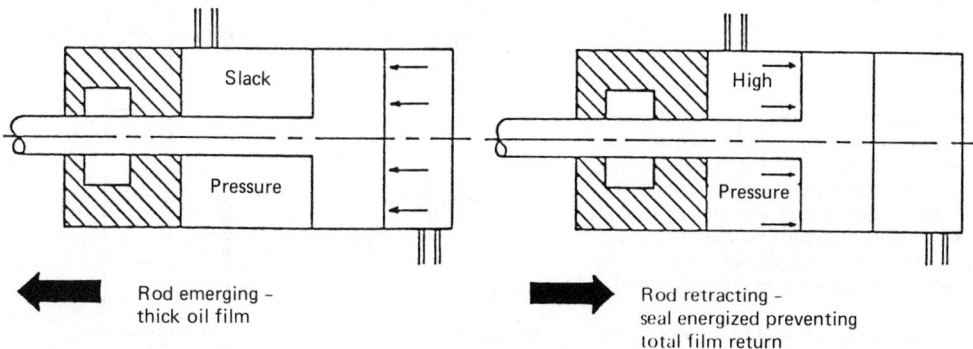

Rod emerging – thick oil film

Rod retracting – seal energized preventing total film return

Figure 15.62 Condition of maximum collected oil film

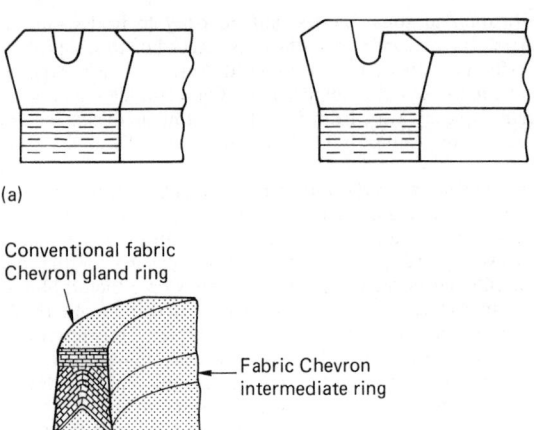

(a)

Conventional fabric
Chevron gland ring

Fabric Chevron
intermediate ring

Lofilm header ring

Composition
gland ring

Fabric Chevron
intermediate ring

Lofilm header ring

(b)

Figure 15.63

Knuckling/offset conditions Experience has shown that many hydraulic cylinders are subjected to a knuckling condition which then causes wear in the bushes and so both the piston and, in particular, the rod seals are required to operate in a misaligned condition. Remembering that leakage typically occurs on the emerging stroke of the rod, often combined with low pressure at the seal, prevention of leakage is then a function of the seal's own retained residual sealing stress. Average lip seals will lose much of their original sealing force fairly quickly.

15.2.2.7 Friction

Many designers of hydraulic equipment will have experienced the reluctance of seal manufacturers to provide even approximate frictional values of their seals. This is not surprising when considering the many variables that will determine the resultant drag, such as: type of oil, degree of resultant sealing stress, whether it be as-moulded, or as the result of subsequent compression, surface finish of the rod/cylinder, direction of rod movement in relation to the pressure on the seal, and the speed of the rod or piston.

Variations in speed above 0.1 m/s (20 ft/min) will not have a noticeable influence on the frictional drag even at speeds of 25 m/s which is encountered in gun recuperator systems.

However, at very slow speeds below approximately 0.02 m/s (4 ft/min) friction can vary and increase significantly – often experienced as 'stick-slip' – during which the actual movement is a constant start–stop sequence involving repeated breakout friction. Depending upon the characteristics of the equipment, this can often occur at a very high frequency resulting in noise and severe vibrations being set up. Whether or not this happens is also directly influenced by the function the hydraulic unit/actuator is performing in relation to the type of external force being exerted to the ram/cylinder. Determining the frictional drag of a seal is further complicated by the fact that many of today's seals incorporate two or three elements of different materials each having a different friction characteristic such as rubber; rubberized fabric; PTFE; nylon, etc.

In general, frictional drag is more of a problem with smaller actuators since the friction loss is proportional to the seal's diameter, whereas the thrust developed by the unit is, of course, directly related to the cylinder/piston rod area. However, when designing small actuators, low-friction seals should ideally be selected and if conventional seals in rubber/fabric materials are necessary the seals should be as short in axial length as possible since seal friction is also approximately a direct function of the seal's effective contact area. This point is shown in Figure 15.64.

Calculation of frictional drag Despite the many variables involved as discussed above, there are certain known factors which have a far greater influence on the result than others, such as fluid pressure, seal diameters and contact length and seal material. After much experimental work it has been established that a reasonable assessment of frictional drag is obtained by the following:

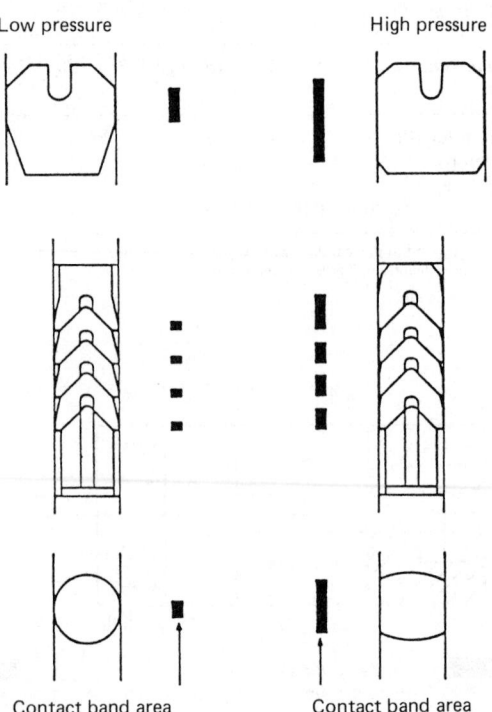

Low pressure   High pressure

Contact band area   Contact band area

Figure 15.64 Seal contact band width

Fabric seals
Dynamic friction $F = DL [0.055 + (0.31R \times 10^{-3})]$

Rubber seals (i.e. 'O' seal or rubber contact seals)
Dynamic friction $F = Dd [0.092 + (0.465R \times 10^{-3})]$

PTFE contact seals
Dynamic friction $F = Dh [0.055 + (0.096R \times 10^{-3})]$

where

F = dynamic friction (kgf)
D = dynamic/sliding seal diameter (mm)
L = effective seal contact length (mm)
R = fluid pressure (bar)
d = 'O' seal cross-section or depth of rubber contact (mm)
h = effective seal contact length (mm)

It must be emphasized that the above gives only an approximate value since it will be appreciated that the result is dependent upon the accuracy of estimation of the contact length of the seal when under pressure. Also, these give 'dynamic' friction values, *not* break-out values, which will often be perhaps 20–30% higher. Break-out values are extremely difficult to predict because quite often the first actual movements will happen before the operating system pressure is reached. When considering seals containing a combination of materials each element must then be considered separately using the above formulae.

Low-friction seals   If friction is a critical consideration then PTFE contact seals must be used. PTFE has many useful characteristics such as its low coefficient of friction but it does not possess any elastomeric properties and does suffer from what is commonly known as cold flow – this being more of a problem with flange gaskets.

For use as an hydraulic seal, regardless of the design of the seal, it will require some mechanical means of maintaining sufficient sealing stress to enable a seal to be maintained until the effect of the hydraulic pressure intensifies the sealing force. This can be achieved by using either springs or an elastomeric energizer. Typical spring-loaded designs are shown in Figure 15.65 and rubber-energized types do exist.

PTFE seals in general give excellent results, but there is always the possibility of a slight steady leakage developing. This is because the sealing face of the seal can be very easily damaged/scratched – perhaps during assembly – and eventually most rods or cylinders develop minute axial scores which can then cause a very small constant leakage.

A typical application where this has been a problem is when a PTFE piston seal is used on the double-acting piston of a down-stroking press typically as shown in Figure 15.66. When in the raised static position any minute leakage across the piston then causes the ram to creep downwards. Clearly, when considering the use of a PTFE contact seal it is useful to remember that these provide excellent dynamic seals. However, their use should be viewed with caution if the design/function of the equipment demands periods when a static pressurized condition exists such that even slight movement/creep of the rod/piston cannot be tolerated. For such

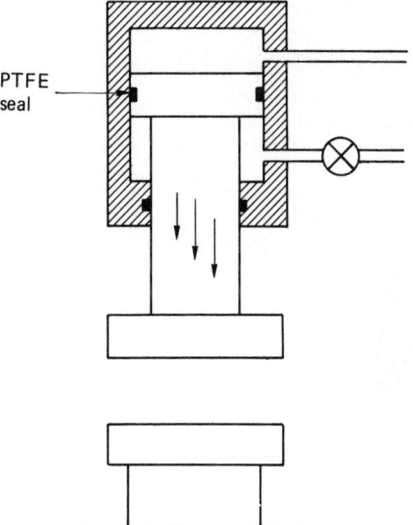

Figure 15.66

applications an elastomeric rubber contact seal must then be considered preferable.

Because of the low friction, PTFE seals are required on equipment which demands very sensitive control. When used as a dynamic double-acting piston seal slight weepage across the seal is often not a problem – it is normally too small to be detectable and assists in lubricating the seal. However, when used as a rod/gland seal any eventual leakage is obviously undesirable. In such applications it is becoming more common to fit a secondary rubber seal outboard of the main PTFE seal with a leakage drain taken from between them. This arrangement is now proving very successful on even large 700 mm diameter ram extrusion presses (a typical arrangement is shown in Figure 15.67).

It will be appreciated that PTFE seals are only efficient in endless form, which is ideal for piston heads but may pose a replacement problem on rod positions, depending upon the overall design and accessibility of the equipment.

15.2.2.8 Piston seals

In the early days of hydraulics it made no difference whether a piston was single- or double-acting since the choice of seal was either a simple 'U' ring type or the multi V-ring Chevron type.

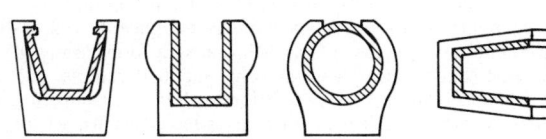

Figure 15.65

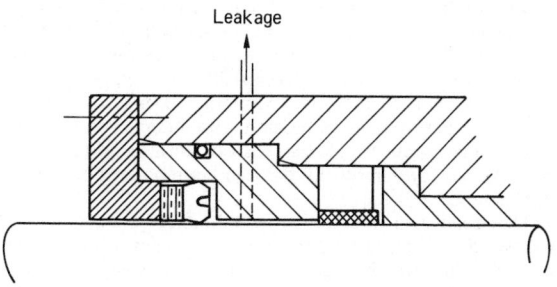

Figure 15.67

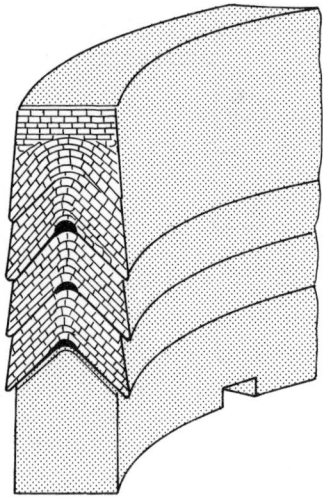

Conventional Chevron

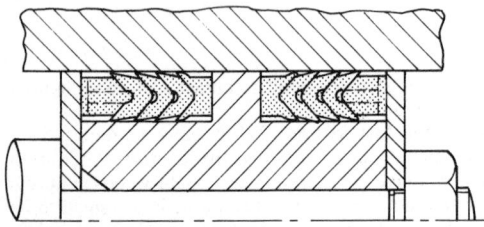

Figure 15.68 A typical double-acting piston seal assembly fitted with conventional Chevron

Single or opposed seals would be fitted as appropriate – typically as in Figure 15.68. Today the approach/choice is rather different. It is extremely rare now even to consider the double-opposed Chevron arrangement, which can cause excessive friction problems and demands a large/long piston which directly influences the overall size of the cylinder and therefore its manufacturing costs, etc.

For a single-acting piston the same type of seals would be considered as for use on a rod or gland position but, for a typical double-acting piston, a single double-acting seal would now normally be preferred of the types shown in Figure 15.60. Alternatively, two opposed single-acting seals could be considered if the minimum possible leakage across the piston was essential.

Most of the double-acting seals have radiused profile sealing areas and when dynamic they do allow a small weepage across the seal. On, say, a 100 mm diameter piston constantly reciprocating, a leakage of perhaps 30 ml h^{-1} may occur but this, in relation to the volume of fluid displaced, is negligible – the performance of the unit is not affected and the seal is well lubricated. These seals also perform a perfect static seal. As already mentioned, this can be important on certain applications.

The trend with new designs is influenced by commercial considerations so as to use the most economical piston design possible. This will obviously be a 'one-piece' component with a simple machined groove into which the seal must be stretched. The seal will usually also be required to incorporate

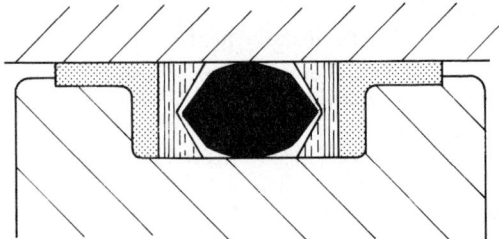

Figure 15.69

integral bearing elements and one such design is shown in Figure 15.69. For such piston designs there are already two recognized standards, i.e. ISO Standard No. 6547 and BS 6241 which provide full housing details.

In general, piston seals can tolerate far more wear before replacement becomes necessary. It is only when their degree of leakage is such that the performance of the cylinder starts to become impaired that the seal is changed, whereas the same degree of wear on a gland seal would have caused far earlier excessive leakage out of the unit, which would be unacceptable.

15.2.2.9 Gland seals

Overall, the sealing of a gland is a more difficult requirement since, unlike a piston seal, no leakage can normally be tolerated. Bearing in mind that to obtain even a reasonable service life the seal must operate on a fluid film, the resultant difference between essential film and acceptable leakage is extremely marginal – the emerging film conditions have already been discussed.

A gland seal is single-acting and therefore demands a different sealing contact profile to the piston seal. With 'U' rings and Chevron V-rings this is obvious but with the change from basic lip seals to the squeeze types the contact profile is somewhat different.

Many of these seals employ what is often termed a 'reverse rake' sealing edge, typically shown in Figure 15.63. To some minds this shape would appear to encourage leakage, but we must recall that most leakages occur at low pressures and what prevents leakage is the radial force the seal exerts on the rod and not the shape of the seal. In this context the lip profile is irrelevant since at pressures too low to energize the seal the lip shape has little or no effect. This point is shown in Figure 15.70. For the same degree of radial interference the rubber seal will normally exert and retain a significantly higher natural sealing force than the lip seals – even though these would be in a far harder material.

All seals which incorporate a rubber sealing element function on the basic 'O' seal principle which demands that, to respond to system pressure, the seal must be exposed to the pressure across its full radial width. The majority of these seals are therefore designed to float axially within the housing, or they may also be provided with ports across the full seal width so permitting the fluid to energize them. All these seals are therefore ideal for non-adjustable fixed-depth housings.

Depending on the installation of the equipment, seal replacement may demand that seals are in split form. Many of the latest shallow-gland seals are only supplied endless, and any seals that can function in split form will be of a multi-ring design requiring a deeper housing space. This aspect of subsequent seal replacement should therefore not be overlooked at the design stage.

effects of high radial loadings (they are also often supplied with a built-in lubricant). Furthermore, within their temperature range of around 100°C maximum they have a significantly higher resistance to shear than, say, nylon or PTFE.

Dirt excluders Ingress of dirt, etc. can cause both seal wear and, in particular, causes scoring of piston rods, hence all gland assemblies should ideally incorporate an efficient wiper. Typical shapes are as seen in Figure 15.71(a) – materials can be rubber, rubberized fabric, polyurethane or nylon.

Bearing elements Many units today, particularly the smaller sizes, employ synthetic bearing elements which can be in nylon, acetal, PTFE or, more often, phenolic resin. Most are simple rectangular section but 'T' section is also used, these being normally in nylon with axial pressure ports as shown in Figure 15.71(b). Regardless of the profile, it is important that these do not restrict the oil from reaching the seal – they must therefore always be split with a positive gap.

15.2.2.10 Seal selection

The ultimate question is: 'What is the best seal for a certain application?' This is a very difficult question and in most cases

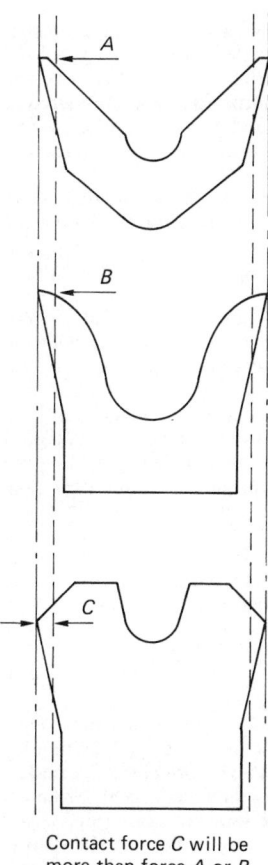

Contact force *C* will be more than force *A* or *B*

Figure 15.70

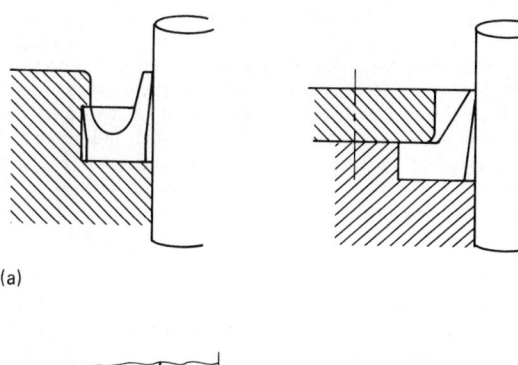

(a)

So far, most of the emphasis has been around low pressures since this is the main cause of leakage. However, the high or operating pressures must not be overlooked. All the seals mentioned here are automatic, whether lip or squeeze type. They will be actuated by the pressure and so the critical element of the seal is its ability to resist extrusion at the side remote from the pressure source. This is achieved by various means – often the latest rubberized fabric compositions are themselves sufficiently strong, or nylon or PTFE heel elements will be used. In many designs the integral bearing elements also double as the anti-extrusion component.

Many seals use nylon/acetal elements that are in split form. This is primarily because of the high thermal expansion and water absorption characteristics of these materials which can significantly influence their sizes. While the change in radial section and depth will not be important, diametral changes can be considerable and a scarf split often allows this expansion to be resolved into a peripheral size change. If these components are not split they will usually have axial ports across them to provide unrestricted passage of the fluid to the seal.

For very high-pressure applications which can often incur expansion of the cylinder, anti-extrusion elements in phenolic resin-based materials are often employed and using a simple triangular shape is extremely successful since the pressure acting through the seal always ensures that this ring is forced directly into the area of the extrusion clearance. These materials are normally used as bearings and so do not suffer any

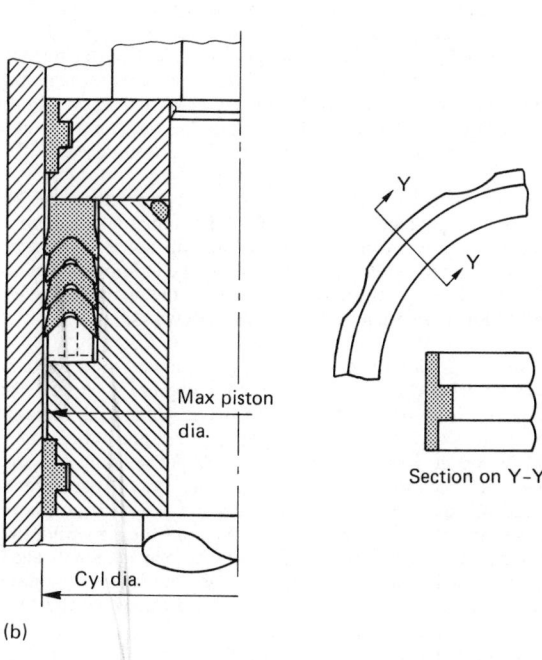

Max piston dia.

Cyl dia.

Section on Y–Y

(b)

Figure 15.71

there will be many seal designs which will apparently meet the given conditions and yet appear significantly different in respect of material, size and price. The fluid power equipment designer can be forgiven for being confused.

The choice will often depend on the user's own experience and preference based on knowledge of the application. Schedule maintenance periods, accessibility, consequence of leakage, initial cost and availability will all play their part in steering the decision towards ultimate security or some other level of cost-effectiveness.

If in doubt or the application/conditions are unusual, then consult the seal manufacturers. If they are a reputable company they will express any reservations rather than offer a standard off-the-shelf product which will possibly experience premature failure. In recent times many applications have demanded a specially designed seal, but not all manufacturers are able to produce something special that is, to them, non-standard.

Having made a choice and being confronted with a range of overall sizes for a given diameter, the largest seal section available should be taken – not the smallest – other requirements allowing. The smaller the seal – whether lip or squeeze – the finer is the working tolerance band of interference and the lower is the capability for absorbing misalignment, vibration, adverse accumulative tolerances, etc. This applies equally to static and dynamic positions.

Even when the correct seal is selected, seals are, of course, prone to eventual failure – prior to failure of mechanical components. Most engineers and indeed end users accept this fact. However, 'correct' seals can fail prematurely through no fault of the seal and this can sometimes be associated with component failures. In many instances such failures can be avoided since the cause lies within the design of the equipment/unit. If the cause is known about in advance it can therefore be designed out of the unit.

15.2.2.11 Causes of seal failure

For many years most seal failures were attributed to either (or a combination of) extrusion of the seal, poor surface finish of the rods/cylinders, dirty fluids containing abrasives, etc. Combined with improved quality of machined metal parts, cleaner fluids are now used and seals are both superior in design and materials. These modes of failure are now far less common, but, with today's higher operating pressures and temperatures less obvious causes of seal failure are often encountered.

Air entrainment Within all hydraulic systems there will be some small percentage of air which, if trapped adjacent to a seal, can cause severe failures. Failure will not always occur and will be dependent upon the operating conditions, particularly the system pressure and the speed of the operating cycle. Seal failures can occur in the following two ways.

Explosive decompression Virtually all rubbers and rubberized fabrics are permeable to various degrees such that air under pressure can be forced into the seal material and will therefore be within the seal at system pressure – the longer the period at pressure, the greater the permeation into the material. The problem occurs when the system pressure is reduced, which will often be from e.g. 350 bar to atmosphere in perhaps a few milliseconds. At this point the air within the seal cannot slowly vent itself out but will expand rapidly such that the seal is severely damaged by splits or blisters. A simple analogy to this is the 'bends' suffered by divers when they are forced to surface too quickly.

The problem is more common with 'O' seals in valves used in multi-phase systems, primarily in the oil industries, but it

has been experienced in hydraulic units. Much development work has taken place in the last few years and specialized elastomers are now available which have largely overcome the problem. These are based on fluorocarbon and are therefore suitable for both mineral oils and phosphate ester fluids. Should this mode of failure be encountered, a solution is now available.

With the rubberized fabric materials they tend to delaminate, but in developing harder materials primarily to resist extrusion these new fabrics are also less permeable, and so at least some of today's seals are less prone to this problem.

Dieseling This is a term often used to describe what happens when a pocket of air/oil mist self-ignites. It can occur when the rise in pressure is fast enough to cause a significant rise in temperature and hence ignition takes place. When this happens in the area of the seal it can cause severe localized damage and even melting of any nylon components. A typical case is in an inclined cylinder as shown in Figure 15.72, where air has become trapped adjacent to the piston seal, and possibly the gland seal, depending upon the position of the connecting ports.

Clearly, in any system serious attention should be given to eliminating the possible problems by trying to ensure that the minimum of air is present and that it can be vented if at all possible.

15.2.2.12 Axial length of seals

All seals fail in basically a similar way in that failure never occurs at the first point of contact with the fluid. With a 'U' ring or Chevron type seal the first lip is normally still in perfect condition since this will always be lubricated, and the failure happens beyond this towards the lower pressure side. This is because the subsequent lips and base of the seal still deform against the rod or cylinder under the hydraulic pressure but is often void of sufficient lubrication, and, obviously, the higher the pressure, the worse are the effects. The more modern squeeze seals do still fail eventually in a very similar way.

Many years ago, for higher-pressure applications the trend was to increase the depth of the seal, i.e. the number of rings, etc. but this is completely wrong, and many of today's most difficult applications are sealed perfectly with a shallow single-element seal. In principle, therefore, although it is sensible to opt for as wide a radial section seal as possible, the axial length must be carefully considered, as too deep a seal can in fact reduce seal life and cause premature failure.

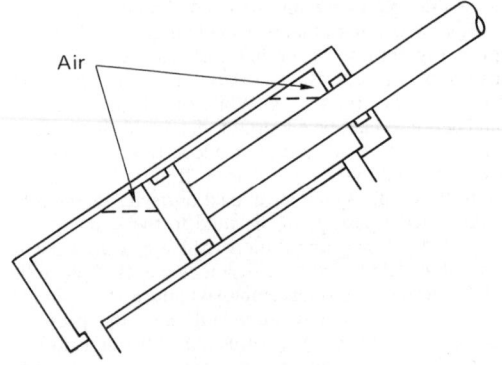

Air

Figure 15.72

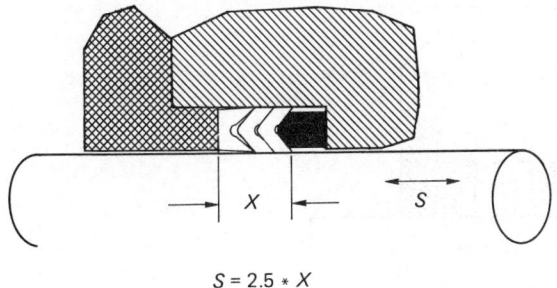

$$S = 2.5 * X$$

Figure 15.73 Short strokes

15.2.2.13 Short strokes

In a similar way, short-stroke applications can cause seal failures. Remembering that the vital fluid film is created by the movement of the rod through the seal, clearly a short stroke may not transport sufficient oil to the complete contact area of the seal. Also, a rod will not carry oil over the full length of its travel. With a seal length of, say, 50 mm a minimum stroke movement of perhaps 150 mm/200 mm will be necessary for oil to reach the full axial length of the seal. If this does not happen the seal may fail simply due to lack of lubrication. A typical sequence is that the 'dry' area towards the low-pressure side of the seal wears and particles of fabric are then dragged towards the forward sealing edge where they effectively become adhered to the seal and so create an uneven contact area and hence leakage. For short-stroke applications it is therefore important to use shallow seals, and a 'rule of thumb' is to ensure that the stroke length is at least two to three times that of the seal's contact area – as indicated in Figure 15.73.

15.2.2.14 Drag flow

This is a term often used to describe the effects of oil within a close clearance between a static and dynamic component. The first time the writer witnessed this was many years ago during tests to establish the friction characteristics of seals using a simple test rig as shown in Figure 15.74, where the right-hand seal had been replaced by a close-fitting bush. During the tests leakage through this bush would be expected, so providing a true friction value for the seal in the left-hand housing. During the tests leakage readily occurred with the rod moving to the right, but when moving to the left, no leakage developed, despite the system pressure of 350 bar!

The effect of this has been shown to be the cause of certain seal failures. A classic case is with a gland seal on a double-acting unit where, quite often, the gland seal will be subjected to exhausting oil when the rod emerges, and often only a relatively low pressure sufficient to return the rod on the inward stroke, i.e. a typical down-stroking press. Under such conditions seals have failed clearly as a result of being subjected to very high pressures, resulting in wear and extrusion, etc., although, according to the system, perhaps only 50 bar is used on the return stroke. With a close-fitting bush as shown in Figure 15.75 the following has been shown to happen.

On the emerging stroke (with low-pressure oil adjacent to the bush) oil is dragged by the movement of the rod towards the seal. On the inward return stroke slightly less oil is dragged back through the bush because of the return system pressure now acting adjacent to the inboard clearance between the rod and the bush. Therefore, during each cycle slightly more oil is transported towards the seal than is returned and, depending upon the frequency of operation, high pressures can be created at the seal. This then causes problems, since the seal is effectively under a constant energizing pressure which results in insufficient oil reaching the rear of the seal contact area. Hence possible failure occurs as described in earlier sections.

This possible mode of failure is easily prevented by simply ensuring that there is a 'hydraulic link' across the bearing

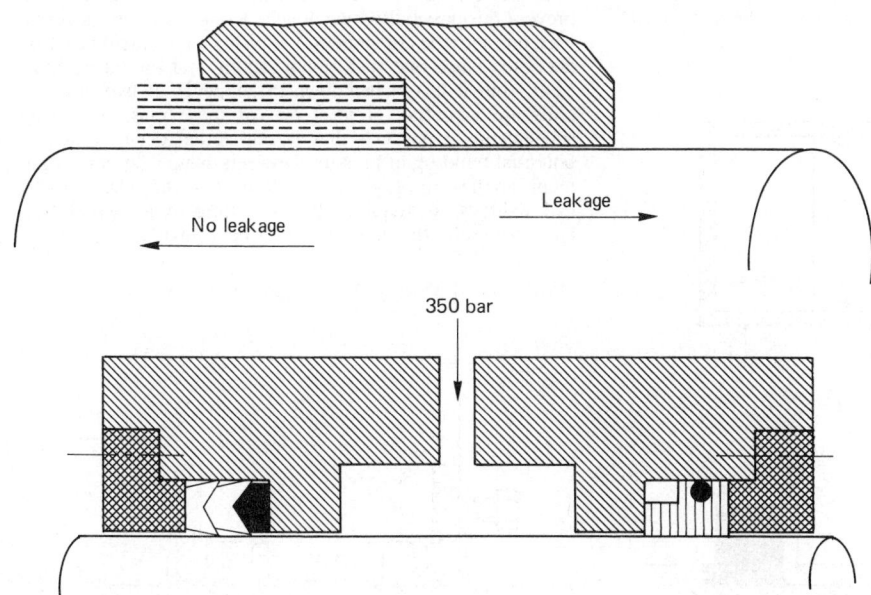

Figure 15.74 Drag flow

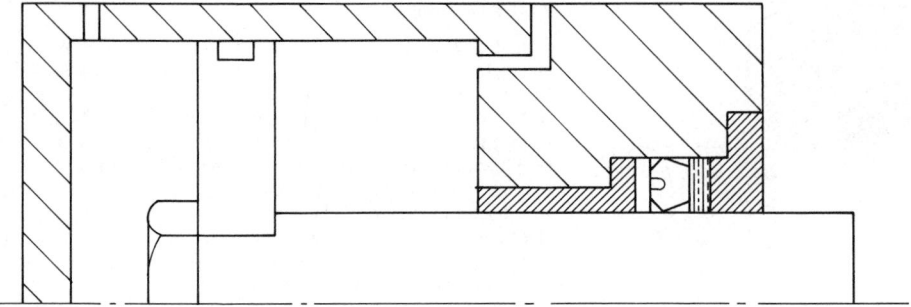

Figure 15.75

element either by holes through it or, more simply, by one or two small axial slots milled down the inside of the bush, or, if preferred, a spirally machined groove – all of which achieve the desired effect.

15.2.2.15 Pressure trapping

This is a situation that can develop between any two seals – it can cause seal failures and/or fracture of metal components of the gland or piston head. It can potentially develop if two seals are used which are of the type that can effectively be double-acting. In general, this applies to most of the more modern squeeze types even though they may appear supposedly as single-acting seals. There are various theories as to how this phenomenon develops but the writer's view is as follows.

Referring to Figure 15.76, i.e. using two simple 'O' rings on a double-acting piston, with the piston moving to the left, seal A is energized by the system pressure but seal B is relaxed and will therefore allow a minute degree of leakage across the seal into the centre space. Even if the centre space is full during this stroke, seal A, because it is pressurized, will not allow the same degree of leakage across it and therefore out from the centre position. On the reverse stroke, exactly the same thing happens, i.e. seal B is energized, but seal A, being relaxed,

will allow a small leakage across it into the centre space. Hence, during each cycle, oil is forced into this centre position but never out of it, and so this space soon becomes permanently pressurized. As this happens the seals try to extrude outwards and, depending upon the seal type and profile, this in itself can encourage further movement of oil across the seals into this space.

In practice, this trapped pressure can become significantly higher than the system pressure such that piston components are under stress from the reverse direction and can fracture as shown in Figure 15.77.

Often there are symptoms revealed that indicate that this phenomenon is starting to develop. Because the *two* seals are now permanently under pressure, excessive friction is occurring. This may then cause the unit to become hot and if the seal incorporates nylon components the cylinders may emit squeaking noises. Also, the speed of the component may slow down and eventually stop. This is more likely the smaller is the diameter of the cylinder such that the developed thrust is lower in relation to the friction of the seals, than will be the case in a large diameter unit.

In cases where seizure has occurred the pistons have often required to be mechanically winched out of the cylinder. To prevent this possibility, two double-acting seals must never be used on the same component, i.e. piston (this should never be necessary since one such double-acting seal should be sufficient – but it has been known to happen!). If two opposed seals are essential, at least one of these must be a true single-acting type such that it can automatically vent any potential build-up of pressure between them. The same argument applies to a gland situation in cases where 100% leak-tightness is required by the fitting of a second (i.e. back-up) seal – the *inner* seal must be a single-acting type.

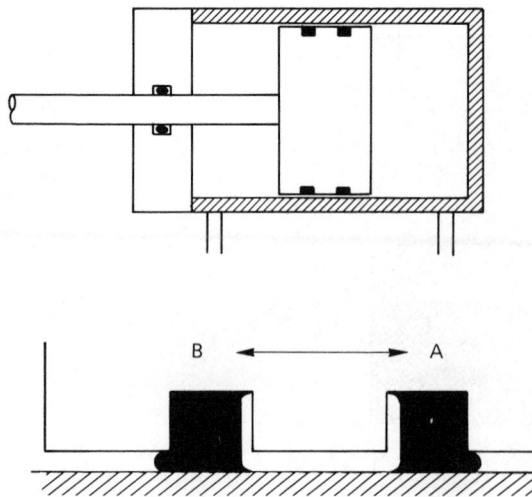

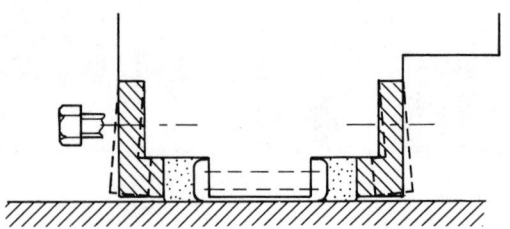

Figure 15.76

Figure 15.77

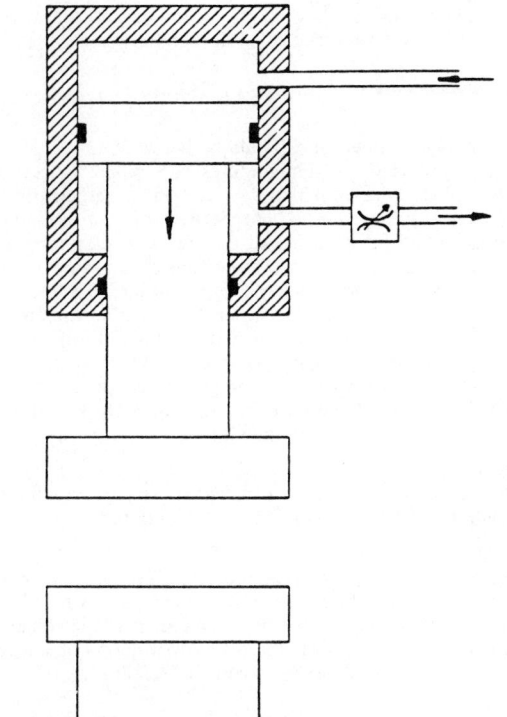

Figure 15.78

15.2.2.16 Exhaust pressures

Exhaust pressures from a double-acting cylinder can cause both seal and gland component failure. The typical way of controlling the speed of a unit is to use a flow control valve in the line so as to restrict the exhausting oil. This achieves a smooth and controlled speed of the unit.

By referring to Figure 15.78 it will be appreciated that during the emerging stroke, depending upon the difference in the diameters of the rod and cylinder, the resultant pressure in the exhausting annulus side can become higher than the inlet pressure to the full piston area. In most instances no apparent problems are encountered but in situations where the system can incorporate gas/oil accumulators, disastrous results have occurred.

In Figure 15.79 it is seen that *without* an accumulator in the system full system pressure will probably not develop until the press actually closes, but with an accumulator, the full system pressure will be applied *during* the closing/approach stroke. In many units the difference between the rod and cylinder diameters is quite small and therefore the developed pressure on the annulus side is equal to the system pressure × [AREA A]/AREA A − AREA B).

Normally the difference in cylinder area to rod area is quite small. Hence pressures during the emerging stroke can be in fact many times the system pressure but in the majority of systems this exhaust pressure is never monitored. Most systems only have gauges showing the system pressure being delivered by the pump. The effect of this extremely high pressure in the annulus side is to cause premature seal failure. This is bad enough, but the more serious effect is that the cylinder and the gland components are being subjected to

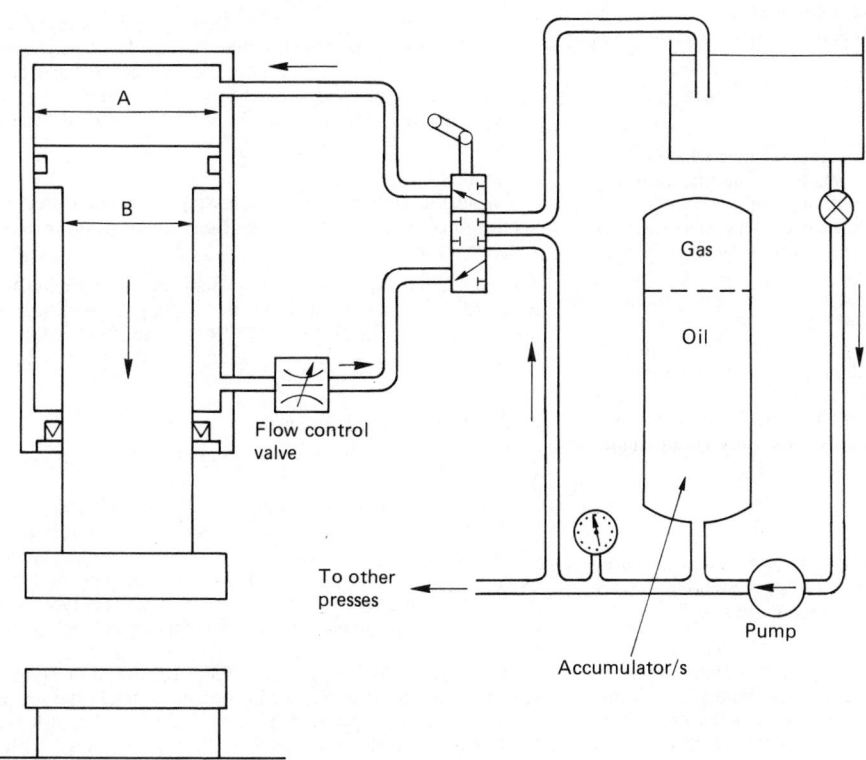

Figure 15.79

pressures far in excess of their design limits, and it is not uncommon to find that all the gland studs have sheared!

This 'unseen' problem can only be overcome by placing some form of flow control on the *inlet* to the piston side of the cylinder which then gives a 'comparable' flow rate in relation to the flow control of the exhaust side – which is often still necessary on larger down-stroking units.

15.2.2.17 Conclusions

This section has not attempted to identify the correct seal for any particular application but it is hoped that it may have at least pointed in the right direction in both the important aspects of the design of the seal housing, etc. and provided sufficient basic information for making a tentative selection of the most suitable seal type to use. It cannot be over-stressed that to consult a seal manufacturer regarding the final seal choice is well worth the cost of a telephone call or time involved in a discussion. It could save both you and your customer considerable time and expense at a later date.

Apart from the selection of the most suitable seal, applying the following rules should assist in obtaining the best out of your unit and may well prevent premature seal and/or component failure:

1. Pay careful attention to seal housing design, i.e. extrusion clearances, etc.
2. Use the widest radial width of seal rather than the narrowest.
3. Endeavour to prevent air entering the system and make provisions for venting – particularly during initial commissioning.
4. Ensure that bearings do not restrict fluid from reaching the seal – provide a positive hydraulic link across/through the bearings.
5. Be cautious when controlling actuator speeds by throttling of the exhausts – check whether excessive pressures are being created.

15.2.2.18 The future

It is unlikely that any radical new seal designs will suddenly emerge. Most so-called new designs are a modification of a well-established sealing concept although minor changes can, of course, give improved performance. The main area of change is in respect of seal materials to give improved properties, i.e. better resistance to extrusion and less friction and wear, etc. In this respect seals are now being manufactured in PTFE-loaded rubbers and rubberized fabric compositions with a PTFE-loaded proofing on the contact surfaces. These early tests are giving excellent results and such seals will become more common in the near future. Also, considerable development is taking place with polyurethane and future materials with a higher-temperature capability could significantly increase its use within the sealing industry.

15.2.2.19 Guidelines on storage

BS 3754: 1963 'Storage of vulcanized rubber' was prepared under the authority of the Rubber Industry Standards Committee and includes the following recommendations:

'Most vulcanized rubbers change in physical properties during storage and ultimately may become unserviceable, for example, because of excessive hardening, softening, cracking, crazing or other surface degradation. These changes may be the result of one particular factor or a combination of factors, namely, the action of oxygen, ozone, light, heat and humidity.

The deleterious effects of these factors may, however, be minimized by careful choice of storage conditions.

Recommendations

Temperature
The storage temperature should be below 25°C and preferably below 15°C. At temperatures exceeding 25°C certain forms of deterioration may be accelerated sufficiently to affect the ultimate service life. Sources of heat in storage rooms should be so arranged that the temperature of no stored article exceeds 25°C. The effects of low temperature are not permanently deleterious to vulcanized rubber articles but they may become stiffer if stored at low temperatures and care should be taken to avoid distorting them during handling at that temperature. When articles are taken from low-temperature storage for immediate use their temperature should be raised to approximately 30°C throughout before they are put into service.

Humidity
Moist conditions should be avoided; storage conditions should be such that condensation does not occur.

Light
Vulcanized rubber should be protected from light, in particular direct sunlight and strong artificial light with a high ultraviolet content. Unless the articles are packed in opaque containers, it is advisable to cover any windows of storage rooms with a red or orange coating or screen.

Oxygen and ozone
Where possible, vulcanized rubber should be protected from circulating air by wrapping, storage in airtight containers, or other suitable means; this particularly applies to articles with large surface area-to-volume ratios, e.g. proofed fabric, cellular rubber.

As ozone is particularly deleterious, storage rooms should not contain any equipment that is capable of generating ozone, such as mercury vapour lamps, high-voltage electrical equipment, electric motors or other equipment which may give rise to electrical sparks or silent electrical discharges.

Deformation
Vulcanized rubber should, wherever possible, be stored in a relaxed condition free from tension, compression or other deformation.

High quality requirements for storage and periodic inspection such as those specified by the Aerospace industry are obtainable from BS 2F.68: 1963, 'Recommendations for the storage and inspection in store of vulcanized rubber items'.

15.2.3 Rotary shaft lip seals

15.2.3.1 Introduction

The spring-loaded rotary shaft lip seal, as shown in Figure 15.80, is widely used to retain operating media/system lubricants and to exclude foreign matter from within mechanical equipment with rotary shafts. This section explains important principles of operation with a review of basic seal types with materials in typical use and fluid compatibility being examined.

The nature of most designs precludes the use of this type of seal from operating at significant pressures unless the sealing lip is adequately supported by a shaped plate. For pressure conditions much in excess of 2.0–3.0 bar, combined with a rotary shaft, it would be preferable to consider the use of either a compression packing or a radial face mechanical seal.

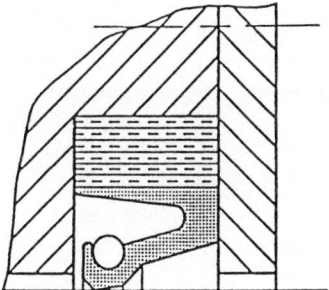

Figure 15.80 Standard fabric back seal with retaining plate

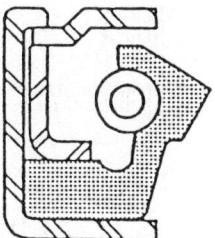

Figure 15.82 Typical metal-encased seal

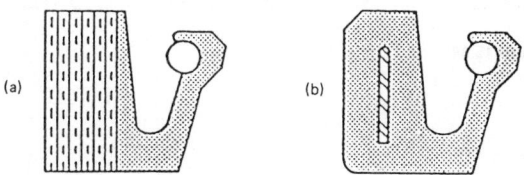

Figure 15.83 (a) Standard fabric back seal; (b) moulded-in flexible steel band self-retaining seal

15.2.3.2 Method of operation

The lip seal as we know it today has changed little in component parts and shape from its pre-war concepts. Originally, leather was the boundary-lubricated contact material but now elastomers predominate in a range of products that are intended for liquid/gas interfaces. The lip profile can be produced in as-moulded, ground or knife-cut condition – the last being demonstrably superior and the most common form of finishing used in contemporary products. In simple terms, the integrity of the seal leakage performance relies on the meniscus effect shown in Figure 15.81, which was first researched by Dr Ernest Jagger of the George Angus Company in Wallsend.

It should be remembered that this stable hydrodynamic film characteristic takes time to establish on start-up and diminishes to a boundary condition again when shaft rotation is being reduced towards stop. Most aggravated wear problems occur in the boundary regime and elastomers must be formulated to suit.

15.2.3.3 Design

A wide variety of lip seal designs is available in terms of overall construction, material and lip profile. Some are intended for fitting in housings which have no separate cover plate and may be supplied with a rigid metal case to which the seal is bonded – the unit being a force fit in the housing (Figure 15.82). Others have the advantage of flexibility by virtue of having a proofed fabric back (Figure 15.83) while requiring the provision of a retaining plate to nip axially the back of the seal to prevent leakage and obviate rotation. More recent developments combine the merits of flexibility and self-retention in a metal-supported all-rubber configuration (Figure 15.83(b)).

One of the several advantages of the fabric back seal is the facility with which split seals can be fitted without reducing

performance which, in many cases, is of the same order of efficiency as an endless ring. A rubber inlay is frequently moulded into the back of such a seal through which the split is effected. The rubber abutment which results ensures good sealing across the split portion. Where shaft dynamics are encountered, this split seal may not provide the degree of sealability required – for such aplications on-site joining is necessary and one method is described later in this section.

The profile of the lip contact area is subject to each manufacturer's design philosophy. Some have a knife-edge contact band and rely on heavy as-moulded lip interference. Others depend on the spring tension to urge the lip into intimate shaft contact. In practice, a careful balance of interference, spring characteristics and contact band width must be allied to knowledge of the type of material being used, bearing type and condition and other environmental considerations such as temperature, fluid and, if any, pressure. A typical range of seal profiles is shown in Figure 15.84, which also illustrates the expedient used when insufficient space is available to accommodate a pair of seals, i.e. the so-called dust lip design.

Finite Element Analysis is extensively employed for the optimization of seal lip designs. This eliminates much costly physical testing and many sophisticated techniques are available for examining both hydrodynamic fluid film characteristics under given conditions and the non-linear properties of

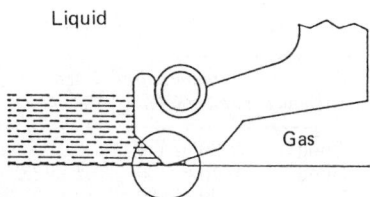
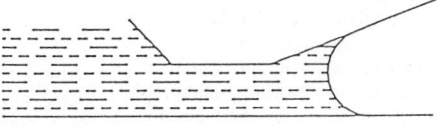

Figure 15.81

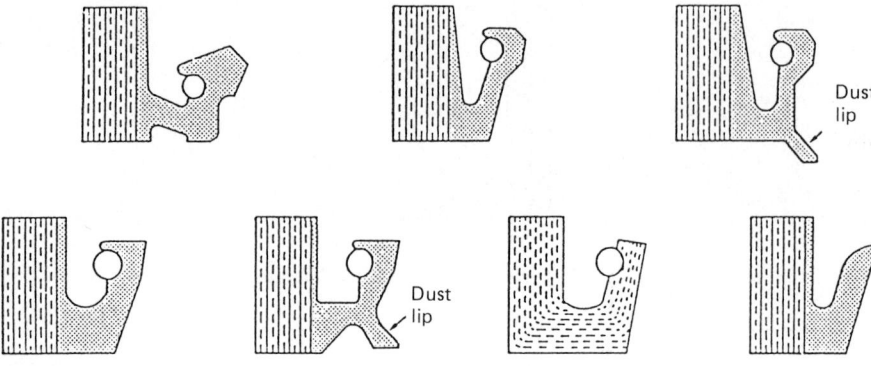

Figure 15.84

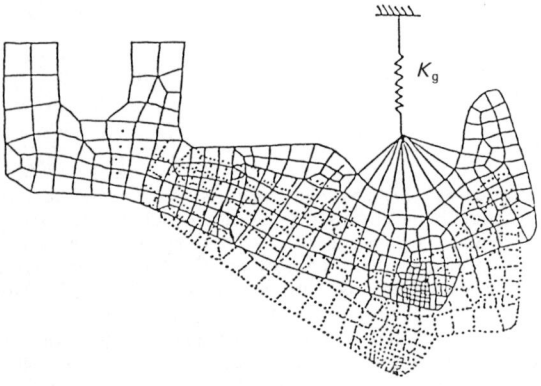

Figure 15.85

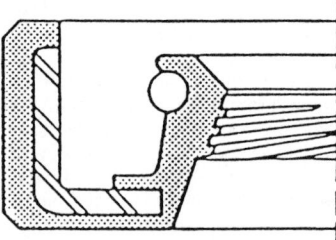

Figure 15.86

complex elastomer formulations which can contain sophisticated additive packages (Figure 15.85).

15.2.3.4 Material

Nitrile-base synthetic rubber compounds are widely employed as standard materials by reason of their compatibility with most lubricating oils and greases and their resistance to wear. There are, however, certain fluids which may cause excessive swelling or shrinkage of such polymers and a fluorocarbon rubber (e.g. Viton) may be a necessary selection. This is particularly true of some palm-oil solutions and other soluble types of oil.

Consideration for operational temperature is important and, in this respect, it is the condition at the lip of the seal which is paramount – not necessarily the environmental temperature. However, it may be said that nitriles are frequently used with success for intermittent service temperatures of 150°C and are continuously rated for 120°C in lubricated conditions. Above these values acrylic rubbers, fluorocarbon or silicone compounds would require investigation.

Many silicone and nitrile rubber shaft seals are fitted in automotive applications for crankshaft and gearbox sealing, in which areas much use is being made of grooved sealing surfaces that are designed to produce a hydrodynamic effect in the oil film being sealed, effectively causing the oil to be pumped away from the seal lip (see Figure 15.86).

Table 15.15 gives a broad indication of cost to temperature and fluid performance. In the past, user selection judgement has been driven totally by cost considerations where an NBR seal would, for example, be selected for continuous running at 120°C. The right choice for much wider performance safety margins would have been a commercial FKM product rated to 180°C where the extra cost would be well justified in terms of operational lifetime and sustained reliability.

Another point worth flagging concerns low-temperature flexibility which can frequently be the Achilles' heel of a wide range of equipment involving working capabilities below −10°C. As temperatures approach the glass transition point of elastomers, stiffening of the material can produce significant changes in seal hysteresis effects in relation to shaft dynamics. Out-of-phase frequencies that result can often lead to aggravated leakage conditions. Grades of VMQ (silicone) and low-nitrile versions of NBR elastomers are available for such duties. Low-temperature versions of FKM rubbers have become established in more recent times.

Table 15.15

Material	Oil resistance	Temperature range (°C)	Price indicator
NBR (nitrile)	Good	−20 to 120	1.0
ACM (polyacrylic)	Good	−10 to 130	1.5 to 2.0
VMQ (silicone)	Fair	−40 to 180	3.0
FKM (fluorocarbon)	Excellent	−25 to 180	5.0 to 10.0

In the overall sense, temperature has a more direct effect on seal materials than oil resistance (excluding fire-resistant fluids of the phosphate ester type). A reduction of even 10°C can produce disproportionate advantages in seal-lifetime terms for any given material. Greater attention to cooling and oil flow rates through equipment can have considerable benefits to actual under-lip temperatures which will frequently be 30–40°C higher than general sump temperatures.

Where systems cannot be changed and overheating has been identified, possible solutions can be found in highly developed liquid surface modifiers which are typically used in a ratio of 1:20 dilution with the specified lubricants. These surface modifiers are preferentially attracted to hot metallic dynamic interfaces and offer durable benefits.

15.2.3.5 Seal lubrication

In common with most other forms of dynamic seal, rotary shaft lip seals depend for their efficient and consistent performance upon the presence of a stable fluid film. The establishment of this film, which will typically be in the order of a few microns' thickness, may not occur immediately and it is not uncommon for a high percentage of wear to be observed on rotary shaft seal lips during this period before steady-state conditions are achieved. For this reason, it is always good practice to apply a lubricant to the seal lip and in the seal cavity before fitting.

Where single seals are housed at each end of the housing there is normally sufficient bearing lubricant in contact to provide adequate lubrication. Again, where two seals are housed together, it is often found that with bearing lubricant on one seal and fluid on the other, further lubrication is unnecessary. In all cases the liberal application of grease to the seals on assembly will ensure lubrication from the beginning, and in some instances this will be found sufficient to last from one fitting to the next.

However, there are many sealing arrangements in which two or more seals are fitted together in the same housing and there is the danger that at least one will run dry unless lubricant is supplied from an external source. This can be best accomplished by drilling a hole through which connects with an annular groove in the back of the seal housing, as shown in Figure 15.87, and using a special type of seal which has a series of radial ports in the base for passage of lubricant to the seal lip. When it is difficult or impossible to machine an annular groove in the back of the seal housing, this groove may be incorporated into the seals themselves as shown in Figure 15.88.

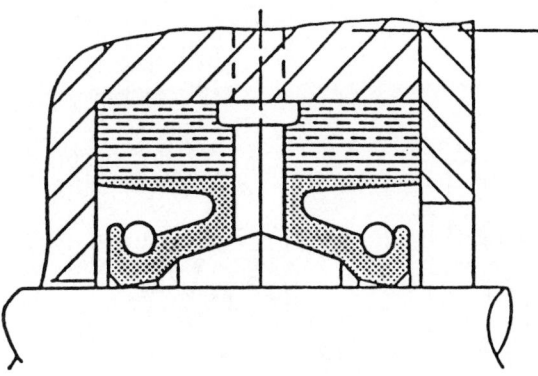

Figure 15.88 Seal lubrication via ports and annular groove in seal base

In general, a good-quality mineral oil or grease is suitable for seal lubrication, but molybdenized lubricants may be used to advantage where application is infrequent. Care should be taken to ensure that the grease or lubricant with which the seal is to come into contact is compatible. For example, where seals are being used with dilute mineral acids, butyl compositions may well be used. This material is likely to swell in contact with mineral oil or grease and an alternative lubricant will be essential.

15.2.3.6 Liquid/liquid interface

In submersible arrangements multiple seal assemblies are required to maintain oil and water separation. Figure 15.89 shows a typical marine stern gland where five seals are used to separate sea water from bearing lubricating oil. In this situation it is difficult to ensure the correct liquid/gas interface necessary for the establishment of stable hydrodynamic film conditions. Where liquid/liquid interfaces develop, the seal will see more contact with the shaft than intended and severe under-lip temperature hikes can result.

If the seal material has absorbed any liquid (usually water) vaporization can occur, leading to the formation of thin-skinned blisters. Rotation of the shaft removes this skin and pockmarking of the running face will result (see Figure 15.90).

Catastrophic failure can follow in the most severe cases. To resist these effects it is necessary to use a highly developed FKM compound which is an oil-industry benchmark standard to resist explosive decompression in 'O' rings and other seals subject to gas mixture absorption and subsequent rapid decompression. The mechanisms of explosive decompression and lip blistering are similar and can be resolved with the same FKM material.

15.2.3.7 Shaft surfaces

The sealing area of the shaft should be a fine ground finish of 0.4–0.8 μm R_a for most applications but, for the higher speed range, it is recommended that the surface finish be improved to 0.2–0.4 μm R_a In all cases it is important that the shaft sealing area be free from machining marks, dents, burrs and scratches. Where fluorocarbon or acrylic rubber seals are being employed, it is also advisable to use the finer level of finish indicated above in order to eliminate pick-up of the seal material. If lubrication is adequate and free from abrasive contact, unhardened mild steel shafts will generally give satisfactory results under normal operating conditions.

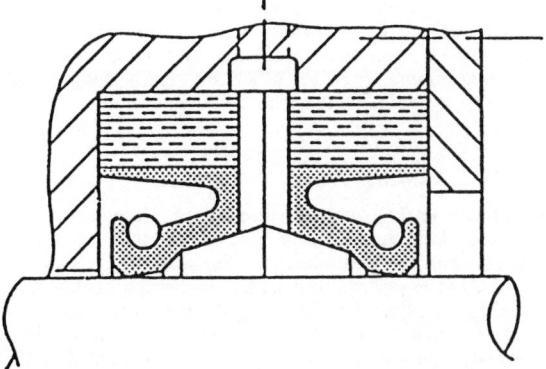

Figure 15.87 Seal lubrication via ports in seal base

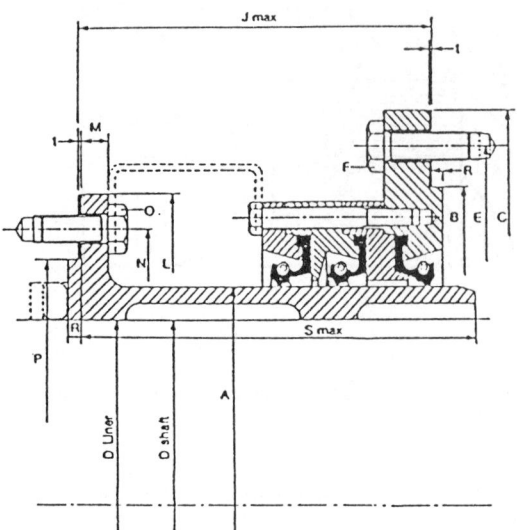
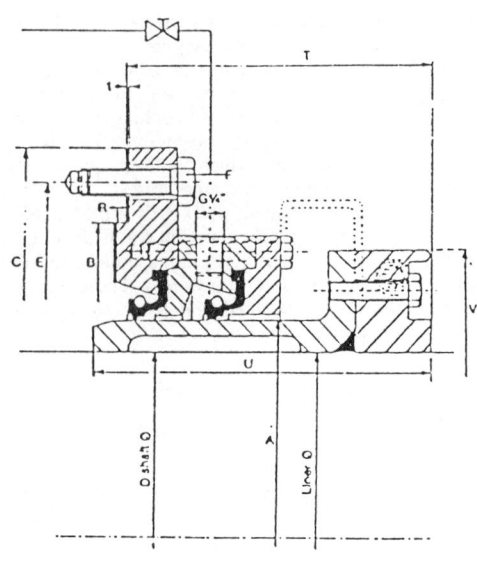

Figure 15.89

Figure 15.90

However, a harder shaft material is to be preferred for applications where lubrication is poor, abrasives are present, or speed and pressure conditions are particularly arduous. A fine machined finish is suitable for the housing bore.

Avoid methods of finishing shafts which produce unidirectional spiral patterns. These have frequently caused external 'pumped' leakage of the fluid being sealed.

15.2.3.8 Friction

The rubbing friction of the seal lip on the shaft material inevitably causes a higher local temperature than is present in the fluid being sealed. This effect will be due to the interference of the seal lip on the shaft material and may be aggravated by inadequate lubrication or speeds in excess of that for which a particular seal was designed. Assuming the system temperature to be well within the bounds of the material capability, a high differential between lip-tip temperature and fluid temperature will be manifested in many cases by hairline cracks in the seal lip co-axial with the shaft. Local carbonization of the rubber compound will also be evident.

Figure 15.91 indicates a set of curves derived by experiment, projecting frictional power loss against seal diameter for a conventional rotary lip seal design operating at 500 rpm in mineral lubricating oil at different system pressures.

15.2.3.9 Speeds

Many of the fabric-backed type of seals together with those of the metal-supported type are working satisfactorily on 400 mm shaft diameters at speeds of up to 25 m s^{-1} over sustained working periods. There are, however, many factors such as surface finish, eccentricity and lubrication which can limit the maximum speed for which any seal is suitable and manufacturers should be consulted in cases of doubt.

15.2.3.10 Eccentricity

Where plain metal or synthetic composition bearings are being employed, it is not uncommon to find that the shaft is not truly concentric with the seal housing owing to bearing clearance and wear. In this event, it is essential for efficient sealing that the seal lip be capable of following all shaft movement and, indeed, on many large comparatively slow-moving shafts, eccentricity values of 2.5 mm have been satisfactorily accommodated. Naturally, the seal performance in terms of eccentricity capability will be speed and shaft diameter dependent.

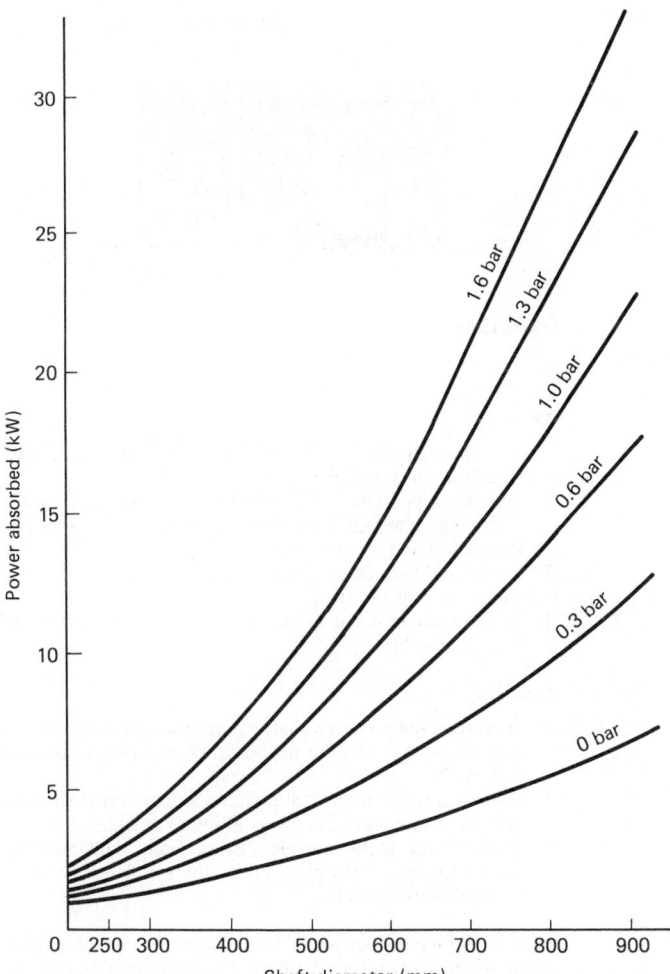

Figure 15.91 Frictional power absorbed rotary shaft lip seals, having conventional lip interference, operating at 500 rpm. (These values are conservative and are given as a worst-case guide)

Where split seals are fitted, then particular consideration is necessary to the problem of shaft eccentricity, since there may be a tendency for a split seal to open at the join. (See Section 15.2.3.14.)

15.2.3.11 Pressure

Although few rotary shaft lip seals are specifically designed as standard components to accept significant pressure, the use of metal supporting plates will extend the usefulness of this type of seal. A typical profile is shown in Figure 15.92. As a result of experimental work on the sealing of oil-filled marine stern glands and manoeuvring thrusters, a seal lip profile has been developed which satisfactorily sustains pressures up to 4.0 bar without the use of a shaped support plate. The base must be fully supported as indicated in Figure 15.93.

Where constant applied pressure is not anticipated it is frequently suggested that grease-lubricated bearings are equipped with relief holes and that oil-lubricated bearings have drains of adequate size. Drains taken from the ends of the bearing near the seals will help to dissipate any localized

pressure build-up. Where possible, steps should be taken in bearing design to prevent escaping high-pressure oil impinging directly upon the seals.

In some cases, where the loss of a small amount of grease is unimportant, a simple relief system may be formed by facing seals away from the bearing and allowing the seal lip to be lifted under the influence of the lubricant pressure.

15.2.3.12 Flexible throttle bush

There is a unique and patented development which can throttle system pressure to lip seal positions in a wide variety of pressurized transmission arrangements. Operating on a simple pressure balance principle that tilts the bushing to give a minimum exit clearance on the fluid film beneath it (see Figure 15.94) the product is used in conjunction with a vertical drain to tank which leaves the lip seal functioning against slack pressure. This advantage has widespread implications to the PV capability of seals in high-speed gear trains at the expense of a small additional requirement in axial length of the total seal assembly. (See Figure 15.95.)

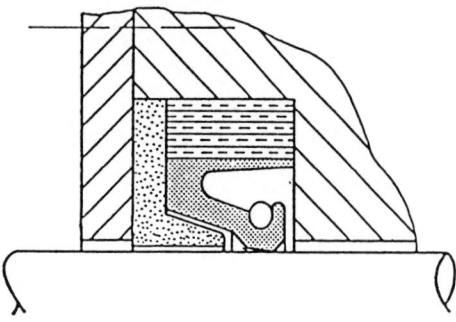

Figure 15.92 Seal with shaped support plate for pressures up to 3.0 bar

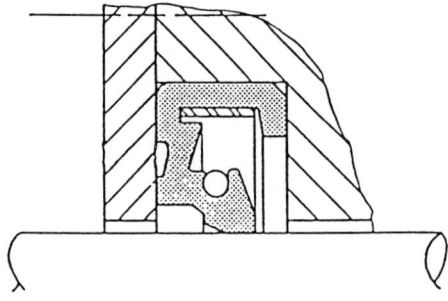

Figure 15.93 Seal development for pressures up to 4.0 bar without shaped support plate

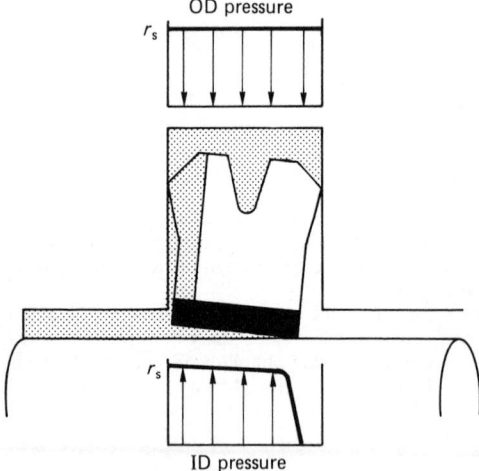

Figure 15.94

15.2.3.13 Care and handling

Fluid seals should be handled at all times with extreme care since the life of bearings or other costly machine parts may depend upon their efficiency. Attention to the following vital points will assist in ensuring trouble-free operation during service.

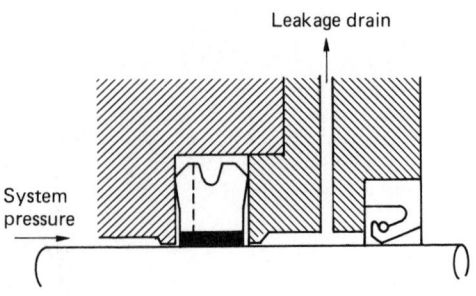

Figure 15.95

Storage

1. The store should have a cool, clean and dry atmosphere, free from dust and grit.
2. Whenever possible, seals should not be removed from the wrapping in which they were supplied as this provides protection and identification.
3. Avoid untidy stacking as the weight may distort the seals at the bottom of the stack.
4. Seals should never be threaded on wire or string as this will damage the lips.

Handling

1. It must be remembered that seal lips are extremely vulnerable to damage and the smallest nick provides a potential leak path.
2. Seal-reinforcing inserts, although adequate for their duty, may deform under adverse handling or stacking.
3. Seals having metal outside surfaces may damage other seals, especially if the metal edges contact the rubber parts of neighbouring seals.

Fitting A high proportion of failures and leakage of oil seals is due to incorrect fitting resulting in damage to both seal and sealing surface. Strict attention to the following matters is essential if best performance is to be obtained:

1. Before fitting, the seal should be examined to ensure that it is clean and undamaged.
2. The sealing lip should be smeared with suitable clean lubricant. Seals used as dust excluders should be packed with a compatible grease.
3. Normally, the sealing lip should face the fluid to be sealed.
4. When fitting, it is important to ensure that the sealing lip is not damaged even by the slightest nick, that the spring is correctly located when in position and that the seal is properly pressed home into the housing recess.
5. Examine the shaft, which should be free from all roughness and sharp edges and avoid passing the sealing lip over keyways, screw threads or shoulders. Shaft edges or shoulders should be well rounded or chamfered and where this is not practicable, a fitting sleeve slightly larger than the shaft with a lead-in taper should be used.
6. According to the type of assembly, it may be necessary either to press the seal first into the housing and subsequently on to the shaft, or alternatively, to pass the seal over the shaft and then press it into the housing. It is preferable first to mount the seal on the shaft where circumstances permit, since this allows observation of the lip during assembly.
7. The assembly should not be allowed to rest for any length of time at an incomplete state of fitting where the weight of

the shaft or housing may be borne by the seal, resulting in damage or distortion to the latter.

8. When pressing the seal into the housing a uniform pressure should be exerted, preferably by means of an arbor press in combination with a suitable tool. The diameter of the tool should be slightly smaller than the diameter of the housing by 0.1–0.4 mm. The outside surface of the seal can be smeared with a suitable lubricant in order to facilitate fitting. Care must be taken to ensure that the seal does not enter the housing recess in a tilted position since this will cause damage to the outer surface.

Figure 15.96 shows a range of typical fitting tools.

15.2.3.14 On-site joining technique

The advantages in terms of maintenance and downtime of having split and abutted lip seals is obvious, since extensive dismantling is avoided. The performance of these fabric backed split lip seals – where the abutment is located through a solid rubber inlay – is frequently acceptable. However, where shaft dynamics are severe, greater leakage may be encountered than with the fully moulded endless product.

Where such leakage is unacceptable but the benefits of assembling a split seal are also required, a new on-site joining method has been developed which precludes the need for costly and cumbersome vulcanizing methods. The kit comprises:

- A split fabric back seal adapted to incorporate a plastic circumferential insert at its join interface.
- Two-part epoxy adhesive in measured quantities. The epoxy cures to a semi-rigid state.
- A joining jig in silicone rubber precision moulded to the exact internal profile of the seal.
- A steel adjustable clamping band.

Fitting tool for use when front of seal is fed into housing

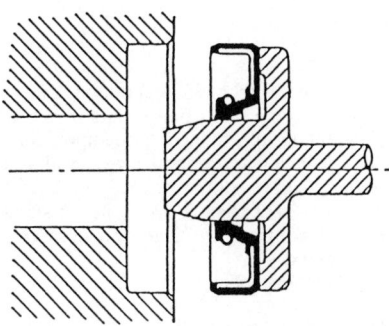

Fitting tool for use when rear of seal is fed into housing.
If necessary, remove spring where possible and refit when seal is assembled

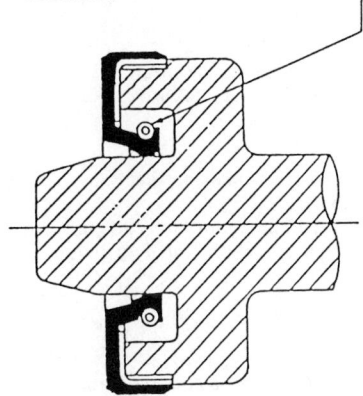

Method of fitting seals

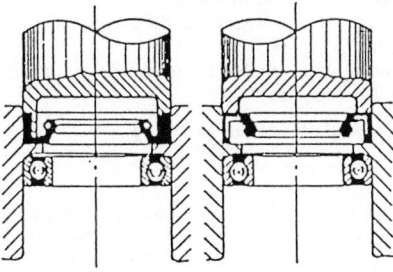

Without shaft

Fitting seal over a splined shaft

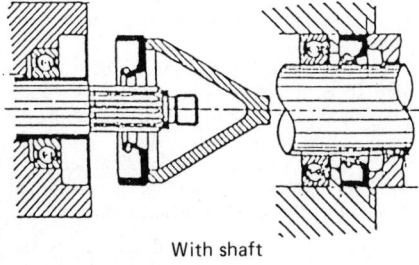

With shaft

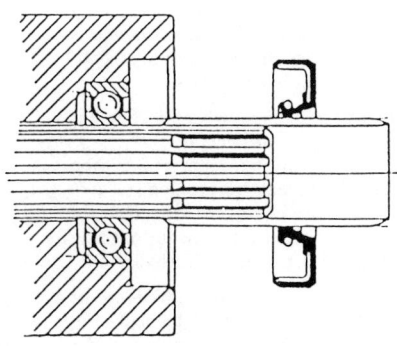

Figure 15.96 Fitting tools for seals

Figure 15.97

- A temperature indicator with self-adhesive backing.
- A cleaning pack comprising degreasing cloth, abrasive stick and self-adhesive backed emergy paper.
- An optional hot air gun.

Full training in the use of the technique is made available by the manufacturers. See Figure 15.97.

The following two sections detail a number of fault-finding procedures which, if taken in sequence, should analyse the reason for a given difficulty.

15.2.3.15 Trouble-tracing

Unacceptable leakage This is almost always associated with oil-lubricating bearings, since grease is not a difficult lubricant to seal. The term 'unacceptable' can have wide interpretation since an occasional drop of oil might be disastrous if it resulted in contamination of the product being handled by the machine concerned in such spheres as the textile, paper or food industries, while it would probably remain unnoticed in a heavy industrial environment.

Cost and consideration for the environment have caused much greater sensitivity to oil losses in any form and leakage rates that hitherto have been ignored are now becoming regarded as unacceptable. When dealing with such complaints it is essential to discover the history of the equipment concerned and this broadly falls into three categories.

I   New equipment recently commissioned, where sealing has been regarded as unsatisfactory from the start;
II Equipment that has been in operation for a period of time and only recently has developed leakage problems;
III Equipment that was satisfactory during its first term of operation but leakage has occurred after fitting replacement seals during routine maintenance or overhaul.

Since trouble tracing is basically a process of eliminating or substantiating suspected faults, the sequence of checks required would vary with each of the above categories.

In order to simplify the procedures and avoid irrelevant investigations, the recommended sequences for each of the above categories are defined by letter symbols to be used in conjunction with the fault-finding chart which follows.

Category I   Full checks in order as A, B, C, D and E until fault is discovered.

Category II   A, B, C(1), D(2), and D(3). If faults as C(1) or D(2) are exhibited, ascertain period of service with seals. This should be calculated in terms of hours of running and related to speed, temperature and other environmental conditions. A moderate speed with good clean lubrication conditions and ambient temperatures would normally anticipate a seal life of around 10 000 hours. High speeds, poor lubrication, elevated temperatures or partially abrasive media could reduce this to as little as 2000 hours. The problem may therefore be simply that of being due for seal replacement.

If faults as in D(3) are in evidence, obtain details of all media in contact with seal (including any cleaning fluids) as a change of lip material may be necessary to obtain compatibility. If seals display no faults check for mechanical defects as E(1) and E(2).

Category III   A, B, C(1), C(3), D(1), D(3), D(4), E(1), E(3).

15.2.3.16 Fault-finding chart

(A) Is leakage actually occurring from the seal or does it stem from such sources as bearing cover flanges and is merely 'collected' by the seal housing, giving a false impression? Check by wiping clean all appropriate areas and run machine to ascertain leakage source.
(B) Is leakage from around seal back or from the lip along the shaft? Check by wiping both clean and observing while machine is running.
(C) If O/D leakage – check the following:
  (1) Is seal a good fit in housing or is it slack? (On split seals a slack seal will display a gap between seal ends.)
  (2) If housing bore is correct size then seal O/D dimension is suspect if slack in housing.
  (3) If seal is good fit in housing, check for damage on housing bore.
  (4) Check housing depth to ensure seal is being axially compressed – if applicable.
(D) If leakage along shaft check the following:
  (1) Shaft size, surface finish, shaft damage at contact area.
  (2) If (1) okay, check condition of sealing lip for hardening and/or cracking. If either in evidence, then speed or temperature conditions are probably incompatible with seal material.
  (3) If lip is soft or swollen this is usually an indication of chemical incompatibility with the media in contact with the seal.
  (4) If (2) and (3) okay, check section width of seal with spring fitted. This should be at least nominal section + 1% immediately on removal and increasing to nominal + 3% after 1 hour in free state. Spring may be shortened by up to 5% of its original length if section appears inadequate. Section measurement should be average of four equidistant readings.
(E) If checks (C) and (D) do not reveal any faults the problem may be due to mechanical conditions and the following should be checked:

(1) Shaft to housing concentricity – check by means of callipers between shaft and housing bore at four points around periphery. Variations of more than 0.3 mm require further investigation.

(2) If smallest calliper measurement occurs between bottom of shaft and housing this may indicate bearing wear with resultant dynamic eccentricity. (Applicable to horizontal shafts only.)

(3) If bearing is okay, then housing offset may be responsible. Seal housings are normally centralized with the shaft by means of a machined register with the bearing housing. Where this feature is not incorporated then it may be possible to centralize the housing by slackening the bolts and repositioning.

(4) If (1), (2) and (3) are blameless then the following requires investigation. If bearings are oil pressure lubricated, is there adequate drainage to prevent pressure build-up against the seal? If equipment operates on a constant oil level principle, are there gear-wheels or ball-journal bearings in close proximity to the seal causing oil turbulence or flooding? Where the latter situation exists, the housing lands should be only marginally larger than shaft diameter in order to form a baffle or, alternatively, a baffle plate fitted between bearing and seal housing. A temporary remedy can be made by using a 1.5 mm semi-rigid gasket at the bottom of the housing, the I/D of which should be shaft diameter plus 0.5 mm maximum. A further gasket of equal thickness should be fitted underneath the retaining plate to restore the correct amount of axial compression in the case of retained seals. (See Figures 15.98 and 15.99.)

15.2.4 Mechanical seals

15.2.4.1 Introduction

Lip seals cannot be used in many rotary shaft sealing applications. This is especially the case for fluids other than oil or when the sealed pressure is greater than 2 bar, but other factors can also be important. The most commonly used high-pressure rotary shaft seal is the mechanical seal, where the dynamic sealing occurs between the faces of a pair of very flat annular rings, one stationary and the other rotating against

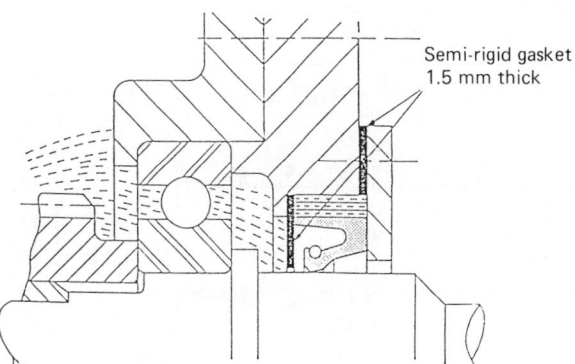

Figure 15.99 How the fitting of gaskets can provide a temporary remedy for the situation in Figure 15.98

it. Mechanical seals are to be found in large numbers worldwide on rotary pumps (ranging from high-duty refinery applications to car-water and washing-machine pumps), compressors, mixers and ship stern-tubes. Mechanical seals have become an industry standard in many areas, frequently replacing other rotary seals such as soft packing and bushings. They have a deserved reputation for low leakage and low maintenance costs compared with other seal types. However, mechanical seals are precision devices which require certain actions on the part of the user to improve the chances of obtaining good performance on any given duty. Chief among these are: correct seal selection, careful installation by trained personnel, and an understanding of the effects of external factors out of the control of the seal manufacturer.

A large number of seal types and variants are commercially available to the user – knowing where to start can be difficult. Most users would probably approach a reputable seal manufacturer, pass on what basic process data were available and let the manufacturer select a seal. Close liaison with seal manufacturers is to be encouraged and is usually the best way to enhance reliability in the field. However, it is also to the benefit of both manufacturer and user for the latter to understand the basic construction, operation, expected performance and requirements of the seal. The manufacturer will also be considerably aided in selection if the user is primed with appropriate data regarding the process, pump, ancillary equipment, installation procedures and any past failures. In particular, an awareness of the API 610 code of practice for centrifugal pumps will be helpful.[23] This document has been compiled by US oil producers for use on oil refineries but is widely regarded as a standard by the mechanical seal manufacturers. API 610 sets out requirements for subjects such as seal cooling, geometric envelopes, piping plans, inspection and testing, and pump data sheets.

It is the intention of the following text to equip the seal user with sufficient knowledge to enable effective liaison with seal manufacturers. Where further information is sought, the reader is referred to more detailed references pertaining to the subject. Reference 24 is particularly useful.

15.2.4.2 Seal design

Seal operation The function of a mechanical seal is to prevent the sealed process fluid escaping from the pump body at the point where the impeller drive-shaft enters. The seal is therefore required to operate under conditions of extreme

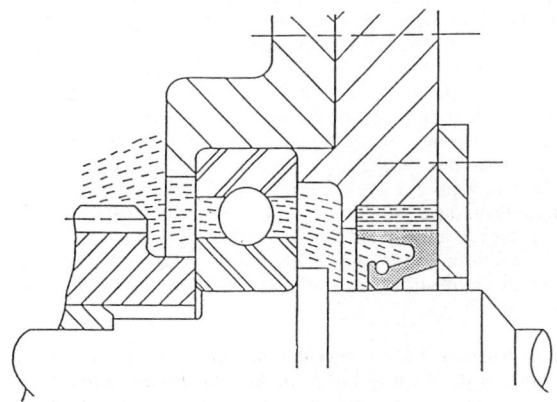

Figure 15.98 How too large a bore diameter of a seal housing land permits high-velocity oil impingement on a seal. A baffle plate between bearing and seal or a close-fitting land seal would reduce risk of oil leakage

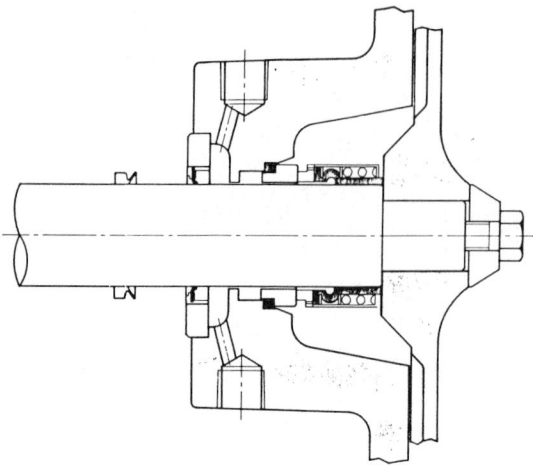

Figure 15.100 Typical mechanical seal installation. (Source: Girdlestone Pumps Ltd)

temperature (both high and low), and run on liquids which are possibly toxic, flammable and corrosive. Sealed pressures can commonly be up to 70 bar and occasionally higher. Often the actual conditions of pressure and temperature in the seal chamber are not known. Understandably, users expect virtually no leakage and reliable operation with predictable life typically in excess of 3 years.

A typical mechanical seal system as installed on a centrifugal pump is shown in Figure 15.100. The mechanical seal is, in essence, a plain annular thrust bearing. Despite the plethora of available seal types, the mechanical seal generally comprises the basic components shown in Figure 15.101. The dynamic sealing takes place in the sliding interface between the stator and the rotor, other leakage paths being sealed with stationary or pseudo-stationary seals. Unlike a conventional bearing, the mechanical seal usually has to run on the pumped fluid which encompasses a wide range of possibilities all with varying degrees of lubricating ability. Moreover, the interface lubricating film is very thin, generally in the order of a few microns. Consequently, the profiles, surface finish and materials of the seal faces are of crucial importance.

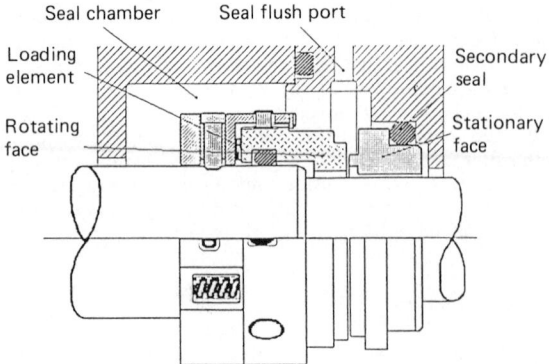

Figure 15.101 Basic features of a mechanical seal. (Copyright BHR Group Ltd)

Sealing interface The stability of the interface fluid film is vital to successful seal operation; should it disappear through vaporization and the seal run dry for any length of time, eventual failure is likely. This will be caused by face material distress, secondary seal degradation due to the high temperatures generated, or high face wear. Consequently, a large amount of research has been concentrated on understanding the behaviour of the sealing interface.

The faces of a mechanical face seal need to be flat to within a few tenths of a micron. This is achieved by lapping and their flatness checked using an optical flat. Surface roughness is normally in the range $0.05–0.20\ \mu m\ R_a$.

The faces are able to move axially and also cone inwards or outwards. The loading element provides a nominal closing force on the face when the seal is stationary. This force is usually designed to be small compared with the hydraulic loading of the seal in operation. Hydraulic forces can be controlled by the seal manufacturer by choice of 'hydraulic balance' or 'balance ratio'. These terms are explained with the aid of the diagrams in Figure 15.102. A seal with a balance ratio of 1 or greater is termed an 'unbalanced seal'. This means that the seal faces experience the full pressure of the pumped fluid. Unbalanced seals are simple and low-cost but cannot operate at pressures much in excess of 10 bar as the high interface loads cause vaporization of the interface film. Most process seals are of the 'balanced' type, that is, they have a balance ratio of less than 1; a ratio of 0.7 being not uncommon. By modifications to the geometry of the seal assembly the net interface area over which the system pressure acts can be reduced by balancing the pressure-induced forces on the seal as shown in Figure 15.102. The added complexity results in more expensive seals but extends the operating pressure envelope of the seal considerably. Choosing the optimum ratio for a particular application can be difficult and requires skill and judgement on the part of the seal manufacturer to obtain the best compromise of face loading and film thickness. Reducing the face loading allows the film thickness to increase and, consequently, the leakage will be greater. Conversely, increasing the face loading leads to thinner films, greater heat generation and an increased dependence on the tribological performance of the seal faces.

The hydraulic force generated by the sealed liquid not only loads the faces axially but also generates a turning moment about the cross-sectional centroid of the seal rings and causes them to cone. Where the sealed liquid is on the outside (the case with most seals) the moment tends to cone the faces inwards to cause contact at the outer edges as shown in Figure 15.103. This is denoted as 'positive' rotation and sets up a face taper which is divergent in the direction of leakage, thereby cutting off the interface film. Seals are also subject to thermal distortions, globally due to bulk warming of the surrounding pumped fluid and also locally due to heat generation at the interface. On most seals, the effect of heating distorts the seal faces to rotate 'negatively' to form a convergent taper as shown in Figure 15.104. Ideally, the pressure-induced and thermally induced rotations should be kept in balance so that the faces run as near parallel as possible. This can be achieved for steady-state running at a particular design condition, but tolerance to off-design or transient conditions is more difficult to engineer.

Face materials Most mechanical seals experience short periods of dry-running and boundary conditions during their lives, due either to the arduous nature of a duty or to transient conditions such as start-ups and process interruptions. The tribological compatibility of the face materials and their ability to survive when in rubbing contact is of paramount importance for long-term seal reliability. In the majority of cases this is

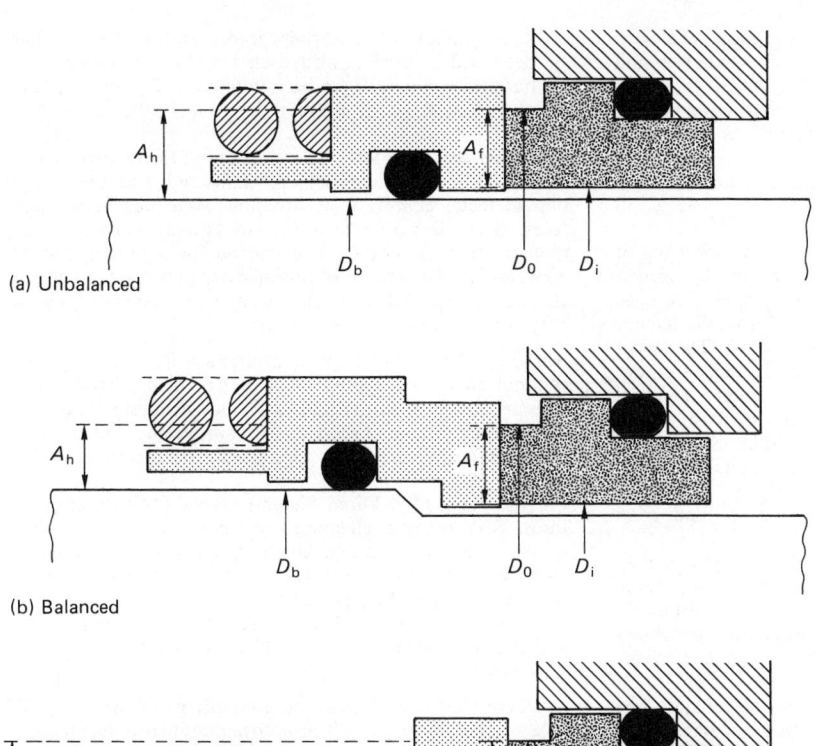

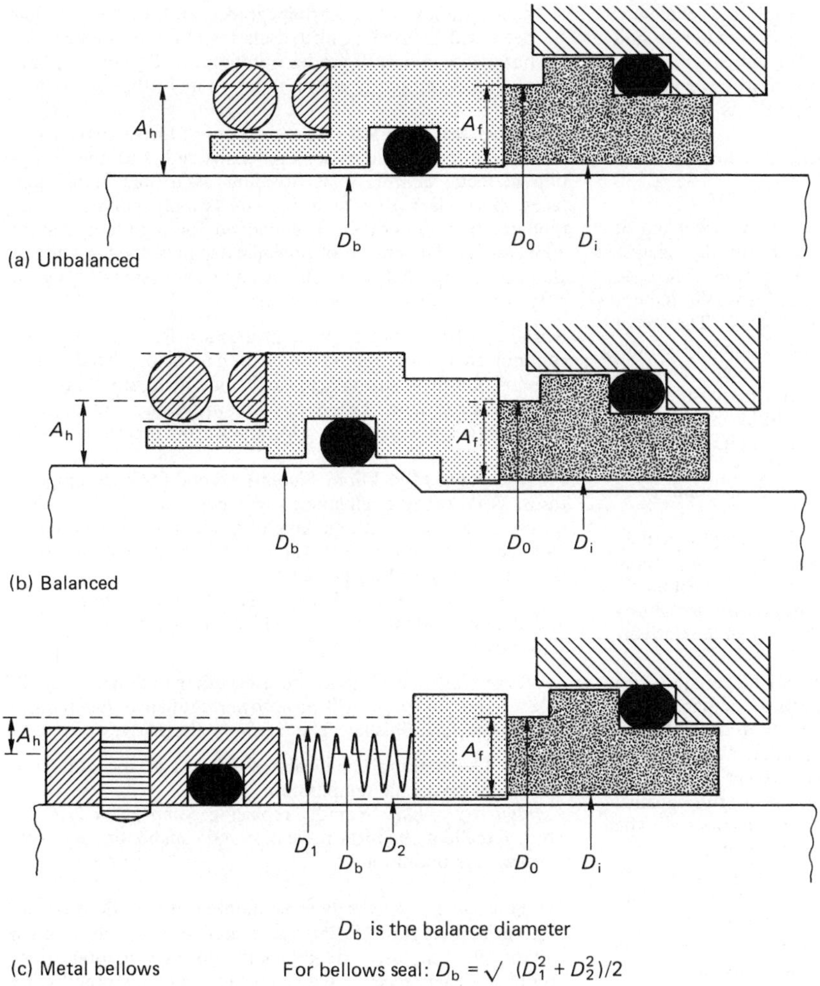

(a) Unbalanced

(b) Balanced

(c) Metal bellows

D_b is the balance diameter

For bellows seal: $D_b = \sqrt{(D_1^2 + D_2^2)/2}$

Figure 15.102   Definition of seal balance. (Copyright BHR Group Ltd)

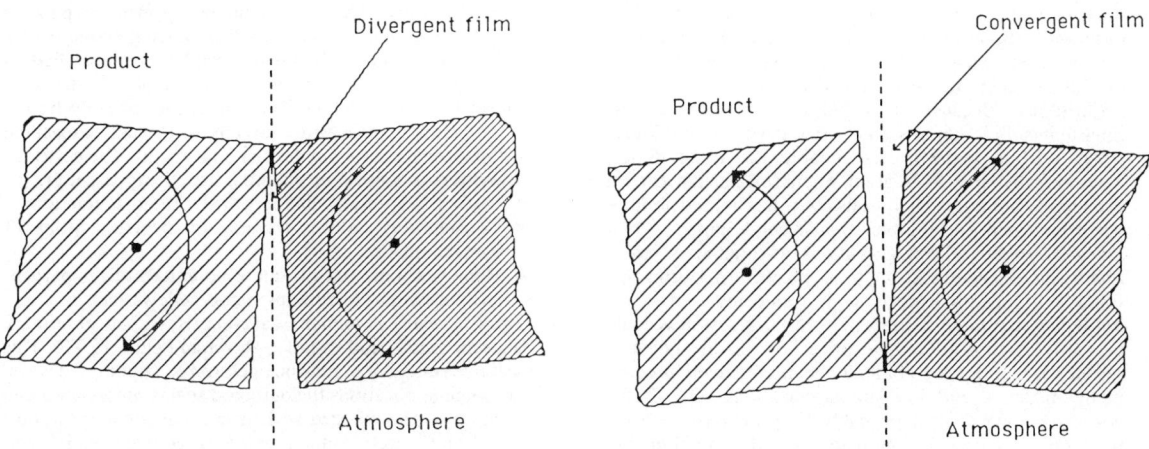

Figure 15.103   Effect of pressure on the sealing interface. (Copyright BHR Group Ltd)

Figure 15.104   Effect of temperature on the sealing interface. (Copyright BHR Group Ltd)

satisfied by using a high-quality carbon-graphite as the 'soft' face. Counterfaces are always harder and stiffer materials and include metals, ceramics and cermets. On duties where the pumped fluid is particularly abrasive, two hard faces may be run against each other, but such combinations are totally dependent on liquid lubrication.

The most commonly encountered seal face materials are summarized as follows:

1. *Carbon-graphite* This terms covers a broad spectrum of different carbon-based materials which, in the sealing industry, are referred to colloquially as 'carbons'. The base grade carbon-graphite has many properties which make it an excellent choice for mechanical seals. The most important are: good lubrication under boundary or dry conditions, tribological compliance, good chemical resistance, wide temperature range and low cost. To be impermeable to fluids, the carbon has to be impregnated – the impregnant chosen depends on the application being considered.
   - Resin-impregnated grades: wear resistant and good chemical resistance.
   - Metal-impregnated grades: better thermal conductivity and stronger than the plain grades. Enhanced running abilities. Metal impregnant limits chemical resistance.
   - Electro-graphite grades: high-temperature capability and good chemical resistance. Weaker and relatively soft.
2. *Ni-Resist* A generic term for particular nickel-alloy cast irons with enhanced chemical properties compared with cast iron. Ni-Resist material is popularly used in low- to medium-duty seals for which it is well suited. It is relatively low cost, easy to machine and fairly tolerant of transient and dry-running conditions. It is best run against resin-impregnated carbon. Ni-Resist is not a material to be used for physically or chemically arduous conditions.
3. *Alumina* Alumina ceramics are also widely used on low and some medium duties (typically at pressures less than 1.5 MPa) particularly where the duty involves aggressive or mildly abrasive or corrosive substances. It has very good wear resistance and performs well on aqueous duties. It is, however, rather intolerant of shock, transients and dry running. Its thermal conductivity is low which is a major drawback for more arduous and critical duties.
4. *Tungsten carbide* A cermet material known for its good performance under arduous and boundary conditions, tungsten carbide has been employed extensively on high-duty applications for many years. Its initial high cost is rapidly offset by its enhanced performance. Improvements in seal life have also encouraged its use on less severe applications. Because it includes a metallic binder, tungsten carbide does suffer from corrosion problems which tend to make it less suitable for acidic duties. Tribologically, it is very good when run against carbon even under boundary conditions but it has limited abilities when run against itself under similar conditions.
5. *Silicon carbide* Silicon carbide is a ceramic which is rapidly increasing in use due to its excellent performance. Initially, its high price precluded it from being used on any but the most arduous duties where cost-effectiveness could be easily demonstrated. The price has steadily fallen in recent years and it has gradually been effectively used on many medium- and low-duty applications. Of all the counterface materials it probably comes closest to being the 'industry standard'. The material itself is anything but standard as it is available in a number of grades, with each material within a grade varying between manufacturers. It does have some limitations particularly on strongly al-

kaline duties where certain grades are attacked. It has been widely used against itself for abrasive applications but again, choice of grades is important. Careful handling of silicon carbide is essential due to its brittleness.

Secondary seal materials The functions of the secondary seal are to seal the secondary leakage paths and, with the exception of metal bellows seals, to allow axial float of the seal faces. Secondary seal materials are usually polymeric, although other materials are employed for high-temperature applications. The choice of polymer depends on thermal and chemical compatibility. Strengths and weaknesses of selected polymeric materials are as follows:

Nitrile (NBR) Good physical properties. Resistant to many common chemicals. Should not be used on halogenated hydrocarbons, ketones, strong acids and some hydraulic fluids.
Ethylene propylene Good physical properties. Particularly good for hot-water duties. Do not use on petroleum-based fluids.
Fluorocarbons (e.g. Viton, Fluorel) Good for high temperatures. Wide range of chemical resistance available depending on the grade – correct matching of grade to duty is essential. Not recommended for hot water, ketones or alcohols. Relatively poor physical properties.
Perfluoroelastomer (e.g. Kalrez) Excellent high-temperature and chemical abilities. Relatively poor physical properties. Expensive.

Where elastomers cannot be used other options are available – generally these will have lower resilience and require superior surface finishes. The common alternatives are:

PTFE Outstanding chemical resistance and the ability to operate at cryogenic temperatures.
Exfoliated graphite Rapidly replacing compressed asbestos fibre. Excellent chemical resistance and capable of very high-temperature operation.

Design variants A considerable number of variations on the basic design (Figure 15.100) are available to the user. Each variant offers the user added benefits for a given application; optimizing the precise design and balancing their strengths and weaknesses is best done by the seal manufacturer. The user should be aware, however, of the principal variants available and their respective generic advantages and disadvantages; these are listed in Table 15.16.

1. *Cartridge seals* These are enjoying growing popularity despite their high initial cost. The cartridge comprises a self-contained mechanical seal which has been installed on the bench by the seal manufacturer; a typical cartridge is shown in Figure 15.109. The seal is set correctly by the manufacturer and is fitted directly to the user's machine without the need for any further adjustment. The cartridge seal offers a number of potential benefits to the user.
   - Option for pre-installation tests by the manufacturer;
   - Considerably reduces the risk of incorrect fitting or damage on-site;
   - Removes some out-of-square potential; this is particularly of benefit to rotating metal bellows seals.
2. *Tandem seals* A tandem seal comprises a pair of primary seals mounted head-to-tail as shown in Figure 15.110. This sealing arrangement may be operated in a number of ways. On some applications the outboard seal is simply used as a back-up to the inboard seal. In normal running the outboard seal merely contains any leakage from the inboard seal, but should the latter fail then the outboard seal will accept the full system pressure and seal the pump until a suitable maintenance window can be scheduled. Alterna-

Table 15.16

Variant	Advantages	Disadvantages
Externally mounted seal (Figure 15.105)	Easy installation and inspection. Minimal components in contact with pumped fluid, hence used on corrosive duties.	Not suited to high pressures. Face cooling more difficult. Leakage containment difficult. Space required outside housing.
Stationary floating seal (Figure 15.106)	Can be used at high speeds. More tolerant of misalignment. Suitable for liquids of higher viscosity.	Greater radial space required. Generally more expensive. Can become clogged by solid particles in product.
Non-metallic bellows seal (Figure 15.107)	Low cost. No hang-up or sleeve wear. PTFE bellows can be used for corrosive duties.	Less robust. Limited pressure capability. Not recommended for high temperatures and speeds.
Metal bellows seal (Figure 15.108)	No sliding secondary seal, hence no hang-up or sleeve wear. Can be used at high temperatures and speeds. Inherently balanced. Compact	Expensive. Less robust. Not generally recommended for highest pressures. More prone to fatigue failure especially with high vibration.

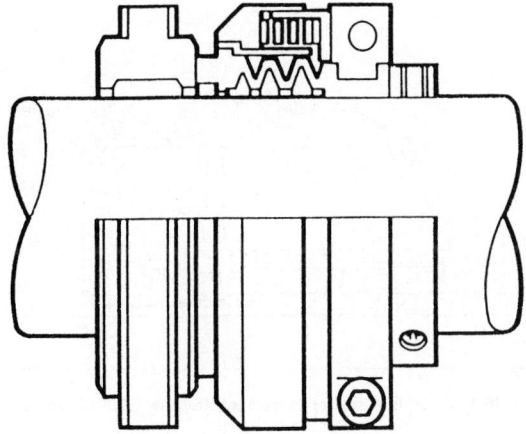

Figure 15.105 Externally mounted seal. (Source: John Crane UK Ltd)

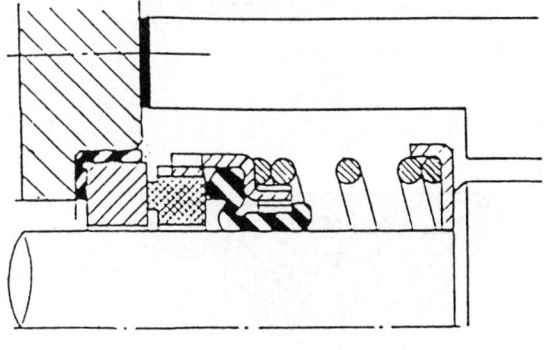

Figure 15.107 Non-metallic bellows seal. (Source: EG & G Ltd – Sealol Division)

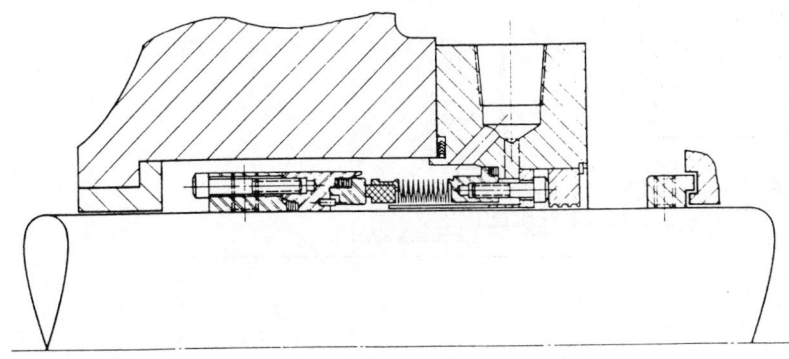

Figure 15.106 Stationary-floating seal. (Source: BW/IP Mechanical Seals Division)

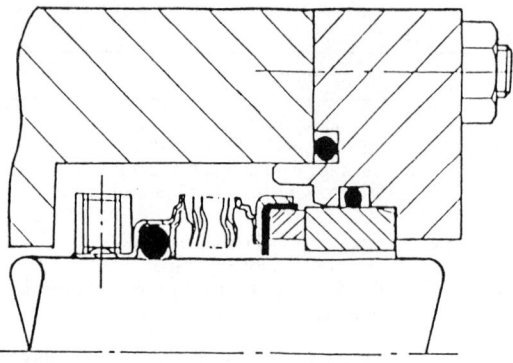

Figure 15.108   Metallic bellows seal. (Source: EG & G Ltd – Sealol Division)

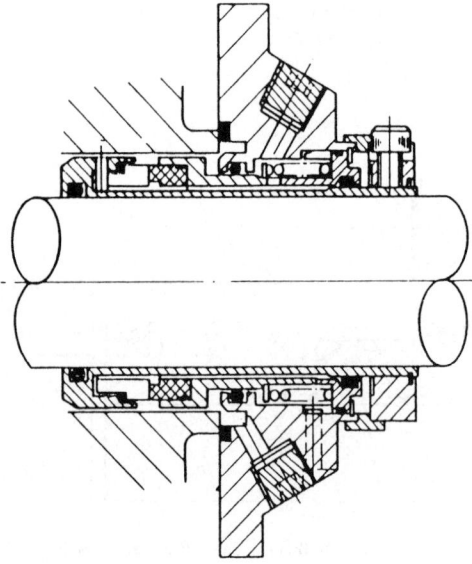

Figure 15.109   Cartridge seal. (Source: Flexibox Ltd)

tively, the tandem seal can be used with a pressurized inter-stage or buffer fluid which reduces the pressure drop across each seal. In this way the seals are sharing the workload, thus permitting the sealing of very high product pressures.

3. *Double seals*   A double seal also utilizes a pair of seals arranged in a back-to-back configuration as shown in Figure 15.111. The appropriately selected buffer fluid is pressurized externally at a pressure just above that of the process and, consequently, there is a slow leakage of buffer fluid into the process. The double seal is used when zero leakage is required; any product leakage into the buffer fluid can be dealt with using appropriate ancillary systems. Double seals offer enhanced reliability as the seal designer is able to select buffer fluids with good lubricating properties for the seals to run on – a rare luxury.

15.2.4.3 Seal performance

The two most important features of seal performance are leakage and life. The type of duty usually dictates the relative importance of the two although every user would ideally wish for low-cost leaktight seals which last, predictably, as long as the pump bearings. In general, this is unlikely to be achieved and it is worth noting that very few seals simply wear out – most 'seal failures' are due to excessive or unacceptable leakage.

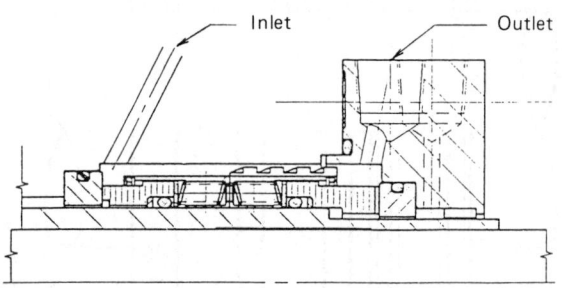

Figure 15.111   Double-seal arrangement. (Source: John Crane UK Ltd)

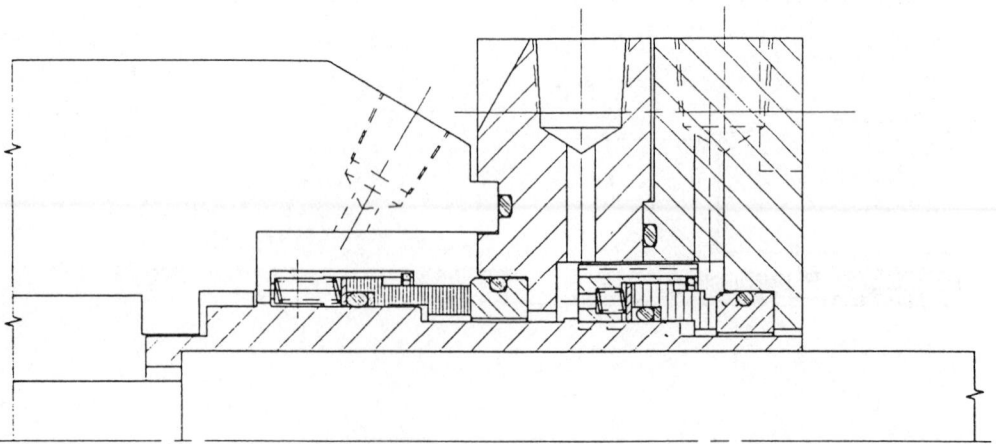

Figure 15.110   Tandem seal arrangement. (Source: John Crane UK Ltd)

Leakage All single mechanical seals leak to some extent, even if the leakage is in the form of invisible vapour. Indeed, vapour leakage can amount to quite high liquid equivalents which would be regarded as unacceptable on, for example, light hydrocarbon duties. Leakage can also be significant when the seal is at rest but still primed, as in the case of a stand-by unit. Typical leakage figures are difficult to give as seal design and size can be influential but, as a rule of thumb, values better than a few millilitres per hour can be achieved for a correctly functioning seal. On duties requiring truly zero leakage a double seal must be installed.

Life Seal life can be notoriously difficult to predict. Lives of thousands of hours are commonplace but cannot be guaranteed. The carbon wear rate on a correctly functioning seal will be of the order of 0.5 mm per year – a remarkable achievement giving the poor lubricating properties of many pumped fluids. There are, however, well-documented examples of a particular seal running successfully for many years and its identical replacement failing after only a few months for no apparent reason. Expected seal life varies depending on the duty, and the seal manufacturer should be able to give an idea of what to expect based on experience elsewhere. It should be emphasized that many factors affecting seal life are outside the direct control of the seal manufacturer. Principally, these are: the sealed fluid, seal cooling arrangements, pump design, and unforeseen process variations or disturbances.

Sealed fluid Apart from double-seal arrangements, the mechanical seal is required to run on the process fluid, however poor its lubricating capabilities. The sealed fluid can exert a major influence on seal performance by its effects on a number of processes inherent to the operation of the seal:

1. *Boundary lubrication of faces* Many process liquids have little or no boundary lubrication ability – the fact that the seal will often run successfully on these fluids is usually due to the careful choice of face materials. This contrasts strongly with bearing lubricants which are developed for optimum boundary performance by incorporating highly effective organic or inorganic additives. Thus, most aqueous duties are considered 'severe' in terms of the tribological performance expected of the seal.
2. *Volatility* In order to maintain a fully liquid film in the sealing interface, the liquid must not be allowed to evaporate. Vaporization of the film will allow solid contact between the faces which may experience thermal distress and high wear. The interface film is subject to local heating due to friction and liquid shear – its temperature is therefore above that of the bulk liquid surrounding the seal. It follows that a seal cannot run with a full liquid film when the temperature of the bulk fluid is within a certain margin of its boiling point or range; this margin is known as the 'Required Temperature Margin' or '$\Delta T_{required}$'. If the interface film is lost the seal can become unstable and will emit spurts of vapour leakage at regular intervals – this phenomenon is known as 'puffing'. A seal behaving in this way will often be obvious to the operator since the puffing is generally both visible and audible.
3. *Other properties* There are many other potential effects arising from properties of the sealed liquid, and the seal manufacturer will be in a position to take these into account. The more common ones are:
   - Dissolved solids can be deposited on the seal faces as the liquid evaporates. These particles can cause severe abrasive damage to the seal faces, particularly the carbon.
   - Liquid leakage can form crystals between the seal faces and on the atmospheric side of the seal faces. Unless

these are continuously removed by a quench, hang-up of the floating member is possible.
   - The light ends of heavy hydrocarbons can evaporate as they reach the atmospheric side of the interface and leave a tarry residue; this must be removed by quenching.
   - Fine suspended solids may be drawn into the seal faces and cause severe abrasive damage.
   - Attack of elastomeric secondary seals.
   - Corrosion of seal components.
   - Solids clogging moving parts such as springs or bellows.

Seal cooling Removing heat from the interface of the seal is vital to its survival. A 50 mm seal running at 3000 rpm and 30 bar can easily generate heat at a rate of one kilowatt in the sealing interface. The volume of liquid in the interface itself is very small and instant vaporization would occur were the heat not removed efficiently. Given that one or both of the seal faces has high thermal conductivity, the fluid flow around the seal must be able to convect this conducted heat away faster than it is being generated. The process of convective heat transfer relies upon a good supply of relatively cool liquid circulating in the vicinity of the sealing interface. Sometimes this can be achieved by the flow behaviour inherent to a particular housing, especially if the duty is dissipating relatively low amounts of heat. This is known as a 'dead-ended' arrangement (API 610 – Plan 01) which may incorporate a water-cooled jacket (API 610 – Plan 02). Unfortunately, the cramped conditions of most seal chambers inhibit the self-flushing process and additional predictable cooling is often provided by flushing the seal chamber.

Flushing arrangements or plans are many and varied; the US petroleum industry pump standard API 610 gives many variants. Basic cooling methods are:

1. *Product recirculation (API 610 – Plan 11)* Product is tapped off from the pump discharge and recirculated through an orifice to the seal chamber, usually entering opposite the seal faces (Figure 15.112). This is a good arrangement providing that the bulk liquid temperature is low compared to its boiling point (i.e. the available ΔT is well in excess of the required ΔT). In addition, the duty should be relatively free of solids which could block the orifice and cause erosion of the softer seal face. The restriction on solids can be obviated in many cases by incorporating a hydro-cyclone in the recirculation line (API 610 – Plan 31).
2. *External injection to seal face (API 610 – Plan 32)* Clean, cool liquid is supplied to the seal faces from an independent external source (Figure 15.113). This allows the seal to operate on more severe applications and makes it less susceptible to process fluctuations. The external flush must be compatible with the pumped liquid; this presents few problems on aqueous duties where water may be used, but may be more difficult for hydrocarbon applications. Some duties, particularly where slurry is being pumped, may experience excessive dilution by external flushing. Removing the flush liquid downstream can be costly. In some industries, sourcing a supply of consistently clean water can be difficult. Finally, it should be borne in mind that external flushing systems are relatively expensive, require careful maintenance, and are more susceptible to operator error than the self-flushing and product recirculation arrangements.

Pump design The effects of pump design are not well understood but circumstantial evidence suggests that seal performance can be impaired by out-of-condition pumps. The seal is the most sensitive part of the pump assembly and any prob-

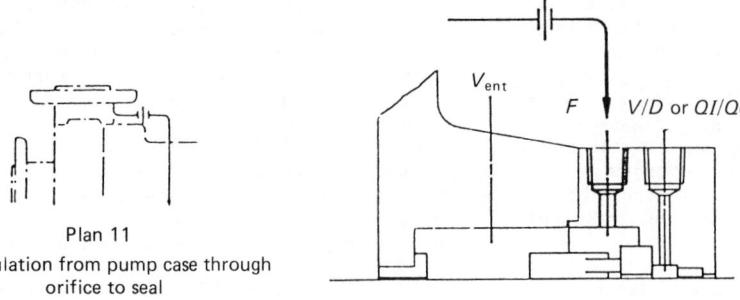

Figure 15.112 Product recirculation. (Source: reference 23)

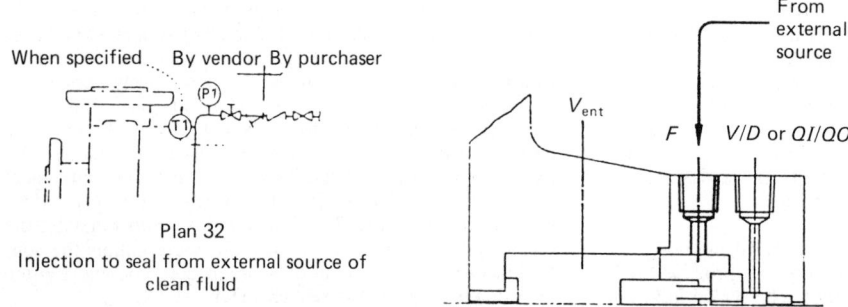

Figure 15.113 External injection. (Source: reference 23)

lems in this area are likely to lead to 'seal failures' even though the root of the problem lies elsewhere. The seal manufacturer should be consulted on the ideal requirements for pump condition and any information available from the user would be of use. In particular, the pump should be checked for:

- Tolerance stack-up which leads to out-of-square seal components or over-compression of the loading member;
- Shaft run-out due to a bent shaft or worn bearings;
- Pump operating away from the best efficiency point (BEP);
- Vibration at the pump due to coupling misalignment, impeller or motor imbalance, hydraulic effects or pipework-borne noise;
- Impeller design and its effect on seal chamber flows and pressures;
- Seal chamber geometry.

Process changes All seal manufacturers recognize that, at some point in the life of the seal, it will experience off-design running. Consequently, seal tolerance is a key area of development which has made significant advances in recent years. For example, it is often commented that the advent of silicon carbide as a counterface to carbon has improved seal reliability noticeably. It is not possible for the manufacturer to foresee, or design for, all fluctuations in the process and the user should be aware of the most common scenarios which can cause particular distress to the seal:

1. *Loss of lubricant* This can arise in a number of ways depending on the sealing arrangement, but usually will be due to the pump de-priming or failure of the seal-flushing system. Many seals are able to withstand short periods of dry running but hot running is inevitable after a while and

this can lead to severe degradation of any polymeric secondary seals. Seal face materials may also experience thermal distress and in extreme cases the carbon may start burning. Many failures occur when the relatively cold process fluid returns as the thermal shock to the brittle seal faces may be sufficient to cause fracture.

2. *Running off BEP* Off-design point running, particularly the low-flow condition, is probably very common and the effects on the seal can be severe. It is most frequently caused by one or more of the following:
   - The over-specification of the pump
   - Changes in the plant throughput
   - The running of two pumps on a duty where one would be adequate.

   Cavitation associated with low flowrates can cause harmful vibrations – these can upset the seal directly by disturbing the interface film and by fatiguing metal bellows and elastomeric components. If the condition persists, the bearings may be damaged and this in turn can lead to eventual seal failure. Cavitation also generates gas or vapour bubbles which tend to be centrifuged onto the seal. In a dead-ended arrangement with a cramped seal chamber the seal may be starved of cooling and lubricating liquid. Another effect of the low flow condition is to deflect the pump shaft – this can be as much as half a millimetre at the seal.

3. *Transients* Process transients are inevitable and many are unavoidable. If particular transients are known to occur the seal manufacturer should be consulted as protective action for the seal may be possible by attention to seal design, face material combination, flushing arrangements, etc. It is worth reviewing whether some transients, e.g.

starting a pump against a closed valve, can be avoided or made less damaging – again the seal manufacturer may be able to help.

15.2.4.4 Seal selection

Selecting the most appropriate sealing system for a specified duty can be a difficult exercise which is best left in the hands of a reputable seal manufacturer unless particular company expertise and experience is available. In addition to the primary seal, secondary containment and ancillary equipment may be required and the manufacturer may also be able to make other suggestions for improving the reliability of the final design. The basic steps involved in seal selection are worth knowing and will aid the liaison betwen manufacturer and user.

The manufacturer will require the following data to make a primary seal selection:

- Precise and complete seal housing dimensions
- Running pressure including the seal chamber pressure, if known
- Running temperature
- Physical and chemical properties of the sealed fluid
- Expected life
- Required leakage

Specification of any secondary containment and ancillary systems will require further information regarding the sealed fluid (e.g. auto-ignition point, toxicity, flammability, tendency to decompose, tendency to crystallize, percentage of solids). Company or other regulations regarding permitted leakage levels should also be consulted. For a more extensive guide to seal selection the reader is referred to reference 24.

15.2.4.5 Seal installation

Training The fitting of mechanical seals is a skilled job and should be carried out by trained personnel. Site surveys indicate that between 25% and 40% of all seal failures may be attributable to incorrect fitting. Cartridge seals reduce the risk of fitting errors significantly and their use is to be encouraged. All reputable mechanical seal manufacturers offer training courses on seal installation and the investment of time and money for making use of these services will inevitably return dividends in reduced premature seal failures in the field.

Handling The rules for handling mechanical seals are:

- Obey any specific instructions in the literature enclosed with the seal.
- Avoid mechanical damage or shock as many seal components are brittle or fragile.
- Do not place the sealing faces down on dirty, unyielding surfaces.
- Unpack the seal carefully; shrunk-wrap packaging should be cut off carefully.
- Check that the seal supplied matches the seal specified for the duty.
- File all relevant technical information supplied by the seal manufacturer.
- Transfer fitting recommendations into company maintenance procedures.

Inspection While detailed checks cannot usually be made on-site, the seal should be inspected prior to installation for any superficial damage:

- Check materials specifications against duty (manufacturers provide references of material codes).

- Check all metal components for physical damage.
- Check seal faces for scratches, nicks or visible imperfections.
- Check secondary seals for cuts, nicks, tears, and chemical attack. Some elastomers are attacked by common fluids such as ozone, water and mineral oils.

Storage Mechanical seals should be stored in the protective packaging supplied by the manufacturer. The packaged seal should be kept in an area free from dirt, excessive moisture, high humidity and extreme cold. Good ventilation is also recommended.

Fitting Seal fitting is best carried out in a clean environment and, if possible, the pump should be removed in entirety to a workshop for stripping down and rebuilding. This practice has a number of attractions:

- The pump components, especially seals and bearings, are not exposed to the elements.
- The existing seal may be removed more carefully and inspected for damage.
- The relatively benign and clean conditions of a workshop are more conducive to good fitting practice.

Prior to fitting the seal the pump should be checked for misalignments which may be harmful to the seal. Checks should include: shaft balance, shaft run-out, seal chamber squareness, and concentricity of the shaft to the seal chamber. The seal manufacturer will be able to give a comprehensive list of recommended checks; alternatively, they are catalogued in Section 9.2 of reference 24.

Most mechanical seals are supplied with detailed fitting instructions and these should be carefully followed. There are some fitting requirements common to all seals:

- Check seal envelope dimensions carefully, particularly the components which dictate the compression length of the loading member. Over-compression of a seal will probably lead to premature failure.
- Avoid twists and kinks in any O-rings. PTFE O-rings should be softened in boiling water immediately prior to fitting.
- Seal faces must be kept clean. Any grease or foreign matter on the faces should be wiped off using lint-free tissue soaked in a suitable solvent such as propanol.
- Check that any ancillary equipment is cleaned and properly commissioned. Piping connections should be inspected for conformance to the seal manufacturer's drawing.
- Ensure adequate coupling alignment – this is very important for long seal life. The seal manufacturer will recommend appropriate tolerances.
- Avoid excessive pipe strains arising from misalignments between the pump flanges and pipework. Coupling alignment should be rechecked after connecting the pipework.
- Check that seal flushing systems are operating correctly and that valves are open.
- If possible, vent the seal chamber at start-up.

15.2.4.6 Seal failures

Defining seal failure is difficult and depends to large extent on the nature of the sealed fluid and the practice of the seal operator. Most seals are removed because of excessive leakage although sometimes it is necessary to inspect if the seal is running hot or squealing. A pump outage caused by a failing seal is obviously irritating to users but they should ensure that vital evidence, which may reveal the reasons for failure, is not lost when the failed seal is removed. Careful records of seal failures are valuable aids to effective troubleshooting and should be made systematically by the trained fitter as the seal is removed. Process conditions prevailing at the time of failure should be logged – many seal failures can be linked to changes causing the seal to experience off-design conditions; for example, low flow rates through the pump may cause cavitation and excessive vibration at the seal.

Leakage may be due to the failure of any of the seals including the secondary seals – these should be carefully inspected as they are removed. Do not handle the rubbing faces before visual inspection. Check the faces for obvious damage, including chemical attack. In addition, note any of the following if evident:

- Thermal distress including surface cracking and discoloration.
- Solids build-up both in the sealing interface and on the sides of the faces.
- Surface pitting and erosion – a magnifying glass can be a useful aid.

A summary of common failure modes and corrective action is given in Table 15.17 and further information can be found in references 25 and 26.

Acknowledgements

The author gratefully acknowledges the British Hydromechanics Research Group Limited for permission to publish the foregoing text.

15.2.5 Clearance seals

15.2.5.1 Introduction

Clearance seals tend to be purpose-designed for use in particular rotary applications where it is not possible to use lip or mechanical seals. Their principal virtues are high reliability and long life compared to other rotary seal types. The price for these advantages is relatively high leakage for pressurized duties even when tight radial and axial tolerances are achieved. The principal areas of use are: steam and gas leakage control, especially for turbines and compressors, large water turbines, grease seals for bearings, high-pressure and/or high-speed reciprocating applications (e.g. diesel fuel injector pumps), and some high-pressure water pumps. Most high-duty clearance seals are built into the piece of machinery by the manufacturer; consequently there are few commercially available units. In terms of geometry the seals range from simple fixed and floating bushings to complex labyrinth and viscoseals. Clearance seals are used frequently as a secondary or back-up device (e.g. throttle bush) to limit leakage in the event of primary seal failure. In these cases the primary seal would often be a mechanical seal.

15.2.5.2 Characteristics

The high reliability and long life of clearance seals are important features. Given a correct initial set-up the life of the seal is usually only limited by wear caused by abrasives in the pumped fluid or contact between the rotating and stationary components due to shaft run-out. In most cases this process is very gradual.

Leakage from clearance seals is usually high compared with contact seals – very tight axial and radial tolerances are required to approach contact seal performance in this regard. However, tight-toleranced seals are more sensitive devices and may be adversely affected by deflections induced hydraulically and thermally in both seal and machine, and also by machine misalignment and external vibration. Tribologically

compatible materials are required since contact is likely at some time. The unit cost is generally higher than more standardized seal types, an exception being grease retainers which are very low unit-cost items.

In general, the choice between individual types of seal will be a balance between cost, life and leakage. Table 15.18 outlines the four main types of clearance seal and their relative merits.

15.2.5.3 Seal types

Fixed bushing The fixed bushing, shown diagrammatically in Figure 15.114, is the simplest design of clearance seal. Its main virtue lies in its low cost. One-piece fixed bushings are mainly used as pump wear rings, as balance drums on multi-stage pumps and throttle bushes as a secondary back-up to other rotary seals. Tolerance requirements can be very tight, raising the true cost of an apparently low-cost seal. This problem is usually overcome by adopting multi-segment designs and/or a floating bush (see below).

The correct choice of materials is very important. Typically, for water duties bushes are segmented and may be manufactured from carbon-graphite and run on a bronze or carbon-steel shaft sleeve. On a clean process liquid the life of the bush may be between 5 and 10 years. If abrasives are present then a flame-hardened or nitrided stainless steel sleeve is recommended and possibly a compatible babbitt bush-lining.

The fixed bushing is characterized by high leakage, which is highly dependent on the radial clearance and relative eccentricity. The amount of leakage can be predicted for both laminar and turbulent flow conditions with compressible and incompressible fluids:

Laminar flow: Theoretical calculation as shown in Table 15.19
Turbulent flow: Empirical data as given in Figures 15.115 and 15.116

Fixed bushings are commonly used as an auxiliary seal on centrifugal pumps to minimize leakage in the event of primary seal failure. The diametral clearance to BS 6836 should be no greater than:

$$\frac{\text{Shaft diameter}}{100} + 0.2$$

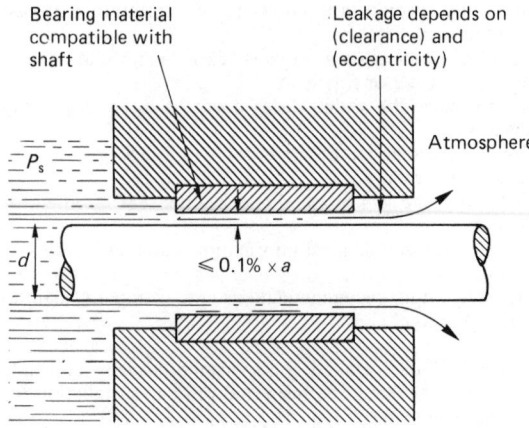

Figure 15.114 Typical fixed bushing seal

Table 15.17 Mechanical seal failure modes (*copyright BHR Group Ltd*)

Special conditions	Symptoms High leak	Over-heat	High wear	Seize	Failure mechanism	Action
Viscosity of fluid high		×	×	×	Excessive frictional heating, film vaporizes	Provide cooling
high	×	×	×	×	Excessive frictional heating, seal distorts	Provide cooling
low		×	×	×	Poor hydrodynamic lubrication, solid contact	Use faces with good boundary lubrication capacity
Low-lubricity fluid	×	×	×	×	Surface seize or 'pick-up'	Use faces with good boundary lubrication capacity. Reduce face loading
Speed of sliding high		×	×	×	Excessive frictional heating, film vaporizes	Provide cooling
high	×				Thermal stress cracking of the face	Use material with higher conductivity or higher tensile strength
high	×	×	×	×	Thermal distortion of seal	Provide cooling
low		×	×	×	Poor hydrodynamic lubrication, solid contact	Use face with good boundary lubrication capacity
Pressure differential low	(×)				Fluid pumped by seal against pressure	Try reversing seal to redirect flow
high		×	×	×	Hydrodynamic film overloaded	Modify area ratio of seal to reduce load
high	×	×	×	×	Seal or housing distorting	Stiffen seal and/or housing
Fluid temperature high	×				Elastomer secondary seal overheated	Use a high-temperature rubber
high	×				Distortion of seal faces	Reduce thermal stresses, e.g. by cooling the seal
high	×				Thermal stress cracking of the faces	Avoid rapid temperature changes or large temperature gradients
Abrasives in fluid	×		×		Solids in interface film	Circulate clean fluid round seal
Crystallizable fluid	×		×		Crystals form at seal face, secondary seal jams	Raise temperature of flush fluid outside seal
Polymerizable fluid	×		×		Solids form at seal face, secondary seal jams	Raise temperature of flush fluid outside seal
Salt solutions Ionic fluid	×	×	×	×	Corrosion damages seal faces	Select resistant materials
Non-Newtonian fluids, suspensions, colloids, etc.	(×)				Fluid behaves unpredictably, leakage may be reversed	Try reversing seal to redirect flow in acceptable direction
Sterilization cycle Cleaning cycle	×				High temp. or solvents incompatible with seal materials, especially elastomer	Use compatible materials
Ozone, radiation exposure to sunlight	×				Seal materials (rubber) fail	Protect seal from exposure, consider other materials
Flushing etc. Auxiliary cooling	×	×	×	×	Stoppage in auxiliary circuit	Overhaul auxiliaries
Double seals	×	×	×	×	Pressure build-up between seals if there is no provision for pressure control	Provide pressure control
Seal faces rough	×	×	×	×	Asperities make solid contact	Lap or grind faces
Bellows type seal	×				Floating seal member vibrates	Fit damping device to bellows
Single-spring seal drive	×				Spring ineffective due to wrong shaft rotation direction	Reverse motor, or change spring
Vibration present	×		×	×	Face separation unstable	Try to reduce vibration, avoid bellows seals, fit damper
Housing flexes due to pressure or temperature changes	×	×	×	×	Seal faces out of alignment, non-uniform wear	Stiffen housing and/or mount seal flexibly
Seal flatness poor	×				Excessive seal gap	Lap faces flatter

Table 15.18 Types of clearance seal (*copyright BHR Group Limited*)

Seal type	Advantages	Disadvantages	Application areas
Labyrinth	High speed, high temperature. 'Zero' wear if shaft located correctly.	High cost. High precision, axially and radially. Static and dynamic leakage. Usually gas seal only.	Gas turbines Gas compressors Steam turbines
Visco-seal	High speed, high temperature. 'Zero' wear. Zero leakage at design speed. Gas or liquid seal.	High cost, high precision, radially. Uni-directional leaks when stopped. Gas ingestion can reduce effectiveness.	Specialized pumps and compressors. e.g., sodium pumps.
Fixed bush	Relatively low cost. Simple design.	High leakage.	Pump wear rings. Very low pressure water seals.
Floating bushing (most commonly segmented)	'Self-adjusting' clearance seal. Will wear in to shaft, so less precision required. Relatively low cost. Large diameters can be accommodated relatively economically.	Pressure/speed limited. Have 'finite' life compared with true clearance seals. Material choice limited to good bearing combinations: seal material generally, carbon, bronzes or babbitt.	Low-duty gas turbines and compressors. Water pumps and turbines (particularly large diameters)

Table 15.19 Leakage from a bushing seal – laminar flow

$q \equiv$ volumetric flow rate/unit pressure gradient/unit periphery $\eta \equiv$ absolute viscosity	Fluid incompressible	Fluid compressible[a]

$$\left(\epsilon = \frac{\delta}{c}\right)$$

Axial bush

$$Q = \frac{2\pi a(P_s - P_a)}{l} \cdot q \quad \text{m}^3\,\text{s}^{-1}$$

b $\quad q = \frac{c^3}{12\eta} \cdot (1 + 1.5\epsilon^2)$

$$q = \frac{c^3}{24\eta} \cdot \frac{(P_s + P_a)}{P_a}$$

Radial bush

$$Q = \frac{2\pi a(P_s - P_a)}{(a - b)} q \quad \text{m}^3\,\text{s}^{-1}$$

$$q = \frac{c^3}{12\eta} \cdot \frac{(a - b)}{a \log_e \frac{a}{b}}$$

$$q = \frac{c^3}{24\eta} \cdot \frac{(a - b)}{a} \cdot \frac{(P_s + P_a)}{P_a}$$

[a] For Mach number < 1.0, i.e. fluid velocity = local velocity of sound.

[b] If shaft rotates, onset of Taylor vortices limits validity of formula to $\frac{V}{v}\sqrt{\frac{c}{a}} < 41.3$ (where V = surface speed and v = kinematic viscosity).

A variation on the fixed clearance seal is the centrifugal liquid barrier seal in which the centrifugal action of the rotating component is sufficient to create a pressure differential to oppose the leakage of sealed fluid. An example of a seal of this type is the hydrodynamic disk seal shown in Figure 15.117 which may be a single or multi-stage device. The performance of this seal type is reported in Merry and Thew.[27]

Floating bushing The floating bushing, shown in Figure 15.118, is a nominally self-aligning version of the fixed bush. It therefore requires less accurate precision in its installation. It is relatively low cost and, in segmented form, is particularly suited to large diameter seals. Its main areas of application are for low-duty gas turbines and compressors, low-pressure water pumps and water turbines. Leakage is relatively high compared with other seals unless very tight tolerances can be held. Estimates of leakage may be made using the figures given for fixed bushings. The floating bush is somewhat more limited in its pressure and speed capabilities and, as for fixed bushings, the choice of materials is vitally important.

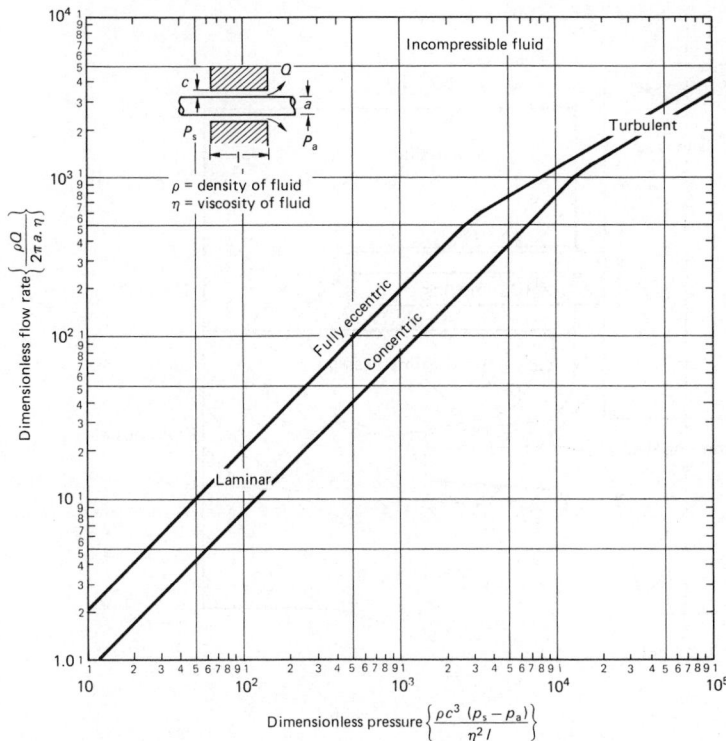

Figure 15.115 Leakage flow in a fixed bushing seal

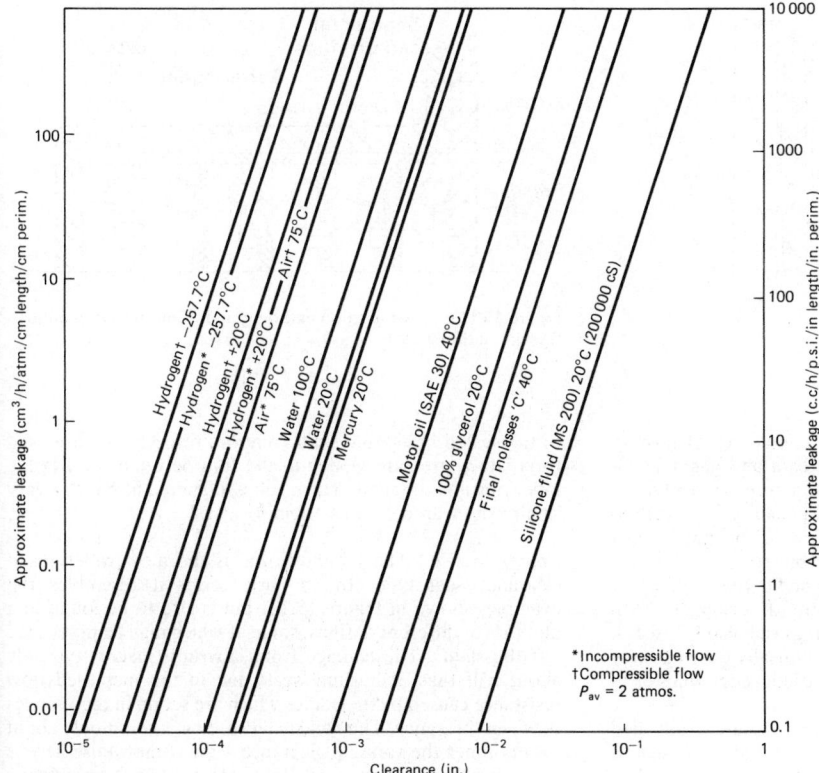

Figure 15.116 Effect of viscosity and clearance on leakage from a bushing seal

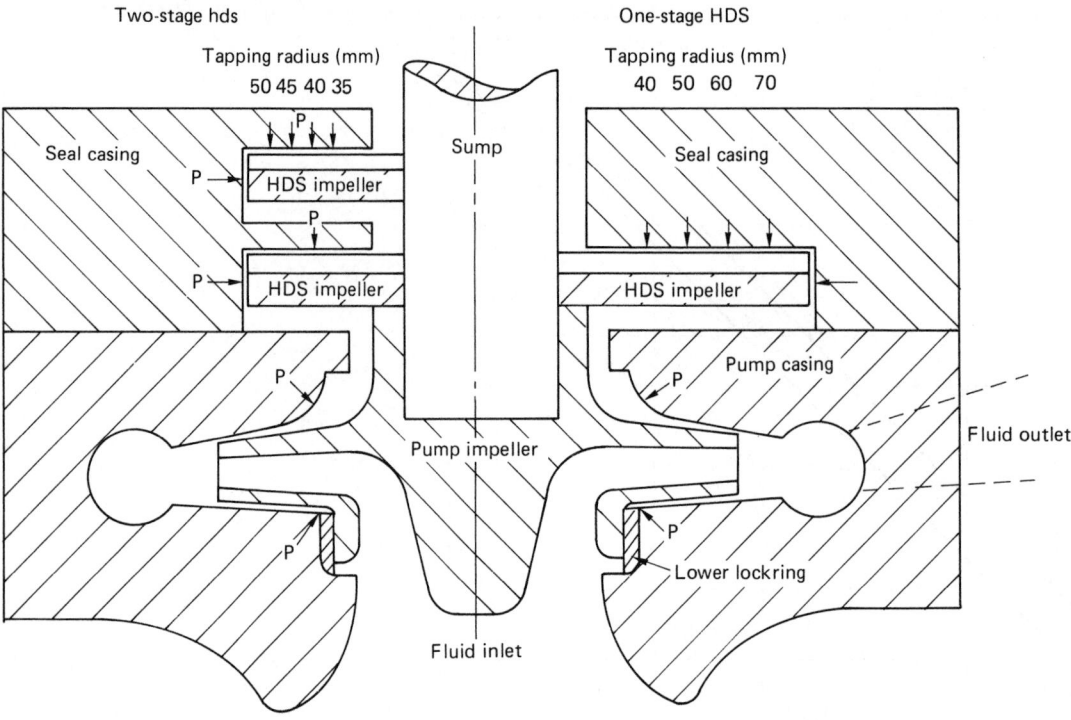

Figure 15.117 Hydrodynamic disk seal. (Copyright BHR Group Ltd)

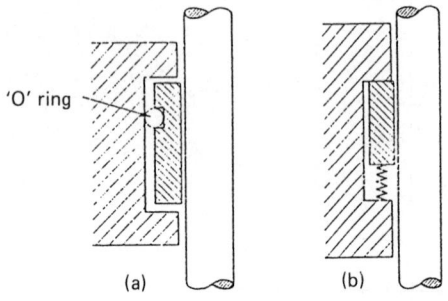

Figure 15.118 Typical floating bushing seals

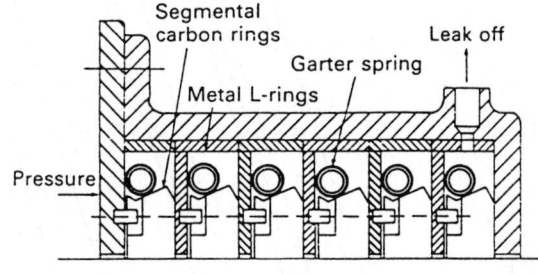

Figure 15.119 Bevel-section carbon gland. (Source: Morganite Special Carbons Ltd)

Floating bushing seals in the non-contacting gland-ring arrangement are commonly used on steam and gas applications. Typically the gland rings are manufactured from carbon-graphite which are often segmented. An example of a carbon gland is shown in Figure 15.119; in this case the rings are of bevel section and spring loaded to enable compensation for ring wear. The bore of the rings is designed to match the shaft diameter at the operating temperature in order to keep leakage to a minimum. Carbon gland rings can also be used in other arrangements for water turbines; typically the rings may be used in a tenon-jointed form, as opposed wedge rings or as hydrostatic radial face seals.

A special type of floating bush seal is the controlled clearance rotary seal shown in Figure 15.120. The seal is designed to be self-energizing and self-compensating. Hydrostatic and hydrodynamic forces resulting from the pressure

of the sealed liquid and shaft rotation, respectively, are used to create a pressure wedge in the film between the flexible inner ring and the shaft. The result is a tapered film with a very small exit clearance, thus effecting a seal.

Labyrinth seal The labyrinth seal is the most widely used clearance seal type. In its basic forms it resembles the examples shown in Figure 15.121 but is also to be found in a plethora of different designs, some of which may be purchased 'off the shelf'. The leakage from labyrinth seals is typically about half that of bushing seals due to the increased flow resistance caused by the eddies which are set up in the grooves between the vanes. The vane axial spacing is commonly about twenty times the vane lip clearance – the compromise here is between achieving sufficiently large grooves to promote strong eddies and limiting the axial length of the seal to manageable

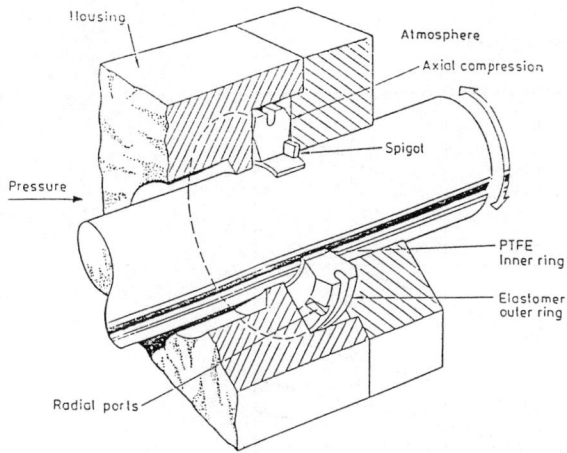

Figure 15.120 Controlled-clearance rotary seal. (Source: James Walker and Co. Ltd)

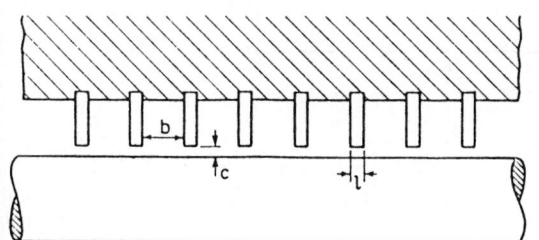

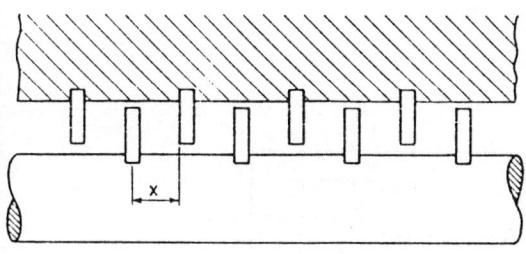

Figure 15.121 Typical labyrinth seals

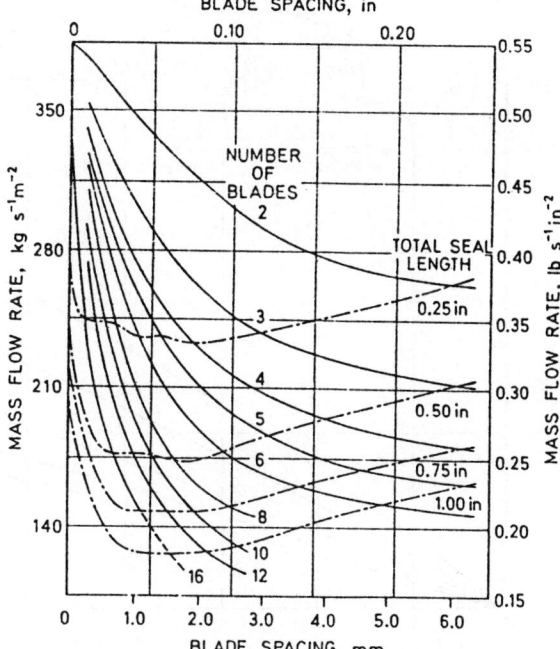

Ambient pressure = 1 atmos.; blade thickness = 0.0055 in; radial clearance = 0.005 in; pressure ratio = 0.551; temperature 310 K

Figure 15.122 Performance of a typical labyrinth seal

dimensions. Typical leakage performance is shown in Figure 15.122, which indicates the influence of blade spacing and the number of blades. For design purposes the user is referred to Figure 15.123.

Labyrinth seals are capable of very high speeds and high temperatures with nominally unlimited life. Consequently, they are used on gas sealing duties, turbines and compressors, and for steam turbines. They are also available as bearing grease seals as an alternative to lip seals. Their main drawbacks are, in general, relatively high cost and the high axial and radial precision required in their installation and operation.

As for other clearance seals, the materials of construction should be tribologically compatible as a precaution against

rubbing contact between the vanes and the rotor. An additional factor may also be creep, especially in high-speed turbines where stresses are large. A convenient material arrangement is to use metal fins and a carbon bush as shown in Figure 15.124. Alternatively, a metal foil honeycomb may be used made, for example, from stainless or Nimonic steels.

Visco-seal The visco-seal or wind-back seal is basically a fixed bush with a helical groove cut either into the shaft or the bore of the bush as shown in Figure 15.125. The effect of the helix is to pump the sealed fluid back into the sealed system as the shaft is rotated. The seal therefore works best for viscous fluids or for high rotational speeds.

The seal is essentially a single speed, uni-directional, device. It is designed to give inward pumping action perfectly matching the leakage flow with the net result of zero leakage at what is known as the *sealing pressure*. Excess pressure will result in leakage while at low pressures the seal runs partially dry and air may be pumped inwards. Typical performance and design criteria are summarized in Figure 15.126. There is no sealing action when the shaft is stationary and it may be necessary to fit an auxiliary static seal. This may lift off automatically when the shaft rotates.

A development of the visco-seal is the barrier visco-seal in which a pair of seals are installed back-to-back as shown in Figure 15.127. The resulting pressure barrier which builds up between the seals may be used to buffer the sealed fluid from atmosphere. Compatibility between sealed fluid (usually a gas) and buffer fluid (usually a liquid) must be established. In tests using grease as the barrier fluid a pressure of 10 bar was sealed by a 13 mm diameter seal running at 1000 rpm.

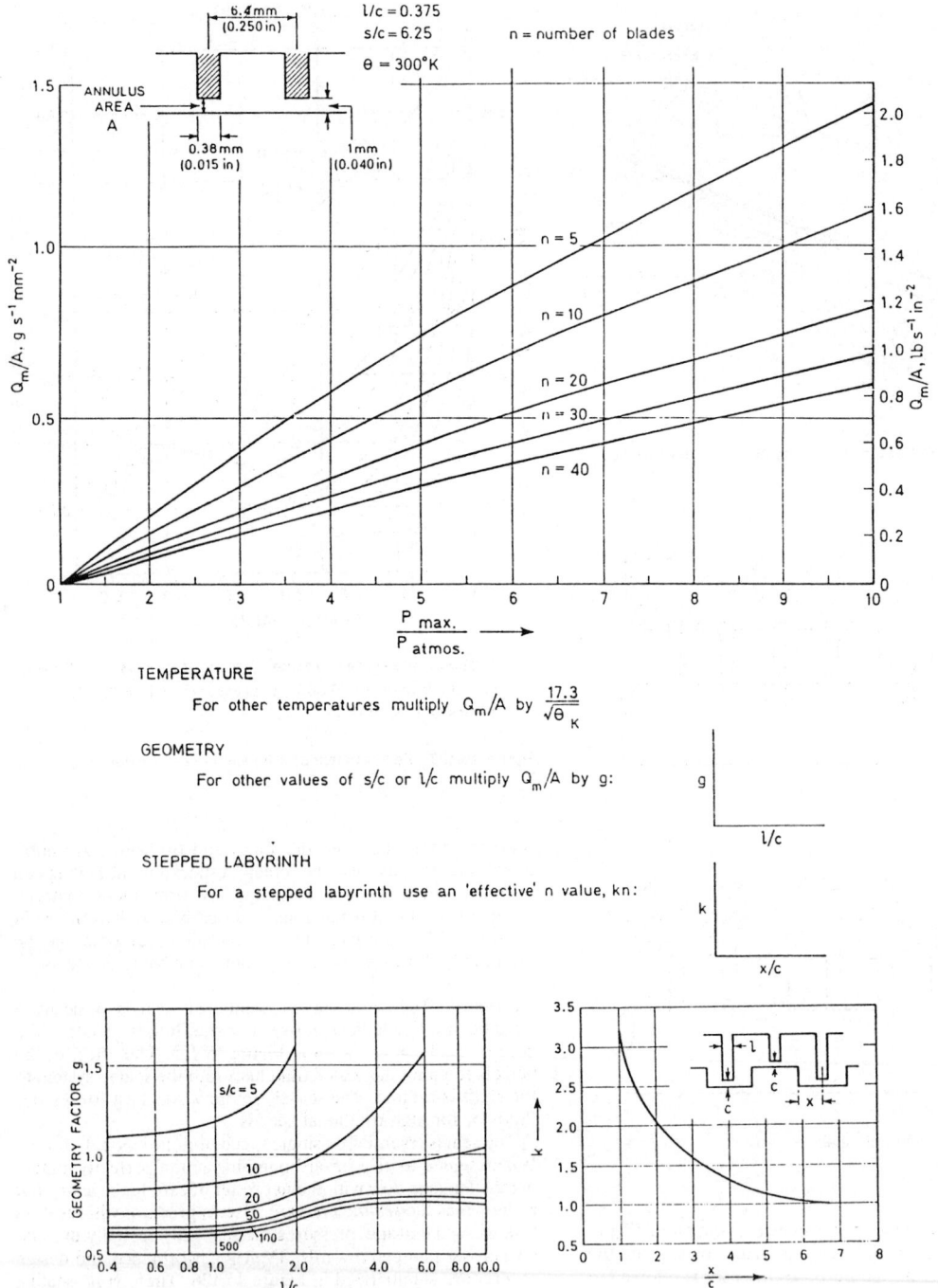

Figure 15.123 Calculation of leakage for a labyrinth seal

The visco-seal is suitable for high-temperature applications where the pressure is low to moderate. However, it requires a high degree of radial and axial precision and its initial cost is relatively high. Its main areas of application are for special-duty very long-life pumps or high-speed rotary compressors. Further information can be found in references 28 and 29.

Acknowledgements

The author gratefully acknowledges the British Hydromechanics Research Group Limited for permission to publish the foregoing text.

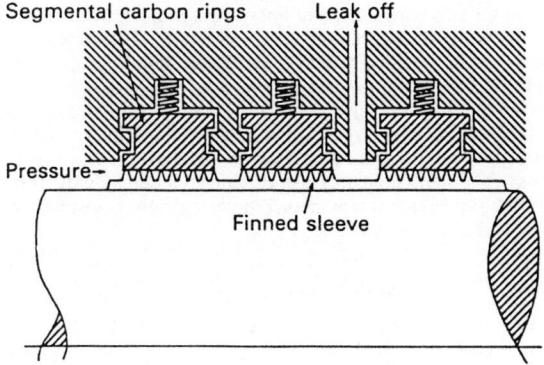

Figure 15.124 Carbon labyrinth gland. (Source: Morganite Special Carbons Ltd)

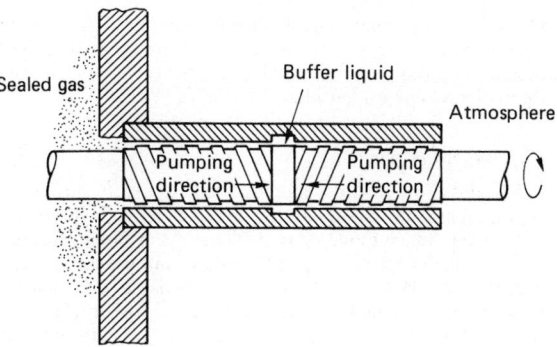

Figure 15.127 Barrier visco-seal

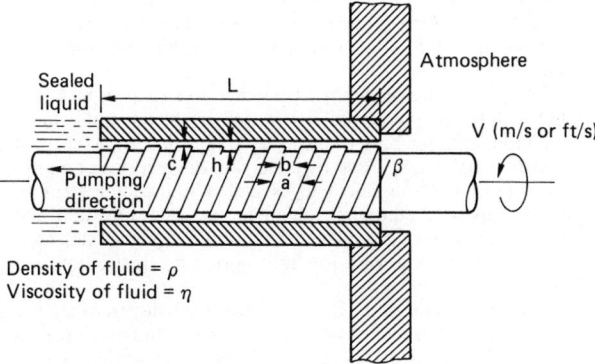

Density of fluid = ρ
Viscosity of fluid = η

Figure 15.125 Visco-seal

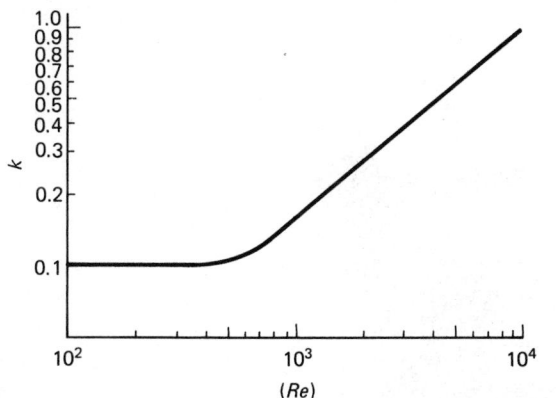

Figure 15.126 Typical visco-seal performance

15.3 Boilers and waste-heat recovery

15.3.1 Types of boilers

This section covers industrial boilers, therefore only units of 500 kg h^{-1} of steam, or equivalent hot water, and above will

be considered. There are eight categories of boiler available. In order of rated output these are:

Cast iron sectional boilers
Steel boilers
Electrode boilers
Steam generators
Vertical shell boilers
Horizontal shell boilers
Water tube boilers
Fluid bed boilers

15.3.1.1 Cast iron sectional boilers

These are used for hot water services with a maximum operating pressure of 5 bar and a maximum output in the order of 1500 kW. Site assembly of the unit is necessary which will consist of a bank of cast iron sections. Each section has internal waterways.

The sections are assembled with screwed or taper nipples at top and bottom for water circulation and sealing between the sections to contain the products of combustion. Tie rods compress the sections together. A standard section may be used to give a range of outputs dependent upon the number of sections used. After assembly of the sections the mountings, insulation and combustion appliance are fitted. This system makes them suitable for locations where it is impractical to deliver a package unit. Models are available for use with liquid, gaseous and solid fuels.

15.3.1.2 Steel boilers

These are similar in rated outputs to the cast iron sectional boiler. Construction is of rolled steel annular drums for the pressure vessel. They may be in either a vertical or a horizontal configuration depending upon the manufacturer.

15.3.1.3 Electrode boilers

These are available for steam raising up to 3600 kg h^{-1}. Normal working pressure would be 10 bar but higher pressures are available. Construction is a vertical pattern pressure shell containing the electrodes.

The length of the electrodes controls the maximum and minimum water level. The electrical resistance of the water allows a current to flow through the water which in turn boils and releases steam. Since water has to be present within the electrode system, lack of water cannot burn out the boiler. The main advantage with these units is that they may be located at the point where steam is required and as no combustion fumes are produced, no chimney is required. Steam may also be raised relatively quickly as there is little thermal stressing to consider.

15.3.1.4 Steam generators

While the term 'steam generator' may apply to any vessel raising steam, this section is intended to cover coil type boilers in the evaporative range up to 3600 kg h^{-1} of steam. Because of the steam pressure being contained within the tubular coil, pressures of 35 bar and above are available although the majority are supplied to operate at up to 10 bar. They are suitable for firing with liquid and gaseous fuels although the use of heavy fuel oil is not common.

The coiled tube is contained within a pressurized combustion chamber and receives both radiant and convected heat. Feed water is pumped through the coil where it converts to steam. As the quantity of water is slightly more than the firing rate in order to protect the coil from damage a steam separator or steam manifold is required to produce an acceptable dryness fraction to the steam. Because there is no stored water in this type of unit they are lighter in weight and therefore suitable for siting on mezzanine or upper floors adjacent to the plant requiring steam. Also, as the water content is minimal, steam raising can be achieved very quickly and can respond to fluctuating demand within the capacity of the generator. Note that close control of suitable water treatment is essential to protect the coil against any build-up of deposits.

15.3.1.5 Vertical shell boilers

This is a cylindrical boiler where the shell axis is vertical to the firing floor. Originally it comprised a chamber at the lower end of the shell which contained the combustion appliance. The gases rose vertically through a flue surrounded by water. Large-diameter (100 mm) cross-tubes were fitted across this flue to help extract heat from the gases which then proceeded to the chimney. Later versions had the vertical flue replaced by one or two banks of small-bore tubes running horizontally before the gases discharged to the chimney. The steam was contained in a hemispherical chamber forming the top of the shell.

The present-day vertical boiler is generally used for heat recovery from exhaust gases from power generation or marine applications. The gases pass through small-bore vertical tube banks. The same shell may also contain an independently fired section to produce steam at such times as there is insufficient or no exhaust gas available.

15.3.1.6 Horizontal shell boilers

This is the most widely used type of boiler in industry. The construction of a single-flue three-pass wetback shell is illustrated in Figure 15.128. As a single-flue design boiler, evaporation rates of up to 16 300 kg h^{-1} F&A 100°C (see Section 15.3.4.7) are normal on oil and gas and 9000 kg h^{-1} F&A 100°C on coal.

In twin-flue design the above figures approximately double. Normal working pressures of 10–17 bar are available with a maximum working pressure for a shell boiler at 27 bar. The outputs of larger boilers will be reduced if high pressures are required.

The boilers are normally despatched to site as a packaged unit with the shell and smokeboxes fully insulated and painted and mounted on a base frame. The combustion appliance and control panel will be fitted together with the feed-water pump, water-level controls and gauges and a full complement of boiler valves. Additional equipment may be specified and incorporated during construction. Larger boilers may have certain items removed for transport, site restrictions or weight.

Some variations of the three-pass wet-back design exist. The most common is the reverse flame boiler and is shown in Figure 15.129. In this design the combustion appliance fires into a thimble-shaped chamber in which the gases reverse back to the front of the boiler around the flame core. The gases are then turned in a front smokebox to travel along a single pass of

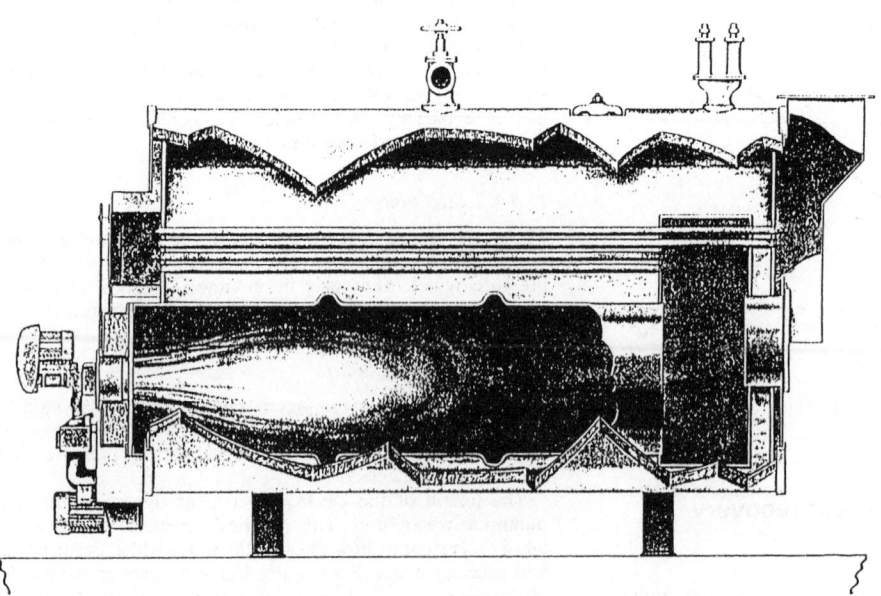

Figure 15.128 Three-pass wet-back shell

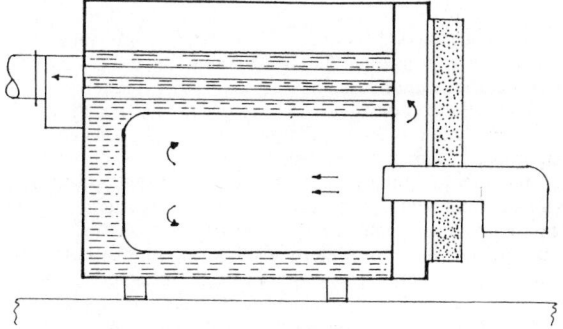

Figure 15.129 Reverse flame shell

smoketubes to the rear of the boiler and then to the chimney. In order to extract maximum heat, gas turbulators or retarders are fitted into these tubes to agitate the gases and help produce the required flue-gas outlet temperature. Evaporative outputs up to 4500 kg h^{-1} F&A 100°C using liquid and gaseous fuels are available.

Other variations of the three-pass wet-back design are the two-pass, where only one pass of smoketubes follows the combustion tube, and the four-pass, where three passes of smoketubes follow the combustion tube. Neither of these are as widely used as the three-pass design.

Dry-back boilers are still occasionally used when a high degree of superheat is required necessitating a rear chamber to house the superheater too large for a semi-wet-back chamber. A water-cooled membrane wall chamber would be an alternative to this.

With twin-flue design boilers it is usual to have completely separate gas passes through the boiler with twin wet-back chambers. It is then possible to operate the boiler on one flue only which effectively doubles its turndown ratio. For

example, a boiler rated at 20 000 kg h^{-1} F&A 100°C may reasonably be expected to operate down to 2500 kg h^{-1} F&A 100°C on oil or gas providing suitable combustion equipment and control is incorporated. It would be good practice to alternate, on a planned time scale, which flue takes the single-flue load if prolonged periods of single-flue operation occur.

Shell boilers are supplied with controls making them suitable for unattended operation although certain operations such as blowdown of controls are called for by the insurance companies to comply with safety recommendations. Oil-, gas- and dual-fired boilers are available with a range of combustion appliances. The smaller units have pressure jet-type burners with a turndown of about 2:1, while larger boilers may have rotary cup, medium-pressure air (MPA) or steam-atomizing burners producing a turndown ratio of between 3:1 and 5:1 depending upon size and fuel. The majority have rotary cup-type burners while steam- or air-atomizing burners are used where it is essential that the burner firing is not interrupted even for the shortest period.

For coal-fired boilers, chain-grate stokers, coking stokers and underfeed stokers are supplied. An alternative to these is the fixed-grate and tipping-grate boiler with coal being fed through a drop tube in the crown of the boiler (Figure 15.130). With the fixed grate de-ashing is manual, while with the tipping grate a micro-sequence controller signals sections of the grate to tip, depositing the ash below the grate where it is removed by a drag-link chain conveyor and then to a suitable ash-disposal system.

It is possible to design a boiler to operate on all or any combination of liquid, gaseous and solid fuel and waste heat sources, although its complexity may outweigh its flexibility in practical and commercial terms.

Access to both water- and fireside surfaces of the boiler is important. All boilers will have an inspection opening or manway on the top of the shell with inspection openings in the lower part. Some larger boilers will have a manway in the lower part of the shell or end plate. With a three-pass wet-back boiler all tube cleaning and maintenance is carried

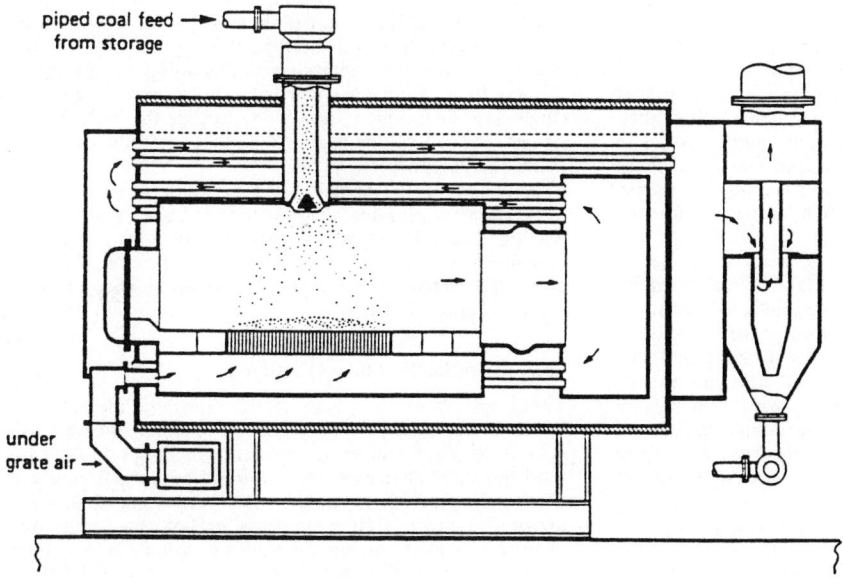

Figure 15.130 Fixed-grate coal-fired boiler

out from the front. The front smokebox doors will be hinged or fitted with davits. On most sizes of boilers bolted-on access panels are sufficient on the rear smokebox. As the majority of shell boilers operate under forced-draught pressurized combustion, steam raising is relatively quick. While good practice would require a cold boiler to come up to pressure over a period of several hours, once it is hot it may be brought up to pressure in minutes, not hours.

For hot water shell boilers the above still applies. The shells would be slightly smaller for equivalent duties due to the absence of steam space. There are three accepted operating bands for hot-water boilers. Low-temperature hot water (LTHW) refers to boilers having a mean water temperature (between flow and return) of below 95°C. Medium-temperature hot water (MTHW) would cover the range 95–150°C. High-temperature hot water (HTHW) covers applications above 150°C.

The flow and return connections will be designed to suit the flow rates and temperature differentials required. The water-return connection will be fitted with either an internal diffuser or a venturi nozzle to assist mixing of the water circulating within the shell and prevent water stratification. The flow connection will incorporate the temperature control stat to signal control of the firing rate for the burner.

Hot-water boilers are potentially more susceptible to gas-side corrosion than steam boilers due to the lower temperatures and pressures encountered on low- and medium-temperature hot-water boilers. With low-temperature hot-water the water-return temperature in particular may drop below the water dewpoint of 50°C, causing vapour in the products of combustion to condense. This, in turn, causes corrosion if it persists for long periods of time. The remedy is to ensure that adequate mixing of the return water maintains the water in the shell above 65°C at all times. Also, if medium or heavy fuel oil is to be used for low- or medium-temperature applications it is desirable to keep the heat transfer surfaces above 130°C, this being the approximate acid dewpoint temperature of the combustion gases. It may therefore be seen how important it is to match the unit or range of unit sizes to the expected load.

15.3.1.7 Water tube boilers

Originally, water tube boilers would have been installed for evaporation of 10 000 kg h^{-1} of steam with pressures as low as 10 bar. At that time this would have been the maximum evaporation expected from a shell-type boiler. Now shell boilers are available at much greater duties and pressures as described in Section 15.3.1.6. It may be appreciated that there will be an area where a water tube boiler is required because of its use on high pressure and then ultimately for the increased duty.

Figure 15.131 illustrates a stoker fired unit. Generally, an output of 60 MW from a single unit may be considered for industrial installations. Higher duties are available if required.

Water tube boilers supplied for national power generation will have outputs up to 900 MW, pressures of 140 bar and final steam temperatures of 500°C. Industrial units are usually supplied with pressures up to 65 bar and with a final steam temperature up to 500°C. This is the maximum temperature and pressure likely to be required for small turbine-driven generating units although turbines are available to operate at much lower pressures of, say, 17 bar.

Construction is a water-cooled wall combustion chamber connected to a steam drum at high level. The bottoms of the walls are connected to headers. Sometimes a bottom or mud drum is incorporated but improved water treatment now available does not always necessitate this.

The chamber is externally insulated and clad. Combustion equipment for solid fuel may be spreader or travelling-grate stokers or by pulverized fuel or fluid bed. Oil and gas burners may be fitted either as main or auxiliary firing equipment. The boilers will incorporate superheaters, economizers and, where necessary, air preheaters and grit arresters and gas-cleaning equipment to meet clean air legislation.

Where water tube boilers are used to recover waste heat (for example, exhaust gases from reciprocating engines) lower gas temperatures may be involved and this, in turn, could obviate the need for water-cooled walls. In this case tube banks may be contained within a gastight insulated chamber.

There are two basic types of water tube boilers: assisted and natural circulation. Assisted circulation might apply where heat is from a convection rather than a radiation source such as a waste-heat application. Natural circulation is more suited where radiant heat and high gas temperatures are present.

Depending upon the required duty and the site, units may be shop assembled or of modular construction. Site-erected units may be designed to have their main components arranged to fit in with the space available.

15.3.1.8 Fluid bed boilers

The name derives from the firebed produced by containing a mixture of silica sand and ash through which air is blown to maintain the particles in suspension. The beds are in three categories: shallow, deep and recirculating. Shallow beds are the most used and are about 150–250 mm in depth in their slumped condition and around twice that when fluidized. Heat is applied to this bed to raise its temperature to around 600°C by auxiliary oil or gas burners. At this temperature coal and/or waste is fed into the bed which is controlled to operate at 800–900°C. Water-cooling surfaces are incorporated into this bed connected to the water system of the boiler. As its name implies, the deep bed is similar to the shallow bed but in this case may be up to 3 m deep in its fluidized state, making it suitable only for large boilers. Similarly, the recirculating fluid bed is only applicable to large water-tube boilers.

Several applications of the shallow bed system are available for industrial boilers, the two most used being the open-bottom shell boiler and the composite boiler. With the open-bottom shell the combustor is sited below the shell and the gases then pass through two banks of horizontal tubes. In the composite boiler the combustion space housing the fluid bed is formed by a water-tube chamber directly connected to a single-pass shell boiler. In order to fluidize the bed, the fan power required will be greater than that with other forms of firing equipment.

To its advantage, the fluid bed may utilize fuels with a high ash content which affect the availability of other systems (see Section 15.3.4.9). It is also possible to control the acid emission by additions to the bed during combustion. They are also less selective in fuels and can cope with a wide range of solid fuel characteristics.

15.3.2 Application and selection

The graph shown in Figure 15.132 illustrates the selective bands for various types of boilers. The operating pressure will govern the steam temperature, except where superheaters are used. For hot-water units the required flow temperature will dictate the operating pressure. It is important that when arriving at the operating pressure for hot-water units due allowance is made for the head of the system, an anti-flash steam margin of 17°C, and a safety valve margin of 1.5 bar.

When deciding to install one or more boilers the following should be considered. The first choice, providing the load is

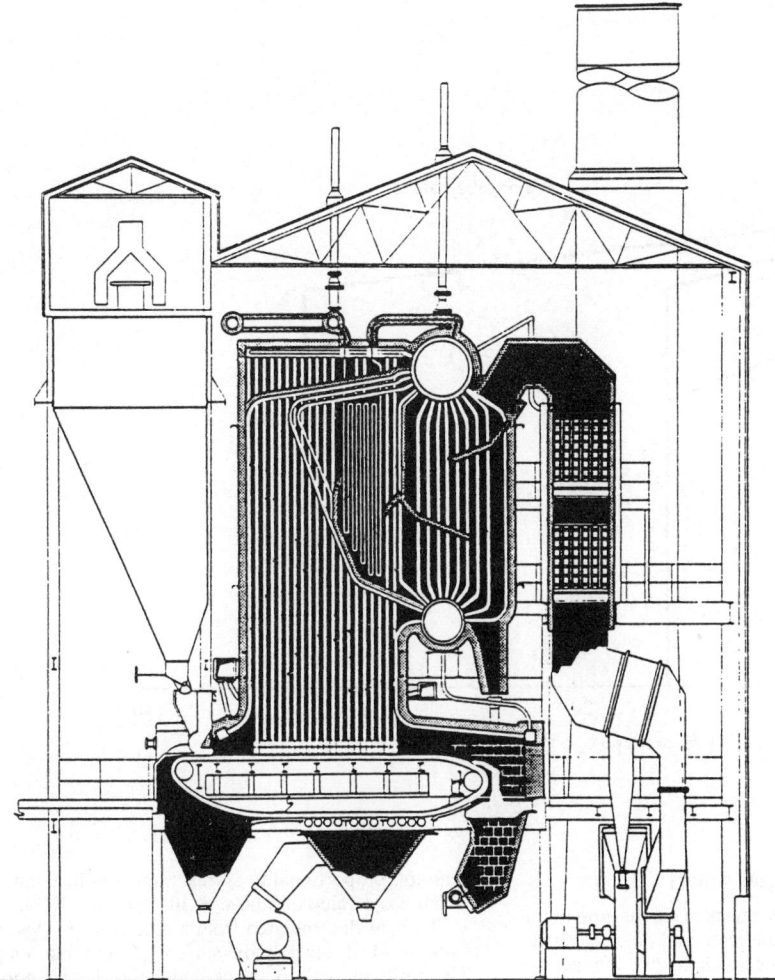

Figure 15.131 Stoker-fired water-tube boiler of 36 300 kg h^{-1} steam capacity at 28 bar and 385°C. (Source: the British Coal publication *Boiler House Design for Solid Fuel* (1980) and with kind permission from the College of Fuel Technology)

within the duty range of the boiler, will be a single unit. This is economically the most attractive providing account is taken of the following:

1. If there is a breakdown on the boiler will services be seriously affected immediately?
2. Will adequate spare parts for the boiler be held in stock or available within an acceptable time and will there be labour to carry out the repair work?
3. Will time be available to service the boiler?
4. The duty will preferably fall within the modulating firing rate of the burner.
5. Prolonged periods of intermittent operation should be avoided.
6. Is there an existing standby unit?

If any or all of these points are not accepted then the next consideration for a shell boiler could be a twin-flue unit suitable for single-flue operation. This has the advantage of using less space than two smaller boilers and having only one set of services.

Moving now to two boilers, the heat load may comprise two elements: one a production process where interruption would cause problems and the other, say, a heating load where any interruption would not be noticed immediately. Assuming that the two elements were of equal duty it would be reasonable to install two boilers each 50% of the total load. One boiler would then be able to cover the process load. An extension to this is to install two boilers each capable of handling the total combined load. Depending upon the boiler size, there may be only a relatively small difference in total capital cost between the above two schemes.

Further options involving three or more boilers must take into account minimum and maximum loads in order to run the plant efficiently. When considering hot water it may be advantageous to consider units in a range of outputs. This will help in operation so that a unit may be brought into duty to match the load and thus avoid low-load conditions and consequent danger of dewpoints. Also, should the plant be fired on solid fuel it will help in maintaining a more even firing rate and a clean stack.

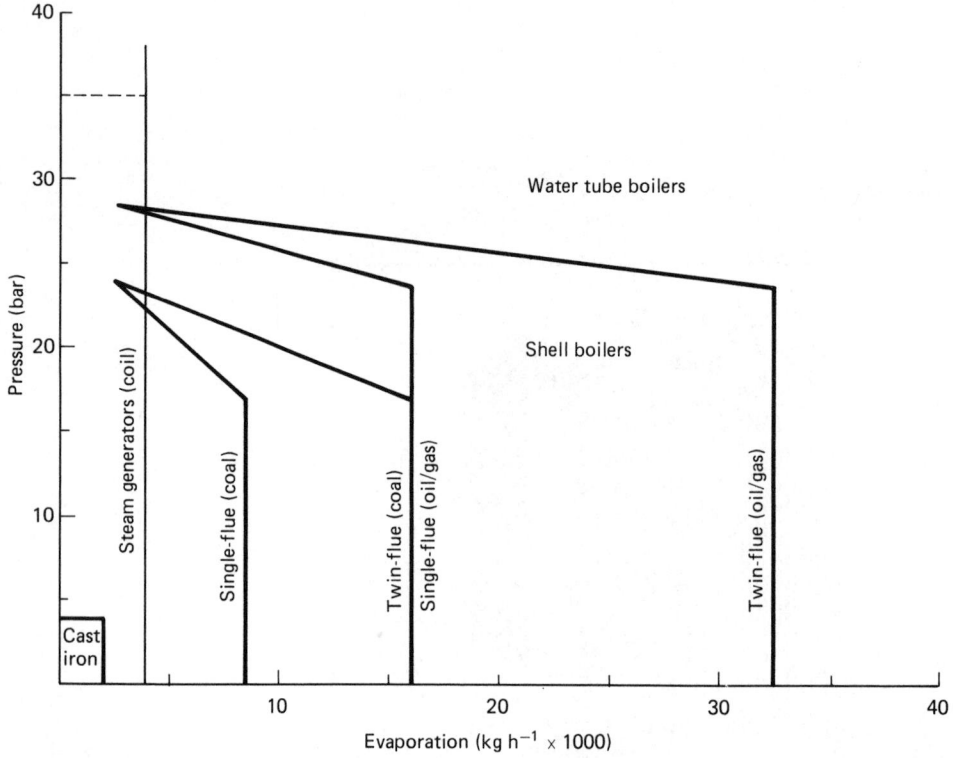

Figure 15.132 Guide to boiler capacities

15.3.3 Heat transfer in industrial boilers

Heat is transferred from the hot products of combustion to the boiler heating surfaces, through the plate and tube walls, and to the water by various mechanisms which involve conduction, radiation and convection.

15.3.3.1 Conduction

The rate at which heat is transferred by conduction through a substance without mass transfer is given by Fourier's Law. This states that the heat flow rate per unit area, or heat flux, is proportional to the temperature gradient in the direction of heat flow. The relationship between heat flux and temperature gradient is characterized by the thermal conductivity, which is a property of the substance. It is temperature dependent and is determined experimentally.

For a plate of area A (m^2), thickness e (m) and with hot and cold face temperatures of T_1 and T_2 (°C), respectively, the normal heat flux ϕ and heat transfer rate Q are given by:

$$\phi = \frac{Q}{A} = \frac{K(T_1 - T_2)}{e} \quad (\text{W m}^{-2})$$

where K = thermal conductivity (W m^{-1} K^{-1}).

15.3.3.2 Radiation

Thermal radiation takes place by the emission of electromagnetic waves, at the velocity of light, from all bodies at temperatures above absolute zero. The heat flux from an ideal or 'black-body' radiating surface is proportional to the fourth power of the absolute temperature of the surface. The con-

stant of proportionality is the Stefan–Boltzmann constant, which has a value of 5.6696×10^{-8} (W m^{-2} K^{-4}).

The heat flux radiated from a real surface is less than that from an ideal 'black-body' surface at the same temperature. The ratio of real to 'black-body' flux is the normal total emissivity. Emissivity, like thermal conductivity, is a property which must be determined experimentally.

Although the rate of emission from a surface is independent of the condition of the surroundings, the net overall exchange of radiant heat between surfaces at different temperatures depends on a number of factors. The continuous interchange of energy is a result of the reciprocal processes of radiation and absorption, and these are dependent on geometrical relationships, emissivity differences and the presence of any absorbing and emitting gases in the intervening space.

15.3.3.3 Convection

Convective heat transmission occurs within a fluid, and between a fluid and a surface, by virtue of relative movement of the fluid particles, that is, by mass transfer. Heat exchange between fluid particles in mixing and between fluid particles and a surface is by conduction. The overall rate of heat transfer in convection is, however, also dependent on the capacity of the fluid for energy storage and on its resistance to flow in mixing.

The fluid properties which characterize convective heat transfer are thus thermal conductivity, specific heat capacity and dynamic viscosity. Convection is classified according to the motivating flow. When the flow takes place as a result of density variations caused by temperature gradients the motion is called natural convection. When the flow is caused by an external agency such as a pump or a fan the process is called forced convection.

At a convection heat transfer surface the heat flux (heat transfer rate per unit area) is related to the temperature difference betweeen fluid and surface by a heat transfer coefficient. This is defined by Newton's law of cooling:

$$\phi = \frac{Q}{A} = h_c \Delta T_m$$

where

ϕ = heat flux (W m^{-2})
Q = heat transfer rate (W)
A = surface area (m^2)
ΔT_m = mean temperature difference between fluid and surface (K)
h_c = convective heat transfer coefficient (W m^{-2} K^{-1})

The heat transfer coefficient is correlated experimentally with the fluid transport properties (specific heat, viscosity, thermal conductivity and density), fluid velocity and the geometrical relationship between surface and fluid flow.

15.3.3.4 Furnace heat transfer

Heat transfer in the furnace is mainly by radiation, from the incandescent particles in the flame, and from hot radiating gases such as carbon dioxide and water vapour. The detailed theoretical prediction of overall radiation exchange is complicated by a number of factors such as carbon particle and dust distributions, and temperature variations in three-dimensional mixing. This is overcome by the use of simplified mathematical models or empirical relationships in various fields of application.

For industrial boilers the mean gas temperature at the furnace exit, or at the entrance to the convection section of the boiler, may be calculated using the relationship

$$T = k(H/A)^{0.25}$$

where

T = gas temperature (°C)
H = heat input rate (W) based on the net calorific value of the fuel
A = effective (projected) water-cooled absorption surface area (m^2)
k = a constant which depends on the fuel and the excess air in the combustion products.

The value of k is determined experimentally by gas-temperature measurement. The measurement error of a simple pyrometer can be 250–300 K, due to re-radiation to water-cooled surroundings, and the values given below are based on measurement by a 'Land' multi-shielded high-velocity suction pyrometer.

Typical values for normal excess air at or near full boiler load are:

natural gas $\quad k = 52.4$
gas oil $\qquad\quad k = 49.1$
heavy fuel oil $k = 48.3$
coal $\qquad\quad\;\; k = 40.3$

In calculating the smoke tube inlet gas temperature of a shell boiler, A includes the effective water-cooled surface in the reversal chamber. In coal-fired boilers any water-cooled surface below the grate is excluded from A.

The total furnace heat absorption may be estimated by using the calculated furnace exit gas temperature and analysis to determine the enthalpy (excluding the latent heat of water

vapour) and thus deducting the heat-rejection rate from the net heat input rate.

15.3.3.5 Boiler tube convection heat transfer

The radiant section of an industrial boiler may typically contain only 10% of the total heating surface yet, because of the large temperature difference, it can absorb 30–50% of the total heat exchange. The mean temperature difference available for heat transfer in the convective section is much smaller. To achieve a thermally efficient yet commercially viable design it is necessary to make full use of forced convection within the constraint of an acceptable pressure drop.

Forced convection heat transfer has been measured under widely differing conditions and correlation of the experimental results is made by using the dimensionless groups:

Nusselt number $\qquad\qquad Nu = \dfrac{h_c D}{k}$

Reynolds number $\qquad\qquad Re = \dfrac{GD}{\mu}$

Prandtl number $\qquad\qquad Pr = \dfrac{C_p \mu}{k}$

where

h_c = heat transfer coefficient (W m^{-2} K^{-1})
D = characteristic dimension (m)
K = thermal conductivity (W m^{-1} K^{-1})
G = gas mass velocity (kg m^{-2} s^{-1})
μ = dynamic viscosity (kg m^{-1} s^{-1})
C_p = specific heat at constant pressure (J kg^{-1} K^{-1})

In applying the correlations use is made of the concept of logarithmic mean temperature difference across the boundary layer. For a boiler section, or pass, this is given by:

$$\Delta T_m = \frac{(T_1 - T_w) - (T_2 - T_w)}{\log_n((T_1 - T_w)/(T_2 - T_w))} \;\; (\text{K})$$

where

T_1 = inlet gas temperature (°C)

T_2 = outlet gas temperature (°C)

T_w = tube wall temperature (°C)

The difference in temperature between the tube wall and the water is small, typically less than 10 K in the convective section. Therefore little error is introduced by using the water temperature as T_w in the evaluation of the gas transport properties.

The representative gas temperatures used in the correlations are the bulk temperature and the film temperature. These are defined as:

Bulk temperature $\qquad T_b = T_w + \Delta T_m$
Film temperature $\qquad T_f = (T_b + T_w)/2$

For longitudinal flow in the tubes of shell boilers the mean heat transfer coefficient may be determined from:

$$Nu = 0.023 Re^{0.8}\, Pr^{0.4}\, (1 + (D/L)^{0.7})$$

where D/L is the tube inside diameter to length ratio and the characteristic dimension in Nu and Re is the tube inside diameter. Gas properties are evaluated at the film temperature.

Correlations for forced convection over tubes in cross-flow are complicated by the effect of the tube bank arrangement.

For the range of Reynolds numbers likely to be encountered in industrial boilers the following equations may be used:

In-line arrays $\qquad Nu = 0.211Re^{0.651} Pr^{0.34} F_1 F_2$
Staggered arrays $\qquad Nu = 0.273Re^{0.635} Pr^{0.34} F_1 F_2$

In these cases gas properties are evaluated at the bulk temperature, the characteristic dimension in Nu and Re is the tube outside diameter, and the Reynolds number is based on the mass velocity through the minimum area for flow between tubes. F_1 is a correction factor for wall-to-bulk property variation which can be calculated from the relationship:

$$F_1 = \frac{(Pr_b)^{0.26}}{(Pr_w)}$$

where Pr_b and Pr_w are Prandtl numbers at the bulk and wall temperatures, respectively. F_2 is a correction factor for the depth of the tube bank in the direction of flow. For bank depths of 10 rows or more $F_2 = 1$. For smaller bank depths the following values of F_2 may be used:

No. of rows	1	2	3	4	5
In-line tubes	0.64	0.8	0.87	0.9	0.92
Staggered tubes	0.68	0.75	0.83	0.89	0.92

No. of rows	6	7	8	9
In-line tubes	0.94	0.96	0.98	0.99
Staggered tubes	0.95	0.97	0.98	0.99

Gas transport properties for the products of combustion of the common fuels, fired at normal excess air at or near full boiler load, may be obtained from Tables 15.20–15.23.

Non-luminous gas radiation has a small overall effect in the convective section, typically 2–5% of total convection. It may therefore be neglected for a conservative calculation.

15.3.3.6 Waterside conditions

In the radiant section of a boiler the fourth power of the wall temperature is typically less than 2% of the fourth power of the mean flame and gas temperature. The effect of waterside conditions and wall thickness on the heat transfer rate are therefore negligible. Even the presence of a dangerous layer of waterside scale reduces the heat flux only by a few per cent. Although this means that scale has little effect on radiant

Table 15.20 Transport properties: natural gas products of combustion

Temp. (°C)	Spec. heat (J kg^{-1} K^{-1})	Viscosity (kg m^{-1} s^{-1} × 10^6)	Conductivity (W m^{-1} K^{-1} × 10^3)	Sp. vol. (m^3 kg^{-1})
100	1098	20.01	27.27	1.1
200	1133	23.97	34.45	1.395
300	1166	27.55	41.34	1.69
400	1198	30.83	47.94	1.985
500	1227	33.89	54.25	2.28
600	1255	36.74	60.29	2.575
700	1281	39.44	66.09	2.87
800	1305	41.99	71.61	3.164
900	1328	44.43	76.86	3.459
1000	1348	46.75	81.86	3.754
1100	1367	48.98	86.6	4.049
1200	1384	51.13	91.08	4.344
1300	1400	53.2	95.31	4.639
1400	1413	55.2	99.25	4.934

Table 15.21 Transport properties: gas oil products of combustion

Temp. (°C)	Spec. heat (J kg^{-1} K^{-1})	Viscosity (kg m^{-1} s^{-1} × 10^6)	Conductivity (W m^{-1} K^{-1} × 10^3)	Sp. vol. (m^3 kg^{-1})
100	1061	20.32	27.24	1.058
200	1096	24.29	34.4	1.342
300	1128	27.88	41.22	1.625
400	1159	31.16	47.73	1.909
500	1188	34.2	53.92	2.192
600	1215	37.05	59.81	2.476
700	1240	39.72	65.42	2.76
800	1263	42.26	70.71	3.043
900	1284	44.67	75.73	3.327
1000	1303	46.98	80.46	3.61
1100	1320	49.19	84.89	3.894
1200	1336	51.32	89.02	4.177
1300	1349	53.37	92.88	4.461
1400	1361	55.35	96.43	4.745

Table 15.22 Transport properties: heavy fuel oil products of combustion

Temp. (°C)	Spec. heat (J kg^{-1} K^{-1})	Viscosity (kg m^{-1} s^{-1} × 10^6)	Conductivity (W m^{-1} K^{-1} × 10^3)	Sp. vol. (m^3 kg^{-1})
100	1054	20.37	27.22	1.05
200	1088	24.34	34.37	1.332
300	1121	27.93	41.17	1.613
400	1152	31.21	47.66	1.895
500	1181	34.25	53.82	2.176
600	1207	37.09	59.69	2.458
700	1232	39.44	66.09	2.87
800	1255	42.3	70.51	3.02
900	1276	44.71	75.47	3.302
1000	1294	47.01	80.15	3.583
1100	1311	49.22	84.51	3.865
1200	1326	51.35	88.59	4.146
1300	1339	53.40	92.38	4.428
1400	1351	55.38	95.86	4.709

Table 15.23 Transport properties: bit coal products of combustion

Temp. (°C)	Spec. heat (J kg^{-1} K^{-1})	Viscosity (kg m^{-1} s^{-1} × 10^6)	Conductivity (W m^{-1} K^{-1} × 10^3)	Sp. vol. (m^3 kg^{-1})
100	1031	20.82	27.43	1.034
200	1065	24.83	34.63	1.312
300	1096	28.44	41.39	1.589
400	1125	31.73	47.78	1.866
500	1152	34.78	53.8	2.143
600	1177	37.63	59.5	2.421
700	1201	40.3	64.88	2.698
800	1222	42.83	69.93	2.975
900	1242	45.24	74.68	3.252
1000	1259	47.55	79.11	3.53
1100	1275	49.75	83.23	3.807
1200	1289	51.87	87.05	4.084
1300	1301	53.92	90.56	4.361
1400	1311	55.89	93.77	4.638

section performance, it also indicates that the metal temperature escalation due to the presence of scale is not self-limiting but is almost proportional to scale thickness.

The thermal conductivity of an average boiler scale is $2.2 (W\ m^{-1}\ K^{-1})$ and that of complex silicate scales is $0.2–0.7 (W\ m^{-1}\ K^{-1})$. Since the furnace peak wall flux can be over $300\,000\ W\ m^{-2}$ it may readily be seen that a small thickness of scale can raise the metal temperature into the creep region, resulting in very expensive repairs.

In the convective section the gas-side heat transfer coefficient controls the heat flux distribution since the waterside coefficient and the thermal conductance of the tube walls are very large in comparison. For this reason it is usually satisfactory to make an allowance by adding 10 K to the water temperature in steam boilers. In hot-water generators the allowance should be about 20 K because sub-cooled nucleate boiling generally takes place only on the radiant walls and, in shell boilers, on the reversal chamber tubeplate. Waterside heat transfer on the major part of the convective heating surface in these units is by convection without boiling.

15.3.3.7 Further reading

A good introduction to the vast literature on the science and technology of heat transfer, with 87 further references, is given in Rose, J. W. and Cooper, J. R., *Technical Data on Fuel*, 7th edn, British National Committee, World Energy Conference, London, p. 48 (1977).

15.3.4 Terminology

The following explain some of the more fundamental terms encountered when considering boilers.

15.3.4.1 Shell boiler

A boiler in which the products of combustion or hot gases pass through a series of tubes surrounded by water. All are contained in an outer shell.

15.3.4.2 Water tube boiler

A boiler where water circulates through small-bore tubes constructed in banks and connected to drums or headers. The external surfaces of the tubes are exposed to the products of combustion or hot gases.

15.3.4.3 Dry-back boiler

A horizontal shell boiler where the gas-reversal chamber from the combustion tube to the first pass of tubes is external to the rear tube plate and is formed by a refractory-lined steel chamber.

15.3.4.4 Wet-back boiler

A horizontal shell boiler where the gas-reversal chamber from the combustion tube to the first pass of tubes is integral within the boiler shell and is surrounded by water.

15.3.4.5 Economic boiler

A term applied to the early free-standing shell boilers of two- and three-pass construction. Originally they were dry-back and later wet-back. These boilers superseded the brickset Cornish and Lancashire boilers. The earliest economic boilers

were also brickset, the gases from the front smokebox returning across the lower external part of the shell contained within the brick setting to form a third pass.

15.3.4.6 Packaged boiler

A concept of a factory-built and assembled shell boiler complete with its combustion appliance, feed-water pump and controls, valves, base frame and insulation. Before this the economic boiler was delivered to site as a bare shell and assembled *in situ*.

Originally in the early 1960s packaged boilers were designed to be as compact as possible, resulting in some inherent faults. Since then design criteria have greatly improved and the present packaged boiler is constructed to acceptable commercial standards.

15.3.4.7 Evaporation

This is the quantity of steam produced by the boiler at temperature and pressure. It may be stated as equivalent evaporation 'from and at 100°C', usually expressed 'F&A 100°C', or 'actual evaporation'.

Evaporation F&A 100°C is a figure taken for design purposes and is based on the amount of heat required to convert water at 100°C to steam at the same temperature. Actual evaporation is the amount of steam passing the crown valve of the boiler. As boilers operate with differing working pressures and feed-water temperatures the above may be compared by using the 'factor of evaporation'.

15.3.4.8 Factor of evaporation

This is the figure obtained by dividing the total heat of steam at working condition by the latent heat of steam at atmospheric condition (i.e. $2256\ kJ\ kg^{-1}$). Then

$$\text{Factor of evaporation} = \frac{H - T}{2256}$$

where

H = total heat in 1 kg of steam at working pressure above 0°C taken from steam tables $(kJ\ kg^{-1})$

T = heat in feed water $(kJ\ kg^{-1})$

2256 = the latent heat of steam at atmospheric conditions.

15.3.4.9 Availability

This is the period of time that a boiler may be expected or required to operate before being shut down for cleaning or maintenance. It will vary with the type of boiler, the fuel being used and the operating load on the boiler.

15.3.4.10 Priming

This is when the water surface in the boiler shell becomes unstable. Vigorous surging will occur and this may cause the boiler to go to low water and cut out or possibly lock out. This, in turn, will exacerbate the condition.

There are two possible causes. The first could be incorrect control of water treatment and blowdown. This can result in excessive levels of suspended solids in the boiler water, organic matter in the boiler water or high alkalinity. The second can be mechanical.

If the boiler is operated below its designed working pressure it will increase the efflux velocity of the steam leaving the water surface area to a point where it may lift the water surface and drop the water level. It is important therefore to give due consideration to the steam load required from the boiler.

15.3.4.11 Thermal storage

A method of supplying a steam load in excess of the maximum continuous rating of the boiler for limited periods.

15.3.4.12 Cavitation

This is a condition which occurs when the feed-water pump is unable to deliver feed water to the boiler although the feed tank has water available. The temperature of the feed water coupled with the possible suction effect from the feed-water pump in the line between the feed tank and the pump effectively drops the pressure, causing the feed water to flash to steam. The pump then loses its water supply. In most cases this condition may be avoided by arranging a sufficient head of water and by correct sizing of the feed-water pipework.

15.3.4.13 Turndown

Turndown, or modulation range, refers to the firing range of the combustion appliance and therefore the output of the boiler. It is the range between the maximum continuous firing rate (MCR) and cut-out to on/off operation. It may be restricted by either the design of the combustion appliance or the boiler. For example, the combustion gas exit temperature from the boiler should not be below steam saturation temperature or fall into dewpoint areas.

15.3.5 Waste-heat boilers

A waste-heat boiler will always be designed to suit its particular installation. Occasionally it may be possible to offer a standard boiler shell for certain applications, but this will only be after careful appraisal by the boiler designer and manufacturer. The sources of waste gases vary widely and could be furnaces, incinerators, gas turbines, diesel exhausts and process plants such as cement or chemical works.

The prime requirement is that the waste gases must contain sufficient useful heat to produce steam or hot water at the condition required. In most processes there is a practical limit on the minimum gas temperature from the waste-heat boiler in order to prevent dewpoint occurring with its associated corrosion problems.

Waste-heat boilers may be horizontal or vertical shell or water tube type. The limitations between shell and water tube are similar to those indicated in Figure 15.132 although now it is also necessary to take account of the mass flow of the gases. These may produce a velocity too high for a shell-type boiler although it is within its evaporative and pressure range.

Shell waste-heat boilers will normally be of single- or two-pass design with one or two banks of tubes. This is acceptable with most gases up to 1200°C. If the waste gas temperature exceeds this, as it may from incineration, then an additional pass similar to a combustion tube will be incorporated to reduce the gas temperature before it enters the tube bank.

Supplementary firing equipment may also be included if a standby heat load is to be met and the waste gas source is intermittent. Waste-heat boilers may be designed to use either radiant or convected heat sources. In some cases problems may arise due to the source of waste heat and due consideration must be taken of this. Examples are plastic content in waste being burned in incinerators, carry-over from some type of furnaces causing strongly bonded deposits and carbon from heavy oil-fired engines. Some may be dealt with by maintain-

ing gas exit temperatures at a predetermined level to prevent dewpoint being reached and others by soot blowing. Currently there is a strong interest in small combined heat and power (CHP) stations and these will normally incorporate a waste-heat boiler.

15.3.6 Economizers

Economizers are installed in the exhaust gas flow from the boiler. They take heat from the flue gases which they transfer via extended surface elements to the feed water immediately prior to the water entering the boiler. They therefore increase the efficiency of the boiler and have the added advantage of reducing thermal shock.

In water-tube boilers they may be incorporated within the structure of the boiler or supplied as a free-standing unit. With shell boilers they will be separate units fitted between the boiler flue gas outlet and the chimney.

Figure 15.133 is a schematic illustration of such a unit. It is desirable for each boiler to have its own economizer. Where one economizer is installed to take the exhaust gases from more than one boiler special considerations must be taken into account. These will include gastight isolation dampers, consideration of flue gas pressures at varying loads, maximum and minimum combined heat load to match economizer and a pumped feed-water ring-main. Economizers may be used for both forced- and induced-draught boilers and in both cases the pressure drop through the economizer must be taken into account when sizing the fans.

Economizers are fitted to most water-tube boilers, an exception being on a waste-heat application. Here it may be desirable due to the nature of the products being burned to maintain a relatively high gas outlet temperature to prevent corrosive damage to the boiler outlet, ductwork and chimney.

With water-tube boilers economizers may be used when burning coal, oil or gas. The material for the economizer will depend on the fuel. They may be all steel, all cast iron or cast iron protected steel. An all-steel construction would be used for non-corrosive fuel gases from the burning of natural gas, light oil and coal. Cast iron may be used where the feed-water condition is uncertain and may attack the tube bore. Fuels may be heavy fuel oil or coal and there is a likelihood of metal temperatures falling below the acid dewpoint. Cast iron protected steel is used when heavy fuel oil or solid fuel firing is required and feed-water conditions are suitably controlled. As cast iron can withstand a degree of acid attack these units have the advantage of being able to operate without a gas bypass where interruptible natural gas supplies are used with oil as standby.

With shell boilers, economizers will generally only be fitted if natural gas is used as main fuel, and then only on larger units. It would be unlikely that a reasonable economic case could be made for boilers of less than 4000 kg h^{-1} F&A 100°C evaporative capacity. The economizer will incorporate a flue-gas bypass with isolating dampers to cover for periods when oil is used and for maintenance. The dampers require electric interlocks to the selected fuel.

As the majority of shell boilers operate in the pressure range 7–10 bar the flue gas outlet temperature will be in the range of 190–250°C. It may be appreciated from this that the boiler needs to operate at 50–100% of its maximum continuous rating for most of the working day to produce an economic return.

Where an economizer is installed it is essential to have water passing through the unit at all times when the burners are firing to prevent boiling. Therefore boilers fitted with economizers will have modulating feed-water control. Even then it is possible that the water flow requirement can become out of

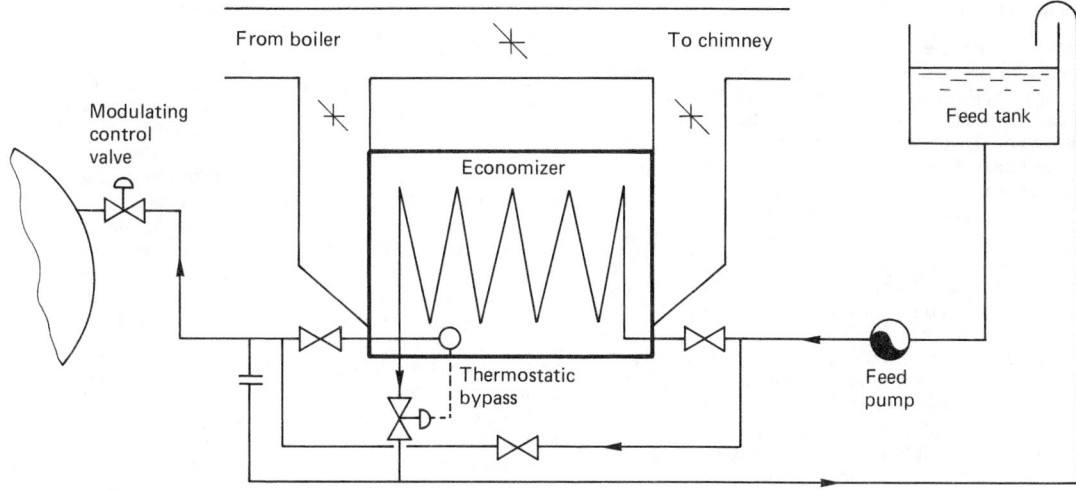

Figure 15.133 Schematic illustration of an installed economizer

phase with the burner firing rate. To prevent damage, a temperature-controlled valve allows a spillage of water back to the feed-water tank thus maintaining a flow of water through the unit. Each economizer will be fitted with a pressure-relief safety valve.

Due to the amount of water vapour produced when natural gas is burned it is important not to allow the exhaust gas temperature to fall below 80°C otherwise the water dewpoint will be reached. Not only the economizer but also the duct-work and chimney must be considered and provision incorporated for drainage.

In the event of a separate use for low-grade hot water being required it is sometimes practical to install a secondary condensing economizer. With this the material of which the economizer is constructed allows for condensate to form and drain away without excessive attack from corrosion.

A recent development in heat recovery has been the heat tube. This is a sealed metal tube which has been evacuated of air and which contains a small quantity of liquid which for boiler applications could be water. When heat from the flue gases is applied to one end of the heat pipes the water in the tube boils, turns to steam, and absorbs the latent heat of evaporation. The steam travels to the opposite end of the tube which is surrounded by water where it gives up its latent heat, condenses and returns to the heated end of the tube. Batteries of these tubes can be arranged to form units usually as a water jacket around a section of flue.

15.3.7 Superheaters

Steam produced from a boiler is referred to as dry saturated and its temperature will correspond with the working pressure of the boiler. In some instances, particularly with shell boilers, this is perfectly acceptable. There are occasions, however, where it is desirable to increase the temperature of the steam without increasing the pressure. This function is performed by a superheater.

Superheated steam may be required where steam-distribution pipework in a plant is over extended distances resulting in a loss of heat and an increase in wetness of the steam. Another case may be where a process requires a temperature above the working pressure of the plant. The

third case is where steam is used for turbines. Here it improves the performance of the turbine where for every 6°C increase in steam temperature it can produce about a 1% reduction in steam consumption.

Steam from the drum or shell of the boiler is passed through a bank of tubes whose external surfaces are exposed to the combustion gases, thus heating the steam while not increasing the pressure. Where a superheater is fitted the boiler working pressure must be increased to allow for the pressure drop through the elements. This will be between 0.3 and 1.0 bar.

In a water-tube boiler the superheater is a separate bank of tubes or elements installed in the area at the rear or outlet of the combustion chamber. Saturated steam temperature may be increased by 200°C with final steam temperature of up to 540°C.

For shell boilers superheaters may be one of three types depending upon the degree of superheat required. The first and simplest is the pendant superheater installed in the front smokebox. The maximum degree of superheat available from this would be around 45°C. The second pattern is again installed in the front smokebox but with this the elements are horizontal 'U' tubes which extend into the boiler smoketubes. The degree of superheat from this pattern is around 55°C. Third, a superheater may be installed in the reversal chamber of the boiler. A wet-back chamber presents problems with lack of space and therefore either a semi-wet-back, dry-back or water-cooled wall chamber may be considered. Maximum degree of superheat would be around 100°C.

Superheater elements are connected to inlet and outlet headers. The inlet header receives dry saturated steam from the steam drum of a water-tube boiler or the shell of a horizontal boiler. This steam passes through the elements where its temperature is raised and to the outlet header which is connected to the services. A thermometer or temperature recorder is fitted to the outlet header.

It should be appreciated that a steam flow must be maintained through the elements at all times to prevent them burning away. If a single boiler is used then provision to flood the superheater during start-up periods may be required.

Superheaters may also be supplied as independently fired units. These may be used when either the amount of super-heated steam required is much less than the boiler evaporation or is required only on an intermittent basis.

15.3.8 Steam storage

Most boilers built now together with their combustion equipment are quick to respond to load fluctuations. Occasionally where very rapid load changes occur, the firing rate of a gas or oil burner can be virtually instantaneous by the use of special control equipment. This control will have to work in conjunction with the boiler and therefore the boiler should have adequate steam space and water surface area to help accommodate the rapid changes in steam demand. Good water treatment is especially important here in order to reduce the risk of priming during peak draw-off periods. A boiler with a large shell will have an advantage over one with a smaller shell, assuming equal heating surfaces, but it will give no more than a slight buffer against severe loads.

Most boiler plants can be installed using one or more boilers which can accommodate minimum to maximum loads. Occasionally heavy peak loads occur for only relatively short periods and here there may be an advantage, on economic running grounds, to install boilers whose firing rate will not meet these peaks. In these cases there are two methods which may be used. One is thermal storage and the other is an accumulator.

15.3.8.1 Thermal storage

The principle is based on a special feed-water control system which allows a volume of water already at temperature and pressure within the shell to convert to steam to meet a load in excess of the maximum firing rate. Conversely, during periods of low steam demand the control system allows the water level to re-establish itself. This is achieved using a constant burner firing rate which should match the average steam demand, thus allowing maximum efficiency. It is claimed that it is possible with this system to control the limits of boiler working pressure to within ±0.07 bar.

15.3.8.2 The accumulator

Unlike thermal storage, this depends upon differential pressures. It is suited to a situation where both high- and low-pressure steam systems are required (for example, 17 bar and 7 bar). Alternatively, if no high-pressure steam is needed then the boilers must be designed to operate at a higher pressure with all steam supplies going to the process through a pressure-reducing station. Any high-pressure surplus then goes to the accumulator to help meet peak loads. Figure 15.134 shows the diagrammatic layout of an accumulator.

The storage vessel is filled to around 90% of its volume with water. The overflow valve is controlled by the pressure of the boilers. On rising steam pressure indicating that the boilers are producing more steam than the process requires a signal to the overflow valve allows all surplus high-pressure steam to flow into the accumulator via a non-return valve and internal distribution header. Here it is condensed and its thermal energy stored. If a peak load develops on the high-pressure system then controls will close the overflow valve and allow steam to discharge from the accumulator through the pressure-reducing valve set to meet the low-pressure steam requirement. Similarly, if the peak develops on the low-pressure system then high-pressure steam may pass directly to the pressure-reducing set to supplement steam from the accumulator.

Every accumulator will be designed to meet its specified duty. It will be appreciated that the greater the differential pressure, the smaller the vessel will need to be.

15.3.9 Water-level control

Water-level controls continuously monitor the level of water in a steam boiler in order to control the flow of feed water into the boiler and to protect against a low-water condition which may expose the heating surfaces with consequent damage. The controls may be either float operated or conductivity probes.

With water tube boilers the control of the water level needs to be precise and sensitive to fluctuating loads due to the high evaporative rates and relatively small steam drums and small water content. Control will be within ±10 mm on the working water level and will be two- or three-element control. Two-element control will comprise modulating feed-water control with the first low-water alarm and high–low control with

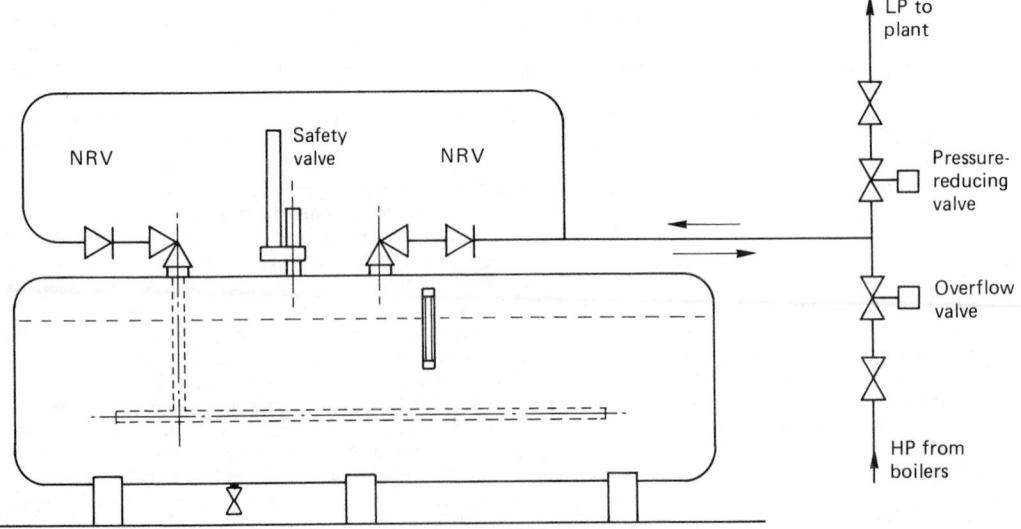

Figure 15.134 Diagrammatic layout of a steam accumulator

low-water cut-out and alarm. The second element will be monitoring of the steam flow to give early indication of any increase in steam demand. This signal may then be linked to the firing rate of the burners and the feed-water modulating valve. The third element senses a drop in feed-water demand which would signal the firing rate of the burners to modulate down.

Shell boilers will have two external level controls each independently attached to the shell. Boilers up to about 9000 kg h^{-1} F&A 100°C will have a dual control and either a single or high–low control. The dual control instigates the feed-water pump which operates on an on–off cycle over a water-level band of ±15 mm and also operates the first low-water alarm. The single or high–low control will incorporate a second low-water alarm with burner lock-out, and with the high–low control also an indication of high water which may be linked to shut down the feed-water pump with automatic restart when the water level drops to normal.

Boilers of larger evaporations will have modulating and high–low control. The modulating level control monitors the working water level in the shell and operates a control valve in the feed-water line allowing water to enter the boiler from a continuously running feed pump. It will also incorporate the first low-water alarm. The high–low control operates as before.

The advantage of modulating control is that it maintains a constant working water level and therefore the boiler is always in its best condition to supply steam for peak loading. These controls may also be fitted to boilers below 9000 kg h^{-1} F&A 100°C if severe loads are present or when the working pressure is above 10 bar.

With water-level controls it is important to check that they are functioning correctly and that they will be operated daily to simulate low-water condition. Shell boilers will also be subject to a weekly evaporation test to prove low-water controls. Blowdown of external level controls is invariably manual but may be motorized on water tube boilers.

Shell boilers may alternatively be fitted with internal level controls. These will have an electronic testing facility operating automatically and arranged to take the boiler to lock-out in the event of a fault.

For fully automatic unstaffed operation three independent level controls are required. These will be first low water and burner cut-out, second low water and burner lock-out and the third level control will be for the feed-water pump and high-water alarm.

With external level controls the sequencing blowdown valve will be motorized and for internal level controls they will be subject to continuous electronic monitoring.

Comprehensive information on Automatically Controlled Steam and Hot Water Boilers is given in the Health and Safety Executive Guidance Note PM5 (Automatically Controlled Steam and Hot Water Boilers).

15.3.10 Automatic control

15.3.10.1 Automatic control of boilers

Whether the boiler is fired on oil, gas or solid fuel, it may be expected to operate automatically. When boiler plant is not run continuously initial start-up may be manual, time clock or through an energy-management system. Manual attendance may be limited to maintenance functions dictated by the size and type of plant.

Automatic controls will cover combustion appliance, water level and blowdown. Requirements to comply are detailed in the Health and Safety Executive Guidance Note PM5. Any degree of automation to the boiler plant should be imple-

mented only after consultation with the covering insurance company.

15.3.10.2 Automatic boiler start

In order to control the operating times of a boiler it is a simple matter to fit each with time-clock control. Alternatively, they may be controlled through a central energy-management system. Either way, a boiler or boilers may be shut down at the end of each day and programmed to restart the following day or when required. Special considerations need to be made if standing periods are extended allowing a boiler to go cold.

With hot-water units time-clock control can operate satisfactorily as automatic bypass valves built into the distribution system will help the heater to achieve its working temperature quickly. With steam boilers it is important that the boiler achieves a reasonable working pressure before steam is allowed into the distribution system.

For example, if boilers are left open to a system for an extended length of time while not firing they will quickly lose their pressure. This is not only wasteful of energy but eventually creates a problem on start-up. To start a boiler on a zero pressure system with all valves open will undoubtedly cause the boiler to prime and go to lock-out condition but not before condensate has, at least in part, flooded the system. Therefore where a time clock is incorporated it is recommended that the crown valve(s) be closed at the end of each working day and opened only after the boiler has reached working pressure the next time it is required. This operation can be automated by the use of motorized or similar valves.

These valves may be fitted to each steam supply line from a manifold adjacent to the boilers and providing adequate safeguards are incorporated to protect the boilers the on-line boiler(s) may be left open to the manifold. Alternatively, each boiler may have its own automatic motorized start-up valve.

Each valve would have a control panel incorporating a timer. This may initially be set to a 'crack' position timed to open after the boiler has started to fire and is already building up pressure. This will allow gentle warm-up of the system while allowing the boiler to achieve working pressure. After this the valve may be set to open in timed adjustable steps to its fully open position. At the end of a timed period, coinciding with the time clock fitted to the boiler, the valve automatically closes at the end of the working cycle.

Where multiple valves are used their control may be incorporated into a single panel or, alternatively, become part of an energy-management system.

15.3.10.3 The automatic boiler house

As described in Section 15.3.10.1, the boiler will operate automatically and may be programmed to operate to suit various cycles (Section 15.3.10.2). There are, however, other areas within the boiler house which still require consideration. The first of these would be the feed-water pump. Each boiler may be fitted with duplicate pumps and these can be arranged for automatic changeover in the event of a failure of one pump by use of a pressure switch and motorized valves. Other areas would be oil-circulating pumps, gas boosters and water-treatment plant and pumps.

There must be an economic limit as to the extent of duplication and on a multi-boiler installation it would be expected to have a degree of reserve capacity if even for a limited period. An energy-management system would probably form part of the automatic boiler house and as such would have the facility to bring on any standby capacity.

15.3.11 Feed-water requirements

Poor or unsuitable water can be a major factor where failure in a boiler occurs. There are four problem areas for which feed water needs suitable treatment and control. These are sludge, foam, scale and corrosion.

Boiler feed water may be from various supplies. If it is from a mains water supply further filtering prior to treatment is unlikely but for other supplies such as boreholes, lakes, rivers and canals filters may be required. Impurities in water may be classed as dissolved solids, dissolved gases and suspended matter and suitable treatment is required.

Table 15.24 indicates the recommended water characteristics for shell boilers and Table 15.25 the water quality guidelines for industrial water tube boilers. Due to the wide parameters encountered in the quality of feed water it is not possible to be specific and define which treatment suits a particular type and size of boiler. The quality of make-up and percentage of condensate returns in a system will both have to be taken into consideration.

For some small boilers it may be possible to supply internal dosing subject to a suitable water supply and other conditions being favourable. However, for anything other than very small installations external treatment is recommended. For shell boiler installations a simplex or duplex base exchange system with suitable dosing is usual, although on larger installations or if the water is excessively hard and there is little condensate return then a de-alkalization plant may be used.

Table 15.24 Recommended water characteristics for shell boilers

For pressures up to 25 bar[a]			
Total hardness in feed water, mg/l in terms of $CaCo_3$ max.	2	20	40

Feed water			
pH value	7.5–9.5	7.5–9.5	7.5–9.5
Oxygen	b	b	b
Total solids, alkalinity, silica	b	b	b
Organic matter	b	b	b

Boiler water			
Total hardness, mg/l in terms of $CaCo_3$ max.	ND[c]	ND	ND
Sodium phosphate, mg/l as Na_3PO_4[d]	50–100	50–100	50–100
Caustic alkalinity, mg/l in terms of $CaCo_3$ min.	350	300	200
Total alkalinity, mg/l in terms of $CaCo_3$ max.	1200	700	700
Silica, mg/l as SiO_2 max.	Less than 0.4 of the caustic alkalinity.		
Sodium sulphite, mg/l as Na_2SO_3 or	30–70	30–70	30–70
Hydrazine, mg/l as N_2H_4	0.1–1.0	0.1–1.0	0.1–1.0
Suspended solids, mg/l max.	50	200	300
Dissolved solids, mg/l max.	3500	3000	2000

[a] 1 bar = 10^5 N/m^2 = 100 kPa = 14.5 lb/in^2.
[b] Numerical values depend upon circumstances but the comments are relevant.
[c] ND: not detectable.
[d] Phosphate is usually added as sodium phosphate but determined as phosphate (PO_4^{3-}): $Na_3PO_4 = 1.73 \times PO_4^{3-}$.

Based on Table 2 of BS 2486: 1978 by permission of BSI.

Table 15.25 Water-quality guidelines recommended for reliable, continuous operation of modern industrial water tube boilers

Boiler feed water			
Drum pressure (psig)	Iron (ppm Fe)	Copper (ppm Cu)	Total hardness, (ppm $CaCo_3$)
0–300	0.100	0.050	0.300
301–450	0.050	0.025	0.300
451–600	0.030	0.020	0.200
601–750	0.025	0.020	0.200
751–900	0.020	0.015	0.100
901–1000	0.020	0.015	0.050
1001–1500	0.010	0.010	ND[d]
1501–2000	0.010	0.010	ND[d]

Boiler water			
Drum pressure (psig)	Silica (ppm SiO_2)	Total alkalinity[a] (ppm $CaCO_3$)	Specific conductance, (μmho/cm)
0–300	150	350[b]	3500
301–450	90	300[b]	3000
451–600	40	250[b]	2500
601–750	30	200[b]	2000
751–900	20	150[b]	1500
901–1000	8	100[b]	1000
1001–1500	2	NS[c]	150
1501–2000	1	NS[c]	100

[a] Minimum level of hydroxide alkalinity in boilers below 1000 psi must be individually specified with regard to silica solubility and other components of internal treatment.
[b] Maximum total alkalinity consistent with acceptable steam purity. If necessary, the limitation on total alkalinity should override conductance as the control parameter. If make-up is demineralized water at 600–1000 psig, boiler water alkalinity and conductance should be shown in the table for the 1001–1500 psig range.
[c] NS (not specified) in these cases refers to free sodium- or potassium-hydroxide alkalinity. Some small variable amount of total alkalinity will be present and measurable with the assumed congruent control or volatile treatment employed at these high-pressure ranges.
[d] None detectable.

For water tube boilers base exchange or de-alkalization may be used providing the water quality is suitable and the boilers are not operating at pressures in excess of 30–35 bar. With modern water tube boilers demineralized water is recommended. Where boilers are operating at high pressures or are used for power generation it is essential to use demineralized water in order to prevent build-up of deposits, particularly silica, on turbine blades.

With hot-water installations it is equally important that water suitably treated for hardness and corrosion should be used. Even when cleaning or flushing a new or modified system care must be taken to prevent premature corrosion occurring by the addition of a suitable treatment. Few (if any) hot-water systems are completely sealed and provision should be designed into the system to treat all make-up water. Draw-off of hot water directly from the system should never be done and a calorifier always used. Analysis of the water in the boiler and system should be carried out at least monthly and more frequently during the commissioning period of a new installation or where an existing system has ben refilled.

Where steam or hot-water boilers are not required to operate for a period of time it is important that suitable measures are taken to prevent waterside corrosion. For periods of a few days the water may be left at its normal level but daily testing must be carried out as if the boiler were in use and corrective treatment added as necessary. If the period is for several months then the boiler should be fully flooded to exclude all air and the water treated. Regular testing of this water should be carried out and corrective treatment used. For longer periods boilers should be drained completely and thoroughly dried out. The boiler may then either be left vented with the addition of a small electric heater inside or sealed and trays of moisture absorbing chemicals such as hydrated lime or silica gel laid inside. In potentially humid atmospheres such as near sea coasts the dry method is preferred, as keeping the boiler full of cold water will cause condensation to be continuously present on the fireside, giving rise to surface corrosion.

15.3.12 Blowdown requirements control and tanks

In order to maintain the level of dissolved and suspended solids within the boiler as recommended in Section 15.3.11 it is necessary for the boiler to be blown down. This is an operation where a quantity of water is drained from the boiler while the boiler is operating at pressure. This may be achieved by various methods.

The simplest, and that applied to small boilers, is for the main bottom blowdown valve to be opened for a set period of time at regular intervals (e.g. 20 seconds every 8 hours). This method may also extend to larger boilers where conditions are such that there is little build-up of solids. Such conditions could be high-condense returns and good-quality make-up feed water.

The second method could be automatic intermittent blowdown. With this a timer-controlled valve is installed at the bottom of the boiler prior to the main blowdown valve. A programme is then designed to operate this valve in short bursts which disperses any sludge and controls the levels of solids. This method is preferred for boilers having internal treatment.

The third method would be continuous blowdown through a regulating or micrometer valve. The take-off position for this should preferably be about 250 mm below the working water level and may either be on the side of the shell or on the crown with a dip pipe down to the correct level. If a connection is not available it is possible to install the valve on the bottom connection prior to the main blowdown valve.

All these methods will require careful monitoring initially to set up and determine the correct rate of blowdown once the plant is operating. In order to take the necessary sample from the boiler the boiler(s) should be fitted with a sample cooler. To automate the continuous blowdown a conductivity-controlled system may be installed. Here a controller continuously compares the boiler water electrical conductivity with a value set in the controller. Depending if this is above or below the set rate it will automatically adjust the blowdown flow rate.

While the above methods control the level of dissolved and suspended solids in the boiler it will still be an insurance requirement to operate the main blowdown valve periodically.

The minimum amount of blowdown may be calculated as a percentage of the evaporation rate by the following formula:

$$\text{Blowdown rate} = \frac{F}{B - F} \times 100\%$$

where F = the total dissolved solids content of the feed in parts per million allowing for the mixture of make-up and condensate plus any chemical treatment and B = the maxi-

mum recommended solids content for boiler water in parts per million.

While vitally necessary, blowdown can be expensive in terms of lost heat. Therefore a point will be reached when it is economical to install a blowdown heat-recovery system. Generally, the heat content in the blowdown water for a shell boiler will represent only about 25% of the heat content in the same percentage of steam. Therefore if a blowdown rate of 10% is required this represents an approximate heat loss of 2.5% from the boiler capacity. This differential reduces and eventually becomes insignificant on high-pressure water tube boilers. The blowdown from the boiler(s) will be run to a flash steam vessel mounted adjacent to the feed tank. Flash steam will be introduced into the feed tank through a dip pipe terminating in a distribution manifold. The drain from the flash vessel may then be taken to a residual blowdown heat exchanger. Any remaining heat is then transferred to the make-up water to the tank before the blowdown runs to drain.

Blowdown from the boiler(s) should always be taken to either a blowdown sump or blowdown vessel before discharging into drains. Both should be adequately sized to give cooling by dilution and be fitted with vent pipes to dissipate pressure safely. The boiler(s) should have independent drain lines for the main manually operated blowdown valve and the drains from a continuous blowdown system. This is set out in Clause 8.7 (Boiler Blowdown and Drain Mountings) in BS 2790: 1989.

15.3.13 Clean Air Act requirement for chimneys and flue designs

15.3.13.1 Introduction

The function of a chimney is to discharge in a manner to give adequate dispersal to the products of combustion in accordance with the third edition of the 1956 Clean Air Act Memorandum on Chimney Heights. The scope of the memorandum is as follows:

1. The publication provides for the use of local authorities, industry and others who may need to determine the height appropriate for certain new chimneys a relatively simple method of calculating the appropriate height desirable in normal circumstances.
2. Heights determined by these methods should be regarded as a guide rather than as a mathematically precise decision on chimney height. The conclusions may need to be modified in the light of particular local circumstances such as valleys, hills and other topographical features.
3. The advice given is applicable only to chimneys of fuel burning plant with a gross heat input of between 0.15 MW and 150 MW, including stationary diesel generators. It does not deal with direct-fired heating systems which discharge into the space being heated, gas turbines or incinerators (which require separate treatment, depending on the pollutants emitted).
4. The main changes from the second edition are the inclusion of a method dealing with very low-sulphur fuel, the extension of the method for taking into account the height of nearby buildings and the extension of the range of the size of furnace included.

15.3.13.2 Gas velocity

In order to maximize the chimney height the efflux velocity of the gases leaving the chimney should be designed on 12 m s^{-1} at maximum continuous rating (MCR) of the boiler. On some very small boilers this may be impractical to achieve but a target velocity of not less than 6 m s^{-1} at MCR should be attempted. With boilers at the top end of the range a velocity of 15 m s^{-1} at MCR is required. Some inner-city authorities may stipulate higher efflux velocities and some plants have been installed with gas velocities of 22 m s^{-1}.

15.3.13.3 Chimney height

Originally the height of the chimney was designed to produce a draught sufficient to produce induced-draught air for combustion. With modern boiler plant forced-draught and/or induced-draught fans are used. This allows for the greater degree of control of the air to be designed into the combustion appliance. The chimney is therefore required only to disperse the gases.

When using gaseous fuel it is normally sufficient to terminate the chimney 3 m above the boiler house roof level subject to there being no higher buildings adjacent to the boiler house. In such cases these buildings may need to be considered.

On medium-size boiler plant where gas is to be the main fuel it may have oil as a secondary standby fuel. In this case the chimney height must be based on the grade of fuel oil capable of being burned.

The methods of calculating proposed chimney height are clearly laid out in the Clean Air Act Memorandum and will be based on:

1. Quantity of fuel burned
2. Sulphur content of fuel burned
3. District Category
4. Adjacent buildings
5. Any adjacent existing emissions.

Application for approval of the proposed chimney height should be made to the appropriate authority at an early stage of a project in order to ascertain their approval or other height they may require. Failure to do this can result in an embarrassing situation where insufficient finance has been allocated due to their requiring a larger chimney than was included in the planned costings.

Where waste products are being incinerated special consideration may have to be given to the resulting flue gases. This may involve having to arrive at a chimney height in conjunction with HM Inspectorate of Factories for Pollution.

15.3.13.4 Grit and dust emissions

Solids emissions from solid and liquid fuel-fired plant are covered in the HMSO publication *Grit and Dust – The measurement of emissions from boiler and furnace chimneys*. This states levels of emissions which should be achieved in existing plant and which should be specified for new plant. Suitable sampling connections should be incorporated into the flue ducting for the use of test equipment if permanent monitoring is not installed.

15.3.14 Energy conservation

Energy conservation in the boiler house can be considered in two areas. One is the selection and installation of suitable equipment and the second is good operation and management.

15.3.14.1 Plant installation

The boiler, flues and chimney, pipework and hotwell where installed should all be insulated to adequate standards and finish. Valves should be enclosed in insulated boxes, although on small installations this can prove disproportionately expensive. The boilers may be fitted with either inlet or outlet air-sealing dampers. These will prevent the flow of ambient air through the boiler during off-load and standby periods thus helping to maintain the heat already in the boiler.

Economizers may be installed particularly if gas is the main fuel. It is unlikely that an economic case can be made for a single boiler if less than 4000 kg h^{-1} evaporation. An economizer can produce fuel savings of 4–5% but it must be remembered that this will be at MCR and if the load factor of the installation is lower then the savings will also be proportionately lower.

Combustion controls such as oxygen trim help to maintain optimum operating conditions especially on gaseous fuels. Instrumentation can give continuous visual and recorded information of selected boiler and plant functions. To be effective, it must be maintained, the data assessed and any required action taken before the information is stored.

Energy-management systems will form an important part of a multiboiler installation whether on steam or hot water. Boiler(s) for base load will be selected and further boilers brought on-line or taken off-line as required. The important feature of these systems is that the selection of boilers coming either on- or off-line will be ahead of the load and programmed to anticipate rising or falling demands.

Computer monitoring and control systems have been recently introduced. These are designed to operate in place of conventional instrumentation. Using intelligent interface outstations connected to a desktop computer, many plant functions may be programmed into the computer and controlled centrally.

15.3.14.2 Operation and maintenance

As most boiler plants installed today are designed for unattended operation it is even more important that early action is taken in the event of the boiler requiring adjustment of combustion or other maintenance. If full instrumentation is not installed then a portable test kit should be used and the plant checked and logged daily or weekly. Perhaps the most obvious waste to look for after steam leaks is a rise in the flue-gas outlet temperature. The boiler will progressively have deposits adhere to its heating surfaces but at an increase in temperature of no more than 16°C above its design outlet temperature it should be cleaned. The time period between cleaning will vary according to the type of fuel and operational load.

15.3.15 Design Standards for pressure vessels, pipes and flanges

BS 779   Cast Iron Boilers for Central Heating and Indirect Hot Water Supply (44 kW and above)

BS 855   Welded Steel Boilers for Central Heating and Indirect Hot Water Supply (44 kW to 3 MW)

BS 2790   Design and Manufacture for Shell Type Boilers

BS 1113   Design and Manufacture for Water Tube Steam Generators (Including superheaters, reheaters, and steel tube economizers)

BS 5500   Design and Manufacture for Unfired Fusion Welded Pressure Vessels

BS 3601   Specification for Carbon Steel Pipe and Tube

BS 3602   Specification for Seamless and ERW Steel Tubes

BS 4504   Circular Flanges for Pipe, Valves and Fittings

BS 4882   Bolting for Flanges and Pressure Containing Purposes

ASME 1989 Part 1   Power Boilers
ASME 1989 Part 2   Material Specification
ASME 1989 Part 8   Pressure Vessel Division 1 Design Code

15.4 Heating, ventilation and air conditioning

15.4.1 Heating

15.4.1.1 Statutory heating regulations

Except for some defined types of accommodation, the use of fuel or electricity to heat premises above a temperature of 19°C is prohibited by the Fuel and Electricity (Heating) (Control) Order 1980. The current Order is an amendment to an earlier Regulation, which limited the temperature to a maximum of 20°C, and although 19°C is generally taken to refer to air temperature the Order does not specify this. The minimum temperature was laid down in the Factories Act 1961 and should be reached one hour after the commencement of occupation.

15.4.1.2 Building regulations

Unfortunately, the optimum results in cutting down space heating energy usage can often be obtained only when a building is at the design stage. Insulation, draught exclusion and the best possible heating system can then be built in at minimum cost. It is usually more expensive to add to (or modify) an existing building. Space heating is probably the largest usage of energy in buildings, so this section considers what can be done to improve insulation and other thermal properties. When energy was relatively cheap, little thought was given to conservation, and these omissions now have to be rectified.

In 1957 the Thermal Insulation (Industrial Buildings) Act laid down standards of insulation for roofs of new buildings. In 1978, Amendments to the Building Regulations specified standards for walls and windows. At this point it is necessary to define the term 'U value', or the insulation characteristic of the building material. This measures the rate at which energy flows through the material when there is a temperature difference of 1°C between the inside and outside faces, and this value is measured in watts (the unit of energy) per square metre of surface area, i.e. W/m^2 °C or W/m^2 K.

Symbol 'K' = °C temperature difference.

The amendments can briefly be summarized in Table 15.26.

Table 15.26

Industrial and commercial buildings
External walls of building enclosing heated spaces, internal walls exposed to unheated ventilated spaces, floors where the undersurface is exposed to outside air or an unheated ventilated space, and roofs over heated spaces (including the cases of ceilings with an unheated ventilated space above them).

Maximum average U value
For factories and storage buildings, such as warehouses, the U value is laid down to be 0.7. For shops, offices, institutional buildings and places of assembly, such as meeting halls, theatres, etc., the maximum average U value is to be 0.6.

The U values for walls, roofs and floor are intended as average figures, so it is permissible to have some areas of the structure underinsulated (i.e. with higher U values) providing other areas have sufficient extra insulation to bring the average of all areas down to (or below) the Regulation values.

Limits are also imposed on window areas and apply to all buildings above 30 m^2 floor area. For the first group, industrial and commercial buildings, these limits apply both to rooflights and to windows in the walls. These percentages for windows or rooflights assume single glazing, and somewhat larger values can be used if double or triple glazing is to be fitted. However, calculations must be produced to show that the total heat loss from such units would be no greater than the single-glazed unit complying with the set limits (Table 15.27).

In most single- and two-storey buildings the largest proportion of heat loss from the building structure is usually through the roof. (In buildings of three storeys or more the losses through walls and windows may overtake the roof loss.)

New Building Regulations for the Conservation of Fuel and Power for England and Wales came into operation on 1 April 1990. The new maximum U values of the elements (W/m^2 K) are shown in Table 15.29.

Table 15.28 gives some of the insulation properties for various building materials. The property given is for the rate at which energy would pass through a unit area of the material. In the standard units it becomes the number of watts that would be transferred through a square metre of the material of normal thickness in the form it would be used, if the air at either side of the material shows a temperature difference of 1°C. In SI units this becomes W/m^2 °C, which, in this case, is commonly known as the U value. The larger the U value, the more energy it will transfer, so the worse are its insulation properties.

The U values are given in W/m^2 °C for various building material under normal weather conditions. There will always be slight variations around these values, dependent on particular manufacturers of the materials. With any insulation which is being fitted, advice should be sought regarding the fire risk and condensation problems.

15.4.1.3 Estimation of heat losses from buildings

The normal procedure in estimating the heat loss from any building is as follows:

1. Decide upon the internal air temperature to be maintained at the given external air temperature.

Table 15.27

Type of building	Maximum permitted glazed area	
	In walls as percentage of wall area	As rooflights as percentage of roof area
Factories and storage	15	20
Offices, shops and places of assembly	35	20
Institutional, including residential	25	20

Note: Where figures for both rooflights and windows in walls are given, these really apply as a combined total. If the full wall window allowance is not used the balance can be reallocated to rooflight areas and vice versa. For example, a factory with only 10% of wall area as windows could add the other 5% of wall area as an increase to the permitted 20% of roof area that could be rooflights.

Table 15.28 *U* values

Roofs

Pitched covered with slates or tiles, roofing felt underlay, foil-backed plasterboard ceiling	1.5
Pitched covered with slates or tiles and roofing felt underlay, foil-backed plasterboard ceiling with 100 mm glass-fibre insulation between joists	0.35
Corrugated steel or asbestos cement roofing sheets	6.1–6.7
Corrugated steel or asbestos cement cladding with 75 mm fibreglass lightweight liner	0.38
Corrugated steel or asbestos cement roofing sheets with cavity and aluminium foil-backed 10 mm plasterboard lining	1.9–2.0
Corrugated double-skin asbestos cement sheeting with 25 mm glass-fibre insulation between with cavity and aluminium foil-backed 10 mm plasterboard lining; ventilated air space	0.8
Steel or asbestos cement roofing sheets, no lining with rigid insulating lining board 75 mm	0.4
Asphalt 19 mm thick or felt/bitumen layer on solid concrete 150 mm thick	3.5
Asphalt 19 mm thick or felt/bitumen layer on 150 mm autoclaved aerated concrete roof slabs	0.9
Flat roof, three layers of felt on chipboard or plasterboard	1.54
Flat roof, three layers of felt on rigid insulating board 100 mm thick	0.29
Timber roof with zinc or lead covering and 25 mm plaster ceiling	0.96

Walls

Steel or asbestos cement cladding	5.3–5.7
Steel or asbestos cement cladding 75 mm fibre glass lightweight liner	0.37
Steel or asbestos cement cladding with plasterboard lining and 100 mm fibre insulating roll	0.4
Solid brick wall unplastered 105 mm	3.3
Solid brick wall unplastered 335 mm	1.7
Solid brick wall 220 mm thick with 16 mm lightweight plaster on inside face	1.9
Brick/cavity/brick (260 mm total thickness)	1.4
260 mm brick/mineral fibre-filled cavity/brick	0.5
260 mm brick/cavity/load-density block	1.0–1.1
Brick/expanded polystyrene board in cavity/low-density block/inside face plastered	0.5
Weather boarding on timber framing with 10 mm plasterboard lining, 50 mm glass-fibre insulation in the cavity and building paper behind the boarding	0.62

Glazing

Single glazing	Wood frame	4.3
	Metal frame	5.6
Double glazing	Wood frame	2.5
	Metal frame	3.2
Triple glazing		2.0
Roof skylights		6.6

Floors

20 mm intermediate wood floor on 100 mm × 50 mm joists 10 mm plasterboard ceiling allowed for 10% bridging by joists	1.5
150 mm concrete intermediate floor with 150 mm screed and 20 mm wood flooring	1.8

Table 15.28 *(cont'd)*

The heat loss through floors in contact with the earth is dependent upon the size of the floor and the amount of edge insulation. Insulating the edge of a floor to a depth of 1 m can reduce the *U* value by 35%. Following are some typical *U* values for ground floors. Effectively, most of the heat loss is around the perimeter of the floor.

Solid floor in contact with the earth with four exposed edges:

150 m × 50 m	0.11
60 m × 60 m	0.15
15 m × 60 m	0.32
15 m × 15 m	0.45
7.5 m × 15 m	0.62
3 m × 3 m	1.47

Suspended timber floors directly above ground. Bare or with linoleum, plastic or rubber tiles:

150 m × 60 m	0.14
60 m × 60 m	0.16
15 m × 60 m	0.37
15 m × 15 m	0.45
7.5 m × 15 m	0.61
3 m × 3 m	1.05

Suspended timber floors directly above ground with carpet or cork tiles:

150 m × 60 m	0.14
60 m × 60 m	0.16
15 m × 60 m	0.34
15 m × 15 m	0.44
7.5 m × 15 m	0.59
3 m × 3 m	0.99

Table 15.29

Building type	Ground floors	Exposed walls and floors	Semi-exposed walls and floors	Roofs
Industrial storage and other buildings, excluding dwellings	0.45	0.45	0.60	0.45

Note: An exposed element is exposed to the outside air; a semi-exposed element separates a heated space from a space having one or more elements which are not insulated to the levels in the table.
   Maximum window areas for single glazing in buildings other than dwellings will be unchanged.

2. Decide the heat transmission coefficient (*U* values) for the outside walls and glass, roof and bottom floor, and the inside walls, ceilings, or of heated spaces adjacent to non-heated spaces.
3. Measure up the area of each type of surface and compute the loss through each surface by multiplying the transmission coefficient by the measured area by the difference between the inside and the outside temperatures.
4. Calculate the cubic contents of each room and, using the appropriate air change rate, the amount of heat required

to warm the air to the desired temperature by multiplying the volume of air by the difference between the inside and outside temperatures and the specific heat of air.

The above calculations will give the heat losses after the building has been heated. Under conditions in which the heating system will operate continuously, satisfactory results will be obtained if the heating system is designed to provide heat equivalent to the amount calculated above. Suitable allowance must be made for losses from mains.

When, however, operation is intermittent, safety margins are necessary. These are, of course, speculative, but the following suggestion has frequently proved satisfactory. When it is necessary to operate after a long period of vacancy, as may happen in certain types of substantially built buildings, it is necessary to add up to 30% to the 'steady state' heat transmissions. In buildings of light construction this margin may be reduced.

In selecting the appropriate U values we must pay due regard to the exposure and aspect of the room. It appears reasonable to make allowance for the height of a room, bearing in mind that warm air rises towards the ceiling. Thus in a room designed to keep a comfortable temperature in the lower $1\frac{1}{2}$ or 2 m, a higher temperature must exist nearer the ceiling, which will inevitably cause greater losses through the upper parts of windows, walls and roof. This effect is greatest with a convective system, i.e. one which relies on the warming of the air in the room for the conveyance of heat. This would occur in the case of conventional radiators, convectors and warm air systems. In the case of radiant heated rooms, this does not occur, and a much more uniform temperature exists from floor to ceiling.

15.4.1.4 Allowance for height of space

In heat loss calculations a uniform temperature throughout the height of the heated space is assumed, although certain modes of heating cause vertical temperature gradients which lead to increased heat losses, particularly through the roof. These gradients need to be taken into account when sizing appliances. Attention is also drawn to the means of reducing the effect of temperature stratification, discussed in Section 15.4.1.9.

15.4.1.5 Characteristics of heat emitters

Designers will need to decide whether it is necessary to add a margin to the output of heat emitters. During the warm-up cycle with intermittently operated heating systems, emitter output will be higher than design because space temperatures are lower. Also, boost system temperatures may be used to provide an emission margin during warm-up. The need for heat emitter margins to meet extreme weather conditions will depend on the design parameters used in determining heat losses.

In summary, although the addition of a modest margin to heat emitter output would add little to the overall system cost and a margin on the heat generator or boiler output can only be utilized if the appropriate emitter capacity is available, the decision should be based on careful discrimination rather than using an arbitrary percentage allowance. In general, for buildings of traditional construction and for the incidence of design weather in normal winters in the UK an emitter margin in excess of, say, 5% or 10% is unlikely to be justified. However, for well-insulated buildings the heat loss reduces in significance relative to the heat stored in or needed to warm up the structure. For such applications a larger heating system

margin is required, and the emitter margin provided would need to be considered accordingly.

15.4.1.6 Central plant size

In estimating the required duty of a central plant for a building it should be remembered that the total net infiltration of outdoor air is about half the sum of the rates for the separate rooms. This is because, at any one time, infiltration of outdoor air takes place only on the windward part of the building, the flow in the remainder being *outwards*.

When intermittent heating is to be practised the pre-heating periods for all rooms in a building will generally be coincident. The central plant rating is then the sum of the individual room heat demands, modified to take account of the *net* infiltration.

If heating is to be continuous some diversity between the several room heating loads can be expected. When mechanical ventilation is combined with heating, the heating and the ventilation plant may have different hours of use, and the peak loads on the two sections of the plant will often occur at different times.

The central plant may also be required to provide a domestic hot water supply and/or heat for process purposes. These loads may have to be added to the net heating load to arrive at the necessary plant duty, but careful design may avoid the occurrence of simultaneous peaks. In large installations the construction of boiler curves may indicate whether savings in boiler rating can be made. In many cases little or no extra capacity may be needed for the hot water supply, its demands being met by 'robbing' the heating circuits for short periods.

15.4.1.7 Selective systems

In some cases the various rooms of a building do not all require heating at the same time of day and here a so-called 'selective system' may be used. The supply of heat is restricted to different parts of the building at different times of the day; the whole building cannot be heated at one time. A typical application is in dwellings where the demands for heat in living spaces and bedrooms do not normally coincide.

In a selective system the individual room appliances must be sized as indicated above, to provide the appropriate output according to heat loss, gains and intermittency. The central plant need only be capable of meeting the greatest simultaneous demands of those room units which are in use at the same time. This will generally lead to a large power being available to meet the demands of those units which form the lesser part of the load. These units may then be operated with a high degree of intermittency.

15.4.1.8 Multiple-boiler installations

Load variation throughout the season is clearly large, and consideration should be given to the number of boilers required in the system. Operation at low loads leads to corrosion and loss in efficiency and should be avoided. On the other hand, a number of smaller boilers gives an increase in capital costs.

It has been shown that when boilers are chosen which have a fairly constant and good efficiency over a working range of 30–100%, then the effects on overall costs (running + capital) of varying the number and relative sizes of boilers in the system is less than 5%. The optimum number depends on the frequency of occurrence of low loads.

Under these circumstances the engineer is free to choose the number of boilers in the system based on practical rather than economic considerations.

Table 15.30 Design water temperatures for warm and hot water heating systems

Category	System design water temperatures (°C)
Warm	40–70
LTHW	70–100
MTHW	100–120
HTHW	Over 120

Note: Account must be taken of the margin necessary between the maximum system operating temperature and saturation temperature at the system operating pressure.

15.4.1.9 Heating systems

Warm and hot water heating systems Warm water or low-, medium- or high-temperature hot water systems are categorized in Table 15.30. Warm water systems may use heat pumps, fully condensing boilers or similar generators, or reclaimed heat. In many cases the system design may incorporate an alternative heat generator for standby purposes or for extreme weather operation. Under such circumstances the system may continue to function at warm water temperatures or could operate at more conventional LTHW ones.

LTHW systems are usually under a pressure of static head only, with an open expansion tank, in which case the design operating temperature should not exceed 83°C. Where MTHW systems operating above 110°C are pressurized by means of a head tank, an expansion vessel should be incorporated into the feed and expansion pipe. This vessel should be adequately sized to take the volume of expansion of the whole system so that boiling will not occur in the upper part of the feed pipe. On no account should an open vent be provided for this type of system.

MTHW and HTHW systems require pressurization such that the saturation temperature at operating pressure at all points in the circuit exceeds the maximum system flow temperature required. A margin of 17 K (minimum) is recommended and is based on the use of conventional automatic boiler plant and includes an allowance for tolerances on temperature set points for the automatic control of heat-generation output. A check must be made on actual tolerance used in the design of a control system to ensure that this allowance is adequate.

When selecting the operating pressure, allowance must be made for the effect of static head reduction at the highest point of the system and velocity head reduction at the circulating pump section, to ensure that all parts of the system are above saturation pressure within an adequate anti-flash temperature margin. Additionally, the margin on the set point of the high-temperature cut-out control should be 6 K, except for boilers fired with solid fuel automatic stokers, where it should be at least 10 K.

Medium- and high-temperature systems should be fully pressurized before the operating temperature is achieved and remain fully pressurized until the temperature has dropped to a safe level. In all systems the heat generator or boiler must be mechanically suitable to withstand the temperature differentials, and the return temperature to the boiler must be kept high enough to minimize corrosion. Automatic controls may be used to achieve this.

Design water flow temperature For low-temperature heating systems using natural convective or radiant appliances the normal design water flow temperature to the system is 83°C

(see also Table 15.30). Boost temperatures may be used on modulated-temperature systems because of the changes in heat output characteristics with varying temperatures. Additionally, comfort aspects must be borne in mind, as forced convective emitters operating on modulated temperature systems can deliver airstreams at unacceptably low temperatures.

For MTHW and HTHW systems heat emitters may be as for LTHW systems, except that, for safety reasons, units with accessible surfaces at water temperature would not normally be employed. Embedded panel coils may be used in conjunction with a MTHW or HTHW distribution system, with insulating sleeves around the coil piping to reduce the heat flow. Alternatively, the coils can be operated as reduced temperature secondary systems by allowing only a small, carefully controlled proportion of flow temperature water to be mixed with the water circulating in the coils. Design arrangements for reduced-temperature secondary systems (sometimes referred to as injection circuits) include fixed provisions for minimum dilution rates. Conventional system-balancing devices with three-port automatic modulating valves to regulate mixed water temperatures and, hence, heat output are used. Automatic safety controls must prevent excessive temperatures occurring in the coil circuits, as floor fabrics or finishes could be damaged very rapidly.

Maximum water velocity The maximum water velocity in pipework systems is limited by noise generation and erosion/corrosion considerations. Noise is caused by the free air present in the water, sudden pressure drops (which, in turn, cause cavitation or the flashing of water into steam), turbulence or a combination of these. Noise will therefore be generated at valves and fittings where turbulence and local velocities are high, rather than in straight pipe lengths.

A particular noise problem can arise where branch circuits are close to a pump and where the regulating valve used for flow-rate balancing may give rise to considerable pressure differences. Oversizing regulating valves should be avoided, as this will result in poor regulation characteristics; the valve operating in an almost shut position and creating a very high local velocity.

High water velocities can result in erosion or corrosion due to the abrasive action of particles in the water and the breakdown of the protective film which normally forms on the inside surface of the pipe. Erosion can also result from the formation of flash steam and from cavitation caused by turbulence.

Minimum water velocity Minimum water velocities should be maintained in the upper floors of high-rise buildings where air may tend to come out of solution because of reduced pressures. High velocities should be used in down-return mains feeding into air-separation units located at a low level in the system.

System temperature drop British practice on LTHW systems uses a typical system temperature drop of 11 K and a maximum system temperature of 17 K. Continental practice has tended to use higher drops (up to 40 K). An advantage of a higher system temperature drop is the reduction in water flow rates. This will result in reduced pipe sizes with savings in capital cost and distribution heat losses and a reduced pump duty, with savings in running costs. A disadvantage of higher system temperature drops is the need for larger and consequently more expensive heat emitters. However, if it is possible to raise the system flow temperature so that the mean water temperature remains the same, then with certain types of emitter only a small increase in size is required. With large system temperature drops the average water temperature in a radiator tends to fall below the mean of flow and return temperature and, thus, a larger surface is needed. Furthermore, on one-pipe circuits the progressive reduction in temperature around the circuit may lead to excessively large heat emitters.

Higher system temperature drops can be used with MHTW and HTHW systems since the mean temperature of the heat emitters will be correspondingly higher. Additionally, these media are well suited to use for primary distribution systems, conveying heat over long distances.

Precautions should be taken to prevent the danger of injury from contact with hot surfaces. The safe temperature for prolonged contact is relatively low and reference should be made to BS 4086 and other sources.

Use of temperature-limiting valves on emitters On some group and district heating schemes, outlet limiting valves which permit flow only when the water temperature has dropped to a specified low level are used. This procedure minimizes the water quantity to be pumped and permits indicative heat metering by water quantity alone. In such cases care must be taken to size emitters to suit the available water temperatures. The effect of low water velocities through the emitter must also be taken into consideration, since the heat output of some convective appliances is greatly reduced under such conditions.

Miscellaneous components Data regarding relief valves, feed and expansion cisterns are available in Table 15.31.

Distribution system design The design of pipework distribution systems must allow for the following:

1. Future extensions, where required, by the provision of valved, plugged or capped tee connections.
2. Provision for isolation for maintenance. Where it is necessary to carry out maintenance on a 'live' system, valves must be lockable and may need to be installed in tandem.
3. Thermal expansion.
4. Provision for distribution flow rate balancing for initial commissioning or rebalancing to meet changed operational requirements. Typical provisions for balancing comprise the following:
   (a) A measuring station – which may be an orifice plate, a venturi, an orifice valve or other proprietary device – provided with a pair of tappings to permit the measurement of upstream and downstream system dynamic pressures.
   (b) An associated regulating valve – preferably a double-regulating valve or other arrangement which permits the required setting to remain undisturbed by closure.
5. Provision for drainage, including drainage after precommission flushing; water circulation during flushing must be in excess of design flow rates and, in order to discharge the flushing effluent effectively, drainage connections must be full diameter.
6. Removal of air from the system by provision of:
   (a) Air separators, one form of which uses the principle of centrifugal force to separate the heavier constituent (water) from the lighter one (non-condensable gases). Best results are achieved by locating the separator at the highest temperature point of the system where air has a greater tendency to come out of solution. The velocity of the medium requires to be above the minimum stated by the manufacturer (usually about 0.25 m/s).

   (b) Automatic air vents for systems operating at temperatures below atmospheric boiling point.
   (c) Air bottles with manually operated needle valves to release accumulated air, for systems operating at temperatures in excess of atmospheric boiling point.
7. Provision of test points for sensing temperature and pressure at selected locations.

Sealed heating systems Pressurization of medium- and high-temperature hot water sealed heating systems referred to above may take the following forms:

1. *Pressurization by expansion of water* The simplest form of pressurization uses the expansion of the water content of the system to create a sufficient pressure in an expansion vessel to provide an anti-flash margin of, say, 17°C at the lowest pressure (highest point) of the system. The main disadvantage of a naturally pressurized expansion vessel is the ability of water to absorb air and the consequent risk of oxygen corrosion.

   A diaphragm expansion vessel is divided into two compartments by a special membrane or diaphragm of rubber or rubber composition which prevents the water coming into contact with the air. On one side of the diaphragm the vessel is filled with air or nitrogen at the required pressure. The other section of the vessel is connected directly to the water system. A correctly positioned air separator will assist in de-aerating the water in the system.

2. *Pressurization of elevated header tanks* Given very careful attention to design, installation and commissioning, MTHW systems may be operated with the necessary system pressure provided by an elevated feed and expansion tank. Where the system operating temperature exceeds 110°C an expansion vessel should be sized to absorb the volume of expansion for the complete system, thus preventing water at operating temperatures entering the feed and expansion tank and causing boiling. On no account should an open vent be provided for this type of system.

3. *Gas pressurization with spill tank* This form consists of a pressure cylinder maintained partly filled with water and partly with gas (usually nitrogen) which is topped up from pressure bottles. Water expansion is usually arranged to discharge from the system through a pressure-control valve into a spill tank open to atmosphere or to a closed cylinder lightly pressurized with nitrogen. A pump is provided to take water from the spill tank and return it under pressure to the system as cooling-down results in a pressure drop. The pump operation is regulated by a system presure sensor.

4. *Hydraulic pressurization with spill tank* In this form the pressure is maintained by a continuously running centrifugal pump. A second pump under the control of a pressure switch is provided to come into operation at a predetermined pressure differential and as an automatic standby to the duty pump. Surplus water is delivered to or taken from a spill tank or cylinder as described previously.

5. *Example of pressure differential*

   Assume system flow temperature of 120°C
   Allow 17 K anti-flash margin – 137°C
   Corresponding absolute pressure   3.4 bar
   Assume static absolute pressure on system   2.0 bar
   Minimum absolute pressure at cylinder   5.4 bar
   Allow operating differential on pressure cylinder,
   say –   0.5 bar
   Minimum operating absolute pressure of system 5.9 bar

6. *Example of water expansion*

Assume water capacity of system   200 000 l
Assume ambient temperature of 10°C
Assume system maximum flow temperature of 120°C
Assume system minimum return temperature of 65°C
Increase in volume from 10°C to 65°C

$$200\ 000\ \frac{(999.7 - 980.5)}{980.5} \quad = \quad 3916\ l$$

Increase in volume from 65°C to 120°C

$$200\ 000\ \frac{(980.5 - 943.1)}{943.1} \quad = \quad 7931\ l$$

Total increase in volume   =   11 847 l

Maintenance of water heating systems A common practice in many hot water heating installations is to drain the complete system during summer months. This practice, involving a complete change of raw water every year, is to be deprecated. It introduces additional hardness salts and oxygen to the system, resulting in very significant increases in scaling and corrosion. Where it is necessary to drain the boiler or heat generator or other parts of the system for inspection or maintenance purposes, isolating valves or other arrangements should be used to ensure that the section drained is kept to a minimum.

Steam heating systems These are designed to use the latent heat of steam at the heat emitter. Control of heat output is generally by variation of the steam saturation pressure within the emitter. For heating applications with emitters in occupied areas low absolute pressures may be necessary in order to reduce the saturation temperature to safe levels.

The presence of non-condensable gases in steam systems (e.g. air and CO_2) will reduce the partial pressure of the steam, and hence its temperature, thus affecting the output of the appliance. A further adverse effect is the presence of a non-condensable gas at the inside surface of a heat emitter. This impedes condensation and, hence, heat output. It is therefore imperative that suitable means are provided to prevent formation of CO_2 and to evacuate all gases from the system.

Superheat, which must be dissipated before condensation occurs, can be used to reduce condensation in the distribution mains.

On–off control of steam systems can result in the formation of a partial vacuum, leading to condensate locking or back feeding, and infiltration of air which subsequently reduces the heat transfer.

When using modulating valves for steam, heat emitter output must be based on the steam pressure downstream of the valve, which often has a high-pressure drop across it, even when fully open.

Steam traps must be sized to cope with the maximum rate of condensation (which may be on start-up) but must perform effectively over the whole operational range, minimizing the escape of live steam.

Partial waterlogging of heater batteries can lead to early failure due to differential thermal expansion. Steam trap selection should take account of this.

Where high temperatures are required (e.g. for process work) and lower temperatures for space heating, it is desirable to use flash steam recovery from the high-temperature condensate to feed into the low-temperature system, augmented as required by reduced pressure live steam.

Steam as a medium for heating is now seldom used. Hot water, with its flexibility to meet variable weather conditions and its simplicity, has supplanted it in new commercial buildings. Steam is, however, often used for the heating of industrial buildings where steam-raising plant occurs for process or other purposes. It is also employed as a primary conveyor of heat to calorifiers such as in hospitals, where again steam boiler plant may be required for sundry duties such as in kitchens, laundry and for sterilizing. Heating is then by hot water served from calorifiers.

High-temperature thermal fluid systems Where high operating temperatures are required, high-temperature thermal fluid systems may be used instead of pressurized water or steam systems. These systems operate at atmospheric pressure using non-toxic media such as petroleum oil for temperatures up to 300°C or synthetic chemical mixtures where temperatures in excess of this are required (up to 400°C). Some advantages and disadvantages of thermal fluid or heat transfer oil systems are listed below.

Advantages

No corrosion problems.
Statutory inspections of boilers/pressure vessels not required.
No scale deposits.
No need for frost protection of system.
Cost of heat exchangers/heat emitters less, as only atmospheric pressures are involved.
Better energy efficiency than steam systems.
Operating temperature can be increased subsequent to design without increasing operating pressure.

Disadvantages

Medium more expensive than water (but no treatment costs).
Medium is flammable under certain conditions.
Heat transfer coefficient is inferior to that of water.
Care necessary in commissioning and in heat-up rates due to viscosity changes in medium.
Circulating pump necessary (not required for steam systems).
Air must be excluded from the system.
In the event of leakage the medium presents more problems than water.

Warm air heating systems These may be provided with electric or indirect oil- or gas-fired heaters or with a hot water heater or steam battery supplied from a central source. Because the radiant heat output of warm-air systems is negligible, the space air temperature will generally need to be higher for equivalent comfort standards than for a system with some radiant output. This will increase energy use, and legislative standards for limiting space temperatures should be considered. Attention is drawn to the vertical temperature gradient with convective systems and, when used for cellular accommodation, the likelihood of some spaces being overheated due to the difficulty of controlling such systems on a room-by-room basis.

With the advent of natural gas, direct-fired warm air systems are used where the heat and products of combustion, diluted by fresh air introduced into the system, are distributed to the heated spaces. In designing such installations account must be taken of the requirements of the Building Regulations 1985, Part J, and of the Regional Gas Authority. Care must also be taken in design and application to ensure that the moisture in the products of combustion will not create condensation problems. Direct-fired systems are more suited to large, single-space low-occupancy applications such as warehouses and hangars and should not be used to serve sleeping accommodation.

Reducing the effect of temperature stratification As with all convective systems, warm air heating installations produce large temperature gradients in the spaces they serve. This results in the inefficient use of heat and high heat losses from roofs and upper wall areas. To improve the energy efficiency of warm air systems, pendant-type punkah fans or similar devices may be installed at roof level in the heated space. During the operational hours of the heating system these fans work either continuously or under the control of a roof-level thermostat and return the stratified warm air down to occupied levels.

The energy effectiveness of these fans should be assessed, taking into account the cost of the electricity used to operate them. The following factors should also be borne in mind:

1. The necessary mounting height of fans to minimize draughts;
2. The effect of the spacing of fans and the distance of the impeller from the roof soffit;
3. Any risk to occupants from stroboscopic effects of blade movements;
4. The availability of multi- or variable-speed units.

Punkah fans may also be operated during summer months to provide air movement and offer a measure of convective cooling for occupants.

High-temperature high-velocity warm air heating systems
These systems, best suited to heating large, single spaces, may use indirect heating by gas or oil or direct gas heating. Relatively small volumes of air are distributed at high temperature (up to 235°C) and high velocity (30–42.5 m/s from heater unit) through a system of well-insulated conventional ductwork. Air outlets are in the form of truncated conical nozzles discharging from the primary ductwork system into purpose-designed diffuser ducts. The high-velocity discharge induces large volumes of secondary air to boost the outlet volume and reduces the outlet temperature delivered to the space, thereby reducing stratification. Most of the ductwork thermal expansion is absorbed by allowing free movement and long, drop-rod hangers are used for this purpose. Light, flexible, axial-bellows with very low thrust loads can also be employed where free expansion movement is not possible. System design and installation is generally handled as a package deal by specialist manufacturers.

15.4.1.10 Heating equipment – attributes and applications

Water system heating equipment The range of heat emitters may be divided into three generic groups:

1. Radiant
2. Natural convective
3. Forced convective

Table 15.31 lists the principal types of appliance in each group, together with descriptive notes. Typical emission ranges are quoted for each type over its normal span of working temperatures. These are intended as a guide only and manufacturers' catalogues should be consulted for detailed performance values.

Electric heating equipment Where electric heating equipment is installed within the space to be heated the total electrical input is converted into useful heat. There are two categories of electric heating equipment, direct acting and storage heating. The two types of electric heating can be used independently or to complement one another to meet particular heating requirements.

Gas- and oil-fired heating equipment Where gas or oil appliances are used for heating and installed within the heated space, between 70% and 90% of the total energy content of the fuel input will be converted into useful heat.

15.4.2 Ventilation

15.4.2.1 Introduction

Ventilation systems are defined as systems providing air movement through a space without artificially heating or cooling the air. It must be said, however, that, in practice, there is often a large degree of overlap, since office ventilation systems often provide heating in winter and complex ducted ventilation systems share much equipment and design procedures with air-conditioning systems.

This section is intended to provide guidance towards defining needs, assessing whether ventilation is the correct solution and selecting equipment and systems to match these requirements in as economic a manner as possible.

Reasons for ventilation Ventilation is used to maintain a satisfactory environment within enclosed spaces. The environmental criteria controlled may be:

Temperature – relief from overheating
Humidity – prevention of condensation or fogging
Odour – dilution of odour from smoking, body odour, processes, etc.
Contamination – dilution or removal of dangerous or unpleasant fumes and dust

The required values for these criteria will depend upon the reason the space is being ventilated. It may be for the benefit of people, processes, equipment, materials, livestock, horticulture, building preservation or any combination of these. Guidance on selection of these values is provided by CIBSE[30] and ASHRAE.[33]

Definitions

Aerodynamic area – The effective theoretical open area of an opening. It is related to the measured area by the coefficient of entry or discharge (C_d).
Air-handling unit – A self-contained package incorporating all equipment needed to move and treat air, requiring only connection to ductwork and services to provide a complete ventilation system.
Coefficient (entry or discharge) – The ratio of aerodynamic (effective) area to the measured area of an opening. The value for a square-edged hole of 0.61 is used for most building openings.
Capture velocity – The air velocity needed to capture a contaminant at source, overcoming any opposing air currents.
Automatic fire ventilation – See *Smoke ventilation.*
Dilution ventilation – A ventilation strategy whereby contaminants are allowed to escape into the ventilated space and are then diluted to an acceptable level by means of the ventilation system.
Industrial ventilation – A term used to cover any ventilation system designed to remove contaminants. Its use is sometimes restricted to local extract systems.
Maximum Exposure Limit (MEL) – Maximum limits of concentration of airborne toxic contaminants, listed by the Health and Safety Executive[47] which must not be exceeded.
Occupational Exposure Standards (OES) – Limits of concentration of airborne toxic contaminants, listed by the Health and Safety Executive[47] which are regarded as safe for prolonged exposure for 8 hours per day.

Table 15.31   Characteristics of water system heating equipment

Type	Advantages	Disadvantages	Emission range
Radiant Radiant panel	No moving parts, hence little maintenance required; may be mounted at considerable height or, in low-temperature applications, set flush into building structure.	Slow response to control; must be mounted high enough to avoid local high intensities of radiation (e.g. onto head).	350 W/m² to 15 kW/m² of which up to 60% may be radiant.
Radiant strip	No moving parts, hence little maintenance required; may be mounted at considerable height or, in low-temperature applications, set flush into building structure.	Slow response to control; must be mounted high enough to avoid local high intensities of radiation (e.g. onto head).	150 W/m to 5 kW/m of which radiant emission may be up to 65% of total.
Natural convective Radiators	Cheap to install; little maintenance required.	Fairly slow response to control. With steel panel radiators there is a risk of corrosive attack in areas having aggressive water, which may be accentuated by copper swarf left in the radiator. This leads to rapid failure unless a suitable inhibitor is used. Not suitable for high-temperature water or steam.	450–750 W/m².
Natural convectors	May be used on high-temperature hot water or low-pressure steam without casing temperature becoming dangerously high: fairly rapid response to control.	Take up more floor space than radiators. Likelihood of fairly high-temperature gradients when using high-temperature heating media.	200 W to 20 kW.
Continuous convectors	Take up relatively little space; give even distribution of heat in room. May be used with medium-temperature hot water or low-pressure steam without casing temperatures becoming dangerously high. Return pipework may be concealed within casing.	May produce large temperature gradients on high-temperature heating media if poorly sited.	500 W/m to 4 kW/m.
Skirting heating	May be used on water or low-pressure steam. Gives low-temperature gradients in the room. All pipework concealed.	Relatively low output per metre of wall. More work involved when installing in existing building as existing skirting has to be removed.	300 W/m to 1.3 kW/m.
Forced convective Far convectors	Rapid response to control by individual thermostat. By use of variable speed motors rapid warm-up available in intermittent systems; filtered fresh air inlet facility.	Electric supply required to each individual unit.	2 to 25 kW.
Unit heaters	Rapid response to control by individual thermostat; by use of multi-speed motors rapid warm-up available on intermittent systems; filtered fresh air inlet facility.	Electric supply required for each individual unit.	3 to 300 kW.

Infiltration – Movement of air through a space with no specific ventilation openings by natural forces.

Local extract – A ventilation strategy whereby heat, steam or contaminants are captured at source and ducted to discharge outside the space.

Mechanical ventilation – See *Powered ventilation*.

Natural ventilation – A ventilation system in which air movement is produced through purpose-designed openings by natural forces (wind and thermal buoyancy).

Powered ventilation – A ventilation system in which air movement is induced by mechanical means – almost invariably a fan.

Smoke logging – The filling of a space with smoke in the event of fire.

Smoke ventilation – A ventilation system designed to remove smoke and heat in the event of fire to prevent or delay smoke logging allowing personnel to escape and firefighters to attack the fire.

Spot cooling – A ventilation strategy whereby the space temperature is allowed to rise and air movement is induced locally to provide comfort conditions within a limited area.

Threshold Limit Value (TLV) – Maximum values of concentrations of airborne toxic contaminants, listed by the American Conference of Governmental Industrial Hygienists[34] (ACGIH), regarded to be safe for 8 hours per day exposure.

Transport velocity – The air velocity required in a duct to transport a contaminant without it falling out of suspension.

15.4.2.2 Ventilation systems and controls

How natural ventilation works Natural ventilation operates by means of airflows generated by pressure differences across the fabric of the building. An airflow will occur wherever there is a crack, hole or porous surface and a pressure difference.

For the relatively large openings in which we are interested the flow rate can be found from the velocity or airflow generated through the aerodynamic area of the opening from the formulae:

$$V = \sqrt{\frac{2\Delta P}{\rho}} \tag{15.38}$$

where V = velocity (m/s),
ΔP = pressure difference (Pa),
ρ = density (kg/m^3).

Then flow rate:

$$\dot{V} = AC_d V \tag{15.39}$$

where $\dot{V}$ = volumetric flow rate (m^3/s),
A = measured area of opening (m^2),
C_d = coefficient of opening.

For purpose-built ventilators the manufacturer will be able to provide values of C_d. For other openings it is conventional to use the value for a sharp-edged square orifice of 0.61.

The pressure can be generated by three mechanisms:

1. Powered ventilation equipment;
2. Buoyancy (temperature difference);
3. Wind.

In still air conditions the source of pressure difference to drive ventilation is buoyancy due to the decrease in density of heated air. In any occupied building there will be a higher temperature inside than outside due to heat gains from people, plant and solar radiation. The lighter heated air will try to rise, causing an increase in internal pressure at high level and a reduction at low level with a neutral plane between the two conditions. Any opening above the neutral plane will therefore exhaust air and any opening below the neutral plane will provide inlet air. Under steady heat load conditions a balance will be achieved with a throughput of air dependent upon the heat load and the size and location of the openings. Conditions at this balance point can be readily calculated using one of the following formulae:

For more than one opening (inlets all at one height, exhausts all at one height)

$$\dot{V} = A_e C_e \sqrt{\frac{2gH\Delta t}{\bar{T}}} \tag{15.40}$$

For a single opening

$$\dot{V} = \frac{AC_d}{3} \sqrt{\frac{gh\Delta t}{\bar{T}}} \tag{15.41}$$

where g = acceleration due to gravity (m/s^2),
H = height between centre lines of inlet and outlet openings (m),
Δt = temperature difference between inside and outside (°C),
$\bar{T}$ = average of inside and outside temperatures (absolute) (K),
h = height of single opening (m),
$C_e A_e$ = overall effective opening size calculated from

$$\frac{1}{(C_e A_e)^2} = \frac{1}{(\Sigma\, C_i A_i)^2} + \frac{1}{(\Sigma\, C_v A_v)^2}$$

(subscript i denotes inlet opening, subscript v exhaust opening).

Under wind conditions a complex system of pressures is set up on the external surfaces of the building which will vary with wind speed and direction. Pressure coefficients C_p[35,36] define the relationship according to the formula:

$$\dot{V} = A_e C_e U_r \sqrt{(\Delta C_p)} \tag{15.42}$$

where U_r = reference wind speed,
ΔC_p = difference between coefficients at ventilation openings.

The coefficients C_p will vary across each surface of the building and, except for very simple shapes, can only be found by model or full-scale test. Since the coefficients will change with wind direction, complete calculation of wind-induced ventilation is very unwieldy, needing computer analysis.

When both wind and temperature difference act on ventilation openings the result is very complex, but a reasonable approximation of flow rate is made by taking the higher of the two individual flow rates. This means that we can, for ventilation design purposes, generally ignore wind effects and design on temperature difference only, since wind effects can be assumed only to increase the ventilation rate.

Advantages and disadvantages

Advantages	Disadvantages
Quiet	Variable flow rate and
Virtually no running cost	direction dependent upon
Self-regulation (flow rate	wind conditions
increases with heat load)	Filtration is generally
Low maintenance cost	impractical
Provides daylight when	Limited ducting can be
open (roof vent)	tolerated
Psychological appeal of	Effectiveness depends on
clear sky (roof vent)	height and temperature
Easy installation	difference

When to use natural ventilation Natural ventilation is used in a number of situations:

1. Shallow-plan offices – by opening windows to remove heat and odour;
2. Large single-storey spaces (factories, warehouses, sports halls, etc.) – by roof and wall ventilators – to remove heat, contaminants, smoke, steam;
3. Plant rooms.

It is not suitable in situations where:

1. Dust, toxic or noxious contaminants must be removed at source;
2. Unfavourable external conditions exist requiring treatment to incoming air – e.g. noise, dust, pollution;
3. A steady controlled flow rate is required – e.g. hospitals, commercial kitchens;
4. Existing mechanical ventilation will affect the flow adversely;
5. Abnormal wind effects can be anticipated due to surrounding higher buildings;
6. The space is enclosed so as to have no suitable source of inlet air.

In many of these situations a system of natural inlet/powered exhaust or powered inlet/natural exhaust will be the best option.

Control Low-level ventilation openings, whether windows, doors or ventilators, are generally manually operated for simplicity and economy, allowing personnel to control their own environment. High-level openings can also be manually controlled by means of rod or cable operation, although this has generally lost favour (except in the case of simple windows) and automatic operation is preferred.

Automatic operation may be by means of compressed air, operating a pneumatic cylinder, or electricity. Pneumatics are generally favoured for industrial applications and electricity for commercial premises. Economy of installation is normally the deciding factor, since running costs are low for either system.

Automatic control allows a number of options to be considered to provide the best form of control for the circumstances. Generally available controls offer the following features:

1. Local control by personnel;
2. Automatic thermostatic control (single or multiple stage);
3. Fire override to open ventilators automatically by means of a connection to the fire-detection system or fireman's switch. This normally overrides all other control settings;
4. Timeswitch control to shut ventilators during unoccupied periods;
5. Weather override to close ventilators during rain or snow;
6. Wind override to shut high-level exhaust ventilators on windward walls (mainly used for smoke ventilation).

How powered (mechanical) ventilation works[38] By definition, a powered ventilation system includes a mechanical means of inducing an airflow using an external power source. This is invariably an electrically driven fan. When a fan blade rotates it does work on the air around it, creating both a static pressure increase (P_s) and an airflow across the fan. The airflow has a velocity pressure associated with it, defined as $P_V = \frac{1}{2}\rho V^2$, and the fan can be described as producing a total pressure $P_T = P_s + P_V$. The pressure generated is used to overcome pressure losses (resistances) within the ventilation system.

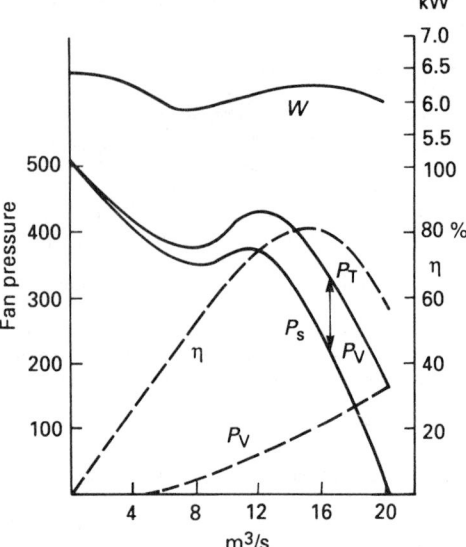

Figure 15.135 Typical fan curve for an axial fan

Each fan has a unique set of characteristics which are normally defined by means of a fan curve produced by the manufacturer which specifies the relationship between airflow, pressure generation, power input, efficiency and noise level (see Figure 15.135). For geometrically similar fans the performance can be predicted for other sizes, speeds, gas densities, etc. from one fan curve using the 'fan laws' set out below.

For a given size of fan and fluid density:

1. $\dfrac{\dot{V}_1}{\dot{V}_2} = \dfrac{N_1}{N_2}$ Volume flow is directly proportional to fan speed

2. $\dfrac{P_1}{P_2} = \left(\dfrac{N_1}{N_2}\right)^2$ Total pressure and static pressure are directly proportional to the square of the fan speed

3. $\dfrac{W_1}{W_2} = \left(\dfrac{N_1}{N_2}\right)^3$ Air power and impeller power are directly proportional to the cube of the fan speed

For changes in density:

4. $\dfrac{P_1}{P_2} = \dfrac{W_1}{W_2} = \dfrac{\rho_1}{\rho_2}$ Pressure and power are directly proportional to density and therefore for a given gas are inversely proportional to absolute temperature

For geometrically similar fans operating at constant speed and efficiency with constant fluid density:

5. $\dfrac{\dot{V}_1}{\dot{V}_2} = \left(\dfrac{D_1}{D_2}\right)^3$ Volume flow is directly proportional to the cube of fan size

6. $\dfrac{P_1}{P_2} = \left(\dfrac{D_1}{D_2}\right)^2$ Total pressure and static pressure are directly proportional to the square of fan size

7. $\dfrac{W_1}{W_2} = \left(\dfrac{D_1}{D_2}\right)^5$ Air power and impeller power are directly proportional to the fifth power of fan size

where

$\dot{V}$ = volumetric flow (m³/s),
P = pressure (kN/m²),
W = power (W),
D = size parameter (diameter) (mm).

In passing through the fans gases are compressed slightly due to the increase in pressure. For absolute accuracy note should be taken of this effect using the gas compressibility factor which will affect flow rate, static and total pressure and power. However, in most fan systems the effect is very small, since the pressure increase through the fan is insignificant compared to atmospheric pressure. By convention, compressibility effects are therefore normally ignored.

Since the pressure generated by most fans is far in excess of pressure differences due to buoyancy and wind, the performance of a powered ventilation system is effectively independent of these, and flow rates and directions can be confidently predicted and will be constant regardless of conditions. The high-pressure generation also allows resistive components such as heater batteries, filters and attenuators to be used within the system. Advantages and disadvantages are:

Advantages	Disadvantages
Weatherproof	Fixed air flow – not self-regulating
Predictable constant performance	Running costs (electrical and maintenance)
Air treatment can be incorporated	Noise
Fresh air can be delivered at optimum volume, velocity and temperature	

When to use powered ventilation Powered ventilation is essential in some instances:

1. Local extract;
2. When pre-treatment of incoming air is required;
3. When a steady controlled airflow is required;
4. When there are no suitable external walls or roof for natural ventilation;
5. In deep-plan offices or large industrial spaces to provide positive air movement in central zones.

It can also be used in any situation where natural ventilation is suitable, generally becoming more economic as the roof height lowers, subject to noise levels being acceptable.

Control Simple systems are normally controlled by a starter or contactor with manual push-button or thermostatic operation to start and stop the fan. More complex systems incorporating other components needing control or monitoring are normally operated from purpose-built central control panels. The most common functions provided are fan motor stop, start and speed control, damper control, filter-condition indication and heater battery control. For optimum control the system should be automatically controlled from thermostats or other sensors and a timeswitch.

15.4.2.3 Powered ventilation equipment

This falls into two basic groupings: supply air systems and extract systems. The equipment used for both is similar, comprising, as a minimum, a fan and weather-proof cowl, plus ducting, air-treatment equipment and grilles as required.

Fans Five main types of fan are used in ventilation systems as described below.

1. *Centrifugal* The airflow changes direction through 90°, which can make this type of fan difficult to use within a ducted system. Two blade types are used, backward curved providing high-pressure at low volume flow and forward curved providing medium-pressure and volume flow. Typical static efficiencies are 70–75% and 80–85%, respectively.
2. *Axial* Duties are usually high- to medium-volume flow rates at medium to low pressures. In its simplest form there is an impeller and its drive motor only mounted within the cylindrical casing, and the discharge flow usually contains a fairly pronounced element of rotational swirl which may, if not corrected, materially increase the resistance of the downstream part of the system. More sophisticated versions include either downstream or upstream guide vanes to correct the swirl. Typical static efficiencies are 60–65% or 70–75% with guide vanes.
3. *Propeller* This is really a simple form of axial fan but with its impeller mounted in a ring or diaphragm which permits it to discharge air with both axial and radial components. Duties covered are high volume and low pressure. Static efficiency is normally under 40%.
4. *Mixed flow* This is a fan in which the air path through the impeller is intermediate between the axial and centrifugal types giving the benefit of increased pressures but capable of being constructed to provide either axial or radial discharge. Static efficiency is typically 70–75%.
5. *Cross flow* This type normally has a long cylindrical impeller having a relatively large number of shallow forward-curved blades. Due to the shape of the casing surrounding this impeller, air enters all along one side of the cylindrical surface of the impeller and leaves on another side. Static efficiency is typically 40–50%.

In general, axial fans are used for roof extract units and small ducted systems and centrifugal fans for large-ducted systems.

Roof extract units These are the most commonly used powered ventilators in large open buildings such as factories, warehouses and sports halls. Mounted directly onto the roof or wall, they comprise an axial fan, a safety grille and a weatherproof casing. Two forms are normally available, the vertical-discharge type which tends to have a complex casing arrangement but which throws the exhaust clear of the building, and the low-discharge type which has a simple casing but directs the exhaust onto the roof of the building. Vertical discharge is essential when smoke or fumes are being exhausted.

These ventilators can normally be used with limited ducting or accessories. A variation, fitted with a centrifugal fan, is available for more extensive ducting and is often used for duties such as toilet extract in commercial buildings.

Roof inlet units This specialized form of supply air system is often used in large open industrial spaces. It comprises a modular system of components which can be built up into simple systems. A typical system might have a roof inlet cowl, a recirculation damper, a heater battery, a fan, one or two outlet grilles and short sections of connecting ductwork, and would handle airflows up to 3–4 m³/s, depending on size. A number of individual systems would be used to provide the total airflow required in the space (Figure 15.136). Systems are normally manufactured with aluminium casings to reduce the roof load.

Ducted systems[39,40] Larger ducted ventilation systems, as used in offices and commercial premises using a central air-handling unit and fabricated distribution ductwork, are

Figure 15.136 Roof inlet system

akin to air-conditioning systems but with less treatment to the air at the AHU (see Section 15.4.3).

Local extract systems[41] Local extract systems are designed specifically to remove fumes, dust, mists, heat, etc. at source from machinery and fume cupboards. The main design considerations are capture of the contaminant which will normally involve special hoods or cabins, and extract at sufficient velocity to satisfactorily transport the contaminant. Ductwork must be manufactured to resist abrasion or corrosion and sufficiently well sealed to prevent leakage. Welded ductwork is often needed. The fan may also need protection and the motor may need to be flameproof or out of airstream if the contaminant is flammable or corrosive. Treatment of the exhaust may be required to reduce pollution and nuisance and to comply with legislation.

Air cleaners A wide range of types of air cleaners are available to match the number of contaminants needing removal from air. Figure 15.137 shows typical particle size ranges and the range of operation of each type of air cleaner.

15.4.2.4 Natural ventilation equipment

The natural ventilation equipment commercially available can be split into two basic groups, fixed and controllable. The fixed ventilation, normally a weathered louvre system, is used in applications where ventilation is constantly needed, winter and summer, with no need for control to maintain human comfort conditions. Typical examples are plant rooms and building block drying rooms. Louvre systems are also often used for inlet or exhaust to powered ventilation or air-conditioning systems and for architectural cladding. Controllable ventilation, which may be weatherproof if required, is used in buildings which are normally occupied and where waste process heat is not sufficient to heat a ventilated building in winter (i.e. most buildings).

Fixed ventilation Fixed ventilation has to be weatherproof to some degree. If it were not, then a simple hole in the wall or roof would be sufficient. The normal form of fixed ventilator is

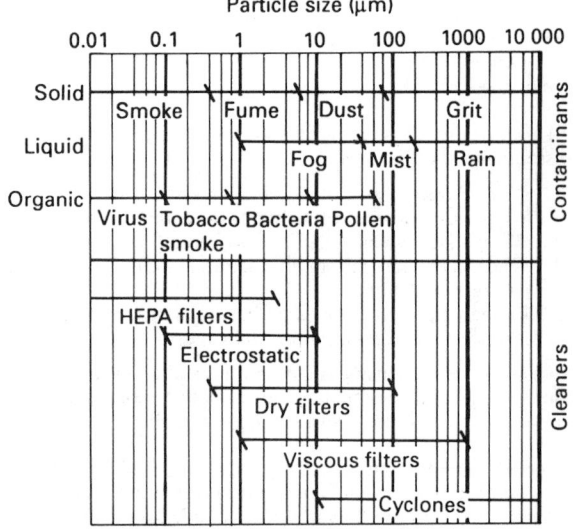

Figure 15.137 Typical particle sizes

the louvre panel. At its simplest, this may be a number of slats of wood mounted at 45–60° in a door or wall opening, and at its most sophisticated an aerodynamically designed two- or three-bank 'chevron' of roll-formed or extruded aluminium.

With all louvres there is a balance between their weatherproofing qualities and their airflow, and it is important to select a louvre with the correct balance for each application. No louvre can be guaranteed to be 100% waterproof under all conditions, but the best ones approach this standard under normal flow and wind conditions. A bird or insect guard (as applicable) is normally mounted behind the louvre.

Controllable ventilation Controllable ventilation is normally provided by one of three types of ventilator: louvred, opening flap and weathered. The most common is the louvred ventilator which has a number of centre hinged louvre blades controlled from a pneumatic ram or electric actuator which can be fully opened when ventilation is needed and fully closed to a weatherproof condition when it is not. These may normally be roof or wall mounted.

Opening flap ventilators have one or two hinged flaps (opaque or glazed) which are normally held closed but which may be opened to between 45° and 90°, depending upon design. Control is normally pneumatic or electric. Specific designs are available for roof or wall mounting.

Weathered ventilators need to have a complex air path in order to prevent rain entry so they therefore provide a restricted airflow. To overcome this, multifunctional ventilators are often used providing a direct airpath during dry weather and a restricted one during rainfall. These ventilators are normally only roof mounted.

Most ventilators are powered in one direction with spring return. Thus a pneumatic ventilator might be described as 'pressure to open' or 'pressure to close'. With the pressure to open type the ventilator will fail to the closed position under the influence of the return spring, ensuring the building remains weatherproof. Where ventilators are installed mainly as smoke ventilators it is important that they fail to the open position, so pressure to close ventilators should be used. A fusible link is normally fitted into the controls so that in the event of fire, affected ventilators will open automatically to release smoke and heat.

A limited range of accessories may be provided, normally bird guard, insect guard or attenuators, but since airflow through the ventilators is driven by very small pressure differences (perhaps 5 Pa), great care has to be taken to ensure the ventilation remains effective.

15.4.2.5 System design

Overheating Overheating can be due to a number of causes, but is usually from solar gain and machinery heat losses. If ventilating to provide thermal comfort in a space for humans or animals there are a number of considerations apart from simply air temperature:

1. Air temperature
2. Mean radiant temperature
3. Radiant temperature asymmetry
4. Air velocity
5. Humidity

Thus someone working near a window will receive hot solar radiation through the window in summer and cold radiation from the cold window surface in winter, causing uncomfortable radiant asymmetry even if the room temperature is perfect for personnel working away from the window. Ventilation will not be the correct solution for this problem – sun shading and double glazing would be more effective.

It is therefore important when considering ventilation design to consider the whole thermal environment and not to simply assume that reduction of air temperature to 20–22°C will cure all problems.

Fume dilution Dilution of fumes in factories, warehouses, etc. is generally required for one of two reasons: either to reduce the level of harmful (toxic or irritant) fumes to a safe level, normally below the OES[47] (Occupational Exposure Standard) or to dilute offensive odours. Care must be taken with the latter to ensure that the problem is not merely passed on to neighbours. If it is, then a local extract with air-cleaning equipment will be preferred if it is practical. Indeed, a local extract system is always preferable, since it removes the problem at source, resulting in a cleaner environment within the building, and may be mandatory (see 15.4.2.6).

The starting point for design of a dilution ventilation scheme is normally a hygiene survey in which levels of pollutants and the ventilation rate are measured under worst conditions.

The ventilation should be designed so that airflow is directed from clean to dirty areas to keep the majority of the building as clean as possible. However, this form of ventilation does little for operatives working at the source of contamination.

Normal ventilation equipment can be used unless the fumes are corrosive or flammable. Powered equipment is normally employed to ensure that a steady airflow is provided.

Since the ventilation will be needed both summer and winter, there will be an energy penalty for the high level of airflow, and either the inlet airflow must be pre-heated or additional space heating will be required.

In offices and public spaces fume dilution is normally required to reduce body odour and tobacco odours. CIBSE[30] gives current UK recommendations for fresh air requirements for various rooms to overcome these, varying from 5 to 25 l/s per person. These figures give a balance between energy loss and air purity, and higher levels of ventilation may be needed if a clean atmosphere is of overriding importance.

In most older offices the natural ventilation achieved by infiltration is sufficient to provide reasonable conditions, but areas of high occupancy such as conference rooms, theatres, bars and restaurants normally require mechanical ventilation. Sufficient ventilation must be provided to match the highest demand, whether for odour dilution or summer overheating.

An alternative to ventilation to clear tobacco smoke is the use of electrostatic air cleaners which clean and recirculate air within a room. These provide smoke dilution without the energy penalty of extra ventilation but require frequent cleaning to keep operating at maximum efficiency.

Prevention of condensation Condensation occurs whenever moist air comes into contact with a surface which is colder than the 'dewpoint' of the air. The dewpoint is the minimum temperature at which the air can contain the amount of moisture within it, and it will vary with moisture content. Condensation can also occur in the air when warm moist air meets cold air, when it is known as 'fogging'.

Condensation will appear on the inside surface of porous or impervious materials, forming first on the worst-insulated surfaces (normally glazing or steelwork). On porous surfaces condensation can occur within the material or at an internal boundary. This is known as interstitial condensation, and it is especially dangerous, since it is often not known about until it has caused noticeable damage.

Condensation can be avoided or reduced by insulation, reduction of moisture emission, ventilation or dehumidification. Calculation for condensation problems is complex but is covered in the CIBSE Guide[30] in some depth.

Local extract Local extract is used to remove contaminants directly from a process to the exterior without passing through personnel breathing zones. It thus provides a high degree of safety and because small volumes of air are extracted relative to a dilution ventilation system it is energy efficient.

Where dusts, grits, sawdust or other large particles are being extracted it is normally mandatory to include an air-cleaning device before the air is exhausted. Unless they are controlled under the Environmental Protection Act, fumes can normally be exhausted into the atmosphere at a suitable location well away from anywhere with normal personnel access.

The design of a system can be broken into three main areas: capture, transport and cleansing. Capture of the contaminant is of paramount importance. Depending upon the source of the contaminant, capture may be via a hood, slot, booth or enclosure, with the airflow designed to take the contaminant from the source into the duct system without passing through the operatives' breathing zone. Some typical examples of good design are shown in Figure 15.138. To capture the contaminant, a minimum air velocity, referred to as the 'capture velocity', is needed at the source. Full guidance in the design of capture systems is given by the ACGIH.[41]

Once captured, the contaminant has to be carried along a duct system. If the duct velocity is too low, particles will tend to drop out of suspension and collect in or fall back down the duct. It is therefore essential that a suitable minimum duct velocity, referred to as the transport velocity, is maintained. Nothing is gained by velocities far in excess of the recommended transport velocity, and in some cases as much is lost, since abrasion of particles on the duct can cause premature erosion and failure.

Removal of particles from the airstream is generally carried out in a cyclone, positioned outside the building and taking particles from all sources in the building. Care should be taken to position cyclones away from noise-sensitive areas since both ducts and cyclones can be noisy, especially if grits or chips are being carried.

Selection of fans and ducting for local extract must be more rigorous than for other systems. The fan must be capable of

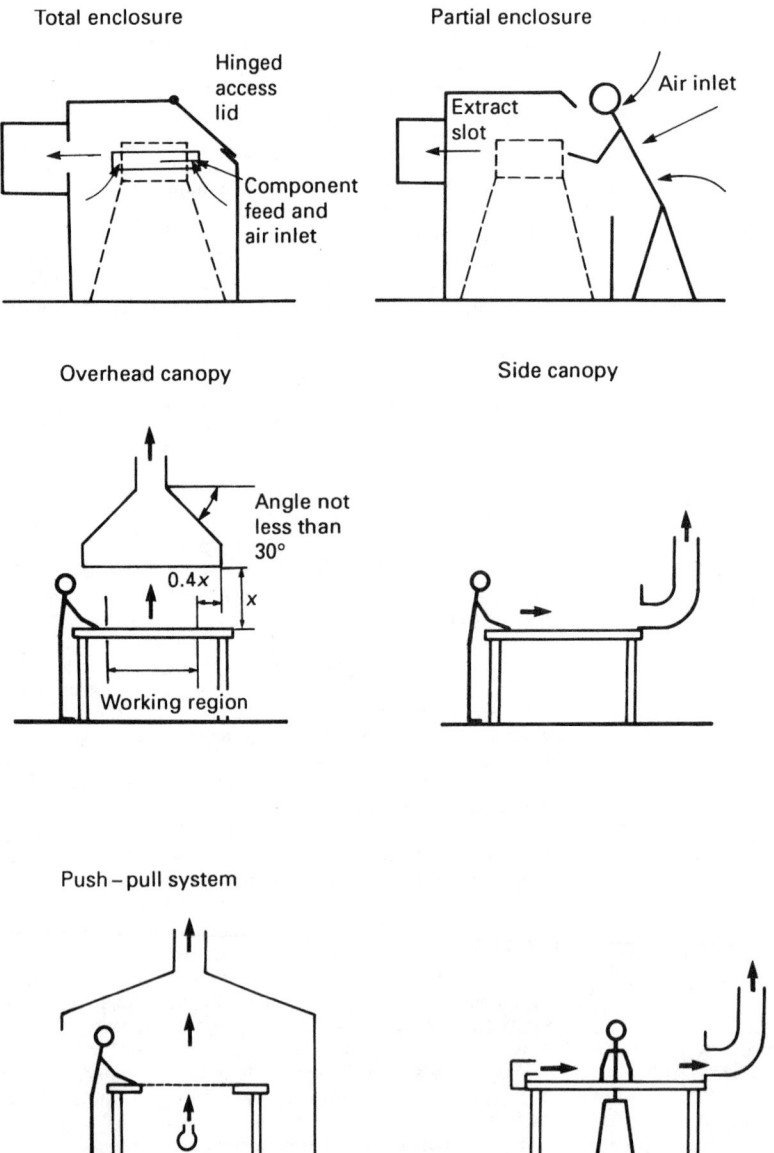

Figure 15.138 Typical local extract hoods and enclosures

withstanding abrasion or corrosion from the contaminants, and if they are flammable must have a flameproof or out-of-airstream motor. The ducting also must be able to withstand abrasion or corrosion and must be fully sealed to prevent escape of contaminants within the building. For specialist applications such as fume cupboard extract, ducting and fans are often of plastic construction.

Smoke ventilation Smoke-ventilation systems are designed to clear smoke and heat from a building in the event of fire. In large open spaces it is impossible to 'smother' a fire (as is often recommended in domestic situations) by closing doors and windows. The aim of smoke ventilation is to minimize damage due to smoke staining and heat and to assist evacuation and firefighting by providing a layer of clear air below the smoke. Without smoke ventilation a space can become 'smoke logged' from ceiling to floor in only a few minutes.

The actual design of a smoke-ventilation system is very complex, and although much published guidance is available[43–46] it should be left to experts.

In discussing design or vetting tenders, there are a number of important points to consider:

1. *Design fire size* – An accurate assessment is needed of the maximum size of fire which is likely and which can be designed for. Since the whole design is based upon this value it is essential that this be carefully considered.
2. *Interaction with sprinklers* – The likely size of fire will be smaller in sprinklered buildings and the temperature of the smoke produced will be lower since the smoke is cooled by the sprinkler flow.
3. *Fire detection* – For efficient operation, control of the ventilation must be linked to a fire-detection system or sprinkler flow switch to ensure that the ventilators are operated as early as possible. Natural ventilators must incorporate a fusible link or bulk as a back-up fail-safe device, but this should not be considered as the main form of emergency operation.
4. *Fire resistance* – Fans, motors, cabling and controls which are expected to operate under fire conditions must be suitably rated for the temperature expected.
5. *Air inlet* – A suitable low-level inlet ventilation area must be provided for the expected air flow rate. Where personnel escape routes will be used for inlet, the inlet velocity must be low enough not to impede progress.

Smoke ventilation is not intended to replace other forms of fire prevention and control but to work as an important component in an overall scheme.

15.4.2.6 Legislation and codes of practice

Legislation Most legislation regarding ventilation has been aimed at controlling the environment within the workplace, and until very recently there was a plethora of assorted and outdated regulations in force. Most of these have now been repealed under COSHH.

Legislation regarding the state of the air exhausted into the external environment is limited, the Environmental Protection Act being the main Act in force. The major pieces of existing legislation are discussed briefly below:

1. *Health and Safety at Work etc. Act 1974 (HASAWA)* This does not directly relate to ventilation but places duties upon employers and employees regarding health and safety and is an enabling Act for further Regulations.

2. *Control of Substances Hazardous to Health Regulations 1988 (COSHH)* Enabled under HASAWA, these Regulations provide a requirement for adequate control of hazardous substances and therefore directly cover use of ventilation to keep the atmosphere clean within the workplace. Ventilation equipment installed to comply with COSHH must be regularly inspected and tested at no more than 14-month intervals, and monitoring may be required to ensure its continued effectiveness.

3. *Workplace (Health, Safety and Welfare) Regulations 1992* Regulations 6 and 7 require effective and suitable ventilation in enclosed workplaces and the temperature in all workplaces inside buildings during working hours to be reasonable.

4. *Highly Flammable Liquids and Liquefied Petroleum Gases Regulations 1972* These require mechanical ventilation to be provided, preferably by local extract from cabinets or enclosures, to avoid dangerous concentrations of vapours occurring in workrooms.

5. *Woodworking Machines Regulations 1974* These Regulations require local extract to be provided at designated machines to remove sawdust, chips, etc. and for all solid particles collected to be discharged into suitable receptacles.

6. *Environmental Protection Act 1990* This controls emission of pollutants from premises in two ways. HMIP control emissions of prescribed substances and from prescribed processes under the Integrated Pollution Control mechanism – the scope of this is steadily widening. Other emissions are controlled as statutory nuisances by local council environmental health departments.

7. *Control of Pollution Act 1974* This Act allows local authorities to require the occupier of any premises to provide estimates or other information as specified concerning emission of pollutants.

   Note: These synopses are the author's interpretation only. Neither the author nor the publishers can take any responsibility for any result of any actions taken as a result of reference to this section.

Codes of Practice

1. *BS 5925: 1980: Design of buildings: ventilation principles and designing for natural ventilation* This British Standard gives recommendations on the principles which should be observed when designing natural ventilation of buildings for human occupation and provides a basis for choice between natural and powered systems.

2. *BS 5720: 1979: Code of Practice for mechanical ventilation and air conditioning in buildings* This deals with design, planning, installation, testing and maintenance of systems.

3. *BS 6540: Part 1: 1985: Methods of test for atmospheric dust spot efficiency and synthetic dust weight arrestance* This gives standard test methods for filters used in ventilation and air-conditioning systems.

4. *ISO 7730: Moderate thermal environments – determination of the PMV and PPD indices and specification of the conditions for thermal comfort* This gives a method of measuring and evaluating moderate thermal environments to which people are exposed.

5. *EH22 (Revised May 1988): Ventilation of the workplace* This Health and Safety Executive Guidance Note provides information on standards of general ventilation and fresh air requirements in the workplace.

6. *EH40/93: Occupational Exposure Limits 1993* This gives advice on limits to which exposure to airborne substances

hazardous to health should be controlled in workplaces. It is revised annually.

15.4.2.7 After installation

Commissioning and testing Once any system has been installed it is important that it is properly commissioned to ensure that everything is working satisfactorily and to specification. A guide to commissioning ducted systems is available from BSRIA. Commissioning of other systems should be in accordance with the manufacturers' recommendations.

Maintenance Most simple ventilation systems require only annual maintenance unless some form of air cleaning or filtration is incorporated, although where systems are pneumatically operated the compressor will need weekly checking. Air cleaners or filters will need regular emptying, cleaning or replacement to maintain efficiency and prevent clogging up. This can either be carried out on a regular schedule (based on the manufacturer's guidance and site experience) or when indicated by a pressure differential gauge or alarm. The cost of a gauge is easily repaid by savings in maintenance costs by maximizing intervals between cleaning or replacement. The task of maintenance can be made easier by taking care in design to provide good access.

Compressors will need a weekly oil level and receiver auto drain check and oil changes and filter cleaning at (typically) 500-hour intervals, although for compressors used only to operate ventilation this can be only an annual task due to the limited usage. Manufacturers' recommendations should be followed.

Running costs Running costs of ventilation systems can be broken down into three main areas:

1. *Maintenance costs* Regular service checks such as filter cleaning and compressor oil level are normally carried out in-house and time can be allocated for these tasks once some experience has ben gathered. Annual maintenance may be carried out in-house or by specialist service engineers employed either by manufacturers or HVAC service companies. A service contract can often include breakdown cover, which has the advantage of reducing risk of unexpected bills and ensuring that prompt repairs are effected at the cost of a higher annual premium.

2. *Electrical costs* The only significant electrical cost involved in ventilation systems is operation of fans. Other electrical equipment such as dampers, compressors, etc. generally run for such short periods that costs are negligible.

   Electrical costs for fans can be estimated from the following formula:

   $$C = V \times \phi \times A \times \text{fuel cost} \times \text{hours run} \qquad (15.43)$$

   where C = cost per annum (£),
   V = motor voltage per phase (V),
   ϕ = number of phases,
   A = operating current (A),
   Fuel cost = cost of electricity (in £/k Wh),
   Hours run = total running hours per annum.

3. *Heat loss* Where a ventilation system is required to run during periods when heating is provided then there is an energy cost associated with the heated air being exhausted from the building. This is related to the extra heat input needed from the heating system to balance the heat loss through the extra ventilation. A calculation method is available in section B18 of the CIBSE Guide.[31]

15.4.3 Air conditioning

15.4.3.1 Basic principles and terms

Abbreviations   In addition to the abbreviations used in SI, the following are employed in air conditioning work:

db	Dry bulb temperature
wb	Wet bulb temperature
dp	Dewpoint temperature
rh	Relative humidity
kg/kg	Kilograms water vapour per kilogram dry air (absolute moisture content)
TH	Total heat
SH	Sensible heat
LH	Latent heat
SHR	Sensible heat ratio
ON OFF	The condition of air or water entering or leaving a coil or heat exchanger
TR	Tons of refrigeration capacity
TRE	Tons of refrigeration capacity extracted
TRR	Tons refrigeration rejected (at final cooler)
HP	High pressure (refrigerant)
LP	Low pressure (refrigerant)
DX	Direct expansion cooling
ΔT	Temperature difference
ach	Air changes (room volumes) per hour
ahu	Air-handling unit
swg	Static water gauge
NR	Noise rating. One of a series of curves relating noise level and frequency to speech inteference
NC	Noise criteria. Similar to NR, but differing, particularly at the low-frequency end

Terms

1. *Mechanical ventilation* The movement of air by fan, conveying outside air into the room or expelling air or both. Filtration, heating and control of the distribution pattern may be included. It is not cooling in the sense of temperature reduction but can be used to limit temperature rise when the outside air is below that of the space being treated.

2. *Full air conditioning* This necessitates plant capable of control of temperature by being able to add or subtract heat from the air and control of humidity by being able to add or subtract moisture. The system also comprises fan(s), filtration, a distribution system and may include noise control. Other terms such as 'cooling' or 'comfort cooling' may be met and these can be taken to mean an ability to lower the temperature of the air by refrigeration but without full control of humidity. Moisture may be removed as an incidental characteristic of the cooling coil. The term 'air conditioning' is sometimes used where control of humidity is not included. It is essential to employ clear specifications of performance.

3. *Air* Atmospheric air is a mixture of gases, mainly nitrogen and oxygen together with water vapour. It normally carries many millions of dust particles per cubic metre.

4. *Temperature* A measure of the average energy of the molecules of a substance. The heat intensity.

5. *Heat* A form of energy which, when given to a body, raises its temperature or changes its state from solid to liquid or liquid to gas.

6. *Heat flow* Heat flows from a body at one temperature to a body at a lower temperature. Materials have the property of resistance to the rate of heat flow. It differs from material to material.

7. *Sensible heat* The heat energy causing a change in temperature, as in raising a kettle of water from cold to boiling point.
8. *Latent heat* The heat necessary to change the state of a substance from solid to liquid or from liquid to gas, or the heat given up during the reverse process. There is no change in temperature during these processes. For example, continuing to boil a kettle of water previously raised to 100°C to steam requires the addition of latent heat, but there is no change in temperature if the pressure remains constant.
9. *Total heat* The sensible heat plus latent heat in such a mixture as moist air. In air-conditioning work it is referred to a base a little below 0°C, not absolute zero.
10. *Sensible heat ratio* Sensible heat flow divided by the total heat flow.
11. *Enthalpy* The heat content of a substance per unit mass.
12. *Dry bulb temperature* The temperature of air as indicated by a dry sensing element such as a mercury-in-glass thermometer.
13. *Psychrometrics* The study of moist air. The psychrometric chart shows the relationship between the various properties of moist air in graphical form and can be used for the solution of problems.
14. *Wet bulb temperature* The temperature of air as indicated by a thermometer when its bulb is enclosed by a water-wet wick. If the surrounding air is not saturated water will evaporate, taking the necessary latent heat from the thermometer bulb which then gives a lower reading than a dry bulb in the same air. The depression in wet bulb temperature is proportional to the amount of moisture in the air. Normal practice is to arrange a flow of air over the wick by using a sling (whirling) or fan-assisted instrument. If the thermometer is stationary an area of higher saturation builds up around the wick but the reading may be referred to tables for screen instead of sling readings.
15. *Partial pressure* The contribution by each constituent gas to the total air pressure. Standard air pressure is 1013 mbar.
16. *Vapour* A gas which is below its critical temperature and which can therefore be turned to liquid by an increase in pressure.
17. *Saturation* There is a limit to the amount of water vapour air can hold. It is higher at higher dry bulb temperatures. At the limit, air is said to be saturated.
18. *Relative humidity* This compares the amount of moisture in a sample of air with the amount it would contain if saturated. More accurately, relative humidty is the partial pressure of vapour present divided by saturation vapour pressure × 100%. Saturation = 100% relative humidity.
19. *Dewpoint* The temperature to which a sample of air has to be reduced to bring it to saturation. It is fixed by the moisture content of the air sample.
20. *Absolute humidity* This measures the quantity of water in a sample of air in kg moisture per kg air. The relative humidity then depends on the air dry bulb temperature. Air at 25°C containing 0.01 kg/kg is at 50% relative humidity (rh). If now cooled to 14°C the air would be at its dewpoint (i.e. saturated). If cooled further, moisture is condensed out, the sample remaining saturated as it cools. If now reheated back to 25°C its rh would be lower than 50%. If cooling had not been continued to condense moisture its rh would return to 50% at 25°C.
21. *System resistance* The resistance to air flow which causes a static pressure drop. It is similar to electrical resistance and voltage drop (see Section 15.4.3.4). The term 'resistance' is often used erroneously when pressure drop is meant.
22. *Upstream, downstream* Used to denote positions earlier or later in the system relative to the direction of air flow.
23. *Condensing unit* A refrigeration compressor and condenser on one chassis complete with controls.
24. *Split system* As above but with a remote condenser.
25. *Chiller* A compressor, water-chilling evaporator and condenser on one chassis.

The plant These divide broadly into two types:

1. The direct expansion plant where the air-cooling coil is fed with cold refrigerant;
2. The chilled water plant where the cold refrigerant first chills water (or other liquid) which is fed to the air-cooling coil.

A block diagram of the DX system is shown in Figure 15.139. It has two main circuits – the air circuit and the refrigerant circuit. In the chilled-water system there are additional circuits:

1. Of chilled water between the refrigerant and the air-cooling coil;
2. Of water carrying heat from the refrigerant to the heat-rejecting device.

Figure 15.140 shows a block diagram of the system.

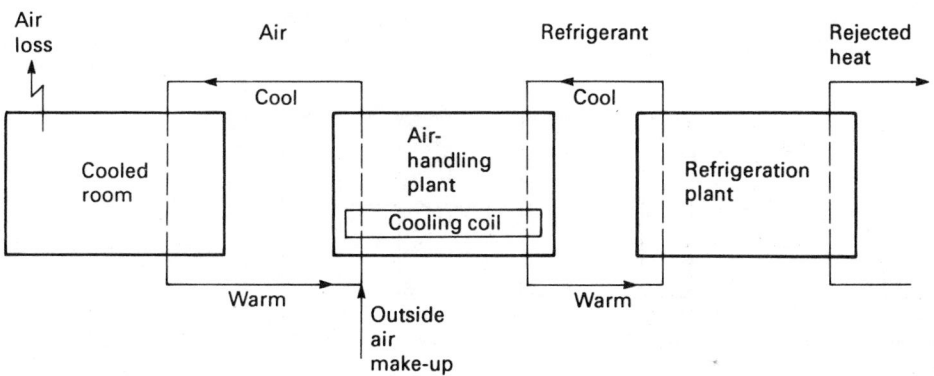

Figure 15.139 Heat flow paths in a direct expansion system

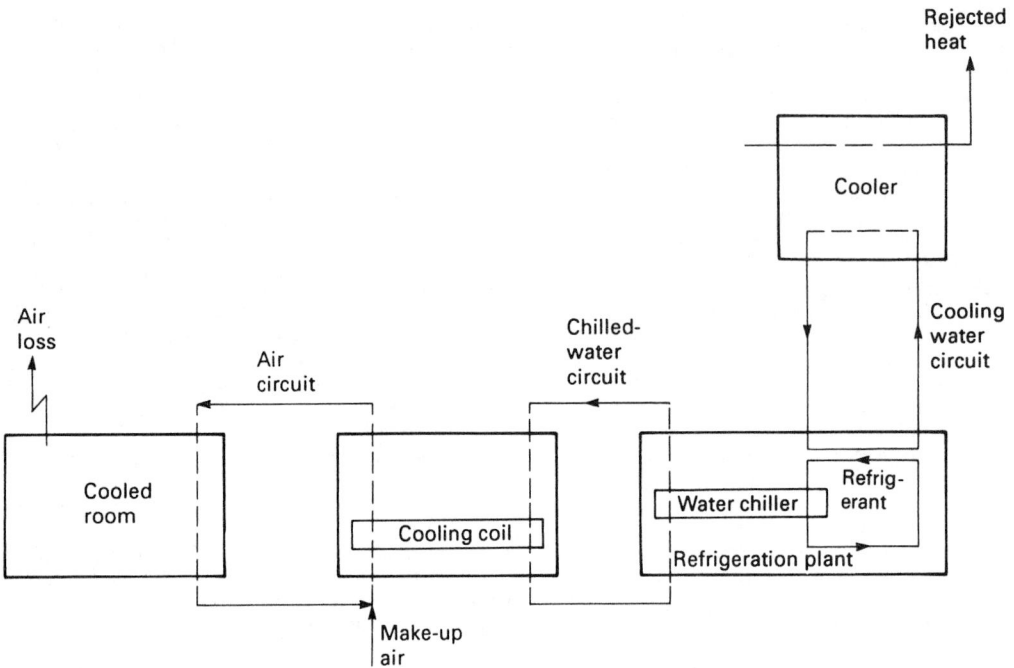

Figure 15.140 Heat flow paths in a chilled-water system

A temperature difference must exist betwen each stage to cause a heat flow from one to the other. The air and refrigeration side are described in further detail below.

The air-handling plant   The plant may comprise one or more complete factory-made units or may be built up on-site from sub-assemblies. There can be variations from the arrangements discussed below and shown in Figure 15.141. The condition or quantity of air input to the conditioned space (referred to below as the room) must be varied such that after it has gained or lost heat or moisture by the applied load its condition and therefore the room condition is as specified.

1. *Outside air intake*   The quantity is discussed in Section 15.4.3.2. Its purpose is to keep the room fresh and to pressurize it against the ingress of unconditioned air. Its psychrometric condition during most of the year will differ from that required. It can be introduced into a chamber, mixing it for treatment with return air from the room. Alternatively, it can be treated in a separate plant before being introduced to the system, but care then has to be exercised in design, since the sometimes small quantity of air has to control the full humidity load. An advantage is that corrosive wet processes are kept out of the main plant.

2. *Mixing chamber*   This is where the outside air intake and recirculated air are brought together before proceeding to the next stage of treatment. As shown in Figure 15.141, it is a low-pressure area which will induce outside air without the use of another fan if the route has low resistance.

When two samples of air are brought together the condition of the mixture may be arrived at arithmetically by adding the heat flow of each and dividing by the total

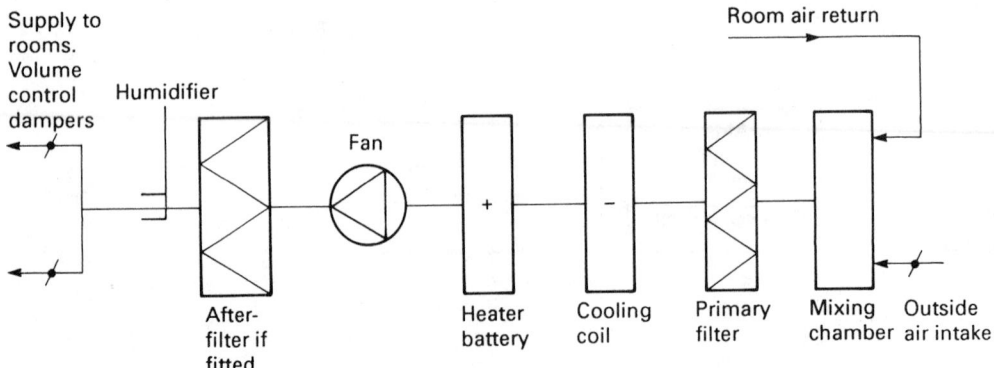

Figure 15.141 Air flow path in an air-conditioning system

mass flow; and similarly for the moisture flow. Alternatively, plot the condition of each onto a psychrometric chart. The mixed condition lies on a straight line between the two in a position proportional to the two quantities.

3. *Pre-filter* Where a high degree of cleanliness is not required it could be the only filter in the system. The subject is covered in more detail in Section 15.4.3.6.

4. *The cooling coil* This is the exchanger where heat flows from the room return or mixed air to cold refrigerant or to chilled water. It is an arrangement of finned tubes normally of aluminium fins on copper tubes, but copper fins can be specified for corrosive atmosphere. Performance characteristics are controlled by fin and tube spacing. If the room rh is high, dehumidification may be brought into use by operating the coil or one of a number of parallel coils at a low temperature. If the room's sensible heat load is low reheat must be allowed to operate at the same time. Dehumidification can be achieved by partially bypassing the coil such that the remaining air travels through the coil at low velocity. This can also be inherent in the full-load design operation of the coil.

5. *Heater battery* This is used when (1) the room needs heating instead of cooling or (2) for reheat as described above. It is vital in close control systems that its capacity is sufficient to maintain room temperature under these conditions, otherwise the system may fall into a loop, with the controls continuing to see high rh due to low temperature. Using only part of the cooling coil for dehumidification will alleviate this situation. A heater capacity of the sensible heat extracted during dehumidification plus half the peak winter fabric loss is recommended where the room load could be nil in winter such as a start-up situation. Separate reheat batteries may be placed in branch ducts where one plant supplies both a main area calling for cooling and an auxiliary room without heat load. Correct rh in the auxiliary rooms results (only) if it is correctly controlled in the main room and they require the same dry bulb temperature. While wasteful of energy, it simplifies the plant design and may be found to use less resources.

6. *The fan* This drives the air around the system against its resistance (see Section 15.4.3.5).

7. *Humidifier* The humidifier is a means of increasing the absolute humidity of the air although usually controlled from a relative humidity sensor. It should be positioned where shown so that it can correct any over-dehumidification by the cooling coil (see Section 15.4.3.7).

8. *Air flow* The quantity supplied must be matched to the load on the plant (see Section 15.4.3.2). After leaving the plant it is distributed to match loads of the rooms or zones to be served.

The refrigeration plant The basic circuit is shown in Figure 15.142 and the principal items are described below:

1. *Evaporator* This is the device where the air or water being cooled gives up its heat to provide the latent heat of evaporation to the refrigerant. Superheat is also added to the refrigerant at this point to prevent damaging liquid forming on the way to the compressor.

2. *Compressor* A compressor circulates the refrigerant around the system, raising its pressure such that the refrigerant can be condensed by removal of latent heat. It may also be considered as raising the temperature of the refrigerant above that of the final cooling medium to which heat is rejected. Lubricating oil is contained in the crankcase but, being miscible in the refrigerant, is carried around the system and returned.

3. *Condenser* This is the vessel where the refrigerant rejects its heat to waste or reclaim, turning back to liquid in the process. Sub-cooling is practised by the removal of further heat. This prevents liquid flashing back to vapour on return to the evaporator.

4. *Expansion valve* A reduction in pressure and hence in temperature takes place across this item before the refrigerant re-enters the cooling coil via distributor pipes.

Controls

1. *Room condition*
   (a) *Dry bulb temperature* This is sensed by a thermostat in the conditioned space or in the return duct. Where underfloor air return is practised it is strongly recommended that the sensor be placed close under a return grille to prevent changes of condition occurring

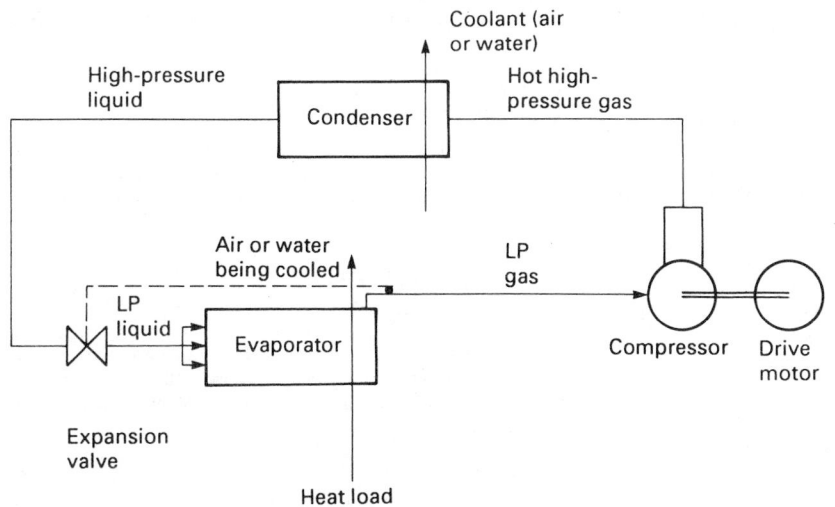

Figure 15.142 Refrigerant flow path in a cooling system

between the room and the sensing position. In large rooms separate thermostats can be arranged to give an average signal but individual zones of control each separately treated are much better. Dry bulb sensors may be bi-metal strips, thermistors or refrigerant-filled phials or bellows responding to pressure differences caused by temperature change. These, in turn, provide an electrical or mechanical signal. The mechanical items are used to alter the value of potentiometers or make-or-break contacts. The signals are transmitted to amplifiers which respond to the degree of error. An important feature is the proportional band of temperature over which the controls call for up to full plant capacity.

(b) *Humidity* Humidity sensors may be animal or plastic skins varying in length with changes in rh or lithium chloride coating changing in electrical resistance. The former are prone to lose calibration. Other comments above apply equally to rh control.

(c) *Control at the evaporator* A phial senses the temperature of the outlet sunction line to the compressor and controls the expansion valve opening to maintain a constant temperature in the coil.

2. *Control at chilled water coil* A three-way motorized valve is modulated between full flow to the coil and full bypass to satisfy the room thermostat.

3. *Control of water-chilling compressors* Being large multi-cylinder machines, the chilling capacity is controlled in steps by rendering cylinder valves inoperative. Control is initiated by sensing water temperature in a storage buffer tank or by sensing return water temperature from the air-cooling coil. Small (DX) compressors are run on a start–stop basis on call from the room thermostat. Safety devices associated with compressors include:

High (gas) pressure cut-out;
Low (gas) pressure cut-out;
Oil differential cut-out (oil feed pressure to be above the crankcase pressure);
Freeze thermostat (low water temperature limit);
Single-phase protection (preventing attempts to start with loss of one phase of a three-phase electrical supply);
Time out (preventing too-frequent starting and motor burn).

Until all the switches in the interlock train are satisfied and closed the compressor will not start or will not continue to run.

The load on the plant The unit of heat is the watt. However, the imperial unit should be understood as it will still be met, particularly outside Europe. The ton of refrigeration is derived from an ability to remove sufficient heat from a short ton (2000 lb) of water at 32°F to turn it to ice at the same temperature in the course of 24 h. This amounts to a heat extraction rate of 3.517 kW.

The load presented by the room is first transmitted to the room air which, in turn, passes it to the cooling coil. Other gains will occur as detailed below. The total load is best considered in two parts – sensible and latent.

Machine heat All electrical energy fed to the room will appear as heat. This presents a load to the plant unless power is conveyed out of the room by cable or hot items are physically removed. Parts of some large computers are cooled by a direct supply of chilled water presenting a load on the refrigeration plant but reducing the load on the air side. Cold outside conditions will result in some of the internal load being met by fabric loss.

Fabric losses or gains In winter in cold climates heat will be lost through fabric of the building. This will be advantageous at times of high internal load but will need to be considered as a heater battery duty at times of low internal load. In winter there can be considerable loads for humidifiers if the structure is not adequately vapour sealed. Weather-tightness is insufficient. For example, a computer room in the UK held at normal conditions may experience up to 100 mm water gauge vapour pressure difference, forcing moisture out. During high summer in temperate and hot climates the external water vapour pressure will be higher than in the conditioned space.

Personnel People give off both sensible and latent heat. During light work such as in a computer suite or laboratory they emit 110 W sensible 30 W latent and while seated, 90 W sensible 20 W latent.

Air ducts A duct carrying cool air through a warm space such as a loft will gain heat before entering the conditioned room, contributing to the load on the plant. Ducts passing through the conditioned space do not add to the load. Similarly, there will be losses in winter if they carry warm air through cold spaces and moisture gains and losses if leaky. Insulation and vapour-tight joints are necessary.

Fan heat The power fed to the fan shaft (or the total electrical power fed to the motor if within the duct) appears as sensible heat in the system. During a heating cycle this is useful. As a rule of thumb, fan shaft power is $17 \times$ flow $m^3 \times$ mm swg watts. This commonly lies between 2% of the cooling duty for small systems to 10% for large systems.

Compressor heat Friction (or, if within the refrigerant path, as in hermetically sealed or semi-hermetic machines, the whole of the input to the motor) increases the amount of heat to be rejected by the final cooler or available for heat recovery. It is uneconomic to operate refrigeration plant at unnecessarily high temperatures in order to assist heat recovery.

Pump heat In chilled-water systems pump shaft power adds to the heat of the circulating water. Similarly, if the chiller has a water-cooled condenser pump heat is added to that handled by the final cooler. Power is proportional to the flow and pressure:

$$\frac{102 \times l/s \times \text{metres head}}{\text{Efficiency}} \text{ watts}$$

Efficiency can vary between 35% for very small pumps and 80% for large ones, and may be found to lie between 3% for small heating circuits and 10% of the heat conveyed for large systems.

15.4.3.2 The air quantity required

Air change rate Change rates (room volumes per hour) can be used to calculate the quantity to be supplied or extracted by a mechanical ventilation system. These figures also apply to parts of an air-conditioning system where stale air must not be recirculated. This is discussed in Section 15.4.2.

Volume to be supplied For the design of close control systems or where large amounts of heat are to be removed the mass flow to be employed must be calculated. Use is made of the specific heat of air which for normal room conditions may be taken as 1.02 kJ/kg°C. 1.02 kW raises a flow of 1 kg/s by 1°C and pro rata. The volume of a given mass of air varies with change in temperature but supply volume is often more convenient to consider. Taking the specific volume to be 0.82 m³/kg then 0.1 m³/s will convey 1 kW with a temperature rise of 8°C and pro rata. A subsequent fall of 8°C across the cooling coil is suitable for areas controlled to 21°C 50% rh. A smaller air quantity would be too close to the limiting temperature to hold the necessary moisture when leaving the coil. Greater temperature differences may be used if the rh is to be controlled to a lower level or not controlled at all. A maximum temperature difference of 10°C supply to room is recommended where occupants are close to supply points. For comfort, larger volumes are preferable to low temperature.

In ventilation systems the temperature rise calculated on the basis of specific heat alone will be pessimistic by one or two degrees because of the effect of building mass. Unless an extract is specifically designed to remove heat from hot spots or lights, the extract and room temperature can be taken to be the same.

Outside air intake When temperature limitation is more important than close control of conditions, a considerable economy of refrigeration plant operation results from arranging to draw in outside air when it is sufficiently cool and rejecting this back to outside after having gained heat from the room. However, where close control of temperature and humidity is required accuracy and economy ensue from minimizing the outside air intake. The quantity may be based on the number of occupants using 0.008 m³/s minimum per person. Alternatively, if larger, use 0.002 m³/s/m² floor area, which is sufficient in a good quality building to keep the room pressurized to one or two mm swg. Because air loss is a function of the building surface, this is preferred to the basis of a proportion of the supply air volume, which is a function of load. If the latter is used for applications such as computer rooms it is seldom necessary to use more than 2.5%.

General considerations Outlets should not be provided in constantly recirculating systems, particularly where close control of humidity is required. The overpressure developed is far less than that exerted by the wind, and for this reason any system which does have both intake and discharge ducts should have them on the same face of the building. While care is necessary to prevent short-circuiting, this alleviates problems arising from the considerable wind pressure difference that can develop on opposite sides of a building.

Air-lock entries should be used where close control of conditions is required and for clean rooms, but conventional doors with close fit and self-closures are sufficient.

Fire authorities may stipulate pressurization of certain areas such as stairways and may require smoke-extract systems to be brought into operation automatically in the event of a fire. They should be consulted at an early stage of the design.

15.4.3.3 Heat losses and gains

Heat losses Heat transfer through a partition is a function of resistance to heat flow, the temperature difference driving the heat through and the surface area. The function of heating systems is to provide the heat lost in maintaining the temperature difference. Thus

Heat flow (watts) is $U \times TD \times$ area

where U is a coefficient for the partition in watts per m²°C. For each room served the loss through each wall, ceiling and floor should be calculated and for each part of those surfaces where differences occur, such as windows in walls or cantilevered structures of upper floors.

The U values of many partitions or composite constructions can be found in standard references but others may have to be calculated where no data are available or changes are contemplated. The resistivity (r) of each element of the partition encountered by the heat in passing through must be found and multiplied by the thickness in metres. Manufacturers are usually able to give resistivities. The sum (R total) of all the elements, including the inner and outer surface resistances and the resistance of any interleaf air gap, is found. The U value is the reciprocal of R total. For example, for a wall:

Outer surface (normal exposure)	0.055
112 mm brick skin 0.12×0.112	0.01344
50 mm uninsulated air gap	0.180
150 mm lightweight block skin 5.88×0.15	0.882
15 mm rendering 2.5×0.015	0.0375
Inner surface	0.123
R total =	1.29094
$U = 1/R =$	0.775

If this is insufficient for the purpose (e.g. the external wall of a dwelling house) the problem could be reworked using insulation in the interleaf gap.

The temperature at any intermediate point is proportional to the R total to that point. This may be used to decide whether that point is above or below the dewpoint of penetrating air.

In maritime climates such as the UK the lowest external temperatures are not sustained for long periods. The mass of the structure has a slugging effect, and it is safe to use a relatively high external design temperature (e.g. -1°C in the south of the UK and -4°C in the north). These figures do not apply to an outside air intake where the full effect of low temperature is felt immediately nor to lightweight structures. When choosing a design temperature difference, one must take into account that adjacent rooms may not always be heated to their design temperature. Heat bridges, which are weak points in the insulation, must also be considered in proportion to their areas.

Cold outside air introduced to the system by infiltration or by design of the plant will require heat to be added by the plant or directly in the room to maintain the room temperature. The heat required is

kg/s $\times 1.02 \times$ °C TD kW

For infiltration it is convenient to use m³/h $\times 0.33 \times$ °C TD watts, but where air conditioning is employed infiltration should not be allowed. Air lost from the space by pressurization carries heat away but this is not an additional load beyond that mentioned above.

The available heating capacity should exceed the calculated figure, this being sufficient only to balance the losses under steady conditions. A 25–50% excess capacity is recommended to provide warm-up from cold and good response to controls but without excess overshoot. It is important in close control air-conditioning design to have sufficient heating capacity to raise the room temperature from cold following plant stoppage. If the room is cold the controls will see high rh and call for dehumidification upon restarting. Unless this is countered by sufficient reheat the control of conditions will not recover automatically.

Internal gains Any heat liberated within the room reduces the heating effort required by the plant. At any time these gains exceed the loss more than marginally, cooling is required. Heat sources are:

1. The total electrical input to the room unless power is carried away by cable or heat by pipe. This includes lighting and it should be remembered that the input to fluorescent fittings is greater than the tube ratings. Some luminaires are designed as air-extract fittings, in which case their heat is a load on cooling plant, but does not contribute to room heating unless the plant is in a non-cooling recycling mode;
2. Heat from other processes unless carried away by pipe or items are taken out of the room;
3. Heat from personnel (see above);
4. There should be no space heating within air-conditioned rooms but some gains may arise from hot water pipes passing through to other areas;
5. Gains from adjacent spaces held at a higher temperature;
6. Heat gains into ducts where these pass through warm areas *en route*;
7. The shaft power of fans will appear in the system as heat. Where the drive motor is in the airstream the whole of the motor input will appear as heat.

Heat gains The calculation of heat gains through the building fabric is more complicated than for heat losses, taking into account the gains from both the air-temperature difference and from solar intensity. The gain varies during the day with the movement of the sun and changes in air temperature. Heavily glazed buildings are susceptible to large gains from low sun elevations at all times of year and here building orientation can have a considerable effect on plant loads.

A structure with a large mass will result in the peak gain appearing on the inside some hours after the external peak, and the gain will be attenuated since the outside condition will be reduced before that time. The time of peak gain will differ for each of the enclosing surfaces, so it is necessary to calculate each for several hours to find the peak for one room and to repeat this for each room to find the peak load on a plant serving several rooms. Computer packages of varying merit are available to undertake this laborious task.

To avoid temperature shock and for economy, the control of comfort air conditioning may be allowed to drift with extremes of external temperature. However, where the design is to maintain specified internal conditions under virtually all external conditions data may need to be adjusted upwards.

Heat of outside air intake In close control air conditioning the condition of the outside air is rarely as required for passing forward to the controlled space. Therefore it contributes to the cooling, heating, humidification or dehumidification loads on the plant. Whether computer rooms, laboratories, etc. with their sparse population are considered or auditoria where there is a larger ventilation requirement, the peak latent heat load caused by treating the outside air is greater than that from the personnel. The heat from personnel is given above.

The calculation of the heat of intake air is in two parts:

Mass flow kg/s × change in latent heat kJ/kg (from tables or psychrometric chart) (kW); plus
Mass flow kg/s × change in sensible heat kJ/kg (kW).

15.4.3.4 Air distribution and system resistance

Duct sizing Ducts convey conditioning air from point to point at a variety of speeds. Slow speeds result in large ducts, costly in themselves and in building space. High speeds result in noise and the need for high fan powers. A good basis for air conditioning is 6–7 m/s adjacent to the plant but, as discussed below, less at distant points.

When we refer here to static pressure we mean the difference between internal and external pressure causing air to tend to flow into or out of ducts. Velocity pressure is that due to the air's forward movement. The sum of the two is total pressure.

Duct design Simple runs of a few metres may be designed for a constant velocity. A supply duct is thus reduced in steps at each outlet or a return or exhaust duct similarly increased in section in the direction of flow.

A large system with branches, several inlets or outlets and some tens of metres long will be more easily controlled at the end distant from the plant if velocities are reduced as we progress down the length of the duct. If air is slowed in a controlled way with the duct sides diverging at not more than 15° included angle its velocity pressure will reduce and (ideally, without loss) its static pressure will rise to maintain a constant total head to compensate for pressure loss as the air progresses down the duct. However, at least by manual methods, design for static regain is laborious and the duct shape unconventional. It is seldom practised.

A method commonly used is that of equal surface friction per unit run. If rectangular ducts are being considered their equivalent circular diameter must be found. This may be obtained by

$$1.3 \times [(w \times d)^{.625}]/[(w + d^{.25})]$$

Here w and d are width and depth and the units may be metres or millimetres. Alternatively, the diameter may be found from published charts.

Where practicable, the large surface area of wide shallow ducts should be avoided to keep pressure gradients to a minimum. There can be no hard-and-fast rule, 4:1 being a suggested limit. The system may, of course, be designed to use only round-section ducts.

System pressure drop The point of interest is the path of highest pressure drop or index leg. Other parallel branches can be designed of appropriate size to pass the required amount of air, those of lower resistance than the index leg being throttled by dampers. The pressure drop is the sum of the drops caused by the following and is calculated to determine the pressure against which the fan must operate:

Surface friction of duct as discussed above (Pa/m × length);
Changes in section;
Bends;
Branches;
Obstructions;
Grilles, meshes, etc.

Plant resistance

In the design of tailor-made plant is is necessary to calculate the above as they occur within the plant and add the pressure drop of all other items such as air filters at their dirty conditions, coils, etc. In the case of proprietary units it is normal for the manufacturers to quote an external pressure against which they will deliver the specified air quantity.

Duct resistance

The basis of good duct design is to arrange gradual changes to section and direction. It is sometimes necessary to construct a 90° elbow with no inside radius. The pressure drop and noise generation can be greatly reduced by incorporating turn vanes which split the air into a number of near-parallel paths.

Ventilated ceilings   A ventilated ceiling is an alternative to ducts, diffusers and grilles as a means of distributing air within a room. It is suspended below the structural ceiling forming a shallow void. One type consists of perforated metal trays or tiles. Each is supplied with a bagged acoustic pad which is removed from those trays or tiles which are to ventilate. Another consists of tiles with single-row slot openings at intervals between runs of tiles, the openings being controlled by dampers integral with the slots. Any type of tile which can shed dust or is combustible should be avoided. Rising hot air and descending cold air mix a few hundred millimetres below the ceiling, giving a near-uniform condition in the occupied levels. The availability of very small areas of control can be advantageous where electronic apparatus with mild chimney effect would have its natural cooling upset by strong down-draughts, by confining air supply to gangways.

The method is useful where large quantities of air are to be introduced without draughts; where the distribution of heat load is unknown at the time of design or is likely to alter; and where, as a measure of reliability, it is required to redistribute the remaining air in the original proportions in the event of partial failure of the plant.

15.4.3.5 Fans

Fan selection   Fans propel the air through the system, and must be chosen to be capable of delivering the required volume flow against the calculated system static pressure, advisedly with a small margin.

Individual fans are capable of operating over a range of volumes and pressures which are interrelated, the performance being shown in manufacturers' tables and curves. Venturing outside the recommended area of operation may, depending on the type of fan, result in motor overload, motor undercooling, vibration or stall. Selection should be made for minimum power input, which is also likely to be the quietest fan for the duty. Performance is usually quoted for a standard condition of $1.2 \text{ m}^3/\text{kg}$. Calculations of system resistance are best carried out at the same condition. The user may find only the static pressure quoted. If total or velocity pressure are also quoted or the outlet velocity can be calculated the designer can calculate how much pressure can be recovered after the exit. Any mismatch due to difficulty in calculating system resistance will cause the volume to rise or fall, to settle on the fan characteristic curve.

Types   The broad range of types used in air conditioning is:

1. Centrifugal
2. Axial
3. Propeller
4. Mixed Flow

These are discussed in Section 15.4.2.3.

Multiple arrangements   Fans may be operated in parallel but are best of similar characteristic to avoid stalling. In the event of failure of one the effect of reduced pressure drop in the system is to give a flow of about two thirds that of two. Non-return dampers should be fitted.

Fans may also be operated in series. Axial-flow fans are available having two (contra-rotating) impellers and motors in the one casing. Further staging may be arranged. If one stage fails or is switched off it will idle round with loss, but since similar losses in centrifugal fans are much higher, these are not usually operated in series. However, supply and extract fans in a recirculating system effectively in series, may differ.

Fan laws   See Section 15.4.2.2.

Volume regulation   This falls under two headings:

1. Where a permanent change is found to be needed during commissioning or change of duty. The volume may be changed:
   (a) By variation of system resistance by damper action. Care is needed to prevent axial fan motors overheating or reaching the stall point;
   (b) By changing the speed of the drive. In centrifugal fans the common practice is to change or adjust belt drive pulleys, taking care not to exceed the power capability of the motor. In axial fans the pitch of the blades may be adjustable by swivel mountings on the hub.
2. Where frequent change is required in the normal operation of the plant:
   (a) By switching motors in multi-stage fans;
   (b) By speed control of motors electrically (e.g. by pole changing), varying the circuit resistance of wound rotor motors, or thyristor part-cycle disconnection;
   (c) By damper modulation;
   (d) By bypass (fan recirculation) dampers;
   (e) By axial fans with pitch variable while running.

Mounting   It is common practice to support the fan on anti-vibration mountings and connect the fan casing to the duct by short lengths of flexible non-combustible material. Care should be taken during installation to see that these are aligned to prevent entry turbulence and noise generation.

15.4.3.6 Dust control and filtration

Dust control   Dust is continually being introduced to the conditioned space by clothing fibres, skin particles, shoe dirt, room processes and the outside air make-up. Dust particle sizes range from $1 \mu\text{m}$ to $75 \mu\text{m}$, smaller particles being described as smoke. In an apparently clean office there may be as many as 30 million particles per cubic metre.

Control is by filtration in the plant but smoke can be removed by local recirculation through fan filter units.

For Clean Rooms (rooms of a very high standard) dust count per unit volume will be specified, but other specifications for room cleanliness are usually in terms of filtration performance against a standard test dust. Other important features are resistance to air flow and dust-holding capacity, leading to the fan energy required and filter life.

By number count the great majority of particles in the outside air are likely to be less than $1 \mu\text{m}$. By weight these small particles will account for a very small proportion of the sample. A filter with a high efficiency measured by weight of particles trapped may be almost transparent to the small ones. Very high counts can be found in rural areas from pollen or agricultural activities.

Where a high cleaning efficiency is required it is sound practice to install a filter of lower performance upstream to trap the larger particles and prolong the life of the more expensive High Efficiency Particulate (HEPA) Air Filter. Two filters of equal merit placed in series will not be materially more efficient than one.

The action of most filters is mechanical. These are normally scrapped when fully loaded but may, for lower efficiencies, be washable. Another type uses electrostatic charges.

Filter life   Filter life varies with make and type, and may be limited by the ability of the fan to operate against pressure drop. It is a function of the dirtiness of the air and the amount of material packed into the filter bank. Life may be quoted in terms of dust held in g/m^2 face area. For fan selection a knowledge of pressure drop is required. Typically, a panel filter might be quoted as operating from, say, 75 Pa initially at 2.5 m/s face velocity to 250 Pa when loaded and a HEPA might operate up to 700 Pa. The pressure drop across a bank of filters is kept within bounds by changing a proportion in rotation.

15.4.3.7 Humidification

The ability to add moisture to the air to raise humidity is an essential part of close control air conditioning. The need may be brought about by a change in the condition of the outside air, fabric losses or a change in the gains within the room or in plant operation.

Humidifier capacity   The humidifier is sized to meet the load of the outside air intake quantity. It is necessary to know the rh to be achieved, outside air intake rate (kg/s) and its lowest winter moisture content (kg/kg). This can be taken as 0.0027 kg/kg in the south UK and 0.002 for the north, corresponding to $-4°$ and $-7°C$ saturated, respectively. Since the capacity will be reduced by blowdown time, allowance is necessary for control, and there will be leaks from the plant and building fabric, an addition of about 30% above the calculated value is recommended.

Types of humidifier   These may be Direct (i.e. installed within the conditioned room) or Indirect (i.e. installed in the conditioning plant). Water may be introduced (a) as a spray or mist or (b) as steam generated separately or by a device within the plant.

In most air-conditioning work the humidifier will be indirect. The psychrometric operation of various types is described below.

Pan humidifier   The steam is generated in a pan of water by electric elements. High-temperature hot water or steam coils could also be used. About 30% of the input appears as waste sensible heat, giving a sensible heat ratio of 0.3.

Where the pan is in the airstream the condition downstream of the pan has an increased moisture content (kg/kg) found from the air flow and moisture input. On a psychrometric chart this will lie on a line of sensible-to-total heat ratio of 0.3. Thus the psychrometric plot shows a steep rise in moisture content with a small rise in dry bulb temperature. The latter is a disadvantage when cooling is required. Regular blowdown is needed, preferably controlled by a timer, to prevent furring when mains water is used, and open pans should not be utilized where the conditioned space is to be dust controlled. A small water volume is an aid to quick response.

Steam jet   Where it is available the source can be a separate boiler plant, but common practice is to employ purpose-made electrode boilers within or adjacent to the plant. The latter reduces sensible gains to the plant but, being essentially saturated steam, condensate return pipes are required. In addition to the rise in moisture content of the air (kg/kg) being dependent on air-flow and steam-injection rates, there is a very small increase in dry bulb temperature by the cooling of the vapour to the air temperature. The rise in total heat is: total heat of steam (kJ/kg) × quantity supplied per kg air.

Proprietary units are supplied with automatic blowdown cycles and can be matched to the broad water analysis. Cylinders have a limited life.

Spray humidifiers   Water is injected into the airstream in a fine mist by pumped jets or spinning disk. For practical purposes, the psychrometric plot follows a wet bulb line. The air provides the latent heat of evaporation, resulting in a fall in dry bulb temperature. If water were to be supplied at up to 100°C the humidified condition would be at a correspondingly higher total heat of 420 kJ per kg water supplied.

Where dust control is important the system should only be used with a supply of demineralized water to avoid solids being passed into the conditioned space. The temperature of the air must be sufficient to hold the quantity of moisture being supplied, any excess being deposited in the duct. Unless drained away, this can give rise to corrosion and to incorrect control by re-evaporation when the humidifier is switched off under control.

Air washers   Banks of sprays discharge water into the airstream with the object of achieving saturation of the whole air flow. Excess water falls into the base tank of the washer from which it is pumped back to the sprays. Downstream eliminator plates entrap any remaining free moisture, acting best within a specified velocity range. Cleanliness is essential to avoid bacterial growth. A constant bleed and make-up is normally arranged to control the accumulation of waterborne solids, but this, in turn, dilutes bacteriocides and inhibitors. The washer does not fully wash the air in the normal sense but does have the cleaning efficiency of a low-grade filter.

If the temperature of the water is not controlled it will come to the wet bulb temperature of the air passing through. Ignoring pump heat, the process is adiabatic. The psychrometric plot follows a wet bulb line.

If heat is added to the water the condition for 100% saturation takes the new wet bulb temperature of the incoming air.

If the water is chilled, cooling of the supply air takes place together with control of dewpoint and hence humidity of the room treated. Saturation efficiency is given by:

$$\frac{\text{Entering air db} - \text{leaving air db}}{\text{Entering air db} - \text{entering air wb}} \times 100\%$$

Due to bypass, a single bank of sprays might achieve 60% saturation and a capillary washer (one where the air passes through a wetted mat) might achieve 95%.

Humidifier run time   Humidification is an expensive process and it is useful to be able to assess energy costs when considering its inclusion in a plant. It is possible from meteorological records for any particular area to find the time in hours per year and extent to which the external moisture content is, on average, at or below a required absolute value. As an example, in the southern UK humidification would be required in varying degrees for 6700 h per year if 21°C 50% rh was to be maintained. If the characteristic of cooling coils is to dehumidify when only sensible cooling is required then the humidifier load will be greater.

15.5 Refrigeration

Two basic types of refrigerator are considered here: the 'vapour compression refrigerator' and the 'gas refrigerator'. The former consists of a compressor followed by a condenser where the refrigerant is liquefied at high pressure. It is then expanded in a 'throttle valve' to a lower pressure and temperature and finally evaporated in an 'evaporator' before re-entry into the compressor. The cycle is similar to the Rankine cycle in reverse.

The gas cycle is the reverse of a closed gas-turbine cycle, i.e. the constant pressure or Joule cycle.

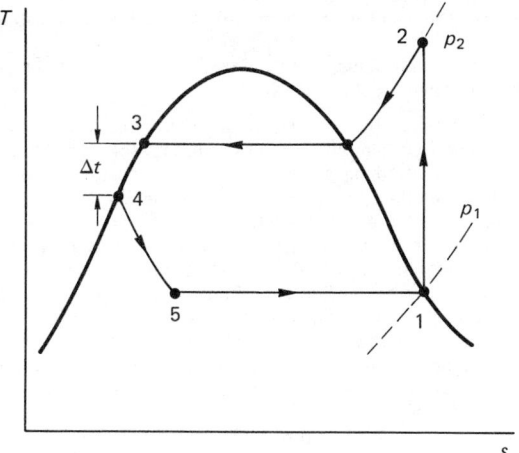

Figure 15.143   Vapour compression cycle

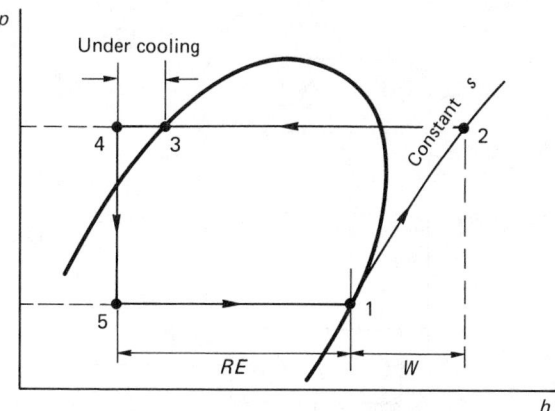

Figure 15.145   Pressure–entropy diagram for a refrigerator

If p_1, p_2 and the undercooling temperature T_4 are known, the diagram can be easily drawn and RE and W scaled off as shown.

15.5.3   Gas refrigeration cycle

Referring to the T–s diagram (Figures 15.146 and 15.147):

Refrigeration effect $RE = c_p(T_1 - T_3) + c_p\eta_t(T_3 - T_4)$

Work in $W = c_p \dfrac{(T_2 - T_1)}{\eta_c} - c_p\eta_t(T_3 - T_4)$

Coefficient of performance COP $= RE/W$

where: $\dfrac{T_1}{T_2} = \dfrac{T_4}{T_3} = \left(\dfrac{p_1}{p_2}\right)\left(\dfrac{\gamma - 1}{\gamma}\right)$ an index should be:

$$\left(\dfrac{p_1}{p_2}\right)^{\left(\frac{\gamma - 1}{\gamma}\right)}$$

and η_t = turbine isentropic efficiency
η_c = compressor isentropic efficiency

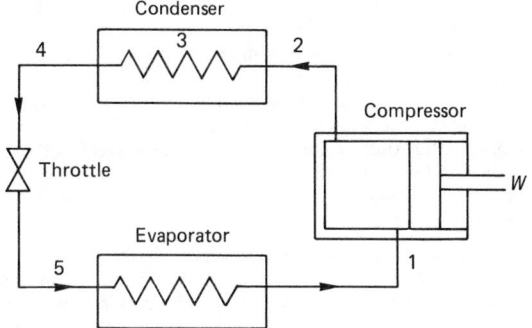

Figure 15.144   Vapour compression refrigerator

15.5.1   Vapour compression cycle

The process can be shown on the temperature entropy (T–s) chart (see Figures 15.143 and 15.144) for the appropriate refrigerant, e.g. ammonia, Freon 12, etc.

1. *Compression*   Work in $W = h_2 - h_1$
   where $h_1 = h_g$ at p_1, h_2 = enthalpy at p_2
   $s_2 = s_1$ (since isentropic compression)
2. *Condensation at constant pressure* p_2
3. *Under-cooling* from T_3 $(= T_s$ at $p_2)$ to T_4
   degree of undercooling $\Delta T = T_3 - T_4$
4. *Throttling* from 4 to 5
   therefore $h_5 = h_4$ and $h_4 = h_f$ at T_4
5. *Evaporation* at pressure p_1
   Refrigeration effect $RE = h_1 - h_5$
   Work in $W = h_2 - h_1$
   Coefficient of performance COP $= RE/W$
   Heat removed $Q = \dot{m}\, RE$, where m = mass flow rate of refrigerant.

15.5.2   Pressure–enthalpy chart

The pressure–enthalpy chart (Figure 15.145) is a more convenient way of showing refrigeration cycles. Work in and refrigeration effect can be measured off directly as the length of a line.

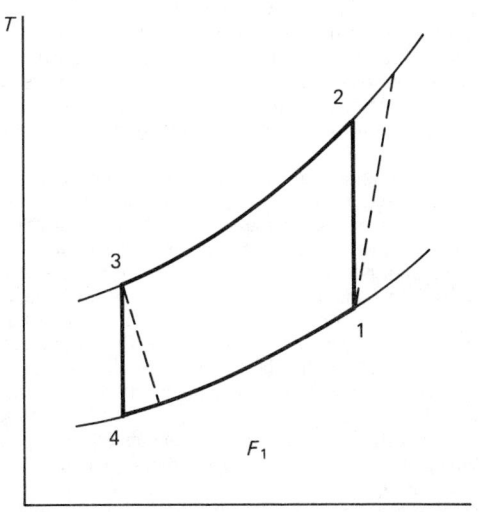

Figure 15.146   Gas refrigeration cycle

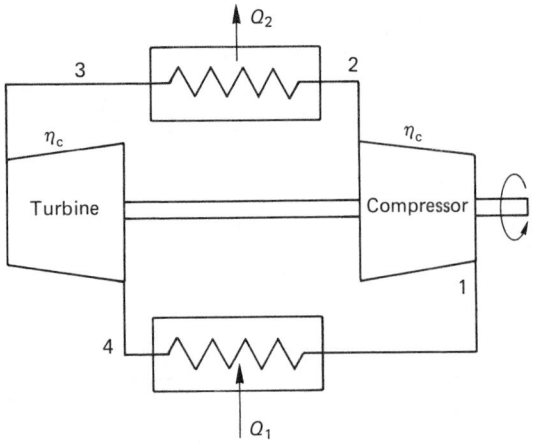

Figure 15.147 Gas refrigeration plant

15.6 Energy management

15.6.1 The energy manager

In 1973 the price of oil increased considerably. At this time about 50% of the UK primary fuel use was oil with 40% coal and 10% gas. One direct result of this price increase was to focus attention on the cost of energy to industry and the concept of an energy manager within a company became established. This is not to suggest that no company was aware of the cost of energy before 1973 but that the serious effects on price inflation were soon apparent and action was needed to contain the problem. In many traditional industries a general knowledge of the amount of energy used was available but the detail of the breakdown of use in particular processes was probably unknown and a knowledge of whether a process was the most energy efficient was certainly not. This latter statement is still true in many situations.

In general terms, we waste about half the energy we obtain from fossil fuels. Much of this waste is due to the widespread use of electricity which is generated at about 35% efficiency and individual companies are not in control of this waste except that they can minimize their consumption of electricity and hence their costs. Other serious causes of waste are poor combustion in boilers, inadequate insulation and poor building environment control, all of which are within individual company control.

It is the duty of an energy manager to know how and why energy is being used within the company and to organize the use of the most cost-effective supply. The nature of this task requires contact with a range of energy users, including both management and workforce, and will require enthusiasm, tact, persuasiveness and patience as well as expert knowledge. It is an ongoing situation in which all changes in energy use due to change of product, new buildings, new equipment, etc. need to be monitored. One key weapon in the manager's armoury is that if everything else remains constant a cut in the energy bill will mean a rise in wages or profits. If this can be achieved with little or no capital cost, but merely by good housekeeping or a change in working practices, then he or she will be in an unassailable position. The energy manager must therefore generate the cooperation of management, unions, shopfloor managers and all the users of energy in the factory, users in the office and users outside in the delivery and sales

vehicles. The manager must be able to communicate at all levels.

Energy management is encouraged by government in at least four ways:

1. Information – the government provides information in the form of free pamphlets, books[48] and newspapers[54] which focus attention on particular aspects of energy use and advise on efficiency. Energy managers should be aware of these valuable sources of information.
2. Grants – from time to time, grants for specific purposes are available. Initially these were mainly for simple one-day surveys by outside consultants who would bring an experienced, impartial eye to look broadly at the way energy was being used and produce a short report with recommended further actions. The availability of grants may change and the Department of Energy should be consulted for current information.
3. Legislation – the government ensures, through the Building Regulations, that adequate insulation standards are applied to buildings, hot fluid pipes and hot fluid stores, that window areas are not excessive and that thermostat control is fitted to heating systems. These regulations are not normally retrospective but they are progressively improved and the energy manager may well see economic sense in applying some to existing situations. The government also sets statutory maximum and minimum temperatures for heated buildings.
4. Audits of industries – detailed analyses have been made of a number of energy-intensive industries. In the present context the Audit of the Engineering Industries[49] is particularly relevant but the other surveys may contain useful information and the energy manager would be well advised to examine the whole series.

The energy manager needs a small library to form a database to assist in the task. Because of the interest in this field in the post-1973 years there is a considerable volume of literature which would take some time to read. Apart from the government publications mentioned above, the books and journals in references 50–53 should enable many questions to be answered.

15.6.2 Energy surveys and audits

It is quite clear that an energy manager must know how energy is being used before any attempt at management is possible. Obviously, the manager must have access to historical information in the form of bills for electricity, gas, coal, oil, petrol, etc. and possibly in daily rates of production. These will show how much was spent and made in any particular period but they will not tell a great deal without further detail. Some points are highlighted below:

1. Unless the firm is very small the bills will not show where the energy was actually used.
2. The bills will not show the total cost of energy. For example, coal and oil are often stored requiring a proportion of the rates, capital value of land and buildings, capital value of the stored fuel and depreciation in quality of some stored fuels, etc.
3. The bills will not show whether the methods of purchase of fuel are the optimum for the business. Contracts for liquid and solid fuels may be renegotiated, gas may be purchased on long-term contract. If electricity is considered as an expensive fuel (only because of the generation efficiency which ensures that it costs about three to four times as much as the raw fossil energy used in the power station) there are a number of tariffs available, depending on how

the electricity is supplied and used. Important considerations for electricity are:

(a) If the electricity is supplied and metered at high voltage to the consumer's own substation or if it is supplied and metered at low voltage;
(b) The amount consumed;
(c) The consumer's power factor;
(d) The consumer's maximum demand, especially in winter when each kilowatt of maximum demand can cost forty times the summer rate.

The supply method is dependent on the capital cost of a substation and decisions will require consultation with the electricity supplier. The supplier may also advise on 'off-peak' electricity. The other three items are within the energy manager's field.[48] The principles to contain these costs are avoiding waste which leads to an excessive consumption, ensuring that the power factor does not fall below 0.9 by fitting power factor-correcting capacitors to plant with inherently low power factors (such as induction motors used for driving machine tools, welders and induction furnaces, etc.) and ensuring that everyone concerned with electrical driven devices understands the maximum demand penalties. The maximum demand charge is related to the generating capacity that has to be available to meet the greatest national demand. The charge is based on twice the maximum number of units of electricity supplied in any single half-hour in the particular month. To avoid maximum demand charges it must be clear that random switching on or off of high-consumption electrical devices (plating plant, furnaces, etc.) must be eliminated. This is a task which requires a *control system* and user cooperation. Ideal control means that the loads are arranged so that the continuous demand is equal to the maximum demand which gives a *load factor* of unity. Computer-controlled demand (Figure 15.148) is one solution but there may well be a need for special intervention in circumstances where a computer power cut to avoid demand charges could cost more in damaged product than is saved.

4. The bills will not relate to external happenings, which do not appear on them. Common examples are special production needs to meet a time penalty or excessively cold weather. The former information may be available internally and the latter in the form of weather statistics. These are published for different areas of the UK as degree days both as a twenty-year average and for the past two years updated monthly in a free Department of Energy publication.[54] The degree days (for heating) in a particular month is the mean number of degrees by which the outside temperature on a given day is less than 15.5°C added up for all the days in the month. Degree days will vary from year to year. For example, in February 1986 Southeast England had 467 degree days rather than the 376 in February 1985, so that to maintain the same internal temperature the energy input would have needed to be about 1.24 times greater in 1986 than in 1985. Although weather is outside the energy managers' control they still face higher electricity bills as both consumption and maximum demand will rise.

5. The bills will not show the efficiency of the plant. However, if all the gas goes to one boiler it may tell the manager something about that piece of plant. Other problems of historical information could be listed but it must be clear that bills are an inadequate method of energy management without further information. A complete and ongoing audit system is required.[48] This can be built up starting with a simple survey of the energy needs

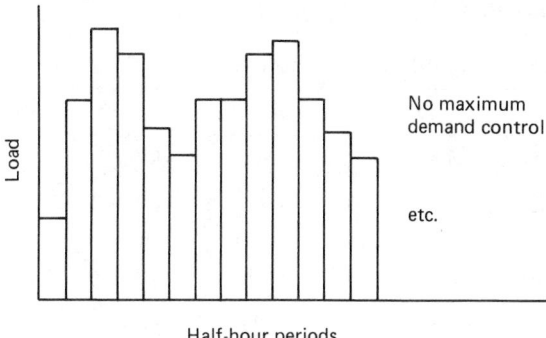

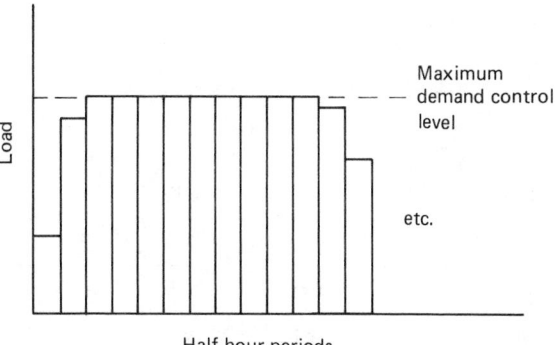

Figure 15.148 In the upper diagram there is no maximum demand control and the maximum demand charge will be 25% higher than that in the lower diagram, where the control has rephased some loads to fit into periods where there is spare capacity. The total consumption in the thirteen periods shown is the same in both diagrams

of all equipment that uses energy. Later it might be desirable to fit individual metering to each piece of equipment and to note consumption, time in operation and goods produced. Between these extremes there could be a compromise. The more instrumentation used, the more complete the picture and the more the use of energy can be controlled either by humans, who are unreliable due to memory lapses, lack of interest, etc., or by computers which are only as succesful as their programs permit.

When an audit has been made the information should tell how, why, when and where energy is being used. This should enable analysis to make sense of the results, detect waste and indicate where savings may be made. Plans can then be made for short-, medium- and long-term actions to minimize energy use. In the short term it might be that boiler efficiency, heat exchanger effectiveness, good housekeeping (closing doors and windows, turning off lights, etc.), lighting levels, draught proofing and stopping leaks of steam or air would pay dividends. In the medium term, insulation, instrumentation, controls and power factor correction might feature and in the long term new boilers, process changes, building use and heat recovery schemes might be considered. Table 15.32 shows possible audit layouts for environment and production energy.

Table 15.32 Simple initial energy audit tables

ENVIRONMENTAL ENERGY AUDIT FOR   19

	Building A	Building B	etc.
Space heating	kWh	kWh	
Heating ratio	$\dfrac{kWh}{m^2}$	$\dfrac{kWh}{m^2}$	
Lighting	kWh	kWh	
Lighting ratio	$\dfrac{kWh}{m^2}$	$\dfrac{kWh}{m^2}$	
Domestic hot water, etc.	kWh	kWh	
Energy cost total	£	£	

Notes

(1) Heating and lighting will vary from month to month with the seasons.
(2) Holidays etc. must be allowed for.
(3) Note the use of the kWh (or any other unit) for *all* cases to enable simple comparisons.

PRODUCTION ENERGY AUDIT FOR   19

	Workshop A	Workshop B	etc.
Electricity	kWh	kWh	
Gas	kWh	kWh	
Other fuels (oil, coal, etc.)	kWh	kWh	
Total energy	kWh	kWh	
Energy cost total	£	£	
Production quantity	units	units	
Cost per unit	£/unit or $\dfrac{kWh}{unit}$	£/unit or $\dfrac{kWh}{unit}$	

Notes

(1) It would be preferable to break down the use of electricity into machine tools, paint drying, air compressors, etc.
(2) The environmental cost of the above table may also be included in the production costs if so desired.
(3) Again note the use of kWh in all cases.

Concurrent with any analysis, long or short term, must be an assessment of the cost of implementing energy-saving measures. This is best determined in terms of the payback period – the time taken to recover the capital cost needed to achieve the savings. Once this period is passed, profits or wages increase. The determination of cost effectiveness when the payback period is lengthy is complex, depending on the accounting system used and will require judgement of the likely long-term benefit. Payback periods of up to three years can enable simple decisions, but beyond this period a decision needs a knowledge of the product stability and the marketing situation. The energy manager can only present findings to the accountancy department who will have conflicting claims for capital expenditure. In the end it is no good being energy efficient if you have no buyer for your product.

Even if a firm does not feel that it wishes to pursue detailed energy management, a person delegated to walk around once a month can save money by monitoring leaks, broken windows, dirty windows and lighting, machine motors idling during lunch periods, vehicle entrances wide open in cold weather, etc. Over a period of time this would lead to noticeable savings in the annual energy expenditure.

15.6.3 Applications

The Audit of the Engineering Industries[49] gives a breakdown of the total primary energy requirement of a range of producers. The word 'primary' implies that the electrical energy purchased has been divided by 0.3 to take generation efficiency into account so that the figures are closer to actual energy costs in financial terms. By averaging the figures listed it is possible to estimate some data which might be considered typical of a very diverse industrial scene. The information is tabulated below and may be seen to fall into three classes:

Building energy
Space heating	50%
Lighting	7%

Production energy
Driving machinery, welding and plating	23%
Heat treatment	10%
Compressed air	2%

Miscellaneous energy
Transport, computing, packing, etc.	8%

Some of these aspects are considered in more detail below.

15.6.3.1 Space heating

Since the average above shows that 50% of industrial energy costs are associated with space heating this is an important area for examination by the energy manager. The buildings in use will have probably been built at various times and may well presently be used for purposes for which they were not originally intended. They will consist of large open-space workshops, possibly with high roofs which allow warm air to convect away from the workforce at floor level, office spaces, controlled environments for inspection or computers, etc. each of which will present different problems. The UK climate is also capricious by nature, presenting quite rapid variations in external conditions which require expensive extra capacity to deal with extremes of heat or cold that may only obtain for a few days each year. Comfort for people (and for computers) is based on both temperature and humidity and it is necessary to ventilate all spaces at an appropriate rate or even air condition, depending on the activity within the space considered. Human beings radiate energy at various rates when working. A seated person at rest radiates at about 115 W and heavy work increases energy release to 440 W. All machines supplied with electrical energy eventually dissipate this energy into the work space as heat so that a machine shop on a hot summer's day is being heated by the motors driving the tools and the operators. Conversely, many spaces have delivery bays for lorries and smaller openings for internal transfers by forklift trucks which may allow considerable ingress or egress of cold or hot air to or from the workplace.

All spaces should be audited to determine the energy balance within them in the worst summer conditions and the worst winter ones. This will allow the largest cooling and heating loads to be determined. The audit should also reveal the heating and cooling methods used and the associated costs, the efficiency of boilers and heaters, the amount of insulation fitted, the building materials in the space, the glazing state, the draughtproofing state, the ventilation rate, the humidity state, the control system used, etc. There will be regulations concerning ventilation, temperature levels and any other health and safety aspects of the particular situation. Fire regulations may also be involved.

A very important consideration for the determination of steady-state heat losses or gains through the building fabric are the U-values of the structural materials. A U-value is a transmittance or overall heat transfer coefficient through a particular wall structure in air measured in $Wm^{-2} K^{-1}$. Thus the heat transfer through a wall is given by

$$\dot{Q} = UA\theta$$

where $\dot{Q}$ is the heat transfer (W), A is the area (m^2) and θ is the temperature difference between the inside air and the outside air (K). (*Note*: a temperature *difference* in K is numerically equal to a temperature *difference* in °C.) The U-value may be found by the addition of the resistances of the air layers inside and outside the wall and of the wall itself. If R is resistance (m^2 KW^{-1}) then

$$U = \frac{1}{\Sigma R}$$

For small heat losses or gains U should be small. This is achieved by suitable *insulation thickness* (Figure 15.149). Tables of data may be found in references 48, 51 and 55. U-values are also used to find heat losses from pipes carrying hot fluids and heat gains to pipes carrying cold fluids, heating effects from room heaters, etc. Examples of the benefits of insulation can be found in a Fuel Efficiency Booklet 16,[48] one of which shows that a 1000 m^2 corrugated steel roof which is uninsulated and has a U-value of 6.7 $Wm^{-2}K^{-1}$ can, with relatively simple insulation, reduce the U-value to 0.7 Wm^{-2} K^{-1} which should save almost £6000 per year in a Midlands situation. This example does not show the cost of the insulation installed but does illustrate a 90% reduction in heat loss costs. There is a danger of overestimating savings since insulation tends to increase inside temperature, and the heating system needs to be adequately controlled to achieve the full benefit.

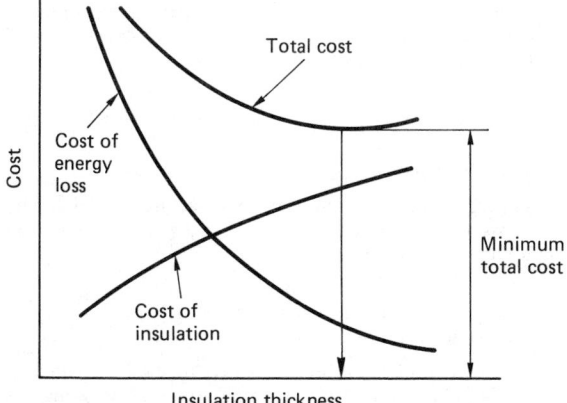

Cost

Total cost

Cost of energy loss

Cost of insulation

Minimum total cost

Insulation thickness

Figure 15.149 This diagram shows how an economic thickness for insulation may be determined. It must be realized that this is drawn for a chosen payback period since the energy loss cost is continuous but the insulation cost is a single occurrence

The methods used to ensure that buildings are heated to the appropriate condition before the workforce arrives and, conversely, that they are not overheated after the workforce leaves also need study. Heating and cooling in these situations are transient problems. Buildings with thick walls store energy within their fabric when the inside air is hot (or due to solar heating) and release it when the inside (or outside) air is cold. The time for a building to warm up or cool down will vary for different structures. A building with thick walls may need to be heated earlier in the morning than a thin-walled structure and the heating may be turned off earlier to compensate.[55]

Overheating a building above the *necessary* temperature, which will lie between the statutory limits, is wasteful. Every degree of excess temperature costs about 5% more in fuel costs and, similarly, every hour of non-essential heating due to the transient effects discussed above may cost about 10% more in an 8-hour day.

Inefficient use of fuel for heating is a waste; a modern boiler should operate with an efficency in excess of 75% at all loads at which it is intended to be used. 'Efficiency' is an imprecise word and should be clearly defined in any discussion. The boiler efficiency has two components: combustion efficiency and heat transfer efficiency, the latter representing the process of transferring energy from the combustion products to the water. This is not perfect, since the exhaust gases carry considerable energy to the atmosphere. This loss is unavoidable, since lowering the exhaust temperature has acid-forming effects. Some modern plant does have lower exhaust temperatures as a design feature. Combustion efficiency is maximized by ensuring that the best air–fuel ratio is used by constant monitoring of the exhaust gas composition (usually CO_2 or O_2 content). Burner nozzles also need cleaning to ensure good combustion. Heat transfer efficiency is maximized by regular blowdown to remove impurities fouling the boiler tubes (blowdown is also necesary to maintain steam quality). Boilers operate best at high loadings, and it may well be economic to have both a summer and a winter boiler to operate effectively. Clearly, the efficient operation of boiler plant is worthwhile.

Space heating and environment control is a complex problem which may well need professional advice to achieve satisfactory results in a modest-sized company, although help can be found in journals[53] and books.[48,50,52,56] Advice is essential when modifying building use or when a new building is being designed. The checklist below may be used but cannot show all considerations in all situations:

Buildings
Structure
Size
Insulation
U-values
Glazing
Draughtproofing
Ventilation
Energy outputs from occupants, machines, lighting, kitchens, furnaces, etc.
Statutory regulations

External environment
Temperature variation (annual and diurnal)
Solar gains
Wind exposure
Dampness
Degree days of location

Internal environment
Statutory regulations
Heating (convective, radiative)
Boilers, fuels and efficiency

Heat pumps
Humidity
Air conditioning

Control
Thermostats
Computers
Climate control
Building management systems

It must be emphasized that, because of the high cost of space heating, all energy managers should regularly monitor temperature levels, heating methods, insulation standards, boiler maintenance, door and window states, the timing of heating systems as the seasons change and the introduction of new machinery which alters loads and of new local heating devices without the energy manager's advice. This should be regarded as prudent housekeeping and should not be an expensive operation.

15.6.3.2 Lighting

Although lighting represents an average of only 7% of industrial energy use it is well worth consideration, because correct illumination is essential to efficient work in both offices and workshops. Savings may be achieved in other costs by better work or less time loss due to sickness as well as in energy costs. The unit by which illumination is measured is the lux (1 lux = 1 lumen m^{-2}). The definition of a lux is that a radiating source of 1 candela at the centre of a sphere of 1 m radius gives an illumination of 1 lux at the surface of the sphere. There are recommended illuminance figures for various grades of work.[48] Some examples are shown below to illustrate the range of levels involved. The symbol for lux is lx and for lumen is lm:

Outdoor stores	30 lx
Corridors	100 lx
Rough bench or machine work	300 lx
Drawing office	750 lx
Fine bench and machine work	1000 lx
Very fine work	1500 lx
Minute detail, engraving	2000 lx

A lighting analysis should decide the level of illumination required and the installation design will require a knowledge of the output of the lamps available. To be energy efficient the lumens per watt of electricity input, known as the lamp efficacy, should be used.[48] Lamp life is also valid information since the lighting system has to be maintained. Typical value of these data are shown below:[48]

Tungsten filament	8–18 lm.W^{-1}	1000–2000 h
Low pressure mercury discharge (tubular fluorescent)	37–90 lm.W^{-1}	5000–10 000 h
High pressure sodium discharge	67–121 lm.W^{-1}	6000–12 000 h
Low pressure sodium discharge	101–175 lm.W^{-1}	6000–12 000 h

There are other lamps available.

From this table it is clear that the most energy-efficient lighting would be sodium lamps, but lamp choice cannot be made solely by lamp efficacy since not all lamps produce the same colour light. For example, the low-pressure sodium lamp produces a monochromatic yellow light so that all colours except yellow appear brown or black. In general, tungsten lamps should be replaced by:

1. Low-pressure mercury types when good colour rendering is required for which there is a choice of type, with differing emphasis on both light colour and quality. Rotating machinery may pose problems as some lamps may cause a flicker effect which could mask rotation at certain speeds. Expert advice is needed for this problem.
2. High-pressure sodium lamps in areas requiring general lighting such as walkways, large workshops, etc.

The design of a lighting installation requires special knowledge[57,58] for the choice and positioning of luminaires is important. When the total load to give the required illumination has been determined it is essential to ensure that it is correctly distributed over the illuminated horizontal, inclined and vertical surfaces without reflection, glare or claustrophobic effects. The positions of windows and the decoration of the spaces also needs to be considered. Window areas may be increased but extra heating cost may result.

The cost of replacing a lighting system is usually recouped within two to three years because the running cost of high-efficiency lamps is about six times the total cost of installation, equipment and maintenance over the lamp life.[52] Once the cost has been recovered the ongoing running costs of the new system should be about a quarter of a tungsten-based system. To be cost effective, a lighting system needs a planned maintenance routine involving cleaning after perhaps 3000 h (including windows) and lamp replacement after 6000 h. By slight overdesign (10–15%) this is found to be cheaper than exact design and lamp replacement on failure[48] (3000 h is 10 h a day, 6 days a week for 50 weeks). Thus even though lighting is a relatively small item in engineering energy costs it can be considerably reduced in many situations.

Even with a good system energy can be wasted by poor use. In a sensible lighting installation attention is paid to switching arrangements. If it is not clear and simple, whole blocks of lamps will be switched on in order to illuminate a single lamp. Individual switching may lead to savings. It is also clear that humans will turn lights on when required but not off when they are not, so that some system of automatic control may be advisable. Time switches or photo-electric devices may be used. Control of lighting for early-morning cleaning, watchmen's rounds and other activities during 'off' periods can also be economic.

The final factors to include in any lighting calculations are:

1. All electricity inputs to lighting systems contribute to the building heating.
2. Lighting loads must be considered in the calculation of maximum demand costs in winter and in the overall power factor considerations in electricity charges.

15.6.3.3 Manufacture

The energy directly consumed by manufacture (i.e. excluding the costs of material, space heating, lighting, etc.) will depend on a number of factors, including:

1. The design of the component
2. The planning of the manufacture
3. The control of the plant used

If energy is considered at all these stages then some changes in the design philosophy, manufacturing processes, working traditions and habits and the plant in use may be considered possible and should be evaluated for their suitability, acceptability and payback time. Some general considerations are discussed below but most industries will have their own particular facets which should appear at the audit stage.

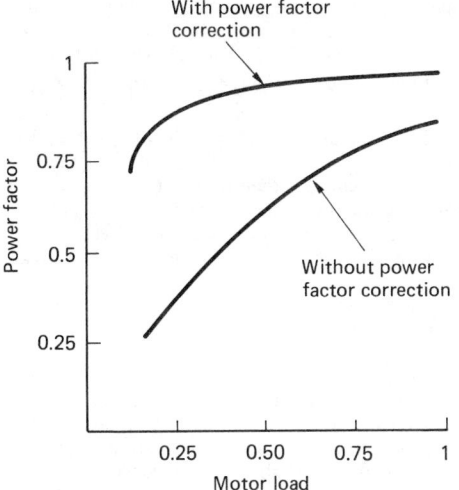

Figure 15.150 The effect of fitting power factor correction to an induction motor is very marked at part-load operation which occurs for a considerable percentage of the motor use

15.6.3.4 Driving machinery, welding and other processes using electricity

Machine tools used for a variety of metal cutting and forming processes are driven by electric motors. Most of these are induction motors which have a poor power factor and in order to achieve their best should run close to their rated power outputs. Induction motors must be fitted with *power factor correction* capacitors (or other devices) which must increase the power factor to more than 0.9 over a wide range of motor loadings. This is particularly valid (Figure 15.150) at part-load running where the power factor is worst in uncorrected motors. The energy audit[49] produced some data showing the distribution of cutting machine loads as working 28%, idling 52% and off 20%.

During cutting the motors operated at 50–80% of their rated power and during idling at 25% of their rated power, the overall result being that 42% of the energy consumed was in idling. It is obvious that there must be idling time between operations but this percentage is high and production planning, including the movement of parts and materials, should be arranged to minimize this waste. In winter a portion of this energy would be useful space heating but in summer it may need cooling plant to remove the excess energy, and even in winter, space heating by idling electric motors is an expensive technique. The planning should include both a time and a power estimation for roughing cuts and finishing cuts in any large-scale repetitive situation, since the energy consumed will depend on both the duration and the rate of consumption of the operation. Reference 49 describes a *measured* range of operating situations for a particular milling machine and a particular lathe. It clearly illustrates that in situations where the power for a machine varies between 23% and 130% of the rated motor power it would be difficult to predict the *maximum demand* in a machine shop. Some control might be possible for heavy power-consuming machines and some statistical data might be used to estimate likely demand in a quasi-steady production environment. A control system to stop some automatic start-up of non-essential plant when the maximum demand is near the limit could save much expensive

idling. Air-compression plant might fall in the category of controlled plant or indeed heating systems.

15.6.3.5 Welding equipment

Welding equipment fed by an a.c. transformer has a very low power factor and such plant must be fitted with *power factor correction*. Design of welded components should minimize the amount of filler required since the electricity consumed is directly proportional to the amount of filler used. Some data are given in reference 49. Pre- and post-heating should, as far as is possible, be localized to avoid waste. Similarly, stress-relieving processes should be considered by choice of method to avoid waste.[49]

15.6.3.6 Heat treatment

Processes such as hardening, tempering and annealing use more energy than any other manufacturing process, and the choice of process and the design of the component will therefore have a marked effect on the cost of the product. Reference 49 shows the relative energy costs for surface hardening of crankshafts (the electrical energy was divided by 0.3 to allow for the cost effect of generation efficiency). It shows that *induction heating* is the most energy efficient method. This technique does not, however, produce the best fatigue resistance nor the hardest surfaces as the other methods, so the design should take account of the proposed production method to be used. Induction heating may be employed for other heat treatment processes such as annealing and normalizing, where it is again found energy efficient.

When furnaces are used for heat treatment, combustion control is important. For example, if a reducing atmosphere is required no oxygen should be present in the gases. Furnaces should be well insulated since the processes are often lengthy, but extra insulation should not be added without consulting the manufacturer, as it may lead to overheating or other unwelcome side-effects. Linings in furnaces should have a low thermal mass to reduce the energy absorbed and wasted on cooling, but this needs analysis of the intended use since low thermal mass ceramic fibre linings are more expensive than traditional refractory bricks. Heat-recovery techniques should be considered for the exhaust gases.

When vats are used for degreasing, cleaning or pickling at elevated temperature there can be considerable loss of heat from the sides of the vat and by evaporation from the liquid surface in the vat. These may be minimized by the vat design, insulation and the use of floating plastic balls on the liquid surface. Other losses due to heating of the work and liquid solvent on the work surfaces are less easily controlled. There will also be an energy loss if the solvent is toxic and the fumes have to be removed by a ventilation system. The design of such a system should minimize the amount of warm room air removed and may include a heat-recovery system.

Paint drying and moisture-evaporation processes should be closely examined to see if the commonly used convection heaters could be replaced by radiant heating which avoids the 'ventilation' losses associated with convective designs. However, the ventilation may be vital with toxic solvents and heat recovery may be possible. It is also possible that the different effects of radiant and convective heating may require some design modifications since radiation penetrates the surface with a different drying effect and may also result in higher product temperature during the process.

15.6.3.7 Compressed air

Many users do not consider the cost of compressed air because, like water, air is erroneously considered to be free. Air-driven equipment should be assessed and the compressor matched to the minimum pressure required. Excess pressure is expensive. Similarly, the compressor should be matched to the demand so that it is running close to the rated output. The electric motor will then be operating at a high power factor. Air is often stored in a receiver and the compressor will operate at times when the receiver dictates. Careful management of the on–off cycle is essential and, if possible, the compressor should recharge the receiver at times chosen to use up spare maximum demand in any half-hour. Any compressor (or other machine) automatically switching on in a high-load half-hour will produce very expensive compressed air (or other product). Matching the compressor to the demand and on–off control are not the only ways to assure energy efficient compressed air supplies. Others include:[48]

1. Avoid leaks by regular inspection of the pipelines and the tools using the air;
2. Compressing air generates heat and both the compressor and the air produced are cooled, therefore an energy-recovery system may be economic;
3. Where tools exhaust air that is still above atmospheric pressure some other use or recycling of the air may be possible;
4. For large installations the use of two-stage compressors with intercooling between the stages will result in lower running costs;
5. Clean filtered *cool* outside air should be compressed rather than warm inside air which increases the energy use for a given delivery amount;
6. Choose the correct pipe sizes to minimize friction loss;
7. Maintain the machinery in accordance with the schedule.

15.6.3.8 Miscellaneous energy use

In any company there will be energy uses specific to the product. It is not possible to cover every aspect here but some comments on the more common, but perhaps unconsidered, peripheral energy uses are given below. *Audits should reveal more.*

15.6.3.9 Transport

Transport embraces a variety of modes: internal movements by battery-powered trucks, road transport of goods, road transport for sales and service staff, other car users, rail transport, sea transport, air freight, etc. The costs of transport may also include any special packaging or containers demanded by the mode. Many factors will decide which method of transport is best. If vehicles are required it is still possible to buy, hire, lease or even to subcontract the work. In spite of all these possibilities, most companies will still own and operate vehicles, and these may be audited to assess their cost and efficiency.

Since vehicles are driven over different routes with different drivers and different loads it will be possible to make comparisons and hence to be able to set reasonable targets in terms of fuel consumption per tonne-km for a particular type of delivery vehicle or litres per 100 km for sales or service duty, and over a period of time to build a picture of what may be reasonably expected. Economy may only be achieved with the cooperation of the workforce and this must also apply to salesforce and management to achieve harmony. If a car is part of the emoluments of the user this should not be debited to the transport bill but to the appropriate fund.

Having established the audit bounds and obtained some ongoing data it should be possible to look for measures to improve the use of energy (fuel). The following list is not exhaustive but indicates fields which might prove to be worth examination:

- Routes should be planned to avoid problems (roadworks, congestion, etc.).
- Telephones may be cheaper than personal visits.
- Loads should use the full vehicle capacity.
- Diesel vehicles are more economic than petrol vehicles.
- Vehicles may no longer be the ideal for their use (too large or too small).
- Speed controllers may be fitted.
- Aerodynamics can be improved with deflectors or load covers.
- Trailers can improve cost effectiveness.
- Vehicle servicing is important, particularly cooling controls for optimum engine temperature, binding brakes, idling speeds, rack or carburettor settings, injectors, ignition, plugs, tyre pressures, wheel alignment, balance, etc.

Multiple small inefficiencies can cost a great deal if vehicles spend most of their time transporting goods and persons. This is what they should be doing, of course, and if vehicles spend excessive time in parking lots then they are not required and hire or contract should prove economic.

15.6.3.10 Refrigeration

Refrigeration may be required for some processes (see also Section 15.5). A refrigerator takes energy from the material being cooled and rejects it to a coolant at a higher temperature. This process requires the input of energy and thus the total energy rejected in the refrigerator condenser is equal to the sum of the energy extracted from the cooled material and the driving energy input. This is likely to be contained in cooling water at a temperature between 20°C and 45°C, depending on the mode of operation of the plant. There is scope for heat recovery here if the plant is large.

15.6.3.11 Computing

The electrical energy supplied to a computer has to be dissipated to avoid overheating. In small installations this energy would be a constituent in the heat input to the space in which the computer is sited. In very large installations there will be a need for air conditioning and associated refrigeration plant from which energy recovery may be possible.

15.6.3.12 Storage, stock, packaging

Space allocated to storage of material for future use, finished product and packaging needs to be heated and lighted (and costs in rates). Careful calculations of stock levels may reduce the energy service cost of this facet of factory operation.

15.6.4 Waste-heat recovery

Waste-heat recovery or energy recovery has been mentioned in various places in this section. The critical factor in the value of waste heat is the temperature at which this energy is available. If the temperature is relatively high then it is not difficult to envisage useful applications of such energy, but it should be realized that each transfer of enegy requires a 10 K temperature difference to drive the heat transfer so that the energy is degraded in value at each transfer. At lower temperatures waste heat is also useful if the quantity is large enough to justify the use of a heat pump. The use of waste heat therefore requires the investigation of heat exchangers[59,60] and heat pumps.[61]

The first consideration for any waste heat must be to decide if it can be used directly. For example, uncontaminated hot water may be used for heating a building in a radiator system. If direct use is not possible due to contamination, toxic effects, etc. then a heat exchanger will be needed to transfer the energy to a suitable medium. Common types of heat exchanger include shell and tube, plate, run-around coils, recuperators and regenerators. The choice will depend on application and the fluids involved.[51]

Three other types of heat-recovery equipment are the rotary regenerator, the heat pump and the heat pipe. Conventional regenerators are cyclic in operation but the rotary design provides continuous recovery. This is achieved by slow rotation of a drum-shaped matrix which transfers the energy gained from the hot waste gas to the cold gas (Figure 15.151). Each of these gases passes continuously through a sector of the rotating matrix. The heat pump (Figure 15.152) is identical in concept to the refrigerator except that the objective is to absorb *low-temperature* waste heat in the evaporator and, by the energy input in the compressor, to deliver the waste heat and the compressor energy input at a *higher temperature*. The

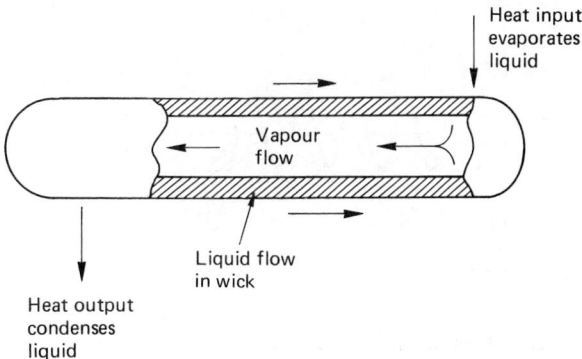

Figure 15.153 The heat pipe in which the energy input evaporates the liquid inside the pipe. At the other end of the pipe the energy is released as the vapour condenses. Various fluids are used inside a heat pipe ranging from nitrogen evaporating at $-196°C$ to lithium evaporating at $1340°C$

heat pipe is a device based on the evaporation of a liquid at one end of the pipe by the waste heat and the recovery of this energy by condensing the vapour at the other end of the pipe (Figure 15.153). Heat pipes may be used to remove energy from small or awkward situations and transmit the energy over distances varying from a few centimetres to a few metres.[60]

Common sources of waste heat include energy in flue gases from driers, boilers and furnaces, condenser coolants, air compressors, refrigerators, engine coolants, process liquids, condensate, flash steam, blowdown, etc.

It is possible to install a plant to produce both electric power and process heat. This combined heat and power plant (CHP) is an energy-efficient way of meeting two demands both on a large scale or as a top-up system in winter when both heat and power needs will probably rise. Maximum demand charges may then be avoided. It is also possible to supply excess electrical output to the grid in some situations. Expert advice should be sought for such schemes.

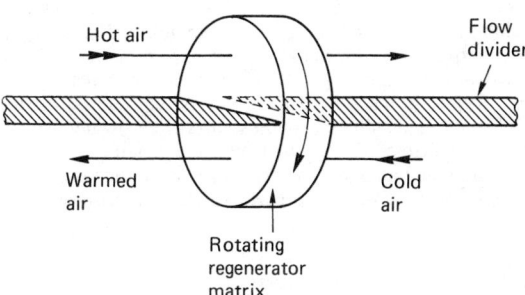

Figure 15.151 The rotary regenerator in which a cold air stream is *continuously* warmed by the waste hot air stream

15.6.5 Control systems

Energy management control systems have made significant advances in recent years. Such systems are essential in any realistic scheme because of the unpredictable, forgetful and habit-forming nature of human beings. Control systems are not inherently perfect but, provided that they have been set or programmed correctly and are adequately maintained, they should continue to operate satisfactorily and save energy. Controls are considered so important that they form part of the Building Regulations and are *mandatory* in new buildings or extensions of appropriate size.

Simple open-loop controls have no feedback of the controlled parameter so that, for example, a control may switch a heater on in a room but without feedback the room is unable to inform the control that it is hot enough and the heater will not be turned off. The addition of a thermostat feedback link from the room will enable the heater to be turned on and off as the temperature falls below or rises above preset limits. The control system is now an energy efficient closed-loop system (Figure 15.154).

From this simple example control systems can be made progressively more complex until we have a Building Automation System[52] where computer-controlled continuous *data*

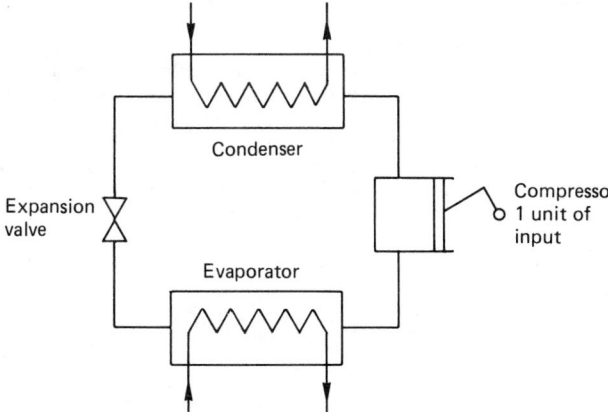

Figure 15.152 A heap pump with a coefficient of performance of 3, which means that for each one unit of 'paid for' energy input to the compressor three units of useful energy are delivered

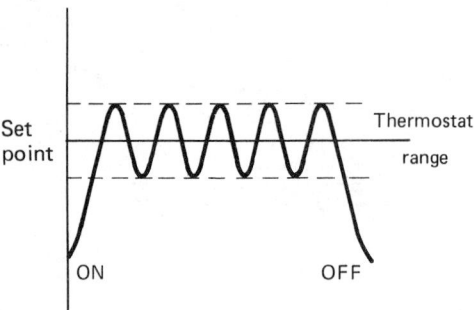

Figure 15.154 Simple thermostat control

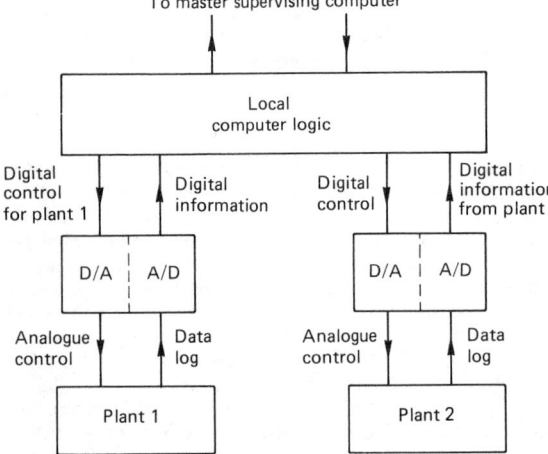

Figure 15.155 A computer control system in which only two plant stations are shown although there will normally be many more. The local computer may control a number of local pieces of plant and there will be several local computers controlling local energy tactics. All the local computers will be supervised by the master computer, which is in charge of the overall energy strategy

logging, assessment and management of the complete building energy demands are achieved (Figure 15.155). This will include heating, cooling, humidity, lighting, ventilation, etc. Other control systems to organize the total electrical load in the whole plant will be interactive with the Building Automation System to ensure that excessive demand is avoided and to make the electrical load factor approach unity. The installation of such complex systems needs expert advice and considerable monitoring at the commissioning stage to ensure that the system cannot cause catastrophic effects due to load shedding to cut demand.

The actuators in a computer-controlled system need regular maintenance. A computer can only data log and issue commands based on the installed program. If the actuators are worn or faulty the system response will be inadequate, leading to incorrect or inefficient operation.

No energy manager can afford to ignore the rapid development of control systems for both large and small factories. Any system that is over 5 years old should be assessed for

updating on a basis of a payback period for possible improvements.

15.6.6 Summary

Key considerations for the energy manager are as follows:

- Building Regulations, Health and Safety, attitude of workforce and management
- Available grants, use of experts
- Audits, instrumentation to give data
- Good housekeeping, insulation, power factor, maximum demand
- Waste-heat recovery
- Control systems
- Continuous assessment of energy consumption against targets

15.7 Condition monitoring

In managing mechanical resources there are various options to deal with the inevitable problem of mechanical breakdown. The first decision is whether to maintain the machine or operate it on a run-to-break basis. Run-to-break is now rarely considered viable in industrial applications as machines are becoming increasingly larger, more complex and expensive. The consequences of mechanical failure at unpredicted times can be catastrophic in terms of finance, loss of production and safety. Run-to-break should thus only be considered in the case of small items of plant (pumps, motor, etc.) where spare parts are available off the shelf and where failure will not damage other components and will be readily noticed. In most machinery some form of preventive maintenance will be required.

15.7.1 Preventive maintenance

There are two available options for preventive maintenance: periodic maintenance or predictive maintenance. Periodic maintenance is outside the scope of this section but, briefly, relies on a statistical background. It is based on the concept that machines will be more likely to fail, the more hours they have operated. While theoretically sound, this assumption does not take into account such factors as load conditions and machine speeds nor, more importantly, production tolerances of the machine, etc.

If high reliability is required (as in machines which will result in expensive downtime if failure occurs) the time interval between services will have to be reduced to well below the shortest time between predicted failures. This, in turn, means a high percentage downtime of the machine and the need to carry large stocks of spares (thus tying up capital). In addition, the increased overhaul frequency leads to greater opportunities for failure caused by human error (due to fault in reassembly, etc.). Also, it may not be practical to prevent failure, by planned maintenance, of certain components which have a relatively long life and are difficult to examine visually at maintenance periods (the inside of roller bearings, for instance).

15.7.2 Predictive preventive maintenance

With the arrival of modern technology, including cheap computer systems, the ideal solution may be found in predictive preventive maintenance. This system involves monitoring the condition of machinery, setting ideal operating limits and dealing with problems that result in these limits being breached. In addition, the method usually allows for diagnostics to be carried out, thus avoiding major strips down. The method has the advantage of much shorter downtime periods and obviously the need to carry large stocks of spare parts. The method is usually capable of predicting failure for some considerable time before it actually occurs. This lends itself to the maintenance being carried out at favourable times (for instance, at periods of minimum production) thus avoiding, as far as possible, production interruption.

15.7.3 Condition monitoring

The benefits of condition monitoring extend beyond maintenance and can be summarized as follows:

Maintenance
Enables repair to be planned, therefore, cheaper labour, part costs, and reduced need to stock excessive amounts of spares
Safety – increased operator safety
Secondary damage – avoided

Production
Reduction in machine downtime
Assured productivity
Improved machine throughput
Reduced scrap levels

Quality Assurance
Early indication of deterioration in machine leads to increase in the quality of the product

Machinery planning
Condition monitoring can be used to assess new machinery prior to it being brought into use

15.7.4 The parameters

The following parameters can be measured and contribute to a condition monitoring system:

Vibration
Bearing temperature
Coolant temperature
Lubricant pressure
Oil and hydraulic fluid analysis (for foreign particles)
Lubricant flow
Electrical parameters (volts and amps)

Vibration is the primary indicator of failure in rotating machinery and all others are secondary indicators. Condition monitoring systems are built up around measuring systems of varying complexity. The most simple system may involve the use of a hand-held vibration meter and moveable accelerometers held manually against the machine. This system will usually involve only overall vibration levels. In order to determine the likely time of failure with the highest order of accuracy this system will not be adequate. Some form of vibration frequency analysis will be necessary – this may have the added benefit of being able to identify the component that is due to fail.

In the case of very large machines, or those which have dangerous rotating components in the areas of interest, it may not be possible to use the simple hand-held meter approach. It is more likely that fixed accelerometers will be installed and cabled to one central data processing point. If a central point approach is chosen more than one input parameter can be analysed simultaneously. This gives advantages in fault diagnosis.

Vibration is perhaps the most costly parameter to measure. In the case of simple machines vibration measurements may

not be cost effective, and one of the other parameters may be used. If this is the case it is likely that some combination of planned and predictive preventive maintenance is likely to have to be adopted. For example, bearing temperature may indicate worn or damaged bearings but will not indicate out-of-balance shafts.

Apart from vibration, the only test for worn or damaged gears in gearboxes is by oil analysis (in itself a complicated procedure). This method can be used to diagnose which part of the assembly has failed (if the materials have distinct characteristics) by the use of spectroscopic techniques. This is a very costly procedure and only suitable for the most expensive machinery and then only on an intermittent basis.

Oil pressure or flow may be used to assess the state of journal bearings but will not be suitable for deciding the state of rolling element bearings. Electrical parameters may be used to indicate excessive loadings on motors, etc. (perhaps due to tight bearings) as well as possible electrical insulation breakdown.

15.7.5 Vibration monitoring for machine condition

The reader is referred to the section of this book on vibration control (Section 15.8) in order that some understanding of the basics of vibration theory and measurement may be gained. This will be necessary in order to fully understand the information that follows.

As previously stated, it is rarely sufficient to measure vibration merely in terms of overall level (RMS). In order to obtain sufficient 'early warning' much more sophisticated techniques will have to be adopted. In describing the various analysis techniques it is first useful to discuss the characteristics of the vibration in a typical device. A gearbox is considered, as it exhibits many types of vibration effects.

15.7.5.1 Low-frequency range

Low-frequency vibration is caused by imbalance, misalignments, bent shafts, etc. and occurs at the shaft rotational speed. Harmonics (i.e. multiples) of this frequency will also be present.

Journal bearings can produce vibration at frequencies below the shaft frequency. This vibration is caused by the centre of gravity of the shaft moving around within the clearance of the bearing. The vibration is caused by 'oil whirl' and seems to depend on the surface of the shaft and the bearing. Typical frequencies are between 40% and 49% of the rotational speed.

Another type of shaft vibration is known as hysteresis whirl. As the shaft passes through a critical frequency the hysteresis whirl will build up and remain at the critical frequency independent of shaft speed.

Mechanical looseness in a system will also be observed. Typically, this may produce interharmonic and subharmonic components (e.g. half-harmonic, one and a half harmonic, etc.).

15.7.5.2 Medium-frequency range

Tooth-meshing frequency will be the next frequency of vibration to be encountered. The frequency of this vibration will depend on the shaft speed multiplied by the number of teeth on the gear. The gear will also produce harmonics of this fundamental gear tooth frequency. Wear on gears produces higher levels at the harmonics (due to an increase in the sliding action of worn gear teeth) rather than a simple increase at the fundamental. General wear can thus be detected by the increase in these harmonics.

A single damaged tooth, on the other hand, produces quite different effects. This will result in a series of pulses in the time-domain signal from the gearbox and will create a line spectrum with each line spaced with the repetition frequency.

Thus damage to a single tooth will show up on the overall spectrum as an increase in the sidebands spaced with rotational speed above as well as below the tooth-meshing frequency. It may be difficult to see this signal as below the tooth-meshing frequency there will be all the low-frequency components (as already discussed) and above it the tooth mesh harmonics will have much more energy. It may, however, just be possible to detect this signal between the tooth meshing harmonics. If more than one tooth is damaged the effect will tend towards an amplitude modulation of the overall time signal. This will result in high-amplitude sidebands around the tooth meshing frequency (and its harmonics) spaced at the rotational frequency.

Another interesting source of gear vibration is 'ghosting'. This arises from errors in the machining of the gear when first produced.

Gears are produced by mounting blanks on a table which is rotated by an index wheel during machining. Inevitably, this index wheel will not itself be perfect and errors will be introduced into the gear being produced. The frequency of vibration produced by the gear in service will correspond to the number of teeth on the index wheel (times the rotational speed of the gear). This ghost frequency will thus be at an integer of the rotational speed of the gear but will not correspond to any tooth meshing frequency of the gearbox. As the gear gradually 'wears-in', the ghost component will gradually disappear (this is one way of determining that the vibration is due to a ghost). Also, the ghost will not be so affected by loading as the main tooth meshing frequency.

15.7.5.3 High-frequency range

At higher frequencies the primary source of vibration is rolling element bearings. Faulty bearings produce a pulse of vibration which repeats periodically dependent upon the shaft speed and the geometry of the bearing. There are four main frequencies associated with a rolling element bearing:

1. Ball-passing frequency outer-race (BPFO) – caused by a fault on the outer race.
2. Ball-passing frequency inner-race (BPFI) – similarly corresponds to the frequency of a ball passing a fault on the inner race.
3. The ball-spin frequency (BSF) – caused by a fault on the rolling element.
4. The fundamental train frequency (FTF) – caused by a fault on the bearing cage.

These frequencies can be determined by:

$$\text{BPFO } (H_z) = \frac{n}{2} f_r \left(1 - \frac{BD}{PD} \cos\beta \right)$$

$$\text{BPFI } (H_z) = \frac{n}{2} f_r \left(1 + \frac{BD}{PD} \cos\beta \right)$$

$$\text{BSF } (H_z) = \frac{PD}{BD} f_r \left[1 - \left(\frac{BD}{PD} \cos\beta \right)^2 \right]$$

$$\text{FTF } (H_z) = \frac{1}{2} f_r \left(1 - \frac{BD}{PD} \cos\beta \right)$$

where

n = number of balls or rollers and f_r = relative rev/s between inner and outer races (see Figure 15.156).

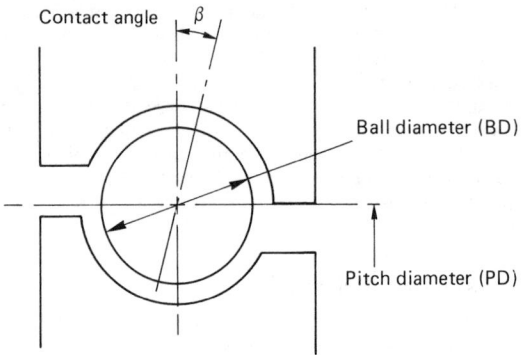

Figure 15.156

In practice, it may be difficult to isolate this ball-passing frequency due to interference from other effects. The pulse of energy transmitted into the race of the bearing will, however, cause it to resonate at its natural frequency. This will typically be at higher frequencies (between 1 kHz and 20 kHz) and will be easier to measure since other sources (as discussed previously) typically occur much lower down the frequency spectrum.

15.7.6 Vibration analysis techniques

15.7.6.1 Overall (RMS) levels

The overall (RMS) level can be measured simply by a portable (hand-held) meter and accelerometer system. This can be carried by an engineer and used on a regular basis to monitor machines. The actual level of vibration measured can be compared with previous readings (at the same location) so that trends in vibration levels may be plotted. In order that a high level of reproducibility is obtained it may be considered necessary to attach special metal disks to the parts of the machine to be monitored and use a magnetically mounted accelerometer. For one-off readings it is possible to gain some insight into machine condition by comparing the reading with ISO standards (ISO 2372/3 and 3945). These give levels of vibration for various classes of machine (in terms of power ratings of machines). The vibration levels are assessed into four bands (good, allowable, just tolerable, not permissible). An increase of 8 dB indicates a change of band. An increase of 20 dB shows that vibration levels have reached the 'danger' zone.

The main disadvantage with using overall (RMS) levels is that the reading is influenced (on the whole) by the single highest frequency 'spike'. In order to influence the overall level a fault signal has to exceed this level. As explained earlier, it is quite possible for a machine to produce high vibration energy at low frequency (perhaps due to slightly out-of-line shafts) and yet operate satisfactorily. Vibration produced by an incipient fault in a bearing will not be detected until its level rises sufficiently to overcome the low-frequency energy, by which time catastrophic failure may be imminent.

The main advantages may thus be classified as follows:

Simple
Cheap
Robust
Portable
Adaptable

Disadvantages are:

Insensitive
Diagnostics not possible

15.7.6.2 Crest Factor

Simple vibration meters often have the facility to measure peak level as well as RMS. The use of this facility, together with the overall (RMS) level, can be used to gain some further insight into the nature of a probable fault.

A typical use for this method is in the detection of faulty rolling element bearings. It may be useful to first filter the signal so as to leave only the range between 1 kHz and 10 kHz. As a bearing operates normally the peak level will remain at a constant figure above the RMS level. As a fault develops, however, the short burst of energy produced by a ball passing a fault increases the peak but does not significantly alter the RMS. The ratio of peak to RMS value therefore increases. This factor is known as the Crest Factor.

As the faults in the bearing progress, more spikes will be produced which will ultimately increase the RMS level. Therefore, towards the end of the bearing life the Crest Factor will fall (perhaps to its original value) while both peak and RMS will be elevated. By observing trends in these data (perhaps by plotting both RMS and peak level on a time graph and obtaining the Crest Factor as the difference between the two curves) the life of a bearing may be predicted.

Advantages and disadvantages are that this technique is very similar to the overall measurement method but it does allow for some diagnostics/trending.

15.7.6.3 Constant percentage bandwidth spectrum comparison

By frequency analysing the signals, using an FFT analyser, an overall 'picture' of the vibration signals is obtained. In order to display the full spectrum it is usual to display the frequency axis in terms of constant percentage bandwidth (i.e. log). A display from 0 Hz to 20 kHz can thus be conveniently displayed on one screen.

Spectrum comparison can be carried out by controlling the FFT analyser with a computer and software package. The computer compares the current displayed spectra with one taken when the machine was in a known 'healthy' condition.

So that small variations in spectra do not trigger alarms a reference 'mask' has to be created around the reference spectra. The mask is created by widening the bands of the reference spectrum, to allow for small speed variations, and increased in height to allow for permissible increases in vibration levels. It is possible to have several 'alarm' levels. Commonly, the maximum permissible increase would be 10× the reference mask. At twice the reference mask it is usual for the software to start to plot the trend in change of level. These trend data can be used in conjunction with data obtained previously to predict the time at which the fault will reach the trend limit (i.e. the 10× level), at which point the machine must be shut down.

These trend data may be used to safely predict that, while a fault has occurred on a particular machine, it will not become terminal before the next available convenient time for shutdown. In practice, it has been found that machines can exceed this suggested trend limit by more than a factor of two before breakdown occurs (see Figure 15.157).

The actual setting of the thresholds will, of course, have to be determined by experience with the actual process, and the above figures should only be used as guides.

As mentioned above, the software should be capable of making compensation for speed changes. If the machine only operates with minor excursions from one set speed, the actual spectral shape may not change significantly, but merely be shifted along the frequency axis. The software will be capable of tracking one characteristic frequency 'spike'. If the machine operates over a wide range of speeds the frequency characteristics may be quite different at the extremes of operating conditions. In this case it will be necessary to break up the operating range into several speed bands and record one reference spectrum (and hence create a mask) for each band. The software will track the signal within one band (as above) but when the speed range goes outside that band it will automatically switch to the next reference spectrum in the memory. This process will repeat depending upon how many separate reference spectra are considered necessary.

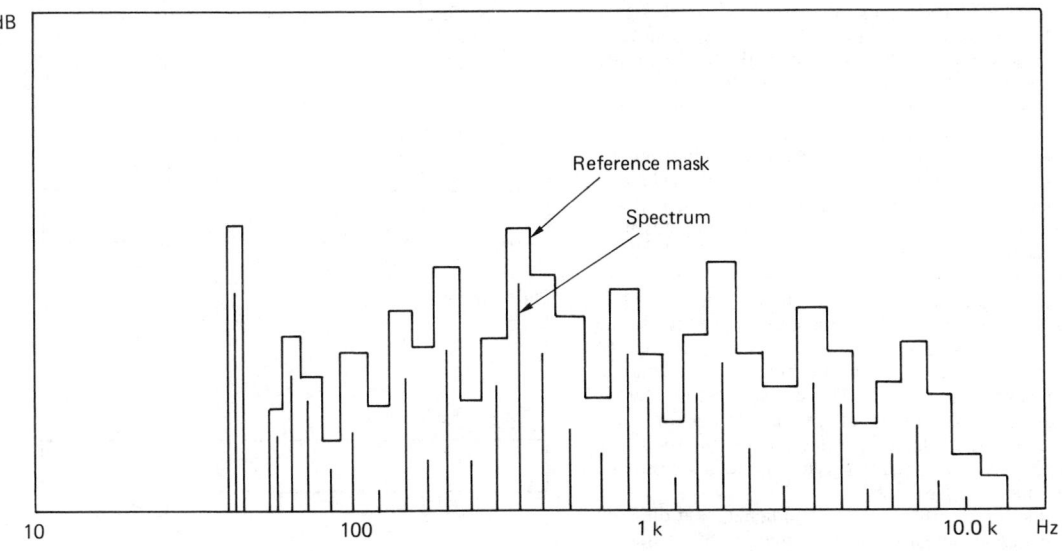

Figure 15.157 Typical reference spectrum and associated mask

15.7.6.4 Narrow band analysis

In order to study a spectrum in more detail it is usual to have the facility to magnify (or zoom) it. The frequency axis is usually expressed as constant bandwidth (as opposed to constant percentage bandwidth).

Narrow band analysis is used as a diagnostic tool (perhaps after an exceeding of a reference mask as in the previous system) and is sensitive enough to isolate the ball passing frequencies of bearings (or perhaps the higher-order harmonics if the fundamental is lost in the background noise 'floor'). This method would be used to analyse sideband build-up as would be necessary in gearbox fault diagnosis (as previously discussed).

15.7.6.5 Envelope analysis

If it is not possible to isolate a fault (in a bearing, for instance) because of interference from other sources this technique can be employed. For example, the bearing passing frequency of a rolling element bearing may be lost in the background noise. The high-frequency element (i.e. the resonance of the bearing race) will contain information concerning the repetition rate of the impacts that cause this resonance. By filtering the time signal so that only the bearing race resonant frequencies remain, the signal will be made clearer. This filtering may be carried out by the use of a high-pass filter or by the analyser's zoom facility. We are thus left with a time signal that contains the resonant frequency amplitude modulated by the impulses in the signal. If we rectify and smooth this signal we will be left with the envelope of the system still containing the impulses. This signal can then be passed through the frequency analyser to display the peak at the fault impact rate and can be identified as being one of fundamental frequencies of the bearing (i.e. BPFO, BPFI, BSF or FTF).

15.7.6.6 Cepstrum analysis

The cepstrum is a spectrum of a logarithmic spectrum (i.e. log amplitude but constant bandwidth frequency). It has the advantage of displaying periodicity in the spectrum. Families of harmonics will be well displayed on a cepstrum and random vibrations will not be displayed.

A spectrum is derived from the time domain signal by a Fourier transform which 'counts' the periodicity. i.e. a signal at 1000 cycles per second in the time domain is represented as a single line at 1 kHz on the spectrum with the amplitude being equal to the RMS time signal amplitude. In the same way, the cepstrum is a Fourier transform of the spectrum and looks for periodicity, i.e. a signal at 100, 200, 300 Hz, etc. (a harmonic series) and represents this as a single line on the frequency scale (the y-axis of the cepstrum). This technique can be very useful in gearbox and turbine fault diagnostics.

In the gearbox, for example, several faults are possible, including eccentricity of gears, single or several tooth faults or general tooth wear. Eccentricity produces an amplitude modulation of the normal tooth meshing frequency with the envelope period corresponding to the rotational frequency. At the same time, the varying contact pressure can give rise to rotational speed fluctuations (shown as frequency modulation of the tooth meshing frequency). This will give rise to a family of sidebands either side of the tooth meshing frequency with a spacing equal to the modulating frequency (i.e. the rotational frequency). Similar sidebands will occur around the harmonics of the tooth meshing frequency.

A single tooth fault will produce a fundamental at the fault rotational frequency and also the harmonics of this frequency (these will all typically be at much lower frequencies than the tooth meshing frequency). General tooth wear will show up as amplitude modulation of the signal (in the same way as eccentricity) and will produce sidebands around the tooth meshing frequencies.

In practice, several of these effects can occur at once and many spikes will be noticed on the spectra. It can thus be almost impossible to locate periodicity by eye alone.

Cepstra can be easily interpreted, as they give a single line corresponding to each sideband spacing with a height equivalent to the average sideband height. The cepstrum is achieved in practice by taking a frequency analysis of a time signal and processing it so as to concentrate on the frequencies of interest (perhaps by the use of the zoom facility). This spectrum is then fed to a personal computer equipped with a specialized software package which controls the frequency analyser. The computer then feeds the spectrum back into the analyser as if it were a time signal. The analyser then produces a further spectrum (which is now the cepstrum of the original signal).

Cepstrum analysis can thus be used as a diagnostic tool and, in addition, may be employed for condition monitoring (by the use of a mask in the same way as spectral comparison). This may be particularly useful if the signal is particularly complex and the component to be monitored is suitable (i.e. one that is likely to produce increased harmonics or sidebands in the event of faults).

15.7.6.7 Phase between transducers

If transducers are mounted on more than one bearing housing along a shaft the phase between the vibrations can give a useful aid to the diagnostics of various faults. Static imbalance produces in-phase vibration at both ends of the shaft (Figure 15.158) and 180° out of phase is likely to be caused by a dynamic imbalance of the shaft (Figure 15.159). By placing accelerometers in more than one plane, further information can be obtained (Figure 15.160). This situation would indicate

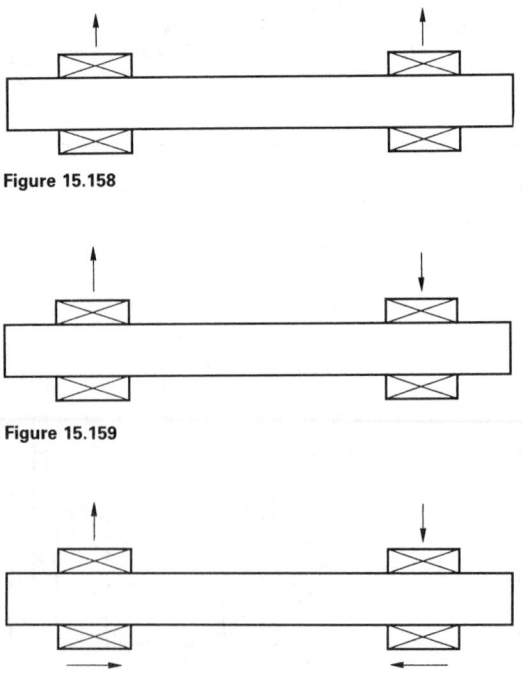

Figure 15.158

Figure 15.159

Figure 15.160

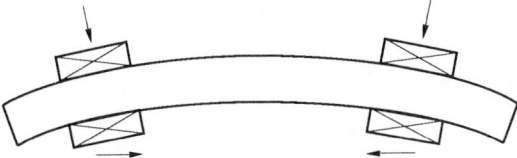

Figure 15.161

a rocking movement of the shaft (which must be caused by an external force) while the above situation would be caused by a bent shaft (Figure 15.161).

15.7.6.8 Synchronous averaging of the time domain signal

In the normal frequency analysis procedure the analyser produces the spectra of the time signal and then averages those to produce a display. The averaging time can be set by the user.

If we imagine a time signal with just two main frequencies (say, 10 Hz and 35 Hz) and their families of harmonics we may wish to analyse just one part without interference from the other. This could be achieved by synchronous time averaging. In this procedure the signals are averaged in the time domain before frequency analysis. If we took the sample rate at 10 Hz, all the 10 Hz signals would be in phase with the sampling rate and would be displayed normally. The 35 Hz signals would, however, progressively come into phase and go completely out of phase with the sample rate and would thus be averaged out to zero.

This procedure can be very useful when analysing the signals from gearboxes with multiple shafts. By choosing the right averaging period the analyser can eliminate most of the signals from the rest of the shafts and focus on the one of interest. In the case of variable-speed devices the triggering of the time domain averaging is done with the use of a trigger attached to the shaft (this can be in the form of an optical device).

If it is not possible to access the shaft in question, it may be possible to access another (perhaps the output shaft) which will be turning at a fixed ratio to the shaft in question. The optical trigger can then be attached to this shaft and fed through a device known as a multiplier. This can be set with the speed ratio of the output shaft to the shaft under examination and produces a signal synchronous with the shaft that is inaccessible.

15.7.6.9 Trending results

Trending of the increase in vibration levels may be a useful indicator in certain cases. The primary use is to predict the time to failure of rolling element bearings and wear on machines in general. Trending may also be applied to changes in oil pressure and other simple indicators, but may not be useful in the case of machines which are subject to one-off (sudden) failures (e.g. compressor turbines losing blades from fans). It may be possible to carry out trending manually; however, by far the most common procedure is to use the computer that is controlling the FFT analyser. If a computer is employed it is usual to trigger the trending procedure when a reference mask is exceeded by a pre-set limit.

The computer generates a line graph (as in Figure 15.162) and 'fits' the points to this line. The correlation (or fit) is expressed as an overall correlation (1.0 = perfect fit to the line). This trend line can then be projected onto the failure point and the lead time derived.

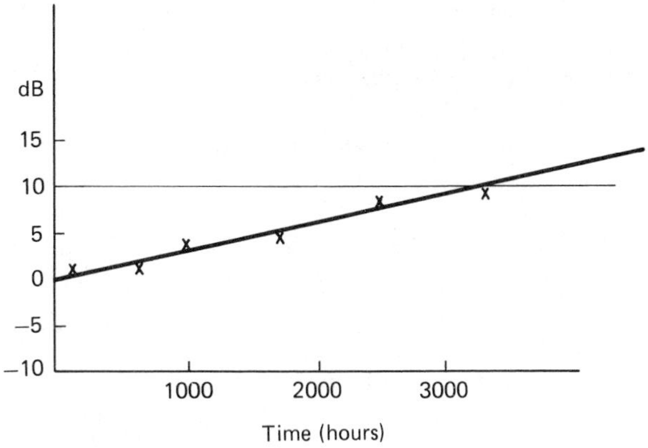

Figure 15.162 Trend curve

The lead time can only be derived accurately if a significant amount of previous testing has been conducted on other machines which have been allowed to fail while being monitored. Even then, it must be remembered that the procedure will allow a degree of statistical error (due to the nature of the actual failure perhaps being slightly different to all other recorded data). A degree of 'safety' must therefore be allowed, and it would be normal to set a shutdown level at perhaps 20 dB above normal running level.

Further information on condition monitoring can be found in references 62 and 63.

15.8 Vibration isolation and limits

15.8.1 Introduction

Vibration may be defined as the oscillatory movement of a mechanical system, and it may be sinusoidal or non-sinusoidal (also known as complex). Vibration can occur in many modes, and the simplest is the single degree of freedom system. If the mass were deflected from its rest position and released the resulting movement would be sinusoidal and decaying with time (as a result of the action of the damper). The actual frequency of vibration will depend on the undamped resonant frequency and the degree of the damping. If no damping is present the mass will always vibrate at the resonant frequency of the system. This frequency is determined by the stiffness of the spring and the mass put on it:

$$f_r = \frac{1}{2\pi} \sqrt{\frac{k}{m}}$$

where
f_r = resonant frequency (Hz)
k = spring stiffness (Nm^{-1})
m = mass (kg)
g = acceleration due to gravity

f_r can also be derived from the amount, d, that the spring is compressed when the weight is put on it (the static deflection) from the following equation:

$$f_r = \frac{1}{2\pi} \sqrt{\frac{g}{d}}$$

The displacement of the object from its rest position can be derived from

$$x = x_{peak} \sin(\omega t)$$

where x = displacement at time t
x_{peak} = peak displacement
ω = angular velocity ($2\pi f$)

Vibration may be expressed in terms of three factors:

Displacement (m)
Velocity (m s^{-1})
Acceleration (m s^{-2})

The relationship between these factors can be expressed by the equation:

$$A_{peak} = +\omega V_{peak} = -\omega^2 x_{peak}$$

For linear systems this can also be expressed as:

$$V(t) = \frac{dx}{dt} \text{ and } A(t) = \frac{d^2x}{dt^2}$$

For sinusoidal vibrations we can therefore conveniently change between displacement, velocity or acceleration units at will, either by the use of a nomograph or, more conveniently, using an integrator fitted to the measurement accelerometer (see below).

There is a phase angle of 90° between displacement and velocity and a further 90° between velocity and acceleration. Vibration may also be expressed in terms of decibels and the standard reference levels used are:

10^{-8} m s^{-1} for velocity
10^{-5} m s^{-2} for acceleration

15.8.2 Damping

The natural resonant frequency of a system is modified by the presence of damping. The damped frequency is

$$f_d = \frac{1}{2\pi} \sqrt{\frac{k}{m} - \frac{(c)^2}{(2m)}}$$

where c = damping force. This can be related to the undamped natural frequency by the following equation:

$$f_d = f_r \sqrt{1 - \frac{(C)^2}{C_c}}$$

where C_c = damping force for critical damping.

15.8.2.1 Critical damping

If the damping force is very high harmonic motion is not possible. Any displacement of the mass from rest will result in it returning exponentially to its rest position. It may be shown that this situation will occur where:

$$\frac{(C)^2}{(2m)} > \frac{k}{m}$$

This is known as supercritical damping. If

$$\frac{(C)^2}{(2m)} < \frac{k}{m}$$

and the mass is displaced and released it will oscillate at the damped natural frequency (see above) and the oscillation will decrease in magnitude with time (decaying more quickly at higher damping rates).

The effect of damping may be shown by Figure 15.163, where

$$\text{Dynamic magnification factor} = \frac{\text{disp. mass}}{\text{disp. base}}$$

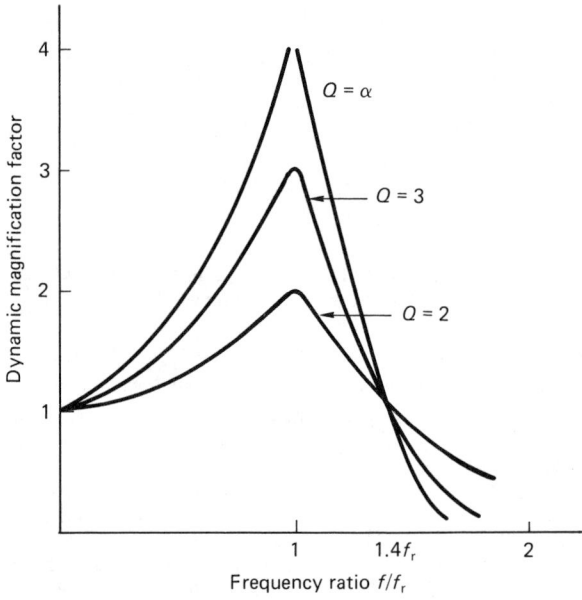

Figure 15.163 Response of damped system (f_r = natural resonant frequency of system)

The Q-factor (the dynamic magnification at resonance) is often used as a measure of damping, i.e. for no damping $Q = \alpha$.

In practice, all mass/spring systems have some natural damping (dependent upon the nature of the spring) and this will tend to reduce the theoretical infinite excursions at resonance.

15.8.3 Multi-degree of freedom systems

Until now we have only concerned ourselves with one mass/spring system. In practice, it is more usual to find multi-degree of freedom systems. These may be considered in terms of Figure 15.164. At low frequencies all the masses will remain coupled as there is low inertia in the system. If the base drives the system all of the masses will move as one with all motion lost in k. As the driving frequency increases to f_{r1} (the resonant frequency of m_1) the masses will vibrate violently again as one. At frequencies above f_{r1}, the masses will be isolated from the base. As f_{r2} approaches (the resonant frequency of m_2) m_2 will oscillate violently and will impose some of its motion on m_1 (dependent on the ratio of masses m_2 to m_1). The graph showing transmissivity will, therefore, be modified by the effect of m_2 and m_3 and dips or bumps in the main transmissivity curve will occur at f_{r2} and f_{r3} (the resonant frequencies of m_2 and m_3).

15.8.4 Vibration isolation

As can be seen from Figure 15.163, above $1.4 \times f_r$ the motion of the suspended mass is less than the motion of the base. The mass is, therefore, said to be vibration isolated.

In practice, isolation systems are chosen so that the forcing frequency is between eight to ten times the resonant frequency of the mounted system. Such a simple system may be suitable for constant-speed machines (which vibrate at a constant frequency). Unfortunately, it is more normal for machines (particularly those involving rotating shafts) to start from a

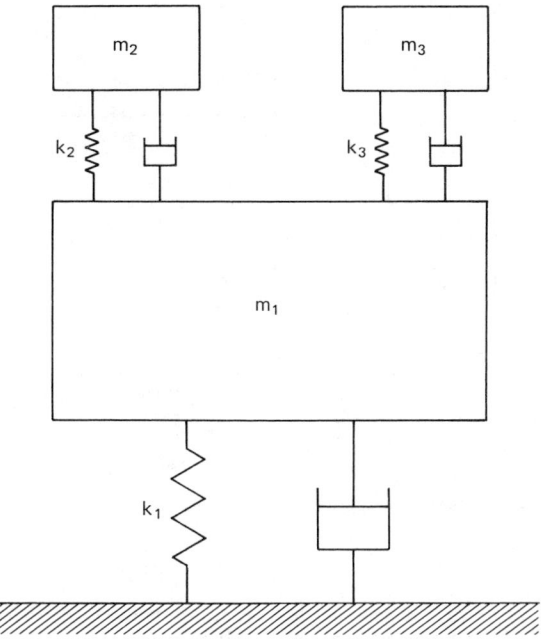

Figure 15.164 Compound mass spring damper system

standstill and accelerate to the normal running speed. For a period, therefore, the machine is likely to be vibrating at the resonant frequency of the system. The excursions of the machine mass may then become unacceptably large. If this were to be the case, damping would have to be applied to reduce the resonance. Unfortunately, this has the side effect of reducing the isolation of the mount under the normal operating conditions.

It may be that the machine will accelerate very rapidly through the resonant frequency and, in this case, the dangerous oscillations may not have sufficient time to build up. The damping could, therefore, be decreased or eliminated altogether (as most spring systems have, at least, some degree of natural damping).

15.8.4.1 Worked example of the vibration isolator selection procedure

Problem A machine has a mass of 20 kg and vibrates predominantly at frequencies between 150 Hz to 1 kHz. This machine is to be isolated from the building by four mounts attached to the four lower corners of the machine.

Solution In order to isolate frequencies greater than 150 Hz the resonant frequency should be chosen as one-tenth of the driving frequency, i.e. 15 Hz. The stiffness of the mounts required to give this resonant frequency can be determined from the formula:

$$k \approx 40 \times m \times (f_r)^2 \text{ N m}^{-1}$$

If four mounts are used the mass on each mount is 5 kg, thus

$$k = 40 \times 5 \times (15)^2 \text{ N m}^{-1}$$
$$k = 45 \times 10^3 \text{ N m}^{-1}$$

Manufacturers specify mounts in terms of the deflection at the maximum permissible imposed weight. The stiffness can be derived from

$$k = \frac{10m}{d} \text{ (N m}^{-1})$$

where m = maximum permissible load (kg) and d = static deflection at maximum permissible load (m).

15.8.4.2 Siting of isolators

It should be remembered that a machine mounted on four springs will exhibit more than one degree of freedom of movement. It is normal to consider six degrees of freedom in such a system (three linear directions and three rotational planes). In order to simplify the calculations and to reduce rocking effects the centre of gravity of the machine must be kept as low as possible. In addition, the mounts must be placed so that they are equidistant from the centre of gravity (they will thus all carry the same load). If this is not possible due to the configuration of the machine two solutions exist.

In the first case a sub-frame is made to fit the machine and this is in turn mounted on the isolators (which may now be mounted symmetrically about the centre of gravity). If this is not an alternative, and different loads have to be carried by each mount, the stiffness of each mount is chosen to be of a different value such that the static deflection of each mount is the same when the weight of the machine is imposed on it.

If the machine is light in weight or the centre of gravity is high, unacceptable excursions may occur. In this case the solution is again to use a sub-frame, but this should incorporate a large mass which will lower the centre of gravity and increase the stiffness of the isolators for a given resonant frequency.

15.8.4.3 Compound vibration isolation

By the use of compound systems (see Figure 15.165) it is possible to increase the attenuation of the system above that possible with simple systems. These systems may also be used where vibration problems exist at one discrete frequency. An example of this type is the inertia damper. In this type the superimposed mass (m_2) is chosen to resonate at the troublesome frequency. This vibration is out of phase with the main machine and will tend to cancel out the energy of the troublesome vibration. The calculation of the exact effect of compound vibration isolation systems is beyond the scope of this work and readers are advised to consult specialist textbooks such as reference 64.

15.8.5 Shock isolation

Shock isolation has many similarities to vibration isolation but there are some key differences and it is necessary to appreciate these prior to designing a shock isolation system.

A shock may be defined as the transmission of kinetic energy to a system which takes place in a relatively short time (compared to the resonant frequency of the system). In order to study shock isolation it is necessary to appreciate the time history of the shock pulse.

Shock pulses may have different time histories and may be explained by Figure 15.166. Frequency analysis of these shocks would reveal that they contain energy spread over all frequencies from zero to infinity, with no distinct frequency 'spikes'. It may thus be seen that the isolator design techniques used for vibration isolators (which depend on frequency measurement) will not be suitable for shock isolators. If the

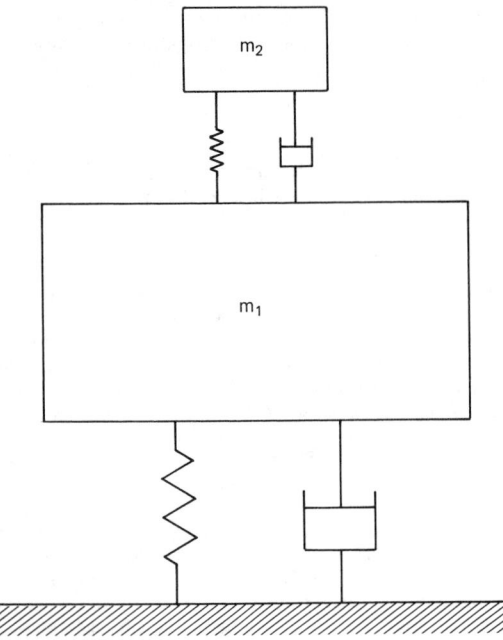

Figure 15.165 Compound system

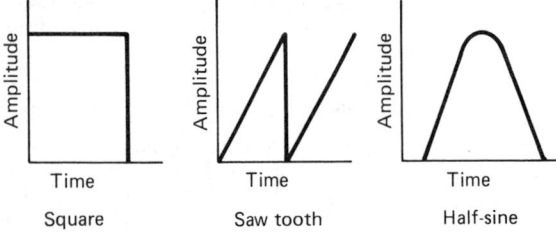

Figure 15.166 Shock pulse time histograms

time of the shock pulse is short compared to the natural frequency of the system the area of the shock pulse (i.e. the area under the amplitude/time graph) is the most important characteristic. Graphs are available for different types of pulse patterns which indicate the shock response spectra.

Knowing the resonant frequency of the system and the force acting on a body the transmitted force may be calculated. The 'force response' is therefore calculated as a fraction of the applied force. This force response will be lower with an increase in the natural period of the system compared with the time duration of the shock pulse. Shock isolation systems should, therefore, be designed with resonant frequency (rad. s⁻¹) in the order of 0.1T. The maximum displacement of the mass can be calculated from $x = F/k$, where F = force response (N) and k = stiffness of spring (N m⁻¹) (see Figure 15.167). Similarly, velocity and acceleration may be determined from the standard relation equations:

$$V = 2\pi f_{res} \frac{F}{k}$$

$$a = 4\pi^2 f_{res}^2 \frac{F}{k}$$

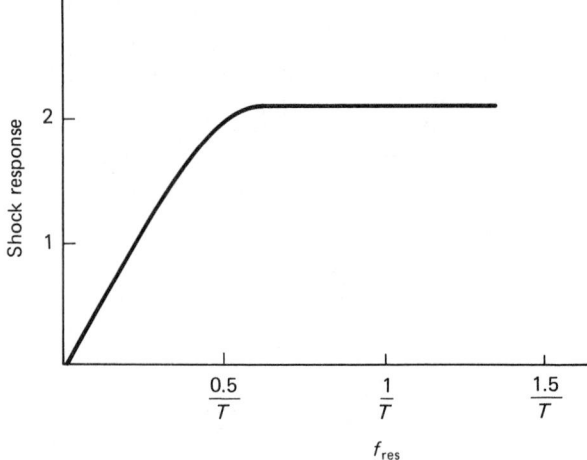

Figure 15.167 Typical shock-response curve

In practice, it may not be easy to assess the likely shock forces on a system mathematically. In this case we may have to resort to the measurement of a test system (see Section 15.8.7) and then produce a 'real' system once the experimental data have been manipulated.

15.8.6 Vibration attenuation

As in noise control, vibration reduction can follow three distinct phases:

1. Reduction of vibration at source
2. Interruption of transmission
3. Protection of affected persons/machines, etc.

15.8.6.1 Reduction of vibration at source

Vibrations may be produced acoustically by out-of-balance forces on rotating components, electrical forces, worn bearings, poor gear meshing and numerous other causes. In the first instance, reduction at source should be considered. This may have the added advantage of improving the efficiency and life of the machine in question. Frequency analysis techniques may be used to isolate the precise source of the vibration (see Section 15.7). Worn or defective components can be dealt with as the first stage of vibration reduction. Out-of-balance forces may be investigated and, if found to be excessive, shafts, etc. should be balanced.

If resonance is found to be a problem, machine rotational speed may be altered in order to take exciting forces away from the resonant frequency range (this is unlikely to be suitable in many cases). This method may be adopted for fans, etc. The flow volume may be maintained by changing the pitch of blades, etc. Finally, damping of the structure may be considered; either by adopting inertia weights (in the form of adhesive backed lead mats, bolt-on weights, etc.) or by stick-on damper materials (felts and rubbers, etc.)

15.8.6.2 Interruption of transmission pathway

This is by far the most common approach to vibration isolation. If all reduction at source has been pursued, the vibrating device can be mounted on vibration isolators. Care should be taken that no 'bridging' occurs or that all services (e.g. power supplies, etc.) are mounted in such a way that they can flex as the machine moves and do not unduly restrict the movement of the machine on the mounts. The nature of the flexible connection should be such as to resist fracture of the service element.

The machine should be sufficiently far from fixed objects so that no collision can occur as the machine deflects on the mounts. Machines may be mounted on a variety of materials and their individual properties are described below.

15.8.6.3 Vibration isolation materials

Steel springs These are suitable for all frequency ranges (as the stiffness may be easily tailored to suit loadings and hence the required resonant frequency). They are, however, most suitable at low frequency. They have the disadvantage of having low internal damping and hence readily transmit high frequencies.

High-frequency transmission is usually overcome by mounting the spring with rubber washers at each end. The steel spring has the advantage of being easy to produce at the required stiffness and is not subject to degradation by oils, etc. It is also tolerant of relatively high temperatures. It is, therefore, the first choice for anti-vibration mounts. Damping is achieved by the use of external damper elements. Steel springs are available as helical compression, helical tension or leaf springs. Leaf springs have the advantage of having higher internal damping. All steel springs can be constructed so as to have a non-linear action.

Rubber mounts Although termed 'rubber' mounts, these are often composed of synthetic rubbers which are not so readily attacked by oils and can operate over a much wider temperature range. The rubber is usually mounted onto a steel base (although it may be used in simple 'mat' form) and may be in compression, shear, or shear-compression (depending upon the configuration of the mount).

Different frequencies may be catered for by changing the hardness of the rubber, but the usual use for rubber mounts is for high-frequency isolation. Damping is better than steel springs but decreases with an increase in the hardness of the rubber.

Mats Machines, etc. may be simply mounted on mats of isolation material and common materials are rubber, cork, foam or felt. These have the disadvantage of having a low static deflection (and hence are only capable of high-frequency attenuation on relatively light machines).

A development of the simple mat is the ribbed rubber mat. In this version higher static deflections are possible and hence lower-frequency control. All mats have the disadvantage of having to be placed under the machine and are hence liable to degradation from oils, etc.

15.8.6.4 Isolation of structures

If it is not possible to mount the machine directly on suspension type mounts the building may be treated as a whole. The isolation of the pathway may start at the machine and continue through isolation of the machine foundations (perhaps by isolating the concrete slab with resilient material) up to complete isolation of a building housing vibrating machines. It is obviously most cost effective to deal with the vibration at the most practicable point near to the machine (as there is less mass to be isolated).

15.8.6.5 Protection of affected persons/machines, etc.

Another approach to vibration isolation is to isolate the affected persons/machines. This will only be practicable if a limited number of positions have to be protected. This method may prove cost effective if the only other alternative would be major structural isolation (if machine isolation was not practical).

Examples of this approach are the suspension in the driver's seat of heavy lorries right up to 'floating' noise refuges in press shops. The vibration isolation principles are exactly the same as vibration source isolation, and similar isolation materials are used.

15.8.7 Measurement of vibration

Vibration measurements are almost universally carried out by the use of accelerometers (these measure acceleration). Other types of vibration transducer are available but are not as adaptable as the accelerometer (at one time it was common to use velocity-sensitive devices). If velocity or displacement measurements are needed it is usual to carry this out electrically on the acceleration signal by means of an integrator.

15.8.7.1 The accelerometer

The majority of accelerometers marketed are of the piezoelectric type. The construction is essentially as shown in Figure 15.168. In this type of accelerometer two disks of piezoelectrical material are trapped between the base of the device and a heavy mass (which is located in position by a spring). When the device is subjected to a vibration the mass exerts a variable force on the disks and generates an electrical output (proportional to the force on the disks). The output is, therefore, proportional to the acceleration of the mass.

The mass will have a resonant frequency which will depend on the weight of the mass and the stiffness of the mounting system. It can be shown that below this resonant frequency (which is commonly in the order of 10–100 kHz) the output of the device will be linear when plotted against frequency.

The actual output (in terms of milli-volts/g) will depend upon several factors, including the piezoelectric properties of the disks and the size of the mass. For a given disk material the output will be primarily governed by the size of the mass (the largest masses giving the largest output).

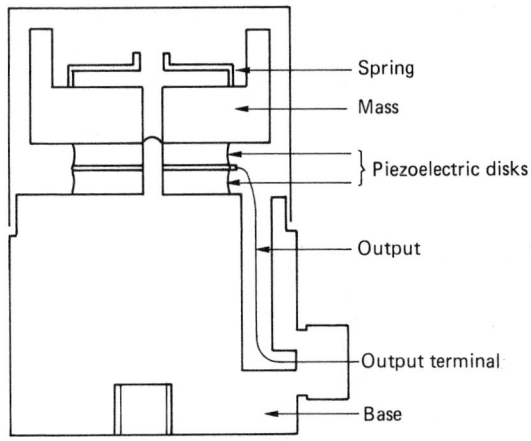

Figure 15.168 Schematic drawing of a piezoelectric accelerometer

Unfortunately, the disadvantage of a high mass transducer may outweigh the voltage gains, as the device may affect the performance of the system to which it is attached (a fundamental design feature of any transducer is that it should not significantly affect the system which it purports to measure). Also the resonant frequency of the accelerometer will be reduced as the mass increases (although this may not be significant at the frequencies we are interested in). Manufacturers supply calibration certificates for each transducer they sell which will give the sensitivity and the frequency at which resonance occurs (usually in graphical as well as numerical form). Graphs are also available so that one may choose the most suitable type of accelerometer for a particular test. An example of such a graph is given in Figure 15.169.

Other factors to consider in choice of accelerometer are:

1. Transverse sensitivity
2. Operating temperature range
3. Acoustic environment
4. Long-term stability

15.8.7.2 Transverse sensitivity

The sensitivity of the accelerometer to vibrations in planes other than in line with the main axis of the device is known as transverse sensitivity. Measurements of this factor are usually carried out perpendicular to the main axis. A good accelerometer should have a low response to this test (one manufacturer quotes less than 3% of the main axial response at low frequencies).

15.8.7.3 Temperature range

Normal operating temperature ranges are up to 250°C and a good-quality device should operate satisfactorily at this temperature (and remain sealed to prevent the ingress of water vapour). Special devices are obtainable which will allow higher operating temperatures (up to 400°C). Temperatures down to −100°C are no problem (except that voltage sensitivity increases slightly).

15.8.7.4 Acoustic environment

It is inevitable that acoustic noise will be present adjacent to all vibrating bodies. The effects of this noise on the device must be considered. Specifications may be given in terms of v/bar. A typical acoustic interference figure may be 1 mV at 140 dB compared to the output at, say, 50 mV g^{-1}). At low vibration levels allowance may have to be made for this effect.

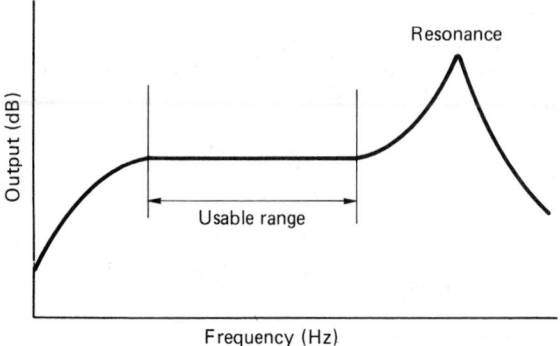

Figure 15.169 Frequency response of an accelerometer

15.8.7.5 Other vibration measurement devices

Devices other than the accelerometer may be used for special applications. These include:

Velocity pick-ups
Displacement transducers
Strain gauges

Velocity pick-ups These are commonly used at low vibration levels at low frequencies. The accelerometer may be at a disadvantage here if information is needed in terms of velocity or displacement as the electronic integration of low-level signals may not be possible. Velocity pick-ups usually operate on either the moving coil or moving magnet principles and have output as high as 350 V m s^{-1}.

Displacement transducers There are two general types: the non-contacting or the contacting. The non-contacting (or proximity sensor) type usually work by capacitive or inductive means. The sensor is brought near to the vibrating object and, as the distance between it and the sensor changes, the capacitance (or inductance) changes and this can be measured electrically. The contacting types actually touch the object which is vibrating and operate an electrical device directly. This type have the disadvantage of damping the object in question and also require very accurate manufacture if they are not to fail early.

Strain gauges A strain gauge measures displacement between two points on the surface of an object (i.e. how much it stretches). The other types of transducer are not suitable for this measurement. These gauges are important in assessing the likely failure of materials when subject to stress. In these devices the active element is stretched by the motion, causing electrical resistance of the material to change. This is measured by 'bridge' amplifiers (which can also be fitted with compensation links to cancel out such effects as temperature change).

15.8.7.6 The measurement system

In vibration assessment it is often important to determine not only the level of the vibration but also the frequency characteristics. The most simple systems consist of a sensing device (often the accelerometer), an amplifier and some form of display (the meter). This system gives no frequency information and is therefore of little use for general application (although it may have specific uses – for example, in machine condition monitoring).

A more practical system involves the use of an analyser (either of the octave/third octave or narrow band type). Such a system can be represented by the schematic shown in Figure 15.170.

For vibration application it is particularly important to study the manufacturer's specifications so that inaccuracies are not introduced. The selection of suitable matching equipment may be made simpler by choosing all equipment from one supplier only. Phase and impedance effects are so crucial at low frequency that the writer would advise all but the most experienced that this is, perhaps, the most significant point to note in all vibration work.

15.8.7.7 Amplifiers

The factors likely to be considered in amplifier choice are:

1. *Frequency response* The response should be linear over the range at which measurements are to be performed.

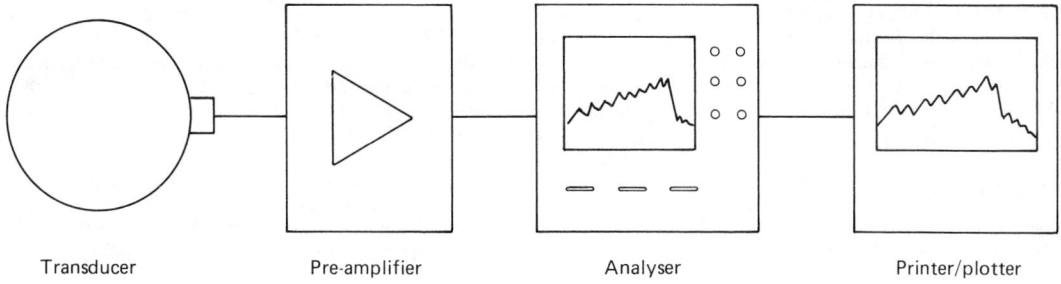

Figure 15.170 Schematic vibration analysis system

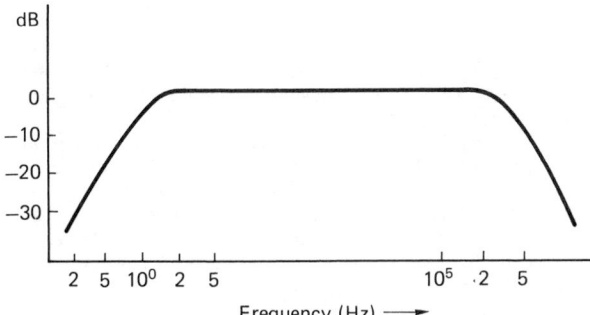

Figure 15.171 Typical amplifier frequency-response curve

The specification may either be given numerically in terms of a 3 dB change in output (e.g. 2 Hz to 200 kHz) or may be displayed as a response graph (Figure 15.171).

2. *Phase* All amplifiers distort phase to a greater or less extent; with complex signals this phase change will cause signal distortion. The phase relationship of periodic sine waves leads to the characteristic shape of the signal (e.g. square wave, saw tooth, etc.) being modified. If the phase of the high- or low-frequency components is changed as compared to the middle frequencies the characteristic shape will be distorted.

15.8.7.8 Harmonic distortion

Amplifiers have a tendency to introduce harmonic distortion (i.e. the generation of multiples or sub-multiples of the original signal). In the case of vibration measurement we may be trying to measure actual frequency harmonics (as in the case of out-of-balance shafts). If these are masked by electronically generated harmonics then actual vibration harmonics may be swamped. Amplifiers with low harmonic distortion should be specified for vibration work and they must never be overloaded.

15.8.7.9 Dynamic range

The dynamic range of an amplifier is the ratio of the lowest to the highest inputs which may be accepted without changing ranges on the controls. The higher the dynamic range, the better, as range-changing may not be possible with rapidly fluctuating signals. Dynamic range should be compared to harmonic distortion figures as high harmonic distortion occurs when overloading (or clipping) occurs. Pure sine waves will be

changed to a flat top (or square wave) shape and extra harmonics rapidly produced.

As a general rule, the ideal operating range of an amplifier (if phase change effects are to be avoided) is ten times the lower-frequency −3 dB point to one-tenth of the high-frequency 3 dB cut-off point.

15.8.7.10 Transient distortion

Transient distortion is that produced when the rise time of the signal is too great for the amplifier to cope with. Again, overload type effects occur and (depending on the level of the signal) distortion or reduction in the response takes place.

15.8.7.11 Frequency analysis

Vibration limits are usually expressed graphically as it is the frequency of the vibration that is often as important as its actual level. It is therefore important that some knowledge of the frequency of the signal is gained. The same frequency analysis equipment as is used for work on acoustic noise is employed, but some special characteristics have to be considered.

The most important point is the likely frequency range to be used. Machinery vibration, in the past, was often at low frequency (perhaps 100 Hz or below) but with modern high-speed machines it is now important to have the ability to measure up to, say, 5 kHz. It is most important, however, that the lower frequencies are accurately measured as it is these that are likely to contain the most energy.

15.8.7.12 Octave analysis

In order to perform octave analysis the signal has to be passed through a series of filters, one at a time. The audible frequency spectrum is divided up into ten bands whose centre frequencies and bandwidths are agreed by international standards. These band centre frequencies are:

32.5, 63, 125, 250, 500, 1k, 2k, 4k, 8k, 16k (Hz)

It should be noted that the centre frequency of each band is twice that of the one below it. Similarly, the upper frequency of each band is twice its lower frequency. The upper and lower frequency of each band can be found from:

Lower frequency = $1/\sqrt{2}$ × centre frequency
Upper frequency = $\sqrt{2}$ × centre frequency

Therefore the band centred on 1 kHz stretches from 707 Hz to 1414 Hz. The 2 kHz band stretches from 1414 Hz to 2828 Hz.

The disadvantage in using octave filters for vibration frequency analysis is that the bandwidth is often too wide for any really useful analysis. Also, the filters have to be used sequentially; this may be of no use when transients are involved as the signal may be 'missed' by one filter. The only option here would be to tape record the signal (onto a precision FM tape recorder) and play the tape back through each filter in turn.

15.8.7.13 One-third octave filters

A development of the octave filter set is the one-third octave filter set. In this progression each centre frequency is

$\sqrt[3]{2}$ × previous centre frequency
Upper frequency = $\sqrt[6]{2}$ × centre frequency
Lower frequency = $1/\sqrt[6]{2}$ × centre frequency

Therefore the band centre on 1 kHz extends from 891 to 1122 Hz.

The third octave filters go some way towards overcoming the limitations of the octave filter set in terms of bandwidth but they exacerbate the problem of having to sequentially filter the signal. The answer to these problems is the use of a frequency analyser.

15.8.7.14 The FFT frequency analyser

A more detailed description of the FFT (Fast Fourier Transform) analyser will be found in Section 15.9.19, but it should be noted that for vibration work this instrument has many advantages compared to the use of filter sets.

The FFT analyser is usually supplied as a self-contained unit or as an add-on unit to a personal computer. It is normal for the machine to operate in the time (as well as the frequency) domain. It is therefore relatively simple to track a time-domain graph to the point of interest (perhaps a transient) and then perform a frequency analysis at this point. This task would be almost impossible without this machine. The FFT analyser may be capable of operating in narrow bands, one-third octaves and octaves at the touch of a switch. All filters are accessed in parallel and hence no sequence is necessary. The results may be either printed as numerical data or plotted graphically.

15.8.8 Vibration limits

Vibration limits are set in order to protect personnel, equipment and buildings from the deleterious effects of the vibration. Numerous sources may be researched for criteria. These include ISO (International Standards Organization),[65,66] DIN (German standards) and scientific journals. The effects of vibrations and some suggested limits are detailed below.

15.8.8.1 Effects of vibration on people

The effect of vibrations on people can be further divided into

1. Vibration-induced noise effects
2. Vibration-induced nuisance
3. Ill health or poor performance caused by vibration

Noise effects Vibration may be transmitted over large distances by structures or objects and then re-radiated as noise at a structure/air interface. The noise limits given in the section on Noise Control then become applicable. Noise reduction is usually achieved by vibration reduction at source or isolation of the transmission pathway.

Vibration-induced nuisance There are no defined limits as to the level of vibration that is likely to constitute a nuisance and

very little case law exists. It should be noted, however, that nuisance levels are probably very little above perceptible levels in the domestic environment. Early work on the effects of vibration on people has led to a set of curves being accepted as reasonably representative of the response of the average person (Figure 15.172).

Ill health or poor performance The body is a complex structure that, when subject to vibration and shock, may exhibit physical and psychological effects. Physical effects may be caused by structural resonance within the body:

At 3–6 Hz the thorax/abdomen system resonates
At 20–30 Hz the head/neck system resonates
At 60–90 Hz the eyeballs resonate
At 100–200 Hz the lower jaw resonates
The skull itself has internal resonances between 300 and 900 Hz.

Apart from physical damage to the body these vibrations cause psychological effects. Again, data is rather limited but some research has been carried out to determine perception, discomfort and pain threshold.

ISO curves are available which give acceptable time periods at differing acceleration levels (Figure 15.173). These curves give the 'fatigue decreased proficiency boundary', i.e. the maximum allowable acceleration with no noticeable decrease in performance. The suggested maximum exposure limits are twice (6 dB upwards) these levels. Perception would be at one-third of these levels.

15.8.8.2 Effects of vibration on machinery

The most common defect in machines caused by excessive vibration is the mechanical failure of a part caused by fatigue. Other failures may be caused by electrical failures due to contacts failing in switches or similar devices, or by physical damage caused by two components being forced into contact with one another.

Stress cracks form in materials which are subject to excessive vibration levels. The exact degree of damage caused will depend on numerous factors but these will include a factor which will depend on the nature of the material, the number of stress reversals and the strain placed upon the material.

If a number of similar samples are subject to the same fatiguing experiments all will not fail at the same time. Some will fail early, some will fail very late, but the majority will fail between these two limits. The actual failure of one sample can be predicted statistically within confidence limits by this experimental process. Other failures of machines can be predicted by experimental or calculus means,[67] although the former method will still be subject to the statistical failure effect.

Failure due to collision of components may be easier to predict and may occur more precisely at the predicted vibration levels (particularly if structural resonance is involved). If complex modes of vibration are encountered it may be possible to model the situation with the use of computer programs.

While on the subject of machine vibration it should be noted that ISO standards exist which grade various types of rotating shaft according to their precision of balance.

15.8.8.3 Effects of vibration on buildings

In order to produce damage to buildings vibration levels will have to be very high compared to nuisance levels. The actual degree and nature of the damage produced will depend upon the building structure – both its type and condition.

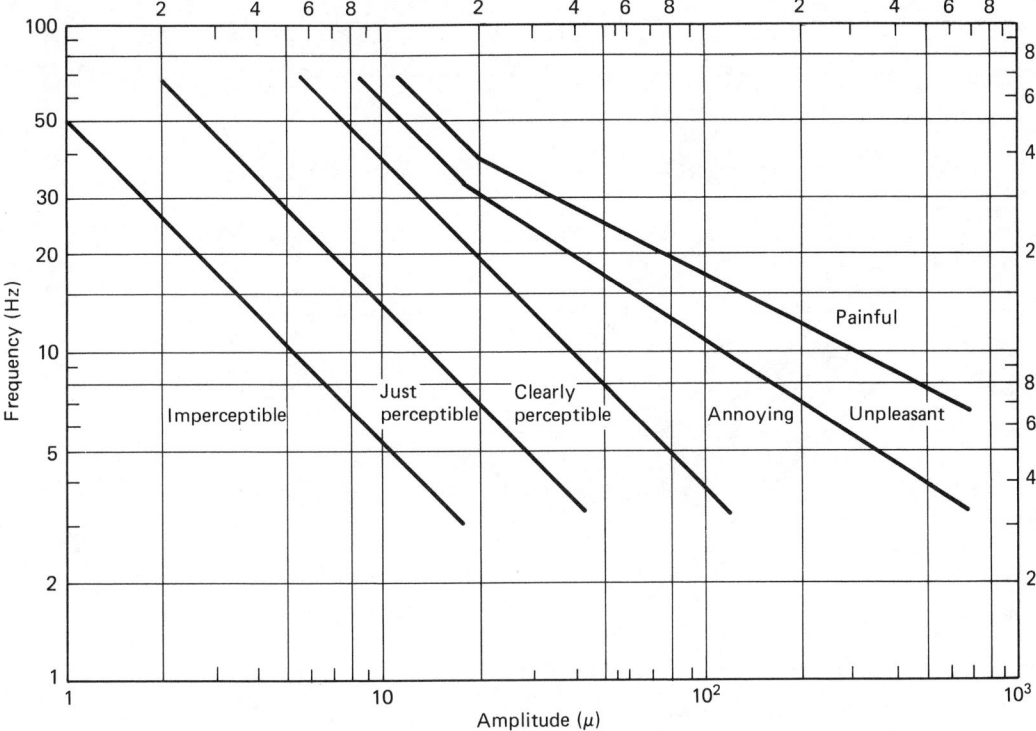

Figure 15.172 Human sensitivity: Reiher–Meister scale (vertical vibration)

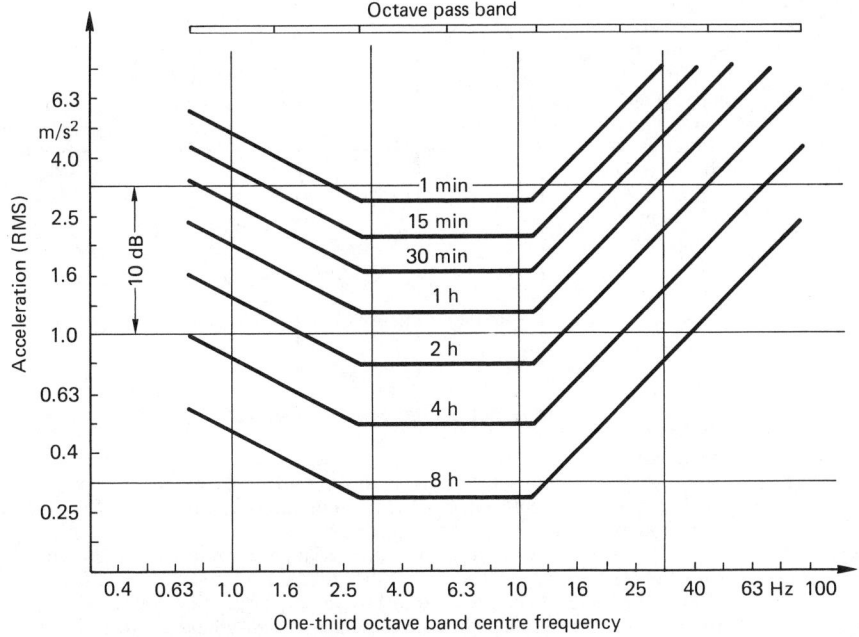

Figure 15.173 ISO vibration criteria for a person in a vertical position

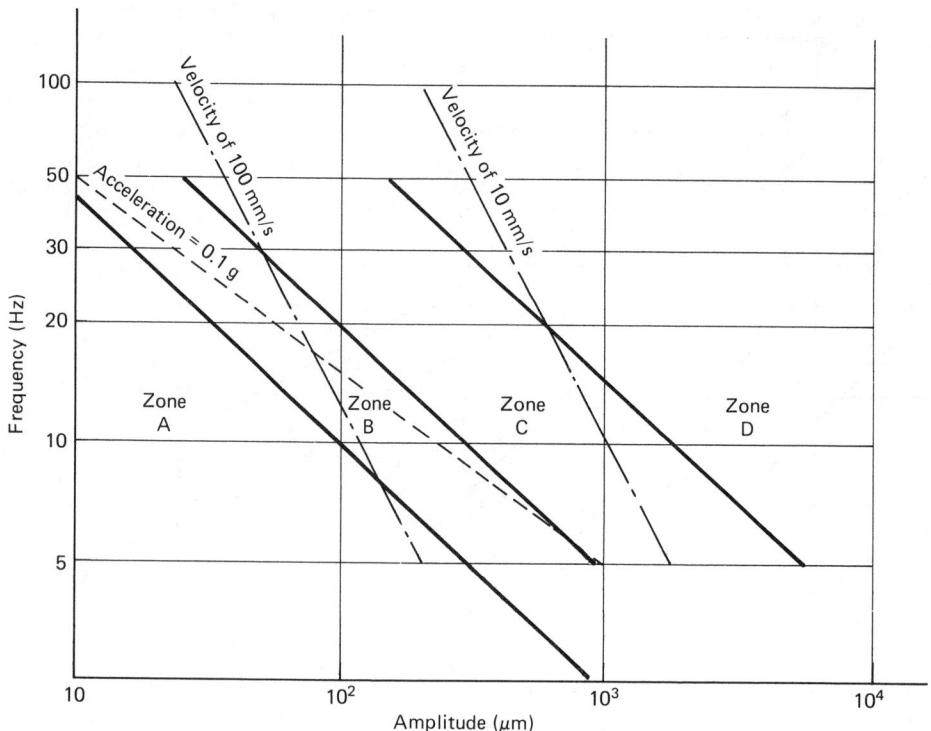

Figure 15.174 Building damage criteria. Zone A – no damage, zone B – plaster cracking possible, zone C – damage to structure, zone D – total destruction

At relatively low vibration levels cracks can occur in plaster (particularly around windows). At higher levels, structural members may crack and ultimately fail. These two types of damage may be easily attributed to mechanical vibration.

Another type of damage may result from building settlement caused by ground-borne vibrations compacting the ground differentially beneath buildings. This type of damage is indistinguishable from settlement caused by other occurrences. Specifications for maximum permissible vibrations may be found in DIN standards which are given in terms of maximum velocity (in mm s^{-1}) which is allowable for different classes of buildings from ruins and historical buildings up to reinforced concrete structures. More accurate criteria may be found in the technical press and HMSO publications. One such type of drawing is shown in Figure 15.174.

15.9 Acoustic noise

15.9.1 Introduction – basic acoustics

Sound can be defined as the sensation in the ear caused by pressure variations in the air. For a pressure variation to be known as sound it must occur much more rapidly than barometric pressure variations. The degree of variation is much less than atmospheric pressure.

Audible sound has a frequency range of approximately 20 Hz to 20 kHz and the pressure ranges from 20×10^{-6} N m^{-2} to 200 N m^{-2}. A pure tone produces the simplest type of wave form, that of a sine wave (Figure 15.175). The average pressure fluctuation is zero. Measurements are thus made in

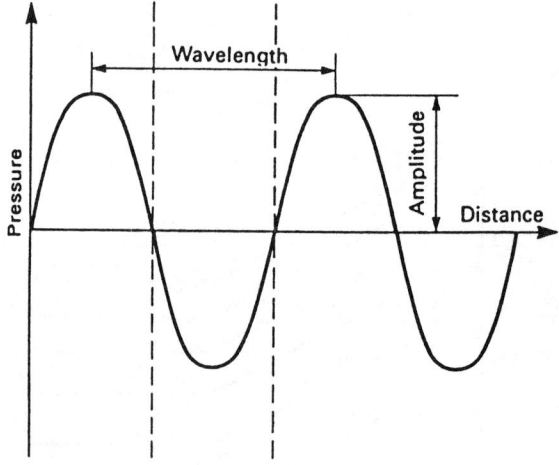

Figure 15.175 Sine wave

terms of the root mean square of the pressure variation (abbreviated to RMS). For the sine wave the RMS is 0.707 times the peak value.

Since RMS pressure variations have to be measured in the range 20×10^{-6} N m^{-2} to 200 N m^{-2} (a range of 10^7) it can be seen that an inconveniently large scale would have to be used if linear measurements were adopted. Additionally, it has been found that the ear responds to the intensity of a sound ($\propto p^2$) in a logarithmic fashion. The unit that has been adopted takes these factors into account and relates the measured sound to a reference level. For convenience, this is taken as the minimum audible sound (i.e. 20×10^{-6} N m^{-2}) at 1 kHz.

The logarithm (to the base 10) of the ratio of the perceived pressure (squared) to the reference pressure (squared) is known as the Bell, i.e.

$$B = \text{Log}_{10} \frac{P^2}{P_{\text{ref}}^2}$$

Since this would give an inconveniently small scale (it would range from approximately 0 to 14 for human response), the Bell is divided numerically by 10 to give the decibel. The equation therefore becomes:

$$\text{dB} = 10 \log_{10} \frac{P^2}{P_{\text{ref}}^2} \equiv 20 \log_{10} \frac{P}{P_{\text{ref}}}$$

15.9.2 Sound intensity

Sound intensity, I, is a measure of energy and its units are watts per metre. Intensity is proportional to the square of pressure. Sound intensity level (SIL) is defined in a similar manner to sound pressure level. In this case the equation is

$$\text{dB (sound intensity level)} = 10 \log_{10}, \frac{I}{I_{\text{ref}}}$$

15.9.3 Sound power

Similarly, the power of a source (measured in watts) can be expressed in terms of decibels (in this case called the sound power level (SWL))

$$\text{dB (sound power level)} = 10 \log_{10}, \frac{W}{W_{\text{ref}}}$$

W_{ref} is taken as 10^{-12}.

It can thus be seen that it is important not only to express the unit but also to state sound pressure level (SPL), sound intensity level (SIL) or sound power level (SWL).

15.9.4 Addition and subtraction of decibels

For coherent sound waves addition of values is possible. It will be apparent that as the scale is logarithmic, values cannot merely be added to one another. Intensities can, however, be added and thus the equation becomes

$$\text{SIL (total)} = 10 \log_{10} \frac{I_1 + I_2}{I_{\text{ref}}}$$

i.e. 70 dB + 73 dB

$$70 = 10 \log_{10} \frac{I_1}{I_{\text{ref}}}$$

$$I_1 = \frac{\text{antilog } 7}{I_{\text{ref}}}$$

$$73 = 10 \log_{10} \frac{I_2}{I_{\text{ref}}}$$

$$I_2 = \frac{\text{antilog } 7.3}{I_{\text{ref}}}$$

$$\begin{aligned}
\text{SIL (total)} &= 10 \log (\text{antilog } 7 + \text{antilog } 7.3) \\
&= 10 \log (2.99526 \times 10^7) \\
&= 74.76 \text{ dB}
\end{aligned}$$

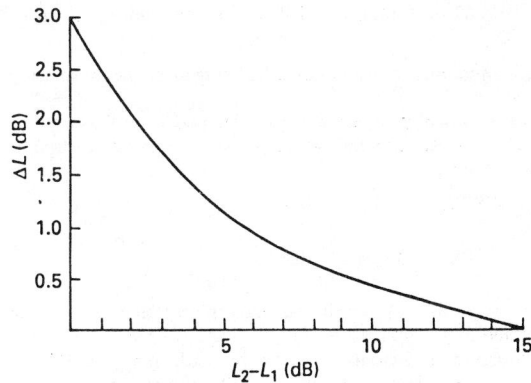

Figure 15.176 Noise-level addition graph

The square of individual pressures must be added and thus the equation in this case must utilize

$$P(\text{total}) = \surd(P_1^2 + P_2^2)$$

15.9.5 Addition of decibels: graph method

It is possible to use a graph to calculate the addition of decibels, even in the case of multiple additions (Figure 15.176). The graph is used in the following way:

1. In the case of the addition of two levels – the difference between the higher and lower levels is plotted on the lower scale of the graph. The correction is then read from the vertical scale by projecting a horizontal line across to this scale from the point on the graph. The correction is added to the highest original level to give the total level.
2. In the case of subtraction of levels – the difference between the total sound level and the one to be subtracted is plotted onto the graph and the correction obtained as above. In this case the correction is subtracted from the total level to give the remaining sound level.
3. In the case of multiple additions – if there are more levels to be added the first two levels are added using the graph and then the third is added to the resultant using the same method.

15.9.6 The relationship between SPL, SIL and SWL

The total acoustic power of a source can be related to the sound pressure level at a distance r by the following equation (assuming spherical propagation):

$$W = P^2 / (\rho c 4 \pi r^2)$$

where ρ = density of the medium and c = velocity of sound in that medium. By substituting this back into the SPL equation we obtain

$$\text{SPL} = \text{SWL} - 20 \log_{10} r - 11 \text{ (spherical propagation)}$$

It is also possible to derive equations for other common situations, i.e.

Point source on a hard reflecting plane
Line source radiating into space
Line source on a hard reflecting plane

These equations are:

$$\text{SPL} = \text{SWL} - 20 \log_{10} r - 8 \text{ (hemispherical propagation)}$$
$$\text{SPL} = \text{SWL} - 10 \log_{10} r - 8 \text{ (line source in space)}$$

$$SPL = SWL - 10 \log_{10} r - 5 \text{ (line source radiating on a plane)}$$

These equations are useful for calculating distance attenuation effects.

If the sound pressure level at distance r_0 is known it is possible to calculate the sound pressure level at position r_1 quite easily:

$$SPL_0 - SPL_1 = 20 \log_{10} r_1 - 20 \log_{10} r_0$$

$$SPL_0 - SPL_1 = 20 \log_{10} \frac{r_1}{r_0} \text{ dB}$$

If r_1 is double r_0 it will be seen that the SPL will be approximately equal to 6 dB ($20_{10} \log 2$). This gives us the principle of a decrease in level by 6 dB per doubling of distance (inverse square law). For the line source the same calculation produces a difference of only 3 dB per doubling of distance.

15.9.7 Frequency weighting and the human response to sound

In practice, noises are not composed of one single pure tone but are usually very complex in nature. It is essential that more than the overall noise level (in dB) is known in order to appreciate the loudness of a noise, as the ear does not respond uniformly to all frequencies.

As previously stated, the ear can respond from 20 Hz to 20 kHz and the response can be demonstrated by equal-loudness contours (Figure 15.177). It can be seen in Figure 15.177 that there is a loss in sensitivity (compared to 2 kHz) of approximately 60 dB at the low-frequency end of the chart. It will also be seen that all the curves are approximately parallel, but there is a tendency to linearity at the higher noise levels.

In order to produce meaningful readings it is therefore important to state the sound pressure level in dB and the frequency of the noise. A weighting can be imposed on noise readings which corresponds to the inverse of the equal-loudness contours. If this weighting is used all readings which are numerically equal will sound equally loud, regardless of frequency.

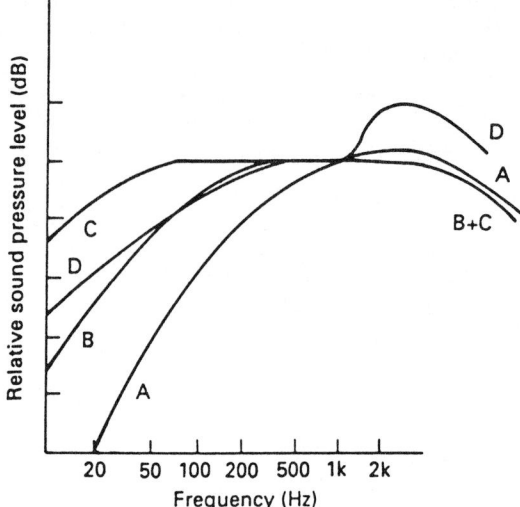

Figure 15.178 Weighting networks

Originally, three networks were proposed (A, B and C) and it was suggested that these be used for low, medium and high noise levels, respectively. It was proved, in practice, that this introduced numerous difficulties particularly with rapidly changing noise levels when a change of filter network was necessary. It was also found that the 'A'-weighting network corresponded very well to annoyance levels at all noise levels (Figure 15.178). It was therefore decided that the 'A' weighting would be used as the norm for noise readings concerning human response. There is another weighting network (the 'D' network) that is used for aircraft noise measurement.

If it is necessary for engineering purposes to know the tonal make-up of a noise, several approaches are possible. The noise can be processed by a bandpass filter. The most common filters are octave band filters and the agreed centre frequencies are as follows

31 63 125 250 500 1k 2k 4k 8k 16k (Hz)

If further resolution is necessary, one-third octave filters can be used, but the number of measurements that are required to be taken is most unwieldy. It may be necessary to record the noise onto tape loops for the repeated re-analysis that is necessary. One-third octave filters are commonly used for building acoustics.

Narrow band real-time analysis can be employed. This is the fastest of the methods and most suitable for transient noise. Narrow band analysis uses a visual display screen to show the graphical results of the fast Fourier transform (FFT) and can also provide octave or one-third octave bar-graph displays.

15.9.8 Noise indices

All the previous discussions have concerned steady-state noise. It will, however, be apparent that most noises change in level with time. It may therefore be necessary to derive indices which describe how noise changes with time. The commonest of these are the percentiles and equivalent continuous noise levels.

Percentiles are expressed as the percentage of time (for the stated period) during which the stated noise level was exceeded, i.e. a 5-minute $L90$ of 80 dB(A) means that for the

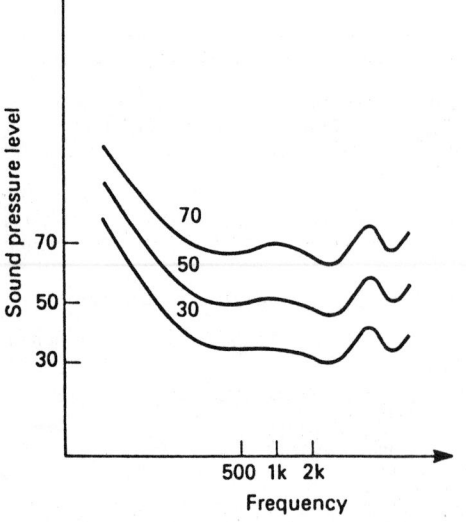

Figure 15.177 Equal-loudness contours

5-minute period of measurement for 90% of the time the noise level exceeded 80 dB(A). Therefore $L0$ is the maximum noise level during any period and $L100$ is the minimum.

The variation of noise levels within a discrete period of time can best be described by a set of Ln results (the more results available, the greater the representation of the noise event). Sound-level meters commonly measure Ln's at seven points (commonly, $L1$, $L2$, $L10$, $L50$, $L90$, $L95$, $L99$). More sophisticated modern machines are capable of being adapted by the user and non-standard Ln's are available.

Leq (the equivalent continuous noise level) is defined as the continuous steady noise level which would contain the same total acoustic energy as the actual fluctuating noise, measured over the same period of time. This concept may be understood by considering electrical power consumption. If a machine uses

4 kW for 1 hour
2 kW for 2 hours
1.5 kW for 4 hours
1 kW for 1 hour

the total usage of power is 15 kW h^{-1}. The equivalent power for the 8-hour period would be 1.875 kW.

If two events are to be added together and the Leq derived we must first convert to intensity units. Addition may then take place directly using the equation:

$$Ieq = \frac{1}{T} (I_1t_1 + I_2t_2 + I_3t_3 \ldots)$$

where T = total time, I_1 = intensity for the first event, t_1 = time for the first event, I_2 = intensity for the second event, t_2 = time for the second event, etc. The total intensity is then converted back to decibel units by

$$dB(A) = 10 \log_{10} \frac{I}{I_{ref}}$$

where I_{ref} = reference intensity. However we usually know the levels in terms of dB(A) rather than intensities, therefore by substitution

$$Leq = 10 \log \frac{1}{T} (t_1 \times 10^{L_1/10} + t_2 \times 10^{L_2/10} + t_3 \times 10^{L_3/10})$$

where L_1 = level 1 in decibels, etc. As noise is often measured on the 'A'-weighted scale Leq is usually expressed in this way. In this case the nomenclature becomes $LAeq$.

A further derivation of equivalent continuous level is the single-event level (SEL), also known as sound-exposure level or Lax. This a special type of Leq used for transient events such as the passage of aircraft, gunshots, etc. The SEL is a one-second Leq and can be defined as the steady level which over one second would contain the same 'A'-weighted energy as the actual event (regardless of its duration). Thus

$$Leq = 10 \log \frac{1}{T} (t_1 \times 10^{L_1/10} + t_2 \times 10^{L_2/10} + \ldots \text{ etc.})$$

$$SEL = 10 \log (t_1 \times 10^{L_1/10} + t_2 \times 10^{L_2/10} + \ldots \text{ etc.})$$

where $t_1 + t_2$ etc. are the durations of levels L_1, L_2, etc. in seconds.

15.9.9 Noise-rating curves

These are a set of graphs that are commonly used as a specification for noise from machinery. They are similar to Noise Criteria Curves (used in the USA to specify noise from ventilation systems). The rating of a noise under investigation

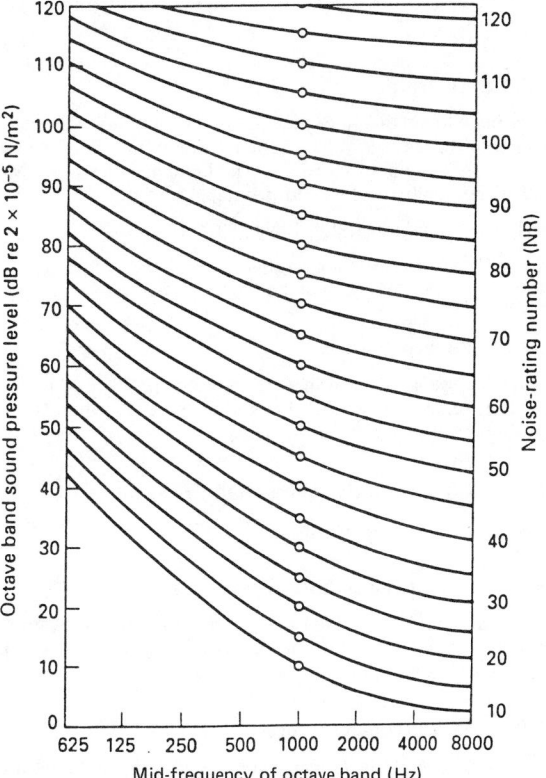

Figure 15.179 Noise-rating curves

is the value of the highest noise-rating curve penetrated by the readings when plotted on the graphs (Figure 15.179).

15.9.10 Community noise units

Noise has been defined as unwanted sound. To quantify noise is therefore much more complicated than to quantify sound itself (which is what we have previously considered). Units have to be derived from these purely acoustic measurements by assessment of experimental psycho-acoustic data. It has been found that the response to different types of aural stimulation cannot be described by one single measurement, and hence a number of different noise measures are used. We now have three distinct classes of measurement:

1. Noise Units – these are the basic physical measurements of sound (i.e. decibel).
2. Noise Scales – these are composed of a combination of physical measurements (usually sound level, time, etc.) (i.e. Ln's, Leq, SEL).
3. Noise Indices – here other factors are used to modify the noise scales in order to more closely relate the noise scale to other factors (annoyance, for instance)

A criterion is a noise index value which is used to describe the reaction of a given percentage of the population.

15.9.11 Road traffic

Road traffic is assessed by an 18-hour $L10$. This is not the percentile for 18 hours but rather the arithmetic average of the

18 one-hour $L10$'s between 6 a.m. and midnight on a normal working day.

15.9.12 Air traffic

It has been found that annoyance caused by aircraft flyovers is related to the average value of the maximum perceived noise levels and the number of events. The index is known as NNI (noise and number index) and is obtained from:

$$NNI = Lpn(\max) + 15 \log(10N) - 80$$

where $Lpn(\max)$ is the logarithmic average of the maxima of the flyovers and N is the number of flyovers.

15.9.13 Railway noise

Railway noise is assessed in Leq units. 65 dB(A) Leq is the usual criterion at which double-glazing is fitted where new housing is built near to railway lines.

15.9.14 Noise from demolition and construction sites

Hourly Leq is used as the index.

15.9.15 Noise from industrial premises

British Standard 4142: 1990 is described in detail in Section 15.9.25 and is derived from the noise measured in Leq compared to a background level measured in Ln.

15.9.16 Measurement of noise

The simplest sound-level meter consists of a microphone, an amplifier and a meter of some type. Sound-level meters are graded according to British and international standards. For most precision work a Type 1 (precision) sound-level meter is used. This has an accuracy of approximately ± 1 dB(A). Type 0 meters (laboratory) grade are rarely encountered. Type 2 (industrial) grade sound meters may be suitable for some initial survey work but may not be sufficiently accurate to comply with legislative requirements at all frequencies. In particular, the lower grade of instruments have poor performance above 10 kHz (the human ear responds to noise at least up to 16 kHz).

15.9.17 Microphones

The microphone is a device for converting pressure fluctuations in the air into an electrical signal. For precision work two types may be chosen. The polarized condensor microphone consists of a very thin metal diaphragm stretched in close proximity to a back plate. This diaphragm is charged to a polarization voltage of 200 V (some are lower). The diaphragm thus forms a condensor with the back plate. Sound causes the diaphragm to move in relation to the back plate, thus changing the charge on the condensor. This can be sensed electrically and used to measure the sound.

The pre-polarized (or electret) microphone is a development of the polarized microphone, the main difference being that the charge across the diaphragm is permanent (or almost) and no polarization is needed (which simplifies the electronics of the pre-amplifiers). The disadvantage of the polarized microphone is that it is very moisture sensitive. Condensation on the diaphragm may result in electrical breakdown which causes sparks. These damage the diaphragm, thus ruining the microphone. The pre-polarized microphone has the disadvantage of slightly reduced long-term stability (although this has now been largely overcome). Other types of microphone have

been used – notably the piezoelectric type – but these are not suitable for anything more than the most basic noise 'survey' meters.

Microphones should be capable of measuring the pressure changes in the air without altering the pressure waves they are trying to measure. This may seem to be a fairly fundamental point but, unfortunately, this is not physically possible. The diagphragm must have sufficient frontal area in order to capture the pressure wave and hence produce a reasonably sensitive output. Some reflections will occur at the diaphragm and hence produce addition and/or cancellation effects with incoming pressure waves. This effect will differ depending upon the angle of incidence of the sound on the diaphragm and the frequency of the pressure fluctuations.

In the past it was necessary to have 25 mm diameter diaphragms in order to get a sensitive response and reflection errors were a significant problem. It is now common to employ 12 mm diameter microphones and these problems are now reduced. There are, however, still many specialized microphones produced but they fall broadly into three types:

1. *Pressure microphones* – used for measuring sound in ducts, etc.;
2. *Free field* – used for measuring sound (usually out of doors) in which the angle of incidence is at 0° to the centre line of the microphone; and
3. *Random incidence* – used for measuring sound (usually indoors) in a reverberant field where the angle of incidence is more random.

Note that most precision sound-level meters are fitted with a switch which can change electronically the response between free field and random response.

For infra-sound (sound below the normal audible range) measurement special microphones may have to be used. Although some ordinary microphones are capable of operating at low frequency, great care has to be exercised in impedance matching if low-frequency cut-off is to be avoided.

15.9.18 The sound-level meter

The precision sound-level meter incorporates the pre-amplifier in the nose of the meter (usually in the stem that the microphone fixes on to). The main amplifier is contained within the body of the meter and may either be auto ranging or may have one or more user-adjusted ranges. In older instruments the range had to be adjusted in 10 dB steps (which was very awkward to use with rapidly changing noise levels). Simple sound-level meters merely display the output of this amplifier onto an analogue meter (Figure 15.180).

Modern sound-level meters are equipped with internal filters and intergrating circuits and can produce outputs in terms of percentiles, Leq and frequency spectra. Some sound-level meters have a computer-controlled circuitry that is addressable from a ROM cartridge which is inserted to load a program and then removed. These sound-level meters can then perform many functions as several cartridges are available. The sound-level meter thereby becomes dedicated to one particular type of task.

Memory power of sound-level meters is increasing daily and it is now common to hold many sets of data (for instance, percentiles) in the sound-level meter memory and download later (perhaps in a kinder environment) either to a printer directly or to a personal computer. If the PC option is chosen the data can be introduced to a graphics program and results displayed in a chosen graphics format which can produce elegant displays.

Digital outputs are available on most sound-level meters which will enable connection to portable computers if much

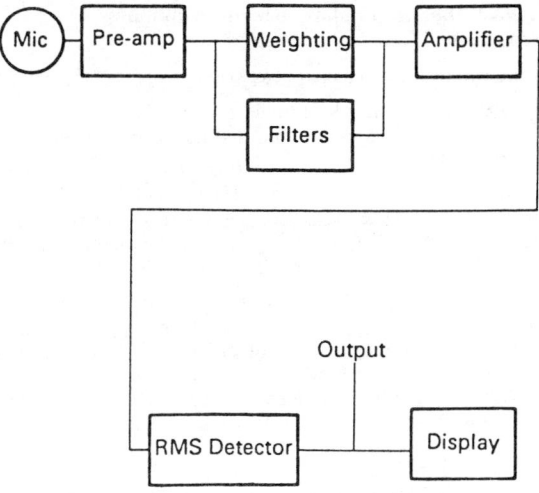

Figure 15.180 Schematic diagram of a sound-level meter

greater memory is required (or if on-site processing is chosen). Sound-level meters are also equipped with a.c. or d.c. outputs which will enable the connection of tape recorders, etc.

Ruggedized sound-level meters are available which are designed for leaving out of doors. These devices (often referred to as environmental noise analysers) are fitted into steel weathertight cases and have a large battery capacity (and the provision for external battery connection). They are fitted with their own printers. Battery and paper life is in the order of six days. Longer life may be obtained by the use of external batteries and minimizing the amount of data being printed to the paper roll.

15.9.19 Digital signal analysis

While analogue filtering of signals may be of some use, as previously described, if detailed information is needed inevitably digital processing is called for. The principle of frequency analysis is known as Fourier Analysis. The Fourier series states that any complex signal can be represented as a series of sine waves of various frequencies, magnitudes and relative phase angles.

An example of this is the square wave. This signal may be represented by the series of sine waves composed of the fundamental frequency – a sine wave at three times the fundamental and one-third of the amplitude, a sine wave at five times the frequency and one-fifth the amplitude, etc., with the progression carrying on to infinity. Electrically, this process is known as FFT (Fast Fourier Transform) analysis. The narrow band FFT analyser displays this signal graphically (as a display with frequency on the x-axis and amplitude on the y-axis).

Octave or one-third octave analysers usually employ digital filters which are arranged such that real-time analysis is possible (where the whole of a signal is analysed rather than merely a snapshot). The sophistication of the machine and the required upper frequency will determine whether real-time operation is possible or not. Both types of analyser have digital outputs which will enable downloading to larger computers for further manipulation or to allow long-term storage.

It is now possible to obtain add-on hardware and software systems for existing personal computers which will enable them to be used both as statistical (Ln and Leq, etc.) and frequency analysers (both narrow and octave band, etc.).

These units are now available in laptop computers. They are not, at present, being produced by the major instrumentation companies, who continue with their dedicated machinery. It has to be said, however, that the add-on units are not as fully developed as they might be.

Current developments include the provision of amplifiers and power supplies to enable microphone connection directly, and if these prove successful the end of the dedicated sound-level meter may be in sight.

15.9.20 Noise control

Noise is capable of causing psychological, physiological and pathological reactions as well as physical damage to plant, machinery and building structures. The need for the control of noise is recognized in many statutes for the protection of both workers and members of the public in their homes.

15.9.21 Noise nuisance

Section 80 of the Environmental Protection Act 1990 gives local authorities the power to serve a notice where certain classes of nuisance have occurred or may occur. The expression 'nuisance' is not defined in the Act or indeed in any other. The use of the expression 'nuisance' can be traced back to legal action as far as the thirteenth century and its meaning is now well understood.

Nuisance describes anti-social un-neighbourly behaviour, and has been taken to mean the interference with one's neighbours in their day-to-day-activities. Noise nuisance can therefore be a statutory nuisance (by virtue of the Environmental Protection Act), a private nuisance (actionable at common law as a tort) or a public nuisance (a crime). For a noise to be a statutory nuisance it must also be a common law nuisance and hence a private or public nuisance.

The concept of private nuisance is now well developed. Private nuisance is a land owner's tort and is a complaint that the use or enjoyment of his or her land has been interfered with. The nuisance only applies to the occupier of the land and not his or her family or sub-tenants.

There are two types of private nuisance. The first concerns rights attached to land (for instance, right of way) and the second to enjoyment of the land (which does have relevance to noise control). This class of nuisance is described as 'where a person is unlawfully annoyed, prejudiced or disturbed in the enjoyment of land or with his health, comfort or convenience as an occupier'. The interference must be substantial and the duration, nature and level of the noise must be considered. A single event may not therefore constitute a nuisance.

The area affected by the nuisance must therefore be considered. One often-quoted remark is taken from the case of *Sturgess* v. *Bridgam* (1879), in which Theiseger, L. J., said 'What would be a nuisance in Belgrave Square would not necessarily be so in Bermondsey'. However, care must be taken if it is to be assumed that because an area is already noisy extra noise will not constitute a nuisance. In one case another printing press in Fleet Street proved to be a nuisance (1907).

Two other legal precedents should be considered at this stage. The first concerns sensitivity. In the case of *Walter* v. *Selfe* (1851) the expression 'ought this inconvenience to be considered – not merely according to elegant or dainty modes of habit or living, but according to plain and sober and simple notions amongst the English people' was quoted. This forms a cornerstone of nuisance law and gives rise to the question of reasonableness of a nuisance. Special sensitivities are not therefore to be considered when the question of nuisance arises. This may have relevance to shift workers, for instance, who while they might expect their daytime sleep to be protected by law, may be disappointed to find that the law will only protect their property against noise that would affect the enjoyment of the average person (i.e. one who is not sleeping during the day). The second precedent concerns the case of the aggrieved person who moves next to a noise source and hence suffers a nuisance.

The law of prescription concerns private nuisances (but not public) and states that if things are done which affect your neighbour (with his or her knowledge) and continue for 20 years, you obtain the right to continue. However, this does not translate well to noise nuisance. If, for example, the noise has continued for more than 20 years but no one has been affected by it, there has been no noise nuisance and hence there can be no prescriptive right.

This can be illustrated by the case of *Sturgess* v. *Bridgman* (1879). The plaintiff was a doctor who built a consulting room at the bottom of his garden against a neighbouring property and was affected by the noise of machinery from that property. The judge ruled that as the doctor had not known about the noise until he built his consulting room no prescriptive right accrued. Therefore in the common case of a complainant moving next door to a factory the normal rules of nuisance will apply, despite the factory occupier's insistence that 'they were there first'.

15.9.22 Health effects

Exposure to noise has been shown (in clinical experiments) to cause nausea, headache, irritability, instability, argumentativeness, reduction in sexual drive, anxiety, nervousness, insomnia, abnormal somnolence, and loss of appetite, as well as the more well-known hearing loss. Generally these health effects were shown to occur at noise levels greater than 85 dB(A).

In the case of hearing damage, numerous experiments have been conducted with the aim of arriving at a safe exposure to noise. It has been found that some individuals are much more susceptible to hearing damage than others. Some people may suffer permanent damage over a few months' exposure while others may take years to develop the same damage (at the same noise levels).

Physical injury occurs at sound pressure levels in excess of 140 dB (at this level there is a risk of rupture of the tympanic membrane) while levels greater than 130 dB result in acute pain. Statistical studies on workers exposed to noise levels between 75 dB(A) and 95 dB(A) lead to the following conclusions:

1. For a 40-year working life a daily Leq of less than 75 dB(A) will lead to negligible risk.
2. The experimental data would indicate that for higher noise levels, and corresponding shorter time periods, the risk to hearing damage is the same. For example, 78 dB(A) for an 8-hour period is the same as 81 dB(A) for a 4-hour period.
3. Above 75 dB(A) 8-hour Leq the risk of hearing damage increases proportionately with the rise in levels.
4. Most countries have legislation which restricts noise levels to 85 dB(A) ±5 dB(A) with a tendency to reducing acceptable levels. It should be noted that at the UK's limit of 90 dB(A) there is some risk of hearing damage.
5. Infra-sound (sound below the normal human audible range) is capable of causing health effects. More recent research indicates an effect similar to excess alcohol consumption and indeed a synergistic effect with alcohol has been noted. It may be that in certain cases infra-sound is capable of causing an increase in accident rates. High infra-sound levels are noted in the foundry industry and in drivers' cabs in large vehicles.

15.9.23 Damage to plant/machinery/building structures

Noise can lead to damage in two ways:

1. Directly – as a result of induced vibrations
2. Indirectly – as a result of interference with the operative's normal function

Direct damage includes vibration fractures of electrical components (particularly switch contacts), structural panels, etc. Damage to buildings occurs particularly around windows (infra-sound is particularly troublesome in this effect).

Indirect damage is probably the greatest effect of noise levels. Operator performance is affected by fatigue and also the inability to hear potential problems with the machine (that might ordinarily be attended to with no significant damage resulting). In addition, the inability to hear shouted warnings may result in accidents and further plant damage.

15.9.24 Legislation concerning the control of noise

15.9.24.1 Environmental Protection Act 1990, Section 80

A notice may be served where a nuisance has occurred or the Local Authority think a nuisance may occur. Noise nuisance is not defined as such, but includes vibration. The notice may not be specific and may merely require the abatement of the nuisance. A notice may, however, require the carrying out of works or specify permissible noise levels. The time period for compliance is not specified in the Act, but must be reasonable.

Appeals against a Section 80 notice must be made to the magistrate's court within 21 days of the serving of the notice. The grounds of appeal are given in the Statutory Nuisance (Appeals) Regulations 1990 and are as follows:

1. That the notice is not justified by the terms of Section 80. The most common reason for this defence is that the nuisance had not already occurred, and that the Local Authority did not have reasonable grounds to believe that the nuisance was likely to occur.
2. That there had been some informality, defect or error in, or in connection with, the notice. It may be that the notice was addressed to the wrong person or contained other faulty wording.
3. That the Authority have refused unreasonably to accept compliance with alternative requirements, or that the requirements of the notice are otherwise unreasonable in character or extent, or are unnecessary. This defence is self-explanatory.
   The Local Authority are only permitted to ask for works that will abate the noise nuisance. Other works (perhaps to comply with other legislation) should not be specified in the notice. They may, however, be contained in a letter separate from the notice. An example of this would be where food hygiene requirements were breached by the fitting of acoustic enclosures to food-manufacturing machines. Readily cleanable enclosures may be a requirement of the Food Hygiene Regulations, but it should not be contained in a Section 80 Environmental Protection Act notice.
4. That the time (or, where more than one time is specified, any of the times) within which the requirements of the notice are to be complied with is not reasonably sufficient for the purpose.
5. Where the noise to which the notice relates is that caused by carrying out a trade or business, that the best practicable means have been used for preventing or for counteracting the effects of the noise. 'Best practicable means'

incorporates both technical and financial possibility. The latter may be related to the turnover of a company. Therefore a solution that may be the best practicable means for one company may not be so for another.

6. That the requirements imposed by the notice are more onerous than those for the time being in force in relation to the noise to which the notice relates of
    (a) Any notice under Sections 60 or 66 of the Control of Pollution Act 1974, or
    (b) Any consent given under Sections 61 or 65, or
    (c) Any determination made under Section 67.
    Section 60 relates to a construction site notice. Section 61 is a consent for construction works. Sections 65–67 relate to noise-abatement zones (see below).

7. That the notice might lawfully have been served on some person instead of the appellant, being the person responsible for the noise.

8. That the notice might lawfully have been served on some person instead of, or in addition to, the appellant, being the owner or occupier of the premises from which the noise is emitted or would be emitted, and that it would have been equitable for it to have been so served.

9. That the notice might lawfully have been served on some person in addition to the appellant, being a person also responsible for the noise, and that it would have been equitable for it to have been so served.

15.9.25 British Standard 4142: 1990

This British Standard is a revision of a standard first published in 1967 and was revised in 1975, 1980, 1982 and 1990. The standard purports to rate noises of an industrial nature affecting persons living in the vicinity. It gives a method of determining a noise level, together with procedures for assessing whether the noise in question is likely to give rise to complaints. It does make the point that while there is a correlation between the incidence of complaints and general community annoyance, quantitive assessment of the latter is beyond the scope of the document, as is the assessment of nuisance.

The previous document has been used extensively as a guide to the assessment of nuisance in various circumstances (certainly outside the scope of the document) and has gained a status that outweighs its original intention. Unfortunately, the early document was very flawed in its methodology (as is the current one) and resulted in numerous difficult legal decisions when it was produced in court as the definitive guide to noise nuisance. In particular, the old BS 4142 had a method for obtaining a 'notional background level' where the actual background level (i.e. that level which exists when the noise in question was suppressed) could not be measured, which was widely discredited as being grossly inaccurate.

The new BS 4142 rates noise in terms of Leq over a measured time interval (one hour in the daytime and 5 minutes at night) and compares this level with a background measured in terms of the $L90$ of the ambient. If a noise has a duration shorter than the measurement period in question, an 'on-time' correction is applied by the use of the following equation:

$$LAeq\ T_r = LAeq\ T_m + 10 \log_{10} \frac{T_{on}}{T_r}$$

where

$LAeq\ T_r = Leq$ for reference period
$LAeq\ T_m = $ measured Leq for the event
$\quad T_{on} = $ time on
$\quad T_r = $ reference time period (5 or 60 minutes)

Table 15.33 Corrections to noise level readings

Noise level reading $LAeq\ T$ minus background $LA90,\ T$ (dB)	Correction subtract from noise level reading (dB)
6–9	1
4–5	2
3	3
<3	Make measurements closer to source and back-calculate theoretical noise level in isolation from background

A further correction may need to be applied if the specific noise does not exceed the background by more than 10 dB. A simplified correction table is used (Table 15.33).

Finally a correction is applied dependent upon the nature of the noise. If the noise contains a distinguishable, discrete, continuous note (whine, hiss, screech, hum, etc.) or if there are distinct impulses in the noise (bangs, clicks, clatters or thumps), or if the noise is irregular in character enough to attract attention, add 5 dB to the specific noise level to obtain the rating level. The assessment for complaint purposes is then made by comparing this rating level with the background noise level. If the rating level exceeds the background by 5 dB the standard states that the result is marginal, and if the rated level exceeds the background by 10 dB or more, complaints are 'likely'.

This background noise level is one of the main criticisms of the document as it is intended to include any existing noise sources in the area. The new noise source is therefore compared against the existing noise climate, even if most of this is produced by the same factory. The example given in the British Standard further reinforces this point by considering premises which produce 40 dB(A) at the nearest house when operating normally and yet the ambient falls to 29 dB(A) during a factory shutdown. Thus the existing contribution is already 11 dB. A new source is assessed which adds 4 dB to the existing (40 dB) ambient, and the result is determined to be marginal! If this situation were to continue the background sound level (as defined in the standard) would 'creep' upwards – obviously an undesirable situation and one that is addressed in a planning circular (Circular 10/73) that deals with planning and noise. This circular particularly addresses creeping ambients and states that 'the introduction of a new noise source into an area is liable to result in a creeping growth of ambient noise level, and consequent deterioration in the quality of the environment, even though each of the new sources, considered separately, would not be liable to give rise to complaints'. This point alone is sufficient for the method to be discredited by Environmental Health Officers when investigating nuisances, and the standard is unlikely to be used by them as a definitive guide. Consequently, operators of industrial premises should not use the information given in this British Standard as evidence when arguing (in legal situations) that they are not causing a statutory nuisance.

Further, this British Standard takes very poor account of the effects of discrete frequency components. It is quite possible for a narrow band component to cause a serious nuisance while being almost unmeasurable on an 'A'-weighted scale. Consequently, more detailed narrow band analysis would be necessary and it is essential to compare the actual noise with the background noise within that narrow band (usually octave or one-third octave). The British Standard makes no mention of such a situation.

15.9.26 Noise-abatement zones

Local Authorities are empowered by the Control of Pollution Act 1974 to designate areas as noise-abatement zones. Within these areas noise levels are measured and entered onto a register. It is an offence to increase noise levels beyond register levels unless a consent is obtained. If the Local Authority are of the opinion that existing noise levels are too high, noise-reduction notices can be served.

In the case of new premises the Local Authority will determine noise levels which it considers acceptable, and these will be entered into the noise level register. Appeals against notices or decisions are made to the Secretary of State.

15.9.27 Planning application conditions

Local Authorities are empowered to impose conditions on planning applications to protect environmental amenities of neighbours. Noise is commonly controlled by conditions. Local Authorities may ask for more onerous controls on planning conditions than the mere avoidance of nuisance. Planning conditions are designed to avoid reduction in amenity of neighbours. This may mean that a process has to be almost inaudible (particularly in the case of Light Industrial Consents). Appeals against planning conditions are made to the Secretary of State.

15.9.28 The Health and Safety at Work etc. Act 1974

Section 2 of this Act imposes a general duty on employers to ensure, so far as is reasonably practicable, the health, safety and welfare at work of all his or her employees. This general section will include the acoustic environment. It should be noted that there are specific regulations made under this Act (The Noise at Work Regulations 1989 – dealt with later) that control noise (primarily to protect hearing) but they do not completely satisfy the overall requirements of Section 2. For instance, if high noise levels mask an audible alarm such that a risk of injury is caused, a breach of Section 2 would be likely (despite the fact that the noise levels in The Noise at Work Regulations have not been exceeded). Section 3 of the Act deals with an employer's duties to non-employees and again imposes a duty of care.

The Noise at Work Regulations have no relevance to members of the public as they apply only to persons at work. Section 3 of the Act would, therefore, control noise exposure to non-employees who would be likely to suffer risks to health and safety. Noise nuisance is controlled by other legislation (The Environmental Protection Act 1990 and others).

Section 4 of the Act imposes the general duty on employers to care for the health and safety of persons, who not being his or her employees, nevertheless have resort to premises under his or her control. This duty is designed to protect subcontractors, etc.

Section 6 concerns articles manufactured for use at work and will control the acoustic output of machines, etc. The control of the manufacture of noisy machines is carried further by The Noise at Work Regulations.

15.9.29 The Noise at Work Regulations 1989

These Regulations came into force on 1 January 1990 and control the exposure to noise of persons at work. They establish three noise levels known as the first, second and peak action levels. Different regulations are applicable as each action level is exceeded. The unit of measurement is known as equivalent continuous sound level and may be defined as 'that notional continuous steady level which would have the same

'A'-weighted acoustic energy as the real fluctuating noise measured over the same period of time'. For the purposes of the Regulations an 8-hour time period is used and the 8-hour equivalent continuous sound level is abbreviated to L_{EP}, d:

The first action level is 85 dB(A) L_{EP},d
The second action level is 90 dB(A) L_{EP},d
The peak action level is 200 Pascals (equivalent to 140 dB)

Damage to the hair cells in the inner ear is proportional to the noise energy received. This is a dose concept comprising the product of noise level and exposure duration. It follows, therefore, that the same amount of deafness will follow from the exposure to a very intense sound for a short period as to a lower level for a proportionally longer period.

It has been shown that the exposure time has to be halved for each 3 dB(A) increase in the noise levels. 3 dB(A) represents a doubling of sound energy, hence this rule has become known as the equal energy damage risk criterion. It follows that 93 dB(A) for 4 hours is also 100% of the permitted exposure for a day; similarly, 2 hours at 93 dB(A) would be 50% of the permitted exposure. Where an employee is likely to be exposed to above the first action level the employer shall ensure that a competent person makes an assessment of the noise levels which is adequate for the purposes:

1. Of identifying which of his or her employees are so exposed, and
2. Of providing him or her with such information with regard to the noise to which those employees may be exposed as will facilitate compliance with his or her duties under the Regulations, specifically:

    (a) Reduction of noise exposure
    (b) Ear protection
    (c) Ear protection zones
    (d) Provision of information to employees

15.9.29.1 The requirements of the Regulations

An employer must:

1. Carry out an assessment when an L_{EP},d of 85 dB(A) is likely to be exceeded.
2. Review the assessment if changes necessitate this.
3. Record the exposure and keep records.
4. Reduce the risk of damage to hearing to the lowest level reasonably practicable.
5. Every employer shall, when any of his or her employees is likely to be exposed to the second action level or above or to the peak action level or above, reduce, so far as is reasonably practicable (other than by the provision of personal ear protectors) the exposure to noise of that employee.
6. If an employee is exposed to greater than the first action level and less than the second action level the employer shall provide hearing protection if so required by the employee.
7. If an employee is exposed to greater than the second action level or greater than the peak action level the employer shall provide hearing protection which, when properly worn, will reduce the risk of hearing damage to below that arising from exposure to the second action level or, as the case may be, to the peak action level.
8. Ear-protection zones (i.e. areas where the second action level is likely to be exceeded) shall be established. Employees must wear ear protection in this zone. The employer shall erect suitable signs.
9. Information, instruction and training shall be provided for employees where exposure is likely to exceed the first

action level or the peak action level. This information shall include:

(a) The risk of damage to an employee's hearing that such exposure may cause;
(b) What steps an employee can take to minimize that risk;
(c) The steps that an employee must take in order to obtain the personal ear protectors which the employer must provide.
(d) The employer's obligation under the Regulations.

15.9.30 Noise control engineering

Before attempting noise control it is important to consider the nature of the problem. The first (and usually the most cost-effective) approach is to silence the noise at source. In order to appreciate likely noise sources and the methods used to reduce their emission, we will consider the ordinary reciprocating piston engine.

The first source is the crankcase wall 'ringing' under the reciprocating forces of the combustion. Excess noise will be produced if the frequency of the combustion pulses is at the resonant frequency of part of the engine. The solution is to de-tune the block by stiffening (which may have an added mass-law effect) or by the addition of damping materials.

The next source to consider is the crankshaft and bearings. Most shafts will be out of balance to some degree and will have a resonant point. A well-designed engine should not run at this resonant frequency. Bearings are two main types – sliding (or plain) bearings and roller. Excess clearance can give rise to bearing knock and the solution here is to replace bearings. Poorly designed systems can give rise to bearing knock if the shaft has a bending mode within the engine's operating range.

Plain (oilite) type bearings can produce screech on start-up from cold (these bearings are used in electric motors, starter motors and alternators, etc.). Replacement may make matters better.

Roller bearings are generally quieter than plain bearings but can produce considerable noise if damaged. Frequency analysis of the noise can be used to assess the source but there are many modes involved in a bearing 'click'. Again the solution is replacement, with care being taken to locate the new bearing without causing damage (by only inserting the bearing with force on the outer race) and by careful checks of the dimensions of the bearing housing.

Gearboxes give rise to noise as teeth contact each other. If excess clearance (and/or poor lubrication) is present, noise will be exacerbated. Again, frequency analysis can help to locate the source of trouble by comparing the dominant frequency of the noise with the gear teeth meshing frequencies. It should also be noted that gears themselves can ring (particularly if they are free and not cast as part of a layshaft, etc.). The dominant frequency may be at the damaged gear's resonant frequency.

Finally, hydraulics can give rise to noise due to the intermittent force pulses produced by pumps. The solution would be to introduce some flexibility into the receiver system to damp out the very high intermittent pressures produced by the incompressibility of the fluid.

Other examples of machines which lend themselves to noise reduction at source are:

1. Presses – is the degree of impact necessary? Can it be adjusted? Can the press operate by pressure alone?
2. Air discharge – use of air tools and nozzles. The turbulence in the boundary layer of air between the rapidly moving airstream and the atmosphere is heard as noise. Can the airstream be diffused (silencers fitted to the exhaust)? Nozzles used for cleaning can have devices fitted which give a gradual transition from the rapidly moving air to atmosphere by the use of an annular ring of small nozzles round the central nozzle. These silence with very little loss in efficiency.
3. Reciprocating compressors – these give rise to very high noise levels at low frequency (below 250 Hz typically). These low-frequency noises are very difficult to attenuate. The most popular solution is to use rotary (vane type) compressors instead. These are inherently quieter and have the further advantage that the noise they generate is at high frequency (typically above 1 kHz) and is, therefore, easy to attenuate.
4. Cutting machines – modifications to the method of restraining material being cut to reduce 'ringing'. Reducing free length of material.

If reduction of noise at source is not possible (or does not provide sufficient reduction) the transmission pathway must be considered:

Source of sound energy → Transmission pathway → Receiver

15.9.31 Noise-reduction principles

There are many thousands of noise control devices. All, however, rely on one or more of the three basic noise control principles. These are: insulation, absorption and isolation.

15.9.31.1 Insulation

The simplest insulator is a sheet of material placed in the sound-transmission pathways. Sound energy reaches the surface in the form of a pressure wave. Some energy passes into the partition and the rest is reflected.

Energy that passes into a partition may be partially absorbed and transformed into heat. This is likely to be very small in a plain partition. The remainder of the energy will then pass through the partition by displacement of molecules and pass as sound in the same way that sound travels in air. This can then pass to the edge of the partition and be re-radiated as sound from other elements of the structure – this is known as flanking transmission. By far the greatest amount of energy, in a thin partition, will pass through the partition by actually causing the partition to vibrate in sympathy with the incident sound and, hence, re-radiating the sound on the opposite side. The amount of sound transmitted through a partition is represented by the ratio of the incident energy to the transmitted energy. This factor when expressed as decibels is known as the sound-reduction index:

$$SRI = 10 \log_{10} \frac{1}{\text{Transmission coefficient}} \text{ (dB)}$$

The movement of the panel (and hence its resistance to the passage of sound) is controlled by a number of factors:

1. The surface mass affects the inertia of the panel. Greater mass causes a corresponding greater inertia and hence more resistance to movement. At high frequencies this becomes even more significant. The mass law can be expressed:

$$SRI = 20 \log_{10} mf - 43 \text{ dB}$$

where m is the superficial weight (kg m^{-2}) and f is the frequency (Hz).

2. Stiffness – at very low frequencies the movement of the panel will be controlled by the stiffness, as inertia is a dynamic force and cannot come into effect until the panel has measurable velocity. Stiffness controls the perfor-

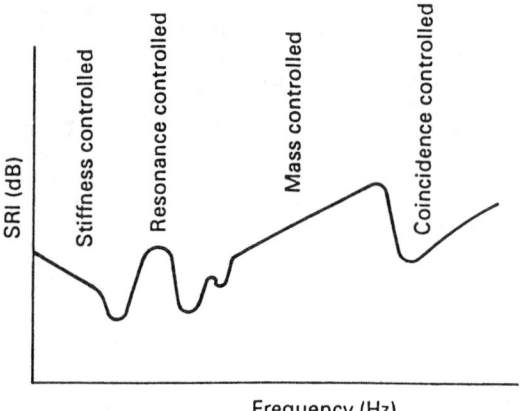

Figure 15.181 Typical insulation characteristics of a partition

mance of the panel at low freqencies until resonance occurs. As the driving frequency increases, the resonance zone is passed and we enter the mass-controlled area. The increase in sound-reduction index with frequency is approximately linear at this point and can be represented by Figure 15.181.
3. Coincidence – a panel will have a bending mode when a wave travels along the length of the sheet of material. The frequency of this bending mode is known as the critical frequency. This mode of bending will be introduced by sound incident at angles greater than 0°. At the critical frequency coincidence will only occur for a sound wave with a grazing incidence (90°). At greater frequencies the partition will still be driven, but in this case by progressively lower angles of incidence. The coincidence dip is not, therefore, a single dip but will result in a loss of sound-reduction index at progressively higher frequencies. The desirable insulation panel will, therefore, be massive but will not be stiff.

15.9.32 Absorbers

15.9.32.1 Porous absorbers

As sound passes through a porous material, energy is lost by friction within the material. The material is usually employed by fixing it to the surfaces in a room. The absorber will have the highest efficiency when positioned where the air molecules are moving the fastest (and hence more energy is absorbed). At the wall surface the molecules are stationary. If we plot a single-frequency graph we find that the maximum particulate velocity occurs at $\lambda/4$ (one-quarter wavelength) from the surface. In practice, incident sound is rarely of single frequency. But the principle can be observed that the absorber must be one-quarter of the wavelength away from the wall (for the frequency of the sound to be absorbed). This can be arranged either by having a thickness greater than $\lambda/4$ for the lowest frequency to be absorbed or, alternatively, to mount the absorber on a frame some distance away from the wall such that the centre of the absorber is at $\lambda/4$ for the frequency to be absorbed.

15.9.32.2 Resonant absorbers

The simplest resonant absorber is known as the Helmholtz resonator. This device consists of a chamber connected to the

duct (or whatever area is to be controlled) by a narrow neck. The volume of air in the chamber will resonate at a frequency, F_{res}, determined by the volume of the chamber, the length of the neck and the cross-sectional area of the neck.

$$F_{res} = 55 \frac{S}{IV} \text{ (Hz)}$$

where S = cross-sectional area (m²), I = length of neck (m) and V = volume of enclosure (m³). As the chamber resonates, air is forced through the narrow neck and hence energy is absorbed in overcoming the resistance.

The degree of attenuation at the critical frequency can be very large, but this type of silencer has a very narrow bandwidth. This device may be suitable when the machine being dealt with emits sound predominantly of a single wavelength. The absorption bandwidth of a Helmholtz resonator can be expanded by lining the chamber with absorbers but this has the effect of reducing the efficiency.

The perforated absorber which forms the basis of many acoustic enclosures and silencers is a development of the resonator principle. As stated previously, the bandwidth may be broadened by packing the chamber with an absorber, but this lowers efficiency. This may be overcome by using multiple absorbers in the sound path. It can be arranged by placing a perforated sheet some distance away from the rigid outer wall of the enclosure and filling the cavity with absorber. It is not necessary to use cross-walls between the 'chambers' so formed. In this case the equation becomes:

$$F_{res} = 5000 \frac{P}{I(t + 0.8d)} \text{ (Hz)}$$

where

I is the depth of airspace (mm)
t = thickness of panel (mm)
d = diameter of holes (mm)
p = percentage open area of panel

15.9.32.3 Panel absorber

This absorber is basically a panel attached to the structural wall which is designed to absorb energy. It is therefore frequency dependent and has an absorption peak at its resonant frequency. This type of absorber is not commonly used.

15.9.33 Vibration isolation

Vibration in machinery or plant can be induced in a number of ways, including:

- Out-of-balance forces on shafts
- Magnetic forces in electrical apparatus
- Frictional forces in sliding objects

The first course of action in vibration isolation is the reduction at source. This may be achieved by balancer shafts in engines, stiffer coils in electrical apparatus or better lubrication between adjacent sliding surfaces.

When all possible vibration reduction has been achieved the machine must be isolated from the structure by some form of spring mounting. Spring mounts have a resonant frequency depending upon the stiffness of the spring and the weight of the object placed upon it. It will be apparent that the static deflection of the spring will also be proportional to the resonant frequency.

As the driving force of the mass/spring increases from zero up to the resonant frequency the amount of transmission of the vibration increases until resonance is reached and the

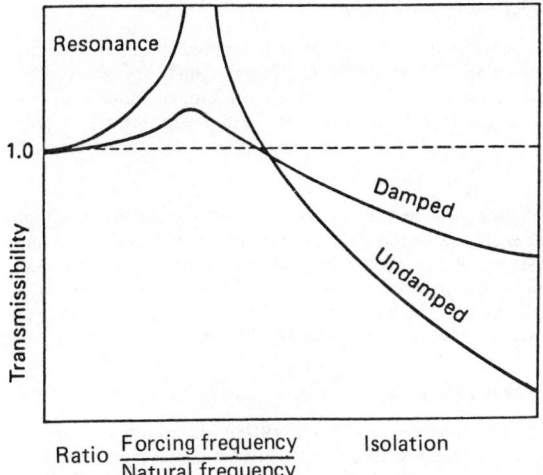

Figure 15.182 Performance of anti-vibration mounting

transmission becomes infinitely large. As the resonant point is passed, the transmission begins to reduce until at some point the transmissibility falls below one (see Figure 15.182), i.e. isolation occurs.

In practice, however, spring systems have some inbuilt damping and this will have the effect of reducing the amplitude of the resonance below infinity. This is very necessary in real systems to avoid excessive excursions of mounted machinery. A damped mounting will follow the second curve in Figure 15.182 and it will be noted that the vibration isolation at high-frequency ratios is less than that for undamped systems. It is important, therefore, to use the lowest degree of damping that is necessary.

15.9.34 Practical applications

15.9.34.1 Acoustic enclosures

Panels of multi-resonator material are made from perforated plate sandwiched with solid plate with an intermediate absorber layer. These panels can be built up into enclosures taking care to seal all junctions adequately. Typically, these enclosures are made to surround small machines (e.g. compressors). They may be fitted together with spring catches to allow for dismantling for maintenance purposes.

Ventilation may be a problem but can be dealt with in several ways:

1. Acoustic louvres – louvres are constructed of the absorbent panel material (suitable for small degree of noise reduction only);
2. Silencer fitted to ventilation duct (see later);
3. Baffled enclosures to ventilation duct.

15.9.34.2 Building insulation

Single-panel insulators have been described earlier. In building insulation it is usual to provide double insulation. In theory, if the insulation panels have no interconnection it should be possible to arithmetically add the sound reduction of the two elements of the structure. In practice, it will be found that there is bridging, either by the structure, wall ties or flanking transmission or by the air between the two elements

acting as a spring. If the two elements of the wall were in rigid connection the insulation would be 3 dB more than the single element alone (mass law) and if totally separated it would be the sum of the figures. In practice, a cavity wall with ties and a 50 mm cavity gives approximately 10 dB more reduction compared to a single-skin wall of half the surface mass.

Double-glazed windows work on the same principle. It is important to avoid the coincidence of the resonant frequencies of the two elements and hence it is usual to arrange for the glazing panels to have different thicknesses (and hence a different resonant frequency). This is not necessary if one element is subdivided by glazing bars to give different size panes from its opposing element. The reveals of a double-glazed window should be lined with acoustically absorbent material to damp the sound within the cavity. The width of the cavity should not be less than 150 mm.

If insulation panels are not of uniform construction, as in the case of a wall containing a window, the average sound-insulation value has to be derived for use in calculations. The total transmission coefficient for the composite panel will equal the sum of the individual coefficients multiplied by their respective areas and divided by the total area. Thus:

$$t_{av} = \frac{t_1 s_1 + t_2 s_2 + t_3 s_3 + \ldots \text{ etc.}}{s_{tot}}$$

and the SRI of the total panel derived from

$$SRI = 10 \log_{10} \left(\frac{1}{t_{av}} \right) \text{ (dB)}$$

15.9.34.3 Control of noise in ducts

Fans produce the least noise when operating at their maximum efficiency. It is, therefore, important to select the correct fan for the airflow and pressure characteristics required. It is also important to remember the noise generated within the system (as opposed to at the fan) depends on the air velocity, and, hence, for a required airflow rate, a larger cross-section duct (with a correspondingly lower velocity) will give quieter results. It will also give other advantages when it comes to providing extra noise attenuation and fitting of silencers.

It is most important in the design of systems to eliminate as much turbulence as possible. To achieve this, the fans should be mounted some distance away from bends (at least one and a half duct diameters). Junctions between pipes and connectors should present a smooth internal profile and inlets to systems should be tapered and not plain. Outlet grilles should be of larger diameter than ducts and have aerodynamically smooth profiles where possible.

If it is necessary to add extra attenuation to a duct it is essential to decide on the required amount. If only a relatively small degree of absorption is required the first course of action is to line part of the duct with absorber. The length of duct to be lined will be determined by the degree of attenuation required and the thickness will be determined by the frequency of the noise. The data for these factors are available from many sources and are usually published as tables by manufacturers. For further attenuation it is necessary to provide a centre-pod type attenuator (Figure 15.183). This increases the area of the absorber and also aids low-frequency attenuation. For further low-frequency attenuation an in-line splitter silencer is employed (Figure 15.184).

These are capable of providing a high degree of attenuation, depending on the width between the elements. The smaller the gap, the higher the attenuation. Again, tables of performance are published by the major manufacturers. It is necessary, in order to decide on the design of the silencer to be

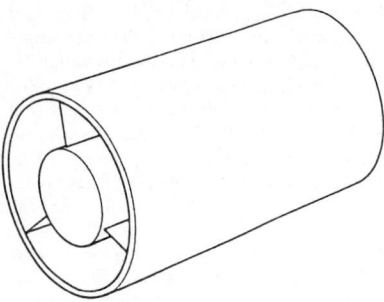

Figure 15.183 Centre-pod silencer

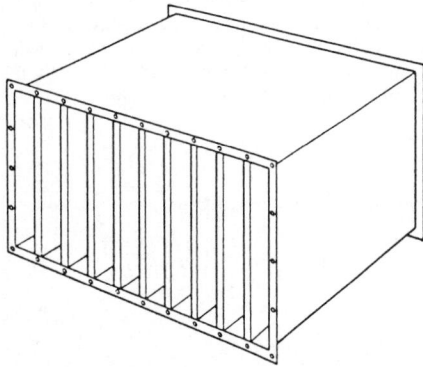

Figure 15.184 Splitter silencer

installed, to know the required attenuation (and something about the frequency/noise level profile) and the permitted pressure loss in the system.

Manufacturers' data can then be consulted. Splitter silencers are also available, made into bent shapes, and these can provide even higher degrees of attenuation as well as aiding installation. Silencers should ideally be fitted in systems as near to the noise source as possible to avoid noise break-out from the duct. Other obstructions in the duct must be considered, however, as they may generate further aerodynamic noise which, if it occurs after the silencer, will not be attenuated.

15.9.34.4 Anti-vibration machinery mounts – in practice

Again, the characteristics of the system need to be considered. The weight of the machine and the frequency will determine the static and dynamic deflections of the mounts and hence the material of which the mount is to be constructed. At very high frequencies, mats may be placed under machinery. These may consist of rubber, cork or foam. At middle frequencies it is usual to use rubber in-shear mounts. At low frequencies metal spring mounts are used.

15.9.34.5 Mats

Anti-vibration mats are very useful for frequencies about 25 Hz. They have the disadvantage of being liable to attack by oils and if they become saturated or deteriorated they will compress and lose their efficiency.

15.9.34.6 Rubber mounts

Although these are loosely termed 'rubber' mounts, they are often composed of synthetic rubbers which are not readily attacked by oils and can operate over a much wider temperature range. Typical maximum static deflections are 12.5 mm.

15.9.34.7 Steel spring mounts

Steel springs have the disadvantage of transmitting the high frequencies along the length of the spring. It is usual to mount the spring with a rubber or neoprene washer under its base. Steel spring mounts are also most vulnerable to resonance problems, and the solution is to build in a damper device. This has the disadvantage of reducing the isolator's efficiency.

15.9.34.8 Positioning of anti-vibration mounts

Machinery must be positioned so that all mounts are equally loaded, and failure to do so will result in the possibility of a rocking motion developing. This may require mounting the machine on a subframe. If this is not possible the load should be assessed at each mounting point and mounts of different stiffness used.

15.9.34.9 Installation

Mounts should be installed so that the whole machine is isolated from the structure. Services (e.g. power, hydraulics, etc.), should be mounted flexibly. Bridging is the most common fault when providing vibration isolation to machines and building structures, and should be carefully avoided. Services should be designed to withstand the degree of movement permitted by the anti-vibration mounts without suffering damage.

Further information on acoustic noise and its control can be found in references 68–70.

References

1. Turton, R. K., *Principles of Turbomachinery*, S & FN Spon, London (1984)
2. Horlock, J. H., *Axial Flow Compressors*, Butterworths, London (1958)
3. Addison, H., *Centrifugal and other Rotodynamic Pumps*, 2nd edn, Chapman & Hall, London (1955)
4. Wislicenus, G., *Fluid Mechanics of Turbomachinery*, Vols 1 and 2, Dover Press, London (1965)
5. Karrasik, I. *et al.*, *Pump Handbook*, McGraw-Hill, New York (1976)
6. Sterling, L., 'Selection of pump type to match systems', 5th BPMA Tech. Conf., Bath (1977)
7. IMechE Conference, 'Part Load Pumping Operation, Control and Behaviour' (1988)
8. Balje, O. E., *Turbomachines*, John Wiley, Chichester (1981)
9. Csanady, G. T., *Theory of Turbomachines*, McGraw-Hill, New York (1964)
10. API 610, Centrifugal Pumps for Refinery Service, 7th edn (1989)
11. ISO 5199, Centrifugal Pumps: Class II (1986)
12. BS 5316, Acceptance Tests for Centrifugal Mixed Flow and Axial Pumps. Part 1 Class C Tests
13. American Hydraulic Institute Standards (1983)
14. BS 848, Fans for General Purposes Part 1: Methods of Testing Performance (1980): Part 2: Methods of Noise Testing (1985)
15. BS 2009, Code for Acceptance Tests for Turbotype Compressors and Exhausters (1953)
16. BS 1571, Testing of Positive Displacement Compressors and Exhausters. Methods for Acceptance Testing (1987): Part 2:

Methods for Simplified Testing of Air Compressors and Exhausters

17 Klay, H. R. and Reich, B., 'Gas compressors – a brief survey', *Sulzer Technical Review*, 2/1988

18 Final Report of the Advisory Committee on Asbestos (Vols 1 and 2) October 1979, HMSO

19 Asbestos Regulations 1969 Statutory Instrument No. 690, HMSO

20 Health and Safety Executive Guidance Note: Environmental Hygiene: 10 Asbestos

21 Department of the Environment, Waste Management Papers: Number 18: Asbestos Wastes Number 23: Special Wastes (Chapter 4), HMSO

22 'Material Health & Safety Data Sheets', Asbestos Information Centre

23 *A Mechanical Seal Guide to API 610 Standard*, 7th edn, John Crane Inc., New York (1990)

24 Summers-Smith, J. D. (ed.), *Mechanical Seal Practice for Improved Performance*, Mechanical Engineering Publications, London (1988)

25 Flitney, R. K., Nau, B. S. and Reddy, D., *The Seal User's Handbook*, BHRA, Cranfield (1984)

26 Fern, A. G. and Nau, B. S., *Seals*, Engineering Design Guide 15, Oxford University Press/Design Council, Oxford (1976)

27 Merry, S. L. and Thew, M. T., 'Comparison between a hydrodynamic disc seal and neckrings for a small process pump running in water and in mercury', *9th BHRA International Conference on Fluid Sealing*, 1981, Paper H2, p. 333

28 Neale, M. J. (ed.), *Tribology Handbook*, Butterworths, London (1973)

29 Warring, R. H., *Seals and Packings*, Trade & Technical Press Limited, London (1967)

30 CIBSE Guide, Volume A, *Design Data* (1986)

31 CIBSE Guide, Volume B, *Installation and Equipment Data* (1986)

32 ASHRAE Handbook, *Fundamentals* (1989)

33 ASHRAE Handbook, *HVAC Systems and Applications* (1987)

34 *Threshold Limit Values and Biological Exposure Indices for 1988–1989*. American Conference of Governmental Industrial Hygienists

35 BRE Digest 119, *Assessment of Wind Loads* (July 1970)

36 BRE Digest 210, *Principles of Natural Ventilation* (B. B. Daly) (February 1978)

37 BRE Digest 346 (7 parts), *The Assessment of Wind Loads* (1989)

38 Daly, B. B., *Woods Practical Guide to Fan Engineering*, Woods of Colchester Ltd (1978)

39 *Ductwork Specification DW142*, HVCA (Heating and Ventilation Contractors Association)

40 CIBSE Technical Memorandum TM8, *Design Notes for Ductwork Industrial Ventilation, A Manual of Recommended Practice*, 20th edn, American Conference of Governmental Industrial Hygienists (1984)

42 ISO 7730: 1984, Modern thermal environments – determination of the PMV and PPD indices and specification of the conditions for thermal comfort

43 Fire Paper 7, 'Investigations into the flow of hot gases in roof venting', HMSO (now available from Colt International Ltd) (1963)

44 Fire Paper 10, 'Design of roof venting systems for single storey buildings', HMSO (now available from Colt International Ltd) (1964)

45 BS 7346, Components for smoke and heat control systems. Part 1: Specification for natural smoke and heat exhaust ventilators

46 BS 7346, Components for smoke and heat control systems. Part 2: Specification for powered smoke and heat exhaust ventilators

47 EH 40/89 Occupational Exposure Limits, Health and Safety Commission (1989)

48 *Fuel Efficiency Booklets*, Department of Energy (1984–1986):
   1. Energy audits
   2. Steam
   3. Economic use of fired space heaters for industry and commerce
   4. Compressed air and energy use
   5. *
   6. *
   7. Degree days
   8. The economic thickness of insulation for hot pipes
   9. Economic use of electricity
   10. Controls and energy savings
   11. The economic use of refrigeration plant
   12. Energy management and good lighting practices
   13. The recovery of waste heat from industrial processes
   14. Economic use of oil-fired boiler plant
   15. Economic use of gas-fired boiler plant
   16. Economic thickness of insulation for existing industrial buildings
   17. Economic use of coal-fired boiler plant
   18. *
   19. Process plant insulation and fuel efficiency
   20. Energy efficiency in road transport
*New titles in preparation

49 *Energy Audit Series, No. 18 The Engineering Industries*, Department of Energy (1984)

50 Payne, G. A. *The Energy Managers' Handbook*, Westbury House, London (1980)

51 NIFES, *Energy Managers' Handbook*, Graham and Trotman, London (1985)

52 *Energy Manager's Workbook*, Energy Publications, Cambridge (1982)

53 *Energy Manager*, Maclaren Publishing, Croydon (monthly)

54 *Energy Management*, Department of Energy (monthly)

55 *CIBS Guide*, Chartered Institution of Building Services, London
   A3 Thermal Properties of Building Structures
   A4 Thermal Response of Buildings

56 Murphy, W. R. and McKay, G. *Energy Management*, Butterworths, London (1982)

57 *CIBS Code for Interior Lighting*, Chartered Institution of Building Services, London

58 Lyons, S. L. *Handbook of Industrial Lighting*, Butterworths, London (1981)

59 Boyen, J. L. *Thermal Energy Recovery*, Wiley, Chichester (1980)

60 Reay, D. A. *Industrial Energy Conservation*, Pergamon Press, Oxford (1977)

61 Von Cube, H. L. and Steimle, F. *Heat Pump Technology*, Butterworths, London (1981)

62 Randall, R. B., *Application of B&K Equipment to Frequency Analysis*, Bruel & Kjaer, Denmark

63 Angelo, M., *Vibration Monitoring of Machines*, Technical Review No. 1, 1987, Bruel & Kjaer, Denmark

64 Broch, J. T., *Mechanical Vibration and Shock Measurement*, Bruel & Kjaer, Denmark (1976)

65 International Standards Organization, ISO 2372/3, Vibration Severity Standards

66 International Standards Organization, ISO 3945, Vibration Severity Standards

67 Thomson, W. T. and Rankin, D., 'Case histories of on-line rotor cage fault diagnosis', Conf. on Condition Monitoring, 1987

68 Beronek, L. L. (ed.), *Noise and Vibration Control*, McGraw-Hill, New York (1971)

69 Burns, W., *Noise and Man*, John Murray, London (1968)

70 Kerse, C. S., *Noise*, Oyez Publishing, London (1975)

16 Manufacturing methods

Donald B. Richardson
(Section 16.1)

Tadeusz Z. Blazynski
(Section 16.2)

Edward N. Gregory
(Section 16.3)

Allan R. Hutchinson
(Section 16.4)

Leslie M. Wyatt
(Section 16.5)

Contents

16.1 Large-chip metal removal 16/3
  16.1.1 Large-chip processes 16/3
  16.1.2 Cutting-tool geometry 16/6
  16.1.3 Cutting-tool materials 16/7
  16.1.4 Cutting fluids 16/8
  16.1.5 Forces and power in metal cutting 16/9
  16.1.6 Surface-finish considerations 16/10
  16.1.7 Tool-life assessment 16/10
  16.1.8 Economics of metal cutting 16/10

16.2 Metal forming 16/12
  16.2.1 Introduction 16/12
  16.2.2 Classification of processes 16/12
  16.2.3 Characteristics of the basic groups of processes 16/15
  16.2.4 Rolling processes and products 16/18
  16.2.5 Forging operations 16/26
  16.2.6 Extrusion 16/33
  16.2.7 Cold drawing of wire and tube 16/40
  16.2.8 Sheet-metal forming 16/43
  16.2.9 High-energy-rate operations 16/47
  16.2.10 Superplastic and mashy state forming 16/52

16.3 Welding, soldering and brazing 16/54
  16.3.1 Welding 16/54
  16.3.2 Soldering and brazing 16/67
  16.3.3 Productivity and welding economics 16/68

16.4 Adhesives 16/70
  16.4.1 General comments 16/70
  16.4.2 Definitions and terms 16/70
  16.4.3 Adhesives 16/70
  16.4.4 Adhesion and surface pretreatment 16/73
  16.4.5 Joint design 16/77
  16.4.6 Fabrication and assembly 16/83
  16.4.7 Quality control and non-destructive testing 16/84

16.5 Casting and foundry practice 16/85
  16.5.1 Introduction 16/85
  16.5.2 Problems inherent in casting technology 16/85
  16.5.3 Ingot, billet and slab casting 16/88
  16.5.4 Sand casting 16/93
  16.5.5 Permanent mould or 'die' casting 16/103
  16.5.6 Centrifugal casting processes 16/108

References 16/109

Further reading 16/111

16.1 Large-chip metal removal

16.1.1 Large-chip processes

All the large-chip processes use cutting tools of defined geometry which are applied in a controlled manner to remove metal at a predetermined rate. The processes could be classified in many ways, but it is convenient to consider them in terms of the kinematics of the machine tools. With this in mind, they have been separated into four main machine groups:

1. Turning (rotating work)
2. Shaping (reciprocating tool or work)
3. Milling (rotating tool)
4. Drilling and boring (rotating tool)

Turning machines embrace the wide variety of lathes and vertical boring machines which can be controlled manually or automatically. Automatic control can be achieved using cams, sequential controllers, hydraulic copying devices or numerical programming. All machines in this group are capable of performing six basic operations as shown in Figure 16.1. In addition, copying lathes and numerically controlled lathes can generate non-parallel forms by traversing the tool simultaneously in two planes.

Most turning processes use tools with a single cutting edge where the cutting action is characterized by a relatively uniform section of material being presented to the cutting zone, resulting in a continuous chip when cutting ductile materials or a repetitive form of short discontinuous chips when cutting brittle materials. Although the production of continuous chips indicates an efficient cutting action the chip

streamer itself presents disposal problems, frequently wrapping itself round the workpiece, the cutter or parts of the machine tool, creating a hazard to both the process and the operator.

Chip breakers are extensively used to induce continuous chips to break into short lengths which are relatively safe and can be easily disposed of. These push against the underside of the chip and cause it to curl into a tight spiral, the free end of which strikes against the tool, and the resulting bending stress causes fracture. The earliest form of chip breaker, still extensively used with flat-top tools, consists of a hard wedge-shaped block of sintered carbide clamped to the rake face of the tool about 2 or 3 mm from the cutting edge (Figure 16.2(a)). The introduction of disposable sintered carbide inserts has allowed more complicated rake-face geometries to be used which act as built-in chip breakers (Figure 16.2(b)). Effective chip breaking is largely a matter of trial and error, being influenced by the feed, tool bluntness and cutting speed as well as by the material being machined.

The development of new and improved cutting materials has enabled a hundredfold reduction in cutting time to be achieved since the beginning of the twentieth century. Unfortunately, the reduction in idle time, caused by the need for tool adjustment, and in the tool approach and retraction times before and after cutting has not been of a comparable order. In achieving lower production costs the emphasis has now rightly moved away from further reducing cutting time to attacking the disproportionately large amount of non-cutting time.

Disposable sintered cutting inserts are made to a high level of precision which allows them to be indexed or replaced in tool holders in a few seconds, usually without the need for sizing cuts. When all the cutting edges are worn the inserts are discarded, obviating the need for time-consuming regrinding which is common practice when using high-speed steel tools. When resetting lathes between work batches a substantial time saving can now be achieved by using preset tools mounted in holders which can be replaced as cartridges in the tool post. No doubt further improvements will be achieved in the development of cutting tool materials, but reduction of manufacturing time in future will be determined mainly by reducing idle time, both by better tool changing mechanisms and by improved machine tool design to facilitate chip disposal and to reduce the tool approach and retraction times.

The shaping group of machine tools produce chips by a relative linear motion between the cutting tool and the work. It includes shaping machines, planing machines and slotting machines, all of which are used mainly for tool manufacture or maintenance work and have little application in modern production. They operate on a reciprocating principle, cutting

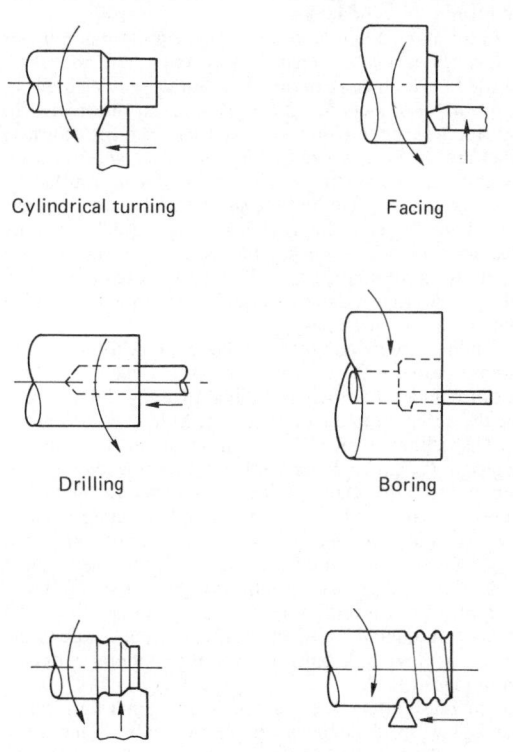

Cylindrical turning   Facing

Drilling   Boring

Forming   Screwcutting

Figure 16.1 Basic lathe operations

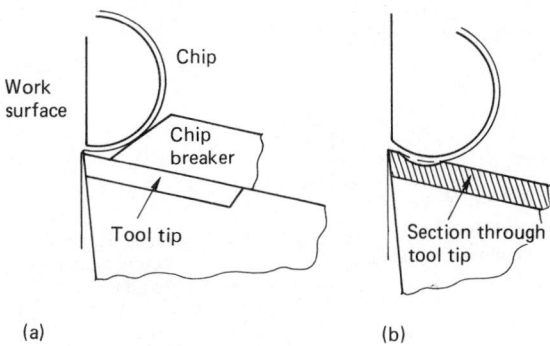

Figure 16.2 Chip breakers: (a) clamped, (b) built-in

on the forward stroke and idling on the return stroke. Although they have quick-return mechanisms the cutting time is only in the order of half the reciprocating cycle time. Swarf disposal is usually no great problem due to the intermittent nature of the cut. Figure 16.3 shows typical configurations of these three machine types, which have changed little in recent years.

Other machines in the shaping group are gear shapers and gear planers, outlines of which are shown in Figure 16.4. In gear shaping the cutter resembles a side-relieved spur gear, the involute profile being generated by rapid reciprocation of the cutter while slowly revolving the cutter and gear blank in synchronism. Helical gears can be generated using a cutter with helical teeth and applying an appropriate helical motion to the spindles. Gear shaping is used for producing gears when hobbing would be impossible due, for instance, to a turned shoulder close to the involute profile. Gear planers have little modern use, particularly in a production environment. The cutter is in the form of a straight-tooth rack, suitably relieved, and the gear is generated by reciprocating the cutter and moving the gear blank and the cutter at a constant speed. To enable a short rack to be used it can be removed from the cut and indexed back at intervals.

Broaching machines also belong to the shaping group, but these produce the required form in a single pass. Internal broaching is for opening circular holes to produce non-circular forms. The cutter is a broach which has a number of cutting edges along its length and which is usually drawn (but some-

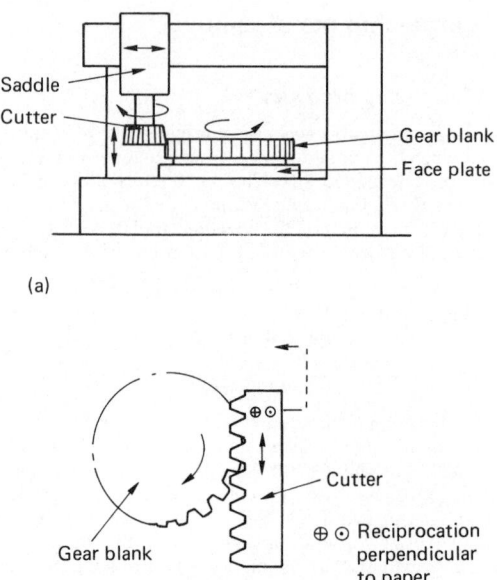

(a)

(b)

Figure 16.4 Kinematics of gear-generating machines: (a) gear shaper, (b) gear planer

times pushed) through the hole by means of hydraulic pressure. Each cutting edge is larger than its predecessor by about 0.05–0.08 mm, so the number of cutting teeth is determined by the form to be produced.

Push broaching is limited to broaches with a small number of teeth which have a low length/cross-section ratio and which would not buckle under compression. Surface broaching is of more recent introduction, and is used as an alternative to milling for the production of external surfaces. Surface broaches are rigidly clamped to a machine slide and traversed against the component being machined, producing a surface in a shorter time than is required for milling, and usually giving a superior finish. Whereas internal broaching is usually the only feasible method for producing the desired shapes, surface broaching is an alternative to milling and is usually justified only if the quantities required are sufficient to absorb the high equipment and tooling costs.

The milling group comprises a large range of manually operated or numerically controlled machines, many of which can perform operations such as drilling, reaming and boring as well as the accepted milling operations. Milling cutters generate surfaces either by means of cutting edges on the periphery or the face of the cutter. Peripheral milling is now seldom used for generating large plane surfaces, its main use being for machining slots or profiles. Although peripheral cutters can be fitted with carbide cutting edges the majority are of high-speed steel and, except when used for machining the more exotic hard materials, will probably continue to be so in the foreseeable future. Frequently they also have shallow teeth on the cutter face, although these teeth usually contribute little to the total metal removed. A range of typical peripheral cutters is shown in Figure 16.5.

Face milling cutters are essentially for generating plane surfaces. They are fastened to the end of stub arbors in the machine spindle and their configuration makes them suited to the use of specially designed carbide inserts (Figure 16.6). The cutting edges of both peripheral and face milling cutters are in contact with the uncut part of the component for, at most, half

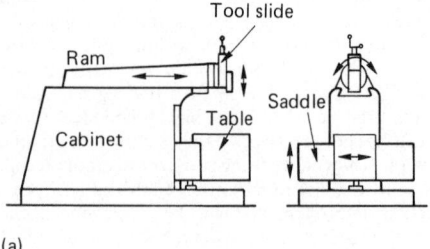

(a)

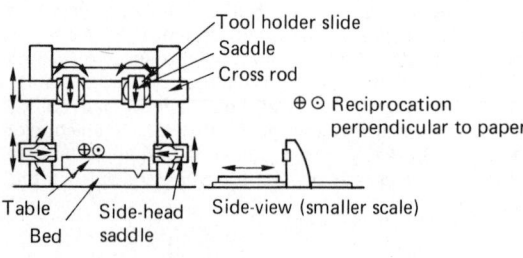

(b)

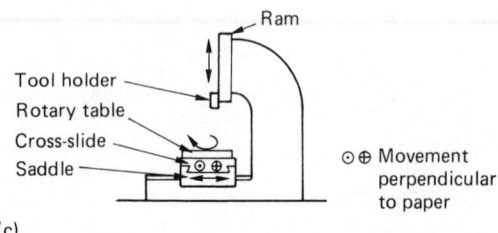

(c)

Figure 16.3 Outlines of reciprocating machines: (a) shaping, (b) planing, (c) slotting

Figure 16.5 Peripheral cutters: (A) high radial rake cutter, (B) helical cutter, (C) side and face cutter, (D) end milling cutter, (E) slot drill

Figure 16.6 Face milling cutters: (A) zero corner angle, (B) 15° corner angle

a revolution. Since the chip length is in the order of one third of the length of the uncut surface the chips have a maximum length approximating to the cutter radius, so chip breaking poses few problems.

Gear hobbing machines also belong to the milling group. Hobs are in the form of a screw with a straight side rack-form thread, gashed to give cutting edges and relieved to provide cutting clearance. Gears are generated by rotating the blank and the hob in synchronism and the hob is fed parallel to the axis of the arbor on which the blank is mounted. When a number of identical gears are required, several blanks can be fitted to the same arbor and machined at a single pass.

The machines in the drilling and boring group can be sub-divided into drill presses, radial drilling machines, jig-boring machines and horizontal boring machines. They have rotating spindles which hold drills, stepped cutters, taps, reamers or single-point boring tools, and the cut is applied by feeding either the spindle or the work table. Typical configurations within this group are shown in Figure 16.7.

Increased metal-removal rates, made possible by the development of new cutting materials, have forced machine-tool manufacturers to design new machines capable of large ranges of spindle speeds and feeds. This has necessitated more powerful motors and structures having high rigidity to resist the increased cutting forces and to reduce the likelihood of self-induced vibrations, giving rise to chatter.

With increased rates of metal removal the problem of swarf disposal has become more acute. The magnitude of this problem can be visualized when it is realized that swarf occupies about one hundred times the volume of the metal removed, so a 10 kW motor running at full power can generate about 1 m^3 of mild steel swarf or about 3 m^3 of aluminium swarf per hour.

Modern high-production metal-cutting machines commonly cost in excess of £100 000. If they are amortized over a period of 5 years the depreciation cost of such a machine when used continuously on a double-shift basis is more than £5 per hour. It follows that such plant requires high utilization, efficient programming to produce at optimal metal-removal rates, an effective system of tool management to reduce non-cutting time and intelligent application of terotechnology to minimize lost time due to maintenance.

Where large-scale production justifies continuous or large-batch manufacture the achievement of these objectives becomes a feasible possibility. Unfortunately, few products are marketed in such large quantities and a large-batch approach usually results in uneconomically high stocks. British manufacturing industry in the 1970s became notorious for its inflexibility and for the disproportionately high stocks which pursuit of large-batch policies entailed. Present policies are directed towards small batches, and this manufacturing philosophy has highlighted the need for rapid change-overs and for manufacturing systems accommodating large numbers of tools which can be called into use in response to the demands of small batches.

16.1.2   Cutting-tool geometry

Most of the research into chip formation has been based on orthogonal cutting, a simplified situation which is seldom met in practice. The tool approaches the work with its cutting edge parallel to the uncut work surface and at right angles to the direction of cutting. To prevent end effects the tool is wider than the work, as shown in Figure 16.8. Orthogonal cutting can be achieved only in a planing or shaping operation, although a close approximation can be obtained when turning on the end of a thin-walled tube.

In practice, the cutting tools usually approach the work obliquely and have rake angles in both directions on the rake face together with a nose radius at the end of the cutting edge. The direction in which the chip flows across the tool surface is determined by this complicated geometry. British Standard 1296: 1972 defines the angles on single-point cutting tools in terms of the normal rake system (Figure 16.9), based on two coordinate rake angles.[1] The back rake or cutting edge inclination λ_s is measured parallel to the cutting edge in the vertical plane and the normal rake γ_n is measured in a plane at right angles to the cutting edge and perpendicular to the rake face. ψ_r is the tool approach angle and κ_r' is the horizontal clearance angle, or the tool minor cutting edge angle. In addition, the tool is relieved to give vertical clearance angles of about 5°.

Other systems of tool nomeclature relate the rake angles to the coordinate axes of the tool shank, or to the cutting edge, measuring the angles in each case in the vertical plane. Although these systems are conceptually simpler, they are of little use in deducing the direction of chip flow. The British Standard relates to single-point tools but it can also be applied to multi-point tools and is generally preferable to the other systems.

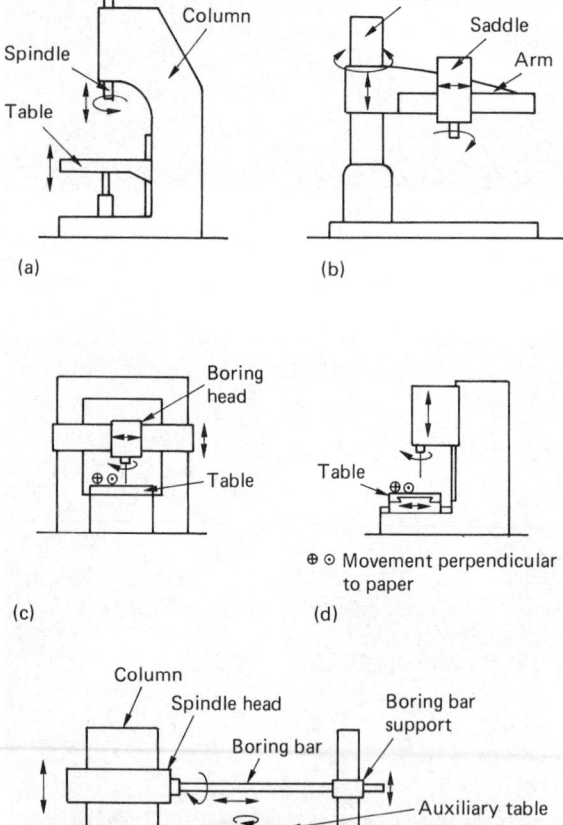

Figure 16.7   Machines in the drilling and boring group: (a) single-spindle drill press, (b) radial drilling machine, (c) gantry-type jig-boring machine, (e) horizontal boring machine

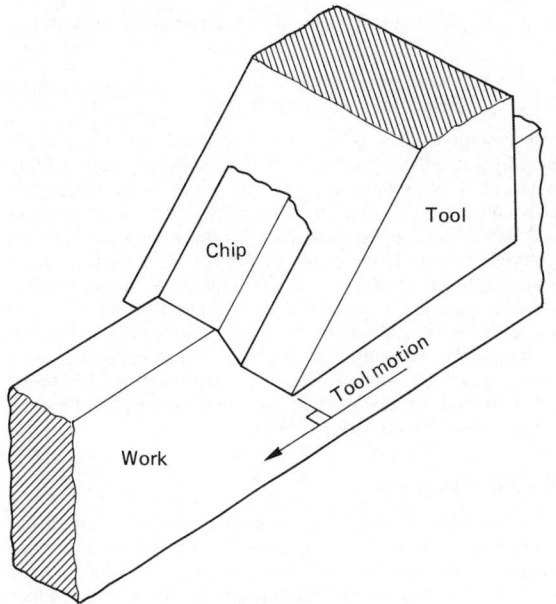

Figure 16.8 Orthogonal machining

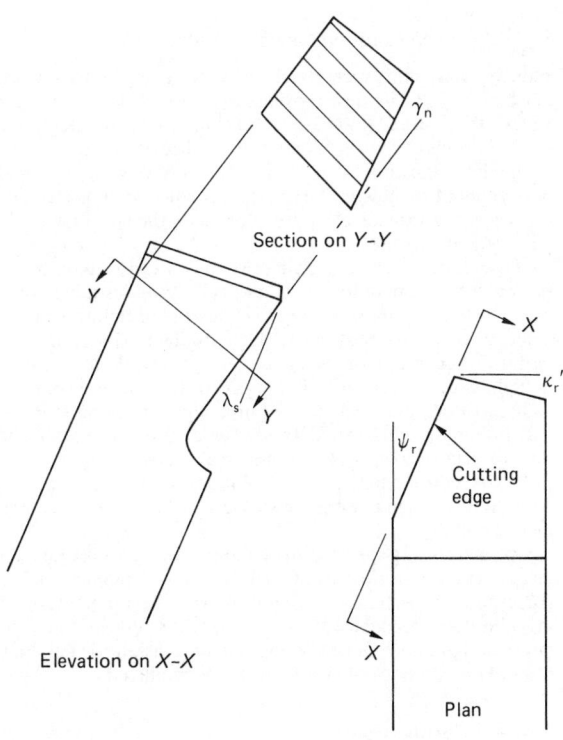

Figure 16.9 Angles in the Normal rake system

16.1.3 Cutting-tool materials

The variety of cutting-tool materials has increased rapidly in recent years due to the development of more difficult-to-machine materials and to the insistent demand for higher productivity. Any attempt to summarize these materials is unlikely to be completely successful due to the rate at which improvements and innovations are occurring. However, this is no excuse for ignoring the published state of the art at the time of writing.

When selecting a tool material for a particular application it is necessary to measure its rating against the following list of properties, some of which are mutually opposed. For instance, in most cases hardness and impact resistance of competing tools tend to be inversely related. The essential properties are:

1. High hardness at elevated temperatures
2. High compressive strength
3. Adequate impact resistance
4. Insusceptibility to violent local temperature variation
5. Chemical inertness at working temperatures
6. Low interface friction
7. Good abrasion resistance

Temperatures at the chip/tool interface can be about 1000°C when machining steel, and considerably higher when machining some of the more exotic materials, particularly if heat is applied to help soften the work material. Since cutting temperature is largely dependent on cutting speed, it follows that pursuit of higher productivity creates increasing demands on the high-temperature properties of the tool materials.

16.1.3.1 High-speed steel

High-speed steels (HSS) are likely to continue to be used in the foreseeable future for many applications such as drilling, reaming, tapping and dieing, forming, broaching and milling due to the ease with which they can be shaped in both the soft and hardened state. Typically, they consist of carbon steel alloyed with tungsten or molybdenum, together with percentages of chromium, vanadium and cobalt. The alloying elements raise the temperature at which tempering occurs, allowing HSS to be used at temperatures up to about 650°C. Their hardness is limited to 750 HV, adequate for machining most of the common metals, including alloy steels in their unhardened forms.

Cutting speeds are necessarily limited to prevent excessive rise in temperature. When machining mild steel, cutting speeds of about 1.5 m/s are possible if a plentiful supply of coolant is provided. A recent development is the coating of HSS drills with a deposit of about 3–5 μm of titanium nitride which allows rotational speeds to be increased, resulting in a 50% increase in penetration rate and longer tool life. With a few exceptions of this sort it is unlikely that HSS will ever again pose a threat to the supremacy of sintered carbides for heavy-metal removal.

Despite their relatively low hardness and susceptibility to softening at high temperatures, high-speed steel tools are tougher than most of the competing materials, enabling them to be used for interrupted cuts without fear of fracture. They can also be reground, giving a number of cutting lives before they must be finally discarded.

16.1.3.2 Cast non-ferrous alloys

These alloys, consisting of cobalt, chromium, tungsten and carbon, although less versatile than high-speed steels, enable cutting to be performed at higher temperatures. Their main use is for drilling, where their superior hardness at elevated temperatures is an advantage when the application of fluids is frequently a problem. Their cutting performance is generally superior to high-speed steel but inferior to sintered carbides, so they are unlikely to grow in popularity.

16.1.3.3 Sintered carbides

The introduction of sintered carbides for cutting has been the most important single contribution to increased productivity during the past 50 years. They are essentially cermets, which consist of hard carbide ceramic particles embedded in a metal matrix. Early carbide tools were usually of tungsten carbide and cobalt. Their brittleness encouraged the use of negative rakes to promote compressive stresses and restricted their use to continuous cutting. Subsequent improvements in the sintering process, and the introduction of alternative ceramic and metallic components, has enabled a range of carbide tools to be produced which withstand the rigours of thermal and mechanical shock, making them suitable for interrupted cutting in turning and milling.

The first generation of carbide tools consisted of sintered tips brazed to steel shanks. This provided a fairly rigid cutting system and the tools could be reground when blunt. More recently, the brazed tip has been almost completely superseded by disposable tips which are mechanically clamped into steel tool holders. These tips are polygonal, with three or more cutting edges which can be indexed when worn to expose new edges. The negative rake varieties can be inverted to double the number of edges. They are a throwaway concept, regrinding being uneconomic. Their introduction has forced a reappraisal of metal-cutting economics, making tool lives in the order of 15 min a desirable objective. Practical cutting speeds are about three times as great as could be achieved with high-speed steel, 5 m/s being a typical maximum when machining mild steel.

16.1.3.4 Ceramics

Sintered ceramic tools based on aluminium oxide (Al_2O_3) have been available for more than 30 years. Their brittleness and poor thermal shock resistance can be improved by additions of zirconium oxide and titanium carbide, but they are generally unsuited to interrupted cutting. They are suitable for machining hardened steels and chilled cast irons, and are similarly suitable for heat-assisted machining of hard materials such as the nimonic alloys and Stellite. However, their brittle behaviour has proved a severe limitation for general-purpose machining of steels, an area in which their high metal-removal potential would have been an advantage.

Mixed ceramics, based on carboxides with dispersed titanium carbide, have achieved better impact resistance without significant loss of hardness. They can be used at high cutting speeds in the order of 15–20 m/s when operated at low feeds, making them suitable for finish turning and finish milling hard materials.

A recent addition to the range of commercially available ceramic cutting tools uses silicon nitride (Si_3N_4) with differing levels of aluminium and oxygen substitution. The silicon nitride ceramics have good resistance to thermal and mechanical shock, enabling them to be used for discontinuous cutting. They can be operated at higher cutting speeds than carbides together with higher feeds than other types of ceramics. Their main applications to date are for rough turning and rough milling grey cast iron and for turning nickel-alloy steels. Due to their high thermal shock resistance they can be used to cut dry or with coolant, the latter method being inadvisable with other ceramics. Chemical action occurs at cutting temperatures, causing rapid wear when machining most steels, so at present there is little likelihood of silicon

nitride supplanting carbides in this important area of manufacture.

16.1.3.5 Cubic boron nitride (CBN)

Polycrystalline cubic boron nitride is another comparatively recently introduced addition to the range of metal-cutting materials. It has a hardness considerably in excess of ceramic tools and retains this hardness at temperatures well in excess of 1000°C. The main application is for machining hard ferrous materials at very high cutting speeds, giving a surface finish comparable to grinding. It has relatively good impact resistance, allowing it to be used for interrupted cutting. It is also of use for hot machining of refractory metals such as Stellite, where a 90% reduction of machining time compared with carbide has been claimed. Due to the high cost of CBN tools, the manufacturers do not claim great cost savings, but the time saving is very significant.

16.1.3.6 Diamonds

Diamond is the hardest material known to humans and it has found a limited cutting application where this is an important attribute. Natural diamonds, brazed to steel holders, have been used for many years for producing fine finishes on copper and aluminium. Being monocrystalline, natural diamonds have planes of weakness which render them unsuitable for anything but fine finishing cuts.

The production of polycrystalline synthetic diamonds (PCD) has extended the usefulness of diamonds by improving their impact reistance. PCD cutting tools are now extensively used to machine abrasive aluminium–silicon alloys, fused silica, and reinforced plastics. They are chemically reactive at high temperatures, so they are of little use for machining ferrous materials.

16.1.3.7 Limitations imposed by machine tools

Industry has, until recently, been very reluctant to replace machine tools while they continue to perform the function for which they were purchased. Machine-tool manufacturers market machines which adequately utilize the cutting tools available at the time of purchase. Inevitably, with the rapid development of new cutting materials, the existing machine tools cease to provide a service which uses the cutting tools in an economic manner.

Self-induced vibration, giving rise to chatter, is undesirable in any cutting operation. It is particularly undesirable when using brittle cutting tools where catastrophic failure can become a very real possibility. The problem of self-induced vibration becomes more acute as metal-removal rates increase at high cutting speeds. Ideally, resonant frequencies should be as high as possible, but this requires high structural stiffness and low mass. Unfortunately, dynamic stiffness tends to be directly related to mass, so a simple scale-related solution does little to reduce chatter. The solution, if it exists, lies in structural redesign to enhance stiffness without a proportional increase in mass.

Optimal cutting speeds using modern cutting tools require a large increase in the rotational speed of spindles and a corresponding increase in input power. Manufacturers of machine tools recognize this need, which is reflected in their latest designs, together with improved provision for handling the greater volumes of swarf which are produced.

16.1.4 Cutting fluids

Cutting fluids are used for three main purposes: as a lubricant at low cutting speeds, to cool the tool and work, and to assist in clearing the swarf. At cutting speeds in excess of about 0.7 m/s there is little noticeable lubricating effect. Below this speed extreme pressure (EP) mineral oils containing sulphur or chlorine additives can be used to reduce friction in the latter stages of chip/tool contact due to the formation of low-shear-strength sulphides or chlorides. The balance of the cutting forces is affected, giving rise to a larger shear angle and reduced contact length, and encouraging the thinner chip to curl, making for more effective chip breaking. High compressive stresses near the tool point prevent lubricant penetrating in this area, so the lubricating effect is limited to the latter part of the chip/tool interface where sliding friction occurs. The chemical reaction giving rise to low-strength compounds is both temperature and time dependent. Hence, at higher cutting speeds the lubricant rapidly loses its effectiveness.

High-speed steel cutting tools start to soften at temperatures above about 650°C. When using these materials the cooling effect of cutting fluids enables the tools to be operated at higher cutting speeds than would be possible when cutting dry. Water-soluble oils, having high specific heat and good metal-wetting properties are better coolants than the mineral oils used as lubricants. With carbides and ceramics the poor resistance to thermal shock makes the use of cutting fluids inadvisable except in special circumstances. Fortunately, these materials can be used satisfactorily at high temperatures, and coolants are therefore not usually required.

The purely mechanical function of using cutting fluid to assist swarf disposal is sometimes of prime importance. An example of this application is in deep-hole drilling, where cutting fluid is pumped to the cutting edges at high pressures of about 6 N/mm^2.

16.1.5 Forces and power in metal cutting

Most lathes and milling machines lack the power to exploit the cutting tools in an economic manner. This shortcoming is usually aggravated by a natural trepidation on the part of operators to run machine tools near their power limits for fear of stalling the drive motor. It is surprising that few machine tools are fitted with wattmeters, so operators usually have no idea how near they are to causing an overload.

Among the more sophisticated numerically controlled machines, very few are fitted with adaptive control devices which cause feed or cutting speed to respond to excessive power demands. The vast majority operate from a predetermined program which has been based on safety considerations where the power requirements are well within the rated output of the motors.

There is an ill-founded belief that the cutting forces, and hence the power required, increase significantly as the tools wear, increases of 40% sometimes being quoted. In fact, cutting power seldom increases by more than 10% over the life of the tools.

At rated power, transmission losses usually account for about 30% of the input power, with a correspondingly greater percentage loss when operating at lower energy levels. Transmission losses are higher when the machine tool is cold, and drop significantly over the first half-hour of operation. It is desirable, therefore, to record the transmission power over the full range of cutting speeds and feeds on a machine tool in both the cold and warmed-up conditions. Only then is it possible to know the available power which can be used for cutting.

16.1.5.1 Forces in turning processes

Although a knowledge of cutting forces is desirable to prevent excessive structural loads, the main reason for wishing to know is as a basis for estimating power. The power in watts is simply the product of the peripheral speed in metres per second and the tangential cutting force measured in newtons.

Due to the formation of a built-up edge when cutting steel at low speeds, forces on the tool vary in an unpredictable manner, but above about 2.5 m/s become relatively constant when built-up edge ceases to have a significant effect. At cutting speeds below about 0.7 m/s the lubricating effect of cutting fluids can do much to inhibit build-up, and on the rare occasions when such low speeds are used the cutting force can often be reduced by this means. When using sintered tools it is usually possible to operate at cutting speeds high enough for the forces to be considered constant.

16.1.5.2 Forces and power in milling operations

Peripheral milling removes metal by means of teeth on the circumference of the cutter. It is seldom used to produce large flat surfaces, which are more effectively generated by face milling. Mostly, peripheral milling cutters are used for end milling slots or for producing slots or stepped surfaces by using one or more horizontally mounted side and face cutters or helical slab milling cutters.

End milling seldom requires the rated power of the drive motor. The limiting factor is usually the maximum recommended feed per cutting tooth which will prevent damage to the cutter. Horizontal peripheral milling, however, can be limited by the power of the drive motor, and it is useful to consider the way in which the cutting parameters affect the power required.

Peripheral cutting can be performed in either the upcut or climb mode (Figure 16.10). In upcut milling the cutting edge must penetrate the previously cut surface before chip generation commences. This causes a high radial force at the commencement of the cut which does not happen with climb milling. Cutters with large radial rakes have a weak tooth form which results in rapid wear when subjected to the high radial forces associated with upcut milling. Radial forces experi-

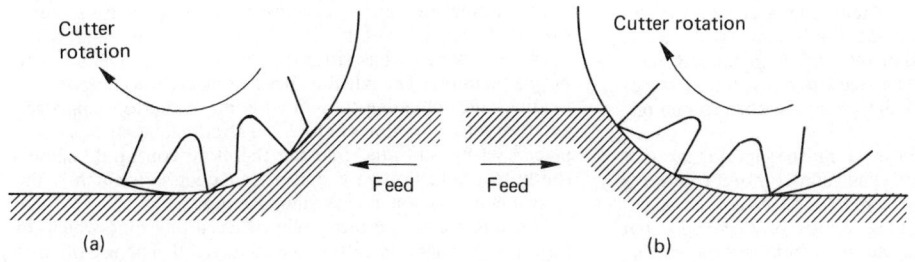

Figure 16.10 Climb and upcut milling: (a) climb, (b) upcut

enced in climb milling are much lower and wear is usually not a severe problem.

Summarizing, the most economic performance in terms of specific power can be achieved when climb milling with a high radial rake cutter operated at large feeds. It is preferable to operate at high feeds rather than high cutting speeds since the index of feed in the cutting power equation is less than unity whereas power is directly proportional to cutting speed.

16.1.5.3 Hot machining

Some work materials pose machining problems which cannot readily be solved by conventional methods. These include the nimonic alloys and Stellite cast alloys. To give some idea of the problems which are encountered, some of the nimonic alloys when machined with carbide tools on a 100 mm diameter bar necessitate cutting speeds as low as 0.15 m/s and tool failure commonly occurs after machining a 100 mm length of the bar. When the surface of the bar is preheated with a gas tungsten arc or a transferred plasma arc struck between the electrode and the work surface and using ceramic or CBN tools, cutting speeds of about 2.5 m/s are possible, and the tools remain serviceable after machining a considerable length of bar.

This technique is not one which would be advised if alternatives are possible, but with some of the more refractory metals now in use the hot-machining process is frequently the only practical solution. The surface preheat temperature is about 600°C, giving such high interface temperatures that carbide tools cannot be used.

16.1.6 Surface-finish considerations

Built-up edge is one of the main factors contributing to poor surface finish. When machining most materials this can be reduced (if not eliminated) by operating at high cutting speeds. Where finish cuts are required, the uncut chip area is relatively small so the cutting power is never likely to be an important consideration, even at very high cutting speeds. If ceramic tools are used the speed limitation is usually that imposed by the available spindle speeds, but when using carbides or high-speed steels the speed constraint is usually that imposed by tool wear.

The theoretical surface roughness in turning is determined by tool plan geometry, a pointed tool operated at a given feed producing a rougher surface than one having a nose radius.

The surface generated by a peripheral milling cutter is directional in property. In the direction of feed the theoretical surface is geometrically similar to that for a turned surface, the cusps having a radius equal to that of the cutter, and the pitch between cusps being equal to the feed per cutting tooth. Due to the almost inevitable lack of straightness of the arbor on a horizontal milling machine, the contour generated by the cutter teeth varies as some teeth take a greater depth of cut than others. In severe cases one tooth may take such a disproportionately deep cut that the surface generated has a periodicity corresponding to the feed per revolution rather than the feed per tooth, and the cusps are correspondingly deeper.

Face milling usually produces a finish superior to that generated by peripheral cutters. The geometry of the cutter is specially designed so that the combination of corner angle, end cutting-edge angle and nose radius produce very flat cusps. In addition, due to the use of carbide cutters, cutting speeds are much higher than those achievable using HSS peripheral cutters.

16.1.7 Tool-life assessment

It is fortunate that, with few exceptions, tool wear occurs in a predictable manner. Although it takes different forms, each is associated with a known cause which happens within particular ranges of feed and cutting speed. The most usual causes are associated with crumbling of the cutting edge, cratering of tungsten carbide, plastic deformation of the tool, thermal and mechanical shock, or attritive wear on the clearance face.

16.1.8 Economics of metal cutting

The main financial objective of a manufacturing company is maximization of return on capital. This implies a knowledge of profitability which, in turn, requires a knowledge of cost and selling price. Production engineers are concerned with the processes needed to make components which are eventually assembled into finished products. Selling price, therefore, is not usually a very useful statistic for the process planner. He or she must settle for sub-objectives such as minimum cost or, sometimes, maximum output which, although not synonymous with profitability, at least contribute to its achievement. When discussing manufacturing economics these are the objectives to which we must address ourselves.

Metal cutting is an intrinsically wasteful operation, involving the removal of large quantities of material. Although there are no reliable figures to support this contention, it is probable that only about 70% of the material purchased is contained in finished parts, the balance being expensively converted to swarf which has a very low resale value. Intelligent design can do much to increase material utilization but material wastage will always be a significant proportion of the total component cost. In spite of the attractiveness of contending production options such as metal forming, it is inevitable that cutting processes will continue to be extensively used. The subsequent analysis assumes that due cognizance has been taken at the design stage of the importance of material utilization, and the cost factors include only the direct cost of manufacture and its associated overhead. The operating cost, taking account of direct labour, machine depreciation and factory overhead, may well be in the order of £20 per hour.

16.1.8.1 Minimum-cost production

The cost of manufacture per component, K, can be divided into five parts:

1. Set-up and idle time cost per component, K_1
2. Machining cost per component, K_2
3. Tool-changing cost per component, K_3
4. Tool-depreciation cost per component, K_4
5. Tool regrinding cost per component, K_5

where $K = K_1 + K_2 + K_3 + K_4 + K_5$. Assuming that disposable inserts are used, there is no regrinding cost, so K_5 can be ignored.

The setting cost can be substantially reduced by using preset tooling. However, with the current trend towards small batch sizes, the setting cost ascribed to each component will increase proportionally. The idle time per cutting cycle is composed of loading and unloading time in addition to the tool approach and tool retraction times before and after machining has taken place. Set-up and idle time can therefore contribute significantly to production cost, and its reduction is frequently the largest single factor in cost minimization.

Machining cost is directly related to cutting time which, in turn, is dependent on cutting speed and feed. The use of large feeds and high cutting speeds reduces machining cost but decreases tool life and, consequently, increases both the unit tool depreciation and unit tool changing costs. It is therefore the minimization of the total of these three costs which determines minimum production cost for any given set-up.

Figure 16.11 illustrates by way of a carpet plot how K typically varies with both feed, f, and cutting speed, v. The main point worth noting is that within the broad range of feeds and cutting speeds selected the minimum cost occurs inside the speed range and at the maximum value of feed.

16.1.8.2 The influence of chatter

A great deal of research has been published on the stability boundary between width of cut and spindle speed.[8] The boundary envelopes vary with the machine tool and the

process parameters, but a typical boundary is shown in Figure 16.12, where the maximum width of cut under stable machining conditions decreases as the spindle speed increases. A large tool approach angle and nose radius increase the effective length of the cutting edge, which is tantamount to increasing the width of cut and decreasing the uncut chip thickness. If chatter is likely to occur it is obvious that tool approach angle and nose radius should be kept as small as possible, a recommendation which conflicts with recommendations as regards tool life.

The influence of feed on the stability boundary is more significant, but this does not appear to have received the attention it deserves. *Figure 16.13* shows how the critical width of cut increases as feed increases, making the case for using large feeds to oppose the onset of chatter.[9]

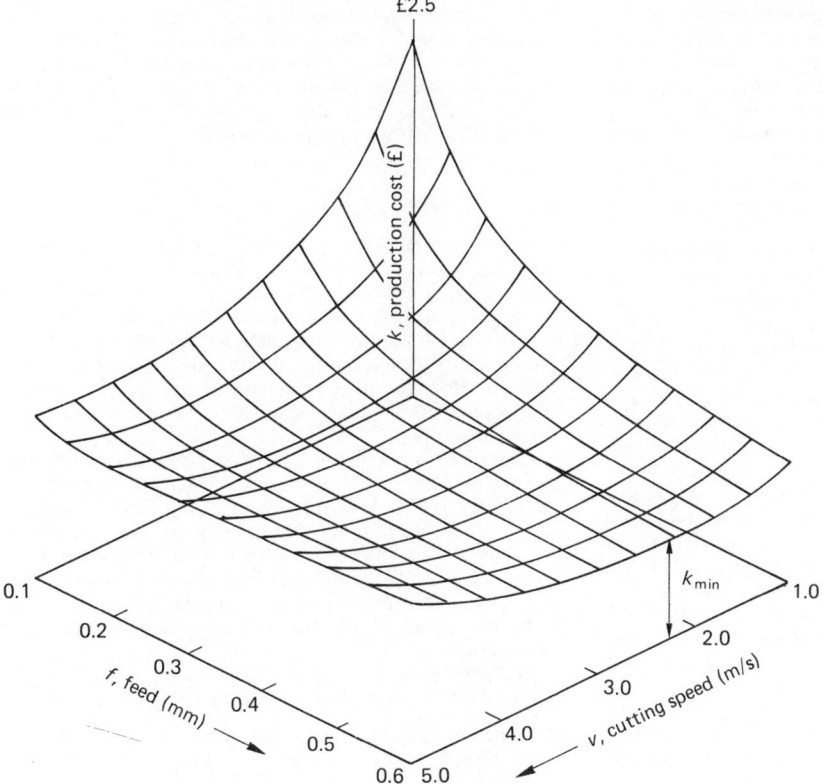

Figure 16.11 Production cost related to cutting speed and feed

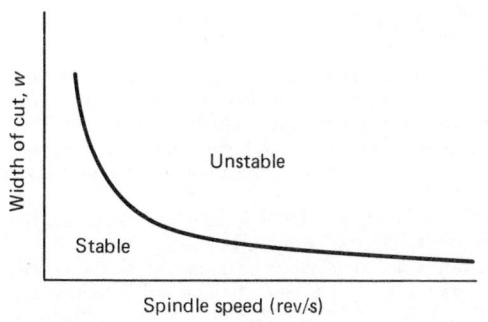

Figure 16.12 Variation of stability threshold with spindle speed

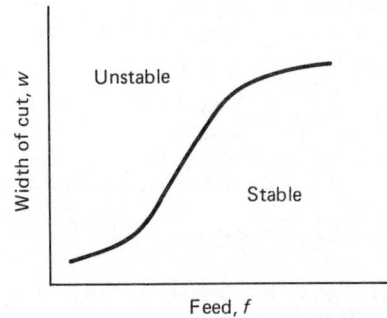

Figure 16.13 Variation of stability threshold with feed

In conclusion, it is pertinent to stress again the importance of large feeds in achieving not only chatter-free performance but also low specific cutting power and minimum-cost machining.

16.2   Metal forming

16.2.1   Introduction

Metal forming, i.e. changing the shape of the material without actually removing any part of it, was practised at least 3000 years ago in Egypt, where hammer forging to produce gold sheet, cut subsequently to make wire, is recorded in the Bible to have taken place. Rolling in wooden mills was employed to manufacture papyrus. Manual swaging and wire drawing were well established in the Middle Ages but, naturally, were limited in scope by the power available. It was only with the advent of the Industrial Revolution that progress was made and processes like extrusion and cross and longitudinal rolling became available. But even here, the restrictions imposed by the low quality of tool materials, lubrication problems and the lack of understanding of the basic precepts of plasticity impeded progress until, in some cases, well into the twentieth century.

The ever-increasing demand for high quality products—often of sophisticated shape in difficult to process materials—economically produced, fabricated or semi-fabricated, combined with the rising cost of metallic engineering alloys has focused attention on metal-forming processes and techniques.

The emphasis here lies on the 'chipless' approach to shaping. This provides an economical, direct means of converting a cast ingot to slab, plate, billet or bloom and then—in another chipless operation—of changing these basic shapes into profiled finished or semi-finished products. The avoidance of the removal of the material during a forming operation enhances the economics of the process by reducing wastage associated with the swarf-producing machining. Whereas the latter has, of course, a very considerable and necessary role to play in the range of manufacturing activities, its indiscriminate use (a feature of the early years of plentiful supply of cheap labour and materials) is no longer acceptable when high tonnage of accurately manufactured product can be obtained at a much lower cost.

In the most simplistic terms, the desired change in shape is effected either in the cold, warm or hot state (the latter below the melting point of the material) by the application of external forces, pressures or torques of sufficient magnitude to induce plastic flow, and thus a permanent set, of the material through the forming pass. Depending on the operation, the material is forced to flow between driven rolls, through (or into) open or closed dies, or between sets of dies and rolls. Solid or hollow sections are thus produced from the initially solid blocks of metal.

The *standard* basic operations are:

1.   rolling (flat, oblique or longitudinal),
2.   extrusion (axisymmetrical or asymmetrical),
3.   drawing (solid or hollow components),
4.   sheet forming (deep drawing, bending, pressing or bulging),
5.   forging (solid and hollow sections), and
6.   cropping (shearing and piercing).

Within the compass of any of these operations, a number of variants exists which reflects not only a variety of manufacturing routes and subroutes, but also the nature, properties and characteristic responses of the processed materials. Modern

metal-forming technology makes use of solid and semi-solid ('mashy' state), and superplastic, as well as explosively pre-welded metallic composites and dynamically compacted particulate matter. Mixtures of metallic and/or ceramic and polymeric materials are formed to manufacture composites of very specific properties. The problem of forming these into desirable shapes presents the engineer with new and often difficult situations to solve. Selection of the appropriate forming process, the tool design, the effects of the pass geometry on the final physical and mechanical properties of the product, the dimensional accuracy, and the achievement of the as near as possible final shape in the minimum of operational stages have to be faced.

The apparently simple sequence of ingot–slab–semifabricate–finished product becomes complex unless there is good understanding of the basic characteristics of the individual processes and an appreciation of the principles of the theory of plasticity, as well as that of the concepts of tool and process design. The bases for and fundamentals of the major processes and technological developments are discussed in the following sections, but detailed treatment of the individual topics is only indicated by reference to the appropriate literature.

16.2.2   Classification of processes

For a given application, the selection of the correct process necessitates the introduction of a criterion of process classification. Since hot working homogenizes and refines the crystallographic structure of the material and thus, ultimately, improves its strength and toughness, whereas cold working increases strength, hardness, dimensional tolerances and improves surface finish, these temperature-induced effects are often used to differentiate between the various manufacturing methods.

Important as the processing temperature is, in some circumstances other criteria of classifying metal-forming processes may well be more appropriate. From a purely manufacturing point of view, quantity and shape may have to be considered, while the likely response of the processed material to the level and/or rate of stressing, as well as the manner of application of the forming load system, may offer a better clue to the desirability or otherwise of using a particular technique or operation.

The parameters that characterize forming operations give rise to the following possible classification systems:

1.   operational temperature (hot, warm or cold forming),
2.   shape effect (bulk or sheet forming),
3.   operational stress system,
4.   operational strain rate,
5.   starting material (ingot, slab, billet, bloom, slurry, or powder).

16.2.2.1   *Operational-temperature criterion*

The idea behind the subdivision into hot, warm and cold processing of materials is not only to indicate the nature of the operation, but also to draw attention to the plant and ancillary equipment needed, to the level of force parameters required, and to the likely metallurgical response of the processed material.

An outline of this classification scheme, including only the basic operations, is given in Table 16.1.

Starting with a cast ingot, the primary hot operations of flat, billet and slab rolling, and slab forging will produce the starting stage for the secondary, further processing of the slab into plate, billet or a large forging. These, in turn, will form the first step in the manufacturing route of a more sophisti-

Table 16.1 Classification of dynamic regimes

Characteristic time (s)	10^6	10^6	10^2	10^0	10^{-2}	10^{-4}	10^{-6}	10^{-8}
Strain rate (s^{-1})	10^{-8}	10^{-6}	10^{-4}	10^{-2}	10^0	10^2	10^4	10^6
	Creep rates		Quasi-static rates		Intermediate strain rates		High strain rates	Very high strain rates
Primary load environment	High or moderate temperatures		Slow deformation rates		Rapid loading or low velocity impact		High velocity impact or loading	Very high velocity or hypervelocity impact
Usual method of loading	Constant load or constant stress machines		Conventional hydraulic or mechanical machines		Fast-acting hydraulic or pneumatic machines, cam plastometers, low impact devices		High velocity impact devices, expanding-ring technique, high-speed metal cutting	Light gas gun or explosively accelerated plate or projectile impact
Dynamic considerations in testing	Strain versus time		Constant-strain-rate test		Machine stiffness wave effects in specimen and testing machine		Elastic–plastic wave propagation	Shock wave propagation, fluid-like behaviour

◄——— Isothermal ———► ◄——— Adiabatic ———►

◄——— Inertia forces neglected ———► ◄——— Inertia forces important ———►

cated, profiled product. Hot operations are carried out at elevated temperatures exceeding annealing and normalizing ranges and, consequently, yield a hot-finished product showing a relatively low level of flow stress. However, the force parameters required match the mechanical properties of the material and are also relatively low. It follows that the rate of wear of the tooling can be kept at an economical level especially if the lubrication problems are well under control.

To improve the mechanical properties of the product, while at the same time keeping the loading at a moderate level, warm processing is used. Here, the temperatures are well above ambient but, equally, well below the hot-processing range, and usually slightly less than for recrystallization. The increased material ductility is sufficient to reduce the power requirement of the plant. Cold-working conditions are confined to ambient temperature and are characterized by a high energy requirement—necessitated by large operational forces and/or torques—but result in very high quality final product displaying both good dimensional tolerances and mechanical properties.

A rough guide to the temperature ranges can be obtained by considering the operational temperature/melting point ratio. On this scale, hot working takes place when the ratio is >0.6, warm working when the ratio is 0.3–0.5 (the latter corresponds to recrystallization conditions), and cold when the ratio is <0.3.

16.2.2.2 Shape-effect criterion

The effect of shape reflects the geometry of both the initial and final component and, consequently, the nature of the change imposed on it by the forming operation.

A process in which a component of a relatively small initial surface area/thickness ratio is deformed in such a way that the ratio is increased, is often classed as a 'bulk deformation operation'. On the other hand, the component of an initially high surface area/thickness ratio, shaped in a process which does not impose any change in the thickness but effects shape changes only, is said to be 'sheet formed'. Any change in the

thickness of such a component can easily lead to tensile plastic instability and incipient, localized yielding.

Bulk processes are those of rolling, extrusion, forging and solid- and/or hollow-section drawing. Bending, pressing, deep drawing, spinning and shearing are the main sheet-forming operations.

16.2.2.3 Operational-stress system

Because of the inherent severity of many forming processes, particularly the rotary ones, a consideration of the type and property of the induced stress field is of primary importance. The success of the operation may well depend on its compatibility with the properties of the processed material.

The presence of tensile and compressive stress fields results in the appearance of shearing stresses which, in turn, lead to the sliding of molecular planes and, eventually, to the yielding and plastic flow of the metal. Stress systems containing these components are most likely to give rise to plastic flow which, if it is controlled, will produce the desired amount of deformation.

Purely compressive or tensile systems create conditions of hydrostatic pressure in a triaxial field (absence of shear), or produce shearing stresses in uni- and bi-axial conditions. Clearly, since it is the configuration of the individual stress system that is indicative of the type of deformation which can be expected, its assessment prior to choosing a forming system is imperative. These various possibilities are illustrated, diagrammatically in Figure 16.14.

As an indication of the incidence of any of the stress systems, the following, non-exhaustive, list can be considered:

Tensile–compressive systems
Biaxial tension/uniaxial compression:

1. under a roll of a two-roll piercer,
2. under a roll of a two- and three-roll piercer, and
3. under a roll in the helical rolling process.

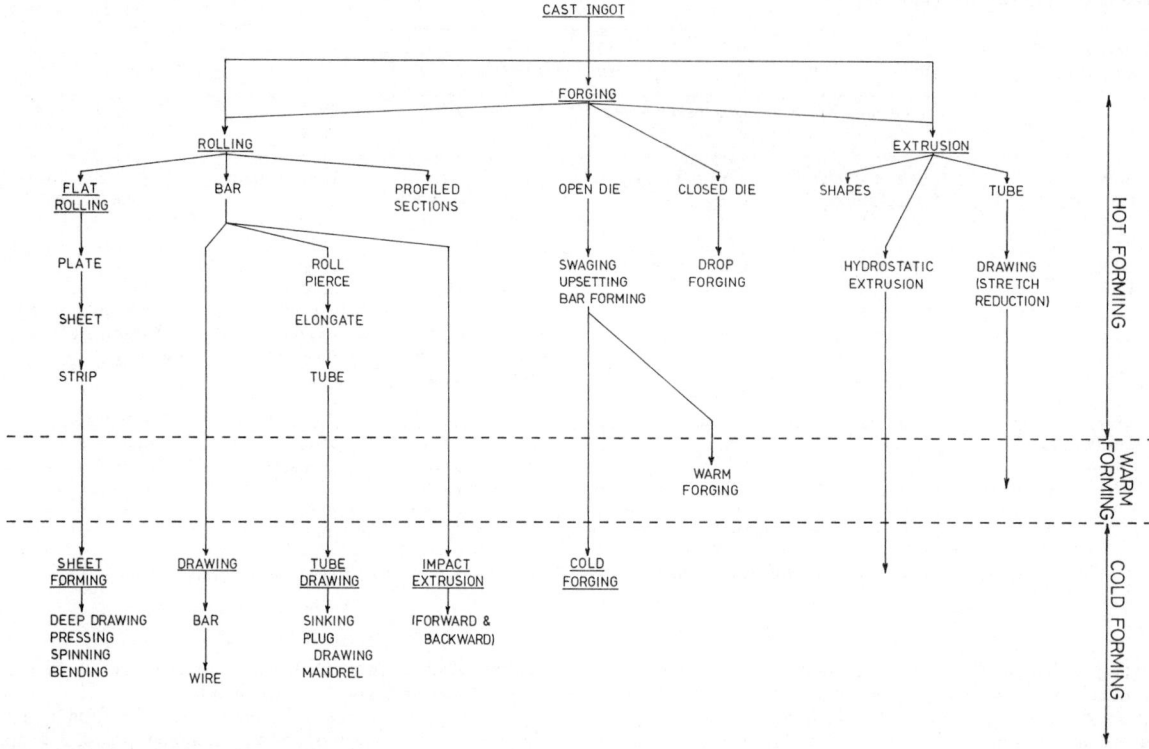

Figure 16.14 Process classification system based on operational temperature

Uniaxial tension/uniaxial compression:

1. between the rolls in roll forming, and
2. in the flange in deep drawing.

Uniaxial tension/biaxial compression:

1. in the drawing die.

Compressive stress systems
Triaxial stresses:

1. In the oblique zone of a three-roll rotary plug piercing mill,
2. in the closed forging die,
3. near the die throat in extrusion of bar, and
4. under the punch in tube extrusion.

Biaxial stress:

1. between the rolls of a longitudinal rolling mill with no front and/or back tension, and
2. in the upsetting, open dies.

Tensile stress system
Biaxial stress:

1. stretch forming, and
2. bulging.

16.2.2.4 Operational strain rate

A number of engineering alloys and even some practically 'pure' materials, e.g. commercially pure aluminium, are susceptible to the changes in the rate of straining. Modern technological techniques have either 'speeded up' conven-

tional processes—for instance, wire can be drawn at some 120 m min^{-1} —or have introduced new ones that operate in truly dynamic conditions. Impact extrusion, explosive forming, welding and compaction, and mechanically and electrically induced discharges of energy producing high strain rates, have all combined to introduce an entirely new field of high-energy rate fabrication, known commonly as HERF.

The range of possibilities arising in this context are listed in Table 16.2 which provides a detailed insight into the effect of different strain rates and the means of producing them in an industrial environment.

Table 16.2 Mass-velocity relationship contributing to total kinetic energy

Forming system	Mass contribution (kg)	Velocity contribution (m s^{-1})	Velocity/mass
Drop hammer	144	122	0.84
Trapper rubber	144	122	0.84
Floating piston	4.54	3050	672
Explosive in gas	0.004	2.75×10^6	687×10^6
Explosive in liquid	0.008	1.52×10^6	190×10^6
Electric discharge	0.008	1.52×10^6	190×10^6

16.2.2.5 Starting material

Since some modern processes do not require bulk solids as starting materials, but utilize particular matter and semi-solid substances, a classification based on the initial physical state of the material offers an interesting alternative to the more conventional approach.

Typical examples of unconventional starting materials are: 'mashy' state processing, leading to conventional rolling of composite sandwich components; the Conform-type extrusion, starting with a powder or granulated material, or an explosive compaction of powders.

16.2.3 Characteristics of the basic groups of processes

Of the major processes listed in Section 16.2.1, forging is the most diverse and cannot therefore be described in more general terms. For this reason, the basic characteristics of only four groups of processes are indicated here and those of forging, sheet forming, cropping, etc., are discussed later.

All rolling processes rely on the forces transmitted through the rolls to the material to effect deformation and on the rigidity of the roll system for the dimensional accuracy of the product.

Sheet and plate are initially obtained from a slab by rolling the slab in a relatively simple system (Figure 16.15). Driven rolls introduce the material into the roll gap, or working zone of the pass, and reduce the thickness. The success of any further processing to obtain strip rather than a sheet or large area of plate, depends on the ability of the system to maintain a constant width of the processed metal and on reduction of the thickness (this being equivalent to the reduction in the cross-sectional area). These requirements call for a plane strain operation which is possible only if the lubrication of the pass is very efficient. Processing in this mode can proceed in either cold or hot conditions.

A much more complex rolling system is that of longitudinal rolling, which is employed in the production of axisymmetrical billets, bars and hollows (Figure 16.16). A train of suitably shaped rolls, mounted on stands (either in pairs or in three-roll configurations) inclined at right angles (between the successive stands) is used, as shown diagrammatically in Figure 16.16(a). A gradual reduction in the cross-sectional area of the material takes place (Figure 16.16(b)) as the specimen moves axially forward through the sets of driven rolls. While fully engaged in the train, the processed material experiences, additionally, axial tensions resulting from a differential distribution of successive stand velocities. The ovality of the early passes is slowly reduced along the train until the last stand is reached. Here, the final, circular cross-section is expected to be achieved.

An alternative to longitudinal rolling is offered by the oblique-rolling system in which a single set of two or three driven rolls produces tractive, frictional forces which propel the specimen axially while, at the same time, causing it to rotate. The motion of an element of the worked material is thus forward, but helical.

Figure 16.17 illustrates, using an example of tube rolling, the basic principle involved. In this case, three profiled, driven rolls, disposed at 120° to each other, and inclined at an angle α (the feed angle) to the horizontal mill axis, and an angle β (the cone angle) in the vertical plane, introduce the bloom (supported internally in the bore by a mandrel) into the forming pass. The bloom is 'sunk' onto the mandrel in the zone AB and has its wall thickness reduced on the roll 'hump' BC. Slight elastic recovery takes place along DE. The bloom is thus elongated and its wall is thinned. The amount of deformation imposed depends on the size of the inter-roll opening or the 'gorge'.

These basic characteristics of oblique rolling operations (the variants of which are discussed later) are common to all operations, as indicated, for instance, in Figure 16.18. This shows, diagrammatically, the operation of the so-called 'secondary piercing', or 'oblique plug rolling' of a tube—a process in which a long cylindrical mandrel is replaced by a short profiled plug.

On the other hand, processes of profiling by rolling can take various forms, one of which is indicated in Figure 16.19 where a stepped shaft, required to acquire a series of specific profiles, can be manufactured by oblique rolling in a single three-roll stand. An operation in which the billet is rotated and fed through a system of driven rolls produces this effect. In another variant of oblique rolling, a two-roll system of helically ribbed rolls (Figure 16.20) will produce metal balls out of a solid cylindrical billet.

These few examples illustrate the versatility of rolling operations, a more detailed discussion of which is given in Section 16.2.4.

When the initial shape of the work piece has been imposed on it by one of the processes described above, there often arises the problem of how to achieve a degree of further deformation leading, possibly, to the final product. Drawing processes answer this need by providing a means of producing either solid (bar, rod or wire) or hollow tubular sections, either circular or non-circular in shape. The drawing operation is carried out in a die—or a set of consecutive dies forming a tandem drawing system—into which the work piece, with a swaged leading end, is introduced (Figure 16.21). An axial force is applied through a gripping device (as indicated by the arrow in Figure 16.21) and the work piece is pulled through the die. In the case of a solid specimen, the outer dimension only is reduced, whereas with a hollow section there is also a change in the wall thickness. Lubrication of the working zone of the pass (the part of the die surface along which the deformation is effected) is of importance from the point of view of the magnitude of both the drawing load and the induced drawing stress, and in view of the surface finish.

Similar results can be obtained in extrusion, a process in which the starting billet (sometimes referred to as the 'slug') is inserted into a cylindrical container and is then pushed mechanically through a suitably profiled die (Figure 16.22). There is a number of variants of this process (see Section 16.2.6), but the two basic operations are those of forward (or direct) and inverse (or backward) extrusion. In the forward extrusion a solid moving ram is brought into direct contact with the billet and activates the latter by moving it axially forward through the die. In inverse extrusion a hollow ram is in contact with a movable die which bears onto the billet, firmly held in the container. When the pressure exerted by the tooling is sufficiently high to exceed the yield stress of the material, plastic flow is initiated and backward extrusion into and through the hollow ram takes place.

Considerable control over the dimensional accuracy can be exercised in such systems but, again, solution of the lubrica-

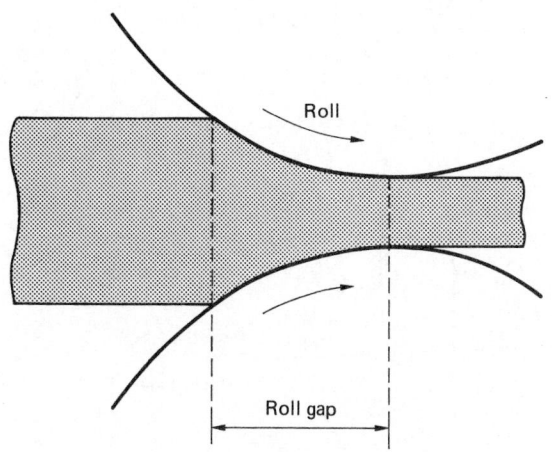

Figure 16.15 The principle of sheet and plate rolling

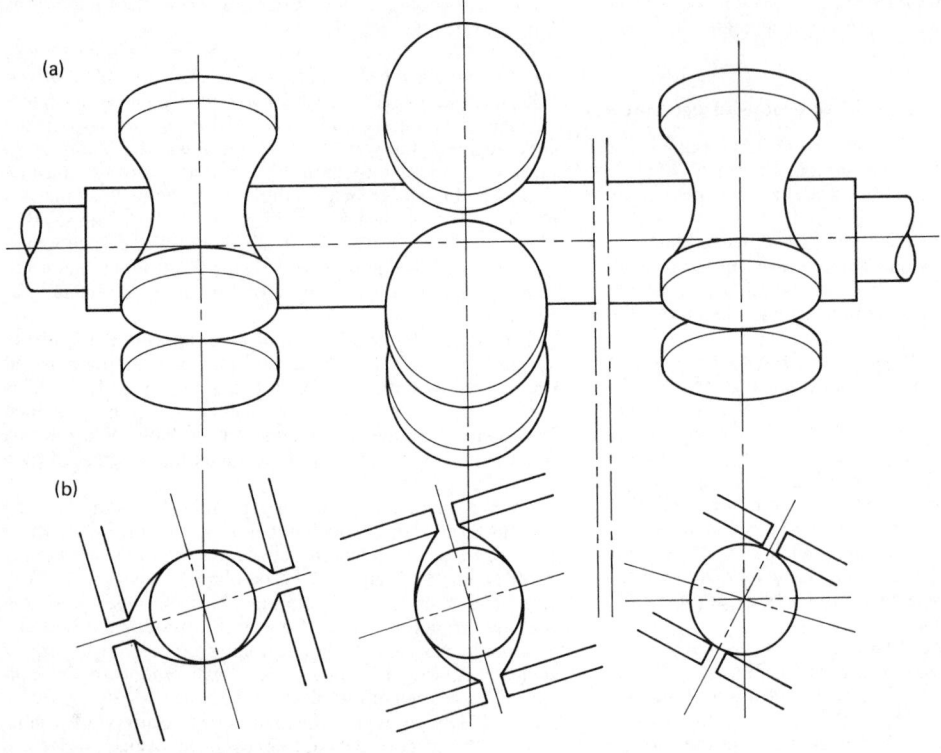

Figure 16.16 Longitudinal rolling: (a) a roll train; (b) successive roll passes

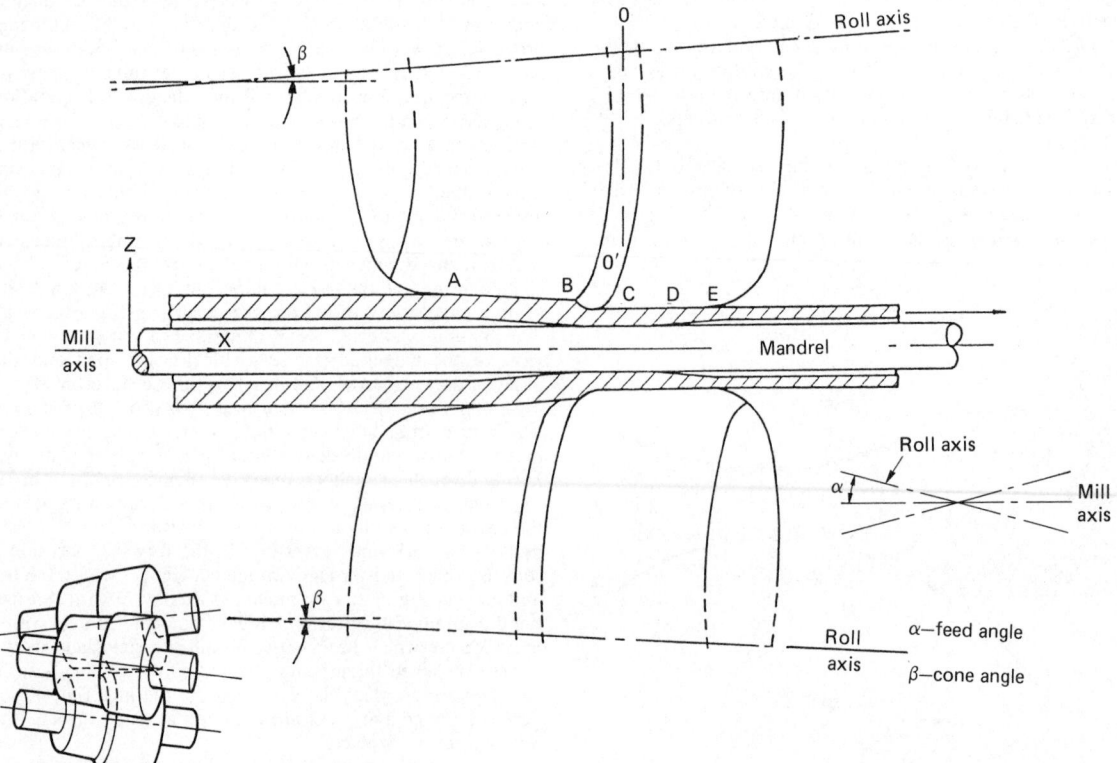

Figure 16.17 Oblique, three-roll tube rolling on a mandrel

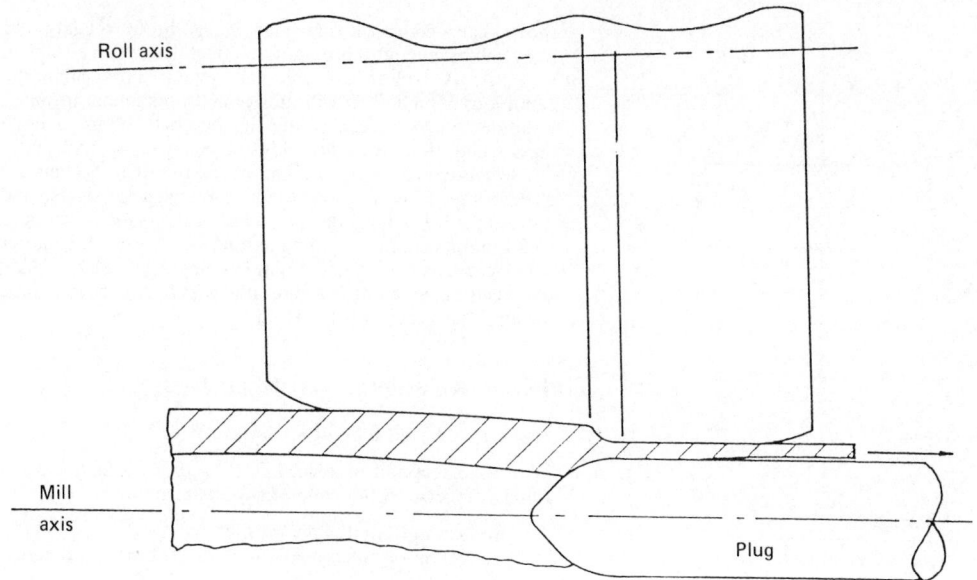

Figure 16.18 Oblique, tube rolling on a plug (secondary piercing)

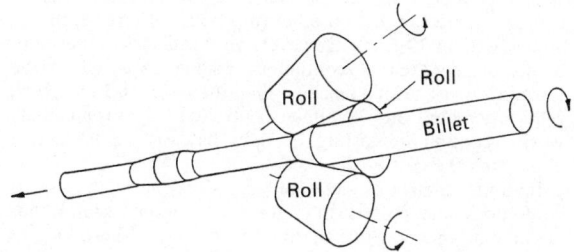

Figure 16.19 Three-roll shaft shaping

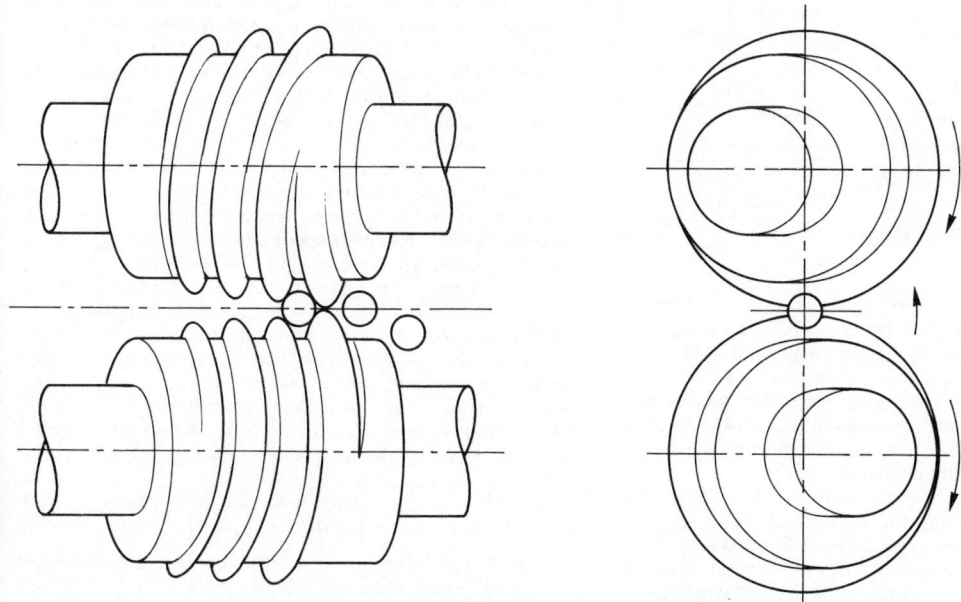

Figure 16.20 Two-roll ball manufacture

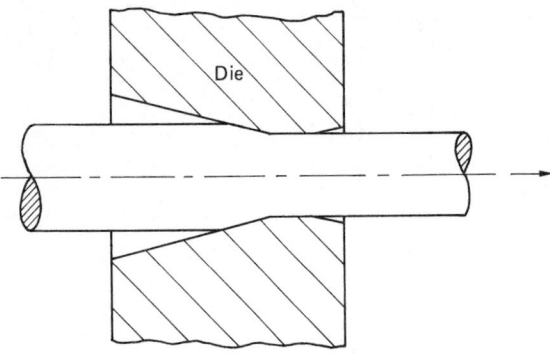

Figure 16.21   Open die drawing

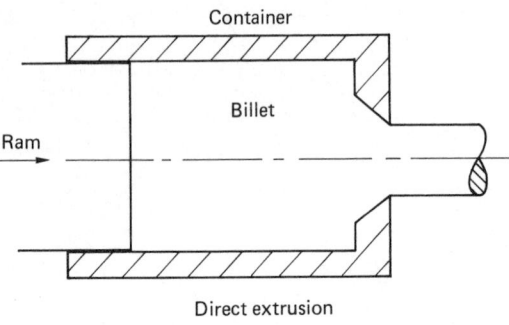

Direct extrusion

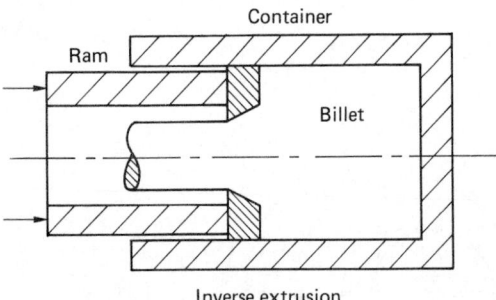

Inverse extrusion

Figure 16.22   Direct and inverse extrusion

tion problem is of importance. In this latter context, hydrostatic extrusion (to be described later) provides an important alternative to the conventional arrangements indicated here.

A large group of 'unorthodox', dynamic processes introduces a number of new elements and opens new operational possibilities of using materials which are sometimes difficult to process and of reducing manufacturing costs by dispensing with heavy plant and equipment.

The high-energy-rate processes stem essentially from the usually overlooked fact that the working of metal requires energy and not merely the application of force, and that, in addition, the rate of dissipation of energy is of importance. A simple consideration of the basic equation for kinetic energy

shows that a comparatively small change in the velocity of a body will have a more pronounced effect than will a change in its mass. A typical conventional system approaching the conditions of high-energy forming, i.e. drop-hammer forming, is limited in its usefulness by the necessity of using large masses and, therefore, unwieldy and costly equipment.

The sources of energy used in the high-velocity systems are chemical explosives, electrostatic and magnetic fields, and pneumatic–mechanical devices. The basic processes are those of forming (shaping), welding and powder/particulate-matter compaction. A variety of forming systems exists, each displaying specific characteristics associated with either sheet or tube forming, for which it is intended.

16.2.4   Rolling processes and products

16.2.4.1   Classification of processes

With the exception of special steels and/or profiled sections rolling processes follow one of two main routes:

1.   production of plate and strip; or
2.   production of billets, bars, rod, sheet, tube or sections.

Traditionally, both routes start with a cast ingot (Figure 16.23) which is then rolled down to slabs (route 1) by cogging. In route 2, cogging again leads to the production of a bloom (a product of over 10 cm^2, or equivalent, in cross-section), and then to either a variety of small flats or large rounds or, through a billet mill, to a billet (a product of cross-sectional area less than 10 cm^2). However, very satisfactory developments in the area of continuous casting have led to the introduction of casting machines into these cycles. In the new, fully automated and computer-controlled, high-productivity works, continuous casting of slabs has to a great extent eliminated the cast ingot.

In route 2, in a modern mill the stress is on the use of continuous billet casters (in preference to bloom casters), thus eliminating one stage of the production line. Where blooms are still required, normal practice is to employ two or three strands of material which are then rolled in two or three passes to produce blooms. With smaller sizes of billet, up to six strands can be cast.

It is clear from *Figure 16.23* that the manufacture of a wide range of either semifabricates or finished products calls for a variety of mills and plant settings. A very brief review of these is provided here but, again, detailed information can only be obtained from the Further Reading at the end of this chapter.

Basically, the process, whether hot or cold, begins with the preparation of stock such as an ingot (in older plant) or continuously cast bloom or billet. In hot operations this is followed by heating in a strictly controlled atmosphere and temperature, and then rolling proper. Finishing of the work piece includes a number of operations such as cutting, cooling and, very often, straightening. In cold operations, which are used to enhance the mechanical properties of the material and improve dimensional accuracy, the ancillary equipment consists of furnaces for heat treatment and plant for surface finishing. Whereas modern plant comprise not only the rolling mill(s) proper, but also a number of pieces of ancillary equipment concerned with the preparation of the material prior to and post rolling, interest centres mainly on the actual mill since the dimensional quality of the product will depend mainly on its performance.

According to their actual functions, rolling mills are subdivided into the following classes:

1.   cogging mills (production of blooms, billets and slabs from ingots, where these are still used);

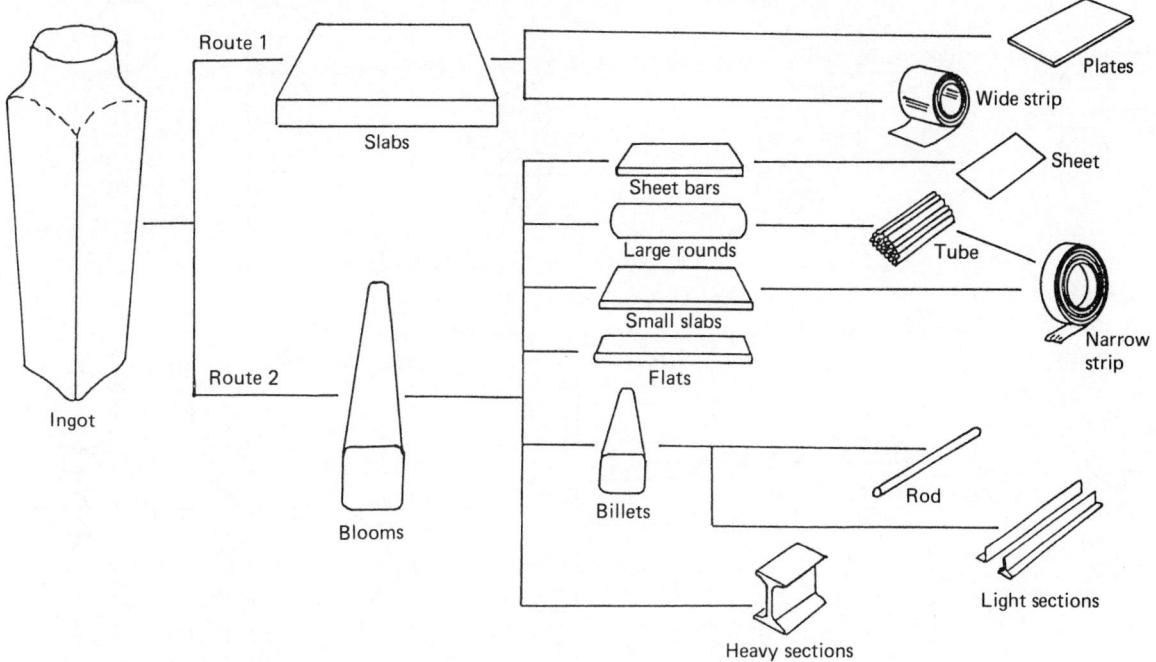

Figure 16.23 Schematic, simplified representation of the sequences of the basic rolling processes. Route 1 leads to the production of plate and wide strip. Route 2 results in sheet, narrow strip, tube and sections

2. plate, strip and sheet mills;
3. tube mills (longitudinal rolling and cross-rolling);
4. section mills (production of profiled sections, rounds, flats and strip); and
5. special mills (production of machine parts).

An individual mill is characterized by the function for which it is intended and also by the type, position and number of rolls. These are located in stands, which in some cases are incorporated in a stand train. Depending on the position of the stands, mills can be further classified as linear (where the stands are in a line and there may be one or two lines driven by a single motor), or continuous (where a number of stands is placed in tandem and, at any given time, rolling takes place in more than one stand).

In each stand the number of rolls involved defines the stiffness of the structure and, therefore, reflects on the dimensional accuracy of the product, whereas the disposition of the rolls specifies the function of the system. Irrespective of the complexity or otherwise of the roll arrangement, the actual work of deformation is normally performed by a pair (or a set) of innermost rolls which are backed up (for the mentioned structural stiffness) by a number of other usually larger diameter sets of rolls. The number of these gives rise to the name of the mill which is described as *n*-high.

A selection of basic arrangements for flat, horizontal rolling is given in Figure 16.24. With reference to the figure, the following types of mill can be distinguished.

1. *Two-high*—either (i) reversing or (ii) non-reversing.
    (i) Used for slabbing, large rail, profiled sections and thick plate rolling.
    (ii) Used for continuous rolling of billets, rods, rounds and plate, linear rolling of sheet, and for dimensional calibration of the product.

2. *Double duo*—used mainly in older mills for rolling small and medium-sized rods, strip and profiled sections.
3. *Three high*—used for rolling medium and large billets, rails, rods and profiled sections.
4. *Three-high with the middle roll oscillating between the two outer rolls*—all the rolls are driven; used for blooming or slabbing of ingots of around 3.5 ton in weight.
5. *Three-high Lauth, with an idle middle roll*—used for rolling thick plate and as a first stage in linear hot rolling of thin sheet.
6. *Four-high*—either (i) reversing or (ii) non-reversing.
    (i) Used for cold rolling thin strip and hot rolling thin and thick plate.
    (ii) Used for continuous hot rolling strip and plate and cold rolling strip.
7. *HC rolling mill (Hitachi)*—used mainly for cold rolling plate and strip.
8. *Taylor mill*—used for precision rolling of thin sheet and strip.
9. *Six-high*—used for cold rolling thin sheet and foil.
10. *Sendzimir twelve-high mill*—used for cold rolling thin sheet and foil.
11. *Twenty-high, planetary Sendzimir mill*—used for cold rolling thin sheet and foil.

The quality of product in terms of material properties depends considerably on whether processing takes place at an elevated or at ambient temperature. Rolling hot is not only intended to minimize the forces and torques necessary to effect deformation, but also to improve the structure through recrystallization. Unless recrystallization takes place, strain hardening (produced by deformation) will adversely alter the structural characteristics of the metal, even at temperatures much higher than ambient. The processing temperature should, therefore, lie above that of recrystallization, in order

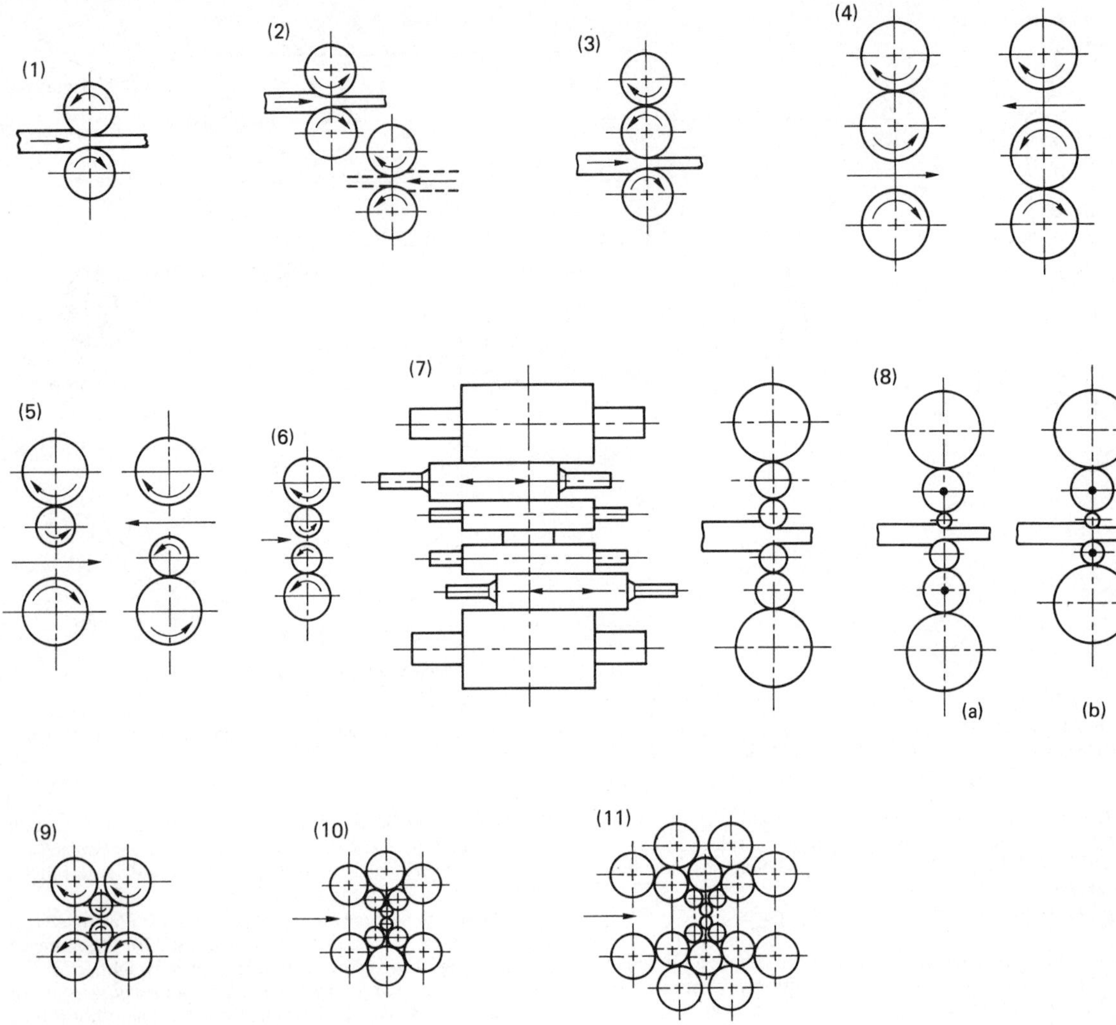

(1)

(2)

(3)

(4)

(5)

(6)

(7)

(8)

(a)   (b)

(9)

(10)

(11)

Figure 16.24 Flat rolling: scheme to show the position of horizontal roll systems. For a description of (1) to (11), see text

to produce new, small grains (Figure 16.25), but below the value at which oxidation takes place and affects the surface of the product.

The main function of cold rolling, as already mentioned, is to improve the mechanical properties of the alloy, to impart a surface finish suitable for the future use of the product, and to enable the manufacture of dimensionally accurate very thin sheet, strip and foil (0.001–0.2 mm thick) and 'medium' thick sheet (1.0–1.5 mm thick). Depending on the quality of the surface, the rolled sheet can be used for stamping when the material displays fine-grain finish, for plating when the surface is smooth, or for further artificial surface coating when the surface is dull.

16.2.4.2   Rolling of plate, strip and sheet

The definition of each of these products is somewhat vague, but it is usually taken that, in hot processes, a rolled item is referred to as 'plate' when its thickness is in the range 4–160 mm and its width is 600–3800 mm. Sheet that is 4–50 mm thick and 200–1500 mm wide is also sometimes called 'universal', whereas 'thin sheet' or 'thin strip' is material

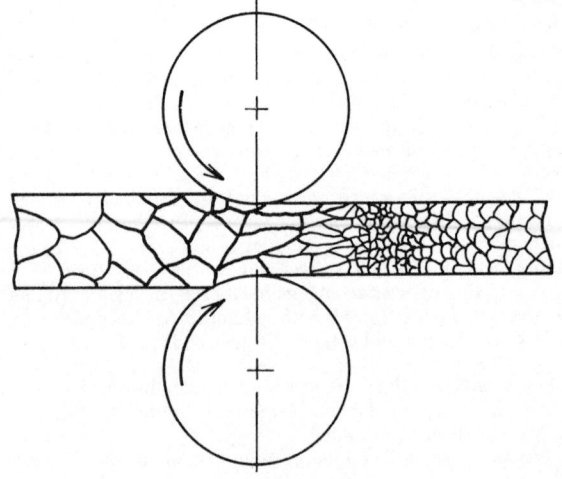

Figure 16.25 Recrystallization in hot rolling

of thickness 0.2 to <4 mm with a width of 600–2400 mm. 'Plate', intended as a constructional material for bridge structures, general building purposes, boilers, ship hulls, car and railway rolling stock bodies, etc., is rolled up to a length of some 36 m, depending on the final use of the product. Rolling is done either from slabs or flat billets. Thicker plates are usually produced in linear, two-stand, two-high reversing systems ((1) in Figure 16.24) followed by a train of four-high finishing stands. Typically, thinner plates may be rolled from slabs in a single stand, Lauth linear system ((5) in Figure 16.24) or, for greater accuracy, in a planetary mill.

More recently, thick and thin products have been rolled in semi-continuous systems employing two universal stands in tandem, followed by a continuous mill arrangement.

Strip and medium heavy sheet are produced in lengths of 4–20 m on either linear or planetary mills; with the latter, in particular, high thickness reductions of up to 95% are obtained.

Thin sheet, intended as structural material, is often rolled, as separate items, to a length of approximately 1800 mm, or alternatively in long lengths in continuous mills and is then coiled. The car and electronic industries, in particular, make considerable use of steel and aluminium alloy sheets and strips coated with thin layers of plastic material. The metallic product used in such applications varies in thickness from some 0.3 to 2 mm.

Cold rolling operations have a two-fold objective: to improve the mechanical properties and dimensional accuracy of the product and to enhance the economic feasibility of the operation (hot rolling of materials to less than about 2 mm thickness is uneconomical). Due to economic considerations hot-rolled strip or sheet is further processed in a variety of mills (very similar to those shown in Figure 16.24), but particularly in planetary or cluster mills if very thin material is required. A typical arrangement employed in the latter case is shown in Figure 16.26. The usual working sequence consists of pickling the hot-rolled material, rolling to the required size, followed, where necessary, by heat treatment and, finally, by general finishing operations.

The range of thicknesses that can be obtained depends, to a degree, on the properties of the starting material, since the roll forces and torques required can be very high. It is often because of this that cluster mills, characterized by their high rigidity, are used. Low carbon steel can be rolled down to a thickness of about 0.2–0.4 mm. Similar strip thicknesses are obtained in Inconel, Monel, nimonic and bimetallic combinations, rolled in Sendzimir mills which use rolls of 450–1300 mm in length. Consecutive rolling operations are employed to produce very thin foils.

16.2.4.3 Rolling of structural sections

Semifinished or hot-finished structural sections are rolled in semicontinuous, continuous, Universal, linear, or H mills. Although a variety of shapes and components can be made on such mills, of particular engineering interest is the rolling of medium and heavy sections. Depending on the actual sizes and weight per length run, these comprise channels, I- and T-beams, rails and angular sections, as well as semiproducts intended for further processing, such as rounds, squares and flats. Without being specific about the type of mill, it is generally understood that 'bar mills' are used for rolling heavy sections and 'rod mills' for light sections.

Channels (50–450 mm high, designated as heavy sections when the height exceeds 200 mm) are usually processed in continuous mills, as are T-sections (30–220 mm high). The method of manufacture of I-beams is less uniform. Smaller sizes (below, say, 200 mm in height) are produced in continuous mills, whereas heavy sections (up to 600 mm and, occasionally, up to 1 m in height, with flanges of 200–400 mm in width) are rolled in universal-type mills. The method of rolling steel rails depends on the weight per metre run. In general, sizes range from 8 to 75 kg m^{-1}, the 40–75 kg m^{-1} sizes representing heavy sections.

In most cases, two-line, two- or three-high, two- to seven-stand linear systems are used but, more recently, special universal and H-type mills have been introduced.

The semifabricates (rounds 6–200 mm in diameter; square sections 8–200 mm side) are produced on semicontinuous and continuous mills, as are rods (5–8 mm in diameter) intended for wire manufacture. These products are rolled in up to 39 stands.

16.2.4.4 Rolling of seamless tubing

Seamless tubing is produced by either longitudinal or cross (oblique) rolling of hot finished rounds. Both manufacturing routes start with pickling, dressing, heating and piercing the billet to provide a roughly made tube which can then be further hot and, if necessary, cold processed. The latter stages are intended to give dimensional tolerances to the product and, in the case of cold processing, to impart both better mechanical properties and, often, better surface finish.

The two routes comprise the following intermediate processes:

1. *Longitudinal rolling* consists of rotary or punch piercing, plug or mandrel rolling. It occasionally may incorporate the Erhardt or push-bench process, followed by the finishing sequences of sinking and stretch reducing.

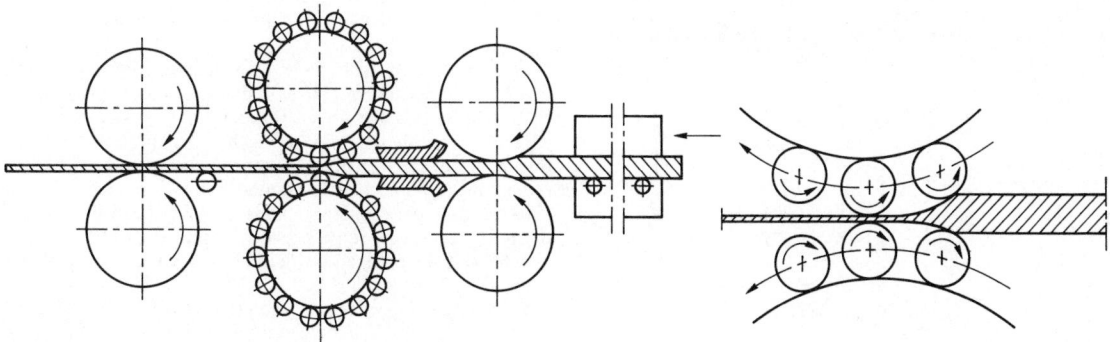

Figure 16.26 System incorporating a planetary mill for cold strip rolling

2. *Cross-rolling* involves rotary piercing, Assel elongating (possibly a combined piercing–elongating process) or Diescher elongating.

Irrespective of the route followed, the piercing stage precedes further operations and consists either of rotary piercing of a solid round or hydraulic punch piercing of a square billet.

Of these two methods, cross-rolling (Figure 16.27) is carried out in a two- or three-roll system in which the rolls are inclined at an angle (feed angle α) to the central axis of the mill, and a plug (situated on a long bar) is inserted between the rolls. The plug bar can either be free to rotate or driven.

With the rolls set obliquely to each other, the billet is drawn into the pass by frictional tractive forces, the axial component of the forces being responsible for the forward movement and the tangential component for the rotation. The physical conditions that develop in the pass are equivalent first to oblique rolling and then to cross-rolling. The cross-rolling zone extends from the plug nose to the exit from the pass, and the billet is worked between the roll and plug surfaces.

The process is not continuous, in the sense that the successive arcs of the billet circumference come, in turn, into contact with the rolls. The set-up clearly tends to develop ovality in the billet, unless horseshoe-type adjustable guides are introduced between the rolls. Up to a point, these guides rectify this undesirable condition and help to preserve circularity of the tube.

Longitudinal rolling The main elongation of the bloom, to produce a hot finished hollow, is carried out in either a plug or a mandrel (continuous) mill and is then completed in one or two finishing stages.

In the plug-mill system, the plug, a short cylindrical, detachable tool, is maintained in position by a plug bar, an arrangement similar to that in the rotary piercing plant, and controls the dimensions of the bore of the hollow. The work piece passes through grooved, oval rolls that are usually positioned in two stands and maintained at 90° to each other. The mill is generally used as a breaking-down stage and operates on tubing of 3–40 mm wall thickness and 20–400 mm diameter.

In the mandrel-mill system the hollow bloom is threaded on a long mandrel and is fed into a continuous train of 7–10 pairs of rolls arranged in tandem. The tube is rolled out on the mandrel without interruption to approximately the finished

thickness. The axes of the rolls are alternately at right angles to one another. The earlier breaking-down passes have considerable ovality. The finishing passes are designed to leave the tube loose on the mandrel and more or less circular so that no reeler is required to enable the mandrel to be withdrawn.

The method is no more continuous than any other seamless process which starts with a billet of finite length and rolls it into tube. The continuous mill is outstanding for producing long lengths of all thickness at a very high rate of output.

The final rolling stage uses a sizing or reducing mill to give the tube its final outside diameter with, preferably, little change in wall thickness. Reducing mills fall into three broad categories:

1. sizing,
2. stretch reducing, and
3. sinking.

Sizing mills are used to produce an accurately sized finished tube without appreciably reducing the outer diameter. They are usually equipped with three to seven stands, each containing two rolls, and are normally designed for an overall diameter reduction of 3–8%. In general, the diameter reduction is accompanied by a thickening of the tube wall.

Sizing mills are usually installed in plants designed to produce the larger tube-size range, but can be used with advantage in the manufacture of smaller tubes when following main seamless processes that are reasonably flexible in the sizes that they can produce.

Reducing sinking mills comprise a continuous train of passes designed to reduce a larger tube to a smaller one without a mandrel or plug to support the bore. The amount of reduction per pass is usually 3–5% according to the type of mill.

A reducing mill is usually employed in conjunction with a main unit to improve the performance of the latter. This may be done by:

1. reducing the number of size changes to be made in the mill;
2. increasing the weight of the billet worked by the mill and, therefore, the rate of output; and
3. increasing the maximum length available, and decreasing the available minimum diameter.

In obtaining these advantages the following disadvantages must be taken into account:

1. a reheating furnace is necessary;
2. an additional process is thus introduced increasing costs of fuel, power, labour and tenance;
3. tubes become thicker during reduction and, therefore, the mill must roll thinner gauge; and
4. the quality of the bore of the tube deteriorates for various reasons and this limits the amount of reduction possible.

These mills are usually designed with up to about 24 two-roll stands, and are capable of overall diameter reductions of up to 55%. The stands are driven through bevel gear boxes from common line shafts, and the behaviour of the tube-wall thickness depends on the relationship between the gearbox reductions. It is normal for sinking mills to operate with no axial tension between the stands so that the wall thickens freely. If, however, some tension is applied, wall thickening can be restricted to about half of that which would occur without tension.

Stretch reducing mills are primarily used for the economic production of small diameter thin-walled tubes. As the name implies, the wall thickness is reduced in the process and, to achieve this condition, considerable tension must be applied between the stands (Figure 16.28).

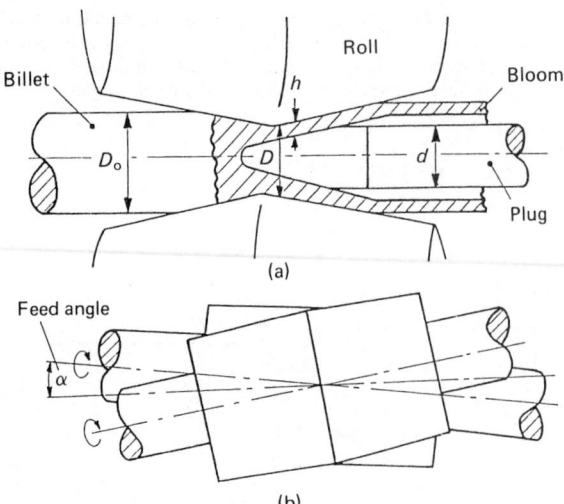

(a)

(b)

Figure 16.27 Rotary piercing of rounds

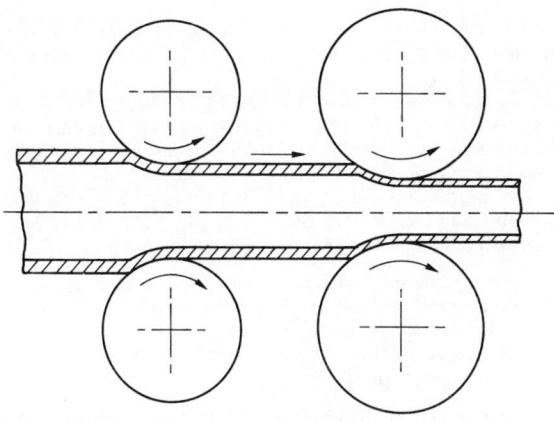

Figure 16.28 Rolling with interstand tension in a stretch-reducing mill

The mills are designed with up to 22 stands, each with two or three rolls, and are capable of overall diameter reductions of more than 75%, and wall reductions of up to 40%. To obtain the control required over the interstand tensions, the mills are designed with either d.c. motor drives to each stand or hydraulically regulated drives.

As the ends of the tube cannot be subjected to full tension in the mill, they are inevitably heavy walled and of a length that is proportional to the elongation, stand spacing and percentage wall reduction. The end crop can have a significant effect on the yield and, therefore, it is necessary to put the longest possible length into the mill to minimize the percentage loss.

In addition to the essential change in the outer diameter of the tube in the reducing operation, there is always some change in wall thickness. The thickness change plays a vital role in the design and operation of reducing mills and depends on the rolling condition.

No-stretch conditions prevail when a tube passes through a single stand, so that its diameter is reduced without any external axial forces being applied to it; the thickness also increases. In this case, where the change in wall thickness occurs purely as a result of the compressive action of rolls on the tube, the reduction is said to take place under the 'no-stretch condition'.

In a mill consisting of a number of stands that are all reducing the tube simultaneously, this condition can arise providing that the roll speeds in successive stands are set such that there is no tension or compression in the length of tube between each pass. Since there is always an increase in the linear speed of a tube as it is reduced, the roll speeds in successive stands must be increased in relation to the elongation that occurs. It follows that if a tube is being continuously processed through a number of reducing mill stands, the volume of flow must be constant. The cross-sectional areas multiplied by the linear speed must, therefore, be the same at all points in the mill, and as the tube area is reduced its speed must correspondingly increase.

Under no-stretch conditions, the whole tube thickens uniformly throughout its length, the amount of thickening being dependent on the diameter to thickness ratio of the in-going tube and the total diameter reduction given. It is unaffected by the reduction given in each stand or by the number of stands working on the tube at any time.

Conversely, in no-stretch conditions, if tension occurs between the stands, wall thickening can be partly or fully compensated for, or the wall thickness can be reduced. Tension is applied by arranging for each pass to run at a higher

speed, relative to the preceding pass, than is required for no-stretch conditions.

With *stretch conditions*, not only is the thickness reduced between the stands but the diameter is also reduced so that the actual reduction in diameter brought about by the compressive action of the rolls is slightly less than that which takes place with no stretch, resulting in slightly less thickening of the tube wall between the rolls. Stretching can occur only on a section of tubing which is between two passes. Consequently, the extreme ends of the tube can never be subjected to stretch nor can an appreciable length at the ends suffer the full stretching action. Therefore, a mill that is designed to stretch or in which stretching occurs produces a tube that is thinner in the centre than at the ends, the thickness at the extreme ends being that which would arise under no-stretch conditions and the centre thickness being that arising from the stretching action. The thickness change is not abrupt but gradual and there is normally an appreciable length of uniform thickness between the thickened ends.

The push-bench process (Figure 16.29) forms an alternative to the above routes. The standard push-bench system incorporates punch-piercing of the hot billet, while retaining a solid base on the leading end, elongating on the bench proper and hot-reducing on an appropriate mill. To retain a reasonable degree of concentricity, the depth of penetration of the piercing punch must not exceed seven times the punch diameter. The 'bottle' which is fed into the bench is relatively short and the process may not be fully economical.

Modern developments have led to the introduction of roller-type dies (the Manfred–Weiss process) and, even more recently, to the incorporation of an elongator in the cycle. The elongator increases the depth of penetration to nine times the punch diameter while retaining good concentricity through the medium of equalization of the wall thickness. Since the elongator will also reduce the diameter of the bottle, the amount of deformation to be obtained on the bench is lower and the quality of the manufactured tube is higher.

Some of the advantages of the elongator are, however, offset by the fact that the gauge correction cannot be carried out at the ends of the bottle and that some reduction in output must therefore be expected.

When the elongator forms an integral part of the cycle, reheating of the bottle, after punch piercing, will be necessary and will, of course, add to the cost of the operation.

The modern push-bench is normally used for the manufacture of tubing of outer diameter of 6–15 cm. Thin-walled tubing can be obtained in lengths of up to 10 m when an elongator is used, and in lengths of up to 6 m without the elongator.

The advantage of the modern roller-die beds of the bench proper over the old ring-die system lies in the reduction of thrust required by about 30%, and a reduction in the size of the die bed by about 25%. Although the provision of three or four rolls in a cluster is more expensive than the installation

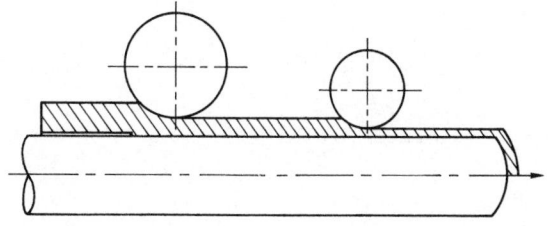

Figure 16.29 Push-bench process

and maintenance of ring dies, the considerable increase in speed of the working cycle, resulting from the possibility of using more passes on a given bed, and from the actual increase in the axial speed of the operation, offsets the apparent economic disadvantage. High speeds of operation are, in fact, required in the case of alloy steels which harden rapidly with a drop in temperature.

Cross rolling The two major rotary production processes employed are those of Mannesmann and Assel. In the Mannesmann cycle a cold centred round billet is heated in a roller-type furnace, pierced in an oblique rotary mill, possibly repierced on another rotary piercer, hot forged, reheated in a continuous furnace and stretch reduced. The piercing and repiercing operations are of primary importance from the point of view of the incidence of redundancy.

In the Assel cycle (Figure 16.30), the billet is heated in a roller-type or rotary-hearth furnace, pierced in an oblique rotary mill, elongated in the Assel mill, reheated in a continuous furnace, and then either sized or sunk on a longitudinal rolling mill. The first two stages, i.e. piercing and elongating, account for most of the redundant shears induced in the worked material.

To increase the range of possible deformations, it is occasionally necessary to follow the piercing stage by a repiercing or secondary piercing operation in which the bore of the pierced bloom is increased while its wall thickness is reduced. A purely elongating operation in which the bore remains unchanged and the wall is further deformed will follow. Secondary piercing operations have been developed from the basic Assel elongating technique and are considered in that context.

An Assel mill is shown diagrammatically in Figure 16.17. In this process, the previously pierced tube, with a cylindrical floating mandrel inserted in the bore, is drawn by frictional forces into the pass defined by the setting of the rolls. In the entry zone of the pass, the diameter of the tube is slightly reduced. The hump of the roll then reduces the wall thickness, and a smoothing operation takes place in the reeling zone of the pass to remove the triangulation of the vertical section of the tube. Final rounding of the tube proceeds in the exit zone of the roll. Elongation occurs primarily as a result of wall thinning, but it is also slightly influenced by the reduction in tube diameter.

Of special interest in tube making is the production of stainless-steel tubing. This, as will be seen later, is generally manufactured in extrusion, but rotary processes are also used, since the rate of yield is likely to be higher.

The production routes used in this case are somewhat complex and reflect both the cost of the material and basic processing difficulties. The main process lines are:

1. punch piercing, Calmes elongator, rotary hot forging;
2. punch piercer, push-bench, polishing mill;
3. disk rotary piercer, plug mill; and
4. Mannesmann rotary piercer, hydraulic push elongator, rotary hot forging.

A possible alternative to asseling is offered by the Diescher elongator. This mill incorporates two oblique barrel rolls and two disks (set at right-angles to the rolls) which replace the horseshoe guides necessary (see piercing) to correct the ovality of the tube. Elongation is carried out between the rolls and a bore-supporting mandrel. The mill improves concentricity of the tube and, by effecting elongation in both forward and backward directions relative to the gorge, imparts a burnishing finish to the tube.

16.2.4.5 *Flow forming*

Practical difficulties of producing large diameter, thin-walled long tubing in a variety of materials, including stainless steel, has introduced a new approach because the ordinary manufacturing techniques could not always cope with the sizes and materials involved and, even where this was possible, the cost of production was very high.

A solution to this problem was first proposed, on a modest scale, in the early 1950s when the horizontal flow-turning lathe was introduced. This was followed by a vertical cold 'floturning' machine in which three stands, carrying five rollers each, were introduced. In this, short thick-walled bloom is supported in the bore by a solid cylinder of hardened tool steel machined to the exact tolerances and to the degree of surface

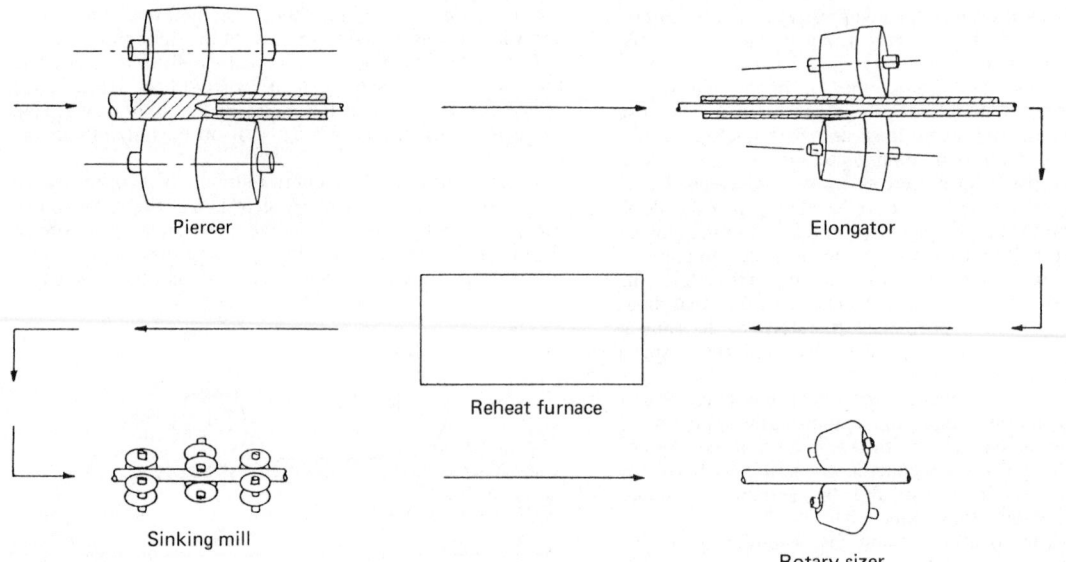

Piercer

Elongator

Reheat furnace

Sinking mill

Rotary sizer

Figure 16.30 The Assel cycle

finish which is required of the bore. Each station, capable of angular adjustment to obtain optimum forming conditions, is powered hydraulically. Rollers are advanced into the forming positions and retracted at the end of the cycle. The wall thickness of the bloom is easily reduced by one-sixth of the total reduction by each roller. As an example of the possibilities, in one pass a steel cylinder of 1.25 m outer diameter, 1.25 mm thickness and 3.5 m in length can be formed from a hollow of 8 mm thickness and about 0.75 m in length. In effect, therefore, the machine reduces the wall thickness of the hollow, elongates the work piece between the rollers, but retains the original diameter of the bore. The control of the mechanical properties of the material can be maintained throughout annealing, and that of the surface finish can be maintained throughout the finish of the roller surfaces.

Some limitations on the length of the tubing obtained on a vertical machine can be expected and this appears to be the reason for a parallel development of horizontal 'floturning' equipment. The floturn process does, in fact, utilize, in general, the horizontal-type machine.

The development of horizontal machines was initially stimulated by the demand for large diameter, long tubing required for the boosters on space vehicles. The equipment designed for this purpose is capable of turning out precise tubing of up to 635 mm in outer diameter and of 30 m length in alloy steels, titanium, Monel, aluminium and stainless steel.

Essentially, the process consists of reducing the diameter of an initially thick-walled, short hollow by threading and clamping the latter on a rotating mandrel, and driving a carriage with the forming rolls along the work piece. The carriage contains three rollers located at 120° to each other which, when driven slowly over the surface of the hollow, will reduce its thickness. This particular process, known as 'par-forming', can retain wall-thickness tolerances to within ±0.51 mm on, say, a 655 mm outer diameter, 2.95 mm thick and 13 m long aluminium tube, and on a 508 mm outer diameter and 2.15 mm thick stainless steel component.

To achieve these high tolerances, a certain amount of preparatory dressing of the hollow is necessary. Initially, the hollow is turned, faced and machined in the bore. The bore is usually honed to remove tool marks and to impose the required tolerances. The tube should fit the mandrel to within 200 μm and, for the final tube sizes stipulated above, it should be about 1.5 m long. Mandrels are made of high quality chromium–molybdenum–tungsten steels and are ground to give tolerances of the order of 51 μm.

The actual forming operation takes place in two stages. Again, for the sizes discussed, something of the order of five passes of the forming rolls over the hollow will be needed, these will be followed by annealing and, finally, by two further finishing passes.

Machines of this type are fully capable of forming tubular components with shoulders, tapers and partial or full closures. The range of tube sizes varies from about 75 to 700 mm in outer diameter, the range of wall thickness being 0.38–3.8 mm. Surface finish can be controlled to within 380 μm on the bore and 760 μm on the outer surface.

Floturning machines are often incorporated in mechanized plant and this facilitates cleaning of tubing prior to adhesive bonding, where required, of two dissimilar metallic components. Cleaning sequences depend on the material processed but, in general, will include emulsion, water and alkaline cleaning for non-ferrous materials (e.g. aluminium) and pickling in nitric hydrofluoric acid for, say, stainless steels.

The capital cost of the equipment involved in these operations is obviously high, as is the cost of hollow dressing, and the cost increases with the increase in the outer diameter of the tube. Clearly, the existing industrial units serve very specific purposes and are not necessarily intended for ordinary commercial purposes. In their own field of application, i.e. large diameters, thin walls, and substantial lengths, they are well advanced compared with other methods.

The possibility, however, of using the same principle for processing small diameter tubing has not been neglected. One of the main problems arising here is that of buckling. In the nose forming of steel tubes, where the components are compressed on a mandrel by conical rollers, the effect of wall thickness on the stability and formability of the tube is noticeable.

16.2.4.6 Rolling of machine parts

Although developed primarily for tube manufacture, cross-rolling processes are now often employed, for reasons of economy, in the production of a variety of machine parts.

These range from a three-roll, oblique rotary grooved shaping of a shaft (Figure 16.19) to the helically two-roll manufactured special hubs (Figure 16.31). Three-roll (inclined at 120° to each other) finned-roll systems are used in the production of finned tubing of high dimensional accuracy, whereas grooved two-roll mills are used to produce ball bearings (Figure 16.20). Bevel and spur gears, as well as bearing races and similar components are manufactured by cross-rolling with a very substantial saving in material as compared with many machining production routes in a variety of rolling systems.

16.2.4.7 Sheet roll bending

An interesting development in the area of cold-rolling is that of using especially profiled rolls to bend metal sheet (usually aluminium, but also hot or cold rolled steel) into simple or even complex shapes that serve as structural components in the car and aircraft industries.

Bending is carried out in machines incorporating roll trains of driven and idle tools, which can produce long length (up to 30 m) suitably profiled components, without changing the initial thickness of the sheet.

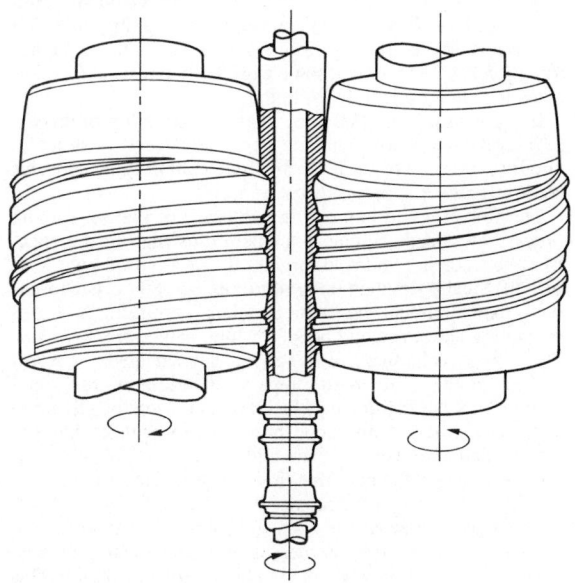

Figure 16.31 Cross-rolling shaped-hub manufacture

16.2.5 Forging operations

16.2.5.1 Basic concepts

Forging, one of the oldest known metal-forming operations, relies entirely on the application of compressive forces to effect the change of shape. In its simplest form, forging is used to preform a billet by changing its dimensions (preparatory to further shaping), and in its more sophisticated form it produces complex shapes to a very high degree of accuracy. Depending on whether the operation is carried out hot, warm or cold, forging affects the structure and properties of the forged component to varying degrees.

Whilst being essentially simple in concept, forging processes, in their many varied forms, are in fact extremely complex. By far the most common group of processes is concerned with the forging of ferrous alloys and a breakdown of the costs involved in producing the average ferrous forging illustrates the reason for the introduction of new methods, techniques and ideas. The cost can be apportioned as follows:

Material in final forging	35%
Material wasted in forging	15%
Labour	10%
Overheads	30%
Tools	10%

It is clear from these figures that material usage is an area in which savings should be introduced and, consequently, innovative preforming processes of powder forging, transverse rolling and cast preform forging, aimed specifically at reducing material usage, must be considered in addition to the standard techniques normally employed. In this context, the idea of forging is of particular interest.

In the hot forging process, the work piece preheat temperature is usually chosen to be as high as possible, consistent with the production of a sound forging, thus exploiting the benefit of minimum flow stress. Any accompanying consideration of economic and technical feasibility must make reference to the work piece preheating costs, corresponding handling costs and the effects of surface oxidation and decarburization. Cold forging of steel, on the other hand, is conducted at room temperature with a consequent saving in heating costs and material wastage, but against this must be set the higher tooling costs. The fact that some work hardening usually takes place is only infrequently exploited.

It is not surprising, therefore, that the possibility of forging at intermediate warm temperatures is advocated in order to obtain the benefits of both hot and cold forging. In this connection, two approaches are adopted.

The first seeks to improve the accuracy of the hot forging process by selecting preheat temperatures high in the spectrum between hot and cold forging. If, for a typical mild (low carbon) steel a forging temperature of, say, 800°C is chosen, then oxidation will be reduced, thus permitting a lower machining allowance and benefit will be gained from the low flow stress. Tolerances of ±0.25 mm are claimed for warm forging in this temperature region. In addition, the small amount of strain-hardening which occurs may be beneficial in some cases. It is important, however, to realize that the heating and attendant costs are still present. As the temperature is reduced further, the flow stress begins to increase rapidly and the process becomes less attractive.

The second approach is from the cold-working end of the spectrum. Due to strain-hardening and strain-ageing effects, lower grade steels may be taken to higher final strengths. This is also in competition with heat treatment over which it has the advantage of being more controllable. A further refinement is to start the process with a quenched and tempered structure

on top of which warm forging will produce even more strength and ductility.

In its classical form, forging is of three main types:

1. open die,
2. impression die, and
3. closed die.

In addition, the already mentioned modern developments have given rise to a whole series of processes of hot and cold rotary forging (including tube making), orbital forging, high-energy-rate (dynamic) forging of both shapes and particulate matter, and cast-preform forging.

The objective of the 'standard' operations is to produce, in stages, machine parts and components such as gears, wheels, compressor and turbine disks, crankshafts and connecting rods, small tooling, screws and bolt heads, and coins and medals, and to assist in the conduct of other operations by providing simple preforms. Punch piercing of billets is a typical example of the latter application of forging.

16.2.5.2 An outline of open-, impression- and closed-die operations

Open-die forging, also known as 'upsetting', is concerned mainly with reducing the height of a cylindrical billet. This is generally done between two flat dies, although the dies can be profiled in a simple manner to impart a specific shape to the ends of the upset specimen.

The outcome of an upsetting operation, in terms of the shape of the preform, depends on the frictional effects that develop between the dies and the faces of the billet. With efficient lubrication, the reduction in height produced by the application of compressive forces is accompanied by an increase in the diameter of the billet, but with the billet retaining its original sharp edges. In an unlubricated or poorly lubricated operation, the deformation becomes inhomogenous and barrelling occurs. The amount of barrelling depends on the value of the width/height (d/h) ratio and the reduction in height r (Figure 16.32).

Although open-die forging offers a simple and relatively inexpensive means of producing small components, it calls for a high degree of manipulative skill on the part of the operator, since acquisition of the basic shape can only be achieved by turning the work piece to different positions between successive blows. The deforming forces are applied either mechanically by means of powered hammers or manually. In either case, the rate of yield of the component is low and the process is unsuitable for mass production. The additional difficulty, which has an additional cost associated with it, is the requirement to machine the specimen in order to obtain both the desired final shape and the required dimensional tolerances, neither of which is likely to result from the forging operation on its own. The machining stage naturally introduces an element of wastage of material and an additional labour requirement.

To introduce a high degree of dimensional accuracy and better material usage, impression and closed-die forging operations are used. These operations are shown in their most basic form in Figure 16.33. The fundamental requirement is that the die cavity, whether that of an impression die (Figure 16.33(a)) or a closed die (Figure 16.33(b)), is completely filled with the forged material and, most importantly, that the material remains structurally sound. The more complex the shape to be forged, the more involved the production line becomes since, clearly, it is impossible to convert a starting cylindrical or square billet to the desired shape in a single blow. Not only are preforming operations required, but several actual forging stages may also be needed.

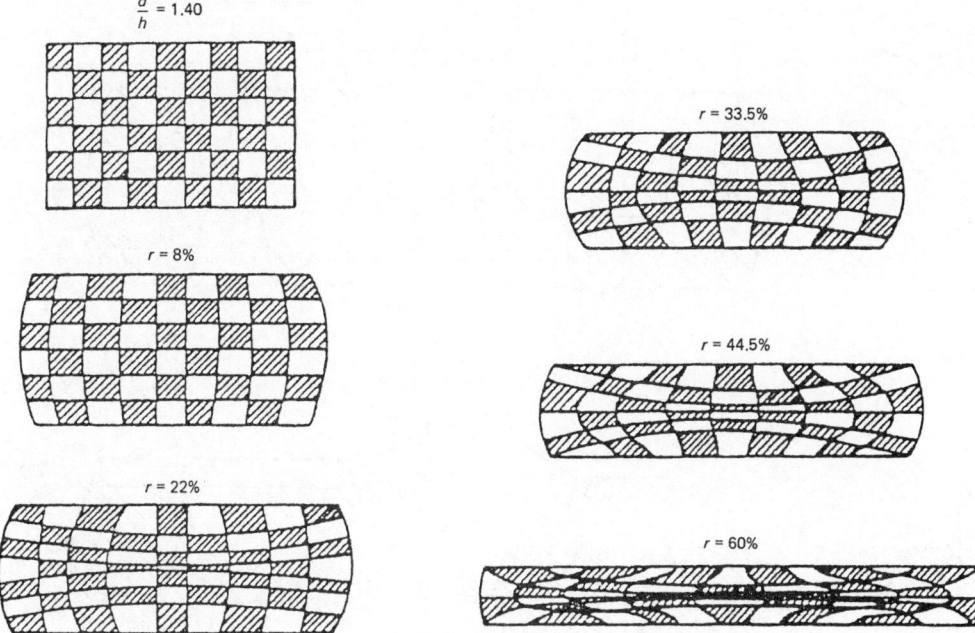

Figure 16.32 Barrelling in unlubricated, open-die forging

The basic characteristics of impression dies is the fact that, when activated by a mechanical hammer or hydraulic press, they do not close completely, and thus some metal is allowed to escape and forms a flash (Figure 16.33(a)) between the flat surfaces. The presence and magnitude of the flash are of considerable importance in impression-die forging because they influence the mechanics of the operation. By the nature of its geometry, the flash is subjected to high pressures and, consequently, it experiences a high degree of frictional res-

istance as it propagates radially outwards. This constrains the tendency of the bulk of the material to flow between the dies and, therefore, creates better conditions for the filling of the die cavity. This action is further assisted by rapid cooling of the flash (as compared with the cooling of the material in the die), which results in a further increase in the resistance to flow. These characteristics form the basis of die design.

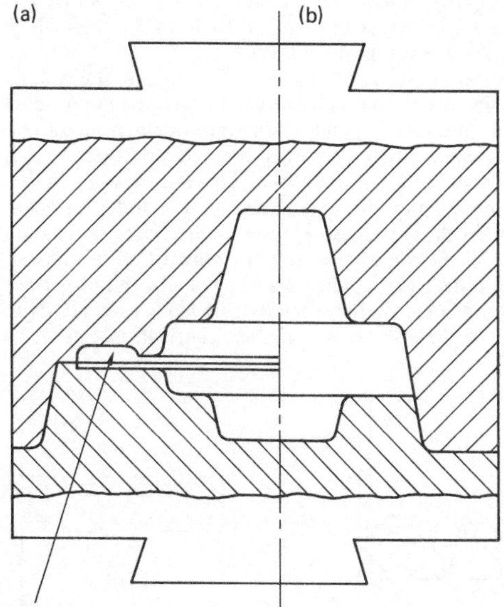

In the normal sequence of operations, preforming will be necessary and may consist of the following operations:

1. simple upsetting,
2. blocking, and
3. rolling.

Upsetting and blocking in blocker dies are the more usual routines, although rolling (see previous sections) of profiled shapes is also used.

A blocker die belongs to the impression-die group, but provides only a general outline of the work piece. The final stage is not achieved using a blocker die, but the material is distributed in a way that ensures a more acceptable uniformity. On trimming the flash, the work piece can undergo further processing in a finishing die that will impart to it its final shape.

A set of impression dies will produce a parting line in the specimen and, unless the die design takes into account the most likely response of the material, in terms of its pattern of flow and formation of fibres in its structure, problems may arise because, on trimming, the fibres are severed.

To facilitate the removal of the forging on completion of the operation, dies are provided with draft angles (a situation similar to that in casting).

Reduction in height, piercing and bulging are the basic operations associated with impression-die forging. However, their sequence and intensity depend on the required shape of the forging. Figure 16.34 illustrates a selection of the possibilities arising in these operations. If the required forging is of simple shape, the arrangement shown in Figure 16.34(a) may

Figure 16.33 (a) Impression-die and (b) closed-die forging

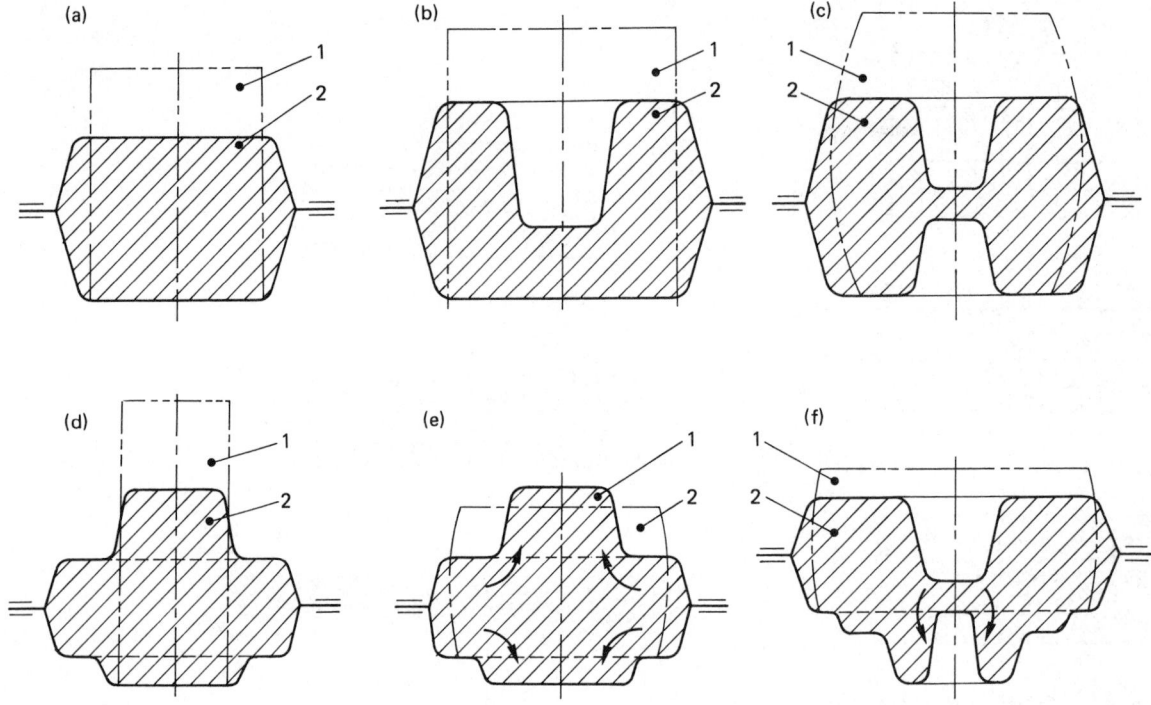

Figure 16.34 Different routes of filling the cavity of a finishing die: 1, starting shape; 2, final shape of forging

be sufficient. If one- or two-sided indentations are needed (Figure 16.34(b) and (c)), the final shape is obtained by height reduction and piercing in the die. Formation of more complex shapes with side bulges, etc., depends on the dimensions of the starting material. If the diameter of the billet is less than that of the elongated part of the axisymmetric forging, then the cavity is filled simply by reducing the billet height (Figure 16.34(d)). On the other hand, when the starting diameter is larger than that of the forging, height reduction and flow into the die cavities are required (Figure 16.34(e)). A more complex operation is required (Figure 16.34(f)) when the forging has both indentations and side bulges, and the starting billet diameter is larger than that of the side of the forging. Height reduction, piercing and bulging must occur to give the final impression.

In true closed-die forging, the operation produces a completely flashless forging (Figure 16.33(b)). To achieve this highly desirable situation, the die/billet relationship must be carefully worked out, since incorrect dimensions of the billet lead to either incomplete filling of the die cavity or damage to the dies and punches. If the forging system is correctly designed and the die (or dies) and the punch are machined to the required degree of accuracy, a precision, or near-net-shaped forging will be obtained.

16.2.5.3 Special applications of closed-die forging

A number of special operations associated with closed-die forging are carried out cold, i.e. at a temperature below the recrystallization temperature of the metal in question. However, some difficulty is experienced in the case of steels, since these alloys recrystallize at temperatures above 600–700°C, temperatures that are too high to be called 'cold'. Consequently, forging at room temperature is often referred

to as 'cold', and forging at elevated, but below recrystallization, temperatures is called 'warm'. Generally, the purpose of cold forging is to produce a finished part with high dimensional accuracy.

Highly ductile materials, such as aluminium, lead and tin, have been cold forged for a long time, mainly in the form of extrudates, but 'proper' cold forging is normally limited to either small parts in low and medium carbon steels (such as bolts and nuts) or larger parts of up to 10 kg in mass which require good dimensional tolerances.

Of the better known applications in this area are the coining and embossing operations (Figure 16.35) used for the production of coins and medals, and for the improvement in dimensional accuracy of other preforms. The coining operation is one that is actually carried out in a closed-die system, sometimes with embossing forming, a stage that includes an open-die system. In these processes, three-dimensional details are reproduced in the material giving not only a faithful impression of the punch and/or die surface, but also good surface finish. It is not surprising that the forging loads used are high. The design of the dies is critical since lubrication cannot be

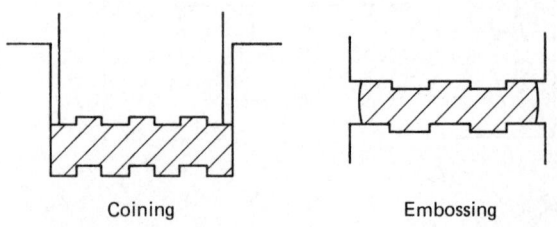

Coining   Embossing

Figure 16.35 Coining and embossing

used in coining operations because of the danger of entrapment and the consequent damage of the forging.

A completely different type of application of closed-die hot forging is that of powder forming, also known as 'sinter forging'. This process, an offshoot of conventional powder metallurgy processes, involves the following sequence of operations. Cold powder is compacted in a press to produce a powder preform. The preform is subsequently sintered in a controlled-atmosphere furnace but, instead of being allowed to cool in that atmosphere (as in conventional powder metallurgy processes), the preform is removed from the furnace whilst still hot and forged in closed dies to produce the final shape. Since only the exact amount of powder required to make the final shape is actually used, there is an obvious elimination of waste.

The same final shape could also be made in the first compaction operation, but the introduction of the forging stage produces a superior product. Improvements in strength, ductility and density are expected, with accuracy increasing to some 0.025 mm. Against these advantages must be balanced the drawbacks of additional processing and the possibility of deleterious oxidation occurring immediately before and after the forging operation.

Commercial applications of powder forging range from forged-powder connecting rods, through material properties approaching those of commercial forged steels, to valve spring cage components. In the latter, for instance, the density (as compared with normally sintered material) increases by some 30%, the elongation quadruples, and the ultimate tensile strength almost trebles.

More recently, the process has been extended to the cold forging of polymeric powders which, although they show some post-forging elastic recovery, can be quite successfully compacted in this way.

16.2.5.4 Subsidiary forging operations

These operations are often associated with preforming and comprise:

1. cogging,
2. fullering,
3. heading, and
4. hubbing.

Cogging or drawing (Figure 16.36) of a rectangular section specimen involves reduction in height by successive blows or bites, leading to a gradual elongation of the original blank. The successive bites reduce the force requirement, but they must be grouped closely together to produce an even surface. The operation will not cause plastic deformation of the bulk material to occur, unless bites of sufficient length are initiated. To avoid possible buckling of the specimen, the height/width (h/w) ratio should be kept below 2.5.

Fullering (Figure 16.37) is another preform operation in which the original bar is shaped by profiled open dies into an outline suitable for more detailed, further forging. Thus the basic function of fullering is to redistribute the bar material along its length, prior to cropping of the individual preforms.

Heading is an operation that combines forging and extrusion. In the forging part an upsetting operation carried out between flat dies will flatten (by bulging) the end of a cylindrical billet, thus forming a bolt head. The remaining portion of the billet can then be extruded to form the bolt.

Hubbing is used as a means of forming die cavities by indenting the material of the blank with suitably profiled punch heads. The operation is thus very similar to in-die punch piercing (discussed above). It generally calls for punch

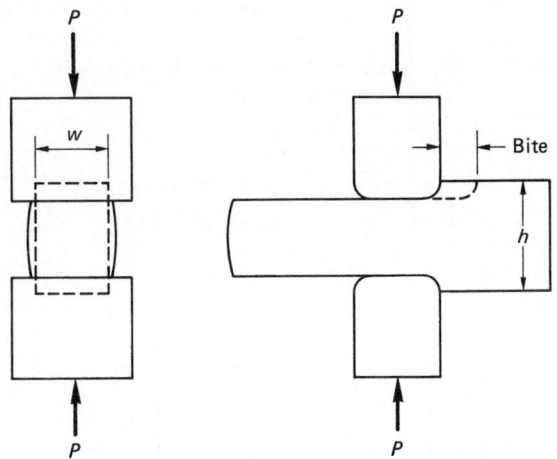

Figure 16.36 Cogging of a rectangular specimen

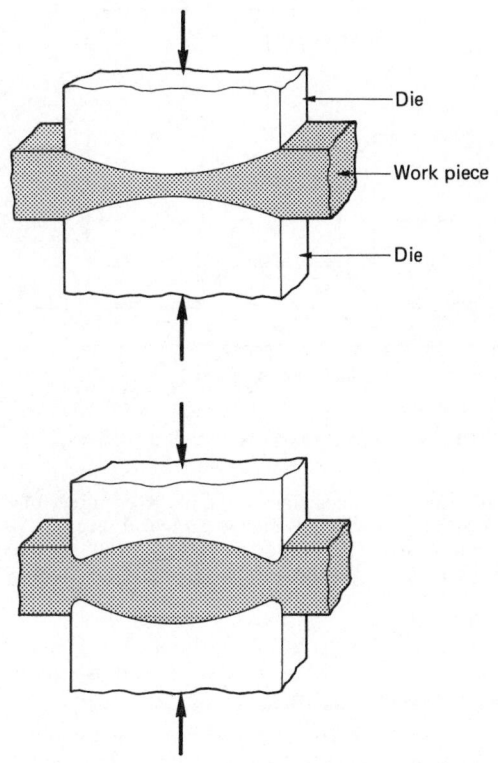

Figure 16.37 Fullering

pressures of about three times the flow stress of the material to be forged.

16.2.5.5 Rotary forging

Rotary forging, also known as 'transverse rolling', is used as a preform operation in which profiled tools are positioned on driven rolls (see Figure 16.38).

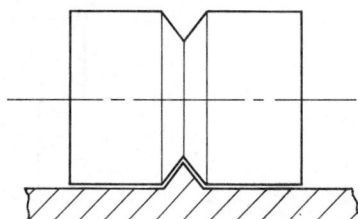

Reduction at position 1

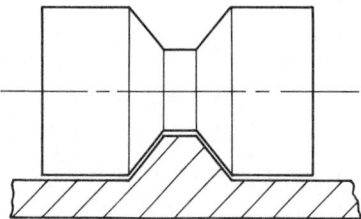

Reduction at position 2

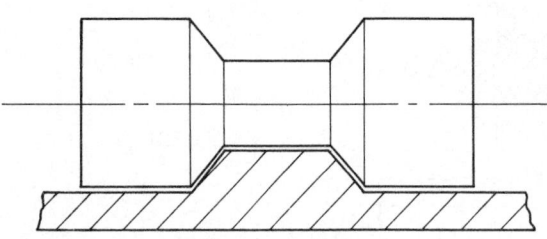

Reduction at position 3

Figure 16.38 An example of transverse roll forging

The bar stock is located transversely between three rolls. The rolls make one revolution and the bar makes several. The wedge impressions on the roll surface (Figure 16.39) are then progressively imparted to the work piece to give the required preform shape. Subsequent flash discards of 15–20% are usually expected.

16.2.5.6 Rotary tube forging

Of very considerable industrial importance is the use of the rotary-forging concept in hot and cold seamless tube manufacture. The operation is carried out on a pilger mill.

In hot pilgering, the already rotary-pierced bloom is threaded onto a cylindrical floating mandrel and is steadily advanced forward and retracted in a reciprocating motion in which the forward stroke always exceeds the length of the backward one (hence the name of the operation, which is derived from the movement of the chorus of pilgrims in an opera) and is executed by two profiled, driven rolls. Each successive movement is accompanied by a 90° turn of the tube–mandrel assembly.

Figure 16.40 shows a sequence of successive 'bites' culminating in a finished tube. Irrespective of whether the working roll surfaces are parabolic, hyperbolic or form logarithmic

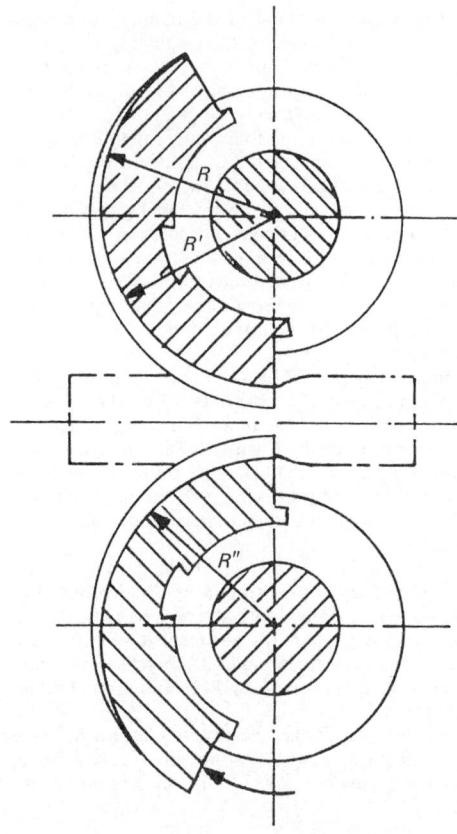

Figure 16.39 Profiled roll system in rotary forging

each roll (being basically a cam) is characterized by three distinct zones. The working zone AB consists of the reducing portion that produces the highest degree of wall reduction. It is this deformation which, in turn, converts a bloom to a tube and is accomplished by a change in roll radius. The finishing zone BC imparts the required dimensions to the tube; the roll radius remaining constant along this part of the profile. The third zone is the idle intercept CA during which reciprocating and rotational motions of the tube–mandrel assembly are performed. To obtain the final geometry of the tube up to six passes may be required, with each successive bite reducing the wall thickness of the bloom further.

To obtain even thinner walled seamless tubing displaying high dimensional accuracy and good mechanical properties, cold pilgering (also known as 'cold reducing') is employed. The cold-reducing process is a step-by-step operation which uses tapered grooved rolls (Figure 16.41). Several dissimilarities exist between hot and cold pilgering. The most obvious of which is that, whilst in the hot process the rolls revolve completely and continuously, in cold reducing they rotate backwards and forwards through approximately 180°. In the hot process the rolls are mounted on a fixed stand, and the work piece reciprocates, whilst in cold reducing the roll stand reciprocates and the work piece remains stationary.

The machine operates as follows. Each time the saddle is at the end of the stroke, so that the large ends of the die grooves are presented to the in-going hollow, the cross-head pushes the hollow forward by a fixed amount (or the feed). The dies then roll the material thus inserted down the cone, simulta-

and the quality of the tubing, with regard to surface or structure, calls for cold deformation. Tube obtained in this manner can either be used directly, in industrial applications, or can form an intermediate stage of production requiring a cold-drawing operation.

16.2.5.7 Orbital forging

The orbital forging process is intended to allow smaller machines to be used for suitable forging operations. This process is based on the simple concept that the axial force required to effect a desired deforming zone between the platens is confined to a small region. The plastic zone is then moved through the work piece, thus resulting in progressive deformation.

The plastically deforming region is formed by initially indenting with a conical platen into the work piece which is supported by a lower platen capable of axial movement only. The plastic region is then moved through the material by the upper conical platen which is capable of rotation about the central axis of the machine and, also, about its own geometric axis. In order to produce a flat top surface on the work piece, the geometric axis of the upper conical platen is inclined to the central axis of the machine.

The process, known also as 'Rotaform', can be applied successfully to the hot, warm and cold forging of a wide range of materials and axisymmetric shapes. The Rotaform is a completely automatic machine provided with automatic feeder and ejection equipment. Figure 16.42 illustrates the principle involved.

The die uses a pair of cooperating dies, one of which is adapted to perform a wobbling motion, relative to the other, about a centre at or near the axial centre-line of the dies. The wobbling die, which has a circular rocking motion without actual rotation about its axis, is actuated at high frequency. The complementary half-die is secured to a hydraulic ram which conveys the work piece to, and presses against, the wobbling die. The cycle of operations commences with the introduction of a billet into the receiving station of the automatic feeder which places the billet accurately into the non-wobbling die. The hydraulic ram raises the non-wobbling die at a speed programmed for the most suitable deformation rate throughout the forging stroke. When the forging has assumed its final shape, the ram retracts rapidly and the forging is automatically ejected and blown clear of the dies.

The cycle time is very short and outputs of 15 forgings per minute have been achieved.

16.2.5.8 Cast-preform forging

The material of all metal components usually originates from a cast and subsequently undergoes a considerable degree of processing before being finally transformed by, say, forging into a component. Since all the intermediate mechanical working improves the properties of the structure, this improvement, to a level necessary for most components, can be achieved by cast-preform forging.

From a technological point of view, the structure of a reasonably homogeneous casting can be transformed to a forged structure by quite small amounts of hot working, amounts which could quite easily be achieved in the forging process itself. Internal cavities can be closed up and welded, providing they are located in regions of high deformation. Due to oxidation, however, cavities extending to the surface are likely to remain as faults. The economics of such a process require some consideration, but it becomes obvious that the operation is viable when the casting can be produced without a lot of waste material in the form of runners and risers. In

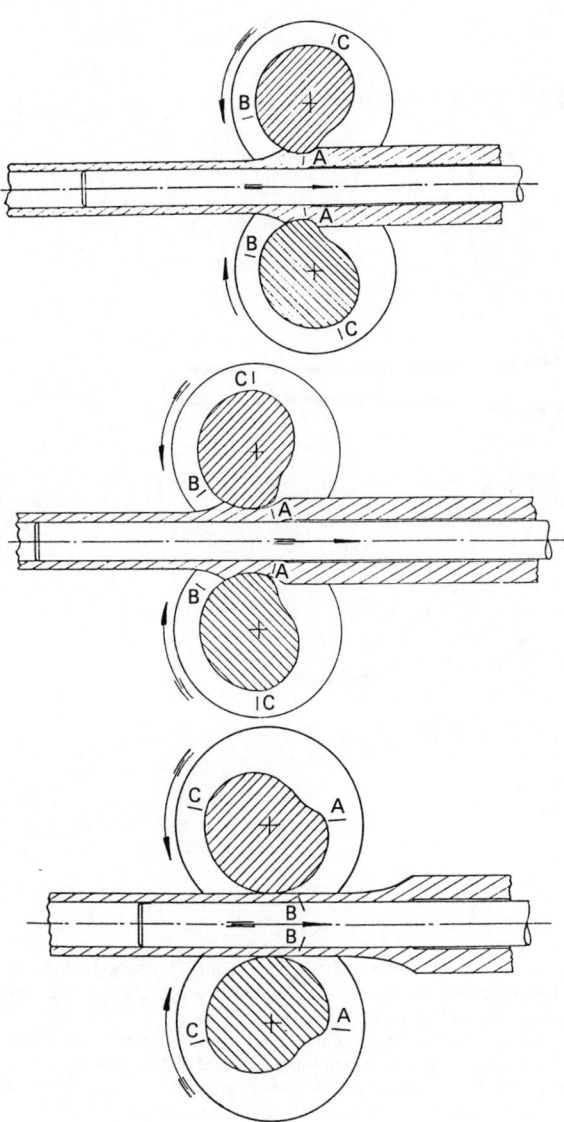

Figure 16.40 Sequence of events in hot pilgering of a seamless tube. AB, working zone; BC, finishing zone; CA, idle zone

neously thinning the wall and reducing the diameter, with the consequent elongation of the tube. At the far end of the saddle stroke, the tube and mandrel are rotated through approximately 60°, and the saddle returns. This is repeated at frequencies of 1–2 Hz, depending on the size and type of machine.

At each end of the saddle stroke, because of the clearance in the dies and also because the rolls turn through slightly more than 180°, there is a short period of time when the dies are not in contact with the tube being rolled. It is during this time that feed has to occur at one end of the stroke and turning of the tube at the other.

In general, this cold forging process is associated with deformation of tubing in the 10–230 mm wall thickness range, and is employed whenever the specified wall-thickness tolerances of the finished product are required to be below ±10%,

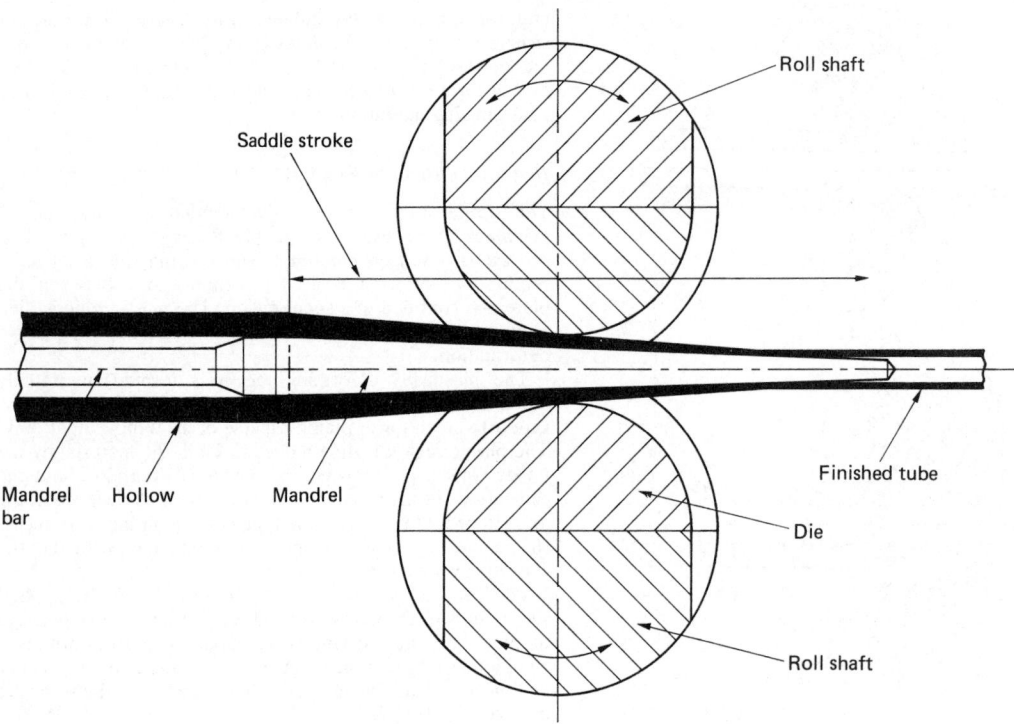

Figure 16.41 Cold pilgering (reducing) tube forging process

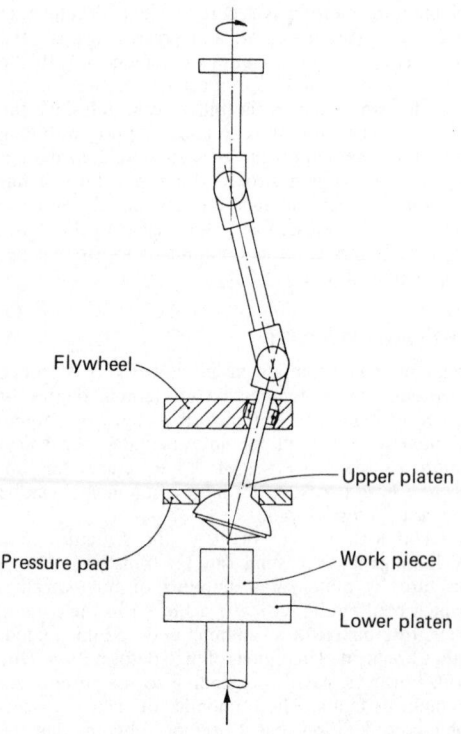

Figure 16.42 Orbital forging arrangement

addition, shapes normally forged from a cropped billet in single impression dies, where material usage is high, are not amenable to this process.

Major economic advantages can be gained if the casting and forging operations are combined. Several techniques are available for this; for example, a conventional steel-casting plant is situated alongside the forging press and reheating is done between the two operations. This technique has been developed as the 'Auto-forge' process which consists of a rotary table with a number of stations at which the various operations are performed. Casting is carried out at the first station and the moulds are water cooled while the table rotates to the next station where the forging operation is performed. The work piece then rotates to the clipping station. Another type of process, the 'Auto Forcast', consists of a melting plant coupled with a continuous casting plant. The various sections of cast steel produced are straightened and hot cropped into billets before being fed into the forging press.

16.2.5.9 Isothermal forging

A more expensive and complicated process is that of isothermal forging in which the dies are heated to the temperature of the blank prior to the operation itself.

The advantages of this technique include low stress fields and, consequently, improved flow of the material which, combined with the absence of a temperature gradient at the tool–work piece interface, reduces the inherent difference in the bulk forging material.

16.2.6 Extrusion

16.2.6.1 General concepts

One of the very widely used metal-forming processes is that of extrusion. In its basic form, the operation involves preparation of a, generally, cylindrical billet (sometimes referred to as the 'slug'), insertion of it into a container (which holds a suitably profiled die), and applying pressure (by means of a punch or a ram) to the trailing end of the billet. The billet is thus pushed, or extruded, through the die.

Since extrusion normally constitutes a more or less direct step from the billet to the finished product, tool design (including its material) and lubrication are of primary importance.

The term 'extrusion' is sometimes applied to forging operations that involve elongation of the product, but it is used here only for the case of 'pure' extrusion described above.

The extrusion process is used in the manufacture of round or profiled bars and tubes, profiled sections, simple and shaped containers, finned or ribbed components and, more recently, spur gears.

Actual extrusion operations are carried out using either mechanical or hydraulic presses; the former being of the vertical type and the latter, as a rule, of the horizontal type. These machines provide extrusion ratios (initial to final cross-section ratios) varying from 10 to 100, with punch speeds of up to 500 mm s^{-1}. Slower speeds are used for most of the lighter non-ferrous alloys of aluminium, magnesium and copper, and higher velocities of deformation are needed when processing ferrous alloys, refractory metals and titanium.

Mechanical presses show a slight advantage over hydraulic machines in terms of speed. The speed of operation precludes almost completely the possibility of using glass as a lubricant which plays an important role in extrusion of stainless steels. Consequently, the process becomes more severe and the billet to hollow ouput can be slightly lower than that for hydraulic presses. A further disadvantage of mechanical presses lies in the fact that the punch velocity varies during the working stroke, and so very considerable difficulties can be created when processing alloys susceptible to strain-rate variations.

In addition, the maximum length of the product is limited by the length of the crankshaft. Hydraulic presses remove these difficulties because the speed of extrusion is controllable and can be maintained almost constant while the length of the extrudate depends on the load capacity of the press and not on its mechanical characteristics.

A distinct, additional and practical advantage of this type of press is the ease of removal of long tubing, or other extrudate, as opposed to the vertical, mechanical presses where a provision for, say, a pit must be made.

It is, however, advisable to remember that, in general, alignment is better on vertical presses than on horizontal ones, and that loading conditions are less severe on the former.

With the great advantages of glass as a lubricant, the number of horizontal presses has inevitably increased, being at present about three times that of vertical presses.

Because of its very wide range of application, the extrusion process is looked upon in a variety of ways that reflect the particular use to which it is put in any specific circumstance. As a consequence of this, the classification of extrusion operations follows diverse courses.

The most basic classification recognizes four extruding techniques:

1. forward or direct extrusion,
2. backward or inverse extrusion,
3. side extrusion, and
4. continuous extrusion.

However, depending on the shape of the product, a further subdivision is often used:

1. solid (circular or non-circular rod) extrusion,
2. hollow (tubular) extrusion, and
3. can extrusion.

In view of the importance of die design and, therefore, the die profile, the process is sometimes classified with reference to the tooling:

1. square (flat faced) die extrusion,
2. conical (linearly converging) die extrusion, and
3. profiled die extrusion.

Naturally, extrusion processes can be carried out at elevated temperatures (hot extrusion), or below the recrystallization temperature of the alloy (warm or cold extrusion).

Square-die processing is used normally in the hot extrusion or profiled light metal (aluminium or copper alloys) components. Continuous profile dies are mainly used for lubricated cold or warm extrusion to enhance material properties and improve surface finish. Streamlined dies are employed because of the higher degree of homogeneity of flow that can be obtained.

Depending on the properties of the starting material and the complexity of the shape of the extrudate, cold extrusion processes (which clearly require high extrusion pressures) involve three distinct techniques which, under specific conditions, ameliorate to some degree their severity:

1. conventional extrusion,
2. impact extrusion, and
3. hydrostatic extrusion.

As in other cold-forming operations, the mechanical properties of the metal are considerably improved if, of course, the increase in temperature associated with the operation does not exceed the temperature of recrystallization. Dimensional tolerances are of a very high order and, as already mentioned, efficient lubrication gives good surface finish, which is also made possible by the absence of oxidizing effects that are present in hot working conditions.

16.2.6.2 Basic extrusion operations

The three most basic extrusion operations of forward, backward and side flow are shown in Figure 16.43. Although, as shown in the figure, these operations refer to axisymmetric solid or hollow components of circular cross-section, profiled or asymmetric shapes can also be obtained in this way. Again, the techniques are equally applicable to hot, warm or cold conventional processing.

In a forward extruding operation involving the manufacture of a solid rod or hollow tube (Figure 16.43(a),(b)) the previously prepared billet is placed in the container and is either directly extruded to form the rod or, if prepierced, it is threaded onto a cylindrical mandrel which supports its bore and is then extruded. The deformation of the tube results in the thinning of its wall and the consequent elongation of the extrudate. If there is no prepiercing, a piercing punch is positioned in the container to provide, on the one hand, the initial hole and, on the other, to act as a mandrel.

In the extrusion of a can (Figure 16.43(a)), a counterpunch in the container acts as a die by causing the material to flow between the container walls and its own outer surface. At the conclusion of the operation and retraction of the punch an ejector removes the formed can from the container.

In the backward extrusion of a rod (Figure 16.43(d)), the billet is placed in the bottom of the container and a hollow punch, of bore diameter corresponding to the outer diameter of the rod to be extruded, is forced into the material, causing it to flow upwards.

A similar arrangement is used when extruding tubular components (Figure 16.43(e)), but here either a prepierced hollow is used or piercing in the container forms the first stage which is followed by extrusion. Again, an ejector is necessary to remove the product.

When extruding a can (Figure 16.43(f)), a reverse of the forward process is effected. The billet is placed in the die and a solid punch is moved axially into the metal. An ejector is necessary to remove the can.

Side extrusion, either one- or two-sided, is limited to rod (Figure 16.43(g)) or tube (Figure 16.43(h)) manufacture. In the first case, the die is situated in the side of the container with the billet positioned at right angles to it, but supported at its lower end. The punch moves axially downwards forcing the metal to flow through the die. When making a tube, the billet must be pierced first, inserted into the container and the

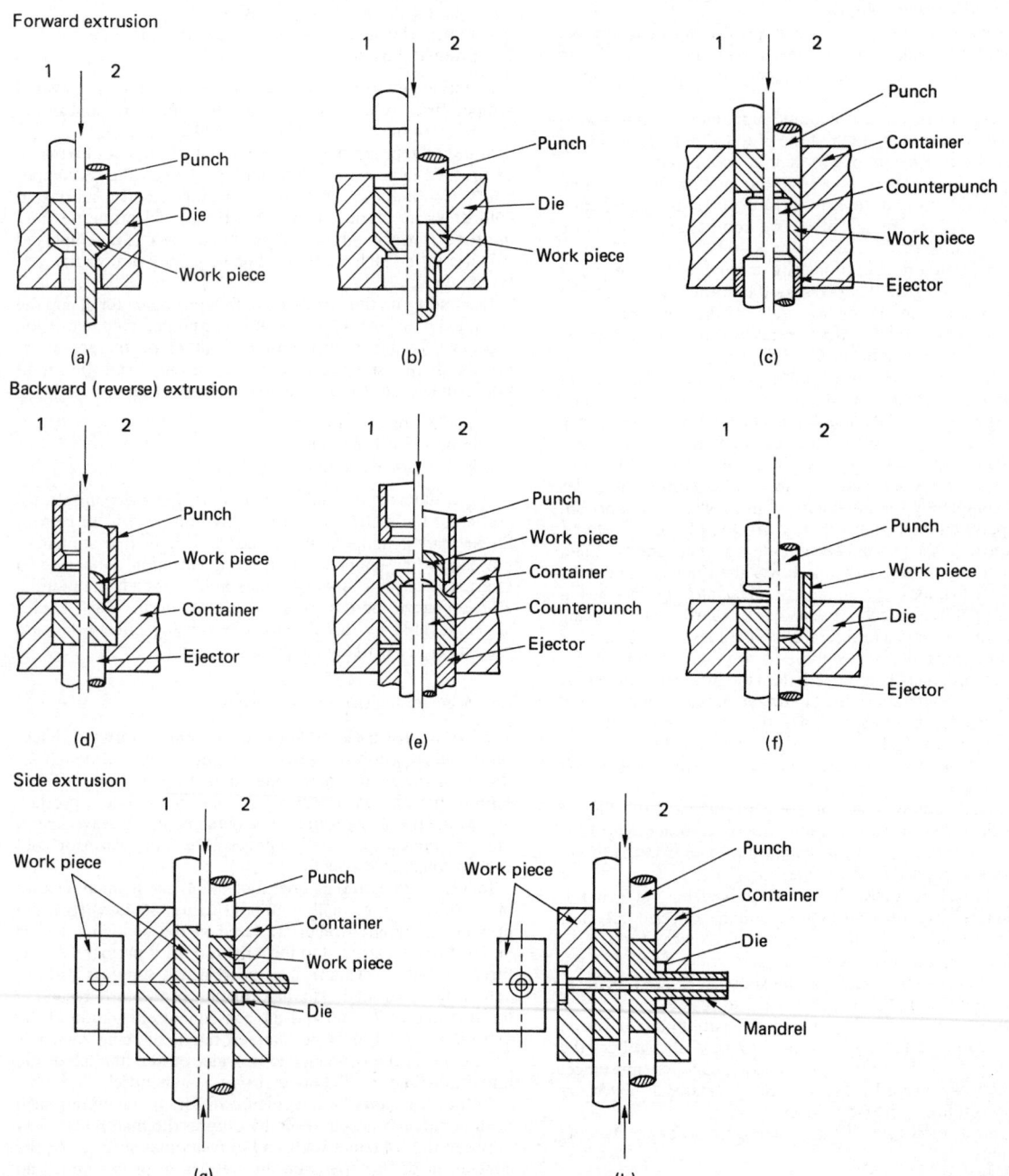

Figure 16.43 Forward extrusion of (a) solid, (b) tube, and (c) can. Backward (reverse) extrusion of (d) solid, (e) tube, and (f) can. Side extrusion of (g) solid, and (h) tube. 1, Pre-extrusion condition; 2, post-extrusion condition

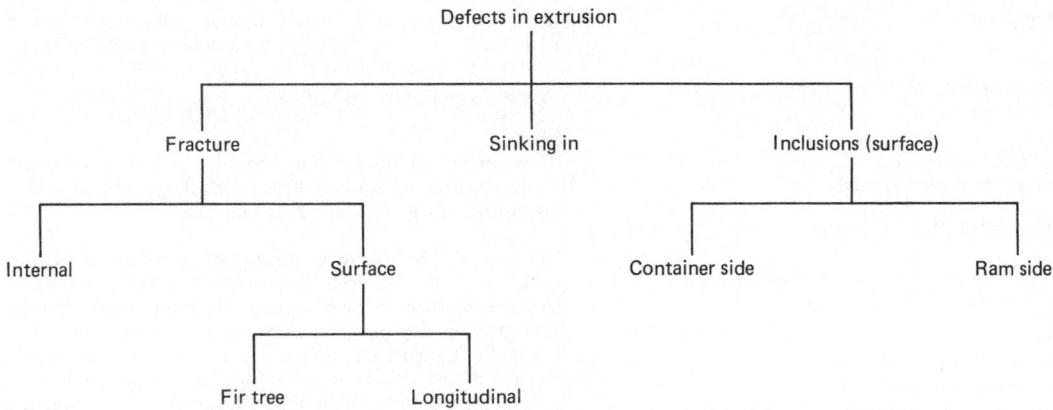

Figure 16.44 Possible defects caused by extrusion

mandrel threaded through it. It is only then that the punch can be actuated and cause plastic flow to commence.

As in other forming processes, extrusion, in any form, produces a number of material defects which are associated either with the characteristics of the process itself, or with the selected geometry of the forming pass. In more general terms, these can be summarized in the form given in Figure 16.44. Defective items in extrusion may reach a proportion as high as 10–15% of the total volume of the product and, although inspection of extrudates can prevent the use of defective components, rejection of parts increases production costs since, in addition to the expense of full processing, the inspection itself is expensive.

Defects that may occur are usually due to any, or a number, of the following:

1. defective billets,
2. defective or unsuitable tooling, and/or
3. processing technique.

Irrespective of their origin, all these defects can be reduced or even eliminated by correct design of the extrusion tooling.

A very characteristic feature of extrusion, especially when using flat dies, is the phenomenon of the 'dead-metal zone'.

It can be seen from Figure 16.45 that in flat dies (90° die semi-angle) part of the billet material becomes trapped in the corner of the die/container space and does not participate in extrusion. The bulk of the material moving through the die shears past the trapped annular ring of stationary metal which thus effectively forms a new, curved die surface that merges with the proper die.

Depending on the point of view adopted, the formation of a dead-metal zone can be regarded either as a defect or as a desirable phenomenon which may enable the material to adopt the optimal flow path.

16.2.6.3 Conventional tube extrusion

One of the most important applications of the extrusion process is the manufacture of seamless tubing and, in particular, the manufacture of the stainless-steel variety. This is because rotary, longitudinal rolling processes often give unsatisfactory results, and operations such as Assel elongating fail altogether. The importance of extrusion becomes obvious in these circumstances, and its applicability to the processing of ferrous alloys cannot be stressed too strongly.

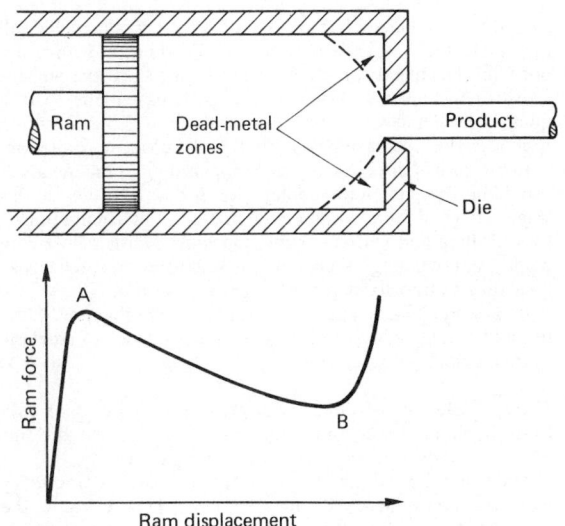

Figure 16.45 Formation of a dead-metal zone. Mechanism of formation and the associated ram-force displacement curve

Although extrusion has been used for non-ferrous metals since about the middle of the last century, serious interest in its application to steels was not shown until the mid-1920s when the first experiments carried out in France, the USA and Germany showed the distinct feasibility of the operation. The problem encountered was the high rate of tool wear, in particular of the dies. The slow development of tool materials and manufacturing techniques prohibited any extensive use of extrusion until almost the outbreak of World War II, but rapid progress was then made in Germany and the USA, whereas in France, Sejournet continued the development of his glass lubricating technique, which became widely accepted in the 1950s.

Since the extrusion of stainless steel offers wide scope for discussion, these materials are used in the following as the basis for a review of techniques and practices.

The current practice of extruding stainless-steel as tubing and sections, either by direct or reverse methods, is based on the following sequence of operations:

1. preparation of billet,
2. heating,
3. lubricating,
4. providing a pilot hole,
5. reheating and relubricating,
6. extruding,
7. removing the lubricant, and
8. straightening whenever required.

Reheating and relubricating (5) are not necessary in some processes.

In general, four main variants of the process are in operation. These involve the following techniques:

1. prepiercing, extruding;
2. drilling the billet, expanding, extruding;
3. drilling the billet, extruding; and
4. staving (dumping) the billet, piercing, extruding.

The choice and application of a given technique are examined in the following detailed discussion of the general sequence of operations.

Preparation of billet Irrespective of the process involved, all billets are faced at least on one end. Furthermore, they are normally machined, or sometimes ground, since the surface finish of the billet determines to a great extent the surface quality of the hollow. A surface finish of about 7.5 μm is often required, thus increasing considerably the cost of production.

In the case of direct extrusion (using drilling as the means of providing the pilot hole), billets are drilled centrally at this stage. Holes of up to 50 mm in diameter are machined out on twist drilling and vertical boring machines. Larger holes are made by trepanning. Trepanning is also used on smaller size holes in an effort to increase the initial length of billets.

Invariably, billets are either radiused or chamfered externally at the leading ends to offset the tendency to cracking in this region during extrusion.

Heating Heating of billets is of great importance, not only from the point of view of the time involved, but also in terms of the economics of the possible descaling operation.

The general trend appears to be towards the low frequency inert gas atmosphere heating of billets. Theoretically, the economic use of a low frequency induction furnace puts severe limitations on the use of smaller diameter billets below, say, 100 mm diameter; the range above 150 mm being considered economical. The advantage of obtaining a scale-free billet may sometimes outweigh the disadvantage of higher heating costs and, consequently, the intermediate range between 100 and 150 mm is occasionally used.

When using induction furnaces, the general practice is towards preheating billets to about 1020°C followed by further induction heating to about 1200°C for staving and prepiercing, and expanding processes.

Other techniques are also in use, e.g. preheating to about 820°C in a gas-fired furnace to avoid heavy scaling, followed by heating in a salt bath to extrusion temperature. A practice widely adopted in the USA consists of preheating in a Selas radiant heat slot furnace to about 1200°C, followed by heating in a barium or sodium chloride bath (to dissolve scale) to extrusion temperature.

A new heating system that makes use of dual fuel furnaces is also in use. In this system, steel is preheated to 900°C in a gas-fired furnace and is brought rapidly to 1250°C in an induction furnace. Scaling appears to be negligible.

In the case of billet drilled for direct extrusion, heating is usually carried out in a salt bath, with the exception of the revolving Balestra type furnace. The latter combines the heating and lubricating operations, being a drum type furnace in which the refractory lining is coated with a thick layer of molten glass. Prior to heating, glass wool is inserted into each end of the hole drilled in the billet in order to prevent oxidation. Equiverse type furnaces, which have proven very successful in the case of low carbon steels, are not often used for stainless steels.

Essentially, the final heating operation is carried out either in an induction- or salt-type furnace. Each of these processes has distinct advantages and disadvantages.

Lubrication The inherent difficulty in extruding steels, and particularly stainless steels, requires more efficient and, to a certain extent, more sophisticated lubricating methods than those used for the processing of ordinary engineering alloys. Originally, graphite type lubricants were widely used mainly because of the possibility of lubricating the mandrel when extruding small-bore hollows. The Ugine–Sejournet process, using glass as lubricant, has changed this situation significantly.

The innovation consists not only in using glass, but also in introducing the novel idea (in hot working) of lubricating only the work piece and not the tools.

However, the possibility of increasing the range of extrudable materials and the amount of deformation in a given operation when using glass can be predicted from a theoretical analysis of the process.

Glass is applied by means of pads of fibre, cloth or pressed powder. The only disadvantage of glass is the slight difficulty in removing it from the extrudate. In a properly controlled process, however, the layer of glass is very thin and on cooling there is a tendency for iron oxides to form. These dissolve glass slightly and facilitate its removal.

Providing a pilot hole An initial hole is made either by drilling or machining, or by hot piercing. Drilling and machining, followed directly by extrusion, are used where the finished bore is small (usually up to about 30 mm). The cost of the waste material and labour involved is considerable, but the lack of eccentricity in the finished tube outweighs to a certain extent this distinct disadvantage. This technique is also used whenever the formed metal is difficult to hot pierce.

To obtain bore sizes of 32–115 mm, small pilot holes are drilled and then expanded. Pilot holes for this operation are 20–25 mm in diameter. For bore sizes larger than 115 mm in diameter, hot piercing followed by extrusion is employed.

Extrusion The techiques of extrusion are discussed in Section 16.2.6.2 but, in the case of stainless steel and the Ugine–Sejournet process, tube sizes and the corresponding required press capacities need careful assessment.

The inherent weakness in the extrusion process lies in the production of discard associated with the formation of the dead-metal zone which, naturally, represents the total loss of material. With the high cost of stainless steels, the problem of discard is more serious than for other materials. The weight of discard is approximately proportional to the cube of the mandrel diameter. The weight is not affected by the introduced variations in the diameter of the die. Although a reduction in the die angle gives some improvement, this is small in comparison with the improvement achieved by completing the piercing–extruding operation in two stages. In this operation, the billet is first pierced against a solid plate which temporarily replaces the die. On completion of piercing, the plate is removed and the extruding die is introduced into its place.

The advantage of reducing the weight of the discard in this way must be set against the cost of the labour involved and the provision of back-plates which, obviously, can be damaged easily.

Removal of lubricant Graphite-based lubricants are easily removable by conventional methods. When glass is used as lubricant, techniques vary slightly but, essentially, they all depend on the application of some type of pickling bath. Thus, glass is removed by pickling in a mixture of 4% hydrofluoric acid and 14% nitric acid, or in a mixture of hydrofluoric acid and sodium sulphate.

Small amounts of the mixture of partly decomposed glass and scale are sometimes removed by sand blasting.

The increase in the use of the extrusion process in the last 25 years has been due partly to economical and partly to technical considerations. Extrusion becomes economical and competitive with rolling processes for a range of sizes of up to 150 mm outside diameter, and for comparatively short lengths and runs. The cost of plant required for the production of bigger hollows, both in terms of presses and ancillary equipment, increases rapidly when this limit is exceeded without there being, at the same time, any possibility of increasing the length. At present, the intermediate range (say, 150–250 mm outer diameter) is manufactured using extrusion presses but the process becomes rather expensive owing to considerable tool wear and increased rate of scrap. The additional, inherent disadvantage of the process is its comparative slowness with, on average, 50–60 extrusions per hour.

A more important reason for the use of extrusion is the fact that, from the point of view of formability, the process is capable of dealing with very difficult materials. This is possible due to the lower incidence of redundant strains than in rotary processes, and the consequent reduction in the severity of the operation.

16.2.6.4   Cold extrusion processes

Conventional cold extrusion operations are based on the same three basic systems of forward, backward and side operations (see Figure 16.43). The range of materials usefully employable in engineering applications is limited to steels and aluminium and copper alloys. Although easily extrudable, materials like tin, lead and magnesium show no benefit from strain hardening and, in any case, are of no great industrial importance.

Although hollow sections are produced by extrusion in non-ferrous alloys, the bulk of the products consists of profiled sections (particularly in aluminium) used extensively in the car and aircraft industries, as well as in domestic situations.

Very high extrusion pressures, combined with extensive frictional effects that occur in the container and in the die, make conventional extrusion unsuitable for processing less ductile materials or composites. Although the individual demand for either is not yet very high, developments in modern technology make it imperative that processes capable of coping with such materials be developed. Two processes fulfilling the demand are hydrostatic and impact extrusion. Both of these rely on high pressure, but reduced friction, forming.

The process of hydrostatic extrusion differs from conventional extrusion in that it employs a pressurized liquid instead of an extrusion ram. A diagram of the type of hydrostatic extrusion systems available is shown in Figure 16.46. The extrusion die has, in this case, a relatively small cone angle, but the high bursting stresses associated with this are offset by the fact that the fluid pressure acts as a containing element around the die circumference.

In practice, this operation has a number of advantages over conventional extrusion, most of which are related to the fact that there is no contact between the billet and the extrusion container wall. This results in:

1. lower extrusion pressure,
2. improved lubrication from the pressurizing liquid,

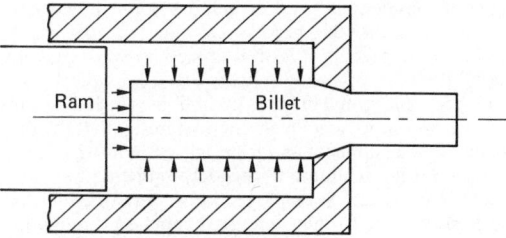

(a) Basic hydrostatic extrusion

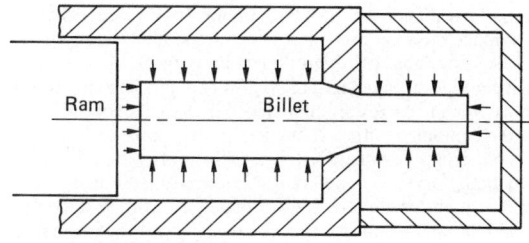

(b) Differential extrusion

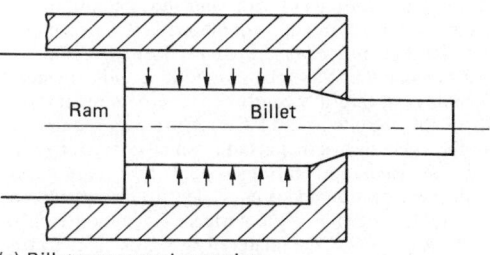

(c) Billet augmented extrusion

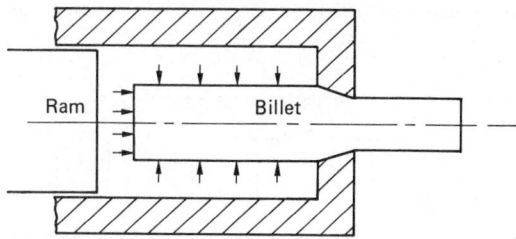

(d) Product augmented extrusion

Figure 16.46 Different hydrostatic extrusion systems

3. a reduction in redundant work since smaller die angles can be used, and
4. the extrusion of longer billets—the limit being the length of the container.

A further advantage of hydrostatic extrusion is associated with the improvement in ductility which most materials undergo when deformed under hydrostatic pressure. In practice, this means that the extrudability of materials is improved when the hydrostatic pressure component is present and under such conditions many nominally brittle materials have been satisfactorily cold extruded.

As in other processes, it is found that for a given amount of deformation and specific frictional conditions, there is an optimum die angle which gives the best balance between friction and inhomogeneous strain and, hence, there is a minimum extrusion pressure.

In spite of considerable research and development work on hydrostatic extrusion, the process has not been as widely accepted as an important industrial process as was originally expected.

Away from the conventional operations, including hydrostatic extrusion, it is the high-energy-rate operations that are steadily finding application in specialized industrial areas, and extrusion is no exception to this development.

Impact extrusion can be employed in the case of either bar or tube forming. The forming energy is supplied through the medium of a ram which, in turn, is actuated by a sudden expansion of gas produced by, say, an explosive charge or a pneumatic–mechanical system.

In the forward extrusion of bar (Figure 16.47), immediately after impact the ram and billet travel together at high speed with the extrusion of the billet taking place in the die. Relatively high strain rates can be attained and in some alloys these will lead to a substantial increase in the value of the yield stress Y, combined with a reduction in the strain to fracture. This is particularly noticeable towards the end of the operation when a high degree of deceleration is reached. It is also at this stage that high tensile stresses are reached at the base of the extruded bar and result in either the necking or breaking off of the product.

The maximum possible deformation depends mainly on the ability of the tool materials to withstand shock conditions.

Impact extrusion of tubular components is more conveniently carried out in a reverse system (Figure 16.48). This employs a ram of either flat or contoured face and a means of applying additional carefully controlled external pressure p to the walls of the extruded product. Pressure p is required to minimize the effect of two major faults which characterize the operation, i.e. cavitation and fish skin. Fish skin occurs throughout the operation and is equivalent to the appearance of circumferential tearing of the surface. Cavitation is associated with the flow of the material upwards, leaving the corners of the container unfilled. Rounded instead of sharp edges are thus produced. The practical usefulness of this particular process depends, to a considerable degree, on

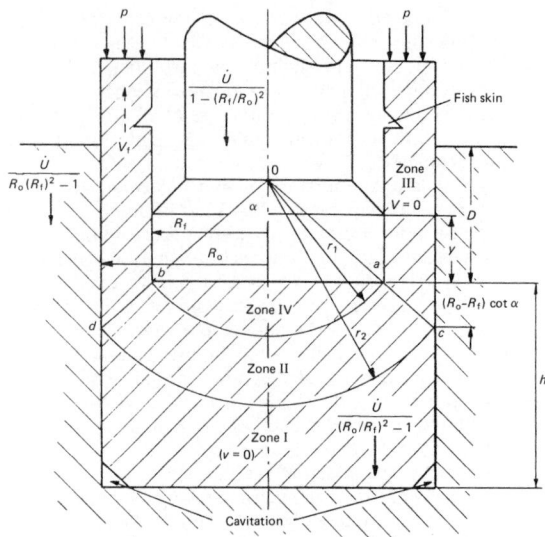

Figure 16.48 Reverse impact tube extrusion

whether the incidence of faults such as break-offs in a bar or cavitation and fish skin in, say, tubular bottle manufacture, can be avoided.

16.2.6.5 Continuous extrusion

Continuous extrusion processes have been explored in some depth, but it appears that, at present, the Conform process is the only one that is used extensively.

The process relies on frictional forces generated between the billet and container. These are sufficiently high to effect

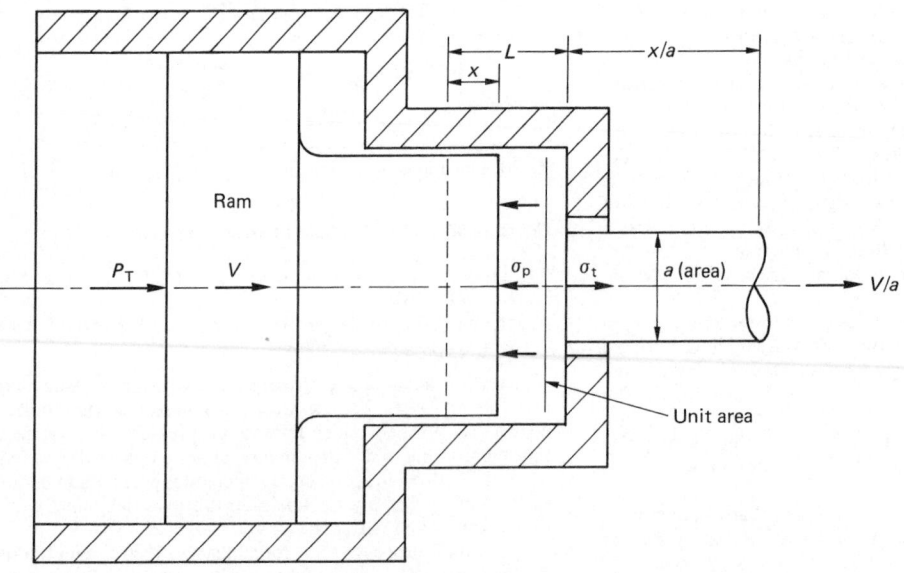

Figure 16.47 Forward impact extrusion of bar

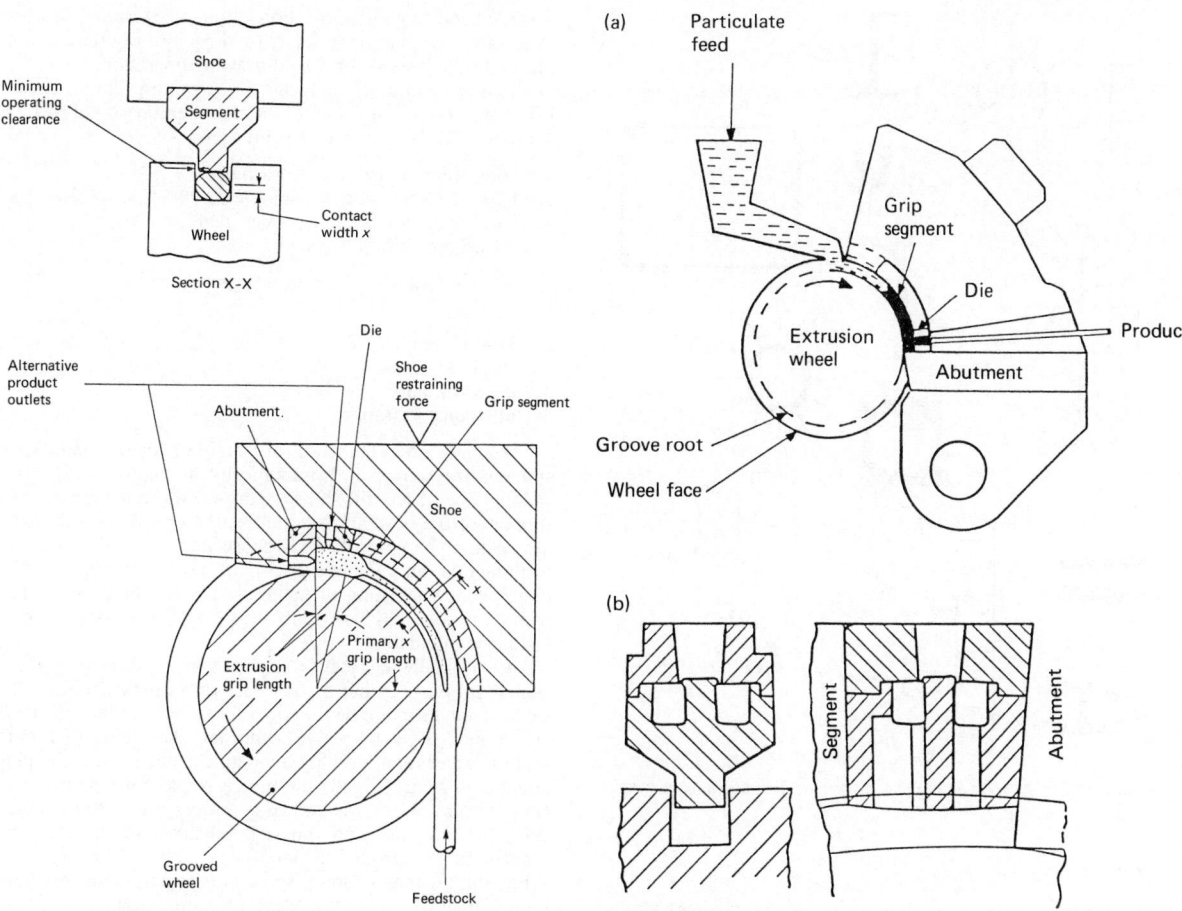

Figure 16.49 Arrangement of a Conform extrusion machine

Figure 16.50 (a) Detail of the Conform machine. (b) Twin-port dies

extrusion through the die. The operational system is illustrated in Figure 16.49. The product can be delivered either axially or radially (as in side extrusion) as shown in Figure 16.50(a).

The tooling consists of a rotating wheel with a circumferential groove, a shoe which overlaps a portion of the wheel surface and includes a grip segment and an abutment containing the die.

Solid and tubular sections can be extruded in aluminium and copper through single- or twin-port dies (Figure 16.50(b)). Both solid and particulate matter can be used as starting material, but very high extrusion ratios of up to 200 produce temperatures of up to 500°C and die pressures of up to 1000 MPa. Consequently, good tool materials must be used and the system must be cooled efficiently. With the groove of the wheel undergoing cyclic thermal stressing, the possibility of fatigue must be considered when designing this part.

16.2.6.6 *Section and sheet extrusion*

Although interest often centres on the extrusion of rod and wire, a large proportion of extrudates in light-metal alloys is manufactured as profiled sections ranging from simple curtain rails to very complex structural shapes a selection of which is shown in Figure 16.51.

In such cases the flow of metal during the operation is far from uniform and the incidence and level of redundancy can be high unless die profiles are designed correctly.

The requirements of modern industry demand more and more composite bi- or tri-metallic components which possess sufficiently good mechanical and physical properties to be suitable for forming to specific shapes. Examples of such applications involve the use of bimetallic sheets in the chemical industry where, say, a thin layer of stainless steel on a low carbon steel base will provide both structural strength and anticorrosive properties, and can be used in the fabrication of containers and pressure vessels, or in the electrical industry for the manufacture of bimetallic strip conductors.

Bimetallic strip is often produced by rolling but, more recently, an extrusion method has been developed for small width and thickness strip in which billets of different metals are extruded simultaneously from two or more containers to form a composite strip.

This manufacturing method provides a means of obtaining good dimensional tolerances and results locally in sufficiently high stresses to produce pressure welding. High deformations exceeding 50% are needed for this to occur.

Extrusion pressures, experienced by the metals, are only slightly higher than the corresponding pressures occurring with similar extrusion ratios in circular sections.

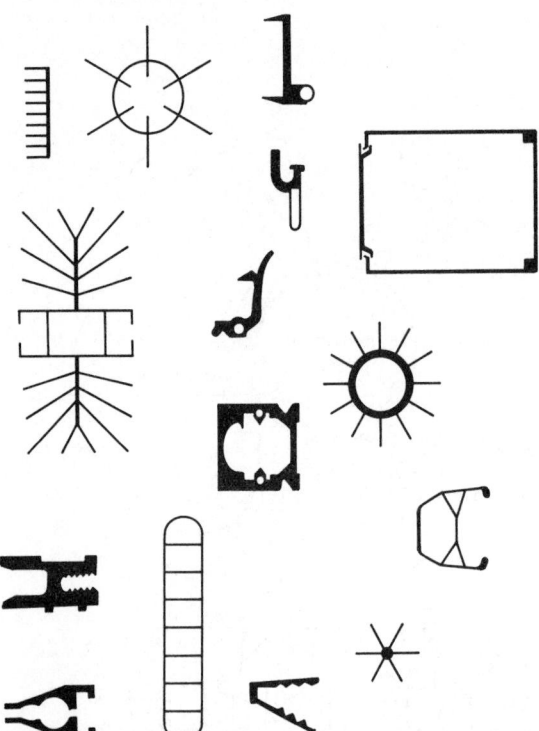

Figure 16.51 A selection of profiled sections extruded in light-metal alloys

16.2.6.7 Die materials

For the hot extrusion of aluminium and similar light alloys, H13 steel dies are normally used. α/β brass extrusions require H10A and H21 steel dies or stellites and cobalt-based alloys. Copper components are extruded through Nimonic 90, or other nickel based alloy dies, whereas steels are usually processed in H13 and H21 steel dies, or in TZM molybdenum alloy. Nickel alloys are invariably extruded through Nimonic 90 dies, and titanium alloys through hot-work refractory oxide coated steel tools.

16.2.7 Cold drawing of wire and tube

16.2.7.1 Introduction

As pointed out earlier, hot processing normally produces low-strength materials and uneven dimensional properties of the product. If high-quality fabricates are required, cold processing must follow the preliminary hot method of shape acquisition. In axisymmetric components, such as wire (rod) and seamless tubing, this is achieved by cold drawing the hot-finished product through a die (or a series of dies). This treatment imparts good mechanical properties and effectively regulates dimensional tolerances.

Basically, a nozzled (in the case of a tube) specimen is inserted in the die and is then gripped by a suitable device which can pull it forward on a mechanical or hydraulic bench. A reduction in the diameter (solid specimens) and wall thickness (tubular specimens) results. Multi-die drawing is necessary in wire production, whereas single-die processing is more usual in the case of a tube although, for special reasons, tandem drawing is sometimes employed.

Coiled rod, obtained in a hot-rolling or extruding process, forms the starting stage of wire drawing, whereas a hot-finished tube is used for a cold-drawing operation.

There are a number of variants of wire-drawing processes. Processes are usually based on the method of lubrication adopted, but the four major routes are:

1. conventional dry or wet drawing,
2. hydrodynamic lubrication system,
3. hydrostatic lubrication system, and
4. ultrasonic vibration system.

The main operational tube drawing processes are:

1. sinking,
2. floating plug,
3. stationary plug,
4. mandrel, and
5. ultrasonic vibration.

The main characteristics of some of these operational systems are shown diagrammatically in Figure 16.52. In a sinking operation, the tube is drawn without any internal support of the bore and, therefore, the change in wall thickness cannot be controlled. Because of the bending and unbending that the wall undergoes in its passage through the die, thinning or thickening of up to 4–11% can take place. The process is generally used only as a preliminary operation to further manipulation.

The normally adopted route is that of drawing over a cylindrical plug. Either a stationary or a floating tool can be used. In the former case the tool is positioned in the die throat and is held there by a rigid plug bar. The tube, of a bore slightly larger than the plug diameter, is pulled over the plug and the plug bar. Initially the tube is 'sunk' onto the plug and is then drawn in what is known as 'close pass' or 'pure draft'. Both the diameter and the wall thickness of the tube are reduced in a controlled manner.

For the drawing of long, small bore, small diameter, thin-walled tubes, floating plug or semi-floating 'captive' processes are used. Here, an initially hot-processed or cold-annealed tube is inserted into a die which contains, in one case, an unsupported conical free plug. The tube is drawn over the plug and is then coiled on a drum for ease of storage and transportation. In the other variant of the operation, a conical plug that is free to float is prevented from moving in one direction by a bar.

The older method of drawing over a long mandrel, which supports the bore, is now seldom used, mainly because the tube drawn in this way clamps tightly onto the mandrel. Thus, to remove the tube after drawing, the tube and mandrel may have to be reeled in a cross-rolling reeler which expands the tube and thus frees it from the mandrel. The reeling operation can impart helical markings to the outer surface of the tube, thus necessitating another drawing operation, and is also likely to affect the uniformity of dimensions along the whole length of the tube. In some cases, where highly polished and dimensionally accurate, but relatively short tubing is needed, mandrel drawing can be used if withdrawal of the tool is possible without reeling. This is the case for tubing with sufficient wall thickness that buckling does not occur when the tube is freed by pulling the mandrel/tube assembly through a 'gate' of a diameter only fractionally larger than that of the mandrel. This results in a stripping operation.

In ultrasonic vibratory systems, used either in wire or tube drawing, the die is vibrated at an appropriate frequency to increase the efficiency of the process by affecting the rate of feed of lubricant and the mechanics of the drawing.

The techniques described above are equally applicable to the manufacture of ferrous and non-ferrous tubing, but con-

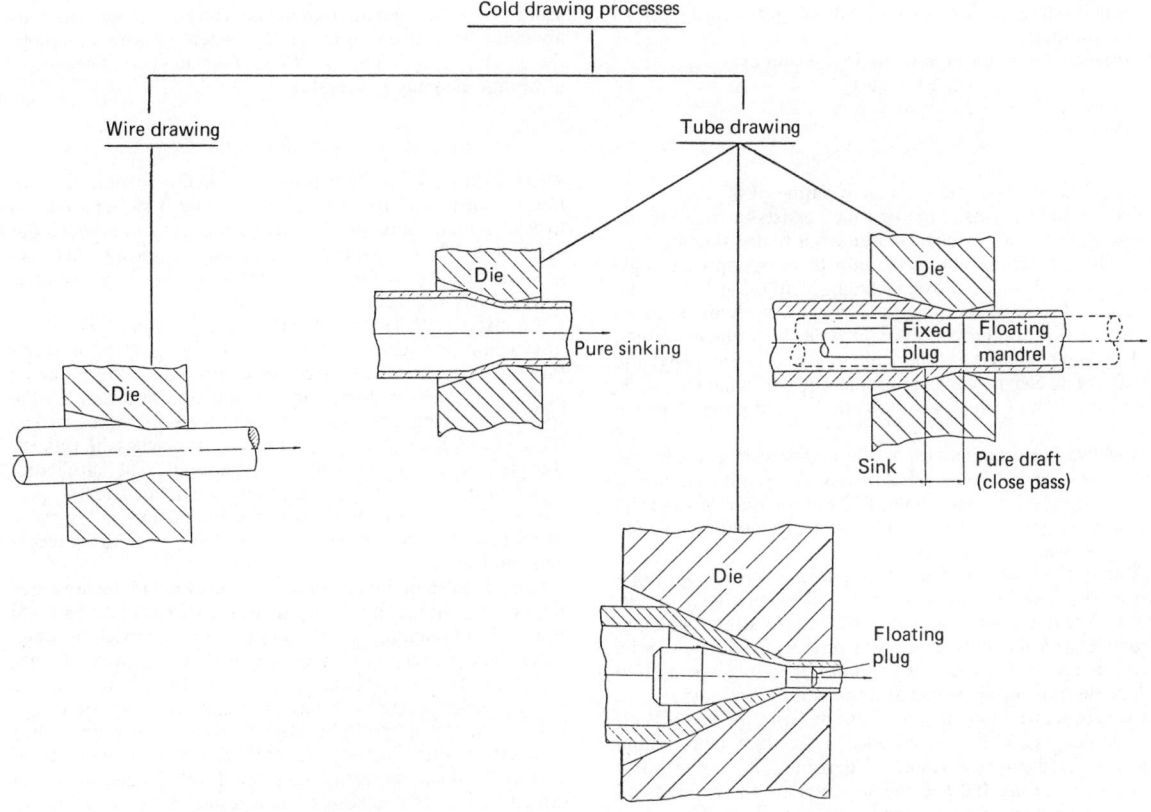

Figure 16.52 Cold drawing of wire and tube

ventional cold and warm drawing of stainless steels and cold drawing of square sections are of particular interest.

In the case of stainless steel, an increase in temperature to about 300°C reduces drawing forces by up to 35% and increases the attainable deformation by about 55%. Warm drawing thus constitutes an important development in this type of processing.

The development of the floating-plug technique, with its saving in tool material, better carry-through of lubricant, and saving of space when coil drawing, has been rapid and, in some cases, has overtaken the use of more conventional fixed-plug operations. Again of particular interest is the ever-increasing use of this process in the coil drawing of stainless and carbon steels. Irrespective of the material being drawn, the rate of production is increased; some 40% saving in time is possible when, for example, coil drawing on 0.9 m diameter drums.

Although the spring back or the final diameter of the drawn coil depends on the material processed, it is easily assessable for aluminium, brass, copper and alloy steels.

Wire drawing, often thought of only in terms of steel and copper, is in fact used to produce satisfactory lengths of rods and wires in a number of more exotic metals and alloys. On the single-metal side, brittle materials such as molybdenum and beryllium can be drawn, whilst tungsten wires must be manufactured at high temperature in order to counteract the unfavourable mechanical properties of this material. Bimetallic wires are manufactured by coaxial drawing of solid cores surrounded by hollow tubular sheaths, including ultrafine composites of niobium and copper, and copper and alumi-

nium. Transformation-induced-plasticity steel has been processed, as has ausformed silicon–chromium steel.

Clearly, the range of application of cold drawing is large and it continues to increase bringing in non-circular section rods and wires.

16.2.7.2 Basic concepts of wire drawing

Although it is customary when considering theoretical aspects of wire drawing to refer to a single pass, it must be made clear that drawing from a coiled rod to the finished product must be carried out in a multistage, normally automated, machine or draw bench. Successive passes call for the correct design of dies and the provision of suitable lubricating conditions. Die design requires minimization of the degree of redundancy and size of the drawing load, leading to the choice of the optimum (for the given conditions) effective die angle, while at the same time allowing a protective lubricating film to develop and be maintained constantly throughout the operation. Failure in lubrication results in impaired quality of the surface finish and an increase in the rate of wear of the die. The cost of remachining or replacement of the tool is then added to the cost due to the loss of dimensional accuracy of the product.

Since successive drawing passes are not able to rectify original surface faults or the presence of scale created by annealing the rod, the preparation of the rod is all important. Here, either 'dry' or 'wet' processing is adopted.

In a typical 'dry' in-line system for rod drawing, the specimen undergoes the following sequence of operations:

1. grit blasting by three or four sets of guns at, say, 90° to each other;
2. grit extraction through filters in a chamber;
3. air blasting to remove the dust;
4. lubricating in an enclosed chamber; and
5. drawing.

A dry blasting operation results in pitting of the surface or, at least, a matt finish, either of which assists in trapping the lubricant and creating local conditions of hydrostatic lubrication. This, in turn, promotes the possibility of applying single heavy passes. Various alloys, including stainless steels, can be treated in this way without any additional surface preparation.

The 'wet' descaling processes include the well-established acid pickling, improved, and sometimes accelerated by, the passage of an electric current of some 7 A cm^{-2} and ultrasonic vibration, both of which help to dislodge and precipitate the scale.

However, environmental considerations weigh against the atmospheric and effluent pollution associated with the use of acids and, therefore, attention has been focused on reducing the original amount of scale and on using molten salt bath heating in place of acid pickling.

A controlled carbon oxide atmosphere is suitable for high-speed, low-scale-forming annealing, and is sometimes preferable to vacuum annealing which may cause strand welding. Depending on the alloy processed, resistance heating can be used for, say, brass annealing, or fluidized beds can be used to reduce the time of operation and the amount of scale.

Electrochemical lubrication, resulting from the use of suitable molten salts of, for instance, potassium and lithium chlorides is the direct result of descaling in a non-acidic environment. Descaled and cleaned rods are then coiled and fed into drawing machines. Drawing of the wire itself is carried out at speeds ranging from 30 to 2500 m min^{-1}, depending on the material, with resulting reductions in cross-sectional area of 15–25% in the case of narrow diameter wires and 20–45% for coarser wires.

The two basic techniques employed are, again, the 'wet' and 'dry' processes. In the former, the entire production line is usually immersed in the lubricating liquid, whereas in the latter the wire 'picks up' the lubricant on passing through a container.

Copper and copper alloy, some aluminium, and very fine diameter wires are normally processed in the wet condition, while ferrous alloys and all other materials tend to be drawn dry.

In consequence, a wide range of lubricants is employed. The lubricant used must account for not only the specific material requirements, but also for the effects of cooling that are now recognized as being of major importance.

For most ferrous materials, the preferred method of lubrication is that of precoating the wire. Ordinary carbon steels are phosphate coated, but stainless steels require either oxalate compounds or borax as a lubricant. Although oxides protect the surface from die damage during drawing, they do not necessarily act as lubricants and so two-stage lubrication is still, reluctantly, used. In addition to oxide films, crystalline lubricants are required, and chlorine and sulphur type additives are used.

Electrochemical deposition of lubricants is still in the development stage but, if proven sufficiently economical, it will provide a relatively easy answer to standard processing lubrication problems in that the deposition and removal (combined with temperature control) of substances such as molybdenum sulphide is relatively simple by this means.

The removal, and possibly re-use (for reasons of economy), of lubricant is of importance. Cleanliness of the lubricant as achieved by, say, filtration of debris, can be high and will thus affect the 'brightness' of the surface when the wire undergoes the final series of passes. When drawing very fine wires, ultrasonic cleaning is necessary.

16.2.7.3 Non-standard wire-drawing techniques

These techniques include principally hydrodynamic lubrication systems and ultrasonic drawing. The basic idea behind hydrodynamic lubrication is to provide a continuous, but sufficiently thick, layer of pressurized lubricant that will separate the tool from the wire. In this way tool wear is considerably reduced.

An effective technique of achieving these conditions is that of drawing through a sealed tube containing oil as lubricant (although occasionally soap can be employed) and terminating in a constriction or nozzle and, eventually, in a die. As the speed of drawing increases (and high speeds are necessary) so does the oil pressure until, theoretically, it reaches the value of the yield stress of the processed material. The difficulties experienced in operating the system are related to the problems of sealing at high pressures and velocities, and to those of precision nozzle design which is 'adjustable' to materials and conditions.

Copper and aluminium wires can be drawn, but the pressures required for 'harder materials' such as, for example, steels, are too high to maintain successfully over a period of time. Nevertheless, since very good lubrication is generated with, consequently, low friction and tool wear, the technique should be considered where economy of operation is important.

In an ultrasonic multidie system, one or more dies when vibrated in the direction of drawing create conditions of back-pull which, in turn, alter the force requirement and directly affect its magnitude. As a result of die vibration, an oscillatory force is induced in the wires, between the consecutive dies and, eventually, the coiler drum. In consequence, greater reductions in cross-sectional area are achievable without either an increase in the degree of plasticity of the material or any decrease in interface friction. The level of this reduction depends on the value of the back-pull exerted and increases with decreasing back-pull factor. The surface finish and mechanical properties of the drawn wire remain unaffected by the oscillatory nature of the force system.

16.2.7.4 Special tube-drawing operations

Two completely diverse drawing systems now in operation are: ultrasonic drawing of ferrous and non-ferrous alloys; and fixed-plug drawing of explosively prewelded bi- or tri-metallic tubing.

Ultrasonic drawing is based on either the volume or surface effects produced by the oscillatory vibration of tools. The volume effect is subdivided into:

1. the superposition mechanism,
2. metallurgical effects, and
3. the swaging effect

The surface effects comprise:

1. the change in friction between the tool and the work piece, and
2. the friction vector.

If only the plug is vibrated at about 20 kHz, friction is reduced by surface effects, i.e. by a thickening of the film of lubricant between the tools and the work piece. This reduces the drawing load and/or eliminates pick-up and chatter. However, practical difficulties are experienced in designing the plug and plug bar that will give the required resonance.

If, however, the die is vibrated ultrasonically in the radial direction, rather than remain stationary as in a conventional drawing operation, a swaging effect is produced in the tube which, although conforming to the subdivision listed above, indicates that, industrially, the division between volume and surface effects is not all that important.

Again, the details of the design of suitable equipment cannot be reproduced here, but information is available in the publications listed in the Further Reading at the end of this chapter.

The process has the following basic characteristics:

1. the reduction in the drawing stress increases with increasing ultrasonic energy density,
2. a reduction in area of more than 54% is achievable,
3. the ultrasonic energy density can be increased by reducing the speed of drawing rather than by increasing the energy input,
4. surface finish is improved when radial die oscillations are used, and
5. pick-up and chatter are low or absent.

The high degree of sophistication required in the rapidly developing technology of the manufacture of petrochemicals, electronic devices (particularly cybernetics, cryogenic systems, and of atomic pile and toxic metals remote control systems) calls for the development of a new range of engineering, tubular composites. Some of these cannot be produced by conventional techniques alone or can only be manufactured at high cost. A typical example of the first group is a multilayer, multimetallic cylindrical pressure vessel, whereas an example of the second is a semiconductor system enclosed permanently in a protective tubular metal sheath. In between these extremes there is a wide range of components such as special bi- or multi-metallic heat exchangers, that combine structural strength with anticorrosive properties and ensure a rate of heat flow as good as that shown by the individual metals of the composite.

Conventional codrawing or coextruding of such assemblies gives less satisfactory results, because even with a high degree of process control it is practically impossible to ensure that no lubricant or debris is trapped on the surfaces. Furthermore, lack of cohesion between the original components of the assembly is likely to lead to very high differential deformation and the associated in-built shearing stress.

Many of these limitations can be eradicated if the integrity of the component is assured *a priori* by, say, explosive welding. In this respect, the manufacture of duplex or triplex, bimetallic or multimetallic cylindrical pressure containers is of practical importance and this has been accomplished successfully. Because of the technical problems involved when long composite cylinders are explosively welded, the usual technique is to weld short, large diameter combinations, and then to obtain the required dimensions by means of cold-plug drawing.

The characteristic features of drawing are the changes effected in the distribution of hardness across the tube section, and in the quality of the weld. While the former simply reflects the effect of strain hardening of the composite associated with the imparted deformation, the latter represents the effect of shearing at interfaces. Unlike hydrostatic extrusion, discussed earlier, where the adhesion of implosively welded elements is improved, in drawing weakening and even failure of the weld can occur. Of course, the failure, often only local, of the weld does not reduce the strength or tightness of the cylinder, since the function of the weld is only to promote more 'homogeneous' drawing conditions.

16.2.8 Sheet-metal forming

16.2.8.1 Forming processes

The considerable success and continuous expansion of the sheet-forming industry are due to a number of unrelated but important developments.

The introduction of the continuous tandem rolling mill has made it possible to produce wide strip which can either be coiled or cut to suitable operational sizes. At the same time, cold forming of sheet ensures greater resistance of the formed section to corrosion, since the preformed material has good surface finish due to treatments such as pickling. Surfaces cleaned in this way form suitable bases for the application of protective layers of, for instance, paint or plastic coating and, furthermore, they can be galvanized and then rolled or drawn.

The good dimensional qualities of preformed sheet make it a good starting material for operations such as bending, piercing, dimpling and drawing, by ensuring general uniformity of thickness. The exception here is the change in sheet thickness on transition sections where thinning is likely to occur. However, in a cold operation, strain hardening accompanied by changes in thickness will tend to counteract the effect of the actual variation in thickness.

The five basic processes effecting a change in shape are:

1. shearing,
2. bending,
3. spinning,
4. stretching, and
5. drawing.

A wide range of sections and shapes, and a varied range of products for industrial, domestic or general use can be produced.

Industrial applications are found in the following areas:

1. computer, nuclear and space industries;
2. the chemical industry;
3. the aircraft industry;
4. the car industry; and
5. the food industry.

Cold-formed sections are used industrially as components, plant, equipment, containers, panels, radiators, body parts, chassis, frames, racks, stiffeners, etc.

Domestic applications include, among others:

1. components,
2. cookers,
3. panels,
4. refrigerator bodies,
5. washing machines, and
6. shower cabinets.

General applications range from building constructions and components through structural elements, to rail and road transport, civil and highway engineering, agricultural machinery and equipment, and architectural and shop fittings. A major part of the output is intended for the electrical heating and ventilating applications represented by conduits, casings, cable supports, electrical appliances, heating panels, and dust-extraction equipment.

16.2.8.2 Shearing operations

The shearing operation constitutes the first stage of any forming process by producing either the starting material (cutting out of a sheet) or preparing an existing work piece by punching a hole or a series of holes before forming.

Basically, the operation involves placing the material between the edges of a shearing tool, which serve as supports, and separating a part of it by the action of a punch. Purely shearing stresses are generated, but the quality of the separating cut depends on the clearance between the tool and the specimen. If the clearance is too small, the cracks produced by the tool do not coalesce and tearing occurs. At the other extreme, too large a clearance results in considerable plastic flow which inhibits formation of the cracks and produces a burr at the upper edge.

16.2.8.3 Bending

Bending is executed either by pushing the strip into a die of the desired profile or, more often, by cold-roll bending (see Section 16.2.4.7). Either method can be used to produce a variety of shapes, a selection of which is shown in Figure 16.53.

Since the final quality depends on the material used, it may well be that the phenomenon limiting the degree of bending allowable in any given case will not be simply that of fracture or instability, but also one that produces an unsatisfactory finish. A selection of possible limiting factors, including the basic fracture and necking, is given in Figure 16.54. The maximum deformation may be limited by indentation caused by the die edges, galling, wrinkling and crease forming.

In addition to actual sheet bending, the operation is often extended to tube or blank flanging. Severe compressive stresses are imposed in these processes and may lead to buckling in a tube.

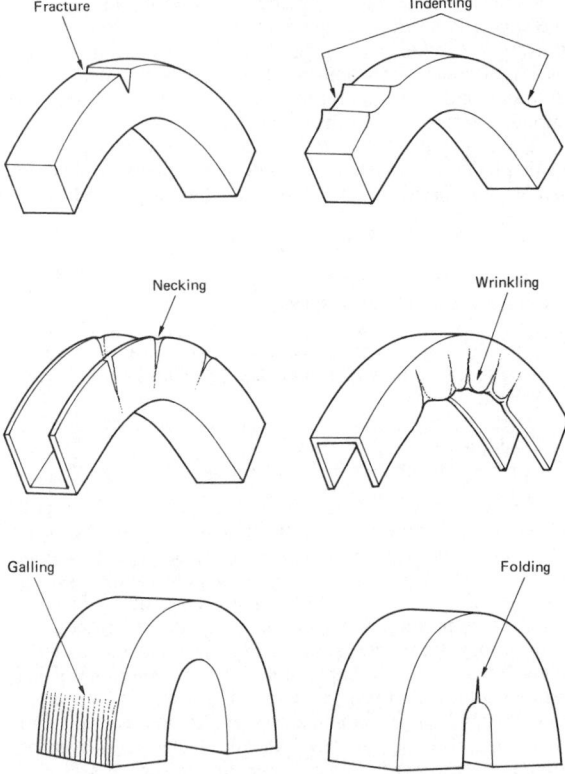

Figure 16.54 Sheet bending limiting product-quality phenomena

16.2.8.4 Spinning

Sheet forming by spinning is applicable to axisymmetric shapes only. The operation involves a driven, rotating system to which the male forming die is attached. The sheet to be formed is held against the die. Either manual or power, cold or hot spinning operations are used. The limit of formability is imposed by the ductility of the material and can be assessed from a simple tensile test.

16.2.8.5 Stretch forming

Stretch forming is often used to produce either simple components, comparable to cylindrical cups, or more complicated pressing-type shapes. Basically, stretch-forming operations depend on the use of conventional punch and die systems or a dieless-punch system. In either case, the rim of the blank is clamped (Figure 16.55) and the stretching, with consequent thinning of the material, can proceed on loading the punch. The thinning is particularly pronounced on the radiused rim of the supporting die or anvil. Instability can set in and lead to failure through localized necking.

Dynamic or explosive stretch forming is also used (see Section 16.2.9.3).

16.2.8.6 Deep drawing

Deep drawing is normally associated with the manufacture of cups, cans and similar containers. The operation is usually divided into two main groups: first-stage drawing, in which a flat circular metal blank is made into a cup; and a redrawing

Figure 16.53 A selection of sheet sections obtained in bending

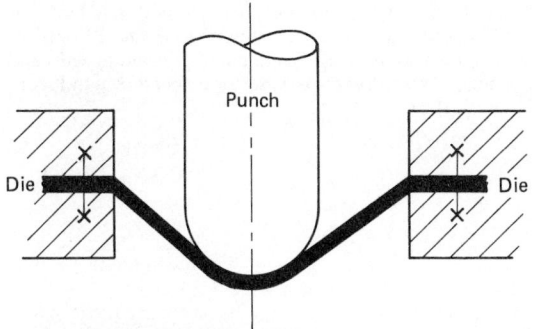

Figure 16.55 Conventional stretch forming

stage (or stages) in which the cup reaches its final size. The latter operation is necessary because first-stage drawing cannot normally produce a higher degree of deformation than that defined by the ratio of the diameter of the blank and the die throat (drawing ratio) of about 2.2, or a cup height/diameter ratio of about 1.

The sequence of operation is as follows. Initially, the specimen held in position by a blank holder, is partly in contact with the die, partly with either the die or punch, and partly with the punch only. The downward movement of the punch initiates drawing. The outer rim of the blank is then subjected to pure radial drawing (i.e. drawing towards the vertical axis of the system) between the die and blank holder. A part of the material bends and slides over the die and is further stretched between the punch and the die, whereas the material initially in the vicinity of the punch head and actually in contact with it bends and slides over the radiused part of the punch and stretches over the punch head.

The redrawing systems often used are shown in Figure 16.56. Parts (a) and (b) in the figure show direct redrawing systems with and without blank holders, respectively, while a reverse system is shown in (c). In (a), the wall of the cup

undergoes double bending and unbending, the severity of which is expected to be high because the respective directions of deformation are at right angles to each other. System (b) shows less severity because of the tapered wall support, although double bending is involved. This system can be used only for relatively low cup diameter/wall thickness ratios which do not require the use of a blank holder. In comparison with the direct methods, system (c), having a generously radiused die profile, tends to reduce the degree of (or with a semicircular profile to eliminate completely) one bending and unbending effect. Whether there is significant advantage to using any system depends on the balance between the reduction in redundancy and practical production considerations.

The definition of 'redundancy' in deep drawing is not easy since redundancy is not necessarily associated with the effects of macroshear. The nature of the processes is such that portions of the blank material undergo some phases of deformation which in themselves induce redundant effects and yet are physically unavoidable if the process is to be completed. It is therefore the degree of severity imposed rather than the avoidance of a certain phase of the operation that matters. In this respect, the process differs significantly from the bulk forming operations discussed previously.

The three main sources of unnecessary strain in and/or distortion of the blank or cup material are flange wrinkling, the already discussed bending and unbending, and, partly, ironing. The latter is used to eliminate the increase in cup wall thickness which can be as much as 30% in the first stage of drawing. If this is followed by a further substantial rise in successive processing stages and is accompanied by wrinkling, an additional drawing operation becomes necessary. As far as redundancy is concerned, ironing is the only operation that brings back the 'standard' features of shearing.

The formability of a material depends on the blank-holder pressure and, consequently, the deep drawing ratio $R = D/d$ may be limited either by wrinkling of the flange, tearing of the cup bottom, or by galling. Figure 16.57 shows diagrammatically the boundaries of these conditions and indicates the presence of a 'safe window' within which deep drawing is likely to be successful.

In determining the drawability, the criterion to be adopted is that relating to the first incidence of any fault.

16.2.8.7 Hydroforming and hydromechanical forming

To increase the depth of the draw, while reducing local stress concentrations, the techniques of fluid-backed or fluid-mechanically augmented processing have been introduced.

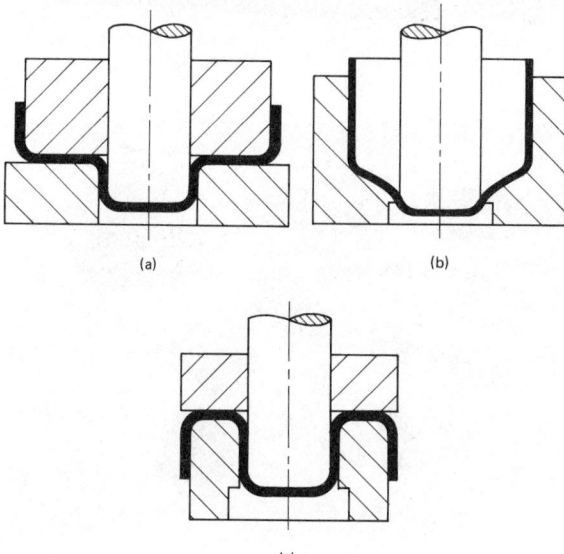

(a) (b)

(c)

Figure 16.56 Direct cup redrawing with (a) and without (b) a blank holder. (c) Reverse drawing

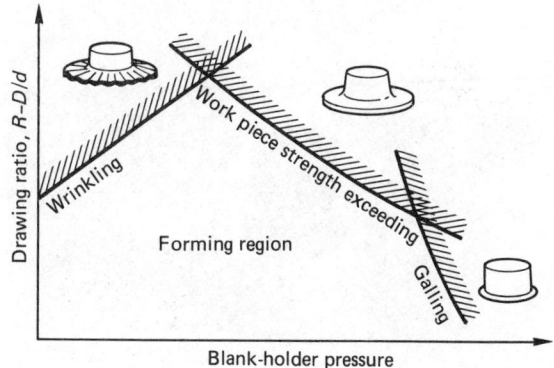

Figure 16.57 The effect of blank-holder pressure on the cup drawability

In a pure hydroforming operation (Figure 16.58(a)), a rubber diaphragm pressurized by the fluid acts as both a blank holder and a flexible tool. The only rigid tool is the punch. Since hydraulic pressure is exerted on the blank, the stressing is uniform and is accurately controlled. With reduced sheet thinning and considerable suppresion of cracks, high draw ratios (up to $R = 3$) are possible. The method is suitable for the forming of both jet-engine components (some 1 m × 1 m in size) and a variety of car components, including side and roof panels. A selection of parts so formed is shown in Figure 16.58((b)–(e)).

(a)

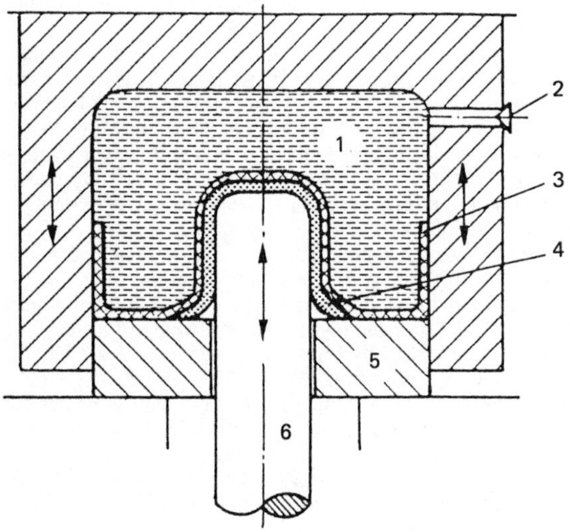

(c)

(d)

(b)

(e)

Figure 16.58 (a) Pure hydroforming operation: 1, hydraulic cushion; 2, pressure regulator; 3, cup diaphragm; 4, blank material; 5, pressure pad; 6, punch. (b, c) A range of jet engine and aircraft components. (d) Side panel of a car body. (e) Car parts. (Parts (b) to (e) courtesy of Saab-Scania AB Car Division, Sweden)

In the Hydromec, or mechanically augmented system (Figure 16.59), the sealing diaphragm is omitted and the blank comes into direct contact with the fluid. The sealing of the fluid is effected through the use of sealing rings. The blank is pressed firmly against the punch and thus possible bulging between the die ring and punch is avoided; a feature of the process that is of particular importance in the deep drawing of parabolic components. Complex parts can be drawn in a single operation, thus reducing operational costs. Sizing tools are claimed to be unnecessary.

16.2.9 High-energy-rate operations

16.2.9.1 Introduction

The last three decades have seen the development of industrial forming techniques based on the utilization of high-energy-rate dissipation. The dynamic aspects and mechanisms of these operations make it possible to manufacture, usually fairly economically, either large components, such as radar dishes, rocket nose cones and pressure vessels, or to produce smaller semifabricates (in sheet or tubular form) to be processed further in conventional operations. Some examples of this approach were given in Sections 16.2.6 and 16.2.7.

The usual sources of energy available are:

1. chemical explosives,
2. electrostatic fields (hydroelectric forming), and
3. magnetic fields (magnetic forming).

These sources of energy are used in:

1. forming (sheet and tube),
2. welding (sheet sandwich semifabricates, and multimetallic tubular components),
3. powder compaction (to form prefabricates),
4. forging, and
5. hardening.

The potential use of any of these operations is conditional on the formability of the given metal in the dynamic condition. An indication of this, based on 1100-0 Al as standard, is given in Figure 16.60.

16.2.9.2 Sources of energy

Explosives The proper utilization of the energy evolved when an explosive charge is detonated depends on the degree of understanding of both the properties of explosives and of the detonation phenomena.

From a practical point of view, commercial chemical explosives will generally fulfil the requirements of the welding, forming and compacting processes. On detonation, the detonating front travels through the explosive converting its mass to a high temperature, high pressure gas which is then used in the generation of a stress wave and, depending on the conditions, in the appearance of shock and release waves.

Chemical explosives are subdivided into 'high' and 'low' (deflagrating) materials. High explosives are characterized by very high detonation pressures and high rates of reaction. Because of this, high explosives are subdivided further into two groups.

1. Primary or detonating explosives which are sensitive and may be detonated by slight impact, flame, static electric charge, or simple ignition. They are normally used in detonators, but seldom as a source of energy in metal working.

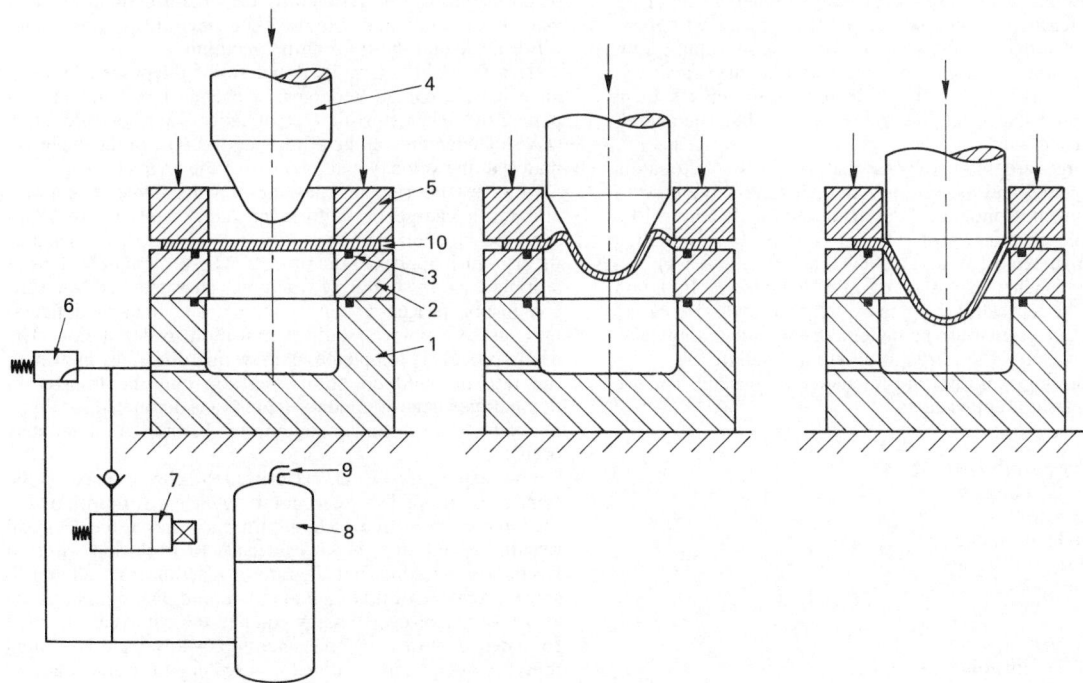

Figure 16.59 Hydromechanical deep drawing (the Hydromec process). 1, Container for hydraulic cushion; 2, die ring; 3, die ring seal; 4, punch; 5, blank holder; 6, pressure regulator; 7, electromagnetic control valve; 8, fluid container; 9, compressed air supply; 10, circular blank

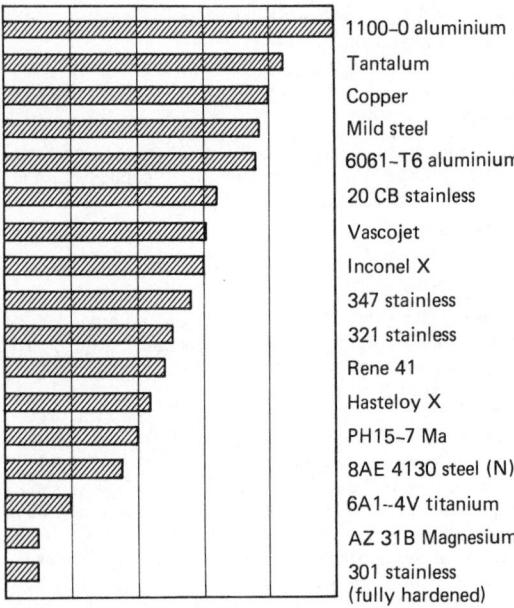

1100–0 aluminium
Tantalum
Copper
Mild steel
6061–T6 aluminium
20 CB stainless
Vascojet
Inconel X
347 stainless
321 stainless
Rene 41
Hasteloy X
PH15–7 Ma
8AE 4130 steel (N)
6A1–4V titanium
AZ 31B Magnesium
301 stainless (fully hardened)

Figure 16.60 An assessment of the formability of metals

2. Secondary high explosives are used mainly in metal working and other industrial applications. They require a detonator to initiate the reaction and sometimes a booster charge to reinforce the detonation wave. They have a higher energy content than primary explosives.

Deflagrating or low explosives burn rather than detonate when the reaction is initiated and produce much lower pressures. They usually contain their own oxygen supply and, therefore, burn easily, but in some materials the reaction is difficult to initiate. Their low rate of burning makes them excellent propellants, but fire risk is considerable when handling such chemicals.

The charges are initiated by detonators. Electric detonators are widely used and are safe to handle if reasonable precautions are taken. Commercial detonators normally consist of a thin metallic container protecting the contents of an initiating primary high explosive, and a small amount (about 1 g) of a sensitive secondary explosive, e.g. PETN or tetryl. Initiation is achieved electrically, using an exploding bridge wire. Some types contain a slow-burning material to provide a time delay when many charges are fired at different time intervals.

The following explosives are commercially available and are used in industrial applications.

Primary high explosives
1. mercury fulminate,
2. lead azide,
3. diazodinitrophenol,
4. lead styphnate, and
5. nitromannite.

Secondary high explosives
1. TNT (trinitrotoluene),
2. tetryl (trinitrophenylmethylnitramine),
3. RDX (cyclotrimethylenetrinitramine),
4. PETN (pentaerythritol tetranitrate),

5. ammonium picrate,
6. picric acid
7. ammonium nitrate,
8. DNT (dinitrotoluene),
9. EDNA (ethylenediaminedinitrate),
10. NG (nitroglycerine), and
11. nitro starch.

Low explosives
1. smokeless powder;
2. nitrocotton;
3. black powder (potassium nitrate, sulphur, charcoal); and
4. DNT (dinitrotoluene ingredient).

Explosive materials are available in different forms and some of them in more than one. Many of them can be melted allowing other explosives to be added in the form of slurries. Powdered, granular, solid, liquid and plastic explosives can be used. One of the most useful types is so-called 'Datasheet' or 'Metabel', which is essentially a PETN explosive combined with other ingredients to form a tough, flexible waterproof sheet that can be cut and shaped to the required size for contact and stand-off operations. It is available in different thicknesses and can be glued together or used as shaping back-up material if a shape charge is required. Another very useful type is powder explosive, particularly various mixtures of TNT with aluminium powder, which can fill a container of any form and then be compacted to attain higher densities. Sealed containers are not required, thus offering an advantage over liquid explosives. Cord explosives, e.g. Cordtex, are also available. These consist of a flexible cord containing a core of explosive. Cords are very useful when continuous long charges are needed and give reasonable accuracy in a number of forming operations.

The effectiveness of a charge depends on the characteristics of the explosive, as reflected by the pressure–time function, velocity of detonation, explosive/specimen mass ratio, stand-off distance and the transmitting medium.

'Detonation' is a term used to describe the process in which an explosive charge undergoes a chemical reaction accompanied by a characteristic type of shock wave (or detonation wave). Depending on the properties and type of the explosive material, the velocity and intensity of the characteristic shock wave varies, but remains constant for a given type of explosive and for a charge of uniform geometry and density. This simplifies the mathematical solution of the hydrodynamic theory which applies to the process. The general behaviour of a primary explosive during reaction is characterized by a slow combustion process at the beginning, and then by deflagration, up to a point of sudden transition to detonation. The whole process is completed in a few microseconds. However, the rate of build up of the reaction and the transfer to detonation when a secondary explosive is detonated without a detonator is much slower and burning before detonation may occur.

Low explosives are characterized by the absence of the transition period. They react at rates which are proportional to the build-up pressure which, in turn, increases as the chemical reaction speeds up. This cycle leads to explosion within a fraction of a second, but the rate of reaction is usually much slower than 1% of that in detonation, and peak pressures are also lower. However, energy comparable with that obtained from high explosives can be generated by low explosives when they are adequately confined or used in sufficient quantity, and pressure distribution is easily controllable.

As far as metal-working processes are concerned, the most important parameters of the detonation process are:

1. the energy released by the detonation,
2. the detonation velocity (i.e the velocity of propagation of the detonation front), and
3. the pressure exerted by the gaseous products of detonation on the specimen.

Other aspects of the process, such as the thermal stability and sensitivity of the explosive, the temperature, the heat generation and the ionization phenomena in the gaseous products, as well as their composition, are of no importance.

To carry out a successful metal-working operation the optimum strain rate for the metal must be ascertained, and the amount of energy dissipated must be adjusted to give the desired rate.

The adjustment of the amount of energy available for a given operation depends on the source used and the method of application. Regardless of the type of source used, the energy can be delivered in one of two ways: through a transmitting medium, or directly to the metal. Apart from some special applications, delivery through a transmitting medium is more usual.

Hydroelectric forming The principle of operation is based on the rapid dissipation and transmission of energy evolved when an electrostatic field is suddenly discharged. Two different techniques of using the energy stored in a bank of condensers are employed: underwater discharge, and exploding wire. In the former case, the discharge across two submerged electrodes produces a shock wave in the transmitting medium, accompanied by heating and vaporizing of the adjacent layers of the medium. The plasma created by the spark expands as a gas bubble, transmitting the force of explosion to the work piece. The efficiency of the operation depends on the conductor material, losses of energy in the circuit, and the geometry and surface conditions of the electrodes.

The second method consists of connecting submerged electrodes with an initiating wire. The transmitted energy vaporizes the wire and converts it into plasma, creating a pressure wave. The increase in volume of the vaporized wire is of the order of 25 000 times its original volume. The exploding-wire method possesses certain distinct advantages over the spark-discharge method in that the process can be more rigidly controlled. The shape of the shock wave can be determined by the shape of the wire, and a long arc discharge can be obtained as opposed to a point source one. The amount of energy can be controlled by the dimensions and material of the wire. For instance, tungsten produces more energy than tantalum, niobium, molybdenum, titanium, nickel or aluminium (in that order).

Electromagnetic forming The principle of electromagnetic forming is, basically, the same as that of an electrohydraulic operation. The energy stored in a bank of condensers is rapidly discharged through a magnetic coil, which surrounds, is placed inside, or is in the proximity of the work piece (Figure 16.61). A high intensity magnetic field is thus created and, providing that the material of the work piece is conductive, electric current is induced in the specimen. The current interacts with the coil field and produces high transient forces. The specimen thus acts as a secondary short-circuited coil. The energy level produced by the magnetic field depends on the conductive properties of the formed metal, its shape and mass and the duration of the initial current pulse. These factors can quite easily be controlled by suitable choice of the materials and geometry of the coils. Materials of low conductivity are sometimes lightly coated with copper. The shape of the impulse wave can be modified by using 'field shapers' which consist of shaped beryllium–copper pieces inserted in the coils. Electrically, shapers help to depress or concentrate the intens-

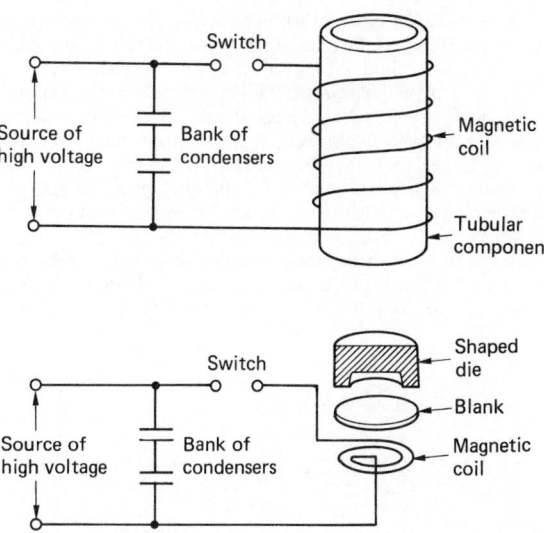

Figure 16.61 Some circuits used for magnetic forming

ity of the magnetic flux in those sections of the work piece which may require a lower or higher degree of deformation.

The life of a coil depends on the magnitude of the force to which it is subjected. This being equal to the force generated on the surface of the work piece, the pressures can be very high. The usual practice consists, therefore, of using one-shot disposable coils for more complicated operations involving only a few parts, and of limiting the pressures to the value of the compressive strength of the coil material for mass-produced parts.

16.2.9.3 Forming systems

Explosive forming Explosives are used primarily for shaping sheet, plate and tubes, and for sizing and flanging, all these operations being of the stand-off type. Close-contact operations are used to a lesser extent and for more sophisticated processes such as extruding, cladding (welding), powder compaction and controlled surface hardening of Hadfield steels. The particular applicability of explosives to forming arises from the fact that virtually any shape in any size can be obtained without recourse to plant or machinery. The forming tool consists of a die which is comparatively light and does not require any foundations since the inertia of the tool mass is sufficient to counteract the applied force. It is essential, however, to realize that explosive forming is not economically viable when a very large number of components is required. Its advantage over the conventional methods lies in the possibility of producing complex parts very accurately and with very little or no machining, but in small quantities. The saving is due to the fact that there is no capital cost of presses, tooling, etc. The dies for stand-off operations are very often made from cheap cast materials such as epoxy or concrete.

The range of metals successfully worked includes: aluminium and its alloys, stainless steels, magnesium and some of its alloys, titanium and its alloys with aluminium, vanadium and manganese, refractory metals, copper and its alloys, and special alloys such as stellite, iron–nickel, nickel–copper, chromium–nickel and cobalt–iron. Carbon and low-alloy steels are less often used because of their low formability in dynamic conditions.

Three essential types of techniques characterize the stand-off operations: (i) free forming (cups, flanging and deep drawing), (ii) cylinder forming, and (iii) bulkhead forming (sheet and plate). These techniques are shown diagrammatically in Figure 16.62. Cylinder shaping is done in an open system using high explosives. A transmitting medium other than air is used to sustain the pressure for a longer period, thus increasing the impulse delivered to the work piece. In general, the forming of metals can be carried out either in a tank sunk in the ground and filled with water as the transmitting medium, or in an empty tank with the water being contained in a polythene bag (in the case of a cylindrical component) or in any suitable, disposable container. The first method is used for large components that require large charges, where the confinement of explosion within a large volume of water serves both as a safety measure and as a means of reducing noise. In such a case, the space between the undeformed metal and the die must be sealed-off and evacuated to enable full deformation to take place. In the second case, it may not be necessary to evacuate the air, providing that the die is fitted with a suitable system of ventilating holes.

A very important aspect of this method of forming is correct die design. The near absence of spring-back depends on the proper balance of tool profile, avoidance of sharp edges and deep narrow grooves and the provision of reasonably smooth transition sections.

Bulk forming of sheet is carried out by using either a single blank subjected, possibly, to a series of shots until the die is filled completely, or by using a mechanically shaped preform which is then given its final accurate dimensions in an explosive operation (Figure 16.63).

Forming without a die is also possible when relatively simple shapes are required, e.g. an impeller. An initially circular specimen can be formed to the shapes shown in Figure 16.64.

Although explosive forming cannot be regarded as a substitute for existing processes, it is an extension of the techniques now in use in the sense that it enables a number of difficult materials to be formed to a high degree of accuracy and at low cost. In the general field of metal shaping, the process is very advantageous when large parts are formed, often in a single operation, as costly heat treatment, tooling and machining are not required. The process is not competitive when parts can be produced in bulk using conventional equipment in a small number of operations. The use of explosives requires special precautions when storing and handling, and for safety reasons forming cannot be done in a congested enclosed area.

Hydroelectric forming In the application of this controlled and repeatable electrical force, the most obvious difference from explosive forming is the design of the forming apparatus. The dies are completely enclosed, since the volume of plasma generated is very small compared with the gas bubble created by a chemical explosive. By closing the die, full advantage can be taken of the pressure wave and, in the case of tubular parts, the shock wave is reflected from the ends, thus giving rise to additional energy. Each die has its own set of electrodes. In either technique the discharge takes place in a matter of microseconds, and is repeatable within 30 s. Operations can be automated and, unlike explosive forming, are eminently suited for mass production. Industrial forming machines are commercially available that have a basic output of, for example, 15 kJ, the amount of energy sufficient to form

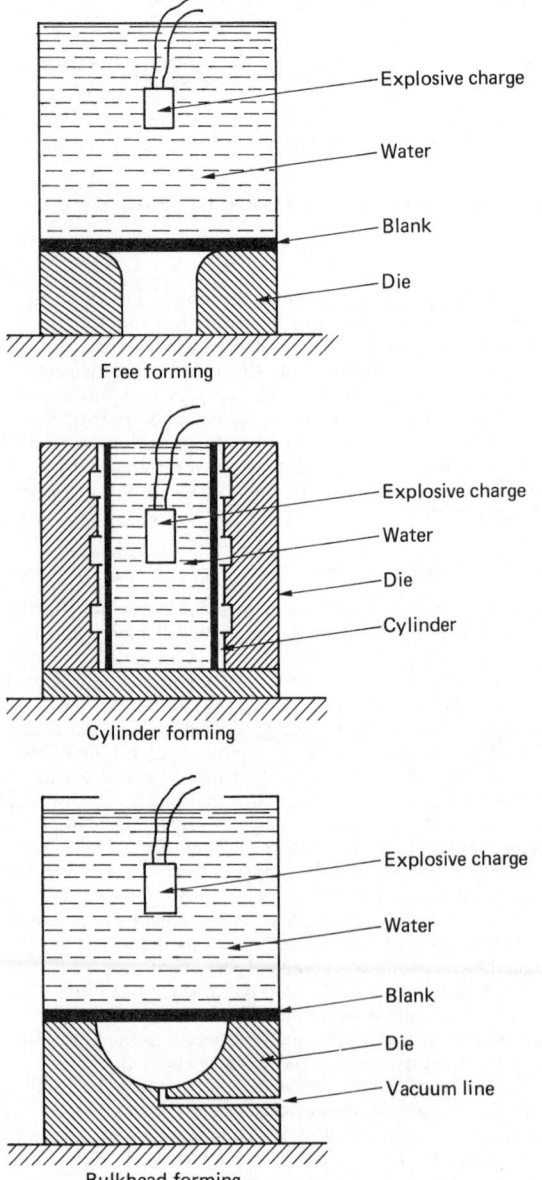

Free forming

Cylinder forming

Bulkhead forming

Figure 16.62 Basic explosive forming systems

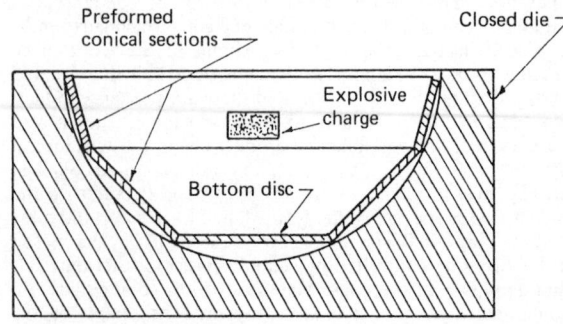

Figure 16.63 Preform bulk forming

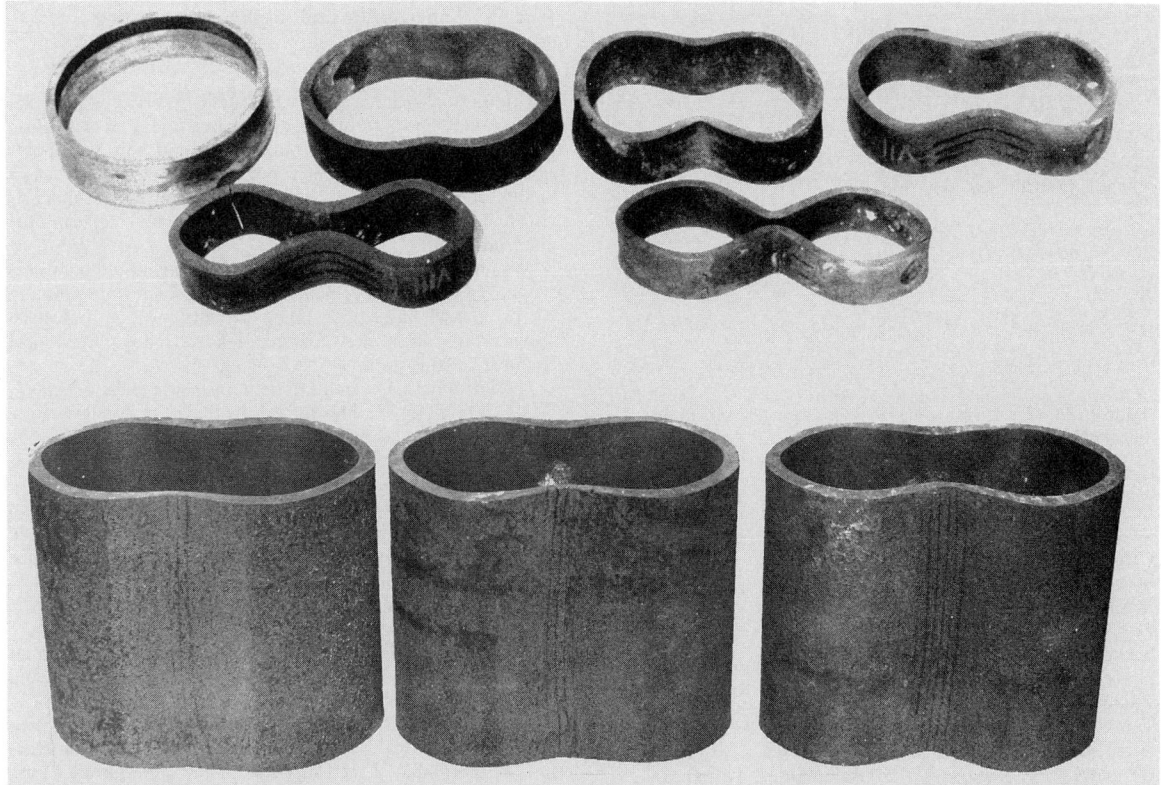

Figure 16.64 Steel impellers formed without using a die

ordinary commercial materials using blanks of 12–100 mm diameter and wall thicknesses of 0.2–2.5 mm. The output can be increased to 60 kJ.

For most purposes, dies used in hydroelectric forming are made of aluminium or cheap castable materials. The serious disadvantage of this technique is the comparatively high cost of the electrical equipment needed.

Electromagnetic forming The dies used in this process are made from cheap materials and, in some cases, are not required at all since the magnetic fields produce uniform pressures sufficient to expand cylindrical parts. The technique is very suitable for shaping sheet and tubular parts, swaging and the production of finned components. The process can be fully automated and forming machines are available. These are capable of producing magnetic flux densities of the order of 3×10^5 G, corresponding to pressures of over 340 MPa. The duration of the pulse is short, because of the permeability of metals, and is usually limited to 10 ms. Magnetic fields are quite sufficient for forming copper, aluminium and similar materials, but for metals like tungsten and for large parts, fields of the order of 5×10^5 to 5×10^6 G are needed.

The major disadvantage of the process is, again, the relatively high cost of the electrical equipment, but this is offset to some extent by the very high reliability, repeatability, speed of production and accuracy obtainable.

16.2.9.4 Explosive welding

The use of explosives for welding is now a well-established practice and provides a means of joining either flat or curved (tubular) surfaces of metallic alloys which are either incompatible (from the point of view of conventional welding) or impossible to join because of the geometry involved.

Explosive welding is a pressure operation in which two initially separate elements are joined together as a result of the metal jetting action which originates at the point of contact between them. The contact is achieved by detonating a charge covering the surface of one component (the flyer) which then gradually collapses onto the other (the base) component. The latter may be rigidly supported on an anvil. The shearing stresses generated at the point or line of contact are too high for either material to sustain and liquefaction takes place. A jet containing the elements of the flyer and the base is created and is propagated along the surface of the base. Since the detonation of the charge continues in a finite, albeit a short, time of microseconds, the flyer collapses in stages, creating new points of contact with the base and thus giving rise to further jetting. An indenting wave is produced (Figure 16.65) which, in addition to molecular welding, also creates a mechanical bond between the two surfaces. Elements welded in this way are usually made of the materials listed in Table 16.3, but more 'exotic' combinations can be produced easily.

Explosive welding is used to produce large sandwich (multi-metallic) plates which can then be processed further by rolling, shearing, bending and/or deep drawing, and to manufacture duplex or triplex cylinders, assemblies of rods and tubes which can be further cold drawn or hydrostatically extruded.

Special welding–forming operations and processes are employed and range from the manufacture of honeycomb panels to multilayered foil cylinders. The latter can be used as light but strong pressure vessels and are made either from layered foils or, if required, can be reinforced by metallic mesh interleaved with foil.

Figure 16.65 Explosively welded stainless steel (lower part) and carbon–steel interface

Table 16.3 Standard cladding combinations

Cladding metals	Base metals
Stainless steels Copper and copper alloys Nickel and nickel alloys Hastelloy Titanium and zirconium Tantalum Aluminium	Carbon and low alloy plates and forgings. Stainless steel
Copper	Aluminium

A cross section of copper/brass non-reinforced multilayer foil cylinder is shown in Figure 16.66.

16.2.9.5   Powder compaction

Various dynamic powder-compaction techniques involving the use of chemical explosives, high pressure gas guns, or electromagnetic or mechanical pneumatic power sources are used to compact metallic powders into semifabricates and finished machine parts. Either cold or preheated powders are used and the products, produced at pressures of 100–1000 kbars, amount to many thousands of tonnes annually. Like explosive-welding operations, explosive powder compacting is an exact and very complex process which requires detailed in depth study.

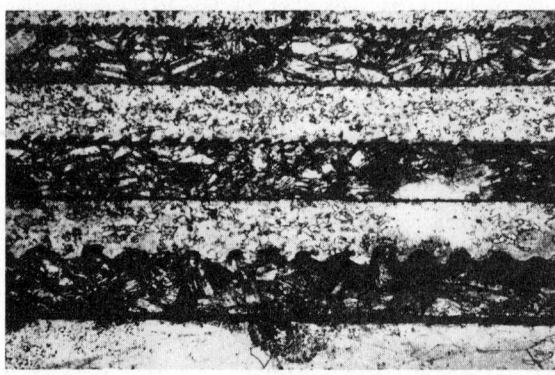

Figure 16.66 Cross-section of an explosively welded and formed multilayer copper/brass foil cylinder. Magnification ×110

16.2.10   Superplastic and mashy state forming

16.2.10.1   Superplasticity

The definition of the term 'superplasticity' is related to the fact that 'normal' alloys rarely show a greater degree of elongation than, at most, 60% when cold worked and only very occasionally exceed 100% when hot worked. In a few cases metallic alloys elongate at elevated temperatures to as much as ten times their original length—these are called 'superplastic' materials.

A number of alloys behave superplastically at temperatures ranging from ambient to 500°C but, because they are prone to corrosion and have poor creep properties, their industrial application is extremely limited. Tin/lead alloys are typical examples of this group.

The introduction of modern manufacturing techniques, e.g. diffusion bonding, enables the production of materials characterized by high strength (e.g. the titanium alloy Ti, 6% Al, 4%V) which are of considerable practical interest, although they require temperatures of up to 900°C.

Although welcome in general terms, the ability of a material to extend by several hundred per cent may not be acceptable from the point of view of the quality of the final product, since it could imply considerable thinning. Consequently, practical considerations limit the extent to which the superplastic tendency of a material can be utilized.

A practical guide defines a desirable material as one that will deform at very low stress, with a high rate sensitivity and freedom from damage, up to an elongation of some 100%.

The usefulness of these materials depends on the grain-growth response to strain rate, because only fine-grain structure is acceptable. Practically acceptable superplastic alloys should, therefore, display the following characteristics:

1. they must be stable at high temperatures (0.3–0.5 T_m);
2. the grain growth must be limited at high strain rates;
3. the flow stresses must be lower than the corresponding stresses in the same, but coarse grain, alloy; and
4. the acceptable properties must be obtainable at normal operational strain rates of up to 10^{-2} s^{-1}.

Since superplastic materials can be formed at low stresses, they are ideal for first-step operations in which blanks are preformed to non-uniform thickness, but are subsequently formed to the required shape exhibiting a very high degree of thickness uniformity.

16.2.10.2   Mashy state operations

When a metal is heated to a temperature higher than its solidus temperature, its structure changes and consists of both solid and liquid elements, the latter generated by partial melting. The 'mashy state' is thus created the properties of which enable easy processing of the material in both the conventional forming processes and composite manufacturing techniques.

In the mashy state, the liquid component is normally present at grain boundaries and separates the solid component consisting of the incompletely melted grains (Figure 16.67). The properties of mashy state metals are as follows:

1. low flow stresses;
2. behaviour like that of a slurry when the solid fraction φ is less than 65%;
3. when φ is low a mashy state metal can be mixed with other metals; and
4. two different mashy state materials can be bonded together.

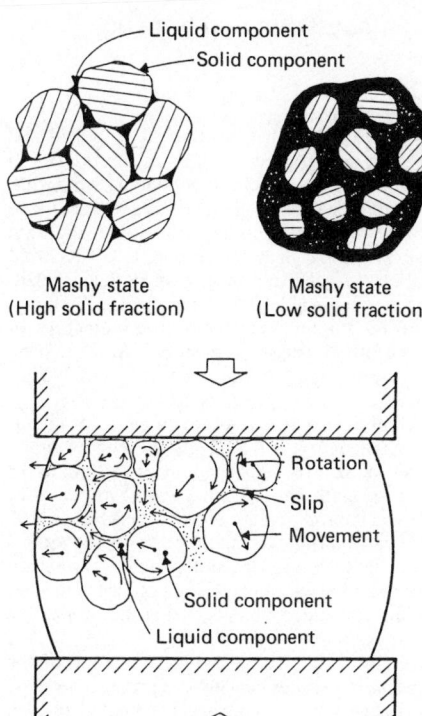

Figure 16.67 Diagrammatic representation of the mashy state

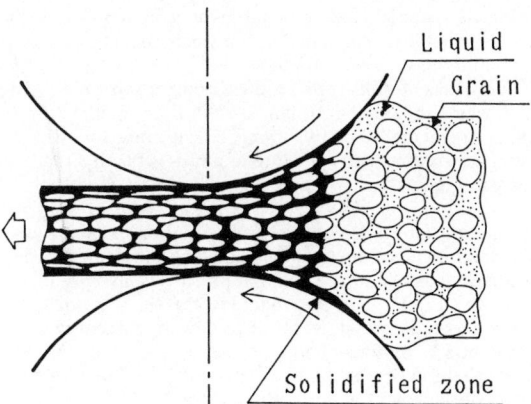

Figure 16.68 Diagrammatic representation of mashy state rolling. (Courtesy of Prof. M. Kiuchi, Tokyo University)

The following operations can be carried out on mashy state metals:

1. extrusion,
2. rolling of sheet metal,
3. production of particle-reinforced metals,
4. production of particle-reinforced cladding metals, and
5. the manufacture of composite sheet.

Extrusion   Extrusion of wires, bars and tubes is carried out in a conventional manner by using an electric furnace and an extrusion press. Billets are heated to the required temperature to obtain solid fractions of at least 70–80% and are extruded using preheated tools.

The characteristics of mashy state extrusion are:

1. a low extrusion pressure (20–25% of that used in hot extrusion),
2. a high extrusion ratio obtainable in a single pass,
3. the liquid component provides lubrication, and
4. complex sections are easy to extrude.

Rolling of sheet metal   The success of this operation depends on steady flow and homogeneous deformation, which can only be achieved if the state of hydrostatic pressure is maintained along the pass. The process has the advantage of enabling metals with complex internal structures to be processed, where a conventional cold- or hot-rolling operation is inappropriate.

A diagrammatic representation of the forming system is given in Figure 16.68. The liquid component is cooled by the rolls, solidifies and is forced into the roll gap, whereas the grains of the solid component are compressed, deformed and

then drawn into the gap. In the gap each grain is deformed and elongated.

The manufacture of particle-reinforced metals   Since additives, in the form of particulate matter, can be mixed effectively with a mashy metal, composites of different metals, or metals and ceramics, can be made and these are particle-reinforced structures. These can then be processed conventionally by extrusion and rolling to produce the desired shapes and dimensions.

Two possible production routes and applications are shown in Figure 16.69. A metallic matrix and, say, reinforcing ceramic particles can be stirred and mixed together and then be extruded directly to form bars, wires or tubes, or be rolled to form sheet. On the other hand, on cooling the metal matrix gradually during stirring, metal powder is obtained. This can then be mixed with reinforcing particulate matter, formed into billets and then be extruded. Both routes are indicated in Figure 16.69.

Particle-reinforced cladding metals   Although particle-reinforced metals possess good hardness and antiwear properties, they are basically brittle and so have low formability. To compensate for this disadvantage, particle-reinforced cladding metals obtained in the mashy state can be used. The particle-reinforced cladding metal consists of a layer of particle-

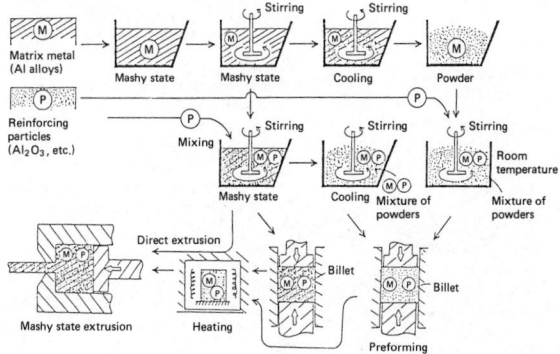

Figure 16.69 Production of particle-reinforced metals in the mashy state. (Courtesy of Prof. M. Kiuchi, Tokyo University)

reinforced material and a metal base which provides the required ductility. The bonding between the two layers is obtained in mashy state forging.

The various manufacturing routes used are shown in Figure 16.70, which shows the possibilities of preforming the reinforcement to sheet followed by forging, cold forming following by forging, or starting with a mixture of powders taken directly to mashy forging.

The manufacture of composite sheet   Either metal/ceramic-fibre reinforced or metal/metal combinations can be produced. An example of the adopted practice is given in Figure 16.71 which refers to the manufacture of a particle- or fibre-reinforced cladding sheet. The new manufacturing processes, briefly outlined here, are beginning to be used in normal industrial routines but, naturally, a great deal of further development work is still necessary.

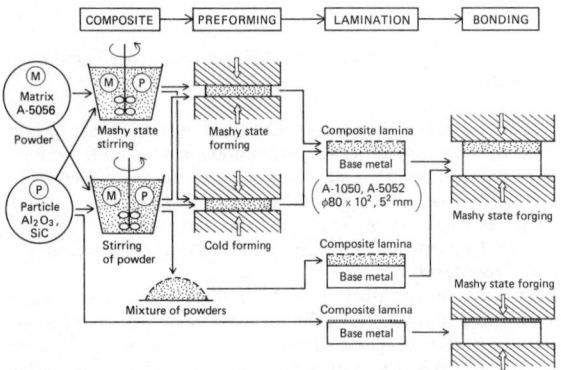

Figure 16.70   The manufacturing techniques for particle-reinforced cladding metals. (Courtesy of Prof. M. Kiuchi, Tokyo University)

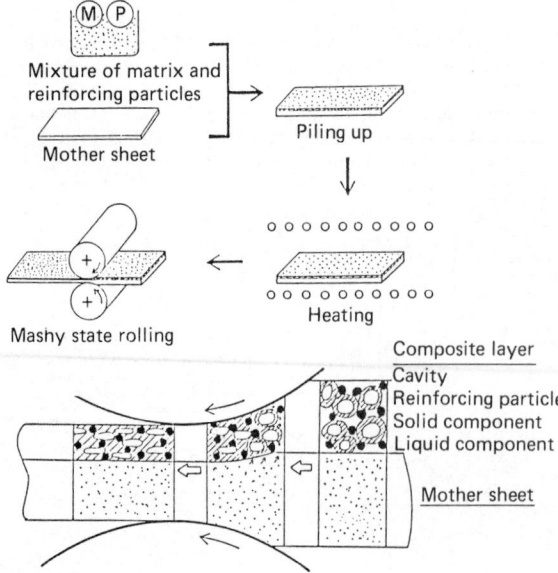

Figure 16.71   The mashy state manufacture of composite sheet. (Courtesy of Prof. M. Kiuchi, Tokyo University)

16.3   Welding, soldering and brazing

16.3.1   Welding

Welding is used for joining metals so that the physical and mechanical properties of the parent metal are reproduced in the joint. The integrity of a welded component, which has metallurgical continuity across the joint, is also characterized by properties such as pressure tightness or heat and corrosion resistance. These properties have contributed to the rapid development, both technical and economic, in all fields including nuclear power, chemical engineering, bridge building, offshore engineering, shipbuilding and the manufacture of cars, railway locomotives and rolling stock, aero engines, domestic appliances, and military hardware from small arms to main battle tanks.

Metallurgical continuity across a joint can cause problems compared with bolted or riveted connections. For example, under certain conditions, cracks originating from fatigue or brittle failure can propagate across welded seams whereas they are arrested by mechanical joints. Nevertheless, extensive research has been carried out over many years on the factors controlling fatigue and brittle failure and guidelines are available in the literature and in various codes of practice which are effective in avoiding cracking problems during the service life of welded structures and components.

A weld is, in effect, a miniature casting and some of the defects that can occur in castings may also be present in welds. The heat affected zone adjacent to a weld is heated to high temperatures during welding and cools rapidly as heat is conducted into the body of the component. This area may be heated again to lower temperatures when more weld metal is deposited on the top of the first weld to build up the joint. The sequence of heating, cooling rapidly, and reheating to a lower temperature is similar to the heat treatment of steel by quenching and tempering.

Some welding processes involve substantial plastic deformation of metal adjacent to the interface between the parts being welded, which is similar in many respects to the forging process used to form metal parts. Thus welding encompasses the whole field of metallurgy, albeit on a small scale, and the application of metallurgical science has successfully solved the major welding problems such as cracking of weld metal or parent plate, porosity and slag inclusions in welds, as well as those that may occur in service such as cracking due to fatigue, brittle fracture or creep, and wear or corrosion.

It will be readily understood that welding technology has developed into a discipline in its own right and this section can only refer to important aspects. However, the engineer must know what methods are available for joining the materials used because the most economical manufacturing route in many cases depends on selection of the most appropriate welding procedure that will produce welded joints having the minimum level of quality fit for their intended purpose. This will require decisions to be taken by an experienced welding engineer who will generally have a professional qualification, but to understand the logic of the welding engineer's decisions a brief knowledge of welding technology is advantageous for any engineer. The purpose of this sub-section is to provide basic information on the principal methods of welding, brazing and soldering and their fields of application. For more detailed information the references at the end of the chapter should be consulted.

16.3.1.1 Types of joint

The types of joint used and their associated weld types are described in detail in BS 499.[10] The commonest joint types are butt, T, corner and lap.

Butt joints These are joints between parts that are generally in line. If two plates are placed in contact (a close square butt joint) they can be welded with full penetration by one run of weld metal deposited by a manual welding process from each side provided that the plate thickness does not exceed approximately 8 mm (Figure 16.72). However, plate above 6 mm thickness is generally bevelled and the Vee edge preparation formed is filled by depositing a number of runs of weld metal. If high-current mechanized welding processes are used penetration of the weld may be at least double the above dimensions and for electron beam welding may be many times as high.

T joints and corner joints The parts may be joined by fillet welds or butt welds made by an arc welding process (Figure 16.73).

Lap joints These are commonly used for sheet metal up to about 3 mm thick in which one sheet is overlapped by another. This type of joint is used for soldering, brazing, resistance spot or seam welding, and for arc spot welding, plug welding, as well as for adhesive bonding. For material of 3 mm or thicker (even up to 10 mm) lap joints are occasionally used and fillet welds are deposited at the plate edges by arc welding.

16.3.1.2 Welding processes

The various welding processes can be used to join the majority of metallic materials, whether in cast or wrought form, in thickness from 1 mm or less up to 1 m or more. A simple classification of welding processes is shown in Figure 16.74. For a complete classification and for definitions of the processes BS 499: Pt 1[10] should be referred to.

A description of the welding processes is as follows.

16.3.1.3 Manual metal arc welding

Manual metal arc welding (referred to in the USA as shielded metal arc welding) is the most widely used process and accounts for approximately 50% of all the welding in the world

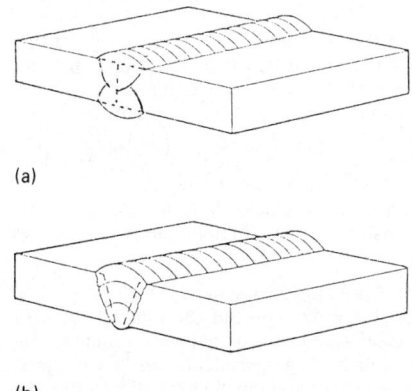

(a)

(b)

Figure 16.72 Edge preparation for butt welds. (a) Square edge; (b) single bevel

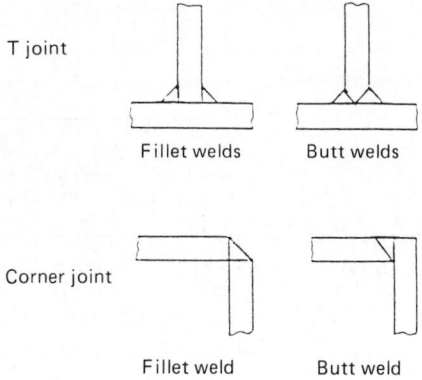

Figure 16.73 Examples of welded joints

today. With this process, welding is carried out with flux-coated electrodes which are connected via an electrode holder and length of cable to one terminal of a welding power source, such as an a.c. transformer or a d.c. generator (Figure 16.75). The other terminal of the power source is connected to the work piece via the earth return or the ground cable, so that when the end of the electrode is placed in contact with the work piece, electric current flows through the circuit. By withdrawing the tip of the electrode to about 3 mm from the work piece an arc will be struck and current will continue to flow in the circuit and pass through the arc which is electrically conductive.

If an arc is maintained between a rod-type electrode and plates to be welded together the tip of the rod becomes molten and so does a portion of the plates (the fusion zone). Gravity causes drops of molten metal to drip onto the plate and form a weld (Figure 16.76).

Apart from gravity, other forces caused by electromagnetic effects propel molten metal globules across the arc and these forces always transfer metal from the rod to the plate, whether a.c. or d.c. is used and whether the polarity is electrode positive or negative. They will also transfer the metal against the force of gravity, so that vertical or overhead welding is possible.

Electrodes These have core wire diameters from 2.4 mm to 10 mm and are 300–450 mm in length. The deposition rate of weld metal, which governs the overall rate of welding, increases with the current and has a maximum value for each electrode length and diameter. Exceeding the maximum current causes overheating of the electrode core wire by resistance heating, which can damage the electrode coating. Welding currents vary from 60 A for the smallest electrodes up to 450 A for the largest. The highest currents and deposition rates can only be used when welding downhand, i.e. in the flat position. Vertical and overhead welding can be used with electrodes having diameters up to 5 mm with maximum currents of approximately 170 A.

A bare wire can be used for welding, but the arc is mechanically unstable and the surface appearance of the weld is rough. The molten weld metal combines with nitrogen and oxygen from the air, resulting in poor mechanical properties.

The above problems can be overcome by coating wires with suitable fluxes and the principal purposes of this are:

1. To facilitate the initiation or striking of an arc and to stabilize it so that it can be easily maintained.
2. To provide a gas shield which protects the molten metal

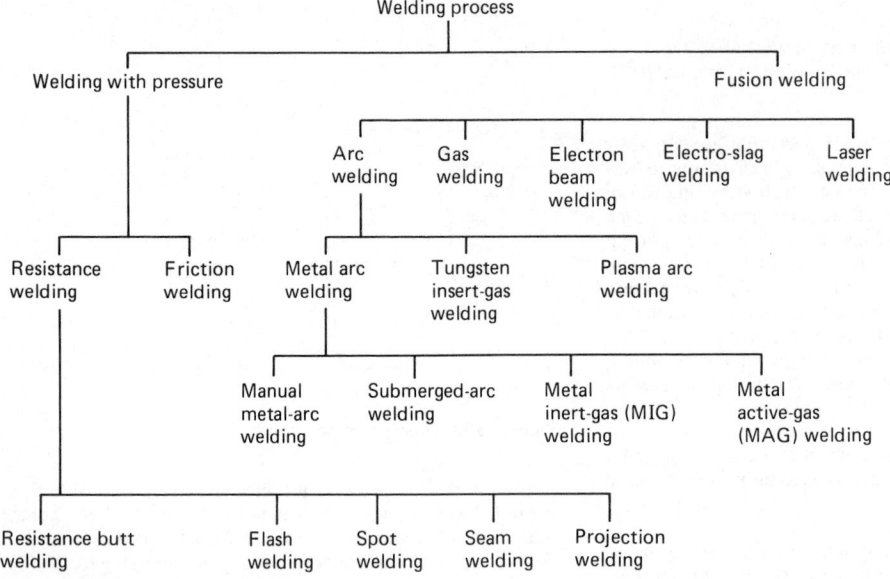

Figure 16.74 Classification of principal welding processes

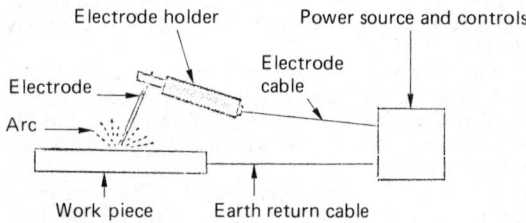

Figure 16.75 Welding circuit for manual metal-arc welding

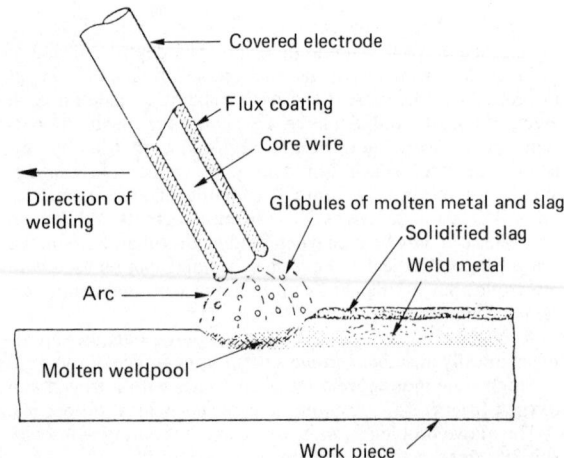

Figure 16.76 Manual metal-arc welding with covered electrode

droplets from oxygen and nitrogen in the air as they are transferred through the arc.

3. To provide a slag which protects the hot, solidifying metal from oxidation. The characteristics of the slag (e.g. melting point, surface tension and viscosity) determine the shape of the weld bead and the suitability of the electrode for positional welding.

4. To supply alloying elements to the weld metal; this means that an inexpensive rimming steel core wire can be used for many different weld metal compositions.

The constituents of the flux covering are mixed together in dry powder form and then binding agents are added. The flux paste is pressed into the form of slugs and loaded into machines which extrude the covering round electrode core wires as they pass at high speed through a die of appropriate size. The electrodes are then dried as they pass through ovens and are stamped with identification marks before being packed.

The classification of flux coverings The development of flux coverings, consisting of mixtures of various minerals, has followed fairly well-defined lines with slight variations between different manufacturers and in different countries. Electrodes can be classified according to their coating types and for a full description BS 639: 1986[11] and the American standard AWS A5.1–81[12] should be consulted.

Steel electrode types are designated by letters in BS 639 and the main characteristics of the different electrodes are as follows:

1. *R (rutile)*: Rutile coverings containing a high proportion of titanium dioxide in the form of the mineral rutile or ilmenite. The electrodes are easy to use but produce weld metal having high hydrogen contents which can cause cracking of the weld or parent metal heat affected zone in heavily restrained joints.

2. *RR (rutile, heavy coated)*: The thick covering enables the electrodes to be used as contact electrodes which can be

held in contact with the parent plate and dragged along the joint at high welding speed. Iron powder is often added to the coating to increase the deposition rate, and the RR type electrodes are not suitable for welding in the vertical and overhead position.

3. *B (basic)*: A basic covering usually has a high content of limestone (calcium carbonate) and fluorspar (calcium fluoride). Basic covered electrodes are often referred to as low hydrogen because they were developed to produce weld metal having a low hydrogen content which reduces any tendency to hydrogen-induced cracking. This covering decomposes to give a gas shield containing a large proportion of carbon dioxide. These electrodes are used extensively because of their ability to weld medium- and high-tensile steels as well as high-sulphur (free-cutting) steels without solidification cracking of the weld metal and also because, by suitable drying treatment, the moisture content of the flux covering can be reduced so that the weld metal hydrogen content will be correspondingly low. This gives insurance against hydrogen-induced cracking of both the weld metal and the heat-affected zone (HAZ). Properly designed basic covered electrodes produce weld metal which has the highest fracture toughness properties, and they have the advantage over other types of electrodes in that high fracture toughness is maintained in all welding positions.

4. *BB (basic, high efficiency)*: These are similar to basic covered electrodes but have iron powder added to the coating so that the quantity of weld metal deposited is at least 130% of the weight of the core wire. The high deposition rates make these electrodes unsuitable for welding in the vertical and overhead positions.

5. *C (cellulosic)*: This designation indicates a covering which has a high content of cellulosic material. These electrodes operate at a high arc voltage, which gives a deep penetrating arc and rapid burn-off. The covering forms a voluminous gas shield, consisting chiefly of carbon monoxide and hydrogen, and a small volume of slag which facilitates work involving changes in welding position such as pipe welding for which these electrodes are particularly suitable. They can also be used for the fast, vertical down welding of vertical seams in storage tanks up to about 12 mm thick. Because of the excellent penetration the root does not require gouging before making a sealing run on the reverse side. In pipe welding the close control of penetration is necessary because the deposition of a sealing run on the inside is generally impossible.

Power sources There are basically three types of power source:

1. A.C. generators
2. D.C. rotary generators
3. D.C. solid state

The choice of power source is described by John and Ellis.[13]

A fourth type of power source of recent development is based on an invertor which is used to convert the mains frequency from 50 Hz to between 5 and 25 kHz. Transformers for currents operating at these high frequencies are much lighter than those used in conventional a.c. generators. In the invertor type power source the a.c. mains input is rectified to give d.c. which is then fed to an invertor which converts it back to a high-frequency a.c. The power is then reduced to the welding voltage by a lightweight transformer and it is rectified again to d.c. for welding.

In Britain transformers have traditionally been the most widely used type of power source for manual metal arc welding because of their relative cheapness, reliability and long life. The development of d.c. solid-state power sources has eliminated the moving parts and maintenance costs associated with rotary generators and has also reduced the capital costs. Therefore the use of d.c. welding is likely to increase in future, particularly in view of the fact that welding is easier because of the more stable arc and ability of d.c. to weld non-ferrous alloys and to produce the highest quality welds in stainless steel.

For use in repair work (for example, in the garage trade and for the DIY market) a number of low-power welding sets are available which operate from the single-phase 220/240 V supply. These power sources can be used with electrodes up to 3.25 mm diameter and both a.c. and d.c. units are available. For the home market, video instruction is available.[14]

Applications The main reasons for the popularity of manual metal arc welding are its versatility, simplicity of operation and the relatively low cost of equipment. The process can be used with equal facility in a workshop or on-site. To weld on a remote part of a structure it is possible, within reason, to lengthen the cables from the power source and when the limits of the extended cables are reached the power sources are readily transported by crane or motor vehicle or by manhandling on level sites. Movement of the power sources is facilitated by their simplicity and robustness.

The range of thicknesses welded varies from less than 2 mm in the fabrication of sheet metal ventilation ducting to 75 mm and above in the production of nuclear containment pressure vessels. These two examples are indicative of the wide range of quality standards that may be required, from general sheet metal work up to the highest possible standards of radiographic soundness and mechanical properties.

Metals that are most commonly fabricated by manual metal arc welding are carbon and carbon manganese steels, low-alloy steels and stainless steels of both the corrosion- and heat-resisting types. By selection of suitable electrodes described in various standards[11, 15–18] the mechanical properties of the weld metal in respect of strength, ductility and toughness match those of the parent plate at ambient temperature and also at elevated or subzero temperatures as required. References to American Welding Society (AWS) specifications are included because of their worldwide use in the oil and petrochemical industries.

Non-ferrous metals such as nickel, copper and aluminium and their alloys are welded much more extensively with the gas shielded processes, although nickel and nickel alloys are readily welded by the manual metal arc process and a wide range of electrodes are available.[19–21] Some tin–bronze (copper–tin), and aluminium–bronze (copper–aluminium) electrodes are manufactured, but their main use is for repair work, particularly of castings (e.g. marine propellers). These can also be used for welding pure copper, because the high conductivity of copper has prevented the successful production of a copper electrode. Pure copper is generally used for its high thermal or electrical conductivity, and therefore the application of copper alloy electrodes is strictly limited to those circumstances where weld metal, having low thermal or electrical conductivity, is satisfactory. Some non-ferrous electrodes based on nickel, nickel–iron or nickel–copper alloys are used for welding the cast irons.[22]

A wide range of electrodes is available for hard-surfacing components to increase their wear resistance under conditions of abrasion, impact, heat or corrosion or various combinations of these factors. Electrodes for hard surfacing are manufactured from core wires of mild steel, carbon and alloy steels, stainless and heat-resisting steels, nickel–chromium and cobalt–tungsten–chromium alloys, and are also made from steel tubes containing granules of refractory metal carbides such as tungsten and chromium carbides.[23]

16.3.1.4 Gravity welding

This is a semi-mechanized welding process which is used principally in shipyards for fillet welding stiffeners to horizontal plates. Covered electrodes of the contact type, typically 600 mm long, are supported by an electrode holder which slides down one arm of a tripod. The other end of the electrode is positioned in the corner of a T-joint to be welded and when the current is switched on an arc is initiated and the electrode moves along the joint line as the electrode holder slides down the arm of the tripod under the force of gravity. One person can operate three gravity welding units simultaneously, thus trebling the rate of manual welding.

16.3.1.5 Open-arc automatic welding

Although no longer used, it is appropriate to mention a mechanized welding process which was employed extensively in the 1950s up to the 1970s in ship building and bridge building. The engineer may find the process referred to in periodic inspection reports of these types of welded structures.

Mechanized welding with covered electrodes was carried out with a continuous coiled electrode having a core wire from 4 to 8 mm diameter. The core wire was wrapped helically with two thin wires about 1 mm diameter, which anchored the extruded flux in place and also acted as a means of conducting electrical current from the jaws of the welding head to the core wire.

Automatic open-arc welding with continuous covered electrodes has now been superseded by the submerged-arc process.

16.3.1.6 Submerged-arc welding

This is the most widely used mechanized welding process (Figure 16.77). A bare wire (1.6–6.3 mm diameter but usually 3.25 or 4 mm) is fed from a coil and an arc is maintained between the end of the wire and the parent metal. As the electrode wire is melted, it is fed into the arc by a servo-controlled motor which matches the wire feed rate to the burn-off rate so that a constant arc length is maintained.

The region of the joint is covered with a layer of granular flux approximately 25 mm thick, fed from a hopper mounted above the welding head. The arc operates beneath this layer of flux (hence the name 'submerged arc'). Some of the flux melts to provide a protective blanket over the weld pool and the

unmelted flux is collected and re-used. The electrode wire, welding head, wire drive assembly and flux hopper are mounted on a traverse system which moves along the work piece as the weld metal is deposited.

The traverse system may consist of a carriage mounted on a boom or it may be a motorized tractor either on rails or running freely with manual adjustment to follow the weld seam. Alternatively, the welding head can remain stationary while the work piece is moved. This method is used for welding the circumferential seams of a pressure vessel while it is rotated under the welding head.

Electrode wires The electrode for submerged-arc welding is a bare wire in coil form usually copper coated. Two types are available – solid wire or tubular wire. The solid wire is widely used for general fabrication of mild and low-alloy steels, stainless steels and non-ferrous metals. For welding mild and low-alloy steels it is either a low-carbon ultra-low-silicon steel or a silicon-killed steel with manganese addition and sometimes low-alloy additions, the selection of either type depending upon the type of flux to be used with it (i.e. a flux with manganese or manganese and alloy additions or a neutral flux, respectively). The tubular wire (made by forming narrow strip into a tube) carries alloy powders which permit the economical production of a wider range of weld compositions than is possible by using the solid wire type. Tubular wires are widely used for hard-facing. Wire compositions for welding carbon steel and medium tensile steel are listed in BS 4165.[25]

With coated manual electrodes, wire and coating are one unit so that such electrodes can be classified according to the type of coating and its effect on weld mechanical properties. In submerged-arc welding, any wire may be used with a number of different fluxes with substantially different results in respect of weld quality and mechanical properties. Consequently, BS 4165 grades wire flux combinations according to the tensile and impact strengths obtained in the weld metal.

A number of tubular wires are available, particularly for surfacing and hard-facing. These contain alloy powders which produce weld metals consisting of low-alloy steels, martensitic and austenitic stainless steels, chromium and tungsten carbides, and various cobalt- and nickel-based heat- and corrosion-resistant alloys. Some corrosion-resistant alloys, including stainless steel, are available in the form of coiled strips from 100 mm to 150 mm wide, 0.5 mm thick for high deposition rate surfacing by a submerged-arc welding process known as strip cladding.

Fluxes Two main types of fluxes are available: fused and agglomerated. Fused fluxes are manufactured by fusing together a mixture of finely ground minerals, followed by solidifying, crushing and sieving the particles to the required grain size. Fused fluxes do not deteriorate during transportation and storage and do not absorb moisture. Agglomerated fluxes are manufactured by mixing finely ground raw materials with bonding agents such as sodium or potassium silicates followed by baking to remove moisture. This type of flux is sensitive to moisture absorption and may require drying before use. Agglomerated fluxes are more prone to mechanical damage which can cause segregation of some of the constituents.

Fluxes are classified as acid, neutral, or basic, the last being subdivided into semi-basic or highly basic. The main characteristics of the fluxes are as follows:

1. *Acid fluxes*: High content of oxides such as silica or alumina. Suitable for high welding currents and fast travel speeds. Resistant to porosity when welding rusty plate. Low notch toughness. Not suitable for multipass welding of thick material.

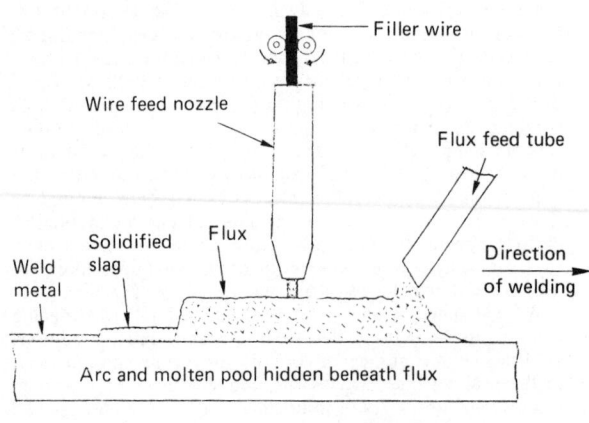

Figure 16.77 Submerged-arc welding

Labels in figure: Filler wire; Wire feed nozzle; Flux feed tube; Solidified slag; Flux; Weld metal; Direction of welding; Arc and molten pool hidden beneath flux; Work piece

2. *Neutral fluxes*: High content of calcium silicate or alumina-rutile. Suitable for fairly high welding currents and travel speeds and also for multipass welding.
3. *Basic fluxes*: High content of chemically basic compounds such as calcium oxide, magnesium oxide and calcium fluoride. Highest weld metal quality in respect of radiographic soundness and impact strength. Lower welding currents and travel speeds are suitable for multipass welding of thick sections. For further information on fluxes references 25 and 26 should be consulted.

Power sources Either d.c. or a.c. may be used. D.C. may be supplied either by a motor generator or by a rectifier, either of which can have flat or sloping current voltage characteristics. These are referred to as constant voltage or constant current types. Generators will deliver up to about 650 A continuous output, rectifiers up to about 1200 A at 100% duty cycle. A.C. is supplied by welding transformers which are designed to give a drooping characteristic. Transformer output may range up to 2000 A.

Power sources are of rugged construction and are designed for the 100% duty cycle. With the constant current type the arc voltage determines the wire feed rate which varies to maintain a constant arc length. The constant voltage type of power source produces a self-adjusting arc in which an increase in arc length or arc voltage causes a decrease in current and burn-off rate so that the original arc length is rapidly obtained. A decrease in arc length increases the current and burn-off rate and this self-adjusting effect occurs with a constant wire feed rate.

Application As the process operates with a continuous coil of electrode wire, butt welds in the flat position requiring multiple runs to fill the joint can be made with minimal stops and starts. Thus circumferential joints in cylindrical bodies such as pressure vessels, pipes, etc. can be made with one stop and start per revolution of the work piece, this stop being necessary to reset the position of the welding head. Consequently, the possibility of stop and start defects is minimized; a most important consideration when reliability in costing is required. Although most widely applied to welding of joints in mild steel, low-alloy high-tensile steel, creep-resisting steels and, to a lesser extent, stainless steels, it is also widely used for building-up work, either for reclamation or replacement of defective parent metal or for hard-surfacing.

Submerged-arc welding is suitable for welding material from 5 mm to 300 mm and even thicker but plates less than about 10 mm thick are generally welded by the gas shielded or flux cored arc welding process. A semi-automatic variant of the process is available in which the welder manually manipulates a welding gun on which is mounted a small hopper containing the flux. Electrode wire is fed to the gun from a coil by a wire feed unit. This process, which is used only to a limited extent, is sometimes referred to as 'squirt' welding.

Other variations of the submerged-arc welding process are mainly concerned with increasing deposition rates and therefore welding speed and productivity. These include:

1. Increasing the electrode extension or stick-out by up to 150 mm by using an insulated guide tube. The resistance of the wire increases the burn-off rate by the I^2R heating effect.[27]
2. Addition of iron powder to the joint which increases the weld volume.
3. The use of multiple wire techniques in which two or more wires are used with two or three separate power sources.[27]
4. Narrow gap welding of plates more than 100 mm thick in which a parallel gap of 14–20 mm between the square edges of the plates is used instead of a V or U groove. The

saving in quantity of weld metal used is considerable, with consequent increase in productivity.[28]

16.3.1.7 Electro-slag welding

Electro-slag welding is an automatic process for welding material 18 mm or thicker in the vertical position. Square edge plate preparation is used and joints are limited to butts and T-butts. Heat for fusion is obtained by the current in the consumable electrode wire which passes through the molten slag formed by melting of a flux formulated to have high electrical resistance in the molten state.

The joint is set up with a wide gap (approximately 25–36 mm, depending on plate thickness), the very large weld pool being contained in the joint by water-cooled copper shoes. One, two or three wires are fed into the joint with or without a reciprocating motion to ensure uniform heat generation, the plate thickness determining the number of wires. One welding head with three wires can weld plate 450 mm thick. The copper shoes rise up the joint to prevent spilling of metal and slag and these shoes form part of the welding head.

A variation of the electro-slag process is known as consumable guide welding. Here the welding head remains stationary and feeds one or more wires down a tube which melts into the pool. The equipment is substantially cheaper than the conventional slag welding machines.

The consumable guide process, while theoretically suitable for very thick sections, is more usually applied to metal up to, say, 50 mm because it is more manageable in these thicknesses. The equipment needed is much simpler than for the conventional process consisting of a constant potential generator, rectifier or transformer with a flat characteristic and a wire feed unit. These are also the essential ingredients of submerged-arc equipment and, by slight modification, they can be adapted for consumable guide welding.

The ultra-slow cooling rate of an electro-slag or consumable guide weld minimizes hydrogen cracking susceptibility in the parent steel and weld metal but produces a large grain size weld. This tends to give a comparatively poor notch impact strength as determined by conventional tests. Consequently, for pressure vessel application, current codes require normalizing after welding.

Fluxes Electro-slag welding fluxes produce complex silicate slags containing SiO_2, MnO, CaO, MgO and Al_2O_3. Calcium fluoride is added to increase electrical conductivity and lower slag viscosity. Slags based on CaF_2–CaO have a strong desulphurizing action which assists the welding of steels higher in carbon than 0.25% without solidification cracking in the weld metal. Fluxes must be kept dry.

Economics of the process On heavy steel plate, for pressure vessels, boiler drums, etc. the actual welding speed (rate of filling the joint) is about twice as fast in 40 mm plate, four times in 90 mm and eight times in 150 mm compared with multi-run submerged arc welding. Welding speed is 1–1.7 m per hour. Plate-edge preparation by bevelling is also avoided. On this evidence it would seem highly attractive economically. However, 'setting up' the machine and selecting the correct welding parameters is a a matter of experience or tests. Therefore the 'setting-up' time must also be considered and the cost of determining the correct procedure included in the cost estimate.

Thus the process gives full economic benefit on repetitive work where set parameters can be used based on experience. An example is the wide use of the process for the longitudinal seams of boiler drums in steel 125–150 mm thick. On one-off applications involving plate thicker than 150 mm the process is economical on each joint and, providing sufficient work of a suitable type is available, the high capital cost is recovered within a reasonable time.

Mechanical properties The weld metal mechanical properties are determined to a considerable degree by the composition and cleanliness of the parent steel. However, proper selection of wire and flux and suitable parent metal confer strength and ductility equal to or better than the parent metal.

In the as-welded condition degradation of notched impact properties in the heat-affected zone results in values lower than the weld metal. Where notched impact requirements must be met, it is necessary to normalize the completed joints. In general, electro-slag welding is an acceptable and economic method of welding thick steel. It finds application for ships' hulls, boiler drums, press frames, nuclear reactors, turbine shafts, rolling mill housings and similar heavy fabrications.

16.3.1.8 Gas-shielded metal arc welding

This general term covers a group of welding processes in which no added flux is used. The molten weld pool is protected by a gas shield which is delivered to the welding gun through a flexible tube at a controlled rate, either from a gas bottle or from a bulk supply. The shielding gas may be inert (e.g. argon or helium or mixtures of these gases) or it may be active (e.g. carbon dioxide (CO_2) or mixtures of CO_2 with other gases such as argon). Sometimes small additions of oxygen or hydrogen are included in the shielding gas.

The wide variety of shielding gases which may be completely inert or non-reactive or may be active (i.e. slightly oxidizing or reducing) has led to the use of the terms MIG (metal inert gas) and MAG (metal active gas) to describe the principal gas-shielded metal arc welding processes.

MIG and MAG welding In MIG or MAG welding, referred to in the USA as gas metal arc welding (GMAW) (Figure 16.78) a small-diameter (0.6–1.6 mm) wire is fed from a coil by a wire feed unit which contains an electric motor, gearbox and grooved drive rolls. The wire is fed to a welding gun that has a trigger which operates the wire feed drive, the current and the flow of shielding gas.

An arc is struck when the wire contacts the work and the arc length depends on the voltage, which is preset by adjustment of a knob on the power source. Welding current is picked up by the wire from a copper contact tube through which the wire passes. The distance between the contact tube where current enters the wire and the end of the wire is usually a maximum

of 25 mm, compared with 300–450 mm for a covered electrode. Therefore overheading is not a problem, particularly as there is no flux coating. Higher currents can be used than those normally employed for manual metal arc welding (e.g. 120–450 A for 1.6 mm diameter wire). Therefore deposition rates are generally higher than for manual metal arc welding. Another advantage of MIG/MAG welding is that the arc is automatically maintained at a length that depends on the arc voltage, which means that the welder has to move the welding gun only along the joint line holding the nozzle of the gun at approximately the same distance from the joint. This is because the arc is self-adjusting because the voltage current characteristic of a MIG/MAG power source is flat or only slightly drooping.

If the welding gun is moved away from the joint the arc length and the arc voltage increase slightly. With a flat or slightly drooping characteristic a small increase in voltage will cause a large decrease in welding current and the wire will burn off at a lower rate. The original arc length will be rapidly attained at which voltage the burn-off rate will once again match the wire feed speed. Similar self-adjustment in the opposite sense will occur when the welding gun is moved towards the work piece.

Filler wires The commonest filler wires are 1.0, 1.2, and 1.6 mm in diameter with 0.6 and 0.8 mm less frequently used. Because of the wide range of current that can be used with each wire it is necessary to stock only one or two diameters. This is in contrast with manual metal arc welding, where a number of different diameters of electrodes and possibly two or more coating types may be required just for welding a single type of material such as carbon–manganese steel.

However, a disadvantage of solid filler wires is the limited range of compositions available because it would be too expensive for a steel maker to produce small quantities of low-alloy or stainless steel wires. Small batches of covered electrodes can readily be produced by introducing alloys in powder form through the coating. This situation is reflected by the number of electrodes and filler wires for welding low-alloy and stainless steels listed in British Standards, which are:

	Low alloy steels	Stainless steels
Covered electrodes	30	39
Solid wires	6	16

Any deoxidizing elements such as silicon or aluminium required to refine or degas the weld pool are contained in the solid wire and compositions of wires available are listed in BS 2901: Parts 1–5[29] for ferritic steels, austenitic stainless steels, copper and copper alloys, aluminium and aluminium alloys and magnesium alloys, and nickel and nickel alloys.

Modes of metal transfer In MIG or MAG welding the operating conditions in terms of current and voltage determine the type of metal transfer which must be suitable for the application. There are four modes of metal transfer.

Short circuiting (dip transfer) This occurs when a low voltage and current are used which causes metal to be transferred from the end of the wire to the work piece by frequent short circuiting of the wire to the weld pool. This technique produces low heat input and a small controllable weld pool essential for welding steel sheet in all positions and thicker steel sections in the vertical and overhead positions. A disadvantage of the process is the production of spatter in the form of globules of metal expelled from the weld pool when each short circuit is broken. Spatter particles which adhere to the work piece can be reduced by fine tuning of the inductance of the power source.

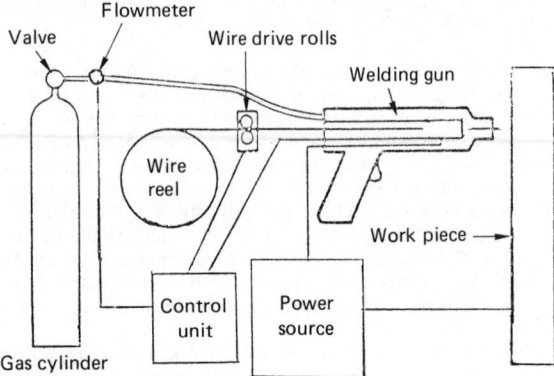

Figure 16.78 Metal inert-gas welding

Globular transfer (semi-shorting) This occurs when somewhat higher currents and voltages are used than for dip transfer welding of steel, but metal transfer still occurs by short circuiting of the filler wire to the weld pool. Because of the large droplet size and the larger weld pool, this mode of welding is not suitable for vertical or overhead welding. The production of spatter still occurs.

Spray transfer Free flight of metal droplets occurs with no short circuiting when the current and voltage are sufficiently high. This give maximum deposition rates and deep penetration welding suitable for flat-position welds in medium and heavy steel plate and for horizontal–vertical fillet welding (e.g. between a vertical and a horizontal plate). Spray transfer is used for welding aluminium and aluminium alloys in all positions because the spray transfer of droplets occurs at much lower welding currents than with other metals. Therefore small weld beads can be deposited which solidify rapidly and enable welding to be carried out in the vertical and overhead positions.

Pulsed transfer This was developed to produce spray transfer at all current levels so that welding of all metal thicknesses in all welding positions could be carried out without the formation of spatter. In pulsed transfer the welding current is switched from a high pulse current to a low background current at a typical frequency of 50 Hz. The background current is sufficient to sustain the arc but it is insufficient for metal transfer. The pulse current is set above the critical level to produce sufficient electromagnetic force with each pulse to transfer one metal droplet from the tip of tthe wire. With the first pulsed arc power supplies the pulse frequency had to be a multiple of mains frequency and setting up welding conditions was difficult to the extent that it hindered the use of the process in industry. The average current, which depended on the background current, the pulse current and the frequency, had to produce a usable burn-off rate at a constant arc length. The process was also sensitive to electrode stick-out (the electrode extension beyond the contact tube) which could disrupt the balance between pulse energy and metal transfer, causing arc extinction and spatter. The full advantages of the pulsed MIG process, including stable low mean current operation particularly when welding aluminium alloys or stainless steel and positional welding capabilities of all metal thicknesses, were made readily available with the development of transistorized power sources referred to in the next subsection.

Power sources MIG and MAG welding are always carried out with d.c. and the principal types of power sources are transformer rectifiers with constant potential or controlled-slope characteristics and motor generators which are used for site work (e.g. welding pipeline). Invertor type power sources, described in the section on manual metal arc welding, are also used for MIG and MAG welding. The main advantage of invertors is the considerable decrease in size and weight compared with conventional transformer rectifiers.

Electronic power control has had a considerable and beneficial influence on MIG/MAG welding, enabling the process parameters to be pre-programmed which eliminates the complicated setting-up operation and enables 'one-knob' control to be achieved. Programmed control for both dip and spray transfer was originally developed in the late 1960s. The relationship between wire feed speed and voltage for any filler wire type and diameter could be programmed into the power source and a single control could be used to vary mean current continuously. The equipment contained preset resistors to store the fixed parameters but microprocessor-controlled units are now available.

A number of different electronic control systems have been developed for MIG/MAG welding in both the dip and spray transfer modes of operation, which overcome the setting-up difficulties mentioned above. These setting-up difficulties were particularly acute with the pulsed arc mode of operation, but they have been overcome by the so-called synergic control technique used in conjunction with a transistorized power source. With synergic control, precise independent regulation of the pulse shape, pulse current time, pulse frequency and background current is obtained. The electronic power source can produce continuously variable 25–250 Hz pulse frequencies. With variable-frequency pulsing the correlation between pulse energy, burn-off rate and arc characteristics can coincide with all electrode wire feed speeds to provide one metal drop transfer per pulse.[30]

With synergic control all the pulse parameters are preprogrammed for a wide range of wire feed speeds. During welding the wire feed rate and pulse frequency automatically adjust together to produce one metal droplet transfer at a constant arc length. The welder only needs to adjust one control – average current.

Modern power supplies have control systems based on microprocessor technology. Memory chips in the control unit store process data and produce the optimum operating parameters if the user presses the appropriate switches to specify the types of filler wire and shielding gas. In some power sources users can load their own operating programs into the control unit. For further information references 31 and 32 should be consulted.

As in manual metal arc welding, there are a number of 'hobby' sets on the market which can be used on the 13 A mains. These have a limited number of current settings for use with 0.6 or 0.8 mm diameter wire and can be used to weld carbon steel, stainless steel and aluminium in thicknesses up to 6 mm. Tuition by video is available.[33]

Applications MIG and MAG welding with the various modes of metal transfer can be used for applications similar to those fabricated by MMA welding. In addition, they are more suitable for welding some of the non-ferrous alloys such as aluminium and copper alloys and are probably equally suitable for welding stainless steel and nickel alloys. For sheet metal thicknesses which are welded by a single run of weld metal, MIG and MAG welding are generally up to 50% faster than MMA welding. In thicker materials, MIG/MAG welding and MMA welding used with the same duty cycle, i.e. the proportion of arcing time to total time, will have approximately the same overall welding rate, provided that full use is made of positioners to enable a large proportion of the welding to be carried out in the flat or horizontal–vertical positions. Claims made in the literature or in suppliers' brochures about the superiority of one process over the other in respect of productivity and economic advantages should be treated with caution, because they may only be valid for a specific application. One great advantage of MIG/MAG welding over MMA welding is the ease with which the process can be mechanized either by fitting the welding gun to a traverse unit or by moving the work piece under a stationary gun either by linear motion or rotation.

Robotic and automated MIG/MAG welding has advanced with the developments in microcomputers and electronic power sources, the latter providing very stable arcing conditions in spite of mains voltage fluctuations. Automated MIG/MAG welding utilizes seam-tracking devices which are necessary to compensate for inaccuracies of the component parts or distortion during welding.

Seam-tracking devices contain contact-type sensors such as probes or guide wheels or the non-contact types such as electromagnetic, ultrasonic or video systems. For further information on robotic or automated welding references 34–36 should be consulted.

16.3.1.9 Flux-cored arc welding

Flux-cored arc welding is similar in many respects to MIG/MAG welding except that in one version of the process no shielding gas is added. In this case the gas shield originates from the decomposition of minerals contained in the tubular core electrode and this version of the process is sometimes referred to as self-shielded welding.

Cored electrodes in coiled form are manufactured from steel strip which is first bent into a U-section as it passes through forming rolls. The U-shaped strip is then filled with a metered quantity of flux and metal powders and the strip is passed through dies to form it into a circular cross-section from 0.9 to 3.2 mm diameter.

Tubular cored electrodes Tubular cored electrodes may be gas shielded with CO_2 or Ar/CO_2 mixtures or they may be self-shielded. Cored electrodes are classified according to the constituents contained in the core which influences the characteristics of the electrode.

For full descriptions of the different carbon and carbon–manganese steel types, BS 7084: 1989,[37] AWS A5.20–79[38] and AWS A5.29–86[39] should be consulted. Many higher-tensile and low-alloy steels can be welded with flux-cored wires having matching strengths. Stainless steel cored wires are available for use either with or without shielding gas and many different types of cored wires are used for hard-facing applications in which a coating is applied to a steel base to confer resistance to wear, corrosion or heat.

Application Flux-cored arc welding is used for applications similar to manual metal arc or MIG/MAG welding and, like MIG/MAG welding, the process can be mechanized. Tubular cored electrodes are available in a wider range of compositions than solid wires because of the ease of introducing alloying elements in powder form. Flux-cored wires, particularly the gas-shielded types, meet the mechanical property requirements of a range of applications and some grades give good low-temperature impact properties. The mechanical properties attainable with self-shielded cored wires is more limited, with maximum weld metal strengths of 700 N mm^{-2}.

Self-shielded wires are particularly useful for site work because unwieldy bottles of shielding gases are not required. Another advantage on-site is that there is no externally added shielding gas which is susceptible to disruption by wind. Flux-cored wires can be used at higher maximum currents than solid wires, resulting in high deposition rates.

16.3.1.10 Gas-shielding tungsten arc (TIG) welding

In this process an arc is established between a tungsten electrode and the parent metal, forming a weld pool into which filler rod is fed, generally by hand (Figure 16.79). Mechanized systems which feed the filler wire are available and movement of the welding head along the joint line can also be mechanized. The tungsten electrode is non-consumable and contamination of the weld pool by air is prevented by an inert shielding gas such as argon, helium or mixtures of these gases. A high level of skill is required by the welder, who can control penetration with great precision. This makes the process particularly suitable for the welding of thin sections and for the deposition of root runs in pipe.

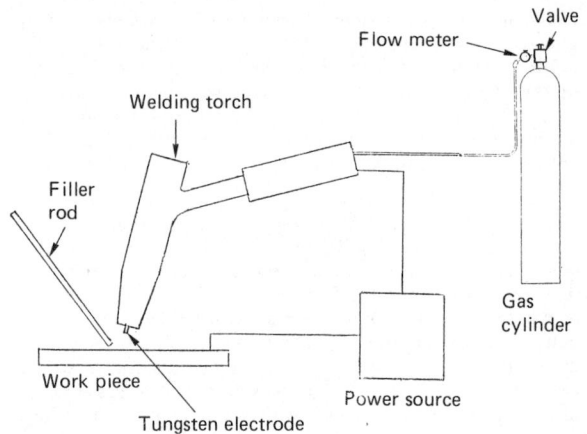

Figure 16.79 Tungsten inert-gas welding

Electrodes and filler rods Pure tungsten electrodes can be used but improved arc initiation and stability are obtained by the use of electrodes containing additions of either thoria (thorium oxide) or zirconia (zirconium oxide). Thoriated electrodes are preferred for d.c. welding and zirconiated electrodes are used for a.c. Electrode diameters vary from 1.2 to 4.8 mm depending on the welding currents used, which can range from 75 to 450 A for thoriated electrodes and from 50 to 200 A for the zirconiated types.

Filler rods which are specified in BS 2981: Parts 1–5[29] have diameters of 1.2–5.0 mm and are available in a wide range of compositions suitable for welding carbon and low-alloy steels, stainless steels, copper and copper alloys, nickel and nickel alloys, aluminium and aluminium alloys, titanium and zirconium.

Power sources An a.c. or d.c. power source with standard generators, rectifiers or transformers is used. For stable operation the power source must have a 'drooping characteristic', so that when variations occur in voltage or arc length the current remains substantially constant. When changes occur in the arc length when the welding torch is manually guided along the joint line the power input remains within $\pm 8\%$ of the preset value.

If the arc is initiated by touching the tungsten electrode onto the parent metal the electrode becomes contaminated and to avoid this, a high-frequency oscillator is incorporated into the power source. Alternatively, a spark starter using a high-voltage coil similar to that in a car-ignition circuit can be used. When the gas in the gap between the electrode and the parent plate is ionized by either the high frequency or the spark discharge the full welding current flows. With d.c. the high frequency is normally turned off automatically after arc initiation but with a.c. it is operated continuously to maintain ionization of the arc path when the arc voltage passes through zero.

Power sources are available for pulsed arc welding which enables a stable arc to be maintained at low currents down to 10 A. In pulsed TIG welding the pulse frequency varies from 10 per second to 1 per second, and each pulse forms a molten pool which solidifies before the next pulse. Pulsed TIG welding can be used to control penetration in thin sheet and in the root runs of pipes and positional welds in plate.

Applications TIG welding is particularly suited to welding light-gauge carbon, alloy and stainless steels and all non-ferrous metals and alloys. A clear, clean weld pool is formed with precise control of heat input and the ability to weld with or without filler metal in all positions makes the process attractive for critical applications where exceptionally high quality is essential. Examples are stainless steel piping for nuclear applications and the wide range of piping compositions used in chemical plant. For such critical applications, fully mechanized orbital welding equipment has been developed in which the welding torch and wire-feeding mechanism rotates round the pipe joint. Thin- and thick-section pipes can be welded with a narrow gap joint preparation and *in-situ* fabrication of nuclear and chemical plant is now possible. Other specialized TIG welding equipment is used for the mechanized welding of tubes to tube plates.

16.3.1.11 Plasma arc welding and cutting

Plasma arc welding was developed from TIG welding by placing a narrow orifice round the arc and supplying a small flow of argon through the orifice (Figure 16.80). The constricted arc dissociates the argon gas into positive and negatively charged electrons to form a plasma. When the plasma gas flows away from the arc column it forms neutral atoms again and gives up its energy in the form of heat.

A low-current pilot arc is initiated between the tungsten electrode and the water-cooled copper orifice. The argon gas flowing through the orifice is ionized and initiates the primary arc between the tungsten electrode and the parent metal when the current is increased. The arc and the weld zone are shielded by a gas flowing through an outer nozzle. The shielding gas consists of argon, helium or gas mixtures of argon with either hydrogen or helium.

A normal tungsten arc has a temperature of approximately 11 000°C but the constricted arc of a plasma torch can reach 20 000°C. The high-temperature ionized gas jet gives up its energy when it contacts the parent metal and thus increases the energy of the tungsten arc. This produces a deep penetration weld with a high depth-to-width ratio with minimum distortion of the parent metal. The term 'keyhole' is used to describe the shape of the hole formed in the parent metal when a close square edge butt joint is welded. As the torch is moved along the joint, molten metal flows round the edges of the hole and solidifies at the rear of the hole. The molten metal at the sides of the hole is held in place by surface tension and the pressure of metal vapour in the hole.

The keyholing welding technique can be used on carbon, low-alloy steels and stainless steels in thicknesses of 2.5–10 mm and in aluminium alloys up to 20 mm. Welding speeds are generally 50–150% higher than those possible with TIG welding.

A low-current version of the process is micro-plasma arc welding, which is used for precision welding of thin sheet from 0.025 to 1.5 mm thick at currents of 0.1–10 A. The plasma arc is much more stable than a TIG arc, which tends to wander from the joint line at low currents.

Plasma cutting If the current and gas flows are increased sufficiently the molten metal formed round the keyhole is ejected at the bottom of the hole and as the plasma torch is traversed along the work piece a cut is formed. Plasma cutting is especially suitable for cutting non-ferrous metals, such as aluminium, copper and nickel, and their alloys which are not easily cut by oxy-fuel gas flames. Most non-ferrous metals are cut using nitrogen, nitrogen–hydrogen mixtures or argon–hydrogen mixtures as the plasma gas. A secondary shielding gas delivered through a nozzle that encircles the plasma gas nozzle is selected according to the material being cut. For mild steel and stainless steel it can be CO_2 and for aluminium it is an argon–hydrogen mixture. Sometimes water is used instead of the ancillary shielding gas and in another variety of the process water is injected round the end of the plasma gas nozzle, which has the effect of concentrating the plasma flame and allowing higher cutting speeds.

Plasma cutting can be used for plate edge preparation (i.e. bevelling) and for shape cutting. The process can be used manually or the torch can be mounted on mechanized cutting equipment identical to that used for oxy-fuel gas cutting. For metal thicknesses up to 75 mm carbon steels can be cut faster by plasma cutting than by oxy-fuel gas, and up to 25 mm thick the cutting speeds can be five times as fast.

An important variation of the process is the use of compressed air for the plasma gas without the provision of any additional shielding gas. The use of compressed air instead of water for cooling enables the torch to be of simplified construction.

Small manual air plasma torches are available which find increasing applications in sheet metal cutting (e.g. motor repair shops). For further information reference 40 should be consulted.

16.3.1.12 Gas welding and cutting

Gas welding is carried out by a flame produced by burning approximately equal volumes of oxygen and acetylene which are delivered at equal pressures from gas bottles to a welding torch. The flame temperature is approximately 3100°C, which is high enough to melt steel and other metals. Filler metal, if required, is added by manually feeding a rod into the front edge of the weld pool while the torch is moved along the joint. The products of combustion provide sufficient protection from the atmosphere when welding steel. When welding other metals such as cast iron, stainless steel, aluminium alloys and copper alloys, fluxes are used to clean and protect the metal from oxidation.

Equipment The welding torch has two knurled control knobs which regulate the flow rates of oxygen and acetylene so that a neutral or slightly oxidizing or reducing flame is obtained, depending on the application. The torch has a screw-in nozzle from a set of nozzles having different diameter holes which produce the appropriate size of flame and therefore the

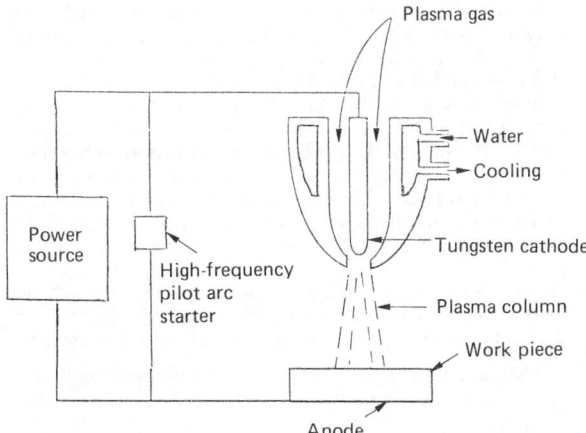

Figure 16.80 Plasma arc welding

required heat input for the particular metal and thickness to be welded. The oxygen and fuel gas hoses are connected between the welding torch and the gas bottles, the gases passing through flashback arresters and pressure regulators. Flashback arresters are safety devices that prevent a flame from travelling back into the cylinders in the case of a backfire. For workshop use the gas bottles are generally mounted in pairs on a trolley which can be moved to where it is required.

Filler metal and fluxes Chemical compositions of filler metals are specified in BS 1453: 1972 and include ferritic steels, cast iron, austenitic stainless steels, copper and copper alloys and aluminium alloys. Ferritic steels do not require the use of a flux but proprietary fluxes are available for other materials.

Applications Gas welding is used mainly for repair and maintenance work, particularly in the repair of car bodies and agricultural implements, although it is slowly being replaced by small TIG and MIG welding equipment. Gas welding is used, to a certain extent, for sheet metal work (i.e. heating and ventilating ducting) and is still employed for making the root runs in pipes, where it is particularly useful for bridging gaps.

Two applications where gas welding has distinct advantages over other processes are in the welding and repair of grey iron casting and in hard-facing with expensive alloys. Grey iron castings can be successfully welded by the use of high preheating temperatures of up to 600°C and gas welding with cast iron filler rods. The deposition of high-cost wear-resistant alloys such as the cobalt–chromium tungsten types or those based on chromium or tungsten carbides can be carried out with minimum melting of the parent metal, so that dilution of the deposited alloy and the consequent decrease in wear resistance is avoided. Gas welding is also successfully applied in jewellery manufacturing with miniature torches and small gas bottles.

Acetylene is the only fuel gas suitable for gas welding because of its favourable flame characteristics of both high temperature and high propagation rates. Other fuel gases, such as propane, propylene or natural gas, produce insufficient heat input for welding but are used for cutting, torch brazing and soldering. They are also used for flame straightening of distorted components and for preheating before welding and post-heating after welding.

Gas cutting Gas cutting, sometimes referred to as flame cutting or oxygen cutting, involves an active exothermic oxidation of the steel being cut when the material has been preheated by an oxy-fuel gas flame to the ignition temperature of around 900°C. The equipment for gas cutting is the same as for welding except that a special cutting nozzle is required. The nozzle has an outer ring of holes through which the preheating gas mixture is delivered and a central hole through which the oxygen jet flows. The exothermic reaction of oxidation of steel forms a fluid slag of iron oxide and after a few seconds, depending on the metal thickness, the section is pierced. Iron oxide and molten metal are expelled from the cut by the oxygen stream. Movement of the cutting torch across the work piece produces a continuous cutting action and the torch can be operated manually or by a motorized carriage. Steel up to 300 mm thick can be cut by this process.

Oxidation-resistant steels such as stainless steel may be cut by specialized methods, including the introduction of iron powder or other proprietary powders into the oxygen stream. These powders react with the refractory chromium oxides and reduce their melting points and increase their fluidity, enabling cutting to take place. For further information reference 40 should be consulted.

Stainless steels and non-ferrous metals and alloys are usually cut by the plasma cutting process, which does not rely on an exothermic reaction.

Manual gas cutting is possible and the accuracy of cutting can be improved by the use of a small wheel mounted on the cutting torch. The wheel may be free-running or motorized. For general cutting and profiling (including cutting bevels on plate edges) mechanized cutting is normally used. For mechanized cutting electronic tracing devices are common and consist of a photoelectric cell that follows the outline of a drawing and guides the cutting nozzle by means of driving motors which regulate the movement of a carriage and cross-arm to which the torch is attached.

Numerically controlled cutting machines are available which use programs stored or punched on magnetic tape which send appropriate signals to the drive motors.

16.3.1.13 Welding and cutting with power beams

Electron beam welding and laser welding utilize high-energy beams which are focused onto a spot of about 0.2 mm diameter on the work piece surface. This intense heat source, which releases its kinetic energy when the beam hits the surface, is radically different from arc welding, in which the arc melts an area of about 5–20 mm in diameter, depending on the welding conditions.

When the power density of an electron or a laser beam at the focused spot is 10 kW mm^{-2} or greater, energy is delivered at a faster rate than can be conducted away in the form of heat in the work piece and the progressive vaporization of metal through the section thickness forms a hole. If the beam is then traversed along the work piece, molten metal flows around the sides of the hole and solidifies at the rear of the hole. Molten metal at the sides of the hole is held in place by surface tension and the presence of metal vapour in the hole in the same manner as that described as the keyholding technique in Section 16.3.1.11.

As with plasma welding, a deep penetration weld with a high depth-to-width ratio is formed with minimum distortion of the parent metal. Maximum penetration depths in steel are approximately 280 mm for electron beam welding and 12 mm for laser welding, although developments in the latter process are likely to increase this to 25 mm or more.

Electron beam welding In this process a finely focused beam of electrons passes from a cathode and travels through a hole in an anode and is focused onto a spot on the work piece 0.2–1 mm diameter by means of a magnetic lens. Deflection coils are used to cause the beam to move in a circular pattern to increase the width of the weld, so that fusing two mating surfaces together in a close square butt joint is possible. The cathode in the electron gun is maintained at a negative potential of 60–150 kV and the gun is contained in a vacuum of 5×10^{-5} torr. In the work chamber a pressure of 5×18^{-3} torr is suitable for welding most metals and for some applications a pressure of 10^{-1} torr or less is used with the advantage of much shorter chamber excavation times, resulting in increased production rates.

High-vacuum welding High-vacuum welding, in which the work chamber is maintained at a pressure of 10^{-3} to 10^{-5} torr depending on the application, has the following features:

1. Maximum weld penetration and minimum weld width and shrinkage, enabling all thicknesses to be welded in a single pass.
2. The highest purity weld metal is produced because of the absence of any contaminating gases such as oxygen or nitrogen.

3. A high vacuum allows a long distance to be maintained between the gun and the work piece, which facilitates observation of the welding process.
4. The pump downtime is lengthy, up to an hour or more depending on the size of the chamber, which would lower the production rate of small jobs with shallow welds but would be insignificant when welds in plate 50 mm or thicker are made with a single pass.

Welding speeds for steel are shown in Table 16.4.

Table 16.4 Penetration depth and welding rates

Welding power (kW)	Plate thickness (mm)	Welding speed (mm/min)
0.25	1	1500
2.0	10	750
4.5	10	750
15	50	150
75	280	50

Medium-vacuum welding In medium-vacuum welding the working chamber is maintained at a pressure generally within the range of 10^{-3} to 10^{-1} torr, although pressures of up to 25 torr are reported in the literature[41] and correspondingly short pump downtimes of a few seconds.

Medium-vacuum welding with small working chambers is used extensively for small repetitive work such as welding of finish machined geartrains and similar high-volume mass-production applications for the motor industry. Many high-precision semi-finished or fully machined components for aircraft engines are also welded by the electron beam process.

Medium-vacuum electron beam welding is not suitable for welding reactive metals and alloys such as titanium and zirconium, which require the high-vacuum process to obtain sound welds.

Out-of-vacuum electron beam welding Provided that the gun-to-work distance is less than about 35 mm to allow for the greater dispersion of the electron beam compared with working in a vacuum, it is possible to weld many materials out of vacuum. With 60 kW non-vacuum equipment single-pass welds can be made in metal thicknesses up to 25 mm. Metals welded out of vacuum include carbon and low-alloy steels, and copper and aluminium alloys. Because of the presence of air, which can cause contamination of the weld metal, it is usually necessary to provide an inert shielding gas to cover the weld zone.

Laser welding and cutting Two types of lasers are used for welding and cutting: the solid-state YAG (yttrium–aluminium–garnet) and the carbon dioxide (CO_2) laser. The term 'laser' is an acronym for Light Amplification by Stimulated Emission of Radiation and a laser is a device for producing monochromatic (single-wavelength) light that is coherent (i.e. all the waves are in the same phase). A laser beam can be transmitted over many metres and can be focused to produce the high-energy density required for welding or cutting.

The solid-state YAG laser The solid-state laser is stimulated to emit coherent radiation by means of the light from one or more powerful flash tubes and the output is in the infrared region around 1.06 μm. Both input and output are generally pulsed and the power output is a maximum of 500 W.

Pulsed laser welding is extensively used in the electronics industry where miniaturization requires very precise positioning of small welds. Typical examples are encapsulation of microelectronic packages and the joining of fire wires by butt, lapped or cross-wire joints. Spot welds can be made between overlapping sheets and seam welds formed by a series of overlapping spot welds. Most metal can be welded, including steel, copper, nickel, aluminium, titanium, niobium, tantalum, and their alloys. Solid-state lasers are also used for precision drilling of holes with very small diameters.

The carbon dioxide laser In the carbon dioxide laser the lasing medium is a gaseous mixture of carbon dioxide, nitrogen and helium at a reduced pressure of 2–50 torr. The output of the carbon dioxide laser is in the mid-infrared (10.6 μm) and CO_2 lasers are available with powers of up to 20 kW. The high power density produced (10^4 W mm^{-2} or more) forms a cavity or keyhole in the work piece which enables a deep penetration weld to be produced similar in appearance to an electron beam weld.

Penetration is less than for an electron beam operating at the same power because of plasma gas formed in the keyhole. The plasma gas escapes from the keyhole and interacts with the laser beam, restricting penetration to approximately 12 mm in mild steel. Methods have been developed for overcoming this problem[41] by the use of either a pulsed output from the laser with pulses of shorter time than that required to generate the plasma or a high-velocity jet of helium to disrupt the plasma above the weld. Such developments are likely to increase the thicknesses that can be laser welded to 25 mm or more.

Carbon dioxide lasers are used in production in some automotive and aerospace applications, particularly for titanium and nickel alloys. The advantage of laser welding over elctron beam welding is that a vacuum is not required, which simplifies the welding operation. Other materials that can be welded by the carbon dioxide laser are steels, copper alloys, zirconium and refractory metals, but aluminium alloys are not readily weldable with the CO_2 laser because of their reflectivity of the laser beam.

Laser cutting Cutting with the CO_2 laser is carried out by a combination of melting and vaporization with an auxiliary jet of gas to blow the molten metal from the cut. Various gases are used for this purpose, including oxygen, compressed air, inert gases and carbon dioxide.

Oxygen produces an exothermic reaction with ferrous metals which increases the efficiency of cutting. The inert gases produce clean unoxidized surfaces – important features when cutting readily oxidized metals such as aluminium and titanium.

16.3.1.14 Resistance welding

Spot, seam and projection welding Spot, seam and projection welding are carried out by electric resistance heating two overlapping metal parts which are pressed together by copper or copper alloy electrodes. Local melting occurs at the faying surfaces and an internal weld nugget is formed. The welding cycle, comprising current, pressure and time, is readily controlled automatically, giving the following advantages:

1. Little skill is required in operation.
2. Welding can readily be built into production lines.
3. Welding can be associated with automatic loading, unloading and transferring of components so that resistance welding is the simplest process for automation or robotic welding.
4. Welding times are short so that output is high.
5. No filler wires or fluxes are used.

6. Distortion is a minimum as heating is confined to a small area.

The high currents required for resistance welding are generally obtained from a single-phase a.c. transformer having a primary winding of several hundred turns. The secondary winding consists of one or two turns of thick copper which may be water cooled. The voltage is stepped down to a value between 4 and 20 V.

Three-phase welding machines are available which are more expensive but they have a better power factor. A three-phase transformer is sometimes used with a d.c. rectifier to carry out resistance welding with d.c. For further information on power sources reference 41 should be consulted.

Spot welding Spot welding (Figure 16.81) is used for the fabrication of sheet metal from 0.6 to 3 mm thick to produce lap joints intermittently welded and therefore not pressure-tight. Typical applications are low-carbon steel components for car bodies, cabinets and general sheet metalwork. Spot welding is also applicable to stainless steel, aluminium and aluminium alloys, and copper alloys.

Because of high electrical conductivity, pure aluminium is difficult to spot weld and its softness results in heavy indentation by the electrodes. The high electrical conductivity of copper makes it unsuitable for spot welding. Stainless and heat-resisting steels are spot welded for aircraft and gas turbine engines.

Seam welding Seam welding (Figure 16.82) is similar in principle to spot welding and uses rotating wheel electrodes which roll the overlapping components between them. The welding current passes intermittently through the electrodes, forming a series of welds that overlap one another. The electrode wheels rotate continuously and at least one is power-driven to move the component along, in addition to carrying the welding current.

Pressure-tight seams can readily be made and the process is faster than spot welding. However, machines are heavier and more costly than spot welding machines. Seam welding is primarily suited to making long straight welds, although curved welds can be made (for example, welds which may occur at the corners when joining two half pressings together to form a fuel tank). Typical applications are pressure-tight seams for oil drums and refrigerator parts.

Projection welding In this process the current is concentrated by the shape of the components themselves, small dimples being formed on one of the sheets to be joined as shown diagrammatically in Figure 16.83. The electrodes are of relatively large area compared with spot welding electrodes and can therefore be made of a hard material having a comparatively high electrical resistance, such as a copper–tungsten alloy. As several welds can be made at the same time, the

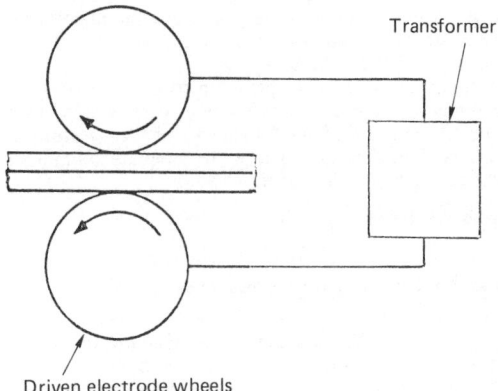

Transformer

Driven electrode wheels

Figure 16.82 Resistance seam welding

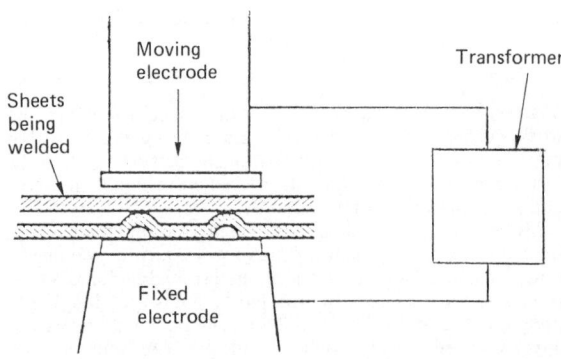

Moving electrode

Transformer

Sheets being welded

Fixed electrode

Figure 16.83 Projection welding

welding machine has to be of high electrical capacity and must also be capable of applying the high total mechanical load to the components during welding. Three projections welded simultaneously give the best results, and up to five projections are common. Applications involving larger numbers of projections are known, but as the number increases it becomes increasingly difficult to ensure uniform and adequate welding of all the projections. The process can also be used for attaching studs, nuts and disks to flat plates.

The main features of the process are as follows:

1. Several welds can be made simultaneously in the time it would take to make a single spot weld.
2. Electrodes have longer lives between dressings compared with spot welding electrodes because of the relatively low current densities.
3. Welds are generally of better external appearance than spot welds due to the absence of indentations.
4. T-joints may be made as well as lap joints.
5. Machines are generally more costly than spot welding machines.
6. The cost of preparing components is greater because of the necessity of forming projections.

16.3.1.15 Resistance butt and flash welding

Resistance butt welding The two ends of the parts to be joined are brought into contact and current is passed across

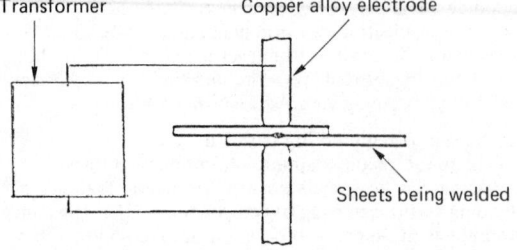

Transformer

Copper alloy electrode

Sheets being welded

Figure 16.81 Resistance spot welding

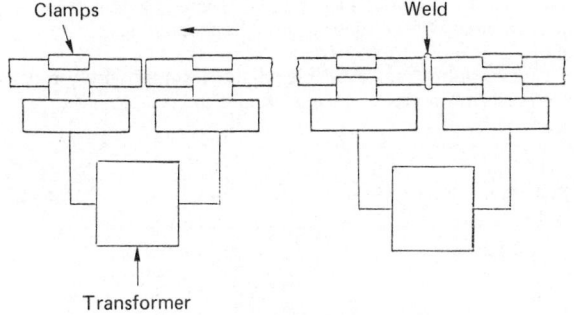

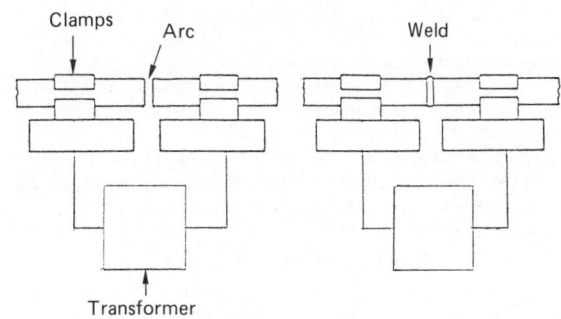

Figure 16.84 Resistance butt welding

Figure 16.85 Flash welding

the joint while a moderate mechanical force is applied to the components (Figure 16.84). As the joint heats by electrical resistance the material softens and welding takes place. The finished joint shows a thickening or upsetting of the components in the joint area, and for some distance away from the joint. Resistance butt welding is essentially a solid-state welding process in which no melting occurs.

Recrystallization of the metal takes place across the faying surfaces and the upsetting action has the effect of removing oxides from the joint. The process has two main applications:

1. Joining two components of the same cross-section end to end (e.g. wire, rod, bar or tubing). Used in rod and wire mills for joining the ends of coils for continuous processing.
2. Continuous welding of longitudinal seams in pipe or tubing formed from flat plate.

The process is applicable to the welding of carbon, alloy and stainless steel, aluminium alloys, copper alloys, nickel alloys, and electrical resistance alloys.

Resistance butt welding machines are generally designed to weld a particular family of alloys because the current densities and pressures as well as the rate of application of the pressure differ widely between, for example, steels and aluminium alloys. The cross-sectional areas to be joined also determine the current capacity of the welding transformer.

The welding machine has two platens on which are mounted clamping dies which grip the components. One platen is stationary and the other moves to produce the pressure required.

Flash welding Flash welding was developed from resistance butt welding and the principle is shown in Figure 16.85. The currrent is switched on before the ends of the components are in contact, with the result that small volumes of metal melt explosively at the points where contact is first made. This process continues as the two components are moved towards one another, causing a large amount of flashing and removal of metal while, at the same time, heating the areas in contact and the material immediately behind. Once suitable temperature conditions have been established, the two ends of the components are forced together by a sudden increase in the load applied and the current is switched off. The effect of this sudden load is to squeeze out from the joint all the overheated and oxidized metal and to form a high-quality pressure weld. The fin or flash of upset metal surrounding the joint is normally removed before the component is put into service.

Typical applications of flash welding are the production of continuous welded railway track, motor-car wheel rims formed from flat steel stock and mitre joints in door and window frames. The materials welded are similar to those that can be resistance butt welded.

16.3.1.16 *Friction welding*

Friction welding is a solid-state joining process, i.e. there is no fused metal involved. The heat to make a forge weld is produced by moving one component relative to a mating component under pressure. The motion is usually rotational, although the linear relative motion can be used and is under development. A friction weld is similar in appearance to a resistance butt or a flash weld with upset metal or flash which is generally removed after welding by a machining operation.

Butt welds can be made in rods or tubes, and rods or tubes can be welded to plates. One of the components is generally circular in cross-section although square rods can be joined by friction welding. A wide range of sizes can be friction welded from small-diameter wires used in the electronics industry up to 150 mm diameter aluminium bus bars.

Typical applications are found where high production rates are required (e.g. the motor industry). Examples are axles, drive shafts, steering shafts and valves where a high-alloy heat-resisting head is welded to a cheaper carbon steel shank. Bar stock can be welded to plates to produce parts that would normally be forged.

Most metals and alloys, with the notable exception of cast iron, can be friction welded. Dissimilar metals can also be welded without the formation of a brittle zone that often occurs with arc welding (e.g. aluminium can be welded to steel).

Friction surfacing is an important development in which a corrosion- or wear-resistant alloy rod is rotated rapidly under pressure against a surface to be clad with the alloy. Under carefully controlled conditions a friction weld can be deposited over the surface and the weld metal is unchanged in composition because there is no dilution of the weld caused by melting of the parent metal as occurs in surfacing by arc welding. For further information on friction welding reference 41 should be consulted.

16.3.2 Soldering and brazing

Brazing and soldering are carried out at temperatures below the melting points of the metals being joined. Brazing filler metals have melting points above 450°C and solders melt below this temperature. In both processes the molten filler metal flows by capillary action between closely fitted surfaces of the parts to be joined.

Unlike welding, soldering and brazing do not generally produce joints having mechanical properties matching those of the parent metal. Typical applications, such as car radiators and electrical connections, require leak-tightness and electrical conductivity, respectively. If high strength or ductility is required correct alloy selection and joint design are essential and for further guidance references 42 and 43 should be consulted.

The low temperatures involved in soldering and brazing have two beneficial effects compared with welding: (1) parent metal properties are less affected, especially by soldering; (2) residual stresses and distortion are lower. For successful joining, the following procedures are required:

1. For wetting of the present metal by the solder or brazing alloy the surfaces must be chemically clean. Degreasing and mechanical cleaning or pickling may be necessary.
2. During the joining operation a flux is generally used to remove surface oxides and to prevent them from reforming. In some brazing processes a reducing or neutral atmosphere or a vacuum is used.
3. Heating of the parent metal must be carefully controlled so that both the mating surfaces reach the melting temperature of the filler metal before it is applied to the joint. When the filler metal is pre-placed in the joint the same requirement applies because the molten filler metal will not wet the surface unless that surface is hot enough.
4. All traces of flux residues which might lead to corrosion must be removed after soldering or brazing.

16.3.2.1 Soldering

Solders   The normal range of soft solders are listed in BS 219: 1977[44] which also indicates typical uses. The commonest solders are the tin–lead alloys, some of which have antimony added. These solders are used in various applications, depending on the composition, including manufacture of food-handling equipment, tin-plated cans, domestic utensils, electronic assemblies, heat exchangers and general engineering work. Tin–antimony solder is used for higher service temperatures (e.g. above 100°C) and in applications for joining stainless steels where lead contamination must be avoided. The alloy is also used for plumbing and refrigeration applications. Tin–silver solder is used for fine instrument work and food applications while tin–lead–silver solder is used for service above 100°C and also below −60°C.

Soldering methods   Many sources of heat can be used for soldering, including the following: heated iron, flame or torch, hot-plate or oven, HF induction, electric resistance, hot gas, as well as dip, wave or flow methods involving molten solder baths. For further information on both manual and mechanized soldering techniques references 43, 45 and 46 should be consulted.

Manual soldering methods using a soldering iron or a torch are fairly slow and require a reasonable degree of skill, although repetitive work carried out by an experienced operator can give high production rates. Dip soldering is faster and the electronics industry uses specially designed equipment for wave soldering of printed circuit boards.

16.3.2.2 Brazing

Brazing alloys are specified in BS 1845[47] under eight groups based on aluminium, silver, copper–phosphorus, copper, copper–zinc, nickel, palladium and gold. The choice of brazing alloy for a particular parent metal can be fairly wide and a decision may be based on metallurgical considerations, including corrosion resistance, maximum temperature to which the parent metal can be heated to avoid deterioration in properties or melting of previously applied brazing alloys. For further

information references 42, 48 and 49 should be consulted. A brief summary of the uses of the eight main groups of brazing alloys is given in Table 16.5.

Brazing alloys are available in the form of wires, rods, powders and inserts of various shapes for pre-placing in the joint.

Table 16.5 Main uses of brazing alloys

Brazing alloy	Materials suitable for brazing
Aluminium	Pure aluminium and some aluminium alloys
Silver	Most ferrous and non-ferrous metals and alloys except aluminium, magnesium and refractory metals and their alloys
Copper–phosphorus	Copper and copper alloys
Copper	Ferrous materials in a protective atmosphere or in a vacuum
Copper–zinc	Stainless steel and other heat- and corrosion-resistant alloys in a protective atmosphere or in a vacuum
Nickel	Stainless steel and nickel–chromium heat-resisting alloys for high-temperature service
Palladium and gold	Metallized ceramics, copper, nickel and ferrous alloys in a protective atmosphere or in a vacuum

Heating methods   A full description of the various heating methods and their advantages and disadvantages are described in BS 1723: Part 2: 1986.[48] The methods comprise hand-torch brazing, mechanized flame brazing, induction brazing, furnace brazing (including protective atmosphere), vacuum and open-furnace brazing, and immersion brazing, including flux bath, dip bath and salt-bath brazing. Special processes referred to are infrared brazing and laser brazing. Reference 48 also contains details of design and location methods for brazing.

Applications   Brazing is applicable to cast irons, steels, galvanized steel, aluminium, copper, magnesium, nickel and their alloys, stainless and heat-resisting steels, titanium, zirconium, ceramics and refractory metals and to dissimilar metal joints. It is selected in preference to other joining processes for the following reasons:

1. Where heating must be restricted to avoid melting or distortion of one of the parts to be joined.
2. Where strength or dimensional accuracy of the assembly would be impaired by heating to a high temperature.
3. Where a soft soldered joint would not be strong enough.
4. Where parts cannot be welded because of the properties of the parent materials involved (e.g. ceramics, refractory metals).
5. Where a number of joints have to be made in a small complicated assembly.
6. Where joints would be inaccessible to the welding processes.

16.3.3 Productivity and welding economics

Productivity, in the broadest sense, is what determines profit. Unfortunately, the cost of welding is one of the least-documented topics in the whole field of welded fabrication. When a company decides to review its fabrication methods there is often a lack of understanding of the principles involved and only limited knowledge of the alternatives available. Reliance may be placed on equipment salespersons who may be fair and have an excellent product to sell but, quite naturally, may be biased in their approach.

There is a lack of information compared with that readily available on machine tools (e.g. cutting speeds, depth of cut which can enable a machine to be integrated into a manufacturing system to give a high utilization or duty cycle and therefore maximize productivity). The same objective should be applied to welding as part of the manufacturing system.

The first task in reviewing fabrication methods is a simple one and concerns design. In the case of arc welding, fabrications should be designed to have the minimum amount of weld metal that will meet the service requirements of the components in terms of mechanical properties.[50] The total quantity of weld metal in a fabrication is, of course, only a very small proportion of the total weight (perhaps 2% or less) but, nevertheless, it is sensible to use the minimum amount necessary. This requires close attention to detail design, shopfloor supervision and inspection to ensure that welded joints are not overdesigned and weld metal is not being wasted.

Sometimes partial-penetration butt joints will be adequate and intermittent fillet welds can replace continuous ones. Fillet weld profiles should not be excessively convex because any weld metal in excess of a mitre fillet does not contribute to the strength of a joint.

For butt joints the plates are usually cut with bevel angles of 30° to given an included angle of 60° when the plates are placed together. With a gap at the root of 1–3 mm this is the optimum edge preparation to give full penetration and room to manipulate a covered electrode or MIG welding gun. If the included angle is increased to 70° the weld volume is increased by 20%, which represents 20% waste. If a fillet weld having a size or leg length of 6 mm is adequate, then a leg length of 8 mm will result in 57% wasted weld metal because of the amount of overwelding. The increase in the cost of wasted consumables may be small but the cost of depositing excess weld metal includes labour costs and overheads, which can increase the overall cost by a considerable amount.

Having ensured that the optimum welded detail design is employed, there may be a desire to increase productivity still further. The next point to consider is the deposition rate, which depends on the welding current whichever arc welding process is used. As a simple example, in manual metal arc welding if the electrode size is increased from 4 mm to 5 mm diameter and the current increased accordingly the cost of welding can be reduced by approximately 25%. A similar increase in welding current has the same effect in other processes such as MIG or submerged-arc welding. The above simple example relates to a particular job where the duty cycle was 30% (the duty cycle being the ratio of arcing time to total elapsed time for the welding operation). A welder is obviously not welding for 100% of the time and typical activities for a manual welder in various industries are shown in Table 16.6.

Duty cycles may be considerably lower than those shown in the table. For example, in repair work where defects have to be laboriously removed by mechanical means, the welding duty cycle may be less than 10%. In contrast, the duty cycle of a robot MIG welding parts of car bodies may have to be 70% or higher to justify the capital investment.

Returning to the manual welder who increases the deposition rate by changing to a larger electrode and increasing the current, this gives a worthwhile cost saving provided that the duty cycle is at least approximately 20%. If the welder only deposits weld metal for 10% of the time, then even if the deposition rate is doubled it will have only a slight effect on productivity. The duty cycle may be low because the welder

Table 16.6 Observed MMA welder activities (percentage breakdown in time)

Industry	Pressure vessel	Site pipework	Engineering Light	Heavy
Activity:				
Arcing (duty cycle)	42	36	62	35
Deslagging, setting up, etc.	22	30	25	32
Other activities: waiting/using crane instructions, etc.	14	23	5	14
Idle	22	21	8	19

has to fetch and carry components and set them up for welding. If a labourer is employed to deliver components and set them up at a second workstation or in a second jig the welder can possibly double or treble the duty cycle and take full advantage of optimizing the deposition rate of the welding process that is being used. It is necessary to carry out an exercise in method study to reveal the true cost of welding and to indicate how improvements in productivity may be accomplished with the equipment currently in use. Costing methods for arc welding must take account of the costs of consumables, weight or volume of weld metal, deposition rates and duty cycles, labour costs and overheads and amortization of equipment. For further information on the calculation of welding costs, references 51–53 should be consulted.

A further important point to consider in economic surveys of welding operations is the welding position. Deposition rates can be maximized by welding in the flat position because a large molten pool of weld metal can be maintained, whereas in the vertical or overhead welding positions the size of the weld pool has to be restricted. Therefore the provision of a welding manipulator or turning rolls can enable work to be rotated so that welding can be carried out in the flat position with maximum deposition rates. In some cases, depending on the shape and size of component, the rate of welding can be doubled or trebled by suitable manipulation of the work piece.

When a welding engineer has considered all the elementary principles of welding engineering and has determined the possible increase in productivity at minimum cost, the next question to be asked is whether this is satisfactory for the company. Management decisions are required on the rate of production required at present and in the foreseeable future, and any plans for the manufacture of different products. New product lines may raise the question of what is technically required in respect of quality of welded joints (for example, static or fatigue strength, fracture toughness, corrosion or wear resistance). With this information, a choice can be made on the most appropriate welding processes, equipment and consumables capable of achieving the quality and production rates required. If a welding engineer is available he or she will obviously be expected to make recommendations on the choice of welding processes and equipment. Otherwise, the company may have to contact outside sources such as equipment suppliers, national advisory bodies or a consultant. In any of these circumstances it is advisable that the manager making investment decisions should have some basic knowledge of welding technology if only to be able to ensure that all the possible alternatives have been considered.

For further information on the choice of manual, mechanized or robotic welding processes coupled with economic considerations, references 36 and 54–56 should be consulted.

16.4 Adhesives

16.4.1 General comments

Adhesives can offer substantial economic advantages over conventional methods of joining. Indeed, it was for this reason that the application of adhesives to metal fabrication, in common with many other technological innovations, was pioneered by the aircraft industry. A realization of the advantages of adhesive bonding over other methods of joining materials has led to a rapid growth of the adhesives industry. This has been accompanied by advances in the science and technology of adhesion and adhesives, with an emphasis on high-performance synthetic products for the mechanical engineering sector. Adhesive bonding, either alone or in combination with other methods of fastening, represents a key enabling technology for the exploitation of new materials and for the development of novel concepts and structural configurations.

The strength of an adhesive bonded joint is determined by the strength of its weakest element; this is not necessarily the adhesive! The correct choice of adhesive and an appreciation of appropriate process procedures are necessary for the satisfactory fabrication of bonded assemblies. Load-bearing joints require proper design. Thus the engineer requires at least a qualitatively correct overall picture of the various factors influencing adhesion and controlling joint performance. This section seeks to present a balanced overview of these factors.

16.4.2 Definitions and terms

16.4.2.1 Adhesives

An adhesive may be defined as any material which, when applied to surfaces, can join them together and resist their separation. Thus adhesive is the general term and may include cement, glue, gum, paste, etc. There are several, largely synthetic, generic groups of adhesives of value to the engineer embracing numerous subgroups, individual formulations and generic combinations.

The adhesive is involved with wetting, adsorption and interdiffusion reactions with the substrate surface before solidification. A schematic cross-section through a bonded interface is given in Figure 16.86.

16.4.2.2 Adhesion

Adhesives join materials primarily by attaching to their surfaces within a layer of molecular dimensions. The term 'adhesion' refers to the attraction between substances whereby, when they are brought into contact, work must be done in order to separate them. Thus adhesion is associated with intermolecular forces acting across an interface, and involves a consideration of surface energies and interfacial tensions.

16.4.2.3 Adherends or substrates

The materials being joined are usually referred to as the adherends or substrates.

16.4.2.4 Adhesive bonding

Bonding means the uniting of similar or dissimilar adherend surfaces with a relatively low modulus interlayer of adhesive material. The resultant properties of the composite made are a function of the bonding, the materials involved and their interaction by stress patterns.

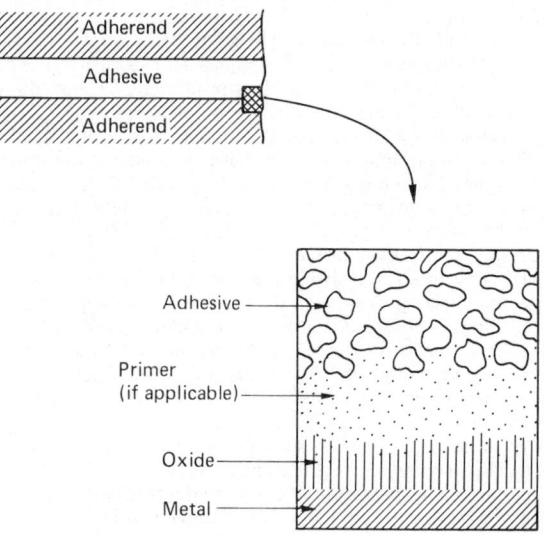

Figure 16.86 Elements of a metal adherend/adhesive interface

16.4.2.5 Advantages and limitations of adhesive bonding

The main advantages and limitations of adhesive bonding as compared with welding or mechanical fastening are given in Table 16.7. The relative importance of individual items naturally depends upon the perspective of different users.

16.4.2.6 Requirements for a satisfactory bonded joint

1. Selection of a suitable adhesive
2. Adequate preparation of the adherend surface
3. Appropriate design of the joint
4. Controlled fabrication of the joint itself
5. Post-bonding quality assurance

16.4.3 Adhesives

16.4.3.1 Introduction

Adhesives may be classified as either organic or inorganic materials in a number of different ways – for example, by origin, by method of bonding, by end-use or by chemical composition. Some classifications employ more than one criterion, such as that offered in Table 16.8.

Natural adhesives such as starch, animal glues and plant resins have been used for centuries, and are still used widely today for packaging and for joining wood. Rubber-based adhesives were introduced in the shoe and tyre industries towards the end of the nineteenth century, but the birth of modern structural adhesives is generally dated from the early twentieth century with the introduction of phenol-formaldehyde resins. Over the past four or five decades the natural adhesives have been improved, and there has been an intense development of synthetic adhesives to meet more technically demanding applications. These synthetic adhesives include thermoplastic and thermosetting types. Thermoplastic adhesives may be softened by heating and rehardened on cooling, and included in this group are relatively low-performance materials such as acrylic polymers, 'hot-melts' and products based upon polyvinyl acetates (PVAs). Thermosetting materials represent thermally stable highly cross-linked

Table 16.7 Advantages and limitations of adhesive bonding

Advantages	Limitations
Ability to join dissimilar materials	Surface pretreatments normally required, particularly with a view to maximum joint strength and durability
Ability to joint thin sheet material efficiently	
More uniform stress distribution in joints which imparts enhanced fatigue resistance	Fairly long curing times frequently involved
	Poor resistance to elevated temperature and fire
Weight savings over mechanical fastening	Structural joints require proper design
Smooth external surfaces are obtained	Brittleness of some products, especially at low temperatures
Corrosion between dissimilar metals may be prevented or reduced	
	Poor creep resistance of flexible products
Glueline acts as a sealing membrane	Poor creep resistance of all products at elevated temperatures
No need for naked flames or high-energy input during joint fabrication	Toxicity and flammability problems with some adhesives
Capital and/or labour costs are often reduced	Equipment and jigging costs may be high
	Long-term durability, especially under severe service conditions, is often uncertain

Table 16.8 Adhesive classification (reprinted from *Polymeric Building Materials* by Feldman, 1989)

Occurrence	Origin	Adhesive
(1) Natural	(a) Animal	Albumen, casein, animal glue, shellac
	(b) Vegetable	Resin: gum arabic, tragacanth, colophony, Canada balsam, natural rubber
		Oils: linseed oil
		Waxes: carnauba wax
		Proteins: soya bean
		Polysaccharides: starch, dextrine
	(c) Mineral	Waxes: paraffin
		Resins: copal, amber
		Other materials: silicates, phosphates, sulphur, magnesia, litharge, bitumen
(2) Synthetic	(a) Thermoplastic	Vinyl polymers and copolymers: poly(vinyl acetate), poly(vinyl alcohol), poly(vinyl acetal)s, polystyrene
		Acrylic polymers: polyacrylates, polymethacrylates, polyacrylamides, poly(cyanoacrylate)s polyamides and saturated polyesters polyurethanes
		Cellulose derivatives: Cellulose acetate, cellulose nitrate, cellulose acetate–butyrate, methyl cellulose, hydroxy ethyl cellulose, carboxy ethyl cellulose, and others
	(b) Thermosetting	Phenolic resins: phenol–formaldehydes, resorcinol formaldehydes
		Amino plastics: urea–formaldehydes, melamine formaldehydes
		Polyepoxides and derivatives: polyepoxides, epoxy–polyamide, epoxy–bitumen, epoxy–polysulphide, etc.
		Unsaturated polyesters
		Polyaromatics: Polybenzimidazole, polyimide, polybenzothiazole

linked structures, and may be regarded as structural adhesives. Included in this group are epoxides and phenolics.

16.4.3.2 Application and setting

During the bonding operation the adhesive must be applied to the substrate in a fluid form, to wet the surface and penetrate its surface irregularities without leaving air voids. The adhesive must then set by changing from a liquid to a solid. The solidification mechanisms used are:

1. Solvent evaporation
2. Cooling from above melting point
3. Chemical reaction

Mechanisms (1) and (2) generally apply to low-performance, thermoplastic, materials while (3) applies to most thermosetting engineering adhesives. The rate of chemical reaction is increased considerably by exposure to heat, and as a rule of thumb is approximately doubled for every 8°C rise in temperature.

16.4.3.3 Selection criteria

Lees[57] states that there exist twelve major family groups of adhesives likely to be of value to the engineer. This poses a major problem in weighing up the pros and cons of different systems for particular applications. However, some computerized processes do exist, and are excellent for newcomers to the technology because they enable the user to gain valuable insight into the criteria which limit selection. Current UK proprietary software includes CATS (Centre for Adhesive Technology), EASeL (Design Council), PAL (Permabond Adhesives) and STICK (Lucas).

The major considerations governing the process of adhesive selection are:

1. Strength requirements (shear, tension, compression, impact, etc.)
2. Deformation characteristics (modulus, strain to failure, creep)
3. Sealing requirements
4. Joint geometry
5. Nature of adherend(s) (mechanical and chemical characteristics)
6. Production processes (method of application, speed of cure, pretreatments, etc.)
7. Service conditions (exposure to heat, moisture, stress)

In general, it will be appreciated that flexible or thin sheet materials should be bonded with relatively flexible adhesives with a high strain to failure, whereas thick sections should be united with a relatively stiff adhesive in order to make a load-bearing joint.

16.4.3.4 Types

It is useful to identify engineering adhesives with four main groups:

1. Epoxy (and toughened variants)
2. Polyurethane
3. Phenolic/resorcinolic
4. Acrylic (including anaerobic, cyanoacrylate and toughened acrylic and tape variants).

Commercial formulations are, in general, complex and sophisticated blends of many components. Fillers, toughening agents, plasticizers, diluents, surfactants and anti-oxidants are among the components which may be added to a 'basic' formulation.

Epoxy Known to most engineers and designers, epoxies represent a group of very strong adhesives. The adhesive comprises an epoxide resin and one of many reactive hardeners. Numerous cold, warm and hot-cured formulations are available to meet particular requirements, and the adhesive can be supplied in paste, film, powder or liquid form. Although often supplied as two components, 'single-part' formulations are also available in which the latent hardener is released for reaction polymerization at temperatures above 100°C. These adhesives may also be toughened by the inclusion of a dispersed rubbery phase to improve peel and impact strength.

Uses are frequently on large components, and where gap-filling, high strength, creep, moisture and heat resistance are required. Many applications in civil engineering (using heavily filled cold-cure formulations), metal plate fabrication, structural engineering, vehicle and general product assembly.

Polyurethane The polyurethane adhesives represent a group of very versatile cold- or warm-cure adhesives, and many formulations are possible. The very active reacting group of the resin is an isocyanate, which is combined with hardeners often similar in nature to those used in epoxies. They may be supplied in paste or liquid form, either as two components or as 'single-part' formulations; moisture is generally the catalyst in the latter case. Unlike epoxies, these materials can harden extremely quickly. However, polyurethanes are generally weaker and more susceptible to moisture attack than other structural adhesives, so that environmental operating conditions must be considered carefully.

Uses are frequently in thin-film form for bonding large sandwich panel assemblies and/or where plastics and timber represent adherend materials. Many applications in vehicle, general product assembly, car repairs, etc., where gap-filling and relatively light-duty connections prevail.

Phenolic/resorcinolic This group represents phenol-formaldehyde and resorcinol-formaldehyde resins, numbered among the earliest structural adhesives to be developed. They are rarely found outside the aircraft or timber and laminating industries. Although good adhesives, they are difficult to use because water is liberated as a by-product of the curing reaction. Heated presses or autoclaves are required to hold the components being bonded together under high pressures, while excess water escapes as steam. They possess outstanding environmental durability.

Uses are in thin-film form for bonding timber components, laminated assemblies and large metal components in the aircraft industry. Their continued use stems from their ability to maintain structural integrity under severe environmental conditions.

Acrylic This group comprises several important cold-curing subgroups, all of which depend upon the activity of the acrylic group for polymerization reactions. The structure of the acrylic backbone is adjusted according to the preferred curing mechanism. Generally, they are used in thin-film form:

1. *Anaerobic*: Sometimes known as sealants and locking compounds, these may be supplied in liquid or paste form. They are single-part adhesives that set characteristically in the presence of metal ions and the absence of atmospheric oxygen; under solely anaerobic conditions the rate of cure is very slow. This cure mechanism is very useful in assembly since hardening of the adhesive occurs only when the joint is closed. Various levels of 'strength' are available, together with toughened variants which enable them to function as true adhesives.

Uses are locking, sealing and retaining closely fitting metal parts – particularly co-axial assemblies. High-viscosity variants are used as gasketing materials.

2. *Cyanoacrylate*: These highly reactive single-component adhesives harden only in thin-film form because surface moisture is used to catalyse them. Traces of (slightly basic) moisture are found on most surfaces, and this slight alkalinity is sufficient to cause polymerization within seconds. Acidic surfaces tend to inhibit curing. Moisture resistance is poor, and there is no gap-filling capability.
   Uses are bonding of small plastic and rubber parts, and even human tissue.

3. *Toughened acrylic*: These represent very versatile two-component adhesives for use in thin-film form. The resin is applied to one surface and an initiator to the other. When the joint is closed the initiator diffuses quickly through the resin and polymerization is complete within minutes – although several hours may be required to achieve full cure. These adhesives bond extremely well to most substrates, and the inclusion of a dispersed rubbery phase confers excellent peel and impact resistance. Recent acrylic derivatives, developed to fill gaps, are true two-part systems similar in nature to epoxy adhesives.
   Uses are bonding of sheet metal, coated metal, plastic parts and some rubbers. Reasonably large structures can be bonded with confidence, especially where surface treatment is minimal.

16.4.3.5 Mechanical characteristics

In general, the properties of adhesive polymers are determined by their internal structure, although the blend of components in many commercial formulations prevents simple chemico-physical relationships being drawn. The types of adhesives of greatest interest to the mechanical engineer range from ductile polyurethane and acrylic formulations through the stiffer epoxies to some stiff (and brittle) anaerobic adhesives. Some stiffer materials still are based upon imide chemistry, and are used in advanced weapons and aerospace structures. The basic property trade-offs for single-phase thermosetting adhesives have been described by Bolger[58] (e.g. Figure 16.87) and it can be seen that high strength is obtained for the price of reduced ductility and toughness. However, the incorporation of rubber toughening into adhesive polymers has meant that reasonably high strength can now be obtained in addition to toughness, at least in acrylates and epoxies. A comparison of the stress/strain characteristics of different adhesives is given in Figure 16.88, showing the range of properties available.

The modulus of polymers arises from the forces required to move molecules with respect to each other. Adhesives therefore possess a relatively low modulus which decreases with increasing temperature. For epoxies, Young's modulus E is typically between 1 and 6 G Nm^{-2}, while the modulus of acrylics and polyurethanes may be a tenth, or even a hundredth, of such values. Variations in temperature can transform materials which are tough and strong at 20°C to ones which are soft and weak at 100°C. The glass transition temperature, T_g, denotes a marked change in the mechanical properties of a polymer. Above T_g the material will be rubbery, and below, glass-like and stiff (see Figure 16.89). The T_g's of sealants are around -140°C, whereas those of room-temperature curing adhesives are around 40–50°C. By warm- or heat-curing adhesives their T_g's are increased, and epoxies cured between 150°C and 200°C will have T_g's in excess of 120°C.

Organic polymers all absorb moisture and one effect is to plasticize the adhesive itself. This modifies its response to mechanical deformation in a manner analogous to a lowering of the T_g. Such effects are, however, reversible, depending upon proximity to moisture – in liquid or vapour form. Thus water and heat have similar effects, and this is illustrated in Figure 16.90 for the case of a cold-cure epoxy formulation.

Finally, adhesive polymers are visco-elastic and the time-dependent component of polymer response is of great importance to the use of adhesives required to sustain either permanent or transient loads; this can be a major determinant of the fatigue life of a bonded joint. Thus low-frequency load cycling and, to a lesser extent, elevated temperatures are very important considerations in adhesive selection.

16.4.3.6 Summary of adhesive considerations

There are a number of sources listing the factors involved in adhesive selection and the performance properties to consider.[57–62] Reference 62 offers a checklist of considerations within the epoxy group, upon which Table 16.9 is based. The choice is clearly a matter of swings and roundabouts.

16.4.4 Adhesion and surface pretreatment

16.4.4.1 Adhesion, or how adhesives stick

Concepts The basic requirements for good adhesion are very simple:

1. Intimate contact between adhesive and substrate
2. Absence of weak layers or contamination at the interface

Adhesives join materials primarily by attaching to their surfaces within a layer of molecular dimensions, i.e. of the order of 0.1–0.5 nm. In joints involving metallic substrates, the adhesive sticks to the surface oxide layer and not to the solid itself. Being liquid, adhesives flow over and into the surface irregularities of a solid, coming into contact with it and, as a result, interact with its atomic forces. The adhesive then solidifies to form the joint.

Interfacial contact Adhesive bonding involves a liquid 'wetting' a solid surface, which implies the formation of a thin film of liquid spreading uniformly without breaking into droplets

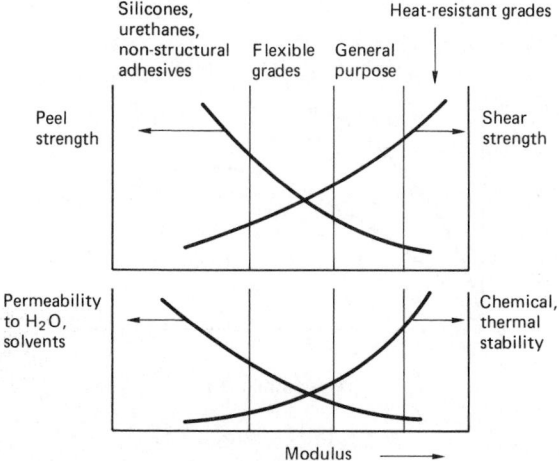

Figure 16.87 Property trade-offs for single-phase thermosetting adhesives. (Reproduced from reference 58, copyright Marcel Dekker)

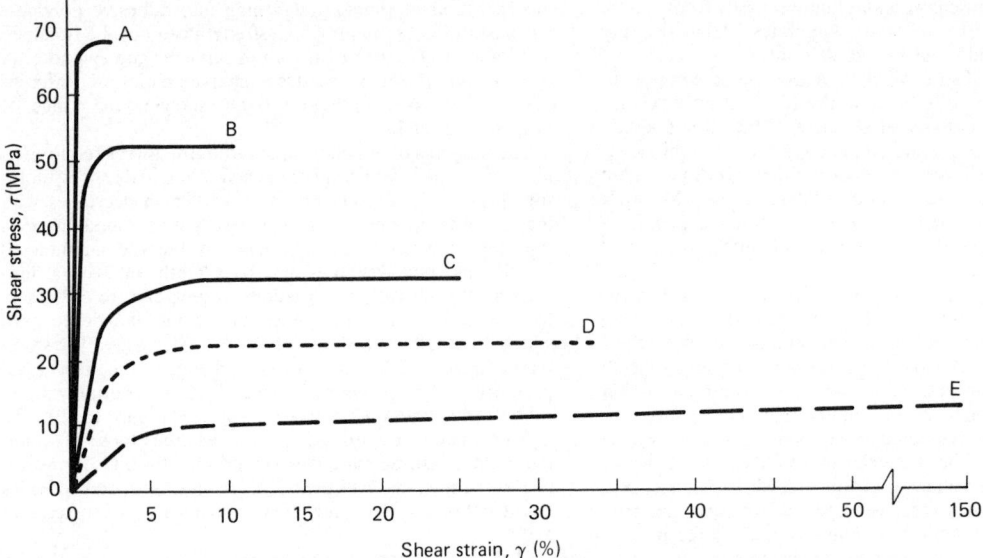

Figure 16.88   Typical stress–strain characteristics of adhesives used for structural and mechanical engineering assembly. (Based upon reference 57, copyright Permabond Adhesives Limited.) A1 – 1 Stiff, heat-resisting, brittle epoxy; B1 – 1 tough, stiff, head-cured, single-part epoxy; C1 – 1 tough, cold-cured, two-part-epoxy; D1 – 1 stiff, cold-cured, polyurethane; E1 – 1 tough, cold-cured ductile acrylic

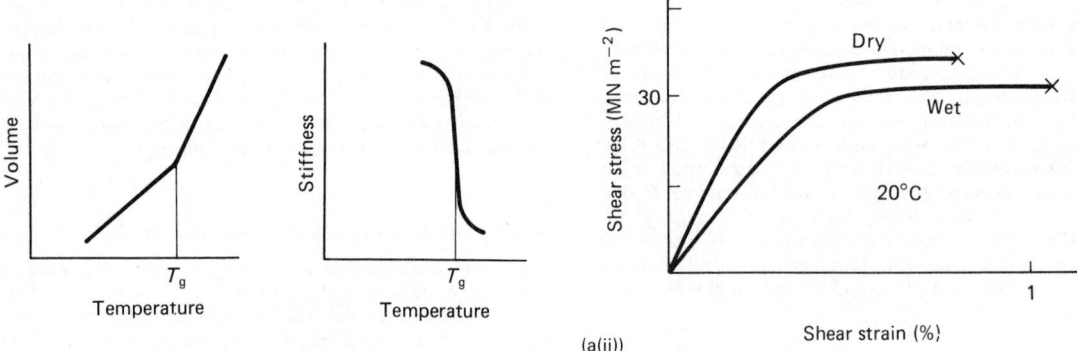

Figure 16.89   The glass transition temperature

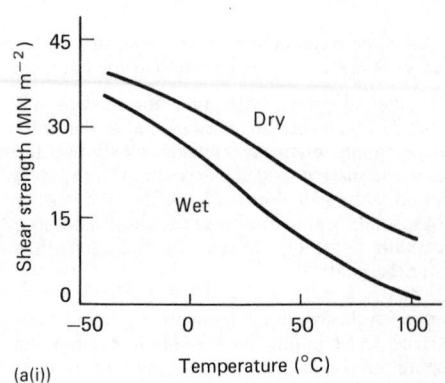

(a(i))

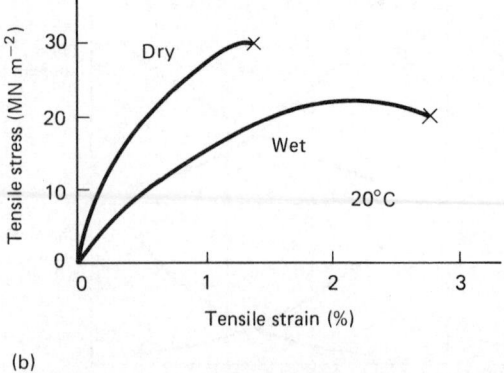

(b)

Figure 16.90   Schematic representation of the effects of temperature and moisture on the mechanical behaviour of an unmodified cold-curing epoxy adhesive. (a) Shear response; (b) tensile response

Table 16.9 Checklist of epoxy adhesive selection considerations

Strength requirements	
Shear	Unlikely to be limiting, but if so use a suitably modified epoxy
Cleavage/tension	Sacrifice shear strength for high toughness and good peel strength with flexibilized or toughened epoxy
Impact	High toughness derived from toughened epoxy; good peel resistance conferred by flexible materials such as epoxy-polysulphide formulations
Deformation characteristics	
Modulus	High for unmodified epoxy
Creep	Use adhesive with T_g well above service temperature; best resistance with unmodified epoxides
Service conditions	
Moisture/temperature	Best resistance with unmodified epoxides and formulations with high T_g's
Stress	Modified, flexibilized and toughened grades to suit function

(Figure 16.91); the water break-free test is the simplest qualitative visual method of assessment. Since adhesion involves intimate contact of two surfaces, it is convenient to think in terms of the free energies of the surfaces involved. Just as liquids have surface tensions or surface energy, so do solid surfaces by virtue of the fact that they are surfaces. However, surface tension of solids tends to go unnoticed because solids are usually too rigid to be visibly distorted by the interatomic, rather than intermolecular, forces holding them together.

Zisman[63] introduced the useful distinction between high- and low-energy surfaces. Most plastics and liquids have surface-free energies below 100 mJ m^{-2}, with organic adhesives such as epoxides having low surface-free energies – usually <50 mJ m^{-2}. Hard solids such as metals and metal oxides, *when atomically clean*, have high surface-free energies typically in excess of 500 mJ m^{-2}. Some important values of surface-free energies are collected in Table 16.10.

An energetic surface will make a wetting liquid spread on it, rather than remain as a discrete drop, so that adhesives should readily spread and wet the oxide layers of metals. There will, however, be a problem with bonding plastics such as polyethylene and polypropylene whose surface-free energies are *less* than those of most organic adhesives. However, the problem with high-energy surfaces is that atmospheric contaminants and moisture are readily adsorbed on them, converting them to surfaces with a free energy barely greater than that of adhesives. This often gives rise to the need for special precautions to be taken with surface pretreatment and process control.

In summary, the ideal conditions for establishing interfacial contact are that the:

1. Surface-free energy of the adherend should be higher than that of the liquid adhesive;
2. Liquid adhesive should exhibit a near-zero contact angle with the adherend(s);
3. Adhesive's viscosity should be relatively low at some time during the bonding operation;
4. Joint should be closed carefully to assist air displacement.

Mechanisms of adhesion Once interfacial contact has been established, the adhesive cures in order to be able to transmit stress. There is some debate regarding the basic nature of the forces then acting across interfaces which prevent them from separating under an applied load. The four main theories of adhesion which have been proposed are: mechanical interlocking, adsorption, diffusion and electrostatic attraction.[63–65] However, the *adsorption* mechanism is generally favoured, with *mechanical keying* also playing an important role. Whichever mechanism or combination of mechanisms are operating, the important surface characteristics are:

1. Surface energy
2. Surface chemistry
3. Surface micro- and macro-morphology

The adsorption theory proposes that adhesive macromolecules are physically adsorbed onto the substrate surface because of the forces acting between the atoms in the two surfaces. In effect, the polar nature of the adhesive molecules acts like a weak magnet and they are attracted toward polar adherend surfaces. The most common interfacial forces are van der Waals' forces, referred to as secondary bonds, although hydrogen bonding and primary bonding are involved in some cases.

Mechanical keying or interlocking of the adhesive into the irregularities or pores of the substrate underlies the instinctive procedure of roughening surfaces to improve adhesion. On porous or fibrous materials, the adhesive certainly flows into

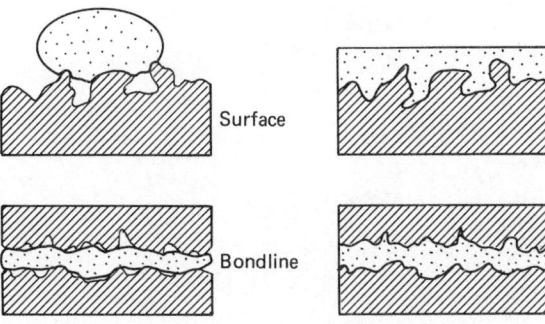

Surface

Bondline

Incomplete wetting   Complete wetting

Figure 16.91 Wetting

Table 16.10   Values of surface-free energies

Surface	Surface-free energy, γ, in vacuo (mJ m^{-2})	Young's modulus, E (G Nm^{-2})	Material class
High-energy surfaces			
C (diamond)	5140	1200	Diamond
Fe$_2$O$_3$ (ferric oxide or red haematite)	1357	210	
Al$_2$O$_3$ (alumina or aluminium oxide)	638	70	Metals
SiO$_2$ (silica or silicon dioxide)	287		
Low-energy surfaces			
H$_2$O (water)[a]	72.2		
CFRP (heavily sanded)[a]	58		
Carbon fibre[a]	51.6		
Epoxide (amine-cured)	~ 42		
Nylon	42	4 (when cured)	Thermosets
PMMA (perspex)	40		
PVC	40		
Cyanoacrylate	~ 33		Thermoplastics
Polyethylene	33	2	
Silicones	25		Elastomers
PTFE	15.5		

[a] Large polar component, γ^p

the structure of the material and is keyed in place. The penetration of adhesive into the *microstructure* of high-alloy metal oxide layers is important, and this micromechanical interlocking at a molecular scale aids the retention of adhesion under severe environmental conditions[64,65] (e.g. Figure 16.92).

16.4.4.2   Surface pretreatment

Adhesives are quite often blamed for 'not sticking', but the general source of the trouble lies with the surface preparation. Surface pretreatments, while greatly affecting bond durability, generally have less effect on initial strength. Inadequate surface pretreatment is usually the main cause of durability problems and of joints failing in service.[66] While adhesives are available which can absorb oily films and a certain amount of surface contamination, some form of pretreatment is generally recommended. The degree of surface pretreatment required

depends upon the nature of the adherends (e.g. Table 16.11), and some effects are summarized in Table 16.12. Surface pretreatment generally involves:

1. Cleaning
2. Removal of weak surface layers
3. Re-cleaning

It can be appreciated from Table 16.12 that surfaces can be made to be very much more 'receptive' towards adhesives in terms of altering the adherend's surface chemistry, energy and morphology. A detailed literature exists giving practical information on pretreatment procedures for a wide range of substrates,[57,60,61,67–71] while a number of surface analytical and optical techniques have been devised for examining surfaces before and after treatment.[65,68] Electron microscopy has been found to be particularly valuable (e.g. Figures 16.92 and 16.93).

The main methods of surface pretreatment fall into four groups:

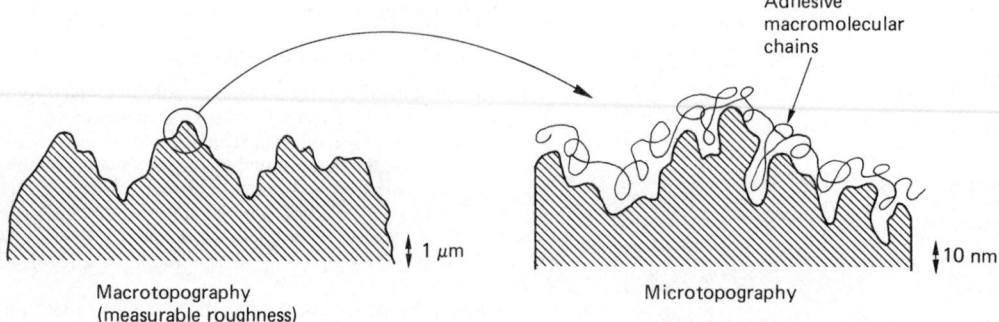

Macrotopography
(measurable roughness)

Microtopography

Adhesive macromolecular chains

1 μm

10 nm

Figure 16.92   Schematic topography of solid surfaces

Table 16.11 Pretreatment requirements

Material	Suitability for bonding	Pretreatment required
Cast iron	*****	Cursory
Steel	****	Straightforward
Stainless steel	***	
Zinc	***	} Quite demanding
Aluminium	***	
Concrete	****	
GRP	****	} Straightforward
CFRP	***	
PVC	**	Rigorous
Polyolefin	*	Complex

Table 16.12 Effects of surface pretreatments

Treatment	Possible effects on surface	Substrate
Solvent etch	Remove weak boundary layer	M, P
	Weaken surface region by plasticization	P
	Increase surface roughness	P
Mechanical	Remove weak boundary layer	M, P
	Increase surface roughness	M, P
Chemical	Remove weak boundary layer	M, P
	Increase, or decrease, surface roughness	M, P
	Alter surface chemistry with consequent changes in the rate and degree of wetting	M, P
Physical (e.g. flame, plasma, corona discharge)	Remove weak boundary layer	M, P
	Weaken surface region by plasticization	P
	Alter surface chemistry	M, P

Substrate abbreviations: M, metal; P, plastic

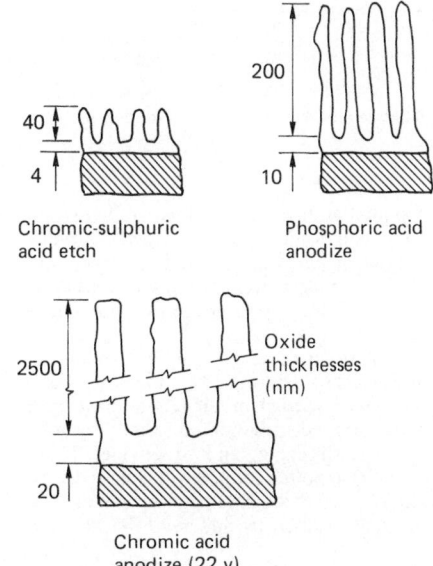

Figure 16.93 Oxide morphology on aluminium alloy following chemical treatments

1. Solvent degreasing – wiping, brushing, dipping or vapour degreasing in organic solvents such as acetone or trichloroethylene. Wiping with MEK or IPA may be suitable for plastics and rubbers. Alkaline cleaners and/or detergent solutions can also be used for metallic substrates.
2. Mechanical abrasion – action of wire brushes, sand and emery papers, abrasive pads, gritblasting, etc. Dust must be removed prior to bonding. The effects on adhesion of such (obvious) roughening are often complex.
3. Chemical treatments – numerous chemical treatment and rinsing procedures exist. Very complex surface changes are brought about in terms of cleaning, altering surface chemistry and morphology (e.g. Figure 16.93).
4. Physical methods – use of ionic bombardment or flaming action to clean and oxidize surfaces. Useful on low-energy thermoplastic materials.

A further possibility lies with the application of an adhesive-compatible primer coating. Generally, the advantages to be gained such as 'pre-wetting' of the substrate surfaces outweigh the disadvantages. Often a more reproducible surface is created and, with the application of some primers (e.g. coupling agents such as silanes), a water-stable interface may result. In the future, it may become commonplace to employ pre-coated metals ready for bonding.

16.4.5 Joint design

Load-bearing bonded joints must be designed properly. It is, for instance, not sufficient simply to substitute adhesive bonding for welding, bolting or riveting. The adhesive, which can be likened to plastic material, represents a low-modulus interlayer and is likely to be the weakest link in a joint, unless very weak or thin-sheet adherends are present. Design considerations generally involve the geometry of the bond, selection of an adhesive, knowledge of the properties of the adhesive and adherend, and analysis of the stresses to which the joint will be subjected.

Adhesive bonded joints should be designed to provide, as far as possible:

1. A large bond area
2. A uniformly thin and continuous bondline
3. Shear (and, ideally, compression) loading on the adhesive
4. Minimum peel loading and avoidance of stress concentrations
5. Self-jigging features to locate and hold the assembly during curing.

It is important to design an assembly correctly to incorporate adhesive bonding. Adhesives generally have good shear strengths but poor peel and cleavage properties, and the joint design should be arranged to eliminate these weak modes of loading – for example, by the use of additional mechanical support (Figures 16.94 and 16.95). Bonds are therefore designed to place the adhesive in shear or, ideally, compression! Other common engineering joints and joint designs are depicted in Figure 16.96.

16.4.5.1 Joint behaviour

The lapped joint is one of the most commonly occurring joints in practice, and is therefore both the most studied and the configuration most often used for testing adhesives.[61,62] However, lapped joints, however fastened, support most of the applied load as stress concentrations at the two ends of the joint. As shown in Figure 16.94(a), the problem is that the loads are not linear, and this causes the joint to rotate. In consequence, the adhesive layer is subjected to shear and tearing stresses at the ends of the joint; the adherends, too, are subjected to shearing, stretching and bending. Double-lap

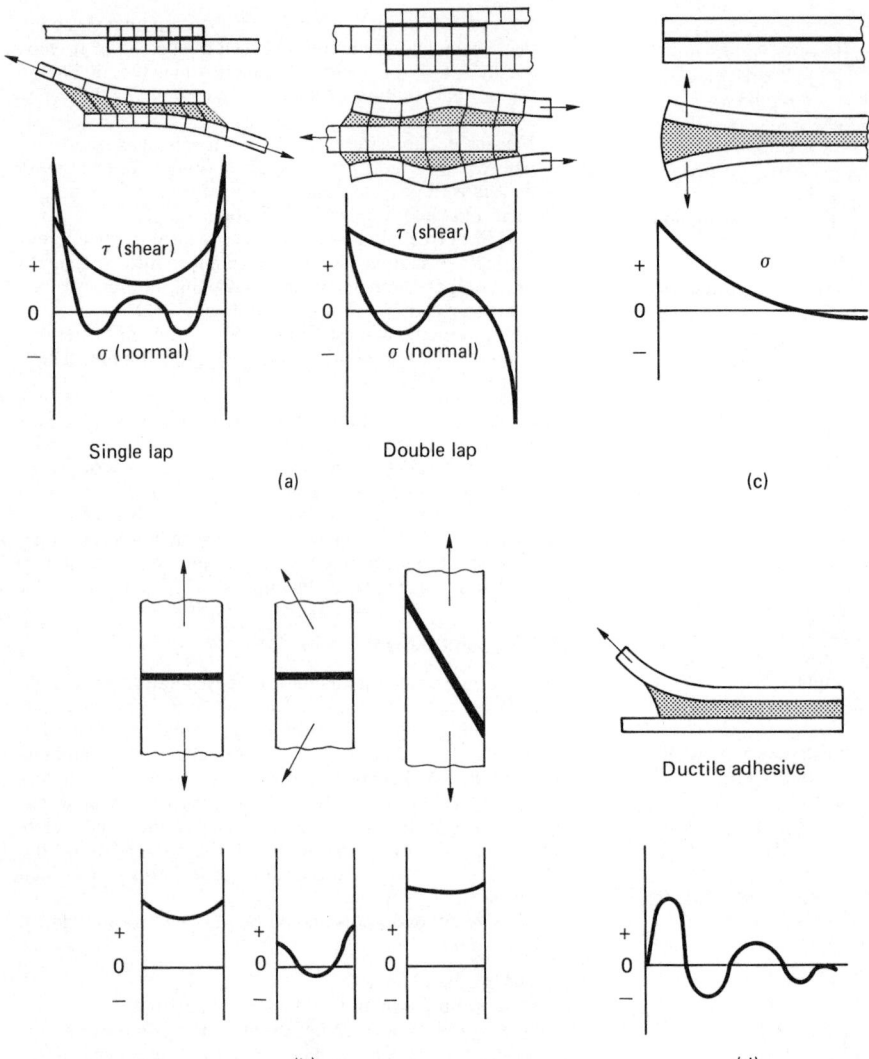

Figure 16.94 Elastic stress distributions in various kinds of bonded joints. (a) Lap shear; (b) butt-tensile and scarf; (c) cleavage; (d) peel

joints, while eliminating gross joint rotation, simply represent a back-to-back arrangement of two single-lap joints and significant tearing stresses still exist.

The stress peaks arising in most forms of bonded joints are at a minimum when stiff adherends and/or flexible adhesives are involved (e.g. acrylics), but can become very large when more flexible substrates or rigid adhesives (e.g. stiff epoxides) are present (Figure 16.97). The factors influencing these peaks are:

1. Adhesive shear modulus, G_a
2. Overlap length, l
3. Adherend moluli, E
4. Adherend thickness(es), h
5. Adhesive layer thickness, t
6. Plastic strain capacity of adhesive

The stress concentration in elastic theory can be described in simple terms by the coefficient $G_a l^2/Eht$. It may therefore be appreciated that lap joint strength depends generally on joint width, and not overlap length (Figure 16.98). However, on the positive side, real bonded joints are inevitably formed with spew fillets of adhesive at their extremities and this generally has the effect of reducing stress concentrations at the end of the overlap length.[62]

Since the strength of structural adhesive joints bonded with relatively stiff adhesives is so determined by the stress peaks induced under load, many design modifications have been introduced to improve stress distribution uniformity by adherend scarfing, tapering, stiffening, and so on (see Figure 16.96). Simple butt joints are often formed with adhesives, but better designs utilize co-axial and rebated concepts in order to subdue large stress peaks caused by adherend deformations (Figure 16.99).

Where very flexible adherends are concerned, flexible adhesives must be used in combination with a joint design that minimizes peel and cleavage stresses in the adhesive. Combi-

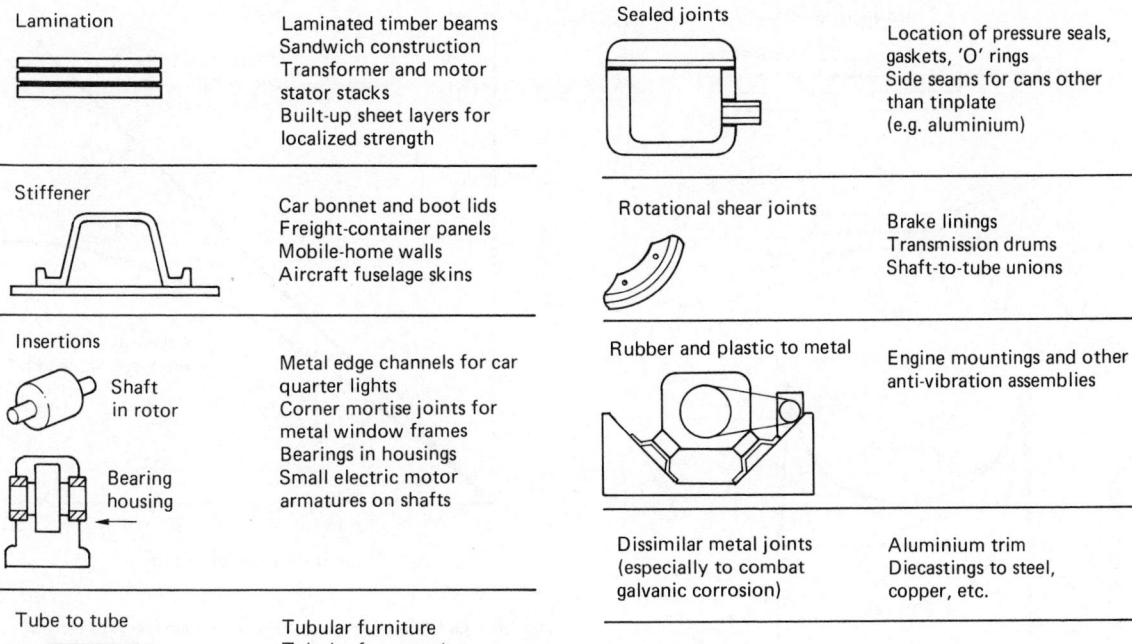

Lamination
- Laminated timber beams
- Sandwich construction
- Transformer and motor stator stacks
- Built-up sheet layers for localized strength

Stiffener
- Car bonnet and boot lids
- Freight-container panels
- Mobile-home walls
- Aircraft fuselage skins

Insertions
- Shaft in rotor
- Bearing housing

Metal edge channels for car quarter lights
Corner mortise joints for metal window frames
Bearings in housings
Small electric motor armatures on shafts

Tube to tube
- Tubular furniture
- Tubular frameworks

Sealed joints
- Location of pressure seals, gaskets, 'O' rings
- Side seams for cans other than tinplate (e.g. aluminium)

Rotational shear joints
- Brake linings
- Transmission drums
- Shaft-to-tube unions

Rubber and plastic to metal
- Engine mountings and other anti-vibration assemblies

Dissimilar metal joints (especially to combat galvanic corrosion)
- Aluminium trim
- Diecastings to steel, copper, etc.

Figure 16.95 Design concepts for bonded structures

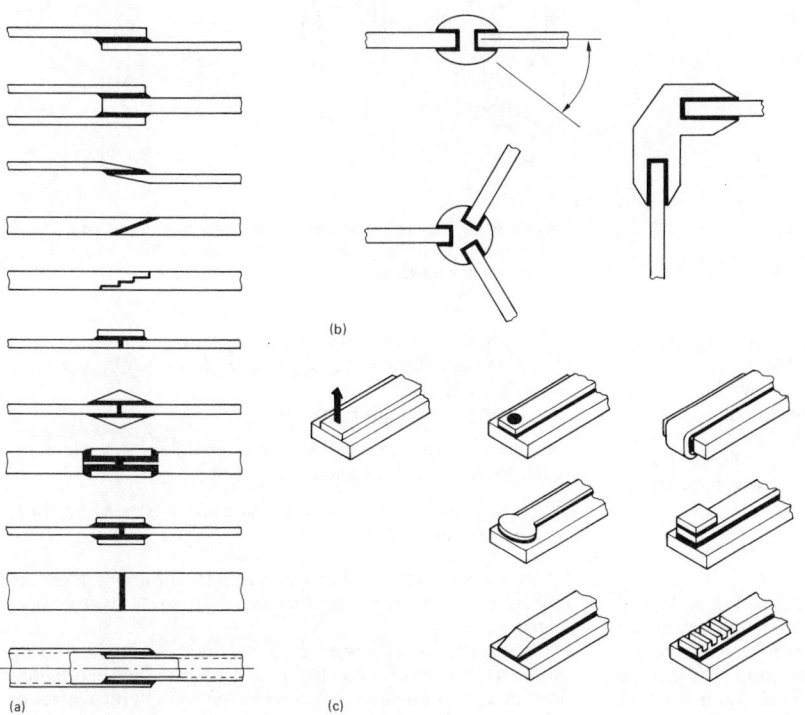

(a)

(b)

(c)

Figure 16.96 Common engineering joints and joint designs. (a) The lap joint and its variants; (b) containment joints for plates, extrusions and pultrusions; (c) ways of minimizing peel in laps, doublers and stiffeners

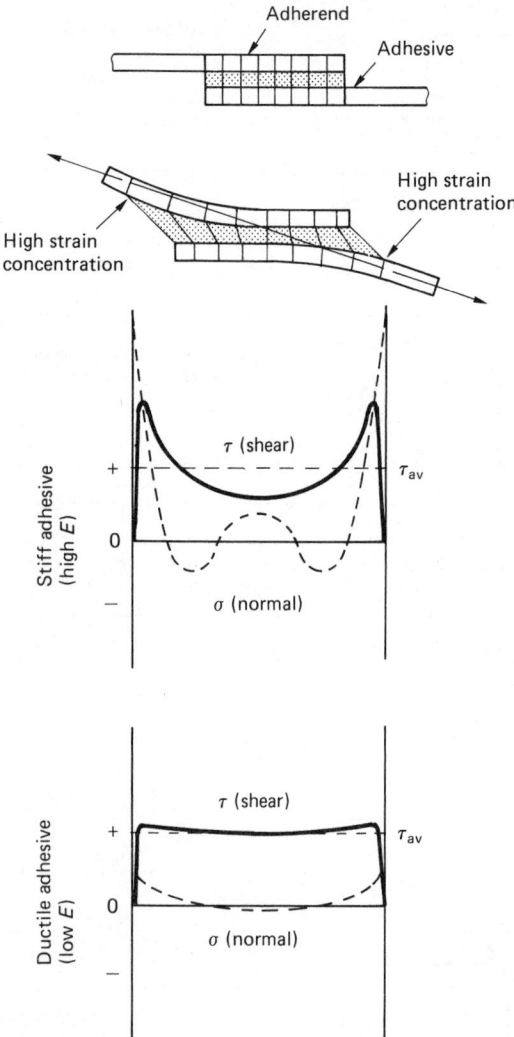

Figure 16.97 Bondline stress distributions in a lap joint under load made with adhesives of low and high modulus

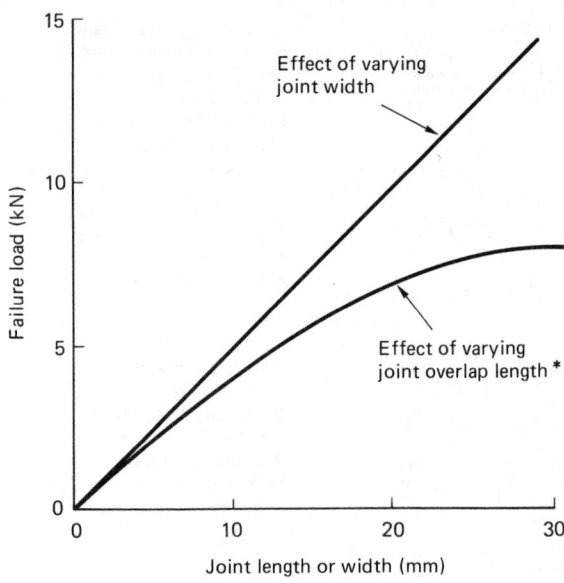

Figure 16.98 Effect of overlap length and width on single lap joint strength. (*The 1.6 mm thick steel adherends begin to yield as overlap length increases beyond, say, 20 mm)

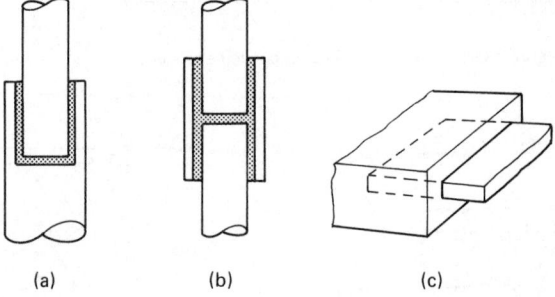

Figure 16.99 Joint configurations suitable for bars, tubes and containment assembly. (a) Bar; (b) bars or tubes with overlapping sleeve; (c) box assembly

nations of mechanical fastening and adhesive bonding are also popular to overcome some of the problems of adhesive bonded-only joints. Thus bolts and rivets can be used to subdue peeling or cleavage forces in the adhesive. Clinching and patent mechanical locking arrangements are also popular, as is spot-welding through the adhesive. These hybrid assemblies are sometimes referred to by the terms 'riv-bond', 'clinch-bond' and 'weld-bond'.

16.4.5.2 Design approaches

Structural adhesive joints are generally designed to be loaded in shear, so that most of the analytical tools so far developed are confined to load transfer through the adhesive in shear. Simple analyses allow only for elastic behaviour, while more sophisticated developments enable plastic behaviour (of both the adhesive and adherends) to be modelled.

All the theoretical models for predicting joint strength require:

1. Adherend tensile modulus (E) and Poisson's ratio (ν)
2. Adhesive shear (G_a) and tensile (E_a) moduli, and Poisson's ratio (ν_a)
3. Ultimate strength properties.

Analyses which allow for adhesive non-linear behaviour require data on ductility such as:

4. Yield stress (strain) and ultimate stress (strain), in shear or tension, or both.

Some very sophisticated analyses require information on the adhesive's coefficients of thermal, and even hygroscopic, expansion.

The empirical approach to the design of simple overlap joints was to construct a correlation diagram between failure load and the joint geometrical ratio, h/l, for a particular set of test conditions. Mathematical treatments of joint analysis, employing differential equations to describe stress (strain) fields, have been developed from the early work of Volkersen and of Goland and Reissner. Sophisticated solutions allowing for gross adherend deformations have been produced by Renton and Vinson, Allman, Hart-Smith, and Grant. These approaches are documented by Adams and Wake[62] and have been adapted and formulated into various software packages (see Section 16.4.5.4).

It is now accepted that non-linear analysis is the key to predicting failure of structural adhesive bonded joints. It is also apparent that joint strength is determined largely by the ultimate stress or the ultimate strain capability of the adhesive in tension. Finite element methods are recognized as being the most useful techniques for analysing real joints, particularly where large non-linearities and/or geometrical complexities can be modelled.[62]

In recent years a fracture mechanics approach to joint failure has been taken, at least for test methods.[65,72,73] The fracture energy (strain energy release rate), G_c, or the critical stress-intensity factor (adhesive fracture toughness), K_c, in tensile Mode I opening (see Figure 16.100) represents the most critical situation in bonded joints. G_c or K_c may be related to a critical crack length in a bonded assembly, such that cracks which are shorter are safe and cracks which are longer are self-propagating and potentially catastrophic. Naturally, the use of rubber-toughened adhesives which inhibit crack propagation is an attractive option.

16.4.5.3 Test procedures

Bulk adhesive and/or bonded joint tests may be used for reasons such as:

1. Comparing the mechanical properties of a group of adhesives
2. As a quality check for a batch of adhesive
3. Checking the effectiveness of surface pretreatment(s)
4. Measurement of environmental effects
5. Determining quantitative mechanical and physical data on adhesives for structural design and analysis
6. Analysis of fracture mechanisms.

A large number of parameters are involved in the fabrication and testing of bulk adhesive specimens and adhesive joints; these must be controlled if meaningful experimental data are to be obtained. Bulk adhesive tests are useful for determining quantitative mechanical and physical properties. Data from joint tests usually reflect the mechanical properties of the adhesive as well as the degree of adhesion and the effectiveness of surface treatments. Many standard test procedures are listed by ASTM, BSI,[74] DIN and other official bodies, and some joint test methods are listed in Table 16.13. Useful commentaries on the relative merits of such methods, and an interpretation of the data generated, are presented by many authors.[61,62,65,75]

Comparative data on adhesive performance is commonly obtained from single-lap shear testing to BS 5350 and/or ASTM D1002. Adhesive formulators generally quote data from such tests, and an enormous data bank now exists worldwide. As discussed previously, the adhesive is subjected to cleavage as well as to shear, in a ratio dependent upon various joint stiffness factors. Nevertheless, comparative properties of different bonding systems and surface treatments may be evaluated over a range of environmental conditions. Cylindrical pin-and-collar assemblies are often used for testing low-viscosity adhesives such as anaerobics. Peel tests are generally employed where flexible substrates are involved, or where a test of adhesion is required.

Quantitative data on adhesive mechanical properties such as moduli, Poisson's ratio, stress and strain to failure, and glass transition temperature are difficult to determine. Bulk tests

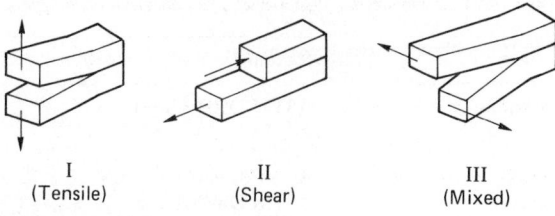

I (Tensile)   II (Shear)   III (Mixed)

Figure 16.100 Principal fracture modes

(e.g. tensile dumb-bells, blocks for compression, strips for torsion pendulum and water sorption experiments) may be conducted on carefully cast specimens. Suitable joint tests include torsional shear or thick adherend shear test methods using appropriate strain measurement. It is usual to conduct such tests over a range of temperatures. A discussion of such methods and procedures is given by the Engineering Sciences Data Unit in London.[76]

Tests for adhesion and surface treatment effects should subject the bonded interface to tensile stresses. Lap shear, peel and cleavage (fracture energy) tests are therefore common. These should be used in conjunction with the results of surface analytical studies to assess surface energy, chemistry and morphology.

Fracture energy tests are usually conducted in tension (Mode I) and/or shear (Mode II). Parallel or tapered cantilever beam configurations are common for investigating G_{IC} and K_{IC}. A thin parallel cantilever beam geometry is used in the 'Boeing Wedge Test' derivative for assessing surface treatments – where crack propagation is monitored along the adherend/adhesive interface, as opposed to extending cohesively within the adhesive layer. Ironically, 'toughened' adhesives can be awkward to test!

Environmental testing Durability assessment represents a very important aspect of adhesive bonding technology. Most test methods are required to provide accelerated ageing. Since the major environmental factors influencing joint degradation are moisture, heat and applied stress, accelerated ageing methods seek to combine some or all of these factors. Suitable test configurations involve small bonded areas and include simple lap joints (sometimes perforated to accelerate moisture accession), peel tests, tensile butt-joints and wedge cleavage specimens. All such joints may be exposed to various natural or aggressive laboratory environments, either unstressed or stressed. Fatigue loading may even be included. Obviously, it is important to note, or monitor, the locus of failure; interfacial failure indicates an unstable surface. A detailed treatment of failure mechanisms and of methods of testing and assessment is given by Kinloch.[66]

Test joints versus real joints Real joints do not consist of simple, separate, elastic materials with a clear mathematical geometry. The adherend surface is often rough, and the thickness and properties of the primer (if applied) and adhesive layer are often difficult to determine. There also remains some debate as to whether the in-bondline, or thin-film form, properties of the adhesive are the same as they are in bulk. In service, the applied loads will generally be much lower than those applied in the laboratory.

There remains a need for the development of appropriate, cheap, simple and quantitative test methods for general engineering applications of adhesives. The test coupon approach to design and quality control has been reviewed by many

Table 16.13 Some standard test methods for adhesive-bonded joints

Test	Standard	Remarks
Definitions	ASTM D907-82(85)	Standard definitions of terms relating to adhesives
Axially loaded butt joints	BS 5350: Part C3: 1978 ASTM D897-78 (83) ASTM D2094-69 (80) and D2095-72 (83) BS 5350: Part C6: 1981	 Specifically for bar- and rod-shaped specimens Bond strength in direct tension in sandwich panels
Lap joints loaded in tension	BS 5350: Part C5: 1976 ASTM D1002-72 (83) ASTM D3528-76 (81) ASTM D3163-73 (84) and D3164-73 (84) ASTM D2295-72 (83) ASTM D2557-72 (83) ASTM D905(86), D906 (82) D2339-82 and D3535-79 (84) ASTM D3983-81 (91) DIN E 54451-77	Single- or double-lap joint test Basic metal-to-metal single lap joint test Double-lap joint test Specifically for polymeric substrates Single-lap joint test for metal-to-metal joints at elevated temperatures As above but at low temperatures Specifically for wooden joints Thick substrates used; shear modulus and strength of adhesive determined
Peel joints	BS 5350: Part C9: 1978 and ASTM D3167-76 (81) BS 5350: Part C10: 1979 and BS 5350: Part C14: 1979 BS 5350: Part C11: 1979 and ASTM D903-49 (83) BS 5350: Part C12: 1979 and ASTM D 1986-72 (83) ASTM D1781-76 (81) BS 5350: Part C13: 1980 ASTM D429-73	Floating-roller test 90° peel test 180° peel test 'T' peel test for flexible-to-flexible assemblies Climbing drum test for skin-sandwich Rubber-to-metal bonding
Shear strength	ASTM D4027-81 (91) ASTM D229-70 (81) ASTM D2182-72 (78) BS 5350: Part C15: 1982 BS 6319: Part 4: 1984 BS5350: Part G2: 1987 ASTM D 3983-81(91) DIN E 54451-77	Modified rail test See Torque strength Disk shear in compression Bond strength in compressive shear Slant shear test, loaded in compression, for resins used in construction; concrete substrates used Collar and pin bonded with anaerobic adhesive and loaded in tension See lap joints loaded in tension
Cleavage strength	BS 5350: Part C1: 1986 ASTM D1062-72(83) ASTM D3433-75(85) ASTM D3762-79(83)	Compact tension specimen Parallel- or tapered double- cantilever-beam joint for determining the adhesive fracture energy, G_{IC} Wedge cleavage test (for aluminium adherends)
Fatigue strength	ASTM D3166-73(79)	Single-lap joint loaded in tension
Flexural strength	ASTM D1184-69(86)	Laminated assemblies
Torque strength	ASTM E229-70(81) ASTM D3658-78(84) BS 5350: Part G1: 1987	For determining pure-shear strength and shear modulus of structural adhesives (napkin-ring specimen) Specifically for ultraviolet light-cured glass–metal joints Anaerobic adhesives on threaded fasteners

Test	Standard	Remarks
Impact resistance	ASTM D3807-79(84) ASTM D950-82 BS 5350: Part C4: 1986	Plastics-to-plastics joints Block shear specimen
Creep resistance	BS 5350: Part C7: 1976 ASTM D1780-72(83) and ASTM D2294-69(80) ASTM D2293-69(80)	Various test geometries permitted Single-lap joint loaded in tension Single-lap joint, having long overlap, and loaded in compression
Environmental resistance	ASTM D2918-71(81) ASTM D2919-84 ASTM D3762-79(83) ASTM D1151-84 ASTM D1183-70(81) ASTM D904-57(82) ASTM D896-84 ASTM D3632-77(82) ASTM D1828-70(81) ASTM D1879-70(81)	Subjected to stress, moisture and temperature; uses peel joint As above, but uses single-lap shear joint loaded in tension As above but uses a wedge test Exposure to moisture and temperature Exposure to cyclic laboratory ageing conditions Exposure to artificial and natural light Exposure to chemical reagents Exposure to oxygen Natural weathering Exposure to high-energy irradiation

authors.[77] A common observation is that problems which are likely to arise in the 'real' structure simply do not arise in the test coupon.

16.4.5.4 The way forward in design

The use of structural adhesives requires both essential choices between the many types available, as well as a considered decision on the design approach appropriate to structures assembled with them. A balance of material requirements and an understanding of their interaction must be achieved for optimum performance. The simplest way forward currently could involve the following steps:

1. Adhesive selection and usage considerations from literature sources and/or proprietary software such as CATS (Centre for Adhesive Technology), EASel (Design Council), PAL (Permabond Adhesives), STICK (Lucas).
2. Structural design using stress analysis programs, e.g. BISEPS and CADEPT (Harwell Laboratory), CATS, ESDU suites,[78,79] PAL, STICK, and software from PERA International. Appropriate factors of safety should be included to allow for adhesive material uncertainties and changes with response to environmental conditions.
3. Experimental work to investigate joint strength and durability, particularly with respect to the range of likely operating conditions.

16.4.6 Fabrication and assembly

Education and training of the personnel involved is essential. They must possess a qualitatively correct overall picture of the importance of the different stages of the bonding operation, and of the health and safety considerations. Control of the working environment (temperature, humidity) may also be important, particularly with regard to cleanliness.

16.4.6.1 Production processes

The adhesive bonding production process generally involves:

1. Surface preparation of the components
2. Application of a primer (if used)
3. Mixing, dispensing and application of the adhesive
4. Curing or setting of the adhesive.

These processes are determined by joint parameters such as the geometry and component fit-up, the materials involved, bond quality and reliability, and production rate. The procedures that need to be followed to ensure high-integrity structural joints can become very expensive in time and equipment.

As indicated in Section 16.4.4, some form of surface pretreatment is recommended – particularly to enhance interfacial stability. This is particularly important where cold-curing epoxy formulations are involved, or where joints will be subjected to hot moist and/or stressed conditions in service. Some adhesives can absorb oily films, but solvent cleaning is very effective in improving joint quality. Depending upon the substrate, mechanical abrasion, chemical etching or physical treatments may additionally be necessary. Such procedures must be well defined.

The time elapsed between surface pretreatment and application of the adhesive should be kept to a minimum in order to minimize subsequent contamination. Priming of the surfaces involved can serve the purposes of protection and sealing (of porous surfaces), as well as providing a more reliable and reproducible surface ready for bonding. Priming may be conducted with, what are essentially, dilute solutions of adhesives themselves, chromate/phosphate solutions, or with chemical coupling agents such as silanes.[80]

Adhesive mixing, dispensing and application must be controlled carefully. The techniques involved should be considered alongside the component design, production rate and adhesive employed. Single-component adhesives can be applied by spraying, brushing, roller-coating, or direct extrusion (for paste-like materials). Two-component systems must be carefully measured and mixed before dispensing, either manually or automatically. In recent years there has been a revolution in adhesive packaging so that weighing or measurement is not required, minimizing errors and mess. For high-volume applications equipment is available for automatic metering, mixing and dispensing, even by robots, of adhesives with a wide range of viscosities and mixing ratios.

Insufficient adhesive will result in joint performance and durability problems, while a gross excess will hamper assembly and necessitate clean-up and extra finishing operations. Joints should also be closed carefully to minimize the inclusion of air. With close-fitting parts over large bonded areas a means of air escape should be provided. A good joint design will assist the positioning and location of the joint parts, but jigs or permanent mechanical fixings may also be required to hold the components while the adhesive cures. Bondline thickness control is required both to confer a uniform design thickness as well as to prevent displacement of the adhesive under pressure from clamps or jigs. It is often desirable to leave fillets of adhesive around the perimeter of a bonded joint, and these can be tooled when the joint has been closed.

Once the adhesive has been applied, the curing and hardening process must take place. This may be within a matter of seconds for some acrylate adhesives or several hours for cold-curing epoxy formulations. The various curing systems employed were discussed in Section 16.4.3, from which it is clear that provision may sometimes have to be made for moisture or solvent evaporation, conversely for providing moisture for polyurethanes or, perhaps, ultraviolet light for some acrylic formulations. The time of reaction can be shortened considerably by the application of heat, and a typical time–temperature curing profile for a single-part epoxy is depicted in Figure 16.101. Ovens, infrared lamps, induction-, radiant- and resistance-heating systems are among the methods for achieving faster curing rates.

16.4.6.2 Health and safety

Safety aspects should partner the measures practised within a quality system. The history of adhesive bonding has shown that accident and health problems associated with the technology are rare. Commonsense precautions such as the use of skin and eye protection are sufficient for many applications.

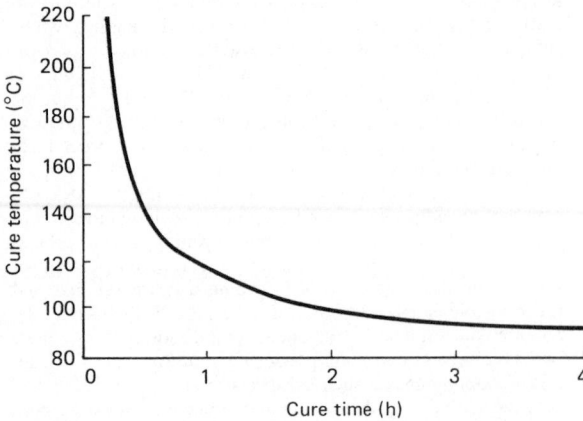

Figure 16.101 Typical time–temperature cure profile for single-part epoxy adhesive in an air circulating overn. (Based upon reference 57, copyright Permabond Adhesives Limited)

Concentration and duration of exposure to hazardous materials generally governs any risks.

Legislation requires the adhesive supplier to label and classify products, as well as including standard risk phrases and safety procedures. Production information sheets must also be supplied in accordance with the Health and Safety at Work Act. Useful guidelines for users of adhesives are documented by the British Adhesives and Sealants Association.[81]

Most engineering adhesives contain no organic solvents and therefore problems such as solvent abuse are not present. However, solvents are sometimes used in considerable volume for surface degreasing processes as well as for cleaning-up operations. Some surface treatment processes utilize acids and other chemicals, for which a hazard is presented both in handling the liquids as well as with disposal of the residues. Mechanical abrasion techniques carry an obvious, albeit limited, risk.

When comparing the risks inherent in other joining processes such as brazing or welding the use of adhesives can significantly improve safety.

16.4.7 Quality control and non-destructive testing

16.4.7.1 Quality control

Test methods for the control of fabrication procedures and non-destructive testing (NDT) are basic requirements for the formation of structural adhesive joints. There are two essential but different quality aspects to be considered. These are the:

1. Adhesion between the adhesive and substrate
2. Cohesive strength and integrity of the cured adhesive layer.

A measurement of potential adhesion only is possible, restricting (1) to an assessment of the adherend surface characteristics prior to bonding through visual inspection, wettability tests, or surface potential difference techniques (for monitoring contamination). Appropriate post-bonding mechanical tests of adhesion include assessments of peel, tension and cleavage.

The quality of the cured adhesive layer can depend on many factors. Adhesive materials may be assessed in the fresh uncured state (e.g. by measurements of viscosity, sag, pot-life, etc.) as well as in their hardened state. In the latter case mechanical measurements and NDT techniques may be used.

Quality variations in the cured adhesive layer may be due to the presence of air voids (trapped during joint assembly), local areas of uncured material, insufficient curing or cracking due to shrinkage on cure. The presence of such defects may or may not be important, depending on their extent, location and so on. In many joints such defects are unimportant unless around the joint perimeter and/or in regions of high stress transfer (e.g. at the ends of a bonded overlap). The nature and significance of defects is reviewed by Adams and Wake.[62]

16.4.7.2 Non-destructive testing (NDT)

NDT represents a large and diverse field in which a number of review papers exist.[62,82,83] The NDT methods available are essentially void detectors, although claims are made that information can additionally be obtained on the density and thickness of the adhesive, and even on the overall structural stiffness of a bonded assembly. Numbered among the techniques are:

1. Sonic methods, such as coin-tapping
2. Ultrasonic methods
3. Acoustic emission

4. Thermography
5. Radiography
6. Holography.

Methods using ultrasound (2), heat (4) and X-rays (5) have been developed quite successfully for monitoring density variations and disbonds in bonded structures, particularly in sandwich panels. Advances in such techniques have been spurred on mainly by aircraft industry requirements for production control and regular inspection of airframe integrity.

16.5 Casting and foundry practice

16.5.1 Introduction

Casting has significant advantages compared with other methods of component manufacture. Castings are generally cheaper than components made in other ways. The casting process in one or other of its forms provides the designer with an unrestricted choice of shape made in a single stage. A casting can usually be made much closer to the chosen design, which provides savings in both material and finishing processes compared with other methods of manufacture. In addition, the cast structure has the highest resistance to deformation at elevated temperatures so that castings have higher creep strengths than wrought and fabricated components. This advantage can be enhanced by modifying and aligning solidification to produce highly creep-resistant structures such as bundles of crystals with one crystallographic axis oriented lengthwise or single crystals (see Chapter 7, Section 7.4 non-ferrous metals). Cast metal may also have superior wear resistance than the equivalent forged metal. These advantages have combined to ensure that casting has become the most important process for the manufacture of components in metals (and in some other materials).

Castings may be cast in one of a variety of 'sand' moulds formed around a pattern and classified as 'sand castings'; in one of a variety of moulds formed around a fusible wax pattern and classified as 'precision' or 'lost wax' castings; in a metal mould and classified as 'die castings'; or they may be formed by centrifugal force and classified as 'centrifugal castings'. 'Splat' casting produces material with the optimum mechanical properties in the form of small flakes.

Even when casting has not been adopted to generate the shape of a component it may well have entered into the process of manufacture at an earlier state having been used to produce the stock for mechanical working, forging, rolling or extrusion by a process such as 'ingot', 'billet', 'continuous' or 'semi-continuous' casting.

The advantages of castings have, in the past, been offset by significant disadvantages compared with wrought products. Castings are considered to be less ductile than the equivalent wrought product, and they have a less consistent performance in fatigue and inferior integrity. The difference in ductility may be more apparent than real. A forged or rolled component may have a higher ductility than a casting in the direction of forging or rolling but a significantly lower transverse ductility. This is a distinct advantage if the longitudinal direction has to resist the principal stress but it is not necessarily a sign of inferiority of the casting.

However, some of the problems concerned with brittleness, lack of integrity and an inferior and less consistent performance under fatigue loading stem from fundamental difficulties in the casting process. These problems will be highlighted and analysed and the way in which modern developments ameliorate and eliminate them will be indicated in the accounts of individual techniques.

16.5.2 Problems inherent in casting technology

The problems inherent in casting technology may be classified under the headings 'Shrinkage', 'Gas' (originating from metal and mould), 'Coarse structure and segregation', 'Inclusions' (originating from metal and mould) and such defects as 'Cold shuts' where metal streams meet but do not unite, or dimensional errors which arise from movement of cores.

16.5.2.1 Shrinkage and contraction

Metals (with the exception of some alloys of antimony, tin and bismuth) contract in volume when they solidify and continue to contract as they cool to room temperature. If, therefore, a substantial body of metal is located alongside thinner sections which solidify earlier so that it cannot be fed from them that body will shrink by forming external sinks, internal shrinkage cavities or possibly interdendritic voids which may communicate through the walls. The influence of shrinkage cavities on the UTS of A357 alloy is shown in Figure 16.102.[84] If the body is constrained excessively by the mould or by the rest of the casting while its metal is in a hot short condition it will crack. (An alloy is 'hot short' when it comprises solid islands or dendrites surrounded by thin bands of molten metal and therefore has no strength or ductility.) If it has reached a temperature at which it is ductile, it will distort and perhaps fail to clean up during subsequent machining.

Design of components to be cast should therefore, wherever possible, avoid the introduction of isolated relatively heavy sections. Where it has not proved possible to achieve this, casting design should ensure either that such sections are heavily chilled and the flow of metal arranged so that they solidify at the same time as the rest of the casting, or that adequate arrangements are made to feed them by liquid metal channelled to them for this purpose. Interdendritic shrinkage cavities may be present even in cast material which may be considered to have been fed satisfactorily.

Cores should, as far as possible, be made weak enough to give against the contraction stresses imposed by the cast metal and should be removed from the casting as soon as possible after the metal has solidified.

16.5.2.2 Gas

The effect of gas varies with the metal cast. Hydrogen dissolved in molten aluminium may generate cavities of its own but generally increases the size of shrinkage cavities. In steel, hydrogen causes a number of very dangerous effects described in Chapter 7, Section 7.3 (Corrosion). The requirement for good-quality castings demands its complete removal from both metals. Fortunately, hydrogen is reasonably easy to remove from molten metal, in the case of aluminium by bubbling gas through the melt before casting. Figure 16.103

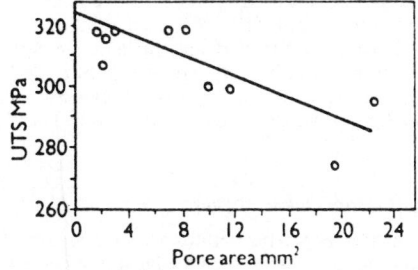

Figure 16.102 Strength of fully heat treateed A357 alloy as a function of maximum observed pore size on the fracture surface

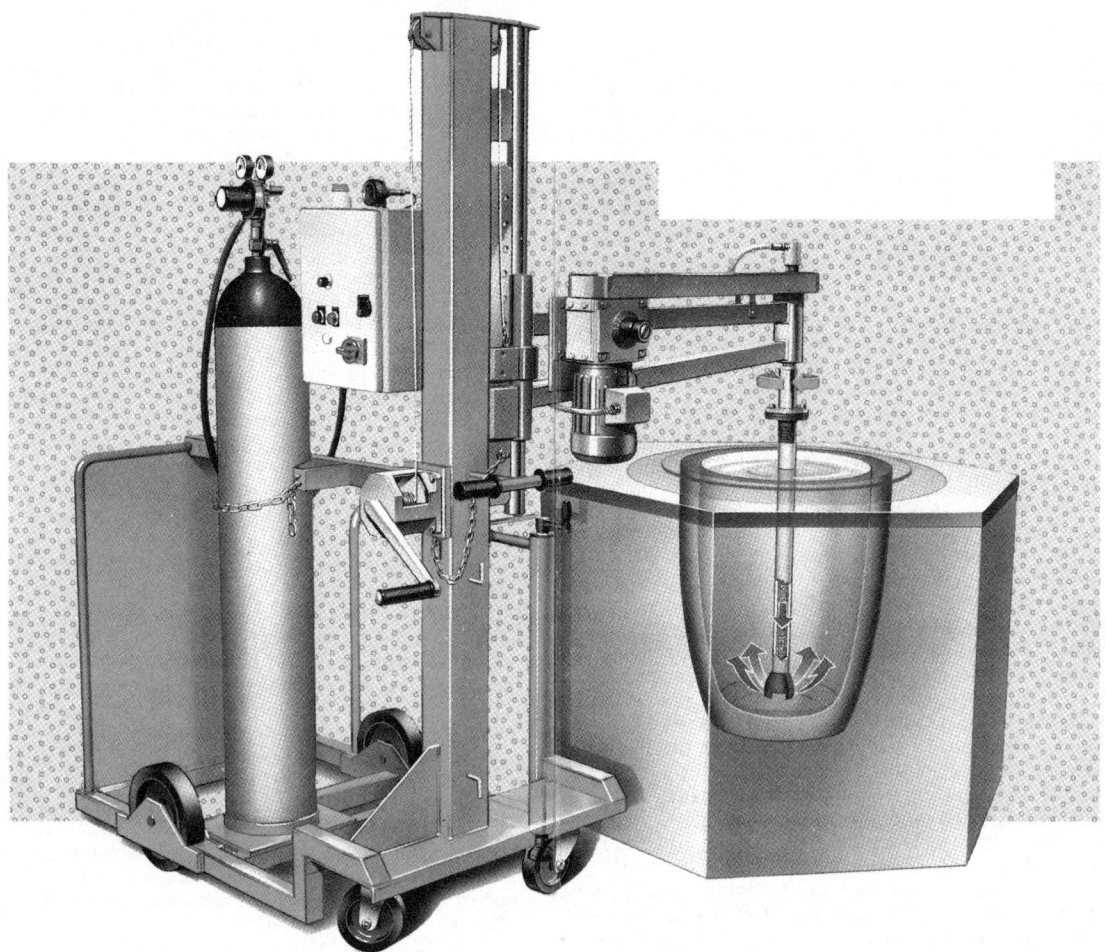

Figure 16.103 Spinning rotor argon lance system for degassing molten aluminium (courtesy of Foseco plc)

shows the arrangement of a spinning rotor argon lance system which, it is claimed, will approximately halve the quantity of hydrogen in 250 kg of molten aluminium in 10 min. Hydrogen is removed from steel by vacuum treatment of the metal in the molten state (see Chapter 7, Section 7.3, Ferrous metals).

Oxygen is removed from molten steel by 'killing' with silicon and/or aluminium. In the case of ingot casting of unkilled or 'rimming' steel advantage is taken of the effect of the bubbles of oxygen evolved when this steel solidifies to neutralize the solidification contraction and eliminate the pipe on the top surface. It therefore becomes possible to forge and roll the whole of the ingot, and not go to the expense of discarding the top 30%, as would have to be done with killed steel.

16.5.2.3 Coarse structure and segregation

Slow cooling of solidifying metal generates a columnar structure of coarse dendrites growing perpendicular to the cooled surface which have (in an alloy with a margin between liquidus and solidus) a lower solute content than the melt. The space between and in advance of the dendrites is therefore enriched

in solute (in the case of steel this includes carbon and other alloying agents) and the result is a coarse segregated structure. The result of variation of cooling rate on tensile properties of A357 alloy is illustrated in Table 16.14[84] and the effect of structural variations on fracture toughness in Figure 16.104.[84] The cooling rate should be the maximum consistent with good feeding and is compared diagrammatically for a selection of casting processes in Table 16.15.

16.5.2.4 Inclusions

Inclusions can originate from oxide formed on the metal surface from slag, flux, from the melting crucible or from the mould and can have just as great an effect on metal properties as pores (see Figure 16.105[84]). Clearly, all possible precautions must be taken during melting to provide as clean a metal as possible. In addition, weirs that trap inclusions should be provided in the runner of the casting and filters which are now available for all types of cast metal can be inserted in the metal stream. Filters not only remove entrained oxide and nonmetallics but will, if strategically placed, eliminate turbulence, the flowing metal on the downstream side emerging as a

Table 16.14 Properties of 357 aluminium alloy obtained with various casting processes

A357	0.2% Proof stress (MPa)	UTS (MPa)	Elong. (%)
Sand cast	200.6	226.6	1.6
Chill cast	248.6	313.6	6.9
Squeeze cast	283	347	9.3
Cosworth	242	312	9.8

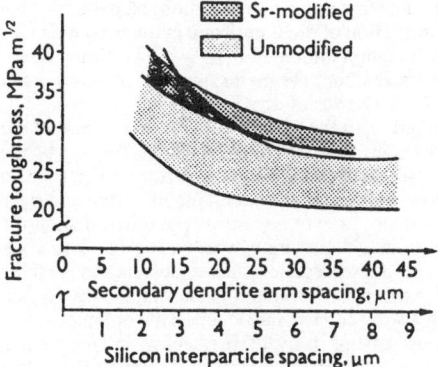

Figure 16.104 Fracture toughness of A357 alloy in castings solidified over a range of cooling rates with and without Sr modification

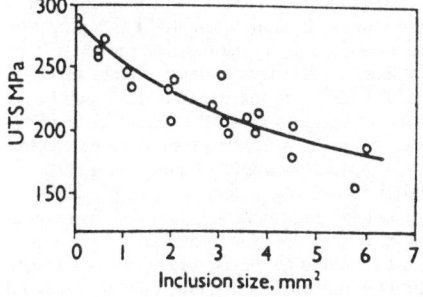

Figure 16.105 Fracture strength of a porosity-free magnesium alloy as a function of inclusion size

smooth coherent flow (see Figure 16.106). Filters restrict metal flows, a filter $50 \times 50 \times 22$ mm gives a flow rate of 6 kg s^{-1} of grey cast iron at 300 mm head. Impingement against the mould wall and consequent entrapment of mould material are avoided.

Upward pouring, as in the low-pressure die casting and Cosworth processes, is a very good solution to this problem because it allows metal to be transferred upwards from below its surface without turbulence, thus minimizing inclusions. Once the casting has been formed the molten metal head is maintained until all parts of the casting have been fed, thus maintaining a reservoir of hot metal as long as required. With top-pouring methods, on the other hand, the top of the feeders may solidify as soon as the casting.

16.5.2.5 Cold shuts

'Cold shuts', i.e. locations where two metal streams meet without uniting, must be avoided by good casting design.

16.5.2.6 Core movement

Cores are normally of lower density than the surrounding metal and may float when submerged or be displaced by the metal flow. Core prints must be accurate in dimensions and adequate in size. Where necessary, core prints may be provided with steel bushes which fit in corresponding bushes in the mould.

16.5.2.7 General

The problems described above have been long appreciated and techniques are being developed which will completely overcome them. These techniques are not necessarily applicable to all metals or to every size and shape of casting, but where they can be applied they will produce castings with consistency in fatigue behaviour adequate for any requirement.

The following sections describe the techniques used to produce castings and stock for mechanical working with particular emphasis on the most recent developments and most advanced processes which provide the highest quality products in the most efficient way. Not every difficulty has been eliminated with every metal and alloy cast, but this account will indicate how castings which will perform fully as well as forgings (and, in some cases, better) can be produced in an increasing proportion of materials.

It is not possible in an account of this length to detail the very large number of variations in casting process which are and have been employed. Where a selection has had to be made the process or variation described is, as far as possible, the most efficient, the most modern, and produces a product

Table 16.15 Spectrum of casting processes represented as a function of cooling rates illustrating some alloys cast by the various techniques

Process	Alloy							Cooling Rate
	Al	Fe	Mg	Ni	Ti	Zn	Cu	
Splat quenching (RSR)	●	●	●	●	●		●	10^6 K/s
Strip casting	●					●	●	
High-pressure die casting	●		●			●	●	
Squeeze casting	●	●	●				●	
Low-pressure die casting	●	●	●					
Gravity die casting	●	●	●			●		
Cosworth and Investment casting	●	●	●		●	●		
Sand casting	●	●	●					10^3 K/day

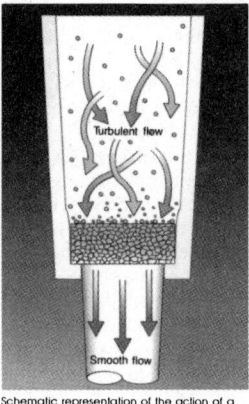

Cut-away of a DYPUR unit.

Schematic representation of the action of a DYPUR unit in filtering and eliminating turbulence in a molten metal stream.

Figure 16.106 'Dypur' filter unit showing how a strategically placed filter can eliminate turbulence in a molten metal stream (courtesy of Foseco plc)

of the highest material quality, mechanical properties, dimensional accuracy and surface finish.

16.5.3 Ingot, billet and slab casting

16.5.3.1 Conventional casting of steel

For many years stock for mechanical working was made by top pouring individual ingots or billets typical of which are the square steel ingots shown in Figure 16.107.[85] During this period great advances have been made in metal treating and handling techniques. Cleaner steelmaking technology, better deoxidation control and secondary steel making in the ladle and the arc furnace have been introduced and furnaces have been designed (for steel and also for aluminium) which yield metal free from both surface and bottom oxides and slag. Hydrogen can be removed from molten steel by vacuum treatment, and from molten aluminium as described above. New types of refractory stoppers, both vertical and sliding,

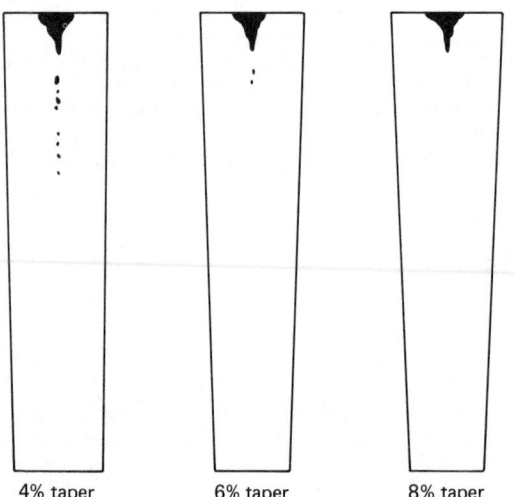

4% taper   6% taper   8% taper

Figure 16.107 Segregation and bridging in typical steel ingots

have made it possible to start, stop and control the rate of flow of the casting stream.

However good the metal quality, however carefully the dimensions, shape and casting parameters of the ingot were chosen, and however well the ingot was fed (and however effective the measures taken to keep the top of the ingot hot) ingots formed from killed steel developed a pipe (which sometimes bridged) in which was located considerable segregation (see Figures 16.107 and 16.108[85]). There was also usually segregation in a band round the ingot where the columnar crystals which advanced from the surface gave way to the equiaxed structure. Segregation also developed if cracks appeared between the columnar crystals at the surface and material of a higher solute concentration bled out of the cracks. (This formation of blebs enriched in impurity elements also occurred in aluminium ingots.) To avoid vertically orientated surface cracks the perimeter section of large ingot moulds consists of a series of arcs intersecting at cusps.

In a top-poured ingot the metal has to fall to the base of the mould from a height at least equal to that of the top of the mould. The resultant splashing caused surface defects which were so serious that stainless steel ingots, where surface quality is paramount, used to be bottom poured in spite of the increase in pipe and segregation which resulted. The combined effects of pipe, segregation surface defects and cracking of individually cast ingots resulted in a substantial proportion of metal being discarded. The introduction of continuous (and semi-continuous) casting proved so effective in overcoming these defects that this process is rapidly displacing individual casting in steel (see Figure 16.109[86]) and other metals.

16.5.3.2 Continuous casting of steel

Billets, blooms, slabs, thin slabs and strip may be continuously cast. Continuous casting eliminates longitudinal segregation and for billet and bloom casting horizontal segregation may be reduced or eliminated by electromagnetic stirring of liquid steel in the mould (M-EMS), in the strand (S-EMS) and/or in the final stage of solidification (F-EMS). (Centreline segregation may, however, be a problem in slab where electromagnetic stirring is difficult to apply.) Defects in quality are therefore restricted to surface defects, including cracking, possible internal cracks and inclusions derived from nonmetallic particles carried in the molten metal.

The practices and techniques devised to improve the efficiency of a continuous slab casting machine may be described with reference to Figure 16.110,[87] which shows a continuous slab casting machine with hot connection facility. Reoxidation of the steel stream from ladle to tundish and from tundish to mould is prevented by using a ladle shroud, a submerged entry nozzle, and by flooding with argon. Flow from the tundish to the mould is controlled by replaceable stoppers. The tundish design (Figure 16.111[87]) has been improved to give longer residence time, to incorporate weirs and dams and to allow bubbling of inert gas through porous plugs located at the bottom to improve flotation and removal of inclusions. Tundish level control has been made automatic and the tundish can be heated for start-up.

Flow control from ladle to tundish and from tundish to mould has been integrated and automated, and level in the mould controlled by a duplex electromagnetic/radioactive system. The metal in the mould is protected and the metal mould interface lubricated by a 50 mm layer of 'black powder'.[88] This is the name given to a series of proprietary fluxes probably containing calcium silicate and fluoride and more than one variety of carbon. The geometry of the particles and

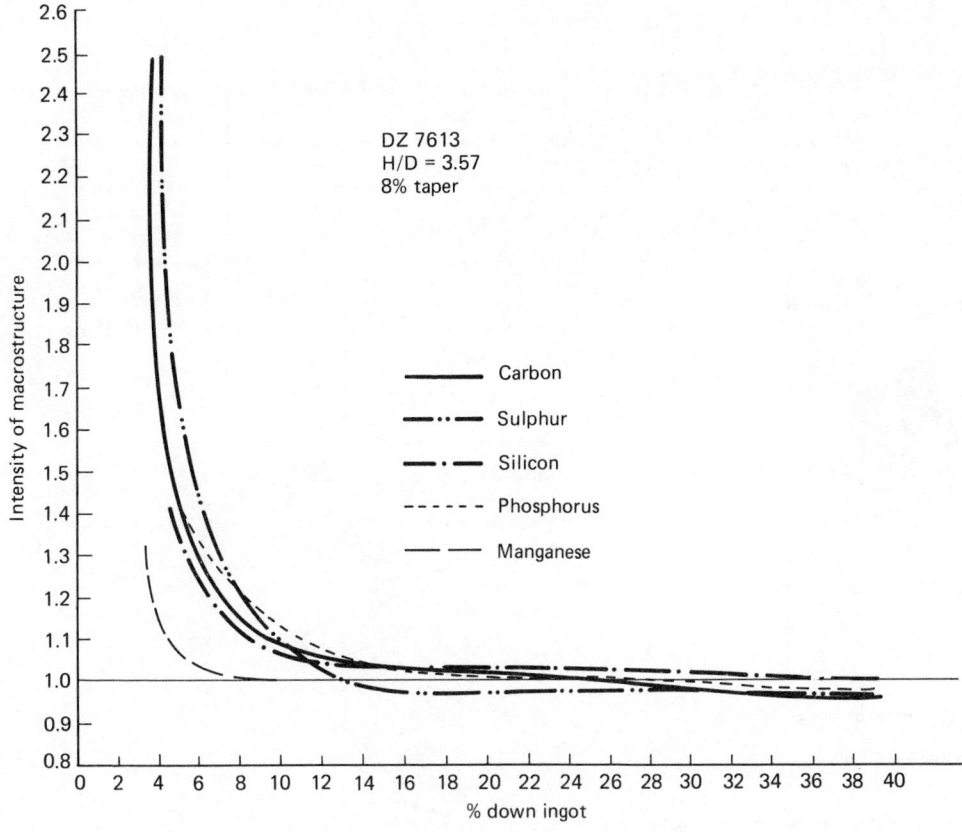

Figure 16.108 Typical macrosegregation of the main solute elements in a wide-end-up ingot

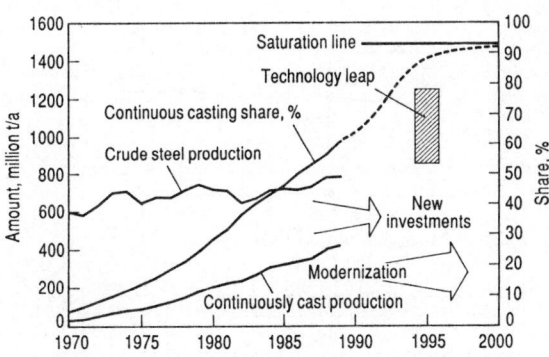

Figure 16.109 Increasing share of continuous casting in production of steel

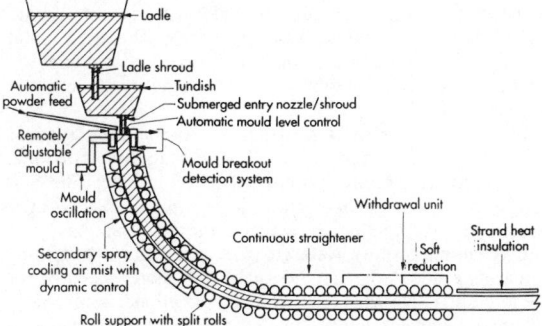

Figure 16.110 Layout of a continuous slab casting machine with hot connection facility

the distribution of their constituents is more critical to performance than their composition. Hollow sphere granules with carbon concentrated on the exterior maintain good insulation until the granules melt and then the flux lubricates the mould/strand interface.

Figure 16.112 shows a flux feeder with automatic gravity feeding. The mould is oscillated vertically to prevent the

strand shell sticking to the copper mould and causing breakout. Oscillation is controlled hydraulically to give a downstroke synchronized with the strand's downward velocity and an up-stroke giving low 'negative strip' time, thus saving powder consumption and improving lubrication. This gives good surface quality and requires low maintenance. The width of the mould can be changed remotely during slab casting

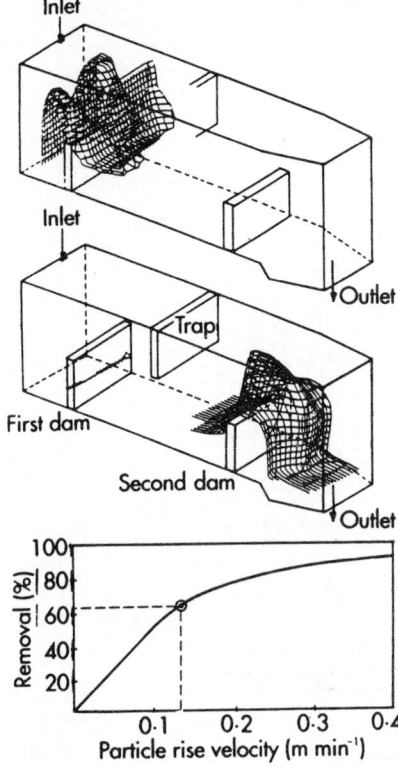

Figure 16.111 Modern tundish design illustrating computer model prediction of particle carry-over. (*Top*) 30% removal by first dam and trap; (*middle*) 65% removal by first dam, trap and second dam

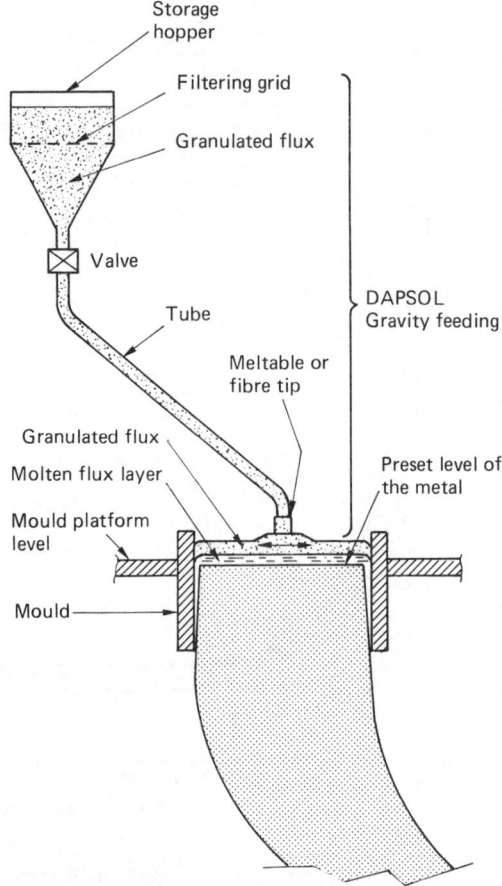

Figure 16.112 'Dapsol' automatic flux feeder for continuous caster showing distribution of flux

without impairing surface quality so that a whole variety of product widths may be cast in one stream. Further operational flexibility is obtained by 'link casting', which allows small batches of differing steel grades to be cast without restranding. When a ladle of one grade has been cast a fabricated metal link is partially immersed in the liquid steel and the exposed part becomes enveloped by the steel of another grade from the following ladle.

Additional flexibility is provided by multi-strand casting. Either dividers are fixed in a single-slab mould or individual slab moulds are mounted within one slab machine. The first stage of solidification takes place in the water-cooled copper mould. Air-mist cooling, in which compressed air has been used to atomize spray water to fine droplets, is becoming increasingly adopted, first for slab and more recently for bloom casting. This has the advantage of greater uniformity, higher heat transfer coefficient, less risk of blockage because of the larger outlet orifice, much higher turndown ratio and greater facility for automatic control than water only.

The principles of computer software for dynamic spray control of secondary cooling devised by one manufacturer are shown diagrammatically in Figure 16.113.[87] The same manufacturer has developed a system of quality assurance in real time that assigns quality values to segments of the strand during casting and enables selection of those slabs suitable for direct rolling from those which have to be cooled for inspection and rectification.

In a machine such as that illustrated in Figure 16.110 slabs are bent, straightened and soft reduced while metal in the

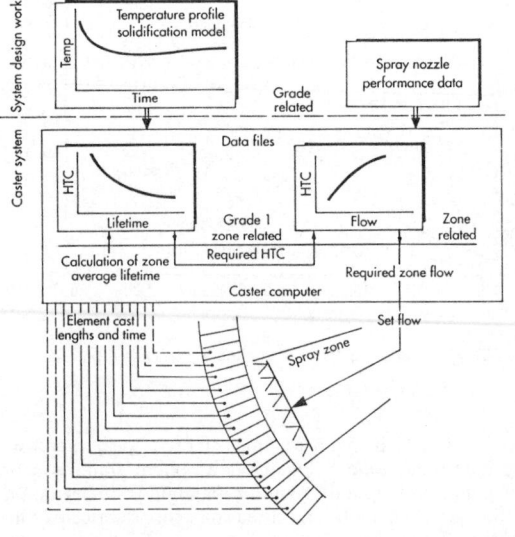

Figure 16.113 Computer software dynamic spray control method of secondary cooling devised by Davy Distington

interior of the strand is still molten. Complex stresses introduced in these processes may result in problems associated with centreline and spot segregation caused by solute enrichment in certain steels for 'offshore-structural' and 'resistance to hydrogen-induced cracking' plate. These problems are overcome by 'continuous straightening', which applies a constant bending moment over the straightening zone, and special roll design and optimized cooling during soft reduction.[89]

There are two further developments. Direct rolling eliminates a reheating furnace, saves energy, reduces scaling and therefore increases yield and, if it can be automated, may save labour. On the other hand, the rolling mill must be sited near the caster, the slab must be transferred rapidly, edge heating is required and scheduling of production is highly complex. Direct rolling is already practised[90] in certain plants but the compromise procedure, hot linking, in which the slab is cut hot into appropriate lengths, insulated and reheated for rolling before it has fully cooled, is more widespread and provides moderate savings in energy and cost.

Low-head casting will enable a continuous casting machine to be sited in an old ingot casting shop, the low ferrostatic head will reduce bulging strains which lead to internal defects and the smaller machine size will reduce weight and therefore cost.

Near-net shape (15–25 mm thick) and final thickness (1–3 mm thick) strip casting have been the subject of considerable study and processes which may include moving-belt moulds and moving casting belts may be realized in the medium or long term.

16.5.3.3 Continuous casting of aluminium

When continuous casting of aluminium alloys was first introduced it provided a very significant improvement in soundness and fineness of structure compared with the 'can cast' method of ingot production, which was the best previously available. A very large number of systems – some vertical, some horizontal, some utilizing fixed moulds while some depend on casting wheels, bands, segmented moulds or rolls – are summarized in Table 16.16.[91]

Table 16.17[91] places the systems in approximate order of cooling rate and the fineness of structure of the material they produce. The twin-roll and stationary mould strip casters will produce 3–7 m dendritic structures in 12.5 mm thick strip.

16.5.3.4 Continuous casting of other metals

Gold, silver with palladium, copper, bronzes, brasses, zirconium copper, chromium copper, platinum, platinum rhodium nickel, solders, cadmium and cobalt are continuously cast in the following size ranges:

Solid rod 2–300 mm dia.
Strip 1.6 mm thick and up to 1000 m wide
Pipe, bore 30 mm (min), OD 350 mm (max), wall 12 mm min

Throughputs 1000–2800 kg h^{-1} (dependent on diameter) can be achieved.

The casting techniques used are those introduced by Propergi, Haslett, Southwire and Ontokampi. Typical examples of plant are shown in Figure 16.114, which illustrates a Technics Guss or Wertli systems caster for strip 2 cm thick and 60 cm wide, and Figure 16.115, a Rautomead RT650 caster for small-diameter copper and alloy rod. Both are horizontal

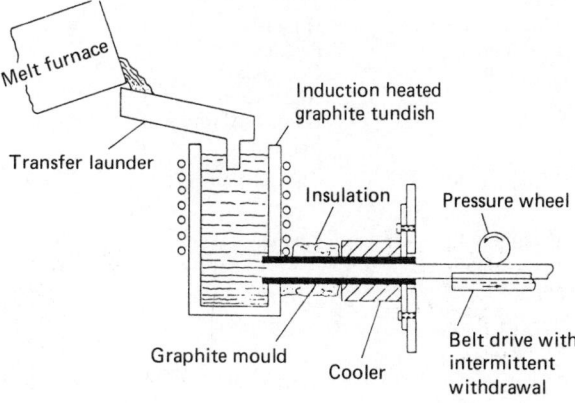

Figure 16.114 Technica Guss or Wertle strip caster for copper base alloys

Table 16.16 Aluminium continuous casting systems

System	Other alloys cast
Ingot, slab, and billet	
Clark single-strand horizontal-casting system	
Wagstaff horizontal-casting machine	
Reynolds horizontal-casting process	
Kaiser aluminium process for horizontal continuous casting	
Alcoa horizontal continuous-casting process	
Sheet, plate, and foil	
Hunter continuous-casting process	
Pechiney 3C process	
Mann rotary strip casting and rolling line	
Alusuisse caster I and caster II	
Hazelett twin-belt caster	Zn, Pb, Cu Steel
Rod, bar, and wire	
Properzi process for continuously cast and rolled rod	Pb, Zn, Cu
Cegedur–Pechiney–Secim continuous casting and rolling process	
Southwire aluminium SCR systems	Cu

Table 16.17   Classification of commercial and experimental casters for aluminium alloys

Class	Type	Section thickness (cm)	Cooling rate (°C/S)	Dendrite cell size (μm)
Moving mould Continuous Two belts Strip	Hazlett Hunter/Douglass	1–2.5	5–7	40–60
"	Alusuisse II	1–2	5	40–70
Moving mould Continuous Casting wheel Metal band Strip	Properzi Rigamonti Pechiney Lubeck	2	10	30–50
Moving mould Continuous Casting wheel/ metal band Rod	Spidem Southwire Mann	2	10	20–40
Moving mould Continuous Two rolls Strip	Hunter Pechiney-Coquillard Alusuisse I Harvey (Martin Marietta)	0.5	10^2–10^3	5–10
Fixed mould Ingot	Chill casting	50–60	0.1	200
Fixed mould Semi-continuous Ingot	Isomet	20–25	1	100–120
" "	Vertical DC Horizontal DC	7–65 5–65	1–5 1–10	30–70 15–30
Fixed mould Continuous Strip	Mortiz/VAW	0.5–1.25	10^2–10^3	2–7

Figure 16.115   Rautomead 650 continuous casting machine for rod

casters and both utilize a graphite tundish or crucible from a hole in the bottom or side of which the molten alloy flows into a high-density polished graphite mould. A protective atmosphere, usually nitrogen, may be maintained around both. Heat is extracted from the mould by a water-cooled copper cooler and the rod or strip is withdrawn intermittently. The key to success is precise control of molten metal temperature, die cooling and withdrawal cycle. Although, unlike in continuous casting of steel, the die is stationary instead of oscillatory, movement of the strand relative to the die is similar, consisting of a rapid pull alternating with a relatively long hold. In the case of the strip caster a pull of 1 cm in 1–2 s alternates with a hold of 2–4 s and overall strip production is 10–15 cm per minute.

Higher casting speeds are possible with the quick withdrawals which can be obtained with a pneumatic-activated pulsed movement via a 'sprag clutch' to feed rolls (see Figure 16.116[92]) and a supercooler heat sink (see Figure 16.117[92]). It is not possible in an account of this length to discuss cooler/strand relationships but Figure 16.118[92,93] gives some idea of the variations in cooling rate. Secondary cooling after the mould may be used to avoid solid-state reactions such as precipitation in alloys where this may occur.

16.5.3.5 Semi-continuous casting

Semi-continuous casting processes ('vacuum arc remelting' and 'electroslag remelting') are employed to produce very high quality stock for mechanical working. In both, one or more cast or forged billets are made into one electrode (where three parallel electrodes are employed three-phase current may be used). The electrode(s) are fed downwards and as they melt, the metal melted by the arc (or by the heat generated by the resistance of a bath of slag) is transferred downwards to the other electrode which is located in a cylindrical water-cooled copper vessel.

In vacuum arc remelting (see Figure 16.119) the arc is contained in a vacuum and the consumable electrode is fed downwards through a seal which retains the vacuum. In electroslag remelting (see Figure 16.120) a bath of fused slag is retained above the other electrode in the copper vessel. In both cases, after the start continuous casting and cooling and feeding conditions exist in the bath of metal which is formed, and climbs continuously up inside the vessel. The resulting billet which is fed and rapidly cooled during solidification is the best stock that can be produced for mechanical working.

Electroslag remelting refines the metal by dissolving such non-metallics as alumina in the flux. Vacuum arc remelting breaks and disperses non-metallic inclusions but does not remove them completely. It does, however, completely remove gaseous impurities such as hydrogen and oxygen from steel.

In general, electroslag remelting is used for steel and nickel and cobalt alloys, vacuum arc remelting for refractory metals, titanium, zirconium, tantalum and molybdenum, but it is also used for steel and nickel alloys.

16.5.4 Sand casting

In sand casting, of which there are many variants, molten metal is poured into a sand 'mould' which defines the exterior shape (including any re-entrant surfaces) of the component. 'Cores' also formed of sand define the interior.

The components of the mould of which there are usually two, the 'cope' and the 'drag', are moulded in sand around 'patterns' and the cores are made in 'core boxes'. Very simple examples of patterns and core box which might be used for hand moulding of the pipe connection (Figure 16.121(a)) are shown in Figures 16.121(b) and (c). Usually each section of the pattern is mounted on a 'moulding board' as shown in Figure 16.122(a). In the moulding board are drilled two holes which register with corresponding holes in the other pattern section assembly and with 'moulding flasks' also shown in Figure 16.122(a). Alignment of the moulding board and pattern is maintained with the flask by means of 'moulding pins' which fit the holes accurately. The flask is filled with moulding sand, which is compacted and levelled off with the top of the flask, which is then turned over to form the 'drag' or bottom section of the mould. The pattern is lifted out vertically and the core, which has been made of sand blown into the core box and hardened, is placed in the mould supported in 'core prints' which are formed by the extensions to the pattern shown in Figure 16.121(b).

The 'cope', the other section of the mould which has been prepared in the same way, is placed on top of the drag and aligned by means of moulding pins. Molten metal is introduced into the mould by means of channels (described later).

After the metal has solidified, the casting is removed from the mould, shaken and brushed to remove sand (which if it should have stuck to the casting must be scraped or ground off) and runners, risers and feeders appended to it for the purposes of casting 'fettled' off. The procedure described imposes certain geometrical constraints on the shape which can be cast.

16.5.4.1 Geometrical considerations

The pattern must be capable of being withdrawn vertically from the mould. Each nominally vertical surface must therefore be given a draft to facilitate this withdrawal and, if a two-part mould only is to be used, no pattern can contain an overhang.

When, as is the case with the three-arm flanged connection shown in Figure 16.123, the design provides for an overhanging flange, whichever way the component is moulded one of the flanges will prevent withdrawal of the pattern. The problem may be overcome by blocking off the overhanging part of

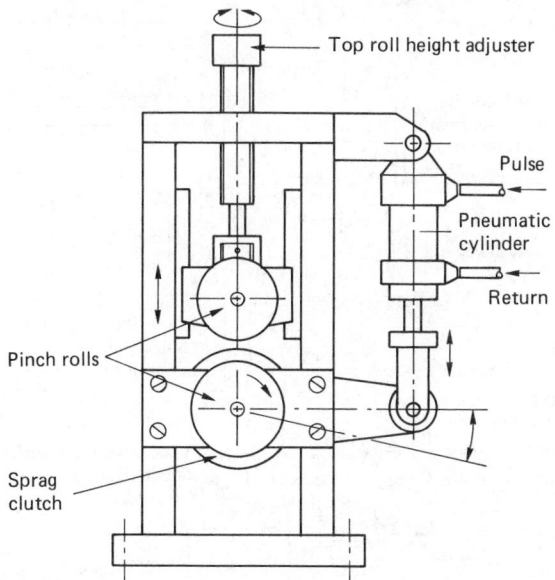

Figure 16.116 Runtomead quick pneumatic withdrawal mechanism providing square/box curve withdrawal

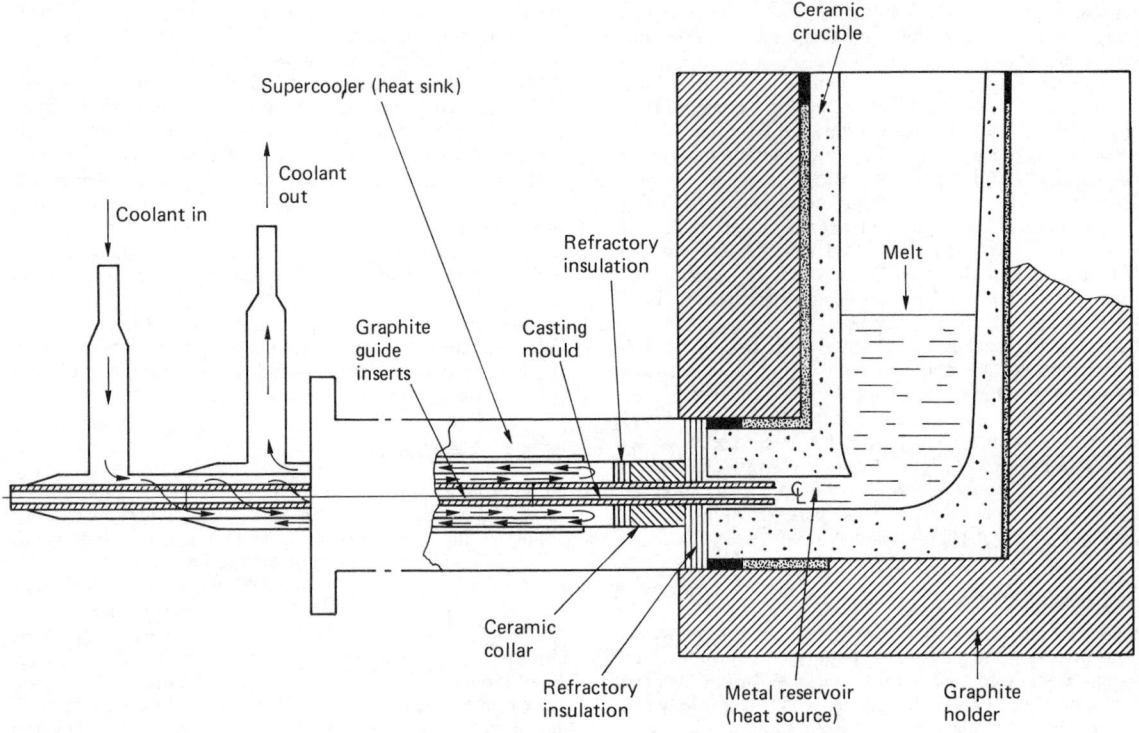

Figure 16.117   Rautomead supercooler and containment for 205 mm diameter casting

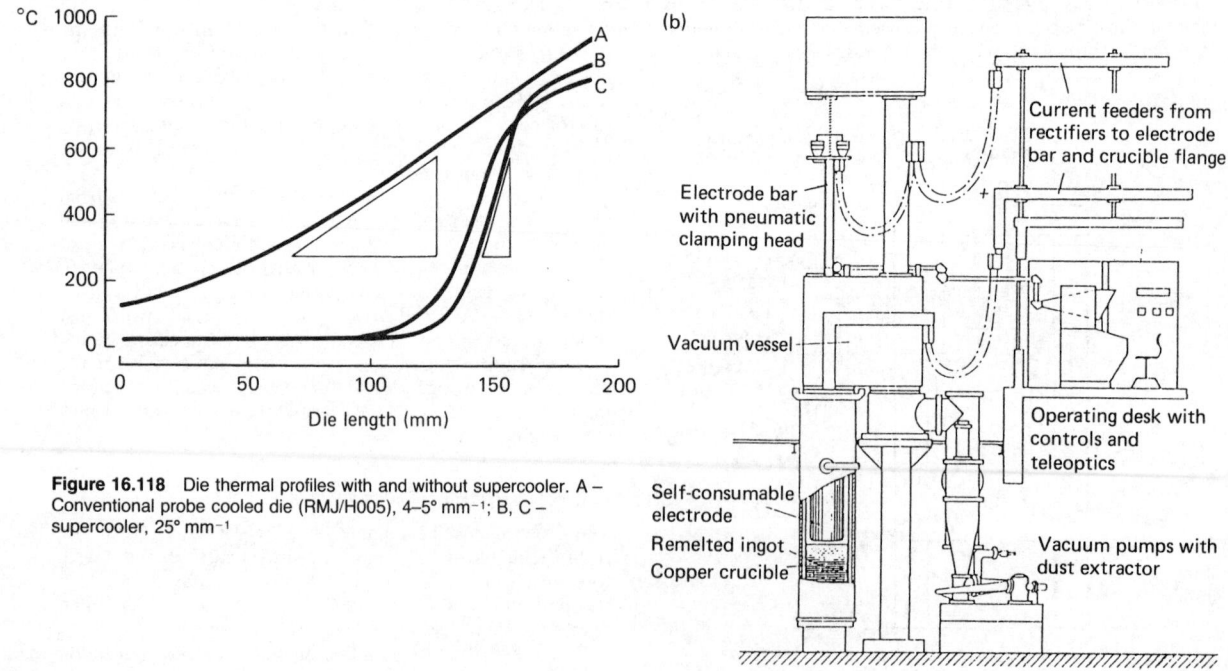

Figure 16.118   Die thermal profiles with and without supercooler. A – Conventional probe cooled die (RMJ/H005), 4–5° mm⁻¹; B, C – supercooler, 25° mm⁻¹

Figure 16.119   Vacuum arc remelting furnace

(a)

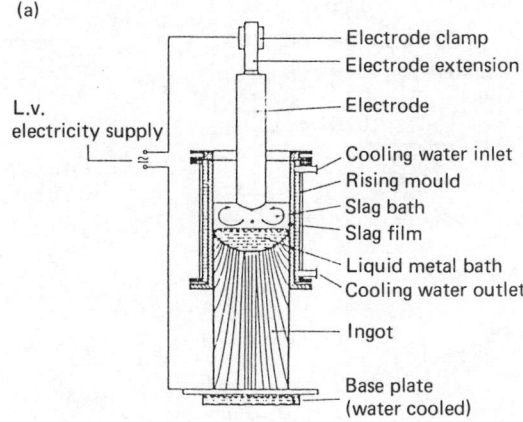

Electrode clamp
Electrode extension
L.v. electricity supply
Electrode
Cooling water inlet
Rising mould
Slag bath
Slag film
Liquid metal bath
Cooling water outlet
Ingot
Base plate (water cooled)

Figure 16.120 Electroslag remelting furnace

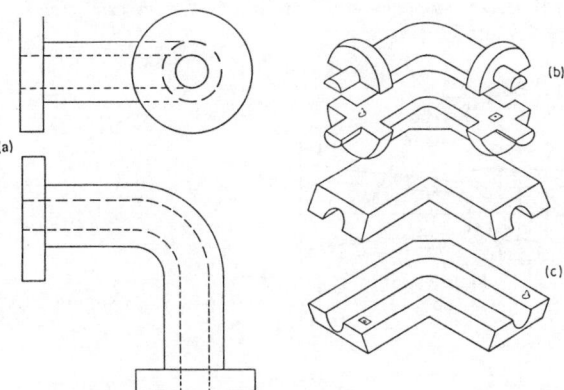

Figure 16.121 Patterns (b) for hand moulding and a core box (c) for moulding the pipe connection elbow (a)

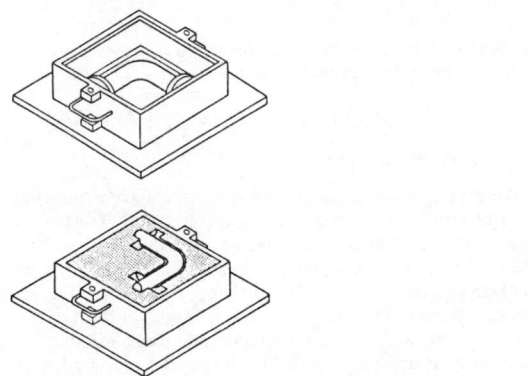

Figure 16.122 Pattern as in *Figure 16.121* mounted on a moulding board with flask and holes for box pins ready for the introduction of sand

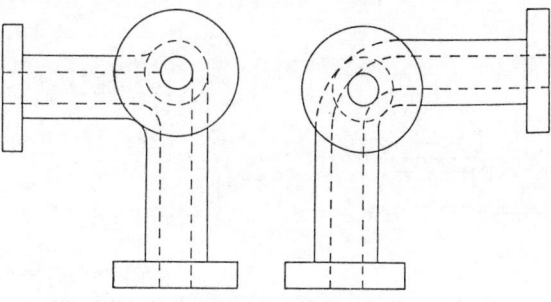

Figure 16.123 Pipe connection elbow with three arms at right angles so that one of the flanges will give difficulty in moulding

the pattern with a 'core print' (see Figure 16.124(a)) and making a core which forms the shape blocked off and itself fits into this core print.

Although simple and neat, this is not a perfect solution. Some draft for which permission must be obtained from the designer must be allowed on the vertical arm of the connection and, as the position stands, with a flat top to the core the top flange will form a sharp angle with the pipe. Such a sharp edge will constitute a stress concentration in operation but, more seriously, may cause cracking or shrinkage cavities as the casting solidifies.

There are other ways of overcoming overhangs including:

1. The use of a three-part pattern and mould.
2. The use of a cover core (Figure 16.125(b)). These methods are to be avoided if possible because they are difficult to mechanize.
3. The incorporation into the pattern of segmented loose pieces (Figure 16.125(c)).

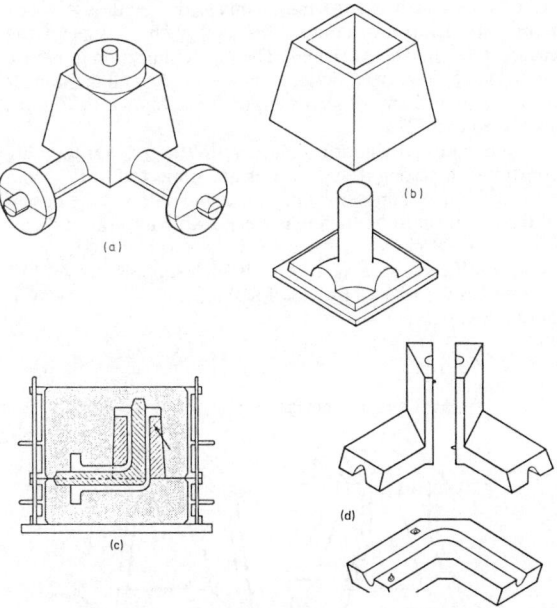

Figure 16.124 Using a core. (a) Pattern with a case joint to overcome difficulty in moulding the flange. (b) Two parts of a simple core box to produce the core. (c) Arrangement of mould and core box. (d) Core box for centre core

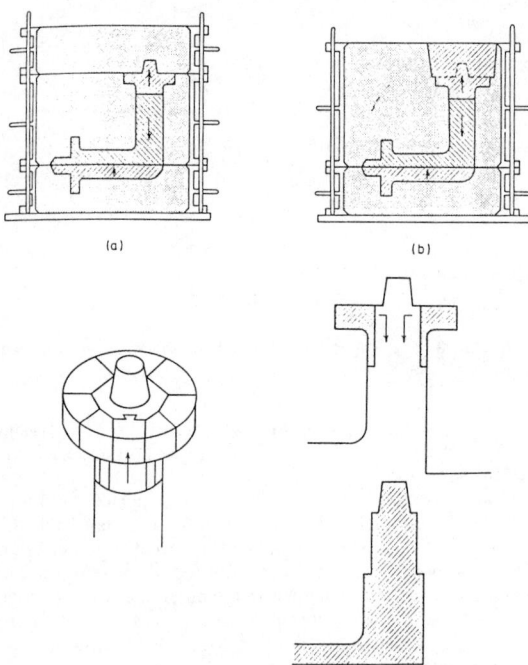

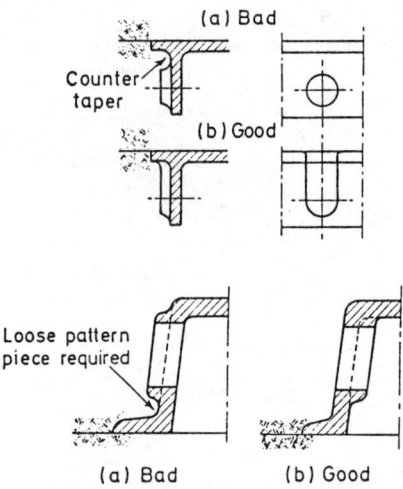

Figure 16.127   Modifications to simplify moulding of small projecting bosses

Figure 16.125   (a) Three-part pattern and mould. The arrows show the direction of pattern part removal. (b) Three-part pattern and two-part mould. Arrows again indicate the direction of removal and the broken lines where the cover core will be replaced

This method introduces a significant amount of hand work into what should be a simple mechanized process. Also, sand compaction under the overhang may be inadequate and the act of removing segments may damage the mould. A much more satisfactory procedure is to modify the design of the casting to eliminate overhangs. The pipe connection in Figure 16.123 might have been designed as a T to align all the flanges in one plane. Some other examples[94] are shown in Figures 16.126 and 16.127.

The need to provide for pattern withdrawal is not the only constraint in casting design. Cores are subjected to flow and buoyancy forces imposed by the molten metal, and the design of the casting must be such that it is possible to support them. Figure 16.128 gives examples of designs which respectively make it difficult and easy to provide such support. Design (b) also makes it easier to provide for the escape of gas evolved during casting.

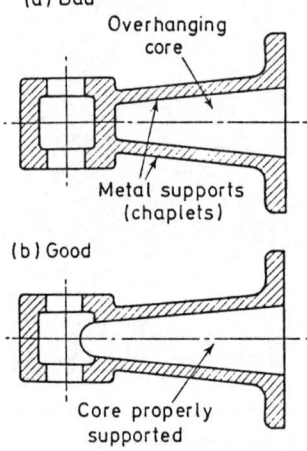

Figure 16.128   Alteration to the design not only gives more support to the core but also improves core venting

16.5.4.2   Running and feeding

In conventional sand casting metal is poured into a 'pouring basin' and enters the casting via a 'sprue' or 'downgate', a runner bar and a number of 'ingates' (see Figure 16.129). The objective is to introduce metal as smoothly and as free from turbulence as possible, so that it fills up and then solidifies uniformly and simultaneously upwards from the bottom of the casting over as wide an area as possible thus minimizing differential contractions. It solidifies last in the feeding heads which therefore contain all the shrinkage cavities. The 'pouring basin' helps to equalize the flow into the downgate and, it is hoped, traps slag or dross.

The 'sprue' or 'downgate' is a component of the runner system necessary to connect the pouring basin (which must be at the top of the cope) to the runner bar (which must usually be at the level of the parting line). It may be tapered from the top to the bottom to allow for the effect of gravity increasing

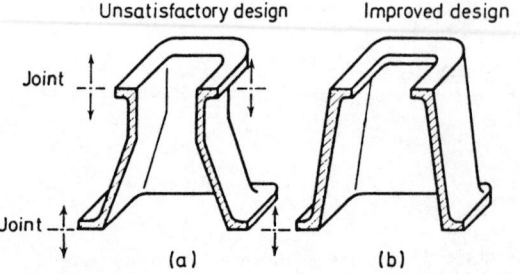

Figure 16.126   Design modification to eliminate a joint

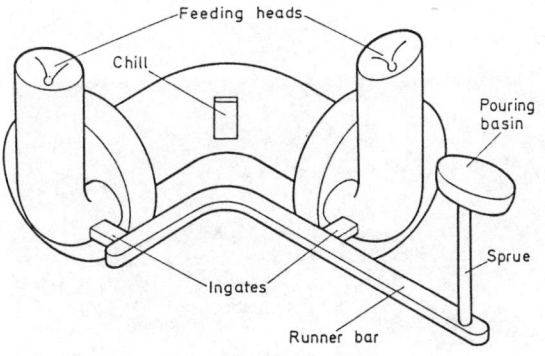

Figure 16.129 Components of a running and feeding system

the rate of flow downwards. Pouring basin and sprue are milled into the mould after moulding is completed.

The 'runner bar' is usually moulded into the face of the cope and the drag, and should extend almost all the way round the casting. It is both larger in section than the sprue (to promote a smooth flow) and the total section of the ingates (to ensure as far as possible that metal flows through them all at the same time). Filters, weirs and spinners are provided for preference between the sprue and the start of the runner bar to trap slag and dross. The correct size and position of the ingates is vital to the success of the casting and may vary according to the metal cast. A recommended design for magnesium is shown in Figure 16.130[94]. This metal has a low heat capacity and solidifies quickly. For this reason, the rule that the cross-section of the ingates is less than that of the runner does not hold for this metal. Filling must be rapid enough to ensure that the casting fills while the metal is molten without resorting to unreasonably high temperature.

Ideally, for all materials (with the exception of a few alloys, which include tin bronze) the aim is to get the whole of the casting to fill with metal while it is liquid. It will then solidify

progressively from the thin sections remote from the feeders and the metal will remain liquid last in the feeder heads at the top.

This can be achieved best by placing the ingates into or next to the base of the feeder heads (as in Figure 16.129). If a flange has been placed in between a feeder and an intermediate section its roots should be chamfered. Otherwise a well of liquid metal heated from the other side of the flange will persist until the metal round it has solidified to form a shrinkage cavity. Feeder heads should be larger than the sections they have to feed, and if they are internal to the mould they should be equipped with exothermic devices to keep them hot.

The majority of castings are poured from small solidification range alloys which tend to form skins of solid metal and shrinkage cavities which have to be 'chased' into the feeders. Those alloys which have large solidification ranges form dispersed porosity. This must be overcome by modifying the temperature gradient by control of the casting temperature, by gating and by the use of chills to produce directional solidification.

However, with some alloys (including tin bronzes) the skin of the casting solidifies with such a steep temperature gradient that it is free from shrinkage, and the microporosity goes to the centre of the section and cannot be chased into a feeder. In this case the foundry worker avoids directional solidification which might concentrate porosity in one part of the casting and encourages an even rate of solidification throughout the casting. In all other cases the design engineer should do his or her best to assist the foundry worker to achieve directional solidification. If possible, section thickness should increase smoothly to some point (probably a flange) which is conveniently fed. Thick sections, isolated in the centre of thinner sections, will cause difficulties. Sharp corners and re-entrant angles will cause shrinkage, puncturing and cracks, and this is accentuated by variation in section.

The way in which various types of junction give rise to increases in mass is illustrated in Figure 16.131,[95] and the way in which they can be lightened and made more flexible by staggering is shown in Figure 16.132.[95] The development of

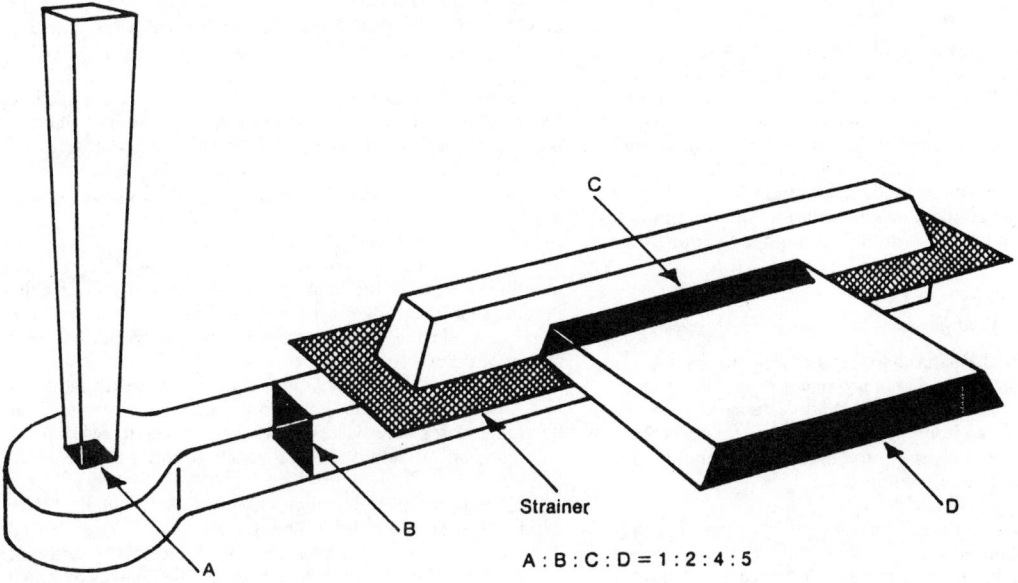

Strainer

A : B : C : D = 1 : 2 : 4 : 5

Figure 16.130 Design of gating system for magnesium

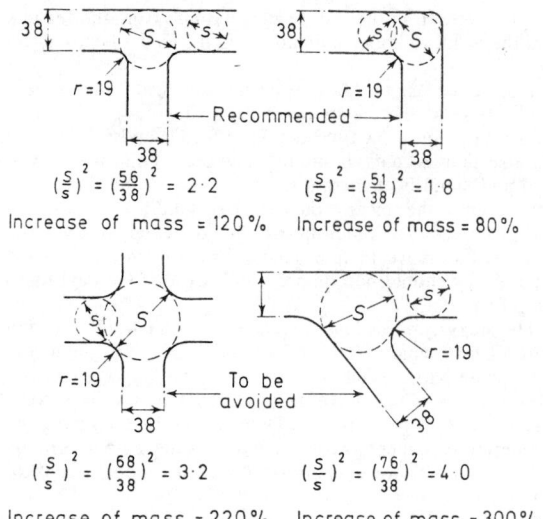

$$\left(\frac{S}{s}\right)^2 = \left(\frac{56}{38}\right)^2 = 2\cdot2 \qquad \left(\frac{S}{s}\right)^2 = \left(\frac{51}{38}\right)^2 = 1\cdot8$$

Increase of mass = 120%   Increase of mass = 80%

$$\left(\frac{S}{s}\right)^2 = \left(\frac{68}{38}\right)^2 = 3\cdot2 \qquad \left(\frac{S}{s}\right)^2 = \left(\frac{76}{38}\right)^2 = 4\cdot0$$

Increase of mass = 220%   Increase of mass = 300%

Dimensions are in millimetres

Figure 16.131   Increase in mass at various types of junction

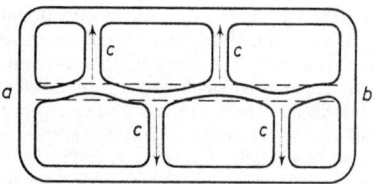

Figure 16.132   Staggering of ribs prevents the increase of section at junctions, reducing shrinkage problems, and also allows easier deformation to reduce the tendency to crack

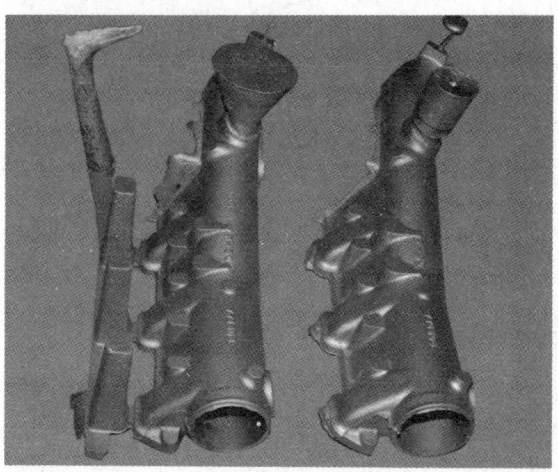

Figure 16.133   Sand cast aluminium alloy inlet manifold. The casting on the left requires a 3 kg running system and a 2 kg riser. On the right the same casting passed through a Dypur unit has a 1 kg riser and needs no running system

the type of filter shown in Figure 16.106 has raised the possibility that the entire gating system may be dispensed with. It is claimed that if a ceramic foam system is placed in a convenient riser as close as possible to the mould cavity, gravity, die and sand castings in aluminium, magnesium and some copper alloys may be direct poured to produce higher-quality casting and save substantial quantities of metal. Figure 16.133 compares an aluminium alloy manifold poured through a conventional running system and through a 'Dypur' filter unit.

16.5.4.3   Modern sand handling and moulding procedures

The traditional mould material for sand casting has been silica sand possibly with a natural binder – new for the facing sand, and recycled to form the backing sand. This is not a well-controlled material and optimum moulding and core-making performance requires precise control of grain size, binder and, where appropriate, water content.

Sand reclamation   The precision material required by a modern foundry requires a reclamation plant that will return pure, dry, uncoated, dust-free particles for mixing and moulding. There are two basic processes: one attrites the sand and burns off residual binders; the other relies on dynamic im-

pulses, grain against grain, that rub off binder coatings and remove them as dust (see Figure 16.134). In either case, 1% of sand only is lost.

The small usage of sand allows an alternative to silica sand to be used. Zircon sand with 99% reclamation gives lower cost than silica sand used conventionally. The heat capacity of the zircon sand produces a finer cast structure giving solidification rates in aluminium equivalent to those in permanent moulds. Zircon sand is more satisfactory for service with higher melting point materials. The thermal expansion of zircon sand is (see Figure 16.135) smaller and more consistent than that of silica, and the lower tolerances which result promote savings in weight of casting and finished component. Metred amounts of various kinds of binder (see below) are added.

Moulding   The processes used for sand moulding include squeeze, jolt, jolt squeeze and sand sling. A method which is claimed to give the most uniform and accurate compaction is the air impact process (see Figure 16.136). This process takes a metered quantity of sand to which bentonite clay has been added.

Modern core-making processes   Core-making differs from moulding because the core must be transferred into the mould, and must therefore be strong enough to be free standing while this process is carried out. (Typical cores are shown in Figure 16.137.)

Originally, cores were of silica sand bonded with molasses and linseed oil with bentonite to provide green strength for handling. Cores were packed in sand and baked to develop dry strength and to remove gases which would be evolved in heating.

The most modern development is to use a small precise addition of resin binder to recovered sand. Figure 16.138 illustrates a typical plant arrangement. A metered quantity of mixed sand is blown into a core box, the two halves of which are clamped together hydraulically. After filling, the binder is cured either by heat from a hot core box or by introducing an

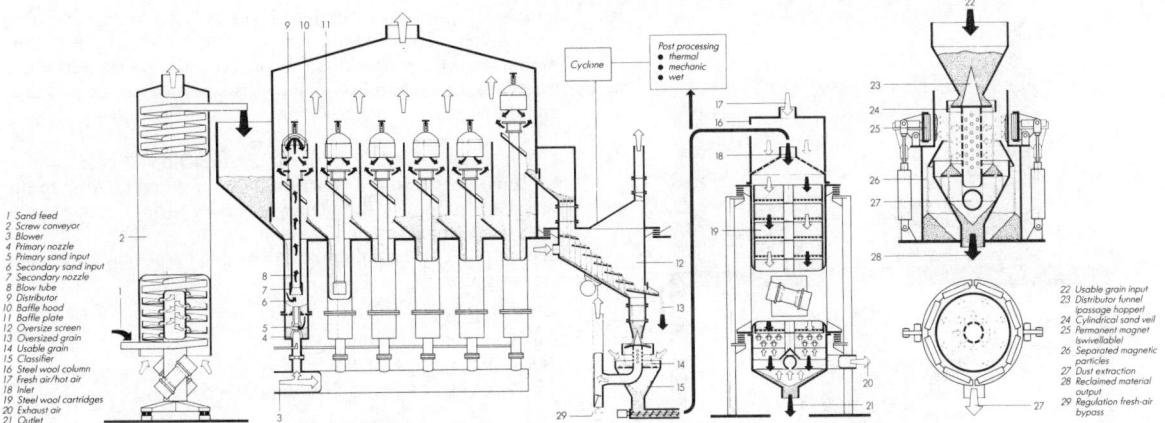

1 Sand feed
2 Screw conveyor
3 Blower
4 Primary nozzle
5 Primary sand input
6 Secondary sand input
7 Secondary nozzle
8 Blow tube
9 Distributor
10 Baffle hood
11 Baffle plate
12 Oversize screen
13 Oversize grain
14 Usable grain
15 Classifier
16 Steel wool column
17 Fresh air/hot air
18 Inlet
19 Steel wool cartridges
20 Exhaust air
21 Outlet

22 Usable grain input
23 Distributor funnel
   (passage hopper)
24 Cylindrical sand veil
25 Permanent magnet
   (swivellable)
26 Separated magnetic
   particles
27 Dust extraction
28 Reclaimed material
   output
29 Regulation fresh-air
   bypass

Figure 16.134 Principle and function of the mechanical/pneumatic reclamation of foundry sand (courtesy of George Fischer)

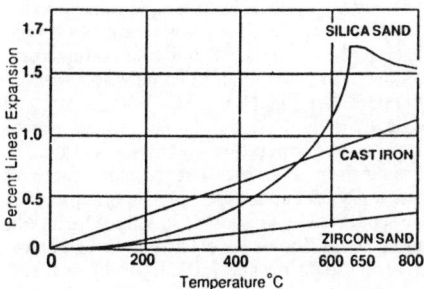

Figure 16.135 Thermal expansion of silica and zircon sands and cast iron

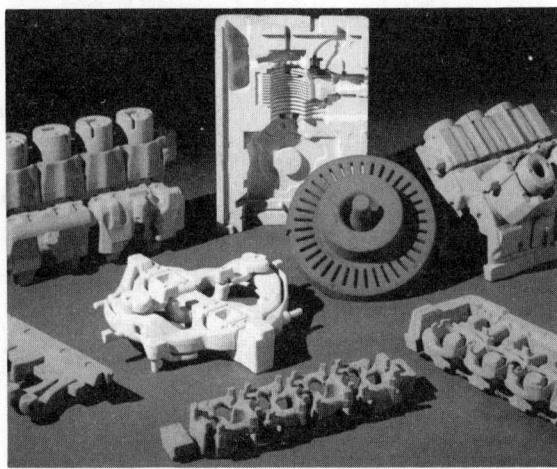

Figure 16.137 Typical cores (courtesy of George Fischer)

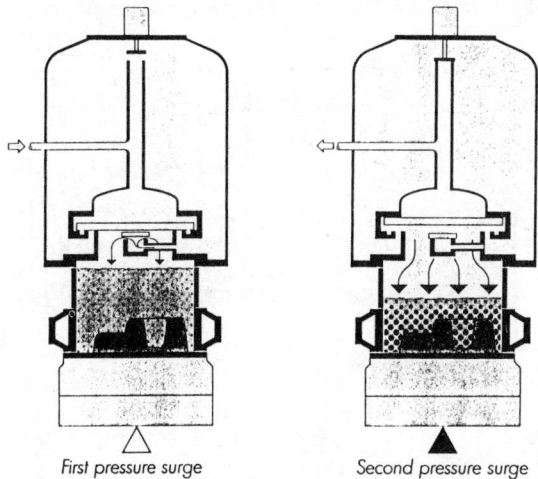

First pressure surge   *Second pressure surge*

Figure 16.136 Air-Impact-Plus moulding process (courtesy of George Fischer)

amine gas accelerator. Both processes are rapid. Up to 80 cores an hour can be made by the cold process on a machine of the type illustrated and the tolerances of the finished castings are the most precise that can be achieved by sand casting (see below). Cores may be located into the moulds either by a steel pin and bush system or by engineered sand to sand locations.

Dressings for mould and core The size of the grains of a moulding sand is a compromise. A large grain size is desirable so that gases evolved near the mould surface are able to diffuse away and not cause blowholes in the casting. A small grain size is required to produce a good surface finish. Where these objectives are incompatible a dressing must be applied to the surface of the mould or core. This can consist essentially of a finer-grain version of the moulding sand but, where a high melting point metal is to be cast, the use of a highly refractory base such as zircon flour has the great advantage of preventing 'burn-on' of the mould onto the casting.

There are a number of proprietary dressings, mostly consisting of suspensions of refractory powders in liquid, which can

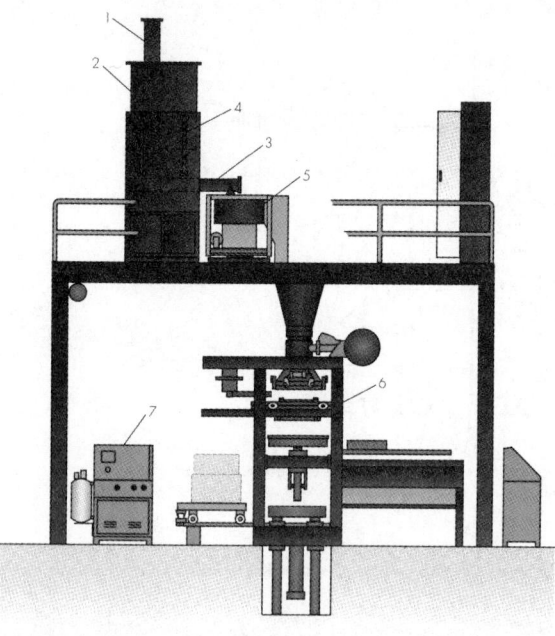

1	INCOMING SAND
2	CORE SAND HOPPER
3	SAND METERING
4	BINDER ADDITION
5	SAND MIXER
6	CORE MACHINE
7	GAS GENERATOR

Figure 16.138   A typical core-making system suitable for hot or cold process (courtesy of George Fischer)

be sprayed or brushed onto the mould. It is essential to ensure that the mould is dried after the dressing has been applied.

A recently developed process which is claimed to give satisfactory results projects a dry powder based on zircon sand towards the mould surface by means of a gun. This charges the particles electrically by contact with its barrel and thereby ensures a uniform adherent coating by electrostatic charge.

16.5.4.4   Specific sand casting processes

Green sand casting   Green sand casting uses a clay bonded sand as described above for the mould, with (where required) a dry sand core. The process is very versatile in the range of materials cast, size and shape of castings, and quantities required. It is also economical and production can start as soon as a pattern becomes available. The sand must be a compromise between the best possible finish, which requires a fine grade, and the possibility of gas evolved from the mould causing blowholes, which requires a relatively coarse grade (unless a mould dressing, which requires an additional, probably manual process, is used) – see above.

Core assembly processes   The core assembly process is similar to green sand but the mould consists of one or more cores made as described above. The 'mould' is stronger and to-

lerates long, thin projections. There is less risk of gas evolution which may cause blowholes than from a green sand mould. However, a number of cores cost more to produce and assemble than a mould which consists only of a cope and drag and more parting line flash may result.

Cold set sand mould   With the introduction of cold catalyst hardening resins the distinction between mould and core has vanished (except in the actual geometry of production). The tolerances which can be worked to are the smallest of any sand casting process. For aluminium cast in zircon sand tolerances are as low as

+0.15 mm up to 100 mm
+0.20 mm up to 300 mm
+0.25 mm up to 800 mm
+0.10 mm across each core joint.

These close tolerances permit much greater precision in design and make possible substantial savings in weight of finished component, in weight of metal cast and also in cost of finishing processes.

The Cosworth Process[96]   The Cosworth Process, developed for aluminium but equally applicable to magnesium and zinc alloys, takes advantage of the close tolerances and high cooling rate conferred by zircon sand and cold setting resin mould and core systems. It uses them in conjunction with an upward pouring system (see Figure 16.139).

Molten metal is held in a large holding furnace which smooths out variations in temperature and analysis. Metal is degassed by bubbling through inert gas at the ingot charging station, and the metal is held under inert gas. Oxide particles and inclusions are separated by floating or sinking. The mould is filled from below by means of an electromagnetic pump. Metal enters the casting smoothly and without turbulence, and the pressure head is maintained during feeding. When feeding is complete, the pressure head is removed and the liquid metal, in what in a conventional casting would be the feeder head, runs back into the bath. This results in a greatly improved metal yield.

To summarize, the advantages claimed for the Cosworth Process compared with the conventional green sand casting process are:

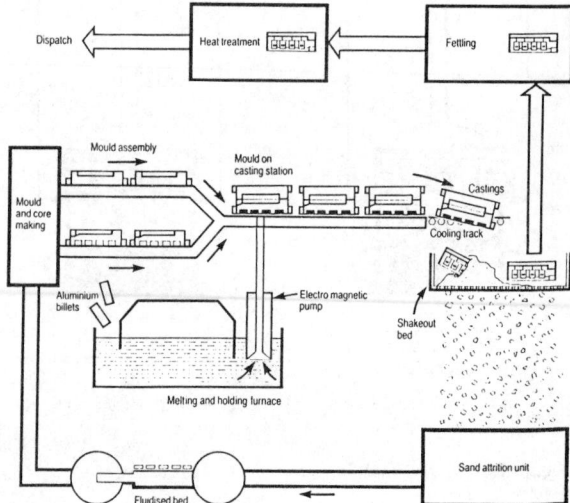

Figure 16.139   The casting unit of the Cosworth Process

- A finer metal structure with tensile strength improved by 30% and elongation almost doubled.
- Improved integrity, pressure-tightness and freedom from porosity inclusions and casting defects.
- Increased dimensional accuracy which results in saving metal, component weight and machining time.
- Additional saving in metal melted because metal in runners and feeders is saved.

Premium-quality castings It has been claimed that metal quality in aluminium castings may be further improved by varying the composition of the material of the mould (so that very high cooling rates are obtained locally) and by optimizing gating and feeding. By these means, castings have been produced with properties equivalent to those of forgings, but they are very expensive.

Other sand casting processes Green sand and core assembly processes produce the vast bulk of sand castings but there are a number of other processes which require mention because they have been, and in some cases still are, extensively used. They are:

1. *Dry sand moulding*: This process starts with a green sand mould which is baked in a stove (or by blowtorches) until it becomes hard. It was used in the past for moulds for very large castings in which the metal pressure might wash away green sand mould material. Drying also makes sure that moisture does not cause blowholes. Dry sand moulding has almost been superseded by the development of high-pressure moulding which confers the requisite strength on green sand.
2. *Loam moulding*: Loam moulding uses very moist green sand which, instead of being formed on a pattern which is relatively expensive and time consuming to manufacture, is formed to a profile against a 'strickle' which is rotated on a spindle. The mould is later dried to remove excess moisture.

Randupson process The Randupson process uses a mixture of sand with cement and water for larger sized castings. The moulds take about 24 hours to harden and the material can be crushed and re-used. However, this causes a substantial dust hazard.

Sodium silicate process This is a cold-hardening process which bonds the sand with sodium silicate that is hardened by passing CO_2 through the mould. The process is used for both moulds and cores and has been developed by adding hardening agents such as glyceryl acetate, cement, ferro-silicon and calcium silicate to the sand immediately before moulding. The process gives higher strength than green sand without the need for storing. Tolerances are better than for green sand, production rate is high and materials are cheap. Steel castings may tend to burn-on, cores are difficult to knock out and there is a problem with dust.

Fluid and castable sand process Sand with sodium silicate and a self-setting agent is suspended in a detergent foam poured around the pattern and allowed to set. No ramming or drying is necessary because of the low water content of the sand, but the process can only be used for simple shaped castings. The mould needs dressing to produce a satisfactory finish and a licence is required to operate the process.

16.5.4.5 Expendable pattern processes

A number of casting processes derive the shape of the finished casting not from a re-usable pattern but from a material which is removed by the action of heat before or during casting. This eliminates the limitations which arise from the need to withdraw a pattern. The mould (or shell, see below) can be made in one piece so that drafts can be eliminated (unless they are needed to make the pattern which may, however, be flexible). Parting-line mismatch and flash are eliminated, and undercuts are easier to deal with.

On the other hand, the operation of making the pattern occurs once for every casting instead of once per component design. This may be satisfactory for prototypes, or one-off castings, for which the expendable pattern may be cheap and easy to produce, and for large runs where the expendable pattern is mass produced. The additional operation may prove costly for intermediate quantities.

The full-mould expendable-pattern process This uses lightweight, foamed plastic patterns moulded or machined and, if required, assembled by cementing sections together. The pattern, incorporating runners and risers, is surrounded by any form of cold setting sand, or by dry unbonded sand, and is removed as a vapour by the heat of the incoming metal. This forms an interface with the evaporating plastic and prevents collapse of the mould even though the sand is unbonded.

16.5.4.6 Shell-moulding techniques

These utilize, instead of a solid mould, a shell a few millimetres thick. This may be either one-piece formed around an expendable pattern, or two-piece formed around a solid pattern.

In the 'Croning' process sand coated with a thermosetting resin is projected onto a metal pattern heated to 250°C. After removal from the pattern, the halves of the shell are fixed together, possibly backed with sand or steel shot, and the metal poured.

In the 'Shaw' process the pattern is dip-coated with a mixture of a fine refractory such as sillimanite and ethyl silicate. The bond produced by this mixture is rubbery when green, which makes stripping from the pattern easy. The mould is flash fired to remove alcohol, fired at 800–1000°C and poured.

Both these processes give good reproduction of details and exceptional accuracy of dimensions, so finishing operations are minimized. Patterns are, however, expensive; size of part is limited; and large runs in the order of 500 castings are needed to make them economic.

16.5.4.7 Casting processes incorporating both shell moulding and expendable patterns

'Investment', 'precision' or 'lost wax' casting[97,98] Lost wax casting is employed to produce components to any conceivable shape in nickel, copper, titanium and aluminium alloys, steel and precious metals. It is not limited by the need to withdraw a pattern. There are no mould parting lines and very fine detail may be reproduced. The surface smoothness of the casting is 1–3 μm rms and dimensioned tolerances as low as ISO 10 IT. For a casting process lost wax is expensive but the cost of a finished component (if indeed it can be made in any other way) is competitive.

The manufacture of weapons, tools and ornaments by lost wax casting continued, with only minor changes, for almost 6000 years up to the introduction, in 1932 by Austenal Laboratories Inc., of hydrolysed ethyl silicate binder for quartz dental moulds. Development was intensified early in the Second World War by the requirements for blades first for aircraft turbosuperchargers and later for aircraft turbines. The high creep strength alloys are difficult to forge and machine and cast blades with a structure of equiaxed crystals, columnar crystals with a preferred orientation, or one single crystal[97] have much superior creep resistance to forged blades.

The investment casting process employs a model or pattern of the object to be cast in wax (or an alternative material such as frozen mercury, a low melting point metal or a polymer) which can easily be removed from the mould which is formed round it. The wax pattern(s) may be made in robot moulding machines with up to 100 tonnes clamping pressure and assembled singly or in clusters with casting gates and/or runners and vents. They are next dipped repeatedly in coating materials consisting of refractory slurry and liquid binder.

There has been several variants of coating materials which use either alkaline silica sol or acidic alcoholic hydrolysed ethyl silicate air dried and ammonia set binders. In some processes these two methods are used for alternate coats. At the time of writing, however, the rapid ammonia set process has been most widely adapted. The filler is fine silica for the primary coat. Subsequent coatings have a higher ratio of liquid binder to solid refractory and the gradings become progressively coarser.

Moulds, typically 8 mm thick, are dried at room temperature and then plunged into furnace chambers at 1050°C. This melts the surface wax before the main body has time to expand and, after dewaxing, cures the ceramic mould.

Some aircraft turbine blades require to be cast hollow to permit the passage of cooling air. These are made by forming the wax round a core of strong ceramic the same shape as the cavity required in the cast component. The completed shell mould may be inserted free-standing in the casting furnace or may be supported by dry granular refractory.

The actual casting may be carried out in normal atmosphere, *in vacuo*, or in a protective atmosphere. The pouring and cooling can be varied according to the material structure required[98, 99] (see Chapter 7, Section 7.4). Centrifugal assistance (see Section 16.5.6.3), or mould heating, may be incorporated to give improved casting detail. The ceramic core of a hollow casting is removed by chemical leaching.

In the Middle Ages the lost wax process was used to cast equestrian statues greater than life size, but the great bulk of modern castings are measured in inches. However, integral turbine wheels and vane rings are now cast in one piece and large structural aerospace components can be made in titanium or nickel base alloys.

The 'Replicast 20, 21' ceramic shell process[100]   The Replicast process resembles the full mould expendable pattern process in using an expanded polystyrene replica of the required article; and the precision casting process in using a ceramic shell. It has the flexibility of the lost wax process (including absence of parting lines, draft or tapers) but is claimed to be more suitable for larger castings because the patterns are light in weight, cheaper and less liable to distortion and fracture as size increases. In addition, the ceramic shells may be produced thinner (4–6 mm) and therefore cheaper because they are subjected to a much lower stress during removal of the polystyrene than is generated by the removal of wax. The shells are positioned on a box supported by loose sand, vibrated to fill all cavities, and then held under vacuum during casting. It is claimed that a casting can be made within 5 hours of the replica being removed from the die.

Dimensional tolerances and surface finish claimed for the Replicast process compared with other casting processes are illustrated in Figures 16.140 and 16.141. It is claimed that the reduction in casting wall section and the casting-in of the bolt holes of the casting shown in Figure 16.142 gave a weight saving in the castings delivered of 23%. Other benefits which derive from the Replicast process as compared with casting in a green sand mould are:

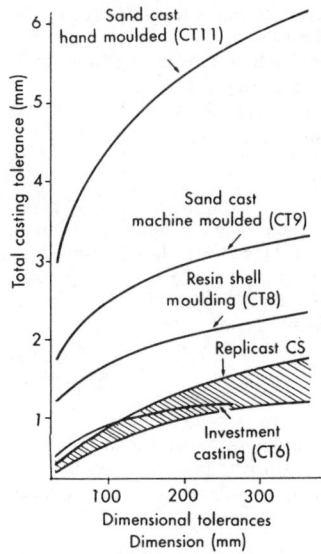

Figure 16.140   Dimensional tolerances achievable with the Replicast process compared with other processes

Method of casting	Roughness average (μm)											
	15.2	14.0	12.7	11.5	9.8	8.9	7.65	6.4	5.0	3.8	2.5	1.3
Investment casting										▬▬		
Replicast ceramic shell							▬▬▬					
Resin shell			▬▬▬									
Sand mould	▬▬											

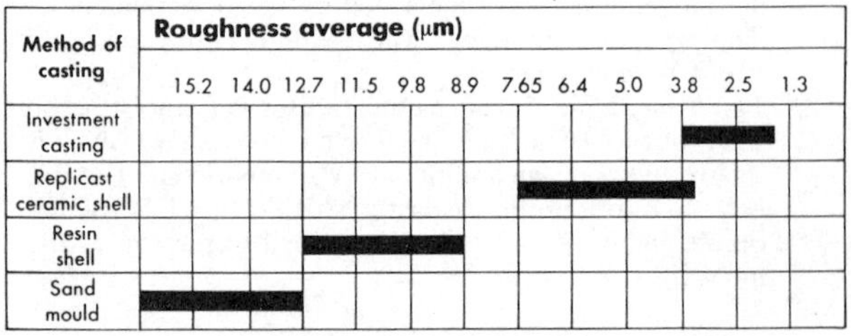

Figure 16.141   Surface finish achievable with the Replicast process compared with other processes

Figure 16.142   Replicast 8-inch ball valve body after shake-out from the mould and fine shot-blasting

1. Improved positioning of feeder heads is possible.
2. The chilling effect of the ceramic enhances feeding and is claimed to eliminate hot tears.
3. The strength and integrity of the ceramic are claimed to eliminate inclusion defects and gas evolution.

The major advantage claimed is, however, the reduction in fettling and machining compared with green sand casting. The limitations and drawbacks are problems in making parts in excess of 250 kg, high labour costs at front-end, higher die costs and a longer lead time for manufacture of new parts. As a consequence, Replicast is best suited to components of fairly complex geometry with stringent acceptance criteria and with a high machining content destined for high-duty applications. The Steel Castings Research and Trade Association (SCRATA) operate a demonstration foundry.

16.5.5 Permanent mould or 'die' casting

Die casting comprises the transfer of molten metal to a die which is a metal block, or assembly of blocks, containing a machined cavity which is the shape of the required casting together with a runner and feeder(s). There may be, inside the cavity, one or more metal or sand cores. Die casting imposes a quicker rate of cooling on the metal as it solidifies than sand casting and produces a finer structure with improved tensile and ductile properties. A dressing is usually applied to the die runner and riser. This may be based on a refractory such as alumina, silica or zirconia where a decreased chill is required, or graphite where a high chill is beneficial. In either case a binder which may be based on sodium silicate is required. The dimensional accuracy of die castings is generally superior to that of sand castings. This leads to savings in metal and machining costs. Most metal alloys can be die cast, but the bulk of die castings are made in metals of medium melting point, mainly alloys of aluminium, zinc and magnesium.

16.5.5.1 Gravity die casting

In gravity die casting molten metal is poured into a die, usually from a hand ladle. Moulds are usually two part split vertically and locked together during casting. Figure 16.143[101] shows one half of a typical die for casting magnesium alloy. The pouring gate is located in the parting line of the die. The junction between the down-sprue and the gate may house either a metal filter or a weir to trap inclusions. The die is filled from the runner and the metal flows in horizontally to the metal in the die cavity to achieve directional solidification. Note that most metals are more fluid than magnesium alloys and will normally require smaller section runners. The feasibility of direct-pouring gravity-lie castings down the feeder using a filter unit such as is shown in Figure 16.106 is worth consideration.

Good mechanical properties significantly superior to those of sand castings can be achieved in well-designed and carefully produced die castings in aluminium, magnesium and zinc. For example, the properties of 'Y alloy' are 275 MPa UTS 1% elongation as compared with 220 MPa UTS 0.5% elongation sand cast, and 400 MPa 15% elongation wrought.

Gravity die castings may also be made in low melting point alloys, copper alloys and cast iron. The process is not normally applied to steel.

Die castings can be made in weights from 150 g to 200 kg but the optimum is from 1 to 50 kg, section thickness from 3 to 50 mm, ISO dimensional tolerances IT12–14, machining allowance roughly 1 mm, and surface smoothness 2.5–25 μm r.m.s.

16.5.5.2 Low-pressure die casting

Low-pressure die casting bears the same relationship to gravity die castings as the Cosworth process does to conventional sand casting. It also has the same advantages. It was developed initially to improve the quality of gravity die casting in aluminium and zinc alloys, and it could be applied to many other metals. Figure 16.144 illustrates the principle of the process.

The mould or die is sited above a bath of molten metal in a gas-tight chamber. A riser tube extends from well below the surface of the metal to the base of the mould cavity. Metal is constrained to rise smoothly and without turbulence until it completely fills the mould by using the pressure of the atmosphere above the melt, usually less than 69 kNm^{-2}. Solidification contraction is fed from the reservoir of molten metal under pressures in the rider tube. The pressure is released when solidification is complete in the mould and the molten metal in the tube drains back into the furnace.

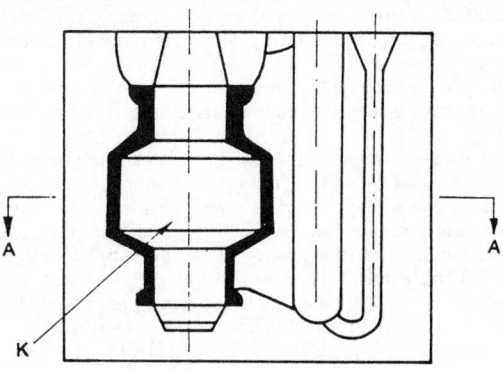

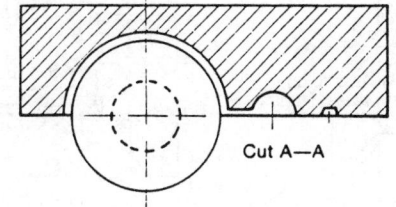

Cut A—A

Figure 16.143 A gravity die half with sand core

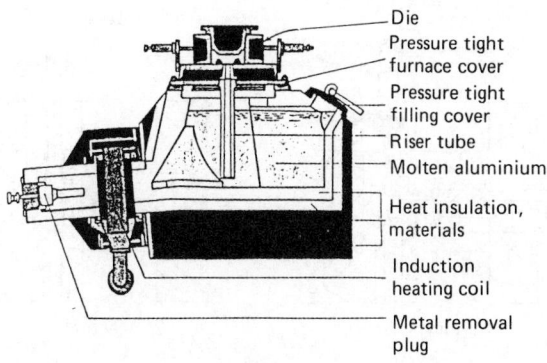

Die

Pressure tight furnace cover

Pressure tight filling cover

Riser tube

Molten aluminium

Heat insulation, materials

Induction heating coil

Metal removal plug

Figure 16.144 The principles of low-pressure die casting

Feeding is greatly improved compared with gravity die casting because of the improved thermal pattern, and this improves strength and pressure-tightness. The absence of turbulence and the metal source from below the surface reduce the chance of inclusions. Yield is very high (90–95%) because only a small butt of metal requires to be trimmed from the gate. This compares with 40–60% for gravity die casting. The process is very suitable for automation using, if required, multi-cavity moulds. The disadvantage compared with gravity die casting is the higher capital cost of the equipment.

16.5.5.3 High-pressure die casting

Alloys of the majority of metals can be high-pressure die cast. In its best-known form this process is known simply as 'die casting' and is used extensively for alloys of zinc, aluminium and magnesium. It is equally applicable to the low melting point metals tin, lead and cadmium.

The process has recently been extended to enable castings up to 3 kg in weight to be produced in carbon (up to 0.5%C) and stainless steels, copper and copper alloys, cobalt and air-melting grades of nickel alloys. There are three variants of the process: high-pressure cold chamber die casting; high-pressure hot chamber die casting; and the GKN 'Ferro Die' process for casting steel and other relatively high melting point metals.

There are two variants of cold chamber diecasting machine: the horizontal and the vertical, the working principles of which are illustrated in Figures 16.145 and 16.146.[101] In both variants, a metered quantity of metal is poured into the shot sleeve and injected into the die at 5–500 M Nm^{-2} pressure by a piston. When the metal has solidified, the die is opened and the casting ejected. It is desirable in the case of the horizontal machine that the piston should move in two stages; slowly at first to expel air, and quickly when the metal reaches the gate.

In hot chamber die casting (Figure 16.147) the shot tube is located below the surface of a reservoir of liquid metal and fills automatically when the plunger is withdrawn. This greatly increases the rate of production, and, presumably, reduces those defects caused by chill in the shot tube.

The hot chamber machine is simple, easy to operate and permits even higher production rates than the cold chamber machines. It operates at lower pressures and is particularly suitable for thin-walled castings in magnesium alloy and the lower aluminium content zinc alloys. Aluminium and the higher aluminium content zinc alloys attack the parts of the machine with which they come in contact so that they are not really suitable for hot chamber high-pressure die casting.

High-pressure die castings can be made in the weight range 0.1 g to 10 kg. They have an extremely good surface finish and a high dimensional accuracy (ISO tolerance IT 11–14). Smaller wall thicknesses may be cast (magnesium > 1.2 mm, aluminium > 0.7 mm, zinc > 0.2 mm) than by any other casting process. Pressure die castings require very little machining and their machining allowance may be as low as 0.25 mm.

Labour costs are low, but die costs and die preparation times high. Dies last up to 100 000 shots before failing (together with the shot tube and piston) by crazing. The process is only economic for large-scale production.

The most serious disadvantage of high-pressure die castings is the large amount of internal porosity. Castings have a so-called 'skin' 0.1–0.2 mm thick, formed by the rapid freezing of the first metal injected into the die cavity. Beneath this skin can be found spherical gas porosity, interdendritic porosity and 'flow line' porosity (see Figure 16.148[102]) which occurs between the skin and the central region of the casting.

The origin of these defects lies in the shot chamber where chilling of the metal can cause solidification to occur and large agglomerates of solid can cause blockages in the gate region. A chart which may be used to determine the possible causes of

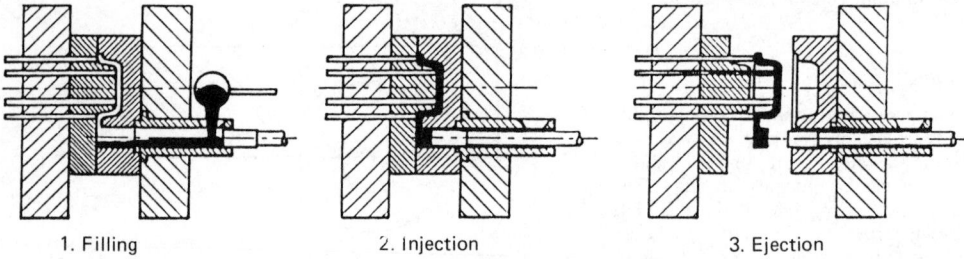

1. Filling   2. Injection   3. Ejection

Figure 16.145 Working principles of horizontal cold chamber high-pressure die casting machine

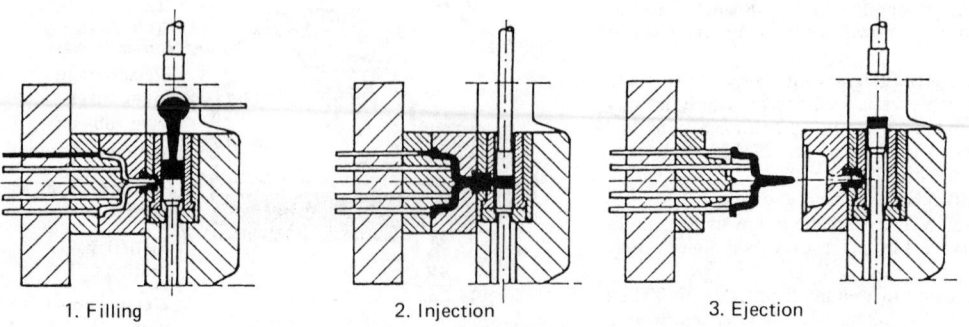

1. Filling   2. Injection   3. Ejection

Figure 16.146 Working principles of vertical cold chamber high-pressure die casting machine

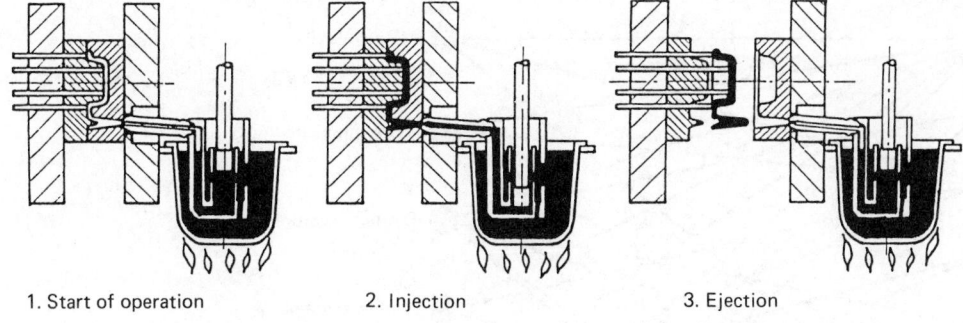

1. Start of operation 2. Injection 3. Ejection

Figure 16.147 Working principles of piston operated hot chamber high-pressure die casting machine

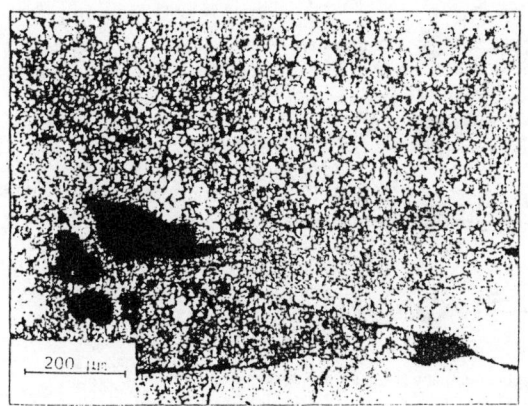

Figure 16.148 Flow line discontinuity with its associated porosity in aluminium alloy pressure die casting

specific defects in magnesium high-pressure die casting is shown in Figure 16.149.[102]

Better properties are achievable by other casting routes and this, and the porosity, limit the useful application of die castings to non-structural components. High-pressure aluminium alloy die castings cannot be heat treated or anodized because the porosity would cause blistering.

In general, high-pressure die castings are used in non-load-bearing situations such as zip fasteners, car door handles and carburettor bodies. Computer printer and other rapidly moving parts are die cast in magnesium alloys.

16.5.5.4 The GKN 'Ferro Die' process

The GKN 'Ferro Die' process is carried out on what is essentially a horizontal cold chamber machine modified by making the parts that come into contact with molten metal (including the die) of molybdenum. The ladle is replaced by a vertical tube furnace heated by an induction coil. A weighed slug of solid metal is placed in the furnace. As soon as the slug melts and falls into the shot tube it is injected into the die. The process is economic for the production of a minimum of 500 castings.

16.5.5.5 Squeeze casting

Squeeze casting gives (pending the development of rheocasting and thixocasting) the most favourable properties obtain-able by any method of casting. Squeeze cast products bear comparison in properties with forgings. It has so far been applied to alloys of aluminium and magnesium to manufacture castings and also metal matrix composites.

In squeeze casting the metal being cast is solidified under pressure between metal dies. In 'direct' squeeze casting a metered quantity of molten metal (which must be clean, free from gas and preferably filtered) is poured into the female die casting or mould and the male die is driven into the metal at a controlled speed so that it forms the required shape in a non-turbulent manner. Pressures between 50 and 150 MPa are maintained on the metal while it solidifies. The dies are coated with a graphite coat and heated. The process is illustrated graphically in Figure 16.150.[103]

Direct squeeze casting has the following advantages over all casting processes described so far:

- The resulting casting is completely free from gas and shrinkage porosity.
- The very high cooling rate which the applied pressure facilitates favours control of microstructure by means of temperature control of mould and liquid metal. Very fine-grain structures may be produced without the use of grain-refining additions.
- No feeders or risers are required.

In the indirect squeeze casting process, metal is injected (as in pressure die casting) into the die cavity by a small-diameter piston. One process arrangement is illustrated in Figure 16.151[103], but other methods have been proposed. Unlike pressure die casting, the pressure remains on throughout solidification, which takes place progressively from the part of the casting away from the punch. Although it is more difficult to obtain such good properties as with the direct process, high quality automobile parts have been produced.

In both direct and indirect versions of squeeze casting there are no internal or external defects which require NDT. Mechanical properties are as good (or sometimes better) than those of forgings. The production cycle times for squeeze castings are less by as much as 66% of those for comparable die castings because of the high heat transfer coefficients.

Many aluminium alloys can be produced with very substantially improved properties by squeeze casting. Table 16.18[103] shows how the properties of 7010 alloy vary with squeezing pressure. The effect of grain size on the fracture toughness of this alloy is shown in Figure 16.152.[103]

Figure 16.153[103] shows how the consistency of the SN curve of 70/mm grain size 7010 compares with that of plate material and also the way in which the fatigue strength varies with the grain size.

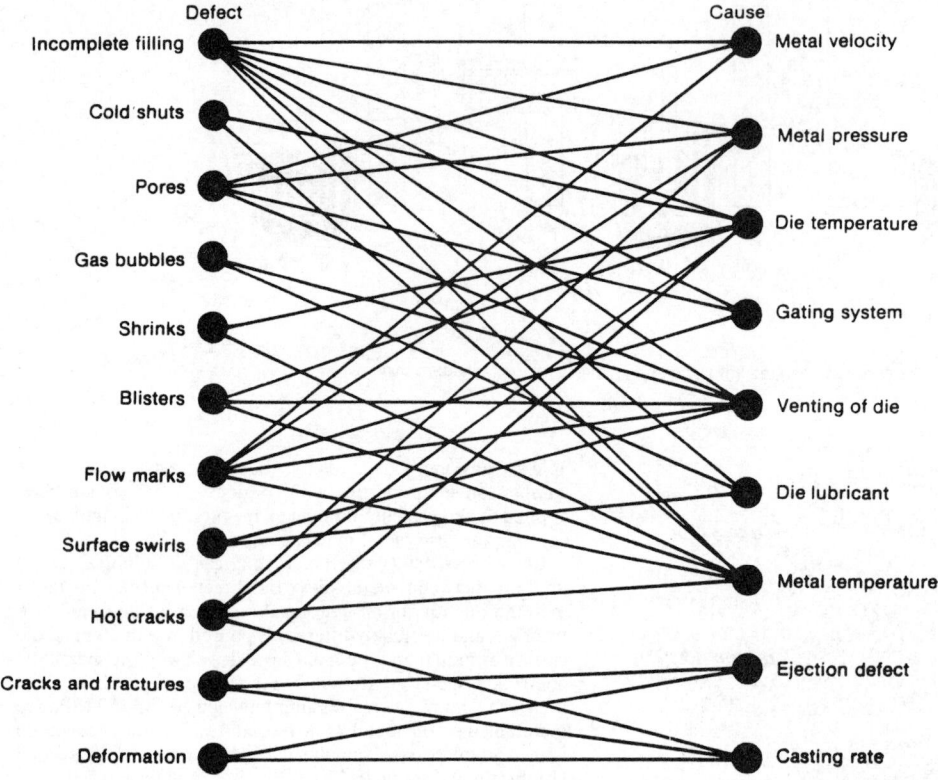

Figure 16.149   A chart which may be used to identify the causes of defects in magnesium alloy pressure die-casting

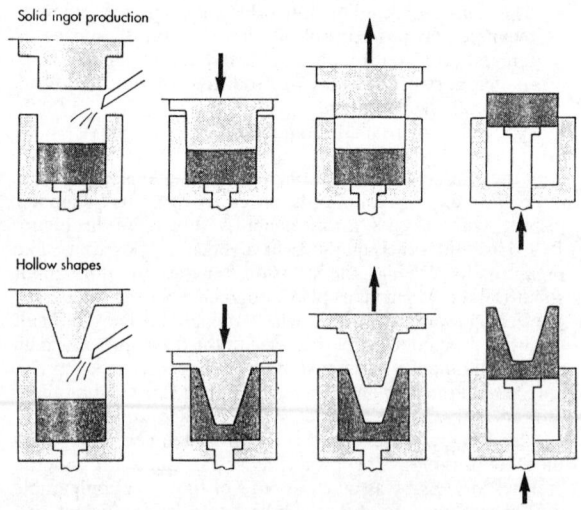

Figure 16.150   Production of solid ingot and hollow shapes by 'direct' squeeze casting

Magnesium alloys can be fabricated by squeeze casting and Figure 16.154 gives a comparison between squeeze and other casting processes in A91 alloy. The squeeze cast properties are

Table 16.18   Effect of casting pressure on the tensile properties of squeeze cast 7010 alloy

Casting pressure (MPa)	0.2% yield (MN m^{-2})	UTS (MN m^{-2})	Elong. (%)
0.1	474	524	5.5
50	480	551	11.1
100	485	551	12.2
300	475	548	11.9
Plate material	(L) 495 (ST) 519	(L) 568 (ST) 563	(L) 15 (ST) 7

better, but in this and in other cases squeeze casting does not appear to provide an advantage comparable with that obtainable with aluminium alloys. More research is required before the full potential of squeeze casting is realized.

16.5.5.6   Rheo-casting and thixo-casting[104, 105]

In both rheo-casting and 'thixo-casting' a slurry (prepared by stirring an alloy while it is in the semi-solid condition between the solidus and liquidus) is cast. Dendrites which may have formed as the alloy is solidifying are broken up and the slurry flows as a thixotropic liquid whose viscosity decreases as it flows but recovers when it is allowed to stand so that the material stiffens.

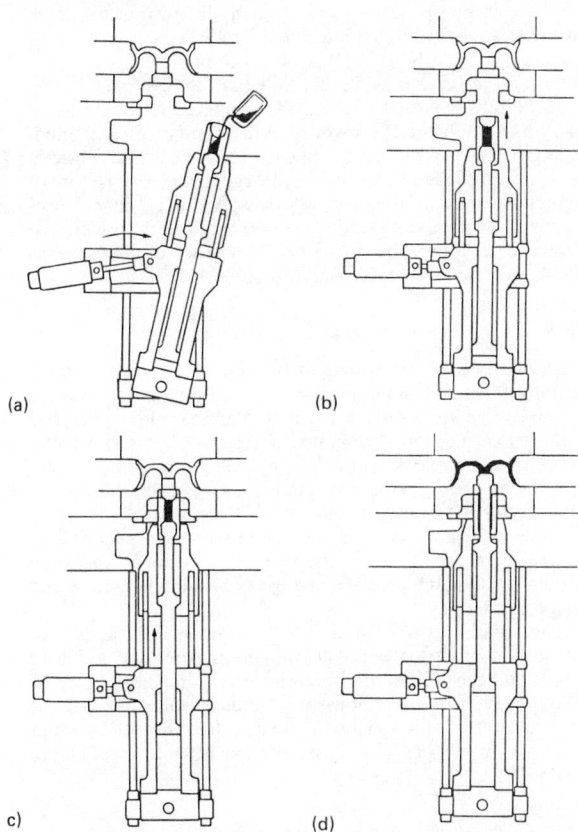

(a)

(b)

c)

(d)

Figure 16.151 An indirect vertical shot squeeze casting system (Ube) (a) Pouring aluminium. (b) Injection cylinder swivels back to vertical. (c) Sleeve holder rises with sleeve and sets in bottom die. (d) Plunger tip rises and injects

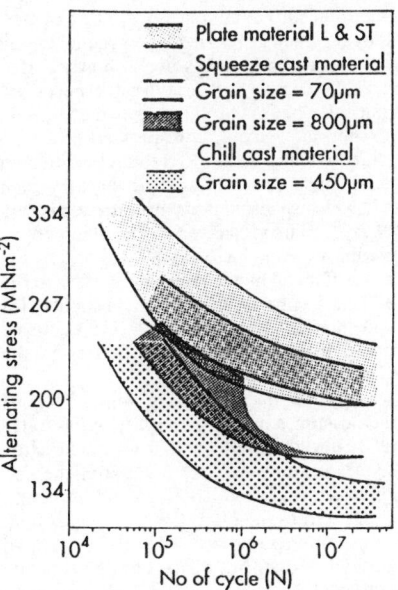

Figure 16.153 *S-N* curves for cast 7010 (peak aged) material and plate material

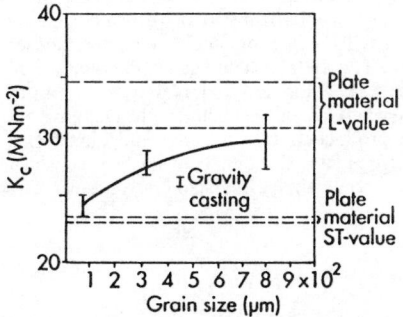

Figure 16.152 Effect of grain size on fracture toughness values (K_{IC}) of squeeze cast 7010

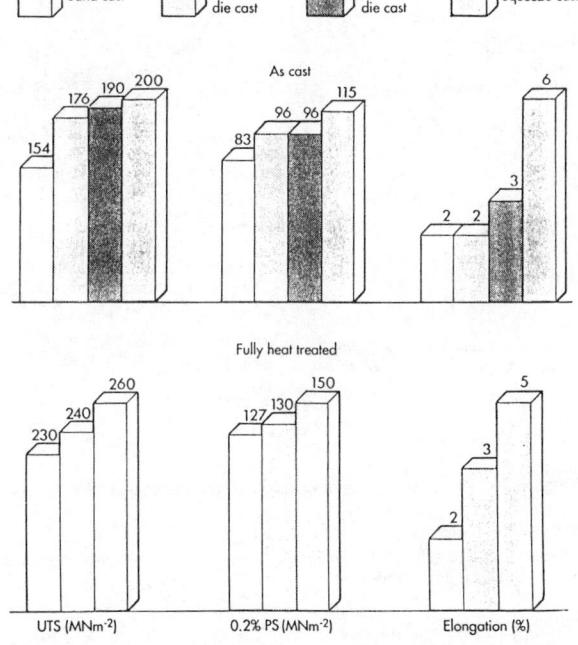

Figure 16.154 Comparison between properties produced by squeeze and other casting processes in A91 alloy

The first slurries produced, containing up to 60% of particles, were formed by mechanical stirring when the alloy had partially solidified. More recently it has been shown[106] that induction electromagnetic stirring will produce the necessary agitation. Slurries may also be prepared by a powder route in which elemental powders are mixed, cold pressed and then heated to a temperature at which an alloy (created by diffusion between the powders) melts. Intermittent stirring may then produce a structure in which spheroidal particles are contained within a liquid matrix.[107] Slurries may also be obtained by deformation, recrystallization and incipient melting and by gas atomizing a fine stream of metal in nitrogen or argon (as in the Osprey process). Whichever process is found to be suitable for the specific alloy will, if successful, produce a fine equiaxed structure which should lead to optimum properties in the resultant casting.

The casting processes are illustrated diagrammatically in Figure 16.155. In rheo-casting a metered quantity of the slurry may be poured directly into the shot chamber of a die-casting machine or into a squeeze casting die. In thixo-casting the slug is allowed to solidify as a billet which is cut up into unit charges. These are reheated to the inter-solidus/liquidus temperature and either die or squeeze cast. (The billet can, of course, be forged instead of cast, but thixo-forging is outside the scope of this section.)

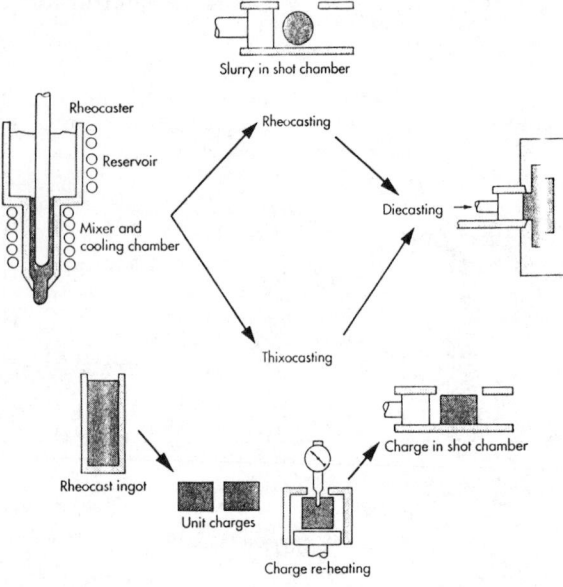

Figure 16.155 Rheo-casting and thixo-casting processes

These casting processes will normally lead to castings with properties significantly better than traditional casting techniques (particularly if thixo-casting is combined with squeeze casting), but there may be additional advantages.

If the powder route is used, the temperature to which the alloy must be heated is lower than if it is completely melted. This greatly eases the duties imposed on furnaces and refractories. This proved to be a great advantage in the manufacture of dental castings in titanium alloys where the temperatures required were about 1250°C as compared with 1650°C for complete melting. The properties of a series of alloys thixo- and conventional-cast are listed in Table 16.19.[108]

16.5.5.7 Casting metal matrix composites

Casting processes are among many which have been used for manufacturing metal matrix composites. Conventional casting is used to manufacture particulate alumina and silicon and tungsten carbides in aluminium, and short-fibre and whisker alumina, silicon carbide and carbon in aluminium. The volume fraction is limited to approximately 20% by the increase in viscosity of the molten metal.

Squeeze casting is used for the Toyota car piston[109] as described in Chapter 7, Section 7.6 (Composites). The main problem is the high pressure involved which limits the size and shape produced.

The liquid pressure forming (LPF) process may be used to produce composites with aluminium, magnesium and lead alloys and the reinforcing materials listed in Table 16.20.[109] The process is based on low-pressure die casting and uses gas pressure to force molten metal into a preform housed in a split steel die which has previously been evacuated (see Figure 16.156[109]).

16.5.6 Centrifugal casting processes

Three types of casting process rely on centrifugal force rather than gravity or pressure to impel molten metal into a mould or die. This has the advantage that a force which can be greater than gravity, and which may be increased or decreased at will, is available to run and feed the casting.

16.5.6.1 Centrifugal casting

Centrifugal casting produces hollow cylinders (generally of cast iron or steel) by pouring molten metal into a launder projecting into the end of a horizontally or vertically rotating cylinder. This is usually made of steel but sometimes of graphite or ceramic. The metal is held against the mould wall by centrifugal force. Feeding is good and dross tends to migrate to the bore where it can be machined off. Dimensional accuracy is good; production rate is rapid and dimensions range from very large pipes and cylinders to small 'pots' which can be machined to form cylinder liners and piston rings. The

Table 16.19 Mechanical properties of thixo-cast and die-cast components in titanium alloys compared

Composition (wt%)	Condition	0.1% PS (MN m^{-2})	UTS (MN m^{-2})	Elong. (%)	Hardness (VPN)
Ti-20Co	Conventional	139	454	1.4	474
Ti-20Co	Thixo-cast	168	486	7.4	480
Ti-20Cu	Conventional	121	162	1.9	350
Ti-20Cu	Thixo-cast	126	170	9.5	375
Ti-17Cu-8Co	Conventional	183	367	1.2	390
Ti-17Cu-8Co	Thixo-cast	212	388	8.8	408

Table 16.20 Range of reinforcing materials and matrix materials used in the LPF process

Continuous fibres	Discontinuous fibres	Particulate	Ceramic
(Max. $V_f = 0.6$)	(Max. $V_f = 0.3$)	(Max. $V_f = 0.6$)	(Max. $V_f = 0.8$)
Alumina	Alumina	SiC	Alumina
SiC (10 μm)	SiC(w)	–	Zirconia
Carbon	–	–	Carbides
Boron (140 μm)	–	–	–
SiC (125 μm)	–	–	–
Glass	–	–	–
Basalt	–	–	–

Matrix materials
Al alloys – 1050, 2024, LM16, LM25, 6061, LM5, LM10
Also Mg alloys and Pb alloys

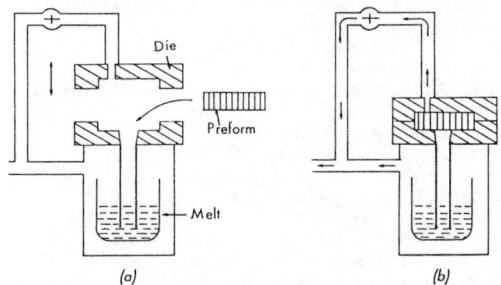

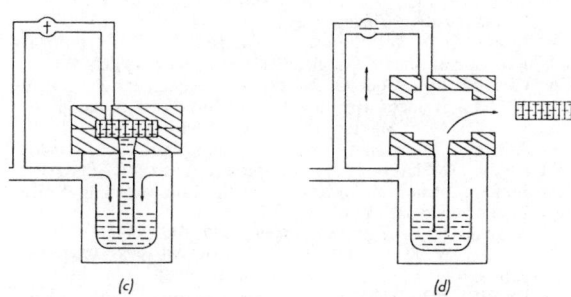

Figure 16.156 The liquid pressure-forming process. (a) Preform placement; (b) evacuation; (c) pressurization and infiltration; (d) ejection

casting shape is, however, limited to cylinders and the equipment is expensive.

16.5.6.2 Semi-centrifugal casting

Semi-centrifugal casting produces rotationally symmetric castings, the moulds for which can be stacked vertically and poured in the common axis while rotating. The centrifugal force available makes this very suitable for pulleys and gear blanks which require the best quality metal on the periphery. This process is very suitable for the production of gears and its use should increase.

16.5.6.3 Centrifugally assisted casting

In this process a number of moulds, often of the investment type, are located symmetrically round a crucible. When the metal has been melted the assembly is rotated at a controlled speed so that the metal runs smoothly into the moulds. When the moulds are filled the speed is increased so that detail is well reproduced and the castings are satisfactorily fed. The process is excellent for casting small articles of jewellery.

References

Most of the British Standards referred to here will have been revised in the early 1990s and will be European EN Standards although the BS numbers will be unchanged.

1 BS 1296: 172. Specifications of single point cutting tools. British Standards Institution, London
2 Stabler, G. V., 'The fundamental geometry of cutting tools', *Proc. Inst. Mech. Engrs*, **165**, 14 (1951)
3 Crookall, J. R. and Richardson, D. B., 'Use of photographed orthogonal grids and mechanical quick stopping techniques in machining research', *Conference on Photography in Engineering*, Institution of Mechanical Engineers, London (1969)
4 Trent, E. M., *Metal Cutting*, Butterworths, London (1977)
5 Taylor, F. W., 'On the art of cutting metals', *Trans. ASME*, **28**, 31 (1907)
6 Brewer, R. C., 'On the economics of the basic turning operation', *Trans. ASME*, **80**, 1479 (1958)
7 Brewer, R. C. and Rueda, R., 'A simplified approach to the optimum selection of machining parameters', *Engineers' Digest*, **24**, 9 (September 1963)
8 Tobias, S. A., *Machine Tool Vibrations*, Blackie, London (1965)
9 Pearce, D. F. and Richardson, D. B., 'Improved stability in metal cutting by control of feed and tool/chip contact length', *Joint Polytechnic Symposium on Manufacturing Engineering*, Leicester (1977)
10 BS 499: Part 1: 1983, Welding Terms and Symbols Part 1. Glossary for welding, brazing and thermal cutting
11 BS 639: 1986, Covered Carbon and Carbon Manganese Steel Electrodes for Manual Metal Arc Welding
12 AWS A.5.1–81, Covered Carbon Steel Arc Welding Electrodes
13 John, R. and Ellis, D. J., 'AC or DC for manual metal arc', *Metal Construction*, **14**, No. 7, 368 (1982)
14 *Manual Metal Arc Welding (mild steel with low-current sets)*, video, The Welding Institute, Abington, Cambridge
15 BS 2493: 1985, Low Alloy Steel Electrodes for Manual Metal Arc Welding

16 AWS A5.5–81, Low Alloy Steel Covered Arc Welding Electrodes
17 BS 2926: 1984, Chromium and Chromium–Nickel Steel Electrodes for Manual Metal Arc Welding
18 AWS A5.4–81, Covered Corrosion-Resisting Chromium and Chromium–Nickel Steel Welding Electrodes
19 AWS A5.11–83, Nickel and Nickel Alloy Covered Welding Electrodes
20 AWS A.5.6–84, Covered Copper and Copper Alloy Arc Welding Electrodes
21 AWS A.5.3–80, Aluminium and Aluminium Alloy Covered Arc Welding Electrodes
22 AWS A.5.15–82, Welding Rods and Covered Electrodes for Cast Iron
23 British Steel Corporation, Corporate Engineering Standard (CES) 23, Part 1, Supplement, Consumables, Weld Deposited Surfaces
24 AWS A5.15–82, Welding Rods and Covered Electrodes for Cast Iron
25 BS 4165: 1984, Electrode Wires and Fluxes for the Submerged Arc Welding of Carbon Steel and Medium-tensile Steel
26 Davis, M. L. E., *An Introduction to Welding Fluxes for Mild and Low Alloy Steels*, The Welding Institute, Abington, Cambridge (1971)
27 Jones, S. B., 'Variations on submerged arc welding', *WI Research Bulletin*, **15**, No. 3, 67 (1974)
28 Kennedy, N. A., 'Narrow gap submerged arc welding of steel, Part 1. Applications', *Metal Construction*, **18**, No. 1, 687 (1986); 'Part 2. Equipment, consumables and metallurgy', *Ibid.*, **18**, No. 12, 765 (1986)
29 BS 2901, Filler Rods and Wires for Gas-shielded Arc Welding
   Part 1: Ferritic Steels
   Part 2: Austenitic Stainless Steels
   Part 3: Copper and Copper Alloys
   Part 4: Aluminium and Aluminium Alloys and Magnesium Alloys
   Part 5: Nickel and Nickel Alloys
30 Lucas, W., 'Synergic pulsed MIG welding – process, equipment and applications', *FWP Journal*, **25**, No. 6, 7 (1985)
31 Lucas, W., 'Microcomputer control in the control of arc welding equipment', *Metal Construction*, **17**, No. 1, 30 (1985)
32 Norrish, J., 'What is synergic MIG?' *Welding and Metal Fabrication*, **55**, No. 5, 227 (1987)
33 *MIG Welding (mild steel with low-current sets)*, video guide, The Welding Institute, Abington, Cambridge
34 Weston, J., 'Arc welding robots – a welding engineering viewpoint', *Proceedings of a Welding Institute Conference on Developments in Mechanical, Automated and Robotic Welding*, London (1980): The Welding Institute, Abington, Cambridge
35 Kennedy, N. A., 'Robotics for welding engineers'
   '1. Introduction', *WI Research Bulletin*, **26**, No. 7, 221 (1985)
   '2a. Programming methods', *Ibid.*, **26**, No. 9, 302 (1985)
   '2b. Programming languages', *Ibid.*, **26**, No. 10, 334 (1985)
   '3a. Hardware for robotic arc welding', *Ibid.*, **26**, No. 12, 412 (1985)
36 Weston, J. (ed.), *Exploiting Robots in Arc Welded Fabrication*, The Welding Institute, Abington, Cambridge (1988)
37 BS 7084: 1989, Carbon and Carbon-Manganese Steel Tubular Cored Welding Electrodes
38 AWS A5.20–79, Specification for Carbon Steel Electrodes for Flux-cored Welding
39 AWS A5.29–80, Specification for Low Alloy Steel Electrodes for Flux-cored Welding
40 *Welding Handbook*, Vol. 2, 7th edn, American Welding Society, p. 592 (1980)
41 *Welding Handbook*, Vol. 3, 7th edn, American Welding Society, p. 459 (1980)
42 Brooker, H. R. and Beatson, E. V., *Industrial Brazing*, 2nd edn, Newnes-Butterworth, London, p. 263 (1975)
43 *Soldering Manual*, American Welding Society, 2nd edn, p. 149 (1977)
44 BS 219: 1977, Soft Solders
45 Lotta, A. J., *Connections in Electronic Assemblies*, Marcel Dekker, New York, p. 277 (1985)
46 Woodgate, R. W., *The Handbook of Machine Soldering*, John Wiley, Chichester, p. 224 (1983)
47 BS 1845: 1984, Specification for Filler Metals for Brazing
48 BS 1723: Part 2: 1986, Guide to Brazing
49 *Brazing Manual*, American Welding Society, p. 309 (1976)
50 Hinkel, J. E., 'Joint designs can be both practical and economical', *Welding Journal*, **49**, No. 6, 449 (1970)
51 Doherty, J., 'Costing methods for arc welding', *WI Research Bulletin*, Nos 1–3, 7, 35 and 73 (1968)
52 Jack, J. T., 'Controlling the cost of welding', *The Australian Welding Journal*, **14**, No. 4, 13 (1970)
53 McMahon, B. P., 'The price of welding', *Welding & Metal Fabrication*, **58**, Nos 1 and 2, 4 and 58 (1970)
54 Reynolds, D. E. H., 'Decreasing welding costs in heavy fabrication', *Welding and Metal Fabrication*, **42**, Nos 3 and 5, 94 and 185 (1974)
55 Linblad, L., 'The economics and methods of automatic welding', *British Welding Journal*, **13**, No. 5, 269 (1966)
56 *Welding Handbook*, Vol. 5, 7th edn, American Welding Society, p. 444 (1984)
57 Lees, W. A. (ed.), *Adhesives and the Engineer*, Mechanical Engineering Publications, London (1989)
58 Bolger, J. C., in Patrick, R. L. (ed.), *Treatise on Adhesion and Adhesives*, Vol. 3, Marcel Dekker, New York, (1973)
59 MacDonald, N. C., 'Codes of practice and specifications', *Design and Assembly with Engineering Adhesives*, Cranfield Institute of Technology, May (1983)
60 Lees, W. A., *Adhesives in Engineering Design*, The Design Council, Springer-Verlag, London (1984)
61 Shields, J., *Adhesives Handbook*, 3rd edn, Newnes-Butterworths, London (1984)
62 Adams, R. D. and Wake, W. C., *Structural Adhesive Joints in Engineering*, Elsevier Applied Science, London (1984)
63 Zisman, W. A., in Lee, L.-H. (ed.), *Polymer Science and Technology*, Vol. 9A, Plenum Press, New York, p. 55 (1975)
64 Wake, W. C., *Adhesion and the Formulation of Adhesives*, 2nd edn, Applied Science, London (1982)
65 Kinloch, A. J., *Adhesion and Adhesives: Science and Technology*, Chapman and Hall, London (1987)
66 Kinloch, A. J., *Durability of Structural Adhesives*, Applied Science, London (1983)
67 Snogren, R. C., *Handbook of Surface Preparation*, Communication Channels, Atlanta (1974)
68 Brewis, D. M. (ed.), *Surface Analysis and Pretreatment of Plastics and Metals*, Applied Science, London (1982)
69 CP 3012: 1972, Code of Practice for Cleaning and Preparation of Metal Surfaces, British Standards Institution, London
70 BS 5350: 1976, Methods of Test for Adhesives, Part A1: Adherend Preparation, British Standards Institution, London
71 ASTM D2651, Preparation of Metal Surfaces for Adhesive Bonding, American Society for Testing and Materials, Vol. 22, Philadelphia (1979)
72 Kinloch, A. J., *J. Materials Science*, **17**, 617 (1982)
73 Bascom, W. D. and Hunston, D. L., in Allen, K. W. (ed.) *Adhesion 6*, Applied Science, London (1982)
74 BS 5350: 1976, Methods of Test for Adhesives, Part C, British Standards Institution, London
75 Skeist, I. (ed.), *Handbook of Adhesives*, 3rd edn, Van Nostrand Reinhold, New York (1990)
76 ESDU 81022, Engineering Sciences Data Unit, London (July 1981)
77 Albericci, P., in Kinloch, A. J. (ed.), *Durability of Structural Adhesives*, Applied Science, London, p. 317 (1983)
78 ESDU 80039, Engineering Sciences Data Unit, London (December 1980)
79 ESDU 79016, Engineering Sciences Data Unit, London (September 1979)
80 Kinloch, A. J. (ed.), *Structural Adhesives: Developments in Resins and Primers*, Elsevier Applied Science, London (1986)
81 Anon., *Safe Handling of Adhesives and Sealants in Industry*, 2nd edn, British Adhesives and Sealants Association (1988)
82 Schliekelmann, R. J., *Int. Conf. on Structural Adhesives in Engineering*, Bristol University, Institution of Mechanical Engineers, London, p. 241 (1986)
83 Guyott, C. C. H., Cawley, P. and Adams, R. D., *J. Adhesion*, **20**, No. 2, 129 (1986)

84 Chadwick, G. A., 'Castings – current practice and future potential', *Metals and Materials*, **2**, No. 11, November, 693–697 (1986)

85 Morgan, P. C., Waterworth, P. W. and Davies, I. G., in *Macrosegregation in Killed Steel Ingots*, Paper 70 'Solidification technology in the foundry and casthouse', The Metals Society (September 1980)

86 Doring, K. and Wiesinger, H., 'Continuous casting and rolling of thin slabs', *MPT Metallurgical Plant and Technology International*, No. 5, pp. 16–29 (1990)

87 Islam, N., 'Developments in the continuous casting of steel illustrated by Davy Distington single slab jumbo slab caster', *Metals and Materials*, **5**, No. 7, July (1989)

88 Branion, R. V., 'Mould fluxes for continuous casting', *ISS 69th Steelmaking Conference*, Washington, DC (1986)

89 Haslio, M., Tozake, Y., Watanabe, T. and Yamashita, Y., 'Improvement in centre line segregation in continuously cast slabs', Concast Workshop on Slab Casting, Kashina (1987)

90 *Continous Thin Slab Casting and Rolling Technology*, Voest-Alpino Industricanlegenbau

91 Merchant, H. D., Kattnais, T. Z. and Moncis, J. G., 'Continuous casting of aluminium alloys', *Proceedings of a symposium sponsored by the Non-Ferrous Metals Committee of the Minerals, Metals and Materials Society Chicago* (1983)

92 Kennedy, C. J. (BSC), *Rapid Cooling and Rapid Haul-off Systems*, Development Engineer Rautomead Ltd (1988)

93 Wilson, R. and Johnson, R. W., 'Continuous casting of high purity small diameter copper rod', *Metallurgia*, **56**, No. 7, July (1989)

94 *Casting Technique*, Normag Magnesium Norsk Hydro, Bygdby Alle, Oslo 2, Norway

95 *The Design and Properties of Steel Castings*, The Steel Castings Research and Trade Association (SCRATA), Sheffield

96 Lavington, M. M., *The Cosworth Process – a new concept in aluminium alloy casting production*. (International Meehanite is the sole world marketer of the Cosworth Process)

97 Taylor, P. R., *Metals and Materials*, No. 11, November (1986)

98 Barnett, S., 'Investment casting: a multi process technology', *Metallurgia*, July (1989)

99 Pratt, D. G., D S and single crystal castings

100 Ashton, M. C., 'The Replicast ceramic shell process', *Metals and Materials*, **7**, No. 1, January (1991)

101 *Casting Technique*, Norsk Hydro, Bygdby Alle 2, Oslo 2, Norway

102 Ghomaschi, M. R. and Chadwick, G. A., 'Cold chamber diecasting of aluminium alloys', *Metals and Materials*, **2**, No. 8, August (1986)

103 Chadwick, G. A. and Yue, T. M., 'Principles and appliciations of squeeze casting', *Metals and Materials*, **5**, No. 1, January (1989)

104 Spencer, D. B., Mehrabian, R. and Flemmings, M. C., *Met. Trans.*, 1925–32 (1972–1973)

105 Joly, P. A. and Mehrabian, R. S., *Mat. Sci.,* **11**, 1393–1418 (1976)

106 UK Patent No. GB 2042386A and European Patent Nos EP 006 3757, EP 0069270 and EP 0071822

107 Rinkwood, D. H. and Kaprianos, P., 'Semi-solid processing of alloys', *Metals and Materials*, **5**, No. 1, January (1989)

108 Wood, J. V. and Tolin, B., *Proc. Conf. on Designing with Titanium* (1986)

109 Nybura, N., 'Casting MMCs by liquid pressure forming', *Metals and Materials*, **7**, No. 1, January (1991)

Further reading

Section 16.2

General

Alting, L., *Manufacturing Engineering Processes*, Marcel Dekker, New York (1982)

Avitzur, B., *Metal Forming: Processes and Analysis*, McGraw-Hill, New York (1968)

Blazynski, T. Z., *Plasticity and Modern Metal-Forming Technology*, Elsevier Applied Science Publishers, London (1989)

Blazynski, T. Z., *Design of Tools for Deformation Processes*, Elsevier/Applied Science, London (1986)

Blazynski, T. Z., *Applied Elasto-Plasticity of Solids*, Macmillan, London (1983)

Blazynski, T. Z., *Metal Forming: Tool Profiles and Flow*, Macmillan, London (1976)

Burke, J. J. and Weiss, V., *Advances in Deformation Processes*, Plenum Press, New York (1978)

Johnson, W. and Mellor, P. B., *Engineering Plasticity*, Van Nostrand Reinhold, London (1973)

Johnson, W., Sowerby, R. and Haddow, J. B., *Plane Strain Slip-Line Fields*, Edward Arnold, London (1977)

Kalpakjian, S., *Manufacturing Processes for Engineering Materials*, Addison-Wesley, London (1984)

Pittman, J. F. T., Wood, R. D., Alexander, J. M. and Zienkiewicz, O. C., *Numerical Methods in Industrial Forming Processes*, Pineridge Press, Swansea (1982)

Rowe, G. W., *Principles of Industrial Metalworking Processes*, Edward Arnold, London (1977)

Schey, J. A., *Introduction to Manufacturing Processes*, McGraw-Hill, New York (1977)

Shey, J. A., *Metalworking Tribology: Friction, Lubrication and Wear*, American Society for Metals, Metals Park, Ohio (1983)

Rolling

Javoronkov, V. A. and Chaturvedi, R. C., *Rolling of Metals*, Yantrik, Bombay (1981)

Larke, E. C., *The Rolling of Strip, Sheet and Plate*, Chapman & Hall, London (1963)

Roberts, W. L., *Cold Rolling of Steel*, Marcel Dekker, New York (1978)

Roberts, W. L., *Hot Rolling of Steel*, Marcel Dekker, New York (1983)

Starling, C. W., *The Theory and Practice of Flat Rolling*, University of London Press, London (1962)

Wusatowski, Z., *Fundamentals of Rolling*, Pergamon Press, New York (1969)

Forging

Altan, T., Oh, S. and Gegel, A., *Metal Forming: Fundamentals and Applications*, American Society of Metals, Metals Park, Ohio (1983)

Feldman, H. D., *Cold Forging of Steel*, Hutchinson, London (1961)

Sabroff, A. M. et al., *Forging Materials and Practices*, Reinhold, New York (1968)

Watkins, M. T., *Metal Forming I, Forging and Related Processes*, Oxford University Press, New York (1975)

Extrusion

Alexander, J. M. and Lengyel, B., *Hydrostatic Extrusion*, Mills & Boon, London (1971)

Lane, K. and Stenger, H., *Extrusion – Processes, Machining, Tooling*, American Society of Metals, Metals Park, Ohio (1981)

Johnson, W. and Kudo, H., *The Mechanics of Metal Extrusion*, Manchester University Press, Manchester (1962)

Pearson, C. E. and Parkins, R. N., *The Extrusion of Metals*, Chapman & Hall, London (1960)

Drawing

Bernhoeft, C. P., *The Fundamentals of Wire Drawing*, The Wire Industry Ltd, London (1962)

Cameron, A., *Principles of Hydrodynamic Lubrication*, Longman, Harlow (1966)

Ibrahim, I. N. and Sansome, D. H., *An Experimental Study of the Mechanics of Ultrasonic Tube Bending*, Conf. Ultrasonics International, Canada (1983)

Sansome, D. H., 'Ultrasonic tube drawing', *Journal of Tube International*, 219, December (1985)

Tassi, O. J., *Non-Ferrous Wire Handbook*, Wire Association Int., Branford, Conn. (1977, 1981)

Sheet forming

Grainger, J. A., *Flow Turning of Metals*, The Machinery Publishing Co., Brighton (1969)

Kostinen, D. P. and Wang, N.-M., *Mechanics of Sheet Metal Forming*, Plenum Press, New York (1978)

Watkins, M. T., *Metal Forming II: Pressing and Related Processes*, Oxford University Press, New York (1975); *Source Book on Forming of Steel Sheet*, American Society for Metals, Metals Park, Ohio, (1976); *Developments in the Drawing of Metals*, Metals Society, London (1983)

High energy-rate forming

Blazynski, T. Z., *Explosive Welding, Forming and Compaction*, Applied Science, London (1983)

Blazynski, T. Z., *High-Energy-Rate Fabrication*, Int. Conf. Proceedings, University of Leeds (1981)

Blazynski, T. Z., *Materials at High Strain-Rates*, Elsevier Applied Science, London (1987)

Crossland, B., *Explosive Welding of Metals and its Application*, Clarendon Press, Oxford (1982)

Schroeder, J. W., and Berman, I., *High Energy Rate-Fabrication '84*, American Society of Mechanical Engineers, New York (1984)

Section 16.4

Brewis, D. M. and Briggs, D. (eds), *Industrial Adhesion Problems*, Orbital Press, Oxford (1985)

Hartshorn, S. R. (ed.), *Structural Adhesives: Chemistry and Technology*, Plenum Press, New York (1986)

Kinloch, A. J. (ed.), *Developments in Adhesives – 2*, Applied Science, London (1981)

Mays, G. C. and Hutchinson, A. R., *Adhesives in Civil Engineering*, Cambridge University Press, Cambridge (1991)

Mittal, K. L. (ed.), *Adhesive Joints*, Plenum Press, New York (1984)

Sadek, M. M. (ed.), *Industrial Applications of Adhesive Bonding*, Elsevier Applied Science, London (1987)

Schneberger, G. L. (ed.), *Adhesives in Manufacturing*, Marcel Dekker, New York (1983)

Wake, W. C. (ed.), *Developments in Adhesives – 1*, Applied Science, London (1977)

Structural Adhesives in Engineering (SAE) Conferences, Bristol University, UK:

SAE I, July 1986. *Proceedings* from Mechanical Engineering Publications, Bury St Edmunds

SAE II, September 1989. *Proceedings* from Butterworth-Heinemann, Oxford

Bonding and Repair of Composites, Seminar Proceedings, July 1989, Butterworth-Heinemann, Oxford

Engineering Applications of Adhesives, Seminar Proceedings, July 1988, Butterworth-Heinemann, Oxford

17 Engineering mathematics

John Barron
(Sections 17.1–17.4)

R. Ken Livesley
(Section 17.5)

Fraidoon Mazda
(Section 17.6)

Contents

17.1   Trigonometric functions and general formulae   17/3
    17.1.1   Mathematical signs and symbols   17/3
    17.1.2   Trigonometric formulae   17/3
    17.1.3   Trigonometric values   17/4
    17.1.4   Approximations for small angles   17/4
    17.1.5   Solution of triangles   17/4
    17.1.6   Spherical triangle   17/4
    17.1.7   Exponential form   17/5
    17.1.8   De Moivre's theorem   17/5
    17.1.9   Euler's relation   17/5
    17.1.10   Hyperbolic functions   17/5
    17.1.11   Complex variable   17/5
    17.1.12   Cauchy–Riemann equations   17/5
    17.1.13   Cauchy's theorem   17/5
    17.1.14   Zeros, poles and residues   17/5
    17.1.15   Some standard forms   17/6
    17.1.16   Coordinate systems   17/6
    17.1.17   Transformation of integrals   17/6
    17.1.18   Laplace's equation   17/6
    17.1.19   Solution of equations   17/7
    17.1.20   Method of least squares   17/8

17.2   Calculus   17/8
    17.2.1   Derivative   17/8
    17.2.2   Maxima and minima   17/8
    17.2.3   Integral   17/8
    17.2.4   Derivatives and integrals   17/8
    17.2.5   Standard substitutions   17/10
    17.2.6   Reduction formulae   17/10
    17.2.7   Numerical integration   17/10
    17.2.8   Vector calculus   17/11

17.3   Series and transforms   17/11
    17.3.1   Arithmetic series   17/11
    17.3.2   Geometric series   17/11
    17.3.3   Binomial series   17/11
    17.3.4   Taylor's series   17/11
    17.3.5   Maclaurin's series   17/11
    17.3.6   Laurent's series   17/12
    17.3.7   Power series for real variables   17/12
    17.3.8   Integer series   17/12
    17.3.9   Fourier series   17/12
    17.3.10   Rectified sine wave   17/12
    17.3.11   Square wave   17/13
    17.3.12   Triangular wave   17/13
    17.3.13   Sawtooth wave   17/13
    17.3.14   Pulse wave   17/13
    17.3.15   Fourier transforms   17/13
    17.3.16   Laplace transforms   17/13

17.4   Matrices and determinants   17/14
    17.4.1   Linear simultaneous equations   17/14
    17.4.2   Matrix arithmetic   17/14
    17.4.3   Eigenvalues and eigenvectors   17/15
    17.4.4   Coordinate transformation   17/15
    17.4.5   Determinants   17/15
    17.4.6   Properties of determinants   17/15
    17.4.7   Numerical solution of linear equations   17/15

17.5   Differential equations   17/16
    17.5.1   Notation and definitions   17/16
    17.5.2   Ordinary differential equations: analytical solutions   17/16
    17.5.3   Ordinary differential equations: approximate solutions   17/18
    17.5.4   Partial differential equations   17/20

17.6   Statistics   17/23
    17.6.1   Introduction   17/23
    17.6.2   Averages   17/23
    17.6.3   Dispersion   17/23
    17.6.4   Skewness   17/24
    17.6.5   Combinations and permutations   17/24
    17.6.6   Regression and correlation   17/25
    17.6.7   Probability   17/25
    17.6.8   Probability distributions   17/26
    17.6.9   Sampling   17/28
    17.6.10   Tests of significance   17/28

Further reading   17/29

17.1 Trigonometric functions and general formulae

17.1.1 Mathematical signs and symbols

Sign, symbol	Quantity
$=$	equal to
$\neq$	not equal to
$\equiv$	identically equal to
$\triangleq$	corresponds to
$\approx$	approximately equal to
$\rightarrow$	approaches
$\simeq$	asymptotically equal to
$\sim$	proportional to
∞	infinity
$<$	smaller than
$>$	larger than
$\leq \leqslant \leqq$	smaller than or equal to
$\geq \geqslant \geqq$	larger than or equal to
$\ll$	much smaller than
$\gg$	much larger than
$+$	plus
$-$	minus
$\times$	multiplied by
$\dfrac{a}{b}$; a/b	a divided by b
$\|a\|$	magnitude of a
a^n	a raised to the power n
$a^{1/2}$; $\sqrt{a}$	square root of a
$a^{1/n}$; $\sqrt[n]{a}$	nth root of a
$\bar{a}$]; a)	mean value of a
$p!$	factorial p, $1 \times 2 \times 3 \times \ldots \times p$
$\dbinom{n}{p}$	binomial coefficient, $\dfrac{n(n-1)\ldots(n-p+1)}{1 \times 2 \times 3 \times \ldots \times p}$
Σ	sum
Π	product
$f(x)$	function f of the variable x
$[f(x)]_a^b$	$f(b) - f(a)$
$\lim\limits_{x \to a} f(x)$; $\lim_{x \to a} f(x)$	the limit to which $f(x)$ tends as x approaches a
Δx	delta x = finite increment of x
δx	delta x = variation of x
$\dfrac{df}{dx}$; df/dx; $f'(x)$	differential coefficient of $f(x)$ with respect to x
$\dfrac{d^n f}{dx^n}$; $f^{(n)}(x)$	differential coefficient of order n of $f(x)$
$\dfrac{\partial f(x,y,\ldots)}{\partial x}$; $\left(\dfrac{\partial f}{\partial x}\right)_{y\ldots}$	partial differential coefficient of $f(x,y,\ldots)$ with respect to x, when $y\ldots$ are held constant
df	the total differential of f
$\int f(x)dx$	indefinite integral of $f(x)$ with respect to x
$\displaystyle\int_a^b f(x)dx$	definite integral of $f(x)$ from $x=a$ to $x=b$
e	base of natural logarithms

Sign, symbol	Quantity
e^x, $\exp x$	e raised to the power x
$\log_a x$	logarithm to the base a of x
$\lg x$; $\log x$; $\log_{10} x$	common (Briggsian) logarithm of x
$\text{lb } x$; $\log_2 x$	binary logarithm of x
$\sin x$	sine of x
$\cos x$	cosine of x
$\tan x$; $\text{tg } x$	tangent of x
$\cot x$; $\text{ctg } x$	cotangent of x

Sign, symbol	Quantity
$\sec x$	secant of x
$\operatorname{cosec} x$	cosecant of x
$\arcsin x$	arc sine of x
$\arccos x$	arc cosine of x
$\arctan x$, $\text{arctg } x$	arc tangent of x
$\operatorname{arccot} x$, $\text{arcctg } x$	arc cotangent of x
$\operatorname{arcsec} x$	arc secant of x
$\operatorname{arcosec} x$	arc cosecant of x
$\sinh x$	hyperbolic sine of x
$\cosh x$	hyperbolic cosine of x
$\tanh x$	hyperbolic tangent of x
$\coth x$	hyperbolic cotangent of x
$\operatorname{sech} x$	hyperbolic secant of x
$\operatorname{cosech} x$	hyperbolic cosecant of x
$\operatorname{arsinh} x$	inverse hyperbolic sine of x
$\operatorname{arcosh} x$	inverse hyperbolic cosine of x
$\operatorname{artanh} x$	inverse hyperbolic tangent of x
$\operatorname{arcoth} x$	inverse hyperbolic cotangent of x
$\operatorname{arsech} x$	inverse hyperbolic secant of x
$\operatorname{arcosech} x$	inverse hyperbolic cosecant of x
i, j	imaginary unity, $i^2 = -1$
$\operatorname{Re} z$	real part of z
$\operatorname{Im} z$	imaginary part of z
$\|z\|$	modulus of z
$\arg z$	argument of z
z^*	conjugate of z, complex conjugate of z
$\bar{A}$, A', A^t	transpose of matrix A
A^*	complex conjugate matrix of matrix A
A^+	Hermitian conjugate matrix of matrix A
$\mathbf{A}$, $\mathbf{a}$	vector
$\|\mathbf{A}\|$, A	magnitude of vector
$\mathbf{A} \cdot \mathbf{B}$	scalar product
$\mathbf{A} \times \mathbf{B}$, $\mathbf{A} \wedge \mathbf{B}$	vector product
∇	differential vector operator
$\nabla\varphi$, $\operatorname{grad} \varphi$	gradient of φ
$\nabla \cdot A$, $\operatorname{div} \mathbf{A}$	divergence of $\mathbf{A}$
$\nabla\times \mathbf{A}$, $\nabla \wedge \mathbf{A}$ curl $\mathbf{A}$, rot $\mathbf{A}$	curl of $\mathbf{A}$
$\nabla^2\varphi$, $\Delta\varphi$	Laplacian of φ

12.1.2 Trigonometric formulae

$$\sin^2 A + \cos^2 A = \sin A \operatorname{cosec} A = 1$$

$$\sin A = \frac{\cos A}{\cot A} = \frac{1}{\operatorname{cosec} A} = (1 - \cos^2 A)^{1/2}$$

$$\cos A = \frac{\sin A}{\tan A} = \frac{1}{\sec A} = (1 - \sin^2 A)^{1/2}$$

$$\tan A = \frac{\sin A}{\cos A} = \frac{1}{\cot A}$$

$$1 + \tan^2 A = \sec^2 A$$

$$1 + \cot^2 A = \text{cosec}^2 A$$

$$1 - \sin A = \text{coversin } A$$

$$1 - \cos A = \text{versin } A$$

$$\tan \tfrac{1}{2}\theta = t; \ \sin \theta = 2t/(1 + t^2); \ \cos \theta = (1 - t^2)/(1 + t^2)$$

$$\cot A = 1/\tan A$$

$$\sec A = 1/\cos A$$

$$\text{cosec } A = 1/\sin A$$

$$\cos (A \pm B) = \cos A \cos B \mp \sin A \sin B$$

$$\sin (A \pm B) = \sin A \cos B \pm \cos A \sin B$$

$$\tan (A \pm B) = \frac{\tan A \pm \tan B}{1 \mp \tan A \tan B}$$

$$\cot(A \pm B) = \frac{\cot A \cot B \mp 1}{\cot B \pm \cot A}$$

$$\sin A \pm \sin B = 2 \sin \tfrac{1}{2}(A \pm B)\cos \tfrac{1}{2}(A \mp B)$$

$$\cos A + \cos B = 2 \cos \tfrac{1}{2}(A + B) \cos \tfrac{1}{2}(A - B)$$

$$\cos A - \cos B = 2 \sin \tfrac{1}{2}(A + B) \sin \tfrac{1}{2}(B - A)$$

$$\tan A \pm \tan B = \frac{\sin (A \pm B)}{\cos A \cos B}$$

$$\cot A \pm \cot B = \frac{\sin (B \pm A)}{\sin A \sin B}$$

$$\sin 2A = 2 \sin A \cos A$$

$$\cos 2A = \cos^2 A - \sin^2 A = 2 \cos^2 A - 1 = 1 - 2 \sin^2 A$$

$$\cos^2 A - \sin^2 B = \cos (A + B) \cos (A - B)$$

$$\tan 2A = 2 \tan A/(1 - \tan^2 A)$$

$$\sin \tfrac{1}{2} A = \pm \left(\frac{1 - \cos A}{2} \right)^{1/2}$$

$$\cos \tfrac{1}{2} A = \pm \left(\frac{1 + \cos A}{2} \right)^{1/2}$$

$$\tan \tfrac{1}{2} A = \frac{\sin A}{1 + \cos A}$$

$$\sin^2 A = \tfrac{1}{2}(1 - \cos 2 A)$$

$$\cos^2 A = \tfrac{1}{2}(1 + \cos 2 A)$$

$$\tan^2 A = \frac{1 - \cos 2 A}{1 + \cos 2 A}$$

$$\tan \tfrac{1}{2}(A \pm B) = \frac{\sin A \pm \sin B}{\cos A + \cos B}$$

$$\cot \tfrac{1}{2}(A \pm B) = \frac{\sin A \pm \sin B}{\cos B - \cos A}$$

17.1.3 Trigonometric values

Angle	0°	30°	45°	60°	90°	180°	270°	360°
Radians	0	$\pi/6$	$\pi/4$	$\pi/3$	$\pi/2$	π	$3\pi/2$	2π
Sine	0	$\tfrac{1}{2}$	$\tfrac{1}{2}\sqrt{2}$	$\tfrac{1}{2}\sqrt{3}$	1	0	-1	0
Cosine	1	$\tfrac{1}{2}\sqrt{3}$	$\tfrac{1}{2}\sqrt{2}$	$\tfrac{1}{2}$	0	-1	0	1
Tangent	0	$\tfrac{1}{3}\sqrt{3}$	1	$\sqrt{3}$	∞	0	∞	0

17.1.4 Approximations for small angles

$$\sin \theta = \theta - \theta^3/6; \quad \cos \theta = 1 - \theta^2/2; \quad \tan \theta = \theta + \theta^3/3;$$
$$(\theta \text{ in radians})$$

17.1.5 Solution of triangles

$$\frac{\sin A}{a} = \frac{\sin B}{b} = \frac{\sin C}{c}; \ \cos A = \frac{b^2 + c^2 - a^2}{2bc}$$

$$\cos B = \frac{c^2 + a^2 - b^2}{2ca}; \ \cos C = \frac{a^2 + b^2 - c^2}{2ab}$$

where A, B, C and a, b, c are shown in Figure 17.1. If $s = \tfrac{1}{2}(a + b + c)$

$$\sin \frac{A}{2} = \sqrt{\frac{(s - b)(s - c)}{bc}}; \ \sin \frac{B}{2} = \sqrt{\frac{(s - c)(s - a)}{ca}}$$

$$\sin \frac{C}{2} = \sqrt{\frac{(s - a)(s - b)}{ab}}$$

$$\cos \frac{A}{2} = \sqrt{\frac{s(s - a)}{bc}}; \ \cos \frac{B}{2} = \sqrt{\frac{(s - b)}{ca}}$$

$$\cos \frac{C}{2} = \sqrt{\frac{s(s - c)}{ab}}$$

$$\tan \frac{A}{2} = \sqrt{\frac{(s - b)(s - c)}{s(s - a)}}; \ \tan \frac{B}{2} = \sqrt{\frac{(s - c)(s - a)}{s(s - b)}}$$

$$\tan \frac{C}{2} = \sqrt{\frac{(s - a)(s - b)}{s(s - c)}}$$

17.1.6 Spherical triangle

$$\frac{\sin A}{\sin a} = \frac{\sin B}{\sin b} = \frac{\sin C}{\sin c}$$

$$\cos a = \cos b \cos c + \sin b \sin c \cos A$$

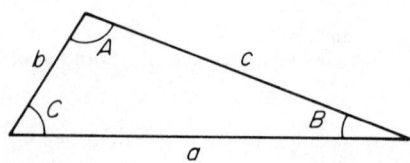

Figure 17.1 Triangle

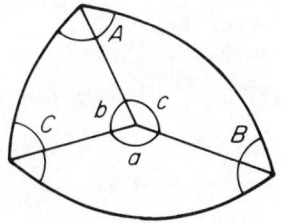

Figure 17.2 Spherical triangle

$\cos b = \cos c \cos a + \sin c \sin a \cos B$

$\cos c = \cos a \cos b + \sin a \sin b \cos C$

where A, B, C and a, b, c are now as in Figure 17.2.

17.1.7 Exponential form

$$\sin \theta = \frac{e^{i\theta} - e^{-i\theta}}{2i} \quad \cos \theta = \frac{e^{i\theta} + e^{+i\theta}}{2}$$

$c^{i\theta} = \cos \theta + i \sin \theta \quad e^{-i\theta} = \cos \theta - i \sin \theta$

17.1.8 De Moivre's theorem

$(\cos A + i \sin A)(\cos B + i \sin B)$
$= \cos(A + B) + i \sin (A + B)$

17.1.9 Euler's relation

$(\cos \theta + i \sin \theta)^n = \cos n\theta + i \sin n\theta = e^{in\theta}$

17.1.10 Hyperbolic functions

$\sinh x = (e^x - e^{-x})/2 \quad \cosh x = (e^x + e^{-x})/2$

$\tanh x = \sinh x \cosh x$

Relations between hyperbolic functions can be obtained from the corresponding relations between trigonometric functions by reversing the sign of any term containing the product or implied product of two sines, e.g.:

$\cosh^2 A - \sinh^2 A = 1$

$\cosh 2A = 2 \cosh^2 A - 1 = 1 + 2 \sinh^2 A$
$\qquad = \cosh^2 A + \sinh^2 A$

$\cosh(A \pm B) = \cosh A \cosh B \pm \sinh A \sinh B$

$\sinh (A \pm B) = \sinh A \cosh B \pm \cosh A \sinh B$

$e^x = \cosh x + \sinh x \quad e^{-x} = \cosh x - \sinh x$

17.1.11 Complex variable

If $z = x + iy$, where x and y are real variables, z is a complex variable and is a function of x and y. z may be represented graphically in an Argand diagram (Figure 17.3).

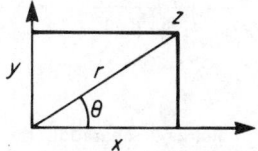

Figure 17.3 Argand diagram

Polar form:

$z = x + iy = |z|(\cos \theta + i \sin \theta)$

$x = r \cos \theta \quad y = r \sin \theta$

where $r = |z|$.

Complex arithmetic:

$z_1 = x_1 + iy_1; z_2 = x_2 + iy_2$

$z_1 \pm z_2 = (x_1 \pm x_2) + i(y_1 \pm y_2)$

$z_1 . z_2 = (x_1 x_2 - y_1 y_2) + i(x_1, y_2 + x_2, y_1)$

Conjugate:

$z^* = x - iy \quad z . z^* = x^2 + y^2 = |z|^2$

Function: another complex variable $w = u + iv$ may be related functionally to z by

$w = u + iv = f(x + iy) = f(z)$

which implies

$u = u(x,y) \quad v = v(x,y)$

e.g.,

$\cosh z = \cosh (x + iy) = \cosh x \cosh iy + \sinh x \sinh iy$
$\qquad\qquad\qquad\qquad = \cosh x \cos y + i \sinh x \sin y$

$u = \cosh x \cos y \qquad v = \sinh x \sin y$

17.1.12 Cauchy–Riemann equations

If $u(x,y)$ and $v(x,y)$ are continuously differentiable with respect to x and y,

$$\frac{\partial u}{\partial x} = \frac{\partial v}{\partial y} \quad \frac{\partial u}{\partial y} = -\frac{\partial v}{\partial x}$$

$w = f(z)$ is continuously differentiable with respect to z and its derivative is

$$f'(z) = \frac{\partial u}{\partial x} + i\frac{\partial v}{\partial x} = \frac{\partial v}{\partial y} - i\frac{\partial u}{\partial y} = \frac{1}{i}\left(\frac{\partial u}{\partial y} + i\frac{\partial v}{\partial y}\right)$$

It is also easy to show that $\nabla^2 u = \nabla^2 v = 0$. Since the transformation from z to w is conformal, the curves $u = $ constant and $v = $ constant intersect each other at right angles, so that one set may be used as equipotentials and the other as field lines in a vector field.

17.1.13 Cauchy's theorem

If $f(z)$ is analytic everywhere inside a region bounded by C and a is a point within C

$$f(a) = \frac{1}{2\pi i} \int_C \frac{f(z)}{z - a} \, dz$$

This formula gives the value of a function at a point in the interior of a closed curve in terms of the values on that curve.

17.1.14 Zeros, poles and residues

If $f(z)$ vanishes at the point z_0 the Taylor series for z in the region of z_0 has its first two terms zero, and perhaps others also: $f(z)$ may then be written

$f(z) = (z - z_0)^n g(z)$

where $g(z_0) \neq 0$. Then $f(z)$ has a *zero* of order n at z_0. The reciprocal

$$q(z) = 1/f(z) = h(z)/(z - z_0)^n$$

where $h(z) = 1/g(z) \neq 0$ at z_0. $q(z)$ becomes infinite at $z = z_0$ and is said to have a *pole* of order n at z_0. $q(z)$ may be expanded in the form

$$q(z) = c_{-n}(z - z_0)^n + \ldots + c_{-1}(z - z_0)^{-1} + c_0 + \ldots$$

where c_{-1} is the *residue* of $q(z)$ at $z = z_0$. From Cauchy's theorem, it may be shown that if a function $f(z)$ is analytic throughout a region enclosed by a curve C except at a finite number of poles, the integral of the function around C has a value of $2\pi i$ times the sum of the residues of the function at its poles within C. This fact can be used to evaluate many definite integrals whose indefinite form cannot be found.

17.1.15 Some standard forms

$$\int_0^{2\pi} e^{\cos\theta} \cos(n\theta - \sin\theta)d\theta = 2\pi/n!$$

$$\int_0^\infty \frac{x^{a-1}}{1 + x} dx = \pi \operatorname{cosec} a\pi$$

$$\int_0^\infty \frac{\sin\theta}{\theta} d\theta = \frac{\pi}{2}$$

$$\int_0^\infty x \exp(-h^2 x^2)dx = \frac{1}{2h^2}$$

$$\int_0^\infty \frac{x^{a-1}}{1 - x} dx = \pi \cot a\pi$$

$$\int_0^\infty \exp(-h^2 x^2)dx = \frac{\sqrt{\pi}}{2h}$$

$$\int_0^\infty x^2 \exp(-h^2 x^2)dx = \frac{\sqrt{\pi}}{4h^3}$$

17.1.16 Coordinate systems

The basic system is the rectangular Cartesian system (x,y,z) to which all other systems are referred. Two other commonly used systems are as follows.

17.1.16.1 Cylindrical coordinates

Coordinates of point P are (x,y,z) or (r,θ,z) (see Figure 17.4), where

$$x = r \cos\theta \quad y = r \sin\theta \quad z = z$$

In these coordinates the volume element is $r\ dr\ d\theta\ dz$.

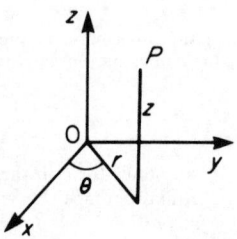

Figure 17.4 Cylindrical coordinates

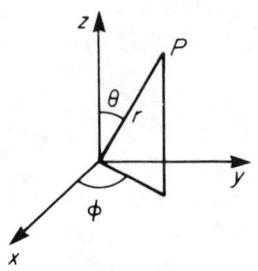

Figure 17.5 Spherical polar coordinates

17.1.16.2 Spherical polar coordinates

Coordinates of point P are (x,y,z) or (r,θ,φ) (see Figure 17.5) where

$$x = r \sin\theta \cos\phi \quad y = r \sin\phi \sin\phi \quad z = r \cos\theta$$

In these coordinates the volume element is $r^2 \sin\theta\ dr\ d\theta\ d\phi$.

17.1.17 Transformation of integrals

$$\iiint f(x,y,z)dx\ dy\ dz = \iiint \varphi(u,v,w)J du\ dv\ dw$$

where

$$J = \begin{vmatrix} \dfrac{\partial x}{\partial u} & \dfrac{\partial y}{\partial u} & \dfrac{\partial z}{\partial u} \\[2mm] \dfrac{\partial x}{\partial v} & \dfrac{\partial y}{\partial v} & \dfrac{\partial z}{\partial v} \\[2mm] \dfrac{\partial x}{\partial w} & \dfrac{\partial y}{\partial w} & \dfrac{\partial z}{\partial w} \end{vmatrix} = \frac{\partial(x,y,z)}{\partial(u,v,w)}$$

is the Jacobian of the transformation of coordinates. For Cartesian to cylindrical coordinates, $J = r$, and for Cartesian to spherical polars, it is $r^2 \sin\theta$.

17.1.18 Laplace's equation

The equation satisfied by the scalar potential from which a vector field may be derived by taking the gradient is Laplace's equation, written as:

$$\nabla^2 \Phi = \frac{\partial^2 \Phi}{\partial x^2} + \frac{\partial^2 \Phi}{\partial y^2} + \frac{\partial^2 \Phi}{\partial z^2} = 0$$

In cylindrical coordinates:

$$\nabla^2 \Phi = \frac{1}{r}\frac{\partial}{\partial r}\left(r\frac{\partial \Phi}{\partial r}\right) + \frac{1}{r^2}\frac{\partial^2 \Phi}{\partial \theta^2} + \frac{\partial^2 \Phi}{\partial z^2}$$

In spherical polars:

$$\nabla^2 \Phi = \frac{1}{r^2}\frac{\partial}{\partial r}\left(r^2\frac{\partial \Phi}{\partial r}\right) + \frac{1}{r^2 \sin\theta}\frac{\partial \Phi}{\partial \theta} + \frac{1}{r^2 \sin^2\theta}\frac{\partial^2 \Phi}{\partial \Phi^2}$$

The equation is solved by setting

$$\Phi = U(u)V(v)W(w)$$

in the appropriate form of the equation, separating the variables and solving separately for the three functions, where (u,v,w) is the coordinate system in use.

In Cartesian coordinates, typically the functions are trigonometric, hyperbolic and exponential; in cylindrical coordinates the function of z is exponential, that of θ trigonometric and that of r is a Bessel function. In spherical polars, typically the function of r is a power of r, that of φ is trigonometric, and that of θ is a Legendre function of $\cos \theta$.

17.1.19 Solution of equations

17.1.19.1 Quadratic equation

$$ax^2 - bx + c = 0$$

$$x = -\frac{b}{2a} \pm \frac{\sqrt{b^2 - 4ac}}{2a}$$

In practical calculations if $b^2 > 4ac$, so that the roots are real and unequal, calculate the root of larger modulus first, using the same sign for both terms in the formula, then use the fact that $x_1 x_2 = c/a$ where x_1 and x_2 are the roots. This avoids the severe cancellation of significant digits which may otherwise occur in calculating the smaller root.

For polynomials other than quadratics, and for other functions, several methods of successive approximation are available.

17.1.19.2 Bisection method

By trial find x_0 and x_1 such that $f(x_0)$ and $f(x_1)$ have opposite signs (see Figure 17.6). Set $x_2 = (x_0 + x_1)/2$ and calculate $f(x_2)$. If $f(x_0)f(x_2)$ is positive, the root lies in the interval (x_1, x_2); if negative in the interval (x_0, x_2); and if zero, x_2 is the root. Continue if necessary using the new interval.

17.1.19.3 Regula falsi

By trial, find x_0 and x_1 as for the bisection method; these two values define two points $(x_0, f(x_0))$ and $(x_1, f(x_1))$. The straight line joining these two points cuts the x-axis at the point (see Figure 17.7)

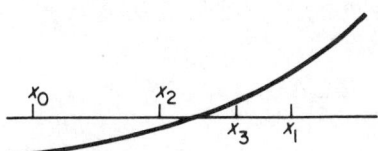

Figure 17.6 Bisection method

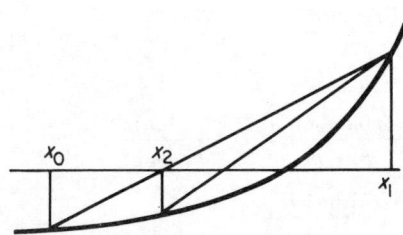

Figure 17.7 Regula falsi

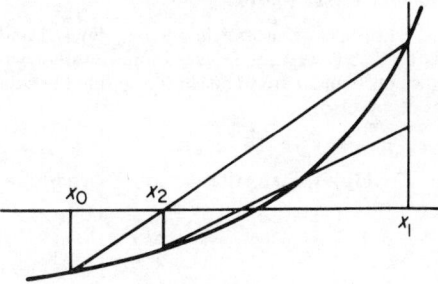

Figure 17.8 Accelerated method

$$x_2 = \frac{x_0 f(x_1) - x_1 f(x_0)}{f(x_1) - f(x_0)}$$

Evaluate $f(x_2)$ and repeat the process for whichever of the intervals (x_0, x_2) or (x_1, x_2) contains the root. This method can be accelerated by halving at each step the function value at the retained end of the interval, as shown in Figure 17.8.

17.1.19.4 Fixed-point iteration

Arrange the equation in the form

$$x = f(x)$$

Choose an initial value of x by trial, and calculate repetitively

$$x_{k+1} = f(x_k)$$

This process will not always converge.

17.1.19.5 Newton's method

Calculate repetitively (Figure 17.9)

$$x_{k+1} = x_k - f(x_k)/f'(x_k)$$

This method will converge unless: (a) x_k is near a point of inflexion of this function; or (b) x_k is near a local minimum; or (c) the root is multiple. If one of these cases arises, most of the trouble can be overcome by checking at each stage that

$$f(x_{k+1}) < f(x_k)$$

and, if not, halving the preceding value of $|x_{k+1} - x_k|$.

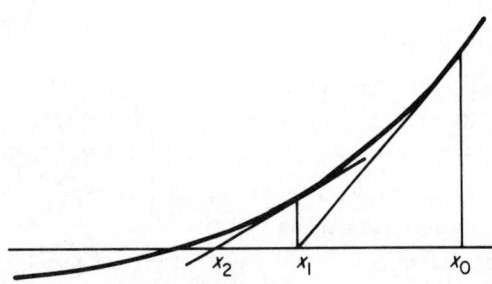

Figure 17.9 Newton's method

17.1.20 Method of least squares

To obtain the best fit between a straight line $ax + by = 1$ and several points $(x_1 y_1)$, (x_2, y_2),..., (x_n, y_n) found by observation, the coefficients a and b are to be chosen so that the sum of the squares of the errors

$$e_i = ax_i + by_i - 1$$

is a minimum. To do this, first write the set of inconsistent equations

$$ax_1 + by_1 - 1 = 0$$
$$ax_2 + by_2 - 1 = 0$$
$$\vdots$$
$$ax_n + by_n - 1 = 0$$

Multiply each equation by the value of x it contains, and add, obtaining

$$a \sum_{i=1}^{n} x_i^2 + b \sum_{i=1}^{n} x_i y_i - \sum_{i=1}^{n} x_i = 0$$

Similarly multiply by y and add, obtaining

$$a \sum_{i=1}^{n} x_i y_i + b \sum_{i=1}^{n} y_i^2 - \sum_{i=1}^{n} y_i = 0$$

Lastly, solve these two equations for a and b, which will be the required values giving the least squares fit.

17.2 Calculus

17.2.1 Derivative

$$f'(x) = \lim_{\delta x \to 0} \frac{f(x + \delta x) - f(x)}{\delta x}$$

If u and v are functions of x,

$$(uv)' = u'v + uv'$$

$$\left(\frac{u}{v}\right)' = \frac{u'v - uv'}{v^2}$$

$$(uv)^n = u^{(n)}v + nu^{(n-1)}v^{(1)} + \ldots + {}^nC_p u^{(n-p)}v^{(p)}$$
$$+ \ldots + uv^{(n)}$$

where

$$^nC_p = \frac{n!}{p!(n - p)!}$$

If $z = f(x)$ and $y = g(z)$, then

$$\frac{dy}{dx} = \frac{dy}{dz}\frac{dz}{dx}$$

17.2.2 Maxima and minima

$f(x)$ has a stationary point wherever $f'(x) = 0$: the point is a maximum, minimum or point of inflexion according as $f''(x) <, >$ or $= 0$.

$f(x, y)$ has a stationary point wherever

$$\frac{\partial f}{\partial x} = \frac{\partial f}{\partial y} = 0$$

Let (a, b) be such a point, and let

$$\frac{\partial^2 f}{\partial x^2} = A, \quad \frac{\partial^2 f}{\partial x \partial y} = H \quad \frac{\partial^2 f}{\partial y^2} = B$$

all at that point, then:

If $H^2 - AB > 0$, $f(x, y)$ has a saddle point at (a, b).
If $H^2 - AB < 0$ and if $A < 0$, $f(x, y)$ has a maximum at (a, b),
    but if $A > 0$, $f(x, y)$ has a minimum at (a, b).
If $H^2 = AB$, higher derivatives need to be considered.

17.2.3 Integral

$$\int_b^a f(x)dx = \lim_{N \to \infty} \sum_{n=0}^{N-1} f\left(a + \frac{n(b - a)}{N}\right)\left(\frac{b-a}{N}\right)$$

$$= \lim_{N \to \infty} \sum_{n=1}^{N} f(a + (n - 1)\,\delta x)\delta x$$

where $\delta x = (b - a)/N$.
If u and v are functions of x, then

$$\int uv'dx = uv - \int u'vdx \quad \text{(integration by parts)}$$

y	$\int ydx$	
$\sin mx \sin nx$	$\dfrac{1}{2}\dfrac{\sin(m - n)x}{m - n} - \dfrac{1}{2}\dfrac{\sin(m + n)x}{m + n}$	$(m \neq n)$
	$\dfrac{1}{2}\left(x - \dfrac{\sin 2mx}{2m}\right)$	$(m = n)$
$\sin mx \cos nx$	$-\dfrac{1}{2}\dfrac{\cos(m + n)x}{m + n} - \dfrac{1}{2}\dfrac{\cos(m - n)x}{m - n}$	$(m \neq n)$
	$-\dfrac{1}{2}\dfrac{\cos 2mx}{2m}$	$(m = n)$
$\cos mx \cos nx$	$\dfrac{1}{2}\dfrac{\sin(m + n)x}{m + n} + \dfrac{1}{2}\dfrac{\sin(m - n)x}{m - n}$	$(m \neq n)$
	$\dfrac{1}{2}\left(x + \dfrac{\sin 2mx}{2m}\right)$	$(m = n)$

17.2.4 Derivatives and integrals

y	$\dfrac{dy}{dx}$	$\displaystyle\int y\,dx$
x^n	nx^{n-1}	$x^{n+1}/(n+1)$
$1/x$	$-1/x^2$	$\ln(x)$
e^{ax}	ae^{ax}	e^{ax}/a
$\ln(x)$	$1/x$	$x[\ln(x)-1]$
$\log_a x$	$\dfrac{1}{x}\log_a e$	$x\log_a\left(\dfrac{x}{e}\right)$
$\sin ax$	$a\cos ax$	$-\dfrac{1}{a}\ln(\cos ax)$
$\cos ax$	$-a\sin ax$	$-\dfrac{1}{a}\sin ax$
$\tan ax$	$a\sec^2 ax$	$-\dfrac{1}{a}\cos ax$
$\cot ax$	$-a\,\mathrm{cosec}^2 ax$	$\dfrac{1}{a}\ln(\sin ax)$
$\sec ax$	$a\tan ax\sec ax$	$\dfrac{1}{a}\ln(\sec ax + \tan ax)$
$\mathrm{cosec}\,ax$	$-a\cot ax\,\mathrm{cosec}\,ax$	$\dfrac{1}{a}\ln(\mathrm{cosec}\,ax - \cot ax)$
$\arcsin(x/a)$	$1/(a^2-x^2)^{1/2}$	$x\arcsin(x/a) + (a^2-x^2)^{1/2}$
$\arccos(x/a)$	$-1/(a^2-x^2)^{1/2}$	$x\arccos(x/a) - (a^2-x^2)^{1/2}$
$\arctan(x/a)$	$a/(a^2+x^2)$	$x\arctan(x/a) - \tfrac{1}{2}a\ln(a^2+x^2)$
$\mathrm{arccot}(x/a)$	$-a/(a^2+x^2)$	$x\,\mathrm{arccot}(x/a) + \tfrac{1}{2}a\ln(a^2+x^2)$
$\mathrm{arcsec}(x/a)$	$a(x^2-a^2)^{-1/2}/x$	$x\,\mathrm{arcsec}(x/a) - a\ln[x+(x^2-a^2)^{1/2}]$
$\mathrm{arccosec}(x/a)$	$-a(x^2-a^2)^{-1/2}/x$	$x\,\mathrm{arccosec}(x/a) + a\ln[x+(x^2-a^2)^{1/2}]$
$\sinh ax$	$a\cosh ax$	$\dfrac{1}{a}\cosh ax$
$\cosh ax$	$a\sinh ax$	$\dfrac{1}{a}\sinh ax$
$\tanh ax$	$a\,\mathrm{sech}^2 ax$	$\dfrac{1}{a}\ln\cosh ax$
$\coth ax$	$-a\,\mathrm{cosech}^2 ax$	$\dfrac{1}{a}\ln(\sinh ax)$
$\mathrm{sech}\,ax$	$-a\tanh ax\,\mathrm{sech}\,ax$	$\dfrac{2}{a}\arctan(e^{ax})$
$\mathrm{cosech}\,ax$	$-a\coth ax\,\mathrm{cosech}\,ax$	$\dfrac{1}{a}\ln\left(\tanh\dfrac{ax}{2}\right)$
$\mathrm{arsinh}(x/a)$	$(x^2+a^2)^{-1/2}$	$x\,\mathrm{arsinh}(x/a) - (x^2+a^2)^{1/2}$
$\mathrm{arcosh}(x/a)$	$(x^2-a^2)^{-1/2}$	$x\,\mathrm{arcosh}(x/a) - (x^2-a^2)^{1/2}$
$\mathrm{artanh}(x/a)$	$a(a^2+x^2)^{-1}$	$x\,\mathrm{artanh}(x/a) - \tfrac{1}{2}a\ln(a^2+x^2)$
$\mathrm{arcoth}(x/a)$	$-a(x^2-a^2)^{-1}$	$x\,\mathrm{arcoth}(x/a) + \tfrac{1}{2}a\ln(x^2-a^2)$

y	$\dfrac{dy}{dx}$	$\displaystyle\int y\,dx$
arsech (x/a)	$-a(a^2 - x^2)^{-1/2}/x$	x arsech $(x/a) + a$ arcsin (x/a)
arcosech (x/a)	$-a(x^2 + a^2)^{-1/2}/x$	x arcosech $(x/a) + a$ arsinh (x/a)
$(x^2 + a^2)^{1/2}$		$\frac{1}{2}x(x^2 + a^2)^{1/2} + \frac{1}{2}a^2$ arsinh (x/a)
$(x^2 - a^2)^{1/2}$		$\frac{1}{2}x(x^2 - a^2)^{1/2} - \frac{1}{2}a^2$ arcosh (x/a)
$(a^2 - x^2)^{1/2}$		$\begin{cases} \frac{1}{2}(x^2 \pm a^2)^{p+1}/(p + 1) & (p \neq -1) \\ \frac{1}{2}\ln(x^2 \pm a^2) & (p = -1) \end{cases}$
$(x^2 \pm a^2)^p x$		
$(a^2 - x^2)^p x$		$\begin{cases} -\frac{1}{2}(a^2 - x^2)^{p+1}/(p + 1) & (p \neq -1) \\ -\frac{1}{2}\ln(a^2 - x^2) & (p = -1) \end{cases}$
$x(ax^2 + b)^p$		$(ax^2 + b)^{p+1}/2a(p + 1) \quad (p \neq -1)$
		$[\ln(ax^2 + b)]/2a \qquad\quad (p = -1)$
$(2ax - x^2)^{-1/2}$		$\arccos\left(\dfrac{a - x}{a}\right)$
$(a^2 \sin^2 x + b^2 \cos^2 x)^{-1}$		$\dfrac{1}{ab} \arctan\left(\dfrac{a}{b} \tan x\right)$
$(a^2 \sin^2 x - b^2 \cos^2 x)^{-1}$		$-\dfrac{1}{ab} \operatorname{artanh}\left(\dfrac{a}{b} \tan x\right)$
$e^{ax} \sin bx$		$e^{ax} \dfrac{a \sin bx - b \cos bx}{a^2 + b^2}$
$e^{ax} \cos bx$		$e^{ax} \dfrac{(a \cos bx + b \sin bx)}{a^2 + b^2}$

17.2.5 Standard substitutions

Integral a function of	Substitute
$a^2 - x^2$	$x = a \sin \theta$ or $x = a \cos \theta$
$a^2 + x^2$	$x = a \tan \theta$ or $x = a \sinh \theta$
$x^2 - a^2$	$x = a \sec \theta$ or $x = a \cosh \theta$

17.2.6 Reduction formulae

$$\int \sin^m x \, dx = -\frac{1}{m} \sin^{m-1} x \cos x + \frac{m - 1}{m} \int \sin^{m-2} x \, dx$$

$$\int \cos^m x \, dx = -\frac{1}{m} \cos^{m-1} x \sin x + \frac{m - 1}{m} \int \cos^{m-2} x \, dx$$

$$\int \sin^m x \cos^n x \, dx = \frac{\sin^{m+1} x \cos^{n-1} x}{m + n}$$

$$+ \frac{n - 1}{m + n} \int \sin^m x \cos^{n-2} x \, dx$$

If the integrand is a rational function of $\sin x$ and/or $\cos x$, substitute $t = \tan \frac{1}{2}x$, then

$$\sin x = \frac{1}{1 + t^2}, \quad \cos x = \frac{1 - t^2}{1 + t^2}, \quad dx = \frac{2\,dt}{1 + t^2}$$

17.2.7 Numerical integration

17.2.7.1 Trapezoidal rule (Figure 17.10)

$$\int_{x_1}^{x_2} y \, dx = \frac{1}{2}h(y_1 + y_2) + O(h^3)$$

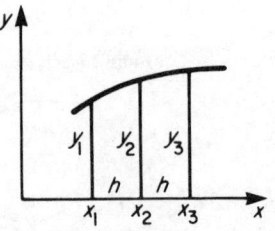

Figure 17.10 Numerical integration

17.2.7.2 Simpson's rule (Figure 17.10)

$$\int_{x_1}^{x_2} y \, dx = 2h(y_1 + 4y_2 + y_3)/6 + O(h^5)$$

17.2.7.3 Change of variable in double integral

$$\iint f(x,y) dx \, dy = \iint F(u,v) |J| du \, dv$$

where

$$J = \frac{\partial(x,y)}{\partial(u,v)} = \begin{vmatrix} \dfrac{\partial x}{\partial u} & \dfrac{\partial x}{\partial v} \\ \dfrac{\partial y}{\partial u} & \dfrac{\partial y}{\partial v} \end{vmatrix} = \begin{vmatrix} \dfrac{\partial x}{\partial u} & \dfrac{\partial y}{\partial u} \\ \dfrac{\partial x}{\partial v} & \dfrac{\partial y}{\partial v} \end{vmatrix}$$

is the Jacobian of the transformation.

17.2.7.4 Differential mean value theorem

$$\frac{f(x+h) - f(x)}{h} = f'(x + \theta h) \; 0 < \theta < 1$$

17.2.7.5 Integral mean value theorem

$$\int_a^b f(x)g(x) dx = g(a + \theta h) \int_a^b f(x) dx$$

$$h = b - a, \, 0 < \theta < 1$$

17.2.8 Vector calculus

Let $s(x, y, z)$ be a scalar function of position and let $\mathbf{v}(x, y, z) = \mathbf{i}v_x(x, y, z) + \mathbf{j}v_y(x, y, z) + \mathbf{k}v_z(x, y, z)$ be a vector function of position. Define

$$\nabla = \mathbf{i}\frac{\partial}{\partial x} + \mathbf{j}\frac{\partial}{\partial y} + \mathbf{k}\frac{\partial}{\partial z}$$

so that

$$\nabla \cdot \nabla = \nabla^2 = \frac{\partial^2}{\partial x^2} + \frac{\partial^2}{\partial y^2} + \frac{\partial^2}{\partial z^2}$$

then

$$\text{grad } s = \nabla s = \mathbf{i}\frac{\partial s}{\partial x} + \mathbf{j}\frac{\partial s}{\partial y} + \mathbf{k}\frac{\partial s}{\partial z}$$

$$\text{div } \mathbf{v} = \nabla \cdot \mathbf{v} = \frac{\partial v_x}{\partial x} + \frac{\partial v_y}{\partial y} + \frac{\partial v_z}{\partial z}$$

$$\text{curl } \mathbf{v} = \nabla \times \mathbf{v} = \mathbf{i}\left(\frac{\partial v_z}{\partial y} - \frac{\partial v_y}{\partial z}\right) + \mathbf{j}\left(\frac{\partial v_x}{\partial z} - \frac{\partial v_z}{\partial x}\right)$$

$$+ \mathbf{k}\left(\frac{\partial v_y}{\partial x} - \frac{\partial v_x}{\partial y}\right)$$

The following identities are then true:

$$\text{div}(s\mathbf{v}) = s \text{ div } \mathbf{v} + (\text{grad } s) \cdot \mathbf{v}$$

$$\text{curl}(s\mathbf{v}) = s \text{ curl } \mathbf{v} + (\text{grad } s) \times \mathbf{v}$$

$$\text{div}(\mathbf{u} \times \mathbf{v}) = \mathbf{v} \cdot \text{curl } \mathbf{u} - \mathbf{u} \cdot \text{curl } \mathbf{v}$$

$$\text{curl}(\mathbf{u} \times \mathbf{v}) = \mathbf{u} \text{ div } \mathbf{v} - \mathbf{v} \text{ div } \mathbf{u} + (\mathbf{v} \cdot \nabla)\mathbf{u} - (\mathbf{u} \cdot \nabla)\mathbf{v}$$

$$\text{div grad } s = \nabla^2 s$$

$$\text{div curl } \mathbf{v} = 0$$

$$\text{curl grad } s = 0$$

$$\text{curl curl } \mathbf{v} = \text{grad}(\text{div } \mathbf{v}) - \nabla^2 \mathbf{v}$$

where ∇^2 operates on each component of $\mathbf{v}$.

$$\mathbf{v} \times \text{curl } \mathbf{v} + (\mathbf{v} \cdot \nabla)\mathbf{v} = \text{grad } \tfrac{1}{2} \mathbf{v}^2$$

potentials:

If curl $\mathbf{v} = 0$, $\mathbf{v} = \text{grad } \phi$ where ϕ is a scalar potential.
If div $\mathbf{v} = 0$, $\mathbf{v} = \text{curl } \mathbf{A}$ where $\mathbf{A}$ is a vector potential.

17.3 Series and transforms

17.3.1 Arithmetic series

Sum of n terms,

$$S_n = a + (a + d) + (a + 2d) + \ldots + [a + (n - 1)d]$$
$$= n[2a + (n - 1)d]/2$$
$$= n(a + l)/2$$

17.3.2 Geometric series

Sum of n terms,

$$S_n = a + ar + ar^2 + \ldots + ar^{n-1} = a(1 - r^n)/(1 - r)$$
$$(\,|\,r\,|\, < 1)$$
$$S_\infty = a/(1 - r)$$

17.3.3 Binomial series

$$(1 + x)^p = 1 + px + \frac{p(p - 1)}{2!} x^2$$

$$+ \frac{p(p - 1)(p - 2)}{3!} x^3 + \ldots$$

If p is a positive integer the series terminates with the term in x^p and is valid for all x; otherwise the series does not terminate, and is valid only for $-1 < x < 1$.

17.3.4 Taylor's series

Infinite form

$$f(x + h) = f(x) + hf'(x) + \frac{h^2}{2!} f''(x) + \ldots$$

$$+ \frac{h^n}{n!} f^{(n)}(x) + \ldots$$

Finite form

$$f(x + h) = f(x) + hf'(x) + \frac{h^2}{2!} f''(x) + \ldots$$

$$+ \frac{h^n}{n!} f^{(n)}(x) + \frac{h^{n+1}}{(n + 1)!} f^{(n+1)}(x + \lambda h)$$

where $0 \leqslant \lambda \leqslant 1$.

17.3.5 Maclaurin's series

$$f(x) = f(0) + xf'(0) + \frac{x^2}{2!} f''(0) + \ldots + \frac{x^n}{n!} f^{(n)}(0) + \ldots$$

Neither of these series is necessarily convergent, but both usually are for appropriate ranges of values of h and x respectively.

17.3.6 Laurent's series

If a function $f(z)$ of a complex variable is analytic on and everywhere between two concentric circles centre a, then at any point in this region

$$f(z) = a_0 + a_1(z - a)$$
$$+ \ldots + b_1/(z - a) + b_2/(z - a)^2 + \ldots$$

This series is often applicable when Taylor's series is not.

17.3.7 Power series for real variables

	Math	Comp
$e^x = 1 + x + \dfrac{x^2}{2!} + \ldots$	all x	$\lvert x \rvert \leqslant 1$
$\ln(1 + x) = x - \dfrac{x^2}{2} + \dfrac{x^3}{3}$ $\quad -\dfrac{x^4}{4} + \ldots$	$-1 < x \leqslant 1$	
$\sin x = x - \dfrac{x^3}{3!} + \dfrac{x^5}{5!} - \dfrac{x^7}{7!} + \ldots$	all x	$\lvert x \rvert \leqslant 1$
$\cos x = 1 - \dfrac{x^2}{2!} + \dfrac{x^4}{4!} - \dfrac{x^6}{6!} + \ldots$	all x	$\lvert x \rvert \leqslant 1$
$\tan x = x + \dfrac{x^3}{3} + \dfrac{2x^5}{15} + \dfrac{17x^7}{315} + \ldots$		$\lvert x \rvert < \dfrac{\pi}{2}$
$\arctan x = x - \dfrac{x^3}{3} + \dfrac{x^5}{5} - \dfrac{x^7}{7} + \ldots$		$\lvert x \rvert \leqslant 1$
$\sinh x = x + \dfrac{x^3}{3!} + \dfrac{x^5}{5!} + \dfrac{x^7}{7!} + \ldots$	all x	$\lvert x \rvert \leqslant 1$
$\cosh x = 1 + \dfrac{x^2}{2!} + \dfrac{x^4}{4!} + \dfrac{x^6}{6!} + \ldots$	all x	$\lvert x \rvert \leqslant 1$

The column headed 'Math' contains the range of values of the variable x for which the series is convergent in the pure mathematical sense. In some cases a different range of values is given in the column headed 'Comp', to reduce the rounding errors which arise when computers are used.

17.3.8 Integer series

$$\sum_{n=1}^{N} n = 1 + 2 + 3 + 4 + \ldots + N = N(N + 1)/2$$

$$\sum_{n=1}^{N} n^2 = 1^2 + 2^2 + 3^2 + 4^2 + \ldots + N^2$$
$$= N(N + 1)(2N + 1)/6$$

$$\sum_{n=1}^{N} n^3 = 1^3 + 2^3 + 3^3 + 4^3 + \ldots + N^3 = N^2(N + 1)^2/4$$

$$\sum_{n=1}^{\infty} \frac{(-1)^{n+1}}{n} = 1 - \frac{1}{2} + \frac{1}{3} - \frac{1}{4} + \ldots = \ln(2)$$

(see $\ln(1 + x)$)

$$\sum_{n=1}^{\infty} \frac{(-1)^{n+1}}{2n - 1} = 1 - \frac{1}{3} + \frac{1}{5} - \frac{1}{7} + \ldots = \frac{\pi}{4} \text{(see arctan } x)$$

$$\sum_{n=1}^{\infty} \frac{1}{n^2} = 1 + \frac{1}{4} + \frac{1}{9} + \frac{1}{16} + \ldots = \frac{\pi^2}{6}$$

$$\sum_{n=1}^{N} n(n + 1)(n + 2) \ldots (n + r)$$
$$= 1 . 2 . 3 \ldots + 2 . 3 . 4 \ldots + 3 . 4 . 5 \ldots + \ldots$$
$$+ N(N + 1)(N + 2) \ldots (N + r)$$
$$= \frac{N(N + 1)(N + 2) \ldots (N + r + 1)}{r + 2}$$

17.3.9 Fourier series

$$f(\theta) = \frac{1}{2} a_0 + \sum_{n=1}^{\infty} (a_n \cos n\theta + b_n \sin n\theta)$$

with

$$a_n = \frac{1}{\pi} \int_0^{2\pi} f(\Theta) \cos n\Theta d\Theta$$

$$b_n = \frac{1}{\pi} \int_0^{2\pi} f(\Theta) \sin n\Theta d\Theta$$

or

$$f(\theta) = \sum_{n=-\infty}^{\infty} c_n \exp(jn\theta)$$

with

$$c_n = \frac{1}{2\pi} \int_0^{2\pi} f(\Theta) \exp(-jn\Theta) d\Theta = \begin{cases} \frac{1}{2}(a_n + jb_n) \ n < 0 \\ \frac{1}{2}(a_n - jb_n) \ n > 0 \end{cases}$$

The above expressions for Fourier series are valid for functions having at most a finite number of discontinuities within the period 0 to 2 of the variable of integration.

17.3.10 Rectified sine wave

$$f(\omega t) = \frac{1}{\pi} + \frac{1}{2} \cos \omega t + \frac{2}{\pi} \sum_{n=1}^{\infty} (-1)^{n+1} \frac{\cos 2n\omega t}{4n^2 - 1}$$

$$f(\omega t) = \frac{\sin(\pi/p)}{\pi/p} + \frac{2p}{\pi} \sin\left(\frac{\pi}{p}\right) \sum_{n=1}^{\infty} (-1)^{n+1} \frac{\cos np\omega t}{p^2 n^2 - 1}$$

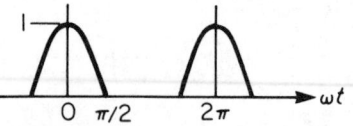

Figure 17.11 Half wave

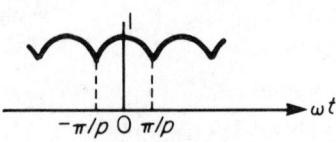

Figure 17.12 p-phase

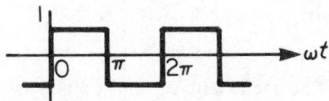

Figure 17.13 Square wave

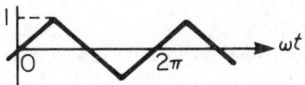

Figure 17.14 Triangular wave

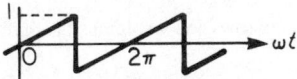

Figure 17.15 Sawtooth wave

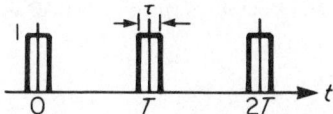

Figure 17.16 Pulse wave

17.3.11 Square wave

$$f(\omega t) = \frac{4}{\pi} \sum_{n=1}^{\infty} \frac{\sin(2n-1)\omega t}{(2n-1)}$$

17.3.12 Triangular wave

$$f(\omega t) = \frac{8}{\pi^2} \sum_{n=1}^{\infty} (-1)^{n+1} \frac{\sin(2n-1)\omega t}{(2n-1)^2}$$

17.3.13 Sawtooth wave

$$f(\omega t) = \frac{2}{\pi} \sum_{n=1}^{\infty} (-1)^{n+1} \frac{\sin n\omega t}{n}$$

17.3.14 Pulse wave

$$f(t) = \frac{\tau}{T} + \frac{2\tau}{T} \sum_{n=1}^{\infty} \frac{\sin(n\omega\tau/T)}{n\pi\tau/T} \cos\left(\frac{2n\pi t}{T}\right)$$

17.3.15 Fourier transforms

Among other applications, these are used for converting from the time domain to the frequency domain.

Basic formulae:

$$\int_{-\infty}^{\infty} U(f)\exp(j2\pi ft)df = u(t) \leftrightarrows U(f) = \int_{-\infty}^{\infty} u(t)\exp(-j2\pi ft)dt$$

Change of sign and complex conjugates:

$$u(-t) \leftrightarrows (U(-f), \; u^*(t) \leftrightarrows U^*(-f)$$

Time and frequency shifts (τ and ϕ constant):

$$u(t-\tau) \leftrightarrows U(f)\exp(-j2\pi f\tau)\exp(j2\pi\phi t)u(t) \leftrightarrows U(f-\phi)$$

Scaling (T constant):

$$u(t/T) \leftrightarrows TU(fT)$$

Products and convolutions:

$$u(t)^*v(t) \leftrightarrows U(f)V(f), \; u(t)v(t) \leftrightarrows U(f)^*V(f)$$

Differentiation:

$$u'(t) \leftrightarrows j2\pi fU(f), \; -j2\pi tu(t) \leftrightarrows U'(f)$$

$$\partial u(t,\alpha)/\partial\alpha \leftrightarrows \partial(U-f,\alpha)/\partial\alpha$$

Integration ($U(0) = 0$, a and b real constants):

$$\int_{-\infty}^{t} u(\tau)d\tau \leftrightarrows U(f)/j2\pi f$$

$$\int_{a}^{b} v(t,\alpha)d\alpha \leftrightarrows \int_{a}^{b} V(f,\alpha)d\alpha$$

Interchange of functions:

$$U(t) \leftrightarrows u(-f)$$

Dirac delta functions:

$$\delta(t) \leftrightarrows 1 \qquad \exp(j2\pi f_0 t) \leftrightarrows \delta(f-f_0)$$

Rect(t) (unit length, unit amplitude pulse, centred on $t = 0$):

$$\text{rect}(t) \leftrightarrows \sin \pi f/\pi f$$

Gaussian distribution:

$$\exp(-\pi t^2) \leftrightarrows \exp(-\pi f^2)$$

Repeated and impulse (delta function) sampled waveforms:

$$\sum_{-\infty}^{\infty} u(t-nT) \leftrightarrows (1/T)U(f) \sum_{-\infty}^{\infty} \delta(f-n/T)$$

$$u(t) \sum_{-\infty}^{\infty} \delta(t-nT) \leftrightarrows (1/T) \sum_{-\infty}^{\infty} U(f-n/T)$$

Parseval's lemma:

$$\int_{-\infty}^{\infty} u(t)v^*(t)dt = \int_{-\infty}^{\infty} U(f)V^*(f)df$$

$$\int_{-\infty}^{\infty} |u(t)|^2 dt = \int_{-\infty}^{\infty} |U(f)|^2 df$$

17.3.16 Laplace transforms

$$\bar{x}_s = \int_{0}^{\infty} x(t)\exp(-st)$$

Function	Transform	Remarks
$e^{-\alpha t}$	$\dfrac{1}{s + \alpha}$	
$\sin \omega t$	$\dfrac{\omega}{s^2 + \omega^2}$	
$\cos \omega t$	$\dfrac{s}{s^2 + \omega^2}$	
$\sinh \omega t$	$\dfrac{\omega}{s^2 - \omega^2}$	
$\cosh \omega t$	$\dfrac{s}{s^2 - \omega^2}$	
t^n	$n!/s^{n+1}$	
$H(t)$	$1/s$	
$H(t - \tau)$	$\dfrac{1}{s} \exp(-s\tau)$	Heaviside step function
$x(t - \tau)H(t - \tau)$	$\exp(-s\tau)\bar{x}(s)$	Shift in t
$\delta(t - \tau)$	$\exp(-s\tau)$	Dirac delta function
$\exp(-\alpha t)x(t)$	$\bar{x}(s + \alpha)$	Shift in s
$\exp(-\alpha t) \sin \omega t$	$\dfrac{\omega}{(s + \alpha)^2 + \omega^2}$	
$\exp(-\alpha t) \cos \omega t$	$\dfrac{(s + \alpha)}{(s + \alpha)^2 + \omega^2}$	
$tx(t)$	$-\dfrac{\mathrm{d}\bar{x}(s)}{\mathrm{d}s}$	
$\dfrac{\mathrm{d}x(t)}{\mathrm{d}t} = x'(t)$	$s\bar{x}(s) - x(0)$	
$\dfrac{\mathrm{d}^2x(t)}{\mathrm{d}t^2} = x''(t)$	$s^2\bar{x}(s - sx(0) - x'(0)$	
$\dfrac{\mathrm{d}^n x(t)}{\mathrm{d}x^n} = x^{(n)}(t)$	$s^n\bar{x}(s) - s^{n-1}x(0) - s^{n-2}x'(0) \dots$ $\qquad\qquad -sx^{(n-2)}(0) - x^{(n-1)}(0)$	

Convolution integral

$$\int_0^t x_1(\sigma)x_2(t - \sigma)\mathrm{d}\sigma \rightarrow \bar{x}_1(s)\bar{x}_2(s)$$

17.4 Matrices and determinants

17.4.1 Linear simultaneous equations

The set of equations

$a_{11}x_1 + a_{12}x_2 + \dots + a_{1n}x_n = b_1$

$a_{21}x_1 + a_{22}x_2 + \dots + a_{2n}x_n = b_1$

$\qquad \dots$

$a_{n1}x_1 + a_{n2}x_2 + \dots + a_{nn}x_n = b_n$

may be written symbolically

$$\mathbf{Ax} = \mathbf{b}$$

in which $\mathbf{A}$ is the *matrix* of the coefficients a_{ij}, and $\mathbf{x}$ and $\mathbf{b}$ are the *column matrices* (or vectors) $(x_1 \dots x_n)$ and $(b_1 \dots b_n)$. In this case the matrix $\mathbf{A}$ is square $(n \times n)$. The equations can be solved unless two or more of them are not independent, in which case

$$\det \mathbf{A} = |\mathbf{A}| = 0$$

and there then exist non-zero solutions x_i only if $\mathbf{b} = 0$. If $\det \mathbf{A} \neq 0$, there exist non-zero solutions only if $\mathbf{b} \neq 0$. When $\det \mathbf{A} = 0$, $\mathbf{A}$ is *singular*.

17.4.2 Matrix arithmetic

If $\mathbf{A}$ and $\mathbf{B}$ are both matrices of m rows and n columns they are *conformable*, and

$$\mathbf{A} \pm \mathbf{B} = \mathbf{C} \text{ where } C_{ij} = A_{ij} \pm B_{ij}$$

17.4.2.1 Product

If $\mathbf{A}$ is an $m \times n$ matrix and $\mathbf{B}$ an $n \times l$, the product $\mathbf{AB}$ is defined by

$$(\mathbf{AB})_{ij} = \sum_{k=1}^{n} (\mathbf{A})_{ik}(\mathbf{B})_{kj}$$

In this case, if $l \neq m$, the product $\mathbf{BA}$ will not exist.

17.4.2.2 Transpose

The transpose of $\mathbf{A}$ is written $\mathbf{A}'$ or $\mathbf{A}^t$ and is the matrix whose rows are the columns of $\mathbf{A}$, i.e.

$$(\mathbf{A}^t)_{ij} = (\mathbf{A})_{ji}$$

A square matrix may be equal to its transpose, and it is then said to be *symmetrical*. If the product $\mathbf{AB}$ exists, then

$$(\mathbf{AB})^t = \mathbf{B}^t\mathbf{A}^t$$

17.4.2.3 Adjoint

The *adjoint* of a square matrix $\mathbf{A}$ is defined as $\mathbf{B}$, where

$$(\mathbf{B})_{ij} = (A)_{ij}$$

and A_{ij} is the *cofactor* of a_{ji} in $\det \mathbf{A}$.

17.4.2.4 Inverse

If $\mathbf{A}$ is non-singular, the *inverse* $\mathbf{A}^{-1}$ is given by

$$\mathbf{A}^{-1} = \mathrm{adj}\,\mathbf{A}/\det \mathbf{A} \text{ and } \mathbf{A}^{-1}\mathbf{A} = \mathbf{A}\mathbf{A}^{-1} = 1$$

the *unit* matrix.

$$(\mathbf{AB})^{-1} = \mathbf{B}^{-1}\mathbf{A}^{-1}$$

if both inverses exist. The original equations $\mathbf{Ax} = \mathbf{b}$ have the solutions $\mathbf{x} = \mathbf{A}^{-1}\mathbf{b}$ if the inverse exists.

17.4.2.5 Orthogonality

A matrix $\mathbf{A}$ is orthogonal if $\mathbf{AA}^t = 1$. If $\mathbf{A}$ is the matrix of a coordinate transformation $\mathbf{X} = \mathbf{AY}$ from variables y_i to variables x_i, then if $\mathbf{A}$ is orthogonal $\mathbf{X}^t\mathbf{X} = \mathbf{Y}^t\mathbf{Y}$, or

$$\sum_{i=1}^{n} x_i^2 = \sum_{i=1}^{n} y_i^2$$

17.4.3 Eigenvalues and eigenvectors

The equation

$$\mathbf{Ax} = \lambda\mathbf{x}$$

where $\mathbf{A}$ is a square matrix, $\mathbf{x}$ a column vector and λ a number (in general complex) has at most n solutions $(\mathbf{x}, \lambda)$. The values of λ are *eigenvalues* and those of $\mathbf{x}$ *eigenvectors* of the matrix $\mathbf{A}$. The relation may be written

$$(\mathbf{A} - \lambda\mathbf{I})\mathbf{x} = 0$$

so that if $\mathbf{x} \neq 0$, the equation $\mathbf{A} - \lambda\mathbf{I} = 0$ gives the eigenvalues. If $\mathbf{A}$ is symmetric and real, the eigenvalues are real. If $\mathbf{A}$ is symmetric, the eigenvectors are orthogonal. If $\mathbf{A}$ is not symmetric, the eigenvalues are complex and the eigenvectors are not orthogonal.

17.4.4 Coordinate transformation

Suppose $\mathbf{x}$ and $\mathbf{y}$ are two vectors related by the equation

$$\mathbf{y} = \mathbf{Ax}$$

when their components are expressed in one orthogonal system, and that a second orthogonal system has unit vectors $\mathbf{u}_1, \mathbf{u}_2, \ldots, \mathbf{u}_n$ expressed in the first system. The components of $\mathbf{x}$ and $\mathbf{y}$ expressed in the new system will be $\mathbf{x}'$ and $\mathbf{y}'$, where

$$\mathbf{x}' = \mathbf{U}^t\mathbf{x}, \quad \mathbf{y}' = \mathbf{U}^t\mathbf{y}$$

and $\mathbf{U}^t$ is the orthogonal matrix whose rows are the unit vectors $\mathbf{u}_1^t$, $\mathbf{u}_2^t$, etc. Then

$$\mathbf{y}' = \mathbf{U}^t\mathbf{y} = \mathbf{U}^t\mathbf{Ax} = \mathbf{U}^t\mathbf{Ax} = \mathbf{U}^t\mathbf{AUx}'$$

or

$$\mathbf{y}' = \mathbf{A}'\mathbf{x}'$$

where

$$\mathbf{A}' = \mathbf{U}^t\mathbf{AU}$$

Matrices $\mathbf{A}$ and $\mathbf{A}'$ are *congruent*.

17.4.5 Determinants

The determinant

$$D = \begin{vmatrix} a_{11} & a_{12} & \cdots & a_{1n} \\ a_{21} & a_{22} & \cdots & a_{2n} \\ \vdots & \vdots & \vdots & \vdots \\ a_{n1} & a_{n2} & \cdots & a_{nn} \end{vmatrix}$$

is defined as follows. The first suffix in a_{rs} refers to the row, the second to the column which contains a_{rs}. Denote by M_{rs} the determinant left by deleting the rth row and sth column from D, then

$$D = \sum_{k=1}^{n} (-1)^{k+1} a_{1k} M_{1k}$$

gives the value of D in terms of determinants of order $n - 1$, hence by repeated application, of the determinant in terms of the elements a_{rs}.

17.4.6 Properties of determinants

If the rows of $|a_{rs}|$ are identical with the columns of $|b_{sr}|$, $a_{rs} = b_{sr}$ and

$$|a_{rs}| = |b_{sr}|$$

that is, the *transposed* determinant is equal to the original.

If two rows or two columns are interchanged, the numerical value of the determinant is unaltered, but the sign will be changed if the permutation of rows or columns is odd.

If two rows or two columns are identical, the determinant is zero.

If each element of one row or one column is multiplied by k, so is the value of the determinant.

If any row or column is zero, so is the determinant.

If each element of the pth row or column of the determinant c_{rs} is equal to the sum of the elements of the same row or column in determinants a_{rs} and b_{rs}, then

$$|c_{rs}| = |a_{rs}| + |b_{rs}|$$

The addition of any multiple of one row (or column) to another row (or column) does not alter the value of the determinant.

17.4.6.1 Minor

If row p and column q are deleted from $|a_{rs}|$, the remaining determinant M_{pq} is called the *minor* of a_{pq}.

17.4.6.2 Cofactor

The *cofactor* of a_{pq} is the minor of a_{pq} prefixed by the sign which the product $M_{pq}a_{pq}$ would have in the expansion of the determinant, and is denoted by A_{pq}:

$$A_{pq} = (-1)^{p+q} M_{pq}$$

A determinant a_{ij} in which $a_{ij} = a_{ji}$ for all i and j is called *symmetric*, whilst if $a_{ij} = -a_{ji}$ for all i and j, the determinant is *skew-symmetric*. It follows that $a_{ii} = 0$ for all i in a skew-symmetric determinant.

17.4.7 Numerical solution of linear equations

Evaluation of a determinant by direct expansion in terms of elements and cofactors is disastrously slow, and other methods are available, usually programmed on any existing computer system.

17.4.7.1 Reduction of determinant or matrix to upper triangular or to diagonal form

The system of equations may be written

$$\begin{bmatrix} a_{11} & a_{12} & \cdots & a_{1n} \\ a_{21} & a_{22} & \cdots & a_{2n} \\ \vdots & \vdots & \vdots & \vdots \\ a_{n1} & a_{n2} & \cdots & a_{nn} \end{bmatrix} \begin{bmatrix} x_1 \\ x_2 \\ \vdots \\ x_n \end{bmatrix} = \begin{bmatrix} b_1 \\ b_2 \\ \vdots \\ b_n \end{bmatrix}$$

The variable x_1 is eliminated from the last $n - 1$ equations by adding a multiple $-a_{i1}/a_{11}$ of the first row to the ith, obtaining

$$\begin{bmatrix} a_{11} & a_{12} & \cdots & a_{1n} \\ 0 & a_{22}' & \cdots & a_{2n} \\ \vdots & \vdots & \cdots & \vdots \\ 0 & 0 & \cdots & a_{nn}'' \end{bmatrix} \begin{bmatrix} x_1 \\ x_2 \\ \vdots \\ x_n \end{bmatrix} = \begin{bmatrix} b_1 \\ b_1' \\ \vdots \\ b_n'' \end{bmatrix}$$

where primes indicate altered coefficients. This process may be continued by eliminating x_2 from rows 3 to n, and so on.

Eventually the form will become

$$
\begin{bmatrix}
a_{11} & a_{12} & \cdots & a_{1n} \\
0 & a'_{22} & \cdots & a'_{2n} \\
\vdots & \vdots & \cdots & \vdots \\
0 & 0 & \cdots & a''_{nn}
\end{bmatrix}
\begin{bmatrix}
x_1 \\
x_2 \\
\vdots \\
x_n
\end{bmatrix}
=
\begin{bmatrix}
b_1 \\
b'_2 \\
\vdots \\
b''_n
\end{bmatrix}
$$

x_n can now be found from the nth equation, substituted in the $(n-1)$th to obtain x_{n-1} and so on.

Alternatively the process may be applied to the system of equations in the form

Ax = Ib

where **I** is the unit matrix, and the same operations carried out upon **I** as upon **A**. If the process is continued after reaching the upper triangular form, the matrix **A** can eventually be reduced to diagonal form. Finally, each equation is divided by the corresponding diagonal element of **A**, thus reducing **A** to the unit matrix. The system is now in the form

Ix = Bb

and evidently $\mathbf{B} = \mathbf{A}^{-1}$. The total number of operations required is $O(n^3)$.

17.5 Differential equations

A differential equation is an equation involving a *dependent* variable and its derivatives with respect to one or more *independent* variables. An *ordinary* differential equation is one in which there is only one independent variable – conventionally x or t. A *partial* differential equation is one in which there are several independent variables.

17.5.1 Notation and definitions

An ordinary differential equation with y as *dependent* variable and x as *independent* variable has the general form

$$
f\left\{x; y, \frac{dy}{dx}, \frac{d^2y}{dx^2}, \ldots\right\} = 0
$$

where $f\{\ \}$ represents some specified function of the arguments. *Solving* a differential equation involves obtaining an explicit expression for y as a known function of x.

The *order* of a differential equation is the order of the highest derivative appearing in it. Thus

$$
\frac{d^2y}{dx^2} + 3\frac{dy}{dx} + 6y = 6
$$

is a second-order equation. A differential equation of order n has a *general* solution containing n *arbitrary constants*. Specified values of the *dependent* variable and/or its derivatives which allow these arbitrary constants to be determined are called *boundary conditions* or (when the independent variable is t and the values are given at $t = 0$) *initial conditions*. Boundary conditions in which the dependent variable or its derivatives are assigned zero values are called *homogeneous* boundary conditions. A solution in which the arbitrary constants take definite values is called a *particular* solution.

A *linear* differential equation is one which is linear in the *dependent* variable and its derivatives, having the general form

$$
p_n(x)\frac{d^ny}{dx^n} + \ldots + p_1(x)\frac{dy}{dx} + p_0(x)y = f(x) \tag{17.1}
$$

where $p_0(x) \ldots p_n(x)$ and $f(x)$ are specified functions of x. If $f(x) \neq 0$ the differential equation is said to be *inhomogeneous*. If $f(x) = 0$, so that

$$
p_n(x)\frac{d^ny}{dx^n} + \ldots + p_1(x)\frac{dy}{dx} + p_0(x)y = 0 \tag{17.2}
$$

the differential equation is said to be *homogeneous*.

In a partial differential equation the independent variables are normally variables defining spatial position plus (possibly) time. A particular solution of a partial differential equation requires the definition of a *solution region* with a bounding curve or bounding surface, together with the specification of suitable boundary conditions on that curve or surface. A partial differential equation, like an ordinary differential equation, may be linear or non-linear, and a linear partial differential equation may be homogeneous or inhomogeneous. Boundary conditions, specifying values of the dependent variable and/or its derivatives, may also be homogeneous or inhomogeneous.

17.5.2 Ordinary differential equations: analytical solutions

Simple analytical solutions exist for first-order linear differential equations and for linear equations of higher order with constant coefficients.

17.5.2.1 First-order linear equations

A *first-order* linear differential equation has the general form $p_1(x)(dy/dx) + p_0(x)y = f(x)$, which can be written as

$$
\frac{dy}{dx} + P(x)y = Q(x) \tag{17.3}
$$

This equation has the general solution

$$
ye^{\int P(x)dx} = \int Q(x)e^{\int P(x)dx}dx + C \tag{17.4}
$$

where C is an arbitrary constant. The function $e^{\int P(x)dx}$ is known as the *integrating factor*.

17.5.2.2 Linear equations with constant coefficients

Homogeneous equations A second-order homogeneous linear differential equation with constant coefficients has the general form

$$
a\frac{d^2y}{dx^2} + b\frac{dy}{dx} + cy = 0 \tag{17.5}
$$

The general solution is

$$
y = C_1e^{\lambda_1x} + C_2e^{\lambda_2x} \tag{17.6}
$$

where λ_1, λ_2 are the roots of the *auxiliary equation* $a\lambda^2 + b\lambda + c = 0$ and C_1, C_2 are arbitrary constants.

If the roots of the auxiliary equation are complex, with values $\lambda_1 = \alpha + \int\beta$, $\lambda_2 = \alpha - i\beta$, it is more convenient to write the general solution of the differential equation in the form

$$
y = e^{\alpha x}(C_1 \cos \beta x + C_2 \sin \beta x) \tag{17.7}
$$

If the roots are equal, i.e. $\lambda_1 = \lambda_2 = \lambda$, say, then the general solution is

$$y = e^{\lambda x}(C_1 + C_2 x) \tag{17.8}$$

where again C_1, C_2 are arbitrary constants.

The solution of third- and higher-order homogeneous equations follows a similar pattern, the auxiliary equation being a polynomial equation in λ of appropriate degree.

Inhomogeneous equations   A second-order inhomogeneous linear differential equation with constant coefficients has the general form

$$a\frac{d^2 y}{dx^2} + b\frac{dy}{dx} + cy = f(x) \tag{17.9}$$

where $f(x)$ is a specified function. The general solution of equation (17.9) is the general solution of the homogeneous equation (17.5) containing two arbitrary constants (this solution is called the *complementary function*) plus a function (called the *particular integral*) which, when substituted into equation (17.9), gives the correct function $f(x)$ on the right-hand side.

For many simple right-hand sides the particular integral can be found by replacing y in the differential equation by a 'trial solution' containing one or more unknown parameters, here written as α, β, etc.

Right-hand side: $f(x)$	Trial solution: $y(x)$
constant	α
$x^n (n \text{ integral})$	$\alpha x^n + \beta x^{n-1} + \ldots$
e^{kx}	αe^{kx}
xe^{kx}	$(\alpha x + \beta)e^{kx}$
$x^n e^{kx}$	$(\alpha x^n + \beta x^{n-1} + \ldots)e^{kx}$

$\left.\begin{array}{l} \sin kx \\ \\ \cos kx \end{array}\right\}$ $\left\{\begin{array}{l} \alpha \sin kx + \beta \cos kx. \text{ (If only even differential co-} \\ \text{efficients occur in the differential equation then} \\ \alpha \sin kx \text{ or } \beta \cos kx \text{ is sufficient.)} \end{array}\right.$

$\left.\begin{array}{l} e^{k_1 x} \sin k_2 x \\ \\ e^{k_1 x} \cos k_2 x \end{array}\right\}$ $\quad e^{k_1 x}(\alpha \sin k_2 x + \beta \cos k_2 x)$

Equating the coefficients of the functions on the two sides of the equation gives the values of the parameters. This technique can also be used to solve equations of third and higher orders.

If $f(x)$ has the same form as one of the terms in the complementary function then the substitution $y = uf(x)$ should be made, where u is an unknown function of x. This substitution generates a simple differential equation for $u(x)$.

Simultaneous linear differential equations   The analysis of a linear mechanical or electrical system with several degrees of freedom may require the solution of a set of simultaneous linear differential equations, in which there is one independent variable (normally time) and several dependent variables. In cases where the equations have constant coefficients, as in the example

$$\frac{du}{dt} + 3\frac{dv}{dt} + u - v = t^2$$

$$2\frac{du}{dt} - \frac{dv}{dt} - 2u + 3v = 0$$

the equations can be solved by a procedure very similar to the elimination method for solving sets of linear algebraic equations. This procedure generates a linear differential equation (with order equal to the sum of the orders of the original equations) for one of the dependent variables: after solution of this equation the other dependent variables can be obtained by back-substitution.

Inserting the initial or boundary conditions   A linear differential equation of order n has a general solution

$$y = \xi_0(x) + C_1\xi_1(x) + C_2\xi_2(x) + \ldots + C_n\xi_n(x) \tag{17.10}$$

where $\xi_0(x)$ is the *particular integral* and $C_1\xi_1(x) + C_2\xi_2(x) + \ldots + C_n\xi_n(x)$ is the *complementary function*. Once this general solution has been found, the values of the n constants $C_1, \ldots, C_n$ can be obtained by imposing n boundary or initial conditions, i.e. n values of y and/or its derivatives at particular values of x. If all the boundary conditions are specified at a single value of x the problem is referred to as a *one-point boundary-value* problem or, if the independent variable is t and the conditions are specified at $t = 0$, as an *initial-value* problem. Initial value problems can also be solved by the use of Laplace transforms (see Section 17.3.16). The Laplace transform method determines a particular solution of a differential equation, with the initial conditions inserted, rather than the general solution (17.10).

Impulse and frequency responses: the convolution integral   The solution of the differential equation

$$a_n\frac{d^n y}{dt^n} + \ldots + a_1\frac{dy}{dt} + a_0 y = f(t) \tag{17.11}$$

for a general function of time $f(t)$ with homogeneous initial conditions

$$\frac{d^{n-1}y}{dt^{n-1}} = \frac{d^{n-2}y}{dt^{n-2}} = \ldots = \frac{dy}{dt} = y = 0 \text{ at } t = 0$$

can be obtained from the *impulse response* $g(t)$, which is the solution of the differential equation with the same initial conditions when $f(t) = \delta(t)$. ($\delta(t)$ is the Dirac δ-function, defined by the equations

$$\int_{-\infty}^{\infty} \delta(t)dt = 1; \ \delta(t) = 0 \text{ if } t \neq 0.)$$

The impulse response can be obtained by solving the homogeneous equation

$$a_n\frac{d^n y}{dt^n} + \ldots + a_1\frac{dy}{dt} + a_0 y = 0 \tag{17.12}$$

with initial conditions $(d^{n-1}y)/(dt^{n-1}) = 1/a_n$, $(d^{n-2})/(dt^{n-2}) = \ldots = dy/dt = y = 0$ at $t = 0$. Alternatively, it can be found by the use of Laplace transforms.

The solution of equation (17.11) for an arbitrary right-hand side $f(t)$ is given in terms of the impulse response $g(t)$ by the *convolution integral*

$$y(t) = \int_0^t g(\tau)f(t - \tau)d\tau \tag{17.13}$$

This integral is symmetric in the functions g and f, and can therefore be written in the alternative form

$$y(t) = \int_0^t f(\tau)g(t - \tau)d\tau \tag{17.14}$$

If $f(t) = e^{i\omega t}$ and equation (17.11) represents a stable system (i.e. the complementary function has no exponential terms with positive real part) then as $t \to \infty$ the solution tends to the 'steady state' form $y(t) = G(\omega)e^{i\omega t}$. The complex function

$G(\omega)$ is called the *frequency response* of the system. It may be obtained from the differential equation by substituting the trial solution $y = \alpha e^{i\omega t}$ or from the impulse response by the use of equation (17.13). The latter derivation gives the result

$$G(\omega) = \int_{-\infty}^{\infty} g(\tau)e^{-i\omega\tau}d\tau \qquad (17.15)$$

This equation states that the frequency response $G(\omega)$ is the *Fourier transform* of the impulse response $g(t)$ (see Section 17.3.15).

17.5.2.3 Linear equations with variable coefficients

Second- and higher-order linear equations with variable coefficients do not, in general, have solutions which are expressible in terms of elementary functions. However, there are a number of second-order equations which occur frequently in applied mathematics and for which tables of solutions exist. Sub-routines for generating these solutions are available on most scientific computers. Two of the most important of these equations are

Bessel's equation: $\quad x^2\dfrac{d^2y}{dx^2} + x\dfrac{dy}{dx} + (\lambda^2x^2 - n^2)y = 0$

$$\qquad (17.16)$$

Legendre's equation: $\quad (1 - x^2)\dfrac{d^2y}{dx^2} - 2x\dfrac{dy}{dx}$

$$+ n(n + 1)y = 0 \qquad (17.17)$$

In certain other cases an equation with variable coefficients can be converted into one with constant coefficients by means of a change of variable. In general, however, solutions of linear differential equations with variable coefficients can only be obtained by approximate methods.

17.5.3 Ordinary differential equations: approximate solutions

Approximate solutions of differential equations can be obtained by graphical, numerical or analytical methods.

17.5.3.1 A graphical method for first-order equations

A graphical solution of the general first-order equation $dy/dx = f(x,y)$ can be obtained as follows. A series of curves $f(x,y) = c_1, c_2, \ldots, c_i, \ldots$ (termed *isoclines*) are drawn in the x, y plane, where the c's are suitable constants. On each isocline line-segments are drawn with slope equal to the associated value of c_i: these segments give the direction of the solutions as they cross the isocline. The general form of these solutions can be obtained by joining up the segments to form continuous curves.

A simple example is shown in Figure 17.17, which illustrates the solution of the differential equation $dy/dx = -x/y$. The isoclines $-x/y = c_1, c_2, \ldots, c_i, \ldots$ are straight lines through the origin, and the segments which form part of the solutions are always perpendicular to the isoclines. It is clear from the figure that the solutions are circles centred on the origin: this is easily verified analytically.

17.5.3.2 Approximate numerical methods

Derivatives and differences If a continuous function $y(x)$ is sampled at a series of equally spaced points $x_0, \ldots, x_n, \ldots, x_N$ to give a set of values $y_0, \ldots, y_n, \ldots, y_N$ then it follows from the definition of a differential coefficient that

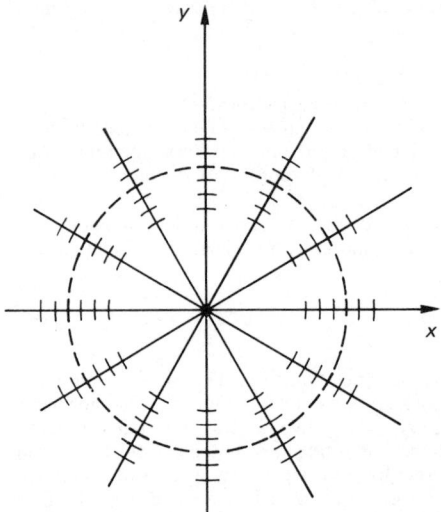

Figure 17.17 Isoclines for the differential equation $dy/dx = -x/y$

$$\left(\frac{dy}{dx}\right)_{n+1/2} \approx (y_{n+1} - y_n)/h, \quad \left(\frac{dy}{dx}\right)_{n-1/2} \approx (y_n - y_{n-1})/h$$

$$\qquad (17.18)$$

or alternatively

$$\left(\frac{dy}{dx}\right)_n \approx (y_{n+1} - y_{n-1})/2h \qquad (17.19)$$

where h is the sampling inverval, as shown in Figure 17.18. Taking the difference of the two equations (17.18) and dividing by h gives

$$\left(\frac{d^2y}{dx^2}\right)_n \approx (y_{n+1} - 2y_n + y_{n-1})/h^2 \qquad (17.20)$$

and the process can be continued in a similar way to give approximations to $(d^3y/dx^3)_{n+1/2}$, etc. The quantities $(y_1 - y_0), \ldots (y_{n+1} - y_n), \ldots (y_N - y_{N-1})$ are termed the *first*

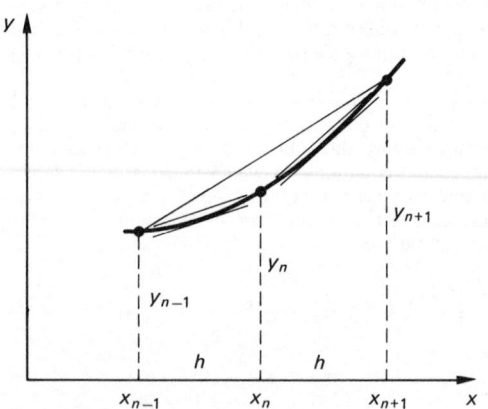

Figure 17.18 Approximate representations of dy/dx

differences of the set of values y_n, the quantities . . . $(y_{n+1} - 2y_n + y_{n-1})$, . . . the *second differences*, and so on. The role of differences in numerical analysis is similar to that of differential coefficients in calculus.

Two-point boundary-value problems An approximate solution of the second-order linear differential equation

$$p_2(x)\frac{d^2y}{dx^2} + p_1(x)\frac{dy}{dx} + p_0(x)y = f(x) \qquad (17.21)$$

with boundary conditions $y = y_0$ at $x = 0$, $y = y_N$ at $x = a$ can be found by dividing the solution range $0 \le x \le a$ into N equal intervals and replacing the continuous function $y(x)$ by a set of $N + 1$ quantities $y_n = y(x_n)$, $(n = 0, \ldots, N)$, where $x_n = nh$ and $h = a/N$. Replacing the differential coefficients in equation (17.21) by the approximations (17.19) and (17.20) gives

$$p_2(x_n)(y_{n+1} - 2y_n + y_{n-1}) + hp_1(x_n)(y_{n+1} - y_{n-1})/2$$
$$+ h^2p_0(x_n)y_n = f(x_n)$$
$$(n = 1, \ldots, N - 1) \qquad (17.22)$$

Setting up an equation of this form at each of the points $x_1, \ldots, x_{N-1}$ produces a set of $n - 1$ simultaneous linear algebraic equations which can be solved for the unknown function values $y_1, \ldots, y_{N-1}$ (the values of y_0 and y_N which appear in these equations are known from the boundary conditions). Intermediate values of $y(x)$ can be found subsequently by interpolation.

Initial-value problems The general first-order differential equation

$$\frac{dy}{dt} = f(t,y) \qquad (17.23)$$

with initial condition $y = y_0$ at $t = t_0$ can be solved by a step-by-step procedure in which approximate function values $y_1, y_2, \ldots$ are computed successively at $t = t_1, t_2, \ldots$ The simplest step-by-step procedure is due to Euler and involves the replacement of the differential equation (17.23) by the approximation

$$y_{n+1} = y_n + hf(t_n, y_n) \ (n = 0, 1, 2, \ldots) \qquad (17.24)$$

where h is equal to the interval $t_{n+1} - t_n$. As shown in Figure 17.19 this procedure takes the tangent at each solution point as the solution over the next interval. The truncation error in a single step is $O(h^2)$. If the step-length h is kept constant over a given range $0 \le t \le T$ the number of steps is T/h, so that the trauncation error over the range is $O(h)$. (The round-off error increases with the number of steps, so that there is an optimum value of h which minimizes the total error.)

The accuracy of the Euler procedure can be improved by using equation (17.24) as a 'predictor' to obtain an approximate value y_{n+1}^*, which is then inserted in a suitable 'corrector' formula to generate a more accurate value of y_{n+1}. A simple predictor/corrector pair is

Predictor $\quad y_{n+1}^* = y_n + hf(t_n, y_n) \qquad (17.25)$
Corrector $\quad y_{n+1} = y_n + h\{f(t_n, y_n) + f(t_{n+1}, y_{n+1}^*)\}/2$

One of the most popular predictor–corrector procedures is the Runge–Kutta. A single step of the procedure involves four evaluations of $f(t,y)$ in accordance with the formulae

$$a_1 = hf(t_n, y_n), \qquad\qquad a_2 = hf(t_n + h/2, y_n + a_1/2)$$
$$a_3 = hf(t_n + h/2, y_n + a_2/2), \ a_4 = hf(t_n + h, y_n + a_3)$$

the final value of y_{n+1} being

$$y_{n+1} = y_n + \{a_1 + 2a_2 + 2a_3 + a_4\}/6 \qquad (17.26)$$

The error per step is $O(h^5)$, so that the error over a given range of t is $O(h^4)$. A computer sub-routine for the Runge–Kutta procedure normally requires a user-supplied sub-routine to evaluate $f(t,y)$ for specified values of t and y.

An initial-value problem involving a differential equation of second or higher order can be solved by reducing the differential equation to a set of first-order equations. For example, the third-order non-linear equation

$$\frac{d^3y}{dt^3} + t\frac{d^2y}{dt^2} - 2t^2\left(\frac{dy}{dt}\right)^2 + y^2 = 10t$$

can be solved by introducing the additional variables u and v and writing the equation as

$$\frac{dy}{dt} = u; \ \frac{du}{dt} = v; \ \frac{dv}{dt} = 10t - tv + 2t^2u^2 - y^2$$

This set of first-order equations for the three variables u, v and y can be solved by any of the methods described above, the step-by-step procedure being carried forward simultaneously for each of the variables.

17.5.3.3 *Approximate analytical methods*

An approximate solution of a linear differential equation can also be obtained by choosing a set of M *basis functions* $B_m(x)$ and expressing the unknown solution $y(x)$ as

$$y(x) \approx c_1B_1(x) + \ldots + c_MB_M(x) = \sum_{m=1}^{M} c_mB_m(x) \quad (17.27)$$

There are a number of methods based on this approach. They may be classified according to the choice of basis functions $B_m(x)$ and the procedure used to find the constants c_m. The most important sets of basis functions are the integral powers of x (which generate power-series approximations) and the harmonic functions $\sin mx$ and $\cos mx$ (which generate Fourier approximations). In the following account the equation to be solved is written as

$$\mathcal{L}y = w(x) \qquad (17.28)$$

where $\mathcal{L}$ represents a specified linear differential operator and $w(x)$ is a specified function of x. It is assumed that a solution is required in an interval $p \le x \le q$ and that sufficient homoge-

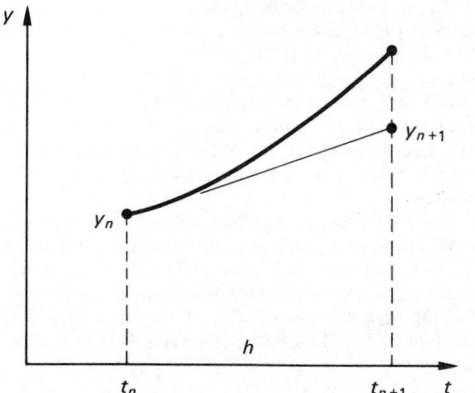

Figure 17.19 Euler's approximate integration procedure

neous boundary conditions are specified at $x = p$ and $x = q$ to make the solution unique. It is further assumed that each of the approximating functions $B_1(x), \ldots, B_M(x)$ satisfies these boundary conditions.

In general the approximation (17.27) will not be capable of satisfying the differential equation (17.28) exactly, whatever values are assigned to the constants c_i: there will be an error function

$$\epsilon(x) = \mathcal{L}\left\{ \sum_{m=1}^{M} c_m B_m(x) \right\} - w(x)$$

which can be written in the form

$$\epsilon(x) = \sum_{m=1}^{M} c_m b_m(x) - w(x) \tag{17.29}$$

where $b_m(x) = \mathcal{L}\{B_m(x)\}$.

Two procedures for finding sets of constants which make the error $\epsilon(x)$ 'small' are *Collocation* and *Galerkin's method*. In the Collocation method the constants c_m are obtained by making $\epsilon(x)$ zero at a selected set of points x_k $(k = 1, \ldots, M)$ in the interval $p \leqslant x \leqslant q$. This generates a set of M simultaneous equations

$$\sum_{m=1}^{M} b_m(x_k) c_m = w(x_k) \; (k = 1, \ldots, M) \tag{17.30}$$

which can be solved for the M constants. In Galerkin's method the constants c_m are obtained by making $\epsilon(x)$ orthogonal to the M basis functions $B(x)$,

$$\int_p^q B_k(x)\epsilon(x)\mathrm{d}x = \int_p^q B_k(x)\left\{ \sum_{m=1}^{M} c_m b_m(x) - w(x) \right\}\mathrm{d}x = 0$$

$(k = 1, \ldots, M)$

These equations can be written in the form

$$\sum_{m=1}^{M} \left\{ \int_p^q B_k(x)b_m(x)\mathrm{d}x \right\} c_m = \int_p^q B_k(x)w(x)\mathrm{d}x$$

$(k = 1, \ldots, M) \tag{17.31}$

Equation (17.31), like equation (17.30), represents a set of M linear algebraic equations for the unknown constants c_m. If the differential operator $\mathcal{L}$ is self-adjoint (a condition satisfied in most practical applications of the method) the coefficients

$$\int_p^q B_k(x)b_m(x)\mathrm{d}x$$

form a symmetric matrix. If, in addition, the functions $B_m(x)$ are chosen to be the normalized eigenfunctions of the differential operator $\mathcal{L}$, so that $\mathcal{L}\{B_m(x)\} = b_m(x) = \lambda_m B_m(x)$, then equation (17.31) takes the simpler form

$$c_k = \int_p^q B_k(x)w(x)\mathrm{d}x/\lambda_k \quad (k = 1, \ldots, M) \tag{17.32}$$

with each constant c_k depending only on the corresponding function $B_k(x)$.

17.5.4 Partial differential equations

Linear partial differential equations can be classified as elliptic, hyperbolic or parabolic. An elliptic differential equation is one in which the boundary conditions imposed on each segment of the boundary affect the solution at *all* points in the solution region or, conversely, one in which the solution at any point depends on the boundary conditions over the *whole*

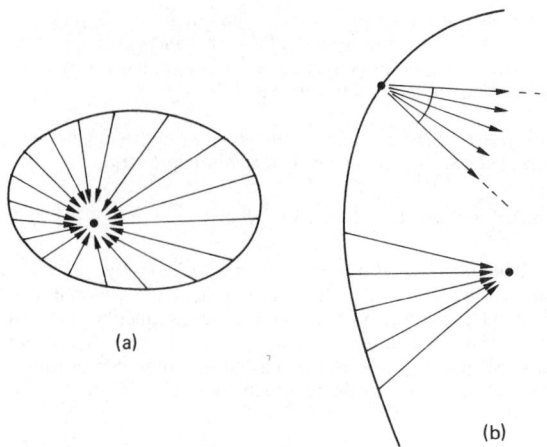

(a)

(b)

Figure 17.20 Partial differential equation types: (a) elliptic, (b) hyperbolic

boundary, as shown in Figure 17.20(a). The commonest elliptic equation is Laplace's equation

$$\frac{\partial^2 \phi}{\partial x^2} + \frac{\partial^2 \phi}{\partial y^2} + \frac{\partial^2 \phi}{\partial z^2} = 0 \tag{17.33}$$

which is the equation governing gravitational fields in free space, steady heat and electrical conduction, seepage flow in soils, etc. The inhomogeneous form of Laplace's equation is Poisson's equation

$$\frac{\partial^2 \phi}{\partial x^2} + \frac{\partial^2 \phi}{\partial y^2} + \frac{\partial^2 \phi}{\partial z^2} = -\sigma \tag{17.34}$$

where σ is a known function of position. This equation governs gravitational fields in regions containing distributed matter, heat conduction in the presence of distributed heat sources, etc.

Another elliptic differential equation of interest to mechanical engineers is the bi-harmonic equation governing the bending of an initially flat plate:

$$\frac{\partial^4 \phi}{\partial x^4} + 2\frac{\partial^4 \phi}{\partial x^2 \partial y^2} + \frac{\partial^4 \phi}{\partial y^4} = -q/D \tag{17.35}$$

where ϕ is the transverse displacement of the plate, q is the known distribution of transverse load and D is a constant representing the stiffness of the plate.

Equations (17.33)–(17.35) can also be written in the more general form $\nabla^2 \phi = 0$, $\nabla^2 \phi = -\sigma$, $\nabla^4 \phi = -q/D$, where ∇^2 is the Laplacian operator of vector calculus. This operator takes various forms, depending on the coordinate system (Cartesian, cylindrical polar, spherical polar, etc.) used to define the solution region.

A hyperbolic differential equation is one in which the boundary conditions on a segment of the boundary only affect a *part* of the solution region or, conversely, one in which the solution at any point only depends on the boundary conditions over *part* of the boundary, as shown in Figure 17.20(b). The commonest hyperbolic differential equation is the wave equation

$$\frac{\partial^2 \phi}{\partial x^2} = \frac{1}{a}\frac{\partial^2 \phi}{\partial t^2} \text{ or, more generally, } \nabla^2 \phi = \frac{1}{a^2}\frac{\partial^2 \phi}{\partial 7 t^2} \tag{17.36}$$

which governs the propagation of sound and other waves in both fluids and solids.

Another common partial differential equation is the diffusion equation

$$\frac{\partial^2 \phi}{\partial x^2} = \frac{1}{a^\delta} \frac{\partial \phi}{\partial t} \quad \text{or, more generally,} \quad \nabla^2 \phi = \frac{1}{a} \frac{\partial \phi}{\partial t} \qquad (17.37)$$

which governs, for example, the unsteady flow of heat in solids. The diffusion equation is an example of a parabolic differential equation. Such equations can be thought of as lying on the borderline between elliptic and hyperbolic forms.

17.5.4.1 Analytical solutions: separation of variables

Simple analytical solutions exist for linear partial differential equations with constant coefficients. For example, Laplace's equation in two dimensions is satisfied by both the real and imaginary parts of any analytic function $f(z)$, where z is the complex variable $x + \mathrm{j}y$. This fact allows many two-dimensional field problems to be solved by a technique known as *conformal mapping*. Similarly, the one-dimensional wave equation

$$\frac{\partial^2 \phi}{\partial x^2} = \frac{1}{a^2} \frac{\partial^2 \phi}{\partial t^2}$$

has solutions of the form $f(x \pm at)$, where f is an arbitrary differentiable function. These solutions represent waves of arbitrary shape travelling along the x axis.

Analytical solutions of linear partial differential equations can be obtained by using the method of *separation of variables*. For a differential equation whose dependent variable is ϕ and whose independent variables are x and y this method involves assuming a solution of the form $\phi = X(x)Y(y)$, where X is an unknown function of x only and Y is an unknown function of y only. Substitution of this solution into the differential equation yields ordinary differential equations for the functions X and Y, which can be solved by methods described in Section 17.5.2.2.

Typical examples of separable solutions are the function

$$\phi = \begin{Bmatrix} \cos \\ \sin \end{Bmatrix} \lambda x \begin{Bmatrix} \cosh \\ \sinh \end{Bmatrix} \lambda y$$

which satisfies both the two-dimensional Laplace equation and the homogeneous plate bending equation and the function

$$\phi = \mathrm{e}^{-a\lambda^2 t} \begin{Bmatrix} \cos \\ \sin \end{Bmatrix} \lambda x$$

which satisfies the one-dimensional diffusion equation.

Separable solutions always contain an arbitrary parameter λ called the *separation constant*. The imposition of boundary conditions on a solution may result in only certain values of λ being permissible. In such cases more general solutions can often be built up by combining a number of basic solutions involving these values of λ. For example, the solution of the one-dimensional diffusion equation given above implies the existence of a more general solution

$$\phi = \sum_{n=1}^{\infty} \mathrm{e}^{-a\lambda_n^2 t} (A_n \cos \lambda_n x + B_n \sin \lambda_n x)$$

which can be made to fit a variety of boundary conditions by suitable choice of the constants A_n and B_n.

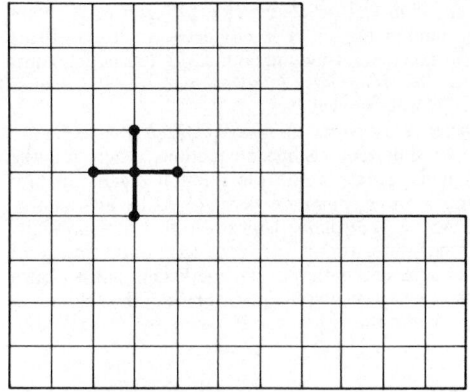

Figure 17.21 A finite-difference mesh

17.5.4.2 Numerical solutions: the finite-difference method

The finite-difference method for solving partial differential equations is similar to the numerical technique for solving ordinary differential equations with two-point boundary conditions described in Section 17.5.3.3. The following example shows how the method can be used to find the steady-state distribution of temperature within the L-shaped region shown in Figure 17.21 when the temperature variation on the boundary of the region is given. In this problem the temperature ϕ satisfies the two-dimensional Laplace equation

$$\frac{\partial^2 \phi}{\partial x^2} + \frac{\partial^2 \phi}{\partial y^2} = 0$$

with appropriate values of ϕ specified on the boundary.

The region is first covered with a uniform grid of squares, as shown in the figure. The intersections of the grid lines within the solution region are called *nodal points* and the values of ϕ at these points are called *nodal values*: it is these values which are determined by the method. At each nodal point the partial derivatives which make up the differential equation are replaced by differences, using an appropriately amended version of equation (17.20). This operation converts the partial differential equation into a linear algebraic equation involving the nodal values at the chosen nodal point and its four nearest neighbours. If these points are labelled as shown in Figure 17.22 then the linear equation associated with the point p is

$$(\phi_q + \phi_r + \phi_s + \phi_t - 4\phi_p)/h^2 = 0 \qquad (17.38)$$

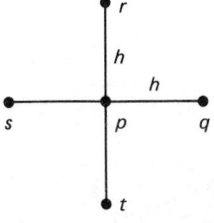

Figure 17.22 Nodal points associated with the difference equation for node p

A similar equation can be constructed for each nodal point within the solution region (it is not necessary to construct equations for nodal points on the boundary). For nodal points adjacent to the boundary at least one of the values $\phi_q, \ldots, \phi_t$ will be known.

This procedure converts the partial differential equation into a set of n simultaneous linear equations, where n is the number of nodal points within the solution region. In pre-computer days these equations were solved by an iterative process known as *relaxation*. Nowadays they are normally solved by a computer routine designed to take advantage of the sparse and banded nature of the coefficient matrix. Once the nodal values have been obtained, values of the solution at other points within the region can be found by interpolation.

17.5.4.3 Numerical solutions: the finite-element method

In recent years the finite-element method has largely replaced the finite-difference method as the standard numerical technique for solving problems of heat conduction and stress analysis in solid bodies. To assist in a comparison of the two approaches the following account considers the heat-conduction problem solved by finite differences in Section 17.5.4.2.

The finite-element method also begins with the construction of a 'mesh' covering the solution region. This mesh is commonly formed from triangles (the 'elements' of the method) although quadrilaterals can also be used. The mesh need not be uniform – indeed, it is standard practice to grade the mesh so that it is finer in regions where the solution is likely to vary rapidly, as shown in Figure 17.23.

The finite-element method, like the finite-difference method, changes the problem of solving a partial differential equation into that of solving a system of linear algebraic equations for a set of nodal values. However, in contrast to the finite-difference method, in which the value of the solution is only defined at the nodal points, the finite-element method replaces the actual solution by an approximation which is linear (or, more generally, a low-order polynomial) within each element.

The first stage of the solution procedure involves the determination of the properties of each individual element. A typical triangular element with nodal values ϕ_1, ϕ_2, ϕ_3 is shown in Figure 17.24. If the temperature ϕ within the

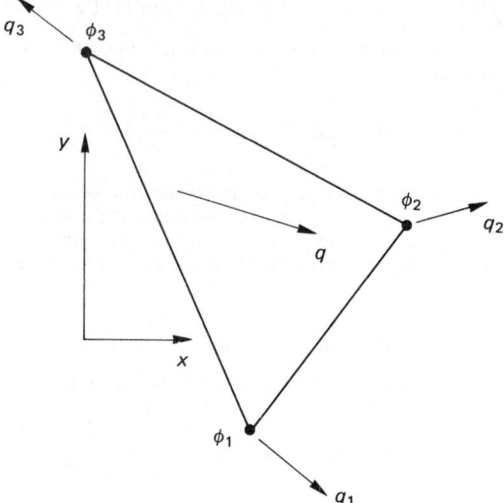

Figure 17.24 A typical finite element

element is assumed to vary linearly with position then it can be expressed in terms of the nodal values as

$$\phi = \phi_1 n_1(x,y) + \phi_2 n_2(x,y) + \phi_3 n_3(x,y) \qquad (17.39)$$

where n_1, n_2, n_3 are simple linear functions of x and y called *shape functions*. From this expression for ϕ it is straightforward to obtain the density of heat flow $\mathbf{q}$ (constant within the element) and the amount of heat flowing across each side of the triangle as linear functions of the nodal temperatures ϕ_1, ϕ_2, ϕ_3. In preparation for the next part of the procedure these distributed boundary flows are replaced by 'equivalent' concentrated flows q_1, q_2, q_3 at the vertices of the element, as shown in Figure 17.24, these concentrated flows being expressed as linear functions of the nodal temperatures ϕ_1, ϕ_2, ϕ_3.

The second stage of the solution procedure involves joining the elements together to form the solution region. This has two consequences. First, it imposes conditions of continuity on the temperature ϕ. If two elements have nodes p and q in common, as shown in Figure 17.25, then they share the same

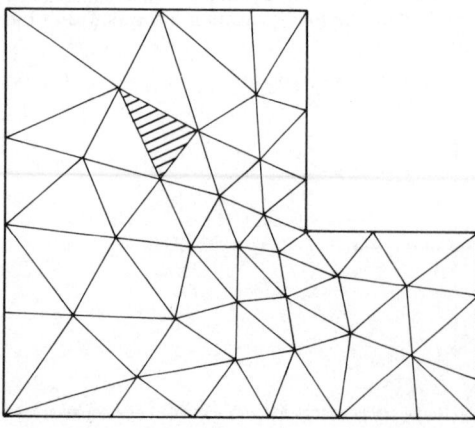

Figure 17.23 A finite-element mesh

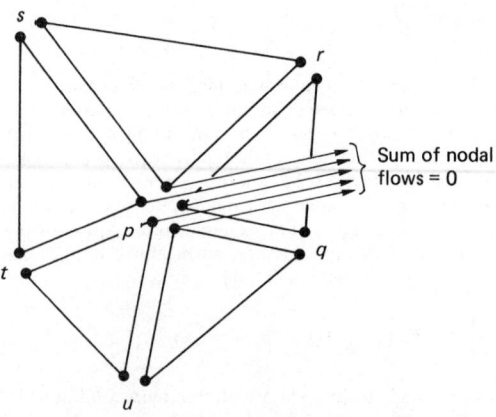

Figure 17.25 Adding up the nodal flows in a finite-element mesh

nodal values ϕ_p and ϕ_q. Furthermore, since ϕ is linear within each element, ϕ is also continuous on their common boundary pq. Second, it imposes conditions on the nodal heat flows. In the exact solution of a steady heat-flow problem the net outflow from any infinitesimal area within the solution region must be zero. In the finite-element method this condition is replaced by the condition that at each node within the solution region the equivalent concentrated nodal flows associated with the node must add up to zero. Since these nodal flows are known linear functions of the associated nodal values, this condition generates a linear equation which relates the nodal values at a group of neighbouring nodes. Thus for the elements shown in Figure 17.25 the condition of zero net outflow at node p generates a linear equation involving ϕ_p and $\phi_q, \ldots, \phi_u$. There is one such equation for each node within the solution region.

The final stage of the method is the solution of the nodal-flow equations for the nodal values. As with the finite-difference method, the coefficient matrix for these equations is both sparse and banded. After the equations have been solved, values of ϕ at points within elements can be found, if required, from equation (17.39).

This example has introduced the simplest form of finite element – the three-node triangle, within which the dependent variable varies *linearly*. Adding three additional nodes, one on each side of the triangle, allows a *quadratic* variation of the dependent variable within the triangle, giving improved accuracy. Four- and eight-noded quadrilaterals are also popular elements in the analysis of two-dimensional problems. In three-dimensional analyses the corresponding elements are tetrahedra, 'bricks' and 'wedges'. Nowadays it is common for applications of the method to involve meshes with tens of thousands of nodes.

The application of the finite-element method to stress analysis follows similar lines, with (vector) displacements replacing (scalar) temperatures and (tensor) stresses replacing (vector) heat flow densities. Many commercial computer programs are now available for solving a wide range of stress and thermal analaysis problems. The method can also be applied to fluid flow and electromagnetic field problems.

17.6 Statistics

17.6.1 Introduction

Data are available in vast quantities in all branches of engineering. This chapter presents the more commonly used techniques for presenting and manipulating data to obtain meaningful results.

17.6.2 Averages

17.6.2.1 Arithmetic mean

The arithmetic mean of n numbers $x_1, x_2, x_3, \ldots, x_n$ is given by

$$\bar{x} = \frac{x_1 + x_2 + x_3 + \ldots + x_n}{n}$$

or

$$\bar{x} = \frac{\sum_{r=1}^{n} x_r}{n} \tag{17.40}$$

The arithmetic mean is easy to calculate and it takes into account all the figures. Its disadvantages are that it is

influenced unduly by extreme values and the final result may not be a whole number, which can be absurd at times, e.g. a mean of $2\frac{1}{2}$ people.

17.6.2.2 Median and mode

Median or 'middle on' is found by placing all the figures in order and choosing the one in the middle, or if there are an even number of items, the mean of the two central numbers. It is a useful technique for finding the average of items which cannot be expressed in figures, e.g. shades of a colour. It is also not influenced by extreme values. However, the median is not representative of all the figures.

The mode is the most 'fashionable' item, that is, the one which appears the most frequently.

17.6.2.3 Geometric mean

The geometric mean of n numbers $x_1, x_2, x_3, \ldots, x_n$ is given by

$$x_g = \sqrt[n]{(x_1 \times x_2 \times x_3 \times \ldots \times x_n)} \tag{17.41}$$

This technique is used to find the average of quantities which follow a geometric progression or exponential law, such as rates of changes. Its advantage is that it takes into account all the numbers, but is not unduly influenced by extreme values.

17.6.2.4 Harmonic mean

The harmonic mean of n numbers $x_1, x_2, x_3, \ldots, x_n$ is given by

$$x_h = \frac{n}{\sum_{r=1}^{n} (1/x_r)} \tag{17.42}$$

This averaging method is used when dealing with rates or speeds or prices. As a rule when dealing with items such as A per B, if the figures are for equal As then use the harmonic mean but if they are for equal Bs use the arithmetic mean. So if a plane flies over three equal distances at speeds of 5 m/s, 10 m/s and 15 m/s the mean speed is given by the harmonic mean as

$$\frac{3}{\frac{1}{5} + \frac{1}{10} + \frac{1}{15}} = 8.18 \text{ m/s}$$

If, however, the plane were to fly for three equal times, of say, 20 seconds at speeds of 5 m/s, 10 m/s and 15 m/s, then the mean speed would be given by the arithmetic mean as $(5 + 10 + 15)/3 = 10$ m/s.

17.6.3 Dispersion

17.6.3.1 Range and quartiles

The average represents the central figure of a series of numbers or items. It does not give any indication of the spread of the figures, in the series, from the average. Therefore, in Figure 17.26, both curves, A and B, have the same average but B has a wider deviation from the average than curve A.

There are several ways of stating by how much the individual numbers, in the series, differ from the average. The range is the difference between the smallest and largest values. The series can also be divided into four quartiles and the dispersion stated as the interquartile range, which is the difference between the first and third quartile numbers, or the quartile deviation which is half this value.

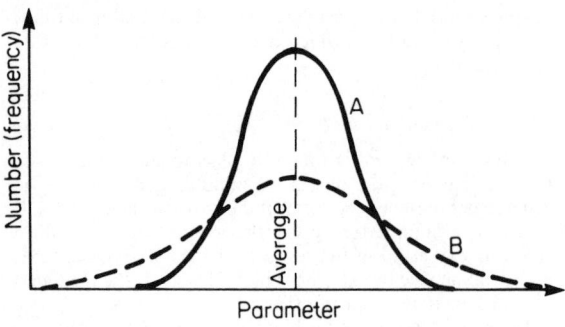

Figure 17.26 Illustration of deviation from the average

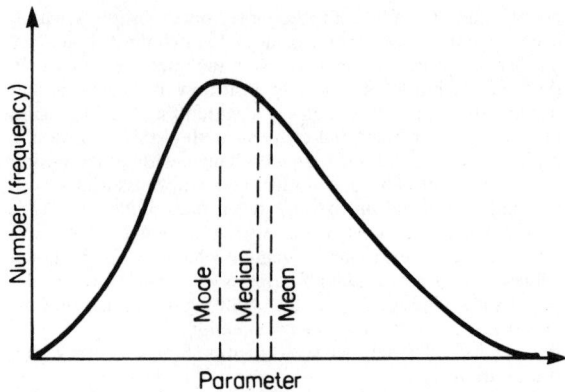

Figure 17.27 Illustration of skewness

The quartile deviation is easy to use and is not influenced by extreme values. However, it gives no indication of distribution between quartiles and covers only half the values in a series.

17.6.3.2 Mean deviation

This is found by taking the mean of the differences between each individual number in the series and the arithmetic mean, or median, of the series. Negative signs are ignored.

For a series of n numbers $x_1, x_2, x_3, \ldots, x_n$ having an arithmetic mean of $\bar{x}$ the mean deviation of the series is given by

$$\frac{\sum_{r=1}^{n} |x_r - \bar{x}|}{n} \tag{17.43}$$

The mean deviation takes into account all the items in the series. But it is not very suitable since it ignores signs.

17.6.3.3 Standard deviation

This is the most common measure of dispersion. For this the arithmetic mean must be used and not the median. It is calculated by squaring deviations from the mean, so eliminating their sign, adding the numbers together and then taking their mean and then the square root of the mean. Therefore, for the series in Section 17.6.3.2 the standard deviation is given by

$$\sigma = \left(\frac{\sum_{r=1}^{n}(x_r - \bar{x})^2}{n} \right)^{1/2} \tag{17.44}$$

The unit of the standard deviation is that of the original series. So if the series consists of the heights of a group of children in metres, then the mean and standard deviation are in metres. To compare two series having different units, such as the height of children and their weights, the coefficient of variation is used, which is unitless:

$$\text{coefficient of variation} = \frac{\sigma}{\bar{x}} \times 100 \tag{17.45}$$

17.6.4 Skewness

The distribution shown in Figure 17.26 is symmetrical since the mean, median and mode all coincide. Figure 17.27 shows a skewed distribution. It has positive skewness although if it bulges the other way, the skewness is said to be negative.

There are several mathematical ways for expressing skewness. They all give a measure of the deviation between the mean, median and mode and they are usually stated in relative terms, for ease of comparison between series of different units. The Pearson coefficient of skewness is given by

$$P_k = \frac{\text{mean} - \text{mode}}{\text{standard deviation}} \tag{17.46}$$

Since the mode is sometimes difficult to measure this can also be stated as

$$P_k = \frac{3(\text{mean} - \text{median})}{\text{standard deviation}} \tag{17.47}$$

17.6.5 Combinations and permutations

17.6.5.1 Combinations

Combinations are the numbers of ways in which a proportion can be chosen from a group. Therefore the number of ways in which two letters can be chosen from a group of four letters A, B, C, D is equal to 6, i.e. AB, AC, AD, BC, BD, CD. This is written as

$$^4C_2 = 6$$

The factorial expansion is frequently used in combination calculations where

$$n! = n \times (n - 1) \times (n - 2) \times \ldots \times 3 \times 2 \times 1$$

Using this the number of combinations of n items from a group of n is given by

$$^nC_r = \frac{n!}{r!(n - r)!} \tag{17.48}$$

17.6.5.2 Permutations

Combinations do not indicate any sequencing. When sequencing within each combination is involved the result is known as a permutation. Therefore the number of permutations of two letters out of four letters A, B, C, D is 12, i.e. AB, BA, AC, CA, AD, DA, BC, CB, BD, DB, CD, DC. The number of permutations of r items from a group of n is given by

$$^nP_r = \frac{n!}{(n - r)!} \tag{17.49}$$

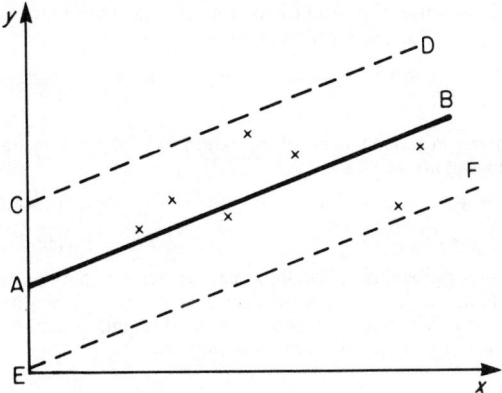

Figure 17.28 A scatter diagram

17.6.6 Regression and correlation

17.6.6.1 Regression

Regression is a method of establishing a mathematical relationship between two variables. Several equations may be used to establish this relationship, the most common being that of a straight line. Figure 17.28 shows the plot of seven readings. This is called a scatter diagram. The points can be seen to lie approximately on the straight line AB.

The equation of a straight line is given by

$$y = mx + c \qquad (17.50)$$

where x is the independent variable, y the dependent variable, m is the slope of the line and c its interception on the y-axis. c is negative if the line intercepts the y-axis on its negative part and m is negative if the line slopes the other way to that shown in Figure 17.28.

The best straight line to fit a set of points is found by the method of least squares as

$$m = \frac{\Sigma xy - (\Sigma x \Sigma y)/n}{\Sigma x^2 - (\Sigma x)^2/n} \qquad (17.51)$$

and

$$c = \frac{\Sigma x \Sigma xy - \Sigma y \Sigma x^2}{(\Sigma x)^2 - n\Sigma x^2} \qquad (17.52)$$

where n is the number of points. The line passes through the mean values of x and y, i.e. $\bar{x}$ and $\bar{y}$.

17.6.6.2 Correlation

Correlation is a technique for establishing the strength of the relationship between variables. In Figure 17.28 the individual figures are scattered on either side of a straight line and although one can approximate them by a straight line it may be required to establish if there is correlation between the x- and y-readings.

Several correlation coefficients exist. The product moment correlation coefficient (r) is given by

$$r = \frac{\Sigma (x - \bar{x})(y - \bar{y})}{n\sigma_x \sigma_y} \qquad (17.53)$$

or

$$r = \frac{\Sigma (x - \bar{x})(y - \bar{y})}{[\Sigma (x - \bar{x})^2 \Sigma (y - \bar{y})^2]^{1/2}} \qquad (17.54)$$

The value of r varies from $+1$, when all the points lie on a straight line and y increases with x, to -1, when all the points lie on a straight line but y decreases with x. When $r = 0$ the points are widely scattered and there is said to be no correlation between x and y.

The standard error of estimation in r is given by

$$S_y = \sigma_y (1 - r^2)^{1/2} \qquad (17.55)$$

In about 95% of cases, the actual values will lie between plus or minus twice the standard error of estimated values given by the regression equation. This is shown by lines CD and EF in Figure 17.28. Almost all the values will be within plus or minus three times the standard error of estimated values.

It should be noted that σ_y is the variability of the y-values, whereas S_y is a measure of the variability of the y-values as they differ from the regression which exists between x and y. If there is no regression then $r = 0$ and $\sigma_y = S_y$.

It is often necessary to draw conclusions from the order in which items are ranked. For example, two judges may rank contestants in a contest and we need to know if there is any correlation between their rankings. This may be done by using the Rank correlation coefficient (R) given by

$$R = 1 - \frac{6\Sigma d^2}{n^3 - n} \qquad (17.56)$$

where d is the difference between the two ranks for each item and n is the number of items. The value of R will vary from $+1$ when the two ranks are identical to -1 when they are exactly reversed.

17.6.7 Probability

If an event A occurs n times out of a total of m cases then the probability of occurrence is stated to be

$$P(A) = n/m \qquad (17.57)$$

Probability varies between 0 and 1. If $P(A)$ is the probability of occurrence then $1 - P(A)$ is the probability that event A will not occur and it can be written as $P(\bar{A})$.

If A and B are two events then the probability that either may occur is given by

$$P(A \text{ or } B) = P(A) + P(B) - P(A \text{ and } B) \qquad (17.58)$$

A special case of this probability law is when events are mutually exclusive, i.e. the occurrence of one event prevents the other from happening. Then

$$P(A \text{ or } B) = P(A) + P(B) \qquad (17.59)$$

If A and B are two events then the probability that they may occur together is given by

$$P(A \text{ and } B) = P(A) \times P(B|A) \qquad (17.60)$$

or

$$P(A \text{ and } B) = P(B) \times P(A|B) \qquad (17.61)$$

$P(B|A)$ is the probability that event B will occur assuming that event A has already occurred and $P(A|B)$ is the probability that event A will occur assuming that event B has already occurred. A special case of this probability law is when A and B are independent events, i.e. the occurrence of one event has no influence on the probability of the other event occurring.

Then

$$P(A \text{ and } B) = P(A) \times P(B) \tag{17.62}$$

Bayes' theorem on probability may be stated as

$$P(A|B) = \frac{P(A)P(B|A)}{P(A)P(B|A) + P(\bar{A})P(B|\bar{A})} \tag{17.63}$$

As an example of the use of Bayes' theorem suppose that a company discovers that 80% of those who bought its product in a year had been on the company's training course. 30% of those who bought a competitor's product had also been on the same training course. During that year the company had 20% of the market. The company wishes to know what percentage of buyers actually went on its training course, in order to discover the effectiveness of this course.

If B denotes that a person bought the company's product and T that they went on the training course then the problem is to find $P(B|T)$. From the data $P(B) = 0.2$, $P(\bar{B}) = 0.8$, $P(T|B) = 0.8$, $P(T|\bar{B}) = 0.3$. Then from equation (17.63)

$$P(B|T) = \frac{0.2 \times 0.8}{0.2 \times 0.8 + 0.8 \times 0.3} = 0.4$$

17.6.8 Probability distributions

There are several mathematical formulae with well-defined characteristics and these are known as probability distributions. If a problem can be made to fit one of these distributions then its solution is simplified. Distributions can be discrete when the characteristic can only take certain specific values, such as 0, 1, 2, etc., or they can be continuous where the characteristic can take any value.

17.6.8.1 Binomial distribution

The binomial probability distribution is given by

$$(p + q)^n = q^n + {}^nC_1pq^{n-1} + {}^nC_2p^2q^{n-2}$$
$$+ \ldots + {}^nC_xp^xq^{n-x} + \ldots + p^n \tag{17.64}$$

where p is the probability of an event occurring, $q(= 1 - p)$ is the probability of an event not occurring and n is the number of selections.

The probability of an event occurring m successive times is given by the binomial distribution as

$$p(m) = {}^nC_mp^mq^{n-m} \tag{17.65}$$

The binomial distribution is used for discrete events and is applicable if the probability of occurrence p of an event is constant on each trial. The mean of the distribution $B(M)$ and the standard deviation $B(S)$ are given by

$$B(M) = np \tag{17.66}$$

$$B(S) = (npq)^{1/2} \tag{17.67}$$

17.6.8.2 Poisson distribution

The Poisson distribution is used for discrete events and, like the binomial distribution, it applies to mutually independent events. It is used in cases where p and q cannot both be defined. For example, one can state the number of goals which were scored in a football match, but not the goals which were not scored.

The Poisson distribution may be considered to be the limiting case of the binomial when n is large and p is small.

The probability of an event occurring m successive times is given by the Poisson distribution as

$$p(m) = (np)^m \frac{e^{-np}}{m!} \tag{17.68}$$

The mean $P(M)$ and standard deviation $P(S)$ of the Poisson distribution are given by

$$P(M) = np \tag{17.69}$$

$$P(S) = (np)^{1/2} \tag{17.70}$$

Poisson probability calculations can be done by the use of probability charts as shown in Figure 17.29. This shows the probability that an event will occur at least m times when the mean (or expected) value np is known.

17.6.8.3 Normal distribution

The normal distribution represents continuous events and is shown plotted in Figure 17.30. The x-axis gives the event and the y-axis the probability of the event occurring. The curve shows that most of the events occur close to the mean value and this is usually the case in nature. The equation of the normal curve is given by

$$y = \frac{1}{\sigma(2\pi)^{1/2}} e^{-(x - \bar{x})^2/(2\sigma^2)} \tag{17.71}$$

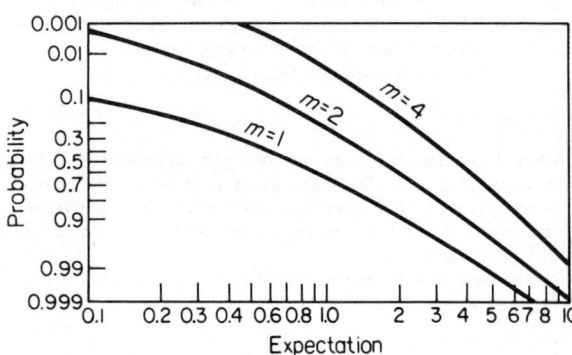

Figure 17.29 Poisson probability paper

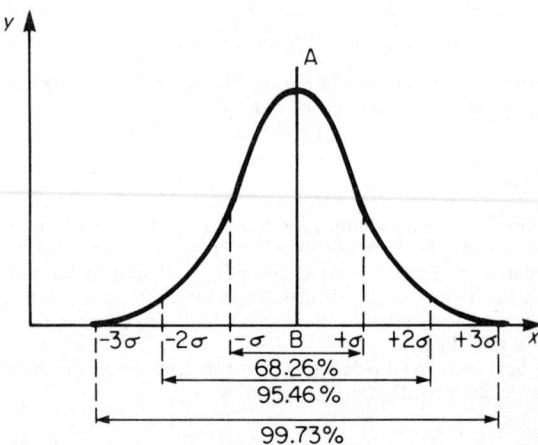

Figure 17.30 The normal curve

where $\bar{x}$ is the mean of the values making up the curve and σ is their standard deviation.

Different distributions will have varying mean and standard deviations but if they are distributed normally then their curves will all follow equation (17.71). These distributions can all be normalized to a standard form by moving the origin of their normal curve to their mean value, shown as B in Figure 17.30. The deviation from the mean is now represented on a new scale of units given by

$$\omega = \frac{x - \bar{x}}{\sigma} \qquad (17.72)$$

The equation for the standardized normal curve now becomes

$$y = \frac{1}{(2\pi)^{1/2}} e^{-\omega^2/2} \qquad (17.73)$$

The total area under the standardized normal curve is unity and the area between any two values of ω is the probability of an item from the distribution falling between these values. The normal curve extends infinitely in either direction but 68.26% of its values (area) fall between $\pm\sigma$, 95.46% between $\pm 2\sigma$, 99.73% between $\pm 3\sigma$ and 99.994% between $\pm 4\sigma$.

Table 17.1 gives the area under the normal curve for different values of ω. Since the normal curve is symmetrical the area from $+\omega$ to $+\infty$ is the same as from $-\omega$ to $-\infty$. As an example of the use of this table, suppose that 5000 street lamps have been installed in a city and that the lamps have a mean life of 1000 hours with a standard deviation of 100 hours.

How many lamps will fail in the first 800 hours? from equation (17.72)

$$\omega = (800 - 1000)/100 = -2$$

Ignoring the negative sign, Table 17.1 gives the probability of lamps not failing as 0.977 so that the probability of failure is $1 - 0.977$ or 0.023. Therefore 5000×0.023 or 115 lamps are expected to fail after 800 hours.

17.6.8.4 Exponential distribution

The exponential probability distribution is a continuous distribution and is shown in Figure 17.31. It has the equation

$$y = \frac{1}{\bar{x}} e^{-x/\bar{x}} \qquad (17.74)$$

where $\bar{x}$ is the mean of the distribution. Whereas in the normal distribution the mean value divides the population in half, for the exponential distribution 36.8% of the population is above the average and 63.2% below the average. Table 17.2 shows the area under the exponential curve for different values of the ratio $K = x/\bar{x}$, this area being shown shaded in Figure 17.31.

As an example suppose that the time between failures of a piece of equipment is found to vary exponentially. If results indicate that the mean time between failures is 1000 hours, then what is the probability that the equipment will work for 700 hours or more without a failure? Calculating K as $700/1000 = 0.7$ then from Table 17.2 the area beyond 0.7 is 0.497 which is the probability that the equipment will still be working after 700 hours.

Table 17.1 Area under the normal curve from $-\infty$ to ω

ω	0.00	0.02	0.04	0.06	0.08
0.0	0.500	0.508	0.516	0.524	0.532
0.1	0.540	0.548	0.556	0.564	0.571
0.2	0.579	0.587	0.595	0.603	0.610
0.3	0.618	0.626	0.633	0.640	0.648
0.4	0.655	0.663	0.670	0.677	0.684
0.5	0.692	0.700	0.705	0.712	0.719
0.6	0.726	0.732	0.739	0.745	0.752
0.7	0.758	0.764	0.770	0.776	0.782
0.8	0.788	0.794	0.800	0.805	0.811
0.9	0.816	0.821	0.826	0.832	0.837
1.0	0.841	0.846	0.851	0.855	0.860
1.1	0.864	0.869	0.873	0.877	0.881
1.2	0.885	0.889	0.893	0.896	0.900
1.3	0.903	0.907	0.910	0.913	0.916
1.4	0.919	0.922	0.925	0.928	0.931
1.5	0.933	0.936	0.938	0.941	0.943
1.6	0.945	0.947	0.950	0.952	0.954
1.7	0.955	0.957	0.959	0.961	0.963
1.8	0.964	0.966	0.967	0.969	0.970
1.9	0.971	0.973	0.974	0.975	0.976
2.0	0.977	0.978	0.979	0.980	0.981
2.1	0.982	0.983	0.984	0.985	0.985
2.2	0.986	0.987	0.988	0.988	0.989
2.3	0.989	0.990	0.990	0.991	0.991
2.4	0.992	0.992	0.993	0.993	0.993
2.5	0.994	0.994	0.995	0.995	0.995
2.6	0.995	0.996	0.996	0.996	0.996
2.7	0.997	0.997	0.997	0.997	0.997
2.8	0.997	0.998	0.998	0.998	0.998
2.9	0.998	0.998	0.998	0.998	0.999
3.0	0.999	0.999	0.999	0.999	0.999

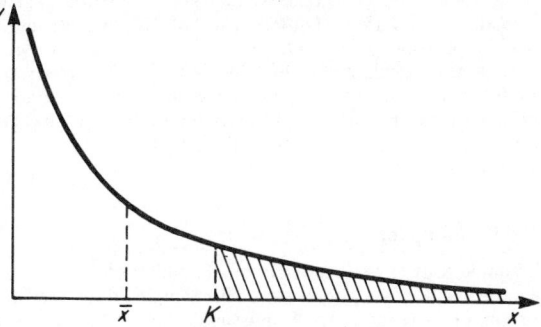

Figure 17.31 The exponential curve

Table 17.2 Area under the exponential curve from K to $+\infty$

K	0.00	0.02	0.04	0.06	0.08
0.0	1.000	0.980	0.961	0.942	0.923
0.1	0.905	0.886	0.869	0.852	0.835
0.2	0.819	0.803	0.787	0.771	0.776
0.3	0.741	0.726	0.712	0.698	0.684
0.4	0.670	0.657	0.644	0.631	0.619
0.5	0.607	0.595	0.583	0.571	0.560
0.6	0.549	0.538	0.527	0.517	0.507
0.7	0.497	0.487	0.477	0.468	0.458
0.8	0.449	0.440	0.432	0.423	0.415
0.9	0.407	0.399	0.391	0.383	0.375

Column 1 lists the ordinal values of ω or K and the corresponding values of area are presented in column 2. Interpolation between ordinal values can be achieved in steps of 0.02 by using the remaining 4 columns.

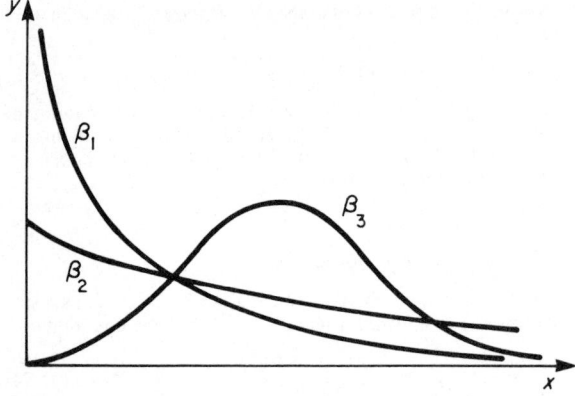

Figure 17.32 Weibull curves ($\alpha = 1$)

17.6.8.5   Weibull distribution

This is a continuous probability distribution and its equation is given by

$$y = \alpha\beta(x - \gamma)^{\beta-1}e^{-\alpha(x-\gamma)^{\beta}} \tag{17.75}$$

where α is called the scale factor, β the shape factor and γ the location factor.

The shape of the Weibull curve varies depending on the value of its factors. β is the most important, as shown in Figure 17.32, and the Weibull curve varies from an exponential ($\beta = 1.0$) to a normal distribution ($\beta = 3.5$). In practice β varies from about $\frac{1}{3}$ to 5. Because the Weibull distribution can be made to fit a variety of different sets of data, it is popularly used for probability distributions.

Analytical calculations using the Weibull distribution are cumbersome. Usually predictions are made using Weibull probability paper. The data are plotted on this paper and the probability predictions read from the graph.

17.6.9   Sampling

A sample consists of a relatively small number of items drawn from a much larger population. This sample is analysed for certain attributes and it is then assumed that these attributes apply to the total population, within a certain tolerance of error.

Sampling is usually associated with the normal probability distribution and, based on this distribution, the errors which arise due to sampling can be estimated. Suppose a sample of n_s items is taken from a population of n_p items which are distributed normally. If the sample is found to have a mean of μ_s with a standard deviation of σ_s then the mean μ_p of the population can be estimated to be within a certain tolerance of μ_s. It is given by

$$\mu_p = \mu_s \pm \frac{\gamma\sigma_s}{n_s^{1/2}} \tag{17.76}$$

γ is found from the normal curve depending on the level of confidence we need in specifying μ_p. For $\gamma = 1$ this level is 68.26%; for $\gamma = 2$ it is 95.46% and for $\gamma = 3$ it is 99.73%.

The standard error of mean σ_e is often defined as

$$\sigma_e = \frac{\sigma_s}{n_s^{1/2}} \tag{17.77}$$

so equation (17.76) can be rewritten as

$$\mu_p = \mu_s \pm \gamma\sigma_e \tag{17.78}$$

As an example suppose that a sample of 100 items, selected at random from a much larger population, gives their mean weight as 20 kg with a standard deviation of 100 g. The standard error of the mean is therefore $100/(100)^{1/2} = 10$ g and one can say with 99.73% confidence that the mean value of the population lies between $20 \pm 3 \times 0.01$ or 20.03 kg and 19.97 kg.

If in a sample of n_s items the probability of occurrence of a particular attribute is p_s, then the standard error of probability p_e is defined as

$$p_e = \left(\frac{p_s q_s}{n_s}\right)^{1/2} \tag{17.79}$$

where $q_s = 1 - p_s$.

The probability of occurrence of the attribute in the population is then given by

$$p_p = p_s \pm \gamma p_e \tag{17.80}$$

where γ is again chosen to cover a certain confidence level.

As an example suppose a sample of 500 items shows that 50 are defective. Then the probability of occurrence of the defect in the sample is $50/500 = 0.1$. The standard error of probability is $(0.1 \times 0.9/500)^{1/2}$ or 0.0134. Therefore we can state with 95.46% confidence that the population from which the sample was drawn has a defect probability of $0.1 \pm 2 \times 0.0134$, i.e. 0.0732 to 0.1268; or we can state with 99.73% confidence that this value will lie between $0.1 \pm 3 \times 0.0134$, i.e. 0.0598 to 0.1402.

If two samples have been taken from the same population and these give standard deviations of σ_{s1} and σ_{s2} for sample sizes of n_{s1} and n_{s2} then equation (17.77) can be modified to give the standard error of the difference between means as

$$\sigma_{de} = \left(\frac{\sigma_{s1}^2}{n_{s1}} + \frac{\sigma_{s2}^2}{n_{s2}}\right)^{1/2} \tag{17.81}$$

Similarly equation (17.79) can be modified to give the standard error of the difference between probabilities of two samples from the same population as

$$p_{de} = \left(\frac{p_{s1} q_{s1}}{n_{s1}} + \frac{p_{s2} q_{s2}}{n_{s2}}\right)^{1/2} \tag{17.82}$$

17.6.10   Tests of significance

In taking samples we often obtain results which deviate from the expected. Tests of significance are then used to determine if this deviation is real or if it could have arisen due to sampling error.

17.6.10.1   Hypothesis testing

In this system a hypothesis is formulated and is then tested at a given confidence level. For example, suppose a coin is tossed 100 times and it comes up heads 60 times. Is the coin biased or is it likely that this falls within a reasonable sampling error? The hypothesis is set up that the coin is not biased. Therefore one would expect that the probability of heads is 0.5, i.e. $p_s = 0.5$. The probability of tails, q_s, is also 0.5. Using equation (17.79) the standard error of probability is given by $p_e = (0.5 \times 0.5/100)^{1/2}$ or 0.05. Therefore from equation (17.80) the population probability at the 95.45% confidence level of getting heads is $0.5 + 2 \times 0.05 = 0.6$. Therefore it is highly likely that the coin is not biased and the results are due to sampling error.

The results of any significance test are not conclusive. For example, is 95.45% too high a confidence level to require? The higher the confidence level the greater the risk of rejecting a true hypothesis, and the lower the level the greater the risk of accepting a false hypothesis.

Suppose now that a sample of 100 items of production shows that five are defective. A second sample of 100 items its taken from the same production a few months later and gives two defectives. Does this show that the production quality is improving? Using equation (17.82) the standard error of the difference between probabilities is given by $(0.5 \times 0.95/100 + 0.02 \times 0.98/100)^{1/2} = 0.0259$. This is less than twice the difference between the two probabilities, i.e. $0.05 - 0.02 = 0.03$, therefore the difference is very likely to have arisen due to sampling error and it does not necessarily indicate an improvement in quality.

17.6.10.2 Chi-square test

This is written as χ^2. If O is an observed result and E is the expected result then

$$\chi^2 = \sum \frac{(O - E)^2}{E} \tag{17.83}$$

The χ^2 distribution is given by tables such as Table 17.3, from which the probability can be determined. The number of degrees of freedom is the number of classes whose frequency can be assigned independently. If the data are presented in the form of a table having V vertical columns and H horizontal rows then the degrees of freedom are usually found as $(V - 1)(H - 1)$.

Returning to the earlier example, suppose a coin is tossed 100 times and it comes up heads 60 times and tails 40 times. Is the coin biased? The expected values for heads and tails are 50 each so that

$$\chi^2 = \frac{(60 - 50)^2}{50} + \frac{(40 - 50)^2}{50} = 4$$

The number of degrees of freedom is one since once we have fixed the frequency for heads that for tails is defined. There-

Table 17.3 The chi-square distribution

Degrees of freedom	Probability level				
	0.100	0.050	0.025	0.010	0.005
1	2.71	3.84	5.02	6.63	7.88
2	4.61	5.99	7.38	9.21	10.60
3	6.25	7.81	9.35	11.34	12.84
4	7.78	9.49	11.14	13.28	14.86
5	9.24	11.07	12.83	15.09	16.75
6	10.64	12.59	14.45	16.81	18.55
7	12.02	14.07	16.01	18.48	20.28
8	13.36	15.51	17.53	20.09	21.96
9	14.68	16.92	19.02	21.67	23.59
10	15.99	18.31	20.48	23.21	25.19
12	18.55	21.03	23.34	26.22	28.30
14	21.06	23.68	26.12	29.14	31.32
16	23.54	26.30	28.85	32.00	34.27
18	25.99	28.87	31.53	34.81	37.16
20	28.41	31.41	34.17	37.57	40.00
30	40.26	43.77	46.98	50.89	53.67
40	51.81	55.76	59.34	63.69	66.77

Table 17.4 Frequency distribution of accidents in a factory during 24 hours

Time (24 hour clock)	Number of accidents
0–6	9
6–12	3
12–18	2
18–24	6

fore entering Table 17.3 with one degree of freedom the probability level for $\chi^2 = 4$ is seen to be above 0.025 ($= 2.5\%$), i.e. there is a strong probability that the difference in the two results arose by chance and the coin is not biased.

As a further example suppose that over a 24-hour period the average number of accidents which occur in a factory is seen to be as in Table 17.4. Does this indicate that most of the accidents occur during the late night and early morning periods? Applying the χ^2 tests, the expected value, (if there was no difference between the time periods) would be the mean of the number of accidents, i.e. 5. Therefore from equation (17.83)

$$\chi^2 = \frac{(9 - 5)^2}{5} + \frac{(3 - 5)^2}{5} + \frac{(2 - 5)^2}{5} + \frac{(6 - 5)^2}{5} = 6$$

There are three degrees of freedom, therefore from Table 17.3 the probability of occurrence of the result shown in Table 17.4 is seen to be greater than 10%. The conclusion would be that although there is a trend, as yet there are not enough data to show if this trend is significant or not. For example, if the number of accidents were each three times as large, i.e. 27, 9, 6, 18 respectively, then χ^2 would be calculated as 20.67 and from Table 17.3 it is seen that the results are highly significant since there is a very low probability, less than 0.5% that it can arise by chance.

17.6.10.3 Significance of correlation

The significance of the product moment correlation coefficient of equations (17.53) or (17.54) can be tested at any confidence level by means of the standard error of estimation given by equation (17.55). An alternative method is to use the Student t test of significance. This is given by

$$t = \frac{r(n - 2)^{1/2}}{(1 - r^2)^{1/2}} \tag{17.84}$$

where r is the correlation coefficient and n the number of items. Tables are then used, similar to Table 17.3, which give the probability level for $(n - 2)$ degrees of freedom.

The Student t for the rank correlation coefficient is given by

$$t = R[(n - 2)/(1 - R^2)]^{1/2} \tag{17.85}$$

and the same Student t tables are used to check the significance of R.

Further reading

Besterfield, D. H., *Quality Control*, Prentice-Hall, Englewood Cliffs, NJ (1979)
Boyce, W. E. and DiPrima, R. C., *Elementary Differential Equations and Boundary Values*, 4th edn, Wiley, Chichester (1986)
Caplen, R. H., *A Practical Approach to Quality Control*, Business Books, London (1982)
Chalk, G. O. and Stick, A. W., *Statistics for the Engineer*, Butterworths, London (1975)

Cohen, S. S., *Practical Statistics*, Edward Arnold (1988)

David, H. A., *Order Statistics*, Wiley, Chichester (1981)

Dunn, R. A. and Ramsing, K. D., *Management Science, a Practical Approach to Decision Making*, Macmillan, London (1981)

Fitzsimmons, J. A., *Service Operations Management*, McGraw-Hill, New York (1982)

Grant, E. L. and Leavenworth, R. S., *Statistical Quality Control*, McGraw-Hill, New York (1980)

Hahn, W. C., *Modern Statistical Methods*, Butterworths, London (1979)

Jones, M. E. M., *Statistics*, Schofield & Sims (1988)

Kreyszig, E., *Advanced Engineering Mathematics*, 5th edn, Wiley, Chichester (1983)

Livesley, R. K., *Finite Elements: an Introduction for Engineers*, Cambridge University Press, Cambridge (1983)

Lyons, S., *Handbook of Industrial Mathematics*, Cambridge University Press, Cambridge (1978)

Mazda, F. F., *Quantitative Techniques in Business*, Gee & Co. (1979)

Pudewicz, E. J., and Mishra, S. N., *Modern Mathematical Statistics*, Wiley (1988)

Siegel, A. F., *Statistics and Data Analysis*, Wiley (1988)

18

Health and safety

Roger C. Webster

Contents

18.1 Health and safety in the European Community   18/3
   18.1.1   Denmark   18/3
   18.1.2   Belgium   18/3
   18.1.3   France   18/3
   18.1.4   The Netherlands   18/3
   18.1.5   Portugal   18/3
   18.1.6   Spain   18/3
   18.1.7   Germany   18/3
   18.1.8   Greece   18/3
   18.1.9   Irish Republic   18/3
   18.1.10   Italy   18/3

18.2 Health and safety at work – law and administration in the USA   18/4

18.3 UK legislation and guidance   18/4

18.4 The Health and Safety at Work etc. Act 1974   18/4
   18.4.1   Duties of employers   18/4
   18.4.2   Duties to others   18/4
   18.4.3   Duties to non-employees on works premises   18/4
   18.4.4   Duties of manufacturers   18/4
   18.4.5   Duties of employees   18/4
   18.4.6   Implementation of the Act   18/4

18.5 The Health and Safety Executive   18/5

18.6 Local Authorities   18/5

18.7 Enforcement Notices   18/5
   18.7.1   Improvement Notice   18/5
   18.7.2   Prohibition Notice   18/5
   18.7.3   Prosecution   18/5
   18.7.4   Other powers   18/5

18.8 Control of Substances Hazardous to Health Regulations 1988   18/5
   18.8.1   Duties under the Regulations   18/6
   18.8.2   The assessment   18/6
   18.8.3   Control measures   18/6

18.9 Asbestos   18/6

18.10 Control of lead at work   18/6

18.11 The Electricity at Work Regulations 1989   18/7
   18.11.1   The relationship to IEE Regulations   18/7
   18.11.2   The Regulations – definitions   18/7
   18.11.3   The Regulations – requirements   18/8

18.12 The Noise at Work Regulations 1989   18/8
   18.12.1   The requirements of the Regulations   18/9

18.13 Safety of machinery   18/9
   18.13.1   Identification of hazard   18/9
   18.13.2   Types of hazard   18/9
   18.13.3   Assessment   18/9
   18.13.4   Safety by design   18/9
   18.13.5   Guarding   18/10

18.14 Personal protective equipment   18/10
   18.14.1   Existing PPE legislation   18/10
   18.14.2   The new Regulations   18/11

18.15 Manual handling   18/11

Further reading   18/12

The regulations governing health and safety are surveyed in this chapter. Obviously, these regulations differ from country to country, but a general trend towards harmonization of European legislation through the European Community is bound to lead to more common legislation worldwide.

A brief section dealing with the legislation and administration of health and safety law in several countries follows. As a country with highly developed legislation and regulation, the UK has been chosen for the subsequent more detailed examination. This gives an illustration of various themes that run through the health and safety regulations of any country. The UK also has a system that is fairly similar to those of most European countries.

18.1 Health and safety in the European Community

The legislation and administrative organization of the European Community with respect to health and safety varies widely. At one extreme are countries like the UK and Denmark, who have highly developed legislation and, at the other, is Portugal, with no specific regulations.

18.1.1 Denmark

Health and safety in Denmark is controlled by the Health and Safety at Work Act 1985. Specific regulations are made under this enabling Act and are enforced by an inspectorate under the Minister of Labour. Safety committees have to be formed by companies with more than 20 employees.

18.1.2 Belgium

Health and safety is controlled in Belgium by File 2 of the General Regulations, for the protection of employment. These detail requirements including use of machinery, handling of materials, fire risks, hygiene of premises, temperature and sanitation. Safety officers have to be appointed by companies. Safety committees are necessary in companies employing more than 50 people. There is no government inspectorate but routine inspections are carried out by authorized private firms.

18.1.3 France

Labour codes control health and safety practices in France. Under labour code 236 companies employing more than 50 people are required to form safety committees:

Code 232-1 requires workplaces to be hygienic.
Code 233-1 protects workers against falls, collisions and suffocation.
Codes 231-6 and 231-7 list dangerous substances and control labelling and carrying instructions.

18.1.4 The Netherlands

Health and safety is controlled in the Netherland by the Working Conditions Act 1980. This Act, which was introduced in phases, requires employers to promote health and safety to employees. Employers are obliged to reveal to workers all information concerning risks to health and are required to assign tasks according to workers' physical and mental capacities.

There is an overriding obligation to organize activities to ensure the best possible working conditions and for companies with more than 100 employees there is a requirement to report annually on these conditions. Works councils must be set up to oversee and promote health and safety. For companies with more than 500 employees safety departments have to be organized and staffed by specialist personnel. The legislation is enforced by an inspectorate employed by the Labour Ministry.

18.1.5 Portugal

There is no formal legislation controlling health and safety in Portugal. Draft documents exist and these have been the subject of discussion since 1990. The regulations that do apply to factories are enforced by the Geneal Inspectorate of Labour. The Director General of Occupational Safety and Health advises on standards and carries out research.

18.1.6 Spain

Spanish codes of practice enforced by the labour inspectorate form the basis of legislative control. Firms employing more than 100 employees have to form health and safety committees. Staff representatives have a role to play in health and safety enforcement. If three-quarters of the safety representatives decide that there is a grave risk of an accident they are entitled to stop work at the premises. A ruling from the labour inspectorate is then made within 24 hours. Fines for breaches of regulations vary between £165 and £1650.

18.1.7 Germany

In Germany the Safety at Work Act 1973 controls health and safety. Inspection of premises is again entrusted to private authorized firms (for example, the Accident Prevention Institute). Employees have a duty to support these inspection agencies and companies must appoint a specialist safety officer if they have more than 20 employees.

18.1.8 Greece

Under safety law 1568/85 firms are required in Greece to form safety committees and a 'safety technician' has to be employed if there are more than 50 employees. A factory doctor is also required. Inspection of premises is carried out by a labour inspectorate under the Minister of Labour.

18.1.9 Irish Republic

Health and safety in the Irish Republic is controlled by the Health Safety and Welfare at Work Act 1989. This statute requires firms that employ more than 10 staff to produce a safety statement. If the statement reveals unsatisfactory situations the firms are obliged to deal with them. An inspectorate is employed which can inspect and require changes to be made to safety statements and, consequently, changes to work practices.

18.1.10 Italy

Under Article 2087 of the Civil Code companies in Italy are responsible for employees' physical and mental wellbeing. Article 41 deals with safety, freedom and human dignity in employment. Under Article 9 of the workers' statute workers have a collective right to verify measures to protect staff and prevent accidents and work-related diseases. Local health authorities can monitor situations with respect to space, light, cleanliness, toilet provision, protective clothing, etc. Regulations exist which control the employment of pregnant women.

18.2 Health and safety at work–law and administration in the USA

In the USA health and safety standards are set by the Occupational Safety and Health Administration (OSHA). During 1991 OSHA conducted 82 484 State and 42 113 Federal inspections and maximum penalties for breaches of legislation were increased by sevenfold during that year. A system of penalties for breaches of legislation is enacted by OSHA.

During 1991 a total of $9.7 million in federal penalties and $32.6 million in state penalties were imposed. Recently, OSHA introduced two new and three proposed standards. The new standards related to the construction industry and detailed safe practices in the use of the 'lift slab' construction method and one on the use of ladders and stairways in construction. The proposed new standards included:

1. A new limit value for formaldehyde
2. Indoor air quality
3. Cadmium

18.3 UK legislation and guidance

Health and safety at work in the UK is controlled by a large number of Statutory Instruments of varying importance and it is necessary to appreciate their individual significance and relationship to others.

The principal way that statute law is implemented is by Act of Parliament. The statute first passes through the House of Commons, after initial drafting and consultation phases, and then to the House of Lords. It passes again through the Commons where any amendments are considered. It will finally receive the Royal Seal of Approval and become an Act of Parliament.

Regulations may be made by persons (usually Ministers) authorized in Acts of Parliament (for instance, by Section 15 of The Health and Safety at Work etc. Act 1974). These Regulations do not pass through Parliament in the same way as an Act (although Parliament does maintain overall control). Regulations carry the same force of law as Acts of Parliament.

Approved Codes of Practice are also commonly employed in health and safety matters. These do not carry the same authority as Acts or Regulations but rather serve to amplify and add detail to requirements imposed by Acts or Regulations. Non-compliance with approved Codes of Practice is not (in itself) an offence, but is likely to result in an offence being committed under the enabling Act (or Regulations).

Guidance notes are also issued. These are of a technical nature and serve to give examples of good practice that will enable compliance with statutes. Again, non-compliance with a guidance note is not, in itself, an offence.

18.4 The Health and Safety at Work etc. Act 1974

This Act is intended to secure the health, safety and welfare of persons at work. In addition, the Act also protects persons (other than those at work) against risks to health and safety arising out of or in connection with the activities of persons at work. The Act also controls the keeping and use of explosives or highly flammable or otherwise dangerous substances and the emission into the atmosphere of noxious or offensive substances.

18.4.1 Duties of employers

Section 2 of the Act applies to employers and imposes a general duty to ensure the health, safety and welfare of all employees. This includes the provision and maintenance of safe plant, arrangements for handling and storage and transport of substances in a safe manner. The provision of instruction, information and training is also a requirement.

The work premises are required to be safe and without risks to health, including safe means of access and egress. The working environment shall similarly be safe and without risks to health, and adequate in facilities and arrangements for welfare at work. All these actions are required to be taken 'as far as is reasonably practicable' – an expression that involves both technical feasibility as well as an element of financial practicality.

Section 2 also requires the writing of safety policies (except as may be prescribed) which, currently, have to be prepared for companies with five or more employees. Statements must be reviewed at regular intervals and brought to the attention of employees.

18.4.2 Duties to others

Section 3 of the Act requires employers and the self-employed to take steps to protect persons not in their employment from work-related activities. In addition, this section requires employers and the self-employed (in prescribed circumstances) to provide certain information to persons who may be affected by their work activities concerning the way in which those work activities may affect those persons' health or safety. In addition, the self-employed have a duty of self-care concerning risks to their own health and safety.

18.4.3 Duties to non-employees on works premises

Persons having a control of premises (not including domestic) where persons (not being their employees) work or use plant or substances provided for their use have a duty of care to persons on their premises.

18.4.4 Duties of manufacturers

Persons who design, manufacture, import or supply articles for use at work have a duty of care to users such that the article is safe and without risks to health when properly used. For this purpose, they should carry out such tests and examinations as may be necessary. They must also make available to users adequate information about the use for which the article was designed and give information concerning any conditions necessary to ensure the safe use of that object.

18.4.5 Duties of employees

Employees (while at work) have a duty to take reasonable care of the health and safety of themselves and other persons who may be affected by their acts and omissions. In addition, they must cooperate with employers or other persons who have duties under this Act to ensure that that duty is performed or complied with.

18.4.6 Implementation of the Act

The Health and Safety Commission is the body that generates policies concerning health and safety. It consists of representatives from employers' and employees' organizations and is chaired by a person appointed by the Secretary of State. In addition, there may be representatives from Local Authority organizations and other professional bodies.

Enforcement of the Act falls to two Authorities (depending upon the nature of the use of the premises); the Health and Safety Executive and Local Authorities.

18.5   The Health and Safety Executive

The Health and Safety Executive has a dual role. One function is to generate guidance on aspects of health and safety, including guidance notes and liaison documents for Local Authorities. The second function is that of enforcement agency. Classes of premises are divided between the Health and Safety Executive and Local Authorities by the Allocation Regulations. Broadly, these Regulations allocate non-industrial premises (e.g. offices, warehouses, shops, places of entertainment) to Local Authorities and the rest falls to the Health and Safety Executive. Where more than one use takes place within one curtilage (for example, a cardboard carton-manufacturing factory which is within a large warehouse and only supplies that warehouse) several tests are applied concerning which use predominates. If the 'factory' use predominates, the Health and Safety Executive enforce, if the non-industrial use predominates, the Local Authority will enforce. Thus, for instance, a food warehouse with the cardboard carton-manufacturing plant used only for the purpose of boxing the food stored within the warehouse would be enforced by the Local Authority, whereas a large cardboard carton manufacturer who supplied other outlets and had a warehouse only to store its own cardboard boxes would be enforced by the Health and Safety Executive.

The test here is the prime purpose of the user. If the purpose is to manufacture and sell boxes the user is a factory (notwithstanding the fact that the warehouse may occupy a larger floor area than the box-manufacturing plant). The Health and Safety Executive is sub-divided into classes of inspectorate, as follows:

HM Factory Inspectorate
HM Agricultural Inspectorate
HM Explosives Inspectorate
HM Mines and Quarries Inspectorate
HM Nuclear Installations Inspectorate

The Health and Safety Executive operates from area offices located around the country, and, in addition, specialist offices are located in certain cities which deal with just one or two specialized industries (which may be prevalent in that particular area).

18.6   Local Authorities

Local Authorities appoint authorized officers to carry out their functions under The Health and Safety at Work etc. Act and these are usually Environmental Health Officers. Local Authority officers have exactly the same powers as the Health and Safety Executive enforcement officers and possess the same degree of expertise. As can be seen from the examples of allocations given above, the types of premises visited by Local Authorities may be essentially the same as those visited by the Health and Safety Executive. If, for instance, a large warehouse operation has an ancillary factory employing, say, one hundred operatives, the Local Authority would be responsible for enforcement, whereas the Health and Safety Executive would be responsible for a much smaller factory if the manufacturing use predominates.

18.7   Enforcement Notices

Breaches of the Health and Safety at Work etc. Act or its Regulations can be dealt with in one of three ways:

1. Improvement Notice
2. Prohibition Notice
3. Prosecution

18.7.1   Improvement Notice

This Notice may be served if the inspector believes that a contravention of one or more of the statutory provisions has taken place and that it is likely that the contravention will be repeated, or there is an ongoing breach of a statutory provision. In this case, the inspector may issue a Notice requiring works to be carried out. The time period for compliance is variable but should not be less than that allowed for appeal against the Notice (under Section 24).

18.7.2   Prohibition Notice

If the activities being carried out will involve a risk of serious personal injury the inspector may serve a Prohibition Notice. This may take immediate effect if the inspector feels that the risk of serious personal injury is imminent or may be deferred to a later date, if this is not the case.

18.7.3   Prosecution

An inspector may also prosecute any person who breaches a statutory provision in addition to (or instead of) the serving of a statutory notice.

18.7.4   Other powers

In addition to exercising the above powers an inspector may also physically render safe or seize any harmful substance or article. Inspectors have the power to enter premises (at any time if they believe that there is a dangerous situation), and may take a police officer with them if they have reason to believe that they will be obstructed. They may take photographs, samples, etc. and may order that premises be left undisturbed while investigations take place.

18.8   Control of Substances Hazardous to Health Regulations 1988

These Regulations (of which the majority came into force on 1 October 1989) lay down essential requirements for the control of hazardous substances and to protect persons exposed to them, and apply to the following:

1. Those substances that have been classified as being very toxic, toxic, harmful, corrosive or irritant under The Classification, Packaging and Labelling of Dangerous Substances Regulations 1984;
2. Those substances which have maximum exposure limits or occupational exposure standards (i.e. listed in HSE document EH/40);
3. Substances that have chronic or delayed effects, i.e. carcinogenic, mutagenic or teratogenic.

A substance should be regarded as hazardous to health if it is hazardous in the form in which it occurs in the work activity, whether or not its mode of action has been identified.

A substance hazardous to health is not just a single chemical compound but also includes mixtures of compounds, microoganisms, allergens, etc. When considering whether a substance is likely to be hazardous to health the following points are likely to be taken into account:

1. The form of the substance, i.e. particle size – respirable dust may be more hazardous than solid;
2. Impurities – contaminants in innocuous materials may render that material hazardous;
3. Fibres – fibrous form may be more hazardous;
4. Synergistic effects of more than one substance;
5. Microorganisms may arise from the work directly or indirectly (e.g. *legionella* – the causative organism of Legionnaires' disease which may be present in wet cooling towers).

18.8.1 Duties under the Regulations

The Regulations require employers to take steps to assess the risk to health arising from work with these hazardous substances and to decide on the precautions that are necessary. The employer then has to take the appropriate measures to prevent or control the risk (where it is not reasonably practicable to prevent it) and must then ensure that these control measures are used and that equipment is properly maintained and procedures observed.

Monitoring the health of workers is necessary in certain circumstances. Finally, instruction, training and the dissemination of information concerning the risks of working with the substances and the precautions to be taken are a requirement of the Act.

18.8.2 The assessment

The duty to carry out the assessment is placed on the employer, and it is acknowledged that the employer is likely to seek the services of an outside consultant (and this is most certainly the recommended course of action). However, the responsibility to make the actual assessment rests with the employer and cannot be delegated as such.

The Regulations require that after 1 January 1990 no work which is liable to expose anyone to substances hazardous to health shall be carried on unless an assessment has been made which need only relate to that part of the work liable to expose people to hazardous substances. If the conclusion is drawn that (while the substances covered are being used) there is no potential for exposure then no further assessment is needed.

The main points of the assessment are as follows:

1. Gather information.
2. Evaluate the risks.
3. Decide what precautions are necessssary.
4. Record the assessment.
5. Decide on the review period. (Review will also be necessary when there are any changes in the working practices or the substances used.)

18.8.3 Control measures

The Regulations require the prevention of exposure to the substance but acknowledge that this may not be reasonably practicable. In these cases the substance should be adequately controlled. The requirement is for adequate control, so far as is reasonably practicable, by means other than the provision of personal protective equipment. This means that engineering solutions, local exhaust ventilation, etc. should be the first considerations. If adequate control is not possible by these means then personal protective equipment should be used in addition to engineering solutions so that adequate control of the substance can be obtained.

18.9 Asbestos

Asbestos means any of the following minerals: crocidolite, amosite, chrysotile, fibrous actinolite, fibrous anthophyllite, fibrous tremolite and any mixture containing any of these minerals. There are various parts of legislation concerning the use, manufacture and removal of asbestos (or materials containing it) and reference must be made to these statutes prior to starting work:

The Asbestos (Licensing) Regulations 1983 prohibit work (subject to the Regulations) without a licence.
The Asbestos (Prohibition) Regulations 1985 (amended 1988), among other things, prohibit the application of asbestos by spraying, the use of asbestos materials for thermal and acoustic insulation, and the supply or application of any paint or varnish containing asbestos.
The Control of Asbestos at Work Regulations 1987 cover all work with asbestos.

18.10 Control of lead at work

The use of lead at work is controlled by The Control of Lead at Work Regulations 1980. An approved Code of Practice has been issued entitled Approved Code of Practice – Lead at Work (revised June 1985). The Regulations apply to work from which lead arises:

1. In the form of lead dust, fume or vapour which is liable to be inhaled;
2. In any form in which it is liable to be ingested (e.g. lead powder, dust, paint or paste);
3. In the form of lead compounds which are liable to be absorbed through the skin (e.g. concentrated lead alkyls).

The Regulations are thus not applicable to work with substances which, while containing lead, are not liable to be inhaled, ingested or absorbed by persons. An example of such a process given in the guidance notes is the handling of finished articles containing lead (e.g. pottery).

The extent to which the Regulations apply to a particular work activity is controlled by the probable nature and degree of exposure to lead and reference levels are set in the Regulations. A standard (known as the lead-in-air standard) is set at the following figures:

Lead (except for tetra-ethyl lead) (as Pb) 0.15 mg/m^3 of air
Tetra-ethyl lead (as Pb) 0.10 mg/m^3 of air

These limits are for 8-hour time-weighted average concentrations. The Regulations do allow some deviation from these standards as it is acknowledged that the absorption of lead is controlled by numerous factors including composition, solubility, particle size and period of exposure.

The degree of variance allowed is as follows:

The 8-hour time-weighted average concentration does not exceed three times the above figures.
The 40-hour time-weighted average does not exceed 0.15 mg/m^3 of air, if there is sufficient information available from biological test results to indicate that the degree of lead absorption is at an acceptable level.

The Regulations use the expression 'significant exposure' to apply or disapply certain Regulations. This is defined as:

1. An exposure of at least half the lead-in-air standard;
2. Where there is a significant risk of ingesting lead;
3. Where there is a risk of skin contact with concentrated lead alkyls.

Where biological monitoring is carried out as part of an assessment if the blood levels are greater than 40 μg per 100 ml or, in the case of workers exposed to lead alkyls, a urinary lead level of greater than 120 μg/l, the worker should be regarded as being significantly exposed to lead.

In the case of intermittent exposure (e.g. a few hours per week) if the lead-in-air standard is not exceeded and the weekly (i.e. 40-hour) average does not exceed one half of the 8-hour standard the exposure is not classed as significant. The main requirements of the Regulations are as follows:

Regulation 4 – where persons are exposed to lead the employer shall make an assessment.

Regulation 5 – employers shall ensure adequate information, instruction and training of their employees.

Regulation 6 – employers shall control the exposure of their employees to lead as far as is reasonably practicable otherwise than by the use of respiratory protective equipment. Control measures are considered adequate when they effectively control the exposure of employees to:

1. Lead-in-air concentrations not exceeding the lead-in-air standard;
2. Lead which can be ingested;
3. Lead which can be absorbed through the skin.

Regulation 7 – where it is not possible to control lead to the above standards the employer shall provide any employee liable to be exposed to airborne lead with respiratory protective equipment.

Regulation 8 – the employer shall provide protective clothing unless the exposure to lead is not 'significant' (see previous definition).

Regulation 9 – washing, changing and clothes storage facilities must be provided.

Regulation 10 – employers shall ensure that employees do not eat, drink or smoke in contaminated areas and that suitable areas are set aside for these activities.

Regulation 11 – the employer shall ensure the cleanliness of the workplace, etc.

Regulation 12 – the employer has a duty to ensure that contamination does not spread from the workplace so as to avoid exposure to persons not engaged in work with lead (e.g. the families of the lead workers).

Regulation 13 – control measures. If control measures (e.g. ventilation, respiratory protection, etc.) are necessary the employer shall ensure that they are properly used. The employer also has a duty under this section to properly use such facilities and to report any defects in such equipment, etc.

Regulation 14 – requires that employers shall adequately maintain control measures.

Regulation 15 – requires air monitoring to be carried out unless the exposure is not significant.

Regulation 16 – requires medical surveillance to be carried out if the exposure to lead is significant or if the Employment Medical Advisory Service-appointed doctor certifies that the employee should be under surveillance. The EMAS doctor can also require that an employer ceases to expose an employee to lead by way of a certificate that may also allow for some partial exposure (under specified circumstances).

Regulation 17 – requires the keeping of records of the assessments, maintenance, air tests and medical surveillance as required by previous Regulations.

18.11   The Electricity at Work Regulations 1989

The Electricity at Work Regulations came into force on 1 April 1990. These require precautions to be taken against the risk of death or personal injury from electricity in work activities. The Regulations apply to all electrical systems and equipment whenever manufactured, purchased or installed even if the use, etc. predates the Regulations.

18.11.1   The relationship to IEE Regulations

The Institution of Electrical Engineers Regulations are non-statutory Regulations and relate principally to the design, selection, erection, inspection and testing of electrical installations (whether permanent or temporary) in or about buildings generally and to agricultural and horticultural buildings, construction sites and caravans and their sites. There are thus many situations to which the IEE Regulations do not apply. In particular, they do not relate to systems operating at 1000 V a.c. or more. Compliance with IEE Regulations will mean that it is highly likely that a system will comply with The Electricity at Work Regulations.

18.11.2   The Regulations – definitions

The term 'system' is used to describe an electrical installation and is defined as including all the constituent parts of the system (e.g. conductors and electrical equipment) and is not a reference solely to the functional circuit as a whole. Where circuits are connected by inductance (e.g. in the windings of a transformer). Even if the circuits are galvanically separated, the two circuits are to be regarded as one system.

Some 'systems' may thus extend over a large geographical area if the circuits are connected in any way. Several persons may thus have control over one system. The Regulations require control over the system only insofar as they have control over the premises.

Electrical equipment is defined as every type of electrical equipment from a battery-powered torch up to a 400 kV overhead line. The Regulations apply where 'danger' may arise. Explosion risks are considered relevant. Thus low-voltage applicances (which present no danger of shock) may give rise to sparks which may present a danger.

Conductors mean any material capable of conducting electricity and include metals and other materials. The definition covers materials and structures not specifically designed for the transmission of electricity and may include, for example, salt water and ionized gases.

Circuit conductors are conductors whose normal function is to carry load currents or to be energized.

Danger is defined as the risk of injury.

Injury is defined as death or personal injury from electric shock, electric burn, electrical explosion or arcing or from fire or explosion initiated by electrical energy, where any such death or injury is associated with the generation, provision, transmission, tranformation, rectification, conversion, conduction, distribution, control, storage, measurement or use of electrical energy. The Regulations require that operatives prevent danger in some circumstances and prevent injury in others. This distinction is important. In some operations it is impossible to prevent danger (i.e. the risk of injury) but it is possible to prevent injury. Injury means death or injury caused by: electric shock, electric burn, fires of electrical origin, electric arcing and explosions initiated or caused by electricity.

Live – means that the conductor is at a voltage produced by a source of electricity.

Charged – means that the item has acquired a charge either because it is live or because is has become charged by static or induction charging.

18.11.3 The Regulations – requirements

The Regulations apply to employers and employees (including the self-employed). All systems shall be constructed so as to prevent danger (as far as is reasonably practicable). Every work activity should be carried out so as to avoid danger (again as far as is reasonably practicable).

The Regulations control both the way in which maintenance is carried out and the need for regular maintenance. The overriding preference for work on electrical systems is that they are made dead before work starts. Safe systems of work are considered most important in this case (particularly to avoid the system becoming inadvertently energized). The Regulations recognize that, under certain circumstances, it may not be possible to isolate circuits and particular requirements are made in terms of protective equipment which may need to be provided. The strength and capability of electrical systems is considered. Systems should take into account possible transient situations as well as normal conditions. Systems should be capable of operating without causing danger.

Adverse or hazardous environments are considered and it is a requirement that systems take account of mechanical damage, weather, wet, dirty or corrosive conditions as well as the presence of any flammable or explosive substances. The requirement again is to avoid danger.

Insulation is required to conductors which may give rise to danger or an alternative is permitted in that conductors may be made safe by position (this may also need the back-up of strictly controlled working practices). Earthing is required when conductors (other than circuit conductors) may become charged so as to cause danger. Alternative means of preventing this danger are also permitted (for instance, double insulation).

No device shall be placed in a reference conductor which is designed to connect to earth that might reasonably be expected to interrupt that conductor in such a way as may result in danger. Connections shall be mechanically and electrically suitable for use so as to prevent danger. Means for protecting from excess of current in any circuit are also to be provided. Where necesssary to prevent danger, a means of cutting off the supply and for isolation shall be supplied. Adequate precautions shall be taken to prevent equipment which has been made dead for the purposes of carrying out works from becoming electrically charged during work if danger may arise.

Work on or near live conductors, if liable to cause danger, is prohibited unless it is unreasonable in all the circumstances for it to be dead and it is reasonable in all the circumstances for the worker to be at work on or near it while it is live and suitable precautions (including, where necessary, the provision of suitable protective equipment) are taken to prevent injury. Working space, access and lighting are considered and the Regulations require that adequate means of access and lighting and adequate working space be provided when work is being carried out in circumstances which may give rise to danger. Regulation 17 of the Electricity (Factories Act) Special Regulations 1908 and 1944 gives dimensions for minimum passageways near switchboards, etc. and these are used as guidance for situations where circuit voltages do not exceed 3000 V.

Competent persons The Regulations require that all persons who work in any activity where technical knowledge or experience is necessary to prevent danger (or injury) are trained so as to possess that knowledge or experience unless under such a degree of supervision as appropriate considering the nature of the work.

Exemption certificates It is possible to obtain exemption from any of the Regulations by writing to the Health and Safety Executive. These exemptions will not be granted unless the HSE are satisfied that the health and safety of persons who are likely to be affected will not be prejudiced in consequence of it. The HSE may impose conditions or other requirements when granting exemption certificates. The HSE is also given the power to issue general exemptions or special exemptions. This power is written in to take account of unforeseen circumstances (as at the time of drafting) and is unlikely to be used in practice.

18.12 The Noise at Work Regulations 1989

These Regulations came into force on 1 January 1990 and control the exposure to noise of persons at work. They establish three noise levels, known as the first action level, the second action level and the peak action level. Different regulations are applicable as each action level is exceeded.

The unit of measurement is known as the equivalent continuous sound level and may be defined as 'that notional continuous steady level which would have the same weighted acoustic energy as the real fluctuating noise measured over the same period of time'. For the purposes of the Regulations an 8-hour time period is used and the 8-hour equivalent continuous sound level is abbreviated to L_{EP},d.

The first action level is 85 dB(A) L_{EP},d.
The second action level is 90 dB(A) L_{EP},d.
The peak action level is 200 Pascals (equivalent to 140 dB).

Damage to the hair cells in the inner ear is proportional to the noise energy received. This is a dose concept comprising the product of noise level and exposure duration. It follows, therefore, that the same amount of deafness will follow from the exposure to a very intense sound for a short period as to a lower level for a proportionally longer one.

It has been shown that the exposure time must be halved for each 3 dB(A) increase in the noise levels. 3 dB(A) represents a doubling of sound energy, hence this rule has become known as the equal energy damage risk criterion. It follows that 93 dB(A) for 4 hours is also 100% of the permitted exposure for a day. Similarly, 2 hours at 93 dB(a) would be 50% of the permitted exposure. Where an employee is likely to be exposed to above the first action level the employer shall ensure that a competent person makes an assessment of the noise levels which is adequate for the purposes:

1. Of identifying which employees are so exposed; and
2. Of providing the employer with such information with regard to the noise to which those employees may be exposed as will facilitate compliance with the employer's duties under the Regulations, specifically:
   (a) Reduction of noise exposure:
   (b) Ear protection;
   (c) Ear protection zones;
   (d) Provision of information to employees.

18.12.1 The requirements of the Regulations

An employer must:

1. Carry out an assessment when an L_{EP},d of 85 dB(A) is likely to be exceeded.
2. Review the assessment if changes necessitate this.
3. Record the exposure and keep records.
4. Reduce the risk of damage to hearing to the lowest level that is reasonably practicable.
5. Every employer shall, when any employee is likely to be exposed to the second action level or above or to the peak action level or above, reduce, so far as is reasonably practicable (other than by the provision of personal ear protectors) the exposure to noise of that employee.
6. If an employee is exposed to greater than the first action level and less than the second action level the employer shall provide hearing protection if so required by the employee.
7. If an employee is exposed to greater than the second action level or greater than the peak action level the employer shall provide hearing protection which, when properly worn, will reduce the risk of hearing damage to below that arising from exposure to the second action level or, as the case may be, to the peak action level.
8. Ear protection zones (i.e. areas where the second action level is likely to be exceeded) shall be established. Employees must wear ear protection in this zone. The employer shall erect suitable signs.
9. Information, instruction and training shall be provided for employees where exposure is likely to exceed the first action level or the peak action level. This information shall include:
    (a) The risk of damage to an employee's hearing that such exposure may cause;
    (b) What steps an employee can take to minimize that risk;
    (c) The steps that an employee must take in order to obtain the personal ear protectors which the employer must provide;
    (d) The employer's obligation under the Regulations.

18.13 Safety of machinery

18.13.1 Identification of hazard

It is first essential to consider all of the phases of a machine's life (i.e. construction through to dismantling). These will include:

Construction
Transport
Installation
Commissioning
Operation
Setting
Adjustment
Maintenance
Decommissioning

The hazard at each point of this list must be assessed. It may be that the most hazardous point will vary with the nature of the operation (i.e. tool setting may be more hazardous than actually operating the machine if the operation is carried out remotely).

18.13.2 Types of hazard

Machines injure in one or more of the following ways:

Entanglement
Trapping
Contact with machine or material in motion
Being struck by objects or parts of machine being ejected
Non-mechanical (electrical, chemical, heat, etc.)

18.13.3 Assessment

Two factors are normally considered, when carrying out risk assessment; the probability of injury and severity of injury produced. In assessing probability of injury the following points should be considered:

1. Frequency of access needed to danger areas;
2. What actions are likely when in danger areas.

An assessment of the two factors can lead to a measure of the proportion of these dangerous accesses likely to result in injury.

The nature of any likely injuries may be used when deciding on the degree of protection necessary. Clearly, an event capable of causing death requires the most stringent safety precautions, while a dangerous machine capable of, say, causing minor bruising might not warrant the same expenditure.

The overall risk is derived from consideration of the likelihood of injury and the probable outcome over all the phases of the machine's life.

18.13.4 Safety by design

Where possible, designers should ensure that new machines do not contain dangerous parts or that these are enclosed by the design of the machine. This is a much more desirable alternative to the fitting of guards to an established machine design. Designers should also pay serious attention to ergonomics. It is important particularly to avoid operator fatigue as much as possible.

The design of safeguards to prevent physical injury can also be used as an opportunity to protect the operator from other hazards – for instance, noise, heat, etc. Examples of typical construction details to reduce the risk of injury may include:

1. Avoidance of shear traps by filling gaps between static and moving parts of machines such that gaps are eliminated or reduced to such an extent that parts of the body cannot enter. Data are available which give information concerning dimensions suitable for the avoidance of trapping various parts of the 'typical' body. An alternative would be to widen a gap sufficiently to prevent body parts being trapped.
2. Drawing-in – consideration of surface roughness, in-running nips, speed or distance of movement, force, torque and inertia.

A consideration of all of the typical hazards to the body should be made and the design of the machine should be such as to eliminate them as far as is reasonably practicable.

If it is not possible to make the machine safe by design other considerations will have to be taken into account, concerning controls, proximity devices and guards. While on the subject of design, other factors to be considered should include:

1. Design of controls to avoid unexpected start-up and/or movement in unexpected directions;
2. Elimination of hazards due to failure of machinery (e.g.

falls of platerns due to hydraulic failure – trapping may be avoided by the use of scotches (props) which are only removed mechanically when guards are closed);

3. Stability – design such that stability is not prejudiced by, for instance, overloading material feed hoppers;
4. Lighting – one of several environmental considerations which are pertinent to risk elimination.

18.13.5 Guarding

If safety by design is not possible the next alternative must be guarding. Within this generic expression we may also include electrical and mechanical interlocking, proximity devices and two-hand switching, etc. There are two main classes of hazardous machinery to be considered:

1. Those involving hazardous parts to which access is not required during normal operation. Access may be needed during maintenance and some form of 'inching' device may be fitted;
2. Those machines to which access is required during normal operations.

The guards for the first type could include the following:

- Fixed enclosed guard;
- Fixed distance guard (barrier or tunnel of sufficient size/length to prevent parts of body from reaching danger area), interlocked guards;
- Trip device (e.g. photoelectric cells, pressure mats, etc.).

In the case of machines to which access is normally required, different types of guards may be necessary, and these would include:

- Interlocked guards (machine disconnected when guard not in position);
- Automatic guard (pushes operator away from danger area);
- Trip device (photocells, etc.);
- Adjustable guards – a poor option, but may be necessary in certain cases where the gap between guard and danger point cannot be completely eliminated (e.g. woodworking machines – these guards tend to be left in the widest gap position);
- Self-adjusting guards – the gap is adjusted by the workpiece itself, thereby reducing the risks of the adjustable guards;
- Two-hand control devices – the operator has to operate two on-switches which are situated to prevent spanning by one hand. The switches must be operated within 0.5 seconds of each other. This device will only protect one operator and is open to abuse if two persons use the machine;
- Hold-to run control – the control is placed out of the danger area and the operative must remain in contact with the switch all the time the machine is operating

18.13.5.1 Installation and practical considerations of guards

Hygiene Guards for use on food machines should be readily cleansable and completely removable from machines – hinged guards are difficult to clean. Removable guards should be adequately interlocked. If fixed guards are necessary they should be mounted on spacers away from the machine in order to permit cleaning, but gaps should not be sufficient to permit parts of the body to reach danger areas.

Corrosion The materials used should be suitable for any corrosive risk from the environment or cleaning materials

used. Stainless steel is the most suitable for the food industry.

Visibility The guard should be designed, if necessary, so that the operative can see the moving parts of the machinery. The use of clear sheet, mesh or grills can be considered.

Strength and durability The materials used should be suitable for the use (or abuse!) to which the guard may be put.

Maintenance Guards should be subject to routine inspection. Routine replacement of moving parts should take place after the end of their design life, particular attention being given to interlock switches, hydraulic valves, etc. Testing of safeguards must be carried out only by properly trained personnel.

Operation without guards There are some circumstances in which machinery must be operated without guards (so as to lubricate, maintain or adjust). It is permissible to carry out these operations. However, they are controlled by The Factories Act 1961, Section 15 (where applicable), and operatives must be properly trained and operations must be specified in writing. Any other operations of machinery without guards is likely to result in a breach of The Health and Safety at Work etc. Act, which could result in the employee and/or the employer being prosecuted as well as the risk of injury.

18.14 Personal protective equipment

Personal protective equipment (PPE) can be defined as:

all equipment designed to be worn or held by a person at work to protect him against one or more risks, and any addition or accessory designed to meet this objective, other than

(a) ordinary working clothes and uniforms not specifically designed to protect the health and safety of the wearer;
(b) personal protective equipment used for protection while travelling on a road within the meaning (in England and Wales) of section 192 (10 of The Road Traffic Act 1988, or (in Scotland) section 151 of The Roads (Scotland) Act 1984;
(c) equipment used during the playing of competitive sports;
(d) self defence or deterrent equipment;
(e) portable devices for detecting and signalling risks and nuisances.

'Risk' means any risk to the health and safety of a person and includes wet or extreme temperature caused by adverse weather or otherwise. This definition is taken from a draft set of regulations published by the Health and Safety Commission. The regulations are intended to implement the requirements of The Health and Safety EC framework directive (which must be implemented by 31 December 1992).

18.14.1 Existing PPE legislation

Section 2, Health and Safety at Work etc. Act 1974
Construction (Head Protection) Regulations 1989
Control of Asbestos at Work Regulations 1987
Control of Lead at Work Regulations 1980
Control of Substances Hazardous to Health Regulations 1988
Ionizing Radiation Regulations 1985
Noise at Work Regulations 1989
Also various pre-Health and Safety at Work etc. Act Regulations.

18.14.2 The new Regulations

Section 2 of The Health and Safety at Work etc. Act 1974 is not sufficiently explicit to comply with the terms of the framework directive. Consequently, new Regulations will have to be made (and may be implemented by the time this book is published). They will be made under Section 15 of The Health and Safety at Work etc. Act and will cover all situations of work as yet not covered by existing Regulations.

Most pre-health and safety at work law concerning PPE is to be revoked as it is not considered 'consistent with modern practice or selection, use and maintenance of PPE'. Some older legislation will remain but will need alteration – primarily to take account of the new European approval method (in place of the old HSE approval). These are Section 30(6) of The Factories Act 1961, Regulations 50, 51 and 60 of The Ship Building and Ship Repairing Regulations 1960 and paragraph 24 to Regulation 18 of The Approval Code of Practice 'Safety in Docks'. These pieces of legislation deal with entry into confined spaces. The scope of the new Regulations would be that of The Health and Safety at Work etc. Act 1974 to include mining, quarrying and offshore work.

The use of PPE on means of transport would be dealt with by Regulations to be made by the Department of Transport (which is why it is specifically excluded from the definition of PPE). Guidance on the selection, maintenance and use of PPE is to be published along with the new Regulations. The Regulations will not apply to the areas covered by existing (post-HSW) Regulations (as listed above) but, rather, these Regulations will be modified so as to be in line with the new general Regulations.

The duty to provide PPE Existing legislation requires risks to be controlled at source and only stipulates the use of PPE if the risk cannot be adequately controlled. This duty is carried on in the new Regulations. The duty is placed on the self-employed as well as employers.

Suitability Regulation 4(3) of the new Regulations defines suitability. The overriding requirement is that the PPE is suitable for the degree of risk 'so far as is reasonably practicable'. After 30 June 1992 all PPE must carry the European mark of approval (the 'CE' mark). Regulations to be published by the DTI (which will implement the EC PPE Product Directive) will require manufacturers and suppliers of new PPE to ensure that their products comply with the basic safety requirements of those Regulations. HSE approvals would cease once the new Regulations apply. Any existing PPE can continue to be used if it complies with Regulation 4(3).

Assessment Regulation 5 requires the assessment of PPE, and this shall comprise:

1. An assessment of any risk or risks which have not been avoided by other means;
2. The definition of the characteristics which PPE must have in order to be effective against the risk referred to in (1) above, taking into account any risks which the equipment itself may create;
3. Comparison of the characteristics of the PPE available with the characteristics referred to in (2) above.

Accommodation for PPE Regulation 7 requires accommodation for PPE, and this is considered necessary to ensure proper maintenance and to enable equipment to be kept clean.

Use of PPE Regulation 9 requires that employers who provide PPE shall take all reasonable steps to ensure that it is properly used. In addition, there is a requirement on employees and the self-employed to make full and proper use of PPE provided under these Regulations and to take all reasonable steps to see that it is returned to the accommodation provided for it, after use.

Charging for the use of PPE An employer is not permitted to charge an employee for the use of PPE at work. There is a provision in the Regulations to allow an employer to charge for the private use (i.e. outside of work) of PPE by the employee. This charge must be reasonable and in line with the cost to the employer resulting from the use of the PPE outside of work. For the purpose of implementing this Regulation, Section 9 of The Health and Safety at Work etc. Act 1974, is disapplied (this section prohibited employers from making any charge) in cases subject to this Regulation.

18.15 Manual handling

More than a quarter of accidents reported to authorities each year are a result of manual handling. While it may be noted that fatalities are rare, major injuries (for instance, those involving major limb fractures) as a result of manual handling constituted 7% of those reported in 1988/1989.

The great majority of lost work days as a result of accidents involving handling were in connection with sprains or strains (approximately 64%). The most likely affected area is the back (40% of all 'over 3-day' injuries).

It should also be noted that it is not the traditional 'heavy' industries that result in the most accidents. The incidence of manual handling injury is widespread. For example, the incidence of manual handling injury in the construction industry is about 37% while that for the medical, veterinary and other health services is 51%.

The latest thinking in manual handling is known as the 'ergonomic approach'. This takes into account the nature of the task, the load, the working environment and the individual's capability. The old legislation approach of simple lifting-weight limits has now fallen into disfavour as being too simplistic and likely to lead to erroneous conclusions concerning an individual's capability.

Draft Regulations concerning manual handling have been proposed but have not yet been brought into force. It is suggested that the Regulations will require an assessment to be made of handling operations likely to result in injury.

In order to avoid carrying out assessments on all manual handling operations (which would be an impossible task), guidelines are given as to the area of manual handling that is not likely to result in injury. These guidelines are not intended to be rigid and detail lifting and lowering of weights. The maximum capability of a person to lift a given weight is represented by the use of a diagram. The greatest lifting power is for weights held close to the body and for those lifts which do not involve lifting above the shoulder or below the knee. As an example, 25 kg may be safely lifted closed to the body, between the upper thigh and the waist, while the capability of lifting a weight at arm's length from the shoulder to above the head is only 5 kg.

The guidelines for carrying are essentially similar to those for lifting with the proviso that no carrying will take place with weights lower than knuckle height. If weights are carried on the shoulder detailed assessment may show that greater weights may be carried.

The guideline for pushing and pulling is a load of 250 Newtons to start or stop the push and 100 Newtons continually.

If lifting from a seated position, the maximum that can be lifted (without further assessment) is 5 kg, if only lifted from waist height to shoulder height near to the body.

It should be noted that all of these weight limits are designed for safe lifting by 95% of all men and between one half and two-thirds of women. If the same degree of protection is required for 95% of women, the weight limits should be reduced by one-third.

In carrying out an assessment the following points should be considered:

The task
Reduce risk of injury
Improve layout
Use the body more effectively (e.g. no twisting)
Rest periods flexibly arranged
Posture
No handling while seated
The load
Make it lighter (subdivide packages)
Make it small
Easier to grasp
More stable
Less dangerous to hold (not oily, corrosive, dirty, etc.)
The working environment
Space constraints

Floors (condition and nature)
Working at different levels
Thermal environment
Lighting
Individual capabilities
Personal capability – injuries, pregnancy, back problems, etc.
Knowledge and training – as a complement to safe systems of work
Training to recognize loads which might cause injury – care with unfamiliar loads

Further reading

BS EN 292, *Safety of Machines*, British Standards Institution, London
Control of Substances Hazardous to Health Regulations, HMSO, London (1988)
Fife, I. and Machin, E. A., *Redgrave, Fife and Machin's Health and Safety*, Butterworths, London (1990)
Health and Safety at Work, Croner Publications, London (looseleaf publication)
Kletz, T. A., *Critical Aspects of Safety and Loss Prevention*, Butterworths, London (1990)
Lees, F. P., *Loss Prevention in the Process Industries*, Butterworths, London (1980)
Ridley, J., *Safety at Work*, 3rd edn, Butterworth–Heinemann, Oxford (1990)

19 Units, symbols and constants

Duncan S. T. Enright

Contents

19.1 SI units   19/3
   19.1.1   Derived units   19/3
   19.1.2   Gravitational and absolute systems   19/3
   19.1.3   Expressing magnitudes of SI units   19/3
   19.1.4   Rules for use of SI units and the decimal multiples and submultiples   19/4
   19.1.5   SI quantities, units and symbols   19/5

19.2 Conversion of existing imperial terms   19/13

19.3 Abbreviations   19/13

19.4 Physical and chemical constants   19/17
   19.4.1   Atomic number   19/17
   19.4.2   Atomic weight   19/17
   19.4.3   Density   19/21
   19.4.4   Melting point   19/21
   19.4.5   Linear coefficient of expansion   19/21
   19.4.6   Heat conductivity   19/21
   19.4.7   Electrical resistivity   19/21

Further reading   19/21

19.1 SI units

The Système International d'Unités (SI) has been adopted and is defined by ISO 1000. Here the system is described and conversions to other commonly used systems are given.

SI comprises seven basic units from which a wide range of quantities can be derived in the form of products and quotients of these units which are shown in Table 19.1. The definitions of these units are as follows.

Metre (m). The metre is the length equal to 1 650 763.73 wavelengths in vacuum of the radiation corresponding to the transition between the levels $2p_{10}$ and $5d_5$ of the krypton-86 atom.

Kilogram (kg). The kilogram is the unit of mass; it is equal to the mass of the international prototype of the kilogram.

Second (s). The second is the duration of 9 192 631 770 periods of the radiation corresponding to the transition between the two hyperfine levels of the ground state of the caesium-133 atom.

Ampere (A). The ampere is that constant current which, if maintained in two straight parallel conductors of infinite length, of negligible circular cross-section, and placed 1 m apart in vacuum, would produce between these conductors a force equal to 2×10^{-7} newtons per metre of length.

Kelvin (K). The kelvin, unit of thermodynamic temperature, is the fraction 1/273.16 of the thermodynamic temperature of the triple point of water.

Candela (cd). The candela is the luminous intensity, in the perpendicular direction, of a surface of $1/600\,000\ m^2$ of a black body at the temperature of freezing platinum under a pressure of 101 325 newtons per square metre.

The supplementary base units are defined as follows:

Plane angle (radian). The angle subtended at the centre of a circle of radius 1 m by an arc of length 1 m along the circumference.

Solid angle (steradian). The solid angle subtended at the centre of a sphere of radius 1 m by an area of $1\ m^2$ on the surface.

Mole (mol) is the amount of substance of a system which contains as many elementary entities as there are atoms in 0.012 kg of carbon-12. The elementary entities must be specified and can be atom molecules, ion electrons, other particles or specified groups of such particles.

19.1.1 Derived units

SI is a rationalized and coherent system because, for any one physical quantity, it admits of only one measurement unit with

Table 19.1 Basic SI units

Quantity	Name of unit	Unit symbol
Length	metre	m
Mass	kilogram	kg
Time	second	s
Electric current	ampere	A
Thermodynamic temperature[a]	kelvin	K
Luminous intensity	candela	cd
Amount of substance	mole	mol

[a] Temperature difference is commonly expressed in degrees Celsius instead of degrees Kelvin. The unit of the temperature interval for these scales is the same: 0 K = −273.15°C; 273.15°K = 0°C.

Table 19.2 Derived units

Physical quantity	SI unit	Unit symbol
Force	newton	$N = kg\ ms^{-2}$
Work, energy quantity of heat	joule	$J = N\ m = kg\ m^2\ s^{-2}$
Power	watt	$W = J\ s^{-1} = kg\ m^2\ s^{-3}$
Electric charge	coulomb	$C = A\ s$
Electric potential	volt	$V = W\ A^{-1} = kg\ m^2\ A^{-1}\ s^{-3}$
Electric capacitance	farad	$F = A\ s\ V^{-1} = A^2 s^4\ kg^{-1}\ m^{-2}$
Electric resistance	ohm	$\Omega = V\ A^{-1} = kg\ m^2\ A^{-2}\ s^{-3}$
Frequency	hertz	$Hz = s^{-1}$
Magnetic flux	weber	$Wb = kg\ m^2\ A^{-1} s^{-2}$
Magnetic flux density	tesla	$T = Wb\ m^{-2} = kg\ A^{-1}\ s^{-2}$
Inductance	henry	$H = kg\ m^2\ A^{-2}\ s^{-2}$
Luminous flux	lumen	$lm = cd\ sr^a$
Illumination	lux	$lx = lm\ m^{-2}$

[a] One steradian (sr) is the solid angle which, having its vertex at the centre of a sphere, cuts off an area of the surface of the sphere equal to that of a square with sides of length equal to the radius of the sphere. The SI unit of electric dipole moment (A s m) is usually expressed as a coulomb metre (C m).

its entire structure derived from no more than seven arbitrarily defined basic units. It is coherent because the derived units are always the products or quotients of two or more of these basic units. Thus the SI unit for velocity is m s^{-1} (metre per second) and for acceleration is m s^{-2} (metre per second every second). Special names (Table 19.2) have been given to some derived units as an aid to communication.

Although SI is complete in itself, certain non-SI units are recognized for use in conjunction with it where, for traditional, commercial or practical purposes, it is difficult to discard them. For example, it is impracticable to disregard the minute (in SI 60 seconds) and the hour (in SI 3600 seconds) which are non-coherent units.

19.1.2 Gravitational and absolute systems

There may be some difficulty in understanding the difference between SI and the Metric Technical System of units which has been used principally in Europe. The main difference is that while mass is expressed in kg in both systems, weight (representing a force) is expressed as kgf (a gravitational unit) in the MKSA system and as N in SI. An absolute unit of force differs from a gravitational unit of force because it induces unit acceleration in a unit mass whereas a gravitational unit imparts gravitational acceleration to a unit mass.

A comparison of the more commonly known systems and SI is shown in Table 19.3. It should be noted in particular how all energy and power, whether from a mechanical, electrical or heat source, share a common derived unit in the SI.

19.1.3 Expressing magnitudes of SI units

To express magnitudes of a unit, decimal multiples and submultiples are formed using the prefixes shown in Table 19.4. This method of expressing magnitudes ensures complete adherence to a decimal system.

Table 19.3   Commonly used units of measurement

	SI (absolute)	*FPS (gravitational)*	*FPS (absolute)*	*cgs (absolute)*	*Metric technical units (gravitational)*
Length	metre (m)	ft	ft	cm	metre
Force	newton (N)	lbf	poundal (pdl)	dyne	kgf
Mass	kg	lb or slug	lb	gram	kg
Time	s	sec	sec	sec	sec
Temperature	°C K	°F	°F °R	°C K	°C K
Energy { mech. / heat	joule[a]	ft lbf / Btu	ft pdl / Btu	dyne cm = erg / calorie	kgf m / k cal.
Power { mech. / elec.	watt	hp / watt	hp / watt }	ergs	metric hp / watt
Electric current	amp	amp	amp	amp	amp
Pressure	N m^{-2}	lbf ft^{-2}	pdl ft^{-2}	dyne cm^{-2}	kgf cm^{-2}

[a] 1 joule = 1 newton metre or 1 watt second.

Table 19.4   The internationally agreed multiples and submultiples

Factor by which the unit is multiplied		*Prefix*	*Symbol*	*Common examples*
One million million	10^{12}	tera	T	
One thousand million	10^{9}	giga	G	gigahertz (GHz)
One million	10^{6}	mega	M	megawatt (MW)
One thousand	10^{3}	kilo	k	kilometre (km)
One hundred	10^{2}	hecto[a]	h	
Ten	10^{1}	deca[a]	da	decagram (dag)
UNITY	1			
One tenth	10^{-1}	deci[a]	d	decimetre (dm)
One hundredth	10^{-2}	centi[a]	c	centimetre (cm)
One thousandth	10^{-3}	milli	m	milligram (mg)
One millionth	10^{-6}	micro	μ	microsecond (μs)
One thousand millionth	10^{-9}	nano	n	nanosecond (ns)
One million millionth	10^{-12}	pico	p	picofarad (pF)
One thousand million millionth	10^{-15}	femto	f	
One million million millionth	10^{-18}	atto	a	

[a] To be avoided wherever possible.

19.1.4   Rules for use of SI units and the decimal multiples and submultiples

1. The SI units are preferred but it is impracticable to limit usage to these, therefore their decimal multiples and submultiples are also required. (For example, it is cumbersome to measure road distances or the breadth of a human hair in metres.)
2. In order to avoid errors in calculations it is preferable to use coherent units. Therefore, it is strongly recommended that in calculations only SI units themselves are used and not their decimal multiples and submultiples. (Example: use N m^{-2} × 10^{6} not MN m^{-2} or N mm^{-2} in a calculation.)
3. The use of prefixes representing 10 raised to a power which is a multiple of 3 is especially recommended. (Example: for length, km . . . m . . . mm . . . μm. Thus hm; dam; dm; cm are non-preferred.)
4. When expressing a quantity by a numerical value of a unit it is helpful to use quantities resulting in numerical values between 0 and 1000. Examples:

12 kN = 12 × 10^{3} N instead of 12 000 N
3.94 mm = 3.94 × 10^{-3} m instead of 0.00394 m
14.01 kN m^{-2} = 14.01 × 10^{3} N m^{-2} instead of 14 010 N m^{-2}

5. Compound prefixes are not used. (Example: write nm not mμm.) Where, however, a name has been given to a product or a quotient of a basic SI unit (for example, the bar (10^{5} N m^{-2})) it is correct practice to apply the prefix to the name (for example, millibar (10^{-3} bar)).
6. In forming decimal multiples and submultiples of a derived SI unit preferably only one prefix is used. The prefix should be attached to the unit in the numerator. (Example: MW m^{-2} not W mm^{-2}.) The exception is stress, where BSI recommend the use of N mm^{-2}.
7. Multiplying prefixes are printed immediately adjacent to the SI unit symbol with which they are associated. The multiplication of symbols is usually indicated by leaving a small gap between them. (Example: mN = millinewton. If written as m N this would indicate a metre newton.)

19.1.5 SI quantities, units and symbols

Tables 19.5 to 19.13 contain lists of SI units used in certain fields together with other units or names of units which may be used and their values in SI units. The tables also show symbols typically used to describe quantities in the same fields.

The units shown in Tables 19.5 to 19.13 generally obey the above guide rules. However, it is expected that a practical attitude should prevail in the interpretation of these recommendations, particularly as certain countries still use Metric Technical Units which include such units as kgf cm^{-2}.

Table 19.5(a) Space and time: units

Quantity	SI units	Conversion factors and remarks on non-SI units
Plane angle[a]	radian (rad)	$1° = 0.017\ 453\ 3$ rad $1' = 2.908\ 88 \times 10^{-4}$ rad $1'' = 4.848\ 14 \times 10^{-6}$ rad 1 right angle $= \Pi/2$ rad $= 90°$
Solid angle	steradian (sr)	
Length[b]	metre (m)	1 inch $= 25.4$ mm 1 foot $= 0.3048$ m 1 yard $= 0.9144$ m 1 mile $= 1.609\ 344$ km 1 UK nautical mile $= 1.853\ 18$ km 1 angstrom (Å) $= 10^{-10}$ metre or 10^{-1} nm
Area	square metre (m^2)	1 sq. in $= 645.16$ mm^2 $\quad\quad\quad\ = 6.4516$ cm^2 1 sq. ft $= 0.092\ 903$ m^2 1 sq. yd $= 0.836\ 127$ m^2 1 acre $= 0.404\ 686$ ha 1 sq. mile $= 258.999$ ha ha $=$ hectare a $=$ are
Volume[c]	cubic metre (m^3)	1 in$^3 = 16.3871$ cm^3 1 ft$^3 = 0.028\ 3168$ m^3 1 yd$^3 = 0.764\ 555$ m^3 1 UK fl. oz $= 28.4131$ mlitre 1 gal $= 4.5461$ litre
Time	second (s)	1 day $= 7.344$ Ms 1 hour $= 3600$ s 1 minute $= 60$ s
Angular velocity	radian per second (rad s^{-1})	rev min$^{-1} = 0.104\ 720$ rad s^{-1} rev s$^{-1} = 6.283\ 19$ rad s^{-1} 1 degree s$^{-1} = 0.017\ 4533$ rad s^{-1}
Velocity	metre per second (m s^{-1})	1 km h$^{-1} = 0.277\ 778$ m s^{-1} 1 ft s$^{-1} = 0.3048$ m s^{-1} 1 mile h$^{-1} = 0.447\ 04$ m s^{-1} or $\quad$ 1.609 34 km h^{-1} knot (kn) $= 0.514\ 444$ m s^{-1} 1 UK knot $= 0.514\ 773$ m s^{-1}
Acceleration	metre per second squared (m s^{-2})	1 ft s$^{-2} = 0.3048$ m s^{-2}
Angular acceleration	radian per second squared (rad s^{-2})	1 in s$^{-2} = 0.0254$ m s^{-2}

[a] *Plane angle.* The Sumerian division of the circle in 360° (hence degrees) is retained for geometry although dynamicists use the radian.
[b] *Length — the centimetre.* In many engineering disciplines the use of the centimetre is non-preferred. It has been adopted as the basic unit of measurement by primary schools and for commercial purposes. Sometimes the centimetre raised to a power (e.g. cm^2; cm^3; cm^4) is used to maintain a sensible range of numerical values in front of the unit. An instance of this concerns steel sections where the moduli of sections and moment of section may be given in steel tables of cm^3 and cm^4 respectively.
Where accuracy to the nearest millimetre is unwarranted, the centimetres can be used to imply a less precise dimension.
[c] *Volume and capacity – the litre.* Before 1964 the 1901 litre was equal to 1.000 028 dm^3. At the XII International CGPM meeting on units (1964) the litre was redefined to equate exactly to 1 cubic decimetre. The same conference agreed that the litre should not be used to express the results of precise measurements, so as to make sure that where high precision was involved (say, greater than 1 part in 20 000) the possibility of confusion between the former (1901) litre and the new (1964) litre would be eliminated.
It is recommended that the results of precise measurements of volume should be given only in terms of m^3, dm^3, cm^3, mm^3, etc. even though the millilitre (ex. cm^3) and litre (ex dm^3) will still be used for operational and commercial purposes.
Because of the possible confusion of the symbol for the litre 'l' with the figure '1', it is strongly recommended that the unit name be spelt in full.
Centilitre is sometimes used for arbitrary quantities implying a greater degree of tolerance.

Table 19.5(b) Space and time: symbols

Symbol	Quantity
θ (α, β, etc.)	plane angle
l	length
A	area
V	volume
t, T	time
ω	angular velocity
v, $\dot{s}$, ds/dt	velocity
a, $\ddot{s}$, d^{2}s/dt^2	acceleration
α, $\dot{\omega}$, dω/dt,	angular acceleration
s, x, y, z	distance or displacement

Table 19.6(a) Periodic and related phenomena: units

Quantity	SI units	Units other than SI	Conversion factors and remarks on non-SI units
Frequency	hertz (Hz)		1 c/s (or c.p.s.) $= 1$ Hz
Rotational frequency[a]	reciprocal second (s^{-1})	rev min^{-1}	1 rev min$^{-1} = 0.016\ 667$ s^{-1}
Wavelength	metre (m)	Å[b]	1 Å $= 10^{-10}$ m

[a] *Rotational frequency (rev/min).* The quantity rev/min is favoured for rotating machinery.
[b] Electromagnetic radiative wavelengths are often quoted in angstroms (Å).

Table 19.6(b) Periodic and related phenomena: symbols

Symbol	Quantity
T	periodic time
τ, (T)	time constant of an exponentially varying quantity
f, v	frequency
η	rotational frequency
ω	angular frequency
λ	wavelength
$\sigma(\bar{v})$	wavenumber
k	circular wavenumber
$\log_e (A_1/A_2)'$	natural logarithm of the ratio of two amplitudes
$10 \log_{10}(P_1/P_2)$	ten times the common logarithm of the ratio of two powers
δ	damping coefficient
Λ	logarithmic decrement
α	attenuation coefficient
β	phase coefficient
γ	propagation coefficient

19.1.5.1 Mass and weight

Confusion sometimes arises over the measuring of the terms 'mass' and 'weight'. Commonly, and in many branches of engineering, it has been the custom to refer to quantities of mass as weights, e.g. weight of coal in kilograms.

Weight, however, is dependent upon the gravitational force acting upon the mass. Thus for a mass (M), weight (W) $= Mg$, where g is the local acceleration due to gravity which varies slightly from point to point on the earth's surface. For practical purposes, an approximated figure of 9.81 or 9.807 metres per second squared (m s^{-2}) is used for g.

The force unit in SI is the newton (N) and by using consistent units becomes the force applied to unit mass (kg) to

Table 19.7(a) Mechanics: units

Quantity	SI units	Conversion factors and remarks on non-SI units
Mass	kilogram (kg)	1 ton = 1016.05 kg or 1.016 05 tonne 1 cwt = 50.8023 kg 1 lb = 0.453 592 37 kg 1 oz = 28.3495 g (avoir)
Mass density	kilogram per cubic metre (kg m^{-3})	1 lb/ft^3 = 16.0185 kg m^{-3} 1 lb/in^3 = 27.6799 g cm^{-3}
Specific volume	cubic metre per kilogram (m^3 kg^{-1})	1 ft^3/lb = 0.062 428 m^3 kg^{-1} 1 cm^3/g = 10^{-3} m^3 kg^{-1}
Momentum	kilogram metre per second (kg m s^{-1})	1 lb ft/s = 0.138 255 kg ms^{-1} 1 g cm/s = 10^{-5} kg m s^{-1}
Angular momentum	kilogram square metre per second (kg m^2 s^{-1})	1 lb ft^2/s = 0.042 140 1 kg m^2 s^{-1} 1 g cm^2/s = 10^{-7} kg m^2 s^{-1}
Moment of inertia	kilogram square metre (kg m^2)	1 lb ft^2 = 0.042 140 1 kg m^2 1 g cm^2 = 10^{-7} kg m^2
Force	newton (N)	1 tonf = 9.964 02 kN 1 lbf = 4.448 22 N 1 ozf = 0.278 014 N 1 pdl = 0.138 255 N 1 dyne = 10^{-5} N 1 kgf or kilopound = 9.806 65 N
Moment of force (torque)	newton metre (N m)	1 tonf ft = 3.037 03 kN m 1 lbf ft = 1.355 82 N m 1 pdl ft = 0.042 140 1 N m 1 lbf in = 0.112 985 N m 1 dyne cm = 10^{-7} N m 1 kgf m = 9.806 65 N m
Mass per unit length	kilogram per metre (kg m^{-1})	1 ton/1000 yds = 1.111 16 kg m^{-1} 1 ton/mile = 0.631 342 kg m^{-1} 1 lb/in = 17.8580 kg m^{-1} 1 lb/ft = 1.488 16 kg m^{-1} 1 lb/yd = 0.496 055 kg m^{-1}
Mass per unit area	kilogram per square metre (kg m^{-2})	1 lb/acre = 1.120 85 × 10^{-4} kg m^{-2} 1 ton/sq. mile = 3.922 98 × 10^{-4} kg m^{-2} 1 lb/1000 ft^2 = 4.882 43 kg m^{-2}
Mass rate of flow	kilogram per second (kg s^{-1})	1 lb/s = 0.453 592 kg s^{-1} 1 lb/h = 1.259 98 × 10^{-4} kg s^{-1} 1 UK ton/h = 0.282 235 kg s^{-1} or 1.016 05 tonne h^{-1}
Volume rate of flow	cubic metre per second (m^3 s^{-1})	1 ft^3/s (cusec) = 28.3168 × 10^{-3} m^3 s^{-1} 1 gal/s = 4.546 09 × 10^{-3} m^3 s^{-1}
Mass flow rate per unit area	kilogram per square metre second (kg m^{-2} s^{-1})	1 lb/ft^{2}h = 1.356 23 × 10^{-3} kg m^{-2} s^{-1}
Pressure and stress	newton per square metre (N m^{-2}) or pascal (Pa)	1 bar = 10^5 N m^{-2} 1 N/m^2 = 1 Pa 1 lbf/in^2 = 6.894 76 kN m^{-2} 1 torr = 1.333 22 mbar = 133.322 N m^{-2} 1 in Hg = 3386.39 N m^{-2} 1 in W.G. = 2.490 89 mbar 1 kgf/cm^2 = 0.980 665 bar = 98.0665 kN m^{-2} 1 tonf/in^2 = 15.4443 N mm^{-2} 1 pieze = 10^3 N m^{-2} 1 std atmosphere = 1013.25 × 10^2 N m^{-2} = 1.033 23 kgf cm^{-2} = 14.695 lbf in^{-2} = 760 torr+ = 29.921 3 in Hg

Table 19.7(a) Cont'd

Quantity	SI units	Conversion factors and remarks on non-SI units
Second moment of area	metre to the power of four (m^4)	$1\ in^4 = 41.6231\ cm^4$ $1\ ft^4 = 863\ 097\ cm^4$
Section modulus	cubic metre (m^3)	$1\ in^3 = 16.3871\ cm^3$
Dynamic viscosity	newton second per square metre ($N\ s m^{-2}$)	$1\ P(poise) = 10^{-1}\ Ns\ m^{-2}$ $1\ lbf\ s/ft^2 = 47.8803\ N\ s\ m^{-2}$ or $47\ 880.3$ cP $1\ pdl\ s/ft^2$ } $= 1.488\ 16\ N\ s\ m^{-2}$ or lb/ft s $\quad$ } $\quad$ or 1488.16 cP
Kinematic viscosity	square metre per second ($m^2 s^{-1}$)	$1\ St\ (stokes) = 10^{-4}\ m^2\ s^{-1}$ $1\ ft^2/h = 2.580\ 64 \times 10^{-5}\ m^2\ s^{-1}$ or 25.806 cSt $1\ ft^2/s = 0.092\ 903\ m^2\ s^{-1}$ or 9.2903×10^4 cSt
Surface tension	newton per metre ($N\ m^{-1}$)	$1\ lbf/ft = 14.5939\ N\ m^{-1}$ $1\ dyne/cm = 10^{-3}\ N\ m^{-1}$ $1\ ft\ lbf = 1.355\ 82\ J$ $1\ erg = 10^{-7}\ J$
Energy, work	joule ($J = N\ m$)	$1\ kgf\ m = 9.806\ 65\ J$ $1\ ft\ pdl = 0.042\ 140\ 1\ J$ $1\ hp\ h = 2.684\ 52 \times 10^6\ J$ $1\ kcal = 4186.8\ J$ $1\ Btu = 1055.06\ J$ $1\ eV\ (electron\ volt) = (1.602\ 10 \pm 0.00007) \times 10^{-19}\ J$ $1\ kWh = 3.6 \times 10^6\ J$
Power	watt ($W = J s^{-1}$)	$1\ hp = 745.7\ W$ $1\ ft\ lbf/s = 1.355\ 82\ W$ $1\ metric\ hp = 735.499\ W$ $1\ kg\ m/s = 9.806\ 65\ W$ $1\ erg/s = 10^{-7}\ W$
Impact strength	joule per square metre ($J\ m^{-2}$)	
Fuel consumption		$1\ gal/mile = 2.825\ litre\ km^{-1}$ $1\ mile/gal = 0.354\ km\ litre^{-1}$
Specific fuel consumption	kilogram per joule ($kg J^{-1}$)	$1\ lb/hp\ h = 0.168\ 97\ kg\ MJ^{-1}$
	cubic metre per joule ($m^3 J^{-1}$)	$1\ pint/hp\ h = 0.211\ 68\ litre\ MJ^{-1}$

impart unit acceleration ($m\ s^{-2}$) to the mass (as distinct from gravitational acceleration, which equals $9.806\ 65\ m\ s^{-2}$). Thus it can be more readily understood by comparing the SI system with other systems for mass, weights and measures as shown in Table 19.8.

19.1.5.2 Megagram

The tonne and kilogram are generally accepted as replacement units for ton and pounds. In particular, most lifting equipment already marked in tons can be considered as adequate for lifting the same number of tonnes because of the small excess (1.6%) of the ton over the tonne.

However, in soil mechanics, the megagram (Mg) rather than the tonne is recommended. This is because with large masses involved in work on soil mechanics, confusion between the ton and the tonne could prove very expensive.

19.1.5.3 Pressure and stress

The SI derived unit for force per unit area is the newton per square metre ($N\ m^{-2}$), referred to as the pascal (Pa), and this unit with suitable multiples is favoured as the unit for stress. There are differences of opinion regarding the unit for pressure, but although some flexibility will have to be allowed in the expression of pressure values, the following practice should be adopted.

1. For the statement of stress property, use, without deviation, $N\ m^{-2}$ and appropriate multiples of it (e.g. $MN\ m^{-2}$) or this, if preferred, expressed as $N\ mm^{-2}$ or, if essential for non-metallic materials, $kN\ m^{-2}$.
2. For pressure statements use either $N\ m^{-2}$ (and suitable multiples and submultiples of it) or bar or mbar. In such cases the conversion $1\ bar = 10^5\ N\ m^{-2}$ will always be quoted for reference.

Table 19.7(b)   Mechanics: symbols

Symbol	Quantity
m	mass
e, ρ	density (mass density)
d	relative density
v	specific volume
p	momentum
b, p_0, p_θ	moment of momentum (angular momentum)
I, J	moment of inertia (dynamic moment of inertia)
F	force
$G(P, W)$	weight
γ	specific weight (weight density)
M	moment of force
M	bending moment
T	torque, moment of a couple
p	pressure
σ	normal stress
τ	shear stress
e, ε	linear strain (relative elongation)
γ	shear strain (shear angle)
$\Theta\theta$	volume strain (bulk strain)
μ, ν	{ Poisson's ratio / Poisson's number
E	Young's modulus (modulus of elasticity)
G	shear modulus (modulus of rigidity)
K	bulk modulus (modulus of compression)
x, κ	compressibility (bulk compressibility)
I, I_a	second moment of area (second axial moment of area)
I_p, J	second polar moment of area
$Z, W\left(\dfrac{I}{v}\right)$	section modulus
$\mu(f)$	coefficient of friction (factor of friction)
$\eta(\mu)$	viscosity (dynamic viscosity)
γ	kinematic viscosity
$\sigma(\gamma)$	surface tension
A, W	work
E, W	energy
E_p, U, V, Φ	potential energy
E_k, K, T	kinetic energy
p	power

3. Pressures or pressure differences measured by manometer tube may often conveniently be expressed as a height of a column of fluid, the nature of the fluid being stated. Such readings must be converted to terms of N m^{-2} if they are to be used in calculations of flow, etc. On the other hand, manometers are sometimes used merely as indicators that a prescribed operating condition has been met. Judgement is therefore required as to when it can be of advantage to use mm H$_2$O, mm Hg, etc. or when it is of advantage to calibrate and read manometers in a suitable multiple of N m^{-2} or in mbar. It is understood that manometers calibrated in mbar are becoming increasingly available and it is recommended that pressures expressed as a height of a column of fluid should progressively give place to a suitable multiple of the SI unit or to the millibar.

4. Pressure units themselves are often not modified to indicate whether the pressure value is 'absolute' (i.e. above zero) or 'gauge' (i.e. above atmospheric pressure). If, therefore, the context leaves any doubt as to which is meant, the word 'pressure' must be qualified appropriately.

   e.g. '. . . at a gauge pressure of 12.5 bar'
   or   '. . . at a gauge pressure of 1.25 MN m^{-2}'
   or   '. . . at an absolute pressure of 2.34 bar'
   or   '. . . at an absolute pressure of 234 kN m^{-2}'

Table 19.9 illustrates some of these practices. Note that this table works on a gauge pressure basis (atmosphere = 0), thus vacuum is shown measured in negative millibars. This continues the custom of associating the higher numerical readings with greater vacuum.

Notwithstanding previous practices of referring to pump performances in terms of pressure, the pump total head should be specified in linear measure (metres).

Table 19.8   Systems of weights and measures

Quantity	Foot pound second	Metric, technical	SI
Mass	1 lb	1 kg	1 kg
Length	1 ft	1 m	1 m
Force	1 lbf	1 kgf	1 N
Definition of force	$lbf = \dfrac{lb \times ft\ s^{-2}}{g}$	$kgf = \dfrac{kg \times m\ s^{-2}}{g}$	$N = kg \times m\ s^{-2}$
Definition of weight (gravitational force)	1 lbf per 1 lb	1 kgf per 1 kg	9.806 65 N per 1 kg

Table 19.9   Gauge vacuum and pressure

	Vacuum			Gauge pressure			
Ins. Hg	30 in Hg	20 in Hg	10 in Hg	10 lbf/in²	100 lbf/in²	1000 lbf/in²	2000 lbf/in²
kN/m²	−101.3	−67.73	−33.86	68.94	689.4	6894	13789.5
	−1.01325	−0.6773	−0.3386				
bar	(−1013.25 mbar)	(−677.3 mbar)	(−338.6 mbar)	0.6894	6.894	68.94	137.895
MN/m²	−0.1013	−0.0677	−0.0338	0.0689	0.6894	6.894	13.7895

Table 19.10(a) Heat: units

Quantity	SI units	Conversion factors and remarks on non-SI units
Absolute temperature	kelvin (K)	K = °C + 273.15 K = 1.8°R (Rankine)
Customary temperature		°C = 5/9 (°F − 32)
Temperature interval	kelvin (K)	1°C = 1 K = 1.8°F (alternative form 1 deg C = 1 deg K = 1.8 deg F)
Temperature coefficient (linear or volumetric)	$(1\ K^{-1})$	
Heat, quantity of heat, internal energy, enthalpy	joule (J)	1 Btu = 1055.06 J 1 cal (IT) = 4.1868 J 1 CHU = 1899.2 J 1 kWh = 3.6 MJ 1 therm = 105.506 MJ 1 erg = 10^{-7} J
Heat flow rate	watt (W)	1 Btu/h = 0.293 071 W 1 kcal/h = 1.163 W 1 cal/s = 4.1868 W 1 frigorie = 4.186 W
Density of heat flow rate	watt per square metre $(W\ m^{-2})$	1 Btu/ft^2 h = 3.154 59 W m^{-2} 1 cal/cm^2 s = 41 868 W m^{-2} 1 CHU/ft^2 h = 5.678 W m^{-2}
Thermal conductivity	watt per metre kelvin $(W\ m^{-1}\ K^{-1})$	1 Btu/ft h °F = 1.730 73 W m^{-1} °C^{-1} 1 kcal/m h °C = 1.163 W m^{-1} °C^{-1} known as k value
Coefficient of heat transfer	watt per square metre kelvin $(W\ m^{-2}\ K)$	1 Btu/ft^2 h °F = 5.678 26 W m^{-2} °C^{-1} 1 cal/cm^2 s °C = 41 868 W m^{-2} °C^{-1} 1 kcal/m^2 h °C = 1.163 W m^{-2} °C^{-1} known as U value
Heat capacity	joule per kelvin $(J\ K^{-1})$	1 Btu=deg R = 1899.11 J °C^{-1} 1 cal/g °C = 4.1868 J K^{-1}
Specific heat capacity	joule per kilogram kelvin $(J\ kg^{-1}\ K^{-1})$	1 Btu/lb °F = 4.1868 kJ kg^{-1} °C^{-1} 1 cal/g °C = 4.1868 J kg^{-1} °C^{-1}
Entropy	joule per kelvin $(J\ K^{-1})$	1 Btu/°R = 1899.11 J K^{-1}
Specific entropy	joule per kilogram per kelvin $(J\ kg^{-1}\ K^{-1})$	1 Btu/lb °F = 4.1868 kJ kg^{-1} K^{-1} 1 cal/g K = 4.1868 kJ kg^{-1} K^{-1}
Specific energy	joule per kilogram $(J\ kg^{-1})$	1 Btu/lb = 2.326 kJ kg^{-1} 1 cal/g = 4.1868 kJ kg^{-1}
Specific enthalpy, specific latent heat	joule per kilogram $(J\ kg^{-1})$	1 Btu/lb = 2.326 kJ kg^{-1}
Specific heat content (i) Mass basis	joule per kilogram $(J\ kg^{-1})$	1 kcal kg^{-1} = 4.1868 kJ kg^{-1} 1 Btu/lb = 2.326 kJ kg^{-1} 1 CHU/lb = 4.186 816 kJ kg^{-1} 1 therm/ton = 103.84 kJ kg^{-1}
(ii) Volume basis	joule per cubic metre $(J\ m^{-3})$	1 Btu/ft^3 = 37.2592 kJ m^{-3} 1 Btu/gal = 0.232 08 kJ litre^{-1} 1 therm/UK gal = 23.208 GJ m^{-3} 1 cal/cm^3 = 4.1868 MJ m^{-3} 1 kcal/m^3 = 4.1868 kJ m^{-3}
Heat release rate	watt per cubic metre $(W\ m^{-3})$	1 Btu/ft^2 s = 37.2589 kW m^{-3} 1 cal/cm^3 h = 1.163 kW m^{-3}

Table 19.10(b) Heat: symbols

Symbol	Quantity
T, Θ	thermodynamic temperature
	absolute temperature
$t, \theta\Theta$	customary temperature
α, λ	linear expansion coefficient
α, β, γ	cubic expansion coefficient
β	pressure coefficient
Q	heat, quantity of heat
$\Phi(q)$	heat flow rate
$q(\phi)$	density of heat flow rate
$\lambda(k)$	thermal conductivity
h, k, U, α	coefficient of heat transfer
$\sigma(\alpha, x, k)$	thermal diffusivity
C	heat capacity
c	specific heat capacity
c_p	specific heat capacity at constant pressure
c_V	specific heat capacity at constant volume
γ, x, k	ratio of the specific heat capacities
S	entropy
s	specific entropy
$U(E)$	internal energy
$H(I)$	enthalpy
F	free energy
G	Gibbs function
$u(e)$	specific internal energy
$h(i)$	specific enthalpy
f	specific free energy
g	specific Gibbs function
L	latent heat
l	specific latent heat

19.1.5.4 Viscosity

The recognized derived SI units for Dynamic and Kinematic Viscosity are $Ns\ m^{-2}$ and $m^2\ s^{-1}$, respectively. However, the existing units, centipoise (cP) and centistoke (cSt) are so well established internationally, particularly for oils, that the operational use of these units will continue.

19.1.5.5 Energy

The choice of a suitable commercial energy unit common to all energy-producing concerns has still to be resolved. The SI unit is the joule and its multiples. However, electrical interests favour the adoption of the kWh ($3.6\ MJ = 1\ kWh$). The following are the probable commercial field quantities:

Coal	tonne
Electricity	kWh
Gas	100 MJ (the therm)
Oil	$\begin{cases} \text{litre; } m^3 \\ \text{kg; tonne} \end{cases}$

19.1.5.6 Hardness values

The hardness unit kgf is used to express the load applied by the indenter and this ensures that most hardness-testing machines and empirically based formulae are not made obsolete. Thus the Rockwell, Vickers and Brinell hardness numbers are used. This number is arbitrary and dimensionless and is dependent upon the resistance offered by the material under test to a definite load.

Table 19.11(a) Electricity and magnetism: units

Quantity	SI units	Conversion factors and remarks on non-SI units
Electric current	ampere (A)	$1\ emu = 10\ A$ $1\ esu = 1/3 \times 10^{-9}\ A$
Electric charge	coulomb (C)	$1\ Ah = 3600\ C$ $C = A\ s$
Charge density	coulomb per cubic metre (C m^{-3})	$1\ emu = 10^7\ C\ m^{-3}$ $1\ esu = 1/3 \times 10^{-3}\ C\ m^{-3}$
Surface density of charge	coulomb per square metre (C m^{-2})	
Electric field strength	coulomb per square metre (C m^{-2})	
	volt per metre (V m^{-1})	
Electric potential	volt (V)	
Displacement	coulomb per square metre (C m^{-2})	$C\ m^{-2} = A\ s\ m^{-2}$
Electric flux	coulomb (C)	
Capacitance	farad (F)	$F = A\ s\ V^{-1} = C\ V^{-1}$
Permittivity	farad per metre (F/m)	$\varepsilon_0 = 8.854 \times 10^{-12}\ F\ m^{-1}$
Electric polarization	coulomb per square metre (C m^{-2})	

Table 19.11(a) Cont'd

Quantity	SI units	Conversion factors and remarks on non-SI units
Electric dipole moment	coulomb metre (C m)	
Current density	ampere per square metre (A m^{-2})	
Linear current density	ampere per metre (A m^{-1})	
Magnetic field strength	ampere per metre (A m^{-1})	1 oersted = $10^3/4\pi$ A m^{-1}
Magnetic potential difference	(A)	1 gilbert = $10/4\pi$ A
Magnetic flux density	tesla (T)	Wb m^{-2} = T 1 gauss = 10^{-4} T
Magnetic flux	weber (Wb)	V s = Wb 1 maxwell = 10^{-8} Wb
Magnetic vector potential	weber per metre (Wb m^{-1})	1 maxwell cm^{-1} = 10^{-6} Wb m^{-1}
Mutual inductance, self-inductance	henry (H)	H = V s A^{-1}
Permeability	henry per metre (H m^{-1})	$\mu_0 = 4\pi \times 10^{-7}$ H m^{-1}
Magnetic moment	ampere square metre (A m^2)	
Magnetization	ampere per metre (A m^{-1})	1 oersted = $10^3/4\pi$ A m^{-1}
Magnetic polarization	tesla (T)	1 gauss = 10^{-4} T
Magnetic dipole moment	newton square metre per ampere (N m^2 A^{-1})	
Resistance	ohm (Ω)	
Conductance	reciprocal ohm (1/Ω)	S = mho = Siemen
Resistivity	ohm metre (Ωm)	
Conductivity	reciprocal ohm metre (1 Ω^{-1} m^{-1})	
Reluctance	reciprocal henry (1 H^{-1})	
Permeance	henry (H)	
Impedance Reactance	ohm (Ω)	
Conductance	reciprocal ohm (1 Ω^{-1})	
Active power	watt (W)	
Apparent power	volt ampere (VA)	
Reactive power	var	
Electric stress	volt per metre (V m^{-1})	1 kV in^{-1} = 0.039 370 1 kV mm^{-1} or 1 V mil^{-1} = 39.370 1 kV m^{-1}

Table 19.11(b) Electricity and magnetism: symbols

Symbol	Quantity
I	electric current
Q	electric charge, quantity of electricity
e	volume density of charge, charge density
σ	surface density of charge
$E, (K)$	electric field strength
V, ϕ	electric potential
$U, (V)$	potential difference, tension
E	electromotive force
D	displacement (rationalized displacement)
D'	non-rationalized displacement
Ψ	electric flux, flux of displacement (flux of rationalized displacement)
Ψ'	flux of non-rationalized displacement
C	capacitance
ε	permittivity
ε_0	permittivity of vacuum
ε'	non-rationalized permittivity
ε_0'	non-rationalized permittivity of vacuum
ε_r	relative permittivity
χ_e	electric susceptibility
χ_e'	non-rationalized electric susceptibility
P	electric polarisation
$p, (p_e)$	electric dipole moment
$J, (S)$	current density
$A, (\alpha)$	linear current density
H	magnetic field strength
H'	non-rationalized magnetic field strength
U_m	magnetic potential difference
F, F_m	magnetomotive force
B	magnetic flux density, magnetic induction
Φ	magnetic flux
A	magnetic vector potential
L	self-inductance
M, L_{12}	mutual inductance
$k, (x, k)$	coupling coefficient
σ	leakage coefficient
μ	permeability
μ_0	permeability of vacuum
μ'	non-rationalized permeability
μ_0'	non-rationalized permeability of vacuum
μ_r	relative permeability
x, k	magnetic suceptibility
x', k'	non-rationalized magnetic susceptibility
m	electromagnetic moment (magnetic moment)
$H_i, (M)$	magnetization
$B_i, (J)$	magnetic polarization
J'	non-rationalized magnetic polarization
ω	electromagnetic energy density
S	Poynting vector
c	velocity of propagation of electromagnetic waves in vacuo
R	resistance (to direct current)
G	conductance (to direct current)
ρ	resistivity
y, σ	conductivity
R, R_m	reluctance
$A, (P)$	permeance
N	number of turns in winding
m	number of phases
p	number of pairs of poles
ϕ	phase displacement
Z	impedance (complex impedance)
$[Z]$	modulus of impedance (impedance)
X	reactance
R	resistance
Q	quality factor
Y	admittance (complex admittance)
$[Y]$	modulus of admittance (admittance)
B	susceptance
G	conductance
P	active power
$S, (P_s)$	apparent power
$Q, (P_q)$	reactive power

19.1.5.7 Concentration

Concentration should preferably be expressed on a mass/mass basis (i.e. $kg\ kg^{-1}$; $mg\ kg^{-1}$) or a volume/volume basis (i.e. $m^3\ m^{-3}$; litre m^{-3}; millilitre m^{-3}). It may also be expressed in parts per million (ppm) or as a percentage 'by mass' or 'by volume', respectively.

Table 19.12(a) Light: units

Quantity	SI units	Conversion factors and remarks on non-SI units
Luminous intensity	candela (cd)	
Luminous flux	lumen (lm)	lm = cd sr (candela steradian)
Illuminance	lux (lx)	lx = lm m^{-2}
Luminance	candela per square metre (cd m^{-2})	stilb = 1 cd cm^{-2} apostilb = π^{-1} cd m^{-2} 1 cd/in^2 = 1550 cd m^{-2} 1 foot lambert = 3426 cd m^{-2} 1 lambert = 3183 cd m^{-2}

Table 19.12(b) Light: symbols

Symbol	Quantity
E	illuminance
r	reflectance factor
V	Munsell value
I	luminous intensity
n	refractive index
ϕ	luminous flux
w	beamwidth

19.1.5.8 pH scale

This is a number based on the logarithm, to the base 10 of the reciprocal of the concentration of hydrogen ions in aqueous solution. It is used as a method of expressing small differences in the acidity or alkalinity of nearly neutral solutions in biological and electrolytic processes.

19.1.5.9 Quantities and units of light

The following definitions are based on the International Lighting Vocabulary.

Luminous flux (symbol ϕ): The light emitted by a source such as a lamp or received by a surface, irrespective of direction.
Lumen (abbreviation lm): The SI unit of luminous flux used in describing the total light emitted by a source or received by a surface. (A 100-watt incandescent lamp emits about 1200 lumens.)
Illumination: The process of lighting an object.
Illumination value (symbol E): The luminous flux incident on a surface, per unit area.
Lux (abbreviation lx): The SI unit of illumination value; it is equal to one lumen per square metre.
Lumen per square foot (abbreviation lm ft^{-2}): A non-metric unit of illumination value, equal to 10.76 lux. (Previously called the foot-candle, a term still used in some countries.)
Service value of illumination: The mean value of illumination throughout the life of an installation and averaged over the working area.
Initial value of illumination: The mean value of illumination averaged over the working area before depreciation has started, i.e. when the lamps and fittings are new and clean and when the room is freshly decorated.
Mean spherical illumination (scalar illumination): The average illumination over the surface of a small sphere centred at a given point; more precisely, it is the flux incident on the surface of the sphere divided by the area of the sphere. The

Table 19.13(a) Sound: units

Quantity	SI units	Conversion factors and remarks on non-SI units
Sound intensity	watt per square metre (W m^{-2})	$1 \text{ erg s}^{-1} \text{ cm}^{-2} = 10^{-3} \text{ W m}^{-2}$
Sound intensity (logarithmic)		$1 \text{ decibel (dB)} = 20 \log_{10}(P/P_0)$ where P = measured sound pressure and P_0 = reference sound pressure of $2 \times 10^{-5} \text{ N m}^{-2}$
Loudness		1 phon
Attenuation		neper per metre (np m^{-1})

Table 19.13(b) Sound: symbols

Symbol	Quantity
T	period, periodic time
f, v	frequency, frequency interval
ω	angular frequency, circular frequency
λ	wavelength
k	circular wave number
ρ	density (mass density)
P_s	static pressure
p	(instantaneous) sound pressure
$\varepsilon, (x)$	(instantaneous) sound particle displacement
u, v	(instantaneous) sound particle velocity
a	(instantaneous) sound particle acceleration
q, U	(instantaneous) volume velocity
c	velocity of sound
E	sound energy density
$P, (N, W)$	sound energy flux, sound power
I, J	sound intensity
$Z_s, (W)$	specific acoustic impedance
$Z_a, (Z)$	acoustic impedance
$Z_m, (w)$	mechanical impedance
$L_p, (L_N, L_w)$	sound power level
$L_p, (L)$	sound pressure level
δ	damping coefficient
Λ	logarithmic decrement
α	attenuation coefficient
β	phase coefficient
γ	propagation coefficient
δ	dissipation coefficient
r, τ	reflection coefficient
γ	transmission coefficient
$\alpha, (\alpha_a)$	acoustic absorption coefficient
R	sound reduction index / sound transmission loss
A	equivalent absorption area of a surface or object
T	reverberation time
$L_N, (\Lambda)$	loudness level
N	loudness

term 'scalar' illumination is the lux: care is needed to avoid confusing the unit with the illumination on a plane which is measured in the same unit.

Illumination vector: A term used to describe the flow of light. It has both magnitude and direction. The magnitude is defined as the maximum difference in the value of illumination at diametrically opposed surface elements of a small sphere centred at the point under consideration. The direction of the vector is that of the diameter joining the brighter to the darker element.

Luminous intensity: The quantity which describes the illuminating power of a source in a particular direction. More precisely, it is the luminous flux emitted within a very narrow cone containing that direction divided by the solid angle of the cone.

Candela (abbreviation cd): The SI unit of luminous intensity. The term 'candle power' designates a luminous intensity expressed in candelas.

19.2 Conversion of existing imperial terms

If it is necessary to convert existing imperial terms to a metric equivalent, care should be taken to ensure that the converted value implies the same degree of accuracy. The conversion factor must convey the same order of precision as the original value. Thus to translate 1 in as 25.4 mm or 1000 ft as 304.8 m conveys a tolerance which, in most cases, would be too precise.

Particular care is needed when converting machined tolerances. With a simple dimension such as 0.836 in, it is reasonable to assume that this dimension can be met because an imperial micrometer can measure to 0.001 in. The conversion factor for 0.836 in is 21.2344 mm. Thus 0.004 mm in our conversion represents an accuracy of 0.000 016 in, which is beyond the scope of most toolroom measuring devices. In such cases it should be borne in mind that a metric micrometer can measure to 0.01 mm and with a vernier attachment to 0.002 mm. Thus for the greatest possible accuracy our converted readings should be 21.234 mm. Table 19.14 gives some metric to other unit conversion factors.

19.3 Abbreviations

Table 19.15 gives an alphabetical list of commonly used abbreviations of units. Obsolete and rarely found units are also included. For definitions of these units see Tables 19.5–19.13 and other reference works listed at the end of this chapter.

Table 19.14 Metric conversion factors

SI units	Other units
SPACE AND TIME	
Length:	
1 μm (micron)	= 39.37 × 10^{-6} in
1 mm	= 0.039 370 1 in
1 cm	= 0.393 701 in
1 m	= 3.280 84 ft
1 m	= 1.093 61 yd
1 km	= 0.621 371 mile
Area:	
1 mm^2	= 1.550 × 10^{-3} in^2
1 cm^2	= 0.1550 in^2
1 m^2	= 10.7639 ft^2
1 m^2	= 1.195 99 yd^2
1 ha	= 2.471 05 acre
Volume:	
1 mm^3	= 61.0237 × 10^{-6} in^3
1 cm^3	= 61.0237 × 10^{-3} in^3
1 m^3	= 35.3147 ft^3
1 m^3	= 1.307 95 yd^3
Capacity:	
10^6 m^3	= 219.969 × 10^6 gal
1 m^3	= 219.969 gal
1 litre (l)	$\begin{cases} = 0.219\ 969\ \text{gal} \\ = 1.759\ 80\ \text{pint} \end{cases}$
Capacity flow:	
10^3 m^{-3} s^{-1}	= 791.9 × 10^6 gal h^{-1}
1 m^3 s^{-1}	= 13.20 × 10^3 gal min^{-1}
1 litre s^{-1}	= 13.20 gal min^{-1}
1 m^3 k^{-1} W h	= 219.969 gal k^{-1} W h
1 m^3 s^{-1}	= 35.3147 ft^3 s^{-1} (cusecs)
1 litre s^{-1}	= 0.588 58 × 10^{-3} ft^3 min^{-1} (cfm)
Velocity:	
1 m s^{-1}	= 3.280 84 ft s^{-1} = 2.236 94 mile h^{-1}
1 km h^{-1}	= 0.621 371 mile h^{-1}
Acceleration:	
1 m s^{-2}	= 3.280 84 ft s^{-2}
MECHANICS	
Mass:	
1 g	= 0.035 274 oz
1 kg	= 2.204 62 lb
1 t	= 0.984 207 ton = 19.6841 cwt
Mass flow:	
1 kg s^{-1}	= 2.204 62 lb s^{-1} = 7.936 64 klb h^{-1}
Mass density:	
1 kg m^{-3}	= 0.062 428 lb ft^{-3}
1 kg litre^{-1}	= 10.022 119 lb gal^{-1}
Mass per unit length:	
1 kg m^{-1}	= 0.671 969 lb ft^{-1} = 2.015 91 lb yd^{-1}
Mass per unit area:	
1 kg m^{-2}	= 0.204 816 lb ft^{-2}
Specific volume:	
1 m^3 kg^{-1}	= 16.0185 ft^3 lb^{-1}
1 litre tonne^{-1}	= 0.223 495 gal ton^{-1}
Momentum:	
1 kg m s^{-1}	= 7.233 01 lbft s^{-1}
Angular momentum:	
1 kg m^2 s^{-1}	= 23.7304 $lbft^2$ s^{-1}
Moment of inertia:	
1 kg m^2	= 23.7304 $lbft^2$
Force:	
1 N	= 0.224 809 lbf
Weight (force) per unit length:	
1 N m^{-1}	= 0.068 521 8 lb ft^{-1} = 0.205 566 lbf yd^{-1}

Table 19.14 Cont'd

SI units	Other units
Moment of force (or torque):	
1 Nm	= 0.737 562 lbf ft
Weight (force) per unit area:	
1 N m^{-2}	= 0.020 885 lbf ft^{-2}
Pressure:	
1 N m^{-2}	= 1.450 38 × 10^{-4} lbf in^{-2}
1 bar	= 14.5038 lbf in^{-2}
1 bar	= 0.986 923 atmosphere
1 mbar	= 0.401 463 in H$_2$O
	= 0.029 53 in Hg
Stress:	
1 N mm^{-2}	= 6.474 90 × 10^{-2} tonf in^{-2}
1 MN m^{-2}	= 6.474 90 × 10^{-2} tonf in^{-2}
1 hbar	= 0.647 490 tonf in^{-2}
Second moment of area:	
1 cm^4	= 0.024 025 in^4
Section modulus:	
1 m^3	= 61 023.7 in^3
1 cm^3	= 0.061 023 7 in^3
Kinematic viscosity:	
1 m^2 s^{-1}	= 10.762 75 ft^2 s^{-1} = 10^6 cSt
1 cSt	= 0.038 75 ft^2 h^{-1}
Energy, work:	
1 J	= 0.737 562 ft lbf
1 MJ	= 0.3725 hph
1 MJ	= 0.277 78 kW h
Power:	
1 W	= 0.737 562 ft lbf s^{-1}
1 kW	= 1.3410 hp = 737.562 ft lbf s^{-1}
Fluid mass:	
(Ordinary) 1 kg s^{-1}	= 2.204 62 lb s^{-1} = 7936.64 lb h^{-1}
(Velocity) 1 kg m^{-2} s	= 0.204 815 lb ft^{-2} s
HEAT	
Temperature:	
(Interval) 1 degK	= 9/5 deg R (Rankine)
1 degC	= 9/5 deg F
(Coefficient) 1 degR^{-1}	= 1 deg F^{-1} = 5/9 deg C
1 degC^{-1}	= 5/9 deg F^{-1}
Quantity of heat:	
1 J	= 9.478 17 × 10^{-4} Btu
1 J	= 0.238 846 cal
1 kJ	= 947.817 Btu
1 GJ	= 947.817 × 10^3 Btu
1 kJ	= 526.565 CHU
1 GJ	= 526.565 × 10^3 CHU
1 GJ	= 9.478 17 therm
Heat flow rate:	
1 W(J s^{-1})	= 3.412 14 Btu h^{-1}
1 W m^{-2}	= 0.316 998 Btu ft^{-2} h^{-1}
Thermal conductivity:	
1 W m^{-1} °C^{-1}	= 6.933 47 Btu in ft^{-2} h^{-1} °F^{-1}
Heat transfer coefficient:	
1 W m^{-3} °C^{-1}	= 0.176 110 Btu ft^{-2} h^{-1} °F^{-1}
Heat capacity:	
1 J ° g^{-1} C^{-1}	= 0.526 57 × 10^{-3} Btu °R^{-1}
Specific heat capacity:	
1 J °C^{-1}	= 0.238 846 Btu lb^{-1} °F^{-1}
1 kJ kg^{-1} °C^{-1}	= 0.238 846 Btu lb^{-1} °F^{-1}
Entropy:	
1 J K^{-1}	= 0.526 57 × 10^{-3} Btu °R^{-1}

Table 19.14 Cont'd

SI units	*Other units*
Specific entropy:	
1 J kg^{-1} degC^{-1}	= 0.238 846 × 10^{-3} Btu lb^{-1} °F^{-1}
1 J kg^{-1} degK^{-1}	= 0.238 846 × 10^{-3} Btu lb^{-1} °R^{-1}
Specific energy/specific latent heat:	
1 J g^{-1}	= 0.429 923 Btu lb^{-1}
1 J kg^{-1}	= 0.429 923 × 10^{-3} Btu lb^{-1}
Calorific value:	
1 kJ kg^{-1}	= 0.429 923 Btu lb^{-1}
1 kJ kg^{-1}	= 0.773 861 4 CHU lb^{-1}
1 J m^{-3}	= 0.026 839 2 × 10^{-3} Btu ft^{-3}
1 kJ m^{-3}	= 0.026 839 2 Btu ft^{-3}
1 kg litre^{-1}	= 4.308 86 Btu gal^{-1}
1 kJ kg^{-1}	= 0.009 630 2 therm ton^{-1}
ELECTRICITY	
Permeability:	
1 H m^{-1}	= 10^7/4Π μ_o
Magnetic flux density:	
1 tesla	= 10^4 gauss = 1 Wb m^{-2}
Conductivity:	
1 mho	= 1 reciprocal ohm
1 Siemen	= 1 reciprocal ohm
Electric stress:	
1 kV mm^{-1}	= 25.4 kV in^{-1}
1 kV m^{-1}	= 0.0254 kV in^{-1}

Table 19.15 Common abbreviations of units

a	year (p.a. = per annum)
Å	ångstrom
A	ampere
asb	apostilb
AU	astronomical unit
AT	assay ton
b	barn
bar	bar
Bi	Biot (unit of current in electromagnetic CGS system)
Btu } BthU }	British thermal unit
c	curie
C	coulomb
°C	degree Celsius
cal	calorie
cc	cubic centimetre
cd	candela
CHU	Centigrade heat unit
Ci	curie
cl	centilitre
cm	centimetre
CM	carat
cP	centipoise
c/s	cycle per second
cSt	centistoke
ct	carat
cu. cm	cubic centimetre
cu ft	cubic foot
cu in	cubic inch
cusec	cubic foot per second
cwt	hundredweight

Table 19.15 Cont'd

d	day
dB	decibel
dm	decimetre
dwt	pennyweight
dyn	dyne
e unit } E unit }	X-ray dosage
erg	erg
eV	electronvolt
f	force
F	farad
°F	degree Fahrenheit
fc	foot candle
ft	foot
ft L	foot Lambert
ft lb	foot pound
g	gram
G	gauss
gal	gallon
Gb	gilbert
g cal	gram calorie
gl	gill
gm	gram
g.p.m.	gallons per minute
g.p.s.	gallons per second
gr	grain
Gs	gauss
h	hour
H	henry
ha	hectare
hp	horsepower
hp hr	horsepower hour

Table 19.15 Cont'd

Hz	hertz
in	inch
in Hg	inch of mercury
J	joule
K	kelvin
kc	kilocycle
kcal	kilocalorie
kc/s	kilocycle per second
kg	kilogram
kgf	kilogram force
km	kilometre
kn kt	knot
kV	kilovolt
kVA	kilovolt ampere
kW	kilowatt
kW h	kilowatt hour
L	lambert
l	litre
lb	pound
lbf	pound force
lea	league
lm	lumen
ly	light year
lx	lux
m	metre
m	molality molal concentration
M	molar concentration
mA	milliampere
mbar	millibar
mcps	mega cycles per second
MeV	mega electron volt
mF	millifarad
micron	length – 10^{-6} metre pressure – 10^{-3} mm Hg
mil	angular – 1/1000 rt. angle length 1/1000 inch volume – millilitre
min	minute (time)
mks	metre kilogram second
ml	millilitre
mL	millilambert
mm	millimetre
mm fd	micromicrofarad
mm Hg	millimetre of mercury
mmm	millimicrons
mol	mole (amount of substance)
mpg	miles per gallon
mpm	metres per minute
m/s mps	metres per second
mt	metric ton
mV	millivolt
mW h	megawatt hour
Mx	maxwell
N	newton
n. mile nm	nautical mile
Np	neper
nt	nit
ntm	net ton mile
n unit	neutron dose
Oe	oersted
oz	ounce (avoirdupois)

Table 19.15 Cont'd

oz. t	ounce (troy)
p	perch
P	poise
P	phon
Pa	pascal
pc	parsec
pdl	poundal
ph	phot
psi	pounds per square inch
pwt	pennyweight
q ql	quintal
qts	quart
r R	Röntgen
R	Réaumier
°R	degree Rankine
rad	radian
rpm	revolutions per minute
rps	revolutions per second
s	second (time)
S	Siemen
S St	stokes
sb	stilb
sn	sthéne
sr	steradian
T	tesla
t	tonne
th	thermie
V	volt
VA	volt ampere
W	watt
Wb	weber
yd	yard

19.4 Physical and chemical constants

The following tables provide physcial and chemical constants most of use in engineering disciplines. Table 19.16 lists universal constants. Table 19.17 is a table of elements. Table 19.18 presents the principal elements in order of valency. Table 19.19 gives the surface tension of some common metals, and Table 19.20 lists specific heats, melting points and densities of selected elements. Table 19.21 reports the sectional properties of metals.

For further information on the properties of engineering materials, including common metals and alloys, see Chapter 7.

19.4.1 Atomic number

Radon has an atomic number of 86 (International Atomic Weights Commission). Isotopes of radon will have the same atomic number but different atomic weights.

19.4.2 Atomic weight

Weights are related to an arbitrary reference value of 16 for oxygen. Atomic weights vary because of natural variation in composition (different isotopes in mixture affect the weight average) and because of relativistic effects.

Table 19.16 Universal constants

Constant	Symbol	Numerical value	SI unit
Speed of light in vacuum	c	2.997 925(1)	10^8 m s^{-1}
Gravitational constant	G	6.670(5)	10^{-11} N m^2 kg^2
Elementary charge	e	1.602 10(2)	10^{-19} C
Avogadro constant	N_A	6.022 52(9)	10^{26} kmol^{-1}
Mass unit	u	1.660 43(2)	10^{-27} kg
Electron rest mass	m_e	9.109 08(13)	10^{-31} kg
		5.485 97(3)	10^{-44} u
Proton rest mass	m_p	1.672 52(3)	10^{-27} kg
		1.007 276 63(8)	u
Neutron rest mass	m_n	1.674 82(3)	10^{-27} kg
		1.008 665 4(4)	u
Faraday constant	F	9.684 70(5)	10^4 C mol^{-1}
Planck constant	h	6.625 59(16)	10^{-34} J s
	$h/2\pi$	1.054 494(25)	10^{-34} J s
Fine-structure constant	α	7.297 20(3)	10^{-3}
	$1/\alpha$	137.038 8(6)	
Charge-to-mass ratio for electron	e/m_e	1.758 796(6)	10^{11} C kg^{-1}
Quantum of magnetic flux	hc/e	4.135 56(4)	10^{-11} Wb
Rydberg constant	R_∞	1.097 373 1(1)	10^7 m^{-1}
Bohr radius	a_0	5.291 67(2)	10^{-11} m
Compton wavelength of electron	h/m_ec	2.426 21(2)	10^{-12} m
	$\lambda C/2\pi$	3.861 44(3)	10^{-13} m
Electron radius	$e^2/m_ec^2 = r_e$	2.817 77(4)	10^{-15} m
Thomson cross-section	$8\pi r_e^2/3$	6.651 6(2)	10^{-29} m^2
Compton wavelength of proton	$\lambda_{C,p}$	1.321 398(13)	10^{-15} m
	$\lambda_{C,p}/2\pi$	2.103 07(2)	10^{-16} m
Gyromagnetic ratio of proton	γ	2.675 192(7)	10^8 rad (s T)$^{-1}$
	$\gamma/2\pi$	4.257 70(1)	10^7 Hz T^{-1}
(uncorrected for diamagnetism of H$_2$O)	γ'	2.675 123(7)	10^8 rad (s T)$^{-1}$
	$\gamma'/2\pi$	4.257 59(1)	10^7 Hz T^{-1}
Bohr magneton	μ_B	9.273 2(2)	10^{-24} J T^{-1}
Nuclear magneton	μ_N	5.050 50(13)	10^{-27} J T^{-1}
Proton magnetic moment	μ_p	1.410 49(4)	10^{-26} J T^{-1}
	μ_p/μ_N	2.792 76(2)	
(uncorrected for diamagnetism in H$_2$O sample)	μ_p'/μ_N	2.792 68(2)	
Gas constant	R_0	8.314 34(35)	J K^{-1} mol
Boltzmann constant	k	1.380 54(6)	10^{-23} J K^{-1}
First radiation constant $(2\pi hc^2)$	c_1	3.741 50(9)	10^{-16} W m^{-2}
Second radiation constant (hc/k)	c_2	1.438 79(6)	10^{-2} m K
Stefan–Boltzmann constant	σ	5.669 7(10)	10^{-8} W m^{-2} K^4

Table 19.17 Table of elements

Element	Symbol	Atomic no.	Atomic weight	Density (kg m^{-3})	Melting point (°C)	Linear coefficient of expansion/°C at normal temp. ($\times 10^{-6}$)	Heat conductivity (W m^{-1} °C^{-1})	Electric resistivity (microhm cm)
Actinium	Ac	89	227.00	10 100	1 230	—	—	—
Aluminium	Al	13	26.98	2 700	657	24	217.7	2.655
Antimony	Sb	51	121.75	6 619	630.5	11.29	18.59	39
Argon	Ar	18	39.944	1.663 (liquid)	−187.9	—	0.017	—
Arsenic	As	33	74.92	5 733	813.8	3.86	—	35
Barium	Ba	56	137.34	3 500	710	—	—	60
Beryllium	Be	4	9.012	1 822	1 285	12.3	161	2.85
Bismuth	Bi	83	209.00	9 802	271	13.45	8.37	115
Boron	B	5	10.81	2 300	2 030	2	—	1.8×10^{12}
Bromine	Br	35	79.9	3 119	−28.3	—	—	35
Cadmium	Cd	48	112.40	8 652	321	29.8	90.86	7.59
Caesium	Cs	55	132.91	1 899	26	97	—	20
Calcium	Ca	20	40.08	1 550	851.3	25	—	4.6
Californium	Cf	98	251	—	—	—	—	—
Carbon (Graphite)	C	6	12.01	2 220	3 500	1.2	23.87	1 000
Cerium	Ce	58	140.12	6 901	775.2	—	—	78
Chlorine	Cl	17	35.457	1 560 (liquid)	−101	11.44	0.007 2	10×10^{15}

Table 19.17 Cont'd

Element	Symbol	Atomic no.	Atomic weight	Density (kg m⁻³)	Melting point (°C)	Linear coefficient of expansion/°C at normal temp. ($\times 10^{-6}$)	Heat conductivity (W m⁻¹ °C⁻¹)	Electric resistivity (microhm cm)
Chromium	Cr	24	52.01	7 139	1 900	8.1	69.08	13.1
Cobalt	Co	27	58.94	8 904	1 490	12.08	69.08	9.7
Copper	Cu	29	63.54	8 941	1 082	16.42	386.5	1.682
Curium	Cm	96	247	—	—	—	—	—
Dysprosium	Dy	66	162.46	8 500	1 500	—	—	89
Erbium	Er	68	167.20	9 000	1 525	—	—	81
Europium	Eu	63	152.00	5 200	830	—	—	—
Fermium	Fm	100	257	—	—	—	—	—
Fluorine	Fl	9	19.00	1 100 (liquid)	−223	—	—	—
Gadolinium	Gd	64	156.90	7 900	1 320	—	—	126
Gallium	Ga	31	69.72	5 910	29.79	18.3	—	57.1
Germanium	Ge	32	72.60	5 363	958.3	—	—	89×10^3
Gold	Au	79	197.20	19 310	1 063	14.4	296.1	2.42
Hafnium	Hf	72	178.60	11 400	1 700	—	—	29.6
Helium	He	2	4.003	0.166 3	2 000	—	0.139	—
Holmium	Ho	67	163.5	8 800	1 500	—	—	77
Hydrogen	H	1	1.008 1	0.083 8	−253	—	0.170	—
Indium	In	49	114.76	7 308	161.2	33	23.87	9
Iodine	I	53	126.92	4 927	113.5	93	0.043 5	1.3×10^{15}
Iridium	Ir	77	193.10	22 400	2 409	6.41	59	6.08
Iron	Fe	26	55.84	7 861	1 536	11.9	79.56	9.8
Krypton	Kr	36	83.70	2 160 (liquid)	−170.5	—	0.009	—
Lanthanum	La	57	138.92	6 146	826.4	—	—	59
Lead	Pb	82	207.21	11 320	327.3	29.5	34.75	20.65
Lithium	Li	3	6.94	534.3	186	56.0	71.14	8.5
Lutetium	Lu	71	175	9 870	1 700	—	—	54
Magnesium	Mg	12	24.32	1 739	651.3	25.7	154.9	4.46
Manganese	Mn	25	54.93	7 418	1 243	23	—	—
Mercury	Hg	80	200.61	13 540	−38.9	—	8.37	95.8
Molybdenum	Mo	42	95.95	10 190	2 620	5.49	146.6	4.77
Neodymium	Nd	60	144.27	7 058	1 024	—	—	79
Neon	Ne	10	20.183	0.839	−248.5	—	0.046	—
Neptunium	Np	93	239	—	1 900	—	—	—
Nickel	Ni	28	58.70	8 915	1 452	13.7	58.61	6.9
Niobium	Nb	41	92.91	8 571	2 420	7.2	52	15.2
Nitrogen	N	7	14.008	1 165	−209.5	—	0.025	—
Osmium	Os	76	190.2	22 480	3 000	5.7	—	9
Oxygen	O	8	16.000	1.332	−218	—	0.025	—
Palladium	Pd	46	106.7	11 990	1 555	11.60	67.41	10
Phosphorus	P	15	31.02	1 819 (yellow)	44.12	11.25	—	10^{17}
Platinum	Pt	78	195.23	21 420	1 774	8.8	69.5	9.83
Plutonium	Pu	94	239	19 800	640	—	—	150
Polonium	Po	84	209	9 320	254	—	—	—
Potassium	K	19	39.096	858.4	62.29	—	99.2	7
Praseodymium	Pr	59	140.92	6 616	940.1	8.3	—	88
Promethium	Pm	61	147	—	—	—	—	—
Protactinium	Pa	91	231	15 400	1 000	—	—	—
Radium	Ra	88	226.05	5 005	960	—	—	—
Radon	Rn	86	222	4 400	−71	—	—	—
Rhenium	Re	75	186.31	20 000	3 000	—	—	21
Rhodium	Rh	45	102.91	12 430	1 966	8.9	89.15	4.93
Rubidium	Rb	37	85.48	1 531	38.3	90.0	—	12.5
Ruthenium	Ru	44	101.7	12 210	2 300	8.5	—	10
Samarium	Sm	62	150.4	7 752	1 050	—	—	91.4
Scandium	Sc	21	45.10	2 434	1 204	—	—	50.5
Selenium	Se	34	78.96	4 816	220.1	87.0	—	8×10^6
Silicon	Si	14	28.08	2 408	1 427	—	83.74	85×10^3
Silver	Ag	47	107.88	10 520	960.5	18.9	407.9	1.62
Sodium	Na	11	22.997	969	97.52	71.0	135.1	4.6
Strontium	Sr	38	87.63	2 602	77.1	—	—	22.76
Sulphur	S	16	32.06	2 076	113	67.48	0.263 7	1.9×10^{17}
Tantalum	Ta	73	180.88	16 620	3 017	6.5	54.43	15.5
Technetium	Tc	43	99	11 400	2 100	—	—	—
Tellurium	Te	52	127.61	6 200	452.3	16.8	6.016	1.6×10^5
Terbium	Tb	65	159.2	11 850	310	—	—	—
Thallium	Tl	81	204.39	11 850	303.8	28.0	38.9	18.1
Thorium	Th	90	232.12	11 520	1 700	12.3	—	18
Thulium	Tm	69	169.4	9 330	1 600	—	—	—
Tin	Sn	50	118.70	7 308	232	21	65.73	11.5
Titanium	Ti	22	47.90	4 512	1 680	7.14	—	—
Tungsten	W	74	183.92	19 320	3 370	4.0	199.3	5.48
Uranium	U	92	238.07	18 710	1 133	—	—	60
Vanadium	V	23	50.95	5 675	1 920	8	—	26
Xenon	Xe	54	131.3	152.7	−112	—	519.2	—
Ytterbium	Yb	70	173.04	6 900	824	—	—	27.7
Yttrium	Y	39	88.905	5 509	1 482	—	—	80
Zinc	Zn	30	65.38	7 142	419.5	30	112	5.5
Zirconium	Zr	40	91.22	6 366	1 850	6.3	—	41

Table 19.18 Table of principal elements arranged in order of valency

Valency	Name	Symbol	Atomic weight
1 ((Monovalent)	Bromine	Br	79.9
	Chlorine	Cl	35.457
	Fluorine	Fl	19.00
	Hydrogen	H	1.008 1
	Iodine	I	126.92
	Potassium	K	39.096
	Silver	Ag	107.88
	Sodium	Na	22.997
2 (Divalent)	Barium	Ba	137.34
	Cadmium	Cd	112.40
	Calcium	Ca	40.08
	Copper	Cu	63.54
	Magnesium	Mg	54.93
	Mercury	Hg	200.61
	Oxygen	O	16.000
	Zinc	Zn	65.38
3 (Trivalent)	Aluminium	Al	26.98
	Bismuth	Bi	209.00
	Boron	B	10.81
	Cobalt	Co	58.94
	Gold	Au	197.20
	Iron	Fe	55.84
	Nickel	Ni	58.70
4 (Tetravalent)	Lead	Pb	207.21
	Platinum	Pt	195.23
	Silicon	Si	28.08
	Tin	Sn	118.70
5 (Pentavalent)	Antimony	Sb	121.75
	Arsenic	As	74.92
	Nitrogen	N	14.008
	Phosphorus	P	31.02
6 (Hexavalent)	Chromium	Cr	52.01
	Manganese	Mn	54.93
	Sulphur	S	32.06

Table 19.19 Surface tension of some common metals

Metal	Temperature (°C)	Surface tension (mN m^{-1})
Antimony	640	350
Bismuth	269	378
Cadmium	320	630
Copper	1131	1103
Gold	1120	1128
Iron (acc. to C content)	1300–1420	1150–1500
Lead	327	452
Mercury	20	465
Silver	998	923
Tin	232	526
Zinc	419	758

Surface tension of water is 78.5 mN m^{-1} at 0°C and decreases by 0.152 mN m^{-1} for each degree rise of temperature.

Table 19.20 Specific heats, melting points and densities

Material	Specific heat (kJ kg^{-1} °C^{-1})	Melting point (°C)	Density (kg m^{-3}) (approx.)
Air (20°C)	0.996 8	—	1.2
Aluminium	0.914 9	657	2 700
Brass	0.395 3	850–950	8 400
Carbon	0.854 3	—	2 220
Cobalt	0.448	1 480	8 904
Copper	0.393 6	1 083	8 941
German silver	0.398	—	8 400
Graphite	0.837 4	—	2 300
Iron, cast	0.46–0.67	1 200	7 000
Lead (solid)	0.131 3	327.4	11 320
Lead (liquid)	0.197 2	—	—
Mercury	0.134	−38.9	13 540
Molybdenum	0.275 9	2 620	10 190
Nickel	0.454	1 452	8 915
Paraffin (solid)	2.6–2.9	38–56	900
Paraffin (liquid)	29.7	—	800
Pitch	—	—	1 100
Platinum	0.138 1	1 774	21 420
Silver	0.232 7	960.8	10 520
Solder	—	205–185	8 300
Tin (solid)	0.234 5	232	7 308
Tin (liquid)	0.268	—	—
Tungsten	0.146 6	3 370	19 320
Type metal	0.163 3	—	—
Water (20°C)	4.186 8	—	—
Zinc (cast)	0.39–0.50	419.5	7 142

Table 19.21 Sectional properties of metals

Substance	E (MN m^{-2})	G (MN m^{-2})	σ	K (MN m^{-2})	Tensile strength (MN m^{-2})
Aluminium	70 300	26 100	0.345	75 500	90–150
Brass	101 000	37 300	0.350	111 800	280–730
Copper	129 800	48 300	0.343	137 800	120–400
Iron (cast)	152 000	60 000	0.270	109 000	100–230
Iron (Wrought)	211 400	81 000	0.293	170 000	260–450
Lead	16 100	5 600	0.440	45 700	12–17
Magnesium	44 700	17 000	0.291	25 600	60–190
Silver	82 700	30 200	0.366	103 600	300
Platinum	168 000	61 000	0.377	228 000	330–370
Tantalum	185 700	69 200	0.342	196 300	800–1 100
Tin	49 900	18 400	0.357	58 200	20–35
Tungsten	411 000	160 000	0.280	311 500	1 500–3 500
Steel (mild)	211 900	82 200	0.291	169 200	430–690
Steel (hardened)	210 400	77 800	0.295	165 200	1 800–2 300

Notes: E is known as Young's modulus or longitudinal elasticity; G is known as Shear or Rigidity modulus; σ is known as Poisson's ratio; K is known as Bulk modulus.

19.4.3 Density

Values are taken at 293 K.

19.4.4 Melting point

Values are for 1 atm pressure (1.01325×10^5 Pa).

19.4.5 Linear coefficient of expansion

This is commonly known as the α value. It is worked out as a mm expansion over 100 m length using the equation $L = l(1 + \alpha t)$ for expansion and $l = L/(1 + \alpha t)$ for a contraction; where l = increase or decrease in length, L = original length, t = temperature change and α = linear coefficient of expansion.

19.4.6 Heat conductivity

This is sometimes called 'thermal conductivity' and is known as the k factor. Values are taken at 293 K.

19.4.7 Electrical resistivity

Values are given at 293 K.

Further reading

Avallone, E. A. and Baumeister III, T., *Marks' Standard Handbook for Mechanical Engineers*, 9th edition, McGraw-Hill, New York (1986)

Blake, L. S. (ed.), *Civil Engineer's Reference Book*, 4th edition, Butterworths, London (1989)

Doran, D. K. (ed.), *Construction Materials Reference Book*, Butterworth-Heinemann, Oxford (1992)

Koshal, D. (ed.), *Manufacturing Engineer's Reference Book*, Butterworth-Heinemann, Oxford (1993)

Laughton, M. A. and Say, M. G., *Electrical Engineer's Reference Book*, 14th edition, Butterworths, London (1985)

Mazda, F. F., *Electronic Engineer's Reference Book*, 6th edition, Butterworths, London (1989)

Nayler, G. H. F., *Dictionary of Mechanical Engineering*, Butterworths, London (1985)

Noltingk, B. E. (ed.), Instrumentation Reference Book, Butterworths, London (1988)

Oberg, E., Jones, F. D. and Horton, H. L., *Machinery's Handbook*, 23rd edition, Industrial Press, New York (1988)

Snow, D. A., *Plant Engineer's Reference Book*, Butterworth-Heinemann, Oxford (1991)

Tool and Manufacturing Engineer's Handbook, Society of Manufacturing Engineers, Dearborn, MI (1983–1988)

Index

Abbott bearing area curve, **9**/121
Abbreviations, unit, **19**/16–**19**/17
Abrasion, **9**/71
Abrasive friction/wear, **9**/4
ABS (Acrilonitrile-Butadiene-Styrene), **7**/117
Absolute system (units), **19**/3
Absorbers, sound, **15**/148
Acceleration, **1**/5
    angular, **1**/5
    constant, **1**/6
    Coriolis, **1**/7
    in mechanisms, **1**/7
    variable, **1**/6
Accelerometer, **15**/133–**15**/134
Access time, disk, **4**/20
Accumulator:
    pulsation reduction, **15**/8
    steam-storage, **15**/86
Accuracy of measurement, **3**/21–**3**/22
Acetal resins, **7**/122
ACGIH, capture systems, **15**/104
Acoustic enclosures, **15**/149
Acoustic sound, **15**/138–**15**/139
Acoustic testing, **7**/176–**7**/177
Acrylic adhesives, **16**/72–**16**/73
Acrylics, **7**/120
ADA language, **4**/44
Adaptive control, **3**/54
ADC (Analogue-to-Digital Converter),
    3/12–**3**/14
ADCCP protocol, **4**/29
Addressing modes, computer, **4**/8
Adhesion, **9**/71
    joint design, **16**/77–**16**/83
    principles, **16**/73–**16**/76
    production processes, **16**/83–**16**/84
    surface pretreatment, **16**/76–**16**/77
Adhesive friction/wear, **9**/4
Adhesives, **6**/20, **16**/70–**16**/72
    types, **16**/72–**16**/73
Adiabatic processes, **1**/31, **1**/36–**1**/37
Advanced Gas-cooled Reactors (AGR), **13**/4,
    13/8
Air change rate, **15**/110
Air cleaners, **15**/102
Air conditioning:
    controls, **15**/109–**15**/110
    plant, **15**/107–**15**/109
    sizing, **15**/110–**15**/113
    terms, **15**/106–**15**/107
Air gauging, *see* Pneumatic surface-
    measurement methods
Air traffic noise, **15**/142
Air washers, **15**/114
Air-handling plant/unit, **15**/97, **15**/108–**15**/109
AISI/SAE number (steel references), **7**/11, **7**/31,
    7/39–**7**/40
Aliasing, **5**/34

Alkyds, **7**/128
All-metal membrane couplings, **10**/40–**10**/42
Alloying elements, steel, **7**/30
Allylics, **7**/128
Alpha-particles, **13**/3, **13**/11
Alternating current, rectification, **2**/14,
    2/41–**2**/42
Alternating current (a.c.), **2**/10–**2**/11
    computerized power control, **3**/11
    motor speed control, **3**/64
    recification, **2**/14, **2**/41–**2**/42
Alternating quantities, electrical, **2**/9
Alternative energy sources, **12**/3
Alternator, **2**/15
    three-phase, **2**/24–**2**/25
Alumina ceramics, **7**/148, **15**/62
Aluminium casting, **7**/63, **16**/91, **16**/92
Aluminium and its alloys, **7**/62–**7**/66
    bronzes, **7**/58
Aluminizing, **9**/75–**9**/76
Anodizing, **7**/165, **9**/79
American Conference of Governmental
    Industrial Hygienists (ACGIH), Threshold
    Limit Value, **15**/99, **15**/104
American Electronics Industry Association
    (EIA) standards, **3**/7
    systems specifications, **4**/27
American Gear Manufacturers Association
    (AGMA) standards, gear oils, **9**/47
American Hydraulics Institute Standards, **15**/5
American National Lubricating Grease
    Institute (NLGI), **9**/6
American National Standards Institute
    (ANSI), **4**/29, **4**/31
American Petroleum Industry (API),
    standards, **15**/13
American Petroleum Institute (API), oil service
    classifications, **9**/9–**9**/10
American Society for Non-Destructive testing
    (ASNT), **13**/6
Aminos, **7**/128
Ampere (unit), **2**/5
Amplifiers, **2**/37–**2**/38, **3**/18–**3**/19
    differential, **2**/41
    effect of feedback, **2**/38
    frequency response, **2**/38
    instrumentation, **2**/41
    noise and distortion, **2**/38
    operational, *see* Operational amplifiers
    vibration, **15**/134–**15**/135
Amplitude distribution, moments of, **9**/120
Analogue input/output, **4**/23
Analogue interfacing, **3**/12–**3**/15
Anemometer, hot wire, **1**/33
Aneroid barometer, **1**/20
Angular displacement measurement, **3**/26–**3**/27
Aniline Point, **9**/12
Annealing, **7**/29
Anodizing, **7**/165, **9**/79

ANSI, *see* American National Standards
    Institute (ANSI)
ANSI Standard X3T9.5, **4**/31
Anthropometric data, **6**/23–**6**/28
Anti-corrosion additives, **9**/6, **9**/48
Anti-foam additives, **9**/6, **9**/20, **9**/48
Anti-oxidant additives, **9**/6, **9**/8, **9**/48
Anti-vibration mats, **15**/150
Anti-vibration mounts, **15**/150
Anti-wear additives, **9**/6, **9**/49
APL (A Programming Language), **4**/44
Application Specific Integrated Circuits
    (ASICs), **2**/55
Aquifers, **12**/28
Aramid fibres, **7**/120
    other uses, **7**/99
    as packing material, **15**/28
    as reinforcement, **7**/93
Arc of control correction factor, **10**/20
Archibald's Equation (fluid flow), **9**/36
Archiving, **4**/34
Argand diagram, **17**/5
Arithmetic hardware, computer, **4**/12
Arithmetic and Logic Unit (ALU), **3**/4, **4**/4
Arithmetic mean, **17**/23
Arithmetic series, **17**/11
Armature, **2**/15
    alternator, **2**/25
    coils, **2**/17
    e.m.f., **2**/17
    winding, **2**/17
Armature reaction, **2**/18
Array processors, **4**/13
Artificial Intelligence (AI), **5**/10, **5**/18
Asbestos:
    control of use of, **18**/6
    health hazards, **15**/30
    removal, **15**/32
ASCII code (American Standard Code for
    Information Interchange), **3**/3, **4**/24, **4**/25
Assel cycle, **16**/24
Assembler/assembly language, **4**/41
Assembly stage, control at, **7**/168–**7**/169
ASTM standards, **9**/20
    fatigue testing, **8**/19, **8**/31, **8**/37
    oil, **9**/20
Astrel, **7**/123
Atmosphere, **1**/18
Atomic number, **19**/17, **19**/18
Atomic weights, **19**/17, **19**/18
Atria, **12**/10
Attenuation, sound, **15**/149
Attributes (for display devices), **5**/34
Austempering, **7**/29
    ductile iron, **7**/54
Austenite, **7**/21, **7**/22, **7**/25
Auto Forcast process, **16**/32
Auto-forge process, **16**/32

Auto-transformers, **2**/33
  as starter, **2**/27
Autocorrelation function (ACF), random
  process, **1**/16, **9**/122–**9**/123
Autocovariance function, **9**/122
Automatically programmed tools (APT), **5**/15
Availability function, **1**/37
Averages, **17**/23

B-rep (Boundary representation) models, **5**/11
B-spline curve fitting, **5**/10
Babbitts, *see* Bearing metals
Back e.m.f., **2**/17
Background noise levels, **15**/145
Bainite, **7**/24
Balance of Systems (BOS) costs, **12**/15
Balancing:
  dynamic, **1**/8
  rotating masses, **1**/8–**1**/9
  static, **1**/8
Ball bearings, *see* Bearings, rolling element
Bandpass filters, **3**/21
Bandwidth, **2**/13
  video, **5**/30
Bar code readers, **3**/39–**3**/40
Base, transistor, **2**/33–**2**/34
BASIC (Beginner's All-purpose Symbolic
  Instruction Code) language, **3**/61, **4**/43
Basic flux covering, **16**/57
Batch processing, **4**/37
Baud (unit), **3**/3, **4**/24
Baudot code, **4**/24
Bayes theorem (probability), **17**/26
Bazin formula (flow velocity), **1**/28
Beams, transverse vibration, **1**/9–**1**/10
Bearing loads, **10**/34–**10**/35
Bearing metals, **7**/90
Bearing ratio, **9**/120–**9**/121
Bearings, **9**/14, **9**/27
  continuous motion, **9**/27–**9**/28
  failure of, **9**/59–**9**/60
  friction limits, **9**/54
  hydrodynamic, **9**/30
    principles, **9**/30–**9**/31
  journal, *see* Journal bearings, design
  life of, **9**/52–**9**/54
  lubrication of, **9**/14, **9**/35–**9**/36
  materials for, **9**/37
  as noise source, **15**/147
  rolling element, **9**/49
    applications, **9**/55–**9**/56
    lubrication, **9**/56–**9**/58
    maintenance, **9**/59
    mounting, **9**/58
    types, **9**/50–**9**/52
  selection of, **9**/28–**9**/29
  self-contained, **9**/39
  speed limits, **9**/55
  thrust, **9**/39–**9**/41
  tilted pad, **9**/40–**9**/41
  vibration, **15**/126, **15**/127
Bearsafe (analysis program), **8**/26
Behavioural science, **6**/5
Belfast Buoy, **12**/34
Belgium, health and safety standards, **18**/3
Bell (unit), **15**/139
Belt drives, **10**/15–**10**/20
  flat, **10**/15–**10**/18
  manufacturers, **10**/25
  miscellaneous, **10**/25
  synchronous, **10**/22–**10**/25
  V-, **10**/18–**10**/22
Bending equation, **1**/3
Bending moment, **1**/3
Bending operation, sheet metal, **16**/44
Bernoulli equation, 135, **1**/23, **1**/24, **1**/25

Beryllia ceramics, **7**/149
Beryllium, **7**/92
Bessel's equation, **17**/18
Beta-particles, **13**/3, **13**/11
Betz limit, **12**/24
Bevel gears, **10**/30
Binary notation, **3**/3
Binomial distribution, **17**/26
Binomial series, **17**/11
Bio-engineering, and surface topography, **9**/107
Biodeterioration, **7**/163–**7**/165
Biogas systems, **12**/38–**12**/39, **12**/40
Biological effects, nuclear, **13**/11
Biomass, **12**/3, **12**/36
  conversion processes, **12**/36–**12**/37, **12**/40
Bisection method, **17**/7
Bisync protocol, **4**/29
Bit stuffing, **4**/29
Bits (BInary digiTs), **3**/3
Black-body radiating surface, **15**/80–**15**/81
Blowdown, boiler, **15**/89
Blowers, **9**/19
Bode plots, **2**/38, **3**/47
Boiler house, automatic, **15**/87
Boiler scale, **15**/82–**15**/83
Boiler tube, heat transfer, **15**/81
Boilers:
  automatic control, **15**/87, **15**/90
  blowdown, **15**/89
  heat transfer, **15**/80–**15**/83
  priming, **15**/83
  selection, **15**/78–**15**/79
  starting, **15**/87
  terminology, **15**/83–**15**/84
  types of, **15**/75–**15**/78, **15**/83, **15**/84
Boiling Water Reactors (BWR), **13**/5
Bolts:
  load-sensing, **6**/17–**6**/18
  threadlocking, **6**/19–**6**/20
Boolean algebra, **2**/43–**2**/44
  logical function application, **4**/10
  minimization principle, **2**/50
Bootstrapping/booting, **4**/7
Boriding, **9**/75
Boring processes, metal, **16**/6
Boron carbide, **7**/150
Boron fibres, **7**/93, **7**/99
Boron nitrides, **7**/150
Boundary layer flow, **1**/28–**1**/30
  laminar, **1**/29
  turbulent, **1**/30
Bourdon tube, **1**/19, **3**/29
Bourneville Solar Village, **12**/9
BPCL (computer language), **4**/44
Brakes, **10**/49
Brasses, **7**/56–**7**/57
Brazing, **16**/67–**16**/68
Breeder reactors, **13**/14, **13**/7–**13**/8
Bresenham line drawing algorithm, **5**/32, **5**/33,
  5/43
Bridge measurements (instrumentation), **3**/22
Brinell hardness, **7**/3, **7**/4
British Gear Association, **10**/26
British Hydrodynamics Research Association
  (BHRA), Fluid Engineering Centre, **15**/40
British Industrial Fasteners Federation (BIFF),
  6/17
British Standards (BS), **6**/3–**6**/4
  BS 4500 (limits and fits), **6**/7–**6**/8, **6**/13, **6**/14
  BS 5750 (quality assessment), **6**/30, **6**/33
Brittle-coating technique, **8**/19
Broaching processes, metal, **16**/4
Bronzes, **7**/57–**7**/58
  for sliding bearings, **9**/69
Brown coal, *see* Lignite
Bubble memory, **4**/8
Buffer (register), **4**/24

Building Automation System (energy
  management), **15**/124–**15**/125
Building regulations, heating requirements,
  125/91
Buildings:
  insulation, **15**/149
  vibration effects on, **15**/136, **15**/138
Bulk deformation operations, **16**/13
Bulk Modulus, **7**/3
  table, **19**/20
Buoyancy, **1**/20–**1**/21
Bussing (bus) systems, **3**/4, **4**/5, **4**/13–**4**/14
Buying specifications, **15**/13
Bytes, **3**/3

C (language), **3**/61, **4**/44, **5**/9
Cache memory, **4**/11–**4**/12
CAD/CAM (Computer-Aided Design/
  Computer-Aided Manufacturing), **5**/3–**5**/4,
  6/6
  applications, **5**/12–**5**/16
  command languages, **5**/9
  future for, **5**/18
  graphics displays, **5**/28
  hardware, **5**/4–**5**/7
  implementation, **5**/16–**5**/17
  modelling techniques, **5**/10
  software, **5**/7–**5**/9
Calculus, vector, **17**/11
Calenderizing, **7**/112
Calorific value, **11**/5
Capacitance, **2**/4
  measurements, **3**/25
Capacitor, **2**/5
  motors, **2**/28
  reservoir, **2**/42
Carbide diffusion treatment, **9**/76
Carbides, sintered, **16**/8
Carbon fibre reinforced plastic (CFRP), **7**/98,
  7/99
Carbon fibres, **7**/93
Carbon graphites, **9**/67–**9**/68
  for sealing, **15**/62
Carbon products, **7**/150
Carbon steel, **7**/12
Carbon-dating, **13**/10
Carbon/carbon composites, **7**/101
Carbon/carbon manganese steels, **7**/25–**7**/27
Carbonitriding, **7**/31, **9**/75
Carborundum, *see* Silicon carbide
Carburizing, *see* Case hardening
Carnot cycle, **10**/3
Carnot efficiency, **1**/36
Carrier depletion layer, **2**/14
Cartesian coordinate system, **1**/6, **5**/32, **17**/6,
  17/7
Cartridge disks, **4**/21
Cartridge seals, **15**/62
Cascade control, **3**/54
Case-depth, **9**/87
Case-hardening, **7**/31, **8**/38, **9**/74–**9**/75
  stels, **7**/12, **7**/31
Cast iron, **7**/52–**7**/55
  tensile testing, **7**/3
Cast non-ferrous alloys, **16**/7
Cast steel, **7**/49–**7**/52, **16**/88–**16**/91
Cast-preform forging, **16**/31
Casting, **16**/85
  Aluminium, **7**/63, **7**/65
  basic problems, **16**/85–**16**/87
  gas problem, **16**/85–**16**/86
  plastic, **7**/112, **7**/113
  processes:
    centrifugal, **16**/108–**16**/109
    continuous, **16**/18, **16**/88–**16**/93
    conventional, **16**/88

die, *see* die casting
  rheo, **16**/106–**16**/108
  sand, **16**/93, **16**/95–**16**/103
  semi-continuous, **16**/93
  steel, **7**/49–**7**/52, **16**/88–**16**/91
Cathode Ray Tube (CRT), **5**/28–**5**/30, **6**/6
Cathodic protection, **7**/165
Cauchy-Riemann equations, **1**/35, **17**/5
Cauchy's theorem, **17**/5
Cavitation, **15**/8, **15**/9
  boiler feed, **15**/84
  damage by, **7**/157
CCITT, *see* Comite Consultatif International
  de Telegraphie et Telephonique
CCMC lubricant ratings, **9**/10
Cellulosic polymers, **7**/120
  flux covering, **16**/57
Cementite, **7**/21
Cemfil, **7**/93
Centrifugal casting, **16**/108–**16**/109
Centistoke (unit), **9**/5
Central processor unit (CPU), **2**/37, **3**/4,
    4/4–**4**/5, **4**/7
  enhancements, **4**/11–**4**/12
  implementation, **4**/11
  instruction set, **4**/9–**4**/10
Centre of buoyancy, **1**/20, **1**/21
Centre of gravity, **1**/3
Centre of pressure, **1**/20
Centrifugal reaction, **1**/7
Centrifuge, **9**/24–**9**/25
Centripetal force, **1**/7
Cepstrum analysis, **15**/128
Ceramics:
  clay-based, **7**/147–**7**/148
  corrosion of, **7**/164
  cutting-tool, **16**/8
  engineering, **7**/147
  future prospects, **7**/153
  hardness, **7**/4
  manufacturing procedures, **7**/153
  mechanical properties, **7**/151–**7**/152
  non-oxide, **7**/147–**7**/150
  oxide-based, **7**/148–**7**/149
  tensile testing, **7**/3
Certification, offshore structures, **14**/33–**14**/34
Cetane Number, **11**/17
CFRP (carbon fibre reinforced plastic), **7**/98,
    7/99
Chain-drives, **10**/35–**10**/36
Character recognition devices, **4**/19–**4**/20
Character stuffing, **4**/29
Characteristic number, pump, **15**/5–**15**/6
Charge, electrical, **2**/3
Charpy beam testing, **7**/4
Chatter, metal-cutting, **16**/11
Chemical constants, **19**/17–**19**/20
Chemical dilution flow measurement, **1**/27
Chemical heat storage, **12**/18
Chemical vapour deposition (CVD), **9**/81
Chemical wear, machinery, **9**/8
Chernobyl accident, **13**/8
Chezy equation, **1**/27–**1**/28
Chi-square test, **17**/29
Chilled water system, **15**/107
Chimneys:
  design, **15**/89–**15**/90
  height of, **15**/90
Chip breakers, metal, **16**/3
Chloroplast, **12**/36
Chopper control, **2**/22
Chromating, **7**/165, **9**/82
Chromium, **7**/91
Chromizing, **9**/75–**9**/76, **9**/81
Chromogenic materials, **12**/17
Circuit breakers, **2**/56
Circuits:

magnetic, **2**/8
pulse, **2**/54–**2**/55
RC, **2**/9–**2**/10
RL and RC, **2**/9
RLC, **2**/11
simple capacitive, **2**/4
simple resistive, **2**/3
Circular motion, **1**/7
Circulation, **1**/22
CISC (Complex Instruction Set Computer),
    4/12
Cladding processes (for wear resistance), **9**/83
Clearn Air Act 1956, **15**/89–**15**/90
Clean Rooms standards, 15, 113
Clearance ratio, **9**/32
Clearance seals, **15**/68, **15**/70, **15**/72–**15**/74
Clipping operation (computer graphics), **5**/37,
    5/38
Closed loop transfer function, **3**/73, **3**/75
Closed-circuit test, transformer, **2**/31
Closed-loop control system, **3**/41
Cloud point, **11**/12
Clustering, *see* Computer systems, loosely
    coupled
Clutch, **10**/43–**10**/47
CMOS devices, **2**/44, **4**/8
Coal Rank Number, **11**/13
Coal tar fuels, **11**/12
Coals, **11**/12
Coating processes, **7**/165
  adhesion testing, **9**/88
  porosity, **9**/88
  protective, **7**/165
  substrate topography, **9**/107
  thickness of, **9**/87–**9**/88
Cobalt and its alloys, **7**/90
COBOL (Common Business-Oriented
    Language), **4**/43, **4**/44
Cockerell Raft, **12**/34
Codasyl standards, **4**/41, **4**/43
Codes of Practice, **6**/3–**6**/4
Coefficient of friction, **1**/6
Coefficient of variation, **17**/24
Coercive force, **2**/8
Coffin-Mansion relationship (fatigue), **8**/37
Cogged bolts, **10**/22
Cogging, **16**/29
Coil steel, **7**/26
Coining operation, **16**/28
Cokes, **11**/13
Cold drawing processes, **16**/40–**16**/43
Cold forging, **16**/28, **16**/30, **16**/31
Cold rolling, **16**/20, **16**/21
Cold shuts, **16**/87
Cold working, **9**/74
Cold-reducing process, **16**/30
Colebrook and White equation (friction), **1**/24
Collector, transistor, **2**/33
Collocation procedure, **17**/20
Colour Look-Up Tables, **5**/31
Combinations (mathematics), **17**/24
Combined heat and power (CHP) plant, **15**/84,
    15/123
Combustion process, **1**/37, **11**/14–**11**/19
  products, **9**/8
  temperature, **11**/15
Comite Consultatitif International de
    Telegraphie et Telephonique (CCITT),
    standards, **4**/27, **4**/29, **4**/36
Command languages, CAD/CAM, **5**/9
Common-emitter (transistor), **2**/34–**2**/35
Common-mode rejection ratio, **2**/41
Communication standards/protocols, **3**/7–**3**/8,
    4/27, **4**/29–**4**/31
Community noise units, **15**/141
Commutator, **2**/16
Compilers, **4**/43

Complementary function, **17**/17
Complex variable, **17**/5
Composite materials, fibre-reinforced, *see*
    Fibre-reinforced composite materials
Composite sheet manufacture, **16**/54
Composites, **7**/92
  elastomeric, *see* Elastomeric composites
  filamentary polymer, **7**/95–**7**/98, **7**/99
  laminar, **7**/105
  particulate, **7**/105
Compounding, rubber, **7**/133
Compressed air plant, energy survey, **15**/122
Compression moulding, **7**/112
Compression packing, **15**/18
  fault-finding, **15**/27
  materials, **15**/21–**15**/22
    development, **15**/28–**15**/31
    selection, **15**/30–**15**/31
  operating principles, **15**/19–**15**/20
  pumps, **15**/19
  standardization, **15**/27–**15**/28
  valves, **15**/19
  wear, **15**/22–**15**/23
Compressors, **9**/18
  control of, **15**/122
  design principles, **15**/3–**15**/12
  lubrication of, **9**/18–**9**/20
  machine selection, **15**/13–**15**/14
  as noise source, **15**/147
  performance, **15**/14, **15**/18
Computer graphics systems, **5**/27, **5**/28
  applications, **5**/32–**5**/38
  displays, **5**/28–**5**/30
  generation and manipulation, **5**/30–**5**/31
  human-machine interface, **5**/31–**5**/32
  workstations, **5**/4–**5**/5, **5**/40–**5**/41
Computer hardware, finishes on, **9**/107
Computer languages, **3**/61, **4**/43–**4**/45
  see also under specific names
Computer programmes, heat transfer
    problems, **1**/47–**1**/48
Computer systems, **4**/3, **4**/4–**4**/7
  central processor unit, *see* Central processor
    unit
  digital, **2**/55
  loosely coupled, **4**/6–**4**/7
  networks, *see* Networks, computer
  output devices, **4**/16
  peripheral devices, **4**/15–**4**/16
  personal, **4**/5, **5**/5
  terminals, **4**/16–**4**/19
Computer-Aided Engineering (CAE), **5**/3
Computer-Aided Process Planning (CAPP),
    5/15
Computing installations, energy survey, **15**/122
Concentration units, **19**/12
Concrete:
  protection of, **7**/166
  reinforced, *see* Reinforced concrete
Concrete gravity structures, **14**/9–**14**/10
Condensation, control of, **15**/103
Condition monitoring, **9**/26, **15**/125
  parameters, **15**/125
Conductance:
  electrical, **2**/3
  thermal, **1**/41, **7**/5, **7**/143, **19**/18, **19**/21
Conduction, **1**/39, **1**/40, **1**/43–**1**/44, **15**/80
Conform process, **16**/38–**16**/39
Conformal gears, **10**/26
Conformal mapping, **17**/21
Conservation of mass, **1**/22
Conservation of momentum, **1**/5
Constant of gravitation, **1**/4
Constant head equation, *see* Bernouilli
    equation
Constant-pressure cycle, **10**/3
  Diesel, **10**/4

Constant-volume cycle, **10**/4
Construction site noise, **15**/142
Constructive solid geometry (CSG), **5**/11
Contact mechanics, **9**/105–**9**/106
Contaminants, air-borne, **15**/103–**15**/105
Continuity equation, **1**/22
Continuous casting process:
    aluminium, **16**/91, **16**/92
    other metals, **16**/91, **16**/93
    steel, **16**/88–**16**/91
Continuous-combustion heaters, **11**/18
Continuum, **1**/22
Contraction coefficient, **1**/24
Control engineering, **3**/40–**3**/42
Control of Pollution Act 1974, **15**/105
    noise-abatement zones, **15**/146
Control registers, **3**/6
Control strategies, **3**/49–**3**/58
Control of Substances Hazardous to Health
    (COSHH) Regulations 1988, **15**/105,
    18/5–**18**/6
Control systems, **3**/40–**3**/43
    hydraulic, **3**/68–**3**/69
    microprocessor-based, **3**/58
        hardware, **3**/59–**3**/60
        software, **3**/60–**3**/62
    multi-layer, **3**/57
    multi-level, **3**/58
    PC-based, **3**/65
    pneumatic, **3**/68–**3**/69
    self-tuning, **3**/56
Convection, **1**/39, **1**/40, **1**/44–**1**/45, **15**/80
    boiler tube, **15**/81
Conversion coatings, **7**/165
Convolution integral, **17**/17
Cooking methods, **12**/37
Coolants, reactor, **13**/5
Coons patch, **5**/10
Coordinate system:
    cylindrical, **17**/6
    spherical polar, **17**/7
    transformation, **17**/15
Copper and its alloys, **7**/55–**7**/62
    sliding bearings, **9**/69
Cordierites, **7**/148
Core memory, **4**/8
Core movement, casting, **16**/87
Coriolis acceleration, **1**/7
Correlation, **17**/25
    significance, **17**/29
Corrosion, **9**/71
    in air, **7**/155
    in aqueous solutions, **7**/156
    of ceramics, **7**/164
    crevice, **7**/158
    electrochemical, **7**/155
    and erosion, **7**/157, **7**/158
    failure identification, **7**/166
    galvanic, **7**/157
    of glasses, **7**/164
    intergranular, **7**/159
    liquid metal, **7**/163
    localised, **7**/157
    of metals, **7**/153–**7**/163
    of plastics, **7**/164–**7**/165
    prevention of, **7**/165–**7**/166, **9**/22–**9**/23, **15**/88
Corrosion fatigue, **7**/162
Corrosion fretting, **7**/162–**7**/163
Corrosion jacking, **7**/156
Corrosion resistance, **7**/7
Cosworth Process, **7**/65, **16**/100–**16**/101
Coulomb (unit), **2**/3
Counters:
    computer, **2**/53, **4**/13
    PLC, **3**/68
Couple, **1**/5
Crack growth, elastomers, **7**/139, **7**/142

Crack-tip opening displacement method, **8**/22
CRC (Cyclic Redundancy Check), **4**/28, **4**/29
Creep, **8**/3, **8**/23–**8**/25
    interpretation of data, **8**/25
    pressure vessel steels, **8**/26, **8**/29
    resistance, **7**/12, **8**/26, **8**/30
    testing, **8**/23, **8**/25, **8**/30
Creep range, **8**/23
    design in, **8**/26, **8**/29
Creep rates, elastomer, **7**/137
Creep rupture, **7**/4
Creep-resisting steels, **7**/12
Crest factor, **15**/127
Croning process, **16**/101
Cross rolling, **16**/24
Cross-assemblers, **3**/62
Cryogenic applications, steel, **7**/45
CSMA/CD concept, **4**/32
Cubic boron nitride (CBN), cutting-tool, **16**/8
Current, diffusion, **2**/14
Current meter, fluid, **1**/27
Cursor key, **5**/31
Customer feedback, **6**/30
Cutting fluids, **16**/8–**16**/9
Cutting process:
    gas, **16**/64
    plasma, **16**/63
Cutting tools:
    geometry, **16**/6
    materials, **16**/7–**16**/8
Cyanoacrylate adhesives, **16**/73

D type connectors (computer), **4**/30
DAC (Digital-to-Analogue Convertor), **3**/9
DAIP (diallyl isophthalate), **7**/128
Damping, **3**/77
    calculations, **3**/43–**3**/44
    elastomers, **7**/142
    vibration, **1**/10–**1**/11, **15**/129, **15**/130
Danbury Park Management Centre, **6**/32
DAP (diallyl phthalate), **7**/128
D'Arcy equation (head loss), **1**/24
    modified, *see* Chezy Equation; Manning
        Equation
Darlington Driver, **3**/6, **3**/8
Darrieus windmill, **12**/25
Data communications, **4**/23
    concepts, **4**/24
    development of, **4**/23–**4**/24
    error detection, **4**/28–**4**/29
    multiplexing, **4**/26
    software management, **4**/39–**4**/40
    terminal equipment, **4**/36
    transfer of, **4**/14
    transmissions, **4**/27–**4**/28
Data registers, **3**/6
Database management, **4**/40–**4**/41
    CAD/CAM, **5**/9
Dating, nuclear, **13**/10
DDC (Direct Digital Control), **3**/40, **3**/59
DDCMP protocol, **4**/29
De Moirre's theorem, **17**/5
De Morgan's theorem, **2**/44
De-stressing notches, **9**/101
Dead-band characteristic, **3**/80
Dead-metal zone, extrusion, **16**/35
Decibels (sound power level), **15**/139
Decimal notation, **3**/3
DECNET, *see* DNA
Dedendum circle (gears), **10**/29
Deep drawing, **16**/44–**16**/45
Defects, shape of fracture, **8**/19
Deflagrating explosives, **16**/48
Deformation mechanisms, **8**/23–**8**/25
Degreasing, **9**/23
Degree days (heating), **15**/117

Degrees of freedom, **1**/9
    multi-, **15**/130
Delamination theory, **9**/105
Delsun process, **9**/79
Delta connection, **2**/23
Demulsibility, lubricant, **9**/20
Denmark, health and safety standards, **18**/3
Density:
    fluid, **1**/18
    fuel, **11**/3–**11**/5
    tables, **19**/18, **19**/20, **19**/21
Derivative, calculus, **17**/8, **17**/9–**17**/10
Design:
    computer-aided, *see* CAD/CAM
    safety considerations, **18**/10
Detecting, nuclear applications for, **13**/9
Detergent/dispersant additives, **9**/7, **9**/8
Determinants, **17**/15
Deterministic response, **1**/15
Detonation Gun process, **9**/82
Detonators, **16**/48
Detoxification, solar, **12**/18
Deuterium-tritium reaction, **13**/4
Deutsche Industrie Normen (DIN) standards,
    gear oils, **9**/10, **9**/47
Deviation:
    mean, **17**/24
    standard, **17**/24
Dezincification, **7**/57
Dial guages, **1**/19
Diamonds, **7**/150
    cutting-tool, **16**/8
    polycrystalline synthetic (PCD), **16**/8
Diaphragm gauges, **1**/19
Die casting, **16**/103
    GKN 'Ferro Die', **16**/105
    gravity, **16**/103
    high-pressure, **16**/104–**16**/105
    low-pressure, **16**/103–**16**/104
    rheo-, **16**/106, **16**/108
    squeeze, **16**/105, **16**/108
    thixo-, **16**/106, **16**/108
Dielectric breakdown, **7**/5
Dielectric loss, **7**/5
Dielectric strength, **2**/5
Diescher elongator, **16**/24
Diesel cycle, **10**/4
Diesel knock, **11**/17
Dieseling (seal failure), **15**/46
Differential equations, **17**/16
    analytical solutions, **17**/16–**17**/18
    approximate solutions, **17**/18–**17**/20
    partial, **17**/20–**17**/23
Differentiation mean value theorem, **17**/11
Diffraction theory, wave forces, **14**/18–**14**/19,
    14/31–**14**/32
Diffuse beam (roughness measurement), **9**/115
Diffusion coatings, **7**/165
Diffusion current, **2**/14
Digital control loop, **3**/71
    z-transform, **3**/72–**3**/73
Digital data-acquisition systems, **9**/111–**9**/112
Digital input/output, **4**/23
Digital logic systems, **2**/42, **2**/44
    interfacing, **3**/8
    positive/negative, **2**/51
    tri-state, **2**/51
Digital sampling, **3**/62
Digital signal analysis, **15**/143
Digitizers, **5**/6, **5**/31
Digitizing pen, **5**/6
Dimensional measurements, **3**/23–**3**/27
    standards, **5**/14–**5**/15
Dimensionless load, **9**/32
Diode, **2**/14
    Zener, **2**/15
Dirac δ-function, **17**/17

Direct expansion plant, air-conditioning, **15**/107
Direct stiffness analysis method, **1**/19–**14**/20
Direct-current generators, **2**/18
  efficiency, **2**/23
Direct-current motors, **2**/19–**2**/20
  efficiency, **2**/23
Disk pack, **4**/21
Disks, computer, **4**/20–**4**/22
Dispersion (mathematical), **17**/23–**17**/24
Displacement force, **1**/21
Displacement thickness, **1**/29
Displacement transducers, **15**/134
Distillation, solar, **12**/12
Distortion, amplifier, **2**/38
Disturbance sensitivity, **3**/53–**3**/54
Diverter (electrical), **2**/21
DLE (Data Link Escape), **4**/29
DMA (Direct Memory Access) device, **3**/6, **4**/14
DNA (Digital Network Architecture), **4**/35
Dom equation (steady-state creep), **8**/24–**8**/25
Doppler techniques, **3**/34
Double integration, change of variable, **17**/11
Double seals, **15**/64
Double-glazed windows, **15**/149
Doublet, **1**/35
Dough-moulding compounds (DMC), **7**/99
Drag flow, **15**/47
Drag forces, **1**/29, **14**/15–**14**/17
  coefficients, **1**/29, **14**/18
Draughting, computer-aided, **5**/14–**5**/15
Drawing operations, sheet metal, **7**/21, **16**/15, **16**/44–**16**/45
Drawings, role of, **6**/4
Drift current (electrical), **2**/14
Drilling processes, metal, **16**/6
Dual-combustion cycle, **10**/4
Dual-in-line (DIL) package, **2**/37
Ducted ventilation systems, **15**/101
Ductility, **7**/3
Ducts:
  gas flow in, **1**/32
  noise control in, **15**/149–**15**/150
Dunkerley's empirical method (loaded beams), **1**/9
Duplex communication, **4**/28
Duplex stainless steel, **7**/40
Dust control, **15**/113
DVST (Direct View Storage Tubes), **5**/5
Dynamic braking, **2**/27
Dynamic friction, **15**/43
Dynamic magnification factor (damping measure), **15**/130
Dynamic magnifier (vibration), **1**/12
Dynamic positioning (mooring system), **14**/6–**14**/7
Dynamic range, amplifiers, **15**/135
Dynamics:
  balancing, **10**/41–**10**/42
  rigid body, **1**/4–**1**/9

E-glass fibres, **7**/93
E-system (reference lines), **9**/119
Earth, composition of, **12**/27
Earth-leakage protection, **2**/56
Earthing:
  electrical, **2**/57
  virtual, **2**/39
EBCDIC (Extended Binary Coded Decimal Interchange Code), **4**/24, **4**/25
Economisers, **15**/84–**15**/85, **15**/90
Eddy current loss, **2**/8
EEPROM (Electrically Erasable Read Only Memory) chip, **3**/5
Efficiency:
  heating systems, **15**/119

machine, **15**/4
EIA/CCITT standards, *see* American Electronic Industries Association
Eigen values, **17**/15
Eigen vectors, **17**/15
Elastic Modulus, *see* Modulus of Elasticity
Elasticity, **8**/3
  linear, **8**/4
  rubber, **7**/134–**7**/137
Elastohydrodynamic lubrication, **9**/3
Elastomeric seals, **15**/32, **15**/34
Elastomers, **7**/107, **7**/109, **7**/132–**7**/133
  component design, **7**/138
  composites, **7**/138–**7**/139
  example applications, **7**/145–**7**/147
  filler effects, **7**/137
  liquid effects on, **7**/143
  processing, **7**/133–**7**/134
  properties, **7**/134–**7**/137, **7**/139–**7**/143
  selection guidelines, **7**/143–**7**/145
  thermoplastic (TPE), **7**/134, **7**/137–**7**/138
Electric burn, **2**/56
Electric heating equipment, **15**/97
Electric shock, **2**/56
Electrical alloys, **7**/85–**7**/86
Electrical material-testing techniques, **7**/173–**7**/174
Electrical motors, *see* Motors, electrical
Electrical properties, **7**/5
Electrical Research Association (ERA), **12**/26
Electrical steels, **7**/12, **7**/28
Electricity at Work Regulation 1989, **2**/55–**2**/56, **18**/7
Electricity generation, **2**/15
  geothermal sources, **12**/29
  hydroelectric power, **12**/19–**12**/23
  nuclear power, **13**/4
  tidal power, **12**/31, **12**/32
  wave power, **12**/34
  wind power, **12**/24–**12**/26
Electricity supply survey, **15**/116–**15**/118
Electricity units and symbols, **19**/10–**19**/12
Electrochromic windows, **12**/17
Electrode boilers, **15**/75
Electrodes, manual welding, **16**/55
Electrogalvanizing, **7**/88
Electroless coatings, **7**/165, **9**/81
Electrolytic coatings, **9**/81
Electromagnetic forming, **16**/49, **16**/51
Electromagnetic systems, **2**/5
Electromotive force (emf), **2**/3
  self-induced, **2**/7
Electron beam hardening, **9**/74
Electron microscopy, **9**/116–**9**/117
Electronic Data Interchange (EDI), **4**/24
Electronic elements, finishes on, **9**/107
Electroplating, **7**/165
Electrostatic systems, **2**/4
  measurement techniques, **9**/112–**9**/113
Elements, table of, **19**/18–**19**/19, **19**/20
Ellipsometry, *see* Polarization measurements
Elongation operations, **16**/24
Embossing operation, **16**/28
Emissions control, **11**/18–**11**/19
Emissivity, energy, **1**/42
Emitter, transistor, **2**/33, **2**/34–**2**/35
Emulator, **4**/35
EN Standards, *see* European Normalen
End-quench (Jominy) test, **7**/30
Energy, **1**/5
  conservation of, **1**/36, **15**/90
  flow of, **12**/3–**12**/4
  in fluid mechanics, **1**/22
  process transfers, **1**/35–**1**/36
Energy crops, **12**/39, **12**/40
  fuel-alcohol systems, **12**/39
  marine sources, **12**/39–**12**/40

short-rotation forestry, **12**/39
Energy management, **15**/87, **15**/90, **15**/116
  control systems, **15**/123
  survey/audits, **15**/116–**15**/118
Energy manager, **15**/116
Energy recovery, **15**/122–**15**/123
Enforcement Notes, Health & Safety, **18**/5
Engine lubricants, *see* Lubricants, engine
Engineering design, **5**/3, **5**/12–**5**/13
  civil and architectural, **5**/14
  computer-aided, *see* CAD/CAM
  electronic, **5**/14
  mechanical, **5**/12–**5**/14
  organization of, **6**/5
Engineering Industries, Audit of Industries, **15**/116, **15**/118
Engineering Industries Training Board (EITB), **6**/32
Engines:
  compression-ignition, **10**/6, **11**/17
  continuous-combustion, **11**/17–**11**/18
  discontinuous-combustion, **11**/16–**11**/17
  spark-ignition, **10**/5–**10**/7, **11**/16, **11**/17
Ensemble (random process), **1**/5
Enthalpy, **1**/22, **1**/36
Entropy, **1**/22, **1**/37
Envelope analysis, **15**/128
Environment:
  effect on fretting, **9**/95
  effect on instruments, **3**/22–**3**/23
  resistance to corrosion, **7**/7
Environmental control, disposal of oil products, **9**/27
Environmental Health Officers (RHO), **18**/5
  noise level nuisances, **15**/145
Environmental Protection Act 1990, **15**/105
  noise nuisance, **15**/143, **15**/144–**15**/145
Environmental resistance (corrosion), **7**/7
Epicyclic units, **10**/32
Epoxide resins, **7**/128–**7**/130
Epoxy adhesives, **16**/72
EPROM (Erasable and Programmable Read Only Memory), **3**/5
  emulation, **3**/61
Equations:
  differential, **17**/16–**17**/23
  of motion
    Morison-equation approach, **14**/30
    multiple degrees of freedom, **14**/29
  simultaneous, **17**/14
  solution of, **17**/7
Equilibrium, floating body, **1**/21
Equipment selection, compressors, fans and pumps, **15**/13–**15**/14
Ergodic process, **1**/16
Ergonomics, **6**/23–**6**/29
  human-machine interface, **5**/31–**5**/32
  for manual handling, **18**/11
Erosion, **9**/71
ETHERNET standard, **3**/70, **4**/30, **4**/32, **5**/7
Euler equation, forms of, **15**/3–**15**/4
Euler procedure/relation, **1**/23, **17**/5, **17**/19
European Industrial Services (company), **6**/15
European Normalen (EN), **6**/4
European Space Research Institute (ESRIN), **6**/6
Eutectoid, **7**/21, **7**/22, **7**/23
Evaporation:
  boiler, **15**/83
  factor of, **15**/83
Exhaust pressures (seal failure), **15**/49
Expendable pattern casting processes, **16**/101
Expert systems technology, **6**/6
Exploration drilling, offshore, **14**/3, **14**/5–**14**/8
Explosive forming, **16**/49–**16**/50
Explosive welding, **16**/51
Explosives, types of, **16**/47–**16**/48

Exponential distribution, **17**/27
Extract systems, local, **15**/102, **15**/103–**15**/105
Extreme pressure (EP) oils, **9**/44–**9**/45, **9**/49
  additives, **9**/7
Extreme-value parameters, **9**/120
Extrusion processes, **16**/15, **16**/33–**16**/35
  cold, **16**/37–**16**/38
  continuous, **16**/38–**16**/39
  conventional tube, **16**/35–**16**/37
  die materials, **16**/40
  hydrostatic, **16**/37
  impact, **16**/38
  mashy state, **16**/53
  metal, **7**/20
  plastic, **7**/111–**7**/112
  section and sheet, **16**/39

Factories Act 1961, **15**/105
Failure asssessment diagram, **8**/22–**8**/23
False brinelling, **9**/89
Fans:
  air-conditioning, **15**/113
  design principles, **15**/3–**15**/12
  machine selection, **15**/13–**15**/14
  performance of, **15**/12, **15**/14, **15**/18
  ventilation, **15**/100, **15**/101
Farad (unit), **2**/4
Faraday's Law, **2**/6–**2**/7, **2**/8
Fast Fourier Transform (FFT) frequency
  analyser, **15**/136, **15**/143
Fast reactors, **13**/5, **13**/7–**13**/8
Fasteners, **6**/14, **6**/20
  automatic insertion, **6**/15–**6**/16
  plastic, **6**/21
  self-sealing, **6**/21
  spring steel, **6**/21
  suppliers of, **6**/21–**6**/22
  threaded, **6**/17
  see also Adhesives; Bolts; Rivets
Fatigue, **8**/3, **8**/31
  analysis, **14**/22–**14**/23
  design, **8**/37–**8**/38
  elastomers, **7**/142
  endurance, **7**/3
  and environment, **8**/39
  failure, **8**/32
  fretting, **9**/89
  parameters and symbols, **7**/4, **8**/40
  performance, **8**/32, **8**/37
  static, **8**/23
  testing, **8**/37
  wear, **9**/72
Fatigue cracks:
  growth calculation, **8**/39–**8**/40
  initiation, **8**/31–**8**/32
  propagation of, **8**/32
Fatigue life, **14**/22–**14**/23
FDDI (Fibre Distributed Data Interface), **4**/31
Feature modelling, computer-generated, **5**/11
Feed-water, **15**/88
Feedback signal effects, **2**/38, **2**/39
Feedforward control, **3**/53–**3**/54
Ferrites, **7**/21, **7**/22, **7**/149
Ferritic steels, **7**/21–**7**/25
Ferrography, **9**/72
Fibre-optic cable, **4**/27
Fibre-reinforced composite material, **8**/8–**8**/9
  discontinuous, **7**/99, **7**/100
  filamentary, **7**/95–**7**/98, **7**/99
  thermal conductivity, **1**/44
Fibre-reinforced glasses, *see* Glasses, fibre-
  reinforced
Fibre-reinforced metals, *see* Metals, fibre-
  reinforced
Field effect transistor (FET), **2**/36
Field regulator, **2**/21

Filamentary polymer composites, **7**/95–**7**/98,
  7/99
File organization, **4**/39–**4**/40
File transfer, **4**/34
File-server, **3**/70
Film bearings, *see* Bearings, hydrodynamic
Film thickness, lubricant, **9**/34
Filter, types of, **9**/24
Filtering, noise, **3**/19–**3**/20
Filtration:
  air, **15**/113–**15**/114
  oil, **9**/13, **9**/23–**9**/24
Fingerprinting (plant surveillance), **7**/169
Finite difference technique, **1**/43–**1**/44, **1**/47,
  17/21
Finite-element analysis:
  elastomeric components, **7**/138
  lip seal design, **15**/51
  structural, **5**/11, **14**/19–**14**/20, **17**/22–**17**/23
Finned surface theory (heat transfer), **1**/46–**1**/47
Fire-resistant hydraulic fluids, **9**/12
Fission process, **13**/4
  in reactor, **13**/4
Fits:
  conditions of, **6**/7
  selection of, **6**/13
Fixed-point iteration process, **17**/7
Fixed-point processor, **4**/12
Flame control, **11**/15–**11**/16
Flame cutting, *see* Gas cutting
Flame hardening, **9**/74
Flammability, **11**/15
Flash point, **9**/5–**9**/6, **11**/10–**11**/11, **11**/15
Flash welding, **16**/67
Flaws, identification of, **7**/171
Flexible Manufacturing Systems (FMS), **5**/16,
  6/29
Flexible throttle bush, **15**/55
Flexural strength, **7**/3
Flip-flops (memory elements):
  D (Data) type, **2**/53
  JK type, **2**/52
  SR (Set/Reset) type, **2**/51–**2**/52
  T (Trigger) type, **2**/52
Floating-point processor, **4**/12
Floats, measurement, **1**/27
Floppy disks, **4**/21
Flow coefficient, pump, **15**/5
Flow forming, **16**/24–**16**/25
Flow measurement, **3**/31–**3**/34
  computerized control, **3**/11
  gas, **1**/33
Flow patterns, **1**/35
Flowmeters, **3**/33–**3**/34
Flues, *see* Chimneys, design
Fluid bed boilers, **15**/78
Fluid drive couplings, **10**/48
Fluid Engineering Centre, **15**/40
Fluid flow, **1**/18, **1**/22–**1**/24
  ideal, **1**/33–**1**/35
  measurement, **1**/25–**1**/35
  and surface roughness, **9**/106
Fluid seals:
  fault-finding, **15**/58–**15**/59
  handling and care, **15**/56–**15**/57
  see also Hydraulic seals; Lip seals
Fluid statics, **1**/18–**1**/21
Flumes, **1**/27
Fluoroplastics, **7**/125
Flux:
  field, **2**/3
  leakage losses, **2**/31
  magnetic, **2**/5–**2**/6
Flux coverings, welding electrode, **16**/55–**16**/56
Fluxes, welding, **16**/58–**16**/59
Focus feedback methods, **9**/115
Foil gauges, **8**/13

Force, **1**/5
  measurement, **3**/28–**3**/29
Force microscope, **9**/117
Force response, shock, **15**/132
Ford, quality control, **6**/33
Forestry, short-rotation, **12**/39
Forez process, **9**/78
Forging operations, **7**/18–**7**/20, **16**/26
  cast-preform, **16**/31–**16**/32
  closed-die, **16**/26, **16**/28
    special applications, **16**/28–**16**/29
  impression-die, **16**/26–**16**/27
  isothermal, **16**/32
  open-die, **16**/26
  orbital, **16**/31
  rotary, **16**/29–**16**/30
  subsidiary, **16**/29
Forming processes, metal, **16**/12–**16**/18,
  16/43–**16**/47
Forsterites, **7**/148
FORTH (language), **3**/61
Fortin barometer, **1**/20
FORTRAN (FORmula TRANslation)
  language, **4**/43, **5**/9
Foundation design, offshore jackets,
  14/21–**14**/22
Four-stroke engine, **10**/5–**10**/6
Fourier's law (for conduction), **1**/40
Fourier's series (mathematical), **9**/123, **17**/12
  transforms, **17**/13, **17**/18
Fourth-generation languages, **4**/44–**4**/45
Fracture mechanics:
  defect shape, **8**/19
  linear elastic, **8**/19
  post-yield, **8**/21–**8**/23
  toughness, **7**/4, **8**/19, **8**/21
Frame buffer, **5**/30
Frame grabber, **5**/30
France, health and safety standards, **18**/3
Francis formula (fluid flow), **1**/27
Francis turbine, **12**/20
Free-body diagram, **1**/3
Free-cutting steels, **7**/30–**7**/31
Freewheels, **10**/43
Freezing point, fuel, **11**/11–**11**/12
Frequency, **2**/9
  response, **17**/18
  vibration, **1**/10, **15**/129, **15**/130
    analysis, **15**/135
Frequency analysis procedure, **15**/129
Frequency-domain analysis, **3**/44–**3**/46
  stability criteria, **3**/47–**3**/49
Fretting, **9**/71, **9**/88–**8**/89
  alleviation of fatigue, **9**/101–**9**/102
  characteristics, **9**/90–**9**/91
  design modifications, **9**/98–**9**/99
  factors affecting, **9**/91–**9**/95
  preventative measures, **9**/98–**9**/102
  relative movement sources, **9**/89–**9**/90
  theory of, **9**/95–**9**/98
  wear evaluation, **9**/98
Friction, **1**/6
  data, **9**/83–**9**/85
  effect on seal lip, **15**/54
  loss, **1**/23
  principles, **9**/4
  on seals, **15**/42
  wear, **9**/8
Friction coefficient, **1**/24
Friction welding, **16**/67
Frictional drag, **15**/42–**15**/43
Fringes, polarised light, **8**/16
Froude-Krylov force (fluid flow), **14**/17, **14**/18,
  14/23, **14**/25
Froude's number (fluid flow), **1**/28
Fuel-alcohol plant, **12**/39
Fuels:

alternative, 11/19–11/20, 12/17
calorific values, 11/5
density of, 11/3–11/5
gaseous, 11/5–11/8
liquid, 11/8–11/12
nuclear, 13/5
    post-reactor processing, 13/8
    provision of initial charge, 13/4–13/5
solid, *see* Solid fuels
supply survey, 15/116–15/118
types of, 11/3
Fullering, 16/29
Fume dilution, 15/103
Furan resin, 7/128
Furnace heat transfer, 15/81
Fuses, 2/56
Fusion process:
nuclear, 13/3–13/4
in reactor, 13/4

Gain scheduling method, 3/54, 3/56
Galerkin's method (finding constants), 17/20
Galling, 9/71
Galvanic corrosion, *see* Corrosion, galvanic
Galvanizing, 7/88, 7/165
Gamma-radiation, 7/174, 13/3, 13/11, 13/12
Ganguillet and Kutter equation (flow velocity), 1/28
Gap meter, *see* Rotameter
Gas Constant, 1/31
Gas content, pump performance, 15/12
Gas cutting, 16/64
Gas exploration, 14/3
Gas flow, 1/31–1/33
measurement of, 1/33
Gas nitrocarburizing, 9/76
Gas refrigeration cycle, 15/115
Gas turbines, 10/13–10/15
cycle, 1/37, 1/39
fuel behaviour in, 11/17–11/18
lubrication, 9/20
nickel alloys for, 7/81, 7/83–7/85
Gas welding, 16/63–16/64
Gaseous fuels, 11/5
manufactured, 11/6, 11/8
natural, 11/6
Gate source voltage, 2/36
Gates:
FET, 2/36
logic, 2/44–2/48
    symbols, 2/44
Gauge factor, 3/22, 8/14
Gauging, nuclear application, 13/9
Gaussian process, 1/16
Gear hobbers, 16/6
Gear oils, 9/44–9/45
Gear planers, 16/4
Gear shapers, 16/4
Gear wheels/gearing, 9/41, 10/26, 10/27, 10/29
bevel, 10/30
conformal, 10/26
heat dissipation in, 9/45–9/47
helical, 10/30
involute profile, 10/26–10/29
lubrication of, 9/14–9/15, 9/42–9/44, 9/47–9/49, 10/34
materials for, 10/32
securing to shaft, 10/32–10/33
tooth profile, 10/26
trains, 10/32
worm drive, 10/30–10/31
Gearboxes, 10/33
as noise source, 15/147
Generative planning, 5/15
Generators, 2/15
compound-wound, 2/19

direct-current, 2/15–2/17
separately-excited, 2/17–2/18
series-wound, 2/19
shunt-wound, 2/19
steam, *see* Steam generators
Geometric mean, 17/23
Geometric series, 17/11
Geometrical measurements, 3/23–3/27
Geothermal energy, *see* Terrestrial energy
Gerber's Parabolic Law (fatigue), 8/38
Germanium, 7/92
Germany, health and safety standards, 18/3
GFRP (glass fibre reinforced plastic), 7/99
Ghosting (gear machining), 15/126
GKN 'Ferro Die' process, 16/105
4GL (language), 5/18
Gland:
adjustable, 15/18–15/19
adjustment, 15/30
design, 15/20–15/21, 15/36
fitting, 15/26–15/27
Gland seals, 15/44–15/45
Glass ceramics, 7/151
Glass fibre-reinforced plastics (GRP), bolted joints, 6/18
Glass fibres, 7/93, 7/103
Glasses, 7/147, 7/150–7/151
corrosion of, 7/164
Glossometer, 9/115
GMAW welding, *see* MIG/MAG welding
Gold, 7/91
Goodman Law (fatigue), 8/38
Graphics, CAD/CAM, 5/9
Graphics tablet, 5/31
Graphite, 7/150, 9/69, 11/13
fibres, 7/93
as packing material, 15/28–15/29, 15/31
Grashof number (heat convection), 1/45
Gravitation, 1/4, 19/5
Gravitational system of units, 19/3
Gravity waves, 14/14
Gray binary code, 2/49, 3/3
Gray (unit), 13/10
Gray-King assay test, 11/13
Gray-scale (computer graphics), 5/28
Grease, 9/21–9/22, 9/56–9/58
drop point, 9/6
penetration, 9/6
see also Lubricants
Greece, health and safety standards, 18/3
Green sand casting, 16/100
Greenhouse effect, 11/18, 11/20
Green's function (velocity potential), 14/31
Grey cast iron, 7/52
Grey-body radiation, 1/45–1/46
Grey-scale (computer graphics), 5/28
Grids, 5/34
Griffith's (catastrophic) failure, 13/5
Group Technology (GT), 5/15
Guarding, machinery, 18/10
Gunmetals, 7/57
Gyroscopic effects, 1/7–1/8

Hafnium, 7/92
Half-life, 13/3
Hard-facing materials, 7/87
Hardenability, steel, 7/29–7/30
Hardened steels, 7/12, 7/28–7/30
Hardening, surface, 9/73–9/74
Hardness, 7/3–7/4
Harmonic distortion, amplifiers, 15/135
Harmonic mean, 17/23
Haskind relations, 14/31
Hastelloys, 7/81
Hazard identification, 18/9
emissions, 11/18–11/19
Hazen-Williams equation (water flow), 1/24

HDLC protocol, 4/29
Head coefficient, pump, 15/5
Head loss, 1/24
Heading operation (forging), 16/29
Health & Safety at Work Act, 18/4
Health physics, 13/10–13/12
Health and safety:
in European Community, 18/3
noise effects, 15/144
petroleum products, 9/26–9/27
in UK, 18/4
in USA, 18/4
Health and Safety at Work &c. Act 1974 (HASAWA), 15/105
acoustic environment, 15/146
Health and Safety Executive (HSE), 18/5, 18/8
boiler control Guidance Note, 15/87
Maximum Exposure Limit, 15/97
Occupational Exposure Standards, 15/97
radiation limits, 13/11
robot safety booklet, 5/26
ventilation Guidance Notes, 15/105
Heat conductivity, *see* Conductance, thermal
Heat emitters, 15/93, 15/97
Heat engines, 1/36, 10/3
cycle, 10/3–10/5
Heat exchangers, 1/47, 15/123
Heat losses/gains, 15/110–15/112
building, 15/91–15/93
Heat pipes, 15/123
Heat recovery, 15/84–15/85
Heat transfer, 1/37–1/46
boiler tube, 15/81
in boilers, 15/80–15/83
coefficients
    overall, 1/41
    surface, 1/40, 1/42
furnace, 15/81
Heat treatment processes, energy survey, 15/121
Heat tube, 15/85
Heat units and symbols, 19/9–19/10
Heat-pump system, 15/123
thermochemical, 12/18
Heat-resisting steels, 7/42–7/44
Heaters, continuous-combustion, 11/18
Heating applications, geothermal, 12/28
Heating regulations, statutory, 15/91
Heating systems:
central plant sizing, 15/93
hot-water, 15/94–15/95
multiple-boilers, 15/93
types, 15/94–15/97
Heatserter, 6/16
Heave motion:
offshore
    on cylindrical buoys, 14/23–14/25
    on semi-submersible platforms, 14/25–14/26
    on tensioned buoyant platforms, 14/26
Helical gears, 10/30
Heliohydroelectric power generation, 12/11–12/12
Heliostats, 12/11
Helmholtz resonator, 15/148
Henry (unit), 2/7
Hertz (unit), 2/9
Hexadecimal (hex) notation, 3/3
Hi-bolt, 6/17
Hidden-line removal (computer graphics), 5/42
Hierarchical control, 3/57
High Efficiency Particulate Air (HEPA) Filter, 15/113
High explosives, 16/47–16/48
High spot count, 9/121
High Torque Drives (HTD), 10/24–10/25

High-energy-rate fabrication operations (HERF), 16/14, 16/18, 16/47
  energy sources, 16/47–16/49
High-speed steels (HSS), 16/7
High-strength low-alloy (HSLA) steels, 7/12, 7/27–7/28
Highly Flammable Liquids and Liquefied Petroleum Gases Regulations 1972, 15/105
Highpass filters, 3/20
Holography, 8/18, 9/115
Holtzer method (vibrations), 1/14
Hooke coupling, 10/39
Hooke's Law, 8/4, 8/7
Hot machining, 16/10
Hot working:
  metal, 7/17
  pressing, 7/153
Hot-dip coating, 7/165
Hot-water heating systems, 15/94–15/95
Hottel charts, 1/45
Hottel-Whillier-Bliss equation (solar radiation), 12/7
Hubbing, 16/29
Human engineering, see Ergonomics
Human response:
  sound, 15/140
    indices, 15/140–15/141
  vibration, 15/136
Human sensitivity graph (vibrations), 15/137
Human-machine interfaces, see Ergonomics
Humidification, 15/114
HWBHF (high water-based hydraulic fluids), 9/13
Hydraulic fluids, 9/10–9/13
Hydraulic leap, see Standing wave
Hydraulic mean depth, 1/28
Hydraulic mean diameter, 1/24
Hydraulic oils, 9/11
Hydraulic seals:
  axial length, 15/46
  designs, 15/36
  development, 15/50
  failure, 15/46–15/49
  fluids, 15/38–15/39
  housings, 15/36–15/37
  low-friction, 15/43
  materials, 1/32, 1/34–1/36
  piston, 15/43–15/44
  reciprocating, 15/32
    materials, 15/32–15/36
  selection, 15/45–15/46, 15/50
  short-stroke applications, 15/47
Hydraulic systems, controlling, 3/68–3/69
Hydrodynamic bearing theory, 9/31
Hydrodynamic loading:
  offshore
    on large bodies, 14/18–14/19
    on slender structures, 14/17–14/18
Hydrodynamic lubrication, 9/3
Hydroelectric forming, 16/49, 16/50–16/51
Hydroelectric power, 12/3, 12/12, 12/19
  economics, 12/22–12/23
  environmental issues, 12/22
  potential for, 12/20–12/21
  small-scale, 12/21–12/22
Hydroforming, 16/45–16/46
Hydrogen, as fuel, 12/17
Hydrogen cracking, 7/161
Hydrogen degradation, 7/161
Hydromec system, 16/47
Hydromechanical forming, 16/47
Hyper-eutectoid steel, 7/23
Hyperbolic functions, 17/5
Hypoid gears, 10/30
Hypothesis testing, 17/28–17/29
Hysteresis, 2/8
  instrument, 3/22

Hysteresis whirl, 15/125

ICI (EPOS) polymer selection program, 7/8
Ideal fluid flow, 1/33–1/35
Ideal gas law, 1/31
IEE Regulations, see under Institution of Electrical Engineers
IEEE, see Institute of Electrical and Electronic Engineers
Ignitability, 11/16
IIR (isobutyl-isoprene), 7/144
Iliums, 7/81
Image construction:
  computer, 5/32–5/35
  processing, 5/28
Immersed surface, pressure on, 1/20
Impact testing, 7/4
Impedance, 2/11
Improvement Notices, 18/5
Impulse, 1/5
  angular, 1/6
In Circuit Emulation, 3/61
Inclusions, casting, 16/86–16/87
Incoloys, 7/81
Inconels, 7/81
Indexed Sequential Access Method (ISAM), 4/40
Inductance:
  mutual, 2/7–2/8
  self-, 2/7
Induction hardening, 9/73
Induction heating, 15/121
Induction motors, 2/26
  braking, 2/27
  single phase, 2/28
  speed control, 2/27–2/28
  starting, 2/26–2/27
Inductor, 2/7
Industrial controllers, 3/60
Industrial noise, 15/142
Industrial vision systems, 5/26–5/27
  applications, 5/27
Inert Atmosphere Plasma Spraying, 9/82
Inertia coefficients, 14/18
Influence quantities (instrumentation), 3/22
Ingot casting, 7/17
Injection moulding, 7/111
Input/output:
  computer, 3/5–3/6, 4/5, 4/11
  bus, 4/13–4/14
  digital and analogue, 4/23
  direct, 4/19–4/20
  transaction types, 4/14
Inserts, threaded, 6/20
Inspection process, 6/30, 7/170, 7/172
Inspection Validation Centre (IVC), 13/5
Inspectors, Health & Safety, 18/5
Institute of Electrical and Electronic Engineers (IEEE), 3/7
  IEEE-488 interface bus, 3/8
  IEEE-488 standard (GPIB), 3/65
  RS422 interface standard, 3/7
  RS423 interface standard, 3/7
  see also RS232 interface standard
Institution of Electrical Engineers (IEE):
  802.5 Standard, 4/31
  Regulations, 18/7
Instrument location, importance of, 6/28–6/29
Insulation:
  building, 15/91, 15/149
  electrical, 2/56
  sound, 15/147–15/148
  thermal
    benefits of, 15/119
    transparent materials, 12/18
Integer series, 17/12

Integrals, 17/8–17/11
  transformations, 17/6
Integrated circuits, metals for, 7/92
Integrating factor, 17/16
Interactive processing, 4/37–4/38
Interfacing, 3/8
  analogue, 3/12–3/15
  controller output, 3/9–3/11
  devices for, 4/24, 4/26
  digital, 3/8
Interference microscopy/interferometers, 9/114–9/115
Internal flaws, 7/171
Internal mean value theorem, 17/11
International Atmosphere, 1/18
International Commission for Radiation Protection (ICRP), 13/12
International Graphics Exchange Standard (IGES), 5/17, 6/6
International Lighting Vocabulary, 19/12
International Standards Organization (ISO), 6/3–6/4
  limits and fits, 6/7–6/8, 6/13
Interrupts, computer, 4/14–4/15
Investment casting, 16/102
Involute curve profile, 10/26–10/29
Ion implantation, 9/79
  steel, 7/35
Ion plating, 9/80
Ionizing radiations, 17/10
Ionizing Radiations (Sealed Sources) Regulations 1969, 7/176
IRHD (International rubber hardness) scale, 7/134
Irish Republic, health and safety standards, 18/3
Irradiance, see Solar energy
Isentropic efficiency, 1/37, 1/39
Isentropic processes, 1/31
ISO Standards, see International Standards Organization
Isoclines, 17/18
  method of, 3/78–3/82
Isolation:
  electrical, 2/56–2/57
  double, 2/57
  shock, 15/131–15/132
  vibration, 15/130–15/131
Isothermal transformation, 7/24–7/25
Italy, health and safety standards, 18/3
Izod cantilever testing, 7/4

J integral method, 8/21–8/22
Jack-up rigs, 14/3
Jackets:
  offshore, 14/8–14/9
    certification, 14/33–14/34
    concrete, 14/9–14/10
    design procedures, 14/32–14/33
Jacobian transformation, 17/6
JCL (Job Control Language), 4/37
JEL Building, Stockport (low energy), 12/9–12/10
Jet Coat process, 9/82
Joint design:
  adhesive, 16/77–16//78, 16/80–16/81
    testing, 16/81–16/83, 16/84
  engineering, 16/77, 16/79
Jominy curves, 7/30
Joule cycle, 1/37, 1/39
Journal bearings:
  design, 9/32–9/39
  monitoring vibration, 15/125
Joystick, 5/31
Junction field effect transistor (JFET), 2/36
Just-In-Time (JIT) technique, 6/30

Kaplan turbine, **12**/20
Kapton, **7**/125
Karnaugh map, **2**/49–**2**/51
Kennedy key, **10**/32
Keulegan-Carpenter number, **14**/17, **14**/18
Kevlar, **7**/93, **7**/120
  fibres, **7**/93
Keyboard, **5**/6
Keypad, **5**/6
Kinetic energy, **1**/5
  angular, **1**/6
Kirchhoff's Laws, **2**/3, **2**/8
Knife-line attack, **7**/47–**7**/48, **7**/157
Knocking, spark, **11**/16
Knoop hardness, **7**/3
Kurtosis, **9**/120

Labyrinth seal, **15**/72–**15**/73
Ladder logic programming, **3**/67–**3**/68
Lagging (a.c.), **2**/10, **2**/12
Lambert's Law (radiation), **1**/45
Laminar flow, **1**/23, **1**/24
  sublayer, **1**/29, **1**/30
Laminates:
  behaviour of, **8**/9, **8**/10
  composite, **7**/95–**7**/96
Lamps:
  efficacy of, **15**/120
  types of, **15**/120
Landfill gas, **12**/38
Language Sensitive Editor (LSE), **4**/38
Language translator, **4**/41
Languages:
  computer, *see* Computer languages
  programming, **3**/61
Lantern ring, **15**/19–**15**/20
Lap windings, armature, **2**/17
Laplace equations (fluid flow), **1**/35, **17**/6, **17**/20
Laplace transform notation, **3**/44, **3**/46, **3**/71
Laptop computer, **4**/5
Large-chip processes (metal removal),
  16/3–**16**/6
Larson-Miller method (creep rupture), **7**/4, **8**/25
Laser alloying, **9**/83
Laser cladding, **9**/83
Laser cutting, **16**/65
Laser glove, **4**/16
Laser hardening, **9**/74
Laser transmissions, **4**/27
Laser welding, **16**/65
Latches (electronic), **2**/48
Laurent's series, **17**/12
Lay-up (of composites), **7**/95
Leaching selective, **7**/157
Lead, control of use of, **18**/6–**18**/7
Lead and its alloys, **7**/88, **7**/90
Leading (a.c.), **2**/10, **2**/12
Least squares method, **17**/8
Legendre's equation (linear equations), **17**/18
Lenz's law (induced e.m.f.), **2**/6
Level measurement, **3**/27–**3**/28
Lever operation, **6**/28
Life-fraction rules, **8**/30
Light sectioning process, **9**/113
Light units and symbols, **19**/12–**19**/13
Light Water-cooled Reactors (LWR), **13**/5
Light-pen, **5**/6, **5**/31
Lighting systems, energy survey, **15**/120
Lignite, **11**/12
Limit cycles, **3**/81
Limit of proportionality, **7**/3
Limit switches, **3**/67
Line voltages/currents, **2**/24
Linear coefficient of expansion, **19**/18–**19**/19,
  19/21
Linear energy transfer, **13**/11

Linear equations, solution of, **17**/15–**17**/16
Lip seals, **15**/36
  on-site joining, **15**/57
  rotary-shaft, **15**/50–**15**/51
    applied pressure, **15**/55
    design, **15**/51
    handling care, **15**/56–**15**/57
    lubrication, **15**/53
    materials, **15**/52
Liquid Crystal Display (LCD), **5**/30
Liquid fuels:
  coal tar, **11**/12
  natural gas as, **11**/9
  natural gasoline, **11**/9
  petroleum fractions, **11**/9–**11**/12
Liquid pressure forming process, **16**/108
Liquid-Metal Fast Breeder Reactor (LMFBR),
  13/6
Liquid/liquid interfaces, sealing, **15**/53
Load angle, **2**/25
Load cells, **3**/29, **8**/15–**8**/16
Load characteristic generator, **2**/18
Local Area Network (LAN), **3**/70, **4**/16, **4**/31,
  4/32, **5**/7
Local Authorities, health and safety
  enforcement, **18**/5
Locking compounds, **16**/72
Logarithmic decrement, **1**/11
Logic systems, **2**/44–**2**/48
  sequential circuits, **2**/51–**2**/52
Logical channels, computer, **4**/35
Loss-lubrication systems, **9**/15
Lost-wax casting, **16**/101–**16**/102
Louvres, **15**/102
Low Pressure Plasma Spraying, **9**/82
Low voltage supplies, **2**/53
Lowpass filters, **3**/20
LPG, *see* Petroleum gases
LRC (Longitudinal redundancy check), **4**/28
Lubricants:
  additives, **9**/6–**9**/7
  applications, **9**/7–**9**/10
  contamination of, **9**/19
  demulsibility, **9**/20
  engine, **9**/7–**9**/10
  for fretting limitation, **9**/99
  gear, **9**/47–**9**/49
  performance ratings, **9**/9–**9**/10
  physical characteristics, **9**/5–**9**/6
  rationalizing requirements, **9**/15
  service life, **9**/49
  solid, **9**/68–**9**/69, **9**/99
  spray, **9**/23
  storage of, **9**/25–**9**/26
  viscosity of, **9**/5
Lubrication:
  bearings, **9**/35–**9**/36
  centralized, **9**/25
  fretting control, **9**/95
  gearing, *see* Gear wheels/gearing, lubrication
  one-shot, **9**/25
  and planned maintenance, **9**/26
  principles, **9**/3
  process fluid, **9**/39
  rolling element bearings, **9**/56–**9**/58
  and surface roughness, **9**/104–**9**/105
  systems, **9**/15
Lubrication and maintenance, planned, **9**/26
Lumped capacity system, **1**/42
Lux (unit), **15**/120
LVDT (Linear Variable Differential
  Transformer), **3**/24

M-system (reference lines), **9**/119
Mach number, **1**/32, **15**/5
Mach Stream process, **9**/82

Machine code (computer), **4**/37
Machine parts:
  rolling, **16**/25
  standardization of, **6**/3
Machine tools, **9**/13–**9**/15, **16**/3–**16**/8
  control interfaces, **3**/15
  energy survey, **15**/121
Machinery, vibration effects on, **15**/136
Machinery oils, **9**/7
Machining operations:
  costs, **16**/10–**16**/11
  hot, **16**/10
  and surface topography, **9**/108
Maclaurin's series, **17**/11–**17**/12
McLeod gauge, **3**/31
Magne particle clutches, **10**/47
Magnesia ceramics, **7**/149
Magnesium and its alloys, **7**/74–**7**/75, **7**/76–**7**/80
Magnetic alloys/materials, **7**/85, **7**/87, **7**/90
Magnetic circuit, **2**/8
Magnetic field, of a toroid, **2**/5–**2**/6
Magnetic Particle Inspection (MPI), **7**/173
Magnetism units and symbols, **19**/11–**19**/12
Magnetomotive force, **2**/5
Magnox gas-cooled reactor, **13**/5
  can development, **7**/9–**7**/10
Magslip, *see* Synchro
Mainframe computer, **4**/6
Maintenance management, **9**/26
  preventive, **15**/124–**15**/125
Malleable iron, **7**/54
Mannesmann cycle, **16**/24
Manning equation (fluid flow), **1**/28
Manometer, **1**/18–**1**/19
Manual handling, safety considerations,
  18/11–**18**/12
Manufacture, energy survey, **15**/120
Manufacturing Automation Protocol (MAP),
  3/70, **5**/7
Maraging steels, **7**/12, **7**/45
Mari Advanced Microelectronics Ltd., QA
  certification, **6**/33
Marine energy sources, **12**/39
Markov process, **9**/122
Martempering, **7**/29
Martensite, **7**/23
Mashy state forming, **16**/14, **16**/52–**16**/54
Masonry, protection of, **7**/166
Mass, **1**/4, **19**/5, **19**/7
Mass flow rate, **1**/32
Material Design Specification, **7**/8
Material/manufacturing costing, **7**/7
Materials:
  choice of, **7**/5
    Classical Procedure, **7**/6–**7**/7, **7**/9–**7**/10
    Comparative Procedure, **7**/8
    computerization, **7**/8–**7**/9
    Imitative Procedure, **7**/7–**7**/8
    techniques, **7**/6
  compression packing, **15**/21–**15**/22
  counterface, **9**/62, **9**/67
  hydraulic seals, **1**/32, **1**/34–**1**/36
  information sources, **7**/8
  polymeric, for seals, **15**/62
  requirement checklist, **7**/6–**7**/7
  sorting, **7**/169
Materials control, **7**/168
  requirement checklist, **7**/6–**7**/7
  sorting, **7**/169
Mathematical signs and symbols, **17**/3
Matrix arithmetic, **17**/14
  transformation representations, **5**/35, **5**/37
Matrix representations, **5**/35
Mats, isolating, **15**/133
Maxima/minima, calculus, **17**/8
Maximum continuous firing rate (MCR), **15**/84
Maximum Exposure Limit (MEL), **15**/97

Mean peak radius, profile, 9/122
Mean slope, profile, 9/121
Measurands, 3/21
Measurement systems, 3/21–3/23
Mechanical seals, 15/59–15/60
 cooling arrangements, 15/65
 design variants, 15/62–15/64
 face materials, 15/60, 15/62
 failures, 15/67–15/68
 installation, 15/67
 performance, 15/64–15/66
Mechanical ventilation, see Powered ventilation
Mechanics units and symbols, 19/6–19/8
Mechatronics, 3/58
Median, 17/23
Medical applications, nuclear, 13/10
Megagram (unit), 19/7
Melamine, 7/128
Melt processing, 7/112
Melting points tables, 19/18, 19/20, 19/21
Memory (computer):
 addressing, 4/8, 4/9
 management of, 4/9
 organization, 4/7–4/8
 protection, 4/9
 technology, 4/8–4/15
 virtual, 4/9
Memory devices, 3/5
Memory map, system, 3/6–3/7
Menus, hierarchical, 5/9
Mersey Barrage, proposed, 12/32
Mesh analysis, 2/4
Mesh connection, see Delta connection
Metacentre, 1/21
Metal-cutting:
 economics, 16/10–16/11
 forces and power in, 16/9–16/10
 machines for, 16/3–16/8
Metal forming, 16/12–16/18, 16/43–16/47
Metal matrix composites, casting, 16/108
Metal Oxide Semiconductor Field Effect
  Transistor (MOSFET), 2/36
Metal rolling, see Rolling processes
Metal spraying, 7/165
Metal-cooled reactor, 13/5
Metal-wetting, 9/48
Metallic alloys and composites (for sliding
  bearings), 9/69–9/70
Metals:
 corrosion protection, 7/165
 fibre-reinforced, 7/101–7/102
 mechanical working, 7/17–7/21
 properties tables, 19/20
Methane, reforming of, 12/17
Metric Technical System (of units), 19/3, 19/5
Metric units, see S.I. units
Metrology:
 automated, 7/169
 optical, 7/172
Microcomputer, 4/5
 graphics systems, 5/27
 or PLC alternative, 3/66–3/67
Microfilm, 6/6
Microphones, 15/142
Microprocessors, 2/55, 3/4–3/7
 control by, see Control, microprocessor-
  based
Microscopy techniques, 9/116–9/117
Microwave transmissions, 4/27
MIG/MAG welding, 16/60–16/61
MIL lubricant ratings, 9/10
Milling processes:
 forces in, 16/9
 metal, 16/4, 16/6
Mindel, 7/123
Mini-OTEC (ocean thermal energy) project,
  12/12

Minicomputer, 4/6
Mitsubishi microcontrollers, 3/60
Mode, 17/23
Modelling techniques, 5/10
 geometric, 5/10–5/11
 schematic, 5/12
 for stress-analysis, 8/17–8/18
Modem, 4/27
Moderator, reactor, 13/5
Modular design, 6/3
Modulation range, boiler, 15/84
Modulus of elasticity, 7/3, 8/3, 8/4, 8/7
 table, 19/20
Moh hardness, 7/3
Mohorovicic seismic discontinuity (Moho),
  12/27
Mohr's stress circle, 8/5
Moiré fringes, 3/25
Molybdenum, 7/91
Molybdenum disulfide, 9/69
Moment of inertia, 1/5
Momentum, 1/5
 angular, 1/6
Momentum thickness, 1/29
Monels, 7/81
Mooney-Rivlin equation, 7/136
Morison's equation (wave loading),
  14/14–14/15, 14/17, 14/18
 for slender structures, 14/30
Motion, curvilinear, 1/6
Motorola microcontrollers, 3/60
Motors:
 electrical, 2/15
 brushless, 2/30
 capacitor, 2/28
 direct current, 2/19–2/23
  speed control, 2/21, 3/63–3/64
 induction, 2/26–2/28
 permanent magnet, 2/28–2/29
 shaded-pole, 2/28
 stepper, 2/29–2/30, 3/64
 synchronous, 2/25–2/26
 universal, 2/28
Moulding, rubber, 7/134
Mounts, anti-vibration, 15/150
Mouse (input method), 5/6, 5/31
Moving coordinate system, 1/7
MOX fuel, 13/5
Multi-degree of freedom systems, 15/130
Multi-layer control, 3/57
Multi-level control, 3/57–3/58
Multigrade oils, 9/8–9/9
Multiplexing (MUX), 3/15, 4/26
Multiprogramming, 4/9, 4/38–4/39
Mutual inductance, 2/7–2/8

Narrow band analysis, spectrum, 15/128
Narrow-band analysis (frequency), 1/17
National Centre of Tribology, 9/100
National Engineering Laboratory, 12/26
National Grid system, 2/30
National Instruments 'lab windows', 3/65
National Lubricating Grease Institute (NLGI),
  9/56–9/58
National Wind Turbine Centre, 12/26
Natural gas, 11/6
Natural ventilation, 15/99–15/100
 equipment, 15/102–15/103
Navier-Stokes equation (fluid flow), 1/23, 9/104
NBR (crylonitrile-butadiene), 7/144
Net energy ratio concept, 12/39
Net positive suction head (NPSH), 15/9–15/10
Netherlands, health and safety standards, 18/3
Nettlefolds Gold Seal fasteners, 6/15
Network theorems, electrical, 2/3
Networks:

CAD/CAM, 5/7
computer, 3/70, 4/16, 4/23, 4/30, 4/31, 5/7
 architecture, 4/36
 concepts, 4/34
 design, 4/35
 types of, 4/32–4/34
Neutrons, 13/3
Newton's Law of Gravitation, 1/4
Newton's Laws of Motion, 1/4, 1/23
Newton's method of successive approximation,
  17/7
Ni-Resist, 15/62
Nibbles, 3/3
Nickel and its alloys, 7/75, 7/80–7/87
Nickel-silvers, 7/59
Niobium, 7/91
Nitriding, 7/35, 9/76–9/77
Nitriles, 7/144
Nitrocarburizing, 9/77
Nitrotec process, 9/78
No-stretch conditions, rolling process, 16/23
Nodal analysis, 2/4
Nodal points/values, 17/21
Nodular graphite (SG) iron, 7/52
Noise:
 amplifier, 2/38
 control of, 15/143
 damage by, 15/144
 filtering, 3/19–3/20
 legislation, 15/144–15/145
 measurement of, 15/142–15/143
 nuisance by, 15/143
 road traffic, 15/141
 and surface roughness, 9/106
Noise at Work Regulations 1989, 15/146, 18/8
Noise control engineering, 15/147
 reduction devices, 3/20, 15/147–15/149
Noise Criteria Curves, 15/141
Noise indices, 15/140–15/141
Noise level:
 background, 15/145
 equivalent continuous, 15/141
 single-event, 15/141
Noise units, Community, 15/141
Noise-abatement zones, 15/146
Noise-rating curves, 15/141
Nomex, 7/120
Non-destructive testing (NDT), 7/166,
  13/5–13/6, 16/84–16/85
 application areas, 7/168
 methods, 7/169
  choice of, 7/177–7/178
 techniques, 7/171–7/177
Non-ferrous materials, surface treatments, 9/79
Non-ferrous metals, as alloying elements,
  7/91–7/92
Non-linearities, accounting for, 3/78
Normal distribution, 17/26
Normal random process, 1/16
Norton's theorem (networks), 2/4
Norwave Tapchan device, 12/35
Noryl, 7/122
Noskuff treatment, 9/75
Notches, flow over, 1/27
Notebook computer, see Laptop computer
Nozzle, 1/25
 gas flow in, 1/32
Nuclear fuel, 13/4
 MOX, 13/5
Nuclear metals, 7/91–7/92
Nuclear radiation, 13/3, 13/9
 limits, 13/11–13/12
Nuisance, definition, 15/143–15/144
Null modems, 4/30
Number systems, 3/3
Numerically-controlled (NC) systems,
  5/15–5/16

direct (DNC), **5**/7, **5**/16
Nusselt number, **1**/44
Nuts:
  free-spinning, **6**/18
  stiff, **6**/20
Nylons, **7**/120
Nyquist plots, **3**/47, **3**/48

O-rings, **5**/32, **15**/36
Occupational Exposure Standards (OES),
  15/97, **15**/103
Ocean thermal energy, **12**/12
Octane Number, **11**/17
Octave analysis, **15**/135–**15**/136
Offices, Shops and Railway Premises Act 1963,
  15/105
Offshore developments:
  current, **14**/3–**14**/11
  future, **14**/11–**14**/13
  structure certification, **14**/14, **14**/33–**14**/34
Ohm (unit), **2**/3
Ohm's Law, **2**/3
Oil:
  cleaning of, **9**/23–**9**/24
  gear, **9**/47–**9**/48
  hydraulic, **9**/11
  mist feeds, **9**/25
  multi-grade, **9**/8–**9**/9
  reconditioning, **9**/26
  tramp, **19**/15
  transformer, **9**/20–**9**/21
  turbine, **9**/20
  see also Lubricants
Oil exploration, **14**/3
Oil film whirl, **9**/38, **15**/125
Oil stores, **9**/25–**9**/26
Oldham coupling, **10**/39
ON/OFF control strategy, **3**/49–**3**/50
Open channel flow, **3**/33
Open System Interconnection Model (OSI),
  3/70, **4**/31, **4**/32, **5**/7
Open-channel flow, **1**/27
  measurement, **1**/27
Open-circuit characteristics, **2**/18
Open-circuit test, transformer, **2**/31
Open-loop control system, **3**/41
Openness, surface, **9**/121
Operating systems:
  computer, **4**/37
  common concepts, **4**/38–**4**/40
Operational amplifiers (op-amp), **2**/39–**2**/41 ,
  3/18–**3**/19
Operational strain rate, metal-forming process,
  16/14
Operational-stress system, metal-forming
  process, **16**/13–**16**/14
Optical Character Recognition (OCR), **4**/20,
  6/6
Optical components, finishes on, **9**/107
Optical interference, **3**/25
Optical properties, **7**/5
Optical surface-measurement methods,
  9/113–**9**/116
Orbital forging, **16**/31
Organic energy resources, **12**/36
Orifice plate, **1**/25, **3**/33
Orthotropic material, **8**/8–**8**/10
Oscillating loads, lubrication for, **9**/36
Oscillation, floating body, **1**/21
Osprey process, **16**/108
Otto cycle, *see* Constant-volume cycle
Output devices, computer, **4**/16–**4**/19
Overall (RMS) levels, **15**/125, **15**/126, **15**/127
Overheating, **15**/103, **15**/119
Oxidation corrosion process, **7**/155
Oxide treatments (for hardening), **9**/78

P + I control strategy, **3**/74–**3**/75, **3**/77–**3**/78
p-n junctions, **7**/92
Package assembler/disassembler (PAD), **4**/36
Packing, length-form, **15**/31
Painting, corrosion protection by, **7**/165
Palette, *see* Colour Look-Up Tables
Palmgren-Miner rule (fatigue damage), **14**/22
Parallel axis theorem, **1**/5
Parallel communication, **3**/7–**3**/8
Parceval's lemma, **17**/13
Parsons' turbine, **10**/10, **10**/12
Part-punching, **6**/17
Particle-reinforced metals, **16**/53–**16**/54
Particular integral, **17**/17
Parts referencing, **6**/4
PASCAL (language), **4**/44
Pascal (unit), **19**/7
Passive solar design, **12**/8
  UK applications, **12**/9
  use of atria, **12**/10
Patenting (steel), **7**/49
Path control, robotic, **5**/20–**5**/21
Peak inverse voltage, **2**/14
Pearlite, **7**/21, **7**/23
Pearson coefficient (of skewness), **17**/24
Peat, **11**/12
PEEK (polyether-ether ketone) composites,
  7/95, **7**/123
Pelton wheel, **12**/20
Pemserter, **6**/16
Penetrant testing, **7**/173
Penetration cracking, **7**/159
People, vibration effects on, **15**/136
Period (time interval), **2**/9
Periodic phenomena units and symbols, **19**/5
Peripheral devices, **4**/5, **4**/15–**4**/16
PERITUS knowledge-based system, **7**/9
Permanent mould casting, *see* die casting
Permeability, **2**/6
Permeance, **2**/6
Permittivity, **2**/4, **7**/5
Permutations, **17**/24
Perpendicular axis theorem, **1**/5
Personal Certification in Non-Destructive
  Testing, **13**/5
Personal computers (PCs), **3**/58
  laptop, **4**/5
  use as controllers, **3**/65
Personal protective equipment, **18**/10
  legislation, **18**/10–**18**/11
Petroff equation (oil requirement), **9**/46
Petroleum fractions, **11**/9–**11**/12
Petroleum gases, **11**/8, **11**/9
Petrov's Law (fluid films), **9**/31, **9**/36
pH scale, **19**/12
Pharmaceutical industry, and surface
  topography, **9**/107
Phase diagram, iron-carbide, **7**/21–**7**/25
Phasor diagrams, **2**/11
Phenolic/resorcinolic adhesives, **16**/72
Phenolics, **7**/127–**7**/128
Phosphating, **7**/165, **9**/81–**9**/82
Photochemical processes, **12**/16
Photoelasticity, **8**/16–**8**/18
Photoelectrochemical cell, **12**/16
Photogrammetric methods, **9**/117
Photosynthesis, **12**/4, **12**/36
Photosynthetically active radiation (PAR),
  12/36
Photovoltaic systems, **12**/3
  energy conversion, **12**/14, **12**/15
Physical constants, **19**/17–**19**/20
Physical vapour deposition (PVD), **9**/80
PID (Proportional Integral and Derivative)
  control, **3**/51, **3**/52, **3**/60, **3**/74
  digital control algorithm, **3**/62–**3**/63
Piezoelectric transducers, **11**/19

Pilgering, **16**/30
Pipe flow:
  change of section losses, **1**/24
  effect of roughness, **1**/24
  friction losses, **1**/23
  measurement of, **1**/25–**1**/27
  pressure Design Standards, **15**/90–**15**/91
Pipe networks, **1**/24
Pipework distribution systems, heating, **15**/95
Pirani thermal conductivity instrument, **3**/31
Piston seals, **15**/43–**15**/44
Pitot tube, **1**/26, **3**/34
Pitot-static tube, **1**/26
Pitting, **7**/158, **9**/104–**9**/105
  steel, **7**/42
Pixel, **5**/27–**5**/28
PL1 (Programming Language 1), **4**/44
Planned lubrication maintenance (PLM)
  schemes, **9**/26
Planning Application conditions, **15**/146
Plant surveillance, **7**/169
Plasma carburizing, **9**/76
Plasma coating, **7**/165
Plasma cutting, **16**/63
Plasticity, **8**/3, **8**/10–**8**/11
  collapse mechanism, **8**/11–**8**/12
Plastics, **7**/107
  component design, **7**/113–**7**/116, **7**/130, **7**/132
  corrosion of, **7**/164–**7**/165
  protection of, **7**/166
Plate Inspection Steering Committee, **13**/5
Plate production, **16**/20, **16**/21
Plating/diffusion treatments (for hardening),
  9/78
Platinum group metals, **7**/91
Plotters, **4**/19, **5**/6–**5**/7, **5**/30
Plugging (motor braking), **2**/27
Plutonium, **7**/92
  as fuel, **13**/4
  isotopes, **13**/4
Plywood, **7**/106
pn junction diode, **2**/13–**2**/14
Pneumatic instruments, **3**/26
Pneumatic surface-measurement methods,
  9/113
Pneumatic systems, controlling, **3**/68–**3**/69
Pneurop Oxidation Test (POT), **9**/18
Pocket computer, **4**/5
Poiseuille's equation (fluid flow), **1**/24
Poisson distribution, **17**/26
Poisson's equation, **2**/6
Poisson's Ratio, **7**/3, **8**/3, **8**/7
  table, **19**/20
Polar coordinates, **1**/6
Polariscope photoelastic, **8**/17
Polarization measurements (surface roughness),
  9/116
Pole (mathematics), **17**/6
Polyamide-imide, **7**/125
Polyamides:
  aliphatic, *see* Nylons
  aromatic, **7**/120
Polybutylene telephthalate (PBT), **7**/122
Polycarbonates, **7**/122
Polyester resins, **7**/128–**7**/129
Polyesters, **7**/122
Polyether ketones (PEK), **7**/123
Polyether sulphones, **7**/123
Polyetherimide, **7**/125
Polyethylene, **7**/117
Polyethylene terephthalate (PETP), **7**/122
Polyimide, **7**/125
Polymer matrices, **7**/93, **7**/94
Polymers, **7**/106–**7**/108, **9**/62–**9**/67
  adhesive, **16**/73
  classification, **7**/109–**7**/111
  for coatings, **7**/165

Polymers – *continued*
  processing, **7**/111–**7**/113
  use of, **9**/100
  see also under specific type names
Polymethyl pentene (TPX), **7**/117
Polymides, thermosetting, **7**/128
Polyphenylene oxide, modified, **7**/122
Polyphenylene sulphide, **7**/122
Polypropylene, **7**/117
Polystyrene, **7**/117
Polysulphone, **7**/123
Polytetrafluoroethylene (PTFE), *see* PTFE
Polytropic process, **1**/37
Polyurethanes, **7**/144
  as adhesives, **16**/72
Polyvinyl chloride (PVC), *see* PVC
Porcelains, *see* Ceramics, clay-based
Porous absorbers, sound, **15**/148
Porous metals, *see* Self-lubricating porous
  metals
Portugal, health and safety standards, **18**/3
Positron measurements, optical, **3**/25–**3**/26
Positive displacement machines, **15**/3
  principles, **15**/6–**15**/8
Position annihilation technique, **7**/175
Post-processing systems, computer, **5**/16
Potential difference, **2**/3
Potential energy, **1**/5
Pour point (oil viscosity), **9**/5
Powder compaction techniques, **16**/52
Powder forming, **16**/29
Power, **1**/5
  belt-drive ratings, **10**/20
  compact sources, **13**/9
  electrical, **2**/3
Power beam welding, **16**/64
Power factor, **2**/12
Power factor correction, welding equipment,
  15/121
Power fasteners, **6**/17
Power series, **17**/12
Power sources, welding, **16**/57, **16**/59
Power spectral density function, **9**/123
Power transmission, **10**/15
  shafts, **10**/34
Power triangle, **2**/12
Power-correction factors, **10**/20
PPI (Programmable Peripheral Interface), **3**/6
Prandtl numbers, **1**/30, **1**/44, **15**/82
Pre-ignition, **11**/17
Pre-production control, **7**/170
Precession, **1**/8
Precious metals, **7**/91
Preferred sizes, modular design, **6**/3
Pressure:
  fluid, **1**/18
    measurement, **1**/18–**1**/20, **3**/29–**3**/31
Pressure ratio, critical, **1**/32
Pressure transducer, **8**/16
Pressure transients, **1**/30–**1**/31
Pressure trapping (seal failure), **15**/48
Pressure units, **19**/7
  absolute, **19**/8
  gauge, **19**/8
Pressure vessels:
  Design Standards, **15**/90–**15**/91
  steels, **7**/26
    ferritic, **8**/26, **8**/29
Pressure wave progress, **1**/31
Pressure-enthalpy chart, refrigeration cycles,
  15/115
Pressurization, heating systems, **15**/95
Pressurized Water-cooled Reactor (PWR),
  13/5, **13**/9
Prestressed concrete, **7**/104
Preventive maintenance, **15**/124
  predictive, **15**/124–**15**/125

Priming, boiler, **15**/83
Printers:
  computer
    dot matrix, **4**/16, **5**/30
    high-speed, **4**/18–**4**/19
    laser, **4**/19, **5**/30
    letter-quality, **4**/17
Probabilistic quantity, **1**/15
Probability, **17**/25–**17**/26
  distribution, **1**/15–**1**/16, **17**/26–**17**/28
    density function, **1**/15
Process control, **3**/41, **7**/170
Process efficiency, **1**/36
Process laws, thermodynamic, **1**/37
Process planning, **5**/15
Product Design Exchange Specification
  (PDES), **5**/18
Product Design Specification, **7**/8
Production platforms:
  offshore, **14**/8–**14**/11
  future, **14**/11–**14**/13
Profile parameters, **9**/122
Programmable Logic Controllers (PLCs), **3**/65
  in automation systems, **3**/65–**3**/66
  ladder diagram, **3**/67–**3**/68
  or microcomputer alternative, **3**/66–**3**/67
  networking, **3**/69–**3**/70
  safety procedures, **3**/69
Programmable Logic Devices (PLD), **2**/55
Programming, numerical control, **5**/15
Programs, computer, **4**/36–**4**/37
Prohibition Notices, **18**/5
Projection welding, **16**/66
Proof stress, **7**/3
Proportional control, **3**/73–**3**/74
Prosecutions, health and safety, **18**/5
Protection, electrical, **2**/56
Proving ring, **3**/29
PTFE (polytetrafluoroethylene), **7**/125
  contact seals, **15**/43
  in O-rings, **15**/35
  as packing material, **15**/28, **15**/29
  use of, **9**/84, **9**/100
  for sliding bearings, **9**/62, **9**/64, **9**/66, **9**/69
Public Packet-Switched Network (PPSN), **4**/36
Pulleys:
  belt-drive, **10**/18–**10**/19, **10**/20
  timing, **10**/24
Pulse circuits, **2**/54–**2**/55
Pulse transfer function, **3**/72–**3**/73
Pulse wave, **17**/13
Pulse Width Modulation (PWM), **3**/10–**3**/11
Pulsed Video Thermography (PVT), **7**/173
Pultrusion, **7**/95
Pumped storage systems, **12**/21
Pumps:
  centrifugal, **15**/9
  characteristics, **15**/5–**15**/6
  design principles, **15**/3–**15**/12
  external gear, **15**/7
  machine selection, **15**/13–**15**/14
  metering, **15**/9, **15**/10
  performance, **15**/14, **15**/18
    limitations, **15**/8–**15**/9
  reciprocating, seals for, **15**/23
Push-bench process, **16**/23
PV (pressure velocity) factors, **9**/61, **9**/63
PVC (Polyvinyl chloride), **7**/117
Pyrolysis, **7**/153
Pyrometers, *see* Thermometers, radiation

Q factor (quality), **2**/13
Q-factor, *see* Dynamic magnification factor
Quacorr RP100A, **7**/128
Quadratic equation, **17**/7
Quality, **6**/29–**6**/33

Quality Assurance (QA), **13**/9
  surface treatments, **9**/87
Quality control, **7**/170
  adhesive joints, **16**/84
  nuclear plants, **13**/8–**13**/9
  surface treatments, **9**/87
Quality system, health and safety, use of
  adhesives, **16**/84
Quartiles, **17**/23–**17**/24
Quartz fibres, **7**/93
Quenching process, **1**/42, **7**/28, **8**/38

R6 method, CEGB, **8**/22–**8**/23
Radel, **7**/123
Radiation:
  nuclear
    limits, **13**/11–**13**/12
    units, **13**/10–**13**/11
  protection from, **7**/176
  solar-energy, **12**/4–**12**/5
  thermal, **1**/39, **1**/42, **1**/45
Radio-toxicity, **13**/5
Radioactive Substances Act 1960, **7**/175
Radioactive Substances (Carriage by Road)
  GB Regulations 1940, **7**/176
Radioactive Substances (Road Transport
  Workers) GB Regulations 1970, **7**/176
Radioactivity, **7**/175–**7**/176, **13**/3–**13**/4
  see also Radiation, nuclear
Radiography, **7**/174
  dynamic, **7**/175
  electron, **7**/175
  flash, **7**/175
  legislation for, **7**/175–**7**/176
  low-voltage, **7**/175
  panoramic, **7**/175
  real-time, **7**/175
Radiometry, **7**/175
Radiosity, **1**/45
Railway noise, **15**/142
RAM (Random Access Memory) chip, **3**/5, **4**/7,
  4/8
Ramp input (control system), **3**/42
Rance Barrage, **12**/31
Random logic, **4**/11
Random processes, **1**/15
Random variable, **1**/15
Randupson process, **16**/101
Range, series, **17**/23
Rank correlation coefficient, **17**/25
Rankine cycle, **10**/9
Rankine-Hugoneot relationship, **1**/33
RAPRA Technology, **6**/32
Rarefaction, **1**/31
Raster-scan, **5**/30
  screen, **5**/5
Ray-tracing, **5**/42
Rayleigh number (fluid flow), **1**/45
Rayleigh's method (vibrating system), **1**/9
RC circuits, **2**/9–**2**/10
Reactance:
  capacitive, **2**/11
  inductive, **2**/10
Reaction, compressor, **15**/4
Reaction bonding, **7**/153
Reaction injection moulding, **7**/113
Reactive power, **2**/12
Reactor, nuclear:
  control of, **13**/6–**13**/7
  types of, **13**/5
Read-only memory (ROM), **4**/7–**4**/8
Real time, **4**/38
Reciprocating piston engine, noise sources
  from, **15**/147
Recirculation, pump performance, **15**/11
Rectification, a.c., **2**/14, **2**/41–**2**/42

Rectified sine wave, **17**/12
Reducing mills, **16**/22
Reduction formulae, calculus, **17**/10
Reference spectrum, **15**/127
Referred values, transformer, **2**/31–**2**/32
Reflection, **5**/35
Refractive index, **7**/5
Refrigeration, **15**/114–**15**/115
  solar, **12**/12–**12**/14
Refrigeration plant:
  air-conditioning, **15**/109
  energy survey, **15**/122
Refrigerator compressors, **9**/19
Refuse-derived fuels, **11**/13, **12**/37
Regenerative cycle, steam, **10**/9
Regenerators, energy management, **15**/123
Registers:
  computer, **4**/8
  flip-flop, **2**/53
Regression, **17**/25
Regula falsi method (successive
  approximation), **17**/7
Reiher-Meister scale, **15**/137
Reinforced concrete, **7**/103–**7**/105
  polymer additives, **7**/104
Reinforced thermoplastic materials (RTP), **7**/99
Reinforcement:
  elastomer composite, **7**/138–**7**/139
  fibre, **7**/93, **7**/94
Relative file organization, **4**/39
Relative permittivity, **7**/5
Relays, control, **3**/68
Reliability, instrument, **3**/22
Reluctance, **2**/6
Remanent flux density/remanence, **2**/8
  life assessment, **8**/30
Renard series, **6**/3
Replicast ceramic shell casting process, **16**/102
Report Program Generator (RPG), **4**/44
Residue (mathematical), **17**/6
Resistance:
  electrical, **2**/3
  measurements, **3**/24–**3**/25
Resistance welding, **16**/65–**16**/66
Resistivity, electrical, **2**/3, **19**/18–**19**/19, **19**/21
Resolution, computer graphics, **5**/28, **5**/29
Resolvent matrix, **3**/86
Resonance condition (electrical), **2**/13
Resonant absorbers (sound), **15**/148
Resource sharing, computer, **4**/35
Response initiate time, **6**/28
Reynolds number, **1**/23, **1**/44, **9**/5, **15**/5
  boiler flow, **15**/82
Reynold's equation (hydrodynamic bearing
  theory), **9**/31, **9**/104
Rheo-casting, **16**/106, **16**/108
Right-hand screw rule, **2**/5
Rigid column theory, **1**/30
Ring pattern, network, **3**/70
Ripple factor, **2**/42
RISC (reduced instruction set computer), **4**/12
Rivets, **6**/21
RL circuits, **2**/9
RLC circuits, **2**/11
Road traffic noise, **15**/141
Robotics:
  industrial, **5**/16, **5**/19
    construction, **5**/19–**5**/20
    drive and control systems, **3**/15–**3**/18, **5**/20,
      5/22
    geometry, **5**/20
    programming methods, **5**/22–**5**/23
    safety with, **5**/25–**5**/26
    selection of, **5**/24
    uses of, **5**/23
Rockets, **11**/18
Rockwell hardness, **7**/3, **7**/4

Rockwell microcomputer, **3**/60
Roll-cladding, **9**/83
Roller bearings, *see* Bearings, rolling element
Rolling mills, **16**/18–**16**/19
Rolling processes, **7**/18, **16**/18–**16**/20
  flow forming, **16**/24–**16**/25
  general, **16**/5
  machine parts, **16**/25
  mashy state, **16**/53
  plate strip and sheet, **16**/20–**16**/21
  seamless tubing, **16**/21–**16**/24
  sheet roll bending, **16**/25
  structural sections, **16**/21
ROM (Read Only Memory) chip, **3**/5, **4**/7, **4**/8
Roof extract units, **15**/101
Roof inlet units, **15**/101
Root mean square values, **2**/9
RotaBolt, **6**/17
Rotaform process, **16**/31
Rotameter, **1**/25, **3**/33
Rotary forging, **16**/29–**16**/30
Rotary-Compressor Oxidation Test (ROCOT),
  9/18
Rotation process (mathematical), **5**/35
Rotodynamic machine principles, **15**/3
Roughness, surface, **6**/14
Routh-Hurwits criterion, **3**/49
RS 232C interface standard, **3**/7, **4**/27,
  4/29–**4**/30
Rubber mounts, **15**/133
  anti-vibration, **15**/150
Rubbers, *see* Elastomers
Runge-Kutta predictor/corrector procedure,
  17/19
Running-in phase, **9**/105
Rupture, *see* Stress rupture testing
Rutile flux covering, **16**/56

SAE viscosity system, *see* Society of
  Automotive Engineers
Safety:
  by design, **18**/9
  electrical, **2**/55–**2**/56
  industrial robots, **5**/25–**5**/26
  machinery, **18**/9–**18**/10
  see also Health and Safety
Salt bath process, *see* Nitrocarburizing
Salter Duck, **12**/34
Sampled-data systems, **3**/56, **3**/63–**3**/64
Sampling (digital), **3**/62
Sampling (mathematical), **17**/28
SAN (styrene/acrilonitrile polymer), **7**/117
Sand casting processes, **16**/93–**16**/98,
  16/100–**16**/103
  sand handling, **16**/98–**16**/100
Sandwik cutting tool program, **7**/8
Saturation characteristics, amplifier, **3**/79–**3**/80
Sawtooth wave, **17**/13
Scaling Laws, **16**/6
Scaling process (graphics), **5**/35
Scanner, **5**/6, **5**/30
Scanning tunnelling microscope, **9**/117
Scatter diagram, **17**/25
Scattering measurements (surface roughness),
  9/115
Schaeffler Diagram, **7**/46, **7**/48
Schematic modelling, **5**/12
Schlerscope hardness, **7**/4
Schmaltz technique, *see* Light sectioning
  process
Schmidt method, **1**/43
Schottky circuits, **2**/44
Scotch-Grip fastening system, **6**/20
Screens, CAD/CAM, **5**/5
Screws, self-tapping, **6**/20
Scuffing, **9**/71, **9**/105

SDLC protocol, **4**/29
Sea waves, *see* Wave power
Seal Compatability Index (SCI), **9**/12
Sealants, **16**/72
Sealed heating systems, **15**/95–**15**/96
Seals, **15**/18
  contact band width, **15**/42
  design of, **9**/106
  hydraulic, *see* Hydraulic seals
  multi-lip, **15**/23
  see also Clearance seals; Mechanical seals
Seam welding, **16**/66
Seamless tubing production, **16**/21
Second moment of area, *see* Moment of inertia
Sectional boilers, **15**/75
Sectional properties table, **19**/20
Segments (graphics), **5**/38
Seismic excitation, **1**/12, **1**/14
Self-induced e.m.f., **2**/7
Self-lubricating porous metals, **9**/69–**9**/70
Self-tuning control, **3**/56
Selsyn, *see* Synchro
Semi-conductor, **2**/13
  diode, **2**/14, **2**/15
  doped, **2**/13
  use in solar cell, **12**/14
Semi-submersible platforms, **14**/3, **14**/5–**14**/6
  future, **14**/12
Separation constant (mathematical), **17**/21
Separation load, **10**/35
Separation of variables method, **17**/21
Sequential control systems, **3**/40
Serial communication, **3**/7
Series (mathematical), **17**/11–**17**/12
Servomechanisms, **3**/40
Severn Barrage (proposed), **12**/32, **12**/33
Shading operation, **5**/42
Shaft:
  critical frequency, **9**/38–**9**/39
  eccentricity, **15**/54
  imbalance, **15**/128–**15**/129
  surface finish, **15**/53
  whirling, **1**/8
Shaft couplings, **10**/36
  all-metal membrane, **10**/40–**10**/42
  flexible ranges, **10**/42–**10**/43
  torque and misalignment, **10**/42
  types of, **10**/37–**10**/40, **10**/43
Shaft seals, *see* Lip seals, rotary-shaft
Shannon's sampling theorem, **3**/62
Shape factor (radiation), **1**/45
Shaping processes, metal, **16**/3–**16**/4
Shaw process, **16**/101
Shear force, **1**/3
Shear transformation process, **5**/35
Shear/Rigidity modulus table, **19**/20
Shearing operation, sheet metal, **16**/43
Sheet steel, **7**/26
  forming, **16**/13, **16**/43–**16**/47
  production, **16**/20–**16**/21
  roll-bending, **16**/25
Sheet-moulding compounds (SMC), **7**/99
Shell boilers, **15**/76–**15**/78
Shell-moulding techniques, **16**/101
Sherardizing, **7**/88
Shielding, nuclear, **13**/11, **13**/12
Shock isolation, **15**/131–**15**/132
Shock noise, **9**/106
Shock pulses, **15**/131–**15**/132
Shock waves, **1**/33
Shore A scale (hardness), **7**/3, **7**/134
Shot peening, **8**/38, **9**/74, **9**/101
Shrinkage/contraction, casting, **16**/85
Shunt-wound generator, **2**/19
SI units, **19**/3
  abbreviations, **19**/13–**19**/17
  conversion factors, **19**/13–**19**/17

SI units – *continued*
 derived units, **19**/3
 listing, **19**/5–**19**/16
 magnitudes/multiples, **19**/3, **19**/4
 rules for, **19**/4
Sialons, **7**/150
Siemens (unit), **2**/3
Sievert (radiation unit), **13**/10
Sigma Ltd DataMyte 769 system, **6**/33
Signal conditioning, **3**/18–**3**/19
Silicon, **7**/92
Silicon bronze, **7**/58
Silicon carbide, **7**/149, **15**/62
 fibres, **7**/93
Silicon cell, *see* Solar cell
Silicon nitride, **7**/149–**7**/150
Silicon-controlled rectifier, *see* Thyristor
Silicone resins, **7**/128
Silicones, **7**/125
Siliconizing, **9**/75–**9**/76
Silver, **7**/91
Simple harmonic motion, **1**/10
Simplex communication, **4**/28
Simpson's Rule (integration), **17**/11
Simulation, robot, **5**/22–**5**/23
Simulation techniques, **5**/22–**5**/23
Simultaneous equations, linear, **17**/14
Single-chip microcontrollers (SCMs), **3**/58, **3**/60
Sinking mills, **16**/22
Sinter forging, *see* Powder forming
Sintering, **7**/153
Sinusoidal input (control system), **3**/42
Sizing mills, **16**/22
Skewness, **9**/120, **17**/24
Skin friction, **9**/106
Slideways, lubrication of, **9**/14
Sliding:
 friction/wear, **9**/4
 unlubricated, **9**/60
  performance rating, **9**/60–**9**/62
Slip amplitude, **9**/92
Slip frequency, **9**/94
Small angles approximations, **17**/4
Smith-Putnam windmill, **12**/23
Smog, **11**/18
Smoke ventilation, **15**/105
Smoothness, depth of, **9**/120
Society of Automotive Engineers (SAE),
 viscosity system, **9**/8
Software:
 computer, **4**/36–**4**/37
  quality maintenance, **6**/32
Solar cell, **12**/14–**12**/15
Solar chemistry, **12**/15
 detoxification, **12**/18
 photochemical processes, **12**/16
 surfaces transformation, **12**/189
Solar distillation, **12**/12
Solar energy, **12**/3
 flat-plate collectors, **12**/6–**12**/7
 heat collectors, **12**/10–**12**/11
 radiation, **12**/4–**12**/5
 space heating, **12**/8
Solar glazing, **12**/17
Solar houses, **12**/8
 direct gain system, **12**/8
 passive system, **12**/8–**12**/10
Solar ponds, **12**/11
Solar refrigeration/cooling, **12**/12–**12**/14
Solar-powered engine, **12**/10
Soldering, **16**/67–**16**/68
Solders, **7**/90
Solid fuels, **11**/12
 cokes, **11**/13
 graphite, **11**/13
 peat to anthracite, **11**/12–**11**/13
 wastes as, **11**/13

 wood and straw, **11**/12
Solid modelling, **5**/10–**5**/11
Solid solutions, **7**/21
Solid-state diffusion, **7**/163
Sommerfeld Number/Reciprocal, **9**/32
Sonic booms, **1**/33
Soot formation, **11**/19
Sound intensity, **15**/139
Sound power, **15**/139
Sound pressure, **15**/139
Sound units and symbols, **19**/13
Sound-level meter, **15**/142–**15**/143
Sound-reduction index (SRI), **15**/147, **15**/149
Space heating, energy survey, **15**/118–**15**/120
Space units, **19**/5
Spain, health and safety standards, **18**/3
Spark-ignition internal combustion engines,
 10/5–**10**/6
 fuel behaviour, **11**/17
 knocking, **11**/16
 performance curves, **10**/7
 timing diagrams, **10**/6
SPATE (Stress Pattern Analysis by Thermal
 Emission), **7**/172
Specific heat, **7**/4–**7**/5
 ratio, **1**/31
 Table, **17**/20
Specific speed, pump, **15**/5
Specific wear rate, **9**/91
Speckle pattern measurements (surface
 roughness), **9**/116
Spectral absorption coefficient, **7**/5
Spectral density, **1**/17, **9**/123
Specular beam, **9**/115
Speed control:
 a.c. motors, **3**/64
 d.c. motors, **2**/21–**2**/22, **3**/63–**3**/64
 electric drives, **3**/63–**3**/64
 induction motors, **2**/27–**2**/28
 stepper motors, **3**/64
Spinning, sheet metal, **16**/44
Spiroid gears, **10**/30
Spontaneous ignition temperature, **11**/16
Spot welding, **16**/65, **16**/66
Sprag clutch, **10**/44
Sprayed coatings, **9**/82
Spring steels, **7**/12, **7**/49
Sprites, **5**/38
Spur-gear device, **15**/7
Sputtering process, **7**/165, **9**/80
Square wave, **17**/13
Squeeze casting, **16**/105–**16**/106, **16**/108
Squeeze seals, **15**/36
Squeeze-film effect, **9**/36
Stability:
 control systems, **3**/47–**3**/49
 floating body, **1**/21
Stagnation conditions (gas flow), **1**/32
Stainless steels, **7**/12, **7**/35–**7**/40
 corrosion resistance, **7**/40–**7**/42
 welding of, **7**/46
Stanal process, **9**/78
Standard deviation, **1**/16
Standard error of mean, **17**/28
Standardization, compression packing,
 15/27–**15**/28
Standardization, design, **6**/3–**6**/4
Standing wave, **17**/13
Stanton number (heat flow), **1**/44
Star connection, **2**/23
Star network, **4**/32
Star pattern, network, **3**/70
Star-delta starter, **2**/26
State transition matrix, **3**/86
State variable modelling, **3**/76–**3**/86
Statically-determinate systems, **1**/3
Statics, rigid bodies, **1**/3

Stationary process, **1**/16
Statutory Nuisances, **15**/105
Steady-flow energy equation, **1**/22
Steam generators, **15**/76
Steam heating systems, **15**/96
Steam plant, **10**/9
Steam storage, **15**/86
Steam turbines:
 impulse, **10**/10
 impulse-reaction, **10**/11–**10**/12
 lubrication, **9**/20
Steatites, **7**/147–**7**/148
Steel springs, isolating, **15**/133
Steel-making, **7**/12–**7**/17
Steels:
 casting, **7**/49–**7**/52, **16**/88–**16**/91
 creep-resistant, **8**/26, **8**/30
 fatigue resistance, **8**/37
 heat-resisting, **7**/42–**7**/44
 pressure vessel, **8**/26, **8**/29
 stainless, *see* Stainless steels
 toughness of, **7**/44–**7**/45
 weldability, **7**/45–**7**/48
 wrought, **7**/11, **7**/12
Stefan-Boltzmann constant (energy emission),
 1/42
Step input (control system), **3**/42
Stepper motors, **2**/29–**2**/30, **3**/64
Sterilization, nuclear, **13**/10
Stirling engine, **12**/18–**12**/19
Stoichiometry, **11**/14
Storage and stock, energy survey, **15**/122
Strain, **8**/3
Strain gauges, **3**/23, **3**/29, **8**/12, **15**/134
 arrangements of, **8**/14–**8**/16
 principles, **8**/13
Strainer, **9**/24
Straw, use of, **12**/37
Stream function, **1**/33
Stream lines, **1**/22
Stream tubes, **1**/22
Strength of materials, **1**/3–**1**/4
Stress, **8**/3, **8**/4
 concentration, **8**/38
  factor, **8**/7–**8**/8
 distribution, under polarized light, **8**/17–**8**/18
 freezing, **8**/18
 impact, **8**/8
 relaxation, **7**/137, **8**/23
 rupture, **7**/4
 testing, **8**/23, **8**/25, **8**/30
Stress corrosion cracking, **7**/42, **7**/160
Stress systems:
 isotropic materials, **8**/4
 plane, **8**/4–**8**/7
Stress units, **17**/7–**17**/8
Stress waves, **8**/8
Stress-pattern analysis by thermal emission
 (SPATE), **8**/19
Stretch conditions, rolling process, **16**/23
Stretch forming, sheet metal, **16**/44
Strip steel, **7**/26
 production, **16**/21
Strobe input, **2**/48
Structural analysis techniques, **14**/19–**14**/22
Structural failure mechanisms, jacket,
 14/22–**14**/23
Structural isolation, **15**/133
Structural steels, **7**/26
 sections production, **16**/21
Structural variability, **7**/171–**7**/172
Structure function, **9**/123
Student *t* test (of significance), **17**/29
Stuffing box, **15**/18–**15**/19
Stylus measurement instruments, **9**/109–**9**/112
Substitutions, standard calculus, **17**/10
Subtractor, *see* Amplifier, differential

Sulf-BT process, 9/78–9/79
Sulfinuz process, 9/77, 9/100
Superheaters, 15/85
Superminicomputer, 4/6
Superplastic forming, 16/52
Superposition theorem, 2/3
Supply air systems, 15/101
Surface chemistry changes, 9/74
Surface engineering, 9/70, 9/88
Surface features, 9/102, 9/103, 9/121–9/123
Surface finishes, 6/14, 9/107, 9/108
  generation of, 16/10
  monitoring of, 9/88
Surface flaws, 7/171
Surface hardening, steel, 7/31
Surface measurements, characterization of,
    9/117–9/124
Surface modelling, 5/10
Surface mount technology, 4/4
Surface replication technique, 8/30
Surface tension, 1/18
  table of, 19/20
Surface topography, 9/124
  effects of, 9/102–9/109
  measurements, 9/109–9/117
    characterization of, 9/117–9/124
  production factors, 9/107–9/108
Surface treatments/coatings, 9/73–9/83,
    9/85–9/87, 9/100
Surge, compressor performance, 15/12
Sursulf process, 9/77–9/78
Suspension processing, 7/113
Swageform screws, 6/20
Swarf disposal, 16/6
Switchgear, 9/20
  oils for, 9/20–9/21
Switching operations, thyristor, 2/37
Synchro (for angle measurement), 3/26
Synchronous motors, 2/25–2/26
Sysgen (system generation), 4/39
Systems Network Architecture (SNA), 4/35

T-T-T diagrams, 7/24
Tactile test, friction feel, 9/112
Tandem seals, 15/62
Tantalum, 7/91
Tape:
  magnetic, 4/22
    streamer unit, 4/22
Taptite screws, 6/20
Task-to-task communications, 4/35
Taylor's series, 17/5–17/6, 17/11
Technical and Office protocols (TOP), 5/7
Tedlar, 7/125
Teflon, 7/125
Tefzel, 7/125
Temper embrittlement, 7/44
Temperature, 1/18
  measurement, 3/34–3/39
Temperature-limiting values, 15/95
Tempering, 7/29
  steels, 7/12, 7/28–7/30
Tensile ductility, 7/3
Tensile testing, parameters, 7/3
Tensioned buoyant platform, 14/26–14/27
Terminal equipment, 4/36
Terneplate, 7/90
Terrestrial energy, 12/3, 12/27
  applications, 12/28
  developments, 12/29–12/30
  electricity generation, 12/29
  hot dry rocks, 12/29
  resources, 12/27
Terylene, 7/122
Tesla (unit), 2/6
Tests of significance, 17/28–17/29

Thermal efficiency, 1/36
Thermal insulation, building, 15/91
Thermal power, 12/10
  chemical energy storage, 12/18
Thermal properties, 7/4–7/5
  conductivity, see Conductance, thermal
  diffusivity, 7/5
  expansion, 3/34, 7/5
  resistance, 1/40, 1/41
Thermal storage, 15/86
Thermal-fluid heating systems, high-
    temperature, 15/96–15/97
Thermaspray adhesives, 6/20
Thermistors, 3/38
Thermo-couples, 3/34–3/37
Thermo-elastic analysis, 8/18–8/19
Thermochemical heatpump, 12/18
Thermodynamics, 1/35–1/36
  laws of, 1/36–1/37
Thermoeconomics, 1/37
Thermographic techniques, 7/172
Thermometers, 3/34
  gas/vapour, 3/39
  radiation, 3/38–3/39
  resistance, 3/37–3/38
Thermoplastic polymers, 7/95, 7/109–7/111,
    7/117–7/122
  high-temperature, 7/123–7/125
Thermosetting polymers, 7/95, 7/111,
    7/126–7/130
Thevenin's theorem (networks), 2/4
Third-generation languages (3GLs), 4/43–4/44
Thixo-casting, 16/106, 16/108
Thoma cavitation number, 15/9
Thorium fuel, 7/92, 13/4
  enrichment process, 13/4
Threadlocking, 6/19–6/20
Three Mile Island accident, 13/8–13/9
Three-dimensional concepts, 5/41–5/43
  motion, 1/7–1/8
Three-phase circuits, 2/23–2/24
  alternators, 2/24–2/25
  power in, 2/24
Threshold Limit Value (TLV), 15/99
Thyristor, 2/37, 2/42, 3/8
  for speed control, 2/22–2/23
TI Index (information system), 6/5
Tidal power, 12/3, 12/30
  economics, 12/33
  impact of, 12/32–12/33
  principles, 12/30–12/31
  schemes, 12/31
  sites, 12/31–12/32
Tiduram process, 9/79
TIG welding, 16/62–16/63
Tightly coupled computer systems, 4/6
Tilting pad bearing, 9/40
Time averaging, synchronous, 15/129
Time delay effect, 3/49
Time sharing, 6/37
Time units and symbols, 19/5
Time-division multiplexing (TDM), 4/76
Time-domain analysis, 3/42–3/44
Time-series modelling, 9/123–9/124
Timers:
  computer, 4/13
  flip-flop circuits, 2/54–2/55
  PLC, 3/68
  programmable, 3/6
Timing belt, 10/23–10/24
Tin and its alloys, 7/88, 7/90
Tinplate, 7/90
Titania ceramics, 7/149
Titanites, 7/149
Titanium and its alloys, 7/66–7/74
Titanium nitride, use of, 9/80–9/81
Tolerances, 6/7–6/8

and dimensioning, 6/13–6/14
and surface conditions, 6/14
Tomography, 7/175
Tonne (unit), 19/7
Tool steels, 7/12, 7/48–7/49
Tooth load, 10/35
Tooth profile, 10/26
Toothmeshing frequency, 15/125
Topografiner, 9/117
Topothesy, 9/118
Torlon, 7/125
Torque, 1/5
  power due to, 1/6
  work done by, 1/6
Torsion equation, 1/4
TOST 'life' (of oil), 9/20
Total Quality Management (TQM), 6/32
Touch screen, 5/32
Toughness:
  steel, 7/44–7/45
    fracture testing, 8/19
Toyota Diffusion Process, 9/76
TPX, see Polymethyl pentene
Tracing, nuclear, 13/10
Tracker ball, 5/31
Tramp oil, 9/15
Transaction processing, 4/38
Transducers, 3/23–3/25
  phase imbalance, 15/128
Transferred Plasma-Arc process, 9/82
Transformation ratio, 2/31
Transformation-hardening, 9/73
Transformations, two-dimensional, 5/34–5/37
Transformer, 2/30–2/33, 9/20
  closed-circuit test, 2/31
  efficiency, 2/32
  losses, 2/31
  oils for, 9/20–9/21
  three-phase, 2/32–2/33
  see also Auto-transformer
Transient switching conditions, 2/9–2/10
Transistor, 2/33–2/35
  collector, 2/33
  use in circuits, 2/35–2/36
  see also Field effect transistor
Transistor/Transistor Logic (TTL), 3/8
Translation, image, 5/34–5/35
Transmission Control Protocol/Internal
    Protocol (TCP/IP), 5/7
Transmission shaft, 10/34
Transport, energy survey, 15/122
Transverse rolling, see Rotary forging
Trapezoidal rule, 17/10
Trend limit, 15/127
Trending procedure, vibration increase, 15/129
Triac, 2/37
Triangles:
  solution of, 17/4
  spherical, 12/4–12/5
Triangular wave, 17/13
Tribology, seal behaviour, 15/40–15/42
Trigonometric formulae, 17/3–17/4
Tritium, 13/4
Trombe wall, 12/8
Truth table (logical functions), 2/43
TTL circuits, 2/44
Tube extrusion processes, 16/35–16/37
Tube-drawing processes, 16/40–16/41,
    16/42–16/43
Tungsten, 7/91
Tungsten carbide, 15/62
Turbine meter, 3/31
Turbine oil, 9/20
Turbines, see Gas turbines; Steam turbines;
    Water turbines
Turbulent flow, 1/23
Turndown, boiler, 15/84

Turning processes:
  forces in, **16**/9
  metal, **16**/3
Two-stroke engine, **10**/6
U-value (insulation), **1**/41, **1**/46, **15**/91, **15**/93, **15**/111, **15**/119
Udel, **7**/123
Ugine-Sejournet process, **16**/36
Ultem, **7**/125
Ultimate tensile stress, **7**/3
Ultrasonics, use of, **6**/20, **7**/176–**7**/177
UNC (Unified coarse) thread, **6**/17
UNF (Unified fine) thread, **6**/17
UNIGRAPHICS 11 software, **6**/32
United Kingdom, health and safety standards, **18**/4
United States, health and safety standards, **18**/4
Universal constants, **19**/18
Universal gas constant, **1**/31
Universal joint, **10**/39
UNIX operating system, **4**/44, **5**/41
Upsetting, *see* Forging operations, open-die
Uranium, **7**/92
  as fuel, **13**/4
  isotopes, **13**/4
Urea formaldehyde, **7**/128
Urethanes, **7**/144

V.24 interface standard, **4**/29–**4**/30
V-belts, **10**/18–**10**/22
Vacuum evaporation coating, **7**/165
Vacuum measurements, **3**/29, **3**/31
Vacuum welding, **16**/64–**16**/65
Vacuum-forming process, **7**/113
Valency values table, **19**/20
Valisys software, **6**/32–**6**/33
Valve closure rates, effect of, **1**/30
Valve steels, **7**/12, **7**/44
Valve stem seals, **15**/30
Van Karmann's momentum integral, **1**/29
Vapour compression cycle, **15**/115
Variable-amplitude loading, effect of, **8**/38
Variance, **1**/16
VAX system, **4**/10, **4**/13, **4**/14, **4**/15, **5**/41
Vector calculus, **17**/11
Vector scan, **5**/30
Velocity, **1**/5
  angular, **1**/5
  measurement, **3**/27
    fluid, **3**/34
  in mechanisms, **1**/7
Velocity diagrams, **15**/4
Velocity meter, **1**/26
  open-channel, **1**/27
Velocity pick-ups, **15**/134
Velocity potential, **1**/34
Velocity potential lines, **1**/34
Vena contracta, **1**/24
Ventilation:
  Codes of Practice, **1**/105
  legislation, **15**/105
Ventilation systems, **15**/97
  commissioning, **15**/106
  definitions, **15**/97, **15**/99
  design factors, **15**/103–**15**/105
  maintenance, **15**/106
  natural, **15**/99–**15**/100
  powered, **15**/100-**15**/102
  running costs, **15**/106
Venturi flumes, *see* Flumes
Venturi meter, **1**/25
Venturi throat, **3**/33
VIA (Versatile Interface Adaptor), **3**/6
Vibration:
  attenuation, **15**/132–**15**/133
  definition, **15**/129–**15**/130

effects, **15**/136, **15**/138
isolation of, **15**/130–**15**/131, **15**/148–**15**/149
  materials, **15**/133
limits, **15**/136, **15**/138
measurement of, **3**/28, **15**/133–**15**/136
multi-degree of freedom, **1**/14
random, **1**/15–**1**/17
reduction of, **15**/132
single-degree of freedom, **1**/9–**1**/14
and surface roughness, **9**/106
torsional, **1**/10
Vibration monitoring, **15**/125–**15**/126
  analysis techniques, **15**/126–**15**/129
  high-frequency range, **15**/126
  low-frequency range, **15**/125
  medium-frequency range, **15**/125–**15**/126
Vickers hardness, **7**/3, **7**/4
Victrex, **7**/123
Virtual terminals, **4**/35
Visco-seal, **15**/73
Viscosity, **1**/18, **9**/31–**9**/32
  fuel, **11**/11
  gear oils, **9**/47–**9**/48
  hydraulic oils, **9**/11
  of lubricants, **9**/5
  pump performance, **15**/10
Viscosity Index (VI), **9**/5, **9**/8, **9**/11
  improvers, **9**/7
Vision systems, industrial, *see* Industrial vision systems
Visual Display Unit (VDU), **4**/17–**4**/18
Visual indicators, locating, **6**/29
Vitrification, **7**/153
VLSI (very-large scale integration), **2**/37, **3**/4, **4**/4
VMS systems, **5**/41
Voltage ratio, **2**/13–**2**/13
Voltage regulation, **2**/32
Voltage/current relationship, **2**/9
Volume measurement, **3**/27
Volume resistivity, **7**/5
Vortex, **1**/35
Vorticity, **1**/22
VRC (Vertical Redundancy Check), **4**/28
Vulcanization, **7**/133
Vulcanized rubber, storage of, **15**/50

Walsh functions, **9**/123
Wankel engine, **11**/16
Ward Leonard drive, **2**/21–**2**/22
Warm air heating systems, **15**/96–**15**/97
Warm forging, **16**/28
Washers, **6**/21
Waste, solid, energy from, **11**/13, **12**/37–**12**/39
Waste lubricating oil, **9**/27
Waste storage, nuclear, **13**/8
Waste-heat boilers, **15**/84
Waste-heat recovery, **15**/122–**15**/123
Water hammer, *see* Pressure transients
Water tube boilers, **15**/78
Water turbines, **12**/19–**12**/20
Water-level control, boiler, **15**/86–**15**/87
Waterside conditions, **15**/82–**15**/83
Wave forces, **14**/14–**14**/17
  on large bodies, **14**/18–**14**/20
  on slender structures, **14**/17–**14**/18
Wave forms, **17**/12–**17**/13
Wave power, **12**/3, **12**/33, **12**/35
  energy convertors, **12**/34
  principles, **12**/33–**12**/34
  shore-based systems, **12**/35
Wave windings, armature, **2**/17
Wavelength:
  average, **9**/121
  root mean square, **9**/121
Wear:

data on, **9**/83–**9**/85
fretting, **9**/89
process identification, **9**/72
sliding, **9**/60–**9**/61, **9**/64–**9**/66
types, **9**/71–**9**/72
Wear equations, **9**/4–**9**/5
Wear Factor, **9**/4
Wear resistance, **7**/7
Wear-resistance surface treatments, choice of, **9**/85–**9**/87
Weber (unit), **2**/5
Wedge belts, *see* V-belts
Weeds, aquatic, **12**/40
Weibull function (probability), **7**/152, **12**/24, **17**/28
Weight, **1**/3, **19**/5, **19**/7
Weight measurement, **3**/28–**3**/29
Weights and measures systems, **19**/8
Weirs, **1**/27
Weld Decay, **7**/47, **7**/159
Welding, **16**/54
  economics, **16**/68–**16**/69
  energy survey, **15**/121
  joint types, **16**/55
  steels for, **7**/45–**7**/48
  for wear resistance, **9**/83
Welding processes:
  consumable guide, **16**/59
  electro-slag, **16**/59–**16**/60
  electron beam, **16**/64
  explosive, **16**/51
  flash, **16**/67
  flux-cored arc, **16**/62
  friction, **16**/67
  gas, **16**/63–**16**/64
  gas-shielded metal arc. **16**/60–**16**/61
  gas-shielding tungsten arc, **16**/62–**16**/63
  gravity, **16**/58
  laser, **16**/64, **16**/65
  manual metal arc, **16**/55–**16**/57
  mechanized, **16**/58
  open-arc automatic, **16**/58
  plasma arc, **16**/62–**16**/63
  resistance, **16**/65–**16**/67
  resistance butt, **16**/66–**16**/67
  submerged arc, **16**/58–**16**/59
  ultrasonic, **6**/20
  vacuum, **16**/64–**16**/65
Wells turbine, **12**/34, **12**/35
Westland Helicopters Ltd., **10**/26
Wheatstone bridge circuit, **8**/14–**8**/15
Whirling (of shafts), **1**/10
White cast iron, **7**/55
White noise, **1**/17
Wide area network (WAN), **4**/16
Wide-band process, **1**/17
Wiggings Teape Basingstoke headquarters, use of atrium, **12**/10
Winchester drive, **4**/21–**4**/22
Wind Energy Group, **12**/26
Wind power, **12**/3, **12**/23, **12**/27
  characteristics, **12**/23–**12**/24
  environmental issues, **12**/26
  potential, **12**/23–**12**/24
Wind turbine, **12**/25–**12**/26
Windchargers, **12**/25
Windings, transformer, **2**/31
Windmills, **12**/24–**12**/25
Window, solar, **12**/17
Windowing operation (graphics), **5**/37–**5**/38
Windows operating system (computer), **4**/5
Windscale accident, **13**/10
Wire-drawing, **7**/27, **16**/41–**16**/43
Wire-frame:
  drawing, **5**/42
  modelling, **5**/10

WLF (Williams, Landel, Ferry) equation, 7/142–7/143
Wobbe number (fuels), 11/6
Wood, 7/105
    protection of, 7/166
    resin-impregnated, 7/106
Woodfuel, 12/37
Woodworking Machines Regulations 1974, 15/105
Work, 1/5
Work transfer, 1/37
Works tests:
    compressors, 15/18
    fans, 15/18
    pumps, 15/14

Workstations, 4/5, 4/18, 5/4–5/5, 5/40–5/41
Worm drive, 10/30–10/31
Worthington plot, 15/5
Writing tablet, 4/20
Wye connection, see Star connection

X-ray fluorescence, 9/87
X-rays, 7/174
    for stress determination, 8/19
Xeroradiography, 7/175

Yield point, 8/3
Yield Stress, 7/3
Young's Modulus, see Modulus of elasticity

z-transform, 3/56–3/57, 3/70–3/75
Zalutite, 7/88
Zener diodes, 2/14, 2/15, 2/42
Zener effect, 2/34
Zener voltage, 2/14
Zero, 17/6
Zero-order-hold device, 3/56, 3/71, 3/72
Zinal treatment, 9/79, 9/84
Zinc coating, 7/88
Zinc and its alloys, 7/87–7/89
Zinc spraying, 7/165
Zirconia ceramics, 7/149
Zirconium, 7/92

DATE DUE

Demco, Inc. 38-293